Other books in the GRANTS Series:

**Directory of Biomedical and Health Care Grants**
**Directory of Grants in the Humanities**
**Funding Sources for Children and Youth Programs**
**Funding Sources for Community and Economic Development**
**Funding Sources for K-12 Education**
**Operating Grants for Nonprofit Organizations**



Littleberry
Press
West Lafayette, Indiana

GRANTS Series

# DIRECTORY OF RESEARCH GRANTS

43rd Edition

Edited by
Louis S. Schafer, Ed.S.

Littleberry Press
West Lafayette, Indiana

Published by Littleberry Press LLC
1048 B Sagamore Pkwy W, #48
West Lafayette, IN 47906
www.littleberrypress.com
765-237-3390
info@littleberrypress.com

Distributed by Ingram Content Group
1 Ingram Blvd.
La Vergne, TN 37086 USA
(615) 793-5000



Published 1975, 43rd Edition, 2020
ISBN 978-1-94075505-0-7

Editor: Louis S. Schafer, Ed.S.
Book and cover design: Anita Schafer
Production: Littleberry Press and IngramSpark
Contributors (in alphabetical order): Kira Axsiom, Andrea Axsom, Zahraa Ouyoun,
Anita Schafer, Louis S. Schafer

Printed and bound in the United States of America

At Littleberry Press, we publish a collection of books that reflect on real events and people and books that help educators and nonprofit personnel achieve their goals.

# CONTENTS

# INTRODUCTION

The entry into the world of grantseeking and grantsmanship is not usually a direct one. Many researchers, students, and nonprofit workers acquire those skills along the way, funding projects and education by learning-by-doing and writing grant proposals. What is considered to be a "soft" skill actually helps to keep the doors open, the lights on, and programs running. Every grant writer, new and experienced, learns quickly that all the grant writing proficiency in the world does not matter unless you have someone or some organization to send your proposal. In order to ask for the money, you must have someone to ask.

For four decades the GRANTS Database and its print complements, including the annual *Directory of Research Grants*, have provided the research community with current, accurate information regarding funding for research-related programs and projects, scholarships, fellowships, conferences, and internships. To meet increased information needs, the GRANTS Database, researched and edited by members of the GrantSelect editorial staff, has grown significantly. This ##st edition of the annual *Directory of Research Grants* offers factual and concise descriptions of more than 4,500 research funding programs. With this new edition both experienced and novice grantseekers will be better able to target only the most applicable programs for their research needs.

**Directory of Research Grants**

The *Directory of Research Grants* features listings of programs that offer non-repayable research funding for projects in medicine, the physical and social sciences, the arts and humanities, education, community and economic development, and children and youth programs. Listings in the main section of the *Directory* contain annotations describing each program's focuses and goals, program requirements explaining eligibility, funding amounts, deadlines, sponsor name and address, contact information, and the sponsor's web address. Grantseekers with access to the Internet can use the addresses to locate further information about the organizations and their application procedures. Internet addresses are also provided, when available, within listings. Some of the grant programs listed in the main section have geographic restrictions for applicants.

**Indexes.** The Subject Index of this *Directory* lists all program titles—with accession numbers—under their applicable subject terms. Other indexes follow, including the Grants by Program Type Index, which lists 44 program categories, such as Basic Research, Fellowships, Travel Grants, etc., with the grants that fall within their scope; and the Geographic Index, which lists programs that have state, regional, or international focus. See "How to Use This Directory" on pages ## for sample index entries.

**Using the *Directory* for Grantseeking**

By using the *Directory of Research Grants*, grantseekers can match the needs of their particular programs with those sponsors offering funding in the researchers' area of interest. The information listed here is meant to eliminate the costs incurred by both grantseekers and grantmakers when inappropriate proposals are submitted. However, because the GRANTS database is updated daily basis, with program listings continually added, deleted, and revised, grantseekers using this *Directory* may also search the GRANTS database online through *GrantSelect* (www.grantselect.com).

All new and revised information within the *Directory* has been taken from (1) sponsors' updates of previously published program statements, (2) questionnaires sent to sponsors whose programs were not listed in previous editions, or (3) other materials published by the sponsor and furnished to Littleberry Press. Updated information for U.S. government programs includes new and revised program information published in the *Federal Register*; the *NIH Guide*, published by the National Institutes of Health; and the *NSF E-Bulletin*, a monthly publication of the National Science Foundation. Included in this edition are identifying document numbers from the NIH and NSF publications. Located at the ends of the program descriptions in certain entries, the numbers indicate the ongoing NIH program number (PA) or the request for applications number (RFA). For programs of the National Science Foundation, the *NSF Bulletin* number appears. This information will help users identify the programs when seeking additional information from program staff.

While Littleberry Press has made every effort to verify that all information is both accurate and current within the confines of format and scope, the publisher does not assume and hereby disclaims any liability to any party for loss or damages caused by errors or omissions in this *Directory*, whether such errors or omissions result from accident, negligence, or any other cause. Questions should be directed to the Editor at Littleberry Press via email at editor@littleberrypress.com.

# HOW TO USE THIS DIRECTORY

The *Directory of Research Grants* is designed to allow the user quick and easy access to information regarding funding programs in a researcher's specific area of interest. This *Directory* is composed of a main section, Grant Programs, which lists grant programs in alphabetical order, and three indexes: the Subject Index, the Grants by Program Type Index, and the Geographic Index.

## GRANT PROGRAMS

Each listing in this section consists of the following elements: an annotation describing each program's focus and goals, requirements explaining eligibility, funding amounts, application and renewal dates, sponsor information, contact information, and Internet address.

GRANT TITLE — **Paul G. Allen Family Foundation Grants** **4298** — ACCESSION NUMBER

The single foundation, created through the consolidation of Allen's six previous foundations (The Allen Foundation for the Arts, The Paul G. Allen Charitable Foundation, The Paul G. Allen Foundation for Medical Research, The Paul G. Allen Forest Protection Foundation, The Allen Foundation for Music, and The Paul G. Allen Virtual Education Foundation), will continue to focus on the Allen family's philanthropic interests in the areas of arts and culture, youth engagement, community development and social change, and scientific and technological innovation. The Arts and Culture Program fosters creativity and promotes critical thinking by helping strong arts organizations become sustainable and supporting projects that feature innovative and diverse artistic forms. — GRANT DESCRIPTION
The Youth Engagement Program improves the way young people learn by supporting organizations that use innovative teaching strategies and provide opportunities for children to address issues relevant to their lives. The Community Development and Social Change Program promotes individual and community development by supporting initiatives and organizations that provide access to resources and opportunities. The Scientific and Technological Innovation Program advances promising scientific and technology research that has the potential to enhance understanding and stewardship of the world in which we live. Organizations may only receive one grant per year. Organizations must not have any delinquent final reports due to any of the Paul G. Allen Foundations for previous grants. Grantseekers are encouraged to apply through the online application process, where basic organizational and project information will be requested. Guidelines are available online.

REQUIREMENTS — *Requirements* 501(c)3 tax-exempt organizations, status from the Internal Revenue government entities, and IRS-recognized tribes are eligible. Eligible organizations must be located in, or serving populations of, the Pacific Northwest, which includes Alaska, Idaho, Montana, Oregon, and Washington.

RESTRICTIONS — *Restrictions* In general, the foundation will not consider requests for general fund drives, annual appeals, or federated campaigns; special events or sponsorships; direct grants, scholarships, or loans for the benefit of specific individuals; projects of organizations whose policies or practices discriminate on the basis of race, ethnic origin, sex, creed, or sexual orientation; contributions to sectarian or religious organizations whose principle activity is for the benefit of their own members or adherents; loans or debt retirement; projects that will benefit the students of a single school; general operating support for ongoing activities; or projects not aligned with the foundation's specified program areas. 509(a) private foundations are ineligible.

APPLICATION/ DUE DATE — *Date(s) Application Is Due* Mar 31; Sep 30.

*Contact* Grants Administrator, (206) 342-2030; fax: (206) 342-3030; email: info@pgafamilyfoundation.org — CONTACT

INTERNET ADDRESS — *Internet* http://www.pgafamilyfoundation.org

*Sponsor* Paul G. Allen Family Foundation — SPONSOR INFORMATION
505 Fifth Ave S, Ste 900
Seattle, WA 98104

## SUBJECT INDEX

The most effective way to access specific funding programs is through the Subject Index. This index lists the subject terms with applicable grants program titles – and their accession numbers – alphabetically under each term. Terms were assigned to target the specific area of research designated in the description of each program. Cross-references are used to link subjects and assist the user in finding specific grant information.

Following are general guidelines that can make your search of this index more successful. First, check under the specific topic of interest rather than a more general term. For instance, if you are interested in chemical engineering, look under "Chemical Engineering" rather than "Engineering." Items indexed under "Engineering" indicate funding in broad areas of engineering.

Use general headings when you want grants covering broader areas or if you can't find a specific topic. For example, many grants list funding for humanities research, health programs, or science and technology. To find these grants use such headings as "Humanities," "Medical Programs," "Science," or "Technology." For additional grant information on more specific humanities research opportunities, such as in American History or Cultural Anthropology, also check under the topics "United States History" and "Anthropology, Cultural."

Many of the grant programs provide funding for research-related scholarships, faculty fellowships, dissertations, undergraduate education, conferences, or internships. If grant funds are designated for specific disciplines, you will find the items under the specific subject. Scholarships and fellowships are also listed under the terms "Native American Education," "African Americans (Student Support)," "Hispanic Education," "Minority Education," and "Women's Education."

Grants concerning study of a particular country are listed under the name of the country. Grants concerning the history, literature, art and language of a country are listed under the name of the country also, e.g., "Chinese Art" and "Chinese Language/ Literature."

**SUBJECT TERM** ——— **Education**

A.L. Mailman Family Foundation Grants, 5
AAAS Science & Technology Policy Fellowships - Health, Education & Human Services, 22
AARP Andrus Foundation Grants, 104
Abbott Laboratories Fund Grants, 125
ACIE Host University Edmund S. Muskie/Freedom Support Act Graduate Fellowships, 156
ACT Awards, 189
Akonadi Foundation Anti-Racism Grants, 351 ——— **PROGRAM TITLE**
Akron Community Foundation Grants, 352
Albert and Margaret Alkek Foundation Grants, 370
Albuquerque Community Foundation Grants, 379
Alcoa Foundation Grants, 380
Alcon Foundation Grants Program, 382

---

## GEOGRAPHIC INDEX

This index lists programs that have state, regional, or international geographic focus. The Geographic Index is arranged by state, followed by Canadian programs, then by international programs by country, and lists grant program titles and their corresponding accession numbers.

**COUNTRY** ——— **United States**

**Alabama** ——— **STATE**

3M Fndn Grants, 2
Alabama Humanities Fndn Grants Program, 368
Arkema Inc. Fndn Science Teachers Program, 705
CDC Injury Control Research Centers Grants, 1198
**PROGRAM TITLE** ——— DOE Experimental Program to Stimulate Competitive Research (EPSCoR), 1653
Hill Crest Fndn Grants, 2293
Linn-Henley Charitable Trust Grants, 2777
NOAA Community-Based Restoration Program (CRP) Grants, 3843
Southern Company's Longleaf Pine Reforestation Fund, 4726

## PROGRAM TYPE INDEX

This index is broken into 44 categories according to the type of program funded:

- Adult Basic Education
- Adult/Family Literacy Training
- Awards/Prizes
- Basic Research
- Building Construction and/or Renovation
- Capital Campaigns
- Centers: Research/Demonstration/Service
- Citizenship Instruction
- Community Development
- Consulting/Visiting Personnel
- Cultural Outreach
- Curriculum Development/Teacher Training
- Demonstration Grants
- Development (Institutional/Departmental)
- Dissertation/Thesis Research Support
- Educational Programs
- Emergency Programs
- Endowments
- Environmental Programs
- Exchange Programs
- Exhibitions, Collections, Performances, Video/Film Production
- Faculty/Professional Development
- Fellowships
- General Operating Support
- Graduate Assistantships
- Grants to Individuals
- International Exchange Programs
- International Grants
- Job Training/Adult Vocational Programs
- Land Acquisition
- Matching/Challenge Funds
- Materials/Equipment Acquisition (Computers, Books, Tapes, etc.)
- Preservation/Restoration
- Professorships
- Publishing/Editing/Translating
- Religious Programs
- Scholarships
- Seed Grants
- Service Delivery Programs
- Symposia, Conferences, Workshops, Seminars
- Technical Assistance
- Training Programs/Internships
- Travel Grants
- Vocational Education

# GOVERNMENT & ORGANIZATION ACRONYMS

**AAAAI** American Academy of Allergy Asthma and Immunology
**AAAS** American Association for the Advancement of Science
**AACAP** American Academy of Child and Adolescent Psychiatry
**AACN** American Association of Critical Care Nurses
**AACR** American Association of Cancer Research
**AAFCS** American Association of Family and Consumer Sciences
**AAF** American Architectural Foundation
**AAFP** American Academy of Family Physicians Foundation
**AAFPRS** American Academy of Facial Plastic and Reconstructive Surgery
**AAP** American Academy of Pediatrics
**AAR** American Academy in Rome
**AAS** American Antiquarian Society
**AASL** American Association of School Libraries
**AAUW** American Association of University Women
**ABA** American Bar Association
**ACC** Asian Cultural Council
ACF Administration on Children, Youth and Families
**ACLS** American Council of Learned Societies
**ACM** Association for Computing Machinery
**ACE** American Council on Education
ACMP Amateur Chamber Music Players
**ACS** American Cancer Society
**ADA** American Diabetes Association
**ADHF** American Digestive Health Foundation
**AF** Arthritis Foundation
**AFAR** American Federation for Aging Research
**AFOSR** Air Force Office of Scientific Research
**AFUD** American Foundation for Urologic Disease
**AFUW** Australian Federation of University Women
**AGS** American Geriatrics Society
**AHA** American Heart Association
**AHAF** American Health Assistance Foundation
**AFHMR** Alberta Heritage Foundation for Medical Research
**AHRQ** Agency for Healthcare Research and Quality
**AICR** American Institute for Cancer Research
**AIIS** American Institute for Indian Studies
**AJA** American Jewish Archives
**AJL** American Jewish Libraries
**ALA** American Library Association
**ALISE** Association for Library and Information Science Education
**AMNH** American Museum of Natural History
**AMS** American Musicological Society
**ANL** Argonne National Library
**ANS** American Numismatic Society
**AOA** American Osteopathic Association
**AOCS** American Oil Chemists' Society
**APA** American Psychological Association
**APAP** Association of Performing Arts Presenters
**APEAL** Asian Pacific Partners for Empowerment and Leadership
**APS** Arizona Public Service
**APSA** American Political Science Association
**ARIT** American Research Institute in Turkey
**ARO** Army Research Office
**ASA** American Statistical Association
**ASCSA** American School of Classical Studies at Athens
**ASECS** American Society for Eighteenth-Century Studies
**ASF** American-Scandinavian Foundation
**ASHA** American Speech-Language-Hearing Association
**ASHRAE** American Society of Heating, Refrigerating, and Air Conditioning Engineers
**ASME** American Society of Mechanical Engineers
**ASNS** American Society for Nutritional Sciences
**ASPRS** American Society of Photogrammetry and Remote Sensing
**ASTA** American String Teachers Association
**ATA** Alberta Teachers Association
**AWHONN** Association of Women's Health, Obstetric, and Neonatal Nurses
**AWU** Associated Western Universities
**AWWA** American Water Works Association
**BA** British Academy
**BBF** Barbara Bush Foundation
**BCBS** Blue Cross Blue Shield
**BCBSM** Blue Cross Blue Shield of Michigan
**BCBSNC** Blue Cross Blue Shield of North Carolina
**BWF** Burroughs Wellcome Fund
**CBIE** Canadian Bureau for International Education
**CCF** Catholic Community Foundation
**CCF** Common Council Foundation
**CCFF** Canadian Cystic Fibrosis Foundation
**CCFF** Christopher Columbus Fellowship Foundation
**CDC** Centers for Disease Control and Prevention
**CDECD** Connecticut Department of Economic and Community Development
**CDI** Children's Discovery Institute
**CEC** Council for Exceptional Children
**CEF** Chemical Educational Foundation
**CES** Council for European Studies
**CF** The Commonwealth Fund
**CFF** Cystic Fibrosis Foundation
**CFFVR** Community Foundation for the Fox Valley Region
**CFKF** Classic for Kids Foundation
**CFNCR** Community Foundation for the National Capital Region
**CFPC** College of Family Physicians of Canada

| | |
|---|---|
| **CFUW** | Canadian Federation of University Women |
| **CHCF** | California Health Care Foundation |
| **CHEA** | Canadian Home Economics Association |
| **CICF** | Central Indiana Community Foundation |
| **CIES** | Council for International Exchange of Scholars |
| **CIUS** | Canadian Institute of Ukrainian Studies |
| **CLA** | Canadian Lung Association |
| **CLF** | Canadian Liver Foundation |
| **CMS** | Centers for Medicare and Medicaid Services |
| **CNCS** | Corporation for National and Community Service |
| **CRI** | Cancer Research Institute |
| **CTCNet** | Community Technology Centers Network |
| **DAAD** | Deutscher Akademische Austauschdienst (German Academic Exchange Service) |
| **DHHS** | Department of Health and Human Services |
| **DOA** | Department of Agriculture |
| **DOC** | Department of Commerce |
| **DOD** | Department of Defense |
| **DOE** | Department of Energy |
| **DOI** | Department of the Interior |
| **DOJ** | Department of Justice |
| **DOL** | Department of Labor |
| **DOS** | Department of State |
| **DOT** | Department of Transportation |
| **EFA** | Epilepsy Foundation of America |
| **EIF** | Entertainment Industry Foundation |
| **EPA** | Environmental Protection Agency |
| **ESF** | European Science Foundation |
| **ETS** | Educational Testing Service |
| **FCAR** | Formation de Chercheurs et L'Aide a la Recherche |
| **FCD** | Foundation for Child Development |
| **FDA** | Food and Drug Administration |
| **FIC** | Fogarty International Center |
| **GAAC** | German American Academic Council |
| **GCA** | Garden Club of America |
| **GEF** | Green Education Foundation |
| **GNOF** | Greater New Orleans Foundation |
| **HAF** | Humboldt Area Foundation |
| **HBF** | Herb Block Foundation |
| **HHS** | Health and Human Services |
| **HHMI** | Howard Hughes Medical Institute |
| **HRSA** | Health Resources and Services Administration |
| **HUD** | Department of Housing and Urban Development |
| **ICC** | Indiana Campus Compact |
| **IIE** | Institute of International Education |
| **IRA** | International Reading Association |
| **IRC** | International Rescue Committee |
| **IREX** | International Research and Exchanges Board |
| **IUCP** | Indiana University Center on Philanthropy |
| **IYI** | Indiana Youth Institute |
| **JDF** | Juvenile Diabetes Foundation International |
| **JMO** | John M. Olin Foundation |
| **JSPS** | Japan Society for the Promotion of Science |
| **KFC** | Kidney Foundation of Canada |
| **LISC** | Local Initiatives Support Corporation |
| **LSA** | Leukemia Society of America |
| **MFRI** | Military Family Research Institute |
| **MHRC** | Manitoba Health Research Council |
| **MLA** | Medical Library Association |
| **MLB** | Major League Baseball |
| **MMA** | Metropolitan Museum of Art |
| **MMS** | Massachusetts Medical Society |
| **MSSC** | Multiple Sclerosis Society of Canada |
| **NAA** | Newspaper Association of America |
| **NAACP** | National Association for the Advancement of Colored People |
| **NAGC** | National Association for Gifted Children |
| **NAPNAP** | National Association of Pediatric Nurse Associates and Practitioners |
| **NARSAD** | National Alliance for Research on Schizophrenia and Depression |
| **NASA** | National Aeronautics and Space Administration |
| **NASE** | National Association for the Self-Employed |
| **NASM** | National Air and Space Museum |
| **NATO** | North Atlantic Treaty Organization |
| **NCCAM** | National Center for Complementary and Alternative Medicine |
| **NCFL** | National Center for Family Literacy |
| **NCI** | National Cancer Institute |
| **NCIC** | National Cancer Institute of Canada |
| **NCRR** | National Center for Research Resources |
| **NCSS** | National Council for the Social Studies |
| **NEA** | National Education Association |
| **NEH** | National Endowment for the Humanities |
| **NEI** | National Eye Institute |
| **NFID** | National Foundation for Infectious Diseases |
| **NFL** | National Football League |
| **NFWF** | National Fish and Wildlife Foundation |
| **NGA** | National Gardening Association |
| **NHGRI** | National Human Genome Research Institute |
| **NHLBI** | National Heart, Lung and Blood Institute |
| **NHSCA** | New Hampshire State Council on the Arts |
| **NIA** | National Institute on Aging |
| **NIAF** | National Italian American Foundation |
| **NIAAA** | National Institute on Alcohol Abuse and Alcoholism |
| **NIAF** | National Italian American Foundation |
| **NIAID** | National Institute of Allergy and Infectious Diseases |
| **NIAMS** | National Institute of Arthritis and Musculoskeletal Skin Diseases |

| | |
|---|---|
| **NICHD** | National Institute of Child Health and Human Development |
| **NIDA** | National Institute on Drug Abuse |
| **NIDCD** | National Institute on Deafness and Other Communication Disorders |
| **NIDCR** | National Institute of Dental and Craniofacial Research |
| **NIDDK** | National Institute of Diabetes, and Digestive and Kidney Diseases |
| **NIDRR** | National Institute on Disability and Rehabilitation Research |
| **NIEHS** | National Institute of Environmental Health Sciences |
| **NIGMS** | National Institute of General Medical Sciences |
| **NIH** | National Institutes of Health |
| **NIJ** | National Institute of Justice |
| **NIMH** | National Institute of Mental Health |
| **NINDS** | National Institute of Neurological Disorders and Strokes |
| **NINR** | National Institute of Nursing Research |
| **NIOSH** | National Institute for Occupational Safety and Health |
| **NIST** | National Institute of Standards and Technology |
| **NJSCA** | New Jersey State Council on the Arts |
| **NKF** | National Kidney Foundation |
| **NL** | Newberry Library |
| **NLM** | National Library of Medicine |
| **NMF** | National Medical Fellowships, Inc. |
| **NMSS** | National Multiple Sclerosis Society |
| **NNEDVF** | National Network to End Domestic Violence Fund |
| **NOAA** | National Oceanic and Atmospheric Administration |
| **NRA** | National Rifle Association |
| **NRC** | National Research Council |
| **NSERC** | Natural Sciences and Engineering Research Council of Canada |
| **NSF** | National Science Foundation |
| **NSTA** | National Science Teachers Association |
| **NYCH** | New York Council for the Humanities |
| **NYCT** | New York Community Trust |
| **NYFA** | New York Foundation for the Arts |
| **NYSCA** | New York State Council on the Arts |
| **OAH** | Organization of American Historians |
| **ODKF** | Outrigger Duke Kahanamoku Foundation |
| **OJJDP** | Office of Juvenile Justice and Delinquency Prevention |
| **ONF** | Oncology Nursing Foundation |
| **ONR** | Office of Naval Research |
| **OREF** | Orthopaedic Research and Education Foundation |
| **ORISE** | Oak Ridge Institute for Science and Education |
| **OSF** | Open Society Foundation |
| **PAS** | Percussive Arts Society |
| **PCA** | Pennsylvania Council on the Arts |
| **PDF** | Peace Development Fund |
| **PDF** | Parkinson's Disease Foundation |
| **PhRMA** | Pharmaceutical Research and Manufacturers of American Foundation |
| **PHSC** | The Photographic Historical Society of Canada |
| **PSEG** | Public Service Enterprise Group |
| **RCF** | Richland County Foundation |
| **RCPSC** | Royal College of Physicians and Surgeons of Canada |
| **RSC** | Royal Society of Canada |
| **RWJF** | Robert Wood Johnson Foundation |
| **SAMHSA** | Substance Abuse and Mental Health Services Administration |
| **SLA** | Special Libraries Association |
| **SME** | Society of Manufacturing Engineers |
| **SOCFOC** | Sisters of Charity Foundation of Cleveland |
| **SORP** | Society of Biological Psychiatry |
| **SSHRC** | Social Sciences and Humanities Research Council of Canada |
| **SSRC** | Social Science Research Council |
| **STTI** | Sigma Theta Tau International |
| **SVP** | Social Venture Partners |
| **SWE** | Society of Women Engineers |
| **TAC** | Tennessee Arts Commission |
| **TOMF** | Tucson Osteopathic Medical Foundation |
| **TRCF** | Three Rivers Community Fund |
| **TSYSF** | Teemu Selanne Youth Sports Foundation |
| **UPS** | United Parcel Service |
| **USHMM** | United States Holocaust Memorial Museum Research Institute |
| **USIA** | United States Information Agency |
| **USAID** | United States Agency for International Development |
| **USDA** | United States Department of Agriculture |
| **USFA** | United States Fencing Association |
| **USGA** | United States Golf Association |
| **USIP** | United States Institute of Peace |
| **USTA** | United States Tennis Association |
| **UUA** | Unitarian Universalist Association |
| **WAWH** | Western Association of Woman Historians |
| **WHO** | Women Helping Others |

# GRANT PROGRAMS

**1 in 9: Long Island Breast Cancer Action Coalition Grants** 1

The Coalition's mission is to promote awareness of the breast cancer epidemic through education, outreach, advocacy, and direct support of research which is being done to find the causes of and cures for breast cancer and other related cancers. The Coalition's major goals are to: raise awareness of the epidemic of breast cancer and keep the disease in the forefront; obtain funding for research; investigate the link between pesticides and breast cancer since Long Island is a farm area; increase detection of breast cancer; and find the causes, a cure and how to prevent breast cancer.

*Geographic Focus:* New Jersey, New York, Pennsylvania

*Contact:* Geri Barish, President; (516) 374-3190; fax (516) 569-1894; info@1in9.org

*Internet:* www.1in9.org/

*Sponsor:* Long Island Breast Cancer Action Coalition
86 East Rockaway Road
Hewlett, NY 11557

**1st Touch Foundation Grants** 2

Established by Derek L. Lee in 2006 in an effort to help his daughter, Jada, giving is centered around the testing and researching of Leber's Congenital Amaurosis (LCA), an extremely rare disease which causes severe vision loss and blindness. The Foundation's primary long-range mission is to help researchers find a cure and eradicate this disease. It will also continue to provide grants for higher education, and help children and youth organizations within its local communities. There are no specific guidelines, application forms, or deadlines with which to adhere, so applicants should contact the Foundation office directly.

*Geographic Focus:* All States

*Contact:* Mark Martella, (415) 421-0535 or (818) 501-4421; Mmartella@1sttouch.org

*Internet:* 1sttouch.org/

*Sponsor:* 1st Touch Foundation
5098 Foothills BOulevard, Suite 3, #492
Roseville, CA 95747

**3M Canada National Teaching Fellowships** 3

In 1986, the Society for Teaching and Learning in Higher Education (STLHE) and 3M Canada joined to reward exceptional contributions to teaching and learning at Canadian universities. Nominations of faculty members teaching at Canadian universities for one-year teaching fellowships are invited. The program embodies the highest ideals of teaching excellence and scholarship with a commitment to enhance the educational experience of every learner. Up to ten awards, presented annually at the STLHE conference in June, recognize exemplary contributions to educational and teaching excellence in Canadian universities. The award includes lifetime membership in the Society. 3M Fellowship winners will attend an all expense paid 3-day retreat at the Fairmont Le Chateau Montebello, to celebrate and share exceptional achievements in teaching and provide an opportunity for outstanding teachers to share experiences and ideas.

*Requirements:* Awards are open to all individuals currently teaching at a Canadian university, regardless of discipline or level of appointment.

*Geographic Focus:* Canada

*Date(s) Application is Due:* Aug 31

*Contact:* Sylvia Avery, Administrator; (905) 525-9140, ext. 20130; riselays@mcmaster.ca
Ron Marken, Program Coordinator; (306) 934-0792; ron.marken@usask.ca

*Internet:* www.mcmaster.ca/3Mteachingfellowships/index.html

*Sponsor:* Society for Teaching and Learning in Higher Education (STLHE)
1280 Main Street West
Hamilton, ON L8S 4L6 Canada

**3M Company Founation Higher Education Grants** 4

Parallel to its Foundation, the 3M Company makes charitable contributions to higher education organizations directly. Support is given on a national basis. Most giving is initiated through a Request for Proposal process that allows the company to focus our giving and maximize results. The Company invests primarily in technical, engineering and business schools, and efforts that further teaching and learning. Its grants make a difference for students and their success through support of new curriculum, scholarship and retention programs. The overall goal is to help students prepare for future opportunities in an ever-changing world. Areas of interest for Higher Education include: to advance teaching and learning in science, engineering and business; to increase participation by underrepresented people in these fields; and to establish a link with K-12 education through professional development for teachers and college readiness for students. Types of support include: building and renovation; capital campaigns; donated equipment; donated land; donated products; employee volunteer services; general operating support; in-kind gifts; internship funds; program development; seed money; technical assistance; and use of facilities.

*Restrictions:* No support is offered for: political, fraternal, social, veterans, or military organizations; propaganda or lobbying organizations; religious organizations not of direct benefit to the entire community; animal-related organizations; or disease-specific organizations. No grants are given to individuals, or for electronic media promotion or sponsorships, athletic events, non-3M equipment, endowments, emergency needs, conferences, seminars, workshops, symposia, fund raising or testimonial events, travel, or film or video production; no cause-related marketing.

*Geographic Focus:* Alabama, Alaska, Arkansas, California, Connecticut, Georgia, Hawaii, Illinois, Indiana, Iowa, Kentucky, Massachusetts, Michigan, Minnesota, Missouri, Nebraska, New Jersey, New York, Ohio, South Carolina, South Dakota, Texas, Utah, Wisconsin

*Contact:* Cynthia F. Kleven, Manager of Corporate Contributions; (651) 733-0144 or (651) 736-8146; fax (651) 737-3061; cfkleven@mmm.com

*Internet:* www.3m.com/3M/en_US/gives-us/education/

*Sponsor:* 3M Company Foundation
3M Center, Building 225-01-S-23
Saint Paul, MN 55144-1000

**3M Company Foundation Community Giving Grants** 5

The 3M Company Foundation supports organizations involved with arts and culture, K-12 education, higher education, the environment, and health and human services. Special emphasis is directed toward programs designed to help prepare individuals and families for success. Fields of interest are: arts; arts education; business education; disaster relief, preparedness, and services; economics; elementary and secondary education; employment training; engineering; environmental causes; family services; federated giving programs; health care; higher education; human services; mathematics; minorities; science programs; youth development; and youth, services. Types of support include: building construction and renovation, capital campaigns, curriculum development, employee matching gifts, general operating support, in-kind gifts, program development, and scholarship funds. The foundation utilizes an invitational Request For Proposal (RFP) process for organizations located in Minneapolis and St. Paul, Minnesota, and Austin, Texas. Application forms are not required.

*Requirements:* Established 501(c)3 tax-exempt organizations in all 3M communities are eligible.

*Restrictions:* The 3M Company Foundation does not accept unsolicited proposals in St. Paul/Minneapolis, Minnesota, and Austin, Texas. No support for religious organizations, conduit agencies, political groups, fraternal organizations, social groups, or veteran organizations. No funding for hospitals, K-12 schools, military organizations, animal-related organizations, or disease-specific organizations. No grants to individuals, or for endowments, emergency operating support, advocacy and lobbying efforts, fundraising events and associated advertising, travel, publications, start-up needs, non -equipment, debt reduction, conferences, athletic events, or film or video production; no loans or investments.

*Geographic Focus:* Alabama, Alaska, Arkansas, California, Connecticut, Georgia, Hawaii, Illinois, Indiana, Iowa, Kentucky, Massachusetts, Michigan, Minnesota, Missouri, Nebraska, New York, Ohio, South Carolina, South Dakota, Texas, Utah, Wisconsin

*Contact:* Cynthia F. Kleven, Manager of Corporate Contributions; (651) 733-0144 or (651) 736-8146; fax (651) 737-3061; cfkleven@mmm.com

*Internet:* www.3m.com/3M/en_US/gives-us/community/

*Sponsor:* 3M Company Foundation
3M Center, Building 225-01-S-23
Saint Paul, MN 55144-1000

**A-T Children's Project Grants** 6

The A-T Children's Project strives to assist respected scientists in developing a clearer understanding of ataxia-telangiectasia. The Project is determined to find a timely cure, or life-improving treatments, for this serious disease. Grant awards are made through a careful and detailed selection process. The members of the Scientific Advisory Board examine each proposal and make their independent recommendations to its Board of Directors which then votes on each proposed project. Proposals from junior investigators, from scientists in related disciplines, and from individuals with innovative new ideas for A-T research are particularly encouraged, as are laboratories and teams working together from industry as well as teaching universities. The Project provides competitive grant awards for basic and translational research grants related to A-T. One- and two-year projects are funded up to a maximum total direct cost of $75,000 per year. Grants of $75,000 per year, however, are rare; grants in the $25,000 to 50,000 per year range are much more common. Budgets for up to $150,000 for a two-year project are acceptable.

*Requirements:* Proposals from junior investigators, from scientists in related disciplines, and from individuals with innovative new ideas for A-T research are particularly encouraged, as are laboratories and teams working together from industry as well as teaching universities. A Letter of Intent is not required. However, prior to submission of a full-length proposal, applicants seeking Scientific Advisory Board input may submit a Letter of Intent directly to the A-TCP Science Coordinator. This letter (not to exceed two pages) should include a brief abstract describing the proposed research, specific aims and an estimated budget. Applicants must submit an electronic copy of their Proposal in either MSWord or PDF formats to grants@atcp.org.

*Restrictions:* The sponsor does not pay for administrative overhead and indirect costs. The sponsor does not pay for institutional construction or renovation; purchase of major capital equipment other than directly needed for proposed experiments; office equipment or furniture; travel (except as required to perform the project); tuition fees; journal subscriptions; dues or memberships; and printing or publishing costs.

*Geographic Focus:* All States

*Date(s) Application is Due:* Mar 1; Sep 1
*Amount of Grant:* Up to 150,000 USD
*Contact:* Cynthia Rothblum-Oviatt, PhD, Science Coordinator; (703) 765-1223 or (954) 481-6611; fax (954) 725-1153; cynthia@atcp.org or grants@atcp.org
*Internet:* www.communityatcp.org/page.aspx?pid=3538
*Sponsor:* A-T Children's Project
5300 W. Hillsboro Boulevard, Suite 105
Coconut Creek, FL 33073

**A-T Children's Project Post Doctoral Fellowships** 7
The A-T Children's Project strives to assist respected scientists in developing a clearer understanding of ataxia-telangiectasia. Post doctorals with one year experience or less post degrees are eligible and must be nominated for this award by their principal investigator (PI). Any interested PI who would like to nominate a new post doc for the A-T Post Doctoral Fellowship Award must send a Letter of Intent to the A-T Children's Project. This letter (not to exceed two pages) should include a brief abstract describing the proposed research, specific aims and an estimated budget. Fellowship award applications are subject to the same conditions and guidelines as the Project's regular investigator initiated grants. However, the level of funding will be in the range of $30,000 to $40,000 per year for two years.
*Restrictions:* The sponsor does not pay for administrative overhead and indirect costs. The sponsor does not pay for institutional construction or renovation; purchase of major capital equipment other than directly needed for proposed experiments; office equipment or furniture; travel (except as required to perform the project); tuition fees; journal subscriptions; dues or memberships; and printing or publishing costs.
*Geographic Focus:* All States
*Date(s) Application is Due:* Mar 1; Sep 1
*Amount of Grant:* 30,000 - 40,000 USD
*Contact:* Cynthia Rothblum-Oviatt, Science Coordinator; (703) 765-1223 or (954) 481-6611; fax (954) 725-1153; cynthia@atcp.org or grants@atcp.org
*Internet:* www.communityatcp.org/page.aspx?pid=3539
*Sponsor:* A-T Children's Project
5300 W. Hillsboro Boulevard, Suite 105
Coconut Creek, FL 33073

**A.L. Mailman Family Foundation Grants** 8
The A.L. Mailman Family Foundation awards grants on the state, regional, and national levels with a focus on childcare, especially on issues of quality, caregiver training, teacher/student ratios, and connecting quality childcare to later achievements. The foundation also supports efforts to stimulate moral growth and the development of social responsibility and for research in and refinement of developmental, individualized education of young children. Types of support include curriculum development, matching/challenge grants, program development, publication, research grants, seed grants, and technical assistance. Prospective applicants should write a letter describing the proposed project. The letter should include a project summary, contact information, fit with Foundation objectives, and an estimated budget and timeframe. The Foundation board meets in April and October to authorize grants. Proposals should be sent to the foundation by June 15 for fall review and by January 15 for spring review. Letters of inquiry should be submitted by May 1 for fall proposals and by December 1 for spring proposals.
*Restrictions:* Grants do not support direct service organizations, childcare centers and schools, locally focused organizations, individuals, capital campaigns and endowments, scholarships, general operating expenses or deficit reduction, or organizations or projects outside of the United States.
*Geographic Focus:* All States
*Date(s) Application is Due:* Jan 15; May 1; Jun 15; Dec 1
*Amount of Grant:* 5,000 - 50,000 USD
*Samples:* Algebra Project for Math Talk. Cambridge, Massachusetts, $10,000 - general operating support; Americans for Responsible Solutions, Washington, D.C., $10,000 - general operating support; College Track, Oakland, California, $2,500 - general operating support.
*Contact:* Bobbie Enright, Grant Administrator; (914) 683-8089; fax (914) 686-5519; info@mailman.org
*Sponsor:* A.L. Mailman Family Foundation
707 Westchester Avenue
White Plains, NY 10604

**A.O. Smith Foundation Community Grants** 9
The A.O. Smith Foundation supports nonprofit organizations in communities where the A.O. Smith Corporation has facilities. The foundation offers funding to: elementary and secondary school projects that focus on the quality of educational programs and curriculum development; civic, cultural, and social welfare of communities; and medical research and improved local health services. Types of support include continuing support, annual campaigns, building and renovation, scholarship funds, and employee matching gifts. Proposals should describe the project's benefits and constituency, budget information including other sources of funding, and how results will be reported. Proposals to be considered for the following year's foundation budget must be received by October 30.
*Requirements:* Nonprofit organizations in company areas may apply.
*Restrictions:* Grants do not support political organizations or organizations whose chief purpose is to influence legislation.
*Geographic Focus:* All States
*Date(s) Application is Due:* Oct 30
*Amount of Grant:* 500 - 20,000 USD
*Contact:* Rita Schwalbach, Grants Officer; (414) 359-4000; fax (414) 359-4064
*Internet:* www.aosmith.com/Foundation/
*Sponsor:* A.O. Smith Foundation
P.O. Box 245008
Milwaukee, WI 53224-9510

**AAAAI ARTrust Mini Grants for Allied Health** 10
Grants are awarded to AAAAI members for research in allergy and immunology. Awards and deadlines vary according to availability. Applicants should refer to the website for the most current information and application process.
*Requirements:* Applicants must be members of the AAAAI.
*Geographic Focus:* All States
*Contact:* Jerome Schultz, Contact; (414) 272-6071; fax (414) 276-6070; jschultz@aaaai.org
*Internet:* www.aaaai.org/professional-education-and-training/grants-awards.aspx
*Sponsor:* American Academy of Allergy, Asthma, and Immunology
555 East Wells Street, Suite 1100
Milwaukee, WI 53202-3823

**AAAAI ARTrust Mini Grants for Clinical Research** 11
The ARTrust invites AAAAI members to apply for mini-grants for for projects to advance the knowledge and/or treatment of allergy, asthma, and immunology for: small clinical or research projects; new educational initiatives for professional and public audiences; and other research including equipment needs, supplemental funding, and FIT projects. Funding is available for up to $15,000 for one year.
*Requirements:* Applicants must be U.S. citizens or permanent residents and AAAAI members in good standing. Those with a membership status of Fellow, Member, or In-training must be ABAI certified or ABAI eligible. Applicants with a membership status of Allied Health may apply. Applicants must submit a one page demographic/CV, budget, and a 1-2 page proposal with the following as section headers: hypothesis (if applicable); rational; methods; and expected outcome. Proposals are submitted online and must not exceed two pages.
*Restrictions:* Funding is not available for indirect costs, or the salary and benefits for the project's principal investigator. Individuals currently receiving an ARTrust grant are not eligible.
*Geographic Focus:* All States
*Date(s) Application is Due:* Nov 12
*Amount of Grant:* Up to 15,000 USD
*Contact:* Jerome Schultz, Contact; (414) 272-6071; fax (414) 276-6070; jschultz@aaaai.org
*Internet:* www.aaaai.org/Aaaai/media/MediaLibrary/PDF%20Documents/Education%20and%20Training/Awards/2013-ARTrust-Mini-Grants.pdf
*Sponsor:* American Academy of Allergy, Asthma, and Immunology
555 East Wells Street, Suite 1100
Milwaukee, WI 53202-3823

**AAAAI ARTrust Mini Grants in Faculty Development** 12
The Faculty Development Grant offers an individual three years of financial support to concentrate his/her research efforts in either basic or clinical allergy and immunology. The objectives of this funding are to recognize and support young researchers and promote the specialty of allergy/immunology. The recipient is expected to participate in a seminar at the AAAAI annual meeting, where awardees present their research.
*Requirements:* The award is open to researchers of junior faculty status—either instructor, assistant professor, or the equivalent grade. Funding is directed toward MDs, but in unusual circumstances, PhDs will be considered. Applicants must be U.S. or Canadian citizens or permanent residents who are AAAAI members (or have applied for membership by September 1).
*Restrictions:* Past recipients are ineligible.
*Geographic Focus:* All States
*Date(s) Application is Due:* Nov 15
*Amount of Grant:* 150,000 USD
*Contact:* Jerome Schultz, Contact; (414) 272-6071; fax (414) 276-6070; jschultz@aaaai.org
*Internet:* www.aaaai.org/global/ARTrust/award-recipients.aspx
*Sponsor:* American Academy of Allergy, Asthma, and Immunology
555 East Wells Street, Suite 1100
Milwaukee, WI 53202-3823

**AAAAI Distinguished Clinician Award** 13
The Distinguished Clinician Award is given to a person or persons whose clinically relevant activities have made a significant constructive impact on the field of allergy and clinical immunology. The awardee(s) should have a primary commitment to patient care of clinical investigation focused on diagnosis, treatment and prevention of asthma and allergic disease. The individual(s) should have a record of substantial clinical research as evidenced by papers published in referenced journals (such as JACI); evidence of scholarship by presentations on the programs of national or international meetings, published books, editorship of supplements to referenced journals or equivalent activity. Especially desirable, but not mandatory, the individual should show dedicated service to the field of allergy and clinical immunology, e.g. service to the American Board of Allergy/Immunology, a voluntary foundation or government agency, or public education, outreach, or community activities in the field of allergy\immunology. Alternatively, the recipient(s) could have worked actively in the AAAAI or similar professional society, serving on major committees, participating in workshops, and/or serving as an officer in the AAAAI.
*Requirements:* Nomination forms are available at the AAAAI website. The nomination letter, along with a copy of the nominee's CV and two letters of recommendation, should be mailed to the Awards Subcommittee at the AAAAI corporate office.
*Geographic Focus:* All States
*Date(s) Application is Due:* Jan 14

*Contact:* Dr. Stuart L. Abramson, Chair; (414) 272-6071; fax (414) 276-6070
*Internet:* www.aaaai.org/professional-education-and-training/grants—-awards/past-aaaail-honorary-award-recipients.aspx#dca
*Sponsor:* American Academy of Allergy, Asthma, and Immunology
555 East Wells Street, Suite 1100
Milwaukee, WI 53202-3823

**AAAAI Distinguished Layperson Award** 14

The Distinguished Layperson Award is granted once a year to the citizen who has made the most substantial contribution to the field of allergy. The AAAAI recognizes that many scientists and physicians who are not clinical allergists have made contributions to the understanding and practice of the specialty. Non-medical citizens have worked in various ways to promote the field. Many are members of voluntary health organizations dedicated to public education and fundraising activities to support research. The nomination instructions are available at the AAAAI website.
*Requirements:* The recipient should be a non-medical person who has made a substantial contribution in service or substance to education, research, or to the practice. The contribution need not be directly related to programs sponsored by the AAAAI. To be eligible, a candidate for the award should still be living at the time the award is made. At least one member of the AAAAI should have first-hand knowledge of the candidate's contribution.
*Restrictions:* Employees of physicians' or health institutions shall not be eligible for the award unless their contribution has been made outside of their duties of employment.
*Geographic Focus:* All States
*Date(s) Application is Due:* Jan 14
*Contact:* Dr. Stuart L. Abramson, Chair; (414) 272-6071; fax (414) 276-3349
*Internet:* www.aaaai.org/professional-education-and-training/grants—-awards/past-aaaail-honorary-award-recipients.aspx#dla
*Sponsor:* American Academy of Allergy, Asthma, and Immunology
555 East Wells Street, Suite 1100
Milwaukee, WI 53202-3823

**AAAAI Distinguished Scientist Award** 15

The Distinguished Scientist Award is presented to recognize scientific contributions to the field of allergy/immunology that have advanced research in the specialty, in addition to leadership contributions to the specialty. Nomination forms are available at the AAAAI website.
*Requirements:* Nominations should include a letter of nomination, with a copy of the nominee's CV and two letters of recommendation.
*Geographic Focus:* All States
*Date(s) Application is Due:* Jan 14
*Contact:* Dr. Stuart L. Abramson, Chair; (414) 272-6071; fax (414) 276-3349
*Internet:* www.aaaai.org/professional-education-and-training/grants—-awards/past-aaaail-honorary-award-recipients.aspx#dsca
*Sponsor:* American Academy of Allergy, Asthma, and Immunology
555 East Wells Street, Suite 1100
Milwaukee, WI 53202-3823

**AAAAI Distinguished Service Award** 16

The Distinguished Service Award recognizes long and distinguished service to the AAAAI rather than primarily for scientific achievement. Past-Presidents are eligible for this award. The number of Distinguished Service Awards should be limited to one per year.
*Requirements:* Nomination forms are available at the AAAAI website. Nomination letters must include the nominee's CV and two letters of recommendation.
*Geographic Focus:* All States
*Date(s) Application is Due:* Jan 14
*Contact:* Dr. Stuart L. Abramson, Chair; (414) 272-6071; fax (414) 276-3349
*Internet:* www.aaaai.org/professional-education-and-training/grants—-awards/past-aaaail-honorary-award-recipients.aspx#dsa
*Sponsor:* American Academy of Allergy, Asthma, and Immunology
555 East Wells Street, Suite 1100
Milwaukee, WI 53202-3823

**AAAAI Fellows-in-Training Abstract Award** 17

Fellows-in-Training are eligible for the Abstract Award. Accepted abstracts are presented at the Annual Meeting. Authors of accepted abstracts will receive a reduced registration rate for the National Conference. Additional information is available at the website, including instructions for online abstract submission.
*Requirements:* Abstracts must be submitted using the online abstract submission process. Submission by members and non-members are welcome, and participation is open to health professionals in any field. However, some sections require sponsors for papers whose authors do not include a member of the respective section. The text of each abstract is limited to 450 words. Images and tables are permitted. Images will not count toward the word limit, but any words or numbers within an included table will be counted accordingly. All authors will be required to complete conflict of interest disclosures. Authors whose abstracts are accepted for presentation will be expected to attend the meeting and make their presentation.
*Restrictions:* An abstract may be submitted to one section or council only.
*Geographic Focus:* All States
*Date(s) Application is Due:* Sep 12
*Contact:* Mariana Duran, Contact; (414) 272-6071; fax (414) 276-6070; mduran@aaaai.org
*Internet:* www.aaaai.org/professional-education-and-training/fellows-in-training.aspx
*Sponsor:* American Academy of Allergy, Asthma, and Immunology
555 East Wells Street, Suite 1100
Milwaukee, WI 53202-3823

**AAAAI Fellows-in-Training Grants** 18

Fellowship status in the AAAAI recognizes the professional achievements in a member's A/I specialty and provides several benefits available only to Fellows. Benefits include: voting, proposing motions, and holding office on the Board of Directors; recognition as a leader in the specialty's premier A/I society; first opportunity to register for seminars, workshops, and courses; complimentary subscriptions to professional journals and magazines; and networking opportunities with other A/I professionals around the world. Applications are accepted year round and reviewed in March, June, and November of each year. Applicants are notified of their status within four weeks after the meeting. The application is available at the corporate website or www.aaaai.org/Aaaai/media/MediaLibrary/PDF%20Documents/Member%20Applications/Fellowship-Application.pdf. No application fee is required for current members.
*Requirements:* All applications must be accompanied by a letter of intent and CV from the applicant. Applicants must also be sponsored by one Fellow of the AAAAI. Applicants should meet the following criteria for consideration: full membership with AAAAI for at least three years; certification by a subspeciality board, the American Board of Allergy and Immunology, or its foreign equivalent; and continuing training and experience in the field of A/I.
*Geographic Focus:* All States, All Countries
*Contact:* Mariana Duran, Fellows-in-Training Contact; (414) 272-6071; fax (414) 276-6070; mduran@aaaai.org or membership@aaaai.org
*Internet:* www.aaaai.org/professional-education-and-training/fellows-in-training.aspx
*Sponsor:* American Academy of Allergy, Asthma, and Immunology
555 East Wells Street, Suite 1100
Milwaukee, WI 53202-3823

**AAAAI Fellows-in-Training Travel Grants** 19

The AAAAI will provide funding to assist Fellows-in-Training (FITs) and non-faculty PhD Post Doc FITS in Allergy/Immunology (A/I) to attend AAAAI's Annual Meeting. Grant awards vary according to the level of the applicant. First year in-training FITs without an accepted abstract for presentation will receive up to $800. Third or fourth year in-training FITs without an accepted abstract will receive up to $650. In-Training or Post Doc FITs who are first author and present their accepted abstract will receive up to $1,100. Candidates will apply through the online application process at the AAAAI website.
*Requirements:* First, second, third, or fourth year FITS enrolled in an A/I program approved by the ACGME or by the Royal College of Physicians and Surgeons of Canada may apply. Accepted abstracts are not required (although funding will typically be higher for accepted abstracts). Post-Doctoral Research FITs may apply if their training has taken place in the U.S. or Canada, and involves A/I related research. Abstracts must be submitted and accepted for presentation at the annual meeting. Fellows-in-Training in a medical or surgical specialty other A/I may apply if they have submitted an abstract and it has been accepted for presentation.
*Restrictions:* Only AAAAI members may apply. A maximum of three FIT awards can be received during the fellowship training years.
*Geographic Focus:* All States
*Date(s) Application is Due:* Nov 2
*Amount of Grant:* 650 - 1,100 USD
*Contact:* Mariana Duran, Fellows-in Training Contact; (414) 272-6071; fax (414) 276-6070; mduran@aaaai.org
*Internet:* www.aaaai.org/professional-education-and-training/fellows-in-training.aspx
*Sponsor:* American Academy of Allergy, Asthma, and Immunology
555 East Wells Street, Suite 1100
Milwaukee, WI 53202-3823

**AAA Foundation for Traffic Safety Grants** 20

The foundation sponsors research that not only identifies critical traffic safety problems, but also searches for underlying causes and possible solutions, as well as identifying early trends and offering solutions before the general public is aware of the problem. The foundation continues research into ways to help the growing numbers of teen drivers and older drivers evaluate and improve their driving performance. A sampling of recent research includes drunk driving, seeking additional solutions; seated for safety; headlight glare countermeasures; supplemental transportation programs for seniors; distracted driving, phase I; and longer combination vehicle safety study. Prior to submitting unsolicited proposals, applicants may email a one-page preproposal to gauge the foundation's general interest in the proposed topic.
*Requirements:* Research institutions and nonprofit organizations with 501(c)3 status are eligible.
*Restrictions:* The foundation does not fund research to develop new devices. No grants are made for community action initiatives or other purely local traffic safety programs.
*Geographic Focus:* All States
*Samples:* Guidebook for Observing Occupant Restraint System Use; Older Driver Involvement in Injury Crashes; Drunk Driving—Seeking Additional Solutions; Children Seated for Safety
*Contact:* Fairley Mahlum, Communications Director; (202) 638-5944; fax (202) 638-5943; info@aaafoundation.org
*Internet:* www.aaafoundation.org/about_us/index.cfm
*Sponsor:* AAA Foundation for Traffic Safety
607 14th Street NW, Suite 201
Washington, D.C. 20005

**AAA Minority Dissertation Fellowships** 21

The American Anthropological Association invites minority doctoral candidates in anthropology to apply for a nonrenewable dissertation writing fellowship of $10,000. Students of any subfield or specialty in anthropology will receive equal consideration. Dissertation topics in all areas of the discipline are welcome. Doctoral students who require financial assistance to complete the write-up phase of the dissertation are urged to apply. The dissertation proposal must be approved by the applicant's committee prior to application.
*Requirements:* An applicant must be a U.S. citizen; a member of an historically underrepresented ethnic minority group; enrolled in a full-time academic program leading to a doctoral degree in anthropology at the time of application; admitted to degree candidacy before the dissertation fellowship is awarded; and a member of the American Anthropological Association.
*Geographic Focus:* All States
*Date(s) Application is Due:* Feb 15
*Amount of Grant:* 10,000 USD
*Contact:* Resat Mursaloglu, Staff Accountant; (703) 528-1902, ext. 1160; fax (703) 528-3546; rmursaloglu@americananthro.org
*Internet:* www.americananthro.org/ParticipateAndAdvocate/Content.aspx?ItemNumber=1621
*Sponsor:* American Anthropological Association
2300 Clarendon Boulevard, Suite 1301
Arlington, VA 22201

**AAAS Award for Public Engagement with Science** 22

AAAS invites nominations for this annual award, which honors working scientists and engineers who make outstanding contributions to the popularization of science. It is intended to encourage technical professionals to communicate with the public by popularizing their own work. Such efforts can include publications or broadcasts, lectures, museum presentations or exhibit designs, or any other outreach project. The Award shall be given annually to scientists or engineers who, while working in their fields, have also contributed substantially to public understanding of science and technology. A monetary prize of $5,000, a commemorative plaque, complimentary registration, and reimbursement for reasonable travel and hotel expenses to attend the AAAS Annual Meeting to receive the prize are given to the recipient. Nominators are encouraged to identify candidates whose contributions reach broad audiences that include women, minorities, disabled persons, and senior citizens.
*Requirements:* Eligible individuals include scientists and engineers (individual or a small group) from all disciplines (including social sciences and medicine) engaged in research, teaching, and related activities that have contributed substantially to the public's understanding of science or technology. Nominations should include name, position, institution, professional address and phone, and home address and phone of the candidate; name, position, institution, and professional address and phone of the nominator; a summary of the action or actions that form the basis for the nomination (about 250 words); a longer statement (no more than three pages) providing additional details of the actions for which the candidate is nominated, including one paragraph that describes the nominee's impact on the general public; the candidate's vita (three-page maximum); two letters of support; and at least one representative sample that illustrates the nominee's contribution. Books, videotapes, brochures, magazine articles, or other materials are welcome.
*Restrictions:* Only materials produced for general audiences, as opposed to professional or trade audiences, will be considered. Employees of the AAAS are not eligible. All materials become the property of AAAS.
*Geographic Focus:* All States
*Date(s) Application is Due:* Aug 1
*Amount of Grant:* 5,000 USD
*Contact:* Rebekah Corlew, (202) 326-6445; fax (202) 371-9849; rcorlew@aaas.org
*Internet:* https://www.aaas.org/page/aaas-award-public-engagement-science
*Sponsor:* American Association for the Advancement of Science
1200 New York Avenue NW
Washington, D.C. 20005-3920

**AAAS Award for Science Diplomacy** 23

Through the Award for Science Diplomacy, AAAS, in collaboration with its affiliated organizations, seeks to recognize an individual or a limited number of individuals working together in the scientific or engineering community for making an outstanding contribution to furthering international cooperation in science and engineering. Nominations for the award may be originated by anyone. CAIP administers the screening and selection process for the awarding of the AAAS Award for Science Diplomacy. The ISC award is presented each February at the AAAS Annual Meeting. A monetary prize of $5,000, a commemorative plaque, complimentary registration, and reimbursement for reasonable travel and hotel expenses to attend the AAAS annual meeting are given to the recipient.
*Requirements:* Nominations should be typed and should include the following information: the nominator's name, address, and phone number; the nominee's name and title, institutional affiliation, and address; a summary of the actions that form the basis for the nomination (about 250 words); a longer statement (no more than three pages) providing additional details of the actions for which the candidate is nominated; at least two letters of support with addresses and phone numbers; and the candidate's vita (no more than three pages) containing professional positions held. Any documentation—books, articles, etc.—that illuminates the significance of the nominee's achievements may also be submitted.
*Restrictions:* The award is open to all regardless of nationality or citizenship. Nominees must be living at the time of their nomination. Any individual or small group in the scientific and engineering community that has contributed substantially to the understanding or development of science or engineering across national boundaries is eligible for this award.
*Geographic Focus:* All States, All Countries
*Date(s) Application is Due:* Sep 1
*Amount of Grant:* 5,000 USD
*Contact:* Marga Gual Soler, Senior Project Director; (202) 326-6659 or (202) 326-6650; fax (202) 289-4958; mgsoler@aaas.org or diplomacy@aaas.org
*Internet:* www.aaas.org/aboutaaas/awards/int/
*Sponsor:* American Association for the Advancement of Science
1200 New York Avenue NW
Washington, D.C. 20005-3920

**AAAS Award for Scientific Freedom and Responsibility** 24

The Award for Scientific Freedom and Responsibility is presented to honor scientists and engineers whose exemplary actions have served to foster scientific freedom and responsibility and recognizes scientists and engineers who have: acted to protect the public's health, safety, or welfare; or focused public attention on important potential impacts of science and technology on society by their responsible participation in public policy debates; or established important new precedents in carrying out the social responsibilities or in defending the professional freedom of scientists and engineers. Nominations should include nominator's name, address, and phone number; and the name and address of the nominee; a summary of the action or actions that form the basis for the nomination (about 250 words); a longer statement (no more than three pages) providing additional details of the actions for which the candidate is nominated; at least two letters of support with addresses and phone numbers; and the candidate's vita (no more than three pages). Any documentation—books, articles, etc.—that illuminates the significance of the nominee's achievements may also be submitted.
*Requirements:* The award is open to all regardless of nationality or citizenship. Eligible nominees are scientists and engineers who have acted to protect the public's health, safety, or welfare; or to focus public attention on potentially serious impacts of science and technology on society by their responsible participation in public policy debates; or to establish important new precedents in social responsibility or in defending the professional freedom of scientists and engineers.
*Restrictions:* All materials become the property of AAAS.
*Geographic Focus:* All States, All Countries
*Date(s) Application is Due:* Aug 1
*Amount of Grant:* 5,000 USD
*Contact:* Deborah Runkle, (202) 326-6794; fax (202) 289-4950; drunkle@aaas.org
*Internet:* https://www.aaas.org/page/aaas-award-scientific-freedom-and-responsibility
*Sponsor:* American Association for the Advancement of Science
1200 New York Avenue NW
Washington, D.C. 20005-3920

**AAAS Community Engagement Fellowships** 25

The AAAS Community Engagement Fellows Program, supported by the Alfred P. Sloan Foundation, is designed to improve online collaboration and community-building in science. The program has three broad goals: professionalize and institutionalize the role of community management within scientific organizations; provide professional development resources to individuals who manage communities and collaborations in research organizations and scientific associations; and collect and disseminate knowledge about building strong collaborations and communities, including the first landscape survey of scientific community engagement professionals. Community engagement is an emerging profession, especially in science. The individuals who carry out these roles may be doing so in informal ways, sometimes in addition to other responsibilities. They may not even recognize themselves as community managers. Frequently they are the only person in the role in their organization, which can leave them feeling isolated and unsupported. The AAAS Community Engagement Fellows Program will provide training, peer support, and a shared knowledge base for Fellows. Training will include a week-long introductory course in January, followed by monthly webinars and mid-year and end-of-year meetings.
*Geographic Focus* All States, District of Columbia, Guam, Marshall Islands, Northern Mariana Islands, Puerto Rico, U.S. Virgin Islands, American Samoa
*Contact:* Kira E. Mock, Program Manager; (202) 326-6700; fax (202) 289-4950; kmock@aaas.org or fellowships@aaas.org
*Internet:* https://www.aaas.org/cefp/about
*Sponsor:* American Association for the Advancement of Science
1200 New York Avenue NW
Washington, D.C. 20005-3920

**AAAS Early Career Award for Public Engagement with Science** 26

The AAAS Early Career Award for Public Engagement with Science, established in 2010, recognizes early-career scientists and engineers who demonstrate excellence in their contribution to public engagement with science activities. A monetary prize of $5,000, a commemorative plaque, complimentary registration to the AAAS Annual Meeting, and reimbursement for reasonable hotel and travel expenses to attend the AAAS Annual Meeting to receive the prize are given to the recipient. For the purposes of this award, public engagement activities are defined as the individual's active participation in efforts to engage with the public on science- and technology-related issues and promote meaningful dialogue between science and society, as highlighted in this video. The award will be given at the AAAS Annual Meeting.
*Requirements:* Nominee must be an early-career scientist or engineer in academia, government or industry actively conducting research in any scientific discipline (including social sciences and medicine). Groups or institutions will not be considered

for this award. AAAS employees are ineligible. One scientist or engineer will be chosen to receive the award on an annual basis. Nominee will have demonstrated excellence in his/her contribution to public engagement with science activities, with a focus on interactive dialogue between the individual and a non-scientific, public audience(s).
*Geographic Focus:* All States
*Date(s) Application is Due:* Aug 1
*Amount of Grant:* 5,000 USD
*Contact:* Rebekah Corlew, (202) 326-6656; fax (202) 289-4950; rcorlew@aaas.org
*Internet:* https://www.aaas.org/page/aaas-early-career-award-public-engagement-science
*Sponsor:* American Association for the Advancement of Science
1200 New York Avenue NW
Washington, D.C. 20005-3920

**AAAS Entry Point Internships** 27

Entry Point is a program of the American Association for the Advancement of Science (AAAS), offering outstanding internship opportunities for students with disabilities in science, engineering, mathematics, computer science, and some fields of business. To meet the challenge of the competitive global economy in the new millennium, private industry and government research agencies must expand the pool of technical talent. The American Association for the Advancement of Science (AAAS) has developed unique partnerships with IBM, NASA, Merck, NOAA, Google, Lockheed Martin, CVS, NAVAIR, Pfizer, Infosys, and university science laboratories to meet their human resources needs. Working with its partners, AAAS identifies and screens undergraduate and graduate students with disabilities who are pursuing degrees in science, engineering, mathematics, computer science, and some fields of business, and places them in paid summer internships.
*Requirements:* Students must: be full-time undergraduate or graduate student, with a disability, majoring in a science or engineering field; maintain a B average or better; and be a U.S. citizen or have a right-to-work permit.
*Geographic Focus* All States, District of Columbia, Guam, Marshall Islands, Northern Mariana Islands, Puerto Rico, U.S. Virgin Islands, American Samoa
*Date(s) Application is Due:* Feb 15
*Contact:* Laureen Summers, Project Director; (202) 326-6649; fax (202) 371-9849; lsummers@aaas.org or entrypoint@aaas.org
*Internet:* https://www.aaas.org/program/entrypoint
*Sponsor:* American Association for the Advancement of Science
1200 New York Avenue NW
Washington, D.C. 20005-3920

**AAAS Eppendorf and Science Prize for Neurobiology** 28

The Eppendorf and Science Prize for Neurobiology acknowledges the increasingly active and important role of neurobiology in advancing our understanding of the functioning of the brain and the nervous system — a quest that seems destined for dramatic expansion in the coming decades. This international prize, established in 2002, encourages the work of promising young neurobiologists by providing support in the early stages of their careers. It is awarded annually for the most outstanding neurobiological research by a young scientist, as described in a 1,000-word essay based on research performed during the past three years. The winner is awarded money and publication of his or her essay in 'Science'. The essay and those of up to three finalists are also published on 'Science Online'. The award is announced and presented at a ceremony at the annual meeting of the Society for Neuroscience. Eppendorf provides financial support to help enable the grand prize winner to attend the meeting. The annual deadline for application submission is June 15.
*Requirements:* Entrants must be a neurobiologist with an advanced degree and not older than 35 years. The research described in the entrant's essay must be based on the methods of molecular and cell biology. The entrant must have performed or directed the work described in the essay. The research must have been performed during the previous three years.
*Restrictions:* Employees of Eppendorf AG, its subsidiaries, Science and AAAS, and their relatives are not eligible for the prize.
*Geographic Focus* All States, District of Columbia, Guam, Marshall Islands, Northern Mariana Islands, Puerto Rico, U.S. Virgin Islands, American Samoa
*Date(s) Application is Due:* Jun 15
*Amount of Grant:* 25,000 USD
*Contact:* Executive Editor; (202) 326-6550; fax (202) 289-7562; eppendorfscienceprize@aaas.org
*Internet:* www.sciencemag.org/eppendorf-science-prize-neurobiology
*Sponsor:* American Association for the Advancement of Science
1200 New York Avenue NW
Washington, D.C. 20005-3920

**AAAS Kavli Science Journalism Awards** 29

The Kavli Science Journalism awards, which are cosponsored by AAAS and Johnson and Johnson Pharmaceutical Research & Development, recognize outstanding reporting for a general audience and honor individuals (rather than institutions, publishers or employers) for their coverage of the sciences, engineering and mathematics. The winning journalists have helped to foster the public's understanding and appreciation of science. Independent screening and judging committees select the winning journalists and their entries based on scientific accuracy, initiative, originality, clarity of interpretation and value in fostering a better understanding of science by the public. One award in each of six categories is given annually: newspapers with daily circulation more than 100,000, newspapers with circulation of less than 100,000, general circulation magazines, radio, and television, and online. Additional International category - Children's Science News (Provide English translations of foreign-language entries whenever possible).
*Requirements:* Persons other than the author may submit entries in accordance with the rules of the competition. The contest is open to journalists who have published or broadcast in media outlets accessible to the general public. Except for the Children's Science News award, which is open to international media outlets, the submitted materials must have been published, broadcast or posted online within the United States by a U.S.-based news organization. Print articles must be available to the public by subscription or newsstand sales.
*Restrictions:* Items must be: exclusively concerning health or medical treatment; published originally in AAAS publicaitons or produced by AAAS; and by employees of AAAS or Johnson and Johnson Pharmaceutical Research Development. Winners from the previous year are not eligible. Individuals winning three times are no longer eligible.
*Geographic Focus* All States, District of Columbia, Guam, Marshall Islands, Northern Mariana Islands, Puerto Rico, U.S. Virgin Islands, American Samoa
*Date(s) Application is Due:* Aug 1
*Amount of Grant:* 3,000 USD
*Contact:* Awards Coordinator; (202) 326-6434; fax (202) 789-0455; media@aaas.org
*Internet:* www.aaas.org/aboutaaas/awards/sja/
*Sponsor:* American Association for the Advancement of Science
1200 New York Avenue NW
Washington, D.C. 20005-3920

**AAAS L'Oréal USA Fellowships for Women in Science** 30

The L'Oréal USA Fellowships for Women in Science program is a national awards program that annually recognizes and rewards five U.S.-based women researchers at the beginning of their scientific careers. Recipients each receive up to $60,000 that must be put towards their postdoctoral research. Launched in 2003 as the U.S. component of the L'Oréal-UNESCO for Women in Science International Fellowship program, the U.S. Fellowships aim to: raise awareness of the contribution of women to the sciences; and identify exceptional female researchers in the U.S. to serve as role models for younger generations. The application period ends on February 2.
*Requirements:* All applicants must: have a conferred Ph.D. and have started in their postdoctoral position by February 2 of the application year; maintain the status of postdoctoral researcher throughout the fellowship year; be American born, naturalized citizen or permanent resident; be affiliated with a U.S. based academic or research institution; plan to conduct their postdoctoral studies and research primarily in the U.S; be involved in basic research in the life and physical/material sciences, engineering and technology, computer science and mathematics; commit to at least twenty hours of activity in support of women and girls in science (e.g. mentoring, classroom visits, media, events); and be available in late October for the For Women in Science Awards Week.
*Geographic Focus* All States, District of Columbia, Guam, Marshall Islands, Northern Mariana Islands, Puerto Rico, U.S. Virgin Islands, American Samoa
*Date(s) Application is Due:* Feb 2
*Contact:* Kira E. Mock, Program Manager; (202) 326-6700; fax (202) 289-4950; kmock@aaas.org or fellowships@aaas.org
*Internet:* https://www.aaas.org/page/l-or-al-usa-fellowships-women-science
*Sponsor:* American Association for the Advancement of Science
1200 New York Avenue NW
Washington, D.C. 20005-3920

**AAAS Martin and Rose Wachtel Cancer Research Award** 31

AAAS and Science Translational Medicine invite applications for the annual AAAS Martin and Rose Wachtel Cancer Research Award. This award, funded by an endowment established through a generous bequest from Martin L. Wachtel, will honor an early-career investigator who has performed outstanding work in the field of cancer research. The award winner will be invited to deliver a public lecture on his or her research and will receive a cash award of $25,000. The award winner's Essay will be published as a Perspective in Science Translational Medicine. The deadline for application is February 15.
*Requirements:* Entrants must be a researcher in the field of cancer, with an advanced degree (Ph.D., M.D., D.V.M.) received within the last 10 years. The research must have been performed during the previous 10 years. The entrant must have performed or personally directed the work described in the Essay.
*Restrictions:* Employees of AAAS and their relatives are not eligible for the prize.
*Geographic Focus* All States, District of Columbia, Guam, Marshall Islands, Northern Mariana Islands, Puerto Rico, U.S. Virgin Islands, American Samoa
*Date(s) Application is Due:* Feb 15
*Amount of Grant:* 25,000 USD
*Contact:* Prize Coordinator; (202) 326-6617 or (202) 326-6700; fax (202) 289-4950; wachtelprize@aaas.org
*Internet:* https://www.aaas.org/page/aaas-martin-and-rose-wachtel-cancer-research-award
*Sponsor:* American Association for the Advancement of Science
1200 New York Avenue NW
Washington, D.C. 20005-3920

**AAAS Mass Media Science and Engineering Fellowships** 32

The highly competitive Mass Media Science and Engineering Fellowship program strengthens the connections between scientists and journalists by placing advanced undergraduate, graduate, and post-graduate level scientists, engineers and mathematicians at media organizations nationwide. Fellows have worked as reporters, editors, researchers, and production assistants at such media outlets as the Los Angeles Times, National Public Radio, The Washington Post, WIRED, and Scientific American. The Mass Media Fellows use their academic training in the sciences as they research, write and report today's headlines, sharpening their abilities to communicate complex scientific issues to non-specialists. Participants come in knowing the importance of translating their work for the public, but they leave with the tools and the know-how to accomplish this important goal. For 10 weeks during the summer, the AAAS Mass

Media Science & Engineering Fellows collaborate with media professionals at radio and television stations, newspapers, and magazines. As part of their job, the student-scientists and their host-journalists strive to make science news easy for the public to understand. The fellowship program is designed to enhance coverage of science-related issues in the media in order to improve public understanding and appreciation of science and technology. Fellows have the opportunity to observe and participate in the process by which events and ideas become news, improve their communication skills by learning to describe complex technical subjects in a manner understandable to the lay public, and increase their understanding of editorial decision making and the way in which information is effectively disseminated. Fellows are provided a weekly stipend of $500 as well as travel expenses to and from AAAS and their sites.
*Requirements:* Applicants must be enrolled as students (upper level undergraduate or graduate) or postdoctoral trainees at a university — or within one year of a completed degree — in the life, physical, health, engineering, computer, or social sciences or mathematics and related fields. Applicants must be U.S. citizens or already hold visas that allow them to receive payment for work during the summer. Successful applicants are required to attend an orientation at AAAS headquarters at the beginning of the summer (early June) and a wrap-up session at the end of the summer (mid-August).
*Restrictions:* Students enrolled in English, journalism, science journalism, or other non-technical fields are not eligible for the AAAS Mass Media Fellowship.
*Geographic Focus* All States, District of Columbia, Guam, Marshall Islands, Northern Mariana Islands, Puerto Rico, U.S. Virgin Islands, American Samoa
*Amount of Grant:* 5,000 USD
*Contact:* Rebekah Corlew, (202) 326-6794 or (202) 326-6700; fax (202) 289-4950; rcorlew@aaas.org
*Internet:* https://www.aaas.org/page/about-1
*Sponsor:* American Association for the Advancement of Science
1200 New York Avenue NW
Washington, D.C. 20005-3920

**AAAS Mentor Awards** 33

The two categories of the AAAS Mentor Awards (Lifetime Mentor Award and Mentor Award) both honor individuals who during their careers demonstrate extraordinary leadership to increase the participation of underrepresented groups in science and engineering fields and careers. These groups include: women of all racial or ethnic groups; African American, Native American, and Hispanic men; and people with disabilities. Both awards recognize an individual who has mentored and guided significant numbers of students from underrepresented groups to the completion of doctoral studies or who has impacted the climate of a department, college, or institution to significantly increase the diversity of students pursuing and completing doctoral studies. Nominations should include name, position, institution, professional address and phone, and home address and phone of the candidate; name, position, institution, and professional address and phone of the nominator; a summary of the actions that form the basis for the nomination (about 250 words); a letter of nomination that enumerates the ways in which the person reflects the purpose of this award; the candidate's vita (three-page maximum); the total number of students the candidate mentored at the bachelor's or master's level who went on to the doctoral level at other institutions, and the total number of underrepresented students the candidate mentored at the doctoral level; a maximum of five supporting letters from students and three supporting letters from colleagues representative of the different spheres in which the candidate has demonstrated effort, results, and commitment.
*Requirements:* The award is open to all regardless of nationality or citizenship. Nominees must be living at the time of their nomination. Commitment and extraordinary effort may be demonstrated by the number and diversity of students mentored; assisting students to present and publish their work, to find financial aid, and to provide career guidance; providing psychological support, encouragement, and essential strategies for life in the scholarly community; and continued interest in the individual's professional advancement. The Lifetime Mentor award winner will have served in the role of mentor for 25 or more years. The Mentor Award winner will have served in the role of mentor for less than 25 years.
*Geographic Focus:* All States, All Countries
*Date(s) Application is Due:* Aug 15
*Amount of Grant:* 5,000 USD
*Contact:* Ysabel Woody, Mentor Awards Coordinator; (202) 326-6670; fax (202) 371-9849; ywoody@aaas.org
*Internet:* https://www.aaas.org/page/aaas-mentor-awards
*Sponsor:* American Association for the Advancement of Science
1200 New York Avenue NW
Washington, D.C. 20005-3920

**AAAS Newcomb Cleveland Prize** 34

The Association's oldest award, the AAAS Newcomb Cleveland Prize, supported by The Fodor Family Trust, was established in 1923 with funds donated by Newcomb Cleveland of New York City and was originally called the AAAS Thousand Dollar Prize. It is now known as the AAAS Newcomb Cleveland Prize, and its value has been raised to $25,000. The winner also receives a commemorative plaque, complimentary registration and reimbursement for reasonable travel and hotel expenses to attend the AAAS Annual Meeting. The prize is awarded to the author or authors of an outstanding paper published in the Research Articles or Reports sections of Science. Each annual contest starts with the first issue of June and ends with the last issue of the following May.
*Requirements:* An eligible paper is one that includes original research data, theory, or synthesis; is a fundamental contribution to basic knowledge or is a technical achievement of far-reaching consequence; and is a first-time publication of the author's own work. Reference to pertinent earlier work by the author may be included to give perspective. Readers of 'Science' are invited to nominate papers appearing in the Articles, Research Articles, or Reports sections. Nominations must be typed and provide the following information: title of paper, author's name, date of issue in which it was published, page number, and a brief statement of justification for nomination.
*Restrictions:* Self-nominations will not be accepted.
*Geographic Focus* All States, District of Columbia, Guam, Marshall Islands, Northern Mariana Islands, Puerto Rico, U.S. Virgin Islands, American Samoa
*Date(s) Application is Due:* Jun 30
*Amount of Grant:* 25,000 USD
*Contact:* Science Editorial Office; (202) 326-6560 or (202) 326-6675; fax (202) 289-7562; acrawfor@aaas.org
*Internet:* www.aaas.org/aboutaaas/awards/newcomb/index.shtml
*Sponsor:* American Association for the Advancement of Science
1200 New York Avenue NW
Washington, D.C. 20005-3920

**AAAS Philip Hauge Abelson Award** 35

The Philip Hauge Abelson Award was established by the AAAS Board of Directors and is presented at the AAAS Annual Meeting. It is awarded annually to either a public servant, in recognition of sustained exceptional contributions to advancing science, or to a scientist whose career has been distinguished both for scientific achievement and for other notable services to the scientific community. The annual nomination deadline is August 1.
*Requirements:* The award is open to all regardless of nationality or citizenship. Nominees must be living at the time of their nomination. Any individual or small group in the scientific and engineering community that has contributed substantially to the understanding or development of science or engineering across national boundaries is eligible for this award.
*Restrictions:* Each nomination must be seconded by at least two other AAAS members. The winner will be selected by a five-member judging panel.
*Geographic Focus* All States, District of Columbia, Guam, Marshall Islands, Northern Mariana Islands, Puerto Rico, U.S. Virgin Islands, American Samoa
*Date(s) Application is Due:* Aug 1
*Amount of Grant:* 5,000 USD
*Contact:* Nicolle Rutledge; (202) 326-6600; fax (202) 289-4950; nrutledg@aaas.org
*Internet:* www.aaas.org/aboutaaas/awards/abelson/index.shtml
*Sponsor:* American Association for the Advancement of Science
1200 New York Avenue NW
Washington, D.C. 20005-3920

**AAAS Roger Revelle Fellowship in Global Stewardship** 36

The Fellowship is designed to provide a unique opportunity for an accomplished scientist to address global stewardship issues by applying his or her broad, multidisciplinary background toward solutions to these important societal problems. The Revelle Fellow will engage with professionals in the public policy arena to strengthen his or her understanding of the intricacies of environmental policymaking and make practical contributions to the more effective use of scientific and technical knowledge in federal decision-making. The fellowship provides the unique opportunity to address sustainability challenges across multiple disciplines. The focus will be on human interaction with ecosystems, which includes, but is not limited to, ecosystem, population, sustainable development, food, oceans, global climate change, and related environmental concerns.
*Requirements:* Applications are invited from candidates in any physical, biological, or social science; any field of engineering; or any relevant interdisciplinary field. All applicants must have a PhD or an equivalent doctoral-level degree and at least three years of post-degree professional experience by the application deadline. Individuals with a master's degree in engineering and six years of post-degree, professional experience may apply. All applicants must be U.S. citizens.
*Restrictions:* Federal employees are not eligible.
*Geographic Focus* All States, District of Columbia, Guam, Marshall Islands, Northern Mariana Islands, Puerto Rico, U.S. Virgin Islands, American Samoa
*Date(s) Application is Due:* Dec 5
*Amount of Grant:* 64,000 USD
*Contact:* Kira E. Mock, Program Manager; (202) 326-6700; fax (202) 289-4950; kmock@aaas.org or fellowships@aaas.org
*Internet:* https://www.aaas.org/story-topics/roger-revelle-fellowship-global-stewardship
*Sponsor:* American Association for the Advancement of Science
1200 New York Avenue NW
Washington, D.C. 20005-3920

**AAAS Science and Technology Policy Fellowships** 37

AAAS Science and Technology Policy Fellowships provide opportunities for scientists and engineers to learn first-hand about policymaking while contributing their knowledge and analytical skills in the federal policy realm. Fellows chosen for this program engage in projects, programs, policies and outreach initiatives to protect environmental and human health, tackle energy challenges and opportunities, and to safeguard our air, water, land, and natural resources. Placement opportunities are available with the national Oceanic Atmospheric Association, National Science Foundation, U.S. Department of Agriculture, USDA Forest Service, U.S. Department of Energy, and U.S. Environmental Protection Agency. Fellowships are awarded to highly qualified individuals interested in learning about the science-policy interface while applying their scientific and technical knowledge and analytical skills to the federal policy realm. Stipends range from $55,000 to more than $85,000. The annual deadline is November 1.
*Requirements:* Prospective fellows must have a PhD or equivalent doctoral level degree. Persons with a master's degree in engineering and at least three years of post-

degree professional experience may apply. Candidates must demonstrate exceptional competence in some area of science or engineering, and an interest in applying their expertise to the economic and technical assessment of problems relating to human health, agriculture, and the environment. Applications are invited from individuals in any physical, biological, or social science; any field of engineering; or any relevant interdisciplinary field. Persons with a DVM, MD, or a PhD in the natural sciences or economics are especially encouraged to apply. Applicants must be U.S. citizens.
*Restrictions:* Federal employees are ineligible.
*Geographic Focus* All States, District of Columbia, Guam, Marshall Islands, Northern Mariana Islands, Puerto Rico, U.S. Virgin Islands, American Samoa
*Date(s) Application is Due:* Nov 1
*Amount of Grant:* 55,000 - 85,000 USD
*Contact:* Kira E. Mock, Program Manager; (202) 326-6700; fax (202) 289-4950; kmock@aaas.org or fellowships@aaas.org
*Internet:* www.aaas.org/program/science-technology-policy-fellowships
*Sponsor:* American Association for the Advancement of Science
1200 New York Avenue NW
Washington, D.C. 20005-3920

**AAAS Science and Technology Policy Fellowships: Congressional** 38

Congressional fellows spend one year working as special legislative assistants on the staffs of members of Congress or congressional committees, beginning in September. The program includes an orientation on congressional and executive branch operations, and a year-long seminar program on issues involving science and public policy. The fellowships are designed to provide a unique public policy learning experience, to demonstrate the value of science-government interaction, and to bring technical backgrounds and external perspectives to the decision-making process in government. AAAS will select and sponsor two Congressional Fellows. Annual stipends are approximately $70,000, with allowances for health insurance, relocation, and professional travel.
*Requirements:* Prospective Fellows must demonstrate exceptional competence in some area of science or engineering; have a good scientific and technical background; be cognizant of and demonstrate sensitivity toward political and social issues; and, perhaps most importantly, have a strong interest and some experience in applying personal knowledge toward the solution of societal problems. Applications are invited from individuals in any physical, biological, or social science, any field of engineering, or any relevant interdisciplinary field. All applicants must have a Ph.D. or an equivalent doctoral-level degree by the application deadline (December 10). Individuals with a master's degree in engineering and at least three years of post-degree professional experience may apply. Persons from underrepresented minority groups and persons with disabilities are encouraged to apply.
*Restrictions:* Applicants must be U.S. citizens. Federal employees are not eligible. Will accept only online application submissions. Applications with fewer than three recommendation letters will not be considered.
*Geographic Focus* All States, District of Columbia, Guam, Marshall Islands, Northern Mariana Islands, Puerto Rico, U.S. Virgin Islands, American Samoa
*Date(s) Application is Due:* Dec 5
*Amount of Grant:* 70,000 USD
*Contact:* Kira E. Mock, Program Manager; (202) 326-6700; fax (202) 289-4950; kmock@aaas.org or fellowships@aaas.org
*Internet:* https://www.aaas.org/program/science-technology-policy-fellowships
*Sponsor:* American Association for the Advancement of Science
1200 New York Avenue NW
Washington, D.C. 20005-3920

**AAAS Science and Technology Policy Fellowships: Diplomacy, Security, and Development** 39

Selected fellows for this program apply their scientific and technical expertise into policy development and program planning, implementation and evaluation in the areas of: foreign policy and international trade, treaty engagement, and multilateral cooperation; disaster preparedness and response; infrastructure, environmental, cyber and health security, terrorism and warfare prevention, and nonproliferation; and international aid, capacity building, and development assistance. Placement opportunities are available at Fogerty International Center, National Institutes of Health; Foreign Agriculture Service, U.S. Department of Agriculture; U.S. Agency for International Development; or U.S. Department of State.
*Requirements:* Applicants must have a PhD or equivalent doctoral level degree at the time of application. Persons with a master's degree in engineering and at least three years of postdegree professional experience will also be considered. All applicants must be U.S. citizens. Prospective fellows must demonstrate exceptional competence in some area of science and engineering (multidisciplinary expertise is especially valued); be cognizant of the ways in which science and technology affect a broad range of international development and foreign policy issues; communicate and work effectively with decision makers and others outside of the scientific and engineering communities; exhibit a willingness and flexibility to tackle problems in a number of nonscientific areas; demonstrate sensitivity toward political, economic, and social issues; and have some experience and/or strong interest in applying knowledge toward the solution of problems in the area of foreign affairs or international development. Applications are invited from individuals in any physical, biological, or social science; any field of engineering; or any relevant interdisciplinary field. Some program areas and agencies seek additional qualifications. See the website for more specific guidelines.
*Restrictions:* Federal employees are not eligible.
*Geographic Focus:* All States
*Date(s) Application is Due:* Dec 5
*Amount of Grant:* 55,000 - 85,000 USD
*Contact:* Christine Rovner, Program Manager; (202) 326-6700; fax (202) 289-4950; crovner@aaas.org or fellowships@aaas.org
*Internet:* https://www.aaas.org/taxonomy/term/3275/all/feed
*Sponsor:* American Association for the Advancement of Science
1200 New York Avenue NW
Washington, D.C. 20005-3920

**AAAS Science and Technology Policy Fellowships: Global Health and Development** 40

The AAAS Science and Technology Policy Fellowships have partnered with the Bill and Melinda Gates Foundation to create the Global Health and Development (GHD) Fellowships. This fellowship opportunity places Fellows at the Gates Foundation in Seattle, Washington, for one year, with the potential to renew for a second year. The GHD Fellowships provide opportunities for Fellows to learn from and contribute to the international development initiatives of the Foundation, by helping to identify and support strategies and programs in which science and technology can be harnessed to advance development in designated countries and/or regions.
*Requirements:* This fellowship opportunity is open to current and alumni AAAS Science & Technology Policy Fellows. All applicants must have a doctoral-level degree (PhD, MD, DVM, DSc, PharmD, and other terminal degrees), in any physical, biological, health/medical or social science, any field of engineering, or any relevant interdisciplinary field (individuals with a master's degree in engineering and at least three years of post-degree professional experience also may apply). Note: All requirements for the degree must be completed by the application deadline. A prospective fellow must demonstrate exceptional competence in their specialty appropriate to their career stage, and have the strong endorsement of three references. Show an understanding of the opportunities for science and engineering to support a broad range of non-scientific issues, and display a commitment to apply their scientific or technical expertise to serve society; Exhibit awareness and sensitivity to the political, economic and social issues that influence policy; Be articulate communicators, both verbally and in writing, to decision-makers and non-scientific audiences, and have the ability to work effectively with individuals and groups outside the scientific community; Demonstrate integrity and good judgment, and the flexibility and willingness to address policy issues outside their scientific realm.
*Restrictions:* Federal employees are ineligible. All applicants must be U.S. citizens (dual citizenship from the United States and another country is acceptable). Some program areas and host agencies seek additional qualifications. Holding a Title 42 position is considered federal employment and therefore renders an applicant ineligible for a AAAS Science and Technology Policy Fellowship. Individuals working in postdoctoral positions at government labs or agencies, for private contractors, and in state government posts are not federal employees and therefore are eligible for the AAAS Fellowships.
*Geographic Focus* All States, District of Columbia, Guam, Marshall Islands, Northern Mariana Islands, Rhode Island, U.S. Virgin Islands, American Samoa
*Date(s) Application is Due:* Dec 5
*Amount of Grant:* 70,000 - 90,000 USD
*Contact:* Christine Rovner, Program Manager; (202) 326-6700; fax (202) 289-4950; crovner@aaas.org or fellowships@aaas.org
*Internet:* https://www.aaas.org/program/science-technology-policy-fellowships
*Sponsor:* American Association for the Advancement of Science
1200 New York Avenue NW
Washington, D.C. 20005-3920

**AAAS Science and Technology Policy Fellowships: Health, Education and Human Services** 41

Fellows chosen for this program focus their experience to support improved programs, policies, planning, and risk analysis for initiatives in: preventive and community health, disease identification and response, and medical research; individual, family and community social services, systems and support; food, processing and distribution safety; and science education, research, and innovation. Placement opportunities are available with the National Institutes of Health, National Science Foundation, USDA Food Safety Inspection Services and U.S. Department of Health & Human Services. Fellowships are awarded to highly qualified individuals interested in learning about the science-policy interface while applying their scientific and technical knowledge and analytical skills to the federal policy realm.
*Requirements:* Prospective fellows must have a PhD or equivalent doctoral level degree. Persons with a master's degree in engineering and at least three years of post-degree professional experience may apply. Candidates must demonstrate exceptional competence in some area of science or engineering, and an interest in applying their expertise to the economic and technical assessment of problems relating to human health, agriculture, and the environment. Applications are invited from individuals in any physical, biological, or social science; any field of engineering; or any relevant interdisciplinary field. Persons with a DVM, MD, or a PhD in the natural sciences or economics are especially encouraged to apply. Applicants must be U.S. citizens.
*Restrictions:* Federal employees are not eligible.
*Geographic Focus* All States, District of Columbia, Guam, Marshall Islands, Northern Mariana Islands, Puerto Rico, U.S. Virgin Islands, American Samoa
*Date(s) Application is Due:* Dec 5
*Amount of Grant:* 55,000 - 85,000 USD
*Contact:* Robert Harper, Program Manager; (202) 326-6700; fax (202) 289-4950; rharper@aaas.org or fellowships@aaas.org
*Internet:* https://www.aaas.org/page/stpf/executive-branch-host-agencies-health-education-human-services-program
*Sponsor:* American Association for the Advancement of Science
1200 New York Avenue NW
Washington, D.C. 20005-3920

**AAAS Science and Technology Policy Overseas Fellowships at USAID Missions** 42
Re-launched in 2011, the AAAS Overseas Fellowships offer Fellows the experience of living and working overseas at a U.S. Embassy or U.S. Agency for International Development (USAID) Mission, as well as the opportunity to expand both the service and professional development opportunities of the S&T Policy Fellowship experience. AAAS supports Fellows to pursue international assignments via the Overseas Fellowships. AAAS Overseas Fellowships are currently available only at USAID Missions. Overseas Fellowships offer one- to two-year international assignments. The Fellow's primary responsibility will be to help the Mission identify key opportunities to develop strategies and programs that contribute to development in science and technology and areas where S&T can be harnessed to advance development in the host country and/or region. Anticipated placement opportunities will vary from year to year.
*Requirements:* This fellowship opportunity is limited to current and alumni AAAS Science & Technology Policy Fellows. Fellows may have served in any program area.
*Restrictions:* Applicants must be U.S. citizens. Federal employees are not eligible. Will accept only online application submissions. Applications with fewer than three recommendation letters will not be considered.
*Geographic Focus* All States, District of Columbia, Guam, Marshall Islands, Northern Mariana Islands, Puerto Rico, U.S. Virgin Islands, American Samoa
*Date(s) Application is Due:* Dec 5
*Amount of Grant:* 55,000 - 85,000 USD
*Contact:* Christine Rovner, Program Manager; (202) 326-6700; fax (202) 289-4950; crovner@aaas.org or fellowships@aaas.org
*Internet:* https://www.aaas.org/page/stpf/overview
*Sponsor:* American Association for the Advancement of Science
1200 New York Avenue NW
Washington, D.C. 20005-3920

**AABB Dale A. Smith Memorial Award** 43
The AABB Dale A. Smith Memorial Award, created in 2002 and funded by Fenwal, Incorporated, honors Dale A. Smith, a long-time Baxter Healthcare executive who was responsible for establishing the Fenwal Division of Baxter. This award will be presented to an individual or institution to recognize groundbreaking work performed in the application of technology to the practice of transfusion medicine or cellular therapies. The recipient will receive a $7,500 honorarium. The Award is presented annually.
*Requirements:* Award recipients are not required to be members of AABB. A nomination form must be completed for each nomination submission.
*Geographic Focus:* All States
*Amount of Grant:* 7,500 USD
*Contact:* Awards Coordinator; (301) 907-6977; fax (301) 907-6895; aabb@aabb.org
*Internet:* www.aabb.org/about/awards/Pages/descriptions.aspx#smith
*Sponsor:* American Association of Blood Banks
4550 Montgomery Avenue, Suite 700, North Tower
Bethesda, MD 20814

**AABB Emily Cooley Memorial Award and Lectureship** 44
The AABB Emily Cooley Memorial Award and Lectureship began as a lectureship in 1963, and was designated as a Memorial Award in 1983. The person who receives this award shall have demonstrated teaching ability and have made a major contribution to the field of transfusion medicine or cellular therapies. Acceptance of the award requires attendance at the Annual Meeting and the presentation of a lecture. Recipient will receive a $5,000 honorarium. The Award and Lectureship is presented annually and the recipient is chosen one year in advance by a Joint Committee composed of leaders from the Cellular Therapies Section coordinating committee and Transfusion Medicine Section coordinating committee.
*Geographic Focus:* All States
*Amount of Grant:* 5,000 USD
*Contact:* Awards Coordinator; (301) 907-6977; fax (301) 907-6895; aabb@aabb.org
*Internet:* www.aabb.org/about/awards/Pages/descriptions.aspx#cooley
*Sponsor:* American Association of Blood Banks
4550 Montgomery Avenue, Suite 700, North Tower
Bethesda, MD 20814

**AAB Berlin Prize Fellowships** 45
The American Academy in Berlin was established in 1994 with the primary goal of fostering greater understanding and dialogue between the people of the United States and the people of Germany. The Academy offers residential fellowships at its Hans Arnhold Center to American scholars, writers, policymakers, and artists who wish to engage in independent study in Berlin. Around two dozen of these various Berlin Prize Fellowships are conferred annually. The academy welcomes younger as well as established scholars, artists, and professionals. Fellowships have been awarded in a variety of disciplines, including history, political science, economics, German studies, art history, musicology, anthropology, law, and linguistics, as well as to writers and poets, public policy experts, journalists, composers, and artists. Fellows are in residence for one academic semester or occasionally for a full academic year (Bosch Fellows in Public Policy usually stay for six to eight weeks). Fellowship benefits include round-trip airfare, housing at the Academy, partial board, and a stipend of $5,000 per month. Furnished apartments at the Hans Arnhold Center are available for individuals and couples; limited accommodations are also available for families with children. Fellows are encouraged to work in association with a Berlin institution such as a museum, library, archive, university, government agency, studio, or media organization. Current guidelines and applications are made available at the Academy's website in mid-June. Application deadlines may vary from year to year. Candidates should verify current deadline dates at the Academy's website.
*Requirements:* All candidates should be permanently based in the United States and must be either U.S. citizens or permanent residents. They must have completed their doctorate or equivalent professional degree. Writers must have published at least one book at the time of application. Candidates need not work on German topics, but their project descriptions should explain how a residency in Berlin will contribute to further professional development. Academy fellows are comprised of established and emerging scholars, writers, and professionals who wish to engage in independent study. Berlin Prizes have been awarded to historians, economists, filmmakers, art historians, journalists, legal scholars, musicologists, public-policy experts, and writers, among others. They are appointed by an independent selection committee. The annual deadline for applications is September 28.
*Restrictions:* The Academy does not accept project proposals in mathematics and the hard sciences. The Academy does not accept applications from visual artists. Instead a small number of candidates are contacted individually by members of the Academy's Visual Arts Jury. Only one fellowship in music composition will be awarded per selection cycle. Applications in music composition are due in the Academy's New York office on October 1. Music composition applicants should use the Application for Music Composition, not the general application for fellowships. Music composition applicants may not apply online.
*Geographic Focus:* All States
*Date(s) Application is Due:* Sep 28
*Amount of Grant:* 7,500 - 60,000 USD
*Contact:* Emma Lo, Junior Coordinator, Fellows Selection; +49 (30) 804 83-153 or +49 (30) 804 83-830; fax +49 (30) 804 83-111; el@americanacademy.de
Carol Scherer, Manager, Fellows Selection; +49 (30) 804 83-106 or +49 (30) 804 83-830; fax +49 (30) 804 83-111; cs@americanacademy.de or mailbox@americanacademy.de
*Internet:* www.americanacademy.de/apply/apply-for-a-fellowship/
*Sponsor:* American Academy in Berlin
Am Sandwerder 17-19
Berlin, 14109 Germany

**AABB Fenwal Transfusion Medicine Fellow Scholarships** 46
AABB Fenwal Transfusion Medicine Fellow Scholarships program was created in 1982 and funded by Fenwal, Inc., now being funded by Fresenius Kabi, to recognize individuals who are enrolled, accepted for enrollment in, or who have recently completed training (within12 months) leading to SBB certification from an AABB-accredited institution. Recipients will receive a $1,500 honorarium. The AABB-Fenwal SBB Scholarship Awards are presented annually as distinguished candidates are identified.
*Geographic Focus:* All States
*Amount of Grant:* 1,500 USD
*Contact:* Administrator; (301) 215-6526 or (301) 215-6493; standards@aabb.org
*Internet:* www.aabb.org/development/scholarships/Pages/default.aspx
*Sponsor:* American Association of Blood Banks
4550 Montgomery Avenue, Suite 700, North Tower
Bethesda, MD 20814

**AABB Fenwal Transfusion Medicine Fellow Scholarships** 47
The AABB Fenwal Transfusion Medicine Fellow Scholarship was created in 1982 and funded by Fenwal, Inc., to recognize professionals with a Doctor of Medicine or Doctor of Osteopathic Medicine degree who are current fellows in a transfusion medicine program or training program that includes at least one continuous year in transfusion medicine training. Recipients will receive a $1,500 honorarium. The AABB-Fenwal TMF Scholarship Awards are presented annually as distinguished candidates are identified.
*Geographic Focus:* All States
*Contact:* Administrator; (301) 215-6526 or (301) 215-6493; standards@aabb.org
*Internet:* www.aabb.org/development/scholarships/Pages/default.aspx
*Sponsor:* American Association of Blood Banks
4550 Montgomery Avenue, Suite 700, North Tower
Bethesda, MD 20814

**AABB Hemphill-Jordan Leadership Award** 48
The AABB Hemphill-Jordan Leadership Award, renamed in 2005 after Bernice Hemphill, W. Quinn Jordan, and Joel Solomon, honors leaders from the transfusion medicine and cellular therapy community. The award recognizes an individual who made significant contributions in the areas of administration, quality programs, law and/or government affairs. The individual shall have demonstrated leadership qualities and a consistent willingness to lend his/her expertise to his/her peers. It may recognize one particular act or an accumulation of years of contributions. Acceptance of the award requires attendance at the Annual Meeting and the presentation of a lecture. The recipient will receive a $5,000 honorarium. The Award is presented annually and the recipient is chosen one year in advance.
*Requirements:* Award recipients are not required to be members of AABB. A nomination form must be completed for each nomination submission.
*Geographic Focus:* All States
*Amount of Grant:* 5,000 USD
*Contact:* Awards Coordinator; (301) 907-6977; fax (301) 907-6895; aabb@aabb.org
*Internet:* www.aabb.org/about/awards/Pages/descriptions.aspx#hemphill
*Sponsor:* American Association of Blood Banks
4550 Montgomery Avenue, Suite 700, North Tower
Bethesda, MD 20814

**AABB John Elliott Memorial Award** 49

The AABB John Elliott Memorial Award, established in 1956, recognizes an individual who has given outstanding service to AABB by demonstrating a willingness to lend his/her expertise to the association through work on committees, the AABB Board of Directors and other areas. The recipient will receive a $1,000 honorarium. The Award is presented in odd years only, and the recipient is selected by a Joint Committee composed of leaders from the Cellular Therapies Section coordinating committee and Transfusion Medicine Section coordinating committee.

*Requirements:* Award recipients are not required to be members of AABB. A nomination form must be completed for each nomination submission.

*Geographic Focus:* All States

*Amount of Grant:* 1,000 USD

*Contact:* Awards Coordinator; (301) 907-6977; fax (301) 907-6895; aabb@aabb.org

*Internet:* www.aabb.org/about/awards/Pages/descriptions.aspx#elliott

*Sponsor:* American Association of Blood Banks

4550 Montgomery Avenue, Suite 700, North Tower

Bethesda, MD 20814

**AABB Karl Landsteiner Memorial Award and Lectureship** 50

Initiated in 1954 to honor Karl Landsteiner, MD, whose lifetime research laid the foundation for modern blood transfusion therapy, the AABB Karl Landsteiner Memorial Award and Lectureship recognizes a scientist whose original research resulted in an important contribution to the body of scientific knowledge. The scientist who receives the award shall have an international reputation in transfusion medicine or cellular therapies. The recipient must agree to lecture at the AABB Annual Meeting. The recipient will receive a $7,500 honorarium. The Award and Lectureship will be awarded as distinguished candidates are identified. The recipient is chosen one year in advance by AABB's Board of Directors.

*Requirements:* Award recipients are not required to be members of AABB. A nomination form must be completed for each nomination submission.

*Geographic Focus:* All States

*Amount of Grant:* 7,500 USD

*Contact:* Awards Coordinator; (301) 907-6977; fax (301) 907-6895; aabb@aabb.org

*Internet:* www.aabb.org/about/awards/Pages/descriptions.aspx#landsteiner

*Sponsor:* American Association of Blood Banks

4550 Montgomery Avenue, Suite 700, North Tower

Bethesda, MD 20814

**AABB Sally Frank Memorial Award and Lectureship** 51

The Sally Frank Memorial Award and Lectureship was established in 1982 in memory of Ms. Sally Frank and her dedication to red cell serology and education. This award recognizes an individual who is, or has been, a medical technologist involved with these fields and has demonstrated quality research, teaching and/or service abilities in the technical aspects of immunohematology. Recipient will receive a $1,000 honorarium. The Award and Lectureship is presented annually and the recipient is chosen one year in advance by AABB's Transfusion Medicine Section Coordinating Committee with Formal Approval by AABB's Board of Directors.

*Requirements:* Award recipients are not required to be members of AABB. A nomination form must be completed for each nomination submission.

*Geographic Focus:* All States

*Amount of Grant:* 1,000 USD

*Contact:* Awards Coordinator; (301) 907-6977; fax (301) 907-6895; aabb@aabb.org

*Internet:* www.aabb.org/about/awards/Pages/descriptions.aspx#frank

*Sponsor:* American Association of Blood Banks

4550 Montgomery Avenue, Suite 700, North Tower

Bethesda, MD 20814

**AABB Tibor Greenwalt Memorial Award and Lectureship** 52

The AABB Tibor Greenwalt Memorial Award and Lectureship honors Tibor Greenwalt, MD, who was the first registrant at the first AABB Annual Meeting and founding editor of Transfusion. The award recognizes an individual who made major scientific or clinical contributions to hematology, transfusion medicine or cellular therapies, and succinctly communicated these advances. Acceptance of this award requires attendance at the AABB Annual Meeting and the presentation of a lecture on a current topic of interest. Recipient will receive a $1,000 honorarium. The Award and Lectureship is presented annually and the recipient is chosen one year in advance by a Joint Committee Composed of Leaders from the Cellular Therapies Section Coordinating Committee and Transfusion Medicine Section Coordinating Committee.

*Requirements:* Award recipients are not required to be members of AABB. A nomination form must be completed for each nomination submission.

*Geographic Focus:* All States

*Amount of Grant:* 1,000 USD

*Contact:* Awards Coordinator; (301) 907-6977; fax (301) 907-6895; aabb@aabb.org

*Internet:* www.aabb.org/about/awards/Pages/descriptions.aspx#greenwalt

*Sponsor:* American Association of Blood Banks

4550 Montgomery Avenue, Suite 700, North Tower

Bethesda, MD 20814

**AACAP Educational Outreach Program for Child and Adolescent Psychiatry Residents** 53

The Educational Outreach Program (EOP) funding provides the opportunity for up to 50 child and adolescent psychiatry residents to receive a formal overview to the field of child and adolescent psychiatry, establish child and adolescent psychiatrists as mentors and experience the AACAP Annual Meeting. Participants will be exposed to the breadth and depth of the field of child and adolescent psychiatry, including research opportunities, access to mentors, and various networking opportunities. Participation in this program provides up to $1,000 for travel expenses, which includes airfare, hotel, and meals (maximum $75/day).

*Requirements:* Applicants must: be child and adolescent psychiatry residents at the time of the AACAP Annual Meeting; be currently enrolled in a residency program in the United States; be residents in their first or second year of child fellowship training are eligible (Triple Boarders in their fourth or fifth year of training in their triple board programs are eligible); either be members of the AACAP or have a membership application pending at the time of application; and attend all AACAP Annual Meeting events specified by AACAP.

*Geographic Focus:* All States

*Date(s) Application is Due:* Jul 12

*Amount of Grant:* Up to 1,000 USD

*Contact:* Ashley Partner, (202) 966-7300, ext. 117; fax (202) 966-2891; apartners@aacap.org

*Internet:* www.aacap.org/cs/residents/eop-capfellows

*Sponsor:* American Academy of Child and Adolescent Psychiatry

3615 Wisconsin Avenue, NW

Washington, D.C. 20016-3007

**AACAP Educational Outreach Program for General Psychiatry Residents** 54

The Educational Outreach Program (EOP) for General Psychiatry residents provides the opportunity for up to 20 general psychiatry residents to receive a formal overview to the field of child and adolescent psychiatry, establish child and adolescent psychiatrists as mentors and experience the AACAP Annual Meeting. Participants will be exposed to the breadth and depth of the field of child and adolescent psychiatry, including research opportunities, access to mentors, and various networking opportunities. Participation in this program provides up to $1,500 for travel expenses to the AACAP Annual Meeting.

*Requirements:* Applicants must: be general psychiatry residents at the time of the AACAP Annual Meeting; be currently enrolled in a residency program in the United States; be residents in their first, second or third year of general psychiatry training (Triple Boarders in their first, second or third year of training in their triple board programs are eligible); either be members of the AACAP or have a membership application pending at the time of application; and attend all AACAP Annual Meeting events specified by AACAP.

*Geographic Focus:* All States

*Date(s) Application is Due:* Jul 12

*Amount of Grant:* Up to 1,500 USD

*Contact:* Ashley Partner, (202) 966-7300, ext. 117 or (202) 587-9663; fax (202) 966-2891; apartner@aacap.org or training@aacap.org

*Internet:* www.aacap.org/cs/residents/eop-generalresidents

*Sponsor:* American Academy of Child and Adolescent Psychiatry

3615 Wisconsin Avenue, NW

Washington, D.C. 20016-3007

**AACAP Educational Outreach Program for Residents in Alcohol Research** 55

The Educational Outreach Program (EOP) for Residents in Alcohol Research provides the opportunity for up to 10 child and adolescent psychiatry and general psychiatry residents to pursue advanced knowledge in adolescent alcohol research. The program also allows residents to expand their knowledge of adolescent alcohol-related issues by attending the AACAP Annual Meeting. EOP recipients will meet with an assigned alcohol research mentor and with a representative from NIAAA at the AACAP Annual Meeting to encourage mentorship connections and career development. Participants will be exposed to the field of child and adolescent psychiatry, including research opportunities, access to mentors, and various networking opportunities. Participation in this program provides up to $1,000 for travel expenses to the AACAP Annual Meeting.

*Requirements:* Applicants must be child and adolescent psychiatry or general psychiatry residents at the time of the AACAP Annual Meeting. Participants must: be currently enrolled in a child and adolescent psychiatry, general psychiatry, or triple board residency program in the United States; either be members of the AACAP or have a membership application pending at the time of application; and attend all Annual Meeting events specified by AACAP.

*Geographic Focus:* All States

*Date(s) Application is Due:* Jul 12

*Amount of Grant:* Up to 1,000 USD

*Contact:* Ashley Partner, (202) 966-7300, ext. 117 or (202) 587-9663; fax (202) 966-2891; apartner@aacap.org or training@aacap.org

*Internet:* www.aacap.org/cs/award_opportunities/details/aacap_educational_outreach_program_for_residents_in_alcohol_research

*Sponsor:* American Academy of Child and Adolescent Psychiatry

3615 Wisconsin Avenue, NW

Washington, D.C. 20016-3007

**AACAP George Tarjan Award for Contributions in Developmental Disabilities** 56

The AACAP George Tarjan Award for Contributions in Developmental Disabilities recognizes a child and adolescent psychiatrist and Academy member who has made significant contributions in a lifetime career or single seminal work to the understanding or care of those with mental retardation and developmental disabilities. These contributions must have national and/or international stature and clearly demonstrate lasting effects. The contributions may be in areas of teaching, research, program development, direct clinical service, advocacy or administrative commitment. A cash prize of up to $1,000 will be awarded. The award winner will be recognized at a Distinguished Awards Luncheon and make an Honors Presentation about his or her work during the AACAP Annual Meeting.

*Requirements:* Nomination letters must be accompanied by a CV for the individual nominated.

*Geographic Focus:* All States
*Date(s) Application is Due:* Apr 30
*Amount of Grant:* 1,000 USD
*Contact:* Bryan King, Chair; (202) 966-7300; fax (202) 966-2891; bking@aacap.org or clinical@aacap.org
*Internet:* www.aacap.org/cs/distinguishedmembers
*Sponsor:* American Academy of Child and Adolescent Psychiatry
3615 Wisconsin Avenue, NW
Washington, D.C. 20016-3007

**AACAP Irving Philips Award for Prevention** 57

The AACAP Irving Philips Award for Prevention recognizes a child and adolescent psychiatrist and Academy member who has made significant contributions in a lifetime career or single seminal work to the prevention of mental illness in children and adolescents. These contributions must have national and/or international stature and clearly demonstrate lasting effects. The contributions may be in the areas of teaching, research, program development, direct clinical service, advocacy or administrative commitment. The award pays $2,500 to the winner and a $2,000 donation to a prevention program or center of the awardee's choice. The award winner will be recognized at a Distinguished Awards Luncheon and make an Honors Presentation about his or her work during the AACAP Annual Meeting.
*Requirements:* Nomination letters must be accompanied by a CV for the individual nominated.
*Geographic Focus:* All States
*Date(s) Application is Due:* Apr 30
*Amount of Grant:* 4,500 USD
*Contact:* James Hudziak, Chair; (202) 966-7300; fax (202) 966-2891; jhudziak@aacap.org or clinical@aacap.org
*Internet:* www.aacap.org/cs/distinguishedmembers
*Sponsor:* American Academy of Child and Adolescent Psychiatry
3615 Wisconsin Avenue, NW
Washington, D.C. 20016-3007

**AACAP Jeanne Spurlock Lecture and Award on Diversity and Culture** 58

The Jeanne Spurlock Lecture and Award on Diversity and Culture reflects the spirit of the nomination for outstanding contributions to diversity and culture and in some way encourages individuals from diverse cultural backgrounds to become child and adolescent psychiatrists. Nominees should be individuals who have made contributions in the areas of social awareness including: civil rights; spirituality and/or religion; social welfare; public information; scientific research; education and mentoring; and the arts (literature, theatre, music, painting, sculpture or photography). The award includes an honorarium of $2,500, with the award winner being recognized at a Distinguished Awards Luncheon where he or she will make an honors award presentation about his or her work during the AACAP Annual Meeting.
*Geographic Focus:* All States
*Date(s) Application is Due:* Apr 30
*Amount of Grant:* 2,500 USD
*Contact:* Adriano Boccanelli, Clinical Practice Manager; (202) 966-7300, ext. 133; fax (202) 966-9518; aboccanelli@aacap.org or clinical@aacap.org
*Internet:* www.aacap.org/cs/distinguishedmembers
*Sponsor:* American Academy of Child and Adolescent Psychiatry
3615 Wisconsin Avenue, NW
Washington, D.C. 20016-3007

**AACAP Jeanne Spurlock Research Fellowship in Substance Abuse and Addiction for Minority Medical Students** 59

The Jeanne Spurlock Research Fellowship in Substance Abuse and Addiction for Minority Medical Students offers a unique opportunity for minority medical students to explore a research career in substance abuse in relation to child and adolescent psychiatry, gain valuable work experience, and meet leaders in the child and adolescent psychiatry field. The fellowship opportunity provides up to $4,000 for 12 weeks of summer research under a child and adolescent psychiatrist researcher/mentor. The research training plan must provide for significant contact between the student and the mentor and for exposure to state-of-the-art drug abuse and addiction research. The plan should include program planning discussions, instruction in research planning and implementation, regular meetings with the mentor, laboratory director, and the research group, and assigned readings. Research assignments may include responsibility for part of the observation or evaluation, developing specific aspects of the research mechanisms, conducting interviews or tests, use of rating scales, and psychological or cognitive testing of subjects.
*Requirements:* All fellowship participants must attend the AACAP Annual Meeting. Applications are considered from African-American, Native American, Alaskan Native, Mexican American, Hispanic, Asian, and Pacific Islander students in accredited U.S. medical schools.
*Geographic Focus:* All States
*Date(s) Application is Due:* Feb 15
*Amount of Grant:* Up to 4,000 USD
*Contact:* Ashley Partner, (202) 966-7300, ext. 117; fax (202) 966-2891; apartner@aacap.org or training@aacap.org
*Internet:* www.aacap.org/cs/students/opportunities/SpurlockResearch
*Sponsor:* American Academy of Child and Adolescent Psychiatry
3615 Wisconsin Avenue, NW
Washington, D.C. 20016-3007

**AACAP Junior Investigator Awards** 60

The AACAP Junior Investigator Award offers two awards of up to $30,000 a year for two years for child and adolescent psychiatry junior faculty (assistant professor level or equivalent). The program is intended to facilitate innovative research. The research may be basic or clinical in nature but must be relevant to our understanding, treatment and prevention of child and adolescent mental health disorders. The award also includes the cost of attending the AACAP Annual Meeting for five days.
*Requirements:* Recipients are required to submit a poster or oral presentation on his or her research for the AACAP's Annual Meeting. Applicants must: be board eligible or certified in child and adolescent psychiatry; have a doctoral level degree and be in a faculty or independent research position; and have an on-site mentor who has had experience in the type of research that is being proposed that will normally include work with children and adolescents. Candidates must either be AACAP members or have a membership application pending (not paid by the award).
*Restrictions:* Applicants who have served as Principal Investigator on an NIH R01 grant are not eligible.
*Geographic Focus:* All States
*Date(s) Application is Due:* Mar 1
*Amount of Grant:* Up to 30,000 USD
*Contact:* Alyssa Sommer, Research Coordinator; (202) 966-7300, ext. 157; fax (202) 364-5925; asommer@aacap.org or research@accap.org
*Internet:* www.aacap.org/cs/root/research_and_training_awards/aacap_junior_investigator_awards
*Sponsor:* American Academy of Child and Adolescent Psychiatry
3615 Wisconsin Avenue, NW
Washington, D.C. 20016-3007

**AACAP Life Members Mentorship Grants for Medical Students** 61

The Life Members Mentorship Grants for Medical Students (MGM) provides the opportunity for seven medical students to attend the AACAP Annual Meeting and receive an introduction into the field of child and adolescent psychiatry through the AACAP Mentorship Program. MGM recipients also participate in programs sponsored by the Life Members, which is a group of the oldest and most distinguished members of AACAP, all having been members for at least 30 years. Many of those in this group served as AACAP leadership and also pioneered many of the significant discoveries and developments in the field of child and adolescent psychiatry. Partnered with the Mentorship Program, this program provides participants with networking opportunities, exposure to varying specialties, and interaction with Life Members. Participation in this program provides up to $1,000 for travel expenses to the AACAP Annual Meeting, which includes airfare, hotel, and meals (maximum $75/day).
*Requirements:* Applicants must be enrolled in a medical school in the United States at the time of the AACAP Annual Meeting. Recipients are required to: share the AACAP recruitment video with fellow medical students in their program within the six months following the Annual Meeting; and write a follow-up report on the experience that they would be encouraged to submit to their program's listserve and/or website and share a copy of this with AACAP. Participants must: be in good standing at their medical school; and attend all AACAP Annual Meeting events specified by AACAP.
*Geographic Focus:* All States
*Date(s) Application is Due:* Jul 12
*Amount of Grant:* Up to 1,000 USD
*Contact:* Ashley Partner, (202) 966-7300, ext. 117 or (202) 587-9663; fax (202) 966-2891; apartner@aacap.org or training@aacap.org
*Internet:* www.aacap.org/cs/students/opportunities/life_members_mentorship_grants_for_medical_students
*Sponsor:* American Academy of Child and Adolescent Psychiatry
3615 Wisconsin Avenue, NW
Washington, D.C. 20016-3007

**AACAP Mary Crosby Congressional Fellowships** 62

The Mary Crosby Congressional Fellowships program is designed to educate policy makers and Congressional staff about child and adolescent psychiatry, and to foster awareness of children's mental health issues. This experience will allow the Fellow to develop a keen understanding of how public policy affects patient care, education, and health insurance issues. The AACAP Mary Crosby Congressional Fellow will be placed with a Congressional office or Committee where they will gain invaluable experience as they assist in the development of legislative and public policy initiatives. The award includes an $85,000 stipend, up to $3,000 in relocation or moving expenses, and health insurance for the length of the fellowship program.
*Requirements:* Successful applicants must: be an AACAP member; be a PGY-4 psychiatry resident or beyond; have completed training in or be scheduled to complete training in one of the child and adolescent psychiatry programs (CAP training program, Combined General Psychiatry/CAP Program, or Triple Board Program); have a demonstrated interest in public policy issues; exhibit an interest in applying medical and scientific knowledge to the resolution of public policy issues; and be a U.S. citizen or permanent resident.
*Geographic Focus:* All States
*Amount of Grant:* 85,000 - 88,000 USD
*Contact:* Ashley Partner, (202) 966-7300, ext. 117; fax (202) 966-2891; apartner@aacap.org or training@aacap.org
*Internet:* www.aacap.org/cs/root/legislative_action/2008_congressional_fellowship_program
*Sponsor:* American Academy of Child and Adolescent Psychiatry
3615 Wisconsin Avenue, NW
Washington, D.C. 20016-3007

**AACAP Pilot Research Award for Attention-Deficit Disorder** 63

The AACAP Pilot Research Award for Attention Disorders, supported by The Elaine Schlosser Lewis Fund, offers $15,000 for child and adolescent psychiatry residents and junior faculty who have an interest in beginning a career in child and adolescent mental health research. By providing one award to a child and adolescent psychiatry junior faculty member or resident for pilot research on attention disorders, the AACAP supports a young investigator at a critical stage, encouraging a future career in child and adolescent psychiatry research. The recipient has the opportunity to submit a poster presentation on his or her research for AACAP's Annual Meeting, and present at the Elaine Schlosser Lewis Luncheon.

*Requirements:* Candidates must be board eligible, certified in child and adolescent psychiatry, or enrolled in a child psychiatry residency or fellowship program. Candidates must also have a faculty appointment in an accredited medical school or be in a fully accredited child and adolescent psychiatry clinical research or training program.

*Restrictions:* At the time of application, candidates may not have more than two years experience following graduation from residency/fellowship training. Candidates must not have any previous significant, individual research funding in the field of child and adolescent mental health. These include the following: NIMH/NIH Funding (Small Grants, T Award or R-01) or similar foundation or industry research funding.

*Geographic Focus:* All States

*Date(s) Application is Due:* Apr 29

*Amount of Grant:* 15,000 USD

*Contact:* Ashley Partner, Research Coordinator; (202) 966-7300, ext. 117; fax (202) 364-5925; apartner@aacap.org

*Internet:* www.aacap.org/cs/root/research_and_training_awards/pilot_research_award_for_attention_disorder_for_a_junior_faculty_or_child_psychiatry_resident_supported_by_the_elaine_schlosser_lewis_fund

*Sponsor:* American Academy of Child and Adolescent Psychiatry
3615 Wisconsin Avenue, NW
Washington, D.C. 20016-3007

**AACAP Rieger Psychodynamic Psychotherapy Award** 64

The AACAP Rieger Psychodynamic Psychotherapy Award recognizes the best published or unpublished paper, written by an AACAP member that addresses the use of psychodynamic psychotherapy in clinical practice and fosters development, teaching, and practice of psychodynamic psychotherapy within child and adolescent psychiatry. Papers that express a novel hypothesis, raise questions about existing theory, or integrate new neuroscience and developmental psychotherapy research with psychodynamic principles may be nominated. Unpublished, new papers and papers published within the last three years may be submitted by their authors. Published papers may be nominated by any member of the AACAP. Authors may be senior or junior faculty members or residents. Delivery of the winning paper will be made at the AACAP Annual Meeting Honors Presentation.

*Geographic Focus:* All States

*Date(s) Application is Due:* Apr 30

*Amount of Grant:* 4,500 USD

*Contact:* Tim Dugan, Co-Chair; (202) 966-7300; fax (202) 966-2891; tdugan@aacap.org or clinical@aacap.org

Efrain Bleiberg, Co-Chair; (202) 966-7300; fax (202) 966-2891; ebleiberg@menninger.edu or clinical@aacap.org

*Internet:* www.aacap.org/cs/awards/riegerpsychotherapy

*Sponsor:* American Academy of Child and Adolescent Psychiatry
3615 Wisconsin Avenue, NW
Washington, D.C. 20016-3007

**AACAP Rieger Service Program Award for Excellence** 65

The AACAP Rieger Service Program Award for Excellence recognizes innovative programs that address prevention, diagnosis, or treatment of mental illnesses in children and adolescents, and serve as model programs to the community. This award of $4,500 is shared among the awardee and his or her service program. The award winner will be recognized at a Distinguished Awards Luncheon and make an Honors Presentation about his or her work during the AACAP Annual Meeting.

*Requirements:* Nomination letters must be accompanied by a CV and any support materials for the individual or organization nominated.

*Geographic Focus:* All States

*Date(s) Application is Due:* Apr 30

*Amount of Grant:* 4,500 USD

*Contact:* Kaye McGinty, Co-Chair; (202) 966-7300; fax (202) 966-2891; kmcginty@aacap.org or clinical@aacap.org

Mark Chenven, Co-Chair; (202) 966-7300; fax (202) 966-2891; mchenven@aacap.org or clinical@aacap.org

*Internet:* www.aacap.org/cs/distinguishedmembers

*Sponsor:* American Academy of Child and Adolescent Psychiatry
3615 Wisconsin Avenue, NW
Washington, D.C. 20016-3007

**AACAP Robert Cancro Academic Leadership Award** 66

The AACAP Cancro Academic Leadership Award recognizes a currently serving General Psychiatry Training Director, Medical School Dean, CEO of a Training Institution, Chair of a Department of Pediatrics or Chair of a Department of Psychiatry for his or her contributions to the promotion of child and adolescent psychiatry. Named in honor of Robert Cancro, M.D., Chairman at New York University, this award offers a $2,000 honorarium to the awardee. The award is presented in even-numbered years. Nominations for the award may be made by Child and Adolescent Training Directors or Division Directors and must include a CV for the individual nominated. The recipient of this award will receive a plaque and be recognized at the AACAP Annual Meeting in San Francisco, CA. The award recipient will be honored at the Distinguished Awards Luncheon, Residency Program Directors Luncheon, and will provide an Honors Presentation on his or her work during the Annual Meeting.

*Requirements:* Nominations must include a CV for the individual nominated

*Geographic Focus:* All States

*Date(s) Application is Due:* Apr 30

*Amount of Grant:* 2,000 USD

*Contact:* Jeffrey Hunt, Co-Chair; (202) 966-7300; fax (202) 966-2891; jhunt@aacap.org or training@aacap.org

Howard Liu, Co-Chair; (202) 966-7300; fax (202) 966-2891; hliu@aacap.org or clinical@aacap.org

*Internet:* www.aacap.org/cs/distinguishedmembers

*Sponsor:* American Academy of Child and Adolescent Psychiatry
3615 Wisconsin Avenue, NW
Washington, D.C. 20016-3007

**AACAP Sidney Berman Award for the School-Based Study and Intervention for Learning Disorders and Mental Ilness** 67

The AACAP Sidney Berman Award for the School-Based Study and Treatment of Learning Disorders and Mental Illness recognizes an individual or program that has shown outstanding achievement in the school-based study or delivery of intervention for learning disorders and mental illness. A cash prize of $4,500 will be awarded. The award winner will be recognized at a Distinguished Awards Luncheon and make an Honors Presentation about his or her work during the AACAP Annual Meeting.

*Requirements:* Nomination letters must be accompanied by a CV for the individual nominated or program information.

*Geographic Focus:* All States

*Date(s) Application is Due:* Apr 30

*Amount of Grant:* 4,500 USD

*Contact:* Sheryl Kataoka, Co-Chair; (202) 966-7300; fax (202) 966-2891; skataoka@aacap.org or clinical@aacap.org

Shashank Joshi, Co-Chair; (202) 966-7300; fax (202) 966-2891; sjoshi@aacap.org or clinical@aacap.org

*Internet:* www.aacap.org/cs/distinguishedmembers

*Sponsor:* American Academy of Child and Adolescent Psychiatry
3615 Wisconsin Avenue, NW
Washington, D.C. 20016-3007

**AACAP Simon Wile Leadership in Consultation Award** 68

The Simon Wile Leadership in Consultation Award recognizes innovative programs that address prevention, diagnosis, or treatment of mental illnesses in children and adolescents, and serve as model programs to the community. This award of $4,500 is shared among the awardee and his or her service program. Nomination letters must be accompanied by a CV and any support materials for the individual or organization nominated. The recipient will be recognized at a Distinguished Awards Luncheon and make an Honors Presentation about his or her work during the AACAP Annual Meeting.

*Requirements:* Practicing child and adolescent psychiatrists are eligible for nomination.

*Geographic Focus:* All States

*Date(s) Application is Due:* Apr 30

*Amount of Grant:* 500 USD

*Contact:* Kaye McGinty, Co-Chair; (202) 966-7300; fax (202) 966-2891; kmcginty@aacap.org or clinical@aacap.org

Mark Chenven, Co-Chair; (202) 966-7300; fax (202) 966-2891; mchenven@aacap.org or clinical@aacap.org

*Internet:* www.aacap.org/cs/distinguishedmembers

*Sponsor:* American Academy of Child and Adolescent Psychiatry
3615 Wisconsin Avenue, NW
Washington, D.C. 20016-3007

**AACAP Summer Medical Student Fellowships** 69

AACAP Summer Medical Student Fellowships offer a chance for medical students to explore a career in child and adolescent psychiatry, gain valuable work experience, and meet leaders in the child and adolescent psychiatry field. The fellowship opportunity provides up to $3,500 for 12 weeks of clinical or research training under a child and adolescent psychiatrist mentor. Fellowship stipends will be distributed in two installments. Upon receipt of the fellowship, $2,500 will be sent to the student's mentor to be disbursed at the onset of the summer fellowship. Upon meeting all program requirements, the last installment will be sent directly to the recipient and will be pro-rated according to the total amount of time spent completing the fellowship. Fellowships lasting the full 12 weeks will receive the maximum amount. Participants are required to attend the AACAP Annual Meeting. The application deadline for this fellowship is February 15.

*Requirements:* Applicants must be students enrolled in accredited U.S. medical schools.

*Geographic Focus:* All States

*Date(s) Application is Due:* Feb 15

*Amount of Grant:* 3,500 USD

*Contact:* Ashley Partner, (202) 966-7300, ext. 117; fax (202) 966-2891; apartner@aacap.org or training@aacap.org

*Internet:* www.aacap.org/cs/students/summerfellowship

*Sponsor:* American Academy of Child and Adolescent Psychiatry
3615 Wisconsin Avenue, NW
Washington, D.C. 20016-3007

**AACAP Systems of Care Special Program Scholarships** 70
The American Academy of Child and Adolescent Psychiatry (AACAP) is pleased to announce the opportunity for child and adolescent psychiatry residents to apply for the 2012 Systems of Care Special Program Scholarship. The Special Program is a day-long event that explores the psychological, psychosocial and psycho-educational needs of children and adolescents as manifest in their experiences in the schools. Policy and best practice considerations will be reviewed as regards health promotion, screening, assessment and treatment intervention for students and student groups, with the goal of advancing the knowledge, skills and attitudes of child psychiatrists and other mental health providers so that they can better function as service providers and systems consultants for youth and their school systems. The scholarship includes $750 for travel expenses to the AACAP's 59th Annual Meeting. This includes airfare, hotel, and meals, maximum $75 per day.
*Geographic Focus:* All States
*Date(s) Application is Due:* Jul 12
*Amount of Grant:* 750 USD
*Contact:* Adriano Boccanelli, Clinical Practice Assistant; (202) 966-7300, ext. 133 or (202) 587-9671; fax (202) 966-2891; aboccanelli@aacap.org or clinical@aacap.org
*Internet:* www.aacap.org/cs/root/medical_students_and_residents/residents/2012_systems_of_care_special_program_scholarship
*Sponsor:* American Academy of Child and Adolescent Psychiatry
3615 Wisconsin Avenue, NW
Washington, D.C. 20016-3007

**AACC International Alsberg-French-Schoch Memorial Lectureship Awards** 71
The Alsberg-French-Schoch Memorial Lectureship Award, sponsored by the Corn Refiners Association, is the organization's highest award given every two to four years in recognition of scientists who have made significant and superior contributions to fundamental starch science. The award is given every two to four years and includes a $2,000 honorarium and a plaque, as well as the opportunity to present a lecture on some phase of starch science before an annual meeting of the Association. To be nominated for an award, contact information for both the person suggesting the nominee and the individual being nominated, as well as a summary of the nominee's career, accomplishments, and contributions should be forwarded directly to AACC International headquarters. Nominations should address the activities most relevant to the award, with a clear statement of the impact the nominee has had in his/her specific area of expertise. Nomination forms are available at the AACC website.
*Requirements:* Nominee must be an AACC International member. All members are strongly encouraged to submit nominations for these awards at any time. All nominations are considered for three years.
*Geographic Focus:* All States, All Countries
*Date(s) Application is Due:* Jan 1
*Amount of Grant:* 2,000 USD
*Contact:* Linda Schmitt, AACC International Award Chairperson; (651) 994-3828 or (651) 454-7250; fax (651) 454-0766; lschmitt@scisoc.org or aacc@scisoc.org
*Internet:* www.aaccnet.org/membership/awards/Pages/AlsbergFrenchSchochMemorialLectureship.aspx
*Sponsor:* American Association of Cereal Chemists International
3340 Pilot Knob Road
St. Paul, MN 55121

**AACC International Best Student Paper and Best Student Poster Award** 72
The Protein Division selects the best protein chemistry papers presented by a student at the AACC International Annual Meeting each year. The Protein Division Best Student Paper Awards consists of an engraved plaque, an honorarium, and recognition at the annual meeting. The Best Student Poster Award consists of an engraved plaque, an honorarium, and recognition at the annual meeting.
*Geographic Focus:* All States, All Countries
*Date(s) Application is Due:* Jan 1
*Contact:* Bin X. Fu, Chair; (800) 853-6705 or (204) 983-2770; fax (204) 983-2751
*Internet:* www.aaccnet.org/membership/divisions/Pages/Protein.aspx
*Sponsor:* American Association of Cereal Chemists International
3340 Pilot Knob Road
St. Paul, MN 55121

**AACC International Bruce Wasserman Young Investigator Award** 73
The purpose of the Bruce Wasserman Young Investigator Award is to recognize young scientists who have made outstanding contributions to the field of cereal biotechnology. The research recognized by this award should be relevant to the broad aims of the AACC International. The work can be either basic (e.g., improved understanding of structure-function relationships) or applied (e.g., new approaches for cereal production, breeding or utilization). For the purposes of this award, 'Cereal Biotechnology' is broadly defined, and encompasses any significant body of research using plants, microbes, genes, proteins or other biomolecules. Contributions in the disciplines of genetics, molecular biology, biochemistry, microbiology and fermentation engineering are all included. The research will be evaluated by its impact as measured as either enhanced knowledge of cereal functionality or improved cereal production or utilization. Achievements will be considered in relation to the age and experience of the nominee. AACC International membership is not required.
*Requirements:* Nominees must be no older than 40 by July 1, but nominations of younger scientists are particularly encouraged.
*Geographic Focus:* All States, All Countries
*Date(s) Application is Due:* Jan 15
*Amount of Grant:* 1,000 USD
*Contact:* Tandace A. Bell, Committee Chair; (816) 891-0459 or (651) 454-7250; fax (651) 454-0766; tandace.a.bell@usda.gov
*Internet:* www.aaccnet.org/membership/divisions/Pages/Biotechnology.aspx
*Sponsor:* American Association of Cereal Chemists International
3340 Pilot Knob Road
St. Paul, MN 55121

**AACC International Carl Wilhelm Brabender Award** 74
The Carl Wilhelm Brabender Award was established to honor the distinguished contributions to the application of rheology in milling and baking. Every three years, the Award provides a plaque and a travel grant to enable the recipient to visit scientists and laboratories in other parts of the world to increase the Awardee's knowledge and experience in the field. The only qualification is to be professionally active in rheology, specifically those engaged in the milling and baking industry. The Awardee shall present an address at the AACC International Annual Meeting or at one of the three annual meetings of the AG at which the Award is given. The nomination form is available at the AACC website.
*Requirements:* All members are strongly encouraged to submit nominations for these Awards at any time. All Awards approved by the Board of Directors will be presented at the annual meeting. All nominations for AACCI Awards are considered for three years if the nominee meets the Award criteria.
*Geographic Focus:* All States, All Countries
*Date(s) Application is Due:* Jan 1
*Contact:* Linda Schmitt, AACC International Award Chairperson; (651) 994-3828 or (651) 454-7250; fax (651) 454-0766; lschmitt@scisoc.org or aacc@scisoc.org
*Internet:* www.aaccnet.org/membership/awards/Pages/Brabender.aspx
*Sponsor:* American Association of Cereal Chemists International
3340 Pilot Knob Road
St. Paul, MN 55121

**AACC International Cecil F. Pinney Travel Awards** 75
The Cecil F. Pinney Travel Awards are presented to individuals establishing a career in grain science and technology. The candidate can apply for the travel award or can be nominated. The application process consists of an application form, nomination letter (or letter of context if self nominating), and the candidate's resume. Achievements and participation in AACCI activities will be considered in relation to the age and experience of the candidate. Up to two $750 awards are available. In addition, a one-year membership to the division of their choice will be awarded to the recipient(s) and the registration fee to the annual meeting will be waived, as well as admission to a division function at the annual meeting. The application is available at the AACCI website, and must also include a cover sheet, resume, and supporting data and descriptions of previous AACCI involvement, along with activities in grain science and technology.
*Requirements:* Candidates must be AACCI members interested in some aspect of grain science and technology. This award is for new professionals who have been in their career less than eight years as of March 1 of the year of application. Preference will be given to nominees or applicants with a demonstrated need for the support to attend the AACCI Annual Meeting. Previous Cecil F. Pinney Travel Award recipients are eligible to apply for future awards, but they are limited to receiving this award three times within the eight-year time frame.
*Geographic Focus:* All States, All Countries
*Date(s) Application is Due:* Mar 1
*Amount of Grant:* 750 USD
*Contact:* Linda Schmitt, Pinney Award Coordinatorr; (651) 994-3828 or (651) 454-7250; fax (651) 454-0766; lschmitt@scisoc.org or aacc@scisoc.org
*Internet:* www.aaccnet.org/membership/foundation/awards/Pages/PinneyTravelAward.aspx
*Sponsor:* American Association of Cereal Chemists International
3340 Pilot Knob Road
St. Paul, MN 55121

**AACC International Edith A. Christensen Award** 76
In recognition of the central role that analytical methodology and approved methods have played in the history of AACC International and their future importance, the Edith A. Christensen Award for outstanding contributions in analytical methodology was created to recognize scientific contributions in analytical methodology, and service to the society through approved methods leadership. The award consists of a suitably inscribed plaque and a $1,500 honorarium.
*Requirements:* The recipient must have contributed to the development of new analytical technologies, and/or the application of new analytical technologies to cereal grains and products, and/or the application of existing analytical technologies to solving detection and measurement problems in the field of cereal science. Further, the recipient must have demonstrated involvement and leadership in approved methods activity within AACC International. Nomination must be made by letter of nomination detailing the nominee's accomplishments in analytical methodology and the approved methods process, and two additional letters of support for the nominee. Awardees will have the option to present a lecture during an annual meeting of the Association when the Award is presented.
*Geographic Focus:* All States, All Countries
*Date(s) Application is Due:* Jan 1
*Amount of Grant:* 1,500 USD
*Contact:* Linda Schmitt, AACC International Award Chairperson; (651) 994-3828 or (651) 454-7250; fax (651) 454-0766; lschmitt@scisoc.org or aacc@scisoc.org
*Internet:* www.aaccnet.org/membership/awards/Pages/EdithAChristensen.aspx
*Sponsor:* American Association of Cereal Chemists International
3340 Pilot Knob Road
St. Paul, MN 55121

**AACC International Engineering and Processing Division's Stanley Watson Award 77**
The Stanley Watson Award recognizes outstanding AACC International members who, through the application of engineering principles, have made significant contributions in the area of cereal/grain processing. The Award is presented every two years, and includes a $1,000 honorarium, engraved plaque, and a lecture presentation at the division luncheon. Nominations are accepted through June 1.
*Requirements:* Nominations must include: name of nominee; position; complete address and contact information; area of research; and a nominating statement detailing the nominee's significant contributions in the area of cereal/grain processing via application of engineering principles.
*Geographic Focus:* All States, All Countries
*Date(s) Application is Due:* Jun 1
*Amount of Grant:* 1,000 USD
*Contact:* Heather Maskus, AACC Division Chair; (204) 984-3139 or (204) 983-5344; fax (204) 983-2642; hmaskus@cigi.ca
*Internet:* www.aaccnet.org/divisions/divisionsdetail.cfm?CODE=ENG
*Sponsor:* American Association of Cereal Chemists International
3340 Pilot Knob Road
St. Paul, MN 55121

**AACC International Excellence in Teaching Award** 78
The Excellence in Teaching Award is to be presented to an AACC International member and current teacher who has made significant contributions through teaching in the broad field of Cereal Science and Technology. The Award consists of an honorarium of $1,500 and a suitably inscribed plaque. To be nominated for an Award, contact information for both the person suggesting the nominee and the individual being nominated, as well as a summary of the nominee's career, accomplishments, and contributions should be forwarded directly to AACC International headquarters. Nominations should address the activities most relevant to the Award, with a clear statement of the impact the nominee has had in his/her specific area of expertise. Nomination forms are available at the AACC website.
*Requirements:* Nominee must be an AACC International member. All members are strongly encouraged to submit nominations for these awards at any time.
*Geographic Focus:* All States, All Countries
*Date(s) Application is Due:* Jan 1
*Amount of Grant:* 1,500 USD
*Contact:* Linda Schmitt, AACC International Award Chairperson; (651) 994-3828 or (651) 454-7250; fax (651) 454-0766; lschmitt@scisoc.org or aacc@scisoc.org
*Internet:* www.aaccnet.org/membership/awards/Pages/ExcellenceinTeaching.aspx
*Sponsor:* American Association of Cereal Chemists International
3340 Pilot Knob Road
St. Paul, MN 55121

**AACC International George W. Scott Blair Lecture Award** 79
The Rheology Division of AACC International sponsors several awards to scientists and students working in the area of rheology, in order to recognize their accomplishments, subsidize their efforts and publicize the field of rheology in general. The purpose of the George W. Scott Blair Lecture Award is to recognize scientists who have demonstrated outstanding ability in research in the area of rheology and texture as related to cereal based products. The Award consists of an engraved plaque, $500 honorarium and travel expenses (if necessary) to attend the AACC International Annual Meeting. This Award is presented in even years.
*Geographic Focus:* All States, All Countries
*Amount of Grant:* 500 USD
*Contact:* Hulya Dogan, Chair; (785) 532-2628; fax (785) 532-7010; dogan@ksu.edu
*Internet:* www.aaccnet.org/membership/divisions/Pages/Rheology.aspx
*Sponsor:* American Association of Cereal Chemists International
3340 Pilot Knob Road
St. Paul, MN 55121

**AACC International Graduate Fellowships** 80
Graduate fellowships of between $2,000 and $3,000 will be available, supported by endowment funds of the AACCI Foundation, and from annual contributions by companies and AACCI divisions. The purpose of the fellowship program is to encourage graduate research in grain-based food science and technology. Graduate awards are paid directly to the winners in September after AACCI receives verification of registration for the fall semester or quarter from the appropriate department head. If the student is already receiving an assistantship or fellowship, they will still receive the amount of the AAEE Fellowship. The application and additional guidelines are available at the AAEE website.
*Requirements:* Applicants must be a current AACCI Student Division member. The candidate must be enrolled in graduate studies by the time the Fellowship becomes effective, or be a current graduate student pursuing a course of study leading to an M.S. or a Ph.D. degree. The educational institution in which the recipient is enrolled must be conducting fundamental investigations for the advancement of cereal science and technology, including oilseeds. The recipient must be enrolled in an academic schedule that meets the minimum requirements of the university involved, for full-time graduate study.
*Restrictions:* Research in such disciplines as genetics, horticulture, nutrition, microbiology, biochemistry, engineering, or chemistry is only eligible if it is directly related to the student's primary research program dealing with a grain-based food science and technology related areas.
*Geographic Focus:* All States, All Countries
*Date(s) Application is Due:* Mar 5
*Amount of Grant:* 2,000 - 3,000 USD
*Contact:* Linda Schmitt, AACC International Award Chairperson; (651) 994-3828 or (651) 454-7250; fax (651) 454-0766; lschmitt@scisoc.org or aacc@scisoc.org
*Internet:* www.aaccnet.org/membership/foundation/awards/Pages/GraduateProgram.aspx
*Sponsor:* American Association of Cereal Chemists International
3340 Pilot Knob Road
St. Paul, MN 55121

**AACC International Milling & Baking Division's John C. Halverson Memorial Lectureship** 81
The John C. Halverson Memorial Lectureship is an award presented by the Milling and Baking Division to honor an individual for their contributions to the technology of milling and baking. The award consists of an honorarium and suitably inscribed certificate. The recipient is given the opportunity to present a technical paper detailing their findings and accomplishments.
*Geographic Focus:* All States, All Countries
*Contact:* Barbara B. Heidolph, Chair; (314) 983-7500; fax (314) 983-7636
*Internet:* www.aaccnet.org/membership/divisions/Pages/MillingBaking.aspx
*Sponsor:* American Association of Cereal Chemists International
3340 Pilot Knob Road
St. Paul, MN 55121

**AACC International Milling & Baking Division's M. Rella Dwyer Graduate Fellowship** 82
The M. Rella Dwyer Graduate Fellowship is offered annually to a deserving student preparing a career in milling and baking. Applications for AACCI Scholarships and Fellowships are solicited from appropriate University department heads and other sources by AACCI. Candidates may review the AACC website for additional information.
*Geographic Focus:* All States, All Countries
*Amount of Grant:* 3,000 USD
*Contact:* Barbara B. Heidolph, Chair; (314) 983-7500; fax (314) 983-7636
*Internet:* www.aaccnet.org/membership/divisions/Pages/MillingBaking.aspx
*Sponsor:* American Association of Cereal Chemists International
3340 Pilot Knob Road
St. Paul, MN 55121

**AACC International Phil Williams Applied Research Award** 83
The AACC International Phil Williams Applied Research Award is presented for a significant body of distinguished contributions to the application of science in the cereals area. Of primary interest is the development of novel machinery, processes, products, patents, tests, methodologies, or procedures taking advantage of the wealth of accumulated knowledge from basic research as available in the literature together with basic/applied research performed by the applicant in developing the product/patent. The methodologies, applications, products and processes must have a clear and sustained record of adoption and/or commercialization and must have contributed positively to the development of science and knowledge of relevance to the cereal/food industry. The Award will consist of a $2,000 honorarium and a suitably inscribed medal in the form of a plaque that will be given to an individual scientist or a team of scientists. Recipients of the Award are also accorded the status of AACC International Fellow for their contributions leading to this Award. Awardees will have the opportunity to present a lecture during an annual meeting of the Association.
*Requirements:* Nominee must be an AACC International member. All members are strongly encouraged to submit nominations for these awards at any time.
*Geographic Focus:* All States, All Countries
*Date(s) Application is Due:* Jan 1
*Amount of Grant:* 2,000 USD
*Contact:* Linda Schmitt, AACC International Award Chairperson; (651) 994-3828 or (651) 454-7250; fax (651) 454-0766; lschmitt@scisoc.org or aacc@scisoc.org
*Internet:* www.aaccnet.org/membership/awards/Pages/PhilWilliamsAppliedResearch.aspx
*Sponsor:* American Association of Cereal Chemists International
3340 Pilot Knob Road
St. Paul, MN 55121

**AACC International Student Division's Best Research Paper Awards** 84
The objectives of the Student Division's Best Research Paper Award competition are to challenge students to demonstrate superior presentation skills, encourage students to present their research work at AACC International, highlight the best research conducted and presented by students, and offer an opportunity for students to interact with the AACC International community at an early stage in their career. The competition will be judged in two stages. The Best Student Research Paper Competition jury will review the nomination forms, abstracts, and initial presentations. Based on the judge's scores, six students will be chosen to participate in the competition. Winners of the competition will be announced at the annual meeting where the competition took place. Each of the six finalists will receive a $650 travel grant for attending the the annual meeting. First place awards include $1,250 cash award, $650 travel grant, and the opportunity to publish their paper in Cereal Chemistry, Cereal Foods World, or the Cereal Science Knowledge Database. Second place awards include $750 cash award, $650 travel grant, and the opportunity to publish their paper in Cereal Chemistry, Cereal Foods World, or the Cereal Science Knowledge Database. Third place award includes $500 cash award, $650 travel grant, and the opportunity to publish their paper in Cereal Chemistry, Cereal Foods World, or the Cereal Science Knowledge Database. Academic departments may submit one nominee each for the competition. More than one student may be nominated from a university as long as the nomination comes from different departments. It is up to the department to determine which student will be chosen for the competition.

Submitters need not be AACCI members. Additional information and guidelines are available at the AACC website.
*Requirements:* Applicants must be a full-time student at the the time of submission date. Applicants must provide proof of student status by way of department chair signature on the nomination form. The student presenting the paper must be an author of the presentation, must have made a major contribution to the research, must present it, and must take primary responsibility for its content.
*Geographic Focus:* All States, All Countries
*Date(s) Application is Due:* Feb 28
*Amount of Grant:* 650 - 1,250 USD
*Contact:* Cindy Scheller, Student Award Contact; (651) 454-7250; fax (651) 454-0766; cscheller@scisoc.org
*Internet:* www.aaccnet.org/membership/divisions/Pages/AACCIBestStudentResearchPaperCompetition.aspx
*Sponsor:* American Association of Cereal Chemists International
3340 Pilot Knob Road
St. Paul, MN 55121

**AACC International Student Division's Product Development Awards** **85**
The AACC International Student Division Product Development Competition (PD.C.) is organized by students and sponsored by food companies who are members of AACC International. The PD.C. provides an opportunity for students to apply the knowledge and skills acquired in school to create new food products professionally and intelligently in the real business world. The competition is designed to encourage innovative thinking of students in the cereal science area, expose students to a professional experience, and promote networking with food industry representatives. Students can work individually or in a team to develop a novel food product containing at least one major cereal/legume ingredient. A total of three prizes will be awarded: first place ($2,550); second place ($1,700); and third place ($850). All finalists receive $300 per team for travel expenses. The entry form and additional guidelines are available at the AACC website.
*Requirements:* All participating team members must be student members of AACC International. Each participant must be enrolled at a four-year accredited college or a graduate student during the spring semester.
*Geographic Focus:* All States, All Countries
*Date(s) Application is Due:* Jun 8
*Amount of Grant:* 850 - 2,550 USD
*Contact:* Tom Jondiko, Chair; (651) 454-7250; fax (651) 454-0766; jondiko@yahoo.com
*Internet:* www.aaccnet.org/membership/divisions/Pages/AACCIProductDevelopmentCompetition.aspx
*Sponsor:* American Association of Cereal Chemists International
3340 Pilot Knob Road
St. Paul, MN 55121

**AACC International Student Travel Grant** **86**
AACC International's Protein Division awards a student travel grant of $250 to a student attending the annual meeting. Students presenting a poster or paper in the area of protein chemistry are eligible to apply.
*Geographic Focus:* All States, All Countries
*Date(s) Application is Due:* Jan 1
*Amount of Grant:* 250 USD
*Contact:* Bin X. Fu, Division Chair; (800) 853-6705 or (204) 983-2770; fax (204) 983-2751
*Internet:* www.aaccnet.org/membership/divisions/Pages/Protein.aspx
*Sponsor:* American Association of Cereal Chemists International
3340 Pilot Knob Road
St. Paul, MN 55121

**AACC International Thomas Burr Osborne Medal** **87**
The Thomas Burr Osborne Medal was established in 1926, and is awarded to an individual whose research in the field of cereal chemistry has contributed significantly to the progress of the science. The award consists of an honorarium of $2,000 and a suitably inscribed medal in the form of a plaque. The recipient shall present an address at the AACC International Annual Meeting at which the medal is presented. Recipients of the Medal are also accorded the status of AACC International Fellow.
*Requirements:* AACCI Membership is required.
*Geographic Focus:* All States, All Countries
*Date(s) Application is Due:* Jan 1
*Amount of Grant:* 2,000 USD
*Contact:* Linda Schmitt, AACC International Award Chairperson; (651) 994-3828 or (651) 454-7250; fax (651) 454-0766; lschmitt@scisoc.org or aacc@scisoc.org
*Internet:* www.aaccnet.org/membership/awards/Pages/ThomasBurrOsborneMedal.aspx
*Sponsor:* American Association of Cereal Chemists International
3340 Pilot Knob Road
St. Paul, MN 55121

**AACC International Walter Bushuk Graduate Research Award** **88**
The AACC International Walter Bushuk Graduate Research Award in Cereal Protein Chemistry for Ph.D. students is presented to an individual for outstanding contributions in basic and/or applied research in cereal protein chemistry. This Award recognizes research in relevant to the broad aims of AACC International. The research can be either fundamental/basic (such as improved understanding of structure-function relationships) or applied (such as development of new products, processes, techniques). The Awardee will receive an honorarium of $750 and a plaque, and will be required to present a lecture at the AACC International annual meeting of the year in which the award is given. Nomination forms are available at the AACC website, and are due January 1.
*Geographic Focus:* All States, All Countries
*Date(s) Application is Due:* Jan 1
*Amount of Grant:* 750 USD
*Contact:* Bin X. Fu, Chair; (800) 853-6705 or (204) 983-2770; fax (204) 983-2751
*Internet:* www.aaccnet.org/membership/divisions/Pages/Protein.aspx
*Sponsor:* American Association of Cereal Chemists International
3340 Pilot Knob Road
St. Paul, MN 55121

**AACC International Young Scientist Research Award** **89**
The Award is presented to an individual for outstanding contributions in basic and applied research to cereal science with the expectation that contributions will continue. This Award recognizes research relevant to the broad aims and interests of AACC International. The research can be either basic (such as improved understanding of structure-function relationships) or applied (such as development of new products, processes, or techniques) and nominees may be from government, academia or industry. Evidence of accomplishments can include awards, publications, invitations to lecture, patents and commercialized foods, processes and/or testing procedures. The research will be evaluated based on impact on advancing cereal science knowledge or improved production or utilization. Achievements will be considered in relation to the age and experience of the nominee. The recipient will receive a $1,000 honorarium, a plaque, and will be encouraged to present a lecture at the AACC International meeting of the year in which the award is given.
*Requirements:* Nominees must be no older than 40 years by June 1 of the year the award is sought, but nominations of younger scientists are particularly encouraged. AACC International membership is required.
*Geographic Focus:* All States, All Countries
*Date(s) Application is Due:* Jan 1
*Amount of Grant:* 1,000 USD
*Contact:* Linda Schmitt, AACC International Award Chairperson; (651) 994-3828 or (651) 454-7250; fax (651) 454-0766; lschmitt@scisoc.org or aacc@scisoc.org
*Internet:* www.aaccnet.org/membership/awards/Pages/YoungScientistResearch.aspx
*Sponsor:* American Association of Cereal Chemists International
3340 Pilot Knob Road
St. Paul, MN 55121

**AACC Texture Technologies Quality Research Award** **90**
Established in 2012, the Texture Technologies Quality Research Award is awarded annually to two authors who write and present high quality research which broadly involves instrumental texture analysis. The objective of the award is to increase awareness among professors and students of the importance of publishing impacting research in sufficient high quality and detail that fellow researchers can replicate and build upon the work. One award will be presented to an author whose paper was published in Cereal Chemistry or Cereal Foods World the previous year. The second award will be presented to an oral or poster during the upcoming AACCI Annual Meeting.
*Requirements:* Award recipients must be either Professional or Student AACCI members. Submitted papers and posters will be judged by the Texture Award Committee.
*Geographic Focus:* All States, All Countries
*Date(s) Application is Due:* Jan 1
*Contact:* Linda Schmitt, AACC International awards Chairperson; (651) 994-3828 or (651) 454-7250; fax (651) 454-0766; lschmitt@scisoc.org or aacc@scisoc.org
*Internet:* www.aaccnet.org/membership/awards/Pages/TextureTechnologiesQualityResearchAward.aspx
*Sponsor:* American Association of Cereal Chemists International
3340 Pilot Knob Road
St. Paul, MN 55121

**AAF Richard Morris Hunt Prize** **91**
The Hunt Prize is named for Richard Morris Hunt, the first American architect to study at the École des Beaux-Arts in Paris, France. Hunt is one of the most renowned 19th-century American architects and helped to formalize architecture as a profession in the U.S. and promote urbanism. The award, co-sponsored by the American Architectural Foundation and the French Heritage Society as part of a commitment to stewardship of American and French heritage, is awarded to architects pursuing a career in historic preservation. The Prize permits young architects from France and the United States to observe and practice the latest preservation technologies and techniques in each other's countries. Inaugurated in 1990 as part of the organizations' commitment to foster stewardship of architectural heritage, the recipients represent the best their countries have to offer in the field of historic preservation. The six-month fellowship alternates each year between a French and an American architect and carries a stipend of $20,000. Since 2011, the Richard Morris Hunt Prize Jury gives an additional opportunity through the Hunt Scholarship Prize to an architect with the characteristics outlined above to spend five weeks in France or America with the assistance of the Managing Teams.
*Geographic Focus:* All States, France
*Amount of Grant:* 20,000 USD
*Contact:* Vanessa Ofwono; (202) 787-1018 or (202) 787-1001; fax (202) 787-1002; vofwono@archfoundation.org or info@archfoundation.org
*Internet:* www.archfoundation.org/2011/11/richard-morris-hunt-prize-overview/
*Sponsor:* American Architectural Foundation
740 15th Street NW
Washington, D.C. 20005

**AAR Italian Fellowships in Humanities** 92

The American Academy in Rome offers up to six fellowships for Italian artists and scholars. The Italian Fellows are an integral part of the AAR community. Italian artists and scholars live and work in the Academy community, pursuing their own projects participating in its collaborative, interdisciplinary environment and helping make connections between Rome Prize Fellows and their Italian peers. In the humanities, the Academy invites proposals in archaeology, classics, history, the history of art and architecture, literature, religious studies, anthropology, and musicology. Any approach to these disciplines or combination of approaches to these areas will be given consideration, as well as historical approaches that overlap with the social sciences. Projects that consider multiple chronological periods listed are welcome. Up to two postdoctoral fellowships in the Humanities will be awarded. Each fellowship will run for about 4½ months. Projects do not need to address topics specific to Rome or to Italy. Candidates should, however, address how they plan to make use of resources in Rome and at the Academy. The annual deadline for application submission is February 1.
*Requirements:* Candidates must: have Italian citizenship (dual citizenship is acceptable); possess a PhD from an Italian university during the past five years; and have excellent spoken and written English, which will be verified by interview.
*Geographic Focus:* All States, Italy
*Date(s) Application is Due:* Feb 1
*Contact:* Shawn Miller, Programs Director; (212) 751-7200; fax (212) 751-7220; shawn.miller@aarome.org or info@aarome.org
*Internet:* www.aarome.org/apply/italian-fellowships
*Sponsor:* American Academy in Rome
7 East 60th Street
New York, NY 10022-1001

**AAR Italian Fellowships in the Arts** 93

The American Academy in Rome offers up to six fellowships for Italian artists and scholars. The Italian Fellows are an integral part of the AAR community. Italian artists and scholars live and work in the Academy community, pursuing their own projects participating in its collaborative, interdisciplinary environment and helping make connections between Rome Prize Fellows and their Italian peers. In the arts, the Academy invites proposals in: architecture (including architecture, landscape architecture and urban planning); design (jewelry and fashion, furniture, set and lighting design, graphic design, and other related design fields); literature; and the visual arts. Any approach to these disciplines or combination of approaches to these areas will be given consideration, as well as historical approaches that overlap with the social sciences. Projects that consider multiple chronological periods listed are welcome. Up to four Italian Fellowships in the Arts will be awarded, with each fellowship running approximately 3½ months.
*Requirements:* All candidates must be nominated; unsolicited applications will not be accepted. Candidates must: have Italian citizenship (dual citizenship is acceptable); possess a PhD from an Italian university during the past five years; and have excellent spoken and written English, which will be verified by interview.
*Geographic Focus:* All States, Italy
*Date(s) Application is Due:* Feb 1
*Contact:* Shawn Miller, Programs Director; (212) 751-7200; fax (212) 751-7220; shawn.miller@aarome.org or info@aarome.org
*Internet:* www.aarome.org/apply/italian-fellowships
*Sponsor:* American Academy in Rome
7 East 60th Street
New York, NY 10022-1001

**AAR Oscar Broneer Traveling Fellowship** 94

In addition to Rome Prize and Italian Fellows, the American Academy in Rome hosts recipients of fellowships awarded by educational and cultural organizations around the world. The Academy and the American School of Classical Studies at Athens award the Oscar Broneer Traveling Fellowship to encourage the study of the Greco-Roman world. The Fellowship will be awarded for research in Greece and Italy in alternate years. It is expected that the Fellow will use either the American Academy in Rome (AAR) or the American School of Classical Studies at Athens (ASCSA) as a base from which to pursue work through trips to sites, museums, or repositories of materials relevant to the Fellow's research. The Fellowship is awarded to one individual per year. It may be held at any time during the academic year for a minimum of three and a maximum of six months. The award is for a maximum of $30,000 and is meant to cover expenses including housing at the host institution, travel (only one round-trip excursion from home base in the United States or Europe to Rome and travel within Italy will be funded), and living expenses. Support from the host institution includes access to research facilities, assistance with obtaining permissions, passes to state museums and monuments, letters of introduction, invitations to various scholarly and other events, and access to other activities offered by the host institution. The annual deadline for application submission is March 15.
*Requirements:* Individuals who have spent a year as a Fellow of the ASCSA are eligible for this Fellowship based at the AAR and for research based in Rome and Italy. Candidates must have an approved dissertation proposal or, if they already hold the PhD, they should be at the beginning of their teaching career and without tenure. Projects may focus on any period of study in the humanities, although preference will be given to topics dealing with classical antiquity.
*Geographic Focus:* All States, All Countries
*Date(s) Application is Due:* Mar 15
*Amount of Grant:* Up to 30,000 USD
*Contact:* Shawn Miller, Programs Director; (212) 751-7200; fax (212) 751-7220; shawn.miller@aarome.org or s.miller@aarome.org
*Internet:* www.aarome.org/apply/affiliated-fellowships
*Sponsor:* American Academy in Rome
7 East 60th Street
New York, NY 10022-1001

**AAR Rome Prize Fellowships** 95

The American Academy in Rome is a leading American overseas center for independent studies and advanced research in the fine arts and humanities. It began as a collaborative effort in 1893 at the World's Columbian Exhibition when a small group, including architects Charles Follen McKim and Daniel Burham, painters John LaFarge and Francis Millet, and sculptors Augustus Saint-Gaudens and Daniel Chester French resolved to create a center to study art amid the classical tradition of ancient Rome. Today the academy awards up to thirty Rome Prize Fellowships to emerging artists and scholars in the early or middle stages of their careers. Fellows are chosen from the disciplines of architecture, design, historic preservation and conservation, landscape architecture, literature, musical composition, visual arts, ancient studies, medieval studies, renaissance and early modern studies, and modern Italian studies. Prize recipients are invited to Rome for six months, eleven months, or two years to immerse themselves in the Academy community. Each Rome Prize winner is provided with a stipend, meals, a bedroom with private bath, and a study or studio. Recipients with a spouse receive a double room. Recipients with children under eighteen live in partially subsidized apartments nearby. Stipends for six-month and eleven-month fellowships are $16,000 and $28,000 respectively. Winners of two-year fellowships receive $28,000 annually. Eleven-month fellowships generally begin in early-September and end at the beginning of August. Winners of six-month fellowships may choose to begin in early September and end in early March or begin in early February and end in early August. The deadline for applications is November 1 but applications will be accepted through November 15 for a higher application fee. Application fees are based upon date of application receipt and number of disciplines for which an applicant applies. The deadline for receipt of letters of recommendation (when required) is November 30. The Academy makes application forms available on their website in early September of the current funding cycle.
*Requirements:* Applicants for the Rome Prize fellowships must be U.S. citizens at the time of application, with the exception of those competing for postdoctoral fellowships sponsored by the National Endowment for the Humanities. U.S. citizens and those foreign nationals who have lived in the United States for the three years immediately preceding the application deadline may apply for the NEH postdoctoral fellowships. Applicants applying solely for NEH post-doctoral fellowships are not required to submit application fees.
*Restrictions:* Graduate students in the humanities may apply only for pre-doctoral fellowships. Previous winners of the Rome Prize are not eligible to reapply. Undergraduate students are not eligible for Rome Prize fellowships. The Rome Prize for Literature Fellowship is awarded only by nomination through the American Academy of Arts and Letters. Rome Prize recipients may not hold full-time jobs while at the Academy.
*Geographic Focus:* All States
*Date(s) Application is Due:* Nov 1; Nov 15
*Amount of Grant:* 16,000 - 54,000 USD
*Contact:* Shawn Miller, Program Director; (212) 751-7200; fax (212) 751-7220; shawn.miller@aarome.org or info@aarome.org
*Internet:* www.aarome.org/apply
*Sponsor:* American Academy in Rome
7 East 60th Street
New York, NY 10022-1001

**AASF Lydia Donaldson Tutt-Jones Memorial Research Grant** 96

The African American Success Foundation (AASF) is a 501(c)3 organization with a mission to promote the positive image of African Americans by funding research and collecting and disseminating information that sheds light on the many successes being attained by African Americans. Holder of a bachelor of science degree in nursing and a masters of science degree in human resources development, Ms. Tutt-Jones (for whom this AASF research grant is named) was committed to excellence and achievement in her life. She dedicated many years of her professional career traveling around the country to recruit highly talented and motivated teachers to work in the Fort Lauderdale public-school system. The Tutt-Jones grant provides financial support to students and professionals who conduct research activities to study African-American success particularly in the area of education. The research can focus upon early childhood education or student performance in the elementary, middle, or high-school years as well as the attitudes and behaviors that help individuals finish college and graduate school. The research can also focus upon parental behaviors and home and community variables that contribute to student success and promote excellence in educational performance. The foundation's interest is to increase the body of knowledge about African-American students who are excelling in school to aid in the replication of that success for others. The duration of the Tutt-Jones grant is for one year; former applicants may re-apply. No special application form, recommended format, or length of material is required. Interested individuals should submit a letter of interest, curriculum vita, a description of the proposed research project (including a timeline and a letter of recommendation from their faculty mentor, if students, or from their department chairperson, agency head, or officer of their professional association, if professionals). An original and five copies of the application materials should be submitted, along with a stamped, self-addressed postcard that will be used to acknowledge receipt of the application. Deadlines may vary. All proposals must be received by 5 pm on the deadline date. Detailed guidelines are available on the website.
*Requirements:* Graduate students and professionals who reside in the United States are eligible to apply. Recipients must prepare a publishable caliber paper upon the completion of the project and provide a copy to the foundation. Student applicants must

be recommended by a faculty mentor who agrees to oversee the project and submission of a publishable caliber paper.
*Geographic Focus:* All States
*Date(s) Application is Due:* Jun 1
*Amount of Grant:* 5,000 USD
*Contact:* Cynthia Wilson, Ph.D., Chairperson of the Grant Selection Committee; (954) 792-1117; fax (954) 792-9191; info@blacksuccessfoundation.org
*Internet:* https://blacksuccessfoundation.org/the-grant/
*Sponsor:* African American Success Foundation
7027 West Broward Boudlevard, #313
Fort Lauderdale, FL 33317

**AASL Research Grant** 97
The AASL Research Grants are given to up to two school librarians, library educators, library information science, or education professors to conduct innovative research aimed at measuring and evaluating the impact of school library programs on learning and education. The study should have the potential to serve as a model for future school library research, and researchers should furnish documentation of the results of their work. Up to two grants of $2,500 each are given every year. Recipients of the Research Grant will be invited to present the results of their projects at a concurrent session during the AASL National Conference. Recipients may also be invited to publish a summary of their findings in School Library Research and/or Knowledge Quest. Winners are announced in the spring.
*Requirements:* Applications are evaluated on the following criteria: potential to measure and/or evaluate the impact of school library programs on learning and education; originality of the research project and methodology; potential for replication; demonstrated ability of the applicant(s) to undertake and successfully complete the project; evidence that sufficient time and resources have been allocated to the effort; support and commitment (financial and otherwise) by institutions and organizations to the project; clarity and completeness of the proposal; personal resume of individual applicant attached; and personal membership in ALA/AASL at the time of application by the applicant.
*Geographic Focus:* All States
*Date(s) Application is Due:* Feb 1
*Amount of Grant:* Up to 2,500 USD
*Contact:* Shannon Carter, Program Coordinator; (312) 280-4381 or (800) 545-2433, ext. 4381; fax (312) 440-9374; scarter@ala.org
*Internet:* www.ala.org/awardsgrants/research-grant-0
*Sponsor:* American Association of School Librarians
50 East Huron Street
Chicago, IL 60611-2795

**AAUW American Dissertation Fellowships** 98
AAUW advances equality for women and girls through advocacy, education, philanthropy, and research. The American Fellowships program has been in existence since 1888, making it the oldest non-institutional source of graduate funding for women in the United States. The program provides fellowships for women writing their dissertations and those pursuing postdoctoral research. Research publication grants are also available to enable scholars to complete manuscripts for publication. The purpose of the Dissertation Fellowship is to offset a scholar's living expenses while she completes her dissertation. Award amount is $20,000, with a November 15 deadline for online submissions of applications.
*Requirements:* Applicants must be U.S. citizens or permanent residents by the application deadline. When comparing proposals of equal merit, special consideration will be given to women holding junior academic appointments who are seeking research leave, women who have held the doctorate for at least three years, and women whose educational careers have been interrupted.
*Restrictions:* Candidates may apply for only one of the awards described. Former recipients of these awards are not eligible to apply for additional American Fellowships or publication grants. Students holding any fellowship for writing a dissertation in the year prior to the AAUW Educational Foundation fellowship year are not eligible.
*Geographic Focus* All States, District of Columbia, Guam, Marshall Islands, Northern Mariana Islands, Puerto Rico, U.S. Virgin Islands, American Samoa
*Date(s) Application is Due:* Nov 15
*Amount of Grant:* 20,000 USD
*Contact:* Gloria Blackwell; (202) 785-7700 or (800) 326-2289; fax (202) 872-1425; fellowships@aauw.org or blackwellg@aauw.org
*Internet:* https://www.aauw.org/what-we-do/educational-funding-and-awards/american-fellowships/af-dissertation-application/
*Sponsor:* American Association of University Women
1310 L Street NW, Suite 1000
Washington, D.C. 20005

**AAUW American Postdoctoral Research Leave Fellowships** 99
AAUW advances equality for women and girls through advocacy, education, philanthropy, and research. Postdoctoral Research Leave Fellowships are designed to assist scholars in obtaining tenure and other promotions by enabling them to spend a year pursuing independent research. The primary purpose of the fellowship is to increase the number of women in tenure-track faculty positions and to promote equality for women in higher education. The award amount is $30,000, with an annual deadline date for submission of online applications being November 15.
*Requirements:* Applicants must be U.S. citizens or permanent residents by the application deadline. Candidates are evaluated on the basis of scholarly excellence; the quality and originality of project design; and active commitment to helping women and girls through service in their communities, professions, or fields of research.
*Restrictions:* Candidates may apply for only one of the awards described. Former recipients of these awards are not eligible to apply for additional American Fellowships or publication grants. Tenured professors are not eligible.
*Geographic Focus* All States, District of Columbia, Guam, Marshall Islands, Northern Mariana Islands, Puerto Rico, U.S. Virgin Islands, American Samoa
*Date(s) Application is Due:* Nov 15
*Amount of Grant:* 30,000 USD
*Contact:* Gloria Blackwell; (202) 785-7700 or (800) 326-2289; fax (202) 872-1425; fellowships@aauw.org or blackwellg@aauw.org
*Internet:* https://www.aauw.org/what-we-do/educational-funding-and-awards/american-fellowships/af-postdoctoral-research-application/
*Sponsor:* American Association of University Women
1310 L Street NW, Suite 1000
Washington, D.C. 20005

**AAUW American Summer/Short-Term Research Publication Grants** 100
AAUW advances equality for women and girls through advocacy, education, philanthropy, and research. Summer/Short-Term Research Publication Grants provide funds for women college and university faculty and independent researchers to prepare research for publication. Time must be available for eight consecutive weeks of final writing and editing in response to issues raised in critical reviews. These grants can be awarded to both tenure-track and part-time faculty, and new and established researchers. The grants are designed to assist the candidate in obtaining tenure and other promotions. The annual online application deadline is November 15, with grant awards of $6,000 available for an eight-week program.
*Requirements:* Applicants must be U.S. citizens or permanent residents by the application deadline. Applicants must have received their doctorates by the application deadline.
*Restrictions:* Funds are not available for research. Candidates may apply for only one of the awards described. Former recipients of these awards are not eligible to apply for additional American Fellowships or publication grants. Tenured professors are not eligible.
*Geographic Focus* All States, District of Columbia, Guam, Marshall Islands, Northern Mariana Islands, Puerto Rico, U.S. Virgin Islands, American Samoa
*Date(s) Application is Due:* Nov 15
*Amount of Grant:* 6,000 USD
*Contact:* Gloria Blackwell; (202) 785-7700 or (800) 326-2289; fax (202) 872-1425; fellowships@aauw.org or blackwellg@aauw.org
*Internet:* https://www.aauw.org/what-we-do/educational-funding-and-awards/american-fellowships/af-research-publication-grants-application/
*Sponsor:* American Association of University Women
1310 L Street NW, Suite 1000
Washington, D.C. 20005

**AAUW Community Action Grants** 101
Community Action grants provide funds to individuals, AAUW branches and AAUW state organizations as well as local community-based nonprofit organizations for innovative programs or non-degree research projects that promote education and equality for women and girls. One-year grants from $2,000 to $7,000 provide funding for community-based projects. Two-year grants from $5,000 to $10,000 provide start-up funds for new projects that address the particular needs of the community and develop girls' sense of efficacy through leadership or advocacy opportunities. Topic areas are unrestricted, but should include a clearly defined activity that promotes education and equality for women and girls. Special consideration is given to projects focused on K-12 and community college girls' and women's achievements in science, technology, engineering, or math. The annual deadline for application is January 15.
*Requirements:* Applicants must be women who are U.S. citizens or permanent residents. Grant projects must have direct public impact, be nonpartisan, and take place within the U.S. or its territories.
*Geographic Focus:* All States
*Date(s) Application is Due:* Jan 15
*Amount of Grant:* 2,000 - 10,000 USD
*Contact:* Gloria Blackwell; (202) 785-7700 or (800) 326-2289; fax (202) 872-1425; fellowships@aauw.org or blackwellg@aauw.org
*Internet:* https://www.aauw.org/what-we-do/educational-funding-and-awards/community-action-grants/
*Sponsor:* American Association of University Women
1310 L Street NW, Suite 1000
Washington, D.C. 20005

**AAUW International Doctoral Fellowships** 102
International Fellowships are awarded for full-time study or research in the United States to women who are not U.S. citizens or permanent residents. Both graduate and postgraduate studies at accredited U.S. institutions are supported. Funding may be renewable for a second year. The annual application deadline is December 1, with an award amount for the International Doctoral Fellowships set at $20,000.
*Requirements:* Candidates must have an academic degree equipment to the U.S. bachelor's degree; a specific plan of study or research that will advance the applicant's professional competence; the intention to return home to pursue a professional career; devote full time to graduate work; and must have satisfactory English proficiency. Applicants whose native language is not English must submit up-to-date scores on one of four language proficiency examinations. Preference will be given to applicants who have specific positions to return to in their own countries.
*Geographic Focus:* All Countries
*Date(s) Application is Due:* Dec 1

*Amount of Grant:* 20,000 USD
*Contact:* Gloria Blackwell; (202) 785-7700 or (800) 326-2289; fax (202) 872-1425; fellowships@aauw.org or blackwellg@aauw.org
*Internet:* aauw-international.scholarsapply.org/
*Sponsor:* American Association of University Women
1310 L Street NW, Suite 1000
Washington, D.C. 20005

**AAUW International Master's/First Professional Degree Fellowships** 103
International Fellowships are awarded for full-time study or research in the United States to women who are not U.S. citizens or permanent residents. Both graduate and postgraduate studies at accredited U.S. institutions are supported. Funding may be renewable for a second year. The annual application deadline is December 1, with an award amount established at $18,000.
*Requirements:* Candidates must have an academic degree equivalent to the U.S. bachelor's degree; a specific plan of study or research that will advance the applicant's professional competence; the intention to return home to pursue a professional career; must devote full time to graduate work; and must have satisfactory English proficiency. Applicants whose native language is not English must submit up-to-date scores on one of four language proficiency examinations. Preference will be given to applicants who have specific positions to return to in their own countries.
*Restrictions:* Previous and current recipients of AAUW fellowships are not eligible to apply.
*Geographic Focus:* All Countries
*Date(s) Application is Due:* Dec 1
*Amount of Grant:* 18,000 USD
*Contact:* Gloria Blackwell; (202) 785-7700 or (800) 326-2289; fax (202) 872-1425; fellowships@aauw.org or blackwellg@aauw.org
*Internet:* https://www.aauw.org/what-we-do/educational-funding-and-awards/international-fellowships/
*Sponsor:* American Association of University Women
1310 L Street NW, Suite 1000
Washington, D.C. 20005

**AAUW International Postdoctoral Fellowships** 104
AAUW is one of the nation's largest sources of private funding exclusively for educational programs that directly benefit women and girls. International Fellowships are awarded for full-time study or research in the United States to women who are not U.S. citizens or permanent residents. Funding may be renewable for a second year. The application and additional information are available at the website. The deadline for International Postdoctoral Fellowships is December 1, with an award amount established at $30,000.
*Requirements:* Candidates must have an academic degree equivalent to the U.S. bachelor's degree; a specific plan of study or research that will advance the applicant's professional competence; the intention to return home to pursue a professional career; devote full time to the work pursued with the fellowship; and must have satisfactory English proficiency. Applicants whose native language is not English must submit up-to-date scores on one of four language proficiency examinations. Preference will be given to applicants who have specific positions to return to in their own countries.
*Restrictions:* Previous and current recipients of AAUW fellowships are not eligible to apply.
*Geographic Focus:* All Countries
*Date(s) Application is Due:* Dec 1
*Amount of Grant:* 30,000 USD
*Contact:* Gloria Blackwell; (202) 785-7700 or (800) 326-2289; fax (202) 872-1425; fellowships@aauw.org or blackwellg@aauw.org
*Internet:* https://www.aauw.org/what-we-do/educational-funding-and-awards/international-fellowships/
*Sponsor:* American Association of University Women
1310 L Street NW, Suite 1000
Washington, D.C. 20005

**AAUW Selected Professions Fellowships** 105
Selected Professions Fellowships are awarded to women who intend to pursue a full-time course of study at accredited U.S. institutions during the fellowship year in one of the designated degree programs where women's participation traditionally has been low (architecture, computer/information sciences, engineering, and mathematics/statistics). Fellowships range from $5,000 to $18,000, with January 10 being the annual deadline for applications.
*Requirements:* Applicants must be U.S. citizens or permanent residents of the U.S. Fellowships in the following degree programs are restricted to women of color, who have been underrepresented in these fields: Master's in business administration — applicants may apply for second year of study only; Law — applicants may apply for third year of study only; and Doctorate in medicine — applicants may apply for third or fourth year of study only.
*Restrictions:* Fellowships in business administration, law, and medicine are restricted to women of color, who have been underrepresented in these fields.
*Geographic Focus* All States, District of Columbia, Guam, Marshall Islands, Northern Mariana Islands, Puerto Rico, U.S. Virgin Islands, American Samoa
*Date(s) Application is Due:* Jan 10
*Amount of Grant:* 5,000 - 18,000 USD
*Contact:* Gloria Blackwell; (202) 785-7700 or (800) 326-2289; fax (202) 872-1425; fellowships@aauw.org or blackwellg@aauw.org
*Internet:* https://www.aauw.org/what-we-do/educational-funding-and-awards/selected-professions-fellowships/
*Sponsor:* American Association of University Women
1310 L Street NW, Suite 1000
Washington, D.C. 20005

**Abbott Fund Global AIDS Care Grants** 106
Abbott has been a significant contributor to the fight against HIV/AIDS for more than two decades. Since 2000, Abbott and the Abbott Fund have invested $225 million in grants and product donations targeted to resource poor countries most impacted by HIV/AIDS. The focus of these efforts includes expanding access to care, testing and treatment; strengthening HIV/AIDS health care systems; preventing mother-to-child transmission; and supporting children and families affected by HIV/AIDS. In addition, Abbott and the Abbott Fund have helped pioneer innovative model programs to combat the disease. For example, we helped build the Baylor International Pediatric AIDS Initiative's first pediatric outpatient clinic in Romania in 2001 that has served an average of 600 patients per year. The clinic has reduced pediatric HIV mortality rates by more than 90 percent. Today Baylor has replicated the Romania model in additional clinics throughout Africa and now serves nearly 60,000 children and young people with HIV – including those at two clinics built by and supported by the Abbott Fund in Malawi and Tanzania.
*Requirements:* Grants are made to tax-exempt organizations supporting global HIV/AIDS programs.
*Restrictions:* Contributions will not be made to individuals; for-profit entities; purely social organizations; political parties or candidates; sectarian religious organizations; advertising; symposia, conferences, and meetings; ticket purchases; memberships; business-related purposes; volunteer efforts of non-Abbott employees; or marketing sponsorships.
*Geographic Focus:* Arizona, California, Illinois, Kansas, Massachusetts, Michigan, New Jersey, New York, North Carolina, Ohio, Puerto Rico, Texas, Utah, Virginia
*Amount of Grant:* 10,000 - 100,000 USD
*Contact:* Cindy Schwab, Vice President; (847) 937-7075; fax (847) 935-5051; cindy.schwab@abbott.com
*Internet:* www.abbottfund.org/projects
*Sponsor:* Abbott Fund
100 Abbott Park Road, Department 379, Building 6D
Abbott Park, IL 60064-3500

**Abbott Fund Science Education Grants** 107
The world urgently needs people who are well-trained in science and technology, and the Abbott Fund is committed to doing its part to address this challenge. Serving as a catalyst by stimulating community investment and engagement, Abbott's investment in science education: engages and inspires students, families and teachers in scientific exploration in out-of-school informal settings; encourages young people to be more proficient in science and attracts more scientists to the field; and builds strong partnerships that are systemic, replicable and sustainable for multiple years and multiple locations. Complete and submit the Abbott Fund online grant application, posted when available. It should include your organization's Federal Tax ID. At times when the Abbott Fund is accepting unsolicited grant applications, it will acknowledge receipt of an online application via email. The Fund will notify an applicant of its decision on a funding request within six to eight weeks.
*Requirements:* Grants are made to tax-exempt organizations supporting company operating areas in Arizona, California, Illinois, Kansas, Massachusetts, Michigan, New Jersey, New York, North Carolina, Ohio, Puerto Rico, Texas, and Virginia, and Utah. It also supports the communities of: Abingdon, England; Brockville, Canada; Campoverde, Italy; Clonmel, Ireland; Cootehill, Ireland; Delkenheim, Germany; Kanata, Canada; Katsuyama, Japan; Ludwigshafen, Germany; Queenborough, England; Rio de Janeiro, Brazil; Sligo, Ireland; and Zwolle, the Netherlands.
*Restrictions:* Contributions will not be made to individuals; for-profit entities; purely social organizations; political parties or candidates; sectarian religious organizations; advertising; symposia, conferences, and meetings; ticket purchases; memberships; business-related purposes; volunteer efforts of non-Abbott employees; or marketing sponsorships.
*Geographic Focus:* Arizona, California, Illinois, Kansas, Massachusetts, Michigan, New Jersey, New York, North Carolina, Ohio, Puerto Rico, Texas, Utah, Virginia, Brazil, Canada, Germany, Great Britain, Ireland, Italy, Japan, Netherlands
*Amount of Grant:* 20,000 - 100,000 USD
*Contact:* Cindy Schwab, Vice President; (847) 937-7075; fax (847) 935-5051; cindy.schwab@abbott.com
*Internet:* www.abbottfund.org/tags/science/1
*Sponsor:* Abbott Fund
100 Abbott Park Road, Department 379, Building 6D
Abbott Park, IL 60064-3500

**Abdus Salam ICTP-ELETTRA Users Grants** 108
The ICTP-ELETTRA Users Program is offering access to the synchrotron radiation facility ELETTRA in Trieste in the years 2008-2013 to scientists who are citizens of developing countries and work in those countries. A minimum of 1500 hours of beamtime can be made available within this program for measurement and applications at any of the existing ELETTRA beamlines. The program offers a limited number of grants to cover travel and living expenses of individuals and small groups who will use beamtime shift at ELETTRA. The number of scientists per group who receive support depends on the number of allocated shifts and available funds: up to 2 scientists for experiments which received less than 12 shifts; and up to 3 scientists for experiments which received 12 shifts or more. In order to allow experiment preparation and/or sample disposal, support can be offered towards stays up to 3 days before the beginning date of the experiment and 1 day after its completion. In exceptional cases, training on a specific topic for longer periods can be arranged if it is meant to improve the expertise of relevance to a project. In that case a quota is deducted from the total of annual hours.
*Geographic Focus:* All Countries
*Date(s) Application is Due:* Mar 15; Sep 15

*Contact:* Patrizia Passarella, Program Assistant; (+39) 040 2240 389; fax (+39) 040 2240 388; itlabs@ictp.trieste.it
*Internet:* www.elettra.trieste.it/userarea/ictp-elettra-users-programme.html
*Sponsor:* Abdus Salam International Centre for Theoretical Physics
Strada Costiera 11
Trieste, I - 34014 Italy

**Abdus Salam ICTP-OEA Affiliated Centre Grants** **109**
An ICTP Affiliated Centre (ICAC) is an institute or university department of physics or mathematics that agrees to carry out a specific long-term research project on a definite subject with well-defined purposes. An ICAC is expected to have a regional character and to be strongly supported by the local authorities and the hosting Institution. The proposer is expected to submit a scientific project to the ICTP, in collaboration with the leading scientists of that department, and possibly, of other departments in the country. Many physicists participate in the ICAC activities. As a result of such collaboration, an ICAC scientific advisory board is normally formed. The scientific project must consist of: a general program for activities planned over a period of three years; and a detailed project and itemized budget for the first year. Funding will be determined by scope of the project.
*Requirements:* An institute or university department of physics or mathematics in a developing country that agrees to carry out a specific long-term research project on a definite subject with well defined purposes may apply.
*Geographic Focus:* All Countries
*Contact:* Patrizia Passarella, Program Assistant; (+39) 040 2240 323 or (+39) 040 2240 322; fax (+39) 040 2240 443; itlabs@ictp.trieste.it or oea@ictp.it
*Internet:* oea.ictp.it/
*Sponsor:* Abdus Salam International Centre for Theoretical Physics
Strada Costiera 11
Trieste, I - 34014 Italy

**Abdus Salam ICTP-OEA Projects Grants** **110**
The OEA promotes initiatives aimed at "training for research" of young scientists. A Project offers scientific advice and financial support towards the establishment of new Ph.D. courses in physics or mathematics especially at universities in the less developed countries. A Project may also provide support towards fellowships for young scientists from less developed countries to work in institutions in more advanced developing countries in their region. These are often organized on the basis of cooperation agreements between the ICTP and those institutions. In general a Project is meant to undertake actions which are likely to develop into a long term research program. Thus a Project could also be the preliminary phase leading eventually to an Affiliated Center or to the formation of research groups.
*Requirements:* Developing country university physics or mathematics departments, especially those in less developed countries may apply.
*Geographic Focus:* All Countries
*Contact:* Patrizia Passarella, Program Assistant; (+39) 040 2240 323 or (+39) 040 2240 322; fax (+39) 040 2240 433; itlabs@ictp.trieste.it or oea@ictp.it
*Internet:* oea.ictp.it/oea/programmes/projects/
*Sponsor:* Abdus Salam International Centre for Theoretical Physics
Strada Costiera 11
Trieste, I - 34014 Italy

**Abdus Salam ICTP-OEA Scientific Meetings Grants** **111**
The Abdus Salam ICTP Office of External Activities (O.E.A.) encourages the organization of international and regional scientific meetings in developing countries by offering financial assistance to the organizers of conferences, workshops, and schools held in O.E.A. supported countries. Grants are offered for meetings in the fields of physics and mathematics. A similar program sponsored by the Third World Academy of Sciences offers support for meetings in the following fields of natural sciences: agricultural, biological, chemical, engineering, geological and medical sciences.
*Requirements:* Institutions in developing countries that want to organize international and regional scientific meetings in developing countries should apply. The application should include a scientific program and list of invited speakers with the topics indicated in detail together with title of lectures. Importance will be given to applications with most of the key speakers already confirmed.
*Geographic Focus:* All Countries
*Contact:* Eleonora Crotta, (+39) 040 2240 323 or (+39) 040 2240 322; fax (+39) 040 2240 443; crotta@ictp.it or oea@ictp.it
*Internet:* oea.ictp.it/oea/programmes/sm/
*Sponsor:* Abdus Salam International Centre for Theoretical Physics
Strada Costiera 11
Trieste, I - 34014 Italy

**Abdus Salam ICTP-OEA Visiting Scholars/Consultants Grants** **112**
The Visiting Scholar/Consultant Program provides assistance to Institutes or research groups in a developing country requiring expert advice to enhance existing activities or to initiate a new research program. The objective of the program is to establish a permanent link between an outside expert in a particular field of physics or mathematics and the applying Institute on a long-term basis. The Visiting Scientist is an established researcher whose visits will provide a clear benefit to the research group of the applicant in particular and to the host institution in general. To ensure a long-term scientific collaboration the visiting scientist will be expected to deliver a series of topical lectures/seminars and to interact closely with the hosting scientists. The visiting scientist should be the same throughout the project. Only rarely may different scientists carry out the same project, provided they are from the same scientific group as the visiting scientist. It is important to note that the grant is awarded to the proposer and not to the visiting scientist. Round trip travel costs of the visiting scientist are allowed. The amount provided depends on the application, but usually does not exceed Euro 2,500 for each visit.
*Requirements:* The visiting scientist should make a commitment for two or three visits of about one month each over a period of three years. The Assignments will automatically terminate three years after they are granted.
*Geographic Focus:* All Countries
*Date(s) Application is Due:* Sep 30
*Amount of Grant:* Up to 7,500 EUR
*Contact:* Patrizia Passarella, Program Assistant; (+39) 040 2240 323 or (+39) 040 2240 322; fax (+39) 040 2240 443; itlabs@ictp.trieste.it or oea@ictp.it
*Internet:* https://oea.ictp.it/oea/programmes/vs/
*Sponsor:* Abdus Salam International Centre for Theoretical Physics
Strada Costiera 11
Trieste, I - 34014 Italy

**Abdus Salam ICTP/IAEA Sandwich Training Education Grants** **113**
The Sandwich Training Educational Program (STEP) aims at offering fellowship opportunities to Ph.D. students from developing countries in the fields of physics and mathematics. The goal of the STEP is to strengthen the scientific capability of young scientists and researchers from developing countries to better contribute and serve the scientific, technical, and economic development of their home countries. The STEP offers fellowships to candidates from the International Atomic Energy Agency (IAEA) member states in the following fields: atomic and nuclear physics; nuclear, isotope and laser techniques applied to human health, agriculture, biotechnology and biophotonics, and water resources and environmental studies; synchrotron radiation and applications; and medical radiation physics. Fellowships are also provided by the ICTP, within its scientific and technical competence, to candidates from developing countries in the fields listed below: condensed matter and statistical physics; high energy physics; mathematics; earth system physics; and applied physics, including micro- and nano-electronics, fluid dynamics, and dense plasma source physics. Applications are expected from students in developing countries who are enrolled in a PhD course at their home university. PhD students who are supervised by an ICTP Associate are encouraged to apply. The program aims at an equitable participation of fellows, and therefore especially welcomes applications from woman scientists. The STEP fellowship includes a stipend of Euro 1,400 per month during each stay at the host institute, plus travel costs and medical insurance. Additional information is available at the ICTP website.
*Geographic Focus:* All Countries
*Date(s) Application is Due:* Feb 15; Aug 15
*Contact:* Patrizia Passarella, Program Assistant; (+39) 040 2240 389; fax (+39) 040 2240 388; itlabs@ictp.trieste.it or calligar@ictp.it
*Internet:* portal.ictp.it/step/
*Sponsor:* Abdus Salam International Centre for Theoretical Physics
Strada Costiera 11
Trieste, I - 34014 Italy

**Abdus Salam ICTP Federation Arrangement Scheme Grants** **114**
Federation Arrangements are agreements signed by the ICTP and a scientific institute (Federated Institute) in developing countries. This enables the Institute to send junior scientists - up to the age of 40 - to the ICTP for a total of 60 to 150 days, during the validity of the contract (i.e. three years), on a cost-sharing basis. While living expenses are always covered by the Center, travel expenses are either covered by the Federated Institute or shared with the Center. The Federation Arrangement Scheme gives the opportunity to junior scientists from developing countries to participate in the Center's scientific programs. Their participation is proposed by the Federated Institute and approved by the ICTP. Presently there are five different standard types of Federation Arrangements depending upon geographical areas. While the junior scientists' living expenses are always covered by the Centre, travel expenses are either covered by the Federated Institute or shared with the Center.
*Requirements:* Scientific institutes from developing countries (to send junior scientists to ICTP) are eligible to apply.
*Geographic Focus:* All Countries
*Contact:* Patrizia Passarella, Program Assistant; (+39) 040 2240 389; fax (+39) 040 2240 388; itlabs@ictp.trieste.it or diploma@ictp.it
*Internet:* assoc.ictp.it/federation-scheme/description.html
*Sponsor:* Abdus Salam International Centre for Theoretical Physics
Strada Costiera 11
Trieste, I - 34014 Italy

**Abdus Salam ICTP Group Associates Grants** **115**
The ICTP Associateship Scheme was created at the very inception of the Centre. It represents one of the means through which the vocation of the ICTP for the promotion and the development of scientific knowledge in Third World Countries has concretely been applied to reality. In the context of the fundamental goal of the ICTP of promoting research in developing countries it has become apparent that, in many countries, due in part to historical reasons combined with budgetary restrictions, research is hampered not only by isolation but even more by the atomization of research interests. For this reason, research collaboration groups rather than individual awards were established. Group Association awards last six years and are composed of a certain number of Associate awards, according to age and therefore, its members are either Junior, Regular or Senior Associates. Each Group will avail of a number of Associateship privileges depending on the number of components of the Group such as to allow each member at least one visit

at ICTP during the allocation period. Depending on composition of group, each member receives entitlements as either Junior, Regular or Senior Associates.
*Requirements:* Research Groups must be operating in developing countries and have people working and living in the same or in nearby institutes. Each member of the group must fill in an individual online application form, and specify that they are applying for a group association.
*Geographic Focus:* All Countries
*Date(s) Application is Due:* Sep 30
*Contact:* Patrizia Passarella, Program Assistant; (+39) 040 2240 389; fax (+39) 040 2240 388; itlabs@ictp.trieste.it or diploma@ictp.it
Daniela Valentini; (+39) 040 2240 551; fax (+39) 040 2240 511; valentin@ictp.it
*Internet:* assoc.ictp.it/associateship-scheme/group.html
*Sponsor:* Abdus Salam International Centre for Theoretical Physics
Strada Costiera 11
Trieste, I - 34014 Italy

**Abdus Salam ICTP Joint Laurea Magistralis Grants** **116**
There are a limited number of scholarships available to students from developing countries selected to follow the training track in Condensed Matter Physics. Each scholarship consists of: 800 Euro a month, travel grant, insurance and tuition fees. ICTP cooperates with the University of Trieste's Faculty of Sciences in providing physics education to graduate students from developing countries. This international study programme for the "Laurea Magistralis Interateneo" (Universities of Trieste and Udine) in physics, includes curricula in: microphysics and condensed matter physics; theoretical physics; astrophysics and cosmology; and earth and environmental physics. The Programme covers two academic years and is designed to provide young promising students in physics from developing countries with graduate training suitable for pursuing teaching and research in physics. Courses are held in English and each are followed by an examination. A course on Italian language will be organized during the first year. After having passed all exams, participants prepare and defend a dissertation. Upon successful completion of the entire study plan, the students will obtain the "Laurea Magistralis" from the University of Trieste. The participants may also be required to take part in ongoing activities of ICTP.
*Requirements:* The Program is open to young, qualified graduates from all countries that are members of the United Nations, UNESCO or IAEA. The minimum qualification for applicants is a degree equivalent to an M.Sc. (or an exceptionally good B.Sc.) in physics. The selection of candidates will be based on their university record and academic recommendations. Fluency in both spoken and written English is required.
*Geographic Focus:* All Countries
*Date(s) Application is Due:* May 31
*Amount of Grant:* 9,600 EUR
*Contact:* Patrizia Passarella, Program Assistant; (+39) 040 2240 389; fax (+39) 040 2240 388; itlabs@ictp.trieste.it or diploma@ictp.it
*Internet:* www.ictp.it/programmes/joint-laurea-description.aspx
*Sponsor:* Abdus Salam International Centre for Theoretical Physics
Strada Costiera 11
Trieste, I - 34014 Italy

**Abdus Salam ICTP Joint PhD Programme in Environmental Fluid Mechanics Grants** **117**
The Doctorate School in Environmental and Industrial Fluid Mechanics trains students in the field of Fluid Dynamics with special emphasis on basic fluid dynamics, mathematical methods, the physics of large scale flows and application to industrial problems. Due to the interdisciplinarity of the program, students have different backgrounds in the fields of engineering, physics and mathematics, and acquire skills useful both for the academic environment and for a career in high-tech companies. The PhD school is based at the University of Trieste in collaboration with the following institutes: International Center for Theoretical Physics; Istituto Nazionale di Oceanografia e Geofisica Sperimentale; Agenzia Regionale Protezione Ambientale - Friuli Venezia Giulia; and Istituto di Scienze Marine.
*Geographic Focus:* All Countries
*Date(s) Application is Due:* Oct 31
*Contact:* Patrizia Passarella, Program Assistant; (+39) 040 2240 389; fax (+39) 040 2240 388; itlabs@ictp.trieste.it or diploma@ictp.it
Vincenzo Armenio; (+39) 040 2240 389; fax (+39) 040 2240 388; armenio@dica.units.it
*Internet:* phdfluidmechanics.appspot.com/
*Sponsor:* Abdus Salam International Centre for Theoretical Physics
Strada Costiera 11
Trieste, I - 34014 Italy

**Abdus Salam ICTP Junior Associates Grants** **118**
The ICTP Associateship Scheme was created at the very inception of the Centre. It represents one of the means through which the vocation of the ICTP for the promotion and the development of scientific knowledge in Third World Countries has concretely been applied to reality. Junior Associates are intended for promising young scientists up to the age of 35. The award lasts for six years and entitles them to spend from 35 to 70 days at the Center, three times during the validity of the award, with fare and living allowance covered by the Associateship. After the three visits are completed, should the Associate Member still have a certain number of days remaining from his 210 day allocation, then he/she is entitled to visit the ICTP with living expenses covered.
*Requirements:* Scientists up to age 35 from and working in developing countries are eligible.
*Geographic Focus:* All Countries
*Date(s) Application is Due:* Sep 30
*Contact:* Patrizia Passarella, Program Assistant; (+39) 040 2240 389; fax (+39) 040 2240 388; itlabs@ictp.trieste.it or diploma@ictp.it
Daniela Valentini; (+39) 040 2240 551; fax (+39) 040 2240 511; valentin@ictp.it
*Internet:* assoc.ictp.it/associateship-scheme/junior.html
*Sponsor:* Abdus Salam International Centre for Theoretical Physics
Strada Costiera 11
Trieste, I - 34014 Italy

**Abdus Salam ICTP Mathematical Research Fellowships** **119**
The Abdus Salam International Centre for Theoretical Physics (ICTP) invites applications for a small number of fellowships for outstanding mathematicians from developing countries. There are two types: Visiting Fellowships are open to mathematicians of any age and fund visits of between 1 and 12 months to ICTP during the next year; Post-Doctoral Fellowships are intended for younger mathematicians who have obtained a PhD in mathematics after January 1. Fellowships have a 12 month duration with a possibility to be renewed for a further 12 months. Preference is given to candidates who will benefit most from the time spent at ICTP, in the sense of pursuing their own research, using the ICTP facilities, interacting with other mathematicians, and ultimately turning the fellowship into a positive opportunity for their home institution or country as well as for themselves. Candidates should apply using the online application.
*Geographic Focus:* All Countries
*Date(s) Application is Due:* Jan 13
*Contact:* Patrizia Passarella; (+39) 040 2240 201; fax (+39) 040 2240 490; math@ictp.it
*Internet:* www.ictp.it/research/math/visiting-fellowships.aspx
*Sponsor:* Abdus Salam International Centre for Theoretical Physics
Strada Costiera 11
Trieste, I - 34014 Italy

**Abdus Salam ICTP Networks Grants** **120**
A Network is a system of research groups in an entire region, or among different regions, that pursues a common scientific project over an extended period. The network program represents an efficient approach to overcoming the problem of isolation and counteracting the "brain-drain" of scientists in the Third World. Scientific network projects receive support in the following ways: as institutes or individuals with common scientific interests who agree upon a well defined joint project that emphasizes collaboration and the sharing of expertise and facilities. The proposer of the network is expected to act as coordinator and his/her Institution as headquarters. A network may organize seminars or workshops, arrange for visiting scientists and maintain joint links with the rest of the international scientific community. The ICTP contributes towards the travel expenses of scientists who belong to the network to visit other member Institutes. Grants can be offered to young scientists to work for extended periods at institutes that are part of the network. The fields of study are physics or mathematics.
*Requirements:* A network project proposal should include: a detailed scientific and financial plan for one year; the list of member institutions/individuals; a description of the facilities of the member institutions that will be involved in the network; a justification for the proposed network; and the commitment of the member institutions to contribute substantially to the expenses of the project, in particular, to offer hospitality and local services. Academic institutions, nonprofit organizations, and Ph.D./M.D./ other professionals may apply.
*Geographic Focus:* All Countries
*Contact:* Patrizia Passarella, Program Assistant; (+39) 040 2240 323 or (+39) 040 2240 322; fax (+39) 040 2240 433; itlabs@ictp.trieste.it or oea@ictp.it
*Internet:* oea.ictp.it/programmes/net/index_html
*Sponsor:* Abdus Salam International Centre for Theoretical Physics
Strada Costiera 11
Trieste, I - 34014 Italy

**Abdus Salam ICTP Postdoctoral Positions in Theoretical High Energy Physics** **121**
Up to four new postdoctoral positions will be available in the Abdus Salam ICTP High Energy, Cosmology and Astroparticle Physics (HECAP) section for the upcoming academic years. The award amount is unspecified. Appointments will be made initially for two years, with the possibility of an extension for a third year. Members of the section are currently working on: string theory, including D-branes, gauge/string theory duality, phenomenological string theory/M theory; particle phenomenology including grand unification, supersymmetry, extra dimensions, LHC/collider physics; cosmology including large scale structures, physics of the early universe, present acceleration; topological field theories; and neutrino physics and astrophysics.
*Geographic Focus:* All Countries
*Date(s) Application is Due:* Dec 1
*Contact:* Patrizia Passarella, Program Assistant; (+39) 040 2240 323 or (+39) 040 2240 322; fax (+39) 040 2240 443; itlabs@ictp.trieste.it or rosanna@ictp.it
*Internet:* www.ictp.it/about-ictp/media-centre/news/postdoctoral-positions-in-theoretical-high-energy-physics.aspx
*Sponsor:* Abdus Salam International Centre for Theoretical Physics
Strada Costiera 11
Trieste, I - 34014 Italy

**Abdus Salam ICTP Postgraduate Diploma Grants** **122**
Talented young science students in developing countries are sometimes limited in achieving their full potential by the absence of advanced training at a postgraduate level, and up to standard international criteria. The late Professor Abdus Salam, Nobel Laureate and former Director of ICTP instituted an intense 12-month program in the main research fields of the center, with the aim of filling this gap. Among the projects on capacity building that ICTP has undertaken over time is the so-called Postgraduate Diploma Programme. The Programme consists of a rigorous, one-year long, pre-doctoral

course work, with a small part devoted to independent projects. The Program initially had three streams: High Energy Physics, Condensed Matter Physics and Mathematics. A fourth stream has recently been added, Earth System Physics. The goal in these programs is to take good students from the least developed countries generally – and for the Physics Program students from only sub-Saharan Africa – and educate them so they can compete favorably for graduate studies in any center of learning in the world.
*Geographic Focus:* All Countries
*Date(s) Application is Due:* Dec 31
*Contact:* Patrizia Passarella, Program Assistant; (+39) 040 2240 389; fax (+39) 040 2240 388; itlabs@ictp.trieste.it or diploma@ictp.it
*Internet:* diploma.ictp.it/
*Sponsor:* Abdus Salam International Centre for Theoretical Physics
Strada Costiera 11
Trieste, I - 34014 Italy

**Abdus Salam ICTP Ramanujan Prize for Young Mathematicians** 123

ICTP has created the Ramanujan Prize for young mathematicians from developing countries. The Prize is awarded annually to a researcher from a developing country less than 45 years of age on 31 December of the year of the award, who has conducted outstanding research in a developing country. Researchers working in any branch of the mathematical sciences are eligible. The Prize carries a $15,000 cash award and travel and subsistence allowance to visit ICTP for a meeting where the winner will be required to deliver a lecture. The Selection Committee will take into account not only the scientific quality of the research, but also the background of the candidate and the environment in which the work was carried out. The Committee will in particular favor candidates who have overcome adversity to achieve distinction in mathematics. The Prize is usually awarded to one person, but may be shared equally among recipients who have contributed to the same body of work.
*Requirements:* Nominations should include a CV and a list of publications, as well as a letter of recommendation. Additional supporting letters are encouraged.
*Restrictions:* Self nominations are not eligible.
*Geographic Focus:* All Countries
*Date(s) Application is Due:* Feb 1
*Amount of Grant:* 15,000 USD
*Contact:* Patrizia Passarella, Program Assistant; (+39) 040 2240 323 or (+39) 040 2240 322; fax (+39) 040 2240 443; director@ictp.it
*Internet:* www.ictp.it/about-ictp/prizes-awards/the-ramanujan-prize.aspx
*Sponsor:* Abdus Salam International Centre for Theoretical Physics
Strada Costiera 11
Trieste, I - 34014 Italy

**Abdus Salam ICTP Regular Associates Grants** 124

The ICTP Associateship Scheme represents one of the means through which the promotion and the development of scientific knowledge in Third World countries has concretely been applied to reality. Regular Associates between the age of 36-45 are entitled to visit for 35 to 70 days, three times in a period of six years. For three visits the Centre provides their fare and a living allowance. After the three visits are completed, should the Associate Member still have a certain number of days remaining from his 210 day allocation, then he/she is entitled to visit the ICTP with living expenses covered.
*Requirements:* Scientists up to age 35 from and working in developing countries are eligible.
*Geographic Focus:* All Countries
*Date(s) Application is Due:* Sep 30
*Contact:* Patrizia Passarella, Program Assistant; (+39) 040 2240 389; fax (+39) 040 2240 388; itlabs@ictp.trieste.it or diploma@ictp.it
Daniela Valentini, Senior Assistant; (+39) 040 2240 551; fax (+39) 040 2240 511; valentin@ictp.it
*Internet:* assoc.ictp.it/associateship-scheme/regular.html
*Sponsor:* Abdus Salam International Centre for Theoretical Physics
Strada Costiera 11
Trieste, I - 34014 Italy

**Abdus Salam ICTP Schools, Conferences, Workshops and Extended Research Workshops Grants** 125

Each year, the ICTP organizes training and research activities in all areas of physica; sciences including mathematics. The subjects cover theory, simulations, and experiments. Proposals may be areas where there is in-house interest or research at ICTP, as follows: high energy, cosmology and astroparticle physics; condensed matter physics including statistical physics; fundamental and applied mathematics; earth system physics including geophysics, ocean cynamics, weather and climate modeling, upper atmosphere; and applied physics including fluid dynamics, plasma physics, nuclear data and applications, laser physics, energy systems, medical physics, ecological and environmental economics, machine learning, computer algorithms, soft matter, biophysics and quantitative aspects of life sciences and radio propagation. Proposals on web technologies and distance education, information and communication technologies, open source for science and development are also welcome. Subjects are not restricted to theory and the activities are selected on the basis of scientific novelty, the impact on the international community with special emphasis on bringing together scientists from South and North, basic training for younger scientists, and hands-on training for computer-intensive subjects. Where appropriate, proposals that combine pedagogical material with advanced research results, or those that spur new results in an emerging or multidisciplinary field, or those that resolve an outstanding or controversial issue will be preferred.
*Requirements:* Proposals are welcome at any time. Decisions concerning the program for next year are taken around March or April. Since the decision process involves several stages, early submission of proposals with a lead time of a few months is strongly encouraged. Proposals arriving after February 15 will as a rule not be processed for inclusion in the next year program, and might be considered for subsequent years only if there is room available.
*Geographic Focus:* All Countries
*Date(s) Application is Due:* Feb 15
*Contact:* Patrizia Passarella, Program Assistant; (+39) 040 2240 389; fax (+39) 040 2240 388; itlabs@ictp.trieste.it or diploma@ictp.it
Claudio Arezzo; (+39) 040 2240 389; fax (+39) 040 2240 388; arezzo@ictp.it
Stefano Luzzatto; (+39) 040 2240 389; fax (+39) 040 2240 388; luzzatto@ictp.it
*Internet:* www.ictp.it/call-for-proposals.aspx
*Sponsor:* Abdus Salam International Centre for Theoretical Physics
Strada Costiera 11
Trieste, I - 34014 Italy

**Abdus Salam ICTP Senior Associates Grants** 126

The ICTP Associateship Scheme represents one of the means through which the vocation of the ICTP for the promotion and the development of scientific knowledge in Third World countries has concretely been applied to reality. All candidates over the age of 45 will automatically be considered for a Senior Associateship award. However, a Senior Associateship may be granted to candidates below the age of 45 who opt only to apply for a Senior Associateship award by stating this very clearly on their application form. The upper age limit for Senior Associates is 62 at the time of award. During the six years, Senior Associate Members may apply to visit the Center as often and for as long as they wish, until the allocation is exhausted, although the maximum foreseen duration of any visit is 70 days. Payment is Euros 8,000, which must be used to cover travel and living expenses during the 6-year appointment.
*Geographic Focus:* All Countries
*Date(s) Application is Due:* Sep 30
*Amount of Grant:* 8,000 EUR
*Contact:* Patrizia Passarella, Program Assistant; (+39) 040 2240 389; fax (+39) 040 2240 388; itlabs@ictp.trieste.it or diploma@ictp.it
Daniela Valentini; (+39) 040 2240 551; fax (+39) 040 2240 511; valentin@ictp.it
*Internet:* assoc.ictp.it/associateship-scheme/senior.html
*Sponsor:* Abdus Salam International Centre for Theoretical Physics
Strada Costiera 11
Trieste, I - 34014 Italy

**Abdus Salam ICTP Training and Research in Italian Laboratories Grants** 127

The ICTP Programme gives scientists from Developing Countries in Central and Eastern Europe the opportunity to spend periods ranging from a few months to one year at Italian research laboratories of universities, governmental and private institutions, which operate in different branches of Physics. This includes academic studies as well as practical applications and industrial projects. Follow up visits and collaboration programs are also envisioned. Agreements of collaboration have been established with Italian research institutions, i.e. CNR (National Research Council), INFN (National Institute for Nuclear Physics), ENEA (Italian Commission for New Technologies, Energy and Environment) and INFM (National Institute for the Physics of Matter), APAT (Agency for Environmental Protection and Technical Services), Sincrotrone Trieste (Elettra Synchrotron Light Source), National Institute of Oceanography and Experimental Geophysics (OGS) and others. They offer hospitality and a special financial contribution. The fields covered, which reflect activities held at the ICTP, are presently: physics of condensed matter - solid state physics and materials science, computational methods, atomic and molecular physics, laser spectroscopy, semiconductor physics and technology; physics and energy - plasma physics and fusion, non-conventional energy sources, applied nuclear physics; physics and technology - optical physics and lasers, communication physics and technology, optical fibers, microprocessors and informatics, synchrotron radiation applications; earth and environmental sciences - geophysics and seismology, oceanography, physics of the atmosphere and aeronomy, meteorology and climatology, soil physics, environmental control, mathematical ecology, modelling of environmental systems; physics of the living state - biophysics, medical physics, neurophysics; and miscellaneous - instrumentation for nuclear and subnuclear physics, space physics, topics at the interface with chemistry, engineering, or biology. The minimum amount of these grants will be Euro 1,300 per month. Limited funds for travel expenses will be available in case these are not covered by other sources. Health insurance coverage will be provided for visitors and their accompanying family members. A monthly family allowance for those having dependent family members will be granted. The average length of a visit is 6 to 12 months. Additional information, including the application form, is available at the ICTP website.
*Requirements:* Less developed countries in Central and Eastern Europe may apply.
*Geographic Focus:* All Countries
*Amount of Grant:* Up to 15,600 EUR
*Contact:* Patrizia Passarella, Program Assistant; (+39) 040 2240 389; fax (+39) 040 2240 388; itlabs@ictp.trieste.it or diploma@ictp.it
*Internet:* users.ictp.it/www_users/ItaLab/index.html
*Sponsor:* Abdus Salam International Centre for Theoretical Physics
Strada Costiera 11
Trieste, I - 34014 Italy

**Abdus Salam ITCP Postgraduate Diploma Programme in General Physics** 128

In each Postgraduate Diploma Programme, scholarships and travel grants will be awarded to those 10 students selected from developing countries. There are no course fees. The scholarships cover living expenses for the entire year of study (around 800 Euro per month). A limited number of qualified candidates who meet the same selection

criteria as others, may attend the courses at their own cost. They must supply written evidence from the funding agency, or banking institution that funds are available for their entire year's stay, and are equal to those offered by the Centre.
*Requirements:* B.Sc. (but a M.Sc. is preferred) candidates in Physics or Mathematics are preferred. Applicants under 26 are also preferred from less developed countries, particularly sub-Sahara Africa.
*Geographic Focus:* All States, Angola, Benin, Botswana, Burkina Faso, Burundi, Cameroon, Cape Verde, Central African Republic, Chad, Comoros, Congo, Congo, Democratic Republic of, Cote d' Ivoire (Ivory Coast), Djibouti, Equatorial Guinea, Eritrea, Ethiopia, Gabon, Gambia, Ghana, Guinea, Guinea-Bissau, Kenya, Lesotho, Liberia, Madagascar, Malawi, Mali, Mauritania, Mozambique, Namibia, Niger, Nigeria, Reunion, Rwanda, Sao Tome & Principe, Senegal, Seychelles, Sierra Leone, Somalia, South Africa, Sudan, Swaziland, Tanzania, Togo, Uganda, Western Sahara, Zambia, Zimbabwe
*Date(s) Application is Due:* Dec 31
*Amount of Grant:* Up to 9,600 EUR
*Contact:* Patrizia Passarella, Program Assistant; (+39) 040 2240 389; fax (+39) 040 2240 388; itlabs@ictp.trieste.it or diploma@ictp.it
*Internet:* www.ictp.it/programmes.aspx
*Sponsor:* Abdus Salam International Centre for Theoretical Physics
Strada Costiera 11
Trieste, I - 34014 Italy

**Abdus Salam Joint ICTP/SISSA PhD Programme in Physics and Mathematics Grants** 129

The Abdus Salam International Centre for Theoretical Physics (ICTP) and the International School for Advanced Studies (SISSA) offer the opportunity for carrying out Ph.D. studies in the following fields: Applied Mathematics; Mathematical Analysis; Geometry; Mathematical Physics; Astroparticle Physics; Astrophysics; Elementary Particles; Physics and Chemistry of Biological Systems; Statistical Physics; and Theory and Numerical Simulation of the Condensed Matter. Admitted candidates will be awarded a fellowship of the amount of Euro 13,638.47 gross per year (approx. Euro 1030,00 net per month) for the entire duration of the course (3 years extendable to a fourth, subject to availability of funds and approval by ICTP/SISSA Governing bodies). A contribution towards living expenses of the amount of Euro 100,00 gross per month will be awarded to those who will subscribe a registered rent contract. A contribution towards the expenses to be enrolled in the health insurance system up to Euro 198.77 per year will be granted.
*Requirements:* Applicants interested in participating in this selection must be possess an "ICTP Diploma" or be enrolled in the last year of the "ICTP diploma Program". All the admitted candidates must hold the "ICTP Diploma" by the date of the beginning of the courses (3 October for Applied Mathematics, Astroparticle Physics, Geometry, Mathematical Analysis, Mathematical Physics and Theory of Elementary Particles, 7 November for all other courses). Admitted candidates will have to pass a qualifying examination within the first year of study.
*Geographic Focus:* All Countries
*Date(s) Application is Due:* Mar 31
*Amount of Grant:* 14,838 - 15,037 EUR
*Contact:* Patrizia Passarella, Program Assistant; (+39) 040 2240 389; fax (+39) 040 2240 388; itlabs@ictp.trieste.it or diploma@ictp.it
*Internet:* www.ictp.it/programmes/degree-programmes.aspx
*Sponsor:* Abdus Salam International Centre for Theoretical Physics
Strada Costiera 11
Trieste, I - 34014 Italy

**Abington Foundation Grants** 130

The Abington Foundation awards grants to Ohio nonprofit organizations serving Cuyahoga County in its areas of interest, including pre-primary and higher education, geriatric healthcare and nursing, the promotion or sustenance of individual and family economic independence, and cultural activities. Priority is given to funding requests for specific programs or projects that represent critical periods in a child's development from birth to age five, and to early adolescence, ages ten through fifteen. Requests for endowment or general operating support are discouraged. Foundation staff is available for consultation during the proposal preparation process.
*Requirements:* Ohio nonprofit organizations serving Cuyahoga County are eligible.
*Geographic Focus:* Ohio
*Date(s) Application is Due:* May 1; Sep 1; Dec 1
*Amount of Grant:* 5,000 - 50,000 USD
*Samples:* Cleveland Public Theatre, Cleveland, Ohio, $15,000 - to support the Brick City after school and summer arts program (2018); Montclair Art Museum, Cleveland, Ohio, $30,000 - to support the Pilot an Artist in Residence program (2018); University Circle Incorporated, Cleveland, Ohio, $50,000 - for the development of the Cozad-Bates House Interpretive Center (2018).
*Contact:* Cristin Slesh, Consultant; (216) 621-2901 or (216) 621-2632; fax (216) 621-8198; cslesh@fmscleveland.com
*Internet:* www.fmscleveland.com/abington/grant-philosophy.cfm
*Sponsor:* Abington Foundation
1422 Euclid Avenue, Suite 966
Cleveland, OH 44115-1952

**Abracadabra Foundation Grants** 131

The Abracadabra Foundation was established in Oregon in 1997 to provide funding support for Attention Deficit Disorder (ADD) and Attention Deficit Hyperactivity Disorder (ADHD) research in the Portland, Oregon, region. Its primary fields of interest, therefore, are medical research and education. Types of funding generally include internships and research funds. A specific application form is not required, and interested parties should forward a letter directly to the Foundation office. There are no annual deadlines. Most recently, awards have ranged from $2,500 up to $200,000.
*Requirements:* 501(c)3 non-profits and research facilities, primarily in the Portland, Oregon, metropolitan area, should apply.
*Geographic Focus:* Oregon
*Samples:* OHSU Foundation, Portland, Oregon, $200,000 - ADHD research funding; Stanford University, Stanford, California, $2,500 - general donation.
*Contact:* Steven J. Sharp, President; (503) 615-9408
*Sponsor:* Abracadabra Foundation
3605 NW Bliss Road
Vancouver, WA 98685-1512

**Academy of American Franciscan History Dissertation Fellowships** 132

The Academy of American Franciscan History is accepting applications for four dissertation fellowships, each worth $10,000. As many as two of these fellowships will be awarded for projects dealing with some aspect of the history of the Franciscan Family in Latin America, including the United States Borderlands, Mexico, Central and South America. Up to another two fellowships will be awarded to support projects dealing with some aspect of the history of the Franciscan Family in the rest of the United States and Canada. Projects may deal with any aspect of the history of the Franciscan Family, including any of the branches of the Family, male, female, tertiary, Capuchin, etc. The deadline for applications is February 1 each year.
*Requirements:* The recipient must be engaged in full-time research during the period of the fellowship. The applicant must be a doctoral candidate at a University in the Americas, and the bulk of the research should be conducted in the Americas.
*Geographic Focus:* All States, Argentina, Belize, Bolivia, Brazil, Canada, Chile, Colombia, Costa Rica, Ecuador, El Salvador, Falkland Islands (Malvinas), French Guiana, Guatemala, Guyana, Honduras, Mexico, Nicaragua, Panama, Paraguay, Peru, Suriname, Uruguay, Venezuela
*Date(s) Application is Due:* Feb 1
*Amount of Grant:* 10,000 USD
*Contact:* Dr. Jeffrey M. Burns; (510) 548-1755; acadafh@aol.com or acadafh@fst.edu
*Internet:* aafh.org/Scholarships.html
*Sponsor:* Academy of American Franciscan History
1712 Euclid Avenue
Berkley, CA 94709-1208

**ACCP Infectious Diseases Research Fellowships** 133

The purpose of ACCP fellowships is to support the development of Clinical Scientists through postgraduate fellowship experiences, or through the support of graduate students who have at least completed their qualifying examinations. Research activities must relate directly to the stated areas of emphasis. Fellowships support a research training experience and not a clinical pharmacy residency program. A pharmacy fellowship is a directed, highly individualized postgraduate program designed to prepare the participant to become an independent researcher. Fellowships exist primarily to develop competency in the scientific research process. The ACCP Research Institute believes that clinical pharmacy fellowships should be at least two years in length, and will give strong preference to funding applications that describe a multi-year training program. This grant will provide the fellow's stipend for either the first, second, or third year. Preceptors will have the option to receive the fellowship's entire funding over one year, or half funding for each of two years.
*Requirements:* Preceptor: An ACCP Full Member at the time application is made. Fellowship applications submitted concurrently with an application for Full Membership will not be considered. Fellow: An ACCP Member, Associate Member, or applicant for member. If a graduate student, (a) preference will be given to supporting students who have received or are concurrently enrolled in a Pharm.D. degree program; and (b) the Fellow must at least have completed his/her qualifying examinations.
*Restrictions:* Ineligible applicants include: ACCP Board of Regent members; ACCP Research Institute Board of Trustee members; Grants and Fellowship Selection Committee members; preceptor of an ACCP Research Institute fellowship within the previous year.
*Geographic Focus:* All States
*Date(s) Application is Due:* Mar 23
*Amount of Grant:* 40,500 USD
*Contact:* Carla Scarborough; (913) 492-3311; fax (913) 492-0088; cscarborough@accp.com
*Internet:* www.accp.com/frontiers/research.php#resfel
*Sponsor:* American College of Clinical Pharmacy
13000 W 87th Street Parkway
Lenexa, KS 66215-4530

**ACF Center for Research on Hispanic Children and Families Grants** 134

The Office of Planning, Research and Evaluation (OPRE) within the Administration for Children and Families (ACF) is considering soliciting applications for a Center for Research on Hispanic Children and Families, to lead and support investigation of the needs of Hispanic populations served by ACF and promising approaches to promote social and economic well-being among low-income Hispanic families. This research center (hereafter referred to as "Center") would ideally bring together a diverse, interdisciplinary team of academic and organizational partners to provide leadership in culturally competent research that can inform policies concerning low-income Hispanic families and to foster significant scholarship regarding the needs and experiences of the diverse Hispanic populations throughout the nation. Through a combination of research, research capacity building, and communication activities, this Center will ideally address a range of relevant issues including, but not limited to: a) definition and measurement

of significant characteristics and experiences of Hispanic individuals, families, and communities; b) identification of barriers and facilitators of service utilization by Hispanic populations; c) examination of strategies to promote the social and economic well-being of Hispanic children and families and to strengthen Hispanic families; d) evaluation of promising approaches to serving Hispanic populations. The Center will be expected to develop research products, resources, and a comprehensive communication plan that aims to build research capacity in the field and to improve understanding of Hispanic populations in order to inform policy development and programmatic responses. The Center will focus broadly on marriage and relationships, fatherhood, child care and early education, and related issues such as employment and economic mobility of low-income families and children's healthy development and learning. Key topics of interest will be identified in the Funding Opportunity Announcement and will include, but are not limited to, research regarding mental health; substance abuse prevention and treatment; coordinated services; program evaluation; and rural populations. This award would be for a 60-month project period with five 12-month budget periods (i.e., October 2018 - September 2023). For more information about the current cooperative agreement, please visit https://www.acf.hhs.gov/opre/research/project/center-for-research-on-hispanic-children-families. The annual application due date is July 11.
*Requirements:* Eligible applicants include: public and state controlled institutions of higher education; state governments; nonprofits having a 501(c)3 status with the IRS, other than institutions of higher education; Native American tribal governments (federally recognized); nonprofits that do not have a 501(c)3 status with the IRS, other than institutions of higher education; public housing authorities/Indian housing authorities; special district governments; county governments; private institutions of higher education; Native American tribal organizations (other than federally recognized tribal governments); independent school districts; and city or township governments. Faith-based and community organizations that meet the eligibility requirements are also eligible to receive awards under this funding opportunity announcement.
*Geographic Focus:* All States
*Date(s) Application is Due:* Jul 11
*Contact:* Ann Rivera, Program Manager; (202) 401-5506 or (202) 401-9220; fax (202) 205-3598; ann.rivera@acf.hhs.gov
*Internet:* https://ami.grantsolutions.gov/index.cfm?switch=foa&fon=HHS-2018-ACF-OPRE-PH-1378
*Sponsor:* Administration for Children and Families
330 C Street, SW
Washington, D.C. 20201

**ACF Child Care and Development Block Grant Implementation Research and Evaluation Grants** 135

In May 2017, the Office of Planning, Research and Evaluation (OPRE) funded three cooperative agreements, under HHS-2017-ACF-OPRE-YE-1217 (Child Care and Development Block Grant Implementation Research and Evaluation Planning Grants: Phase I/Cohort 2), for 18-months to Child Care and Development Fund (CCDF) Lead Agencies. Those grants provided funding for Lead Agencies to develop a research-based evaluation plan of the implementation of policies and initiatives in response to the goals of the CCDBG Act of 2014. Under this funding opportunity announcement (FOA), all grantees awarded planning grants (Phase I/Cohort 2) could be funded to implement their research plan (up to three awards may be made). Awards will be on a competitive basis according to the evaluation criteria in the FOA. During the period of this grant, CCDF Lead Agencies will refine their research and evaluation plan with their research partners (either within their organization or with an outside partner) and implement their research project. Grantees are also expected to clearly document the policies of interest that will be evaluated, document how they will measure the policy implementation or change in response to the goals of the CCDBG Act of 2014, and articulate how these findings will be used in future policy and planning decisions. These awards will be for a 48-month project period with four 12-month budget periods. The application due date is August 14.
*Requirements:* Eligible applicants include: public and state controlled institutions of higher education; state governments; nonprofits having a 501(c)3 status with the IRS, other than institutions of higher education; Native American tribal governments (federally recognized); nonprofits that do not have a 501(c)3 status with the IRS, other than institutions of higher education; public housing authorities/Indian housing authorities; special district governments; county governments; private institutions of higher education; Native American tribal organizations (other than federally recognized tribal governments); independent school districts; and city or township governments. Faith-based and community organizations that meet the eligibility requirements are also eligible to receive awards under this funding opportunity announcement.
*Geographic Focus:* All States
*Date(s) Application is Due:* Aug 14
*Contact:* Meryl Barofsky, Program Officer; (202) 401-5871 or (202) 401-9220; fax (202) 205-3598; meryl.barofsky@acf.hhs.gov
*Internet:* https://ami.grantsolutions.gov/index.cfm?switch=foa&fon=HHS-2019-ACF-OPRE-YE-1380
*Sponsor:* Administration for Children and Families
330 C Street, SW
Washington, D.C. 20201

**ACF Early Care and Education Research Scholars: Child Care Research Scholars** 136

Funds for Child Care Research Scholars grants are available to support dissertation research on child care policy issues in partnership with State Child Care and Development Fund (CCDF) lead agencies. Since 2000, Congress has appropriated about $10 million per year of CCDF discretionary funds to be used for child care research and evaluation. These funds have supported projects that add to our knowledge about the efficacy of child care subsidy policies and programs in supporting employment and self-sufficiency outcomes for parents, and providing positive learning and school readiness outcomes for children. Previously funded Child Care Research Scholars have made significant contributions to the child care policy research field. To ensure that research is responsive to the changing needs of low-income families, partnerships between the graduate student, their mentor and the State CCDF lead agency are essential. This partnership ensures the research will be policy-relevant and is the foundation that fosters skills necessary to build the graduate student's career trajectory of successful partnership-building and contributions to the policy and scientific communities. An estimated six (6) grants will be awarded, ranging from $20,000 to $25,000. Electronically submitted applications must be submitted no later than 11:59 p.m., ET, on April 10.
*Requirements:* Eligible applicants include: public and state controlled institutions of higher education; state governments; nonprofits having a 501(c)3 status with the IRS, other than institutions of higher education; Native American tribal governments (federally recognized); nonprofits that do not have a 501(c)3 status with the IRS, other than institutions of higher education; public housing authorities/Indian housing authorities; special district governments; county governments; private institutions of higher education; Native American tribal organizations (other than federally recognized tribal governments); independent school districts; and city or township governments. Faith-based and community organizations that meet the eligibility requirements are also eligible to receive awards under this funding opportunity announcement.
*Geographic Focus:* All States
*Date(s) Application is Due:* Apr 10
*Amount of Grant:* 20,000 - 25,000 USD
*Contact:* Ann Rivera, Program Officer; (202) 401-5506 or (202) 401-9220; fax (202) 205-3598; ann.rivera@acf.hhs.gov
*Internet:* https://www.acf.hhs.gov/opre/research/project/child-care-research-scholars
*Sponsor:* Administration for Children and Families
330 C Street, SW
Washington, D.C. 20201

**ACF Early Care and Education Research Scholars: Head Start Graduate Student Research Grants** 137

The Head Start Graduate Student Research grant program is designed to build research capacity in and knowledge of effective early childhood interventions with low-income children and families. The grant program does this by providing support for dissertation research conducted by graduate students working in partnership with local Head Start or Early Head Start programs. Many former grantees have become leading researchers who continue to conduct research that informs and improves Head Start/Early Head Start, other early childhood intervention practices, and our understanding of low-income populations. The immediate goals of the grant program are to: support the completion of high quality research projects directed at the current concerns of Head Start and Early Head Start programs and policy makers; encourage research with Head Start and Early Head Start populations; promote mentor-student relationships that provide project supervision in the field and support students' professional development; emphasize the importance of developing working research partnerships with Head Start and Early Head Start programs, thereby increasing the effectiveness of the applied research work; and support active communication, networking and collaboration among the group of grantee graduate students, their mentors and other prominent researchers in the field. An estimated six (6) grants will be awarded, ranging from $20,000 to $25,000. Electronically submitted applications must be submitted no later than 11:59 p.m., ET, on April 21.
*Requirements:* Eligible applicants include: public and state controlled institutions of higher education; state governments; nonprofits having a 501(c)3 status with the IRS, other than institutions of higher education; Native American tribal governments (federally recognized); nonprofits that do not have a 501(c)3 status with the IRS, other than institutions of higher education; public housing authorities/Indian housing authorities; special district governments; county governments; private institutions of higher education; Native American tribal organizations (other than federally recognized tribal governments); independent school districts; and city or township governments. Faith-based and community organizations that meet the eligibility requirements are also eligible to receive awards under this funding opportunity announcement.
*Geographic Focus:* All States
*Date(s) Application is Due:* Apr 21
*Amount of Grant:* 20,000 - 25,000 USD
*Contact:* Wendy DeCourcey, Federal Project Officer; (202) 260-2039 or (202) 401-9220; fax (202) 205-3598; wendy.decourcey@acf.hhs.gov
*Internet:* https://www.acf.hhs.gov/opre/research/project/head-start-graduate-student-research-program
*Sponsor:* Administration for Children and Families
330 C Street, SW
Washington, D.C. 20201

**ACF Family Strengthening Scholars Grants** 138

The Family Strengthening Scholars grant program is designed to build research capacity in the healthy marriage/responsible fatherhood field (HM/RF). These grants are to support dissertation research on HM/RF policy issues and are meant to build capacity in the research field to focus research on questions that have direct implications for HM/RF decision-making and program administration. They are intended to focus particularly on underserved/understudied populations, such as low-income families and minority population, utilize rigorous methodology and help inform the development of future intervention research. The specific goals of the Family Strengthening Scholars grants are: to address issues of significant relevant to policy decisions related to low-income families in order further the HM/RF field; to directly support graduate students' engagement

in HM/RF research; to foster mentoring relationships between faculty members and graduate students who are pursuing doctoral-level research in the HM/RF field; and to encourage active communication, networking and collaboration among graduate students, their mentors, and other senior family strengthening researchers. An estimated three (3) grants will be awarded, ranging from $10,000 to $25,000. Electronically submitted applications must be submitted no later than 11:59 p.m., ET, on May 10.
*Requirements:* Eligible applicants include: private institutions of higher education; and public and State controlled institutions of higher education.
*Geographic Focus:* All States
*Date(s) Application is Due:* May 20
*Amount of Grant:* 10,000 - 25,000 USD
*Contact:* Kathleen Mccoy, Program Officer; (202) 401-5436 or (202) 401-9220; fax (202) 205-3598; kathleen.mccoy@acf.hhs.gov
*Internet:* https://www.acf.hhs.gov/opre/research/project/family-strengthening-scholars
*Sponsor:* Administration for Children and Families
330 C Street, SW
Washington, D.C. 20201

**A Charitable Foundation Grants** 139

Founded in 1997 in Nevada and funded by the Tactical Investment Management Corporation, A Charitable Foundation offers support primarily in California, Hawaii, and New York, as well as Brazil. Its major fields of interest include: educational research; the environment; food banks; human services; and international development. Types of support include annual campaigns, building and renovation projects, emergency assistance, basic research, and grants to individuals. Applications should include a detailed description of the project, purpose of the research (if applicable), and a detailed budget. There are no annual deadlines, and grants typically range from $2000 to $35,000.
*Geographic Focus:* California, Hawaii, New York, Brazil
*Amount of Grant:* 250 - 30,000 USD
*Samples:* Self-Realization Fellowship, Los Angeles, California, $5,000 - support of a fellowship program; Wailua United Church of Christ, Wailua, Hawaii, $18,000 - for public charity work; North Shore Community Land Trust, Haleiwa, Hawaii, $750 - services for public charities.
*Contact:* David S. Druz, President; (702) 248-8184; ddruz@aol.com
*Sponsor:* A Charitable Foundation
2657 Windmill Parkway, Suite 220
Henderson, NV 89014

**Achelis Foundation Grants** 140

Elisabeth Achelis was born in Brooklyn Heights in 1880. She used her inheritance from her father Fritz Achelis, who was President of the American Hard Rubber Company, to establish the Achelis Foundation in 1940 to aid and contribute to charitable, benevolent, educational and religious uses and purposes for the moral, ethical, physical, mental and intellectual well-being and progress of mankind; to aid and contribute to methods for the peaceful settlement of international differences; to aid and contribute to the furtherance of the objects and purposes of any charitable, benevolent, educational or religious institution or agency; and to establish and maintain charitable, benevolent and educational institutions and agencies. The Achelis Foundation shares trustees, staff, office space, and even a website with the Bodman Foundation which has a similar mission and geographic area of concentration (both foundations give in New York City, while the Bodman Foundation also gives in New Jersey). Funding is concentrated in six program areas: arts and culture; education; employment; health; public policy; and youth and families. Most recent awards have ranged from $10,000 to $200,000, though typical awards average $20,000 to $50,000.
*Requirements:* 501(c)3 organizations based in New York City that fall within the foundation's areas of interest are welcome to submit an inquiry or proposal letter by regular mail (initial inquiries by email or fax are not accepted, nor are CDs, DVDs, computer discs, or video tapes). An initial inquiry to the foundation should include only the following items: a proposal letter that briefly summarizes the history of the project, need, objectives, time period, key staff, project budget, and evaluation plan; the applicant's latest annual report and complete set of audited financial statements; and the applicant's IRS 501(c)3 tax-exemption letter. Applications may be submitted at any time during the year. Each request is reviewed by staff and will usually receive a written response within thirty days. Those requests deemed consistent with the interests and resources of the foundation will be evaluated further and more information will be requested. Foundation staff may request a site visit, conference call, or meeting. All grants are reviewed and approved by the Trustees at one of their three board meetings in May, September, or December.
*Restrictions:* The foundation generally does not make grants for the following purposes or program areas: nonprofit organizations outside of New York; annual appeals, dinner functions, and fundraising events; endowments and capital campaigns; loans and deficit financing; direct grants to individuals; individual day-care and after-school programs; housing; organizations or projects based outside the U.S; films or video projects; small art, dance, music, and theater groups; individual K-12 schools (except charter schools); national health and mental health organizations; and government agencies or nonprofit organizations significantly funded or reimbursed by government agencies. Limited resources prevent the foundations from funding the same organization on an ongoing annual basis.
*Geographic Focus:* New York
*Amount of Grant:* 10,000 - 200,000 USD
*Samples:* American Ballet Theatre, New York, New York, $25,000 - general operating support; East Harlem School at Exodus House, New York, New York, $25,000 - general operating support (2014; International Center for the Disabled, New York, New York, $25,000 - to support the ReadyNow program.
*Contact:* John B. Krieger, Executive Director; (212) 644-0322; fax (212) 759-6510; main@achelis-bodman-fnds.org
*Internet:* www.achelis-bodman-fnds.org/guidelines.html
*Sponsor:* Achelis Foundation
767 Third Avenue, 4th Floor
New York, NY 10017-2023

**Acid Maltase Deficiency Association Helen Walker Research Grant** 141

The Helen Walker Research Grant, offered through the Acid Maltase Deficiency Association, is an annual award aimed at promoting research and education about Pompe disease. The Acid Maltase Deficiency Association was established in 1995 to assist in funding research and to promote public awareness of the affliction. Pompe disease is one of a family of 49 rare genetic disorders known as Lysosomal Storage Diseases, or LSDs. Pompe disease is also known as Acid Maltase Deficiency or Glycogen Storage Disease type II, and it affects an estimated 5,000 to 10,000 people in the developed world. The research grant is in the amount of $140,000. The annual deadline for application is September 4.
*Geographic Focus:* All States
*Date(s) Application is Due:* Sep 4
*Amount of Grant:* 140,000 USD
*Contact:* Tiffany L. House, AMDA President; (210) 494-6144; fax (210) 490-7161; TiffanyLHouse@aol.com
*Internet:* www.amda-pompe.org/index.php/main/news/2015_helen_walker_research_grant_for_pompe_disease_now_accepting_applicatio
*Sponsor:* Acid Maltase Deficiency Association
P.O. Box 700248
San Antonio, TX 78270

**ACIE Title VIII Combined Research and Language Training Fellowships** 142

Funded by the U.S. Department of State's Program for Research and Training on Eastern Europe and the Independent States of the Former Soviet Union (Title VIII), the American Councils Combined Research and Language Training (CRLT) Program serves graduate students and scholars who, in addition to support for research in the independent states of the former Soviet Union, require supplemental language instruction. The synergistic nature of the research and language training aspects of the Combined Research and Language Training Program promises both more productive research and efficient language acquisition. The total value of Title VIII CRLT fellowships, administered by American Councils, ranges from $5,000 to $25,000. Typical awards include: international airfare from the scholar's home city to his/her host city overseas; academic affiliation at a leading local university; visa(s) arranged by American Councils in direct collaboration with academic host institutions in order to facilitate archive access and guarantee timely visa registration; a housing stipend; a monthly living stipend; health insurance of up to $100,000 per accident or illness; and ongoing logistical support from American Councils offices throughout the region, including in-country orientation and 24-hour emergency aid. The application deadline is October 1.
*Requirements:* Applicants must conduct research and language study for a minimum of three months in the field to a maximum of nine months in the field. Participants must have attained at least an intermediate level of language proficiency to apply. Applicants must be U.S. citizens. The competition for funding is open and merit-based. All applications will receive consideration without regard to any factor such as race, color religion, sex, sexual orientation, national origin, martial status, political affiliation, or disability.
*Geographic Focus:* All States
*Amount of Grant:* 5,000 - 25,000 USD
*Contact:* Liliya Kholod, Program Officer, Teacher Programs; (202) 833-7522; fax (202) 833-7523; lkholod@americancouncils.org
*Internet:* researchfellowships.americancouncils.org/crlt
*Sponsor:* American Councils for International Education
1828 L Street, NW, Suite 1200
Washington, D.C. 20036

**ACIE Title VIII Research Scholar Fellowships** 143

Funded by the U.S. Department of State's Program for Research and Training on Eastern Europe and the Independent States of the Former Soviet Union (Title VIII), the American Councils Research Scholar Program provides full support for graduate students, faculty, and independent scholars seeking to conduct in-country, independent research for three consecutive months to nine consecutive months in Central Asia, Moldova, Russia, the South Caucasus, Southeast Europe, and Ukraine. The total value of Title VIII Research Scholar fellowships, administered by American Councils, ranges from $5,000 to $25,000. Typical awards include: international airfare from the scholar's home city to his/her host city overseas; academic affiliation at a leading local university; visa(s) arranged by American Councils in direct collaboration with academic host institutions in order to facilitate archive access and guarantee timely visa registration; a housing stipend; a monthly living stipend; health insurance of up to $100,000 per accident or illness; and ongoing logistical support from American Councils offices throughout the region, including in-country orientation and 24-hour emergency aid. The application deadline is October 1.
*Requirements:* Scholars in the social sciences and humanities are eligible to apply for the program. While a wide-range of topics receive support each year, all funded research must contribute to a body of knowledge enabling the U.S. to better understand the region and formulate effective policies within it. All applicants should clearly describe the policy-relevance of their work, be it in anthropology, history, international relations, political science, or some other field.
*Geographic Focus:* All States
*Date(s) Application is Due:* Oct 1

*Amount of Grant:* 5,000 - 25,000 USD
*Contact:* Liliya Kholod, Program Officer; (202) 833-7522; fax (202) 833-7523; lkholod@americancouncils.org
*Internet:* researchfellowships.americancouncils.org/researchscholar
*Sponsor:* American Councils for International Education
1828 L Street, NW, Suite 1200
Washington, D.C. 20036

**ACL Field Initiated Projects Program: Minority-Serving Institution Development Grants** 144

The purpose of the Field Initiated (FI) Projects program is to develop methods, procedures, and rehabilitation technology that maximize the full inclusion and integration into society, employment, independent living, family support, and economic and social self-sufficiency of individuals with disabilities, especially individuals with the most severe disabilities. Another purpose of the FI Projects program is to improve the effectiveness of services authorized under the Rehabilitation Act of 1973, as amended. In carrying out a development project under a FI Projects development grant, a grantee must use knowledge and understanding gained from research to create materials, devices, systems, or methods beneficial to the target population, including design and development of prototypes and processes. It is expected that one award of $200,000 will be given annually. Letters of Intent are due each year by May 6, with the annual deadline for application submission being June 14.
*Requirements:* Minority entities are the only applicants that are eligible to apply for this grant opportunity.
*Geographic Focus:* All States
*Date(s) Application is Due:* Jun 14
*Contact:* Marlene Spencer, Program Officer; (202) 245-7532 or (202) 401-4634; fax (202) 245-7323; marlene.spencer@acl.hhs.gov or aclinfo@acl.hhs.gov
*Internet:* www.acl.gov/Funding_Opportunities/Announcements/Index.aspx
*Sponsor:* U.S. Department of Health and Human Services
330 C Street SW
Washington, D.C. 20201

**ACL Field Initiated Projects Program: Minority-Serving Institution Research Grants** 145

The purpose of the Field Initiated (FI) Projects program is to develop methods, procedures, and rehabilitation technology that maximize the full inclusion and integration into society, employment, independent living, family support, and economic and social self-sufficiency of individuals with disabilities, especially individuals with the most severe disabilities. Another purpose of the FI Projects program is to improve the effectiveness of services authorized under the Rehabilitation Act of 1973, as amended. In carrying out a development project under a FI Projects development grant, a grantee must use knowledge and understanding gained from research to create materials, devices, systems, or methods beneficial to the target population, including design and development of prototypes and processes. It is expected that one award of $200,000 will be given annually. Letters of Intent are due each year by May 6, with the annual deadline for application submission being June 14.
*Requirements:* Minority entities are the only applicants that are eligible to apply for this grant opportunity.
*Geographic Focus:* All States
*Date(s) Application is Due:* Jun 14
*Amount of Grant:* 200,000 USD
*Contact:* Marlene Spencer, Project Officer; (202) 245-7532 or (202) 401-4634; fax (202) 245-7323; marlene.spencer@acl.hhs.gov or aclinfo@acl.hhs.gov
*Internet:* www.acl.gov/Funding_Opportunities/Announcements/Index.aspx
*Sponsor:* U.S. Department of Health and Human Services
330 C Street SW
Washington, D.C. 20201

**ACLS/New York Public Library Fellowships** 146

The American Council of Learned Societies and The New York Public Library offer a collaborative program to support up to five residential fellowships per year at the Library's Dorothy and Lewis B. Cullman Center for Scholars and Writers. The Center provides opportunities for up to five Fellows to explore the rich, diverse collections in the Library's Stephen A. Schwarzman Building. The Center also serves as a forum for the exchange of ideas among Fellows, invited guests, the wider academic and cultural communities, and the interested public. It provides individual office space and common areas in the Library building. Fellows are required to be in residence from September through May, and to participate in Center activities such as lunches, panel discussions, public conversations, symposiums, and interviews. Specific guidelines, the online ACLS application, and link to the Cullman Center application (where candidates must also apply) is available at the website. The annual deadline for application submission is September 23.
*Requirements:* Applicants must be U.S. citizens or permanent residents by the application date. They must also have earned a Ph.D. at least two years before the application deadline. An applicant's most recent supported research leave must have been two years before the application date.
*Geographic Focus:* All States
*Date(s) Application is Due:* Sep 23
*Contact:* Valerie Popp; (212) 697-1505, ext. 153; fax (212) 949-8058; vpopp@acls.org
*Internet:* https://www.acls.org/Competitions-and-Deadlines/ACLS-Fellowships
*Sponsor:* American Council of Learned Societies
633 Third Avenue, 8th Floor
New York, NY 10017-6795

**ACLS/Robert H.N. Ho Family Foundation Grants for Critical Editions and Scholarly Translations** 147

Critical editions are of crucial importance to the expanding field of Buddhist studies. Translation has been at the core of Buddhism since the Buddha's instruction to his monks to teach the dharma in many languages. These grants support a broad range of endeavor, from the creation of critical editions (with full scholarly apparatus), to translation of canonical texts into modern vernaculars, to the translation of scholarly works on Buddhism from one modern language into another. Collaborative projects will receive priority. Ideally, they will involve scholars who are native speakers of the relevant source and target languages, and/or scholars experienced in research and study in those languages. Award funds can be used as stipends for work performed (e.g., to secure release time), for travel, and for related office costs, including reproduction or digitization of images, according to a proposed budget. Pending confirmation of funding, ACLS anticipates offering publication subventions to recipients of these grants who complete manuscripts. The subventions will be available through a separate competition to be announced at a later time. Institutions of higher education worldwide are eligible to apply for grants up to $320,000 (to be expended over four years) in support of new teaching positions in Buddhist studies. Completed applications should be submitted through the ACLS online fellowship and grant administration system by the annual deadline of November 13.
*Requirements:* Individual applicants and leaders of collaborative teams must have PhD degrees conferred by an accredited university. This is a global competition. There are no restrictions as to the location of work proposed, or the citizenship/residence of applicants. Although the final products of research can be in any language, all applications must be submitted in English.
*Restrictions:* There are no restrictions as to the language of the final product prepared for publication.
*Geographic Focus:* All States, All Countries
*Date(s) Application is Due:* Nov 13
*Amount of Grant:* 80,000 - 320,000 USD
*Contact:* Valerie Popp, Senior Program Officer; (212) 697-1505, ext. 153; fax (212) 949-8058; vpopp@acls.org
*Internet:* https://www.acls.org/Competitions-and-Deadlines/The-Robert-H-N-Ho-Family-Foundation-Program-in-Buddhist-Studies#crit
*Sponsor:* American Council of Learned Societies
633 Third Avenue, 8th Floor
New York, NY 10017-6795

**ACLS/Robert H.N. Ho Family Foundation Postdoctoral Fellowships in Buddhist Studies** 148

The American Council of Learned Societies (ACLS) announces an initiative supporting research and teaching in Buddhist studies, funded by the Robert H.N. Ho Family Foundation. In cooperation with the Foundation, ACLS offers an integrated set of fellowship and grant competitions supporting work that will expand the understanding and interpretation of Buddhist thought in scholarship and society, strengthen international networks of Buddhist studies, and increase the visibility of innovative currents in those studies. Postdoctoral Fellowships in Buddhist Studies provide two years of funding to recent recipients of the PhD for residence at a university for the purpose of revising the dissertation into a publishable manuscript or for beginning the first new project after completion of the PhD degree. The teaching of one course per year is encouraged. Priority is given to residence at universities or colleges providing a collegial atmosphere and working conditions enabling the postdoctoral Fellow's entry into an academic career in Buddhist studies. An annual stipend of $55,000 is offered. In addition, if required, a relocation and health-insurance allowance of $6,000 is also available. Completed applications should be submitted through the ACLS online fellowship and grant administration system by the annual deadline of November 13.
*Requirements:* This is a global competition. There are no restrictions as to the location of work proposed, or the citizenship/residence of applicants. Although the final products of research can be in any language, all applications must be submitted in English.
*Geographic Focus:* All States, All Countries
*Date(s) Application is Due:* Nov 13
*Amount of Grant:* 55,000 - 61,000 USD
*Contact:* Valerie Popp; (212) 697-1505, ext. 153; fax (212) 949-8058; vpopp@acls.org
*Internet:* https://www.acls.org/Competitions-and-Deadlines/The-Robert-H-N-Ho-Family-Foundation-Program-in-Buddhist-Studies#postdoc
*Sponsor:* American Council of Learned Societies
633 Third Avenue, 8th Floor
New York, NY 10017-6795

**ACLS/Robert H.N. Ho Family Foundation Research Fellowships in Buddhist Studies** 149

Research fellowships offer support for research and writing in Buddhist studies for scholars who hold a PhD degree, with no restrictions on time from the PhD. These fellowships provide scholars time free from teaching and other responsibilities to devote full-time to research and writing on the project proposed. The fellowship period may last up to nine months, during which time no teaching, commissioned research on other topics, or administrative duties are allowed. The fellowship may be separated into two periods, each of which must be a minimum of three months. If the entire duration is less than a total of nine months (minimum of six months), the stipend will be prorated. There are no restrictions as to the location of the work conducted. Applicants must identify a significant scholarly product (monograph, series of journal articles, etc.) that will result from the fellowship. They will be asked for a final report at the end of the fellowship tenure describing progress made, and will be asked to report on the products resulting

from it when they are published. Tenure of the fellowship may begin no earlier than July 1. A stipend of up to $70,000, to provide release from other duties for devoting full time to the proposed project is available. The exact amount will be calculated based on the Fellows' current academic salary and other fellowships, grants, sabbatical salary to be received during fellowship tenure. Completed applications must be submitted through the ACLS online fellowship and grant administration system by the November 13 deadline.
*Requirements:* This is a global competition. There are no restrictions as to the location of work proposed, or the citizenship/residence of applicants. Although the final products of research can be in any language, all applications must be submitted in English.
*Restrictions:* A Research Fellowship may be held concurrently with other fellowships and grants and any sabbatical pay, but the total received cannot exceed 125% of the Fellow's academic annual salary.
*Geographic Focus:* All States, All Countries
*Date(s) Application is Due:* Nov 14
*Amount of Grant:* Up to 70,000 USD
*Contact:* Valerie Popp; (212) 697-1505, ext. 153; fax (212) 949-8058; vpopp@acls.org
*Internet:* https://www.acls.org/Competitions-and-Deadlines/The-Robert-H-N-Ho-Family-Foundation-Program-in-Buddhist-Studies#research
*Sponsor:* American Council of Learned Societies
633 Third Avenue, 8th Floor
New York, NY 10017-6795

**ACLS African Humanities Postdoctoral Fellowships** 150
African Humanities Postdoctoral (AHP) Fellowship stipends allow recipients an academic year free from teaching and other duties for revising the dissertation for publication or for the first major research project after the PhD. Fellows are also eligible for additional benefits such as residential stays for writing, manuscript development workshops, and publication support. Each fellow may request a residential stay at an African institute for advanced study. Residencies have proved to be extremely popular and productive, granting fellows time and space to concentrate on writing. Because residencies must be taken at an institute outside the home country, they foster international communication. Currently AHP Fellows may take residencies at six institutes from South Africa to Senegal, Ghana to Tanzania. Fellows who have completed their fellowship tenures may apply to attend a Manuscript-Development Workshop to discuss their manuscripts with AHP mentors and other fellows in a weeklong, intensive retreat. Many authors use these discussions to guide their final revisions before submitting manuscripts for publication. AHP invites fellows to submit their manuscripts to the AHP Series. The rigorous processes of peer-review and developmental editing of AHP publications is overseen by the Series Editors, Fred Hendricks, Rhodes University, South Africa and Adigun Agbaje, University of Ibadan, Nigeria.
*Requirements:* Applicants must be hold a Ph.D., and be U.S. citizens or permanent residents who have lived in the U.S. at least three years prior to application. These fellows must also submit a final report to the National Endowment for the Humanities and the American Council of Learned Societies.
*Geographic Focus* All States, District of Columbia, Guam, Marshall Islands, Northern Mariana Islands, Puerto Rico, U.S. Virgin Islands, American Samoa
*Date(s) Application is Due:* Nov 20
*Contact:* Valerie Popp; (212) 697-1505, ext. 153; fax (212) 949-8058; vpopp@acls.org
*Internet:* https://www.acls.org/Competitions-and-Deadlines/African-Humanities-Program
*Sponsor:* American Council of Learned Societies
633 Third Avenue, 8th Floor
New York, NY 10017-6795

**ACLS Collaborative Research Fellowships** 151
The American Council of Learned Societies (ACLS) Collaborative Research Fellowships support collaborative research in the humanities and related social sciences. Collaborations need not be interdisciplinary or inter-institutional, but must involve at least two scholars. Applicants at the same institution must demonstrate why local funding is insufficient to support the project. It is hoped that projects of successful applicants will help demonstrate the range and value of both collaborative research and inquiry in the humanities, and model how such collaboration may be carried out successfully. Collaborations that involve the participation of assistant and associate faculty members, or that of scholars at different kinds of institutions, are particularly encouraged. Award amounts will range from $60,000 to $201,000 in total, depending on the nature and duration of the collaboration, the kinds of expenses projected to carry out the research, and the number of collaborators. Salary replacement stipends are based on academic rank: up to $35,000 for Assistant Professor and career equivalent; up to $40,000 for Associate Professor and career equivalent; and up to $60,000 for full Professor and career equivalent, as well as up to $20,000 in collaboration funds (which may be used for travel, materials, or research assistance). Applications must be filled out through the online system. Specific guidelines, a sample budget and application, and advice on how to write ACLS proposals are located at the website. This fellowship is currently on hiatus.
*Requirements:* A collaborative project consists of at least two scholars who are each seeking salary-replacement stipends for six to twelve continuous months of supported research leave to pursue full-time collaborative research during the fellowship tenure. The Project Coordinator must have an appointment at a U.S.-based institution of higher education. Other project members may be at institutions outside the United States or may be independent scholars. All project collaborators must hold a Ph.D. degree or its equivalent in publications and professional experience at the time of application.
*Restrictions:* ACLS does not fund creative work (e.g., novels or films), textbooks, straightforward translation, or pedagogical projects. The Fellowship does not fund the following: large research clusters that do not produce any collaborative-authored publications; collaborative projects that result in an anthology or edited volume of secondary scholarship; projects that are not primarily focused on research; projects whose primary aim is to transform existing research results into digital format; and projects whose primary emphasis is on organization of events (workshops, lectures, exhibitions).
*Geographic Focus* All States, District of Columbia, Guam, Marshall Islands, Northern Mariana Islands, Puerto Rico, U.S. Virgin Islands, American Samoa
*Amount of Grant:* 60,000 - 201,000 USD
*Contact:* Valerie Popp, Director of Fellowship Programs; (212) 697-1505, ext. 153; fax (212) 949-8058; vpopp@acls.org
*Internet:* https://www.acls.org/Competitions-and-Deadlines/ACLS-Collaborative-Research-Fellowships
*Sponsor:* American Council of Learned Societies
633 Third Avenue, 8th Floor
New York, NY 10017-6795

**ACLS Comparative Perspectives on Chinese Culture and Society Grants** 152
The Comparative Perspectives on Chinese Culture and Society Grants support projects that bridge disciplinary or geographic boundaries, engage new kinds of information, develop fresh approaches to traditional materials and issues, or bring innovative perspectives to the study of Chinese culture and society. The program aims to provide opportunities for interchange among scholars who may not otherwise have chances to work with one another. The grants support planning meetings, workshops, and/or conferences leading to publication of scholarly volumes. Proposals are expected to be empirically grounded, theoretically informed, and methodologically explicit. The program will support collaborative work of three types: formal research conferences intended to produce significant new research published in a conference volume; workshops or seminars, designed to less formally facilitate new research on newly available or inadequately researched problems, data, or texts; and planning meetings to organizers of the above-described types of projects. Grants up to $6,000 will be offered for one-day meetings to plan conferences or workshops, or for less structured explorations, e.g., brainstorming sessions. Grants of $10,000 to $15,000 will be offered for workshops designed to promote discussion and the exchange of ideas on newly available or inadequately researched data or texts in a collegial, seminar-like setting. Grants up to $25,000 will be offered for formal research conferences intended to produce significant new research that will be published in a conference volume. There are no application forms, but the submitted materials must include the following: the online application information sheet; a description of the project and its purposes; a budget for the proposed event; a short bibliography of relevant sources; a two page CV of the principal organizer; a list of those invited with their CVs; and, for conferences only, an appendix containing abstracts for each paper to be presented at the conference. Currently, this program is on hiatus
*Requirements:* The principal organizer must be affiliated with a university or research institution and must hold a Ph.D. Proposals must include at least one scholar from Taiwan as a participant. There are no restrictions to citizenship of participants or location of the project; however, it is expected the scholars from academic institutions in Taiwan will participate in conferences, workshops, and planning meetings.
*Restrictions:* The program will not support regularly scheduled meetings or conventions. Funds awarded in these competitions cannot be used to cover the following expenses: direct research expenses, such as those of field work, obtaining research materials, or analysis of data; salaries or released time for organizers or participants; honoraria or speaker fees; purchase of equipment; or institutional overhead (although direct administrative costs are allowed).
*Geographic Focus* All States, District of Columbia, Guam, Marshall Islands, Northern Mariana Islands, Puerto Rico, U.S. Virgin Islands, American Samoa
*Amount of Grant:* 6,000 - 25,000 USD
*Contact:* Valerie Popp, Senior Program Officer; (212) 697-1505, ext. 153; fax (212) 949-8058; vpopp@acls.org
*Internet:* https://www.acls.org/Competitions-and-Deadlines/Comparative-Perspectives-on-Chinese-Culture-and-Society
*Sponsor:* American Council of Learned Societies
633 Third Avenue, 8th Floor
New York, NY 10017-6795

**ACLS Digital Extension Grants** 153
The American Council of Learned Societies (ACLS) Digital Extension Grants support digitally based research projects in all disciplines of the humanities and humanities-related social sciences. It is hoped that projects of successful applicants will help advance digital humanistic scholarship by broadening understanding of its nature and exemplifying the robust infrastructure necessary for creating such works. The grants are intended to support a twelve- to eighteen-month period dedicated to work on a major scholarly project that takes a digital form. Up to $150,000 is available to cover salary replacement, staffing, equipment and other costs. Projects may: address a consequential scholarly question through new research methods, new ways of representing the knowledge produced by research, or both; create new digital research resources; increase the scholarly utility of existing digital resources by developing new means of aggregating, navigating, searching, or analyzing those resources; propose to analyze and reflect upon the new forms of knowledge creation and representation made possible by the digital transformation of scholarship. All applications must include the endorsement of a senior administrator of the applicant's institution. This endorsement should include discussion of how the institution's existing cyber infrastructure complements and supports the technologies to be developed for the specified project. Completed applications must be submitted through the ACLS online application system. Specific guidelines, FAQs, a sample application, and advice on how to write an ACLS proposal are available at the website. The annual deadline for submissions is January 8.

*Requirements:* Projects must be hosted by an institution of higher education in the United States. A project's principal investigator must: be a scholar in a field of the humanities and the humanistic social sciences; and have a PhD degree conferred prior to the application deadline.
*Restrictions:* ACLS does not support creative works (e.g., novels or films), textbooks, straightforward translations, or purely pedagogical projects.
*Geographic Focus* All States, District of Columbia, Guam, Marshall Islands, Northern Mariana Islands, Puerto Rico, U.S. Virgin Islands, American Samoa
*Date(s) Application is Due:* Jan 8
*Amount of Grant:* Up to 150,000 USD
*Contact:* Valerie Popp; (212) 697-1505, ext. 153; fax (212) 949-8058; vpopp@acls.org
*Internet:* https://www.acls.org/Competitions-and-Deadlines/ACLS-Digital-Extension-Grants
*Sponsor:* American Council of Learned Societies
633 Third Avenue, 8th Floor
New York, NY 10017-6795

**ACL Self-Advocacy Resource Center Grants** 154

The Self-Advocacy Resource Center, guided by an advisory committee, will provide information to several entities including self advocacy organizations. The advisory committee members will be selected by the grantee and 75% of its members will be people with I/DD. The Center will have several activities including: documenting and disseminating tools; collecting success stories and best practices concerning self-advocacy; research activities, policy development, training and technical assistance; small leadership grants; and creating and maintaining a website. It is expected that one award of $400,000 will be given annually. Letters of Intent are due each year by June 14, with the annual deadline for application submission being July 25.
*Requirements:* Eligible applicants include: nonprofits that do not have a 501(c)3 status with the IRS, other than institutions of higher education; and nonprofits having a 501(c)3 status with the IRS, other than institutions of higher education.
*Geographic Focus:* All States
*Date(s) Application is Due:* Jul 25
*Amount of Grant:* 400,000 USD
*Contact:* Katherine Cargill-Willis, Project Officer; (202) 357-3436 or (202) 401-4634; katherine.cargill-willis@acl.hhs.gov or aclinfo@acl.hhs.gov
*Internet:* www.grants.gov/web/grants/view-opportunity.html?oppId=284144
*Sponsor:* U.S. Department of Health and Human Services
330 C Street SW
Washington, D.C. 20201

**ACLS Fellowships** 155

The American Council of Learned Societies Fellowships fund research in all disciplines of the humanities and humanities-related social sciences. Appropriate fields of specialization include but are not limited to: anthropology, archeology, art history, economics, geography, history, languages and literatures, law, linguistics, musicology, philosophy, political science, psychology, religion, and sociology. Proposals in the social science fields listed are eligible only if they employ predominantly humanistic approaches (e.g., economic history, law and literature, political philosophy). Proposals in interdisciplinary and cross-disciplinary studies are welcome, as are proposals focused on any geographic region or on any cultural or linguistic group. The Fellowships are intended as salary replacement and may be held concurrently with other fellowships and grants and any sabbatical pay to reach that goal. Specific guidelines, the online application, a sample application, a suggestions for writing ACLS proposals are located at the website.
*Requirements:* Applicants must have a Ph.D. degree completed two years prior to the application deadline. The applicant must have U.S. citizenship or permanent resident status as of the application deadline. The most recent supported research leave must have concluded two years prior to July 1. (Supported research leave is defined as the equivalent of one semester or more of time free from teaching or other employment to pursue scholarly research or writing supported by sabbatical pay or other institutional funding, fellowships and grants, or a combination of these.) This definition will apply to independent scholars as well as those with institutional affiliations.
*Restrictions:* ACLS does not fund creative work (e.g., novels or films), textbooks, straightforward translation, or pedagogical projects.
*Geographic Focus:* All States
*Date(s) Application is Due:* Sep 23
*Amount of Grant:* 40,000 - 75,000 USD
*Contact:* Valerie Popp; (212) 697-1505, ext. 153; fax (212) 949-8058; vpopp@acls.org
*Internet:* https://www.acls.org/Competitions-and-Deadlines/ACLS-Fellowships
*Sponsor:* American Council of Learned Societies
633 Third Avenue, 8th Floor
New York, NY 10017-6795

**ACLS Frederick Burkhardt Residential Fellowships for Recently Tenured Scholars** 156

The American Council of Learned Societies (ACLS) Burkhardt Fellowships support long-term, unusually ambitious projects in the humanities and related social sciences. The ultimate goal of the project should be a major piece of scholarly work by the applicant. Burkhardt Fellowships are intended to support an academic year (nine months) of residence at any one of the national residential research centers (listed on the website) participating in the program. Such an environment, beyond providing free time, encourages exchanges across disciplinary lines that can be especially helpful to expanding the significance of projects in the humanities and related social sciences. Appropriate fields of specialization include but are not limited to: anthropology, archaeology, art history, economics, geography, history, languages and literature, law, linguistics, musicology, philosophy, political science, psychology, religion, and sociology. Proposals in the social science fields listed above are eligible only if they employ predominantly humanistic approaches (e.g., economic history, law and literature, political philosophy). Proposals in interdisciplinary and cross-disciplinary studies are welcome, as are proposals focused on any geographic region or on any cultural or linguistic group. Fellowships are in the amount of $95,000, plus funds for research costs and related scholarly activities of up to $7,500 and for relocation up to $3,000. Applications must be submitted online. Specific guidelines also are available at the website.
*Requirements:* The Burkhardt Fellowship Program is open to recently tenured humanists scholars who will have begun their first tenured contracts by the application deadline but began their first tenured contracts no earlier than the fall 2002 semester or quarter. An applicant must be employed in a tenured position at a degree-granting academic institution in the U.S., remaining so for the duration of the fellowship. American citizenship or permanent residency is not required, and previous supported research leaves do not affect eligibility for a Burkhardt Fellowship.
*Restrictions:* ACLS does not fund creative work (e.g., novels or films), textbooks, straightforward translation, or pedagogical projects.
*Geographic Focus* All States, District of Columbia, Guam, Marshall Islands, Northern Mariana Islands, Puerto Rico, U.S. Virgin Islands, All Countries, American Samoa
*Date(s) Application is Due:* Sep 23
*Amount of Grant:* 95,000 - 105,500 USD
*Contact:* Valerie Popp; (212) 697-1505, ext. 153; fax (212) 949-8058; vpopp@acls.org
*Internet:* https://www.acls.org/Competitions-and-Deadlines/Frederick-Burkhardt-Residential-Fellowships-for-Recently-Tenured-Scholars
*Sponsor:* American Council of Learned Societies
633 Third Avenue, 8th Floor
New York, NY 10017-6795

**ACRL Instruction Section Innovation Award** 157

This annual award recognizes a project that demonstrates creative, innovative, or unique approaches to information literacy instruction or programming. Projects nominated for the award should demonstrate recognized creativity, quality, and innovation within the context of national trends in information literacy instruction or programming. A certificate and a $3,000 award, donated by ProQuest, is given. Recipients(s) are recognized during the Instruction Section Program at the ALA Annual Conference. The award is administered by the Instruction Section of ACRL.
*Requirements:* Academic librarians or academic project teams that include an academic librarian are eligible to receive the award. Recipients must have implemented their project in an academic or research library or through the aegis of a professional library organization no more than two years prior to the nomination submission deadline.
*Geographic Focus:* All States
*Date(s) Application is Due:* Dec 7
*Amount of Grant:* 3,000 USD
*Contact:* Robin Kear, IS Innovation Award Co-Chair; (412) 648-7728; rlk25@pitt.edu
Erin L. Ellis, IS Innovation Award Co-Chair; (785) 864-5589; eellis@ku.edu
*Internet:* www.ala.org/acrl/awards/achievementawards/innovationaward
*Sponsor:* Association of College and Research Libraries
50 East Huron Street
Chicago, IL 60611-2795

**ACSE J. Waldo Smith Hydraulic Fellowship** 158

The J. Waldo Smith Hydraulic Fellowship is restricted to research in the field of experimental hydraulics as distinguished from research concentrating on purely theoretical hydraulics. Emphasis is to be placed on practical experiments designed and executed for the purpose of advancing knowledge with respect to the laws of hydraulic flow, rather than to the type of research that proceeds on the theory of mathematical analysis based on assumptions of unknown validity. The purpose of the research is to test the assumptions that are currently made, and also to develop a better understanding of fluid flow. The fellowship is offered every third year, and runs for one full academic year. Administration is in part through the institution that invites cooperation through its engineering faculty. Application packets are available through the student chapter advisor or from ASCE.
*Requirements:* Fellowships are awarded to graduate students, who are members of the society, who give promise of best fulfilling the ideals of the fellowship. There is a brief application form. Each application shall include a statement in general terms of the purposes for which the funds are expected to be used.
*Geographic Focus:* All States
*Date(s) Application is Due:* Feb 9
*Amount of Grant:* Up to 5,000 USD
*Contact:* John Casazza, Managing Director, Continuing Education; (703) 295-6155; fax (703) 295-6132; jcasazza@asce.org
*Internet:* www.asce.org/Student-Organizations/Scholarships-and-Fellowships/J—Waldo-Smith-Hydraulic-Fellowship/
*Sponsor:* American Society of Civil Engineers
1801 Alexander Bell Drive
Reston, VA 20191-4400

**ACSM Carl V. Gisolfi Memorial Fund Grant** 159

Carl V. Gisolfi, Ph.D., FACSM, was one of the most influential leaders of ACSM in its 50+ year history. With Gatorade, ACSM created the Gisolfi Memorial Fund to honor his many contributions to ACSM and the exercise science field. This grant is designated to encourage research in thermoregulation, exercise, and hydration. One award will be dispersed to ACSM doctoral students in the amount of $5,000 each for a one-year period.

*Requirements:* Applicants for student research grants must have graduate student status during the term of the grant to be considered for funding. Applicants must be current members of ACSM at the time of submitting an application to be funded. Grants are open to all ACSM members, including international members.
*Geographic Focus:* All States, All Countries
*Date(s) Application is Due:* Jan 15
*Amount of Grant:* 5,000 USD
*Contact:* Jane Senior, (317) 637-9200, ext. 143; fax (317) 634-7817; jsenior@acsm.org
*Internet:* acsm.org/find-continuing-education/awards-grants/research-grants/carl-v.-gisolfi-memorial-fund
*Sponsor:* American College of Sports Medicine
401 West Michigan Street
Indianapolis, IN 46202-3233

**ACSM Charles M. Tipton Student Research Award** 160
ACSM invests in the future of sports medicine and exercise science by giving to its student members. Student members are eligible for grants and awards to offset the rising costs of tuition and travel. The Charles M. Tipton Student Research Award is presented annually to the student with the most outstanding research project of the year. Criteria for the awards include: the quality of the research project, which is evaluated via; and the extent of the students participation in the project.
*Requirements:* At the time of application, the applicant must be presently enrolled/accepted into a graduate or professional program in the areas of basic of clinical exercise science or sports medicine at an accredited university or have graduated from such a program within one year prior to this award's application date. In the latter case, only work accomplished before graduation should be submitted. The applicant must be the first author of the abstract submitted and accepted as a free communication for presentation at the ACSM Annual Meeting. The abstract must be original research in exercise science and should include the following: statement of the program; methods and results and conclusions. The work reported in the abstract must have been completed for academic credit toward completion of a graduate degree. Finally, the applicant must be a current member of the American College of Sports Medicine or one of its Regional Chapters at the time of application and at the time the award is presented.
*Geographic Focus:* All States
*Date(s) Application is Due:* Feb 1
*Contact:* Jane Senior, Research Administration and Programs Department; (317) 637-9200, ext. 143 or (317) 637-9200, ext. 117; fax (317) 634-7817; jsenior@acsm.org or education@acsm.org
*Internet:* acsm.org/find-continuing-education/awards-grants/student-awards/charles-m.-tipton-student-research-awards
*Sponsor:* American College of Sports Medicine
401 West Michigan Street
Indianapolis, IN 46202-3233

**ACSM Coca-Cola Company Doctoral Student Grant on Behavior Research** 161
For research addressing behavioral strategies and techniques to promote the adoption and maintenance of physical activity/exercise. Studies that focus on innovative strategies to improve individual-level adoption and/or maintenance of physical activity will be given first priority. The proposed study can focus on youth, adolescents, adults, or older adults 55 and over. One award of $5,000 is available.
*Requirements:* Applicants for student research grants must have graduate student status during the term of the grant to be considered for funding. Applicants must be current members of ACSM at the time of submitting an application to be funded. Grants are open to all ACSM members, including international members. However, the NASA initiative is open to U.S. residents only.
*Geographic Focus:* All States, All Countries
*Date(s) Application is Due:* Jan 17
*Amount of Grant:* 5,000 USD
*Contact:* Jane Senior or Michael F. Dell, (317) 637-9200, ext. 143; fax (317) 634-7817; jsenior@acsm.org or mdell@acsm.org
*Internet:* acsm.org/find-continuing-education/awards-grants/research-grants/2011/08/16/the-coca-cola-company-doctoral-student-grant-on-behavior-research
*Sponsor:* American College of Sports Medicine
401 West Michigan Street
Indianapolis, IN 46202-3233

**ACSM Dr. Raymond A. Weiss Research Endowment Grant** 162
Wishing to advance research into the health benefits - physical, mental and emotional - of physical activity and sports, the Dr. Raymond A. Weiss Research Endowment has been established for ACSM students studying in this field. One project will be funded for applied rather than for basic research, with the intent of applying the results to programs involving physical activity and sports. The psychological and emotional benefits of physical activity are of particular importance to the benefactors of this endowment, Drs. Raymond and Rosalee Weiss, and proposals addressing those issues will be given priority. One reward of $1,500 is available.
*Requirements:* Applicants for student research grants must have graduate student status during the term of the grant to be considered for funding. Applicants must be current members of ACSM at the time of submitting an application to be funded. Grants are open to all ACSM members, including international members. However, the NASA initiative is open to U.S. residents only.
*Geographic Focus:* All States, All Countries
*Date(s) Application is Due:* Jan 15
*Amount of Grant:* 1,500 USD
*Contact:* Jane Senior, (317) 637-9200, ext. 143; fax (317) 634-7817; jsenior@acsm.org
*Internet:* acsm.org/find-continuing-education/awards-grants/research-grants/dr.-raymond-a.-weiss-research-endowment
*Sponsor:* American College of Sports Medicine
401 West Michigan Street
Indianapolis, IN 46202-3233

**ACSM Foundation Clinical Sports Medicine Endowment Grants** 163
ACSM's Clinical Sports Medicine Endowment will fund an annual award of $5,000 for research that is directly related to clinical sports medicine. The purpose of the award is to stimulate clinical research in sports medicine and research in clinical sports medicine. Clinical research in sports medicine must be derived from active care of athletes, such as prevention strategies, treatment protocols, or treatment outcomes. Research in clinical sports medicine must be directly related to issues of active care of athletes, such as pathogenesis, pathophysiology, biomechanics, or environmental issues. The proposals must demonstrate the study's direct relationship to one of these purposes. The active patient care must be provided to physically active recreational, competitive or elite athletes. It is anticipated that the completed research will be presented at a highlighted session at the ACSM Annual Meeting.
*Requirements:* Grants are open to all ACSM members, including international members. This clinical research award is open to MDs, DOs, PTs, ATCs, and other medical professionals (ACSM members only) involved in the conduct of patient-based clinical research. Applicants for student research grants must have graduate student status during the term of the grant to be considered for funding. Applicants must be current members of ACSM at the time of submitting an application to be funded.
*Geographic Focus:* All States, All Countries
*Date(s) Application is Due:* Jan 15
*Amount of Grant:* 5,000 USD
*Contact:* Jane Senior, (317) 637-9200, ext. 143; fax (317) 634-7817; jsenior@acsm.org
*Internet:* acsm.org/find-continuing-education/awards-grants/research-grants/clinical-sports-medicine-endowment
*Sponsor:* American College of Sports Medicine
401 West Michigan Street
Indianapolis, IN 46202-3233

**ACSM Foundation Doctoral Student Research Grants** 164
ACSM Foundation Research Grants for doctoral students awards up to $5,000 for a one-year period. The awards are to be used for experimental subjects, supplies, and small equipment needs. There are two parts to the application. Both Part A and Part B should be submitted as one complete PDF application.
*Requirements:* Doctoral students enrolled in full-time programs are eligible to apply. Applicants for student research grants must have graduate student status during the term of the grant to be considered for funding. Applicants must be current members of ACSM at the time of submitting an application to be funded. Grants are open to all ACSM members, including international members. However, the NASA initiative is open to U.S. residents only.
*Geographic Focus:* All States, All Countries
*Date(s) Application is Due:* Jan 15
*Amount of Grant:* 5,000 USD
*Contact:* Jane Senior, (317) 637-9200, ext. 143; fax (317) 634-7817; jsenior@acsm.org
*Internet:* acsm.org/find-continuing-education/awards-grants/research-grants/acsm-foundation-doctoral-student-research-grant
*Sponsor:* American College of Sports Medicine
401 West Michigan Street
Indianapolis, IN 46202-3233

**ACSM Foundation Research Endowment Grants** 165
The research endowment has been made possible by charitable funds and individual donations derived from the annual campaign, specifically targeted for research. The intent of the research endowment is to use a portion of the interest derived from these funds to support basic and applied research in exercise science. The primary goal is to fund mechanistic, hypothesis driven, basic and applied research. A $10,000 grant is available and is applicable to all relevant fields of exercise science.
*Requirements:* Funding is primarily targeted for new or junior investigators, within seven years of attaining a terminal degree (e.g., Ph.D., Ed.D.). It is the intent of this grant to provide seed money support after which further funding would be sought from other sources. Only one application per person is allowed. Applicants for student research grants must have graduate student status during the term of the grant to be considered for funding. Applicants must be current members of ACSM at the time of submitting an application to be funded. Grants are open to all ACSM members, including international members. However, the NASA initiative is open to U.S. residents only.
*Geographic Focus:* All States, All Countries
*Date(s) Application is Due:* Jan 15
*Amount of Grant:* 10,000 USD
*Contact:* Jane Senior, (317) 637-9200, ext. 143; fax (317) 634-7817; jsenior@acsm.org
*Internet:* acsm.org/find-continuing-education/awards-grants/research-grants/research-endowment
*Sponsor:* American College of Sports Medicine
401 West Michigan Street
Indianapolis, IN 46202-3233

**ACSM International Student Award** 166

The American College of Sports Medicine invests in the future of sports medicine and exercise science by giving to its student members. Student members are eligible for grants and awards to offset the rising costs of tuition and travel. The International Student Award, in the amount of $1,000, honors students from countries outside of the United States and helps fund travel expenses for those traveling to the ACSM Annual Meeting to present their scholarly work. Submission of an abstract for presentation at the ACSM Annual Meeting is due November 1, with the completed application deadline three months after, on February 1.
*Requirements:* Applicants must be a student at the time of the ACSM Annual Meeting and must not hold a completed doctoral degree of any type (Ph.D., Ed.D., M.D., DPH, etc.) at the time of application. Applicants must: be ACSM members in good standing; have been accepted to present research at the ACSM Annual Meeting; be the first author of the accepted abstract and should plan to personally present the research; not be residents of the United States.
*Restrictions:* Previous International Student Award winners are not eligible to apply.
*Geographic Focus:* All States
*Date(s) Application is Due:* Feb 1
*Amount of Grant:* 1,000 USD
*Contact:* Jane Senior, Research Administration and Programs Department; (317) 637-9200, ext. 143 or (317) 637-9200, ext. 117; fax (317) 634-7817; jsenior@acsm.org or education@acsm.org
*Internet:* acsm.org/find-continuing-education/awards-grants/student-awards/international-student-award
*Sponsor:* American College of Sports Medicine
401 West Michigan Street
Indianapolis, IN 46202-3233

**ACSM International Student Awards** 167

ACSM international awards provide funding for professionals and students to participate in clinical and research exchange opportunities. The International Student Award honors students from countries outside of the United States and helps fund travel expenses for those traveling to the ACSM Annual Meeting to present their scholarly work. Awards will be determined by the quality of research to be presented at the ACSM Annual Meeting. Award recipients will receive $1,000 USD to assist with travel costs for attending ACSM's Annual Meeting.
*Requirements:* Applicants may not be residents of the United States. Applicants must be a student at the time of the ACSM Annual Meeting and must not hold a completed doctoral degree of any type (Ph.D., Ed.D., M.D., DPH, etc.) at the time of application. Applicants must be ACSM members in good standing and must have been accepted to present research at the ACSM Annual Meeting. Applicants should be the first author of the accepted abstract and should plan to personally present the research.
*Restrictions:* Previous International Student Award winners are not eligible to apply.
*Geographic Focus:* All Countries
*Date(s) Application is Due:* Feb 1
*Amount of Grant:* 1,000 USD
*Contact:* Jane Senior or Michael F. Dell, (317) 637-9200, ext. 143; fax (317) 634-7817; jsenior@acsm.org or mdell@acsm.org
*Internet:* acsm.org/find-continuing-education/awards-grants/international-awards/2011/08/16/international-student-award
*Sponsor:* American College of Sports Medicine
401 West Michigan Street
Indianapolis, IN 46202-3233

**ACSM Michael L. Pollock Student Scholarship** 168

The American College of Sports Medicine invests in the future of sports medicine and exercise science by giving to its student members. Student members are eligible for grants and awards to offset the rising costs of tuition and travel. The Michael L. Pollock Memorial Fund was established to help pay travel expenses for graduate students traveling to the ACSM Annual Meeting to present their scholarly work. Each scholarship, in the amount of $200, is given to an applicant whose research focuses on health-and-fitness or clinical exercise physiology. The annual deadline for applications is February 1.
*Requirements:* Applicants must: have been a graduate student in good standing within one year of submitting an application; and be the first author on an accepted research project (free communication or poster).
*Geographic Focus:* All States
*Date(s) Application is Due:* Feb 1
*Amount of Grant:* 200 USD
*Contact:* Jane Senior; (317) 637-9200, ext. 143 or (317) 637-9200, ext. 117; fax (317) 634-7817; jsenior@acsm.org or education@acsm.org
*Internet:* acsm.org/find-continuing-education/awards-grants/student-awards/michael-l.-pollock-student-scholarship
*Sponsor:* American College of Sports Medicine
401 West Michigan Street
Indianapolis, IN 46202-3233

**ACSM NASA Space Physiology Research Grants** 169

ACSM's Foundation Research Grant Program started in 1989 with one research initiative in basic and applied science, and it awarded $50,000 among five doctoral research students. Through the National Aeronautics and Space Administration (NASA), $10,000 is available for research grants in the area of exercise, weightlessness, and musculoskeletal physiology. Doctoral students enrolled in full-time programs are eligible to apply. Grants can range up to $5,000 and are available for a one-year period. There are two parts to the application. Both Part A and Part B should be submitted as one complete PDF application. The annual deadline is January 15.
*Requirements:* Doctoral students (U.S. residents only) enrolled in full-time programs are eligible to apply. Applicants for student research grants must have graduate student status during the term of the grant to be considered for funding. Applicants must be current members of ACSM at the time of submitting an application to be funded. Grants are open to all ACSM members, including international members. However, the NASA initiative is open to U.S. residents only.
*Geographic Focus:* All States
*Date(s) Application is Due:* Jan 15
*Amount of Grant:* Up to 5,000 USD
*Contact:* Jane Senior, (317) 637-9200, ext. 143; fax (317) 634-7817; jsenior@acsm.org or foundation@acsm.org
*Internet:* https://www.acsm.org/find-continuing-education/awards-grants/research-grants/nasa-space-physiology-research-grant
*Sponsor:* American College of Sports Medicine
401 West Michigan Street
Indianapolis, IN 46202-3233

**ACSM Oded Bar-Or International Scholar Awards** 170

The Oded Bar-Or International Scholar Award allows professionals to gain technical expertise and/or scientific knowledge through an international exchange program. Award recipients from the United States and Canada are required to travel to an institution outside of the United States or Canada. Similarly, award recipients from countries other than Canada and the United States are required to travel to an institution within the United States or Canada. Membership in ACSM is not a requirement. Awards will be determined by the likelihood that the activities will open avenues for professional growth, the intrinsic quality of the proposed research activity and evidence the project can be completed. Award recipients will receive round-trip air transportation to the host institution, a lodging and living stipend, medical health insurance for two months, complimentary registration and lodging for the ACSM Annual Meeting.
*Requirements:* The award is open to all ACSM members, including international members. Applicants must: be professionals in a discipline recognized by ACSM (minimum of a bachelor's degree is required; plan to engage in a professional activity that will advance the professional standing and reputation of scientists (Though many proposals will be for research-based activities, requests for other types of support will be considered; choose a host scientist who is a member of ACSM (ACSM will not match applicants with potential host scientists).
*Geographic Focus:* All States, Canada
*Date(s) Application is Due:* Feb 1
*Contact:* Jane Senior or Michael F. Dell, (317) 637-9200, ext. 143; fax (317) 634-7817; jsenior@acsm.org or mdell@acsm.org
*Internet:* acsm.org/find-continuing-education/awards-grants/research-grants/2011/08/16/oded-bar-or-international-scholar-award
*Sponsor:* American College of Sports Medicine
401 West Michigan Street
Indianapolis, IN 46202-3233

**ACSM Paffenbarger-Blair Fund for Epidemiological Research on Physical Activity Grants** 171

Dr. Ralph S. Paffenbarger, Jr., a co-recipient of the first Olympic Prize in Exercise Science and Sports Medicine, made an extraordinarily generous gift to the American College of Sports Medicine by donating his share of the award to the College's capital campaign with the monies added to the College's epidemiology endowment. The interest on this contribution makes this grant possible. The intent of this award is to encourage researchers early in their career to become involved with physical activity epidemiology. The applications may focus on observational studies of physical activity and health outcomes, or on randomized controlled trials that are clearly focused on physical activity and important public health issues. Applicants are expected to apply within two (2) years of receiving a postgraduate degree or completion of clinical training. The award will be in the amount of $10,000 for a one year period.
*Requirements:* Applicants for student research grants must have graduate student status during the term of the grant to be considered for funding. Applicants must be current members of ACSM at the time of submitting an application to be funded. Grants are open to all ACSM members, including international members. However, the NASA initiative is open to U.S. residents only.
*Geographic Focus:* All States, All Countries
*Date(s) Application is Due:* Jan 15
*Amount of Grant:* 10,000 USD
*Contact:* Jane Senior, (317) 637-9200, ext. 143; fax (317) 634-7817; jsenior@acsm.org
*Internet:* acsm.org/find-continuing-education/awards-grants/research-grants/paffenbarger-blair-fund-for-epidemiological-research-on-physical-activity
*Sponsor:* American College of Sports Medicine
401 West Michigan Street
Indianapolis, IN 46202-3233

**ACSM Steven M. Horvath Travel Award** 172

ACSM invests in the future of sports medicine and exercise science by giving to its student members. Student members are eligible for grants and awards to offset the rising costs of tuition and travel. The Steven M. Horvath Travel Award, in the amount of $500, funds travel expenses for underrepresented minority graduate students traveling to the ACSM Annual Meeting to present their scholarly work. The annual deadline for applications is February 1.

*Requirements:* Award requirements include: must be an underrepresented minority (as defined by NIH) graduate student in good standing with a research focus related to sports medicine and/or exercise sciences; and must be presenting a research project at the ACSM Annual Meeting.
*Geographic Focus:* All States
*Date(s) Application is Due:* Feb 1
*Contact:* Jane Senior, Research Administration and Programs Department; (317) 637-9200, ext. 143; fax (317) 634-7817; jsenior@acsm.org or foundation@acsm.org
*Internet:* acsm.org/find-continuing-education/awards-grants/student-awards/steven-m.-horvath-travel-award
*Sponsor:* American College of Sports Medicine
401 West Michigan Street
Indianapolis, IN 46202-3233

**Active Living Research, University of California, San Diego, Grants** 173
Active Living Research supports research examining how environments and policies impact physical activity, especially among ethnic minorities and children living in low-income communities. Findings are expected to inform environmental and policy changes that will promote active living among children and families. the Foundation encourages experts in fields such as public health, public administration, law, economics, transportation, medicine, recreation, urban planning, architecture, the behavioral sciences, child care and education to work together to identify promising approaches for increasing physical activity and preventing obesity among children and families.
*Requirements:* Requirements vary from one Call for Proposals to the next.
*Restrictions:* The program does not provide support for international activities, or for institutions in other countries.
*Geographic Focus:* All States
*Contact:* James F. Sallis, Program Director; (619) 260-5534; fax (619) 260-1510; ALR@projects.sdsu.edu
*Internet:* activelivingresearch.org/grantsearch/granteeresources
*Sponsor:* Active Living Research, University of California, San Diego
3900 Fifth Avenue, Suite 310
San Diego, CA 92103

**Acuity Charitable Foundation Grants** 174
The Acuity Charitable Foundation was established in 2003 with a major donation from the Acuity Mutual Insurance Company. The foundation primarily supports Sheboygan, Wisconsin, organizations involved with arts and culture, elementary and secondary education, engineering, health care, cancer, and youth development. Types of support include general operating support, program development, and sponsorships. Though formal application forms are not required, proposals should be submitted using the organization's letterhead. There are no deadlines, and applicants should submit the entire proposal, which should include: need for the project, group being served, evaluation process, and necessary funding.
*Requirements:* 501(c)3 organizations based in or serving residents of Sheboygan, Wisconsin, can apply.
*Geographic Focus:* Wisconsin
*Amount of Grant:* Up to 125,000 USD
*Samples:* Aurora Health Foundation, Aurora, Wisconsin, $125,000 - general operating support; Feeding America, Milwaukee, Wisconsin, $68,200 - general operating support; Mental Health America, Sheboygan, Wisconsin, $83,810 - general operating support.
*Contact:* Carli R. Miller, (920) 458-9131
*Sponsor:* Acuity Charitable Foundation
2800 South Taylor Drive, P.O. Box 58
Sheboygan, WI 53082-0058

**ADA Accelerator Award** 175
The American Diabetes Association's Pathway to Stop Diabetes program is a bold, innovative initiative designed to radically transform diabetes research. The Vision is simple yet revolutionary: find a new generation of brilliant scientists at the peak of their creativity, and provide them with freedom, autonomy, and the financial and professional resources to set them on the road to breakthrough discoveries. This initiative enables scientists to advance not only their research, but also their careers. Pathway scientists have access to the Mentor Advisory Group, an annual Pathway symposia, select speaking opportunities, and innovative technologies, all designed to foster interactions and collaboration that will enable them to accelerate their research progress. The Accelerator Award is designed to support exceptional, independent early-career researchers who have distinguished themselves as promising investigators and are in the beginning stages of establishing successful, sustainable diabetes research programs. Candidates must be identified through institutional nomination; applications will be accepted only from individuals with the appropriate institution support. Awards provide five years of research support at a maximum of $325,000 per year (including 30% indirect costs), for a total of $1,625,000. The annual deadline is July 2.
*Requirements:* Awards are available to early-career diabetes investigators proposing innovative and ambitious diabetes-related research programs. Applicants must hold faculty positions and have demonstrated independent productivity in diabetes research. Applicants may currently hold independent NIH funding (K, U or R awards, including an initial R01/U01) but must not have applied for, or received, an R01/U01 renewal or a second R01/U01 award.
*Geographic Focus:* All States
*Date(s) Application is Due:* Jul 2
*Amount of Grant:* Up to 1,625,000 USD
*Contact:* Research Staff; (800) 342-2383; pathway@diabetes.org or askada@diabetes.org
*Internet:* https://professional.diabetes.org/meetings/pathway-stop-diabetes%C2%AE
*Sponsor:* American Diabetes Association
2451 Crystal Drive, Suite 900
Arlington, VA 22202

**ADA Foundation Dentsply International Research Fellowships** 176
The American Dental Association Foundation's Dentsply International Research Fellowships may be applied for by finalists in the Annual Dental Student Research Competition conducted by Dentsply International. Fellowship recipients receive a ten-week research opportunity, a $4,000 stipend, and room and board expenses in order to work in the Foundation's Paffenbarger Research Center located in Gaithersburg, Maryland. Applications are available at the dental school research advisor and/or dean's office.
*Geographic Focus:* All States
*Amount of Grant:* 4,000 USD
*Contact:* Dr. Clifton M. Carey, Director of Administration; (301) 975-6806; fax (301) 926-9143; clif.carey@nist.gov
*Internet:* www.ada.org/applyforassistance.aspx#research
*Sponsor:* American Dental Association Foundation
211 East Chicago Avenue
Chicago, IL 60611

**ADA Foundation George C. Paffenbarger Student Research Awards** 177
The ADA Foundation George C. Paffenbarger Student Research Awards are offered in collaboration with the Academy of Operative Dentistry. Awards honors George C. Paffenbarger, an internationally recognized authority on dental materials and a pioneer in the development of specifications and standards for testing materials. The Foundation's Paffenbarger Research Center is named in his honor. Awards are open to dental students at all levels who wish to undertake novel research relevant to contemporary operative dentistry. The budget for the proposed research may not exceed $6,000. The awardee will also receive up to $1,000 to attend and present the research findings at the annual meeting. Applications must be submitted electronically to the Research Committee Chair.
*Requirements:* The application should take the form of a protocol outlining the background, aims and hypothesis of the proposed research, the methodology to be employed, and the anticipated work on the clinical practice of operative dentistry. Application information, including deadline and submission format, can be found at the Foundation's website.
*Geographic Focus:* All States
*Date(s) Application is Due:* Jan 13
*Amount of Grant:* Up to 7,000 USD
*Contact:* Cristina Garcia, Grants Program Coordinator; (312) 440-2763; fax (312) 440-3526; garciac@ada.org or adaf@ada.org
Dr. Tilly Peters, Research Committee Chair/Application Contact; (734) 763-3366; mcpete@umich.edu
*Internet:* www.ada.org/applyforassistance.aspx#tarrson
*Sponsor:* American Dental Association Foundation
211 East Chicago Avenue
Chicago, IL 60611

**ADA Foundation Scientist in Training Fellowship** 178
The American Dental Association Foundation's Scientist in Training Fellowships are conducted at the Foundation's Paffenbarger Research Center (PRC) in Gaithersburg, Maryland. The program includes a full-time fellow working in conjunction with the PRC's scientific research staff. Fellowships have been made possible by the Great-West Life and Annuity Insurance Company. Individuals interested in learning more about the program, application process or deadlines are invited to contact Paffenbarger Research Center.
*Geographic Focus:* All States
*Contact:* Dr. Clifton M. Carey, Director of Administration; (301) 975-6806; fax (301) 926-9143; clif.carey@nist.gov or prc@ada.org
*Internet:* www.ada.org/applyforassistance.aspx#grants
*Sponsor:* American Dental Association Foundation
211 East Chicago Avenue
Chicago, IL 60611

**ADA Foundation Summer Scholars Fellowships** 179
The ADA Foundation's Paffenbarger Research Center annually appoints two young investigators to its summer scholars program. The program brings industrial and dental researchers together in an environment outside the dental school. This program is made possible by Dentsply International. Individuals interested in learning more about the program, including the application process and deadlines, are invited to contact the Program Coordinator at the Paffenbarger Research Center.
*Geographic Focus:* All States
*Contact:* Gretchen Duppins, Summer Scholars Program Coordinator; (301) 975-6806; fax (301) 926-9143; gretchen@dupins.nist.gov
*Internet:* www.ada.org/applyforassistance.aspx
*Sponsor:* American Dental Association Foundation
211 East Chicago Avenue
Chicago, IL 60611

**ADA Initiator Award** 180
The American Diabetes Association's Pathway to Stop Diabetes program is a bold, innovative initiative designed to radically transform diabetes research. The Vision is simple yet revolutionary: find a new generation of brilliant scientists at the peak of their creativity, and provide them with freedom, autonomy, and the financial and professional resources to set them on the road to breakthrough discoveries. This initiative enables scientists to

advance not only their research, but also their careers. Pathway scientists have access to the Mentor Advisory Group, an annual Pathway symposia, select speaking opportunities, and innovative technologies, all designed to foster interactions and collaboration that will enable them to accelerate their research progress. The Initiator Award is designed to support the transition of scientists from mentored training to independent research faculty. Candidates must be identified through institutional nomination; applications will be accepted only from individuals with the appropriate institution support. Awards provide two distinct phases of research support: Phase 1 provides up to two years of support for mentored training at a maximum of $100,000 per year (including 10% indirect costs), Phase 2 provides up to five years of support for independent research at a maximum of $325,000 per year (including 30% indirect costs). Maximum combined support for Phase 1 and Phase 2 is $1,625,000. The annual deadline is July 2.
*Requirements:* Eligible applicants must currently be in research training positions (post-doctoral fellow, research fellowship) and have no more than seven years of research training following terminal doctoral degree.
*Restrictions:* Applicants cannot concurrently hold an NIH K99/R00 grant.
*Geographic Focus:* All States
*Date(s) Application is Due:* Jul 2
*Amount of Grant:* Up to 1,625,000 USD
*Contact:* Research Staff; (800) 342-2383; pathway@diabetes.org or askada@diabetes.org
*Internet:* https://professional.diabetes.org/meetings/pathway-stop-diabetes%C2%AE
*Sponsor:* American Diabetes Association
2451 Crystal Drive, Suite 900
Arlington, VA 22202

**ADA Innovative Basic Science Awards** 181
The American Diabetes Association's Core Research Program funds research with novel and innovative hypotheses in any area relevant to the etiology or pathophysiology of diabetes and its complications that hold significant promise for advancing the prevention, cure or treatment of diabetes. Basic, clinical, translational and epidemiological or health services research approaches are encouraged. Innovative Basic Science Awards support basic research with novel and innovative hypotheses in any area relevant to the etiology or pathophysiology of diabetes and its complications that holds significant promise for advancing the prevention, cure or treatment of diabetes. Applications proposing high-risk projects with the potential for high-impact results are encouraged, as are studies that may not be sufficiently developed for traditional funding sources. Awards provide up to $115,000 per year for up to three years, including indirect costs. Up to 20% of total costs may be requested for PI salary support. The annual application deadline is April 16.
*Requirements:* Applicants must be authorized to work in the United States or U.S. possessions. Any level faculty is eligible to apply.
*Geographic Focus:* All States
*Date(s) Application is Due:* Apr 16
*Amount of Grant:* Up to 345,000 USD
*Contact:* Research Department; (800) 342-2383; grantquestions@diabetes.org
*Internet:* https://professional.diabetes.org/meetings/core-program
*Sponsor:* American Diabetes Association
2451 Crystal Drive, Suite 900
Arlington, VA 22202

**ADA Innovative Clinical or Translational Science Grants** 182
The American Diabetes Association's Core Research Program funds research with novel and innovative hypotheses in any area relevant to the etiology or pathophysiology of diabetes and its complications that hold significant promise for advancing the prevention, cure or treatment of diabetes. Basic, clinical, translational and epidemiological or health services research approaches are encouraged. Innovative Clinical or Translational Science Grant awards support research with novel and innovative hypotheses, performed in human subjects, or research approaches to accelerate the transition of scientific discoveries into clinical application. Studies supported with these awards must directly involve human subjects, human samples and/or data, and offer considerable promise for advancing the cure, prevention or treatment of diabetes. Applications proposing high-risk projects with the potential for high-impact results are encouraged, as are studies that may not be sufficiently developed for traditional funding sources. Awards provide up to $200,000 per year for up to three years, including indirect costs. Up to 20% of total costs may be requested for PI salary support. The annual application deadline is April 16.
*Requirements:* Applicants must be authorized to work in the United States or U.S. possessions. Any level faculty is eligible to apply.
*Geographic Focus:* All States
*Date(s) Application is Due:* Apr 16
*Amount of Grant:* Up to 600,000 USD
*Contact:* Research Department; (800) 342-2383; grantquestions@diabetes.org or research@diabetes.org
*Internet:* https://professional.diabetes.org/meetings/core-program
*Sponsor:* American Diabetes Association
2451 Crystal Drive, Suite 900
Arlington, VA 22202

**ADA Junior Faculty Development Awards** 183
The American Diabetes Association's Core Research Program funds research with novel and innovative hypotheses in any area relevant to the etiology or pathophysiology of diabetes and its complications that hold significant promise for advancing the prevention, cure or treatment of diabetes. Basic, clinical, translational and epidemiological or health services research approaches are encouraged. Junior Faculty Development Awards support early investigators as they establish independence as diabetes researchers. Awards are up to $138,000 per year for up to four years including indirect costs.PI's salary support cannot exceed $75,000 per year. Additionally, applicants can request up to $10,000 per year towards the repayment of the principal on loans for a doctoral degree (MD, PhD, PharmD, DPM or DO). The annual application deadline is April 16.
*Requirements:* Eligible applicants must be full-time independent faculty with less than 10 years of research experience since their terminal degree who do not have previous or current NIH support (R00, R01, U01 or equivalent). Applicants may not hold concurrent awards of similar intent (NIH K awards, foundation or non-profit career development awards), however they can have completed previous NIH K or other career development funding. Applicants must dedicate at least 75% total effort to research activities. Applicants must be authorized to work in the United States or U.S. possessions.
*Geographic Focus:* All States
*Date(s) Application is Due:* Apr 16
*Amount of Grant:* Up to 592,000 USD
*Contact:* Research Department; (800) 342-2383; grantquestions@diabetes.org or research@diabetes.org
*Internet:* https://professional.diabetes.org/meetings/core-program
*Sponsor:* American Diabetes Association
2451 Crystal Drive, Suite 900
Arlington, VA 22202

**ADA Minority Postdoctoral Fellowships** 184
The American Diabetes Association's Core Research Program funds research with novel and innovative hypotheses in any area relevant to the etiology or pathophysiology of diabetes and its complications that hold significant promise for advancing the prevention, cure or treatment of diabetes. Basic, clinical, translational and epidemiological or health services research approaches are encouraged. Minority Postdoctoral Fellowships are available to minority postdoctoral researchers (MD, MD/PhD, PhD, DVM, or equivalent) who are underrepresented in biomedical research. Fellowships provide support for high quality training in disciplines and topics relevant to diabetes, in an environment conducive to beginning a career in diabetes research. Awards are between $47,484 and $55,581 peryear for up to three years, plus yearly $5,000 training and $5,000 fringe benefit allowances.
*Requirements:* Researchers (PhD, MD, PharmD, DO, DPM, or equivalent) with confirmed postdoctoral position by award start date are eligible to apply. Applicants must be from racial/ethnic groups that are underrepresented in biomedical research, and may not have more than ten years research experience beyond conferral of their doctoral degree. Applicants must be authorized to work in the United States or U.S. possessions.
*Geographic Focus:* All States
*Date(s) Application is Due:* Apr 16
*Amount of Grant:* 47,484 - 197,523 USD
*Contact:* Research Department; (800) 342-2383; grantquestions@diabetes.org or research@diabetes.org
*Internet:* https://professional.diabetes.org/meetings/core-program
*Sponsor:* American Diabetes Association
2451 Crystal Drive, Suite 900
Arlington, VA 22202

**ADA Minority Undergraduate Internships** 185
The American Diabetes Association's Core Research Program funds research with novel and innovative hypotheses in any area relevant to the etiology or pathophysiology of diabetes and its complications that hold significant promise for advancing the prevention, cure or treatment of diabetes. Basic, clinical, translational and epidemiological or health services research approaches are encouraged. Minority Undergraduate Internships are designed to provide a mentored diabetes research experience to undergraduate minority students who are underrepresented in biomedical research. The award is equal to a one-time $3,000 award. The annual application deadline is April 16.
*Requirements:* Current Association-funded investigators may apply for a stipend to support the training of a minority undergraduate student working in their laboratory. Applicants must be authorized to work in the United States or U.S. possessions.
*Geographic Focus:* All States
*Date(s) Application is Due:* Apr 16
*Amount of Grant:* 3,000 USD
*Contact:* Research Department; (800) 342-2383; grantquestions@diabetes.org or research@diabetes.org
*Internet:* https://professional.diabetes.org/meetings/core-program
*Sponsor:* American Diabetes Association
2451 Crystal Drive, Suite 900
Arlington, VA 22202

**Adam Richter Charitable Trust Grants** 186
The Trust, which is administered by the Bank of America in Dallas, Texas, was established in California in 1994. Offering grants on a national basis, its primary fields of interest include: the arts; animal welfare; Christian agencies and churches; the environment; health organizations; higher education; hospitals; and human services. There are no deadlines, though a formal application is required. Interested applicants should begin by sending a letter of inquiry to the Trust office.
*Restrictions:* Funding is not available for grants to individuals.
*Geographic Focus:* California
*Amount of Grant:* 3,000 - 20,000 USD
*Samples:* Educational Media Foundation, Rocklin, California, forr unrestricted contribution, payable over one year, $20,000; Habitat for Humanity of Sonoma Country, Santa Rosa, California, for unrestricted contribution, payable over one year, $6,000; Idol

Gives Back Foundation, Los Angeles, California, for unrestricted contribution, payable over one year, $13,000.
*Contact:* Michael Schlebach, Specialty Asset Manager; (866) 461-7281
*Sponsor:* Adam Richter Charitable Trust
c/o Bank of America, N.A.
Dallas, TX 75201-3115

**ADA Postdoctoral Fellowships** 187
The American Diabetes Association's Core Research Program funds research with novel and innovative hypotheses in any area relevant to the etiology or pathophysiology of diabetes and its complications that hold significant promise for advancing the prevention, cure or treatment of diabetes. Basic, clinical, translational and epidemiological or health services research approaches are encouraged. Postdoctoral Fellowships are available to postdoctoral researchers (MD, MD/PhD, PhD, DVM, or equivalent) to provide support for high quality training in disciplines and topics relevant to diabetes, in an environment conducive to beginning a career in diabetes research. Awards are between $47,484 and $55,581 per year for up to three years, plus yearly $5,000 training and $5,000 fringe benefit allowances. The annual application deadline is April 16.
*Requirements:* Researchers (PhD, MD, PharmD, DO, DPM, or equivalent) with confirmed postdoctoral position by award start date are eligible. Applicants must be authorized to work in the United States or U.S. possessions.
*Geographic Focus:* All States
*Date(s) Application is Due:* Apr 16
*Amount of Grant:* 47,484 - 196,743 USD
*Contact:* Research Department; (800) 342-2383; grantquestions@diabetes.org or research@diabetes.org
*Internet:* https://professional.diabetes.org/meetings/core-program
*Sponsor:* American Diabetes Association
2451 Crystal Drive, Suite 900
Arlington, VA 22202

**ADA Targeted Program Funding Grants** 188
The American Diabetes Association's Core Research Program funds research with novel and innovative hypotheses in any area relevant to the etiology or pathophysiology of diabetes and its complications that hold significant promise for advancing the prevention, cure or treatment of diabetes. Basic, clinical, translational and epidemiological or health services research approaches are encouraged. Targeted Awards provide support for research focused on a specific topic area. Targeted Requests for Applications (RFAs) can be any award type (research, development or training award). Annual application deadlines vary, with amounts typically ranging up to a maximum of $100,000.
*Requirements:* Applicant must be secondary teachers from the Near East, South Asia and Southeast Asia
*Geographic Focus:* New Independent States
*Contact:* Research Department; (800) 342-2383; grantquestions@diabetes.org
*Internet:* https://professional.diabetes.org/content/targeted-rfas
*Sponsor:* American Diabetes Association
2451 Crystal Drive, Suite 900
Arlington, VA 22202

**ADA Visionary Awards** 189
The American Diabetes Association's Pathway to Stop Diabetes program is a bold, innovative initiative designed to radically transform diabetes research. The Vision is simple yet revolutionary: find a new generation of brilliant scientists at the peak of their creativity, and provide them with freedom, autonomy, and the financial and professional resources to set them on the road to breakthrough discoveries. This initiative enables scientists to advance not only their research, but also their careers. Pathway scientists have access to the Mentor Advisory Group, an annual Pathway symposia, select speaking opportunities, and innovative technologies, all designed to foster interactions and collaboration that will enable them to accelerate their research progress. The Visionary Awards are designed to support established, experienced investigators with strong records of outstanding productivity in fields outside of diabetes who are interested in applying their considerable skills and expertise to diabetes research. These awards are highly competitive and intended to support particularly innovative and transformational ideas that have the potential to have an exceptional impact in diabetes. Candidates must be identified through institutional nomination; applications will be accepted only from individuals with the appropriate institution support. Awards provide two distinct phases of research support: Phase 1 provides up to three years of support at a maximum of $325,000 per year (including 30% indirect costs), Phase 2 provides up to two years of support at a maximum of $325,000 per year (including 30% indirect costs). Phase 2 is contingent upon demonstration of significant contributions to the field of diabetes research in Phase 1. Maximum combined support for Phase 1 and Phase 2 is $1,625,000. The annual deadline is July 2.
*Requirements:* Applicants must hold independent faculty positions and have demonstrated significant productivity in their current field of research. Applicants cannot have previously received national grant support (NIH, NIDDK, ADA, JDRF, etc.), as either a PI or Co-PI, in diabetes or diabetes-related research.
*Geographic Focus:* All States
*Date(s) Application is Due:* Jul 2
*Amount of Grant:* Up to 1,625,000 USD
*Contact:* Research Staff; (800) 342-2383; pathway@diabetes.org or askada@diabetes.org
*Internet:* https://professional.diabetes.org/meetings/pathway-stop-diabetes%C2%AE
*Sponsor:* American Diabetes Association
2451 Crystal Drive, Suite 900
Arlington, VA 22202

**AERA-MET Minority Dissertation Fellowships** 190
The American Educational Research Association (AERA) with funding support from the Bill and Melinda Gates Foundation announces a dissertation fellowship program to support graduate students in education research to conduct secondary data analysis using the Measures of Effective Teaching (MET) Longitudinal Database. The AERA-MET Dissertation Fellowship provides funding and professional development and training to dissertation stage graduate students who use the MET data to address research questions and examine issues that will contribute to knowledge about teaching and learning. The program supports high-quality science undertaken by the education research field through dissertation research on topics related to teaching and instruction, the effects of the classroom and school climate, student achievement, children and youth, and other educational issues. This is a field-initiated grants competition. The MET data allow researchers to explore many possible topics using a variety of research methods. Some examples of projects that might be conducted using MET data are listed below (but many other topics of research are equally appropriate to study). AERA will award each fellow up to a $20,000 stipend to study education, teaching, learning, or other education research topic using the MET Longitudinal Database. The fellowship funds can be used for research related expenses such as tuition, living expenses, travel to scholarly conferences, books, computer equipment, and other expenses that are directly related to conducting this research. As part of the proposal, applicants provide a budget that outlines anticipated research related expenses. AERA encourages cost sharing from universities in the form of tuition assistance, office space, university fees, and other expenses. Institutions cannot charge overhead or indirect costs to administer the fellowship funds. In addition to the funding, fellows will be paired with an Advisory Committee member who will provide mentoring, monitor fellows' progress, and help guide their research. The annual deadline for applications is May 16.
*Requirements:* Eligible graduate students for the AERA-MET Dissertation Fellowship Program will be at the dissertation stage in an accredited graduate program in education research or another social or behavioral science disciplinary or interdisciplinary field, such as sociology, economics, psychology, or political science. Applicants must be U.S. citizens or U.S. permanent residents enrolled in a doctoral program. Underrepresented racial and ethnic minority researchers and women are strongly encouraged to apply. Applicants should be familiar with the principles and framework of the MET study and the instruments and methods used to collect the data.
*Geographic Focus:* All States
*Date(s) Application is Due:* May 16
*Amount of Grant:* 20,000 USD
*Contact:* Felice J. Levine, Programs Manager; (202) 238-3200, ext. 201; fax (202) 238-3250; flevine@aera.net or fellowships@aera.net
*Internet:* www.aera.net/Professional-Opportunities-Funding/AERA-Funding-Opportunities/AERA-MET-Dissertation-Fellowship
*Sponsor:* American Educational Research Association
1430 K Street, NW, Suite 1200
Washington, D.C. 20005

**AERA Dissertation Grants** 191
AERA invites education-related dissertation proposals using NCES, NSF, and other federal databases. Dissertation Grants are available for advanced doctoral students and are intended to support the student while writing the doctoral dissertation. Applications are encouraged from a variety of disciplines, such as but not limited to, education, sociology, economics, psychology, demography, statistics, and psychometrics. Awards for Dissertation Grants are up to $20,000 for one-year projects. In addition to the dissertation grant award, grantees will participate in a two-day conference in Washington, D.C. The conference will provide unique professional development experiences for grantees, including highly qualified speakers on topics of education policy and career development, presentations of dissertation research by former grantees, and interaction with the Governing Board and federal agency staff. This conference is specifically for AERA grantees, and travel expenses will be paid by AERA.
*Requirements:* Applicants may be U.S. citizens or U.S. permanent residents enrolled in a doctoral program. Non-U.S. citizens enrolled in a doctoral program at a U.S. institution are also eligible to apply. Applicants should be advanced doctoral students at the dissertation writing stage. Underrepresented racial and ethnic minority researchers are strongly encouraged to apply. Grantees will present their research at a poster session during the AERA Annual Meeting and will participate in a one-day capstone workshop. Grantees must include travel funds (up to $1,000) in their grant budget to attend the AERA Annual Meeting held in Spring.
*Restrictions:* Grants are not renewable. In accordance with AERA's agreement with the funding agency, institutions may not charge indirect costs on these awards.
*Geographic Focus:* All States
*Date(s) Application is Due:* Jan 14
*Amount of Grant:* Up to 20,000 USD
*Contact:* Felice J. Levine, Program Manager; (202) 238-3200, ext. 201; fax (202) 238-3250; flevine@aera.net or grantsprogram@aera.net
*Internet:* www.aera.net/Professional-Opportunities-Funding/AERA-Funding-Opportunities/Grants-Program/Dissertation-Grants
*Sponsor:* American Educational Research Association
1430 K Street, NW, Suite 1200
Washington, D.C. 20005

**AERA Minority Dissertation Fellowships** 192
In 1991, the Council of the American Educational Research Association (AERA) established the AERA Minority Dissertation Fellowship in Education Research to provide support for doctoral dissertation research. The purposes of the program are to

advance education research by outstanding minority graduate students and to improve the quality and diversity of university faculties. This program offers doctoral fellowships to enhance the competitiveness of outstanding minority scholars for academic appointments at major research universities. It supports fellows conducting education research and provides mentoring and guidance toward the completion of their doctoral studies. The annual application deadline is November 1.
*Requirements:* This program is open to U.S. citizens and permanent residents. Fellows will be required to provide proof of advancement to candidacy at the beginning of the grant period. In the letter of recommendation from the applicant's advisor, the dissertation advisor will attest to the proposal constituting research for the doctoral degree. Applicants must work full-time on their dissertations and course requirements. This program is targeted for members of racial and ethnic groups historically underrepresented in higher education (e.g., African Americans, Alaskan Natives, American Indians, Asian Americans, Hispanics or Latinos, and Native Hawaiian or Pacific Islanders). Each fellowship award is for one year, beginning July 1 or later, and is nonrenewable. Fellowships are awarded for doctoral dissertation research conducted under faculty sponsorship in any accredited university in the United States.
*Geographic Focus:* All States
*Date(s) Application is Due:* Nov 1
*Contact:* Felice J. Levine, Programs Manager; (202) 238-3200, ext. 201; fax (202) 238-3250; flevine@aera.net or fellowships@aera.net
*Internet:* www.aera.net/Professional-Opportunities-Funding/AERA-Funding-Opportunities/Minority-Dissertation-Fellowship-Program
*Sponsor:* American Educational Research Association
1430 K Street, NW, Suite 1200
Washington, D.C. 20005

**AERA Research Grants** 193
The AERA Grants Program, with support from the National Science Foundation (NSF), announces calls for proposals each year for its Research Grants competition. The program seeks to stimulate research on U.S. education issues using data from the large-scale, national and international data sets supported by the National Center for Education Statistics (NCES), NSF, and other federal agencies, and to increase the number of education researchers using these data sets. The program supports research projects that are quantitative in nature, include the analysis of existing data from NCES, NSF or other federal agencies, and have U.S. education policy relevance. AERA invites education-related research proposals using NCES, NSF, and other federal databases. Research Grants are available for faculty at institutions of higher education, postdoctoral researchers, and other doctoral-level scholars. Applications are encouraged from a variety of disciplines, such as but not limited to, education, sociology, economics, psychology, demography, statistics, and psychometrics. Awards for Research Grants are up to $20,000 for one-year projects, or up to $35,000 for two-year projects.
*Requirements:* Applicants may be U.S. citizens or U.S. permanent residents. Non-U.S. citizens working at a U.S. institution are also eligible to apply. Applicants must have received the doctoral degree by the start date of the grant. Underrepresented racial and ethnic minority researchers are strongly encouraged to apply. Applicants are encouraged to submit proposals that: develop or benefit from new quantitative measures or methodological approaches for addressing education issues; include interdisciplinary teams with subject matter expertise, especially when studying science, technology, engineering and mathematics (STEM) learning; analyze TIMSS, PISA, or other international data resources; include the integration and analysis of more than one data set. Proposals for Research Grants will be reviewed twice a year, once in the fall and once in the winter.
*Restrictions:* In accordance with AERA's agreement with the funding agencies, institutions may not charge indirect costs or overhead on these awards.
*Geographic Focus:* All States
*Date(s) Application is Due:* Jan 24; Oct 1
*Amount of Grant:* Up to 35,000 USD
*Contact:* Felice J. Levine, Program Manager; (202) 238-3200, ext. 201; fax (202) 238-3250; flevine@aera.net or grantsprogram@aera.net
*Internet:* www.aera.net/Professional-Opportunities-Funding/AERA-Funding-Opportunities/Grants-Program/Research-Grants
*Sponsor:* American Educational Research Association
1430 K Street, NW, Suite 1200
Washington, D.C. 20005

**AES-Grass Young Investigator Travel Awards** 194
To recognize outstanding young investigators conducting research in basic or clinical neuroscience related to epilepsy, the Grass Foundation and the American Epilepsy Society (AES) have partnered to present the Young Investigator Travel Awards for up to eight candidates to help support travel costs to present their research at the annual AES meeting. The AES will announce the awardees and their talk/poster titles in an award ceremony during the annual program. Following the initial review of application abstracts, as part of the application process, a recommendation may be required from one of the applicant's trainers/mentors. This will be solicited by the AES review committee should the application be deemed competitive for the award. This recommendation should include the applicant's research focus, role in the abstract, promise as a future investigator, and any other pertinent information.
*Requirements:* Candidates will be investigators early in their careers, including graduate students, postdoctoral fellows, residents, clinical fellows or junior faculty (M.D., M.D./Ph.D., Ph.D., or the equivalent), who are conducting basic or clinical research relating to epilepsy. Candidates must also be no more than five years out of postgraduate training. Applicants should also be first authors on an abstract submitted to the AES. Applicants can self-nominate, or nominations from second parties will be accepted. Applicants submit an abstract through the AES online submission program as first author, then indicate the Junior Investigator Travel Award box. Candidates must be present at the AES Annual Meeting.
*Geographic Focus:* All States
*Date(s) Application is Due:* Jun 15
*Amount of Grant:* 1,000 USD
*Contact:* Cheryl-Ann Tubby, Assistant Executive Director; (860) 586-7505, ext. 542; fax (860) 586-7550; ctubby@aesnet.org
*Internet:* www.aesnet.org/research/aes-sponsored-grant/grass-travel-award/
*Sponsor:* American Epilepsy Society
342 North Main Street
West Hartford, CT 06117-2507

**AES Robert S. Morison Fellowship** 195
The American Epilepsy Society, in partnership with the Grass Foundation, offers the Robert S. Morison Fellowship for the training of academic clinician scientists in epilepsy. This two-year post-doctoral Fellowship will be awarded to a promising young investigator possessing an M.D. degree who intends to continue training in basic science in an epilepsy research laboratory. The Fellowship was created to honor the contributions of Dr. Morison, one of the founding Trustees of The Grass Foundation. The Fellowship funds $40,000 per year for two years of salary with an additional $10,000 (maximum) per year to be used for institutional fringe benefits plus $1,000 for the Fellow to travel to the AES Annual Meeting to present his/her results. The award is announced and presented during the AES Annual Meeting.
*Requirements:* Applicants must apply at the Epilepsy Foundation website for either a Post-doctoral Fellowship or a Research and Training Fellowship for Clinicians. Candidates should note in the application their desire to be considered specifically for the Morison Fellowship. Applicants for this Fellowship can at the same time be considered for a Fellowship in the category under which he/she applied.
*Geographic Focus:* All States, All Countries
*Amount of Grant:* Up to 50,000 USD
*Contact:* Cheryl-Ann Tubby, Assistant Executive Director; (860) 586-7505, ext. 542; fax (860) 586-7550; ctubby@aesnet.org
*Internet:* www.aesnet.org/go/research/aes-sponsored-grant-program/grass-foundation/aes-robert-s-morison-fellowship
*Sponsor:* American Epilepsy Society
342 North Main Street
West Hartford, CT 06117-2507

**AES Susan S. Spencer Clinical Research Epilepsy Training Fellowship** 196
The Spencer Fellowship is a two-year fellowship to support clinical research training in the field of epilepsy. The Fellowship is supported by the American Academy of Neurology Foundation (AAN), the American Epilepsy Society, and the Epilepsy Foundation. Each Fellowship will consist of $55,000 per year for two years, plus $10,000 per year for tuition to support formal education in clinical research methodology at the applicant's institution or elsewhere. Supplementation of the stipend with other grants or by the fellowship institution is permissible, but Fellows may not accept other Fellowships, similar awards, or have another source of support for more than 50% of their research salary. Only direct costs will be funded by this award. Detailed information about the application and its process is available at the AES website. All materials must be submitted online.
*Requirements:* For the purpose of this Fellowship, clinical research is defined as "patient-oriented research conducted with human subjects, or translational research specifically designed to develop treatments or enhance diagnosis of neurological disease." These areas of research include epidemiologic or behavioral studies, clinical trials, studies of disease mechanisms, the development of new technologies, and health services and outcomes research." Disease related studies not directly involving humans or human tissue also are encouraged if the primary goal is the development of therapies, diagnostic tests, or other tools to prevent or mitigate neurological diseases. Applicants must be AAN members interested in an academic career in clinical research who have completed a residency or a post-doctoral Fellowship (for a Ph.D.) within the past five years. Those early in their clinical research careers will be given priority.
*Geographic Focus:* All States
*Date(s) Application is Due:* Oct 1
*Amount of Grant:* 65,000 USD
*Contact:* Terry Heinz, Grants Administrator; (651) 695-2746; theinz@aan.com
*Internet:* www.aesnet.org/go/research/aes-sponsored-grant-program
*Sponsor:* American Epilepsy Society
342 North Main Street
West Hartford, CT 06117-2507

**Aetna Foundation Integrated Health Care Grants** 197
The Aetna Foundation is advancing an integrated health care agenda that focuses on care coordination and builds on a foundation of strong primary care. The Aetna Foundation seeks to support projects that promote evidence-based models of care coordination that can lead to high-quality, patient-centered health care services, improve health outcomes and lower costs. The Foundation's goal is to demonstrate the key components, best practices and benefits of care coordination that is centered on strong primary care. Examples of grants supported include projects or studies that: develop standards and metrics of care coordination in ambulatory care settings; evaluate models of care coordination that enhance providers' communication with each other and with their patients and lead to improved patient outcomes and experiences with their care; evaluate methods of care coordination for engaging patients as partners in their care, particularly in the management of chronic conditions; identify best practices to align financial and

other incentives for achieving well-coordinated care; and identify and assess models of care coordination that reduce the cost of care while improving patient health outcomes.
*Requirements:* Grants will be made only to nonprofit organizations with evidence of IRS 501(c)3 designation or de facto tax-exempt status.
*Restrictions:* The Foundation generally does not fund: endowment or capital costs, including construction, renovation, or equipment; direct delivery of reimbursable health care services; basic biomedical research; grants or scholarships to individuals; work for which results and impact cannot be measured; advertising; golf tournaments; advocacy, political causes or events; sacramental or theological functions of religious organizations; operational expenses; or existing deficits.
*Geographic Focus:* All States
*Amount of Grant:* 25,000 - 100,000 USD
*Contact:* Melenie O. Magnotta, Grants Manager; (860) 273-1012 or (860) 273-0123; fax (860) 273-7764; aetnafoundation@aetna.com
*Internet:* www.aetna-foundation.org/foundation/aetna-foundation-programs/integrated-healthcare/index.html
*Sponsor:* Aetna Foundation
151 Farmington Avenue
Hartford, CT 06156-3180

**Aetna Foundation Minority Scholars Grants** 198
Diversifying the next generation of health care professionals and researchers is an important part of the Aetna Foundation's efforts to support high-quality health care for all population groups. Founded in 2010, the AcademyHealth/Aetna Foundation Minority Scholars Program provides professional development, mentoring and networking opportunities for graduate-level students, post-doctoral fellows and other researchers to attract more men and women from underrepresented groups to the field of racial and ethnic disparities research in health outcomes and access to health care.
*Restrictions:* The Foundation generally does not fund: endowment or capital costs, including construction, renovation, or equipment; direct delivery of reimbursable health care services; basic biomedical research; grants or scholarships to individuals; work for which results and impact cannot be measured; advertising; golf tournaments; advocacy, political causes or events; sacramental or theological functions of religious organizations; operational expenses; or existing deficits.
*Geographic Focus:* All States
*Contact:* Melenie O. Magnotta, Grants Manager; (860) 273-1012 or (860) 273-0123; fax (860) 273-7764; aetnafoundation@aetna.com
*Internet:* www.aetna-foundation.org/foundation/aetna-foundation-programs/scholars/index.html
*Sponsor:* Aetna Foundation
151 Farmington Avenue
Hartford, CT 06156-3180

**Aetna Foundation National Medical Fellowship in Healthcare Leadership** 199
Diversifying the next generation of health care professionals and researchers is an important part of the Aetna Foundation's efforts to support high-quality health care for all population groups. In 2011, in partnership with National Medical Fellowships (NMF), the Aetna Foundation established the Aetna Foundation/NMF Healthcare Leadership program to provide scholarships to second- and third-year medical students from underrepresented minority groups. Recipients commit to practice medicine in medically underserved communities and are distinguished by their community service and leadership potential.
*Restrictions:* The Foundation generally does not fund: endowment or capital costs, including construction, renovation, or equipment; direct delivery of reimbursable health care services; basic biomedical research; grants or scholarships to individuals; work for which results and impact cannot be measured; advertising; golf tournaments; advocacy, political causes or events; sacramental or theological functions of religious organizations; operational expenses; or existing deficits.
*Geographic Focus:* All States
*Contact:* Melenie O. Magnotta, Grants Manager; (860) 273-1012 or (860) 273-0123; fax (860) 273-7764; aetnafoundation@aetna.com
*Internet:* www.aetna-foundation.org/foundation/aetna-foundation-programs/scholars/index.html
*Sponsor:* Aetna Foundation
151 Farmington Avenue
Hartford, CT 06156-3180

**AFP Research Grants** 200
Research grants up to $10,000 annually are available through the Association of Fundraising Professionals Research Council to members who want to explore new aspects, trends or issues related to fundraising. Major directions for the Research Council in the next two to three years include: salary survey, especially incentive compensation; cost of fundraising and fundraising effectiveness projects; trends and gaps in knowledge; international issues; and demographics of fundraisers and donors. Grants are awarded on a rolling basis twice a year, with deadlines of February 1 and September 1.
*Requirements:* Fundraising professionals who wish to submit proposals are encouraged to seek advice from or partner with academic institutions in developing the research design and to attend the AFP International Conference information session on how to submit a grant application to AFP. Applicants will also be asked to submit documentation of compliance with ethical guidelines for any research involving human subjects.
*Geographic Focus:* All States
*Date(s) Application is Due:* Feb 1; Sep 1
*Amount of Grant:* Up to 10,000 USD
*Contact:* Cathlene Williams, Awards Administrator; (703) 684-0410 or (800) 666-3863; fax (703) 684-0540; cwilliams@afpnet.org or profadv@afpnet.org
*Internet:* www.afpnet.org/Audiences/ReportsResearchDetail.cfm?ItemNumber=780
*Sponsor:* Association of Fundraising Professionals
4300 Wilson Boulevard, Suite 300
Arlington, VA 22203

**AFP Skystone Ryan Prize for Research on Fundraising and Philanthropy** 201
Each year, the Association of Fundraising Professionals Research Council awards the Skystone Partners Prize for Research to the author of a book that contributes substantially to the knowledge and understanding of fundraising or philanthropic behavior. The Prize for Research is made possible by an endowment from Skystone Partners to encourage advanced research that extends the knowledge of fundraising and philanthropy. Authors of published works on research in fundraising and philanthropy are invited to submit books or monographs for the Skystone Partners Prize for Research on Fundraising and Philanthropy. Other individuals or organizations can nominate appropriate publications. The annual application/nomination submission deadline is November 1. Prize winners receive a cash award of $3,000 and are honored at the AFP International Conference on Fundraising.
*Requirements:* To be considered for the Prize, works must be: a book or monograph of fifty or more pages; published by a commercial publishing house or a professional organization during the 23 months preceding the current deadline; based on either applied or basic research; reflect a standard publisher selection process without regard to the source and sponsorship of the research.
*Restrictions:* Articles, directories, op-ed pieces, and self-published or unpublished works will not be considered.
*Geographic Focus:* All States
*Date(s) Application is Due:* Nov 1
*Amount of Grant:* 3,000 USD
*Samples:* Dr. Lilya Wagner, book entitled Diversity and Philanthropy: Expanding the Circle of Giving; Marilyn Taylor, Robert Strom, and David Renz, book entitled Handbook of Research on Entrepreneurs' Engagement in Philanthropy: Perspectives; Penelope Cagney, CFRE and Bernard Ross, book entitled Global Fundraising: How the World is Changing the Rules of Philanthropy.
*Contact:* Cathlene Williams, Awards Administrator; (800) 666-3863 or (703) 684-0410; fax (703) 684-0540; cwilliams@afpnet.org or profadv@afpnet.org
*Internet:* www.afpnet.org/About/AwardsDetailcfm?ItemNumber=783&navItem Number=13283
*Sponsor:* Association of Fundraising Professionals
4300 Wilson Boulevard, Suite 300
Arlington, VA 22203

**AGA Foundation Research Scholar Award** 202
The AGA Foundation Research Scholar Award provides $90,000 per year for three years (total $270,000) to a young investigator, instructor, research associate or equivalent working toward an independent career in gastroenterology, hepatology or related areas. The objective of the Award is to support young gastroenterologists working toward independent and productive research careers in digestive diseases by ensuring that a major proportion of their time is protected for research (i.e. a minimum of 75 percent effort dedicated to the proposed project). The award will support young faculty (not fellows) who have demonstrated exceptional promise and have some record of accomplishment in research. Up to five awards will be given annually. The deadline for applications is September 8, with a start date of July 1 the following year.
*Requirements:* Candidates for this award must: hold an MD, PhD or equivalent degree (e.g. MB, ChB, MBBS, DO) and a full-time faculty position at a North American institution by the start date of this award; and be an AGA member in good standing at the time of application submission. AGA membership must be maintained throughout the three-year duration of the award. Women and under-represented minorities are encouraged to apply.
*Geographic Focus:* All States
*Date(s) Application is Due:* Sep 8
*Amount of Grant:* 270,000 USD
*Contact:* Harmony Excellent, Senior Manager of Development; (301) 272-1602 or (301) 654-2055; hexcellent@gastro.org or awards@gastro.org
*Internet:* www.gastro.org/grants/research-scholar-award-rsa
*Sponsor:* American Gastroenterological Association Foundation
4930 Del Ray Avenue
Bethesda, MD 20814

**AGA Foundation Takeda Pharmaceuticals Research Scholar Award in Inflammatory Bowel Disease** 203
The AGA-Takeda Pharmaceuticals Research Scholar Award in Inflammatory Bowel Disease provides $90,000 per year for three years (total $270,000) to a young investigator, instructor, research associate or equivalent working toward independent careers in gastroenterology, hepatology or related areas. The objective of the Award is to support young gastroenterologists working toward independent and productive research careers in IBD by ensuring that a major proportion of their time is protected for research (i.e. a minimum of 75 percent effort dedicated to the proposed project). The award will support young faculty (not fellows) who have demonstrated exceptional promise and have some record of accomplishment in research. One award will be given annually. The deadline for applications is September 8, with a start date of July 1 the following year.
*Requirements:* Candidates for this award must: hold an MD, PhD or equivalent degree (e.g. MB, ChB, MBBS, DO) and a full-time faculty position at a North American institution by the start date of this award; and be an AGA member in good standing at the time of application submission. AGA membership must be maintained throughout the three-year duration of the award. Women and under-represented minorities are

encouraged to apply. Physician-scientist investigators and candidates interested in translational research are especially encouraged to apply for this award.
*Geographic Focus:* All States
*Date(s) Application is Due:* Sep 8
*Amount of Grant:* 270,000 USD
*Contact:* Harmony Excellent, Senior Manager of Development; (301) 272-1602 or (301) 654-2055; hexcellent@gastro.org or awards@gastro.org
*Internet:* www.gastro.org/grants/aga-takeda-pharmaceuticals-research-scholar-award-in-inflammatory-bowel-disease
*Sponsor:* American Gastroenterological Association Foundation
4930 Del Ray Avenue
Bethesda, MD 20814

**AGA Rady Children's Institute for Genomic Medicine Research Scholar Award in Pediatric Genomics** 204

The AGA-Rady Children's Institute for Genomic Medicine Research Scholar Award in Pediatric Genomics provides $90,000 per year for three years (totaling $270,000) to a young investigator, instructor, research associate or equivalent working toward independent careers in gastroenterology, hepatology, or related areas. The proposed research may be basic, translational or clinical and must use genomics as an approach to enhance understanding of pediatric digestive diseases toward prevention, treatment, and/or cure of such diseases. The objective of the Award is to support young gastroenterologists working toward independent and productive research careers in digestive diseases by ensuring that a major proportion of their time is protected for research (i.e. a minimum of 75 percent effort dedicated to the proposed project). The award will support early-career faculty (not fellows) who have demonstrated exceptional promise and have a record of accomplishment in research. The proposed research may be basic, translational or clinical and must incorporate genomics as an approach to enhance understanding of pediatric digestive diseases toward prevention, treatment, and/or cure of such diseases. One award will be given annually. The deadline for applications is September 8, with a start date of July 1 the following year.
*Requirements:* The funded research must be conducted full-time at the Rady Children's Institute for Genomic Medicine in San Diego, California, or at Rady Children's Hospital in San Diego. Candidates must: hold a full-time faculty position at a North American institution by the start date of this award; identify a mentor amongst faculty based at the Rady Children's Institute for Genomic Medicine; hold an MD, PhD, and/or equivalent degree (e.g. MB, ChB, MBBS, DO) by the start date of this award per the following guidelines; and be an AGA member in good standing at the time of application submission. AGA membership must be maintained throughout the three-year duration of the award.
*Geographic Focus:* All States
*Date(s) Application is Due:* Sep 8
*Amount of Grant:* 270,000 USD
*Contact:* Harmony Excellent, Senior Manager of Development; (301) 272-1602 or (301) 654-2055; hexcellent@gastro.org or awards@gastro.org
*Internet:* www.gastro.org/grants/aga-rady-children-s-institute-for-genomic-medicine-research-scholar-award-in-pediatric-genomics
*Sponsor:* American Gastroenterological Association Foundation
4930 Del Ray Avenue
Bethesda, MD 20814

**AGA Research Foundation Caroline Craig Augustyn and Damian Augustyn Award in Digestive Cancer** 205

The AGA-Caroline Craig Augustyn and Damian Augustyn Award in Digestive Cancer awards $40,000 for one year to an early career investigator who currently holds a federal or non-federal career development award devoted to conducting research related to digestive cancer. The primarey objective is to support early career investigators conducting research relevant to the pathogenesis, prevention, diagnosis, or treatment of digestive cancer. The award is intended to supplement existing career development funding. The deadline for applications is January 12, with a start date of July 1 the same year.
*Requirements:* Candidates for this award must: hold an MD, PhD or equivalent degree (e.g. MB, ChB, MBBS, DO) and a full-time faculty position at an accredited North American institution; conduct research at an accredited North American institution; and hold an NIH K series (e.g. K01, K08, K23, K99, R00) or other federal or non-federal career development award of at least four years' duration. The proposed research must focus on the pathogenesis, prevention, diagnosis, or treatment of digestive cancer.
*Geographic Focus:* All States
*Date(s) Application is Due:* Jan 12
*Amount of Grant:* 40,000 USD
*Contact:* Harmony Excellent, Senior Manager of Development; (301) 272-1602 or (301) 654-2055; hexcellent@gastro.org or awards@gastro.org
*Internet:* www.gastro.org/grants/aga-caroline-craig-augustyn-damian-augustyn-award-in-digestive-cancer
*Sponsor:* American Gastroenterological Association Foundation
4930 Del Ray Avenue
Bethesda, MD 20814

**AGA Research Foundation Elsevier Gut Microbiome Pilot Research Award** 206

The AGA-Elsevier Gut Microbiome Pilot Research Award provides $25,000 for one year to a recipient at any career stage researching projects on the relationship of the gut microbiota to digestive health and disease. The objective of this pilot research award is to provide funds for early career investigators to help establish their research careers or to support projects that represent new research directions for established investigators. Projects must focus on the relationship of the gut microbiota to digestive health and disease. A single award will be given annually. The deadline for applications is January 12, with a start date of July 1 the same year.
*Requirements:* Candidates for this award must hold an MD, PhD, or equivalent degree (e.g., MBChB, MBBS, DO) and a full-time faculty position at an accredited North American institution. AGA membership is required at the time of application submission. Women and minorities are strongly encouraged to apply.
*Restrictions:* Applicants may not submit the same proposal to multiple AGA pilot awards offered in the same year.
*Geographic Focus:* All States
*Date(s) Application is Due:* Jan 12
*Amount of Grant:* 25,000 USD
*Contact:* Harmony Excellent, Senior Manager of Development; (301) 272-1602 or (301) 654-2055; hexcellent@gastro.org or awards@gastro.org
*Internet:* www.gastro.org/grants/aga-elsevier-gut-microbiome-pilot-research-award
*Sponsor:* American Gastroenterological Association Foundation
4930 Del Ray Avenue
Bethesda, MD 20814

**AGA Research Foundation Elsevier Pilot Research Award** 207

The AGA-Elsevier Pilot Research Award provides $25,000 for one year to a recipient at any career stage performing research in gastroenterology- or hepatology-related areas. The objective of this pilot research award is to provide funds to early career investigators to help establish their research careers, or to support projects that represent new research directions for established investigators. Up to two awards will be given annually. The deadline for applications is January 12, with a start date of July 1 the same year.
*Requirements:* Candidates for this award must hold an MD, PhD, or equivalent degree (e.g., MBChB, MBBS, DO) and a full-time faculty position at an accredited North American institution. AGA membership is required at the time of application submission. Women and minorities are strongly encouraged to apply.
*Restrictions:* Applicants may not submit the same proposal to multiple AGA pilot research awards offered in the same year.
*Geographic Focus:* All States
*Date(s) Application is Due:* Jan 12
*Amount of Grant:* 25,000 USD
*Contact:* Harmony Excellent, Senior Manager of Development; (301) 272-1602 or (301) 654-2055; hexcellent@gastro.org or awards@gastro.org
*Internet:* www.gastro.org/grants/aga-elsevier-pilot-research-award
*Sponsor:* American Gastroenterological Association Foundation
4930 Del Ray Avenue
Bethesda, MD 20814

**AGA Research Foundation Pfizer Young Investigator Pilot Research Award in Inflammatory Bowel Disease** 208

The AGA-Pfizer Young Investigator Pilot Research Award in Inflammatory Bowel Disease provides $30,000 for one year to recipients at any career stage researching new directions focused on improving the diagnosis and treatment of inflammatory bowel disease (IBD). The objective of this pilot research award is to provide funds for early career investigators to help establish their research careers or to support projects that represent new research directions for established investigators. Projects must focus on new research areas that could improve the diagnosis and treatment of IBD. Up to three awards will be given annually. The deadline for applications is January 12, with a start date of July 1 the same year.
*Requirements:* The proposed research must focus on new research areas that could improve the diagnosis and treatment of IBD. Candidates for this award must hold an MD, PhD, or equivalent degree (e.g., MBChB, MBBS, DO) and a full-time faculty position at an accredited North American institution. AGA membership is required at the time of application submission. Women and minorities are strongly encouraged to apply.
*Restrictions:* Applicants may not submit the same proposal to multiple AGA pilot awards offered in the same year.
*Geographic Focus:* All States
*Date(s) Application is Due:* Jan 12
*Amount of Grant:* 30,000 USD
*Contact:* Harmony Excellent, Senior Manager of Development; (301) 272-1602 or (301) 654-2055; hexcellent@gastro.org or awards@gastro.org
*Internet:* www.gastro.org/grants/aga-pfizer-young-investigator-pilot-research-award-in-inflammatory-bowel-disease
*Sponsor:* American Gastroenterological Association Foundation
4930 Del Ray Avenue
Bethesda, MD 20814

**AGA Research Foundation Rome Foundation Functional GI and Motility Disorders Pilot Research Award** 209

The AGA-Rome Foundation Functional GI and Motility Disorders Pilot Research Award is one-year, $30,000 research grant offered to a recipient at any career stage to support pilot research projects pertaining to functional GI and motility disorders. The award is sponsored by the Rome Foundation, and the primary objective of this award is to provide investigators funds to help establish their research careers or support pilot projects that represent new research directions. The intent of the award is to stimulate research in the areas of functional GI and motility disorders by providing time for investigators to obtain new data that can ultimately lead to subsequent grant applications for more substantial funding and duration. Up to two awards will be given annually. The deadline for applications is January 12, with a start date of July 1 the same year.

*Requirements:* International candidates may apply for this award. Candidates for this award must hold an MD, PhD or equivalent degree (e.g. MB, ChB, MBBS, DO) and a full-time faculty position at an accredited institution. AGA membership is required at the time of application submission. Women and minorities are strongly encouraged to apply.
*Restrictions:* Rome Foundation board members are ineligible.
*Geographic Focus:* All States, All Countries
*Date(s) Application is Due:* Jan 12
*Amount of Grant:* 30,000 USD
*Contact:* Harmony Excellent, Senior Manager of Development; (301) 272-1602 or (301) 654-2055; hexcellent@gastro.org or awards@gastro.org
*Internet:* www.gastro.org/grants/aga-rome-foundation-functional-gi-and-motility-disorders-pilot-research-award
*Sponsor:* American Gastroenterological Association Foundation
4930 Del Ray Avenue
Bethesda, MD 20814

**AGHE Graduate Scholarships and Fellowships** **210**
The association awards graduate scholarships and fellowships for research on aging. Gerontology and aging studies faculty are invited to nominate qualified students. This program is funded by the AARP Andrus Foundation. Faculty nomination and student application forms are available online. Annual deadline dates may vary; contact program staff for exact dates.
*Requirements:* Gerontology and aging studies faculty are invited to nominate qualified students for these awards.
*Geographic Focus:* All States
*Amount of Grant:* 2,000 - 15,000 USD
*Contact:* M. Angela Baker, Director; (202) 289-9806, ext. 125; fax (202) 289-9824; abaker@aghe.org
*Internet:* www.aghe.org/677858
*Sponsor:* Association for Gerontology in Higher Education
1220 L Street, NW, Suite 901
Washington, D.C. 20005-4015

**AHIMA Research Awards** **211**
Since their inception in 1994, the AHIMA Triumph Awards have honored those professionals responsible for advancing the field of health information management (HIM). Each year, AHIMA members nominate those that have demonstrated extraordinary leadership, volunteerism and talent. As part of a long and cherished tradition, the Triumph Awards continue to be the highest recognition of excellence, dedication and service, and are presented annually at the AHIMA Convention and Exhibit. Specifically, the AHIMA Research Award honors those individuals who have made an outstanding contribution to original and applied HIIM (Health Information and Informatics Management) research. Nominations are due each year by June 3.
*Requirements:* Qualified nominees are those HIM professionals or groups who exemplify vision and leadership in understanding HIM philosophy and conduct and support research demonstrating the application of these concepts to healthcare, and presented or published the research, etc., findings in a peer review format (i.e, presentations, Journal of AHIMA, etc.). Required nomination criteria for this award include the following: showed evidence of sustained body of research that has advanced theory and significantly impacted health information, health informatics, or related bodies of knowledge; obtained external peer-reviewed funding for research submitted for consideration for this award; performed as the Principal Investigator (PI) or co-Principal Investigator (co-PI) for the research projects submitted for consideration for this award; and submitted research for consideration for this award has been published in a refereed journal or manuscript(s) accepted for publication.
*Geographic Focus:* All States
*Date(s) Application is Due:* Jun 3
*Contact:* Bonnie Aguda, Grants and Sponsored Programs; (312) 233-1175 or (312) 233-1100; fax (312) 233-1537; bonnie.aguda@ahimafoundation.org or fore@ahima.org
*Internet:* www.ahima.org/about/recognition/triumph/award-categories
*Sponsor:* Foundation of Research and Education of the American Health Information Management Association
233 North Michigan Avenue, 21st Floor
Chicago, IL 60601-5809

**AHL Grace Charity Foundation Research Fellowship** **212**
The AHL (Art, Humanity, and Love) Foundation will accept annual applications for the Grace Charity Foundation Research Fellowship. The archive of Korean Artists in America (AKAA) was established in 2013 to bolster the AHL Foundation's mission in recognizing and supporting the accomplishments of talented Korean artists working in the United States. The AKAA Fellow will be expected to devote at least one to two days a week researching, collecting and digitizing archival materials for online access; identifying artists for inclusion in the archive; and conducting artist interviews to be published online and in print. Depending on interest and ability there may be additional opportunities to be involved with other programs such as curating exhibitions or leading workshops/lectures. A stipend of $5,000 will be available for the Research Fellow. The annual deadline for applications is June 30.
*Requirements:* The annual one-year fellowship is open to all young-in-career art historian or curator with an interest in researching and archiving artists of Korean heritage and their activities. Knowledge of Korean or Asian art and culture preferred. An ideal candidate would be someone who is in the process of completing his/her master's or Ph.D. in art history or cultural studies with a strong interest in Asian Contemporary Art. Candidates must have excellent written and verbal communication skills in English. Some basic knowledge of Korean culture or language is preferred. Postdoctoral researchers and scholars are welcome to contribute to the archive as well and can contact the AHL Foundation for other opportunities.
*Geographic Focus:* All States
*Date(s) Application is Due:* Jun 30
*Amount of Grant:* 5,000 USD
*Contact:* Jain Kwak, Programs Manager; (516) 983-3935; info@ahlfoundation.org or applications@ahlfoundation.org
*Internet:* www.ahlfoundation.org/ahl-grace-charity-foundation-research-fellowship-2018/
*Sponsor:* AHL Foundation
420 West 23rd Street, Suite 7A
New York, NY 10011

**AIAR Educational and Cultural Affairs Junior Research Fellowships** **213**
The Educational and Cultural Affairs Junior Research Fellowship program offers $31,000 for up to three awards from four-and-one-half to nine months. The award is equal to $2138 per month. Room and half-board at the Institute is $1,200 per month and the remainder is stipend. Fellowships are open to students and scholars in Near Eastern studies from prehistory through the early Islamic period, including the fields of archaeology, anthropology, art history, Bible, epigraphy, historical geography, history, language, literature, philology and religion and related disciplines. Contact the Director with any quests related to the duties of the position. The annual deadline for applications is October 15.
*Requirements:* The Fellowship is open to post-doctoral scholars, who are U.S. citizens. The research period should be continuous, without frequent trips outside the country. Residence at the Albright is required. The option to accommodate dependents is subject to space available at the Albright.
*Geographic Focus:* All States
*Date(s) Application is Due:* Oct 15
*Amount of Grant:* 9,621 - 19,242 USD
*Contact:* Dr. Joan R. Branham, Chair, Albright Fellowship Committee; (401) 865-1789; fax (401) 865-2410; jbranham@providence.edu
*Internet:* www.aiar.org/available-fellowships/
*Sponsor:* W.F. Albright Institute of Archaeological Research
Department of Art and Art History, Providence College
Providence, RI 02918

**AIAR George A. Barton Fellowship** **214**
The George A. Barton Fellowship provides a $3,000 award for one month. Room and half-board at the Institute is $1,200 per month and the remainder is stipend. Fellowships are open to students and scholars in Near Eastern studies from prehistory through the early Islamic period, including the fields of archaeology, anthropology, art history, Bible, epigraphy, historical geography, history, language, literature, philology and religion and related disciplines. Contact the Director with any quests related to the duties of the position. The annual deadline for applications is October 15.
*Requirements:* The Fellowship is open to doctoral students and recent Ph.D. recipients of all nationalities. The research period should be continuous, without frequent trips outside the country. Residence at the Albright is required. The option to accommodate dependents is subject to space available at the Albright.
*Geographic Focus:* All States, All Countries
*Date(s) Application is Due:* Oct 15
*Amount of Grant:* 3,000 USD
*Contact:* Dr. Joan R. Branham, Chair, Albright Fellowship Committee; (401) 865-1789; fax (401) 865-2410; jbranham@providence.edu
*Internet:* www.aiar.org/available-fellowships/
*Sponsor:* W.F. Albright Institute of Archaeological Research
Department of Art and Art History, Providence College
Providence, RI 02918

**AIAR National Endowment for the Humanities Fellowships** **215**
National Endowment for the Humanities Fellowships makes $58,800 available for up to three awards from four to nine months. The award is $4,200 per month. Room and half-board at the Institute is $1,200 per month and the remainder is stipend. Fellowships are open to students and scholars in Near Eastern studies from prehistory through the early Islamic period, including the fields of archaeology, anthropology, art history, Bible, epigraphy, historical geography, history, language, literature, philology and religion and related disciplines. The annual deadline for applications is October 15.
*Requirements:* The Fellowship is open to post-doctoral scholars who are U.S. citizens (or alien residents for at least three years). Residence at the Albright is required (except for the Frerichs Fellow/Program Coordinator). The option to accommodate dependents is subject to space available at the Albright.
*Geographic Focus:* All States
*Date(s) Application is Due:* Oct 15
*Amount of Grant:* 16,800 - 37,800 USD
*Contact:* Dr. Joan R. Branham, Chair, Albright Fellowship Committee; (401) 865-1789; fax (401) 865-2410; jbranham@providence.edu
*Internet:* www.aiar.org/available-fellowships/
*Sponsor:* W.F. Albright Institute of Archaeological Research
Department of Art and Art History, Providence College
Providence, RI 02918

**AIAR Noble Group Fellowships for Chinese Students and Scholars** 216

The Noble Group Fellowships for Chinese Students and Scholars makes $75,000 available for up to five awards with a maximum grant of $25,000 for nine months and $12,500 for four-and-one-half months. Room and half-board at the Institute is $1,200 per month and the remainder is stipend. Fellowships are open to Chinese students and scholars in Near Eastern studies from prehistory through the early Islamic period, including the fields of archaeology, anthropology, art history, Bible, epigraphy, historical geography, history, language, literature, philology and religion and related disciplines. Contact the Director with any quests related to the duties of the position. The annual deadline for applications is January 15.

*Requirements:* The Fellowship is open to Chinese students who are doctoral scholars. The research period should be continuous, without frequent trips outside the country. Residence at the Albright is required. The option to accommodate dependents is subject to space available at the Albright.

*Geographic Focus:* China

*Date(s) Application is Due:* Jan 15

*Amount of Grant:* 12,500 - 25,000 USD

*Contact:* Dr. Joan R. Branham, Chair, Albright Fellowship Committee; (401) 865-1789; fax (401) 865-2410; jbranham@providence.edu

*Internet:* www.aiar.org/available-fellowships/

*Sponsor:* W.F. Albright Institute of Archaeological Research
Department of Art and Art History, Providence College
Providence, RI 02918

**Aid to Scholarly Publications (ASPP)** 217

The ASPP is designed to assist the publication of works of advanced scholarship which make an important contribution to knowledge, but which are unlikely to be self-supporting. Manuscripts in the various disciplines in the humanities and social sciences are considered for the ASPP process if they meet ASPP eligibility criteria. An award from the ASPP is designed to help offset costs of marketing and promoting a scholarly work. All funds are paid directly to a book's publisher. Also, five subventions per year will be devoted to translations. Successful candidates applying for a translation subvention will also be eligible for a grant of up to $4,000 to help offset the costs of translating the work.

*Requirements:* With few exceptions, ASPP-funded authors must be Canadian citizens or landed immigrants. In general, only book-length manuscripts (at least 100 typescript pages) written by Canadian citizens or landed immigrants / permanent residents, and to be published in Canada by approved Canadian scholarly presses, are eligible for consideration. Authors wishing to work with overseas presses should consult page 3 of these Guidelines, item 3, Submitting Party. Authors seeking ASPP support are strongly advised not to sign a contract with a non-Canadian publisher before obtaining ASPP approval of that publisher. The only modern languages in which submissions will be considered are English and French. Classical-language works may also be admissible. Guidelines and registrations forms are available at the sponsor's website.

*Restrictions:* The ASPP does not award research grants. The ASPP does not provide grants for the publication of: text books; technical reports; original works of poetry, fiction and drama; scholarly journals and articles; or, conference proceedings. Not all publishers are eligible to work with the ASPP (please see List of Eligible Publishers available at the website).

*Geographic Focus:* All States, Canada

*Amount of Grant:* 8,000 CAD

*Samples:* A searchable database of authors and titles is available at www.fedcan.ca/english/aspp/titles/database/.

*Contact:* Secretariat; (613) 238-6112, ext. 350; fax (613) 238-6114; secaspp@fedcan.ca

*Internet:* www.fedcan.ca/english/aspp/about/general/

*Sponsor:* Canadian Federation for the Humanities and Social Sciences
151 Slater Street, Suite 415
Ottawa, ON K1P 5H3 Canada

**AIHS/Mitacs Health Pilot Partnership Internship Grants** 218

The AIHS/Mitacs Health Pilot Partnership is a partnership between AIHS, Mitacs and not-for-profit organizations. This is a pilot program created to address the existing gap and lack of funding available to engage highly qualified graduate students in the public and not-for-profit sector, specifically in health-related research. The health pilot partnership will fund up to three internships for graduate and postgraduate health research and innovation trainees in Alberta. The AIHS/Mitacs Health Pilot Partnership provides a grant of $15,000 for a four-month internship administered by Mitacs-Accelerate and a partner university. The Partnership will meet the following objectives: connect not-for-profit organizations with peer-reviewed research expertise, ensuring high quality research that meets the organization's needs; apply knowledge by interns to real-world issues to create solutions to challenging problems; and provide research labs with a unique opportunity to train graduate and postgraduate trainees in new and applied skills that can be brought back to the lab. Interns may work in the areas of health economics; remote health monitoring; health policy; e-health; chronic disease management; or knowledge transfer/translation. A minimum $10,000 stipend will be provided to the intern, and the remaining $5,000 will be available for other project-related expenses. A longer eight-month double internship can be developed if there is strong justification to accommodate larger projects. Interns will spend approximately 50% of their time with the partnered not-for-profit organization, with the remaining time spent at the university, advancing the research under the guidance of their faculty supervisor.

*Requirements:* An applicant must be officially accepted into, or currently enrolled in, graduate or postgraduate studies at the University of Alberta or the University of Calgary; able to participate in a four-month single internship or an eight-month double internship with 50% of the intern's time being spent outside of the lab; and have a defined research project that they will undertake with the partnered not-for-profit organization.

*Geographic Focus:* All States, Canada

*Date(s) Application is Due:* May 15

*Amount of Grant:* 15,000 USD

*Contact:* Dr. Ryan Perry, Director, Programs; (780) 423-5727; fax (780) 429-3509; ryan.perry@albertainnovates.ca

*Internet:* www.aihealthsolutions.ca/grants/training-and-early-career-development-programs/mitacs/

*Sponsor:* Alberta Innovates - Health Solutions
10104-103 Avenue, Suite 1500
Edmonton, AB T5J 4A7 Canada

**AIHS Alberta/Pfizer Translational Research Grants** 219

The Alberta/Pfizer Translational Research Fund Opportunity is a partnership between Pfizer Canada Inc. (Pfizer), Alberta Innovates - Health Solutions (AIHS), Alberta's Ministry of Enterprise and Advanced Education, and Western Economic Diversification Canada. This partnership will provide opportunities to focus on the development and commercialization of innovations in health. The Alberta/Pfizer Translational Research Fund Opportunity will support innovative translational research projects in areas of unmet health or health system need that have a strong likelihood for technology transfer and commercialization within the next two to five years. Grants will support the development and commercialization of innovations in priority areas identified by Pfizer and the province of Alberta. Those areas include the following: neuroscience/ Alzheimer's Disease/ neurodegeneration/ neurological diseases; pain and sensory disorders; immunology and autoimmunity; inflammation and remodeling; diabetes and cardiovascular or other chronic disease issues as they relate to the consequences of diabetes; and orphan and genetic diseases. Preference will be given to research projects focused on the development, use and/or application of innovative technological platforms. These platforms include, but are not limited to: metabolomics and other 'omics' platforms that facilitate personalized medicine strategies; drug discovery; information and communication technologies; nanotechnology; and health-systems-based research approaches in the creation of new knowledge and its potential use. Translational research grants will provide up to $200,000 for a period of up to 18 months. Eligible expenses include salaries of research personnel, purchase and/or leasing of equipment, and research material required for the project. Payments will be based on agreed upon milestones and deliverables, and the submission of periodic reports outlining progress.

*Requirements:* Organizations will submit a letter of intent and full application. Additional guidelines are available at the AIHS website.

*Geographic Focus:* All States, Canada

*Amount of Grant:* Up to 200,000 USD

*Contact:* Tara McCarthy, Grant Coordinator, Programs; (780) 423-5727; fax (780) 429-3509; tara.mccarthy@albertainnovates.ca

*Internet:* www.aihealthsolutions.ca/grants/industry-partnered-translational-fund/pfizer/

*Sponsor:* Alberta Innovates - Health Solutions
10104-103 Avenue, Suite 1500
Edmonton, AB T5J 4A7 Canada

**AIHS Clinical Fellowships** 220

The AIHS Clinical Fellowships program is designed for highly qualified individuals who hold an MD or DDS degree, and who anticipate undertaking a career in health related or clinical research in Alberta. The awards consist of a stipend and a research allowance. A Clinical Fellowship award is normally tenable for a maximum of three years. However, if the trainee is registered in a graduate program, the term may be extended.

*Requirements:* Normally, an award will be held within the Province of Alberta; however, candidates who are Canadian citizens or permanent residents with records of outstanding performance in postgraduate training may seek research training elsewhere, if sponsored by an Alberta faculty with an expressed interest in future recruitment of the candidate. Candidates must hold an MD or DDS degree and have received a significant portion of their postgraduate training in Alberta.

*Restrictions:* Clinical fellows are expected to commit to a minimum of two years of full time research.

*Geographic Focus:* Canada

*Date(s) Application is Due:* Mar 1; Oct 1

*Amount of Grant:* 20,000 - 50,000 USD

*Contact:* Pamela Valentine, Vice President of Research and Innovation; (780) 423-5727, ext. 230; fax (780) 429-3509; pamela.valentine@albertainnovates.ca or grants.health@albertainnovates.ca

*Internet:* www.aihealthsolutions.ca/grants/Clin-fellow.php

*Sponsor:* Alberta Innovates - Health Solutions
10104-103 Avenue, Suite 1500
Edmonton, AB T5J 4A7 Canada

**AIHS Collaborative Research and Innovation Grants - Collaborative Program** 221

The AIHS Collaborative Research and Innovation Opportunities (CRIO) aims to bring together experts in different disciplines, fields, and areas to tackle health research problems in areas of strategic priority that would benefit from an interdisciplinary approach. The objectives of the CRIO are to: catalyze and support collaborative, interdisciplinary, multi-sectoral and/or multi-institutional research with a focus on achieving solutions that address complex health problems or issues; support areas of research that are aligned with the thematic priority areas defined in Alberta's Health Research and Innovation Strategy; engage end users in the process to enhance impact on the health of Albertans and/or the health care system; provide opportunities for

interdisciplinary research training and mentorship; and encourage the use of the AIHS collaborative opportunities to connect Alberta to national and international initiatives and leverage additional opportunities. A maximum of up to $500,000 per year for five years may be awarded. The Collaborative Program involves three to five principal collaborators who may already have an informal arrangement or shared funding to support interdisciplinary work, or have specific plans to develop such arrangements. The Collaborative Program may be considered for a one time renewal, based on agreed-upon deliverables set up in the management plan for years 1 through 5.
*Requirements:* The application process involves registration, a letter of intent, and full application. Each CRIO must involve Collaborative Lead(s) who are established individual(s) with demonstrated leadership. Collaborative Leads can lead more than one CRIO Project, Program and/or Team simultaneously. There are no limits on the number of applications a Collaborative Lead can submit. Collaborative members may be defined as needed to address the research objectives, and may include individuals from outside the Alberta area. Interdisciplinary, multi-institutional activity is encouraged. The focus is on a complex program of health research and innovation. A full list of eligibility requirements is available at the AIHS website.
*Geographic Focus:* All States, Canada
*Date(s) Application is Due:* Jan 18
*Amount of Grant:* Up to 500,000 USD
*Contact:* Kathy Morrison, Grants Coordinator, Programs; (780) 423-5727; fax (780) 429-3509; kathy.morrison@albertainnovates.ca
*Internet:* www.aihealthsolutions.ca/grants/crio/#project
*Sponsor:* Alberta Innovates - Health Solutions
10104-103 Avenue, Suite 1500
Edmonton, AB T5J 4A7 Canada

**AIHS Collaborative Research and Innovation Grants - Collaborative Project 222**
The AIHS Collaborative Research and Innovation Opportunities (CRIO) aims to bring together experts in different disciplines, fields, and areas to tackle health research problems in areas of strategic priority that would benefit from an interdisciplinary approach. The objectives of the CRIO are to: catalyze and support collaborative, interdisciplinary, multi-sectoral and/or multi-institutional research with a focus on achieving solutions that address complex health problems or issues; support areas of research that are aligned with the thematic priority areas defined in Alberta's Health Research and Innovation Strategy; engage end users in the process to enhance impact on the health of Albertans and/or the health care system; provide opportunities for interdisciplinary research training and mentorship; and encourage the use of the AIHS collaborative opportunities to connect Alberta to national and international initiatives and leverage additional opportunities. The CRIO Collaborative Project allows approximately three collaborators to come together to complete a defined project. There is no expectation that the collaboration will continue beyond the duration of this non-renewable grant. The opportunity also provides a framework for collaborators who choose to evolve their project into a more complex Collaborative Program or Collaborative Team. The application process involves registration, a letter of intent, and full application. Up to $250,000 per year may be funded for three years.
*Requirements:* The Collaborative Lead must be an individual with an established career and proven leadership skills and experience. Collaborative members will have an extensive record of success, be creative and original in their approach to research and its translation, and have experience working in teams. Members may also be located outside the Alberta area. Co-funding from philanthropic, national or international organizations is encouraged and should be declared as part of the submission. A full list of eligibility requirements is available at the AIHS website.
*Geographic Focus:* Canada
*Date(s) Application is Due:* Aug 31
*Amount of Grant:* Up to 250,000 USD
*Contact:* Kathy Morrison, Grants Coordinator, Programs; (780) 423-5727; fax (780) 429-3509; kathy.morrison@albertainnovates.ca
*Internet:* www.aihealthsolutions.ca/grants/crio/
*Sponsor:* Alberta Innovates - Health Solutions
10104-103 Avenue, Suite 1500
Edmonton, AB T5J 4A7 Canada

**AIHS Collaborative Research and Innovation Grants - Collaborative Team 223**
The AIHS Collaborative Research and Innovation Opportunities (CRIO) aims to bring together experts in different disciplines, fields, and areas to tackle health research problems in areas of strategic priority that would benefit from an interdisciplinary approach. The objectives of the CRIO are to: catalyze and support collaborative, interdisciplinary, multi-sectoral and/or multi-institutional research with a focus on achieving solutions that address complex health problems or issues; support areas of research that are aligned with the thematic priority areas defined in Alberta's Health Research and Innovation Strategy; engage end users in the process to enhance impact on the health of Albertans and/or the health care system; provide opportunities for interdisciplinary research training and mentorship; and encourage the use of the AIHS collaborative opportunities to connect Alberta to national and international initiatives and leverage additional opportunities. The Collaborative Team fosters and support a large interdisciplinary collaborative team or network with national or international stature. The complexity of the health issues of interest will require the involvement of many stakeholder groups. Up to $1,000,000 per year for five years may be funded.
*Requirements:* The application process involves registration, a letter of intent, and a full application. Each CRIO must involve a Collaborative Lead() who are established individuals with demonstrated leadership. Collaborative members may be defined as needed to address the research objectives and may include individuals from outside Alberta's knowledge or end users. Interdisciplinary, multi-institutional activity is encouraged. A full list of eligibility requirements is available at the website.
*Geographic Focus:* All States, All Countries
*Date(s) Application is Due:* Jan 18
*Amount of Grant:* Up to 1,000,000 USD
*Contact:* Tara McCarthy, Grants Coordinator, Programs; (780) 423-5727; fax (780) 429-3509; tara.mccarthy@albertainnovates.ca
*Internet:* www.aihealthsolutions.ca/grants/crio/#project
*Sponsor:* Alberta Innovates - Health Solutions
10104-103 Avenue, Suite 1500
Edmonton, AB T5J 4A7 Canada

**AIHS Fast-Track Fellowships 224**
AIHS will accept, at any time, applications from Alberta-based institutions for full-time Fellowship awards for outstanding candidates currently training outside of Alberta. The fast-track program is intended to assist these institutions in recruiting highly qualified trainees to the province. Up to a maximum of 10 awards per calendar year will be awarded through the fast-track review process in each category. Applications to the program will be reviewed on a first come-first serve basis. Fast-track awards are tenable for up to 12 months.
*Requirements:* Applications must include a letter from the Dean/designate of the faculty indicating that the proposal for a fast-track award has been reviewed and stating why a fast-track application should be considered for the candidate in question. Please refer to the sections on Full-Time Fellowships or Full-Time Studentships for further details regarding eligibility for training awards.
*Geographic Focus:* Canada
*Amount of Grant:* 20,000 - 50,000 USD
*Contact:* Pamela Valentine, Vice President of Research and Innovation; (780) 423-5727, ext. 230; fax (780) 429-3509; pamela.valentine@albertainnovates.ca or grants.health@albertainnovates.ca
*Internet:* www.aihealthsolutions.ca/grants/fasttrack.php
*Sponsor:* Alberta Innovates - Health Solutions
10104-103 Avenue, Suite 1500
Edmonton, AB T5J 4A7 Canada

**AIHS ForeFront Executive-in-Residence Award 225**
ForeFront supports career, industry and innovation development. Alberta's universities are integral to the innovation system and the creation of the medical and health industries through research, technology transfer, and spin-off company creation. The Executive-in-Residence Award enables the technology transfer offices to recruit senior-level management who have relevant industry experience and appropriate networks to assist university spin-off companies with the commercialization of technologies. The goal is to advance technologies and build better companies that will both improve health care and diversify and contribute to the economy.
*Geographic Focus:* All States, Canada
*Contact:* Carla Weyland, Manager, Training and Development Programs; (780) 423-5727; fax (780) 429-3509; carla.weyland@albertainnovates.ca
*Internet:* www.aihealthsolutions.ca/forefront/EIR.php
*Sponsor:* Alberta Innovates - Health Solutions
10104-103 Avenue, Suite 1500
Edmonton, AB T5J 4A7 Canada

**AIHS ForeFront Internships 226**
The AIHS Internship Program addresses the need for highly trained staff and management of the Alberta-based health, medical products and biotechnology industries. This program is intended to support technology commercialization training and experience for individuals with an appropriate background in science and/or business. Although the work plan should identify the specific area of focus, the intern should receive a broad exposure to all of the following areas: product development; intellectual property strategy; regulatory requirements; market research and strategy; and business planning and development. Interns are awarded a $5,000 training allowance, which may be used for expenses related to educational workshops/seminars/courses necessary for relevant commercialization training, plus a $45,000 to $50,000 stipend. The application and additional guidelines are available at the AIHS website.
*Requirements:* The applicant must hold, or be in the final year of a degree in business, management, or science. The applicants should have some relevant experience in the business or private sector. Applicants must be sponsored by an Alberta organization actively engaged in the commercialization of medical or health related technology. The organization must have the experience and resources to provide an appropriate training environment, and designate one person within the organization who will be the primary supervisor for the intern. The organization must agree to provide direct and continuous supervision.
*Geographic Focus:* All States, Canada
*Amount of Grant:* 45,000 - 50,000 USD
*Contact:* Carla Weyland, ForeFront Officer; (780) 423-5727; fax (780) 429-3509; carla.weyland@albertainnovates.ca
*Internet:* www.aihealthsolutions.ca/forefront/internship.php
*Sponsor:* Alberta Innovates - Health Solutions
10104-103 Avenue, Suite 1500
Edmonton, AB T5J 4A7 Canada

**AIHS ForeFront MBA Studentship Award 227**
This award will enable students with a background in the medical/health/life sciences to apply their knowledge and pursue a career in management in the medical/health industry by supporting their education and hands-on training in the University of Alberta Master

of Business Administration program, specializing in Technology Commercialization. The Tech Com MBA builds and delivers a tool kit that emphasizes the importance of capital, management practices, infrastructure, information, and networks. The goal of the Tech Com MBA is to position graduates for a leadership role in today's leading edge, high technology industries. The main objectives are to: encourage Albertans with a background in science to explore careers that will give them the opportunity to apply their knowledge to commercializing medical discoveries that will lead to improved health; and build capacity of knowledgeable, highly-skilled technology managers to meet the needs of the Alberta medical/health industry. The MBA Studentship Award includes a contribution towards tuition fees and a stipend of $2,000 per month for the duration of the student's full-time MBA studies. The MBA Program may be completed in two years on a full time basis. Students have a maximum of six years to complete the course requirements. Any extensions to the two-year full-time term must be approved by AIHS. Additional guidelines and terms of the award are available at the AIHS website.
*Requirements:* Students must hold a graduate degree in medical/health/life sciences (graduated within the last two years) and meet the minimum eligibility criteria for admission to the MBA program.
*Geographic Focus:* All States, Canada
*Contact:* Carla Weyland, Manager, Training and Development Programs; (780) 423-5727; fax (780) 429-3509; carla.weyland@albertainnovates.ca
*Internet:* www.aihealthsolutions.ca/forefront/mba.php
*Sponsor:* Alberta Innovates - Health Solutions
10104-103 Avenue, Suite 1500
Edmonton, AB T5J 4A7 Canada

**AIHS ForeFront MBT Studentship Awards** **228**

The award will enable students to pursue a career in the medical, health, and biotechnology industries by supporting their education and hands-on training in the Master of Biomedical Technology (MBT) Program at the University of Calgary. The MBT Program takes a multi-disciplinary approach and exposes students to a broad range of courses and cutting-edge technologies that will ensure the students acquire practical competencies and extensive knowledge of concepts in biomedical and bioinformatics disciplines, including the application of this knowledge through commercialization of medical and health-related innovations. The main objectives are to: encourage individuals with a background in science to explore careers that will give them the opportunity to apply their knowledge to commercializing medical discoveries that will lead to improved health; and build capacity of knowledgeable, highly-skilled technology managers to meet the needs of the Alberta medical/health industry. The MBT Studentship Award includes a contribution towards tuition fees and a stipend for the practicum (up to a maximum of $3,000). The MBT Program may be completed in one year on a full time basis. Any extensions to the one-year full-time term must be approved by AIHS. Application information is available at the AIHS website.
*Requirements:* Candidates must hold at least an undergraduate degree in an area of relevance to the objectives of Alberta Innovates – Health Solutions (AIHS).
*Geographic Focus:* All States, Canada
*Contact:* Kathy Morrison, Grants Coordinator; (780) 423-5727; fax (780) 429-3509; kathy.morrison@albertainnovates.ca
*Internet:* www.aihealthsolutions.ca/forefront/mbt.php
*Sponsor:* Alberta Innovates - Health Solutions
10104-103 Avenue, Suite 1500
Edmonton, AB T5J 4A7 Canada

**AIHS Full-Time Fellowships** **229**

The AIHS Full-Time Fellowships program is designed to enable highly qualified doctoral graduates to prepare for careers in medical or health research as independent investigators. A fellowship award will provide to the host institution funding for one year's stipend, its associated benefits, and a research allowance. Fellows may engage in teaching activities related to their research discipline a maximum of 20 percent of their time. The maximum term is three years
*Requirements:* Candidates must have a Ph.D., M.D., D.D.S., D.V.M. or D.Pharm. degree. Normally, support will not be provided beyond 6 years after receipt of the Ph.D. degree, or beyond 8 years after receipt of the M.D., D.D.S., D.V.M. or D.Pharm. degrees.
*Geographic Focus:* Canada
*Date(s) Application is Due:* Mar 1; Oct 1
*Amount of Grant:* 35,000 - 50,000 USD
*Contact:* Pamela Valentine, Vice President of Research and Innovation; (780) 423-5727, ext. 230; fax (780) 429-3509; pamela.valentine@albertainnovates.ca or grants.health@albertainnovates.ca
*Internet:* www.aihealthsolutions.ca/grants/FT-fellow.php
*Sponsor:* Alberta Innovates - Health Solutions
10104-103 Avenue, Suite 1500
Edmonton, AB T5J 4A7 Canada

**AIHS Full-Time Health Research Studentships** **230**

The Health Research Studentship program enables academically superior students to undertake full-time training in health research. The award supports training in: research on the organization and delivery of health care; technology assessment; community health; health promotion; disease prevention; and related disciplines. The Health Research Studentship consists of a stipend and a research allowance. The term of the award is a maximum of five years (three years at the graduate level).
*Requirements:* Candidates must have been accepted into, or be currently studying in, a full-time, thesis-based, graduate program at an Alberta-based university in a health-related discipline leading to a Master's or doctoral degree. Normally, support will not be provided beyond six years of enrollment in graduate school. Candidates who have interrupted their training for parenting or other reasons and who have consequently exceeded this time limit, may apply for studentship support. In such cases, however, the candidate is advised to clearly explain the nature of their particular circumstances at the time of application.
*Restrictions:* The award is not available to students registered in a course-based program.
*Geographic Focus:* Canada
*Date(s) Application is Due:* Mar 1; Oct 1
*Amount of Grant:* 20,000 USD
*Contact:* Pamela Valentine; (780) 423-5727, ext. 230; fax (780) 429-3509; pamela.valentine@albertainnovates.ca or grants.health@albertainnovates.ca
*Internet:* www.aihealthsolutions.ca/grants/FT-student.php#HRS
*Sponsor:* Alberta Innovates - Health Solutions
10104-103 Avenue, Suite 1500
Edmonton, AB T5J 4A7 Canada

**AIHS Full-Time M.D./Ph.D. Studentships** **231**

The AIHS Full-Time M.D./Ph.D. Studentships are intended to provide an opportunity for exceptional candidates, who wish to pursue careers as Clinical Investigators, and study simultaneously for the M.D. and the Ph.D. degrees. Support is complementary to the formal M.D./Ph.D. programs at the University of Alberta and the University of Calgary. Awards are tenable only at an Alberta-based university. The term of the award is a maximum of six years.
*Requirements:* Applicants are limited to those students that formally enter the MD/PhD program no later than the start of their second year of PhD studies at either the University of Alberta or the University of Calgary. AIHS's offer of support is contingent on the granting of complementary stipend support from the Faculty of not less than 15% of the value of the AIHS award, for the duration of the award. Furthermore, it is expected that the universities will undertake to provide administrative support to the offices of the coordinators of the M.D./Ph.D. programs.
*Geographic Focus:* Canada
*Date(s) Application is Due:* Mar 1; Oct 1
*Amount of Grant:* 20,000 USD
*Contact:* Pamela Valentine, Vice President of Research and Innovation; (780) 423-5727, ext. 230; fax (780) 429-3509; pamela.valentine@albertainnovates.ca or grants.health@albertainnovates.ca
*Internet:* www.aihealthsolutions.ca/grants/FT-student.php#MPS
*Sponsor:* Alberta Innovates - Health Solutions
10104-103 Avenue, Suite 1500
Edmonton, AB T5J 4A7 Canada

**AIHS Full-Time Studentships** **232**

AIHS Full-Time Studentships enable academically superior students to undertake full-time research training in the basic biomedical sciences or in clinical research. The award consists of a stipend and a research allowance. Approved uses of the research allowance include: the purchase of scientific materials, supplies and expendables; the purchase of minor equipment; computer software programs; costs for the use of libraries, or computers; costs associated with the publication of research results; travel expenses to attend scientific meetings; and purchase of books, periodicals and journals. The term of the award is a maximum of five years (three years at the graduate level).
*Requirements:* Candidates must have been accepted into, or be currently engaged in, a full-time, thesis-based, graduate program at an Alberta-based university in a health-related discipline leading to a Master's or doctoral degree.
*Restrictions:* This award is not available to students registered in a course-based program. Normally, support will not be provided beyond 6 years of enrollment in graduate school. Candidates who have interrupted their training for parenting or other reasons and who have consequently exceeded this time limit, may apply for studentship support. In such cases, however, the candidate is advised to clearly explain the nature of their particular circumstances at the time of application.
*Geographic Focus:* Canada
*Date(s) Application is Due:* Mar 1; Oct 1
*Amount of Grant:* 20,000 USD
*Contact:* Pamela Valentine, Vice President of Research & Innovation; (780) 423-5727, ext. 230; fax (780) 429-3509; pamela.valentine@albertainnovates.ca or grants.health@albertainnovates.ca
*Internet:* www.aihealthsolutions.ca/grants/FT-student.php#fts_description
*Sponsor:* Alberta Innovates - Health Solutions
10104-103 Avenue, Suite 1500
Edmonton, AB T5J 4A7 Canada

**AIHS Heritage Youth Researcher Summer Science Program** **233**

The Heritage Youth Researcher Summer Program (HYRS) is an intensive six-week summer science program for high school student in Alberta. The program is designed to give exemplary students from around the province first-hand biomedical and health research experience at a university campus, and introduce them to the many educational and career options available in this field. Students work on health research projects in a variety of work environments of researchers who are on faculty at the University of Alberta, the University of Calgary, and the University of Lethbridge. Schedules vary slightly at each University, but are generally from July 4 through August 15. Students will be supervised by a faculty member or by a member of their research team. Research projects could be in cardiology, cell biology, genetics, biomedical engineering, population health, epidemiology, or other health research areas. The application and additional guidelines are available at the AIHS website.

*Requirements:* Applicants require at least an 85% standing in each of math 20-1 or math 20-2, biology 20, and one other grade 11 science class, which is the minimum requirement to work in a university laboratory. Applicants must have two reference forms completed by two of their science teachers, or one math and one science teacher. They must also submit a reference form completed by an adult who is not a relative and is from outside their school and who has known the student for at least one year. Additional guidelines are available at the website.
*Restrictions:* Students must be willing to commit to a full six weeks. Any time taken off will disqualify the student for the program.
*Geographic Focus:* All States, Canada
*Date(s) Application is Due:* Mar 23
*Contact:* Danica Wolkow, HYRS Program Provincial Coordinator; (780) 423-5727, ext 237; fax (780) 429-3509; hyrs@albertainnovates.ca
*Internet:* www.aihealthsolutions.ca/HYRS/
*Sponsor:* Alberta Innovates - Health Solutions
10104-103 Avenue, Suite 1500
Edmonton, AB T5J 4A7 Canada

**AIHS Interdisciplinary Team Grants** 234
The AIHS Interdisciplinary Team Grants Program provides opportunities for high-quality, internationally recognized teams of investigators to complete research initiatives with defined health outcomes. Funds available in this competition are to support collaborative, interdisciplinary and multi-institutional teams that address important research questions, health problems or issues in defined areas of research that are aligned with strategic research priorities of the AIHS and Alberta. The Program will provide up to $1 million per year per team for up to five years. The area of research to be pursued must be relevant to one of a number of areas of special interest identified by AIHS in collaboration with other stakeholders in Alberta. Those areas are: maternal, fetal and child health; mental health, mental illness and addictions; health system sustainability; modern lifestyles and health; health issues in rural and remote environments; health, genes and the environment; health and injury; modern techniques and technologies (including IT) and health; health and behavior: disease prevention; health and infectious diseases; and food and health. The LOI and full application must clearly identify under which of these theme areas the proposed program of research falls and how the proposed research is relevant to the theme. More than one theme can be identified as appropriate. Applicants are asked to make special reference as to how their program of research will be relevant to health issues facing vulnerable populations.
*Requirements:* Each eligible Interdisciplinary Team Grant application will include: A Team Leader—the Team Leader must be an established researcher with proven leadership skills and experience who will act as research program director and who will assume administrative responsibility for the grant. It is expected that the Team Leader will devote a significant and appropriate portion of their time to these tasks. The Team Leader will have their primary academic appointment at an Alberta-based university; at least two additional independent investigators with established research track records. Teams with a nucleus of experienced investigators are encouraged to include promising new investigators as part of their group; team members who collectively have an extensive record of success, are creative and original in their approach to research and its translation, and who have experience working in research teams. The specific contribution of each team member and end-user partner, where applicable, must be described; representation from more than one research discipline and from more than one "research pillar" (i.e. biomedical; clinical science; health systems and services; and the social, cultural and other factors that affect the health of populations.); and representation from more than one Alberta-based university. Multi-institutional collaboration is strongly encouraged in this Program. In addition, the members of the Team may pursue other research in addition to their commitment to the Interdisciplinary Team Grant. However, each individual investigator must contribute sufficient time to the Team Grant research program to ensure the achievement of its research and translational objectives. Collaborators from outside of Alberta who make a substantial intellectual contribution to the research program may be listed as team members. Specific and justified requests that AIHS funding be used for work performed outside of Alberta will be considered. However, it is expected that the great majority of the funds will remain in Alberta.
*Geographic Focus:* All States, Canada
*Amount of Grant:* Up to 1,000,000 USD
*Contact:* Pamela Valentine, Vice President of Research and Innovation; (780) 423-5727; fax (780) 429-3509; pamela.valentine@albertainnovates.ca
*Internet:* www.aihealthsolutions.ca/grants/team_guidelines.php
*Sponsor:* Alberta Innovates - Health Solutions
10104-103 Avenue, Suite 1500
Edmonton, AB T5J 4A7 Canada

**AIHS Knowledge Exchange - Visiting Professorships** 235
Visiting Professorships are awarded to particularly distinguished scientists to contribute to the education and training of students and fellows, and promote the progress of medical or health research in Alberta. Visiting Professorships may be awarded in order to assist the Alberta-based universities in planning or initiating new research directions, or for the purpose of evaluating AIHS-funded programs. The award includes an honorarium and per diem, plus travel costs. The length of the award is expected to be no longer than seven days.
*Requirements:* Applications should be submitted directly to AIHS at least three months prior to the proposed visit.
*Geographic Focus:* All States, Canada
*Contact:* Donna Angus, Director, Knowledge Transfer; (780) 423-5727, ext 208; fax (780) 429-3509; donna.angus@albertainnovates.ca
*Internet:* www.aihealthsolutions.ca/grants/knowledge.php
*Sponsor:* Alberta Innovates - Health Solutions
10104-103 Avenue, Suite 1500
Edmonton, AB T5J 4A7 Canada

**AIHS Knowledge Exchange - Visiting Scientists** 236
The Visiting Scientist program is intended to bring productive scientists with expertise not currently available in provincial institutions to Alberta. This expertise may consist of special knowledge, new approaches to research, and the management of new technology. The program is also intended to allow Alberta-based institutions to sponsor Alberta scientists to visit major medical or health research centers, and acquire knowledge abut new research concepts or technology. The visit must result in important contributions to medical or health related science in Alberta, and an increase in the visitor's research productivity. The sponsoring institution must assure the return of the Alberta scientist to their regular position at the completion of the award. Visiting Scientists will receive travel assistance and a research allowance to cover some of the costs of materials, supplies, and expenses critical for their participation in the program. The length of the visit ranges from one month to 12 months.
*Requirements:* The sponsoring institution must identify the candidate's special expertise, as well as the expected long-term research benefits of the visit. This applies to both visiting scientists to and visiting scientists from Alberta. Applications demonstrating that arrangements have been made for the visitor to meet with scientists in other Alberta universities or institutions will receive preference.
*Geographic Focus:* All States, Canada
*Contact:* Donna Angus, Director, Knowledge Transfer; (780) 423-5727, ext 208; fax (780) 429-3509; donna.angus@albertainnovates.ca
*Internet:* www.aihealthsolutions.ca/grants/knowledge.php
*Sponsor:* Alberta Innovates - Health Solutions
10104-103 Avenue, Suite 1500
Edmonton, AB T5J 4A7 Canada

**AIHS Media Summer Fellowship** 237
The AIHS Media Fellowship offers University of Alberta and University of Calgary students with strong biomedical science backgrounds the opportunity to spend 12 weeks during the summer (May through August) working with CBC Radio. Media Fellows will work as reporters, researchers and productions assistants with CBC Radio in either Edmonton or Calgary. Through the Media Fellowship, students will have an opportunity to impact the discussion of biomedical science in popular media and will be contributing to the translation of knowledge from the scientists to the media and the general public. The Media Fellowship will strengthen the relationship between scientists and the media; provide young scientists, at a critical stage in their careers, the opportunity to observe and participate in how events and ideas become news; improve the Fellow's skills in communicating complex technical subjects to the public; increase the Fellow's understanding of editorial decision-making and how news is communicated to the public; enhance the coverage of science and technology by the media; and improve public understanding and appreciation of science and medical research. The Media Fellowship provides a stipend to successful applicants, to be determined at the time of acceptance.
*Requirements:* An applicant must be an undergraduate- or graduate-level university student at either the University of Alberta or the University of Calgary with a strong background in biomedical science; continuing university studies in the fall in any discipline; demonstrate excellent writing skills; and work well independently as well as part of a team.
*Geographic Focus:* All States, Canada
*Date(s) Application is Due:* Mar 23
*Contact:* Dwayne Brunner, Interim Media Relations Manager; (780) 423-5727, 224; fax (780) 429-3509; dwayne.brunner@albertainnovates.ca
*Internet:* www.aihealthsolutions.ca/grants/training-and-early-career-development-programs/media-fellowship/
*Sponsor:* Alberta Innovates - Health Solutions
10104-103 Avenue, Suite 1500
Edmonton, AB T5J 4A7 Canada

**AIHS Part-Time Fellowships** 238
The AIHS Part-Time Fellowships enable highly qualified doctoral graduates to prepare for careers in medical or health research as independent investigators doing part-time research training during the regular academic year while enrolled in a full-time professional degree program in Alberta. Part-time awards will consist of a stipend only and will be pro-rated to the amount of time the candidate is prepared to commit to research. The full-time fellowship rate will be used as the basis for this calculation. The maximum term for the award is three years.
*Requirements:* Candidates must hold a PhD and must be engaged in a professional degree program (usually the MD program).
*Restrictions:* Part-time fellowship awards are tenable only at an Alberta-based university.
*Geographic Focus:* Canada
*Date(s) Application is Due:* Mar 1; Oct 1
*Amount of Grant:* 5,000 - 30,000 USD
*Contact:* Pamela Valentine, Vice President of Research and Innovation; (780) 423-5727, ext. 230; fax (780) 429-3509; pamela.valentine@albertainnovates.ca or grants.health@albertainnovates.ca
*Internet:* www.aihealthsolutions.ca/grants/PT-fellow.php
*Sponsor:* Alberta Innovates - Health Solutions
10104-103 Avenue, Suite 1500
Edmonton, AB T5J 4A7 Canada

**AIHS Part-Time Studentships** 239

The Alberta Innovates - Health Solutions (AIHS) Part-Time Studentships enable students enrolled in a full-time professional degree program (usually the M.D. program) in Alberta to engage in part-time research training during the regular academic year. The award consists of a stipend only. Part-time students may elect to engage in full-time research during the summer; these students should contact AIHS to request that full-time summer trainee status be implemented.
*Requirements:* Candidates must normally have been accepted into, or be currently engaged in, a full-time graduate program at an Alberta-based university in a health-related discipline leading to a Master's or doctoral degree. It is expected that the amount of time the candidate will commit to research will not be less than the equivalent of one day per week, taken either as one full day or two half days.
*Restrictions:* The award is tenable at an Alberta-based university only.
*Geographic Focus:* Canada
*Date(s) Application is Due:* Mar 1; Oct 1
*Contact:* Pamela Valentine, Vice President of Research and Innovation; (780) 423-5727, ext. 230; fax (780) 429-3509; pamela.valentine@albertainnovates.ca or grants.health@albertainnovates.ca
*Internet:* www.aihealthsolutions.ca/grants/PT-student.php
*Sponsor:* Alberta Innovates - Health Solutions
10104-103 Avenue, Suite 1500
Edmonton, AB T5J 4A7 Canada

**AIHS Proposals for Special Initiatives** 240

The AIHS welcomes proposals for innovative initiatives that accelerate the achievement of its overall objectives. Institutions recruiting internationally-recognized researchers to lead new initiatives should consult AIHS, if current programs are inappropriate. AIHS will also consider proposals for pilot projects in patient or population-based research. These projects must have the potential to develop information, and the investigators must have the expertise to attract grants from other agencies. The review process and financial support for special initiatives may be shared by AIHS and other agencies.
*Restrictions:* Applications that are normally eligible for regular operating grants from other agencies will not be considered under this program.
*Geographic Focus:* All States, Canada
*Contact:* Jacques Magnan, Vice President, Research; (780) 423-5727; fax (780) 429-3509; jmagnan@albertainnovates.ca
*Internet:* www.aihealthsolutions.ca/grants/initiatives.php
*Sponsor:* Alberta Innovates - Health Solutions
10104-103 Avenue, Suite 1500
Edmonton, AB T5J 4A7 Canada

**AIHS Research Prize** 241

The AIHS Research Prize is intended to maintain and improve AIHS Personnel Awards as a potent vehicle for the recruitment and retention of highly qualified, internationally competitive investigators at Alberta institutions. The Prize will be made to every eligible investigator supported by AIHS (Population Health Investigator, Clinical Investigator, Scholar, Senior Scholar, Scientist). The Research Prize is not determined by institutional rank and salary scale. It is provided in recognition of the individual investigator's research accomplishments, and of his/her success in obtaining an AIHS Independent Investigator award. The Research Prize is a personal prize to the recipient. No application is required.
*Requirements:* To be eligible, the investigator must be currently supported by an unconditional AIHS investigator award. Investigators receiving AIHS terminal support or those on leaves of absence are not eligible to receive the Research Prize.
*Geographic Focus:* All States, Canada
*Amount of Grant:* 10,000 - 20,000 USD
*Contact:* Pamela Valentine, Vice President of Research and Innovation; (780) 423-5727; fax (780) 429-3509; pamela.valentine@albertainnovates.ca
*Internet:* www.aihealthsolutions.ca/grants/resprize.php
*Sponsor:* Alberta Innovates - Health Solutions
10104-103 Avenue, Suite 1500
Edmonton, AB T5J 4A7 Canada

**AIHS Summer Studentships** 242

The AIHS Summer Studentships offer motivated students with exceptional academic records an opportunity to participate in medical or health research in Alberta during the summer months. The award is meant to encourage students to consider pursuing formal training and a career in health research. The award is tenable for a minimum of two months, and a maximum of four months during the period of May to August. The award consists of a stipend of $1300 per month. Applications are available at the AIHS website.
*Requirements:* Candidates must meet one of the following criteria to be eligible to apply: registered in an Alberta-based undergraduate degree program in a medical or health-related field; registered in an undergraduate degree program outside of Alberta, and desiring to engage in research during the summer at an Alberta institution; registered in an M.D. program, and who may also hold an undergraduate or graduate degree; exceptional high school students with records of participation in the health care system, and a clear interest in pursuing a health research career; or currently in the last term of their undergraduate degrees and who have applied either to medical school or to a graduate program that would start in the coming fall semester.
*Restrictions:* A faculty supervisor possessing both a record of productive health-oriented research and sufficient resources to ensure the satisfactory conduct of the research must sponsor the candidate. The supervisor may sponsor no more than two summer students per year, and each student must have a separate project. These awards are not renewable, but students may reapply in subsequent years.
*Geographic Focus:* All States, Canada
*Contact:* Pamela Valentine, Vice President of Research and Innovation; (780) 423-5727; fax (780) 429-3509; pamela.valentine@albertainnovates.ca
*Internet:* www.aihealthsolutions.ca/grants/Sum-student.php
*Sponsor:* Alberta Innovates - Health Solutions
10104-103 Avenue, Suite 1500
Edmonton, AB T5J 4A7 Canada

**AIHS Sustainability Grants** 243

The Sustainability Funding Opportunity provides bridge funding for Alberta health researchers who have submitted renewal applications that, although highly rated, were not funded in the CIHR Open Operating Grants or NSERC Discovery Grant competitions. The objective of these bridge grants is to ensure the maintenance of excellent research programs without loss of momentum, staff, or trainees until the Principal Investigators have the opportunity to resubmit their research proposals. A webinar discussing the grants is available at the AIHS website.
*Requirements:* The grant recipient must be a full-time university-based independent investigator. The recipient must also have applied to a Tri-Council agency for a peer-reviewed, individual research grant renewal in the CIHR Open Operating Grant or NSERC Discovery Grant competitions within the prior year, and was deemed excellent but was not funded. The applicant must have at least one more opportunity for reapplication of grant renewal with the external agency. The initial renewal application to the Tri-Council agency's operating grant competition was internally peer-reviewed according to processes defined at the institution prior to submission to the Tri-Council agency. Recipients must provide specific plans for a future application to the Vice President Research of their university.
*Restrictions:* The Sustainability Funding Program will not supplement existing grants or cover shortfalls in approved grant support. Each institution will work through their Office of the Vice President Research for the review of applications and selection of health researchers to be offered support under the Sustainability Funding Opportunity according to the defined eligibility criteria. Applications will be evaluated by the university receiving a Sustainability Funding allocation, not by AIHS.
*Geographic Focus:* All States, Canada
*Amount of Grant:* Up to 50,000 USD
*Contact:* Pamela Valentine, Vice President of Research and Innovation; (780) 423-5727, ext 230; fax (780) 429-3509; pamela.valentine@albertainnovates.ca
*Internet:* www.aihealthsolutions.ca/grants/sustainability.php
*Sponsor:* Alberta Innovates - Health Solutions
10104-103 Avenue, Suite 1500
Edmonton, AB T5J 4A7 Canada

**AINSE Honours Scholarships** 244

The Australian Institute of Nuclear Science and Engineering will award ten Honours scholarships annually. Eligible students will be enrolled in an Honours or Masters program, and their university supervisor will hold a AINSE Research Award. AINSE Honour Scholars receive a $5,000 stipend.
*Geographic Focus:* Australia
*Date(s) Application is Due:* Apr 15
*Amount of Grant:* 5,000 AUD
*Contact:* Frank Bruhn, Managing Director; +61 02 9717 3388 or +61 02 9717 3376; fax +61 02 9717 9268; frankb@ainse.edu.au or ainse@ansto.gov.au
*Internet:* www.ainse.edu.au/grad_students2/honours_scholarships
*Sponsor:* Australian Institute of Nuclear Science and Engineering
Locked Mail Bag 2001
Kirrawee, NSW 2232 Australia

**AINSE International Conference Travel Scholarships** 245

Travel scholarships will be awarded to students and post doctoral fellows from AINSE member universities who wish to present their AINSE supported research at an international meeting which is being conducted outside Australia. Successful applicants will be awarded up to $900. More support will be given when you are able to indicate an intention to publish the results being presented in a prestigious journal.
*Requirements:* Applicants should provide their personal details, name, university, supervisor, etc. as well as the following information: full name of the conference; location of the conference; dates of the conference; are you presenting an oral or poster? If so, provide evidence of the status of your presentation oral/poster, refereed/not refereed, accepted/not accepted; details of (or plans for) publication of the results; note that the committee looks more favorably on applications with an intention to publish in high impact journals; the AINSE Award number(s) which relate to the research being presented; abstract of your presentation; and, your budget including any additional support you have secured for the conference. Applications should be sent by email to ainse@ansto.gov.au.
*Geographic Focus:* Australia
*Amount of Grant:* Up to 900 AUD
*Contact:* Frank Bruhn, Managing Director; +61 2 9717 3388 or +61 02 9717 3376; fax +61 02 9717 9268; frankb@ainse.edu.au or ainse@ansto.gov.au
*Internet:* www.ainse.edu.au/grad_students2/travel_scholarships
*Sponsor:* Australian Institute of Nuclear Science and Engineering
Locked Mail Bag 2001
Kirrawee, NSW 2232 Australia

**AINSE John Ferris Memorial Award** 246

Dr. John Ferris was a cherished identity at ANSTO. He graduated from the University of Adelaide in South Australia with a B.Sc (Hons) and from the University of Tasmania in 1986 with a Ph.D in Limnology. John joined ANSTO in 1991, following post-doctoral work with the Australian Antarctic Division and the Centre for Water Research at the University of Western Australia. He brought considerable expertise in ecosystem studies involving algae which was to be the basis of both his research contributions within ANSTO's Institute for Environmental Research and more recently in environmental management in safety and radiation services. John became involved in bio monitoring and environmental management at ANSTO and was responsible for ANSTO's Annual Environmental Report for many years. He was a significant contributor to ANSTO's ISO 14001 certification. In more recent times, John's research interests focused on radiation effects on non-human biota and his work contributed to international thinking in this area. A John Ferris Memorial Award has been established to honour John's life and work. The award is designed assist in supporting the best post-graduate student whose application is in environmental sciences.
*Geographic Focus:* Australia
*Date(s) Application is Due:* Apr 15
*Contact:* Frank Bruhn, Managing Director; +61 2 9717 3388 or +61 02 9717 3376; fax +61 02 9717 9268; frankb@ainse.edu.au or ainse@ansto.gov.au
*Internet:* www.ainse.edu.au/grad_students2/john_ferris_memorial_award
*Sponsor:* Australian Institute of Nuclear Science and Engineering
Locked Mail Bag 2001
Kirrawee, NSW 2232 Australia

**AINSE Postgraduate Research Awards** 247

Awards are offered by the Australian Institute of Nuclear Science and Engineering (AINSE) for suitably qualified persons wishing to undertake studies in AINSE's field of interest for a higher degree at a member university of AINSE. Applications for PGRAs will also be considered, as in past years, where the research aligns with the broad spectrum of research foci supported by AINSE. A full list of over 100 major and minor facilities facilities is available on request. Research Supplements ($7,500 pa) are the principal form of award and are offered to scholars who are (or will be) in receipt of an Australian Postgraduate Award (APA) or equivalent scholarship. The award also provides $10,000 pa for costs involved in using the facilities and services at Lucas Heights. Certain travel and accommodation costs to enable students to work at Lucas Heights are also provided.
*Requirements:* Candidates for an AINSE Postgraduate Research Supplement must be nominated by the Australian or New Zealand university where PhD enrollment is held or proposed for tenure commencing before 30 June of the year of application. The university must be a member of AINSE. Guidelines and applications can be found at the website.
*Geographic Focus:* Australia
*Date(s) Application is Due:* Apr 15
*Amount of Grant:* 7,500 - 17,500 AUD
*Contact:* Frank Bruhn, Managing Director; +61 02 9717 3388 or +61 02 9717 3376; fax +61 02 9717 9268; frankb@ainse.edu.au or ainse@ansto.gov.au
*Internet:* www.ainse.edu.au/grad_students2/postgraduate_awards
*Sponsor:* Australian Institute of Nuclear Science and Engineering
Locked Mail Bag 2001
Kirrawee, NSW 2232 Australia

**AINSE Research Fellowships** 248

AINSE Research Fellowship are offered at academic level B for outstanding and enthusiastic scientists with three to eight years of postdoctoral experience to undertake research in areas of science and technology requiring use of the research facilities at the Australian Nuclear Science and Technology Organisation (ANSTO). Fellows will be expected to build a research group in one of the AINSE member universities and be attached to ANSTO for experiments on the new, state of the art facilities on the OPAL research reactor or other facilities at ANSTO.
*Requirements:* AINSE is highlighting three research directions in calling for applications for research fellowships: neutron scattering; radiopharmaceuticals; and, high resolution climate records using nuclear techniques. Applications will also be considered which fall outside these topics which but still require use of ANSTO facilities and which align with the joint research interests of ANSTO and AINSE member universities.
*Geographic Focus:* Australia
*Date(s) Application is Due:* May 31
*Contact:* Frank Bruhn, Managing Director; +61 2 9717 3388 or +61 02 9717 3376; fax +61 02 9717 9268; frankb@ainse.edu.au or ainse@ansto.gov.au
*Internet:* www.ainse.edu.au/academic_researchers2
*Sponsor:* Australian Institute of Nuclear Science and Engineering
Locked Mail Bag 2001
Kirrawee, NSW 2232 Australia

**AINSE Research Fellowships with Institute for Science and International Security Support** 249

AINSE Research Fellowship are offered at academic level B for outstanding and enthusiastic scientists with 3 - 8 years of postdoctoral experience to undertake research in areas of science and technology requiring use of the research facilities at the Australian Nuclear Science and Technology Organisation (ANSTO). Fellows will be expected to build a research group in one of the AINSE member universities and be attached to ANSTO for experiments on the new, state of the art facilities on the OPAL research reactor or other facilities at ANSTO. Since the demise of the Access to Major Research Facilities Program (AMRFP), there is no mechanism for Australian researchers to gain funds to travel to the United Kingdom. AINSE has received a number of requests from researchers needing funding for international travel. Following discussion, AINSE will now support travel for researchers to attend ISIS, on the same basis that the AMRF program allocated funding (one trip per individual per 6 months). ANSTO research scientists will also receive the same level of support as university researchers. Maximum grant available is $5,000 for any one visit.
*Requirements:* AINSE is highlighting three research directions in calling for applications for research fellowships: neutron scattering; radiopharmaceuticals; and, high resolution climate records using nuclear techniques. Applications will also be considered which fall outside these topics which but still require use of ANSTO facilities and which align with the joint research interests of ANSTO and AINSE member universities.
*Geographic Focus:* All States
*Date(s) Application is Due:* May 31
*Contact:* Frank Bruhn, Managing Director; +61 02 9717 3388 or +61 02 9717 3376; fax +61 02 9717 9268; frankb@ainse.edu.au or ainse@ansto.gov.au
Jorden Lickiss, +61 02 9717 3376; fax +61 02 9717 9268; jlk@ainse.edu.au or ainse@ansto.gov.au
*Internet:* www.ainse.edu.au/academic_researchers2/isis_support
*Sponsor:* Australian Institute of Nuclear Science and Engineering
Locked Mail Bag 2001
Kirrawee, NSW 2232 Australia

**ALA/ACRL Academic/Research Librarian of the Year Award** 250

This award is presented by the Association of College and Research Libraries. Funded by the GOBI Library Solutions from EBSCO, it is intended to recognize an individual member of the library profession who is making an outstanding national or international contribution to academic or research librarianship and library development. Individuals nominated should have demonstrated achievements in such areas as service to the organized profession through ACRL and related organizations, significant and influential research on academic or research library service, publication of a body of scholarly and/or theoretical writing contributing to academic or research library development, or planning and implementing a library program of such exemplary quality that it has served as a model for others. The award, in the amount of $5,000, has an annual deadline of December 7.
*Requirements:* Individuals may nominate themselves or others. Electronic submissions are required. Applicants must submit a nomination form and supporting documents via email.
*Geographic Focus:* All States, All Countries
*Date(s) Application is Due:* Dec 7
*Amount of Grant:* 5,000 USD
*Contact:* Chase Ollis, Staff Liaison; (312) 280-2521 or (312) 944-6780; fax (312) 440-9374; collis@ala.org or ala@ala.org
*Internet:* www.ala.org/awardsgrants/awards/206/apply
*Sponsor:* American Library Association
50 East Huron Street
Chicago, IL 60611-2788

**ALA Achievement in Library Diversity Research Award** 251

ALA Achievement in Library Diversity Research Award provides support of the propagation of library-based diversity research. The award is given annually, and provides a plaque and complimentary ALA Annual Conference registration. Awardees typically moderate the presentation of the year's Diversity Research Grant award winners at ALA's annual conference. The annual deadline for nominations is April 30.
*Requirements:* Individuals may nominate themselves or others. Electronic submissions are required. Applicants must submit a nomination form and supporting documents via email.
*Geographic Focus:* All States
*Date(s) Application is Due:* Apr 30
*Contact:* Jody Gray, Staff Liaison; (312) 280-5048 or (312) 944-6780; fax (312) 280-3256; jgray@ala.org or diversity@ala.org
*Internet:* www.ala.org/awardsgrants/awards/159/apply
*Sponsor:* American Library Association
50 East Huron Street
Chicago, IL 60611-2788

**Alabama Humanities Foundation Major Grants** 252

Alabama Humanities Foundation Major Grants fund a broad range of projects. Formats include public discussion programs, such as workshops, reading and discussion, lecture and discussion, and other community forums which encourage scholar and audience dialogue. They may be supplemented by other formats such as media items or exhibitions. The Foundation funds a wide array of project types, including: portable exhibit rentals; exhibition design and production; book discussions; humanities radio broadcasts; lecture and discussion programs; dramas based on oral histories and dramatic historical interpretation; interpretive readings of poetry, prose, or song; historical figures in costume and site interpretations; television & film documentaries and study guides to stimulate discussion on issues; community forums or town meetings to encourage public debate and discussion; and teachers seminars, institutes, workshops and more. Major Grants also support: exhibitions, including wall-mounted or portable interpretive exhibitions for displays at public sites such as libraries, museums, and government buildings; and teacher programs in a variety of formats focusing on critical reading and discussion, analytical writing, and discourse on the humanities currently in the Alabama public school curriculum. Final annual application submission deadlines are June 16 and September 15. Preliminary Grant Applications must be submitted via email at least four weeks prior to the final full application deadline. Major grants will receive notice of award by August 10 and October 31. Up to a maximum of $7,500 in outright funds are available.

*Requirements:* In order for a proposal to be considered for funding, a Preliminary Grant Application must be submitted via email to the Grants Director at least four weeks in advance of the application deadline. The Preliminary Grant Application must consist of a two-page Summary Narrative outlining project goals, humanities focus and scholarship, intended audience, key project personnel, and publicity and marketing strategy, and a Summary Budget detailing proposed grant funds and proposed applicant cost-share, which may include in-kind. A one-to-one cost-share match, which may include in-kind contributions, is required.
*Restrictions:* Grants are not made to individuals. AHF does not fund projects that: promote a course of action or advocacy; discriminate against persons or groups; support individual research or scholarship; support permanent planning, construction, or restoration; result in permanent acquisition of equipment; or realize a profit or result in a free-standing publication.
*Geographic Focus:* Alabama
*Date(s) Application is Due:* Jun 16; Sep 15
*Amount of Grant:* 1,500 - 7,500 USD
*Contact:* Thomas Bryant, Grants Director; (205) 558-3997 or (205) 558-3980; fax (205) 558-3981; tbryant@alabamahumanities.org or info@alabamahumanities.org
*Internet:* www.alabamahumanities.org/grants/categories/
*Sponsor:* Alabama Humanities Foundation
1100 Ireland Way, Suite 202
Birmingham, AL 35205

**Alabama Humanities Foundation Mini Grants** 253
Designed for flexibility, Alabama Humanities Foundation Mini Grants support a wide variety of projects from workshops to radio programs and reading/discussion series. Up to $1,500 in outright is available, though a 1:1 cost share match is required. Final annual application submission deadlines are June 16 and September 15. Preliminary Grant Applications must be submitted at least six weeks prior to the final application deadline. Mini grants will receive notice of award by July 15 and October 15, respectively.
*Requirements:* Applications must be submitted three months prior to start date of the program. Grants are made to nonprofit public or private organizations. Funds are to be used in state only.
*Restrictions:* Grants are not made to individuals.
*Geographic Focus:* Alabama
*Date(s) Application is Due:* Jun 16; Sep 15
*Amount of Grant:* Up to 2,000 USD
*Contact:* Thomas Bryant, Grants Director; (205) 558-3997 or (205) 558-3980; fax (205) 558-3981; tbryant@alabamahumanities.org or info@alabamahumanities.org
*Internet:* www.ahf.net/programs/grantsProgram.html
*Sponsor:* Alabama Humanities Foundation
1100 Ireland Way, Suite 202
Birmingham, AL 35205

**Alabama Library Association CUS Research Grants** 254
Sponsored by the Research Committee of the College, University, and Special Libraries (CUS) Division, Alabama Library Association, research supported through this program must relate to college, university, or special libraries. The Committee looks for the following: originality, validity, relevance to the profession, and applicant's need for funding support. Grant funds may be used for expenses such as duplicating, mailing, travel, office supplies, consulting services, and computer services. The deadline for submitting an application is normally December 1.
*Requirements:* Competition for the grant is open to CU.S. members employed in Alabama college, university, and special libraries, and to CU.S. members attending an ALA-accredited library school.
*Geographic Focus:* Alabama
*Date(s) Application is Due:* Dec 1
*Amount of Grant:* 250 USD
*Contact:* Kathy Wheeler, Research Chair; (251) 460-7938; kwheeler@southalabama.edu
*Internet:* www.allanet.org/?page=CUSMaterials
*Sponsor:* Alabama Library Association
6030 Monticello Drive
Montgomery, AL 36117

**ALA Carroll Preston Baber Research Grant** 255
The Carroll Preston Baber Research Grant is given annually to one or more librarians or library educators who will conduct innovative research that could lead to an improvement in services to any specified group(s) of people. The project should aim to answer a question that is of vital importance to the library community and the researchers should plan to provide documentation of the results of their work. Up to $3,000 is available. Half of the amount will be paid within one month of the Annual Conference. The remaining half will be provided approximately six months later after a satisfactory progress report. The application cover sheet and information about additional materials required are available at the website. The annual deadline is February 11.
*Requirements:* Any ALA member may apply. The jury would welcome projects that involve both a practicing librarian and a researcher. Applications must be clearly defined, with the potential to address a currently important issue in library service that is national in scope. Applications must also demonstrate the ability to undertake and successfully complete the project, with a reasonable dissemination plan.
*Restrictions:* Institutional overhead is not an acceptable budget item within the parameters of the Baber Grant, nor should overhead be listed as institutional support. If travel for dissemination of results is in the budget, it must be less than 10% of the total.
*Geographic Focus:* All States
*Date(s) Application is Due:* Feb 11
*Amount of Grant:* Up to 3,000 USD
*Contact:* Courtney Jones, Awards Coordinator; (800) 545-2433, ext. 5839 or (312) 944-6780; fax (312) 440-9374; cjones@ala.org or ala@ala.org
*Internet:* www.ala.org/awardsgrants/awards/55/apply
*Sponsor:* American Library Association
50 East Huron Street
Chicago, IL 60611-2788

**ALA Donald G. Davis Article Award** 256
The Donald G. Davis Article Award is presented by the Library History Round Table of the American Library Association every second year to recognize the best article written in English in the field of United States and Canadian library history including the history of libraries, librarianship, and book culture. Entries are judged on quality of scholarship, clarity of style, depth of research, and ability to place research findings in a broad social, cultural, and political context. One award will be given every second year unless the jury does not find a suitable candidate for that biennial period. Any member of the Library History Round Table may nominate one or more articles by sending a recommendation to the Chair of the Davis Award Committee. Also, the editor of the LHRT Newsletter Library History Bibliography will submit a list of candidates. The winner will be announced in a press release on or about June 1st of the award year. Certificates honoring the author(s) and journal will be presented at a Library History Round Table awards ceremony during the American Library Association Annual Conference in the year of the award. The annual deadline for applications is January 22.
*Requirements:* Entries for each biennial award must have been published between January 1 and December 31 of the two years preceding the award year.
*Restrictions:* Papers that have won the Justin Winsor or Jesse Shera Awards are not eligible for consideration.
*Geographic Focus:* All States
*Date(s) Application is Due:* Jan 22
*Contact:* Kelsey Henke, Staff Liaison; (312) 280-4283 or (800) 545-2433, ext. 4283; fax (312) 280-4392; khenke@ala.org or ala@ala.org
*Internet:* www.ala.org/awardsgrants/awards/138/apply
*Sponsor:* American Library Association
50 East Huron Street
Chicago, IL 60611-2788

**ALA Emerald Research Grant Awards** 257
The Emerald Research Grant is awarded to individuals seeking support to conduct research in business librarianship. Two awards of $5,000 each are given each year. The awards (check and citation) will be presented at the Reference and User Services Association (RUSA) Awards Ceremony at the annual conferences each year. Recipients will be required to attend the awards ceremony. Recipients may be asked to present their findings at a Business Reference and Services Section (BRASS) event within two years of receiving the award (at the discretion of the BRASS Executive Committee). Recipients will also be required to acknowledge the Emerald Research Grant when publishing or presenting their research. The annual deadline for application submissions is January 6.
*Requirements:* Proposals will be accepted from both individual researchers and those working collaboratively. Applicants must be members of ALA. Candidates must submit a detailed proposal outlining their proposed research project; methodology, scope and timetable; how this project fits into the existing literature; and projected outcomes, including a statement outlining how this research will benefit the library profession.
*Geographic Focus:* All States
*Date(s) Application is Due:* Jan 6
*Amount of Grant:* 5,000 USD
*Contact:* Leighann Wood, Staff Liaison; (312) 280-5837 or (312) 944-6780; fax (312) 440-9374; lwood@ala.org or ala@ala.org
*Internet:* www.ala.org/awardsgrants/awards/132/apply
*Sponsor:* American Library Association
50 East Huron Street
Chicago, IL 60611-2788

**ALA Excellence in Academic Libraries Award** 258
The Excellence in Academic Libraries Award recognizes academic librarians and staff who work together to develop academic libraries that are outstanding in furthering the educational missions of their insitutions. The program will support three annual awards of $3,000 each by type of academic library (community, college, and university). Awards are judged on the following criteria: how academic librarians and staff work together to develop an academic library that is outstanding in furthering the educational mission of its parent institution; creativity and innovation in meeting the needs of their academic community; leadership in developing and implementing exemplary programs that other libraries can emulate; and substantial and productive relationships with classroom faculty and students. Awards will be made by an ACRL representative at a presentation at each institution receiving the award. Further recognition of award recipients will be made at the ACRL President's Program at the annual ALA conference. Funding will cover the outright awards and provide $3,000 for travel to presentation ceremonies. Applications are kept in the award pool for three years. Applications may be updated each year, as long as they are received by the award deadline. Application forms are available at the website.
*Requirements:* Academic and research libraries may apply. The award winners will agree to take responsibility for organizing and funding a recognition ceremony on campus to receive the award.
*Geographic Focus:* All States
*Date(s) Application is Due:* Dec 2
*Amount of Grant:* 6,000 USD

*Contact:* Megan Griffin; (800) 545-2433, ext. 2514; fax (312) 440-9374; mgriffin@ala.org
Lori Goetsch, Chair; (785) 532-7400; fax (785) 532-7415; lgoetsch@ksu.edu
*Internet:* www.ala.org/awardsgrants/awards/234/apply
*Sponsor:* American Library Association
50 East Huron Street
Chicago, IL 60611-2788

**ALA Frederick G. Kilgour Award for Research in Library and Information Technology** 259

The Frederick G. Kilgour Award for Research in Library and Information Technology is sponsored by OCLC Online Computer Center, Inc. and the Library and Information Technology Association (LITA), a division of the American Library Association. The purpose of this award is to bring attention to research relevant to the development of information technologies, especially work which shows promise of having a positive and substantive impact on any aspect of the publication, storage, retrieval and dissemination of information, or the processes by which information and data is manipulated and managed. The award recipient receives $2,000 cash and an expense paid trip to the ALA conference (airfare and two nights lodging).
*Requirements:* Nominations will be accepted from any member of the American Library Association. Nominating letters should briefly address: how the research is relevant to libraries; is creative in its design or methodology; builds on existing research or enhances potential for future exploration; and/or solves an important current problem in the provision of information resources. In addition, a complete curriculum vita and a copy of several seminal publications by the nominee must be included. Nominees may be asked to submit additional information to the Kilgour Committee. Preference will be given to completed research over works in progress.
*Restrictions:* Currently-serving officers and elected officials of LITA, members of Kilgour Committee and OCLC employees and their immediate family members are not eligible.
*Geographic Focus:* All States
*Date(s) Application is Due:* Dec 31
*Amount of Grant:* 2,000 USD
*Contact:* Melissa Prentice; (312) 280-4268; fax (312) 280-3257; mprentice@ala.org
Valerie Edmonds-Merritt; (800) 280-4269; fax (312) 280-3257; vedmonds@ala.org
*Internet:* www.ala.org/awardsgrants/awards/227/apply
*Sponsor:* American Library Association
50 East Huron Street
Chicago, IL 60611-2788

**ALA Gale Cengage History Research and Innovation Award** 260

The Gale Cengage History Research and Innovation Award will be granted to an MLS degreed librarian from an ALA accredited school to facilitate and further research relating to history and history librarianship. The winner will receive a citation and up to $2,500. The annual deadline for nomination submission is December 15.
*Requirements:* All applicants should provide persuasive evidence that the proposed project has practical and/or scholarly value to the history librarianship community. The proposal should be as explicit as possible about the current state of knowledge of the subject, state the proposed objective(s), and tell what accomplishment of the project will achieve. The research design should be outlined as specifically as possible and should demonstrate why (and where) travel is necessary, if that is proposed.
*Geographic Focus:* All States
*Date(s) Application is Due:* Dec 15
*Amount of Grant:* Up to 2,500 USD
*Contact:* Leighann Wood, Staff Liaison; (312) 280-5837 or (312) 944-6780; fax (312) 440-9374; lwood@ala.org or ala@ala.org
*Internet:* www.ala.org/awardsgrants/awards/32913/apply
*Sponsor:* American Library Association
50 East Huron Street
Chicago, IL 60611-2788

**ALA Gale Cengage History Research and Innovation Award** 261

The Gale Cengage History Research and Innovation Award consists of up to $2,500 and a citation recognizing an MLS degreed librarian to further research relating to history and history librarianship. The project will be evaluated with the following criteria: value of the project to the history librarianship community; quality, depth, comprehensiveness of the literature reviewed; feasibility of the project, with a combination of the appropriateness of the research method/research design and the reasonableness of the timeline; the applicant's qualifications; and the raasonableness of proposed expenditures. The nomination form is available at the website. The annual deadline for application submissions is January 6.
*Requirements:* Librarians with an ALA accredited degree are encouraged to apply. Historical research must consist of an emphasis in an area reflected by the History Section's subject-oriented committees: American history; genealogy, local history, instruction and research services, and historical materials. Candidates must be members of the RUSA History section by the time the award is presented.
*Geographic Focus:* All States
*Date(s) Application is Due:* Jan 6
*Amount of Grant:* Up to 2,500 USD
*Contact:* Shannon Carter, Program Coordinator; (312) 280-4381 or (800) 545-2433, ext. 4381; fax (312) 440-9374; scarter@ala.org
*Internet:* www.ala.org/awardsgrants/awards/31876/apply
*Sponsor:* American Library Association
50 East Huron Street
Chicago, IL 60611-2788

**ALA Innovation Award** 262

This annual award recognizes a project that demonstrates creative, innovative, or unique approaches within the context of national trends to information literacy instruction or programming. The winner receives $3,000 and a certificate.
*Requirements:* Academic librarians or academic project teams that include an academic librarian are eligible to receive the award. Recipients must have implemented their project in an academic or research library, or through the sponsorship of a professional library organization no more than two years prior to the nomination submission deadline. Electronic submissions are required. Nominations must include a description of the innovative project, a letter supporting the project, and sufficient supporting documentation for the committee to understand its purpose, content, impact, and innovative aspects.
*Geographic Focus:* All States
*Date(s) Application is Due:* Dec 3
*Amount of Grant:* 3,000 USD
*Contact:* Casey Kinson; (800) 545-2433, ext. 2511; fax (312) 440-9374; ckinson@ala.org
Erin Ellis, IS Awards Co-Chair; (785) 864-5589; eellis@ku.edu
Robin Ewing; (320) 308-2093 or (320) 308-0121; rlewing@stcloudstate.edu
*Internet:* www.ala.org/awardsgrants/awards/244/apply
*Sponsor:* American Library Association
50 East Huron Street
Chicago, IL 60611-2788

**ALA Jan Merrill-Oldham Professional Development Grant** 263

The Award is established by the Preservation and Reformatting Section (PARS) of the Association for Library Collections & Technical Services (ALCTS) to honor the career and influence of Jan Merrill-Oldham, distinguished leader, author, and mentor in the field of library and archives preservation.Over the past three decades Ms. Merrill-Oldham has been instrumental in the development of standards and best practices, writing more than forty publications including co-authoring the Guide to the ANSI/NISO/LBI Library Binding Standard, a document used by almost all libraries and commercial library binders. Ms. Merrill-Oldham has served on key committees within ALCTS, ALA, the Association of Research Libraries, the Council on Library and Information Resources, the National Information Standards Organization and many others. She has educated and mentored countless preservation librarians and conservators and her support for students and dedication to the field serves as a model to all of us. In September of 2010, Ms. Merrill-Oldham announced her retirement after a long and notable career in library and archives preservation. In December 2010, she was named the recipient of the ALCTS Ross Atkinson Lifetime Achievement Award. This award recognizes Jan Merrill-Oldham's wide ranging contributions, deep commitment to the field of library and archives preservation, and her undying support of young professionals by supporting participation in an ALA Annual Conference.It is important however to encourage new and young preservation librarians and staff to further pursue their professional development. A significant part of that involvement is to attend the ALA Annual Conference to network and learn from their colleagues. Ms. Merrill-Oldham over her career has dedicated herself to mentoring young professionals and it is in recognition of that service that an award that supports professional development and involvement by librarians and para-professionals new to the preservation field be established.The Jan Merrill-Oldham Professional Development Grant is awarded by the ALCTS Preservation and Reformatting Section to provide librarians and paraprofessionals new to the preservation field with the opportunity to attend a professional conference and encourages professional development through active participation at the national level. The grant is to be used for airfare, lodging, and registration fees to attend the ALA Annual Conference.This annual award consists of a $1,250 grant donated by the Library Binding Institute, and a citation. The grant is applicable toward airfare, lodging, and registration fees related to the ALA Annual Conference attendance. Application instructions are available at the website.
*Requirements:* Applicants must have no more than five years of experience in the field of library and archives preservation, and no previous attendance to an ALA Annual Conference. They should currently work as a librarian or para-professional within a library or archives preservation department, or have preservation responsibilities within their institution, or be currently enrolled in a preservation-related graduate program. They must be willing to participate in designated conference events, and submit a summary of their conference experience to the ALCTS Newsletter Online no later than 30 days after the conference.
*Restrictions:* Previous winners are not eligible.
*Geographic Focus:* All States
*Date(s) Application is Due:* Dec 1
*Amount of Grant:* 1,250 USD
*Contact:* Keri Cascio, Staff Liaison; (312) 280-5030 or (312) 944-6780; fax (312) 440-9374; kcascio@ala.org or ala@ala.org
*Internet:* www.ala.org/awardsgrants/awards/31536/apply
*Sponsor:* American Library Association
50 East Huron Street
Chicago, IL 60611-2788

**ALA Jesse H. Shera Award for the Support of Dissertation Research** 264

The Jesse H. Shera Award for the Support of Dissertation Research is intended to provide recognition for dissertation research employing exemplary research design and methods. The Award is given to provide recognition and monetary support for dissertation research employing exemplary research design and methods. Any research design is acceptable, but the topic of the research must relate in at least a general way to library and information studies. Candidates for this award should submit a two-part paper based on

their approved dissertation proposal. Part one should be a brief problem statement not to exceed 500 words. Part two should be a summary of the design/methodology portion of their dissertation proposal, not to exceed 2,500 words. The approval of the research by the doctoral candidate's dissertation committee should be supported by a brief statement from the dissertation committee chair. A certificate and a $250 award will be given out annually, with a submission deadline of January 31.
*Requirements:* The award is restricted to research designs that have been approved by the doctoral candidate's dissertation committee and that are about to be employed or are in the initial stage of use. Completed research is not eligible for consideration.
*Geographic Focus:* All States
*Date(s) Application is Due:* Jan 31
*Amount of Grant:* 250 USD
*Contact:* Kelsey Henke, Staff Liaison; (312) 280-2788 or (312) 944-6780; fax (312) 440-9374; khenke@ala.org or ala@ala.org
*Internet:* www.ala.org/awardsgrants/awards/188/apply
*Sponsor:* American Library Association
50 East Huron Street
Chicago, IL 60611-2788

**ALA LBI George Cunha and Susan Swartzburg Award** **265**

The LBI George Cunha and Susan Swartzburg Award honors the memory of George Cunha and Susan Swartzburg, early leaders in cooperative preservation programming and strong advocates for collaboration in the field of preservation. The award acknowledges and supports cooperative preservation projects and/or rewards individuals or groups that foster collaboration for preservation goals. Recipients of the award demonstrate vision, endorse cooperation, and advocate for the preservation of published and primary source resources that capture the richness of our cultural patrimony. The award recognizes the leadership and initiative required to build collaborative networks designed to achieve specific preservation goals. Since collaboration, cooperation, advocacy and outreach are key strategies that epitomize preservation, the award promotes cooperative efforts and supports equitable preservation among all libraries, archives and historical institutions. Given out on an annual basis, the award consists of $1,250 and a citation. The annual deadline for application submissions is December 1.
*Requirements:* Any person or group is eligible for this award. Membership in the ALA organization is not required.
*Restrictions:* Employees of LBI are not eligible.
*Geographic Focus:* All States
*Date(s) Application is Due:* Dec 1
*Amount of Grant:* 1,250 USD
*Contact:* Keri Cascio, Staff Liaison; (800) 545-2433, ext. 5030; fax (312) 440-9374; kcascio@ala.org or ala@ala.org
*Internet:* www.ala.org/awardsgrants/awards/182/apply
*Sponsor:* American Library Association
50 East Huron Street
Chicago, IL 60611-2788

**ALA Margaret T. Lane/Virginia F. Saunders Memorial Research Award** **266**

The award is given annually to an author or shared among collaborative authors of an outstanding research article in which government documents, either published or archival in nature, form a substantial part of the documented research. Preference may be given to articles published in library literature and those that appeal to a broader audience. A plaque and $2,000 is awarded to the recipient(s). The nomination form is available at the website. Awards are announced following the ALA Midwinter meeting and presented at the Government Documents Round Table (GODORT) reception at the ALA Annual Conference.
*Requirements:* The article must be published, either in print or online, no more than 18 months prior to the due date for nominations. The award is not restricted to articles in library journals.
*Restrictions:* Only articles are eligible for the award. Book chapters, books, and dissertations are not eligible.
*Geographic Focus:* All States
*Date(s) Application is Due:* Dec 3
*Amount of Grant:* 2,000 USD
*Contact:* Andrea Sevetson, Award Coordinator; (541) 992-5461; asevetson@hotmail.com
*Internet:* www.ala.org/awardsgrants/awards/31624/apply
*Sponsor:* American Library Association
50 East Huron Street
Chicago, IL 60611-2788

**ALA Marta Lange/CQ Press Award** **267**

This annual award, supported by Congressional Quarterly, recognizes an academic or law librarian who, through research, other creative activity, or service to the profession, makes distinguished contributions to bibliography and information service in law or political science. Nominees should have achieved distinction in one or more of the following areas: planning and implementing a model bibliography/information services program in a law or political science library; history of contributions to the field through research, publications, and other activities displaying active participation in the advancement of law/political science librarianship; service to the profession through ACRL or related regional and national organizations; and promotion or development of an education program for law and political science librarianship that has served as a model for other courses and programs.
*Requirements:* There are no restrictions on who may apply. Individuals may nominate themselves or others. Electronic submissions are required. Nominators should submit a letter of nomination including the name, address, and phone number of both the nominating party and the nominee, as well as a narrative supporting the nomination, and a current vita.
*Geographic Focus:* All States
*Date(s) Application is Due:* Dec 2
*Amount of Grant:* 1,000 USD
*Contact:* Casey Kinson; (312) 280-2511 or (312) 280-2523; fax (312) 280-2520; ckinson@ala.org
Nadine R. Hoffman, Award Committee Chair; (403) 220-8392; nadine.hoffman@calgary.ca or nhoffman@ucalgary.ca
Megan Griffin, Staff Liaison; (312) 280-2514; fax (312) 440-9374; mgriffin@ala.org
*Internet:* www.ala.org/awardsgrants/awards/73/apply
*Sponsor:* American Library Association
50 East Huron Street
Chicago, IL 60611-2788

**ALA Miriam Dudley Instruction Librarian Award** **268**

The Dudley Award recognizes a significant contribution by a librarian to the advancement of instruction in a college or research library environment. Nominees should have achieved distinction in one or more of the following areas: planning and implementation of an academic instruction program in a library environment that has served as a model for other programs nationally or regionally; production of a body of research and publication that has a demonstrable impact on the concepts and methods of teaching and information-seeking strategies in a college or research institution; sustained participation in organizations, at the national or regional level, devoted to the promotion and enhancement of academic instruction in a library environment; or promotion, development, and integration of education for instruction in ALA-accredited library schools or professional continuing education programs that have served as models for other courses and programs. The $1,000 award is given annually. The annual deadline is December 1.
*Restrictions:* Nominees will be judged on an individual basis; this award cannot be given to a pair or group of persons.
*Geographic Focus:* All States
*Date(s) Application is Due:* Dec 1
*Amount of Grant:* 1,000 USD
*Contact:* Chase Ollis, Staff Liaison; (312) 280-2521 or (800) 545-2433, ext. 2521; fax (312) 440-9374; collis@ala.org or ala@ala.org
*Internet:* www.ala.org/awardsgrants/awards/355/apply
*Sponsor:* American Library Association
50 East Huron Street
Chicago, IL 60611-2788

**ALA NewsBank/Readex Catharine J. Reynolds Award** **269**

The NewsBank/Readex Catharine J. Reynolds Award provides funding for research in the field of documents librarianship or in a related area that would benefit the individual's performance as a documents librarian or make a contribution to the field. This award is named for Catharine J. Reynolds, former Head of Government Publications at the University of Colorado. It is supported by an annual contribution from the NewsBank/Readex Corporation. Up to three grants a year are awarded. Recipients are selected on the basis of proven or potential ability, promise of future usefulness and permanence in the profession, financial need, and benefit of the project to the profession as well as the individual. The application is available at the website. Award recipient(s) will be announced as soon as all recipients are informed following the ALA Midwinter meeting. Awards are presented at the Government Documents Round Table (GODORT) reception at the ALA Annual Conference.
*Requirements:* Any documents librarian may apply. The awards support research in all areas of government information, U.S., foreign, international, state, regional, or local. The grant can be used to finance research costs, such as computer time or travel.
*Restrictions:* The Awards Committee may designate the amount of each award at its discretion.
*Geographic Focus:* All States
*Date(s) Application is Due:* Dec 2
*Amount of Grant:* 2,000 USD
*Contact:* Beth Clausen, Awards Committee Chair; (847) 491-2891; fax (847) 491-4655; b-clausen@northwestern.edu
*Internet:* www.ala.org/awardsgrants/awards/207/apply
*Sponsor:* American Library Association
50 East Huron Street
Chicago, IL 60611-2788

**ALA Phyllis Dain Library History Dissertation Award** **270**

The award, named in honor of a library historian widely known as a supportive advisor and mentor as well as a rigorous scholar and thinker, recognizes outstanding dissertations in the general area of library history. An award of $500 is given for a work that embodies original research on a significant topic relating to the history of books, libraries, librarianship or information science. The award is given in the odd numbered years. Dissertations completed and accepted during the preceding two academic years are eligible. The annual deadline for nominations is January 20.
*Requirements:* Entries are judged on: clear definition of the research questions and/or hypotheses; use of appropriate source materials; depth of research; superior quality of writing; ability to place the subject within its broader historical context; and significance of the conclusions. Four copies of the dissertation and a letter of support from the doctoral advisor or from another faculty member at the degree-granting institution are required.
*Geographic Focus:* All States
*Date(s) Application is Due:* Jan 20
*Amount of Grant:* 500 USD

*Contact:* Kelsey Henke, Staff Liaison; (800) 545-2433, ext. 4283 or (312) 280-4283; fax (312) 440-9374; khenke@ala.org or ala@ala.org
*Internet:* www.ala.org/awardsgrants/awards/208/apply
*Sponsor:* American Library Association
50 East Huron Street
Chicago, IL 60611-2788

**ALA ProQuest/GODORT Documents to the People Award** 271
The ProQuest/GODORT Documents to the People Award is a tribute to an individual, library, institution, or other non-commercial group that has most effectively encouraged the use of government documents in support of library service. A framed plaque with the individual's name and achievement plus a $3,000 monetary prize to be assigned to a project of the recipient's choice is given annually. The nomination form is available at the website, and the annual deadline for submission is December 2.
*Restrictions:* The recipient is responsible for any travel expenses to attend the ALA Annual Conference.
*Geographic Focus:* All States
*Date(s) Application is Due:* Dec 2
*Amount of Grant:* 3,000 USD
*Contact:* Beth Clausen, Awards Committee Chair; (847) 491-2891 or (847) 491-7627; fax (847) 491-4655; b-clausen@northwestern.edu
*Internet:* www.ala.org/awardsgrants/awards/435/apply
*Sponsor:* American Library Association
50 East Huron Street
Chicago, IL 60611-2788

**Alaska Airlines Corporate Giving Medical Emergency and Research Grants** 272
Alaska Airlines and Horizon Air Corporate Giving support health and human services, arts and cultural programs, as well as education, environmental, and civic organizations. The program focuses on communities served now or in the near future, and where a significant number of its employees live or work. When considering requests, the corporation favors organizations and efforts that are most likely to enhance a community's cultural and economic vitality and improve the quality of life for its citizens. Objectives of the Corporate Giving Program are to: provide support to charitable and cultural organizations within our communities; and maintain a reasonable balance of contributions supporting a wide range of organizations. In the areas of Medical Emergency and Research, the corporation regularly responds to the need for health and human services relief, especially in the state of Alaska, where transportation infrastructure limits access to medical facilities. Through in-kind contributions, the giving program assists hundreds of individual emergency and medical transportation needs. Its partnerships with Angel Flight West and the Shriners' Hospitals provide well over 1,000 complimentary passenger seats each year for those needing treatment in another city.
*Requirements:* Giving is on a national and international basis in areas of company operations, with emphasis on Alaska, Oregon, and Washington; some further giving occurs in Canada and Mexico.
*Restrictions:* Alaska Airlines or Horizon Air do not provide support for: loans and grants to individuals or private business; groups that discriminate; endowments; religious organizations (if for sacramental purposes); pageants; capital projects; multi-year commitment or automatic renewal grants; general operating expenses; publicly or privately funded educational institutions; or organizations whose prime purpose is to influence legislation.
*Geographic Focus:* Alaska, Oregon, Washington, Canada, Mexico
*Contact:* Tim R. Thompson, Manager, Public Affairs; (907) 266-7230; fax (907) 266-7229; tim.thompson@alaskaair.com
Barbara J. Johnson, Manager; (206) 392-5383; barb.johnson@alaskaair.com
*Internet:* https://www.alaskaair.com/content/about-us/social-responsibility/corporate-giving.aspx
*Sponsor:* Alaska Airlines
4750 Old International Airport Road
Anchorage, AK 99502

**Alaska Conservation Foundation Watchable Wildlife Conservation Trust Grants** 273
The Watchable Wildlife Conservation Trust is a donor-advised fund developed in cooperation with the Alaska Department of Fish and Game. The purpose of the Trust is to fund those projects that emphasize non-consumptive use of wildlife, expand wildlife conservation efforts, and broaden public support for conserving Alaska's wildlife resources. Areas of interest include: interpretation (e.g., signage, viewing guides, media programming); education (e.g., curriculum development or publication development and distribution); facilities construction; applied ecosystem research; and projects involving a state wildlife sanctuary. The Trust makes grants once per calendar year. The annual deadline is October 1. Prospective grantees should contact ACF to determine if the project fits within the grant making program and when to submit their proposal. The Trust generally funds projects in the $1,000 to $4,000 range.
*Requirements:* As a 501(c)3 public foundation, ACF accepts requests for funding from incorporated, tax-exempt organizations, non-incorporated organizations, and individuals. Applicants must meet the following criteria: be a non-profit organization; be based in Alaska, or have an Alaska-based program; focus operations within Alaska; and, if a current ACF grantee, be in good standing with required grant reports. Eligible projects and costs include: interpretation (e.g. signage, viewing guides, media programming); wildlife education (e.g. curriculum development or publication development and distribution); trail construction and maintenance; facilities construction; research (e.g. ecosystem research or research on the social and economic values of watching wildlife); and acquisition of critical wildlife habitats.
*Restrictions:* Ineligible requests include: grants to individuals; projects outside the state of Alaska; general operating support; litigation or lobbying; endowments; and scholarships.
*Geographic Focus:* Alaska
*Date(s) Application is Due:* Oct 1
*Amount of Grant:* 1,000 - 4,000 USD
*Contact:* Michael Barber, Program Officer; (907) 276-1917, ext. 205; fax (907) 274-4145; mbarber@alaskaconservation.org or grants@alaskaconservation.org
Anna Dalton, Grants Manager; (907) 276-1917; fax (907) 274-4145; adalton@alaskaconservation.org or grants@alaskaconservation.org
*Internet:* alaskaconservation.org/grant-opportunities/watchablewildlife/
*Sponsor:* Alaska Conservation Foundation
911 West 8th Avenue, Suite 300
Anchorage, AK 99501

**ALA STS Innovation in Science and Technology Librarianship Award** 274
The purpose of the award is to recognize creative, innovative approaches to solving problems or improving products and services in science and technology librarianship. The award recognizes and promotes leadership and initiative that demonstrate impact on the academic and research community. Projects nominated for the award should inspire information professionals to overcome obstacles in the profession, and should demonstrate creativity, quality and innovation within the context of national trends in science and technology librarianship. The award is administered by the ACRL Science and Technology Section and is presented at the ALA Annual Conference each year.
*Requirements:* Any library-related project is eligible to be nominated for the award. Nominations must include a description of the innovative project, and sufficient supporting documentation for the committee to understand the purpose, content, impact, and innovative aspects of the program. Electronic submissions are required.
*Restrictions:* Projects from the award sponsors are ineligible for consideration. Projects of the STS Chair/Vice-Chair while those persons are in office are not eligible to apply.
*Geographic Focus:* All States
*Date(s) Application is Due:* Dec 2
*Amount of Grant:* 3,000 USD
*Contact:* Casey Kinson, Program Coordinator; (312) 280-2511 or (312) 280-2523; fax (312) 280-2520; ckinson@ala.org
Elizabeth Berman, Award Co-Chair; (802) 656-8130; fax (802) 656-4038; elizabeth.berman@uvm.edu
Megan Griffin; (800) 545-2433, ext. 2514; fax (312) 440-9374; mgriffin@ala.org
*Internet:* www.ala.org/awardsgrants/awards/411/apply
*Sponsor:* Association of College and Research Libraries
50 East Huron Street
Chicago, IL 60611-2795

**ALA Ulrich's Serials Librarianship Award** 275
This award, donated by the CSA/Ulrich Company, is given for contributions to serials librarianship in areas of professional association participation, library education, serials literature, research, or development of tools leading to better understanding of the field of serials. This annual award consists of $1,500 and a citation. The winner is notified after the Midwinter meeting.
*Requirements:* Applicants must be members of the Association for Library Collections and Technical Services (ALCTS). They must show a record of outstanding contributions to serials and continuing resources librarianship over the course of a professional career. Nominations should include a resume and letters of recommendation.
*Restrictions:* Employees of the CSA/Ulrich Company are not eligible to apply.
*Geographic Focus:* All States
*Date(s) Application is Due:* Dec 1
*Amount of Grant:* 1,500 USD
*Contact:* Charles Wilt, Staff Liaison; (312) 280-5030; fax (312) 280-5033; cwilt@ala.org
*Internet:* www.ala.org/awardsgrants/awards/216/apply
*Sponsor:* American Library Association
50 East Huron Street
Chicago, IL 60611-2788

**Alavi Foundation Education Grants** 276
The Alavi Foundation is a private not-for-profit organization devoted to the promotion and support of Islamic culture and Persian language, literature and civilization. Since its inception in 1973, the Foundation has been an independent charitable organization based in the United States. The Foundation has been promoting projects for over 30 years by financially supporting charitable and philanthropic causes through educational, religious, and cultural programs. It supports nine (9) core areas: grants to colleges and universities; donations to Persian schools; donations and loans to Islamic organizations; free distribution of Islamic books; donations for disaster relief; support for the arts; scholarly research; interfaith dialogue and religious pluralism; and student loans.
*Requirements:* The Foundation's contributions are limited only to public charities that are tax exempt under IRS ruling.
*Geographic Focus:* All States
*Contact:* Contact; (212) 944-8333; fax (646) 619-4272; info@alavifoundation.org
*Internet:* alavifoundation.us/grantprograms/
*Sponsor:* Alavi Foundation
650 5th Avenue, Suite 2406
New York, NY 10019-6108

**ALA WESS De Gruyter European Librarianship Study Grant** 277

The De Gruyter Study Grant supports research in European studies with an emphasis on librarianship, the book trade, resource documentation, and similar information-science related topics. The grant funds travel to and from Europe, in addition to transportation, room, and board in Europe for up to thirty consecutive days. The purpose of the grant is to support research on the acquisition, organization, or use of library resources from or relating to Europe. Proposals are reviewed with the following in mind: the work to be accomplished; the need for and value of the proposal research; the research methodology; whether the work can be accomplished within the time frame proposed; and whether the applicant's qualifications are sufficient to carry out the research. Additional application information is available at the website.

*Requirements:* Each applicant must be a member of ACRL and employed as a librarian or information professional in a university, college, community college, or research library in the year prior to application for the award. The application must include a proposal; tentative travel itinerary of up to thirty days; travel budget; and current curriculum vitae. Applications and supporting materials must be emailed.

*Geographic Focus:* All States

*Date(s) Application is Due:* Dec 1

*Amount of Grant:* 2,500 EUR

*Contact:* Chase Ollis, Staff Liaison; (312) 280-2521 or (312) 944-6780; fax (312) 440-9374; collis@ala.org or ala@ala.org

*Internet:* www.ala.org/awardsgrants/awards/426/apply

*Sponsor:* American Library Association

50 East Huron Street

Chicago, IL 60611-2788

**Albert and Mary Lasker Foundation Awards** 278

The major purpose of the awards is to honor the individual(s) who have made significant contributions in basic or clinical research in diseases that are the main cause of death and disability. Four categories of awards are made. The Albert Lasker Basic Medical Research Award honors the scientist or scientists who have made fundamental investigations that open new areas of biomedical science. The Lasker-DeBakey Clinical Medical Research Award honors the scientist(s) whose contributions, directly or indirectly, have led to the improvement of the clinical management or treatment of patients and to the alleviation or elimination or one of the major medical causes of disability or death. The Lasker-Koshland Special Achievement Award in Medical Science honors a scientist whose contributions to research are of unique magnitude and immeasurable influence on the course of science, health, or medicine, and whose professional career has engendered extreme respect within the biomedical community. And the Lasker~Bloomberg Public Service Award honors men and women who have helped make possible the federal legislation and funding that supports research, and who have created public communication, public health, and advocacy programs of major importance. Nominations must be received by the listed application deadline.

*Geographic Focus:* All States

*Date(s) Application is Due:* Feb 1

*Amount of Grant:* 25,000 - 50,000 USD

*Samples:* Pierre Chambon, Institute of Genetics and Molecular and Cellular Biology (Strasbourg, France)—for the 2004 Lasker Award for Basic Medical Research, $50,000; Charles Kelman, International Retinal Research Foundation (Birmingham, AL)—for the 2004 Lasker Award for Clinical Medical Research, $50,000; Matthew Meselson, Harvard U (Cambridge, MA)—for the 2004 Lasker Award for Special Achievement in Medical Science, $25,000.

*Contact:* David Keegan, Senior Program Director; (212) 286-0222; fax (212) 286-0924; dkeegan@laskerfoundation.org

*Internet:* www.laskerfoundation.org/awards/index.htm

*Sponsor:* Albert and Mary Lasker Foundation

110 East 42nd Street, Suite 1300

New York, NY 10017

**Albert and Mary Lasker Foundation Clinical Research Scholars** 279

The Lasker Foundation and the National Institutes of Health have joined together in an innovative partnership to nurture the next generation of clinical researchers. The Lasker Clinical Research Scholars Program is an NIH-funded initiative providing up to 11 years of support for medical doctors to bridge the widening gap between cutting-edge research and improved patient care. Under the program, NIH Institutes will specify clinical research topics that merit support and accept applications each year from early-career investigators in those areas. Applications will be reviewed for scientific merit by a peer review group of scientists, convened by the NIH Center for Scientific Review, and make recommendations to the NIH scientific and clinical directors who will annually select outstanding scholars for placement on the NIH tenure-track. The NIH expects to appoint at least three Lasker Scholars in the first year of the program, and to ultimately reach a steady-state of 20-30 active awardees. The Lasker Foundation will assist with providing mentors for the scholars, fund travel costs for NIH lecturers and the attendance by scholars at the Lasker Awards program. The Foundation will also support annual meetings for presentations and discussions on topics related to the program. Scholars' work will receive regular evaluation at NIH. After five to seven years, each scholar will be reviewed again by a group of distinguished clinical investigators. By mutual agreement between the NIH and the candidate, a decision will be made as to whether the scholar will stay on as a tenured senior investigator at the NIH, or whether the candidate will be eligible for outside research grant support worth $500,000 a year plus overhead for four years.

*Geographic Focus:* All States

*Contact:* David Keegan, Senior Program Director; (212) 286-0222; fax (212) 286-0924; dkeegan@laskerfoundation.org

*Internet:* www.laskerfoundation.org/programs/index.htm

*Sponsor:* Albert and Mary Lasker Foundation

110 East 42nd Street, Suite 1300

New York, NY 10017

**Alcatel-Lucent Technologies Foundation Grants** 280

The Alcatel-Lucent Foundation is the philanthropic arm of Alcatel-Lucent and it leads the company's charitable activities. With a focus on volunteerism, the Foundation's mission is to support the commitment of Alcatel-Lucent to social responsibility by serving and enhancing the communities where its employees and customers live and work. To accomplish its mission, the Foundation manages grants and employee volunteerism on a global level. It receives its income from the corporation - Alcatel-Lucent - whose name it bears. However, legally the Foundation is an independent, charitable, non-profit and private entity and is governed by its own board of trustees that is separate from the corporate board of directors. Global Foundation grants are dedicated to the main focus areas of the Foundation and are managed by the Foundation.

*Requirements:* Giving is on an international basis.

*Geographic Focus:* All States, All Countries

*Amount of Grant:* 500 - 5,000 USD

*Contact:* Bishalakhi Ghosh, Executive Director; +91-99-58418547 or +91-22-66798700; fax +91-22-26598542; bishalakhi.ghosh@alcatel-lucent.com

*Internet:* www.alcatel-lucent.com/wps/portal/foundation

*Sponsor:* Alcatel-Lucent Technologies Foundation

600 Mountain Avenue, Room 6F4

Murray Hill, NJ 07974-2008

**Alcon Foundation Grants** 281

The foundation supports organizations in the fields of health care, leadership programs, research, education and community responsibility. Programs that advance the education and skill levels of eye care professionals are given special consideration. General operating grants are given to organizations and institutions improving education and research in the areas of specialization of Alcon Laboratories ophthalmology and vision care. Grants are also awarded to community activities that benefit company employees. Applications may be submitted at any time.

*Requirements:* The Alcon Foundation awards grants to qualified, U.S.-based tax-exempt public 501(c)3 charities that provide high-quality programs and services, well-defined goals, a program commitment to maximizing available resources, and a reputation for meeting objectives and reporting measured results.

*Restrictions:* The Alcon Foundation does not support the following: non-501(c)3 designated organizations; individual or family requests for scholarships, fellowship assistance or other types of support; religious-based activities for the purpose of furthering religious doctrine; fraternal, labor, political or veterans organizations; organizations that discriminate on the basis of race, religion, creed, gender or national origin; matching gifts; university administrative, management or indirect fees; golf tournaments, athletic events, or league or team sponsorships; school-affiliated orchestra, band, or choir student trips or tours; public and private K-12 schools; unrestricted grants; books, research papers or articles in professional journals; travel expenses; or fundraising activities or advertising sponsorships.

*Geographic Focus:* All States

*Amount of Grant:* 100 - 50,000 USD

*Contact:* Melissa Thompson, Interim President; (800) 222-8103; fax (817) 615-3811

*Internet:* www.alconlabs.com/corporate-responsibility/alcon-foundation.asp

*Sponsor:* Alcon Foundation

1 Health Plaza, Building 200

East Hanover, NJ 07936-1016

**Aldo and Jeanne Scaglione Prize for a Translation of a Literary Work** 282

The Committee on Honors and Awards of the Modern Language Association invites translators and publishers to compete for the sixteenth Aldo and Jeanne Scaglione Prize for a Translation of a Literary Work. The prize, established by the Aldo and Jeanne Scaglione Endowment Fund, is awarded for an outstanding translation into English of a book-length literary work. For the competition held in this year, translations published in the previous year are eligible. Translators need not be members of the association. To enter a translation into the competition, authors or publishers should send six copies along with brief biographical and professional data about the translator; a total of 12–15 pages of the text in its original language taken from the beginning, middle, and end of the work; and a letter identifying the work. In the case of a born-digital publication, send the relevant media or URL and a letter identifying the work to the email listed. Publishers may submit more than one title. Works submitted for this prize will also be automatically considered for the Lois Roth Award. Definitions and criteria are the same for both awards. Separate submissions are not required. Entries will be accepted until May 1 each year. The winning author will be contacted in September, and a public announcement will be made in early December. The prize, which consists of a cash award and a certificate, will be presented to the winning author at the association's annual convention in January.

*Geographic Focus:* All States

*Date(s) Application is Due:* May 1

*Amount of Grant:* 2,000 USD

*Samples:* Linda Coverdale, Brooklyn, New York, for Slave Old Man, $2,000 (2018); Alissa Valles, Boston University, for Our Life Grows, $2,000; Jeffrey Angles, Western Michigan University, for The Book of the Dead, $2,000.

*Contact:* Dennis Looney, Director of Programs; (646) 576-5000; fax (646) 458-0030; awards@mla.org

Annie M. Reiser, Coordinator of Book Prizes and Administrative Assistant; (646) 576-5141 or (646) 576-5000; fax (646) 458-0030; awards@mla.org
*Internet:* https://www.mla.org/Resources/Career/MLA-Grants-and-Awards/Award-Submissions-and-Nominations/Competitions-for-MLA-Publication-Awards/Annual-Prizes-with-Competitions-in-2020/Aldo-and-Jeanne-Scaglione-Prize-for-a-Translation-of-a-Literary-Work
*Sponsor:* Modern Language Association
85 Broad Street, Suite 500
New York, NY 10004-2434

**Alexander von Humboldt Foundation Anneliese Maier Award** 283

Nominations for the Anneliese Maier Award can be made for researchers from abroad in the humanities and social sciences. Researcher achievements should be internationally recognized, along with research collaboration with specialist colleagues in Germany. A sustainable contribution is expected by these researchers toward the further internationalization of the humanities and social sciences in Germany. The Award is also aimed at researchers who are not yet so advanced in their scientific careers, but who are already known in the shaping of the humanities and social sciences in Germany. Award winners are expected to spend up to five years cooperating on a long-term research project with the nominator and/or specialist colleagues at a research institution in Germany. Up to eight Maier Research Awards are given annually, and are valued at 250,000 EUR each. Additional information is available at the website.
*Requirements:* Nominations may be submitted by established academics in Germany. The nominees must be recognized as established leaders in their subject. Researchers who are not yet advanced, but internationally established may also be recognized. One international expert and one German expert review by collaborative partners and/or academics are necessary.
*Restrictions:* Direct applications are not accepted.
*Geographic Focus:* All States, All Countries
*Date(s) Application is Due:* Apr 30
*Amount of Grant:* 250,000 EUR
*Contact:* Dr. Gerhard Lauer; (+49) 0228-833-0; fax (+49) 0228-833-212; info@avh.de
*Internet:* www.humboldt-foundation.de/web/anneliese-maier-award.html
*Sponsor:* Alexander von Humboldt Foundation
Jean-Paul-Street 12
Bonn, D-53173 Germany

**Alexander von Humboldt Foundation CAPES-Humboldt Fellowships for Experienced Researchers** 284

The CAPES-Humboldt Fellowships for Experienced Researchers enable highly qualified scientists from Brazil to spend extended periods of research (6-18 months) in Germany. Research projects are carried out in cooperation with academic hosts at research institutions in Germany. Candidates are expected to have their own, clearly defined research profile. The Fellowship is worth 3,100 EUR per month and includes the following benefits: an allowance for research costs; language fellowship for the Fellow and marital partner at selected language institutes in Germany immediately prior to the actual period of research; family and child allowances; a Europe allowance for a research stay at a research institute if necessary to ensure the success of the research project; and an extensive alumni sponsorship once the successful research stay has come to an end. A candidate's academic eligibility is assessed on the following criteria: academic record and performance; quality of key publications; quality of research plan; and the academic's future potential. Additional information is available at the website.
*Requirements:* Candidates should be working at the level of assistant professor, junior research group leader, or be able to document independent research work over a number of years. Scientists and scholars from all disciplines may apply. Applicants must hold a doctorate from Brazil or abroad with an official revalidation in Brazil completed less than twelve years prior to the application date. They must also hold Brazilian nationality or permanent residence visa in Brazil, with proven teaching or research activity in Brazil. Natural scientists and engineers must have a good knowledge of German or English. Physicians and researchers from humanities and social sciences must have a good knowledge of German if is necessary to carry out the project (otherwise a good knowledge of English is required). Eligibility requirements also include confirmation that research facilities are available, with a mentoring agreement and detailed expert's report by the academic host.
*Restrictions:* Research already funded in other programs cannot be financed again within the CAPES Fellowships. Potential applicants who have been in Germany for more than six months at the time of application are not eligible.
*Geographic Focus:* Brazil
*Date(s) Application is Due:* Jul 31; Dec 31
*Amount of Grant:* 37,200 EUR
*Contact:* Denise de Menezes Neddermeyer, Directorate of International Relations; (61) 2022-6900; fax (61) 2022-6902; dri@capes.gov.br
*Internet:* www.humboldt-foundation.de/web/capes-humboldt-fellowship.html
*Sponsor:* Alexander von Humboldt Foundation
Jean-Paul-Street 12
Bonn, D-53173 Germany

**Alexander von Humboldt Foundation CAPES-Humboldt Fellowships for Postdoctoral Researchers** 285

The CAPES-Humboldt Fellowships for postdoctoral researchers enable highly qualified scientists from Brazil to spend extended periods of research (6-24 months) in Germany. The Fellowship is worth 2,600 EUR per month. Additional benefits include an allowance for research costs, a language stipend for the Fellow and marital partner at selected language institutes in Germany, family and child allowances, and extensive alumni sponsorship. Scientists and scholars from all disciplines may apply, and there are no quotas for fields of disciplines. Candidate selection is based on academic record and performance; quality of key publications; quality of research plan; and the applicant's future potential and career prospects. The application documents and additional information are available at the website.
*Requirements:* Postdoctoral researchers must be scientists or scholars from Brazil at the beginning of their career with above average qualifications, who have earned a doctoral degree within the last four years. Candidates choose their own research projects and their host in Germany, and prepare their own research plan. Details of the research project and time schedule must be agreed upon with the prospective host in advance. They must also have an academic publication file, with proven teaching or research activity in Brazil, along with Brazilian nationality or permanent residence visa in the country. Applicants must confirm that research facilities are available. A mentoring agreement with a detailed expert's report and proposed timetable should be discussed within the application. Natural scientists and engineers must have a good knowledge of German or English; physicians and researchers in the humanities and social sciences must have a good knowledge of German if it is necessary to carry out the project successfully; otherwise a good knowledge of English is required.
*Restrictions:* Short term visits for study and training purposes or for attending conferences are not eligible. Potential applicants who have been in Germany for more than six months are not eligible to apply. Research already funded in other programs cannot be financed again with the CAPES Fellowship.
*Geographic Focus:* Brazil
*Date(s) Application is Due:* Jul 31; Dec 31
*Amount of Grant:* 31,200 EUR
*Contact:* Denise de Menezes Neddermeyer, Directorate of International Relations; (61) 2022-6900; fax (61) 2022-6902; dri@capes.gov.br or humboldt@capes.gov.br
*Internet:* www.humboldt-foundation.de/web/capes-humboldt-fellowship.html
*Sponsor:* Alexander von Humboldt Foundation
Jean-Paul-Street 12
Bonn, D-53173 Germany

**Alexander von Humboldt Foundation Feodor-Lynen Research Fellowships for Experienced Researchers** 286

The Feodor Lynen Research Fellowships for experienced researchers allows a researcher to carry out a long-term project (6 to 18 months) selected in cooperation with an academic host at a research institution in Germany. Candidates choose their own research projects and their host abroad and prepare their own research plan. The fellowship is flexible and can be divided into as many as three stays within three years. The Foundation tries to involve the host in financing the research fellowship. Scientists and scholars from all disciplines may apply to the Foundation directly at any time. The provisions of the fellowship comprise a basic monthly sum and a monthly foreign allowance which may differ according to location and marital status. Marital and child allowances may be paid in addition. Additional benefits may include: return travel expenses; a reintegration allowance; a return fellowship; and an alumni sponsorship. Applicants can apply if he/she has been living in Germany continuously for at least five years at the time of application and Germany is their principal country of residence and work. The Foundation awards up to 100 Fellowships for postdoctoral researchers and experienced researchers annually. There is no closing date for submitting applications.
*Requirements:* Applicants must furnish proof of work published in recognized academic journals; submit a research plan agreed with the foreign host, including confirmation that research facilities can be made available; and possess a good knowledge of the host country's language or at least a very good knowledge of English, provided they can carry out their research project in that language. The host must be an academic working abroad who has already been sponsored by the Humboldt Foundation.
*Geographic Focus:* All States, Germany
*Amount of Grant:* 33,600 - 52,800 EUR
*Contact:* Dr. Anne Bohnenkamp-Renken, Chairperson; (+49) 0228-833-0; fax (+49) 0228-833-212; info@avh.de
*Internet:* www.humboldt-foundation.de/web/lynen-fellowship.html
*Sponsor:* Alexander von Humboldt Foundation
Jean-Paul-Street 12
Bonn, D-53173 Germany

**Alexander von Humboldt Foundation Feodor-Lynen Research Fellowships for Postdoctoral Researchers** 287

Feodor Lynen Research Fellowships for postdoctoral researchers are the instrument with which the Alexander von Humboldt Foundation enables highly-qualified scientists and scholars from Germany who are just embarking on their academic careers and who completed their doctorates less than four years agol to spend extended periods of research (6-24 months) abroad. Applicants can apply if he/she has been living in Germany continuously for at least five years at the time of application and Germany is the candidate's principal country of residence and work. Scientists and scholars from all disciplines may apply for any target country abroad. Candidates choose their own research projects and their host abroad and prepare their own research plan. The provisions of the fellowship comprise a basic monthly sum and a monthly foreign allowance which may differ according to location and marital status. Marital and child allowances may be paid in addition. Additional benefits may include: return travel expenses; reintegration allowance; a return fellowship; and alumni sponsorship. There is no closing date for submitting applications.
*Requirements:* Applicants must be less than 38 years old; have successfully completed their doctorate with very good or good ratings; furnish proof of work published in recognized academic journals; submit a research plan agreed with the foreign host, including confirmation that research facilities can be made available; and possess a good

knowledge of the host country's language or at least a very good knowledge of English, provided they can carry out their research project in that language. The academic host must be an academic working abroad who has already been sponsored by the Humboldt Foundation. Details of the research project and the time schedule must be agreed upon with the prospective host in advance.
*Restrictions:* Short-term visits for study and training purposes or for attending conferences are not eligible for sponsorship.
*Geographic Focus:* All States, Germany
*Amount of Grant:* 32,600 - 52,800 EUR
*Contact:* Dr. Anne Bohnenkamp-Renken, Chairperson; (+49) 0228-833-0; fax (+49) 0228-833-212; info@avh.de
*Internet:* www.humboldt-foundation.de/web/lynen-fellowship.html
*Sponsor:* Alexander von Humboldt Foundation
Jean-Paul-Street 12
Bonn, D-53173 Germany

**Alexander von Humboldt Foundation Fraunhofer-Bessel Research Awards 288**
The Fraunhofer-Gesellschaft for the Promotion of Applied Research and the Alexander von Humboldt Foundation grant up to three research awards annually to internationally renowned academics from all non-European countries in recognition of their outstanding academic performance in applied research to date. Scientists and scholars from all non-European countries, who have completed their doctorates less than twelve years ago, who are internationally recognized for their achievements in the field of applied research, and who in future are expected to continue producing cutting-edge achievements which will have a seminal influence on their discipline beyond their immediate field of work, are eligible to be nominated for an Award. Winners are invited to spend a period of up to one year cooperating on a long-term research project with specialist colleagues at one of the Fraunhofer Institutes in Germany. The stay may be divided up into blocks. The Foundation and the Fraunhofer Society for the Advancement of Applied Research jointly grant up to three Awards annually. The award is valued at 45,000 EUR.
*Requirements:* Academics of all disciplines can be nominated for the Fraunhofer-Bessel Research Award, provided that they have received their doctoral degree within the past 18 years. The Foundation particularly encourages the nomination of qualified female academics. At the time of nomination, the candidate must have been predominantly living and working abroad for at least five years. Nominations may be submitted by the heads of institutes or senior academics at institutes belonging to the Fraunhofer Society. Direct applications are not accepted.
*Restrictions:* The Fraunhofer-Bessel Research Award can only be granted once to any one academic. Researchers who have already received an award for their academic achievement from the Humboldt Foundation are not eligible to be nominated. Furthermore, the nominee should not have recently won any awards or fellowships in Germany for his or her academic achievements.
*Geographic Focus:* All States, Germany
*Amount of Grant:* 45,000 EUR
*Contact:* Dr. Martin Carrier, Chairperson; (+49) 0891-205-4710 or (+44) 0207-882-2218; fax (+49) 0228-833-212; info@avh.de or m.j.carrier@qmul.ac.uk
*Internet:* www.humboldt-foundation.de/web/fraunhofer-bessel-award.html
*Sponsor:* Alexander von Humboldt Foundation
Jean-Paul-Street 12
Bonn, D-53173 Germany

**Alexander von Humboldt Foundation Friedrich Wilhelm Bessel Research Awards 289**
The Alexander von Humboldt Foundation awards annually about 25 Friedrich Wilhelm Bessel Research Awards, funded by the Federal Ministry for Education and Research, to internationally renowned scientists and scholars from abroad in recognition of their outstanding accomplishments in research to date and their exceptional promise for the future. Scientists and scholars of all disciplines or research areas, regardless of their nationality, may be nominated for an Award, provided they have received their doctoral degree within the past 12 years. The Foundation particularly encourages the nomination of female scholars. In addition, awardees are invited to conduct an original research project of their own design in close collaboration with an appropriate colleague in Germany over a period of six to twelve months, which may be divided into segments. The award is valued at 45,000 Euro, and nominations may be submitted by regular mail at any time..
*Requirements:* A candidate must be nominated by distinguished scientists/scholars employed by any university or research institution in Germany. Direct applications are not accepted. Scientists and scholars who are nominated must have attained international reputation through research in their chosen field; they must also show promise that their further research will remain on the cutting edge and have impact beyond their area of specialization.
*Geographic Focus:* All States, All Countries, Germany
*Amount of Grant:* 45,000 EUR
*Contact:* Dr. Martin Carrier; (+49) 0228-833-0; fax (+49) 0228-833-212; info@avh.de
*Internet:* www.humboldt-foundation.de/web/bessel-award.html
*Sponsor:* Alexander von Humboldt Foundation
Jean-Paul-Street 12
Bonn, D-53173 Germany

**Alexander von Humboldt Foundation Georg Forster Fellowships for Experienced Researchers 290**
In providing Georg Forster Research Fellowships for experienced researchers, the Alexander von Humboldt Foundation enables highly-qualified scientists and scholars from abroad, who completed their doctorates less than twelve years ago to spend extended periods of research (6-18 months; may be divided up into a maximum of three blocks) in Germany. Candidates are expected to have their own, clearly defined research profile. This means they should usually be working at least at the level of Assistant Professor or Junior Research Group Leader or be able to document independent research work over a number of years. The fellowship is worth 3,150 EUR per month. This includes a mobility lump sum and a contribution towards health and liability insurance. Additional benefits may include: return travel expenses; language fellowship; family allowances; additional extension; a subsidy towards research costs; a Europe allowance; and an extensive alumni sponsorship. There is no closing date for submitting applications. Applications are processed as part of an ongoing procedure.
*Requirements:* Scientists and scholars from all disciplines from developing countries, emerging economies and transition states may apply. Applicants must have completed their doctorates less than twelve years ago. Scholars in the humanities or social sciences and physicians must have a good knowledge of German if it is necessary to carry out the project successfully; otherwise a good knowledge of English; scientists and engineers must have a good knowledge of German or English.
*Restrictions:* Short-term study tours, participation in conferences, or educational visits cannot be funded. Applications for extension of research stays already commenced in Germany cannot be considered.
*Geographic Focus:* All Countries
*Date(s) Application is Due:* Jun 15
*Amount of Grant:* Up to 37,800 EUR
*Contact:* Dr. Bernhard Fleischer, Chairperson; (+49) 0228-833-0 or (+49) 4042-818-401; fax (+49) 0228-833-212; info@avh.de or bernhard.fleischer@t-online.de
*Internet:* www.humboldt-foundation.de/web/georg-forster-fellowship-experienced.html
*Sponsor:* Alexander von Humboldt Foundation
Jean-Paul-Street 12
Bonn, D-53173 Germany

**Alexander von Humboldt Foundation Georg Forster Fellowships for Postdoctoral Researchers 291**
Fellowships for postdoctoral researchers are the instrument with which the Foundation enables highly-qualified scientists and scholars who are just embarking on their academic careers and who completed their doctorates less than four years ago to spend extended periods of research (6-24 months) in Germany. The research proposal should deal with issues of major relevance to the future development of the candidate's country of origin and, in this context, particularly suited to transferring knowledge and methods to developing and threshold countries. Research projects are carried out in cooperation with academic hosts at research institutions in Germany. Candidates choose their own research projects and their host in Germany and prepare their own research plan. The fellowship is worth 2,650 euro per month. This includes a mobility lump sum and a contribution towards health and liability insurance. There is no closing date for submitting applications. Applications are processed as part of an ongoing procedure.
*Requirements:* Scientists and scholars from all disciplines from developing and threshold countries (excluding People's Republic of China, India and Turkey) may apply.
*Geographic Focus:* All States
*Date(s) Application is Due:* Jun 15
*Amount of Grant:* 31,800 EUR
*Contact:* Dr. Bernhard Fleischer, Chairperson; (+49) 0228-833-0 or (+49) 4042-818-401; fax (+49) 0228-833-212; info@avh.de or bernhard.fleischer@t-online.de
*Internet:* www.humboldt-foundation.de/web/georg-forster-research-fellowship-postdoc.html
*Sponsor:* Alexander von Humboldt Foundation
Jean-Paul-Street 12
Bonn, D-53173 Germany

**Alexander von Humboldt Foundation Georg Forster Research Award 292**
The Georg Forster Research Award is granted in recognition of a researcher's entire achievements to academics of all disciplines whose fundamental discoveries, new theories, or insights have had a significant impact on their own discipline and beyond, and who are expected to continue developing research-based solutions to the specific challenges facing transition and developing countries. Up to four awards are offered every year at 60,000 EUR each. Award winners are invited to conduct a research project of their own choosing in Germany in close collaboration with a specialist colleague. The project duration of about six to twelve months may be divided into segments. In addition, further expenditures (such as travel or language courses) are covered in the context of the invitation to stay in Germany for research purposes. To support the collaboration, the Alexander von Humboldt Foundation may grant additional funding of up to 25,000 EUR, particularly for participating in scientific conferences, additional material resources, such as specialist literature and scientific equipment for the nominee's own institute, as well as for involving junior researchers. An established academic employed by a university or research institution in Germany must take the initiative to nominate a candidate for the Georg Forster Research Award. Nominations may also be initiated by award winners of the Humboldt Foundation working abroad, provided that the nomination is made jointly with a colleague working in Germany. Nomination information is posted on the website.
*Requirements:* Nominees must be nationals of a developing or transition country (a full list of qualifying countries is available at the website). At the time of nomination, they must have had their main residence and place of work in one of these countries for at least five years. They must be recognized internationally, and continue developing research based solutions for the challenges facing transition and developing countries.
*Restrictions:* Self nomination is not acceptable.
*Geographic Focus:* All States, All Countries
*Date(s) Application is Due:* Jan 15
*Amount of Grant:* Up to 85,000 EUR
*Contact:* Dr. Bernhard Fleischer; (+49) 0228-833-0; fax (+49) 0228-833-212; info@avh.de

*Internet:* www.humboldt-foundation.de/web/georg-forster-award.html
*Sponsor:* Alexander von Humboldt Foundation
Jean-Paul-Street 12
Bonn, D-53173 Germany

**Alexander von Humboldt Foundation German Chancellor Fellowships for Prospective Leaders from the People's Republic of China** 293

The Alexander von Humboldt Foundation's German Chancellor Fellowships are for university graduates from the People's Republic of China with an interest in international issues and demonstrated leadership potential. Up to ten Fellowships are given annually. Fellows are mentored by hosts in Germany chosen by the Fellows on the basis of their expertise in the respective areas. The program is targeted at accomplished young professionals who are likely to become decision-makers, thought leaders, and influential voice in their respective fields. During their time in Germany, Fellows conduct independent projects at their host institutions. Projects should be relevant to modern societies, have a long term and visible impact, and help to advance Fellow's careers and professional development. It should also involve original exploration of a topic or issue or research in the Fellow's field of interest. Through their experience in Germany, Fellows will acquire greater knowledge of their fields, gain new international experience, and strengthen their intercultural competence. In addition to a stipend, Fellows also receive extensive benefits which are discussed in detail in the guidelines at the website. The application is also available at the website.
*Requirements:* Applicants may apply from a broad range of areas such as politics and public policy, law, media, business, the non-governmental sector, and the arts. Candidates will spend one year in Germany where they will network with other prospective leaders from abroad to explore new solutions to global issues. Candidates must meet the following *Requirements:* citizenship from the People's Republic of China; a bachelor's or comparable academic degree completed less than twelve years prior to the Fellowship's starting date; demonstrated leadership potential; either English or German language skills; a project plan detailing the project's subject matter, goals, and a time plan; a detailed host's statement, including a confirmation of support; and two to three expert recommendations from those qualified to comment on the candidate's professional, personal, and/or academic background.
*Geographic Focus:* China
*Date(s) Application is Due:* Sep 15
*Amount of Grant:* 25,800 - 33,000 EUR
*Contact:* Jianhua Zhu, Selection Committee Chairperson; (+49) 0228-833-0; fax (+49) 0228-833-212; chancellor-chn.select@avh.de or buka@avh.de
*Internet:* www.humboldt-foundation.de/web/4074.html
*Sponsor:* Alexander von Humboldt Foundation
Jean-Paul-Street 12
Bonn, D-53173 Germany

**Alexander von Humboldt Foundation German Chancellor Fellowships for Prospective Leaders from the Russian Federation** 294

The Alexander von Humboldt Foundation's German Chancellor Fellowships are for university graduates from the Russian Federation with an interest in international issues and demonstrated leadership potential. Up to ten Fellowships are given annually. Fellows are mentored by hosts in Germany, whom the Fellows have chosen on the basis of their expertise in their respective areas. The program is targeted at accomplished young professionals who are likely to become decision-makers, thought leaders, and influential voices in their respective fields. During their time in Germany, Federal Chancellor Fellows conduct independent projects at their host institutions. Projects should be of relevance to modern societies, have a long-term and visible impact, and help to advance Fellows' careers and professional development. It should also involve original exploration of a topic or issue, or research in the Fellow's respective field of interest. Through their experience in Germany, Fellows will acquire greater knowledge of their fields, gain new international experience, and strengthen their intercultural competence. In addition to a stipend, Fellows also receive extensive benefits which are discussed in detail in the guidelines at the website. The application is also available at the website.
*Requirements:* Applicants may apply from a broad range of areas such as politics and public policy, law, media, business, the non-governmental sector, and the arts. Candidates will spend one year in Germany where they will network with other prospective leaders from abroad to explore new solutions to global issues. Candidates must meet the following *Requirements:* citizenship from the Russian Federation; a bachelor's or comparable academic degree, completed less than twelve years prior to the starting date of the Fellowship; demonstrated leadership potential; either English or German language skills; a project plan detailing the project's subject matter, goals and a time plan; a detailed host's statement, including a confirmation of support; and two or three expert recommendations from those qualified to comment on the candidate's professional, personal, and/or academic background.
*Geographic Focus:* Russian Federation
*Date(s) Application is Due:* Sep 15
*Amount of Grant:* 25,800 - 33,000 EUR
*Contact:* Leonid Zhmud, Selection Committee Chairperson; (+49) 0228-833-0; fax (+49) 0228-833-212; info@avh.de
*Internet:* www.humboldt-foundation.de/web/4074.html
*Sponsor:* Alexander von Humboldt Foundation
Jean-Paul-Street 12
Bonn, D-53173 Germany

**Alexander von Humboldt Foundation Hezekiah Wardwell Fellowships** 295

Up to ten Hezekiah Wardwell Fellowships are offered annually to young, highly-gifted Spanish musicians and musicologists for further training or advanced studies, primarily in Germany. The Fellowship is aimed at candidates from Spain in transition to their professional career. Fellows are supervised by recognized music teachers at colleges of music or conservatories, or by specialist scholars at universities. The Foundation expects candidates to establish contact with supervisors prior to submitting applications, and to obtain binding clarification for acceptance and academic supervision. Fellowships offer EUR 800 per month. In addition, the Foundation pays a single lump sum on arrival as well as travel expenses to Fellows traveling directly from Spain. Initially, applications can be made for a Fellowship lasting two academic semesters or ten months. Additional information is available at the website.
*Requirements:* Applicants should have a good knowledge of German. The Fellowship period must begin no later than three years from the end of the applicant's graduation. Preference will be given to candidates who have previously had no opportunity to continue their music studies in Germany.
*Geographic Focus:* Spain
*Date(s) Application is Due:* Jan 10
*Amount of Grant:* 8,250 - 16,250 EUR
*Contact:* Paulo Sergio Guimares Alvares, Fellowship Committee Chairperson; (+49) 0228-833-0; fax (+49) 0228-833-199; info@avh.de
*Internet:* www.humboldt-foundation.de/web/wardwell-es.html
*Sponsor:* Alexander von Humboldt Foundation
Jean-Paul-Street 12
Bonn, D-53173 Germany

**Alexander von Humboldt Foundation International Climate Protection Fellowships** 296

The International Climate Protection Fellowships enable prospective leaders to conduct a research-related project of their own choice during a one-year stay in Germany. Up to 15 Fellowships are awarded. Any prospective leader from a non-European or developing country working the climate protection and resource conservation in academia, business, or administration may apply. A complete list of countries is available at the website (Africa, East Asia and the Pacific, Eurasia, Near East, South and Central Asia, Latin America, and the Caribbean). The Fellowship will enable the recipients to conduct a research-related project of their own choice with hosts in Germany that they have chosen. The program includes a preparatory intensive language course in Germany, a several week introductory seminar, a several day training course, and a final meeting in Berlin. These activities will enable Fellows to gain additional insights into academic, social, cultural, economic and political life in Germany. Applicants who have been trained in the natural and engineering sciences, and candidates who have been engaged in legal, economic and societal issues relating to climate change are encouraged to apply. The application and additional information, including specific information about the Fellowship stipend and funding allowances, are available at the website.
*Requirements:* Applicants must provide a clearly visible leadership potential either by experience in a first leadership position or be able to provide appropriate references. They must also have completed their first university degree less than 12 years prior to the start of the Fellowship with outstanding results. They must also have extensive professional experience in a leadership role (at least 48 months at the time of application) or hold a further academic or professional qualification. Fellows are expected to have gained initial practical experience through involvement in projects related to climate protection and resource conservation.
*Geographic Focus:* All Countries
*Date(s) Application is Due:* Dec 1
*Contact:* Francois Claude Buscot, Selection Committe Chairperson; (+49) 0228-833-0; fax (+49) 0228-833-199; info@avh.de
*Internet:* www.humboldt-foundation.de/web/icf.html
*Sponsor:* Alexander von Humboldt Foundation
Jean-Paul-Street 12
Bonn, D-53173 Germany

**Alexander von Humboldt Foundation JSPS Research Fellowships** 297

The Alexander von Humboldt Foundation JSPS Research Fellowships assist promising, highly-qualified, young foreign researchers from Germany who hold a doctoral degree. These Fellowships allow researchers to pursue collaborative research under the leadership of Japanese host researchers in Japanese universities and research institutes. This allows them to advance their own research while promoting scientific advancement in Japan and the counterpart countries through close collaboration in scientific activities. JSPS offers two funding lines for a stay in Japan. JSPS Postdoctoral Fellowships (short-term) provide support for doctoral candidates and post-doctoral researchers of all disciplines to spend 1-12 months working on research in Japan. For sponsorship periods exceeding six months, the Humboldt Foundation preselects post-doctoral candidates on behalf of JSPS. Up to 20 JSPS Postdoctoral Fellowships are granted annually. Research stays lasting between 12 and 24 months are funded in the framework of JSPS Postdoctoral Fellowships (standard). Pre-selection of candidates for this funding line is conducted exclusively by the Alexander von Humboldt Foundation. Up to 16 JSPS Postdoctoral Fellowships (standard) are granted annually. Fellows receive 362,000 EUR per month, plus a 200,000 EUR relocation allowance. The application and additional detailed guidelines are available at the website.
*Requirements:* All fields of the humanities, social sciences, and natural sciences may apply. Each candidate for the Fellowship must also meet the following requirements to be eligible: possess either the nationality, citizenship or permanent residency of the nominating country; hold a doctorate degree which must have been received within six

years prior to April 1 of the Fellowship year; and have arranged in advance a research plan with his/her Japanese host.
*Restrictions:* Short-term study visits, participation in congresses and training courses are not eligible for funding. Medical doctors without a Ph.D. are not eligible. Persons of Japanese nationality are not eligible to apply, nor are those of dual nationality if one is Japanese.
*Geographic Focus:* Germany
*Amount of Grant:* 362,000 EUR
*Contact:* Matthias Kleiner, Selection Committee Chairperson; (+49) 0228-833-0; fax (+49) 0228-833-199; info@avh.de
*Internet:* www.humboldt-foundation.de/web/jsps-fellowship-postdoc.html
*Sponsor:* Alexander von Humboldt Foundation
Jean-Paul-Street 12
Bonn, D-53173 Germany

**Alexander von Humboldt Foundation Konrad Adenauer Research Award 298**
The Konrad Adenauer Research Award is made annually to one Canadian scholar in the humanities or social sciences. The aim of the Award is to promote academic relations between Canada and Germany. The Award is given to a highly qualified Canadian scholar whose research work in the humanities or social sciences has brought international recognition, and who belongs to a group of leading scholars in the respective area of specialization. The Awardee will carry out research work at German research institutes for up to one year. The winner is included in the Foundation's sponsorship program, and will be invited to take part in all functions arranged by the Foundation. The amount of the Award includes travel for Awardee and family members. Additional information is available at the website.
*Requirements:* Candidates are nominated by their universities for this award. Nomination should include a letter from the candidate's institution regarding the candidate's academic qualifications; a brief statement of the candidate's research proposal; three references who can provide information on the candidate's academic qualifications; and the names of German scholars who would provide guidance to the scholar during his/her stay in Germany.
*Restrictions:* Self-nominations will not be accepted. Applications will not be considered if the candidate holds or has held another award from the Humboldt Foundation within the previous three years.
*Geographic Focus:* Canada
*Date(s) Application is Due:* Jan 31
*Amount of Grant:* Up to 50,000 EUR
*Contact:* Barbara Konig; (+49) 0228-833-0; fax (+49) 0228-833-199; info@avh.de
*Internet:* www.humboldt-foundation.de/web/31764.html
*Sponsor:* Alexander von Humboldt Foundation
Jean-Paul-Street 12
Bonn, D-53173 Germany

**Alexander von Humboldt Foundation Max Planck Research Award 299**
The Max Planck Award is conferred upon exceptionally highly-qualified international scientists. The Award is given to one international researcher and one researcher working in Germany every other year. The researchers are expected to have already achieved international recognition and to continue to produce outstanding academic results in international collaboration. On an annually-alternating basis, the call for nominations addresses areas within the natural and engineering sciences, the life sciences, and the humanities. Recipients of the Award will be granted funding of up to 750,000 EUR over a period of three to a maximum of five years in order to pursue research of their own choosing. Nomination information is available at the website.
*Geographic Focus:* All States, All Countries
*Date(s) Application is Due:* Jan 31
*Amount of Grant:* Up to 750,000 EUR
*Contact:* Michaela Kreilos, Selection Department Coordinator; (+49) 0228-833-0; fax (+49) 0228-833-212; michaela.kreilos@avh.de
*Internet:* www.humboldt-foundation.de/web/6674.html
*Sponsor:* Alexander von Humboldt Foundation
Jean-Paul-Street 12
Bonn, D-53173 Germany

**Alexander von Humboldt Foundation NSC Taiwan Fellowships 300**
The NSC Taiwan Fellowships assist promising, highly-qualified, young foreign researchers who have held a doctoral degree for less than twelve years. This program allows researchers to pursue collaborative research at Taiwan universities and research institutes. This activity allows them to advance their own research while promoting scientific advancement in Taiwan and the counterpart countries through close collaboration in scientific activities. All fields of humanities, social sciences, and natural sciences are eligible. Up to two Fellowships are granted each year. The application and additional guidelines are available at the website.
*Requirements:* Applicants must be a post-doctoral researcher from Germany. They must also have completed their doctorate within the last twelve years. Candidates must have a good knowledge of Chinese or English, and have academic publications. They must also have confirmation that research facilities are available, with a detailed expert's report from an academic host from a Taiwan institute.
*Geographic Focus:* Germany
*Contact:* Program Coordinator; (+49) 0228-833-0; fax (+49) 0228-833-199; info@avh.de
*Internet:* www.humboldt-foundation.de/web/3626.html
*Sponsor:* Alexander von Humboldt Foundation
Jean-Paul-Street 12
Bonn, D-53173 Germany

**Alexander von Humboldt Foundation Philipp Franz von Siebold Award for Japanese Researchers 301**
The Philipp Franz von Siebold Award is given annually to a Japanese academic for outstanding services to enhancing mutual understanding of culture and society in Germany and Japan. The award winner will be invited to spend up to a year researching in Germany. The prize is valued at 50,000 EUR. Nominations are accepted from presidents of Japanese universities, former Siebold award winners, the heads of the Goethe Institutes in Japan, and the German Institute for Japanese Studies, as well as the German Ambassador in Tokyo and the German Consul General in Osaka. Humboldt Fellows and award winners may make nominations via the Vice Chancellor of a Japanese university. A decision on the award is made by a special committee under the chairmanship of the German Ambassador in Japan.
*Requirements:* The scientist or scholar should be under 50 years of age, and have a working knowledge of German.
*Restrictions:* Self nominations are not accepted.
*Geographic Focus:* Japan
*Amount of Grant:* 50,000 EUR
*Contact:* Barbara Konig; (+49) 0228-833-0; fax (+49) 0228-833-199
DAAD Tokyo Office; (03) 3582-5962; fax (03) 3582-5554; daad-tokyo@daadjp.com
*Internet:* www.humboldt-foundation.de/web/31771.html
*Sponsor:* Alexander von Humboldt Foundation
Jean-Paul-Street 12
Bonn, D-53173 Germany

**Alexander von Humboldt Foundation Professorships 302**
Academics of all disciplines from abroad, who are internationally recognized as leaders in their field and who are expected to contribute to enhancing Germany's sustained international competitiveness as a research location in consequence of the award, are eligible to be nominated. The Professorship, which is being financed by the Federal Ministry of Education and Research through the International Research Fund for Germany, enables award winners to carry out long-term and ground-breaking research at universities and research institutions in Germany. The prize money, totaling 5 million Euro for academics in experimental disciplines and 3.5 million Euro for researchers in theoretical disciplines, is being made available for a period of five years.
*Requirements:* Nominations may be made by German universities; non-university research institutions may also submit nominations jointly with a German university. The nominee must be: recognized internationally as an outstandingly qualified academic; be established as an academic abroad; and be nominated by an institution entitled to make nominations.
*Geographic Focus:* All States, All Countries, Germany
*Date(s) Application is Due:* Apr 15; Oct 15
*Contact:* Contact; (+49) 0228-833-0; fax (+49) 0228-833-199; avh-professur@avh.de
*Internet:* www.humboldt-foundation.de/en/programme/preise/ahp.htm
*Sponsor:* Alexander von Humboldt Foundation
Jean-Paul-Street 12
Bonn, D-53173 Germany

**Alexander von Humboldt Foundation Reimar Lust Award 303**
The Award recognizes the contribution of outstanding humanities scholars and social scientists from abroad to promoting cultural and academic relations between Germany and their own countries. Up to two Awards are given each year. Nominations for the Award may be made on behalf of highly respected academics who, as multipliers in and through the field of academic studies, have made an exceptional contribution to the enduring promotion of bilateral relations between Germany and their own countries. These services include the transfer of traditions of and approaches to academic studies from Germany to other countries and from other countries to Germany, translation work, or the promotion of institutional cooperation. In order to strengthen and extend bilateral cooperative relations to their partners and other specialist colleagues in Germany on a sustainable basis, the award winners are invited to spend up to a year in Germany, which may be divided up into blocks. The Award is valued at 50,000 EUR. Additional information is available at the website.
*Restrictions:* Academics who have already been granted an award by the Alexander von Humboldt Foundation are not eligible to be nominated for the Award.
*Geographic Focus:* All States, All Countries, Germany
*Date(s) Application is Due:* Oct 31
*Amount of Grant:* 50,000 EUR
*Contact:* Gerhard Lauer, Selection Committee Chairperson; (+49) 0228-833-0; fax (+49) 0228-833-199; info@avh.de
*Internet:* www.humboldt-foundation.de/web/luest-award.html
*Sponsor:* Alexander von Humboldt Foundation
Jean-Paul-Street 12
Bonn, D-53173 Germany

**Alexander von Humboldt Foundation Research Awards 304**
Each year the Alexander von Humboldt Research Awards honor up to 100 internationally renowned scientists and scholars from abroad. These awards are based primarily on the scientist's/scholar's entire academic record. Scientists and scholars who are nominated for an Award must have contributed fundamental discoveries, new theories, or insights that significantly impact their own research area and also reach beyond their specialized field of research. Nominees should be active researchers who are likely to achieve the same amount high level of achievement in the future. A candidate must be nominated by distinguished scientists/scholars employed by any university or research institution

in Germany. The Award amounts to 60,000 EUR. Additional information about the nomination process is available at the website.
*Requirements:* Scientists and scholars of all disciplines or research areas, regardless of their nationality, may be nominated.
*Restrictions:* Direct applications are not accepted.
*Geographic Focus:* All States, All Countries
*Amount of Grant:* 60,000 EUR
*Contact:* Barbara Konig, Selection Committee Chairperson; (+49) 0228-833-0; fax (+49) 0228-833-199; info@avh.de
*Internet:* www.humboldt-foundation.de/web/6446.html
*Sponsor:* Alexander von Humboldt Foundation
Jean-Paul-Street 12
Bonn, D-53173 Germany

**Alexander von Humboldt Foundation Research Awards for Top Researchers in Germany** 305

Research awards for Top Researchers in Germany are granted by partner organizations of the von Humboldt Foundation. Foreign researchers who wish to nominate a colleague in Germany in their field for a research award must do so through the relevant programs of the respective partner organization in their country. The respective partner organizations are independent from the Humboldt Foundation and follow their own procedures concerning the selection and final decision on the awards. As a rule, the award sum is conferred in conjunction with an invitation to spend up to one year working on research in the host country, which may be divided up into several blocks. Additional information about the agreements for each country is available at the website.
*Requirements:* The nominator should contact the partner organization posted on the website, since nomination procedures and regulations vary from country to country.
*Geographic Focus:* Germany
*Contact:* Barbara Konig, Selection Committee Chairperson; (+49) 0228-833-0; fax (+49) 0228-833-199; info@avh.de
*Internet:* www.humboldt-foundation.de/web/award-researchers-in-germany.html
*Sponsor:* Alexander von Humboldt Foundation
Jean-Paul-Street 12
Bonn, D-53173 Germany

**Alexander von Humboldt Foundation Research Fellowships for Postdoctoral Researchers** 306

The Research Fellowships for Postdoctoral Researchers are the instrument with which the Foundation enables highly-qualified scientists and scholars from abroad who are just embarking on their academic careers and who completed their doctorates less than four years ago to spend extended periods of research (6-24 months) in Germany. Scientists and scholars from all disciplines and countries may apply. Projects are carried out in cooperation with academic hosts at research institutions in Germany. Candidates choose their own research projects and their host in Germany and prepare their own research plan. Details of the research project and the time schedule must be agreed upon with the prospective host in advance. Candidates are selected solely on the basis of their academic record. There are no quotas for individual disciplines or countries. The Fellowship is worth 2,600 EUR per month. This includes a mobility lump sum and a contribution towards health and liability insurance. The application and additional guidelines are available at the website. Applications may be submitted at any time.
*Requirements:* Application requirements include the following: a doctorate or comparable academic degree (Ph.D., C.Sc. or equivalent), completed less than twelve years prior to the date of application; the candidate's own research profile documented by a comprehensive list of academic publications reviewed according to international standards and printed in journals and/or by publishing houses; and a good knowledge of German or English if it is necessary to carry out the project successfully.
*Restrictions:* Short-term visits for study and training purposes or for attending conferences are not eligible for sponsorship.
*Geographic Focus:* All States, All Countries
*Amount of Grant:* 31,200 EUR
*Contact:* Matthias Kleiner, Selection Committee Chairperson; (+49) 0228-833-0; fax (+49) 0228-833-212; info@avh.de
*Internet:* www.humboldt-foundation.de/web/humboldt-fellowship-postdoc.html
*Sponsor:* Alexander von Humboldt Foundation
Jean-Paul-Street 12
Bonn, D-53173 Germany

**Alexander von Humboldt Foundation Sofja Kovalevskaja Awards** 307

Up to eight Sofja Kovalevskaja Awards are granted in recognition of the outstanding academic achievements of exceptionally promising junior researchers from abroad. It enables them to establish their own groups of junior researchers at research establishments in Germany. Scientists and scholars from abroad, whose research records to date have already qualified them to be recognized as top-level junior researchers, and who are expected to continue producing outstanding results as recipients of the Award, are eligible to apply. The award winners are able to concentrate on high-level, innovative research work of their own choice in Germany without restraint and strengthen the internationalization of research in Germany. The Award allows the winners to finance their own working group at a university or non-university research establishment of their own choice in Germany. It also covers their living expenses. The application and additional information is available at the website.
*Requirements:* Application is open to scientists and scholars from all countries and disciplines who have completed their doctorates with distinction within the past six years, and have published work in prestigious international journals or publishing houses. Foreign academics staying in Germany are eligible to apply if they have not been working in an academic capacity in Germany for more than two years at the time of application. German academics are eligible to apply if they have been working in an academic capacity abroad for at least five years at the time of application.
*Geographic Focus:* All States, All Countries
*Date(s) Application is Due:* Jul 31
*Amount of Grant:* Up to 1,650,000 EUR
*Contact:* Barbara Konig, Selection Committee Chairperson; (+49) 0228-833-0; fax (+49) 0228-833-199; info@avh.de
*Internet:* www.humboldt-foundation.de/web/7360.html
*Sponsor:* Alexander von Humboldt Foundation
Jean-Paul-Street 12
Bonn, D-53173 Germany

**Alex C. Walker Educational and Charitable Foundation Grants** 308

The Alex C. Walker Educational and Charitable Foundation awards grants to investigate the causes of economic imbalances; investigate the effect of the monetary system in fostering a sustainable economy; investigate causes tending to destroy or impair the free-market system; explore and develop free-market solutions; and disseminate information on these issues. The foundation funds projects dealing with the research and development of innovative ideas, education of market principles, the application of ideas developed through seed grants and research, and ecological economics with a free-market orientation. Projects with the following qualities are favored: innovation, impact, longevity, collaboration, solution oriented, trackability, objectivity, and sustainability. Grant requests must be submitted on a formal application form through this website. Annual deadlines for submission are April 1 and October 1.
*Requirements:* United States 501(c)3 tax-exempt organizations are eligible.
*Restrictions:* Ideological or political activities and general educational programs are not funded. Grants do not support endowment, building funds, or individuals.
*Geographic Focus:* All States
*Date(s) Application is Due:* Apr 1; Oct 1
*Amount of Grant:* 5,000 - 100,000 USD
*Contact:* Barrett P. Walker, Grants Administrator; (404) 378-2752; fax (404) 378-2751; barrettwalker@mac.com
*Internet:* walker-foundation.org/grant-guidelines
*Sponsor:* Alex C. Walker Educational and Charitable Foundation
1729 Coventry Place
Decatur, GA 30030

**Alex Stern Family Foundation Grants** 309

The Alex Stern Family Foundation awards grants to North Dakota and Minnesota nonprofits in its areas of interest, which include: arts and culture; child welfare; the elderly; alcohol abuse; community affairs; family and social services; education; minorities; hospices; and cancer research. Types of support include: general operating support; continuing support; annual campaigns; building construction and renovation; equipment acquisition; emergency funds; program development; scholarship funds; research; and matching funds. The annual deadlines for submission are March 31 and August 31. Applications are reviewed in June and November. Most recent awards have ranged from $1,000 up to $75,000.
*Requirements:* Moorhead, Minnesota, and Fargo, North Dakota, nonprofit organizations are eligible to apply.
*Restrictions:* Grants are not awarded to individuals or for endowments.
*Geographic Focus:* Minnesota, North Dakota
*Date(s) Application is Due:* Mar 31; Aug 31
*Amount of Grant:* 1,000 - 75,000 USD
*Samples:* Dakota Medical Foundation, Fargo, North Dakota, $75,000 - capital campaign; Blue Stem Performing Center, Noorhead, Minnesota, $20,000 - general operating support; Jamestown College, Jamestown, North Dakota, $7,500 - scholarship funding.
*Contact:* Donald L. Scott, Executive Director; (701) 237-0170; fax (701) 271-0408; donlscott@yahoo.com
*Sponsor:* Alex Suitern Family Foundation
4141 28th Avenue South
Fargo, ND 58104-8468

**Alfred P. Sloan Foundation Behavioral Economics and Household Finance Grants** 310

This program supports research on the structure, behavior, and performance of the U.S. economy with the goal of providing objective and nonpartisan insights that can inform and strengthen critical decisions facing leaders, policymakers, and the public. Projects in the Behavioral Economics and Household Finance sub-program study individuals and households, specifically with regard to the quality of their economic decision-making. Appropriate research topics include the annuity paradox; the energy efficiency paradox; insurance markets; risk-taking, savings, and personal bankruptcy; cognitive biases; public understanding of economics and markets for financial advice. Successful grant applications to the Foundation's program in Economic Institutions, Behavior and Performance have several features in common. Typically, they are: empirical and hypothesis-driven; policy-relevant, but neither policy research nor advocacy; motivated by non-ideological questions rather than preconceived answers; engaged with fundamental puzzles but using fresh approaches; unbiased, statistically significant, and replicable; careful about baselines, controls, and econometrics; savvy about markets, institutions, transaction costs, regulatory arbitrage, etc; contributors to research infrastructure, datasets, or resources for general use; generators of highly-cited results in high-quality

journals; and ultimately concerned with the quality of life in the United States. The Foundation encourages projects by multidisciplinary teams where appropriate.
*Requirements:* Concise, well-organized proposals are preferred. In no case should the body of the proposal exceed 20 double-spaced pages.
*Restrictions:* The Foundation does not make grants: to individuals; to for-profit institutions (except in rare instances); in religion, medical research or the humanities; to projects aimed at pre-college students; to projects in the creative or performing arts, except when those projects are related to educating the public about science, technology or economics; for endowments, buildings or equipment, or fundraising drives, including fundraising dinners; or to political campaigns, to support political activities or to lobby for or against particular pieces of legislation.
*Geographic Focus:* All States
*Contact:* Daniel L. Goroff, Program Director; (212) 649-1676 or (212) 649-1649; fax (212) 757-5117; goroff@sloan.org
*Internet:* www.sloan.org/major-program-areas/economic-performance-and-quality-of-life/economic-institutions-behavior-and-performance/
*Sponsor:* Alfred P. Sloan Foundation
630 Fifth Avenue, Suite 2200
New York, NY 10111-0242

**Alfred P. Sloan Foundation Civic Initiatives Grants** **311**

The goal of the Program is to make a contribution to the Foundation's home area, New York City. There are two directions to the Program: to respond to special opportunities in New York City; and to fund high-leverage projects in New York City that are related to other parts of our program. Interested readers should refer to the descriptions of other program areas in the Foundation website. Grant requests can be made at any time for support of activities related to Foundation program areas and interests. The Foundation is generally limited to supporting tax-exempt organizations.
*Requirements:* Concise, well-organized proposals are preferred. In no case should the body of the proposal exceed 20 double-spaced pages.
*Restrictions:* The Foundation's activities do not normally extend to religion, the creative or performing arts, elementary or secondary education, medical research or health care, the humanities or to activities outside the United States. Grants are not made for endowments or for buildings or equipment.
*Geographic Focus:* New York
*Contact:* Paula J. Olsiewski, Program Director; (212) 649-1658 or (212) 649-1649; fax (212) 757-5117; olsiewski@sloan.org
*Internet:* www.sloan.org/major-program-areas/select-national-issues/civic-initiatives/
*Sponsor:* Alfred P. Sloan Foundation
630 Fifth Avenue, Suite 2200
New York, NY 10111-0242

**Alfred P. Sloan Foundation Data and Computational Research Grants** **312**

From the natural sciences to the social sciences to the humanities to the arts, the availability of more data and cheaper computing is transforming research. As costs for sensors, sequencing, and other forms of data collection decline, researchers can generate data at greater and greater scale, relying on parallel increases in computational power to make sense of it all and allowing the investigation of phenomena too large or complex for conventional observation. Grants in this sub-program aim to help researchers develop tools, establish norms, and build the institutional and social infrastructure needed to take full advantage of these important developments in data-driven, computation-intensive research. Emphasis is placed on projects that encourage access to and sharing of scholarly data, that promote the development of standards and taxonomies necessary for the interoperability of datasets, that enable the replication of computational research, and that investigate models of how researchers might deal with the increasingly central role played by data management and curation.
*Requirements:* Concise, well-organized proposals are preferred. In no case should the body of the proposal exceed 20 double-spaced pages.
*Restrictions:* The Foundation's activities do not normally extend to religion, the creative or performing arts, elementary or secondary education, medical research or health care, the humanities or to activities outside the United States. Grants are not made for endowments or for buildings or equipment.
*Geographic Focus:* All States
*Contact:* Joshua M. Greenberg, Program Director; (212) 649-1633 or (212) 649-1649; fax (212) 757-5117; greenberg@sloan.org
*Internet:* www.sloan.org/major-program-areas/digital-information-technology/data-and-computational-research/
*Sponsor:* Alfred P. Sloan Foundation
630 Fifth Avenue, Suite 2200
New York, NY 10111-0242

**Alfred P. Sloan Foundation Deep Carbon Observatory Grants** **313**

The relatively new Deep Carbon Observatory program aims to revolutionaize our understanding of the carbon deep in the Earth, including its connections to the origins of life and to the origins, distribution, and abundance of fossile fuels, through a multidisciplinary international network that develops and applies new instruments, takes observations, and performs analyses. A three-year 2009 grant to the Carnegie Institution of Washington supported the initial phase of the program, which was focused on developing instruments to meet the severe technical challenges associated with probing the high-pressure, high-temperature rocesses in Earth's deep interior and on building an organizational infrastructure to set strategic priorities, engage a network of researchers, and secure funding commitments from institutional partners.
*Requirements:* Concise, well-organized proposals are preferred. In no case should the body of the proposal exceed 20 double-spaced pages.
*Restrictions:* The Foundation's activities do not normally extend to religion, the creative or performing arts, elementary or secondary education, medical research or health care, the humanities or to activities outside the United States. Grants are not made for endowments or for buildings or equipment.
*Geographic Focus:* All States
*Contact:* Daniel L. Goroff, Program Director; (212) 649-1676 or (212) 649-1649; fax (212) 757-5117; goroff@sloan.org
*Internet:* www.sloan.org/major-program-areas/basic-research/deep-carbon-observatory/
*Sponsor:* Alfred P. Sloan Foundation
630 Fifth Avenue, Suite 2200
New York, NY 10111-0242

**Alfred P. Sloan Foundation Economic Analysis of Science and Technology Grants** **314**

This program supports research on the structure, behavior, and performance of the U.S. economy with the goal of providing objective and nonpartisan insights that can inform and strengthen critical decisions facing leaders, policymakers, and the public. Projects in the Economic Analysis of Science and Technology sub-program study universities and groundbreaking industries, specifically regarding human capital development and applications of information technology. Appropriate research topics include labor markets for scientists and engineers; high-skilled immigration; patterns of scientific publication, collaboration, and intellectual property protection; the economics of digitization; and the international distribution of returns on high-tech investments. Successful grant applications to the Foundation's program in Economic Institutions, Behavior and Performance have several features in common. Typically, they are: empirical and hypothesis-driven; policy-relevant, but neither policy research nor advocacy; motivated by non-ideological questions rather than preconceived answers; engaged with fundamental puzzles but using fresh approaches; unbiased, statistically significant, and replicable; careful about baselines, controls, and econometrics; savvy about markets, institutions, transaction costs, regulatory arbitrage, etc; contributors to research infrastructure, datasets, or resources for general use; generators of highly-cited results in high-quality journals; and ultimately concerned with the quality of life in the United States. The Foundation encourages projects by multidisciplinary teams where appropriate.
*Requirements:* Concise, well-organized proposals are preferred. In no case should the body of the proposal exceed 20 double-spaced pages.
*Restrictions:* The Foundation does not make grants: to individuals; to for-profit institutions (except in rare instances); in religion, medical research or the humanities; to projects aimed at pre-college students; to projects in the creative or performing arts, except when those projects are related to educating the public about science, technology or economics; for endowments, buildings or equipment, or fundraising drives, including fundraising dinners; or to political campaigns, to support political activities or to lobby for or against particular pieces of legislation.
*Geographic Focus:* All States
*Contact:* Daniel L. Goroff, Program Director; (212) 649-1676 or (212) 649-1649; fax (212) 757-5117; goroff@sloan.org
*Internet:* www.sloan.org/major-program-areas/economic-performance-and-quality-of-life/economic-institutions-behavior-and-performance/
*Sponsor:* Alfred P. Sloan Foundation
630 Fifth Avenue, Suite 2200
New York, NY 10111-0242

**Alfred P. Sloan Foundation Economic Implications of the Great Recession Grants** **315**

This program supports research on the structure, behavior, and performance of the U.S. economy with the goal of providing objective and nonpartisan insights that can inform and strengthen critical decisions facing leaders, policymakers, and the public. Projects in Economic Implications of the Great Recession sub-program study markets and governments, specifically with regard to lessons we can draw from the recent financial crisis and Great Recession. Appropriate research topics include systemic stability; international regulatory coordination; risk measurement, capital requirements, and credit ratings; labor market recovery rates and liquidity; and dataset and model development concerning labor trends. Successful grant applications to the Foundation's program in Economic Institutions, Behavior and Performance have several features in common. Typically, they are: empirical and hypothesis-driven; policy-relevant, but neither policy research nor advocacy; motivated by non-ideological questions rather than preconceived answers; engaged with fundamental puzzles but using fresh approaches; unbiased, statistically significant, and replicable; careful about baselines, controls, and econometrics; savvy about markets, institutions, transaction costs, regulatory arbitrage, etc; contributors to research infrastructure, datasets, or resources for general use; generators of highly-cited results in high-quality journals; and ultimately concerned with the quality of life in the United States. The Foundation encourages projects by multidisciplinary teams where appropriate.
*Requirements:* Concise, well-organized proposals are preferred. In no case should the body of the proposal exceed 20 double-spaced pages.
*Restrictions:* The Foundation does not make grants: to individuals; to for-profit institutions (except in rare instances); in religion, medical research or the humanities; to projects aimed at pre-college students; to projects in the creative or performing arts, except when those projects are related to educating the public about science, technology or economics; for endowments, buildings or equipment, or fundraising drives, including fundraising dinners; or to political campaigns, to support political activities or to lobby for or against particular pieces of legislation.

*Geographic Focus:* All States
*Contact:* Daniel L. Goroff, Program Director; (212) 649-1676 or (212) 649-1649; fax (212) 757-5117; goroff@sloan.org
*Internet:* www.sloan.org/major-program-areas/economic-performance-and-quality-of-life/economic-institutions-behavior-and-performance/
*Sponsor:* Alfred P. Sloan Foundation
630 Fifth Avenue, Suite 2200
New York, NY 10111-0242

**Alfred P. Sloan Foundation Empirical Economic Research Enablers Grants 316**

This program supports research on the structure, behavior, and performance of the U.S. economy with the goal of providing objective and nonpartisan insights that can inform and strengthen critical decisions facing leaders, policymakers, and the public. Projects in the Empirical Economic Research Enablers sub-program study economic researchers, specifically with regard to their needs, opportunities, incentives, and professional practices. Appropriate research topics include legal entity identifiers; data citation standards; identification and tracking systems for scholars; federal statistics; smart disclosure platforms for obfuscated markets; data and metadata management protocols; the replicability of empirical research; and the economics of knowledge contribution and distribution. Successful grant applications to the Foundation's program in Economic Institutions, Behavior and Performance have several features in common. Typically, they are: empirical and hypothesis-driven; policy-relevant, but neither policy research nor advocacy; motivated by non-ideological questions rather than preconceived answers; engaged with fundamental puzzles but using fresh approaches; unbiased, statistically significant, and replicable; careful about baselines, controls, and econometrics; savvy about markets, institutions, transaction costs, regulatory arbitrage, etc; contributors to research infrastructure, datasets, or resources for general use; generators of highly-cited results in high-quality journals; and ultimately concerned with the quality of life in the United States. The Foundation encourages projects by multidisciplinary teams where appropriate.
*Requirements:* Concise, well-organized proposals are preferred. In no case should the body of the proposal exceed 20 double-spaced pages.
*Restrictions:* The Foundation does not make grants: to individuals; to for-profit institutions (except in rare instances); in religion, medical research or the humanities; to projects aimed at pre-college students; to projects in the creative or performing arts, except when those projects are related to educating the public about science, technology or economics; for endowments, buildings or equipment, or fundraising drives, including fundraising dinners; or to political campaigns, to support political activities or to lobby for or against particular pieces of legislation.
*Geographic Focus:* All States
*Contact:* Daniel L. Goroff, Program Director; (212) 649-1676 or (212) 649-1649; fax (212) 757-5117; goroff@sloan.org
*Internet:* www.sloan.org/major-program-areas/economic-performance-and-quality-of-life/economic-institutions-behavior-and-performance/
*Sponsor:* Alfred P. Sloan Foundation
630 Fifth Avenue, Suite 2200
New York, NY 10111-0242

**Alfred P. Sloan Foundation Energy and Environment Grants 317**

Grantmaking in this small interdisciplinary program looks for unique opportunities to expand our understanding of the economic, technological, organizational, regulatory, national security, and environmental consequences of energy production and consumption. Past grantmaking in this program has led to the publication of the influential MIT reports, The Future of Nuclear Power, and The Future of Coal and the Foundation is currently supporting a similar project which focuses on economic, technological, and institutional issues associated with the use of solar energy. Other recent Foundation grants in this area support the exploration of strategies for the safe, responsible expansion of nuclear power around the world, a project to examine the feasibility of extending the life of existing nuclear plants, a project to disseminate safety guidelines for the responsible use of radioactive materials at academic, industrial and medical institutions, a project on potential improvements to the measurement of carbon sequestered in forests, and two projects on natural gas: one examining the economics of state-of-the-art natural gas extraction based on detailed analyses of the geological and economic attributes of five important U.S. shale gas deposits, and one examining environmental and regulatory issues surrounding shale gas exploration and extraction. Due to the significant funding available from both public and private sources for energy and environmental research, the Foundation is very selective in the grants it makes in this area, supporting only projects for which non-partisan funding is not readily available, projects related to other Foundation programs or priorities, or projects where Foundation support could be leveraged to significantly raise the chances of the project's success.
*Requirements:* Concise, well-organized proposals are preferred. In no case should the body of the proposal exceed 20 double-spaced pages.
*Restrictions:* The Foundation's activities do not normally extend to religion, the creative or performing arts, elementary or secondary education, medical research or health care, the humanities or to activities outside the United States. Grants are not made for endowments or for buildings or equipment.
*Geographic Focus:* All States
*Contact:* Evan S. Michelson, Program Officer; (212) 649-1623 or (212) 649-1649; fax (212) 757-5117; michelson@sloan.org
*Internet:* www.sloan.org/major-program-areas/select-national-issues/energy-and-environment/
*Sponsor:* Alfred P. Sloan Foundation
630 Fifth Avenue, Suite 2200
New York, NY 10111-0242

**Alfred P. Sloan Foundation International Science Engagement Grants 318**

The International Science Engagement program is an early stage program currently focused on South Asia that seeks to bring scientists and engineers in conflict regions together to collaborate on subjects and projects of mutual interest. A pilot grant was made to the Lee Kuan Yew School of Public Policy/National University of Singapore for a preliminary meeting on rice science involving scientists from eight countries including: India, Pakistan, Bangladesh, Sri Lanka, Nepal, Bhutan, the Philippines, and China. Other grants have supported the Institute for International Education's Scholar Rescue Fund which works to relocate endangered scientists, engineers and mathematicians to research institutes abroad.
*Requirements:* Concise, well-organized proposals are preferred. In no case should the body of the proposal exceed 20 double-spaced pages.
*Restrictions:* The Foundation's activities do not normally extend to religion, the creative or performing arts, elementary or secondary education, medical research or health care, the humanities or to activities outside the United States. Grants are not made for endowments or for buildings or equipment.
*Geographic Focus:* All States, All Countries
*Contact:* Doron Weber, Program Director; (212) 649-1652 or (212) 649-1649; fax (212) 757-5117; weber@sloan.org
*Internet:* www.sloan.org/major-program-areas/select-issues/international-science-engagement/
*Sponsor:* Alfred P. Sloan Foundation
630 Fifth Avenue, Suite 2200
New York, NY 10111-0242

**Alfred P. Sloan Foundation Microbiology of the Built Environment Grants 319**

Formerly known as the Indoor Environment Grant, the goal of the Alfred P. Sloan Foundation Microbiology of the Built Environment program is to understand the human indoor environment at the microbial level. The Foundation believes that this understanding may ultimately help to make this artificial environment more hospitable to human life or resistant to biological attacks. The Foundation plan has three parts. First, the Foundation wants to establish a basic understanding of the indoor microbial environment. Second, it wants to help develop tools for understanding and probing the indoor environment. And lastly, it wants to apply the knowledge and tools to various human indoor environments, such as hospitals. Grant requests can be made at any time for support of activities related to Foundation program areas and interests. The Foundation is generally limited to supporting tax-exempt organizations. The Foundation has no deadlines or standard forms. The Foundation accepts proposals sent by email. A brief letter of inquiry, rather than a fully developed proposal, is an advisable first step for an applicant, conserving his or her time and allowing for a preliminary response regarding the possibility of support.
*Requirements:* Concise, well-organized proposals are preferred. In no case should the body of the proposal exceed 20 double-spaced pages.
*Restrictions:* The Foundation's activities do not normally extend to religion, the creative or performing arts, elementary or secondary education, medical research or health care, the humanities or to activities outside the United States. Grants are not made for endowments or for buildings or equipment.
*Geographic Focus:* All States, Canada
*Contact:* Paula J. Olsiewski, Program Director; (212) 649-1658 or (212) 649-1649; fax (212) 757-5117; olsiewski@sloan.org
*Internet:* www.sloan.org/major-program-areas/basic-research/mobe/
*Sponsor:* Alfred P. Sloan Foundation
630 Fifth Avenue, Suite 2200
New York, NY 10111-0242

**Alfred P. Sloan Foundation Public Understanding of Science and Technology: Book Grants 320**

The Alfred P. Sloan Foundation supports a wide range of books aimed at public understanding of science and technology. Popular categories include books that elucidate the scientific basis of issues that are often confusing or controversial or unnecessarily obscure. The Foundation also supports books that profile scientific and technological figures from varying angles, but with an emphasis on the human story. Books about the relevance of technology to daily life and about the relationship between women and technology also find occasional support. Inventing America: A History of the United States, a popular college textbook supported by the Foundation, explores the central role of science, technology and business in the nation's development. The Foundation's Book program aims to reach a wide, lay audience.
*Requirements:* Concise, well-organized proposals are preferred. In no case should the body of the proposal exceed 20 double-spaced pages.
*Restrictions:* The Foundation's activities do not normally extend to religion, the creative or performing arts, elementary or secondary education, medical research or health care, the humanities or to activities outside the United States. Grants are not made for endowments or for buildings or equipment.
*Geographic Focus:* All States
*Contact:* Doron Weber, Program Director; (212) 649-1652 or (212) 649-1649; fax (212) 757-5117; weber@sloan.org
*Internet:* www.sloan.org/major-program-areas/public-understanding-of-science-and-technology/books/
*Sponsor:* Alfred P. Sloan Foundation
630 Fifth Avenue, Suite 2200
New York, NY 10111-0242

**Alfred P. Sloan Foundation Public Understanding of Science and Technology: New Media Grants** 321

The Alfred P. Sloan Foundation occasionally sponsors innovative efforts using a range of media and other platforms to reach a broad, cross-cultural audience or to target specific segments of the public to enhance public understanding. These efforts may take the form of conferences, multimedia events, performances, ebooks, interactive games, science and arts festivals, and more. The Foundation makes grants year-round. Interested grant seekers with a relevant project idea should email a one page letter of inquiry to Program Director.

*Requirements:* Concise, well-organized proposals are preferred. In no case should the body of the proposal exceed 20 double-spaced pages.

*Restrictions:* The Foundation's activities do not normally extend to religion, the creative or performing arts, elementary or secondary education, medical research or health care, the humanities or to activities outside the United States. Grants are not made for endowments or for buildings or equipment. Telephone inquiries are not accepted.

*Geographic Focus:* All States

*Contact:* Doron Weber, Vice President, Programs; (212) 649-1652 or (212) 649-1649; fax (212) 757-5117; weber@sloan.org

*Internet:* www.sloan.org/major-program-areas/public-understanding-of-science-and-technology/new-media/

*Sponsor:* Alfred P. Sloan Foundation
630 Fifth Avenue, Suite 2200
New York, NY 10111-0242

**Alfred P. Sloan Foundation Public Understanding of Science and Technology: Radio Grants** 322

The Alfred P. Sloan Foundation supports a range of radio programs that deal with science and technology and seeks to increase both the quantity and the quality of science and technology coverage. Grant requests can be made at any time for support of activities related to Foundation program areas and interests. The Foundation is generally limited to supporting tax-exempt organizations. The Foundation has no deadlines or standard forms. The Foundation accepts proposals sent by email. To begin the application process, send a one page letter of inquiry to program director.

*Requirements:* Concise, well-organized proposals are preferred. In no case should the body of the proposal exceed 20 double-spaced pages.

*Restrictions:* The Foundation's activities do not normally extend to religion, the creative or performing arts, elementary or secondary education, medical research or health care, the humanities or to activities outside the United States. Grants are not made for endowments or for buildings or equipment. Telephone inquiries are not accepted.

*Geographic Focus:* All States

*Samples:* Daniel Charles, Washington, D.C.—to research & broadcast half a dozen segments on the technology of mapmaking for NPR; Public Radio International, Minneapolis, MN—planning grant for science and technology component of a satirical news program on public radio; L.A. Theatre Works, Los Angeles, CA—to record 4 new science radio plays, including 2 Sloan commissions, and to broadcast the complete 17-science-play series on public radio and in libraries and schools.

*Contact:* Doron Weber, Program Director; (212) 649-1652 or (212) 649-1649; fax (212) 757-5117; weber@sloan.org

*Internet:* www.sloan.org/major-program-areas/public-understanding-of-science-and-technology/radio/

*Sponsor:* Alfred P. Sloan Foundation
630 Fifth Avenue, Suite 2200
New York, NY 10111-0242

**Alfred P. Sloan Foundation Public Understanding of Science and Technology: Television Grants** 324

The Foundation is a major supporter of public television documentaries, docudramas and dramas about science, technology and the lives of the men and women involved in scientific and technological pursuit. Foundation-supported shows highlight the role of engineering and technology in society or broaden our view of the nation's history and of the central role of science and technology in the country's narrative. Some programs celebrate scientific and technological breakthroughs and milestones, while others reveal just how little we know, and how far we still have to go. The Foundation has a longtime interest in the underappreciated role of women and minorities in science and technology. The Foundation also supports television programs based on projects it has sponsored in other media, such as books and plays. The Foundation also continues to support commercial television by developing various film projects with cable and broadcast networks.

*Requirements:* Concise, well-organized proposals are preferred. In no case should the body of the proposal exceed 20 double-spaced pages. Grant requests can be made at any time for support of activities related to Foundation program areas and interests. The Foundation is generally limited to supporting tax-exempt organizations. The Foundation has no deadlines or standard forms. The Foundation accepts proposals sent by email. A brief letter of inquiry, rather than a fully developed proposal, is an advisable first step for an applicant, conserving his or her time and allowing for a preliminary response regarding the possibility of support. Interested grant seekers with a relevant project idea should email a one page letter of inquiry to Program Director Doron Weber.

*Restrictions:* The Foundation's activities do not normally extend to religion, the creative or performing arts, elementary or secondary education, medical research or health care, the humanities or to activities outside the United States. Grants are not made for endowments or for buildings or equipment. Telephone inquiries are not accepted.

*Geographic Focus:* All States

*Contact:* Doron Weber, Vice President, Programs; (212) 649-1652 or (212) 649-1649; fax (212) 757-5117; weber@sloan.org

*Internet:* www.sloan.org/major-program-areas/public-understanding-of-science-and-technology/television/?L=imbzkscxn

*Sponsor:* Alfred P. Sloan Foundation
630 Fifth Avenue, Suite 2200
New York, NY 10111-0242

**Alfred P. Sloan Foundation Public Understanding of Science and Technology: Theater Grants** 324

In this program, the Foundation makes grants to encourage playwrights and theater companies to write and produce new plays about scientists, engineers and mathematicians that will break down the barrier between the two cultures. The Foundation's major partners include: the Ensemble Studio Theatre; the Manhattan Theatre Club and; Playwrights Horizons. The Foundation has been a leading force in commissioning and producing new science and technology plays. In this role, it has helped to stimulate numerous playwrights to tackle these subjects while also creating a more hospitable environment among the theater-going public for such fare, heralding an important cultural shift.

*Restrictions:* The Foundation does not make grants to: individuals (except through its Books program); for-profit institutions; religion, medical research, or for research in the humanities; projects aimed at pre-college students (except through its Civic Initiatives program); projects in the creative or performing arts (except when those projects are related to educating the public about science, technology, or economics); endowments or fundraising drives, including fundraising dinners; support buildings, laboratories, equipment, or instruments unless it can be demonstrated that such capital expenditures are essential to the success of a Foundation-supported research project, research program or educational initiative; or political campaigns, to support political activities, or to lobby for or against particular pieces of legislation.

*Geographic Focus:* All States

*Contact:* Doron Weber, Program Director; (212) 649-1652 or (212) 649-1649; fax (212) 757-5117; weber@sloan.org

*Internet:* www.sloan.org/major-program-areas/public-understanding-of-science-and-technology/theater/

*Sponsor:* Alfred P. Sloan Foundation
630 Fifth Avenue, Suite 2200
New York, NY 10111-0242

**Alfred P. Sloan Foundation Research Fellowships** 325

Alfred P. Sloan Foundation Research Fellowships stimulate fundamental research by young scholars of outstanding promise at a time in their careers when their creative abilities are especially high and when government or other support is difficult to obtain. Fellows are free to pursue whatever lines of inquiry that are of the most compelling interest to them. Funds are awarded directly to the fellow's institution and may be used by the fellow for such purposes as equipment, technical assistance, professional travel, trainee support, or activities directly related to the fellow's research. The foundation welcomes nominations of all candidates who meet the traditional high standards of this program and strongly encourages the participation of women and members of underrepresented minority groups. Awards are made in the fields of physics, chemistry, neuroscience, economics, pure mathematics, applied mathematics, computer science, and computational and evolutionary molecular biology. Fellowships are awarded for a two-year period, with possible extension for another two years. Each year 126 fellows are selected from the nominations received. Nominations are due September 15 for awards to begin the following September. Nomination forms are available online.

*Requirements:* Candidates must hold a PhD or equivalent in chemistry, physics, mathematics, computer science, economics, neuroscience or computational and evolutionary molecular biology, or in a related interdisciplinary field, and must also be a faculty member at a U.S. or Canadian university. Candidates must be no more than six years from the completion of the most recent PhD or equivalent as of the year of their nomination.

*Restrictions:* Direct applications are not accepted. Candidates must be nominated by department heads or other senior scholars.

*Geographic Focus:* All States, Canada

*Date(s) Application is Due:* Sep 15

*Amount of Grant:* 50,000 USD

*Contact:* Daniel L. Goroff, Program Director; (212) 649-1676 or (212) 649-1649; fax (212) 757-5117; goroff@sloan.org

*Internet:* www.sloan.org/sloan-research-fellowships/

*Sponsor:* Alfred P. Sloan Foundation
630 Fifth Avenue, Suite 2200
New York, NY 10111-0242

**Alfred P. Sloan Foundation Scholarly Communication Grants** 326

The shift to digitally-mediated forms of scholarship has been characterized by a substantial growth in channels for and diversity of scholarly work. We see this in the flourishing of content in preprint servers and rapid-publication channels like arXiv, PLoS ONE, and the Social Science Research Network alongside unconventional forms of scholarly communication like research blogs and personal websites, all of which enable scholars to put their work out for broad access. Grants in this sub-program aim to ease this transition by supporting the development of new models of filtering and curating online scholarly materials and by engaging the emerging community of stakeholders and practitioners tackling similar issues in widely divergent disciplinary contexts. Grantseekers with general inquiries or with a relevant project idea regarding the Data and Computational Research or Scholarly Communications sub-programs should send a one-page letter of inquiry to Program Director.

*Requirements:* Concise, well-organized proposals are preferred. In no case should the body of the proposal exceed 20 double-spaced pages.

*Restrictions:* The Foundation does not make grants: to individuals; to for-profit institutions (except in rare instances); in religion, medical research or the humanities; to projects aimed at pre-college students; to projects in the creative or performing arts, except when those projects are related to educating the public about science, technology or economics; for endowments, buildings or equipment, or fundraising drives, including fundraising dinners; or to political campaigns, to support political activities or to lobby for or against particular pieces of legislation.
*Geographic Focus:* All States
*Contact:* Joshua M. Greenberg, Program Director; (212) 649-1633 or (212) 649-1649; fax (212) 757-5117; greenberg@sloan.org
*Internet:* www.sloan.org/major-program-areas/digital-information-technology/scholarly-communication/
*Sponsor:* Alfred P. Sloan Foundation
630 Fifth Avenue, Suite 2200
New York, NY 10111-0242

**Alfred P. Sloan Foundation Science of Learning STEM Grants** 327
Grantmaking in the Science of Learning STEM program aims to improve the quality of higher education in STEM fields through the support of original, high-quality research on the factors affecting undergraduate and graduate student learning and retention in STEM fields. Grants primarily support consortia of colleges, universities, and other educational institutions with plans to develop and to study the impact and effectiveness of new approaches to STEM pedagogy, especially in "gateway" courses, with an explicit commitment to institutionalize successful initiatives. Successful proposals are expected to be hypothesis-driven, sensitive to the heterogeneity of STEM disciplines, attentive to differences in student motivations to choose STEM majors and persist in STEM careers, and concerned with the dissemination and portability of results to other institutions.
*Restrictions:* The Foundation does not make grants to: individuals (except through its Books program); for-profit institutions; religion, medical research, or for research in the humanities; projects aimed at pre-college students (except through its Civic Initiatives program); projects in the creative or performing arts (except when those projects are related to educating the public about science, technology, or economics); endowments or fundraising drives, including fundraising dinners; support buildings, laboratories, equipment, or instruments unless it can be demonstrated that such capital expenditures are essential to the success of a Foundation-supported research project, research program or educational initiative; or political campaigns, to support political activities, or to lobby for or against particular pieces of legislation.
*Geographic Focus:* All States
*Contact:* Elizabeth Boylan, Program Director; (212) 649-1634 or (212) 649-1649; fax (212) 757-5117; boylan@sloan.org
*Internet:* www.sloan.org/major-program-areas/stem-higher-education/the-science-of-learning-stem/
*Sponsor:* Alfred P. Sloan Foundation
630 Fifth Avenue, Suite 2200
New York, NY 10111-0242

**Alfred P. Sloan Foundation Sloan Digital Sky Survey Grants** 328
The Sloan Digital Sky Survey (SDSS) program is expanding our understanding of the evolution and structure of the universe by providing the largest uniform, detailed archive of objects in the skies that has ever existed. In cooperation with the Astrophysical Research Consortium, the Foundation has helped build and operate a specially designed telescope to observe and archive galaxies, quasars and other cosmological phenomena. Data collected through the SDSS is providing an increasingly detailed picture of the universe, including data bearing on the curvature of the universe itself, on the existence of dark energy, and on the features of the Milky Way. Grant requests can be made at any time for support of activities related to Foundation program areas and interests. The Foundation is generally limited to supporting tax-exempt organizations. The Foundation has no deadlines or standard forms. The Foundation accepts proposals sent by email. A brief letter of inquiry, rather than a fully developed proposal, is an advisable first step for an applicant, conserving his or her time and allowing for a preliminary response regarding the possibility of support.
*Requirements:* Concise, well-organized proposals are preferred. In no case should the body of the proposal exceed 20 double-spaced pages.
*Restrictions:* The Foundation's activities do not normally extend to religion, the creative or performing arts, elementary or secondary education, medical research or health care, the humanities or to activities outside the United States. Grants are not made for endowments or for buildings or equipment.
*Geographic Focus:* All States
*Contact:* Evan S. Michelson, Program Officer; (212) 649-1623 or (212) 649-1649; fax (212) 757-5117; michelson@sloan.org
*Internet:* www.sloan.org/major-program-areas/basic-research/sloan-digital-sky-survey/
*Sponsor:* Alfred P. Sloan Foundation
630 Fifth Avenue, Suite 2200
New York, NY 10111-0242

**Alfred P. Sloan Foundation Working Longer Grants** 329
Americans are working later in life, delaying retirement for a variety of reasons. While some continue working by choice, others needto remain in the workforce for financial reasons. Regardless of their motivation, the result is that people in the United States are working beyond what conventionally is thought of as retirement age. While most studies related to this issue have focused on how individually based factors such as health status, pension plans, and financial incentives influence the decision to retire, relatively little is known about employment patterns, obstacles to employment, or the ensuing economic impact. The Foundation's Working Longer program began grant making in 2010 to expand and deepen our understanding of aging Americans' work patterns. The goal is to understand; employer practices by industry and sector; obstacles to continued employment of older Americans; and the economic consequences for both individuals and for the federal budget.
*Requirements:* Concise, well-organized proposals are preferred. In no case should the body of the proposal exceed 20 double-spaced pages.
*Restrictions:* The Foundation does not make grants: to individuals; to for-profit institutions (except in rare instances); in religion, medical research or the humanities; to projects aimed at pre-college students; to projects in the creative or performing arts, except when those projects are related to educating the public about science, technology or economics; for endowments, buildings or equipment, or fundraising drives, including fundraising dinners; or to political campaigns, to support political activities or to lobby for or against particular pieces of legislation.
*Geographic Focus:* All States
*Contact:* Kathleen E. Christensen, Program Director; (212) 649-1628 or (212) 649-1649; fax (212) 757-5117; christensen@sloan.org
*Internet:* www.sloan.org/major-program-areas/economic-performance-and-quality-of-life/working-longer/
*Sponsor:* Alfred P. Sloan Foundation
630 Fifth Avenue, Suite 2200
New York, NY 10111-0242

**Allen Foundation Educational Nutrition Grants** 330
Established in 1975 by agricultural chemist William Webster Allen and based in Midland, Michigan, the Allen Foundation makes grants to projects that benefit human nutrition in the areas of education, training, and research. Priorities include training programs for children and young adults to improve their health and development; training programs for educators and demonstrators concerned with good nutritional practices; programs for the education and training of mothers during pregnancy and after the birth of their children, so that good nutritional habits can be formed at an early age; and programs that aid in the dissemination of information regarding healthful nutrition practices. The Allen Foundation will consider requests from the following: hospitals or medical clinics; social, religious, fraternal, or community organizations; private foundations; and K-12 public, parochial or private schools. Preference may be given to proposals that include matching funds from the institution or other partners including in-kind contribution. Third party contribution to matching funds such as computer or software donated from a company may be included. Interested organizations should begin by taking the online eligibility quiz. Applications are also available online. The annual deadline is December 31, though proposals are reviewed throughout the year.
*Requirements:* 501(c)3 tax-exempt organizations nationwide may apply. In certain circumstances, the foundation will consider requests from the following: hospitals or medical clinics; social, religious, fraternal, or community organizations; private foundations; and K-12 public, parochial, or private schools.
*Geographic Focus:* All States
*Date(s) Application is Due:* Dec 31
*Amount of Grant:* 5,000 - 250,000 USD
*Contact:* Dale Baum, Secretary; (989) 832-5678 or (979) 695-1132; fax (989) 832-8842; dbaum@allenfoundation.org or Lucille@allenfoundation.org
*Internet:* www.allenfoundation.org/commoninfo/aboutus.asp
*Sponsor:* Allen Foundation
P.O. Box 1606
Midland, MI 48641-1606

**Allstate Foundation Safe and Vital Communities Grants** 331
Established in 1952, The Allstate Foundation is an independent, charitable organization made possible by subsidiaries of The Allstate Corporation. The Foundation is dedicated to fostering safe and vital communities where people live, work and raise families. We strive to foster communities that are economically strong, crime-free, and give residents a sense of belonging and commitment. Our Safe and Vital Communities funding priorities are: teen safe driving - helping to save young lives and instill a lifetime of safe driving attitudes and behaviors; catastrophe response - rebuilding lives after a natural disaster strikes; and neighborhood revitalization - nurturing safe, strong, and healthy communities. Proposals are accepted throughout the year, though the regional grant proposal deadline is May 31. The average grant amount ranges from $5,000 to $20,000 for regional grants.
*Requirements:* The Allstate Foundation makes grants to nonprofit, tax-exempt organizations under Section 501(c)3 of the Internal Revenue Code, or be a municipal, state or federal government entity.
*Restrictions:* The Foundation does not support the following: individuals; fundraising events, sponsorships; capital and endowment campaigns; equipment purchase unless part of a community outreach program; athletic events; memorial grants; athletic teams, bands, and choirs; organizations that advocate religious beliefs or restrict participation on the basis of religion; groups or organizations that will re-grant the Foundation's gift to other organizations or individuals; scouting groups; private secondary schools; requests to support travel; grant requests for production of audio, film, or video; multiyear pledge requests; or non-domestic (international) causes.
*Geographic Focus:* All States
*Date(s) Application is Due:* May 31
*Amount of Grant:* 5,000 - 20,000 USD
*Contact:* Vicky Dinges, Vice President of Corporate Social Responsibility; (847) 402-5600 or (847) 402-7893; fax (847) 326-7517; allfound@allstate.com
*Internet:* www.allstatefoundation.org/grant-focus-areas
*Sponsor:* Allstate Foundation
2775 Sanders Road, Suite F4
Northbrook, IL 60062-6127

**Alpha Omega Foundation Grants** 332

The Foundation was established to advance the dental profession through research, grants, and scholarships. Fields of interest include: dental care, dental education, and medical research. Grants are given to individuals, program development, and higher education institutions. The Foundation was responsible for dispersing almost $250,000 this past year for projects in Israel and worldwide. The application must be typed and submitted via email by September 1 each year. Notification will be completed by the following December.
*Geographic Focus:* All States, All Countries
*Date(s) Application is Due:* Sep 1
*Amount of Grant:* 1,000 - 50,000 USD
*Contact:* Heidi Weber, Association Director; (301) 738-6400; fax (301) 738-6403; hweber@ao.org or foundation@ao.org
*Internet:* www.ao.org/index.php?option=com_content&task=view&id=80&Itemid=147
*Sponsor:* Alpha Omega Foundation
50 W Edmonston Drive, Suite 303
Rockville, MD 20852-1274

**ALSAM Foundation Grants** 333

The ALSAM Foundation was named in honor of well-known Utah businessman and benefactor L.S. Sam Skaggs, who shunned the spotlight while quietly helping pioneer the development of a retail superstore. He built a retail food-and-drug empire from his Salt Lake City headquarters that operated under the American Stores name. During his lengthy career, his Foundation has funneled millions of dollars to charitable causes. Today, the Foundation continues to award grants in its areas of interest, including: agriculture; Christian agencies and churches; higher education; health care; medical research; human services; minorities; and the economically disadvantaged. Types of support include building construction and renovation, general operating support, and scholarships. The board meets in January and October. Contact the office for application forms.
*Requirements:* Higher education institutions, nonprofit organizations, religious organizations, and research institutions are eligible to apply.
*Restrictions:* No grants to are given directly to individuals.
*Geographic Focus:* All States
*Amount of Grant:* 5,000 - 50,000 USD
*Contact:* Ronny Cutshall, President; (801) 266-4950
*Sponsor:* ALSAM Foundation
6190 Moffat Farm Lane
Salt Lake City, UT 84121

**Alston Bannerman Sabbatical Fellowships** 334

The Center for Social Inclusion (CSI) recognizes that working for social change can lead to pressures that result in a loss of creative and critical leadership. Because of this, Sabbatical Fellows receive a $25,000 award to take three months off for reflection and renewal. Sabbaticals strengthen organizational planning and leadership development, encourage innovation and collaboration, promote the value of reflection, and foster connections between organizers and organizations. By reinvigorating experienced grassroots organizers and encouraging younger ones to see organizing as a long-term career, the Fellowship helps build the necessary infrastructure for systemic and sustainable change. Fellows are selected through an annual application process. The Selection Committee is comprised of leaders from around the country working in or for communities of color. Fellows are required to attend pre- and post-sabbatical retreats; begin their sabbaticals within one year of the awarding of the Fellowship; and report on their sabbaticals within four months of their completion to describe what they did, what they learned that will improve their leadership, and what impact had on their work. The application is available at the CSI website.
*Requirements:* A Fellowship applicant must be a person of color; have more than ten years of community organizing experience; be committed to social change work in communities of color; and live in the United States, Puerto Rico, Guam, American Samoa, or the U.S. Virgin Islands. Beyond the basic eligibility, the Alston Bannerman Program seeks applicants whose work attacks root causes of inequity by organizing those affected to take collective strategic action; challenges the systems that perpetrate injustice and effects institutional and structural change; builds community capacity for democratic participation and develops grassroots leadership; acknowledges the cultural values of the community; creates accountable participatory structures in which community members have decision-making power; and contributes to building a movement for social change by making connections between issues, developing alliances with other constituencies, and collaborating with other organizations.
*Restrictions:* Awards are not made to individuals who exclusively provide services, such as substance abuse counseling, after school programs, HIV-AIDS outreach or shelter for the homeless; or who advocate on behalf of a community without directly involving the members of that community in asserting their own interests and choosing their own leadership.
*Geographic Focus* All States, Guam, Puerto Rico, U.S. Virgin Islands, American Samoa
*Amount of Grant:* 25,000 USD
*Contact:* Madeleine Adamson; (212) 248-2785; fax (212) 248-6409; info@thecsi.org
*Internet:* www.centerforsocialinclusion.org/leadership/alston-bannerman-sabbatical-fellows/
*Sponsor:* Center for Social Inclusion
150 Broadway, Suite 303
New York, NY 10038

**Alzheimer's Association Conference Grants** 335

The Alzheimer's Association has a long history of supporting scientific conferences that advance research on Alzheimer's disease. One of the principal goals of the Association from its inception has been to increase public awareness and to facilitate the exchange of information through the scientific and clinical communities. The support of conferences, workshops and meetings has been a key vehicle in achieving this goal. The objectives for conference support are to: facilitate and speed the exchange of information relevant to Alzheimer's disease research; convene experts to address emerging issues in Alzheimer's disease research; offer opportunities for new investigators and graduate students to participate in scientific meetings; facilitate the creation of networks among investigators in related areas; and increase visibility of the research interests and programs of the Alzheimer's Association. At this time, support requests must be limited to no more than $10,000 per conference. Most awarded conference support requests have been in the range of $2000 to $5000.
*Requirements:* Requests for conference support may be submitted at any time. It is recommended that requests be submitted at least three months before the conference.
*Geographic Focus:* All States
*Date(s) Application is Due:* Jan 4; Mar 6; May 1; Jun 7; Aug 7; Oct 2; Nov 1
*Amount of Grant:* Up to 10,000 USD
*Contact:* Angela Worlds, (312) 335-5807; fax (312) 335-4034; angela.worlds@alz.org
*Internet:* www.alz.org/research/alzheimers_grants/types_of_grants.asp
*Sponsor:* Alzheimer's Association
225 North Michigan Avenue, Suite 1700
Chicago, IL 60601-7633

**Alzheimer's Association Development of New Cognitive and Functional Instruments Grants** 336

The Alzheimer's Association is launching a new initiative to stimulate the scholarly investigation and development of cognitive or functional evaluation instruments that can capture the earliest changes in the disease, are sensitive to change over time, and/or could be used in clinical trials. This RFA is designed to enable pilot research or proof-of-principle studies that can provide preliminary data for subsequent inquiry. Allowable costs under this award include: purchase and care of laboratory animals; small pieces of laboratory equipment and laboratory supplies; computer equipment if used strictly for data collection; travel (up to $1,000 per year); and salary for the principal investigator, scientific (including post-doctoral fellows) and technical staff (including laboratory technicians and administrative support staff whose work is directly related to the funded project). The Association anticipates funding up to 2 approaches addressing needs in cognitive and functional instrument development. Each award is limited to $400,000 (direct and indirect costs) for two to three years. Requests in any given year may not exceed $200,000 (direct and indirect costs). Indirect costs are capped at 10 percent (rent for laboratory or office space is expected to be covered by indirect costs paid to the institution). Letters of Intent (LOIs) must be received by December 20, with completed applications due by February 7.
*Requirements:* Researchers with full-time staff or faculty appointments are encouraged to apply.
*Restrictions:* Applications from post-doctoral candidates will not be accepted. Costs not allowed under this award include: tuition; computer hardware or software for investigators; rent for laboratory or office spaces; or construction or renovation costs.
*Geographic Focus:* All States
*Date(s) Application is Due:* Feb 7
*Amount of Grant:* Up to 400,000 USD
*Contact:* Nico Stanculescu, Program Administrator; (312) 335-5747 or (312) 335-5862; fax (312) 335-4034; nico.stanculescu@alz.org or grantsapp@alz.org
*Internet:* www.alz.org/research/alzheimers_grants/types_of_grants.asp
*Sponsor:* Alzheimer's Association
225 North Michigan Avenue, Suite 1700
Chicago, IL 60601-7633

**Alzheimer's Association Everyday Technologies for Alzheimer Care Grants** 337

Everyday Technologies for Alzheimer Care (ETAC) is a cooperative research funding initiative sponsored by the Alzheimer's Association and Intel Corporation. ETAC seeks proposals on personalized diagnostics, preventive tools and interventions for adults coping with the spectrum of cognitive aging and neurodegenerative disease, particularly Alzheimer's disease. ETAC is designed to support exploratory multidisciplinary research that would not typically be funded by national health and science granting foundations. Minor iterations in testing plans or populations will not be considered for funding. Collaboration between social science/medical/public health and computer science/engineering researchers is valued. Mobile computing, high bandwidth sensing, robotics, imaging, face recognition, natural language processing, statistical modeling and a host of other technology advances allow unprecedented opportunities to study disease progression and therapeutic strategies in the context of everyday life. ETAC supports research that integrates such emerging technology capabilities with leading directions in behavioral science and biomedical research. Grants that merely create Internet-based versions of existing services or paper tools will not be considered. Submissions must be original ideas, not continuations of previously funded ETAC projects. Please see links provided below for examples of studies that have been funded by ETAC. Letters of Intent (LOIs) must be received by December 20, with completed applications due by February 7. The Association anticipates funding two awards under this program.
*Requirements:* Researchers with full-time staff or faculty appointments are encouraged to apply.
*Restrictions:* The Alzheimer's Association will not accept new research grant applications from currently funded Alzheimer's disease investigators who are delinquent in submitting interim/final scientific or interim/final financial reports on active grants. ETAC applications from post-doctoral candidates will not be accepted. Each total award is limited to $200,000 (direct and indirect costs) for up to three years. Requests in any given year may not exceed $90,000 (direct and indirect costs). Indirect costs are capped at 10 percent (rent for laboratory/office space is expected to be covered by indirect costs paid to the institution).
*Geographic Focus:* All States
*Date(s) Application is Due:* Feb 7

*Amount of Grant:* Up to 200,000 USD
*Contact:* Nico Stanculescu, Program Administrator; (312) 335-5747 or (312) 335-5862; fax (312) 335-4034; nico.stanculescu@alz.org or grantsapp@alz.org
*Internet:* www.alz.org/research/alzheimers_grants/types_of_grants.asp
*Sponsor:* Alzheimer's Association
225 North Michigan Avenue, Suite 1700
Chicago, IL 60601-7633

**Alzheimer's Association Investigator-Initiated Research Grants 338**

The program is structured to provide one to three years of sustained project support for independent, ongoing research. Proposals are solicited for basic, clinical, and social/behavioral research relevant to degenerative brain diseases such as Alzheimer's disease. Allowable costs for this award include the purchase and care of laboratory animals; small pieces of laboratory equipment and laboratory supplies; and salary for the principal investigator, scientific (including postdoctoral fellows) and technical staff (including laboratory technicians and modest secretarial support). It is required that most of the funds awarded under this program be used for direct research support. Allowable costs under this award include: purchase and care of laboratory animals; small pieces of laboratory equipment and laboratory supplies; computer software if used strictly for data collection; salary for the principal investigator, scientific (including post-doctoral fellows) and technical staff (including laboratory technicians and administrative support related directly to the funded project); and travel to scientific and professional meetings, not to exceed $1,000 per year. Letters of Intent (LOIs) must be received by December 20, with completed applications due by February 7.
*Requirements:* Public, private, domestic and foreign research laboratories, medical centers and hospitals, and universities are eligible to apply. Investigators from all stages of their research career development are encouraged to apply.
*Restrictions:* Your budget must not exceed the maximum amount of the award, $240,000 or $100,000 per year. Costs not allowed under this award include: tuition; computer hardware or standard software (e.g., Microsoft Office) for investigators; rent for laboratory or office space; or construction or renovation costs.
*Geographic Focus:* All States
*Date(s) Application is Due:* Feb 7
*Amount of Grant:* 100,000 - 240,000 USD
*Contact:* Nico Stanculescu, Program Administrator; (312) 335-5747 or (312) 335-5862; fax (312) 335-4034; nico.stanculescu@alz.org or grantsapp@alz.org
*Internet:* www.alz.org/research/alzheimers_grants/types_of_grants.asp
*Sponsor:* Alzheimer's Association
225 North Michigan Avenue, Suite 1700
Chicago, IL 60601-7633

**Alzheimer's Association Mentored New Investigator Research Grants to Promote Diversity 339**

The Mentored New Investigator Research Grant to Promote Diversity (MNIGD) is a three-year award intended to be a research-based and mentoring investment in the process of closing the health disparities gap between diverse and non-diverse investigator populations. The Alzheimer's Association feels strongly that the mentoring and involvement of diverse researchers in independently funded Alzheimer's research is a pressing need. The MNIRGD is intended to enhance the capacity of diverse and non-diverse scientists to conduct basic, clinical and social/behavioral research. Each MNIRGD award is limited to $170,000. A total of $150,000 will be awarded for costs related to the proposed research for up to three years (direct and indirect costs). Allowable costs under this award include: purchase and care of laboratory animals; small pieces of laboratory equipment and laboratory supplies; computer equipment if used strictly for data collection; travel (up to $1,000 per year); and salary for the principal investigator, scientific (including post-doctoral fellows) and technical staff (including laboratory technicians and administrative support related directly to the funded project) Letters of Intent (LOIs) must be received by December 20, with completed applications due by February 7.
*Requirements:* Eligibility for this grant competition is restricted to investigators who have less than 10 years of research experience after receipt of their terminal degree.
*Restrictions:* Each MNIRGD award is limited to $170,000. A total of $150,000 will be awarded for costs related to the proposed research for up to three years (direct and indirect costs). Requests in any given year may not exceed $60,000 (direct and indirect costs). Indirect costs are capped at 10 percent (rent for laboratory and/or office space is expected to be covered by indirect costs paid to the institution). Costs not allowed under this award include: tuition; computer hardware or software for investigators; rent for laboratory/office space; or construction or renovation costs.
*Geographic Focus:* All States
*Date(s) Application is Due:* Feb 7
*Amount of Grant:* Up to 170,000 USD
*Contact:* Nico Stanculescu, Program Administrator; (312) 335-5747 or (312) 335-5862; fax (312) 335-4034; nico.stanculescu@alz.org or grantsapp@alz.org
*Internet:* www.alz.org/research/alzheimers_grants/types_of_grants.asp
*Sponsor:* Alzheimer's Association
225 North Michigan Avenue, Suite 1700
Chicago, IL 60601-7633

**Alzheimer's Association Neuronal Hyper Excitability and Seizures in Alzheimer's Disease Grants 340**

The Alzheimer's Association is launching a new initiative to stimulate the development of new pharmacological strategies to prevent or treat seizures and abnormal neural network activity in Alzheimer's disease (AD). The Association's Request for Applications (RFAs) is aimed at the identification, screening and development of therapeutic strategies to reduce seizures and other types of abnormal neural network activity and at the evaluation of drug safety and efficacy at the preclinical and clinical levels. The RFA is designed to enable preliminary pilot research or proof-of-principle studies that can provide data for further research support by other funding agencies. Allowable costs under this award include: purchase and care of laboratory animals; small pieces of laboratory equipment and laboratory supplies; computer equipment if used strictly for data collection; travel (up to $1,000 per year); and salary for the principal investigator, scientific (including post-doctoral fellows) and technical staff (including laboratory technicians and administrative support staff whose work is directly related to the funded project). The Association anticipates funding up to two (2) Neuronal Hyper Excitability and Seizures in AD awards. Each award is limited to $400,000 (direct and indirect costs) for two to three years. Requests in any given year may not exceed $200,000 (direct and indirect costs). Indirect costs are capped at 10 percent (rent for laboratory/office space is expected to be covered by indirect costs paid to the institution). Letters of Intent (LOIs) must be received by December 20, with completed applications due by February 7.
*Requirements:* Researchers with full-time staff or faculty appointments are encouraged to apply.
*Restrictions:* Applications from post-doctoral candidates will not be accepted. Costs not allowed under this award include: tuition; computer hardware or software for investigators; rent for laboratory/office spaces; or construction or renovation costs.
*Geographic Focus:* All States
*Date(s) Application is Due:* Feb 7
*Amount of Grant:* Up to 400,000 USD
*Contact:* Nico Stanculescu, Program Administrator; (312) 335-5747 or (312) 335-5862; fax (312) 335-4034; nico.stanculescu@alz.org or grantsapp@alz.org
*Internet:* www.alz.org/research/alzheimers_grants/types_of_grants.asp
*Sponsor:* Alzheimer's Association
225 North Michigan Avenue, Suite 1700
Chicago, IL 60601-7633

**Alzheimer's Association New Investigator Research Grants 341**

The purpose of this program is to provide new investigators with funding that will allow them to develop preliminary or pilot data, to test procedures, and develop hypotheses which will then underpin the preparation of research grant applications to NIH, NSF, and other funding agencies and groups, including the Alzheimer's Association. All applications submitted to the program must focus on a question or questions in interventions for Alzheimer's disease to be considered responsive to the program announcement. Thirty awards will be made under this program. Letters of Intent (LOIs) must be received by December 20, with completed applications due by February 7.
*Requirements:* Public, private, domestic and foreign research laboratories, medical centers and hospitals, and universities are eligible to apply. Eligibility is restricted to investigators who have less than 10 years of research experience, including postdoctoral fellowships or residencies, after receipt of the doctoral degree. Applications from graduate and doctoral students for research projects, which will be used for the thesis or dissertation, will be accepted.
*Restrictions:* The Association anticipates funding 45 awards under this competition. Each total award is limited to $100,000 (direct and indirect costs) for up to two years. Requests in any given year may not exceed $60,000 (direct and indirect costs). Indirect costs are capped at 10 percent (rent for laboratory/office space is expected to be covered by indirect costs paid to the institution).
*Geographic Focus:* All States
*Date(s) Application is Due:* Feb 7
*Amount of Grant:* Up to 100,000 USD
*Contact:* Nico Stanculescu, Program Administrator; (312) 335-5747 or (312) 335-5862; fax (312) 335-4034; nico.stanculescu@alz.org or grantsapp@alz.org
*Internet:* www.alz.org/research/alzheimers_grants/types_of_grants.asp
*Sponsor:* Alzheimer's Association
225 North Michigan Avenue, Suite 1700
Chicago, IL 60601-7633

**Alzheimer's Association New Investigator Research Grants to Promote Diversity 342**

The New Investigator Research Grant to Promote Diversity (NIRGD) in Alzheimer's research is a two-year award to investigators who are currently underrepresented at academic institutions in Alzheimer's or related dementias research. The objective of this award is to increase the number of highly trained investigators from diverse backgrounds whose basic, clinical and social/behavioral research interests are grounded in the advanced methods and experimental approaches needed to solve problems related to Alzheimer's and related dementias in general and in health disparities populations. Each NIRGD award is limited to $100,000 (direct and indirect costs) for up to two years. Requests in any given year may not exceed $60,000 (direct and indirect costs). Letters of Intent (LOIs) must be received by December 20, with completed applications due by February 7.
*Requirements:* Eligibility to apply for this grant competition is restricted to investigators who have less than 10 years of research experience after receipt of their terminal degree.
*Restrictions:* Costs not allowed under this award include: tuition; computer hardware or software for investigators; rent for laboratory/office space; or construction or renovation costs. Your budget must not exceed the maximum amount of the award, $100,000 or $60,000 per year.
*Geographic Focus:* All States
*Date(s) Application is Due:* Feb 7
*Amount of Grant:* Up to 100,000 USD
*Contact:* Nico Stanculescu, Program Administrator; (312) 335-5747 or (312) 335-5862; fax (312) 335-4034; nico.stanculescu@alz.org or grantsapp@alz.org
*Internet:* www.alz.org/research/alzheimers_grants/types_of_grants.asp
*Sponsor:* Alzheimer's Association
225 North Michigan Avenue, Suite 1700
Chicago, IL 60601-7633

**Alzheimer's Association U.S.-U.K. Young Investigator Exchange Fellowships** 343

The U.S.-U.K. Young Investigator Exchange Fellowship provides a three-year grant to fund quality scientific research into the causes, diagnosis and treatment of Alzheimer's disease. The fellowship aims to address important research questions as well as to help and encourage promising scientists as they establish their careers within Alzheimer's research internationally. It is also hoped that by supporting meaningful scientific collaboration between scientists in the United Kingdom and the United States, there will be mutual benefit to the research output of both countries. In addition to an exchange of ideas, an important aim is to promote the learning of new experimental techniques and methodologies. The purpose of this fellowship is to provide new investigators with funding that will allow them to develop preliminary or pilot data, to test procedures and to develop hypotheses on an international level. The intent is to support international early-career development that will lay the groundwork for future research grant applications to other international funding agencies and groups. Allowable costs under this award include: purchase and care of laboratory animals; small pieces of laboratory equipment and laboratory supplies; computer equipment if used strictly for data collection; travel (up to $1,000 per year); and salary for the principal investigator, scientific (including post-doctoral fellows) and technical staff (including laboratory technicians and administrative support staff whose work is directly related to the funded project). The Alzheimer's Association and the Alzheimer's Reearch United Kingdom anticipate funding up to four awards (two originating from each country) under this competition. Each total award is limited to $300,000 USD. A total of $260,000 USD will be awarded for costs related to the proposed research for up to three years (direct and indirect costs). Requests in any given year may not exceed $90,000 USD (direct and indirect). Indirect costs are capped at 10 percent (rent for laboratory/office space is expected to be covered by indirect costs paid to the institution). The PI must commit to a 75 percent effort toward the proposed project over the funding period. Letters of Intent (LOIs) must be received by December 20, with completed applications due by February 7.
*Requirements:* Researchers with full-time staff or faculty appointments are encouraged to apply.
*Restrictions:* Applications from post-doctoral candidates will not be accepted. Costs not allowed under this award include: tuition; computer hardware or software for investigators; rent for laboratory/office spaces; or construction or renovation costs.
*Geographic Focus:* All States, United Kingdom
*Date(s) Application is Due:* Feb 7
*Amount of Grant:* Up to 300,000 USD
*Contact:* Nico Stanculescu, Program Director; (312) 335-5747 or (312) 335-5862; fax (312) 335-4034; nico.stanculescu@alz.org or grantsapp@alz.org
*Internet:* www.alz.org/research/alzheimers_grants/types_of_grants.asp
*Sponsor:* Alzheimer's Association
225 North Michigan Avenue, Suite 1700
Chicago, IL 60601-7633

**Alzheimer's Association Zenith Fellows Awards** 344

The Zenith Fellows award was initiated in 1991 to provide a vehicle for research support for donors with a substantial personal commitment to the advancement of Alzheimer's disease research. The objective of the Zenith Fellows Awards competition is to provide major support for investigators who have: contributed significantly to the field of Alzheimer's disease research or made significant contributions to other areas of science and are now beginning to focus more directly on problems related to Alzheimer's disease; and are likely to make substantial contributions in the future. We anticipate funding up to four awards under this competition. Each award is limited to $450,000 (direct and indirect costs) for three years. Requests in any given year may not exceed $250,000 (direct and indirect costs). Indirect costs are capped at 10 percent (rent for laboratory/office space is expected to be covered by indirect costs paid to the institution).
*Requirements:* Only established independent investigators are eligible as evidenced by: academic appointment; major, peer-reviewed, external multi-year grant support on which the applicant is the principal investigator (PI); independent laboratory operation; and quality and independence of publication record.
*Restrictions:* Previous recipients of Zenith Awards, Alzheimer's Disease Center Directors (P50 and P30), Medical and Scientific Advisory Council, and members of the National Board of the Alzheimer's Association are not eligible to apply.
*Geographic Focus:* All States
*Date(s) Application is Due:* Feb 7
*Amount of Grant:* Up to 450,000 USD
*Contact:* Nico Stanculescu, Program Administrator; (312) 335-5747 or (312) 335-5862; fax (312) 335-4034; nico.stanculescu@alz.org or grantsapp@alz.org
*Internet:* www.alz.org/research/alzheimers_grants/types_of_grants.asp
*Sponsor:* Alzheimer's Association
225 North Michigan Avenue, Suite 1700
Chicago, IL 60601-7633

**Ama OluKai Foundation Grants** 345

Preserving, protecting and celebrating the cultural heritage and Aloha spirit in Hawai'i is the mission of the Ama OluKai Foundation. The Ama OluKai Foundation targets programs that are directly designed to educate and serve Hawaiian communities. The Foundation's goal is to partner with organizations that promote the Hawaiian culture and heritage from its ancestral past to present day. Primary funding interests include: Supporting endemic organizations whose efforts are directed towards the development of quality communities and individuals in Hawai'i; Preserving the rich traditions and lessons from the past, with emphasis on the arts, sciences, language, customs, morals, physical activities and educational outreach; Supporting organizations that perpetuate the Aloha spirit.
*Requirements:* The Foundation offers grants to 501(c)3 non-profit organizations located in Hawai'i. Grants will be awarded to organizations participating directly with the community.
*Restrictions:* Organizations that re-grant, outsource or redistribute to other non-profits or beneficiaries are not eligible.
*Geographic Focus:* Hawaii
*Contact:* Dan McInerny; fax (808) 748-0835; Info@amaolukaifoundation.org
*Internet:* www.amaolukaifoundation.org/grant-request/
*Sponsor:* Ama OluKai Foundation
3023 B Loomis Street
Honolulu, 96822

**AMDA Foundation Quality Improvement Award** 346

The AMDA Foundation Quality Improvement Award is a program designed to encourage the development of innovative projects that will help to make a distinct impact on the quality of long term care. The Awards support initiatives that focus on facility staff education, quality improvement programs, research on interventions and treatment, and health literacy to directly enhance the quality of care provided to patients in long term care settings. One project that supports a Quality Improvement program will be selected for the $5,000 award. The winner will present their project results at a future educational session at AMDA's annual symposium. The AMDA Foundation Quality Improvement application should be completed and submitted online by the annual November 9 deadline.
*Requirements:* The program is open to all: AMDA members (including associate members); residents or fellows in an accredited training program; and mid-career and junior career faculty members of schools of medicine, osteopathy, nursing, or other health-related academic institutions. If the applicant is not an AMDA member the submission must be sponsored by an AMDA member.
*Geographic Focus:* All States
*Date(s) Application is Due:* Nov 9
*Amount of Grant:* 5,000 USD
*Contact:* Christine Ewing, Program Manager; (410) 992-3134; fax (410) 740-1318; cewing@amdafoundation.org
*Internet:* www.amdafoundation.org/index.php/our-work/recognize-awards/amda-foundation-qi-awards
*Sponsor:* American Medical Directors Association Foundation
11000 Broken Land Parkway, Suite 405
Columbia, MD 21044

**American Academy of Advertising Distinguished Service Award** 347

The American Academy of Advertising (AAA), founded in the late 1950's, is an organization of advertising scholars and professionals. The academy fosters research that is relevant to the advertising field and provides a forum for the exchange of ideas among its academic and professional members. The Distinguished Service Award is given by AAA to individuals who have rendered distinguished service to the organization and/or to advertising education. Nominees should be champions of advertising education, active in providing opportunities for the professional development of students and faculty, and should unselfishly perform the service for which they are nominated. Nominations and supporting documentation should be submitted via email to the Chair of the Awards Committee on or before the submission deadline. Deadlines may vary from year to year; Academy members who wish to nominate are encouraged to visit AAA's website to confirm current deadline dates and to obtain complete guidelines for submission of nominations.
*Requirements:* A nominee need not be a member of the AAA to be considered for the Distinguished Service Award; however, only AAA members may submit a nomination.
*Geographic Focus:* All States, All Countries
*Date(s) Application is Due:* Nov 15
*Contact:* Dr. Jef I. Richards, Chair, AAA Awards Committee; (512) 471-1101; fax (512) 471-7018; jef@mail.utexas.edu
*Internet:* www.aaasite.org/Award_DistinguishedService.html
*Sponsor:* American Academy of Advertising
24710 Shaker Boulevard
Beachwood, OH 44122

**American Academy of Advertising Doctoral Dissertation Competition Awards** 348

The American Academy of Advertising (AAA), founded in the late 1950's, is an organization of advertising scholars and professionals. The academy fosters research that is relevant to the advertising field and provides a forum for the exchange of ideas among its academic and professional members. Members' research findings and scholarly contributions are disseminated through the academy's annual conference proceedings and journals (the Journal of Advertising, the Journal of Current Issues and Research in Advertising, and the Journal of Interactive Advertising). The purpose of the academy's annual Doctoral Dissertation Competition is to promote doctoral research in advertising. Awards are based on a competitive review of dissertation proposals; any topic in advertising may be addressed. Each award is in the range of $1,000 to $2,500. In addition to the standard awards, the competition gives the Dunn Award for outstanding proposals in the area of international advertising. Recipients receive half their award at the time of selection and half when their dissertation has been successfully defended. All submissions must reach the Chair of the Research Committee electronically by 5 p.m. on the deadline date. Complete guidelines for submission and format of proposals are on the academy's website. Application deadlines may vary from year to year and should be verified at the academy's website.
*Requirements:* Only members of the American Academy of Advertising working on their dissertation at the time of proposal submission are eligible for these awards. Award recipients must maintain their membership until they complete their project. Applicants must be currently enrolled in a graduate program.

*Restrictions:* Submissions based on completed or near completed dissertations are not eligible; submissions should be in the proposal stage. Recipients have three years from the time of the award to complete their dissertation.
*Geographic Focus:* All States, All Countries
*Date(s) Application is Due:* Nov 2
*Amount of Grant:* 1,000 - 2,500 USD
*Contact:* Dr. Eric Haley, Chair; (865) 974-5102; haley@utk.edu
*Internet:* www.aaasite.org/Doctoral_Dissertation_Competition.html
*Sponsor:* American Academy of Advertising
24710 Shaker Boulevard
Beachwood, OH 44122

**American Academy of Advertising Ivan Preston Outstanding Contribution to Research on Advertising Award** 349

The American Academy of Advertising (AAA), founded in the late 1950's, is an organization of advertising scholars and professionals. AAA fosters research that is relevant to the advertising field and provides a forum for the exchange of ideas among its academic and professional members. AAA's Ivan Preston Outstanding Contribution to Research on Advertising Award is designed to honor an individual who has made an outstanding contribution to the discipline of advertising through a systematic and sustained program of published research. The award carries a $1,000 prize. Nominations must be submitted electronically to the AAA Research Committee Chair by the deadline date; nominations should include a letter of nomination, at least two supporting letters from leading scholars, the nominee's curriculum vitae, and copies of the nominee's five most influential works (in the case of books, a representative chapter my be submitted). The winner will be notified prior to the AAA annual conference and is required to attend the conference to receive the award. Deadlines may vary from year to year; nominators are encouraged to check the AAA website for current deadline dates and complete guidelines for submitting a nomination.
*Requirements:* To be eligible for the award, a nominee must satisfy the following *Requirements:* have received a doctorate more than 20 years ago; have been promoted to rank of Full Professor prior to nomination; and have been an active member of AAA for the last five consecutive years. Active membership is defined as having participated in AAA annual conference activities as a reviewer, a presenter, a track chair, a session chair, a discussant, and/or a special-topics-session participant. Additionally active membership should include service in at least one, but preferably two of the following ways: as an AAA committee member or chair; as an AAA officer; as a Webmaster; as a newsletter editor; or as a Journal of Advertising reviewer or Review Board member.
*Restrictions:* The Ivan Preston award will be given only to active and contributing AAA members for exemplary research on advertising and is not necessarily awarded every year.
*Geographic Focus:* All States, All Countries
*Date(s) Application is Due:* Nov 2
*Amount of Grant:* 1,000 USD
*Contact:* Dr. Eric Haley, Chair; (865) 974-5102; haley@utk.edu
*Internet:* www.aaasite.org/IPAward_Research.html
*Sponsor:* American Academy of Advertising
24710 Shaker Boulevard
Beachwood, OH 44122

**American Academy of Advertising Kim Rotzoll Award for Advertising Ethics and Social Responsibility** 350

The American Academy of Advertising (AAA), founded in the late 1950's, is an organization of advertising scholars and professionals. The academy fosters research that is relevant to the advertising field and provides a forum for the exchange of ideas among its academic and professional members. The Kim Rotzoll Award was established by AAA in 2004 and recognizes individuals or organizations who have demonstrated an outstanding commitment to advertising ethics and social responsibility. Nominations and supporting documentation should be submitted via email to the Chair of the Awards Committee on or before the submission deadline. Deadlines may vary from year to year; Academy members who wish to nominate are encouraged to visit AAA's website to confirm current deadline dates and to obtain complete guidelines for submission of nominations.
*Requirements:* Nominees need not be members of AAA, but only AAA members may submit a nomination. Nominees must evidence lifetime achievement in the fields of advertising ethics or social responsibility. For academics, such evidence may be in the form of scholarly publications; for advertising practitioners or organizations, the results of specific programs or activities that have had a positive impact on society may serve as evidence.
*Restrictions:* The Rotzoll Award is not to be considered an annual award and will be given only to individuals or organizations who have demonstrated an outstanding commitment to advertising ethics and social responsibility.
*Geographic Focus:* All States, All Countries
*Date(s) Application is Due:* Nov 15
*Contact:* Dr. Jef I. Richards, Chair, AAA Awards Committee; (512) 471-1101; fax (512) 471-7018; jef@mail.utexas.edu
*Internet:* www.aaasite.org/Award_KimRotzoll.html
*Sponsor:* American Academy of Advertising
24710 Shaker Boulevard
Beachwood, OH 44122

**American Academy of Advertising Research Fellowships** 351

The American Academy of Advertising (AAA), founded in the late 1950's, is an organization of advertising scholars and professionals. The academy fosters research that is relevant to the advertising field and provides a forum for the exchange of ideas among its academic and professional members. The academy's Research Fellowship Competition promotes the continued scholarship of professors and advertising professionals who have completed their education and are doing research in advertising. Awards are based on a competitive review of research proposals; any topic that is appropriate for potential publication in the academy's Journal of Advertising may be addressed. Each award is typically in the range of $1,000 to $3,000. Fellows receive half their award at time of selection and half when the first completed paper is submitted to the Journal of Advertising. Complete guidelines for submission and format of proposals are at the academy's website. All submissions must reach the Chair of the Research Committee electronically by 5 p.m. on the deadline date. Winners will be announced at the academy's annual conference and communicated to the membership through the AAA newsletter. Application deadlines may vary from year to year and should be verified at the academy's website.
*Requirements:* Applicants must comply with the following *Requirements:* be current members of the academy; maintain membership until they complete the project; grant the Journal of Advertising first right of refusal on any papers resulting from the supported research; and acknowledge the AAA Fellowship in all publications resulting from the project. Winners are asked to publicize the award on their campuses and in their communities as widely as possible. Researchers who previously have received an AAA Research Fellowship are eligible for the award upon completion or termination of the previous award. Researchers who have submitted proposals that have not been awarded support in the past may resubmit proposals for similar projects.
*Restrictions:* While members may submit as many single or joint proposals as desired within a given year, authors cannot receive funding for more than one proposal; should more than one proposal containing an author's name receive high rankings from the reviewers, only the highest ranking proposal will be funded. Winners have three years from the time of the award to complete their project and are not eligible for additional AAA research fellowships while the award is active. Doctoral students are not eligible for these fellowships. Joint research with doctoral students is likewise excluded unless the research will not be part of the student's dissertation.
*Geographic Focus:* All States, All Countries
*Date(s) Application is Due:* Sep 14
*Amount of Grant:* 1,000 - 3,000 USD
*Contact:* Dr. Eric Haley, Chair; (865) 974-5102; haley@utk.edu
*Internet:* aaasite.org/Award_ResearchFellowship.html
*Sponsor:* American Academy of Advertising
24710 Shaker Boulevard
Beachwood, OH 44122

**American Academy of Arts and Sciences Hellman Fellowships** 352

As part of the Initiative for Science, Engineering, and Technology, the American Academy has established the the Hellman Fellowship in Science and Technology Policy for an early-career professional with training in science or engineering who is interested in transitioning to a career in public policy and administration. While in residence, the Hellman Fellow will work with senior scientists and policy experts on critical national and international policy issues related to science, engineering, and technology. The focus of the work will be on one or more of the ongoing projects under the Academy's Initiative for Science, Engineering, and Technology to which the Hellman Fellow will contribute substantively. The mission of the Initiative is to examine, in broad terms, how the world of science and technology is evolving, how to help the public understand these changes, and how society can better adapt. The Initiative brings together scientists and public policy experts in a neutral setting, outside of the constraints of the political process.
*Requirements:* Applicants must have a Ph.D. in an area of science or engineering and have some experience or a demonstrated interest in an area related to science and technology policy. Masters degrees may be considered in the fields of engineering and computer science. Strong writing and organizational skills are desired. Candidates must be U.S. citizens, permanent residents, or current employees of an academic or professional organization in the United States.
*Geographic Focus:* All States
*Date(s) Application is Due:* Jan 17
*Contact:* Hillary Dobel, Program Coordinator; (617) 576-5024; fax (617) 576-5050; hdobel@amacad.org or vsp@amacad.org
*Internet:* www.amacad.org/hellman.aspx
*Sponsor:* American Academy of Arts and Sciences
136 Irving Street
Cambridge, MA 02143

**American Academy of Arts and Sciences Visiting Scholars** 353

The Academy's Visiting Scholars Program provides residential fellowships for junior faculty members and postdoctoral scholars in the humanities and social sciences. The fellowship program offers scholars a year for research and writing free from teaching and administrative duties, a collaborative work environment, and the opportunity to interact with Academy members. It also creates a national network for these scholars, assisting them in their research and professional development. The Academy seeks proposals in the humanities and social sciences relating to American history, culture, and public policy from the founding period to the present. The academy also welcomes proposals that explore the impact of scientific and technological advances over the past two centuries on American institutions, humanities and culture in America, American foreign policy, and global security. Proposals should take into account the academy's emphasis on interdisciplinary work, as well as its interest in broadening public understanding of important intellectual trends and contemporary policy choices. In addition to pursuing individual projects, scholars participate in activities of the academy; collaboration with academy fellows is encouraged. Visiting scholars remain in residence at the academy throughout the academic year (September-May), with some fellowships eligible for a one-year renewal. Faculty remain in residence for one academic year, although one-semester fellowships may be considered. Scholars receive office space, computer services,

library privileges, and assistance in locating housing. Health benefits can be arranged. Applicants are responsible for contacting three references and ensuring that their letters are mailed to the academy by the postmark deadline. Faculty can receive up to $65,000 (not to exceed one-half of current salary). Postdoctoral students receive an annual stipend of up to $45,000. The Academy provides office space, computer services, library privileges, and information on locating housing.
*Requirements:* Preference is given to untenured junior faculty but the program is also open to qualified postdocs. Candidates must be U.S. citizens, permanent residents, or current employees of an academic or professional organization in the United States. The Ph.D., J.D, or equivalent professional training (e.g., public policy) should have been completed within the last 10 years (although exceptional circumstances will be taken into consideration). Graduate student applicants must complete all degree requirements by August 1.
*Restrictions:* Research trips, interviews, attendance at scholarly meetings or speaking engagements must be limited to no more than twenty days during the fellowship term, and scholars must not accept teaching appointments or other major commitments during their fellowship. Applications must be submitted electronically.
*Geographic Focus:* All States
*Date(s) Application is Due:* Oct 22
*Amount of Grant:* 45,000 - 65,000 USD
*Contact:* Hilary Dobel, Program Coordinator; (617) 576-5024; fax (617) 576-5050; hdobel@amacad.org or vsp@amacad.org
*Internet:* https://www.amacad.org/content/about/about.aspx?d=363&t=4&s=0
*Sponsor:* American Academy of Arts and Sciences
136 Irving Street
Cambridge, MA 02143

**American Automatic Control Council Control Practice Award** 354

The Control Engineering Practice Award is given to one individual or one team to be selected from those nominated for significant contribution to the advancement of control practice. The primary criterion for selection will be for the application and implementation of innovative control concepts, methodology, and technology, for the planning, design, manufacture, and operation of control systems. Achievement and usefulness will be evidenced by the benefit to society and by the degree of acceptance by those who use control as a tool. The work on which the nomination is based must have been performed while the nominated individual or at least one member of the team was a resident of the USA. The award consists of a certificate and an honorarium. In the event that the winner is a team, each member of the team will receive a certificate and the honorarium will be divided equally among the team members. Nomination forms as well as additional information may be obtained from the AACC secretariat.
*Requirements:* Nomination packages should consist of a letter of nomination, three to five reference letters, current resume, current list of publications and patents, any other supporting material, and a nomination form. The nomination letter should clearly identify the primary reason that the nominee should receive the award, as well as four ancillary reasons. All materials should be submitted in a single package.
*Geographic Focus:* All States
*Date(s) Application is Due:* Dec 15
*Contact:* B. Wayne Bequette, AACC Secretariat; (518) 276-6683; fax (518) 276-4030
*Internet:* a2c2.org/awards/control-engineering-practice-award
*Sponsor:* American Automatic Control Council
2145 Sheridan Road
Evanston, IL 60208-3118

**American Automatic Control Council Donald P. Eckman Award** 355

The Donald P. Eckman Award is given for outstanding accomplishments by a young engineer in the field of automatic control. Nominees must be younger than 35 years on January 1 of the year of the award and have made contributions to the field while residing in the United States. Contributions may be technical or scientific publications, theses, patents, inventions, or combinations or the above in the field of automatic control. The award consists of a certificate and honorarium, which are presented at the annual conference. Nomination forms as well as additional information may be obtained from the AACC secretary.
*Requirements:* Nomination packages should consist of letter of nomination, three to five letters of reference, current resume, current list of publications and patents, any other supporting material, and nomination form. The nomination letter should clearly identify the primary reason that the nominee should receive the award, as well as up to four ancillary reasons. Supporting documentation must include a full endorsement by at least one responsible supervisor.
*Geographic Focus:* All States
*Contact:* B. Wayne Bequette, AACC Secretariat; (518) 276-6683; fax (518) 276-4030
*Internet:* a2c2.org/awards/donald-p-eckman-award
*Sponsor:* American Automatic Control Council
2145 Sheridan Road
Evanston, IL 60208-3118

**American Automatic Control Council Hugo Schuck Award** 356

The Hugo Schuck Award is given to recognize the best two papers presented at the previous ACC. One award is for a paper emphasizing contributions to theory and the other emphasizing significant or innovative applications. The papers must have been presented by the awardee or a coauthor. Criteria for selection include the quality of the written and oral presentation, the technical contribution, timeliness, and practicality. Additional information may be obtained from the AACC secretariat.
*Geographic Focus:* All States
*Date(s) Application is Due:* Dec 15
*Contact:* B. Wayne Bequette, AACC Secretariat; (518) 276-6683; fax (518) 276-4030
*Internet:* a2c2.org/awards/o-hugo-schuck-best-paper-award
*Sponsor:* American Automatic Control Council
2145 Sheridan Road
Evanston, IL 60208-3118

**American Automatic Control Council John R. Ragazzini Education Award** 357

The John R. Ragazzini Education Award is given to recognize outstanding contributions to automatic control education in any form. These contributions can be from any source and in any media, i.e., electronic, publications, courses, etc. Nomination forms as well as additional information may be obtained from the AACC secretariat.
*Requirements:* Nomination packages should consist of a letter of nomination, three to five reference letters, current resume, current list of publications and patents, any other supporting material, and a nomination form. The nomination letter should clearly identify the primary reason that the nominee should receive the award, as well as four ancillary reasons. All materials should be submitted in a single package.
*Geographic Focus:* All States
*Date(s) Application is Due:* Dec 15
*Contact:* B. Wayne Bequette, AACC Secretariat; (518) 276-6683; fax (518) 276-4030
*Internet:* a2c2.org/awards/john-r-ragazzini-education-award
*Sponsor:* American Automatic Control Council
2145 Sheridan Road
Evanston, IL 60208-3118

**American Automatic Control Council Richard E. Bellman Control Heritage Award** 358

The Richard E. Bellman Control Heritage Award recognizes distinguished career contributions to the theory or applications of automatic control. It is the highest recognition of professional achievement for U.S. control systems engineers and scientists. The nominee must have spent a significant part of his/her career in the United States. The award consists of a certificate and honorarium, which are presented at the council's annual conference. The awardee is expected to make a short acceptance speech at the AACC Awards Ceremonies. Contact the AACC secretary for the nomination form and further information.
*Requirements:* Nomination packages should consist of a letter of nomination, three to five reference letters, current resume, current list of publications and patents, any other supporting material, and a nomination form. The nomination letter should clearly identify the primary reason that the nominee should receive the award, as well as four ancillary reasons. All materials should be submitted in a single package.
*Geographic Focus:* All States
*Contact:* B. Wayne Bequette, AACC Secretariat; (518) 276-6683; fax (518) 276-4030
*Internet:* a2c2.org/awards/richard-e-bellman-control-heritage-award
*Sponsor:* American Automatic Control Council
2145 Sheridan Road
Evanston, IL 60208-3118

**American Board of Emergency Medicine Fellowships** 359

The overall purpose of the American Board of Emergency Medicine (ABEM) Fellowship is to provide talented, early-career health science scholars in emergency medicine with the opportunity to experience and participate in evidence-based healthcare or public health studies that improve the care and access to care of patients in domestic and global health care systems. The program especially welcomes nominations of under-represented minority candidates. Diplomates or Active Candidates for Certification by ABEM, who hold non-tenured faculty positions in any university, are eligible for the program. The ABEM Fellowship is awarded for a two-year period. During this time, the fellow is expected to continue their work at his/her primary academic/research post, while being assigned to a board of the IOM. The boards are: Children, Youth, and Families; Food and Nutrition; Global Health; Health Care Services; Health of Select Populations; Health Sciences Policy; and Population Health and Public Health Practice. A flexible research stipend of $25,000 will be awarded to each ABEM Fellow. The stipend will be administered through the appropriate department in the Fellow's home institution. Stipends are not intended for use as salary offsets. The Call for Nominations begins on April 1 each year, with the deadline for submission being June 8.
*Requirements:* The fellowship requires a ten-to-twenty percent commitment of time for two years, and includes attendance at a one-week orientation to health policy in October, the Fall IOM Annual Meeting and the meetings of the assigned board. Additionally, the fellow will participate actively in the work of an appropriate expert study committee or round table, including contributing to its reports or other products. Studies will be identified that would enable the potential for the fellow to contribute to the value of integrating medicine and public health.
*Geographic Focus:* All States
*Date(s) Application is Due:* Jun 8
*Amount of Grant:* 25,000 USD
*Contact:* Marie E. Michnich; (202) 334-1506; fax (202) 334-3862; mmichnich@nas.edu
*Internet:* www.nationalacademies.org/hmd/Activities/Education/EmergencyMedicine.aspx
*Sponsor:* National Academies
500 5th Street, NW, Room 508
Washington, D.C. 20001

**American Chemical Society AGFD Advancement of the Application of Agricultural and Food Chemistry Award** 360

The purpose of the American Chemical Society Division of Agricultural and Food Chemistry (AGFD) Advancement of the Application of Agricultural and Food Chemistry Award is to recognize and encourage outstanding contributions to pure or applied agricultural and food chemistry. The Award consists of $3,000, a plaque, and

meeting travel expenses, and is co-sponsored by the AGFD and International Flavors and Fragrances. The annual deadline is November 1.
*Requirements:* Applicants must have made an outstanding application of chemistry or chemical technology to the solution of agricultural or food problems of importance to the nourishment and health of mankind or outstanding contributions to the advancement of pure or applied agricultural and food chemistry, special consideration given to independence of thought and the originality shown, or to the importance of the work when applied to public welfare or the needs of humanity, must be actively engaged in the line of work, granted without regard to sex, creed, or nationality.
*Geographic Focus:* All States
*Date(s) Application is Due:* Nov 1
*Amount of Grant:* 3,000 - 4,500 USD
*Contact:* Neil Da Costa, AGFD Chair; (732) 335-2110; neil.dacosta@iff.com
*Internet:* webapps.acs.org/findawards/detail.jsp?ContentId=WPCP_012439
*Sponsor:* American Chemical Society
1155 Sixteenth Street, NW
Washington, D.C. 20036-4801

**American Chemical Society AGFD Donald A. Withycombe Graduate Fellowship Award in Food Chemistry** 361
The purpose of the Division of Agricultural and Food Chemistry (AGFD) Donald A. Withycombe Graduate Fellowship Award in Food Chemistry is to support and encourage graduate students who display outstanding potential early in their program. One prize will be awarded comprising of a cash award of $1000. Travel expenses will be paid up to $1,000. The annual deadline for applications is October 31.
*Requirements:* Applicants must have outstanding and beginning graduate grade point averages (GPA) and show promise of an excellent graduate research career.
*Geographic Focus:* All States
*Date(s) Application is Due:* Oct 31
*Amount of Grant:* 1,000 - 2,000 USD
*Contact:* Neil Da Costa, AGFD Chair; (732) 335-2110; neil.dacosta@iff.com
*Internet:* webapps.acs.org/findawards/detail.jsp?ContentId=WPCP_012447
*Sponsor:* American Chemical Society
1155 Sixteenth Street, NW
Washington, D.C. 20036-4801

**American Chemical Society AGFD Teranishi Graduate Fellowship Award** 362
The purpose of the Division of Agricultural and Food Chemistry (AGFD) Teranishi Graduate Fellowship Award is to recognize an AGFD member who has made outstanding scientific contributions to the field of agricultural and food chemistry. The award includes $2,500 and an appropriate plaque. The annual deadline for applications is February 1.
*Requirements:* Only AGFD member may apply.
*Geographic Focus:* All States
*Date(s) Application is Due:* Feb 1
*Amount of Grant:* 2,000 USD
*Contact:* Neil Da Costa, AGFD Chair; (732) 335-2110; neil.dacosta@iff.com
*Internet:* webapps.acs.org/findawards/detail.jsp?ContentId=WPCP_012454
*Sponsor:* American Chemical Society
1155 Sixteenth Street, NW
Washington, D.C. 20036-4801

**American Chemical Society AGFD Undergraduate Research Symposium Travel Grants** 363
American Chemical Society AGFD Undergraduate Research Symposium Travel Grants are offered to those students chosen to participate in the symposium, and each will receive travel expenses up to $1,000 to attend and present their research at the Spring meeting of the ACS. The AGFD Advisory Board will choose the students who will participate in the symposium. Two prizes will be awarded. First prize will receive a cash award of $750. Second place will receive a cash award of $250. The annual deadline for applications is October 31.
*Geographic Focus:* All States
*Date(s) Application is Due:* Oct 31
*Contact:* Neil Da Costa, AGFD Chair; (732) 335-2110; neil.dacosta@iff.com
*Internet:* agfd.sites.acs.org/awards.htm
*Sponsor:* American Chemical Society
1155 Sixteenth Street, NW
Washington, D.C. 20036-4801

**American Chemical Society AGFD Young Scientist Award** 364
The primary purpose of the American Chemical Society AGFD Young Scientist Award is to recognize outstanding research in agricultural and food chemistry by a young scientist. The winner receives a $1000 honorarium, an engraved plaque and paid expenses (up to $1500) to the ACS meeting to present a paper and to receive the award.
*Requirements:* Each applicant must be an AGFD member and have earned last degree within eight years of October 1 of the presentation year.
*Geographic Focus:* All States
*Amount of Grant:* 1,000 USD
*Contact:* Neil Da Costa, Chair; (732) 335-2110; neil.dacosta@iff.com
*Internet:* webapps.acs.org/findawards/detail.jsp?ContentId=WPCP_012462
*Sponsor:* American Chemical Society
1155 Sixteenth Street, NW
Washington, D.C. 20036-4801

**American Chemical Society AGRO Award for Innovation in Chemistry of Agriculture** 365
The American Chemical Society AGRO Award for Innovation in Chemistry of Agriculture is given to an active researcher working in North America for a chemical innovation that significantly enhances agricultural or veterinary pest management and productivity. The awardee will be asked to give an award address at the National ACS meeting. Nominations must be received by the committee chair by September 30 of each year. Balloting will be conducted in November/December and results will be announced shortly thereafter.
*Geographic Focus:* All States
*Date(s) Application is Due:* Sep 30
*Contact:* Dr. James Seiber, Awards Committee Chair; (530) 752-1465 or (530) 752-1141; jnseiber@ucdavis.edu
*Internet:* www.agrodiv.org/awards/
*Sponsor:* American Chemical Society
1155 Sixteenth Street, NW
Washington, D.C. 20036-4801

**American Chemical Society AGRO Education Awards** 366
The AGRO Division has established an endowment fund in collaboration with Bayer CropScience that will be used to promote an understanding of the role of chemistry in agriculture. The purpose of the American Chemical Society AGRO Education Awards is to promote an understanding of the role of chemistry in agriculture as embraced in the following areas related to pest management chemistry: synthesis; metabolism; regulatory; biotechnology; delivery; risk assessment; resistance; residues; mode of action; and fate and/or behavior. To apply, a graduate student should submit the application materials to be received no later than March 1.
*Requirements:* An applicant must be a graduate student.
*Geographic Focus:* All States
*Date(s) Application is Due:* Mar 1
*Contact:* Dr. James Seiber; (530) 752-1465 or (530) 752-1141; jnseiber@ucdavis.edu
*Internet:* webapps.acs.org/findawards/detail.jsp?ContentId=WPCP_012468
*Sponsor:* American Chemical Society
1155 Sixteenth Street, NW
Washington, D.C. 20036-4801

**American Chemical Society AGRO Fellow Awards** 367
The American Chemical Society Division of Agrochemicals has established the Fellow Awards to recognize its members whose dedicated and enthusiastic service has kept the Division moving forward. Nomination criteria shall be continued and substantial contributions of time, talents, and service to the AGRO Division of ACS, and to the agrochemical science over a period of at least six years. Nominations include a letter, noting the contributions to the Division, and a current curriculum vitae. Deadline for submitting nominations is May 31 of each year.
*Geographic Focus:* All States
*Date(s) Application is Due:* May 31
*Contact:* Dr. James Seiber; (530) 752-1465 or (530) 752-2490; jnseiber@ucdavis.edu
*Internet:* www.agrodiv.org/fellow-awards/
*Sponsor:* American Chemical Society
1155 Sixteenth Street, NW
Washington, D.C. 20036-4801

**American Chemical Society AGRO International Award for Research in Agrochemicals** 368
The ACS International Award for Research in Agrochemicals is given to a scientist who has made outstanding contributions to the field of agrochemicals at the international level. Their vision and sustained contributions will have opened new horizons for other investigators in their field and beyond. Nominations are due by December 31.
*Requirements:* The award is given to an individual without regard to nationality for outstanding contributions to research in agrochemicals.
*Geographic Focus:* All States
*Date(s) Application is Due:* Dec 31
*Contact:* Dr. James Seiber; (530) 752-1465 or (530) 752-1141; jnseiber@ucdavis.edu
*Internet:* www.agrodiv.org/awards/
*Sponsor:* American Chemical Society
1155 Sixteenth Street, NW
Washington, D.C. 20036-4801

**American Chemical Society AGRO Sterling B. Hendricks Memorial Lectureship** 369
The USDA-Agricultural Research Service (ARS) is seeking nominations for the Sterling B. Hendricks Memorial Lectureship Award. This Lectureship was established in 1981 by ARS to honor the memory of Sterling B. Hendricks and to recognize scientists who have made outstanding contributions to the chemical science of agriculture. Dr. Hendricks contributed to many diverse scientific disciplines, including soil science, mineralogy, agronomy, plant physiology, geology, and chemistry. He is most frequently remembered for discovering phytochrome, the light-activated molecule that regulates many plant processes. The lecture should address a scientific topic, trend, or policy issue related to agriculture. The lectureship is presented at an AGFD symposium in even-numbered years, and in an AGRO symposium in odd-numbered years. The award includes an honorarium of $2000, a bronze medallion, and expenses to attend the meeting. The annual deadline for nominations is November 5.
*Requirements:* Nominees will be outstanding senior scientists in industry, universities, consulting, or government positions.

*Restrictions:* Current ARS employees are not eligible.
*Geographic Focus:* All States
*Date(s) Application is Due:* Nov 5
*Amount of Grant:* 2,000 - 3,500 USD
*Contact:* Dr. James Seiber; (530) 752-1465 or (530) 752-1141; jnseiber@ucdavis.edu
Kim Kaplan, Lecture Coordinator; (301) 504-1637
*Internet:* webapps.acs.org/findawards/detail.jsp?ContentId=WPCP_012471
*Sponsor:* American Chemical Society
1155 Sixteenth Street, NW
Washington, D.C. 20036-4801

**American Chemical Society AGRO Young Scientist Research Awards 370**
AGRO, a Division of the American Chemical Society, brings together a worldwide community of scientists and stakeholders to advance knowledge and promote innovative solutions for the protection of agricultural productivity, public health and the environment. The purpose of the American Chemical Society AGRO Young Scientist Research Awards is to recognize outstanding agrochemical research by pre- or postdoctoral students.
*Requirements:* Eligibility includes only pre- and postdoctoral students whose work has been conducted primarily at U.S. institutions.
*Geographic Focus:* All States
*Contact:* Dr. James Seiber; (530) 752-1465 or (530) 752-1141; jnseiber@ucdavis.edu
*Internet:* webapps.acs.org/findawards/detail.jsp?ContentId=WPCP_012474
*Sponsor:* American Chemical Society
1155 Sixteenth Street, NW
Washington, D.C. 20036-4801

**American Chemical Society Alfred Bader Award in BioInorganic or BioOrganic Chemistry 371**
This Alfred Bader Award in BioInorganic or BioOrganic Chemistry is given annually to recognize outstanding contributions to bioorganic or bioinorganic chemistry. The award will be granted for outstanding research accomplishments without regard to age or nationality. The award is intended to recognize significant accomplishments that are at the interface between biology and organic or inorganic chemistry. Special consideration will be given to applications of the fundamental principles and experimental methodology of chemistry to areas of biological significance. The award consists of $5,000 and a certificate. Up to $2,500 for travel expenses to the meeting at which the award will be presented will be reimbursed.
*Requirements:* Any individual, except a member of the awards committee, may submit one nomination or seconding letter for the award in any given year. The nominating documents consist of a letter of not more than 1000 words containing an evaluation of the nominee's accomplishments and a specific identification of the work to be recognized, a biographical sketch including date of birth, and a list of publications and patents authored by the nominee.
*Restrictions:* Self-nominations are not accepted.
*Geographic Focus:* All States
*Date(s) Application is Due:* Nov 1
*Amount of Grant:* 5,000 - 7,500 USD
*Contact:* Awards Administrator; (800) 227-5558 or (202) 872-4408; fax (202) 776-8008; awards@acs.org
*Internet:* webapps.acs.org/findawards/detail.jsp?ContentId=CTP_004491
*Sponsor:* American Chemical Society
1155 Sixteenth Street, NW
Washington, D.C. 20036-4801

**American Chemical Society Alfred Burger Award in Medicinal Chemistry 372**
This American Chemical Society Alfred Burger Award in Medicinal Chemistry, supported by GlaxoSmithKline in 1978, is awarded biennially in even-numbered years to recognize outstanding contributions to research in medicinal chemistry. The award is granted without regard to age or nationality, and the recipient presents an award address at the spring meeting of the ACS Division of Medicinal Chemistry. Applications are accepted in odd-numbered years. The award consists of $3,000 and a certificate. Up to $2,500 for travel expenses to the meeting at which the award will be presented will be reimbursed. The award is presented biennially in even-numbered years, and the recipient will present an award address at the spring meeting of the ACS Division of Medicinal Chemistry.
*Requirements:* Any individual, except a member of the award committee, may submit one nomination or seconding letter for the award in any given year. The nominating documents consist of a letter of not more than 1000 words containing an evaluation of the nominee's accomplishments and a specific identification of the work to be recognized, a biographical sketch including date of birth, and a list of publications and patents authored by the nominee.
*Restrictions:* Self-nominations are not accepted.
*Geographic Focus:* All States
*Date(s) Application is Due:* Nov 1
*Amount of Grant:* 3,000 - 5,500 USD
*Contact:* Awards Administrator; (800) 227-5558 or (202) 872-4408; fax (202) 776-8008; awards@acs.org
*Internet:* webapps.acs.org/findawards/detail.jsp?ContentId=CTP_004491
*Sponsor:* American Chemical Society
1155 Sixteenth Street, NW
Washington, D.C. 20036-4801

**American Chemical Society ANYL Arthur F. Findeis Award for Achievements by a Young Analytical Scientist 373**
The purpose of the American Chemical Society ANYL Arthur F. Findeis Award for Achievements by a Young Analytical Scientist is to recognize and encourage outstanding contributions to the fields of analytical chemistry by a young analytical scientist. Evidence shall be presented for one or more of the following outstanding accomplishments: conceptualization and development of unique instrumentation that has had an enabling impact upon analytical chemistry and has substantively advanced the field; development of novel and important analytical methods or methodologies that have found significant beneficial applications in the chemical sciences; elucidation of fundamental events or processes involved in or important to analytical chemistry; authorship of books, patents, and/or research papers that have had an influential role in the development of analytical chemistry; or other significant contributions to the furtherance of analytical chemical sciences role in the use of chemical instrumentation. The Award consists of a plaque and $2,500.
*Requirements:* The awardee must have earned his or her highest degree within ten years of January 1 of the year of the award. Both the nationality of the young analytical scientist and the arena (e.g., academic, industrial, national laboratory) in which the contributions of the young analytical scientist have been made are unrestricted.
*Geographic Focus:* All States
*Date(s) Application is Due:* Nov 1
*Amount of Grant:* 2,500 USD
*Contact:* Miquela Sena, (505) 820-0443; fax (505) 989-1073; office@analyticalsciences.org
*Internet:* www.analyticalsciences.org/awards.php#Findeis
*Sponsor:* American Chemical Society
2019 Galisteo Street, Building I-1
Santa Fe, NM 87505

**American Chemical Society ANYL Award for Distinguished Service in the Advancement of Analytical Chemistry 374**
The purpose of the American Chemical Society ANYL Award for Distinguished Service in the Advancement of Analytical Chemistry is to honor an individual who through professional service in activities such as teaching, writing, research, and administration has substantially and uniquely enhanced the field of analytical chemistry. Nominations should present evidence for one or more of the following outstanding accomplishments: enhance the positive perception of analytical chemistry in the public eye; foster the development of analytical chemistry research in academic institutions, government laboratories, or in private industries; develop and implement programs that benefit the analytical community (these can be but are not limited to efforts within the Division of Analytical Chemistry or within the American Chemical Society; advance and promote the careers of analytical chemists in any area of employment; and play a central role in improving the way analytical chemistry is practiced role in the use of chemical instrumentation. The Award consists of a plaque and $2,500.
*Geographic Focus:* All States
*Date(s) Application is Due:* Nov 1
*Amount of Grant:* 2,500 USD
*Contact:* Miquela Sena, (505) 820-0443; fax (505) 989-1073; office@analyticalsciences.org
*Internet:* www.analyticalsciences.org/awards.php#Service
*Sponsor:* American Chemical Society
2019 Galisteo Street, Building I-1
Santa Fe, NM 87505

**American Chemical Society ANYL Award in Chemical Instrumentation 375**
The purpose of the American Chemical Society ANYL Award in Chemical Instrumentation is to recognize and encourage achievement in the origination or improvement of analytical methods of instrumentation. Advancing the field of chemical instrumentation through achievement of one or more of the following: conceptualization and development of unique instrumentation that has made a significant impact on the field; demonstration of innovative use of instrumentation in chemical measurement; stimulation of other researchers to use instrumentation in chemical measurement; and authorship of research papers or books that have had an influential role in the use of chemical instrumentation. This Award consists of a plaque and $2,500.
*Requirements:* Candidates must be an ANYL member and must have made original and eminent contributions to chemical instrumentation.
*Geographic Focus:* All States
*Date(s) Application is Due:* Nov 1
*Amount of Grant:* 2,500 USD
*Contact:* Miquela Sena, (505) 820-0443; fax (505) 989-1073; office@analyticalsciences.org
*Internet:* www.analyticalsciences.org/awards.php#Instrumentation
*Sponsor:* American Chemical Society
2019 Galisteo Street, Building I-1
Santa Fe, NM 87505

**American Chemical Society ANYL Award in Electrochemistry 376**
The purpose of the American Chemical Society ANYL Award in Electrochemistry is to recognize an individual who through scholarly activity has definitely and uniquely advanced the field of electrochemistry. Advancing the field of electrochemical analysis in one or more of the following: conceptualization and development of unique instrumentation that has made a significant impact on the field; development of novel and important instrumentation; elucidation of fundamental events or processes important to the field; and authorship of important research papers and/or books that have had an influential role in the development of the field of chemical instrumentation. This Award consists of a plaque and $2,500.
*Geographic Focus:* All States

*Date(s) Application is Due:* Nov 1
*Amount of Grant:* 2,500 USD
*Contact:* Miquela Sena, (505) 820-0443; fax (505) 989-1073; office@analyticalsciences.org
*Internet:* www.analyticalsciences.org/awards.php#Electrochemistry
*Sponsor:* American Chemical Society
2019 Galisteo Street, Building I-1
Santa Fe, NM 87505

**American Chemical Society ANYL Award in Spectrochemical Analysis** 377

The purpose of the American Chemical Society ANYL Award in Spectrochemical Analysis is to recognize an individual who through scholarly activity has definitely and uniquely advanced the field of spectrochemical analysis and optical spectrometry. Advancing the fields of spectrochemical analysis and optical spectrometry in one or more of the following: conceptualization and development of unique instrumentation that has made a significant impact on the field; development of novel and important instrumentation; elucidation of fundamental events or processes important to the field; and authorship of important research papers and/or books that have had an influential role in the development of the field of chemical instrumentation. This Award consists of a plaque and $2,500.
*Geographic Focus:* All States
*Date(s) Application is Due:* Nov 1
*Amount of Grant:* 2,500 USD
*Contact:* Miquela Sena, (505) 820-0443; fax (505) 989-1073; office@analyticalsciences.org
*Internet:* www.analyticalsciences.org/awards.php#Spectrochemical
*Sponsor:* American Chemical Society
2019 Galisteo Street, Building I-1
Santa Fe, NM 87505

**American Chemical Society ANYL I.M. Kolthoff Enrichment Awards for Undergraduate Students** 378

The purpose American Chemical Society ANYL I.M. Kolthoff Enrichment Awards are given annually to deserving undergraduates who have done undergraduate research in the field of analytical chemistry, broadly defined, in celebration of the life and accomplishments of the late Professor Izaak Maurits Kolthoff, viewed by many as the father of modern analytical chemistry. The Kolthoff awards provide travel funding up to $750 for undergraduate students to present a poster on their undergraduate research this year
*Requirements:* Only U.S. citizens and permanent residents are eligible, and preference will be given to those who have not made a previous presentation at a national scientific meeting.
*Restrictions:* Prior recipients are not eligible.
*Geographic Focus:* All States
*Date(s) Application is Due:* Nov 1
*Amount of Grant:* 750 USD
*Contact:* Miquela Sena, (505) 820-0443; fax (505) 989-1073; office@analyticalsciences.org
*Internet:* www.analyticalsciences.org/awards.php?action=kolthoff
*Sponsor:* American Chemical Society
2019 Galisteo Street, Building I-1
Santa Fe, NM 87505

**American Chemical Society ANYL J. Calvin Giddings Award for Excellence in Education** 379

The purpose of the American Chemical Society ANYL J. Calvin Giddings Award for Excellence in Education is to recognize the enhancement of the personal and professional development of students in the study of analytical chemistry. Enhancing the personal and professional development of students in the study of analytical chemistry in one or more of the following ways: authorship of an influential textbook for an analytical chemistry course; design and implementation of a successful new approach to teaching analytical chemistry; stimulation through teaching or research mentorship a significant number of students to become analytical chemists; development and publication of innovative experiments; design of improved equipment for teaching labs; and publication of widely quoted articles on teaching analytical chemistry. This Award consists of a plaque and $2,500.
*Geographic Focus:* All States
*Date(s) Application is Due:* Nov 1
*Amount of Grant:* 2,500 USD
*Contact:* Miquela Sena, (505) 820-0443; fax (505) 989-1073; office@analyticalsciences.org
*Internet:* www.analyticalsciences.org/awards.php#Teaching
*Sponsor:* American Chemical Society
2019 Galisteo Street, Building I-1
Santa Fe, NM 87505

**American Chemical Society ANYL Pfizer Graduate Travel Awards in Analytical Chemistry** 380

The purpose of the American Chemical Society ANYL Pfizer Graduate Travel Awards in Analytical Chemistry is to provide funding for graduate students to travel to an ACS national meeting and to present a poster. The award maximum is $750.
*Requirements:* Only U.S. citizens and permanent residents are eligible, and preference will be given to those who have not made a previous presentation at a national scientific meeting.
*Geographic Focus:* All States
*Date(s) Application is Due:* Nov 1
*Amount of Grant:* 750 USD
*Contact:* Miquela Sena, (505) 820-0443; fax (505) 989-1073; office@analyticalsciences.org
*Internet:* webapps.acs.org/findawards/detail.jsp?ContentId=WPCP_012500
*Sponsor:* American Chemical Society
2019 Galisteo Street, Building I-1
Santa Fe, NM 87505

**American Chemical Society Arthur C. Cope Award** 381

The American Chemical Society Arthur C. Cope Award is given to recognize outstanding achievement in the field of organic chemistry, the significance of which has become apparent within the five years preceding the year in which the award will be considered. The award consists of $25,000, a medallion with a presentation box, and a certificate. Up to $2,500 for travel expenses to the spring national meeting at which the award will be presented and to the fall national meeting to deliver an award address will be reimbursed. In addition to the award to the recipient, an unrestricted grant in aid for $150,000 for research in organic chemistry, under the direction of the recipient, will be made to any university or nonprofit institution selected by the recipient. The recipient may choose to assign the Arthur C. Cope Fund Grant to an institution for use by others for research or education in organic chemistry.
*Requirements:* Any individual, except a member of the award committee, may submit one nomination or seconding letter for the award in any given year. The nominating documents consist of a letter of not more than 1000 words containing an evaluation of the nominee's accomplishments and a specific identification of the work to be recognized, a biographical sketch including date of birth, and a list of publications and patents authored by the nominee.
*Restrictions:* Self-nominations are not accepted.
*Geographic Focus:* All States
*Date(s) Application is Due:* Nov 1
*Amount of Grant:* 175,000 - 177,500 USD
*Contact:* Awards Administrator; (800) 227-5558 or (202) 872-4408; fax (202) 776-8008; awards@acs.org
*Internet:* webapps.acs.org/findawards/detail.jsp?ContentId=CTP_004495
*Sponsor:* American Chemical Society
1155 Sixteenth Street, NW
Washington, D.C. 20036-4801

**American Chemical Society Arthur C. Cope Scholar Awards** 382

The American Chemical Society Arthur C. Cope Scholar Awards are given to recognize and encourage excellence in organic chemistry. Ten (10) Arthur C. Cope Scholars will be named annually: four between the ages of thirty-six and forty-nine, four age fifty or older, and two thirty-five years old and younger all inclusive before November 1 of the year in which the award is presented. The award consists of $5,000, a certificate and an unrestricted research grant in the amount of $40,000 to be assigned by the recipient to any university or nonprofit institution. The recipient is required to deliver a lecture at the annual Arthur C. Cope Symposium; traveling expenses incidental to participation in the symposium will also be paid.
*Requirements:* Any individual qualified to evaluate the nominee's accomplishments may submit one nomination for the award in any year; nomination should include a letter containing an evaluation of the nominee's contributions to organic chemistry with identification of the work to be recognized including literature and/or patent references; a biographical sketch of the nominee including date of birth and record of employment; and a list of publications and patents authored by the nominee.
*Restrictions:* Self-nominations are not accepted. An individual may not receive the award a second time. Recipients of the Arthur C. Cope Award are ineligible to be named Arthur C. Cope Scholars.
*Geographic Focus:* All States
*Date(s) Application is Due:* Nov 1
*Amount of Grant:* 45,000 - 47,500 USD
*Contact:* Awards Administrator; (800) 227-5558 or (202) 872-4408; fax (202) 776-8008; awards@acs.org
*Internet:* webapps.acs.org/findawards/detail.jsp?ContentId=CTP_004496
*Sponsor:* American Chemical Society
1155 Sixteenth Street, NW
Washington, D.C. 20036-4801

**American Chemical Society Award for Achievement in Research for the Teaching and Learning of Chemistry** 383

The award honors qualitative, quantitative, experimental research or case study research that investigates some aspect of the teaching and learning of chemistry. The research contributions may include the testing of research hypotheses, basing experiments on existing theories, incorporating what is known in the literature into the research, using an experimental design or recognized data-collection procedure, analyzing quantitative or qualitative data, and drawing valid conclusions from the observations and data. Experimental research should include one or more recognized techniques and be designed as control-group designs with random assignments of subjects or the use of in-tact class sections; factorial designs, bi-variate or multivariate correlation studies, etc. The research must show independence of thought and originality. The award consists of $5,000 and a certificate. Up to $2,500 for travel expenses to the meeting at which the award will be presented will be reimbursed. This is an international award.
*Requirements:* Any individual, except a member of the award committee, may submit one nomination or seconding letter for the award in any given year. The nominating documents consist of a letter of not more than 1000 words containing an evaluation of the nominee's accomplishments and a specific identification of the work to be recognized, a biographical sketch including date of birth, and a list of publications and patents authored by the nominee.
*Restrictions:* Self-nominations are not accepted.
*Geographic Focus:* All States
*Date(s) Application is Due:* Nov 1
*Amount of Grant:* 5,000 - 7,500 USD

*Samples:* Loretta L. Jones, $5,000; David F. Treagust, $5,000; Michael R. Abraham, $5,000; Alexander Henry Johnstone, $5,000.
*Contact:* Awards Administrator; (202) 872-4575 or (202) 872-4408; fax (202) 776-8008; awards@acs.org
*Internet:* webapps.acs.org/findawards/detail.jsp?ContentId=CTP_004484
*Sponsor:* American Chemical Society
1155 Sixteenth Street, NW
Washington, D.C. 20036-4801

**American Chemical Society Award for Affordable Green Chemistry** 384

The purposes of the Award are: to recognize outstanding scientific discoveries that lay the foundation for environmentally friendly products or manufacturing processes at a cost comparable to or less than that of current technologies, or discoveries that deliver new applications with compelling cost/benefit profiles; and to identify and recognize discovery of new eco-friendly chemistries with the potential to enable products or manufacturing processes that are less expensive than existing alternatives. Nominees may include: individuals; research teams (maximum 3 individuals); representatives from an organization (company, university or national lab); or professors or other academic-based teams whose work has been commercialized. The award will consist of $5,000 and a certificate. Up to $2,500 for travel expenses to the meeting at which the award will be presented will be reimbursed.
*Requirements:* Nominee eligibility is neither contingent on the timeframe in which the technology was developed, nor on demonstrated commercial success. Discoveries must have significant commercial potential, however, to be considered for award. Any individual, except a member of the award committee, may submit one nomination or seconding letter for the award in any given year. The nominating documents consist of a letter of not more than 1000 words containing an evaluation of the nominee's accomplishments and a specific identification of the work to be recognized, a biographical sketch including date of birth, and a list of publications and patents authored by the nominee.
*Restrictions:* Self-nominations are not accepted.
*Geographic Focus:* All States
*Date(s) Application is Due:* Nov 1
*Amount of Grant:* 5,000 - 7,500 USD
*Samples:* William J. Kruper, $5,000; Richard P. Wool, $5,000; Vincent J. D'Amico, Emiel H. van Broekhoven, and Juha J. Jakkula, $5,000.
*Contact:* Awards Administrator; (202) 872-4575 or (202) 872-4408; fax (202) 776-8008; awards@acs.org
*Internet:* webapps.acs.org/findawards/detail.jsp?ContentId=CTP_004487
*Sponsor:* American Chemical Society
1155 Sixteenth Street, NW
Washington, D.C. 20036-4801

**American Chemical Society Award for Computers in Chemical or Pharmaceutical Research** 385

The purpose of the Award is to recognize and encourage the use of computers in the advancement of the chemical and biological sciences. The award consists of $5,000 and a certificate. Up to $1,000 for travel expenses to the meeting at which the award will be presented will be reimbursed. The annual deadline is November 1.
*Requirements:* Any individual, except a member of the award committee, may submit one nomination or seconding letter for the award in any given year. The nominating documents consist of a letter of not more than 1000 words containing an evaluation of the nominee's accomplishments and a specific identification of the work to be recognized, a biographical sketch including date of birth, and a list of publications and patents authored by the nominee.
*Restrictions:* Self-nominations are not accepted.
*Geographic Focus:* All States
*Date(s) Application is Due:* Nov 1
*Amount of Grant:* 5,000 - 6,000 USD
*Contact:* Awards Administrator; (202) 872-4575 or (202) 872-4408; fax (202) 776-8008; awards@acs.org
*Internet:* webapps.acs.org/findawards/detail.jsp?ContentId=CTP_004503
*Sponsor:* American Chemical Society
1155 Sixteenth Street, NW
Washington, D.C. 20036-4801

**American Chemical Society Award for Creative Advances in Environmental Science and Technology** 386

The purpose of the Award is to encourage creativity in research and technology or methods of analysis to provide a scientific basis for informed environmental control decision-making processes, or to provide practical technologies that will reduce health risk factors. The award consists of $5,000 and a certificate. Up to $2,500 for travel expenses to the meeting at which the award will be presented will be reimbursed.
*Requirements:* Any individual, except a member of the award committee, may submit one nomination or seconding letter for the award in any given year. The nominating documents consist of a letter of not more than 1000 words containing an evaluation of the nominee's accomplishments and a specific identification of the work to be recognized, a biographical sketch including date of birth, and a list of publications and patents authored by the nominee.
*Restrictions:* The award will be granted without regard to age or nationality. Self-nominations are not accepted.
*Geographic Focus:* All States
*Date(s) Application is Due:* Nov 1
*Amount of Grant:* 5,000 - 7,500 USD
*Contact:* Awards Administrator; (202) 872-4575 or (202) 872-4408; fax (202) 776-8008; awards@acs.org
*Internet:* webapps.acs.org/findawards/detail.jsp?ContentId=CTP_004504
*Sponsor:* American Chemical Society
1155 Sixteenth Street, NW
Washington, D.C. 20036-4801

**American Chemical Society Award for Creative Invention** 387

The purpose of the Award is to recognize a single inventor for the successful application of research in chemistry and/or chemical engineering that contributes to the material prosperity and happiness of people. The award consists of $5,000 and a certificate. Up to $1,000 for travel expenses to the meeting at which the award will be presented will be reimbursed.
*Requirements:* An inventor/nominee must be a resident of the United States or Canada. A patent must have been granted for the work to be recognized, with the nominee listed as an inventor, and it will have been developed during the 17 years ending January 1 of the award year. A copy of the patent must be submitted with the nominating documents. Any individual, except a member of the award committee, may submit one nomination or seconding letter for the award in any given year. The nominating documents consist of a letter of not more than 1000 words containing an evaluation of the nominee's accomplishments and a specific identification of the work to be recognized, a biographical sketch including date of birth, and a list of publications and patents authored by the nominee.
*Restrictions:* Self-nominations are not accepted.
*Geographic Focus:* All States
*Date(s) Application is Due:* Nov 1
*Amount of Grant:* 5,000 - 6,000 USD
*Contact:* Awards Administrator; (202) 872-4575 or (202) 872-4408; fax (202) 776-8008; awards@acs.org
*Internet:* webapps.acs.org/findawards/detail.jsp?ContentId=CTP_004506
*Sponsor:* American Chemical Society
1155 Sixteenth Street, NW
Washington, D.C. 20036-4801

**American Chemical Society Award for Creative Research and Applications of Iodine Chemistry** 388

The purpose of the Award is to support, promote, and motivate global research of iodine chemistry and develop its use and knowledge through applications. Applications may include but are not limited to its uses in medicine, catalysis, food, and photography. The award consists of $10,000 and a certificate. Up to $1,000 for travel expenses to the meeting at which the award will be presented will also be reimbursed. The award is presented biennially in odd-numbered years.
*Requirements:* A nominee must have performed outstanding and creative research related to iodine chemistry or its applications. Any individual, except a member of the award committee, may submit one nomination or seconding letter for the award in any given year. The nominating documents consist of a letter of not more than 1000 words containing an evaluation of the nominee's accomplishments and a specific identification of the work to be recognized, a biographical sketch including date of birth, and a list of publications and patents authored by the nominee.
*Restrictions:* Self-nominations are not accepted.
*Geographic Focus:* All States
*Date(s) Application is Due:* Nov 1
*Amount of Grant:* 10,000 - 11,000 USD
*Contact:* Awards Administrator; (202) 872-4575 or (202) 872-4408; fax (202) 776-8008; awards@acs.org
*Internet:* webapps.acs.org/findawards/detail.jsp?ContentId=CTP_004507
*Sponsor:* American Chemical Society
1155 Sixteenth Street, NW
Washington, D.C. 20036-4801

**American Chemical Society Award for Creative Work in Fluorine Chemistry** 389

The purpose of the Award is to recognize outstanding contributions to the advancement of the chemistry of fluorine. The award consists of $5,000 and a certificate. Up to $2,500 for travel expenses to the meeting at which the award will be presented will be reimbursed. The award is presented in odd-numbered years at the Biennial Winter Fluorine Conference. The award is presented in even-numbered years during an ACS national meeting at which the ACS Division of Fluorine Chemistry meets.
*Requirements:* A nominee must have made an outstanding contribution or contributions to the field of fluorine chemistry. Any individual, except a member of the award committee, may submit one nomination or seconding letter for each award in any given year. The nominating documents consist of a letter of not more than 1000 words containing an evaluation of the nominee's accomplishments and a specific identification of the work to be recognized, a biographical sketch including date of birth, and a list of publications and patents authored by the nominee.
*Restrictions:* Self-nominations are not accepted.
*Geographic Focus:* All States
*Date(s) Application is Due:* Nov 1
*Amount of Grant:* 5,000 - 7,500 USD
*Contact:* Awards Administrator; (202) 872-4575 or (202) 872-4408; fax (202) 776-8008; awards@acs.org
*Internet:* webapps.acs.org/findawards/detail.jsp?ContentId=CTP_004505
*Sponsor:* American Chemical Society
1155 Sixteenth Street, NW
Washington, D.C. 20036-4801

**American Chemical Society Award for Creative Work in Synthetic Organic Chemistry** 390

The purpose of the Award is to recognize and encourage creative work in synthetic organic chemistry. The award consists of $5,000 and a certificate. Up to $1,000 for travel expenses to the meeting at which the award will be presented will be reimbursed.
*Requirements:* A nominee must have accomplished outstanding creative work in synthetic organic chemistry that has been published. Any individual, except a member of the award committee, may submit one nomination or seconding letter for the award in any given year. The nominating documents consist of a letter of not more than 1000 words containing an evaluation of the nominee's accomplishments and a specific identification of the work to be recognized, a biographical sketch including date of birth, and a list of publications and patents authored by the nominee.
*Restrictions:* Self-nominations are not accepted.
*Geographic Focus:* All States
*Date(s) Application is Due:* Nov 1
*Amount of Grant:* 5,000 - 6,000 USD
*Contact:* Awards Administrator; (202) 872-4575 or (202) 872-4408; fax (202) 776-8008; awards@acs.org
*Internet:* webapps.acs.org/findawards/detail.jsp?ContentId=CTP_004508
*Sponsor:* American Chemical Society
1155 Sixteenth Street, NW
Washington, D.C. 20036-4801

**American Chemical Society Award for Distinguished Service in the Advancement of Inorganic Chemistry** 391

The purpose of the Award is to recognize individuals who advanced inorganic chemistry by significant service in addition to performance of outstanding research. Activities recognized by the award may include such fields as teaching, writing, research, and administration. The award consists of $5,000 and a certificate. Up to $1,000 for travel expenses to the meeting at which the award will be presented will be reimbursed. The annual deadline is November 1.
*Requirements:* A nominee must have demonstrated extensive contributions to the advancement of inorganic chemistry. Any individual, except a member of the award committee, may submit one nomination or seconding letter for the award in any given year. The nominating documents consist of a letter of not more than 1000 words containing an evaluation of the nominee's accomplishments and a specific identification of the work to be recognized, a biographical sketch including date of birth, and a list of publications and patents authored by the nominee.
*Restrictions:* Self-nominations are not accepted.
*Geographic Focus:* All States
*Date(s) Application is Due:* Nov 1
*Amount of Grant:* 5,000 - 6,000 USD
*Contact:* Awards Administrator; (202) 872-4575 or (202) 872-4408; fax (202) 776-8008; awards@acs.org
*Internet:* webapps.acs.org/findawards/detail.jsp?ContentId=CTP_004486
*Sponsor:* American Chemical Society
1155 Sixteenth Street, NW
Washington, D.C. 20036-4801

**American Chemical Society Award for Encouraging Disadvantaged Students into Careers in the Chemical Sciences** 392

The purpose of this Award is to recognize significant accomplishments by individuals in stimulating students, underrepresented in the profession, to elect careers in the chemical sciences and engineering. The award consists of $5,000 and a certificate. A grant of $10,000 will be made to an academic institution, designated by the recipient, to strengthen its activities in meeting the objectives of the award. Up to $1,500 for travel expenses to the meeting at which the award will be presented will be reimbursed. The annual deadline for nominations is November 1.
*Requirements:* Nominees for the award may come from any professional setting: academia, industry, government, or other independent facility. The award is intended to recognize significant accomplishments in the United States by individuals in stimulating students, especially those currently underrepresented in the profession, to elect careers in the chemical sciences and engineering, and in generating a broader appreciation of chemistry as the central science.
*Restrictions:* Self-nominations are not accepted.
*Geographic Focus:* All States
*Date(s) Application is Due:* Nov 1
*Amount of Grant:* Up to 16,500 USD
*Contact:* Awards Administrator; (202) 872-4575 or (202) 872-4408; fax (202) 776-8008; awards@acs.org
*Internet:* webapps.acs.org/findawards/detail.jsp?ContentId=CTP_004509
*Sponsor:* American Chemical Society
1155 Sixteenth Street, NW
Washington, D.C. 20036-4801

**American Chemical Society Award for Encouraging Women into Careers in the Chemical Sciences** 393

The purpose of the Award is to recognize individuals who have significantly stimulated or fostered the interest of women in chemistry, thereby promoting their professional development as chemists or chemical engineers, and/or increasing their appreciation of chemistry as the central science. Nominees for the award may come from any professional setting: academia, industry, government, or other independent facility. The award is intended to recognize significant accomplishments by individuals in stimulating women to elect careers in the chemical sciences and engineering, and in generating a broader appreciation of chemistry as the central science. The award consists of $5,000 and a certificate. A grant of $10,000 will be made to an academic institution, designated by the recipient, to strengthen its activities in meeting the objectives of the award. Up to $1,500 for travel expenses to the meeting at which the award will be presented will be reimbursed. The annual deadline is November 1.
*Requirements:* Any individual, except a member of the award committee, may submit one nomination or seconding letter for the award in any given year. Nominating documents consist of a letter of not more than 1000 words containing an evaluation of the nominee's accomplishments and a specific identification of the work to be recognized, a biographical sketch including date of birth, and a list of publications and patents authored by the nominee.
*Restrictions:* Self-nominations are not accepted.
*Geographic Focus:* All States
*Date(s) Application is Due:* Nov 1
*Amount of Grant:* 15,000 - 16,500 USD
*Contact:* Awards Administrator; (202) 872-4575 or (202) 872-4408; fax (202) 776-8008; awards@acs.org
*Internet:* webapps.acs.org/findawards/detail.jsp?ContentId=CTP_004517
*Sponsor:* American Chemical Society
1155 Sixteenth Street, NW
Washington, D.C. 20036-4801

**American Chemical Society Award for Research at an Undergraduate Institution** 394

The purpose of the Award is to recognize the importance of research with undergraduates. The award will honor a chemistry faculty member whose research in an undergraduate setting has achieved wide recognition and contributed significantly to chemistry and to the professional development of undergraduate students. This award recognizes fundamental research that constitutes advances in science as evidenced by refereed publications with undergraduate coauthors in leading scientific research journals, external research grant support, and the subsequent professional development of students who have participated in the research program. The award will be given for significant work over a long period of time rather than for a specific, limited project. The award consists of $5,000 and a certificate. Up to $2,500 for travel expenses to the meeting at which the award will be presented will be reimbursed. The program will also provide a grant of $5,000 directly to the recipient's institution.
*Requirements:* The nominee's department may offer work leading to the master's degree but cannot have a doctoral program. A nominee must be a tenured faculty member of a predominantly undergraduate institution. Any individual, except a member of the award committee, may submit one nomination or seconding letter for the award in any given year. The nominating documents consist of a letter of not more than 1000 words containing an evaluation of the nominee's accomplishments and a specific identification of the work to be recognized, a biographical sketch including date of birth, and a list of publications and patents authored by the nominee.
*Restrictions:* Self-nominations are not accepted.
*Geographic Focus:* All States
*Date(s) Application is Due:* Nov 1
*Contact:* Awards Administrator; (202) 872-4575 or (202) 872-4408; fax (202) 776-8008; awards@acs.org
*Internet:* webapps.acs.org/findawards/detail.jsp?ContentId=CTP_004549
*Sponsor:* American Chemical Society
1155 Sixteenth Street, NW
Washington, D.C. 20036-4801

**American Chemical Society Award for Team Innovation** 395

This award was established to highlight the value and importance of technical teams and teamwork to the chemical and allied industries by recognizing a multidisciplinary team for successfully moving an innovative idea to a product now in commercial use. For each team member, the award consists of $3,000 and a certificate. Up to $1,000 for travel expenses to the meeting at which the award will be presented will be reimbursed to each team member. A certificate will also be provided to the employers of the team.
*Requirements:* The team shall be multidisciplinary in nature and consist of not fewer than two nor more than five members. The team's work leading to this award must have been carried out primarily in the United States and the technical accomplishments of the team must be documented in the technical literature as a publication(s) or a patent(s). The output of the team must also demonstrate innovation, commercialization of a product or process, be of a special value to society, and of a nature that could only be achieved by professionals working together effectively. The rate of commercialization of the team's output will also be considered in the selection of the recipients of this award.
*Restrictions:* Application or self-nomination is not acceptable.
*Geographic Focus:* All States
*Date(s) Application is Due:* Nov 1
*Contact:* Awards Administrator; (202) 872-4575 or (202) 872-4408; fax (202) 776-8008; awards@acs.org
*Internet:* webapps.acs.org/findawards/detail.jsp?ContentId=CTP_004554
*Sponsor:* American Chemical Society
1155 Sixteenth Street, NW
Washington, D.C. 20036-4801

**American Chemical Society Award in Analytical Chemistry** 396

This award, sponsored by Battelle Memorial Institute and administered by the American Chemical Society, is given annually to recognize and encourage outstanding contributions to the science of analytical chemistry, pure or applied, carried out in the United States or Canada. Special consideration will be given to the independence of

thought and the originality shown, or to the importance of the work when applied to public welfare, economics, or the needs and desires of humanity. The award consists of $5,000 and a certificate. Up to $2,500 for travel expenses to the meeting at which the award will be presented will be reimbursed.
*Requirements:* A nominee must be a resident of the United States or Canada and must have made an outstanding contribution to analytical chemistry. Any individual, except a member of the award committee, may submit one nomination or seconding letter for the award in any given year. The nominating documents consist of a letter of not more than 1000 words containing an evaluation of the nominee's accomplishments and a specific identification of the work to be recognized, a biographical sketch including date of birth, and a list of publications and patents authored by the nominee.
*Restrictions:* Self-nominations are not accepted.
*Geographic Focus:* All States
*Date(s) Application is Due:* Nov 1
*Amount of Grant:* 5,000 - 7,500 USD
*Contact:* Awards Administrator; (202) 872-4575 or (202) 872-4408; fax (202) 776-8008; awards@acs.org
*Internet:* webapps.acs.org/findawards/detail.jsp?ContentId=CTP_004492
*Sponsor:* American Chemical Society
1155 Sixteenth Street, NW
Washington, D.C. 20036-4801

**American Chemical Society Award in Applied Polymer Science** **397**
This award, sponsored by Eastman Chemical Company and administered by the American Chemical Society, is given annually to recognize and encourage outstanding achievements in the science or technology of plastics, coatings, polymer composites, adhesives, and related fields. The recipient will be selected primarily on the basis of scientific contributions made to the specified fields during the 10-year period preceding date of selection. To avoid repeating specific areas of technology, preference will be given whenever recognized by the granting of this award in the two preceding years. The award consists of $5,000 and a certificate. Up to $2,500 for travel expenses to the meeting at which the award will be presented will be reimbursed.
*Requirements:* Any individual, except a member of the award committee, may submit one nomination or seconding letter for the award in any given year. The nominating documents consist of a letter of not more than 1000 words containing an evaluation of the nominee's accomplishments and a specific identification of the work to be recognized, a biographical sketch including date of birth, and a list of publications and patents authored by the nominee.
*Restrictions:* Self-nominations are not accepted.
*Geographic Focus:* All States
*Date(s) Application is Due:* Nov 1
*Amount of Grant:* 5,000 - 7,500 USD
*Contact:* Awards Administrator; (202) 872-4575 or (202) 872-4408; fax (202) 776-8008; awards@acs.org
*Internet:* webapps.acs.org/findawards/detail.jsp?ContentId=CTP_004493
*Sponsor:* American Chemical Society
1155 Sixteenth Street, NW
Washington, D.C. 20036-4801

**American Chemical Society Award in Chromatography** **398**
The purpose of this Award is to recognize outstanding contributions to the fields of chromatography. The award consists of $5,000 and a certificate. Up to $2,500 for travel expenses to the meeting at which the award will be presented will be reimbursed. The annual deadline is November 1.
*Requirements:* A nominee must have made an outstanding contribution to the field with particular consideration given to developments of new methods. Any individual, except a member of the award committee, may submit one nomination or seconding letter for the award in any given year. The nominating documents consist of a letter of not more than 1000 words containing an evaluation of the nominee's accomplishments and a specific identification of the work to be recognized, a biographical sketch including date of birth, and a list of publications and patents authored by the nominee.
*Restrictions:* Self-nominations are not accepted.
*Geographic Focus:* All States
*Date(s) Application is Due:* Nov 1
*Amount of Grant:* 5,000 - 7,500 USD
*Contact:* Awards Administrator; (202) 872-4575 or (202) 872-4408; fax (202) 776-8008; awards@acs.org
*Internet:* webapps.acs.org/findawards/detail.jsp?ContentId=CTP_004500
*Sponsor:* American Chemical Society
1155 Sixteenth Street, NW
Washington, D.C. 20036-4801

**American Chemical Society Award in Colloid and Surface Chemistry** **399**
The award, sponsored by the Procter and Gamble Company, is given annually to recognize and encourage outstanding scientific contributions to colloid or surface chemistry in the United States or Canada. Recognition will also be given to originality and independence of thought, and to the technological impact of the nominee's contribution. In odd-numbered years, awards are made for contributions in surface chemistry. In even-numbered years, awards are made for advances in colloid chemistry. The award consists of $5,000 and a certificate. Up to $2,500 for travel expenses to the meeting at which the award will be presented will be reimbursed.
*Requirements:* The nominee must be a resident of the United States or Canada and must have made outstanding scientific contributions to colloid or surface chemistry. Any individual, except a member of the award committee, may submit one nomination or seconding letter for the award in any given year. The nominating documents consist of a letter of not more than 1000 words containing an evaluation of the nominee's accomplishments and a specific identification of the work to be recognized, a biographical sketch including date of birth, and a list of publications and patents authored by the nominee.
*Restrictions:* Self-nominations are not accepted.
*Geographic Focus:* All States
*Date(s) Application is Due:* Nov 1
*Amount of Grant:* 5,000 - 7,500 USD
*Contact:* Awards Administrator; (202) 872-4575 or (202) 872-4408; fax (202) 776-8008; awards@acs.org
*Internet:* webapps.acs.org/findawards/detail.jsp?ContentId=CTP_004502
*Sponsor:* American Chemical Society
1155 Sixteenth Street, NW
Washington, D.C. 20036-4801

**American Chemical Society Award in Industrial Chemistry** **400**
The award, sponsored by ACS Division of Business Development and Management, is given annually to recognize outstanding contributions to industrial chemistry resulting in the commercialization of an economically significant new product or process. Any field of chemical, chemical engineering, or biochemical research is appropriate if it is of general interest and reflects the concerns of modern society. Any chemical research, whether industrial, governmental, or academic, is eligible, provided the work was done in North America and yielded significant commercial results for a period of more than one year. The award consists of $5,000 and a certificate. Up to $2,500 for travel expenses to the meeting at which the award will be presented will be reimbursed.
*Requirements:* Any individual, except a member of the award committee, may submit one nomination or seconding letter for the award in any given year. The nominating documents consist of a letter of not more than 1000 words containing an evaluation of the nominee's accomplishments and a specific identification of the work to be recognized, a biographical sketch including date of birth, and a list of publications and patents authored by the nominee.
*Restrictions:* Self-nominations are not accepted.
*Geographic Focus:* All States
*Date(s) Application is Due:* Nov 1
*Amount of Grant:* 5,000 - 7,500 USD
*Contact:* Awards Administrator; (202) 872-4575 or (202) 872-4408; fax (202) 776-8008; awards@acs.org
*Internet:* webapps.acs.org/findawards/detail.jsp?ContentId=CTP_004531
*Sponsor:* American Chemical Society
1155 Sixteenth Street, NW
Washington, D.C. 20036-4801

**American Chemical Society Award in Inorganic Chemistry** **401**
This award, sponsored by the Aldrich Chemical Company and administered by the American Chemical Society, is given annually to recognize and encourage fundamental research in the field of inorganic chemistry. A nominee must have accomplished outstanding research in the preparation, properties, reactions, or structure of inorganic substances. Special consideration shall be given to the independence of thought and originality shown. The award consists of $5,000 and a certificate. Up to $2,500 for travel expenses to the meeting at which the award will be presented will be reimbursed.
*Requirements:* Any individual, except a member of the award committee, may submit one nomination or seconding letter for the award in any given year. The nominating documents consist of a letter of not more than 1000 words containing an evaluation of the nominee's accomplishments and a specific identification of the work to be recognized, a biographical sketch including date of birth, and a list of publications and patents authored by the nominee.
*Restrictions:* Self-nominations are not accepted.
*Geographic Focus:* All States
*Date(s) Application is Due:* Nov 1
*Amount of Grant:* 5,000 - 7,500 USD
*Contact:* Awards Administrator; (202) 872-4575 or (202) 872-4408; fax (202) 776-8008; awards@acs.org
*Internet:* webapps.acs.org/findawards/detail.jsp?ContentId=CTP_004532
*Sponsor:* American Chemical Society
1155 Sixteenth Street, NW
Washington, D.C. 20036-4801

**American Chemical Society Award in Organometallic Chemistry** **402**
The Award, sponsored by the Dow Chemical Company Foundation and administered by the American Chemical Society, is given annually to recognize a recent advancement that is having a major impact on research in organometallic chemistry. A nominee must have shown outstanding research in the preparation, reactions, properties, or structure of organometallic substances. Special consideration will be given to demonstrated creativity and independence of thought. The award consists of $5,000 and a certificate. Up to $1,000 for travel expenses to the meeting at which the award will be presented will be reimbursed.
*Requirements:* Preference will be given to U.S. citizens. Any individual, except a member of the award committee, may submit one nomination or seconding letter for the award in any given year. The nominating documents consist of a letter of not more than 1000 words containing an evaluation of the nominee's accomplishments and a specific identification of the work to be recognized, a biographical sketch including date of birth, and a list of publications and patents authored by the nominee.
*Restrictions:* Self-nominations are not accepted.

*Geographic Focus:* All States
*Date(s) Application is Due:* Nov 1
*Amount of Grant:* 5,000 - 6,000 USD
*Contact:* Awards Administrator; (202) 872-4575 or (202) 872-4408; fax (202) 776-8008; awards@acs.org
*Internet:* webapps.acs.org/findawards/detail.jsp?ContentId=CTP_004542
*Sponsor:* American Chemical Society
1155 Sixteenth Street, NW
Washington, D.C. 20036-4801

**American Chemical Society Award in Polymer Chemistry** 403

The Award, supported by the ExxonMobil Chemical Company, is intended to recognize outstanding contributions to polymer chemistry. This award is specifically intended to acknowledge the impact in basic science of a researcher's body of work in the broad field of polymer chemistry on our global needs for advanced polymeric systems and materials. The award consists of $5,000 and a certificate. Up to $1,000 for travel expenses to the meeting at which the award will be presented will be reimbursed. The annual deadline is November 1.
*Requirements:* The nominees for this award are expected to have demonstrated pioneering research accomplishments in synthetic methods of unique polymeric materials and/or achieved profound discoveries in polymer physical chemistry leading to recognized advancements in our understanding of complex macromolecular behavior.
*Restrictions:* Self-nominations are not accepted.
*Geographic Focus:* All States
*Date(s) Application is Due:* Nov 1
*Amount of Grant:* 5,000 - 6,000 USD
*Contact:* Awards Administrator; (202) 872-4575 or (202) 872-4408; fax (202) 776-8008; awards@acs.org
*Internet:* webapps.acs.org/findawards/detail.jsp?ContentId=CTP_004544
*Sponsor:* American Chemical Society
1155 Sixteenth Street, NW
Washington, D.C. 20036-4801

**American Chemical Society Award in Pure Chemistry** 404

This award, sponsored by Alpha Chi Sigma Fraternity and the Alpha Chi Sigma Educational Award Foundation, is given annually to recognize and encourage fundamental research in pure chemistry carried out in North America by young men and women. Special consideration is given to independence of thought and the originality shown in the research. The award consists of $5,000 and a certificate. Up to $1,000 for travel expenses to the meeting at which the award will be presented will be reimbursed. The annual deadline is November 1.
*Requirements:* A nominee must have been born after April 30, 1978 and must have accomplished research of unusual merit for an individual on the threshold of her or his career. Any individual, except a member of the award committee, may submit one nomination or seconding letter for the award in any given year. The nominating documents consist of a letter of not more than 1000 words containing an evaluation of the nominee's accomplishments and a specific identification of the work to be recognized, a biographical sketch including date of birth, and a list of publications and patents authored by the nominee.
*Restrictions:* Self-nominations are not accepted.
*Geographic Focus:* All States
*Date(s) Application is Due:* Nov 1
*Amount of Grant:* 5,000 - 6,000 USD
*Contact:* Awards Administrator; (202) 872-4575 or (202) 872-4408; fax (202) 776-8008; awards@acs.org
*Internet:* webapps.acs.org/findawards/detail.jsp?ContentId=CTP_004546
*Sponsor:* American Chemical Society
1155 Sixteenth Street, NW
Washington, D.C. 20036-4801

**American Chemical Society Award in Separations Science and Technology** 405

This award is given annually to recognize outstanding accomplishments in fundamental or applied research directed to separations science and technology. The award shall be granted to an individual without regard to age or nationality. The scope of the award is to be as broad as possible, covering all fields where separations science and technology is practiced, including but not limited to biology, chemistry, engineering, geology, and medicine. The award consists of $5,000 and a certificate. Up to $2,500 for travel expenses to the meeting at which the award will be presented will be reimbursed. The recipient will deliver a lecture at the annual Division of Industrial and Engineering Chemistry Separations Science and Technology Symposium.
*Requirements:* Any individual, except a member of the award committee, may submit one nomination or seconding letter for the award in any given year. The nominating documents consist of a letter of not more than 1000 words containing an evaluation of the nominee's accomplishments and a specific identification of the work to be recognized, a biographical sketch including date of birth, and a list of publications and patents authored by the nominee.
*Restrictions:* Self-nominations are not accepted.
*Geographic Focus:* All States
*Date(s) Application is Due:* Nov 1
*Amount of Grant:* 5,000 - 7,500 USD
*Samples:* Abraham M. Lenhoff—award recipient, $5,000; Allan S. Myerson—award winner, $5,000; William H. Pirkle—award winner, $5,000.
*Contact:* Awards Administrator; (800) 227-4575 or (202) 872-4408; fax (202) 776-8008; awards@acs.org
*Internet:* webapps.acs.org/findawards/detail.jsp?ContentId=CTP_004552
*Sponsor:* American Chemical Society
1155 Sixteenth Street, NW
Washington, D.C. 20036-4801

**American Chemical Society Award in the Chemistry of Materials** 406

This award, supported by E.I. du Pont de Nemours and Company, is given annually to recognize and encourage creative work in the chemistry of materials. A nominee must have made outstanding contributions to the chemistry of materials. Particular emphasis will be placed on research relating to materials of actual or potential technological importance, where a fundamental understanding of the chemistry associated with materials preparation, processing, or use is critical. The award consists of $5,000 and a certificate. Up to $2,500 for travel expenses to the meeting at which the award will be presented will be reimbursed. The annual deadline is November 1.
*Requirements:* Any individual, except a member of the award committee, may submit one nomination or seconding letter for the award in any given year. The nominating documents consist of a letter of not more than 1000 words containing an evaluation of the nominee's accomplishments and a specific identification of the work to be recognized, a biographical sketch including date of birth, and a list of publications and patents authored by the nominee.
*Restrictions:* Self-nominations are not accepted.
*Geographic Focus:* All States
*Date(s) Application is Due:* Nov 1
*Amount of Grant:* 5,000 USD
*Contact:* Awards Administrator; (202) 872-4575 or (202) 872-4408; fax (202) 776-8008; awards@acs.org
*Internet:* webapps.acs.org/findawards/detail.jsp?ContentId=CTP_004499
*Sponsor:* American Chemical Society
1155 Sixteenth Street, NW
Washington, D.C. 20036-4801

**American Chemical Society Award in Theoretical Chemistry** 407

The purpose of this annual award, which is sponsored by Dell Incorporated, is to recognize innovative research in theoretical chemistry that either advances theoretical methodology or contributes to new discoveries about chemical systems. Emphasis in the selection process will be on work characterized by depth, originality, and scientific significance. The award is given without regard to age or nationality, and consists of $5,000 and a certificate, with an additional maximum of $2,500 in travel expenses. The annual deadline is November 1.
*Requirements:* Any individual, except a member of the award committee, may submit one nomination or seconding letter for each award in any given year. The letter of nomination may not exceed 1000 words and should contain an evaluation of the nominee's accomplishments and a specific identification of the work to be recognized.
*Restrictions:* Self-nominations are not accepted.
*Geographic Focus:* All States
*Date(s) Application is Due:* Nov 1
*Amount of Grant:* 5,000 - 7,500 USD
*Contact:* Awards Administrator; (202) 872-4575 or (202) 872-4408; fax (202) 776-8008; awards@acs.org
*Internet:* webapps.acs.org/findawards/detail.jsp?ContentId=CTP_004555
*Sponsor:* American Chemical Society
1155 Sixteenth Street, NW
Washington, D.C. 20036-4801

**American Chemical Society GCI Pharmaceutical Roundtable Research Grants** 408

The Roundtable is a partnership with the ACS GCI (Green Chemistry Institute) and pharmaceutical corporations united by a shared commitment to integrate the principles of green chemistry and engineering into the business of drug discovery and production. The program was created to support research in the key green chemistry research areas. (Ten chemical transformations and two process related operations were identified and subsequently published - download the list from the sponsor's website). The primary purpose of this grant is to publish research to make information publicly available. One grant is planned to be awarded. The total award is limited to $150,000 for a grant period of 12 to 24 months.
*Requirements:* Proposals will only be accepted from public and private institutions of higher education. The grant is not limited to institutions in the United States. Proposals must be submitted through the appropriate institutional office for external funding. Proposed research must focus on the key targeted areas identified by the Roundtable. Proposals must clearly describe how the proposed research will provide improvements over existing technology that are consistent with the goals of Green Chemistry including improved atom economy, use of less hazardous reagents, more energy efficient processes, elimination of by-products, reduced waste generation, improved environmental, health, and safety and life cycle impacts, catalytic versus stoichiometric methods, and development of inherently safer transformations. Illustrations of the relevance of the proposed technology to synthetic and process related problems within the pharmaceutical industry are encouraged. Complete guidelines can be downloaded at the sponsor's website.
*Restrictions:* Applicants may have only one research grant with the ACS GCI Pharmaceutical Roundtable at a time. Current research grant holders may not apply for a new grant until their active grant or no-cost time extension officially expires.
*Geographic Focus:* All States
*Date(s) Application is Due:* Sep 14
*Amount of Grant:* Up to 150,000 USD
*Contact:* Gayle Peterman, Finance and Grants Manager; (202) 872-6092 or (202) 872-4481; fax (202) 872-6319; g_peterman@acs.org or gcipr@acs.org

*Internet:* portal.acs.org/portal/acs/corg/content?_nfpb=true&_pageLabel=PP_TRANSITIONMAIN&node_id=1456&use_sec=false&sec_url_var=region1
*Sponsor:* American Chemical Society
1155 Sixteenth Street, NW
Washington, D.C. 20036-4801

**American Chemical Society Green Chemistry Chapter Award for Student Affiliates** 409

The Education Division of the American Chemical Society (ACS) sponsors the award and the ACS Green Chemistry Institute assists with evaluating the green chemistry activities. The award's purpose is to recognize Student Affiliates Chapters of the ACS that engage in at least three green chemistry activities during the academic year. The Student Affiliates of the American Chemical Society is the official ACS organization for undergraduate chemical science majors. Each chapter receiving the green chemistry chapter award receives a plaque during the Student Affiliates Awards Ceremony, held annually during the Spring ACS National Meeting.
*Geographic Focus:* All States
*Contact:* Ron Siatkowski; (202) 872-6091; fax (202) 872-6319; r_siatkowski@acs.org
*Internet:* webapps.acs.org/findawards/detail.jsp?ContentId=WPCP_012552
*Sponsor:* American Chemical Society
1155 Sixteenth Street, NW
Washington, D.C. 20036-4801

**American Chemical Society Green Chemistry Grants** 410

The American Chemical Society Green Chemistry Institute sponsors a green chemistry grant program by promoting funding, increasing opportunities, and developing information on the benefits of green chemistry. Its efforts span the spectrum from theory and basic research through application and commercialization of science and technology. The program also sponsors a number of research fellowships and scientific exchanges through its grants programs.
*Geographic Focus:* All States
*Contact:* Gayle Peterman, Finance and Grants Manager; (202) 872-6092 or (202) 872-4481; fax (202) 872-6319; g_peterman@acs.org
*Internet:* portal.acs.org/portal/acs/corg/content?_nfpb=true&_pageLabel=PP_TRANSITIONMAIN&node_id=1456&use_sec=false&sec_url_var=region1&__uuid=c14e3223-89d3-4263-8072-51ccffe52007
*Sponsor:* American Chemical Society
1155 Sixteenth Street, NW
Washington, D.C. 20036-4801

**American Chemical Society Joseph Breen Memorial Fellowship in Green Chemistry** 411

The American Chemical Society (ACS) established the Joseph Breen Memorial Fund through the ACS International Endowment Fund in 2000. This fund commemorates Joe Breen's commitment and accomplishments for the advancement of green chemistry. The purpose of the Joseph Breen Memorial Fellowship in Green Chemistry is to sponsor a young international green chemistry scholar to participate in an international green chemistry technical meeting, conference or training program. The award is a travel reimbursement up to a limit decided by the award judges. The amount of the award is based on the estimated cost of travel, accommodation, and fees associated with the event.
*Geographic Focus:* All States
*Contact:* Ron Siatkowski, Director, Office of Research Grants; (202) 872-6091; fax (202) 872-6319; r_siatkowski@acs.org
*Internet:* webapps.acs.org/findawards/detail.jsp?ContentId=WPCP_012550
*Sponsor:* American Chemical Society
1155 Sixteenth Street, NW
Washington, D.C. 20036-4801

**American Chemical Society Kenneth G. Hancock Memorial Student Award in Green Chemistry** 412

The ACS Division of Environmental Chemistry and the National Institute of Standards and Technology (NIST) sponsor the Kenneth G. Hancock Memorial Student Award in Green Chemistry. The award is administered by the ACS Green Chemistry Institute, and its purpose is to recognize and honor outstanding student contributions to furthering the goals of green chemistry through research and/or studies. The award provides national recognition for outstanding student contributions to furthering the goals of green chemistry (i.e., the research, development, and implementation of fundamental and innovative chemical technologies that incorporate the principles of green chemistry into chemical design, manufacture, and use, and that have the potential to be utilized in achieving national pollution prevention goals). The recipient of the award receives $1000. The award is presented in conjunction with the annual ceremony for the Presidential Green Chemistry Challenge Awards, held in Washington, D.C. in June.
*Requirements:* The award is open to undergraduate and graduate students, regardless of citizenship or country of study. Students who completed their degree prior to May 1 of the year before the application year are not eligible for the award.
*Geographic Focus:* All States
*Date(s) Application is Due:* Feb 1
*Amount of Grant:* 1,000 USD
*Contact:* Ron Siatkowski; (202) 872-6091; fax (202) 872-6319; r_siatkowski@acs.org
*Internet:* webapps.acs.org/findawards/detail.jsp?ContentId=WPCP_012551
*Sponsor:* American Chemical Society
1155 Sixteenth Street, NW
Washington, D.C. 20036-4801

**American Chemical Society Overcoming Challenges Award** 413

The Women Chemists Committee of the American Chemical Society established this award to recognize a woman undergraduate from a two-year or four-year institution for her efforts in overcoming hardship to achieve success in chemistry. One award is offered annually. The awardee receives a monetary prize of $250 and up to $1,000 for travel expenses to attend the fall ACS national meeting. The awardee is presented with a plaque and recognized at the WCC Luncheon at the fall meeting. The current application deadline is April 1.
*Requirements:* To be considered a candidate, students must meet the following criteria: applicant must be a female undergraduate currently enrolled in a two-year chemistry-related program or pursuing a major or minor in a four-year chemistry program at a school that does not grant doctorates in chemistry; applicant must have completed one semester of college-level chemistry; applicant must demonstrate that she has overcome hardships (economic, personal, or academic) in pursuit of her education; and applicants are reviewed on: improvement, initiative, successes and grades from the previous two semesters (not cumulative grade point average).
*Geographic Focus:* All States
*Date(s) Application is Due:* Apr 1
*Amount of Grant:* 250 - 1,000 USD
*Contact:* Ron Siatkowski; (202) 872-6091; fax (202) 872-6319; r_siatkowski@acs.org
*Internet:* webapps.acs.org/findawards/detail.jsp?ContentId=WPCP_010370
*Sponsor:* American Chemical Society
1155 Sixteenth Street, NW
Washington, D.C. 20036-4801

**American Chemical Society PCD Award in Experimental Physical Chemistry** 414

The purpose of the Physical Chemistry Division's Award in Experimental Physical Chemistry is to recognize outstanding contributions in experimental physical chemistry. The recipient will receive a $2,000 honorarium, a plaque, and will be invited to present his/her work at the fall ACS meeting that immediately follows the announcement of the award. At that meeting, he/she will present an invited talk and receive the award at the annual reception sponsored by the PHYS Division and the Journal of Physical Chemistry. The Division will pay up to $1,500 to cover the travel, lodging, and meal expenses for the award winner while at that ACS meeting. The annual deadline is November 1.
*Requirements:* Eligibility is restricted to Physical Chemistry Division members who, at the time of the nomination, have not yet won a national award from a scientific society that is based on the nominee's scientific accomplishments. Members of the National Academy of Science are also ineligible, but fellowship in a professional society is not considered a national award in this context. Current serving members of the PHYS Division Executive Committee or Executive Committees of any of the Subdivisions of the PHYS Division are ineligible for nomination for this award until after their term of service.
*Geographic Focus:* All States
*Date(s) Application is Due:* Nov 1
*Amount of Grant:* 2,000 - 3,500 USD
*Contact:* Awards Administrator; (202) 872-4575; fax (202) 776-8008; awards@acs.org
*Internet:* webapps.acs.org/findawards/detail.jsp?ContentId=CNBP_030731
*Sponsor:* American Chemical Society
1155 Sixteenth Street, NW
Washington, D.C. 20036-4801

**American Chemical Society PCD Award in Theoretical Chemistry** 415

Sponsored by the Physical Chemistry Division, Telluride School on Theoretical Chemistry, the Award in Theoretical Chemistry's purpose is to recognize outstanding contributions in theoretical chemistry. The recipient will receive a $2,000 honorarium, a plaque, and will be invited to the next Telluride School on Theoretical Chemistry (TSTC), which are held every other summer. At that meeting, he/she will present a plenary lecture and receive the award. The TSTC will pay up to $1,500 to cover travel, lodging, and meal expenses for the Award winner while at the TSTC. The recipient will also be invited to present his/her work at the fall ACS meeting that immediately follows the announcement of the award. At that meeting, he/she will also be recognized at the annual reception sponsored by the PHYS Division and the Journal of Physical Chemistry. The Division will pay up to $1,500 to cover the travel, lodging, and meal expenses for the award winner while at that ACS meeting. The annual deadline is November 1.
*Requirements:* Eligibility is restricted to Physical Chemistry Division members who, at the time of the nomination, have not yet won a national award from a scientific society that is based on the nominee's scientific accomplishments. Members of the National Academy of Science are also ineligible, but fellowship in a professional society is not considered a national award in this context. Currently serving members of the PHYS Division Executive Committee or Executive Committees of any of the Subdivisions of the PHYS Division are ineligible for nomination for this award until after their term of service.
*Geographic Focus:* All States
*Date(s) Application is Due:* Nov 1
*Contact:* Awards Administrator; (202) 872-4575; fax (202) 776-8008; awards@acs.org
*Internet:* webapps.acs.org/findawards/detail.jsp?ContentId=CNBP_030732
*Sponsor:* American Chemical Society
1155 Sixteenth Street, NW
Washington, D.C. 20036-4801

**American Chemical Society PCD Early Career Awards** 416

Established in 2012, the purpose of the Physical Chemistry Division's Early Career Awards is to recognize outstanding contributions in experimental and theoretical physical chemistry by young investigators. Two awards will be given each year – one each in theoretical and experimental physical chemistry – to honor achievements by investigators in the early stages of their careers. Each recipient will receive a plaque and

will be invited to present a lecture at the divisional award symposium held at the fall ACS meeting that immediately follows the announcement of the award. Award winners will be introduced at the annual reception sponsored by the PHYS Division and the Journal of Physical Chemistry. The Division will pay up to $1,500 to cover the travel, lodging, and meal expenses for the award winner while at that ACS meeting. Winners will also be invited to the Sunday evening dinner following the PHYS Executive Committee meeting. The annual deadline is November 1.
*Requirements:* Eligibility is restricted to Physical Chemistry Division members who, at the time of the nomination, are in the first ten years of their careers as independent researchers.
*Geographic Focus:* All States
*Date(s) Application is Due:* Nov 1
*Amount of Grant:* Up to 1,500 USD
*Contact:* Awards Administrator; (202) 872-4575 or (202) 872-4408; fax (202) 776-8008; awards@acs.org
*Internet:* webapps.acs.org/findawards/detail.jsp?ContentId=CNBP_030736
*Sponsor:* American Chemical Society
1155 Sixteenth Street, NW
Washington, D.C. 20036-4801

**American Chemical Society Petroleum Research Fund Doctoral New Investigator Grants** 417

The goals of the American Chemical Society Petroleum Research Fund are: the support of fundamental research in the petroleum and energy fields; and development of the next generation of engineers and scientists through support of advanced scientific education. The emphasis of the DNI grants is focused on providing start-up funding for scientists and engineers who are within the first three years of their first academic appointment, those who have limited or no preliminary results for a research project they wish to pursue, with the intention of using the preliminary results obtained to seek continuation funding from other agencies. An estimated 90 awards will be given each year. Proposals are considered at three advisory board meetings per year (early fall, mid-winter, and late spring), which are coordinated with the ACS Grants and Awards board meetings. The amount of grants will be a maximum of $100,000 over two years, and the average number of awards is 90 annually.
*Requirements:* Regularly-appointed faculty members at U.S. academic institutions are eligible to apply. Research or adjunct faculty holding non-tenured, long-term scientific appointments also may be eligible. Current eligibility criteria are: the non-profit institution submitting the DNI proposal must certify that the individual listed as a principal investigator on the cover page qualifies as a principal investigator under the institution's policies; each principal investigator must be eligible to serve as the formal, official supervisor of the students for whom he/she is seeking support; the terms of appointment of the principal investigator must promise reasonable continuity of service. The appointment should continue at least through the period of funding requested in the proposal.
*Restrictions:* Although a PI may send the same proposal to more than one agency, PRF will not support a project having overlap, or partial overlap, with research funded by another agency. New investigators may submit only one research proposal for consideration per meeting and may not hold more than one active PRF research grant at a time. New investigators may have only three (3) DNI grant applications considered in their career. Thereafter, the new investigator may apply to the New Directions (ND) Grant Program. A principal investigator with an active PRF research grant, including a grant on time extension, may not submit an application for a new grant. No overhead costs may be charged, which includes secretarial and/or administrative salaries. Funds may not be used to support laboratory technicians, contractors, consultants, or visiting faculty.
*Geographic Focus:* All States
*Date(s) Application is Due:* Mar 16
*Amount of Grant:* 100,000 USD
*Contact:* Ron Siatkowski, Director, Office of Research Grants; (202) 872-6091; fax (202) 872-6319; r_siatkowski@acs.org
*Internet:* portal.acs.org/portal/acs/corg/content?_nfpb=true&_pageLabel=PP_SUPERARTICLE&node_id=1264&use_sec=false&sec_url_var=region1&__uuid=95d2e453-63db-4752-b1bc-abbd6a0ff8df
*Sponsor:* American Chemical Society
1155 Sixteenth Street, NW
Washington, D.C. 20036-4801

**American Chemical Society Petroleum Research Fund New Directions Grants** 418

The New Directions (ND) grants program provides funds to scientists and engineers with limited—or even no—preliminary results for a research project they wish to pursue, who intend to use the PRF-driven preliminary results to seek continuation funding from other agencies. ND grants are to be used to illustrate proof of concept/feasibility. Accordingly, they are to be viewed as seed money for new research ventures. The amount of grants will be a maximum of $100,000 over two years, and the average number of awards is 100 annually.
*Requirements:* The lead principal investigator must demonstrate that this is an entirely new research direction from what the lead principal investigator has done to date. Proposals that are considered to not be a new direction of research will be denied without external review, or may be rejected. Regularly-appointed faculty members at academic institutions in countries where ACS PRF can administer grants are eligible. In addition, certain other long-term scientific appointments may be eligible. Current eligibility criteria are: (a) The non-profit institution submitting the ND proposal must certify that each individual listed as a principal investigator on the cover page qualifies as a principal investigator under the institution's Principal Investigator policies. (b) In view of the long-standing policy of The Petroleum Research Fund to give priority to support of students (undergraduate, graduate, or postdoctoral), each principal investigator (lead and co-PI) must be eligible to serve as the formal, official supervisor of the students for whom he/she is seeking support. (c) The terms of appointment of the principal investigator (and if appropriate the co-PI) must promise reasonable continuity of service through the life of the grant. Proposals are due by 5:00 pm of the published deadline date.
*Restrictions:* Principal investigators (lead or co-) may submit only one research proposal per meeting. Principal investigators may not hold more than one active PRF research grant at a time. Principal investigators with an active PRF research grant, including a grant on time extension, may not submit an application for a new grant. Principal investigators may have only one ND grant application considered in a 12-month period. Therefore, a principal investigator who has an ND proposal denied may not submit another ND proposal until the Advisory Board meeting one year later. No overhead costs may be charged, which includes secretarial and/or administrative salaries. Funds may not be used to support laboratory technicians, contractors, consultants, or visiting faculty.
*Geographic Focus:* All States
*Date(s) Application is Due:* Mar 16
*Amount of Grant:* 100,000 USD
*Contact:* Ron Siatkowski, Director, Office of Research Grants; (202) 872-6091; fax (202) 872-6319; r_siatkowski@acs.org
*Internet:* portal.acs.org/portal/acs/corg/content?_nfpb=true&_pageLabel=PP_SUPERARTICLE&node_id=1262&use_sec=false&sec_url_var=region1&__uuid=64a4c12d-a548-4e81-a299-89007bd0aeb3
*Sponsor:* American Chemical Society
1155 Sixteenth Street, NW
Washington, D.C. 20036-4801

**American Chemical Society Petroleum Research Fund Scientific Education Grants** 419

The program assists with the travel expenses of foreign (defined as non-U.S. and non-Canadian) scientists who participate in major symposia in this country (or in Canada). The symposium topic must be fundamental research in the petroleum field to be eligible for support. The PRF Advisory Board makes recommendations on Type SE proposals during meetings three times per year: September, January, and May. Funding for foreign travel assistance is limited to $1,500 per foreign speaker with a maximum of $4,500 per symposium (three or more visiting speakers). Download the application packet from the sponsor's website.
*Requirements:* Researchers at colleges and universities are eligible. Type SE proposals for ACS symposia should be written on behalf of an ACS division, rather than the principal investigator's institutional affiliation; proposals for support of non-ACS symposia should be written on behalf of the institution organizing the symposium. In order to be eligible, researchers must meet the following *Requirements:* the symposium topic must be 'fundamental research in the petroleum field'; each eligible speaker must be from a foreign (non-U.S. and non-Canadian) non-profit institution, and must give a petroleum-relevant or alternate-energy presentation; the sponsor of the symposium must be a non-profit scientific or educational institution; the meeting or symposium must be accessible and open; symposia whose attendance is by invitation only, or where facilities sharply limit attendance have not been favored for support by the PRF Advisory Board; and the results of the meeting or symposium must be publishable in peer-reviewed literature.
*Restrictions:* The program does not support travel of American scientists to similar meetings in foreign countries, on the rationale that more American scientists derive benefit from having foreign speakers travel to North America, and that support helps American organizations to enhance the value of their programs and, in some instances, to meet their obligations as hosts for international meetings.
*Geographic Focus:* All States
*Amount of Grant:* 1,500 - 4,500 USD
*Contact:* Ron Siatkowski, Director, Office of Research Grants; (202) 872-6091; fax (202) 872-6319; r_siatkowski@acs.org
*Internet:* portal.acs.org/portal/acs/corg/content?_nfpb=true&_pageLabel=PP_SUPERARTICLE&node_id=1265&use_sec=false&sec_url_var=region1&__uuid=7e98ac5c-1769-493b-ba5b-086690c3c85d
*Sponsor:* American Chemical Society
1155 Sixteenth Street, NW
Washington, D.C. 20036-4801

**American Chemical Society Petroleum Research Fund Undergraduate Faculty Sabbatical Grants** 420

The Undergraduate Faculty Sabbatical grant program is designed to provide year-long, full-time, research sabbaticals for qualified faculty performing fundamental research in the petroleum field during a sabbatical. This is a matching grants program that can provide up to $45,000 in faculty salary and qualified benefits, and up to $5,000 in relocation and research expenses for a faculty member's sabbatical year. The maximum individual award is $50,000. Applications are considered each year at the September PRF Advisory Board meeting.
*Requirements:* An eligible sabbatical must be at least one academic year (i.e., 9 months or more) and must be carried out in its entirety at a location other than the applicant's home institution. Sabbaticals may be taken anywhere in the world. Academic, national lab, and industrial sabbaticals are eligible, as are sabbaticals that involve more than one institution. Faculty with tenured appointments in departments that do not offer a doctoral degree and faculty at two- and four-year institutions are eligible to apply. Current ACS PRF grant recipients, as well as those with pending PRF proposals are also eligible. Complete eligibility requirements and application packet can be downloaded from the sponsor's website.
*Restrictions:* Fundamental research is required. ACS PRF cannot support applied research or methods development. Note also that the proposed research need not be directly connected to petroleum, but rather may provide a basis for subsequent research directly connected with the petroleum field. No overhead or indirect costs may be charged to the grant.
*Geographic Focus:* All States

*Amount of Grant:* Up to 50,000 USD
*Contact:* Ron Siatkowski, Director, Office of Research Grants; (202) 872-6091; fax (202) 872-6319; r_siatkowski@acs.org
*Internet:* portal.acs.org/portal/acs/corg/content?_nfpb=true&_pageLabel=PP_SUPERARTICLE&node_id=1266&use_sec=false&sec_url_var=region1&__uuid=b9dc3d16-4e42-41c6-b235-86d242b3a346
*Sponsor:* American Chemical Society
1155 Sixteenth Street, NW
Washington, D.C. 20036-4801

**American Chemical Society Petroleum Research Fund Undergraduate New Investigator Grants** 421

The goals of the American Chemical Society Petroleum Research Fund are: to support fundamental research in the petroleum field; and to develop the next generation of engineers and scientists through support of advanced scientific education. The Undergraduate New Investigator Grants program is intended to initiate the research program of new scientists and engineers who are faculty members at undergraduate research institutions and to provide financial incentives for students at those institutions to become involved in research activities leading to employment or continued study in graduate school. These grants provide funds for scientists and engineers who are beginning their independent careers in academia and have limited or no preliminary results for a research project they wish to pursue. The UNI grants are to be used to illustrate proof of principle, i.e., feasibility, and accordingly, are to be viewed as seed money for generating preliminary results that can be used to apply for continuation funding from other agencies. The amount of grants will be a maximum of $50,000 over two years, and the average number of awards is 45 annually.
*Requirements:* Eligibility for a UNI grant requires that a PI is in a department without a doctoral program, and that the students receiving stipends for the work to be done are undergraduates (M.S.-level students can also be supported IF one or more undergraduates are also supported from this grant). Accordingly, the research being proposed need not be high risk but it should be of publishable quality. The research opportunities afforded must be of the highest caliber, and provide a compelling educational experience for the student. To be eligible as a principal investigator for a UNI grant, the applicant must be a member of the faculty of a college or university within the United States; be within the first three years of their first academic appointment as a regular faculty member; have completed all requirements for the Ph.D. (however, an application may be submitted before the degree has been awarded); and be appointed at the rank of Assistant Professor or the equivalent. In addition, applicants must meet the following three criteria: (a) The non-profit institution submitting the UNI proposal must certify that the person listed as a principal investigator on the cover page qualifies as a principal investigator under the institution's policies. (b) In view of the long-standing policy of The Petroleum Research Fund to give priority to support of students, the principal investigator must be eligible to serve as the formal, official supervisor of the students for whom he/she is seeking support. (c) The term of appointment of the principal investigator must promise reasonable continuity of service. The appointment should continue at least through the period of funding requested in the UNI proposal.
*Restrictions:* A principal investigator may submit only one research proposal per meeting. Only one UNI grant may be held during an investigator's career.
*Geographic Focus:* All States
*Date(s) Application is Due:* Mar 16
*Amount of Grant:* 50,000 USD
*Contact:* Ron Siatkowski; (202) 872-6091; fax (202) 872-6319; r_siatkowski@acs.org
*Internet:* portal.acs.org/portal/acs/corg/content?_nfpb=true&_pageLabel=PP_SUPERARTICLE&node_id=1796&use_sec=false&sec_url_var=region1&__uuid=30190564-11ef-4b06-a76b-71cdf4d4c22c
*Sponsor:* American Chemical Society
1155 Sixteenth Street, NW
Washington, D.C. 20036-4801

**American Chemical Society Petroleum Research Fund Undergraduate Research Grants** 422

The goals of the American Chemical Society Petroleum Research Fund are: to support fundamental research in the petroleum field; and to develop the next generation of engineers and scientists through support of advanced scientific education. The Undergraduate Research (UR) grants program supports the research programs of established scientists and engineers at non-doctoral departments and provides financial support for students at those institutions to become involved in advanced research activities, in preparation for continued study in graduate school or employment. Demonstration of productivity is important, but a UR grant may also be used for a project with limited or no preliminary results in a new research area the PI wishes to pursue, with the intention of using the preliminary results obtained to seek continuation funding from other agencies. The amount of each grant over a three-year period is $65,000, and the estimated number of awards is 45 annually.
*Requirements:* Eligibility for a UR grant requires that a PI is in a department without a doctoral program, and that undergraduates are involved in the project. Investigators from Master's degree-granting departments are eligible, and support can be provided for M.S. level research, but undergraduates must be included in the project. Accordingly, the research being proposed need not be comparable in volume and scope of research in doctoral settings, but it should be of publishable quality, and the research opportunities afforded the students must be of the highest caliber.
*Restrictions:* A principal investigator may have only one research proposal active or under consideration per meeting. This limitation extends to those who have co-PI status on an application. A PI must close an active grant, after the full grant period has been completed, before submitting another proposal.
*Geographic Focus:* All States
*Date(s) Application is Due:* Mar 16
*Amount of Grant:* Up to 65,000 USD
*Contact:* Ron Siatkowski; (202) 872-6091; fax (202) 872-6319; r_siatkowski@acs.org
*Internet:* portal.acs.org/portal/acs/corg/content?_nfpb=true&_pageLabel=PP_SUPERARTICLE&node_id=1263&use_sec=false&sec_url_var=region1&__uuid=b499d85c-9121-440b-bbec-7a0976442e2e
*Sponsor:* American Chemical Society
1155 Sixteenth Street, NW
Washington, D.C. 20036-4801

**American Chemical Society Presidential Green Chemistry Challenge Awards** 423

The Office of Pollution Prevention and Toxics of the United States Environmental Protection Agency sponsors and administers these awards and the American Chemical Society (ACS Green Chemistry Institute) convenes the panel of technical experts to evaluate the nominations. The Award's purpose is to recognize outstanding chemical technologies that incorporate the principles of green chemistry into chemical design, manufacture, and use. The program invites nominations that describe the science and innovation of a green chemistry technology, its human health and environmental benefits, and its applicability and impact. The five award categories are: Greener Synthetic Pathways; Greener Reaction Conditions; Designing Greener Chemicals; Academic; and Small Business. The recipient of the award is recognized at an awards ceremony in Washington, D.C., typically held during June. A crystal sculpture will be presented to the primary sponsor(s) of the winning green chemistry technology in each of the five award categories. Certificates will be presented to individuals (as identified by the primary sponsor) who contributed to the research, development, or implementation of the chemistry.
*Requirements:* Nominations are accepted from individuals, groups, non-profit and for-profit organizations, academia, small business, and industry. The nominated green chemistry technology must have reached a significant milestone within the past five years (i.e., it must been researched, demonstrated, implemented, applied, patented, etc.). It must also have a significant component within the United States.
*Geographic Focus:* All States
*Date(s) Application is Due:* Dec 31
*Contact:* Ron Siatkowski; (202) 872-6091; fax (202) 872-6319; r_siatkowski@acs.org
*Internet:* webapps.acs.org/findawards/detail.jsp?ContentId=WPCP_012553
*Sponsor:* American Chemical Society
1155 Sixteenth Street, NW
Washington, D.C. 20036-4801

**American Chemical Society WCC Rising Star Award** 424

The ACS Women Chemists Committee (WCC) serves the membership of the American Chemical Society with its mission to be leaders in attracting, retaining, developing, promoting, and advocating for women in the chemical sciences. Established in 2011, the WCC Rising Star Award recognizes exceptional early to mid-career women chemists across all areas of chemistry on a national level. To be given annually, this award is also intended to help promote retention of women in science. The Award includes the following: an opportunity to present contributions at a WCC-sponsored symposium at the spring ACS national meeting; a $1,000 stipend to cover spring national meeting travel expenses; and visibility and networking opportunities at the spring national meeting—awardees are recognized at the WCC Just Cocktails reception and at the WCC Luncheon, where awardees are seated at the head tables along with ACS governance, WCC members, and fellow award winners. The deadline for receipt of nominations is June 15.
*Requirements:* The award is open to all female ACS members in chemistry and chemical engineering working in academic, industrial, government, non-profit or other employment sectors. Appropriate candidates will typically be no more than 15 years from receipt of their terminal scientific degree and have demonstrated outstanding promise for contributions to their respective fields. Applicants can either be self-nominated or can be nominated by another individual for this award.
*Restrictions:* The following are not eligible: those who are already widely acknowledged and recognized in their area of chemistry or chemical engineering and are at the pinnacle of their careers; ACS national award winners; those who have received a prior award under this program; or members of the Women Chemists Committee.
*Geographic Focus:* All States
*Date(s) Application is Due:* Jun 15
*Contact:* Ron Siatkowski; (202) 872-6091; fax (202) 872-6319; r_siatkowski@acs.org
*Internet:* webapps.acs.org/findawards/detail.jsp?ContentId=CNBP_029245
*Sponsor:* American Chemical Society
1155 Sixteenth Street, NW
Washington, D.C. 20036-4801

**American College of Bankruptcy Foundation Grants** 425

The American College of Bankruptcy Foundation was established to promote the ends of justice through educational and charitable activities. Its missions, focused in the areas of bankruptcy and insolvency, are: sponsorship and encouragement of legal research, publications and forums; establishment of scholarships; providing for the collection and maintenance of data and documents for scholarly research; and fostering the institution and maintenance of legal aid facilities for the indigent. In moving toward these goals, the American College of Bankruptcy and American College of Bankruptcy Foundation have joined together to award grants to legal services organizations whose activities are consistent with the objective of the respective missions of the College and Foundation. Grant awards are made by the College and Foundation's Pro Bono Committee.

Applications for grants during a calendar year must be received on or before July 1 of the calendar year. Typically, grants have ranged from $1,000 to $10,000.
*Geographic Focus:* All States
*Date(s) Application is Due:* Jul 1
*Amount of Grant:* 1,000 - 10,000 USD
*Contact:* Shari Bedker, Executive Director; (434) 939-6005; fax (434) 939-6030; sbedker@acbfoundation.org or info@acbfoundation.org
*Internet:* www.acbfoundation.org/
*Sponsor:* American College of Bankruptcy Foundation
P.O. Box 249
Stanardsville, VA 22973

**American Foodservice Charitable Trust Grants** 426
American Foodservice Charitable Trust provides funding in the following areas of interest: Christian agencies & churches; community/economic development; higher education; hospitals (general) and; youth development. The types of support available are: general/operating support; program development; scholarship funds. Giving is primarily available in the areas of: King of Prussia, Pennsylvania; Thomasville, Georgia; and Fort Worth, Texas.
*Requirements:* Qualifying IRA 501(c)3 nonprofit organizations are eligible to apply. There's no application form nor is there a deadline date to adhere to when submitting a proposal to the Foundation.
*Restrictions:* No grants to individuals.
*Geographic Focus:* Georgia, Pennsylvania, Texas
*Amount of Grant:* 200 - 8,000 USD
*Samples:* Catholic Social Services, West Chester, PA, $200—social services grant; American Red Cross, Philadelphia, PA, $1,000—disaster relief grant; Police Atheletic League, Philadelphia, PA, $2,000—youth program grant.
*Contact:* Richard S. Downs, Trustee; (610) 933-9792
*Sponsor:* American Foodservice Charitable Trust
860 First Avenue, Suite 9A
King of Prussia, PA 19406-1404

**American Historical Association Albert B. Corey Prize in Canadian-American Relations** 427
The prize is awarded biennially in even-numbered years to an American or Canadian resident for the best book on the history of Canadian-American relations or on the history of both countries. The prize is awarded jointly by the Canadian Historical Association and the American Historical Association. Books bearing an imprint during the previous two years prior to the prize year are eligible.
*Requirements:* Books bearing an imprint of the previous two years are eligible for the prize.
*Geographic Focus:* All States
*Date(s) Application is Due:* May 15
*Contact:* Julia Brookins, Special Projects Coordinator; (202) 544-2422, ext. 119; fax (202) 544-8307; jbrookins@theaha.org or aha@theaha.org
*Internet:* www.historians.org/prizes/index.cfm?PrizeAbbrev=COREY
*Sponsor:* American Historical Association
400 A Street SE
Washington, D.C. 20003

**American Historical Association Albert J. Beveridge Award** 428
The Award is given annually for the best book in English on the history of the United States, Latin America, or Canada from 1492 to the present. Books that employ new methodological or conceptual tools or that constitute a significant reinterpretation of an important historical problem are given preference in the awarding of this prize. Biographies, monographs, and works of synthesis or interpretation are eligible.
*Requirements:* U.S. and Canadian citizens and permanent residents are eligible. Books published after May 1 of the previous year and before April 30 of the award year are eligible.
*Restrictions:* Translations, anthologies, and collections of documents are not eligible.
*Geographic Focus:* All States
*Date(s) Application is Due:* May 15
*Contact:* Julia Brookins, Special Projects Coordinator; (202) 544-2422, ext. 119; fax (202) 544-8307; jbrookins@theaha.org or aha@theaha.org
*Internet:* www.historians.org/prizes/index.cfm?PrizeAbbrev=Beveridge
*Sponsor:* American Historical Association
400 A Street SE
Washington, D.C. 20003

**American Historical Association Albert J. Beveridge Grants for Research in the History of the Western Hemisphere** 429
The Association announces the availability of modest grants to support research in the history of the Western hemisphere (United States, Canada, and Latin America). The funds for this program come from the earnings of the Albert J. Beveridge Memorial Fund. The grants are intended to further research in progress and may be used for travel to a library or archive, for microfilms, photographs, or photocopying-a list of purposes that is meant to be merely illustrative, not exhaustive (other expenses, such as child care, can be included). Of particular interest are those applicants with specific research needs, such as the completion of a project or completion of a discrete segment thereof. The annual deadline is February 15.
*Requirements:* Preference will be given to Ph.D. candidates and junior scholars.
*Restrictions:* Only members of the Association are eligible. Within a five-year period, no individual is eligible to receive more than a combined total of $1,000 from the Beveridge, Kraus, and Littleton-Griswold grant programs.
*Geographic Focus:* All States
*Date(s) Application is Due:* Feb 15
*Amount of Grant:* Up to 1,000 USD
*Contact:* Julia Brookins, Special Projects Coordinator; (202) 544-2422, ext. 119; fax (202) 544-8307; jbrookins@theaha.org or aha@theaha.org
*Internet:* www.historians.org/prizes/BeveridgeGrantInfo.htm
*Sponsor:* American Historical Association
400 A Street SE
Washington, D.C. 20003

**American Historical Association Award for Scholarly Distinction** 430
In 1984 the AHA Council established the American Historical Association Award for Scholarly Distinction to honor senior historians in the United States. According to the selection criteria of this Award, recipients must be senior historians of the highest distinction who have spent the bulk of their professional careers in the United States. Generally, they must also be of emeritus rank, if from academic life, or equivalent standing otherwise. Under normal circumstances the award is not intended to go to former presidents of the Association; rather, the intent is to honor persons not otherwise recognized by the profession to an extent commensurate with their contributions.
*Geographic Focus:* All States
*Date(s) Application is Due:* Apr 30
*Contact:* Sharon K. Tune, Assistant Director; (202) 544-2422, ext. 101; fax (202) 544-8307; stune@theaha.org or aha@theaha.org
*Internet:* www.historians.org/prizes/ScholarlyDistinction.htm
*Sponsor:* American Historical Association
400 A Street SE
Washington, D.C. 20003

**American Historical Association Bernadotte E. Schmitt Research Grants in European, African, or Asian History** 431
Modest grants support research in the history of Europe, Africa, and Asia. The funds for this program come from the earnings of a bequest from Bernadotte E. Schmitt, president of the Association in 1960. The grants are intended to further research in progress and may be used for: travel to a library or archive; microfilms, photographs, or photocopying; coding and keypunching; and other approved research-related activities.
*Requirements:* Preference will be given to those with specific research needs, such as the completion of a project or completion of a discrete segment thereof. Preference will be given to Ph.D. candidates and junior scholars.
*Restrictions:* Only members of the Association are eligible to apply.
*Geographic Focus:* All States
*Date(s) Application is Due:* Feb 15
*Amount of Grant:* Up to 1,000 USD
*Contact:* Julia Brookins, Special Projects Coordinator; (202) 544-2422, ext. 119; fax (202) 544-8307; jbrookins@theaha.org or aha@theaha.org
*Internet:* www.historians.org/prizes/SchmittGrantInfo.htm
*Sponsor:* American Historical Association
400 A Street SE
Washington, D.C. 20003

**American Historical Association Clarence H. Haring Prize** 432
The Clarence H. Haring Prize is awarded every five years to a Latin American who, in the opinion of the committee, has published the most outstanding book on Latin American history during the preceding five years ending June 1 of the award year. There is no language limitation on works submitted. There is no language limitation on works submitted.
*Requirements:* Books published between May 1, 2011, and April 30, 2016, will be eligible for the prize in 2016.
*Geographic Focus:* All States
*Date(s) Application is Due:* May 16
*Contact:* Julia Brookins, Special Projects Coordinator; (202) 544-2422, ext. 119; fax (202) 544-8307; jbrookins@theaha.org or aha@theaha.org
*Internet:* www.historians.org/prizes/index.cfm?PrizeAbbrev=HARING
*Sponsor:* American Historical Association
400 A Street SE
Washington, D.C. 20003

**American Historical Association Fellowship in Aerospace History** 433
The American Historical Association will fund one Fellow to undertake a research project related to aerospace history. It will provide a Fellow with an opportunity to engage in significant and sustained advanced research in all aspects of the history of aerospace from the earliest human interest in flight to the present, including cultural and intellectual history, economic history, history of law and public policy, and the history of science, engineering, and management. The Fellowship term is for a period of at least six months, but not more than one year. The Fellow will be expected to devote the term entirely to the proposed research project. The Fellow will have (and be encouraged to take advantage of) the opportunity to use the documentary resources of the National Aeronautics and Space Administration, and may also spend the Fellowship in residence at the NASA headquarters or one of the NASA centers. The stipend is $20,000 for a 6 to 9 month fellowship. This amount is adjustable to the length of the fellowship term. Applications and letters of recommendation must be postmarked by April 1.
*Requirements:* Applicants must possess a doctorate degree in history or in a closely related field, or be enrolled as a student (having completed all coursework) in a doctoral degree-granting program. The applicant must complete an application form and offer a specific and detailed research proposal that will be the basis of the Fellow's research during the

term. At the term's conclusion, the Fellow will be expected to write a report, and to present a paper or a public lecture on the Fellowship experience.
*Restrictions:* Funds may not be used to support tuition or fees. A Fellow may not hold other major fellowships or grants during the fellowship term, except sabbatical and supplemental grants from their own institutions, and small grants from other sources for specific research expenses. Sources of anticipated support must be listed in the application form.
*Geographic Focus:* All States
*Date(s) Application is Due:* Apr 1
*Contact:* Julia Brookins, Special Projects Coordinator; (202) 544-2422, ext. 119; fax (202) 544-8307; jbrookins@theaha.org or aha@theaha.org
*Internet:* www.historians.org/prizes/NASA.htm
*Sponsor:* American Historical Association
400 A Street SE
Washington, D.C. 20003

**American Historical Association Folger Shakespeare Library Fellowship** 434
The American Historical Association/Folger Shakespeare Library Fellowship is sponsored jointly by the AHA and the Folger Shakespeare Library. It is awarded for research on seventeenth- and eighteenth-century western European history; the fellow will be awarded a one-month fellowship to be taken at the Folger Shakespeare Library. Apply directly to the Folger by March 1.
*Requirements:* Applicants must hold the PhD at the time of application and must be a member in good standing of the AHA.
*Geographic Focus:* All States
*Date(s) Application is Due:* Mar 1
*Contact:* Julia Brookins, Special Projects Coordinator; (202) 544-2422, ext. 119; fax (202) 544-8307; jbrookins@theaha.org or aha@theaha.org
*Internet:* www.historians.org/prizes/Folger.cfm
*Sponsor:* American Historical Association
400 A Street SE
Washington, D.C. 20003

**American Historical Association George L. Mosse Prize** 435
The American Historical Association awards the George L. Mosse Prize annually for an outstanding major work of extraordinary scholarly distinction, creativity, and originality in the intellectual and cultural history of Europe since the Renaissance. This prize was established with funds donated by former students, colleagues, and friends of Dr. Mosse. Only books of a high scholarly distinction should be submitted. Research accuracy, originality, and literary merit are important selection factors.
*Requirements:* Books published after May 1 of the previous year and before April 30 of the award year are eligible.
*Geographic Focus:* All States
*Date(s) Application is Due:* May 15
*Contact:* Julia Brookins, Special Projects Coordinator; (202) 544-2422, ext. 119; fax (202) 544-8307; jbrookins@theaha.org or aha@theaha.org
*Internet:* www.historians.org/prizes/index.cfm?PrizeAbbrev=MOSSE
*Sponsor:* American Historical Association
400 A Street SE
Washington, D.C. 20003

**American Historical Association George Louis Beer Prize** 436
The American Historical Association offers the George Louis Beer Prize in recognition of outstanding historical writing in European international history since 1895. This prize was established in accordance with the terms of a bequest by George Louis Beer (d. 1920), historian of the British colonial system before 1765, to be awarded annually for the best work on any phase of European international history since the year 1895 that is submitted by a scholar who is a United States citizen or permanent resident. The phrase "European international history since the year 1895" may be understood to mean any study of international history since the year 1895 with a significant European dimension. Only books of a high scholarly historical nature should be submitted. Research accuracy, originality, and literary merit are important factors. Entries must be postmarked by or on May 15.
*Requirements:* Books published after May 1 of the previous year and before April 30 of the award year are eligible. Only books of a high scholarly historical nature should be submitted.
*Geographic Focus:* All States
*Date(s) Application is Due:* May 15
*Contact:* Julia Brookins, Special Projects Coordinator; (202) 544-2422, ext. 119; fax (202) 544-8307; jbrookins@theaha.org or aha@theaha.org
*Internet:* www.historians.org/prizes/index.cfm?PrizeAbbrev=BEER
*Sponsor:* American Historical Association
400 A Street SE
Washington, D.C. 20003

**American Historical Association Helen and Howard R. Marraro Prizes** 437
The Helen and Howard R. Marraro Prizes are given annually by the American Historical Association, the American Catholic Historical Association, and the Society for Italian Historical Studies for the best works on any epoch of Italian cultural history or on Italian-American relations written by American and Canadian resident citizens. The associations appoint a joint selection committee. One copy of each work, together with a brief curriculum vita and bibliography of the author, must be submitted to designated committee members. The American Historical Association will send instructions upon request.
*Requirements:* Books published after May 1 of the previous year and before April 30 of the award year are eligible. Entries must first have been published in English by a historian whose usual residence is North America.
*Geographic Focus:* All States
*Date(s) Application is Due:* May 15
*Contact:* Julia Brookins, Special Projects Coordinator; (202) 544-2422, ext. 119; fax (202) 544-8307; jbrookins@theaha.org or aha@theaha.org
*Internet:* www.historians.org/prizes/index.cfm?PrizeAbbrev=MARRARO
*Sponsor:* American Historical Association
400 A Street SE
Washington, D.C. 20003

**American Historical Association Herbert Baxter Adams Prize** 438
The American Historical Association offers the Herbert Baxter Adams Prize annually for a distinguished book by an American author in the field of European history. The prize was established in memory of the first secretary of the Association, Herbert Adams of Johns Hopkins University, who was also one of the Association's founders. Together with the Leo Gershoy Award, the Adams Prize is the most important distinction bestowed by the profession in the field of European history and thus has considerable prestige.
*Requirements:* The author must be a citizen or permanent resident of the United States or Canada. Since the prize is designed especially to encourage scholars who have not yet obtained an established reputation, the entry must be the author's first substantial book.
*Restrictions:* Textbooks in the strict sense of the word are not eligible, but a work of wide scope which interprets a major period or area would certainly qualify. Pamphlets, anthologies, edited works, and other small-scale efforts will not qualify.
*Geographic Focus:* All States
*Date(s) Application is Due:* May 15
*Contact:* Julia Brookins, Special Projects Coordinator; (202) 544-2422, ext. 119; fax (202) 544-8307; jbrookins@theaha.org or aha@theaha.org
*Internet:* www.historians.org/prizes/index.cfm?PrizeAbbrev=ADAMS
*Sponsor:* American Historical Association
400 A Street SE
Washington, D.C. 20003

**American Historical Association Herbert Feis Award** 439
Established in 1984, the Herbert Feis Award is offered annually to recognize the scholarly interests of historians outside academe and the importance of the work of independent scholars in the United States. The terms of the award define both contribution and public history broadly. Contributions could include work as the administrator of a public history group or agency (such as a historical society, a historic site, or a community history project) or as the creator or producer of a public history product or products (such as a museum exhibit, radio script, web site, oral history collection, or film). The contribution may be the result of years of effort in the field, but the prize might also recognize a singular contribution of major importance such as a path breaking museum exhibit. Public history is defined as work primarily directed at non-academic, non-school-based audiences. Those audiences could be very broad (e.g., television viewers) or highly specialized (e.g., policymakers). Although the audience should be primarily outside of academia, the recipient of the award could be employed at a university. Nominations must be postmarked by or on August 1.
*Requirements:* Works published or issued in-house between May 1 of the previous year and April 30 of the upcoming year are eligible.
*Geographic Focus:* All States
*Date(s) Application is Due:* Aug 1
*Contact:* Julia Brookins, Special Projects Coordinator; (202) 544-2422, ext. 119; fax (202) 544-8307; jbrookins@theaha.org or aha@theaha.org
*Internet:* www.historians.org/prizes/index.cfm?PrizeAbbrev=FEIS
*Sponsor:* American Historical Association
400 A Street SE
Washington, D.C. 20003

**American Historical Association J. Franklin Jameson Award In Outstanding Editorial Achievement** 440
Commencing in 1980, this prize is awarded biennially in odd-numbered years for outstanding achievement in the editing of historical sources. The prize is named in honor of J. Franklin Jameson, a founder of the Association and longtime managing editor of the American Historical Review.
*Requirements:* To be eligible for consideration, works must be of a scholarly, historical nature; review/journal editing is ineligible. Books published after May 1 of the previous year and before April 30 of the award year are eligible.
*Geographic Focus:* All States
*Date(s) Application is Due:* May 15
*Contact:* Julia Brookins, Special Projects Coordinator; (202) 544-2422, ext. 119; fax (202) 544-8307; jbrookins@theaha.org or aha@theaha.org
*Internet:* www.historians.org/prizes/index.cfm?PrizeAbbrev=Jameson
*Sponsor:* American Historical Association
400 A Street SE
Washington, D.C. 20003

**American Historical Association J. Franklin Jameson Fellowship** 441
The Franklin Jameson Fellowship in American History is offered annually by the Library of Congress and the American Historical Association to support significant scholarly research for one semester in the collections of the Library of Congress by scholars at an early stage in their careers in history. The fellowship is named in honor of J. Franklin Jameson, a founder of the Association, longtime managing editor of the American Historical Review, formerly Chief of the Manuscript Division of the Library of Congress, and the first incumbent of the Library's Chair of American History. It is designed to assist scholars early in their careers.

The applicant's project in American history must be one for which the general and special collections of the Library of Congress offer unique research support. Applicants should include a statement substantiating this relationship. The fellowship will be awarded for 2 to 3 months, as the Jameson Fellow desires, to spend in full-time residence at the Library of Congress. Working space will be provided by the Library of Congress, but the Jameson Fellow is required to spend at least two months in residence. The total stipend may vary from year to year. Currently the amount will be $5,000. This will not be prorated according to the proposed length of tenure of the fellowship.
*Requirements:* At the time of application, applicants must hold the Ph.D. degree or equivalent; must have received this degree within the past seven years, and must not have published or had accepted for publication a book-length historical work.
*Restrictions:* The fellowship will not be awarded to permit completion of a doctoral dissertation.
*Geographic Focus:* All States
*Date(s) Application is Due:* Mar 15
*Amount of Grant:* 5,000 USD
*Contact:* Julia Brookins, Special Projects Coordinator; (202) 544-2422, ext. 119; fax (202) 544-8307; jbrookins@theaha.org or aha@theaha.org
*Internet:* www.historians.org/prizes/Jameson_fellowship.htm
*Sponsor:* American Historical Association
400 A Street SE
Washington, D.C. 20003

**American Historical Association J. Russell Major Prize** 442
The American Historical Association awards the J. Russell Major Prize annually for the best work in English on any aspect of French history. The prize was established in memory of J. Russell Major, the distinguished scholar of French history who died on December 12, 1998 at the age of 77. Major served on the history faculty at Emory University from 1949 until his retirement in 1990, and wrote 10 books, including Representative Government in Early Modern France and From Renaissance Monarchy to Absolute Monarchy: French Kings, Nobles and Estates.
*Requirements:* Books published after May 1 of the previous year and before April 30 of the award year are eligible.
*Geographic Focus:* All States
*Date(s) Application is Due:* May 15
*Contact:* Julia Brookins, Special Projects Coordinator; (202) 544-2422, ext. 119; fax (202) 544-8307; jbrookins@theaha.org or aha@theaha.org
*Internet:* www.historians.org/prizes/index.cfm?PrizeAbbrev=MAJOR
*Sponsor:* American Historical Association
400 A Street SE
Washington, D.C. 20003

**American Historical Association James A. Rawley Prize in Atlantic History** 443
The American Historical Association announced the establishment of the James A. Rawley Prize in Atlantic History in recognition of outstanding historical writing that explores aspects of integration of Atlantic worlds before the 20th century. The prize was established in accordance with the terms of a gift from James A. Rawley, Carl Adolph Happold Professor of History Emeritus at the University of Nebraska at Lincoln. Only books of high scholarly and literary merit will be considered. Research accuracy and originality also will be important factors in the evaluation of the books.
*Requirements:* Books published after May 1 of the previous year and before April 30 of the award year are eligible. Only books of a high scholarly historical nature should be submitted.
*Geographic Focus:* All States
*Date(s) Application is Due:* May 15
*Contact:* Julia Brookins, Special Projects Coordinator; (202) 544-2422, ext. 119; fax (202) 544-8307; jbrookins@theaha.org or aha@theaha.org
*Internet:* www.historians.org/prizes/index.cfm?PrizeAbbrev=Rawley
*Sponsor:* American Historical Association
400 A Street SE
Washington, D.C. 20003

**American Historical Association James Harvey Robinson Prize** 444
The James Harvey Robinson Prize was established by the AHA Council in 1978 and is awarded biennially for the teaching aid which has made the most outstanding contribution to the teaching and learning of history in any field for public or educational purposes. Teaching aid encompasses textbooks, source and reference materials, audiovisuals, computer-assisted instruction, and public history or museum materials. Monographs and revisions will not be considered. The winner will receive a one-year membership in the Association. The work should: have the potential to influence history education; demonstrate recent and good historical scholarship; and be well written and attractively presented.
*Requirements:* Monographs and revisions will not be considered. Books published after May 1 two years prior and before April 30 of the award year are eligible.
*Restrictions:* Monographs and revisions will not be considered.
*Geographic Focus:* All States
*Date(s) Application is Due:* May 15
*Contact:* Julia Brookins, Special Projects Coordinator; (202) 544-2422, ext. 119; fax (202) 544-8307; jbrookins@theaha.org or aha@theaha.org
*Internet:* www.historians.org/prizes/index.cfm?PrizeAbbrev=ROBINSON
*Sponsor:* American Historical Association
400 A Street SE
Washington, D.C. 20003

**American Historical Association James Henry Breasted Prize** 445
Established in 1985, this prize, named in honor of James Henry Breasted, a pioneer in ancient Egyptian and Near Eastern history and president of the Association in 1928, is offered for the best book in English in any field of history prior to CE 1000 and rotates annually among the following geographical areas: Near East and Egypt; Far East and South Asia; Africa, North America, and Latin America; and Europe. Only works of high scholarly and literary merit will be considered. One copy of each entry must be submitted to designated committee members.
*Requirements:* Books published after May 1 of the previous year and before April 30 of the award year are eligible.
*Geographic Focus:* All States
*Date(s) Application is Due:* May 15
*Contact:* Julia Brookins, Special Projects Coordinator; (202) 544-2422, ext. 119; fax (202) 544-8307; jbrookins@theaha.org or aha@theaha.org
*Internet:* www.historians.org/prizes/index.cfm?PrizeAbbrev=BREASTED
*Sponsor:* American Historical Association
400 A Street SE
Washington, D.C. 20003

**American Historical Association Joan Kelly Memorial Prize in Women's History** 446
Established in 1984, this prize, named in memory of Joan Kelly (1928-82), is awarded annually for the book in women's history and/or feminist theory that best reflects the high intellectual and scholarly ideals exemplified by the life and work of Joan Kelly. The prize was established by the Coordinating Committee on Women in the Historical Profession and the Conference Group on Women's History (now the Coordinating Council for Women in History), and is administered by the American Historical Association.
*Requirements:* Books in any chronological period, any geographical location, or in an area of feminist theory that incorporates a historical perspective are eligible for consideration. Books should demonstrate originality of research, creativity of insight, graceful stylistic presentation, analytical skills, and a recognition of the important role of sex and gender in the historical process. The inter-relationship between women and the historical process should be addressed. Books published after May 1 of the previous year and before April 30 of the award year are eligible.
*Geographic Focus:* All States
*Date(s) Application is Due:* May 15
*Contact:* Julia Brookins, Special Projects Coordinator; (202) 544-2422, ext. 119; fax (202) 544-8307; jbrookins@theaha.org or aha@theaha.org
*Internet:* www.historians.org/prizes/index.cfm?PrizeAbbrev=KELLY
*Sponsor:* American Historical Association
400 A Street SE
Washington, D.C. 20003

**American Historical Association John F. Richards Prize in South Asian History** 447
The John F. Richards Prize in South Asian History recognizes the most distinguished work of scholarship on South Asian history published in English during the previous calendar year. South Asia is defined as the geographic area included in the modern states of Afghanistan, Bangladesh, Bhutan, Burma, India, the Maldives, Nepal, Pakistan, and Sri Lanka. Eligibility will be defined quite broadly, including books on any period or field of South Asian historical studies and works which integrate South Asian history with broader global issues and movements. In making its selection, the prize committee will pay particular attention to depth of research, methodological innovation, conceptual originality and literary excellence. Works that reinterpret old themes or develop new theoretical perspectives are welcome.
*Requirements:* Books published after May 1 of the previous year and before April 30 of the award year are eligible. Only books of a high scholarly historical nature should be submitted.
*Restrictions:* Anthologies, encyclopedias and other edited volumes will not be considered.
*Geographic Focus:* All States
*Date(s) Application is Due:* May 15
*Contact:* Julia Brookins, Special Projects Coordinator; (202) 544-2422, ext. 119; fax (202) 544-8307; jbrookins@theaha.org or aha@theaha.org
*Internet:* www.historians.org/prizes/index.cfm?PrizeAbbrev=Richards
*Sponsor:* American Historical Association
400 A Street SE
Washington, D.C. 20003

**American Historical Association John H. Dunning Prize** 448
In recognition of outstanding historical writing in United States history, the American Historical Association offers the John H. Dunning Prize. Established by a bequest from Mathilde Dunning in 1927 to be awarded for an outstanding monograph on any subject relating to United States history, this prize is offered biennially in odd-numbered years. To be eligible for consideration, an entry must be of a scholarly historical nature.
*Requirements:* An entry must be the author's first or second book published or completed. Books published after May 1 of the previous year and before April 30 of the deadline year are eligible.
*Geographic Focus:* All States
*Date(s) Application is Due:* May 15
*Contact:* Julia Brookins, Special Projects Coordinator; (202) 544-2422, ext. 119; fax (202) 544-8307; jbrookins@theaha.org or aha@theaha.org
*Internet:* www.historians.org/prizes/index.cfm?PrizeAbbrev=DUNNING
*Sponsor:* American Historical Association
400 A Street SE
Washington, D.C. 20003

**American Historical Association John K. Fairbank Prize in East Asian History** 449
The John K. Fairbank Prize in East Asian History is awarded annually for an outstanding book on the history of China proper, Vietnam, Chinese Central Asia, Mongolia, Manchuria, Korea, or Japan substantially since the year 1800. Only books of high scholarly and literary merit will be considered. One copy of each entry must be submitted to each committee member. Application forms are available from the association.
*Requirements:* Books published after May 1 of the previous year and before April 30 of the award year are eligible.
*Restrictions:* Anthologies, edited works, and pamphlets are ineligible for the competition.
*Geographic Focus:* All States
*Date(s) Application is Due:* May 15
*Contact:* Julia Brookins, Special Projects Coordinator; (202) 544-2422, ext. 119; fax (202) 544-8307; jbrookins@theaha.org or aha@theaha.org
*Internet:* www.historians.org/prizes/index.cfm?PrizeAbbrev=FAIRBANK
*Sponsor:* American Historical Association
400 A Street SE
Washington, D.C. 20003

**American Historical Association Leo Gershoy Award** 450
In recognition of outstanding historical writing, the American Historical Association offers the Leo Gershoy Award in 17th- and 18th-century western European history. This award was established in 1975 by a gift from Mrs. Ida Gershoy in memory of her late husband, a specialist in European history. Professor Gershoy was associated with the graduate faculty of New York University for more than 25 years. The prize is awarded annually to the author of the most outstanding work published in English on any aspect of the fields of 17th- and 18th-century western European history. Only books of high scholarly and literary merit will be considered.
*Requirements:* Books published after May 1 of the previous year and before April 30 of the award year are eligible.
*Geographic Focus:* All States
*Date(s) Application is Due:* May 15
*Contact:* Julia Brookins, Special Projects Coordinator; (202) 544-2422, ext. 119; fax (202) 544-8307; jbrookins@theaha.org or aha@theaha.org
*Internet:* www.historians.org/prizes/index.cfm?PrizeAbbrev=GERSHOY
*Sponsor:* American Historical Association
400 A Street SE
Washington, D.C. 20003

**American Historical Association Littleton-Griswold Prize in American Law and Society** 451
Established in 1985, the Littleton-Griswold Prize is offered annually for the best book in any subject on the history of American law and society. The $1,000 prize is administered by a joint committee of the American Historical Association and the American Society for Legal History. Only books of high scholarly and literary merit will be considered. Recipients are announced at the association's annual meeting. Submission guidelines must be obtained from the association.
*Requirements:* Books published between May 1 of the previous year and April 30 of the award year are eligible.
*Geographic Focus:* All States
*Date(s) Application is Due:* May 15
*Amount of Grant:* 1,000 USD
*Contact:* Julia Brookins, Special Projects Coordinator; (202) 544-2422, ext. 119; fax (202) 544-8307; jbrookins@theaha.org or aha@theaha.org
*Internet:* www.historians.org/prizes/index.cfm?PrizeAbbrev=LITTLETON%2DGRISWOLD
*Sponsor:* American Historical Association
400 A Street SE
Washington, D.C. 20003

**American Historical Association Littleton-Griswold Research Grants in U.S. Legal History** 452
The Littleton-Griswold Research Grants in U.S. Legal History are given annually for research in U.S. legal history and in the general field of law and society. The funds for this program come from the earnings of the Littleton-Griswold Fund. The grants are intended to further research in progress and may be used for travel to a library or archive, for microfilms, photographs, or photocopying-a list of purposes that is meant to be merely illustrative, not exhaustive (other expenses, such as child care, can be included).
*Requirements:* Preference will be given to those with specific research needs, such as the completion of a project or completion of a discrete segment thereof. Preference will be given to Ph.D. candidates and junior scholars.
*Restrictions:* Only members of the Association are eligible to apply. Specifically excluded from funding are use of research assistants except in cases of technical skills such as data entry, coding of data, and other procedures relating to the preparation of data for machine computation; help with typing except for typing of camera-ready copy for an article, dissertation, or book, which would qualify as a technical expense; and use of funds for partial salary replacement.
*Geographic Focus:* All States
*Date(s) Application is Due:* Feb 15
*Amount of Grant:* Up to 1,000 USD
*Contact:* Julia Brookins, Special Projects Coordinator; (202) 544-2422, ext. 119; fax (202) 544-8307; jbrookins@theaha.org or aha@theaha.org
*Internet:* www.historians.org/prizes/Littleton-GriswaldGrantInfo.htm
*Sponsor:* American Historical Association
400 A Street SE
Washington, D.C. 20003

**American Historical Association Martin A. Klein Prize in African History** 453
The Martin A. Klein Prize in African History will recognize the most distinguished work of scholarship on African history published in English during the previous calendar year. The books must focus primarily on continental Africa (including those islands usually treated as countries of Africa). Eligibility will otherwise be defined quite broadly, to include books on any period of African history and from any disciplinary field that incorporates an historical perspective. In making its selection, the prize committee will pay particular attention to methodological innovation, conceptual originality and literary excellence. Works that reinterpret old themes or develop new theoretical perspectives are welcome. The prize is named for Martin A. Klein, who is currently professor of history at the University of Toronto. Funding for the prize was completed thanks to a substantial donation from Dr. Mougo Nyaggah of California State University at Fullerton and his wife Dr. Lynette Nyaggah. Mougo Nyaggah was Klein's first graduate student at the University of California Berkeley. Nyaggah credits the completion of his doctorate to Klein's mentoring, guidance, enthusiasm, and commitment to the research and teaching of African history. He observed that, "There are many Martins who have or will mentor and inspire many Africanist students in American universities. Those mentors will be honored by this prize for their human and scholarly contribution." Books may be submitted by authors or publishers, by sending one copy to each of the members of the committee. Anthologies, encyclopedias and other edited volumes will not be considered.
*Requirements:* Books published after May 1 of the previous year and before April 30 of the award year are eligible.
*Geographic Focus:* All States
*Date(s) Application is Due:* May 15
*Contact:* Julia Brookins, Special Projects Coordinator; (202) 544-2422, ext. 119; fax (202) 544-8307; jbrookins@theaha.org or aha@theaha.org
*Internet:* www.historians.org/prizes/index.cfm?PrizeAbbrev=Klein
*Sponsor:* American Historical Association
400 A Street SE
Washington, D.C. 20003

**American Historical Association Michael Kraus Research Grant in American Colonial History** 454
The American Historical Association announces the availability of the Michael Kraus Research Grant to recognize the most deserving proposal relating to work in progress on a research project in American colonial history, with particular reference to the intercultural aspects of American and European relations. Only members of the Association are eligible. The grants are intended to further research in progress and may be used for travel to a library or archive, for microfilms, photographs, or photocopying—a list of purposes that is meant to be merely illustrative, not exhaustive.
*Requirements:* Applicants must be members of the association. Preference will be given to those with specific research needs such as the completion of a project or a discrete segment thereof. Preference will be given to Ph.D. candidates and junior scholars.
*Restrictions:* Specifically excluded from funding are use of research assistants except in cases of technical skills such as keypunching, coding of data, and other procedures relating to the preparation of data for machine computation; help with typing except for typing of camera-ready copy for an article, dissertation, or book, which would qualify as a technical expense; and use of funds for partial salary replacements. Within a five-year period, no individual is eligible to receive more than a combined total of $1,000 from the Beveridge, Kraus, and Littleton-Griswold grant programs.
*Geographic Focus:* All States
*Date(s) Application is Due:* Feb 15
*Amount of Grant:* Up to 800 USD
*Contact:* Julia Brookins, Special Projects Coordinator; (202) 544-2422, ext. 119; fax (202) 544-8307; jbrookins@theaha.org or aha@theaha.org
*Internet:* www.historians.org/prizes/KrausGrantInfo.htm
*Sponsor:* American Historical Association
400 A Street SE
Washington, D.C. 20003

**American Historical Association Morris D. Forkosch Prize** 455
The American Historical Association offers the Morris D. Forkosch Prize annually in recognition of the best book in English in the field of British, British Imperial, or British Commonwealth history since 1485. Submission of books relating to the shared common law heritage of the English-speaking world are particularly encouraged in memory of the late Professor Forkosch's contributions to the field of legal studies and legal history.
*Requirements:* Books published after May 1 of the previous year and before April 30 of the award year are eligible.
*Geographic Focus:* All States
*Date(s) Application is Due:* May 15
*Contact:* Julia Brookins, Special Projects Coordinator; (202) 544-2422, ext. 119; fax (202) 544-8307; jbrookins@theaha.org or aha@theaha.org
*Internet:* www.historians.org/prizes/index.cfm?PrizeAbbrev=FORKOSCH
*Sponsor:* American Historical Association
400 A Street SE
Washington, D.C. 20003

**American Historical Association Paul Birdsall Prize in European Military and Strategic History** 456
Commencing in 1986 this prize, endowed by Hans Gatzke, Yale University and named in honor of the late Paul Birdsall of Williams College (d. 1970), is offered biennially for a major work in European military and strategic history since 1870. Preference will be given to the international aspects of military history (military/diplomatic), but the

political, economic, and social impact of technological developments, strategic planning, and military events on society will also qualify.
*Requirements:* Authors must be citizens of the United States or Canada. Preference will be given to younger academics, but older scholars and nonacademic candidates will not be excluded. Books published after May 1 of the previous year and before April 30 of the award year are eligible.
*Geographic Focus:* All States
*Date(s) Application is Due:* May 15
*Contact:* Julia Brookins, Special Projects Coordinator; (202) 544-2422, ext. 119; fax (202) 544-8307; jbrookins@theaha.org or aha@theaha.org
*Internet:* www.historians.org/prizes/index.cfm?PrizeAbbrev=BIRDSALL
*Sponsor:* American Historical Association
400 A Street SE
Washington, D.C. 20003

**American Historical Association Premio Del Rey** 457
The American Historical Association awards the Premio Del Rey biennially for a distinguished book in English in the field of early Spanish history. It was endowed by a gift of Robert I. Burns S.J., from his Llull and Catalonia prizes and covers the medieval period in Spain's history and culture CE 500-1516. The terms of the prize include works on Hispanic history and culture, including the Islamic and Jewish communities of Medieval Spain as well as early New World topics prior to 1516. Only books of a high scholarly historical nature should be submitted. Research accuracy, originality, and literary merit are important factors.
*Requirements:* Books published after May 1 of the previous year and before April 30 of the award year are eligible.
*Geographic Focus:* All States
*Date(s) Application is Due:* May 15
*Contact:* Julia Brookins, Special Projects Coordinator; (202) 544-2422, ext. 119; fax (202) 544-8307; jbrookins@theaha.org or aha@theaha.org
*Internet:* www.historians.org/prizes/index.cfm?PrizeAbbrev=DEL-REY
*Sponsor:* American Historical Association
400 A Street SE
Washington, D.C. 20003

**American Historical Association Raymond J. Cunningham Prize** 458
The Raymond J. Cunningham Prize is given for the best article published in a history department journal written by an undergraduate student. The prize was established in memory of Raymond J. Cunningham who was an associate professor of history at Fordham University. He was an authority on American Historian Herbert Baxter Adams. The deadline is May 15.
*Geographic Focus:* All States
*Date(s) Application is Due:* May 15
*Contact:* Julia Brookins, Special Projects Coordinator; (202) 544-2422, ext. 119; fax (202) 544-8307; jbrookins@theaha.org or aha@theaha.org
*Internet:* www.historians.org/prizes/index.cfm?PrizeAbbrev=Cunningham
*Sponsor:* American Historical Association
400 A Street SE
Washington, D.C. 20003

**American Historical Association Waldo J. Leland Prize** 459
The prize is awarded every five years (next in 2016) by the American Historical Association to an American resident for the most outstanding reference tool in the field of history. The award is honorific. The term reference tool encompasses bibliographies, indexes, encyclopedias, and other scholarly apparatus.
*Requirements:* Books published after May 1 of the previous award year and before April 30 of the current award year are eligible.
*Geographic Focus:* All States
*Date(s) Application is Due:* May 16
*Contact:* Julia Brookins, Special Projects Coordinator; (202) 544-2422, ext. 119; fax (202) 544-8307; jbrookins@theaha.org or aha@theaha.org
*Internet:* www.historians.org/prizes/index.cfm?PrizeAbbrev=LELAND
*Sponsor:* American Historical Association
400 A Street SE
Washington, D.C. 20003

**American Historical Association Wesley-Logan Prize** 460
The Wesley-Logan Prize in African diaspora history is jointly sponsored by the American Historical Association and the Association for the Study of African American Life & History. The prize is awarded annually for an outstanding book in African diaspora history. The AHA Committee on Minority Historians established the prize in 1992 in memory of two early pioneers in the field, Charles H. Wesley and Rayford W. Logan. The prize is offered for a book on some aspect of the history of the dispersion, settlement and adjustment, and/or return of peoples originally from Africa. Eligible for consideration are books in any chronological period and any geographical location.
*Requirements:* Books published between May 1 of the previous year and April 30 of the current year are eligible. Only books of high scholarly and literary merit will be considered.
*Geographic Focus:* All States
*Date(s) Application is Due:* May 15
*Contact:* Julia Brookins, Special Projects Coordinator; (202) 544-2422, ext. 119; fax (202) 544-8307; jbrookins@theaha.org or aha@theaha.org
*Internet:* www.historians.org/prizes/index.cfm?PrizeAbbrev=WESLEY%2DLOGAN
*Sponsor:* American Historical Association
400 A Street SE
Washington, D.C. 20003

**American Mathematical Society AAAS Congressional Fellowships** 461
The American Mathematical Society (AMS), in conjunction with the American Association for the Advancement of Science (AAAS), sponsors a Congressional Fellow each year who spends the year working on the staff of a Member of Congress or a congressional committee, working as a special legislative assistant in legislative and policy areas requiring scientific and technical input. The program includes an orientation on congressional and executive branch operations, and a year-long seminar series on issues involving science, technology and public policy. The fellowship stipend for the Fellowship from September of next year to August of the following year is $74,872 for the fellowship period, with allowances for relocation and professional travel and a contribution towards health insurance.
*Requirements:* Prospective Fellows must demonstrate expertise in some area of the mathematical sciences; have a good scientific and technical background; be cognizant of and demonstrate sensitivity toward political and social issues; and, most importantly, have a strong interest and some experience in applying personal knowledge toward the solution of societal problems. Eligible applicants are individuals in the mathematical sciences with a Ph.D. or an equivalent doctoral-level degree in the mathematical sciences by the application deadline of February 15.
*Geographic Focus:* All States
*Contact:* Dr. Donald McClure, Executive Director; (401) 455-4105 or (401) 455-4000; fax (401) 331-3842; ams@ams.org
*Internet:* www.ams.org/programs/ams-fellowships/ams-aaas/ams-aaas-congressional-fellowship
*Sponsor:* American Mathematical Society
201 Charles Street
Providence, RI 02904-2294

**American Mathematical Society AAAS Mass Media Fellowships** 462
The American Mathematical Society (AMS) sponsors a Mass Media Fellow each summer through the Mass Media Science & Engineering Fellowship program organized by the American Association for the Advancement of Science (AAAS). It is a highly competitive program designed to improve public understanding of science and technology by placing advanced science, mathematics and engineering students in newsrooms nationwide. Fellows work with media professionals to improve their communication skills and increase their understanding of the editorial process by which events and ideas become news. Fellows receive a weekly stipend of $450, plus travel expenses, to work for ten weeks during the summer months as reporters, researchers, and production assistants in media organizations. They observe and participate in the process by which events and ideas become news, improve their ability to communicate about complex technical subjects in a manner understandable to the public, and increase their understanding of editorial decision making and of how information is effectively disseminated. Eligible candidates should review information on the application procedures and submit their application by the January 15 deadline.
*Requirements:* The program is available to college or university students (in their senior year, or in any graduate or post graduate level) in the natural, physical, health, engineering, computer or social sciences or mathematics with outstanding written and oral communication skills and a strong interest in learning about the media.
*Geographic Focus:* All States
*Amount of Grant:* 4,500 USD
*Contact:* Dr. Donald McClure, Executive Director; (401) 455-4105 or (401) 455-4000; fax (401) 331-3842; ams@ams.org
*Internet:* www.ams.org/programs/ams-fellowships/media-fellow/massmediafellow
*Sponsor:* American Mathematical Society
201 Charles Street
Providence, RI 02904-2294

**American Mathematical Society Centennial Fellowships** 463
The AMS Centennial Research Fellowship Program makes awards annually to outstanding mathematicians to help further their careers in research. From 1997-2001, the fellowship program was aimed at recent PhDs. Recently, the AMS Council approved changes in the rules for the fellowships. The primary selection criterion for the Centennial Fellowship is the excellence of the candidate's research. Applications will be accepted from those currently holding a tenured, tenure track, post-doctoral, or comparable (at the discretion of the selection committee) position at an institution in North America. The stipend for most recent fellowships awarded was $82,000, with an additional expense allowance of about $8,200. Applications will be accepted from October 1 through December 1.
*Requirements:* Preference will be given to candidates who have not had extensive fellowship support in the past. A recipient of the fellowship shall have held his or her doctoral degree for at least three years and not more than twelve years at the inception of the award.
*Restrictions:* Recipients may not hold the Centennial Fellowship concurrently with another major research award such as a Sloan fellowship, NSF Postdoctoral fellowship, or CAREER award. Under normal circumstances, the fellowship cannot be deferred.
*Geographic Focus:* All States
*Date(s) Application is Due:* Dec 1
*Amount of Grant:* 90,200 USD
*Contact:* Dr. Donald McClure, Executive Director; (401) 455-4105 or (401) 455-4000; fax (401) 331-3842; ams@ams.org
*Internet:* www.ams.org/programs/ams-fellowships/centennial-fellow/emp-centflyer
*Sponsor:* American Mathematical Society
201 Charles Street
Providence, RI 02904-2294

**American Mathematical Society China Exchange Travel Grants** 464
The purpose of these American Mathematical Society China Exchange Travel Grants is to support travel of mathematical scientists at institutions in the U.S. or Canada on visits to research colleagues in China, and to support travel of (generally, less senior) mathematical scientists from China to the U.S. or Canada for visits with U.S. or Canadian research colleagues at U.S. or Canadian institutions. In the case of American or Canadian travelers, preference will be given to senior mathematicians whose visit will have a positive impact on the mathematical research community in China. In the case of Chinese travelers, preference will be given to scientists in the beginning stages of their research careers. For U.S. or Canadian institutions applying, US$1000 will be set aside from the total award and given to the home institution (in China) of the visitor (in the case of multiple visitors, the amount should be divided equally), in lieu of honorarium. The annual deadline is March 15 for travel that takes place the following academic year (between September and August).
*Requirements:* It is expected that the host institution will provide some type of additional support for the travel or subsistence of the visitor to ensure a stay of suitable length.
*Restrictions:* Funds from this program may not be used specifically to support conference attendance, but trips may sometimes be combined with conference attendance.
*Geographic Focus:* All States, Canada
*Date(s) Application is Due:* Mar 15
*Amount of Grant:* 1,000 USD
*Contact:* Steven Ferrucci, (401) 455-4113 or (401) 455-4000; fax (401) 331-3842; ams@ams.org
*Internet:* www.ams.org/programs/travel-grants/china-exchange/china-exchange
*Sponsor:* American Mathematical Society
201 Charles Street
Providence, RI 02904-2294

**American Mathematical Society Graduate Student Travel Grants** 465
The AMS, with support from a private gift, accepts applications for partial support for two types of AMS meetings: Joint Mathematics Meetings - available to selected doctoral students in mathematics who are in their last year of study (that is, have not received the PhD before the application deadline, but do expect to receive the PhD degree within the twelve month period following the application deadline); and Sectional Meetings - program accepts applications from current full time doctoral students. The deadline for Joint Mathematics Meetings applications is September 26, while the deadlines for Sectional Meetings are July 18 and November 14. Support level for the January JMM is $500. Support level for sectional meetings is $250.
*Requirements:* The current institution of the applicant must be more than 50 miles from the site of the meeting per google maps. A representative of the university must submit a statement verifying the expected date of graduation. Each awardee's university is strongly encouraged, but not required, to provide matching support.
*Geographic Focus:* All States
*Date(s) Application is Due:* Jul 18; Sep 26; Nov 14
*Amount of Grant:* 250 - 500 USD
*Contact:* Steven Ferrucci, (401) 455-4113 or (401) 455-4000; fax (401) 331-3842; ams@ams.org
*Internet:* www.ams.org/programs/travel-grants/grad-students/emp-student-JMM
*Sponsor:* American Mathematical Society
201 Charles Street
Providence, RI 02904-2294

**American Mathematical Society International Congress of Mathematicians Travel Grants** 466
The American Mathematical Society has applied to the National Science Foundation (NSF) for funds to permit partial travel support for U.S. mathematicians attending the International Congress of Mathematicians. The application period will be September 1 through November 15. This program is open to U.S. mathematicians (those who are U.S. citizens or affiliated with a U.S. institution).
*Requirements:* Early career mathematicians (those within six years of their doctorate), women and underrepresented minorities will be especially encouraged to apply. Early career mathematicians will be required to supply one letter from an individual who has made a commitment to serve as a mentor to the applicant during the period leading up to the International Congress, advising him or her about the appropriate sessions and activities in which to participate at the ICM.
*Restrictions:* Mathematicians accepting grants for partial support of the travel to ICM may not supplement them with any other NSF funds.
*Geographic Focus:* All States
*Contact:* Steven Ferrucci, (401) 455-4113 or (401) 455-4000; fax (401) 331-3842; ams@ams.org
*Internet:* www.ams.org/programs/travel-grants/icm
*Sponsor:* American Mathematical Society
201 Charles Street
Providence, RI 02904-2294

**American Mathematical Society Simons Travel Grants** 467
Each AMS-Simons Travel Grant provides an early-career mathematician with $2,000 per year for two years to be used for research-related travel. Sixty awards will be made. Each applicant will choose a mentor with whom to interact during the two-year period of the grant. The ideal mentor is someone who is an active senior participant in the applicant's research area and or a trusted senior colleague in the applicant's department. The mentor will be expected to interact with the awardee on a regular basis throughout the two-year period of the grant. In particular, the mentor will consider the appropriateness of proposed travel with the awardee. The awardee must receive approval from his or her mentor before each trip.
*Requirements:* Applicants must be located in the United States (or be U.S. citizens employed outside the U.S.) and must have completed the PhD within the last four years.
*Geographic Focus:* All States
*Amount of Grant:* Up to 2,000 USD
*Contact:* Steven Ferrucci, (401) 455-4113 or (401) 455-4000; fax (401) 331-3842; ams@ams.org
*Internet:* www.ams.org/programs/travel-grants/AMS-SimonsTG
*Sponsor:* American Mathematical Society
201 Charles Street
Providence, RI 02904-2294

**American Mathematical Society Travel Grant for the Mathematical Congress of the Americas** 468
The American Mathematical Society Travel Grant Program for the Mathematical Congress of the Americas is intended to provide partial travel support for up to sixty U.S.-based mathematicians to attend the inaugural Mathematical Congress of the Americas. It is expected that this travel grant program will provide travel support for both U.S.-based invited speakers (senior mathematicians) and early career mathematians; in addition, it is expected that the standard award amount will be up to $1,600 for invited speakers (senior mathematicians) and up to $1,800 for early career mathematicians. Applications will be evaluated by a panel of mathematical scientists under the terms of a proposal submitted to the National Science Foundation (NSF) by the Society.
*Requirements:* Support is available for U.S. mathematicians who are in one of the following categories: early career mathematicians within six years of their doctorate; MCA Invited Speakers from U.S. institutions who desire funding.
*Restrictions:* Mathematicians accepting grants for partial support of the travel to MCA may not supplement them with any other NSF funds.
*Geographic Focus:* All States
*Amount of Grant:* Up to 1,800 USD
*Contact:* Steven Ferrucci, (401) 455-4113 or (401) 455-4000; fax (401) 331-3842; ams@ams.org
*Internet:* www.ams.org/programs/travel-grants/mca
*Sponsor:* American Mathematical Society
201 Charles Street
Providence, RI 02904-2294

**American Philosophical Society Center for Native American and Indigenous Research Fellowship** 469
The American Philosophical Society Library in Philadelphia seeks applicants for one-year residential fellowships to assist in the completion of doctoral dissertation research. One fellowship will be awarded in each of three areas of study. The Center for Native American and Indigenous Research Fellowship research pertains to Native American history, anthropology, linguistics, cultural revitalization, and related fields. This 12-month fellowship is intended for advanced Ph.D. students working toward the completion of the dissertation. $25,000 for twelve months will be awarded to all successful applicants, paid in two installments after arrival in the Library. The annual application deadline is February 1, with notification of the successful applicant by April 15.
*Requirements:* The successful applicant must be affiliated with CNAIR at the American Philosophical Society.
*Restrictions:* The fellowship cannot be used to supplement external awards of similar purposes. Institutions are not eligible to apply.
*Geographic Focus:* All States
*Date(s) Application is Due:* Feb 1
*Amount of Grant:* 25,000 USD
*Contact:* Linda Musumeci, Director of Grants and Fellowships; (215) 440-3429; lmusumeci@amphilsoc.org
*Internet:* https://amphilsoc.org/library/fellowships/long-term-pre-docs
*Sponsor:* American Philosophical Society
104 South Fifth Street
Philadelphia, PA 19106-3387

**American Philosophical Society Daland Fellowships in Clinical Investigation** 470
Through the American Philosophical Society Daland Fellowships in Clinical Investigation, limited number of fellowships are awarded annually for research in clinical medicine, including the fields of: internal medicine; neurology; psychiatry; pediatrics; and surgery. Patient-oriented research is emphasized. Essentially 100 percent of the fellow's time will be devoted to research. Teaching or clinical service of a limited amount is permitted. Additional salary may be granted by the institution at which the fellow is located, from another fellowship, or from a similar award during the tenure of the fellowship. The term of the fellowship is one year, with possible renewal for another year. Applications are available on the Web site, or may be obtained by written request, stating when the MD or MD/PhD degree was awarded; include a self-addressed mailing label. The term of the fellowship is one year, with renewal for one year if satisfactory progress is demonstrated. Stipends for the fellowship are $40,000 for the first year and $40,000 for the second year. Applications for first-year fellowships are due no later than September 15.
*Requirements:* These fellowships are designed for qualified persons who have held an MD, or MD/PhD degree for less than eight years. The fellowship is generally intended to be the first post-clinical fellowship; but each case will be decided on its merits. Preference is generally given to candidates who have not more than two years of post-doctoral training and research. Applicants must expect to perform their research at an institution in the United States, under the supervision of a scientific adviser. Candidates are to be nominated by their department chairman, in a letter providing assurance that the nominee will work with the guidance of a scientific adviser of established reputation who has guaranteed adequate space, supplies, etc. for the Fellow. The adviser need not be a member of the department nominating the Fellow, nor need the activities of the Fellow

be limited to the nominating department. As a general rule, no more than one fellowship will be awarded to a given institution in the same year of competition.
*Restrictions:* The society does not provide funds for institutional overhead.
*Geographic Focus:* All States
*Date(s) Application is Due:* Sep 15
*Amount of Grant:* Up to 80,000 USD
*Contact:* Linda Musumeci; (215) 440-3429; lmusumeci@amphilsoc.org
*Internet:* https://amphilsoc.org/grants/daland
*Sponsor:* American Philosophical Society
104 South Fifth Street
Philadelphia, PA 19106-3387

**American Philosophical Society Franklin Research Grants** 471
Through the Franklin Research program, the American Philosophical Society awards small grants to scholars to support the cost of research leading to publication in all areas of knowledge. The program is particularly designed to help meet the costs of travel to libraries and archives for research purposes; the purchase of microfilm, photocopies, or equivalent research materials; the costs associated with fieldwork; or laboratory research expenses. Franklin grants are made for noncommercial research. They are not intended to meet the expenses of attending conferences or the costs of publication. The society is particularly interested in supporting the work of young scholars who have recently received their PhDs. Applicants who have received Franklin grants may reapply after an interval of two years. Funding is offered up to a maximum of $6,000. The annual application deadlines are October 1 and December 1.
*Requirements:* Applicants are expected to have a doctorate, or to have published work of doctoral character and quality. American citizens and residents of the United States may use their Franklin awards at home or abroad. Foreign nationals must use their Franklin awards for research in the United States.
*Restrictions:* Predoctoral graduate students are not eligible. The society does not pay overhead or indirect costs to any institution. Grants will not be made to replace salary during a leave of absence or earnings from summer teaching; pay living expenses while working at home; cover the costs of consultants or research assistants; or purchase permanent equipment such as computers, cameras, tape recorders, or laboratory apparatus.
*Geographic Focus:* All States
*Date(s) Application is Due:* Oct 1; Dec 1
*Amount of Grant:* Up to 6,000 USD
*Contact:* Linda Musumeci; (215) 440-3429; lmusumeci@amphilsoc.org
*Internet:* https://amphilsoc.org/grants/franklin
*Sponsor:* American Philosophical Society
104 South Fifth Street
Philadelphia, PA 19106-3387

**American Philosophical Society Friends of the APS Fellowship in Early American History** 472
The American Philosophical Society Library in Philadelphia seeks applicants for one-year residential fellowships to assist in the completion of doctoral dissertation research. One fellowship will be awarded in each of three areas of study. The successful applicant will receive an appointment as a Research Associate at the McNeil Center for Early American Studies, which will provide library and computer privileges at the University of Pennsylvania to those who agree to participate regularly in the McNeil Center's seminars and other programming. This 12-month fellowship is intended for advanced Ph.D. students working toward the completion of the dissertation. $25,000 for twelve months will be awarded to all successful applicants, paid in two installments after arrival in the Library. The annual application deadline is February 1, with notification of the successful applicant by April 15.
*Restrictions:* The fellowship cannot be used to supplement external awards of similar purposes. Institutions are not eligible to apply.
*Geographic Focus:* All States
*Date(s) Application is Due:* Feb 1
*Amount of Grant:* 25,000 USD
*Contact:* Linda Musumeci, Director of Grants and Fellowships; (215) 440-3429 or (215) 440-3400; lmusumeci@amphilsoc.org
*Internet:* https://amphilsoc.org/library/fellowships/long-term-pre-docs
*Sponsor:* American Philosophical Society
104 South Fifth Street
Philadelphia, PA 19106-3387

**American Philosophical Society Jacques Barzun Prize in Cultural History** 473
The Jacques Barzun Prize honors historian and cultural critic Jacques Barzun, University Professor Emeritus of Columbia University, a member of the American Philosophical Society since 1984. A gift from Roger Williams, Distinguished Professor of History Emeritus at the University of Wyoming, a former student of Professor Barzun, funds this prize. It is awarded annually to the author or authors whose book exhibits distinguished work in American or European cultural history. Authors and publishers are invited to submit entries for the past year prize by January 31 each succeeding year.
*Requirements:* Authors must be citizens or permanent residents of the United States. Books must be written in English, and published in the United States between January 1 and December 31 of the most recent past year. Books must be single-authored volumes, not collections of articles or edited texts. The subject must be American or European cultural history.
*Restrictions:* There is strict a 3-book limit per publisher, whether books are submitted by the publisher or directly by an author. If more than 3 books are submitted by a publisher, all of them will be dropped from consideration.
*Geographic Focus:* All States
*Date(s) Application is Due:* Jan 31
*Contact:* Linda Musumeci, Director of Grants and Fellowships; (215) 440-3429 or (215) 440-3400; lmusumeci@amphilsoc.org
*Internet:* https://amphilsoc.org/prizes/barzun
*Sponsor:* American Philosophical Society
104 South Fifth Street
Philadelphia, PA 19106-3387

**American Philosophical Society John Hope Franklin Dissertation Fellowship** 474
The John Hope Franklin Dissertation Fellowship is designed to support an outstanding African-American graduate student attending any PhD-granting institution in the United States, in any field of knowledge. There is no residential requirement. The stipend for this fellowship is $25,000 for a twelve-month period, plus $5,000 to support the cost of residency in Philadelphia, for a total award of $30,000. The twelve-month period is flexible.
*Requirements:* Candidates must have completed all course work and examinations preliminary to the doctoral dissertation and be prepared to devote full time for 12 months - with no teaching obligations - to research on their dissertation projects or the writing of their dissertations.
*Geographic Focus:* All States
*Date(s) Application is Due:* Apr 1
*Amount of Grant:* 30,000 USD
*Contact:* Linda Musumeci; (215) 440-3429; lmusumeci@amphilsoc.org
*Internet:* https://www.amphilsoc.org/grants/johnhopefranklin
*Sponsor:* American Philosophical Society
104 South Fifth Street
Philadelphia, PA 19106-3387

**American Philosophical Society Judson Daland Prize** 475
The American Philosophical Society began awarding medals in 1790. Since then, it has continued by awarding medals and prizes for outstanding achievement in the sciences, humanities, arts, professions, and public service. The $50,000 Judson Daland Prize recognizes outstanding achievement in patient-oriented research. The prize will be awarded at the meeting of the American Philosophical Society in November of each year. The Society reserves the right to award a shared prize. Nominations and all three letters of support must be received by March 15.
*Requirements:* Nominees must have done their work in an institution in the United States. Nominees need not, however, be citizens of the United States. Nominees should be no more than 15 years beyond receipt of the M.D. degree.
*Geographic Focus:* All States
*Date(s) Application is Due:* Mar 15
*Amount of Grant:* 50,000 USD
*Contact:* Linda Musumeci, Director of Grants and Fellowships; (215) 440-3429 or (215) 440-3400; lmusumeci@amphilsoc.org
*Internet:* www.amphilsoc.org/prizes/daland.htm
*Sponsor:* American Philosophical Society
104 South Fifth Street
Philadelphia, PA 19106-3387

**American Philosophical Society Lewis and Clark Fund for Exploration and Field Research** 476
The American Philosophical Society Lewis and Clark Fund for Exploration and Field Research program, established in 2004 in honor of the Lewis and Clark Bicentennial, encourages exploratory field studies for the collection of specimens and data and to provide the imaginative stimulus that accompanies direct observation. Applications are invited from disciplines with a large dependence on field studies, such as archeology, anthropology, biology, ecology, geography, geology, linguistics, and paleontology, but grants will not be restricted to these fields. Budgets should be limited to travel and related expenses, including personal field equipment. Amounts will depend on travel costs but will ordinarily be in the range of several hundred dollars up to about $5,000. The annual application deadline is February 1.
*Requirements:* Grants are available to graduate students, postdoctoral students, junior and senior scientists, and social scientists who wish to participate in field studies for their theses or for other purposes. A graduate student applicant should ask his or her academic supervisor or field trip leader to write one of the two letters of recommendation, specifying the role of the student in the field trip and the educational contribution of the trip. The competition is open to U.S. residents wishing to carry out research anywhere in the world. Foreign applicants must either be based at a U.S. institution or plan to carry out their work in the United States.
*Restrictions:* Undergraduates are not eligible.
*Geographic Focus:* All States
*Date(s) Application is Due:* Feb 1
*Amount of Grant:* Up to 5,000 USD
*Contact:* Linda Musumeci; (215) 440-3429; lmusumeci@amphilsoc.org
*Internet:* https://amphilsoc.org/grants/lewisandclark
*Sponsor:* American Philosophical Society
104 South Fifth Street
Philadelphia, PA 19106-3387

**American Philosophical Society Lewis and Clark Fund for Exploration and Field Research in Astrobiology** 477
The Lewis and Clark Fund for Exploration and Field Research in Astrobiology is open to field studies in any area of interest to astrobiology. Astrobiology is the study of the origin, evolution, distribution, and future of life on Earth and in the universe. It encompasses research in, among others, the fields of astronomy, chemistry, evolutionary biology, field and population biology, geology, microbiology, molecular biology, oceanography,

paleontology, and planetary science. Astrobiology includes investigations of the geologic and fossil record to understand the conditions of the early Earth when life arose. Its scope also includes research of contemporary locations on Earth that might be similar to early earth and to environments elsewhere in our Solar System (such as on Mars, Europa, and Titan), which may be, or have been in the past, suitable for life. Astrobiology is also about understanding the characteristics of life, which requires investigations into extreme natural environments on Earth and, eventually, elsewhere. Budgets should be limited to travel and related expenses, including personal field equipment. Amounts will depend on travel costs, but will ordinarily be in the range of several hundred dollars up to about $5,000. The deadline for applications and two letters of support is February 1, with notification in May, for work in June and beyond.
*Requirements:* Grants will be available to graduate students and post-doctoral and junior scientists who wish to participate in field studies for their theses or for other purposes. Eligibility for applicants with doctorates is limited to those five years or fewer beyond their Ph.D. or equivalent degree, although, rarely, exceptions may be made. The competition is open to those affiliated with a U.S.-based institution, who may carry out research anywhere in the world.
*Geographic Focus:* All States
*Date(s) Application is Due:* Feb 1
*Amount of Grant:* Up to 5,000 USD
*Contact:* Linda Musumeci, Research Administrator; (215) 440-3429; lmusumeci@amphilsoc.org or applications@amphilsoc.org
*Internet:* https://amphilsoc.org/grants/astrobiology
*Sponsor:* American Philosophical Society
104 South Fifth Street
Philadelphia, PA 19106-3387

**American Philosophical Society Library Digital Humanities Fellowships** 478
The American Philosophical Society (APS) Library in Philadelphia seeks applicants for its Digital Humanities Fellowship. This fellowship, for up to two (2) months, is open to scholars who are comfortable creating tools and visualizations, as well as those interested in working collaboratively with the APS technology team. Examples include (but are not limited to) projects that incorporate timelines, text analytics, network graphs, and maps. $3,000 per month will be awarded to the successful applicant after their arrival at the Library. The annual deadline for applications is March 1.
*Requirements:* Scholars, including graduate students, at any stage of their career may apply. Special consideration will be given to proposals that present APS Library holdings in new and engaging ways.
*Restrictions:* Grants will not be made for journalistic or other writing for general readership; the preparation of textbooks, casebooks, anthologies, or other materials for classroom use by students; nor for the work of creative and performing artists. Grants are rarely made to persons who have had the doctorate less than a year and never for predoctoral study or research. The committee will seldom approve more than two grants for the same project within any five-year period.
*Geographic Focus:* All States
*Date(s) Application is Due:* Oct 1
*Amount of Grant:* 3,000 - 6,000 GBP
*Contact:* Linda Musumeci; (215) 440-3429; lmusumeci@amphilsoc.org
*Internet:* https://amphilsoc.org/library/fellowships/digital-humanities
*Sponsor:* American Philosophical Society
104 South Fifth Street
Philadelphia, PA 19106-3387

**American Philosophical Society Library Short-Term Resident Research Fellowships** 479
Short-term residential fellowships for up to three months encourage research in the library's collections. Outstanding historical collections and subject areas include the papers of Benjamin Franklin; the American Revolution; 18th- and 19th-century natural history; western scientific expeditions and travel including the journals of Lewis and Clark; polar exploration; the papers of Charles Wilson Peale, including family and descendants; history of genetics, eugenics, and evolution; history of biochemistry, physiology, and biophysics; 20th-century medical research; and history of physics. A stipend of $3,000 per month is awarded for a minimum of one month and a maximum of three months, paid after arriving in the Library to work on their fellowship. Application procedures are entirely online. The annual deadline is March 1.
*Requirements:* U.S. citizens and foreign nationals who hold the PhD or the equivalent, PhD candidates who have passed their preliminary exams, and independent scholars are eligible to apply. The fellowships are intended to encourage research in the library collections by scholars who reside beyond a 75-mile radius of Philadelphia.
*Geographic Focus:* All States, All Countries
*Date(s) Application is Due:* Mar 1
*Amount of Grant:* 3,000 - 9,000 USD
*Contact:* Linda Musumeci, Director of Grants and Fellowships; (215) 440-3429 or (215) 440-3443; fax (215) 440-3423; lmusumeci@amphilsoc.org or libfellows@amphilsoc.org
*Internet:* https://amphilsoc.org/library/fellowships/short-term-fellowships
*Sponsor:* American Philosophical Society
104 South Fifth Street
Philadelphia, PA 19106-3387

**American Philosophical Society Phillips Fund Grants for Native American Research** 480
The Phillips Fund of the American Philosophical Society provides grants for research in Native American linguistics, ethnohistory, and the history of studies of Native Americans, in the continental United States and Canada. Grants are not made for projects in archaeology, ethnography, psycholinguistics, or for the preparation of pedagogical materials. The committee distinguishes ethnohistory from contemporary ethnography as the study of cultures and culture change through time. The grants are intended for such costs as travel, tapes, films, and consultants' fees but not for the purchase of books or permanent equipment. The average award is about $3,200; grants do not exceed $3,500. The annual deadline for applications is March 1.
*Requirements:* Applicants may be graduate students who have passed their qualifying examinations for either the master's or doctorate degrees; postdoctoral applicants are eligible.
*Restrictions:* Grants are not made for projects in archaeology, ethnography, or psycholinguistics, nor for the preparation of pedagogical materials.
*Geographic Focus:* All States
*Date(s) Application is Due:* Mar 1
*Amount of Grant:* Up to 3,500 USD
*Contact:* Linda Musumeci; (215) 440-3429; lmusumeci@amphilsoc.org
*Internet:* https://amphilsoc.org/grants/phillips
*Sponsor:* American Philosophical Society
104 South Fifth Street
Philadelphia, PA 19106-3387

**American Psychiatric Association/Kempf Fund Award for Research Development in Psychobiological Psychiatry** 481
This award recognizes a senior researcher who has made a significant contribution to research on the causes and treatment of schizophrenia as both a researcher and a mentor. A stipend will support the research career development of a young research psychiatrist working in a mentor-trainee relationship with the award winner on further research in this field. Submissions will be judged on the excellence of the nominee's overall contribution to the body of research in schizophrenia, including submission of the nominee's most significant paper or book; a description of the nominee's role as a mentor to younger colleagues in the field; and a description of the qualifications of the young research psychiatrist, as well as a detailed description of his or her career development plan.
*Geographic Focus:* All States
*Date(s) Application is Due:* Oct 14
*Amount of Grant:* 1,500 - 20,000 USD
*Contact:* Ernesto Guerra, (703) 907-7300; fax (703) 907-1085; eguerra@psych.org
*Internet:* psych.org/MainMenu/Research/ResearchTrainingandFunding/ResearchFellowships/KempfFundAwardforResearchDevelopmentinPsychobiologicalPsychiatry.aspx
*Sponsor:* American Psychiatric Association
1000 Wilson Boulevard, Suite 1825
Arlington, VA 22209-3901

**American Psychiatric Foundation Call for Proposals** 482
APF support allows organizations across the country to make a difference through unique educational, informational and outreach initiatives that promote the early recognition and treatment of mental illness and encourage leadership in the field of psychiatry. The Call for Proposals program is intended to fund public education programs, as well as information and outreach initiatives that promote the early recognition and treatment of mental illness. Up to $750,000 in grant funds are available over the course of three years. Average grants are in the $50,000 range. Grant making under this program has been temporarily suspended while a new Call for Proposals is being developed. Contact the office for further updates.
*Requirements:* Organizations that have been in existence for at least two years and currently maintain a 501(c)3 charitable status and American Psychiatric Association District Branches and subsidiaries. Organizations need not be mental health programs.
*Geographic Focus:* All States
*Amount of Grant:* Up to 100,000 USD
*Contact:* Paul T. Burke, Executive Director; (703) 907-8518 or (703) 907-8512; fax (703) 907-7851; pburke@psych.org or apf@psych.org
*Internet:* www.psychfoundation.org/GrantAndAwards/Grants/CallforProposals.aspx
*Sponsor:* American Psychiatric Foundation
1000 Wilson Boulevard, Suite 1825
Arlington, VA 22209

**American Psychiatric Foundation James H. Scully Jr., M.D., Educational Fund Grants** 483
The fund supports education and research within the American Psychiatric Association (APA), its subsidiary organizations, or affiliated organizations, including district branches, state associations, or other psychiatric organizations that are in partnership with the APA. The fund can be used to support activities that improve the quality of care for patients with mental illness, advance the prevention of mental illness, raise awareness of mental illness and its treatments, and increase access to mental health services. Requests for funding should be sent in writing to the foundation and include the following: identification and significance of need; description of program/activity to address that need; expected outcomes and evaluation process; budget and time line; and background of key project personnel. Funding decisions are made by the foundations board of directors.
*Restrictions:* The fund will not support public policy or lobbying activities.
*Geographic Focus:* All States
*Contact:* Paul T. Burke, Executive Director; (703) 907-8518 or (703) 907-8512; fax (703) 907-7851; pburke@psych.org or apf@psych.org
*Internet:* www.psychfoundation.org/GrantAndAwards/Grants/JamesHScullyEducationalFund.aspx
*Sponsor:* American Psychiatric Foundation
1000 Wilson Boulevard, Suite 1825
Arlington, VA 22209

**American Psychiatric Foundation Jeanne Spurlock Congressional Fellowship** 484
The Jeanne Spurlock Congressional Fellowship provides general psychiatry and child psychiatry residents an opportunity to work in a congressional office or committee on federal health policy, particularly policy related to child and minority issues. The recipient will serve a ten-month fellowship in Washington, D.C.. The fellow will be introduced to the structure and development of federal and congressional health policy procedures, with a focus on mental health issues affecting minorities and under-served populations, including children. Fellows traditionally help develop legislative proposals, track and analyze legislative initiatives, arrange hearings, and brief Members of Congress and their staff, and interact with their constituents. During the fellowship, recipients have opportunities to interact with health policymakers and advocacy/professional groups, including the APA.
*Requirements:* PGY-II and III general psychiatry residents, child psychiatry residents, or child psychiatrists who will be out of training for less than one year at the time of the fellowship may apply. Applicants must be U.S. citizens or permanent residents. The recipient will be required to submit a written summary of the Fellowship experience at the end of the fellowship and may make recommendations or suggestions for improving the fellowship.
*Geographic Focus:* All States
*Date(s) Application is Due:* Mar 14
*Amount of Grant:* 35,000 - 37,500 USD
*Contact:* Paul T. Burke, Executive Director; (703) 907-8518 or (703) 907-8512; fax (703) 837-5451; pburke@psych.org or apf@psych.org
*Internet:* www.psychfoundation.org/GrantAndAwards/Fellowships/SpurlockFellowship.aspx
*Sponsor:* American Psychiatric Foundation
1000 Wilson Boulevard, Suite 1825
Arlington, VA 22209

**American Psychiatric Foundation Minority Fellowships** 485
The Minority Fellowships Program (MFP) endeavors to eliminate racial and ethnic disparities in mental health and substance abuse care by providing specialized training to psychiatry residents and medical students interested in serving minority communities. The MFP is designed to: provide fellowship recipients with enriching training experiences through participation in the APA September and Annual Meetings; provide recipients with resources to support activities that enhance culturally relevant aspects of their training program; stimulate their interest in pursuing training in areas of psychiatry where minority groups are underrepresented, such as research, child psychiatry, and addiction psychiatry; and develop leadership to improve the quality of mental health care for the following federally recognized ethnic minority groups: American Indians, Native Alaskans, Asian Americans, Native Hawaiians, Native Pacific Islanders, African Americans and Hispanics/Latinos.
*Geographic Focus:* All States
*Contact:* Paul T. Burke, Executive Director; (703) 907-8518 or (703) 907-8512; fax (703) 837-5451; pburke@psych.org or apf@psych.org
*Internet:* www.psychfoundation.org/GrantAndAwards/Fellowships/MFP.aspx
*Sponsor:* American Psychiatric Foundation
1000 Wilson Boulevard, Suite 1825
Arlington, VA 22209

**American Rivers Anthony A. Lapham River Conservation Fellowship** 486
The Anthony A. Lapham River Conservation Fellowship at American Rivers provides an excellent professional development opportunity for talented post-graduates pursuing careers as leaders in the field of conservation advocacy. The program will develop the next generation of skilled conservation leaders who can promote practical solutions that benefit natural and human communities. Recent master's degree graduates will focus on an applied research project that will make a tangible contribution to AR's mission. The 12-month Fellowship will be supported by a team of conservation staff and members of our Scientific and Technical Advisory Committee, which includes some of the nation's foremost experts on freshwater and other conservation science and policy.
*Geographic Focus:* All States
*Date(s) Application is Due:* Feb 15
*Amount of Grant:* 35,000 USD
*Contact:* Ann C. Mills, (202) 347-7550; fax (202) 347-9240; amills@americanrivers.org or rivergrants@americanrivers.org
Garrett Russo, (202) 347-7073; fax (202) 347-9240; grusso@americanrivers.org or rivergrants@americanrivers.org
*Internet:* www.americanrivers.org/newsroom/press-releases/2007/new-lapham-fellowship.html
*Sponsor:* American Rivers
1101 14th Street NW, Suite 1400
Washington, D.C. 20005

**American Schlafhorst Foundation Grants** 487
The American Schlafhorst Foundation, which was established in 1987, awards grants to eligible North Carolina nonprofit organizations in its areas of interest, including: arts and culture; botanical gardens; children and youth; community college education; the elderly; elementary and secondary education; employment; health care; higher education; hospital care; housing development; museums; science; and social services delivery. Types of support include: building construction and renovation; equipment acquisition; general operating support; research; scholarships; and seed money funding. There are no specific annual application deadlines, so interested parties should begin by contacting the Foundation directly with a proposed concept. Most recently, awards have ranged from $10,000 to $25,000.
*Requirements:* 501(c)3 tax-exempt organizations serving the greater Charlotte, NC, area are eligible.
*Restrictions:* Individuals are not eligible.
*Geographic Focus:* North Carolina
*Amount of Grant:* 10,000 - 25,000 USD
*Contact:* Dan W. Loftis, Wxecutive Director; (704) 554-0800; info@schlafhorst.com
*Sponsor:* American Schlafhorst Foundation
8801 South Boulevard, P.O. Box 240828
Charlotte, NC 28224-0828

**American Society for Yad Vashem Grants** 488
Established in the State of New York in 1981, the American Society for Yad Vashem gives on an international basis with emphasis on Israel. The society is the development arm of Yad Vashem, Jerusalem, which serves to remember and honor all victims of the Holocaust. Its primary fields of interest include: historical activities; genealogy; historic preservation; support of historical societies; and Jewish agencies and synagogues. There are no specified application forms or annual deadlines, and applicants should begin by forwarding a letter of application to the Society.
*Geographic Focus:* All States, Israel
*Contact:* Eli Zborowski, Chairperson; (212) 220-4304; fax (212) 220-4308; ezborowski@yadvashemusa.org
*Internet:* www.yadvashemusa.org/history.html
*Sponsor:* American Society for Yad Vashem
500 5th Avenue, 42nd Floor
New York, NY 10110-1699

**AMS Alfred Einstein Award** 489
The AMS Alfred Einstein Award is given annually to the author of the article on a musicological subject selected by a committee of scholars to be the most significant published in a periodical during the preceding calendar year. Nominations should include the name of the author, the title of the article, and the name and year of the periodical or other collection in which it was published. Additionally, the award committee will solicit the curriculum vita of each nominee.
*Requirements:* Eligible are Canadian and U.S. scholars in the early stages of their careers.
*Geographic Focus:* All States
*Date(s) Application is Due:* May 1
*Amount of Grant:* 400 USD
*Contact:* Robert Judd, Executive Director; (212) 992-6340 or (877) 679-7648; fax (877) 679-7648; rjudd@amsmusicology.org or ams@amsmusicology.org
*Internet:* https://www.amsmusicology.org/page/einstein
*Sponsor:* American Musicological Society
20 Cooper Square, Floor 2
New York, NY 10003

**ANA/Grass Foundation Award in Neuroscience** 490
The ANA/Grass Foundation Award was established to honor outstanding young investigators doing research in basic or clinical neuroscience. The awardee receives a $1,000 honorarium; inclusion in the annual meeting as either a poster or platform presenter; complimentary registration; a commemorative plaque; and up to $1,500 travel reimbursement.
*Requirements:* Eligible candidates include: physician-scientist neurology faculty members early in their careers (M.D. or M.D./Ph.D.); those who are five years or less out of postgraduate or post fellowship training; and those conducting research in neuroscience. Nominators may send a letter of recommendation and the nominee's CV and biography. Self nominations and second party nominations are accepted.
*Geographic Focus:* All States
*Date(s) Application is Due:* Feb 15
*Amount of Grant:* 2,500 USD
*Contact:* Laurie Dixon; (952) 545-6284; fax (952) 545-6073; ana@llmsi.com
*Internet:* www.aneuroa.org/i4a/pages/index.cfm?pageid=3360
*Sponsor:* American Neurological Association
5841 Cedar Lake Road, Suite 204
Minneapolis, MN 55416

**ANA Wolfe Neuropathy Research Prize** 491
The Wolfe Neuropathy Research Prize was designed to honor an outstanding investigator who has identified a new cause or treatment of axonal peripheral neuropathy. The recipient is awarded $2,000 and a plaque.
*Geographic Focus:* All States
*Amount of Grant:* 2,000 USD
*Contact:* Linda Scher; (952) 545-6284; fax (952) 545-6073; lindascher@llmsi.com
*Internet:* www.aneuroa.org/i4a/pages/index.cfm?pageid=3358
*Sponsor:* American Neurological Association
5841 Cedar Lake Road, Suite 204
Minneapolis, MN 55416

**Anderson Foundation Grants** 492
Established in Ohio in 1949, the Anderson Family Foundation Trust gives grants in the greater Toledo, Ohio, region, including Maumee and Columbus. Giving also to organizations located within the areas of the Anderson plants in the following states: Champaign, Illinois; Delphi, Lafayette, and Dunkirk, Indiana; and Albion, Potterville, Webberville, and White Pigeon, Michigan. The Foundation's primary fields of interest include: agriculture; arts; children and youth services; education; the environment; government and public administration; higher education; human services; religion; secondary school education; and federated giving programs. Types of support include: annual campaigns; building and renovation programs; capital campaigns; conferences and seminars; emergency funds; general operating support; matching/challenge grants; program development; publications; research; scholarship funds; and seed money.

Specific application forms are not required, and board decisions are made on the 3rd Monday of the month in March, June, September, and December. Grant amounts range up to $100,000 on occasion, though most average between $5,000 and $10,000.
*Requirements:* Nonprofit 501(c)3 agencies in Ohio, Illinois, Indiana, and Michigan serving areas of company operation are eligible to apply.
*Restrictions:* No support is available for private foundations, public high schools, or elementary schools. No grants are given to individuals, or for endowment funds, travel, or building or operating funds for churches or elementary schools.
*Geographic Focus:* Illinois, Indiana, Michigan, Ohio
*Date(s) Application is Due:* Feb 28; May 30; Aug 30; Nov 30
*Amount of Grant:* Up to 100,000 USD
*Samples:* Chance for Change Foundation, Toledo, Ohio, $5,000 - human services; DeKalb County Fair Association, DeKalb, Indiana, $5,000 - agriculture ($5,000); Camp Tecumseh, Brookston, Indiana, $5,000 - human services.
*Contact:* Fredi Heywood, Chairman; (419) 243-1706 or (419) 893-5050; fax (419) 242-5549; fredi@toledocf.org
*Sponsor:* Anderson Foundation
480 W. Dussel Drive, P.O. Box 119
Maumee, OH 43537-0119

**Andres Memorial Trust Grants** 493

The Andres Memorial Trust was established in both the Matilda Andres Trust Agreement and the Leonard Andres Trust Agreement on August 5, 1968. The Trust is intended to augment the efforts and funds of its beneficiaries and to inspire and encourage the leaders of the organizations benefited, and the residents of Floyd County, Iowa to achieve superior accomplishments in the areas of public service benefited by this trust. The Andres Memorial Trust is comprised of two trusts, one established by Leonard L. Andres, better known as L. L. Andres, and the other established by his wife, Matilda Andres. Program areas of interest include: health; human services; and the public/society benefit. Applications must be submitted through the online grant application form or alternative accessible application designed for assistive technology users. Grant applications should be submitted only for capital needs for buildings, facilities and equipment. The average grant funded is $20,000, with six awards given each year. Applications are accepted year-round. Applications must be submitted by June 30 to be reviewed at the annual grant meeting that occurs in July.
*Requirements:* To be eligible, organizations must qualify as exempt governmental units or charitable organizations under Section 501(c)3 of the Internal Revenue Code. Grants are limited to organizations located in Floyd County, Iowa, and are generally made to the following organizations: Charles City Public Library; City Improvement Association of Charles City, Iowa; the Floyd County Historical Society; the St. Charles Cemetery Association of Charles City, Iowa; hospitals; facilities for medical research; facilities for the treatment of diseases or afflictions of aging persons; homes for the aged; and homes for the indigent.
*Restrictions:* The foundation normally does not fund requests for: churches; educational institutions; operating expenses; or salaries.
*Geographic Focus:* Iowa
*Date(s) Application is Due:* Jun 30
*Amount of Grant:* Up to 25,000 USD
*Contact:* Sangita Karra, Vice President; (609) 799-3108; fax (609) 799-4189; grantadministration@wellsfargo.com
*Internet:* https://www.wellsfargo.com/private-foundations/andres-memorial-trust
*Sponsor:* Andres Memorial Trust
Greeley Square, Post Office Box 20459
New York, NY 10001-0459

**Andrew Family Foundation Opportunity Knocks Grants** 494

The Opportunity Knocks program is a specific initiative created by the Foundation to focus on educational and mentoring opportunities for disadvantaged youth and families that would not otherwise be available. The Andrew Family Foundation seeks to provide expansion capital to an established organization(s) to build a new program or to enhance an existing program that addresses the issue of literacy and focuses on teaching young children in grades K-3 how to read. The program must positively affect and enhance the lives of disadvantaged youth ("program participants") and enable them to read at grade level by the end of 3rd Grade. In addition to financial support, the board members of the Andrew Family Foundation may desire to provide expertise and resources necessary to ensure the success of the program, including the pursuit of collaborative efforts to replicate ideas and to implement proven solutions to expand the program to other geographical locations. Accordingly, providing opportunities for involvement may be a plus. The program's objectives should include a plan: to develop early childhood literacy among the program participants by properly teaching them to read and develop reading skills for the future; to maximize the impact on the greatest number of program participants; and to become self-sustaining in three to five years. Preference will be given to organizations that focus on collaborating with other organizations to replicate proven solutions. The Foundation will award grant(s) of up to $250,000 over a three-year period.
*Requirements:* The organization must meet the following criteria for eligibility: must be a U.S. nonprofit (federally tax-exempt) organization as defined under Section 501(c)3 of the Internal Revenue Code; must focus on literacy development specifically for grades K-3; government funding must not exceed 50% of the organization's total budget; geographic focus must remain in the United States; must have two or more years of operating experience; must have an established board of directors; must be able to provide informal progress reports on a quarterly basis; and must have ability to measure results, which reflect the success of the program. Preference will be given to those organizations and/or programs with budgets of less than $2 million.
*Restrictions:* The Andrew Family Foundation will not consider programs that directly impact public policy or provide funding of endowments or capital grants. The financial support will be restricted to a specific purpose or project and can include operational support.
*Geographic Focus:* All States
*Date(s) Application is Due:* Sep 15
*Amount of Grant:* Up to 250,000 USD
*Contact:* Connor Humphrey, Grants Administrator; (602) 828-8471 or (708) 460-1288; fax (602) 385-3267; Connor.Humphrey@GenSpring.com or aff@inlignwealth.com
*Internet:* https://online.foundationsource.com/andrew/ok.htm
*Sponsor:* Andrew Family Foundation
14628 John Humphrey Drive
Orland Park, IL 60462

**Andy Warhol Foundation for the Visual Arts Curatorial Research Fellowships** 495

Scholarly research undertaken in the field of contemporary art is funded through Andy Warhol Foundation for the Visual Arts Curatorial Research Fellowships. It is assumed that research will lead to a significant exhibition, though this is not a requirement. Grants to curators do not preclude separate proposals from sponsoring institutions in any given grant round. The postmark deadlines for proposals are March 1 with notification on July 1 and September 1 with notification on January 1.
*Requirements:* Institutionally-affiliated curators at any stage of their careers are eligible to apply and must have the formal support of their director.
*Geographic Focus:* All States
*Date(s) Application is Due:* Mar 1; Sep 1
*Amount of Grant:* 1,000 - 100,000 USD
*Contact:* Rachel Bers, Program Director; (212) 387-7555; fax (212) 387-7560; info@warholfoundation.org
*Internet:* warholfoundation.org/grant/curatorial_fellowships.html
*Sponsor:* Andy Warhol Foundation for the Visual Arts
65 Bleecker Street, Seventh Floor
New York, NY 10012

**Animal Behavior Society Amy R. Samuels Cetacean Behavior and Conservation Awards** 496

The Animal Behavior Society (ABS) was founded in 1964 to promote the study of animal behavior. ABS members come from all over the world, but primarily from North, Central, and South America. Current members' research activities span the invertebrates and vertebrates, both in the field and in the laboratory, and include experimental psychology, behavioral ecology, neuroscience, zoology, biology, applied ethology, and human ethology as well as many other specialized areas. The ABS offers several first-time research awards to graduate-student members of the society. The Amy R. Samuels Cetacean Behavior and Conservation Grant was established to provide financial support for graduate students studying cetacean behavior and/or cetacean conservation in natural environments. The goal of the award is to encourage original and significant research that focuses on endangered populations or species, or on critically important problems in cetacean behavioral biology. A maximum of two awards up to $2,000 each will be awarded to successful applicants. The annual deadline for submission is December 14.
*Requirements:* Applicants for student grants must be enrolled in a graduate program and must be active members of the Animal Behavior Society.
*Restrictions:* Individuals may receive only one ABS research grant during their lifetime. The award must be used for field research; it cannot be used for laboratory research. The award shall be given only if there are proposals deemed worthy by the judges in any given year. Thus, there may be years in which no award is made.
*Geographic Focus:* All States, All Countries
*Date(s) Application is Due:* Dec 14
*Amount of Grant:* 2,000 USD
*Contact:* Melissa Dawn Hughes, Student Research Committee Chair; (850) 644-588 or (504) 202-6026; fax (850) 644-6402; mh11c@my.fsu.edu or studentresearchgrants@animalbehaviorsociety.org
*Internet:* www.animalbehaviorsociety.org/web/awards-cetacean.php
*Sponsor:* Animal Behavior Society
402 North Park Street
Bloomington, IN 47408

**Animal Behavior Society Career Awards** 497

The Animal Behavior Society (ABS) was founded in 1964 to promote the study of animal behavior. ABS members come from all over the world, but primarily from North, Central, and South America. Current members' research activities span the invertebrates and vertebrates, both in the field and in the laboratory, and include experimental psychology, behavioral ecology, neuroscience, zoology, biology, applied ethology, and human ethology as well as many other specialized areas. The ABS has a series of Career Awards which include the following: Distinguished Animal Behaviorist, for outstanding lifetime achievement in animal behavior; Exemplar Award, for major long-term contribution to animal behavior; Outstanding New Investigator, for outstanding contribution by a new investigator; Quest Award, for outstanding seminal contribution; Exceptional Service Award, for sustained service contributions to the Animal Behavior Society; and Penny Bernstein Distinguished Teaching Award, which recognizes a sustained record of excellent teaching of animal behavior in the classroom or informal education setting. All members of the society are encouraged to submit nominations for the above awards. The annual deadline for nominations is November 30.
*Requirements:* To aid the Selection Committee and to help codify the procedures involved, the following items must be submitted for a nomination: a letter of nomination indicating the award for which the nominee is being proposed which should provide details on

the reasons the nominee should be considered for that award; a curriculum vitae of the nominee; and additional supporting letters from colleagues solicited by the nominator.
*Geographic Focus* All States, District of Columbia, Guam, Marshall Islands, Northern Mariana Islands, Puerto Rico, U.S. Virgin Islands, All Countries, American Samoa
*Date(s) Application is Due:* Nov 30
*Contact:* Danielle J. Whittaker, Public Affairs Committee Chair; (517) 884-2561 or (517) 355-1855; publicaffairs@animalbehaviorsociety.org
*Internet:* www.animalbehaviorsociety.org/web/awards-career.php
*Sponsor:* Animal Behavior Society
402 North Park Street
Bloomington, IN 47408

**Animal Behavior Society Caregiver Travel Grants** **498**
The Animal Behavior Society was founded in 1964 to promote the study of animal behavior. ABS members come from all over the world, but primarily from North, Central, and South America. Current members' research activities span the invertebrates and vertebrates, both in the field and in the laboratory, and include experimental psychology, behavioral ecology, neuroscience, zoology, biology, applied ethology, and human ethology as well as many other specialized areas. The Caregiver Travel Grant award is intended to provide financial assistance for active ABS members whose caregiving responsibilities for dependent persons would otherwise limit their ability to attend and present at the annual ABS meeting. Dependents include but are not limited to young children and aging parents. Receiving this award does not preclude ABS members from applying for additional funding from ABS. Candidates are prioritized according to their need, abstract and CVs. Preference will be given to applicants who are presenting their work. The award money may be spent in any of a variety of ways to support your conference participation. Examples include daycare at home or at the conference venue, costs of travel for the dependent or a second caregiver so that a child may accompany you to the conference, or costs of additional care at home while you are away. Funds should be used to support dependent care rather than cover other costs associated with meeting attendance, for example conference registration or travel for the applicant. The annual deadline for applications is the last day of abstract submissions on April 1.
*Requirements:* Applicants must both be a resident of and conducting research in a developing country to be eligible for this award. Caregiver Travel Grant Awards are for ABS members who will be attending the annual ABS meeting. Receiving this award does not preclude ABS members from applying for additional funding from ABS.
*Geographic Focus* All States, District of Columbia, Guam, Marshall Islands, Northern Mariana Islands, Puerto Rico, U.S. Virgin Islands, All Countries, American Samoa
*Date(s) Application is Due:* Apr 1
*Contact:* Alexander Ophir, Animal Care Committee Chair; (607) 255-3714 or (607) 254-4636; fax (607) 255-4644; ago25@cornell.edu or animalcare@animalbehaviorsociety.org
*Internet:* www.animalbehaviorsociety.org/web/awards-care.php
*Sponsor:* Animal Behavior Society
402 North Park Street
Bloomington, IN 47408

**Animal Behavior Society Charles H. Turner Award** **499**
The Animal Behavior Society was founded in 1964 to promote the study of animal behavior. ABS members come from all over the world, but primarily from North, Central, and South America. Current members' research activities span the invertebrates and vertebrates, both in the field and in the laboratory, and include experimental psychology, behavioral ecology, neuroscience, zoology, biology, applied ethology, and human ethology as well as many other specialized areas. Charles Turner was one of the very first African-American researchers in animal behavior. Among other things, his research focused on insect sensory biology and task learning. By naming the undergraduate program after him, the ABS Diversity Committee emphasizes its goal to increase the diversity of its membership by encouraging researchers of all ages, levels, and ethnic groups to participate in the annual meetings. In 2002 the Animal Behavior Society created the Charles H. Turner Award to encourage undergraduate participation at the annual Society meetings. With generous support from the National Science Foundation, the ABS Diversity committee brings a group of about ten undergraduates to the meetings each year, covering all expenses, and providing mentors and a full program of mentoring events, including a pre-meeting workshop. Charles Turner was one of the very first African-American researchers in animal behavior. Among other things, his research showed that insects can hear and exhibit trial-and-error learning. By naming the undergraduate program after him, the ABS Diversity Committee emphasizes its goal to increase the diversity of its membership by encouraging researchers of all ages, levels, and ethnic groups to participate in the annual meetings. Applications are submitted via an online form available from the Turner-Award webpage at the ABS website. The form is generally available from January through May. Submission deadlines may vary from year to year.
*Requirements:* Participants must be citizens or permanent residents of the United States. Students must attend the entire meeting (generally in June), and must not have completed their undergraduate degree the prior December. Students who are members of groups traditionally under-represented in the sciences (women, ethnic and racial minorities, those with disabilities, etc.) are particularly urged to apply. Students with special needs are also encouraged to apply but should discuss their specific requirements with the organizers as part of the application process.
*Geographic Focus* All States, District of Columbia, Guam, Marshall Islands, Northern Mariana Islands, Puerto Rico, U.S. Virgin Islands, American Samoa
*Date(s) Application is Due:* Apr 30
*Contact:* Daniel R. Howard, Diversity Committee Chair; (603) 862-0242 or (603) 862-1453; daniel.howard@unh.edu or diversity@animalbehaviorsociety.org
*Internet:* www.animalbehaviorsociety.org/web/awards-turner.php
*Sponsor:* Animal Behavior Society
402 North Park Street
Bloomington, IN 47408

**Animal Behavior Society David S. Tuber Award** **500**
The Animal Behavior Society (ABS) was founded in 1964 to promote the study of animal behavior. ABS members come from all over the world, but primarily from North, Central, and South America. Current members' research activities span the invertebrates and vertebrates, both in the field and in the laboratory, and include experimental psychology, behavioral ecology, neuroscience, zoology, biology, applied ethology, and human ethology as well as many other specialized areas. The ABS offers several first-time research awards to graduate-student members of the society. The ABS David S. Tuber award was created to promote excellence in research into Applied Animal Behavior. Named in honor of David S. Tuber (1942-1995), this award was created by Victoria L. Voith and Peter L. Borchelt, editors of the book "Readings in Companion Animal Behaviour" for his seminal contribution to the field of clinical animal behavior. David S. Tuber was one of the first behaviorists to apply the principles of animal behavior to the behavioral problems of companion animals. The annual deadline for submission is December 14.
*Requirements:* Applicants for student grants must be enrolled in a graduate program and must be active members of the Animal Behavior Society.
*Restrictions:* Individuals may receive only one ABS research grant during their lifetime.
*Geographic Focus:* All States, All Countries
*Date(s) Application is Due:* Dec 14
*Contact:* Melissa Dawn Hughes, Student Research Committee Chair; (850) 644-588 or (504) 202-6026; fax (850) 644-6402; mh11c@my.fsu.edu or studentresearchgrants@animalbehaviorsociety.org
*Internet:* www.animalbehaviorsociety.org/web/awards-tuber.php
*Sponsor:* Animal Behavior Society
402 North Park Street
Bloomington, IN 47408

**Animal Behavior Society Developing Nations Research Grants** **501**
The Animal Behavior Society was founded in 1964 to promote the study of animal behavior. ABS members come from all over the world, but primarily from North, Central, and South America. Current members' research activities span the invertebrates and vertebrates, both in the field and in the laboratory, and include experimental psychology, behavioral ecology, neuroscience, zoology, biology, applied ethology, and human ethology as well as many other specialized areas. The goal of the Developing Nations Research Grants is to support animal behaviorists in developing countries who may have difficulty securing financial resources to fund their research. Awards are provided for research to be conducted within a 1-year period from the date of the award. The number of awards varies annually, and amounts range from $500 to $800, depending on need and funding availability. Applicants must use ABS's online submission site to submit their applications. The site generally opens in November and stops accepting submissions in late December or early January. Please note that membership dues must be paid in full and the ABS Central Office contacted to activate access to the on-line application system. The account activation cut-off date occurs several days prior to the actual submission deadline. Detailed submission guidelines are provided at the website. The annual deadline for submission is December 14.
*Requirements:* Only members of the Animal Behavior Society who are residents of developing nations and who are conducting research at an institution in a developing nation are eligible to apply. Applicants may be graduate students as well as more established members of the research community. Eligible candidates may receive one award as a student and one additional award as an established researcher.
*Restrictions:* The following nations shall not be considered as developing nations: the United States, Canada, Israel, Japan, Australia, New Zealand, Iceland, Norway, Sweden, Finland, Denmark, Germany, the Netherlands, Belgium, Luxembourg, Ireland, Switzerland, Austria, Italy, France, the United Kingdom, Spain, and Portugal.
*Geographic Focus:* All States, All Countries
*Date(s) Application is Due:* Dec 14
*Amount of Grant:* 500 - 800 USD
*Contact:* Melissa Dawn Hughes, Student Research Committee Chair; (850) 644-5888 or (504) 202-6026; fax (850) 644-6402; mh11c@my.fsu.edu or studentresearchgrants@animalbehaviorsociety.org
*Internet:* www.animalbehaviorsociety.org/web/awards-developing.php
*Sponsor:* Animal Behavior Society
402 North Park Street
Bloomington, IN 47408

**Animal Behavior Society Diversity Fund Student Registration Fee Awards** **502**
The Animal Behavior Society was founded in 1964 to promote the study of animal behavior. ABS members come from all over the world, but primarily from North, Central, and South America. Current members' research activities span the invertebrates and vertebrates, both in the field and in the laboratory, and include experimental psychology, behavioral ecology, neuroscience, zoology, biology, applied ethology, and human ethology as well as many other specialized areas. Each year, the ABS offers a limited number of awards from the Diversity Fund to cover registration fees for minority graduate students attending the annual ABS Meeting, with the goal of broadening minority and ethnic representation. To apply for this award, applicants should submit a web-based form that includes their email address, full name, qualifying status, name and address of their school, the title of their talk for the meeting, and their preliminary abstract. The link to the form is on the society's Diversity Grant webpage. The annual deadline for applications is the last day of abstract submissions on April 1.

*Requirements:* Applicants must be enrolled in a graduate program at the time of application and be a member of an under-represented minority, including African Americans, Native Americans, Asian Americans, and Hispanic Americans who live in North America or are citizens of Central or South America.
*Geographic Focus* All States, District of Columbia, Guam, Marshall Islands, Northern Mariana Islands, Puerto Rico, U.S. Virgin Islands, All Countries, American Samoa
*Date(s) Application is Due:* Apr 1
*Contact:* Daniel R. Howard, (603) 862-0242 or (603) 862-1453; daniel.howard@unh.edu or diversity@animalbehaviorsociety.org
*Internet:* www.animalbehaviorsociety.org/web/awards-diversity.php
*Sponsor:* Animal Behavior Society
402 North Park Street
Bloomington, IN 47408

**Animal Behavior Society Edward O. Wilson Conservation Award** **503**
The Animal Behavior Society (ABS) was founded in 1964 to promote the study of animal behavior. ABS members come from all over the world, but primarily from North, Central, and South America. Current members' research activities span the invertebrates and vertebrates, both in the field and in the laboratory, and include experimental psychology, behavioral ecology, neuroscience, zoology, biology, applied ethology, and human ethology as well as many other specialized areas. The ABS offers several first-time research awards to graduate-student members of the society. E.O. Wilson, professor at Harvard University, who in 2002 received the ABS Distinguished Animal Behaviorist Award, is one of the world's most eminent scientists and pioneers in biodiversity conservation. In his long career, he has transformed his field of research—the behavior of ants—and applied his scientific perspective and experience to illuminate the human circumstance, including human origins, human nature, and human interactions. Wilson has also been a pioneer in spearheading efforts to preserve and protect the biodiversity of this planet. Beginning with his unusual childhood in Alabama, "E.O. Wilson—Of Ants and Men" chronicles the famed biologist and Pulitzer Prize-winning author's lifelong love for the natural world and the groundbreaking research that would establish him as the foremost authority on ants. The Award named in his honor seeks to encourage graduate students of animal behavior to participate in meaningful conservation-related research. This single award of up to $2,000 is given in recognition of a research proposal considered meritorious for its integration of behavior and conservation. The annual deadline for submission is December 14.
*Requirements:* Applicants for student grants must be enrolled in a graduate program and must be active members of the Animal Behavior Society.
*Restrictions:* Individuals may receive only one ABS research grant during their lifetime.
*Geographic Focus:* All States, All Countries
*Date(s) Application is Due:* Dec 14
*Amount of Grant:* 2,000 USD
*Contact:* Melissa Dawn Hughes, Student Research Committee Chair; (850) 644-5888 or (504) 202-6026; fax (850) 644-6402; mh11c@my.fsu.edu or studentresearchgrants@animalbehaviorsociety.org
*Internet:* www.animalbehaviorsociety.org/web/awards-wilson.php
*Sponsor:* Animal Behavior Society
402 North Park Street
Bloomington, IN 47408

**Animal Behavior Society Founders Memorial Poster Paper Award** **504**
The Animal Behavior Society was founded in 1964 to promote the study of animal behavior. ABS members come from all over the world, but primarily from North, Central, and South America. Current members' research activities span the invertebrates and vertebrates, both in the field and in the laboratory, and include experimental psychology, behavioral ecology, neuroscience, zoology, biology, applied ethology, and human ethology as well as many other specialized areas. The Founders Memorial Poster Paper Award is in honor of the founders of the Animal Behavior Society. It is for outstanding posters presented at the ABS annual meeting. The annual deadline for applications is the last day of abstract submissions on April 1.
*Requirements:* Any graduate student or postdoctoral member of the society in good standing is eligible to enter the competition. Applicants can indicate their intention to enter the competition by marking the appropriate box on the Meeting Abstract submission form. All Founders presenters will be expected to provide a single page preview of their poster for the judges.
*Geographic Focus* All States, District of Columbia, Guam, Marshall Islands, Northern Mariana Islands, Puerto Rico, U.S. Virgin Islands, All Countries, American Samoa
*Date(s) Application is Due:* Apr 1
*Contact:* Daniel R. Howard, Diversity Committee Chair; (603) 862-0242 or (603) 862-1453; daniel.howard@unh.edu or diversity@animalbehaviorsociety.org
*Internet:* www.animalbehaviorsociety.org/web/awards-founders.php
*Sponsor:* Animal Behavior Society
402 North Park Street
Bloomington, IN 47408

**Animal Behavior Society Genesis Award for the Outstanding Undergraduate Poster Presentation** **505**
The Animal Behavior Society was founded in 1964 to promote the study of animal behavior. ABS members come from all over the world, but primarily from North, Central, and South America. Current members' research activities span the invertebrates and vertebrates, both in the field and in the laboratory, and include experimental psychology, behavioral ecology, neuroscience, zoology, biology, applied ethology, and human ethology as well as many other specialized areas. First presented at the 2000 ABS Annual Meeting the Genesis Award honors the best undergraduate poster presentation at the meeting. To encourage quality presentations the following guidelines and suggestions are available for viewing. The annual deadline for applications is the last day of abstract submissions on April 1.
*Requirements:* To judge each project judges will: consider each poster presentation on its own merits; identify the strengths and weaknesses to determine the best student poster presentation; and offer encouragement and constructive criticism.
*Geographic Focus* All States, District of Columbia, Guam, Marshall Islands, Northern Mariana Islands, Puerto Rico, U.S. Virgin Islands, All Countries, American Samoa
*Date(s) Application is Due:* Apr 1
*Contact:* Daniel R. Howard; (603) 862-0242 or (603) 862-1453; daniel.howard@unh.edu Matthew Wund, Genesis Sub-committee Chair; (609) 771-2897 or (609) 771-2356; fax (609) 637-5118; wundm@tcnj.edu or biology@tcnj.edu
*Internet:* www.animalbehaviorsociety.org/web/awards-genesis.php
*Sponsor:* Animal Behavior Society
402 North Park Street
Bloomington, IN 47408

**Animal Behavior Society George W. Barlow Award** **506**
The Animal Behavior Society (ABS) was founded in 1964 to promote the study of animal behavior. ABS members come from all over the world, but primarily from North, Central, and South America. Current members' research activities span the invertebrates and vertebrates, both in the field and in the laboratory, and include experimental psychology, behavioral ecology, neuroscience, zoology, biology, applied ethology, and human ethology as well as many other specialized areas. The ABS offers several first-time research awards to graduate-student members of the society. The ABS George W. Barlow Award's purpose is to encourage excellence in graduate student research in the field of animal behavior. The Barlow Award will be given annually to one top-ranked proposal in the Student Research Grant cycle in accordance with the most recent ABS Student Research Grant competition rules. Applications will be accepted through the online Grant Submission System. The annual deadline for submission is December 14.
*Requirements:* Applicants for student grants must be enrolled in a graduate program and must be active members of the Animal Behavior Society.
*Restrictions:* Individuals may receive only one ABS research grant during their lifetime.
*Geographic Focus:* All States, All Countries
*Date(s) Application is Due:* Dec 14
*Contact:* Melissa Dawn Hughes, Student Research Committee Chair; (850) 644-5888 or (504) 202-6026; fax (850) 644-6402; mh11c@my.fsu.edu or studentresearchgrants@animalbehaviorsociety.org
*Internet:* www.animalbehaviorsociety.org/web/awards-barlow.php
*Sponsor:* Animal Behavior Society
402 North Park Street
Bloomington, IN 47408

**Animal Behavior Society Graduate Student Travel Grants** **507**
The Animal Behavior Society (ABS) was founded in 1964 to promote the study of animal behavior. ABS members come from all over the world, but primarily from North, Central, and South America. Current members' research activities span the invertebrates and vertebrates, both in the field and in the laboratory, and include experimental psychology, behavioral ecology, neuroscience, zoology, biology, applied ethology, and human ethology as well as many other specialized areas. The ABS offers several first-time research awards to graduate-student members of the society. The ABS Graduate Student Travel Grant is intended to provide financial support for graduate students to attend and present at the annual Animal Behavior Society meeting by helping to defray the costs of travel, housing and/or meals at the meetings. The deadline for applications is the last day of abstract submissions. Priority will be given to Ph.D. students who are attending an ABS meeting for the first time. Award values vary according to funds available. Candidates are prioritized according to the quality of their abstracts and CVs. Applications consist of: a submitted abstract, including the title and indicate whether talk or poster; a curriculum vita (CV); and justification for request for support, including information about whether support for attendance is covered by other sources. Applicants should indicate whether or not they have attended or presented at a previous ABS meeting and, if so, provide a description. Applicants will be able to apply via the abstract submission system after they register for the conference.
*Restrictions:* Graduate Student Travel Awards are restricted to graduate students who will be presenting a talk/poster at the annual ABS meeting and who are active members of ABS. Receiving this award does not preclude students from applying for additional funding from ABS.
*Geographic Focus:* All States, All Countries
*Contact:* Melissa Dawn Hughes, Student Research Committee Chair; (850) 644-5888 or (504) 202-6026; fax (850) 644-6402; mh11c@my.fsu.edu or studentresearchgrants@animalbehaviorsociety.org
*Internet:* www.animalbehaviorsociety.org/web/awards-student.php
*Sponsor:* Animal Behavior Society
402 North Park Street
Bloomington, IN 47408

**Animal Behavior Society Latin American Travel Awards** **508**
The Animal Behavior Society was founded in 1964 to promote the study of animal behavior. ABS members come from all over the world, but primarily from North, Central, and South America. Current members' research activities span the invertebrates and vertebrates, both in the field and in the laboratory, and include experimental psychology, behavioral ecology, neuroscience, zoology, biology, applied ethology, and human ethology as well as many other specialized areas. The Latin American Travel

Awards are intended to encourage greater participation by Latin-American researchers in the Animal Behavior Society (ABS)'s annual meetings by helping to defray the costs of international travel, housing, and/or meals. Prospective applicants who have successfully submitted an abstract for their talk or poster via the ABS annual-meeting webpage should also email their submitted abstract (including the title) and curriculum vita (CV) to the second ABS member-at-large. Each document should be attached to the email message as a separate pdf file, and each labeled with the applicant's last name, initials, and the contents of each file (i.e. abstract or CV). A list of ABS Officers with their contact information is posted at the ABS website. Checks in the amount of the award will be sent directly to recipients following announcement of the results. The current deadline for submissions is April 1.
*Requirements:* Priority will be given to Ph.D. students who are attending an ABS meeting for the first time or who have not previously received a Latin American Travel Award. Award values vary according to funds available. Candidates are prioritized according to the quality of their abstracts and CVs.
*Restrictions:* Latin American Travel Awards are restricted to Latin-American graduate students enrolled in Latin-American institutions, who will be presenting a talk/poster at the annual ABS meeting.
*Geographic Focus:* All Countries
*Date(s) Application is Due:* Apr 1
*Contact:* Lilian Tonelli Manica, Latin American Affairs Committee Chair; +55 (41) 3360-5000; gabo@fcien.edu.uy or latinamericanaffairs@animalbehaviorsociety.org
*Internet:* www.animalbehaviorsociety.org/web/awards-latin.php
*Sponsor:* Animal Behavior Society
402 North Park Street
Bloomington, IN 47408

**Animal Behavior Society Student Research Grants** **509**
The Animal Behavior Society (ABS) was founded in 1964 to promote the study of animal behavior. ABS members come from all over the world, but primarily from North, Central, and South America. Current members' research activities span the invertebrates and vertebrates, both in the field and in the laboratory, and include experimental psychology, behavioral ecology, neuroscience, zoology, biology, applied ethology, and human ethology as well as many other specialized areas. The ABS offers several first-time research awards to graduate-student members of the society: the George W. Barlow Award, the E.O. Wilson Conservation Award, and the Amy R. Samuels Cetacean Behavior and Conservation (CBC) Award (among others). A brief description of each award follows: the George W. Barlow Award's purpose is to encourage excellence in graduate-student research in the field of animal behavior and will be given to one top-ranked proposal in ABS's annual Student-Research Grant cycle; the E.O. Wilson Conservation Award annually recognizes one research proposal considered meritorious for its integration of animal behavior and conservation; the CBC Award provides financial support for graduate students studying cetacean behavior and/or cetacean conservation in natural environments. Specifically, the CBC award encourages original and significant research that focuses on endangered populations or species, or on critically important problems in cetacean behavioral biology. The award is given only if there are proposals deemed worthy by the judges in any given year; up to two awards of a maximum of $2000 each will be made to successful applicants. The award must be used for field research and cannot be used for laboratory research. Applicants apply for all three grants through a single-application process, usually at the end of the year. All eligible applicants will be considered for all three grants, but may receive only one of them. Applicants must use ABS's online submission site to submit their applications. The site generally opens around the end of October and stops accepting submissions in early January. Please note that membership dues must be paid in full and the ABS Central Office contacted to activate access to the on-line application system. The account-activation cut-off date occurs several days prior to the actual submission deadline.
*Requirements:* Applicants for student grants must be enrolled in a graduate program and must be active members of the Animal Behavior Society. Members of the Association for the Study of Animal Behavior are eligible for the Cetacean Behavior Conservation grant.
*Restrictions:* Applicants may receive only one of these grants in a lifetime.
*Geographic Focus:* All States, All Countries
*Date(s) Application is Due:* Dec 14
*Amount of Grant:* 500 - 2,000 USD
*Contact:* Melissa Dawn Hughes, Student Research Committee Chair; (850) 644-5888 or (504) 202-6026; fax (850) 644-6402; mh11c@my.fsu.edu or studentresearchgrants@animalbehaviorsociety.org
*Internet:* www.animalbehaviorsociety.org/web/awards-student-grants.php
*Sponsor:* Animal Behavior Society
402 North Park Street
Bloomington, IN 47408

**Animal Behavior Society Warder Clyde Allee Competition** **510**
The Animal Behavior Society was founded in 1964 to promote the study of animal behavior. ABS members come from all over the world, but primarily from North, Central, and South America. Current members' research activities span the invertebrates and vertebrates, both in the field and in the laboratory, and include experimental psychology, behavioral ecology, neuroscience, zoology, biology, applied ethology, and human ethology as well as many other specialized areas. The Warder Clyde Allee Session for Best Student Paper Competition will be held during the annual meeting of the Animal Behavior Society, typically scheduled in mid- to late-July. All eligible students and recent graduates are encouraged to participate. The annual deadline for applications is the last day of abstract submissions on April 1.
*Requirements:* Any independent graduate student research (including, but not limited to, the doctoral dissertation) is suitable for the application. The work presented may be part of a larger collaborative effort, but the student must be first author and have principal responsibility for conceptualization and design of the research, collection and analysis of the data, and interpretation of the results. Interested candidates must: check the appropriate box in the online abstract submission system for the annual meeting; submit a cover letter and an electronic version of their paper as specified below; confirm that they meet all eligibility requirements; present a spoken version during the Annual Meeting; and attend both the Allee welcoming dinner and the awards ceremony and closing banquet (complimanetary banquet ticket provided) during the Annual Meeting
*Restrictions:* The entrant cannot have been awarded the Ph.D. degree before the start of the preceding ABS annual meeting. An individual can enter the session only once per lifetime.
*Geographic Focus* All States, District of Columbia, Guam, Marshall Islands, Northern Mariana Islands, Puerto Rico, U.S. Virgin Islands, All Countries, American Samoa
*Date(s) Application is Due:* Apr 1
*Contact:* Daniel R. Howard, Diversity Committee Chair; (603) 862-0242 or (603) 862-1453; daniel.howard@unh.edu or diversity@animalbehaviorsociety.org
*Internet:* www.animalbehaviorsociety.org/web/awards-allee.php
*Sponsor:* Animal Behavior Society
402 North Park Street
Bloomington, IN 47408

**AOCS A.R. Baldwin Distinguished Service Award** **511**
The A. Richard Baldwin Award was given to Baldwin in recognition of his lengthy and distinguished service to the AOCS. It is the Society's highest service award. The award includes a $2,000 honorarium, a plaque, and up to $1,500 in travel expenses to the Annual Meeting. Nomination materials should include a letter of nomination describing the candidate's contributions and service to the Society, in addition to up to two letters supporting the nomination, submitted from individuals familiar with the candidate's Society service. Self nominations are welcomed. The annual deadline for nominations is October 1.
*Requirements:* Candidates must have made outstanding and extraordinary contributions to the Society or supported the activities of the Society through officer and/or committee participation over a substantial period of time. Candidates must also be current Society members in good standing.
*Geographic Focus:* All States, All Countries
*Date(s) Application is Due:* Oct 1
*Amount of Grant:* 2,000 - 3,500 USD
*Contact:* Moghis U. Ahmad, Awards Chairperson; (217) 693-4804 or (217) 359-2344; fax (217) 693-4849; moghis@jinapharma.com or awards@aocs.org
*Internet:* www.aocs.org/Membership/content.cfm?ItemNumber=782
*Sponsor:* American Oil Chemists' Society
2710 South Boulder
Urbana, IL 61802-6996

**AOCS ACI/NBB Innovation Research Award** **512**
Sponsored by the American Cleaning Institute (ACI) and the National Biodiesel Board (NBB), the Glycerine Innovation Research Award recognizes outstanding achievement for research into new applications for glycerine with particular emphases on commercial viability. The Award includes a $5,000 honorarium, a plaque for the recipient, and an award lecture at the Annual Meeting. Nominations for this award are accepted from members of the Society, colleagues, or directly from individual candidates. The nomination form is available at the website.
*Requirements:* The research of candidates must reflect particular emphasis to existing or potential commercial importance. Candidate materials must include the following: a two-page letter of nomination, describing original research work in new applications and uses of glycerine; up to two letter of support from individuals familiar with the candidate's accomplishments; and any published works relating to the research.
*Geographic Focus:* All States, All Countries
*Date(s) Application is Due:* Nov 1
*Amount of Grant:* 5,000 USD
*Contact:* Barb Semeraro, Awards Program Coordinator; (217) 693-4804; fax (217) 693-4849; barbs@aocs.org
*Internet:* www.aocs.org/Membership/content.cfm?ItemNumber=804
*Sponsor:* American Oil Chemists' Society
2710 South Boulder
Urbana, IL 61802-6996

**AOCS Alton E. Bailey Award** **513**
The Alton E. Bailey Award recognizes outstanding research and exceptional service in the field of lipids and associated products. This prestigious award commemorates Alton E. Bailey's outstanding contributions to the field of fats and oils as a researcher, an author of several standard books in the field and as a leader in the work of the Society. The recipient receives a $750 honorarium, a plaque, and award lecture at the Annual Meeting. The application is available at the website. Self nominations are welcomed.
*Requirements:* Candidates must be AOCS members in good standing, with outstanding leadership to the Society. They must also have made significant contributions to the science and technology of lipids.
*Geographic Focus:* All States, All Countries
*Date(s) Application is Due:* Oct 15
*Amount of Grant:* 750 USD
*Contact:* Barb Semeraro, ; (217) 693-4804; fax (217) 693-4849; barbs@aocs.org

Silvana Martini, Awards Administration Chairperson; (435) 797-8136; fax (435) 797-2379; silvana.martini@usa.edu
*Internet:* www.aocs.org/Membership/content.cfm?ItemNumber=781
*Sponsor:* American Oil Chemists' Society
2710 South Boulder
Urbana, IL 61802-6996

**AOCS Analytical Division Student Award** 514

The Analytical Division Student Award is given to graduate students in the filed of lipid analytical chemistry. Graduate students at any institution of higher learning who are doing research toward a degree are eligible. The Award involves a certificate, $250 honorarium, a $500 travel allowance to present at the Annual Meeting, and an award lecture or poster presentation. Nominations should consist of the completed application, abstract tracking number and title, professor ranking form, and the professor letter of recommendation. Self-nominations are welcomed. The annual deadline for nominations is November 1.
*Requirements:* This award is open to graduate students presenting within the Analytical Division's technical program at the AOCS Annual Meeting.
*Geographic Focus:* All States, All Countries
*Date(s) Application is Due:* Nov 1
*Amount of Grant:* 750 USD
*Contact:* Moghis Ahmad, Awards Chairperson; (217) 693-4804 or (217) 359-2344; fax (217) 693-4849; moghis@jinapharma.com or awards@aocs.org
*Internet:* https://www.aocs.org/network-and-connect/awards-x1888#analytical-division
*Sponsor:* American Oil Chemists' Society
2710 South Boulder
Urbana, IL 61802-6996

**AOCS Australasian Award for Scientific Excellence in Lipid Research** 515

The AOCS Award for Scientific Excellence in Lipid Research is given to a scientist from within the Australasian region that has made a significant research contribution toward fats and oil research, either cumulative or one major advancement. The Award is offered every two years with the Section's biannual conference. The scientist receives a plaque, $1,500 AU towards expenses to attend the AOCS Annual Meeting, complimentary registration for the biannual AAOCS meeting where the award is presented, and presentation at that meeting. Candidates may nominate themselves or others. All nomination material must be emailed to the AAOCS Award Coordinator.
*Requirements:* Candidates must meet the following criteria to be eligible: make a significant contribution in basic or applied research, with that knowledge having a significant effect on advances in the field or hold substantial promise for its effect in the near future; and hold membership with the AAOCS during the year they are applying for the Award. Candidate materials must include the following: letter of nomination with summary of the nominee's research accomplishments; letters of support from at least two other scientists; and professional biographical data.
*Geographic Focus:* Australia
*Date(s) Application is Due:* May 17
*Amount of Grant:* 1,500 ARA
*Contact:* Amy Logan, AAOCS Award Coordinator; +61 3 9731 3478; fax +61 3 9545 2175; Amy.Logan@csiro.au
*Internet:* www.aocs.org/Membership/content.cfm?ItemNumber=18316
*Sponsor:* American Oil Chemists' Society
2710 South Boulder
Urbana, IL 61802-6996

**AOCS Biotechnology Division Student Awards** 516

The Biotechnology Division Student Awards are given to students presenting an outstanding paper in the field of biotechnology at the Annual Meeting. The Awards are given in three categories: $300 (1st place); $200 (2nd place); and $100 (3rd place).
*Geographic Focus:* All States, All Countries
*Date(s) Application is Due:* Oct 15
*Amount of Grant:* 600 USD
*Contact:* Barb Semeraro, Awards Program Coordinator; (217) 693-4804; fax (217) 693-4849; barbs@aocs.org
Silvana Martini, Awards Administrative Chairperson; (435) 797-8136; fax (435) 797-2379; silvana.martini@usa.edu
*Internet:* www.aocs.org/Membership/content.cfm?ItemNumber=786
*Sponsor:* American Humane Association
1400 16th Street NW, Suite 360
Washington, D.C. 20036

**AOCS Edible Applications Technology Division Outstanding Achievement Award** 517

The Edible Applications Technology Division Outstanding Achievement Award recognizes a scientist, technologist, or leader making contributions to the advancement of edible oils and/or the Division. The Award includes a plaque for the recipient, $500 cash prize, and a complimentary Division dinner ticket.
*Requirements:* Preference is given to AOCS members involved in the Edible Applications Technology Division, Society, or related organizations over a substantial amount of time. Candidates must also meet the following criteria: significant contribution to the knowledge, technology, and history of the industry; demonstrated character, ethics and a reputation that puts the individual above his or her peers; and service as a career and technical mentor.
*Geographic Focus:* All States, All Countries
*Date(s) Application is Due:* Oct 15
*Amount of Grant:* 500 USD
*Contact:* Barb Semeraro, Awards Program Coordinator; (217) 693-4804; fax (217) 693-4849; barbs@aocs.org
Silvana Martini, Awards Administration Chairperson; (435) 797-8136; fax (435) 797-2379; silvana.martini@usa.edu
*Internet:* www.aocs.org/Membership/content.cfm?ItemNumber=18315
*Sponsor:* American Oil Chemists' Society
2710 South Boulder
Urbana, IL 61802-6996

**AOCS Edible Applications Technology Division Student Award** 518

The Edible Applications Technology Division Student Award is given to a graduate student presenting within the Edible Applications Technology program at the Annual Meeting. The Award includes a certificate for the recipient, an award lecture or poster presentation at the Annual Meeting, and $500 in travel funds to attend the Meeting.
*Geographic Focus:* All States, All Countries
*Date(s) Application is Due:* Oct 15
*Amount of Grant:* 500 USD
*Contact:* Barb Semeraro, Awards Program Coordinator; (217) 693-4804; fax (217) 693-4849; barbs@aocs.org
Silvana Martini, Awards Administration Chairperson; (435) 797-8136; fax (435) 797-2379; silvana.martini@usa.edu
*Internet:* www.aocs.org/Membership/content.cfm?ItemNumber=796
*Sponsor:* American Oil Chemists' Society
2710 South Boulder
Urbana, IL 61802-6996

**AOCS George Schroepfer Medal** 519

The George Schroepfer Medal was established to honor the memory of George J. Schroepfer, Jr., a leader in the sterol and lipid field for more than 40 years. The award aims to foster Schroeper's ideals of personal integrity, high scientific standards, perseverance, and a strong spirit of survival, tempered by charm and wit. The award recognizes major advances in the steroid or sterol field, and is presented every even numbered year. It includes a bronze medal for the recipient, an award lecture at the Annual Meeting, and an honorarium determined by the endowed fund investment return. Self nominations are welcomed.
*Requirements:* The candidate's work may represent a single major achievement or a series of significant accomplishments, but should be based upon sound primary experimental data. Preference is given to accomplishments in the following: biochemistry and physiology with biomedical application; and interdisciplinary research in which rigorous chemical and analytical methods were applied to elucidate the physiological roles of steroids in animals, plants, or microorganisms. Fundamental advances that are primarily chemical, pharmacological, or analytical will also be considered
*Geographic Focus:* All States, All Countries
*Date(s) Application is Due:* Oct 15
*Contact:* Barb Semeraro, Awards Program Coordinator; (217) 693-4804; fax (217) 693-4849; barbs@aocs.org
Silvana Martini, Awards Administration Chairperson; (435) 797-8136; fax (435) 797-2379; silvana.martini@usa.edu
*Internet:* www.aocs.org/Membership/content.cfm?ItemNumber=934
*Sponsor:* American Oil Chemists' Society
2710 South Boulder
Urbana, IL 61802-6996

**AOCS Hans Kaunitz Award** 520

The Hans Kaunitz Award is given to a student doing research in the sciences relating to fats and oils and detergent technology. The Award includes a certificate, $1,000 honorarium, $500 travel allowance to the Annual Meeting, and a speaking engagement at the section Annual Meeting.
*Requirements:* The applicant must be scholastically outstanding, and a graduate student attending a post-secondary school in the U.S. The applicant must be actively performing research toward a Master's or Doctorate degree. This research must be related to fats, oils, protein co-products and/or detergents, and be interested in the areas of science and technology fostered by AOCS.
*Geographic Focus:* All States
*Date(s) Application is Due:* Oct 15
*Amount of Grant:* 1,500 USD
*Contact:* Barb Semeraro, Awards Program Coordinator; (217) 693-4804; fax (217) 693-4849; barbs@aocs.org
Silvana Martini, Awards Administration Chairperson; (435) 797-8136; fax (435) 797-2379; silvana.martini@usa.edu
*Internet:* www.aocs.org/Membership/content.cfm?ItemNumber=923
*Sponsor:* American Oil Chemists' Society
2710 South Boulder
Urbana, IL 61802-6996

**AOCS Health and Nutrition Division Student Award** 521

The Health and Nutrition Division of AOCS promotes and facilitates communication and cooperation among professionals whose interests in lipid biochemistry and physiology relate to all aspects of dietary fats and health. The Division encompasses the technical areas of dietary fats and general health. The Health and Nutrition Division Student Award is given to a graduate student in the health and nutrition field. The Award includes a $500 honorarium, certificate, and award lecture or poster presentation at the Annual Meeting. Self nominations are welcomed.

*Requirements:* Graduate students from any institution of higher learning who are doing research toward a degree are eligible.
*Geographic Focus:* All States, All Countries
*Date(s) Application is Due:* Oct 15
*Amount of Grant:* 500 USD
*Contact:* Barb Semeraro, Awards Program Coordinator; (217) 693-4804; fax (217) 693-4849; barbs@aocs.org
Silvana Martini, Awards Administration Chairperson; (435) 797-8136; fax (435) 797-2379; silvana.martini@usa.edu
*Internet:* www.aocs.org/Membership/content.cfm?ItemNumber=2048
*Sponsor:* American Oil Chemists' Society
2710 South Boulder
Urbana, IL 61802-6996

**AOCS Health and Nutrition Poster Competition** **522**
The Health and Nutrition Division promotes and facilitates communication and cooperation among professionals whose interests in lipid biochemistry and physiology relate to all aspects of dietary fats and health; encompasses the technical areas of dietary fats and general health. The Health and Nutrition Division Poster Competition was created to stimulate additional interest in the submission of quality poster student presentations at the Annual Meeting. Awards are given for best overall poster ($300), 1st place ($300), 2nd place ($200), and 3rd place ($100). All awardees also receive certificates, complimentary division dinner tickets, and award ribbons for their poster display.
*Requirements:* All poster abstracts submitted and selected for the Health and Nutrition Division Poster session are eligible for the competition.
*Geographic Focus:* All States, All Countries
*Date(s) Application is Due:* Oct 15
*Amount of Grant:* 900 USD
*Contact:* Barb Semeraro, Awards Program Coordinator; (217) 693-4804; fax (217) 693-4849; barbs@aocs.org
*Internet:* www.aocs.org/Membership/content.cfm?ItemNumber=18280
*Sponsor:* American Oil Chemists' Society
2710 South Boulder
Urbana, IL 61802-6996

**AOCS Herbert J. Dutton Award** **523**
The AOCS Analytical Division promotes professional interest, communication, and expert competency in the analytical chemistry field. The Herbert J. Dutton Award recognizes significant contributions to analytical chemistry. The Award includes a plaque for the recipient, $1,000 honorarium, and an award lecture at the Annual Meeting.
*Requirements:* Eligible candidates include those who have fulfilled one of the following criteria: significant contributions to the analysis of fats, oil, and related products (contributions reflected in current analytical methodologies); or those whose work has resulted in major advances in the understanding of processes utilized in the fats and oils industry.
*Geographic Focus:* All States, All Countries
*Date(s) Application is Due:* Oct 15
*Amount of Grant:* 1,000 USD
*Contact:* Barb Semeraro, Awards Program Coordinator; (217) 693-4804; fax (217) 693-4849; barbs@aocs.org
Silvana Martini, Awards Administration Chairperson; (435) 797-8136; fax (435) 797-2379; silvana.martini@usa.edu
*Internet:* www.aocs.org/Membership/content.cfm?ItemNumber=948
*Sponsor:* American Oil Chemists' Society
2710 South Boulder
Urbana, IL 61802-6996

**AOCS Honored Student Award** **524**
The American Oil Chemists' Society Honored Student Award is designed to encourage graduate student research in lipid chemistry and greater interest in AOCS. This award is sponsored by the AOCS Foundation. The award consists of a certificate, a $500 travel allowance, and complimentary lodging and registration for the recipient to attend the AOCS Annual Meeting, where they will present an award lecture.
*Requirements:* Applicants must be graduate students at any institution of higher learning, in any area of science and technology fostered by the AOCS, who are doing research toward an advanced degree.
*Restrictions:* Candidates must have not already received a graduate degree, or begun their career employment prior to May of the presentation year. An individual may receive an Honored Student Award only once.
*Geographic Focus:* All States, All Countries
*Date(s) Application is Due:* Nov 1
*Amount of Grant:* 500 USD
*Contact:* Moghis U. Ahmad, Awards Chairperson; (217) 693-4804 or (217) 359-2344; fax (217) 693-4849; moghis@jinapharma.com or awards@aocs.org
*Internet:* https://www.aocs.org/network-and-connect/awards-x1888#honored-students
*Sponsor:* American Oil Chemists' Society
2710 South Boulder
Urbana, IL 61802-6996

**AOCS Industrial Oil Products Division Student Award** **525**
The AOCS Industrial Oil Products Division supports professionals involved in research, development, engineering, marketing, and testing of industrial products and co-products from fats and oil, including glycerine and its derivatives. The Division's Student Award is given to a graduate student for travel to the Annual Meeting to present a paper. The Award includes a certificate for the recipient, an award lecture or poster presentation at the Annual Meeting, and $500 in travel funds for that meeting.
*Requirements:* Graduate students presenting within the Industrial Oil Products Division technical program at the Annual Meeting are eligible.
*Geographic Focus:* All States, All Countries
*Date(s) Application is Due:* Oct 15
*Amount of Grant:* 500 USD
*Contact:* Barb Semeraro, Awards Program Coordinator; (217) 693-4804; fax (217) 693-4849; barbs@aocs.org
*Internet:* www.aocs.org/Membership/content.cfm?ItemNumber=810
*Sponsor:* American Oil Chemists' Society
2710 South Boulder
Urbana, IL 61802-6996

**AOCS Lipid Oxidation and Quality Division Poster Competition** **526**
The Lipid Oxidation and Quality Division supports professionals in lipid oxidation with a major focus in food applications. The Lipid Oxidation and Quality Division Poster Competition is awarded to a student for travel to present a poster at the Annual Meeting. All poster abstracts submitted by students and selected for the Lipid Oxidation and Quality Poster session are considered for the competition. The Award includes a certificate, Award ribbon for poster display, complimentary division dinner ticket, and $300.
*Geographic Focus:* All States, All Countries
*Date(s) Application is Due:* Oct 15
*Amount of Grant:* 300 USD
*Contact:* Barb Semeraro, Awards Program Coordinator; (217) 693-4804; fax (217) 693-4849; barbs@aocs.org
*Internet:* www.aocs.org/Membership/content.cfm?ItemNumber=18281
*Sponsor:* American Oil Chemists' Society
2710 South Boulder
Urbana, IL 61802-6996

**AOCS Manuchehr Eijadi Award** **527**
The Manuchehr Eijadi Award is given to a graduate student doing research toward an advanced degree. The Award includes a $1,000 in addition to a plaque and certificate.
*Requirements:* Applicants must be previous recipients of the AOCS Honored Student Award, with graduate research focused on chemical engineering.
*Geographic Focus:* All States, All Countries
*Date(s) Application is Due:* Oct 15
*Amount of Grant:* 1,000 USD
*Contact:* Barb Semeraro, Awards Program Coordinator; (217) 693-4804; fax (217) 693-4849; barbs@aocs.org
*Internet:* www.aocs.org/Membership/content.cfm?ItemNumber=797
*Sponsor:* American Oil Chemists' Society
2710 South Boulder
Urbana, IL 61802-6996

**AOCS Peter and Claire G. Kalustian Award** **528**
The Peter and Claire G. Kalustian Award is given to a graduate student doing research toward an advanced degree. The Award includes $1,000 and a certificate.
*Requirements:* Applicants must be graduate students in any area of science and technology fostered by AOCS, who are doing research toward an advanced degree. Students must be previous AOCS Honored Students. Preference is given to students of Armenian descent.
*Geographic Focus:* All States, All Countries
*Date(s) Application is Due:* Oct 15
*Amount of Grant:* 1,000 USD
*Contact:* Barb Semeraro, Awards Program Coordinator; (217) 693-4804; fax (217) 693-4849; barbs@aocs.org
Silvana Martini, Awards Administration Chairperson; (435) 797-8136; fax (435) 797-2379; silvana.martini@usa.edu
*Internet:* www.aocs.org/Membership/content.cfm?ItemNumber=812
*Sponsor:* American Oil Chemists' Society
2710 South Boulder
Urbana, IL 61802-6996

**AOCS Phospholipid Best Paper Award** **529**
The Phospholipid Best Paper Award recognizes an outstanding published paper or monograph in the area of phospholipid research or applications for phospholipids. The Award includes a plaque for the main author, certificates for each author, funding the attend the Annual Meeting, and an award address at that meeting. Self nominations are welcomed.
*Requirements:* Papers eligible for nomination include those relating to phospholipids that have been published in the past year from any journal or monograph. Papers do not need to be from an AOCS publication.
*Geographic Focus:* All States, All Countries
*Date(s) Application is Due:* Oct 15
*Contact:* Barb Semeraro, Awards Program Coordinator; (217) 693-4804; fax (217) 693-4849; barbs@aocs.org
Silvana Martini, Awards Administration Chairperson; (435) 797-8136; fax (435) 797-2379; silvana.martini@usa.edu
*Internet:* www.aocs.org/Membership/content.cfm?ItemNumber=904
*Sponsor:* American Oil Chemists' Society
2710 South Boulder
Urbana, IL 61802-6996

**AOCS Processing Distinguished Service Award** 530

The AOCS Processing Distinguished Service Award recognizes and honors outstanding meritorious service to the oilseed processing industry. The Award includes a certificate and travel expenses to the Annual Meeting.
*Requirements:* Preference is given to AOCS members involved in Division, Society, or related organizations over a substantial amount of time. Candidates must meet the following criteria: significant contributions to the knowledge, technology, and history of the industry; demonstrated character, ethics and a reputation that puts the candidate above his or her peers; and service as a career and technical mentor.
*Geographic Focus:* All States, All Countries
*Date(s) Application is Due:* Oct 15
*Contact:* Barb Semeraro, Awards Program Coordinator; (217) 693-4804; fax (217) 693-4849; barbs@aocs.org
*Internet:* www.aocs.org/Membership/content.cfm?ItemNumber=922
*Sponsor:* American Oil Chemists' Society
2710 South Boulder
Urbana, IL 61802-6996

**AOCS Processing Division Student Award** 531

The AOCS Processing Division advances the processing knowledge and managerial skills by providing a forum of technical information and networking opportunities, to promote and facilitate communication and cooperation between members, and to mentor young professionals. The Processing Division Student Award is given to a graduate student researching oilseed handling preparation and extraction, refining and processing, oil products and packaging, feed ingredients, by-products utilization, safety and health, and environmental concerns. The Award includes a certificate for the recipient, an award lecture or poster presentation at the Annual Meeting, and a $500 honorarium.
*Requirements:* Graduate students presenting within the Processing Division's technical program at the Annual Meeting are eligible.
*Geographic Focus:* All States, All Countries
*Date(s) Application is Due:* Oct 15
*Amount of Grant:* 500 USD
*Contact:* Barb Semeraro, Awards Program Coordinator; (217) 693-4804; fax (217) 693-4849; barbs@aocs.org
*Internet:* www.aocs.org/Membership/content.cfm?ItemNumber=924
*Sponsor:* American Oil Chemists' Society
2710 South Boulder
Urbana, IL 61802-6996

**AOCS Protein and Co-Products Division Student Poster Competition** 532

The Protein and Co-Products Division Student Poster Competition is awarded to a student presenting a poster at the AOCS Annual Meeting. Awards are given to a first place winner for $300, and a second place winner for $200. Each student also receives a certificate, an award ribbon for their poster display, and a complimentary division dinner ticket.
*Requirements:* All poster abstracts submitted by students and selected for the Protein and Co-Products Poster session are considered for the competition.
*Geographic Focus:* All States, All Countries
*Date(s) Application is Due:* Oct 15
*Amount of Grant:* 500 USD
*Contact:* Barb Semeraro, Awards Program Coordinator; (217) 693-4804; fax (217) 693-4849; barbs@aocs.org
*Internet:* www.aocs.org/Membership/content.cfm?ItemNumber=18314
*Sponsor:* American Oil Chemists' Society
2710 South Boulder
Urbana, IL 61802-6996

**AOCS Ralph H. Potts Memorial Fellowship** 533

The Ralph H. Potts Memorial Fellowship is awarded to a graduate student doing research in fats and oils. The award is named for Ralph Potts in recognition for his work in fats, fatty acids, and derivatives, and the personal esteem of his colleagues. The recipient receives a plaque, a travel allowanced for the AOCS Annual Meeting, and a $2,000 honorarium. The application is available at the website.
*Requirements:* The applicant must be a graduate student at a North American university, with research involving fatty acids or their derivatives such as long chain alcohols, amines, or other nitrogen compounds. The applicant's research should fall in the category of synthesis, processing utilization, or characterization.
*Geographic Focus:* All States, Puerto Rico, Anguilla, Antigua & Barbuda, Aruba, Bahamas, Barbados, Belize, Bermuda, British Virgin Islands, Canada, Cayman Islands, Costa Rica, Cuba, Dominica, Dominican Republic, El Salvador, Grenada, Guatemala, Haiti, Honduras, Jamaica, Martinique, Mexico, Montserrat, Nicaragua, Panama, Saint Kitts And Nevis, St. Lucia, St. Vincent and the Grenadines, Trinidad and Tobago, Turks and Caicos Islands, Virgin Islands (U.S.)
*Date(s) Application is Due:* Oct 15
*Amount of Grant:* 2,000 USD
*Contact:* Barb Semeraro, Awards Program Coordinator; (217) 693-4804; fax (217) 693-4849; barbs@aocs.org
Silvana Martini, Awards Administration Chairperson; (435) 797-8136; fax (435) 797-2379; silvana.martini@usa.edu
*Internet:* www.aocs.org/Membership/content.cfm?ItemNumber=906
*Sponsor:* American Oil Chemists' Society
2710 South Boulder
Urbana, IL 61802-6996

**AOCS Ralph Holman Lifetime Achievement Award** 534

The AOCS Health and Nutrition Division promotes and facilitates communication and cooperation among professionals whose interests in lipid biochemistry and physiology relate to all aspects of dietary fats and health. The Ralph Holman Lifetime Achievement Award recognizes outstanding performance and meritorious contributions. The Award is named for Ralph Holman in recognition of his lifetime service to the study of essential fatty acids and their impact on health. The Award includes a framed orchid photograph, $1,000 honorarium, and an award lecture at the Annual Meeting.
*Requirements:* Candidates must have made significant lifetime and meritorious achievements in areas of interest to the Health and Nutrition Division of AOCS.
*Geographic Focus:* All States, All Countries
*Date(s) Application is Due:* Oct 15
*Amount of Grant:* 1,000 USD
*Contact:* Barb Semeraro, Awards Program Coordinator; (217) 693-4804; fax (217) 693-4849; barbs@aocs.org
Silvana Martini, Awards Administration Chairperson; (435) 797-8136; fax (435) 797-2379; silvana.martini@usa.edu
*Internet:* www.aocs.org/Membership/content.cfm?ItemNumber=807
*Sponsor:* American Oil Chemists' Society
2710 South Boulder
Urbana, IL 61802-6996

**AOCS Samuel Rosen Memorial Award** 535

The Samuel Rosen Memorial Award recognizes accomplishments in surfactant chemistry. The Award is named for Samuel Rosen, who worked for more than 40 years as an industrial chemist on the formulation of printing inks. The Award includes a plaque for the recipient, a $2,000 honorarium, and an award lecture at the Annual Meeting.
*Requirements:* Candidates must meet the following criteria for eligibility: significant advancements, cumulative advancements, or application of the principles of surfactant chemistry; and a scientist or technician who has worked the majority of his or her career in the industry, preferable within the last three years of the award recognition.
*Geographic Focus:* All States, All Countries
*Date(s) Application is Due:* Oct 15
*Amount of Grant:* 2,000 USD
*Contact:* Barb Semeraro, Awards Program Coordinator; (217) 693-4804; fax (217) 693-4849; barbs@aocs.org
Silvana Martini, Awards Administration Chairperson; (435) 797-8136; fax (435) 797-2379; silvana.martini@usa.edu
*Internet:* www.aocs.org/Membership/content.cfm?ItemNumber=931
*Sponsor:* American Oil Chemists' Society
2710 South Boulder
Urbana, IL 61802-6996

**AOCS Stephen S. Chang Award** 536

The Stephen S. Change Award is given annually to recognize a scientist or technologist who has made decisive accomplishments in research for the improvement or development of products related to lipids. The award includes a lecture at the Annual Meeting, a jade sculpture of a galloping horse, and an honorarium determined by the endowed fund investment return. Self nominations are welcomed.
*Requirements:* Candidates must have made significant and distinguished accomplishments in basic research that have been utilized by industries for the development or improvement of products related to lipids. Applicants may be recognized for either one major breakthrough or by an accumulation of publications.
*Geographic Focus:* All States, All Countries
*Date(s) Application is Due:* Oct 15
*Contact:* Barb Semeraro, Awards Program Coordinator; (217) 693-4804; fax (217) 693-4849; barbs@aocs.org
*Internet:* www.aocs.org/Membership/content.cfm?ItemNumber=788
*Sponsor:* American Oil Chemists' Society
2710 South Boulder
Urbana, IL 61802-6996

**AOCS Supelco/Nicholas Pelick-AOCS Research Award** 537

The Supelco/Nicholas Pelick-AOCS Research Award is given to a qualified scientist for outstanding original research on fats, oils, lipid chemistry, and/or biochemistry, the results of which have been published in technical papers of high quality. The recipient receives a plaque and a $10,000 honorarium. The award winner also gives a lecture at the AOCS Annual Meeting. The application is available at the website. Self nominations are welcomed.
*Requirements:* Candidates must have published technical papers of high quality Preference is given to individuals who are actively associated with research and have made influential discoveries in their fields.
*Geographic Focus:* All States, All Countries
*Date(s) Application is Due:* Oct 15
*Amount of Grant:* 10,000 USD
*Contact:* Barb Semeraro, Awards Program Coordinator; (217) 693-4804; fax (217) 693-4849; barbs@aocs.org
Silvana Martini, Awards Administration Chairperson; (435) 797-8136; fax (435) 797-2379; silvana.martini@usa.edu
*Internet:* www.aocs.org/Membership/content.cfm?ItemNumber=940
*Sponsor:* American Oil Chemists' Society
2710 South Boulder
Urbana, IL 61802-6996

**AOCS Surfactants and Detergents Division Student Award** 538

The AOCS Surfactants and Detergents Division encompasses the sciences and technologies associated with the development and production of surfactants, detergents, and soaps. The Surfactants and Detergents Division Student Award is given to a graduate student to help in their travel to present a paper at the Annual Meeting. The Award includes a certificate for the recipient(s), an award lecture or poster presentation at the Annual Meeting, and $500 in travel funds. Self nominations are welcomed.
*Requirements:* Graduate students presenting within the Surfactants and Detergents Division technical program at the AOCS Annual Meeting are eligible.
*Geographic Focus:* All States, All Countries
*Date(s) Application is Due:* Oct 15
*Amount of Grant:* 500 USD
*Contact:* Barb Semeraro, Awards Program Coordinator; (217) 693-4804; fax (217) 693-4849; barbs@aocs.org
Silvana Martini, Awards Administration Chairperson; (435) 797-8136; fax (435) 797-2379; silvana.martini@usa.edu
*Internet:* www.aocs.org/Membership/content.cfm?ItemNumber=942
*Sponsor:* American Oil Chemists' Society
2710 South Boulder
Urbana, IL 61802-6996

**AOCS Thomas H. Smouse Memorial Fellowship** 539

The Thomas H. Smouse Memorial Fellowship is awarded to a graduate student doing research in areas of interest to AOCS. The award includes a $10,000 stipend, plus $5,000 for travel to professional meetings and research expenditures related to the student's graduate program. Self nominations are welcomed.
*Requirements:* Individuals must be scholastically outstanding with interests and involvement outside the academic discipline.
*Geographic Focus:* All States, All Countries
*Date(s) Application is Due:* Oct 15
*Amount of Grant:* 15,000 USD
*Contact:* Barb Semeraro, Awards Program Coordinator; (217) 693-4804; fax (217) 693-4849; barbs@aocs.org
Silvana Martini, Awards Administration Chairperson; (435) 797-8136; fax (435) 797-2379; silvana.martini@usa.edu
*Internet:* www.aocs.org/Membership/content.cfm?ItemNumber=938
*Sponsor:* American Oil Chemists' Society
2710 South Boulder
Urbana, IL 61802-6996

**AOCS Timothy L. Mounts Award** 540

The Edible Applications Technology Division is comprised of professionals in the processing and utilization of lipids industry with primary focus on food applications. The Timothy L. Mounts Award recognizes research relating to the science and technology of edible oils, basic or applied in nature. The Award includes a plaque for the recipient, a $500 honorarium, and an award lecture at the Annual Meeting.
*Requirements:* Mid-career scientists engaged in edible oil research are eligible.
*Geographic Focus:* All States, All Countries
*Date(s) Application is Due:* Oct 15
*Amount of Grant:* 500 USD
*Contact:* Barb Semeraro, Awards Program Coordinator; (217) 693-4804; fax (217) 693-4849; barbs@aocs.org
Silvana Martini, Awards Administration Chairperson; (435) 797-8136; fax (435) 797-2379; silvana.martini@usa.edu
*Internet:* www.aocs.org/Membership/content.cfm?ItemNumber=831
*Sponsor:* American Oil Chemists' Society
2710 South Boulder
Urbana, IL 61802-6996

**AOCS Uses of Soybean Oil Award** 541

The AOCS Uses of Soybean Oil Award is sponsored by the United Soybean Board. The Award recognizes outstanding research into new industrial applications or uses for soybean oil. The Award includes a plaque for the recipient, a $3,000 honorarium, and award lecture at the Annual Meeting. Nomination information is available at the website.
*Geographic Focus:* All States, All Countries
*Date(s) Application is Due:* Oct 15
*Amount of Grant:* 3,000 USD
*Contact:* Barb Semeraro, Awards Program Coordinator; (217) 693-4804; fax (217) 693-4849; barbs@aocs.org
*Internet:* www.aocs.org/Membership/content.cfm?ItemNumber=943
*Sponsor:* American Oil Chemists' Society
2710 South Boulder
Urbana, IL 61802-6996

**AOCS Young Scientist Research Award** 542

The Young Scientist Research Award recognizes substantial research contributions in one of the following areas: agricultural microscopy; biotechnology; edible applications technology; health and nutrition; industrial oil products; lipid oxidation and quality; phospholipid; protein and co-products; and surfactants and detergents. The Award includes a plaque for the recipient, a $1,000 honorarium, travel and registration for the Annual Meeting, and award lecture at the Annual Meeting. Self nominations are welcomed.
*Requirements:* The recipient must be younger than 36 years of age by June 1 of the year of the presentation, or have received his or her highest degree within the previous ten years. The candidate must also have significant contributions to advances in basic or applied research, or hold substantial promise for its effect in the near future.
*Geographic Focus:* All States, All Countries
*Date(s) Application is Due:* Oct 15
*Amount of Grant:* 1,000 USD
*Contact:* Barb Semeraro, Awards Program Coordinator; (217) 693-4804; fax (217) 693-4849; barbs@aocs.org
Silvana Martini, Awards Administration Chairperson; (435) 797-8136; fax (435) 797-2379; silvana.martini@usa.edu
*Internet:* www.aocs.org/Membership/content.cfm?ItemNumber=944
*Sponsor:* American Oil Chemists' Society
2710 South Boulder
Urbana, IL 61802-6996

**APHL Emerging Infectious Diseases Fellowships** 543

The Emerging Infectious Diseases (EID) Laboratory Fellowship Program, sponsored by the Association of Public Health Laboratories (APHL) and the Centers for Disease Control and Prevention (CD.C.), trains and prepares scientists for careers in public health laboratories and supports public health initiatives related to infectious disease research. Two types of fellowships are offered: the EID Advanced Laboratory Training Fellowship is a one-year program designed for bachelor's or master's level scientists, with emphasis on the practical application of technologies, methodologies and practices related to emerging infectious diseases; the EID Laboratory Research Fellowship is a two-year program designed for doctoral level (Ph.D., M.D. or D.V.M.) scientists to conduct high-priority research in infectious diseases. Up to 30 pre-doctoral one-year and 15 post-doctoral two-year fellowships will be awarded. Fellows are placed in local, state or federal (CD.C.) public health laboratories throughout the US. Fellows are provided with a stipend, medical insurance, travel to the host laboratory and a professional development allowance. The fellowship program offers a wide variety of training and research experiences. Examples of projects include research in molecular genetics, pathogenesis, epidemiology and cell biology. Fellows may be trained in specific laboratory techniques such as real-time PCR, DNA sequencing and mass spectrometry. Fellows may receive specialized training and experience with a specific pathogen, or generalized training in influenza or STD surveillance, newborn screening or chemical and biological warfare agents. Other work may include diagnostic virology, bioanalytical chemistry, foodborne disease research including Pulsed Field Gel Electrophoresis (PFGE), clinical and environmental diagnostics and epidemiologic studies of disease outbreaks. Fellows also have opportunities to participate in research seminars, teleconferences, regional and national meetings and other continuing education opportunities. All fellows participate in an orientation session at CD.C. in Atlanta to gain a general understanding of the public health laboratory system and how it relates to infectious disease surveillance, prevention, research and control. All applicants are required to submit one original single-sided application to APHL by the application deadline. The application may be downloaded from the EID fellowship website in writeable pdf format. All required supporting documents (listed in the application) should be collected by the applicant and forwarded with the application to APHL. Deadline dates may vary from year to year. Applicants are encouraged to check the EID fellowship website or contact the Fellowship Manager to verify current the deadline date. APHL member laboratories may apply to host one or more EID Fellows. The link to the Host Laboratory Instructions and Application page is available from the EID Fellowship website.
*Requirements:* EID Advanced Laboratory Training Fellowship applicants must have completed a bachelor's or master's degree in microbiology, virology or a related discipline and/or completed an accredited medical technologist program prior to the start of the fellowship. EID Laboratory Research Fellowship applicants must have received a Ph.D., M.D., or D.V.M., or have completed all requirements for such a degree prior to the start of the fellowship. Applicants without a degree in a physical science will be considered if they have significant laboratory science and/or public health coursework or experience. All bachelor's and master's level candidates must have completed their degree by the time the fellowship begins (mid-September). Postdoctoral candidates may delay their start date to February 2012. Degree requirements must be completed by by this time. All appointees must attend the August Orientation session in Atlanta.
Restrictions: All applicants must be U.S. citizens.
*Geographic Focus:* All States
*Date(s) Application is Due:* Feb 11
*Amount of Grant:* 32,733 - 89,780 USD
*Contact:* Heather Roney, Manager, Fellowship Program; (240) 485-2778 or (240) 485-2745; fax (240) 485-2700; fellowships@aphl.org or heather.roney@aphl.org
*Internet:* www.aphl.org/mycareer/fellowships/eid/pages/default.aspx
*Sponsor:* Association of Public Health Laboratories
8515 Georgia Avenue, Suite 700
Silver Spring, MD 20910

**Appalachian Regional Commission Health Care Grants** 544

Access to comprehensive, affordable health care is vital to social and economic growth in the Appalachian Region. The Appalachian Regional Commission's health projects focus on community-based efforts to encourage health-promotion and disease-prevention activities. Strategies include: using best practices in public health to develop targeted approaches to wellness and disease prevention; supporting partnerships that educate children and families about basic health risks; using telecommunications and other technology to reduce the high cost of health-care services; and encouraging the development and expansion of health professional education services within the Region. ARC health care grants have helped provide equipment for hospitals and rural clinics, training for health care professionals, and support for community-based health education

activities. ARC also works with other organizations to address the high incidence of life-threatening diseases in the Region, as in its ongoing partnership with the Centers for Disease Control and Prevention in diabetes and cancer education, prevention, and treatment programs in the Region's distressed counties.
*Requirements:* States, and through states, public bodies and private nonprofit organizations are eligible to apply.
*Restrictions:* Generally, ARC grants are limited to 50% of project costs.
*Geographic Focus:* Alabama, Georgia, Kentucky, Maryland, Mississippi, New York, North Carolina, Ohio, Pennsylvania, South Carolina, Tennessee, Virginia, West Virginia
*Contact:* Jill Wilmoth, Budget and Program Specialist; (202) 884-7668 or (202) 884-7700; fax (202) 884-7691; jwilmoth@arc.gov
*Internet:* https://www.arc.gov/program_areas/index.asp?PROGRAM_AREA_ID=16
*Sponsor:* Appalachian Regional Commission
1666 Connecticut Avenue NW, Suite 700
Washington, D.C. 20009-1068

**APS/British Academy Fellowship for Research in London** **545**
In collaboration with the British Academy, the American Philosophical Society offers an exchange postdoctoral fellowship for a minimum of one and a maximum of two months' research in the archives and libraries of London during the upcoming year. This award includes travel expenses between the United States and the United Kingdom, as well as a monthly subsistence paid by the APS. Candidates should specify that they are asking for the British Academy Fellowship and apply by October 1; applicants not selected for the British Academy Fellowship will be considered for a Franklin Research Grant. Independent scholars and faculty members at all four-year and two-year research and non-research institutions are welcome to apply. Funding is offered up to a maximum of $6,000.
*Requirements:* Applicants are expected to have a doctorate or to have published work of doctoral character and quality. American citizens and residents of the United States may use their Franklin awards at home or abroad. Foreign nationals not affiliated with a U.S. institution must use their Franklin awards for research in the United States. Applicants who have previously received a Franklin grant may reapply after an interval of two years.
*Restrictions:* Ph.D. candidates are not eligible to apply, but the Society is particularly interested in supporting the work of young scholars who have recently received the doctorate. This Fellowship is not intended to meet the expenses of attending conferences or the costs of publication. The Society does not pay overhead or indirect costs to any institution, and grant funds are not to be used to pay income tax on the award. Grants will not be made to replace salary during a leave of absence or earnings from summer teaching; pay living expenses while working at home; cover the costs of consultants or research assistants; or purchase permanent equipment such as computers, cameras, tape recorders, or laboratory apparatus.
*Geographic Focus:* All States, All Countries
*Date(s) Application is Due:* Oct 1
*Amount of Grant:* Up to 6,000 USD
*Contact:* Deborah Mikula; (215) 440-3429 or (215) 440-3400; lmusumeci@amphilsoc.org
*Internet:* https://amphilsoc.org/grants/franklin
*Sponsor:* American Philosophical Society
104 South Fifth Street
Philadelphia, PA 19106-3387

**APS/Institute for Advanced Studies in the Humanities Fellowship for Research in Edinburgh** **546**
In collaboration with the Institute for Advanced Studies in the Humanities (IASH) at the University of Edinburgh, the APS offers a visiting fellowship of between two and four months for research in Edinburgh in the upcoming calendar year in any aspect of the humanities and social sciences. To maximize the benefits of the fellowship, applicants are strongly encouraged to schedule their visit to overlap with one of the two main teaching semesters (January through March and September through December). This award includes travel expenses between the United States and the United Kingdom, a private office, library and research facilities at the IASH, and a monthly subsistence paid by the APS. Travel expenses and the monthly subsistence amount will not exceed a maximum of $6,000. Candidates should specify that they are asking for the IASH Fellowship and apply by October 1; applicants not selected for the IASH Fellowship will be considered for a Franklin Research Grant.
*Requirements:* Applicants are expected to have a doctorate or to have published work of doctoral character and quality. American citizens and residents of the United States may use their Franklin awards at home or abroad. Foreign nationals not affiliated with a U.S. institution must use their Franklin awards for research in the United States. Applicants who have previously received a Franklin grant may reapply after an interval of two years.
*Restrictions:* Ph.D. candidates are not eligible to apply, but the Society is particularly interested in supporting the work of young scholars who have recently received the doctorate. This Fellowship is not intended to meet the expenses of attending conferences or the costs of publication. The Society does not pay overhead or indirect costs to any institution, and grant funds are not to be used to pay income tax on the award. Grants will not be made to replace salary during a leave of absence or earnings from summer teaching; pay living expenses while working at home; cover the costs of consultants or research assistants; or purchase permanent equipment such as computers, cameras, tape recorders, or laboratory apparatus.
*Geographic Focus:* All States
*Date(s) Application is Due:* Oct 1
*Amount of Grant:* Up to 6,000 USD
*Contact:* Linda Musumeci, Director of Grants and Fellowships; (215) 440-3429 or (215) 440-3400; lmusumeci@amphilsoc.org
*Internet:* https://amphilsoc.org/grants/franklin
*Sponsor:* American Philosophical Society
104 South Fifth Street
Philadelphia, PA 19106-3387

**APS/Phoenix Suns Energy and Environmental Minigrants** **547**
The program supports innovative, student-based educational projects that enhance learning with the objective of improving student achievement. Projects should have the ability of being replicated or adapted by other schools. Projects addressing energy, the environment, reading, mathematics or character education will be considered. Annual deadline dates may vary; contact program staff for exact dates.
*Requirements:* The program is open to educators who teach in public schools in the Arizona Public Service territory. Funds may be applied to equipment, materials and field trips that would not normally be provided by the school or school district.
*Geographic Focus:* Arizona
*Date(s) Application is Due:* Oct 15
*Amount of Grant:* 2,500 USD
*Contact:* Louise Moskowitz, (602) 250-2291; fax (602) 250-2113; louise.moskowitz@aps.com
*Internet:* www.aps.com/my_community/EducationTools/LearningCenter_1.html
*Sponsor:* Arizona Public Service Corporation
P.O. Box 53999, MS 8010
Phoenix, AZ 85072-3999

**APSA Congressional Fellowships for Communications Scholars and Journalists** **548**
The purpose of this fellowship is to give early- to mid-career scholars and journalists an opportunity to learn more about Congress and the legislative process through direct participation. Ideal candidates should have an analytical interest in politics, communications, and public policy, and show promise of making a significant contribution to the public understanding of the political process. A comprehensive orientation begins in November. Office assignments as full-time legislative aides in the House of Representatives and/or Senate run from December to August. Fellows receive a stipend of $43,000, plus a small travel allowance.
*Requirements:* All candidates must be U.S. citizens or permanent U.S. residents and demonstrate an interest in communications and the policymaking process. Scholars must have a Ph.D completed within the last 15 years or will have defended their dissertation by November of the fellowship year. Print journalists must have a bachelor's degree and two to ten years of continuous, full-time professional experience in newspaper or magazine at the time of application (background may be in reporting or editing). Broadcast journalists must have a bachelor's degree and two to ten years of continuous, full-time professional experience in radio or television at the time of application (background may be in reporting, producing, directing or writing). Preference is given to candidates without extensive Washington experience.
*Geographic Focus:* All States
*Date(s) Application is Due:* Dec 1
*Amount of Grant:* 43,000 USD
*Contact:* Laura Leddy, Program Director; (202) 483-2512; fax (202) 483-2657; lleddy@apsanet.org or cfp@apsanet.org
*Internet:* www.apsanet.org/cfp
*Sponsor:* American Political Science Association
1527 New Hampshire Avenue, NW
Washington, D.C. 20036-1206

**APSA Congressional Fellowships for Federal Executives** **549**
The purpose of this fellowship is to give senior-level federal executives an opportunity to learn more about Congress and the legislative process through direct participation. A comprehensive orientation begins in November. Office assignments as full-time legislative aides in the House of Representatives and/or Senate run from December to August. Fellows continue to receive their regular salaries from their agencies during the fellowship. Agencies with substantial activity with foreign governments are additionally invited to nominate candidates to participate in the Foreign Affairs Seminar. Foreign Affairs Fellows attend an intensive eight-week course at the Johns Hopkins University School of Advanced International Studies in Washington from September to November. They then join the rest of the Fellows for the November general orientation.
*Requirements:* All candidates must have a minimum grade of GS-13 or equivalent at time of application, at least two years of federal service in the executive branch, and long-term career goals relevant to a congressional experience. Additionally, candidates must demonstrate: a commitment to public service; adaptability to new, diverse working environments; and an interest in the legislative process and public affairs.
*Geographic Focus:* All States
*Date(s) Application is Due:* Mar 31
*Contact:* Laura Leddy, Program Director; (202) 483-2512; fax (202) 483-2657; lleddy@apsanet.org or cfp@apsanet.org
*Internet:* www.apsanet.org/cfp
*Sponsor:* American Political Science Association
1527 New Hampshire Avenue, NW
Washington, D.C. 20036-1206

**APSA Congressional Fellowships for International Scholars** **550**
The Congressional Fellowships for International Scholars program is the nation's oldest and most prestigious congressional fellowship. More than fifty years later, the program remains devoted to its original objective of expanding knowledge and awareness of Congress. For nine months, select political scientists, journalists, doctors, federal executives and international scholars gain hands on understanding of the legislative process by serving on congressional staffs. Through this unique opportunity, the Association enhances public understanding of policy-making and improves the quality of scholarship, teaching and reporting on American national politics. The CFP is currently investigating the possibilities of expanding the program. Presently, however, opportunities are limited to: German Fellows—selected by the German Marshall Fund

of the United States; Asian Fellows—selected by the Asia Foundation; Senior Fulbright Scholars—selected by the State Department Bureau of Education and Cultural Affairs and the Council for International Exchange of Scholars; and Australian Fellows—selected by the American Australian Association.
*Geographic Focus:* All States
*Date(s) Application is Due:* Jan 31
*Contact:* Laura Leddy, Program Director; (202) 483-2512; fax (202) 483-2657; lleddy@apsanet.org or cfp@apsanet.org
*Internet:* www.apsanet.org/cfp
*Sponsor:* American Political Science Association
1527 New Hampshire Avenue, NW
Washington, D.C. 20036-1206

**APSA Congressional Fellowships for Journalists** 551
The purpose of this fellowship is to give early- to mid-career journalists an opportunity to learn more about Congress and the legislative process through direct participation. A comprehensive orientation begins in November. Office assignments as full-time legislative aides in the House of Representatives and/or Senate run from December to August. Fellows receive a stipend of $43,000, plus a small travel allowance.
*Requirements:* All candidates must be U.S. citizens or permanent U.S. residents and hold a bachelor's degree. Print journalists must have two to ten years of continuous, full-time professional experience in newspaper or magazine at the time of application (background may be in reporting or editing). Broadcast journalists must have two to ten years of continuous, full-time professional experience in radio or television at the time of application (background may be in reporting, producing, directing or writing). Preference is given to candidates with background in political reporting but without extensive Washington experience.
*Geographic Focus:* All States
*Date(s) Application is Due:* Dec 1
*Amount of Grant:* 43,000 USD
*Contact:* Laura Leddy; (202) 483-2512; fax (202) 483-2657; lleddy@apsanet.org
*Internet:* www.apsanet.org/cfp
*Sponsor:* American Political Science Association
1527 New Hampshire Avenue, NW
Washington, D.C. 20036-1206

**APSA Congressional Fellowships for Political Scientists** 552
The purpose of Congressional Fellowships for Political Scientists is to give early- to mid-career political scientists an opportunity to learn more about Congress and the legislative process through direct participation. Office assignments as full-time legislative aides in the House of Representatives and/or Senate run from December to August. Fellows are selected in a nationwide competition. Candidates must first submit their applications for a paper review before being invited before an outside selection panel. Interviews take place at the APSA offices in Washington, D.C. APSA reimburses travel to and from interviews and arranges for hotel accommodations. Applicants not currently in the United States must fund their own transportation, should they be selected for an interview. Interviews will be announced in January. All finalists will be notified in February. Further application guidelines are available online.
*Requirements:* Eligibility qualifications include: U.S. citizenship or permanent U.S. residency; a Ph.D completed within the last 15 years or a dissertation near completion (ABD); and a scholarly interest in Congress and the policy-making process.
*Geographic Focus:* All States
*Date(s) Application is Due:* Dec 1
*Amount of Grant:* 43,000 USD
*Contact:* Laura Leddy; (202) 483-2512; fax (202) 483-2657; lleddy@apsanet.org
*Internet:* www.apsanet.org/cfp
*Sponsor:* American Political Science Association
1527 New Hampshire Avenue, NW
Washington, D.C. 20036-1206

**APSA Congressional Health and Aging Policy Fellowships** 553
Supported by the Atlantic Philanthropies and administered by Columbia University, this national program seeks to provide professions in health and aging with the experience and skills necessary to make a positive contribution to the development and implementation of health policies that affect older Americans. The program will offer two different tracks for individual placement: a residential track that includes a nine-to-12-month placement in Washington, D.C. or at a state agency (as a legislative assistant in Congress, a professional staff member in an executive agency or in a policy organization; and a non-residential track that includes a health policy project and brief placement(s) throughout the year at relevant sites. Core program components focused on career development and professional enrichment are provided for fellows in both tracks.
*Requirements:* The program is open to physicians, nurses and social workers at all career stages (early, mid, and late) with demonstrated commitment to health and aging issues and a desire to be involved in health policy at the federal, state or local level. Other professionals with clinical backgrounds (e.g. pharmacists, dentists, clinical psychologists) working in the field of health and aging are also eligible to apply.
*Geographic Focus:* All States
*Contact:* Jeffrey R. Biggs, Program Director; (202) 483-2512, ext. 521; fax (202) 483-2657; jbiggs@apsanet.org or cfp@apsanet.org
*Internet:* www.apsanet.org/content_50835.cfm
*Sponsor:* American Political Science Association
1527 New Hampshire Avenue, NW
Washington, D.C. 20036-1206

**APSA E.E. Schattschneider Award** 554
The E.E. Schattschneider prize is awarded annually for the best doctoral dissertation completed and accepted during that year or the previous year in the field of American government. This award fund was established in 1971. It carries a prize of $750. This award was set up in honor of Elmer Eric Schattschneider, a former APSA president, and widely published and respected political scientist.
*Requirements:* Political science departments are invited to nominate not more than one doctoral dissertation for this award. The dissertations must have been completed and accepted during the two calendar years preceding the date of the prize.
*Geographic Focus:* All States
*Date(s) Application is Due:* Jan 15
*Contact:* Lauren Barry, Awards Coordinator; (202) 483-2512, ext. 631; fax (202) 483-2657; lbarry@apsanet.org or awards@apsanet.org
*Internet:* www.apsanet.org/content_4125.cfm
*Sponsor:* American Political Science Association
1527 New Hampshire Avenue, NW
Washington, D.C. 20036-1206

**APSA Edward S. Corwin Award** 555
The Edward S. Corwin prize is awarded annually for the best dissertation in the field of public law. It carries a prize of $750. The Corwin award is for the best doctoral dissertation completed and accepted during that year or the previous year in the field of public law, broadly defined to include the judicial process, judicial behavior, judicial biography, courts, law, legal systems, the American constitutional system, civil liberties, or any other substantial area, or any work which deals in a significant fashion with a topic related to or having substantial impact on the American Constitution. Edward S. Corwin was a former Association president who is nationally known and widely published. He consulted with many other academics as well as politicians involved with constitutional issues, most notably when he publicly supported Franklin D. Roosevelt's Supreme Court reorganization (court packing) plan. Nominations should be submitted each year by the February 13 deadline.
*Requirements:* Political science departments are invited to nominate not more than one doctoral dissertation for this award. The dissertations must have been completed and accepted during the two calendar years preceding the date of the prize.
*Geographic Focus:* All States
*Date(s) Application is Due:* Feb 13
*Amount of Grant:* 750 USD
*Contact:* Betsy Super, Deputy Director; (202) 483-2512; fax (202) 483-2657; bsuper@apsanet.org or awards@apsanet.org
*Internet:* www.apsanet.org/PROGRAMS/APSA-Awards/Edward-S-Corwin-Award
*Sponsor:* American Political Science Association
1527 New Hampshire Avenue, NW
Washington, D.C. 20036-1206

**APSA Frank J. Goodnow Award** 556
The Frank Johnson Goodnow Award was established by the APSA Council in 1996 to honor service to the community of teachers, researchers, and public servants who work in the many fields of politics. Frank J. Goodnow, the first president of the American Political Science Association, a pioneer in the development of judicial politics, and former president of Johns Hopkins University, is an exemplar of the public service and volunteerism that this award represents. Each year the Association solicits nominations from members of the Council, former winners, and APSA members. Nominations are then considered by the Goodnow Committee. One or more recipients may be selected, and in some years no recipients may be named. Recipients are approved by the APSA Administrative Committee. Nominations should be submitted each year by the February 13 deadline.
*Geographic Focus:* All States
*Date(s) Application is Due:* Feb 13
*Contact:* Betsy Super, Deputy Director; (202) 483-2512; fax (202) 483-2657; bsuper@apsanet.org or awards@apsanet.org
*Internet:* www.apsanet.org/PROGRAMS/APSA-Awards/Frank-J-Goodnow-Award
*Sponsor:* American Political Science Association
1527 New Hampshire Avenue, NW
Washington, D.C. 20036-1206

**APSA Gabriel A. Almond Award** 557
The Gabriel A. Almond prize is awarded annually for the best dissertation in the field of comparative politics. It carries a prize of $750. The award was created in recognition of Gabriel Almond's contributions to the discipline, profession, and Association. He was a long time faculty member at Stanford University and former APSA President (1966). Almond's scholarly work contributed directly to the development of theory in comparative politics and brought together work on the developing areas and Western Europe that prevented splintering into an array of disparate areas studies. Nominations should be submitted each year by the February 13 deadline.
*Requirements:* Political science departments are invited to nominate not more than one doctoral dissertation for this award. The dissertations must have been completed and accepted during the two calendar years preceding the date of the prize.
*Geographic Focus:* All States
*Date(s) Application is Due:* Feb 13
*Amount of Grant:* 750 USD
*Contact:* Betsy Super; (202) 483-2512; fax (202) 483-2657; bsuper@apsanet.org
*Internet:* www.apsanet.org/almondaward
*Sponsor:* American Political Science Association
1527 New Hampshire Avenue, NW
Washington, D.C. 20036-1206

**APSA Harold D. Lasswell Award** 558

Harold D. Lasswell's extensive and wide-ranging books and essays are extraordinarily rich sources of ideas, methods, and topics for the study of political behavior. Whether and how the legacy of his writings is used by contemporary political scientists and theorists is reported here by way of an investigation of references to his work appearing in mainstream political science journals (available through JSTOR) for the 17 years following the end of his academic career. We find that most references to Lasswell are superficial (perfunctory, suggestive, deferential), although a few are more substantial (critical, extending). We conclude that Lasswell's legacy is today undervalued and underused, to the discipline's detriment. The Harold D. Lasswell prize is awarded annually for the best dissertation in the field of public policy. The award is co-sponsored by the Policy Studies Organization. It carries a prize of $1,000. Nominations should be submitted each year by the February 13 deadline.
*Requirements:* Political science departments are invited to nominate not more than one doctoral dissertation for this award. The dissertations must have been completed and accepted during the two calendar years preceding the date of the prize.
*Geographic Focus:* All States
*Date(s) Application is Due:* Feb 13
*Amount of Grant:* 1,000 USD
*Contact:* Betsy Super, Deputy Director; (202) 483-2512; fax (202) 483-2657; bsuper@apsanet.org or awards@apsanet.org
*Internet:* www.apsanet.org/PROGRAMS/APSA-Awards/Harold-D-Lasswell-Award
*Sponsor:* American Political Science Association
1527 New Hampshire Avenue, NW
Washington, D.C. 20036-1206

**APSA Leonard D. White Award** 559

Leonard Dupee White (17 January 1891 - 23 February 1958) was a historian of the field of public administration in the United States.[1] His technique was to study administration in the context of grouped U.S. presidential periods. An important founder of the field, White worked at the University of Chicago after service in the administrations of Franklin D. Roosevelt. The Leonard D. White prize is awarded annually for the best dissertation successfully defended during the previous two years in the field of public administration. The award is supported by the University of Chicago. It carries a prize of $750.
*Requirements:* Political science departments are invited to nominate not more than one doctoral dissertation for this award. The dissertations must have been completed and accepted during the two calendar years preceding the date of the prize.
*Geographic Focus:* All States
*Date(s) Application is Due:* Jan 15
*Amount of Grant:* 750 USD
*Contact:* Lauren Barry, Awards Coordinator; (202) 483-2512, ext. 631; fax (202) 483-2657; lbarry@apsanet.org or awards@apsanet.org
*Internet:* www.apsanet.org/content_4127.cfm
*Sponsor:* American Political Science Association
1527 New Hampshire Avenue, NW
Washington, D.C. 20036-1206

**APSA Leo Strauss Award** 560

The Leo Strauss prize is awarded annually for the best dissertation in the field of political philosophy. It carries a prize of $750. The Fund was developed by former students of Strauss' who sought to recognize his extraordinary influence on generations of students and his contributions to the field of political philosophy. He was a major figure in the department of political science at the University of Chicago, where he taught from 1949 to 1967.
*Requirements:* Political science departments are invited to nominate not more than one doctoral dissertation for this award. The dissertations must have been completed and accepted during the two calendar years preceding the date of the prize.
*Geographic Focus:* All States
*Date(s) Application is Due:* Jan 15
*Amount of Grant:* 750 USD
*Contact:* Lauren Barry, Awards Coordinator; (202) 483-2512, ext. 631; fax (202) 483-2657; lbarry@apsanet.org or awards@apsanet.org
*Internet:* www.apsanet.org/content_4126.cfm
*Sponsor:* American Political Science Association
1527 New Hampshire Avenue, NW
Washington, D.C. 20036-1206

**APSA Merze Tate Award** 561

In 1956, Helen Dwight Reid, a political scientist who taught at Bryn Mawr and the University of Buffalo, established a nonprofit foundation to support projects in education, international affairs, and the sciences. The Helen Dwight Reid prize was awarded annually for the best dissertation successfully defended during the previous two years in the field of international relations, law, and politics. In 2010, the Helen Dwight Reid Foundation ceased to exist, and sponsorship of the award ended. The award was recently renamed the Merze Tate Award, after the first African American woman studying international relations to receive a doctoral degree in Government. In choosing to name the award after Merze Tate, the committee noted that "her perseverance in the face of significant structural obstacles is inspiring and particularly meaningful for a dissertation award." The Award carries a prize of $750. Nominations should be submitted each yea the February 13 deadline.
*Requirements:* Political science departments are invited to nominate not more than one doctoral dissertation for this award. The dissertations must have been completed and accepted during the two calendar years preceding the date of the prize.
*Geographic Focus:* All States
*Date(s) Application is Due:* Feb 13
*Amount of Grant:* 750 USD
*Contact:* Betsy Super, Deputy Director; (202) 483-2512; fax (202) 483-2657; bsuper@apsanet.org or awards@apsanet.org
*Internet:* www.apsanet.org/awards/merze-tate
*Sponsor:* American Political Science Association
1527 New Hampshire Avenue, NW
Washington, D.C. 20036-1206

**APSA Minority Fellowships** 562

The program is designed primarily for minority students applying to enter a doctoral program in political science for the first time. The association has re-focused and increased its efforts to assist minority students in completing their doctorates by concentrating not only on the recruitment of minorities, but also on the retention of these groups within the profession. The program designates six stipend minority fellows each year. Additional applicants who do not receive funds from the association may also be recognized and recommended for admission and financial support to graduate political science programs. Fellows with stipends receive a fellowship that is disbursed in two payments - one at the end of their first graduate year and one at the end of their second - provided that they remain in good academic standing. Awards are based on students' undergraduate course work, GPA, extracurricular activities, GRE scores, and recommendations from faculty. Guidelines are available online.
*Requirements:* Applicants must: be members of one of the following racial/ethnic minority groups - African Americans, Latinos/as, and Native Americans (federal and state recognized tribes); demonstrate an interest in teaching and potential for research in political science; be a U.S. citizen at time of award; and demonstrate financial need.
*Geographic Focus:* All States
*Date(s) Application is Due:* Oct 2
*Amount of Grant:* 4,000 USD
*Contact:* Jeffrey R. Biggs, Program Director; (202) 483-2512, ext. 521; fax (202) 483-2657; jbiggs@apsanet.org or cfp@apsanet.org
Shaunda Ragland, Program Associate; (202) 483-2512, ext. 362; fax (202) 483-2657; sragland@apsanet.org
*Internet:* www.apsanet.org/content_3284.cfm
*Sponsor:* American Political Science Association
1527 New Hampshire Avenue, NW
Washington, D.C. 20036-1206

**APSA Presidency Research Fund Fellowships** 563

The Presidency Research Fund Fellowship is designed to support research on the American presidency. With an interest in explaining relationships, institutions, and environment surrounding the President, the Fund will support the work of people whose scholarly research brings them to the Washington area. Whether scholars plan to conduct archival work, library research, or interviews with key officials, the fellowship will provide a place at the Centennial Center where people are supported as they conduct their work. Applications will be reviewed, and selection made by, a three-member committee appointed by the APSA president.
*Geographic Focus:* All States
*Contact:* Jeffrey R. Biggs, Program Director; (202) 483-2512, ext. 521; fax (202) 483-2657; jbiggs@apsanet.org
*Internet:* www.apsanet.org/content_6335.cfm
*Sponsor:* American Political Science Association
1527 New Hampshire Avenue, NW
Washington, D.C. 20036-1206

**APSA Ralph J. Bunche Award** 564

Ralph Johnson Bunche was an American political scientist, academic, and diplomat who received the 1950 Nobel Peace Prize for his late 1940s mediation in Israel. He was the first African American to be so honored in the history of the prize. He was involved in the formation and administration of the United Nations. In 1963, he was awarded the Medal of Freedom by President John F. Kennedy. For more than two decades, Bunche served as chair of the Department of Political Science at Howard University, where he also taught generations of students. He served as a member of the Board of Overseers of his alma mater, Harvard University, as a member of the board of the Institute of International Education, and as a trustee of Oberlin College, Lincoln University, and New Lincoln School. In August 2008, the United States National Archives and Records Administration made public the fact that Ralph Bunche had joined the U.S. Office of Strategic Services (O.S.S.) – the precursor organization to the Central Intelligence Agency – during World War II. The Ralph Bunche Award, named in his honor, is given annually for the best scholarly work in political science that explores the phenomenon of ethnic and cultural pluralism. The Award carries a prize of $1,000, and nominations should be submitted each year by the February 13 deadline.
*Geographic Focus:* All States
*Date(s) Application is Due:* Feb 13
*Amount of Grant:* 1,000 USD
*Contact:* Betsy Super, Deputy Director; (202) 483-2512; fax (202) 483-2657; bsuper@apsanet.org or awards@apsanet.org
*Internet:* www.apsanet.org/PROGRAMS/APSA-Awards/Ralph-J-Bunche-Award
*Sponsor:* American Political Science Association
1527 New Hampshire Avenue, NW
Washington, D.C. 20036-1206

**APSA Robert Wood Johnson Foundation Health Policy Fellowships** 565
The purpose of this fellowship is to give outstanding mid-career health professionals in academic and community-based settings an opportunity to learn more about the legislative process through direct participation. Fellows actively contribute to the formulation of national health policies and accelerate their careers as leaders in health policy. Initiated in 1973, the program is funded by The Robert Wood Johnson Foundation (RWJF)and conducted by the Institute of Medicine (IOM) of the National Academy of Sciences. The application process typically opens in September and the application deadline typically occurs in mid-November of each year. Fellows must commit 100 percent of their time to program activities during the first 12 months. Up to six grants of up to $165,000 each will be made.
*Geographic Focus:* All States
*Date(s) Application is Due:* Nov 14
*Amount of Grant:* Up to 165,000 USD
*Contact:* Jeffrey R. Biggs, Program Director; (202) 334-1506 or (202) 483-2512, ext. 521; fax (202) 483-2657; jbiggs@apsanet.org or info@healthpolicyfellows.org
*Internet:* www.apsanet.org/content_3544.cfm
*Sponsor:* American Political Science Association
1527 New Hampshire Avenue, NW
Washington, D.C. 20036-1206

**APSA Small Research Grants** 566
The APSA Small Research Grant Program supports research in all fields of political science. The intent of these grants is to support the research of political scientists who are not employed at PhD granting department in the field and to help further the careers of these scholars. Grant recipients in the last five years have published over 12 books, 16 refereed articles, 8 book chapters, many working papers, and a large number of conference presentations. Others are still working on book, article, and conference projects and many recipients report great benefits for their students. Students have co-authored projects, worked as research assistants, and benefited through the professors' classroom use of their research. Many recipients also were able to use the APSA grant as seed money to get additional funding.
*Requirements:* Applicants must be APSA members at the time of application and at the time funds are dispersed. In addition, the principal investigator and any co-author must be one of the following: a faculty member at a college or university that does not award a PhD in Political Science, Public Administration, Public Policy, International Relations, Government, or Politics and whose primary appointment is in one of these departments; or a political scientist not affiliated with an academic institution and is either: unemployed; or working in a research organization such as a think tank.
*Restrictions:* This grant does not provide support for dissertation research or writing and graduate students are not eligible to apply.
*Geographic Focus:* All States
*Date(s) Application is Due:* Feb 3
*Amount of Grant:* 2,500 USD
*Contact:* Sean Twombly, Director, Member Services and Development; (202) 483-2512, ext. 513; fax (202) 483-2657; twombly@apsanet.org
*Internet:* www.apsanet.org/content_9222.cfm
*Sponsor:* American Political Science Association
1527 New Hampshire Avenue, NW
Washington, D.C. 20036-1206

**APSA Special Fund for the Study of Women and Politics** 567
The Special Fund for the Study of Women and Politics is designed to provide scholars conducting research on issues of women and politics with supplemental funding. Monies awarded through this fund may support resident or non-resident stays at the Centennial Center. If applying as a Visiting Scholar, a separate application file is not necessary. Interest in this assistance may be noted on the Centennial Center Visiting Scholar Application. Non-resident scholars interested in funding should complete the Non-Resident Scholar Application Form.
*Geographic Focus:* All States, Florida
*Contact:* Jeffrey R. Biggs, Program Director; (202) 483-2512, ext. 521; fax (202) 483-2657; jbiggs@apsanet.org
*Internet:* www.apsanet.org/content_6334.cfm
*Sponsor:* American Political Science Association
1527 New Hampshire Avenue, NW
Washington, D.C. 20036-1206

**APSA Travel Grants** 568
APSA provides travel grants to assist some members in attending the Annual Meeting. U.S. graduate students, international graduate students studying in the U.S., unemployed APSA members, and international scholars are encouraged to apply. Priority will be given to presenters at the Annual Meeting, first-time applicants, or those who have not received a travel grant for three years. Application forms are available upon request.
*Requirements:* Only applicants who are APSA members and pre-registered for the APSA Annual Meeting are eligible to receive travel grants.
*Restrictions:* Travel grants will be awarded as cash only; hotel rooms will no longer be included as part of the award.
*Geographic Focus:* All States
*Date(s) Application is Due:* Jun 3
*Amount of Grant:* Up to 700 USD
*Contact:* Jeffrey R. Biggs, Program Director; (202) 483-2512, ext. 521; fax (202) 483-2657; jbiggs@apsanet.org or travelgrants@apsanet.org
*Internet:* www.apsanet.org/content_11228.cfm
*Sponsor:* American Political Science Association
1527 New Hampshire Avenue, NW
Washington, D.C. 20036-1206

**APSA Warren E. Miller Fellowship in Electoral Politics** 569
Warren E. Miller was an American political scientist in the field of American political behavior. Best known as a co-author of the seminal book, The American Voter, alongside Angus Campbell, Philip Converse and Donald Stokes, which provided the basis for the social-psychological Michigan school of thought in American political behavior, Miller had a full and impressive career as a political scientist. Miller was a professor of political science at the University of Michigan from 1951 until 1981 when he left for Arizona State University where he worked until his death in 1999. The Warren E. Miller Fellowship honors the outstanding contributions of Miller to the study of electoral politics and other areas of political science inquiry. The fellowship will support research residencies in national and comparative electoral politics. The Miller fellowships will support scholars for a period one to six months at the Centennial Center.
*Geographic Focus:* All States
*Contact:* Laura Leddy, Program Director; (202) 483-2512; fax (202) 483-2657; lleddy@apsanet.org or cfp@apsanet.org
*Internet:* www.apsanet.org/cfp
*Sponsor:* American Political Science Association
1527 New Hampshire Avenue, NW
Washington, D.C. 20036-1206

**APSA William Anderson Award** 570
The William Anderson prize is awarded annually for the best dissertation in the general field of federalism or intergovernmental relations, state and local politics. It carries a prize of $750. The award was set up in honor of William Anderson, former APSA president, who was a leading American authority in the areas of local government, public administration, intergovernmental relations and the history of political science. He did much to shape teaching and research in these fields not only at his own university, but throughout the country. Nominations should be submitted each year by the February 13 deadline.
*Requirements:* Political science departments are invited to nominate not more than one doctoral dissertation for this award. The dissertations must have been completed and accepted during the two calendar years preceding the date of the prize.
*Geographic Focus:* All States
*Date(s) Application is Due:* Feb 13
*Amount of Grant:* 750 USD
*Contact:* Betsy Super, Deputy Director; (202) 483-2512; fax (202) 483-2657; bsuper@apsanet.org or awards@apsanet.org
*Internet:* www.apsanet.org/PROGRAMS/APSA-Awards/William-Anderson-Award
*Sponsor:* American Political Science Association
1527 New Hampshire Avenue, NW
Washington, D.C. 20036-1206

**Archer Daniels Midland Foundation Grants** 571
The foundation prefers to fund programs that will directly impact the communities where operating units are located. Nearly a third of the funding is directed toward educational institutions, including elementary, secondary, and higher education. The advocacy category covers world affairs and foreign relations groups that deal with international trade; also supported are projects stressing free enterprise and assistance for women. Support of social services goes to minority group development, cultural activities, and hospital and youth agencies. Health service funding supports hospital and disease association programs. Conservation funding supports programs for protecting the environment and beautification as well as conservation. Religion grants support Christian, Jewish, and Roman Catholic organizations, such as churches, colleges and universities, international ministries/missions, Jewish welfare, and the Salvation Army. Applications are accepted at any time; initial contact by letter of inquiry is encouraged.
*Requirements:* Tax-exempt organizations are eligible. Current United Way recipients and national organizations are also eligible.
*Geographic Focus:* Arkansas, California, Georgia, Illinois, Indiana, Iowa, Kansas, Kentucky, Massachusetts, Michigan, Minnesota, Mississippi, Missouri, Montana, Nebraska, New Jersey, New York, North Carolina, North Dakota, Ohio, Oklahoma, Pennsylvania, Puerto Rico, South Carolina, Tennessee, Texas, Washington, Wisconsin
*Amount of Grant:* 10,000 USD
*Contact:* Brian Peterson, Program Contact; (217) 424-5413 or (217) 424-5200; corpaffairs@admworld.com
*Internet:* www.adm.com/en-US/responsibility/social_investing/Pages/default.aspx
*Sponsor:* Archer Daniels Midland Foundation
4666 Faries Parkway, P.O. Box 1470
Decatur, IL 62526-5630

**Arizona Commission on the Arts Research and Development Grants** 572
Artist Research and Development Grants are designed to support individual artists from all disciplines. The purpose of this grant is to aid in the development of artistic work, support the advancement of artistic research and recognize the contributions individual artists make to Arizona's communities. Artist Research and Development Grants are made to practicing artists who demonstrate strong, original work that impacts both their own artistic practice and the broader community, as defined by the artist. Artist Research and Development Grant applicants may include, but are not limited to, artists who are: developing new techniques or methodologies; experimenting with new materials or technologies; creating new works; or advancing new strategies for engagement. The deadline for applications is September 17

*Requirements:* This grant is for individual artists. At the time of application, an applicant must be: 18 years of age or older; a resident of Arizona; and not enrolled in more than three (3) credit hours at a college or university.
*Restrictions:* Organizations are not eligible to apply.
*Geographic Focus:* Arizona
*Date(s) Application is Due:* Sep 17
*Amount of Grant:* 3,000 - 5,000 USD
*Contact:* Casandra Hernandez, Artist Programs Coordinator; (602) 771-6501 or 602) 771-6530; fax (602) 256-0282; chernandez@azarts.gov or info@azarts.gov
*Internet:* azarts.gov/grant/artist-research-and-development/
*Sponsor:* Arizona Commission on the Arts
417 West Roosevelt Street
Phoenix, AZ 85003-6501

**Arkansas Arts Council AIE Arts for Lifelong Learning Mini-Grants** **573**
The Arts for Lifelong Learning Mini-Grants program will enrich the lives of adults of all experience levels and abilities through quality arts learning experiences. This new program provides funding for eligible institutions that wish to host a short-term (up to 10 days) residency program with professional artists who are listed on the Arts in Education Artist Roster. The Arts in Education Artist Roster is a listing of Arkansas artists of all disciplines who have been selected by an independent panel to participate in Arts in Education residency programs. Arts activities that include multi-generational populations are especially encouraged. Community-based or governmental organizations and institutions, whether they be facilities and/or organizations for elders, hospitals, counseling or rehabilitation centers, social service agencies, prisons or detention centers, or organizations that provide facilities and/or guidance for disabled adults, are encouraged to participate in this program. Institutions of higher education, as well as museums, libraries, and community centers that offer educational programming for adult populations, are also encouraged to apply. Applications are available for programs occurring from October 1 to August 31 of the following year. The maximum amount to be requested is $1,000, with a one-to-one match. The annual deadline for applications is June 30.
*Requirements:* An applicant must meet the following criteria: be a certified 501(c)3 tax-exempt organization that does not receive operational support from the Arkansas Arts Council; or be a federal, state, or local government or governmental unit, church or convention/association of churches, hospital, hospital service organization or medical research affiliate, public school of institution of higher learning.
*Geographic Focus:* Arkansas
*Date(s) Application is Due:* Jun 30
*Amount of Grant:* Up to 1,000 USD
*Contact:* Kim Davis, Arts in Education Program Manager; (501) 324-9759 or (501) 324-9150; fax (501) 324-9207; kimberly.davis@arkansas.gov
*Internet:* www.arkansasarts.org/aac-grants/arts-in-education-arts-for-lifelong-learning-mini-grants
*Sponsor:* Arkansas Arts Council
1100 North Street
Little Rock, AR 72201

**Arkansas Arts Council Individual Artists Fellowships** **574**
The Arkansas Arts Council's Individual Artist Fellowships are unconditional, non-matching awards made directly to individual artists in recognition of their artistic abilities. Up to nine fellowships of $4,000 each may be awarded. Awards are based on the creative excellence of works submitted for review, and are awarded to a limited number of artists of outstanding talent. Artist categories for funding vary from year to year (literary arts, performing arts, visual arts, etc.). The annual application deadline is April 13.
*Requirements:* Applicants must be no younger than 25 and residents of Arkansas for at least one year prior to application. They must also be legal U.S. citizens or prove eligibility for full-time, permanent employment. Work submitted must be original and created within the past five years. If the creation involves collaboration, the applicant must be the primary creator and identify his or her role and the amount and type of participation contributed in creating the work. In all categories, it is best to submit documentation materials that show a concentration rather than a range of work.
*Restrictions:* An application may submit only one individual artist application each year. Previous individual fellowship winners are not eligible to apply. Current Arkansas Arts Council advisory board members, Department of Arkansas Heritage staff, or an immediate family members are also not eligible.
*Geographic Focus:* Arkansas
*Date(s) Application is Due:* Apr 13
*Amount of Grant:* 4,000 USD
*Contact:* Robin Muse McClea, Artist Services Program Manager; (501) 324-9348 or (501) 324-9150; fax (501) 324-9207; robin.mcclea@arkansas.gov
*Internet:* www.arkansasarts.org/grants/iaf.aspx
*Sponsor:* Arkansas Arts Council
1100 North Street
Little Rock, AR 72201

**Armstrong McDonald Foundation Health Grants** **575**
The Armstrong McDonald Foundation was incorporated in the State of Nebraska in 1986. The mission of the Foundation is to continue the philanthropic ideals and goals of James M. McDonald, Sr. through prudent and impartial review of all qualifying grant requests received annually to insure that awards are made to soundly conceived and operated non-profit organizations. For its Health category, the Foundation supports a great variety of projects related to health issues ranging from lab supplies for teaching hospitals to multi-drug resistance research. Other examples of projects falling within this category are the need for medical related equipment, renovating areas of hospitals for specialized treatment, sending kids with cancer to camp, recovery programs for burn victims, outpatient therapy for abused and neglected children, programs related to fighting eating disorders, and costs of providing onsite facilities at hospitals providing specialized care for children. The Foundation provides a formal downloadable application which is to be completed by all applicants approved for submission of a grant request. All required application materials need to be received on or before September 15.
*Requirements:* The Armstrong McDonald Foundation will only accept unsolicited grant requests from those IRS approved non-profits listed on the Pre-Approved for Grant Submission List. Please note that this list will be updated annually in December with additions and/or deletions. All other IRS approved non-profits desiring to submit a grant request to this foundation must meet the following three qualifications: be incorporated in either the State of Arizona or Nebraska; have a physical office located in their state of incorporation; and spend any awarded grant funds within their state of incorporation.
*Restrictions:* Grants do not support advocacy organizations, individuals, international organizations, political organizations, or state and local government agencies. The foundation does not fund capital campaigns, salaries/stipends, and multi-year projects. Organization must have received a grant from the foundation within the last five years unless it is located within the states of Arizona or Nebraska. No organization east of the Mississippi is eligible for a grant.
*Geographic Focus:* Arizona, Nebraska
*Date(s) Application is Due:* Sep 15
*Amount of Grant:* 1,000 - 80,000 USD
*Contact:* Laurie L. Bouchard, President; (520) 878-9627; fax (520) 797-3866; info@ArmstrongMcDonaldFoundation.org
*Internet:* www.armstrongmcdonaldfoundation.org/cat.html
*Sponsor:* Armstrong McDonald Foundation
P.O. Box 70110
Tucson, AZ 85737-0110

**Arnold and Mabel Beckman Foundation Beckman-Argyros Award in Vision Research** **576**
The Beckman-Argyros Award in Vision Research is an annual award established in 2013 to honor and celebrate a decades-long friendship of the two remarkable men, Dr. Arnold O. Beckman and Ambassador George L. Argyros, and to continue their commitment, dedication and shared vision to make the world a better place. The Beckman-Argyros Award honors Ambassador Argyros for his 22 years of service as Chairman of the Board of the Arnold and Mabel Beckman Foundation and recognizes the special and unique friendship he shared with Arnold O. Beckman for over forty years. The Beckman-Argyros Award in Vision Research is intended to reward individuals who are making significant transformative breakthroughs in vision research; this may include those whose contributions to science in general, or through the development of an innovative technology or fundamental scientific breakthrough have been applied to, aided and/or improved the vision sciences. The recipient of the Beckman-Argyros Award will receive a total of $500,000, as outlined below, along with a solid gold commemorative medallion: $100,000 prize to award recipient; $400,000 grant to support award recipient's research at a university or other qualifying public 501(c)3 charity specifying its designated use.
*Geographic Focus:* All States
*Amount of Grant:* 500,000 USD
*Contact:* Catrina Bryant, Program Administrator; (949 721-2246 or (949 721-2222; fax (949) 721-2225; beckmanargyros@beckman-foundation.com or administration@beckman-foundation.com
*Internet:* www.beckman-foundation.org/programs/beckman-argyros-vision-program-information
*Sponsor:* Arnold and Mabel Beckman Foundation
100 Academy Drive
Irvine, CA 92617

**Arnold and Mabel Beckman Foundation Postdoctoral Fellowships** **577**
The Arnold O. Beckman Postdoctoral Fellows Award Program supports postdoctoral scholars with the highest potential for success in an independent academic career in chemistry and the life sciences, to assist in their transition from mentored, yet independent, postdoctoral projects to an independent, tenure-track position. These individuals are expected to become the next generation of leaders and innovators in science, engineering, and technology. Fellows will be appointed for two years, with possibility for renewal for a third and final year. Each fellow will begin as early as September of their application year but no later than December 31. The award amount will range from $63,300 to $77,000, based upon years of postdoctoral research experience, with 75% of the award going towards the stipend, and 25% used towards research expenditures. The annual online application submission deadline is September 19.
*Requirements:* Applicants must be at time of application: a U.S. college, university or institution which offers postdoctoral fellowships and holds a 501(c)3, or similarly qualifying, IRS designation; a U.S. citizen or permanent resident of the United States; granted and/or conferred PhD, MD, or MD/PhD; and eligible for a tenure track position at a scientific research institution at the end of the Arnold O. Beckman Postdoctoral Fellowship. The criteria for a sponsoring mentor at time of application must be: a PhD, MD, or MD/PhD; a full-time tenured or tenure track; an ctive investigator in the applicant's area of research; and only sponsor one applicant for consideration.
*Geographic Focus:* All States
*Date(s) Application is Due:* Sep 19
*Amount of Grant:* 63,300 - 77,000 USD
*Contact:* Catrina Bryant, Program Administrator; (949) 721-2246 or (949 721-2222; fax (949) 721-2225; admin@beckman-foundation.org or aobpostdoc@beckman-foundation.org

*Internet:* www.beckman-foundation.org/programs/beckman-postdoctoral-fellows
*Sponsor:* Arnold and Mabel Beckman Foundation
100 Academy Drive
Irvine, CA 92617

**Arnold and Mabel Beckman Foundation Scholars Grants** 578
The Arnold and Mabel Beckman Foundation Scholars program awards are institutional, university or college awards. Each year, the Foundation selects a number of research, doctoral, masters and baccalaureate universities and colleges to be invited to submit applications for the Scholars program. Each institution may submit one application for consideration for an award. If an institution is interested in the Beckman Scholars Program, but has not been invited to participate, it should review the guidelines for eligibility. The amount given for each award is $26,000 for two summers and one academic year.
*Restrictions:* The Beckman Scholars program does not make awards directly to individuals.
*Geographic Focus:* All States
*Amount of Grant:* 26,000 USD
*Contact:* Sarah Marinkovich, BSP Program Administrator; (949) 721-2222; fax (949) 721-2225; bsp@beckman-foundation.org or administration@beckman-foundation.com
*Internet:* www.beckman-foundation.org/programs/beckman-scholars-program-information
*Sponsor:* Arnold and Mabel Beckman Foundation
100 Academy Drive
Irvine, CA 92617

**Arnold and Mabel Beckman Foundation Young Investigators Grants** 579
The Arnold and Mabel Beckman Foundation makes Young Investigators grants to nonprofit research institutions to promote research in chemistry and the life sciences, broadly interpreted, and particularly to foster the invention of methods, instruments, and materials that will open up new avenues of research in science. The program is intended to provide research support to the most promising young faculty members in the early stages of academic careers in the chemical and life sciences. The program is open to persons with tenure-track appointments in academic and nonprofit institutions that conduct fundamental research in the chemical and life sciences. The program is intended primarily for U.S. institutions. Only proposals of exceptional merit from foreign institutions will receive consideration. Projects are normally funded for a period of two years. When extraordinary circumstances warrant, support may be provided over a one-year or three-year period. All versions of the required Letter of Intent must be submitted online prior to August 15. Potential candidates for application submission will be notified in early December, and awardees will be announced in the following July.
*Requirements:* To be eligible, an applicant should not have completed more than three full years in his or her tenure-track or other comparable independent research appointment on the anniversary date of initial appointment in the year in which application is to be made. Candidates must be citizens or permanent residents of the United States at the time of application.
*Restrictions:* Funding will not be considered for general institutional expenses or general fund-raising campaign expenses such as dinners and mass mailings. The foundation does not provide funds for overhead or indirect costs. Persons who have applied for permanent residency but have not received their government documentation by the time of application are not eligible. Individuals may not apply more than three times. Applicants from institutions with a currently funded Beckman Young Investigator are not eligible to apply.
*Geographic Focus:* All States
*Date(s) Application is Due:* Aug 15
*Amount of Grant:* Up to 750,000 USD
*Contact:* Nicole Forino; (949) 721-2222; fax (949) 721-2225; younginvestigators@beckman-foundation.com or administration@beckman-foundation.com
*Internet:* www.beckman-foundation.org/programs/beckman-young-investigators-program-information
*Sponsor:* Arnold and Mabel Beckman Foundation
100 Academy Drive
Irvine, CA 92617

**Arthur F. and Alice E. Adams Charitable Foundation Grants** 580
The Arthur F. and Alice E. Adams Foundation was established by Arthur and Alice Adams out of their desire to support various charitable initiatives. Arthur Adams was a business entrepreneur who managed the Gary Group of telephone interests in the Midwest. Under his leadership as president and CEO, the Strowger Automatic Electric Company of Chicago developed and patented the dial telephone. Alice ("Elise") Adams was born in Istanbul, Turkey. Her family relocated to Egypt during her childhood, and she spent much of her time in private school in Switzerland. She became proficient in the use of foreign languages and eventually mastered eleven of them. In the late 1930s, Arthur met and married Elise. They maintained several homes and eventually settled in Miami Beach, Florida. Following Arthur's death in 1958, Elise spent much of her time supporting philanthropic causes until she passed away in 1987. Established in Florida in 1985, the Foundation is managed by a board of governors with the Wells Fargo Philanthropic Services acting as corporate governor. The Foundation offers support in the arts, education, and human services. Primary fields of interest include: performing arts centers, opera, and medical research. Grants range from $5,000 to $250,000, with giving in the areas around Miami, New York City, and Memphis. The annual deadline is February 15, though board members meet in both May and November. Applicants should forward a letter describing their need and budget. Awards up to a maximum of $50,000 are typical.
*Requirements:* There are no geographic limitations; however, the foundation has a practice of primarily supporting organizations located in Memphis, Miami, or New York City areas.
*Restrictions:* No grants are awarded to individuals.
*Geographic Focus:* All States
*Date(s) Application is Due:* Feb 15
*Amount of Grant:* 5,000 - 250,000 USD
*Samples:* Circuit Playhouse on the Square, Memphis, Tennessee, $125,000 - capital campaign; Preservation League of New York, Albany, New York, $50,000 - general operating support; Florida International University Foundation, Miami, Florida, $50,000 - ongoing support.
*Contact:* Sangita Karra, Vice President; (609) 799-3108; fax (609) 799-4189; grantadministration@wellsfargo.com
*Internet:* https://www.wellsfargo.com/private-foundations/adams-charitable-foundation
*Sponsor:* Arthur F. and Alice E. Adams Charitable Foundation
Greeley Square, Post Office Box 20459
New York, NY 10001-0459

**Arthur W. Adamson Award for Distinguished Service in the Advancement of Surface Chemistry** 581
This annual award, supported by Occidental Petroleum Corporation, recognizes distinguished services in the advancement of surface chemistry. Activities recognized by the award may include such fields as teaching, writing, research, and administration. The award consists of $5,000, a medallion with a presentation box, and a certificate.
*Requirements:* Any individual, except a member of the award committee, may submit one nomination or seconding letter for each award in any given year. The letter of nomination should contain an evaluation of the nominee's accomplishments and a specific identification of the work to be recognized, and may not exceed 1000 words in length.
*Restrictions:* Self-nominations are not accepted.
*Geographic Focus:* All States
*Date(s) Application is Due:* Nov 1
*Amount of Grant:* 6,000 USD
*Samples:* Steven J. Sibener—award recipient, $5,000; Patricia A. Thiel—award recipient, $5,000; David A. King—award recipient, $5,000.
*Contact:* Felicia Dixon, Awards Administrator; (202) 872-6283 or (202) 872-4575; fax (202) 776-8008; f_dixon@acs.org or awards@acs.org
*Internet:* portal.acs.org/portal/acs/corg/content?_nfpb=true&_pageLabel=PP_ARTICLEMAIN&node_id=1319&content_id=CTP_004494&use_sec=true&sec_url_var=region1
*Sponsor:* American Chemical Society
1155 Sixteenth Street, NW
Washington, D.C. 20036-4801

**ASCE Arthur S. Tuttle Memorial Scholarships** 582
Arthur S. Tuttle Memorial Scholarships are for tuition assistance to students during their first year of graduate studies in civil engineering. Any student in good standing who is also a national student member of the society may apply; the funds may be applied only for tuition expenses during the first year of formal graduate civil engineering education in an accredited educational institution. Financial need and educational standing will be considered in selecting recipients. The total sum for awards available will be determined annually based on earnings available from the endowment.
*Requirements:* Application consists of a completed application form; applicant's justification for applying; plans for continuing formal education; official transcripts, recommendation forms, and supporting statements from the student chapter faculty advisor and two other members of the engineering school faculty; and a detailed financial plan outlining both income and expenses anticipated for the award year. Applicants must be a member of the society, in any grade, and must be in good standing at the time of application and award. Membership applications may be submitted with scholarship application.
*Geographic Focus:* All States
*Date(s) Application is Due:* Feb 9
*Amount of Grant:* 2,000 - 5,000 USD
*Contact:* John Casazza, Managing Director, Continuing Education; (703) 295-6155; fax (703) 295-6132; jcasazza@asce.org
*Internet:* www.asce.org/Student-Organizations/Scholarships-and-Fellowships/Arthur-S—Tuttle-Memorial-Scholarship/
*Sponsor:* American Society of Civil Engineers
1801 Alexander Bell Drive
Reston, VA 20191-4400

**ASCE Freeman Fellowships** 583
Freeman Fellowships to aid and encourage young engineers, especially in research work, are made to fund expenses for experiments, observations, and compilations to discover new and accurate data that will be useful in engineering. The fellowships also recognize underwriting fully or in part some of the loss that may be sustained in the publication of meritorious books, papers, or translations pertaining to hydraulic science and art that might, except for some such assistance, remain mostly inaccessible. Traveling scholarships are available to members younger than 45 years of age in recognition of achievement or promise for the purpose of aiding the candidate to visit engineering works in the United States or any other part of the world where there are good prospects of obtaining information useful to engineers. The grant may be used for assisting in the translation or publication in English of papers or books in foreign languages pertaining to hydraulics. The grant is available annually. In a given year there may be more than one recipient, or no recipient if no meritorious applications are received. Stipend is based on funds available annually from the endowment.
*Requirements:* Application consists of a completed form and a statement in general terms of the purposes for which the funds are expected to be used. Any member of ASCE is eligible to apply.
*Geographic Focus:* All States
*Date(s) Application is Due:* Feb 9
*Amount of Grant:* 2,000 - 5,000 USD

*Contact:* John Casazza, Managing Director, Continuing Education; (703) 295-6155; fax (703) 295-6132; jcasazza@asce.org
*Internet:* www.asce.org/Student-Organizations/Scholarships-and-Fellowships/Freeman-Fellowship/
*Sponsor:* American Society of Civil Engineers
1801 Alexander Bell Drive
Reston, VA 20191-4400

**ASCE Jack E. Leisch Memorial National Graduate Fellowship** **584**
The Jack E. Leisch Memorial National Graduate Fellowship is a memorial to the outstanding professional accomplishments and contributions of Jack E. Leisch, M. ASCE to the fields of geometric design, traffic engineering and transportation planning. The fellowship is supported by the income from donations in Mr. Leisch's honor. Any full-time student who is enrolled in or admitted to a transportation/traffic engineering graduate degree program (master or doctoral) in Civil Engineering at a qualified university may apply.
*Requirements:* Applicants must be a member of the Society (Membership applications may be submitted with scholarship applications), and must be in good standing at the time of the award. Applicants must be enrolled at a qualified university (those whose undergraduate Civil Engineering program is accredited by the Accreditation Board for Engineering and Technology).
*Restrictions:* Previous Jack E. Leisch Scholarship recipients are not eligible to reapply.
*Geographic Focus:* All States
*Date(s) Application is Due:* Feb 9
*Amount of Grant:* 4,000 USD
*Contact:* John Casazza, Managing Director, Continuing Education; (703) 295-6155; fax (703) 295-6132; jcasazza@asce.org
Andrea C. Baker, Manager; abaker@asce.org
*Internet:* www.asce.org/Content.aspx?id=25021
*Sponsor:* American Society of Civil Engineers
1801 Alexander Bell Drive
Reston, VA 20191-4400

**ASCE O.H. Ammann Research Fellowship in Structural Engineering** **585**
Nominations are being solicited for the annual O.H. Ammann Research Fellowship in Structural Engineering, which is bestowed annually to a member for the purpose of encouraging the creation of new knowledge in the field of structural design and construction. The recipient may apply for a renewal, additional stipend to be determined annually. Applications must be postmarked by the listed application deadline date of November 1, with notification being made in May for a fellowship beginning that same year.
*Requirements:* Applicants must be members of the society or applicants for membership; citizens of countries other than the United States may apply if they are eligible under this rule. Each application should include completed application form, official transcripts, a description of proposed research and its objectives, a statement from the institution at which the research is to be done that the applicant and proposed research are acceptable to that institution, and a statement in general terms of the purposes for which the funds are expected to be used.
*Restrictions:* During the tenure of the fellowship, the recipient may not work on research projects other than that for which this fellowship has been granted, but recipients may accept other awards if the conditions of such awards are the same as those for this fellowship.
*Geographic Focus:* All States
*Date(s) Application is Due:* Nov 1
*Amount of Grant:* 5,000 USD
*Contact:* Susan Reid; (703) 295-6299; fax (703) 295-6132; sreid@asce.org
*Internet:* content.seinstitute.org/inside/ammann.html
*Sponsor:* American Society of Civil Engineers
1801 Alexander Bell Drive
Reston, VA 20191-4400

**ASCE Trent R. Dames and William W. Moore Engineering Fellowship** **586**
The Trent R. Dames and William W. Moore Engineering Fellowship is available to practicing engineers, earth scientists, professors, or graduate students for engineering research. Preference is given to research that will contribute to the solution of conflicting engineering issues involving social, economic, environmental, or political concerns or compromises. Other selection criteria include, but are not limited to, relevant professional experience, previous research or course of study either as a practitioner or in academia, and prior contribution to the engineering profession and/or the applied earth sciences. The fellowship may be awarded to co-researchers in a single project or divided among multiple projects.
*Requirements:* Application must include the completed application form and attachments; description of proposed research and its objectives, including a statement from the institution at which the research is to be done that the applicant and proposed research are acceptable to the institution; and total budget. Applicants must be practicing engineers or earth scientists, professors, or graduate students.
*Geographic Focus:* All States
*Date(s) Application is Due:* Feb 9
*Amount of Grant:* 2,000 - 6,000 USD
*Contact:* John Casazza, Managing Director, Continuing Education; (703) 295-6155; fax (703) 295-6132; jcasazza@asce.org
*Internet:* www.asce.org/Student-Organizations/Scholarships-and-Fellowships/Trent-R—Dames-and-William-W—Moore-Fellowship/
*Sponsor:* American Society of Civil Engineers
1801 Alexander Bell Drive
Reston, VA 20191-4400

**ASECS/IASECS Maria Salgado Student Travel Grant** **587**
Maria Salgado was born in Toro in the province of Zamora, Spain. From the beginning she assimilated and became familiar with the basis and sources of traditional folk music. In Segovia, the great master of Castilian music, Agapito Marazuela, revealed to her the essence of genuine traditional interpretation. In fact, on the basis of his teachings she made a great recording of songs taken from the Cancionero de Castilla, which Marazuela painstakingly recovered at the beginning of the 20th century. In Canciones De Amor y Trabajo, Maria rearranged old melodies and included the most modern musical instruments. This she did in such a way that these vigorous songs lose none of their true spirit. The ASECS/IASECS Maria Salgado Student Travel Grant offers up to a maximum of $400 to graduate students from North American universities who will be presenting a paper focused on eighteenth-century Spain, Portugal, or Ibero-America, at the American Society for Eighteenth Century Studies (ASECS) national annual meeting. The annual deadline for application submissions is February 15.
*Geographic Focus:* All States
*Date(s) Application is Due:* Feb 15
*Amount of Grant:* Up to 400 USD
*Contact:* Lisa Berglund, Executive Director; (716) 878-3405 or (716) 878-5417; fax (716) 878-4939; berglul@buffalostate.edu or asecsoffice@gmail.com
Enid Valle, Professor of Romance Languages and Literature; (269) 337-7121; valle@kzoo.edu
*Internet:* https://asecs.press.jhu.edu/general%20site/asecsawards.html
*Sponsor:* American Society for Eighteenth-Century Studies
1300 Elmwood Avenue, Ketcham Hall 213
Buffalo, NY 14222

**ASECS/IASECS Pilar Saenz Annual Student Essay Prize** **588**
The ASECS/IASECS Pilar Saenz Annual Student Essay Prize is awarded to the best essay dealing with eighteenth-century Spain, Portugal or Ibero-America. The prize is open to graduate students enrolled in a North American university. Advanced undergraduate work could be considered provided it is accompanied by the recommendation of their professor. All entries should be received by the annual deadline date of February 15. The Ibero-American Society for Eighteenth Century Studies (IASECS) Essay Prize Committee will announce the award at the ASECS annual meeting. The winner will receive $250; a one-year subscription to the journal Dieciocho; and a two-year membership in the IASECS.
*Requirements:* An applicant must be a member in good standing of ASECS, be a postdoctoral scholar no more than 10 years beyond receipt of the PhD or equivalent degree at the time of application, and be working on a project concerning the 18th century. Contact should be made to the specific library of interest; complete contact information for each library is available on the ASECS Web site.
*Geographic Focus:* All States
*Date(s) Application is Due:* Feb 15
*Amount of Grant:* 250 USD
*Contact:* Lisa Berglund, Executive Director; (716) 878-3405 or (716) 878-5417; fax (716) 878-4939; berglul@buffalostate.edu or asecsoffice@gmail.com
Enid Valle, Professor of Romance Languages and Literature; (269) 337-7121; valle@kzoo.edu
*Internet:* https://asecs.press.jhu.edu/general%20site/asecsawards.html
*Sponsor:* American Society for Eighteenth-Century Studies
1300 Elmwood Avenue, Ketcham Hall 213
Buffalo, NY 14222

**ASECS A.C. Elias, Jr. Irish-American Research Travel Fellowship** **589**
The ASECS A.C. Elias, Jr. Irish-American Research Travel Fellowship, with a $2,500 funding eligibility, is given annually to support documentary research in Irish repositories, both in the Republic of Ireland and the North, by U.S.-based scholars of Ireland in the period between the Treaty of Limerick (1691) and the Act of Union (1800). In alternate years, the award will go to Irish-based scholars seeking to travel to North America. The fellowship is restricted to documentary scholars whose research centers on primary sources from the 18th century (printed matter, manuscripts, buildings, works of art, or other artifacts), rather than on extant secondary literature. Applications for the next Elias fellowship are due on November 15.
*Requirements:* Applicants must be members of the society at the time of the award and must be residents of North America. The fellowship is restricted to documentary scholars, whose research centers on primary sources from the 18th century (printed matter, manuscripts, buildings, works of art, or other artifacts), rather than on the secondary literature already extant.
*Geographic Focus:* All States
*Date(s) Application is Due:* Nov 15
*Amount of Grant:* 2,500 USD
*Contact:* Lisa Berglund, Executive Director; (716) 878-3405 or (716) 878-5417; fax (716) 878-4939; berglul@buffalostate.edu or asecsoffice@gmail.com
Dr. Jason McElligott, Director (the Keeper); (353) 1 4543511; jason.mcelligott@marshlibrary.ie
*Internet:* https://asecs.press.jhu.edu/general%20site/travelgr.html
*Sponsor:* American Society for Eighteenth-Century Studies
1300 Elmwood Avenue, Ketcham Hall 213
Buffalo, NY 14222

**ASECS Annibel Jenkins Biography Prize** **590**
The biennial Annibel Jenkins Prize is given to the author of the best book-length biography of a late seventeenth-century or eighteenth-century subject and carries an award of $1,000. The prize has been named in honor of Annibel Jenkins, Professor of English (Emerita) at the Georgia Institute of Technology. A founding member of the

Southeastern American Society for Eighteenth-Century Studies, she was an outstanding teacher and scholar who has been for many years one of the most active and encouraging members of the academic community in America. The annual deadline for application submissions is November 15.
*Requirements:* Applicants must be members of the society at the time of the award.
*Geographic Focus:* All States
*Date(s) Application is Due:* Nov 15
*Amount of Grant:* 1,000 USD
*Contact:* Lisa Berglund, Executive Director; (716) 878-3405 or (716) 878-5417; fax (716) 878-4939; berglul@buffalostate.edu or asecsoffice@gmail.com
*Internet:* https://asecs.press.jhu.edu/Weekly%20Announcements/jenkins1.html
*Sponsor:* American Society for Eighteenth-Century Studies
1300 Elmwood Avenue, Ketcham Hall 213
Buffalo, NY 14222

**ASECS Aubrey L. Williams Research Travel Fellowship** **591**
Aubrey Lake Williams, Jr., was a retired English professor. Before finishing his education, Dr. Williams was a military pilot in the World War II Army Air Corps with the rank of first lieutenant. He then graduated Louisiana State University and Yale University where he earned his Ph.D. in English. He taught at Yale University, Rice University and the University of Florida, specializing in eighteenth-century English literature. He authored several books and mentored many graduate students who now teach throughout the United States. He donated many rare books to both LSU and the University of Florida, and has been honored in many ways, particularly by his students starting a research travel fellowship in his name to advance the studies of eighteenth-century English literature. The ASECS Aubrey L. Williams Research Travel Fellowship of $1,500 is awarded annually to support documentary research in 18th-century English literature by US-based scholars. The fellowship is restricted to doctoral students working on the dissertation in the field of 18th-century English literature. Applicants must be members of the society at the time of the award and must be residents of North America. The annual deadline for application submission is January 1.
*Requirements:* The fellowship is restricted to doctoral students at work on a dissertation in the field of eighteenth-century English literature.
*Geographic Focus:* All States
*Date(s) Application is Due:* Jan 1
*Amount of Grant:* 1,500 USD
*Contact:* Lisa Berglund, Executive Director; (716) 878-3405 or (716) 878-5417; fax (716) 878-4939; berglul@buffalostate.edu or asecsoffice@gmail.com
*Internet:* https://asecs.press.jhu.edu/general%20site/travelgr.html
*Sponsor:* American Society for Eighteenth-Century Studies
1300 Elmwood Avenue, Ketcham Hall 213
Buffalo, NY 14222

**ASECS Award for Excellence in Graduate Student Mentorship** **592**
The ASECS Women's Caucus Editing and Translation Fellowship, is an annual award of $1000 to support an editing or a translation work in progress of an eighteenth-century primary text on a feminist or a Women's Studies subject. Editing and translation work of eighteenth-century texts in languages other than English are eligible. The award is meant to fund works in progress, commensurate in scope with a scholarly article-length project to a longer scholarly and/or a classroom edition with a strong scholarly basis for which research and work is well under way, rather than work that is already completed. Projects typically fall within the period from 1660 to 1820. Submissions must be received by January 15.
*Requirements:* The award is open to all members of ASECS who have received a PhD. Proposals from Emeritae/i faculty that do not already have professional support for the project will also be considered. To be eligible for the prize, projects must translate and/or edit works by eighteenth-century women writers or works that significantly advance our understanding of women's experiences in the eighteenth century or offer a feminist analysis of any aspect of eighteenth-century culture and/or society.
*Geographic Focus:* All States
*Date(s) Application is Due:* Jan 15
*Amount of Grant:* 1,000 USD
*Contact:* Lisa Berglund, Executive Director; (716) 878-3405 or (716) 878-5417; fax (716) 878-4939; berglul@buffalostate.edu or asecsoffice@gmail.com
Byron R. Wells; (336) 727-4694; fax (336)- 727-4697; asecs@wfu.edu
*Internet:* Women's Caucus Editing and Translation Fellowship
*Sponsor:* American Society for Eighteenth-Century Studies
1300 Elmwood Avenue, Ketcham Hall 213
Buffalo, NY 14222

**ASECS Daiches-Manning Memorial Fellowship in 18th-Century Scottish Studies** **593**
Co-sponsored by ASECS and the Eighteenth-Century Scottish Studies Society (ECSSS), the $3,000 Daiches-Manning Memorial Fellowship in 18th-Century Scottish Studies honors the memory of David Daiches (1912–2005) and Susan Manning (1953–2013). Each was an eminent eighteenth-century literary and cultural historians of Scotland and America, recipients of the ECSSS Lifetime Achievement Award, and directors of the Institute for Advanced Studies in the Humanities (IASH) at the University of Edinburgh. The Fellowship supports two to six months of research on eighteenth-century Scotland as a Fellow in residence at IASH. The annual deadline for application submission is February 28.
*Requirements:* Post-doctoral scholars without regard to nationality, residence, or academic discipline are eligible to apply. Fellows must be members of ASECS and ECSSS at the time the fellowship is awarded.
*Geographic Focus:* All States, United Kingdom
*Date(s) Application is Due:* Feb 28
*Amount of Grant:* 3,000 USD
*Contact:* Lisa Berglund, Executive Director; (716) 878-3405 or (716) 878-5417; fax (716) 878-4939; berglul@buffalostate.edu or asecsoffice@gmail.com
*Internet:* https://asecs.press.jhu.edu/general%20site/travelgr.html
*Sponsor:* American Society for Eighteenth-Century Studies
1300 Elmwood Avenue, Ketcham Hall 213
Buffalo, NY 14222

**ASECS Dora Wiebenson Graduate Student Prize** **594**
Dora Louise Wiebenson was born in Cleveland, Ohio, in 1926, the daughter of a building contractor, Edward Ralph Wiebenson, and his wife Jeannette Rodier (Wiebenson). After graduating from Vassar College with an A. B. in 1946 she entered Harvard University, securing a second Bachelor's degree in Architecture in 1951. She moved to New York where she worked as a designer and draftsperson in various architectural offices in New York City between 1951 and 1958. That year she was awarded an A.M. from New York University, continuing for her doctorate. While working on her Ph. D., she acted as an independent architectural consultant in New York City. She was appointed a lecturer in architecture at Columbia University in 1966. Wiebenson moved to the University of Maryland, College Park as an associate professor in 1968. In 1972, she accepted the position of professor of architectural history at University of Virginia, Charlottesville. She was director of the Society of Architectural Historians between 1974 and 1977. She was promoted to professor of architectural history and chair of division in 1977. The following year, her book on landscape architecture in France, The Picturesque Garden in France, appeared. She retired from the University in 1992. The Historians of Eighteenth-Century Art and Architecture offers the Dora Wiebenson Graduate Student Prize in her honor for the best graduate student paper presented during the previous calendar year at a scholarly conference or as a sponsored lecture. Honorable mention is an option for papers of distinction that do not win the prize. The annual deadline for application submission is February 15.
*Requirements:* Winners must be HECAA members in good standing.
*Geographic Focus:* All States
*Date(s) Application is Due:* Feb 15
*Contact:* Lisa Berglund, Executive Director; (716) 878-3405 or (716) 878-5417; fax (716) 878-4939; berglul@buffalostate.edu or asecsoffice@gmail.com
*Internet:* https://asecs.press.jhu.edu/Weekly%20Announcements/wiebenson1.html
*Sponsor:* American Society for Eighteenth-Century Studies
1300 Elmwood Avenue, Ketcham Hall 213
Buffalo, NY 14222

**ASECS Graduate Student Research Paper Award** **595**
Recognizing the pioneering research contributions of the next generation of scholars of eighteenth-century studies, this $200 Graduate Student Research Paper Award will recognize an outstanding research essay of 15 to 30 pages, which has not been previously published. Copies of the submission as well as a letter of endorsement from a mentoring professor, which outlines the originality and contributions that the essay makes to the field of eighteenth-century studies should be sent. The deadline for applications is January 1, and the announcement of the winner will be made at the annual meeting.
*Geographic Focus:* All States
*Date(s) Application is Due:* Jan 1
*Amount of Grant:* 200 USD
*Contact:* Lisa Berglund, Executive Director; (716) 878-3405 or (716) 878-5417; fax (716) 878-4939; berglul@buffalostate.edu or asecsoffice@gmail.com
*Internet:* https://asecs.press.jhu.edu/Weekly%20Announcements/gsresearch.html
*Sponsor:* American Society for Eighteenth-Century Studies
1300 Elmwood Avenue, Ketcham Hall 213
Buffalo, NY 14222

**ASECS Gwin J. and Ruth Kolb Research Travel Fellowship** **596**
The Gwin J. and Ruth Kolb Research Travel Fellowship of $500 is awarded annually to supplement costs for 18th-century scholars to travel to distant collections in North America and abroad. There are no restrictions based on applicants' age, sex, race, religion, or academic rank or discipline. Advanced doctoral candidates with demonstrable need for specific collections necessary for their dissertations also are encouraged to apply. The annual deadline for applications is January 1.
*Requirements:* Applicants must be members of ASECS at the time of application and faculty or independent scholars within the first five years of receipt of their PhD.
*Geographic Focus:* All States
*Date(s) Application is Due:* Jan 1
*Amount of Grant:* 500 USD
*Contact:* Lisa Berglund, Executive Director; (716) 878-3405 or (716) 878-5417; fax (716) 878-4939; berglul@buffalostate.edu or asecsoffice@gmail.com
*Internet:* https://asecs.press.jhu.edu/general%20site/travelgr.html
*Sponsor:* American Society for Eighteenth-Century Studies
1300 Elmwood Avenue, Ketcham Hall 213
Buffalo, NY 14222

**ASECS Innovative Course Design Competition** **597**
To encourage excellence in undergraduate teaching of the eighteenth century, the Society invites proposals from members in any of its constituent disciplines. Proposals should be for a new approach to teaching a unit within a course on the eighteenth century, covering perhaps one to four weeks of instruction, or for an entire new course. For example, participants may offer a new approach to a specific work or theme, a comparison of two

related works from different fields (music and history, art and theology), an interdisciplinary approach to a particular social or historical event, new uses of instructional technology (e.g., web sites, internet resources and activities), or a new course that has never been taught or has been taught only very recently for the first time. Participants are encourage to include why books and topics were selected and how they worked. Applicants should submit a 3-5 page proposal (double-spaced) as an email attachment and should focus sharply on the leading ideas distinguishing the unit to be developed. A syllabus draft of the course should also be provided. The annual submission deadline is October 1. The Committee will select the top three proposals by November 15. A major criterion for judging the proposals is how specific they are in relation to design, readings, pedagogy, and/or activities. The authors will be asked to develop a brief presentation for delivery in the Teaching Competition seminar at the Annual Meeting. A distinguished teacher-scholar will be invited to moderate the session. A $500 award will be presented to each of the participants, and they will be invited to submit a twelve-page account of the unit or course, with a syllabus or other supplementary materials for publication on the website.
*Requirements:* Only submissions by ASECS members will be accepted.
*Geographic Focus:* All States
*Date(s) Application is Due:* Oct 1
*Amount of Grant:* 500 USD
*Contact:* Lisa Berglund, Executive Director; (716) 878-3405 or (716) 878-5417; fax (716) 878-4939; berglul@buffalostate.edu or asecsoffice@gmail.com
*Internet:* https://asecs.press.jhu.edu/Weekly%20Announcements/innovativecourse.html
*Sponsor:* American Society for Eighteenth-Century Studies
1300 Elmwood Avenue, Ketcham Hall 213
Buffalo, NY 14222

**ASECS James L. Clifford Prize** **598**
The James L. Clifford Prize, which includes a cash award of $500, is awarded annually for an outstanding article appearing in a journal, festschrift, or other serial publication; the article must be an outstanding study of some aspect of 18th-century culture interesting to any 18th-century specialist, regardless of discipline, and no longer than 15,000 words. Nominations must be accompanied by eight copies of the article. The annual deadline for submissions is January 1.
*Requirements:* The article may be nominated by any member of the society, by its author, or by an editor of the publishing journal. The author must be a member of the society at the time of presentation of the award.
*Geographic Focus:* All States
*Date(s) Application is Due:* Jan 1
*Amount of Grant:* 500 USD
*Contact:* Lisa Berglund, Executive Director; (716) 878-3405 or (716) 878-5417; fax (716) 878-4939; berglul@buffalostate.edu or asecsoffice@gmail.com
*Internet:* https://asecs.press.jhu.edu/Weekly%20Announcements/clifford1.html
*Sponsor:* American Society for Eighteenth-Century Studies
1300 Elmwood Avenue, Ketcham Hall 213
Buffalo, NY 14222

**ASECS Louis Gottschalk Prize** **599**
Louis Gottschalk (1899-1975) second President of ASECS, President of the American Historical Association, and for many years Distinguished Service Professor at the University of Chicago, exemplified in his scholarship the humanistic ideals that this award is meant to encourage. The Prize named in his honor is awarded annually for an outstanding historical or critical study on a subject of 18th-century interest and carries an award of $1,000. Books, which may be written in any modern language, may be commentaries, critical studies, biographies, or critical editions.
*Requirements:* Applicants must be members of the society at the time of the award. Submission must be made by the publisher and received at ASECS by the listed application deadline. All scholarly books, including commentaries, critical studies, biographies, collections of essays by a single author and critical editions, written in any modern language are eligible. Books that are primarily translations and multiauthored collections of essays are not eligible.
*Restrictions:* Books that are primarily translations are ineligible.
*Geographic Focus:* All States
*Date(s) Application is Due:* Nov 15
*Amount of Grant:* 1,000 USD
*Contact:* Lisa Berglund, Executive Director; (716) 878-3405 or (716) 878-5417; fax (716) 878-4939; berglul@buffalostate.edu or asecsoffice@gmail.com
*Internet:* https://asecs.press.jhu.edu/Weekly%20Announcements/gottschalk1.html
*Sponsor:* American Society for Eighteenth-Century Studies
1300 Elmwood Avenue, Ketcham Hall 213
Buffalo, NY 14222

**ASECS Mary D. Sheriff Research and Travel Award** **600**
Mary Diana Lee Sheriff was an American art historian, and a Distinguished Professor of Art History at University of North Carolina at Chapel Hill who specialized in Eighteenth century French art, decorative arts, gender studies, and material culture. Sheriff was a leading scholar on 18th- and 19th-century French art and culture and the Rococo period. Her work focused on issues of creativity, sexuality, gender, and travel and culture exchange. Her research included traditional facets of visual culture such as painting and sculpture, while also incorporating gardens, book illustration, material culture, performance, and the graphic arts. The Mary Sheriff Travel and Research Award, in the amount of $2,000, supports feminist topics in eighteenth-century art history and visual culture. Initially, the Award will be given every other year, beginning in 2018, and continuing in even years from that point forward. The annual deadline for application submission is January 15. The Award will be presented at the Historians of Eighteenth Century Art and Architecture (HECAA) annual luncheon and business meeting at ASECS.
*Requirements:* Doctoral candidates and early career scholars are eligible to apply.
*Geographic Focus:* All States
*Date(s) Application is Due:* Jan 15
*Amount of Grant:* 2,000 USD
*Contact:* Lisa Berglund, Executive Director; (716) 878-3405 or (716) 878-5417; fax (716) 878-4939; berglul@buffalostate.edu or asecsoffice@gmail.com
*Internet:* https://asecs.press.jhu.edu/general%20site/asecsawards.html
*Sponsor:* American Society for Eighteenth-Century Studies
1300 Elmwood Avenue, Ketcham Hall 213
Buffalo, NY 14222

**ASECS Paula Backscheider Archival Fellowship** **601**
Paula Backscheider is a Philpott-Stevens Eminent Scholar employed by the Department of English at Auburn University. She earned her Ph.D. at Purdue University in West Lafayette, Indiana, and now believes that teaching eighteenth-century literature has never been easy, and it certainly is not going to become more so. "There's the language, especially for the period before 1740," claims Backscheifder, "and it has always seemed to to require the most interdisciplinary knowledge and skills of any period. Politics, economics, social issues, historical events, legal change, medical advances, the arts, and philosophical currents, some from other countries such as France and the Netherlands, are routinely mentioned or discussed in depth in every kind of literature from belle letters to chapbooks, and researchers must be conversant with them. However, a great attraction of the literature is that it was written in a time when people thought literature and printed texts mattered and actually impacted the world and the people around them." The Paula Backscheider Archival Fellowship, named in her honor, is a $1,000 funding award that supports those researchers whose projects necessitate work in archives, repositories, and special collections (public and private) in foreign countries and/or in the United States. The annual deadline for application submissions is January 1.
*Requirements:* Only members of ASECS in good standing are eligible to apply.
*Geographic Focus:* All States
*Date(s) Application is Due:* Jan 1
*Amount of Grant:* 1,000 USD
*Contact:* Lisa Berglund, Executive Director; (716) 878-3405 or (716) 878-5417; fax (716) 878-4939; berglul@buffalostate.edu or asecsoffice@gmail.com
*Internet:* https://asecs.press.jhu.edu/general%20site/travelgr.html
*Sponsor:* American Society for Eighteenth-Century Studies
1300 Elmwood Avenue, Ketcham Hall 213
Buffalo, NY 14222

**ASECS Richard H. Popkin Research Travel Fellowship** **602**
Richard Henry Popkin was an academic philosopher who specialized in the history of enlightenment philosophy and early modern anti-dogmatism. His 1960 work, entitled The History of Scepticism from Erasmus to Descartes, introduced one previously unrecognized influence on Western thought in the seventeenth century, the Pyrrhonian Scepticism of Sextus Empiricus. Popkin also was an internationally acclaimed scholar on Jewish and Christian millenarianism and messianism. The ASECS Richard H. Popkin Research Travel Fellowship is a $250 award intended to supplement costs for scholars to travel to collections in North America and abroad. The annual deadline for submissions is January 1.
*Requirements:* Only members of ASECS in good standing are eligible to apply.
*Geographic Focus:* All States
*Date(s) Application is Due:* Jan 1
*Amount of Grant:* 250 USD
*Contact:* Lisa Berglund, Executive Director; (716) 878-3405 or (716) 878-5417; fax (716) 878-4939; berglul@buffalostate.edu or asecsoffice@gmail.com
*Internet:* https://asecs.press.jhu.edu/general%20site/travelgr.html
*Sponsor:* American Society for Eighteenth-Century Studies
1300 Elmwood Avenue, Ketcham Hall 213
Buffalo, NY 14222

**ASECS Robert R. Palmer Research Travel Fellowship** **603**
Robert Roswell Palmer, commonly known as R. R. Palmer, was a distinguished American historian at Princeton and Yale universities, who specialized in eighteenth-century France. His most influential work of scholarship, The Age of the Democratic Revolution: A Political History of Europe and America, 1760–1800, examined an age of democratic revolution that swept the Atlantic civilization between 1760 and 1800. He was awarded the Bancroft Prize in History for the first volume. Palmer also achieved distinction as a history text writer. The ASECS Robert R. Palmer Research Travel Fellowship of $500 is awarded annually to support documentary research related primarily to the history and culture of France. Applicants must be members of the society at the time of application. There are no restrictions based on age, sex, race, religion, or academic rank. The annual deadline for applications is January 1.
*Requirements:* All members of ASECS are eligible to apply.
*Geographic Focus:* All States
*Date(s) Application is Due:* Jan 1
*Amount of Grant:* 500 USD
*Contact:* Lisa Berglund, Executive Director; (716) 878-3405 or (716) 878-5417; fax (716) 878-4939; berglul@buffalostate.edu or asecsoffice@gmail.com
*Internet:* https://asecs.press.jhu.edu/general%20site/travelgr.html
*Sponsor:* American Society for Eighteenth-Century Studies
1300 Elmwood Avenue, Ketcham Hall 213
Buffalo, NY 14222

**ASECS Srinivas Aravamudan Prize** 604
The Srinivas Aravamudan Prize goes to the author of an article published in the previous year that pushes the boundaries, geographical and conceptual, of eighteenth-century studies, especially by using a transnational, comparative, or cosmopolitan approach. It carries an award of $250. The article may be nominated by a member of the society, by its author, or by an editor of the publishing journal. The annual deadline for nominations is January 1.
*Requirements:* The article should be no longer than 15,000 words. The article must have been published in a journal or any peer-reviewed publication between July and June of the previous fiscal year period.
*Geographic Focus:* All States
*Date(s) Application is Due:* Jan 1
*Amount of Grant:* 250 USD
*Contact:* Lisa Berglund, Executive Director; (716) 878-3405 or (716) 878-5417; fax (716) 878-4939; berglul@buffalostate.edu or asecsoffice@gmail.com
*Internet:* https://asecs.press.jhu.edu/general%20site/asecsawards.html
*Sponsor:* American Society for Eighteenth-Century Studies
1300 Elmwood Avenue, Ketcham Hall 213
Buffalo, NY 14222

**ASECS Theodore E.D. Braun Research Travel Fellowship** 605
Theodore E.D. Braun was born in Brooklyn, the youngest of six children, twenty minutes after his twin brother, Larry. Life was tough for the Braun family, who lived in extreme poverty in what was then called a cold-water flat. But they all survived and, fortunately Braun listened to his mother when she told him that the way to a better life was through education. He won a full-tuition scholarship at Teacher's College of St. John's University, which was then located in Brooklyn, and he received a B.A. degree in French Education in 1955. After a year in Valence - a small city on the Rhone River in France, about 100 km south of Lyon, he became an assistant d'anglais at the Lycee Emile Loubet before he was drafted in the army (his serial number began with US, which he always claimed stood for Unwilling Soldier). The ASECS Theodore E.D. Braun Research Travel Fellowship, in the amount of $1,000, is intended to support researchers, regardless of rank, who are working in French literary studies. The annual application deadline for submissions is January 1.
*Requirements:* Only members of ASECS in good standing are eligible to apply.
*Geographic Focus:* All States
*Date(s) Application is Due:* Jan 1
*Amount of Grant:* 1,000 USD
*Contact:* Lisa Berglund, Executive Director; (716) 878-3405 or (716) 878-5417; fax (716) 878-4939; berglul@buffalostate.edu or asecsoffice@gmail.com
*Internet:* https://asecs.press.jhu.edu/general%20site/travelgr.html
*Sponsor:* American Society for Eighteenth-Century Studies
1300 Elmwood Avenue, Ketcham Hall 213
Buffalo, NY 14222

**ASECS Travelling Jam-Pot: Fund for Graduate Students Awards** 606
The Travelling Jam-Pot: Fund for Graduate Student award recipients are ABDs and PhDs within a year after receiving doctoral degree. An award of up to $300 will be given toward the cost of attending the ASECS annual meeting. The annual application deadline is November 1.
*Requirements:* Four copies of the following information must be submitted by each applicant: a statement of need; identity of other sources of funding sought; budget; endorsement from a faculty member (one copy is sufficient); and number of professional conferences attended in the past year.
*Geographic Focus:* All States
*Date(s) Application is Due:* Nov 1
*Contact:* Lisa Berglund, Executive Director; (716) 878-3405 or (716) 878-5417; fax (716) 878-4939; berglul@buffalostate.edu or asecsoffice@gmail.com
*Internet:* https://asecs.press.jhu.edu/Weekly%20Announcements/jampot.html
*Sponsor:* American Society for Eighteenth-Century Studies
1300 Elmwood Avenue, Ketcham Hall 213
Buffalo, NY 14222

**ASGH Arno Motulsky-Barton Childs Award for Excellence in Human Genetics Education** 607
The American Society of Human Genetics (ASGH) established the Award for Excellence in Human Genetics Education (renamed the Arno Motulsky-Barton Childs Award for Excellence in Human Genetics Education in 2015) to recognize outstanding contributions to human genetics education. Nominees should have long-standing involvement in genetics education, contributions in more than one area, and contributions of substantive influence on individuals and/or organizations. Examples of the types of contributions that might qualify a nominee include: producing a set of writings that has had a major influence on human genetics education, developing a curriculum or innovative teaching program that is widely emulated, writing a book that has been adopted by many universities, developing an educational website, or directing a fellowship program that has consistently produced exemplary graduates. Nominees may be individuals or groups.
*Requirements:* Nominees must have made contributions that are recognized nationally or internationally as being of exceptional quality and great importance to human genetics education. All nominees and winners must be current ASHG members. If a group is nominated, at least one nominee in the group must be a current member. A monetary award and plaque will be presented to the recipient(s) at the Society's Annual Meeting. Nominees for the ASHG Arno Motulsky-Barton Childs Award for Excellence in Human Genetics Education may emerge from submissions from the membership, from the Awards Committee, or from the Information and Education Committee.
*Geographic Focus:* All States
*Amount of Grant:* 10,000 USD
*Contact:* Mona Miller, Executive Director; (866) 486-4363 or (301) 634-7300; fax (301) 634-7079; mmiller@ashg.org
*Internet:* www.ashg.org/awards/index.shtml
*Sponsor:* American Society of Human Genetics
6120 Executive Boulevard, Suite 500
Rockville, MD 20852

**ASGH C.W. Cotterman Award** 608
Each September, the editorial board of The American Journal of Human Genetics selects two articles published in the journal in the previous year that best represent outstanding scientific contributions to the field of human genetics. Two Cotterman Awards will be given annually and a monetary award of $1,000 and a plaque will be presented to the recipients for the top two papers published in the Journal during the previous year on which the first author was either a pre- or post- doctoral trainee and an ASHG member.
*Geographic Focus:* All States
*Amount of Grant:* 1,000 USD
*Contact:* Mona Miller, Executive Director; (866) 486-4363 or (301) 634-7300; fax (301) 634-7079; mmiller@ashg.org
*Internet:* www.ashg.org/awards/index.shtml
*Sponsor:* American Society of Human Genetics
6120 Executive Boulevard, Suite 500
Rockville, MD 20852

**ASGH Charles J. Epstein Trainee Awards for Excellence in Human Genetics Research** 609
The American Society of Human Genetics (ASGH) honors excellence in research conducted by predoctoral and postdoctoral trainees (including genetic counseling trainees) through merit-based awards that recognize highly competitive abstracts submitted and presented at the ASHG Annual Meeting. These awards were renamed in 2012 to honor the late Dr. Charles Epstein, who was a past president of ASHG, former editor of AJHG, and winner of both the William Allan Award and the McKusick Leadership Award. The Program Committee and Awards Committee, in consultation with the ASHG Board of Directors, determine the number of awards, the categories for which they are given, and the prize amounts of each award (currently $750 plus complimentary registration for approximately 60 semifinalists, plus an additional $250 to 18 finalists). An additional $1,000 is given to the winners. Abstract scoring is completed by the Program Committee, and the top-ranking abstracts (semifinalists) are submitted to the Awards Committee, which selects 18 finalists prior to the Annual Meeting. Six winners are chosen after the committee members judge the finalists' presentations. The total value of ASHG's Trainee Awards is approximately $70,000 annually.
*Geographic Focus:* All States
*Amount of Grant:* 750 - 2,000 USD
*Contact:* Mona Miller, Executive Director; (866) 486-4363 or (301) 634-7300; fax (301) 634-7079; mmiller@ashg.org
*Internet:* www.ashg.org/awards/index.shtml
*Sponsor:* American Society of Human Genetics
6120 Executive Boulevard, Suite 500
Rockville, MD 20852

**ASGH Curt Stern Award** 610
The American Society of Human Genetics (ASGH) Curt Stern Award honors the memory of Curt Stern (1902-1981) as an outstanding pioneering human geneticist. This award is presented yearly for outstanding scientific achievements in human genetics that occurred in the last 10 years. The work could be a single major discovery or may be a series of contributions on a similar or related topic. A plaque and a monetary award of $10,000 are presented to the recipient(s) at the Society's Annual Meeting. Nominations for the Stern Award will be solicited from the ASHG Awards Committee and the general membership.
*Geographic Focus:* All States
*Amount of Grant:* 10,000 USD
*Contact:* Mona Miller, Executive Director; (866) 486-4363 or (301) 634-7300; fax (301) 634-7079; mmiller@ashg.org
*Internet:* www.ashg.org/awards/index.shtml
*Sponsor:* American Society of Human Genetics
6120 Executive Boulevard, Suite 500
Rockville, MD 20852

**ASGH Early-Career Award** 611
The Early-Career Award recognizes scientists who are in the early stages of their careers as independent investigators. The candidate must not have been an independent investigator for more than ten years and the work submitted for consideration must have been primarily the product of the candidate's independent lab. A plaque and a $10,000 prize will be presented at the ASHG Annual Meeting.
*Geographic Focus:* All States
*Amount of Grant:* 10,000 USD
*Contact:* Mona Miller, Executive Director; (866) 486-4363 or (301) 634-7300; fax (301) 634-7079; mmiller@ashg.org
*Internet:* www.ashg.org/awards/index.shtml
*Sponsor:* American Society of Human Genetics
6120 Executive Boulevard, Suite 500
Rockville, MD 20852

**ASGH Victor A. McKusick Leadership Award** 612

The prestigious Victor A. McKusick Leadership Award, established by the American Society of Human Genetics (ASHG) in honor of the late Dr. Victor A. McKusick, is presented on behalf of the Society to an individual whose professional achievements have fostered and enriched the development of various human genetics disciplines. Potential recipients should exemplify the enduring leadership and vision required to ensure that the field of human genetics will flourish and successfully assimilate into the broader context of science, medicine, and health. They also may have made major contributions to awareness or understanding of human genetics by policy makers or by the general public. A plaque and $10,000 prize will be presented in honor of the awardee at the ASHG Annual Meeting.
*Geographic Focus:* All States
*Amount of Grant:* 10,000 USD
*Contact:* Mona Miller, Executive Director; (866) 486-4363 or (301) 634-7300; fax (301) 634-7079; mmiller@ashg.org
*Internet:* www.ashg.org/awards/index.shtml
*Sponsor:* American Society of Human Genetics
6120 Executive Boulevard, Suite 500
Rockville, MD 20852

**ASGH William Allan Award** 613

The William Allan Award is the top prize given by the American Society of Human Genetics (ASHG); it was established in 1961 in memory of William Allan (1881-1943), who was one of the first American physicians to conduct extensive research in human genetics. The Allan Award is presented annually to recognize substantial and far-reaching scientific contributions to human genetics, carried out over a sustained period of scientific inquiry and productivity. The recipient is presented with an engraved medal and a monetary award of $25,000 at the ASHG Annual Meeting. The Allan Award winner is also invited to present a 30-45 minute address at the ASHG Annual Meeting, and it is customary to publish a manuscript of the presentation in The American Journal of Human Genetics. This award is given on a yearly basis, but can be omitted in any given year at the discretion of the Board. Nominations will be solicited from the ASHG Awards Committee and the general membership.
*Geographic Focus:* All States
*Amount of Grant:* 25,000 USD
*Contact:* Mona Miller, Executive Director; (866) 486-4363 or (301) 634-7300; fax (301) 634-7079; mmiller@ashg.org
*Internet:* www.ashg.org/awards/index.shtml
*Sponsor:* American Society of Human Genetics
6120 Executive Boulevard, Suite 500
Rockville, MD 20852

**ASHG Advocacy Award** 614

The ASHG Advocacy Award honors individuals or groups of individuals who have exhibited excellence and achievement in promoting the science of human genetics and its application for the common good. Areas of advocacy may include, but are not limited to, the following: promoting the importance of genetics and an understanding of genetically based healthcare in society at large; increasing public awareness of ethical, legal, and policy-related issues in genetics; promoting the importance of funding for biomedical research; spearheading/shepherding/implementing genetics policy or legislation; and improving familiarity with genetics research and genetic testing among a broad range of audiences. The award recipient receives a plaque and a monetary award of $10,000 at the Society's Annual Meeting. The award is presented annually, but may be omitted in any given year at the discretion of the Nominating Committee and Board.
*Restrictions:* Nominees need not be members of ASHG, but nominations must come from members of the Society.
*Geographic Focus:* All States
*Amount of Grant:* 10,000 USD
*Contact:* Mona Miller, Executive Director; (866) 486-4363 or (301) 634-7300; fax (301) 238-3709; mmiller@ashg.org
*Internet:* www.ashg.org/awards/index.shtml
*Sponsor:* American Society of Human Genetics
6120 Executive Boulevard, Suite 500
Rockville, MD 20852

**ASHG Mentorship Award** 615

The ASHG Mentorship Award honors ASHG members who have significant records of accomplishment as mentors. The award is open to individuals at all academic ranks, but eligible candidates must have demonstrated a sustained pattern of exemplary mentorship at the graduate student, postdoctoral, residency, or fellowship level. Candidates can be nominated by a colleague or a mentee. The nomination letter should describe the mentorship accomplishments of the candidate in sufficient detail for the selection committee to evaluate the candidacy. While it is not necessary to provide a list of all individuals trained by the nominee, the selection committee will consider information regarding mentorship style and its profound personal and/or professional impact on representative trainees, including their career decisions, trajectories and accomplishments. A second supporting letter is required. Either or both letters can contain testimonial support from other trainees with their expressed permission. At least one of these letters must come from a mentee, and at least one of the two individuals writing letters must be a member of ASHG. In addition, the nominee's CV must accompany the letters of nomination. The awardee will receive a plaque and a monetary award of $10,000 at the Society's Annual Meeting.
*Geographic Focus:* All States
*Amount of Grant:* 10,000 USD
*Contact:* Mona Miller, Executive Director; (866) 486-4363 or (301) 634-7300; fax (301) 634-7079; mmiller@ashg.org
*Internet:* www.ashg.org/awards/index.shtml
*Sponsor:* American Society of Human Genetics
6120 Executive Boulevard, Suite 500
Rockville, MD 20852

**ASHG Rosalind Franklin Young Investigator Award** 616

In recognition of the groundbreaking contributions of Dr. Rosalind Franklin, this career development research award is intended to inspire and support new generations of women in the field of genetics. Successful candidates are women in the first one to three years of an independent faculty-level position whose work displays originality and scientific creativity, making seminal discoveries within their fields. The Rosalind Franklin Young Investigator Award is funded by The Gruber Foundation, and is administered by a joint committee appointed by the Genetics Society of America and the American Society of Human Genetics. Every three years, two women are chosen to receive the Rosalind Franklin Award, and each awardee receives $75,000 over three years ($25,000 per year). One award funds genetics research in human and non-human mammals, and another award funds genetics research in model organisms.
*Geographic Focus:* All States
*Amount of Grant:* 75,000 USD
*Contact:* Mona Miller, Executive Director; (866) 486-4363 or (301) 634-7300; fax (301) 634-7079; mmiller@ashg.org
*Internet:* www.ashg.org/awards/index.shtml
*Sponsor:* American Society of Human Genetics
6120 Executive Boulevard, Suite 500
Rockville, MD 20852

**Ashland Corporate Contributions Grants** 617

The foundation awards grants to eligible Kentucky, Ohio, and West Virginia nonprofit organizations in its areas of interest, including education, arts, communities/civic, disaster relief, environment, and health and human services. Types of support include seed money grants, project grants, endowments, matching gifts, and employee volunteers. The foundation does not provide an application. Most giving is centered on programs that best address the needs of Ashland's employees, stockholders, customers, and other constituencies, and the communities in which they live. Funding requests are not solicited but will be considered. The primary focus is on education, with an emphasis on mentoring, literacy and/or diversity.
*Requirements:* 501(c)3 tax-exempt organizations in Boyd and Greenup Counties, KY, Lawrence County, OH, and Cabell and Wayne Counties, WV, are eligible. Charitable groups within church or religious organizations are eligible.
*Restrictions:* The foundation does not support individuals, capital campaigns for building or equipment, endowments, travel, film or video production, tickets, religious or political activities, or goodwill advertising.
*Geographic Focus:* Kentucky, Ohio, West Virginia
*Samples:* University of Kentucky, Lexington, KY—for the William T. Young Library Endowment, $1 million.
*Contact:* Program Contact; (859) 815-3333; community@ashland.com
*Internet:* www.ashland.com/commitments/contributions.asp
*Sponsor:* Ashland
50 East Rivercenter Boulevard, P.O. Box 391
Covington, KY 41012-0391

**ASID Foundation Irene Winifred Eno Grant** 618

The Irene Winifred Eno Grant award in the amount of $5,000 is intended to provide financial assistance to individuals or groups engaged in the creation of an educational program or an interior design research project dedicated to health, safety and welfare. The grant will be awarded on the basis of the project description, break down of potential use of funds and the marketing plan for the use/distribution of the end product of the project. The annual deadline for applications is April 18.
*Requirements:* The grant is open to students, educators, interior design practitioners, institutions or other interior design related group.
*Geographic Focus:* All States
*Date(s) Application is Due:* Apr 18
*Amount of Grant:* 5,000 USD
*Contact:* Matthew DeGeeter, Director; (202) 675-2379 or (202) 546-3480; fax (202) 546-3240; mdegeeter@asid.org
*Internet:* www.asidfoundation.org/SCHOLARSHIPS.html
*Sponsor:* American Society of Interior Designers Foundation
718 7th Street NW, 4th Floor
Washington, D.C. 20001

**ASID Foundation Joel Polsky Academic Achievement Award** 619

This Joel Polsky Academic Achievement Award in the amount of $5,000 is given annually to recognize an outstanding interior design research or thesis project by an undergraduate or graduate student. Entries should address the needs of the public, designers, and students on such topics as educational research, behavioral science, business practice, design process, theory, or other technical subjects. Material will be judged on bibliography/references, breadth of material, comprehensive coverage of topic, graphic presentation and organization, and innovative subject matter. Entries will be returned upon request with a $10 postage and handling fee. The annual deadline is April 18.
*Requirements:* Application requirements include: an abstract (maximum of 250 words); a digital copy of thesis, dissertation, or research project; a biographical statement; and a headshot.

*Geographic Focus:* All States
*Date(s) Application is Due:* Apr 18
*Amount of Grant:* 5,000 USD
*Contact:* Matthew DeGeeter, Director; (202) 675-2379 or (202) 546-3480; fax (202) 546-3240; mdegeeter@asid.org
*Internet:* www.asidfoundation.org/SCHOLARSHIPS.html
*Sponsor:* American Society of Interior Designers Foundation
718 7th Street NW, 4th Floor
Washington, D.C. 20001

**ASID Foundation Joel Polsky Prize** **620**
The Joel Polsky Prize in the amount of $5,000 is given by the ASID Foundation to recognize outstanding academic contributions to the discipline of interior design through literature or visual communication. Entries should address the needs of the public, designers and students on such topics as educational research, behavioral science, business practice, design process, theory or other technical subjects. Material will be judged on innovative subject matter, comprehensive coverage of topic, organization, graphic presentation, bibliography and references. The annual deadline for applications is April 18.
*Requirements:* Application requirements include: description of the publication or visual communication (maximum of 250 words); copy of the publication or visual communication; a biographical statement; and a headshot.
*Geographic Focus:* All States
*Date(s) Application is Due:* Apr 18
*Amount of Grant:* 5,000 USD
*Contact:* Matthew DeGeeter, Director; (202) 675-2379 or (202) 546-3480; fax (202) 546-3240; mdegeeter@asid.org
*Internet:* www.asidfoundation.org/SCHOLARSHIPS.html
*Sponsor:* American Society of Interior Designers Foundation
718 7th Street NW, 4th Floor
Washington, D.C. 20001

**ASID Foundation Legacy Scholarship for Graduate Students** **621**
The Legacy Scholarship for Graduate Students will be awarded on the basis of academic accomplishment, creative achievement, and personal statement. Applicants should write an essay on the following subject: as an interior designer, how will you embrace and advance the professional trend to positively impact health and well-being in your anticipated area of design practice? The award is equal to $4,000, and the application deadline is April 18.
*Requirements:* The ASID Foundation Legacy Scholarship is open to students who are enrolled in or have been accepted for admission to a graduate-level interior design program at an accredited university.
*Geographic Focus:* All States
*Date(s) Application is Due:* Apr 18
*Amount of Grant:* 4,000 USD
*Contact:* Matthew DeGeeter, Director; (202) 675-2379 or (202) 546-3480; fax (202) 546-3240; mdegeeter@asid.org
*Internet:* www.asidfoundation.org/SCHOLARSHIPS.html
*Sponsor:* American Society of Interior Designers Foundation
718 7th Street NW, 4th Floor
Washington, D.C. 20001

**ASM/CDC Fellowships in Infectious Disease and Public Health Microbiology** **622**
The ASM/CD.C. Postdoctoral Research Fellowship Program is sponsored by the American Society for Microbiology (ASM) and the Centers of Disease Control and Prevention (CD.C.). Founded in 1899, the American Society for Microbiology (ASM) promotes the microbiological sciences and their application for the common good. The Society accomplishes its mission by disseminating information, stimulating research, promoting education, and advancing the profession. ASM's headquarters office is located in Washington, D.C. The Centers for Disease Control and Prevention (CD.C.) is the agency of the United States Public Health Service (PHS) responsible for developing and applying disease control and prevention methods and improving the health of the citizens of the United States. The CD.C. has been a leader in efforts to prevent and control infectious diseases for nearly half a century. CD.C. supports surveillance, research, prevention efforts, and training in the area of infectious diseases. The goal of the two-year ASM/CD.C. Fellowships is to support the development of new approaches, methodologies and knowledge in infectious disease prevention and control in areas within the public health mission of the CD.C.. The program awards up to ten fellowships annually. The fellowships are similar to those at the postdoctoral level in universities, industry, and other government laboratories and allow bright, highly motivated, recent doctoral graduates to perform research-in-residence while headquartered at a CD.C. location. Participating CD.C. organizational units are located in: Atlanta, Georgia; Fort Collins, Colorado; and San Juan, Puerto Rico and are comprised of: the Division of Bacterial Diseases; the Division of Foodborne, Bacterial and Mycotic Diseases; the Division of Healthcare Quality Promotion; the Division of HIV/AIDS Prevention; the Division of Influenza; the Division of Viral Hepatitis; the Division of Parasitic Diseases; the Division of Tuberculosis Elimination; the Divisions of Vector-Borne Infectious Diseases; the Division of Viral and Rickettsial Diseases; the Division of Viral Diseases; and the Division of Viral Hepatitus. The fellowship appointments are for two years assuming satisfactory progress. A stipend (up to $45,243 annually), health benefits (up to $3000 annually), relocation benefits (up to $500), and professional development funds (up to $2000 annually) are provided. Fellows are classified as fellows and are not employees of either CD.C. or ASM but rather have the status of a visiting scientist. Applicants must use the ASM online application system to apply for the fellowship. The link to the online system is available at the program website on the top right corner of the screen. Applicants are encouraged to review the "Full Program Catalog" to obtain complete information about the program, research opportunities, and the application process. The link to the catalog is available at the program website. The ASM/CD.C. Postdoctoral Research Fellowship Program is sponsored by the U.S. Department of Health and Human Services.
*Requirements:* The program is intended for individuals who either earned their doctorate degree or completed a primary residency within three years of their proposed start date (considerations will be given to individuals with more experience if there are compelling reasons). Non-U.S. nationals whose primary language is not English must demonstrate command of the English language through the Test of English as a Foreign Language (TOEFL). These applicants may also be subject to an oral interview. Non-U.S. nationals who are offered awards must have valid visas throughout their tenure. Fellows will be responsible for managing all their own visa matters. Fellows must devote their full-time effort to the approved research project and must be in residence at the sponsoring CD.C. laboratory for the entire two-year period. Fellows must begin the program between October 1 and December 31 during the year in which they are accepted. All fellows must undergo security clearance to work at CD.C.. Support after the first year of the Program is contingent upon satisfactory progress in the proposed research. Interim and final reports and/or presentations are required. Fellows are expected to publish results of their studies in the scientific literature and to present findings at national meetings. Fellows are subject to the general regulations of the laboratory and are expected to conduct research according to the highest scientific and ethical standards and in compliance with all applicable laws, regulations, and policies regarding protection of human research subjects, humane care and use of laboratory animals, and laboratory safety. Participation in CD.C. seminar programs is also expected. After completion of the two-year program, fellows are required to keep ASM informed of their current position, affiliation, and continued professional development for a minimum of ten years. This arrangement will ensure proper assessment and longitudinal studies regarding the effectiveness of such training.
*Restrictions:* Applicants may not have a faculty position or be enrolled in a graduate degree program during the fellowship. Monies from other appointments, fellowships, private employment, or consulting or contract work are not permitted during the two-year appointment.
*Geographic Focus:* All States, All Countries
*Date(s) Application is Due:* Jan 15
*Amount of Grant:* Up to 100,986 USD
*Contact:* Tiffani Fonseca, Coordinator, Fellowship Programs; (202) 942-9295 or (202) 942-9283; fax (202) 942-9329; fellowships@asmusa.org or tfonseca@asmusa.org
*Internet:* www.asm.org/asm/index.php/education/asmcdc-program-in-infectious-disease-and-public-health-microbiology.html
*Sponsor:* American Society for Microbiology
1752 N Street, N.W.
Washington, D.C. 20036-2904

**ASME Adaptive Structures and Material Systems Award** **623**
The American Society of Mechanical Engineers Adaptive Structures and Material Systems Award recognizes significant contributions to the sciences and technologies associated with adaptive structures and/or materials systems. The award is intended to honor a lifetime of achievement and sustained impact in the field and is given only to a senior researcher. The awardee will receive $2,000, the Vermeil Medal, and a certificate. The annual deadline for nominations is January 31.
*Requirements:* The nominee must be a senior researcher who has made a significant contribution to the field of adaptive structures and material systems. The General Awards Committee screens nominations and makes recommendations to the Committee on Honors. It consists of Members or Fellows representing the Districts. Special Award Committees are usually associated with a particular honor or award. They screen nominations and make recommendations to the other two committees.
*Restrictions:* Except for the Melville Medal, an individual will receive only one honor in recognition of the same achievement. The receipt of one ASME honor shall not bar the recipient from another ASME honor provided it is for different accomplishments. Honors are not awarded posthumously except if a nominee's death occurs after the nomination has been received at ASME Headquarters.
*Geographic Focus:* All States, All Countries
*Date(s) Application is Due:* Jan 31
*Amount of Grant:* 2,000 USD
*Contact:* Ralph C. Smith, Committee Chair; (919) 515-7552 or (800) 843-2763; fax (212) 705-7739; rsmith@ncsu.edu
*Internet:* https://www.asme.org/about-asme/honors-awards/achievement-awards/adaptive-structures-material-systems-award
*Sponsor:* American Society of Mechanical Engineers
Two Park Avenue
New York, NY 10016-5990

**ASME Allan Kraus Thermal Management Medal** **624**
The ASME Allan Kraus Thermal Management Medal was established in 1994 by the Electronic and Photonic Packaging Division and operated as a divisional award until 2009, when it was elevated to a Society Award and renamed the Allan Kraus Thermal Management Medal. The award recognizes an individual who has demonstrated outstanding achievements in thermal management of electronic systems and his or her commitment to the field of thermal science and engineering. The award includes a $1,000 stipend, bronze medal, a certificate, and up to $750 in travel expenses to attend the award presentation. The annual deadline for nominations is February 1.
*Requirements:* The nominee for the award should have significant contributions in thermal management of electronic systems demonstrated by successful product development,

seminal papers, filed patents and/or leadership of research and development programs. The award may be made to one recipient of any age who need not be a member of ASME. The General Awards Committee screens nominations and makes recommendations to the Committee on Honors. It consists of Members or Fellows representing the Districts. Special Award Committees are usually associated with a particular honor or award. They screen nominations and make recommendations to the other two committees.
*Restrictions:* Services performed within the context of an individual's normal employment are not eligible. Except for the Melville Medal, an individual will receive only one honor in recognition of the same achievement. The receipt of one ASME honor shall not bar the recipient from another ASME honor provided it is for different accomplishments. Honors are not awarded posthumously except if a nominee's death occurs after the nomination has been received at ASME Headquarters.
*Geographic Focus:* All States, All Countries
*Date(s) Application is Due:* Feb 1
*Amount of Grant:* 1,000 - 1,750 USD
*Contact:* Ravi V. Mahajan, Committee Chair; (480) 554-3715 or (800) 843-2763; ravi.v.mahajan@intel.com
*Internet:* https://www.asme.org/about-asme/participate/honors-awards/achievement-awards/allan-kraus-thermal-management-medal
*Sponsor:* American Society of Mechanical Engineers
Two Park Avenue
New York, NY 10016-5990

**ASME Arthur L. Williston Medal** 625

The ASME Arthur L. Williston Medal is presented for the best paper or thesis submitted by an undergraduate or junior engineer fostering a spirit of civil service. Candidates must be student members or associate members with an ASME sponsor for the award. First place recipients are awarded a bronze medal, certificate, $1,000, and up to $750 in travel expenses to attend the award presentation. Second place award winners receive a certificate and $500, while third place winners receive a certificate and $250. The annual deadline for nominations is March 1.
*Requirements:* Applicant must be a student member or an associate member of ASME who has not graduated more than two years prior.
*Restrictions:* Services performed within the context of an individual's normal employment are not eligible. Except for the Melville Medal, an individual will receive only one honor in recognition of the same achievement. The receipt of one ASME honor shall not bar the recipient from another ASME honor provided it is for different accomplishments. Honors are not awarded posthumously except if a nominee's death occurs after the nomination has been received at ASME Headquarters.
*Geographic Focus:* All States, All Countries
*Date(s) Application is Due:* Mar 1
*Amount of Grant:* 250 - 1,750 USD
*Contact:* Fran McKivor, Committee on Honors; (212) 591-7094 or (800) 843-2763; mckivorf@asme.org or awards@asme.org
*Internet:* https://www.asme.org/about-asme/get-involved/honors-awards/literature-awards/arthur-l-williston-medal
*Sponsor:* American Society of Mechanical Engineers
Two Park Avenue
New York, NY 10016-5990

**ASME Barnett-Uzgiris Product Safety Design Award** 626

The American Society of Mechanical Engineers Barnett-Uzgiris Safety Design Award recognizes individuals who have made significant contributions to the safe design of products through teaching, research, and professional accomplishments. The award was established as the Triodyne Safety Award by the Design Engineering Division and operated as a division award until 2008 when it was elevated to a Society award and renamed. The award includes a bronze medal, $1,000, a certificate, and travel expenses of a maximum of $750 to attend the award presentation. Nomination forms and instructions are available at the website. The nomination deadline is February 1.
*Requirements:* The General Awards Committee screens nominations and makes recommendations to the Committee on Honors. It consists of Members or Fellows representing the Districts. Special Award Committees are usually associated with a particular honor or award. They screen nominations and make recommendations to the other two committees.
*Restrictions:* Except for the Melville Medal, an individual will receive only one honor in recognition of the same achievement. The receipt of one ASME honor shall not bar the recipient from another ASME honor provided it is for different accomplishments. Honors are not awarded posthumously except if a nominee's death occurs after the nomination has been received at ASME Headquarters.
*Geographic Focus:* All States, All Countries
*Date(s) Application is Due:* Feb 1
*Amount of Grant:* 1,000 - 1,750 USD
*Contact:* J.R. Mountain; (281) 283-3846 or (800) 843-2763; mountain@uhcl.edu
*Internet:* https://www.asme.org/about-asme/participate/honors-awards/achievement-awards/barnett-uzgiris-product-safety-award
*Sponsor:* American Society of Mechanical Engineers
Two Park Avenue
New York, NY 10016-5990

**ASME Ben C. Sparks Medal** 627

The ASME Ben C. Sparks Medal is presented to a mechanical engineer for recognition of eminent service to mechanical engineering technology education through contributions toward ASME; higher education professional service; and the teaching and/or research facility roles in mechanical engineering and mechanical engineering technology. The award includes a bronze metal, a certificate, $1,000, and up to $750 in travel expenses to attend the award presentation. The annual deadline for nominations is September 1.
*Requirements:* Candidates must have a record of outstanding performance in all of these areas: a broad range of contributions to ASME; dedicated efforts and a record of achievements in the advancement of engineering education and engineering technology education; and contributions to the positive support, recognition, or interaction of mechanical engineering technology and mechanical engineering programs, their faculties, or students. Recipients must hold current academic appointments in ABET-accredited departments where mechanical engineering and/or mechanical engineering technology is taught.
*Restrictions:* Services performed within the context of an individual's normal employment are not eligible. Except for the Melville Medal, an individual will receive only one honor in recognition of the same achievement. The receipt of one ASME honor shall not bar the recipient from another ASME honor provided it is for different accomplishments. Honors are not awarded posthumously except if a nominee's death occurs after the nomination has been received at ASME Headquarters.
*Geographic Focus:* All States, All Countries
*Date(s) Application is Due:* Sep 1
*Amount of Grant:* 1,000 - 1,750 USD
*Contact:* Kenneth Ball, Committee on Honors; (571) 488-8961 or (800) 843-2763; ball@gmu.edu or awards@asme.org
*Internet:* https://www.asme.org/about-asme/participate/honors-awards/achievement-awards/ben-c-sparks-medal
*Sponsor:* American Society of Mechanical Engineers
Two Park Avenue
New York, NY 10016-5990

**ASME Bergles-Rohsenow Young Investigator Award in Heat Transfer** 628

The American Society of Mechanical Engineers Bergles-Rohsenow Young Investigator Award in Heat Transfer is given to a young engineer no older than 35 years of age who has received a Ph.D. or an equivalent degree in engineering. The individual must be committed to pursuing research and making significant contributions in heat transfer. Such contributions may include analytical/numerical methods, equipment/instrumentation, or experimentation, which should lead to peer-reviewed publications. The award includes a bronze metal, a certificate, and $1,000. Nomination forms and instructions are available at the website. The annual nomination deadline is October 1.
*Requirements:* The General Awards Committee screens nominations and makes recommendations to the Committee on Honors. It consists of Members or Fellows representing the Districts. Special Award Committees are usually associated with a particular honor or award. They screen nominations and make recommendations to the other two committees.
*Restrictions:* Except for the Melville Medal, an individual will receive only one honor in recognition of the same achievement. The receipt of one ASME honor shall not bar the recipient from another ASME honor provided it is for different accomplishments. Honors are not awarded posthumously except if a nominee's death occurs after the nomination has been received at ASME Headquarters.
*Geographic Focus:* All States, All Countries
*Date(s) Application is Due:* Oct 1
*Amount of Grant:* 1,000 USD
*Contact:* Webb Marmer; (626) 825-5785 or (800) 843-2763; marner@ucla.edu
*Internet:* https://www.asme.org/about-asme/participate/honors-awards/achievement-awards/bergles-rohsenow-young-investigator-award-in-heat
*Sponsor:* American Society of Mechanical Engineers
Two Park Avenue
New York, NY 10016-5990

**ASME Bernard F. Langer Nuclear Codes and Standards Award** 629

The ASME Bernard F. Langer Award established in 1977 in honor of B.F. Langer who was instrumental in the development of the rules for nuclear vessels. It recognizes individuals who have contributed to the nuclear power plant industry through the development and promotion of ASME nuclear codes and standards or the ASME nuclear certification program. The award includes $1,000, a crystal oracle, and a certificate. The annual deadline for nominations is February 1.
*Requirements:* The award may be made to one recipient of any age who need not be a member of ASME. The General Awards Committee screens nominations and makes recommendations to the Committee on Honors. It consists of Members or Fellows representing the Districts. Special Award Committees are usually associated with a particular honor or award. They screen nominations and make recommendations to the other two committees.
*Restrictions:* Services performed within the context of an individual's normal employment are not eligible. Except for the Melville Medal, an individual will receive only one honor in recognition of the same achievement. The receipt of one ASME honor shall not bar the recipient from another ASME honor provided it is for different accomplishments. Honors are not awarded posthumously except if a nominee's death occurs after the nomination has been received at ASME Headquarters.
*Geographic Focus:* All States, All Countries
*Date(s) Application is Due:* Feb 1
*Amount of Grant:* 1,000 USD
*Contact:* Tom Vogan; (941) 753-9875 or (800) 843-2763; tomvogan45@gmail.com
*Internet:* https://www.asme.org/about-asme/participate/honors-awards/achievement-awards/bernard-f-langer-nuclear-codes-standards-award
*Sponsor:* American Society of Mechanical Engineers
Two Park Avenue
New York, NY 10016-5990

**ASME Blackall Machine Tool and Gage Award** 630

The ASME Blackall Machine Tool and Gage Award is given annually for the best paper or papers clearly concerned with or related to the design or application of machine tools or dimensional measuring instruments, submitted to ASME for presentation and publication. Awardees receive $1,000 and a plaque. Papers by multiple authors are eligible. The award is made annually if warranted. The annual deadline for nominations is December 15.

*Requirements:* Any individual member, group of members, or committee may nominate candidates for this award. The General Awards Committee screens nominations and makes recommendations to the Committee on Honors. It consists of Members or Fellows representing the Districts. Special Award Committees are usually associated with a particular honor or award. They screen nominations and make recommendations to the other two committees.

*Restrictions:* Services performed within the context of an individual's normal employment are not eligible. Except for the Melville Medal, an individual will receive only one honor in recognition of the same achievement. The receipt of one ASME honor shall not bar the recipient from another ASME honor provided it is for different accomplishments. Honors are not awarded posthumously except if a nominee's death occurs after the nomination has been received at ASME Headquarters.

*Geographic Focus:* All States, All Countries

*Date(s) Application is Due:* Dec 15

*Amount of Grant:* 1,000 USD

*Contact:* Yuebin Guo, Committee on Honors; (205) 348-2615 or (800) 843-2763; Yguo@eng.ua.edu or awards@asme.org

*Internet:* https://www.asme.org/about-asme/participate/honors-awards/literature-awards/blackall-machine-tool-and-gage-award

*Sponsor:* American Society of Mechanical Engineers
Two Park Avenue
New York, NY 10016-5990

**ASME Burt L. Newkirk Award** 631

The ASME Newkirk Award is given to candidates who have made a notable contribution to the field of tribology in research or development as evidenced by important tribology publications. Tribology for this award is defined as pertaining to the science and technology associated with surfaces in contact and relative motion with each other. It covers all the fundamentals associated with the field of friction, lubrication, and wear. It embraces all the technological aspects of bearings, brakes, clutches, and gears. The award winner is given a certificate and $1,000. The annual deadline for nominations is February 1.

*Requirements:* Nominees must be under the age of 40 and members of ASME for at least five years. The General Awards Committee screens nominations and makes recommendations to the Committee on Honors. It consists of Members or Fellows representing the Districts. Special Award Committees are usually associated with a particular honor or award. They screen nominations and make recommendations to the other two committees.

*Restrictions:* Services performed within the context of an individual's normal employment are not eligible. Except for the Melville Medal, an individual will receive only one honor in recognition of the same achievement. The receipt of one ASME honor shall not bar the recipient from another ASME honor provided it is for different accomplishments. Honors are not awarded posthumously except if a nominee's death occurs after the nomination has been received at ASME Headquarters.

*Geographic Focus:* All States, All Countries

*Date(s) Application is Due:* Feb 1

*Amount of Grant:* 1,000 USD

*Contact:* Benyebka Bou-Saïd, Committee on Honors; +33 478890980 or (800) 843-2763; benyebka.bou-said@insa-lyon.fr

*Internet:* https://www.asme.org/about-asme/participate/honors-awards/achievement-awards/burt-l-newkirk-award

*Sponsor:* American Society of Mechanical Engineers
Two Park Avenue
New York, NY 10016-5990

**ASME Calvin W. Rice Lecture Award** 632

The ASME Calvin W. Rice Lecture honors Calvin Rice, who served as Secretary of the Society from 1906 to 1934. It is intended to continue the effort for which Rice was known, to increase understanding among engineers of various countries. The Subcommittee looks for a lecturer and topic to fit the purpose of the lecture. It is also important that the quality and prestige of the lecture continue on a high level. Preference is given to a lecturer whose accomplishments have been achieved overseas. The award includes a certificate, a lifetime membership with ASME, and a $4,000 honorarium to offset expenses incurred to present the lecture. Nomination forms and instructions are available at the website.The annual deadline for nominations is February 1.

*Requirements:* The nomination package should include the following: a letter from the principal nominator addressing how the nominee meets the criteria for the award; supporting letters from a minimum of three references that show why the nominee should be considered for the award; curriculum vitae of the nominee listing educational and professional background, awards, publications, and other information pertinent to the nomination.

*Geographic Focus:* All States, All Countries

*Date(s) Application is Due:* Feb 1

*Amount of Grant:* 4,000 USD

*Contact:* Fran McKivor, Award Administrator; (212) 591-7094 or (800) 843-2763; mckivorf@asme.org or awards@asme.org

*Internet:* www.asme.org/about-asme/honors-awards/unit-awards/calvin-w—rice-lecture-award

*Sponsor:* American Society of Mechanical Engineers
Two Park Avenue
New York, NY 10016-5990

**ASME Charles Russ Richards Memorial Award** 633

The ASME Charles Russ Richards Memorial Award, established in 1944, is jointly sponsored by Pi Tau Sigma Honorary Mechanical Engineering Fraternity and ASME. It is given annually to the engineering graduate who has demonstrated outstanding achievement in mechanical engineering 20 years or more after graduation from the regular engineering course of a recognized college or university. Achievement shall be all or in part in any field including industrial, educational, political, research, civic, and artistic. The candidate's achievements will be examined for an application of basic engineering methods or principles. Nomination forms and instructions are available at the website. The award includes a certificate, $1,000, and up to $750 in travel expenses to attend the award presentation. The annual deadline for nominations is February 15.

*Requirements:* The General Awards Committee screens nominations and makes recommendations to the Committee on Honors. It consists of Members or Fellows representing the Districts. Special Award Committees are usually associated with a particular honor or award. They screen nominations and make recommendations to the other two committees.

*Restrictions:* Services performed within the context of an individual's normal employment are not eligible. Except for the Melville Medal, an individual will receive only one honor in recognition of the same achievement. The receipt of one ASME honor shall not bar the recipient from another ASME honor provided it is for different accomplishments. Honors are not awarded posthumously except if a nominee's death occurs after the nomination has been received at ASME Headquarters.

*Geographic Focus:* All States, All Countries

*Date(s) Application is Due:* Feb 15

*Amount of Grant:* 1,000 - 1,750 USD

*Contact:* Darryl James, Committee Chair; (806) 834-3386 or (800) 843-2763; darryl.james@ttu.edu or awards@asme.org

*Internet:* https://www.asme.org/about-asme/participate/honors-awards/achievement-awards/charles-russ-richards-memorial-award

*Sponsor:* American Society of Mechanical Engineers
Two Park Avenue
New York, NY 10016-5990

**ASME Charles T. Main Awards** 634

The purpose of the ASME Charles T. Main Awards is to encourage student members and young engineers to become active in public service. The awards recognize ASME student members whose leadership and service qualities have contributed to the program and operation of a Student Section of the ASME. Award consideration is given to local, regional, and/or national levels of involvement. Candidates must have been members for more than a year. First place award winners receive a gold medal, certificate, $3,000, and travel expenses to attend the award presentation. Second place winners receive a silver medal, certificate, $2,000, and travel expenses to attend the award presentation. Up to eight $500 honorable mentions will be presented to qualified candidates. The annual deadline for nominations is March 15.

*Requirements:* Any individual member, group of members, or committee may nominate candidates for this award. The General Awards Committee screens nominations and makes recommendations to the Committee on Honors. It consists of Members or Fellows representing the Districts. Special Award Committees are usually associated with a particular honor or award. They screen nominations and make recommendations to the other two committees.

*Restrictions:* Services performed within the context of an individual's normal employment are not eligible. Except for the Melville Medal, an individual will receive only one honor in recognition of the same achievement. The receipt of one ASME honor shall not bar the recipient from another ASME honor provided it is for different accomplishments. Honors are not awarded posthumously except if a nominee's death occurs after the nomination has been received at ASME Headquarters.

*Geographic Focus:* All States, All Countries

*Date(s) Application is Due:* Mar 15

*Amount of Grant:* 500 - 3,000 USD

*Contact:* Fran McKivor, Committee Contact; (212) 591-7094 or (800) 843-2763; mckivorf@asme.org or awards@asme.org

*Internet:* https://www.asme.org/about-asme/participate/honors-awards/achievement-awards/charles-t-main-student-section-awards

*Sponsor:* American Society of Mechanical Engineers
Two Park Avenue
New York, NY 10016-5990

**ASME Daniel C. Drucker Medal** 635

The American Society of Mechanical Engineers Daniel C. Drucker Medal is given in recognition of distinguished contributions to the field of applied mechanics and mechanical engineering through research, teaching and service to the community over a substantial period of time. The award includes a bronze medal, a certificate, $2,000, and up to $750 in travel expenses to attend the award presentation. Nomination forms and instructions are available at the website. The annual deadline for nominations is September 15.

*Requirements:* The General Awards Committee screens nominations and makes recommendations to the Committee on Honors. It consists of Members or Fellows representing the Districts. Special Award Committees are usually associated with a particular honor or award. They screen nominations and make recommendations to the other two committees.

*Restrictions:* Except for the Melville Medal, an individual will receive only one honor in recognition of the same achievement. The receipt of one ASME honor shall not bar the recipient from another ASME honor provided it is for different accomplishments. Honors are not awarded posthumously except if a nominee's death occurs after the nomination has been received at ASME Headquarters.

*Geographic Focus:* All States, All Countries
*Date(s) Application is Due:* Sep 15
*Amount of Grant:* 2,000 - 2,750 USD
*Contact:* Balakumar Balachandran; (301) 405-4747 or (800) 843-2763; balab@umd.edu
*Internet:* https://www.asme.org/about-asme/participate/honors-awards/achievement-awards/daniel-c-drucker-medal
*Sponsor:* American Society of Mechanical Engineers
Two Park Avenue
New York, NY 10016-5990

**ASME Dedicated Service Award** 636

In 1983, the ASME Board of Governors approved the establishment of the ASME Dedicated Service Award (DSA). It honors unusual dedicated voluntary service to the Society marked by outstanding performance, demonstrated effective leadership, prolonged and committed service, devotion, enthusiasm and faithfulness. The award may be presented to selected individuals who have served the Society for at least ten years in one or more of the following areas: Standards and Certification; Public Affairs and Outreach; Technical Events and Content; Student and Early Career Development; Board of Governors; ASME Foundation; and the ASME Auxiliary. The award includes a plaque, certificate and lapel pin. The annual deadline for nominations is December 1.
*Requirements:* Any individual member, group of members, or committee may nominate candidates for this award. The General Awards Committee screens nominations and makes recommendations to the Committee on Honors. It consists of Members or Fellows representing the Districts. Special Award Committees are usually associated with a particular honor or award. They screen nominations and make recommendations to the other two committees.
*Geographic Focus:* All States
*Date(s) Application is Due:* Dec 1
*Contact:* Fran McKivor, Committee on Honors; (212) 591-7094 or (800) 843-2763; mckivorf@asme.org or awards@asme.org
*Internet:* https://www.asme.org/about-asme/participate/honors-awards/service-awards/dedicated-service-award
*Sponsor:* American Society of Mechanical Engineers
Two Park Avenue
New York, NY 10016-5990

**ASME Dixy Lee Ray Award** 637

The ASME Dixy Lee Ray Award recognizes significant achievements and contributions in the broad field of environmental protection. Achievement in the following areas will be recognized: environmental engineering, including environmental technology and related topics; environmental health, environmental sciences, environmental management and policy. The award include a bronze metal, a certificate, $1,000, and up to $750 in travel expenses to attend the award presentation. The annual deadline for nominations is February 1.
*Requirements:* Any individual member, group of members, or committee may nominate candidates for this award. The General Awards Committee screens nominations and makes recommendations to the Committee on Honors. It consists of Members or Fellows representing the Districts. Special Award Committees are usually associated with a particular honor or award. They screen nominations and make recommendations to the other two committees.
*Restrictions:* Services performed within the context of an individual's normal employment are not eligible. Except for the Melville Medal, an individual will receive only one honor in recognition of the same achievement. The receipt of one ASME honor shall not bar the recipient from another ASME honor provided it is for different accomplishments. Honors are not awarded posthumously except if a nominee's death occurs after the nomination has been received at ASME Headquarters.
*Geographic Focus:* All States, All Countries
*Date(s) Application is Due:* Feb 1
*Amount of Grant:* 1,000 - 1,750 USD
*Contact:* Steven M. Unikewicz, Committee on Honors; (703) 439-7133 or (800) 843-2763; unikewiczs@asme.org or awards@asme.org
*Internet:* https://www.asme.org/about-asme/participate/honors-awards/achievement-awards/dixy-lee-ray-award
*Sponsor:* American Society of Mechanical Engineers
Two Park Avenue
New York, NY 10016-5990

**ASME Edward F. Obert Award** 638

The ASME Edward F. Obert Award recognizes an outstanding paper on thermodynamics authored during the preceding two calendar years. The award shall be made for a paper presented at the Advanced Energy Systems Division's International Mechanical Engineering Congress symposium on energy systems analysis (so long as the symposium continues). The award shall be presented at the following ASME Congress. The award includes a certificate, $5,000, and up to $750 in travel expenses to attend the award presentation. Nomination are accepted until the annual March 1 deadline.
*Requirements:* The paper nominated must have been written during the preceding two years of the presentation. Any individual member, group of members, or committee may nominate candidates for this award. The General Awards Committee screens nominations and makes recommendations to the Committee on Honors. It consists of Members or Fellows representing the Districts. Special Award Committees are usually associated with a particular honor or award. They screen nominations and make recommendations to the other two committees.
*Restrictions:* Services performed within the context of an individual's normal employment are not eligible. Except for the Melville Medal, an individual will receive only one honor in recognition of the same achievement. The receipt of one ASME honor shall not bar the recipient from another ASME honor provided it is for different accomplishments. Honors are not awarded posthumously except if a nominee's death occurs after the nomination has been received at ASME Headquarters.
*Geographic Focus:* All States, All Countries
*Date(s) Application is Due:* Mar 1
*Amount of Grant:* 5,000 - 5,750 USD
*Contact:* Howard N. Shapiro, Committee on Honors; (313) 310-2629 or (800) 843-2763; hshapiro513@gmail.com or awards@asme.org
*Internet:* https://www.asme.org/about-asme/participate/honors-awards/literature-awards/edward-f-obert-award
*Sponsor:* American Society of Mechanical Engineers
Two Park Avenue
New York, NY 10016-5990

**ASME Edwin F. Church Medal** 639

The American Society of Mechanical Engineers Edwin F. Church Medal is given annually to an individual who has rendered eminent service in increasing the value, importance, and attractiveness of mechanical engineering education. The award includes $1,000, a bronze medal, a certificate, and travel expenses up to $750 to attend the award presentation. Mechanical engineering is used here in its broadest sense of preparation for any aspect or level of mechanical engineering through any appropriate mechanism including universities, technical institutes, professional societies and private groups, in-house professional development programs of industrial concerns and governmental agencies, or programmed learning and self-instruction systems. The nomination instructions and form are available at the website. The annual deadline for nominations is September 1.
*Requirements:* The award may be made to one recipient of any age who need not be a member of ASME. The General Awards Committee screens nominations and makes recommendations to the Committee on Honors. It consists of Members or Fellows representing the Districts. Special Award Committees are usually associated with a particular honor or award. They screen nominations and make recommendations to the other two committees.
*Restrictions:* Services performed within the context of an individual's normal employment are not eligible. Except for the Melville Medal, an individual will receive only one honor in recognition of the same achievement. The receipt of one ASME honor shall not bar the recipient from another ASME honor provided it is for different accomplishments. Honors are not awarded posthumously except if a nominee's death occurs after the nomination has been received at ASME Headquarters.
*Geographic Focus:* All States, All Countries
*Date(s) Application is Due:* Sep 1
*Amount of Grant:* 2,500 - 3,250 USD
*Contact:* Ken Ball; (571) 488-8961 or (800) 843-2763; ball@gmu.edu or awards@asme.org
*Internet:* https://www.asme.org/about-asme/participate/honors-awards/achievement-awards/edwin-f-church-medal
*Sponsor:* American Society of Mechanical Engineers
Two Park Avenue
New York, NY 10016-5990

**ASME Engineer-Historian Award** 640

The ASME History and Heritage Award Committee recognizes outstanding published work or works by an engineer dealing with the history of mechanical engineering through its Engineer-Historian Award. Preference is given to those publications dealing with the art and science of mechanical engineering. The award is intended to encourage the active interest by mechanical engineers in the history of their profession. The award with its honorarium of $1,000 and certificate is presented annually, if merited, at the International Mechanical Engineering Congress. Nominations for the Engineer-Historian Award are accepted year-round for consideration. Candidates may remain under consideration in following years.
*Requirements:* Candidates can be nominated by any ASME member, with submission accompanied by 10 photocopies of at least one published work to be distributed to Committee members. The names of these candidates as well as previously nominated candidates are considered during the regular meeting agenda. Deadline for submission of a nomination is one month prior to the Spring meeting (approximately April 1), for the following meeting's annual presentation. Criteria includes subject matter, depth of treatment, technical level of material, and the historical significance of the subject matter. Candidate must be or have been active in the practice of the art and science of mechanical engineering, including managers, teachers, and museum professionals directly concerned with mechanical engineering historical artifacts (provided they have been trained as and practiced as mechanical engineers). Multiple co-authors are permitted as long as all authors fit the above definition. Self nominations are accepted. The award is international in scope, and open to members and nonmembers.
*Restrictions:* History and Heritage Committee members are not eligible. Submissions must be in English or translated.
*Geographic Focus:* All States, All Countries
*Date(s) Application is Due:* Mar 1
*Amount of Grant:* 1,000 USD
*Contact:* Wil Haywood, Award Contact; (212) 591-7094 or (800) 843-2763; haywoodw@asme.org or awards@asme.org
*Internet:* www.asme.org/about-asme/honors-awards/unit-awards/engineer-historian-award-%281%29
*Sponsor:* American Society of Mechanical Engineers
Two Park Avenue
New York, NY 10016-5990

**ASME Fluids Engineering Award** 641
The American Society of Mechanical Engineers Fluids Engineering Award is given for outstanding contributions over a period of years to the engineering profession and especially to the field of fluids engineering through research, practice, and/or teaching. The award includes a bronze medal, a certificate, and $1,000. Nomination forms and instructions are available at the website. The annual deadline for nominations is November 30.
*Requirements:* The General Awards Committee screens nominations and makes recommendations to the Committee on Honors. It consists of Members or Fellows representing the Districts. Special Award Committees are usually associated with a particular honor or award. They screen nominations and make recommendations to the other two committees.
*Restrictions:* Except for the Melville Medal, an individual will receive only one honor in recognition of the same achievement. The receipt of one ASME honor shall not bar the recipient from another ASME honor provided it is for different accomplishments. Honors are not awarded posthumously except if a nominee's death occurs after the nomination has been received at ASME Headquarters.
*Geographic Focus:* All States, All Countries
*Date(s) Application is Due:* Nov 30
*Contact:* Ning Zhang; (785) 475-5873 or (800) 843-2763; nzhang@mcneese.edu
*Internet:* https://www.asme.org/about-asme/participate/honors-awards/achievement-awards/fluids-engineering-award
*Sponsor:* American Society of Mechanical Engineers
Two Park Avenue
New York, NY 10016-5990

**ASME Frank Kreith Energy Award** 642
The ASME Frank Kreith Energy Award, established in 2005, honors individuals for significant contributions to a secure energy future with particular emphasis on innovations in conservation and/or renewable energy. Contributions may be through research, education, practice, or significant service to society that will lead to a sustainable energy future. The award includes a plaque, a certificate, $2,000, and up to $750 in travel expenses to attend the award presentation. The annual deadline for nominations is December 1.
*Requirements:* The award may be made to one recipient of any age who need not be a member of ASME. The General Awards Committee screens nominations and makes recommendations to the Committee on Honors. It consists of Members or Fellows representing the Districts. Special Award Committees are usually associated with a particular honor or award. They screen nominations and make recommendations to the other two committees.
*Restrictions:* Services performed within the context of an individual's normal employment are not eligible. Except for the Melville Medal, an individual will receive only one honor in recognition of the same achievement. The receipt of one ASME honor shall not bar the recipient from another ASME honor provided it is for different accomplishments. Honors are not awarded posthumously except if a nominee's death occurs after the nomination has been received at ASME Headquarters.
*Geographic Focus:* All States, All Countries
*Date(s) Application is Due:* Dec 1
*Contact:* Andy Walker, ; (303) 384-7531 or (800) 843-2763; andy.walker@nrel.gov
*Internet:* https://www.asme.org/about-asme/participate/honors-awards/achievement-awards/frank-kreith-energy-award
*Sponsor:* American Society of Mechanical Engineers
Two Park Avenue
New York, NY 10016-5990

**ASME Frank von Flue Award** 643
The Frank von Flue Award recognizes significant contributions by an individual or a group to the promotion of lifelong learning for mechanical engineers. Lifelong learning is defined as the continued process of education and/or professional development, following a post-secondary education degree, to maintain the competence and flexibility of mechanical engineers throughout their career. Contributions may include assessment, development, distribution or promotion of materials for lifelong learning via appropriate mechanisms, such as professional development or university extension courses and programmed or self-paced learning. The award includes a bronze medal, a certificate, $1,000, and up to $750 in travel expenses to attend the award presentation. The annual deadline for nominations is February 1.
*Requirements:* Any individual member, group of members, or committee may nominate candidates for this award. The General Awards Committee screens nominations and makes recommendations to the Committee on Honors. It consists of Members or Fellows representing the Districts. Special Award Committees are usually associated with a particular honor or award. They screen nominations and make recommendations to the other two committees.
*Restrictions:* Services performed within the context of an individual's normal employment are not eligible. Except for the Melville Medal, an individual will receive only one honor in recognition of the same achievement. The receipt of one ASME honor shall not bar the recipient from another ASME honor provided it is for different accomplishments. Honors are not awarded posthumously except if a nominee's death occurs after the nomination has been received at ASME Headquarters.
*Geographic Focus:* All States, All Countries
*Date(s) Application is Due:* Feb 1
*Amount of Grant:* 1,000 - 1,750 USD
*Contact:* Fran McKivor, Committee on Honors; (212) 591-7094 or (800) 843-2763; mckivorf@asme.org or awards@asme.org
*Internet:* https://www.asme.org/about-asme/get-involved/honors-awards/achievement-awards/frank-von-flue-award
*Sponsor:* American Society of Mechanical Engineers
Two Park Avenue
New York, NY 10016-5990

**ASME Freeman Scholar Award** 644
The ASME Freeman Scholar Award is given annually to a person with extensive experience in fluids engineering. The applicant is expected to review a coherent topic in their specialty and suggest key future research needs. The awardee is given $10,000, a certificate, and a travel supplement not to exceed $750. The results will be presented at the winter meeting and published in the ASME Journal of Fluids Engineering. Deadline date for applications and nominations is September 1.
*Requirements:* The recipient may be from industry, government, education, or private professional practice, and need not be an ASME member.
*Geographic Focus:* All States, All Countries
*Date(s) Application is Due:* Sep 1
*Amount of Grant:* 10,000 - 10,750 USD
*Contact:* Fran McKivor, Committee on Honors; (212) 591-7094 or (800) 843-2763; mckivorf@asme.org or awards@asme.org
*Internet:* www.asme.org/about-asme/honors-awards/literature-awards/freeman-scholar-award
*Sponsor:* American Society of Mechanical Engineers
Two Park Avenue
New York, NY 10016-5990

**ASME Gas Turbine Award** 645
The ASME Gas Turbine Award is given in recognition of an outstanding individual or multiple-author contribution to the literature of combustion gas turbines or gas turbines thermally combined with nuclear or steam power plants. The paper may be devoted to the design aspects of overall gas turbines or individual components and/or systems such as compressors, combustion systems, turbines, controls and accessories, bearings, regenerators, inlet air filters, silencers, etc. It may cover topics specifically related to gas turbines, such as high-temperature materials or fuel considerations including erosion and corrosion complications. It also can be devoted to application or operational aspects of gas turbines for aircraft propulsion and ground power units, or automotive, electric utility, gas pipeline pumping, locomotive, marine, oil field pumping, petrochemical, space power, steel, and similar uses. The award includes a plaque and $1,000. The annual deadline for nominations is June 1.
*Requirements:* Papers published anywhere in the world are eligible; authors are not restricted by nationality, age, profession, or membership in any engineering society or other organizations. Any individual member, group of members, or committee may nominate candidates for this award. The General Awards Committee screens nominations and makes recommendations to the Committee on Honors. It consists of Members or Fellows representing the Districts. Special Award Committees are usually associated with a particular honor or award. They screen nominations and make recommendations to the other two committees.
*Restrictions:* Services performed within the context of an individual's normal employment are not eligible. Except for the Melville Medal, an individual will receive only one honor in recognition of the same achievement. The receipt of one ASME honor shall not bar the recipient from another ASME honor provided it is for different accomplishments. Honors are not awarded posthumously except if a nominee's death occurs after the nomination has been received at ASME Headquarters.
*Geographic Focus:* All States, All Countries
*Date(s) Application is Due:* Jun 1
*Amount of Grant:* 1,000 USD
*Contact:* John Blanton, Committee on Honors; (864) 382-0626 or (800) 843-2763; blantonj@asme.org or awards@asme.org
*Internet:* www.asme.org/about-asme/honors-awards/literature-awards/gas-turbine-award
*Sponsor:* American Society of Mechanical Engineers
Two Park Avenue
New York, NY 10016-5990

**ASME George Westinghouse Medals** 646
The ASME George Westinghouse Medals are bestowed annually for eminent achievement or distinguished service in the power field of mechanical engineering. Considering power in the broad sense, the basis of the awards shall include contributions of utilization, application, design, development, research, and the organization of such activities in the power field. Any individual member, group of members, or committee may nominate candidates for the awards. Awardees receive $1,500, a vermeil medal, a certificate, a travel expenses to attend the award presentation. A silver medal, certificate, $1,000, and up to $750 in travel expenses are awarded to those under 45 years of age. The annual deadline for nominations is February 1.
*Requirements:* Candidates are not restricted by profession nor by membership in any engineering society or organization. Any individual member, group of members, or committee may nominate candidates for this award. The General Awards Committee screens nominations and makes recommendations to the Committee on Honors. It consists of Members or Fellows representing the Districts. Special Award Committees are usually associated with a particular honor or award. They screen nominations and make recommendations to the other two committees.
*Restrictions:* Services performed within the context of an individual's normal employment are not eligible. Except for the Melville Medal, an individual will receive only one honor in recognition of the same achievement. The receipt of one ASME honor shall not bar the recipient from another ASME honor provided it is for different accomplishments. Honors are not awarded posthumously except if a nominee's death occurs after the nomination has been received at ASME Headquarters.
*Geographic Focus:* All States, All Countries
*Date(s) Application is Due:* Feb 1
*Amount of Grant:* 1,500 - 2,250 USD

*Contact:* Ashwani K. Gupta, Committee on Honors; (301) 474-2294 or (800) 843-2763; akgupta@umd.edu or awards@asme.org
*Internet:* https://www.asme.org/about-asme/participate/honors-awards/achievement-awards/george-westinghouse-medals
*Sponsor:* American Society of Mechanical Engineers
Two Park Avenue
New York, NY 10016-5990

**ASME Graduate Teaching Fellowships** **647**
The ASME Graduate Teaching Fellowships are a collaboration between ASME and mechanical engineering departments to encourage outstanding doctoral candidates in mechanical engineering education (and related engineering fields), particularly women and minorities. Fellowship awards of $5,000 a year are made for a maximum of two years. Fellows are selected or renewed annually by the ASME Board on Education. The current application is posted on the website.
*Requirements:* Applicants must be U.S. citizens or permanent residents, Ph.D. students in mechanical engineering with an interest in a teaching career, and members or student members of ASME. Applicants must submit undergraduate grade point averages (GPA), GRE scores, a current resume or vitae, two letters of recommendation from faculty or M.S. committee, transcripts of all academic work, and a statement about a faculty career. A Master's degree or passage of qualifying exam is required. A lecture-responsibility teaching assistantship commitment from the applicant's department is also required.
*Geographic Focus:* All States
*Date(s) Application is Due:* Feb 15
*Amount of Grant:* 5,000 USD
*Contact:* Beth Lefever, Centers Administrator; (212) 591-7790 or (212) 591-7094; fax (212) 591-7143; lefeverb@asme.org
*Internet:* www.asme.org/about-asme/scholarship-and-loans/graduate-teaching-fellowships
*Sponsor:* American Society of Mechanical Engineers
Two Park Avenue
New York, NY 10016-5990

**ASME Gustus L. Larson Memorial Award** **648**
The Gustus L. Larson Memorial Award, established in 1974, honors Gustus L. Larson, ASME Fellow and founder of Pi Tau Sigma at the University of Wisconsin. He was a recognized leader in heating, ventilating, and air conditioning and was a former president of the ASHVE. The award is sponsored jointly by the Pi Tau Sigma Honorary Mechanical Engineering Fraternity and ASME, and is given annually to the engineering graduate who has demonstrated outstanding achievement in mechanical engineering within 10 to 20 years after graduating from a regular engineering course of a recognized college or university. Achievement shall be all or in part in any field including industrial, educational, political, research, civic, and artistic. The candidate's achievements will be examined for an application of basic engineering methods or principles. Awardees are given a certificate, $1,000, and up to $750 in travel expenses to attend the award presentation. The annual deadline for nominations is February 15.
*Requirements:* The nominee must be a person who has received a baccalaureate degree in Mechanical Engineering or other related filed within ten to twenty calendar years from the year of the award presentation. The award may be made to one recipient of any age who need not be a member of ASME. The General Awards Committee screens nominations and makes recommendations to the Committee on Honors. It consists of Members or Fellows representing the Districts. Special Award Committees are usually associated with a particular honor or award. They screen nominations and make recommendations to the other two committees.
*Restrictions:* Services performed within the context of an individual's normal employment are not eligible. Except for the Melville Medal, an individual will receive only one honor in recognition of the same achievement. The receipt of one ASME honor shall not bar the recipient from another ASME honor provided it is for different accomplishments. Honors are not awarded posthumously except if a nominee's death occurs after the nomination has been received at ASME Headquarters.
*Geographic Focus:* All States, All Countries
*Date(s) Application is Due:* Feb 15
*Amount of Grant:* 1,000 - 1,750 USD
*Contact:* Darryl James, Committee on Honors; (806) 834-3386 or (800) 843-2763; darryl.james@ttu.edu or awards@asme.org
*Internet:* https://www.asme.org/about-asme/participate/honors-awards/achievement-awards/gustus-l-larson-memorial-award
*Sponsor:* American Society of Mechanical Engineers
Two Park Avenue
New York, NY 10016-5990

**ASME H.R. Lissner Award** **649**
The ASME H.R. Lissner Award was established in honor of H.R. Lissner for his pioneering contributions in biomechanics. The award is bestowed annually for outstanding accomplishments in the area of bioengineering. These achievements may be in the form of: significant research contributions in bioengineering; development of new methods of measuring in bioengineering; design of new equipment and instrumentation in bioengineering; educational impact in the training of bioengineers; and/or service to the bioengineering community, in general, and to the Bioengineering Division of ASME, in particular. The award includes a bronze medal, certificate, $1,000, and up to $750 in travel expenses to attend the award presentation. Nomination forms and instructions are available at the website. Nomination packets will be held and considered for three years; a new nomination packet must be submitted after that time. The annual deadline for nominations is September 1.
*Requirements:* Candidate must be an active member of the bioengineering division. The General Awards Committee screens nominations and makes recommendations to the Committee on Honors. It consists of Members or Fellows representing the Districts. Special Award Committees are usually associated with a particular honor or award. They screen nominations and make recommendations to the other two committees.
*Restrictions:* Services performed within the context of an individual's normal employment are not eligible. Except for the Melville Medal, an individual will receive only one honor in recognition of the same achievement. The receipt of one ASME honor shall not bar the recipient from another ASME honor provided it is for different accomplishments. Honors are not awarded posthumously except if a nominee's death occurs after the nomination has been received at ASME Headquarters.
*Geographic Focus:* All States
*Date(s) Application is Due:* Sep 1
*Amount of Grant:* 1,000 - 1,750 USD
*Contact:* Noshir A. Langrana, Committee on Honors; (732) 445-4500 or (800) 843-2763; langrana@rutgers.edu
*Internet:* https://www.asme.org/about-asme/participate/honors-awards/achievement-awards/h-r-lissner-medal
*Sponsor:* American Society of Mechanical Engineers
Two Park Avenue
New York, NY 10016-5990

**ASME Heat Transfer Memorial Award** **650**
The American Society of Mechanical Engineers Heat Transfer Memorial Award was established by the Heat Transfer Division in 1959 and operated as a division award until 1974 when it was elevated to a Society award. It is bestowed on individuals who have made outstanding contributions to the field of heat transfer through teaching, research, design, or publications. Each award is based on papers in an area of heat transfer or on a paper dealing with the science or art of heat transfer. Awards are made annually in each of the following categories: the science of heat transfer; the art of heat transfer; and the general subject of heat transfer. The awardees receive a plaque, along with a certificate and $1,000. The annual deadline for nominations is October 1.
*Requirements:* There are no restrictions by age, nationality, or society membership. The General Awards Committee screens nominations and makes recommendations to the Committee on Honors. It consists of Members or Fellows representing the Districts. Special Award Committees are usually associated with a particular honor or award. They screen nominations and make recommendations to the other two committees.
*Restrictions:* Except for the Melville Medal, an individual will receive only one honor in recognition of the same achievement. The receipt of one ASME honor shall not bar the recipient from another ASME honor provided it is for different accomplishments. Honors are not awarded posthumously except if a nominee's death occurs after the nomination has been received at ASME Headquarters.
*Geographic Focus:* All States, All Countries
*Date(s) Application is Due:* Oct 1
*Amount of Grant:* 1,000 USD
*Contact:* Webb Marmer, Committee Chair; (626) 825-5785 or (800) 843-2763; marner@ucla.edu or awards@asme.org
*Internet:* https://www.asme.org/about-asme/participate/honors-awards/achievement-awards/heat-transfer-memorial-award
*Sponsor:* American Society of Mechanical Engineers
Two Park Avenue
New York, NY 10016-5990

**ASME Henry Hess Early Career Publication Award** **651**
The ASME Henry Hess Early Career Publication Award is given for an original technical paper submitted to ASME for presentation and publication. Joint authorship is permissible provided all authors meet the requirements. The paper shall be specifically recommended for the award by a review committee or a qualified individual. The award includes a $2500 honorarium, a certificate, and a travel supplement not to exceed $750. The annual deadline for nominations is March 1.
*Requirements:* Author must not be older than 35 at the time of submission and must be an ASME student member or associate member to be eligible to apply. Any individual member, group of members, or committee may nominate candidates for this award. The General Awards Committee screens nominations and makes recommendations to the Committee on Honors. It consists of Members or Fellows representing the Districts. Special Award Committees are usually associated with a particular honor or award. They screen nominations and make recommendations to the other two committees.
*Restrictions:* Services performed within the context of an individual's normal employment are not eligible. Except for the Melville Medal, an individual will receive only one honor in recognition of the same achievement. The receipt of one ASME honor shall not bar the recipient from another ASME honor provided it is for different accomplishments. Honors are not awarded posthumously except if a nominee's death occurs after the nomination has been received at ASME Headquarters.
*Geographic Focus:* All States, All Countries
*Date(s) Application is Due:* Mar 1
*Amount of Grant:* 2,500 - 3,250 USD
*Contact:* Fran McKivor, Committee on Honors; (212) 591-7094 or (800) 843-2763; mckivorf@asme.org or awards@asme.org
*Internet:* www.asme.org/about-asme/honors-awards/literature-awards/henry-hess-award
*Sponsor:* American Society of Mechanical Engineers
Two Park Avenue
New York, NY 10016-5990

**ASME Henry Laurence Gantt Medal** 652

The American Society of Mechanical Engineers Henry Laurence Gantt Medal is given for distinguished achievement in management and for service to the community. The medal was established in honor of Henry Laurence Gantt, a prolific writer and one of the first leaders in the scientific management movement to express concern for the human element of productivity. His enduring legacy is the philosophy that seeks to turn the potential of industry into a broad contribution of service to society. He believed that managers should "view their activities from the vantage point of the larger communities business serves, thus dedicating themselves to the doctrine of service." The award includes a bronze medal, a certificate, and $1,000. Nomination forms and instructions are available at the website. The annual deadline for nominations is February 1.
*Geographic Focus:* All States, All Countries
*Date(s) Application is Due:* Feb 1
*Amount of Grant:* 1,000 USD
*Contact:* Steve H. Long, Committee Chair; (281) 425-8770 or (800) 843-2763; Steve.Long@snclavalin.com
*Internet:* https://www.asme.org/about-asme/participate/honors-awards/achievement-awards/henry-laurence-gantt-medal
*Sponsor:* American Society of Mechanical Engineers
Two Park Avenue
New York, NY 10016-5990

**ASME Henry R. Worthington Medal** 653

The ASME Henry R. Worthington Medal is given annually for eminent achievement in the field of pumping machinery, systems and concepts. Such achievement may be in the areas of research, development, design, innovation, management, education, or literature. Awardees are presented with a Bronze Medal, certificate, $5,000, and up to $750 in travel expenses to attend the award presentation. Any individual member, group of members, or committee may nominate candidates for the award. The annual deadline for nominations is February 1.
*Requirements:* Any individual member, group of members, or committee may nominate candidates for this award. The General Awards Committee screens nominations and makes recommendations to the Committee on Honors. It consists of Members or Fellows representing the Districts. Special Award Committees are usually associated with a particular honor or award. They screen nominations and make recommendations to the other two committees.
*Restrictions:* Services performed within the context of an individual's normal employment are not eligible. Except for the Melville Medal, an individual will receive only one honor in recognition of the same achievement. The receipt of one ASME honor shall not bar the recipient from another ASME honor provided it is for different accomplishments. Honors are not awarded posthumously except if a nominee's death occurs after the nomination has been received at ASME Headquarters.
*Geographic Focus:* All States, All Countries
*Date(s) Application is Due:* Feb 1
*Amount of Grant:* 5,000 - 5,750 USD
*Contact:* Judith A. Bamberger, Committee on Honors; (509) 375-3898 or (800) 843-2763; bambergerj@asme.org or awards@asme.org
*Internet:* https://www.asme.org/about-asme/participate/honors-awards/achievement-awards/henry-r-worthington-medal
*Sponsor:* American Society of Mechanical Engineers
Two Park Avenue
New York, NY 10016-5990

**ASME Holley Medal** 654

The American Society of Mechanical Engineers Holley Medal is given to an individual who has contributed a unique engineering benefit for the public good. Nominees are not limited by education, societal or organizational membership, or an official position. The achievement should be worthy of public gratitude and admiration of engineers. More than one individual may be recognized for a single achievement provided that each individual made equal or comparable contribution. Nomination forms and instructions are available at the website. Awardees are presented with a vermeil medal, certificate, and $1,000. The annual deadline for nominations is March 1.
*Requirements:* Any individual member, group of members, or committee may nominate candidates for this award. The General Awards Committee screens nominations and makes recommendations to the Committee on Honors. It consists of Members or Fellows representing the Districts. Special Award Committees are usually associated with a particular honor or award. They screen nominations and make recommendations to the other two committees.
*Restrictions:* Except for the Melville Medal, an individual will receive only one honor in recognition of the same achievement. The receipt of one ASME honor shall not bar the recipient from another ASME honor provided it is for different accomplishments. Honors are not awarded posthumously except if a nominee's death occurs after the nomination has been received at ASME Headquarters.
*Geographic Focus:* All States, All Countries
*Date(s) Application is Due:* Mar 1
*Amount of Grant:* 1,000 USD
*Contact:* Fran McKivor, Committee Chair; (212) 591-7094 or (800) 843-2763; fax (212) 705-7739; mckivorf@asme.org or awards@asme.org
*Internet:* https://www.asme.org/about-asme/participate/honors-awards/achievement-awards/holley-medal
*Sponsor:* American Society of Mechanical Engineers
Two Park Avenue
New York, NY 10016-5990

**ASME Honorary Member Award** 655

The American Society of Mechanical Engineers Honorary Member Award is given to a person who has made distinctive contributions to engineering, science, industry, research, public service, or other pursuits allied with and beneficial to the engineering profession. Honorary Membership was first awarded in 1880, the founding year of the Society. The roster of Honorary Members contains the names of leaders of world renown who have been selected under carefully drawn procedures rigorously maintained by the Society over the years. The Award consists of a silver medal, a certificate, and up to a $750 travel expense supplement to attend the award presentation. The annual deadline for nominations is March 1.
*Requirements:* The General Awards Committee screens nominations and makes recommendations to the Committee on Honors. It consists of Members or Fellows representing the Districts. Special Award Committees are usually associated with a particular honor or award. They screen nominations and make recommendations to the other two committees.
*Restrictions:* Except for the Melville Medal, an individual will receive only one honor in recognition of the same achievement. The receipt of one ASME honor shall not bar the recipient from another ASME honor provided it is for different accomplishments. Honors are not awarded posthumously except if a nominee's death occurs after the nomination has been received at ASME Headquarters.
*Geographic Focus:* All States, All Countries
*Date(s) Application is Due:* Mar 1
*Amount of Grant:* Up to 750 USD
*Contact:* Fran McKivor, (212) 591-7094 or (800) 218-9946; mckivorf@asme.org
*Internet:* https://www.asme.org/about-asme/participate/honors-awards/achievement-awards/honorary-member
*Sponsor:* American Society of Mechanical Engineers
Two Park Avenue
New York, NY 10016-5990

**ASME Internal Combustion Engine Award** 656

The ASME Internal Combustion Engine Award is given in recognition of eminent achievement or distinguished contribution over a substantial period of time, which may result from research, innovation, or education in advancing the art of engineering in the field of internal combustion engines or in directing the efforts and accomplishments of those engaged in engineering practice in the design, development, application, and operation of internal combustion engines. The awardee receives a plaque and $1,000. The annual deadline for nominations is February 1.
*Geographic Focus:* All States, All Countries
*Date(s) Application is Due:* Feb 1
*Amount of Grant:* 1,000 USD
*Contact:* Robert M Wagner, Committee Chair; (865) 946-1239 or (800) 843-2763; wagnerrm@ornl.gov or awards@asme.org
*Internet:* https://www.asme.org/about-asme/participate/honors-awards/achievement-awards/internal-combustion-engine-award
*Sponsor:* American Society of Mechanical Engineers
Two Park Avenue
New York, NY 10016-5990

**ASME J. Hall Taylor Medal** 657

The ASME J. Hall Taylor Medal is presented for distinguished service or eminent achievement in codes and standards pertaining to the fields of piping and pressure vessels that are sponsored or undertaken by ASME. The scope shall include contributions to technical advancement and administration. The awardee receives a Bronze Medal, certificate, and $1,000. The award was established to commemorate the pioneering work of J. Hall Taylor in the field of standardization of industrial products and safety codes for their usage. The annual deadline for nominations is November 15.
*Requirements:* Any individual member, group of members, or committee may nominate candidates for this award. The General Awards Committee screens nominations and makes recommendations to the Committee on Honors. It consists of Members or Fellows representing the Districts. Special Award Committees are usually associated with a particular honor or award. They screen nominations and make recommendations to the other two committees.
*Restrictions:* Services performed within the context of an individual's normal employment are not eligible. Except for the Melville Medal, an individual will receive only one honor in recognition of the same achievement. The receipt of one ASME honor shall not bar the recipient from another ASME honor provided it is for different accomplishments. Honors are not awarded posthumously except if a nominee's death occurs after the nomination has been received at ASME Headquarters.
*Geographic Focus:* All States, All Countries
*Date(s) Application is Due:* Nov 15
*Amount of Grant:* 1,000 USD
*Contact:* Jon S. Batey, Committee on Honors; (979) 480-9428 or (800) 843-2763; jbatey1953@gmail.com or awards@asme.org
*Internet:* https://www.asme.org/about-asme/participate/honors-awards/achievement-awards/j-hall-taylor-medal
*Sponsor:* American Society of Mechanical Engineers
Two Park Avenue
New York, NY 10016-5990

**ASME J.P. Den Hartog Award** 658

The American Society of Mechanical Engineers J.P. Den Hartog Award was established by the Design Engineering Division in 1987 and operated as a division award until 2010 when it was elevated to a Society award. It recognizes lifetime contributions to the teaching and practice of vibration engineering. The award includes a bronze medal, certificate,

and $3,000, and is given in odd-numbered years. Nomination forms and instructions are available at the website. The bi-annual deadline for nominations is July 31 of even years.
*Requirements:* The General Awards Committee screens nominations and makes recommendations to the Committee on Honors. It consists of Members or Fellows representing the Districts. Special Award Committees are usually associated with a particular honor or award. They screen nominations and make recommendations to the other two committees.
*Restrictions:* Except for the Melville Medal, an individual will receive only one honor in recognition of the same achievement. The receipt of one ASME honor shall not bar the recipient from another ASME honor provided it is for different accomplishments. Honors are not awarded posthumously except if a nominee's death occurs after the nomination has been received at ASME Headquarters.
*Geographic Focus:* All States, All Countries
*Date(s) Application is Due:* Jul 31
*Amount of Grant:* 3,000 USD
*Contact:* Massimo Ruzzene, Committee Chair; (404) 894-3078 or (800) 843-2763; massimo.ruzzene@aerospace.gatech.edu
*Internet:* https://www.asme.org/about-asme/participate/honors-awards/achievement-awards/j-p-den-hartog-award
*Sponsor:* American Society of Mechanical Engineers
Two Park Avenue
New York, NY 10016-5990

**ASME James Harry Potter Gold Medal** **659**
The ASME James Harry Potter Gold Medal is awarded in recognition of eminent achievement or distinguished service in the appreciation of the science of thermodynamics in mechanical engineering. The basis of the award shall include contributions involving the teaching, appreciation, or utilization of thermodynamic principles in research, development, and design in mechanical engineering. The award includes $2,000, a vermeil medal, and a certificate. Any individual member, group of members, or committee may nominate candidates for consideration of this award. The annual deadline for nominations is March 1.
*Requirements:* Any individual member, group of members, or committee may nominate candidates for this award. The General Awards Committee screens nominations and makes recommendations to the Committee on Honors. It consists of Members or Fellows representing the Districts. Special Award Committees are usually associated with a particular honor or award. They screen nominations and make recommendations to the other two committees.
*Restrictions:* Services performed within the context of an individual's normal employment are not eligible. Except for the Melville Medal, an individual will receive only one honor in recognition of the same achievement. The receipt of one ASME honor shall not bar the recipient from another ASME honor provided it is for different accomplishments. Honors are not awarded posthumously except if a nominee's death occurs after the nomination has been received at ASME Headquarters.
*Geographic Focus:* All States, All Countries
*Date(s) Application is Due:* Mar 1
*Amount of Grant:* 2,000 USD
*Contact:* Tiruvadi RaviGururajan, Committee on Honors; (316) 978-6370 or (800) 843-2763; ts.ravi@wichita.edu or awards@asme.org
*Internet:* https://www.asme.org/about-asme/participate/honors-awards/achievement-awards/james-harry-potter-gold-medal
*Sponsor:* American Society of Mechanical Engineers
Two Park Avenue
New York, NY 10016-5990

**ASME James N. Landis Medal** **660**
The ASME James N. Landis Medal is given for outstanding personal performance related to designing, constructing, or managing the operation of major steam-powered electric stations using nuclear or fossil fuels, coupled with personal leadership in humanitarian pursuits. This may include committee activity, ASME section leadership, or the broad nontechnical professional activity of the nominee's engineering society. The award is presented preferably to a member of ASME, and includes $7,500, bronze medal, certificate, and up to $750 in travel expenses to attend the award presentation. The annual deadline for nominations is February 1.
*Requirements:* The award may be made to one recipient of any age who need not be a member of ASME. The General Awards Committee screens nominations and makes recommendations to the Committee on Honors. It consists of Members or Fellows representing the Districts. Special Award Committees are usually associated with a particular honor or award. They screen nominations and make recommendations to the other two committees.
*Restrictions:* Services performed within the context of an individual's normal employment are not eligible. Except for the Melville Medal, an individual will receive only one honor in recognition of the same achievement. The receipt of one ASME honor shall not bar the recipient from another ASME honor provided it is for different accomplishments. Honors are not awarded posthumously except if a nominee's death occurs after the nomination has been received at ASME Headquarters.
*Geographic Focus:* All States, All Countries
*Date(s) Application is Due:* Feb 1
*Amount of Grant:* 7,500 - 8,250 USD
*Contact:* Jovica R. Riznic, Committee Chair; (613) 943-0132 or (800) 843-2763; jovica.riznic@cnsc-ccsn.gc.ca or awards@asme.org
*Internet:* https://www.asme.org/about-asme/participate/honors-awards/achievement-awards/james-n-landis-medal
*Sponsor:* American Society of Mechanical Engineers
Two Park Avenue
New York, NY 10016-5990

**ASME M. Eugene Merchant Manufacturing Medal** **661**
The ASME M. Eugene Merchant Manufacturing Medal of ASME/SME is awarded to an individual who has had significant influence and responsibility for improving the productivity and efficiency (either by research or by implementation of research) of the manufacturing operation(s). The award includes a vermeil medal, a certificate, and $1,500. The annual deadline for nominations is February 1.
*Requirements:* The General Awards Committee screens nominations and makes recommendations to the Committee on Honors. It consists of Members or Fellows representing the Districts. Special Award Committees are usually associated with a particular honor or award. They screen nominations and make recommendations to the other two committees.
*Restrictions:* Services performed within the context of an individual's normal employment are not eligible. Except for the Melville Medal, an individual will receive only one honor in recognition of the same achievement. The receipt of one ASME honor shall not bar the recipient from another ASME honor provided it is for different accomplishments. Honors are not awarded posthumously except if a nominee's death occurs after the nomination has been received at ASME Headquarters.
*Geographic Focus:* All States, All Countries
*Date(s) Application is Due:* Feb 1
*Amount of Grant:* 1,500 USD
*Contact:* John W. Sutherland, Committee on Honors; (765) 496-9697 or (800) 843-2763; jwsuther@purdue.edu
*Internet:* https://www.asme.org/about-asme/participate/honors-awards/achievement-awards/m-eugene-merchant-manufacturing-medal-of-asme-sme
*Sponsor:* American Society of Mechanical Engineers
Two Park Avenue
New York, NY 10016-5990

**ASME Machine Design Award** **662**
The ASME Machine Design Award is made to recognize eminent achievement or distinguished service in the field of machine design, which includes application, research, development, or teaching of machine design. Recipients receive $1,000, a plaque, and a certificate. The annual deadline for nominations is February 1.
*Requirements:* Any individual member, group of members, or committee may nominate candidates for this award. The General Awards Committee screens nominations and makes recommendations to the Committee on Honors. It consists of Members or Fellows representing the Districts. Special Award Committees are usually associated with a particular honor or award. They screen nominations and make recommendations to the other two committees.
*Restrictions:* Services performed within the context of an individual's normal employment are not eligible. Except for the Melville Medal, an individual will receive only one honor in recognition of the same achievement. The receipt of one ASME honor shall not bar the recipient from another ASME honor provided it is for different accomplishments. Honors are not awarded posthumously except if a nominee's death occurs after the nomination has been received at ASME Headquarters.
*Geographic Focus:* All States, All Countries
*Date(s) Application is Due:* Feb 1
*Amount of Grant:* 1,000 USD
*Contact:* Jeffrey R. Mountain, Committee on Honors; (802) 485-2814 or (800) 843-2763; mountain@uhcl.edu or awards@asme.org
*Internet:* https://www.asme.org/about-asme/participate/honors-awards/achievement-awards/machine-design-award
*Sponsor:* American Society of Mechanical Engineers
Two Park Avenue
New York, NY 10016-5990

**ASME Marshall B. Peterson Award** **663**
The ASME Marshall B. Peterson Award is given in recognition of early career achievements in tribology research, as demonstrated by papers published in ASME scientific journals. It is preferred, but not required, that the awardee's research emphasis be on materials aspects of tribology, such as wear, friction materials, and solid lubricants. The award includes a certificate and $2,500. The even year bi-annual deadline for nominations is February 1.
*Requirements:* Any individual member, group of members, or committee may nominate candidates for this award. The General Awards Committee screens nominations and makes recommendations to the Committee on Honors. It consists of Members or Fellows representing the Districts. Special Award Committees are usually associated with a particular honor or award. They screen nominations and make recommendations to the other two committees.
*Restrictions:* Services performed within the context of an individual's normal employment are not eligible. Except for the Melville Medal, an individual will receive only one honor in recognition of the same achievement. The receipt of one ASME honor shall not bar the recipient from another ASME honor provided it is for different accomplishments. Honors are not awarded posthumously except if a nominee's death occurs after the nomination has been received at ASME Headquarters.
*Geographic Focus:* All States, All Countries
*Date(s) Application is Due:* Feb 1
*Amount of Grant:* 2,500 USD
*Contact:* Benyebka Bou-Said, Committee Chair; +33 478890980 or (800) 843-2763; benyebka.bou-said@insa-lyon.fr or awards@asme.org
*Internet:* https://www.asme.org/about-asme/participate/honors-awards/achievement-awards/marshall-b-peterson-award
*Sponsor:* American Society of Mechanical Engineers
Two Park Avenue
New York, NY 10016-5990

**ASME Mayo D. Hersey Award** 664

The American Society of Mechanical Engineers Mayo D. Hersey Award is bestowed on an individual in recognition of distinguished contributions over a substantial period of time to the advancement of the science and engineering of tribology. Distinguished contributions may result from significant original research in one or more of the many scientific disciplines related to lubrication, from excellence and creativity in lubrication engineering practice, or from sustained efforts and dissemination of information on the theory and practice of lubrication. The award includes a plaque and $2,000. Nomination forms and instructions are available at the website. The annual deadline for nominations is February 1.
*Requirements:* The General Awards Committee screens nominations and makes recommendations to the Committee on Honors. It consists of Members or Fellows representing the Districts. Special Award Committees are usually associated with a particular honor or award. They screen nominations and make recommendations to the other two committees.
*Restrictions:* Except for the Melville Medal, an individual will receive only one honor in recognition of the same achievement. The receipt of one ASME honor shall not bar the recipient from another ASME honor provided it is for different accomplishments. Honors are not awarded posthumously except if a nominee's death occurs after the nomination has been received at ASME Headquarters.
*Geographic Focus:* All States, All Countries
*Date(s) Application is Due:* Feb 1
*Amount of Grant:* 2,000 USD
*Contact:* Benyebka Bou-Said, Committee Chair; +33 478890980; fax (404) 894-8336; benyebka.bou-said@insa-lyon.fr
*Internet:* www.asme.org/about-asme/honors-awards/achievement-awards/mayo-d—hersey-award
*Sponsor:* American Society of Mechanical Engineers
Two Park Avenue
New York, NY 10016-5990

**ASME McDonald Mentoring Award** 665

The ASME McDonald Mentoring Award, established in 2007, recognizes the outstanding mentoring of other professionals by an engineer in industry, government, education or private practice. The award includes a bronze medal, a certificate, $1,000, and up to $750 in travel expenses to attend the award presentation. The annual deadline for nominations is March 1.
*Requirements:* Candidates for this award must have attained a record of outstanding performance in the following areas: contribute significantly to the engineering community, and serve as an effective advocate and guide, exhibiting true concern for mentees; demonstrate a portion of their mentoring or advising activities within the past five years; demonstrate the impact of practices, processes, and programs established by the nominee; current memberships with ASME, or another ICOMES member society for at least five years; possess a baccalaureate or equivalent degree in recognized field of engineering or engineering science, but not necessary in mechanical engineering.
*Restrictions:* Services performed within the context of an individual's normal employment are not eligible. Except for the Melville Medal, an individual will receive only one honor in recognition of the same achievement. The receipt of one ASME honor shall not bar the recipient from another ASME honor provided it is for different accomplishments. Honors are not awarded posthumously except if a nominee's death occurs after the nomination has been received at ASME Headquarters.
*Geographic Focus:* All States, All Countries
*Date(s) Application is Due:* Mar 1
*Amount of Grant:* 1,000 - 1,750 USD
*Contact:* Sonya T. Smith, Committee on Honors; (202) 806-4837 or (800) 843-2763; ssmith@howard.edu
*Internet:* https://www.asme.org/about-asme/participate/honors-awards/achievement-awards/mcdonald-mentoring-award
*Sponsor:* American Society of Mechanical Engineers
Two Park Avenue
New York, NY 10016-5990

**ASME Medal** 666

The American Society of Mechanical Engineers (ASME) Medal is the highest award that the Society can bestow and is to recognize eminently distinguished engineering achievement. Only one ASME Medal is be awarded annually. Although some individuals have been honored by both the ASME Medal and Honorary Membership, each award has been made on the basis of different accomplishments. Awardees receive $15,000, a gold medal, certificate, and travel expenses to the attend the award presentation. Nomination forms and instructions are available at the website.
*Requirements:* The General Awards Committee screens nominations and makes recommendations to the Committee on Honors. It consists of Members or Fellows representing the Districts. Special Award Committees are usually associated with a particular honor or award. They screen nominations and make recommendations to the other two committees.
*Restrictions:* Except for the Melville Medal, an individual will receive only one honor in recognition of the same achievement. Honors are not awarded posthumously except if a nominee's death occurs after the nomination has been received at ASME Headquarters.
*Geographic Focus:* All States, All Countries
*Date(s) Application is Due:* Mar 1
*Amount of Grant:* 15,000 USD
*Contact:* Y.Y. Bayazitoglu, Committee Chair; (212) 591-7094 or (800) 843-2763; fax (212) 705-7739; mckivorf@asme.org or awards@asme.org
*Internet:* https://www.asme.org/about-asme/participate/honors-awards/achievement-awards/asme-medal
*Sponsor:* American Society of Mechanical Engineers
Two Park Avenue
New York, NY 10016-5990

**ASME Melville Medal** 667

The ASME Melville Medal is the highest ASME honor for the best current original paper, not published elsewhere, which has been presented before ASME or approved for publication by ASME during the two previous calendar years. The paper may have more than one author but one of the authors must be an ASME corporate member (fellow, member, or associate member). The paper must be specifically recommended for the medal by a review committee or qualified individual. The medal may be awarded for a paper that has been selected for some other best paper award in the society if specifically recommended by the committee responsible for that specific award. The award includes a $2,000 honorarium, a bronze medal, a certificate, and a travel supplement not to exceed $750. The annual deadline for nominations is March 1.
*Requirements:* Any individual member, group of members, or committee may nominate candidates for this award. The General Awards Committee screens nominations and makes recommendations to the Committee on Honors. It consists of Members or Fellows representing the Districts. Special Award Committees are usually associated with a particular honor or award. They screen nominations and make recommendations to the other two committees.
*Restrictions:* Services performed within the context of an individual's normal employment are not eligible. Except for the Melville Medal, an individual will receive only one honor in recognition of the same achievement. The receipt of one ASME honor shall not bar the recipient from another ASME honor provided it is for different accomplishments. Honors are not awarded posthumously except if a nominee's death occurs after the nomination has been received at ASME Headquarters.
*Geographic Focus:* All States, All Countries
*Date(s) Application is Due:* Mar 1
*Amount of Grant:* 2,000 - 2,750 USD
*Contact:* Fran McKivor, Committee on Honors; (212) 591-7094 or (800) 843-2763; mckivorf@asme.org or awards@asme.org
*Internet:* https://www.asme.org/about-asme/participate/honors-awards/literature-awards/melville-medal
*Sponsor:* American Society of Mechanical Engineers
Two Park Avenue
New York, NY 10016-5990

**ASME Melvin R. Green Codes and Standards Medal** 668

The American Society of Mechanical Engineers Melvin R. Green Codes and Standards Medal was established in 1976 as the Codes and Standards Medal and renamed the Melvin R. Green Codes and Standards Medal in 1996 to honor the memory and extraordinary contributions of Melvin R. Green, an ardent supporter of industrial standards. The award recognizes outstanding contributions to: the development, promulgation, or management of documents, objects, or devices used in ASME programs of technical codification, standardization, and conformity assessment; or the acceptance of ASME Codes and Standards within the United States or internationally. The award includes a bronze medal, a certificate, and $1,500. Nomination forms and instructions are available at the website. The annual deadline for nominations is January 1.
*Requirements:* The recipient may be from industry, government, education, or private professional practice, and need not be an ASME member. The General Awards Committee screens nominations and makes recommendations to the Committee on Honors. It consists of Members or Fellows representing the Districts. Special Award Committees are usually associated with a particular honor or award. They screen nominations and make recommendations to the other two committees.
*Restrictions:* Except for the Melville Medal, an individual will receive only one honor in recognition of the same achievement. The receipt of one ASME honor shall not bar the recipient from another ASME honor provided it is for different accomplishments. Honors are not awarded posthumously except if a nominee's death occurs after the nomination has been received at ASME Headquarters.
*Geographic Focus:* All States, All Countries
*Date(s) Application is Due:* Jan 1
*Amount of Grant:* 1,500 USD
*Contact:* Mohinder L. Nayyar, Committee Chair; (301) 774-3806 or (800) 843-2763; nayyar1222@gmail.com or awards@asme.org
*Internet:* https://www.asme.org/about-asme/participate/honors-awards/achievement-awards/melvin-r-green-codes-standards-medal
*Sponsor:* American Society of Mechanical Engineers
Two Park Avenue
New York, NY 10016-5990

**ASME Milton C. Shaw Manufacturing Research Medal** 669

The ASME Milton C. Shaw Manufacturing Research Medal recognizes significant fundamental contributions to the science and technology of manufacturing processes. The award includes a medal, certificate, and $1,500. The annual deadline for nominations is February 1.
*Requirements:* Any individual member, group of members, or committee may nominate candidates for this award. The General Awards Committee screens nominations and makes recommendations to the Committee on Honors. It consists of Members or Fellows representing the Districts. Special Award Committees are usually associated with a particular honor or award. They screen nominations and make recommendations to the other two committees.
*Restrictions:* Services performed within the context of an individual's normal employment are not eligible. Except for the Melville Medal, an individual will receive only one honor

in recognition of the same achievement. The receipt of one ASME honor shall not bar the recipient from another ASME honor provided it is for different accomplishments. Honors are not awarded posthumously except if a nominee's death occurs after the nomination has been received at ASME Headquarters.
*Geographic Focus:* All States, All Countries
*Date(s) Application is Due:* Feb 1
*Amount of Grant:* 1,500 USD
*Contact:* Yong Huang, Committee on Honors; (352) 392-5520 or (800) 843-2763; fax (352) 392-7303; yongh@ufl.edu or awards@asme.org
*Internet:* https://www.asme.org/about-asme/participate/honors-awards/achievement-awards/milton-c-shaw-manufacturing-research-medal
*Sponsor:* American Society of Mechanical Engineers
Two Park Avenue
New York, NY 10016-5990

**ASME Nadai Medal** 670
The ASME Nadai Medal was established in 1975 on the proposal of the Materials Division to honor Arpad L. Nadai, who was a pioneer in the field of engineering materials, contributing particularly to the area of plasticity. His perspective also enabled him to give strong impetus to development in fatigue and high temperature behavior. The award is awarded in recognition of significant contributions and outstanding achievements which broaden the field of materials engineering. Such achievements may be in the areas of education, research, development, and service to the field and profession. The award includes a bronze medal, a certificate, $1,000, and up to $750 in travel expenses. The annual deadline for nominations is February 1.
*Requirements:* The General Awards Committee screens nominations and makes recommendations to the Committee on Honors. It consists of Members or Fellows representing the Districts. Special Award Committees are usually associated with a particular honor or award. They screen nominations and make recommendations to the other two committees.
*Restrictions:* Services performed within the context of an individual's normal employment are not eligible. Except for the Melville Medal, an individual will receive only one honor in recognition of the same achievement. The receipt of one ASME honor shall not bar the recipient from another ASME honor provided it is for different accomplishments. Honors are not awarded posthumously except if a nominee's death occurs after the nomination has been received at ASME Headquarters.
*Geographic Focus:* All States, All Countries
*Date(s) Application is Due:* Feb 1
*Amount of Grant:* 1,000 - 1,750 USD
*Contact:* Yong Zhu; (919) 513-7735 or (800) 843-2763; yzhu7@ncsu.edu
*Internet:* https://www.asme.org/about-asme/participate/honors-awards/achievement-awards/nadai-medal
*Sponsor:* American Society of Mechanical Engineers
Two Park Avenue
New York, NY 10016-5990

**ASME Nancy DeLoye Fitzroy and Roland V. Fitzroy Medal** 671
The American Society of Mechanical Engineers Nancy DeLoye Fitzroy and Roland V. Fitzroy Medal recognizes pioneering contribution to the frontiers of engineering leading to a breakthrough(s) in existing technology or leading to new applications or new areas of engineering endeavor. The award include a bronze metal, a certificate, $3,000, and up to $750 in travel expenses to attend the award presentation. The nomination forms and instructions are available at the website. The annual deadline for nominations is February 1.
*Requirements:* The General Awards Committee screens nominations and makes recommendations to the Committee on Honors. It consists of Members or Fellows representing the Districts. Special Award Committees are usually associated with a particular honor or award. They screen nominations and make recommendations to the other two committees.
*Restrictions:* Except for the Melville Medal, an individual will receive only one honor in recognition of the same achievement. The receipt of one ASME honor shall not bar the recipient from another ASME honor provided it is for different accomplishments. Honors are not awarded posthumously except if a nominee's death occurs after the nomination has been received at ASME Headquarters.
*Geographic Focus:* All States, All Countries
*Date(s) Application is Due:* Feb 1
*Amount of Grant:* 3,000 - 3,750 USD
*Contact:* Cristina Amon; (416) 978-3131 or (800) 843-2763; cristina.amon@utoronto.ca
*Internet:* https://www.asme.org/about-asme/participate/honors-awards/achievement-awards/nancy-deloye-fitzroy-and-roland-v-fitzroy-medal
*Sponsor:* American Society of Mechanical Engineers
Two Park Avenue
New York, NY 10016-5990

**ASME Old Guard Early Career Award** 672
The ASME Old Guard Early Career Award recognizes outstanding engineers who have advanced quickly in their professional career and education, shown leadership in ASME activities, and volunteered in their communities. The Old Guard is made up of ASME members who have reached the age of 65 and retired. These members continue to contribute to the Society, and their contributions are used to support the ASME Student and Early Career Engineer Development. The first place award includes a plaque, prepaid ASME membership for life, and $5,000. The second place award includes $2,000, a plaque and prepaid ASME membership for life. The annual deadline for nominations is February 1.
*Requirements:* The General Awards Committee screens nominations and makes recommendations to the Committee on Honors. It consists of Members or Fellows representing the Districts. Special Award Committees are usually associated with a particular honor or award. They screen nominations and make recommendations to the other two committees.
*Restrictions:* Services performed within the context of an individual's normal employment are not eligible. Except for the Melville Medal, an individual will receive only one honor in recognition of the same achievement. The receipt of one ASME honor shall not bar the recipient from another ASME honor provided it is for different accomplishments. Honors are not awarded posthumously except if a nominee's death occurs after the nomination has been received at ASME Headquarters.
*Geographic Focus:* All States, All Countries
*Date(s) Application is Due:* Feb 1
*Amount of Grant:* 2,000 - 5,000 USD
*Contact:* Dennis A. Armstrong, Committee on Honors; (509) 943-4193 or (800) 843-2763; daa3@msn.com or armstrongd@asme.org
*Internet:* https://www.asme.org/about-asme/participate/honors-awards/achievement-awards/old-guard-early-career-award
*Sponsor:* American Society of Mechanical Engineers
Two Park Avenue
New York, NY 10016-5990

**ASME Old Guard Prizes** 673
The Old Guard Prizes are awarded for the best four contenders at the Old Guard Oral Presentation Competition during the ASME International Mechanical Engineering Congress and Exposition (IMECE). These competitors are student members who won first prize at the district level competitions held at ASME Student Professional Development Conferences. The district level first prizes also include partial travel support to participate in the IMECE. The Old Guard is made up of ASME dues exempt members, those who have reached the age of 65 and have retired. They continue to contribute to the Society and their contributions are used to support the "Old Guard" Prizes and similar Society activities related to Student and Early Career members. Awards include the following: 1st prize: $2,000; 2nd prize: $1,500; 3rd prize: $1,000; and 4th prize: $500. Further information including specific rules, criteria, and current deadline for the competition is available at the website.
*Requirements:* The General Awards Committee screens nominations and makes recommendations to the Committee on Honors. It consists of Members or Fellows representing the Districts. Special Award Committees are usually associated with a particular honor or award. They screen nominations and make recommendations to the other two committees.
*Restrictions:* Services performed within the context of an individual's normal employment are not eligible. Except for the Melville Medal, an individual will receive only one honor in recognition of the same achievement. The receipt of one ASME honor shall not bar the recipient from another ASME honor provided it is for different accomplishments. Honors are not awarded posthumously except if a nominee's death occurs after the nomination has been received at ASME Headquarters.
*Geographic Focus:* All States, All Countries
*Amount of Grant:* 500 - 2,000 USD
*Contact:* Dennis A. Armstrong, Committee on Honors; (509) 943-4193 or (800) 843-2763; daa3@msn.com or armstrongd@asme.org
*Internet:* https://www.asme.org/about-asme/participate/honors-awards/achievement-awards/old-guard-prizes
*Sponsor:* American Society of Mechanical Engineers
Two Park Avenue
New York, NY 10016-5990

**ASME Outstanding Student Section Advisor Award** 674
The Outstanding Student Section Advisor Award recognizes the leadership and service qualities of a student who has completed at least three academic years as a Student Section Advisor prior to nomination for the award. A nomination for this award is expected to substantiate the leadership and service qualities of the nominee by reference to accomplishments in the following areas: encouraging engineering students to become Student Members as the initial step in their program of professional development; providing guidance to the Student Section on the activities and programs that will stimulate interest in the profession; and creating a professional awareness in the Student Members which will inspire them to upgrade to Member before graduation and to maintain a continuous active membership in ASME. The award includes a silver medal, certificate, $1,500, and up to $750 in travel expenses to attend the award presentation. The annual deadline for nominations is March 15.
*Requirements:* Any individual member, group of members, or committee may nominate candidates for this award. The General Awards Committee screens nominations and makes recommendations to the Committee on Honors. It consists of Members or Fellows representing the Districts. Special Award Committees are usually associated with a particular honor or award. They screen nominations and make recommendations to the other two committees.
*Restrictions:* Services performed within the context of an individual's normal employment are not eligible. Except for the Melville Medal, an individual will receive only one honor in recognition of the same achievement. The receipt of one ASME honor shall not bar the recipient from another ASME honor provided it is for different accomplishments. Honors are not awarded posthumously except if a nominee's death occurs after the nomination has been received at ASME Headquarters.
*Geographic Focus:* All States, All Countries
*Date(s) Application is Due:* Mar 15
*Amount of Grant:* 1,500 - 2,250 USD

*Contact:* Fran McKivor, Committee on Honors; (212) 591-7094 or (800) 843-2763; mckivorf@asme.org or awards@asme.org
*Internet:* www.asme.org/about-asme/honors-awards/service-awards/student-section-advisor-award
*Sponsor:* American Society of Mechanical Engineers
Two Park Avenue
New York, NY 10016-5990

**ASME Patrick J. Higgins Award** 675
The American Society of Mechanical Engineers Patrick J. Higgins Award recognizes an individual who has contributed to the enhancement of standardization through the development and promotion of ASME Codes and Standards or Conformity Assessment Programs in the following areas: plumbing materials and equipment; screw threads; tools (machine, cutting and hand); fasteners, (e.g, bolts, nuts, rivets, screws, washers); chain attachments and sprockets for power transmission and conveying; metal and metal alloy wrought mill product nominal sizes; pressure and temperature instruments and accessories; classification and designation of surface qualities; gage blanks; chemical standard pumps; dimensional metrology; gas turbine procurement; overhead hoists; measurement of fluid flow in closed conduits; steel stacks; and engineering drawing and related documentation practices. The award includes a bronze metal, a certificate, and $1,000. Nomination forms and instructions are available at the website. The annual deadline for nominations is September 1.
*Requirements:* The General Awards Committee screens nominations and makes recommendations to the Committee on Honors. It consists of Members or Fellows representing the Districts. Special Award Committees are usually associated with a particular honor or award. They screen nominations and make recommendations to the other two committees.
*Restrictions:* Except for the Melville Medal, an individual will receive only one honor in recognition of the same achievement. The receipt of one ASME honor shall not bar the recipient from another ASME honor provided it is for different accomplishments. Honors are not awarded posthumously except if a nominee's death occurs after the nomination has been received at ASME Headquarters.
*Geographic Focus:* All States, All Countries
*Date(s) Application is Due:* Sep 1
*Amount of Grant:* 1,000 USD
*Contact:* Peter V. DeMarco, Committee Chair; (732) 329-1237 or (800) 843-2763; pete.demarco@iapmo.org
*Internet:* https://www.asme.org/about-asme/participate/honors-awards/achievement-awards/patrick-j-higgins-award
*Sponsor:* American Society of Mechanical Engineers
Two Park Avenue
New York, NY 10016-5990

**ASME Per Bruel Gold Medal for Noise Control and Acoustics** 676
The American Society of Mechanical Engineers Per Bruel Gold Medal for Noise Control and Acoustics honors Dr. Per Bruel, who pioneered the development of sophisticated noise and vibration measuring and processing equipment. The Medal is given annually in recognition of eminent achievement and extraordinary merit in the field of noise control and acoustics. The achievement must include useful applications of the principles of noise control and acoustics to the art and science of mechanical engineering. The award includes $1,000, a vermeil medal, certificate, and travel expenses up to a maximum of $750 to attend the award presentation. Nomination instructions and forms are located on the website. The annual nomination deadline is February 1.
*Requirements:* The General Awards Committee screens nominations and makes recommendations to the Committee on Honors. It consists of Members or Fellows representing the Districts. Special Award Committees are usually associated with a particular honor or award. They screen nominations and make recommendations to the other two committees.
*Restrictions:* Except for the Melville Medal, an individual will receive only one honor in recognition of the same achievement. The receipt of one ASME honor shall not bar the recipient from another ASME honor provided it is for different accomplishments. Honors are not awarded posthumously except if a nominee's death occurs after the nomination has been received at ASME Headquarters.
*Geographic Focus:* All States, All Countries
*Date(s) Application is Due:* Feb 1
*Amount of Grant:* 1,000 - 1,750 USD
*Contact:* Andrew Norris, Committee Chair; (908) 789-8767 or (800) 843-2763; norris@rutgers.edu or awards@asme.org
*Internet:* https://www.asme.org/about-asme/participate/honors-awards/achievement-awards/per-bruel-gold-medal-for-noise-control-and
*Sponsor:* American Society of Mechanical Engineers
Two Park Avenue
New York, NY 10016-5990

**ASME Performance Test Codes Medal** 677
The ASME Performance Test Codes Medal was established to recognize the first voluntary codes and standards activity in the Society and in the United States. The codes are recognized throughout the world for their excellence, providing industry and the engineering community with the technology that promotes the philosophy of accurate and reliable performance evaluation. The Medal is awarded to an individual (or individuals in exceptional circumstances) who has made an outstanding contribution to the development and promotion of ASME performance test codes including the supplements on instruments and apparatus. The award includes a vermeil medal, a certificate, and $1,000. The annual deadline for nominations is January 1.
*Requirements:* Any individual member, group of members, or committee may nominate candidates for this award. The General Awards Committee screens nominations and makes recommendations to the Committee on Honors. It consists of Members or Fellows representing the Districts. Special Award Committees are usually associated with a particular honor or award. They screen nominations and make recommendations to the other two committees.
*Restrictions:* Services performed within the context of an individual's normal employment are not eligible. Except for the Melville Medal, an individual will receive only one honor in recognition of the same achievement. The receipt of one ASME honor shall not bar the recipient from another ASME honor provided it is for different accomplishments. Honors are not awarded posthumously except if a nominee's death occurs after the nomination has been received at ASME Headquarters.
*Geographic Focus:* All States, All Countries
*Date(s) Application is Due:* Jan 1
*Amount of Grant:* 1,000 USD
*Contact:* David R. Keyser, Committee on Honors; (410) 586-8940 or (800) 843-2763; dave.keyser@survice.com or awards@asme.org
*Internet:* https://www.asme.org/about-asme/participate/honors-awards/achievement-awards/performance-test-codes-medal
*Sponsor:* American Society of Mechanical Engineers
Two Park Avenue
New York, NY 10016-5990

**ASME Pi Tau Sigma Gold Medal** 678
The ASME Pi Tau Sigma Gold Medal, sponsored jointly by the Pi Tau Sigma Honorary Mechanical Engineering Fraternity and ASME, is awarded annually to a young engineering graduate who has demonstrated outstanding achievement in mechanical engineering no more than ten years after receiving their baccalaureate degree. Achievement shall be all or in part in any field, including industrial, educational, political, research, civic, or artistic. The candidate's achievements will be examined for an application of basic engineering methods or principles. Special nomination form must be obtained from ASME. The award includes gold medal, certificate, $1,000, and up to $750 in travel expenses to attend the award presentation. The annual deadline for nominations is February 15.
*Requirements:* The qualifications for the recipients of each of the three awards are identical except for the period of recognized achievements. In each case the recipient shall have received a baccalaureate degree from a regular engineering program of a recognized college or university and shall have attained outstanding achievement within the period stated for each award. Achievement shall be all or in part in any field, including industrial, educational, political, research, civic and artistic. The candidate's achievements will be examined for an application of basic engineering methods or principles. The three awards are administered by a Joint Board of Award appointed by Pi Tau Sigma and ASME. The Chair is a Pi Tau Sigma appointee. The Board of Award provides forms and invites ASME Sections and other units to nominate for the awards. The Board of Award submits the names and records of the persons selected for the three awards to the ASME Committee on Honors for the formal selection of the recipients.
*Restrictions:* Services performed within the context of an individual's normal employment are not eligible. Except for the Melville Medal, an individual will receive only one honor in recognition of the same achievement. The receipt of one ASME honor shall not bar the recipient from another ASME honor provided it is for different accomplishments. Honors are not awarded posthumously except if a nominee's death occurs after the nomination has been received at ASME Headquarters.
*Geographic Focus:* All States, All Countries
*Date(s) Application is Due:* Feb 15
*Amount of Grant:* 1,000 - 1,750 USD
*Contact:* Darryl James, Committee on Honors; (806) 834-3386 or (800) 843-2763; darryl.james@ttu.edu or awards@asme.org
*Internet:* https://www.asme.org/about-asme/participate/honors-awards/achievement-awards/pi-tau-sigma-gold-medal
*Sponsor:* American Society of Mechanical Engineers
Two Park Avenue
New York, NY 10016-5990

**ASME Prime Movers Committee Award** 679
The ASME Prime Movers Committee Award recognizes outstanding contributions to the literature of thermal electric station practice or equipment which are available through public presentation and publication. Those papers approved by the appropriate papers review committee as meeting ASME standards and available in printed form may be considered for this award. Papers need not be current and may be written by a single author or multiple authors. The award includes a certificate and $1,000. The annual deadline for nominations is February 1.
*Requirements:* Any individual member, group of members, or committee may nominate candidates for this award. The General Awards Committee screens nominations and makes recommendations to the Committee on Honors. It consists of Members or Fellows representing the Districts. Special Award Committees are usually associated with a particular honor or award. They screen nominations and make recommendations to the other two committees.
*Restrictions:* Services performed within the context of an individual's normal employment are not eligible. Except for the Melville Medal, an individual will receive only one honor in recognition of the same achievement. The receipt of one ASME honor shall not bar the recipient from another ASME honor provided it is for different accomplishments. Honors are not awarded posthumously except if a nominee's death occurs after the nomination has been received at ASME Headquarters.
*Geographic Focus:* All States, All Countries
*Date(s) Application is Due:* Feb 1

*Amount of Grant:* 1,000 USD
*Contact:* S.A. Huff, Committee on Honors; (314) 682-1642 or (800) 843-2763; shuff@burnsmcd.com or awards@asme.org
*Internet:* www.asme.org/about-asme/honors-awards/literature-awards/prime-movers-committee-award
*Sponsor:* American Society of Mechanical Engineers
Two Park Avenue
New York, NY 10016-5990

**ASME R. Tom Sawyer Award** **680**
The R. Tom Sawyer Award is given to an individual who has made important contributions to advance the purpose of the gas turbine industry and to the International Gas Turbine Institute over a substantial period of time. The contribution may be in any area of institute activity, but must be marked by sustained efforts. The award includes a plaque, certificate, and $1,000. The annual deadline for nominations is August 30.
*Requirements:* Any individual member, group of members, or committee may nominate candidates for this award. The General Awards Committee screens nominations and makes recommendations to the Committee on Honors. It consists of Members or Fellows representing the Districts. Special Award Committees are usually associated with a particular honor or award. They screen nominations and make recommendations to the other two committees.
*Restrictions:* Services performed within the context of an individual's normal employment are not eligible. Except for the Melville Medal, an individual will receive only one honor in recognition of the same achievement. The receipt of one ASME honor shall not bar the recipient from another ASME honor provided it is for different accomplishments. Honors are not awarded posthumously except if a nominee's death occurs after the nomination has been received at ASME Headquarters.
*Geographic Focus:* All States, All Countries
*Date(s) Application is Due:* Aug 30
*Amount of Grant:* 1,000 USD
*Contact:* Jaroslaw Szwedowicz, Committee on Honors; +562 0548-8000 or (800) 843-2763; jaroslaw.szwedowicz@ge.com or awards@asme.org
*Internet:* https://www.asme.org/about-asme/participate/honors-awards/achievement-awards/r-tom-sawyer-award
*Sponsor:* American Society of Mechanical Engineers
Two Park Avenue
New York, NY 10016-5990

**ASME Ralph Coats Roe Medal** **681**
The ASME Ralph Coats Roe Medal is awarded annually to an individual selected by the society for a significant contribution to a better public understanding and appreciation of the engineer's worth to contemporary society. The successful candidate is expected to have the attributes that qualify him/her as an authoritative lecturer on his/her contribution at a general session during the winter meeting. Candidates are not restricted by profession nor by membership in any engineering society or organization. Award includes a gold medal, $12,000, a certificate, and up to $750 in travel expenses to attend the award presentation. The annual deadline for nominations is September 15.
*Requirements:* Any individual member, group of members, or committee may nominate candidates for this award. The General Awards Committee screens nominations and makes recommendations to the Committee on Honors. It consists of Members or Fellows representing the Districts. Special Award Committees are usually associated with a particular honor or award. They screen nominations and make recommendations to the other two committees.
*Restrictions:* Services performed within the context of an individual's normal employment are not eligible. Except for the Melville Medal, an individual will receive only one honor in recognition of the same achievement. The receipt of one ASME honor shall not bar the recipient from another ASME honor provided it is for different accomplishments. Honors are not awarded posthumously except if a nominee's death occurs after the nomination has been received at ASME Headquarters.
*Geographic Focus:* All States, All Countries
*Date(s) Application is Due:* Sep 15
*Amount of Grant:* 12,000 - 12,750 USD
*Contact:* Betty Bowersox, Committee Chair; (636) 651-1025 or (800) 843-2763; betty@bowersox.org or awards@asme.org
*Internet:* www.asme.org/about-asme/honors-awards/achievement-awards/ralph-coats-roe-medal
*Sponsor:* American Society of Mechanical Engineers
Two Park Avenue
New York, NY 10016-5990

**ASME Robert E. Koski Medal** **682**
The ASME Robert E. Koski Medal recognizes individuals who have advanced the art and practice of fluid power motion and control through education and/or innovation. The honoree does not have to be a member of ASME. The Medal was established by the Fluid Power Systems and Technology Division to honor Koski's contributions to the field of Design Engineering. The award includes a vermeil medal, a certificate, $10,000, and up to $750 in travel expenses to attend the award presentation. Nomination forms and instructions are available at the website, and the annual deadline is February 11.
*Geographic Focus:* All States, All Countries
*Date(s) Application is Due:* Feb 11
*Amount of Grant:* 10,000 USD
*Contact:* Nariman Sepehri, Committee Chair; (204) 474-6834 or (800) 843-2763; deankim@bradley.edu
*Internet:* https://www.asme.org/about-asme/participate/honors-awards/achievement-awards/robert-e-koski-medal
*Sponsor:* American Society of Mechanical Engineers
Two Park Avenue
New York, NY 10016-5990

**ASME Robert Henry Thurston Lecture Award** **683**
The ASME Robert Henry Thurston Lecture, presented annually at the International Mechanical Engineering Congress, provides an outstanding leader in pure or applied science or engineering with the honor of presenting a lecture to the Society that encourages stimulating thinking on a subject of broad technical interest to engineers. In addition to the lecture, the awardee receive a plaque, $500, and up to $500 in travel expenses to attend the award presentation. The annual deadline for nominations is February 15.
*Requirements:* Any individual member, group of members, or committee may nominate candidates for this award. The General Awards Committee screens nominations and makes recommendations to the Committee on Honors. It consists of Members or Fellows representing the Districts. Special Award Committees are usually associated with a particular honor or award. They screen nominations and make recommendations to the other two committees.
*Restrictions:* Services performed within the context of an individual's normal employment are not eligible. Except for the Melville Medal, an individual will receive only one honor in recognition of the same achievement. The receipt of one ASME honor shall not bar the recipient from another ASME honor provided it is for different accomplishments. Honors are not awarded posthumously except if a nominee's death occurs after the nomination has been received at ASME Headquarters.
*Geographic Focus:* All States, All Countries
*Date(s) Application is Due:* Feb 15
*Amount of Grant:* 500 - 1,000 USD
*Contact:* Arunachalam M. Rajendran, Committee on Honors; (662) 915-5770 or (800) 843-2763; raj@olemiss.edu or awards@asme.org
*Internet:* https://www.asme.org/about-asme/participate/honors-awards/achievement-awards/robert-henry-thurston-lecture-award
*Sponsor:* American Society of Mechanical Engineers
Two Park Avenue
New York, NY 10016-5990

**ASME Robert M. Nerem Education and Mentorship Medal** **684**
The American Society of Mechanical Engineers Robert M. Nerem Education and Mentorship Medal was established by the Bioengineering Division in 2017, and is given to an individual who has demonstrated a sustained level of outstanding achievement in education and mentoring of trainees. Examples of meritorious activities include leadership within the nominee's institution, mentoring activities that are above and beyond those expected from others employed in similar positions, mentoring activities tailored to meet the needs of the trainees, and innovative mentoring activities. The award includes $1,000, bronze medal, certificate and a travel supplement not to exceed $750. The annual deadline for nominations is September 1.
*Requirements:* The award may be made to one recipient of any age who need not be a member of ASME. The General Awards Committee screens nominations and makes recommendations to the Committee on Honors. It consists of Members or Fellows representing the Districts. Special Award Committees are usually associated with a particular honor or award. They screen nominations and make recommendations to the other two committees.
*Restrictions:* Services performed within the context of an individual's normal employment are not eligible. Except for the Melville Medal, an individual will receive only one honor in recognition of the same achievement. The receipt of one ASME honor shall not bar the recipient from another ASME honor provided it is for different accomplishments. Honors are not awarded posthumously except if a nominee's death occurs after the nomination has been received at ASME Headquarters.
*Geographic Focus:* All States, All Countries
*Date(s) Application is Due:* Sep 1
*Amount of Grant:* 1,000 - 1,750 USD
*Contact:* Sara E. Wilson; (785) 864-2103 or (800) 843-2763; sewilson@ku.edu
*Internet:* https://www.asme.org/about-asme/honors-awards/achievement-awards/robert-nerem-education-mentorship-medal
*Sponsor:* American Society of Mechanical Engineers
Two Park Avenue
New York, NY 10016-5990

**ASME Rufus Oldenburger Medal** **685**
The ASME Rufus Oldenburger Medal recognizes significant contributions and outstanding achievements in the field of automatic control. Such achievements may be in the areas of education, research, development, innovation, and service to the field and profession. The award includes a bronze medal, a certificate, and $2,000. The annual deadline for nominations is February 1.
*Requirements:* The General Awards Committee screens nominations and makes recommendations to the Committee on Honors. It consists of Members or Fellows representing the Districts. Special Award Committees are usually associated with a particular honor or award. They screen nominations and make recommendations to the other two committees.
*Restrictions:* Services performed within the context of an individual's normal employment are not eligible. Except for the Melville Medal, an individual will receive only one honor in recognition of the same achievement. The receipt of one ASME honor shall not bar the recipient from another ASME honor provided it is for different accomplishments.

Honors are not awarded posthumously except if a nominee's death occurs after the nomination has been received at ASME Headquarters.
*Geographic Focus:* All States, All Countries
*Date(s) Application is Due:* Feb 1
*Amount of Grant:* 2,000 USD
*Contact:* Santosh Devasia; (206) 685 3401 or (800) 843-2763; devasia@u.washington.edu
*Internet:* https://www.asme.org/about-asme/participate/honors-awards/achievement-awards/rufus-oldenburger-medal
*Sponsor:* American Society of Mechanical Engineers
Two Park Avenue
New York, NY 10016-5990

**ASME Ruth and Joel Spira Outstanding Design Educator Award** 686
The ASME Ruth and Joel Spira Outstanding Design Educator Award was established to recognize a person who exemplifies the best in furthering engineering design education through interacting with students and industry, and serving as a role model for other educators to emulate. The award includes a vermeil medal, a certificate, and $1,000. The annual deadline for nominations is February 1.
*Requirements:* Any individual member, group of members, or committee may nominate candidates for this award. The General Awards Committee screens nominations and makes recommendations to the Committee on Honors. It consists of Members or Fellows representing the Districts. Special Award Committees are usually associated with a particular honor or award. They screen nominations and make recommendations to the other two committees.
*Restrictions:* Services performed within the context of an individual's normal employment are not eligible. Except for the Melville Medal, an individual will receive only one honor in recognition of the same achievement. The receipt of one ASME honor shall not bar the recipient from another ASME honor provided it is for different accomplishments. Honors are not awarded posthumously except if a nominee's death occurs after the nomination has been received at ASME Headquarters.
*Geographic Focus:* All States, All Countries
*Date(s) Application is Due:* Feb 1
*Amount of Grant:* 1,000 USD
*Contact:* Jeffrey R. Mountain, Committee on Honors; (802) 485-2814 or (800) 843-2763; mountain@uhcl.edu or awards@asme.org
*Internet:* https://www.asme.org/about-asme/participate/honors-awards/achievement-awards/ruth-and-joel-spira-outstanding-design-educator
*Sponsor:* American Society of Mechanical Engineers
Two Park Avenue
New York, NY 10016-5990

**ASME S.Y. Zamrik PVP Medal** 687
The ASME S.Y. Zamrik PVP Medal is given for outstanding contributions in the field of pressure vessel and piping technology including, but not limited to, research, development, teaching, and significant advancements of the state-of-the-art. The award includes a bronze medal, a certificate, and $4,000. The annual deadline for nominations is June 15.
*Requirements:* Any individual member, group of members, or committee may nominate candidates for this award. The General Awards Committee screens nominations and makes recommendations to the Committee on Honors. It consists of Members or Fellows representing the Districts. Special Award Committees are usually associated with a particular honor or award. They screen nominations and make recommendations to the other two committees.
*Restrictions:* Services performed within the context of an individual's normal employment are not eligible. Except for the Melville Medal, an individual will receive only one honor in recognition of the same achievement. The receipt of one ASME honor shall not bar the recipient from another ASME honor provided it is for different accomplishments. Honors are not awarded posthumously except if a nominee's death occurs after the nomination has been received at ASME Headquarters.
*Geographic Focus:* All States, All Countries
*Date(s) Application is Due:* Jun 15
*Amount of Grant:* 4,000 USD
*Contact:* Clay Rodery, Committee on Honors; (832) 661-0625 or (804) 541-0812; roderyc@asme.org or awards@asme.org
*Internet:* https://www.asme.org/about-asme/participate/honors-awards/achievement-awards/s-y-zamrik-pvp-medal
*Sponsor:* American Society of Mechanical Engineers
Two Park Avenue
New York, NY 10016-5990

**ASME Safety Codes and Standards Medal** 688
The American Society of Mechanical Engineers Safety Codes and Standards Medal was established in 1986 by the Council on Codes and Standards, and is presented to one or more individuals who have contributed to the enhancement of public safety through the development and promotion of ASME codes and standards or the ASME safety accreditation activity. The award includes $1,500, a bronze medal, a certificate, and travel expenses up to $750 to attend the award presentation. The annual deadline for nominations is January 1.
*Requirements:* The award may be made to one recipient of any age who need not be a member of ASME. The General Awards Committee screens nominations and makes recommendations to the Committee on Honors. It consists of Members or Fellows representing the Districts. Special Award Committees are usually associated with a particular honor or award. They screen nominations and make recommendations to the other two committees.
*Restrictions:* Services performed within the context of an individual's normal employment are not eligible. Except for the Melville Medal, an individual will receive only one honor in recognition of the same achievement. The receipt of one ASME honor shall not bar the recipient from another ASME honor provided it is for different accomplishments. Honors are not awarded posthumously except if a nominee's death occurs after the nomination has been received at ASME Headquarters.
*Geographic Focus:* All States, All Countries
*Date(s) Application is Due:* Jan 1
*Amount of Grant:* 1,500 - 2,250 USD
*Contact:* Thomas F. Conry, Committee Chair; (217) 369-1501 or (800) 843-2763; tconry@illinois.edu or awards@asme.org
*Internet:* https://www.asme.org/about-asme/participate/honors-awards/achievement-awards/safety-codes-standards-medal
*Sponsor:* American Society of Mechanical Engineers
Two Park Avenue
New York, NY 10016-5990

**ASME Savio L-Y. Woo Translational Biomechanics Medal** 689
The American Society of Mechanical Engineers Savio L-Y. Woo Translational Biomechanics Medal is bestowed upon an individual who has translated meritorious bioengineering science to clinical practice through research, education, professional development, and with service to the bioengineering community. Examples of meritorious activity might be: basic bioengineering science that translates into a medical device or equipment; contributions to new approaches of disease treatment; or establishment of new injury treatment modalities. The award includes he award includes $1000, bronze medal, certificate and a travel supplement not to exceed $750. The annual deadline for nominations is September 1.
*Requirements:* The award may be made to one recipient of any age who need not be a member of ASME. The General Awards Committee screens nominations and makes recommendations to the Committee on Honors. It consists of Members or Fellows representing the Districts. Special Award Committees are usually associated with a particular honor or award. They screen nominations and make recommendations to the other two committees.
*Restrictions:* Services performed within the context of an individual's normal employment are not eligible. Except for the Melville Medal, an individual will receive only one honor in recognition of the same achievement. The receipt of one ASME honor shall not bar the recipient from another ASME honor provided it is for different accomplishments. Honors are not awarded posthumously except if a nominee's death occurs after the nomination has been received at ASME Headquarters.
*Geographic Focus:* All States, All Countries
*Date(s) Application is Due:* Sep 1
*Amount of Grant:* 1,000 - 1,750 USD
*Contact:* Baruch Barry Lieber, Committee Chair; (631) 444-1278 or (800) 843-2763; baruch.lieber@stonybrook.edu or awards@asme.org
*Internet:* https://www.asme.org/about-asme/honors-awards/achievement-awards/savio-ly-woo-translational-biomechanics-medal
*Sponsor:* American Society of Mechanical Engineers
Two Park Avenue
New York, NY 10016-5990

**ASME Sia Nemat-Nasser Early Career Award** 690
The ASME Sia Nemat-Nasser Early Career Award recognizes early career research excellence in the areas of experimental, computational, and theoretical mechanics and materials by young investigators who are within 10 years after their Ph.D. degree, with special emphasis placed on under-represented groups. The award was established by the by the Materials Division in 2008 and operated as a division award until 2012 when it was elevated to a Society award. The award includes a bronze medal, a certificate, $3,000, and up to $750 in travel expenses. The annual deadline for nominations is February 1.
*Requirements:* The General Awards Committee screens nominations and makes recommendations to the Committee on Honors. It consists of Members or Fellows representing the Districts. Special Award Committees are usually associated with a particular honor or award. They screen nominations and make recommendations to the other two committees.
*Restrictions:* Services performed within the context of an individual's normal employment are not eligible. Except for the Melville Medal, an individual will receive only one honor in recognition of the same achievement. The receipt of one ASME honor shall not bar the recipient from another ASME honor provided it is for different accomplishments. Honors are not awarded posthumously except if a nominee's death occurs after the nomination has been received at ASME Headquarters.
*Geographic Focus:* All States, All Countries
*Date(s) Application is Due:* Feb 1
*Amount of Grant:* 3,000 - 3,750 USD
*Contact:* Yong Zhu; (919) 513-7735 or (800) 843-2763; yzhu7@ncsu.edu
*Internet:* https://www.asme.org/about-asme/get-involved/honors-awards/achievement-awards/sia-nematnasser-early-career-award
*Sponsor:* American Society of Mechanical Engineers
Two Park Avenue
New York, NY 10016-5990

**ASME Soichiro Honda Medal** 691
The American Society of Mechanical Engineers Soichiro Honda Medal recognizes an individual for an outstanding achievement or a series of significant engineering contributions in developing improvement in the field of personal transportation. Attention shall be concentrated on the brilliance of the achievement or on the overall effect of a series of contributions, not on the individual. The achievement should be of such public importance as to be worthy of the gratitude of the nation and to call forth the admiration of engineers. The award includes a gold medal, a certificate, $7,500, and

up to $750 in travel expenses to attend the award ceremony. Nomination forms and instructions are available at the website.
*Requirements:* There are no restrictions as to nominee's age, nationality, society membership, degree of education, employment, or official position. The General Awards Committee screens nominations and makes recommendations to the Committee on Honors. It consists of Members or Fellows representing the Districts. Special Award Committees are usually associated with a particular honor or award. They screen nominations and make recommendations to the other two committees. The annual deadline for nominations is October 1.
*Restrictions:* Except for the Melville Medal, an individual will receive only one honor in recognition of the same achievement. The receipt of one ASME honor shall not bar the recipient from another ASME honor provided it is for different accomplishments. Honors are not awarded posthumously except if a nominee's death occurs after the nomination has been received at ASME Headquarters.
*Geographic Focus:* All States, All Countries
*Date(s) Application is Due:* Oct 1
*Amount of Grant:* 7,500 - 8,250 USD
*Contact:* Ryoichi S. Amano; (414) 229-2345 or (800) 843-2763; amano@uwm.edu
*Internet:* https://www.asme.org/about-asme/participate/honors-awards/achievement-awards/soichiro-honda-medal
*Sponsor:* American Society of Mechanical Engineers
Two Park Avenue
New York, NY 10016-5990

**ASME Spirit of St. Louis Medal** **692**
The ASME Spirit of St. Louis Award is given for meritorious service in the advancement of aeronautics and astronautics. The award is not limited to members of the ASME or the engineering profession. The award includes a vermeil medal, a certificate, and $1,000. The annual deadline for nominations is January 20.
*Requirements:* Any individual member, group of members, or committee may nominate candidates for this award. The General Awards Committee screens nominations and makes recommendations to the Committee on Honors. It consists of Members or Fellows representing the Districts. Special Award Committees are usually associated with a particular honor or award. They screen nominations and make recommendations to the other two committees.
*Restrictions:* Services performed within the context of an individual's normal employment are not eligible. Except for the Melville Medal, an individual will receive only one honor in recognition of the same achievement. The receipt of one ASME honor shall not bar the recipient from another ASME honor provided it is for different accomplishments. Honors are not awarded posthumously except if a nominee's death occurs after the nomination has been received at ASME Headquarters.
*Geographic Focus:* All States, All Countries
*Date(s) Application is Due:* Jan 20
*Amount of Grant:* 1,000 USD
*Contact:* Anthony M. Waas, Committee on Honors; (206) 221-2569 or (800) 843-2763; awaas@aa.washington.edu or awards@asme.org
*Internet:* https://www.asme.org/about-asme/participate/honors-awards/achievement-awards/spirit-of-st-louis-medal
*Sponsor:* American Society of Mechanical Engineers
Two Park Avenue
New York, NY 10016-5990

**ASME Thomas A. Edison Patent Award** **693**
The American Society of Mechanical Engineers Edison Patent Award recognizes creativity of a patented device or process that has the potential of significantly enhancing some aspect of mechanical engineering. Nominations for this award are expected to demonstrate the significant impact of the nominee's invention to the practice of mechanical engineering or the significant impact that the invention had on at least one industrial/commercial sector. Such an impact might include: creating a new business or industry; substantially improving the efficiency, method of operation, or productivity of an existing machine or process; dramatically reducing the size, weight, or energy requirements of an existing machine or device; improving the quality of life for a definable segment or all of humanity; and financially impacting an existing business or industry. The nomination should focus on a single patent rather than the inventor's entire body of work. The award includes a bronze plaque, a certificate, and up to $750 in travel expenses to attend the award presentation. Nomination forms and instructions are available at the website. The annual deadline for nominations is February 1.
*Requirements:* The patent must have been registered in the U.S. The General Awards Committee screens nominations and makes recommendations to the Committee on Honors. It consists of Members or Fellows representing the Districts. Special Award Committees are usually associated with a particular honor or award. They screen nominations and make recommendations to the other two committees.
*Restrictions:* Except for the Melville Medal, an individual will receive only one honor in recognition of the same achievement. The receipt of one ASME honor shall not bar the recipient from another ASME honor provided it is for different accomplishments. Honors are not awarded posthumously except if a nominee's death occurs after the nomination has been received at ASME Headquarters.
*Geographic Focus:* All States, All Countries
*Date(s) Application is Due:* Feb 1
*Amount of Grant:* Up to 750 USD
*Contact:* Jeffrey R. Mountain; (281) 283-3846 or (800) 843-2763; mountain@uhcl.edu
*Internet:* https://www.asme.org/about-asme/participate/honors-awards/achievement-awards/thomas-a-edison-patent-award
*Sponsor:* American Society of Mechanical Engineers
Two Park Avenue
New York, NY 10016-5990

**ASME Timoshenko Medal** **694**
The ASME Timoshenko Medal is given in recognition of Stephen Timoshenko, a world-renowned authority in the field of applied mechanics. The award include a bronze medal, a certificate, $2,500, and up to $750 in travel expenses to attend the award presentation. The annual deadline for nominations is September 15.
*Requirements:* Any individual member, group of members, or committee may nominate candidates for this award. The General Awards Committee screens nominations and makes recommendations to the Committee on Honors. It consists of Members or Fellows representing the Districts. Special Award Committees are usually associated with a particular honor or award. They screen nominations and make recommendations to the other two committees.
*Restrictions:* Services performed within the context of an individual's normal employment are not eligible. Except for the Melville Medal, an individual will receive only one honor in recognition of the same achievement. The receipt of one ASME honor shall not bar the recipient from another ASME honor provided it is for different accomplishments. Honors are not awarded posthumously except if a nominee's death occurs after the nomination has been received at ASME Headquarters.
*Geographic Focus:* All States, All Countries
*Date(s) Application is Due:* Sep 15
*Amount of Grant:* 2,500 - 3,250 USD
*Contact:* Balakumar Balachandran, Committee on Honors; (301) 563-6526 or (800) 843-2763; balab@umd.edu or awards@asme.org
*Internet:* https://www.asme.org/about-asme/participate/honors-awards/achievement-awards/timoshenko-medal
*Sponsor:* American Society of Mechanical Engineers
Two Park Avenue
New York, NY 10016-5990

**ASME Van C. Mow Medal** **695**
The ASME Van C. Mow Medal is given to an individual who has demonstrated meritorious contributions to the field of bioengineering through research, education, professional development, leadership in the development of the profession, mentorship to young bioengineers, and service to the bioengineering community. The award includes a bronze metal, a certificate, $1,000, and up to $750 in travel expenses to attend the award presentation. Nomination forms and instructions are available at the website. Nomination packets are held and considered for three years. A new nomination packet must be submitted after that time. The annual deadline for nominations is September 1.
*Requirements:* Candidates must have earned a Ph.D. or equivalent degree in any field of engineering, physics, medicine, or life sciences. Nominees must have earned their degree between ten and twenty years prior to June 1 of the year of the award. Nominees must also be active members of the bioengineering division of ASME.
*Restrictions:* Services performed within the context of an individual's normal employment are not eligible. Except for the Melville Medal, an individual will receive only one honor in recognition of the same achievement. The receipt of one ASME honor shall not bar the recipient from another ASME honor provided it is for different accomplishments. Honors are not awarded posthumously except if a nominee's death occurs after the nomination has been received at ASME Headquarters.
*Geographic Focus:* All States, All Countries
*Date(s) Application is Due:* Sep 1
*Amount of Grant:* 1,000 - 1,750 USD
*Contact:* John Bischof; (612) 625-5513 or (800) 843-2763; bischof@umn.edu
*Internet:* https://www.asme.org/about-asme/participate/honors-awards/achievement-awards/van-c-mow-medal
*Sponsor:* American Society of Mechanical Engineers
Two Park Avenue
New York, NY 10016-5990

**ASME Warner T. Koiter Medal** **696**
The ASME Warner T. Koiter Medal was funded by the Technical University of Delft, The Netherlands, to honor Warner T. Koiter for his fundamental work in nonlinear stability of structures, his diligence in the effective application of these theories, his international leadership in mechanics, and his effectiveness as a teacher and researcher. The Koiter Medal recognizes distinguished contributions to the field of solid mechanics with special emphasis on the effective blending of theoretical and applied elements of the discipline, and on a high degree of leadership in the international solid mechanics community. The award includes a bronze metal, a certificate, $2,000, and up to $750 in travel expenses to attend the award presentation. Nomination forms and instructions are available at the website, and the annual deadline is September 15.
*Geographic Focus:* All States, All Countries
*Date(s) Application is Due:* Sep 15
*Amount of Grant:* 2,000 - 2,750 USD
*Contact:* Balakumar Balachandran; (301) 405-4747 or (800) 843-2763; balab@umd.edu
*Internet:* https://www.asme.org/about-asme/participate/honors-awards/achievement-awards/warner-t-koiter-medal
*Sponsor:* American Society of Mechanical Engineers
Two Park Avenue
New York, NY 10016-5990

**ASME Wilfred C. LaRochelle Conformity Assessment Award** 697

The American Society of Mechanical Engineers Wilfred C. LaRochelle Conformity Assessment Award, established in 2017, recognizes distinguished service in the area of conformity assessment including, but not limited to, the establishment, advancement and promotion of the Society's product and personnel certification and accreditation programs. The award, established in 1977, recognizes an individual who has contributed to the nuclear power plant industry through the development and promotion of ASME nuclear codes and standards or the ASME Nuclear Certification Program. During his 35 year career at HSB GS, LaRochelle held key supervisory and management positions, all related to safety inspections during the fabrication or operation of boilers, pressure vessels and nuclear components in accordance with ASME codes. The award includes $1,000, a bronze medal, a certificate, and travel expenses up to $750 to attend the award presentation. The annual deadline for nominations is February 1.
*Requirements:* The award may be made to one recipient of any age who need not be a member of ASME. The General Awards Committee screens nominations and makes recommendations to the Committee on Honors. It consists of Members or Fellows representing the Districts. Special Award Committees are usually associated with a particular honor or award. They screen nominations and make recommendations to the other two committees.
*Restrictions:* Services performed within the context of an individual's normal employment are not eligible. Except for the Melville Medal, an individual will receive only one honor in recognition of the same achievement. The receipt of one ASME honor shall not bar the recipient from another ASME honor provided it is for different accomplishments. Honors are not awarded posthumously except if a nominee's death occurs after the nomination has been received at ASME Headquarters.
*Geographic Focus:* All States
*Date(s) Application is Due:* Feb 1
*Amount of Grant:* 1,000 - 1,750 USD
*Contact:* R. Stevenson, Committee Chair; (781) 615-1597 or (800) 843-2763; stevenrr@westinghouse.com or awards@asme.org
*Internet:* https://www.asme.org/about-asme/honors-awards/achievement-awards/wilfred-larochelle-conformity-assessment-award
*Sponsor:* American Society of Mechanical Engineers
Two Park Avenue
New York, NY 10016-5990

**ASME William T. Ennor Manufacturing Technology Award** 698

The American Society of Mechanical Engineers William T. Ennor Manufacturing Technology Award was established by the Production Engineering Division (now the Manufacturing Engineering Division) in conjunction with the Alcoa Company in 1990. It is presented to an individual or team of individuals for developing or contributing significantly to an innovative manufacturing technology that contributes substantial economic and/or societal benefits. The award includes a vermeil medal, a certificate, and $1,000. Nomination forms and instructions are available at the website. The annual deadline for nominations is February 1.
*Requirements:* The General Awards Committee screens nominations and makes recommendations to the Committee on Honors. It consists of Members or Fellows representing the Districts. Special Award Committees are usually associated with a particular honor or award. They screen nominations and make recommendations to the other two committees.
*Restrictions:* Except for the Melville Medal, an individual will receive only one honor in recognition of the same achievement. The receipt of one ASME honor shall not bar the recipient from another ASME honor provided it is for different accomplishments. Honors are not awarded posthumously except if a nominee's death occurs after the nomination has been received at ASME Headquarters.
*Geographic Focus:* All States, All Countries
*Date(s) Application is Due:* Feb 1
*Amount of Grant:* 1,000 USD
*Contact:* Yuebin Guo; (205) 348-2615 or (800) 843-2763; fax (205) 348-6419; Yguo@eng.ua.edu
*Internet:* https://www.asme.org/about-asme/participate/honors-awards/achievement-awards/william-t-ennor-manufacturing-technology-award
*Sponsor:* American Society of Mechanical Engineers
Two Park Avenue
New York, NY 10016-5990

**ASME Worcester Reed Warner Medal** 699

The ASME Worcester Reed Warner Medal is awarded annually for outstanding contributions to the permanent literature of engineering. Contributions may be single papers, treatises or books, or a series of papers. These contributions should be progressive ideas relating to engineering, scientific, and industrial research associated with mechanical engineering; the design and operation of mechanical and associated equipment; industrial engineering or management, organization, operation, and the concomitants of each; or other subjects closely associated with the foregoing. A certificate, vermeil medal, and $2,000 is included with the award. The annual deadline for nominations is March 1.
*Requirements:* To qualify as having permanent value, any paper or treatise must be at least five years old. Recipients may be nonmembers of ASME. Any individual member, group of members, or committee may nominate candidates for this award. The General Awards Committee screens nominations and makes recommendations to the Committee on Honors. It consists of Members or Fellows representing the Districts. Special Award Committees are usually associated with a particular honor or award. They screen nominations and make recommendations to the other two committees.
*Restrictions:* Services performed within the context of an individual's normal employment are not eligible. Except for the Melville Medal, an individual will receive only one honor in recognition of the same achievement. The receipt of one ASME honor shall not bar the recipient from another ASME honor provided it is for different accomplishments. Honors are not awarded posthumously except if a nominee's death occurs after the nomination has been received at ASME Headquarters.
*Geographic Focus:* All States, All Countries
*Date(s) Application is Due:* Mar 1
*Amount of Grant:* 2,000 USD
*Contact:* Fran McKivor, Committee on Honors; (212) 591-7094 or (800) 843-2763; mckivorf@asme.org or awards@asme.org
*Internet:* https://www.asme.org/about-asme/participate/honors-awards/literature-awards/worcester-reed-warner-medal
*Sponsor:* American Society of Mechanical Engineers
Two Park Avenue
New York, NY 10016-5990

**ASME Y.C. Fung Young Investigator Award** 700

The American Society of Mechanical Engineers Y.C. Fung Young Investigator Award was established to recognize young investigators who are committed to pursuing research and have demonstrated significant potential to make substantial contributions to the field of bioengineering. Such accomplishments may take the form of design or development of new methods, equipment or instrumentation in bioengineering, and research publications in peer-reviewed journals. The award includes a bronze medal, certificate, $1,000, and up to $750 in travel expenses to attend the award presentation. Nomination forms and instructions are available at the website. The annual deadline for nominations is September 1.
*Requirements:* Candidates must have earned a Ph.D. or equivalent degree in any field of engineering, physics, medicine or life sciences. Candidates must have received their Ph.D. degree within seven years of their award nomination and must be no more than 35 years of age on June 1st of the year in which they are nominated. Candidate must be an active member of the Bioengineering Division. The General Awards Committee screens nominations and makes recommendations to the Committee on Honors. It consists of Members or Fellows representing the Districts. Special Award Committees are usually associated with a particular honor or award. They screen nominations and make recommendations to the other two committees.
*Restrictions:* Except for the Melville Medal, an individual will receive only one honor in recognition of the same achievement. The receipt of one ASME honor shall not bar the recipient from another ASME honor provided it is for different accomplishments. Honors are not awarded posthumously except if a nominee's death occurs after the nomination has been received at ASME Headquarters.
*Geographic Focus:* All States, All Countries
*Date(s) Application is Due:* Sep 1
*Amount of Grant:* 1,000 - 1,750 USD
*Contact:* Naomi C. Chesler; (608) 265-8920 or (800) 843-2763; chesler@engr.wisc.edu
*Internet:* https://www.asme.org/about-asme/participate/honors-awards/achievement-awards/y-c-fung-young-investigator-award
*Sponsor:* American Society of Mechanical Engineers
Two Park Avenue
New York, NY 10016-5990

**ASME Yeram S. Touloukian Award** 701

Yeram Sarkas Touloukian was a professor of mechanical engineering and director of the Thermophysical Properties Research Center (now known as CINDAS) at Purdue University in West Lafayette, Indiana. He was world-renowned for his work in thermophysics and his name has become synonymous with the field of thermophysical properties. The American Society of Mechanical Engineers Yeram S. Touloukian Award, established in 1997, is bestowed triennially to recognize outstanding technical contributions in the field of thermophysical properties. An individual who is internationally recognized for major contributions in the thermophysical properties field is eligible to receive this award. Fields recognized by the award include, but are not limited to, mechanical engineering, chemical engineering, physics and chemistry. The award includes a bronze medal, a certificate, and a travel supplement not to exceed $750. The deadline for nominations is September 1 in the year before the Symposium on Thermophysical Properties.
*Requirements:* The award may be made to one recipient of any age who need not be a member of ASME. The General Awards Committee screens nominations and makes recommendations to the Committee on Honors. It consists of Members or Fellows representing the Districts. Special Award Committees are usually associated with a particular honor or award. They screen nominations and make recommendations to the other two committees.
*Restrictions:* Services performed within the context of an individual's normal employment are not eligible. Except for the Melville Medal, an individual will receive only one honor in recognition of the same achievement. The receipt of one ASME honor shall not bar the recipient from another ASME honor provided it is for different accomplishments. Honors are not awarded posthumously except if a nominee's death occurs after the nomination has been received at ASME Headquarters.
*Geographic Focus:* All States
*Date(s) Application is Due:* Sep 1
*Amount of Grant:* Up to 750 USD
*Contact:* Daniel Friend; (303) 497-5424 or (800) 843-2763; daniel.friend@nist.gov
*Internet:* https://www.asme.org/about-asme/participate/honors-awards/achievement-awards/yeram-s-touloukian-award
*Sponsor:* American Society of Mechanical Engineers
Two Park Avenue
New York, NY 10016-5990

**Aspen Institute William Randolph Hearst Endowed Fellowship for Minority Students** 702

The Aspen Institute Program on Philanthropy and Social Innovation (PSI) in Washington, D.C., offers the William Randolph Hearst Endowed Fellowship three times annually. The fellowship, which is based on academic excellence and need, is open to both undergraduate and graduate students of color. The Hearst Fellow serves as an intern with PSI. Through this fellowship, PSI seeks to introduce a diverse group of students to issues and challenges affecting philanthropy, social enterprise, nonprofit organizations, and other actors in the social sector. Recipients may arrange with their colleges or universities to receive academic credit for this experience. In his or her internship, the Hearst Fellow undertakes research, writing, logistical, and administrative support for PSI's leadership initiatives, public programs, and convenings. The student must be able to intern for 12-15 weeks at the Washington, D.C., office of the Aspen Institute. Fall and Spring internships will be part-time (15-20 hours per week) and Summer internships will be full-time. A fellowship grant of approximately $2,000 will be awarded to the Fall and Spring fellows and approximately $4,000 will be awarded to the Summer fellow.

*Requirements:* The ideal candidate for this fellowship is a highly motivated continuing graduate or undergraduate student from an underrepresented community of color. She or he should have an excellent academic record and also have the following: demonstrated interest or experience in nonprofit organizations, philanthropy, and the social sector; excellent research and writing skills; demonstrated financial need; and U.S. citizenship or U.S. permanent residency. There is an online application form for this fellowship.

*Restrictions:* Travel and housing costs are not covered by the fellowship grant.

*Geographic Focus:* All States

*Date(s) Application is Due:* Feb 28; Jul 18; Nov 22

*Amount of Grant:* 2,000 - 4,000 USD

*Contact:* Fellowship Director; (202) 736-2500; fax (202) 467-0790; lauren.stebbins@aspeninstitute.org

*Internet:* www.aspeninstitute.org/policy-work/nonprofit-philanthropy/leadership-initiatives/hearst

*Sponsor:* Aspen Institute
1 Dupont Circle NW, Suite 700
Washington, D.C. 20036

**ASPEN Rhoads Research Foundation Abbott Nutrition Research Grants** 703

The ASPEN Rhoads Research Foundation provides annual grant support to nutrition researchers at all stages of their careers. This grant, funded by a donation from the Abbott Foundation, is intended to assist a nutrition investigator by providing preliminary funding for promising new research in the field of nutrition and metabolic support and related areas of clinical nutrition. Persons applying for the grant must be in a post-training position and commit at least 20% of their time to research. Corporate employees are eligible to receive grants if the research project is not part of their normal duties and they have at least a part-time academic appointment or are associated with a professional institute that conducts nutrition research. In making choices among otherwise high-quality applicants, the reviewers will take into account the geographic distribution of grant winners; their ethnic background and gender, should the applicant choose to supply such information; their professional discipline; and other criteria as appropriate to ensure an equitable system. Funds may be used only for technician salary, equipment, supplies, animals, clinical research costs or other expenses directly related to the conduct of the proposed research. Grants can be renewed for a second year of funding, assuming satisfactory progress.

*Requirements:* Applicants must submit a letter from their supervisor or department head at the institution confirming his/her commitment to the project. If the project involves human subjects, a letter pledging support in recruiting patients from the primary care provider and the institutional review board overseeing human studies is required. For junior faculty and applicants who do not hold faculty positions, three letters of reference are required.

*Restrictions:* Individuals who have received other research grants in excess of $25,000 are ineligible to apply, as are those who completed training more than 5 years prior to the start of the grant period. This grant is not intended to support pursuit of an additional degree. Grant funds may not be used for: comparison of commercial products; indirect costs or overhead; costs of patient care; constructing or renovating facilities; furniture or office equipment; secretarial services; honoraria or membership dues; textbooks or periodicals; repair or service contract costs on institutional equipment; entertainment; travel; or salary support for the Principal Investigator.

*Geographic Focus:* All States

*Date(s) Application is Due:* Sep 8

*Amount of Grant:* Up to 25,000 USD

*Contact:* Michelle Spangenburg, Director of Education and Research; (301) 587-6315, ext. 127 or (301) 920-9132; fax (301) 587-2365; michelles@nutritioncare.org

*Internet:* www.nutritioncare.org/Research/ARRF/A_S_P_E_N__Rhoads_Research_Foundation_Grants/

*Sponsor:* American Society for Parenteral and Enteral Nutrition Rhoads Research Foundation
8630 Fenton Street, Suite 412
Silver Spring, MD 20910

**ASPEN Rhoads Research Foundation Baxter Parenteral Nutrition Research Grant** 704

The ASPEN Rhoads Research Foundation provides annual grant support to nutrition researchers at all stages of their careers. The Baxter Parenteral Nutrition Research grant is intended to assist a nutrition investigator by providing preliminary funding for promising new research in the field of parenteral nutrition and metabolic support and related areas of clinical nutrition. Priority consideration is given to applications that involve a multidisciplinary team or that investigate the efficacy of parenteral or enteral nutrition. Applicants still in training positions may apply for the Fleming Grant. Corporate employees are eligible to receive grants if the research project is not part of their normal duties and they have at least a part-time academic appointment or are associated with a professional institute that conducts nutrition research. In making choices among otherwise high-quality applicants, the reviewers will take into account the geographic distribution of grant winners; their ethnic background and gender, should the applicant choose to supply such information; their professional discipline; and other criteria as appropriate to ensure an equitable system. Funds may be used only for technician salary, equipment, supplies, animals, clinical research costs or other expenses directly related to the conduct of the proposed research. The grant can be renewed for a second year of funding, assuming satisfactory progress.

*Requirements:* Applicants must submit a letter from their supervisor or department head at the institution confirming his/her commitment to the project. If the project involves human subjects, a letter pledging support in recruiting patients from the primary care provider and the institutional review board overseeing human studies is required. For junior faculty and applicants who do not hold faculty positions, three letters of reference are required.

*Restrictions:* Individuals who have received other research grants in excess of $25,000 are ineligible to apply, as are those who completed training more than five years prior to the start of the grant period. This grant is not intended to support pursuit of an additional degree. Grant funds may not be used for: comparison of commercial products; indirect costs or overhead; costs of patient care; constructing or renovating facilities; furniture or office equipment; secretarial services; honoraria or membership dues; textbooks or periodicals; repair or service contract costs on institutional equipment; entertainment; travel; or salary support for the Principal Investigator.

*Geographic Focus:* All States

*Date(s) Application is Due:* Sep 8

*Amount of Grant:* Up to 50,000 USD

*Contact:* Michelle Spangenburg, Director of Education and Research; (301) 587-6315, ext. 127 or (301) 920-9132; fax (301) 587-2365; michelles@nutritioncare.org

*Internet:* www.nutritioncare.org/Research/ARRF/A_S_P_E_N__Rhoads_Research_Foundation_Grants/

*Sponsor:* American Society for Parenteral and Enteral Nutrition Rhoads Research Foundation
8630 Fenton Street, Suite 412
Silver Spring, MD 20910

**ASPEN Rhoads Research Foundation C. Richard Fleming Grant** 705

The ASPEN Rhoads Research Foundation provides annual grant support to nutrition researchers at all stages of their careers. This grant, funded by the donations of ASPEN members, is intended to assist a nutrition investigator by providing preliminary funding for promising new research in the field of nutrition and metabolic support and related areas of clinical nutrition. Applicants still in training positions may apply. Corporate employees are eligible to receive grants if the research project is not part of their normal duties and they have at least a part-time academic appointment or are associated with a professional institute that conducts nutrition research. In making choices among otherwise high-quality applicants, the reviewers will take into account the geographic distribution of grant winners; their ethnic background and gender, should the applicant choose to supply such information; their professional discipline; and other criteria as appropriate to ensure an equitable system. Funds may be used only for technician salary, equipment, supplies, animals, clinical research costs or other expenses directly related to the conduct of the proposed research.

*Requirements:* Applicants must submit a letter from their supervisor or department head at the institution confirming his/her commitment to the project. If the project involves human subjects, a letter pledging support in recruiting patients from the primary care provider and the institutional review board overseeing human studies is required. For junior faculty and applicants who do not hold faculty positions, three letters of reference are required.

*Restrictions:* This grant is not intended to support pursuit of an additional degree. Grant funds may not be used for: comparison of commercial products; indirect costs or overhead; costs of patient care; constructing or renovating facilities; furniture or office equipment; secretarial services; honoraria or membership dues; textbooks or periodicals; repair or service contract costs on institutional equipment; entertainment; travel; or salary support for the Principal Investigator.

*Geographic Focus:* All States

*Date(s) Application is Due:* Sep 8

*Amount of Grant:* Up to 10,000 USD

*Contact:* Michelle Spangenburg, Director of Education and Research; (301) 587-6315, ext. 127 or (301) 920-9132; fax (301) 587-2365; michelles@nutritioncare.org

*Internet:* www.nutritioncare.org/Research/ARRF/A_S_P_E_N__Rhoads_Research_Foundation_Grants/

*Sponsor:* American Society for Parenteral and Enteral Nutrition Rhoads Research Foundation
8630 Fenton Street, Suite 412
Silver Spring, MD 20910

**ASPEN Rhoads Research Foundation Maurice Shils Grant** 706

The ASPEN Rhoads Research Foundation provides annual grant support to nutrition researchers at all stages of their careers. This grant, funded by a donation from Baxter Health Care and Nestle Clinical Nutrition, is intended to assist a nutrition investigator by providing preliminary funding for promising new research in the field of nutrition and metabolic support and related areas of clinical nutrition. Persons applying for the grant must be in a post-training position and commit at least 20% of their time to research. Corporate employees are eligible to receive grants if the research project is not part of their normal duties and they have at least a part-time academic appointment or are associated with a professional institute that conducts nutrition research. In making choices among otherwise high-quality applicants, the reviewers will take into account the geographic distribution of grant winners; their ethnic background and gender, should the applicant choose to supply such information; their professional discipline; and

other criteria as appropriate to ensure an equitable system. Funds may be used only for technician salary, equipment, supplies, animals, clinical research costs or other expenses directly related to the conduct of the proposed research. Grants can be renewed for a second year of funding, assuming satisfactory progress.
*Requirements:* Applicants must submit a letter from their supervisor or department head at the institution confirming his/her commitment to the project. If the project involves human subjects, a letter pledging support in recruiting patients from the primary care provider and the institutional review board overseeing human studies is required. For junior faculty and applicants who do not hold faculty positions, three letters of reference are required.
*Restrictions:* Individuals who have received other research grants in excess of $25,000 are ineligible to apply, as are those who completed training more than 5 years prior to the start of the grant period. This grant is not intended to support pursuit of an additional degree. Grant funds may not be used for: comparison of commercial products; indirect costs or overhead; costs of patient care; constructing or renovating facilities; furniture or office equipment; secretarial services; honoraria or membership dues; textbooks or periodicals; repair or service contract costs on institutional equipment; entertainment; travel; or salary support for the Principal Investigator.
*Geographic Focus:* All States
*Date(s) Application is Due:* Sep 8
*Amount of Grant:* Up to 25,000 USD
*Contact:* Michelle Spangenburg, Director of Education and Research; (301) 587-6315, ext. 127 or (301) 920-9132; fax (301) 587-2365; michelles@nutritioncare.org
*Internet:* www.nutritioncare.org/Research/ARRF/A_S_P_E_N__Rhoads_Research_Foundation_Grants/
*Sponsor:* American Society for Parenteral and Enteral Nutrition Rhoads Research Foundation
8630 Fenton Street, Suite 412
Silver Spring, MD 20910

**ASPEN Rhoads Research Foundation Norman Yoshimura Grant** **707**
The ASPEN Rhoads Research Foundation provides annual grant support to nutrition researchers at all stages of their careers. This grant, supported by B. Braun Medical funding, is intended to assist a nutrition investigator by providing preliminary funding for promising new research in the field of nutrition and metabolic support and related areas of clinical nutrition. Applicants still in training positions may apply. Corporate employees are eligible to receive grants if the research project is not part of their normal duties and they have at least a part-time academic appointment or are associated with a professional institute that conducts nutrition research. In making choices among otherwise high-quality applicants, the reviewers will take into account the geographic distribution of grant winners; their ethnic background and gender, should the applicant choose to supply such information; their professional discipline; and other criteria as appropriate to ensure an equitable system. Funds may be used only for technician salary, equipment, supplies, animals, clinical research costs or other expenses directly related to the conduct of the proposed research.
*Requirements:* Applicants must submit a letter from their supervisor or department head at the institution confirming his/her commitment to the project. If the project involves human subjects, a letter pledging support in recruiting patients from the primary care provider and the institutional review board overseeing human studies is required. For junior faculty and applicants who do not hold faculty positions, three letters of reference are required.
*Restrictions:* This grant is not intended to support pursuit of an additional degree. Grant funds may not be used for: comparison of commercial products; indirect costs or overhead; costs of patient care; constructing or renovating facilities; furniture or office equipment; secretarial services; honoraria or membership dues; textbooks or periodicals; repair or service contract costs on institutional equipment; entertainment; travel; or salary support for the Principal Investigator.
*Geographic Focus:* All States
*Date(s) Application is Due:* Sep 8
*Amount of Grant:* Up to 25,000 USD
*Contact:* Michelle Spangenburg, Director of Education and Research; (301) 587-6315, ext. 127 or (301) 920-9132; fax (301) 587-2365; michelles@nutritioncare.org
*Internet:* www.nutritioncare.org/Research/ARRF/A_S_P_E_N__Rhoads_Research_Foundation_Grants/
*Sponsor:* American Society for Parenteral and Enteral Nutrition Rhoads Research Foundation
8630 Fenton Street, Suite 412
Silver Spring, MD 20910

**ASPET-Pharmacia Award for Experimental Therapeutics** **708**
The ASPET-Pharmacia Award in Experimental Therapeutics is given annually to recognize and stimulate outstanding research in pharmacology and experimental therapeutics - basic laboratory or clinical research that has had, or potentially will have, a major impact on the pharmacological treatment of disease. The winner will receive a $2,500 honorarium, a plaque, hotel and economy airfare for the winner and spouse to the award ceremony at the ASPET annual meeting. The annual deadline for nominations is September 14.
*Requirements:* Nominations must be made by ASPET members; however, the nominee need not be a member. No restrictions exist as to age or institutional affiliation of nominee.
*Geographic Focus:* All States
*Date(s) Application is Due:* Sep 14
*Amount of Grant:* 2,500 USD
*Contact:* Carol Paronis, Awards Committee Chairperson; (301) 634-7060; fax (301) 634-7061; cparonis@aspet.org or awards@aspet.org
*Internet:* https://www.aspet.org/aspet/meetings-awards/aspet-awards/aspet-scientific-achievement-awards/pharmacia-aspet-award-for-experimental-therapeutics
*Sponsor:* American Society for Pharmacology and Experimental Therapeutics
9650 Rockville Pike, Suite E220
Bethesda, MD 20814-3995

**ASPET Benedict R. Lucchesi Award in Cardiac Pharmacology** **709**
The Benedict R. Lucchesi Award in Cardiac Pharmacology was established to honor Dr. Lucchesi's lifelong scientific contributions to our better understanding and appreciation of pharmacological treatment and prevention of cardiovascular disease and for his mentoring of countless prominent functional (in vivo) cardiovascular pharmacologists. The Benedict R. Lucchesi Award is a biennial award, consisting an honorarium of $1,000, a custom-designed crystal bowl depicting the named Lectureship, and up to $2,000 travel expenses including registration to the annual spring ASPET meeting. A recipient will be selected and invited to deliver a state-of-the-art lecture on recent advances in the field of cardiac and electropharmacology at the spring ASPET meeting (Division's programming session). The presentation of his/her research should be of broad interest and contribute to the growth of the Cardiovascular Pharmacology Division. The bi-annual deadline for nominations is September 15.
*Requirements:* There are no restrictions on institutional affiliation, nationality, or age of the candidate, but the recipient must be a member of the ASPET. Nominations must be made by a member of the ASPET, and no member may nominate more than one candidate per year.
*Geographic Focus:* All States
*Date(s) Application is Due:* Sep 15
*Amount of Grant:* 1,000 - 3,000 USD
*Contact:* Carol Paronis, Awards Committee Chairperson; (301) 634-7060; fax (301) 634-7061; cparonis@aspet.org or awards@aspet.org
*Internet:* https://www.aspet.org/aspet/membership-community/divisions/division-for-cardiovascular-pharmacology-(cvp)/division-awards/benedict-r-lucchesi-young-scientist-travel-award-in-cardiac-pharmacology
*Sponsor:* American Society for Pharmacology and Experimental Therapeutics
9650 Rockville Pike, Suite E220
Bethesda, MD 20814-3995

**ASPET Bernard B. Brodie Award in Drug Metabolism** **710**
This biennial award (even-numbered years), established by CIBA-GEIGY Corporation to honor the fundamental contributions of Bernard B. Brodie, is presented to recognize outstanding original research contributions in drug metabolism and disposition, particularly those having a major impact on future research in the field. Only one nominator is necessary although more are acceptable. The award consists of a $2,000 honorarium, a commemorative medal, hotel and economy airfare to the award ceremony at the annual meeting. Nominations must be received by the deadline of September 14 prior to the award year.
*Requirements:* Nominees must be members of ASPET; nominators need not be members of ASPET.
*Restrictions:* Supporting research accomplishments must not have been used to win any other major award.
*Geographic Focus:* All States
*Date(s) Application is Due:* Sep 14
*Amount of Grant:* 2,000 USD
*Contact:* Carol Paronis, Awards Committee Chairperson; (301) 634-7060; fax (301) 634-7061; cparonis@aspet.org or awards@aspet.org
*Internet:* https://www.aspet.org/aspet/membership-community/divisions/division-for-drug-metabolism-and-disposition-(dmdd)/bernard-b-brodie-award-in-drug-metabolism
*Sponsor:* American Society for Pharmacology and Experimental Therapeutics
9650 Rockville Pike, Suite E220
Bethesda, MD 20814-3995

**ASPET David Lehr Award** **711**
The David Lehr Research Award is intended to extend funding for preclinical or clinical research directed towards improving human health. This award is made possible by an endowment to ASPET from Mrs. Lisa Lehr in honor of her husband, the late Dr. David Lehr, former chair of the Department of Pharmacology for New York Medical College. The following investigators are encouraged to apply: one who requires bridging funds to enhance a grant application for competitive renewal; one who has a research project currently supported by a no cost extension and desires additional funds to enhance a grant re-submission; or one who desires funds to acquire preliminary data in support of a new grant submission. The award may be used to support the salaries of the principal investigator (PI), postdoctoral fellows, or research staff whose efforts are directed toward the aims of the proposal. The award is presented biennially in odd-numbered years, at the award ceremony during the ASPET Annual Meeting. The award consists of: $50,000 per year for two years ($100,000 total); a commemorative plaque; and complimentary registration and travel expenses for the winner and spouse to attend the award presentation at the ASPET Annual Meeting at Experimental Biology.
*Requirements:* Applicants for this award must be: at least 45 years of age by September 14 (nomination deadline) of the year in which they are applying; a current ASPET member in good standing; and the principal investigator (PI) of the award proposal submitted. The awarded funds must be used to support research activities for which there is a gap in funding or for the purpose of acquiring preliminary data to support the submission of a new grant proposal.
*Restrictions:* The applicant may only submit one (1) application per award cycle. The awarded funds will not support facilities and administrative costs or travel.
*Geographic Focus:* All States
*Date(s) Application is Due:* Sep 14
*Amount of Grant:* 100,000 USD
*Contact:* Carol Paronis, Awards Committee Chairperson; (301) 634-7060; fax (301) 634-7061; cparonis@aspet.org or awards@aspet.org
*Internet:* https://www.aspet.org/aspet/meetings-awards/aspet-awards/aspet-scientific-achievement-awards/the-david-lehr-award

*Sponsor:* American Society for Pharmacology and Experimental Therapeutics
9650 Rockville Pike, Suite E220
Bethesda, MD 20814-3995

**ASPET Division for Toxicology Career Award** 712

The ASPET Division for Toxicology Career Award has been established to recognize outstanding original research contributions to Toxicology by an established investigator. The award is presented annually and consists of $500, a plaque, and complimentary registration plus partial travel expenses (to a maximum of $500) for the winner to attend the awards ceremony at the annual meeting. The primary criterion for the award is the level of excellence and originality of the research conducted by the candidate in the field of toxicology during their career. The annual application due date is September 14.
*Requirements:* Candidates must be a member of ASPET, and can be self-nominated or nominated by another ASPET member. Selection will be made by the Executive Committee of the Division for Toxicology. The award winner is expected to submit a symposia proposal for the Experimental Biology meeting.
*Geographic Focus:* All States
*Date(s) Application is Due:* Sep 14
*Amount of Grant:* 500 - 1,000 USD
*Contact:* Carol Paronis, Awards Committee Chairperson; (301) 634-7060; fax (301) 634-7061; cparonis@aspet.org or awards@aspet.org
*Internet:* https://www.aspet.org/aspet/membership-community/divisions/division-for-toxicology-(tox)/division-awards#career
*Sponsor:* American Society for Pharmacology and Experimental Therapeutics
9650 Rockville Pike, Suite E220
Bethesda, MD 20814-3995

**ASPET Division for Toxicology Junior Investigator Award** 713

The ASPET Division for Toxicology Junior Investigator Award has been established to recognize excellent original research by early career investigators in the area of Toxicology. The award is presented annually and consists of $500, a plaque, and complimentary registration plus partial travel expenses (to a maximum of $500) for the winner to attend the awards ceremony at the annual meeting. The primary criterion for the award is the excellence and originality of research conducted by the candidate in the field of toxicology. Candidates can be self-nominated or nominated by another ASPET member. Nominations are due by the annual deadline date of September 14.
*Requirements:* Candidates for this award must have a doctoral degree (e.g., Ph.D., M.D., Pharm.D., D.V.M.), be within 8 to 15 years of having received their final degree, and a member of ASPET.
*Geographic Focus:* All States
*Date(s) Application is Due:* Sep 14
*Amount of Grant:* 500 - 1,000 USD
*Contact:* Carol Paronis, Awards Committee Chairperson; (301) 634-7060; fax (301) 634-7061; cparonis@aspet.org or awards@aspet.org
*Internet:* https://www.aspet.org/aspet/membership-community/divisions/division-for-toxicology-(tox)/division-awards#career
*Sponsor:* American Society for Pharmacology and Experimental Therapeutics
9650 Rockville Pike, Suite E220
Bethesda, MD 20814-3995

**ASPET Division for Toxicology New Investigator Award** 714

The ASPET Division for Toxicology New Investigator Award has been established to recognize excellent original research by early career investigators in the area of Toxicology. The award consists of $500, a plaque, and complimentary registration plus partial travel expenses (to a maximum of $500) for the winner to attend the awards ceremony at the annual meeting. The primary criterion for the award is the excellence of research conducted by the candidate in the field of toxicology. Candidates can be self-nominated or nominated by another ASPET member. The annual nomination deadline is September 14.
*Requirements:* Candidates for this award must have a doctoral degree (e.g., Ph.D., M.D., Pharm.D., D.V.M.), be within 7 years of having received their final degree, no longer a trainee (postdoc, resident, fellow, etc), and a member of ASPET.
*Geographic Focus:* All States
*Date(s) Application is Due:* Sep 14
*Amount of Grant:* 500 - 1,000 USD
*Contact:* Carol Paronis, Awards Committee Chairperson; (301) 634-7060; fax (301) 634-7061; cparonis@aspet.org or awards@aspet.org
*Internet:* https://www.aspet.org/aspet/membership-community/divisions/division-for-toxicology-(tox)/division-awards#career
*Sponsor:* American Society for Pharmacology and Experimental Therapeutics
9650 Rockville Pike, Suite E220
Bethesda, MD 20814-3995

**ASPET Early Career Achievement Award in Drug Metabolism and Disposition** 715

The ASPET Division for Drug Metabolism Early Career Achievement Award has been established to recognize excellent original research by early career investigators in the area of drug metabolism and disposition. The award is presented biennially in odd-numbered years. The award consists of $1,000, a plaque, and complimentary registration plus travel expenses (to a maximum of $1,000) for the winner to attend the award ceremony at the annual meeting. The awardee will deliver a lecture at the annual meeting describing their relevant research accomplishments. The awardee will be invited to publish a review article on the subject matter of the award lecture in Drug Metabolism and Disposition. The bi-annual deadline is September 15.
*Requirements:* Nominees for this award must have a doctoral degree (e.g. Ph.D., M.D., Pharm. D., D.V.M.) and must be within 15 years of having received their final degree, as of December 31 of the year of the award. There are no restrictions on institutional affiliation and a candidate need not be a member of ASPET.
*Geographic Focus:* All States
*Date(s) Application is Due:* Sep 15
*Amount of Grant:* 1,000 - 2,000 USD
*Contact:* Carol Paronis; (301) 634-7060; fax (301) 634-7061; cparonis@aspet.org
*Internet:* https://www.aspet.org/aspet/membership-community/divisions/division-for-drug-metabolism-and-disposition-(dmdd)/early-career-achievement-award-in-drug-metabolism-and-disposition
*Sponsor:* American Society for Pharmacology and Experimental Therapeutics
9650 Rockville Pike, Suite E220
Bethesda, MD 20814-3995

**ASPET Goodman and Gilman Award in Drug Receptor Pharmacology** 716

The Louis S. Goodman and Alfred Gilman Award in Receptor Pharmacology was established in 1980 to recognize and stimulate outstanding research in pharmacology of biological receptors. Such research might provide a better understanding of the mechanisms of biological processes and potentially provide the basis for the discovery of drugs useful in the treatment of diseases. The Award is given biennially (even-numbered years) to recognize and stimulate outstanding research in the pharmacology of biological receptors. It is the hope that such research might provide a better understanding of the mechanism of biological processes and potentially provide the basis for the discovery of drugs useful in the treatment of diseases. The award includes a monetary prize of $2,500, an engraved plaque, an invitation to deliver the Goodman and Gilman Award Lecture at the ASPET Annual Meeting at Experimental Biology, and travel support for winner and spouse to attend the award ceremony. Nominations must be received September 14 of the year prior to the year in which the award will be made.
*Requirements:* Nominations must be made by ASPET members; however, there are no restrictions as to age, sex, nationality, or institutional affiliation. Nominee need not be a member of ASPET.
*Geographic Focus:* All States
*Date(s) Application is Due:* Sep 14
*Amount of Grant:* 2,500 USD
*Contact:* Carol Paronis, Awards Committee Chairperson; (301) 634-7060; fax (301) 634-7061; cparonis@aspet.org or awards@aspet.org
*Internet:* https://www.aspet.org/aspet/meetings-awards/aspet-awards/aspet-scientific-achievement-awards/the-goodman-and-gilman-award-in-receptor-pharmacology
*Sponsor:* American Society for Pharmacology and Experimental Therapeutics
9650 Rockville Pike, Suite E220
Bethesda, MD 20814-3995

**ASPET John J. Abel Award in Pharmacology** 717

The John J. Abel Award in Pharmacology, named after the founder of ASPET, was established in 1946 to stimulate fundamental research in pharmacology and experimental therapeutics by young investigators. The Award shall be made for original, outstanding research in the field of pharmacology and/or experimental therapeutics. Independence of thought, originality of approach, clarity, and excellence of data presentation are important criteria. The award is presented annually, at the award ceremony during the ASPET Annual Meeting. The award consists of: an invitation to deliver the John J. Abel Award Lecture at the ASPET Annual Meeting at Experimental Biology; a $5,000 honorarium; an engraved plaque; and complimentary registration and travel expenses for the winner and spouse to attend the ASPET Annual Meeting at Experimental Biology. The annual nomination deadline is September 14.
*Requirements:* Candidates may not have passed their 45th birthday as of as the nomination deadline, or previously received an award from the sponsor for the same technical accomplishment. Candidates need not be members of the society; however, nominations must be made by ASPET members.
*Geographic Focus:* All States
*Date(s) Application is Due:* Sep 14
*Amount of Grant:* 5,000 USD
*Contact:* Carol Paronis, Awards Committee Chairperson; (301) 634-7060; fax (301) 634-7061; awards@aspet.org or cparonis@aspet.org
*Internet:* https://www.aspet.org/aspet/meetings-awards/aspet-awards/aspet-scientific-achievement-awards/john-j-abel-award
*Sponsor:* American Society for Pharmacology and Experimental Therapeutics
9650 Rockville Pike, Suite E220
Bethesda, MD 20814-3995

**ASPET Julius Axelrod Award in Pharmacology** 718

The Julius Axelrod Award in Pharmacology was established in 1991 to honor the memory of the eminent American pharmacologist who shaped the fields of neuroscience, drug metabolism, and biochemistry and who served as a mentor for numerous eminent pharmacologists around the world. The Award is presented annually for significant contributions to understanding the biochemical mechanisms underlying the pharmacological actions of drugs and for contributions to mentoring other pharmacologists. The award consists of an honorarium of $5,000, a medal, hotel and economy airfare for the winner and spouse to the annual meeting. The formal presentation of this award and medal will be made at the annual meeting of ASPET. The recipient will be invited by the President of the Society to deliver the Julius Axelrod Lecture and organize the Julius Axelrod Symposium at the annual meeting a year hence. The recipient will also be invited by the Catecholamine Club to give a less formal

presentation at its annual dinner meeting that year of the award. The annual deadline for nominations is September 14.
*Requirements:* No candidate may be nominated for more than one major ASPET award in any given year.
*Restrictions:* There are no restrictions on nominees for this award.
*Geographic Focus:* All States
*Date(s) Application is Due:* Sep 14
*Amount of Grant:* 5,000 USD
*Contact:* Carol Paronis, Awards Committee Chairperson; (301) 634-7060; fax (301) 634-7061; cparonis@aspet.org or awards@aspet.org
*Internet:* https://www.aspet.org/aspet/meetings-awards/aspet-awards/aspet-scientific-achievement-awards/julius-axelrod-award-in-pharmacology
*Sponsor:* American Society for Pharmacology and Experimental Therapeutics
9650 Rockville Pike, Suite E220
Bethesda, MD 20814-3995

**ASPET Neuropharmacology Early Career Independent Investigator Award** **719**
The Early Career Independent Investigator Award recognizes and honors a young investigator who is working in any area of Neuropharmacology. The award is open to early career stage investigators from all types of organizations, including academia, industry, private or government institutes who is a primary member of ASPET's Division for Neuropharmacology. An early career stage investigator is defined as someone who at the time of nomination is under 40 years of age or, if over 40 has been in an independent position for no more than 5 years. An independent position is considered to be one that is responsible for securing and administering their own budgets for research (traditionally a faculty position, or a team leader in a non-university setting). The award consists of: a commemorative plaque; complimentary registration and travel expenses (up to $1,000) for the winner to attend the ASPET Annual Meeting at Experimental Biology; membership on that year's Division for Neuropharmacology program committee, with the expectation that they will put together an Early Career Award Symposium, on a theme of their choosing; and membership on the Division for Neuropharmacology Executive Committee for two years. Candidates can be self-nominated or nominated by other ASPET members. The annual deadline for nominations is September 14.
*Requirements:* The award is open to early career stage investigators from all types of organizations, including academia, industry, private or government institutes. The candidate must be a member of ASPET.
*Geographic Focus:* All States
*Date(s) Application is Due:* Sep 14
*Amount of Grant:* Up to 1,000 USD
*Contact:* Carol Paronis, Awards Committee Chairperson; (301) 634-7060; fax (301) 634-7061; cparonis@aspet.org or awards@aspet.org
*Internet:* https://www.aspet.org/aspet/membership-community/divisions/division-for-neuropharmacology-(neu)/neuropharmacology-early-career-independent-investigator-award
*Sponsor:* American Society for Pharmacology and Experimental Therapeutics
9650 Rockville Pike, Suite E220
Bethesda, MD 20814-3995

**ASPET Otto Krayer Award in Pharmacology** **720**
The Otto Krayer Award in Pharmacology is given biennially in even years to commemorate the enduring legacy of Otto Krayer's personal qualities: his ethical behavior, his commitment to teaching, his high standards of scientific scholarship, publication and editorship, his promotion of interdisciplinary research to reveal the actions of drugs or other chemicals, and his guidance and support of younger scientists. The purpose of the award is to recognize an individual whose character and career contributions to pharmacology are in accord with those exemplified by Otto Krayer. This award was originally established in 1985, suspended in 2003, and became active again with a presentation at the 2017 ASPET Annual Meeting at Experimental Biology. The award consists of: a $2,500 honorarium; a commemorative plaque; and complimentary registration and travel expenses for the winner and spouse to attend the ASPET Annual Meeting at Experimental Biology. The awardee may also be invited to deliver a lecture at the ASPET Annual Meeting at Experimental Biology. The annual deadline for nominations is September 14.
*Requirements:* The candidate must be a member of ASPET, and the nomination must be made by an ASPET member.
*Restrictions:* No candidate may be nominated for more than one major ASPET award in any given year.
*Geographic Focus:* All States
*Date(s) Application is Due:* Sep 14
*Amount of Grant:* 2,500 USD
*Contact:* Carol Paronis, Awards Committee Chairperson; (301) 634-7060; fax (301) 634-7061; awards@aspet.org or cparonis@aspet.org
*Internet:* https://www.aspet.org/aspet/meetings-awards/aspet-awards/aspet-scientific-achievement-awards/otto-krayer-award-in-pharmacology
*Sponsor:* American Society for Pharmacology and Experimental Therapeutics
9650 Rockville Pike, Suite E220
Bethesda, MD 20814-3995

**ASPET P.B. Dews Lifetime Achievement Award for Research in Behavioral Pharmacology** **721**
ASPET's Division of Behavioral Pharmacology sponsors the P. B. Dews Award for Research in Behavioral Pharmacology to recognize outstanding lifetime achievements in research, teaching and professional service in the field of Behavioral Pharmacology and to honor Peter Dews for his seminal contributions to the development of behavioral pharmacology as a discipline. The Award consists of $1,000, a plaque, and partial travel expenses (up to $500) to the award ceremony at the ASPET annual meeting. The recipient will be invited by the Chair of the Division of Behavioral Pharmacology to deliver a special lecture on this occasion. The lecture will be published subsequently in an appropriate ASPET-sponsored publication. The annual deadline for nominations is September 14.
*Requirements:* Nominations may be made by members of ASPET or of any relevant scientific society.
*Restrictions:* There are no restrictions on nominees for this award.
*Geographic Focus:* All States
*Date(s) Application is Due:* Sep 14
*Amount of Grant:* 1,000 - 1,500 USD
*Contact:* Carol Paronis, Awards Committee Chairperson; (301) 634-7060; fax (301) 634-7061; cparonis@aspet.org or awards@aspet.org
*Internet:* https://www.aspet.org/aspet/membership-community/divisions/division-for-behavioral-pharmacology-(beh)/p-b-dews-lifetime-achievement-award-for-research-in-behavioral-pharmacology
*Sponsor:* American Society for Pharmacology and Experimental Therapeutics
9650 Rockville Pike, Suite E220
Bethesda, MD 20814-3995

**ASPET Paul M. Vanhoutte Distinguished Lectureship in Vascular Pharmacology** **722**
The Paul M. Vanhoutte Award in Vascular Pharmacology was established to honor Dr. Vanhoutte's lifelong scientific contributions to our better understanding and appreciation of the importance of endothelial cells and vascular smooth muscle function in health and disease and for his mentoring of countless prominent endothelial and vascular biologists and pharmacologists. This is a biennial award, consisting an honorarium of $1,000, a custom-designed crystal bowl depicting the named Lectureship, and up to $2,000 travel expenses including registration to the annual spring ASPET meeting. The annual deadline for nominations is September 14.
*Requirements:* There are no restrictions on institutional affiliation, nationality, or age of the candidate, but the recipient must be a member of the ASPET.
*Geographic Focus:* All States
*Date(s) Application is Due:* Sep 14
*Amount of Grant:* 1,000 - 3,000 USD
*Contact:* Carol Paronis, Awards Committee Chairperson; (301) 634-7060; fax (301) 634-7061; cparonis@aspet.org or awards@aspet.org
*Internet:* https://www.aspet.org/aspet/membership-community/divisions/division-for-cardiovascular-pharmacology-(cvp)/division-awards/paul-m-vanhoutte-distinguished-lectureship-in-vascular-pharmacology
*Sponsor:* American Society for Pharmacology and Experimental Therapeutics
9650 Rockville Pike, Suite E220
Bethesda, MD 20814-3995

**ASPET Reynold Spector Award in Clinical Pharmacology** **723**
The Reynold Spector Award in Clinical Pharmacology was established in 2014 by ASPET in recognition of Dr. Spector's dedication and contributions to clinical pharmacology. The Award recognizes excellence in research and/or teaching in clinical pharmacology. This award is made possible by an endowment to ASPET from Dr. Reynold and Mrs. Michiko Spector. The award is presented biennially in odd numbered years, at the award ceremony during the ASPET Annual Meeting. The award consists of: an invitation to deliver the Reynold Spector Lecture at the ASPET Annual Meeting at Experimental Biology; an $8,000 honorarium; a commemorative plaque; and complimentary registration and travel expenses for the winner and spouse to attend the ASPET Annual Meeting at Experimental Biology. The bi-annual deadline for nominations is September 13.
*Requirements:* There are no restrictions on nominees for this award, however, the nomination must be made by a member of ASPET.
*Restrictions:* No candidate may be nominated for more than one major ASPET award in any given year.
*Geographic Focus:* All States
*Date(s) Application is Due:* Sep 13
*Amount of Grant:* 8,000 USD
*Contact:* Carol Paronis, Awards Committee Chairperson; (301) 634-7060; fax (301) 634-7061; cparonis@aspet.org or awards@aspet.org
*Internet:* https://www.aspet.org/aspet/meetings-awards/aspet-awards/aspet-scientific-achievement-awards/the-reynold-spector-award-in-clinical-pharmacology
*Sponsor:* American Society for Pharmacology and Experimental Therapeutics
9650 Rockville Pike, Suite E220
Bethesda, MD 20814-3995

**ASPET Robert R. Ruffolo Career Achievement Award in Pharmacology** **724**
The Robert R. Ruffolo Career Achievement Award in Pharmacology was established in 2011 in recognition of the contributions made to drug discovery and development by Dr. Ruffolo. The Award recognizes the scientific achievements of scientists who are at the height of their careers (typically mid- to late-career) and who have made significant contributions to any area of pharmacology. The award is presented annually, at the award ceremony during the ASPET Annual Meeting. The award consists of: a $2,500 honorarium; a commemorative medal; and complimentary registration and travel expenses for the winner and spouse to attend the ASPET Annual Meeting at Experimental Biology. The annual deadline for nominations is September 14.
*Requirements:* No candidate may be nominated for more than one major ASPET award in any given year.
*Restrictions:* There are no restrictions on nominees for this award. However, the nomination must be made by a member of ASPET.

*Geographic Focus:* All States
*Date(s) Application is Due:* Sep 14
*Amount of Grant:* 2,500 USD
*Contact:* Carol Paronis, Awards Committee Chairperson; (301) 634-7060; fax (301) 634-7061; cparonis@aspet.org or awards@aspet.org
*Internet:* https://www.aspet.org/aspet/meetings-awards/aspet-awards/aspet-scientific-achievement-awards/the-robert-r-ruffolo-career-achievement-award-in-pharmacology
*Sponsor:* American Society for Pharmacology and Experimental Therapeutics
9650 Rockville Pike, Suite E220
Bethesda, MD 20814-3995

**ASPET Summer Undergraduate Research Fellowships (SURF)** 725

The purpose of the Fellowship program is to introduce undergraduate students to pharmacology research in order to heighten interest in science as a career, with an emphasis on pharmacology graduate training to increase the number of young scientists entering the research discipline of pharmacology. A group of at least five ASPET Regular members from one institution may wish to apply for support of up to $9,000 for an undergraduate fellowship program, to include up to five students stipends. Students are to receive at least a $2,800 stipend for a minimum of ten weeks participation. It is anticipated that the request for funding will include matching funds of at least $5,000 from local resources. Awards are normally made for three (3) years.
*Restrictions:* No indirect costs will be provided. All funds shall be used for student stipends/wages only.
*Geographic Focus:* All States
*Date(s) Application is Due:* Mar 1; Oct 1
*Amount of Grant:* 2,800 - 9,000 USD
*Contact:* Carol Paronis, Awards Committee Chairperson; (301) 634-7060; fax (301) 634-7061; cparonis@aspet.org or awards@aspet.org
*Internet:* https://www.aspet.org/aspet/education-careers/aspet-programs/summer-undergraduate-research-fellow-surf-awards
*Sponsor:* American Society for Pharmacology and Experimental Therapeutics
9650 Rockville Pike, Suite E220
Bethesda, MD 20814-3995

**Assisi Foundation of Memphis General Grants** 726

The foundation supports organizations in its areas of interest, including health and human services—promote the health and well-being of the Mid-South community and help the health care system respond more effectively to community needs; education and literacy—projects/programs that build organizational capacity of provider agencies, provide professional development to service providers, promote collaboration among provider agencies, and leverage resources (local, state, and federal); social justice/ethics—projects/programs that strengthen ethical values among Mid-South citizens and promote social justice leading to a better understanding of and a more effective response to economic or social threats to the community; and cultural enrichment and the arts—projects/programs that foster an appreciation of the arts in the Greater Memphis community. Religious organizations seeking funding for religious programs also are eligible. Deadlines are set to coordinate with quarterly meetings of the Board of Directors of the Foundation.
*Requirements:* The Foundation makes grants only to organizations that are classified as tax-exempt under Section 501(c)3 of the Internal Revenue Code and as public charities under Section 509(a) of that Code. The Foundation uses its resources for charitable endeavors that advance the well-being of people and institutions located in Shelby, Fayette, and Tipton Counties in Tennessee; Crittenden County, Arkansas; and Desoto County, Mississippi.
*Restrictions:* Grants are not made for individuals, national fundraising drives, projects that address the needs of only one congregation, tickets for benefits, political organizations or candidates for public office, lobbying activities, recurring budget deficits, or tournament fees and/or travel for athletic competitions.
*Geographic Focus:* Arkansas, Mississippi, Tennessee
*Date(s) Application is Due:* Feb 15; May 17; Aug 16; Nov 15
*Amount of Grant:* Up to 20,000 USD
*Contact:* Jan Young, Executive Director; (901) 684-1564; jyoung@assisifoundation.org
*Internet:* www.assisifoundation.org/generalgrants.html
*Sponsor:* Assisi Foundation of Memphis
515 Erin Drive
Memphis, TN 38117

**ASTA Academic Scholarships** 727

ASTA's Tourism Cares refers to its scholarships for undergraduate or graduate level students as "Academic" Scholarships. Tourism Cares is offering 59 Academic Scholarships, ranging from $1,000 from $5,000. The Academic Scholarship Program benefits undergraduate or graduate students of travel- and-tourism-related, or hospitality-related, programs of study, at accredited schools. Each undergraduate scholarship award can be used to cover expenses for tuition, books, and educational fees only. Each graduate scholarship award can be used to cover expenses for tuition, books, educational fees, and research-related expenses only. Each graduate research scholarship award can be used to cover expenses for tuition, books, educational fees, and research-related expenses only. Scholarship offerings (for Graduating High School Seniors or Undergraduate Students include: ASTA Alaska Airlines Scholarship (one at $1,000); ASTA America Express Travel Scholarship (one at $1,000); ASTA Arizona Scholarship (one at $2,000); ASTA Holland America Line Undergraduate Scholarship (one at $1,500); ASTA Northern California Chapter Richard Epping Scholarship (one at $2,000); ASTA Pacific Northwest Chapter - William Hunt Scholarship (two at $1,500); ASTA Princess Cruises Scholarship (one at $2,500); ASTA Stan and Leone Pollard Scholarship (one at $1,000). Scholarship offerings for Graduate Students or Graduate Research Students include: ASTA Arnold Rigby Scholarship ( two at $2,500); ASTA Avis Budget Group Scholarship (one at $2,000); ASTA Healy Scholarship (one at $1,000); ASTA Holland America Line Graduate Research Scholarship (one at $4,000); ASTA Joseph R. Stone Scholarship (three at $2,500). Scholarship offering for an Undergraduate or Graduate Internship (at ASTA's Research Department in the Washington, D.C. area fall semester): ASTA David J. Hallissey Memorial Internship (one at $2,000). Scholarships for Entering, or Presently Working, Travel Professionals include: ASTA A. J. Spielman Scholarship (one at $1,000); ASTA George Reinke Scholarship (two at $1,500). Specific ASTA Chapters also have scholarship offerings of their own, though many of these have resident requirements.
*Requirements:* Tourism Cares scholarship awards are merit-based, meaning applicants are evaluated on: academic performance, extracurricular and volunteer activities and experiences, motivation, and leadership potential. In general, all applicants must be enrolled in school for the upcoming fall semester or term to be eligible to receive the scholarship award. Scholarship eligibility criteria are determined by each fund donor contact. Applicants may only apply for the Academic Scholarship Program through the on-line application.
*Restrictions:* A student may win only one scholarship award from Tourism Cares per calendar year.
*Geographic Focus:* All States
*Amount of Grant:* 1,000 - 4,000 USD
*Contact:* Danielle Belanger, Student Programs Coordinator; 781-821-5990 x 215; scholarships@tourismcares.org or danielleb@tourismcares.org
*Internet:* www.asta.org/Education/content.cfm?ItemNumber=2552&navItemNumber=614
*Sponsor:* American Society of Travel Agents
1101 King Street, Suite 200
Alexandria, VA 22314

**Astronomical Society of the Pacific Amateur Achievement Award** 728

The Amateur Achievement Award is given annually to recognize significant contributions to astronomy or amateur astronomy by those not employed in the field of astronomy in a professional capacity. Nominations may be submitted by any individual or group. All nominations will remain active for three years of competition. Additional support letters or nomination updates are accepted at any time.
*Restrictions:* Self nominations or nomination by a family member will not be accepted.
*Geographic Focus:* All States
*Date(s) Application is Due:* Jan 1
*Amount of Grant:* 500 USD
*Contact:* Awards Coordinator; (415) 337-1100; fax (415) 337-5205
*Internet:* www.astrosociety.org/membership/awards/amateur.html
*Sponsor:* Astronomical Society of the Pacific
390 Ashton Avenue
San Francisco, CA 94112

**Astronomical Society of the Pacific Dorothea Klumpke-Roberts Award** 729

The Dorothea Klumpke-Roberts Award is annually for outstanding contributions to public understanding and appreciation of astronomy. Nominees can include a broad range of individuals involved in science, education, writing/publishing, broadcasting, astronomy popularization, the arts, or other pursuits that advance public understanding and appreciation of astronomy. Letters of nomination should highlight how the nominee or nominees have distinguished themselves in this endeavor and/or cite exceptional achievement. Nominations are solicited from anyone familiar with or knowledgeable of the achievements of the nominee or nominees. The Award recipient will be selected by the Awards Committee appointed by the Board of Directors, with a plaque and monetary award given at the annual awards banquet. If none of the nominees are judged to be sufficiently outstanding, the Award will not be given for the year in question. In February or March, the secretary of the Society will communicate with the nominee and, if the Award is accepted, it will be announced in the spring of award year. Nominations received for the current year's competition remain under consideration for the two years following. The nomination form and a list of previous recipients are available at the website.
*Geographic Focus:* All States
*Date(s) Application is Due:* Jan 1
*Amount of Grant:* 500 USD
*Contact:* Awards Coordinator; (415) 337-1100; fax (415) 337-5205
*Internet:* www.astrosociety.org/membership/awards/klumpke.html
*Sponsor:* Astronomical Society of the Pacific
390 Ashton Avenue
San Francisco, CA 94112

**Astronomical Society of the Pacific Las Cumbres Amateur Outreach Award** 730

The Las Cumbres Amateur Outreach Award honors outstanding educational outreach by an amateur astronomer to K-12 children and the interested lay public. "Amateur"is understood to mean someone who does not receive compensation (other than expenses) for their activity and does not receive the majority of their income from a profession in astronomy. The Award consists of a recognition plaque and $500 presented at the annual meeting awards banquet. The Award recipient will be selected by the Awards Committee appointed by the Board of Directors. If, in the opinion of the Committee, none of the nominees is sufficiently outstanding, the Award will not be given for the year in question. In February or March, the secretary of the Society will communicate with the nominee and, if the Award is accepted, it will be announced in the spring of award year.
*Requirements:* Nominations are made by members of the astronomical community. Nomination forms are available at the ASP website.
*Geographic Focus:* All States

*Date(s) Application is Due:* Jan 1
*Amount of Grant:* 500 USD
*Contact:* Awards Coordinator; (415) 337-1100; fax (415) 337-5205
*Internet:* www.astrosociety.org/membership/awards/cumbres.html
*Sponsor:* Astronomical Society of the Pacific
390 Ashton Avenue
San Francisco, CA 94112

**Astronomical Society of the Pacific Maria and Eric Muhlmann Award** 731

The Maria and Eric Muhlmann Award of the ASP is given annually for recent significant observational results made possible by innovative advances in astronomical instrumentation, software, or observational infrastructure. The Award will be made to the scientist(s) who acquired the results, to the builder of the instrument/software (if a different individual), or to both as appropriate. The Award recipient(s) will be selected by the Awards Committee appointed by the Board of Directors of the ASP. Once accepted, the Award will be announced in the spring. The Award includes a $500 cash award and a plaque to be presented at the Annual Meeting Awards Banquet. Nomination forms are available at the website.
*Requirements:* Individuals as well as small teams of individuals are eligible. The Award will be granted for significant advances in any areas of astronomy, without restriction to wavelength or space/ground-based observations. Nominations are solicited from major observatories and departments of astronomy worldwide. Nominations will be accepted from any member of the astronomical community. The nominator(s) should submit an original and seven copies of a letter (no more than two typed pages in length) giving strong support to the choice. The letter should discuss the significance of the scientific results and clearly delineate the importance of the instrumentation/software/infrastructure advancement that enabled its achievement. Appropriate journal citations to the scientific results must also be included.
*Geographic Focus:* All States
*Amount of Grant:* 500 USD
*Contact:* Awards Coordinator; (415) 337-1100; fax (415) 337-5205
*Internet:* www.astrosociety.org/membership/awards/muhlmann.html
*Sponsor:* Astronomical Society of the Pacific
390 Ashton Avenue
San Francisco, CA 94112

**Astronomical Society of the Pacific Priscilla and Bart Bok Awards** 732

The Priscilla and Bart Bok Awards are given jointly by the American Astronomical Society (AAS) and the Astronomical Society of the Pacific (ASP). The main criterion for selecting the two Bok Awards is scientific merit. Observational, instrumental, theoretical and interdisciplinary projects involving physics, mathematics, computer sciences, and engineering, are eligible. The awarded funds are intended to be used by the recipients to further their education and research efforts. There is a first prize award of $1,000, and a second prize award of $500. Applicants may contact the Society for further information about the award and how to apply.
*Geographic Focus:* All States
*Contact:* Awards Coordinator; (415) 337-1100; fax (415) 337-5205
*Internet:* www.astrosociety.org/membership/awards/bok.html
*Sponsor:* Astronomical Society of the Pacific
390 Ashton Avenue
San Francisco, CA 94112

**Astronomical Society of the Pacific Richard H. Emmons Award for Excellence in College Astronomy Teaching** 733

The Richard A. Emmons Award of the ASP is awarded annually to an individual demonstrating outstanding achievement in the teaching of college-level introductory astronomy for non-science majors. Nominees must be teachers of college-level introductory astronomy for non-science majors. The award consists of a recognition plaque and a check for $500 which is presented at the annual meeting awards banquet. Letters of nomination should highlight how the nominee has distinguished himself or herself by citing exceptional achievement. Nomination forms are available at the ASP website.
*Requirements:* Nominations are solicited from colleagues, students, or others familiar with or knowledgeable of the nominee's achievements. Nominations received for the current year's competition remain under consideration for the two years following.
*Geographic Focus:* All States
*Date(s) Application is Due:* Jan 1
*Amount of Grant:* 500 USD
*Contact:* Awards Coordinator; (415) 337-1100; fax (415) 337-5205
*Internet:* www.astrosociety.org/membership/awards/emmons.html
*Sponsor:* Astronomical Society of the Pacific
390 Ashton Avenue
San Francisco, CA 94112

**Astronomical Society of the Pacific Robert J. Trumpler Award** 734

The Trumpler Award is given each year to a recent recipient of the PhD degree in North America whose research is considered unusually important to astronomy. Nominations may be made by chairpersons of the astronomy and physics departments of these institutions. A plaque is included with the monetary award. Nomination forms and instructions for additional information are available at the website.
*Requirements:* To be considered for the Trumpler Award, a nominee's PhD must have been awarded within two years of the time of the nomination and most or all of the thesis must have been published in a recognized journal. The nominee remains eligible for two awards cycles — the cycle for which the nomination was initially submitted and for the cycle following. The department chairman should submit an original and seven copies of a letter of nomination, supporting the selection and discussing the extent of independence exhibited by the nominee in the choice and execution of the thesis research. Eight reprints of the published articles must accompany the nomination.
*Geographic Focus:* All States
*Date(s) Application is Due:* Jan 1
*Amount of Grant:* 500 USD
*Contact:* Awards Coordinator; (415) 337-1100; fax (415) 337-5205
*Internet:* www.astrosociety.org/membership/awards/trumpler.html
*Sponsor:* Astronomical Society of the Pacific
390 Ashton Avenue
San Francisco, CA 94112

**ASU Graduate College Completion Fellowships** 735

The Graduate College Completion Fellowships are intended to support the degree completion of doctoral and MFA in the arts, humanities, social sciences, and education during the semester. Nominations should be made on the basis of demonstrated ability to complete research/projects of high quality and the assurance that a semester of full-time effort will allow the student to complete their dissertation/project. The Completion Fellowship provides a spring or fall stipend of $9,500. It will not include tuition and health insurance. It is expected that recipients will have completed all course work for their degree and will only need to register for one credit hour. The application is available at the graduate college website. The deadline for spring semester is October 31, while the fall deadline is March 30.
*Requirements:* Applicants must be officially advanced to candidacy at the time of nomination; in good academic standing with the department and the Graduate College; and finishing his or her doctoral or MFA degree by the end of the semester. Applicants must provide a copy of their doctoral candidacy letter; a letter from their advisor describing the substance of the dissertation/practical project and its significance and completion date for the dissertation/final project; and a document from the applicant with a detailed outline of the dissertation/project; a timeline for its completion; a one page dissertation summary in non-technical language; and a copy of the latest chapter completed or other substantive work as evidence of dissertation progress.
*Restrictions:* Each Fellows is personally responsible for individual taxes related to this award. Additional information regarding tax issues related to awards can be found at the Fellowship website or by contacting the ASU financial aid office. Recipients holding these awards will not be allowed t hold a research or teaching assistantship, or any kind of employment during the semester awarded. International awardees may request exemption to the full-course of study. MFA student needing to register for more than one credit hour must justify this at the time of the application.
*Geographic Focus:* All States
*Date(s) Application is Due:* Mar 30; Oct 31
*Amount of Grant:* 9,500 USD
*Contact:* Pat Saldivar, Fellowship Contact; (480) 965-5906; psaldivar@asu.edu
*Internet:* graduate.asu.edu/completion
*Sponsor:* Arizona State University Graduate College
Interdisciplinary Building, B-Wing
Tempe, AZ 85287-1003

**ASU Graduate College Dean's Fellowships** 736

Graduate College Dean's Fellowships provide up to $10,000 of support to regularly admitted first year graduate degree students who are Arizona residents or underrepresented in their discipline, and who demonstrate academic excellence. This can include under-represented minorities, such as Hispanics, African Americans and Native Americans in all disciplines and under-represented students in particular disciplines, such as Asian Americans in the Humanities and Social Sciences, women in Mathematics, Science, Engineering and some Agricultural and Life Sciences programs, men in Nursing, Public Health or Women's Studies. The award is disbursed in two equal payments each semester. Award amount will depend on the student's demonstrated financial need. Other fellowships, teaching or research assistantships, or other scholarships can be used to supplement this award. The total amount of the offer (including this fellowship), should add up to the typical offer made to comparable students. The application is available at the Fellowship website.
*Requirements:* Applicants must enroll for a minimum of six graduate level credit hours each semester. Students must be: newly admitted for spring, summer, or fall of the academic year; regularly admitted to a graduate degree program, with a minimum 3.0 GPA; in their first year of study; and either a resident of Arizona or underrepresented in their discipline.
*Restrictions:* International students are not eligible for this support. Each fellow is personally responsible for individual taxes related to this award. Additional information regarding tax issues related to awards can be found at the graduate college financial aid website.
*Geographic Focus:* Arizona
*Amount of Grant:* Up to 10,000 USD
*Contact:* Catherine Lebihan, Fellowship Contact; (480) 965-6916; clebihan@asu.edu
*Internet:* graduate.asu.edu/financing/fellowships/graduate-college-fellowships/deans
*Sponsor:* Arizona State University Graduate College
Interdisciplinary Building, B-Wing
Tempe, AZ 85287-1003

**ASU Graduate College Dissertation Fellowships** 737

Dissertation Fellowships are competitive university-wide fellowships designed to support highly meritorious doctoral students who are in the final stages of post-candidacy doctoral work. The Fellowship provides a $17,000 award plus tuition during the final year of study. At the time of the award, the student must be in good academic standing as defined by the department and the Graduate College. The academic unit nominating the

student provides health insurance. Awards will be made across the following five areas: arts, humanities, and social sciences; natural sciences and mathematics; engineering; professional programs and education; and interdisciplinary research. The application is available at the fellowship website, along with a list of required documentation.
*Requirements:* The nominees must have demonstrated the ability to carry out original research at a high level and advanced to doctoral candidacy at time of application; be prepared to devote full time to writing the dissertation with no other employment, including university assistantships; within one year of completion of the dissertation upon receipt of the award; and maintain a minimum of six graduate level credit hours each semester. Nine graduate level credit hours is highly recommended.
*Restrictions:* Each Fellow is personally responsible for individual taxes related to this award. Additional information regarding tax issues related to awards can be found at the financial aid website.
*Geographic Focus:* All States
*Date(s) Application is Due:* Feb 1
*Amount of Grant:* 17,000 USD
*Contact:* Pat Saldivar, Fellowship Contact; (480) 965-5906; psaldivar@asu.edu
*Internet:* graduate.asu.edu/dissfellowship
*Sponsor:* Arizona State University Graduate College
Interdisciplinary Building, B-Wing
Tempe, AZ 85287-1003

**ASU Graduate College Doctoral Enrichment Fellowships** **738**
Doctoral Enrichment Fellowships support new and regularly-admitted first-year doctoral Doctor of Philosophy (Ph.D.) graduate students who demonstrate academic excellence and are underrepresented in their field of study. The fellowship provides a $17,000 award plus tuition during the first year of study. In the second year, the academic unit will provide at least a 50% TA or RA position (at the department standard program rate) or similar funding assuming satisfactory academic progress. Student Health Insurance is provided by the academic unit for the first year.
*Requirements:* New students must be nominated by their admitting academic unit. The nominee must: be a new Ph.D. degree seeking student who will be admitted and start the degree program the next academic year; demonstrate academic excellence and contribute to diversity in the field of study and the profession; and be a U.S. citizen or assigned an IP visa.
*Restrictions:* Student Health Insurance (provided by the academic unit for the first year)
*Geographic Focus:* All States
*Amount of Grant:* 17,000 USD
*Contact:* Jennifer Cason, (480) 965-6113; fax (480) 727-6615; Jennifer.Cason@asu.edu
*Internet:* graduate.asu.edu/financing/fellowships/graduate-college-fellowships/doctoral_enrichment
*Sponsor:* Arizona State University Graduate College
Interdisciplinary Building, B-Wing
Tempe, AZ 85287-1003

**ASU Graduate College Earl and Lenore Tripke Fellowship** **739**
One nomination of a doctoral student per school in the College of Public Programs will be accepted for consideration of the Earl A. and Lenore H. Tripke Fellowship award. This Fellowship is awarded to one student for the spring semester from one of the following schools at ASU Graduate College: Community Resources and Development; Criminology and Criminal Justice; Public Affairs; and Social Work. A detailed list of the contents of the academic nomination packet are available at the Fellowship website.
*Requirements:* Nominees must be regularly admitted to an ASU graduate degree program; enrolled in spring semester with a minimum of six graduate-level credit hours (or more if required); and in good academic standing, making satisfactory academic progress. International students must have been enrolled in an ASU graduate program for one academic year immediately prior to the award period.
*Geographic Focus:* All States
*Date(s) Application is Due:* Oct 1
*Amount of Grant:* 4,000 USD
*Contact:* Pat Saldivar, Fellowship Contact; (480) 965-5906; psaldivar@asu.edu
*Internet:* graduate.asu.edu/tripke
*Sponsor:* Arizona State University Graduate College
Interdisciplinary Building, B-Wing
Tempe, AZ 85287-1003

**ASU Graduate College Herman E. Demund Memorial Fellowships** **740**
The Herman E. DeMund Memorial Fellowship awards a $2,000 fellowship to be split between fall and spring semesters. The Fellowship covers tuition for fall and spring semesters. The award will be made using the following criteria: academic accomplishments including, but not limited to GPA, test scores, publications, exhibitions, presentations; contribution to the discipline including clarity and importance of the applicant's research questions or creative endeavors, as demonstrated in the letter of nomination and vita; progress towards the degree, with typical time-to-degree in the academic program clarified in the letter of nomination; and contributions to the academic community (e.g., teaching awards or other recommendations, substantial service activities). The nomination form is available at the Fellowship website. Nomination packets must also include a detailed letter of nomination and transcripts.
*Requirements:* Nominees must be regularly admitted to a graduate degree program with a minimum of six graduate level credit hours to be eligible. International students must have been enrolled for one academic year immediately prior to the award period to be eligible.
*Restrictions:* Each fellow is personally responsible for individual taxes related to this award. Additional information regarding tax issues related to awards can be found at the university financial aid office.
*Geographic Focus:* All States
*Amount of Grant:* 2,000 USD
*Contact:* Pat Saldivar, Fellowship Contact; (480) 965-5906; psaldivar@asu.edu
*Internet:* graduate.asu.edu/financing/fellowships/graduate-college-fellowships/demund
*Sponsor:* Arizona State University Graduate College
Interdisciplinary Building, B-Wing
Tempe, AZ 85287-1003

**ASU Graduate College Lattie and Elva Coor Building Great Communities Graduate Fellowship** **741**
The Lattie and Elva Coor Fellowship is a unique opportunity created to support graduate students doing community engaged scholarly work. There is no specified disciplinary focus, but the student's work must include active engagement with communities in Arizona and include research or activities directed towards strengthening neighborhoods and/or improving the lives of Arizonans. The Fellow candidate's work should contribute to sustainable associations between ASU and one or more communities throughout the state. Projects that lead to enhanced civic participation in Arizona, improved Arizona public policy, or reducing domestic violence in Arizona are of particular interest. Successful applicants will demonstrate a clear integration of community work with scholarly work and pragmatic application of their discipline to promote enlightenment, understanding, and solutions to current community conditions. During the Fellowship, students must satisfy the following *Requirements:* at least one briefing with Office of Public Affairs staff to review project progress; a project presentation to Lattie and Elva Coor and the Office of Public Affairs staff; and at least one presentation to a community organization or other entity demonstrating the applicability of the research region's issues and concerns. The application is available at the Fellowship website.
*Geographic Focus:* All States
*Date(s) Application is Due:* Sep 28
*Amount of Grant:* 3,000 USD
*Contact:* Kenja Hassan, Fellowship Contact; (480) 965-0695; kenja.hassan@asu.edu
*Internet:* graduate.asu.edu/financing/fellowships/graduate-college-fellowships/coor
*Sponsor:* Arizona State University Graduate College
Interdisciplinary Building, B-Wing
Tempe, AZ 85287-1003

**ASU Graduate College Martha E. Bernal Memorial Award** **742**
The Martha E. Bernal Memorial Award was established to recognize the efforts of a Ph.D. student who is doing work similar to the career accomplishments of Professor Bernal, which focused on minority mental health issues, broadly defined. Dr. Bernal was one of the leading researchers on understanding minority mental and health issues, ethnic identity development and training of ethnic minority graduate students in psychology. The Department of Psychology is responsible for the selection of the award recipient. Nominations will be sought by emailing graduate students and faculty within the Psychology Department. In addition, the directors of schools in the Social Science Division of the College of Liberal Arts and Sciences, the director of the Hispanic Research Center, and the director of the Applied Linguistics Program will be contacted. Finally, information will be posted on the Graduate College's website. The selection committee will consist of the members of the Graduate Studies Committee in the Department of Psychology. The application is available at the Award website. Applicants should also include a brief statement of purpose, a current vita, and a letter of recommendation from an ASU faculty member.
*Requirements:* The recipient shall be a graduate student pursuing a Ph.D. with formal standing affiliated with the Department of Psychology, a school within the Social Science Division of the College of Liberal Arts and Sciences, the Hispanic Research Center, Applied Linguistics Program, or any other discipline congruent with the spirit of Martha Bernal's work and relevant to minority mental health issues, as broadly defined. The applicant should be enrolled as a full time student, making acceptable and timely progress towards the Ph.D.
*Restrictions:* Each Fellow is personally responsible for individual taxes related to this award. Additional information regarding tax issues related to awards can be found at the University's financial aid office. The award is not renewable, but current recipients may be nominated for the award in the future. There is no residency or financial need requirement associated with this award.
*Geographic Focus:* All States
*Date(s) Application is Due:* Sep 30
*Amount of Grant:* 600 USD
*Contact:* Kathryn Lemery-Chalfant, Fellowship Contact; klemery@asu.edu
*Internet:* graduate.asu.edu/financing/fellowships/graduate-college-fellowships/bernal
*Sponsor:* Arizona State University Graduate College
Interdisciplinary Building, B-Wing
Tempe, AZ 85287-1003

**ASU Graduate College Reach for the Stars Fellowship** **743**
Reach for the Stars Fellowship support new and regularly-admitted first-year students who demonstrate academic excellence and are underrepresented in their field of study. The Fellowship provides a $15,000 award for the first academic year plus tuition. In the second year, the academic unit will provide at least a 50% teacher assistant or research assistant position (at the department standard program rate) or similar funding assuming satisfactory academic progress. The nomination form is available at the Fellowship website.
*Requirements:* The nominee must be regularly admitted into a Master's degree program at the time the award is disbursed; demonstrate academic excellence and contribute to diversity in the field of study and the profession; and be a U.S. citizen or assigned an IP visa. All Fellowship recipients are required to register and participate in the one credit

Interdisciplinary Research Colloquium; enroll for a minimum of nine graduate-level credit hours during fall and spring semesters during the award period; meet regularly with a faculty mentor to discuss program goals, objectives and progress toward degree completion; meet regularly with a peer mentor to discuss getting acclimated to graduate school and student life; maintain satisfactory academic progress as defined by the academic unit and the Graduate College; not be employed as a TA, RA or in any other position at ASU during the year of this award; and submit a progress report signed by the academic mentor at the end of each semester.
*Restrictions:* Students must be nominated by their admitting academic unit. Mandatory fees, program fee, or any other fees are not covered as part of the fellowship.
*Geographic Focus:* All States
*Date(s) Application is Due:* May 15
*Amount of Grant:* 15,000 USD
*Contact:* Jennifer Cason; (480) 965-6113; fax (480) 727-6615; Jennifer.Cason@asu.edu
*Internet:* graduate.asu.edu/financing/fellowships/graduate-college-fellowships/stars
*Sponsor:* Arizona State University Graduate College
Interdisciplinary Building, B-Wing
Tempe, AZ 85287-1003

**ASU Graduate College Science Foundation Arizona Bisgrove Postdoctoral Scholars** **744**

The Bisgrove Postdoctoral Scholars Program (sponsored by Science Foundation Arizona) is designed to attract the nation's best early career scientists and engineers who exhibit the potential for outstanding competence and creativity in their research areas, strong communication skills, a passion for communicating the importance of their research to society, and a keen interest in educational science outreach to the community. Bisgrove Scholars will receive an annual stipend of approximately $60,000, benefits and additional funding for research expenses. The Bisgrove appointment is renewable on a year-to-year basis for a maximum initial term of two years, depending on the availability of funds. The ASU Graduate College Bisgrove Leadership Academy has been launched to provide training and guidance for Bisgrove Postdoctoral Scholars in areas of collaborative interdisciplinary research, user-inspired research, laboratory management and leadership, together with related resources, that will ensure their success as they develop as scientists in a multidisciplinary setting. The major activity of the Academy will include Strategies for Success workshops and seminars led by experts in a range of fields, with participating speakers from ASU, industry and the community. Bisgrove Scholars will also be invited to participate in the Preparing Future Faculty (PFF) program. Specific details about the application package are available at the Fellowship website.
*Requirements:* United States citizens or permanent residents of the U.S. should have completed a Ph.D. by the time of appointment, with no prior post-doctoral experience. They should also demonstrate research training and potential to transform ideas into value for society and the interest to work at the convergence of several disciplines. Scholars should be prepared to commit time to mentor undergraduate and graduate students in related disciplines; meet regularly with a faculty member to discuss research goals and objectives; submit a progress report every 90 days, summarizing their goals and evidence of progress; and co-write at least one federal grant application with the main research faculty mentor during the second year of the award, assuming satisfactory progress.
*Geographic Focus:* All States
*Date(s) Application is Due:* Jan 14
*Amount of Grant:* 60,000 USD
*Contact:* Jennifer Cason; (480) 965-8968; Jennifer.Cason@asu.edu
*Internet:* graduate.asu.edu/bisgrove
*Sponsor:* Arizona State University Graduate College
Interdisciplinary Building, B-Wing
Tempe, AZ 85287-1003

**ASU Graduate College Science Foundation Arizona Fellowships in Science and Engineering** **745**

Arizona State University's commitment to research aimed at solving specific societal challenges is being supported by graduate fellowship funding from Science Foundation Arizona. These graduate fellowships will be awarded to biologists, chemists, physicists, bioengineers, chemical engineers, electrical engineers, materials engineers, computer scientists, social scientists and individuals with related degrees interested in addressing one of the following critical needs: diagnosis and prevention of disease; sustainable energy and the environment; and information and communications technologies at the human interface. Candidates must be nominated by the degree-granting academic unit. Science Foundation Fellowships offer up to two years of competitive support. Each year of the award carries a $25,000 personal stipend plus a $5,000 supplemental allowance to the SFAz Fellow to advance the Fellow's research and/or to offset relocation expenses. Full tuition is covered. Second year support depends upon evidence of outstanding performance and progress toward the degree.
*Requirements:* Nominee must be admitted into a graduate program or related degrees interested in addressing one of the following critical needs: diagnosis and prevention of disease; sustainable energy and the environment; and information and communications technologies at the human interface. Detailed information about each of these needs is available at the Fellowship website.
*Geographic Focus:* All States
*Amount of Grant:* 30,000 USD
*Contact:* Dr. Andrew Webber; (480) 965-5906; andrew.webber@asu.edu
*Internet:* graduate.asu.edu/financing/fellowships/graduate-college-fellowships/sfaz
*Sponsor:* Arizona State University Graduate College
Interdisciplinary Building, B-Wing
Tempe, AZ 85287-1003

**ASU Institute for Humanities Research Fellowships** **746**

The purpose of the 2012-2013 Institute for Humanities Research Fellowship at Arizona State University is to encourage serious critical explorations of home as a term that resonates on cultural, emotional, intellectual, religious, philosophical, political and spiritual levels—as a place, a space, a myth, a source of identity, a promised land, a state of being, a war zone, an impossibility, and/or an inalienable right. The program invites humanities scholars from various disciplines to apply for the IHR fellowship to conduct research projects from multiple disciplinary and transdisciplinary perspectives that illuminate and enrich our understanding of how the concept of home informs and is informed by geography and history, identities and ideologies, imagined and lived experiences. Projects may explore what it means to create a home or to be without home. For example, are diasporic communities without a home, or is home more real and vital because it resides in culture, history, and memory? And while the homeless, are obviously without a physical home, what about those who reside in literal and figurative borderlands, for whom feeling at home is always contingent, always at risk? The IHR ASU Fellows program provides funding for either individual tenured or tenure-track faculty or research teams (two to three faculty) to engage in a year of research related to the annual theme, share their research with the academic community (via lectures, a conference or symposium), and produce a strong application for a large external grant.
*Geographic Focus:* All States
*Date(s) Application is Due:* Mar 4
*Amount of Grant:* Up to 20,000 USD
*Contact:* Sally Kitch, Director; (480) 965-3000; fax (480) 965-4300; sally.kitch@asu.edu
*Internet:* ihr.asu.edu/funding/fellows
*Sponsor:* Arizona State University
P.O. Box 876505
Tempe, AZ 85287-6505

**ASU Institute for Humanities Research Seed Grants** **747**

In fall and spring competitions each year, projects are selected for Seed Grant awards of up to $7,500 for individual researchers or up to $12,000 per team. The IHR seed grant program is designed to provide support for projects that advance the transdisciplinary, collaborative, and issue-focused mission of the Institute and are intended for sufficient development to be competitive for national grants.These grants further advance faculty research and often improve the quality of grant proposals to external funding agencies. The Institute supports those projects that best address its mission and that have strong prospects of receiving external funding. Outcomes include: conferences and symposia; invited guest scholars; publications; external grant support; and curricular impacts.
*Requirements:* Seed grant funds are available for junior and senior humanities faculty members and/or collaborative teams involving non-humanists. All IHR projects should incorporate humanities methodologies and involve at least one humanities scholar.
*Geographic Focus:* All States
*Date(s) Application is Due:* Oct 22
*Amount of Grant:* Up to 12,000 USD
*Contact:* Sally Kitch, Director; (480) 965-3000; fax (480) 965-4300; sally.kitch@asu.edu
*Internet:* ihr.asu.edu/funding/seed-grants
*Sponsor:* Arizona State University
P.O. Box 876505
Tempe, AZ 85287-6505

**ASU Institute for Humanities Research Visiting Fellowships** **748**

The Visiting Fellowship program is for tenured and tenure-track scholars from other institutions of higher education in the United States and abroad to spend spring semester in residence at the Institute for Humanities Research (IHR), participating in the intellectual life of the IHR and the university community. The Fellowship provides the opportunity to conduct research, collaborate with ASU faculty, and write. The Fellowship also promotes an exchange of ideas among visitors and ASU Fellows also working on the annual theme, which this year will be the Humanities and Home. Visitors will participate in weekly fellows meetings and give public lectures and seminars on their research topics while in residence at the IHR.
*Geographic Focus:* All States, All Countries
*Date(s) Application is Due:* Feb 18
*Amount of Grant:* Up to 20,000 USD
*Contact:* Sally Kitch, Director; (480) 965-3000; fax (480) 965-4300; sally.kitch@asu.edu
*Internet:* ihr.asu.edu/funding/fellows
*Sponsor:* Arizona State University
P.O. Box 876505
Tempe, AZ 85287-6505

**As You Sow** **749**

As You Sow was founded in 1992 and has grown into two programs that strive to increase corporate accountability. The Environmental Enforcement Program seeks to reduce and remove carcinogenic exposures by pursuing compliance with California's Safe Drinking Water and Toxic Enforcement Act. In its consideration of grant requests, As You Sow will give priority to proposals in the following areas: education of consumers about the risks of toxic exposure and their right to know what goes into consumer products; toxics education and reduction in minority communities; enforcement of local, state, or federal consumer and environment-related laws; education of youth on environmental issues; and related areas will be considered on a case-by-case basis but the proposal must generally have an environmental enforcement or education component. The grants are generally in the range of $5,000 to $10,000.
*Requirements:* As You Sow gives priority to proposals that will substantially benefit citizens of California through neighborhood, regional, or statewide projects. As You

Sow will also give consideration to proposals from organizations outside the state if the project or organization has a California emphasis or component.
*Geographic Focus:* All States
*Amount of Grant:* 5,000 - 10,000 USD
*Contact:* Conrad MacKerron; (415) 391-3212; fax (415) 391-3245; mack@asyousow.org
*Internet:* www.asyousow.org/grantmaking/guidelines.shtml
*Sponsor:* As You Sow
311 California Street, Suite 510
San Francisco, CA 94104

**AT&T Foundation Health and Welfare Grants** **750**
The AT&T Foundation is pivotal in enriching and strengthening the diverse communities it serves. The Foundation supports nonprofit organizations and programs that foster inclusion and create opportunities for diverse populations. Through its Health and Welfare program, approximately $35 million is contributed annually to support: economically challenged; health and health care access; urban families; and ethnic and racial minority groups. In addition, AT&T is committed to supporting underserved populations including women, seniors, youth, people with disabilities and low-income families. Interested parties should begin by forwarding a letter of inquiry to the Foundation office, outlining the proposed program or project in a brief one- to two-page letter.
*Requirements:* The Foundation makes grants to 501(c)3 tax-exempt, nonprofit organizations nationally, with emphasis on California, District of Columbia, Florida, Minnesota, and Texas.
*Restrictions:* No support is offered for: religious organizations not of direct benefit to the entire community; for political, discriminatory, or disease-specific organizations; medical clinics; or research. No grants are given: directly to individuals (except for employee-related disaster grants). Funding is not available for: capital campaigns; endowment funds; goodwill ads; ticket or dinner purchases; sports programs or events; cause-related marketing; or product donations.
*Geographic Focus:* All States
*Contact:* Nicole Anderson, President; (210) 821-4105; foundation@att.com
*Internet:* about.att.com/content/csr/home/frequently-requested-info/social.html
*Sponsor:* AT&T Foundation
208 S. Akard Street, Suite 100
Dallas, TX 75202

**ATA Doctoral Fellowships in Education** **751**
Each year, the Association awards two fellowships, valued at $15,000 each, to ATA members who have been accepted into, or have already embarked upon, full-time study (as specified by the institution) in a doctoral program in education at a recognized Alberta public university. The fellowship program is intended to recognize academic excellence and to help defray the financial costs of university study. In selecting fellowship recipients, the Association takes into account the applicant's academic standing, contribution to the Association and public education, and exemplary teaching practices. The Association names a selection committee, which meets in early spring to review the applications and to select potential recipients. The committee forwards its recommendations to Provincial Executive Council, which makes the final decision at its April meeting. Applications are available at the ATA website.
*Requirements:* Applicants must meet the following conditions: hold a permanent Alberta teaching certificate and have completed at least five years of successful teaching in Alberta; hold the highest level of membership in ATA, with membership at that level for at least five years; entering his or her first year of full time doctoral study; and intend to continue a career in education in Alberta. Applicants cannot be previous fellowship winners and must be ATA members in good standing.
*Restrictions:* Fellowships are paid to recipients in two equal installments, beginning in the fall. Recipients are required to file a copy of their research with the Association.
*Geographic Focus:* Canada
*Date(s) Application is Due:* Mar 1
*Amount of Grant:* 15,000 CAD
*Contact:* Corinne Anderson, Administrative Officer; (780) 447-9470 or (800) 232-7208; corinne.anderson@ata.ab.ca
*Internet:* www.teachers.ab.ca/Professional+Development/Grants+and+Scholarships/Fellowships+and+Scholarships.htm
*Sponsor:* Alberta Teachers Association
11010 142nd Street NW
Edmonton, AB T5N 2R1 Canada

**ATA Educational Trust Grants - Curriculum Resource or Research Project** **752**
Each year, the Trust awards a number of project grants of up to $3,000 to help individuals or groups conduct research or develop practical classroom materials for Alberta teachers. The materials and resources produced through these grants are placed in the ATA library where they can be borrowed by interested teachers. Grants are payable against receipts for expenses. Applications are available at the ATA website, but must be mailed. Faxed or emailed applications are not accepted.
*Requirements:* Any individual or group involved in education and planning to undertake such a project during the next school year is eligible to apply. Submissions in the areas of special education, global and environmental education, languages, Francophone education, native education and early childhood education are of particular interest to the Trust.
*Geographic Focus:* Canada
*Date(s) Application is Due:* May 1
*Amount of Grant:* Up to 3,000 CAD
*Contact:* Educational Trust Contact; (800) 232-7208 or (780) 447-9494
*Internet:* www.teachers.ab.ca/For%20Members/Programs%20and%20Services/Grants%20Awards%20and%20Scholarships/Pages/The%20ATA%20Educational%20Trust.aspx
*Sponsor:* Alberta Teachers Association
11010 142nd Street NW
Edmonton, AB T5N 2R1 Canada

**ATA Education Research Award** **753**
The ATA Education Research Award is presented annually to a faculty member or sessional lecturer at an Alberta university who has undertaken high-quality research on classroom teaching and learning. Research must be directly related to school and classroom practice; be focused on school teaching and/or learning; be current (either ongoing or completed within the last two years); be related to critical issues; have involved classroom teachers and/or students; be applicable to the Alberta context and must be of practical benefit to teachers in improving their professional practice, and be of high quality in terms of purpose, approach, methodology, originality and clarity. The researcher(s) must also be prepared to present the research to Association subgroups. The application is available at the ATA website.
*Requirements:* To qualify for an award, the research must meet the following criteria: directly relate to school and classroom practice; focus on school teaching and/or learning; have current value and relate to critical issues; involve classroom teachers and/or students; apply to the Alberta context; hold practical benefit to teachers in improving their professional practice, with high quality in terms of purpose, methodology and originality.
*Geographic Focus:* Canada
*Date(s) Application is Due:* May 11
*Amount of Grant:* 5,000 CAD
*Contact:* J.C. Couture; (780) 447-9462 or (800) 232-7208; jc.couture@ata.ab.ca
*Internet:* www.teachers.ab.ca/For%20Members/Programs%20and%20Services/Grants%20Awards%20and%20Scholarships/Pages/Educational%20Research%20Award.aspx
*Sponsor:* Alberta Teachers Association
11010 142nd Street NW
Edmonton, AB T5N 2R1 Canada

**ATA Nadene M. Thomas Graduate Research Bursary** **754**
The Nadene M. Thomas Graduate Research Bursary is offered annually to a member of the Alberta Teachers Association who is enrolled in a graduate program in a specialty education at a recognized Canadian university. In selecting bursary recipients, the Association takes into account these factors: the applicant's academic standing; the applicant's contribution to the Association and public education; exemplary teaching practice; and the applicability of the research on health issues affecting teachers and/or teachers' working conditions. The selected recipient is granted the bursary in two equal installments, the first upon receipt of the award and the second upon receipt of the completed research. Recipients are required to file a copy of their research with the Association. The application is available at the ATA website.
*Requirements:* Applicants must meet the following conditions: hold a permanent Alberta teaching certificate and have completed at least five years of successful teaching in Alberta; be at the highest level of membership possible and at the highest possible level of membership, in good standing, for at least five years; intend to continue a career in the K-12 public education system in Alberta; and be registered as a graduate student in a specialty education at a recognized Canadian university. Research must focus on health issues affecting teachers and/or on teachers' working conditions.
*Restrictions:* The applicant must not have received an ATA fellowship or bursary in the past.
*Geographic Focus:* Canada
*Date(s) Application is Due:* Mar 1
*Amount of Grant:* 5,000 CAD
*Contact:* Corinne Anderson, Administrative Officer; (780) 447-9470 or (800) 232-7208; corrine.anderson@ata.ab.ca
*Internet:* www.teachers.ab.ca/For%20Members/Programs%20and%20Services/Grants%20Awards%20and%20Scholarships/Pages/Nadene%20M%20Thomas%20Graduate%20Research%20Bursary.aspx
*Sponsor:* Alberta Teachers Association
11010 142nd Street NW
Edmonton, AB T5N 2R1 Canada

**Auburn Foundation Grants** **755**
The purpose of the Foundation is to stimulate giving and cooperative leadership among the citizens of Auburn; help improve the lives of all community residents, especially those who are most vulnerable; and enrich the cultural environment and community life. Of special interest are projects that bring together all ages and sections of the town, or that contribute to healthy, active living. Application is available online.
*Requirements:* Any nonprofit organization that serves residents of Auburn is invited to apply.
*Restrictions:* Grants will not be awarded to for-profit businesses or expenses already incurred by the applicant.
*Geographic Focus:* Massachusetts
*Date(s) Application is Due:* Apr 15
*Amount of Grant:* 5,000 USD
*Samples:* Bancroft School (Worcester, MA)—for general support, $45,000; New England Science Ctr, Rutland House (Worcester, MA)—for repairs, $20,000; Tower Hill Botanic Garden (Boylston, MA)—for general support, $30,000.
*Contact:* Lois Smith; (508) 755-0980, ext. 107; lsmith@greaterworcester.org
*Internet:* www.greaterworcester.org/grants/Auburn.htm
*Sponsor:* Auburn Foundation
370 Main Street, Suite 650
Worcester, MA 01608-1738

**AUPHA Corris Boyd Scholarships** 756

The Association of University Programs in Health Administration offers the Corris Boyd Scholarship in honor of a healthcare leader who dedicated his life to diversity and excellence in leadership, especially among people of color and women. The program currently offers two $40,000 scholarships to deserving students of color entering a graduate program in healthcare management. The award will be paid in four installments, paid in August and December of the first year, then August and December of the second year. The annual deadline for applicants is May 3.
*Requirements:* At the time of award, applicants must meet the following eligibility *Requirements:* have applied and been accepted into an AUPHA full-member masters degree program for start in the upcoming Fall semester (not before August 1); be an under-represented minority; have a minimum 3.0 GPA (out of 4.0) in undergraduate coursework; and have U.S. citizenship or permanent resident status.
*Geographic Focus:* All States
*Date(s) Application is Due:* May 3
*Amount of Grant:* 40,000 USD
*Contact:* Lydia Middleton, President and CEO; (703) 894-0940, ext. 115; fax (703) 894-0941; lmiddleton@aupha.org or aupha@aupha.org
*Internet:* www.aupha.org/i4a/pages/Index.cfm?pageID=3541
*Sponsor:* Association of University Programs in Health Administration
2000 14th Street North, Suite 780
Arlington, VA 22201

**Australasian Institute of Judicial Administration Seed Funding Grants** 757

The Australasian Institute of Judicial Administration (AIJA) is a research and educational institute associated with Monash University. It is funded by the Standing Council on Law and Justice (SCLJ) and also from subscription income from its membership. The principal objectives of the Institute include research into judicial administration and the development and conduct of educational programmes for judicial officers, court administrators and members of the legal profession in relation to court administration and judicial systems. The AIJA is interested in working with academics to develop proposals for research projects. It is offering financial assistance to provide an incentive for academics to prepare ARC grant applications in areas of AIJA interest. Broad areas of research identified as relevant to judicial administration are: the proportional use of judicial resources for optimum practical effectiveness and efficiency; and public perceptions and understanding of the justice system. Included in these broad areas are: the examination of the relationship between the principles of adjudicatory independence and administrative accountability; a comparative analysis of reforms in the justice system in other jurisdictions; performance measurement, including the creation of best practice models for the introduction of bench-marking and yardstick competition models in the courts; and proportionality, namely, the proportion of the value of claims expended on legal costs and the expenditure of court resources to meet the instances and areas of greatest need and access to justice.
*Geographic Focus:* All States, All Countries
*Contact:* Gregory Reinhardt, Executive Director; +61 3 9600 1311; fax +61 3 9606 0366; Gregory.Reinhardt@monash.edu
*Internet:* www.aija.org.au/index.php/research/seed-funding
*Sponsor:* Australasian Institute of Judicial Adminstration
555 Lonsdale Street, Ground Floor
Melbourne, VIC 3000 Australia

**Avon Products Foundation Grants** 758

The foundation's two-fold focus is to support education, community and social services, and arts organizations and programs that provide economic opportunities for women and girls; and to support breast cancer and other women's health organizations and programs. The foundation awards grants in cities and regions with a large concentration of representatives and business operations, with the majority of funds going to US-based institutions. National, international, and New York metropolitan area programs are administered through the foundation's headquarters in New York. For regional funding support, refer to the Avon Foundation website.
*Requirements:* Applying organizations must be tax-exempt; national and municipal organizations are eligible. Request the guidelines brochure prior to submitting a formal proposal.
*Restrictions:* Grants do not support individuals; memberships; lobbying organizations; political activities and organizations; religious, veteran, or fraternal organizations; fundraising events; and journal advertisements.
*Geographic Focus:* All States
*Amount of Grant:* Up to 23,000,000 USD
*Contact:* Grants Administrator; (866) 505-2866; info@avonfoundation.org
*Internet:* www.avoncompany.com/women/avonfoundation
*Sponsor:* Avon Products Foundation
1345 Avenue of the Americas
New York, NY 10105

**Axe-Houghton Foundation Grants** 759

The Axe-Houghton Foundation operates in the New York City metropolitan area exclusively for charitable, educational, and scientific purposes to foster and encourage an appreciation of the English language, with major emphasis on the spoken language. Priority is given to projects for the improvement of speech and its uses in the areas of public affairs, education, theater, poetry, debate, and the oral interpretation of literature. A portion of available funds is devoted to speech remediation and to scientific research pertaining to speech. Types of support include program grants and seed money grants. Applications are accepted only from organizations who have been invited to apply. Interested parties can contact the Foundation for information on how to secure an invitation to apply. Funds are limited and awards rarely exceed $5,000.
*Requirements:* To be eligible to receive a grant from the Foundation, the proposed grantee must be either: an organization which has been determined by the Internal Revenue Service to be a tax-exempt organization described in section 501(c)3 of the Internal Revenue Code, and not a private foundation within the meaning of Section 509(a) of the Internal Revenue Code; a governmental unit as referred to in Section 170(c)1 of the Internal Revenue Code; or an agency or instrumentality of a governmental unit, including, without limitation, a state college or university within the meaning of Section 511(a)2 of the Internal Revenue Code. Initial inquiries should be by email or phone. Invited applicants should include a brief statement of objective, duration of time required to complete this objective, qualifications of personnel responsible for conducting the project, and the general amounts of funds needed. The proposal should also indicate the degree of internal support that would be committed.
*Restrictions:* Grants are made only to tax-exempt institutions and not to individuals, private foundations or organizations outside of the United States. Grants are not made for scholarships, fellowships, capital, or general support programs.
*Geographic Focus:* New York
*Date(s) Application is Due:* Sep 1
*Amount of Grant:* 1,000 - 7,500 USD
*Contact:* John Johnson, Client Services; (800) 839-1754; fax (800) 421-6579; jjohnson@foundationsource.com
*Internet:* fdnweb.org/axehoughton/guidelines/
*Sponsor:* Axe-Houghton Foundation
55 Walls Drive, 3rd Floor
Fairfield, CT 06824

**Bainum Family Foundation Grants** 760

Stewart and Jane Bainum founded what is now the Bainum Family Foundation in 1968. It was first known as The College Foundation and later as the Commonweal Foundation. Today, the Bainum Family Foundation continues to combine proven expertise with a passion for helping the whole child by going beyond academics to help them thrive. The Foundation awards grants to eligible nonprofit organizations serving disadvantaged youth in the area comprising the corridor between Baltimore, MD, and Washington, D.C. It also offers some assistance to elementary education, educational research, and some health care causes. Programs include Partners in Learning (address low literacy skills of economically disadvantaged children), community assistance grants (i.e., after-school tutoring, parenting classes, and shelters for the homeless or abused women), and learning disabilities support program (special education services to economically disadvantaged children). Contact the office for application procedures.
*Requirements:* 501(c)3 tax-exempt organizations serving disadvantaged youth in the corridor between Baltimore, Maryland, and Washington, D.C., are eligible to apply.
*Geographic Focus:* District of Columbia, Maryland, Virginia
*Date(s) Application is Due:* Feb 1; May 1; Aug 1; Nov 1
*Amount of Grant:* 10,000 - 25,000 USD
*Contact:* Barbara Bainum, CEO and Director; (240) 450-0000; fax (240) 450-4115
*Internet:* www.commonweal-foundation.org/index.html
*Sponsor:* Bainum Family Foundation
7735 Old Georgetown Road, #1000
Bethesda, MD 20814

**Ball Brothers Foundation General Grants** 761

Founded in the name of Edmund B. Ball and his brothers, the Foundation seeks to build and sustain a high quality of life in Indiana by awarding grants to nonprofit organizations in broad subject areas, including elementary, secondary, higher, and adult basic education and literacy skills; cultural activities; community betterment; the environment; and health and human services. Usually, Muncie and Delaware Counties receive a higher priority for funding than requests from across the state. Types of support include general operations, annual campaigns, capital campaigns, building construction/renovation, program development, conferences and seminars, professorships, publication, curriculum development, research, fellowships, matching funds, seed grants, and technical assistance. Preference will be given to catalytic grants that will stimulate others to participate in problem solving or in matching fund programs and to innovative approaches for addressing either traditional or emerging community needs. Applications are reviewed by the board of directors in January, May, and September of each calendar year. Proposals are encouraged to be submitted from February to May. Grant seekers may send a preliminary proposal, complete proposal, or ask for a personal visit to discuss a potential grant request.
*Requirements:* Indiana 501(c)3 nonprofit institutions and organizations are eligible.
*Restrictions:* The Foundation will not support: direct assistance to individuals or scholarships; applications coming from outside of Indiana; booster organizations; on-going salary requests of staff personnel to support an organization; services that the community-at-large should normally underwrite (i.e. roads, bus transportation, etc.); capital building projects; research projects (except for philanthropic studies); or unsolicited proposals (all requests must begin with a preliminary proposal).
*Geographic Focus:* Indiana
*Date(s) Application is Due:* Apr 1; Sep 1
*Amount of Grant:* Up to 100,000 USD
*Samples:* Open Door Health Center, Muncie, Indiana, $90,000 - construction of a environmentally-friendly facility; Indiana Council for Economic Education, West Lafayette, Indiana, $5,000 - economic education and financial literacy; Delaware County Soil and Water, Muncie, Indiana, $150,00 - wetlands preserve.
*Contact:* Donna Munchel, Executive Assistant; (765) 741-5500; fax (765) 741-5518; donna.munchel@ballfdn.org or info@ballfdn.org

*Internet:* www.ballfdn.org
*Sponsor:* Ball Brothers Foundation
222 South Mulberry Street
Muncie, IN 47305

**Ball Brothers Foundation Rapid Grants** 762
Ball Brothers Foundation offers funding opportunities for a limited number of Ball Rapid Grants that are designed to provide funding to organizations requiring immediate funding for the following types of needs: to continue or complete a project; to provide professional development; to buy equipment or materials for a project; for travel to meet representatives to advance ideas for a current or future project; to formulate a project idea; to carry out a mandated law or event; or for seed money to begin a new project. Generally requests up to $5,000 are considered. Ball Rapid Grants do not fall within the normal granting period. Requests may be submitted at any time between February 1 and November 30; decisions will be made within four business days and awards sent within 7 to 10 business days.
*Requirements:* Ball Brothers Foundation is restricted by its charter to grants to nonprofit institutions and organizations within Indiana.
*Restrictions:* The Foundation will not support: direct assistance to individuals or scholarships; applications coming from outside of Indiana; booster organizations; on-going salary requests of staff personnel to support an organization; services that the community-at-large should normally underwrite (i.e. roads, bus transportation, etc.0; capital building projects; research projects (except for philanthropic studies); or unsolicited proposals (all requests must begin with a preliminary proposal).
*Geographic Focus:* Indiana
*Date(s) Application is Due:* Nov 30
*Amount of Grant:* Up to 5,000 USD
*Contact:* Donna Munchel, Executive Assistant; (765) 741-5500; fax (765) 741-5518; donna.munchel@ballfdn.org or info@ballfdn.org
*Internet:* www.ballfdn.org/index/staff/generalinfo.asp
*Sponsor:* Ball Brothers Foundation
222 South Mulberry Street
Muncie, IN 47305

**Baltimore Washington Center for Psychotherapy and Psychoanalysis Adult Fellowships** 763
The Adult Psychoanalytic Fellowship Program is a one-year program designed for individuals who have a significant interest in psychoanalysis as a body of knowledge and as a framework with which to understand and carry out therapeutic efforts. Intended for advanced graduate students and residents as well as recent graduates in psychiatry, psychology and social work, the program consists of monthly seminars presented by psychoanalysts on clinically-related topics. Fellows are also welcome to attend Psychoanalytic Forums and other conferences offered by the Center. The annual deadline for applications is July 15. Fellowships will be announced in early September.
*Geographic Focus:* All States
*Date(s) Application is Due:* Jul 15
*Contact:* Elizabeth Manne, Executive Director; (301) 470-3635 or (410) 792-8060; fax (410) 792-4912; admin@bwanalysis.org or info@bwanalysis.org
*Internet:* bwanalysis.org/adult-psychoanalytic-fellowship-program.html
*Sponsor:* Baltimore Washington Center for Psychotherapy and Psychoanalysis
14900 Sweitzer Lane, Suite 106
Laurel, MD 20707

**Baltimore Washington Center for Psychotherapy and Psychoanalysis Child Fellowships** 764
The Child Fellowship Program is a one-year program specifically designed for child mental health professionals who are interested in deepening their understanding of working with children from a psychoanalytic or psychodynamic perspective. The program is tuition-free. Detailed discussions of actual clinical work with children and reviews of relevant psychoanalytic literature will be used as a framework to enhance understanding and ability to carry out therapeutic work. Topics covered will include: issues in diagnostic understanding and treatment planning with children and adolescents; psychoanalytic psychotherapy with children; professional and ethical issues in child psychotherapy; work with children who are traumatized, learning disabled, and from diverse cultures; work with parents and families; and case discussions. The annual deadline for applications is July 15.
*Requirements:* Professionals, including psychiatrists, psychologists, social workers, and counselors currently working with children in psychotherapy should apply. Recent graduates of training programs should also apply.
*Geographic Focus:* All States
*Date(s) Application is Due:* Jul 15
*Contact:* Elizabeth Manne, Executive Director; (301) 470-3635 or (410) 792-8060; fax (410) 792-4912; admin@bwanalysis.org
*Internet:* bwanalysis.org/child-fellowship-program.html
*Sponsor:* Baltimore Washington Center for Psychotherapy and Psychoanalysis
14900 Sweitzer Lane, Suite 106
Laurel, MD 20707

**Banfield Charitable Trust Grants** 765
This trust offers grants annually to nonprofit organizations and educational institutions that make life better for Pets and their families. Special considerations will be given to collaborative programs with Pet-related organizations and Banfield team members working together towards a shared goal. At this time, Banfield Charitable Trust is pursuing the following funding priorities: promotion of preventative healthcare for Pets; educating children about veterinary medicine and the Pets they love; programs based on the human-Pet bond and how this relates to longer, healthier lives for Pets and people; and veterinary education programs. The Trust does not generally fund an entire project, but expects to be one of multiple funding sources. Typically, the Trust will not fund more than 50% of the project's entire budget.
*Requirements:* 501(c)3 tax-exempt organizations and educational institutions are eligible.
*Restrictions:* Funding will not be considered for the following purposes: Spay and neuter or adoption programs; General operating expenses, deficit reduction, or general administrative overhead expenses; Fundraising campaigns, including special events; Grants to individuals, or to provide support for business enterprise.
*Geographic Focus:* All States
*Date(s) Application is Due:* Jun 30; Nov 30
*Samples:* Portland Children's Museum, Portland, OR, $15,000—this grant will help the museum to host the national traveling exhibit Clifford the Big Red Dog, based on the highly acclaimed children's book series and PBS television show. The exhibit and accompanying programs will teach children how to care for their pets and show them the benefits and joys of the pet-owner relationship; Sarge's Animal Rescue Foundation, Waynesville, NC, $3,000—this grant will help to cover the cost of providing senior cats to senior citizens at no cost. The program benefits both the cats who are adopted into loving homes and the seniors who get a companion along with the proven health benefits of pet ownership; Humane Society of Saint Joseph County, Mishawaka, IN, $10,000—this grant will help support their Humane Education Learning Program (H.E.L.P.) which includes classroom presentations about animal care to their community.
*Contact:* Darlene Schwartz, Program Coordinator; (503) 922-5801; fax (503) 922-6801; info@banfieldcharitabletrust.org
*Internet:* www.banfieldcharitabletrust.org/Grants
*Sponsor:* Banfield Charitable Trust
8000 North East Tillamook Sreet, P.O. Box 13998
Portland, OR 97213

**Bank of America Charitable Foundation Matching Gifts** 766
The Bank of America Charitable Foundation Matching Gifts program encourages employees to contribute to qualifying charitable organizations. This program supports employee giving by offering a way to double – up to $5,000 per person each calendar year – employees' cash or securities contributions to their favorite charitable organizations and thus improve their communities. Annually, the Bank of America Charitable Foundation provides more than $25 million in matching gifts on behalf of employee donations.
*Requirements:* Charitable organizations in the United States must be tax-exempt under section 501(c)3 of the Internal Revenue Code and not be classified as a private foundation. Charitable Organizations located in England or Wales must be registered with the Charity Commission. Charitable organizations outside of the United States, England or Wales must be qualified as eligible for donations from CAFAmerica.
*Restrictions:* The Bank of America Charitable Foundation does not: match charitable gifts to private, family or donor advised funds, or gifts to political or fraternal organizations; or match charitable gifts that benefit students directly or that result in an employee receiving a benefit, including tuition or sponsorships.
*Geographic Focus* All States, District of Columbia, Guam, Marshall Islands, Northern Mariana Islands, Puerto Rico, U.S. Virgin Islands, American Samoa, Canada, United Kingdom
*Date(s) Application is Due:* Mar 31
*Amount of Grant:* Up to 5,000 USD
*Contact:* Anne M. Finucane, Foundation Chairperson/Chief Marketing Officer; (617) 434-9410 or (800) 218-9946; anne.m.finucane@bankofamerica.com
*Internet:* about.bankofamerica.com/en-us/global-impact/matching-gifts-features-and-eligibility.html#fbid=wq9wE7VdEpC
*Sponsor:* Bank of America Charitable Foundation
100 North Tryon Street
Charlotte, NC 28255

**Baxter International Foundation William B. Graham Prize for Health Services Research** 767
Since 1986, The Baxter International Foundation and the Association of University Programs in Health Administration have awarded the William B. Graham Prize for Health Services Research to recognize researchers who have made major contributions to the health of the public through innovative research in health services. The Prize honors the late William B. Graham, longtime chairman and CEO of Baxter International Inc., and is internationally regarded as the premier recognition for individuals conducting health services research. The Prize recognizes individuals who have had a significant impact on the health of the public in one of three primary focus areas: Health Services Management, Health Policy Development and Healthcare Delivery. The Prize includes an award of $25,000 to the individual and $25,000 to a not-for-profit institution that supports the winner's work.
*Geographic Focus:* All States, All Countries
*Amount of Grant:* 50,000 USD
*Contact:* Prize Coordinator; (847) 948-4605; fdninfo@baxter.com
*Internet:* www.baxter.com/about_baxter/sustainability/international_foundation/index.html
*Sponsor:* Baxter International Foundation
One Baxter Parkway
Deerfield, IL 60015-4633

**Bay and Paul Foundations Research and Initiatives for Ecosystem Integrity Grants** 768
The Bay and Paul Foundations' overall mission is to foster and accelerate initiatives that prepare agents of change working to strengthen our social compact and develop authentic solutions to the challenges of this pivotal century. With the Research and Initiatives for

Ecosystem Integrity grant program, the Foundation supports scientific research and community-based initiatives to preserve or repair biodiversity and ecosystem integrity, often emphasizing culturally significant natural heritage conservation with economic and social sustainability goals. The Foundation no longer accepts unsolicited proposals; all full proposals will be by invitation only and are considered stage two of a two-part process. Applicants may submit an online funding inquiry (OFI) related to one or more of its current program areas. Annual funding cycles for OFI's are from January 19 through February 19 and July 19 through August 19.
*Requirements:* Nonprofits in Connecticut, Massachusetts, Maine, New Hampshire, New Jersey, New York, Rhode Island, and Vermont are eligible.
*Restrictions:* Grants do not support requests for endowments, building campaigns, building construction or maintenance, sectarian religious programs, books or studies, individual scholarships or fellowships, loans, travel, film, television or video productions, programs consisting primarily of conferences, for annual fund appeals, or to other than publicly recognized charities. First time grants for K-12 arts-in-education programs and K-12 science and math programs are currently geographically restricted to the New York City metropolitan area.
*Geographic Focus:* All States
*Date(s) Application is Due:* Feb 19; Aug 19
*Contact:* Danielle Reda,; (212) 663-1115; fax (212) 932-0316; info@bayandpaul.org
*Internet:* bayandpaulfoundations.org/areas-of-focus/research-initiatives-for-ecosystem-integrity/index.html
*Sponsor:* Bay and Paul Foundations
17 West 94th Street, 1st Floor
New York, NY 10025

**Bay Area Community Foundation Michigan Nonprofits Count Campaign Grants 769**
The Bay Area Community Foundation offers several opportunities for grant funding for a registered 501(c)3 organization or government entity. Getting an accurate count for Bay County residents in the 2020 Census is crucial for our community. Each person left uncounted equals approximately $1,800 a year in loss of government funding for ten years. Partnering with Midland Area Community Foundation, Saginaw Community Foundation, and Mt. Pleasant Community Foundation through Michigan Nonprofits Association, the Bay Area Community Foundation has a census hub for the region and has grant funding available for Bay County nonprofits looking to help ensure an accurate count for our community. The Michigan Nonprofits Count Campaign is a collaborative, coordinated, statewide effort to encourage participation in the 2020 census in communities that are at significant risk of being under-counted. Led by the Michigan Nonprofit Association, the campaign will mobilize nonprofits to encourage participation in the 2020 census by: providing trainings and tools for nonprofits on effective outreach tactics; assisting nonprofits in identifying historically under-counted communities; awarding grants to local nonprofits; and working with government officials to avoid duplication of efforts and enhance government's communication and outreach efforts to support a fair and accurate census count. With support from the Michigan Nonprofit Association and local census hub, grant recipients will: educate and encourage historically under-counted communities to complete the 2020 census; provide opportunities for those served/represented to complete the 2020 census; and address community concerns in relation to the 2020 census.
*Requirements:* Required basic eligibility criteria includes: emerging and established organizations at all stages of work are encouraged to apply for funding; organizations must be a legally incorporated entity in Michigan with501(c)3 status or a project sponsored by a government agency with a clear charitable purpose; organizations with a fiscal sponsor are also eligible to apply. Organizations must satisfy all reporting requirements including narrative and financial reports to local hub administrators.
*Restrictions:* Campaign support cannot be used for partisan efforts.
*Geographic Focus:* Michigan
*Contact:* Joni King, Program Officer; (989) 893-4438; jonik@bayfoundation.org or bacfnd@bayfoundation.org
*Internet:* www.bayfoundation.org/index.php/nonprofits-side/grant-programs
*Sponsor:* Bay Area Community Foundation
1000 Adams Street, Suite 200
Bay City, MI 48708

**Bayer Foundation Grants 770**
The Bayer Foundation supports programs that enhance the quality of life, provide unique and enriching opportunities that connect diverse groups and ensure preparedness for tomorrow's leaders. The Foundation welcomes proposals from 501(c)3 organizations whose programming matches at least one of the following areas: civic and social service programs; education; workforce development; arts and culture; and health and human services. The application submission process is online. Applicants will be asked to provide a brief history of the organization, contact information, an executive summary of the proposed project, the requested amount, a proposed budget, program time frame, a letter from the IRS that verifies the organization's 501(c)3 status, a W-9 form, and a listing of other funding sources.
*Requirements:* 501(c)3 nonprofit organizations in Bayer operating communities are eligible. Submit proposals to regional offices in: Berkley and northern California; Kansas City, Missouri; northern New Jersey; Pittsburgh, Pennsylvania; Raleigh-Durham, North Carolina; Houston and Baytown, Texas; and Shawnee, Kansas. Additionally, the foundation considers proposals from STEM education programs that have a national focus.
*Restrictions:* The Foundation will not fund: for-profit organizations or those without Internal Revenue Service code 501(c)3 nonprofit, tax-exempt status; organizations that discriminate on the basis of race, color, creed, gender, sexual orientation or national origin; general operating support for United Way affiliated agencies (only support for special projects will be considered); organizations or programs designed to influence legislation or elect candidates to public office; general endowment funds; Deficit reduction or operating reserves; religious organizations; charitable dinners, events or sponsorships; community or event advertising; individuals; student trips or exchange programs; athletic sponsorships or scholarships; telephone solicitations; or organizations outside of the United States or its territories.
*Geographic Focus:* California, Kansas, New Jersey, North Carolina, Pennsylvania, Texas
*Contact:* Sarah Toulouse, Executive Director; (412) 777-5725; bayerusafoundation@bayer.com
*Internet:* www.bayerus.com/about/community/i_foundation.html
*Sponsor:* Bayer USA Foundation
100 Bayer Road, Building 4
Pittsburgh, PA 15205-9741

**BBF Fellowships 771**
Barbara Bush Fellows originated at TCALL in 2007, and are funded by The Barbara Bush Texas Fund for Family Literacy. The Barbara Bush Fellows are doctoral students at Texas A&M University-College Station who are engaged in research, writing, and development of resources and presentations in topics relevant to the field of family literacy education. These fellowships consist of a one-year award, which may be used to offset the cost of travel expenses to conferences, dissertation research, and/or tuition, fees, and maintenance while a student is enrolled at Texas A&M University. During their fellowship year, students will have the opportunity to present their work at national family literacy conferences and publish their research in peer-reviewed journals. Up to four Barbara Bush Fellowships of $25,000 each will be awarded for research on topics relevant to Family Literacy.
*Requirements:* Applicants for the Barbara Bush Fellowship at TCALL must: be admitted on a full-time basis to Texas A&M University's main College Station campus. You are not required to reside in the Bryan-College Station area, but you must be willing to make yourself available for meetings as needed; be enrolled in a doctoral program. All qualified applicants are encouraged to apply regardless of the discipline or program of study; and submit the complete application packet to TCALL by the deadline.
*Geographic Focus:* All States
*Date(s) Application is Due:* Feb 6
*Amount of Grant:* 25,000 USD
*Contact:* Ken Appelt; (979) 845-6615; fax (979) 845-0952; kappelt@tamu.edu
*Internet:* www-tcall.tamu.edu/bbfellows/index.html
*Sponsor:* Barbara Bush Foundation for Family Literacy
1201 15th Street NW, Suite 420
Washington, D.C. 20005

**BCBSM Foundation Investigator Initiated Research Grants 772**
The program provides grants for applied research that focuses on quality, cost and appropriate access to health care in Michigan. Grants average $75,000 for one year. Multi-year grants or grants in excess of $75,000 are awarded for exemplary projects. Projects that focus on the quality and cost of health care and appropriate access are considered priority and include: Organization and delivery of health care services; Evaluation of new methods or approaches to containing health care costs; Evaluation of new methods or approaches to providing access to high quality health care; Assessment and assurance of quality care; Identification and validation of clinical protocols and evidence-based practice guidelines. Applications are accepted at any time.
*Requirements:* Applications will be accepted from medical- and doctoral-level researchers based in hospitals, university settings, non-profit health care organizations, health systems, medical or nursing schools, schools of public health, or other relevant academic disciplines such as psychology, sociology and urban studies.
*Restrictions:* This program does not support basic science or biomedical research, including drug studies or studies using animals, costs of equipment (e.g., personal computers), hardware, software or on-going operating or developmental expenses. The BCBSM Foundation does not provide support to for-profit organizations or individuals associated with organizations not located in Michigan. Blue Cross and Blue Shield of Michigan employees, members of their immediate families, and employees and immediate family members of any Blue Cross and Blue Shield of Michigan affiliate or subsidiary are not eligible to receive BCBSM Foundation grants.
*Geographic Focus:* Michigan
*Amount of Grant:* Up to 100,000 USD
*Samples:* Wei Du, Ph.D., Children's Hospital of Michigan, Department of Pediatrics, $74,994 - Diagnosing Adverse Drug Reactions in Pediatric ICU; Teresa Wehrwein, Ph.D., R.N., Michigan State University – College of Nursing, $82,290 - Preparing Nurses for Roles in Patient-Centered Medical Homes: Case Management and Quality/Safety Management; Faith Hopp, Ph.D., Hospice of Michigan, $89,646 - Evaluation of @HOMe Support Program.
*Contact:* Nora Maloy; (313) 225-8706; fax (313) 225-7730; foundation@bcbsm.com
*Internet:* www.bcbsm.com/foundation/grant.shtml
*Sponsor:* Blue Cross Blue Shield of Michigan Foundation
600 East Lafayette Boulevard
Detroit, MI 48226-2998

**BCBSM Foundation Physician Investigator Research Awards 773**
The program provides seed money to physicians to explore the merits of a particular research idea for further study. Grants are offered of up to $10,000 for projects that include pilot, feasibility or small research studies in clinical or health services research.
*Requirements:* Applicants must be a physician interested in research, who is licensed and domiciled in the state of Michigan. Applicants may include physicians working in research environments such as medical schools or university-affiliated hospitals, health care systems, or nonprofit agencies. An application may be submitted at any time.
*Geographic Focus:* Michigan

*Amount of Grant:* Up to 10,000 USD
*Contact:* Nora Maloy, Dr.P.H., Senior Program Officer; (313) 225-8706; fax (313) 225-7730; foundation@bcbsm.com
*Internet:* www.bcbsm.com/foundation/grant.shtml
*Sponsor:* Blue Cross Blue Shield of Michigan Foundation
600 East Lafayette Boulevard
Detroit, MI 48226-2998

**BCF Leonard and Helen R. Stulman Charitable Foundation Grants** **774**
The Leonard and Helen R. Stulman Charitable Foundation supports work in four areas of interest. Ninety five percent (95%) of funding from The Foundation is reserved for programs in greater Baltimore and the State of Maryland. The foundation will make a limited number of new grants each year to projects that address one or more of the following: research as to causes and treatment of mental illness and services for people with mental illness and their families; initiatives for an older adult and aging population and programs for geriatric medicine; programs engaged in health care treatment and prevention as well as those involved in medical research; and higher educational institutions. To apply to the Foundation for funding, applicants should submit a two-page letter of inquiry, with basic background on your organization, the identified needs your project proposes to address, an overview of the proposed project, and the amount you intend to request. Awards range from $25,000 to $150,000. There is a rolling deadline.
*Requirements:* Organizations (or their fiscal agents) that qualify as public charities under section 501(c)3 of the Internal Revenue Code and do not discriminate on the basis of race, creed, national origin, color, physical handicap, gender or sexual orientation are eligible to apply.
*Geographic Focus:* Maryland
*Amount of Grant:* 25,000 - 150,000 USD
*Contact:* Cathy Brill, Stulman Foundation Executive Director; (410) 332-4172 or (410) 332-4171; fax (410) 837-4701; cbrill@bcf.org or grants@bcf.org
*Internet:* www.bcf.org/For-Grant-Scholarship-Seekers/Grants/Leonard-Helen-R-Stulman-Charitable-Foundation
*Sponsor:* Baltimore Community Foundation
2 East Read Street, 9th Floor
Baltimore, MD 21202

**Beckman Coulter Foundation Grants** **775**
Beckman Coulter has a long-standing commitment to its Community Relations program. Within the program, Beckman Coulter associates strive to improve its local communities through a wide range of engagement activities that includes (but is not limited to): service projects; toy drives; blood drives; and walks, runs and bike rides. The Community Relations program also serves as a vehicle that encourages associates to work toward fulfilling the corporate vision of advancing healthcare for every person through our associate-driven fundraising activities that supports science, science education and healthcare-related research to help fight diseases. Since the Foundation is not currently accepting unsolicited applications, interested organization should begin by contacting the Foundation with a concept. If invited, grant applications are reviewed by the Foundation's staff on a regular basis. Each request is evaluated to determine its eligibility and conformity to the Beckman Coulter Foundation's giving guidelines.
*Requirements:* Organizations seeking funding should submit a letter of concept providing the following information: full legal name of the organization; brief description of the organization (including a mission statement and services provided); amount requested; and a description of how the funds will be used. Additional attachments should include: history of previous support by Beckman Coulter; description of any involvement by Beckman Coulter employees; statement as to why you consider Beckman Coulter an appropriate donor; and a list of governing board members.
*Restrictions:* The Foundation does not support the following: advertising; awards and recognition programs; beauty or talent contests; capital or building campaigns, including new construction or renovations; for-profit organizations; fraternal, labor, or veteran's organizations and activities; fund-raisers; galas, banquets, or dinners; golf tournaments; indirect costs, media productions (e.g. radio, T.V., film, webcast, publications); meetings, conferences, workshops, forums, summits and symposiums; organizations that discriminate on the basis of race, color, sex, sexual orientation, marital status, religion, age, national origin, veteran's status, or disability; private foundations; political organizations, campaigns and activities; religious organizations or groups; or sports affiliated activities.
*Geographic Focus:* All States
*Contact:* Marci Raudez, Senior Specialist, Foundation and Community Relations; (714) 961-6672 or (714) 961-6338; mfraudez@beckman.com
*Internet:* www.beckmancoulterfoundation.org/
*Sponsor:* Beckman Coulter Foundation
250 South Kraemer Boulevard, E1.SE.03
Brea, CA 92821

**Bernard G. Berkman and Nancy J. Berkman Foundation Grants** **776**
The Bernard G. Berkman and Nancy J. Berkman Foundation was established in Boston, Massachusetts, in 1986. It was founded by an investment and auction firm named Bernard G. Berkman Associates, and has a geographic focus on its home state. The Foundation's primary interest areas of support include: higher education; philanthropy; and religion. Typical funding strategies include program development, research, and research evaluation. Most recent awards have ranged from $100 to $5,000. There are no specific application deadlines or formats.
*Restrictions:* Grants are never given directly to individuals.
*Geographic Focus:* Massachusetts
*Amount of Grant:* 100 - 5,000 USD
*Samples:* Babson College Leonard Schlesinger Endowment Fund, Wellesley, Massachusetts, $5,000 - general support for education; Beth Israel Deaconess Medical Center, Boston, Massachusetts, $1,000 - general research,; Brigham Women's Hospital, Boston, Massachusetts, $1,000 - general research and evaluation.
*Contact:* Bernard G. Berkman, Trustee; (617) 566-5212 or (617) 266-6100
*Sponsor:* Bernard G. Berkman and Nancy J. Berkman Foundation
842A Beacon Street
Boston, MA 02215

**Better Way Foundation Grants** **777**
Formerly Alpha Omega Foundation, the Better Way Foundation was established in Florida in 1994. Giving is centered geographically in California, Indiana, Minnesota, Washington, and Tanzania. For the most part, the Foundation supports programs designed to provide holistic and cost-effective development opportunities to young children and families. Special emphasis is directed toward programs designed to improve early childhood outcomes. Its primary fields of interest include: Catholic agencies and churches; early childhood education; family services; health care; higher education; human services; and nutrition. Types of funding support includes: capital campaigns; general operating support; program development; research; and scholarship funding. Application forms are not required, and there are no annual deadlines. Funding amounts range up to $200,000.
*Requirements:* Unsolicited full proposals are not accepted. Organizations interested in presenting an idea for funding must submit a brief letter of inquiry.
*Geographic Focus:* California, Indiana, Minnesota, Washington, Tanzania
*Amount of Grant:* Up to 200,000 USD
*Contact:* Matthew Rauenhorst, (952) 656-4597 or (952) 656-4806; info@betterwayfoundation.org
*Sponsor:* Better Way Foundation
10350 Bren Road West
Minnetonka, MN 55343-9014

**Bikes Belong Foundation Research Grants** **778**
The Bikes Belong Foundation, launched in 2006, is a separate, complementary organization to the Bikes Belong Coalition, an organization sponsored by the U.S. bicycle industry. The foundation's focus is on bicycle-safety projects and children's bicycle programs. The foundation offers three distinct grant programs: the Paul David Clark Bicycling Safety Fund; the REI Grant Program; and Bikes Belong Research Grants. The latter focuses on two priority areas of research: projects that examine the economic impact of additional or improved bicycling facilities or bike-related events; and innovative or unique research on the benefits of bicycling (special opportunities). Applicants interested in the special opportunities research area are asked to contact the Research Analyst (see contact information on this page) before submitting an application. Examples of research that has particular interest for Bikes Belong are as follows: economic benefits of bicycling facilities; economic benefits of removing parking and adding bicycle infrastructure; economic benefits of Sunday Streets/Ciclovia events; Quantitative trails to sales studies (bike businesses only); savings for businesses that promote bicycling among employees; academic performance of children who bike to school; and state, regional, or city-level studies of the economic impact of bicycling. Bikes Belong Foundation reviews research applications twice per year. Application guidelines and downloadable applications are available at the website. Organizations should submit their completed applications along with a cover letter and any required supporting materials in one pdf file via email by the deadline date. See application guidelines for more detailed information. Deadlines may vary from year to year. Prospective applicants are encouraged to verify current year deadlines at the grant website.
*Requirements:* Applicants are limited to U.S. colleges, universities, or other institutions of higher education and non-profit research organizations.
*Restrictions:* Individuals will not be considered for funding.
*Geographic Focus:* All States
*Date(s) Application is Due:* Mar 30; Sep 28
*Amount of Grant:* 5,000 - 10,000 USD
*Samples:* Furman University, Greenville, South Carolina, $10,000 - to measure the impact of the Swamp Rabbit Trail development on communities surrounding it; University of Minnesota, Minneapolis, Minnesota, $7,500 - "Bike-Share Station Local Economic Activity Study"; Iowa Bicycle Coalition in partnership with the University of Northern Iowa, Coralville, Iowa, $6,000 - to study the statewide economic impacts of bicycling by looking at retail sales, tourism, and bicycling-related events and to use the results to guide policy development and spur investments in bicycle transportation and infrastructure.
*Contact:* Kate Powlison, Research Analyst and Communications Coordinator; (303) 449-4893 ext. 7; fax (303) 442-2936; kate@bikesbelong.org or grants@bikesbelong.org
Zoe Kircos, Grants Manager; (303) 449-4893 ext. 5; fax (303) 442-2936; zoe@bikesbelong.org or grants@bikesbelong.org
*Internet:* www.bikesbelong.org/bikes-belong-foundation/foundation-grants/research-grants
*Sponsor:* Bikes Belong Foundation
207 Canyon Boulevard Suite 202
Boulder, CO 80302

**Bill and Melinda Gates Foundation Agricultural Development Grants** **779**
The Agricultural Development Program supports projects that enable small farmers in developing countries to break the cycle of hunger and poverty, to sell what they grow or raise, increase their incomes, and make their farms more productive and sustainable. Previously funded initiatives include projects that employ a collaborative and comprehensive approach to agricultural development; provide small farmers with the supplies and support they need to succeed; address the needs of women farmers; help small farmers profit from their crops; use science and technology to develop

crops that can thrive; gather and analyze data to improve decision-making; encourage greater investment and involvement in agricultural development; and encourage policy and advocacy efforts that accelerate progress against the world's most acute poverty. Additional information on each agricultural development initiative can be found on the website. New proposals are considered, as well as expansion of existing initiatives currently funded by the Foundation.
*Requirements:* Proposals should aim to help the world's poorest people lift themselves out of hunger and poverty. The Foundation seeks proposals that: are able to produce measurable results; use preventive approaches; promise significant and long-lasting change; leverage support from other sources; and accelerate or are in accordance with work the Foundation already supports.
*Restrictions:* The majority of funding is made to organizations that are independently identified by Foundation staff. Unsolicited proposals are not accepted. Proposals must be made through 501(c)3 or other tax-exempt organizations. The Foundation is unable to make grants directly to individuals. The Foundation will not fund projects addressing health problems in developed countries; political campaigns and legislative lobbying efforts; building or capital campaigns; or projects that exclusively serve religious purposes.
*Geographic Focus:* All Countries
*Contact:* Sam Dryden, Director, Agricultural Development; (206) 709-3400 or (206) 709-3140; info@gatesfoundation.org
*Internet:* www.gatesfoundation.org/agriculturaldevelopment/Pages/default.aspx
*Sponsor:* Bill and Melinda Gates Foundation
P.O. Box 23350
Seattle, WA 98102

**Bill and Melinda Gates Foundation Financial Services for the Poor Grants** **780**
The Gates Foundation seeks to deliver reliable access to a range of safe, affordable financial tools and services to help the world's poorest households build better, healthier lives. The Foundation currently supports distribution channels, saving products, financial systems, and complementary financial systems. To ensure these services benefit the poorest populations will require new models and innovative approaches. New proposals are considered, as well as expansion of existing initiatives currently funded under the Gates Foundation.
*Requirements:* New technologies and innovative partnerships make it possible to create a "next-generation" banking system. The Foundation seeks to partner with banks, governments, mobile phone companies, retail store chains, and others to make financial services and technology accessible to billions of people throughout the world. The Foundation seeks proposals that are able to produce measurable results; use preventive approaches; promise significant and long-lasting change; leverage support from other sources; and accelerate or are in accordance with work the Foundation already supports.
*Restrictions:* The majority of funding is made to organizations that are independently identified by Foundation staff. Unsolicited proposals are not accepted. Proposals must be made through 501(c)3 or other tax-exempt organizations. The Foundation is unable to make grants directly to individuals. The Foundation will not fund projects addressing health problems in developed countries; political campaigns and legislative lobbying efforts; building or capital campaigns; or projects that exclusively serve religious purposes.
*Geographic Focus:* All States, All Countries
*Contact:* Rodger Vorhies, Director, Financial Services for the Poor; (206) 709-3140; info@gatesfoundation.org
*Internet:* www.gatesfoundation.org/financialservicesforthepoor/Pages/default.aspx
*Sponsor:* Bill and Melinda Gates Foundation
P.O. Box 23350
Seattle, WA 98102

**Bill and Melinda Gates Foundation Water, Sanitation and Hygiene Grants** **781**
Poor sanitation causes severe diarrhea, which kills 1.5 million children each year. Smart investments in sanitation can reduce disease, increase family incomes, keep girls in school, help preserve the environment, and enhance human dignity. The Foundation is looking to work with partners in an effort to expand affordable access to sanitation. Detailed information is available at the Foundation website.
*Requirements:* The Gates Foundation seeks proposals that are able to produce measurable results; use preventive approaches; promise significant and long-lasting change; leverage support from other sources; and accelerate or are in accordance with work the foundation already supports.
*Restrictions:* The majority of funding is made to organizations that are independently identified by Foundation staff. Unsolicited proposals are not accepted. Proposals must be made through 501(c)3 or other tax-exempt organizations. The Foundation is unable to make grants directly to individuals. The Gates Foundation will not fund projects addressing health problems in developed countries; political campaigns and legislative lobbying efforts; building or capital campaigns; or projects that exclusively serve religious purposes.
*Geographic Focus:* All States, All Countries
*Contact:* Kellie Sloan, Interim Director of Water, Sanitation, and Hygiene; (206) 709-3140; info@gatesfoundation.org
*Internet:* www.gatesfoundation.org/watersanitationhygiene/Pages/home.aspx
*Sponsor:* Bill and Melinda Gates Foundation
P.O. Box 23350
Seattle, WA 98102

**Biogen Foundation Fellowships** **782**
The Biogen Foundation's mission is to improve the quality of people's lives and contribute to the vitality of the communities in which Biogen operates. It is committed to sparking a passion for science and discovery, supporting innovative STEM (science, technology, engineering and math) initiatives, and strengthening efforts to make science accessible to diverse populations. The purpose of Biogen's fellowship program is to support the career development of fellows striving to become specialized in the areas of neurology or hematology. Three distinct fellowships are offered for distinct regions of the world (Europe and Canada, worldwide, and the United States), all of which are committed to improving the lives of people with Multiple Sclerosis (MS) through clinical research. Each fellowship will award up to 75,000 Euro for a single academic year in support,
*Geographic Focus:* All States, All Countries
*Amount of Grant:* Up to 75,000 EUR
*Contact:* Kathryn Bloom, Senior Director, Public Affairs; (617) 914-1299 or (866) 840-1146; fax (617) 679-2617; grantsoffice@biogen.com
*Sponsor:* Biogen Foundation
225 Binney Street
Cambridge, MA 02142

**Biogen Foundation Scientific Research Grants** **783**
Biogen supports non-accredited medical education activities that enhance and expand medical knowledge and skills for researchers. Biogen corporate philanthropic activities are closely aligned with its mission to increase disease awareness, improve patient access to care, and help patients with unmet medical needs. When evaluating a request Biogen will consider: adherence to all relevant regulations, guidelines, and policies; alignment with Biogen areas of interest; robust and quality needs assessment; measurable educational objectives; educational design that incorporates evidence-based adult learning principles and promotes continuous improvement and performance measurement for the targeted audience during the educational intervention; and innovative and cost-effective approaches in designing quality educational activities.
*Requirements:* Scientific research grants may be requested by independent accredited educational providers or organizations tax exempt under 501(c)3 of the Internal Service Revenue Code.
*Restrictions:* Biogen will not offer an Scientific Research grant with the intent of, directly or indirectly, implicitly or explicitly, influencing or encouraging the recipient to purchase, prescribe, refer, sell, arrange for the purchase or sale, or recommend any Biogen product.
*Geographic Focus:* All States
*Contact:* Kathryn Bloom, Senior Director, Public Affairs; (617) 914-1299 or (866) 840-1146; fax (617) 679-2617; grantsoffice@biogen.com
*Internet:* grantsoffice.biogen.com/scientific-programs
*Sponsor:* Biogen Foundation
225 Binney Street
Cambridge, MA 02142

**BLM National Interagency Fire Center (NIFC) Joint Fire Science Program Graduate Research Innovation (GRIN) Grants** **784**
BLM National Interagency Fire Center (NIFC) Joint Fire Science Program (JFSP) Graduate Research Innovation (GRIN) Grants award is for current master and doctoral students enrolled at colleges or universities within the United States in the field of wildland fire and related physical, biological, and social sciences, to enhance student exposure to the management and policy relevance of their research to achieve beneficial outcomes of funded work.
*Requirements:* Eligibility to apply for these funds is unrestricted.
*Geographic Focus* All States, District of Columbia, Guam, Marshall Islands, Northern Mariana Islands, Puerto Rico, U.S. Virgin Islands, American Samoa
*Date(s) Application is Due:* May 16
*Amount of Grant:* 25,000 - 500,000 USD
*Contact:* Becky Jenison, Program Coordinator; (208) 387-5958; bjenison@blm.gov
*Internet:* https://www.grants.gov/web/grants/search-grants.html
*Sponsor:* U.S. Department of the Interior
1849 C Street NW, Room 5665
Washington, D.C. 20240

**Blum-Kovler Foundation Grants** **785**
The Blum-Kovler Foundation was established in 1985 after Everett Kovler retired from his position as President of James Beam Distilling Company. The Foundation awards general operating grants to eligible nonprofit organizations in its areas of interest, including social services, Jewish welfare funds, higher education, health services and medical research, and cultural programs. The foundation also supports youth- and child-welfare agencies and public-interest and civic-affairs groups. Grants are awarded primarily in the Chicago metropolitan area and in the Washington, D.C. area, though funding is also provided nationally. There are no application forms. Applicants should submit a one to two page written proposal with a copy of their IRS determination letter by mid-November to considered for the current year. Typical grant awarded is between $1,000-$5,000, though a few are as much as $100,000.
*Requirements:* Illinois and District of Columbia nonprofit organizations are eligible.
*Geographic Focus:* All States, District of Columbia, Illinois
*Date(s) Application is Due:* Nov 15
*Amount of Grant:* 1,000 - 1,000,000 USD
*Samples:* Children's Hospital Foundation, Silver Spring, Maryland, $32,500 - general operating support; East Village Youth Program, Chicago, Illinois, $5,000 - general operating support; Illinois Holocaust Museum, Skokie, Illinois, $2,500 - general operating support,
*Contact:* Jonathan Kovler, President; (312) 664-5050
Peter Kovler, Chairperson and Vice President; (312) 664-5050
*Sponsor:* Blum-Kovler Foundation
875 North Michigan Avenue, Suite 3400
Chicago, IL 60611-1958

**Blum Family Foundation Grants** 786

The Blum Family Foundation (formerly known as the Samuel Blum Foundation) was established in Baltimore, Maryland, with an expressed interest in supporting: Higher education; elementary and secondary education; human services; and Judaism. Types of support typically include: general operating support; presentations and productions; program development; research and evaluation; and publications. Most recent awards have ranged from $50,000 to as much as $150,000. The geographic focus is Maryland, New York, and Washington, D.C. An application form is required, and can be secured by contacting the Foundation office. There are no specified annual deadlines.

*Requirements:* 501(c)3 organizations supporting the residents of the states of Maryland and New York, as well as the Washington, D.C. area, are eligible to apply.

*Geographic Focus:* District of Columbia, Maryland, New York

*Amount of Grant:* 50,000 - 150,000 USD

*Samples:* American Associates Ben-Gurion University, Bethesda, Maryland, $50,000 - general operating support; American Friends of Hebrew University, Washington, D.C., $75,000 - general operating support; American Technion Society, New York, New York, $102,000 - general operating support.

*Contact:* Marc P. Blum, President; (410) 685-4606

*Sponsor:* Blum Family Foundation

233 E. Redwood Street, Garrett Building, Suite 100

Baltimore, MD 21202-3332

**BMW of North America Charitable Contributions** 787

BMW of North America funds charitable programs that benefit society in the areas of education, road-traffic safety, and the environment. The corporation supports education at all levels and specifically focuses on the following: intercultural learning for K-12 students and their teachers; automotive technology, mechanics, and career and repair programs in high schools, technical schools, and community colleges; and research in the areas of safety design, ergonomics, and new materials. In the area of road traffic safety, the corporation supports driver-education programs geared at teenagers and new drivers; basic auto-maintenance programs for women; consumer education on general road-safety issues; and programs to promote the safety of children and young people on the road. In the area of the environment, BMW is committed to sustainable development and focuses grant making on the following: conservation/preservation of natural resources, in particular park lands and waterways; research and promotion of alternative fuels; and environmental education for K-12 students. In general, grants are awarded for specific projects rather than for general operating support, although some operating and capital grants are given consideration. Interested organizations may download application instructions and guidelines at the grant website. Organizations wishing to be considered for a grant must submit an application; telephone solicitations will not considered.

*Requirements:* 501(c)3 charities or 501(c)9 organizations are eligible to apply.

*Restrictions:* The corporate giving program does not support non-tax-exempt organizations; individuals; religious organizations for religious purposes; political candidates or lobbying organizations; organizations with a limited constituency, such as fraternal, labor, or veterans groups; travel by groups or individuals; national or local chapters of disease-specific organizations; national conferences, sports events, and other one-time, short-term events; sponsorships or advertising; anti-business groups; team sponsorships or athletic scholarships; or organizations outside the United States or its territories.

*Geographic Focus:* All States

*Contact:* Grants Coordinator; (201) 307-4000; fax (201) 307 3607

*Internet:* www.bmwgroupna.com/philanthropy.htm

*Sponsor:* BMW of North America

300 Chestnut Ridge Road

Woodcliff Lake, NJ 07677-7731

**Bob Herrington Student Travel Grant** 788

The Gopher Tortoise Council was formed in 1978 by a group of southeastern biologists and other citizens concerned with the decline of the gopher tortoise. Bob Herrington, a long-time member of the Gopher Tortoise Council, who served as chair of the GTC's J. Larry Landers Student Research Award Committee for more than 20 years, passed away in April 2016. In recognition of Bob's contributions to the Council and his dedication to 30 years of teaching and mentoring students at Georgia Southwestern State University, the student travel award was renamed in his honor. This award provides financial assistance to students wishing to present their research at the Annual Gopher Tortoise Council meeting and is funded through GTC membership fees.

*Requirements:* Students must reside greater than 60 miles from the meeting venue to qualify. Recipients will be selected randomly from among those who submit applications with abstract submissions. Awards will be presented at the meeting.

*Geographic Focus:* All States

*Date(s) Application is Due:* Sep 1

*Amount of Grant:* Up to 100 USD

*Contact:* Keri Landry, klandry@wlf.la.gov

*Internet:* www.gophertortoisecouncil.org/who/grants.php

*Sponsor:* Gopher Tortoise Council

303 Main Street #241

Safety Harbor, FL 34695

**Bodenwein Public Benevolent Foundation Grants** 789

The Bodenwein Public Benevolent Foundation was established in 1938 under the will of Theodore Bodenwein, owner and publisher of The Day newspaper, to support and promote quality educational, cultural, human-services, and health-care programming for underserved populations. The Foundation specifically serves the people of Greater New London County where The Day has a substantial circulation. The majority of grants from the Bodenwein Public Benevolent Foundation are one year in duration; on occasion, multi-year support is awarded. Applicants must apply online at the grant website. Applicants are strongly encouraged to do the following before applying: review the downloadable state application procedures for additional helpful information, requirements, and restrictions; review the downloadable online-application guidelines at the grant website; review the foundation's funding history (link is available from the grant website); review the online application questions in advance; and review the list of required attachments. These will generally include: a list of board members, financial statements (audited, reviewed, or compiled by independent auditor); an organization summary; a list of other funding sources; an IRS Determination letter; and other required documents. All attachments must be uploaded in the online application as PDF, Word, or Excel files. The annual application deadline is 11:59 p.m. on November 15. Applicants will be notified of grant decisions by letter within three to four months after the proposal deadline.

*Requirements:* Nonprofit organizations serving Greater New London County (East Lyme, Groton, Ledyard, Lyme, Montville, Mystic, New London, North Stonington, Old Lyme, Salem, Stonington, and Waterford, Connecticut) are eligible to apply. A breakdown of number/percentage of people served by specific towns will be required in the online application.

*Restrictions:* The foundation does not support requests from individuals, organizations attempting to influence policy through direct lobbying, or any political campaigns.

*Geographic Focus:* Connecticut

*Date(s) Application is Due:* Nov 15

*Amount of Grant:* Up to 5,000 USD

*Samples:* New London Homeless Hospitality Center, New London, Connecticut, $5,000 - guest financial support; Center for Hospice Care Southeast Connecticut, Norwich, Connecticut, $3,000 - support of the Special Care Fund; Waterford Country School, Quaker Hill, Connecticut, $2,000 - support of special education field trips.

*Contact:* Amy R. Lynch; (860) 244-4870; amy.r.lynch@ustrust.com

*Internet:* https://www.bankofamerica.com/philanthropic/foundation.go?fnId=50

*Sponsor:* Bodenwein Public Benevolent Foundation

200 Glastonbury Boulevard, Suite # 200

Glastonbury, CT 06033-4056

**Bodman Foundation Grants** 790

The Bodman Foundation was established by George M. Bodman and his wife Louise Clarke Bodman in 1945. George was born in Toledo, Ohio, in 1882 and died in 1950. Mrs. Bodman was born in Chicago in 1893 and died in 1955. The Bodmans lived for much of their lives in Red Bank, New Jersey, and in New York City, where George Bodman was a senior partner at the investment banking firm of Cyrus J. Lawrence and Sons. The Bodmans were generous supporters of numerous cultural, civic, and service organizations. During World War I Mr. Bodman headed the Intelligence Service of the War Trade Board. During World War II, he served as executive assistant to the Red Cross Commissioner for Great Britain and was regional director in charge of American Red Cross Club operations in England, Scotland, and Ireland. The Bodman Foundation's Certificate of Incorporation states that its funds are to be used for the aid, support or benefit of religious, educational, charitable, and benevolent objects and purposes for the moral, ethical and physical well-being and progress of mankind." The Bodman Foundation shares trustees, staff, office space, and even a website with the Achelis Foundation which has a similar mission and geographic area of concentration (both foundations give in New York City, while the Bodman Foundation also gives in New Jersey). Funding is concentrated in six program areas: arts and culture, education, employment, health, public policy, and youth and families. Most recently, awards have ranged from $15,000 to $200,000.

*Requirements:* 501(c)3 organizations based in New York City that fall within the foundation's areas of interest are welcome to submit an inquiry or proposal letter by regular mail (initial inquiries by email or fax are not accepted, nor are CDs, DVDs, computer discs, or video tapes). An initial inquiry to the foundation should include only the following items: a proposal letter that briefly summarizes the history of the project, need, objectives, time period, key staff, project budget, and evaluation plan; the applicant's latest annual report and complete set of audited financial statements; and the applicant's IRS 501(c)3 tax-exemption letter. Applications may be submitted at any time during the year. Each request is reviewed by staff and will usually receive a written response within thirty days. Those requests deemed consistent with the interests and resources of the foundation will be evaluated further and more information will be requested. Foundation staff may request a site visit, conference call, or meeting. All grants are reviewed and approved by the Trustees at one of their three board meetings in May, September, or December.

*Restrictions:* The foundation generally does not make grants for the following purposes or program areas: nonprofit organizations outside of New York; annual appeals, dinner functions, and fundraising events; endowments and capital campaigns; loans and deficit financing; direct grants to individuals; individual day-care and after-school programs; housing; organizations or projects based outside the U.S; films or video projects; small art, dance, music, and theater groups; individual K-12 schools (except charter schools); national health and mental health organizations; and government agencies or nonprofit organizations significantly funded or reimbursed by government agencies. Limited resources prevent the foundations from funding the same organization on an ongoing annual basis.

*Geographic Focus:* New Jersey, New York

*Amount of Grant:* 15,000 - 200,000 USD

*Samples:* Chamber Music Society of Lincoln Center, New York, New York, $75,000 - program support; Scholarship Fund for Inner-City Children, Newark, New Jersey, $50,000 - support the Newark Program of the Children's Scholarship Fund; Rockefeller University, New York, New York, $200,000 - support the Bio-Imaging Resource Center.

*Contact:* John B. Krieger, Executive Director; (212) 644-0322; fax (212) 759-6510; main@achelis-bodman-fnds.org
*Internet:* www.achelis-bodman-fnds.org/guidelines.html
*Sponsor:* Bodman Foundation
767 Third Avenue, 4th Floor
New York, NY 10017-2023

**Boeing Company Contributions Grants** 791

The Boeing U.S. contributions program welcomes applications in five focus areas: education; health and human services; arts and culture; civic; and the environment. Primary fields of interest include: arts; elementary and secondary education; the environment; family services, prevention of domestic violence; health care; public affairs; public safety; substance abuse programs; and general human services. The largest single block of charitable contributions goes toward supporting programs and projects related to education. Boeing also looks for innovative initiatives that promote the economic well-being of the community and neighborhood revitalization. Boeing invests in programs that promote participation in arts and cultural activities and experiences, programs that increase public understanding of and engagement in the processes and issues that affect communities and programs that protect and conserve the natural environment. Boeing accepts applications for cash grants, in-kind donations, and services.
*Requirements:* To apply for support you must be a U.S. based IRS 501(c)3 qualified charitable or educational organization or an accredited K-12 educational institution. U.S. grant guidelines and applications are available online.
*Restrictions:* Grants do not support: an individual person or families; adoption services; political candidates or organizations; religious activities, in whole or in part, for the purpose of further religious doctrine; memorials and endowments; travel expenses; nonprofit and school sponsored walk-a-thons, athletic events and athletic group sponsorships other than Special Olympics; door prizes or raffles; U.S. hospitals and medical research; school-affiliated orchestras, bands, choirs, trips, athletic teams, drama groups, yearbooks and class parties; general operating expenses for programs within the United States; organizations that do not follow our application procedures; follow-on applications from past grantees that have not met our reporting requirements or satisfactorily completed the terms of past grants; fundraising events, annual funds, galas and other special-event fundraising activities; advertising, t-shirts, giveaways and promotional items; documentary films, books, etc; debt reduction; dissertations and student research projects; loans, scholarships, fellowships and grants to individuals; for-profit businesses; gifts, honoraria, gratuities; capital improvements to rental properties.
*Geographic Focus:* Alabama, Arizona, California, Colorado, District of Columbia, Florida, Georgia, Hawaii, Illinois, Kansas, Maryland, Missouri, Nevada, New Mexico, Ohio, Oklahoma, Oregon, Pennsylvania, South Carolina, Texas, Utah, Washington, Australia, Canada
*Contact:* Antoinette Bailey, (312) 544-2000; fax (312) 544 - 2082
*Internet:* www.boeing.com/companyoffices/aboutus/community/charitable.htm
*Sponsor:* Boeing Company Contributions
100 North Riverside
Chicago, IL 60606-1596

**BOEM Alaska Coastal Marine Institute Cooperative Agreement** 792

The Environmental Studies Program (ESP) of the BOEM is offering a cooperative agreement opportunity through the University of Alaska Coastal Marine Institute (CMI) to units within the University of Alaska system that have the ability to conduct research in topics that serve the public interest of safe and environmentally sound energy production and meet the goals of the BOEM. The general types of studies that may be funded include: scientific studies for better understanding marine, coastal or human environments affected or potentially affected by offshore oil and gas or other mineral exploration and extraction on the outer continental shelf (OCS); modeling studies of environmental, social, economic, or cultural processes related to OCS oil and gas activities in order to improve scientific predictive capabilities; experimental scientific studies for better understanding of environmental processes, or the causes and effects of OCS activities; projects which design or establish mechanisms or protocols for sharing data or information regarding marine or coastal resources, or human activities to support prudent management of oil, gas and marine mineral resources; and synthesis studies of scientific environmental or socioeconomic information relevant to the OCS oil and gas program. The most relevant geographic areas are potential OCS oil and gas lease areas off Alaska, including the Beaufort Sea, the Chukchi Sea, Cook Inlet, and the northern Gulf of Alaska. Following the submission of letters of intent (LOI) from the University of Alaska, the BOEM, in discussion with representatives from the State of Alaska, have selected the LOI topics listed below for development of full proposals. This request is not a guarantee of funding. Proposals that do not target these selected research areas will not be considered: CMI Administration 2019-2023; Western Beaufort and Chukchi Sea surface current analysis; Evaluating novel assessment approaches for coastal ice seal haulout areas and behavior in the Alaskan Beaufort Sea; Tracking fish and marine mammal responses to Arctic change: a community-based biological observing program in Arctic Alaska; Kelp Restoration in the Boulder Patch; Changing relationships among climate variables and cumulative climate stress on the Gulf of Alaska ecosystem; Utilization of the under-ice habitat by Arctic Cod in the western Arctic Ocean: a multidisciplinary collaborative study; and Investigating marine movement and behavior of steelhead trout in and near OCS oil and gas lease areas. Awards will range from $50,000 to $400,000. The annual application deadline is April 25.
*Requirements:* Public and State controlled institutions of higher education are eligible.
*Geographic Focus:* Alaska
*Date(s) Application is Due:* Apr 25
*Amount of Grant:* 50,000 - 400,000 USD
*Contact:* Christy Tardiff, Program Officer; (703) 787-1367; Christy.Tardiff@bsee.gov
*Internet:* https://www.grants.gov/web/grants/search-grants.html
*Sponsor:* U.S. Department of the Interior
1849 C Street, NW
Washington, D.C. 20240

**BOEM Environmental Studies Cooperative Agreement** 793

The Pacific Outer Continental Shelf (OCS) Region of the Bureau of Ocean Energy Management (BOEM) is offering a cooperative agreement for the study of Understanding Biological Connectivity Among Offshore Structures and Natural Reefs to test hypotheses on biological connectivity among artificial and natural habitats using genetic markers. This award will utilize genetic analyses to test hypotheses of connectivity and understand how connectivity may influence non-indigenous species (NIS). The total anticipated amount of available funding is approximately $800,000 subject to the availability of funds. The anticipated period of performance will be from fiscal year (FY) 2019 through FY 2024. BOEM anticipates making one cooperative agreement award for one proposal submitted under this program announcement. Work performance under these awards must start before October 1, and the proposed work must be completed no later than 48 months from the start date. This announcement is specifically to announce intent to undertake the following project(s): Understanding Biological Connectivity Among Offshore Structures and Natural Reefs. This involves substantial involvement by BOEM scientists in various aspects of study development and/or study conduct. The application deadline is May 7.
*Requirements:* Public and State controlled institutions of higher education are eligible.
*Geographic Focus:* All States
*Date(s) Application is Due:* May 7
*Amount of Grant:* 350,000 - 800,000 USD
*Contact:* Dominique Bruce, Program Officer; (703) 787-1342; Dominique.Bruce-Morton@bsee.gov
*Internet:* https://www.grants.gov/web/grants/search-grants.html
*Sponsor:* U.S. Department of the Interior
1849 C Street, NW
Washington, D.C. 20240

**Boyd and Evelyn Mullen Foundation Grants** 794

Evelyn Mullen and her husband Boyd owned a car dealership in Pasadena, Texas. Throughout their lives, the Mullens were involved in their community and strove to meet its changing needs. Evelyn dreamed of going to college to study nursing, but while she did not have an opportunity to do so, she became a strong advocate for education during her lifetime. The Boyd and Evelyn Mullen Foundation was established to support charitable organizations in Texas, with a priority for the Houston area, endeavoring to advance education, literary studies, science, and religion, thereby improving the welfare of those in need. Specific areas of funding interest include: education via college scholarships; disaster relief; enriching the lives of the blind and visually impaired; meeting the needs of homeless men, women and children; and medical research with a focus on nutrition. These areas of focus may be adjusted from time to time to address the most pressing challenges facing residents of Houston and surrounding areas. Current priority funding areas include: higher education; literary studies and humanities; disaster preparedness and relief; human services; medical research; science research; and religion. Grants are made to support primarily operating and program or project needs, but requests for one-time capital improvements may also be considered. Grants typically range between $5,000 and $30,000. The annual deadline for application submissions is July 31.
*Requirements:* 501(c)3 organizations serving the residents of Texas are eligible to apply. Priority is given to the Houston metro area.
*Restrictions:* Grants for individuals or for endowment campaigns are not considered. Only one request will be accepted from an organization during a twelve-month period.
*Geographic Focus:* Texas
*Date(s) Application is Due:* Jul 31
*Amount of Grant:* 5,000 - 30,000 USD
*Contact:* Debra Goldstein Phares, Philanthropic Client Director; (214) 209-1830 or (800) 357-7094; debra.g.phares@ustrust.com or tx.philanthropic@ustrust.com
*Internet:* https://www.bankofamerica.com/philanthropic/foundation.go?fnId=175
*Sponsor:* Boyd and Evelyn Mullen Foundation
901 Main Street, 19th Floor
Dallas, TX 75202-3714

**Boyle Foundation Grants** 795

The Boyle Foundation was established in Massachusetts in 1990, and works diligently to support its primary fields of interest, including: animal welfare and wildlife; basic and emergency aid; biodiversity; diseases and conditions; elementary and secondary education; health and health care; higher education; human services; international development; out-patient medical care;p science research; and shelter and residential care. Most often, funding comes in the form of general operating support and program development. An application form is required, though interested parties should begin by forwarding a letter of interest to the office. That letter should include a brief history of the organization and description of its mission, along with a detailed description of the project and the amount of funding requested. Most recently, grant awards have ranged from as little as $750 to a maximum of $10,000. There are no specified annual application submission deadlines.
*Requirements:* Applicants must either be located in, or support the residents of, the states of Massachusetts, New Hampshire, New York, Maine, Vermont, Maryland, Georgia, Minnesota, Arizona, and Oregon.

*Geographic Focus:* Arizona, Georgia, Maine, Maryland, Massachusetts, Minnesota, New Hampshire, New York, Oregon, Vermont
*Amount of Grant:* 750 - 10,000 USD
*Contact:* Robert W Scharar, Trustee; (508) 349-7955
*Sponsor:* Boyle Foundation
P.O. Box 786
Truro, MA 02666-0786

**Brach Family Foundation Grants** 796
Established in New York in 1991, the Brach Family Foundation offers funding primarily in Brooklyn, New York, to Jewish agencies, temples, and schools. Types of support include research and general operations. Since an application form is not required, applicants should submit the name, address, and phone number of the organization, as well as and argument for financial need. There are no annual deadlines, and grant amounts range from $250 to $10,000.
*Geographic Focus:* New York
*Contact:* Zigmond Brach; (718) 236-8000 or (800) 444-5671; fax (718) 236-2400
*Sponsor:* Brach Family Foundation
1600 63rd Street
Brooklyn, NY 11204-2713

**Brain and Behavior Research Foundation Colvin Prize for Outstanding Achievement in Mood Disorders Research** 797
The Colvin Prize for Outstanding Achievement in Mood Disorders Research, is an award of $50,000 given to an outstanding scientist carrying out work on the causes, pathophysiology, treatment, or prevention of affective disorders. The scientist to be recognized is one who gives particular promise for advancing the understanding of affective illness or its basic brain mechanisms that will lead to new treatment approaches. The Prize is given at the annual New York City Awards Dinner held each October. Nominations are due by June 1.
*Geographic Focus:* All States
*Date(s) Application is Due:* Jun 1
*Amount of Grant:* 50,000 USD
*Contact:* Sho Tin Chen, Grants Manager; (800) 829-8289 or (646) 681-4888; schen@bbrfoundation.org or grants@bbrfoundation.org
*Internet:* https://bbrfoundation.org/annual-prizes#Bipolar
*Sponsor:* Brain and Behavior Research Foundation
90 Park Avenue, 16th Floor
New York, NY 10016

**Brain and Behavior Research Foundation Goldman-Rakic Prize for Cognitive Neuroscience** 798
The Goldman-Rakic Prize for Cognitive Neuroscience Research is given in recognition of a research scientist who has made distinguished contributions to the understanding of cognitive neuroscience. The Prize includes $40,000 and an honorary lecture at Yale University to honor the memory and accomplishments of Patricia Goldman-Rakic. The Prize is specifically for excellence in neurobiological research at the cellular, physiological, or behavioral levels that may lead to a greater understanding of underlying psychiatric or neurological disease. The Prize is given at the annual New York City Awards Dinner held each October. Nominations for the Award are due by June 1.
*Geographic Focus:* All States
*Date(s) Application is Due:* Jun 1
*Amount of Grant:* 40,000 USD
*Contact:* Sho Tin Chen, Grants Manager; (800) 829-8289 or (646) 681-4888; schen@bbrfoundation.org or grants@bbrfoundation.org
*Internet:* https://bbrfoundation.org/annual-prizes#Goldman
*Sponsor:* Brain and Behavior Research Foundation
90 Park Avenue, 16th Floor
New York, NY 10016

**Brain and Behavior Research Foundation Lieber Prize for Outstanding Achievement in Schizophrenic Research** 799
The Lieber Prize for Schizophrenia Research is given in recognition of a research scientist who has made distinguished contributions to the understanding of schizophrenia. The $50,000 cash rewards past achievement and provides further incentive for an outstanding working scientist to continue to do exceptional research into the causes, prevention, and treatment of schizophrenia. The prize is bestowed at the annual New York City Awards Dinner held each October. The presentation of the prize helps build public understanding of the importance of schizophrenia research, and acts as a goal towards which researchers might strive. Nominations are due by June 1. The Lieber Prize winner will select the winner of the the Sidney R. Baer, Jr. Prize for Schizophrenia Research, worth $40,000. The winner is typically a NARSAD Young Investigator at the Lieber Prize winner's institution, conducting work in schizophrenia research.
*Geographic Focus:* All States
*Date(s) Application is Due:* Jun 1
*Amount of Grant:* 50,000 USD
*Contact:* Sho Tin Chen, Grants Manager; (516) 829-0091 or (800) 829-8289; fax (516) 487-6930; schen@bbrfoundation.org or info@bbrfoundation.org
*Internet:* https://bbrfoundation.org/annual-prizes#Lieber
*Sponsor:* Brain and Behavior Research Foundation
90 Park Avenue, 16th Floor
New York, NY 10016

**Brain and Behavior Research Foundation Maltz Prize for Outstanding Achievement in Schizophrenia Research** 800
Established in 2005, the prize was renamed in 2016 in honor of Board Members Milton and Tamar Maltz. The Maltz Prize provides $40,000 to an investigator who has undertaken innovative and promising research in schizophrenia. Winners of this prize are selected by the Lieber Prize recipient(s) of the same year. The prize is bestowed at the annual New York City Awards Dinner held each October. The presentation of the prize is intended to stimulate the career of the investigator and further the field of schizophrenia research. Nominations for the Award are due by June 1.
*Geographic Focus:* All States
*Date(s) Application is Due:* Jun 1
*Amount of Grant:* 40,000 USD
*Contact:* Sho Tin Chen, Grants Manager; (800) 829-8289 or (646) 681-4888; schen@bbrfoundation.org or grants@bbrfoundation.org
*Internet:* https://bbrfoundation.org/annual-prizes#Maltz
*Sponsor:* Brain and Behavior Research Foundation
90 Park Avenue, 16th Floor
New York, NY 10016

**Brain and Behavior Research Foundation NARSAD Distinguished Investigator Grants** 801
The Brain and Behavior Research Foundation (formerly National Alliance for Research on Schizophrenia and Depression, or NARSAD) distributes funds for psychiatric brain and behavior disorder research. NARSAD Distinguished Investigator Grants provide support for experienced investigators (full professor or equivalent) conducting neurobiological and behavioral research. Areas of particular interest include patient populations with unique or unusual characteristics and central nervous system developments. A one-year grant of up to $100,000 is provided for established scientists pursuing innovative projects in diverse areas of neurobiological research.
*Requirements:* Applicants must be a full professor (or equivalent), and maintain peer reviewed competitively funded scientific programs. Guidelines and application deadlines are at the Foundation's website.
*Restrictions:* Previous grantees may not receive the grant for a second time until five years have elapsed since the beginning date of the prior award. Grant requirements from all previous awards must be met. Only one applicant may apply per application. Applicants may submit only one application per cycle.
*Geographic Focus:* All States
*Date(s) Application is Due:* Jul 15
*Amount of Grant:* Up to 100,000 USD
*Contact:* Sho Tin Chen, Grants Manager; (800) 829-8289 or (646) 681-4888; schen@bbrfoundation.org or grants@bbrfoundation.org
*Internet:* https://bbrfoundation.org/di
*Sponsor:* Brain and Behavior Research Foundation
90 Park Avenue, 16th Floor
New York, NY 10016

**Brain and Behavior Research Foundation NARSAD Independent Investigator Grants** 802
The Brain and Behavior Research Foundation (formerly National Alliance for Research on Schizophrenia and Depression, or NARSAD) distributes funds for psychiatric brain and behavior disorder research. The NARSAD Independent Investigator Grants provide support for investigators during the critical period between the initiation of research and the receipt of sustained funding. Basic and/or clinical investigators are supported, but research must be relevant to schizophrenia, major affective disorders, or other serious mental illnesses. A two-year grant up to $100,000, or $50,000 per year, is provided to scientists at the associate professor level or equivalent, who are clearly independent and have won national competitive support as a principal investigator.
*Requirements:* Applicants must have a doctoral-level degree, be an associate professor (or equivalent), and have received, as Principal Investigator, competitive research support at a national level. Independent Investigators are eligible for a maximum of two grants. Previous awardees at the associate professor level are encouraged to apply for a second grant. Applications must be submitted through the online application system. Additional guidelines can be found at the Foundation's website.
*Restrictions:* The total budget request may not exceed $100,000. Awards cannot overlap in time, all prior grant requirements (including NARSAD Young Investigator Grants) must be met, and an applicant cannot submit more than one application per cycle. Investigators that have been promoted to associate professor, but have currently active NARSAD Young Investigator Grants, are not eligible.
*Geographic Focus:* All States
*Date(s) Application is Due:* Dec 15
*Amount of Grant:* Up to 50,000 USD
*Contact:* Sho Tin Chen, Grants Manager; (800) 829-8289 or (646) 681-4888; schen@bbrfoundation.org or grants@bbrfoundation.org
*Internet:* https://bbrfoundation.org/narsad-independent-investigator-grants
*Sponsor:* Brain and Behavior Research Foundation
90 Park Avenue, 16th Floor
New York, NY 10016

**Brain and Behavior Research Foundation NARSAD Young Investigator Grants** 803
The Brain and Behavior Research Foundation (formerly National Alliance for Research on Schizophrenia and Depression, or NARSAD) distributes funds for psychiatric brain and behavior disorder research. The Foundation's NARSAD Young Investigator Grants offer up to $35,000 a year for up to two years to enable promising investigators to either

extend their research fellowship training or to begin careers as independent research faculty. Grants are intended to facilitate innovative research opportunities and support basic, as well as translational and/or clinical investigators. Research must be relevant to the understanding, treatment and prevention of serious psychiatric disorders such as schizophrenia, bipolar, mood and anxiety disorders, or child and adolescent psychiatric disorders. Awardees become eligible for the Freedman Prize for Outstanding Research, with a cash prize of $1,000. They are also eligible for the Klerman Prize for Outstanding Clinical Research, also with a cash prize of $1,000. Young Investigator winners also become eligible for the Baer Prize for Schizophrenia Research, worth $40,000.
*Requirements:* Applicants must have a doctoral level degree and already be employed in research training, or be in a faculty or independent research position. Applicants must have an on-site mentor or senior collaborator who is an established investigator in areas relevant to psychiatric disorders. Applications must be submitted online from the Foundation's website.
*Restrictions:* Pre-doctoral students or investigators at the rank of associate professor or equivalent are not eligible. Institutional overhead is not allowable.
*Geographic Focus:* All States
*Date(s) Application is Due:* Feb 15
*Amount of Grant:* Up to 70,000 USD
*Contact:* Sho Tin Chen, Grants Manager; (800) 829-8289 or (646) 681-4888; schen@bbrfoundation.org or grants@bbrfoundation.org
*Internet:* https://bbrfoundation.org/yi
*Sponsor:* Brain and Behavior Research Foundation
90 Park Avenue, 16th Floor
New York, NY 10016

**Brain and Behavior Research Foundation Ruane Prize for Child and Adolescent Psychiatric Research** 804
The Ruane Prize is an award of $50,000 given to an outstanding scientist carrying out research on the causes, pathophysiology, treatment, or prevention of severe mental illness in children. The scientist to be recognized is one who gives particular promise for advancing the understanding of psychotic, affective or other severe brain and behavior disorders having their onset in childhood or adolescence. Contributions may be for clinical research or relevant basic science. The Prize is given at the annual New York City Awards Dinner held each October. Nominations are due by June 1.
*Geographic Focus:* All States
*Date(s) Application is Due:* Jun 1
*Amount of Grant:* 50,000 USD
*Contact:* Sho Tin Chen, Grants Manager; (800) 829-8289 or (646) 681-4888; schen@bbrfoundation.org or grants@bbrfoundation.org
*Internet:* https://bbrfoundation.org/annual-prizes#Ruane
*Sponsor:* Brain and Behavior Research Foundation
90 Park Avenue, 16th Floor
New York, NY 10016

**Brainerd Foundation Conservation Capacity Grants** 805
The Brainerd Foundation is dedicated to protecting the environmental quality of the Pacific Northwest, including Washington, Oregon, Idaho, Montana, Alaska, and British Columbia. Specifically, the Conservation Capacity program focuses on three initiatives that build upon the foundation's historic investments; inspire the next generation of conservation philanthropists; strengthen the capacity of emerging conservation leaders and activists; and support a culture of innovation within conservation advocacy organizations. Grants are made in the following areas: endangered ecosystems—conservation biology, conservation assessment, and mining reform and roadless areas; and communications and capacity building—organizational development, and allied voices. These grants are awarded to cover costs associated with activities such as public education and grassroots outreach, media strategies, litigation, scientific and economic studies, computer networking, and building organizational capacity. Additional types of support include: general operating support; continuing support; equipment acquisition; conferences and seminars; seed money; research; technical assistance; and employee matching gifts. Typical conservation capacity grants range from $10,000 to $50,000 and can be awarded for multiple years. Inquiries are accepted on an ongoing basis. Grants are awarded at one of three board meetings held each year. Proposals are accepted by invitation only. Applications are available online.
*Requirements:* Nonprofit organizations in the Pacific Northwest are eligible to apply.
*Restrictions:* The Foundation does not favor proposals for school education programs, land acquisition, endowments, capital campaigns, projects sponsored by government agencies, basic research, fellowships, or books or videos that are not part of a broader strategy.
*Geographic Focus:* Alaska, Idaho, Montana, Oregon, Washington, Canada
*Amount of Grant:* 10,000 - 50,000 USD
*Contact:* Hattie Cambridge; (206) 448-0676; fax (206) 448-7222; hattiec@brainerd.org
*Internet:* www.brainerd.org/funding/conservation-capacity.php
*Sponsor:* Brainerd Foundation
1601 Second Avenue, Suite 610
Seattle, WA 98101-1541

**Brian G. Dyson Foundation Grants** 806
The Foundation, established in Atlanta in 1994 by former Coca-Cola Bottling executive, Brian G. Dyson, offers funding to community foundations, higher education, and federated programs in the Atlanta region and throughout Georgia. Grants typically range up to $20,000, and funding supports general operating costs. There are no specific guidelines, application formats, or deadlines with which to adhere, and initial contact should be made in writing.
*Requirements:* Applicants must be colleges, public or private schools, or other non-profit organizations located within the state of Georgia.
*Geographic Focus:* Georgia
*Amount of Grant:* Up to 20,000 USD
*Contact:* Brian G. Dyson, Director; (404) 364-2940
*Sponsor:* Brian G. Dyson Foundation
3060 Peachtree Road NW, Suite 1465
Atlanta, GA 30305-2241

**Brinson Foundation Grants** 807
The foundation supports education, public health, and scientific research programs that engage, inform, and inspire committed citizens to confront the challenges that face humanity. Grantmaking priorities are education—awareness and outreach, democracy and citizenship, economically disadvantaged, and libraries and literacy; public health—awareness and outreach, and economically disadvantaged; and scientific research—astrophysics, cosmology, geophysics, medical research (i.e., Alzheimer's disease, cancer, Lou Gehrig's Disease (ALS), and stroke). Types of support include general operating grants and project grants. The foundation does not accept unsolicited grant applications. Grantseekers are asked to review the foundation's mission, vision, beliefs, priorities (accessed from the Who We Are link), and guidelines. If a grantseeker believes the request would match one or more of the foundation's grantmaking priorities, they can make an inquiry by completing the online Grantseeker Information Form. The completed form should be emailed to the office. Further application is by invitation.
*Requirements:* The foundation will consider inviting grant applications from organizations: whose request matches one or more of the Foundation's grantmaking priorities; located in the United States of America that are exempt from tax under Section 501(c)3 of the Internal Revenue Code and are defined as charitable organizations as described in Section 509(a)1, 2 or 3 or 170(b)1A; located outside the United States of America provided they produce a written legal opinion stating that they are a charitable equivalency to a qualifying U.S. organization and/or a written affidavit containing sufficient information for the Foundation to make a reasonable judgment that the organization is charitable. The Foundation's education and public health grants are generally made to organizations that serve individuals and communities in the greater Chicago area. It also considers leading U.S.-based programs that reach broader populations across the U.S. and internationally or have the potential to have a meaningful impact on best practices at the national or international level. The Foundation's physical science research grants are made to leading organizations across the United States. In this priority area, the location of the program is less critical than the match with the Foundation's grantmaking priorities.
*Restrictions:* The Foundation will not consider grant inquiries from organizations that: discriminate on the basis of race, gender, religion, ethnicity or sexual orientation. The Foundation will not consider grant inquiries that request funding for: activities that attempt to influence public elections; voter registration; political activity; lobbying efforts; promotion of a specific religious faith; medical research involving human cloning. The Foundation discourages grant inquiries requesting funds for: capital improvements; endowments; fundraising events.
*Geographic Focus:* All States
*Samples:* Adler Planetarium and Astronomy Museum, Chicago, Illinois, $80,000 - for cosmology and astrophysics research; Lincoln Park Zoo, Chicago, Illinois, $55,000 - for general support; Institute for Humane Studies, Arlington, Virginia, $20,000 - summer seminars for college students.
*Contact:* Cheryl A. Heads; (312) 799-4500; fax (312) 799-4310; mail@brinsonfoundation.org
*Internet:* www.brinsonfoundation.org/grant_seekers/grantseekers.shtml
*Sponsor:* Brinson Foundation
737 N Michigan Avenue, Suite 1850
Chicago, IL 60611

**Bristol-Myers Squibb / Meade Johnson Award for Distinguished Achievement in Nutrition Research** 808
The Award for Distinguished Achievement in Nutrition Research was first presented in 1978. The Award includes a $50,000 cash prize and a silver commemorative medallion. The winners are selected by an independent peer-review selection committee whose members are grant administrators of current Bristol-Myers Squibb/Mead Johnson Unrestricted Nutrition Research Grants. The annual application deadline is in late April.
*Geographic Focus:* All States
*Date(s) Application is Due:* Apr 24
*Amount of Grant:* 50,000 USD
*Contact:* John L. Damonti; (212) 546-4000 or (800) 332-2056; fax (212) 546-9574
*Internet:* www.bms.com/aboutbms/grants/data/nutrit.html
*Sponsor:* Bristol-Myers Squibb Foundation
345 Park Avenue, Suite 4364
New York, NY 10154

**Bristol-Myers Squibb Clinical Outcomes and Research Grants** 809
Bristol-Myers Squibb's mission is to extend and enhance human life. To help achieve that mission, the Company has established programs to support Investigator Sponsored Trials (ISTs). ISTs must be medically appropriate and scientifically valid. While the Company will consider requests for clinical research trials in all clinical and therapeutic areas, it currently gives priority to proposals in the following therapeutic areas: Cardiovascular/Metabolics, Infectious Diseases, Neuroscience, Oncology, Immunology, and Virology. Bristol-Myers Squibb maintains a strict policy of not exercising any influence or control over the design of any investigator initiated clinical research trial supported by BMS.

*Requirements:* Individuals in the following settings are eligible for support: private practice, hospitals, community health centers, cooperative groups, physician networks, and academic medical centers and universities.
*Geographic Focus:* All States
*Contact:* Amit Duggal, (212) 546-4000; fax (212) 546-9574; amit.duggal@bms.com
*Internet:* www.bms.com/responsibility/building_our_communities/Pages/default.aspx
*Sponsor:* Bristol-Myers Squibb Company
777 Scudders Mill Road
Plainsboro, NJ 08536

**Bristol-Myers Squibb Foundation Fellowships** 810

The Program provides support for fellows to gain experience in epidemiological and clinical research as it relates to the care of individuals infected with HIV/AIDS and/or HBV. Goals of the Program include: support of studies that will further strengthen the science and knowledge of HIV/AIDS and HBV; development of a foundation for future prospective and retrospective studies; to provide a forum to share research findings; and support of the development of future clinical researchers. Up to 18 fellows will be selected annually, with grant awards up to $20,000 to support research-related expenses for a one-year research period. Applications must be submitted online.
*Requirements:* Candidates must: be an active Fellow in good standing in an ACGME-accredited Fellows training program; be a senior Fellow not in the first years of the fellowship; desire to enhance knowledge and skill development in the area of HIV/AIDS, HBV, or Oncology research; and identify a faculty member to serve as the project mentor.
*Restrictions:* Total grant amount is inclusive of indirect costs and associated IRB fees and is not permitted for use towards travel to conferences or for durable equipment.
*Geographic Focus:* All States
*Date(s) Application is Due:* Feb 26
*Amount of Grant:* Up to 20,000 USD
*Contact:* John L. Damonti; (212) 546-4000 or (800) 332-2056; fax (212) 546-9574
*Internet:* www.bms.com/responsibility/grantsandgiving/Corporate_Giving/Pages/fellowships_scholarships_awards.aspx
*Sponsor:* Bristol-Myers Squibb Foundation
345 Park Avenue, Suite 4364
New York, NY 10154

**Bristol-Myers Squibb Foundation Global HIV/AIDS Initiative Grants** 811

The intent of this initiative is to develop new models in awareness, in medical care, in community development and in prevention and treatment in poor and resource-limited areas of the world, where the need for all such efforts is greatest. Health care infrastructures must be developed and enhanced, stigmatization must be overcome, health care worker capacity must be built and preserved and local people must be empowered to generate and sustain local solutions to this global problem.
*Requirements:* The Bristol-Myers Squibb Foundation considers requests for support only from tax-exempt organizations that satisfy the requirements of section 501(c)3 of the U.S. Internal Revenue Code.
*Restrictions:* The Foundation does not award funds to: individuals; political, fraternal, social or veterans' organizations; religious or sectarian organizations unless engaged in a significant project benefiting the entire community; organizations receiving support through United Way or other federated campaigns; endowments; courtesy advertising; or conferences/special events/videos.
*Geographic Focus:* All States
*Contact:* John L. Damonti; (212) 546-4000 or (800) 332-2056; fax (212) 546-9574
*Internet:* www.bms.com/responsibility/grantsandgiving/Pages/default.aspx
*Sponsor:* Bristol-Myers Squibb Foundation
345 Park Avenue, Suite 4364
New York, NY 10154

**Bristol-Myers Squibb Foundation Health Disparities Grants** 812

One mission of the Bristol-Myers Squibb Foundation is to reduce health disparities by strengthening community-based health care worker capacity, integrating medical care and community-based supportive services, and mobilizing communities to fight disease. To this end, this Foundation attempts to address health disparities in four strategic disease areas representing major public health burdens and in four highly affected geographies: hepatitis in Asia, HIV/AIDS in Africa, serious mental illness in the U.S., and cancer in Europe. Additional areas of concern include: metabolic diseases, infectious diseases; rheumatoid arthritis; cardiovascular diseases; substance abuse; women's health issues; and overall health care giving.
*Requirements:* Nonprofit organizations in communities where Bristol-Myers Squibb maintains a facility should submit their requests for company contributions directly to that location. Contact persons are listed at the company website.
*Restrictions:* The foundation does not support individuals; conferences, special events, or videos; political, fraternal, social, or veterans organizations; religious or sectarian activities, unless they benefit the entire community; organizations funded through federated campaigns; endowments; or courtesy advertising.
*Geographic Focus:* Connecticut, Indiana, Massachusetts, New Jersey, New York
*Contact:* John Damonti; (212) 546-4000 or (800) 332-2056; fax (212) 546-9574
*Internet:* www.bms.com/foundation/reducing_health_disparities/Pages/default.aspx
*Sponsor:* Bristol-Myers Squibb Foundation
345 Park Avenue, Suite 4364
New York, NY 10154

**Bristol-Myers Squibb Foundation Health Education Grants** 813

The Program supports the development and replication of novel approaches aimed at individuals in the community to learn more about their health and wellbeing so that they can become more informed decision makers about health care and they can actively participate in disease prevention and management. In many of the grants, the Foundation seeks to be a catalyzing force for changes in government or social policies, supporting programs that can demonstrate the value of new approaches to health care through educational efforts. Nearly all the education programs focus on taking on major diseases, like hepatitis B or cancer and creating model programs for either preventing them or helping treat aspects of them in community settings.
*Requirements:* The Bristol-Myers Squibb Foundation considers requests for support only from tax-exempt organizations that satisfy the requirements of section 501(c)3 of the U.S. Internal Revenue Code.
*Restrictions:* The Foundation does not award funds to: individuals; political, fraternal, social or veterans' organizations; religious or sectarian organizations unless engaged in a significant project benefiting the entire community; organizations receiving support through United Way or other federated campaigns; endowments; courtesy advertising; or conferences/special events/videos.
*Geographic Focus:* All States
*Contact:* John L. Damonti; (212) 546-4000 or (800) 332-2056; fax (212) 546-9574
*Internet:* www.bms.com/responsibility/grantsandgiving/Pages/default.aspx
*Sponsor:* Bristol-Myers Squibb Foundation
345 Park Avenue, Suite 4364
New York, NY 10154

**Bristol-Myers Squibb Foundation Women's Health Grants** 814

Bristol-Myers Squibb Women's Health grants support projects that enhance women's health with strategies that improve education, prevention, diagnosis, treatment and access to care for women worldwide. Support has been given to projects that test innovative outreach programs, cultivate multi-sectoral partnerships and add new information to the existing body of knowledge to help define and achieve improved health for women around the world. The goal of the program is to generate initiatives that will help enhance women's health through novel interdisciplinary strategies that improve education, prevention, diagnosis, treatment and access to care for women worldwide. Since its inception, significant resources have been invested in programs that educate women about diseases and conditions that particularly threaten them as women.
*Requirements:* The Bristol-Myers Squibb Foundation considers requests for support only from tax-exempt organizations that satisfy the requirements of section 501(c)3 of the U.S. Internal Revenue Code.
*Restrictions:* The Foundation does not award funds to: individuals; political, fraternal, social or veterans' organizations; religious or sectarian organizations unless engaged in a significant project benefiting the entire community; organizations receiving support through United Way or other federated campaigns; endowments; courtesy advertising; or conferences/special events/videos.
*Geographic Focus:* All States
*Contact:* John L. Damonti; (212) 546-4000 or (800) 332-2056; fax (212) 546-9574
*Internet:* www.bms.com/Documents/foundation/women2002.pdf
*Sponsor:* Bristol-Myers Squibb Foundation
345 Park Avenue, Suite 4364
New York, NY 10154

**Broadcasting Board of Governors Internships** 815

Broadcasting Board of Governors Internships provide real world experience in a diverse and dynamic international broadcast organization. Typical areas available include: public affairs; engineering; and the Office of Research and Assessment. These educational experiences, which are predominantly unpaid, last between six and twelve weeks. In the area of public affairs, duties include cooperating and communicating with key strategic audiences, event planning, and promoting different Broadcasting Board of Governors projects, creating content for BBG web properties and drafting press releases, multimedia production. In the field of engineering, duties are tailored to the skill and interest level of the intern, including technical (engineering design or computer programming), analytical (data base research), or business support (international market research) projects (nb. TSI manages transmission, program delivery, and technical support). And in the Office of Research and Assessment, interns may data analysis in SPSS and producing charts, graphs, and analytics memos or presentations, reviewing qualitative transcripts and composing analytical reports on the findings, performing data checks on datasets, assisting in instrument design for qualitative and quantitative studies, performing desk research on media environment issues, perform market research.
*Requirements:* Applicants must have current enrollment at an accredited college or university, though exceptions are occasionally made for outstanding high-school students.
*Restrictions:* International students are not eligible to apply.
*Geographic Focus:* All States
*Contact:* Cheryl Peters, Director, Office of Contracts; (202) 382-7866 or (202) 203-4000; fax (202) 382-7877; cpeters@bbg.gov or publicaffairs@bbg.gov
*Internet:* https://www.bbg.gov/careers/bbg-intern-and-student-programs/
*Sponsor:* Broadcasting Board of Governors
330 Independence Avenue, SW
Washington, D.C. 20237

**Brookdale Foundation Leadership in Aging Fellowships** 816

The Leadership in Aging Fellowships provide two years of support to junior academics to focus on a project that will help establish them in an area of aging research. Fellowships are open to a broad range of disciplines including, but not limited to, medical, biological

and basic sciences, nursing, social sciences, and the arts and humanities. The Fellowship is paid to the candidate's sponsoring institution in support of the candidate's research project. The Fellowship amount of up to $125,000 each year is intended to cover 75% of the fellow's time, base salary and fringe benefits. The award could also be used to include the support of a graduate assistant if necessary as long as the total amount does not exceed $125,000. Additional information is available at the website. The annual deadline for applications is November 1.
*Requirements:* Fellowships are open to all professionals in the field of aging. Candidates must meet the following criteria: leadership potential; ongoing commitment to a career in aging; a mentor (or mentors); and a willingness to commit at least 75% of his or her time for career development during each of the two years of the Fellowship.
*Geographic Focus:* All States
*Date(s) Application is Due:* Nov 1
*Amount of Grant:* Up to 125,000 USD
*Contact:* Valerie Hall, Grants Coordinator; (201) 836-4602; fax (201) 836-4382; vah@brookdalefoundation.org
*Internet:* www.brookdalefoundation.net/Leadership/Fellows/fellows.html
*Sponsor:* Brookdale Foundation
300 Frank W. Burr Boulevard, Suite 13
Teaneck, NJ 07666

**Brown Advisory Charitable Foundation Grants** 817
Established in Maryland in 2009, the Brown Advisory Charitable Foundation offers grant funding primarily throughout the State of Maryland. The Foundation has identified the following fields of interest: the arts; primary and secondary education; higher education; human services; and federated giving programs. Typically, funding takes the form of either general operating support or program development. Most recently, grants have ranged from $100 to $40,000. A formal application is required, and can be secured by contacting the Foundation office. There are no annual application submission deadlines.
*Requirements:* Applicants must be 501(c)3 organizations located in, or serving the residents of, the State of Maryland.
*Geographic Focus:* Maryland
*Amount of Grant:* 100 - 40,000 USD
*Samples:* CollegeTracks, Bethesda, Maryland, $3,000 - general operating support; Maryland Zoo in Baltimore, Baltimore, Maryland, $12,500 - general operating support; Baltimore School for the Arts, Baltimore, Maryland, $40,000 - general operating support.
*Contact:* Irene Alisa Stesch, Secretary; (410) 537-5503 or (410) 537-5400
*Internet:* www.brownadvisory.com/community/tabid/79/Default.aspx
*Sponsor:* Brown Advisory Charitable Foundation
901 Bond Street, Suite 400
Baltimore, MD 21231-3340

**Bupa Foundation Healthy Futures Grants** 818
The Bupa Foundation is an independent medical research charity that funds medical research to prevent, relieve and cure sickness and ill health. The Foundation aims to produce long-term benefits that have the potential to improve the health of individuals and populations worldwide. The Foundation funds practical projects that tackle specific challenges in health and social care, and make a direct impact on people's health and well being. Its funding programs are developed around clearly defined themes, with specific priorities and criteria outlined for each theme when a program is announced. Before an applicant begin preparing a proposal, it is suggested that he/she read the funding priorities and criteria for specific funding programs carefully to make sure that the work fits the priorities and criteria outlined. In the area of Healthy Futures, the Foundation focuses on supporting and empowering young adults aged 18 to 25 living with ongoing health challenges. Young adults are particularly vulnerable as they leave secondary education and move towards becoming independent. For young adults with health concerns this period can be even more difficult. Poor transitions, changing support networks and gaps in provision, along with new responsibilities all have the potential to lead to poorer outcomes and experiences. Over £380,000 has been awarded in grants within the Healthy Futures funding program, helping a variety of organizations to deliver a wide range of projects that will make an impact on the health and well being of young adults living with health challenges, empowering them to live life to the fullest.
*Restrictions:* There are a number of areas which are not eligible for funding, including: work that does not fall within the scope of the Foundation's charitable purposes; work that is not clearly aligned with the stated funding priorities of or criteria for a specific funding program; projects delivered outside the United Kingdom; long-term projects and initiatives; work being delivered by local authorities or housing associations; work which might reasonably be eligible for funding from statutory bodies; general awareness and information campaigns; sponsorship of or attendance at meetings, events and conferences; fundraising appeals, including requests for contributions to capital or equipment costs; unrestricted funding for charities or other organizations; educational bursaries or grants for university or postgraduate education, school trips or projects, gap year or elective year projects; and academic research, including funding for educational or research posts.
*Geographic Focus:* United Kingdom
*Contact:* Lucy Delaney, Foundation Coordinator; 020 7656 2591; lucy.delaney@bupa.com or bupaukfoundation@bupa.com
*Internet:* https://www.bupaukfoundation.org/page/programmehealthy-futures/
*Sponsor:* Bupa Foundation
Battle Bridge House
London, GLONDON WC1X 8DU United Kingdom

**Bupa Foundation Mid-Life Mental Health Grants** 819
The Bupa Foundation is an independent medical research charity that funds medical research to prevent, relieve and cure sickness and ill health. The Foundation aims to produce long-term benefits that have the potential to improve the health of individuals and populations worldwide. The Foundation funds practical projects that tackle specific challenges in health and social care, and make a direct impact on people's health and well being. Its funding programs are developed around clearly defined themes, with specific priorities and criteria outlined for each theme when a program is announced. Before an applicant begin preparing a proposal, it is suggested that he/she read the funding priorities and criteria for specific funding programs carefully to make sure that the work fits the priorities and criteria outlined. In the area of Mid-Life Mental Health, the Foundation has awarded a total of £830,000 in grants to 24 projects across the UK, from Cornwall to Newcastle. These projects will support and improve people's mental health in many different ways by: piloting new interventions and services; supporting people in crisis; breaking down barriers and taboos; developing skills and employment opportunities; and creating communities for mutual support.
*Requirements:* All projects should develop, apply or test research-based approaches in community settings - with clear intended impacts on public health. The areas for funding as detailed above are not intended to be exclusive of each other. Health professionals working for public or private organizations may apply for research grants for UK-based projects.
*Restrictions:* There are a number of areas which are not eligible for funding, including: work that does not fall within the scope of the Foundation's charitable purposes; work that is not clearly aligned with the stated funding priorities of or criteria for a specific funding program; projects delivered outside the United Kingdom; long-term projects and initiatives; work being delivered by local authorities or housing associations; work which might reasonably be eligible for funding from statutory bodies; general awareness and information campaigns; sponsorship of or attendance at meetings, events and conferences; fundraising appeals, including requests for contributions to capital or equipment costs; unrestricted funding for charities or other organizations; educational bursaries or grants for university or postgraduate education, school trips or projects, gap year or elective year projects; and academic research, including funding for educational or research posts.
*Geographic Focus:* All States, All Countries
*Date(s) Application is Due:* Nov 30
*Contact:* Lucy Delaney, Foundation Coordinator; 020 7656 2591; lucy.delaney@bupa.com or bupaukfoundation@bupa.com
*Internet:* https://www.bupaukfoundation.org/page/programmemental-health/
*Sponsor:* Bupa Foundation
Battle Bridge House
London, GLONDON WC1X 8DU United Kingdom

**Bushrod H. Campbell and Adah F. Hall Charity Fund Grants** 820
Established in Massachusetts in 1956, the Bushrod H. Campbell and Adah F. Hall Charity Fund awards grants to organizations in the Boston area devoted to basic needs for the elderly, projects relating to medicine and medical research, health care, hospitals, the blind and deaf, and certain discretionary projects. Grants are also awarded countrywide for projects addressing population control. Types of support include: capital grants; general operating grants; program grants' and research grants. There are four quarterly deadlines each year for online application submission: January 1, April 1, August 1, and October 1.
*Requirements:* Tax-exempt organizations located within Boston and neighboring communities and U.S. tax-exempt organizations devoted to population control are eligible.
*Restrictions:* Grants are not awarded to individuals. The Fund does not typically support direct medical care or public policy advocacy.
*Geographic Focus:* Massachusetts
*Date(s) Application is Due:* Jan 1; Apr 1; Aug 1; Oct 1
*Amount of Grant:* 3,000 - 7,500 USD
*Samples:* Strong Women, Strong Girls, Pittsburgh, Pennsylvania, $7,000 - general operating support; Boston Health Care Homeless Program, Boston, Massachusetts, $10,000 - general operating support; Adolescent Consultation Services, Cambridge, Massachusetts, $5,000 - general operating support.
*Contact:* Rita Goldberg, Grants Manager; (617) 557-9766 or (617) 557-9766; fax (617) 227-7940, ext. 775; rgoldberg@hembar.com
Gioia C. Perugini; (617) 557-9777 or (617) 557-9766; gperugini@hembar.com
*Internet:* https://hembar.com/client_services/bushrod-h-campbell-and-adah-f-hall-charity-fund
*Sponsor:* Bushrod H. Campbell and Adah F. Hall Charity Fund
75 State Street, 16th Floor
Boston, MA 02109

**BWF Ad Hoc Grants** 821
To complement the competitive award programs, the Burroughs Wellcome Fund makes modest grants on an ad hoc basis to nonprofit organizations conducting activities intended to improve the general environment for science. These noncompetitive grants are for activities closely related to the Foundation's focus areas. The Foundation places special priority on working with non-profit organizations, including government agencies, to leverage financial support for targeted areas of research, and on encouraging other foundations to support biomedical research. Proposals should be brief and include a description of the focus of the activity, the expected outcomes, qualifications of the organization or individuals involved; provide certification of the sponsor's Internal Revenue Service tax-exempt status; and give the total budget for the activity, including any financial support obtained or promised. Proposals are given careful preliminary review, and those deemed appropriate are presented for consideration by BWF's board of directors.
*Geographic Focus:* All States, Canada

*Contact:* Rolly L. Simpson, Jr., Senior Program Officer; (919) 991-5100; fax (919) 991-5160; info@bwfund.org or rsimpson@bwfund.org
*Internet:* www.bwfund.org/grant-programs/ad-hoc-grants
*Sponsor:* Burroughs Wellcome Fund
21 TW Alexander Drive, P.O. Box 13901
Research Triangle Park, NC 27709-3901

**BWF Career Awards at the Scientific Interface** 822
BWF's Career Awards at the Scientific Interface (CASI) provide $500,000 over five years to bridge advanced postdoctoral training and the first three years of faculty service. These awards are intended to foster the early career development of researchers who have transitioned or are transitioning from undergraduate and/or graduate work in the physical/mathematical/computational sciences or engineering into postdoctoral work in the biological sciences, and who are dedicated to pursuing a career in academic research. Scientific advances such as genomics, quantitative structural biology, imaging techniques, and modeling of complex systems have created opportunities for exciting research careers at the interface between the physical/computational sciences and the biological sciences. Tackling key problems in biology will require scientists trained in areas such as chemistry, physics, applied mathematics, computer science, and engineering. These awards are open to U.S. and Canadian citizens or permanent residents as well as to U.S. temporary residents. Candidates are expected to draw from their training in a scientific field other than biology to propose innovative approaches to answer important questions in the biological sciences. Examples of approaches include, but are not limited to, physical measurement of biological phenomena, computer simulation of complex processes in physiological systems, mathematical modeling of self-organizing behavior, building probabilistic tools for medical diagnosis, developing novel imaging tools or biosensors, developing or applying nanotechnology to manipulate cellular systems, predicting cellular responses to topological clues and mechanical forces, and developing a new conceptual understanding of the complexity of living organisms. Proposals that include experimental validation of theoretical models are particularly encouraged.
*Requirements:* Citizens and non-citizen permanent residents of the U.S. and Canada are eligible. Candidates must: hold a Ph.D. degree in one of the fields of mathematics, physics, chemistry, computer science, statistics, or engineering (this includes related areas of physical, mathematical, computational, theoretical, and engineering science); demonstrate that their work is truly interdisciplinary; have completed at least 12 months but not more than 48 months of postdoctoral research by the date of the full invited application deadline; be committed to a full-time career in research as an independent investigator at a North American degree-granting institution; have at least one first-author publication; including papers on which first authorship is shared; and be based at a non-profit institution in the U.S. or Canada.
*Restrictions:* Candidates cannot hold nor have accepted, either in writing or verbally, a faculty appointment as a tenure-track assistant professor at the time of application - both pre-proposal and full application. Candidates must not hold nor have accepted a K99 award from the U.S. National Institutes of Health. Temporary residents of Canada are not eligible.
*Geographic Focus:* All States, Canada
*Date(s) Application is Due:* Jan 8
*Amount of Grant:* Up to 500,000 USD
*Contact:* Rusty Kelley, Ph.D., Senior Program Officer; (919) 991-5120 or (919) 991-5100; fax (919) 991-5160; rkelley@bwfund.org or info@bwfund.org
Tiffania Taylor, Senior Program Associate; (919) 991-5116 or (919) 991-5100; fax (919) 991-5160; ttaylor@bwfund.org or info@bwfund.org
*Internet:* www.bwfund.org/grant-programs/interfaces-science/career-awards-scientific-interface
*Sponsor:* Burroughs Wellcome Fund
21 TW Alexander Drive, P.O. Box 13901
Research Triangle Park, NC 27709-3901

**BWF Career Awards for Medical Scientists (CAMS)** 823
The Career Awards for Medical Scientists (CAMS) is a highly competitive program that provides $700,000 awards over five years for physician-scientists, who are committed to an academic career, to bridge advanced postdoctoral/fellowship training and the early years of faculty service. Awards are made to degree granting institutions in the U.S. or Canada on behalf of the awardee. The postdoctoral/fellowship portion of the award may last up to two years, and allows the awardee the use of up to $95,000 per year ($65,000 per year salary support). BWF will make up to two additional awards to clinically trained psychiatrists who focus on research at the interface between neuroscience and psychiatry. These proposals must clearly demonstrate evidence of integration of neuroscience and psychiatry in project design.
*Requirements:* The ideal candidate will be two years away from becoming an independent investigator, have at least two years or more of postdoctoral research experience, and have a significant publication record. Candidates must be nominated by their dean or department chair at the degree-granting institution where they will conduct the postdoctoral/fellowship training under the award. Applications must be approved by an official responsible for sponsored programs (generally from the grants office, office of research, or office of sponsored programs) at the degree-granting institution. Candidates must hold an M.D., D.D.S., or D.V.M. degree. Proposals must be in the area of basic biomedical, disease-oriented, or translational research. Proposals in health services research or involving large-scale clinical trials are ineligible. Candidates must be a fellow, resident, or a postdoctoral researcher and have at least two years of postdoctoral research experience at the time of application.
*Restrictions:* Proposals in health services research or involving large-scale clinical trials are ineligible.
*Geographic Focus:* All States, Canada
*Date(s) Application is Due:* Oct 1
*Amount of Grant:* 700,000 USD
*Contact:* Rolly L. Simpson, Jr., Senior Program Officer; (919) 991-5100; fax (919) 991-5160; info@bwfund.org or rsimpson@bwfund.org
Debra Holmes, Senior Program Associate; (919) 991-5134 or (919) 991-5100; fax (919) 991-5160; dholmes@bwfund.orf or info@bwfund.org
*Internet:* www.bwfund.org/grant-programs/biomedical-sciences/career-awards-medical-scientists-2
*Sponsor:* Burroughs Wellcome Fund
21 TW Alexander Drive, P.O. Box 13901
Research Triangle Park, NC 27709-3901

**BWF Collaborative Research Travel Grants** 824
The Collaborative Research Travel Grant (CRTG) program provides up to $15,000 in support for researchers from degree-granting institutions to travel either domestically or internationally to a laboratory to acquire a new research technique or to facilitate a collaboration. To capitalize on what appears to be an opportunity to provide relatively unrestricted travel funds to academic scientists and trainees and to provide a stimulus for those working or contemplating working at the interface of science, this program provides travel grants that can be used both internationally and domestically to acquire new research techniques, and to promote collaborations.
*Requirements:* Postdoctoral fellows or faculty at degree-granting institutions in the U.S. or Canada are eligible to apply. Applicants must hold a Ph.D. in mathematics, physics, chemistry, computer science, statistics, or engineering at the time of application; and interested in investigating research opportunities in the biological sciences. Biologists holding a doctorate degree at the time of application who are interested in working with physical scientists, mathematicians, engineers, chemists, statisticians, or computer scientists to incorporate their ideas and approaches to answering biological questions are eligible to apply. Applicants must be citizens or permanent residents of the U.S. or Canada at the time of application.
*Restrictions:* Grants cannot be used for travel to domestic or international conferences, to cover salary support or purchase computers or equipment. Laboratory or lecture courses are not eligible for this grant. No indirect costs or overhead charges can be applied to this grant. This grant is not bi-directional or reciprocal for the collaborator. Public health, health policy, or epidemiology proposals will not be accepted. This grant is not open to researchers at NIH institutions. Previous travel grant recipients are not eligible.
*Geographic Focus:* All States, Canada
*Date(s) Application is Due:* Feb 3
*Amount of Grant:* Up to 15,000 USD
*Contact:* Rolly L. Simpson, Jr., Senior Program Officer; (919) 991-5110 or (919) 991-5100; fax (919) 991-5160; rsimpson@bwfund.org or info@bwfund.org
Debra Holmes, Senior Program Associate; (919) 991-5134 or (919) 991-5100; fax (919) 991-5160; dholmes@bwfund.orf or info@bwfund.org
*Internet:* www.bwfund.org/grant-programs/biomedical-sciences/collaborative-research-travel-grants
*Sponsor:* Burroughs Wellcome Fund
21 TW Alexander Drive, P.O. Box 13901
Research Triangle Park, NC 27709-3901

**BWF Innovation in Regulatory Science Grants** 825
BWF's Innovation in Regulatory Science Awards provide up to $500,000 over five years to academic investigators developing new methodologies or innovative approaches in regulatory science that will ultimately inform the regulatory decisions the Food and Drug Administration (FDA) and others make. These awards are open to U.S. and Canadian citizens or permanent residents who have a faculty or adjunct faculty appointment at a North American degree-granting institution. Awards are made to degree-granting institutions in the U.S. or Canada on behalf of the awardee. The application process consists of two phases: a preproposal followed by a full proposal invitation. Preproposal applicants selected by the Advisory Committee deemed to meet the goals of this initiative will be invited to submit full proposals.
*Requirements:* Candidates must: hold an M.D., Ph.D., D.V.M., D.D.S., or M.D.-Ph.D. degree; be based at a non-profit institution [501(c)3 or equivalent] in the U.S. or Canada; hold a faculty position at an accredited, degree-granting institution in the United States or Canada; be an investigator at the adjunct, assistant, associate, or full professor level; be a citizen or permanent resident of the U.S. or Canada at the time of application.
*Restrictions:* Ineligible applicants include: persons who have applied for permanent resident status but have not received their government documentation by the time of the preproposal application deadline; temporary residents of the U.S. or Canada; postdoctoral students; and candidates who currently hold a BWF award.
*Geographic Focus:* All States, Canada
*Date(s) Application is Due:* Apr 1
*Amount of Grant:* 500,000 USD
*Contact:* Rusty Kelley, Senior Program Officer; (919) 991-5120 or (919) 991-5100; fax (919) 991-5160; rkelley@bwfund.org or info@bwfund.org
Tiffanie Taylor, Senior Program Associate; (919) 991-5116 or (919) 991-5100; fax (919) 991-5160; ttaylor@bwfund.ord or info@bwfund.org
*Internet:* www.bwfund.org/grant-programs/regulatory-science/innovation-regulatory-science
*Sponsor:* Burroughs Wellcome Fund
21 TW Alexander Drive, P.O. Box 13901
Research Triangle Park, NC 27709-3901

**BWF Institutional Program Unifying Population and Laboratory Based Sciences Grants** 826
The Institutional Program Unifying Population and Laboratory Based Sciences program offers five-year institutional training awards of $500,000 per year to bridge the gap between the population and computational sciences and the laboratory-based biological sciences. The award will support the training of researchers between existing concentrations of

research strength in population approaches to human health and in basic biological sciences. The goal is to establish training programs by partnering researchers working in schools of medicine and schools (or academic divisions) of public health. The programs supported by these awards will develop young researchers who will be equally at home with the ideas, approaches, and insights generated at the molecular scale and at the population scale. Proposals that cross institutional boundaries are encouraged.
*Requirements:* Degree-granting institutions in the U.S. or Canada may submit applications. Proposals must be driven by core components within medical and public health schools, but beyond those required components, departments or centers located within non-medical parts of a university, existing inter-institutional collaboratives, research museums, free-standing research institutes, and other non-profit institutions that provide advanced-level training are all acceptable as potential additional partners. Dental, osteopathic, and veterinary medical schools are appropriate applicants.
*Restrictions:* For-profit companies may not participate in the application, but could be valuable partners in such training programs. Research groups working at national laboratories and within the federal government are allowable as partners.
*Geographic Focus:* All States, Canada
*Date(s) Application is Due:* May 19
*Amount of Grant:* 500,000 USD
*Contact:* Victoria McGovern, Ph.D., Senior Program Officer; (919) 991-5112 or (919) 991-5100; fax (919) 991-5160; vmcgovern@bwfund.org or info@bwfund.org
Jean Kramarik, Senior Program Associate; (919) 991-5122 or (919) 991-5100; fax (919) 991-5160; jkramarik@bwfund.org or info@bwfund.org
*Internet:* www.bwfund.org/grant-programs/population-sciences/institutional-program-unifying-population-and-laboratory-based
*Sponsor:* Burroughs Wellcome Fund
21 TW Alexander Drive, P.O. Box 13901
Research Triangle Park, NC 27709-3901

**BWF Investigators in the Pathogenesis of Infectious Disease Awards** **827**
The Investigators in the Pathogenesis of Infectious Disease Award provides $500,000 over a period of five years to support accomplished investigators at the assistant professor level to study pathogenesis, with a focus on the interplay between human and microbial biology, shedding light on how human and microbial systems are affected by their encounters. The awards are intended to give recipients the freedom and flexibility to pursue new avenues of inquiry and higher-risk research projects that hold potential for significantly advancing the biochemical, pharmacological, immunological, and molecular biological understanding of how microbes and the human body interact. The goal of the program is to provide opportunities for accomplished investigators still early in their careers to study what happens at the points where human and microbial systems connect. The program supports research that sheds light on the fundamentals that affect the outcomes of this encounter: how colonization, infection, commensalism and other relationships play out at levels ranging from molecular interactions to systemic ones.
*Requirements:* Candidates must: generally have an M.D., D.V.M., or Ph.D. degree; have an established record of independent research and hold a tenure-track position as an assistant professor or equivalent (at the time of application) at a degree-granting institution; and be nominated by accredited, degree-granting institutions in the United States or Canada.
*Restrictions:* Applications from non-tenure track investigators at tenure-offering, degree-granting institutions will not be accepted.
*Geographic Focus:* All States, Canada
*Date(s) Application is Due:* Nov 3
*Amount of Grant:* 500,000 USD
*Contact:* Victoria McGovern, Ph.D., Senior Program Officer; (919) 991-5112 or (919) 991-5100; fax (919) 991-5160; vmcgovern@bwfund.org or info@bwfund.org
Jean Kramarik, Senior Program Associate; (919) 991-5122 or (919) 991-5100; fax (919) 991-5160; jkramarik@bwfund.org or info@bwfund.org
*Internet:* www.bwfund.org/grant-programs/infectious-diseases-0
*Sponsor:* Burroughs Wellcome Fund
21 TW Alexander Drive, P.O. Box 13901
Research Triangle Park, NC 27709-3901

**BWF Preterm Birth Initiative Grants** **828**
The Preterm Birth Initiative was created to increase the understanding of the biological mechanisms underlying parturition and spontaneous preterm birth and will provide up to $600,000 over a four-year period ($150,000 per year). The initiative is designed to stimulate both creative individual scientists and multi-investigator teams to approach the problem of preterm birth using creative basic and translation science methods. Postdoctoral fellows nearing their transition to independent investigator status through senior established investigators are encouraged to apply. Molecular and computational approaches such genetics/genomics, immunology, microbiology, evolutionary biology, mathematics, engineering, and other basic sciences hold enormous potential for new insights independently or in conjunction with more traditional areas of parturition research such as maternal fetal medicine, obstetrics, and pediatrics. The formation of new connections between reproductive scientists and investigators who are involved in other areas will give preterm birth research a fresh and unique look, and stimulate a new work face to tackle this challenge. Proposals can be submitted by individual investigators or research teams designating a contact principal investigator. Proposals that cross institutional boundaries (partnerships between multiple universities or collaborations within larger universities) are welcomed.
*Requirements:* Proposals must be submitted from degree-granting institutions in the United States or Canada. The principal investigator must be a postdoctoral fellow in the final 1-2 years of postdoctoral training or hold faculty appointment at a degree-granting institution in the U.S. or Canada. The principal investigator must be a citizen or permanent resident of the U.S. or Canada.
*Restrictions:* Prior preterm birth full research grant recipients are not eligible to reapply.
*Geographic Focus:* All States, Canada
*Date(s) Application is Due:* Dec 1
*Amount of Grant:* Up to 600,000 USD
*Contact:* Rolly L. Simpson, Jr., Senior Program Officer; (919) 991-5110 or (919) 991-5100; fax (919) 991-5160; rsimpson@bwfund.org or info@bwfund.org
Debra Holmes, Senior Program Associate; (919) 991-5134 or (919) 991-5100; fax (919) 991-5160; dholmes@bwfund.orf or info@bwfund.org
*Internet:* www.bwfund.org/grant-programs/reproductive-sciences/preterm-birth-initiative
*Sponsor:* Burroughs Wellcome Fund
21 TW Alexander Drive, P.O. Box 13901
Research Triangle Park, NC 27709-3901

**Bydale Foundation Grants** **829**
The Bydale Foundation emphasizes international understanding, public policy research, environmental quality, cultural programs, the law and civil rights, social services, higher education, and economics. Funding includes support for: conference/seminars, continuing support, general operating support, matching/challenging support, program development, publication, research, and seed money. An application form is not required. Submit an initial approach in the form of a letter or proposal. Inquiry application forms are due in the Bank of America office by August 1. When this deadline falls on a weekend, proposals are due on the previous business day.
*Requirements:* U.S. 501(c)3 nonprofits are eligible. The Bydale Foundation's domestic geographic areas of interest are Connecticut, Westchester County (N.Y.), North Carolina, or those with a nationwide focus.
*Geographic Focus:* All States
*Date(s) Application is Due:* Aug 1
*Contact:* Christine O'Donnell, Bank of America; (646) 855-1011; bydale@ustrust.com
*Internet:* fdnweb.org/bydale/
*Sponsor:* Bydale Foundation
114 West 47th Street, 10th Floor
New York, NY 10036

**CAA Professional Development Fellowships for Graduate Students** **830**
CAA's Professional-Development Fellowships support promising artists and art historians who are enrolled in MFA and PhD programs nationwide. Fellows are honored with $5,000 grants to help them with various aspects of their work, whether it be for job-search expenses or purchasing materials for the studio. Fellows also receive a free one-year CAA membership and complimentary registration to the Annual Conference. Honorable mentions, given at the discretion of the jury, also earn a free one-year CAA membership and complimentary conference registration. In Fellowships in the Visual Arts, five student artists in MFA programs will each receive a one-time award of $5,000. In Fellowships in Art History, two doctoral candidates in art history will each receive a one-time award of $5,000.
*Requirements:* CAA seeks applications from students who are current members; are citizens or permanent residents of the United States; will receive their MFA or PhD degree in the calendar year following the year of application; and have outstanding capabilities and demonstrate distinction in approach, technique, or perspective in their contribution to art history and the visual arts. A jury of artists, curators, and other professionals will review all applications in the fall and announce the recipients in January of the following year. All applications materials must be sent by mail and postmarked by the deadline date.
*Geographic Focus:* All States
*Date(s) Application is Due:* Oct 1
*Amount of Grant:* 5,000 USD
*Contact:* Michael Fahlund, Deputy Director; (212) 392-4401; fax (212) 627-2381; mfahlund@collegeart.org
*Internet:* www.collegeart.org/fellowships/
*Sponsor:* College Art Association
50 Broadway, 21st Floor
New York, NY 10004

**CAA Wyeth Publication Grant** **831**
Since 2005, the Wyeth Foundation for American Art has supported the publication of books on American art through the Wyeth Foundation for American Art Publication Grant, administered by CAA. For this grant program, "American art" is defined as art created in the United States, Canada, and Mexico through 1970. Eligible for the grant are book-length scholarly manuscripts in the history of American art, visual studies, and related subjects that have been accepted by a publisher on their merits but cannot be published in the most desirable form without a subsidy.
*Requirements:* Application is by the publisher, not the author. Applications are considered only for book-length scholarly manuscripts in the history of American art that are under contract for publication. Awards are open to publishers of all nations; authors need not live or work in the United States. Commercial, university, and museum presses are all eligible. Applicant authors and presses must be institutional CAA members, although on a case-by-case basis CAA may waive that requirement upon request for non-US-based independent presses. Excluded from consideration are excavation or other technical reports, articles, previously published works (including collections of previously published essays), and congress proceedings. Museum exhibition or collection catalogues containing substantial scholarship are eligible. High scholarly and intellectual merit is the sine qua non for an award; however, the jury is also attentive to the following criteria: Topics on art through 1970; Topics with a naturally small market or unusually high expenses; Works by disadvantaged scholars, including those at the earlier stages of a career, or by younger scholars or curators; or issued by smaller museums; or by or about

underserved constituencies; Books that break new ground, contribute new scholarship, or publish important primary-source material; and, Beautiful books that increase the audience for American art .
*Restrictions:* A project that has been rejected for grant may not be resubmitted to the same grant, except in a rare case where substantial revision has been made to the material, and the publisher has so noted in the application. At its discretion, the jury may decline to review the resubmitted application.
*Geographic Focus:* All States, All Countries
*Date(s) Application is Due:* Sep 15
*Samples:* Katherine A. Bussard, Unfamiliar Streets: Photographs by Richard Avedon, Charles Moore, Martha Rosler, and Philip-Lorca DiCorcia, Yale University Press. Melissa Dabakis, The American Corinnes: Women Sculptors and the Eternal City, 1850–1876, Pennsylvania State University Press.
*Contact:* Alex Gershuny, Editorial Associate; (212) 392-4424; fax (212) 627-2381; agershuny@collegeart.org
*Internet:* www.collegeart.org/wyeth/
*Sponsor:* College Art Association
50 Broadway, 21st Floor
New York, NY 10004

**Cabot Corporation Foundation Grants** **832**
The goal of Cabot Corporation Foundation is to support community outreach objectives, with priority given to science and technology, education, and community and civic improvement efforts in the communities where the company has major facilities or operations. Types of support include capital grants, challenge grants, employee matching gifts, fellowships, general support, professorships, project support, research, scholarships, and seed money. The board meets in January, April, July, and October to consider requests. Applications must be received at least 30 days before a board meeting.
*Requirements:* The Foundation supports only nonprofit 501(c)3 tax-exempt organizations in areas of company operation. Modest support is available for international organizations that qualify under U.S. tax regulations.
*Restrictions:* Contributions are not made to individuals; fraternal, political, athletic or veterans organizations; religious institutions; capital and endowment campaigns; sponsorships of local groups/individuals to participate in regional, national, or international competitions, conferences or events; advertising sponsorships; or Tickets or tables at fundraising events.
*Geographic Focus:* Georgia, Illinois, Louisiana, Massachusetts, New Mexico, Pennsylvania, Texas, West Virginia, Belgium, Canada, China, Switzerland, United Kingdom
*Amount of Grant:* 2,000 - 75,000 USD
*Contact:* Cynthia L. Gullotti, Program Manager; (617) 345-0100; fax (617) 342-6312; Cynthia_Gullotti@cabot-corp.com or cabot.corporation.foundation@cabotcorp.com
*Internet:* www.cabot-corp.com/About-Cabot/Corporate-Giving
*Sponsor:* Cabot Corporation Foundation
Two Seaport Lane, Suite 1300
Boston, MA 02210-2019

**Caesars Foundation Grants** **833**
Founded as Harrah's Foundation in Nevada in 2002, giving is in the area of company operations. The foundation supports programs designed to help older individuals live longer, healthier, and more fulfilling lives; promote a safe and clean environment; and improve the quality of life in communities where Caesars operates. Fields of interest include: aging centers and services; Alzheimer's disease; developmentally disabled services; the environment; food distribution programs; food services; health care; patient services; higher education; hospitals; human services; mental health services; public affairs; public safety; nutrition; and youth services. Types of support being offered include: building and renovation; capital campaigns; continuing support; general operating support; program development; research; scholarship funding; and sponsorships. There are no specific deadlines or application forms. Caesars Foundation Trustees meet on a quarterly basis, typically around the second week of each quarter. Check with your nearest Caesars Entertainment property for deadlines. The foundation generally funds programs and projects of $10,000 or more.
*Requirements:* Eligible organizations are 501(c)3 nonprofits operating programs in the communities where Caesars employees and their families live and work. Applying organizations must also: demonstrate diversity by providing services and volunteer opportunities to all without regard to race, ethnicity, gender, religion, sexual orientation, identity or disability; illustrate strong leadership that will significantly strengthen communities in which Caesars operates; show sound administrative and financial condition; provide opportunities for Caesars staff involvement as volunteers and/or opportunity to serve on Board of Directors; and, provide branding opportunities and openly support Caesars Foundation in a public forum.
*Restrictions:* The Foundation is not designed to react to last-minute requests or event sponsorships—plan the timing of your proposal accordingly. Caesars Foundation does not accept requests for in-kind contributions.
*Geographic Focus:* Arizona, California, Illinois, Indiana, Iowa, Louisiana, Mississippi, Missouri, Nevada, New Jersey, North Carolina, Pennsylvania
*Amount of Grant:* 10,000 USD
*Samples:* Atlantic Cape Community College Foundation, Atlantic City, New Jersey, $75,000 - general operating support; Calumet College of St. Joseph, Whiting, Indiana, $60,000 - construction of new student center building; University of Chicago Medical Center, Chicago, Illinois, $50,000 - research and development.
*Contact:* Gwen Migita, Community Affairs; (702) 880-4728 or (702) 407-6358; fax (702) 407-6520; caesarsfoundation@caesars.com
*Internet:* www.caesarsfoundation.com/
*Sponsor:* Caesars Foundation
1 Caesars Palace Drive
Las Vegas, NV 89109-8969

**Callaway Golf Company Foundation Grants** **834**
The foundation strives to support initiatives in communities where Callaway Golf Company employees live and work. The geographic area of focus is primarily North San Diego County, California. The foundation offers support in the form of: matching funds, special projects, and general operating budgets. Areas of interest include but are not limited to: children and youth; biomedical research, with a special interest in the field of cancer; golf; education; boys & girls clubs of America; drug prevention; American Red Cross; housing; veterans; youth programs; food banks; social services; emergency programs; scholarship program for dependents of Callaway Golf employees; grants for training, competition and equipment needs. The foundation does not require the completion of a formal application document but requests a description of the organization and its history, the project at issue including goals and time lines, the qualifications of the leadership personnel involved in the project, and a detailed project budget. Grants are awarded semiannually. The Callaway Golf Company Foundation does not accept unsolicited requests for grants.
*Requirements:* IRS 501(c)3 nonprofit organizations in California are eligible.
*Restrictions:* The foundation will not fund applicants that illegally discriminate on the basis of gender, race, color, religion, national origin, ancestry, age, marital status, medical condition, or physical disability, either in the services they provide or in the hiring of staff; or promote political or particular religious doctrines.
*Geographic Focus:* California
*Amount of Grant:* 500 - 10,000 USD
*Contact:* Paul Thompson, Executive Director; (760) 930-8686; fax (760) 930-5021; cgcfoundation@callawaygolf.com
*Internet:* www.callawaygolf.com/Global/en-US/Corporate/CallawayGolfFoundation.html
*Sponsor:* Callaway Golf Company Foundation
2180 Rutherford Road
Carlsbad, CA 92008-7328

**Camille Dreyfus Teacher-Scholar Awards** **835**
The Camille Dreyfus Teacher-Scholar Awards Program supports the research and teaching careers of talented young faculty in the chemical sciences. Based on institutional nominations, the program provides discretionary funding to faculty at an early stage in their careers. Criteria for selection include an independent body of scholarship attained within the first five years of their appointment as independent researchers, and a demonstrated commitment to education, signaling the promise of continuing outstanding contributions to both research and teaching. The Camille Dreyfus Teacher-Scholar Awards Program provides an unrestricted research grant of $75,000. Of the total amount, $7,500 is for departmental expenses associated with research and education. Guidelines and required information are available online.
*Requirements:* The Camille Dreyfus Teacher-Scholar Awards Program is open to academic institutions in the States, Districts, and Territories of the United States of America that grant a bachelor's or higher degree in the chemical sciences, including biochemistry, materials chemistry, and chemical engineering. Nominees must hold a full-time tenure-track academic appointment, and are normally expected to have been appointed no earlier than mid-year 2006. Awardees are from Ph.D. granting departments in which scholarly research is a principal activity. Undergraduate education is an important component of the nominee's activities. Institutions may submit only one Camille Dreyfus nomination annually.
*Restrictions:* Defrayal of academic-year salary is not permitted.
*Geographic Focus:* All States
*Date(s) Application is Due:* Feb 9
*Amount of Grant:* 75,000 USD
*Contact:* Mark J. Cordillo, Executive Director; (212) 753-1760; fax (212) 593-2256; mcardillo@dreyfus.org or admin@dreyfus.org
*Internet:* www.dreyfus.org/awards/camille_dreyfus_teacher_award.shtml
*Sponsor:* Camille and Henry Dreyfus Foundation
555 Madison Avenue, 20th Floor
New York, NY 10022-3301

**Campbell Foundation HIV/AIDS Research Grants** **836**
The Campbell Foundation awards grants for research projects on ways to prevent and treat HIV, AIDS, and related conditions or illnesses. Proposals focusing on alternative, nontraditional avenues of study receive preference. Most grants support clinical research, although other types of work are considered. After initial screening, requests are reviewed by a peer review board of noted HIV/AIDS physicians and then sent to the board for final decisions. There are no application deadlines, but as all grants are handled chronologically, the foundation suggests submitting grant requests as early as possible. Typical awards range from $20,000 to $100,000.
*Requirements:* U.S. nonprofit organizations are eligible to apply.
*Geographic Focus:* All States
*Amount of Grant:* 20,000 - 100,000 USD
*Samples:* Massachusetts General Hospital, Boston, Massachusetts, $77,495 - support of brain inflammation and cognitive decline in HIV patients with dysbiotic microbiome; Northwestern University, Evanston, Illinois, $80,000 - controlling HIV release from cells with first-in-class drugs targeting virus budding; Scripps Research Institute, Jupiter, Florida, $79,151 - targeting HIV-1 latent infection with a potent tat inhibitor.
*Contact:* Ken Rapkin, Executive Director; (954) 493-8822; fax (954) 493-8801; ken@campbellfoundation.net or info@campbellfoundation.net
*Internet:* https://www.campbellfoundation.net/grant-requests
*Sponsor:* Campbell Foundation
5975 North Federal Highway, Suite 126
Fort Lauderdale, FL 33308

**Campbell Soup Foundation Grants** 837

Since 1953, the Campbell Soup Foundation has provided financial support to local champions that inspire positive change in communities throughout the United States where Campbell Soup Company employees live and work. The Foundation places particular emphasis on Camden, New Jersey, birthplace of Campbell's flagship soup business and world headquarters. The Campbell Soup Foundation focuses its giving on four key areas: hunger relief-supporting food bank organizations in the communities of operation; wellness-addressing the health of consumers in the communities where they live; education-leveraging the Campbell brand portfolio to support educational programs; community revitalization-enhancing the quality of life in the communities that Campbell operates in. The Foundation only considers applications that meet the following criteria: the proposal must fit one of the key focus areas; the organization must display strong and effective leadership; the proposed plan must be clear and compelling, with measurable and sustainable commitments expressed in terms of real results; the proposed activity must be sufficiently visible to leverage additional support from other funding sources. The Campbell Soup Foundation now hosts only one funding cycle per year, in the fall. The funding cycle is typically four months in length, and runs July through October.

*Requirements:* The Foundation limits grants to nonprofit organizations which are tax-exempt under Section 501(c)3 of the Internal Revenue Code. Grants are made to institutions that serve: Camden and East Brunswick, New Jersey; Bakersfield, California; Dixon, California; Stockton, California; Bloomfield and Norwalk, Connecticut; Downers Grove, Illinois; Ferndale and Grand Rapids, Michigan; Maxton, North Carolina; Napoleon and Willard, Ohio; Denver and Downingtown, Pennsylvania; Paris, Texas; Richmond, Utah; Everett, Washington; and Milwaukee, Wisconsin. Organizations do not need to be located in these communities in order to qualify for funding. However, the programs to be funded must serve these communities. Proposals must be submitted electronically. Proposals should be prepared in a concise, narrative form, without extensive documentation.

*Restrictions:* Grants are not made to the following: organizations that are based outside the United States and its territories; individuals; organizations that limit their services to members of one religious group or whose services propagate religious faith or creed; political organizations and those having the primary purpose of influencing legislation of/or promoting a particular ideological point of view; units of government; events and sponsorships; sports related events, activities and sponsorships. Organizations may not submit the same or similar proposals more than once in a Foundation fiscal year (July 1 through June 30). Proposals submitted via regular mail will not be reviewed.

*Geographic Focus:* California, Connecticut, Florida, Illinois, Michigan, New Jersey, North Carolina, Ohio, Pennsylvania, Texas, Utah, Washington, Wisconsin

*Date(s) Application is Due:* Nov 3

*Contact:* Grant Administrator; (856) 342-6423 or (800) 257-8443; fax (856) 541-8185; community_relations@campbellsoup.com

*Internet:* https://www.campbellsoupcompany.com/about-campbell/corporate-responsibility/campbell-soup-foundation/

*Sponsor:* Campbell Soup Foundation
1 Campbell Place
Camden, NJ 08103-1701

**Carl C. Icahn Foundation Grants** 838

The foundation awards grants to New York and New Jersey nonprofits in the areas of education, arts and culture, health care, child welfare, and Jewish temples and organizations. Types of support include general operating support, annual campaigns, building construction/renovation, and matching funds. There are no application deadlines or forms.

*Requirements:* New York and New Jersey nonprofits are eligible to apply.

*Restrictions:* No grants are provided to individuals.

*Geographic Focus:* New Jersey, New York

*Amount of Grant:* 500 - 1,600,000 USD

*Samples:* Randall's Island Sports Foundation, New York, NY, $1,666,000—to support program/general purposes; Wildlife Conservation Society, Pelham, NY, $10,000—to support program/general purposes; The Ladies' Village Improvement Society, East Hampton, NY, $500—to support program/general purposes.

*Contact:* Gail Golden-Icahn, Vice Presicent; (212) 702-4300; fax (212) 750-5815

*Sponsor:* Carl C. Icahn Foundation
767 5th Avenue, 47th Floor
New York, NY 10153-0023

**Carl M. Freeman Foundation Grants** 839

The Carl M. Freeman Foundation has historically emphasized it's support in the following communities: Montgomery County, Maryland; Sussex County, Delaware; and the Eastern Panhandle of West Virginia . Funding is available for a wide variety of community organizations, having supported everything from arts organizations and hunger centers to educational and health related organizations. To simplify the application process, funding applications are available in six areas of interest: arts and culture; education; environment; health and human services; housing; other (this may include anything you feel does not fit in the above categories, for example spaying cats and dogs). Additional guidelines and applications are available at the website.

*Requirements:* 501(c)3 tax-exempt organizations in Maryland, Delaware and West Virginia are eligible. Nonsectarian religious programs also are eligible.

*Restrictions:* Grants will not be distributed to: individuals; political associations or candidates; organizations that would disperse the funding to others; organizations that discriminate by race, creed, gender, sexual orientation, age, religion, disability or national origin.

*Geographic Focus:* Delaware, Maryland, West Virginia

*Amount of Grant:* 5,000 - 30,000 USD

*Contact:* Melissa Rizer, Grants Administrator; (302) 436-3555; melissa@freemanfoundation.org

*Internet:* www.freemanfoundation.org/carl/CarlMFreemanFoundation/Grants/GrantGuidelines/tabid/181/Default.aspx

*Sponsor:* Carl M. Freeman Foundation
31556 Winterberry Parkway
Selbyville, DE 19975

**Carnegie Corporation of New York Grants** 840

The Carnegie Corporation of New York provides research, study, and support for projects to improve government at all levels, to increase public understanding of social policy issues, to equalize opportunities for minorities and women, and to increase participation in political and civic life. Also supported are projects that promote electoral reform; education reform from early childhood through higher education; early childhood development; and urban school reform. The foundation will also fund research on the increasing availability and success of after-school and extended service programs for children and teenagers, particularly those in urban areas, that promote high academic achievement. Dissemination of best practices in teacher education will also be emphasized. There is no formal procedure for submitting a proposal. To apply under any of the corporation's grantmaking programs, applicants should submit a full proposal that describes the project's aims, duration, methods, amount of financial support required, and key personnel. The board meets four times a year, in October, February, April, and June.

*Requirements:* Only full proposals that have been invited for submission will be considered. After a letter of inquiry has been reviewed, applicants may be invited via email to submit a full proposal.

*Restrictions:* Grants are not made for construction or maintenance of facilities or endowments. The Corporation does not generally make grants to individuals except through the Carnegie Scholars Program, that supports the work of select scholars and experts conducting research in the foundation's fields of interest.

*Geographic Focus:* All States

*Amount of Grant:* Up to 4,000,000 USD

*Samples:* Massachusetts Institute of Technology, Cambridge, Massachusetts, $1,000,000 - international peace and security; Center for Better Schools, Portsmouth, Rhode Island, $650,000 - strengthening teaching and human capital; Citizen Schools, Boston, Massachusetts, $500,000 - strengthening education.

*Contact:* Nicole Howe Buggs, Grants Manager; (212) 371-3200; fax (212) 754-4073; externalaffairs@carnegie.org

*Internet:* https://www.carnegie.org/grants/grantseekers/

*Sponsor:* Carnegie Corporation of New York
437 Madison Avenue
New York, NY 10022

**Carnegie Endowment for International Peace Junior Fellowships** 841

Each year the endowment offers eight to ten one-year fellowships to uniquely qualified graduating college seniors and individuals who have graduated during the past academic year. They are selected from a pool of nominees from close to 400 colleges (a list of participating colleges available online). Fellows work as research assistants to the Endowment's senior associates. Junior Fellows have the opportunity to conduct research for books, co-author journal articles and policy papers, participate in meetings with high-level officials, contribute to congressional testimony and organize briefings attended by scholars, journalists and government officials.

*Requirements:* The Carnegie Endowment accepts applications only through participating universities via designated nominating officials. Applicants should contact their university for deadlines and a current application.

*Restrictions:* Students who have started graduate studies are not eligible for nomination. Students need not be U.S. citizens; however, all applicants must be eligible to work in the United States from August 1 through July 31 following graduation. Due to the nominations process and varying on-campus application procedures, the Endowment is not able to send applications directly to students.

*Geographic Focus:* All States, Canada, Ireland, Spain, United Kingdom

*Date(s) Application is Due:* Jan 15

*Amount of Grant:* 35,000 USD

*Contact:* Fellowship Administrator; (202) 483-7600; fax (202) 483-1840; jrfellowinfo@CarnegieEndowment.org

*Internet:* www.carnegieendowment.org/about/index.cfm?fa=jrFellows

*Sponsor:* Carnegie Endowment for International Peace
1779 Massachusetts Avenue NW
Washington, D.C. 20036-2103

**Carter G. Woodson Institute Research Fellowships** 842

The Carter G. Woodson Institute for Afro-American and African Studies, located at the University of Virginia, sponsors residential pre- and postdoctoral research fellowships. All fellowships are for the completion of research on race, ethnicity, and society in Africa and the Atlantic world (broadly defined as the African diaspora). Preference will be given to applicants whose research is substantially complete. Tenure for the fellowships begins in August. Postdoctoral fellowships are for one year; pre-doctoral fellowships are for two years. Application guidelines are available on the Web site. Annual deadline dates may vary; contact program staff for exact dates.

*Requirements:* Applicants for the fellowships must have been awarded their PhD by the time of application or must furnish proof that it will have been received prior to August 1. The competition is open to qualified candidates without restriction as to citizenship or current residence.

*Restrictions:* Employees of the University of Virginia and former university employees whose termination dates are less than one year prior to the time of application are not eligible.
*Geographic Focus:* All States
*Date(s) Application is Due:* Dec 1
*Amount of Grant:* 15,000 - 25,000 USD
*Contact:* Deborah E. McDowell, Director of the Fellowship Program; (434) 924-8892 or (434) 924-3109; fax (434) 924-8820; dem8z@virginia.edu
*Internet:* woodson.virginia.edu/fellowship-program
*Sponsor:* Carter G. Woodson Institute for Afro-American and African Studies
108 Minor Hall, P.O. Box 400162
Charlottesville, VA 22904-4162

**Casler Foundation Grants** **843**
The Casler Foundation was established in Rockford, Illinois, in 1988, with primary fields of interest to support: human services; alleviation of diseases; religious causes; medical research; and science. Target populations include: children and youth; economically disadvantaged; the homeless; low-income; destitute; and students. Types of funding most often provided comes in the form of general operating support, program development, research, evaluation of programs, and re-granting. Grant amounts typically range from $500 to $4,000, with approximately fifteen to twenty awards given annually. There are no annual deadlines specifies, so interested organizations should begin the process by contacting the Foundation directly.
*Requirements:* Any 501(c)3 supporting the residents of Illinois is eligible to apply.
*Geographic Focus:* Illinois
*Amount of Grant:* 500 - 4,000 USD
*Samples:* Crusaders Health Foundation, Rockford, Illinois, $4,000 - research and education support; Heartland Church, Rockford, Illinois, $4,000 - religious outreach support; Hope for Christian Church, Rockford, Illinois, $4,000 - religious outreach support.
*Contact:* N. Jean Cassens, Trustee; (815) 877-8449 or (815) 997-3791
*Sponsor:* Casler Foundation
10959 Hamborg Road
Roscoe, IL 61073

**Catherine Holmes Wilkins Foundation Grants** **844**
The Catherine Holmes Wilkins Foundation awards grants in Washington's Puget Sound region to nonprofit organizations in its areas of interest. Funding priorities include medical research and education (medical and academic centers conducting research and training in areas such as cancer, heart disease, and mental illness); physically handicapped and mentally ill (community nonprofit agencies providing direct social services to people with physical disabilities or mental illness); and services for the needy (community-based programs providing immediate support to the needy, with particular emphasis on services for abused women and children). Preference will be given to project support, rather than ongoing operating expenses. Between forty and fifty grants are awarded annually, ranging from $2,000, to $5,000. Proposals are accepted throughout the year, with annual deadlines designated as the second Monday in March, the second Friday in October, and the fourth Friday in June. Grants are awarded three times per year.
*Requirements:* 501(c)3 nonprofit organizations that operate within or significantly affect the residents of the greater Seattle region (Tacoma to Everett, Seattle, and the Eastside) are eligible.
*Restrictions:* Grants are not made for multi-year projects, debt retirement, operational deficits, or to individuals or for scholarships.
*Geographic Focus:* Washington
*Date(s) Application is Due:* Mar 13; Jun 23; Oct 13
*Amount of Grant:* 2,000 - 5,000 USD
*Samples:* Amara Parenting and Adoption Services, Seattle, Washington, $5,000 - support of the Emergency Sanctuary for Vulnerable Children; Pike Market Senior Center, Seattle, Washington, $5,000 - support of the Food Bank Capacity Growth; University Street Ministry, Seattle, Washington, $4,000 - support of the Teen Feed Meal program.
*Contact:* Keri Healey, Grants Consultant; (206) 604-5374; keri@kerihealey.com
*Internet:* https://www.bankofamerica.com/philanthropic/foundation.go?fnId=137
*Sponsor:* Catherine Holmes Wilkins Foundation
800 5th Avenue, P.O. Box 24565, WA1-501-33-23
Seattle, WA 98124

**CBA Viscount Bennett Fellowship** **845**
The Viscount Bennett Fellowship is made possible by a donation from the Right Honorable Viscount Bennett, a CBA Past President (1929-30) who in 1943 established a trust fund to provide income to be used to encourage a high standard of legal education, training and ethics. One fellowship per year is granted for study in the field of legal education, training, and ethics tenable at an institution in Canada, the United States, England, or France. Fellowships are awarded for one year and are not renewable. Application forms may be obtained from the association. The award is in the amount of $25,000, and the annual deadline for application submissions is November 15. The Award will be paid as a fellowship for post-graduate studies at a university approved by the Committee.
*Requirements:* Canadian citizens who have graduated from an approved law school in Canada or who are in the final year of a law school program at time of application are eligible. All applicants must be Canadian Bar Association members in good standing.
*Geographic Focus:* Canada
*Date(s) Application is Due:* Nov 15
*Amount of Grant:* 25,000 CAD
*Samples:* Anthony Sangiuliano, Maple, Ontario, $25,000 - Viscount Bennett Fellowship (2018); Daniel Del Gobbo, Toronto, Ontario, $25,000 - Viscount Bennett Fellowship; François Goyer, Deux-Montagnes, Quebec, $25,000 - Viscount Bennett Fellowship.
*Contact:* Crystal Lam, Liaison Officer; (613) 237-2925, ext. 133 or (613) 237-1988; fax (613) 237-0185; crystall@cba.org
*Internet:* https://www.cba.org/Who-We-Are/About-us/Awards-and-Recognition/Search-Awards/Students/Viscount-Bennett-Fellowship
*Sponsor:* Canadian Bar Association
500 - 865 Carling Avenue, Suite 500
Ottawa, ON K1S 5S8 Canada

**CCA Alberta Creative Development Initiative: Grants to Individuals** **846**
The Alberta Creative Development Initiative (ACDI) is a partnership between the Alberta Foundation for the Arts and the Canada Council for the Arts, in collaboration with the Calgary Arts Development Authority and the Edmonton Arts Council. The Grant is designed support projects by Alberta professional artists, arts administrators and collectives. The objectives are to: support the development of professional artists and arts organizations in Alberta; foster a dynamic, vibrant arts community in Alberta; and enable the success of Alberta artists and arts organizations locally, provincially, nationally and internationally.
*Requirements:* Eligible applicants must be: Alberta-based professional artists, arts administrators or collectives working in all artistic disciplines and artistic practices currently supported by the Council or the Alberta Foundation for the Arts. Individuals must be a Canadian citizen or permanent resident, have maintained a regular residence in Alberta in the past year and have been defined as an Alberta resident on their last tax return. Eligible applicants can request funding for projects that will advance their artistic practice and/or professional development. Eligible projects may be research, creation, production/and or dissemination projects or professional development projects including residencies. Aboriginal artists and artists of diverse cultural and regional communities of Canada are eligible.
*Restrictions:* Full-time undergraduate students at a school, college or university are ineligible. Graduate or post-graduate students are eligible only by meeting the definition of a professional artist or arts administrator, and the proposed project cannot be related to the program of study and cannot be under the direct supervision of a university professor.
*Geographic Focus:* Canada
*Date(s) Application is Due:* Nov 1
*Amount of Grant:* 15,000 CAD
*Contact:* Marion Vitrac, Program Officer; (800) 263-5588 or (613) 566-4414, ext. 4414; fax (613) 566-4409
*Internet:* www.canadacouncil.ca/grants/acdi_idca/tx128351161983158103.htm
*Sponsor:* Canada Council for the Arts
150 Elgin Street, 2nd Floor, P.O. Box 1047
Ottawa, ON K2P 1L4 Canada

**CCA Diefenbaker Award to update** **847**
This annual award enables a distinguished German scholar to spend up to 12 months in Canada, which may include brief periods in the United States. While research must be the primary activity during the tenure of the award, the recipient will also be encouraged to participate in the activities of the host institution and to interact more extensively with the research communities in Canada and the United States by visiting other institutions. Candidates may not apply for this award, but instead must be nominated by a host university department or research institute in Canada.
*Requirements:* Candidates must be German citizens and must have a sound working knowledge of one of Canada's two official languages. German scholars who have demonstrated outstanding ability, especially through a substantial publication record over several years, are eligible to apply.
*Geographic Focus:* All States, Germany
*Date(s) Application is Due:* Dec 1
*Amount of Grant:* 95,000 CAD
*Contact:* Nathalie Lauzon, Administrative Assistant; (800) 263-5588 or (613) 566-4414, ext. 4083; fax (613) 566-4430; endowments.prizes@canadacouncil.ca
*Internet:* www.canadacouncil.ca/prizes/john_g_diefenbaker/hn127223038493593750.htm
*Sponsor:* Canada Council for the Arts
150 Elgin Street, 2nd Floor, P.O. Box 1047
Ottawa, ON K2P 1L4 Canada

**CCFF Chairman's Distinguished Life Sciences Award** **848**
The $10,000 Chairmen's Distinguished Life Sciences Award will be presented to a scientist or researcher who is making or has recently made a significant and positive contribution toward the development of a cutting edge innovation in the field of life sciences. The Chairmen's Award will provide incentive for continuing the research. Individuals will compete and their accomplishments will be judged on how they exemplify excellence in life sciences. Winners must attend the awards ceremony in April at Washington, D.C., in order to receive the award.
*Requirements:* All Nominees must be living United States citizens. The Nomination must demonstrate a program of ongoing work with a specific outcome suggesting that an important achievement will result. The Nomination's proposed achievement must benefit human health in a unique and effective manner, demonstrating originality from other work in the same or related field(s). The proposed achievement must correlate to a tangible outcome to advance human health.
*Restrictions:* Nominees must be permitted, by their employer or any other relevant authority, to accept a monetary award bestowed by the Foundation, a Federal government agency. Anyone who cannot attend the award ceremony will automatically be disqualified.
*Geographic Focus:* All States
*Date(s) Application is Due:* Feb 26
*Amount of Grant:* 10,000 USD

*Contact:* Judith M. Shellenberger, Executive Director; (315) 258-0090; fax (315) 258-0093; judithmscolumbus@cs.com
*Internet:* www.columbusfdn.org/lifesciences/
*Sponsor:* Christopher Columbus Fellowship Foundation
110 Genesee Street, Suite 390
Auburn, NY 13021

**CCFF Distinguished Agriscience Scientist Award** 849

The Christopher Columbus Fellowship Foundation, a Federal government agency, and the American Farm Bureau Federation, have formed a partnership to present four monetary awards and research funding to three groups representing new innovative approaches in the field of agriscience around the country. One $10,000 Distinguished Agriscience Scientist Award will be presented to a scientist who is making or has recently made significant and positive contributions in the field of agriscience.
*Requirements:* All Candidates must be living United States citizens. The Nomination must demonstrate a program of ongoing work with a specific outcome suggesting that an important achievement will result. The Nomination's proposed achievement must benefit the field of agriscience in a unique and effective manner, demonstrating originality from other work in the same or related field(s). The proposed achievement must correlate to a tangible outcome to advance the field of agriscience.
*Geographic Focus:* All States
*Date(s) Application is Due:* Feb 12
*Amount of Grant:* 10,000 USD
*Contact:* Judith M. Shellenberger, Executive Director; (315) 258-0090; fax (315) 258-0093; judithmscolumbus@cs.com
*Internet:* www.columbusfdn.org/agriscience/
*Sponsor:* Christopher Columbus Fellowship Foundation
110 Genesee Street, Suite 390
Auburn, NY 13021

**CCH Documentary Project Research and Development Grants** 850

The California Documentary Project (CDP) is a competitive grant program of the California Council for the Humanities (CCH). CDP supports film, radio, and new media projects in which the humanities are used to provide context, depth, and perspective. The humanities spring from a fundamental interest in understanding the values and practices that inform our lives; the need to reflect on the past and the present to make critical choices; and the desire to be moved, delighted, and make sense of the world in which we live. Humanities disciplines include, but are not limited to: history, literature, philosophy, folklore, ethnic studies, religious studies, ethics, jurisprudence, and qualitative approaches in the social sciences. They emphasize analysis, interpretation, and the exchange of ideas. CDP Research and Development grants are designed to strengthen the humanities content and approach of documentary media productions in their earliest stages. Projects must actively involve at least three humanities advisors to help frame and contextualize subject matter throughout the research and development phase. Eligible projects may apply for funding up to $7,000.
*Requirements:* Eligible applicant organizations/project directors must: have tax-exempt organizational status or a tax-exempt organization as fiscal sponsor; be in good standing with CCH (e.g., without unfulfilled reporting requirements), if a previous grantee; and not have an open grant with CCH, unless a CCH-approved sponsoring organization (see FAQ for more information). Film, radio, and new media projects must meet all of the following criteria: be currently in the research and development stage; document the California experience and explore issues of significance to Californians; approach the subject matter from a humanities perspective—for example, making use of existing or new scholarship and research on the topic, incorporating a variety of perspectives, seeking to foster critical reflection and thoughtful analysis on the part of the audience; actively involve at least three humanities advisors throughout the research and development phase; and be conducted by experienced film, radio, or new media producers and directors. A previously completed work sample must be submitted.
*Geographic Focus:* California
*Date(s) Application is Due:* Oct 3
*Amount of Grant:* Up to 7,000 USD
*Contact:* Vanessa Whang; (415) 391-1474; fax (415) 391-1312; vwhang@calhum.org
*Internet:* www.calhum.org/guidelines/cdp_2011_RandD_guidelines.htm
*Sponsor:* California Council for the Humanities
312 Sutter Street, Suite 601
San Francisco, CA 94108-4371

**CDC Experience Applied Epidemiology Fellowships** 851

The goal of the CD.C. Experience Applied Epidemiology Fellowship is to increase the number of physicians with a population health perspective by providing medical students with an understanding of applied epidemiology, the role of epidemiology in medicine and health, and the role of physicians in the public health system. The fellowship is held at the Centers of Disease Control and Prevention (CD.C.) headquarters in Atlanta, Georgia. Fellows start in August and work for ten or twelve months. While at CD.C., fellows will: gain training and work experience in applied epidemiology and public health; perform epidemiologic analyses and research; participate in public health field experiences; report findings through written and oral presentations at scientific meetings; write scientific manuscripts for publication; participate in monthly seminars and journal clubs; and attend the annual Epidemic Intelligence Service conference. CD.C. Program areas available to applicants are chronic diseases, emergency preparedness, environmental health, global health, infectious diseases, injury prevention, and reproductive health. Fellows receive a stipend for living expenses. Applicants must apply online. Supporting documents are required and must be postmarked by the deadline date. Applicants should check the CD.C. Experience website for further details. Annual deadline dates may vary; applicants are encouraged to check the website or contact program staff to verify current deadline dates. The CD.C. Experience Applied Epidemiology Fellowship is made possible by a public/private partnership supported by a grant to the CD.C. Foundation from External Medical Affairs, Pfizer Inc., and is considered a valuable experience to those those seeking career advancement in the areas of clinical medicine, clinical epidemiology, health services research, preventive medicine, and public health.
*Requirements:* Applicants must: be medical students in their 2nd or 3rd year when they apply; be enrolled in a school accredited by the Liaison Committee on Medical Education or the American Osteopathic Association; be a U.S. citizen or permanent resident; and be willing to relocate to Atlanta, Georgia for the duration of the fellowship. Awardees may start the fellowship while in their 3rd or 4th year.
*Geographic Focus:* All States
*Date(s) Application is Due:* Dec 2
*Contact:* Virginia Watson, MA, Program Contact; (404) 498-6151; VWatson1@cdc.gov or cdcexperience@cdcfoundation.org
Internet: www.cdc.gov/CD.C.ExperienceFellowship/index.html
*Sponsor:* Centers for Disease Control and Prevention
1600 Clifton Road
Atlanta, GA 30333

**CDC Foundation Atlanta International Health Fellowships** 852

The Atlanta International Health Fellowships (AIHF) were created in 1984 by Drs. Bob Chen and Katy Irwin to provide financial assistance to international participants in CD.C.'s varied public health training courses. The two young physicians enlisted help from other CD.C. employees and retirees to raise funds and then formed a partnership with Emory University, which provided free tuition to one individual each year to attend Emory's International Course in Applied Epidemiology. Drs. Chen and Irwin also formed a partnership with Villa International to provide free housing to the AIHF recipient. In 1997, a grant from the Tull Charitable Foundation was added to existing funds, and the Atlanta International Health Fellowship became the first endowed fund at the CD.C. Foundation. Since the first fellowships were awarded in 1991, 25 individuals from over 22 countries have received stipends to cover costs of travel, tuition and lodging as they enrich their public health expertise at CD.C.. Interested applicants are encouraged to contact the CD.C. Foundation or Emory University for information on how to apply for the AIHF fellowship.
*Geographic Focus:* All States, All Countries
*Contact:* Verla S. Neslund, J.D., Vice President for Programs; (888) 880-4232 or (404) 653-0790; fax (404) 653-0330; vneslund@cdcfoundation.org
Pia Vaerliano, B.S.C., M.B.A, Co-Director of Special Programs; (404) 727-3485; fax (404) 727-4590; pvaleri@sph.emory.edu
*Internet:* www.cdcfoundation.org/what/program/atlanta-international-health-fellowship-endowment
Sponsor: Centers for Disease Control and Prevention (CD.C.) Foundation
55 Park Place NE, Suite 400
Atlanta, GA 30303

**CDC Foundation Tobacco Network Lab Fellowship** 853

According to a Centers for Disease Control and Prevention (CD.C.) morbidity and mortality report, tobacco use is the world's leading single preventable cause of death. Worldwide tobacco-related deaths currently exceed 5 million a year and are expected to exceed 8 million a year by 2030 if left unchecked. In the United States, tobacco use is the single leading preventable cause of disease, disability, and death. In 2003, the World Health Organization (WHO) created the Tobacco Laboratory Network (TobLabNet) of government, academic, and independent laboratories worldwide to strengthen national and regional capacity for testing and research of the contents and emissions of tobacco products. The Network's stated goals are: to establish global capacity to test tobacco products for regulatory compliance; to research and develop harmonized standards for contents and emissions testing; to share tobacco research and testing standards and results; to inform risk assessment activities related to the use of tobacco products; and to develop harmonized reporting of results so that data can be transformed into meaningful trend information that can be compared across countries and over time. The Tobacco Network Lab Fellowship is a collaboration between CD.C. and WHO to place a scientist in CD.C.'s Division of Laboratory Sciences to help TobLabNet address testing and research of tobacco products at the global level in accordance with the tobacco product regulation provisions outlined by the WHO Framework Convention on Tobacco Control. A link to brief information about the fellowship is at the grant website under Chronic Disease & Birth Defects Programs. Interested applicants are encouraged to contact the CD.C. Foundation or the World Health Organization for more information on how to apply for the fellowship.
*Geographic Focus:* All States, All Countries
*Contact:* Verla S. Neslund, J.D., Vice President for Programs; (888) 880-4232 or (404) 653-0790; fax (404) 653-0330; vneslund@cdcfoundation.org
Dr Haik Nikogosian, Convention Secretariat; 41 22 791 5043; fax 41 22 791 5830; fctcsecretariat@who.int
*Internet:* www.cdcfoundation.org/what/programs/list#program-518
Sponsor: Centers for Disease Control and Prevention (CD.C.) Foundation
55 Park Place NE, Suite 400
Atlanta, GA 30303

**CDC Preventive Medicine Residency and Fellowship** 854

The Preventive Medicine Residency and Fellowship (PMR/F) is an umbrella that includes two preventive medicine programs that provide hands-on experiences in public health agencies at the federal, state, and local levels. The programs consist of a

residency (PMR) for physicians only and a fellowship (PMF) for physicians and other health professionals. The PMR is a 24-month nationally accredited graduate medical education program for physicians. It is accredited by the Accreditation Council for Graduate Medical Education (ACGME) and meets the residency requirement of the American Board of Preventive Medicine (ABPM) in the specialty of Public Health and General Preventive Medicine. Select candidates can be considered for one year of residency training. The PMF is a 12-month program open to veterinarians, dentists, registered nurses, physician assistants, and international medical graduates (physicians) who have no ACGME- or American Osteopathic Association (AOA)- accredited postgraduate clinical training. The residency (PMR) and the fellowship (PMF) both provide service, training, and supervised on-the-job learning in public health and preventive medicine practice. PMR/F activities include: designing, implementing, and evaluating public health programs; developing or analyzing health policy; managing public health and preventive medicine projects; bridging medical and public health sectors to improve wellness; and developing and applying leadership skills. The PMR/F program goals are to prepare clinicians for leadership roles in public health and general preventive medicine at the federal, state or local levels. Residents and fellows will receive mentoring by experienced public health and preventive medicine practitioners; hands-on experience in public health and preventive medicine practice; classroom training to enhance professional growth; and M.P.H. coursework for select physicians in the PMR program. Residents and fellows receive a salary and benefits, moving expenses, travel to PMR/F-sponsored training, and tuition support for an M.P.H. degree for residents (as needed). (Qualified applicants interested in the American Board of Preventive Medicine "complementary pathway to board exam certification" must have completed an ACGME accredited residency in another specialty and have an M.P.H. or equivalent degree per ACGME.) Interested applicants may apply online between July 1 and September 15. Additional supporting documents are required. Applications are due by midnight eastern daylight time on the due date. Supporting documents must also be received on the application due date. Applicants are encouraged to mail supporting documents early. All mail delivery to the CD.C. goes to a central mail facility before being delivered to the PMR/F office. Applicants are encouraged to click on the "Application Information" link at the PMR/F website for additional instructions, guidelines, and information on how to apply and submit the supporting documents.
*Requirements:* Applicants must: commit to a 1- or 2-year full-time training period (depending on the program) starting in mid-June; be willing to relocate for duration of training; meet the professional and licensing requirements for hiring by the U.S. government according to the U.S. Office of Personnel Management (link provided at the PMR/F website); and have trained in the Epidemic Intelligence Service (EIS) program or have comparable applied epidemiology experience. PMR applicants must be physicians who have completed at least 12 months of ACGME-accredited postgraduate clinical training involving at least 11 months of direct patient care; have a current, full, and unrestricted medical license from a U.S. licensing jurisdiction; and have a Master of Public Health (M.P.H.) or equivalent accredited degree, per ACGME requirements. See the PMR/F website for more information about these requirements. There are some exceptions to the M.P.H. degree requirement. Physicians may be sponsored during the residency to obtain an M.P.H. degree. The completion of necessary courses to earn the degree will be integrated into the training program. M.P.H. sponsored applicants must be Commissioned Corps officers in the U.S. Public Health Service at the time of application with a commitment to an additional 2-year service obligation in the Commissioned Corps following the residency and a willingness to attend the academic institution for the M.P.H. selected by the residency (be willing to relocate). PMF applicants must be veterinarians, dentists, registered nurses, physician assistants, or international medical graduates (physicians who have no ACGME- or American Osteopathic Association- accredited postgraduate clinical training) who have an M.P.H. or equivalent degree from an accredited institution; have a current, full, and unrestricted license from a U.S. licensing jurisdiction in their qualifying clinical specialty; and demonstrate ability in applied epidemiology, public health practice, and written and oral communication for scientific audiences.
*Restrictions:* Applicants must be U.S. citizens or permanent residents. Physicians who have graduated from a medical school outside the U.S. and do not have a current, full, and unrestricted license from a U.S. licensing jurisdiction may apply.
*Geographic Focus:* All States
*Date(s) Application is Due:* Sep 15
*Contact:* PMR/F Coordinator; (404) 498-6140; fax (404) 498-6135; PrevMed@cdc.gov
*Internet:* www.cdc.gov/PrevMed/index.html
*Sponsor:* Centers for Disease Control and Prevention
1600 Clifton Road
Atlanta, GA 30333

**CDC Steven M. Teutsch Prevention Effectiveness Fellowships** **855**

The Prevention Effectiveness Fellowship (PEF) is a two-year, highly-selective research fellowship for recent doctoral graduates with a background in economics, policy analysis, operations research, decision sciences, and other quantitative areas, and was established to address demand in the field of public health for economics-based inquiry, quantitative policy analysis, and integrative-health-services research. The PEF allows applicants to apply their academic training in quantitative methods to the science of health protection, health promotion, and disease prevention. Prevention effectiveness uses economic evaluation, policy analysis, and decision modeling to systematically assess the impact of public health interventions, policies, programs, and practices on health outcomes. Econometric, decision, simulation, and operations modeling and methods are employed to determine the efficacy, effectiveness, efficiency, and cost of public health services, policies, and interventions in order to facilitate better decision-making about the allocation of resources around programmatic priorities. At CD.C., prevention effectiveness methods are used to explain: health-related burden and costs; effectiveness and efficiency of health protection, health promotion, and disease prevention programs; health program prioritization based on optimization modeling; disease transmission and intervention modeling; and health system performance. Fellows have the opportunity to work with leading researchers on issues of critical importance to create the expertise, information, and tools that enable people and communities to protect their health through: health promotion; prevention of disease, injury and disability; and preparedness for new health threats. Fellows are assigned to various CD.C. Centers, Institute, and Offices (CIOs) where they assess the public health impact of policies, programs, and practices by determining their cost, effectiveness, and overall impact. The PEF provides initial classroom training and hands-on experience to support the development of valuable research skills. After an introduction to basic public health sciences and prevention-effectiveness methods, fellows are assigned to a CD.C. program area where they take a lead role in prevention-effectiveness studies. Assignment activities include: designing and conducting studies; working closely with national and international experts in public health; presenting and publishing results; and providing technical assistance and service on specific projects or methods. In addition, professional support and career development are provided through didactic training, workshops, research seminars, and peer discussion groups. Fellows are full-time CD.C. employees during the fellowship and will receive the same benefits (such as medical insurance and vacation) as other full-time CD.C. employees. During the first year, fellows are paid the equivalent of a GS 12, Step 1 salary. In the second year, they are eligible for salary equivalent to a GS 12, Step 2 with geographic adjustments. Applicants must apply online at the PEF website (click on "Apply Now" on the left hand side of the screen). Applications are due on January 21 before midnight. All required supporting documents must be postmarked by January 19. Deadlines may vary from year to year; applicants should check the PEF website to verify current submission deadlines for proposals and supporting documents. Further essential details on application and supporting document requirements are available from the PEF website.
*Requirements:* Applicants must hold a doctoral degree in economics or applied economics, decision sciences, health services research or related health sciences, industrial engineering or operations research, public policy or policy analysis, or a related quantitatively-oriented field. A medical degree with additional relevant training is acceptable. Dissertations must be defended by June 17 and degree requirements must be met by July 15 prior to the start date of the fellowship. Written and spoken proficiency in English is essential. Relocation to Atlanta, Georgia is required. As stated in the FAQ at the PEF website, up to $3,000 is provided to cover relocation expenses. Non-U.S. citizens must be legal permanent residents or eligible for J-1 status prior to the program's start date.
Restrictions: Citizens of countries that have been determined by the U.S. Secretary of State to be State Sponsors of Terrorism will not be able to get security clearance to work at CD.C. and should not apply to the PE Fellowship Program. A list of these may be found at the U.S. Department of State website.
*Geographic Focus:* All States, All Countries
*Date(s) Application is Due:* Jan 11
*Amount of Grant:* 149,198 USD
*Contact:* Adam G. Skelton, Ph.D., M.P.H., Fellowship Coordinator; (404) 498-6324 or (404) 498-6786; PEF@cdc.gov or adam.skelton@cdc.hhs.gov
*Internet:* www.cdc.gov/PEF/index.html
*Sponsor:* Centers for Disease Control and Prevention
1600 Clifton Road
Atlanta, GA 30333

**CDECD Elizabeth L. Mahaffey Arts Administration Fellowship** **856**

The Elizabeth L. Mahaffey Arts Administration Fellowship annually recognizes one exceptionally accomplished Connecticut arts administrator with a $2,500 award to be used for professional development activities. The application is available at the CDECD website.
*Requirements:* Administrators must be full time state residents employed by Connecticut nonprofit arts organizations, or involved in ongoing arts programs of nonprofit organizations, municipalities, colleges and universities. Applicants must have at least seven years of professional arts administration experience, including three with their current organization.
*Restrictions:* Administrators employed as consultants, contractors or subcontractors are not eligible to apply. Unpaid and volunteer staff members are also ineligible.
*Geographic Focus:* Connecticut
*Date(s) Application is Due:* Feb 1
*Amount of Grant:* 2,500 USD
*Contact:* Rhonda Olisky, Special Projects Coordinator; (860) 256-2752; fax (860) 256-2811; rhonda.olisky@ct.gov
*Internet:* www.cultureandtourism.org/cct/cwp/view.asp?a=3933&q=293768
*Sponsor:* Connecticut Department of Economic and Community Development
One Constitution Plaza, 2nd Floor
Hartford, CT 06103

**CDI Postdoctoral Fellowships** **857**

The Children's Discovery Institute (CDI) is a world-class center for pediatric research and innovation created to encourage researchers to ask bold questions and take bold risks to uncover answers. By funding the work of creative scientists and clinicians in collaborative, multi-disciplinary research aimed at some of the most devastating childhood diseases and disorders, CDI will accelerate the realization of better treatments, cures, and preventions. Up to $30,000 per year for up to two years is available for especially promising postdoctoral trainees to complete a research project under the direction of a mentor on an issue relevant to the goals of the CDI.
*Requirements:* Proposals will be accepted from Washington University faculty and postdoctoral trainees. Preference will be given to faculty members prior to achieving tenure, to teams of investigators from multiple disciplines, and to multi-investigator projects to develop interactive research groups. Preference will also be given to projects

with potential for acquisition of new knowledge/translational impact and scientific experience; and to programs synergistic with Washington University and St. Louis Children's Hospital. Letters of intent must be submitted by August 1. For investigators invited to submit a proposal, applications will be due on October 15.
*Geographic Focus:* Missouri
*Date(s) Application is Due:* Oct 15
*Amount of Grant:* Up to 60,000 USD
*Contact:* Angela Mayer (Corless); (314) 286-2711; cdi@kids.wustl.edu
*Internet:* www.childrensdiscovery.org/Grants/FundingMechanismsDueDates.aspx
*Sponsor:* Children's Discovery Institute
660 S. Euclid Avenue
St. Louis, MO 63110

**Cedar Tree Foundation David H. Smith Conservation Research Fellowship** **858**
The David H. Smith Conservation Research Fellowship program identifies and supports early-career scientists who will shape the field of applied conservation biology. David H. Smith Conservation Research Fellowships are available to post-doctoral researchers (of any nationality) affiliated with a United States institution. In 1998, to help address the need for post-doctoral opportunities for conservation biology graduates, the foundation and founding partner The Nature Conservancy established the David H. Smith Conservation Research Fellowship Program, devoted exclusively to applied conservation research problems. By fostering the development of promising conservation scientists, the Smith Fellowship Program helps encourage this rapidly expanding field of scientific inquiry and link it to the practice of conservation. Each Fellow will receive an annual salary of $50,000 plus benefits, with the post-doctoral position expected to run for two consecutive years. In addition to the stipend, each Fellow receives a travel budget of $8,000 and a research fund of $32,000 over the 2-year fellowship period.
*Requirements:* To be eligible individuals must have completed their doctorate within the past five years or by the time the award is made.
*Geographic Focus:* All States
*Date(s) Application is Due:* Sep 20
*Amount of Grant:* 140,000 USD
*Contact:* Mike Dombeck, Executive Director; (202) 234-4133, ext. 307; info@cedartreefound.org or loi@cedartreefound.org
*Internet:* www.conbio.org/mini-sites/smith-fellows
*Sponsor:* Cedar Tree Foundation
100 Franklin Street, Suite 704
Boston, MA 02110-1401

**CEE Research Grants** **859**
The Center for Education and Ethics announces a new grant program. The Center will make awards of up to $40,000 for research projects in philosophy as it relates to educational policy and practice. It encourages applicants to understand educational policy and practice in broad terms, including issues that directly relate to K-12 schools and higher education institutions, but also concerning policies that influence children's growth and development in the family and other institutions. The Center also encourage diverse kinds of philosophical research ranging from the highly abstract to the highly applied. Proposals might concern any of the following topics: the proper content of moral education and of the rights of parents to choose its content; the place of religion in schools; justice and efficiency in the allocation of public funds across schools and school districts; the content of the curriculum; the commercialization of schools and childhoods generally; the obligations to students with special educational needs; the proper content of sex education in particular and education for living more generally (concerning e.g., parenting, financial self-management) and the extent to which it is right for schools to defer to parental preferences regarding these matters; the moral rights of school students to privacy, to freedom of expression, to freedom of association; the rights and obligations of teachers with respect to abusive or violent children; should schools cultivate the virtues needed to sustain a democratic society, and if so, what are they and how is this best done given the other values schools should realize and pursue; and ethical considerations in college admissions and enrollment. The annual deadline for applications is November 6.
*Requirements:* To apply, applicants will need to complete a short online form and upload a proposal narrative, budget, and curriculum vita.
*Geographic Focus:* All States
*Date(s) Application is Due:* Nov 6
*Amount of Grant:* 40,000 USD
*Contact:* Harry Brighouse, Director; (608) 263-2709; cee@wcer.wisc.edu
*Internet:* ethicsandeducation.wceruw.org/research-grants.html
*Sponsor:* Center for Ethics and Education
1025 West Johnson Street
Madison, WI 53706

**Center for the Study of Democratic Politics Fellowships** **860**
The Center for the Study of Democratic Politics (CSDP) was created as a research program within the Woodrow Wilson School of Public and International Affairs. The Center supports empirical research on democratic political processes and institutions. The CSDP's research program focuses on the relationship between democratic ideals and democratic practice. Its goal is to encourage rigorous social scientific analysis that informs and is informed by normative theories of democracy. The Center seeks visiting scholars and/or post-doctoral research scholars for research on democratic political processes and institutions. Applications are welcome from political scientists, in addition to related social science disciplines. Each scholar will pursue research and contribute to the intellectual life of the Center, the Department of Politics, and the Woodrow Wilson School of Public and International Affairs. Successful candidates will receive a competitive salary, applicable Princeton University benefits, a taxable moving allowance, and a $5,000 research fund. The application is available at the CSDP website.
*Requirements:* Center scholars must be in residence at Princeton at least four days each week during the academic terms, and are expected to participate in weekly CSDP seminars in the Department of Politics, as well as occasional lectures, conferences, and other events.
*Geographic Focus:* All States
*Date(s) Application is Due:* Nov 30
*Amount of Grant:* 5,000 USD
*Contact:* Helene Wood, CSDP Program Assistant; (609) 258-6493 or (609) 258-6980; fax (609) 258-5014; hwood@princeton.edu
*Internet:* www.princeton.edu/csdp/apply/
*Sponsor:* Princeton University
Center for the Study of Democratic Politics
Princeton, NJ 08544-1013

**CFF-NIH Funding Grants** **861**
The Cystic Fibrosis Foundation (CFF) has developed the overall research grant program to complement the awarding mechanism of the NIH. Support from CFF, through various mechanisms, is intended to provide for the development of sufficient preliminary data to make CF-related grant applications highly competitive in the NIH review process. However, as a result of funding constraints on the NIH, coupled with the growing interest in CF research, occasions arise in which highly meritorious grant applications are submitted to the NIH but are not funded. In an effort to assure that all meritorious CF-related research is supported, CFF has developed the CFF/NIH-unfunded Award mechanism to provide funding. The objective of this award is to support excellent CF-related research projects that have been submitted to and approved by the National Institutes of Health (NIH), but cannot be supported by available NIH funds. Applications must fall within the upper 40th percentile. CFF support ranges from $75,000 to $125,000 per year for up to two years.
*Restrictions:* Indirect costs are not allowed.
*Geographic Focus:* All States
*Amount of Grant:* 75,000 - 125,000 USD
*Contact:* Office of Grants Management; (800) 344-4823 or (301) 951-4422; fax (301) 951-6378; grants@cff.org
*Internet:* www.cff.org/research/ForResearchers/FundingOpportunities/ResearchGrants/#Pilot_and_Feasibility_Awards
*Sponsor:* Cystic Fibrosis Foundation
6931 Arlington Road, 2nd Floor
Bethesda, MD 20814

**CFF Clinical Research Grants** **862**
Cystic Fibrosis Foundation Therapeutics (CFFT), the nonprofit drug discovery and development affiliate of the Cystic Fibrosis Foundation, offers competitive awards to support clinical research projects directly related to CF treatment and care. Projects may address diagnostic or therapeutic methods related to CF or the pathophysiology of cystic fibrosis, and applicants must demonstrate access to a sufficient number of patients from Foundation-accredited care centers and to appropriate controls. Up to $100,000 per year (plus 8% indirect costs) for a maximum of three years may be requested for single-center clinical research grants. For multi-center clinical research, the potential award is up to $225,000 per year (plus 8% indirect costs) for a maximum of three years.
*Requirements:* Applications will be accepted only after submission and approval of a letter of intent. CFFT strictly adheres to the letter of intent and award application deadlines.
*Geographic Focus:* All States
*Date(s) Application is Due:* Mar 6; Sep 11
*Amount of Grant:* Up to 225,000 USD
*Contact:* Office of Grants Management; (800) 344-4823 or (301) 951-4422; fax (301) 951-6378; grants@cff.org
*Internet:* www.cff.org/research/ForResearchers/FundingOpportunities/ClinicalResearchAwards/
*Sponsor:* Cystic Fibrosis Foundation
6931 Arlington Road, 2nd Floor
Bethesda, MD 20814

**CFF First- and Second-Year Clinical Fellowships** **863**
The Cystic Fibrosis Foundation offers competitive clinical fellowships for up to five years for physicians interested in cystic fibrosis and other chronic pulmonary and gastrointestinal diseases of children, adolescents, and adults. The intent of this award is to encourage specialized training early in a physician's career and to prepare well-qualified candidates for careers in academic medicine. Training must take place in one of the foundation's accredited centers and must provide thorough grounding in diagnostic and therapeutic procedures, comprehensive care, and cystic fibrosis-related research. All first- and second-year programs must commit a significant portion (at least 30 percent over two years) to research training. Fellows funded by other sources for their first year of training may apply to the foundation for subsequent training support. Third-year fellowships are available for additional basic and/or clinical research training. Research fellowships are also available from the foundation to support physicians beyond the fellowship experience; salary will be commensurate with experience.
*Requirements:* Applicants must be U.S. citizens or have permanent U.S. resident visas, and must have completed pediatric training and be eligible for board certification in pediatrics by the time the fellowship begins. Candidates with prior training in internal medicine may also apply but must have completed at least two years of an approved adult pulmonary or GI fellowship; be jointly sponsored by the departments of medicine and

pediatrics; and be willing to commit at least 75 percent of their time to cystic fibrosis and related problems of young adults.
*Geographic Focus:* All States
*Date(s) Application is Due:* Oct 2
*Amount of Grant:* 47,600 - 49,250 USD
*Contact:* Office of Grants Management; (800) 344-4823 or (301) 951-4422; fax (301) 951-6378; grants@cff.org
*Internet:* www.cff.org/research/ForResearchers/FundingOpportunities/TrainingGrants/
*Sponsor:* Cystic Fibrosis Foundation
6931 Arlington Road, 2nd Floor
Bethesda, MD 20814

**CFF Harry Shwachman Clinical Investigator Award** 864

The three-year CFF Harry Shwachman Clinical Investigator Award provides the opportunity for clinically trained physicians to develop into independent biomedical research investigators who have active involvement in cystic fibrosis-related areas. It is also intended to facilitate the transition from postdoctoral training to a career in academic medicine. Support is available for up to $76,000 per year plus $15,000 for supplies. Support is based on a full-time, twelve-month appointment. Applications are due the second Wednesday in September.
*Requirements:* U.S. citizenship or permanent resident status is required.
*Restrictions:* Indirect costs are not allowed.
*Geographic Focus:* All States
*Date(s) Application is Due:* Sep 11
*Amount of Grant:* Up to 91,000 USD
*Contact:* Office of Grants Management; (800) 344-4823 or (301) 951-4422; fax (301) 951-6378; grants@cff.org
*Internet:* www.cff.org/research/ForResearchers/FundingOpportunities/ResearchGrants/#Pilot_and_Feasibility_Awards
*Sponsor:* Cystic Fibrosis Foundation
6931 Arlington Road, 2nd Floor
Bethesda, MD 20814

**CFF Leroy Matthews Physician-Scientist Awards** 865

In honor of Dr. LeRoy Matthews' dedication and commitment to cystic fibrosis (CF) research and care, the Cystic Fibrosis Foundation (CFF) announces the Physician Scientist Award named in his memory. Awards will provide up to six years of support for outstanding newly trained pediatricians and internists (MDs and MD/PhDs) to complete sub-specialty training, develop into independent investigators, and initiate research programs. Institutional and individual grants are available. Support ranges from a $48,000 stipend, plus $10,000 for research and development for year one, up to a $76,000 stipend, plus $15,000 for research and development for year six.
*Requirements:* U.S. citizenship or permanent resident status is required.
*Restrictions:* Indirect costs are not allowed.
*Geographic Focus:* All States
*Date(s) Application is Due:* Sep 11
*Amount of Grant:* 48,000 - 86,000 USD
*Contact:* Office of Grants Management; (800) 344-4823 or (301) 951-4422; fax (301) 951-6378; grants@cff.org
*Internet:* www.cff.org/research/cystic_fibrosis_foundation_grants/research_grants
*Sponsor:* Cystic Fibrosis Foundation
6931 Arlington Road, 2nd Floor
Bethesda, MD 20814

**CFF Pilot and Feasibility Awards** 866

The Cystic Fibrosis Foundation offers feasibility grants for developing and testing new hypotheses and/or new methods, as well as to support promising new investigators as they establish themselves in research areas relevant to cystic fibrosis. Proposed work must be hypothesis driven and must reflect innovative approaches to critical questions in cystic fibrosis research. Funding priority will be placed on those projects proposing to better understand the mechanisms behind disease pathophysiology and to develop strategies to prevent or treat it. Up to $40,000 per year (plus 8% indirect costs) for two years may be requested.
*Requirements:* Electronic submission is required.
*Restrictions:* The award is not meant to support continuation of programs begun under other granting mechanisms.
*Geographic Focus:* All States
*Date(s) Application is Due:* Sep 11
*Amount of Grant:* Up to 40,000 USD
*Contact:* Office of Grants Management; (800) 344-4823 or (301) 951-4422; fax (301) 951-6378; grants@cff.org
*Internet:* www.cff.org/research/cystic_fibrosis_foundation_grants/research_grants
*Sponsor:* Cystic Fibrosis Foundation
6931 Arlington Road, 2nd Floor
Bethesda, MD 20814

**CFF Postdoctoral Research Fellowships** 867

Competitive postdoctoral research fellowships in basic or clinical research are made on an annual basis, renewable for a second year. The fellowships are awarded in areas of basic cellular and metabolic research or in other problem areas pertinent to cystic fibrosis and related chronic and recurrent pulmonary and gastrointestinal diseases of childhood. Preference is shown to recent graduates or those just beginning their investigative careers. A third year is offered on a limited basis to highly qualified candidates. Interested individuals are encouraged to contact the foundation for application procedures and/or to discuss the potential relevance of their work to the objectives of the foundation. Stipends are $39,000 (first year), $40,100 (second year), and $42,300 (optional third year). Research expenses of $3,750 per year are available, as well.
*Requirements:* These awards are offered to MDs, PhDs, and MD/PhDs interested in conducting basic or clinical research related to cystic fibrosis. U.S. citizenship or permanent resident status is required.
*Restrictions:* Electronic submission only. Indirect costs are not allowed.
*Geographic Focus:* All States
*Date(s) Application is Due:* Sep 11
*Amount of Grant:* 39,000 - 46,050 USD
*Contact:* Office of Grants Management; (800) 344-4823 or (301) 951-4422; fax (301) 951-6378; grants@cff.org
*Internet:* www.cff.org/research/ForResearchers/FundingOpportunities/TrainingGrants/
*Sponsor:* Cystic Fibrosis Foundation
6931 Arlington Road, 2nd Floor
Bethesda, MD 20814

**CFF Research Grants** 868

Grants are offered by the Cystic Fibrosis Foundation in support of high-quality research projects ranging from basic cellular and metabolic mechanisms to therapy of cystic fibrosis and related chronic and recurrent pulmonary and gastrointestinal diseases of childhood. These grants are broadly oriented to the support of projects for developing and initially testing new hypotheses and/or new methods, or those being applied to problems of cystic fibrosis for the first time. The intent of this award is to enable the investigator to collect sufficient data to compete successfully for long-term support from NIH. Interested individuals are encouraged to contact the foundation for guidelines and/or to discuss the potential relevance of their work to the objectives of this program. Support is available for $90,000 per year (plus 8% indirect costs) for a period of two years, at which time a grant may be competitively renewed for an additional year of funding.
*Requirements:* Investigators who seek support from the Foundation under these funding mechanisms must submit a Letter of Intent (LOI). Applications may be submitted by established investigators or new investigators starting their independent research careers but may not represent support for continuation of a line of research already established by the applicant.
*Geographic Focus:* All States
*Date(s) Application is Due:* Sep 11
*Amount of Grant:* Up to 90,000 USD
*Contact:* Office of Grants Management; (800) 344-4823 or (301) 951-4422; fax (301) 951-6378; grants@cff.org
*Internet:* www.cff.org/research/ForResearchers/FundingOpportunities/ResearchGrants/
*Sponsor:* Cystic Fibrosis Foundation
6931 Arlington Road, 2nd Floor
Bethesda, MD 20814

**CFF Student Traineeships** 869

Student Traineeships are offered by the Cystic Fibrosis Foundation to qualified students in order to introduce them to cystic fibrosis research. Each recipient must work with a faculty sponsor on a research project related to cystic fibrosis. Interested individuals are encouraged to contact the foundation for application procedures and/or to discuss the potential relevance of their work to the objectives of the program. Applications are accepted throughout the year. A maximum of $300 of the award may be used for laboratory expenses with the remainder used as a $1,500 stipend for the trainee.
*Requirements:* Applicants must be students in or about to enter a doctoral program (MD, PhD, or MD/PhD); senior-level undergraduates planning to pursue graduate training may also apply.
*Geographic Focus:* All States
*Amount of Grant:* 1,500 USD
*Contact:* Office of Grants Management; (800) 344-4823 or (301) 951-4422; fax (301) 951-6378; grants@cff.org
*Internet:* www.cff.org/research/ForResearchers/FundingOpportunities/TrainingGrants/
*Sponsor:* Cystic Fibrosis Foundation
6931 Arlington Road, 2nd Floor
Bethesda, MD 20814

**CFF Third-, Fourth-, and Fifth-Year Clinical Fellowships** 870

The Cystic Fibrosis Foundation offers competitive clinical fellowships for up to five years for physicians interested in cystic fibrosis and other chronic pulmonary and gastrointestinal diseases of children, adolescents, and adults to encourage specialized training early in a physician's career and to prepare well-qualified candidates for careers in academic medicine. The third-, fourth-, and fifth-year fellowship award offers support for additional intense basic and/or clinical research training related to cystic fibrosis. Up to $68,250 may be awarded: $58,250 for stipend and $10,000 for research costs.
*Requirements:* Applicants and sponsors must submit proposals of the research studies to be undertaken and other specialized training that will be offered during this year. Preference will be given to applicants whose training was supported by the foundation.
*Restrictions:* Recipients who do not enter a career of academic medicine will be subject to payback provisions. Indirect costs are not allowed.
*Geographic Focus:* All States
*Date(s) Application is Due:* Sep 11
*Contact:* Office of Grants Management; (800) 344-4823 or (301) 951-4422; fax (301) 951-6378; grants@cff.org
*Internet:* www.cff.org/research/ForResearchers/FundingOpportunities/TrainingGrants/
*Sponsor:* Cystic Fibrosis Foundation
6931 Arlington Road, 2nd Floor
Bethesda, MD 20814

**CFNEM Tobacco Settlement Grants** 871
The Community Foundation for Northeast Michigan Tobacco Settlement Grants focus on programs or projects that address tobacco-related issues, and benefit the residents of the northeast Michigan counties of Alcona, Alpena, Montmorency, and Presque Isle. The maximum amount for these awards is $3,000, and the annual grant application deadline is January 15.
*Requirements:* IRS 501(c)3 nonprofit organizations, schools, churches (for non-sectarian purposes), cities, townships, and other governmental units serving the four-county area of Alcona, Alpena, Montmorency, and Presque Isle in northeast Michigan are eligible to apply.
*Restrictions:* Grants are not given to individuals, except for awards or scholarships from designated donor funds. Grants are made for future projects only. No funding will be given for projects completed before final board approval. Board approval is usually within six (6) weeks of the grant deadline. Grants are not made for routine operating needs or budget deficits.
*Geographic Focus:* Michigan
*Date(s) Application is Due:* Feb 1
*Amount of Grant:* Up to 3,000 USD
*Contact:* Julie Wiesen, Program Director; (989) 354-6881 or (877) 354-6881; fax (989) 356-3319; wiesenj@cfnem.org
*Internet:* www.cfnem.org/grants/tobacco-settlement-grants.html
*Sponsor:* Community Foundation for Northeast Michigan
100 N. Ripley, Suite F, P.O. Box 495
Alpena, MI 49707-0495

**CFR Edward R. Murrow Press Fellowship** 872
The Council on Foreign Relations (CFR) awards one fellowship annually to a distinguished foreign correspondent or editor for the purpose of promoting the quality of responsible and discerning journalism that distinguished the work of radio and television journalist Edwin R. Murrow. The fellowship was initially launched in 1949 with help from the Carnegie Foundation and was subsequently renamed to honor Mr. Murrow in 1965. The Edward R. Murrow Press Fellow spends nine months full-time-in-residence at CFR's headquarters in New York. The program enables the fellow to engage in sustained analysis and writing, expand his or her intellectual and professional horizons, and extensively participate in CFR's active program of meetings and events. The fellow will be part of the David Rockefeller Studies Program (CFR's think tank) alongside the program's full-time, adjunct, and visiting fellows, whose expertise extends across the broad range of significant foreign-policy issues facing the United States and the international community. Interested candidates who meet the program's eligibility requirements can apply at the CFR website between January 1 and March 1 each year. Official selections and announcement of the award will be made in April. The fellowship should preferably begin in September. The program awards a stipend of $65,000 as well as a modest travel grant. The fellow is considered an independent contractor rather than an employee of CFR, and is not eligible for employment benefits, including health insurance. Founded in 1921, the Council on Foreign Relations is an independent, nonpartisan membership organization, think tank, and publisher dedicated to being a resource for its members, government officials, business executives, journalists, educators and students, civic and religious leaders, and other interested citizens in order to help them better understand the world and the foreign policy choices facing the United States and other countries. The CFR takes no institutional positions on matters of policy.
*Requirements:* The program is open only to applicants who have distinguished credentials in the field of journalism and who have covered international news as a working journalist for print, broadcast, or online media widely available in the United States. Applicants are limited to those individuals who are authorized to work in the United States and who will continue to be authorized for the duration of the fellowship. Selection as a Murrow Fellow is based on a combination of the following criteria: depth and breadth of professional experience as a foreign correspondent or editor; firm grounding in foreign policy and international relations; and a clear, creative, and original application statement. With respect to the application statement, candidates must submit a detailed description of their research project as well as their expected CFR participation and any planned fellowship travel. The work plan will be judged on its originality, practicality, potential, likelihood of completion during the fellowship period, and the contribution it will make to the applicant's individual career development as well as to CFR.
*Restrictions:* CFR does not sponsor for visas.
*Geographic Focus:* All States
*Date(s) Application is Due:* Mar 1
*Amount of Grant:* 65,000 USD
*Contact:* Janine Hill, Director, Fellowship Affairs and Studies Strategic Planning; (212) 734-9489 or (212) 434-9740; fax (212) 434-9800; Fellowships@cfr.org
*Internet:* www.cfr.org/thinktank/fellowships/murrow.html
*Sponsor:* Council on Foreign Relations
58 East 68th Street
New York, NY 10065

**CFR International Affairs Fellowships** 873
The Council on Foreign Relations (CFR) awards ten 12-month International Affairs Fellowships (IAF)s annually. The purpose of this fellowship program, founded in 1967, is to strengthen the mid-career development of outstanding individuals by helping them acquire and apply foreign-policy skills beyond the scope of their professional and scholarly achievements. The program's distinctive character lies in the contrasting professional experiences fellows receive; fellows from academia and the private sector spend their tenures in public service and policy-oriented settings, while government officials spend their tenures in a scholarly atmosphere free from operational pressure. CFR's Fellowship Affairs Office assists all fellows in finding a suitable affiliation for the year. In addition to providing the opportunity to carry out research, the IAF Program integrates all fellows into the intellectual life of CFR. Fellows who are not placed at CFR during their tenure are invited to attend and participate in select CFR meetings and events. Alumni of the program stay connected with CFR and its prestigious network of professionals and leaders, and convene at CFR's annual IAF Conference in New York City each spring. Interested candidates who meet the program's eligibility requirements can apply at the CFR website between June 1 and October 1 each year. Candidates who are selected as IAF finalists will be notified between December and January, with finalist interviews scheduled in Washington, D.C., and New York City between January and February. Official selections and announcement of IAF awards will be made between February and March. Fellowships should preferably begin in September. Though deferment is not an option, requests to do so, for up to one year only, will be considered on a case-by-case basis and under special circumstances. The program awards a stipend of $85,000. Fellows are considered independent contractors rather than employees of CFR, and are not eligible for employment benefits, including health insurance. Founded in 1921, the Council on Foreign Relations is an independent, nonpartisan membership organization, think tank, and publisher dedicated to being a resource for its members, government officials, business executives, journalists, educators and students, civic and religious leaders, and other interested citizens in order to help them better understand the world and the foreign policy choices facing the United States and other countries. The CFR takes no institutional positions on matters of policy.
*Requirements:* The IAF Program is open only to U.S. citizens and permanent residents between the ages of twenty-seven and thirty-five who are eligible to work in the United States. While a Ph.D. is not a requirement, selected fellows generally hold an advanced degree and possess a strong record of work experience as well as a firm grounding in the field of foreign policy. Selection as an IAF is based on a combination of the following criteria: scholarly qualifications, achievements, and promise; depth and breadth of professional experience; firm grounding in foreign policy and international relations; and an application proposal that focuses on solutions to identified problems in U.S. foreign policy. The proposal will be judged on the proposed work's originality, practicality, potential, likelihood of completion during the fellowship period, and the contribution it will make to the applicant's individual career development. Applicants are encouraged to plan a systematic approach to assessing the major substantive and process issues of their planned research.
*Restrictions:* The program does not fund pre- or postdoctoral research, work toward a degree, or the completion of projects for which substantial progress has been made prior to the fellowship period. CFR does not sponsor for visas.
*Geographic Focus:* All States
*Date(s) Application is Due:* Oct 1
*Amount of Grant:* 85,000 USD
*Contact:* Janine Hill, Director, Fellowship Affairs and Studies Strategic Planning; (212) 434-9489 or (212) 434-9740; fax (212) 434-9800; Fellowships@cfr.org
*Internet:* www.cfr.org/thinktank/fellowships/iaf.html
*Sponsor:* Council on Foreign Relations
58 East 68th Street
New York, NY 10065

**CFR International Affairs Fellowships in Nuclear Security** 874
The Council on Foreign Relations (CFR) generally awards two International Affairs Fellowships in Nuclear Security (IAF-NS) annually to university-based scholars. The program places academics in U.S. Government positions or with international organizations to work with practitioners. Its goal is to close the gap between research and practice and to enrich the teaching and scholarship of academics while also benefiting policymakers by exposing them to cutting-edge scholarly research. Interested candidates who meet the program's eligibility requirements must submit a cover letter, a CV, and a proposal (maximum 1,000 words in length). Each applicant should arrange to have two letters of recommendation sent that assess the policy relevance of the applicant's proposed project as well as the applicant's qualifications for carrying it out. All application materials must be submitted to the email address given by November 1. The duration of the fellowship is twelve months, preferably beginning in September. Though deferment is not an option, requests to do so, for up to one year only, will be considered on a case-by-case basis and under special circumstances. The program awards a stipend of $125,000. Fellows are considered independent contractors rather than employees of CFR, and are not eligible for employment benefits, including health insurance. The IAF-NS program is supported by the Stanton Foundation. Founded in 1921, CFR is an independent, nonpartisan membership organization, think tank, and publisher dedicated to being a resource for its members, government officials, business executives, journalists, educators and students, civic and religious leaders, and other interested citizens in order to help them better understand the world and the foreign policy choices facing the United States and other countries. The CFR takes no institutional positions on matters of policy.
*Requirements:* The IAF-NS is open only to faculty members with tenure or on tenure-track lines at accredited universities who propose to conduct policy-relevant research on nuclear security issues. Qualified candidates must be U.S. citizens or permanent residents who are eligible to work in the United States and be between the ages of twenty-nine and forty. Selection as an IAF-NS is based on a combination of the following criteria: scholarly qualifications, achievements and promise; depth and breadth of professional experience; firm grounding in foreign policy and international relations; and an application proposal that focuses on nuclear security issues. The proposal will be judged on the proposed work's originality, practicality, potential, likelihood of completion during the fellowship period, and the contribution it will make to the applicant's individual career development.
*Restrictions:* CFR does not sponsor for visas.
*Geographic Focus:* All States
*Date(s) Application is Due:* Nov 1
*Amount of Grant:* 125,000 USD

*Contact:* Janine Hill, Director, Fellowship Affairs and Studies Strategic Planning; (212) 434-9489 or (212) 434-9740; fax (212) 434-9800; fellowships@cfr.org
*Internet:* www.cfr.org/thinktank/fellowships/iaf_nuclear.html
*Sponsor:* Council on Foreign Relations
58 East 68th Street
New York, NY 10065

**CFR International Affairs Fellowships in South Korea** 875
The International Affairs Fellowship in South Korea (IAF-SK), sponsored by the Asan Institute for Policy Studies, was founded in 2010 to strengthen mutual understanding and cooperation between the rising generations of leaders in the United States and South Korea by expanding cultural exchange and enhancing communication between both countries on global issues. The program assists two to three mid-career scholars and professionals from the public and private sectors to advance their analytic capabilities and broaden their foreign-policy experience in South Korea while affiliated with the Asan Institute. Fellows are drawn from academia, business, government, media, NGOs, and think tanks. Interested candidates who meet the program's eligibility requirements must submit an application form (available at the website as a downloadable PDF), a CV, and a proposal (as specified in the application form). Each applicant should arrange to have two letters of recommendation sent that assess the policy relevance of the applicant's proposed project as well as the applicant's qualifications for carrying it out. All application materials must be submitted to the email address given by October 1. Candidates who are selected as IAF-SK finalists will be notified in November, with finalist interviews scheduled in Washington, D.C., and New York City shortly after. Official selections and announcement of IAF-SK awards will be made around December. The duration of the fellowship is between four and twelve months. The program awards a stipend which covers living expenses in Seoul, South Korea plus international travel and necessary research expenses. Fellows are considered independent contractors rather than employees of CFR, and are not eligible for employment benefits, including health insurance. Founded in 1921, the Council on Foreign Relations is an independent, nonpartisan membership organization, think tank, and publisher dedicated to being a resource for its members, government officials, business executives, journalists, educators and students, civic and religious leaders, and other interested citizens in order to help them better understand the world and the foreign policy choices facing the United States and other countries. The CFR takes no institutional positions on matters of policy.
*Requirements:* The IAF-SK is only open to U.S. citizens. Applicants do not have to be Korea specialists, and knowledge of the Korean language is not a requirement. While a Ph.D. is not a requirement, selected fellows generally hold an advanced degree and possess a strong record of work experience as well as a firm grounding in the field of foreign policy. Applicants should have a capacity for independent work and be eager to undertake serious foreign-policy analysis. Selection as an IAF-SK is based on a combination of the following criteria: scholarly qualifications, achievements and promise; depth and breadth of professional experience; firm grounding in foreign policy and international relations; and an application proposal that focuses on U.S.-South Korea foreign policy issues. The proposed work will be judged on originality, practicality, potential, likelihood of completion during the fellowship period, and the contribution it will make to the applicant's individual career development. With respect to the proposal, applicants are encouraged to plan a systematic approach to assessing the major substantive and process issues of their planned research.
*Restrictions:* The program does not fund pre- or postdoctoral research, work toward a degree, or the completion of projects for which substantial progress has been made prior to the fellowship period.
*Geographic Focus:* All States
*Date(s) Application is Due:* Oct 1
*Contact:* Janine Hill, Director, Fellowship Affairs and Studies Strategic Planning; (212) 434-9489 or (212) 434-9740; fax (212) 434-9800; fellowships@cfr.org
*Internet:* www.cfr.org/thinktank/fellowships/iaf_southkorea.html
*Sponsor:* Council on Foreign Relations
58 East 68th Street
New York, NY 10065

**CFR International Affairs in Japan Fellowships** 876
The International Affairs Fellowship in Japan (IAF-J), sponsored by Hitachi, Ltd., was founded in 1997 to strengthen mutual understanding and cooperation between the rising generations of leaders in the United States and Japan. The program provides three to five mid-career U.S. citizens the opportunity to expand their professional horizons by spending a period of research or other professional activity in Japan. Fellows are drawn from academia, business, government, media, NGOs, and think tanks. In cooperation with the Council on Foreign Relations (CFR), Hitachi, Ltd., assists fellows in finding suitable host organizations in Japan. A non-exclusive list of cooperating institutions in Japan can be found on the IAF-J's webpage at the CFR website. Interested candidates who meet the program's eligibility requirements can apply at the CFR website between June 1 and October 1 each year. Candidates who are selected as IAF-J finalists will be notified between December and January, with finalist interviews scheduled in Washington, D.C., and New York City between January and February. Official selections and announcement of IAF-J awards will be made between February and March. The duration of the fellowship is between three and twelve months. The program awards a stipend in yen which covers travel and living expenses in Japan. Fellows are considered independent contractors rather than employees of CFR, and are not eligible for employment benefits, including health insurance. Founded in 1921, the Council on Foreign Relations is an independent, nonpartisan membership organization, think tank, and publisher dedicated to being a resource for its members, government officials, business executives, journalists, educators and students, civic and religious leaders, and other interested citizens in order to help them better understand the world and the foreign policy choices facing the United States and other countries. The CFR takes no institutional positions on matters of policy.
*Requirements:* The IAF-J is open only to U.S. citizens between the ages of twenty-seven and forty-five. The program is intended primarily for those without substantial prior experience in Japan, although the selection committee has made exceptions when it considered that the fellowship would allow an individual to add a significant new dimension to his or her career. Knowledge of the Japanese language is not a requirement. Fellows are selected on the basis of academic and professional accomplishments and promise; the merits and feasibility of their specific research or action proposals; character and personal qualities conducive to promoting cross-cultural communication and cooperation; and the contribution that the proposed research or professional activity will make to the applicant's individual career development.
*Geographic Focus:* All States
*Date(s) Application is Due:* Oct 1
*Contact:* Janine Hill, Director, Fellowship Affairs and Studies Strategic Planning; (212) 434-9489 or (212) 434-9740; fax (212) 434-9800; fellowships@cfr.org
*Internet:* www.cfr.org/thinktank/fellowships/iaf_japan.html
*Sponsor:* Council on Foreign Relations
58 East 68th Street
New York, NY 10065

**CFR Military Fellowships** 877
Each year, the Chiefs of Staff of the Army and the Air Force, the Chief of Naval Operations, and the Commandants of the Coast Guard and the Marine Corps nominate an outstanding officer from their respective services as a candidate for a Council on Foreign Relations (CFR) Military Fellowship. CFR usually awards four to five such fellowships annually. The program enables the selected officers to broaden their understanding of international relations by spending a year in residence at CFR's headquarters in New York. Fellows expand their knowledge of international relations through a program of individual study, research and reflection, extensive participation in CFR's active program of meetings and events, and interaction with the CFR's diverse membership. They also assist in arranging several military trips for CFR members during the year. Founded in 1921, the Council on Foreign Relations is an independent, nonpartisan membership organization, think tank, and publisher dedicated to being a resource for its members, government officials, business executives, journalists, educators and students, civic and religious leaders, and other interested citizens in order to help them better understand the world and the foreign policy choices facing the United States and other countries. The CFR takes no institutional positions on matters of policy.
*Requirements:* Interested candidates from the U.S. Army, U.S. Air Force, U.S. Coast Guard, U.S. Marine Corps, and U.S. Navy should contact their human resources officer to learn more about the fellowship application process.
*Geographic Focus:* All States
*Contact:* Janine Hill, Director, Fellowship Affairs and Studies Strategic Planning; (212) 434-9489 or (212) 434-9740; fax (212) 434-9800; fellowships@cfr.org
*Internet:* www.cfr.org/thinktank/fellowships/mil.html
*Sponsor:* Council on Foreign Relations
58 East 68th Street
New York, NY 10065

**CFR Stanton Nuclear Security Fellowships** 878
The Stanton Nuclear Security Fellowship (SNSF) Program, made possible by a generous grant from the Stanton Foundation, offers younger scholars studying nuclear-security issues the opportunity to spend a twelve-month period at the Council on Foreign Relations' (CFR) offices in New York or Washington, D.C., conducting policy-relevant research. While in residence full-time at CFR, fellows will be expected to lead a project of their own design, conduct original research, and write at least one policy-relevant document as well as participate fully in CFR's intellectual life. SNSF fellows will be mentored by the fellows of CFR's David Rockefeller Studies Program. Interested candidates must submit an application form (available at the website as a downloadable PDF), a curriculum vitae, and a proposal outlining the work proposed to conduct. Additionally, CFR must receive two letters of recommendation that assess the policy relevance of the applicant's proposed project as well as the applicant's qualifications for carrying it out. All application materials must be submitted to the email address given by the deadline date. The program awards a stipend of $100,000 for junior (non-tenured) faculty; $75,000 for post-doctoral; and $50,000 for pre-doctoral fellows. Payment will be made in twelve equal monthly installments. Fellowships should preferably begin in September. Fellows are considered independent contractors rather than employees of CFR, and are not eligible for employment benefits, including health insurance. Founded in 1921, the Council on Foreign Relations is an independent, nonpartisan membership organization, think tank, and publisher dedicated to being a resource for its members, government officials, business executives, journalists, educators and students, civic and religious leaders, and other interested citizens in order to help them better understand the world and the foreign policy choices facing the United States and other countries. The CFR takes no institutional positions on matters of policy.
*Requirements:* The SNSF Program seeks to stimulate the development of the next generation of thought leaders in nuclear security. Qualified candidates must be junior (non-tenured) faculty, post-doctoral fellows, or pre-doctoral candidates from any discipline who are working on a nuclear-security related issue. The program is only open to U.S. citizens and permanent residents who are eligible to work in the United States. The fellowships will be awarded on the basis of academic and professional accomplishments and promise, and on the merits of the specific research projects proposed. Fellows may work on a wide range of issues, including nuclear terrorism, nuclear proliferation, nuclear weapons, nuclear force posture, and, as it relates to nuclear security, nuclear energy.

*Restrictions:* CFR does not sponsor for visas.
*Geographic Focus:* All States
*Date(s) Application is Due:* Dec 14
*Amount of Grant:* 50,000 - 100,000 USD
*Contact:* Janine Hill, Director, Fellowship Affairs and Studies Strategic Planning; (212 434-9489 or (212) 434-9740; fax (212) 434-9800; fellowships@cfr.org
*Internet:* www.cfr.org/thinktank/fellowships/StantonFellowship.html
*Sponsor:* Council on Foreign Relations
58 East 68th Street
New York, NY 10065

**Charles Edison Fund Grants** 879

The Charles Edison Fund is an endowed philanthropic institution, incorporated in Delaware in 1948, dedicated to the support of worthwhile endeavors generally within the areas of medical research, science education and historic preservation. Funding is available on a national level but, the majority of institutions and, organizations assisted are based principally in the New York-New Jersey Metropolitan area. This concentration of interest facilitates the efforts of the Trustees to evaluate the work of recipient groups, frequently accomplished by personal visitations. The Fund is a seed money organization, primarily starting projects which would not otherwise get off the ground. The Fund meets three times a year, usually in February or March, June and December at which time requests which have been submitted at least three weeks prior to the meeting will be considered.
*Requirements:* Public and private schools, universities and colleges, and nonprofits are eligible. The Fund recommends that a grant request be submitted on the requesting organization's letterhead and be signed by an official on behalf of the governing board. The Fund does not require or supply application forms. The request should be detailed, complete and include background information about the organization, a full explanation of the project and its costs, and a financial report, current budget and evidence of tax-exempt status of the requesting organization.
*Geographic Focus:* New Jersey, New York
*Contact:* John P. Keegan, President; (973) 648-0500; fax (973) 648-0400; info@charlesedisonfund.org
*Internet:* www.charlesedisonfund.org/thefund.html
*Sponsor:* Charles Edison Fund
One Riverfront Plaza, 4th Floor
Newark, NJ 07102

**Charles H. Farnsworth Trust Grants** 880

The Charles H. Farnsworth Trust was established in 1930 to assist older adults to live in dignity and with independence. In describing the purpose of his legacy, Mr. Farnsworth made clear his interest in supporting housing, particularly in developing affordable housing options, and in providing support services to older adults. Program interests include: development of housing, especially housing with support services; services for elderly persons (i.e. health care, homemaker assistance, and nutritional support to enable the elderly to continue living in the community); and research, planning, and communication to better inform individuals, institutions, and the community at large on ways to improve the quality and quantity of housing and support services for seniors. Capital grants related to construction or renovation of housing for older adults is of particular interest. While general operating grants are provided, requests for support for new or special projects/programs are preferred. Planning grants investigating strategies for the development of supportive housing for the elderly are encouraged. Capital grants generally range between $25,000 and $250,000. General operating grants are usually no more than $10,000. Applicants must apply online at the grant website. Applicants are strongly encouraged to do the following before applying: review the downloadable state application procedures for additional helpful information and clarifications; review the downloadable online-application guidelines at the grant website; review the trust's funding history (link is available from the grant website); review the online application questions in advance; and review the list of required attachments. These will generally include: a list of board members, financial statements (audited, reviewed, or compiled by independent auditor); an organization summary; a list of other funding sources; an IRS Determination letter; and other required documents. All attachments must be uploaded in the online application as PDF, Word, or Excel files. The Farnsworth Trust has biannual deadlines of February 1 and October 15. Grant applicants for the February deadline will be notified of grant decisions by May 31 and applicants for the October deadline will be notified of grant decisions by December 31.
*Restrictions:* Grant opportunities are restricted to the Greater Boston area.
*Geographic Focus:* Massachusetts
*Date(s) Application is Due:* Feb 1; Oct 15
*Amount of Grant:* 10,000 - 250,000 USD
*Samples:* Hearth, Boston, Massachusetts, $100,000; Greater Boston Chinese Golden Age Center, Boston, Massachusetts, $289,346; Friends of Saint Josephs Food Pantry, Salem, Massachusetts, $10,000.
*Contact:* Michealle Larkins; (866) 778-6859; michealle.larkins@baml.com
*Internet:* https://www.bankofamerica.com/philanthropic/fn_search.action
*Sponsor:* Charles H. Farnsworth Trust
225 Franklin Street, 4th Floor, MA1-225-04-02
Boston, MA 02110

**Charles H. Price II Family Foundation Grants** 881

The Charles H. Price II Family Foundation, established in Kansas City, Missouri, by a prominent American businessman and former Ambassador of the United States, offers support to educational programs and research primarily in Missouri and California (though funding is also occasionally provided outside of these two states). Grants are given for general operating support and scholarships. There are no specific applications or deadlines with which to adhere, and applicants should approach the Foundation in writing via a letter of application.
*Geographic Focus:* California, Missouri
*Amount of Grant:* 500 - 5,000 USD
*Samples:* Baker University, Baldwin, Kansas, $1,000 - scholarship fund; Eisenhower Medical Center, Rancho Mirage, California, $5,000 - general operating support; Nelson-Atkins Museum of Art, Kansas City, Missouri, $2,000 - general operating support.
*Contact:* Charles H. Price II, Director; (816) 360-6174 or (816) 360-6176
*Sponsor:* Charles H. Price II Family Foundation
1 W. Armour Boulevard, Suite 300
Kansas City, MO 64111-2004

**Charles Koch Foundation Afghanistan and Syria Research Grants** 882

The Charles Koch Foundation provides funding for academic and public policy research directed at solving social problems through voluntary action and free enterprise. Recently, President Trump announced his intentions to withdraw troops from Syria and draw down America's military presence in Afghanistan. Given the President's recent directives—and urgent questions raised about when and how U.S. deployments in either country will end—the Charles Koch Foundation will fund a series of grants to support research projects (e.g. papers, conferences, workshops, media, etc.) that explore viable solutions to accomplish stated U.S. objectives in Afghanistan and Syria. To that end, the Foundation is actively soliciting proposals for projects ranging up to $25,000 to consider these imperatives between now and November of this year. The proposal submission deadline is May 30. Until that time, the Foundation is especially interested in research that: addresses the unintended consequences of U.S. military actions abroad and explores the costs and impact of the engagements in Afghanistan and Syria; explores the impact of U.S. military engagement in Afghanistan and Syria upon American society (for instance, the effect of our activities on civil liberties or the health and welfare of veterans); explores the concept of threat inflation and critically examines theories of credibility and reputation in relation to decisions to reduce commitments and withdraw from foreign conflicts; analyzes executive–legislative relations as concern these conflicts, including the constitutional division of war powers; examines the impact of domestic interest groups (businesses, think tanks, and the permanent national security bureaucracy on U.S. foreign policy) as they relate to the wars in Afghanistan and Syria; scrutinizes Pentagon spending, force structure, and the strategic demands of remaining in Afghanistan and Syria, including structural incentives for military expenditures; and examines civil-military relations, especially in the context of U.S. involvement in the Afghanistan and Syrian wars. Application materials should include: a one-to-five-page abstract of the project on behalf of your university, college, think tank, or other 501(c)3 organization (the abstract should provide sufficient detail for reviewers to assess the nature and feasibility of the idea); a CV or resume; and a brief, itemized budget. Funding levels are commensurate with the requirements of the research and the potential for the research to advance an understanding of critical issues. Accepted proposals may also receive support to disseminate the research findings.
*Requirements:* Any college, university, think tank, or 501(c)3 organization in the U.S. is eligible to apply.
*Geographic Focus:* All States
*Date(s) Application is Due:* May 30
*Amount of Grant:* Up to 25,000 USD
*Contact:* Elizabeth B. Koch, Director; (703) 875-1770 or (703) 875-1658; fax (703) 875-1766; grants@charleskochfoundation.org
*Internet:* https://www.charleskochfoundation.org/apply-for-grants/requests-for-proposals/foreign-policy/afghanistan-and-syria-research-grants/
*Sponsor:* Charles Koch Foundation
1320 N. Courthouse Road, Suite 500
Arlington, VA 22201

**Charles Koch Foundation Communities Grants** 883

The Charles Koch Foundation provides funding for academic and public policy research directed at solving social problems through voluntary action and free enterprise. In the area of Communities, the Foundation understands that charities, non-profits, and NGOs stand as a powerful proof of concept of the independent sector's past successes in increasing social mobility, building safer communities, and helping the least fortunate realize their potential. Yet just as society's challenges have evolved, the conditions under which Americans volunteer their time, talents, and resources have changed, leaving questions about how best to support the independent sector as one of the most important assets in the drive to eliminate inequality and injustice. Colleges and universities are uniquely positioned to help identify, study, and disseminate best practices, and the Charles Koch Foundation is interested in supporting scholarly projects that offer real-world impact by contributing to the body of knowledge on which policies and practices best position the independent sector to contribute to strong communities. It is particularly interested in supporting research and initiatives that: provide voluntary associations with the tools necessary to solve community problems using a combination of local knowledge and evidence-driven strategies; advance understanding of where and how civil society produces the greatest value for individuals and communities; define and measure the health of communities, with a particular interest in projects that improve insight into the process of personal transformation; clarify and assess the overlap between private, public, and independent sector solutions to problems such as poverty, addiction, or education, with a particular interest in projects that help identify the cost of unintended consequences; and shed light on the historical, intellectual, and political development of voluntary associations in America, with a particular interest in

case studies that offer insight into contemporary challenges. Application materials should include: a one-to-two-page abstract of the project on behalf of your university, college, think tank, or other 501(c)3 organization (the abstract should provide sufficient detail for reviewers to assess the nature and feasibility of the idea); a CV or resume; and a brief, itemized budget. Funding levels are commensurate with the requirements of the research and the potential for the research to advance an understanding of critical issues. Accepted proposals may also receive support to disseminate the research findings. Proposals will be accepted and evaluated on a rolling basis.
*Requirements:* Any college, university, think tank, or 501(c)3 organization in the U.S. is eligible to apply.
*Geographic Focus:* All States
*Amount of Grant:* Up to 10,000,000 USD
*Contact:* Elizabeth B. Koch, Director; (703) 875-1770 or (703) 875-1658; fax (703) 875-1766; grants@charleskochfoundation.org
*Internet:* https://www.charleskochfoundation.org/apply-for-grants/requests-for-proposals/communities/
*Sponsor:* Charles Koch Foundation
1320 N. Courthouse Road, Suite 500
Arlington, VA 22201

**Charles Koch Foundation Corporate Welfare Reform Research Grants** 884

The Charles Koch Foundation provides funding for academic and public policy research directed at solving social problems through voluntary action and free enterprise. In the area of Corporate Welfare Reform Research, the Foundation invites scholars, researchers, and policy experts to submit proposals on issues related to cronyism and corporate welfare reform. While the Foundation is interested in all papers and larger projects that shed light on the various aspects and consequences of government-granted privilege, it is especially interested in projects that: identify and analyze corporate welfare at the local, state, or national levels (e.g. economic development incentives, tax-increment financing, subsidies for film production and sports stadiums, favoritism in the tax code); measure the size, scope, costs, and impact of corporate welfare at the local and state levels, and develop alternative approaches to spurring job creation, growth, and innovation; identify and analyze levels of transparency and accountability in corporate welfare and develop alternative approaches; analyze the costs and impact of occupational licensing, including effects on quality, safety, employment, and entrepreneurship; evaluate potential institutional reforms to cronyism and corporate welfare at local, state, and national levels; and identify and analyze the underlying causes of government-granted privilege at the local, state, and national levels. Application materials should include: a one-to-two-page abstract of the project on behalf of your university, college, think tank, or other 501(c)3 organization (the abstract should provide sufficient detail for reviewers to assess the nature and feasibility of the idea); a CV or resume; and a brief, itemized budget. Funding levels are commensurate with the requirements of the research and the potential for the research to advance an understanding of critical issues. Accepted proposals may also receive support to disseminate the research findings. Proposals will be accepted and evaluated on a rolling basis.
*Requirements:* Any college, university, think tank, or 501(c)3 organization in the U.S. is eligible to apply.
*Geographic Focus:* All States
*Amount of Grant:* Up to 10,000,000 USD
*Contact:* Elizabeth B. Koch, Director; (703) 875-1770 or (703) 875-1658; fax (703) 875-1766; grants@charleskochfoundation.org
*Internet:* https://www.charleskochfoundation.org/apply-for-grants/requests-for-proposals/corporate-welfare-reform/
*Sponsor:* Charles Koch Foundation
1320 N. Courthouse Road, Suite 500
Arlington, VA 22201

**Charles Koch Foundation Criminal Justice Reform Grants** 885

The Charles Koch Foundation provides funding for academic and public policy research directed at solving social problems through voluntary action and free enterprise. In the area of Criminal Justice Reform, the Foundation is especially interested in research that: analyzes the possible personal and/or non-economic and economic (including employment) effects of various sentencing reform ideas on individuals, families, and communities; examines the causes, costs, and consequences of police militarization, civil asset forfeiture, over-ciminalization and the proliferation of laws that carry criminal penalties, and collateral consequences of a criminal record; collects or creates data sets to fill scholarly gaps related to criminal justice and policing reform issues; assesses the effectiveness of various penalties imposed for crimes and attempts to make victims whole; examines the incentives that lead to growing prison populations; explores issues relating to access to justice and constitutional due process protections such as access to counsel, discovery process, and fines and fees; and explores prosecutorial incentives, including the use of plea bargains and their impact on pretrial detention, length of time to trial, and ultimate outcomes. Application materials should include: a one-to-two-page abstract of the project on behalf of your university, college, think tank, or other 501(c)3 organization (the abstract should provide sufficient detail for reviewers to assess the nature and feasibility of the idea); a CV or resume; and a brief, itemized budget. Funding levels are commensurate with the requirements of the research and the potential for the research to advance an understanding of critical issues. Accepted proposals may also receive support to disseminate the research findings. Proposals will be accepted and evaluated on a rolling basis.
*Requirements:* Any college, university, think tank, or 501(c)3 organization in the U.S. is eligible to apply.
*Geographic Focus:* All States
*Amount of Grant:* Up to 10,000,000 USD
*Contact:* Elizabeth B. Koch, Director; (703) 875-1770 or (703) 875-1658; fax (703) 875-1766; grants@charleskochfoundation.org
*Internet:* https://www.charleskochfoundation.org/apply-for-grants/requests-for-proposals/criminal-justice-policing-reform/
*Sponsor:* Charles Koch Foundation
1320 N. Courthouse Road, Suite 500
Arlington, VA 22201

**Charles Koch Foundation Economic Freedom Research Grants** 886

The Charles Koch Foundation provides funding for academic and public policy research directed at solving social problems through voluntary action and free enterprise. In the area of Economic Freedom Research, the Foundation feels that freedom is a critical component of economic well-being, leading to greater opportunity and an improved quality of life. Application materials should include: a one-to-two-page abstract of the project on behalf of your university, college, think tank, or other 501(c)3 organization (the abstract should provide sufficient detail for reviewers to assess the nature and feasibility of the idea); a CV or resume; and a brief, itemized budget. Final projects should be original and of academic journal quality, and must not have been previously published. Funding levels are commensurate with the requirements of the research and the potential for the research to advance an understanding of critical issues. Accepted proposals may also receive support to disseminate the research findings. Proposals will be accepted and evaluated on a rolling basis.
*Requirements:* Any college, university, think tank, or 501(c)3 organization in the U.S. is eligible to apply.
*Geographic Focus:* All States
*Amount of Grant:* Up to 1,000,000 USD
*Contact:* Elizabeth B. Koch, Director; (703) 875-1770 or (703) 875-1658; fax (703) 875-1766; grants@charleskochfoundation.org
*Internet:* https://www.charleskochfoundation.org/apply-for-grants/requests-for-proposals/economic-freedom/
*Sponsor:* Charles Koch Foundation
1320 N. Courthouse Road, Suite 500
Arlington, VA 22201

**Charles Koch Foundation Entrepreneurship Research and Education Grants** 887

The Charles Koch Foundation provides funding for academic and public policy research directed at solving social problems through voluntary action and free enterprise. In the area of Entrepreneurship Research and Education, the Foundation is building its understanding of the social institutions, theories, and mindsets that promote or limit entrepreneurial activity, which will help capture society's full potential, drive life-improving innovations, and ensure everyone has a chance to pursue a meaningful life. The Foundation requests proposals for research and educational grants that will advance the understanding of entrepreneurial theory, practice and impacts. It is especially interested in entrepreneurship research that: utilizes an interdisciplinary approach to advance understanding of the key institution and policies that inform entrepreneurial decision-making; provides undergraduate students with the opportunity to explore theories of entrepreneurship; supports PhD students and early-career faculty interested in exploring entrepreneurship theory and relevant policies; prepares high school and post-secondary instructors to teach foundational concepts in entrepreneurial theory; advances understanding of the key values and behaviors that distinguish between principled entrepreneurship and rent-seeking behavior; and explore the connection between life satisfaction, entrepreneurship, and whether entrepreneurial ventures enable individuals involved to better fulfill their potential. Application materials should include: a one-to-two-page abstract of the project on behalf of your university, college, think tank, or other 501(c)3 organization (the abstract should provide sufficient detail for reviewers to assess the nature and feasibility of the idea); a CV or resume; and a brief, itemized budget. Funding levels are commensurate with the requirements of the research and the potential for the research to advance an understanding of critical issues. Accepted proposals may also receive support to disseminate the research findings. Proposals will be accepted and evaluated on a rolling basis.
*Requirements:* Any college, university, think tank, or 501(c)3 organization in the U.S. is eligible to apply.
*Geographic Focus:* All States
*Amount of Grant:* Up to 10,000,000 USD
*Contact:* Elizabeth B. Koch, Director; (703) 875-1770 or (703) 875-1658; fax (703) 875-1766; grants@charleskochfoundation.org
*Internet:* https://www.charleskochfoundation.org/apply-for-grants/requests-for-proposals/entrepreneurship/
*Sponsor:* Charles Koch Foundation
1320 N. Courthouse Road, Suite 500
Arlington, VA 22201

**Charles Koch Foundation Foreign Policy Research Grants** 888

The Charles Koch Foundation provides funding for academic and public policy research directed at solving social problems through voluntary action and free enterprise. In recent decades, the United States has pursued an approach to foreign policy known as primacy or liberal hegemony. This foreign policy has required active and aggressive U.S. military engagement around the globe that has undermined our safety and imperiled our prosperity and liberties. In the area of Foreign Policy Research, the Foundation supports scholars and institutions interested in challenging this current approach, providing alternative visions for U.S. foreign policy, and engaging in research that can bridge the gap between ideas and policy. The Foundation is especially interested in foreign policy research projects from political science, international relations, history, or economics.

However, proposals from all fields will be considered on their merits. The Foundation is especially interested in research that: explores topics and issues related to a grand strategy of restraint; examines the role of values and ethics in the formulation of U.S. foreign policy; addresses the unintended consequences of U.S. military actions abroad and explores the costs and impact of engagements in Afghanistan, Iraq, Somalia, Syria, Yemen, and other places where significant counterterrorism efforts have been conducted; considers the impact of U.S. military engagement abroad on American society—for instance, the effect of our foreign policy decisions on civil liberties or the health and welfare of veterans; explores the concept of threat inflation and problematizes theories of credibility and reputation in relation to U.S. foreign policy; analyzes executive–legislative relations in foreign policy, including the constitutional division of war powers; examines the impact of domestic interest groups, businesses, think tanks, and the permanent national security bureaucracy on U.S. foreign policy; explores the growth of the intelligence and national security establishments since 9/11; surveys the costs, risks, and impacts of foreign aid and alliance commitments; studies the costs and benefits of burden-sharing and burden-shifting with current allies; scrutinizes Pentagon spending, force structure, and the strategic demands of U.S. defense policy—including structural incentives for military expenditures; considers the consequences of an increasingly multipolar world, especially as concerns principal geostrategic regions (e.g., Europe, the Middle East and North Africa, and the Western Pacific); or explores how changes in technology (such as A2/AD) that impact the offense-defense balance can be leveraged to support a sound approach to grand strategy. Application materials should include: a one-to-two-page abstract of the project on behalf of your university, college, think tank, or other 501(c)3 organization (the abstract should provide sufficient detail for reviewers to assess the nature and feasibility of the idea); a CV or resume; and a brief, itemized budget. Funding levels are commensurate with the requirements of the research and the potential for the research to advance an understanding of critical issues. Accepted proposals may also receive support to disseminate the research findings. Proposals will be accepted and evaluated on a rolling basis.
*Requirements:* Any college, university, think tank, or 501(c)3 organization in the U.S. is eligible to apply.
*Geographic Focus:* All States
*Amount of Grant:* Up to 10,000,000 USD
*Contact:* Elizabeth B. Koch, Director; (703) 875-1770 or (703) 875-1658; fax (703) 875-1766; grants@charleskochfoundation.org
*Internet:* https://www.charleskochfoundation.org/apply-for-grants/requests-for-proposals/foreign-policy/
*Sponsor:* Charles Koch Foundation
1320 N. Courthouse Road, Suite 500
Arlington, VA 22201

**Charles Koch Foundation Immigration Research Grants** 889

The Charles Koch Foundation provides funding for academic and public policy research directed at solving social problems through voluntary action and free enterprise. Presently, the U.S. immigration system keeps too many from fully participating in the American experience—to the detriment of our society. Reforms that incentivize legal immigration and welcome the majority of prospective immigrants, while preventing those who would come to do us harm, can yield safer communities and a stronger economy. In the area of Immigration Research, the Foundation supports scholars exploring solutions and innovative alternatives to a system burdened by backlogs in family-based immigration and asylum processing and posing countless employment barriers for foreign graduates of U.S. universities. The Foundation is especially interested in research that: examines the costs and benefits of merit- or points-based immigration systems, temporary visa programs, and flexible, provisional, or dual-purpose visa alternatives; measures the impact of immigration on formal and informal institutions, including social capital; explores social and cultural integration mechanisms and their effects; analyzes the consequences of overlapping immigration, law enforcement, and national security policies; tracks entrepreneurship among immigrant populations and their contribution to economic growth; evaluates and proposes improvements on interior immigration and border enforcement practices and the adjudication and processing of refugee and asylum claims; and collects data and produces longitudinal, structured data sets on resident attitudes towards immigration, measures of immigrant integration such as language acquisition, educational attainment, wage assimilation, or inter-generational social mobility, status adjustment, naturalization, and emigration rates, and immigrants' access to, use of, and contribution to welfare and public services. Application materials should include: a one-to-two-page abstract of the project on behalf of your university, college, think tank, or other 501(c)3 organization (the abstract should provide sufficient detail for reviewers to assess the nature and feasibility of the idea); a CV or resume; and a brief, itemized budget. Funding levels are commensurate with the requirements of the research and the potential for the research to advance an understanding of critical issues. Accepted proposals may also receive support to disseminate the research findings. Proposals will be accepted and evaluated on a rolling basis.
*Requirements:* Any college, university, think tank, or 501(c)3 organization in the U.S. is eligible to apply.
*Geographic Focus:* All States
*Amount of Grant:* Up to 10,000,000 USD
*Contact:* Elizabeth B. Koch, Director; (703) 875-1770 or (703) 875-1658; fax (703) 875-1766; grants@charleskochfoundation.org
*Internet:* https://www.charleskochfoundation.org/apply-for-grants/requests-for-proposals/immigration-research-grants/
*Sponsor:* Charles Koch Foundation
1320 N. Courthouse Road, Suite 500
Arlington, VA 22201

**Charles Koch Foundation Philosophy, Politics, and Economics Grants** 890

The Charles Koch Foundation provides funding for academic and public policy research directed at solving social problems through voluntary action and free enterprise. In the area of Philosophy, Politics, and Economics, the Foundation is promoting an interdisciplinary approach that brings together students and scholars spread across the college campus with a variety of perspectives to apply theory and practice facilitates the study of the core concepts of societies that encourage individual fulfillment, happiness, and success. In doing so, the Foundation seeks to support research that looks to break through the barriers to multi-disciplinary and interdisciplinary scholarship and foster real-world applications of the scholarly lineage of these diverse fields. It is particularly interested in supporting research and educational programming that: further develops the methodologies utilized by PPE scholars and illuminates their strengths and limitations; utilizes PPE methodologies to explore keys issues related to toleration, corporate welfare, innovation, or regulatory policy; equips students with the ability to understand and employ interdisciplinary perspectives to solve address pressing issues; and provides doctoral students and early-career faculty with the opportunity to build their analytical skill set. Application materials should include: a one-to-two-page abstract of the project on behalf of your university, college, think tank, or other 501(c)3 organization (the abstract should provide sufficient detail for reviewers to assess the nature and feasibility of the idea); a CV or resume; and a brief, itemized budget. Funding levels are commensurate with the requirements of the research and the potential for the research to advance an understanding of critical issues. Accepted proposals may also receive support to disseminate the research findings. Proposals will be accepted and evaluated on a rolling basis.
*Requirements:* Any college, university, think tank, or 501(c)3 organization in the U.S. is eligible to apply.
*Geographic Focus:* All States
*Amount of Grant:* Up to 10,000,000 USD
*Contact:* Elizabeth B. Koch, Director; (703) 875-1770 or (703) 875-1658; fax (703) 875-1766; grants@charleskochfoundation.org
*Internet:* https://www.charleskochfoundation.org/apply-for-grants/requests-for-proposals/philosophy-politics-economics/
*Sponsor:* Charles Koch Foundation
1320 N. Courthouse Road, Suite 500
Arlington, VA 22201

**Charles Koch Foundation Technology and Innovation Research Grants** 891

The Charles Koch Foundation provides funding for academic and public policy research directed at solving social problems through voluntary action and free enterprise. In the area of Technology and Innovation, the Foundation seeks to support research that helps bridge the divide between theory and practice. The Foundation invites scholars and policy experts to submit proposals for work that sheds light on the various effects of technological progress (policy, cultural, or otherwise) and explores the concept of permissionless innovation. Proposals for projects in law, economics, history, political science, and philosophy are encouraged. Projects in computer science, engineering, and the hard sciences will also be considered. It is especially interested in work that: analyzes the role of innovation in society, including the connection between technological progress, job creation, and the future of work; explores the cultural and/or policy conditions necessary for breakthroughs in knowledge, including historical or contemporary case studies in which well-being was advanced by technological innovation; examines questions related to digital free speech, including design, structure, legal rights of users, privacy from government surveillance, intermediary liability, legal threats to free speech such as SLAPPs, and encryption; examines the legal and regulatory frameworks most conducive to innovation, especially in data, health care, manufacturing, competition, and transportation; and provides mentorship, research support, and networking opportunities for the next generation of researchers who will tackle unforeseen challenges at the intersection of innovation and public policy. Application materials should include: a one-to-two-page abstract of the project on behalf of your university, college, think tank, or other 501(c)3 organization (the abstract should provide sufficient detail for reviewers to assess the nature and feasibility of the idea); a CV or resume; and a brief, itemized budget. Final projects should be original and meet the highest standards of their field, and must not have been previously published. Funding levels are commensurate with the requirements of the research and the potential for the research to advance an understanding of critical issues. Accepted proposals may also receive support to disseminate the research findings. Proposals will be accepted and evaluated on a rolling basis.
*Requirements:* Any college, university, think tank, or 501(c)3 organization in the U.S. is eligible to apply.
*Geographic Focus:* All States
*Contact:* Elizabeth B. Koch, Director; (703) 875-1770 or (703) 875-1658; fax (703) 875-1766; grants@charleskochfoundation.org
*Internet:* https://www.charleskochfoundation.org/apply-for-grants/requests-for-proposals/technology-innovation/
*Sponsor:* Charles Koch Foundation
1320 N. Courthouse Road, Suite 500
Arlington, VA 22201

**Charles Koch Foundation Toleration and Free Speech Research Grants** 892

The Charles Koch Foundation provides funding for academic and public policy research directed at solving social problems through voluntary action and free enterprise. In the area of Toleration and Free Speech Research, the Foundation requests proposals for research, programmatic support, or education activities related to understanding what drives intolerance and the best ways to cure it. The Foundation is particularly interested in research that explores: impediments to free speech and open inquiry in the academy; legal interpretations of the First Amendment and their consequences; exploring the connections between the classical liberal tradition and toleration; exploring the individual

and structural drivers of intolerance and tribalism in society; press freedom, including research that explores freedom of information laws and the connections between press freedom and free societies; digital free speech, including research related to intermediary liability, privacy, or other legal threats to free speech; and undergraduate programming that models civil discourse and educates students about the value of free expression. Application materials should include: a one-to-two-page abstract of the project on behalf of your university, college, think tank, or other 501(c)3 organization (the abstract should provide sufficient detail for reviewers to assess the nature and feasibility of the idea); a CV or resume; and a brief, itemized budget. Funding levels are commensurate with the requirements of the research and the potential for the research to advance an understanding of critical issues. Accepted proposals may also receive support to disseminate the research findings. Proposals will be accepted and evaluated on a rolling basis.
*Requirements:* Any college, university, think tank, or 501(c)3 organization in the U.S. is eligible to apply.
*Geographic Focus:* All States
*Amount of Grant:* Up to 10,000,000 USD
*Contact:* Elizabeth B. Koch, Director; (703) 875-1770 or (703) 875-1658; fax (703) 875-1766; grants@charleskochfoundation.org
*Internet:* https://www.charleskochfoundation.org/apply-for-grants/requests-for-proposals/toleration-free-speech/
*Sponsor:* Charles Koch Foundation
1320 N. Courthouse Road, Suite 500
Arlington, VA 22201

**Charles Koch Foundation Trade Policy Research Grants** **893**
The Charles Koch Foundation provides funding for academic and public policy research directed at solving social problems through voluntary action and free enterprise. Since the end of World War II, countries have dramatically liberalized international trade. Tariffs have been reduced or eliminated, while quotas and other non-tariff barriers have been lowered. These reforms have enabled an expansion in global trade and played an important role in reducing poverty in poor nations and boosting living standards in developed countries. In the area of Trade Policy Research, the Foundation supports scholars and institutions interested in research related to the benefits of global trade and the harm of various protectionist initiatives, in their various forms. It invites proposals for trade policy research projects from scholars in the disciplines of economics, political science, international relations, and history, as well as policy experts. However, proposals from all qualified individuals and fields will be considered on their merits. The Foundation requests proposals on research that bridges the gap between theory and practice and contributes to contemporary trade debates. It is especially interested in non-technical studies that will be broadly accessible to non-specialists and in research that: explores topics and issues related to the economic, cultural, and social benefits of trade; analyzes the impact of trade on consumer welfare, including for different groups and segments of society; generates data on the employment impact of trade liberalization; explores the impact of the World Trade Organization and how it might be improved to deal with current challenges; examines the impact of bilateral and regional Free Trade Agreements; analyzes the economic implications of trade balances, currency exchange rates, and the impact of monetary policy on trade; explores issues related to free trade with China, as well as potential reforms; reviews trade rules with non-market nations; examines the cost and extent of subsidies that impact trade as well as non-tariff barriers; analyzes the political economy of trade wars; examines and analyzes the political economy of protectionism, including the special interests, major players, and key constituencies; explores the historical experience with unilateral liberalization, how it impacted individuals in those countries, and how similar initiatives might work today; generates data on the impact of protectionism; explores the role of mutual recognition for trade agreements; analyzes the interaction of trade balances and capital flows; and/or examines the question of whether and how Congress can play a productive role in trade policy. Application materials should include: a one-to-two-page abstract of the project on behalf of your university, college, think tank, or other 501(c)3 organization (the abstract should provide sufficient detail for reviewers to assess the nature and feasibility of the idea); a CV or resume; and a brief, itemized budget. Funding levels are commensurate with the requirements of the research and the potential for the research to advance an understanding of critical issues. Accepted proposals may also receive support to disseminate the research findings. Proposals will be accepted and evaluated on a rolling basis.
*Requirements:* Any college, university, think tank, or 501(c)3 organization in the U.S. is eligible to apply.
*Geographic Focus:* All States
*Amount of Grant:* Up to 10,000,000 USD
*Contact:* Elizabeth B. Koch, Director; (703) 875-1770 or (703) 875-1658; fax (703) 875-1766; grants@charleskochfoundation.org
*Internet:* https://www.charleskochfoundation.org/apply-for-grants/requests-for-proposals/trade-policy-research-grants/
*Sponsor:* Charles Koch Foundation
1320 N. Courthouse Road, Suite 500
Arlington, VA 22201

**Charles Lafitte Foundation Grants** **894**
The foundation is committed to helping groups and individuals foster lasting improvement on the human condition by providing support to education, children's advocacy, medical research, and the arts. Children's advocacy grants support organizations working to improve the quality of life for children, particularly in relation to child abuse, literacy, foster housing, hunger, and after-school programs. Education grants support innovative programs that work to resolve social service issues, address the needs of students with learning disabilities, provide technology and computer-based education, offer leadership skills education, and support at-risk students. Colleges and universities also receive support for research and conferences. The foundation's medical issues and research grants support healthcare studies, with emphasis on cancer research and treatment, children's health, health education, and promoting healthy living and disease prevention. Art grants support emerging artists and educational art programs.
*Requirements:* 501(c)3 tax-exempt organizations are eligible. The Foundation does not respond to unsolicited submissions, so submit a letter of inquiry prior to preparing an application.
*Restrictions:* Political organizations or religious-based programs are not supported.
*Geographic Focus:* All States
*Samples:* Girl Scouts of the Jersey Shore, $200,000. Bridge of Books Foundation, $2,500. Hand in Hand, $10,000. Jersey Shore University Medical Center, $700,000.
*Contact:* Jennifer Vertetis, President; jennifer@charleslafitte.org
*Internet:* charleslafitte.org/grants/overview/
*Sponsor:* Charles Lafitte Foundation
29520 2nd Ave SW
Federal Way, WA 98023

**Charles M. and Mary D. Grant Foundation Grants** **895**
The Foundation awards grants primarily in the southeast United States in its areas of interest, including community and economic development, health and human services, environment, and education. The Foundation prefers project support, but considers operating support proposals from organizations with budgets of less than $1 million. Grants are made in September. A minimum of three years must elapse between grant awards. Further information is available at the website.
*Requirements:* Southeast U.S. 501(c)3 tax-exempt organizations are eligible. Specific grant guidelines are available online with the application, which must be filed online.
*Restrictions:* No grants are made to individuals or for loans. A minimum of three years must elapse between grant awards.
*Geographic Focus:* Alabama, Florida, Georgia, Kentucky, Mississippi, North Carolina, South Carolina, Tennessee, Virginia, West Virginia
*Date(s) Application is Due:* Apr 30
*Amount of Grant:* 20,000 - 40,000 USD
*Samples:* Collaborative for the 21st Century Appalachia, Inc., Charleston, West Virginia, first installment of a $60,000 grant to launch the e-Resource and Training Center, $30,000; Georgia Legal Services Program, Inc., Atlanta, Georgia, final installment of a $70,000 grant to support affordable housing work in rural Georgia, $30,000; Louisiana Bucket Brigade, New Orleans, Louisiana, final installment of a $40,000 grant for the Environmental Justice Corps, $20,000.
*Contact:* Casey Castaneda, Program Officer; (212) 464-2487; fax (212) 464-2305; casey.b.castaneda@jpmchase.com
*Internet:* fdncenter.org/grantmaker/grant
*Sponsor:* Charles M. and Mary D. Grant Foundation
J.P. Morgan Private Bank, Philanthropic Services
Dallas, TX 75222-7237

**Charles Stewart Mott Foundation Grants** **896**
The focus of the foundation's grant making is organized in four programs: civil society; environment; Flint, Michigan; and poverty. Flexibility to investigate new opportunities is maintained through an exploratory and special projects program. The civil society program promotes and supports civil society in the United States; Central/Eastern Europe and South Africa. The environment program supports efforts to achieve a healthy global environment capable of sustaining all forms of life. The Flint program seeks to strengthen the capacity of local institutions, including schools and school districts, in the foundation's home community of Flint, Michigan, to respond to economic and social needs. The poverty program addresses issues that contribute to improved life outcomes for children, youth, and families in low-income communities. Programs in low-income communities that connect schools and communities through systemic reform, improved teaching, leadership development, networking, technical assistance, and advocacy are also applicable. In all grant making, particular interest will be given to fresh approaches to solving community problems in the defined program areas; approaches that can generate long-term support from other sources and/or can be replicated in other communities; public policy development and research and development activities to further existing programs as well as to explore new fields of interest; and approaches and activities that lead to systemic change. Although proposals may be submitted at any time, applicants are strongly encouraged to submit during the first quarter of the year for which funding is requested. Grant expenditures are determined by September 1 of each year. The review process takes up to four months from the time the proposal is received. Therefore, proposals should be submitted at least four months prior to the start of the proposed grant period. Funding for unsolicited proposals is limited. It is recommended that letters of inquiry be submitted instead of a full proposal.
*Requirements:* Only 501(c)3 organizations are eligible, including schools and school districts.
*Restrictions:* Grants are not made to/for individuals; religious activities or programs that serve, or appear to serve, specific religious groups or denominations; or local projects outside the Flint area unless the projects are part of a national demonstration or foundation-planned network of grants that have clear and significant implications for replication in other communities.
*Geographic Focus:* All States
*Samples:* Regional Foundation for Local Development Zamah, Croatia, $82,000—Civil Society Grant for general purposes; Clean Wisconsin, Wisconsin, $100,000—Environmental Grant; Direct Action and Research Training Center, Florida, $300,000—Pathways Out of Poverty Grant for general purposes.
*Contact:* Mary A. Gailbreath; (810) 238-5651; fax (810) 766-1753; info@mott.org
*Internet:* www.mott.org/about/programs.aspx
*Sponsor:* Charles Stewart Mott Foundation
503 S Saginaw Street, Suite 1200
Flint, MI 48502-1851

**Chicago Community Trust Fellowships** 897

The Chicago Community Trust Fellowship offers professional development opportunities for both emerging and experienced leaders who are open to new learning, have a track record of accomplishments, have potential for significant community impact and demonstrate commitment to building strong nonprofit and public sectors. Fellowship applicants customize their own professional development plans. The Trust offers two fellowship tracks, one for Emerging Leaders and one for Experienced Leaders. Emerging leaders are awarded up to $30,000 and experienced leaders are awarded up to $60,000 to put these plans into practice. The award may fund salary and benefits during a leave of absence from work, costs related to academic courses, coaching or learning from other practitioners.
*Requirements:* Candidates for the Trust Fellowship must have a demonstrated commitment to, and passion for, their field and must be currently employed full-time in the nonprofit or public sector in the metropolitan Chicago region, which includes Cook, DuPage, Kane, Lake, McHenry and Will counties. For candidates pursuing the Emerging Leaders track, the following are required: program, budget and supervisory responsibility appropriate to the scale of the organization; at least five years of professional working experience and at least two years of management experience; track record of accomplishments which demonstrates leadership ability and potential for professional growth; and, endorsement from his/her organization's leader. For candidates pursuing the Experienced Leaders track, the following are required: membership in an organization's senior leadership team with responsibilities for organizational strategic planning, program/budget and personnel management; at least 15 years of professional working experience and 10 years of management experience; working relationship with the board of directors; and, endorsement from his/her organization's leader or board chair.
*Restrictions:* Past recipients of Trust Fellowships, professionals from academic institutions and employees or board members of The Chicago Community Trust and its affiliates are not eligible to apply.
*Geographic Focus:* Illinois
*Date(s) Application is Due:* Jun 13
*Amount of Grant:* Up to 60,000 USD
*Samples:* 2012 Emerging Leaders: Robert Castaneda, Gabriela Fitz, Benjamin Helphand, Darlene M. Oliver, Rebeccah Sanders. 2012 Experienced Leaders: Paul W. Hamann, Marco Jacome, Barbara Otto.
*Contact:* Anne Blanton, (312) 616-8000 or (312) 616-7955; fellowship@cct.org
*Internet:* www.cct.org/about/partnership-initiative/chicago-community-trust-fellowship
*Sponsor:* Chicago Community Trust
111 East Wacker Drive, Suite 1400
Chicago, IL 60601

**Chicago Community Trust Housing Grants: Advancing Affordable Rental Housing** 898

The Trust is committed to support innovation and excellence, and to fund organizations advancing research, policy, advocacy, housing development and renters' rights, making Chicago a national leader in rental solutions for the 21st century. The Trust will directly solicit proposals supporting a coordinated approach to research, policy and advocacy designed to increase resources for rental and supportive housing, remove obstacles to its preservation, rehabilitation and development, and ensure that it meets standards of quality, affordability, location and sustainability across the region; and will directly solicit proposals from groups working to provide information and resources to renters, especially those living in subsidized housing. Finally, the Trust will also issue a request for proposals from nonprofit developers that directly provide quality affordable housing and services to low- and moderate-income families in innovative, cost-effective and impactful ways.
*Requirements:* The Trust funds nonprofit agencies with evidence of tax-exempt status under Section 501(c)3 of the Internal Revenue Code that are not classified as private foundations. The Trust also accepts applications from agencies that operate under a nonprofit fiscal sponsor. Organizations must be located within and/or primarily serving residents of Cook County, except for regional, statewide or national projects or research that may benefit a substantial portion of Cook County residents. Organizations must also be non-discriminatory in the hiring of staff or in providing services on the basis of race, religion, gender, sexual orientation, age, national origin or disability.
*Restrictions:* The Trust will not provide grants for: scholarships; individuals; sectarian purposes (programs that promote or require a religious doctrine); support of single-disease oriented research, treatment or care; the sole purpose of writing, publishing, producing or distributing audio, visual or printed material; the sole purpose of conducting conferences, festivals, exhibitions or meetings; or, reducing operating deficits or liquidating existing debt.
*Geographic Focus:* Illinois
*Date(s) Application is Due:* Jan 5
*Contact:* Juanita Irizarry, Program Officer; juanita@cct.org
*Internet:* www.cct.org/apply/funding-priorities/housing#priority2
*Sponsor:* Chicago Community Trust
111 East Wacker Drive, Suite 1400
Chicago, IL 60601

**Chicago Reporter Data Visualization Fellowship** 899

The Chicago Reporter seeks a part-time fellow to produce data visualizations that supports the Reporter's investigative reporting into race, poverty and income inequality. It is looking for someone who can tell stories by combining design with data. The ideal candidate will not only have experience producing charts, maps and other visuals, but also an ability to grasp journalistic processes and convey complex information in a way readers will easily understand. Qualifications include: a grasp of visual design concepts; and the ability to parse various data formats (e.g. csv and json); be adept with html, css and javascript. Possession of a background in journalism is preferred, but not required. Each year, this is a temporary, part-time position of 20-25 hours per week over three months, based out of the Reporter newsroom in downtown Chicago.
*Requirements:* Applicants must have a bachelor's degree, a minimum of three years reporting experience, excellent news judgment, and a strong interest in urban affairs and investigative reporting. Fluency in Spanish a plus.
*Geographic Focus:* All States
*Contact:* Matt Kiefer, Program Contact; (312) 427-4830, ext. 3864; fax (312) 427-6130; mkiefer@chicagoreporter.com
*Internet:* chicagoreporter.com/about/data-viz/
*Sponsor:* Chicago Reporter
111 W. Jackson Boulevard, Suite 820
Chicago, IL 60604

**Children's Brain Tumor Foundation Research Grants** 900

The Children's Brain Tumor Foundation (CBTF), a national not-for-profit organization, was founded in 1988 by a group of dedicated parents, physicians and friends to improve the treatment, quality of life, and long-term outlook for children with brain and spinal cord tumors through research, support, education, and advocacy. The number one priority of CBTF is the awarding of grants for research into the causes of and effective treatments for pediatric brain and spinal cord tumors. When funding is available, the grant cycle begins in the spring of the year. Investigators from the U.S. and Canada should submit a pre-application form to be reviewed by an expert panel. Based on this initial review, a selected group of applicants will then be invited to submit a full proposal.
*Requirements:* Funding is currently restricted to principal investigators at institutions within the United States.
*Restrictions:* The Foundation does not award grants to individuals or to private foundations, and does not fund debt reduction, capital improvements, or travel expenses. Overhead expenses are not to exceed 10 percent of the total project cost.
*Geographic Focus:* All States
*Amount of Grant:* Up to 150,000 USD
*Contact:* Wade Iwata, Director of Quality of Life Programs and Research; (212) 448-9494 or (866) 228-4673; fax (212) 448-1022; wiwata@cbtf.org or info@cbtf.org
*Internet:* www.cbtf.org/professionals/grants
*Sponsor:* Children's Brain Tumor Foundation
1460 Broadway
New York, NY 10036

**Children's Leukemia Research Association Research Grants** 901

The Children's Leukemia Research Association supports research efforts into the causes and cure of leukemia and gives patient aid to families in need while meeting the expenses incurred in leukemia treatment. A medical advisory committee consisting of prominent internationally known and respected hematologists reviews grant proposals submitted for consideration of projects that are not otherwise funded. Grants are available for start-up funding for laboratory or clinical investigations in leukemia. Although the association prefers to fund new investigators, applications from established investigators for new initiatives are also solicited. Renewal for a second year is considered if other funding for promising projects has not been obtained. Grants are limited to a maximum of $30,000 each. The annual application deadline is June 30.
*Requirements:* Any doctor at the Ph.D. or MD level who is involved in research towards finding the causes and cure for leukemia may apply.
*Restrictions:* Previously funded projects are not usually considered for additional funding.
*Geographic Focus:* All States
*Date(s) Application is Due:* Jun 30
*Amount of Grant:* Up to 30,000 USD
*Contact:* Aileen T. Sullivan, Administrative Assistant; (516) 222-1944; fax (516) 222-0457; info@childrensleukemia.org
*Internet:* www.childrensleukemia.org/researchgrants.html
*Sponsor:* Children's Leukemia Research Association
585 Stewart Avenue, Suite 18
Garden City, NY 11530

**Children's Tumor Foundation Clinical Research Awards** 902

In order to accelerate neurofibromatosis research, the Children's Tumor Foundation launched the Clinical Research Award (CRA) program in 2007. The program's primary purpose is to support early stage pilot clinical trials of candidate therapeutics or interventions for treatment of manifestations of NF1, NF2 and schwannomatosis. In addition, these awards support innovative clinical trial enabling studies. It is hoped that applications for Clinical Research Awards will be broad-thinking, novel ideas. Examples of projects encouraged may include but are not limited to the following: preclinical and clinical collaborative studies (note that all studies must include a clinical element – preclinical-only studies are not eligible); pilot clinical trials (CTF leverages our relationship with industry or larger funding organizations to provide drug at no cost to successful applicants); and clinical enabling studies such as biomarker and patient stratification that may better inform and help accelerate a clinical trial. It is anticipated that the Foundation will fund up to three Clinical Research Awards annually of up to $150,000 each (including overheads and indirect costs). Letters of Intent are required by July 1, with invited full applications due by August 17.
*Geographic Focus:* All States
*Date(s) Application is Due:* Aug 17
*Amount of Grant:* Up to 150,000 USD
*Contact:* Patrice Pancza, Research Program Director; (212) 344-7291 or (212) 344-6633; fax (212) 747-0004; ppancza@ctf.org or grants@ctf.org
*Internet:* www.ctf.org/CTF-Awards-Grants-and-Contracts/Clinical-Research-Awards/

*Sponsor:* Children's Tumor Foundation
120 Wall Street, 16th Floor
New York, NY 10005

**Children's Tumor Foundation Drug Discovery Initiative Awards** 903

Founded in 1978 as the National Neurofibromatosis Foundation, the Children's Tumor Foundation is a non-profit organization committed to identifying effective drug therapies for neurofibromatosis type 1 (NF1), neurofibromatosis type 2 (NF2) and schwannomatosis, and to improving the lives of those living with these disorders. The Drug Discovery Initiative Awards program supports early stage testing of candidate drug therapies for the treatment of neurofibromatosis (NF): NF1, NF2 and schwannomatosis. These Awards have yielded over $5M in follow-on funding from the federal government and other sources, as well as multiple industry collaborations and publications. DDI Awards is a catalyst program that has helped to fuel the drug pipeline with promising leads. Funding is as follows: up to $40,000 DDI in vitro Awards to fund cell-based preclinical drug testing studies; and up to $85,000 DDI in vivo awards to fund animal-based preclinical drug testing studies. Applications are welcomed from both the academic and private sectors. Applications are due by August 25.

*Requirements:* Applicants should have an MD, PhD, or equivalent, and have full access to, or identified collaborators with, all required resources including all in vivo and in vitro models. There are no citizenship requirements.
*Geographic Focus:* All States, All Countries
*Date(s) Application is Due:* Aug 25
*Amount of Grant:* Up to 85,000 USD
*Contact:* Patrice Pancza, Research Program Director; (212) 344-7291 or (212) 344-6633; fax (212) 747-0004; ppancza@ctf.org or grants@ctf.org
*Internet:* www.ctf.org/CTF-Awards-Grants-and-Contracts/Drug-Discovery-Initiative/
*Sponsor:* Children's Tumor Foundation
120 Wall Street, 16th Floor
New York, NY 10005

**Children's Tumor Foundation Schwannomatosis Awards** 904

The Children's Tumor Foundation is focused on ending neurofibromatosis (NF) through research. To date CTF has committed over $35M to NF research, ranging from preclinical drug testing to clinical research and a national NF Clinic Network. Schwannomatosis is the rarest form of NF, affecting 1 in every 40,000 persons, and causing peripheral nerve tumors and unmanageable pain. In 2007 a candidate schwannomatosis gene, INI-1/Smarc-B1/Snf-5 was identified. Harnessing this discovery, over the past several years CTF has: convened a series of Schwannomatosis Workshops to identify priorities for advancing schwannomatosis research; and funded schwannomatosis research totaling over $1.3 million since 2007. Results from this include the first schwannomatosis mouse models, utilizing these for preclinical drug trials, advancing schwannomatosis genetics, and establishing an international Schwannomatosis Database. CTF is offering Schwannomatosis Awards of up to $75,000 each. Schwannomatosis Awards may be requested for research in any area of relevance to advancing schwannomatosis research in the following priority areas: genetics; cell biology and translational research; and clinical research.

*Geographic Focus:* All States
*Contact:* Patrice Pancza, Research Program Director; (212) 344-7291 or (212) 344-6633; fax (212) 747-0004; ppancza@ctf.org or grants@ctf.org
*Internet:* www.ctf.org/Research/Schwannomatosis-Awards.html
*Sponsor:* Children's Tumor Foundation
120 Wall Street, 16th Floor
New York, NY 10005

**Children's Tumor Foundation Young Investigator Awards** 905

Founded in 1978, the Children's Tumor Foundation is a non-profit organization committed to identifying effective drug therapies for neurofibromatosis type 1 (NF1), neurofibromatosis type 2 (NF2) and Schwannomatosis, and to improving the lives of those living with these disorders. The Young Investigator (YIA) Awards program provides two-year awards for young scientists early in their careers, bringing them into the NF field and helping to establish them as independent investigators. Though a number of YIAs have made significant research findings and made notable publications the main function of the YIA program has been as a 'seeding mechanism' for researchers who went on to secure larger grants from NIH and CDMRP NFRP. Fellowship amounts range from $32,000 to $54,000, depending upon: level of training at the time of the application; and the request year (1st or 2nd).

*Requirements:* Applicants must be one of the following: a postdoctoral fellow (MD, PhD, or equivalent) but no more than seven years past the completion of their first doctoral degree; or a graduate student pursuing an MD, PhD, or equivalent.
*Geographic Focus:* All States
*Amount of Grant:* 32,000 - 54,000 USD
*Contact:* Patrice Pancza, Research Program Director; (212) 344-7291 or (212) 344-6633; fax (212) 747-0004; ppancza@ctf.org or grants@ctf.org
*Internet:* www.ctf.org/CTF-Awards-Grants-and-Contracts/CTF-Young-Investigator-Award/
*Sponsor:* Children's Tumor Foundation
120 Wall Street, 16th Floor
New York, NY 10005

**ChLA Faculty Research Grants** 906

ChLA Faculty Research Grants have a combined maximum fund of up to $5,000 per year, and individual awards may range from $500 to $1,500, based on the number and needs of the winning applicants. The grants are awarded for proposals dealing with criticism or original scholarship with the expectation that the undertaking will lead to publication and make a significant contribution to the field of children's literature in the area of scholarship or criticism. In honor of the achievement and dedication of Dr. Margaret P. Esmonde, proposals that deal with critical or original work in the areas of fantasy or science fiction for children or adolescents will be awarded the Margaret P. Esmonde Memorial Grant. Applications will be evaluated based upon the quality of the proposal and the potential of the project to enhance or advance Children's Literature studies. Funds may be used for—but are not restricted to—research-related expenses such as travel to special collections or purchasing materials and supplies. Proposals should be received via email by the annual February 1 deadline, with the application window beginning on January 1.

*Requirements:* Recipients must either be members of the Children's Literature Association or join the association before they receive any funds. Grantees should acknowledge ChLA in any publication resulting from the grant.
*Restrictions:* The awards may not be used for obtaining advanced degrees, for researching or writing a thesis or dissertation, for textbook writing, or for pedagogical projects.
*Geographic Focus:* All States
*Date(s) Application is Due:* Feb 1
*Amount of Grant:* 500 - 1,500 USD
*Contact:* Administrator; (630) 571-4520; fax (708) 876-5598; info@childlitassn.org
*Internet:* www.childlitassn.org/faculty-research-grant
*Sponsor:* Children's Literature Association
1301 W. 22nd Street, Suite 202
Oak Brook, IL 60523

**ChLA Hannah Beiter Diversity Research Grants** 907

The Diversity Research Grant, initiated in 2013 and approved by the board for disbursement beginning in 2014, is a grant to support research related to children's and young adult cultural artifacts (including media, culture, and texts) about populations that have been traditionally underrepresented or marginalized culturally and/or historically. Applications for this grant are to be considered annually and will be awarded as warranted. Awards will range from $500 to $1000, depending upon the winning proposal's projected budgetary considerations. Each grant will be awarded with the expectation that the undertaking will lead to publication and make a significant contribution to the field of children's literature scholarship or criticism. Within two years of receiving the grant, the recipient will be asked to submit a paper proposal based upon the project for presentation at a ChLA annual conference. Funds may be used for - but are not restricted to -research-related expenses such as travel to special collections, subvention funds, or purchasing materials and supplies. The annual deadline for receiving the application is midnight, February 1.

*Requirements:* Applications will be evaluated based upon the quality of the proposal and the potential of the project to enhance or advance Children's Literature studies. Winners must be members of the Children's Literature Association before they receive any funds. Winners must acknowledge ChLA in any publication or other presentation resulting from the grant.
*Restrictions:* The awards may not be used for obtaining advanced degrees, for researching or writing a thesis or dissertation, for textbook writing, or for pedagogical projects.
*Geographic Focus:* All States
*Date(s) Application is Due:* Feb 1
*Amount of Grant:* 500 - 1,000 USD
*Contact:* Administrator; (630) 571-4520; fax (708) 876-5598; info@childlitassn.org
*Internet:* www.childlitassn.org/diversity-research-grant
*Sponsor:* Children's Literature Association
1301 W. 22nd Street, Suite 202
Oak Brook, IL 60523

**ChLA Hannah Beiter Graduate Student Research Grants** 908

The Hannah Beiter Graduate Student Research Grants were established to honor the memory of Dr. Hannah Beiter, a long-time supporter of student participation in the Children's Literature Association. The Beiter Grants have a combined maximum fund of up to $5,000 per year, and individual awards may range from $500 to $1,500, based on the number and needs of the winning applicants. The grants are awarded for proposals of original scholarship with the expectation that the undertaking will lead to publication or a conference presentation and contribute to the field of children's literature criticism. The award may be used to purchase supplies and materials (e.g., books, videos, equipment), as research support (photocopying, etc.), or to underwrite travel to special collections or libraries. Proposals should be received via email by the annual February 1 deadline, with the application window beginning on January 1.

*Requirements:* Award recipients must either be members of the Children's Literature Association or join the association before they receive any funds.
*Restrictions:* Beiter grant funds are not intended as income to assist in the completion of a graduate degree (e.g., applied to tuition), but as support for research that may be related to the dissertation or master's thesis.
*Geographic Focus:* All States
*Date(s) Application is Due:* Feb 1
*Amount of Grant:* 500 - 1,500 USD
*Contact:* Administrator; (630) 571-4520; fax (708) 876-5598; info@childlitassn.org
*Internet:* www.childlitassn.org/beiter-graduate-student-research-grant
*Sponsor:* Children's Literature Association
1301 W. 22nd Street, Suite 202
Oak Brook, IL 60523

**Christensen Fund Regional Grants** 909

The fund (TCF) focuses its grantmaking on maintaining the biological and cultural diversity of the world by focusing on four geographic regions: the greater South West (Southwest United States and Northwest Mexico); Central Asia and Turkey; the African Rift Valley (Ethiopia); and Northern Australia and Melanesia. Grants within these programs are generally directed to organizations based within those regions or, where appropriate, to internationally based organizations working in support of people and institutions on the ground. In general, grants are one year or less; currently grants up to two years are by invitation only.

*Requirements:* 501(c)3 nonprofit organizations and non-USA institutions with nonprofit or equivalent status in their country of origin are eligible. Partnerships or associations with USA-based nonprofit organizations are preferred.

*Restrictions:* The fund does not make grants directly to individuals but rather assists individuals through institutions qualified to receive nonprofit support with which such individuals are affiliated.

*Geographic Focus:* All States

*Amount of Grant:* Up to 200,000 USD

*Contact:* Grants Administrator; (415) 644-1600; fax (415) 644-1601; info@christensenfund.org

*Internet:* www.christensenfund.org/index.html

*Sponsor:* Christensen Fund

260 Townsend Street

San Francisco, CA 94107

**Christine Wilson Award** 910

The Society for the Anthropology of Food and Nutrition (SAFN) invites students to submit papers in competition for the annual Christine Wilson Awards presented to outstanding undergraduate and graduate student research papers on that examine topics within the perspectives in nutrition, food studies and anthropology. Winners will be recognized and presented with an award at the annual AAA meeting and receive a year's membership in SAFN.

*Requirements:* Papers may report on research undertaken in whole or in part by the author. Co–authored work is acceptable, provided that submitting student is first author. Papers must have as their primary focus an anthropological approach to the study of food and/or nutrition and must present original, empirical research; literature reviews are not eligible. Papers that propose a new conceptual framework or outline novel research designs or methodological approaches are especially welcome. Eligibility is further restricted to students (undergraduate or graduate) who are currently enrolled or who were enrolled during in the past academic year.

*Geographic Focus:* All States

*Date(s) Application is Due:* Oct 5

*Contact:* Michael R. McDonald, PhD, (239) 590-7212; mmcdonal@fgcu.edu

*Internet:* foodanthro.com/christine-wilson-award/

*Sponsor:* American Anthropological Association

2200 Wilson Boulevard, Suite 600

Arlington, VA 22201

**Chun Foundation Grants** 911

The Chun Foundation in Memory of Say Chong Chun, Chew Hoong Lum Chun, and Bowman Michael Chun was created in 1994 for the purpose of funding organizations primarily for medical research into the causes and cures of the diseases of children, and secondarily to provide funds for the medical care of such children, particularly the children of Hawaii. The donor, Bowman Michael Chun, established the foundation in honor of his parents and in recognition of his concern for the plight and suffering of children in Hawaii who are afflicted with disease. Proposals will be considered for either research (basic or clinical research) or medical care. Grant awards generally range between $10,000 and $15,000. All applications should be postmarked or delivered by the annual deadline of April 15, with notification of awards typically made by August 31.

*Requirements:* Only proposals for projects submitted by organizations determined to be tax exempt, publicly supported and charitable by the Internal Revenue Service as a 501(c)3 organization, or a unit of government will be reviewed.

*Geographic Focus:* Hawaii

*Amount of Grant:* 10,000 - 15,000 USD

*Contact:* Flora Arney, Assistant Vice President; (808) 694-4393; flora.arney@boh.com

*Internet:* https://www.boh.com/apps/foundations/FoundationDetails.aspx?foundation=14&show=0

*Sponsor:* Chun Foundation

P.O. Box 3170, Dept. 758

Honolulu, 96802-3170

**CICF Efroymson Contemporary Arts Fellowships** 912

The Central Indiana Community Foundation Efroymson Contemporary Arts Fellowships were established to increase public awareness of contemporary art in Indiana, Illinois, Kentucky, and Ohio. The intent of the fellowships is to reward creativity, and to encourage both emerging and established individual artists by supporting their artistic development. Applications are available at the Foundation's website.

*Requirements:* Applicants must be at least 25 years old and residents of select counties in Indiana, Kentucky, Illinois, or Ohio, with proof that they have lived there for at least one year. They must also make a commitment to reside in their county for the duration of the one year fellowship. Selection is based on: quality and skill of the artist's work; creativity and uniqueness of the artist's work; artist's commitment to developing his or her work; and the impact of the award on the artist's career.

*Restrictions:* Applicants should refer to the website for a specific list of each state's counties.

*Geographic Focus:* Illinois, Indiana, Kentucky, Ohio

*Date(s) Application is Due:* Sep 23

*Amount of Grant:* 20,000 USD

*Contact:* Beth Barich; (317) 634-2423, ext. 175; fax (317) 684-0943; bethb@cicf.org

*Internet:* www.cicf.org/efroymson-contemporary-arts-fellowships

*Sponsor:* Central Indiana Community Foundation

615 North Alabama Street, Suite 119

Indianapolis, IN 46204-1498

**CIC Smithsonian Institution Fellowships** 913

The Committee on Institutional Cooperation (CIC) and the Smithsonian Institution (SI) invite fellowship applications for one-year fellowships to support research in residence at Smithsonian Institution facilities. Fellowships carry a stipend of $30,000. All fields of study that are actively pursued by the museums and research organizations of the Smithsonian Institution are eligible. Under this Fellowship, the Smithsonian Institution and the Fellow's university share the cost of support. Students enrolled at CIC member universities apply through their university graduate school, but their application is submitted through the Smithsonian's Online Application System (SOLAA). Each university will nominate one application for the CIC/SI competition. All unfunded applications will be eligible for consideration in the general Smithsonian Institution competition. Fellows are eligible to receive allowances for related expenses and for temporary relocation to the Smithsonian. All applications are submitted through the Smithsonian Online Academic Appointment System, with the link located at the CIC website.

*Requirements:* In order to qualify, students must have completed all course work for their programs, admitted into doctoral candidacy, and satisfied all requirements except completion of the dissertation. Only students enrolled at CIC institutions are eligible to apply (University of Chicago; University of Illinois; Indiana University; University of Iowa; University of Michigan; Michigan State University; University of Minnesota; University of Nebraska-Lincoln; Northwestern University; Ohio State University; Pennsylvania State University; Purdue University; and University of Wisconsin-Madison).

*Geographic Focus:* All States

*Date(s) Application is Due:* Nov 15

*Amount of Grant:* 30,000 USD

*Contact:* Amber Marks, Associate Director; (217) 265-8106; aemarks@staff.cic.net

*Internet:* www.cic.net/Home/Students/SmithsonianFellowship.aspx

*Sponsor:* Committee on Institutional Cooperation

1819 South Neil Street, Suite D

Champaign, IL 61820-7271

**CIC Summer Research Opportunities** 914

The Summer Research Opportunities Program (SROP) is a gateway to graduate education at CIC universities ( University of Chicago; University of Illinois at Urbana-Champaign; Indiana University; University of Iowa; University of Michigan; Michigan State University; University of Minnesota; Northwestern University; Ohio State University; Pennsylvania State University; and Purdue University). The goal of the program is to increase the number of underrepresented students who pursue graduate study and research careers. These summer opportunities help prepare undergraduates for graduate study through intensive research experiences with faculty mentors and enrichment activities. The CIC SROP application is accepted from November 1 through February 10. The application is available at the CIC website, in addition to the application guide and videos about the application process, and comments from former participants in the SROP. Applicants must review the application guidelines before applying, where detailed instructions for all required attachments are discussed. Individual campus profiles with program dates, types of applications, stipends, and whether academic credit is available, are posted on the CIC website.

*Requirements:* Applications must meet all of the following criteria: a cumulative grade point average of 3.0 or higher; citizenship or permanent residency in the U.S; enrolled in a degree-granting program at a college or university in the U.S., Puerto Rico, or other U.S. territory; completed at least two semesters of undergraduate education by the time of the summer experience; at least one semester of undergraduate education remaining after completing the summer research experience; and a strong interest in pursuing a Ph.D. Selected participants are expected to complete 40 hours of research each week. Participants must devote full-time to the program during the eight- to 10-week session. Due to the intense research schedule, SROP participants may not work an additional job during the summer program.

*Restrictions:* The SROP is not for students pursuing professional degrees such as law, medical, or MBA. Students who have complete an undergraduate degree are not eligible.

*Geographic Focus:* All States

*Amount of Grant:* 3,000 - 6,000 USD

*Contact:* Charity Farber, Program Manager; (217) 333-6314; cfarber@staff.cic.net

*Internet:* www.cic.net/Home/Students/SROP/Home.aspx

*Sponsor:* Committee on Institutional Cooperation

1819 South Neil Street, Suite D

Champaign, IL 61820-7271

**CIC Traveling Scholars** 915

Traveling Scholar allows doctoral students to spend up to a full academic year pursuing specialized courses of study, researching unique library collections, and working in advanced laboratories and facilities at other CIC institutions, with no change in registration procedures from their home university or additional tuition. Doctoral-level students interested in becoming a Scholar must first consult their advisor to determine whether the off-campus opportunity is likely to enhance the student's course of study, and to verify that it is not available on their home campus. The instructor or advisor at

the host campus must also approve before a student completes the online application. The application is available at the CIC website with detailed instructions.
*Requirements:* Students must be enrolled in a doctoral program at their home institution. Advisors at the student's home and host campuses must agree to allow the student to enroll in a course via the Traveling Scholar before the student submits the online application. Scholars remain registered at their home universities, paying the regular fees charged when they are in residence, and must be registered at both institutions. Scholars are not billed by the host institution for registration or fees, but must provide evidence of tuition paid to the home institution. Host universities provide access to libraries, laboratories, recreational facilities, housing, etc., on the same basis as they are made available to resident graduate students. Scholars may be assessed fees for certain services or benefits, and may be required to secure their own health and medical coverage. Credit earned by Scholars is accepted at home universities for the equivalent level credit (undergraduate or graduate) upon receipt of grade reports or transcripts from the host university. Credit toward a degree, however, will be at the discretion of the home university or department. Visits of Scholars may not exceed two semesters or three quarters regardless of the number of courses taken. In most cases, Scholars physically attend courses at another CIC institution. Technology facilitated/distance courses are occasionally permitted, but students must contact the CIC office before submitting a Scholar application.
*Restrictions:* Candidates must apply at least six weeks prior to the start of the intended course at the host institution.
*Geographic Focus:* All States
*Contact:* Charity Farber, Program Director; (217) 333-6314 or (217) 333-8475; fax (217) 244-7127; cfarber@staff.cic.net
*Internet:* www.cic.net/Home/Projects/SharedCourses/TScholar/Introduction.aspx
*Sponsor:* Committee on Institutional Cooperation
1819 South Neil Street, Suite D
Champaign, IL 61820-7271

**CIGNA Foundation Grants** 916

The Cigna Foundation has identified four areas for grant consideration: health and human services, education, community and civic affairs, and culture and the arts. Health and education are of primary concern and receive priority. Under education, priority is placed on public secondary education, higher education for minorities, and adult basic education/literacy. The foundation also considers requests from U.S. cultural, educational, and public policy organizations that have international components. Requests are accepted and reviewed throughout the year. Consideration will be given to requests for general operating support, program development, annual campaigns, conferences and seminars, fellowships, scholarship funds and employee-related scholarships, and matching gifts and funds.
*Requirements:* Organizations with 501(c)3 tax-exempt status are eligible.
*Restrictions:* The foundation will not consider applications for grants to individuals, organizations operating to influence legislation or litigation, political organizations, or religious activities. In general, the foundation will not consider applications from organizations receiving substantial support through the United Way or other CIGNA-supported federated funding agencies; hospitals' capital improvements; or research, prevention, and treatment of specific diseases.
*Geographic Focus:* All States
*Amount of Grant:* 5,000 - 50,000 USD
*Contact:* Jill Holliday, Program Contact; (860) 226-2094 or (866) 865-5277; jill.holliday@cigna.com
*Internet:* www.cigna.com/aboutus/cigna-foundation
*Sponsor:* Cigna Foundation
1601 Chestnut Street, TL06B
Philadelphia, PA 19192-1540

**Cincinnati Children's Charlotte R. Schmidlapp Women Scholars Grants** 917

Funded training programs at Cincinnati Children's support our trainees and scholars. Through financial and research support, they advance their skills and research interests, leading to better clinical care and the advancement of medical science. The programs range from private to federal funding sources and provide support for clinicians and researchers in all phases of their careers. The Charlotte R. Schmidlapp Women Scholars program goal is to support the academic career development of women faculty who have demonstrated academic potential and leadership skills. The intention of the award is to provide career support so that more women reach the rank of Associate and Full Professor. All applicants also have the opportunity to receive individual academic mentoring. The award amount is variable according to identified needs; up to $70,000 per year is given to the scholar for use in a manner that she deems appropriate for her research. The deadline for application submission is October 4, with the scholar chosen by January 1.
*Requirements:* This program has been designed for women faculty who are Assistant or Associate Professors in departments / divisions at Cincinnati Children's, as well as receive a paycheck from Cincinnati Children's. In general, successful applicants have been in their current academic rank at Cincinnati Children's for at least 3 years. The eligible candidate should have a record of academic accomplishments, such as grants and publications, but needs additional support to ensure academic success.
*Restrictions:* Funds cannot be used for salary support of the scholar.
*Geographic Focus:* Ohio
*Date(s) Application is Due:* Oct 4
*Amount of Grant:* Up to 70,000 USD
*Contact:* Katherine E. Yutzey, Chairperson; (513) 636-8340 or (513) 636-4200; katherine.yutzey@cchmc.org or lydia.west@cchmc.org
*Internet:* https://www.cincinnatichildrens.org/education/funded/schmidlapp
*Sponsor:* Cincinnati Children's Hospital Medical Center
3333 Burnet Avenue
Cincinnati, OH 45229-3026

**Cincinnati Children's Child Health Research Career Development Awards** 918

The Child Health Research Career Development Award (CHRCDA) Program is a K12 Program grant award. The CHRCDA program is supported by the National Institutes of Health and is designed to stimulate pediatric research over a variety of disciplines through training of productive physician scientists. The CHRCDA program enhances the capacity of pediatric research institutions to provide junior faculty investigators with cutting-edge training in an academic research setting and to foster translational research by more rapidly applying basic science findings to clinical problems. In turn, these forward the goal of the CHRCDA program ? to promote the performance of research and transfer of findings that will benefit the health of children. The Child Health Research Career Development Award (CHRCDA) Scholar position is a junior faculty position. Each year, applications are due by October 1 for December appointments.
*Requirements:* Qualified junior faculty investigators are welcome to apply for the CHRCDA. This junior faculty appointment is for eligible candidates who meet the following qualifications and obligations: must be U.S. citizens or non-citizen nationals or must be able to provide legal proof of lawful admission for permanent residence. Individuals on temporary or student visas are not eligible; must have successfully completed an MD and pediatric residency; are no more than 4 years beyond attaining sub-specialty board eligibility; cannot have been the principal investigator of an NIH research grant or contract (including R, P, and K awards); must be willing to spend a minimum of 75 percent of their full-time professional effort (9 person-months) in research and research career development activities; must be willing, and are encouraged and expected, to seek independent grant support during the award period; and must have the full support of their supervisor and division director for this program.
*Restrictions:* Scholars may not accept or hold any other PHS award that duplicates the provisions of this career award during the period of this award.
*Geographic Focus:* All States
*Date(s) Application is Due:* Oct 1
*Contact:* Louis J. Muglia, CHRCDA Training Director; (513) 803-7902 or (513) 636-4200; louis.muglia@cchmc.org
*Internet:* https://www.cincinnatichildrens.org/education/funded/chrcda
*Sponsor:* Cincinnati Children's Hospital Medical Center
3333 Burnet Avenue
Cincinnati, OH 45229-3026

**Cincinnati Children's Procter Scholar Awards** 919

The goal of the Procter Scholar Program is to support the development of highly skilled faculty members who have a primary appointment at Cincinnati Children's (Departments of Pediatrics, Surgery, Radiology and Anesthesia) and who have a strong commitment to pursuing a career in academic research. Successful Candidates will have a: strong commitment to pursuing a career in academic research; record of accomplishment in research; strong mentorship along with a clear path to independent research; and well-defined scientific research plan. The deadlines for applications are the first Monday in May or October at 5:00 pm.
*Requirements:* Applications should be submitted by Jr. faculty, MD or MD/PhD, who are within 5 years of their initial appointment. Current K08 or K23 Awardees, to provide additional funds to expand research program (preference for those in 3rd or 4th year of funding).
*Geographic Focus:* All States
*Contact:* Stavra Xanthakos, Program Director; (513) 636-4315 or (513) 636-4200; stavra.xanthakos@cchmc.org
*Internet:* https://www.cincinnatichildrens.org/education/funded/procter
*Sponsor:* Cincinnati Children's Hospital Medical Center
3333 Burnet Avenue
Cincinnati, OH 45229-3026

**Cincinnati Children's Trustee Grants** 920

The goal of the Trustee Grant Award is to provide research funds for junior faculty to aid in the rapid achievement of independent, sustained extramural funding for their research program. Applications should be submitted within the first 4 years of initial appointment at Cincinnati Children's by faculty whose rank does not exceed Assistant Professor (Tenure Track) or Research Associate Professor (Non-Tenure Track). Funding of both laboratory and patient-based research projects will be considered, including proposals which are likely to have a direct impact on clinical practice such as phase I and phase II clinical trials, The anticipated outcomes of the Award are the successful transition to an NIH R01 award, R01-equivalent award from foundations or industry, and publications in peer-reviewed journals. The deadline for submission is the first Monday in May or October, at 5:00 pm. A maximum of $100,000 per year for two years can be requested initially. Renewal for year 2 is dependent upon the timely submission of progress report after year 1. Progress reports are due June 15th or December 15th depending on funding cycle.
*Geographic Focus:* All States
*Amount of Grant:* Up to 100,000 USD
*Contact:* Dan Starczynowski, Program Officer; (513) 803-5317 or (513) 636-4200; daniel.starczynowski@cchmc.org
*Internet:* https://www.cincinnatichildrens.org/education/funded/trustee
*Sponsor:* Cincinnati Children's Hospital Medical Center
3333 Burnet Avenue
Cincinnati, OH 45229-3026

**Cintas Foundation Fellowships** 921

Cintas Fellowships acknowledge creative accomplishments and encourage excellence in architecture, literature, music composition and the visual arts. Eligibility is limited to artists of Cuban citizenship or direct descent (having a Cuban parent or grandparent). Applications are reviewed by panels of qualified members from each artistic

discipline. The Board of the Cintas Foundation, Inc., awards fellowships based on the recommendations of the panels. Cintas Fellows are free to pursue their artistic activities as indicated in their application. The fellowships are not awarded for academic study or research, or to performing artists. Fellowships are not awarded more than twice to the same person. Second awards are granted only to candidates demonstrating outstanding artistic development. Fellowships are awarded annually in the amount of $10,000 each and are paid in quarterly stipends, beginning in September, for twelve consecutive months. The award is currently considered taxable income. The Cintas Foundation Fiscal year starts on September 1st and ends on August 31st of the following year.
*Requirements:* Creative artists, living outside of Cuba, of Cuban citizenship or direct lineage (having a Cuban parent or grandparent) are eligible.
*Restrictions:* The fellowships are not awarded for academic study or research, or to performing artists. Fellowships are not awarded more than twice to the same person. Second awards are granted only to candidates demonstrating outstanding artistic development.
*Geographic Focus:* All States
*Date(s) Application is Due:* Aug 1
*Contact:* Jessica Brodsky, (305) 237-7722; fax (305) 237-7711; jbrodsky@mdc.edu
*Internet:* www.cintasfoundation.org/fellowships.htm
*Sponsor:* Cintas Foundation
300 Northeast 2nd Avenue
Miami, FL 33132

**CJ Foundation for SIDS Research Grants** **922**
In support of its mission, the CJ Foundation for SIDS (Sudden Infant Death Syndrome) is strongly committed to providing grant support to best-in-class researchers who work toward the ultimate eradication of SIDS and other sleep-related infant and toddler deaths. Applications currently are by invitation only. Preference is given to past grantees and pilot projects that are gathering data for eventual submission to NIH for funding. Applications are reviewed by the Foundation's Medical Advisory Board. Interested applicants should contact the foundation for further information.
*Geographic Focus:* All States
*Samples:* University of California, San Diego, $202,000; Mayo Clinic, Rochester, Minnesota, $10,000; Children's Hospital, Boston, Massachusetts, $23,000.
*Contact:* Wendy Jacobs, Assistant Executive Director, Programs & Grants; (866) 314-7437 or (551) 996-5111; fax (551) 996-5326; wendy@cjsids.com or info@cjsids.org
*Internet:* www.cjsids.org/grants/grants-overview.html
*Sponsor:* CJ Foundation for SIDS
30 Prospect Avenue
Hackensack, NJ 07601

**Clark and Carolyn Adams Foundation Grants** **923**
Established in 1998 in Florida by attorney, John Clark Adams, the Clark and Carolyn Adams Foundation offers general operating support primarily in Florida for a number of causes, including: community improvement; community recreation; elementary and secondary education; disease research; health care; higher education; human service; legal services; mental health care; and religious groups. A formal application is required, though the initial approach should be in the form of a letter offering a detailed description of the proposed project and funding requested. Most recently, grants have ranged from $1,000 to $10,000. The annual deadline for application submission is June 30.
*Requirements:* Any 501(c)3 organization or school district supporting the residents of Florida may apply.
*Geographic Focus:* Florida
*Date(s) Application is Due:* Jun 30
*Amount of Grant:* 1,000 - 10,000 USD
*Samples:* Moffitt Cancer Center, Tampa, Florida, $1,000 - general operating support; Crisis Center of Tampa Bay, Tampa, Florida, $10,000 - general operating support; Ransom Everglades School, Coconut Grove, Florida, $8,000 - general operating support.
*Contact:* John Clark Adams, Trustee; (305) 448-9022
*Sponsor:* Clark and Carolyn Adams Foundation
540 Biltmore Way
Coral Gables, FL 33134

**CLAS Foreign Language and Area Studies Fellowships** **924**
The Center for Latin American Studies Foreign Language and Area Studies Fellowships supports the study of the less commonly taught languages of the region and related area studies courses through this Fellowship program, which is funded by a grant from the U.S. Department of Education. Students may use the fellowship to study any less commonly taught language of the region, including Brazilian Portuguese, Amerindian languages, and Haitian Kreyole. Priority is given to: in-residence students who demonstrate a strong area focus; students advancing in the language; and, in Portuguese, to students who already demonstrate proficiency equivalent to one year. Fellowships provide up to $5,000 tuition, plus a stipend of $2,500. The annual deadline for application submission is February 8.
*Requirements:* The Fellowship is available for all Uchicago PhD students and MA students in programs that last 2 or more years.
*Geographic Focus:* All States
*Date(s) Application is Due:* Feb 8
*Amount of Grant:* 2,500 - 7,500 USD
*Contact:* Natalie Arsenault, Associate Director; (773) 702-9741; fax (773) 702-1755; ncarsenault@uchicago.edu
*Internet:* https://clas.uchicago.edu/grants/area_studies.shtml
*Sponsor:* University of Chicago
5848 S University Avenue, Kelly Hall 117
Chicago, IL 60637

**CLAS Tinker Field Research Grants** **925**
The Center for Latin American Studies Tinker Field Research Grant supports master's, doctoral, and professional school students conducting pre-dissertation fieldwork in Spanish- and Portuguese-speaking countries of the Western Hemisphere, excluding Puerto Rico, providing an opportunity to establish professional and institutional contacts, assess potential research sites, and refine the dissertation project proposal. The annual application deadline is February 29. The grants provide graduate students with funds for travel to and within Latin America plus a small amount toward local living and research costs.
*Requirements:* Award recipients are selected from among all divisions and professional schools of the University of Chicago. Applicants must be degree-seeking graduate students currently enrolled at the University of Chicago.
*Geographic Focus:* All States
*Date(s) Application is Due:* Feb 29
*Contact:* Jamie Gentry, Program Officer; (773) 702-8420; fax (773) 702-1755; clas@uchicago.edu or jagentry@uchicago.edu
*Internet:* https://clas.uchicago.edu/page/tinker#overlay-context=page/grants
*Sponsor:* University of Chicago
5848 S University Avenue, Kelly Hall 117
Chicago, IL 60637

**Cleveland-Cliffs Foundation Grants** **926**
Contributions are made to nonprofit organizations to enhance the quality of life of Cleveland-Cliffs Inc employees and in recognition of a corporate responsibility toward educational, health, welfare, civic, and cultural matters within the communities where the company operates. The foundation was formed for the purpose of making contributions to groups organized and operated exclusively for religious, charitable, scientific, literary, or educational purposes and for the prevention of cruelty to children or animals. Types of support include general operating support, annual campaigns, capital campaigns, building/renovation, professorships, scholarship funds, research, and employee matching gifts. The foundation's major emphasis is on supporting education through a matching gift program and direct contributions to educational institutions. Requests for support must be in writing.
*Requirements:* Nonprofit organizations in the mining communities in which Cleveland-Cliffs Inc operates, including Alabama, Michigan, Minnesota, Ohio and, West Virginia are eligible.
*Geographic Focus:* Alabama, Michigan, Minnesota, Ohio, West Virginia
*Amount of Grant:* 250 - 50,000 USD
*Samples:* Bell Memorial Hospital, Ishpeming, MI, $400,000; University of Saint Thomas, Minneapolis, MN, $10,000; Michigan Tech Fund, Houghton, MI, $9,550.
*Contact:* Dana W. Byrne, Vice President; (216) 694-5700; fax (216) 694-4880; publicrelations@cleveland-cliffs.com
*Internet:* www.cliffsnaturalresources.com/Development/CommunityRelations/Pages/Cleveland-CliffsFoundation.aspx
*Sponsor:* Cleveland-Cliffs Foundation
1100 Superior Avenue, Suite 1500
Cleveland, OH 44114-2589

**Cleveland Foundation Sally and John Morley Family Fund Grants** **927**
The Sally and John Morley Family fund focuses its support on U.S.-based organizations striving to achieve significant systemic change in the area of mental health and/or at the intersection occurring among two or more of the following areas: education, health, the environment and socioeconomically disadvantaged populations. Focus on organizations providing services or doing research that provide or enable long-lasting, systemic and/or prevention benefits to their target populations (i.e. solutions with long-term impacts rather than providing short-term necessities such as food and clothing) and organizations/areas in which the fund's giving can make an impact. Grants will typically range from $5,000 to $50,000 with the focus on the following geographical areas: Cleveland, Ohio, and Boulder, Colorado. The annual deadline for application submission is August 1.
*Requirements:* The Cleveland Foundation makes most of its grants to tax-exempt (private agencies classified as 501(c)3 organizations) public charities as defined by the Internal Revenue Service. Some grants are also made to government agencies. The foundation requires all potential grant applicants to submit a grant inquiry, outlining basic information about the proposed project. Grant inquiries can be submitted at any time. If it is determined that your project fits the foundation's guidelines, you will be asked within a few weeks to submit a full application. It is recommended to submit one inquiry per proposed project.
*Restrictions:* Grants will not be awarded to/for: individuals; for-profit organizations; endowment campaigns, annual appeals, or membership drives; religious organizations for religious purposes; travel for individuals or groups when travel is the proposal's primary focus; community services such as police and fire protection; staff positions for government agencies; capital projects for hospitals, nursing homes, or institutions of higher education; publications, audiovisual projects, or video productions, however, consideration may be given when they fall within a promising project.
*Geographic Focus:* Colorado, Ohio
*Date(s) Application is Due:* Aug 1
*Amount of Grant:* 5,000 - 50,000 USD
*Contact:* Kristi Andrasik, Program Officer; (216) 615-7254 or (216) 615-7192; kAndrasik@CleveFdn.org
*Internet:* https://www.clevelandfoundation.org/grants/affiliated-funds/
*Sponsor:* Cleveland Foundation
1422 Euclid Avenue, Suite 1300
Cleveland, OH 44115

**CMHC-NSERC Postdoctoral Fellowships** 928

The NSERC and the Canada Mortgage and Housing Corporation (CMHC) have launched the CMHC-NSERC Postdoctoral Fellowship, valued at $45,000 per year for two years. The primary purpose of the initiative is to support fellows conducting research in the priority areas of the National Housing Strategy. Two awards will be offered in each of the six identified priority areas, which include: housing for those in greatest need; community housing sustainability; Indigenous housing; northern housing; sustainable housing and communities; and a balanced supply of housing. The annual deadline for application submissions is October 15.

*Requirements:* An applicant must be a Canadian citizen or a permanent resident of Canada and hold or expect to hold a doctorate in one of the fields of research that NSERC supports. In addition the applicant must: expect to complete all doctoral requirements by January 1 of the year following the award announcement; have completed all requirements of your first doctoral degree no more than two years before the deadline date of the year in which the application is made.

*Geographic Focus:* Canada

*Date(s) Application is Due:* Oct 15

*Amount of Grant:* 45,000 CAD

*Contact:* Coralie Rochefort, Program Officer; (613) 943-0703 or (855) 275-2861; schol@nserc-crsng.gc.ca or coralie.rochefort@nserc-crsng.gc.ca

*Internet:* www.nserc-crsng.gc.ca/Students-Etudiants/PD-NP/PDF-BP_eng.asp

*Sponsor:* Natural Sciences and Engineering Research Council of Canada
350 Albert Street, 16th Floor
Ottawa, ON K1A 1H5 Canada

**CMS Hispanic Health Services Research Grants** 929

The purpose of the Hispanic Health Services Research Grant is to implement Hispanic American health services research activities to meet the needs of diverse CMS beneficiary populations. The program seeks competitive applications for small applied research projects that relate to identifying and evaluating solutions for eliminating health disparities among Hispanic Americans. Funding is available for grants to implement research related to health care delivery and health financing issues affecting Hispanic American communities, including issues of access to health care, utilization of health care services, health outcomes, quality of services, cost of care, health and racial disparities, socioeconomic differences, cultural barriers, managed care systems, and active ties related to health screening, prevention, outreach, and education.

*Requirements:* Investigators should be associated with a university, college, community-based health organization, or a professional association that has a health services research component. Researchers are encouraged to use CMS data as part of their research projects. Researchers are also expected to become involved in the design, implementation, and operation of research projects that address health care issues such as financing, delivery, access, quality, and barriers affecting the Hispanic American community. Applicants must meet one of the following three requirements in order to qualify for funding under this grant program: a health services or disparities researcher at a university or college offering a Ph.D. or Master's Degree program in one or more of the following disciplines Allied Health, Gerontology, Health Care Administration, Health Education, Health Management, Nursing, Nutrition, Pharmacology, Public Health, Public Policy, Social Work; or a member of a community-based health organization with a Hispanic health services research component; or a member of a professional association focusing on Hispanic health services and health disparities issues.

*Restrictions:* Grant funds may not be used for any of the following: to provide direct services to individuals except as explicitly permitted under the grant solicitation; to match any other Federal funds; or, to provide services, equipment, or supports that are already the legal responsibility of another party under Federal law.

*Geographic Focus:* All States

*Date(s) Application is Due:* Jul 21

*Amount of Grant:* 200,000 - 250,000 USD

*Contact:* Linda Gmeiner, Grants Management Specialist; (410) 786-9954 or (410) 786-3000; linda.gmeiner@cms.hhs.gov

*Internet:* https://www.cms.gov/Research-Statistics-Data-and-Systems/Research/ResearchDemoGrantsOpt/Hispanic_Serving_Institution_HSI_Health_Services.html

*Sponsor:* Centers for Medicare and Medicaid Services
7500 Security Boulevard, Room C2-21-15
Baltimore, MD 21244-1850

**CMS Historically Black Colleges and Universities Health Services Research Grants** 930

The purpose of the Historically Black Colleges and Universities (HBCU) Health Services Research Grant program is to support researchers in implementing health services research activities to meet the needs of diverse CMS beneficiary populations. The goals of the grant program are to: encourage HBCU health services researchers to pursue research issues which impact the Medicare, Medicaid, and SCHIP (State Children's Health Insurance Program) programs; assist CMS in implementing its mission focusing on health care quality and improvement for its beneficiaries; assist HBCU researchers by supporting extramural research in health care capacity development activities for the African American communities; increase the pool of HBCU researchers capable of implementing the research, demonstration, and evaluation activities of CMS; and assist in fostering inter-university communication and collaboration regarding African American health disparity issues. Funding is available for grants to implement research related to health care delivery and health financing issues affecting African American communities, including issues of access to health care, utilization of health care services, health outcomes, quality of services, cost of care, health and racial disparities, socioeconomic differences, cultural barriers, managed care systems, and activities related to health screening, prevention, outreach, and education.

*Requirements:* To be eligible for grants under this program, an organization must be an HBCU (Historically Black College or University) and meet one of the following three *Requirements:* offer a Ph.D. or Master's Degree program in one or more of the following disciplines - Allied Health, Gerontology, Health Care Administration, Health Education, Health Management, Nursing, Nutrition, Pharmacology, Public Health, Public Policy, Social Work; have a School of Medicine; or have a member of the National HBCU Network for Health Services and Health Disparities. All proposals should describe research to be conducted with relevance to the CMS Medicare, Medicaid, and SCHIP programs and which area of Healthy People is served by this project. Applications must be submitted electronically.

*Restrictions:* Grant funds may not be used for any of the following: to provide direct services to individuals except as explicitly permitted under the grant solicitation; to match any other Federal funds; to provide services, equipment, or supports that are already the legal responsibility of another party under Federal law.

*Geographic Focus:* All States

*Date(s) Application is Due:* Jul 18

*Amount of Grant:* 200,000 - 250,000 USD

*Contact:* Linda Gmeiner, Grants Management Specialist; (410) 786-9954 or (410) 786-3000; linda.gmeiner@cms.hhs.gov

*Internet:* https://www.cms.gov/Research-Statistics-Data-and-Systems/Research/ResearchDemoGrantsOpt/Historically_Black_Colleges_and_Universities.html

*Sponsor:* Centers for Medicare and Medicaid Services
7500 Security Boulevard, Room C2-21-15
Baltimore, MD 21244-1850

**CMS Research and Demonstration Grants** 931

The general purpose of the Centers for Medicare and Medicaid Services' (CMS) research and demonstration grant program is to conduct and support projects to develop, test, and implement new health care financing and payment policies and to evaluate the impact of the agency's programs on its beneficiaries, providers, States, and other customers and partners. The scope of the agency's activities embraces all areas of health care: costs, access, quality, service delivery models, and financing and payment approaches. Generally, the Agency's extramural research and demonstration activities, such as evaluations, demonstration implementation, and research studies, are funded through contracts. CMS awards grants and cooperative agreements under certain specific, focused programs.

*Requirements:* The following themes represent the agency's current priorities in research. Note that all projects must fall into the agency's statutory authorities to operate and improve Medicare, Medicaid, and other CMS programs and activities: monitoring and avaluating CMS programs; strengthening medicaid, the State Children's Health Insurance Program (SCHIP), and other state programs; expanding beneficiaries' choices and availability of managed care options; developing FFS payment and service delivery systems; improving quality of care and performance under CMS programs; improving the health of our beneficiary population; prescription drugs; and building research capacity. An application kit is available online.

*Restrictions:* Applicants are expected to contribute towards the project costs. Generally five percent of the total project costs is considered acceptable. CMS rarely approves grants or cooperative agreements for research or demonstration projects in which the Federal Government covers 100 percent of the project's costs. The budget may not include costs for construction or remodeling or for project activities that take place before the applicant has received official notification of our approval of the project.

*Geographic Focus:* All States

*Amount of Grant:* 25,000 - 1,000,000 USD

*Contact:* Linda Gmeiner, Grants Management Specialist; (410) 786-9954 or (410) 786-3000; linda.gmeiner@cms.hhs.gov

*Internet:* https://www.cms.gov/Research-Statistics-Data-and-Systems/Research/ResearchDemoGrantsOpt/index.html?redirect=/ResearchDemoGrantsOpt/04_Other_CMS-Grant_Opportunities.asp

*Sponsor:* Centers for Medicare and Medicaid Services
7500 Security Boulevard, Room C2-21-15
Baltimore, MD 21244-1850

**CNIB Baker Applied Research Fund Grants** 932

The CNIB Baker Applied Research Fund supports research that is focused on the social, educational, cultural and rehabilitative needs of Canadians living with vision loss, and the application of assistive devices to meet these needs. This program is intended both to advance the career development of new investigators, and to encourage senior researchers to develop programs of applied research. Proposals are invited from both. Grants are for a one-year period, with a limit of $35000.

*Requirements:* Projects must be completed within 12 months. Funds that are unused after 24 months after the start up of the grant must be returned unless a special request for continuance is approved. Applicants must be residents of Canada and research must be conducted primarily in Canada. The applicant must: have an academic appointment at a Canadian university or occupy a supervisory position at a health care facility, or vision rehab facility. All applications will be evaluated with regard to the following criteria: excellence in the health and social sciences as applied to vision care; feasibility of research results translating into improved interventions, programs, and services for Canadians living with vision loss; enhancement of Canada's outstanding research into the needs of Canadians living with vision loss; and creation of training opportunities for students and service providers.

*Restrictions:* Equipment costs will not generally be covered, but in certain circumstances equipment may be funded on the basis of matching funds from the host facility. Travel and publications costs should not exceed $2,000.

*Geographic Focus:* Canada
*Date(s) Application is Due:* Jan 15
*Amount of Grant:* Up to 35,000 CAD
*Contact:* Shampa Bose, Executive Assistant and Research Coordinator; (416) 486-2500, ext. 7622 or (800) 563-2642; fax (416) 480-7059; shampa.bose@cnib.ca
*Internet:* www.cnib.ca/en/research/funding/eabaker-applied/
*Sponsor:* Canadian National Institute for the Blind
1929 Bayview Avenue
Toronto, M4G 3E8 Canada

**CNIB Baker New Researcher Fund Grants** 933
The CNIB Baker New Researcher Fund provides one-year grants to encourage new investigations that may lead to the prevention of vision loss. It is intended to benefit new investigators (within 5 years after an academic faculty appointment) by giving them experience and results which can assist them in further grant applications and pilot investigations. A new investigator is someone who is affiliated with a research facility through an academic posting and who has been in the field for less than five years. Grants are for one year, in amounts up to $35,000 (CAD). The application must be submitted online no later than January 15 of each year, with grant decisions made by April 1.
*Requirements:* Projects should be completed within 12 months. Funds that are unused after 24 months of the start up of the grant must be returned unless a special request for continuance is approved. Applicants must be residents of Canada and research must be conducted primarily in Canada.
*Restrictions:* Equipment costs will not generally be covered, but in certain circumstances equipment may be funded on the basis of matching funds from the host facility. Travel and publication costs should not exceed $2,000. Applications must be focused and no longer than 10 pages in total.
*Geographic Focus:* Canada
*Date(s) Application is Due:* Jan 15
*Amount of Grant:* Up to 35,000 CAD
*Contact:* Shampa Bose, Executive Assistant and Research Coordinator; (416) 486-2500, ext. 7622 or (800) 563-2642; fax (416) 480-7059; shampa.bose@cnib.ca
*Internet:* www.cnib.ca/en/research/funding/eabaker-researcher/
*Sponsor:* Canadian National Institute for the Blind
1929 Bayview Avenue
Toronto, M4G 3E8 Canada

**CNIB Barbara Tuck MacPhee Award** 934
The CNIB Barbara Tuck MacPhee Award supports researchers in the field of macular degeneration. The Tuck-MacPhee Award provides a one-year grant of up to $35,000. Projects should be completed within 12 months. The investigator will provide CNIB with a progress report six months following the start up of the grant. A second report including a copy of relevant findings of the research is required at the end of the research and prior to releasing the final quarterly payment. All publications and presentations of this research should acknowledge the support of the CNIB Baker Tuck MacPhee Fund. Applicants must apply online no later than January 15 of each year. The application will be reviewed by a multi-disciplinary review committee, and decisions will be finalized by April 1.
*Requirements:* Funds that are unused after 24 months must be returned unless a special request for continuance is approved. Equipment costs will not generally be covered, but in certain circumstances equipment may be funded on the basis of matching funds from the host facility. This competition is open to all researchers; however applicants must be residents of Canada and research must be conducted primarily in Canada.
*Restrictions:* Trainees (residents, fellows, post graduate students) are not eligible to apply.
*Geographic Focus:* All States, All Countries
*Date(s) Application is Due:* Jan 15
*Amount of Grant:* Up to 35,000 CAD
*Contact:* Shampa Bose, Executive Assistant and Research Coordinator; (416) 486-2500, ext. 7622 or (800) 563-2642; fax (416) 480-7059; shampa.bose@cnib.ca
*Internet:* www.cnib.ca/en/research/funding/Tuck-MacPhee/
*Sponsor:* Canadian National Institute for the Blind
1929 Bayview Avenue
Toronto, M4G 3E8 Canada

**CNIB Canada Glaucoma Clinical Research Council Grants** 935
The CNIB-CGCRC Award represents a partnership of CNIB and the Canada Glaucoma Clinical Research Council designed to invest in clinical glaucoma research. This partnership contributes about $225,000 over a three year period. Funding of the CNIB-CGCRC award is via an unrestricted grant from Alcon Canada Inc. The application deadline for the CNIB-CGCRC Award is June 30. There are two levels of funding available: $5,000 or less for investigators with training or practice for less than 5 years and $5000 to $25,000 for investigators with training or practice over 5 years.
*Geographic Focus:* All States
*Amount of Grant:* Up to 25,000 CAD
*Contact:* Shampa Bose, Executive Assistant and Research Coordinator; (416) 486-2500, ext. 7622 or (800) 563-2642; fax (416) 480-7059; shampa.bose@cnib.ca
*Internet:* www.cnib.ca/en/research/funding/cnib-cgrc/
*Sponsor:* Canadian National Institute for the Blind
1929 Bayview Avenue
Toronto, M4G 3E8 Canada

**CNIB E. (Ben) & Mary Hochhausen Access Technology Research Grants** 936
The award was established by Ben and Mary Hochhausen to further their long-time volunteer and financial contribution to CNIB. This international award encourages research in the field of access technology for people living with vision loss. Grants up to $10,000 are made each year at the discretion of the E. (Ben) and Mary Hochhausen Fund trustees. Applications are accepted from any country in the world. Awards may be applied to: research projects; study at centres of excellence in technology; fellowships; development of prototypes; or development costs for bringing important new products to market. Initial application should be received by the E. (Ben) and Mary Hochhausen Fund trustees by September 30.
*Requirements:* Trustees will consider: students enrolled in a recognized College, University or equivalent program; or researchers from industry, academia or applied science.
*Geographic Focus:* All States, All Countries
*Date(s) Application is Due:* Sep 30
*Amount of Grant:* Up to 10,000 CAD
*Contact:* Shampa Bose, Executive Assistant and Research Coordinator; (416) 486-2500, ext. 7622 or (800) 563-2642; fax (416) 480-7059; shampa.bose@cnib.ca
*Internet:* www.cnib.ca/en/research/funding/hochhausen/
*Sponsor:* Canadian National Institute for the Blind
1929 Bayview Avenue
Toronto, M4G 3E8 Canada

**Coca-Cola Foundation Grants** 937
The foundation has established education as its philanthropic focus and set aside most of its funds to support educational initiatives that address pressing needs. To help prepare youth for life, the foundation gives in three areas: higher education—pipeline programs that connect various levels of education and help students stay in school, scholarships, and minority advancement; classroom teaching and learning—innovative K-12 projects, teacher development, and small projects that deal with classroom activities; and global education—projects that encourage international studies, global understanding, and student exchange. Grants are awarded to both public and private institutions at all levels of education: universities, colleges, and secondary and elementary schools. International educational institutions and health care organizations also receive consideration. Types of support include annual campaigns, donated equipment, employee matching gifts, operating budgets, special projects, capital campaigns, continuing support, fellowships, internships, endowment funds, matching funds, and scholarship funds. Proposals may be submitted at any time.
*Requirements:* The following are eligible to apply: IRS 501(c)3 non-profits; a foreign organization that has received a ruling from the IRS that it is a section 501(c)3 tax exempt organization; or a foreign organization that is the equivalent of a U.S. charity.
*Restrictions:* The foundation does not make grants to individuals, religious endeavors, political or fraternal organizations, or organizations without 501(c)3 status.
*Geographic Focus:* All States
*Amount of Grant:* 10,000 - 6,000,000 USD
*Samples:* Holyfield Foundation, Inc., Fairburn, GA, $75,000—contribution to the Giving Disadvantaged Youth a Fighting Chance scholarship program; United Way of Metropolitan Atlanta, Atlanta, GA, $15,000—support for Samaritan House, Impact Group and Atlanta Union Mission; Atlanta Children's Shelter, Inc., Atlanta, GA, $10,000—contribution to support shelter programs.
*Contact:* Helen Smith Price, Executive Director; (404) 676-2568; fax (404) 676-8804
*Internet:* www.thecoca-colacompany.com/citizenship/our_communities.html
*Sponsor:* Coca-Cola Foundation
P.O. Box 1734
Atlanta, GA 30301

**Coleman Foundation Cancer Care Grants** 938
The Foundation has been an advocate for raising the standards of cancer care in the Midwest region and assuring that direct cancer services are available to cancer patients in the Chicago Metro area. Recently, the Foundation has employed an impact framework which provides a basic outline of the goals for the Foundation's grantmaking. For the Cancer program area, the Foundation has identified strategies that can enable health care providers to gain effective tools and resources in providing supportive oncology (care from diagnosis through end of life). The Cancer Impact Plan (see website) is intended to inform potential grantees as to the particular strategies the Foundation seeks to fund. The Foundation welcomes potential grantees to review the Cancer Impact Plan to assess and determine the best possible strategy to meet the Foundation's intended impact.
*Requirements:* Grants are made only to 501(c)3 or 509(a)1 nonprofit organizations that are not private foundations. The Foundation's primary geographic focus is the Midwest region, particularly the State of Illinois and the Chicago metropolitan area. Only programs within the United States will be considered, which excludes all international programs. Applicants should submit a letter of inquiry first; LOIs are accepted throughout the calendar year. The Foundation will advise you if a full proposal should be submitted for further review. Proposals are presented by Foundation staff and approved by the Board at quarterly meetings, usually in February, May, August and November.
*Restrictions:* The program does not fund for-profit businesses, individuals, individual scholarships, advertising books, tickets, equipment purchases (including computer hardware or software), or advertising. General solicitations and annual appeals will not be considered.
*Geographic Focus:* Illinois, Indiana, Iowa, Michigan, Ohio, Wisconsin
*Amount of Grant:* 50,000 - 500,000 USD
*Samples:* University of Chicago Medicine, Chicago, Illinois, $500,000 - matching grant award is to support the Chicago South Side Asthma Center for pediatrics; Ann and Robert H. Lurie Children's Hospital, Chicago, Illinois, $80,000 - funding for improvements in supportive oncology service delivery; Center for Independence through Conductive Education, Countryside, Illinois, $163,000 - to support startup and implementation of the Conductive Education program at a fourth site to serve children and young adults with physical disabilities living in the south and southwest corridor of Chicago.

*Contact:* Rosa Berardi, Program Officer; (312) 902-7120; fax (312) 902-7124; rberardi@colemanfoundation.org or info@colemanfoundation.org
Lisa Torres, Grants Manager; (312) 902-7120, ext. 105; fax (312) 902-7124; ltorres@colemanfoundation.org
*Internet:* https://colemanfoundationorg.presencehost.net/what_we_fund/cancer/
*Sponsor:* Coleman Foundation
651 West Washington Boulevard, Suite 306
Chicago, IL 60661

**College of Human Sciences Scholarships** 939

Human sciences undergraduate and graduate scholarships are available in the College of Human Sciences at Texas Tech University. Programs include human development and family studies, marriage and family therapy, interior design, merchandising, family financial planning, clothing and textiles, restaurant/hotel management, food and nutrition, early childhood, family and consumer sciences education, consumer economics, environmental design, and fashion design. Application deadline is February 1 unless specifically stated otherwise. Check Web site for current details and criteria of available scholarships.
*Requirements:* Scholarships are awarded only to students who have officially applied to Texas Tech University. Unless the scholarship award is for multiple years, students are required to resubmit scholarship applications during the spring semester to be eligible for renewal of their scholarship.
*Geographic Focus:* All States
*Date(s) Application is Due:* Feb 1
*Amount of Grant:* 200 - 2,000 USD
*Contact:* Dr. Karen Alexander, Scholarship Coordinator; (806) 742-3031; fax (806) 742-1849; karen.alexander@ttu.ed
*Internet:* www.depts.ttu.edu/hs/undergraduate_scholarships.php
*Sponsor:* Texas Tech University
Box 41162
Lubbock, TX 79409-1162

**Colonel Stanley R. McNeil Foundation Grants** 940

Colonel Stanley R. McNeil and his wife Merna created the Colonel Stanley R. McNeil Foundation in 1993 to support and promote quality educational, human-services, and health-care programming for underserved populations. Special consideration is given to charitable organizations that serve the needs of children. During their lifetimes, the McNeils were actively involved in their local church and community. Colonel McNeil also served on a number of boards, including Lake Bluff Children's Home and Ravenswood Hospital. Although the McNeils had no children, they were strong supporters of children's causes. The foundation is particularly interested in funding programs or organizations that focus on children's causes, start-up initiatives within the human-services or arts and culture arenas, and healthcare. To better support the capacity of nonprofit organizations, multi-year funding requests are considered. Grant requests for naming opportunities that honor the donors are strongly encouraged. Applicants must apply online at the grant website. Applicants are strongly encouraged to do the following before applying: review the downloadable state application procedures for additional helpful information and clarifications; review the downloadable online-application guidelines at the grant website; review the foundation's funding history (link is available from the grant website); review the online application questions in advance; and review the list of required attachments. These will generally include: a list of board members, financial statements (audited, reviewed, or compiled by independent auditor); an organization summary; a list of other funding sources; an IRS Determination letter; and other required documents. All attachments must be uploaded in the online application as PDF, Word, or Excel files. The Colonel Stanley R. McNeil Foundation has biannual deadlines of February 1 and June 1. Applicants for the February deadline will be notified of grant decisions by June 30, and applicants for the June deadline will be notified by November 30. Typical awards have ranged from $5,000 to $125,000.
*Requirements:* Illinois nonprofit organizations serving the Chicago metropolitan area are eligible.
*Restrictions:* Because requests for support usually exceed available resources, organizations can only apply to either the Lang Burk Fund or the Colonel Stanley McNeil Foundation in the same calendar year. Grant requests to both foundations during the same calendar year will no longer be accepted. The Foundation will consider requests for general operations only if the organization's operating budget is less than $1 million. In general, grant request for individuals, endowment campaigns or capital projects will not be considered.
*Geographic Focus:* Illinois
*Date(s) Application is Due:* Feb 1; Jun 1
*Amount of Grant:* 5,000 - 125,000 USD
*Samples:* Adler University, Chicago, Illinois, $50,000 - general operating support; Metropolis Strategies, Chicago, Illinois, $2550,000 - Illinois Justice Project support; World Sport Chicago, Chicago, Illinois, $25,000 - general operating support.
*Contact:* Srilatha Lakkaraju, Philanthropic Relationship Manager; (312) 828-8166; ilgrantmaking@ustrust.com
*Internet:* https://www.bankofamerica.com/philanthropic/foundation.go?fnId=84
*Sponsor:* Colonel Stanley R. McNeil Foundation
231 South LaSalle Street, IL1-231-13-32
Chicago, IL 60604

**Communities Foundation of Texas Grants** 941

The unrestricted funds of the foundation support programs and projects intended to improve the quality of life for the citizens of the Dallas, TX, area. The funds support education, health and hospitals, social services, youth, and cultural programs. The foundation encourages projects developed in consultation with other agencies and planning groups and that promote coordination, cooperation, and sharing among organizations. Types of support include seed grants, emergency funds, building funds, equipment acquisition, matching funds, technical assistance, research, capital campaigns, and operating budgets. Requests for operating funds generally are not granted.
*Requirements:* 501(c)3 organizations in the Dallas, TX, area may apply.
*Restrictions:* Grants are not made to or for individuals, endowments, sectarian religious purposes, political or lobbying efforts, deficit financing, media projects, or for operational expenses of established organizations.
*Geographic Focus:* Texas
*Samples:* Undermain Theatre, $25,000—to help restore 90 original Frank Lloyd Wright chairs that comprise the theater's current seating arrangement; Learning Center of North Texas, $25,000—to upgrade technology infrastructure, provide technology training for staff, and engage IT support services for one year; Contact Crisis Line, $16,200—to upgrade the computer hardware and software used by volunteers on the crisis help lines; Boy Scouts of America National Council (Circle Ten Council) $50,000—to fund a new shower and restroom facility at Camp Wisdom as part of the organization's $20 million centennial capital campaign.
*Contact:* Brent Christopher, President; (214) 750-4222; fax (214) 750-4210; bchristopher@cftexas.org
*Internet:* www.cftexas.org
*Sponsor:* Communities Foundation of Texas
5500 Caruth Haven Lane
Dallas, TX 75225-8146

**Community Foundation for Greater Atlanta AIDS Fund Grants** 942

The Atlanta AIDS Fund is a collaborative funding partnership between The Community Foundation, United Way of Greater Atlanta and Fashion Cares and the Allen Thornell HIV Care and Service Fund. The Community Foundation was an early supporter of HIV/AIDS initiatives, making its first grant in 1981. At that time the disease was known as GRID – Gay Related Immune Disease. The Atlanta community, the country and the world has made incredible progress since that time, but the challenges of HIV/AIDS are still present, particularly in our Atlanta region. The Atlanta AIDS Fund was created in 1991 to support metro Atlanta's HIV/AIDS advocacy, prevention education and service efforts through funding and leadership. Since that time nearly $11 million in grants have been awarded to organizations in the metro area. Currently, the Foundation is working on five goals: offering training to AIDS service organizations and their case management staff to serve as licensed patient navigators; providing funds to AIDS service organizations and/or community centers to facilitate mergers, consolidations and/or joint programming; providing funds for advocacy efforts to ensure Georgia's Medicaid expansion occurs and includes a full-range of covered services and care for people living with HIV/AIDS (including those at high-risk for infections); providing funds for advocacy efforts to monitor Georgia's AIDS Drug Assistance Program for access issues and challenges; and engaging an evaluator to assess progress and impact. After completion of an online Organizational Profile, selected will be invited to apply by the annual July 24 deadline.
*Requirements:* To apply for a grant organizations must: be located and providing services within the Foundation's 23-county service area; spend funds within the 23-county service area; be classified by the U.S. Internal Revenue Service under Section 501(c)3 of the I.R.S. code as a nonprofit, tax-exempt organization, donations to which are deductible as charitable contributions under Section 170(c)2 and the I.R.S. determination must be current; be registered with the Georgia Secretary of State as a nonprofit; have a minimum two-year operating history after the date of receipt of its 501(c)3 classification; have annual operating expenses greater than $100,000 as reflected in the most recently filed IRS form 990; have at least one full-time paid employee (paid minimum wage or more, working 2,080 hours or more) for the 12 months prior to submitting a Letter of Intent; and have a current written strategic or business plan for the whole organization that covers at least 24 months which includes the organization's entire current fiscal year.
*Restrictions:* The following organizations are not eligible to apply for funding: private and publicly funded schools (K-12) and institutions of higher learning (this does not include nonprofit charter schools); organizations that raise funds for publicly funded schools (K-12), institutions of higher learning and government agencies; organizations that require participation in religious services and/or education as a condition of receiving services; and organizations that have discriminatory policies and/or practices on the basis of race, color, national origin, age, disability, sex/gender, marital status, familial status, parental status, religion, sexual orientation, genetic information or political beliefs. Foundation funds may not be used to support the following: religious services and/or education; fundraising and marketing events; endowment funds; or capital campaign contributions (for building construction or renovation) or use of funds to cover capital campaign feasibility studies or campaign implementation expenses.
*Geographic Focus:* Georgia
*Date(s) Application is Due:* Jul 24
*Contact:* Lita Ugarte Pardi, Director, Resource Deployment; (404) 526-1131 or (404) 688-5525; fax (404) 688-3060; lpardi@cfgreateratlanta.org
*Internet:* cfgreateratlanta.org/nonprofits/available-grants/atlanta-aids-fund/
*Sponsor:* Community Foundation for Greater Atlanta
191 Peachtree Street NE, Suite 1000
Atlanta, GA 30303

**Community Foundation for Greater Buffalo Grants** 943

The Community Foundation for Greater Buffalo (CFGB) is a public charity holding more than 800 different charitable funds, large and small, established by individuals, families, nonprofit agencies and businesses to benefit Western New York. Since 1919, the foundation has served the needs of it's community and the wishes of it's donors through personalized service, financial stewardship, local expertise, and community leadership. The Foundation focuse's on four main areas of interest: strengthen the region as a center for arts and culture; natural, historic, and architectural resources; reduce racial and ethnic

disparities; increase economic self-sufficiency for low-income individuals and families. Special funding will also be available to support the following: AIDS research and its cure ($5,000); programs that serve the visual, speech or hearing impaired ($5,000). Preference will be given to requests that align with one or more of the four focus areas.
*Requirements:* Applicants must be 501(c)3 not-for-profit organizations located within the eight counties of Western New York: Allegany, Cattaraugus, Chautauqua, Erie, Genesee, Niagara, Orleans, and Wyoming.
*Restrictions:* The Foundation will not consider competitive funding for: endowments; religious purposes; schools not registered with the New York State Education Department; attendance at or sponsorship of fundraising events for organizations; annual events or festivals; any partisan political activity. Funds from the foundation cannot be used to support or oppose a candidate for political office. Projects and activities that have occurred. The Foundation will not, except in extraordinary cases, provide payment or reimbursement for expenses incurred prior to the funding decision being communicated to the applicant.
*Geographic Focus:* New York
*Date(s) Application is Due:* Mar 1
*Amount of Grant:* 1,000 - 25,000 USD
*Samples:* Carnegie Art Center, $4,500—to support the continued restoration of the landmark Carnegie Art Center building; Grassroots Gardens, $25,000—to support new community gardens that provide beautification, food and employment; Youth Character Development, $9,100—to aid refugee students in the transition to formal education and encourage college enrollment.
*Contact:* Jean McKeown, Senior Program Officer; (716) 852-2857, ext. 204; fax (716) 852-2861; jeanm@cfgb.org
*Internet:* www.cfgb.org/page17000.cfm
*Sponsor:* Community Foundation for Greater Buffalo
712 Main Street
Buffalo, NY 14202

**Community Foundation for Greater New Haven Quinnipiac River Fund Grants** 944
The Community Foundation for Greater New Haven (CFFGNH) is a philanthropic institution that was established in 1928. The foundation's mission is to create positive and sustainable change in Connecticut's Greater New Haven region by increasing the amount of and enhancing the impact of community philanthropy. The Quinnipiac River Fund established in 1990 was a result of a settlement between the National Resources Defense Council, Connecticut Fund for the Environment, and the Upjohn Corporation to improve the environmental quality of Quinnipiac River, the New Haven Harbor, and their watersheds, and otherwise benefit the environment of these resources. Consideration given to grant proposals that: research methods of reducing pollution, or otherwise improving the Quinnipiac River's environmental health; address means of reducing both non-point and point sources of pollution to the River; research the permitting process and look at the permits themselves; entail environmental advocacy, except litigation; study the ecology of the Quinnipiac River and the New Haven Harbor; provide public education about the Quinnipiac River and its watershed; purchase land on the Quinnipiac River for conservation purposes, or to reduce pollution and improve public access to the River.
*Requirements:* IRS 501(c)3 nonprofit organizations are eligible to apply in the greater New Haven area, which includes: Ansonia; Bethany; Branford; Cheshire; Derby; East Haven; Guilford; Hamden; Madison; Milford; New Haven; North Branford; North Haven; Orange; Oxford; Seymour; Shelton; Wallingford; West Haven and; Woodbridge.
*Restrictions:* Grants are not made to support religious activities, lobbying, or travel.
*Geographic Focus:* Connecticut
*Date(s) Application is Due:* Jan 15
*Amount of Grant:* 5,000 - 18,000 USD
*Samples:* Quinnipiac River Watershed Association, $15,000—to support The Friends of the Quinnipiac Project to remedy the effects of poor land use practiced by managers and owners of riparian properties through outreach, education, training, public recognition and technical assistance; Elm City Park Conservancy, $7,800—to reduce erosion in the Rip-Rap at the edge of the River in Quinnipiac Park; Watershed Partnership Inc., $18,000—to support the Safe Grounds Campaign which helps reduce non-point source pollution from lawn pesticides and synthetic fertilizers in the Quinnipiac River watershed.
*Contact:* Sarah Fabish, Grant/Scholarship Director; (203) 777-2386 or (203) 777-7075; fax (203) 787-6584; sfabish@cfgnh.org
*Internet:* qrwgis.newhaven.edu/QRFprojects.htm
*Sponsor:* Community Foundation for Greater New Haven
70 Audubon Street
New Haven, CT 06510

**Community Foundation for Muskegon County Grants** 945
The community foundation was established to serve the needs of the people of Muskegon County and nearby western Michigan. The foundation's strategic plan for grant making focuses on the prevention of problems rather than the cure; encourages programs that are collaborative, comprehensive, and have the potential to be continuous; encourages leveraging and matching grant opportunities from multiple funders; and supports seed money opportunities for innovative projects. Major grant interest areas include the arts, community development and urban revitalization, education, environment, health/human services, and needs of young children (ages 0-3). Types of support include seed money grants, special projects, matching funds, equipment, scholarship funds, loans, research, publications, conferences and seminars, endowment funds, consulting services, continuing support, emergency funds, internships, professorships, and renovation projects. Although the foundation generally makes one-year grant commitments, it will consider making longer term commitments for new efforts that show strong promise for positive impact. Applications are accepted on specific dates each year; it is recommended that the applicant contact the foundation to discuss their interests and to obtain the next grant application deadline.
*Requirements:* IRS 501(c)3 organizations and institutions in Muskegon County and nearby western Michigan are eligible.
*Restrictions:* Support will not be provided for routine operating expenses; capital equipment, computer hardware and software, and motor vehicles; conferences, publications, videos, films, television, or radio programs; endowment campaigns; special fund-raising events; religious programs that serve specific denominations; existing obligations or debts; individual schools or districts; or individuals.
*Geographic Focus:* Michigan
*Amount of Grant:* 250 - 50,000 USD
*Contact:* Marcy Joy, Program Officer; (231) 722-4538 or (231) 332-4124; fax (231) 722-4616; mjoy@cffmc.org
*Internet:* www.cffmc.org/grantapply.php
*Sponsor:* Community Foundation for Muskegon County
425 West Western Avenue, Suite 200
Muskegon, MI 49440

**Community Foundation of East Central Illinois Grants** 946
The community foundation awards grants from its unrestricted funds to Champaign County, IL, nonprofit organizations to use for charitable purposes. Representative categories include arts and humanities, environmental concerns, education, health, human services, research, urban affairs, church, and youth programs. Types of support include continuing support, annual campaigns, building/renovation, equipment, publication, consulting services, and scholarships.
*Requirements:* Champaign County, IL, nonprofit organizations are eligible.
*Geographic Focus:* Illinois
*Date(s) Application is Due:* Aug 31
*Contact:* Joan Dixon, President & CEO; (217) 359-0125; fax (217) 352-6494; joandixon@cfeci.org or info@cfeci.org
*Internet:* www.cfeci.org/non-profit/grant-application
*Sponsor:* Community Foundation of East Central Illinois
404 West Church Street
Champaign, IL 61820-3411

**Community Foundation of Louisville Belle S. and I. M. Kimmelfield Memorial Fund For Education Grants** 947
The Community Foundation of Louisville is unique, in that it responds to the evolving needs and opportunities in the community. There is no set agenda, and no pre-determined recipients. The Belle S. and I.M. Kimmelfield Memorial Fund For Education, administered by the Community Foundation, was established by attorney Carolyn Balleisen in loving memory of her parents to fund scholarships and educational research in Israel and the United States. Eligible organizations may apply to receive a grant award of up to $20,000. Letters of Intent (LOIs) are due by July 8, and the deadline for invited applications is August 15 at 4:00 pm.
*Requirements:* Eligible organizations include those that are: headquartered in Jefferson County, Kentucky (if based outside of the county, then the organization must demonstrate that a majority of beneficiaries are located in Jefferson County, Kentucky); and classified as a 501(c)3 public charity in good standing (organizations with a pending application for 501(c)3 status may apply with proof of Form 1023 receipt from the IRS).
*Geographic Focus:* Kentucky, Israel
*Date(s) Application is Due:* Aug 15
*Amount of Grant:* 100 - 20,000 USD
*Contact:* Chelsea VanHook, Grants Administrator; (502) 855-6963; fax (502) 585-4649; chelseav@cflouisville.org
*Internet:* https://www.cflouisville.org/grants-partnerships/
*Sponsor:* Community Foundation of Louisville
325 W Main Street, Suite 1110, Waterfront Plaza, West Tower
Louisville, KY 40202

**Community Foundation of Louisville Bobbye M. Robinson Fund Grants** 948
The Community Foundation of Louisville is unique, in that it responds to the evolving needs and opportunities in the community. There is no set agenda, and no pre-determined recipients. Field of Interest Funds support organizations working within a specific geographic region or toward a specific purpose. The Community Foundation makes grants to the most appropriate and effective organizations working in areas such as arts and culture, education, youth, health, and human services. The Bobbye M. Robinson Fund, administered by the Community Foundation, was established by Nancy Klempner Patton, in memory of her mother, to support programs that serve patients with colon cancer. Eligible organizations may apply to receive a grant award of up to $20,000. Letters of Intent (LOIs) are due by July 8, and the deadline for invited applications is August 15 at 4:00 pm.
*Requirements:* Eligible organizations include those that are: headquartered in Jefferson County, Kentucky (if based outside of the county, then the organization must demonstrate that a majority of beneficiaries are located in Jefferson County, Kentucky); and classified as a 501(c)3 public charity in good standing (organizations with a pending application for 501(c)3 status may apply with proof of Form 1023 receipt from the IRS).
*Geographic Focus:* Kentucky
*Date(s) Application is Due:* Aug 15
*Amount of Grant:* 100 - 20,000 USD
*Contact:* Anne McKune, Program Officer; (502) 855-6948 or (502) 585-4649; fax (502) 855-6173; annemc@cflouisville.org
*Internet:* www.cflouisville.org/grants-partnerships/field-of-interest-grants/
*Sponsor:* Community Foundation of Louisville
325 W Main Street, Suite 1110, Waterfront Plaza, West Tower
Louisville, KY 40202

**Community Foundation of Louisville C. E. and S. Endowment for the Parks Fund Grants** 949

The Community Foundation of Louisville is unique, in that it responds to the evolving needs and opportunities in the community. There is no set agenda, and no pre-determined recipients. The C. E. and S. Endowment for the Parks Fund, administered by the Community Foundation, was created to provide support for the maintenance of children's play areas in California and Cherokee Parks, and to support research, development or implementation of other enhancements to parks in Jefferson County, Kentucky. Eligible organizations may apply to receive a grant award of up to $20,000. Letters of Intent (LOIs) are due by July 8, and the deadline for invited applications is August 15 at 4:00 pm.
*Requirements:* Eligible organizations include those that are: headquartered in Jefferson County, Kentucky (if based outside of the county, then the organization must demonstrate that a majority of beneficiaries are located in Jefferson County, Kentucky); and classified as a 501(c)3 public charity in good standing (organizations with a pending application for 501(c)3 status may apply with proof of Form 1023 receipt from the IRS).
*Geographic Focus:* Kentucky
*Date(s) Application is Due:* Aug 15
*Amount of Grant:* 100 - 20,000 USD
*Contact:* Chelsea VanHook, Grants Administrator; (502) 855-6963; fax (502) 585-4649; chelseav@cflouisville.org
*Internet:* https://www.cflouisville.org/grants-partnerships/
*Sponsor:* Community Foundation of Louisville
325 W Main Street, Suite 1110, Waterfront Plaza, West Tower
Louisville, KY 40202

**Community Foundation of Louisville Diller B. and Katherine P. Groff Fund for Pediatric Surgery Grants** 950

The Community Foundation of Louisville is unique, in that it responds to the evolving needs and opportunities in the community. There is no set agenda, and no pre-determined recipients. Field of Interest Funds support organizations working within a specific geographic region or toward a specific purpose. The Community Foundation makes grants to the most appropriate and effective organizations working in areas such as arts and culture, education, youth, health, and human services. The Diller B. and Katherine P. Groff Fund for Pediatric Surgery was established in 1990 in support of needs in the areas of pediatric surgery and surgical care of children, as well as education and research in allied disciplines. Typically, eligible organizations may apply to receive a grant award of up to $20,000. Letters of Intent (LOIs) are due by July 8, and the deadline for invited applications is August 15 at 4:00 pm.
*Requirements:* Eligible organizations include those that are: headquartered in Jefferson County, Kentucky (if based outside of the county, then the organization must demonstrate that a majority of beneficiaries are located in Jefferson County, Kentucky); and classified as a 501(c)3 public charity in good standing (organizations with a pending application for 501(c)3 status may apply with proof of Form 1023 receipt from the IRS).
*Geographic Focus:* Kentucky
*Date(s) Application is Due:* Aug 15
*Amount of Grant:* 100 - 10,000 USD
*Contact:* Anne McKune; (502) 855-6948; fax (502) 585-4649; annemc@cflouisville.org
*Internet:* www.cflouisville.org/grants-partnerships/field-of-interest-grants/
*Sponsor:* Community Foundation of Louisville
325 W Main Street, Suite 1110, Waterfront Plaza, West Tower
Louisville, KY 40202

**Community Foundation of Louisville Dr. W. Barnett Owen Memorial Fund for the Children of Louisville and Jefferson County Grants** 951

The Community Foundation of Louisville is unique, in that it responds to the evolving needs and opportunities in the community. There is no set agenda, and no pre-determined recipients. Field of Interest Funds support organizations working within a specific geographic region or toward a specific purpose. The Community Foundation makes grants to the most appropriate and effective organizations working in areas such as arts and culture, education, youth, health, and human services. The Dr. W. Barnett Owen Memorial Fund for the Children of Louisville and Jefferson County, administered by the Community Foundation, was established by Dr. Albert P. Williams to honor the memory of Dr. Owen, a children's health care activist. Owen was instrumental in establishing the Kosair Children's Hospital in Louisville, when polio was the scourge of adolescence. The Fund supports children and youth health and other programs. Eligible organizations may apply to receive a grant award of up to $20,000. Letters of Intent (LOIs) are due by July 8, and the deadline for invited applications is August 15 at 4:00 pm.
*Requirements:* Eligible organizations include those that are: headquartered in Jefferson County, Kentucky (if based outside of the county, then the organization must demonstrate that a majority of beneficiaries are located in Jefferson County, Kentucky); and classified as a 501(c)3 public charity in good standing (organizations with a pending application for 501(c)3 status may apply with proof of Form 1023 receipt from the IRS).
*Geographic Focus:* Kentucky
*Date(s) Application is Due:* Aug 18
*Amount of Grant:* 100 - 20,000 USD
*Contact:* Anne McKune, Program Officer; (502) 855-6948 or (502) 585-4649; fax (502) 855-6173; annemc@cflouisville.org
*Internet:* www.cflouisville.org/grants-partnerships/field-of-interest-grants/
*Sponsor:* Community Foundation of Louisville
325 W Main Street, Suite 1110, Waterfront Plaza, West Tower
Louisville, KY 40202

**Community Foundation of Louisville Irving B. Klempner Fund Grants** 952

The Community Foundation of Louisville is unique, in that it responds to the evolving needs and opportunities in the community. There is no set agenda, and no pre-determined recipients. Field of Interest Funds support organizations working within a specific geographic region or toward a specific purpose. The Community Foundation makes grants to the most appropriate and effective organizations working in areas such as arts and culture, education, youth, health, and human services. The Irving B. Klempner Fund, administered by the Community Foundation, was established by Nancy Klempner Patton, in memory of her father, to support programs that serve patients with prostate cancer. Eligible organizations may apply to receive a grant award of up to $20,000. Letters of Intent (LOIs) are due by July 8, and the deadline for invited applications is August 15 at 4:00 pm.
*Requirements:* Eligible organizations include those that are: headquartered in Jefferson County, Kentucky (if based outside of the county, then the organization must demonstrate that a majority of beneficiaries are located in Jefferson County, Kentucky); and classified as a 501(c)3 public charity in good standing (organizations with a pending application for 501(c)3 status may apply with proof of Form 1023 receipt from the IRS).
*Geographic Focus:* Kentucky
*Date(s) Application is Due:* Aug 15
*Amount of Grant:* 100 - 20,000 USD
*Contact:* Anne McKune, Program Officer; (502) 855-6948 or (502) 585-4649; fax (502) 855-6173; annemc@cflouisville.org
*Internet:* www.cflouisville.org/grants-partnerships/field-of-interest-grants/
*Sponsor:* Community Foundation of Louisville
325 W Main Street, Suite 1110, Waterfront Plaza, West Tower
Louisville, KY 40202

**Community Foundation of Louisville Lee Look Fund for Spinal Injury Grants** 953

The Community Foundation of Louisville is unique, in that it responds to the evolving needs and opportunities in the community. There is no set agenda, and no pre-determined recipients. The Lee Look Fund for Spinal Injury, administered by the Community Foundation, was established by Lee Look in support of public charities that help people with spinal injuries. In 1999, while living in Los Angeles, Lee injured his neck in an accident. Following an examination and x-rays at a medical center, the diagnosis provided by the physicians was a sprain. Lee left with instructions to "take Advil." A few weeks later, in Louisville, he had his fifth and sixth vertebrae fused with a steel plate and a bone graft from his own hip. He started the Lee Look Fund for Spinal Injury to help the thousands of people dealing with the challenges of living with such an injury. The fund supports research that will hopefully find more effective treatments, therapies and even cures for the types of disabilities that result from spinal injuries. Eligible organizations may apply to receive a grant award of up to $20,000. Letters of Intent (LOIs) are due by July 8, and the deadline for invited applications is August 15 at 4:00 pm.
*Requirements:* Eligible organizations include those that are: headquartered in Jefferson County, Kentucky (if based outside of the county, then the organization must demonstrate that a majority of beneficiaries are located in Jefferson County, Kentucky); and classified as a 501(c)3 public charity in good standing (organizations with a pending application for 501(c)3 status may apply with proof of Form 1023 receipt from the IRS).
*Geographic Focus:* Kentucky
*Date(s) Application is Due:* Aug 15
*Amount of Grant:* 100 - 20,000 USD
*Contact:* Chelsea VanHook, Grants Administrator; (502) 855-6963; fax (502) 585-4649; chelseav@cflouisville.org
*Internet:* https://www.cflouisville.org/grants-partnerships/
*Sponsor:* Community Foundation of Louisville
325 W Main Street, Suite 1110, Waterfront Plaza, West Tower
Louisville, KY 40202

**Community Foundation of Riverside and San Bernardino County Desert Legacy Fund-California Desert Research Grants** 954

The Desert Legacy Fund-California Desert Research grant program supports graduate student scientific research that contributes to the awareness and conservation of desert parks, wilderness, and other natural open spaces in the California desert areas in Riverside, San Bernardino, Inyo and Imperial Counties. Funds were awarded each year in May to six research programs that support the goals of the Fund. The grants are given to the universities sponsoring the graduate students' research projects, with winning grants recommended through a competitive process by a committee, and selected with the final approval of the full Board of Directors of the Community Foundation. The annual application deadline is March 29. Grants ranging from $1,000 to $4,000 are available.
*Requirements:* Grants will be awarded on the basis of: the potential of the research to contribute to the understanding, management, restoration and/or conservation of desert wildernesses; the feasibility of completing research as stated in the application; and the degree to which the project meets the selection criteria and considerations described in the guidelines.
*Geographic Focus:* California
*Date(s) Application is Due:* Mar 29
*Amount of Grant:* 1,000 - 4,000 USD
*Contact:* Celia Cudiamat, Executive Vice President of Grant Programs; (951) 241-7777; fax (951) 684-1911; ccudiamat@thecommunityfoundation.net
*Internet:* www.thecommunityfoundation.net/grants/grants/grant-schedule/15-grants/53-desert-legacy-fund-california-desert-research
*Sponsor:* Community Foundation of Riverside and San Bernardino County
3700 Sixth Street, Suite 200
Riverside, CA 92501

**Conquer Cancer Foundation of ASCO Advanced Clinical Research Award in Colorectal Cancer** 955

The Advanced Clinical Research Award (ACRA) in Colorectal Cancer is designed to fund investigators who are committed to clinical cancer research and who wish to conduct original research not currently funded. The grant totals $450,000 and is paid in 3 annual increments of $150,000 made on or about July 1 of each year of the grant term. The grant funds will be directed to the sponsoring institution and should be used towards salary support, supplies, equipment, travel, etc. necessary for the pursuit of the recipient's research project. Letters of Intent are due by September 30, with full applications due by November 18. Award notifications are given by April 1.
*Requirements:* The applicant must meet the following criteria at the time of grant award: be a physician (MD, DO, or international equivalent) working in any country; be in the fourth to ninth year of a full-time, primary faculty appointment in a clinical department at an academic medical institution at the time of grant submission; have completed productive post-doctoral or post fellowship research and demonstrated the ability to undertake independent investigator-initiated clinical research; be an ASCO member (Full Member or International Corresponding Member) or have submitted a membership application with the grant application; and be able to commit 75% of full-time effort in research (applies to total research, not just the proposed project) during the award period.
*Geographic Focus:* All States
*Date(s) Application is Due:* Nov 18
*Amount of Grant:* 450,000 USD
*Contact:* Eileen Melnick, Director, Grants, Awards, and Foundation Giving; (571) 483-1435 or (571) 483-1700; melnicke@asco.org or grants@conquercancerfoundation.org
*Internet:* https://www.conquercancerfoundation.org/advanced-clinical-research-award
*Sponsor:* Conquer Cancer Foundation
2318 Mill Road, Suite 800
Alexandria, VA 22314

**Conquer Cancer Foundation of ASCO Advanced Clinical Research Award in Glioma** 956

The Conquer Cancer Foundation invites physicians with full-time faculty appointments to apply for an Advanced Clinical Research Award (ACRA) in Glioma. The program funds clinical investigators who are committed to clinical research. Their research must have a direct patient-oriented focus including clinical trials and/or translational research involving human subjects. The Advanced Clinical Research Award is designed to provide funding beyond the Clinical Research Career Development Award (CDA). By continuing to support proven clinical researchers who are post-CDA but at a critical stage in their early career, the foundation hopes to expand the cadre of expert clinical researchers who are developing promising research initiatives. The ACRA is a three-year grant totaling $450,000 to pay for personnel and/or research expenses, and travel to attend the CCF Grants and Awards Ceremony at the ASCO Annual Meeting. Letters of Intent are due by September 30, with full applications due by November 18. Award notifications are given by April 1.
*Requirements:* The applicant must meet the following criteria at the time of grant award: be a physician (MD, DO, or international equivalent) working in any country; be in the fourth to ninth year of a full-time, primary faculty appointment in a clinical department at an academic medical institution at the time of grant submission; have completed productive post-doctoral or post fellowship research and demonstrated the ability to undertake independent investigator-initiated clinical research; be an ASCO member (Full Member or International Corresponding Member) or have submitted a membership application with the grant application; and be able to commit 75% of full-time effort in research (applies to total research, not just the proposed project) during the award period.
*Geographic Focus:* All States
*Date(s) Application is Due:* Nov 18
*Amount of Grant:* 450,000 USD
*Contact:* Eileen Melnick, Director, Grants, Awards, and Foundation Giving; (571) 483-1435 or (571) 483-1700; melnicke@asco.org or grants@conquercancerfoundation.org
*Internet:* https://www.conquercancerfoundation.org/advanced-clinical-research-award
*Sponsor:* Conquer Cancer Foundation
2318 Mill Road, Suite 800
Alexandria, VA 22314

**Conquer Cancer Foundation of ASCO Advanced Clinical Research Award in Lung Cancer** 957

The Conquer Cancer Foundation invites physicians with full-time faculty appointments to apply for an Advanced Clinical Research Award (ACRA) in Lung Cancer. The program funds clinical investigators who are committed to clinical cancer research. Their research must have a direct patient-oriented focus including clinical trials and/or translational research involving human subjects. The Advanced Clinical Research Award is designed to provide funding beyond the Clinical Research Career Development Award (CDA). By continuing to support proven clinical researchers who are post-CDA but at a critical stage in their early career, the foundation hopes to expand the cadre of expert clinical researchers who are developing promising research initiatives. The ACRA is a three-year grant totaling $450,000 to pay for personnel and/or research expenses, and travel to attend the CCF Grants and Awards Ceremony at the ASCO Annual Meeting. Letters of Intent are due by September 30, with full applications due by November 18. Award notifications are given by April 1.
*Requirements:* The applicant must meet the following criteria at the time of grant award: be a physician (MD, DO, or international equivalent) working in any country; be in the fourth to ninth year of a full-time, primary faculty appointment in a clinical department at an academic medical institution at the time of grant submission; have completed productive post-doctoral or post fellowship research and demonstrated the ability to undertake independent investigator-initiated clinical research; be an ASCO member (Full Member or International Corresponding Member) or have submitted a membership application with the grant application; and be able to commit 75% of full-time effort in research (applies to total research, not just the proposed project) during the award period.
*Geographic Focus:* All States
*Date(s) Application is Due:* Nov 18
*Amount of Grant:* 450,000 USD
*Contact:* Eileen Melnick, Director, Grants, Awards, and Foundation Giving; (571) 483-1435 or (571) 483-1700; melnicke@asco.org or grants@conquercancerfoundation.org
*Internet:* https://www.conquercancerfoundation.org/advanced-clinical-research-award
*Sponsor:* Conquer Cancer Foundation
2318 Mill Road, Suite 800
Alexandria, VA 22314

**Conquer Cancer Foundation of ASCO Advanced Clinical Research Awards in Breast Cancer** 958

The Conquer Cancer Foundation invites physicians with full-time faculty appointments to apply for an Advanced Clinical Research Award (ACRA) in Breast Cancer. The program funds clinical investigators who are committed to clinical cancer research. Their research must have a direct patient-oriented focus including clinical trials and/or translational research involving human subjects. The Advanced Clinical Research Award is designed to provide funding beyond the Clinical Research Career Development Award (CDA). By continuing to support proven clinical researchers who are post-CDA but at a critical stage in their early career, the foundation hopes to expand the cadre of expert clinical oncology researchers who are developing promising research initiatives. The ACRA is a three-year grant totaling $450,000 to pay for personnel and/or research expenses, and travel to attend the CCF Grants and Awards Ceremony at the ASCO Annual Meeting. Letters of Intent are due by September 30, with full applications due by November 18. Award notifications are given by April 1.
*Requirements:* The applicant must meet the following criteria at the time of grant award: be a physician (MD, DO, or international equivalent) working in any country; be in the fourth to ninth year of a full-time, primary faculty appointment in a clinical department at an academic medical institution at the time of grant submission; have completed productive post-doctoral or post fellowship research and demonstrated the ability to undertake independent investigator-initiated clinical research; be an ASCO member (Full Member or International Corresponding Member) or have submitted a membership application with the grant application; and be able to commit 75% of full-time effort in research (applies to total research, not just the proposed project) during the award period.
*Geographic Focus:* All States
*Date(s) Application is Due:* Nov 18
*Amount of Grant:* 450,000 USD
*Contact:* Eileen Melnick, Director, Grants, Awards, and Foundation Giving; (571) 483-1435 or (571) 483-1700; melnicke@asco.org or grants@conquercancerfoundation.org
*Internet:* https://www.conquercancerfoundation.org/advanced-clinical-research-award
*Sponsor:* Conquer Cancer Foundation
2318 Mill Road, Suite 800
Alexandria, VA 22314

**Conquer Cancer Foundation of ASCO Advanced Clinical Research Awards in Sarcoma** 959

The Advanced Clinical Research Award (ACRA) in Sarcoma is designed to fund investigators who are committed to clinical cancer research and who wish to conduct original research not currently funded. The award is intended to support proposals with a patient-oriented focus, including a clinical research study and/or translational research involving human subjects. The grant totals $450,000 and is paid in three annual increments of $150,000 made on or about July 1 of each year of the grant term. Preference will be given to proposals involving therapeutic vaccine or immunotherapeutic approaches for sarcoma. Letters of Intent are due by September 30, with full applications due by November 18. Award notifications are given by April 1.
*Requirements:* The applicant must meet the following criteria at the time of grant award: be a physician (MD, DO, or international equivalent) working in any country; be in the fourth to ninth year of a full-time, primary faculty appointment in a clinical department at an academic medical institution at the time of grant submission; have completed productive post-doctoral or post fellowship research and demonstrated the ability to undertake independent investigator-initiated clinical research; be an ASCO member (Full Member or International Corresponding Member) or have submitted a membership application with the grant application; and be able to commit 75% of full-time effort in research (applies to total research, not just the proposed project) during the award period.
*Restrictions:* Proposals with a predominant focus on in vitro or animal studies (even if clinically relevant) are not allowed.
*Geographic Focus:* All States
*Date(s) Application is Due:* Nov 18
*Contact:* Eileen Melnick, Director, Grants, Awards, and Foundation Giving; (571) 483-1435 or (571) 483-1700; melnicke@asco.org or grants@conquercancerfoundation.org
*Internet:* https://www.conquercancerfoundation.org/advanced-clinical-research-award
*Sponsor:* Conquer Cancer Foundation
2318 Mill Road, Suite 800
Alexandria, VA 22314

**Conquer Cancer Foundation of ASCO Career Development Award** 960

The Career Development Award (CDA) is a research grant that provides funding to clinical investigators who have received their initial faculty appointment to establish an independent clinical cancer research program. The cancer research must have a patient-oriented focus, including a clinical research study and/or translational research involving

human subjects. Proposals with a predominant focus on in vitro or animal studies (even if clinically relevant) are not allowed. CDA is a three-year cancer grant totaling $200,000, paid in three annual increments to the awardee's institution.
*Requirements:* The applicant must: be a physician (MD, DO, or international equivalent) who is within the first to third year of a full-time, primary faculty appointment in a clinical department at an academic medical institution at the time of grant submission; have a valid, active medical license at the time of application; have completed productive postdoctoral research and demonstrated the ability to undertake independent investigator-initiated clinical research; be an ASCO member (Full Member or International Corresponding) or have submitted a membership application with the grant application; be able to commit more than 50% of full-time effort in research (applies to total research, not just the proposed project) during the award period; have a mentor from the sponsoring institution who must provide a letter of support. If the mentor is not an ASCO Member, a supporting letter from an ASCO Member from the sponsoring institution must be included; be up-to-date and in compliance with all requirements (e.g. progress reports, final reports, budget summaries, IRB approvals, etc.) of any past grants received from Conquer Cancer Foundation. Applicants holding Instructor/Lecturer appointments are eligible to apply; however, the institutional letter of support must include information about the institution's commitment to support the applicant for the duration of the grant period. The full application must be submitted by 11:59PM EDT on the deadline date.
*Restrictions:* Eligible physicians are allowed to hold only one grant from the Conquer Cancer Foundation at a time. Applicants should not have any current career development awards (i.e., K23, K08, or any other type of career development award) and have not been a Principal Investigator on any large project grants (i.e., R01 or international equivalent, or private foundation grants). Past recipients of training fellowships (i.e., Young Investigator Award or an F32 grant) are eligible. Applicants with institutional KL2/K12 grants are eligible to apply but will be asked to relinquish their institutional grant at the start of the CDA grant period.
*Geographic Focus:* All States
*Date(s) Application is Due:* Sep 22
*Amount of Grant:* 200,000 USD
*Contact:* Eileen Melnick, Director, Grants, Awards, and Foundation Giving; (571) 483-1435 or (571) 483-1700; melnicke@asco.org or grants@conquercancerfoundation.org
*Internet:* https://www.conquercancerfoundation.org/career-development-award
*Sponsor:* Conquer Cancer Foundation
2318 Mill Road, Suite 800
Alexandria, VA 22314

**Conquer Cancer Foundation of ASCO Clinical Trials Participation Award** 961
The Conquer Cancer Foundation Clinical Trials Participation Award (CTPA) is an award that recognizes community-based practices to increase awareness of and participation in clinical trials among physicians. CTPA recipients are recognized at the ASCO Annual Meeting and receive a recognition plaque, complimentary meeting registration, and a $2,000 award. The annual deadline for application submission is January 14, with award notifications made by the following March.
*Requirements:* Practices must meet the following criteria to qualify for the Clinical Trials Participation Award: be U.S. based and community-based; have conducted clinical research for at least five years; have not received the CTPA in the past five years; and have a quality research program with high-quality audit reports and investigators in good standing.
*Geographic Focus* All States, District of Columbia, Guam, Marshall Islands, Northern Mariana Islands, Puerto Rico, U.S. Virgin Islands, American Samoa
*Date(s) Application is Due:* Jan 14
*Amount of Grant:* 2,000,000 USD
*Contact:* Eileen Melnick, Director, Grants, Awards, and Foundation Giving; (571) 483-1435 or (571) 483-1700; fax (571) 366-9554; melnicke@asco.org or grants@conquercancerfoundation.org
*Internet:* https://www.conquercancerfoundation.org/clinical-trials-participation-award
*Sponsor:* Conquer Cancer Foundation
2318 Mill Road, Suite 800
Alexandria, VA 22314

**Conquer Cancer Foundation of ASCO Comparative Effectiveness Research Professorship in Breast Cancer** 962
The Comparative Effectiveness Research in Breast Cancer Professorship is designed to provide flexible funding to outstanding translational researchers who have made, and are continuing to make, significant contributions that have changed the direction of cancer research and who provide mentorship to future translational researchers. The award is intended to support qualified individuals who are dedicated to bringing advances in basic sciences into the clinical arena and to mentoring other translational researchers. The award totals $500,000 and is paid in five annual increments of $100,000 made on or about July 1 of each year of the grant term. The award is not designed to cover the total cost of a research project or the investigator's entire compensation, but can be used at the discretion of the awardee to support their overall research activities. Applications are due annually in September.
*Requirements:* Applicants must: have an MD, DO, PhD or equivalent degree; have the rank of full professor (or equivalent); have a full-time faculty appointment at an academic medical center; have made significant contributions that have changed the direction of breast cancer research; be serving as a research mentor to one or more researcher(s) in training, and must be planning to continue to provide leadership in this area throughout the award period; lead a research team in conducting research on comparative effectiveness in breast cancer; be an active member of ASCO or have submitted a membership application with the grant application; commit to spending 75% of time during the award period dedicated to research (applies to total research, not just the proposed project) including leading a team of researchers (in the lab and/or the clinic) and mentoring physician-scientists. The sponsoring institution must be a not-for-profit institution.
*Geographic Focus* All States, District of Columbia, Guam, Marshall Islands, Northern Mariana Islands, Puerto Rico, U.S. Virgin Islands, American Samoa
*Amount of Grant:* 500,000 USD
*Contact:* Eileen Melnick, Director, Grants, Awards, and Foundation Giving; (571) 483-1435 or (571) 483-1700; fax (571) 366-9554; melnicke@asco.org or grants@conquercancerfoundation.org
*Internet:* https://www.conquercancerfoundation.org/research-professorship
*Sponsor:* Conquer Cancer Foundation
2318 Mill Road, Suite 800
Alexandria, VA 22314

**Conquer Cancer Foundation of ASCO Drug Development Research Professorship** 963
The Drug Development Research Professorship (DDRP) is a five-year award designed to provide flexible cancer research funding to outstanding researchers who have made, and are continuing to make, significant contributions that have changed the direction of cancer research and who provide mentorship to future researchers. This award is intended to support qualified individuals who are dedicated to bringing advances into the clinical arena by exploring new and promising therapeutic compounds that will lead to improved treatments, and to mentoring other researchers in the scientific and regulatory aspects of drug development. The total award amount is $500,000 payable on July 1 in annual increments of $100,000 over five years. The award is not designed to cover the total cost of a research project or the investigator's entire compensation, but can be used at the discretion of the awardee to support their overall research activities. The annual deadline for applications is in September.
*Requirements:* Applicants must meet the following criteria at the time of grant submission: be a physician (MD, DO, or international equivalent) working in any country with a full-time faculty appointment in a clinical department at an academic medical center; currently hold the rank of full professor (or equivalent) at an academic medical center; made significant contributions to the development of new therapies for cancer; be serving as a research mentor to researcher(s) in training, and must be planning to continue to provide leadership in the area of drug development throughout the award period; lead a research team in the conduct of drug development research that includes clinical trials; be a full member of ASCO in good standing or submit a membership application with grant application; expect to spend 75% of time during the award period dedicated to research and drug development activities, leading a team of researchers (both in the lab and/or in the clinic) and mentoring junior faculty and trainees on research and drug development. Individuals with an endowed professorship from their own institution are eligible to apply for this award. Individuals with administrative responsibilities (such as a Division or Department Chair) must clearly document their time commitment for performing research activities and mentoring activities. The Drug Development Research Professorship is open to international applicants.
*Geographic Focus* All States, District of Columbia, Guam, Marshall Islands, Northern Mariana Islands, Puerto Rico, U.S. Virgin Islands, American Samoa
*Amount of Grant:* 500,000 USD
*Contact:* Eileen Melnick, Director, Grants, Awards, and Foundation Giving; (571) 483-1435 or (571) 483-1700; fax (571) 366-9554; melnicke@asco.org or grants@conquercancerfoundation.org
*Internet:* https://www.conquercancerfoundation.org/research-professorship
*Sponsor:* Conquer Cancer Foundation
2318 Mill Road, Suite 800
Alexandria, VA 22314

**Conquer Cancer Foundation of ASCO International Development and Education Award** 964
The International Development and Education Award (IDEA) provides support for early-career oncologists in developing countries and facilitates the sharing of knowledge between these oncologists and ASCO members. The program pairs IDEA Recipients with a leading ASCO member "Mentor" and enables Recipients to attend the ASCO Annual Meeting, participate in a post-meeting visit to their Mentor's institution, and develop long-term relationships to improve cancer care in their countries and inform ASCO programs in developing countries. The award is not intended to completely cover all travel or daily expenses during a recipient's trip but will cover much of the expense.
*Requirements:* Applicants must: have a current passport that does not expire before December; be a current resident of a country classified by the World Bank as Low-Income, Lower-Middle-Income, or Upper-Middle-Income, and have limited resources to attend the ASCO Annual Meeting; be a full Member, Member in Training, or International Corresponding member of ASCO or willing to submit an application for ASCO membership as part of the IDEA application process; be less than ten years past teir oncology program training; be fluent in English (both writing and speaking). Although applicants are not required to submit an abstract for the ASCO Annual Meeting, applicants who do submit an abstract will be given the strongest consideration.
*Restrictions:* Applicants cannot: have completed more than one academic year of "formal training" in a country classified by the World Bank as High-Income; have previously received the IDEA award or its predecessor, the International Travel Grant (ITG); have previously attended more than one ASCO meeting in the past five (5) years.
*Geographic Focus:* All States, All Countries
*Date(s) Application is Due:* Jan 6
*Amount of Grant:* 1,160 USD
*Contact:* Eileen Melnick, Director, Grants, Awards, and Foundation Giving; (571) 483-1435 or (571) 483-1501; fax (571) 366-9554; melnicke@asco.org or grants@conquercancerfoundation.org

*Internet:* https://www.conquercancerfoundation.org/international-development-and-education-award
*Sponsor:* Conquer Cancer Foundation
2318 Mill Road, Suite 800
Alexandria, VA 22314

**Conquer Cancer Foundation of ASCO International Development and Education Award in Palliative Care** 965

The International Development and Education Award in Palliative Care (IDEA-PC) provides medical education in palliative care, assists with career development, and helps establish strong relationships with leading ASCO members in the field of palliative care who serve as scientific Mentors to each recipient. IDEA-PC Recipients are expected to share the knowledge and training they receive through the program with colleagues in their home countries once they return. The IDEA-PC program enables Recipients to attend sessions at the ASCO Annual Meeting that address symptom management, communication, end-of-life care, and other relevant topics. The program also includes a visit to the Mentor's U.S. or Canadian institution to experience palliative care performed in a multidisciplinary environment, and provides an opportunity to cultivate long-term professional relationships that will help improve palliative care in the Recipient's home countries.
*Requirements:* Applicants must: have a current passport that does not expire before December; be a current resident of a country classified by the World Bank as Low-Income, Lower-Middle-Income, or Upper-Middle-Income, and have limited resources to attend the ASCO Annual Meeting; be a full Member, Member in Training, or International Corresponding member of ASCO or willing to submit an application for ASCO membership as part of the IDEA application process; be less than ten years past teir oncology program training; be fluent in English (both writing and speaking). Although applicants are not required to submit an abstract for the ASCO Annual Meeting, applicants who do submit an abstract will be given the strongest consideration.
*Restrictions:* Applicants cannot: have completed more than one academic year of "formal training" in a country classified by the World Bank as High-Income; have previously received the IDEA award or its predecessor, the International Travel Grant (ITG); have previously attended more than one ASCO meeting in the past five (5) years.
*Geographic Focus:* All States, All Countries
*Date(s) Application is Due:* Jan 6
*Amount of Grant:* 1,160 USD
*Contact:* Eileen Melnick, Director, Grants and Awards; (571) 483-1435 or (571) 483-1501; fax (571) 366-9554; melnicke@asco.org or grants@conquercancerfoundation.org
*Internet:* https://www.conquercancerfoundation.org/international-development-and-education-award
*Sponsor:* Conquer Cancer Foundation
2318 Mill Road, Suite 800
Alexandria, VA 22314

**Conquer Cancer Foundation of ASCO International Innovation Grant** 966

The International Innovation Grant provides research funding in support of novel and innovative projects that can have a significant impact on cancer control in low- and middle-income countries. The grant is a one-year research grant of up to $20,000 that is awarded and paid directly to a nonprofit organization or governmental agency in a low-income or middle-income country. The grant may be used by the organization only for the approved, budgeted costs of the research project. Each grant will have a Principal Investigator who is an ASCO member, is affiliated with the Grantee Organization, and is a resident of the low-income or middle-income country. International Innovation Grants are hypothesis-driven research grants that fund a specific research project that may result in the discovery of new knowledge about how to advance cancer control in a low- or middle-income setting. It is anticipated that novel approaches and clinical designs proposed for this grant may differ from what would be considered standard practice within high-income settings. Grantee Organizations and Principal Investigators will be expected to share and disseminate the knowledge gained during their research project.
*Requirements:* The applicant organization must: have a charitable purpose registered as a not-for-profit with the relevant national authority or must be a government agency; be located in a country categorized by the World Bank as Low-Income, Lower-Middle-Income or Upper-Middle Income; have been operating for at least one full year, have an acceptable management structure and processes in place, and be solvent with or without the support of the International Innovation Grant; should have experience in carrying out activities with tangible outcomes. The Principal Investigator must: be a member of ASCO or have submitted a membership application with the grant application; be a citizen or permanent resident of a country defined by the World Bank as low-income or middle-income, and currently residing in that country; be affiliated with the applicant organization.
*Geographic Focus:* All States, All Countries
*Date(s) Application is Due:* Oct 8
*Amount of Grant:* Up to 20,000 USD
*Contact:* Eileen Melnick, Director, Grants and Awards; (571) 483-1435 or (571) 483-1700; fax (571) 366-9554; melnicke@asco.org or grants@conquercancerfoundation.org
*Internet:* https://www.conquercancerfoundation.org/international-innovation-grants3
*Sponsor:* Conquer Cancer Foundation
2318 Mill Road, Suite 800
Alexandria, VA 22314

**Conquer Cancer Foundation of ASCO Long-Term International Fellowships** 967

The Conquer Cancer Foundation of ASCO Long Term International Fellowship (LIFe) provides early-career oncologists in developing nations the support and resources needed to advance their training by deepening their relationship with a U.S. or Canadian colleague and his or her institution. Through a one or two year fellowship the recipient will earn valuable training and experience with which they can affect change in cancer care in their home country. Recipients must return to their home institutions after completing their fellowship and are expected to disseminate the knowledge they have gained. The Foundation will administer fundraising and grant distribution for the fellowships. Funds will be sent from TACF to the host institution in January of the award year. The host institution will then be responsible for financial management of the fellowship. Any funds not spent on the fellowship will be returned to TACF. Recipients must attend the ASCO Annual Meeting that falls during the first year of their award term where they will be honored at The ASCO Cancer Foundation Grants Brunch. Recipients will also be expected to submit their research to ASCO via an abstract or publication submission following their fellowship. Applications are due to ASCO by November 3 for the fellowship to begin anytime in the next calendar year. Applications should be completed and returned jointly by the prospective host and fellow. A decision will be distributed by April 1.
*Requirements:* Hosts must: be a full ASCO member in good standing; be employed at a U.S. or Canadian institution with sufficient infrastructure to provide for a 1 year fellowship; independently identify fellowship applicants; agree to cover any costs in excess of ASCO's contribution; complete the host section of the LIFe Application; complete fellowship evaluations as requested by ASCO. Applicants must: be an ASCO member in good standing; reside in a country identified by the World Bank as income level low, lower middle, and upper middle; have the support of their home institution including a letter of support from their supervisor; be a medical doctor who has finished a fellowship in clinical oncology or the equivalent; have less than 10 years experience in the field of oncology; complete the applicant section of the LIFe application including submission of a CV of up to 3 pages in length; commit to return to country of origin at the completion of their fellowship.
*Geographic Focus:* All States
*Date(s) Application is Due:* Nov 3
*Amount of Grant:* 115,000 USD
*Contact:* Eileen Melnick, Director, Grants, Awards, and Foundation Giving; (571) 483-1435 or (571) 483-1700; fax (571) 366-9554; melnicke@asco.org or grants@conquercancerfoundation.org
*Internet:* https://www.conquercancerfoundation.org/long-term-international-fellowship
*Sponsor:* Conquer Cancer Foundation
2318 Mill Road, Suite 800
Alexandria, VA 22314

**Conquer Cancer Foundation of ASCO Medical Student Rotation Grants** 968

This Medical Student Rotation opportunity designed to facilitate the recruitment and retention of individuals from populations underrepresented in medicine to cancer careers and increase access to quality care for underserved communities. The Medical Student Rotation program was initially developed as part of the Conquer Cancer Foundation/ASCO/Susan G. Komen for the Cure Quality of Care Initiative. The program provides 8- to 10-week clinical or clinical research oncology rotations for U.S. medical students from populations underrepresented in medicine who are interested in pursuing oncology as a career. Recipients receive a $5,000 stipend for the rotation plus $1,500 for future travel to the ASCO Annual Meeting, An additional $2,000 will be provided to support the student's mentor.
*Requirements:* Applicants must be enrolled in an MD or DO, U.S. medical school program and be of an underrepresented population (for the purposes of this program, underrepresented population is defined according to the Association of American Medical Colleges definition, currently specified as those racial and ethnic populations that are underrepresented in the field of medicine relative to their numbers in the general population. Races/ethnicities that have been identified include but are not limited to: American Indian/Alaska Native, Black/African American, Hispanic/Latino, and Native Hawaiian/Pacific Islander). Candidates must be U.S. citizens, U.S. nationals, or permanent residents. Candidates also must demonstrate an interest in pursuing oncology as a career and have a record of good academic standing. Medical students may choose a mentor and rotation setting on their own, or they can request the Conquer Cancer Foundation assistance in pairing them with a mentor. Mentors are responsible for ensuring that the student receives consistent guidance and supervision throughout the rotation, either directly or by another member of the oncology practice or staff able to perform this role. The mentor is responsible for defining the goals and objectives of the rotation for the student, orienting the student to the office or hospital setting, and for providing regular feedback to the student.
*Geographic Focus:* All States
*Date(s) Application is Due:* Dec 7
*Amount of Grant:* Up to 8,500 USD
*Contact:* Eileen Melnick, Director, Grants and Awards; (571) 483-1435 or (571) 483-1700; fax (571) 366-9554; melnicke@asco.org or grants@conquercancerfoundation.org
*Internet:* https://www.conquercancerfoundation.org/medical-student-rotation-award-underrepresented-populations
*Sponsor:* Conquer Cancer Foundation
2318 Mill Road, Suite 800
Alexandria, VA 22314

**Conquer Cancer Foundation of ASCO Merit Awards** 969

The Merit Award is open to oncology trainees who are first authors on abstracts selected for presentation at an ASCO scientific meeting, including the ASCO Annual Meeting or smaller thematic meetings. Based on funding availability at the time of the Award, a limited number of Merit Awards will be awarded to fellows and residents whose research is addressed in high-quality abstracts submitted to an ASCO Meeting and recognized for its scientific merit. Recipients of Merit Awards are honored with the opportunity to present their abstract at the Annual Meeting and ASCO thematic meetings and to

receive between $1,000 to $1,500 in monetary support, complimentary registration for the ASCO Meeting and access to reserved Meeting housing. Merit Awards are also given for the following meetings subject to the availability of funding: ASCO Annual Meeting; Cancer Survivorship Symposium; Gastrointestinal Cancers Symposium; Genitourinary Cancers Symposium; Molecular Markers in Cancer Meeting; Palliative Care in Oncology Symposium; and Quality Care Symposium.
*Requirements:* Merit Award candidates must apply for the award at the time of abstract submission. The first author of an abstract must meet the following requirements to be considered for an Award: hold a doctoral degree (including but not limited to MD, DO, PharmD, or PhD) or be a doctoral degree candidate at the time of abstract submission; be enrolled in an oncology fellowship training program or equivalent training program at the time of abstract submission; work in an oncology laboratory or clinical research setting; agree to present the abstract at the Annual Meeting; provide a letter of support from his or her Training Program Director, indicating eligibility for the award; and provide a two-page curriculum vitae.
*Geographic Focus* All States, District of Columbia, Guam, Marshall Islands, Northern Mariana Islands, Puerto Rico, U.S. Virgin Islands, American Samoa
*Amount of Grant:* 1,000 - 1,500 USD
*Contact:* Eileen Melnick, Director, Grants, Awards, and Foundation Giving; (571) 483-1435 or (571) 483-1700; fax (571) 366-9554; melnicke@asco.org or grants@conquercancerfoundation.org
*Internet:* https://www.conquercancerfoundation.org/merit-award
*Sponsor:* Conquer Cancer Foundation
2318 Mill Road, Suite 800
Alexandria, VA 22314

**Conquer Cancer Foundation of ASCO Resident Travel Award for Underrepresented Populations** 970
The Conquer Cancer Foundation of ASCO Resident Travel Award for Underrepresented Populations (RTA) is designed to facilitate the recruitment and retention of individuals from underrepresented populations in medicine to cancer careers and increase access to quality care for underserved communities. RTA provides financial support for residents from underrepresented populations to attend ASCO's Annual Meeting. The intention of this award is to attract residents from underrepresented populations to a possible career in one of the oncology specialties. This travel award will allow the recipients to travel to the ASCO Annual Scientific Meeting, where they will have an opportunity to meet oncologists and to understand the career possibilities in this area. The Resident Travel Award for Underrepresented Populations includes a $1,500 travel advance, complimentary Annual Meeting registration, and access to the Annual Meeting housing block and ASCO's travel agent.
*Requirements:* Applicants must meet the following criteria to qualify for the Resident Travel Award: be enrolled in an ACGME-accredited residency program required for future training in a cancer related subspecialty (i.e., Internal Medicine considering Medical Oncology, Surgery considering Surgical Oncology, etc.); be a U.S. citizen, U.S. national or permanent resident; and be of an underrepresented population as defined by the program criteria.
*Restrictions:* Residents already enrolled in oncology programs (i.e., Radiation Oncology) are not eligible.
*Geographic Focus* All States, District of Columbia, Guam, Marshall Islands, Northern Mariana Islands, Puerto Rico, U.S. Virgin Islands, American Samoa
*Date(s) Application is Due:* Dec 7
*Amount of Grant:* 1,500 USD
*Contact:* Eileen Melnick, Director, Grants, Awards, and Foundation Giving; (571) 483-1435 or (571) 483-1700; fax (571) 366-9554; melnicke@asco.org or grants@conquercancerfoundation.org
*Internet:* https://www.conquercancerfoundation.org/resident-travel-award-underrepresented-populations
*Sponsor:* Conquer Cancer Foundation
2318 Mill Road, Suite 800
Alexandria, VA 22314

**Conquer Cancer Foundation of ASCO Translational Research Professorships** 971
The Translational Research Professorship is designed to provide flexible funding to outstanding translational researchers who have made and are continuing to make significant contributions that have changed the direction of cancer research and who provide mentorship to future translational researchers. The award is intended to support qualified individuals who are dedicated to bringing advances in basic sciences into the clinical arena and to mentoring other translational researchers. The award totals $500,000 payable on July 1 in annual increments of $100,000 over five years. Applications are due annually in September.
*Requirements:* The applicant must meet the following criteria at the time of grant award: be a physician (MD, DO or international equivalent) with a full-time faculty appointment in a clinical department at an academic medical center; currently hold the rank of full professor (or equivalent) at an academic medical center. If applicant does not hold the rank of full professor, but meets all other eligibility criteria, he/she must provide written explanation of why his/her current rank at their institution should be considered equivalent; made significant contributions that have changed the direction of cancer research; be serving as a research mentor to one or more translational researcher(s) in training, and must be planning to continue to provide leadership in this area throughout the award period; lead a research team in the conduct of translational research; be a member of the American Society of Clinical Oncology (ASCO); expect to spend 75% of time during the award period dedicated to translational research activities, leading a team of translational researchers (both in the lab and in the clinic) and mentoring physician-scientists. Individuals with an endowed professorship from their own institution are eligible to apply for this award. Individuals with administrative responsibilities (such as a Division or Department Chair) must clearly document their time commitment for performing research activities and mentoring activities. The sponsoring institution must be a not-for-profit institution.
*Geographic Focus* All States, District of Columbia, Guam, Marshall Islands, Northern Mariana Islands, Puerto Rico, U.S. Virgin Islands, American Samoa
*Amount of Grant:* 500,000 USD
*Contact:* Eileen Melnick, Director, Grants, Awards, and Foundation Giving; (571) 483-1435 or (571) 483-1700; fax (571) 366-9554; melnicke@asco.org or grants@conquercancerfoundation.org
*Internet:* https://www.conquercancerfoundation.org/research-professorship
*Sponsor:* Conquer Cancer Foundation
2318 Mill Road, Suite 800
Alexandria, VA 22314

**Conquer Cancer Foundation of ASCO Young Investigator Award** 972
The Young Investigator Award provides funding to promising investigators to encourage and promote quality research in clinical oncology. The purpose of this award is to fund physicians during the transition from a fellowship program to a faculty appointment. Priority consideration will be given to proposals that include patient-oriented and, ultimately, clinical research. Since many awards are supported with restricted grants from outside organizations, the ASCO is particularly interested in identifying young researchers working in the following sub-specialties and emerging disciplines: breast cancer; cancer survivorship; geriatric oncology; health disparities; kidney cancer; multiple myeloma; ovarian cancer; pancreatic cancer; sarcoma; survivorship; and young adult cancer. The Award is a one-year grant totaling $50,000 to pay for personnel and/or research expenses, and travel to attend the CCF Grants and Awards Ceremony at the ASCO Annual Meeting. Full applications are due by September 22. However, The ASCO Foundation welcomes application submissions in all oncology subspecialties.
*Requirements:* The recipient must be a physician (MD or DO) who, at the time of grant award is in the final year of his/her final sub-specialty training program or in the first year post his/her final sub-specialty training. The sponsoring facility must be an academic medical institution. The primary mentor must be in the candidate's proposed research field, must assume responsibility, and provide guidance for the research. The applicant must either be a member of ASCO or submit a membership application with the grant application. The applicant should spend at least 60% of his or her time in research during the award period.
*Geographic Focus* All States, District of Columbia, Guam, Marshall Islands, Northern Mariana Islands, Puerto Rico, U.S. Virgin Islands, American Samoa
*Date(s) Application is Due:* Sep 22
*Amount of Grant:* 50,000 USD
*Contact:* Eileen Melnick, Director, Grants, Awards, and Foundation Giving; (571) 483-1435 or (571) 483-1700; fax (571) 366-9554; melnicke@asco.org or grants@conquercancerfoundation.org
*Internet:* https://www.conquercancerfoundation.org/young-investigator-award
*Sponsor:* Conquer Cancer Foundation
2318 Mill Road, Suite 800
Alexandria, VA 22314

**Conseil des arts de Montreal Residency of the Americas** 973
The Residency of the Americas aims to increase the circulation of art and artists across the American continent. It is divided into two parts: a three-month residency for a visual artist and a second, two-month residency for a curator or art critic. Both residencies will enjoy a live-in studio with access to common living areas. They will also have access to high-speed Internet, wood and metal workshops and technical assistance, if needed. The residency includes various activities such as open house events, exhibitions and studio visits by professionals that will enable residents to come in contact with both the public and Montreal's professional arts community. The residencies include: return airfare (max $1,200); a live-in studio for three months; and a stipend of $250 per week. The application and guidelines are available on the Foundry website.
*Requirements:* Applicants must be artists or curators residing in the United States or Canadian artists living outside of Quebec.
*Geographic Focus:* All States, Canada
*Date(s) Application is Due:* Sep 30
*Amount of Grant:* 1,450 CAD
*Contact:* Americas Residency Contact; (514) 392-1544; residence@fonderiedarling.org
*Internet:* www.artsmontreal.org/en/programs/residence-des-ameriques
*Sponsor:* Conseil des arts de Montreal
1210 Sherbrooke Street East
Montreal, QC H2L 1L9 Canada

**Conservation, Food, and Health Foundation Grants for Developing Countries** 974
The Conservation, Food, and Health Foundation's geographic focus is the developing world. Through grants to support research and through targeted grants to help solve specific problems, the foundation helps build capacity within developing countries in three areas of interest: conservation, food, and health. The foundation concentrates its grant-making on research, technical assistance, and training projects of benefit to the Third World; favors grants for pilot projects and special programs that have a potential for replication; prefers to support projects that employ and/or train personnel from the developing world; and favors research concerning problems of importance to the developing world. The foundation has two 4-month funding cycles each year. Concept papers received by January 1 will be considered for eligibility for the March 1 full proposal deadline; concept papers received by July 1 will be considered for eligibility for the September 1 full proposal deadline. Full proposals are by invitation. Concept

applications must be submitted through the foundation's online application system. Detailed guidelines, instructions, and faqs are provided at the website.
*Requirements:* The following organizations are eligible to apply: 501(c)3 organizations that are not private foundations under section 509(a) of the United State Internal Revenue Code; "501(c)3 equivalent" foreign or domestic government units; and nongovernmental foreign organizations which can provide secure evidence of their nongovernmental status and charitable purpose.
*Restrictions:* The foundation does not provide support for buildings, land purchases, or vehicles; quantity purchases of durable medical equipment; endowments or fundraising activities; famine or emergency relief; direct medical care or treatment; feeding or food distribution programs; films, videos, or web-site production; scholarships, fellowships, or travel grants; conferences; re-granting through intermediaries; general operating support; or individuals (however, the foundation may support an individual engaged in research on a problem of significance to the developing world where the research is sponsored by an established, nonprofit organization such as an educational institution and conducted in close partnership with a local nongovernmental organization). The foundation does not consider the states of the former Soviet Union or former Eastern Bloc countries as within its geographic focus.
*Geographic Focus:* All States, All Countries
*Date(s) Application is Due:* Jan 1; Jul 1
*Amount of Grant:* Up to 25,000 USD
*Samples:* Fundacion Natura Bolivia, Santa Cruz, Bolivia, $29,242 - to develop a training program that helps municipal watershed authorities in Bolivia to develop payment schemes that help upstream landowners protect forests valuable to downstream communities; One Acre Fund, Falcon Heights, Minnesota, $25,000 - to refine a training, financing, education, and support program that helps rural Kenyan and Rwandan farmers double their harvests and income; Hohenheim University, Stuttgart, Germany, $23,813 - for a study in collaboration with Veld Products Research & Development in Botswana to assess how strip intercropping wild grass with sorghum-cowpea contributes to the field water balance as well as effects on field-water-use efficiency.
*Contact:* Prentice Zinn, Administrator; (617) 391-3091 or skype: prentice.zinn; pzinn@gmafoundations.com
*Internet:* cfhfoundation.grantsmanagement08.com/?page_id=5
*Sponsor:* Conservation, Food, and Health Foundation
77 Summer Street, 8th Floor
Boston, MA 02110-1006

**Cord Foundation Grants** **975**
The foundation makes grants to a variety of groups in three broad areas: education, social services, and the arts. Grant recipients have included higher education institutions, research groups, youth organizations, community service agencies, religious organizations, and performing and visual arts centers. Types of support include general operating support, building construction/renovation, equipment acquisition, emergency funds, program development, scholarship funds, research, and matching funds. Grants are awarded nationwide, with preference given to requests from northern Nevada. Application forms are not required.
*Requirements:* Nonprofit organizations are eligible but giving is primarily in the northern NV area.
*Restrictions:* The foundation does not support general fund-raising events, memorial campaigns, deficit fundings, conferences, dinners, or mass mailings.
*Geographic Focus:* Nevada
*Amount of Grant:* 10,000 - 250,000 USD
*Samples:* Saint Marys Hospital Foundation, Reno, NV, $200,000— for hospice program; Reno, City of, Reno, NV, $500,000—for purchase of YMCA property; Food Bank of Northern Nevada, Sparks, NV, $200,000—for new distribution center.
*Contact:* William Bradley, Trustee; (775) 323-0373
*Sponsor:* Cord Foundation
418 Flint Street
Reno, NV 89501-2008

**Covidien Advanced MIS Surgical Fellowships** **976**
Covidien sponsors fellowships each year at leading institutions that enable individuals to engage in educational and clinical activities that develop their expertise in minimally invasive procedures and techniques. The Foundation for Surgical Fellowship (FSF) is a 501(c)3 organization that handles all aspects of fellowship administration. The Foundation exists to fund high quality fellowships in various areas of surgery, including but not limited to flexible endoscopy, minimal access surgery, and other related disciplines. Through a blinded application and review process, the Foundation identifies and distributes funds to qualified organizations based on criteria of excellence. The Foundation promotes and supports the continued availability of postgraduate surgical training opportunities of the highest quality. Applicants may apply for the Fellowship through the FSF link located at the Covidien website.
*Geographic Focus:* All States
*Date(s) Application is Due:* May 18
*Contact:* Teresa Hacunda; (508) 261-8000; Teresa.Hacunda@covidien.com
Matt Clark, Project and Communications Coordinator; (310) 424-3332; fax (310) 437-0585; matt@surgicalfellowships.org or info@surgicalfellowships.org
*Internet:* www.covidien.com/covidien/pages.aspx?page=Education/Fellowships
*Sponsor:* Covidien Healthcare Products
15 Hampshire Street
Mansfield, MA 02048

**Covidien Surgical Mini-Fellowships** **977**
The Covidien Surgical Mini-Fellowship is a three to five day comprehensive educational program meant to more thoroughly train surgeons to perform a specified surgical procedure, and to educate surgeons on procedural specifics and patient management strategies. Fellowships may include detailed review/lecture of surgical techniques, potential intra-operative complications; discussion/lecture about surgeon's role in the care of the patient; discussion/lecture of post-operative patient care via observation of hospital rounds, review of treatment for potential post-operative complications; discussion of reimbursement strategies for laparoscopic procedures; observation of and/or participation in several laparoscopic and/or open cases; observation of operating room procedures with review of patient preparation and optimal patient positioning for the various procedures; and dry skills or animate lab. Applicants may call the contact person for current deadlines and information about the Fellowship.
*Requirements:* Applicants must have completed at least one master (CME or non CME) and at least one clinical immersion. Additional information about each of these classes is available at the Covidien.
*Geographic Focus:* All States
*Contact:* Teresa Hacunda; (508) 261-8000; Teresa.Hacunda@covidien.com
*Internet:* www.covidien.com/covidien/pages.aspx?page=Education/AdvancedTraining/Mini-Fellowship
*Sponsor:* Covidien Healthcare Products
15 Hampshire Street
Mansfield, MA 02048

**Crane Foundation General Grants** **978**
Created in 1951, the Crane Foundation is a non-profit corporation organized exclusively to make charitable contributions for religious, educational and scientific purposes, including institutions that qualify as tax exempt under section 501(c)3 of Internal Revenue Code. As part of the Foundation, the Crane Company offers a program for Matching Gifts to educational institutions. Any employee with at least one year of service may receive a 100% match on their donation up to a maximum of $5,000 per year.
*Requirements:* Any 501(c)3 organization in the U.S. may apply.
*Geographic Focus:* All States
*Amount of Grant:* 1,000 - 10,000 USD
*Contact:* Administrator; (201) 585-0888 or (203) 363-7300; crfdn@craneco.com
*Internet:* www.craneco.com/Category/35/Crane-Foundation-Inc.html
*Sponsor:* Crane Foundation
140 Sylvan Avenue, Suite 4
Englewood Cliffs, NJ 07632

**CSRPC/CSGS Joint Dissertation Fellowship** **979**
The Centers for the Study of Race, Politics, and Culture (CSRPC), in coordination with the Center for Gender and Sexuality (CSGS), announce the Dissertation Fellowship competition for the academic year. There will be one fellowship awarded annually. The successful candidates will receive a stipend of $23,000, along with a research fund of up to $1,000. Tuition, student fees, and basic health insurance will be covered by the award. The recipient of this fellowship will also be given a shared office at the Centers for the Study of Race, Politics, and Culture and Gender and Sexuality from September 1, through August 31. The recipient of the fellowship is expected generally to participate in the intellectual life of the two centers. The annual deadline for applications is April 4.
*Requirements:* University of Chicago Ph.D. candidates from all disciplines who are working in the intersection of the topics of race and ethnicity and gender and/or sexuality studies are encouraged to apply. University of Chicago doctoral students who have completed all requirements for the Ph.D. but the dissertation, including formal admission to candidacy, and expect to complete all field work by September 1 of the following year, are eligible to apply. Residency at CSRPC/CSGS during the main three academic quarters is essential. Some travel for research for brief periods is allowed, but this must be discussed with the Faculty Directors in advance.
*Geographic Focus:* All States
*Date(s) Application is Due:* Apr 4
*Amount of Grant:* 23,000 - 24,000 USD
*Contact:* Sarah Tuohey, Student Affairs Administrator; (773) 702-2365; fax (773) 834-2000; stuohey@uchicago.edu or csrpc@uchicago.edu
*Internet:* csrpc.uchicago.edu/fellowships_and_funding/
*Sponsor:* University of Chicago
5733 S. University Avenue
Chicago, IL 60637

**CSRPC Dissertation Fellowships** **980**
The goal of the fellowship is to enable an outstanding doctoral student interested in the study of race and ethnicity to devote his or her full energies to the completion of the dissertation. The fellowship carries a stipend of $23,000 for the academic year; a $1,000 travel and research budget; and will cover advanced residence tuition, fees, and basic university student health insurance, if needed. The successful applicant will be provided with an office and use of a computer at the center. The fellow will be expected to be in residence during the award year, present his or her work at one of the Reproduction of Race and Racial Ideologies Workshop meetings and to actively participate in the workshop and other activities sponsored by the center.
*Requirements:* University of Chicago doctoral students who have completed all requirements for the PhD but the dissertation, including formal admission to candidacy, and expect to complete all field work by September 26 are eligible to apply. Any dissertation that has as its central focus issues related to race or racialized groups will be

considered. Special consideration will be given to projects that attend to the intersection of race or ethnicity with other identities such as gender, class, sexuality, and nationality. *Restrictions:* The fellow may not engage in any remunerative activity, including teaching either on or off campus, while holding the award, and will be ineligible for further internal University funding from any source, including teaching appointments, if the degree is not completed within six months of the end of the CSRPC fellowship tenure.
*Geographic Focus:* All States
*Date(s) Application is Due:* Apr 4
*Amount of Grant:* 23,000 - 30,000 USD
*Contact:* Sarah Tuohey, Student Affairs Administrator; (773) 702-2365; fax (773) 834-2000; stuohey@uchicago.edu or csrpc@uchicago.edu
*Internet:* csrpc.uchicago.edu/fellowships_and_funding/
*Sponsor:* University of Chicago
5733 S. University Avenue
Chicago, IL 60637

**CSRPC Graduate Research and Travel Grants** **981**
The goal of the Graduate Research and Travel grant is to support the research efforts of outstanding graduate students interested in the study of race and ethnicity. The grant may only be used for expenses associated with a clearly defined research project. Activities that will be considered for funding include, but are not limited to, travel-related expenses (transportation and lodging) associated with field or primary archival research, transcription costs, duplication services, or the acquisition of data sets for quantitative analysis. Preference will be given to dissertation-related research, though other projects will also be considered. The annual application deadline is April 4. Grants of up to $2,500 will be awarded to successful candidates. The funds will be available for distribution after July 1.
*Requirements:* Funding is available to University of Chicago graduate or professional students in good standing. Travel must be completed and all research- and travel-related receipts submitted by May 15 of the following year.
*Restrictions:* These grants may be used to supplement other small grants, but cannot be combined to yield support of more than $5,000.
*Geographic Focus:* All States
*Date(s) Application is Due:* Apr 4
*Amount of Grant:* Up to 2,500 USD
*Contact:* Sarah Tuohey, Student Affairs Administrator; (773) 702-2365; fax (773) 834-2000; stuohey@uchicago.edu or csrpc@uchicago.edu
*Internet:* csrpc.uchicago.edu/fellowships_and_funding/
*Sponsor:* University of Chicago
5733 S. University Avenue
Chicago, IL 60637

**CSRPC Residential Fellowships** **982**
The Center for the Study of Race, Politics and Culture at the University of Chicago invites applications for Residential Fellows for the upcoming academic year. Qualified graduate students at the University from all disciplines are welcome. All applications and materials should be received by the annual deadline of April 4. This fellowship provides a research fund of $1,000 along with shared office space at the Center. Recipients of the fellowship are expected to be in residence during the award year, to actively participate in the Reproduction of Race and Racial Ideologies Workshop and other activities sponsored by the Race Center during the academic year, and to acknowledge the support of CSRPC in the dissertation and/or any research publications that may result from this fellowship.
*Requirements:* Applicants are required to be enrolled and attending a graduate program of study.
*Geographic Focus:* All States
*Date(s) Application is Due:* Apr 4
*Amount of Grant:* 1,000 USD
*Contact:* Sarah Tuohey, Student Affairs Administrator; (773) 702-2365; fax (773) 834-2000; stuohey@uchicago.edu or csrpc@uchicago.edu
*Internet:* csrpc.uchicago.edu/fellowships_and_funding/
*Sponsor:* University of Chicago
5733 S. University Avenue
Chicago, IL 60637

**CSTE CDC/CSTE Applied Epidemiology Fellowships** **983**
The Council of State and Territorial Epidemiologists (CSTE), in collaboration with the Centers for Disease Control and Prevention (CD.C.), the Association of Schools of Public Health (ASPH), and the Health Resources and Services Administration (HRSA) has established the Applied Epidemiology Fellowship Program to train recent graduates in the expanding field of applied epidemiology in an effort to meet the nation's ongoing need for applied epidemiology workforce capacity in state and local health departments. In achieving the fellowship's goal which is to provide a high quality training experience and to secure long-term career placement for fellows at the state or local level, the program focuses on balancing three key concepts as identified in the Applied Epidemiology and Training Program (AETP) Development Handbook (McDonnell 2002): providing service to the sponsoring agency; creating and training a core group of public health workers; and strengthening the capacity in applied epidemiology across public health institutions. Participating fellows will receive two years of on-the-job training at a state health agency (starting in June - August of the program year). Fellows are placed under the guidance of highly-trained and experienced primary and secondary mentors and work closely with epidemiologists at the state, local, and federal levels. A primary mentor must have a doctoral-level degree and the secondary mentor is preferred to have doctoral training as well. Mentors must devote four hours per week to spend with the fellow during the first month of the fellowship and two hours per week for the duration. The mentors oversee the training, research and field activities of the fellow, ensure that the fellow is familiar with relevant techniques in a given specialty, help provide day-to-day activities that provide "hands on" epidemiology experience, and encourage the overall professional development of the fellow. Mentors receive an initial orientation conducted by the CSTE National Office prior to the fellow's arrival, and they participate in conference calls at various intervals during the course of the fellowship. Host health agencies are CD.C.- and CSTE-approved, with a demonstrated capacity to provide an Applied Epidemiology Fellow with technical training, research opportunities, and practical experience in the application of epidemiologic methods. Host sites are selected based on: the career interests of the fellow; the scope, quality, and diversity of the experience offered to the fellow; the experience of supervisors in management; mentors' number of years of experience in epidemiology; academic training of the mentors; the availability of office space; and letters of support for the agency seeking a fellow. Fellows and host agencies are also carefully matched based on the program areas of interest of the fellow and available opportunities of the host agency. Program areas include infectious disease, injury, maternal and child health, environmental heath, chronic disease, substance abuse, and occupational health. Host agencies are expected to integrate fellows into their worksites and to treat them as entry-level permanent employees. If employee programs are offered to regular permanent employees, host agencies are expected to provide comparable programs and financial support for the fellow. The fellowship program will provide M.P.H.-degree fellows with a bimonthly stipend of up to $39,000 annually. Doctoral-level candidates will be paid up to $56,000. Stipends will be set according to location, cost of living, and pay structure at the host agency. The fellowship program will make available up to $3,200 per year to supplement the fellow's individual health insurance coverage. Up to $1,000 will be provided for moving-related expenses. $970 per year is provided each fellow for professional development. Host health agencies and mentors are strongly encouraged to provide financial support and opportunities as well for the fellow to participate in other public health activities that will expand the fellow's scope and depth of epidemiologic knowledge and/or expand his/her job-related capabilities. CD.C./CSTE Applied Epidemiology fellows will develop a comprehensive set of core skills through competency-based training. Applicants may view a list of core competencies at the fellowship website by clicking on the competencies tab. Applicants may apply online beginning in November. The link is available from the Application page of the fellowship website (click on "Application" on the left hand side of the screen). Applicants are encouraged to review the fellowship website for complete information on the fellowship and the application process.
*Requirements:* Applicants must have: an M.P.H., M.S.P.H., M.S., or an equivalent degree or advanced degree in a public-health-related field (ie. Ph.D. in epidemiology, biostatistics, or other pertinent field; an M.D. degree; or a D.V.M. degree); have completed coursework in at least three advanced graduate-level epidemiology courses and one graduate level biostatistics course; and a desire to pursue a long-term career at the state or local level.
*Restrictions:* United States citizenship
*Geographic Focus:* All States
*Date(s) Application is Due:* Feb 1
*Amount of Grant:* 39,970 - 56,970 USD
*Contact:* Ashlyn Beavor, Workforce and Fellowship Administrator; (770) 458-3811; fax (770) 454-8516; abeavor@cste.org
*Internet:* www.cste.org/dnn/ProgramsandActivities/FellowshipProgram/tabid/259/Default.aspx
*Sponsor:* Council of State and Territorial Epidemiologists
2872 Woodcock Boulevard Suite 303
Atlanta, GA 30341

**Cudd Foundation Grants** **984**
The Cudd Foundation awards grants to eligible nonprofit organizations in its areas of interest, including arts, culture, performing arts, children and youth, education, environment, health care, historic preservation, and social services. Types of support include annual campaigns, capital campaigns, continuing support, curriculum development, emergency grants, endowments, program development, research, and scholarship funds. The listed application deadline is for letters of interest; full proposals are by invitation only.
*Requirements:* Louisiana, Oklahoma, California, and New Mexico 501(c)3 nonprofit organizations are eligible to apply.
*Geographic Focus:* California, Louisiana, New Mexico, Oklahoma
*Amount of Grant:* 250 - 222,088 USD
*Samples:* Bayou District Foundation, New Orleans, Louisiana, $500 - education and development; Caldera Arts Organization, Agoura Hills, California, $5,000 - art programs; Rio Grande School, Santa Fe, New Mexico, $6,000 - educational support.
*Contact:* Amanda Cudd Stuermer, Trustee; (505) 986-8416; fax (505) 986-8427; cuddfdn@aol.com
*Sponsor:* Cudd Foundation
P.O. Box 1980
El Prado, NM 87529

**Curtis and Edith Munson Foundation Grants** **985**
The Curtis and Edith Munson Foundation began making grants in 1987-88 with an emphasize on partnerships, collaborations and seed funding for new projects. Over 85% of the grant program is dedicated to the conservation of natural resources in North America and the Caribbean Basin with emphasis on the United States. Main programs areas are: marine resource conservation and management with particular interest in fisheries; south Florida ecosystems; Alabama environmental issues; Washington D.C. metro area environmental issues; and population and environmental issues. Grants generally range from $10,000 to $50,000, with an average award size of $15,000 to $20,000. Initial approach should be by letter of inquiry, which is due by the 1st Friday in April or the 2nd Friday in September.

*Requirements:* Non-profits with a IRS 501(c)3 letter and EIN number are eligible. No unsolicited grant proposals excepted, send a 1 - 2 page letter of inquiry and a project summary (form available on the Foundations website) to the Foundation before sending any additional materials. Letter of inquiry deadline dates are April 10th and September 11th. email inquires to info@munsonfdn.org. If invited, send in a proposal by May 15th and October 16, see Foundations website for guidelines.
*Restrictions:* Site specific grants are limited to South Florida and Alabama. No grants to individuals or endowment funds. Most grants are restricted and require a 1:1 match.
*Geographic Focus:* All States
*Date(s) Application is Due:* Apr 5; Sep 13
*Amount of Grant:* 10,000 - 50,000 USD
*Samples:* Blue Climate Solutions, Washington, D.C., $15,000 - to support BCS' work to promote blue carbon solutions; Restore America's Estuaries, Arlington, Virginia, $30,000 - to support the work on including seagrass (with mangroves and coral reefs) in Blue Carbon certification and for the annual TCS conference; Shark Advocates International, Washington, D.C., $15,000 - first of a two-year grant in support of SAI's work towards sustainable management of elasmobranchs, especially cownose and other rays.
*Contact:* Wolcott Henry; (202) 887-8992; fax (202) 887-8987; info@munsonfdn.org
*Internet:* https://www.munsonfdn.org/guidelines.html
*Sponsor:* Curtis and Edith Munson Foundation
1320 19th Street, NW, Suite 500
Washington, D.C. 20036

**Curtis Foundation Grants** 986
The Curtis Foundation, based in Raleigh, North Carolina, supports a wide array of causes, including: abuse prevention; basic and emergency aid; Christianity; diseases and conditions; domesticated animals; equal opportunity in education; graduate and professional education; higher education; human services; international development; Methodism; news and public information; community nonprofits; public policy; reproductive health care; shelter and residential care; special population support; and youth development programs. The Foundation's geographic focus is the State of North Carolina. Applicants should submit the following in letter format: copy of IRS Determination Letter; a brief history of organization and description of its mission; and a detailed description of project and amount of funding requested. Although there are no specified deadlines, applicants should submit their proposal at least six months prior to the requested disbursement date.
*Requirements:* Any 501(c)3 supporting residents of North Carolina are eligible to apply.
*Geographic Focus:* North Carolina
*Amount of Grant:* Up to 100,000 USD
*Samples:* University of North Carolina at Chapel Hill, Chapel Hill, North Carolina, $110,000 - general operating support; Edenton Street United Methodist Church, Raleigh, North Carolina, $76,000 - general operating support; Food Bank of Central and Eastern Nort Carolina, Raleigh, North Carolina, $15,000 - general operating support.
*Contact:* Donald W. Curtis, President; (919) 781-6119
*Sponsor:* Curtis Foundation
P.O. Box 20443
Raleigh, NC 27619-0443

**Cystic Fibrosis Canada Clinical Fellowships** 987
Cystic Fibrosis Canada (CFC) Clinical Fellowships are intended for those physicians who have already obtained their residency training and who wish to pursue additional clinical training in Cystic Fibrosis (CF) care. The purpose is to train physicians to become CF specialists, so that they can provide ongoing clinical care to individuals with CF in Canada. This training would include developing competence to allow them to participate in clinical trials. This fellowship is not intended to train individuals from other countries who will not stay in Canada. Up to two highly-ranked competitive Clinical Fellowships will be offered by the Foundation each year. The value of a CCFF Clinical Fellowship is dependent upon academic qualifications, and/or research experience. Award levels are reviewed on an annual basis, and will correspond with prevailing Canadian rates.
*Requirements:* Canadian citizens or permanent residents who have an M.D. degree, have recently completed their clinical training, are exam eligible, and have obtained medical licensure in Canada, are eligible to apply. Clinical Fellowships are tenable in Canada or abroad and it is intended that this experience will directly benefit the Canadian CF community.
*Restrictions:* Clinical Fellowships will not be awarded to individuals who have not completed residency training.
*Geographic Focus:* Canada
*Date(s) Application is Due:* Oct 1
*Contact:* Administrator; (416) 485-9149; fax (416) 485-0960; info@cysticfibrosis.ca
*Internet:* www.cysticfibrosis.ca/en/treatment/TrainingAwards_ClinicalFellowships.php
*Sponsor:* Cystic Fibrosis Canada
2221 Yonge Street, Suite 601
Toronto, ON M4S 2B4 Canada

**Cystic Fibrosis Canada Clinical Project Grants** 988
Clinical Project grants are intended to bolster the Canadian Cystic Fibrosis Foundation's (CCFF's) commitment to clinical research, and to promote the further development of the Clinical Studies Network. Awarded at the discretion of the leaders of the Medical/Scientific Advisory Committee, in consultation with the Clinical Studies Network, clinical project grants provide a mechanism whereby the most important ideas for clinical studies can be forged into viable protocols, and strategies of urgent clinical relevance can be pursued (i.e. seed money for small, start-up pilot projects). Applications should be made in the form of a letter in which a clear hypothesis is detailed, relevance to CF is demonstrated, and a brief budget is outlined. Applications may be submitted at any time, but the Foundation must be consulted in advance with respect to the availability of funds. The amount of a grant will be determined by the Medical/Scientific Advisory Committee, following a detailed review of the applicant's proposed budget, up to $15,000.
*Geographic Focus:* Canada
*Date(s) Application is Due:* Apr 1
*Amount of Grant:* Up to 15,000 CAD
*Contact:* Administrator; (416) 485-9149; fax (416) 485-0960; info@cysticfibrosis.ca
*Internet:* www.cysticfibrosis.ca/en/research/clinicalprojectgrants.php
*Sponsor:* Cystic Fibrosis Canada
2221 Yonge Street, Suite 601
Toronto, ON M4S 2B4 Canada

**Cystic Fibrosis Canada Clinic Incentive Grants** 989
Clinic Incentive grants are intended to enhance the standard of clinical care available to Canadians with cystic fibrosis, by providing funds to initiate a comprehensive program for cystic fibrosis patient care, research, and teaching; or to strengthen an existing program. The amount of a grant will be determined by the Medical/Scientific Advisory Committee, following a detailed review of the applicant's proposed budget.
*Requirements:* Canadian hospitals and/or medical schools are eligible to apply for Clinic Incentive grants.
*Restrictions:* Basic support for clinical care is the responsibility of provincial governments, and Canadian Cystic Fibrosis Foundation Clinic Incentive grants, which are under the jurisdiction of the cystic fibrosis clinic director, are not intended to supplant public funding.
*Geographic Focus:* Canada
*Date(s) Application is Due:* Oct 1
*Contact:* Administrator; (416) 485-9149; fax (416) 485-0960; info@cysticfibrosis.ca
*Internet:* www.cysticfibrosis.ca/en/treatment/ClinicIncentiveGrants.php
*Sponsor:* Cystic Fibrosis Canada
2221 Yonge Street, Suite 601
Toronto, ON M4S 2B4 Canada

**Cystic Fibrosis Canada Fellowships** 990
A limited number of competitive fellowships are offered by the foundation each year for basic or clinical research training in areas of the biomedical or behavioral sciences pertinent to cystic fibrosis. Fellowships are tenable at approved universities, hospitals, and research institutes in Canada. Initial fellowships are awarded for a two-year period. Fellows may apply for a one-year renewal. No one may receive more than three years of support under a CCFF fellowship. Canadian fellowship applicants of exceptional quality requesting funding to study abroad will be considered. Applicants are expected to demonstrate that comparable training is not available in Canada. Equitable consideration is given to applicants from outside of Canada who intend to return to their own country on completion of the fellowship.
*Requirements:* Individuals who hold MD or PhD degrees are eligible to apply. Medical graduates should have already completed basic residency training, and must be eligible for Canadian licensure.
*Restrictions:* Fellowship (Initial) applicants who will have completed four or more years of training as of the October 1st application deadline, following their PhD or MD (post-MD clinical training), are not eligible for Cystic Fibrosis Canada Fellowships. Applicants who have already completed six or more years of postgraduate study or training are not eligible.
*Geographic Focus:* Canada
*Date(s) Application is Due:* Oct 1
*Contact:* Administrator; (416) 485-9149; fax (416) 485-0960; info@cysticfibrosis.ca
*Internet:* www.cysticfibrosis.ca/en/research/TrainingAwards_Fellowships.php
*Sponsor:* Cystic Fibrosis Canada
2221 Yonge Street, Suite 601
Toronto, ON M4S 2B4 Canada

**Cystic Fibrosis Canada Research Grants** 991
Research grants are intended to facilitate the scientific investigation of all aspects of cystic fibrosis. Generally, the Foundation funds research which is carried out in Canada, and the principal investigator must be based at a Canadian institution. While the Foundation will fund a collaboration from outside of Canada, funds must be directed through a Canadian institution and Canadian investigators intending to collaborate with individuals from outside of Canada should contact the Foundation. Applications may be submitted by groups of individuals who plan to collaborate. Grants are awarded for up to three years.
*Requirements:* A principal investigator should hold a recognized, full-time faculty appointment in a relevant discipline at a Canadian university or hospital. Under exceptional circumstances, and at the discretion of the Research Subcommittee, research grant applications by individuals may be evaluated on a case-by-case basis, with significant emphasis placed on the degree of independence of the applicant, and on the institutional commitments to this individual. Such applications must include a statement from the applicant addressing the issue of salary support during the term of the grant, and the availability of laboratory space; and letters from the Departmental Chair and Dean of Faculty, clarifying the nature and extent of the institutional commitment to the applicant. This same eligibility criterion regarding a faculty appointment applies to any named co-investigator.
*Restrictions:* Research grants do not provide support for graduate students, postdoctoral fellows, or summer students; construction costs; institutional overheads for laboratory facilities; or purchase of equipment in excess of $10,000.
*Geographic Focus:* Canada
*Date(s) Application is Due:* Oct 1
*Contact:* Administrator; (416) 485-9149; fax (416) 485-0960; info@cysticfibrosis.ca
*Internet:* www.cysticfibrosis.ca/en/index.php
*Sponsor:* Cystic Fibrosis Canada
2221 Yonge Street, Suite 601
Toronto, ON M4S 2B4 Canada

**Cystic Fibrosis Canada Scholarships** 992

Cystic Fibrosis Canada Scholarships provide salary support to a limited number of exceptional investigators, who have made outstanding contributions to, and have demonstrated leadership in, cystic fibrosis research. This award offers new investigators an opportunity to develop outstanding cystic fibrosis research programs, unhampered by heavy teaching or clinical loads. The award will be for a three-year period, renewable for an additional two years on receipt of a satisfactory progress report during the third year of the award. In no case will an award be for more than five years. The value of a Cystic Fibrosis Canada Scholarship is $60,000 per year.

*Requirements:* Applications are accepted from candidates who have received their first faculty appointment within the preceding five calendar years.

*Geographic Focus:* Canada

*Date(s) Application is Due:* Oct 1

*Contact:* Administrator; (416) 485-9149; fax (416) 485-0960; info@cysticfibrosis.ca

*Internet:* www.cysticfibrosis.ca/en/research/TrainingAwards_Scholarships.php

*Sponsor:* Cystic Fibrosis Canada
2221 Yonge Street, Suite 601
Toronto, ON M4S 2B4 Canada

**Cystic Fibrosis Canada Senior Scientist Research Training Awards** 993

The Cathleen Morrison Senior Scientist Research Training Award provides support to a limited number of established senior investigators by offering them an opportunity to obtain additional training that will enhance their capacity to conduct research directly relevant to cystic fibrosis. Cathleen served the CF cause in Canada, and internationally, for thirty years, as Cystic Fibrosis Canada's Executive Director, and then Chief Executive Officer, from 1981 until 2011. Cathleen has been the consummate advocate for CF research – and health research in general. She has recruited, encouraged and guided many researchers from across Canada, throughout their careers. This award can be used for sabbatical support for qualified individuals. Senior Scientist Research Training awards are awarded for a minimum of three months to a maximum of one year. Frequency of application should be reasonable; once every five years. The amount of the award, to a maximum of $15,000, will be determined by the Medical/Scientific Advisory Committee, following a review of the application.

*Requirements:* Applicants must have held a recognized, full-time faculty appointment in a relevant discipline at a Canadian university or hospital for at least six years. In addition, applicants must hold major research grant funding.

*Restrictions:* Applicants may hold only one Canadian Cystic Fibrosis Foundation personnel award concurrently.

*Geographic Focus:* Canada

*Date(s) Application is Due:* Apr 1; Oct

*Amount of Grant:* Up to 15,000 CAD

*Samples:* Dr. Robert E.W. Hancock, The University of British Columbia, $29,116 - novel genetic screen employing mouse homozygous knockout ES cells.

*Contact:* Administrator; (416) 485-9149; fax (416) 485-0960; info@cysticfibrosis.ca

*Internet:* www.cysticfibrosis.ca/en/research/TrainingAwards_SeniorScientistResearch.php

*Sponsor:* Cystic Fibrosis Canada
2221 Yonge Street, Suite 601
Toronto, ON M4S 2B4 Canada

**Cystic Fibrosis Canada Small Conference Grants** 994

Small conference grants are offered to support medical/scientific conferences which focus on subjects of direct relevance to cystic fibrosis. Such grants will be up to a maximum of $C2500 and are intended to supplement other funding sources. The small conference grants will also support inter-clinic exchanges which are used to facilitate the exchange of special expertise between larger, university-based CF clinics and smaller, more remote clinics. Requests of this type will involve fewer individuals, and will not normally exceed $C1000. Applications may be submitted at any time, but Cystic Fibrosis Canada should be consulted in advance with respect to the availability of funds. Grants are on a first-come, first-served basis.

*Requirements:* CF clinic directors and CCFF-funded investigators are eligible to apply.

*Geographic Focus:* Canada

*Amount of Grant:* Up to 1,000 CAD

*Contact:* Administrator; (416) 485-9149; fax (416) 485-0960; info@cysticfibrosis.ca

*Internet:* www.cysticfibrosis.ca/en/research/SmallConferenceGrants.php

*Sponsor:* Cystic Fibrosis Canada
2221 Yonge Street, Suite 601
Toronto, ON M4S 2B4 Canada

**Cystic Fibrosis Canada Special Travel Grants For Fellows and Students** 995

Cystic Fibrosis Canada (CFC) encourages supported Fellows and Students to attend and participate in scientific meetings related to cystic fibrosis. In addition, allowances may be utilized to acquire new knowledge (i.e. visit another lab). Special Travel allowances may be awarded to CFC-supported Fellows and Students, for each year of their award, upon written application and pending the availability of funds. No CFC-supported Fellow or Student may receive more than one grant per year. Subject to the availability of funds and under certain circumstances, CFC-supported Scholars, physicians who are not CF clinic directors but who specialize in the clinical care of adults with cystic fibrosis at Canadian CF clinics, and recipients of the Skelly and Lackey Summer Studentship awards, may also be eligible. Special Travel allowances are intended to cover travel costs not to exceed the lowest applicable advance booking airfare, and reasonable accommodation and meals, up to a $1,800. Additional guidelines are available at: www.cysticfibrosis.ca/en/research/TravelAllowancesStudents.php.

*Requirements:* Applications should be made in the form of a letter, and must be submitted prior to the proposed travel. The letter should indicate the conference or meeting for which the allowance is requested and must include a proposed travel and accommodation budget, along with the supervisor's signed endorsement. Proof of active participation (i.e. abstract acceptance, etc.) is also required.

*Restrictions:* An itemized breakdown of expenses incurred and the original receipts must be submitted to the Foundation upon return from the conference or meeting, and any unexpended funds returned to the Foundation.

*Geographic Focus:* Canada

*Amount of Grant:* Up to 1,200 CAD

*Contact:* Grants Administrator; (416) 485-9149; fax (416) 485-0960; ResearchPrograms@cysticfibrosis.ca or info@cysticfibrosis.ca

*Internet:* www.cysticfibrosis.ca/en/research/TravelAllowancesStudents.php

*Sponsor:* Cystic Fibrosis Canada
2221 Yonge Street, Suite 601
Toronto, ON M4S 2B4 Canada

**Cystic Fibrosis Canada Studentships** 996

A limited number of competitive studentships are offered by the foundation each year to highly qualified graduate students who are registered for a higher degree and who are undertaking full-time research training in areas of the biomedical or behavioral sciences relevant to cystic fibrosis. Students are expected to spend at least 75 percent of their time on the research training described in their application. Equitable consideration will be given to Studentship applicants from outside of Canada, who intend to return to their own country on completion of a Studentship. The value of a Cystic Fibrosis Canada Studentship is reviewed on an annual basis, and will correspond with prevailing Canadian rates. Cystic Fibrosis Canada Studentships are salary awards, and the payment, in addition, of benefits to Cystic Fibrosis Canada students is a matter of host institutional policy.

*Requirements:* Awards are tenable only at Canadian universities.

*Geographic Focus:* Canada

*Date(s) Application is Due:* Oct 1

*Contact:* Administrator; (416) 485-9149; fax (416) 485-0960; info@cysticfibrosis.ca

*Internet:* www.cysticfibrosis.ca/en/research/TrainingAwards_Studentships.php

*Sponsor:* Cystic Fibrosis Canada
2221 Yonge Street, Suite 601
Toronto, ON M4S 2B4 Canada

**Cystic Fibrosis Canada Summer Studentships** 997

Cystic Fibrosis Canada Summer studentships are intended to provide support to students engaged in summer research projects in areas of the biomedical or behavioral sciences relevant to cystic fibrosis, under the direction of CF clinic directors or principal investigators. The research project should be attainable within the three-month term of the award. The value of a Cystic Fibrosis Canada Summer Studentship is reviewed on an annual basis, and will correspond with prevailing rates. Most recently, the amount has been established at $5,000 per award.

*Requirements:* Full-time students pursuing an undergraduate degree in an appropriate discipline are eligible to receive this award. Under certain circumstances, other special academic situations, may be considered for a summer studentship.

*Restrictions:* Applications must be made by the proposed supervisor on behalf of a student. Applications directly from students cannot be accepted.

*Geographic Focus:* Canada

*Date(s) Application is Due:* Feb 1

*Amount of Grant:* 5,000 CAD

*Contact:* Administrator; (416) 485-9149; fax (416) 485-0960; info@cysticfibrosis.ca

*Internet:* www.cysticfibrosis.ca/en/research/TrainingAwards_SummerStudentships.php

*Sponsor:* Cystic Fibrosis Canada
2221 Yonge Street, Suite 601
Toronto, ON M4S 2B4 Canada

**Cystic Fibrosis Canada Travel Supplement Grants** 998

The Cystic Fibrosis Canada has a special travel fund to assist a limited number of individuals who plan to play an active role in European Cystic Fibrosis Society (ECFS) conferences. Up to five travel supplement grants for the upcoming conference will be awarded on a competitive basis, the maximum award will be $2,000 per person. The application deadline each year is February 1 (for Basic Science Conference) and April 1 (for ECFS Conference).

*Requirements:* The travel supplements will be available to CFC-funded investigators; Cystic Fibrosis Canada clinic directors or clinicians; and Cystic Fibrosis Canada clinic coordinators and allied health professionals (applications from clinicians, coordinators and health professionals must be endorsed by the clinic director); individuals requesting the supplement must show evidence of active participation at the conference, i.e. documentation indicating that you will be presenting a paper or abstract, or participating in a workshop; application is by letter to the Cystic Fibrosis Canada, with a brief budget included; and frequency of application from any particular individual should be reasonable (i.e. once every four years).

*Restrictions:* An itemized breakdown of expenses incurred and the original receipts must be submitted to the Cystic Fibrosis Canada upon return from the ECFS conference, and any unspent funds returned to the Cystic Fibrosis Canada.

*Geographic Focus:* Canada

*Date(s) Application is Due:* Feb 1; Apr 1

*Amount of Grant:* Up to 2,000 CAD

*Contact:* Grants Administrator; (416) 485-9149; fax (416) 485-0960; ResearchPrograms@cysticfibrosis.ca or info@cysticfibrosis.ca

*Internet:* www.cysticfibrosis.ca/en/research/TravelSupplementECFs.php

*Sponsor:* Cystic Fibrosis Canada
2221 Yonge Street, Suite 601
Toronto, ON M4S 2B4 Canada

**Cystic Fibrosis Canada Visiting Allied Health Professional Awards** 999

A limited number of awards may be made to allied health professionals from abroad who are invited to engage in Cystic Fibrosis clinical observation or activity at a Canadian institution; or Canadian allied health professionals who wish to visit another clinic in Canada or abroad. It is intended that this experience as a Canadian Cystic Fibrosis Foundation Visiting Allied Health Professional will, in some way, benefit Canadian Cystic Fibrosis clinical care. Applications may be submitted at any time, but the Foundation should be consulted in advance with respect to the availability of funds.
*Requirements:* To qualify, an allied health professional must be associated with a recognized Cystic Fibrosis clinic, and be an active member in Cystic Fibrosis clinical care. Applications should be made in the form of a letter, with a brief budget included and accompanied by a supporting letter from the head of the appropriate department of the applicant's institution. A supporting letter signed by the department head of the host institution should also be provided.
*Geographic Focus:* All States, All Countries
*Contact:* Administrator; (416) 485-9149; fax (416) 485-0960; info@cysticfibrosis.ca
*Internet:* www.cysticfibrosis.ca/en/treatment/TrainingAwards_VisitingAlliedAwards.php
*Sponsor:* Cystic Fibrosis Canada
2221 Yonge Street, Suite 601
Toronto, ON M4S 2B4 Canada

**Cystic Fibrosis Canada Visiting Clinician Awards** 1000

Awards may be made to clinicians from abroad who are invited to engage in Cystic Fibrosis clinical observation or activity at a Canadian institution; or Canadian clinicians who wish to visit another clinic in Canada or abroad. Applications may be submitted at any time, but Cystic Fibrosis Canada should be consulted in advance with respect to the availability of funds.
*Requirements:* To qualify, a clinician must be associated with a recognized Cystic Fibrosis clinic, and be an active member in Cystic Fibrosis clinical care. It is intended that this experience as a Canadian Cystic Fibrosis Foundation Visiting Clinician will, in some way, benefit Canadian Cystic Fibrosis clinical care. Applications should be made in the form of a letter, with a brief budget included and accompanied by a supporting letter from the head of the appropriate department of the applicant's institution. A supporting letter signed by the department head of the host institution should also be provided.
*Geographic Focus:* All States, All Countries
*Contact:* Administrator; (416) 485-9149; fax (416) 485-0960; info@cysticfibrosis.ca
*Internet:* www.cysticfibrosis.ca/en/treatment/TrainingAwards_VisitingClinicianAwards.php
*Sponsor:* Cystic Fibrosis Canada
2221 Yonge Street, Suite 601
Toronto, ON M4S 2B4 Canada

**Cystic Fibrosis Canada Visiting Scientist Awards** 1001

Awards are made to senior investigators from abroad who are invited to engage in CF research at a Canadian institution; or junior and senior Canadian investigators who wish to work in another laboratory in Canada or abroad. It is intended that this experience as a CCFF visiting scientist will, in some way, benefit the Canadian CF research effort. Applications may be submitted at any time, but the foundation should be consulted in advance with respect to the availability of funds.
*Requirements:* Senior investigators from abroad or junior and senior Canadian investigators who wish to work in another laboratory in Canada or abroad are eligible to apply.
*Geographic Focus:* Canada
*Contact:* Administrator; (416) 485-9149; fax (416) 485-0960; info@cysticfibrosis.ca
*Internet:* www.cysticfibrosis.ca/en/research/TrainingAwards_VisitingScientist.php
*Sponsor:* Cystic Fibrosis Canada
2221 Yonge Street, Suite 601
Toronto, ON M4S 2B4 Canada

**Cystic Fibrosis Research Elizabeth Nash Fellowships** 1002

Cystic Fibrosis Research, Inc. (CFRI) is currently accepting applications for postdoctoral research fellowships in basic or clinical research related to cystic fibrosis. Under the Elizabeth Nash Memorial Fellowship Program, $40,000 per year grants will be awarded to a research institution in California to support a principal investigator, who in turn appoints a post-doctoral research fellow. Of particular interests are projects that reflect novel and original research in cystic fibrosis. Fellowships are awarded for one year with a second year's funding contingent on progress reviews. Post-doctoral fellows and Principal Investigators are expected to attend the Spring and Summer research symposia in the San Francisco Bay Area. The annual deadline for application is December 31 for an April 1 start date.
*Requirements:* Applicants must be California-based research institutions.
*Geographic Focus:* California
*Date(s) Application is Due:* Dec 31
*Amount of Grant:* 40,000 USD
*Contact:* Carroll Jenkins; (855) 237-4669; fax (650) 404-9981; cjenkins@cfri.org
David Soohoo, Director of Programs and Operations; (650) 404-9445 or (855) 237-4669; fax (650) 404-9981; dsoohoo@cfri.org
*Internet:* www.cfri.org/ElizabethNashFellowship.shtml
*Sponsor:* Cystic Fibrosis Research, Inc.
2672 Bayshore Parkway, Suite 520
Mountain View, CA 94043

**Cystic Fibrosis Research New Horizons Campaign Grants** 1003

Since 1975, Cystic Fibrosis Rresearch, Inc., has supported cystic fibrosis research at well-established medical facilities through the funding of grants. Its investment in new ideas has enabled researchers to bring new perspectives to the study of this disease. Those involved in varied disciplines of research are encouraged to submit grants to advance the current understanding of cystic fibrosis. The New Horizons Campaign funds two to four research projects at well-established cystic fibrosis laboratories, providing up to $75,000 per year for a maximum of two years for each project. The Research Advisory Committee (RAC) submits vetted proposals to leaders in the field of CF research for their independent review. The top candidates are then selected for funding by CFRI.
*Geographic Focus:* All States
*Amount of Grant:* Up to 75,000 USD
*Contact:* Carroll Jenkins; (855) 237-4669; fax (650) 404-9981; cjenkins@cfri.org
David Soohoo, Director of Programs and Operations; (650) 404-9445 or (855) 237-4669; fax (650) 404-9981; dsoohoo@cfri.org
*Internet:* www.cfri.org/research.shtml
*Sponsor:* Cystic Fibrosis Research, Inc.
2672 Bayshore Parkway, Suite 520
Mountain View, CA 94043

**Cystic Fibrosis Trust Research Grants** 1004

The trust was founded in 1964 to find a complete solution for cystic fibrosis and to improve upon current methods of treatment. Current areas of research include: gene therapy; microbiology; transplantation; preventing lung damage in young children; immunology; pharmacology; and basic science Research grants are awarded in the United Kingdom and internationally in the fields of science and medicine.
*Requirements:* Only researchers from recognized academic institutions are eligible to apply.
*Geographic Focus:* All States, All Countries
*Date(s) Application is Due:* Mar 11
*Contact:* Grants and Research Officer; 020 8290 8024 or 020 8464 7211; researchgrants@cftrust.org.uk
*Internet:* www.cftrust.org.uk/scope/page/view.go?layout=cftrust&pageid=54
*Sponsor:* Cystic Fibrosis Trust
11 London Road
Bromley, BR1 1BY United Kingdom

**DAAD Bilateral Exchange of Academics Grants** 1005

These DAAD Bilateral Exchange of Academics Grant program aims to intensify academic collaborative research projects by promoting the project-related exchange of academics. These are bilateral programs agreed between the DAAD in Germany and partner organizations in a number of countries. Support is given to persons participating in specific academic collaborative research projects. Particular importance is attached to the continuing education and training and specialization of young academics working within the framework of research collaboration projects. Financial support is only offered to fund specifically person-related additional costs in as far as these arise through the exchange of participating scientists, scholars, graduates and, for some countries, advanced students. Diploma-level candidates can only be included in the support framework if their final dissertation forms part of the project. The German or respectively the foreign side will pay the travel costs and the stay at the partner institute for the participants of their country on the basis of the generally applicable conditions. Payments by the host institute (for example, for accommodation) will be taken into account. Project-related additional costs (for example, computer time, material purchases, documentation, photocopying, printing) cannot be reimbursed by the DAAD, nor is there any element in the grant to cover bench fees or equipment. To date, programs have been implemented with Argentina, Armenia, Azerbaijan, Belarus, Brazil, Bulgaria, Canada, Chile, China, Croatia, Czech Republic, Finland, France, Georgia, Greece, Hong Kong, Hungary, India, Italy, Kazakhstan, Mexico, Moldova, Norway, Poland, Portugal, Russia Federation, Slovakia, Spain, Taiwan, Thailand, Ukraine, the United Kingdom, and the United States. Country-specific information, dates and deadlines, and application addresses differ from one country to the next. Monthly stipends of: 2,000 Euros are available for assistant teachers, assistant professors, and lecturers; and 2,150 Euros are available for professors. A maximum of three months is allowed.
*Requirements:* Applications must be accompanied by a specific academic or scientific research project proposal. Academics from Germany and from the respective partner country must be collaborating in this project. Primary financing for the collaborative research project (personnel and running costs on both sides) must be provided by other sources.
*Restrictions:* Proposals based purely on study and training are not funded.
*Geographic Focus:* All States, Argentina, Armenia, Azerbaijan, Belarus, Brazil, Bulgaria, Canada, Chile, China, Croatia, Czech Republic, Finland, France, Greece, Hong Kong, Hungary, India, Italy, Kazakhstan, Mexico, Moldova, Norway, Poland, Portugal, Russian Federation, Slovakia, Spain, Taiwan, Thailand, Ukraine, United Kingdom
*Amount of Grant:* 2,000 - 6,450 USD
*Contact:* Dr. Holger Finken, Director Head of Section; +49 (228) 882-334; fax +49 (228) 882-659; finken@daad.de or daadny@daad.org
*Internet:* www.daad.de/go/en/stipa50015508
*Sponsor:* German Academic Exchange Service (DAAD)
871 United Nations Plaza
New York, NY 10017

**DAAD German Studies Research Grants** 1006

The aim of the DAAD German Studies Research Grants research program is to fund and encourage young academics to study subjects relating to German affairs on a long-term basis. It offers up to five grants to highly qualified undergraduate and graduate students who are nominated by their department or program chairs. The grant may be used for short-term research (one to two months) in either North America or Germany. The program is designed to encourage research and promote the study of cultural, political, historical, economic and social aspects of modern and contemporary German affairs from an inter- and multi-disciplinary perspective. Research support ranging in value

from $1,500 to $3,000 is available to individual scholarship recipients and is intended to offset living and travel costs during the active research phase.
*Requirements:* Undergraduates with at least junior standing pursuing a German Studies track or minor may be nominated for the grant by their department and/or program chair. Applicants are expected to have completed two years of college German and a minimum of three courses in German Studies (literature, history, politics or other fields) at the time of nomination. Grants are restricted to citizens of the U.S. who are enrolled full time at the university that nominates them. Applicants must be younger than 32 at the start of the grant period.
*Restrictions:* Support cannot be provided for stays in Germany in the context of study abroad programs.
*Geographic Focus:* All States
*Date(s) Application is Due:* May 1; Nov 1
*Amount of Grant:* 1,500 - 2,500 USD
*Contact:* Dr. Holger Finken, Director Head of Section; +49 (228) 882-334; fax (212) 755-5780; finken@daad.de or daadny@daad.org
*Internet:* daad.de/go/en/stipa10000321
*Sponsor:* German Academic Exchange Service (DAAD)
871 United Nations Plaza
New York, NY 10017

**DAAD Intensive Language Courses Scholarships** **1007**
DAAD offers grants to graduate students at North American universities to attend eight-week intensive language courses at leading institutes in Germany. Extensive descriptions of the institutes, their teaching philosophies, course content, as well as course dates are available on the website of each institute. Scholarships are awarded to students currently enrolled full-time in a graduate or PhD program in all fields of study except English, German, or any other modern language or literature. Applicants will be assessed on the basis of their academic record and statements of projected academic and professional future. The courses run from June through December, with a one-off scholarship payment of 2,300 Euros
*Requirements:* As a rule, applicants must be citizens of the U.S. or Canada. Foreign nationals may be eligible if they have been full-time graduate students at a U.S. or Canadian University for at least one academic year at the time of application. Applicants must have completed three semesters of college German or have achieved an equivalent level of language proficiency. Applicants must be younger than 32 at the start of the grant period.
*Restrictions:* Applicants who have received a DAAD summer language grant in the past three years are not eligible to apply.
*Geographic Focus:* All States, Canada
*Date(s) Application is Due:* Dec 11
*Contact:* Administrator; (212) 758-3223; fax (212) 755-5780; graaff@daad.org
*Internet:* www.daad.de/go/en/stipa50046548
*Sponsor:* German Academic Exchange Service (DAAD)
871 United Nations Plaza
New York, NY 10017

**DAAD Research Stays for University Academics and Scientists** **1008**
These grants and scholarships aim to provide foreign academics and scientists working in higher education or at research institutes with an opportunity to carry out a research project at a state (public) or state-recognized higher education institution or non-university research institute in Germany. Depending on the applicant's work schedule, the research stay will last between one and three months. Depending on the applicant's academic status, the monthly award will amount to 2,000 Euros for assistant lecturers, assistant professors, and young lecturers, and 2,150 Euros for professors. In addition to these payments, the DAAD generally will pay an appropriate flat-rate travel allowance, unless these costs are covered by the home country or by another funding source. Applications for such research grants are decided twice a year. Applications must have been submitted by August 1 or January 15. At the earliest, grants can begin 4 months after the date of application.
*Requirements:* Applications for DAAD research stays are open to excellently-qualified academics and scientists who should generally hold a doctorate/PhD. All applicants must be working in higher education or at a research institute in their home country. Besides their previous academic achievements (for example, recent publications), the most important selection criterion is a convincing and well-planned research project to be completed during the stay in Germany. The application must provide proof of a workplace being provided at the host institute.
*Restrictions:* DAAD support for a research stay can only be awarded once in any three-year period. No travel expenses can be paid from the grant.
*Geographic Focus:* All States
*Date(s) Application is Due:* Jan 15; Aug 1
*Amount of Grant:* 2,000 - 6,450 EUR
*Contact:* Dr. Holger Finken, Director Head of Section; +49 (228) 882-334; finken@daad.de or daadny@daad.org
*Internet:* www.daad.de/go/en/stipa50015456
*Sponsor:* German Academic Exchange Service (DAAD)
871 United Nations Plaza
New York, NY 10017

**Dade Community Foundation Grants** **1009**
The funding for this program is made available through the Foundation's unrestricted and field of interest funds. This program is designed to honor both the donors interests and address significant community issues such as: education; health; human services; arts and culture; environment; economic development; at-risk youth; abused and neglected children; living with HIV/AIDS; homelessness; social justice; care of animals; heart disease and more.
*Requirements:* Eligible applicants include nonprofit tax-exempt organizations, as defined by the Internal Revenue Code, which are serving the residents of Miami-Dade County. Preference will be given to organizations based in Miami-Dade County or if located outside the county, are working in partnership with an organization based in Miami-Dade.
*Restrictions:* The Foundation does not provide grants to individuals, for memberships, fundraising events or memorials. Grants to government agencies are made on a very restricted basis.
*Geographic Focus:* Florida
*Date(s) Application is Due:* Nov 15
*Amount of Grant:* 7,500 USD
*Samples:* Alliance for Musical Arts Productions, $7,500 - to provide after school and summer camp services to youth from low-income households living in Opa-locka; Chai Lifeline, $6,500 - to support Smile S'more, a unique program designed to bring fun and love to children who are being treated for cancer and other life-threatening illnesses on pediatric specialty units of hospitals; Earth Learning, $10,000 - to create an EarthFest network that will form an umbrella to unite Earth Day celebrations across South Florida under one banner.
*Contact:* Charisse Grant, Vice President for Programs; (305) 371-2711; fax (305) 371-5342; charisse.grant@dadecommunityfoundation.org
*Internet:* www.dadecommunityfoundation.org/Site/programs/overview.jsp
*Sponsor:* Dade Community Foundation
200 S Biscayne Boulevard, Suite 505
Miami, FL 33131-5330

**Daimler and Benz Foundation Fellowships** **1010**
This program is addressed to young postdoctoral scientists in Germany, who are in the early phase of their postgraduate research. The program is, in the sense of the foundation's statutes, open for all scientific disciplines and subject areas. Prerequisites are an own research project and an affiliation with a scientific institution. The amount of the fellowship is 20.000 Euros annually, which is granted for a total of two years. The funds are disbursed by the respective research institution's third-party funds administration, where the fellowship holder is doing research. The funds can be used for equipment, especially for computers, for financing scientific assistants, research travel, participation in conferences, as well as for the organization of own conferences. The fellowship can not, however, be used for living expenses, printing costs, or for expendable items (e.g., chemicals, paper, etc.). The Foundation plans to award approximately ten per year scholarships, the exact number of scholarships may vary depending on the funds available. In accordance with the Foundation's statutes, the program is open to all disciplines and subjects.
*Requirements:* Postdoctoral scientists and assistant professors in the early phase of their scientific work as well as young researchers in a similar position, such as independent director of junior research groups, are eligible to apply. The applicant must be affiliated with a German university or research institution and carry out their own research project. Applications may be submitted in German and/or English. Each application is reviewed by a committee. The Commission consists of the Board of Management of Daimler and Benz Foundation, the managing director and evaluators. After the deadline initially there will be a pre-selection. The selected applicants will be invited to a seminar in the Daimler and Benz Foundation in which they present their science project in the presence of representatives of the Foundation as well as peer reviewers personally. Participation in this seminar and the presentation of the research project are a prerequisite for the subsequent award of a scholarship. Self-nomination for the Daimler and Benz postdoctoral program is possible and is done without special application forms.
*Geographic Focus:* Germany
*Date(s) Application is Due:* Oct 15
*Amount of Grant:* 20,000 EUR
*Contact:* Dr. Jörg Klein, Managing Director; 06 203-10 92-0; fax 06 203-10 92-5; info@daimler-benz-stiftung.de
*Internet:* www.daimler-benz-stiftung.de/cms/index.php?page=home_en
*Sponsor:* Daimler and Benz Foundation
Dr.-Carl-Benz-Platz 2
Ladenburg, 68526 Germany

**Dart Center Ochberg Fellowship** **1011**
The Dart Center provides six or more expense-paid fellowships to mid-career journalists who want to apply knowledge of emotional trauma to improving coverage of violent events. Twelve selected fellows will attend an intensive week long program of seminars and discussions held at Columbia University in New York City. Program activities include briefings by prominent interdisciplinary experts in the trauma and mental health fields; conversations with journalist colleagues on issues of ethics, craft and other aspects of professional practice; and a host of other opportunities for intellectual engagement and peer learning. The Ochberg Fellowship covers air or train travel, accommodations in New York City, most meals and other expenses directly related to program participation. The online application deadline is October 1.
*Requirements:* Fellowships are open to print and broadcast reporters, photographers, editors, and producers with at least five years of journalism experience.
*Restrictions:* The program does not cover costs related to travel visas, health insurance or ground transportation in fellows' home cities.
*Geographic Focus:* All States
*Date(s) Application is Due:* Oct 1
*Contact:* Kate Black, Associate Director; (212) 854-8056; kate.black@dartcenter.org
*Internet:* dartcenter.org/ochberg-fellowship-guidelines
*Sponsor:* Dart Center for Journalism and Trauma
2950 Broadway
New York, NY 10027

**Dave Thomas Foundation for Adoption Grants** 1012

The foundation supports advocacy and social service organizations working in the field of adoption. The primary goal of the foundation is to raise public awareness about children in the public welfare system awaiting adoption and to form public and private partnerships to make the adoption process easier and more affordable. Types of support include challenge/matching grants, conferences and seminars, demonstration grants, professorships, program development grants, research grants, seed money grants, and technical assistance. National, regional, or statewide projects will be considered. Priority will be given to projects that request seed money; include matching or other material support from other organizations, government agencies, or funders; coordinate service providers; include a measurement component to evaluate the project's success; and are easily replicable. Proposals are accepted throughout the year. Funding decisions are made four times per year. In order to receive serious consideration, proposals must be received by the submission deadline date for each quarter. Guidelines are available online.
*Requirements:* Eligible applicants are nonprofit 501(c)3 organizations.
*Restrictions:* The foundation will not consider funding individual adoption expenses; operating budgets or budget deficits; endowments or capital campaigns; adoption searches or reunions; scholarships; special events; institutions that discriminate on the basis of race, creed, gender, national origin, age, disability, or sexual orientation; or organizations engaged in sectarian religious activities. Unless solicited by the foundation, the following will not be considered: research, educational or promotional videos, publications, television productions, conferences, or public service announcements.
*Geographic Focus:* All States, Canada
*Amount of Grant:* 35,000 - 50,000 USD
*Contact:* Connie Ackert, Grants Director; (800) 275-3832; fax (614) 766-3871; connie_ackert@davethomasfoundation.org
*Internet:* www.davethomasfoundationforadoption.org
*Sponsor:* Dave Thomas Foundation for Adoption
525 Metro Place North, Suite 220
Dublin, OH 43017

**David Alan and Susan Berkman Rahm Foundation Grants** 1013

The David Alan and Susan Berkman Rahm Foundation was established by a pair of prominent attorneys in New York, New York, in 1997. Today, the Foundation concentrates its efforts in the support of: art and culture; children and youth programs; general operating support; higher education; and religion. Typically, funding is given for general operating support, program development, research, and research evaluation. The geographic focus is New York City, and most recent awards have ranged up to nearly $12,000. There are no specified application deadlines, and interested organizations should forward an email letter outlining a detailed description of project or organization, and the amount of funding requested.
*Geographic Focus:* New York
*Amount of Grant:* Up to 12,000 USD
*Samples:* Film Society of Lincoln Center, New York, New York, $2,500 - general operating support; Food Bank for New York City, New York, New York, $5,000 - general operating support; Manhattan School of Music, New York, New York, $11,780 - general operating support.
*Contact:* Susan B. Rahm, President; (212) 348-9551 or (212) 831-4442
*Sponsor:* David Alan and Susan Berkman Rahm Foundation
1125 Park Avenua
New York, NY 10128-1243

**Daviess County Community Foundation Health Grants** 1014

The Daviess County Community Foundation considers proposals for grants on a yearly cycle, which begins each May. At the start of each cycle, a notice is mailed to nonprofit organizations that have applied for grants in the past, have received grants in the past, or have otherwise requested notification of the start of each cycle. Grants in the area of health include activities that: improve and promote health outcomes; provide general and rehabilitative health services; offer mental health services; provide crisis intervention programs; strengthen associations or services associated with specific diseases, disorders, and medical disciplines; and support medical research. Samples of previously funded projects are posted on the website.
*Requirements:* The Foundation welcomes proposals from nonprofit organizations that are deemed tax-exempt under sections 501(c)3 and 509(a) of the Internal Revenue Code and from governmental agencies serving the County of Daviess, Indiana. Proposals from nonprofit organizations not classified as a 501(c)3 public charity may be considered provided the project is charitable and supports a community need.
*Restrictions:* Funding is not available for the following: religious organizations for strictly religious purposes; political parties or campaigns; endowment creation or debt reduction; operating costs (not directly related to the proposed project or program); capital campaigns; annual appeals or membership contributions; travel requests for groups or individuals such as bands, sports teams, or classes. Not more than 20% of any grant request may be for personnel costs, office supplies, or other operating costs. Operating costs for any organization must be directly related to the project or program for which funding is being requested.
*Geographic Focus:* Indiana
*Contact:* Mary E. Smith, Regional Director; (812) 254-9354; fax (812) 254-9355; mary@daviesscommunityfoundation.org
*Internet:* www.daviesscommunityfoundation.org/program-areas
*Sponsor:* Daviess County Community Foundation
320 East Main Street, P.O. Box 302
Washington, IN 47501

**Delaware Division of the Arts Individual Artist Fellowships** 1015

Delaware Division of the Arts Individual Artist Fellowships provide funding to Delaware creative artists working in the visual, performing, media, folk, and literary arts. Fellowships are designed to enable recipients to purchase equipment and materials, allocate working time, or fulfill other needs that will allow them to advance in their careers. Emerging Professional Fellowships of $3,000 are available to artists who have not yet established reputations in their fields. Established Professional Fellowships of $6,000 are available to artists who have significant achievements in their fields. A Masters Fellowship in the amount of $10,000 is available on a rotating basis, with category applications changing from year to year. The application and additional guidelines are posted at the website, including guidelines for submitting work samples for each category. Applications should be submitted electronically by the annual deadline of August 3.
*Requirements:* Applicants must be at least 18 years of age and residents of Delaware for at least one year at the time of application. Recipients must remain Delaware residents during their fellowship year and not be enrolled in degree or certificate-granting educational programs at the time of application or for the duration of their fellowships. Artists can receive only one fellowship per artistic discipline in each category. Emerging Professional Fellowship recipients may apply in the Established Professional category after a three-year period. Only one fellowship application may be submitted per year. Artists may win only one Masters Fellowship. Visual arts applicants must have a current file in the Delaware Division of the Arts (DDOA) Visual Images Registry.
*Geographic Focus:* Delaware
*Date(s) Application is Due:* Aug 3
*Contact:* Paul Weagraff, Director; (302) 577-8278; paul.weagraff.state.de.us
*Internet:* www.artsdel.org/grants/default.shtml
*Sponsor:* Delaware Division of the Arts
820 North French Street, Carvel State Office Building
Wilmington, DE 19801

**Delta Air Lines Foundation Health and Wellness Grants** 1016

Established in 1968 as Delta's company-managed giving system, the Delta Air Lines Foundation contributes more than $1 million annually in endowed funds to deserving organizations and programs. Delta's Health and Wellness Grants support organizations that dedicate their efforts to the health and well-being of communities with a focus on broad-reaching research for cures and education around diseases that affect all walks of life. Once an application is received, applicants should allow up to three months before review.
*Requirements:* For proposals which meet the foundation's area of focus, priority will be given to: programs meeting compelling needs in communities where Delta has a presence; proposals that exhibit clear, reasonable goals, and measurable outcomes; distinctive projects where the foundation's involvement will leave a legacy; projects that include collaboration or cooperation with other nonprofit organizations; projects that offer opportunities for Delta employee involvement. The foundation Board of Trustees reviews and approves funding in March, June, September, and November. The deadline for receiving completed proposals is the first day of each of these months.
*Restrictions:* The foundation will generally not consider: individual applicant's request for support of personal needs; religious activities; political organizations or campaigns; specialized single-issue health organizations; annual or automatic renewal grants; general operating expenses; endowment campaigns; capital campaigns; multi-year commitments; fraternal organizations, professional associations, or membership groups; fundraising events such as benefits; charitable dinners, or sporting events.
*Geographic Focus:* All States
*Date(s) Application is Due:* Jun 1; Sep 1; Nov 1
*Contact:* Administrator; (404) 715-5487 or (404) 715-2554; fax (404) 715-3267
*Internet:* www.delta.com/about_delta/community_involvement/delta_foundation/
*Sponsor:* Delta Air Lines Foundation
P.O. Box 20706, Department 979
Atlanta, GA 30320-6001

**Descartes Center Junior Fellowships** 1017

The Descartes Centre offers fellowships to promote intellectual exchange and collaboration in the history and/or the philosophy of the sciences and the humanities. A Junior Fellow conducts research at Utrecht University and/or at Dutch museums or archives, and presents the results of his/her work at one of the colloquia of the Descartes Centre. Candidates should have a clear affinity with the history and/or the philosophy of the sciences and the humanities, as shown by their list of publications and research proposal. Furthermore, the application should demonstrate an existing or potential collaboration with one or more members of the Descartes Centre, a collaboration that is supported by a corresponding letter of recommendation. Fellows are expected to participate in the activities of the Descartes Centre whenever possible. The start and end date of a fellowship should make such involvement practically feasible. Fellows are expected to participate in the activities of the Descartes Centre whenever possible. The start and end date of a fellowship should make such involvement practically feasible. Junior Fellows can have their costs (housing etc.) compensated for with a maximum of 1500 euro per month. The annual deadline for applications is May 1.
*Geographic Focus:* All States, Netherlands
*Date(s) Application is Due:* May 1
*Amount of Grant:* Up to 4,500 EUR
*Contact:* Bert Theunissen, Director; +31 (0)30 253 7218 or +31 (0)30 253 8180; l.t.g.theunissen@uu.nl or infodescartescentre@uu.nl
*Internet:* https://www.uu.nl/en/descartes-centre/about-us/fellowships
*Sponsor:* Descartes Center
Drift 6, Rooms 2.04-2.06
Utrecht, UT 3512 BS Netherlands

**Descartes Center Senior Fellowships** 1018

The Descartes Centre offers fellowships to promote intellectual exchange and collaboration in the history and/or the philosophy of the sciences and the humanities. A Senior Fellow not only conducts research but also contributes actively to the Descartes Centre by means of public lectures, research seminars and/or teaching. Candidates should have a clear affinity with the history and/or the philosophy of the sciences and the humanities, as shown by their list of publications and research proposal. Furthermore, the application should demonstrate an existing or potential collaboration with one or more members of the Descartes Centre, a collaboration that is supported by a corresponding letter of recommendation. Fellows are expected to participate in the activities of the Descartes Centre whenever possible. The start and end date of a fellowship should make such involvement practically feasible. Fellows are expected to participate in the activities of the Descartes Centre whenever possible. The start and end date of a fellowship should make such involvement practically feasible. Senior Fellows can have their costs (housing etc.) compensated for with a maximum of 2500 euro per month. The annual deadline for applications is May 1.
*Geographic Focus:* All States, Netherlands
*Date(s) Application is Due:* May 1
*Amount of Grant:* Up to 4,500 EUR
*Contact:* Bert Theunissen, Director; +31 (0)30 253 7218 or +31 (0)30 253 8180; l.t.g.theunissen@uu.nl or infodescartescentre@uu.nl
*Internet:* https://www.uu.nl/en/descartes-centre/about-us/fellowships
*Sponsor:* Descartes Center
Drift 6, Rooms 2.04-2.06
Utrecht, UT 3512 BS Netherlands

**Descartes Center Theme-Group Fellowships** 1019

Descartes Theme-Group Fellowships are awarded to scholars working together as part of the Creating a knowledge society in a globalizing world (1450-1800) project, a long term collaboration between the Descartes Centre for the History and Philosophy of the Sciences and the Humanities (Utrecht University), the Max Planck Institute for the History of Science (Berlin), and the Huygens ING (The Hague). Descartes Theme-groups work on research topics that provide a historical and philosophical perspective on scientific developments that have a major impact on society. Descartes theme-groups are made up of a group of scholars working in flexible constellations at one of the various participating institutes. The first group's project will investigate the historical roots of the global knowledge society to increase understanding of the knowledge society in a global context. Members work at NIAS (Wassenaar), Descartes Centre for the History and Philosophy of the Sciences and the Humanities (Utrecht University), the Max Planck Institute for the History of Science (Berlin) or the Huygens ING (The Hague). Theme-group members at NIAS are offered a residential fellowship.
*Requirements:* Descartes theme-group fellowships are by nomination only.
*Geographic Focus:* All States, Netherlands
*Date(s) Application is Due:* May 1
*Contact:* Bert Theunissen, Director; +31 (0)30 253 7218 or +31 (0)30 253 8180; l.t.g.theunissen@uu.nl or infodescartescentre@uu.nl
*Internet:* www.nias.knaw.nl/fellowships/co-sponsored-fellowships/descartes-theme-group-fellowships
*Sponsor:* Descartes Center
Drift 6, Rooms 2.04-2.06
Utrecht, UT 3512 BS Netherlands

**DHHS Emerging Leaders Program Internships** 1020

The Emerging Leaders Program (ELP) is a competitive, two-year paid, federal internship with the U.S. Department of Health and Human Services (HHS). The Program provides a unique opportunity to develop enhanced leadership skills in one of the largest federal agencies in the nation. Upon successful completion of the ELP, participants will be eligible for non-competitive conversion to a permanent appointment. The ELP offers participants the following: formal competency-based leadership training and professional development; challenging developmental rotational assignments; mentorship; fast-paced and diverse work environment; and promotion to a permanent career track that targets the following mission-critical occupation specialties: scientific, social science, human resources, administrative, information technology, public health, and law enforcement. To begin the application process, applicants apply to one or more career track(s) that match their education and/or experience specialty. Later applicants may apply to the specific positions under their chosen career track(s). At the date of this writing, the website states that the ELP is not currently accepting applications, but encourages those interested to check back for future updates.
*Requirements:* Applicants must qualify for a minimum of one career track to move on to the next phase of the selection process.
*Geographic Focus:* All States
*Contact:* Internship Coordinator; (877) 696-6775; ELP@hhs.gov
*Internet:* https://tod.learning.hhs.gov/elp.asp
*Sponsor:* U.S. Department of Health and Human Services
200 Independence Avenue SW
Washington, D.C. 20201

**Diabetes Research of Western Australia Grants** 1021

The primary purpose of the Diabetes Research of Western Australia is to raise funds for research into the causes of, prevention of and cure for diabetes in all its forms. This includes research that prevents complications from diabetes and improvement in the quality of life for people living with diabetes. Ultimately, the goal is to find a cure. The program offers two grants of $80,000 each year. The first will be for basic or fundamental research, while the second is aimed at clinical/transitional research. Projects that have not received funding in the previous round and which include an association or collaboration with an institution or co-funding from other sources will be considered favorably. The annual deadline for application submissions is July 30.
*Geographic Focus:* Australia
*Date(s) Application is Due:* Jul 30
*Amount of Grant:* 80,000 USD
*Contact:* Sherl Westlund, Executive Director; +61 8 9224 1006; fax +61 8 9224 1008; sherl@diabetesresearchwa.com.au or info@diabetesresearchwa.com.au
*Internet:* https://www.diabetesresearchwa.com.au/research/grants/apply-grant
*Sponsor:* Diabetes Research of Western Australia
Level 5 Ainslie House, RPH Precinct
Perth, WA 6000 Australia

**Dietrich W. Botstiber Foundation Institute for Wildlife Fertility Control** 1022

The Botstiber Institute for Wildlife Fertility Control (BIWFC) seeks public policy research projects that advance the knowledge, acceptance and implementation of wildlife fertility control as a method of wildlife management throughout the world. The BIWFC's goal is to support projects that will help the public and policy makers understand and evaluate existing public policies on wildlife fertility control. The BIWFC will support projects that demonstrate ways to examine current management programs and policies, increase public awareness and stimulate and introduce new ideas into the policy debate. Grants may include support for related seminars, surveys, publications, and documentaries. Grants for tuition will not be considered. A grant of up to $25,000 will be considered to support project costs, travel, or other necessary expenses. Applicants wishing to be considered for a BIWFC grant should submit a letter of intent (LOI). LOIs may be submitted at any time. BIWFC will review all LOIs within six weeks of submission. The BIWFC may request a full proposal after consideration of the LOI.
*Geographic Focus:* All States
*Amount of Grant:* Up to 25,000 USD
*Contact:* Monique Principi, Managing Director; (610) 566-3375; fax 610-566-3376; mprincipi@botstiber.org or biwfc@botstiber.org
*Internet:* https://www.wildlifefertilitycontrol.org/grants/
*Sponsor:* Dietrich W. Botstiber Foundation
200 East State Street, Suite 306-A
Media, PA 19063

**DLR-DAAD Research Fellowships** 1023

The DAAD Research Fellowship program implemented by the German Aerospace Center (DLR) and the German Academic Exchange Service (DAAD). DLR is Germany's national research center for aeronautics and space. Its extensive research and development work in Aeronautics, Space, Transportation and Energy is integrated into national and international cooperative ventures. As Germany's space agency, DLR has been given responsibility for the forward planning and the implementation of the German space program by the German federal government as well as for the international representation of German interests. Approximately 8,000 people are employed in DLR's thirty-three institutes and facilities at twenty locations in Germany. Length of funding depends on the educational status of the applicant, with senior scientists funded for one to three months, postdoctoral students funded for six months to two years, and doctoral students funded for up to three years. The amount of stipend is also dependent upon the status of the applicant, with senior scientists receiving 2,300 Euros per month, postdoctoral students receiving 2,000 Euros per month, and doctoral students receiving 1,365 Euros per month (plus an 102 Euro allowance for materials per month).
*Requirements:* This program is intended for highly-qualified foreign doctoral and postdoctoral students as well as senior scientists. DLR-DAAD Fellowships are defined and awarded on an individual basis. Each fellowship announcement will indicate the specific qualification requirements and terms of visit.
*Geographic Focus:* All States, Germany
*Amount of Grant:* 2,300 - 52,812 EUR
*Contact:* Dr. Holger Finken, Director Head of Section; +49 (228) 882-334; finken@daad.de or daadny@daad.org
*Internet:* www.daad.de/go/en/stipa50019749
*Sponsor:* German Academic Exchange Service (DAAD)
871 United Nations Plaza
New York, NY 10017

**DND/NSERC Discovery Grant Supplements** 1024

The DND/NSERC Discovery Grant Supplements, a joint initiative between the Department of National Defence (DND) and NSERC, aim to support unclassified university-based research dedicated to the development of non-weapon specific technologies with dual-use applications in defence and security target areas. This supplement provides additional resources to researchers whose Discovery Grant applications were successful in the Discovery Grants competition and whose research addresses the DND target areas. Each award is valued at up to $120,000 over three years (up to $40,000 annually). These funds may be used to expand the recipient's research group (e.g., students, postdoctoral fellows, technicians), to purchase or access specialized equipment, or for other initiatives or resources that would accelerate the recipient's research program in the targeted areas. The annual deadline for application submission is November 1.
*Requirements:* In addition to the usual NSERC eligibility criteria for applicants, the DND Supplement is only accessible to researchers who are applying to the current Discovery Grants competition and whose research falls within the DND target areas.
*Restrictions:* Adjunct professors whose primary place of employment is outside the university sector (e.g., government, industry or colleges) are not eligible for the DND Supplement.

*Geographic Focus:* Canada
*Date(s) Application is Due:* Nov 1
*Amount of Grant:* Up to 40,000 USD
*Contact:* Coralie Rochefort, Program Officer; (613) 943-0703 or (855) 275-2861; fax (613) 992-5337; dndsuppmdn@nserc-crsng.gc.ca or coralie.rochefort@nserc-crsng.gc.ca
*Internet:* www.nserc-crsng.gc.ca/Professors-Professeurs/Grants-Subs/DND-NSERC_eng.asp
*Sponsor:* Natural Sciences and Engineering Research Council of Canada
350 Albert Street, 16th Floor
Ottawa, ON K1A 1H5 Canada

**DOE Coal Liquefaction Research Grants** 1025
The objectives of the program are to develop advanced technology and an engineering and operations database sufficient for the production of marketable liquid fuels from coal by the year 2015 at a cost competitive with crude oil, between $25 and $30 per barrel. The implementation strategy is based on catalytic staged direct liquefaction, coprocessing of coal/waste and petroleum resid, slurry-phase indirect liquefaction technology, and advanced processes based on exploratory research. Areas covered by the program include direct/indirect coal liquefaction processes, coprocessing of coal-petroleum mixture/coal-waste mixture, refining and chemicals, and supporting research. Proposals are accepted at any time.
*Requirements:* States, local governments, universities, governmental entities, consortia, nonprofit institutions, commercial corporations, joint federal/industry corporations, U.S. territories, and individuals are eligible to apply.
*Geographic Focus:* All States
*Amount of Grant:* 50,000 - 400,000 USD
*Contact:* Fred Glaser, Assistant Secretary of Fossil Energy; (301) 903-2786
*Internet:* www.fe.doe.gov
*Sponsor:* U.S. Department of Energy
19901 Germantown Road, MS FE-3
Germantown, MD 20874-1290

**DOE Initial H-Prize Competition for Breakthrough Advances in Materials for Hydrogen Storage** 1026
The H-Prize, enacted by Congress, authorized the Secretary of Energy to create a program to competitively award cash prizes that will advance the commercial application of hydrogen energy technologies by dramatizing and incentivizing accelerated research. The H-Prize was originally established by the Energy Independence and Security Act of 2007, in Sec. 654. There are several H-Prize categories, including production, storage, distribution, utilization, and prototypes and transformational technologies. The 2009-11 prize will be awarded in the area of storage materials in mobile systems for light-duty vehicles. The Hydrogen Education Foundation, the charitable, education-focused arm of the National Hydrogen Association, has been chosen to administer the H-Prize. The competition will be a single award for $1 million in the subject area of advanced materials for hydrogen storage—a critical challenge to enable widespread commercialization of hydrogen and fuel cell technologies. (FR Doc. E9-20552)
*Requirements:* The competition is open to participants, defined as individuals, entities or teams, that are organized or incorporated in the United States, and maintain for the duration of the H-Prize Competition a primary place of business in the United States. Teams may be comprised of any combination of individuals or entities. All individuals (whether participating singly or as part of an entity or team) must be a citizen of, or an alien lawfully admitted for permanent residence into, the United States as of the date of registration in the competition and maintain that status for the duration of the competition. The winner must be present at the award presentation in order to win the H-Prize. If a team or entity is the winner, only the team leader must be present.
*Restrictions:* The participant, or any member of a participant, shall not be a Federal entity, a Federal employee acting within the scope of his or her employment, or an employee of a National Laboratory acting within the scope of his or her employment. The participant, or any member of a participant, cannot have received Federal funding for research and development of hydrogen storage materials since October 2008. An individual cannot participate on more than one team or compete with multiple entries. Participants cannot be on more than one team or compete with multiple entries.
*Geographic Focus:* All States
*Date(s) Application is Due:* Feb 15
*Amount of Grant:* 1,000,000 USD
*Contact:* Jeffrey Serfass, Project Director
*Internet:* www.hydrogenprize.org/
*Sponsor:* U.S. Department of Energy
1000 Independence Avenue SW, Forrestal Building
Washington, D.C. 20585-1290

**Dolan Children's Foundation Grants** 1027
The Dolan Children's Foundation was founded in New York in 1886 by Charles F. and Helen A. Dolan. Charles F. Dolan was the founder of the cable network HBO, and is the owner of Cablevision Systems Corporation, a cable television provider in New York City that also owns Madison Square Garden, Radio City Music Hall, the New York Knicks, and the New York Rangers. The Foundation awards community service grants with a focus on Long Island, New York. Areas of interest include: human services, disability services, mental health, schools, rehabilitation, hospitals, health facilities, and Catholic agencies and churches. Types of support include: general operating support, building and renovations; capital campaigns; equipment purchase and rental; land acquisition; program development; research; and matching grants. There are no application deadlines, though an application form is required. Recent awards have ranged from $25,000 to $3,000,000. Guidelines are available upon request.
*Requirements:* Schools and other nonprofits in New York City and Long Island are eligible. Applicants outside New York should call or write prior to submitting proposals.
*Geographic Focus:* New York
*Amount of Grant:* 25,000 - 3,000,000 USD
*Samples:* Chaminade High School, Mineola, New York, $3,000,000 - general operating support; Lesley University, Cambridge, Massachusetts, $250,000 - general operating support; Rising Sun Youth Foundation, Plushing, New York, $50,000 - general operating support.
*Contact:* Robert Vizza, President; (516) 803-9200
*Sponsor:* Dolan Children's Foundation
340 Crossways Park Drive
Woodbury, NY 11797-2050

**Donald and Sylvia Robinson Family Foundation Grants** 1028
The Donald and Sylvia Robinson Family Foundation (formerly known as the Alex and Leona Robinson Family Foundation) awards grants to U.S. nonprofit organizations in its areas of interest, including: animal and wildlife protection; arms control; arts (general, performing arts, and visual arts); environmental conservation and protection; eye diseases and eye research; family planning and human reproductive health; food distribution; international affairs; Israel; Jewish social services; and social service delivery programs. Types of support include annual campaigns, building construction/renovation, capital campaigns, and general operating support. There are no application deadlines or forms, so applicants should forward a letter of interest stating the purpose of the contribution.
*Requirements:* National nonprofit organizations are eligible to apply.
*Restrictions:* Individuals are ineligible.
*Geographic Focus:* Pennsylvania
*Amount of Grant:* 500 - 5,000 USD
*Samples:* Southern Alleghenies Museum of Art, Altoona, Pennsylvania, $10,000 - general operating support; Feeding South Florida, Pembroke Park, Florida, $1,000 - general operating support; Carnegie Library of Pittsburgh, Pittsburgh, Pennsylvania, $10,000 - general operating support.
*Contact:* Donald Robinson, Director; (412) 661-1200; fax (412) 661-4645
*Sponsor:* Donald and Sylvia Robinson Family Foundation
6507 Wilkins Avenue
Pittsburgh, PA 15217

**Donald W. Reynolds Foundation Aging and Quality of Life Grants** 1029
The Foundation launched its Aging and Quality of Life Program in 1996. Its goal remains improving the quality of life for America's elderly by preparing physicians to provide better care for frail older people. Most physicians today lack adequate training to meet the needs of the frail elderly patient. Such patients typically suffer from interacting physical, social and psychological conditions –both acute and chronic – that limit their independence and threaten their capacity to function in daily life.
*Requirements:* Proposals are by invitation only. Periodic requests for proposals are announced on the website. Organizations should contact the Foundation to discuss their project before submitting a proposal.
*Geographic Focus:* All States
*Contact:* Rani Snyder, Senior Program Officer; (702) 804-6000; fax (702) 804-6099; rani.snyder@dwrf.org or GeneralQuestions@dwrf.org
*Internet:* www.dwreynolds.org/Programs/National/Aging/Aging.htm
*Sponsor:* Donald W. Reynolds Foundation
1701 Village Center Circle
Las Vegas, NV 89134-6303

**Doris Duke Charitable Foundation Clinical Research Systems Grants** 1030
The Program is founded on the premise that clinical research requires not only human and research capital, facilities, and equipment, but it also requires robust support systems, including appropriate regulatory organizations and processes, common language, analytical tools, databases and information systems, as well as mechanisms to distribute findings and interface with the public. The Program is interested in supporting activities that seek to identify ways to strengthen the regulatory processes and systems for clinical research, such as the protection of human subjects. The foundation is willing to consider any exceptional opportunities that may arise.
*Geographic Focus:* All States
*Contact:* Elaine Gallin, Program Director for Medical Research; (212) 974-7104; fax (212) 974-7590; egallin@ddcf.org
*Internet:* www.ddcf.org/page.asp?pageId=302
*Sponsor:* Doris Duke Charitable Foundation
650 Fifth Avenue, 19th Floor
New York, NY 10019

**Dorr Foundation Grants** 1031
Dorr Foundation grants are made primarily for programs designed to develop new science curricula from sixth to 12th grade. Support is also given to special education projects for youth relating to conservation and the environment if such projects involve the school's curriculum, equipment purchase, program development, emergency funding, and seed money. In addition, some grants are made available to promote research and disseminate information on chemical, metallurgical, and sanitation engineering. Grants are awarded on a national basis, with emphasis in the Northeast states. Types of support include equipment, emergency funds, program development, seed money, curriculum development, scholarship funds, and research. Initial contact should be a phone call. There is no deadline for application submission. No response can be expected unless

there is interest on the part of the trustees. Applications are accepted at any time. Typical awards range from $5,000 to $30,000.
*Requirements:* 501(c)3 tax-exempt organizations are eligible.
*Restrictions:* Grants are not made to individuals or for operation budgets, continuing support, annual campaigns, deficit financing, endowment funds, or conferences and seminars.
*Geographic Focus:* Connecticut, Maine, Massachusetts, New Hampshire, New York, Vermont
*Amount of Grant:* 5,000 - 30,000 USD
*Samples:* Maritime Museum, Norwalk, Connecticut, $20,000 - building STEM Academy for teachers; Maine Natural History Observatory, Gouldsboro, Maine, $15,000 - population counts of nesting Gulls and cormorants along the coast of Maine; Young Men's Leadership Academy, Kennedy, Texas, $5,561 - Ozone North and South project.
*Contact:* Barbara McMillan, Chairperson; (603) 433-6438
*Sponsor:* Dorr Foundation
84 Hillside Drive
Portsmouth, NH 03801-5328

**Dorr Institute for Arthritis Research and Education** 1032
Established in California in 2000, the Dorr Institute for Arthritis Research and Education promotes medical treatment, research and education concerning the causes and treatment of arthritis and joint implant surgery procedures. Its primary fields of interest are arthritis and arthritis research. Recognized as a foundation, it currently funds research for improvement in total hip replacement and total knee replacement. Annually, they support two types of fellows who participate in the research; a research fellow is dedicated to investigations; and a clinical fellow participates in the care of the patients and also participates in research.
*Geographic Focus:* All States
*Contact:* Jeri Ward, Director; (323) 442-5762 or (213) 977-2280; jeri@drdorr.com
*Internet:* dorrarthritisinstitute.org/research.html
*Sponsor:* Dorr Institute for Arthritis Research and Education
1520 San Pablo Street, Suite 2000
Los Angeles, CA 90033

**Do Something TMI Intern** 1033
TMI is the agency of DoSomething.org that works with companies, not-for-profits, and governments to help them in the areas of technology, millennials and impact. The TMI Intern works with the agency team on both the business development front (creating and pitching new ideas for potential clients) and on the client services side (research, writing, etc.) supporting existing projects. Primary responsibilities include: research and development for potential Do Something clients - everything from global education to local teen health; researching companies and brands that are relevant to a teen audience; brainstorming new youth marketing and youth outreach ideas; assisting in the management of our current client and sponsor relations; helping create weekly internal reports with teen insights and helpful data; some administrative work as necessary; helping to manage TMI's social media channels (FB, Twitter, LinkedIn); and helping to develop strategies for client campaigns. Compensation includes: a $1,000 stipend or school credit; tons of fun; a great reference letter; and a chance to work in a super fun and positive work environment. Fall, winter, and summer semester internships are available.
*Requirements:* General requirements include: ability to travel to the New York City office; commitment to working 12 to 15 hours per week for nine to ten weeks; being a self starter – ability to work well alone and with others; passionate about young people making an impact and committed to social change; member of DoSomething.org; love of the DoSomething.org culture and can work with a quirky, close-knit team. Specific requirements include: outgoing and enthusiastic personality; confident working independently and taking ownership of full projects; excellent communication skills; solid grammar; ability to convey message clearly; good telephone manner; passion for social change; interest in CSR, marketing, and everything data; knowledge of pop culture and the teen trends of the moment; and the ability to produce client-facing documents and attend meetings with clients.
*Geographic Focus:* All States
*Date(s) Application is Due:* Aug 14
*Amount of Grant:* 1,000 USD
*Contact:* Lisa Boyd, Senior TMI Strategist; (212) 254-2390; lboyd@dosomething.org or intern@tmiagency.org
*Internet:* https://www.dosomething.org/about/internships#tmi-intern-the-dosomething.org-agency
*Sponsor:* Do Something
19 West 21st Street, 8th Floor
New York, NY 10010

**Do Something User Experience Research Intern** 1034
The Do Something User Experience Research Intern will help get the research needed to make a clean, simple, and intuitive user experience. Responsibilities include: fighting for the user - uncovering what is in the best interest of young people who are looking for ways they can create social change; interviews - recruit, screen, and schedule members and other young people for user interviews; research - field feedback from members and review usability videos, and then consolidate learnings in a way that determines if more research is needed or readiness to implement changes; data digging - be an active member of the quantitative and quantitatively driven data team, helping to lead and report on small and large scale surveys; customer service - use of organizational skills and love of efficiency to come up with better systems that give quick responses to users who need help, or clearly communicate to other team members (tech team, campaigns, social media, etc.); innovation - follow the industry wide conversation on UX innovations and research methodologies. Compensation includes: a $1,000 stipend or school credit; tons of fun; a great reference letter; and a chance to work in a super fun and positive work environment. Fall, winter, and summer semester internships are available.
*Requirements:* General requirements include: ability to travel to the New York City office; commitment to working 12 to 15 hours per week for nine to ten weeks; being a self starter – ability to work well alone and with others; passionate about young people making an impact and committed to social change; member of DoSomething.org; love of the DoSomething.org culture and can work with a quirky, close-knit team. Specific requirements include: background in a social science (e.g., psychology, sociology, behavioral economics), design (arts), or a related field; curiosity; passion for design or science; and customer service experience.
*Geographic Focus:* All States
*Amount of Grant:* 1,000 USD
*Contact:* Luke Patton, User Experience Designer; (212) 254-2390; internships@dosomething.org or lpatton@dosomething.org
*Internet:* https://www.dosomething.org/about/internships#user-experience-research-intern
*Sponsor:* Do Something
19 West 21st Street, 8th Floor
New York, NY 10010

**Douglass Institute Postdoctoral Fellowships** 1035
The Frederick Douglass Institute (FDI) for African and African-American Studies at the University of Rochester brings together scholars working in the humanities and social sciences in the area of Africa and the African Diasporas. Its mission is to further undergraduate and graduate education, faculty research, and intellectual exchange within the University community. The Postdoctoral Fellowship is awarded to scholars who hold a Ph.D. degree in a field that contributes to African and African-American Studies. It carries an annual stipend of $40,000 and a $3,000 stipend for research-related activities. The Fellow will participate in the intellectual life of the Institute, pursue his or her own scholarship, and teach one course in his or her area of specialization. All Fellows receive office space in the Institute, full access to the University's facilities, and opportunities to interact and collaborate with scholars relevant to their projects within the University. Fellows must be in full-time residence during the tenure of their awards, and they are required to be engaged in scholarly activity on a full-time basis. Fellows are expected to be available for consultation with students and colleagues; make at least one formal presentation based upon their research; and contribute generally to the intellectual life of the Institute. Applicants must have their Ph.D. in hand before the fellowship begins. Applications are accepted from September 1 through December 31, with awards announced in early March.
*Requirements:* Applicants must submit the following: the application form available at the website; a curriculum vitae; a 3-5 page description of the research the applicant plans to carry out during the fellowship year, plus a description of the undergraduate course the applicant would like to teach; a sample of published or unpublished writing on a topic related to the proposal; and three letters of recommendation that comment upon the value and feasibility of the work proposed.
*Geographic Focus:* All States
*Amount of Grant:* 43,000 USD
*Contact:* Ghislaine Radegonde-Eison, Program Manager; (585) 275-9161 or (585) 276-5744; fax (585) 256-2594; fdi@mail.rochester.edu
*Internet:* www.rochester.edu/College/AAS/fellowships/
*Sponsor:* University of Rochester - Douglass Institute
302 Morey Hall
Rochester, NY 14627-0440

**Douglass Institute Predoctoral Fellowships** 1036
The Frederick Douglass Institute for African and African-American Studies at the University of Rochester brings together scholars working in the humanities and social sciences in the area of Africa and the African Diasporas. Its mission is to further undergraduate and graduate education, faculty research, and intellectual exchange within the University community. The Predoctoral Fellowship is awarded to a graduate student of any university whose degree contributes to the scholarship in the field of African and African-American Studies. A principal aim of this Fellowship is to provide a different intellectual environment that will enrich and expedite the completion of the Fellow's dissertation. The Fellow will also gain valuable experience through work with the Institute's director in organizing colloquia, lectures, and other events. There is no teaching obligation. The award carries an annual stipend of $26,000, and the Institute offers research funds to support the Fellow's research. All Fellows receive office space in the Institute, full access to the University's facilities, and opportunities to interact and collaborate with scholars relevant to their projects within the University. Fellows must be in full-time residence during the tenure of their awards and they are required to be engaged in scholarly activity on a full-time basis. Fellows are expected to be available for consultation with students and colleagues; make at least one formal presentation based upon their research; and contribute generally to the intellectual life of the Institute. To qualify for this residential Fellowship, an applicant will need to have completed the following: all required courses; qualifying oral and/or written exams; at least one chapter of the dissertation (this chapter is a required portion of the application package). Application materials and additional information are available at the FDI website. Applications are accepted from September 1 through December 31, with awards announced in early March.
*Requirements:* Applicants should submit a completed FDI fellowship application form; a curriculum vitae; an official transcript showing completion of all preliminary coursework and qualifying examinations; the dissertation prospectus; completed chapter(s) from the dissertation; and three letters of recommendation, including one from the dissertation supervisor that specifically addresses the candidate's prospects for completing the project within a year.
*Geographic Focus:* All States
*Amount of Grant:* 26,000 USD

*Contact:* Ghislaine Radegonde-Eison, Program Manager; (585) 275-9161 or (585) 276-5744; fax (585) 256-2594; fdi@mail.rochester.edu
*Internet:* www.rochester.edu/College/AAS/fellowships/
*Sponsor:* University of Rochester - Douglass Institute
302 Morey Hall
Rochester, NY 14627-0440

**Dreyfus Postdoctoral Program in Environmental Chemistry Grant** **1037**
The Camille and Henry Dreyfus Foundation seeks to further the development of scientific leadership in the field of environmental chemistry with a postdoctoral fellowship program. The Postdoctoral Program in Environmental Chemistry provides a principal investigator with an award of $120,000 over two years to appoint a Postdoctoral Fellow in environmental chemistry. Applications most likely to be of interest should describe innovative fundamental research in the chemical sciences or engineering related to the environment. Examples include but are not limited to the chemistry associated with: the climate, the atmosphere, aquatic or marine settings, toxicology, soil or groundwater. Also of interest are chemistry-related energy research (renewable sources, sequestration, etc.), and new or green approaches to chemical synthesis and processing, with a clearly stated relation to the environment.
*Requirements:* The program is open to all academic and other not-for-profit organizations that have well-established research efforts in environmental science or engineering in the States, Districts, and Territories of the US. These research activities need not be located in traditional departments in the chemical sciences, and collaboration across departments and institutions is encouraged.
*Restrictions:* No part of the award may be used for institutional administrative purposes.
*Geographic Focus:* All States
*Date(s) Application is Due:* Aug 25
*Amount of Grant:* 120,000 USD
*Contact:* Mark J. Cordillo, Executive Director; (212) 753-1760; fax (212) 593-2256; mcardillo@dreyfus.org or admin@dreyfus.org
*Internet:* www.dreyfus.org/awards/postdoctoral_program.shtml
*Sponsor:* Camille and Henry Dreyfus Foundation
555 Madison Avenue, 20th Floor
New York, NY 10022-3301

**Dreyfus Prize in the Chemical Sciences** **1038**
The Dreyfus Prize in the Chemical Sciences, awarded biennially, consists of a citation, a medal, and a monetary award of $250,000. The prize is awarded to an individual in a selected area of chemistry to recognize exceptional and original research that has advanced the field in a major way. The first Dreyfus Prize was awarded to George Whitesides of Harvard University in the field of materials chemistry, honoring the accomplishments of the Dreyfus brothers, Camille and Henry. All nomination materials and letters of support are due at the Foundation office by the deadline March 1.
*Requirements:* Any person may nominate a candidate for the Dreyfus Prize in the Chemical Sciences. Self-nominations are not accepted.
*Restrictions:* The prize is restricted to U.S. residents. Nobel laureates are not eligible. Dreyfus Foundation Advisors and reviewers who serve in the year of the selection are not eligible. There is no restriction on the number of nominees who are at a given institution, nor is institutional approval required.
*Geographic Focus:* All States
*Date(s) Application is Due:* Mar 1
*Amount of Grant:* 250,000 USD
*Contact:* Mark J. Cordillo, Executive Director; (212) 753-1760; fax (212) 593-2256; mcardillo@dreyfus.org or admin@dreyfus.org
*Internet:* www.dreyfus.org/awards/prize.shtml
*Sponsor:* Camille and Henry Dreyfus Foundation
555 Madison Avenue, 20th Floor
New York, NY 10022-3301

**Dreyfus Senior Scientist Mentor Program Grants** **1039**
The Camille and Henry Dreyfus Foundation supports emeritus faculty who maintain active research programs with undergraduates in the chemical sciences. The Senior Scientist Mentor Program provides an award of $20,000 over two years for undergraduate stipends and modest research support. The program enables undergraduates to conduct research in the chemical sciences in a close working relationship with emeritus faculty members. Only one faculty member may apply per application, although more than one application per department or institution is permitted. A letter of support must be sent directly to the Foundation from a colleague, preferably from outside the institution, who is familiar with the applicant's research and teaching and who can speak to the applicant's experience in mentoring and advising undergraduates. Letters of support cannot be accepted via electronic mail or facsimile.
*Requirements:* Faculty with emeritus status on or before January of the current year, and who maintain active research programs in the chemical sciences may apply for one of a limited number of awards that will allow undergraduates to do research under their guidance. Successful applicants, who are expected to be closely engaged in a mentoring relationship with the students will receive grants annually for two years for undergraduate stipends and modest research support.
*Restrictions:* Charges associated with indirect costs or institutional overhead are not allowed. Faculty salary and any expenses associated with graduate students are not permitted.
*Geographic Focus:* All States
*Date(s) Application is Due:* Nov 12
*Amount of Grant:* 20,000 USD
*Contact:* Mark J. Cordillo, Executive Director; (212) 753-1760; fax (212) 593-2256; mcardillo@dreyfus.org or admin@dreyfus.org
*Internet:* www.dreyfus.org/awards/senior_scientist_mentor.shtml
*Sponsor:* Camille and Henry Dreyfus Foundation
555 Madison Avenue, 20th Floor
New York, NY 10022-3301

**Dreyfus Special Grants Program in the Chemical Sciences** **1040**
This program offers support to eligible institutions for projects that propose to advance the science of chemistry in innovative ways. Proposals are invited in any area consistent with the Foundation's broad objective of advancing the chemical sciences. For projects that require sustaining support, note that this program is intended to seed the initial phases, with the expectation that grant recipients will find continuing funding from other sources. The Foundation encourages proposals that are judged likely to significantly advance the chemical sciences. Examples of areas of interest include (but are not limited to): the increase in public awareness, understanding, and appreciation of the chemical sciences; innovative approaches to chemistry education at all levels (K-12, undergraduate, and graduate); and efforts to make chemistry careers more attractive. Research proposals are not customarily considered.
*Requirements:* The Special Grant Program in the Chemical Sciences is open to institutions in the States, Districts, and Territories of the United States of America that have a focus in the chemical sciences. Institutions include schools, colleges and universities, as well as other not-for-profit organizations, such as scientific societies and science museums.
*Restrictions:* Awards are not made directly to individuals, nor, in general, to private foundations.
*Geographic Focus:* All States
*Samples:* Texas Wesleyan U (Fort Worth, TX)—for a software-based program that will enable students to better access chemical-sciences information, $30,000; Cornell U (NY)—for a project on computational analysis in the chemistry curriculum, $25,000; California Institute of Technology (CA)—to support summer undergraduate research fellowships, $15,000; Columbia U (NY)—for a project on computer-assisted problem-solving tools for general chemistry, $25,000.
*Contact:* Mark J. Cordillo, Executive Director; (212) 753-1760; fax (212) 593-2256; mcardillo@dreyfus.org or admin@dreyfus.org
*Internet:* www.dreyfus.org/awards/special_grant_program_chemical.shtml
*Sponsor:* Camille and Henry Dreyfus Foundation
555 Madison Avenue, 20th Floor
New York, NY 10022-3301

**DTE Energy Foundation Health and Human Services Grants** **1041**
The DTE Energy Foundation is at the core of DTE Energy's commitment to the communities and customers it is privileged to serve. The DTE Energy Foundation is dedicated to strengthening the health and human services sector of these communities. Priority will be given to supporting organizations that are in the forefront of addressing the critical, acute human needs brought on by the economic downturn. Grant amounts generally range from $500 to $100,000, and the application process is distinct for each of the following ranges: $500 to $2,000; $2,001 to $10,000; and any amount greater than $10,000. Applications must be submitted electronically by the stated deadlines.
*Requirements:* Eligible applicants must meet all of the following criteria: be located in or provide services to a community in which DTE Energy does business; and be a nonprofit (i.e. be exempt for federal income tax under section 501(c)3 of the Internal Revenue Code and not a private foundation, as defined in Section 509(a) of the Code).
*Restrictions:* The Foundation does not provide support to: individuals; political parties, organizations or activities; religious organizations for religious purposes; organizations that are not able to demonstrate commitment to equality and diversity; student group trips; national or international organizations, unless they are providing benefits directly to our service-area residents; projects that may result in undue personal benefit to a member of the DTE Energy Foundation board, or to any DTE Energy employee; conferences unless they are aligned with DTE Energy's business interests; single purpose health organizations; hospitals, for building or equipment needs.
*Geographic Focus:* Michigan
*Date(s) Application is Due:* Apr 13; Jul 13; Oct 12; Dec 28
*Amount of Grant:* 500 - 100,000 USD
*Samples:* Center for Autism, Ann Arbor, Michigan, $100,000 - general operating support; Girl Scouts of Southeastern Michigan, Detroit, Michigan, $75,000 - general operating costs; Forgotten Harvest, Oak Park, Michigan, $60,000 - general operating costs.
*Contact:* Karla D. Hall; (313) 235-9271 or (313) 235-9416; foundation@dteenergy.com
*Sponsor:* DTE Energy Foundation
One Energy Plaza, 1046 WBC
Detroit, MI 48226-1279

**DTE Energy Foundation Higher Learning Grants** **1042**
In the area of Higher Learning grants, the DTE Energy Foundation's focus is on the following priorities: specific academic departments of engineering, science and business that prepare students to enter the professional workforce; programs and institutions that prepare students for technical and skilled trade careers in the energy industry; and programs that increase student retention and success in engineering, the sciences and business with a focus on women and minorities. Grant amounts generally range from $2,000 to $750,000, and the application process is distinct for each of the following ranges: $500 to $2,000; $2,001 to $10,000; and any amount greater than $10,000. Applications must be submitted electronically by the stated deadlines.
*Requirements:* Eligible applicants must meet all of the following criteria: be located in or provide services to a community in which DTE Energy does business; and be a nonprofit (i.e. be exempt for federal income tax under section 501(c)3 of the Internal Revenue Code and not a private foundation, as defined in Section 509(a) of the Code).

*Restrictions:* The Foundation does not provide support to: individuals; political parties, organizations or activities; religious organizations for religious purposes; organizations that are not able to demonstrate commitment to equality and diversity; student group trips; national or international organizations, unless they are providing benefits directly to our service-area residents; projects that may result in undue personal benefit to a member of the DTE Energy Foundation board, or to any DTE Energy employee; conferences unless they are aligned with DTE Energy's business interests; single purpose health organizations; hospitals, for building or equipment needs.
*Geographic Focus:* Michigan
*Date(s) Application is Due:* Apr 13; Jul 13; Oct 12; Dec 28
*Amount of Grant:* 2,000 - 750,000 USD
*Samples:* University of Michigan, Ann Arbor, Michigan, $260,602 - ongoing costs; Wayne State University, Detroit, Michigan, $120,250 - ongoing costs; Western Michigan University Foundation, Kalamazoo, Michigan, $67,555 - general operating costs.
*Contact:* Karla D. Hall, Vice-President; (313) 235-9271 or (313) 235-9416; foundation@dteenergy.com
*Internet:* www.dteenergy.com/dteEnergyCompany/community/foundation/whatWeSupport.html
*Sponsor:* DTE Energy Foundation
One Energy Plaza, 1046 WBC
Detroit, MI 48226-1279

**DuPont Pioneer Conferences and Meetings Grants** **1043**

The primary purpose of the DuPont Pioneer Conferences and Meetings Grant program is to support research conferences and meetings addressing topics of relevance to DuPont Pioneer. Areas of interest include: plant science research; global agronomy; germplasm preservation; and plant biotechnology. Funding decisions are made by the Research Awards Committee Chair and Research Management, based on input from PHI scientists. For further information, applicants should forward an email to the local DuPont Pioneer Representative.
*Geographic Focus:* All States
*Contact:* Grants Administrator; (515) 535-3200 or (800) 247-6803, ext. 3915; fax (515) 334-4415; David.Bubeck@pioneer.com
*Internet:* https://www.pioneer.com/home/site/about/business/pioneer-giving/community-giving/
*Sponsor:* DuPont Pioneer
P.O. Box 1000
Johnston, IA 50131-0184

**DuPont Pioneer Honorary Plant Breeding Fellowships** **1044**

The Pioneer Hi-Bred Honorary Plant Breeding Fellowships honor the contributions of 10 outstanding Pioneer plant breeders to the progress of science and agriculture. These fellowships are intended to promote graduate education in plant breeding and genetics, and are expected to attract high-quality students whose future careers will continue to advance the science of plant breeding. A Fellowship Committee established by Pioneer will review applications and select recipients based on their qualifications, including academic excellence; proposed study program area, with consideration given to agronomic crops such as corn, soybeans, canola, sorghum, sunflower, wheat and alfalfa; future professional plans; and leadership and other experiences predicting future success. The funding limit is $25,000 per year for the duration of the fellowship.
*Requirements:* Scholarship candidates must be nominated and sponsored by a Pioneer Hi-Bred employee.
*Geographic Focus:* All States
*Amount of Grant:* Up to 25,000 USD
*Contact:* Tabare Abadie, Senior Research Manager; (800) 247-6803, ext. 3915 or (515) 535-3200; fax (515) 334-4415; Tabare.Abadie@pioneer.com or community.investment@pioneer.com
*Internet:* https://www.pioneer.com/home/site/about/template.CONTENT/guid.83F8816B-34BB-067F-958C-B3A109F553FD
*Sponsor:* DuPont Pioneer
P.O. Box 1000
Johnston, IA 50131-0184

**DuPont Pioneer International Scholarships** **1045**

The DuPont Pioneer International Scholarship is designed to contribute to agricultural research and education by providing financial support for outstanding students studying for graduate level degrees in plant breeding overseas. Funding is considered to be an unrestricted gift. Funding decisions will be reviewed annually and any changes will be communicated by September 30 of each year. To learn whether Pioneer scholarships or fellowships are available at your university, contact the head of the appropriate academic department.
*Geographic Focus:* All States
*Contact:* Administrator; (515) 535-3200 or (800) 247-6803, ext. 3915; fax (515) 334-4415
*Internet:* https://www.pioneer.com/home/site/about/template.CONTENT/guid.83F8816B-34BB-067F-958C-B3A109F553FD
*Sponsor:* DuPont Pioneer
P.O. Box 1000
Johnston, IA 50131-0184

**DuPont Pioneer Society Fellowships** **1046**

The DuPont Pioneer Society Fellowship is designed to contribute to agricultural research and education by providing financial support for outstanding students studying for graduate level degrees in plant breeding. Leading societies that support plant improvement graduate programs in crops of interest to Pioneer have been selected by the Fellowship Sub-committee. The institutions are responsible for student selection and administering the program. Funded societies include: Agronomic Science Foundation; American Phytopathological Society; American Society of Plant Biologists; the Consultative Group on International Agricultural Research (CGIAR); and the Entomological Foundation.
*Requirements:* Pioneer prefers this grant to be used solely to fund a student(s) progressing towards a graduate degree in plant breeding from the funded institution.
*Restrictions:* No university overhead will be paid from these funds.
*Geographic Focus:* All States
*Contact:* Administrator; (515) 535-3200 or (800) 247-6803, ext. 3915; fax (515) 334-4415
*Internet:* https://www.pioneer.com/home/site/about/template.CONTENT/guid.83F8816B-34BB-067F-958C-B3A109F553FD
*Sponsor:* DuPont Pioneer
P.O. Box 1000
Johnston, IA 50131-0184

**DuPont Pioneer University Plant Breeding Fellowships** **1047**

The DuPont Pioneer University Plant Breeding Fellowships are designed to contribute to agricultural research and education by providing financial support for outstanding students working toward graduate level degrees in plant breeding. Leading universities that support plant improvement graduate programs in crops of interest to Pioneer have been selected by the Fellowship Subcommittee. The institutions are responsible for student selection and administering the program. Funded institutions currently include: Cornell University; North Carolina State University; University of Illinois; University of Minnesota; and University of Wisconsin-Madison.
*Geographic Focus:* All States
*Contact:* Administrator; (515) 535-3200 or (800) 247-6803, ext. 3915; fax (515) 334-4415
*Internet:* www.pioneer.com/web/site/portal/menuitem.07196fe7964ae318bc0c0a03d10093a0/
*Sponsor:* DuPont Pioneer
P.O. Box 1000
Johnston, IA 50131-0184

**Durfee Foundation Stanton Fellowship** **1048**

The Durfee Foundation Stanton Fellowship provides up to six fellows with $75,000 each over a two-year period to think deeply about the intractable problems in their sector, and to tease out solutions that will improve life for the people of Los Angeles. Possible uses for the funds may include: travel to sister organizations in the U.S. or abroad; apprenticing to experts; writing policy papers; enrollment in training programs; or supplementing salaries of co-workers to allow release time for the fellow to concentrate on the project.
*Requirements:* Individual candidates for this program should: be outstanding leaders who have demonstrated a track record of contribution to the community; work for a nonprofit organization; have worked a minimum of ten years in the nonprofit sector; hold primary or significant responsibility for management of the organization's funds; be willing to participate in a peer network of program alumni; be employed full-time at the applicant organization; and have financial need. The candidate's employing organization should: be located in Los Angeles County; have at least three full-time staff; need financial assistance to underwrite the candidate's leave; and be able to sustain regular operations during the candidate's absence.
*Geographic Focus:* California
*Amount of Grant:* 75,000 USD
*Contact:* Executive Director; (310) 899-5120; fax (310) 899-5121; admin@durfee.org
*Internet:* durfee.org/what-we-support/stanton-fellowship/
*Sponsor:* Durfee Foundation
1453 Third Street, Suite 312
Santa Monica, CA 90401

**E.L. Wiegand Foundation Grants** **1049**

Established in Nevada in 1982, the Foundation provides grants to develop and strengthen programs and projects in Arizona, California, the District of Columbia, Idaho, Nevada, New York, Oregon, Utah, and Washington. Funding is given to educational institutions in the academic areas of science, business, fine arts, and law; and medicine and health organizations in the areas of heart, eye, and cancer surgery, treatment, and research. The Foundation also considers requests for projects that enrich children, communities, public policy, and the arts. Its primary fields of interest include: arts; biology and life sciences; business education; cancer research; chemistry; elementary education; eye research; heart and circulatory diseases and research; higher education; law school education; medical research and institutes; medical school education; museums; performing arts (music and theater); physics; public affairs; secondary education; and visual arts. Grants for educational equipment, including computers and scientific supplies, are also awarded. The board of trustees meets in February, June, and October to choose recipients, but applications may be submitted at any time. Application guidelines are available upon request.
*Requirements:* Nonprofit organizations in Arizona, California, District of Columbia, Idaho, Nevada, New York, Oregon, Utah, and Washington State are eligible.
*Restrictions:* Institutions are to be in existence a minimum of five years. Proposals for endowment, debt reduction, ordinary operations, general fund raising, emergency funding, multi-year funding, productions of documentaries, films or media presentations, direct or indirect loans, or for the benefit of specific individuals are excluded from consideration. The Foundation does not award grants to government agencies or to charitable institutions which derive significant support from public tax-funds or United Way.
*Geographic Focus:* Arizona, California, District of Columbia, Idaho, Nevada, New York, Oregon, Utah, Washington
*Amount of Grant:* 10,000 - 200,000 USD
*Samples:* Nevada Museum of Art, Reno, Nevada, $89,000 - exhibit development and support; American Enterprise Institute, Washington, D.C., $500,829 - capital campaign

in support of a communications center; Saint Albert the Great School, Reno, Nevada, $500,000 - expansion project.
*Contact:* Kristen A. Avansino, Executive Director; (775) 333-0310; fax (775) 333-0314
*Sponsor:* E.L. Wiegand Foundation
165 W Liberty Street, Suite 200
Reno, NV 89501-2902

**E.W. "Al" Thrasher Awards** **1050**
The purpose of the E.W. Al Thrasher Awards is to improve children's health through medical research, with an emphasis on projects that have the potential to translate into clinically meaningful results within a few years. The Fund awards grants three times per year, with no fixed number of awards given in each funding cycle or in each year. There are no deadlines for the awards. Concept papers are accepted on a rolling basis, and are considered by the Fund's Executive Committee approximately once per month. If a full proposal is invited, the applicant will have one year to submit a full proposal. Full proposals recommended by peer reviewers for further consideration are considered by the Fund's Committees in February, June, or October.
*Requirements:* Principal Investigators must be qualified in terms of education and experience to conduct research. A doctoral-level degree is required. There are no citizenship or residency requirements. The Fund is open to applications from institutions both inside and outside the United States. Ideal applications for the E.W. "Al" Thrasher Award address significant health problems that affect children in large numbers and offer the potential for practical solutions to these problems. Such solutions should be innovative and have the potential for rapid and broad applicability with low financial and/or technical barriers to implementation. Hypothesis-driven research is given priority over exploratory, hypothesis-generating research. The grant amount is based on the actual budgetary needs of the project. The duration of the project can be up to three years. Indirect costs of no more than 7% of direct costs will be paid on E.W. Al Thrasher Awards. Investigators in need of salary support for a specific project may apply for modest support (no more than 20 percent, based on a 40-hour work week). The Fund limits the maximum salary upon which an individual can request salary support. The maximum amount will be based on the NIH guidelines for salary support.
*Restrictions:* Requests for purchase of major equipment (items costing more than $4,000) are discouraged and rarely funded. The Fund does not award grants for educational programs; general operating expenses; general bridge funding for incomplete projects; construction or renovation of buildings or facilities; loans, student aid, scholarships, tuition; or, support of other funding organizations. Grants generally are not awarded for conferences, workshops, or symposia. Proposals in the areas of research on human fetal tissue or behavioral science research will not be considered. Proposals requesting Thrasher to commit money to a larger pool of funding with no distinct project proposed to Thrasher will not be funded.
*Geographic Focus:* All States, All Countries
*Contact:* Megan Duncan; (801) 240-4720; DuncanME@thrasherresearch.org
Aaron V. Pontsler, M.S., MBA, Research Manager; (801) 240-6385; PontslerAV@thrasherresearch.org
*Internet:* www.thrasherresearch.org/sites/www_thrasherresearch_org/Default.aspx?page=220
*Sponsor:* Thrasher Research Fund
68 S. Main Street, Suite 400
Salt Lake City, UT 84101

**East-West Center Asian Development Bank Japan Scholarship** **1051**
The Asian Development Bank Japan Scholarship program (ADB-JSP) provides support for graduate (post-baccalaureate) studies in ADB-approved fields of study at the University of Hawaii at Manoa (UHM), and for participation in the educational, residential, and leadership development programs at the East-West Center (EWC). The ADB-JSP was established to encourage and strengthen human resource development in ADB's developing member countries. After completing their studies, scholars are expected to return to their home countries to apply and share their new knowledge and skills to promote socioeconomic development. Fellowships are awarded annually in an international competition. The annual application submission deadline is December 1.
*Requirements:* To apply, individuals must: be a citizen of a developing member country of the Asian Development Bank; have a four-year bachelor's degree or the equivalent of a four-year bachelor's degree, at the time of application; have at least two years of full-time professional work experience after a university degree; and take either the TOEFL or the IELTS English proficiency test, with the minimum acceptable score of 550 for the paper-based TOEFL, a 213 for the computer-based TOEFL, and 79 for the Internet-based TOEFL. A 6.5 on the IELTS will also be sufficient. Results must be from a test taken within the past two years.
*Geographic Focus:* All States, All Countries
*Date(s) Application is Due:* Dec 1
*Contact:* Kim Small, Scholarship Program Specialist; (808) 944-7752 or (808) 944-7769; fax (808) 944-7070; adbjsp@eastwestcenter.org or smallk@eastwestcenter.org
*Internet:* https://www.eastwestcenter.org/education/student-programs/opportunities-study
*Sponsor:* East-West Center
1601 East-West Road
Honolulu, HI 96848-1601

**Educational Ventures International Foundation Grants** **1052**
The foundation funds nonprofits addressing important educational issues in areas where adequate funding from other sources is unavailable. The foundation's areas of interest include grants for educational technology, which provide grants for hardware, software, and peripherals for K-12 and underserved areas; grants for educational resources, which provide funding for resources to enhance teaching and learning environments in underserved areas; grants to encourage the development of new and innovative uses of technology to enhance teaching and learning; and grants for research, which provide funds to support research in math and technology education. Grants fund educational programs that promote race, class, ethnic, and gender equity in education. The foundation also funds mathematics and education research and publishes papers of those studies.
*Requirements:* Nonprofit organizations are eligible.
*Restrictions:* The foundation typically does not fund projects with political or religious affiliations.
*Geographic Focus:* All States
*Contact:* Grants Administrator; (801) 659-2233; educationalventures@yahoo.com
*Sponsor:* Educational Ventures International Foundation
P.O. Box 5887
Carefree, AZ 85377-5887

**Edward N. and Della L. Thome Memorial Foundation Grants** **1053**
The Edward N. and Della L. Thome Memorial Foundation was established in 2002 by Robert P. Thome to honor the memory of his parents, Edward and Della Thome. The mission of the foundation is twofold: to advance the health of older adults through: the support of direct service projects; and medical research on diseases and disorders affecting older adults. The goal of the foundation's direct services program is to support organizations in Maryland, Missouri, and Michigan that provide direct services addressing one or more of the following critical issues facing older adults: health care, housing, family services, neighborhood involvement, workforce opportunities, and aging with dignity at home. Grant requests for general operating support are strongly encouraged. Program support will also be considered. Program-related capital expenses may be included in general operating or program requests. To better support the capacity of nonprofit organizations, multi-year funding requests are encouraged. While medical research grants have no annual deadline, all others must be submitted by June 15. Most recent awards have ranged from $20,000 to $2,000,000.
*Requirements:* The Thome Foundation direct services program is most interested in expanding services for older adults. Requests should clearly state how many previously-unserved older adults will now be served as a result of funding from the Thome Foundation. Applicants must have 501(c)3 tax-exempt status.
*Restrictions:* The foundation does not support requests from individuals, organizations attempting to influence policy through direct lobbying, or any political campaigns.
*Geographic Focus:* Maryland, Michigan, Missouri
*Date(s) Application is Due:* Jun 15
*Amount of Grant:* 20,000 - 2,000,000 USD
*Samples:* United Methodist Retirement Communities, Chelsea, Michigan, $2,000,000 - support of the Cascade PACE program; Washington Hospital Center, Washington, D.C., $2,000,000 - for for the new New Heart Hospital: The Medstar Heart and Vascular Institute; Medical College of Wisconsin, Milwaukee, Wisconsin, $250,000 - general operating support.
*Contact:* Srilatha Lakkaraju; (312) 828-8166; ilgrantmaking@ustrust.com
*Internet:* https://www.bankofamerica.com/philanthropic/foundation.go?fnId=47
*Sponsor:* Edward N. and Della L. Thome Memorial Foundation
231 South LaSalle Street, IL1-231-13-32
Chicago, IL 60604

**Edward W. and Stella C. Van Houten Memorial Fund Grants** **1054**
Stella C. Van Houten resided in Bergen County, New Jersey. This foundation, providing funding for health and human services, education, education of medical professionals, and the care of children, was established in 1978 in memory of her husband and herself. The Van Houten's had a particular fondness for the Valley Hospital of Ridgewood, New Jersey and for the Rollins College in Florida. The Foundation continues to honor their preferences with grants to these two organizations in addition to grants to other organizations. The Foundation's mission is to: supports agencies, institutions and services in Passaic and Bergen Counties, New Jersey, having to do with the care or cure of sick or disabled persons or for the care of orphaned children or aged persons; educates students in the medical profession; support for educational purposes; support for the care of children. A target of 10% of the grants each year is for medical scholarships.
*Requirements:* Passaic and Bergen Counties, New Jersey non-profits are eligible to apply. The application form & guidelines are available online at the Wachovia website. The applications must be submitted by January 31 for a March meeting & August 1 for an October meeting.
*Geographic Focus:* New Jersey
*Date(s) Application is Due:* Jan 31; Aug 1
*Amount of Grant:* 6,000 - 100,000 USD
*Samples:* Therapeutic Learning Center, Ramsey, NJ, $6,000—speech therapy; William Paterson University Foundation, $20,000—scholarships for undergraduate, minority, and/or graduate nursing students; Christian Health Care Center Foundation, $100,000—to build a Great Room at Heritage Manor East.
*Contact:* Trustee, c/o Wachovia Bank; grantinquiries2@wachovia.com
*Internet:* https://www.wellsfargo.com/private-foundations/van-houten-memorial-fund
*Sponsor:* Edward W. and Suitella C. Van Houten Memorial Fund
190 River Road, NJ3132
Summit, NJ 07901

**Elizabeth Glaser International Leadership Awards** **1055**
The International Leadership Award (ILA) is a three-year grant focused on international investigations in preventing mother-to-child transmission (PMTCT) of HIV and HIV care and treatment. Recent awards were granted for research into improving the effectiveness of PMTCT services; strengthening maternal and infant diagnosis of HIV; and examining the impact of antiretroviral drugs on children. The Award identifies and supports internationally based physicians or scientists who have the training and

potential to develop pediatric HIV prevention and/or treatment programs, but lack the resources to do so. Award recipients are required to mentor a minimum of three additional people who would benefit from the experience and could help them achieve their goals. These are individuals who would be hired and trained to work on the project proposed by the applicant. The Award provides up to $105,000 in total costs per year for three years for a total of $315,000. Funding for the second and third year is dependent on progress in year one, and is not guaranteed. Funds support the salary of the applicant (up to 100% effort), salaries for those hired and mentored as part of the project, and direct costs related to the project itself such as travel, supplies, and equipment. Indirect cost is limited to 5% (included in the total). Salary support per individual investigator may be a maximum of $45,000 per year, and must be commensurate with the level of experience of the investigator, the salary structure in the country and the level of commitment to the project. Awardees must agree to spend at least 50% of their time on the project related to the Award. Final determination as to the amount of the Award is made by the Advisory Board depending on the scope of the project. Interested applicants are asked to submit a Letter of Intent. Only those Letters of the highest caliber will be asked to submit a full grant application. Additional guidelines, instructions, and summaries of previously funded projects are posted on the website.
*Requirements:* Applicants must be from a developing country, and must be performing work in a developing country in order to apply. Applicants must be living in and performing work in either Africa or India. The Foundation is especially interested in applicants who have specific training or experience with HIV/AIDS. Applicants must have an MD or PhD, and demonstrate commitment to continuing work in a developing country on HIV/AIDS when the award is complete.
*Restrictions:* U.S. citizens or European citizens living in developing countries are not eligible to apply. Senior Investigators such as full professors are usually not eligible, unless they are changing the direction of their career to include HIV/AIDS research.
*Geographic Focus:* All Countries
*Amount of Grant:* 105,000 USD
*Contact:* Chris Hudnall, Senior Program Coordinator, Research; (310) 314-3154; fax (310) 314-1469; chris@pedaids.org
*Internet:* www.pedaids.org/What-We-re-Doing/research/grant-programs.aspx
*Sponsor:* Elizabeth Glaser Pediatric AIDS Foundation
1140 Connecticut Avenue, NW, Suite 200
Washington, D.C. 20036

**Elizabeth Glaser Scientist Award** **1056**
The Elizabeth Glaser Scientist Award is the Foundation's highest scientific honor, and is an investment in the most promising HIV/AIDS researchers at a critical stage in their careers. These scientists, who represent the best and brightest investigators from the international medical science community, are selected on the basis of their knowledge, innovation, and dedication. By providing research funding over a five-year period, this award enables recipients to focus their long-term efforts on issues specific to pediatric HIV/AIDS. The Award fosters an unprecedented spirit of collaboration among these scientists. Each year, the Elizabeth Glaser Scientists meet with the Foundation's internationally renowned advisory board to stimulate ideas, report on current programs, and plan future research. Since the program's inception, the Foundation has built an invaluable network of scientists whose work in vaccine development, immune response, breast milk transmission, and other critical areas impacts the entire field of HIV/AIDS research. Interested applicants should contact the Foundation to request a letter of intent cover sheet. Full applications are by invitation.
*Requirements:* Candidates must have an MD, PhD, DDS, or DVM degree and be at the assistant professor level or above.
*Restrictions:* The Foundation will not fund large therapeutic trials or the pharmocologic development of drugs.
*Geographic Focus:* All States, All Countries
*Contact:* Doug Horner, Vice President of Awards; (202) 296-9165; fax (202) 296-9185; research@pedaids.org
*Internet:* www.pedaids.org/What-We-re-Doing/research/grant-programs
*Sponsor:* Elizabeth Glaser Pediatric AIDS Foundation
1140 Connecticut Avenue, NW, Suite 200
Washington, D.C. 20036

**Elizabeth Huth Coates Charitable Foundation Grants** **1057**
Elizabeth Maddux was married to oilman George H. Coates from 1943 until his death in 1973. It was said that George provided the means while Elizabeth provided the philanthropic inspiration which she had learned from her parents growing up in San Antonio. A longtime supporter of the arts and education, she established the Elizabeth Huth Coates Charitable Foundation of 1992 to continue her legacy of support within the San Antonio area. The Foundation's primary fields of interest include: the arts; Catholicism; diseases; education; museums; Protestantism; and zoos. Most recently, awards have ranged from $5,000 to $300,000. The annual deadline for online application submission is December 31.
*Requirements:* Grants are directed to 501(c)3 organizations which fall within the principal charitable purposes of the foundation.
*Geographic Focus:* Texas
*Date(s) Application is Due:* Dec 31
*Amount of Grant:* 5,000 - 300,000 USD
*Samples:* Ballet San Antonio, San Antonio, Texas, $20,000 - general operating support; Saint Mary's Hall, San Antonio, Texas, $300,000 - general operating support; San Antonio Museum of Art, San Antonio, Texas, $200,000 - general operating support.
*Contact:* Brian R. Korb, Senior Vice President; (210) 283-6700 or (210) 283-6500; bkorb@broadwaybank.com
*Internet:* www.broadwaybank.com/wealthmanagement/FoundationElizabethHuth Coates.html
*Sponsor:* Elizabeth Huth Coates Charitable Foundation
P.O. Box 17001
San Antonio, TX 78217-0001

**Elizabeth Nash Foundation Summer Research Awards** **1058**
The Elizabeth Nash Foundation is in the sixth year of a partnership with Children's Hospital and Research Center Oakland (CHORI), an internationally renowned biomedical research institute that bridges basic science and clinical research in the prevention and treatment of human disease. The Elizabeth Nash Foundation Summer Research Award is available to high school seniors and undergraduates who have a strong interest in pursuing cystic fibrosis (CF) research. The award is designed to identify and train the next generation of talented CF researchers. Students receive a stipend from the Foundation and CHORI pairs student with a mentor scientist to guide them through a structured CF research project. Previous recipients have co-authored publications in Free Radical Biology and Medicine and Cellular Physiology and Biochemistry involving their summer research. The deadline for applications for summer is 4:00 pm on April 15. The awardee will be provided a one-time stipend of $2,500 for the nine-week period. Announcement of all awards will be made on May 3.
*Geographic Focus:* All States
*Date(s) Application is Due:* Apr 15
*Amount of Grant:* 2,500 USD
*Contact:* Ann Nash, (408) 399-6919; info@elizabethnashfoundation.org
Deborah Ellen, CHORI Summer Research Program; (510) 450-7674; summerstudentprogram@chori.org
*Internet:* www.elizabethnashfoundation.org/CHORI.html
*Sponsor:* Elizabeth Nash Foundation
P.O. Box 1260
Los Gatos, CA 95031-1260

**El Pomar Foundation Leadership Development Fellowships** **1059**
The El Pomar Leadership Development Fellowship program educates young leaders about the role of nonprofit organizations and the foundation within the communities of Colorado. The two-year postgraduate program also focuses on professional development and prepares fellows for positions of leadership in Colorado and the nation. Fellows serve the foundation's operating programs, such as EPYCS, a high school philanthropy initiative. In addition, fellows will be thoroughly educated in the nature of the nonprofit sector and the role of foundations. At a minimum, fellows participate in monthly courses in public speaking, financial and investment management, nonprofit management, fundraising and strategic planning, and an annual Outward Bound course in the central mountain ranges of Colorado. This fellowship has a starting salary of $27,000.
*Requirements:* Applicant must be a graduate of a four-year university or college, have a Colorado affiliation (e.g., be a state resident, have attended an in-state university or college, or have immediate family who are residents or past residents), and demonstrate a strong leadership capability and potential.
*Geographic Focus:* Colorado
*Amount of Grant:* Up to 27,000 USD
*Contact:* Nicole Cook, Grants Administrator; (719) 633-7733 or (800) 554-7711; fax (719) 577-5702; fellowship@elpomar.org
*Internet:* https://www.elpomar.org/fellowship/
*Sponsor:* El Pomar Foundation
10 Lake Circle
Colorado Springs, CO 80906

**Elton John AIDS Foundation Grants** **1060**
The Elton John AIDS Foundation (EJAF) was established in the United States in 1992 by Sir Elton John, now headquartered in New York City. In 1993, Sir Elton also established his Foundation as a registered charity in the United Kingdom, headquartered in London. Both organizations pursue the same mission – to reduce the incidence of HIV/AIDS through innovative HIV prevention programs, eliminate stigma and discrimination associated with HIV/AIDS, and support direct HIV-related care for people living with HIV/AIDS. The U.S. organization's grant-making program focuses on the following targeted areas: gay men's mobilization for health; youth mobilization for health; black community mobilization for health; ending injection-related HIV transmission; access to healthcare for ex-offenders; and scaled-up of government HIV programming. EJAF conducts an annual open application grant cycle through a Request for Proposals process launched on or around May 1.
*Requirements:* Applicants located and conducting work corresponding to EJAF's stated grant-making priorities in the United States, Canada, the Caribbean, and Central and South America are eligible to apply for funding by first submitting an online letter of intent (LOI). After the LOIs have been reviewed, selected applicants will be invited to submit a full online proposal. These proposals are extensively reviewed, and final grant awards are approved by EJAF's Board of Directors.
*Geographic Focus* All States, Puerto Rico, U.S. Virgin Islands, Anguilla, Antigua & Barbuda, Argentina, Aruba, Bahamas, Barbados, Belize, Bolivia, Brazil, British Virgin Islands, Canada, Caribbean, Cayman Islands, Chile, Colombia, Costa Rica, Cuba, Dominica, Dominican Republic, Ecuador, El Salvador, Grenada, Guadeloupe, Guatemala, Guyana, Haiti, Honduras, Jamaica, Martinique, Mexico, Montserrat, Nicaragua, Panama, Paraguay, Peru, Saint Kitts And Nevis, St. Lucia, St. Vincent and the Grenadines, Suriname, Trinidad and Tobago, Turks and Caicos Islands, Uruguay, Venezuela
*Amount of Grant:* Up to 100,000 USD
*Samples:* Association of Nurses in AIDS Care, Reston, Virginia, $50,000 - support of HIV criminalization awareness and advocacy work the Association; The Attic

Youth Center, Philadelphia, Pennsylvania, $50,000 - continued support of the youth development model to engage LGBTQ youth of color who are HIV-positive or at high risk in leadership and career readiness programming; Big Bend Cares, Tallahassee, Florida, $40,000 - continued support for the Black MSM Testing Linkage and Adherence Program started in 2013.
*Contact:* Matt Blinstrubas, Director of Grants; (212) 219-0670; info@ejaf.org
*Internet:* newyork.ejaf.org/ejaf-grants-overview/
*Sponsor:* Elton John AIDS Foundation
584 Broadway, Suite 906
New York, NY 10012

**EMBO Courses and Workshops Grants** **1061**
The Courses and Workshops program, the European Molecular Biology Organization (EMBO) supports and encourages the sharing of scientific knowledge and skills across all career stages and disciplines in the life sciences. EMBO Courses and Workshops are organized by some of the leading scientists in their fields, and enjoy an excellent reputation throughout the world. EMBO funds over ninety events attracting more than 11,000 participants every year, making it the largest funder of scientific meetings in Europe. Funding is available for workshops, practical courses, global exchange lecture courses, EMBO/FEBS lecture courses, keynote lectures, and the India/EMBO symposia. There are two annual deadlines for applications to support workshops, practical courses, global exchange lecture courses: March 1 and August 1. The only deadline for EMBO/FEBS lecture courses is March 1, while February 15 and July 15 are the two deadlines for the India/EMBO symposia. In addition to providing financial support, EMBO assists organizers with the promotion of the meeting, as well as with the creation of a website with registration system and abstract submission facilities and the design of a poster. EMBO Workshops are scientific meetings that bring together scientists to present and discuss their latest discoveries. EMBO Practical Courses offer training in emerging techniques, taught by experts. Practical courses promote the transfer of new methodologies between laboratories.
*Geographic Focus:* All States, All Countries
*Date(s) Application is Due:* Feb 15; Mar 1; Jul 15; Aug 1
*Contact:* Gerlind Wallon, Program Manager; +49 6221 8891 112; fax +49 6221 8891 200; courses_workshops@embo.org or gerlind.wallon@embo.org
*Internet:* www.embo.org/funding-awards/courses-workshops
*Sponsor:* European Molecular Biology Organization
Meyerhofstrasse 1
Heidelberg, 69117 Germany

**EMBO Gold Medal** **1062**
The EMBO Gold Medal is awarded annually to young scientists for outstanding contributions to the life sciences in Europe. Young independent group leaders play a major role in building a strong research environment. EMBO places a focus on supporting the needs of young researchers at different stages of their careers. By awarding the EMBO Gold Medal, EMBO wishes to acknowledge and highlight some of the remarkable achievements of this group. EMBO Members are invited to nominate candidates by the February 1 annual deadline. The winner of the Gold Medal receives a hand-crafted medal and a bursary of 10,000 euros.
*Geographic Focus:* All States, All Countries
*Date(s) Application is Due:* Feb 1
*Amount of Grant:* 10,000 EUR
*Contact:* Gerlind Wallon, Program Manager; +49 6221 8891 112; fax +49 6221 8891 200; goldmedal@embo.org or gerlind.wallon@embo.org
*Internet:* www.embo.org/funding-awards/gold-medal
*Sponsor:* European Molecular Biology Organization
Meyerhofstrasse 1
Heidelberg, 69117 Germany

**EMBO Installation Grants** **1063**
The aim of this scheme is to strengthen science in participating EMBC member states. At present applications are accepted from Croatia, the Czech Republic, Estonia, Hungary, Portugal, Poland and Turkey. Additional member states have expressed an interest to join in the near future. Grants will help scientists to relocate, set up their labs and rapidly establish a reputation in the European scientific community. The program is entirely funded by the participating EMBC member states and successful applicants receive an annual support of Euro $50,000 for three to five years. Applicants can also apply for additional small grants of up to 10,000 euros per year. Each participating country sets an upper limit to the number of grants available per year. Application forms should be completed by the individual scientists and the receiving institute. The annual deadline for application submissions is April 16.
*Requirements:* Eligible scientists should: have an excellent publication record; have spent at least two consecutive years prior to the application deadline, outside the country in which they are applying to establish their lab; and be negotiating a full-time position at an institute/university in participating member states by the date of application. No age limit applies.
*Geographic Focus:* All States, Albania, Andorra, Armenia, Austria, Azerbaijan, Belarus, Belgium, Bosnia & Herzegovina, Bulgaria, Croatia, Cyprus, Czech Republic, Denmark, Estonia, Finland, France, Georgia, Germany, Greece, Hungary, Iceland, Ireland, Italy, Kosovo, Latvia, Liechtenstein, Lithuania, Luxembourg, Macedonia, Malta, Moldova, Monaco, Montenegro, Norway, Poland, Portugal, Romania, Russia, San Marino, Serbia, Slovakia, Slovenia, Spain, Sweden, Switzerland, The Netherlands, Turkey, Ukraine, United Kingdom, Vatican City
*Date(s) Application is Due:* Apr 16
*Amount of Grant:* 180,000 - 280,000 EUR
*Contact:* Gerlind Wallon, Deputy Director; +49 6221 8891 112; fax +49 6221 8891 200; gerlind.wallon@embo.org
*Internet:* www.embo.org/funding-awards/installation-grants
*Sponsor:* European Molecular Biology Organization
Meyerhofstrasse 1
Heidelberg, 69117 Germany

**EMBO Lecture Grants** **1064**
European Molecular Biology Organization (EMBO) meeting organizers can apply for lecture grants to support: keynote lectures by an EMBO Member at major international scientific meetings about their latest research; research lectures given by an EMBO Young Investigator at national or international meetings; a series of lectures at several institutions to enhance collaboration between researchers worldwide; non-scientific lectures at a scientific event addressing the policy implications of science and technology (such as discussions about ethics, economics, politics, and law); non-scientific lectures at an EMBO funded event addressing issues relating to women in science (including discussions about the reasons for the under-representation of women in the higher ranks of science or academia and reports on initiatives that redress that imbalance). There are three annual deadlines for application submissions to support keynote lectures by EMBO members and the EMBO global exchange lecture series, including February 1, June 1, and October 1. Meanwhile, applications are accepted throughout the year in support of EMBO young investigator lectures, science policy lectures, and women in science lectures.
*Geographic Focus* All States, District of Columbia, Guam, Marshall Islands, Northern Mariana Islands, Puerto Rico, U.S. Virgin Islands, All Countries, American Samoa
*Date(s) Application is Due:* Feb 1; Jun 1; Oct 1
*Contact:* Gerlind Wallon, Program Manager; +49 6221 8891 112; fax +49 6221 8891 200; gerlind.wallon@embo.org
*Internet:* www.embo.org/funding-awards/lecture-grants
*Sponsor:* European Molecular Biology Organization
Meyerhofstrasse 1
Heidelberg, 69117 Germany

**EMBO Long-Term Fellowships** **1065**
The European Molecular Biology Organization's Long-Term Fellowships are awarded for prolonged visits (twelve to twenty-four months) and are intended for advanced training through research. As the aim of the EMBO is to promote international research, it follows that mobility is a crucial element in deciding the eligibility of an application. All applications must involve a laboratory of origin or a receiving institute or applicant's nationality from one of the member states. Stipend rates depend on the country being visited, marital status, etc. Any complete applications received by the second Friday of February, 2:00 pm CET will be evaluated in the spring meeting, with interviews conducted in April. Applications received by the second Friday of August, 2:00 pm CEST, will be evaluated in the autumn meeting, with interviews in October.
*Requirements:* Applicants must hold a doctorate degree or equivalent before the start of the fellowship but not necessarily when applying. For clarity, it is deemed that this requirement is fulfilled only when the thesis has been stamped as submitted for final examination by the university. As a minimum, applicants must have at least one first author publication in press or published in an international peer reviewed journal at the time of application. All applications must involve movement between countries.
*Restrictions:* Applications are only considered from candidates with a maximum of 3 years postdoctoral experience. Applications to go to the country in which the PhD was obtained will not be considered. Applications to return to a laboratory where the candidate has worked in the past for more than three months, either in one or in several visits will not be considered. Applications for a Long-Term Fellowship to be held in the laboratory, institute or department in which the candidate worked to obtain his or her PhD, or to work again with the candidate's supervisor, no matter where the latter is located, will not be considered.
*Geographic Focus* All States, District of Columbia, Guam, Marshall Islands, Northern Mariana Islands, Puerto Rico, U.S. Virgin Islands, American Samoa, Germany
*Date(s) Application is Due:* Feb 8; Aug 10
*Contact:* David del Álamo Rodríguez, Program Manager; +49 6221 8891, ext. 122; fax +49 6221 8891 200; fellowships@embo.org or david.delalamo@embo.org
*Internet:* embo.org/funding-awards/fellowships/long-term-fellowships
*Sponsor:* European Molecular Biology Organization
Meyerhofstrasse 1
Heidelberg, 69117 Germany

**EMBO Short-Term Fellowships** **1066**
The European Molecular Biology Organization's short-term fellowships are established to advance molecular biology research by helping scientists to visit another laboratory with a view to applying a technique not available in the home laboratory. Such fellowships are not awarded for exchanges between two laboratories within the same country, and are intended for joint research work rather than consultations. The fellowships cover travel plus subsistence of the fellow only and not of any dependents. These fellowships are intended for visits of one week to three months duration (for non-European applicants the short-term fellowships are intended for a fixed duration of 90 days). EMBO Short-Term Fellowships are awarded for research exchanges between laboratories in EMBC Member States, Associated Member States or Cooperation Partners. Stipends vary widely from country to country. There is no deadline for applications, however it is recommended to apply three months before the proposed starting date.
*Requirements:* Applicants must be active researchers at any stage in their career with a minimum of one year of research experience at the PhD-student level.
*Restrictions:* All applications must involve either a laboratory of origin or a receiving institute from one of the EMBC member states. Short-Term Fellowships are not awarded

for exchanges between two laboratories within the same country. Applications from outside EMBC Member States, Associated Member States or Cooperation Partners will not be considered. Applications for the exclusive purpose of training in a technique rather than as a component of a wider research project, will not be considered for funding. EMBO does not consider applications for Short-Term Fellowships to prolong visits begun under other auspices, or as bridging fellowships between, or prior to, long term stays funded by EMBO or other organizations. Applications for fellowships to attend courses, workshops, or symposia will not be considered.
*Geographic Focus* All States, District of Columbia, Guam, Marshall Islands, Northern Mariana Islands, Puerto Rico, U.S. Virgin Islands, American Samoa, Germany
*Contact:* David del Álamo Rodríguez, Program Manager; +49 6221 8891, ext. 122; fax +49 6221 8891 200; fellowships@embo.org or david.delalamo@embo.org
*Internet:* embo.org/funding-awards/fellowships/short-term-fellowships
*Sponsor:* European Molecular Biology Organization
Meyerhofstrasse 1
Heidelberg, 69117 Germany

**EMBO Travel Grants** **1067**
A limited number of travel grants are available for eligible participants who are selected to attend EMBO workshops, EMBO practical courses, the EMBO/FEBS lecture courses, and EMBO/EMBL symposia. travel grants are allocated on the basis of the quality of the abstract submitted, with priority given to scientists from the countries listed below. Travel grants may be awarded to participants from anywhere only after qualified applicants from priority countries have been considered. All applicants have to justify the reasons for applying for a travel grant. Priority is given to participants of any nationality working in laboratories in Croatia, Czech Republic, Estonia, Greece, Hungary, Italy, Malta, Lithuania, Poland, Portugal, Slovakia, Slovenia, Spain, and Turkey, as well as from lower income economies. Specific travel grants are allocated for participants of any nationality working in laboratories in Chile, India, Singapore and Taiwan.
*Geographic Focus* All States, District of Columbia, Guam, Marshall Islands, Northern Mariana Islands, Puerto Rico, U.S. Virgin Islands, All Countries, American Samoa
*Contact:* Gerlind Wallon, Program Manager; +49 6221 8891 112; fax +49 6221 8891 200; courses_workshops@embo.org or gerlind.wallon@embo.org
*Internet:* www.embo.org/funding-awards/travel-grants
*Sponsor:* European Molecular Biology Organization
Meyerhofstrasse 1
Heidelberg, 69117 Germany

**EMBO Women in Science Award** **1068**
The Women in Science Award highlights major contributions by female scientists to life sciences research. Winners of the award are inspiring role models for future generations of women in science. The award is a joint initiative of the European Molecular Biology Organization (EMBO) and the Federation of European Biochemical Societies (FEBS). Each year the exceptional achievements of one woman working in the life sciences in Europe will be rewarded. The winner will receive 10,000 euro, a bronze statue and the opportunity to give a plenary lecture at the FEBS Congress. Nominees should be women scientists working in an EMBC Member State or FEBS member country, who have made outstanding contributions to life sciences research and significantly advanced our understanding of a particular discipline in the last five years. The award is generally not meant to be for lifetime achievements. Their research can cover any area of the life sciences including agricultural and biomedical research. The annual closing date for nominations is October 1.
*Geographic Focus:* Albania, Andorra, Armenia, Austria, Azerbaijan, Belarus, Belgium, Bosnia & Herzegovina, Bulgaria, Croatia, Cyprus, Czech Republic, Denmark, Estonia, Finland, France, Germany, Greece, Hungary, Iceland, Ireland, Italy, Kazakhstan, Kosovo, Latvia, Liechtenstein, Lithuania, Luxembourg, Macedonia, Malta, Moldova, Monaco, Montenegro, Netherlands, Norway, Poland, Portugal, Romania, Russia, San Marino, Serbia, Slovakia, Slovenia, Spain, Sweden, Switzerland, Turkey, Ukraine, United Kingdom, Vatican City
*Date(s) Application is Due:* Oct 1
*Amount of Grant:* 10,000 EUR
*Contact:* Gerlind Wallon, Program Manager; +49 6221 8891 112; fax +49 6221 8891 200; women@embo.org or gerlind.wallon@embo.org
*Internet:* www.embo.org/funding-awards/women-in-science-award
*Sponsor:* European Molecular Biology Organization
Meyerhofstrasse 1
Heidelberg, 69117 Germany

**EMBO Young Investigator Awards** **1069**
Through the Young Investigator Program, the European Molecular Biology Organization (EMBO) identifies and supports some of the best young researchers in the life sciences. EMBO Young Investigators are group leaders in the early stages of setting up an independent laboratory in EMBC Member or Associate Member States. They receive financial and practical support for a period of three years to help them develop skills and connections that will help them during this career stage. The Young Investigator Program was started in 2000. EMBO Young Investigators are selected annually and the 73 current members, together with 384 former members of the program and the recipients of EMBO Installation Grants, form a network of Europe's next generation of leading life scientists. EMBO Young Investigators receive a financial award of 15,000 euros in their second year. They can also apply for additional small grants of up to 10,000 euros per year. The annual deadline for application submissions is April 3.
*Requirements:* Candidates applying to become a Young Investigator must: have been an independent group leader for at least one year at the pre-application deadline and for less than four years on January 1 in the year of application; have an excellent track record; carry out research in the life sciences in an EMBC Member State, an EMBC Associate Member State (currently India and Singapore) or a in countries/territories covered by a co-operation agreement (currently Taiwan and Chile); have obtained sufficient funding to run their laboratory; have published at least one last author research paper in an international peer reviewed journal from independent work carried out in their own laboratory; be 40 years old or younger on January 1 in the year of application, unless exceptional circumstances have significantly delayed career progression; and be able to attend an interview in Heidelberg, Germany, at the end of October.
*Geographic Focus* All States, District of Columbia, Guam, Marshall Islands, Northern Mariana Islands, Puerto Rico, U.S. Virgin Islands, All Countries, American Samoa
*Date(s) Application is Due:* Apr 3
*Amount of Grant:* 15,000 - 25,000 EUR
*Contact:* Gerlind Wallon, Deputy Director; +49 6221 8891 112; fax +49 6221 8891 200; yip@embo.org or gerlind.wallon@embo.org
*Internet:* www.embo.org/funding-awards/young-investigators/apply
*Sponsor:* European Molecular Biology Organization
Meyerhofstrasse 1
Heidelberg, 69117 Germany

**Emil Buehler Perpetual Trust Grants** **1070**
The Emil Buehler Perpetual Trust was established in 1984 pursuant to Emil Buehler's last will and testament, in which he described his lifelong interest in aeronautical science. Buehler was an aviation visionary, architect, and engineer. The majesty and mystery of flight was more than a private interest; it was a vision to be shared during his lifetime and beyond. This same vision is perpetuated today through the continuing philanthropy of the Buehler Trust. The Trust provides funding to institutions, organizations, and foundations that carry forward the research and development of new and improved aircraft and promote the field of aviation science and technology. Grants are occasionally made for specific capital purchases if a clear need is demonstrated. Priority is given to cost-effective programs that offer the broadest benefit to the community. The Board of Trustees meets monthly to make grant decisions. Applications must be submitted by March 1 for an April meeting, June 1 for a July meeting, September 1 for an October meeting, or December 1 for a January meeting.
*Requirements:* Grants are only made to organizations qualifying for tax exemption under Section 501(c)3 of the Internal Revenue Code. Applications must be sumbitted through the online grant application form or alternative accessible application designed for assistive technology users.
*Restrictions:* The Trust does not provide grants for: general operating support; endowments; individuals; political activities; loans; or fund-raising events including dinners, benefits and athletic events.
*Geographic Focus:* All States
*Date(s) Application is Due:* Mar 1; Jun 1; Sep 1; Dec 1
*Contact:* George Weaver, Special Trustee; (888) 234-1999; fax (877) 746-5889; grantadministration@ wellsfargo.com
*Internet:* https://www.wellsfargo.com/private-foundations/buehler-perpetual-trust
*Sponsor:* Emil Buehler Perpetual Trust
1740 Broadway
Denver, CO 80274

**Emma Barnsley Foundation Grants** **1071**
Emma Elizabeth Barnsley was born on December 10, 1926, in Crane County, Texas, and grew up in the areas of Crane, Midland, and Odessa, Texas. She moved to New York City as a young woman where she spent most of her adult life. Barnsley had a deep love for animals of all kinds. It is because of this passion she established the Emma Barnsley Foundation for the purposes of the prevention of cruelty to animals and the study, care, protection, and preservation of animals, both domestic and wild, and their environment. The Foundation awards grants exclusively for charitable, scientific, and educational purposes as follows: to aid and assist in the prevention of cruelty to animals, both domestic and wild; to aid and assist in the care, protection and preservation of animals, both domestic and wild, and the environment in which such wild animals may live; to provide scholarships for deserving young men and women to assist them in the study of veterinary medicine at any institution of higher learning in the state of Texas; to assist institutions of higher learning in the state of Texas in the operation and maintenance of courses of study in veterinary medicine and in the study of animals and animal life; and to assist in research projects involving ecological, environmental, and wildlife studies, provided no funds shall be given to any organization utilizing animals for research or experimental purposes if such animals are in any way harmed, mutilated, intentionally inflicted with disease or any substance causing disease, or if such animals are killed in the course of such research. Preference may be given to organizations that operate in the Crane/Midland/Odessa, Texas area or the Manhattan, New York area. Applications are accepted year-round. Applications must be submitted by August 31 to be reviewed at the annual grant meeting that occurs each September. The typical grant ranges from $10,000 to $20,000, and approximately nine awards are provided annually.
*Requirements:* To be eligible, organizations must qualify as exempt organizations under Section 501(c)3 of the Internal Revenue Code. Applications must be submitted through the online grant application form or alternative accessible application designed for assistive technology users.
*Restrictions:* The foundation normally does not fund requests for: general operating support; endowments; individuals; political activities; loans; or fundraising events including dinners, benefits, and athletic events.
*Geographic Focus:* Colorado, New York
*Date(s) Application is Due:* Aug 31
*Amount of Grant:* 10,000 - 20,000 USD

*Contact:* Private Client Services; (720) 947-6725 or (888) 234-1999; fax (877) 746-5889; grantadministration@wellsfargo.com
*Internet:* https://www.wellsfargo.com/privatefoundationgrants/barnsley
*Sponsor:* Emma Barnsley Foundation
1740 Broadway
Denver, CO 80274

**Energy Foundation Buildings Grants** 1072

The Energy Foundation's mission is to assist the United States and China in the transition to a sustainable energy future by promoting energy efficiency and renewable energy. To this end, grants in the buildings sector hope to: promote the adoption of increasingly stringent building codes, with effective local enforcement; establish stringent appliance standards and complementary consumer labels; create programs to benchmark and rate the energy consumed by buildings and encourage retrofitting the large stock of existing buildings; and develop financial incentives to buy and lease energy-efficient buildings. The Foundation has no fixed format for proposals. Applicants are urged to use the form that best conveys the strengths of their project.
*Requirements:* The foundation makes grants to nonprofit 501(c)3 charitable organizations. In general, a complete proposal includes the following: the attached application form as a cover sheet; a clear statement of the need(s) or problem(s) to be addressed; strategy; timeline; results you expect from your project; project budget—including a brief explanation of the budget, a list of other sources of actual and potential funding for the project, and a description of plans to secure additional funding; how you will determine whether your project is successful; and a history of organization, including mission and goals.
*Restrictions:* The Foundation is unable to fund: grants directly to individuals; local projects, unless they have been consciously designed for further replication or have broad regional or national implications; candidates for political office, to influence legislation, or to support sectarian or religious purposes; the research and development of technology (e.g., funds to develop hybrid automobiles or commercialization of an invention); demonstration projects (e.g., model solar homes); community energy projects; endowments or debt reduction; annual fund-raising campaigns or capital construction; the planning, renovation, maintenance, retrofit, or purchase of buildings; the purchase of equipment; or the acquisition of land, even if the intent is to save energy; or unsolicited general operating support.
*Geographic Focus:* All States
*Amount of Grant:* 2,000 - 850,000 USD
*Samples:* Mayor's Fund to Advance New York City, New York, New York, $25,000 - to support the implementation of advanced building energy codes in New York City; Global Cool Cities Alliance, San Francisco, California, $54,000 - to support the Global Cool Cities Alliance in its first few months under a permanent executive director; Center for Neighborhood Technology, Chicago, Illinois, $40,000 - to host a meeting on GreenMLS.
*Contact:* Todd Foland; (415) 561-6700; fax (415) 561-6709; energyfund@ef.org
*Internet:* www.ef.org/programs.cfm?program=buildings
*Sponsor:* Energy Foundation
301 Battery Street, 5th Floor
San Francisco, 94111

**Energy Foundation Climate Grants** 1073

The Energy Foundation's mission is to assist the United States and China in the transition to a sustainable energy future by promoting energy efficiency and renewable energy. To this end, grants in the climate sector support work in the following areas: building support for climate protection among diverse allies, such as business associations, veterans' groups, religious organizations, and agricultural interests, among others; policy analysis (such as assessing the effects of climate policy on consumers, business, jobs, and Gross Domestic Product; modeling state-by-state impacts; or evaluating policy design options on issues such as offsets, allowance allocation, or integrating carbon controls with complementary policies for energy efficiency and renewable energy); targeted efforts to educate opinion leaders and stakeholders—nationally or in key states-about climate change and climate policy; supporting implementation of state and regional climate laws and agreements (such as California's AB 32, the Regional Greenhouse Gas Initiative in the Northeast, and the Western Climate Initiative); and improving understanding of the role of the U.S. in an international climate agreement. The Foundation has no fixed deadlines or format for proposals. Applicants are urged to use the form that best conveys the strengths of their project.
*Requirements:* The foundation makes grants to nonprofit 501(c)3 charitable organizations. In general, a complete proposal includes the following: the attached application form as a cover sheet; a clear statement of the need(s) or problem(s) to be addressed; strategy; timeline; results you expect from your project; project budget—including a brief explanation of the budget, a list of other sources of actual and potential funding for the project, and a description of plans to secure additional funding; how you will determine whether your project is successful; and a history of organization, including mission and goals.
*Restrictions:* The Foundation is unable to fund: grants directly to individuals; local projects, unless they have been consciously designed for further replication or have broad regional or national implications; candidates for political office, to influence legislation, or to support sectarian or religious purposes; the research and development of technology (e.g., funds to develop hybrid automobiles or commercialization of an invention); demonstration projects (e.g., model solar homes); community energy projects; endowments or debt reduction; annual fund-raising campaigns or capital construction; the planning, renovation, maintenance, retrofit, or purchase of buildings; the purchase of equipment; or the acquisition of land, even if the intent is to save energy; or unsolicited general operating support.
*Geographic Focus:* All States
*Amount of Grant:* 2,000 - 850,000 USD
*Samples:* New Policy Institute, Washington, D.C., $88,000 - to develop effective messaging and polling around clean energy programs; Southern Alliance for Clean Energy, Knoxville, Tennessee, $160,000 - to support outreach in the Southeast on climate policies and solutions; Resources Legacy Fund, Sacramento, California, $250,000 - to support the Californians for Clean Energy and Jobs Network.
*Contact:* Todd Foland, Grants Administrator; (415) 561-6700; fax (415) 561-6709; energyfund@ef.org
*Internet:* www.ef.org/programs.cfm?program=climate
*Sponsor:* Energy Foundation
301 Battery Street, 5th Floor
San Francisco, 94111

**Energy Foundation Power Grants** 1074

The Energy Foundation's mission is to assist the United States and China in the transition to a sustainable energy future by promoting energy efficiency and renewable energy. To this end, grants in the power sector support work in the following areas: policies that yield large-scale purchases of renewable energy; policies that open the electric system to renewable power and associated transmission while supporting resource conservation; policies that yield substantial investments to improve energy efficiency by consumers of electric power and natural gas; policies that remove market and regulatory barriers to renewables, efficiency, and clean distributed generation; resistance to permitting and regulatory approvals for new conventional coal-fired power plants; campaigns to reduce greenhouse gases by directing new utility investments toward clean energy resources; to secure stricter performance standards for emissions from new and existing coal plants, including greenhouse gases; and regulatory controls for carbon capture and sequestration systems. The Foundation has no fixed format for proposals. Applicants are urged to use the form that best conveys the strengths of their project.
*Requirements:* The foundation makes grants to nonprofit 501(c)3 charitable organizations. In general, a complete proposal includes the following: the attached application form as a cover sheet; a clear statement of the need(s) or problem(s) to be addressed; strategy; timeline; results you expect from your project; project budget—including a brief explanation of the budget, a list of other sources of actual and potential funding for the project, and a description of plans to secure additional funding; how you will determine whether your project is successful; and a history of organization, including mission and goals.
*Restrictions:* The Foundation is unable to fund: grants directly to individuals; local projects, unless they have been consciously designed for further replication or have broad regional or national implications; candidates for political office, to influence legislation, or to support sectarian or religious purposes; the research and development of technology (e.g., funds to develop hybrid automobiles or commercialization of an invention); demonstration projects (e.g., model solar homes); community energy projects; endowments or debt reduction; annual fund-raising campaigns or capital construction; the planning, renovation, maintenance, retrofit, or purchase of buildings; the purchase of equipment; or the acquisition of land, even if the intent is to save energy; or unsolicited general operating support.
*Geographic Focus:* All States
*Amount of Grant:* 2,000 - 850,000 USD
*Samples:* Little River Bottoms Coalition, Dallas, Texas, $50,000 - to oppose the construction of new coal-fired power plants in Arkansas; Environment Northeast, Rockport, Maine, $175,000 - to advance energy efficiency programs in the Northeast; Georgia Watch, Atlanta, Georgia, $150,000 - to support work to decrease reliance on coal-fired power and increase energy efficiency services in Georgia.
*Contact:* Todd Foland, Grants Administrator; (415) 561-6700; fax (415) 561-6709; energyfund@ef.org
*Internet:* www.ef.org/programs.cfm?program=power
*Sponsor:* Energy Foundation
301 Battery Street, 5th Floor
San Francisco, 94111

**Energy Foundation Transportation Grants** 1075

The Energy Foundation's mission is to assist the United States and China in the transition to a sustainable energy future by promoting energy efficiency and renewable energy. To this end, the transportation sector makes grants to promote innovative federal and state policies that reduce global warming pollution from vehicles, encourage low-carbon fuels, and discourage high-carbon fuels. The Foundation focuses on performance-based standards and incentives that guide markets toward cleaner solutions. The Foundation has no fixed deadlines or format for proposals. Applicants are urged to use the form that best conveys the strengths of their project.
*Requirements:* The foundation makes grants to nonprofit 501(c)3 charitable organizations. In general, a complete proposal includes the following: the attached application form as a cover sheet; a clear statement of the need(s) or problem(s) to be addressed; strategy; timeline; results you expect from your project; project budget-including a brief explanation of the budget, a list of other sources of actual and potential funding for the project, and a description of plans to secure additional funding; how you will determine whether your project is successful; and a history of organization, including mission and goals.
*Restrictions:* The Foundation is unable to fund: grants directly to individuals; local projects, unless they have been consciously designed for further replication or have broad regional or national implications; candidates for political office, to influence legislation, or to support sectarian or religious purposes; the research and development of technology (e.g., funds to develop hybrid automobiles or commercialization of an invention); demonstration projects (e.g., model solar homes); community energy projects; endowments or debt reduction; annual fund-raising campaigns or capital construction; the planning, renovation, maintenance, retrofit, or purchase of buildings; the purchase of equipment; or the acquisition of land, even if the intent is to save energy; or unsolicited general operating support.

*Geographic Focus:* All States
*Amount of Grant:* 2,000 - 850,000 USD
*Samples:* Environment Northeast, Rockport, Maine, $125,000 - to support the development of a low carbon fuels standard in the Northeast; Center for Auto Safety, Washington, D.C., $125,000 - to support policymaker education and media outreach on federal and California vehicle standards; Izaak Walton League of America, Gaitherburg, Maryland, $50,000 - to advance low carbon fuels in Minnesota and more broadly in the Midwest.
*Contact:* Todd Foland; (415) 561-6700; fax (415) 561-6709; energyfund@ef.org
*Internet:* www.ef.org/programs.cfm?program=transportation
*Sponsor:* Energy Foundation
301 Battery Street, 5th Floor
San Francisco, 94111

**Enterprise Community Partners Rose Architectural Fellowships** **1076**
The Fellowship honors the late Frederick P. Rose, a prominent developer and philanthropist who believed strongly in the value of quality design and the spirit of public service. Under the program's innovative structure, the Enterprise Rose Architectural Fellows work deeply in communities for three years, forging community ties, developing leadership skills, and expanding the capacity of their local host organizations to execute projects with the best possible designs, meaningful community engagement, and the most advanced green building features. Fellows and host organizations promote the principles and best practices of community architecture in a variety of venues, including teaching, conference presentations, and publications. They have initiated the adoption of improved design standards and best practices by their state and local governments and industry organizations. The annual stipend paid to fellows in most communities is $45,000 ($52,000 in high-cost cities, including Los Angeles) an amount that will remain the same for the three years of the fellowship. High cost area may merit a salary adjustment. Adjustments, if any, are described in the application kit for a specific fellowship.
*Requirements:* Applicants must have a Social Security number or a United States Tax Identification number and must possess a professional degree in architecture from an accredited college or university, or expect to graduate from an accredited school of architecture before the fellowship begins. A professional graduate degree in architecture is preferred. The expected minimum grade point average is "B" or 3.0 on a 4.0 scale for all undergraduate and graduate work. Preference will be given to candidates who have approximately three years professional experience, have completed their Intern Development Program (IDP) requirements, and can work independently within a non-profit environment. However, recent graduates will also be considered, as will candidates who have not yet completed their IDP requirements. Fellows must make a full-time commitment of three years to the fellowship and to the host organization. Applicants with diverse educational, professional, and personal backgrounds, and individuals from disadvantaged communities, as well as minorities are encouraged to apply.
*Restrictions:* A Fellow may not have any other paid employment or independent consulting work, unless prior written permission is given by the director of the fellowship and an authorized representative of the partner organization. Individuals currently employed by their partner organization are not eligible to apply.
*Geographic Focus:* All States
*Date(s) Application is Due:* May 1
*Amount of Grant:* 45,000 - 52,000 USD
*Samples:* Class 2012-2014: Mark Matel, Boston, Massachusetts; Sam Carlsen, St. Paul, Minnesota; Seam Beall, Burlington, Vermont; Nathan Poel, Yakima, Washington; Ceara O'Leary, Detroit, Michigan.
*Contact:* Trinity Simons, Program Officer; (781) 235-2006 or (781) 235-4011; tsimons@enterprisecommunity.org or rosefellowship@enterprisecommunity.org
*Internet:* www.enterprisecommunity.com/solutions-and-innovation/design-leadership/rose-architectural-fellowship
*Sponsor:* Enterprise Community Partners
70 Corporate Center
Columbia, MD 21044

**Enterprise Community Partners Terwilliger Fellowship** **1077**
This fellowship is named for Enterprise Trustee J. Ronald Terwilliger, chairman and CEO of Trammel Crow Residential. This one-year Fellowship plays a leadership role in launching the new Enterprise Innovation Unit, which has primary responsibility for innovation: research, development and planning of potential new Enterprise programs, products and partnerships. The fellow works with the Enterprise executive management team and key partners to develop new solutions to advance Enterprise's mission.
*Geographic Focus:* All States
*Contact:* Dave Epley, Program Director; (410) 772-2411; fax (410) 964-1918; depley@enterprisecommunity.org
*Internet:* www.enterprisecommunity.com/about/fellowships-and-awards
*Sponsor:* Enterprise Community Partners
70 Corporate Center
Columbia, MD 21044

**EPA Brownfields Area-Wide Planning Grants** **1078**
Brownfields Area-Wide Planning is an Environmental Protection Agency grant program which provides funding to recipients to conduct research, technical assistance and training that will result in an area-wide plan and implementation strategy for key brownfield sites, which will help inform the assessment, cleanup and reuse of brownfields properties and promote area-wide revitalization. Funding is directed to specific areas, such as a neighborhood, downtown district, local commercial corridor, or city block, affected by a single large or multiple brownfield sites. An estimated twenty awards are given annually, with the application deadline set as August 10.
*Requirements:* Eligible governmental entities include a general purpose local unit of government; a land clearance authority or other quasi-governmental entity that operates under the supervision and control of, or as an agent of, a general purpose unit of government; a governmental entity created by a state legislature; a regional council or group of general purpose units of local government; a redevelopment agency that is chartered or otherwise sanctioned by a state; a state; an Indian Tribe (other than in Alaska), or an Alaskan Native Regional Corporation and an Alaska Native Village Corporation as those terms are defined in the Alaska Native Claims Settlement Act (43 U.S.C. 1601 and following); and the Metlakatla Indian Community. Eligible nonprofit organizations include any corporation, trust, association, cooperative, or other organization that is operated mainly for scientific, educational, service, charitable, or similar purpose in the public interest; is not organized primarily for profit; and uses net proceeds to maintain, improve, or expand the operation of the organization.
*Geographic Focus* All States, District of Columbia, Guam, Marshall Islands, Northern Mariana Islands, Puerto Rico, U.S. Virgin Islands, American Samoa
*Date(s) Application is Due:* Aug 10
*Amount of Grant:* Up to 200,000 USD
*Contact:* Aimee Storm, Program Officer; (202) 566-0633 or (202) 566-2777; fax (202) 566-2757; storm.aimee@epa.gov
Wendy Thomi, Program Officer; (202) 566-1462 or (202) 566-2777; fax (202) 566-2757; thomi.wendy@epa.gov
*Internet:* https://www.epa.gov/brownfields/types-brownfields-grant-funding#tab-5
*Sponsor:* Environmental Protection Agency
1200 Pennsylvania Avenue, NW
Washington, D.C. 20460

**EPA Children's Health Protection Grants** **1079**
The objectives of this program are to catalyze community-based and regional projects and other actions that enhance public outreach and communication; assist families in evaluating risks to children and in making informed consumer choices; build partnerships that increase a community's long-term capacity to advance the protection of children's environmental health and safety; leverage private and public investments to enhance environmental quality by enabling community efforts to continue past EPA's ability to provide assistance to communities; and promote protection of children from environmental threats through lessons learned. There are no deadline dates.
*Requirements:* Eligible applicants include community groups, public nonprofit institutions/organizations, tribal governments, specialized groups, profit organizations, private nonprofit institutions/organizations, and municipal and local governments. Potential applicants are strongly encouraged to discuss proposed projects with or submit preapplications to program staff prior to the completion of a full proposal.
*Geographic Focus:* All States
*Amount of Grant:* 5,000 - 250,000 USD
*Contact:* Office of Children's Health Protection; (202) 564-2188; fax (202) 564-2733; fletcher.bettina@epa.gov
*Internet:* yosemite.epa.gov/ochp/ochpweb.nsf/content/grants.htm
*Sponsor:* Environmental Protection Agency
1200 Pennsylvania Avenue, NW
Washington, D.C. 20460

**EPA Compliance Assistance Support for Services to the Regulated Community and Other Assistance Providers** **1080**
The EPA has sponsored partnerships with industry, academic institutions, environmental groups, and other agencies to launch sector-specific Compliance Assistance Centers. Each Compliance Assistance Center addresses real world issues in language that is used by the regulated entities. This program's objective is to provide financial assistance to private nonprofit institutions, universities, and public agencies to develop and maintain Compliance Assistance Centers to improve environmental compliance within an identified industrial/government sector or environmental topic. It is expected that common communications channels used by the industry/government will continue to quickly distribute information to the targeted audience. The sector entry pages will serve as portals to relevant information developed by the Centers and other organizations as appropriate (e.g. EPA). Individual sector entry pages will use navigation structures specific to the sector thereby providing easy access to relevant environmental content. EPA funding must be used only for activities specified in the approved scope of work; discretionary funding will not be provided. EPA generally funds grants and cooperative agreements on a 12-month basis. The grant range is $5,000 to $35,000 per year, with an average of $10,000 annually.
*Requirements:* Applicants must be nonprofit organizations. Colleges, universities, and community colleges are also eligible to apply.
*Restrictions:* EPA's Appropriation Act requires that applicants submitting unsolicited research grant proposals share in the cost of conducting research. The amount of the cost share will be based on the mutuality of interest between the Government and the applicant.
*Geographic Focus:* All States
*Amount of Grant:* 5,000 - 35,000 USD
*Contact:* Tracy Back, National Center for Environmental Research and Quality Assurance; (202) 564-7076; fax (202) 564-0037; back.tracy@epa.gov
*Internet:* https://www.cfda.gov/?s=program&mode=form&tab=step1&id=556b71385864efe15d4605c6ab8f0a72
*Sponsor:* Environmental Protection Agency
1200 Pennsylvania Avenue, NW
Washington, D.C. 20460

**EPA Pollution Prevention Grants** 1081

The EPA Pollution Prevention Grant program applies to the Pollution Prevention (P2) and the Pollution Prevention Information Network (PPIN) grant programs. Both programs were enacted under the Pollution Prevention Act of 1990. The P2 grant program funds grants/cooperative agreements that implement pollution prevention technical assistance services for businesses. The PPIN grant program funds grants/cooperative agreements through a national network of Pollution Prevention Resource Exchange (P2Rx) Centers. The Centers work to deliver pollution prevention information, resources and tools to state technical assistance providers, businesses and industry sectors. Under both grant programs, EPA is authorized to support projects that utilize pollution prevention techniques to reduce and/or eliminate pollution from air, water and/or land. The objectives include: helping business and industry incorporate P2 techniques into their day-to-day business operations as a means of reducing or eliminating harmful pollutants from entering the air, water or land; to encourage education and training in P2/source reduction techniques and applications; and support of P2 information sharing among State and local technical assistance programs and businesses.

*Requirements:* Eligible applicants for purposes of funding under this grant program include the 50 states, the District of Columbia, the U.S. Virgin Islands, the Commonwealth of Puerto Rico, and any territory or possession of the United States, any state agency or instrumentality of a state including state colleges and universities, federally-recognized tribes and intertribal consortia. Organizations receiving grant funds through the P2 and PPIN grant programs are required to match federal funds by at least 50 percent.

*Restrictions:* Local governments, private universities, private nonprofit entities, private businesses, and individuals are not eligible to apply.

*Geographic Focus* All States, District of Columbia, Guam, Marshall Islands, Northern Mariana Islands, Puerto Rico, U.S. Virgin Islands, American Samoa

*Amount of Grant:* 20,500 - 115,000 USD

*Contact:* Lee Fiske; (617) 918-1847; fax (617) 918-1810; fiske.lee@epa.gov

*Internet:* https://www.cfda.gov/?s=program&mode=form&tab=step1&id=53dc0b42e4cb304e9a54a7ad59905a84

*Sponsor:* Environmental Protection Agency
1200 Pennsylvania Avenue, NW, Mail Code 7101-M
Washington, D.C. 20460-7101

**EPA Pollution Prevention Grants** 1082

The Environmental Protection Agency's Pollution Prevention program funds grants/cooperative agreements that implement pollution prevention technical assistance services and/or training for businesses and support projects that utilize pollution prevention techniques to reduce and/or eliminate pollution from air, water and/or land. Proposals should address one of the following regulatory criteria to merit further review; make specific technical assistance available to businesses seeking information about source reduction opportunities, including funding for experts to provide on-site technical advice to businesses and to assist in the development of source reduction plans; target assistance to businesses for whom lack of information is an impediment to source reduction; or provide training in source reduction techniques. Approximately forty awards are given each year, ranging from $40,000 to $360,000. The annual deadline for applications is May 9.

*Requirements:* Grantees must contribute matching resources of at least 50 percent. The match may include cash, in-kind goods and services, and third-party contributions. State governments, state programs or departments as well as other state institutions, such as universities as well as all federally recognized Native American Tribes are eligible to apply, along with the Virgin Islands, Puerto Rico, and U.S. territories.

*Geographic Focus* All States, District of Columbia, Guam, Marshall Islands, Northern Mariana Islands, Puerto Rico, U.S. Virgin Islands, American Samoa

*Date(s) Application is Due:* May 9

*Amount of Grant:* 40,000 - 360,000 USD

*Contact:* James Jones; (202) 564-2902; fax (202) 564-0801; jones.james@epa.gov

*Internet:* www.epa.gov/p2/grants/ppis/ppis.htm

*Sponsor:* Environmental Protection Agency
1200 Pennsylvania Avenue, NW, Mail Code 7101-M
Washington, D.C. 20460-7101

**EPA Regional Healthy Indoor Air Projects for Community Outreach and Education** 1083

The objective of this program is to support surveys, studies, research, training, outreach, education, investigations or demonstrations performed by organizations that lead to effective outreach strategies to educate key audiences about indoor air pollutants, their associated health risks and encourage effective mitigation and control strategies. Currently, the Program focuses on several critical aspects of indoor air quality that pose significant risks to public health, and in particular, to children and other disproportionately impacted populations. Program focus areas include, but are not limited to reducing the exposure of children and others with asthma to indoor triggers that worsen their health condition, promoting the adoption of operation and maintenance practices in schools, homes and other buildings throughout the nation to reduce the harmful effects of poor indoor air quality on the health of the occupants, promoting voluntary radon testing by homeowners and building operators to identify elevated levels and fix them when they are found, as well as working with homebuilders to incorporate radon resistant and improved indoor air quality construction features into new and remodeled homes, and encouraging adult smokers to protect their children from the adverse health effects of environmental exposure to secondhand smoke by making a conscious decision to keep their homes and cars smokefree. The EPA generally awards grants ranging in value from $5,000 to $750,000 per fiscal year. The average value of each grant is $50,000 per fiscal year.

*Requirements:* Assistance under this program is generally available to States, local governments, territories, Indian Tribes, and possessions of the U.S., including the District of Columbia, international organizations, public and private universities and colleges, hospitals, laboratories, other public or private nonprofit institutions, which submit applications proposing projects with significant technical merit and relevance to EPA's Office of Air and Radiation's mission.

*Geographic Focus* All States, District of Columbia, Guam, Marshall Islands, Northern Mariana Islands, Puerto Rico, U.S. Virgin Islands, American Samoa

*Amount of Grant:* 5,000 - 750,000 USD

*Contact:* Victoria Drew, (202) 343-9049; drew.victoria@epa.gov

*Internet:* https://www.cfda.gov/?s=program&mode=form&tab=step1&id=d2de8c076d885a787769417ed4813ffe

*Sponsor:* Environmental Protection Agency - 1
Ariel Rios North Building
Washington, D.C. 20460

**EPA Research, Development, Monitoring, Public Education, Training, Demonstrations, and Studies Grants** 1084

Grants are awarded to support Research, Development, Monitoring, Public Education, Training, Aid, Demonstrations, and Studies assistance relating to the protection of public health and the environment from pesticides and potential risk from toxic substances. Grants and cooperative agreements are available to support recipients' allowable direct costs incident to approved scopes of work plus allowable indirect costs, in accordance with established EPA policies and regulations. Funding awarded for research does not include research within the purview of EPA's Office of Research and Development. Typically, grants in this category range from $1,000 to $1,500,000, with the average being $500,000.

*Requirements:* Assistance under this program is generally available to States, U.S. territories or possessions, federally recognized Indian tribal governments and Native American Organizations, public and private universities and colleges, hospitals, laboratories, other public or private nonprofit institutions, local governments, and individuals and international entities.

*Restrictions:* For profit organizations are not eligible.

*Geographic Focus:* All States

*Amount of Grant:* 1,000 - 1,500,000 USD

*Contact:* Sharon Clark, (202) 564- 3810; fax (202) 564-0575; clark.sharon@epa.gov

*Internet:* https://www.cfda.gov/?s=program&mode=form&tab=step1&id=014d33dd7fea130ebab8d1dca2d45280

*Sponsor:* Environmental Protection Agency
1200 Pennsylvania Avenue, NW, Mail Code 7101-M
Washington, D.C. 20460-7101

**EPA Research Training Program for College and University Students** 1085

The Environmental Protection Agency (EPA) seeks applications from eligible entities to enter into cooperative agreements that will provide training opportunities for undergraduate and graduate students on-site at EPA-ORD research facilities located in Cincinnati, Ohio. It is envisioned that the training program will increase both the effectiveness and number of future environmental scientists. The training received under the mentorship of EPA scientists will complement the trainees' academic coursework. The recipient will be responsible for ensuring that the training projects are supportive of the trainees' academic training. Some examples of the fields of study for desirable trainees include, but are not limited to, those majoring in chemistry, engineering, biology, computer science, physical science, life science, ecology, and urban planning. It is expected that two cooperative agreements will be awarded, ranging up to $1,250,000 each. The current closing date for applications is August 16.

*Requirements:* This solicitation is available to each state, territory and possession, and Tribal nation of the U.S., including the District of Columbia, public and private universities and colleges, hospitals, laboratories, state and local government departments, other public or private nonprofit institutions, and foreign entities.

*Restrictions:* Profit-making firms are not eligible to receive assistance agreements from the EPA under this program.

*Geographic Focus* All States, District of Columbia, Guam, Marshall Islands, Northern Mariana Islands, Puerto Rico, U.S. Virgin Islands, American Samoa

*Date(s) Application is Due:* Aug 16

*Amount of Grant:* Up to 1,250,000 USD

*Contact:* Debra M. Jones, Program Officer; (202) 564-7839; jones.debram@epa.gov

*Internet:* https://www.epa.gov/research-grants/research-training-program-college-and-university-students

*Sponsor:* Environmental Protection Agency
109 Alexander Drive
Durham, NC 27711

**EPA Southeastern Multi-Media and Geographic Priority Project Grants** 1086

The objective of the Southeastern Multi-Media and Geographic Priority Projects is to fund projects, including surveys, studies, investigations, demonstrations, training, outreach, human health and the environment to: prevent, reduce and eliminate pollution through the promotion of environmental sustainability; build capacity on environmental sustainability and public health issues; support the development and implementation of sustainability strategies and enhance environmental literacy; and promote coordination and collaboration between and among federal, state, and local governmental and non-governmental organizations as well as build environmental education capacity at institutions of higher education. The projects should be centered around providing a forum for academic advisors and sustainability coordinators to collaborate with industry, college and graduate level students to establish and broaden environmental

learning opportunities that promote capacity building on environmental sustainability and public health issues, including the creation of hands-on instructional learning of sustainable practices including pollution prevention, water and energy conservation, watershed protection, and environmental stewardship. The project should also serve to foster an environmental literacy plan to ensure that students have an understanding of environmental sustainability and how it interconnects with the green economy. Contact the headquarters or regional office, as appropriate, for application deadlines. The funding range is $15,000 to $100,000, with an average of approximately $90,000 per project.
*Requirements:* Projects conducted within the geographic boundaries of EPA's Region 4, which comprises the states and tribal lands of Alabama, Georgia, Florida, Kentucky, Mississippi, North Carolina, South Carolina and Tennessee, are eligible for funding when it is available. Interstate, state, tribal, intrastate, and local government agencies, districts, and councils; and public and private universities and colleges, laboratories, and research facilities; and public or private non-governmental, non-profit institutions are eligible to apply, unless restricted by the authorizing statutes. Non-profit organizations must have documentation of non-profit status from the U.S. Internal Revenue Service or their state of incorporation, except that non-profits organizations as defined in Section 501(c)4 of the Internal Revenue Code that engage in lobbying as defined in Section 3 of the Lobbying Disclosure Act 1995 or superseding legislation are ineligible.
*Geographic Focus:* Alabama, Florida, Georgia, Kentucky, Mississippi, North Carolina, South Carolina, Tennessee
*Amount of Grant:* 15,000 - 100,000 USD
*Contact:* Caroline Robinson, Senior Advisor; (404) 562-8976 or (404) 562-9900; fax (404) 562-8174; robinson.caroline@epa.gov
*Internet:* https://www.cfda.gov/?s=program&mode=form&tab=step1&id=af6215f7 9818f81c11793c94fa2fc592
*Sponsor:* Environmental Protection Agency
Sam Nunn Atlanta Federal Center
Atlanta, GA 30303-8960

**EPA Support to the National Tribal Water Council 1087**

The Environmental Protection Agency's (EPA) Office of Water supports the National Tribal Water Council (NTWC), composed of tribal governmental representatives, to increase communications, promote sharing of technical information and best management practices for Indian country, raise awareness of tribes on priority water issues, and facilitate tribal input on actions to protect water quality and provide safe drinking water in tribal communities. EPA supports the efforts of the National Tribal Water Council through a multi-year cooperative agreement to a recipient who manages and supports the operations of the Council, and assists in developing and completing products in coordination with the NTWC. Eligible recipients interested in managing the NTWC under this cooperative agreement may apply by submitting applications to EPA for this competitive process. One award is expected, with a ceiling of $1,100,000. The application deadline is August 11.
*Requirements:* States, local governments, federally recognized Indian Tribes, inter-tribal consortia, territories, possessions of the U.S. (including the District of Columbia), public and private universities and colleges, hospitals, laboratories, and public or private non-profit institutions are eligible to apply.
*Restrictions:* For-profit organizations and individuals are not eligible to apply.
*Geographic Focus* All States, District of Columbia, Guam, Marshall Islands, Northern Mariana Islands, Puerto Rico, U.S. Virgin Islands, American Samoa
*Date(s) Application is Due:* Aug 11
*Amount of Grant:* Up to 1,100,000 USD
*Contact:* Felicia Wright, Program Officer; (202) 566?1886; wright.felicia@epa.gov
*Internet:* https://www.epa.gov/tribal/funding-opportunity-cooperative-agreement-support-national-tribal-water-council
*Sponsor:* Environmental Protection Agency
1200 Pennsylvania Avenue, NW
Washington, D.C. 20460

**EPA Surveys, Studies, Research, Investigations, Demonstrations, and Special Purpose Activities Relating to the Clean Air Act 1088**

The primary objective of the program is to support Surveys, Studies, Research, Investigations, Demonstrations and Special Purpose assistance relating to the causes, effects (including health and welfare effects), extent, prevention, and control of air pollution to include such topics as air quality, acid deposition, climate change, global programs, indoor environments, radiation, mobile source technology and community-driven approaches to transportation and emissions reduction. Projects should also focus on addressing environmental justice (EJ) concerns in communities. EJ is the fair treatment and meaningful involvement of all people regardless of race, color, national origin, or income with respect to the development, implementation, and enforcement of environmental laws, regulations, and policies. Grants and cooperative agreements are available to support recipients' allowable direct costs incident to approved Surveys, Studies, Research, Investigations, Demonstrations and Special Purpose plus allowable indirect costs, in accordance with established EPA policies and regulations. Assistance agreement awards under this program may involve or relate to geospatial information. A detailed list of acceptable activities is provided at the grant website. EPA generally award grants ranging in value from $5,000 to $750,000 per fiscal year. The average value of each grant is $150,000 per fiscal year. Contact the headquarters or regional office, as appropriate, for application deadlines.
*Requirements:* Assistance under this program is generally available to States, local governments, territories, Indian Tribes, and possessions of the U.S., including the District of Columbia, international organizations, public and private universities and colleges, hospitals, laboratories, other public or private nonprofit institutions, which submit applications proposing projects with significant technical merit and relevance to EPA's Office of Air and Radiation's mission.
*Geographic Focus:* All States
*Amount of Grant:* 5,000 - 750,000 USD
*Contact:* Maureen Hingeley, (202) 564-1306; hingeley.maureen@epa.gov
*Internet:* https://www.cfda.gov/?s=program&mode=form&tab=step1&id=1d601ed2a d50dabb4dd11eaa76bb7764
*Sponsor:* Environmental Protection Agency - 1
Ariel Rios North Building
Washington, D.C. 20460

**EPA Technical Assistance and Support for Improved Protection of Drinking Water Sources 1089**

The Environmental Protection Agency's Technical Assistance and Support for Improved Protection of Drinking Water Sources program announces the availability of funds and solicits applications from eligible applicants for technical assistance and support that will enhance communication and coordination between the EPA and the states to protect drinking water sources. The technical assistance and support should develop and expand the capability of state and tribal Underground Injection Control (UIC) programs (Classes I-VI) and state and tribal Source Water Protection programs to protect drinking water sources. Funds awarded under this announcement may be used by recipients to promote non-federal participation in meetings that improve the ability to protect drinking water sources and to support the travel expenses of non-federal personnel to attend appropriate meetings. One award in the amount of $900,000will be given. The annual deadline for applications is July 15.
*Requirements:* Eligible applicants under this competition are states, local governments, territories, Indian tribes, and possessions of the U.S. (including the District of Columbia), public and private universities and colleges, hospitals, laboratories, and public and private nonprofit institutions.
*Restrictions:* Individuals and for-profit organizations are not eligible to apply.
*Geographic Focus* All States, District of Columbia, Guam, Marshall Islands, Northern Mariana Islands, Puerto Rico, U.S. Virgin Islands, American Samoa
*Date(s) Application is Due:* Jul 15
*Amount of Grant:* 900,000 USD
*Contact:* Suzanne Kelly; (202) 564-3887; fax (202) 564-3753; kelly.suzanne@epa.gov
*Internet:* https://www.epa.gov/uic/technical-assistance-and-support-improved-protection-drinking-water-sources
*Sponsor:* Environmental Protection Agency
1200 Pennsylvania Avenue, NW
Washington, D.C. 20460

**EPA Temporally Integrated Monitoring of Ecosystems and Long-Term Monitoring Grants 1090**

Clean Air Act section 103(b)(3) authorizes EPA to conduct and promote the coordination and acceleration of research, investigations, experiments, demonstrations, surveys, studies relating to the causes, effects (including health and welfare effects), extent, prevention and control of air pollution. Clean Water Act section 104(b)(3) authorizes EPA to award grants to conduct and promote the coordination and acceleration of research, investigations, experiments, training, demonstrations, surveys and studies relating to the causes, effects, extent, prevention, reduction and elimination of pollution. EPA will award cooperative agreements to study the ecological response to pollution reduction and emissions control programs. The programs will assist in understanding biogeochemical changes in sulfur, nitrogen, mercury, and carbon in streams and lakes in relation to changing pollutant emissions and deposition, as well as other factors, such as potential climate change (e.g., temperature, precipitation, timing of storm events). The TIME/LTM programs will support initiatives that demonstrate the potential for tracking and understanding the ecological response to air pollution reduction programs in acid-sensitive regions of the U.S. Funding will be available for applicants that demonstrate the ability to provide consistent, reliable, quality-assured data from lakes and streams for a variety of surface water chemistry parameters (e.g., pH, sulfate and nitrate concentrations, acid neutralizing capacity, aluminum concentrations, base cations). EPA will implement this grant authority by awarding cooperative agreements to state and local governments, institutions of higher education, and quasi-public and non-profit organizations that demonstrate the capability to perform the sampling, measurement, and parameter characterization of the lakes and streams in this long-term ecological system response program. There is no minimum amount of assistance. EPA anticipates awards ranging in total value of $640,000 to $1,400,000.
*Requirements:* Assistance under this program is generally available to States, local governments, territories, Indian Tribes, and possessions of the U.S. (including the District of Columbia); public and private universities and colleges; hospitals; laboratories; public or private nonprofit institutions; intertribal consortia; and individuals.
*Restrictions:* Nonprofit organizations described in Section 501(c)(4) of the Internal Revenue Code that engage in lobbying activities as defined in Section 3 of the Lobbying Disclosure Act of 1995 are not eligible to apply.
*Geographic Focus* All States, District of Columbia, Guam, Marshall Islands, Northern Mariana Islands, Puerto Rico, U.S. Virgin Islands, American Samoa
*Amount of Grant:* 640,000 - 1,400,000 USD
*Contact:* Clara Funk, (202) 343-9622; funk.clara@epa.gov
*Internet:* https://www.cfda.gov/?s=program&mode=form&tab=step1&id=8f43f83e 888f5af0a3153751394d2c93
*Sponsor:* Environmental Protection Agency - 1
Ariel Rios North Building
Washington, D.C. 20460

**Epilepsy Foundation Behavioral Sciences Student Fellowships** 1091

Epilepsy Foundation Behavioral Sciences Student Fellowships stimulate individuals to pursue careers in epilepsy in either research or practice settings. Appropriate fields include sociology, social work, psychology, anthropology, nursing, economics, vocational rehabilitation, counseling, political science, and others relevant to epilepsy research or practice. Annual deadline dates may vary; contact program staff for exact dates.
*Requirements:* Applicant must be actively enrolled in a degree program at the time the fellowship is to be undertaken. The fellowship must be undertaken during a free period in the student's year, e.g., during the summer. Both graduate and undergraduate students are eligible provided all other eligibility criteria are met. The project must be carried out at an approved facility.
*Geographic Focus:* All States
*Date(s) Application is Due:* Mar 1
*Amount of Grant:* 3,000 USD
*Contact:* Sonya B. Dumanis, Staff Contact; (301) 459-3700; fax (301) 577-2684; sdumanis@efa.org or contactus@efa.org
*Internet:* https://www.epilepsy.com/make-difference/research-and-new-therapies/upcoming-grants
*Sponsor:* Epilepsy Foundation
8301 Professional Place East, Suite 200
Landover, MD 20785-2353

**Epilepsy Foundation Health Sciences Student Fellowships** 1092

The Epilepsy Foundation's three-month Health Sciences Student Fellowships are awarded to medical and health science students for work on epilepsy study projects. The project may be carried out at the U.S. institution of the student's choice where there are ongoing programs of research, training, or service in epilepsy. Application forms and guidelines are available upon request. Annual deadline dates may vary; contact program staff for exact dates.
*Requirements:* Predoctoral training students in the Health Sciences may be accepted at any point in their schooling-following acceptance but before beginning the first year, or in the period immediately following their final year. A supervisor or preceptor must accept responsibility for the supervision of the student program and for the submission of a full report to the institute upon completion of the program.
*Geographic Focus:* All States
*Date(s) Application is Due:* Mar 1
*Amount of Grant:* 3,000 USD
*Contact:* Sonya B. Dumanis, Staff Contact; (301) 459-3700; fax (301) 577-2684; sdumanis@efa.org or contactus@efa.org
*Internet:* https://www.epilepsy.com/make-difference/research-and-new-therapies/upcoming-grants
*Sponsor:* Epilepsy Foundation
8301 Professional Place East, Suite 200
Landover, MD 20785-2353

**Epilepsy Foundation New Therapy Commercialization Grants** 1093

Epilepsy Foundation New Therapy Commercialization Grant program's is to drive the development of new therapies for epilepsy, accelerating the advancement of research from the laboratory to the patient. The Program seeks to advance the development of new therapies including but not limited to medicines, biologics or devices. All proposals must demonstrate a clear path from the lab to the patient and are evaluated with consideration for their potential value to patients, likelihood of successful development including regulatory approval and the time frame of development. The Program also encourages 1:1 matching grants to academic and commercial entities to support the commercialization of novel approaches to the treatment of epilepsy. Preference will be given to proposals that already have a commercial partner engaged to assist with development, and to proposals that have committed or matched funding from a sponsoring institution, commercial partner or other third party source. Areas of interest include: novel approaches to treatment; platform technology to advance screening techniques that can be utilized by multiple laboratories, including utility of techniques for early proof-of-concept trials; and adaptation of treatment in development for another therapeutic area to assess utility for epilepsy (while maintaining patent protection). Awards are up to $350,000 per year, over a two-year period, and are given twice a year. Letters of Intent (LOI) are due February 1, with full proposals due by April 20.
*Requirements:* Priorities will be given to: research programs that might not otherwise be funded through traditional sources; and projects that encourage collaboration among scientists and industry. Projects with full intellectual property (IP) rights (composition of matter) are preferred over method of use IP.
*Geographic Focus:* All States
*Date(s) Application is Due:* Apr 20
*Amount of Grant:* Up to 350,000 USD
*Contact:* Sonya B. Dumanis, Staff Contact; (301) 459-3700; fax (301) 577-2684; sdumanis@efa.org or contactus@efa.org
*Internet:* https://www.epilepsy.com/make-difference/research-and-new-therapies/innovation/epilepsy-therapy-project/new-therapy
*Sponsor:* Epilepsy Foundation
8301 Professional Place East, Suite 200
Landover, MD 20785-2353

**Epilepsy Foundation Predoctoral Research Training Fellowship** 1094

The Epilepsy Foundation Predoctoral Research Training Fellowship supports predoctoral students with dissertation research related to epilepsy, thus strengthening their interest in establishing epilepsy research as a career direction. Annual deadline dates may vary; contact program staff for exact dates.
*Requirements:* Graduate students pursuing a PhD degree in neuroscience, physiology, pharmacology, psychology, biochemistry, genetics, nursing, or pharmacy may apply.
*Geographic Focus:* All States
*Date(s) Application is Due:* Sep 1
*Amount of Grant:* Up to 25,000,000 USD
*Contact:* Sonya B. Dumanis, Staff Contact; (301) 459-3700; fax (301) 577-2684; sdumanis@efa.org or contactus@efa.org
*Internet:* www.epilepsyfoundation.org/research/grants.cfm
*Sponsor:* Epilepsy Foundation
8301 Professional Place East, Suite 200
Landover, MD 20785-2353

**Epilepsy Foundation Seizure Gauge Challenge Grants** 1095

The Epilepsy Foundation awards Seizure Gauge Challenge Grants that will allow a person with epilepsy to monitor the likelihood of a seizure on a daily basis. The primary purpose is to identify and better understand the changes in the body that may precede the onset of a seizure, at a time course that may be hours or days before the clinical (observable) seizure. The goal for this initiative is multi-fold: enhance seizure warning and prevention capabilities; improve quality of life and safety; better understand what triggers a seizure; and identify new therapeutic targets to prevent or stop seizures. Letters of Intent (LOI) to apply are due annually on December 2, with full grant proposals due by May 11.
*Geographic Focus:* All States
*Date(s) Application is Due:* May 11
*Contact:* Sonya B. Dumanis, Staff Contact; (301) 459-3700; fax (301) 577-2684; sdumanis@efa.org or contactus@efa.org
*Internet:* https://www.epilepsy.com/make-difference/research-and-new-therapies/innovation/epilepsy-innovation-institute/seizure-gauge
*Sponsor:* Epilepsy Foundation
8301 Professional Place East, Suite 200
Landover, MD 20785-2353

**Epilepsy Foundation Shark Tank Competition Grants** 1096

The Epilepsy Therapy Project, an initiative of the Epilepsy Foundation, is inviting entries that represent the most innovative new ideas in epilepsy treatment and care for its seventh annual Shark Tank Competition. The winner(s) of the Competition will receive international recognition and compete for awards totaling $150,000 to support the development and commercialization of an important new product, technology, or therapeutic concept to help people with epilepsy. The deadline for submitting a Letter of Intent (LOI) is October 16. The Shark Tank review committee will select as many as six finalists. All finalists will be invited to present their product concepts or candidates at the annual Epilepsy Pipeline Conference, scheduled for February.
*Geographic Focus:* All States
*Amount of Grant:* 150,000 USD
*Contact:* Sonya B. Dumanis, Staff Contact; (301) 459-3700; fax (301) 577-2684; sdumanis@efa.org or contactus@efa.org
*Internet:* https://www.epilepsy.com/make-difference/research-and-new-therapies/innovation/epilepsy-therapy-project/shark-tank
*Sponsor:* Epilepsy Foundation
8301 Professional Place East, Suite 200
Landover, MD 20785-2353

**Epilepsy Foundation SUDEP Challenge Initiative Prizes** 1097

Three million people in the U.S. and 65 million people worldwide have epilepsy, a neurological condition that affects the nervous system and causes seizures. Sudden unexpected death in epilepsy (SUDEP) is the leading cause of death in young adults with uncontrolled seizures. One way the Epilepsy Foundation SUDEP Institute is meeting these needs is with a series of four multidisciplinary prize challenges. The first challenge sought creative and viral advocacy campaign(s) to inform and educate people with epilepsy, families, and the health care community about SUDEP. The second challenge asked solvers to develop an intervention to help people with epilepsy comply with treatment plans and decrease their risk of seizures. In the third challenge, solvers proposed a predictive biomarker or panel of biomarkers to identify people at risk for SUDEP. Identifying biomarkers to predict SUDEP represents a major unmet medical need. Although there are several theories about the causes of SUDEP, who is at risk of dying from epilepsy and how to prevent it remains a mystery. Hence, the fourth and final challenge requires a detailed project plan for the proposed solution for finding a predictive biomarker or panel of biomarkers to identify people at risk for SUDEP or seizures that compromise cardiac or respiratory function. Anyone can be a part of the $1 million SUDEP solution. The Epilepsy Foundation is accepting submissions of solution up through October of 2020. An Award up to $800,000 for the first submission that satisfies all criteria will be granted.
*Geographic Focus:* All States
*Amount of Grant:* Up to 800,000 USD
*Contact:* Sonya B. Dumanis, Staff Contact; (301) 459-3700; fax (301) 577-2684; sdumanis@efa.org or contactus@efa.org
*Internet:* https://www.epilepsy.com/living-epilepsy/our-programs/sudep-institute/sudep-challenge-initiative
*Sponsor:* Epilepsy Foundation
8301 Professional Place East, Suite 200
Landover, MD 20785-2353

**ERC Advanced Grants** 1098

ERC Advanced Grants allow exceptional established research leaders of any nationality and any age to pursue ground-breaking, high-risk projects that open new directions in their respective research fields or other domains. Funding targets researchers who have already established themselves as independent research leaders in their own right. ERC will award up to € 2.5 million per grant (in some circumstances up to € 3.5 million per grant) for up to five years.

*Requirements:* Researchers can be of any nationality and any age. Applicants must be scientifically independent and have a recent research track-record and profile which identifies them as leaders in their respective field(s) of research. Any field of science, engineering and scholarship is acceptable for submission. The proposed research must be conducted in a public or private research organization (known as a Host Institution/HI) located in one of the EU Member States or Associated Countries. The sole evaluation criterion will be the scientific excellence of the researcher and research proposal. Applicants with questions should contact their National Contact Point (see erc.europa.eu/national-contact-points) or send queries to the Enquiry Service.

*Geographic Focus:* Albania, Andorra, Armenia, Austria, Azerbaijan, Belarus, Belgium, Bosnia & Herzegovina, Bulgaria, Croatia, Cyprus, Czech Republic, Denmark, Estonia, Finland, France, Georgia, Germany, Greece, Hungary, Iceland, Ireland, Italy, Kosovo, Latvia, Liechtenstein, Lithuania, Luxembourg, Macedonia, Malta, Moldova, Monaco, Montenegro, Norway, Poland, Portugal, Romania, Russia, San Marino, Serbia, Slovakia, Slovenia, Spain, Sweden, Switzerland, The Netherlands, Turkey, Ukraine, United Kingdom, Vatican City

*Date(s) Application is Due:* Oct 21

*Amount of Grant:* Up to 2,500,000 EUR

*Contact:* National Contact Point

*Internet:* erc.europa.eu/advanced-grants

*Sponsor:* European Research Council
ERC Executive Agency, COV2
Brussels, BE-1049 Belgium

**ERC Call for Proposals for Communication Actions** 1099

Through this Call for proposals, the ERC wishes to support an ambitious series of communication actions to promote and raise awareness of ERC-funded projects and results across Europe to as wide an audience as possible, including scientists, students, media, policy-makers, the business community and the general public. The actions will go beyond traditional scientific conferences and take a novel and creative approach. The maximum duration of the project will be 48 months. Up to two proposals will be selected.

*Requirements:* Coordination and Support Actions are open to legal entities established in a Member State or an Associated country as a legal entity created under national law, International European Interest Organizations (such as CERN, EMBL, etc.), the European Commission's Joint Research Centre (JRC) or an entity created under EU law. Legal entities established in countries outside the EU or Associated Countries and international organizations are also eligible.

*Geographic Focus:* All States, All Countries

*Date(s) Application is Due:* Dec 16

*Amount of Grant:* 1,600,000 EUR

*Contact:* National Contact Point

*Internet:* ec.europa.eu/research/participants/portal/desktop/en/opportunities/h2020/calls/erc-2014-support-1.html

*Sponsor:* European Research Council
ERC Executive Agency, COV2
Brussels, BE-1049 Belgium

**ERC Proof of Concept Grants** 1100

The ERC launched the funding initiative, called "Proof of Concept", in March 2011, open to researchers who have already been awarded an ERC grant. ERC grant holders can apply for this additional funding to establish the innovation potential of ideas arising from their ERC-funded frontier research. ERC will award up to € 150000 for up to 18 months. This funding aims at supporting grant-holders during the pre-demonstration phase to prepare a "package" to be presented to venture capitalists or companies that might invest in the new technology and take it through the early commercialization phase. The funding can be used to: establish viability, technical issues and overall direction; clarify intellectual property rights position and strategy; provide feedback for budgeting and other forms of commercial discussion; provide connections to later stage funding; or cover initial expenses for establishing a company.

*Requirements:* The ERC Proof of Concept funding is made available only to those who already have an ERC award to establish proof of concept of an idea that was generated in the course of their ERC-funded projects. The calls for proposals are published once a year (generally in November) with two deadlines. Applicants with questions should contact their National Contact Point (see erc.europa.eu/national-contact-points) or send queries to the Enquiry Service. National Contact Points (NCPs) were set up across Europe (in all Member States and Associated States) to provide information and personalised support to applicants in their native language.

*Geographic Focus:* Marshall Islands, Afghanistan, Albania, Algeria, American Samoa, Angola, Anguilla, Argentina, Armenia, Aruba, Austria, Azerbaijan, Bangladesh, Belarus, Belgium, Belize, Benin, Bermuda, Bhutan, Bolivia, Bosnia & Herzegovina, Botswana, British Virgin Islands, Bulgaria, Burkina Faso, Burundi, Cambodia, Cameroon, Cape Verde, Cayman Islands, Central African Republic, Chad, Chile, Colombia, Comoros, Congo, Congo, Democratic Republic of, Costa Rica, Cote d'Ivoire (Ivory Coast), Croatia, Cuba, Cyprus, Czech Republic, Denmark, Djibouti, Dominica, Dominican Republic, Ecuador, Egypt, El Salvador, Eritrea, Estonia, Ethiopia, Falkland Islands (Malvinas), Fiji, Finland, France, French Polynesia, Gabon, Gambia, Germany, Ghana, Greece, Greenland, Grenada, Guatemala, Guinea, Guinea-Bissau, Guyana, Haiti, Honduras, Hungary, Indonesia, Iran, Iraq, Ireland, Italy, Jamaica, Jordan, Kazakhstan, Kenya, Kiribati, Korea, Kosovo, Kyrgyzstan, Laos, Latvia, Lebanon, Lesotho, Liberia, Libya, Lithuania, Luxembourg, Macedonia, Madagascar, Malawi, Malaysia, Maldives, Mali, Malta, Mauritania, Micronesia, Moldova, Mongolia, Montenegro, Montserrat, Morocco, Mozambique, Myanmar (Burma), Namibia, Nepal, Netherlands, New Caledonia, Nicaragua, Niger, Nigeria, Pakistan, Palau, Palestinian Territory, Panama, Papua New Guinea, Paraguay, Peru, Philippines, Pitcairn, Poland, Portugal, Romania, Rwanda, Sao Tome & Principe, Senegal, Serbia, Seychelles, Sierra Leone, Slovakia, Slovenia, Solomon Islands, Somalia, South Africa, South Sudan, Spain, Sri Lanka, St. Lucia, St. Vincent and the Grenadines, Sudan, Suriname, Swaziland, Sweden, Syrian Arab Republic, Tajikistan, Tanzania, Thailand, Timor-Lester, Togo, Tonga, Tunisia, Turkey, Turkmenistan, Turks and Caicos Islands, Tuvalu, Uganda, Ukraine, United Kingdom, Uruguay, Uzbekistan, Vanuatu, Venezuela, Vietnam, Wallis and Furuna Islands, Yemen, Zambia, Zimbabwe

*Date(s) Application is Due:* Apr 1; Oct 1

*Amount of Grant:* Up to 150,000 EUR

*Contact:* National Contact Point

*Internet:* erc.europa.eu/proof-concept

*Sponsor:* European Research Council
ERC Executive Agency, COV2
Brussels, BE-1049 Belgium

**ERC Starting Grants** 1101

ERC Starting Grants aim to support up-and-coming research leaders who are about to establish a proper research team and to start conducting independent research in Europe. The scheme targets promising researchers who have the proven potential of becoming independent research leaders. It will support the creation of excellent new research teams. Each grant awarded will receive up to € 1.5 million (in some circumstances up to € 2 million) for up to 5 years.

*Requirements:* Eligible applicants include researchers of any nationality with 2-7 years of experience since completion of PhD (or equivalent degree) and scientific track record showing great promise. Research must be conducted in a public or private research organization (known as a Host Institution/HI) located in one of the EU Member State or Associated Countries. The sole evaluation criterion will be the scientific excellence of researcher and research proposal.

*Geographic Focus:* Albania, Andorra, Armenia, Australia, Azerbaijan, Belarus, Belgium, Bosnia & Herzegovina, Bulgaria, Croatia, Cyprus, Czech Republic, Denmark, Estonia, Finland, France, Germany, Greece, Hungary, Iceland, Ireland, Italy, Kosovo, Latvia, Liechtenstein, Lithuania, Luxembourg, Macedonia, Malta, Moldova, Monaco, Montenegro, Norway, Poland, Portugal, Romania, Russia, San Marino, Serbia, Slovakia, Slovenia, Spain, Sweden, Switzerland, The Netherlands, Turkey, Ukraine, United Kingdom, Vatican City

*Date(s) Application is Due:* May 13

*Amount of Grant:* Up to 1,500,000 EUR

*Contact:* National Contact Point

*Internet:* erc.europa.eu/starting-grants

*Sponsor:* European Research Council
ERC Executive Agency, COV2
Brussels, BE-1049 Belgium

**EREF Doctoral Scholarship** 1102

Scholarships are awarded by EREF to recognize excellence in master's, doctoral or post-doctoral waste management research and education. As part of its mission, EREF furthers education in the solid waste management industry. In the past, multiple outstanding candidates have applied for the Fiessinger award in the same year, prompting EREF to periodically utilize foundation funds to award additional scholarships. EREF now regularly awards Ph.D. and postdoctoral scholarships to notable candidates, expanding the scholarship program. At least one EREF Scholar is awarded annually. Award amounts consider the cost of tuition at the recipient's institution and any other funds received. Doctoral scholarships are awarded up to $12,000 per year, paid monthly, and can be extended for up to 3 years from the initial award date.

*Requirements:* Applications will be considered from full-time PhD students who have a clearly demonstrated interest in environmental research. Applications will be considered from students outside the United States or studying abroad.

*Restrictions:* Full scholarships may not be awarded to students who will be receiving full-tuition scholarships from other sources (excluding direct university assistance such as tuition remission, or assistantship income) but may still be considered for partial awards.

*Geographic Focus:* All States, All Countries

*Date(s) Application is Due:* Aug 1

*Amount of Grant:* 12,000 - 36,000 USD

*Contact:* Bryan F. Staley, President; (919) 861-6876, ext. 102; fax (919) 861-6878; bstaley@erefdn.org

*Internet:* erefdn.org/index.php/scholarships/

*Sponsor:* Environmental Research and Education Foundation
3301 Benson Drive, Suite 301
Raleigh, NC 27609

**EREF Fiessinger Doctoral Scholarship** 1103

Scholarships are awarded by EREF to recognize excellence in master's, doctoral or post-doctoral waste management research and education. Awards to doctoral students are given in memory of Francois Fiessinger, P.E., Ph.D., a graduate of Rutgers University, who was a founding director of the Environmental Research and Education Foundation. Francois believed in scholarly research and was one of very few people who had a long-term research and development vision for the solid waste industry. In 1997, based on a

contribution from Lyonnaise des Eaux, the Francois Fiessinger Memorial Scholarship Fund was established to recognize excellence in Ph.D. or postdoctoral environmental research and education. One Fiessinger Scholar is awarded annually. Award amounts consider the cost of tuition at the recipient's institution and any other funds received. Doctoral scholarships are awarded up to $12,000 per year, paid monthly, and can be extended for up to 3 years from the initial award date.
*Requirements:* Applications will be considered from full-time PhD students who have a clearly demonstrated interest in environmental research. Applications will be considered from students outside the United States or studying abroad.
*Restrictions:* Full scholarships may not be awarded to students who will be receiving full-tuition scholarships from other sources (excluding direct university assistance such as tuition remission, or assistantship income) but may still be considered for partial awards.
*Geographic Focus:* All States, All Countries
*Date(s) Application is Due:* Aug 1
*Amount of Grant:* 12,000 - 36,000 USD
*Contact:* Bryan F. Staley; (919) 861-6876, ext. 102; fax (919) 861-6878; bstaley@erefdn.org
*Internet:* erefdn.org/index.php/scholarships/
*Sponsor:* Environmental Research and Education Foundation
3301 Benson Drive, Suite 301
Raleigh, NC 27609

**EREF Solid Waste Research Grants** **1104**
The Environmental Research and Education Foundation is the only private, grant making institution with a national and international scope whose sole mission is to support solid waste research. Its research grants program is led by a Research Council, a body of volunteers consisting of technical experts in industry, academia and consulting. The work of the Council is guided by a long range strategic plan with the goal to achieve greater sustainability, good environmental stewardship, higher process efficiency and increased knowledge. Council recommended projects are then reviewed by the Board's Projects Committee for a final review and funding allocation. EREF awards several grants each year for research or education in topics pertaining to any aspect of solid waste management. There are two annual deadlines for solicited proposal submissions, including January 5 and July 15. There is no formal application, and submissions should include a description of proposed activity, project timeline, budget, and resumes of principal investigators. Previously awarded grants have ranged from $15,000 to over $500,000 with the average grant being amount being $100,000.
*Restrictions:* Grants will not be provided for capital campaigns; political contributions; religious causes; operating funds; loans; or support of lobbying activities. The Foundation will not pay overhead in excess of 25%. Proposals submitted in response to this RFP that do not address a specific topic in the research agenda will not be reviewed.
*Geographic Focus:* All States
*Date(s) Application is Due:* Jan 5; Jul 15
*Amount of Grant:* 15,000 - 500,000 USD
*Contact:* Bryan F. Staley; (919) 861-6876, ext. 102; fax (919) 861-6878; bstaley@erefdn.org
*Internet:* erefdn.org/index.php/grants/
*Sponsor:* Environmental Research and Education Foundation
3301 Benson Drive, Suite 301
Raleigh, NC 27609

**EREF Sustainability Research Grants** **1105**
The Environmental Research and Education Foundation is the only private, grant making institution with a national and international scope whose sole mission is to support solid waste research. Submissions of scientific research proposals related to sustainable solid waste management practices are invited in the following areas: waste minimization; recycling; waste conversion to energy, bio-fuels, chemicals or other useful products, including (but not limited to) the technologies of waste-to-energy, pyrolysis and gasification, oxidation, anaerobic digestion, and composting; and strategies to promote diversion from landfills (e.g. separating organics, market analysis, optimized material management, logistics, etc.). EREF awards several grants each year for research or education in topics pertaining to any aspect of solid waste management. There are two annual deadlines for solicited proposal submissions, including January 5 and July 15. There is no formal application, and submissions should include a description of proposed activity, project timeline, budget, and resumes of principal investigators. Previously awarded grants have ranged from $15,000 to over $500,000 with the average grant being amount being $100,000.
*Restrictions:* Grants will not be provided for capital campaigns; political contributions; religious causes; operating funds; loans; or support of lobbying activities. The Foundation will not pay overhead in excess of 25%. Proposals submitted in response to this RFP that do not address a specific topic in the research agenda will not be reviewed.
*Geographic Focus:* All States
*Date(s) Application is Due:* Jan 5; Jul 15
*Amount of Grant:* 15,000 - 500,000 USD
*Contact:* Bryan F. Staley; (919) 861-6876, ext. 102; fax (919) 861-6878; bstaley@erefdn.org
*Internet:* erefdn.org/index.php/grants/proposal
*Sponsor:* Environmental Research and Education Foundation
3301 Benson Drive, Suite 301
Raleigh, NC 27609

**Evan Frankel Foundation Grants** **1106**
Although it was incorporated in 1978, the Evan Frankel Foundation became an active philanthropic entity according to Evan's wishes upon his death in 1991. The Foundation awards grants in its areas of interest, including higher education, the arts, humanities, health and science, social services, and the environment. Giving is primarily in Manhattan and Suffolk County, New York and Los Angeles, California. Most recent grants have ranged from as little as $500 up to $500,000. Applicants should submit a letter to request, outlining their program and budgetary needs.
*Restrictions:* Individuals are not eligible.
*Geographic Focus:* All States
*Amount of Grant:* 500 - 500,000 USD
*Samples:* A Place Called Home, Los Angeles, California, $117,000 - social services general operating costs; Harlem Educational Activities Fund, New York, New York, $49,000 - general operating support; East Hampton Library, East Hampton, New York, $204,750 - general operating support.
*Contact:* Ernest Frankel; (631) 329-0010; fax (631) 329-7102; asff@hamptons.com
*Internet:* www.evanfrankelfoundation.org/
*Sponsor:* Evan Frankel Foundation
P.O. Box 5026
East Hampton, NY 11937

**Evjue Foundation Grants** **1107**
The Foundation contributes each year to worthy educational, cultural and charitable organizations that contribute to the quality of life in Madison and Dane County. The Foundation's interests include: theater, hunger, troubled youth, and universities. The Foundation looks favorably upon projects responding to overall community needs and priorities that do not duplicate existing services. Application and guidelines are available online.
*Requirements:* Grants are only made to nonprofit organizations.
*Restrictions:* Grants are not made to individuals. All grants for scholarships are given to educational institutions, which have full responsibility for selecting the individual recipients. Grants are not usually made to establish or add to endowment funds. Grants are not ordinarily made to fund specific medical or scientific research inquiries, nor to support operating expenses and general administrative expenses of organizations.
*Geographic Focus:* Wisconsin
*Date(s) Application is Due:* Mar 5
*Contact:* Arlene Hornung, Executive Director; (608) 252-6401; ahornung@madison.com
*Internet:* host.madison.com/ct/about/evjue/article_36a7a66e-a46a-11de-ac74-001cc4c002e0.html
*Sponsor:* Evjue Foundation
1901 Fish Hatchery Road
Madison, WI 53713

**Expect Miracles Foundation Grants** **1108**
The MMLC Funding Award is given on an annual basis on behalf of the Miracle Maker Leadership Council to cancer fighting causes and patient care programs where such support is not always readily available. The proposed program should have the ability to materially impact an organization or a specific program of a qualified organization.
*Requirements:* Funding Award recipients must have 501(c)3 status as defined by the IRS. Applicants must provide service in at least one of the following areas: cancer patient care support programs and initiatives; cancer awareness and educational programs; or cancer prevention/health projects. Priority will be given to initiatives demonstrating sustainable benefits to the communities where the Foundation events are held and where Foundation supporters and the Miracle Maker Leadership Council (MMLC) members live.
*Restrictions:* Although there are no specific guidelines on the size, annual budget, or annual fundraising of a recipient organization, the program is not designed to assist in funding large institutional/national/society entities. The goal of the MMLC is to select programs and causes where funding is not always readily available. MMLC guidelines prohibit funding for the following: private pursuits; political parties, associations and representatives of advocacy groups; organizations that discriminate by race, creed, gender, sexual orientation, age, religion, or national origin; religious causes; advertising, promotion, or sponsorship; and donation to an individual or team fundraising initiative or program.
*Geographic Focus:* California, Connecticut, District of Columbia, Illinois, Maine, Maryland, Massachusetts, New Hampshire, New Jersey, New York, Pennsylvania, Rhode Island, Vermont
*Date(s) Application is Due:* Mar 5
*Amount of Grant:* Up to 27,000 USD
*Contact:* Alana Chin, (617) 391-9235 or (617) 827-7463; achin@expectmiraclesfoundation.org
*Internet:* www.expectmiraclesfoundation.org/
*Sponsor:* Expect Miracles Foundation
6 Quail Run
Hingham, MA 02043

**ExxonMobil Foundation Malaria Initiative Grants** **1109**
As a major employer and investor in Africa, ExxonMobil is committed to working with partners — local institutions, international NGOs and governments — to stop the spread of malaria on the continent. Since 2000, the corporation's funding has reached nearly 83 million people through support exceeding $110 million. Funding supports partners who help improve the delivery and use of prevention tools such as bed nets; train healthcare workers to ensure malaria interventions are reaching those who need them most; provide technical assistance to help countries increase their capacity to control malaria; and facilitate the monitoring and promotion of progress through integrated communications programs.
*Requirements:* Grants are made to tax-exempt organizations within principal company-operating areas.
*Restrictions:* Grants are not made to individuals for scholarships, fellowships, research, or travel. Organizations that are primarily religious in nature are ineligible.
*Geographic Focus:* All States, United Kingdom
*Amount of Grant:* Up to 100,000,000 USD
*Contact:* Suzanne McCarron, President; (972) 444-1100 or (972) 444-1007; fax (972) 444-1405; contributions@exxonmobil.com

*Internet:* www.exxonmobil.com/Corporate/community_malaria_initiative.aspx
*Sponsor:* ExxonMobil Foundation
5959 Las Colinas Boulevard
Irving, TX 75039-2298

**Eye-Bank for Sight Restoration and Fight for Sight Summer Student Research Fellowship** 1110

Beginning in 2011, The Eye-Bank for Sight Restoration and Fight for Sight co-sponsored a Summer Student Research Fellowship. Annual grants of $2,100 will support full-time research conducted from June to August. The goal of this award is to advance the skills needed to initiate and carry out work in a scientific environment.
*Requirements:* Applicants must be currently enrolled undergraduates, medical students or graduate students studying in the United States who wish to explore ophthalmology or eye research as a career. Students are expected to complete a short, independent project during the summer months under the guidance of a senior scientist or clinician. Candidates are invited to apply by visiting www.fightforsight.org for an application.
*Restrictions:* Lab employees are not eligible for this fellowship.
*Geographic Focus:* All States
*Date(s) Application is Due:* Feb 1
*Amount of Grant:* 2,100 - 2,100 USD
*Contact:* Janice Benson; (212) 679-6060; fax (212) 679-4466; janice@fightforsight.org
*Internet:* www.eyedonation.org/scholarship_program.html#fight_sight
*Sponsor:* Eye-Bank for Sight Restoration
120 Wall Street
New York, NY 10005-3902

**Ezra Jack Keats/Kerlan Collection Memorial Fellowship** 1111

This fellowship provides a talented writer and/or illustrator of children's books with funds from the Children's Literature Research Collections at the University of Minnesota-Twin Cities Campus to use the Kerlan Collection for artistic development. Special consideration will be given to someone who would find it difficult to finance the visit to the Kerlan Collection. The fellow will receive transportation and a per diem allotment.
*Requirements:* Grants are awarded to individuals based on need. Candidates may be of any nationality.
*Geographic Focus:* All States
*Date(s) Application is Due:* Dec 31
*Amount of Grant:* 1,500 USD
*Samples:* Kathleen Wong Bishop and Leslie Ann Hayashi; Kelly Baptist; Amanda Sheperd.
*Contact:* Fellowship Committee; (612) 624-4576; fax (612) 625-5525; clrc@tc.umn.edu
*Internet:* special.lib.umn.edu/clrc/kerlan/awards.php
*Sponsor:* University of Minnesota
222 21st Avenue South, 113 Andersen Library
Minneapolis, MN 55455

**Facebook Fellowships** 1112

Every day Facebook confronts the most complex technical problems and believes that close relationships with the academy will enable them to address many of these problems at a fundamental level and solve them. As part of its ongoing commitment to academic relations, Facebook has created the Facebook Fellowship program to support graduate students. Interest is in a wide range of academic topics, including the following topical areas: Internet Economics—auction theory and algorithmic game theory relevant to online advertising auctions; Cloud Computing—storage, databases, and optimization for computing in a massively distributed environment; Social Computing—models, algorithms and systems around social networks, social media, social search and collaborative environments; Data Mining and Machine Learning—learning algorithms, feature generation, and evaluation methods to produce effective online and offline models of behavioral signals; Systems—hardware, operating system, runtime, and language support for fast, scalable, efficient data centers; Information Retrieval—search algorithms, information extraction, question answering, cross-lingual retrieval and multimedia retrieval. Tuition and fees will be paid for the academic year: $30K stipend (paid over 9 months of the academic year); $5K per year toward conference attendance and travel; $5K toward a personal computer. Fellows will have an opportunity to apply for a paid summer internship. Fellowships will begin in August or September, depending on the start of the award recipient's academic year.
*Requirements:* Full-time Ph.D. students in topical areas represented by these fellowships who are currently involved in on-going research are eligible to apply. Non-U.S. citizens are also eligible but must be enrolled at a U.S. university and meet the eligibility requirements and have a valid Visa. Students must be studying Computer Science, Computer Engineering, Electrical Engineering, System Architecture, or a related area. Students must be enrolled during the academic year that the Fellowship is awarded and must be nominated by a faculty member.
*Restrictions:* Recipients are responsible for the taxation on any funds.
*Geographic Focus:* All States
*Date(s) Application is Due:* Feb 15
*Amount of Grant:* 40,000 USD
*Contact:* Greg Badros; (650) 543-4811; fellowships@facebook.com
*Internet:* www.facebook.com/careers/fellowship.php
*Sponsor:* Facebook
1601 South California Avenue
Palo Alto, CA 94304-1111

**FACHE Albert W. Dent Scholarships** 1113

The Albert W. Dent Scholarships have been established to provide financial aid and to increase the enrollment of minority students in health care management graduate programs and to encourage students, through structured, formalized study, to obtain positions in middle and upper levels of health care management. Applications are accepted between January 1 and March 31, and scholarship awards are announced in July. Interested persons should request application forms from the director of their graduate program in health care administration or from the foundation. FACHE awards up to fifteen $5,000 scholarships each year.
*Requirements:* To be eligible, an applicant must be a U.S. or Canadian student associate in good standing in the ACHE, a minority student who has been enrolled for full-time study for the upcoming fall term, which is his/her final year in a health care management graduate program. Financial need also must be demonstrated.
*Restrictions:* Previous recipients are ineligible.
*Geographic Focus:* All States
*Date(s) Application is Due:* Mar 31
*Amount of Grant:* 5,000 USD
*Contact:* Membership Office; (312) 424-2800 or (312) 424-9400; fax (312) 424-0023; membership1@ache.org
*Internet:* www.ache.org/Faculty_Students/dent_scholarship.cfm
*Sponsor:* Foundation of the American College of Healthcare Executives
One North Franklin Street, Suite 1700
Chicago, IL 60606-3491

**FAMRI Clinical Innovator Awards** 1114

The purpose of the Clinical Innovator Awards (CIA) from the Flight Attendant Medical Research Institute (FAMRI) is to stimulate novel medical and clinical scientific research studies on the topic of second hand tobacco smoke. Of high priority to FAMRI are new technologies to detect second hand tobacco-caused illnesses and novel therapies to treat these diseases. FAMRI is interested in clinical and translational research, from basic molecular mechanisms of disease to occupational health. These areas include: studies on the etiology and pathogenesis of tobacco-caused respiratory diseases in humans; therapeutic interventions; clinical trials; and research to establish a protocol for inquiry regarding second hand tobacco smoke exposure by all health care professionals. FAMRI plans to provide awards of up to $100,000 per year for direct costs plus no more than 8.5% for indirect costs, subject to the availability of funds (FAMRI cannot guarantee annual and total levels of support). Awards usually will be made for three years with a possibility of a no-cost extension for one year upon request. Applicants must first submit an online letter of intent by the due date. Upon invitation, they may then submit online proposals. Instructions and links for submitting the letter of intent and proposal are available under "Current and Upcoming Awards" at the FAMRI website. Deadlines may vary from year to year. To be kept up to date of all current and future funding opportunities, interested applicants are encouraged to sign up for the FAMRI ListServ (click on "listserv signup" under the researchers tab at the FAMRI website).
*Requirements:* Applicants must be certain that their institution can agree to the terms of the FAMRI grant agreement. A copy of the agreement is available for review under "Current and Upcoming Awards" at the FAMRI website. Grantees must devote a minimum of 50% of professional time to research during the period of this award; at least 15% of the grantee's time (or a total of 15% of the combined times of the co-grantees) should be devoted to the research project described in the application. Researchers with an advanced degree (MD, PhD, or MD/PhD, or the equivalent) are eligible to apply. Individuals who apply to FAMRI for funding need not be U.S. citizens. They normally apply through and with an institution or organization that is qualified under United States tax laws to receive grants from FAMRI, a US-based private foundation. The applying institution receives and administers FAMRI's funds and supervises the work of the Principal Investigator. If the applying institution is organized under the laws of a country other than the United States, it must provide FAMRI any and all documentation that FAMRI may require, in form and content acceptable to FAMRI, to ensure that any grant from FAMRI shall be a "qualified distribution" under Internal Revenue Service laws and regulations, including but not limited to affiliation with a qualified US-based institution, a current affidavit from the applicant or an opinion of qualified legal counsel stating the facts and circumstances of the applicant's organization and activities and certifying that any grant from FAMRI shall be a "qualified distribution." Research projects that involve human subjects must follow Public Health Service guidelines (or its equivalent in non-U.S. institutions) and must have the approval of the appropriate Institutional Review Boards (IRB) or foreign equivalent prior to the start of the research. The IRB (or foreign equivalent) approval must be maintained and be in effect throughout the research project. All research must be performed in accordance with all relevant institutional and federal policies and guidelines (or foreign equivalent) relating to human subject protection, radiation and environmental health and safety. All FAMRI grantees are encouraged to collaborate with fellow grantees and may be required to participate periodically in FAMRI's collaborative efforts.
*Restrictions:* The Clinical Innovator Award must not exceed $100,000 per year plus 8.5 % indirect costs for a total of $108,500 per year. FAMRI will only support animal research that utilizes mice or rats. Research involving other animals will not be considered unless the animals will not be sacrificed as a result of the research. Pending Association for Assessment of Laboratory Animal Care (AALAC) approval must be submitted prior to funding of a grant involving animal subjects. To avoid all conflicts, any principal investigator or other participating applicant who is currently accepting tobacco industry funding, has recently applied for tobacco funding, or has any other association or consulting position with a tobacco company or affiliate of the tobacco industry is not eligible for FAMRI funding. All prior funding and/or association with tobacco companies must be disclosed and will be considered on a case-by-case basis.

An award will not be made or will be withdrawn if the applicant becomes the Principal Investigator on a peer-reviewed grant with the same general aims as those proposed in the application to FAMRI. FAMRI reserves the right to announce the award publicly in a manner and at a time of its choosing.
*Geographic Focus:* All States, All Countries
*Date(s) Application is Due:* Sep 26
*Amount of Grant:* Up to 325,500 USD
*Samples:* Diego A. Preciado, MD, PhD, Children's National Medical Center, Washington, D.C. - Tobacco Smoke Exposure Effects on Sinonasal Mucosa; Noam A. Cohen, MD, PhD, Philadelphia Research & Education Foundation, Philadelphia, Pennsylvania - Induction of Bacterial Biofilms by Tobacco Smoke; Jean Kim, MD, PhD, Johns Hopkins University, Baltimore, Maryland - Nasal Epithelial Growth Dysfunction by Second Hand Smoke in Chronic Sinusitis
*Contact:* Elizabeth Kress; (305) 379-7007; fax (305) 577-0005; ekress@famri.org
*Internet:* www.famri.org/current_awards/index.php
*Sponsor:* Flight Attendant Medical Research Institute
201 S Biscayne Boulevard, Suite 1310
Miami, FL 33131

**FAMRI Young Clinical Scientist Awards** **1115**
The Young Clinical Scientist Award (YSCA) is intended to prepare and support new clinical investigators with an MD or PhD as they begin their careers as independent researchers in the field of smoking-related disorders. Awards are offered to two groups of scientists: those individuals who are currently faculty members, or who expect to be before the start date of the award (YCSA faculty), and those individuals who do not expect to become faculty until after the start date of the award (YCSA fellow). Applicants must initially submit an online letter of intent (LOI) by 5 p.m. on the LOI due date, and upon subsequent invitation submit a full proposal by 5 p.m. on the proposal due date. Instructions and links for submitting the letter of intent and proposal are available under "Current and Upcoming Awards" at the FAMRI website. YCSA awards can be extended to a total of no more than five years. Renewal forms will be sent to current grantees in early February of the appropriate year and will be due late March. Grantees who are fellows should notify both the Flight Attendant Medical Research Institute (FAMRI) and the American Institute of Biological Sciences (AIBS) of any change in status. Grantees are encouraged to call or email the FAMRI Administrative Assistant to obtain current AIBS contact information. Deadlines may vary from year to year. To be kept up to date of all current and future funding opportunities, interested applicants are encouraged to sign up for the FAMRI ListServ (click on "listserv signup" under the researchers tab at the FAMRI website).
*Requirements:* Applicants must be certain that their institution can agree to the terms of the FAMRI grant agreement. A copy of the agreement is available for review under "Current and Upcoming Awards" at the FAMRI website. Each individual other than the institutional official must submit a tobacco disclosure form. A tobacco disclosure form is also required from the applicant's institution. Tobacco disclosure forms may be downloaded from the FAMRI website and after they are signed, they may be scanned and uploaded. Researchers with an advanced degree (M.D., Ph.D., or M.D./Ph.D., or the equivalent) are eligible to apply. A YCSA grantee must devote a minimum of 75% of professional time to research during the period of this award and at least 50% of the grantee's time should be devoted to the research project described in the application. Individuals who apply to FAMRI for funding need not be U.S. citizens. They normally apply through and with an institution or organization that is qualified under United States tax laws to receive grants from FAMRI, a US-based private foundation. The applying institution receives and administers FAMRI's funds and supervises the work of the Principal Investigator. If the applying institution is organized under the laws of a country other than the United States, it must provide FAMRI any and all documentation that FAMRI may require, in form and content acceptable to FAMRI, to ensure that any grant from FAMRI shall be a "qualified distribution" under Internal Revenue Service laws and regulations, including but not limited to affiliation with a qualified US-based institution, a current affidavit from the applicant or an opinion of qualified legal counsel stating the facts and circumstances of the applicant's organization and activities and certifying that any grant from FAMRI shall be a "qualified distribution." Research projects that involve human subjects must follow Public Health Service guidelines (or its equivalent in non-U.S. institutions) and must have the approval of the appropriate Institutional Review Boards (IRB) or foreign equivalent prior to the start of the research. The IRB (or foreign equivalent) approval must be maintained and be in effect throughout the research project. All research must be performed in accordance with all relevant institutional and federal policies and guidelines (or foreign equivalent) relating to human subject protection, radiation and environmental health and safety. All FAMRI grantees are encouraged to collaborate with fellow grantees and may be required to participate periodically in FAMRI's collaborative efforts.
*Restrictions:* The program is limited to the development of young researchers in smoking-related disorders. The YCSA/fellow award must not exceed $75,000 per year plus 8.5 % indirect costs for a total of $81,375 per year. The YCSA/faculty award must not exceed $100,000 per year plus 8.5 % indirect costs for a total of $108,500 per year. FAMRI will only support animal research that utilizes mice or rats. Research involving other animals will not be considered unless the animals will not be sacrificed as a result of the research. Pending Association for Assessment of Laboratory Animal Care (AALAC) approval must be submitted prior to funding of a grant involving animal subjects. To avoid all conflicts, any principal investigator or other participating applicant who is currently accepting tobacco industry funding, has recently applied for tobacco funding, or has any other association or consulting position with a tobacco company or affiliate of the tobacco industry is not eligible for FAMRI funding. All prior funding and/or association with tobacco companies must be disclosed and will be considered on a case-by-case basis.
An award will not be made or will be withdrawn if the applicant becomes the Principal Investigator on a peer-reviewed grant with the same general aims as those proposed in the application to FAMRI. FAMRI reserves the right to announce the award publicly in a manner and at a time of its choosing.
*Geographic Focus:* All States, All Countries
*Date(s) Application is Due:* Sep 26
*Amount of Grant:* 162,750 - 542,500 USD
*Samples:* Muragappa Ramanathan, Jr., M.D., John Hopkins University, Baltimore, Maryland - "Nrf2 and Tobacco Smoke Exacerbated Chronic Sinusitis"; Monica Goldklang, M.D., Columbia University, New York, New York - "The Role of MMP—13 in Emphysema Development"; Fabienne Gally, Ph.D., National Jewish Health, Denver, Colorado - "Cigarette Smoke Impairs FABP5—Mediated Bacterial Clearance".
*Contact:* Elizabeth Kress, Executive Director; (305) 379-7007; fax (305) 577-0005; ekress@famri.org
Marti Gammon; (305) 379-7007; fax (305) 577-0005; mgammon@famri.org
*Internet:* www.famri.org/current_awards/index.php
*Sponsor:* Flight Attendant Medical Research Institute
201 S Biscayne Boulevard, Suite 1310
Miami, FL 33131

**Fannie and John Hertz Foundation Graduate Fellowships** **1116**
The foundation supports graduate students working towards the PhD degree in the applied physical sciences. Fellows must have proposed fields of graduate study in engineering, applied physics, computer science, math, chemistry, or other area in the applied physical sciences; have excellent previous scholastic performance, typically including at least an A- grade average during the last two years of undergraduate work; be a U.S. citizen or a permanent resident of the United States; and be ready to make themselves, their skills, and abilities available for the common defense in the event of national emergency.
*Requirements:* Applicant must be a U.S. citizen and have a bachelor's degree and propose to complete a program of graduate study leading to an advanced degree, but not in a professional degree program or a joint PhD/professional degree program.
*Restrictions:* Joint-professional degree programs are not funded.
*Geographic Focus:* All States
*Date(s) Application is Due:* Oct 28
*Amount of Grant:* 31,000 - 36,000 USD
*Contact:* Jay Davis, PhD; (510) 373-1642; fax (510) 373-6329; askhertz@aol.com
*Internet:* www.hertzfoundation.org/dx/fellowships/fellowshipaward.aspx
*Sponsor:* Fannie and John Hertz Foundation
2300 First Street, Suite 250
Livermore, CA 94550

**Fannie E. Rippel Foundation Grants** **1117**
The foundation aids, assists, funds, equips, and provides maintenance for corporations, institutions, associations, organizations, or societies maintained for the relief and care of aged women; provides funds for the building, equipping, and maintenance of hospitals; and provides funds for corporations, institutions, and other organizations existing for treatment of and/or research on heart disease or cancer. The foundation gives emphasis to the equipment and programmatic needs of major teaching medical centers and local rural hospitals, particularly where opportunities exist for leveraging the expertise or capabilities of the medical centers/rural hospitals. Programs should reach underserved rural and urban groups, advocate preventive care, present strategies to change behaviors of the people served, and promote humanistic medicine and mind-body-spirit connections in the healing process. Preference also is given to proposed projects where the benefits can be leveraged through challenge grants.
*Requirements:* Organizations, associations, institutions, and hospitals in the Northeast are eligible.
*Restrictions:* Grants are not awarded to individuals.
*Geographic Focus:* Connecticut, Maine, Massachusetts, New Hampshire, Rhode Island, Vermont
*Amount of Grant:* 50,000 - 300,000 USD
*Contact:* Leigh Scherrer, Foundation Associate; (973) 540-0101, ext. 305; fax (973) 540-0404; lscherrer@rippelfoundation.org
*Internet:* rippelfoundation.org/
*Sponsor:* Fannie E. Rippel Foundation
14 Maple Avenue, Suite 200
Morristown, NJ 07960

**FAR Fund Grants** **1118**
The FAR Fund Project is a New Orleans-based program exploring Hurricane Katrina's effects on New Orleans therapists and therapeutic practice. It was designed by and for clinicians. The project's mission is two fold: to offer support and concrete help to local clinicians, and to develop a model for better understanding how shared trauma affects therapists and therapy. Through this project, the FAR Fund wishes to unite and revitalize clinician communities following large-scale disasters wherever they occur, starting in New Orleans.
*Requirements:* The FAR Fund Project is open to New Orleans psychotherapists of all disciplines and theoretical orientations. Their mission is two-fold: to offer support and concrete help to local mental health clinicians, and to develop a psychodynamic model to better understand how shared trauma affects therapists and therapy.
*Geographic Focus:* Louisiana
*Contact:* Shirlee Taylor; (212) 982-8400; fax (212) 982-8477; FARFund@mac.com
*Sponsor:* Far Fund
928 Broadway, Suite 902
New York, NY 10010

**Farmers Insurance Corporate Giving Grants** 1119

The corporate community relations program awards grants in the areas of education, public safety, arts and culture, civic improvement, and health and human services. Education giving focuses on literacy programs, mentoring programs, adopt-a-school programs, employee matching grants, and aid-to-education undergraduate scholarships. Public safety awards support tougher laws against drunk driving, drug- and alcohol-free graduation night parties, neighborhood crime prevention, highway safety, and earthquake relief. Arts and culture funding supports children's programs and public television. Civic improvement focuses on recognizing exemplary youth, voter registration drives, adopt-a-highway programs, and community paint-a-thons. Health and human services giving supports March of Dimes, United Way, aid for families with cancer, aid to migrant farmworkers, and feeding the hungry. There are no application deadlines. Requests for contributions should be in the form of a letter outlining the purpose of the organization or program. The letter also should include the amount requested, its intended use, and a description of how Farmers' support will be recognized. Additional information should include a budget, annual report, proof of tax-exempt status, and a roster of the board of directors.
*Requirements:* 501(c)3 tax-exempt organizations are eligible.
*Restrictions:* Farmers does not make charitable contributions to individuals, political candidates, or religious groups or for sports events, advertising or raffle tickets, construction projects, or international programs.
*Geographic Focus:* All States
*Contact:* Doris Dunn, Director of Community Relations; (888) 327-6335
*Internet:* www.farmers.com/corporate_giving.html
*Sponsor:* Farmers Insurance Group of Companies
4680 Wilshire Boulevard
Los Angeles, CA 90010

**FCD New American Children Grants** 1120

The foundation awards grants to support programs for children, particularly the disadvantaged, and promote their well-being through basic and policy-relevant research about the factors that promote optimal development of children and adolescents; policy analysis, advocacy, services, and public education to enhance the discussion and adoption of social policies that support families in their important child-raising responsibilities; and leadership development activities linked to the programmatic focus of the foundation. Grants focus on the integration of research, policy, and advocacy in two areas: the availability of and access to early childhood education programs and health care for children. Most grants support research, but a small number of direct service grants are made for New York City-based projects that advance the foundation's research and policy analysis efforts. There are no application deadlines; submit a brief letter of inquiry. Full proposals are by invitation only, following a strict pre-proposal process.
*Requirements:* Nonprofit organizations are eligible.
*Restrictions:* The foundation does not consider requests for scholarships or support for individuals, capital campaigns, building purchase or renovation, or equipment purchase. The foundation does not make grants outside the United States.
*Geographic Focus:* All States
*Amount of Grant:* Up to 800,000 USD
*Samples:* New America Foundation, Washington, D.C., $770,000 - continued support of the PreK-3rd Education Reform Initiative and the Federal Education Budget Project; Rutgers University Foundation, New Brunswick, New Jersey, $30,000 - support of the Burns Family Endowment for Teacher Leadership in Early Childhood Education; University of Texas at Austin, Austin, Texas, $349,561 - continued support of the Dual-Generation Strategy Initiative Project.
*Contact:* Dorothy Pflager, Grants Manager; (212) 867-5777; fax (212) 867-5844; dorothy@fcd-us.org or info@fcd-us.org
*Internet:* fcd-us.org/grants
*Sponsor:* Foundation for Child Development
295 Madison Avenue, 40th Floor
New York, NY 10017

**FCD Young Scholars Program Grants** 1121

The Foundation for Child Development has endured for over 100 years as a result of the constancy of its commitment to promoting the well-being of children and the flexibility with which it evolves with the times, embracing new strategies as children's needs, our society, and research knowledge have changed. The Foundation's Young Scholars Program (YSP) is no exception to this process of renewal. The Foundation remains steadfastly committed to this flagship program aimed at developing the next generation of developmental scientists who are equipped to advance policy and practice on behalf of the nation's children. FCD is committed to the following issues in developing the next generation of the YSP: empirical and professional mentoring; recruiting scholars from underrepresented groups, notably scholars of color; creating a community of interdisciplinary scholars and supporting collaboration among them; ensuring that pressing policy and practice challenges inform the work of the scholars; fostering the scholars' capacity to translate their work for policy and practice audiences; and expanding opportunities for the young scholars to engage with other FCD grantees by, in part, creating greater thematic coherence across all of FCD's strategic initiatives. Letters of Intent (LOIs) must be received by June 1, and final applications by October 19.
*Requirements:* Eligible researchers must have received their doctoral degrees (e.g., Ph.D., Ed.D., Psy.D., M.D., J.D., etc) within one to seven years of application submission (ten years for physician applicants). Applicants must hold this degree or its equivalent in one of the behavior or social sciences, or in an allied professional field (i.e., public policy, public health, education, social work, nursing). Applicants must hold a position as a full-time, tenure-track faculty member of a U.S. college or university. The affiliated private non-profit organization must have a minimum operating budget of $2.5 million and a minimum three-year track record in conducting multi-year research projects.
*Restrictions:* The Foundation does not consider requests for: capital campaigns, the purchase, construction or renovation of buildings, grants for projects outside the United States, the direct provision of preschool education or child care, or health care; or under the foundation's health focus, research, policy, or direct-service projects concerned with specific illnesses.
*Geographic Focus:* All States
*Date(s) Application is Due:* Oct 19
*Contact:* Grants Officer; (212) 867-5777; fax (212) 867-5844; ysp@fcd-us.org
*Internet:* fcd-us.org/our-work/young-scholars-program
*Sponsor:* Foundation for Child Development
295 Madison Avenue, 40th Floor
New York, NY 10017

**FDA Advanced Spectroscopy of Chiral Pharmaceutical Fellowship** 1122

The Advanced Spectroscopy of Chiral Pharmaceutical Fellowship is a research opportunity is available in the Office of Pharmaceutical Quality/Office of Testing and Research at the Center for Drug Evaluation and Research (CDER), Food and Drug Administration (FDA). Raman optical activity spectroscopy will be evaluated as quantitative tool for chiral active pharmaceutical ingredient tests. Linearity, range, LOD, LOQ, precision and accuracy will be determined for 3 API by using predetermine mixture of enantiomers. The main outcome for this project will be to establish the limits in which Raman optical activity can be used as quantitative tool. This program, administered by ORAU through its contract with the U.S. Department of Energy to manage the Oak Ridge Institute for Science and Education, was established through an interagency agreement between DOE and FDA. The initial appointment is for three months, but may be renewed upon recommendation of FDA contingent on the availability of funds. The participant will receive a monthly stipend commensurate with educational level and experience. Proof of health insurance is required for participation in this program. The appointment is full-time at FDA in the St Louis, Missouri area. Participants do not become employees of FDA or the program administrator, and there are no fringe benefits paid.
*Requirements:* Applicants must currently be enrolled at an accredited U.S. college or university or have received a bachelors degree in Chemical Engineering, Biochemistry, or Chemistry within the last five years.
*Geographic Focus:* All States
*Contact:* Darshini Satchi, (301) 796-3603 or (888) 463-6332; fax (301) 827-7101; fdarpp@orau.org
*Internet:* https://www.zintellect.com/Posting/Details/2228
*Sponsor:* U.S. Food and Drug Administration
10903 New Hampshire Avenue
Silver Spring, MD 20993

**FFAR National Cover Crop Initiative Grants** 1123

The Foundation for Food and Agriculture Research builds unique partnerships to support innovative science addressing today's food and agriculture challenges. The National Cover Crop Initiative is the first FFAR Research investment in Healthy Soils, Thriving Farms. The focus of the initiative will be to identify cover crop species with the greatest potential to improve soil health and evaluate such species over a broad geography within three groups: small grains (wheat, rye, oat and triticale), annual legumes (hairy vetch, winter peas and clovers), and brassicas (turnips, radishes, kale and mustards). This initiative will take advantage of advanced breeding techniques that have traditionally been limited in application to high-value row crops
*Geographic Focus:* Maryland, Missouri, Nebraska, North Carolina, Oklahoma
*Contact:* Sally Rockey, Executive Director; (202) 624-0700; srockey@foundationfar.org
*Internet:* foundationfar.org/challenge/healthy-soils-thriving-farms/national-cover-crop-initiative/
*Sponsor:* Foundation for Food and Agriculture Research
401 9th Street NW, Suite 620
Washington, D.C. 20004

**FFAR New Innovator in Food and Agriculture Research Award** 1124

The Foundation for Food and Agricultural Research (FFAR) invites current nominees to its New Innovator in Food and Agriculture Research Award to submit applications. With this program, FFAR intends to support and promote the future generation of exceptionally talented and creative new faculty who are conducting critical research and establishing research programs that will lead to expanded availability of food and facilitate the global practice of sustainable agriculture as the world's population grows to more than 9 billion people by the year 2050. With this award, FFAR seeks to promote career advancement of highly creative and promising new scientists who intend to make a long-term career commitment to research in food and agriculture and bring innovative, ground-breaking research initiatives and thinking to bear on problems facing food and agriculture. FFAR anticipates naming no more than a total of ten awards within the following target areas of research: optimizing agricultural water use; transforming soil health; enhancing sustainable farm animal productivity, resilience, and health; improving plant efficiency; achieving a deeper understanding of nutrition and healthful food choices; managing food production systems to enhance nutritional outcomes; and spurring food system innovation. FFAR will award each successful applicant up to $100,000 per year for three years with an additional $100,000 required cash match from a non-federal source, for a total award of up to $600,000. The deadline to submit an application is 11:59 pm EST on April 21.
*Requirements:* Eligible organizations include: U.S. Institutions of higher education, both public and private; U.S. nonprofit research institutions (other than institutions of higher education); and the U.S. Department of Agriculture.

*Geographic Focus:* All States
*Date(s) Application is Due:* Apr 21
*Amount of Grant:* Up to 300,000 USD
*Contact:* Lucyna Kurtyka, Senior Scientific Program Director; (202) 624-0700; lkurtyka@foundationfar.org
*Internet:* foundationfar.org/new-innovator/
*Sponsor:* Foundation for Food and Agriculture Research
401 9th Street NW, Suite 620
Washington, D.C. 20004

**FFAR Pollinator Health Fund Grants** 1125
The Pollinator Health Fund is a major new initiative from the Foundation for Food and Agriculture Research. Established in response to the increasingly visible agricultural issue of declining pollinator health in agricultural systems, this program focuses on applied research that addresses the social and economic realities faced by beekeepers, farmers, ranchers, private businesses and others engaged in working toward addressing this problem. The Fund will bring together major stakeholders from the academic, nonprofit, public and private sectors to explore how to improve pollinator health by: understanding multiple interacting stressors; developing best management practices and their increasing application; accelerating technology transfer; and enhancing outreach and education. Pollinator Health Fund grants will be awarded to projects that significantly advance pollinator health in the United States. The portfolio of successful proposals will address pollinator issues across the country's geographic regions and diverse cropping systems, including pollinator-dependent commodities as well as pollinator-independent commodities with impacts on pollinator health. The Fund will devote considerable resources to high-intensity agricultural areas that present major opportunities for improvement of pollinator health. The Fund is accepting Pre-Proposals through April 17. Full invited proposals will be due no later than August 18., with applicants notified by November 1. Individual funding requests should not be less than $100,000 or exceed $1,500,000.
*Requirements:* Eligible organizations include: public and private institutions of higher education; nonprofit organizations; and for-profit organizations,
*Geographic Focus:* All States
*Date(s) Application is Due:* Aug 18
*Amount of Grant:* 100,000 - 1,500,000 USD
*Contact:* Tawny Mata, Scientific Program Director; (202) 836-9957 or (202) 624-0700; tmata@foundationfar.org
*Internet:* foundationfar.org/pollinator-health-fund/
*Sponsor:* Foundation for Food and Agriculture Research
401 9th Street NW, Suite 620
Washington, D.C. 20004

**FFAR Rapid Outcomes from Agricultural Research (ROAR) Grants** 1126
The Rapid Outcomes from Agricultural Research (ROAR) program, created by FFAR, provides nimble deployment of funds to support research and outreach in response to emerging or unanticipated threats to the nation's food supply or agricultural systems. ROAR participants, including but not limited to university researchers, farmers or producers, commodity groups and government officials, may apply for funds prior to an outbreak for development of diagnostics, monitoring and mitigation strategies, or enter into an agreement with FFAR that enables the quick release of funds should an outbreak occur. In this way, the ROAR program supports pre-outbreak efforts, and in the case of an outbreak, fills the gap until traditional, longer-term funding sources can be secured. Up to $150,000 per one-year grant is available from FFAR, with the requirement that recipients provide equal or greater matching funds from non-U.S. federal sources. The annual proposal deadline is May 1.
*Requirements:* Eligible organizations include: institutions of higher education; non-profits; farmers and producers; and government agencies.
*Geographic Focus:* All States
*Date(s) Application is Due:* May 1
*Amount of Grant:* Up to 150,000 USD
*Contact:* Tim Kurt, Scientific Program Director; (202) 816-1884 or (202) 624-0700; tkurt@foundationfar.org
*Internet:* foundationfar.org/ROAR/
*Sponsor:* Foundation for Food and Agriculture Research
401 9th Street NW, Suite 620
Washington, D.C. 20004

**FFAR Seeding Solution Grants** 1127
To kick off FFARs new direction of embarking on challenge areas, the Foundation is Seeding Solutions, a call to the community to come forward with bold, innovative, and potentially transformative research proposals in its recently launched challenge areas. The seven designated challenge areas are: food waste and loss; protein challenge; water scarcity; innovation pathway to sustainability; healthy soils, thriving farms; urban food systems; and making my plate your plate. Prospective grantees may come forward with a proposal for up to $1,000,000 of FFAR funding, and should secure one-to-one matching funding from a non-Federal source before a grant will be awarded. FFAR anticipates funding at least one meritorious and transformative proposal in each of its designated challenge areas. Each grant will be awarded over a period of one to five years. FFAR will continue to seek and develop additional partnerships and projects in the Challenge areas; this will serve as a first step to bringing innovative ideas to FFAR's doorstep. The annual pre-proposal deadline is January 16, with invited applications due by April 1. Successful applicants will be notified by June 1.
*Requirements:* To be considered, projects must be transformative research that addresses and provide solutions to an intractable problem and/or accelerates innovation within the Challenge Areas.
*Geographic Focus:* All States
*Amount of Grant:* Up to 1,000,000 USD
*Contact:* Lucyna Kurtyka; (202) 624-0700; lkurtyka@foundationfar.org
Tim Kurt; (202) 816-1884 or (202) 624-0700; tkurt@foundationfar.org
*Internet:* foundationfar.org/open-opportunities/seeding-solutions/
*Sponsor:* Foundation for Food and Agriculture Research
401 9th Street NW, Suite 620
Washington, D.C. 20004

**Fidelity Foundation Grants** 1128
Fidelity Investments Chairman Edward C. Johnson 3rd and his father, the founder of the company, established the Fidelity Foundation in 1965 with several operating principles in mind. These principles, still current today, guide the Foundation's decisions and grantmaking. The Foundation grant program was designed to strengthen the long-term effectiveness of nonprofit institutions. The types of projects it funds, and the way in which it funds them, are specifically intended to help nonprofits build the organizational capabilities they need to better fulfill their missions and serve their constituencies. The Fidelity Foundation considers Letters of Inquiry from organizations with current IRS 501(c)3 public charity status only. Grants are made to fund only significant, transformative projects usually budgeted at $100,000 or more. The Foundation's primary philanthropic investments are allocated to the following sectors: community and social services including youth development, providing access to basic human needs such as food, health care, and housing; museums, historical, and other cultural organizations; financial literacy for youth including financial education and capability; and educational institutions. The Foundation considers influential community projects from organizations in regions around Fidelity Investments' employee sites in the listed metropolitan areas. Organizations of national importance and high-impact projects with potential to inform the nonprofit sector are also of interest. The Foundation seeks to work with organizations on strategic, transformative projects that have the potential to substantially increase the grantee's impact, efficiency, or long-term sustainability. These capacity-building projects fall into three areas: capital investments; planning initiatives; and technology projects. Typically the initial Letter of Inquiry (LOI) review process will take between four and six weeks.
*Requirements:* Grants are generally made only to organizations with operating budgets of $1,000,000 or more. The Fidelity Foundation considers projects from organizations of regional or national importance throughout the United States. Proposed projects must be budgeted at $100,000 or more.
*Restrictions:* Grants are not awarded to support sectarian organizations, disease-specific associations, or public school systems. Support does not go to individuals or for scholarships, civic or start-up organizations, corporate memberships, operating support, or participation in benefit events, film, or video projects.
*Geographic Focus:* All States
*Amount of Grant:* 100,000 - 1,000,000 USD
*Contact:* Genie Healy McGowan; (617) 563-6806; info@FidelityFoundation.org
*Internet:* www.fidelityfoundation.org/grantmaking-guidelines/
*Sponsor:* Fidelity Foundation
82 Devonshire Street, S2
Boston, MA 02109

**Fight for Sight-Streilein Foundation for Ocular Immunology Research Award** 1129
Beginning in 2011, Fight for Sight and the Streilein Foundation for Ocular Immunology co-sponsored a new annual Summer Student Research Fellowship, Fight for Sight-Streilein Foundation for Ocular Immunology Research Award, for the study of ocular immunology. Candidates for the award, an annual $2,100 grant for fulltime research conducted from June to August, are invited to apply by visiting www.fightforsight.org for an application.
*Requirements:* The award is open to undergraduates, graduate and medical students studying in the United States. Applications will be reviewed by the Fight for Sight Scientific Review Committee and Streilein Foundation for Ocular Immunology and a grantee will be selected based upon the committee's recommendations to the Boards of both organizations.
*Geographic Focus:* All States
*Date(s) Application is Due:* Feb 1
*Amount of Grant:* 2,100 - 2,100 USD
*Contact:* Janice Benson; (212) 679-6060; fax (212) 679-4466; janice@fightforsight.org
*Internet:* www.streilein-foundation.org/fellowships.html
*Sponsor:* Streilein Foundation for Ocular Immunology
P.O. Box 6104
Boston, MA 02114

**Fight for Sight Grants-in-Aid** 1130
Fight for Sight was founded in 1946 by Mildred Weisenfeld, a young woman with retinitis pigmentosa, to encourage and fund research in ophthalmology, vision and related sciences. The goal of Fight for Sight is to encourage and facilitate research in detection, understanding, prevention, treatment and cures of visual disorders especially those diseases leading to impaired sight or blindness. To that end, FFS funds Grants-in-Aid to academic researchers within the first three years of their faculty appointments. FFS primarily supports new investigators, promoting the development of scientific skills and enabling the development of preliminary scientific findings and pilot studies necessary to successfully apply for more substantial federal and private funding such as that provided by the National Eye Institute and other divisions of the NIH. Support may be used to defray costs of personnel (but not the applicant), equipment and consumable supplies needed for the specific research project. One year awards of $20,000 are provided and may start between July 1 and September 1.
*Requirements:* Applications will only be considered from researchers who have received their first faculty or research appointment in eye/vision within the previous three years.

Fringe benefits are not included and institutional overhead charges are not covered. Travel costs are generally not supported.
*Geographic Focus:* All States
*Date(s) Application is Due:* Jan 1
*Amount of Grant:* 20,000 - 20,000 USD
*Contact:* Janice Benson; (212) 679-6060; fax (212) 679-4466; janice@fightforsight.org
*Internet:* https://www.fightforsight.org/Grants/Research-Award-Types/
*Sponsor:* Fight for Sight (USA)
381 Park Avenue S, Suite 809
New York, NY 10016

**Fight for Sight Post-Doctoral Awards** **1131**
Post-Doctoral Awards support individuals with a doctorate (Ph.D., M.D., O.D., Dr.PH, or D.V.M.) who are interested in academic careers in basic or clinical research in ophthalmology, vision or related sciences. This funding is intended to offer those interested in an academic career the opportunity to spend a year engaged in vision and eye research under the supervision of a senior scientist/clinician mentor. Clinical post-doctoral researchers are required to spend sufficient time on the funded research project to carry out the proposed objectives while basic researchers are expected to work full-time. One year grants of $20,000 are awarded for start dates between July 1 and September 1.
*Requirements:* Applications are considered from individuals who are within three years of their doctoral degrees or clinical residency training and have not received a previous Fight for Sight fellowship award. If at the time of application a doctorate has not yet been obtained, a cover letter must be submitted by the conferring institution advising when such degree is expected to be awarded. Recipients may supplement their awards with institutional or other funds however any anticipated supplemental support must be disclosed at the time of application. Total combined salary support must not exceed the annual stipend level set by the NIH for National Research Service Award recipients.
*Restrictions:* Fringe benefits are not provided by Fight for Sight.
*Geographic Focus:* All States
*Date(s) Application is Due:* Feb 1
*Amount of Grant:* 20,000 - 20,000 USD
*Contact:* Janice Benson; (212) 679-6060; fax (212) 679-4466; janice@fightforsight.org
*Internet:* https://www.fightforsight.org/Grants/Research-Award-Types/
*Sponsor:* Fight for Sight (USA)
381 Park Avenue S, Suite 809
New York, NY 10016

**Fight for Sight Summer Student Fellowships** **1132**
Fight for Sight was founded in 1946 by Mildred Weisenfeld, a young woman with retinitis pigmentosa, to encourage and fund research in ophthalmology, vision and related sciences. The goal of Fight for Sight is to encourage and facilitate research in detection, understanding, prevention, treatment and cures of visual disorders especially those diseases leading to impaired sight or blindness. Unrestricted awards of $2,100 are given for two to three months of full-time research, usually during June-August.
*Requirements:* Fight for Sight's primary mission is to support and encourage promising scientists early in their careers. FFS primarily supports new investigators, promoting the development of scientific skills and enabling the development of preliminary scientific findings and pilot studies necessary to successfully apply for more substantial federal and private funding such as that provided by the National Eye Institute and other divisions of the NIH. Summer Student Fellowships are available to undergraduates, graduate and medical students who are interested in pursuing eye-related clinical or basic research. For most students, this is their first exposure to eye or vision research and the experience has resulted in many students choosing academic ophthalmology or eye research as a full-time career.
*Restrictions:* Students receiving stipends from other sources are generally not eligible.
*Geographic Focus:* All States
*Date(s) Application is Due:* Feb 1
*Amount of Grant:* 2,100 - 2,100 USD
*Contact:* Janice Benson, Assistant Director; (212) 679-6060; fax (212) 679-4466; janice@fightforsight.org
*Internet:* https://www.fightforsight.org/Grants/Research-Award-Types/
*Sponsor:* Fight for Sight (USA)
381 Park Avenue S, Suite 809
New York, NY 10016

**FINRA Investor Education Foundation General Grants** **1133**
The Foundation funds research and educational projects that support its mission of providing underserved Americans with the knowledge, skills and tools necessary for financial success throughout life. The Foundation seeks proposals that advance its mission through high-impact investor education or investor protection projects that are efficiently delivered and achieve significant and tangible benefits for the target audience; research that expands the body of knowledge and offers solutions in the fields of investor education and investor protection; and initiatives that lead with a research element and follow with a high-impact investor education or investor protection project based on the results of the research. Of particular interest are projects that better prepare older Americans for handling their finances during retirement; encourage Americans to implement steps to achieve basic retirement security and avoid investing pitfalls; encourage women to take control of their financial future through investor education; advance practice, policy, and thought in the fields of investor education and investor protection; and encourage investors to check the background of financial professionals prior to doing business with them. There is no set minimum or maximum grant amount. Projects should be completed within 24 months of grant award. Those exceeding 24 months will be considered if justification for the additional time is included in the proposal.
*Requirements:* 501(c)3 tax-exempt organizations are eligible.
*Restrictions:* The foundation will not award grants to individuals; organizations affiliated with a current member of the foundation board of directors or FINRA board of governors; securities firms regulated by FINRA; organizations affiliated with a securities firm or individual regulated by FINRA, such as a foundation established by a securities firm; securities regulators, self-regulatory organizations, or securities industry trade associations; organizations that are termed disqualified persons pursuant to Article III, Section 3(d) of the FINRA by-laws; foreign organizations; or entities that discriminate on the basis of age, color, disability, marital status, national origin, race, religion, sex, sexual orientation, or veteran status. The Foundation will generally not consider proposals to fund: international programs or projects; expenses that are not directly related to the project for which funding is sought; salaries of permanent staff (for example, prorated salaries of administrative and executive personnel, or oversight and coordination activities of a project principal); capital costs, such as building and construction or equipment such as computer hardware and office furniture; pass-through funding (for example, if the 501(c)3 organization plans to turn over the funding to a proprietary organization or consultant); projects with a potential conflict of interest (for example where funded technical support or expertise might be provided by a board member of the 501(c)3 organization); conferences and similar activities that fail to provide a long-term solution or sufficiently broad outreach; distribution methodologies that require ongoing maintenance when the ability to perform upkeep without continued funding is questionable (for example, materials with a short "shelf life" that would require ongoing funding for frequent updating); projects with proprietary elements, such as for-profit activities, use or purchase of copyrighted or trademarked materials, and proprietary research; lobbying, political contributions, fund-raising events, or other similar activities designed to influence legislation or intervene in political campaigns; donations, endowments, challenge grants, matching funds, and other similar programs; or direct or matching payments to members of the public, such as scholarships, assistance with personal and family financial difficulties, registration fees for conferences and training, or similar activities.
*Geographic Focus:* All States
*Amount of Grant:* Up to 750,000 USD
*Samples:* First Nations Development Institute, Longmont, Colorado, $519,194 - New Distribution Channels for Investor Education; North Carolina State University, Raleigh, North Carolina, $454,378 - A Web-Based Fraud Risk Assessment Tool; National Labor College, Silver Spring, Maryland, $618,035 - Investor Education for Working Families.
*Contact:* Susan M. Sarver, (202) 728-6948; susan.sarver@finra.com
*Internet:* www.finrafoundation.org/grants/general/
*Sponsor:* FINRA Investor Education Foundation
1735 K Street, NW
Washington, D.C. 20006-1506

**FishAmerica Foundation Marine and Anadromous Fish Habitat Restoration Grants** **1134**
The FishAmerica Foundation Marine and Anadromous Fish Habitat Restoration program requests proposals for local efforts to accomplish meaningful on-the-ground restoration of marine, estuarine and riparian habitats, including salt marshes, seagrass beds, mangrove forests, and freshwater habitats important to anadromous fish species (fish like salmon and striped bass that spawn in freshwater and migrate to the sea). Emphasis is on using a hands-on, grassroots approach to restore fisheries habitat across coastal America and the Great Lakes.
*Requirements:* Projects must result in the implementation of locally-driven habitat restoration projects that emphasize stewardship and yield ecological and socioeconomic benefits. These projects must clearly demonstrate significant benefits to marine, estuarine or anadromous fisheries resources, particularly sportfish, and should involve community participation through an educational or volunteer component tied to the restoration activities. Nonprofit organizations such as sporting clubs, civic organizations, conservation groups, and to a lesser extent state agencies, in the United States and Canada are eligible. Projects must result in on-the-ground habitat restoration, clearly demonstrate significant benefits to marine, estuarine or anadromous fisheries resources, particularly sportfish, and must involve community participation through an educational or volunteer component tied to the restoration activities.
*Geographic Focus:* All States
*Date(s) Application is Due:* Apr 30
*Amount of Grant:* 10,000 - 75,000 USD
*Contact:* Patrick Egan, Grants Manager; (703) 519-9691, ext. 247; fax (703) 519-1872; pegan@asafishing.org or fishamerica@asafishing.org
*Internet:* www.fishamerica.org/grants.html
*Sponsor:* FishAmerica Foundation
1001 North Fairfax Street, Suite 501
Alexandria, VA 22314

**FishAmerica Foundation Research Grants** **1135**
Overall, the FishAmerica Foundation will fund research projects that further the National Fish Habitat Plan. Specifically, the program funds research projects that have regional or national implication, not local. Grants support research in the following areas: fisheries management, water quality, habitat studies, stock enhancement, economic impact studies related to sport fishing, and tagging.
*Requirements:* Nonprofit organizations such as sporting clubs, civic organizations, conservation groups, and to a lesser extent state agencies, in the United States and Canada are eligible.
*Restrictions:* Grants do not support the following projects or purposes: donations of fishing tackle; salaries, administration, overhead, or travel for conservation projects; individuals; local stream monitoring programs; political activities that attempt to

influence political campaigns or legislation; access projects such as road construction, boat ramps, or fishing piers; loans, endowments, trips, tours, tickets, advertising, or publications; or permitting and related costs.
*Geographic Focus:* All States
*Date(s) Application is Due:* Jul 31
*Amount of Grant:* Up to 25,000 USD
*Contact:* Patrick Egan, Grants Manager; (703) 519-9691, ext. 247; fax (703) 519-1872; pegan@asafishing.org or fishamerica@asafishing.org
*Internet:* www.fishamerica.org/grants.html
*Sponsor:* FishAmerica Foundation
1001 North Fairfax Street, Suite 501
Alexandria, VA 22314

**Fitzpatrick and Francis Family Business Continuity Foundation Grants** 1136
The Family Business Continuity Foundation is dedicated to providing financial support for the training and education of family business advisors either through grants directly to individuals or to institutions that provide education and training to individuals. The Foundation wants to ensure that family businesses receive the best possible advice as they create Family Councils, select a competent board of directors, and plan to successfully transfer ownership and leadership of their enterprise from one generation to the next. Primary fields of interest include: the arts, community and economic development, and education. A formal application is required, though there are no specified annual deadlines. Funding is primarily offered in Kansas, Missouri, and Vancouver, British Columbia, though the Foundation does provide grants outside of its geographic designated areas. Grants range from $500 to $10,000.
*Geographic Focus:* All States, Canada
*Amount of Grant:* 500 - 10,000 USD
*Samples:* University of British Columbia, Washington, D.C., $10,000 - business program support; Grand Valley University, Grand Rapids, Michigan, $5,000 - research and scholarship support; KTWU Public Television, Topeka, Kansas, $2,500 - programming support.
*Contact:* John J. Fitzpatrick; (785) 273-8596 or (785) 273-8500; fax (785) 273-8768
*Internet:* familybizfoundation.org/
*Sponsor:* Fitzpatrick and Francis Family Business Continuity Foundation
P.O. Box 4815
Topeka, KS 66604-0815

**FIU Díaz-Ayala Library Travel Grants** 1137
The Díaz-Ayala Cuban and Latin American Popular Music Collection is the most extensive publicly available collection of Cuban music in the United States. Its approximately 100,000 items span the history of popular Cuban and other Latin musics. Originally valued at nearly one million dollars, the collection features 28,000 LPs; 17,000 78 rpms; 4,500 cassettes containing interviews with composers and musicians, radio programs, music, and other materials; 4,000 pieces of sheet music; 3,000 books; and thousands of CDs, photographs, videocassettes, and paper files. Among the collection's rarest items are early recordings made in prerevolutionary Cuba. The Cuban Research Institute (CRI), the Latin American and Caribbean Center (LACC), and Florida International University Libraries offer the Díaz-Ayala Library Travel Grants to study the special collections related to Cuba and Cuban Americans. These grants provide scholars and graduate students the opportunity to conduct research on Cuba and its diaspora at the FIU Green Library, thereby expanding access to the library's unique holdings and enhancing its value as a national resource. CRI, LACC, and FIU Libraries offer research stipends of up to $1,500 each to offset the costs of a minimum one-week stay in Miami to use the collections.
*Requirements:* Scholars in the humanities and the social sciences whose work will be enhanced by using the resources of the collection are encouraged to apply. Priority is given to scholars who are not previous recipients of the award.
*Geographic Focus:* All States
*Date(s) Application is Due:* Mar 1
*Amount of Grant:* Up to 2 USD
*Contact:* Coordinator; (305) 348-1991; fax (305) 348-7463; cri@fiu.edu or SoundImage@fiu.edu
*Internet:* https://cri.fiu.edu/programs/library-travel-grants/
*Sponsor:* Florida International University
11200 SW 8th Street
Miami, FL 33199

**Florence Foundation Grants** 1138
Fred Farrel Florence, banker and civic leader, was born in New York City on November 5, 1891, to Lithuanian immigrants Mose and Celia (Freedman) Fromowitz. His family moved in 1892 to New Birmingham, Texas, then two years later to Rusk, where he attended public school through the tenth grade, worked in the family store, sold newspapers, and, like his brothers and sister, legally changed his name to Florence. At age fifteen he began his life's work at the First National Bank in Rusk. He advanced to assistant cashier by 1911 and moved on to positions as bookkeeper at the American Exchange Bank in Dallas (1911), cashier at the First State Bank in Ratcliff (1912), and vice president (1912), then president (1915) of the Alto State Bank. Florence enlisted in the Signal Corps Aviation School of the United States Army with the rank of corporal in 1918 and was discharged later that year as a second lieutenant. He then returned to his bank presidency in Alto and later served as a town alderman, then mayor (1919). In 1920 he became first vice president of the Guaranty Bank and Trust Company, which became Republic National Bank and Trust Company in 1922 and was renamed Republic National Bank of Dallas in 1937. Florence was president from 1929 until 1957, when he became chairman of the executive committee with the duties of chief executive officer. Under his direction deposits at his bank increased from $1 million to almost $900 million. In addition to his banking interests, Florence was appointed by governors Ross S. Sterling and Miriam A. Ferguson and President Franklin Roosevelt to serve on various recovery and relief boards during the Great Depression. Florence established the Florence Foundation in December 1956 to support charitable, religious, educational, scientific, and recreational projects for the benefit of humanity.
*Geographic Focus:* Texas
*Date(s) Application is Due:* Mar 1
*Contact:* Debra Goldstein Phares, Philanthropic Client Director; (214) 209-1830 or (800) 357-7094; debra.g.phares@ustrust.com or tx.philanthropic@ustrust.com
*Internet:* https://www.bankofamerica.com/philanthropic/fn_search.action
*Sponsor:* Florence Foundation
901 Main Street, 19th Floor
Dallas, TX 75202-3714

**Florida Division of Cultural Affairs Individual Artist Fellowships** 1139
The Individual Artist Fellowship program is designed to recognize practicing, professional, and creative artists residing in Florida through monetary fellowship awards. The program provides support for artists of exceptional talent and demonstrated ability to improve their artistic skills and enhance their careers. For purposes of eligibility to apply to the program, a professional creative artist is defined as a person who creates, on an ongoing basis, original works of art within an artistic discipline, and is pursuing it as a means of livelihood or for the highest level of professional recognition. Applicants are eligible to submit only one application and only one artist may apply to the Fellowship program even if there is collaboration between one or more artists.
*Requirements:* To be eligible for a fellowship the applicant must: submit a complete application and sample(s) of work on or before the postmark deadline; be a legal resident of Florida, as defined by section 196.015, Florida Statutes or section 22.17, Florida Statutes and agree to maintain Florida residency for the duration of the fellowship period; be a professional creative artist at least 18 years of age; not be enrolled in any undergraduate or graduate degree-seeking program during the fellowship period; not have received a monetary fellowship award from the Florida Department of State, Division of Cultural Affairs during or after fiscal year 2002; not be an individual serving as a fellowship review panelist within the same discipline; not be identified by name on sample(s) of work or Form 3 of the application, if the application is reviewed under a blind review process; submit sample(s) of work completed within the last three years prior to the application deadline.
*Restrictions:* Funds must be spent on activities directly related to the advancement of the fellowship recipient's artistic career.
*Geographic Focus:* Florida
*Date(s) Application is Due:* Jul 1
*Amount of Grant:* 5,000 USD
*Contact:* Morgan Barr Lewis, Arts Administrator; (850) 245-6356; fax (850) 245-6497; morgan.lewis@dos.myflorida.com
*Internet:* www.florida-arts.org/programs/fellowship/
*Sponsor:* Florida Division of Cultural Affairs
500 South Bronough Street, 3rd Floor
Tallahassee, FL 32399-0250

**Foundation Fighting Blindness Marjorie Carr Adams Women's Career Development Awards** 1140
The Program's primary purpose is to support female clinical research scientists of superior dedication and talent to pursue vigorous research programs to find the cures for retinal degenerative diseases research (RDD) i.e. inherited orphan retinal degenerative diseases and dry age-related macular degeneration. The goal of this program is to facilitate advances in the laboratory and clinical research, to elucidate the mechanisms for the etiology and pathogenesis of RDDs and to develop innovative strategies to prevent, treat and cure these diseases. This award provides support for mentored study and research for clinically trained biomedical research professionals who have the potential to assume leadership roles as clinician-scientists in RDD research.
*Requirements:* Female clinical scientists possessing an, M.D., D.O., or recognized equivalent foreign degrees and who are in their first, second or third year of a junior faculty appointment are eligible to apply for the Adams WCDA. Applicants do not have to be U.S. citizens.
*Geographic Focus:* All States
*Date(s) Application is Due:* Nov 14
*Amount of Grant:* 65,000 - 325,000 USD
*Contact:* Cindy Settar, (410) 568-0150 or (800) 683-5555; csettar@Fightblindness.org or grants@Fightblindness.org
*Internet:* www.blindness.org/index.php?option=com_content&view=section&id=9&Itemid=107#career-development
*Sponsor:* Foundation Fighting Blindness
7168 Columbia Gateway Drive, Suite 100
Columbia, MD 21046

**Foundation for Enhancing Communities Grants** 1141
The Foundation awards grants to Pennsylvania nonprofit organizations in the greater Harrisburg area. Areas of interest include art/cultural programs, education, health care and health organizations, human services, and community development. Types of support include program development, publication, seed money, and scholarships. Deadlines vary according to funding availability.
*Requirements:* Tax exempt 501(c)3 organizations in Pennsylvania's Dauphin, Cumberland, Franklin, and Perry Counties are eligible.
*Restrictions:* Grants do not support individuals, operating budgets, or capital expenses.
*Geographic Focus:* Pennsylvania

*Contact:* Jennifer Kuntch; (717) 236-5040; fax (717) 231-4463; jkuntch@tfec.org
*Internet:* www.ghf.org
*Sponsor:* Foundation for Enhancing Communities
200 North 3rd Street, P.O. Box 678
Harrisburg, PA 17108-0678

**Foundation for Evangelism Doctoral Scholarships** **1142**
The Foundation for Evangelism sponsors scholarships for those pursuing doctoral studies in evangelism. Currently these programs are led by professors supported by The Foundation for Evangelism and administered through a ThD program at Duke Divinity School and a PhD program at Asbury Theological Seminary. The overall goal in establishing these scholarships is to equip scholars in the Wesleyan tradition who express an interest in being considered for seminary teaching positions in evangelism. These students will help fulfill the Foundation's vision to raise up generations of leaders who have a passion for evangelism by not only being a leader but also training others to lead.
*Geographic Focus:* All States
*Contact:* Dr. Stephen Gunter, Professorship Director; (828) 454-6800 or (800) 737-8333; fax (828) 454-6810; sgunter@foundationforevangelism.org
*Internet:* foundationforevangelism.org/grants/doctoral-education/
*Sponsor:* Foundation for Evangelism
125 North Lakeshore Drive, P.O. Box 985
Lake Junaluska, NC 28745

**Foundation for Evangelism Harry Denman Fellowships** **1143**
The Foundation for Evangelism currently funds the Harry Denman Fellowships which are given to scholars in the United Methodist tradition who are pursuing doctoral studies in evangelism or related disciplines. The Fellowships are to encourage and equip people who might one day teach evangelism in United Methodist schools of theology or institutions of higher education. Candidates are chosen for their commitment to Christ and the church as well as for their academic excellence. A Foundation for Theological Education (AFTE) administers these fellowships through their John Wesley Fellowship Program.
*Geographic Focus:* All States
*Contact:* Dr. Stephen Gunter, Professorship Director; (828) 454-6800 or (800) 737-8333; fax (828) 454-6810; sgunter@foundationforevangelism.org
*Internet:* foundationforevangelism.org/grants/doctoral-education/
*Sponsor:* Foundation for Evangelism
125 North Lakeshore Drive, P.O. Box 985
Lake Junaluska, NC 28745

**Foundation for Pharmaceutical Sciences Herb and Nina Demuth Grant** **1144**
The purpose of the Foundation for Pharmaceutical Sciences Herb and Nina Demuth grant program is to help alleviate the shortage of skilled medical and pharmaceutical scientists that could hamper the research, development and manufacture of new pharmaceutical and biotech products and delivery technologies. A maximum award of $25,000 is available. To obtain further information concerning the application process, please contact the Foundation office directly via email. Applications should be postmarked by the May 30 annual deadline. emailed or faxed proposals are accepted.
*Requirements:* The Foundation is not affiliated with the Parenteral Drug Association (PDA). Eligible applicants are graduate students in universities, colleges, or medical centers that need assistance to complete their Doctoral thesis. This includes: U.S. citizens, resident, or foreign students studying in U.S. institutions.
*Geographic Focus:* All States
*Date(s) Application is Due:* May 30
*Amount of Grant:* Up to 25,000 USD
*Contact:* Roger Dabbah; (301) 762-9258; fax (301) 762-5356; rdabbah@verizon.net
*Sponsor:* Foundation for Pharmaceutical Sciences
5 Eaglebrook Court
Potomac, MD 20854

**Foundation of Orthopedic Trauma Research Grants** **1145**
The Foundation for Orthopedic Trauma was founded and incorporated in December of 2004 with the mission of influencing and enhancing orthopedic traumatology through education, research, mentorship and research founded on firm clinical grounds. The objective of Research Grant program is to encourage investigators by providing seed and start-up funding for promising research projects in the field of orthopedic trauma surgery through grants of up to $20,000 for a research project extending over a maximum of two years. Up to $30,000 is available for multi-center grants. Both laboratory and clinical projects are suitable, but in either case clinical relevance must be explicitly and clearly described. Completed applications should be submitted via email. The deadline for submission is June 1, with notification of awards to successful applicants mailed by August 30.
*Requirements:* An orthopedic surgeon, FOT member must serve as either the principal or co-principal investigator. Non-trauma/orthopedic surgeon, M.D.'s, Ph.D.'s or D.V.M.'s may serve as the principal or co-principal investigator, as long as they are affiliated with a trauma/orthopedic department with an FOT member orthopedic surgeon as the co-principal investigator.
*Geographic Focus:* All States
*Date(s) Application is Due:* Jun 1
*Amount of Grant:* Up to 30,000 USD
*Contact:* Lee-Ann Finno, Administrator; (212) 305-6392; Administrator@FOTnorthamerica.org or lee.finno@residentswap.org
*Internet:* www.fotnorthamerica.org/research_grants.asp
*Sponsor:* Foundation of Orthopedic Trauma
622 West 168th Street, 11th Floor
New York, NY 10032

**Foundation of the American Thoracic Society Research Grants** **1146**
The American Thoracic Society established the ATS Foundation Research program to foster the development and training of future leaders in science and biomedical research. Through collaborations with non-profit and corporate partners the research program is able to provide research and career development grants and mentored fellowships to encourage young investigators to devote their talents and time to generating novel scientific ideas leading to new life-saving therapies for patients and their families. Since its founding in 2002, the Foundation of the ATS has awarded $10 million in grants to 105 young researchers investigating a wide spectrum of lung diseases, ranging from asthma and COPD to pulmonary fibrosis and alpha-1 antitrypsin. The program has also funded the career development of 24 pulmonary and critical care fellows. The grants amounts vary and range from $5,000 a year for 1 year to $50,000 per year for two years.
*Restrictions:* Grants are not awarded to individuals.
*Geographic Focus:* All States
*Amount of Grant:* 5,000 - 50,000 USD
*Contact:* Alyssa Chase; (212) 315-8600; fax (212) 315-6498; achase@thoracic.org
*Internet:* foundation.thoracic.org/programs/research-program.php
*Sponsor:* Foundation of the American Thoracic Society
25 Broadway, 18th Floor
New York, NY 10004

**Fragile X Syndrome Postdoctoral Research Fellowships** **1147**
FRAXA aims to accelerate research aimed at finding a specific treatment for fragile X syndrome, with a primary goal of bringing practical treatment into current medical practice as quickly as possible. Therefore, preference will be given to research projects that have a clear practical application and the results of which will be shared in a timely fashion. Grants in two categories: investigator-initiated program grants are intended for only the most translationally relevant proposals, with basic science proposals streamlined out without review; postdoctoral fellowship program, though these grants are still awarded with weight given to therapeutic potential. FRAXA will offer eight to twelve postdoctoral fellowships at a fixed rate of $45,000 per year. Successful applicants for these grants will likely be working in established labs with secure overall funding, which have funding for supplies and any required animal handling costs. If the applicant proposes to work with a particular fragile X model (KO mouse, drosophila, human neural stem cells, etc.) s/he should demonstrate that this model system has already been established in-house.
*Requirements:* Individuals are nominated by applicant institutions for the fellowships and should have training and experience at least equal to the PhD or MD level.
*Geographic Focus:* All States
*Date(s) Application is Due:* Feb 1
*Amount of Grant:* 45,000 USD
*Contact:* Michael Tranfaglie, (978) 462-1866; fax (978) 463-9985; mtranfaglia@fraxa.org or info@fraxa.org
Katie Clapp; (978) 462-1866; fax (978) 463-9985; kclapp@fraxa.org or info@fraxa.org
*Internet:* www.fraxa.org/research/apply/
*Sponsor:* FRAXA Research Foundation
10 Prince Place, Suite 203
Newburyport, MA 01950

**France-Merrick Foundation Education Grants** **1148**
The France-Merrick Foundation's education funding is its largest area of focus, representing over 40% of its annual grant-making. The Foundation supports K-12 public education, independent private school education as well as higher education. The Foundation funds independent private schools, for both capital projects and scholarship support. The scholarship support is intended to help increase diversity and inclusion as well as provide positive educational experiences to qualified students, regardless of income. The Foundation also supports projects directed toward improving educational quality and equity in Baltimore City with the understanding that education is a prerequisite to a strong civil society and an employed workforce— two keys to a better Baltimore. Much funding to higher education is made in recognition of the importance of these institutions in creating the next generation of leaders and the positive economic impact of their graduates. Other university funded projects facilitate campus expansion and improve instruction. Types of support include: building and renovation; capital campaigns; equipment purchase; management development; capacity building; matching grants; and technical assistance. Most recently, awards in this area have ranged from $34,000 to $500,000.
*Requirements:* Grants are made to Maryland nonprofit 501(c)3 organizations and institutions primarily located in the Baltimore metropolitan area.
*Geographic Focus:* Maryland
*Amount of Grant:* 34,000 - 500,000 USD
*Samples:* Towson University, Towson, Maryland, $500,000 - to support the Top Ten Percent Program which provides scholarships for Baltimore City and County public school students; University of Maryland, College Park, Maryland, $200,000 - to support the Baltimore Incentives Awards Program; Baltimore Leadership School for Young Women, Baltimore, Maryland, $50,000 - to support outfitting the STEM labs.
*Contact:* Amy M. Gross; (410) 464-2004; fax (410) 464-2001; info@france-merrickfdn.org
*Internet:* www.france-merrickfdn.org/education
*Sponsor:* France-Merrick Foundation
2 Hamill Road, Suite 302
Baltimore, MD 21210

**Francis P. Garvan-John M. Olin Medal** **1149**
This award, sponsored by the the Francis P. Garvan-John M. Olin Medal Endowment and administered by the American Chemical Society, is given annually to recognize distinguished service to chemistry by women chemists who are U.S. citizens. The award

consists of $5,000, a medallion with a presentation box, and a certificate. Up to $2,500 for travel expenses to the meeting at which the award will be presented will be reimbursed.
*Requirements:* A nominee must be a citizen of the United States and have performed distinguished service to chemistry. Any individual, except a member of the awards committee, may submit one nomination or seconding letter for the award in any given year. The nominating documents consist of a letter of not more than 1000 words containing an evaluation of the nominee's accomplishments and a specific identification of the work to be recognized, a biographical sketch including date of birth, and a list of publications and patents authored by the nominee.
*Restrictions:* Self-nominations are not accepted.
*Geographic Focus:* All States
*Date(s) Application is Due:* Nov 1
*Amount of Grant:* 5,000 - 7,500 USD
*Samples:* Elizabeth C. Theil—award winner, $5,000; Laura L. Kiessling—award winner, $5,000; Lila Gierasch—award winner, $5000.
*Contact:* Awards Administrator; (202) 872-4575; fax (202) 776-8008; awards@acs.org
*Internet:* portal.acs.org/portal/acs/corg/content?_nfpb=true&_pageLabel=PP_ARTICLEMAIN&node_id=1319&content_id=CTP_004521&use_sec=true&sec_url_var=region1
*Sponsor:* American Chemical Society
1155 Sixteenth Street, NW
Washington, D.C. 20036-4801

**Franklin Institute Bower Award and Prize for Achievement in Science 1150**
The Bower Award and Prize for Achievement in Science recognizes outstanding achievement in life, physical, and applied sciences; innovation in the sciences; and training of scientists. The award is given without regard for nationality and includes a gold medal and prize of $250,000. The Franklin Institute seeks to honor an individual who has played a seminal role in either the design and construction of robotic systems or the advancement of enabling technologies as related to robotics such as mechanical structure, sensors, and control algorithms.
*Requirements:* This is an international competition for individuals whose work has had a significant impact on the field of Robotics. In cases of equal scientific merit, the factor of current economic value of the discovery or application will weigh favorably on behalf of the candidate. Candidates for the Award must be living, and the winner must participate in the April Awards Ceremony in Philadelphia.
*Restrictions:* This award is for an individual rather than for a group.
*Geographic Focus:* All States
*Date(s) Application is Due:* May 31
*Amount of Grant:* 250,000 USD
*Contact:* Awards Program Director; (215) 448-1329; fax (215) 448-1364; awards@fi.edu
*Internet:* https://www.fi.edu/awards
*Sponsor:* Franklin Institute
222 North 20th Street
Philadelphia, PA 19103-1194

**Franklin L. Burdette/Pi Sigma Alpha Award 1151**
Dr. Franklin Lillard Burdette was a 1935 political science graduate of the University of Nebraska, and was later granted the honorary doctor of laws degree by Marshall College at Huntington, West Virginia. He became a professor of government at the University of Maryland, where he also headed the Department of Government and Politics and the Bureau of Governmental Research. Students remember him as having a "down-home folksy approach to teaching political science." The Franklin L. Burdette/Pi Sigma Alpha prize is awarded annually for the best paper presented at the previous year's annual meeting. The award is supported by Pi Sigma Alpha. It carries a prize of $750. Nominations should be submitted each year by the February 13 deadline.
*Geographic Focus:* All States
*Date(s) Application is Due:* Feb 13
*Amount of Grant:* 750 USD
*Contact:* Betsy Super, Deputy Director; (202) 483-2512; fax (202) 483-2657; bsuper@apsanet.org or awards@apsanet.org
*Internet:* www.apsanet.org/PROGRAMS/APSA-Awards/Franklin-L-Burdette-Pi-Sigma-Alpha-Award
*Sponsor:* American Political Science Association
1527 New Hampshire Avenue, NW
Washington, D.C. 20036-1206

**FRAXA Research Foundation Clinical Trial Grants 1152**
FRAXA aims to accelerate research aimed at finding a specific treatment for fragile X syndrome, with a primary goal of bringing practical treatment into current medical practice as quickly as possible. Therefore, preference will be given to research projects that have a clear practical application and the results of which will be shared in a timely fashion. FRAXA invites investigator-initiated research applications for innovative pilot studies aimed at developing and characterizing new therapeutic approaches for the treatment and ultimate cure of fragile X syndrome. These are flexible grants, but the expectation is that this program will be much more competitive than in the past, since fewer grants will be awarded. Successful applicants will be pursuing advanced translational, preclinical, and clinical research in fragile X. Grants are given based on innovation and translational relevance: clinical trials are our highest priority, followed by preclinical research (i.e. rescue of animal models) and then translational research. Applications are accepted at any time.
*Requirements:* There is no limit to structure of the grant (can fund PI, postdoc, grad student, technician, supplies, etc.) or time-frame (though all grants over one year still need yearly renewal). Furthermore, there is no limit on amount, but please remember that applications are ranked based on relative value, so smaller requests have an advantage.
*Geographic Focus:* All States
*Contact:* Katie Clapp, President and Co-Founder; (978) 452-1866; fax (978) 463-9985; kclapp@fraxa.org or info@fraxa.org
Michael Tranfaglie, (978) 462-1866; fax (978) 463-9985; mtranfaglia@fraxa.org or info@fraxa.org
*Internet:* www.fraxa.org/research/apply/
*Sponsor:* FRAXA Research Foundation
10 Prince Place, Suite 203
Newburyport, MA 01950

**Frederick Douglass Summer Fellowships 1153**
Clarion University has established the Frederick Douglass Summer Fellowships to offer opportunities to three graduate students who are pursuing a career in college teaching and are in their final year of a terminal degree program. This program is part of a Pennsylvania State System of Higher Education Collaborative that is intended to commemorate Douglass' life-long contributions to equality and public service. Fellowships are designed to attract advanced graduate students primarily from HBCU's and HACU's. Selected scholars will team and/or co-teach one course during a five-week Summer Session II, which typically runs from the first week of June through the first week of July. The stipend for classroom instruction, community presentations, and other duties as assigned (for the full duration of Summer Session II) is $3,000. The University will also provide housing assistance, academic support including office space, computer and library privileges, and an allowance of $1,000 for research-related expenses.
*Requirements:* Applicants must be U.S. citizens, and must have completed all doctoral work, except the dissertation, by the end of the next academic year. In addition to the application, candidates must submit two full sets of the following materials in hard copy: full curriculum vitae and letter of interest; graduate school transcript (a copy of the original transcript is acceptable); letter from the candidate's major or dissertation adviser regarding their qualifications for this program; one page summary of the dissertation topic; and at least two reference letters that address the candidate's qualifications.
*Restrictions:* Electronic applications are not accepted.
*Geographic Focus:* All States
*Date(s) Application is Due:* Apr 30
*Amount of Grant:* 4,000 USD
*Contact:* Stacy Boyer-Davis, Office Manager; (814) 393-2109; fax (814) 393-1618; sboyerdavis@clarion.edu
*Internet:* www.clarion.edu/48794/
*Sponsor:* Clarion University - Office of Social Equity
207 Carrier Building
Clarion, PA 16214

**Frederick Douglass Summer Scholars Fellowship 1154**
In keeping with Frederick Douglass life of public service and commitment to equity and access, the Frederick Douglass Institute for Intercultural Research offers teaching opportunities to graduate students pursuing careers as university faculty members and entering their final year of terminal degree and/or doctoral programs. As a Frederick Douglass Fellow, scholars will teach and/or co-teach one course during the summer sessions, continue their research, and contribute to the life of the university. Dates of employment typically run from May through August. Douglass Scholar applicants may also be considered for full- or partial-year appointments at the discretion of individual departments. The Douglass Fellowships provide professional experiences, mentoring, and potential employment opportunities within universities that are strongly committed to cultural diversity. Applicants from diverse cultural backgrounds, especially those from historically underrepresented groups, are encouraged to apply. Scholars will be compensated as adjunct faculty, according to the university's collective bargaining agreement. Additionally, Fellows may be provided housing and travel. Faculty benefits (office space, computer and network access, and library privileges) are also provided. The application is available at the Douglass Institute website.
*Requirements:* Applicant must be U.S. citizens or permanent residents; M.A. degree completed and enrollment in a terminal degree program (Ph.D., M.F.A., Ed.D., etc.); possess academic background in one of the disciplines taught at the university. Teaching experience is preferred, with A.B.D. or doctorate within the last two years. Priority consideration will be given to completed applications postmarked by December 16.
*Restrictions:* The appointment does not cover families, and scholars are responsible for their own transportation.
*Geographic Focus:* All States
*Date(s) Application is Due:* Dec 16
*Contact:* Dr. Veronica Watson, Director; (724) 357-3299 or (724) 357-4940; Veronica.Watson@iup.edu
*Internet:* www.iup.edu/page.aspx?id=39215
*Sponsor:* Indiana University of Pennsylvania - Douglass Institute
Stright Hall, Room 108
Indiana, PA 15705

**Frederick Gardner Cottrell Foundation Grants 1155**
The Research Corporation Technologies (RCT) established the Frederick Gardner Cottrell Foundation in December 1998 to provide financial support for scientific research and educational programs at qualified nonprofit organizations. RCT named the foundation in honor of the university professor and inventor who championed the transfer of academic innovation to public use. The Foundation receives its support from donations made by RCT and is a private, non-operating entity. Since its formation, the Foundation has provided nearly $13 million in support of selected scientific and educational programs throughout the United States. The Foundation does not accept

unsolicited grant requests, so interested parties should begin the process by contacting the office and providing project information. The Foundation's primary areas of interest include: eye research; higher education; marine science; and science.
*Geographic Focus:* All States
*Amount of Grant:* 100,000 - 300,000 USD
*Contact:* Gary Munsinger; (520) 748-4400; fax (520) 748-0025; munsinger@rctech.com
*Internet:* www.rctech.com/about-us/foundation/
*Sponsor:* Frederick Gardner Cottrell Foundation
6440 N. Swan Road, Suite 200
Tucson, AZ 85718

**Freedom of the Press Internships** **1156**
The Reporters Committee offers internships at its Washington, D.C., headquarters on a regular basis. The work is challenging, but rewarding. Right from the start, interns write articles for our digital magazine and website, contribute to our booklets on specific legal topics, and conduct research for other special projects. Interns are called on to provide significant contributions to our publications. Interns report and write on free press and media law issues for our quarterly digital magazine The News Media & The Law, website and guidebooks, and assist our staff and fellows in research. The Reporters Committee also provides its interns with regular opportunities to interact with working journalists, media lawyers and policymakers through seminars around Washington, D.C. Interns may attend congressional hearings, hear arguments in the U.S. Supreme Court or other federal appellate courts, and visit newsrooms and government agencies where they can learn firsthand about free-press issues. Students or recent graduates may intern during the spring, summer or fall semesters. Interns may receive college credit and are eligible to receive a competitive stipend from the Reporters Committee. Some interns may be enrolled in Washington semester study programs through their colleges or universities. Applications should be sent by January 31 for the summer term; August 1 for the fall term; and October 31 for spring term.
*Requirements:* A journalism intern candidate must be a college student or recent graduates. H/she will cover a broad range of issues, primarily writing for the magazine, website, guidebooks and other publications.
*Geographic Focus:* All States
*Date(s) Application is Due:* Jan 31; Aug 1; Oct 31
*Contact:* Internship Coordinator; (202) 795-9300 or (800) 336-4243; fax (703) 807-2109; internships@rcfp.org
*Internet:* www.rcfp.org/about-us/internships
*Sponsor:* Reporters Committee for Freedom of the Press
1156 15th Street, Suite 1250
Washington, D.C. 20005-1779

**Freshwater Future Project Grants** **1157**
The Project Grants program provides grant awards ranging from $500 to $5,000 (USD) to be used for specific project expenses. Projects to protect aquatic habitats in the following strategic program areas will be considered: projects to protect and restore wetlands that will result in ongoing or positive permanent change in the habitat; on-the-ground restoration activities that include an action component to create permanent change (a small amount of funding for this area); advocating for low-impact solutions to community storm water issues; participation in land use planning and zoning; participation in watershed planning; watch-dogging and participating in the development, implementation, and enforcement of local, state, provincial, and federal aquatic habitat protection regulations; non-partisan voter education, voter registration, and candidate forums; advocating for naturally functioning river and coastal ecosystems; reducing polluted runoff; launching special initiatives or creating unique opportunities to strengthen citizen involvement in aquatic habitat protection and restoration; and communicating how local wetlands, lakes and rivers are interconnected in the Great Lakes Basin. The spring cycle deadline for project grants is March 31, while the fall cycle deadline is September 30.
*Geographic Focus:* All States
*Date(s) Application is Due:* Mar 31; Sep 30
*Amount of Grant:* 500 - 5,000 USD
*Contact:* Jill Ryan, Executive Director; (231) 348-8200; jill@freshwaterfuture.org or info@freshwaterfuture.org
*Internet:* www.freshwaterfuture.org/grant-programs/project-grant-program.html
*Sponsor:* Freshwater Future
P.O. Box 2479
Petoskey, MI 49770

**Friedman Foundation for Educational Choice Grants** **1158**
Established in Indianapolis in 1996, the Friedman Foundation for Educational Choice provides funding to promote school choice to improve, through competition, the quality of K-12 education for all. In partnership with the State Policy Network, a $25,000 grant is available to organizations with 501(c)3 status for programs that encourage new and innovative efforts to promote school choice in specific states and to develop long-lasting partnerships with state-based think tanks as they undertake efforts to promote the issue. Types of support include: convening; general operating funds; research and evaluation; and systems reform. There are no specific application deadlines.
*Requirements:* Any U.S. 501(c)3 organization is eligible to apply.
*Geographic Focus:* All States
*Amount of Grant:* Up to 25,000 USD
*Samples:* Americans for Prosperity Foundation, Arlington, Virginia, $25,000 - general operating support; Barry Goldwater Institute for Public Policy, Pheonix, Arizona, $25,000 - general operating support; Empower Mississippi Foundation, Madison, Mississippi, $75,000 - general operating support.
*Contact:* Robert C. Enlow; (317) 681-0745; fax (317) 681-0945; rcenlow@edchoice.org
*Internet:* https://www.edchoice.org/
*Sponsor:* Friedman Foundation for Educational Choice
1 American Square, Suite 2420, Box 82078
Indianapolis, IN 46282-0014

**Fulbright Distinguished Chairs Awards** **1159**
The Fulbright Distinguished Chair Awards comprise approximately forty distinguished lecturing, distinguished research and distinguished lecturing/research awards ranging from three to 12 months. Awards in the Fulbright Distinguished Chairs program are viewed as among the most prestigious appointments in the Fulbright Scholar Program. Candidates should be senior scholars and have a significant publication and teaching record. Applicants should submit hard copies of the Distinguished Chairs Application Form (one page), a letter of interest (about three pages), a curriculum vitae (maximum eight pages) and, if required, a sample syllabus (maximum four pages). Chairs are available in Australia, Austria, Brazil, Canada, Denmark, Finland, France, Germany, Hungary, Ireland, Israel, Italy, Netherlands, Poland, Portugal, Russia, and Sweden. Because an objective of the Fulbright Program is to provide an educational exchange experience for those not previously afforded such an opportunity, preference will usually be given to candidates who have not had substantial recent experience in the country to which they are applying.
*Requirements:* Applicants must meet all of the following eligibility *Requirements:* U.S. citizenship at the time of application (permanent resident status is not sufficient); possess a Ph.D. or equivalent professional terminal degree at the time of application (for professionals and artists outside academe, recognized professional standing and substantial professional accomplishments); have college or university teaching experience at the level and in the field of the proposed lecturing activity as specified in the award description (for Distinguished Chairs awards, candidates should be senior scholars with a significant publication and teaching record); have foreign language proficiency only if specified in the award description or required for the completion of the proposed project; be of sound physical and mental health; and disclose prior conviction or current indictment for commission of a felony.
*Restrictions:* Previous Fulbright scholar grantees are eligible to apply only if five years will have elapsed between the ending date of one scholar award and the beginning date of the new scholar award. This rule does not apply if the previous grant was for less than two months. Employees, spouses or dependent children of the United States Department of State or public and private organizations under contract to the United States Department of State are ineligible to apply for a Fulbright grant until one year after the employee's termination.
*Geographic Focus:* All States
*Date(s) Application is Due:* Aug 1
*Contact:* Jordanna Enrich, Director; (202) 686-6233; jenrich@iie.org
*Internet:* www.cies.org/program/fulbright-distinguished-chair-awards
*Sponsor:* Council for International Exchange of Scholars
3007 Tilden Street NW, Suite 5L
Washington, D.C. 20008-3009

**Fulbright Scholar-in-Residence (SIR) Program Grants** **1160**
Grants strengthen the institutional commitment of U.S. higher education to international education and scholarship by enabling colleges and universities to invite visiting scholars from abroad in the humanities and social sciences or in professional specializations with strong international focus. A Fulbright Scholar in Residence may teach regular courses from a foreign area perspective, serve as a resource person in interdisciplinary courses, assist in developing new courses, or participate in special seminars. Host institutions are expected to share the scholar's expertise with a wide range of departments and with neighboring institutions, involve him/her in community activities and professional organizations, and provide opportunities for the visitor to pursue personal research interests. Detailed program guidelines, proposal forms, and further information are available upon request; materials are available in May prior to deadline date.
*Geographic Focus:* All States
*Date(s) Application is Due:* Oct 15
*Contact:* Alma Ford; (202) 686-6252; fax (202) 362-3442; aford@cies.iie.org
*Internet:* www.cies.org/sir/
*Sponsor:* Council for International Exchange of Scholars
3007 Tilden Street NW, Suite 5L
Washington, D.C. 20008-3009

**Fulbright Specialists Program Grants** **1161**
The Program is designed to provide short-term academic opportunities (two to six weeks) for U.S. faculty and professionals. Shorter grant lengths give specialists greater flexibility to pursue a grant that works best with their current academic or professional commitments. Applications for the Program are accepted on a rolling basis, and peer review of applications is conducted eight times per year. Program goals include: to increase the participation of leading U.S. scholars and professionals in Fulbright academic exchanges; to encourage new activities that go beyond the traditional Fulbright activities of lecturing and research; and to promote increased connections between U.S. and non-U.S. post-secondary academic institutions.
*Requirements:* Applicants must: be U.S. citizen at the time of application (permanent resident status is not sufficient; if a naturalized citizen, applicant must provide actual date of naturalization); possess a Ph.D. or equivalent professional/terminal degree at the time of application; have a minimum of five years of post-doctoral teaching or professional experience in the field in which you are applying (for professionals and artists outside academe, recognized professional standing and substantial professional accomplishments

plus a minimum of five years of professional experience in the field in which you are applying); disclose prior conviction or current indictment for commission of a felony; and be residing in the United States at the time they are approved for a grant and intend to return to their U.S. institution after the grant's completion.
*Restrictions:* Employees, spouses or dependent children of the United States Department of State or public and private organizations under contract to the United States Department of State exchange programs are ineligible to apply for a Fulbright grant until one year after the employee's termination.
*Geographic Focus:* All States
*Date(s) Application is Due:* Jan 28; Feb 24; Mar 25; Apr 21; May 20; Jul 15; Sep 9; Nov 4; Dec 30
*Contact:* Ryan Hathaway; (202) 686-4026; rhathaway@cies.iie.org
*Internet:* www.cies.org/specialists/#top
*Sponsor:* Council for International Exchange of Scholars
3007 Tilden Street NW, Suite 5L
Washington, D.C. 20008-3009

**Furthermore Grants in Publishing** 1162
The program is concerned with nonfiction book publishing about the city; natural and historic resources; art, architecture, and design; cultural history; and civil liberties and other public issues of the day. Grants apply to writing, research, editing, design, indexing, photography, illustration, and printing and binding. Work is sought that appeals to an informed general audience; gives evidence of high standards in editing, design, and production; promises a reasonable shelf life; might not otherwise achieve top quality or even come into being; and represents a contribution without which we would be the poorer. Book proposals to which a university press or trade publisher is already committed and for which there is a feasible distribution plan are usually preferred. In geographical reach, the program is drawn, but in no way limited, to New York City and to New York State and its Hudson Valley.
*Requirements:* 501(c)3 tax-exempt organizations are eligible to apply. Trade publishers and public agencies may apply for Furthermore grants in partnership with an eligible nonprofit project sponsor. Grants from $500 to roughly $15,000 are awarded in spring and fall with March 1 and September 1 deadlines.
*Restrictions:* Applications from individuals cannot be accepted
*Geographic Focus:* All States
*Date(s) Application is Due:* Mar 1; Sep 1
*Amount of Grant:* 500 - 15,000 USD
*Contact:* Joan K. Davidson; (518) 828-8900; fax (518) 828-8901; info@furthermore.org
*Internet:* www.furthermore.org
*Sponsor:* J.M. Kaplan Fund
261 Madison Avenue, 19th Floor
New York, NY 10016

**Gaius Charles Bolin Fellowships for Minority Graduate Students** 1163
Gaius Charles Bolin was the first black graduate of Williams. Fellowships were founded in 1985 on the centennial of his admission to the College. The Gaius Charles Bolin Fellowships are designed to promote diversity on college faculties by encouraging students from underrepresented groups to complete a terminal graduate degree and to pursue careers in college teaching. Bolin Fellowships are two-year residencies, and up to three scholars or artists are appointed each year. Fellows devote the bulk of the first year to the completion of dissertation work, or in the case of Master of Fine Arts (MFA) applicants, building their professional portfolios, while also teaching one course as a faculty member in one of the College's academic departments or programs. The second year of residency (ideally with degree in hand) is spent on academic career development while again teaching just one course. The annual stipend is $35,000 with a $4,000 allowance for research-related expenses.
*Requirements:* Applicants must be United States citizens or permanent residents who intend to pursue a professorial career in the United States. Ph.D. candidates must have completed all doctoral work except the dissertation by the end of the current academic year. MFA candidates must be recent recipients of the degree. Fellowships are awarded to applicants from underrepresented groups, including ethnic minorities, those who are first-generation college graduates, women in predominantly male fields, or disabled scholars.
*Geographic Focus:* All States
*Date(s) Application is Due:* Nov 15
*Amount of Grant:* 35,000 - 35,000 USD
*Contact:* Peter T. Murphy, Dean of Faculty; (413) 597-3131; gburda@williams.edu or communications@williams.edu
*Internet:* dean-faculty.williams.edu/graduate-fellowships/gaius-charles-bolin-dissertation-and-post-mfa-fellowships/
*Sponsor:* Williams College
880 Main Street
Williamstown, MA 01267

**Garland D. Rhoads Foundation** 1164
Garland D. Rhoads established the Foundation in 1971 to promote the extension of religion, education, the alleviation of human suffering and the prevention and control of diseases. The grantor also intended for funds to be utilized for the acquisition, construction, maintenance and beautification of public buildings, grounds and/or works for the encouragement of public and civic betterment. Rhoads also wanted to provide for the relief of the indigent and to aid in scientific research for the betterment of mankind. The Foundation is currently managed by the Bank of America Philanthropic Solutions. The annual deadline for applications is September 1, with award notifications by June 30. The average grant award for a grant cycle has been $5,000.
*Requirements:* Grants are awarded to qualified 501(c)3 charitable organizations exclusively within the state of Texas.
*Geographic Focus:* Texas
*Date(s) Application is Due:* Sep 1
*Amount of Grant:* 2,000 - 8,000 USD
*Contact:* Jenae Guillory, Philanthropic Relationship Manager; (214) 209-1965 or (214) 209-1370; tx.philanthropic@ustrust.com
*Internet:* https://www.bankofamerica.com/philanthropic/grantmaking.go
*Sponsor:* Garland D. Rhoads Foundation
901 Main Street, 19th Floor
Dallas, TX 75202-3714

**General Service Foundation Human Rights and Economic Justice Grants** 1165
The goal of the General Service Foundation's Human Rights and Economic Justice program is to support efforts that protect, promote and create good jobs with living wages for workers, including low-wage workers in the United States and Mexico. The Foundation seeks to: strengthen worker voices; promote public policies that protect labor rights; and democratize corporate power and promote corporate accountability. In assessing potential grantees, the Foundation will apply the following criteria: supporting organizations that work to address needs identified by the underrepresented and low-income communities that area directly impacted, connecting at the local and national level; and concentrating on organizations that take risks and try new ideas to enhance their skills, and reflect the diversity of their constituency. Samples of previously funded projects are posted on the Foundation website.
*Requirements:* All applicants, regardless of their prior grant history with the Foundation, start their application process by submitting a letter of inquiry in the spring and fall. Organizations should determine if their project is an appropriate fit for the Foundation, then send a letter of inquiry via the Foundation's online application. Letters of inquiry should include: a political and situational analysis, statement of the issues to be addressed under the proposed project, the history and goals of the organization, and an explanation of why the organization or coalition is the entity that is most likely to achieve success; a brief summary of the project, short and long term goals and anticipated outcomes; the approximate starting date and duration of the proposed activities; the total amount of funding needed, the amount requested from the Foundation, a budget, and information about other sources of support; and a copy of IRS tax exempt 501(c)3 letter. If the project is judged appropriate, the organization will be offered an application to submit to the Foundation.
*Restrictions:* The Foundation does not fund: organizations based outside the United States or Mexico; projects without significant promise of impact beyond a local or state level; research and publications not directly linked to policy outcomes; direct service delivery; and development or relief projects.
*Geographic Focus:* All States, Mexico
*Date(s) Application is Due:* Feb 1; Sep 1
*Amount of Grant:* 5,000 - 35,000 USD
*Contact:* Holly Bartling, Program Contact-Mexico grants; (202) 232-1005 or (970) 920-6834, ext 4; fax (970) 920-4578; holly@generalservice.org
Mary Estrin; (970) 920-6834; fax (970) 920-4578; maryestrin@gmail.com
*Internet:* www.generalservice.org/International%20Peace.htm
*Sponsor:* General Service Foundation
557 North Mill Street, Suite 201
Aspen, CO 81611

**George A. and Eliza Gardner Howard Humanities Fellowships** 1166
In the early 1950s, Nicea Howard, a Providence resident long active in supporting young artists, established contact with Henry Wriston, President of Brown University with the assistance of her financial advisor, Robert W. Kenyon of the Brown Class of 1936. After a lengthy process of consultation and deliberation, the Howard Foundation was officially established in 1952 and offered its first fellowships in 1954. Currently, the Foundation awards a limited number of fellowships each year for independent projects in selected fields. It targets its support specifically to early mid-career individuals who have achieved recognition for at least one major project. Approximately ten fellowships of $33,000 are given annually in the following fields: creative non-fiction; literary translation into English; film studies; and literary studies. Howard Fellowships are intended primarily to provide artists, scholars, and writers with time to complete their work. The deadline for submission of the online application is November 15, with recipients announced the following April.
*Requirements:* Candidates can establish their eligibility for a Howard Fellowship by considering a series of questions before they submit an application.
*Restrictions:* Fellowships are not intended for publication subsidies, for equipment purchase, for preparation of exhibits, or to support institutional programs.
*Geographic Focus:* All States
*Date(s) Application is Due:* Nov 15
*Amount of Grant:* 33,000 USD
*Contact:* Director; (401) 863-2640; fax (401) 863-6280; howard_foundation@brown.edu
*Internet:* www.brown.edu/initiatives/howard-foundation/
*Sponsor:* George A. and Eliza Gardner Howard Foundation
346 Brook Street, 20 Manning Walk Marston Hall, Room B-9
Providence, RI 02912

**George A Ohl Jr. Foundation Grants** 1167
The purpose of the George A Ohl Jr. Foundation is to improve the well-being of the citizens of the State of New Jersey through science, health, recreation, education and increased good citizenship. Grants are made to organizations engaged in such work whether through research, publications, health, school or college activities. The Foundation's mission is the relief of the poor; the improvement of living conditions; the care of the sick, the young, the

aged, the homeless, the incompetent and the helpless. The foundation will target: 35% of its grants to Community Redevelopment; 35% to Health and Human Services organizations; 15% for Arts and culture; 15% for Educational requests. Application deadlines are: January 22 for a March meeting and, June 15 for an August meeting. Application forms are available online. Applicants will receive notice acknowledging receipt of the grant request, and subsequently be notified of the grant declination or approval.
*Requirements:* New Jersey 501(c)3 nonprofit organizations are eligible to apply. Proposals should be submitted in the following format: completed Common Grant Application Form; an original Proposal Statement; an audited financial report and a current year operating budget; a copy of your official IRS Letter with your tax determination; a listing of your Board of Directors. Proposal Statements (second item in the above Format) should answer these questions: what are the objectives and expected outcomes of this program/project/request; what strategies will be used to accomplish your objective; what is the timeline for completion; if this is part of an on-going program, how long has it been in operation; what criteria will you use to measure success; if the request is not fully funded, what other sources can you engage; an Itemized budget should be included; please describe any collaborative ventures. Prior to the distribution of funds, all approved grantees must sign and return a Grant Agreement Form, stating that the funds will be used for the purpose intended. Progress reports and Completion reports must also be filed as required for your specific grant. All current grantees must be in good standing with required documentation prior to submitting new proposals to any foundation.
*Restrictions:* Grants are not made for political purposes, nor to organizations which discriminate on the basis of race, ethnic origin, sexual or religious preference, age or gender
*Geographic Focus:* New Jersey
*Date(s) Application is Due:* Jan 22; Jun 15
*Amount of Grant:* 3,000 - 35,000 USD
*Samples:* Mountainside Hospital Foundation, $35,000—Modernization of Internal Medicine Residents' Learning Center; Bloomfield College, $10,000—Improving Academic Success Through Technological Support Project; Medical Missions for Children, $5,000—Giggles Theater Programming Support (12 performances).
*Contact:* Wachovia Bank, N.A., Trustee; grantinquiries2@wachovia.com
*Internet:* https://www.wellsfargo.com/private-foundations/ohl-trust
*Sponsor:* George A Ohl Jr. Foundation
190 River Road, NJ3132
Summit, NJ 07901

**George Foundation Grants** **1168**
The George Foundation strive to support organizations and programs that assist in developing strong, stable families across Fort Bend County region of Texas. The Foundation's areas of interest include: family stability; scholarships; foundation initiated programs, current programs include, Youth in Philanthropy (YIP), Leadership Excellence for Non-Profits, Integrated Mental Health Care, and Transportation. The Foundation prefers to fund the following types of grants to support Fort Bend organizations in their delivery of services to the community (listed in order of priority): program/project support; foundation initiated; general operating; capital. Proposal deadlines for making grant applications to the Foundation are January 15, April 15, July 15 and October 15 of each year. All proposals for capital support will be grouped together for review. The deadline for capital proposals will be October 15 of each year.
*Requirements:* Non-profit organizations in Fort Bend County, Texas may submit grant proposals.
*Restrictions:* The Foundation does not fund: grants to organizations that do not have a current 501(c)3 determination letter; churches or other organized religious bodies; grants to another organization that distributes money to recipients of its own selection, i.e., a regranting organization; regional, national or international programs; grants for research or studies; grants for travel, conferences, conventions, group meetings, or seminars; the purchase of event tickets, tables, ads or sponsorships; support to fairs and festivals; religious or private schools; request for funds to develop films, videos, books or other media projects; direct mail campaigns; loans of any kind; grants to individuals; grants to fraternal organizations; political interests of any kind; and institutions that discriminate on the basis of race, creed, gender, national origin, age, disability or sexual orientation in policy or in practice.
*Geographic Focus:* Texas
*Date(s) Application is Due:* Jan 15; Apr 15; Jul 15; Oct 15
*Amount of Grant:* 1,000 - 4,000,000 USD
*Samples:* Wharton County Junior College, $110,000—to provide scholarships for Fort Bend County residents; YMCA of the Greater Houston Area, $250,000—to assist in acquiring facility to offer outdoor education/recreation experiences to Fort Bend families and children; OakBend Medical Center, $4,000,000—four year grant to provide indigent healthcare to Fort Bend County residents.
*Contact:* Dee Koch, Grant Officer; (281) 342-6109; fax (281) 341-7635; dkoch@thegeorgefoundation.org
*Internet:* www.thegeorgefoundation.org
*Sponsor:* George Foundation
310 Morton Street, PMB Suite C
Richmond, TX 77469

**George Gund Foundation Grants** **1169**
The Foundation's long-standing interests include: arts; economic development; community revitalization; education; environment; and human services. The Foundation considers global climate change an urgent issue that cuts across all of the Foundation's programs. The Foundation takes seriously it's own responsibility and wants to hear from grant applicants what they are doing or considering to reduce or to eliminate their organizational impact on climate change. The Foundation also supports special projects grants, which currently include: Retinal Degenerative Diseases research grants, making an annual commitment for research on the causes, nature and prevention of inherited retinal degenerative diseases and; philanthropic services grants, offering support to organizations that strengthen the infrastructure of the nonprofit and philanthropic communities. The George Gund Foundation also supports capital requests but only for projects that are clearly aligned with their program priorities and that meet Green Building Council LEED (Leadership in Energy and Environmental Design) certification. The Foundation's green building policy covers both planning and construction grants. In addition, the Foundation supports opportunities that cross program boundaries and that integrate elements of the Foundation's interests. Although the Foundation's focus is centered in the Greater Cleveland, Ohio region, a portion of their grantmaking will continue to support state and national policy making that bolsters their work. Proposals should be mailed directly to the George Gund Foundation. All proposals are screened and evaluated by the staff before presentation at Trustee Meetings. Receipt of proposals will be acknowledged by mail.
*Requirements:* 501(c)3 nonprofit organizations are eligible to apply for funding, with a special interest in Greater Cleveland, Ohio region. Proposals are accepted three times: March 15, July 15 and November 15. Proposals are due the next business day if a deadline falls on a weekend. The grant application form is available on the George Gund Foundation website. All proposals must include a climate change statement, the Foundation's website includes resources to assist grantees with this task. Applicants also must include a completed Grant Application Cover Sheet, which is signed by the organization's board chair and executive director. Proposals should also include: organizational background; history; mission; types of programs offered; constituencies served; project description; justification of need; specific goals and objectives; activities planned to meet goals and objectives; project time line; qualifications of key personnel; methods of evaluation; project budget; anticipated expenses, including details about how Foundation funds would be used; anticipated income, including information about other sources approached for funding; organizational budget; previous and current year budget and proposed budget for project year(s) showing both income and expenses; the organization's most recent audited financial statement, do not include IRS 990 forms; supporting documents; list of current trustees; letters of support; readily available printed material about organization such as annual reports and brochures; IRS letter confirming Internal Revenue Code 501(c)3 status and classification as a public charity or information confirming status as a government unit or agency. The Foundation also will accept the Ohio Common Grant Form, available at www.ohiograntmakers.org, if organizations are using it to apply to multiple funders. Faxed or electronic proposals are not accepted.
*Restrictions:* Do not submit proposals in notebooks, binders or plastic folders and print proposals on both sides of each sheet of paper. The Foundation normally does not consider grants for endowments. Capital requests must meet the Foundation's program goals and also adhere to green building standards of environmental sustainability. Details on these requirements are available from the Foundation. Grants are not made for debt reduction or to fund benefit events. The Foundation does not make grants to individuals, nor does it administer programs it supports. Grants are limited to organizations located in the United States.
*Geographic Focus:* All States
*Date(s) Application is Due:* Mar 15; Jul 15; Nov 15
*Samples:* Progressive Arts Alliance Incorporated, Cleveland, OH, $20,000—database capacity building project; Cleveland Restoration Society, Incorporated, Cleveland, OH, $20,000—preservation programs; Center for Community Solutions, Cleveland, OH, $360,000—two year grant for policy and fiscal analysis and GroundWork early childhood campaign.
*Contact:* David Abbott; (216) 241-3114; fax (216) 241-6560; info@gundfdn.org
*Internet:* www.gundfdn.org/what.asp
*Sponsor:* George Gund Foundation
1845 Guildhall Building, 45 Prospect Avenue, West
Cleveland, OH 44115

**George H. Hitchings New Investigator Award in Health Research** **1170**
Provides awards to provide flexible support for young researchers in their second year of graduate school. Students planning to follow a career path in academic teaching and research that is closely tied to the health of North Carolinians and the strength of North Carolina Science are particularly encouraged to apply.
*Requirements:* Applications/nominations must be submitted by the institution on behalf of the student(s). Nomination applications are accepted from Duke, NCCU, NCSU, and UNC-Chapel Hill.
*Restrictions:* Only one nomination/application per graduate program/department within the university will be accepted.
*Geographic Focus:* North Carolina
*Date(s) Application is Due:* Mar 5
*Amount of Grant:* 10,000 USD
*Contact:* Libby Long; (919) 474-8370, ext. 134; fax (919) 941-9208; libby@trialglecf.org
*Sponsor:* Triangle Community Foundation
324 Blackwell Street, Suite 1220
Durham, NC 27701

**George Kress Foundation Grants** **1171**
Incorporated in Wisconsin in 1953, the George Kress Foundation awards grants to eligible Wisconsin nonprofit organizations in its areas of interest, including: arts and culture; boys and girls clubs; children and youth services; Christian agencies and churches; community and economic development; education; family services; health organizations; higher education; historic preservation; historical societies; hospitals; human services; libraries; recreation; United Ways and Federated Giving Programs; and YM/YWCAs and YM/YWHAs. Types of support include: annual campaigns; building and renovation; capital campaigns; continuing support; professorships; program

development; and research. Preference is given to nonprofit organizations that benefit the communities of Green Bay and Madison. Most recent awards have ranged from $200 to $50,000. There are no specified annual deadlines for submission.
*Requirements:* Interested applicants should submit a letter of inquiry describing their proposed project.
*Geographic Focus:* Wisconsin
*Amount of Grant:* 200 - 50,000 USD
*Samples:* St. Vincent Health Systems, Little Rock, Arkansas, $25,000 - overall program support; Weidner Center, Green Bay, Wisconsin, $30,000 - overall program support; University of Wisconsin Foundation, Madison, Wisconsin, $50,000 - overall program support.
*Contact:* John Kress, President; (920) 327-5670 or (920) 433-3109
*Sponsor:* George Kress Foundation
1700 North Webster Avenue, P.O. Box 12800
Green Bay, WI 54307-2800

**George W. Codrington Charitable Foundation Grants** **1172**
The George W. Codrington Charitable Foundation gives primarily to nonprofit organizations in Ohio, but may consider other areas. The Foundation funds higher education, hospitals, museums, arts groups and performing arts, and youth programs. Types of support include annual and capital campaigns, continuing support, equipment, general/operating support, program development, and research.
*Requirements:* Application forms are not required. Applicants should submit three copies of the following: their IRS determination letter; a brief history of the organization and description of its mission; the geographic area to be served; a list of the board of directors, trustees, officers, and other key individuals with their affiliations; and a detailed description of the project and amount of funding requested. Proposals should be submitted one month before the board meets in April, June, September, and December. Organizations are notified of funding promptly after the board meeting.
*Restrictions:* Funding is not available for individuals, endowment funds, or loans.
*Geographic Focus:* Ohio
*Amount of Grant:* 1,000 - 50,000 USD
*Contact:* Craig Martahus; (216) 566-8674; tommie.robertston@thomasonhine.com
*Sponsor:* George W. Codrington Charitable Foundation
127 Public Square, 39th Floor
Cleveland, OH 44114-1216

**Georgia-Pacific Foundation Environment Grants** **1173**
Since a healthy environment is crucial to any town, city or neighborhood, nowhere is Georgia-Pacific's commitment to building strong communities better exemplified than in its stewardship of natural resources and the environment. The Foundation believes that adhering to sound science is the best way to keep and protect our natural resources. Consequently, it works with scientific institutions to develop the solutions, expertise and scientific data necessary to sustain our natural resources and help resolve environmental concerns. To support these efforts, the Georgia-Pacific Foundation funds environmental programs, such as the Keystone Institute Science School, Keep America Beautiful, and The Nature Conservancy.
*Requirements:* Nonprofit organizations in Georgia-Pacific communities are eligible to apply for funding. To find out whether a program qualifies for consideration for a Georgia-Pacific Foundation grant, applicants should complete the Eligibility Survey available online. The Georgia-Pacific Foundation will accept proposals for grants and in-kind donations from January 1 through October 31. It is recommended that applicants submit proposals early in the grant cycle.
*Restrictions:* No support is given for discriminatory organizations, political candidates, churches or religious denominations, religious or theological schools, social, labor, veterans', alumni, or fraternal organizations not of direct benefit to the entire community, athletic associations, national organizations with local chapters already receiving support, medical or nursing schools, or pass-through organizations. No grants to individuals (except for scholarships), or for emergency needs for general operating support, political causes, legislative lobbying, or advocacy efforts, goodwill advertising, sporting events, general operating support for United Way member agencies, tickets or tables for testimonials or similar benefit events, named academic chairs, social sciences or health science programs, fundraising events, or trips or tours.
*Geographic Focus:* Alabama, Arkansas, California, Florida, Georgia, Illinois, Indiana, Iowa, Kentucky, Louisiana, Massachusetts, Michigan, Mississippi, New Jersey, New York, North Carolina, Ohio, Oklahoma, Oregon, Pennsylvania, South Carolina, Tennessee, Texas, Virginia, Washington, West Virginia, Wisconsin
*Date(s) Application is Due:* Oct 31
*Contact:* Curley M. Dossman, Jr.; (404) 652-4182; fax (404) 749-2754; cmdossma@gapac.com
*Internet:* gp.com/gpfoundation/environment.html
*Sponsor:* Georgia-Pacific Foundation
133 Peachtree Street, NE, 39th Floor
Atlanta, GA 30303

**Getty/ACLS Postdoctoral Fellowships in the History of Art** **1174**
ACLS invites applications for Getty/ACLS Postdoctoral Fellowships in the History of Art, made possible by the support of the Getty Foundation. These fellowships are intended to support an academic year of research and/or writing by early career scholars for a project that will make a substantial and original contribution to the understanding of art and its history. The ultimate goal of the project should be a major piece of scholarly work by the applicant. ACLS does not fund creative work (e.g., novels or films), textbooks, straightforward translation, or pedagogical projects. ACLS will award ten fellowships, each with a salary-replacement stipend of $60,000, plus $5,000 for research and travel during the award period. The fellowships are portable and are tenable at the fellow's home institution, abroad, or at another appropriate site for the work proposed. Awards also will include a one-week residence at the Getty Research Institute following the fellowship period. The deadline for application submission is October 28.
*Requirements:* Applicants must have a PhD that was conferred during the past six years. Applicants who earned their PhDs in and/or are currently employed in any humanistic field may apply, so long as they demonstrate that their research draws substantially on the materials, methods, and/or findings of art history, and contributes to the field. Scholars may propose new approaches to art historical scholarship and/or explore connections between art history and other humanistic disciplines.
*Restrictions:* Getty/ACLS Postdoctoral Fellowships may not be held concurrently with other fellowships and grants, though they may be combined with sabbatical.
*Geographic Focus* District of Columbia, Guam, Marshall Islands, Northern Mariana Islands, Puerto Rico, U.S. Virgin Islands, American Samoa
*Date(s) Application is Due:* Oct 23
*Amount of Grant:* 60,000 - 65,000 USD
*Contact:* Valerie Popp, Senior Program Officer; (212) 697-1505, ext. 153; fax (212) 949-8058; vpopp@acls.org
*Internet:* https://www.acls.org/Competitions-and-Deadlines/Getty-ACLS-Postdoctoral-Fellowships-in-the-History-of-Art
*Sponsor:* American Council of Learned Societies
633 Third Avenue, 8th Floor
New York, NY 10017-6795

**Gibson County Community Foundation Health Grants** **1175**
The Gibson County Community Foundation considers proposals for grants on a yearly cycle, which begins each May. At the start of each cycle, a notice is mailed to nonprofit organizations that have applied for grants in the past, have received grants in the past, or have otherwise requested notification of the start of each cycle. Grants in the area of health include activities that: improve and promote health outcomes; provide general and rehabilitative health services; offer mental health services; provide crisis intervention programs; strengthen associations or services associated with specific diseases, disorders, and medical disciplines; and support medical research. Samples of previously funded projects are posted on the website.
*Requirements:* The Foundation welcomes proposals from nonprofit organizations that are deemed tax-exempt under sections 501(c)3 and 509(a) of the Internal Revenue Code and from governmental agencies serving the County of Gibson, Indiana. Proposals from nonprofit organizations not classified as a 501(c)3 public charity may be considered provided the project is charitable and supports a community need.
*Restrictions:* Funding is not available for the following: religious organizations for strictly religious purposes; political parties or campaigns; endowment creation or debt reduction; operating costs (not directly related to the proposed project or program); capital campaigns; annual appeals or membership contributions; travel requests for groups or individuals such as bands, sports teams, or classes. Not more than 20% of any grant request may be for personnel costs, office supplies, or other operating costs. Operating costs for any organization must be directly related to the project or program for which funding is being requested.
*Geographic Focus:* Indiana
*Contact:* Sarah Wagner; (812) 386-8082; swagner@communityfoundationalliance.org
*Internet:* https://www.communityfoundationalliance.org/gibson/program-areas/
*Sponsor:* Gibson County Community Foundation
127 North Hart Street, P.O. Box 180
Princeton, IN 47670

**Gil and Dody Weaver Foundation Grants** **1176**
Established in Texas in 1980, the Gil and Dody Weaver Foundation offers support throughout the States of Texas, Oklahoma, and Louisiana, with some emphasis on the Dallas-Fort worth area. The Foundation's primary fields of interest include: cancer; children and youth services; education; health organizations; human services; social services; and recreational camps. Major types of support come in the form of: annual campaigns; general operating/continual support; and scholarship funds. Although no formal application is required, the Foundation does provide specific application guidelines. These guidelines require a history of the organization, detailed information about the proposed project, and budgetary needs. The annual deadline is May 31, with final notifications by September 30. Recent grants have ranged from $1,000 to $20,000, with occasional higher amounts for special circumstances.
*Requirements:* 501(c)3 organization serving the residents of Texas, Oklahoma, and Louisiana, are welcome to apply.
*Restrictions:* No grants are given to individuals. No applications are accepted from organizations located in states other than Texas, Oklahoma, or Louisiana.
*Geographic Focus:* California, Colorado, Louisiana, Mississippi, New Mexico, Oklahoma, Texas
*Date(s) Application is Due:* May 31
*Amount of Grant:* 1,000 - 20,000 USD
*Contact:* William R. Weaver, (214) 999-9497 or (214) 999-9494; fax (214) 999-9496
*Sponsor:* Gil and Dody Weaver Foundation
1845 Woodall Rodgers Freeway, Suite 1275
Dallas, TX 75201-2299

**Gilbert Memorial Fund Grants** **1177**
The J. Roland Gilbert, Mary R. Gilbert and Elizabeth A. Gilbert Memorial Fund was established under the last will of Mary R. Gilbert in memory of herself, her husband, and their daughter. The fund seeks to support organizations performing research in various areas of medicine. The average grant amount is $5,000, and approximately seven awards are made each year. Applications should be submitted by September 30 to be reviewed at the annual grant meeting that occurs in October.

*Requirements:* To be eligible, organizations must qualify as exempt organizations under Section 501(c)3 of the Internal Revenue Code. Applications must be submitted through the online grant application form or alternative accessible application designed for assistive technology users. Except as otherwise specified when a grant is awarded, a progress report must be submitted within six months after receiving funds.
*Geographic Focus:* All States
*Date(s) Application is Due:* Sep 30
*Amount of Grant:* Up to 8,000 USD
*Contact:* Jason Craig; (888) 234-1999; fax (877) 746-5889; grantadministration@wellsfargo.com
*Internet:* https://www.wellsfargo.com/privatefoundationgrants/gilbert2
*Sponsor:* Gilbert Memorial Fund
1740 Broadway
Denver, CO 80274-0001

**Gilbert S. Omenn Fellowships** **1178**
The overall purpose of the Gilbert S. Omenn Fellowship is to enable talented, early-career scholars combining biomedical science and population health to participate actively in the IOM study process, promoting the linkage of public health and medicine - both scientifically and through practice and policy. The Omenn Fellowship aims to foster a cadre of physician-scientists who will integrate biomedical research, population health, and health policy and will expand the nation's capacity for research, leadership, and policy development that advances health. The program especially welcomes nominations of under-represented minority candidates. The Fellowship is awarded for a two-year period. During this time, the fellow is expected to continue work at his/her main academic post, while being assigned to a board of the IOM. The boards are: Children, Youth, and Families; Food and Nutrition; Global Health; Health Care Services; Health of Select Populations; Health Sciences Policy; and Population Health and Public Health Practice. A flexible research stipend of $25,000 is awarded to each Omenn Fellow. The stipend will be administered through the appropriate department in the fellow's home institution. Stipends are not intended for use as salary offsets. The Call for Nominations opens April 1 each year, with the deadline for submission being June 8.
*Requirements:* The fellowship requires a ten-to-twenty percent commitment of time for two years, and includes attendance at a one-week orientation to health policy in October, the Fall IOM Annual Meeting, and the meetings of an assigned board. Additionally, the fellow will participate actively in the work of an appropriate expert study committee or round table, including contributing to its reports or other products. Studies will be identified that would enable the fellow to contribute value to the integration of medicine and public health.
*Geographic Focus:* All States
*Date(s) Application is Due:* Jun 8
*Amount of Grant:* 25,000 USD
*Contact:* Marie E. Michnich; (202) 334-1506; fax (202) 334-3862; mmichnich@nas.edu
*Internet:* www.nationalacademies.org/hmd/Activities/Education/OmennFellowship.aspx
*Sponsor:* National Academies
500 5th Street, NW, Room 508
Washington, D.C. 20001

**GMFUS Black Sea Trust for Regional Cooperation Grants** **1179**
The Black Sea Trust (BST) for Regional Cooperation invites proposals from organizations and institutions in the countries of the Wider Black Sea Region on the Trust's four programs: civic participation, cross-border initiatives, Eastern links, and confidence building. The Trust promotes regional cooperation and good governance in the Wider Black Sea region; accountable, transparent, and open governments; strong, effective civic sectors; and independent and professional media. To respond to the rapid shifts in the region, BST staff regularly consult with regional experts and aim to sharpen the program's grant making strategy in order to more effectively achieve the Trust's goals. With the complexity and diversity of the region, BST priorities are revised regularly to respond to the region's changing needs. Adjustments are made in consultation with the BST Advisory Board, the German Marshall Fund's network of offices and internal expertise, and in coordination with other donors active in the region. The maximum duration of projects is twelve months. As an exception, outstanding projects of a longer duration may be considered. Grants generally range from $5,000 and $75,000, with most grants falling between $15,000 and $25,000. BST can support exceptional multi-year projects, renewable on an annual basis contingent upon satisfactory interim reports and performance. There are no application deadlines, and grant decisions are made monthly.
*Requirements:* Non-governmental organizations, governmental entities, community groups, policy institutes, other associations legally registered in Armenia, Azerbaijan, Bulgaria, Georgia, Moldova, Romania, Turkey, Ukraine and Russia. The projects have to be implemented in one or more of the countries mentioned. An online application and budget form are required, and must be completed in English.
*Restrictions:* Individuals and political parties may not apply. Non-indigenous organizations may not apply. Cooperative projects between indigenous and non-indigenous organizations are considered under the Eastern Links component, yet non-indigenous administrative costs will not receive BST support. The BST does not support scholarly research, academic fellowships and scholarships, one-off events, humanitarian aid, refugee or IDP return, religious activities, or the arts or sciences. BST only supports travel and website creation/maintenance as components of larger activities.
*Geographic Focus:* Armenia, Azerbaijan, Bulgaria, Moldova, Romania, Russia, Turkey, Ukraine
*Date(s) Application is Due:* May 21; Oct 19
*Amount of Grant:* 15,000 - 25,000 USD
*Contact:* Ana-Maria Aelenei, Deputy Director, Black Sea Trust; (202) 683-2650; fax (202) 265-1662; aaelenei@gmfus.org or blackseatrust@gmfus.org
Sergiu Panainte, Senior Program Officer; (202) 683-2650; fax (202) 265-1662; spanainte@gmfus.org or BlackSeaTrust@gmfus.org
*Internet:* www.gmfus.org/bst-grantmaking
*Sponsor:* German Marshall Fund of the United States
1744 R Street NW
Washington, D.C. 20009

**GNOF Gulf Coast Oil Spill Grants** **1180**
Shortly after the tragic explosion of BP's Deepwater Horizon drilling rig in 2010, the Greater New Orleans Foundation (GNOF) responded to the crisis by establishing the Gulf Coast Oil Spill Fund to serve as a conduit between the donor community and GNOF's nonprofit partners. The Gulf Coast Oil Spill Fund (GCOS) supports work that helps fisherman and their families in the parishes most affected by the oil spill: Plaquemines, St. Bernard, lower Jefferson, Terrebonne, and Lafourche parishes. The long-term goal of GCOS is to help strengthen coastal communities against future environmental catastrophes such as the Deepwater spill. Specifically GCOS embraces the following objectives: the creation of a more resilient commercial fishing industry that is better able to respond to future crises (whether natural or man-made) through the introduction of new technologies and practices; the development and implementation of job training opportunities for fishers and charter-boat workers temporarily or permanently displaced from their chosen vocation because of the spill; the development of innovative marketing and distribution plans for local seafood; the development of a more comprehensive and accessible system for monitoring both water quality and seafood quality; a mechanism for more effectively organizing the region's commercial fishers which focuses on transcending geographic, racial, ethnic, and linguistic barriers that have traditionally divided commercial fishing populations; the development of laws and/or policies that will be of direct benefit to commercial fishers and others who derive the majority of their income from water-related activities; the provision of free or reduced-cost legal advice and mental-health services to commercial fishers and individuals whose income has been curtailed or interrupted by the oil disaster; and a mechanism for better coordinating the provision of emergency assistance to aquatic and other wildlife affected by "mini-landfalls" or other contact with lingering oil from the Deep Water Horizon spill. GNOF accepts GCOS applications through its RFP process. Interested organizations are encouraged to subscribe to GNOF's email newsletter so they can be notified of future RFPs. The subscription link is available from GNOF's Apply-for-a-Grant web page which contains a comprehensive listing of GNOF's current and past funding opportunities. Interested organizations are also encouraged to contact the GNOF Director for Regional Initiatives with any questions.
*Requirements:* Applicant organizations must meet all of the following criteria: be classified by the Internal Revenue Sercie as 501(c)3 nonprofit entities or make use of a fiscal agent that has a 501(c)3 designation; be based within GNOF's thirteen-parish service area (Assumption, Jefferson, Lafourche, Orleans, Plaquemines, St. Bernard, St. Charles, St. James, St. John the Baptist, St. Tammany, Tangipahoa, Terrebonne, and Washington); and demonstrate close collaboration with commercial fishing organizations or populations in affected communities in the GNOF service area.
*Restrictions:* The GCOS fund will not support film or documentary projects, art projects, or academic research (except in the case of developing a more comprehensive and accessible system for monitoring both water quality and seafood quality).
*Geographic Focus:* Louisiana
*Samples:* St. Bernard Wetlands Foundation, Meraux, Louisiana, $9,000—to provide opportunities to learn the value of trees in restoring and protecting wetlands; Just the Right Attitude, New Orleans, Louisiana, $170,000—designated for food distribution and hot meals for families negatively impacted by the BP Oil Spill; Louisiana Justice Institute, New Orleans, Louisiana, $30,000—One Spill; One Gulf; One People: Organizing and Legal Advocacy for Gulf Coast Fishers.
*Contact:* Dr. Marco Cocito-Monoc, Director for Regional Initiatives; (504) 598-4663; fax (504) 598-4676; marco@gnof.org
*Internet:* www.gnof.org/gnof-oil-spill-fund/
*Sponsor:* Greater New Orleans Foundation
1055 St. Charles Avenue, Suite 100
New Orleans, LA 70130

**GNOF IMPACT Grants for Education** **1181**
Through the IMPACT Program, the Greater New Orleans Foundation (GNOF) makes grants to organizations serving the Greater New Orleans region. The ultimate goal of the IMPACT program is to create a resilient, sustainable, vibrant, and equitable region in which individuals and families flourish and in which the special character of the New Orleans region and its people is preserved, celebrated, and given the means to develop. Specifically GNOF hopes to accomplish the following objectives through its IMPACT grants: provide a much needed source of financial and other support to nonprofit organizations that are struggling in the current financial environment and that are important to the health and vibrancy of the region; develop a better sense of the nonprofit organizations serving the region so GNOF can more effectively match donor desires with effective charitable work; identify and nurture promising new leaders and initiatives, especially in those communities that are in greatest need; and gain knowledge that will help nonprofit leaders and GNOF staff develop better long-term strategies for addressing regional needs and taking best advantage of important opportunities. IMPACT grants are awarded in four categories: Arts and Culture, Youth Development, Education, and Health and Human Services. In the category of Education support is available to organizations that seek to ensure that all young people in public K-12 schools in GNOF's 13-parish service area attend high-performing schools. Priority will be given to work that has the following goals: to advance public education reforms in Orleans parish and other under served areas in GNOF's service area; to hold public agencies accountable for the success of our public schools; to improve student achievement by training and engaging key stakeholders in education (parents, school leaders, etc.) in the

skills of organizing, policy research and analysis, policy advocacy, and the use of data in making policy recommendations; and conduct research and/or work with school leaders and other financial decision makers to address the financial sustainability issues within the charter school system. Interested organizations must submit a letter of intent along with all attachments via one email by 5 p.m. on July 30. GNOF program staff will review all letters of intent and will contact those organizations that are invited to submit a full application for funding. Awards are announced in November. Deadlines may vary from year to year. Interested organizations should verify the current deadline at the GNOF website where they can also obtain complete guidelines and requirements as well as a downloadable application form and cover sheet. Prospective applicants can also subscribe to GNOF's email newsletter for announcements of future funding opportunities. The subscription link is available from GNOF's Apply-for-a-Grant web page which contains a comprehensive listing of GNOF's current and past funding opportunities.
*Requirements:* Nonprofit, tax-exempt organizations that serve the thirteen parishes of Greater New Orleans are eligible to apply for funding. Organizations that are not tax-exempt but have a fiscal agent relationship with a 501(c)3 organization are also eligible.
*Restrictions:* Through its IMPACT program, the Greater New Orleans Foundation is unable to fund the following types of requests: requests for individual support, either through scholarships or other forms of financial assistance; special events or conferences; programs that promote religious doctrine; endowments; and scientific or medical research.
*Geographic Focus:* All States
*Date(s) Application is Due:* Jul 30
*Amount of Grant:* Up to 20,000 USD
*Samples:* Bureau of Governmental Research, New Orleans, Louisiana, $10,000—to provide general operating support; Loyola University, Chicago, Illinois, $20,000—to support the Place Matters Initiative; The Lens, New Orleans, Louisiana, $10,000—to support Charter School Reporting Corps.
*Contact:* Roy Williams, Program Assistant; (504) 598-4663; fax (504) 598-4676; grants@gnof.org or roy@gnof.org
*Internet:* www.gnof.org/programs/impact/
*Sponsor:* Greater New Orleans Foundation
1055 St. Charles Avenue, Suite 100
New Orleans, LA 70130

**Goldman-Sonnenfeldt Foundation Grants** 1182
The Goldman-Sonnenfeldt Foundation was established in the State of New York, with the expressed purpose of offering support to the fields of: arts and culture; education; the environment; human services; and Judaism. The Foundation's primary interest is to award grants for program development. Applicants should submit a letter that includes a detailed description of project and the amount of funding requested. There are no specified annual deadlines. Most recent grants have ranged up to a maximum of $22,000.
*Requirements:* Any 501(c)3 supporting residents of Connecticut, New York, and Washington, D.C., are welcome to apply.
*Geographic Focus:* Connecticut, District of Columbia, New York
*Amount of Grant:* Up to 22,000 USD
*Samples:* Abraham Joshua Heschel School, New York, New York, $7,130 - program development; American Jewish Committee, New York, New York, $21,800 - program development; Breast Cancer Research, New York, New York, $1,500 - program development and general research.
*Contact:* Michael Sonnenfeldt, President; (212) 873-1472; robert@ganer.com
*Sponsor:* Goldman-Sonnenfeldt Foundation
1995 Broadway, 16 Floor
New York, NY 10023-5882

**Graham Foundation Carter Manny Award** 1183
The Foundation offers the prestigious annual Award in honor of Carter H. Manny and his long and distinguished service to the Graham Foundation. Manny has served the Foundation since its inception in 1956, first as a Trustee, then as the Director from 1971, and since his retirement in 1993, as Director Emeritus. The Foundation offers two Carter Manny Awards: a research award for a student at the research stage of the doctoral dissertation and a writing award for a student at the writing stage of the doctoral dissertation. The research award is acknowledged with up to $15,000 and the writing award is acknowledged with up to $20,000. The Carter Manny Award applications are reviewed by a diverse panel of recognized scholars in the fields of inquiry represented by the award.
*Requirements:* Ph.D. students who are presently candidates for a doctoral degree are eligible to apply. Candidates must be nominated by their departments.
*Geographic Focus:* All States, All Countries
*Date(s) Application is Due:* Nov 15
*Amount of Grant:* 15,000 - 20,000 USD
*Contact:* Sarah Herda, Director; (312) 787-4071; sherda@grahamfoundation.org or info@grahamfoundation.org
*Internet:* www.grahamfoundation.org/grant_programs?mode=award
*Sponsor:* Graham Foundation
4 West Burton Place
Chicago, IL 60610

**Graham Foundation Research and Development Grants for Individuals** 1184
Though the majority of the Graham Foundation's grant making focuses on production and presentation grants, it does recognize that projects may require support at early stages of formation. Research and Development Grants assist individuals with seed money for research-related expenses such as travel, documentation, materials, supplies, and other development costs. Projects must have clearly defined goals, work plans, and budgets. Upon completion of research projects, recipients should complete a research report and provide documentation that can be archived at the Graham Foundation and/or presented on its website. A recipient is eligible to apply for a Production and Presentation Grant for the same project once the first grant has been satisfied, however, future funding is not guaranteed. Research and Development Grants do not exceed $10,000 and are likely to be less. Given the demand for funding, the Graham Foundation is not always able to fund grantees at the full request amount. Such grants must be completed within one year. Applicants should allow sufficient time to plan, implement, close out their project, and, if funded, acknowledge Graham Foundation support in all published media. The Foundation's application is a two-stage process and is open submission. During step one, an applicant should forward a completed Inquiry Form by the deadline of September 15. If given a go-ahead, a complete application should be submitted by January 15.
*Requirements:* Grantmaking focuses on individuals in the United States. However, the Foundation does make a small percentage of international grants. For individuals who apply, consideration is given to those demonstrating mature creative talents who have specific project objectives.
*Restrictions:* There is no direct scholarship aid. The foundation does not award grants to cover costs of endowments, annual operating expenses, capital projects or construction, architectural fees in support of construction or renovation projects, work focused on K-12 curriculum or learning and/or intended for K-12 audiences, or debt or expenses incurred prior to the date of grant request.
*Geographic Focus:* All States, All Countries
*Date(s) Application is Due:* Jan 15
*Amount of Grant:* Up to 10,000 USD
*Contact:* Sarah Herda, Director; (312) 787-4071; sherda@grahamfoundation.org or info@grahamfoundation.org
*Internet:* www.grahamfoundation.org/grant_programs?mode=individual
*Sponsor:* Graham Foundation
4 West Burton Place
Chicago, IL 60610

**Grass Foundation Marine Biological Laboratory Advanced Imaging Fellowships** 1185
The Grass Foundation Advanced Imaging Fellowships at Marine Biological Laboratory will support individuals working in any area of neuroscience at any stage of their career, ranging from graduate student to professor. This opportunity is intended to allow individuals to bring their preparations to the MBL for up to four weeks, learning and utilizing the available microscopy resources, gathering data and interacting with other imaging specialists. Grass Fellows in Advanced Imaging are integrated into the traditional Grass Fellows program during the summer, providing a rich environment and access to many activities at the MBL. As such, Imaging Fellows will receive bench space in the Grass Laboratory for sample preparation, and are expected to attend Grass Laboratory lab meetings. In addition, Imaging Fellows are encouraged to audit the many classes available including the neurobiology course, which offers lectures on the theory and practice of advanced microscopy. Support will include research supplies, travel, housing and meals at the Marine Biological Laboratory for the individual Fellow. Applicant instructions with detailed information of all attachments needed, such as proposal and vitae, are available at the Grass Foundation website.
*Requirements:* Applicants at any stage in their academic career are eligible. Priority is given to applicants with an experimental need to use high-end state of the art imaging systems. Prior research experience at the MBL, including previous Grass Fellowships, is neither required nor disqualifying. IRS regulations require that international Fellows (i.e., not U.S. citizens or resident aliens) hold a J-1 visa for the duration of the Fellowship. The duration of the program is two to four weeks during July. Fellows are expected to be in residence at the MBL the entire time. Up to three Fellows will be accepted.
*Restrictions:* Holders of H1-B visas are not eligible to apply. Funding is available for the individual Fellow only.
*Geographic Focus:* All States, All Countries
*Date(s) Application is Due:* Feb 20
*Contact:* Ann Woolford; (508) 289-7521; fax (508) 289-7931; awoolford@mbl.edu
*Internet:* www.grassfoundation.org/grass-fellowships-in-advanced-imaging
*Sponsor:* Grass Foundation
P.O. Box 241458
Los Angeles, CA 90024

**Grass Foundation Marine Biological Laboratory Fellowships** 1186
Grass Fellowships at the Marine Biological Laboratory (MBL) in Woods Hole, Massachusetts, support investigator-designed, independent research projects by scientists early in their career. Supported approaches include neurophysiology, biophysics, integrative neurobiology, neuroethology, neuroanatomy, neuropharmacology, systems neuroscience, cellular and developmental neurobiology, and computational approaches to neural systems. Grass Fellowships provide research support including laboratory space, animals, equipment, and supplies for one summer at the MBL. Additionally, the investigator, his/her spouse or legal domestic partner, and dependent children are provided housing, a daily meal allowance and round-trip travel to the MBL. Fellows function as an intellectual and social group within the MBL scientific community while sharing space in the Grass Laboratory. In a weekly private seminar series, eminent investigators at the MBL discuss their work with the Fellows. In addition, a yearly Forbes Lecturer will spend a portion of the summer in the Grass Lab interacting with Fellows. From 8 to 10 Fellowships are awarded annually. The Fellowship last from late May to early August. The application and supplemental material are available at the Foundation website.
*Requirements:* Early investigators (late stage predoctoral trainees and beyond) are eligible to apply. This includes applicants with prior experience at MBL or with the Grass Foundation. MBL course alumni are encouraged to apply. Priority is given to applicants

with a demonstrated commitment to pursuing a research career. International Fellows must hold either a J-1 or H-1B visa for the entire duration of the Fellowship.
*Geographic Focus:* All States, All Countries
*Date(s) Application is Due:* Dec 5
*Contact:* Ann Woolford; (508) 289-7521; fax (508) 289-7931; awoolford@mbl.edu
*Internet:* www.grassfoundation.org/grass-fellowship-program
*Sponsor:* Grass Foundation
P.O. Box 241458
Los Angeles, CA 90024

**Greater Saint Louis Community Foundation Grants** **1187**
The Greater Saint Louis Community Foundation (GSLCF) was founded in 1915, one year after the first community foundation was established in Cleveland, Ohio. Currently GSLCF administers over 400 individual charitable funds that total $170 million in assets. These funds annually make over $17 million in grants that shape the greater Saint Louis region, touch communities across the nation, and reach across the globe. Historically, the mission of GSLCF has been two-fold: to serve donors and ensure that their dollars work in line with the goals that are important to them; and to promote charitable giving through community investment in nonprofit organizations capable of addressing community issues in measurable ways. To this end GSLCF maintains an online database YOURGivingLink to help donors find deserving nonprofit organizations that match their giving interests. Nonprofits are encouraged to register their organizations with the database; the link is available at the GSLCF website.
*Requirements:* While the Foundation predominately supports St. Louis area nonprofits, grants are also made to national and international charities.
*Geographic Focus:* Illinois, Missouri
*Contact:* Amy Basore Murphy, Director of Scholarships and Donor Services; (314) 588-8200, ext. 139 or (314) 880-4965; fax (314) 588-8088; amurphy@stlouisgives.org
*Internet:* www.stlouisgives.org/charities/
*Sponsor:* Greater Saint Louis Community Foundation
319 North Fourth Street, Suite 300
Saint Louis, MO 63102-1906

**Great Lakes Fishery Commission Pilot Research Project Grants** **1188**
Great Lakes Fishery Commission Pilot Research Project grants are designed to explore new hypotheses and concepts and to acquire preliminary data to guide the develpment of full research proposals. Pilot project proposals may be submitted at any time and must be less than $10,000. Funding decisions will be made after the March and September SLRB meetings. Deadlines for submission of pilot project proposals are January 15th for consideration in March or July 15th for consideration in September.
*Geographic Focus:* All States
*Date(s) Application is Due:* Jan 15; Jul 15
*Amount of Grant:* Up to 10,000 USD
*Contact:* Mike Siefkes, Sea Lamprey Program Specialist; (734) 662-3209, ext. 22; fax (734) 741-2010; msiefkes@glfc.org or info@glfc.org
Charles C. Krueger, Science Director; (734) 662-3209, ext. 12; fax (734) 741-2010; ckrueger@glfc.org or info@glfc.org
*Internet:* www.glfc.org/research/SRpp.php
*Sponsor:* Great Lakes Fishery Commission
2100 Commonwealth Boulevard, Suite 100
Ann Arbor, MI 48105

**Great Lakes Fishery Commission Sea Lamprey Research Grants** **1189**
The purpose of the Sea Lamprey Research program is to conduct research that will provide the information needed to further the integrated management of sea lampreys, including the development of novel, effective strategies for their control. Biological, ecological, and management-related research on sea lampreys and their effects on Great Lakes fish communities and fisheries is supported. Theme areas include: assessment; barriers and trapping; lampricides; pheromones, and sterile males. Research projects that do not fit into an existing theme area are supported under the category Non-Theme Research. Highly meritorious concepts, not connected to existing theme areas may be funded. Projects range from $10,000 to $100,000 per year. The average award ranges from $35,000 to $50,000 per year. Most projects are two to three years in duration. Annual funding available for Sea Lamprey Research program projects is typically $500,000.
*Geographic Focus:* All States
*Date(s) Application is Due:* Jun 1
*Amount of Grant:* 10,000 - 100,000 USD
*Contact:* Mike Siefkes, Sea Lamprey Program Specialist; (734) 662-3209, ext. 22; fax (734) 741-2010; msiefkes@glfc.org or info@glfc.org
Charles C. Krueger, Science Director; (734) 662-3209, ext. 12; fax (734) 741-2010; ckrueger@glfc.org or info@glfc.org
*Internet:* www.glfc.org/research/sldesc.php
*Sponsor:* Great Lakes Fishery Commission
2100 Commonwealth Boulevard, Suite 100
Ann Arbor, MI 48105

**Great Lakes Fishery Trust Access Grants** **1190**
The Great Lakes Fishery Trust's (GLFT) Access to the Great Lakes Fishery grant category focuses on significantly increasing access to the Great Lakes fishery for shore-based angling and tribal fishing. The goal of this funding category is to improve or create opportunities for shore-based access to fishing for Great Lakes species for use by tribal and/or recreational users. The GLFT supports its access goals through a combination of competitive and funder-directed grantmaking. Supported activities may include: new construction of access sites; upgrades and renovations to existing access sites; engineering and feasibility studies for proposed access sites; land acquisition to support subsequent access site development; and communication and outreach efforts regarding existing shore-based angling opportunities. Fishery Access proposals are due on August 23 each year.
*Requirements:* Organizations that are eligible to apply for GLFT grants include nonprofit organizations with a 501(c)3 designation from the IRS (or nongovernmental organizations that hold charitable status in their country), as well as educational and governmental (including tribal) organizations.
*Geographic Focus:* Illinois, Indiana, Iowa, Michigan, Ohio, Wisconsin
*Date(s) Application is Due:* Jan 25
*Contact:* Jonathon Beard; (517) 371-7468; fax (517) 484-6549; jbeard@glft.org
Amy Rittenhouse, Operations Manager; (517) 371-7468; fax (517) 484-6549; arittenhouse@glft.org or glft@glft.org
*Internet:* www.glft.org/site/fishing_access_grant.shtml
*Sponsor:* Great Lakes Fishery Trust
600 W. Saint Joseph, Suite 10
Lansing, MI 48933-2265

**Great Lakes Fishery Trust Ecological and Biological Fisheries Research to Inform Management Grants** **1191**
The Great Lakes Fishery Trust pursues its fisheries research efforts through competitive grantmaking within established research themes. Fundable activities include hypothesis-driven research as well as tools, resources, and capacity-building efforts that support the development or use and application of research. The GLFT will support research that: builds research capacity and management expertise needed to understand and manage the Great Lakes ecosystem for sustainable production of valuable species; enhances the ability of managers and agencies to respond to changes in the fishery and ecosystem; and/or reduces the cost of fishery management in relation to benefits provided. he main theme areas for the GLFT's gran tmaking in Ecological and Biological Fisheries Research are: fish health; recruitment; lake sturgeon rehabilitation; ecosystem modeling; and emerging issues. Pre-proposals for Ecological and Biological Fisheries research projects are due by January 22 of each year. The online application will be available in November.
*Requirements:* Organizations that are eligible to apply for GLFT grants include nonprofit organizations with a 501(c)3 designation from the IRS (or nongovernmental organizations that hold charitable status in their country), as well as educational and governmental (including tribal) organizations.
*Geographic Focus:* Illinois, Indiana, Iowa, Michigan, Ohio, Wisconsin
*Contact:* Jonathon Beard; (517) 371-7468; fax (517) 484-6549; jbeard@glft.org
Amy Rittenhouse, Operations Manager; (517) 371-7468; fax (517) 484-6549; arittenhouse@glft.org or glft@glft.org
*Internet:* www.glft.org/site/dam_management.shtml
*Sponsor:* Great Lakes Fishery Trust
600 W. Saint Joseph, Suite 10
Lansing, MI 48933-2265

**Great Lakes Protection Fund Grants** **1192**
The Great Lakes Protection Fund welcomes preproposals for projects that enhance the health of the Great Lakes ecosystem. Applicants should propose projects that will return the greatest ecosystem benefits. Current funding interests include preventing biological pollution, restoring natural flow regimes; and using market mechanisms for environmental improvement. Additional projects are sought to add to, and expand fund-supported work in these areas. The fund also welcomes projects that are designed to identify and explore other master variables that if acted upon, will result in tangible improvements to the health of the Great Lakes ecosystem. Proposals may be submitted at any time. The board makes grant decisions at its March, June, September, and December meetings.
*Requirements:* Nonprofit organizations (including environmental organizations, trade associations, and universities), for-profit businesses, government agencies, and individuals are eligible for fund support.
*Restrictions:* The fund does not give to general operating funds, environmental education, and groups from Indiana.
*Geographic Focus:* All States
*Amount of Grant:* 25,000 - 420,000 USD
*Contact:* J. David Rankin, Program Director; (847) 425-8150; fax (847) 424-9832; drankin@glpf.org or preproposals@glpf.org
*Internet:* www.glpf.org/about-the-fund
*Sponsor:* Great Lakes Protection Fund
1560 Sherman Avenue, Suite 880
Evanston, IL 60201

**Green Diamond Charitable Contributions** **1193**
Green Diamond Resource Company is a fifth-generation, family-owned forest-products company that has operations in the states of California and Washington. The mission of Green Diamond's contributions program is to improve the quality of life in communities where the company has a significant number of employees living and working; and to serve as a catalyst for employees to become involved and to provide leadership in their communities. The company has supported a broad range of organizations in its operating communities that address key community needs such as education, health services, economic development, the arts, and more. The company also provides support to environmental and conservation projects, such as environmental education, habitat restoration, and research relevant to the timber industry, as well as other projects that impact the timber business. Contributions are generally made in locations where the company has operations. To the extent possible, contributions will support organizations of interest to, or recommended by, Green Diamond employees. Green Diamond prefers

to make capital contributions that will benefit the operating communities for the long term as opposed to contributing operating funds. Generally, support is committed for one year at a time and in amounts less than $5,000. The company supports the United Way in all of its operating communities. The company accepts applications twice a year, generally in May and in late summer/early fall. Interested applicants should call or mail the contact person given for their state to request application materials. More detailed information about application deadlines will be communicated at that time.
*Requirements:* Organizations that serve Pierce, Mason, Thurston, and Grays Harbor counties in Washington or Del Norte and Humboldt counties in California are eligible to apply. Criteria taken into account in determining the amount of any contributions are as follows: degree of support from company employees; relative size and importance of company operations in the community (balance among Green Diamond communities); needs of organization or program for which funding is requested; amount of previous company contributions to the organization; amount committed by other companies, foundations, and/or governments (projects should demonstrate broad-based community support); and proximity of the requesting organization to Green Diamond operations or administrative offices.
*Geographic Focus:* California, Washington
*Amount of Grant:* Up to 5,000 USD
*Contact:* Jacki Deuschle, Public Affairs Manager; (707) 668-4488; fax (707) 668-4402; jdeuschle@greendiamond.com
Patti Case, Public Affairs Manager; (360) 426-3381; pcase@greendiamond.com
*Internet:* www.greendiamond.com/charitable-giving/
*Sponsor:* Green Diamond Resource Company
1301 Fifth Avenue, Suite 2700
Seattle, WA 98101-2613

**Green Foundation Human Services Grants** 1194
The Foundation's mission is to uncover new opportunities, encourage growth and ultimately effect positive change within those institutions that best reflect the core focus areas and the communities it serves. Within the human services area, the Foundation focuses on institutions that provide hope and support to those least able to help themselves as well as the general community; including children, adolescents, the elderly, the homeless, and families who struggle with domestic abuse.
*Requirements:* 501(c)3 nonprofits (as per the IRS Service Code of 1986) are eligible. Most grant making is limited to institutions that serve the Los Angeles community; however the Foundation will consider requests beyond this geographic boundary for those institutions with the potential to impact communities statewide or nationally.
*Restrictions:* The Foundation does not provide funds for: those with net assets or fund balances of less than $100,000; multi-year commitments; annual meetings, conferences, and/or seminars; religious programs; capital campaigns; direct mail campaigns; conduit institutions, unified funds, fiscal agents, or institutions using grant funds from donors to support other institutions or individuals; private foundations; or individuals.
*Geographic Focus:* All States
*Amount of Grant:* 1,000 - 150,000 USD
*Samples:* Esperanza Community Housing, Los Angeles, California, $25,000; Foothill Family Service, Pasadena, California, $25,000; Interfaith Community Services, Escondido, California, $25,000.
*Contact:* Kylie Wright; (626) 793-6200, ext. 1; fax (626) 793-6201; kylies@ligf.org
Elena Hermanson; (626) 793-6200, ext. 2; fax (626) 793-6201; elenah@ligf.org
*Internet:* www.ligf.org/humanservices.php
*Sponsor:* Green Foundation
225 South Lake Avenue, Suite 1410
Pasadena, CA 91101

**Greenspun Family Foundation Grants** 1195
The Greenspun Family Foundation supports many causes with an emphasis on education, health, children, Jewish issues, and the greater Las Vegas community. Preference is given to requests from the Las Vegas area. There are no application forms or deadlines. Applicants should submit a letter of inquiry. Information about current programs supported is available at the Foundation website.
*Geographic Focus:* Nevada
*Amount of Grant:* 50,000 - 2,000,000 USD
*Samples:* University of Nevada at Las Vegas, Las Vegas, Nevada, $2.2 million, payable over one year; Nevada Cancer Institute, Las Vegas, Nevada, $1,000,000, payable over one year; Clark County Public Education, Las Vegas, Nevada, $50,000, payable over one year.
*Contact:* Dr. Brian Cram, Director; (702) 259-2323 or (702) 259-4023; fax (702) 259-4019; brian.cram@lasvegassun.com
*Internet:* www.thegreenspuncorp.com/philanthropy.php
*Sponsor:* Greenspun Family Foundation
901 North Green Valley Parkway, Suite 210
Henderson, NV 89074

**Greenwall Fellowship in Bioethics** 1196
The overall purpose of the Greenwall Fellowship in Bioethics is to provide talented, early-career bioethics scholars with the opportunity to experience and participate in evidence-based healthcare or public health studies and policy-making process that improve the care and access to care of patients in domestic and global health care systems. The Greenwall Fellowship in Bioethics enables young investigators to participate actively in the work of the IOM and to further their careers as future leaders addressing bioethics issues in clinical care, biomedical research, and public policy. The program especially welcomes nominations of under-represented minority candidates. The Greenwall Fellowship in Bioethics is awarded for a two-year period. During this time, the fellow is expected to continue work at his/her main academic post, while being assigned to a board of the IOM. The boards are: Children, Youth, and Families; Food and Nutrition; Global Health; Health Care Services; Health of Select Populations; Health Sciences Policy; and Population Health and Public Health Practice. A flexible research stipend of $25,000 is awarded to each Greenwall Fellow in Bioethics. The stipend will be administered through the appropriate department in the fellow's home institution. Stipends are not intended for use as salary offsets. The Call for Nominations opens on June 24 each year, with nomination packets due on July 22.
*Requirements:* The fellowship requires a ten-to-twenty percent commitment of time for two years, and includes attendance at a one-week orientation to health policy in October, the Fall IOM Annual Meeting and the meetings of the assigned board. Additionally, the fellow will participate actively in the work of an appropriate expert study committee or round table, including contributing to its reports or other products. Studies will be identified that would enable the potential for the fellow to contribute to the value of integrating medicine and public health.
*Geographic Focus:* All States
*Date(s) Application is Due:* Jul 22
*Amount of Grant:* 25,000 USD
*Contact:* Marie E. Michnich; (202) 334-1506; fax (202) 334-3862; mmichnich@nas.edu
*Internet:* www.nationalacademies.org/hmd/Activities/Education/GreenwallFellowship.aspx
*Sponsor:* National Academies
500 5th Street, NW, Room 508
Washington, D.C. 20001

**Greenwall Foundation Faculty Scholars Program in Bioethics** 1197
The Greenwall Foundation Faculty Scholars Program in Bioethics is a career development program that enables junior faculty members to carry out original research on policy and moral dilemmas at the intersection of ethics and the life sciences. To maximize scholarly development, three years of support are provided, requiring a 50% time commitment (carefully monitored) in each of the three years. Each year the Scholars participate in a learner-centered educational program that includes required attendance at two workshops (spring and fall). Continued involvement in this activity, after each Scholar's support period is over, provides Scholar-alumni/ae with ongoing professional development and engages them in the training of successor cohorts. Letters of Intent are due by November 1, with invited applications due by January 15. Approximately 15 to 20 applicants will be invited to submit full proposals due in February.
*Requirements:* Applicants must be junior faculty members holding at least a 60% appointment at a university or non-profit research institute in the U.S. Priority will be given to applicants who are below the rank of Associate Professor, who have not received a comparable career development award, and whose work will have an impact on public policy or clinical practice. Within this group, priority will be given to applicants whose research addresses innovative and emerging topics.
*Geographic Focus:* All States
*Date(s) Application is Due:* Jan 15
*Contact:* Rebekkah Boxt, (212) 679-7266; fax (212) 679-7269; rboxt@greenwall.org
*Internet:* www.greenwallfsp.org/
*Sponsor:* Greenwall Foundation
One Penn Plaza, 47th Floor
New York, NY 10119

**Greenwall Foundation Making a Difference in Real-World Bioethics Dilemmas Grants** 1198
The Greenwall Foundation's bioethics grants program, Making a Difference in Real-World Bioethics Dilemmas, supports research to help resolve an important emerging or unanswered bioethics problem in clinical care, biomedical research, public health practice, or public policy. The Foundation's aim is to fund innovative projects that will have a real-world, practical impact. Priority for funding will be given to collaborative projects involving a bioethics scholar and persons with on-the-ground experience in organizations where bioethics dilemmas arise, for example, in clinical care; biomedical research; biotechnology, pharmaceutical, big data, and artificial intelligence companies; or public service. Projects may be empirical, conceptual, or normative. Letters of Intent are due by November 1, with invited applications due by January 15. Approximately 15 to 20 applicants will be invited to submit full proposals due in February.
*Restrictions:* Projects with the following characteristics will not be funded under this program: projects for which bioethics is not the main focus; projects that implement or make incremental improvements in established approaches to bioethics problems, build institutional infrastructure, or provide bioethics education, training, or coursework; projects that simply describe or analyze bioethics issues or provide a conceptual framework, without making practical recommendations for resolving the issues; projects that have predetermined conclusions or advocate for predetermined positions; projects whose main goal is to convene or enhance a meeting, unless there is a well-developed plan to produce a peer-reviewed publication with consensus recommendations, guidelines, or best practices with a strong likelihood of real-world implementation. The applicant must have a strong record in convening successful meetings; projects to support or extend ongoing or core activities of an organization; projects with a principal investigator who does not have a PhD, JD, MD, or an equivalent doctoral-level degree; or applications from unaffiliated individuals, or from institutions that do not have tax-exempt status in the United States.
*Geographic Focus:* All States
*Date(s) Application is Due:* Jan 15
*Amount of Grant:* 5,000 - 50,000 USD
*Contact:* Sam Teigen, Grants Administrator; (212) 679-7266; fax (212) 679-7269; steigen@greenwall.org or admin@greenwall.org
*Internet:* greenwall.org/making-a-difference.php
*Sponsor:* Greenwall Foundation
One Penn Plaza, 47th Floor
New York, NY 10119

**Grifols Community Outreach Grants** 1199

At Grifols, the primary mission is to enhance and enrich lives, which extends well beyond the patient communities it serves. Through a robust community outreach program, Grifols helps foster strong, vibrant communities in regions where its employees live and work. Each year, Grifols employees spend countless hours volunteering their time to improve the lives of people in need. Grifols encourages employee participation in community outreach activities by supporting a range of local and national causes through donations, corporate sponsorships and company-wide volunteer initiatives. The Grifols Community Relations program focuses on the following service areas: programs that improve the health of neighbors; science education and workforce development; and civic and community programs.

*Requirements:* 501(c)3 tax-exempt organizations in Grifols communities (San Francisco East Bay Area, including Alameda, Contra Costa, and Solano Counties; Los Angeles; Seattle; and Philadelphia) are eligible.

*Restrictions:* In general, Grifols does not support organizations that do not have 501(c)3 tax status; religious, fraternal, service, or veterans' organizations; civic or cultural organizations that do not serve the areas in which Grifols is located; alumni drives and teacher organizations; memorials; municipal and for-profit hospitals; labor unions; city, municipal, or federal government departments; organizations or causes that do not support the company's commitment to non-discrimination and diversity; projects of national scope or from national organizations not related to health care; matching gifts; individuals, including scholarships (other than those awarded as part of the company-sponsored college scholarship program); travel support; fund-raising activities related to individual sponsorship; and fund-raising dinners other than those for health care or medical research organizations aligned with company research and product interests.

*Geographic Focus:* California, Pennsylvania, Washington

*Amount of Grant:* 5,000 - 75,000 USD

*Contact:* Rebecca Barnes; (919) 316-6590; rebecca.barnes@grifols.com

*Internet:* www.grifols.com/web/eeuu/community_outreach

*Sponsor:* Grifols

2410 Lillyvale Avenue

Los Angeles, CA 90032-3514

**Gruber Foundation Cosmology Prize** 1200

The Gruber Foundation honors and encourages educational excellence, social justice and scientific achievements that better the human condition. Foundation is a private, United States-based philanthropic organization established in 1993 under the 501(c)3 section of U.S. Corporate Law. It is funded entirely by Peter and Patricia Gruber, who serve as its Chairman and President, respectively. The Foundation is headquartered at Yale University. A major focus of the Foundation's philanthropy is its International Prize Program, created to recognize excellence in science and humanities by highlighting five fields that create a better world: Cosmology, Genetics, Neuroscience, Justice, and Women's Rights. The Cosmology Prize, cosponsored by the International Astronomical Union since 2001, honors a leading cosmologist, astronomer, astrophysicist or scientific philosopher for theoretical, analytical, conceptual or observational discoveries leading to fundamental advances in our understanding of the universe. The Prize acknowledges and encourages further exploration in a field that shapes the way we perceive and comprehend our universe. In doing so, The Gruber Foundation seeks to extend the pioneering legacy of, among others, Plato and Aristotle; Ptolemy and Copernicus; Brahe, Kepler, and Galileo; Newton and Halley; Einstein and Hubble. Individuals whose achievements have produced fundamental, advances in our understanding of the structure and evolution of the universe may be nominated. A gold medal and unrestricted $500,000 cash prize are presented annually.

*Requirements:* Nominations for the Cosmology Prize are invited annually from the fields of astronomy, cosmology, mathematics, and the philosophy of science. Individuals, organizations, and institutions that are active in or have an appreciation for contemporary cosmological research and study may submit nominations.

*Geographic Focus:* All States, All Countries

*Date(s) Application is Due:* Dec 15

*Amount of Grant:* 500,000 USD

*Contact:* Owen Jay Gingerich, Research Professor of Astronomy and History of Science; (617) 495-7216; nominations@gruberprizes.org or ogingerich@cfa.harvard.edu

*Internet:* gruber.yale.edu/cosmology

*Sponsor:* Gruber Foundation

157 Church Street, 9th Floor

New Haven, CT 06510

**Gruber Foundation Gentics Prize** 1201

The Gruber Foundation honors and encourages educational excellence, social justice and scientific achievements that better the human condition. Foundation is a private, United States-based philanthropic organization established in 1993 under the 501(c)3 section of U.S. Corporate Law. It is funded entirely by Peter and Patricia Gruber, who serve as its Chairman and President, respectively. The Foundation is headquartered at Yale University. A major focus of the Foundation's philanthropy is its International Prize Program, created to recognize excellence in science and humanities by highlighting five fields that create a better world: Cosmology, Genetics, Neuroscience, Justice, and Women's Rights. The Genetics Prize, established in 2001, is presented to a leading scientist, or up to three, in recognition of groundbreaking contributions to any realm of genetics research. A gold medal and unrestricted $500,000 cash award is awarded for fundamental insights in the field of genetics. These may include original discoveries in genetic function, regulation, transmission, and variation, as well as in genomic organization. Nominations for the Prize close on December 15.

*Requirements:* Nominations may be submitted by individuals, organizations, and institutions that are active in or have an appreciation for contemporary genetic research or problems. Individuals who have made original discoveries in the fields of genetic function, regulation, transmission, or variation or in genomic organization are eligible.

*Geographic Focus:* All States

*Date(s) Application is Due:* Dec 15

*Amount of Grant:* 500,000 USD

*Contact:* Helen Hobbs, Foundation Contact; (214) 645-8300 or (214) 648-3111; nominations@gruberprizes.org or info@gruberprizes.org

*Internet:* gruber.yale.edu/genetics

*Sponsor:* Gruber Foundation

157 Church Street, 9th Floor

New Haven, CT 06510

**Gruber Foundation International Astronomical Union Fellowships** 1202

With the aim to promote the science of cosmology and other branches of astronomy, the Gruber Foundation has created the TGF Fellowship Program. One Fellowship will be awarded every year in the amount of $50,000. The award can cover travel, subsistence, and research expenses during a postdoctoral appointment for a period which is typically of one year duration starting no later than 1st October of the award year, but may be extended to cover up to two years. The Fellowship may be held in conjunction with any other fellowship or award. The Fellowship will be awarded to an extremely promising, young astrophysicist, working in any field of astrophysics, either theoretical, observational or experimental. There are no limitations on nationality, but preference will be given to applicants from countries in difficult economic conditions. The annual deadline for applications is March 31, with notifications by May 31.

*Geographic Focus:* All States, All Countries

*Date(s) Application is Due:* Mar 31

*Amount of Grant:* 50,000 USD

*Contact:* Foundation Contact; (203) 432-6231 or (340) 775-4430; nominations@gruberprizes.org or info@gruberprizes.org

*Internet:* gruber.yale.edu/international-astronomical-union-fellowships

*Sponsor:* Gruber Foundation

157 Church Street, 9th Floor

New Haven, CT 06510

**Gruber Foundation Neuroscience Prize** 1203

The Gruber Foundation Neuroscience Prize, established in 2004, is supported by the Gruber Foundation and honors scientists for major discoveries that have advanced understanding of the nervous system. Recipients receive a $500,000 prize and a gold laureate pin. For the SfN annual meeting, the recipient serves as keynote speaker at The Peter and Patricia Gruber Lecture and receives complimentary registration, transportation (economy air or ground), and two nights hotel accommodations. The SfN president presents the prize at a lecture at the meeting. Nominations for the Prize close on December 15.

*Requirements:* Individuals who have conducted highly distinguished research in the field of the brain, spinal cord, or peripheral nervous system are eligible for the award. Individuals, organizations, and institutions that are active in, or have an appreciation for, contemporary neuroscience research can submit nominations.

*Geographic Focus:* All States, All Countries

*Date(s) Application is Due:* Dec 15

*Amount of Grant:* 500,000 USD

*Contact:* Robert H Wurtz, Ph.D., Chairperson; (301) 496-7170; fax (301) 402-0511; bob@lsr.nei.nih.gov

*Internet:* gruber.yale.edu/neuroscience-prize-nomination-criteria

*Sponsor:* Gruber Foundation

157 Church Street, 9th Floor

New Haven, CT 06510

**Gruber Foundation Prize for Global Justice and Women's Rights** 1204

The Gruber Foundation honors and encourages educational excellence, social justice and scientific achievements that better the human condition. Foundation is a private, United States-based philanthropic organization established in 1993 under the 501(c)3 section of U.S. Corporate Law. It is funded entirely by Peter and Patricia Gruber, who serve as its Chairman and President, respectively. The Foundation is headquartered at Yale University. A major focus of the Foundation's philanthropy is its International Prize Program, created to recognize excellence in science and humanities by highlighting five fields that create a better world: Cosmology, Genetics, Neuroscience, Justice, and Women's Rights. Since 2001, the Justice Prize has honored individuals who have advanced the cause of justice as delivered through the legal system. And, since 2003, the Women's Rights Prize has honored individuals who have made significant contributions to human rights that advance the rights of women and girls around the world. In 2012, the two prizes have been combined into the Prize for Global Justice and Women's Rights. It includes a gold medal and unrestricted $500,000 cash award, and is an international award presented annually. Recipients are selected by a distinguished panel of international legal experts from nominations that are received from around the world. The annual deadline is December 15.

*Geographic Focus:* All States, All Countries

*Date(s) Application is Due:* Dec 15

*Amount of Grant:* 500,000 USD

*Contact:* Foundation Contact; (203) 432-6231 or (340) 775-4430; nominations@gruberprizes.org or info@gruberprizes.org

*Internet:* gruber.yale.edu/program-justice-and-womens-rights

*Sponsor:* Gruber Foundation

157 Church Street, 9th Floor

New Haven, CT 06510

**Gruber Foundation Rosalind Franklin Young Investigator Award** 1205

The Gruber Foundation funds the Rosalind Franklin Young Investigator Award – a career development research award of $75,000 over three years - given to two young woman geneticists from anywhere in the world. The award, which honors the groundbreaking contributions of Dr. Rosalind Franklin, is intended to inspire and support new generations of women in the field of genetics. One award will fund genetics research in human and non-human mammals, and the other will fund genetics research in model organisms. This unrestricted cash award is presented every three years. The nomination deadline is June 5.
*Requirements:* The successful candidate must be in her first one to three years of an independent faculty-level position in any area of genetics.
*Geographic Focus:* All States, All Countries
*Date(s) Application is Due:* Jun 5
*Amount of Grant:* 75,000 USD
*Contact:* Foundation Contact; (203) 432-6231 or (340) 775-4430; nominations@gruberprizes.org or info@gruberprizes.org
*Internet:* gruber.yale.edu/rosalind-franklin-young-investigator-award
*Sponsor:* Gruber Foundation
157 Church Street, 9th Floor
New Haven, CT 06510

**Gruber Foundation Weizmann Institute Awards** 1206

The Weizmann Institute is a center of basic multidisciplinary scientific research and graduate study, addressing crucial problems in technology, medicine and health, energy, agriculture, and the environment. Supported by personal funds donated by Peter and Patricia Gruber, the Gruber Awards are used to fund the research of new scientists at the Weizmann Institute and the scientific travel of selected postdoctoral fellows and doctoral students through an award of $30,000 per year over a three-year period. A faculty committee advises the President of the Institute as to a worthy Gruber Award recipient. The inaugural award was presented to Elad Schneidman,a senior scientist in the Institute's Department of Neurobiology, in May 2007. One new scientist will receive the award annually. The Award provides vital funding for gifted new scientists at the beginning of their research activities, and helps enrich their work.
*Geographic Focus:* All States, All Countries
*Date(s) Application is Due:* Dec 15
*Amount of Grant:* 90,000 USD
*Contact:* Foundation Contact; (203) 432-6231 or (340) 775-4430; nominations@gruberprizes.org or info@gruberprizes.org
*Internet:* gruber.yale.edu/weizmann-institute-awards
*Sponsor:* Gruber Foundation
157 Church Street, 9th Floor
New Haven, CT 06510

**Gulf Research Program Early-Career Research Fellowships** 1207

Early-Career Research Fellowships will recognize professionals at the critical pretenure phase of their careers for exceptional leadership, past performance, and potential for future contributions to improving oil system safety, human health and well being, or environmental protection. These two-year fellowships will be awarded to tenure-track faculty (or equivalent) at colleges, universities, and research institutions. Up to ten Fellows will be selected each year. Areas of research may include social and behavioral sciences, health and medicine, engineering, earth and life sciences, or relevant interdisciplinary fields. The following will be considered as positive factors in choosing successful applicants: evidence of a strong scientific and technical background and superior scholarship; experience working in an interdisciplinary field or participating in interdisciplinary collaborations; demonstrated verbal and written communication skills; demonstrated leadership experience; and membership in one or more groups that are traditionally underrepresented in the applicant's scientific discipline. The application period opens on December 1, with a deadline of February 17. Fellows will receive an award of $76,000 paid to their institution in the form of a two-year grant.
*Requirements:* Applicants must: hold an Assistant Professor position (or equivalent) at a college, university, or research institution; not receive tenure before September 1 of the fellowship year; be a U.S. citizen, U.S. national, lawful permanent resident, or an individual granted deferred action status under the Department of Homeland Security Deferred Action for Childhood Arrivals program; be affiliated with a non-federal U.S. institution that has a valid tax ID number; and identify a senior-level faculty member at their affiliated institution willing to serve as a career mentor.
*Restrictions:* Federal employees are not eligible for the fellowship.
*Geographic Focus:* All States
*Date(s) Application is Due:* Feb 17
*Amount of Grant:* 76,000 USD
*Contact:* Director; (202) 334-2455; fax (202) 334-1667; GulfFellowships@nas.edu
*Internet:* www.nas.edu/gulf/fellowships/index.html
*Sponsor:* National Academies
500 5th Street, NW, Room 508
Washington, D.C. 20001

**Gulf Research Program Science Policy Fellowships** 1208

Science Policy Fellowships will contribute to leadership development and capacity building by providing recipients with a valuable educational experience at the science-policy interface. Fellows will spend one year on the staff of a state environmental, natural resources, oil and gas, or public health agency; or regional offices of relevant federal agencies in the Gulf region. Fellows will participate in and contribute to the state or federal policy-making process. Depending on placement, duties are likely to include performing background research, analyzing and synthesizing scientific information to support host office activities, writing policy memos, and meeting with stakeholders. Areas of study may include social and behavioral sciences, health and medicine, engineering, earth and life sciences, or relevant interdisciplinary fields. The following will be considered as positive factors in choosing successful applicants: evidence of a strong scientific and technical background and superior scholarship; experience working in an interdisciplinary field or participating in interdisciplinary collaborations; demonstrated verbal and written communication skills; demonstrated leadership experience; and membership in one or more groups that are traditionally underrepresented in the applicant's scientific discipline. Up to ten Fellows will be chosen each year. Fellows who have completed an MA, MS, or MPH degree or who are currently enrolled in a doctoral program will receive an annual stipend of $45,000. Fellows who have completed a PhD, ScD, MD, or DVM will receive an annual stipend of $55,000. Additional support, up to $5,000, will be provided on a reimbursable basis to offset tuition and/or fees associated with enrolling in professional development courses or for professional travel related to their host office duties. The application period begins on December 1, with the deadline for submissions being February 17.
*Requirements:* Doctoral students and those who received their graduate degree on or after September 1 five years previous to the fellowship year are eligible to apply. Eligible degrees include MA/MS, PhD, ScD, MPH, MD, or DVM. Applicants must be a U.S. citizen, U.S. national, lawful permanent resident, or an individual granted deferred action status under the Department of Homeland Security Deferred Action for Childhood Arrivals Program. Applicants are not required to have direct public policy experience but should have a demonstrable interest in applying science to public policy.
*Restrictions:* Federal employees are not eligible for the fellowship.
*Geographic Focus:* All States
*Date(s) Application is Due:* Feb 17
*Amount of Grant:* 45,000 - 60,000 USD
*Contact:* Director; (202) 334-2455; fax (202) 334-1667; GulfFellowships@nas.edu
*Internet:* www.nas.edu/gulf/fellowships/index.html
*Sponsor:* National Academies
500 5th Street, NW, Room 508
Washington, D.C. 20001

**H.A. and Mary K. Chapman Charitable Trust Grants** 1209

H. Allen Chapman was born in Colorado in 1919. In 1976, he established the H.A. and Mary K. Chapman Charitable Trust, a perpetual charitable private foundation that maintains endowments to fund charitable grants to public charities. The trustees and staff that administer the foundation also provide public stewardship through service to charitable organizations and causes. A major charitable focus of H. A. Chapman during his life, and the lives of his philanthropic parents, James A. and Leta Chapman, was education and medical research. Though not limited geographically, most grants and public service are within Oklahoma. Grants to human services and civic and community programs and projects are primarily focused in the area of Tulsa. There are two steps in the process of applying for a grant. The first is a Letter of Inquiry from the applicant. This letter is used to determine if the applicant will be invited to take the second step of submitting a formal Grant Proposal.
*Requirements:* IRS 501(c)3 non-profits are eligible to apply.
*Restrictions:* Grant requests for the following purposes are not favored: endowments, except as a limited part of a capital project reserved for maintenance of the facility being constructed; deficit financing and debt retirement; projects or programs for which the Chapman Trusts would be the sole source of financial support; travel, conferences, conventions, group meetings, or seminars; camp programs and other seasonal activities; religious programs of religious organizations; project or program planning; start-up ventures are not excluded, but organizations with a proven strategy and results are preferred; purposes normally funded by taxation or governmental agencies; requests made less than nine months from the declination of a previous request by an applicant, or within nine months of the last payment made on a grant made to an applicant; requests for more than one project.
*Geographic Focus:* Oklahoma
*Amount of Grant:* Up to 300,000 USD
*Samples:* Gilcrease Museum Management, Tulsa, Oklahoma, $250,000 - general operating funds; Tulsa Air and Space Museum, Tulsa, Oklahoma, $35,000 - general operating support; Friends of the Fairgrounds Foundation, Tulsa, Oklahoma, $15,000 - general operating funds.
*Contact:* Andrea Doyle, Program Officer; (918) 496-7882; fax (918) 496-7887; andie@chapmantrusts.com
*Internet:* www.chapmantrusts.org/grants_programs.html
*Sponsor:* H.A. and Mary K. Chapman Charitable Trust
6100 South Yale, Suite 1816
Tulsa, OK 74136

**H.J. Heinz Company Foundation Grants** 1210

The Foundation is committed to promoting the health and nutritional needs of children and families. Priority is given to programs in communities where Heinz operates with a special focus given to southwestern Pennsylvania. The Foundation proactively donates funds to develop and strengthen organizations that are dedicated to nutrition and nutritional education, youth services and education, diversity, healthy children and families, and quality of life. Application information is available online.
*Requirements:* Only organizations that have 501(c)3 tax status under the U.S. Internal Revenue Code are eligible for support domestically. International organizations are encouraged to submit a letter of inquiry prior to preparing a full proposal to determine eligibility. International organizations will need to provide a 501(c)3 determination letter from the United States Internal Revenue Service or sufficient documentation to demonstrate that the non-U.S. grantee is the equivalent of a U.S. public charity.

Documentation should be provided in English. The Foundation will also accept proposals in the Common Grant Application Format. All proposals must be submitted in writing. Electronic submissions are not accepted at this time.
*Restrictions:* The Foundation will not provide grants to individuals nor make multi-year pledges except for major capital or grant campaigns. Generally, the Foundation does not make loans and does not provide grants for individuals, equipment, conferences, travel, general scholarships, religious programs, political campaigns, and unsolicited research projects.
*Geographic Focus:* All States
*Samples:* King Edwards Memorial Hospital, Subiacio, Perth, Australia, $28,000—Micronutrient long-term development study; The Children's Medical & Research Foundation, Dublin, Ireland, $25,000—new adolescent Hematology/Oncology unit; Pittsburgh Public Theater, Pittsburgh, PA, $30,000—program support.
*Contact:* Tammy B. Aupperle, Program Director; (412) 456-5773; fax (412) 442-3227; heinz.foundation@hjheinz.com
*Internet:* www.heinz.com/sustainability.aspx/social/heinz-foundation.aspx
*Sponsor:* H.J. Heinz Company Foundation
P.O. Box 57
Pittsburgh, PA 15230-0057

**HAF Riley Frazel Memorial Scholarship** 1211
The Riley Frazel Memorial Scholarship is offered for graduating high school senior from Arcata or McKinleyville high schools. Recipients may be planning to attend any form of higher education (college, university, vocational or trade school), full or part time, and are not limited on where they may enroll. Any major is eligible. Financial need is considered, but grade point average is not. Community service and extra curricular activities are taken into consideration. The scholarship is worth a one time award of $1,000. The application is available at the website.
*Requirements:* Applicants must be graduating seniors of either Arcata or McKinleyvile high schools. They must also have attended Fieldbrook Elementary School at any time from kindergarten through eighth grade. Candidates will submit the application, a personal statement, their official transcript, and two letter of recommendation.
*Geographic Focus:* California
*Date(s) Application is Due:* Mar 15
*Amount of Grant:* 1,000 USD
*Contact:* Cassandra Wagner, Program Coordinator; (707) 442-2993, ext. 323; fax (707) 442-3811; cassandraw@hafoundation.org
*Internet:* www.hafoundation.org/index.php?option=content&task=view&id=61
*Sponsor:* Humboldt Area Foundation
373 Indianola Road
Bayside, CA 95524

**Hampton Roads Community Foundation Mental Health Research Grants** 1212
The Hampton Roads Community Foundation's mission is to inspire philanthropy and transform the quality of life in southeastern Virginia. Funding for Mental Health Research Grants is provided by the Benjamin R. Brown and Charles G. Brown Funds. The annual deadline for online applications is October 1. Applicants from the Eastern Shore of Virginia should speak with a Program Officer before submitting a proposal. Grant awards typically range from $1,500 to $15,000.
*Requirements:* Nonprofit organizations serving residents of south Hampton Roads (Chesapeake, Franklin, Norfolk, Portsmouth, Suffolk, Virginia Beach and Isle of Wight County) and the Eastern Shore of Virginia are eligible.
*Geographic Focus:* Virginia
*Date(s) Application is Due:* Oct 1
*Amount of Grant:* 1,500 - 15,000 USD
*Contact:* Linda M. Rice, Vice President of Grantmaking; (757) 622-7951; fax (757) 622-1751; lrice@hamptonroadscf.org or grants@hamptonroadscf.org
*Internet:* www.hamptonroadscf.org/nonprofits/specialInterestGrants.html
*Sponsor:* Hampton Roads Community Foundation
101 W. Main Street, Suite 4500
Norfolk, VA 23510

**Harmony Project Grants** 1213
The Harmony Project is an innovative, creative nonprofit foundation established in Ohio in 2001 by Judith Harmony to promote the training of doctors and scientists, as well as research at Children's Hospital in Cincinnati. Currently, its primary mission is to enhance the education and personal development of girls, including self-efficacy, leadership and mother-daughter relationships, and to enrich their cultural and artistic lives. Funding is given for: curriculum development; program development; evaluation; research; and scholarship endowments. That support is provided within the greater Cincinnati area, as well as in northern Kentucky. An application form is required, though an initial approach should be by letter to the Foundation office. There are no specific annual submission deadlines.
*Requirements:* The Foundation accepts written applications from 509(a)1, 509(a)2, and 509(a)3 organizations supporting the residents of greater Cincinnati and northern Kentucky.
*Geographic Focus:* Kentucky, Ohio
*Contact:* Judith Harmony; (513) 861-8490; fax (513) 281-4326; jharmony@fuse.net
*Internet:* www.harmonyproject.org/
*Sponsor:* Harmony Project
3950 Rose Hill Avenue
Cincinnati, OH 45229-1448

**Harold R. Bechtel Charitable Remainder Uni-Trust Grants** 1214
Established in Davenport, Iowa, in 1987, the Harold R. Bechtel Charitable Remainder Uni-Trust supports programs in Scott County, Iowa, with giving primarily for health associations, education, and for children, youth, and social services. The Foundation's primary interest areas include: children and youth services; community and economic development; education; family services; health organizations and associations; heart and circulatory research; higher education; human services; museums, music; and performing arts. Types of support building construction and renovation, as well as program-related funding. There are no specific deadlines, and applicants should contact the office in writing or via telephone to request application forms.
*Requirements:* 501(c)3 organizations serving the residents of Rock County, Iowa, are eligible to apply.
*Restrictions:* No grants are given to individuals, or for endowment funds, debt retirement, past operating deficit, general or continuing support, or for scholarly research in an established discipline.
*Geographic Focus:* Iowa
*Amount of Grant:* Up to 150,000 USD
*Samples:* River Music Experience, Davenport, Iowa, $150,000; Saint Ambrose University, Davenport, Iowa, $150,000; Figge Art Museum, Davenport, Iowa, $125,000.
*Contact:* R. Richard Bittner; (563) 328-3333; fax (563) 328-3352; loseband@blwlaw.com
*Sponsor:* Harold R. Bechtel Charitable Remainder Uni-Trust
201 West 2nd Street, Suite 1000
Davenport, IA 52801-1817

**Harold R. Bechtel Testamentary Charitable Trust Grants** 1215
Established in Davenport, Iowa, in 1989, the Harold R. Bechtel Testamentary Charitable Trust Grants gives primarily to higher education and youth services. Major fields of interest include: children and youth services; community development; economic development; education; human services; and medical research. Applicants should contact the office in writing or via telephone to request application forms.
*Requirements:* 501(c)3 organizations serving the residents of Davenport, Iowa, Rock Island, Illinois, and Moline, Illinois, are eligible to apply.
*Restrictions:* No grants are given to individuals, or for endowment funds, debt retirement, past operating deficit, general or continuing support, or for scholarly research in an established discipline.
*Geographic Focus:* Iowa
*Amount of Grant:* Up to 150,000 USD
*Contact:* R. Richard Bittner; (563) 328-3333; fax (563) 328-3352; loseband@blwlaw.com
*Sponsor:* Harold R. Bechtel Testamentary Charitable Trust
201 West 2nd Street, Suite 1000
Davenport, IA 52801-1817

**Harris and Eliza Kempner Fund Health Grants** 1216
The Harris and Eliza Kempner Fund provides grants primarily in the Galveston, Texas, area to qualifying organizations in the broad areas of the arts and humanities, community development, education, the environment, health, human services, and preservation. In the area of health, the Fund supports: medical education; medical research; mental health care; public health care programs; sports and physical health; and women's wellness. The Fund gives preference to requests for seed money, general operating funds, small capital needs support, and special projects partnering with other funding sources. Applicants should submit proposals through the Kempner Fund's online grant application. The Fund accepts proposals twice a year: March 15 for the program areas of health or human services; and October 15 for the program areas of arts and humanities, community development, the environment, education, and preservation. Directors will consider one proposal per program area per year.
*Requirements:* The Kempner Fund primarily supports U.S.-based 501(c)3 organizations and non-profit, governmental, and faith-based proposals from Galveston-based or Galveston-serving organizations.
*Restrictions:* Funding is not available for: fund-raising benefits; direct mail solicitations; grants to individuals; and grants to non-U.S. based organizations. emailed or printed copies will not be accepted.
*Geographic Focus:* Texas
*Date(s) Application is Due:* Mar 15
*Amount of Grant:* Up to 50,000 USD
*Samples:* Family Service Center of Galveston County, Galveston, Texas, $30,000 - support for preventing and addressing adverse childhood experiences (2018); Galveston County Food Bank, Texas City, Texas, $19,000 - general operating support for food assistance to Galveston Island families (2018); Galveston's Own Farmers Market, Galveston, Texas, $10,000 - general operating support and program expansion (2018).
*Contact:* Harrette N. Howard, Grants Administartor; (409) 762-1603; fax (409) 762-5435; hnhoward@kempnerfund.org or information@kempnerfund.org
*Internet:* www.kempnerfund.org/program-areas/
*Sponsor:* Harris and Eliza Kempner Fund
2201 Market Street, Suite 1250
Galveston, TX 77550-1529

**Harris and Eliza Kempner Jeane B. Kempner Fellowships** 1217
In characterizing his parents' legacy, Isaac (Ike) Herbert Kempner wrote the following in his Recalled Recollections: "We can be be thankful for the larger vision of our parents which bid us try to emulate their wisdom, their benevolence, their reverence for God and devotion to their country." The impetus for starting a foundation in 1946 came from the family's concern for the many local charities they supported. They realized that conditions that typically follow economic depressions and wars could affect their ability

to support charities in time of greatest need. A philanthropic philosophy thus evolved. By terms of the Last Will and Testament of Daniel W. Kempner, the Jeane B. Kempner Fund was established to provide fellowships for graduates of The University of Texas Medical Branch to pursue advanced studies.
*Requirements:* The Harris and Eliza Kempner Fund provides fellowships primarily in the Galveston, Texas, area to qualifying higher education programs in the broad areas of the arts and humanities, community development, education, the environment, health, human services, and preservation research.
*Geographic Focus:* All States
*Amount of Grant:* Up to 300,000 USD
*Samples:* University of Texas Medical Branch, Galveston, Texas, $296,167 - a single fellowship for research (2018).
*Contact:* Harrette N. Howard, Grants Administrator; (409) 762-1603; fax (409) 762-5435; hnhoward@kempnerfund.org or information@kempnerfund.org
*Internet:* www.kempnerfund.org/program-areas/
*Sponsor:* Harris and Eliza Kempner Fund
2201 Market Street, Suite 1250
Galveston, TX 77550-1529

**Harry A. and Margaret D. Towsley Foundation Grants** 1218
In 1959, Margaret Towsley created the Harry A. and Margaret D. Towsley Foundation with an initial gift of $4 Million in Dow Chemical Company stock. While the Foundation's initial goals were typical of general family foundations, its mission later became focused on programs promoting education, health care, shelter, and nutrition for children. As its assets grew, its areas of concentration expanded into college and university education, medical education, planned parenthood, and interdisciplinary programs with the schools of law and social work. These areas reflected Dr. and Mrs. Towsley's common interest in teaching. The foundation currently awards grants to Michigan organizations in its areas of interest, including environment, medical and preschool education, social services, continuing education, and research in the health sciences. Types of support include annual campaigns, building construction and renovation, capital campaigns, continuing support, employee matching gifts, endowments, general operating support, matching/challenge support, professorships, program development, research, and seed grants. There are no application forms; submit a letter of inquiry between January and the listed application deadline.
*Restrictions:* Grants are not awarded to individuals or for travel, scholarships, fellowships, conferences, books, publications, films, tapes, audio-visual or communication media, or loans.
*Geographic Focus:* Michigan
*Date(s) Application is Due:* Mar 31
*Contact:* Lynn Towsley White, President; (989) 837-1100; fax (989) 837-3240
*Sponsor:* Harry A. and Margaret D. Towsley Foundation
140 Ashman Street, P.O. Box 349
Midland, MI 48640

**Harry B. and Jane H. Brock Foundation Grants** 1219
The Harry B. and Jane H. Brock Foundation honors a former Fort Payne, Alabama, native who changed the structure of the banking industry in Alabama. Harry Blackwell Brock, Jr., first worked as a truck driver for the Brook-Martin Oil Company, and later worked as a car salesman with John Thomas Motors of Gadsden. In 1964, he co-founded the Central Bank and Trust Company (later known as Compass Bank, now as BBVA Compass). In 1968, together with other investors, he gained voting control of the State National Bank of Alabama and, in 1981, he merged the two banks. By 1987, Central Bancshares of the South was the first bank in Alabama to own a bank in another state. Though Brock retired from the Board of Directors in 1991, he went on to serve on the Board of Directors of: Marathon Industries; the Daniel International Corporation; and Alabama Alliance of Business and Industry. He was inducted into the Alabama Academy of Honor in 1983 and the Alabama Business Hall of Fame in 1993. Brock died on July 29, 2015, at the age of 89. Today, the Foundation named in his honor awards grants to eligible Alabama organizations to support community development and higher education. Primary areas of interest include community service, community funds, education, the environment, higher education, volunteerism, women's services, social services, and cancer treatment and research. Types of support include: annual campaigns, capital campaigns, endowments, general operating support, program development, and research. Written proposals should describe the project and organization and include a copy of the IRS tax-determination letter. The annual deadline is November 1.
*Requirements:* 501(c)3 tax-exempt organizations serving the residents of Alabama are eligible.
*Restrictions:* No grants to individuals are awarded. Giving primarily in Birmingham and Huntsville, Alabama.
*Geographic Focus:* Alabama
*Date(s) Application is Due:* Nov 1
*Amount of Grant:* 1,000 - 25,000 USD
*Contact:* Harry B. Brock, Jr.; (205) 939-0236 or (205) 918-0833; fax (205) 939-0806
*Sponsor:* Harry B. and Jane H. Brock Foundation
2101 Highland Avenue, Suite 250, P.O. Box 11643
Birmingham, AL 35202-1643

**Harry Frank Guggenheim Foundation Dissertation Fellowships** 1220
The Foundation awards ten or more dissertation fellowships each year to individuals who will complete their dissertations within the award year. These fellowships are designed to help doctoral candidates finish writing their dissertations rather than to support dissertation research. Questions that interest the Foundation concern violence and aggression in relation to social change, intergroup conflict, war, terrorism, crime, and family relationships, among other subjects. Priority will also be given to areas and methodologies not receiving adequate attention and support from other funding sources. Applicants may be citizens of any country and studying at colleges or universities in any country. The application deadline is February 1, with final decisions made by the Board of Directors at its June meeting for funding that fall.
*Requirements:* Applicants must submit the following to apply: title page; abstract page and survey; advisor's letter and abbreviated CV; applicant's CV and graduate school transcript; project description; protection of subjects; and their other support for funding. Applicants should refer to the guidelines for additional information and specific instructions for their abstract page and survey (based on their individual discipline), and contact the Foundation if they have questions.
*Restrictions:* Dissertations with no relevance to understanding human violence and aggression will not be supported. Applicants should only apply if they are entering the dissertation stage of graduate school (fieldwork or other research is complete and writing has begun). Recipients of the dissertation fellowship must submit a copy of the dissertation, approved and accepted by their institution, within six months after the end of the award year. Any papers, books, articles, or other publications based on the research should also be sent to the foundation.
*Geographic Focus:* All States, All Countries
*Date(s) Application is Due:* Feb 1
*Amount of Grant:* 20,000 USD
*Contact:* Program Officer; (646) 428-0971; fax (646) 428-0981; info@hfg.org
*Internet:* www.hfg.org/df/guidelines.htm
*Sponsor:* Harry Frank Guggenheim Foundation
25 West 53rd Street
New York, NY 10019-5401

**Harry Frank Guggenheim Foundation Research Grants** 1221
The Foundation invites proposals in any of the natural sciences, social sciences or humanities that promise to increase understanding of the causes, manifestations, and control of human violence and aggression. Highest priority is given to research that can increase understanding and improve urgent problems of violence and aggression in the modern world. Questions that interest the foundation concern violence and aggression in relation to social change, intergroup conflict, war, terrorism, crime, and family relationships, among other subjects. Priority will also be given to areas and methodologies not receiving adequate attention and support from other funding sources. Applicants may be citizens of any country. Most awards fall within the range of $15,000 to $40,000 per year for periods of one or two years. Applications for larger amounts and longer durations must be very strongly justified. Applications must be received by August 1, with a decision given in December.
*Requirements:* Applicants must mail two copies of the typed application in English to the Foundation. Applications may not be faxed or emailed. Along with two copies of the application, applicants must include the following: title page; abstract and survey; budget and its justification; personnel; research plan; other support; protection of subjects; and referee comments. Applicants should refer to the Foundation website for specific instructions on how to submit each of the attachments, and contact the Foundation if they have any questions.
*Restrictions:* Research with no relevance to understanding human problems will not be supported, nor will proposals to investigate urgent social problems where the Foundation cannot be assured that useful, sound research can be done.While almost all recipients of the Foundation grants possess a Ph.D., M.D., or equivalent degree, there are no formal degree requirements for the grant. The grant, however, may not be used to support research undertaken as part of the requirements for a graduate degree. Applicants need not be affiliated with an institution of higher learning, although most are college or university professors. The foundation awards research grants to individuals (or a few principal investigators at most) for individual projects, but does not award grants to institutions for institutional programs. Individuals who receive research grants may be subject to taxation.
*Geographic Focus:* All States, All Countries
*Date(s) Application is Due:* Aug 1
*Amount of Grant:* 15,000 - 40,000 USD
*Contact:* Program Officer; (646) 428-0971; fax (646) 428-0981; info@hfg.org
*Internet:* www.hfg.org/rg/guidelines.htm
*Sponsor:* Harry Frank Guggenheim Foundation
25 West 53rd Street
New York, NY 10019-5401

**Harry S. Moss Heart Trust Grants** 1222
Harry S. Moss was born in Canada in 1888, moved to Texas in 1906 and started the Moss Petroleum Company in 1920. He lived in Highland Park, but in 1930 he bought a bull at the State Fair and realized he had no place to keep it. So he bought and developed the 400-acre Moss Haven Stock Farm in Lake Highlands. His house was featured in the 1958 issue of Architectural Digest, and that home still stands in the neighborhood today. In the late 1960s and early '70s, the City of Dallas purchased 54 and later another 225 acres of that farm, an acquisition that played a lynchpin role in the origins of today's Dallas trail system. Dallas Arboretum founders briefly considered some of the land for the city's botanical gardens, but ultimately deemed the DeGloyer Estate more apropos. After suffering and recovering from a heart attack in the 1960s, he became convinced of the importance of private philanthropic support of medical research. Moss died in 1970, and the terms of the Harry S. Moss Heart Trust established under his will state that the income is to be paid to organizations whose main purpose and object is the prevention and cure of diseases of the heart. His wife, Florence, died in 1978 adding a substantial portion of her estate to the trust. The annual deadline for application submission is May 1.
*Requirements:* Grants are awarded to 501(c)3 charitable organizations for medical or scientific research for the cause, treatment, prevention, alleviation or cure of diseases of the heart. Preference is given to institutions in Dallas County.
*Geographic Focus:* All States
*Date(s) Application is Due:* May 1

*Amount of Grant:* 50,000 - 750,000 USD
*Samples:* Children's Medical Center of Dallas, Dallas, Texas, $750,000 - support for Phase V of the expansion and consolidation of the Heart Center; American Heart Association, Irving, Texas, $500,000 - support of Cardiovascular Research; Juvenile Diabetes Research Foundation International- Greater Dallas Chapter. Dallas, Texas, $200,000 - research of sphingosine kinases in beta-cell survival and proliferation.
*Contact:* Debra Goldstein Phares, Philanthropic Client Director; (214) 209-1830 or (800) 557-7394; debra.g.phares@ustrust.com or tx.philanthropic@ustrust.com
*Internet:* https://www.bankofamerica.com/philanthropic/foundation.go?fnId=154
*Sponsor:* Harry S. Moss Heart Trust
901 Main Street, 19th Floor
Dallas, TX 75202-3714

**Harry W. Bass, Jr. Foundation Grants** **1223**
The Dallas-based Harry W. Bass, Jr. Foundation seeks to enrich the lives of the citizens of Texas by providing support to qualified organizations in the areas of education, health, human services, civic & community, science, research, arts and culture. Grant applications for specific programs or projects, capital projects or, less often, general operations are considered. Endowment gifts are rare. The Harry W. Bass, Jr. Foundation also considers program-related investments as part of its grant-making activities. The foundation strives to be responsive to the needs of all eligible organizations and considers requests of any amount. There is no formal application form. Grant requests are accepted at any time throughout the year. Each organization is limited to one application within a twelve-month period. Requests are usually processed within three to four months. Electronic grant applications should be submitted to dcalhoun@hbrf.org (no file attachments, please). For more information, contact the Harry W. Bass, Jr. Foundation.
*Requirements:* Texas 501(c)3 tax-exempt organizations are eligible.
*Restrictions:* In general, grants are not made for purposes of: church or seminary construction; annual fundraising events or general sustentation drives; professional conferences and symposia; out-of-state performances or competition expenses; to other private foundations. Unsolicited grant requests are restricted to organizations based in the Greater Dallas area. In the event a grant is approved, the recipient organization will be unable to submit another grant request for two years due to the foundation policy of not providing subsequent year grants.
*Geographic Focus:* Texas
*Samples:* Southwestern Medical Foundation, $3,000,000—Heart, Lung & Vascular Center of Excellence at UT Southwestern University Hospital - St. Paul; Child Protective Services Community Partners, $10,000—Rainbow Room; Health & Human Services Grant; Wilkinson Center, $20,000—CLIMB After-School and Summer Program.
*Contact:* F. David Calhoun; (214) 599-0300; fax (214) 599-0405; dcalhoun@hbrf.org
*Internet:* www.harrybassfoundation.org/about.asp
*Sponsor:* Harry Bass Foundation
4809 Cole Avenue, Suite 252
Dallas, TX 75205

**Harvard University John M. Olin Prizes** **1224**
The Harvard University John M. Olin Prizes, supported by a grant from the John M. Olin Foundation are awarded annually for the best papers written in the area of law and economics. The amount of a prize is $1,000. A professor must submit the paper on behalf of the student with a statement describing why the professor believes the paper is meritorious. Submission should be made before the end of classes in the Spring semester by email to the Program Administrator.
*Geographic Focus:* All States
*Date(s) Application is Due:* Nov 30
*Amount of Grant:* 10,000 USD
*Contact:* Irina Goldina; (617) 496-1670; fax (617) 384-5653; igoldina@law.harvard.edu
*Internet:* www.law.harvard.edu/programs/olin_center/
*Sponsor:* Harvard University
1563 Massachusetts Avenue
Cambridge, MA 02138

**Harvard University Weatherhead Center for International Affairs Fellowships** **1225**
The Harvard University Weatherhead Center for International Affairs Fellows program is the principal forum for the interaction of practitioners with the faculty, students, and visiting scholars of the Weatherhead Center. The program was inaugurated in 1958, simultaneously with the establishment of the Center itself, in the belief that the Center's research activities would benefit from the contributions of non-academic professionals with broad practical experience in international affairs, and that the Center could broaden the practitioners' horizons and deepen their understanding of the contemporary issues in which they are professionally involved. While on campus, Fellows are exposed to social science theories and research intended to place their experiences in an academic perspective. Their primary activity is to work on a research project of their choosing. They may also organize and lead study groups for Harvard undergraduates, participate in seminars at the Center and elsewhere, provide advice to undergraduate and graduate students, and otherwise interact with the faculty, visiting scholars, and other members of the University's community. For all successful applicants, the Center hopes to establish an agreement with a Harvard faculty member associated with the Weatherhead Center that results in the formation of an active scholarly relationship. Past Fellows have been senior diplomats, military officers, politicians, journalists, international civil servants, officials from nongovernmental organizations, and business leaders from around the world. The single characteristic they share is leadership in international affairs, as demonstrated by both a record of past achievement and the promise of future accomplishments.
*Geographic Focus:* All States
*Contact:* Kathleen Molony, Director, Fellows Program; (617) 495-2644 or (617) 495-4420; fax (617) 496-5642; kmolony@wcfia.harvard.edu
*Internet:* wcfia.harvard.edu/programs/fellows
*Sponsor:* Harvard University
61 Kirkland Street, Room 110
Cambridge, MA 02138

**Harvard University Women's Studies in Religion Research Associate and Visiting Faculty Program** **1226**
The Harvard University Women's Studies in Religion Research Associate and Visiting Faculty program is designed to explore women's distinctive religious experience and perspectives in their cultural and racial diversity, and to examine male-centeredness and sexism in religious traditions of the world. The program encourages and guides the development of women's studies in religion to help transform the study of religion so it more adequately reflects the study of human experience. Each year, the school appoints as visiting lecturers and research associates in the program five scholars whose work provides leadership in women's studies in the various fields of religion. These scholars conduct individual research projects on topics related to the history and function of gender in the symbolization of religious traditions; the institutionalization of roles in religious communities; and the dynamics of the interaction between religious systems of belief and the personal, social, and cultural conditions of women. Each is also a member of one of the school's departments and offers in one of the three curricular areas a semester course related to this research project. Full-time residence at Harvard Divinity School is required while carrying out proposed research projects. In the spring semester each associate gives a public lecture on her/his research. Application guidelines and deadlines are available from the Harvard Divinity School. Salary for each position will be $55,000. The appointment is full-time, lasting ten months, and includes health benefits and reimbursement of some expenses.
*Requirements:* Candidates with doctorates in the fields of religion and those with primary competence in other humanities, social science, and public policy fields who have serious investment in religion are eligible. Applicants must have received their PhD by October 1 of the application year.
*Geographic Focus:* All States
*Date(s) Application is Due:* Oct 15
*Amount of Grant:* 55,000 USD
*Contact:* Divinity School Office; (617) 495-5705 or (617) 495-5761; fax (617) 496-8564; admin@wsrp.harvard.edu
*Internet:* www.hds.harvard.edu/wsrp/apply
*Sponsor:* Harvard University
45 Francis Avenue, Andover Hall
Cambridge, MA 02138

**Harvey Randall Wickes Foundation Grants** **1227**
The Foundation awards grants to eligible Michigan nonprofit organizations in its areas of interest, including the arts; children/youth services; education; hospitals; human services; libraries; and recreation. Types of support include annual campaigns, building/renovation, equipment, and seed money. There are no application deadline dates. Applications must be received two weeks prior to board meetings in March, June, September, and December. Award notification is given within two weeks of the board meeting.
*Requirements:* Michigan 501(c)3 tax-exempt organizations in Saginaw County are eligible.
*Restrictions:* Funding is not available for government where support is forthcoming from tax dollars, individuals, endowments, travel, conferences, loans, or film or video projects.
*Geographic Focus:* Michigan
*Amount of Grant:* 2,500 - 30,000 USD
*Samples:* $13,000, Saginaw, Michigan Junior Achievement, payable over one year; $20,000, Saginaw, Michigan Holy Cross Children's Services, payable over one year.
*Contact:* Hugo Braun, Jr.; (989) 799-1850; fax (989) 799-3327; hrwickes@att.net
*Sponsor:* Harvey Randall Wickes Foundation
4800 Fashion Square Boulevard, Plaza N., Suite 472
Saginaw, MI 48604-2677

**Hatton W. Sumners Foundation for the Study and Teaching of Self Government Grants** **1228**
The Foundation's purpose is to encourage the study, teaching and research in the science and art of self-government so that citizens understand the fundamental principles of democracy in shaping governmental policies. Working through qualified, tax-exempt organizations, the Foundation seeks to reach, educate, and motivate the general public and the current and future leaders of American society. The Foundation gives to youth organizations and higher education. Types of support include: conferences and seminars; continuing support; curriculum development; endowments; fellowships; general operating support; internship funds; matching and challenge support; and research. Application information is available at the Foundation website. Grant applications will be accepted from January 1st through August 1st of each year. Final decisions on grant proposals are made by the Trustees in October of each year.
*Requirements:* Grants are made only to 501(c)3 tax-exempt organizations in Texas, New Mexico, Oklahoma, Louisiana, Arkansas, Kansas, Nebraska and Missouri.
*Restrictions:* The Foundation does not fund religious organizations or individual grants.
*Geographic Focus:* Arkansas, Kansas, Louisiana, Missouri, Nebraska, New Mexico, Oklahoma, Texas
*Date(s) Application is Due:* Aug 1
*Samples:* Bill of Rights Institute, Arlington, Virginia: Constitutional seminars in Oklahoma and New Mexico, $23,000; National Center for Policy Analysis, Dallas, Texas: Sumners Distinguished Lecture Series, $250,000; Big Brothers Big Sisters

of North Texas, Irving, Texas: character education and civic mentoring programs, $116,400; YMCA Youth and Government Programs in Texas, Oklahoma, and New Mexico: general operating support, $236,200.
*Contact:* Hugh Akin, Executive Director; (214) 220-2128; fax (214) 953-0737; hugh@hattonsumners.org or info@hattonsumners.org
*Internet:* www.hattonsumners.org
*Sponsor:* Hatton W. Sumners Foundation for the Study and Teaching of Self Government
325 North St. Paul Street, Suite 3920
Dallas, TX 75201

**Haverford College Mellon Postdoctoral Fellowships** 1229
The John B. Hurford '60 Center for the Arts and Humanities of Haverford College invites applications for a two-year Mellon Postdoctoral Fellowship in the Humanities to begin in the Fall Semester. The program seeks scholars interested in the anarchist tradition and its historical and contemporary influence on art, politics, and culture. Women and members of underrepresented minority groups are strongly encouraged to apply. Area of specialization is open, but might include history, history of art, music, political science, political economy, anthropology, literature, sociology, philosophy, and cultural studies, among other possibilities. Scholars with broad historical and interdisciplinary interests are encouraged to apply.
*Requirements:* Candidates who received the PhD within the last five years as of the listed application deadline date are eligible.
*Geographic Focus:* All States
*Date(s) Application is Due:* Jan 31
*Contact:* Emily Carey Cronin; (610) 896-1336 or (610) 896-1000; ecronin@haverford.edu
*Internet:* www.haverford.edu/HCAH/mellon_post-doctoral_program/index.php
*Sponsor:* Haverford College
370 Lancaster Avenue
Haverford, PA 19041

**Hawai'i Community Foundation Leahi Fund for Pulmonary Research - Junior nvestigator Grants** 1230
The Leahi Fund for Pulmonary Research was established at Hawai'i Community Foundation in 1994. The purpose of the Junior Investigator segment is to support programs of research and education in, and the prevention of, pulmonary disease. Highest priority will be given to: research and education with foreseeable clinical application to pulmonary problems that exist in Hawai'i; scientific research in the prevention or treatment of pulmonary diseases and infirmities. Preference is given to supporting new investigators and to new areas of investigation for established researchers. This biennial award is intended for junior faculty members who: are at the University of Hawai'i and other Hawai'i-based universities; are committed to a career in clinical or translational research; are at the beginning of their research career and therefore not yet competitive for career development awards; and seek to complete 3-5 years of funded career development activity, as a bridge to being competitive for major national funding. Grants will be up to $50,000. Researchers will have up to eighteen (18) months to complete their research project.
*Requirements:* Research grants can only be approved for individuals affiliated with tax-exempt organizations classified by the Internal Revenue Services as 501(c)3 type, or with units of government.
*Restrictions:* Funding is not available for: salary for the principal investigator; patient care; travel; major equipment purchases; printing or publication costs.
*Geographic Focus:* Hawaii
*Date(s) Application is Due:* Mar 8
*Amount of Grant:* Up to 50,000 USD
*Contact:* Susan Maltezo, Senior Grants Manager; (808) 537-6333 or (888) 731-3863; fax (808) 521-6286; smaltezo@hcf-hawaii.org or medicalresearch@hawaiicommunityfoundation.org
*Internet:* www.hawaiicommunityfoundation.org/grants/leahi-fund-for-pulmonary-research-junior-investigator-clinical-research-career-development-grants
*Sponsor:* Hawai'i Community Foundation
827 Fort Street Mall
Honolulu, HI 96813-4317

**Hawai'i Community Foundation Leahi Fund for Pulmonary Scientific Research Grants** 1231
The Leahi Fund for Pulmonary Research was established at Hawai'i Community Foundation in 1994. The purpose of the Leahi Fund is to support programs of research and education in, and the prevention of, pulmonary disease. Interested organizations or individuals may submit a full proposal for a research project for consideration. The Leahi Fund Advisory Committee will review all proposals on a competitive basis and make recommendation for grants that are most consistent with the purpose and priorities of the fund. Priorities of the Fund are: research with foreseeable clinical application to pulmonary problems that exist in Hawai'i; the prevention or treatment of pulmonary diseases and infirmities; collaborations between researchers within the State of Hawai'i; and seed funding and to new areas of investigation for established researchers. Investigators or programs that have received funding for two consecutive years will be given a lower priority. Scientific Research grants will be up to $50,000. Researchers will have up to eighteen (18) months to complete their research project. Online submissions must be received by 4:00 p.m. HST on March 7.
*Requirements:* Research grants can only be approved for expenditures and individuals in Hawai'i affiliated with tax-exempt organizations classified by the Internal Revenue Service as 501(c)3, or with units of government. The Principal Investigator must be at least equivalent to an Assistant Professor.
*Restrictions:* It is not the intention of the Leahi Fund to serve as a long-term source of research funding for any particular investigator. Previously funded Leahi Fund investigators, who in the opinion of the Advisory Committee should seek funding opportunities from national sources including the National Institutes of Health, may be ineligible for additional funding from the Leahi Fund. Fellows and post-doctoral researchers are ineligible for this grant.
*Geographic Focus:* Hawaii
*Date(s) Application is Due:* Mar 7
*Amount of Grant:* Up to 50,000 USD
*Contact:* Susan Maltezo, Senior Grants Manager; (808) 537-6333 or (888) 731-3863; fax (808) 521-6286; smaltezo@hcf-hawaii.org
*Internet:* www.hawaiicommunityfoundation.org/grants/leahi-fund-for-pulmonary-research-scientific-research-grants
*Sponsor:* Hawai'i Community Foundation
827 Fort Street Mall
Honolulu, HI 96813-4317

**Hawai'i Community Foundation Medical Research Grants** 1232
The Hawai'i Community Foundation's Medical Research Grant program supports clinical and basic research in a variety of areas determined by the source of funding. Currently, areas of focus for both clinical and basic research include: Alzheimer's disease; mental or physical diseases related to old age; cancer; heart disease; lung disease; immunology; genetics; medicinal uses of Hawaiian plants; prevention of blindness, with particular emphasis on macular degeneration; and type I diabetes (juvenile diabetes). Grants in these areas are up to $50,000. Applications are also being accepted in the area of research infrastructure, which is intended to improve Hawaii's research organizational effectiveness. Grants in this area are up to $25,000. The annual deadline for online applications is March 7.
*Requirements:* Organizations are eligible to apply for a grant if: it is a 501(c)3 organization or a unit of government; the primary investigator (PI) is based in and is conducting their research in Hawaii; or the PI is at least equivalent to an Assistant Professor, including Assistant Research Professors and non-tenure track (clinical) Assistant Professors. A PI with a current grant may apply if the current grant is due to be completed within six (6) months of the RFP deadline date. However, if awarded, a new grant will not be released until the current grant is completed and all reports are received.
*Restrictions:* Lecturers, Instructors, Junior Researchers, or post-doctoral researchers are not eligible. Only one (1) grant proposal submission per PI will be accepted. This includes being identified as a Co-PI.
*Geographic Focus:* Hawaii
*Date(s) Application is Due:* Mar 7
*Amount of Grant:* Up to 50,000 USD
*Contact:* Marc Alexander, Program Officer; (808) 566-5559 or (888) 731-3863; fax (808) 521-6286; malexander@hcf-hawaii.org
*Internet:* www.hawaiicommunityfoundation.org/grants/medical-research-grants
*Sponsor:* Hawai'i Community Foundation
827 Fort Street Mall
Honolulu, HI 96813-4317

**Hearst Foundations Health Grants** 1233
The Hearst Foundations fund non-profit organizations working in the fields of culture, education, health, and social services. The Foundations have two offices, one in New York which manages funding for organizations headquartered east of the Mississippi River and one in San Francisco which manages funding for organizations to the west. About 80% of the Foundations' total funding goes to prior grantees; the Foundations receive approximately 1,200 grant requests annually. The Foundations' health funding comprises 30% of their total giving; 80% of the Foundations' health funding goes to organizations having budgets over ten-million dollars. In the area of health, the Foundations assist leading regional hospitals, medical centers, and specialized medical institutions providing access to high-quality healthcare for low-income populations. In response to the shortage of healthcare professionals necessary to meet the country's evolving needs, the Foundations also fund programs designed to enhance skills and increase the number of practitioners and educators across roles in healthcare. Because the Foundations seek to use their funds to create a broad and enduring impact on the nation's health, support for medical research and the development of young investigators is also considered. Preference is given to the following types of programs; professional development programs; programs improving access to high-quality healthcare for low-income populations; programs developing and providing specialized care for the complex needs of elderly populations; programs scaling innovative healthcare delivery systems to provide efficient, coordinated care; and research, particularly related to finding new cures and treatments for prevalent diseases, such as cancer. The Foundations provide program, capital, and, on a limited basis, endowment support. Requests are accepted year round. These must be submitted via the Foundations' online application portal. Each request goes through an evaluation process that generally spans four to six weeks. The Foundations conduct a site visit of semi-finalists and may also consult with experts in a given field. Applicants will receive an email confirmation of receipt of submission and can follow the status of their request through the online system. Instructions for using the system, guidelines (in the form of an FAQ), and the link to the Foundations' online portal are at the Foundations' website.
*Requirements:* Grants are made only to 501(c)3 organizations.
*Restrictions:* Organizations must wait one year from the date of their notice of decline before the Foundations will consider another request. Grantees must wait a minimum of three years from their grant award date before the Foundations will consider another request. The Foundations do not fund individuals or the following types of requests: those from organizations operating outside the United States; those from organizations

with operating budgets under one million dollars; those from organizations involved in publishing, radio, film, or television; those from local chapters of organizations; those from organizations lacking demonstrable long-term impact on populations served; requests to fund tours, conferences, workshops, or seminars; requests to fund advocacy or public-policy research; requests to fund special events, tickets, tables, or advertising for fundraising events; requests for seed money or to fund start-up projects; and request to fund program-related investments.
*Geographic Focus:* All States
*Amount of Grant:* Up to 600,000 USD
*Samples:* American Geriatrics Society, New York, New York, $600,000—toward the Chief Resident Immersion Training Program (CRIT) as part of the Hearst Foundation's Aging Initiative, toward an uncommitted goal of $2.2 million; Arthur Ashe Institute for Urban Health Inc., Brooklyn, New York, $50,000—to provide minority high-school students with a college-level science curriculum to prepare them to successfully pursue degrees in science and health; Banner Health, Phoenix, Arizona, $100,000—to support the expansion of the Banner MD Anderson Cancer Center through the purchase of a new linear accelerator.
*Contact:* Mason Granger; (212) 649-3750; fax (212) 586-1917; hearst.ny@hearstfdn.org
Annette Hepler; (415) 908-4500; fax (415) 348-0887; hearst.sf@hearstfdn.org
*Internet:* www.hearstfdn.org/funding-priorities/
*Sponsor:* Hearst Foundations
300 West 57th Street, 26th Floor
New York, NY 10019-3741

**Heart Rhythm Society Award for the Fellow with the Highest Scoring Abstract 1234**
The Heart Rhythm Society (HRS) seeks to recognize and promote excellence in the field of cardiac pacing and electrophysiology. HRS offers a variety of awards and scholarships for heart rhythm professionals at every stage of their careers, from fellow-in-training to seasoned professional. The Award for the Fellow with the Highest Scoring Abstract recognizes the fellow-in-training with the highest scoring abstract in all sub-categories, excluding the Allied Professional subcategory. To be considered, an individual should indicate his/her interest in applying for this award during the regular abstract submission period, with the annual deadline being December 1. One award is given annually. Awardees receive: complimentary registration to the current year's Scientific Sessions; a plaque; and $1,500 for the Presenting Author's travel and housing expenses.
*Requirements:* To be considered for this award, an applicant must be a fellow currently in a training program or no more than one year out of fellowship training as of May of the award year. Nominees need not be members of the Heart Rhythm Society.
*Geographic Focus:* All States
*Date(s) Application is Due:* Dec 1
*Amount of Grant:* 1,500 USD
*Contact:* Kristian L. Adams, Manager, Scientific Sessions and Education; (202) 464-3400; fax (202) 464-3401; kadams@hrsonline.org or awards@hrsonline.org
*Internet:* https://www.hrsonline.org/About-HRS/Awards-Scholarships/Fellow-with-the-Highest-Scoring-Abstract
*Sponsor:* Heart Rhythm Society
1325 G Street NW, Suite 400
Washington, D.C. 20005

**Heart Rhythm Society Award for the Highest Scoring Abstract in the Allied Professionals Category 1235**
The Heart Rhythm Society (HRS) seeks to recognize and promote excellence in the field of cardiac pacing and electrophysiology. HRS offers a variety of awards and scholarships for heart rhythm professionals at every stage of their careers, from fellow-in-training to seasoned professional. The Award for the Fellow with the Highest Scoring Abstract in the Allied Professionals Category recognizes the fellow-in-training with the highest scoring abstract in the Allied Professional subcategory. To be considered, an individual should indicate his/her interest in applying for this award during the regular abstract submission period, with the annual deadline being December 1. One award is given annually. Awardees receive: complimentary registration to the current year's Scientific Sessions; a plaque; and $1,500 for the Presenting Author's travel and housing expenses.
*Requirements:* To be considered for this award, an applicant must be a fellow currently in a training program or no more than one year out of fellowship training as of May of the award year. Nominees need not be members of the Heart Rhythm Society.
*Geographic Focus:* All States
*Date(s) Application is Due:* Dec 1
*Amount of Grant:* 1,500 USD
*Contact:* Kristian L. Adams, Manager, Scientific Sessions and Education; (202) 464-3400; fax (202) 464-3401; kadams@hrsonline.org or awards@hrsonline.org
*Internet:* https://www.hrsonline.org/About-HRS/Awards-Scholarships/Highest-Scoring-Abstract-in-the-Allied-Professionals-Category
*Sponsor:* Heart Rhythm Society
1325 G Street NW, Suite 400
Washington, D.C. 20005

**Heart Rhythm Society Douglas P. Zipes Lectureship Award 1236**
The Heart Rhythm Society (HRS) seeks to recognize and promote excellence in the field of cardiac pacing and electrophysiology. HRS offers a variety of awards and scholarships for heart rhythm professionals at every stage of their careers, from fellow-in-training to seasoned professional. The Douglas P. Zipes Lecture is awarded to an individual who has made a significant and unique contribution to the field of cardiac pacing and electrophysiology as a basic scientist. The HRS President and Scientific Sessions Program Committee Chair informally poll the basic science community and present a recommendation for the winner of this award to the President of CEPS to make a final selection. One award is given annually. The award includes complimentary registration to the current year's Scientific Sessions, a lectureship at the Annual Scientific Sessions, a plaque, and $1,000 for the honoree's travel and housing expenses.
*Requirements:* The lectureship is not considered one of HRS's 'Distinguished Awards' and therefore does not follow HRS's formal award nomination process.
*Geographic Focus:* All States
*Amount of Grant:* 1,000 USD
*Contact:* Kristian L. Adams, Manager, Scientific Sessions and Education; (202) 464-3400; fax (202) 464-3401; kadams@hrsonline.org or awards@hrsonline.org
*Internet:* https://www.hrsonline.org/About-HRS/Awards-Scholarships/Douglas-P.-Zipes-Lectureship-Award
*Sponsor:* Heart Rhythm Society
1325 G Street NW, Suite 400
Washington, D.C. 20005

**Heart Rhythm Society Eric N. Prystowsky Early Career Researcher Award 1237**
The Heart Rhythm Society (HRS) seeks to recognize and promote excellence in the field of cardiac pacing and electrophysiology. HRS offers a variety of awards and scholarships for heart rhythm professionals at every stage of their careers, from fellow-in-training to seasoned professional. The Eric N. Prystowsky Early Career Researcher Award promotes excellence in clinical research by recognizing an early career researcher with the highest scoring abstract from select sub-categories in the Clinical Electrophysiology and Catheter Ablation abstract categories. To be considered, an individual should indicate his/her interest in applying for this award during the regular abstract submission period, with the annual deadline being December 1. One award is given annually. Awardees receive: complimentary registration to the current year's Scientific Sessions; a plaque; and $1,500 for the Presenting Author's travel and housing expenses.
*Requirements:* The award is open to eligible persons submitting an abstract as presenting author in the following catheter ablation sub-categories: SVT/AVNRT/WPW/AT; atrial fibrillation and atrial flutter; and ventricular tachycardia and ventricular fibrillation. The award is also open to eligible persons submitting an abstract as presenting author in the following clinical electrophysiology sub-categories: SCA risk assessment; syncope and bradycardia; SVT/AVNRT/WPW/AT; and atrial fibrillation and atrial flutter. Nominees need not be members of the Heart Rhythm Society.
*Geographic Focus:* All States
*Date(s) Application is Due:* Dec 1
*Amount of Grant:* 1,500 USD
*Contact:* Kristian L. Adams, Manager, Scientific Sessions and Education; (202) 464-3400; fax (202) 464-3401; kadams@hrsonline.org or awards@hrsonline.org
*Internet:* https://www.hrsonline.org/About-HRS/Awards-Scholarships/The-Eric-N.-Prystowsky-Early-Career-Researcher-Award
*Sponsor:* Heart Rhythm Society
1325 G Street NW, Suite 400
Washington, D.C. 20005

**Heart Rhythm Society Eric N. Prystowsky Fellows Clinical Research Award 1238**
The Heart Rhythm Society (HRS) seeks to recognize and promote excellence in the field of cardiac pacing and electrophysiology. HRS offers a variety of awards and scholarships for heart rhythm professionals at every stage of their careers, from fellow-in-training to seasoned professional. The Eric N. Prystowsky Fellows Clinical Research Award promotes excellence in clinical research by recognizing the fellow with the highest scoring abstract in the Clinical Electrophysiology and Catheter Ablation abstract categories. To be considered, an individual should indicate his/her interest in applying for this award during the regular abstract submission period, with the annual deadline being December 1. Awardees receive: complimentary registration to the current year's Scientific Sessions; a plaque; and $1,500 for the Presenting Author's travel and housing expenses.
*Requirements:* The award is open to eligible persons submitting an abstract as presenting author in the following catheter ablation sub-categories: SVT/AVNRT/WPW/AT; atrial fibrillation and atrial flutter; and ventricular tachycardia and ventricular fibrillation. The award is also open to eligible persons submitting an abstract as presenting author in the following clinical electrophysiology sub-categories: SCA risk assessment; syncope and bradycardia; SVT/AVNRT/WPW/AT; and atrial fibrillation and atrial flutter. Nominees need not be members of the Heart Rhythm Society.
*Restrictions:* The award is open to fellows currently in a training program or no more than one year out of fellowship training. The research for the eligible abstract must have been substantially completed during the fellowship period and must involve intact human subjects.
*Geographic Focus:* All States
*Date(s) Application is Due:* Dec 1
*Amount of Grant:* 1,500 USD
*Contact:* Kristian L. Adams, Manager, Scientific Sessions and Education; (202) 464-3400; fax (202) 464-3401; kadams@hrsonline.org or awards@hrsonline.org
*Internet:* https://www.hrsonline.org/About-HRS/Awards-Scholarships/The-Eric-N.-Prystowsky-Fellows-Clinical-Research-Award
*Sponsor:* Heart Rhythm Society
1325 G Street NW, Suite 400
Washington, D.C. 20005

**Heart Rhythm Society Eric N. Prystowsky Lectureship Award 1239**
The Heart Rhythm Society (HRS) seeks to recognize and promote excellence in the field of cardiac pacing and electrophysiology. HRS offers a variety of awards and scholarships for heart rhythm professionals at every stage of their careers, from fellow-in-training to seasoned professional. The Eric N. Prystowsky Lectureship Award is awarded to an

individual who has made a significant and unique contribution to the field of clinical electrophysiology as a clinical scientist. Working with the Heart Rhythm Society staff, the Scientific Sessions Program Committee will, on an annual basis, identify a list of possible recipients and choose the format of the lectureship. The Program Committee corresponds over the summer to select the recipient. One award is given annually. The award includes complimentary registration to the current year's Scientific Sessions, a lectureship at the Annual Scientific Sessions, a plaque, and $1,500 for the honoree's travel and housing expenses.
*Requirements:* Nominees need not be members of the Heart Rhythm Society.
*Geographic Focus:* All States
*Amount of Grant:* 1,500 USD
*Contact:* Kristian L. Adams, Manager, Scientific Sessions and Education; (202) 464-3400; fax (202) 464-3401; kadams@hrsonline.org or awards@hrsonline.org
*Internet:* https://www.hrsonline.org/About-HRS/Awards-Scholarships/The-Eric-N.-Prystowsky-Lectureship-Award
*Sponsor:* Heart Rhythm Society
1325 G Street NW, Suite 400
Washington, D.C. 20005

**Heart Rhythm Society Founders Lectureship Award** 1240
The Heart Rhythm Society (HRS) seeks to recognize and promote excellence in the field of cardiac pacing and electrophysiology. HRS offers a variety of awards and scholarships for heart rhythm professionals at every stage of their careers, from fellow-in-training to seasoned professional. The Founders Lectureship Award will be given to a prominent member of the heart rhythm community, either nationally or internationally, who; has made a significant and unique contribution to the field of cardiac pacing and electrophysiology; is capable of delivering a high-quality lecture that will be of interest to a multidisciplinary audience; can attend the Heart Rhythm Society's Annual Scientific Sessions; and will grant permission to the Heart Rhythm Society to audio and video capture the lecture for future enduring products. One award is given annually. Awardees receive: complimentary registration to the current year's Scientific Sessions; a plaque; and $2,500 for the Presenting Author's travel and housing expenses.
*Requirements:* Working with HRS staff, the Committee will, on an annual basis, identify a list of possible recipients and choose the format of the lectureship.
*Geographic Focus:* All States
*Amount of Grant:* 2,500 USD
*Contact:* Kristian L. Adams, Manager, Scientific Sessions and Education; (202) 464-3400; fax (202) 464-3401; kadams@hrsonline.org or awards@hrsonline.org
*Internet:* https://www.hrsonline.org/About-HRS/Awards-Scholarships/Founders-Lectureship-Award
*Sponsor:* Heart Rhythm Society
1325 G Street NW, Suite 400
Washington, D.C. 20005

**Heart Rhythm Society Nancy L. Stephenson Award for Exemplary Industry/ HRS Community Partnership** 1241
The Heart Rhythm Society (HRS) seeks to recognize and promote excellence in the field of cardiac pacing and electrophysiology. HRS offers a variety of awards and scholarships for heart rhythm professionals at every stage of their careers, from fellow-in-training to seasoned professional. The Nancy L. Stephenson Award for Exemplary Industry/HRS Partnership celebrates an Industry professional who has modeled an exemplary Industry/ Heart Rhythm Society Community relationship and has promoted the advancement of the field. One award is given annually. Awardees receive: complimentary registration to Heart Rhythm; travel and hotel accommodations stipend up to $2,500 for Heart Rhythm; recognition on the HRS website; presentation of award by the CEO of Heart Rhythm; and a commemorative plaque.
*Requirements:* This award is not intended to recognize CEOs or career corporate executives or those in sales/marketing positions, nor is it intended as an annual award.
*Geographic Focus:* All States
*Date(s) Application is Due:* Mar 15
*Amount of Grant:* 2,500 USD
*Contact:* Kristian L. Adams, Manager, Scientific Sessions and Education; (202) 464-3400; fax (202) 464-3401; kadams@hrsonline.org or awards@hrsonline.org
*Internet:* https://www.hrsonline.org/About-HRS/Awards-Scholarships/The-Nancy-L.-Stephenson-Award-for-Exemplary-Industry-HRS-Community-Partnership
*Sponsor:* Heart Rhythm Society
1325 G Street NW, Suite 400
Washington, D.C. 20005

**Heart Rhythm Society Pioneer in Cardiac Pacing and Electrophysiology Award** 1242
The Heart Rhythm Society (HRS) seeks to recognize and promote excellence in the field of cardiac pacing and electrophysiology. HRS offers a variety of awards and scholarships for heart rhythm professionals at every stage of their careers, from fellow-in-training to seasoned professional. The Pioneer in Cardiac Pacing and Electrophysiology Award provides recognition for an individual who has been active in cardiac pacing and/or cardiac electrophysiology for many years, and has made significant contributions to the field. One award is given annually. Awardees receive: complimentary registration to Heart Rhythm; coach airfare for one to Heart Rhythm; hotel accommodations for the duration of Heart Rhythm; recognition on the HRS website; presentation of award by colleague of recipient's choice; and a commemorative plaque.
*Requirements:* Nominees need not be members of the Heart Rhythm Society.
*Geographic Focus:* All States
*Date(s) Application is Due:* Mar 15
*Contact:* Kristian L. Adams, Manager, Scientific Sessions and Education; (202) 464-3400; fax (202) 464-3401; kadams@hrsonline.org or awards@hrsonline.org
*Internet:* https://www.hrsonline.org/About-HRS/Awards-Scholarships/Recognition-Awards
*Sponsor:* Heart Rhythm Society
1325 G Street NW, Suite 400
Washington, D.C. 20005

**Heart Rhythm Society Research Fellowships** 1243
Heart Rhythm Society Research Fellowships are available for projects pertaining to cardiac pacing and/or cardiac electrophysiology. All awards are made for one year, renewable in competition. Also available are traveling fellowships, for a one- to three-month duration, to enable a researcher to observe research at other institutions or attend seminars and conferences. One traveling Canadian fellowship is offered; the applicant must be a Canadian citizen or must train at a site in Canada. Annual deadline dates may vary; contact program staff for exact dates.
*Requirements:* Fellowships are open to applicants who have a doctor of medicine, philosophy, osteopathy, science, or equivalent training or experience in the field of biomedical engineering. Applicants must be citizens or permanent residents of a North American country and include cardiac pacing or cardiac electrophysiology as a central part of his/her long-range career goals.
*Geographic Focus:* All States
*Date(s) Application is Due:* Sep 15
*Amount of Grant:* 5,000 - 40,000 USD
*Contact:* Kristian L. Adams, Manager, Scientific Sessions and Education; (202) 464-3400; fax (202) 464-3401; fellowship@hrsonline.org or kadams@hrsonline.org
*Internet:* https://www.hrsonline.org/Career-Resources/Fellowships-Program-Directors
*Sponsor:* Heart Rhythm Society
1325 G Street NW, Suite 400
Washington, D.C. 20005

**Heart Rhythm Society Young Investigator Award** 1244
The Heart Rhythm Society (HRS) seeks to recognize and promote excellence in the field of cardiac pacing and electrophysiology. HRS offers a variety of awards and scholarships for heart rhythm professionals at every stage of their careers, from fellow-in-training to seasoned professional. The Young Investigator Award (YIA) Competition recognizes and encourages the work of young investigators and their parent institutions. The Heart Rhythm Society YIA Subcommittee, which is part of the Scientific Sessions Program Committee and is comprised of senior researchers, clinicians, and scientists, reviews all submissions for the competition. The Subcommittee then selects six finalists to present their work during the Heart Rhythm Annual Scientific Sessions. The annual deadline for applications is January 3. Two awards and four honorees are given annually. All six YIA finalists will receive complimentary registration to that year's Scientific Sessions, a plaque, and reimbursement up to $1,000 per finalist for travel expenses. There will be two first prize awards: $1,000 for basic research and $1,000 for clinical research. The four honorable mentions will each receive $500.
*Requirements:* The YIA Competition is open to any physician or scientist who is presently in a residency, fellowship, or doctoral training program, or who has been in such a program within the past three years (at the time of submission). The competition is also open to students in graduate or medical schools who are working on a problem concerned with cardiac pacing and/or electrophysiology (EP) in the areas of clinical and basic research. The applicant must be the first author of the manuscript submitted for consideration.
*Geographic Focus:* All States
*Date(s) Application is Due:* Jan 3
*Amount of Grant:* 1,500 - 2,000 USD
*Contact:* Kristian L. Adams, Manager, Scientific Sessions and Education; (202) 464-3400; fax (202) 464-3401; kadams@hrsonline.org or awards@hrsonline.org
*Internet:* https://www.hrsonline.org/About-HRS/Awards-Scholarships/Young-Investigator-Award-Competition
*Sponsor:* Heart Rhythm Society
1325 G Street NW, Suite 400
Washington, D.C. 20005

**Helen Bader Foundation Grants** 1245
Throughout her life, Helen Bader sought to help others. She played many roles - student, mother, businesswoman, and social worker - believing that everyone should have the opportunity to reach their fullest potential. Growing up in the railroad town of Aberdeen, South Dakota, Helen learned the value of hard work and self-reliance. The Great Depression and the sacrifices of World War II also taught her the importance of reaching out to those in need. Helen attended Downer College in Milwaukee, earning a degree in botany. She married Alfred Bader, a chemist from Austria, and together they started a family and created a business, the Aldrich Chemical Company. From the 1950s to the 1970s, their hard work helped build one of Wisconsin's most successful start-up enterprises of the era. The Baders' eventual divorce led Helen to again become self-reliant. She subsequently finished her Master of Social Work at the University of Wisconsin-Milwaukee. While doing her field work with the Legal Aid Society of Milwaukee, Helen met and helped many people in need, including single mothers and adults with mental illness. In the process, she gained a deeper appreciation for their everyday struggles. After graduation, she worked at the Milwaukee Jewish Home, where working with older adults brought home the many issues of aging. At a time when Alzheimer's disease was almost a complete mystery, she helped open the resident' minds and hearts through dance and music. Helen felt that the residents' quality of life depended upon the small details, so she was happy to run errands or escort them to the symphony. She found herself touched by

the arts and studied the violin and guitar at the Wisconsin Conservatory of Music. Helen eventually faced cancer. As the illness began to sap her physical strength, she shared a wish with her family: to continue to aid those in need. She died in 1989. After her death, patterns of Helen's quiet style of philanthropy became more apparent. When she had come across an organization that impressed her, she would just pull out her checkbook without a lot of fanfare. In her name, the Helen Bader Foundation (HBF) supports worthy organizations working in key areas affecting the quality of life in Milwaukee, the state of Wisconsin, and Israel. The foundation also seeks to inspire the generosity in others, as every individual can make a difference through gifts of time, talent, and resources. The foundation will consider multiple-year requests with 24 or 36 month terms. Multi-year grants are subject to annual review before funds for subsequent years are released. The application deadline for the online preliminary proposal is January 5. The application deadline for organizations invited to complete a full proposal is February 2. The link to the online application system is available at the grant website. Application deadlines may vary from year to year. Prospective applicants are encouraged to visit the grant website to verify current deadline dates.

*Requirements:* Grants are awarded for projects consistent with one or more of the Helen Bader Foundation's program areas: Alzheimer's and aging (national in scope, with priority given to Wisconsin); economic development (restricted to the city of Milwaukee); community partnerships for youth (restricted to the city of Milwaukee); community initiatives (restricted to greater Milwaukee); arts (restricted to the city of Milwaukee); and directed grants and initiatives such as aid and support to Israel (for which proposals must be staff-solicited). Grants are given only to U.S. organizations which are tax exempt under Section 501(c)3 of the Internal Revenue Code or to government entities; grants will only be approved for foreign entities which meet specific charitable status requirements.

*Restrictions:* The Foundation does not provide direct support for individuals, such as individual scholarships.

*Geographic Focus:* All States, All Countries

*Date(s) Application is Due:* Jan 5; Feb 25

*Amount of Grant:* 10,000 - 100,000 USD

*Samples:* Aging & Disabiity Resource Center of Portage County, Stevens Point, Wisconsin, $12,000, creation of an Early Memory Loss Program to serve central Wisconsin; Arts at Large, Inc., Milwaukee, Wisconsin, $412,000, inclusive arts programming for low-income children attending six Milwaukee Public Schools in grades K-8 and their affiliated Community Learning Centers; Ben-Gurion University of the Negev, Beer Sheva, Israel, $10,000, Summer School Science Camp for 50 seventh grade Bedouin students from Abu Basma schools.

*Contact:* Tamara Hogans, Grants Manager; (414) 224-6464; fax (414) 224-1441; tammy@hbf.org or info@hbf.org

*Internet:* www.hbf.org/apply.htm

*Sponsor:* Helen Bader Foundation
233 North Water Street, Fourth Floor
Milwaukee, WI 53202

**Helen Kellogg Institute for International Studies Residential Fellowships 1246**

The Kellogg Institute awards Faculty Residential Fellowships to enable Kellogg faculty fellows to undertake significant projects consistent with the Institute's research priorities. The Faculty Residential Fellowship, which normally covers the recipient's salary for one semester, allows a fellow the opportunity to apply for an academic leave from his or her department or college. While not a given, leave is usually granted. Consequently, during the fellowship period the fellow is relieved of teaching and/or administrative responsibilities.

*Requirements:* Only members of the regular teaching and research faculty who are fellows of the Kellogg Institute are eligible for this program. Assistant professors may apply for faculty residential fellowships, but awards are contingent on their having passed third-year review before beginning the fellowship.

*Restrictions:* Applicants may not have had a paid leave in the previous two years, or a Kellogg Faculty Residential Fellowship in the preceding six years. Projects funded by this program must be conducted in residence at Notre Dame. Proposals for field research outside the University and/or for projects that would entail other than brief absences of the faculty member during the fellowship period are not eligible.

*Geographic Focus:* All States

*Date(s) Application is Due:* Nov 2

*Samples:* Nora Segura, Universidad Nacional de Colombia (Bogata)—for Work, Family, and Gender in Displaced Populations in Colombia.

*Contact:* Sharon Schierling, Associate Director; (574) 631-8524 or (574) 631-6580; fax (574) 631-6717; sschierl@nd.edu

*Internet:* kellogg.nd.edu/faculty/grants/resfellowship.shtml

*Sponsor:* University of Notre Dame
130 Hesburgh Center
Notre Dame, IN 46556-5677

**Helen V. Brach Foundation Grants 1247**

Established in 1974 in Illinois by the wife of Frank Brach, principal owner of the E.J. Brach and Sons Candy Company of Chicago, the foundation operates to prevent cruelty to animals or to children; for religious, charitable, scientific, literary, and education purposes; and for public safety testing through support of Midwest 501(c)3 tax-exempt organizations carrying out programs and activities in these areas. Brach provides grants nationally and has wide-ranging interests. For example, it supports homeless and women's emergency shelters, teen pregnancy prevention programs, parenting education, summer school for disadvantaged children, job training for welfare mothers, orphanages, and scholarships for economically disadvantaged students. Types of support include annual campaigns, building construction and renovation, equipment, general operating support, publications, research, and special projects. The foundation ordinarily does not make multi-year grants or commitments. Applicants are required to complete in full a brief application form, which may be obtained from the office. The board of directors gives final consideration to all applications received in a given year at the board's meeting, which is usually held in March.

*Requirements:* Although 501(c)3 nonprofits from across the nation are eligible, giving is primarily made in the Midwest, as well as California, Massachusetts, Ohio, Pennsylvania, and South Carolina.

*Restrictions:* Grants are not made to individuals or to organizations outside the United States. Typically grants are not made in excess of 10 percent of a group's operating budget, which automatically excludes start-up grants.

*Geographic Focus:* All States

*Date(s) Application is Due:* Dec 31

*Amount of Grant:* 5,000 - 25,000 USD

*Samples:* After School Matters, Chicago, Illinois, $20,000 - for sports and technology program; Brookfield Zoo, Brookfield, Illinois, $50,000 - research program on animal behavior; Hospice of Tuscarawas County, Dover, Ohio, $50,000 - support for Hospice House end-of-life health care facility.

*Contact:* John P. Hagnell, Associate Director; (312) 372-4417; fax (312) 372-0290

*Sponsor:* Helen V. Brach Foundation
55 W Wacker Drive, Suite 701
Chicago, IL 60601-1609

**Henry Dreyfus Teacher-Scholar Awards 1248**

The Henry Dreyfus Teacher-Scholar Awards Program supports the research and teaching careers of talented young faculty in the chemical sciences. Based on institutional nominations, the program provides discretionary funding to faculty at an early stage in their careers. The Henry Dreyfus Teacher-Scholar Awards Program is based on accomplishment in scholarly research with undergraduates, as well as a compelling commitment to teaching. The Henry Dreyfus Teacher-Scholar Awards Program provides an unrestricted research grant of $60,000.

*Requirements:* The Henry Dreyfus Teacher-Scholar Awards Program is open to academic institutions in the States, Districts, and Territories of the United States of America that grant a bachelor's or master's degree in the chemical sciences, including biochemistry, materials chemistry, and chemical engineering. If the department has a modest Ph.D. program, the nominee's research accomplishments must be almost exclusively with undergraduates. Nominees must hold a full-time tenure-track academic appointment, be between the fourth and twelfth years of their independent academic careers, and be engaged in research and teaching primarily with undergraduates. Awardees are typically in departments that do not grant a doctoral degree. Institutions may submit only one Henry Dreyfus nomination annually and renominations are accepted.

*Restrictions:* Awards are not made directly to individuals.

*Geographic Focus:* All States

*Date(s) Application is Due:* Aug 6

*Amount of Grant:* 60,000 USD

*Contact:* Mark J. Cordillo, Executive Director; (212) 753-1760; fax (212) 593-2256; mcardillo@dreyfus.org or admin@dreyfus.org

*Internet:* www.dreyfus.org/awards/henry_dryfus_teacher_award.shtml

*Sponsor:* Camille and Henry Dreyfus Foundation
555 Madison Avenue, 20th Floor
New York, NY 10022-3301

**Henry L. Guenther Foundation Grants 1249**

The Henry L. Guenther Foundation, established in 1956, awards grants to California nonprofit organizations in its areas of interest, including: creating opportunities for youth; disease research; education; human services; and community services. Grants support hospitals, medical research, and social services. There are two annual deadlines: May 31 and October 31. Application forms and guidelines are available upon request. The board meets in January and July to consider all proposals. Most recent grants have ranged from $5,000 to $500,000.

*Requirements:* California nonprofit 501(c)3 organizations are eligible to apply. Grants are awarded primarily in southern California.

*Restrictions:* Grants do not support government agencies, religious organizations for religious purposes, individuals, or debt-reduction requests.

*Geographic Focus:* California

*Date(s) Application is Due:* May 31; Oct 31

*Amount of Grant:* 5,000 - 500,000 USD

*Samples:* Loma Linda University, Loma Linda, California, $400,000 - for the purchase of a Siemens cyclotron for a new Research Imaging Center to further develop Proton Therapy capabilities in the treatment of neurological diseases; PIH Health Foundation, Whittier, California, $100,000 - to purchase tomosynthesis digital mammography equipment; Sharp Healthcare Foundation, San Diego, California, $250,000 - to build the seventh floor of the Stephen Birch Healthcare Center.

*Contact:* Sarah C. Milliken, President; (310) 785-0658

*Sponsor:* Henry L. Guenther Foundation
2029 Century Park D, Suite 4392
Los Angeles, CA 90067-3029

**Henry W. Bull Foundation Grants 1250**

Established in 1960 in California, the foundation was established by Maud Bull in memory of her husband, Henry W. Bull. The foundation grants awards on a nationwide basis, giving primarily in the areas of: arts/performing arts; education; health and; human services. Types of support include annual campaigns, building construction/renovation, capital campaigns; program development, equipment acquisition, matching/challenge

grants; general operating support, continuing support, and research. There is no formal grant application form. Submit a simple, concise statement of needs and objectives with pertinent supportive data, c/o Janice Gibbons, Santa Barbara Bank & Trust, P.O. Box 2340, Santa Barbara, CA, 93120-2340.
*Requirements:* 501(c)()3 nonprofits are eligible.
*Restrictions:* No grants to individuals or private foundations.
*Geographic Focus:* All States
*Date(s) Application is Due:* Apr 1; Sep 1
*Amount of Grant:* 500 - 20,000 USD
*Samples:* Shiloh Christian Childrens Ranch, Shelbina, MO, $2,000—general support; FamiliesFirst, Davis, CA, $3,000—to help expand child abuse prevention program; Dunn School, Los Olivos, CA, $10,000—purchase of new biology countertops, data collection equipment and laptop computers for Science Department.
*Contact:* Janice Gibson, Vice-President, Santa Barbara Bank & Trust; (805) 899-8405; fax (805) 884-1404
*Sponsor:* Henry W. Bull Foundation
P.O. Box 2340
Santa Barbara, CA 93120-2340

**Herbert A. and Adrian W. Woods Foundation Grants** **1251**
The Herbert A. and Adrian W. Woods Foundation was established on June 9, 1999 upon the death of Adrian W. Woods. Mrs. Woods had a long history of charitable giving in the St. Louis community and wanted to establish this foundation to continue that legacy of giving. The Foundation supports charitable organizations primarily in the greater St. Louis, Missouri, area. The Trustees will consider the donor's past giving history and any special needs of any of the charities she has given to in the past. The Trustees will also consider requests from charitable organizations that fall into one or more of the following categories: abused, neglected, or troubled children; the poor; the Episcopal Church and affiliates, including outreach programs; arts and culture in the Metropolitan St. Louis area; animal welfare (in Missouri); and victims of illness or disability, including research in this area. The following types of requests may be submitted: special projects; capital campaign requests (capital grants are awarded only as a source of support among a broad community of funders); challenge or matching grants; and general operation funding. The application deadline for this foundation is September 1. Final decisions about grant rewards will be made by November 30. Most recent awards have ranged from $9,700 to $15,000.
*Requirements:* Applicants must have 501(c)3 tax-exempt status.
*Restrictions:* The fund will not consider requests for multi-year grants (pledges) or endowment creation and funding. The fund does not support requests from individuals, organizations attempting to influence policy through direct lobbying, or any political campaigns.
*Geographic Focus:* Missouri
*Date(s) Application is Due:* Sep 1
*Amount of Grant:* 9,700 - 15,000 USD
*Samples:* Stray Rescue of Saint Louis, St. Louis, Missouri, $15,000 - purchase of veterinary medical equipment and emergency medical care for abused dogs and cats; Lydias House, St. Louis, Missouri, $9,700 - general operating support; Ride On Saint Louis, Kimmswick, Missouri, $9,700 - general operating support.
*Contact:* Srilatha Lakkaraju; (312) 828-8166; ilgrantmaking@ustrust.com
*Internet:* https://www.bankofamerica.com/philanthropic/foundation.go?fnId=7
*Sponsor:* Herbert A. and Adrian W. Woods Foundation
231 South LaSalle Street, IL1-231-13-32
Chicago, IL 60604

**Herbert H. and Grace A. Dow Foundation Grants** **1252**
The Foundation has charter goals to improve the educational, religious, economic and cultural lives of Michigan's people. Priority is given to organizations that: have clearly stated objectives, strong and purposeful management and are publicly accountable; have needs which are in areas not normally funded by governmental or public financing; are not hesitant to explore, initiate, volunteer, or execute original ideas or concepts; are willing to collaborate with other persons or organizations to give synergy to a common objective or goal; have purposes which tend to advance private enterprise and the preservation of a free, open and self-resourceful society. There is no formal application form, though grant seekers will find detailed submission information online.
*Requirements:* Only organizations in Michigan are eligible to apply.
*Restrictions:* The Foundation does not make grants directly to individuals. It cannot legally support: organizations to which contributions are not tax deductible, according to Internal Revenue Service regulations; organizations that practice discrimination by race, sex, creed, age or national origin; political organizations or organizations whose purposes are to influence legislation.
*Geographic Focus:* Michigan
*Contact:* Macauley Whiting Jr.; (989) 631-3699; grants@hhdowfoundation.org
*Internet:* www.hhdowfdn.org/guidelines.html
*Sponsor:* Herbert H. and Grace A. Dow Foundation
1018 West Main Street
Midland, MI 48640-4292

**Herbert Hoover Presidential Library Association Travel Grants** **1253**
Travel grants are awarded to encourage scholarly use of the Herbert Hoover Presidential Library in West Branch, Iowa. The program is intended to promote the study of subjects of interest and concern to Herbert Hoover, Lou Henry Hoover, their associates, and other public figures as reflected in the library's collections. It is strongly suggested that the applicant contact the archival staff to determine whether library holdings are pertinent to the applicant's research. In recent years, awards have ranged from $500 to $1,500 per applicant. The Association will consider requests for extended research at the library. An independent committee of distinguished scholars from Iowa colleges and universities evaluates the research proposals.
*Requirements:* Current graduate students, post-doctoral scholars, and independent researchers are eligible to apply. Three separately mailed references must be submitted.
*Geographic Focus:* All States
*Date(s) Application is Due:* Mar 1
*Amount of Grant:* 500 - 1,500 USD
*Contact:* Delene McConnaha, Manager of Academic Programs; (800) 828-0475 or (319) 643-5327; fax (319) 643-2391; delene.mcconnaha@hooverassociation.org or info@hooverassoc.org
*Internet:* www.hooverassociation.org/grantsawards/travel_grant.php
*Sponsor:* Herbert Hoover Presidential Library Association
302 Parkside Drive
West Branch, IA 52358

**Herman P. and Sophia Taubman Foundation Grants** **1254**
Herman P. Taubman was a member of the National Campaign Cabinet of the United Jewish Appeal, Southwest chairman of the Israel bond campaign, and a vice-president of the American Committee for the Weizmann Institute. Taubman, who resided in Tulsa, Oklahoma, was also an honorary life member of the Tulsa Jewish Community Council. The Herman P. and Sophia Taubman Foundation was established in Oklahoma in 1955, with giving centered in the states of California, Hawaii, New York, Texas, and Washington. The Foundation's primary fields of interest include: arts and culture; child welfare; children's museums; research of diseases and conditions; elementary and secondary education; higher education; family services; hospital care; human services; Judaism; orphanages; the performing arts; general philanthropy; shelter and residential care; special population support; supportive housing; universities and colleges; and women's services. Types of support offered include: capital campaigns; infrastructure; endowments; financial stability; general operating support; program development; and research. Though there are no annual deadlines for application submission, interested parties should submit a formal request which includes a detailed description of the project or program need, along with an detailed budget.
*Requirements:* 501(c)3 organizations and non-profits in the States of California, Hawaii, New York, Texas, and Washington are eligible to apply.
*Restrictions:* Grants are never given directly to individuals.
*Geographic Focus:* California, Hawaii, New York, Texas, Washington
*Amount of Grant:* Up to 250,000 USD
*Samples:* University of California San Diego, La Jolla, California, $220,000 - general operating funds; New Seed Foundation, Palo Alto, California, $50,000 - general operating funds; Jewish Federation of Greater Dallas, Dallas, Texas, $10,000 - general operating funds.
*Contact:* H. Perry Taubman, Trustee; (918) 588-6407
*Sponsor:* Herman P. and Sophia Taubman Foundation
P.O. Box 1620
Tulsa, OK 74101-1620

**HHMI Gilliam Fellowships for Advanced Study** **1255**
The goal of HHMI's Gilliam Fellowships for Advanced Study program is to ensure that a diverse and highly trained workforce is prepared to assume leadership roles in science, particularly college and university faculty who have the responsibility to develop the next generation of scientists. The training of scientists and engineers has long been a cornerstone of public science policy. HHMI has invested and continues to invest substantially in graduate and medical research education through the direct support of fellowship and training programs. The Gilliam Fellowships for Advanced Study program promotes excellence in the life sciences and related fields by supporting (i) PhD students committed to a career in academic research, and (ii) faculty and training programs dedicated to developing the talents of promising graduate students from underrepresented groups. The Gilliam program facilitates an environment that supports growth of the fellows' research and professional skills, including expanding their professional networks. The fellows' mentors are expected to participate in workshops designed to improve their mentoring skills and sharpen their ability to evaluate themselves as mentors to a diverse and inclusive pool of future scholars. The nomination process, review criteria, and faculty training are intended to encourage participating institutions to adopt activities and policies that promote inclusion and the development of talented scientists from all backgrounds. The primary desired outcome of the fellowship is that fellows will assume leadership roles in science, particularly in academia. Additional activities are designed to facilitate connections and encourage a sense of community among the different cohorts of awardees. Compensation for the current fellowship year is $50,000. The annual deadline for nominations is November 3, with applications due by February 1.
*Requirements:* HHMI's Gilliam Fellowships for Advanced Study are open to eligible pairs comprising thesis advisors and PhD students. Advisor-student pairs from eligible disciplines must be nominated by the HHMI-designated institutional representative. PhD students must be NIGMS training grant eligible, be from racial, ethnic, and other underrepresented groups in the sciences, or Alumni of the HHMI EXROP and be at the appropriate program stage of their PhD training.
*Geographic Focus:* All States
*Date(s) Application is Due:* Feb 1
*Amount of Grant:* 50,000 USD
*Contact:* Megan Katz, Program and Special Projects Coordinator; (301) 215-8511 or (800) 448-4882, ext. 8511; fax (301) 215-8888; gilliam@hhmi.org or katzm@hhmi.org
*Internet:* www.hhmi.org/developing-scientists/gilliam-fellowships-advanced-study
*Sponsor:* Howard Hughes Medical Institute
4000 Jones Bridge Road
Chevy Chase, MD 20815-6789

**HHMI Hanna H. Gray Fellowships** 1256

The Hanna H. Gray Fellows Program honors the contributions of Hanna Holborn Gray, PhD, over her 28 years of service as a Trustee of the Howard Hughes Medical Institute. Gray was one of eight original Trustees appointed to the board in 1984 and went on to serve as chair for a number of years. During her tenure, HHMI made significant changes to its process for identifying and funding scientists, opening the Institute's doors to an ever-increasing pool of applicants. Today, the Hanna H. Gray Fellows program supports early-career life scientists in academic labs across the U.S. The goal is to recruit and retain individuals who are from gender, racial, ethnic, and other groups underrepresented in the life sciences, including those from disadvantaged backgrounds, early in their careers. We believe these promising scientists have the potential to become leaders in academic research and inspire future generations. Fellows will receive funding through their academic institution for postdoctoral training and may continue to receive funding during their early career years as independent faculty. The program includes opportunities for career development, including mentoring and networking with others in HHMI's scientific community. Postdoctoral level fellows will receive an annual stipend of $60,000 and a $20,000 expense allowance. Faculty level fellows will recieve $250,000 per year in research funding and a $20,000 expense allowance. The application process opens annually during the third week of September, with application submissions due by January 10.

*Requirements:* This competition is open to those dedicated to basic research from both doctoral and/or medical training paths in the biomedical and life science disciplines, including plant, evolutionary, chemical and computational biology, as well as biophysics and biomedical engineering. During the duration of the award, Fellows have freedom to change their research focus and follow their own curiosity.

*Geographic Focus* All States, District of Columbia, Guam, Marshall Islands, Northern Mariana Islands, Puerto Rico, U.S. Virgin Islands, American Samoa

*Amount of Grant:* 60,000 - 270,000 USD

*Contact:* Megan Katz, Program and Special Projects Coordinator; (301) 215-8511 or (800) 448-4882, ext. 8511; fax (301) 215-8888; fellows@hhmi.org or katzm@hhmi.org

*Internet:* https://www.hhmi.org/programs/hanna-h-gray-fellows-program

*Sponsor:* Howard Hughes Medical Institute
4000 Jones Bridge Road
Chevy Chase, MD 20815-6789

**HHMI Medical Research Fellowships** 1257

The Medical Research Fellows Program enables students to be immersed in a high-impact, intensive research experience at a critical time in their professional education, before they make plans for residency or postdoctoral training. Unique aspects of the program include national and regional professional development activities, mentorship by physician-scientists, interaction with HHMI investigators, and integration into a community of like-minded peers. Each student selects and works with a mentor to develop a basic, translational, or applied biomedical research proposal. Both mentor selection and the research proposal are key components of the fellowship application; working with an HHMI scientist is encouraged but not required. Mentors can be at any academic or nonprofit research institution in the United States (excluding the National Institutes of Health and any other government agency) or at HHMI's Janelia Research Campus in Ashburn, Virginia. There are special requirements for the Med Fellows program at Janelia, including a pre-screening application deadline in early November. Med Fellows may be supported by the following organizations partnering with HHMI: American Society of Human Genetics (mentor must be ASHG member); Burroughs Wellcome Fund (veterinary students only); Citizens United for Research in Epilepsy; Duchenne Research Fund; Foundation Fighting Blindness (inherited retinal degenerative disease research only); Parkinson's Foundation; and Society of Interventional Radiology Foundation (mentor must be SIR member). Med Fellows are fully supported for one year with the following funding: a $32,000 stipend; a $5,500 allowance for research-related enrichment activities, including travel to Med Fellows meetings and scientific conferences; and a $5,500 allowance for health, dental, and vision insurance and education and moving expenses. The Intent to Apply annual deadline is March 1, with online applications due by April 2.

*Requirements:* Students must be in good standing at a medical, dental, or veterinary school located in the United States. U.S. citizenship is not required, but students must have an appropriate work visa. Previous bench experience is required, with applicants having had one or more substantial research experiences (summer or part-time). For placement at Janelia, students must have an especially strong record of research accomplishment and commitment.

*Restrictions:* Students cannot be enrolled in a combined medical, dental, or veterinary/PhD program or a PhD or ScD program. They cannot have earned a PhD or ScD in a laboratory-based biological science.

*Geographic Focus* All States, District of Columbia, Guam, Marshall Islands, Northern Mariana Islands, Puerto Rico, U.S. Virgin Islands, All Countries, American Samoa

*Date(s) Application is Due:* Apr 2

*Amount of Grant:* 32,000 - 43,000 USD

*Contact:* Melanie Daub, Program Officer; (301) 951-6708 or (800) 424-9924; fax (301) 215-8888; medfellows@hhmi.org or daubm@hhmi.org

*Internet:* www.hhmi.org/developing-scientists/medical-research-fellows-program

*Sponsor:* Howard Hughes Medical Institute
4000 Jones Bridge Road
Chevy Chase, MD 20815-6789

**Hill Crest Foundation Grants** 1258

The Hill Crest Foundation awards grants to Alabama nonprofits, primarily health associations and human services and education, with some funding for the arts. Types of support include: building and renovation; capital campaigns; endowments; equipment; matching and challenge support; professorships; program development; publication; research; scholarship funds; seed money; and technical assistance. There are no deadlines or applications. The Board of Directors meets quarterly.

*Requirements:* Application forms are not required. Before submitting a proposal, organizations should submit a letter, detailing the project and the amount requested, along with descriptive literature about their organization.

*Restrictions:* Grantmaking is limited to Alabama. The Foundation does not fund individuals.

*Geographic Focus:* Alabama

*Amount of Grant:* 20,000 - 150,000 USD

*Samples:* Methodist Homes for the Aging, Birmingham, Alabama, $100,000 payable over one year; Alabama Symphonic Association, Birmingham, Alabama, $30,000 payable over one year; Bell Center for Early Intervention Programs, Birmingham, Alabama, $50,000 payable over one year.

*Contact:* Charles Terry, Chairperson; (205) 425-5800

*Sponsor:* Hill Crest Foundation
P.O. Box 530507
Mountain Brook, AL 35253-0507

**Hilton Hotels Corporate Giving Program Grants** 1259

Hilton makes charitable contributions to nonprofit organizations involved with K-12 education, youth development, public policy, homelessness, and civic affairs. Support is given primarily in areas of company operations, with emphasis on California, including Los Angeles and San Francisco, and Tennessee, including Memphis; giving also to national organizations. Contact Ellen Gonda for additional information at: corporate_communications@hilton.com.

*Requirements:* 501(c)3 tax-exempt organizations are eligible.

*Restrictions:* No support for sport teams, religious organizations not of direct benefit to the entire community, government-supported organizations (over 20 percent of budget), hospitals, private schools, pre-schools, or day care facilities, film and video production, or promotional materials. No grants to individuals (except for employee-related scholarships), or for fellowships, sports activities, debt reduction, capital campaigns or endowments, film or video projects, or promotional merchandise.

*Geographic Focus:* California, Tennessee

*Contact:* Ellen Gonda, Senior Vice President; (310) 278-4321; fax (310) 205-7678; corporate_communications@hilton.com

*Internet:* www.hiltonworldwide.com

*Sponsor:* Hilton Hotels Corporation
9336 Civic Center Drive
Beverly Hills, CA 90210-3604

**Holberg Memorial Fund International Prize** 1260

The Holberg International Memorial Prize is named after Ludvig Holberg, who was born in Bergen in 1684, and held the Chairs of Metaphysics and Logic, Latin Rhetoric and History at the University of Copenhagen. Holberg played an important part in bringing the Enlightenment to the Nordic countries. Through his interdisciplinary and internationally-oriented efforts, Holberg sought to modernize academic subjects and teaching methods. The main purpose of the Holberg Fund is to increase the profile of the arts and humanities, the social sciences, law and theology. The Prize is awarded to scholars who have made outstanding contributions to research in the arts and humanities, social science, law, or theology, either within these fields or through interdisciplinary work. The prize winner must have had a decisive influence on international research. The Prize is worth 4.5 million NOK (approximately EUR 570,000 or USD 800,000 USD). Specific instructions and a current nomination form are available at the website.

*Requirements:* Scholars holding positions at universities and other research institutions, including academies, are entitled to nominate candidates for the Prize. The letter of nomination should be written in English and state the reasons for the nomination in two to three pages. The nomination should also include the candidate's CV and suggest referees who know the scholar's work. The function of the nomination is to make the Holberg Prize Academic Committee aware of the candidate's work.

*Restrictions:* Nominations are strictly confidential. They should not be disclosed to the nominee or to others at any time.

*Geographic Focus:* All States, All Countries

*Date(s) Application is Due:* Sep 15

*Amount of Grant:* 775,000 USD

*Contact:* Trine Kleven, Program Manager; 47 55 58 69 92; Trine.kleven@holbergprisen.no

*Internet:* www.holbergprisen.no/en/holberg-international-memorial-prize.html

*Sponsor:* University of Bergen
Thormohlensgate 51, 2nd Floor, P.O. Box 7800
Bergen, N-5020 Norway

**Holberg Memorial Fund Nils Klim Prize** 1261

The Nils Klim Prize is awarded to a young Nordic researcher in the academic fields of the arts and humanities, social sciences, law, and theology. The Prize is awarded to researchers who have made an outstanding contribution, either within one of the relevant fields alone or of an interdisciplinary nature. Particular importance is attached to the researcher's independence and innovative ability. Consideration is also given to whether the research contribution in question can be deemed particularly outstanding or promising in light of the researcher's age, the nature of the research project, and the research traditions in the academic field in question. The Board of the Holberg Memorial Fund makes the award on the basis of the recommendation of the Nils Klim Academic Committee, which consists of outstanding Nordic scholars in the relevant academic fields. The prize is worth NOK 250,000 (approximately $32,000 EUR or $42,000 USD). Specific guidelines and a current nomination form are available at the Prize website.

*Requirements:* Candidates must be under the age of 35 on the closing date of the nomination. Scholars holding a senior position at universities and other research institutions within the above mentioned academic fields are entitled to nominate candidates. The letter of nomination must be written in English, no more than five pages, and state the reasons for the nomination. The letter should also contain a CV for the candidate including list of publications and copies of the candidate's doctoral thesis and any another significant publications. The letter should also suggest referees who know the researcher and his/her work.
*Geographic Focus:* Denmark, Finland, Iceland, Norway, Sweden
*Date(s) Application is Due:* Sep 15
*Amount of Grant:* 250,000 NOK
*Contact:* Trine Kleven, Project Manager; +47 55 58 69 92; trine.kleven@holbergprisen.no
*Internet:* www.holbergprisen.no/en/nils-klim-prize.html
*Sponsor:* University of Bergen
Thormohlensgate 51, 2nd Floor, P.O. Box 7800
Bergen, N-5020 Norway

**Holberg Memorial Fund School Project** 1262
Every year students at upper secondary schools in Norway are invited to carry out a research project in cooperation with established researchers in an effort to stimulate young people to take an interest in the arts, social sciences, law, and theology. These projects span a wide range of topics such as different views on alternative medicine or clothes as an expression. Three projects are awarded prizes of NOK 15,000, 10,000, and 5,000 respectively. The prize winning students are also invited to attend the award ceremony. The Project also awards a prize of NOK 15,000 to the teachers who present the best educational program. Specific information is available at the website.
*Geographic Focus:* Norway
*Amount of Grant:* 5,000 - 15,000 NOK
*Contact:* Barbara Wendelbo; +47 55 58 90 45; barbara.wendelbo@holbergprisen.no
*Internet:* www.holbergprisen.no/en/holberg-prize-school-project/info.html
*Sponsor:* University of Bergen
Thormohlensgate 51, 2nd Floor, P.O. Box 7800
Bergen, N-5020 Norway

**Honeybee Health Improvement Project Grants** 1263
The mysterious disappearance of bees, called Colony Collapse Disorder (CCD), is a growing threat to Honey Bees, the mainstay of pollination services in agriculture. The North American Pollinator Protection Campaign (NAPPC), a tri-national coalition dedicated to promoting the health of all pollinators partners with different organizations to perform research for improving the health of honey bees and reversing the threats they face. In partnership with Burt's Bees, the Honeybee Health Improvement Project was created. It awards annual research grants to scientists seeking solutions to improve honeybee health—including genetic stock improvements, commercial beekeeping practices, and forage opportunities on public and private land. NAPPC will fund multiple projects up to $8,000 each.
*Requirements:* Review and selection of proposals will be conducted by members of the Honey Bee Health Improvement Task Force (HBH). Funds must be used within a one-year period. Focused, targeted projects with a high likelihood of providing tangible results that can be applied to improving honey bee health are preferred. Proposals providing valuable extensions of previously funded projects will be considered. The HBH Task Force has identified six priority areas for funding, though other areas will be considered as well: (1) Effects of nutrition on honey bee behavior, physiology and/or colony health; (2) Effects of pesticides on honey bee behavior, physiology and/or colony health; (3) Effects of pathogens and pests on honey bee behavior, physiology and/or colony health; including the development of novel methods to mitigate these effects, such as RNAi technology; (4) Development of methods to improve genetic stocks of managed honey bee populations; (5) Effects of climate or environmental variables on: a) plants, especially nectar and pollen quantity and quality; and/or b) honey bee physiology and/or colony health; (6) The development of diagnostics or indicators for the presence of stressors that effect honey bee health, particularly those that can be used by beekeepers. Proposals that address multiple priority areas or have implications for the health of other managed or native bee species are encouraged.
*Restrictions:* As a nonprofit organization, the Pollinator Partnership/NAPPC does not pay overhead on funded research grants.
*Geographic Focus:* All States
*Date(s) Application is Due:* Dec 19
*Amount of Grant:* Up to 8,000 USD
*Contact:* Jennifer Tsang, (510) 643-7430; jt@pollinator.org
*Internet:* www.pollinator.org/honeybee_health.htm
*Sponsor:* North American Pollinator Protection Campaign
423 Washington Street, 5th Floor
San Francisco, CA 94111

**Honeywell Corporation Habitat and Conservation Grants** 1264
In partnership with environmental organizations, Honeywell provides students with unique learning opportunities and educators with new teaching tools to promote green science in the classroom. The Corporation believes that energy, water, waste management and air quality are among the most prominent issues around the world today. Operated in partnership with the New Jersey Audubon Society, Montezuma Audubon Center, Onondaga Audubon Society, and Maryland Science Center, the Honeywell Institute for Ecosystems Education takes science and social studies teachers through an explorative tour of their local ecosystems. Educators learn new teaching techniques and discover creative methods to integrate environmental science into their existing curriculum. Applications are required, and potential applicants should begin by contacting the office by email. There are no specific deadlines.
*Geographic Focus:* All States
*Contact:* Michael Holland, Hometown Solutions Media Contact; (973) 455-2728 or (973) 727-6891; michael.holland@honeywell.com
*Internet:* honeywell.com/Citizenship/Pages/habitat-conservation.aspx
*Sponsor:* Honeywell Corporation
101 Columbia Road
Morristown, NJ 07962

**Honeywell Corporation Sustainable Opportunities Grants** 1265
The Honeywell Corporation is dedicated to protecting the environment with a comprehensive and contemporary commitment to address some of the world's toughest challenges. Giving in this area is on a national basis within areas of company operations, with emphasis on New Jersey and New York. Types of support include: education, program development, sponsorships, and environmental protection programs. Applications are required, and potential applicants should begin by contacting the office by email. There are no specific deadlines.
*Geographic Focus:* All States
*Contact:* Michael Holland, Hometown Solutions Media Contact; (973) 455-2728 or (973) 727-6891; michael.holland@honeywell.com
*Internet:* honeywell.com/Citizenship/Pages/our-commitments.aspx
*Sponsor:* Honeywell Corporation
101 Columbia Road
Morristown, NJ 07962

**Honor the Earth Grants** 1266
As a unique national Native initiative, Honor the Earth works to raise public awareness, and raise and direct funds to grassroots Native environmental groups. Honor the Earth's Board encourages proposals on Native environmental justice, sustainable development, and cultural preservation, with a grant limit of $5,000. The organization's focus will remain on sustainable Indigenous communities support. Their work also focuses on opposition to fossil fuels extraction and destructive mining practices. Additional information, guidelines, and information on previously funded projects is available at the website. Applicants are encouraged to check back periodically for grant funding through the Building Resilience program. Proposals may be submitted in hard copy or by email. Faxed proposals cannot be accepted.
*Requirements:* Grants are awarded solely to organizations that are led and managed by Native peoples. Priority is given to grassroots, community-based organizations and groups with a lack of access to federal and/or tribal funding resources.
*Restrictions:* Funding is not available for individuals.
*Geographic Focus:* All States, Canada
*Date(s) Application is Due:* Nov 9
*Amount of Grant:* 1,000 - 5,000 USD
*Contact:* Winona LaDuke, Executive Director; (218) 375-3200; HonorGrants@honorearth.org or info@honortheearth.org
*Internet:* www.honorearth.org/grantmaking
*Sponsor:* Honor the Earth
607 Main Avenue
Callaway, MN 56521

**Horowitz Foundation for Social Policy Grants** 1267
The Horowitz Foundation approves approximately fifteen grants each year, in the amount of $7,500 per grant - $5,000 initially and an additional $2,500 upon receipt of a final report or a copy of the product of the research. The foundation supports and advances research and understanding in the major fields of the social sciences, specifically in the fields of psychology, anthropology, sociology, economics, urban affairs, area studies, political science, and other disciplines. Grants provide direct assistance to individual scholars worldwide who require small grants to further their research with emphasis on policy-oriented studies. Preference will be given to projects that deal with contemporary issues in the social sciences or issues of policy relevance, as well as to scholars in the initial stages of work. Applicants can be from any country and any university in the world. U.S. citizenship or residency is not required. The annual deadline is January 31.
*Requirements:* Applicants must: be current PhD candidates who are working on their dissertation; not have a PhD (those who do are ineligible); and have defended their dissertation proposal or had their topic approved by their department.
*Geographic Focus:* All States, All Countries
*Date(s) Application is Due:* Jan 31
*Amount of Grant:* 3,000 - 7,500 USD
*Contact:* Mary E. Curtis, Chairman; (732) 445-2280; fax (732) 445-3138; applications@horowitz-foundation.org or info@horowitz-foundation.org
*Internet:* www.horowitz-foundation.org
*Sponsor:* Horowitz Foundation for Social Policy
P.O. Box 7
Rocky Hill, NJ 08553-0007

**Horowitz Foundation for Social Policy Special Awards** 1268
The Horowitz Foundation provides eight special awards that may be granted for certain projects. These often carry an additional modest stipend beyond the customary amounts of a grant. The eight areas are as follows: the Donald R. Cressey Award for work done at the empirical level that has direct implication for changes in criminal justice and penology practices; the Joshua Feigenbaum Award for empirical research on policy aspects of the arts and popular culture, with special reference to mass communication;

the Eli Ginzberg Award for a project involving solutions to major health and welfare problems in urban settings; the Irving Louis Horowitz Higher Education Award for research that supports the development of organizational leadership in higher education (specific to Rutgers University doctoral students); the Harold D. Lasswell Award for policy related projects in international relations and foreign affairs; the Robert K. Merton Award for studies in the relation between social theory and public policy; the Martinus Nijhoff Award for policy implications of scientific, technological and medical research; and the John L. Stanley Award for a work that seeks to expand our understanding of the political and ethical foundations of policy research.
*Geographic Focus:* All States
*Date(s) Application is Due:* Jan 31
*Contact:* Mary E. Curtis, Chairman; (732) 445-2280; fax (732) 445-3138; applications@horowitz-foundation.org or info@horowitz-foundation.org
*Internet:* https://www.horowitz-foundation.org/grant-info
*Sponsor:* Horowitz Foundation for Social Policy
P.O. Box 7
Rocky Hill, NJ 08553-0007

**Hospital Libraries Section / MLA Professional Development Grants** 1269
The purpose of this Medical Libraries Association (MLA) award which is sponsored by the association's Hospital Libraries Section (HLS) is to provide librarians working in hospital and similar clinical settings with the support needed for educational or research activities. These include developing and acquiring the knowledge and skills delineated in "Competencies for Lifelong Learning and Professional Success: The Educational Policy Statement of the Medical Library Association" and "The Research Imperative: The Research Policy Statement of the Medical Library Association." The grant may also be used to support reimbursement for expenses incurred in conducting scientific research, such as professional assistance in survey-research design, statistical analyses, etc. The application form and guidelines are available for download at the website. Applications and supporting documents must be received by MLA via email, fax, or mail by December 1. email applications will be accepted as PDF or MS Word files only. Founded in 1898, the Medical Library Association (MLA) is a nonprofit, educational organization of more than 1,100 institutions and 3,600 individual members in the health sciences information field committed to educating health-information professionals, supporting health-information research, promoting access to the world's health-sciences information, and working to ensure that the best health information is available to all.
*Requirements:* Applicants must have been employed as a health-sciences librarian within the last year in either a hospital or other clinical-care institution. It is preferred that applicants be a member of the Hospital Libraries Section of the MLA.
*Restrictions:* Previous recipients of the HLS/MLA Professional Development Grant are not eligible. Applicants must not have received an MLA grant, scholarship, or other award within the last year. An applicant can only receive one award per year. An award will be made to no more than one employee per institution per year. Awards will not be given to support work toward a degree or certificate program.
*Geographic Focus:* All States, Canada
*Date(s) Application is Due:* Dec 1
*Amount of Grant:* Up to 800 USD
*Contact:* Carla Funk, Executive Director; (312) 419-9094, ext. 14; fax (312) 419-8950; mlapd2@mlahq.org or grants@mlahq.org
*Internet:* www.mlanet.org/awards/grants/index.html
*Sponsor:* Medical Library Association
65 East Wacker Place, Suite 1900
Chicago, IL 60601-7246

**Hospitals Contribution Fund Research Foundation Grants** 1270
The HCF Research Foundation's mission is to encourage examination and improvement of the provision, administration and delivery of health care services to drive better health care for all Australians. The HCF Research Foundation has developed research topics to deliver $1 million in funding to the research community following consultation with Australia's leading medical colleges and professional bodies. The Foundation is seeks Expressions of Interest (EOIs) each year from research teams interested in funding for health services research that will drive improvements in the quality, safety, appropriateness and cost effectiveness of health care in Australia. EOI's are invited for research projects that specifically address the following topics: end of life care, consisting of multidisciplinary programs that assist patients to die with dignity at home; clinical registry data, consisting of a registry data to directly improve patient outcomes (e.g. heart disease, orthopaedics, cancer); and mental health care programs that provide the most efficacious and cost effective outcomes for patients. Recent research grants awarded by the Foundation have ranged in value from $20,000 to $300,000. The annual deadline for application submissions is June 15.
*Requirements:* Applications involving collaborations between investigators from a range of backgrounds, including career researchers, practicing clinicians, policy makers, and the private sector, will be given priority.
*Geographic Focus:* Australia
*Date(s) Application is Due:* Jun 15
*Amount of Grant:* 20,000 - 300,000 AUD
*Contact:* Wayne Adams, Foundation Manager; +61 2 9290 0199 or +61 2 9290 0444; hcf_foundation@hcf.com.au
*Internet:* https://www.hcf.com.au/about-us/hcf-foundation
*Sponsor:* Hospitals Contribution Fund Research Foundation
GP.O. Box 4242
Sydney, NSW 2001 Australia

**HRAMF Alagille Syndrome Accelerator Award** 1271
The Alagille Syndrome Accelerator Award was created to advance ongoing research towards a cure or improved therapeutic treatment of patients with Alagille Syndrome. Projects must contribute to finding effective therapies or cures for Alagille Syndrome, or the discovery and testing of potential drugs directed at relevant pathways in animal models or support of exploratory small clinical studies in human volunteers. A one-year $25,000 award will support an outstanding investigator working in a non-profit academic, medical, or research institution in North America. Funding began on September 15, 2019. The current application deadline is July 23.
*Requirements:* Applicants must hold an M.D. or Ph.D; projects must be conducted at a non-profit institution in North America; and all proposals must address discovery of effective therapies or cures for Alagille Syndrome.
*Geographic Focus:* All States
*Date(s) Application is Due:* Jul 23
*Amount of Grant:* 25,000 USD
*Contact:* Kimberley Lezak, Interim Director; (617) 279-2275 or (617) 451-0049; fax (617) 423-4619; klezak@hria.org
*Internet:* https://hria.org/tmf/alagille/
*Sponsor:* Health Resources in Action Medical Foundation
2 Boylston Street, 4th Floor
Boston, MA 02116

**HRAMF Charles A. King Trust Postdoctoral Research Fellowships** 1272
The Charles A. King Trust was established in 1936 to support and promote the investigation of human disease and the alleviation of human suffering through improved treatment. In keeping with these principles, the King Trust today supports postdoctoral fellows in the basic sciences and clinical/health services research. The program is designed to support postdoctoral scientists in non-profit academic, medical or research institutions in Massachusetts. Two-year grants ranging from $104,008 to $117,760, inclusive of a $2,000 expense allowance, will be awarded each September 1. Awards are made to nonprofit academic, medical, or research institutions in the State of Massachusetts on behalf of the Awards Recipient.
*Requirements:* Applicants must be working in an academic or medical research institution in the state of Massachusetts and have the required minimum/maximum years of experience. Applicants must apply for the fellowships under the guidance of a Mentor who is an established investigator with an active research program. Mentors are expected to be involved in the planning, execution, and supervision of the proposed research. Mentors must confirm that degrees obtained outside the United States are equivalent to the M.D., D.M.D., Ph.D. or other doctoral degree. United States citizenship is not required; visa documentation is not required. The Basic Science area requires that, by September 1 of each funding cycle, applicants must have completed at least three years, but not more than five years, of full-time postdoctoral research experience. The Clinical and/or Health Services Research area requires that, by September 1 of each funding cycle, applicants with clinical responsibilities must have no more then five years of postdoctoral research training since completing residency training. Applicants without clinical responsibilities must have completed at least three years, but not more than five years, of full-time postdoctoral research experience.
*Restrictions:* Only one applicant per Mentor may apply per application cycle; there are not institutional limitations on the number of applicants who may submit applications. Funding begins on July 1. A later start date is permitted but the Award must be activated on or before October 1.
*Geographic Focus:* Massachusetts
*Date(s) Application is Due:* Sep 1
*Amount of Grant:* 104,008 - 117,760 USD
*Contact:* Lindsey Carver; (617) 279-2238; fax (617) 423-4619; lcarver@hria.org
*Internet:* https://hria.org/tmf/king/
*Sponsor:* Health Resources in Action Medical Foundation
2 Boylston Street, 4th Floor
Boston, MA 02116

**HRAMF Charles H. Hood Foundation Child Health Research Awards** 1273
The Charles H. Hood Foundation was incorporated in 1942 to improve the health and quality of life for children through grant support of New England-based pediatric researchers. The intent of the Child Health Research Awards Program is to support newly independent faculty, provide the opportunity to demonstrate creativity, and assist in the transition to other sources of research funding. Two-year grants of $165,000 ($82,500 per year inclusive of 10% indirect costs) are awarded to researchers who are within five years of their first faculty appointment by the funding start date. Application deadlines occur in the fall of each year on October 7.
*Requirements:* Applicants must be working in nonprofit academic, medical or research institutions within the six New England states. Grants support hypothesis-driven clinical, basic science, public health, health services research, and epidemiology projects focused on child health. Applicants must hold a doctoral degree with a demonstrated level of independence confirmed by the Department or Division Chair. The Applicant's potential for a lifetime career as an investigator in pediatric research is also critical in the review process. Applicants are required to devote at least 20% effort to the proposed Hood research project. Applicants who have pending R01s or other large applications to the NIH and other agencies are encouraged to submit proposals to the Hood Foundation. Online submissions must be received by April 8, 12:00 Noon, U.S. Eastern time. Printed copies must be received by April 14, 5:00 pm, U.S. Eastern time.
*Restrictions:* Applicants are ineligible if they have combined federal and non-federal funding totaling $450,000 or more in direct costs at any time during the two years of the Award. This figure refers to external funding only and not an Applicant's start-up

package, other intramural support or the Hood Award. Applicants are ineligible if they are currently or have previously been designated as Principal Investigator or Co-P.I. on an R01, P01, Pioneer Award, New Innovator Award or equivalents from federal agencies such as the National Science Foundation (NSF) or Department of Defense (DOD). Applicants who have completed the R00 phase of a K99/R00 are also ineligible for a Hood Award.
*Geographic Focus:* Connecticut, Maine, Massachusetts, New Hampshire, Rhode Island, Vermont
*Date(s) Application is Due:* Oct 7
*Amount of Grant:* 82,500 - 165,000 USD
*Contact:* Charlene Mancusi, Program Officer; (617) 279-2230 or (617) 451-0049; fax (617) 423-4619; cmancusi@hria.org
*Internet:* https://hria.org/tmf/hood/
*Sponsor:* Health Resources in Action Medical Foundation
2 Boylston Street, 4th Floor
Boston, MA 02116

**HRAMF Deborah Munroe Noonan Memorial Research Grants** 1274
The Deborah Munroe Noonan Memorial Research Fund, established in 1947 by Frank M. Noonan in memory of his mother, continues its proud tradition of supporting improvements in the quality of life for children with disabilities with two-year grants up to $160,000 (inclusive of 20% indirect costs). Recognizing that children's health services and supports are provided in a wide range of community settings as well as hospitals, the Noonan Research program welcomes research proposals from both nonprofit organizations and academic institutions that serve children with physical or developmental disabilities and associated health-related complications. Eligible organizations and target populations must be within the Fund's geographic area of interest of Greater Boston. Neither a faculty position nor an advanced degree is required.
*Requirements:* Applicants must hold a position within a nonprofit institution or organization within the Fund's geographic area of interest. New investigators are encouraged to submit proposals. Project must address the target age range of birth through 23 years old. All application information must be completed online, and the proposal uploaded as a PDF by 12:00 noon U.S. Easter time of the published deadline date.
*Restrictions:* Proposals for basic science research will not be considered nor will applications for capital costs such as buildings, renovations, or major equipment items. The Noonan Fund does not support direct service, primary prevention projects, primary medical conditions such as obesity or device development. Drug trials are rarely supported by the Noonan Research Fund.
*Geographic Focus:* Massachusetts
*Date(s) Application is Due:* Oct 10
*Amount of Grant:* Up to 80,000 USD
*Contact:* Jeanne Brown; (617) 279-2255 or (617) 451-0049; fax (617) 423-4619; jbrown@hria.org
*Internet:* https://hria.org/tmf/noonan/
*Sponsor:* Health Resources in Action Medical Foundation
2 Boylston Street, 4th Floor
Boston, MA 02116

**HRAMF Edward N. and Della L. Thome Memorial Foundation Awards Program in Age-Related Macular Degeneration Research** 1275
The Edward N. and Della L. Thome Memorial Foundation was created in 2002 to advance the health of older adults through the support of direct service projects and medical research on diseases and disorders affecting older adults. In keeping with the Foundation's mission, the goal of the Awards Program is to support translational research that will lead to improved therapies for individuals suffering from age-related macular degeneration (AMD). As steward of the Thome Memorial Foundation, Bank of America N.A., works with The Medical Foundation division's Scientific Review Committees to select the most qualified candidates. Two-year awards of $500,000 ($250,000 per year) will be made to investigators who hold a faculty appointment at a non-profit academic, medical, or research institution in the United States.
*Requirements:* Successful research proposals will extend recent basic research findings regarding the underlying mechanisms of AMD. Examples of funding areas include, but are not limited to, genetically engineered animal models, the discovery and testing of small molecule therapies directed at promising targets including immune- or inflammation-related pathways associated with AMD, local drug delivery systems, and neuroprotective strategies. Basic research and clinical trials are outside the scope of this Program. However, clinical studies that involve the evaluation of new imaging modalities with no potential risks to human subjects are eligible. Eligibility is not limited to those investigators currently working in age-related macular degeneration research; scientists who have conducted research exploring the biologic causes of related disorders and/or similar translational research programs are encouraged to apply. All applicants must hold a faculty appointment at a nonprofit academic, medical, or research institution in the United States. Applicants do not need to be nominated by their institutions. United States citizenship is not required; visa documentation is not required. Applications must be submitted by 12:00 Noon, U.S. Eastern time of the published deadline date.
*Restrictions:* Only one application may be submitted per applicant.
*Geographic Focus:* All States
*Date(s) Application is Due:* Aug 30
*Amount of Grant:* Up to 500,000 USD
*Contact:* Lindsey Carver, Grants Officer; (617) 279-2238 or (617) 451-0049; fax (617) 423-4619; lcarver@hria.org
*Internet:* https://hria.org/tmf/thomeamd/
*Sponsor:* Health Resources in Action Medical Foundation
2 Boylston Street, 4th Floor
Boston, MA 02116

**HRAMF Edward N. and Della L. Thome Memorial Foundation Awards Program in Alzheimer's Disease Drug Discovery Research** 1276
The Edward N. and Della L. Thome Memorial Foundation was created in 2002 to advance the health of older adults through the support of direct service projects and medical research on diseases and disorders affecting older adults. In keeping with the foundation's mission, the goal of the research program is to support pilot studies towards innovative drug discovery research that will lead to improved therapies for individuals suffering from Alzheimer's disease. Examples of funding areas include the design, synthesis, and development of target compounds or the modification of existing compounds to improve drug effectiveness and safety as well as other approaches within the field of medicinal chemistry. Researchers dedicated to the validation and testing of target compounds, small molecule therapies, nanotechnologies, or similar techniques are also encouraged to apply. Two-year awards of up to $500,000 ($250,000 per year) will be made (all awards inclusive of 10% indirect costs). Invited applications must be submitted through the online system and must be received by 12:00 noon, U.S. Eastern time on September 9.
*Requirements:* All applicants must hold a faculty appointment at a non-profit academic, medical, non-governmental or research institution in the United States. U.S. citizenship is not required. Applicants do not need to be nominated by their institutions. Preference will be given to originality of ideas, regardless of faculty seniority.
*Restrictions:* Genetic studies, biomarker research, neuroimaging, clinical studies, basic research and new target discovery are currently outside the scope of this program.
*Geographic Focus:* All States
*Date(s) Application is Due:* Sep 9
*Amount of Grant:* Up to 500,000 USD
*Contact:* Lindsey Carver, Grants Officer; (617) 279-2238 or (617) 451-0049; fax (617) 423-4619; lcarver@hria.org
*Internet:* https://hria.org/tmf/thomead/
*Sponsor:* Health Resources in Action Medical Foundation
2 Boylston Street, 4th Floor
Boston, MA 02116

**HRAMF Harold S. Geneen Charitable Trust Awards for Coronary Heart Disease Research** 1277
The Harold S. Geneen Charitable Trust Awards Program for Coronary Heart Disease Research supports research in the area of the prevention and control of coronary and circulatory failure. The Awards Program seeks to contribute to the prevention of coronary heart disease or circulatory failure, and improving care for patients with these medical conditions. In accordance with Mr. Geneen's directives the Program seeks to establish a more direct and personalized relationship with grant recipients than is normally possible in dealing with the diffuse and bureaucratic administrations through which large organizations are managed, and to support smaller institutions rather than major universities or medical complexes which have a demonstrated capacity to raise funds from the public generally. The award is equal to $300,000 distributed over two years (inclusive of 10% indirect costs). The annual deadline for application submission is July 15 at 12:00 noon.
*Requirements:* Invited institutions may submit one application to the Program which meets the eligibility requirements for the grant cycle. Applicants must be full-time faculty at an invited nonprofit academic, medical, or research institution. United States citizenship is not required. Junior faculty are encouraged to apply.
*Geographic Focus:* All States
*Date(s) Application is Due:* Jul 15
*Amount of Grant:* 300,000 USD
*Contact:* Emily Breen; (617) 391-9467 or (617) 451-0049; fax (617) 423-4619; ebreen@hria.org
*Internet:* https://hria.org/tmf/geneen/
*Sponsor:* Health Resources in Action Medical Foundation
2 Boylston Street, 4th Floor
Boston, MA 02116

**HRAMF Jeffress Trust Awards in Interdisciplinary Research** 1278
The Thomas F. and Kate Miller Jeffress Memorial Trust was established in 1981 to support basic research in chemical, medical, and other scientific research through grants to educational and research organizations in the Commonwealth of Virginia. The trust was established under the will of Mr. Robert M. Jeffress, a business executive and philanthropist of Richmond, Virginia. Grants are made to assist scientists in educational and research institutions to conduct investigations in the natural sciences, generally considered to include chemistry, biology (with the exception of field studies, classification, or other largely observational studies), and the basic medical sciences, such as biochemistry, microbiology and others. The principal evaluation criteria will be the scientific significance of the proposed work and the competence of the investigator. The trust is particularly interested in supporting fundamental research by scientists early in their careers and new areas of research or more speculative projects by established investigators. In general, grants will cover direct-expense items essential to the successful completion of the proposed project. Grants from the Jeffress Memorial Trust are one year in duration. Funding may be renewed at the discretion of the allocations committee. Application materials are available for download at the grant website. Applicants are strongly encouraged to review the application guidelines for additional helpful information and clarifications before applying.
*Requirements:* Eligible research areas for Jeffress support include astronomy, biosciences, chemistry, computer sciences, engineering, environmental sciences, material science, mathematics and physics. Full-time faculty at institutions in Virginia that are within seven years of their first faculty appointment are eligible to apply as Principal Investigators. Student participation is a requirement of the proposed research plan. Online submissions must be received by January 14, 12:00 Noon, U.S. Eastern time. Mailed hard copies must be received by The Medical Foundation division by January 8, 5:00 pm, U.S. Eastern time.

*Restrictions:* Research in social, economic and behavioral sciences, including psychology, are currently outside the scope of Jeffress funding. Faculty from medical schools and schools of Osteopathic Medicine may not apply as Principal Investigators; however, they are encouraged to collaborate as Co-Investigators.
*Geographic Focus:* Virginia
*Date(s) Application is Due:* Feb 7
*Amount of Grant:* 100,000 USD
*Contact:* Jeanne Brown; (617) 279-2240, ext. 709; fax (617) 423-4619; jbrown@hria.org
*Internet:* https://hria.org/tmf/jeffress/
*Sponsor:* Health Resources in Action Medical Foundation
2 Boylston Street, 4th Floor
Boston, MA 02116

**HRAMF Ralph and Marian Falk Medical Research Trust Catalyst Awards 1279**

The Dr. Ralph and Marian Falk Medical Research Trust was created by Marian Falk in 1979 to support biomedical research. Marian Falk sought to fund medical research to improve treatments of the past and eventually find cures for diseases for which no definite cure is known. Bank of America serves as Trustee for the Trust. Beginning with the 2015 grant cycle, the Medical Foundation division of Health Resources in Action began to manage the grant making for both the Falk Catalyst and Transformational programs on behalf of Bank of America. The Catalyst Research Award program provides one year of seed funding to support high-risk, high-reward projects that address critical scientific and therapeutic roadblocks within the program's principal areas of focus. The program is designed to enable planning and development of projects, teams, tools, techniques and management infrastructure necessary to successfully compete for two-year awards through the Transformational Research Award program.
*Requirements:* Applicants must meet the following *Requirements:* hold a full-time faculty appointment; and be an independent investigator with demonstrated institutional support and the specialized space and facilities needed to conduct the proposed research. United States citizenship and visa documentation are not required.
*Restrictions:* Applicants may not have funding for a similar project. A PI may only submit one application.
*Geographic Focus:* All States
*Contact:* Jeanne Brown, Program Officer; (617) 279-2255 or (617) 451-0049; fax (617) 423-4619; jbrown@hria.org
*Internet:* https://hria.org/tmf/falkcap/
*Sponsor:* Health Resources in Action Medical Foundation
2 Boylston Street, 4th Floor
Boston, MA 02116

**HRAMF Ralph and Marian Falk Medical Research Trust Transformational Awards 1280**

The Dr. Ralph and Marian Falk Medical Research Trust was created by Marian Falk in 1979 to support biomedical research. Falk sought to fund medical research to improve treatments of the past and eventually find cures for diseases for which no definite cure is known. Bank of America serves as Trustee for the Falk Medical Research Trust. The Transformational Awards Program provides $1,000,000 for a two year funding cycle to successful Catalyst Award Recipients to continue their work tackling critical scientific and therapeutic roadblocks and thereby opening avenues for treating and curing disease. Transformational Awards will be granted based on both scientific merit and having successfully attained the proposed milestones and benchmarks of the Catalyst Award, thus demonstrating its successful execution within the proposed budget and projected time frame. The program is available to researchers nationally.
*Requirements:* Only Catalyst Award Recipients are eligible to apply for a Transformational Award. Applicants must meet the following *Requirements:* hold a full-time faculty appointment; and be an independent investigator with demonstrated institutional support and the specialized space and facilities needed to conduct the proposed research. United States citizenship and visa documentation are not required.
*Restrictions:* Applicants may not have funding for a similar project. A PI may only submit one application.
*Geographic Focus:* All States
*Amount of Grant:* 1,000,000 USD
*Contact:* Jeanne Brown, Program Officer; (617) 279-2255 or (617) 451-0049; fax (617) 423-4619; jbrown@hria.org
*Internet:* https://hria.org/tmf/falktap/
*Sponsor:* Health Resources in Action Medical Foundation
2 Boylston Street, 4th Floor
Boston, MA 02116

**HRAMF Robert E. Leet and Clara Guthrie Patterson Trust Mentored Research Award 1281**

The Robert E. Leet and Clara Guthrie Patterson Trust was established in 1980 to support and promote the investigation of human disease and the alleviation of human suffering through improved treatment. Two-year Mentored Research Awards of $190,000 ($50,000 per year) will be awarded to applicants working in Connecticut, New Jersey and Rhode Island. The primary goal of the Program is to support research that focuses on the causes of human disease and/or improving treatment. Research Awards will support mentored investigators in the early stages of their research careers across a wide range of disciplines, including clinical research, health services or policy research, health education, or epidemiological studies including comparative outcomes research. The Program also seeks to impact public health by supporting researchers with an interest in and commitment to population health issues. Pilot and preliminary studies are encouraged. The annual due date for applications is July 23.
*Requirements:* Applicants must hold a position within a nonprofit institution or organization within the Fund's geographic area of interest. New investigators are encouraged to submit proposals. Project must address the target age range of birth through 23 years old. All application information must be completed online, and the proposal uploaded as a PDF by 12:00 noon U.S. Easter time of the published deadline date.
*Restrictions:* Proposals for basic science research will not be considered nor will applications for capital costs such as buildings, renovations, or major equipment items. The Fund does not support direct service, primary prevention projects, primary medical conditions such as obesity or device development. Drug trials are rarely supported by the Fund.
*Geographic Focus:* Connecticut, New Jersey, New York
*Date(s) Application is Due:* Jul 23
*Amount of Grant:* 100,000 USD
*Samples:* Yale School of Medicine, New Haven, Connecticut, Section of Endocrinology, $100,000 - Role of Insulin in Endometrial Cancer Pathogenesis: Metabolic Phenotyping of Women with Endometrial Cancer (2019); Yale School of Public Health, New Haven, Connecticut, $100,000 - Role of Internal Migration in the Transmission Dynamics of Multidrug-Resistant Tuberculosis in Urban China (2019); Yale University, New Haven, Connecticut, $100,000 - henotyping Deceased Donor Acute Kidney Injury to Predict Recipient Outcomes (2019).
*Contact:* Jeanne Brown, Program Officer; (617) 279-2255 or (617) 451-0049; fax (617) 423-4619; jbrown@hria.org
*Internet:* https://hria.org/tmf/patterson/
*Sponsor:* Health Resources in Action Medical Foundation
2 Boylston Street, 4th Floor
Boston, MA 02116

**HRAMF Smith Family Awards for Excellence in Biomedical Research 1282**

For the past quarter century, the Smith Family Foundation has been supporting groundbreaking medical research through the Smith Family Awards Program for Excellence in Biomedical Research. Its mission is to launch the careers of newly independent biomedical researchers with the ultimate goal of achieving medical breakthroughs. Applications focus on all fields of basic biomedical science and may also be submitted by investigators in physics, chemistry and engineering. Three year awards in the amount of $300,000 ($100,000 per year inclusive of 5% indirect costs) target new faculty who are within two years of their first independent faculty appointment as of July 1 of the application year.
*Requirements:* Applicants, nominated by their institutions, must be full-time faculty at nonprofit academic, medical or research institutions in Massachusetts, at Brown University or at Yale University. Applications from women scientists are encouraged. Applications must be received by 12:00 Noon, U.S. Eastern time of the published deadline date.
*Geographic Focus:* Massachusetts
*Date(s) Application is Due:* Aug 2
*Amount of Grant:* 300,000 USD
*Samples:* David Breslow, Ph., D.Assistant Professor of Molecular, Cellular & Developmental Biology, Yale University, New Haven, Connecticut, $300,000 - Deconstructing the Cellular Antenna: Primary Cilium Disassembly and Control of Cell Growth (2018); Alan Brown, Ph.D., Assistant Professor, Harvard Medical School, Boston, Massachusetts, $300,000 -Structural Basis of Intraflagellar Transport (2018).
*Contact:* Lindsey Carver, Senior Program Officer; (617) 279-2238 or (617) 451-0049; fax (617) 423-4619; lcarver@hria.org
*Internet:* https://hria.org/tmf/smith/
*Sponsor:* Health Resources in Action Medical Foundation
2 Boylston Street, 4th Floor
Boston, MA 02116

**HRAMF Smith Family Foundation Odyssey Award 1283**

The Smith Family Foundation is committed to effecting permanent positive change in the lives of the residents of Greater Boston, particularly individuals and families in economically disadvantaged communities. In the area of health, the Foundation supports the advancement of biomedical research and the expansion of access to high-quality health care at safety net institutions serving low-income individuals and communities of Greater Boston. The new Smith Family Foundation Odyssey Award was created in 2017 to fuel creativity and innovation in junior investigators in the basic sciences. The Award supports the pursuit of high impact ideas to generate breakthroughs and drive new directions in biomedical research. The awards will fund high-risk, high-reward pilot projects solicited from our brightest junior faculty in the region. Junior investigators that are beyond 4 years from their first independent faculty appointment and have not yet received tenure are eligible to apply for this award. There is also an upper eligibility limit at 9 years past their first independent faculty appointment. Those investigators with tenure, or who are beyond the 9 year limit, are ineligible to apply for this award. Awardees of the Smith Family Awards Program for Excellence in Biomedical Research are eligible to apply for the Smith Family Foundation: Odyssey Award. Up to five $300,000 two-year awards will be made to independent junior faculty in nonprofit academic, medical, and research institutions in Massachusetts as well as at Brown University and Yale University.
*Requirements:* All applicants must be an approved tax-exempt non-profit organization as defined by the Internal Revenue Service. Areas of greatest importance to the Foundation are funded on a national basis. However, a good deal of the Foundation's grant making in other program areas is focused in areas of geographic importance to the donors–the northeastern United States (southern New England & the middle Atlantic states), as well as California. Priority will be given to requests that show specific plans for funding beyond the present. Applicants must: hold a Ph.D., M.D., or comparable degree and work at a non-profit academic, medical or research institutions in the United States; and have a full-time independent faculty appointment and their first faculty appointment must have been received no more than five (5) years prior to the award's funding start date.

*Restrictions:* The Foundation will not allow any funds to be earmarked for indirect costs or institutional overhead in cases where the grant relationship was developed independent of that institution's direct involvement. The Foundation generally will not provide grants to the following: organizations not determined to be tax-exempt under section 501(c)3 of the Internal Revenue Code; individuals; general fundraising drives; endowments; government agencies; or organizations that subsist mainly on third party funding and have demonstrated no ability or expended little effort to attract private funding.
*Geographic Focus:* Massachusetts
*Date(s) Application is Due:* Mar 18
*Amount of Grant:* 300,000 USD
*Contact:* Lindsey Carver, Program Officer; (617) 279-2238 or (617) 451-0049; fax (617) 423-4619; lcarver@hria.org
*Internet:* https://hria.org/tmf/smithodyssey/
*Sponsor:* Health Resources in Action Medical Foundation
2 Boylston Street, 4th Floor
Boston, MA 02116

**HRAMF Taub Foundation Grants for MDS Research** 1284
The Henry and Marilyn Taub Foundation works with The Medical Foundation, a division of Health Resources in Action (HRiA), to select the most qualified applicants. HRiA is a nonprofit organization in Boston that advances public health and medical research. The Taub Foundation Grants Program for Myelodysplastic Syndromes (MDS) Research was created to support high-impact, innovative translational research to understand the underlying causes of MDS and to advance its treatment and prevention. The Program specifically focuses on MDS research, exclusive of AML and MPN. Studies focusing on molecular genetics, epigenetics, splicing factors, stem cells, the microenvironment and novel therapeutic targets relevant to MDS are encouraged. Three-year awards of $600,000 ($200,000 per year, inclusive of 10% indirect costs) will be made to independent investigators working in non-profit, non-governmental academic, medical, or research institutions within the United States. Collaborative efforts are encouraged.
*Requirements:* All applicants must hold a faculty appointment at a nonprofit, governmental academic, medical, or research institution in the United States. Applicants do not need to be nominated by their institutions. United States citizenship is not required; visa documentation is not required. Awards are not restricted to investigators currently working in MDS. Applications from investigators in other fields and collaborative efforts are encouraged. The complete initial online application must be received by The Medical Foundation by August 14.
*Restrictions:* Only one application may be submitted per applicant.
*Geographic Focus:* All States
*Date(s) Application is Due:* Oct 15
*Amount of Grant:* 600,000 USD
*Contact:* Lindsey Carver, Program Officer; (617) 279-2238 or (617) 451-0049; fax (617) 423-4619; lcarver@hria.org
*Internet:* https://hria.org/tmf/taub/
*Sponsor:* Health Resources in Action Medical Foundation
2 Boylston Street, 4th Floor
Boston, MA 02116

**HRF Graduate Fellowships** 1285
The Hudson River Foundation (HRF) will award up to six full-time research Fellowships to advanced graduate students conducting research on the Hudson River system. Fellowships awarded to doctoral students will include a stipend of up to $15,000 for one year, with a research budget of up to $1,000. Fellowships awarded to master's level students will include a stipend of up to $11,000 for one year, and research budget of up to $1,000. In special cases, applicants can apply for a 15-month Fellowship in order to extend the proposed project through an additional summer. Application documents are available at the grant website. Applications must be emailed in a single PDF document by 5:00 p.m. on the deadline date. One original copy, postmarked by the deadline date, must be mailed to the address given. Applications are generally due in March. Decisions will be made in June.
*Requirements:* Applicants must be enrolled in an accredited doctoral or master's program, have a thesis advisor and advisory committee (if appropriate to the institution), and have a thesis research plan approved by the student's institution or department. The student's home university will be expected to cover any indirect costs associated with the project, and to be the primary source of support for materials and expenses required to do the research.
*Restrictions:* The Foundation will not accept applications for extensions of existing fellowships. The award is conditional upon a full tuition waiver or reimbursement by the University.
*Geographic Focus:* All States
*Date(s) Application is Due:* Mar 21
*Amount of Grant:* 12,000 - 16,000 USD
*Contact:* Dr. Dennis Suszkowski, Science Director; (212) 483-7667; fax (212) 924-8325; dennis@hudsonriver.org or info@hudsonriver.org
*Internet:* www.hudsonriver.org/graduate_fellow.htm
*Sponsor:* Hudson River Foundation
17 Battery Place, Suite 915
New York, NY 10004

**HRF Hudson River Improvement Grants** 1286
The HRF Improvement Grants were created to support projects that promote the enhancement of public use and enjoyment of the natural, scenic and cultural resources of the Hudson River and its shores, with an emphasis on physical projects that require capital construction, development or improvement. Examples of projects that may be considered for funding include: development or improvement of facilities that increase public physical or visual access to the Hudson River, including but not limited to docks, boats, piers and shore front access points; repair, restoration or creation of habitat; development or improvement and equipping of facilities suitable for Hudson River education programs, such as interpretive centers, marsh boardwalks or waterfront classrooms. The Improvement Fund encourages proposals that focus on environmentally beneficial shoreline enhancement and education projects. Samples of previously funded projects are available at the website. Application guidelines and a the application form are available at the website. Proposals should be assembled electronically and emailed. Additionally, applicants must mail six complete hard copies (one signed) to the contact person.
*Requirements:* Applicants must include a copy of the organization's tax exemption certificate under Section 501(c)3 of the Internal Revenue Code or an indication that the applicant is a subdivision of a government unit.
*Geographic Focus:* New York
*Date(s) Application is Due:* Nov 19
*Amount of Grant:* Up to 10,000 USD
*Samples:* Albany Rowing Center, Albany, New York, $7,200, to continue upgrading floating docks at the City of Albany Corning Preserve public boat launch; Beczak Environmental Education Center, Sleepy Hollow, New York, $5,000, for equipment and furnishings for Scenic Hudson's new environmental center in Kingland Point Park in the Village of Sleepy Hollow; Cary Institute of Ecosystem Studies, Poughkeepsie, New York, $8,000, to create an LED display on the Walkway Over the Hudson of real-time water conditions collected from the Hudson River Environmental Conditions Observing System.
*Contact:* Dr. Dennis Suszkowski, Science Director; (212) 483-7667; fax (212) 924-8325; dennis@hudsonriver.org or info@hudsonriver.org
*Internet:* www.hudsonriver.org/hrif/
*Sponsor:* Hudson River Foundation
17 Battery Place, Suite 915
New York, NY 10004

**HRF Mark B. Bain Graduate Fellowships** 1287
The Foundation awards up to six full-time research Fellowships to advanced graduate students conducting research on the Hudson River system. Doctoral student Fellowships include a stipend of up to $15,000 for one year, and research budget of up to $1,000. Master's student Fellowships include a stipend of up to $11,000 for one year, and a research budget of up to $1,000. Additional information about the Fellowship and application process can be found at the Call for Proposals attachment posted on the HRF call for proposals site.
*Requirements:* The award is conditional upon a full tuition waiver or reimbursement by the University. Applicants must be enrolled in an accredited doctoral or master's program, must have a thesis advisor and advisory committee (if appropriate to the institution), and must have a thesis research plan approved by the student's institution or department. The student's home university will be expected to be the primary source of support for materials and expenses required to do the thesis research and will also be expected to cover any indirect costs associated with the project. In special cases, applicants can apply for a 15-month Fellowship in order to extend the proposed project through an additional summer.
*Geographic Focus:* All States
*Date(s) Application is Due:* Mar 18
*Amount of Grant:* Up to 28,000 USD
*Contact:* Dennis J. Suszkowski, Science Director; (212) 483-7667; fax (212) 924-8325; dennis@hudsonriver.org or info@hudsonriver.org
*Internet:* www.hudsonriver.org/hrfund/
*Sponsor:* Hudson River Foundation
17 Battery Place, Suite 915
New York, NY 10004

**HRF New York City Environmental Grants for Newton Creek** 1288
The New York City Environmental Fund (NYCEF) was formed to encourage active community stewardship of the waterways, shorelines, parklands and open spaces of the City of New York. NYCEF was created to support projects that (a) protect, preserve, improve or restore the environment; (b) facilitate the public's enjoyment of natural resources; or (c) enhance the public's awareness, knowledge, and understanding of ecological principles, natural resources and other environmental issues. The NYCEF Newtown Creek Fund requests proposals for environmental education, public access and stewardship projects that will benefit communities along Newtown Creek, one of New York City's six designated Maritime and Industrial Zones. In accordance with the settlement agreement creating the Fund, priority will be given to projects in and around Greenpoint, Brooklyn, and along Newtown Creek and the East River (both Brooklyn and Queens). NYCEF will provide financial support for qualifying projects that promote the environmental health and revitalization of natural resources and that increase public understanding of, access to and enjoyment of the environment. The NYCEF Newtown Creek Environmental Benefits Program seeks proposals for projects that protect and improve the environmental health of natural areas and the environment; educate people about the value of natural resources, biodiversity, conservation, and environmental protection; enhance access to natural areas and encourage stewardship activities to protect them; promote community involvement and encourage stewardship (personal and community responsibility and guardianship) for natural areas within the Newton Creek target area; and provide a framework for community stewardship and programming related to broader capital projects and improvements. The typical grant size will be between $5,000 and $25,000.
*Requirements:* Any nonprofit group or organization that qualifies as tax-exempt under Section 501(c)3 of the Internal Revenue Code is eligible to apply for support from the NYCEF Newtown Creek Environmental Benefits Program. Additionally, schools and government agencies with qualifying projects may apply. Groups without 501(c)3 certification may, for the purposes of administration of a grant, associate with an organization that does qualify.
*Geographic Focus:* New York

*Amount of Grant:* 5,000 - 25,000 USD
*Contact:* Dr. Dennis Suszkowski, Science Director; (212) 483-7667; fax (212) 924-8325; dennis@hudsonriver.org or info@hudsonriver.org
Lisa Garrison, Program Director; (212) 483-7667; fax (212) 924-8325; Lisa@hudsonriver.org or info@hudsonriver.org
*Internet:* www.hudsonriver.org/nycef/ncf.html
*Sponsor:* Hudson River Foundation
17 Battery Place, Suite 915
New York, NY 10004

**HRF Tibor T. Polgar Fellowships** 1289
The Polgar Fellowships are conducted jointly by the Hudson River Foundation and the New York State Department of Environmental Conservation. The Fellowship provides summer funding, and funds for up to eight undergraduate and graduate students to conduct research on the Hudson River. Researchers gather important information on all aspects of the river, and train students in conducting estuarine studies and public policy research. Polgar Fellowships may be awarded for studies anywhere within the tidal Hudson estuary from New York Harbor to the Federal Dam at Troy, New York, including the four marshes of the National Estuarine Research Reserve (Stockport Flats, the Tivoli Bays, Iona Island Marsh, and Piermont Marsh).
*Requirements:* Graduate and undergraduate students are eligible. Because of the training and educational aspects of this program, each potential Fellow must be sponsored by a primary advisor. The advisor must be willing to commit sufficient time for supervision of the research, and to attend two meetings (orientation and final reports) with their students. Advisors will receive a $500 stipend.
*Geographic Focus:* All States
*Date(s) Application is Due:* Feb 20
*Amount of Grant:* 3,800 - 4,800 USD
*Contact:* Dr. Dennis Suszkowski, Science Director; (212) 483-7667; fax (212) 924-8325; dennis@hudsonriver.org or info@hudsonriver.org
Helen Andreyko, Administrator; (212) 483-7667; fax (212) 924-8325; info@hudsonriver.org
*Internet:* www.hudsonriver.org/polgar.htm
*Sponsor:* Hudson River Foundation
17 Battery Place, Suite 915
New York, NY 10004

**HRK Foundation Health Grants** 1290
The HRK Foundation is defined by quiet leadership and philanthropy. The Board seeks to improve the fabric of society by promoting healthy families and healthy communities. In the area of Health, the Foundation funds programs that strengthen families and promote healthy lives for children, as well as programs that provide access to health and human services. Within the Saint Croix Valley and Ashland and Bayfield Counties in Wisconsin, the board will consider local community-specific health projects. Types of support include: general operating support; seed funding; community development; matching/challenge grants; equipment acquisition; and annual campaigns. Requests are considered two times each year, with deadlines on March 31 and September 15. Most recent awards in this area have ranged from $20,000 to $25,000.
*Requirements:* The foundation makes grants only to qualified IRS 501(c)3 organizations that specifically benefit people in the area surrounding Bayport and Saint Paul, Minnesota, and Ashland and Bayfield counties in Wisconsin. The Foundation supports all families, both traditional and non-traditional.
*Restrictions:* The foundation does not make loans or provide grants to individuals.
*Geographic Focus:* Minnesota, Wisconsin
*Date(s) Application is Due:* Sep 15
*Amount of Grant:* 20,000 - 25,000 USD
*Samples:* University of Minnesota, Minneapolis, Minnesota, $20,000 - support of the Stem Cell Institute; International Institute of Minnesota, Saint Paul, Minnesota, $25,000 - support for the Mental Health Initiative for New Americans; Medical College of Wisconsin, Milwaukee, Wisconsin, $25,000 - general operating support.
*Contact:* Kathleen Fluegel, Foundation Director; (651) 298-0550 or (866) 342-5475; fax (651) 298-0551; info@HRKFoundation.org
*Internet:* www.hrkfoundation.org/purpose.html
*Sponsor:* HRK Foundation
345 Saint Peter Street, Suite 1200
Saint Paul, MN 55102-1639

**HRSA Nurse Education, Practice, Quality and Retention (NEPQR) Grants** 1291
Through its Nurse Education, Practice, Quality and Retention grants program, the Division of Nursing will solicit three-year cooperative agreements that propose to develop and implement innovative career ladder programs that will increase the enrollment, progression, and graduation of Veterans in Bachelor of Science in Nursing programs. This VBSN program supports HRSA's strategic plan to improve access to quality health care and services; strengthen, the nation's healthcare workforce; build healthy communities; and improve health equity. Awarded VBSN applications will complement the collaborative efforts of the Health Resources and Services Administration (HRSA), the Department of Defense (DoD), and the Department of Veteran¿s Affairs (VA) that seek to: reduce barriers that prevent veterans from transitioning into nursing careers; develop BSN career ladder programs targeted to the unique needs of veterans; explore innovative educational models to award academic credit for prior health career experience/training or other relevant military training; improve employment opportunities for veterans through high demand careers training, and address the growing national demand for BSN prepared Registered Nurses. The intermediate program goals are to facilitate the transition of veterans into the field of professional nursing, while building upon skills, knowledge, and training acquired during their military service in order to increase employment opportunities. Sub-goals include increasing and diversifying the health workforce, and ensuring that healthcare providers are trained to provide high quality care that is culturally and linguistically aligned with the communities they will serve. The VBSN project will: provide program participants with the opportunity to receive academic credit for prior military medical training and experience; provide participants with knowledge, skills, and support(s) needed to successfully matriculate through innovative BSN career ladder training programs; and prepare program participants for the National Council Licensing Examination for Registered Nurses (NCLEX- RN).
*Requirements:* Eligible applicants include accredited schools of nursing as defined in section 801(2), of the Public Health Service Act, a health care facility as defined in section 801(11) of the Public Health Service Act, or partnership of such a school and facility.
*Geographic Focus:* All States
*Date(s) Application is Due:* Feb 18
*Contact:* Rebecca Spitzgo; (301) 443-5794 or (888) 275-4772; fax (301) 443-8586
*Internet:* www.grants.gov/view-opportunity.html?oppId=249594
*Sponsor:* Health Resources and Services Administration
5600 Fishers Lane
Rockville, MD 20857

**HSBC Contributions Community Grants** 1292
HSBC believes that banks play a positive role in the economy and society by providing individuals and businesses with the financial services they need to meet their ambitions. Focusing on education and the environment, HSBC builds strong, enduring relationships with a wide range of organizations, providing resources and tools to sustain long-term results. HSBC supports the communities in which it operates both through the involvement of its employees and monetary contributions. On a limited basis, HSBC provides support for initiatives that fall outside of its focus areas of education and the environment. This support is primarily driven by its Community Reinvestment Act activities and employee volunteer efforts.
*Requirements:* HSBC initiates the majority of its grant relationships. Eligible non-profits (those with 501(c)3 tax-deductible status located in areas in which HSBC serves) will be invited to submit their proposal for funding to HSBC via the online HSBC Contribution Application. Proposals must be received by November 1 of each year; all organizations will receive a response in writing. If approved, the correspondence will outline the grant terms along with any requirements for evaluation and reporting. For Education proposals, the goal is to support initiatives that help give young people hope, a positive attitude, opportunities for success and a better future. Providing access to education, particularly for children from disadvantaged backgrounds or for schools in economically deprived areas, is the central aim of HSBC's focus. HSBC Contributions support primary and secondary education programs for children ages 3-25, including environmental education. Additionally, preference will be given to efforts aimed at raising levels of financial and business literacy in the community, especially among young people. Note that general operating grants are not provided to elementary and secondary schools. For Environment proposals, HSBC supports organizations/programs that focus on environmental policy development, public awareness-building, education of teachers or other adults (i.e. over 25), scientific research and projects to improve the sustainability or resilience of eco-systems. An important component in any contribution is the opportunity for HSBC employees to be engaged and have opportunities to volunteer in relevant environmental projects, most importantly where HSBC has a significant local presence.
*Restrictions:* Support is prohibited to oOrganizations outside of the United States and/or organizations that do not hold 501(c)3 tax deductible status under the Internal Revenue Code; fraternal, veteran, labor, or athletic organizations; for-profit student aid or scholarship programs, aside from those already established by the company; political, lobbying or voter registration programs, or those supporting the candidacy of a particular individual; funds to support travel - group or individual; organizations that might in any way pose a conflict with HSBC's corporate values, products, customers or employees; religious organizations that serve a limited constituency; advertising; or, individuals.
*Geographic Focus:* All States
*Date(s) Application is Due:* Nov 1
*Samples:* Asian Americans for Equality Community Development Fund (AAFECDF), New York, for programs in the areas of pre- and post-purchase homeownership counseling, foreclosure prevention, and to support outreach and translation of materials provided to their constituencies. Brooklyn Botanic Garden, Brooklyn, New York, for the Project Green Reach initiative, an award-winning, supplemental science education program reaching approximately 2,000 K-8 students and 80 teachers in 40 of Brooklyn's Title I schools. National Environmental Education Foundation, Washington, D.C., for the Be Water Wise program.
*Contact:* Jan Garner; (224) 880-7128; janice.e.garner@us.hsbc.com
*Internet:* www.us.hsbc.com/1/2/home/about/corporate-sustainability/community-partners
*Sponsor:* HSBC Bank USA
P.O. Box 2013
Buffalo, NY 14240

**ACL/Robert H. Ho Family Foundation Program in Buddist Studies Grants** 1293
The American Council of Learned Societies (ACLS) announces an initiative supporting research and teaching in Buddhist studies, funded by the Robert H.N. Ho Family Foundation. In cooperation with the Foundation, ACLS offers an integrated set of fellowship and grant competitions supporting work that will expand the understanding and interpretation of Buddhist thought in scholarship and society, strengthen international networks of Buddhist studies, and increase the visibility of innovative currents in those studies. Dissertation Fellowships in Buddhist Studies provide one-year stipends to PhD candidates for full time preparation of dissertations. The fellowship period may be used for fieldwork, archival research, analysis of findings, or for writing after research is complete. A

stipend of $30,000 is given. Completed applications should be submitted through the ACLS online fellowship and grant administration system by the annual deadline of November 13.
*Requirements:* This is a global competition. There are no restrictions as to the location of work proposed, or the citizenship/residence of applicants. Although the final products of research can be in any language, all applications must be submitted in English.
*Geographic Focus:* All States, All Countries
*Date(s) Application is Due:* Nov 13
*Amount of Grant:* 30,000 USD
*Contact:* Valerie Popp, Senior Program Officer; (212) 697-1505, ext. 153; fax (212) 949-8058; vpopp@acls.org or ahp@acls.org
*Internet:* https://www.acls.org/Competitions-and-Deadlines/The-Robert-H-N-Ho-Family-Foundation-Program-in-Buddhist-Studies
*Sponsor:* American Council of Learned Societies
633 Third Avenue, 8th Floor
New York, NY 10017-6795

**Hubbard Foundation Grants** 1294
The Hubbard Foundation was established in California in December of 2006 by Dr. David R. and Arlene P. Hubbard. The Foundation investigates the vascular component of neurological disorders, such as Chronic Cerebrospinal Venous Insufficiency (CCSVI) in Multiple Sclerosis, using functional MRI (fMRI), as well as MRV, flow-quantification, perfusion, and other measures. Our mission is to develop tests for earlier recognition of brain abnormalities, which may lead to neurological diseases. Applications for research in these areas should be submitted by the annual deadline of December 31. Though an initial approach should be by letter, a more formal application is required.
*Requirements:* Any 501(c)3 involved in the research of Chronic Cerebrospinal Venous Insufficiency (CCSVI) in Multiple Sclerosis, using functional MRI (fMRI), as well as MRV, flow-quantification, perfusion, and other measures is eligible to apply.
*Restrictions:* No grants are given to to individuals.
*Geographic Focus:* All States
*Date(s) Application is Due:* Dec 31
*Contact:* David R. Hubbard, Chairman; (858) 668-3380 or (858) 212-8074; davidhubbardmd@gmail.com
*Internet:* www.hubbardfoundation.org/
*Sponsor:* Hubbard Foundation
14168 Donart Drive
Poway, CA 92064-3416

**Huisking Foundation Grants** 1295
In the defense of freedom, Francis Robert Huisking paid the ultimate sacrifice during War II. As co-captain of a B-24 bomber, he and his crew were lost when the bomber went down over Italy in 1944. To honor his sacrifice and to perpetuate and preserve as a living memory the sweet nature, the sterling qualities, the grand character, and the noble attributes he displayed during the short time he was with us, his parents Charles and Catherine Huisking founded the Frank R. Huisking Foundation on October 24, 1946. In addition to Charles and Catherine, the original founding Directors and Officers included five of Francis' seven brothers and sisters; Charles Jr., Evelyn, William, Edward, and Richard. These individuals managed the growth, investments, contributions, and distributions of the Foundation for the next twenty plus years. In 1971, the name of the Foundation was officially changed to the Huisking Foundation. Today, the Foundation's primary fields of interest include: animal welfare; Catholicism; communication media; elementary and secondary education; higher education; historic preservation; hospital care; human services; museums; natural resources; nonprofits; and performing arts. The Foundation is small, and dedicated to the areas outlined above. To assure that funds distributed meet these principles, projects are reviewed and recommended by specific Directors who have knowledge of the project or program and thus supports its goals. This process results in the Board having a greater degree of confidence and assurance that the support being provided clearly meets the principles and intent as set out to honor Frank R. Huisking and his family.
*Geographic Focus:* All States
*Amount of Grant:* 200 - 60,000 USD
*Contact:* Frank Huisking, Treasurer; (203) 426-8618; wwh@huiskingfoundation.org
*Internet:* www.huiskingfoundation.org/
*Sponsor:* Huisking Foundation
291 Peddlers Road
Guilford, CT 06437-2324

**Human Source Foundation Grants** 1296
The Human Source Foundation is organized to support charitable and educational initiatives that improve the human condition. The Foundation provides assistance and financial support for programs involved with education, human services, and youth development. Giving is restricted to the state of Texas, primarily in Denton and Fort Worth counties. There are no specific deadlines or application forms, and applicants should begin by forwarding a letter of request.
*Requirements:* Tax-exempt 501(c)3 organizations serving Texas, primarily Denton and Fort Worth counties, are eligible to apply.
*Geographic Focus:* Texas
*Samples:* Botanical Resource of Texas, Fort Worth, Texas, $4,500; Meals on Wheels, Fort Worth, Texas, $7,700.
*Contact:* Mary G. Palko, President; (817) 926-2799; fax (817) 926-5202; mary@ftw.com
*Sponsor:* Human Source Foundation
2409 Winton Terrace West, P.O. Box 100423
Fort Worth, TX 76185-0423

**Huntington Babara Thom Postdoctoral Fellowships** 1297
The Huntington is an independent research center with holdings in British and American history, literature, art history, and the history of science and medicine. Babara Thom Postdoctoral Fellowships are designed to support non-tenured faculty members who are revising a manuscript for publication. The amount of each award is $50,000 for a nine- to twelve-month tenure. Applications should be received by November 15.
*Requirements:* Applicants must: be pursuing scholarship in a field appropriate to the Huntington's collections; and be non-tenured faculty. Recipients of all fellowships are expected to be in continuous residence at the Huntington and to participate in and make a contribution to its intellectual life.
*Geographic Focus:* All States
*Date(s) Application is Due:* Nov 15
*Amount of Grant:* 50,000 USD
*Contact:* Steve Hindle, Director of Research; (626) 405-3411 or (626) 405-2100; fax (626) 449-5703; shindle@huntington.org or fellowships@huntington.org
*Internet:* huntington.org/WebAssets/Templates/content.aspx?id=566#long
*Sponsor:* Huntington Library, Art Collections, and Botanical Gardens
1151 Oxford Road
San Marino, CA 91108

**Huntington Clark-Huntington Joint Bibliographical Fellowship** 1298
The Huntington is an independent research center with holdings in British and American history, literature, art history, and the history of science and medicine. Sponsored jointly by the Clark and the Huntington Libraries, this two-month fellowship provides support for bibliographical research in early modern British literature and history as well as other areas where the two libraries have common strengths; eligible projects include textual scholarship, analytical/descriptive bibliography, history of printing and/or publishers, and related fields. The Fellowship has a two-month duration, with one month at the William Andrews Clark Library and one month at the Huntington. The amount of the award is $5,500 Applications should be received by November 15.
*Requirements:* Applicant scholars must have the PhD or equivalent degree.
*Geographic Focus:* All States
*Date(s) Application is Due:* Nov 15
*Amount of Grant:* 5,500 USD
*Contact:* Steve Hindle, Director of Research; (626) 405-3411 or (626) 405-2100; fax (626) 449-5703; shindle@huntington.org or fellowships@huntington.org
*Internet:* huntington.org/WebAssets/Templates/content.aspx?id=566#short
*Sponsor:* Huntington Library, Art Collections, and Botanical Gardens
1151 Oxford Road
San Marino, CA 91108

**Huntington Corpus Christi College/Lincoln College/New College/Trinity Hall Travel Stipend** 1299
The Huntington is an independent research center with holdings in British and American history, literature, art history, and the history of science and medicine. Accommodation and hospitality is provided by the college, although the timing of the fellowship may be subject to the availability of housing options and to the rhythms of the academic year. The Huntington will reimburse the fellow for economy round-trip airfare before going to England. The fellow must provide a written report on his or her experience. Applications must be received by November 15.
*Requirements:* Applicant scholars must have the PhD or equivalent degree or be doctoral candidates at the dissertation stage. Recipients of all fellowships are expected to be in continuous residence at the Huntington and to participate in and make a contribution to its intellectual life.
*Geographic Focus:* All States
*Date(s) Application is Due:* Nov 15
*Contact:* Steve Hindle, Director of Research; (626) 405-3411 or (626) 405-2100; fax (626) 449-5703; shindle@huntington.org or fellowships@huntington.org
*Internet:* huntington.org/WebAssets/Templates/content.aspx?id=566#short
*Sponsor:* Huntington Library, Art Collections, and Botanical Gardens
1151 Oxford Road
San Marino, CA 91108

**Huntington Dana and David Dornsife Fellowships** 1300
The Huntington Library and Art Gallery is an independent research center with holdings in British and American history, literature, art history, the history of science, and photography. The Dana and David Dornsife fellowships support researchers pursuing scholarship in a field appropriate to the Huntington's collections. Preference will be given to scholars who have not held major awards in the three years preceding the year of the award. Recipients are expected to be in continuous residence at the Huntington during the nine- to 12-month fellowship tenure and to participate in the intellectual life of the center. There is no specific application form. Application consists of cover sheet, description of the project, curriculum vita, and three letters of recommendation that must be received by the application deadline of November 15. The award is equal to $50,000.
*Requirements:* Application is open to individuals with the PhD or equivalent degree who have received their degree by June of the previous year.
*Geographic Focus:* All States
*Date(s) Application is Due:* Nov 15
*Amount of Grant:* 50,000 USD
*Contact:* Steve Hindle, Director of Research; (626) 405-3411 or (626) 405-2100; fax (626) 449-5703; shindle@huntington.org or fellowships@huntington.org
*Internet:* huntington.org/WebAssets/Templates/content.aspx?id=566#long
*Sponsor:* Huntington Library, Art Collections, and Botanical Gardens
1151 Oxford Road
San Marino, CA 91108

**Huntington Dibner History of Science Long-Term Fellowships** 1301
The Huntington is an independent research center with holdings in British and American history, literature, art history, and the history of science and medicine. The Dibner History of Science Program offers historians of science and technology the opportunity to study in the Burndy Library and the other history of science and technology resources at the Huntington. Long-term fellowship awards are of a nine- to twelve month duration, and are in the amount of $50,000. Applicants can be conducting research or already be at the writing stage and need reference materials only. The application must be received by November 15.
*Requirements:* Eligible candidates must have a PhD or equivalent as of June of the previous year.
*Geographic Focus:* All States
*Date(s) Application is Due:* Nov 15
*Amount of Grant:* 50,000 USD
*Contact:* Steve Hindle, Director of Research; (626) 405-3411 or (626) 405-2100; fax (626) 449-5703; shindle@huntington.org or fellowships@huntington.org
*Internet:* huntington.org/WebAssets/Templates/content.aspx?id=566#dibner
*Sponsor:* Huntington Library, Art Collections, and Botanical Gardens
1151 Oxford Road
San Marino, CA 91108

**Huntington Dibner History of Science Short-Term Fellowships** 1302
The Huntington is an independent research center with holdings in British and American history, literature, art history, and the history of science and medicine. The Dibner program in the History of Science offers historians of science and technology the opportunity to study in the Burndy Library and to work in the other resources for the history of science and technology at the Huntington. Short-term awards are $3,000 per month for a one- to five-month period. Applications must be received by November 15.
*Requirements:* Applicants for short-term awards must: have a PhD or equivalent; or be a doctoral candidate at the dissertation stage.
*Geographic Focus:* All States
*Date(s) Application is Due:* Nov 15
*Amount of Grant:* 3,000 - 15,000 USD
*Contact:* Steve Hindle, Director of Research; (626) 405-3411 or (626) 405-2100; fax (626) 449-5703; shindle@huntington.org or fellowships@huntington.org
*Internet:* huntington.org/WebAssets/Templates/content.aspx?id=566#dibner
*Sponsor:* Huntington Library, Art Collections, and Botanical Gardens
1151 Oxford Road
San Marino, CA 91108

**Huntington Linacre College Fellowship Travel Stipend** 1303
The Huntington is an independent research center with holdings in British and American history, literature, art history, and the history of science and medicine. A stipend of $3,000 is provided by the Huntington to the recipient of the Linacre College Fellowship before traveling to England, along with reimbursement for economy round-trip airfare. Accommodation is provided by the college with the stipulation that the fellowship must be taken up in July; the fellow is responsible for paying for the accommodation. The fellow must provide a written report on his or her experience. Applications must be received by November 15.
*Requirements:* Applicant scholars must have the PhD or equivalent degree or be doctoral candidates at the dissertation stage. Recipients of all fellowships are expected to be in continuous residence at the Huntington and to participate in and make a contribution to its intellectual life.
*Geographic Focus:* All States
*Date(s) Application is Due:* Nov 15
*Amount of Grant:* 3,000 USD
*Contact:* Steve Hindle, Director of Research; (626) 405-3411 or (626) 405-2100; fax (626) 449-5703; shindle@huntington.org or fellowships@huntington.org
*Internet:* huntington.org/WebAssets/Templates/content.aspx?id=566#short
*Sponsor:* Huntington Library, Art Collections, and Botanical Gardens
1151 Oxford Road
San Marino, CA 91108

**Huntington Mellon Fellowships** 1304
The Huntington is an independent research center with holdings in British and American history, literature, art history, and the history of science and medicine. This Mellon Fellowship is designed to support applicants pursuing scholarship in a field appropriate to the Huntington's collections. The amount of each award is $50,000 for a nine- to twelve-month tenure. Applications must be received by November 15.
*Requirements:* Applicants must: be pursuing scholarship in a field appropriate to the Huntington's collections; and have received the PhD in June of the application year. Recipients of all fellowships are expected to be in continuous residence at the Huntington and to participate in and make a contribution to its intellectual life.
*Geographic Focus:* All States
*Date(s) Application is Due:* Nov 15
*Amount of Grant:* 50,000 USD
*Contact:* Steve Hindle, Director of Research; (626) 405-3411 or (626) 405-2100; fax (626) 449-5703; shindle@huntington.org or fellowships@huntington.org
*Internet:* huntington.org/WebAssets/Templates/content.aspx?id=566#long
*Sponsor:* Huntington Library, Art Collections, and Botanical Gardens
1151 Oxford Road
San Marino, CA 91108

**Huntington Molina Fellowship in the History of Medicine and Allied Sciences** 1305
The Huntington is an independent research center with holdings in British and American history, literature, art history, and the history of science and medicine. The Molina Fellowship in the History of Medicine and Allied Sciences provides a $50,000 award to those PhD's pursuing scholarship in the history of medicine and related sciences, including public health. The tenure duration of the Fellowship is nine to twelve months. The application must be received by November 15.
*Requirements:* Applicant scholars must have the PhD or equivalent degree or be doctoral candidates at the dissertation stage. Recipients of all fellowships are expected to be in continuous residence at the Huntington and to participate in and make a contribution to its intellectual life.
*Geographic Focus:* All States
*Date(s) Application is Due:* Nov 15
*Amount of Grant:* 50,000 USD
*Contact:* Steve Hindle, Director of Research; (626) 405-3411 or (626) 405-2100; fax (626) 449-5703; shindle@huntington.org or fellowships@huntington.org
*Internet:* huntington.org/WebAssets/Templates/content.aspx?id=566#long
*Sponsor:* Huntington Library, Art Collections, and Botanical Gardens
1151 Oxford Road
San Marino, CA 91108

**Huntington National Endowment for the Humanities Fellowships** 1306
The Huntington is an independent research center with holdings in British and American history, literature, art history, and the history of science, will award a minimum of three fellowships funded by National Endowment for the Humanities. These fellowships are designed to permit scholars to spend from four to twelve months at the Huntington. Fellowships may be combined with sabbaticals or other stipendiary support. Preference will be given to scholars who have not held major fellowships during the three years preceding the year for which the center is awarding the fellowships. Fellows will be able to pursue their own research and writing while participating in the intellectual life of the Huntington. The amount of each fellowship is $50,000, and the application must be received by November 15.
*Requirements:* Applicants must be established scholars at the postdoctoral level or its equivalent, must be U.S. citizens or foreign nationals who have resided in the United States for at least three years, and must be pursuing scholarship in a field appropriate to the Huntington's collections.
*Geographic Focus:* All States
*Date(s) Application is Due:* Nov 15
*Amount of Grant:* 50,000 USD
*Contact:* Steve Hindle, Director of Research; (626) 405-3411 or (626) 405-2100; fax (626) 449-5703; shindle@huntington.org or fellowships@huntington.org
*Internet:* huntington.org/WebAssets/Templates/content.aspx?id=566#long
*Sponsor:* Huntington Library, Art Collections, and Botanical Gardens
1151 Oxford Road
San Marino, CA 91108

**Huntington Short-Term Fellowships** 1307
The Huntington is an independent research center with holdings in British and American history, literature, art history, and the history of science and medicine. Within this category, The Huntington awards a number of specialized fellowships, including: Francis Bacon Foundation Fellowships in Renaissance England; Reese Fellowship in American Bibliography and the History of the Book in the Americas; Trent R. Dames Civil Engineering History Fellowship; Francis J. Weber Research Fellowship in Roman Catholic History; Cheng Fellowship in the History of the Asian American Experience in the U.S; and Christopher Isherwood Foundation Fellowships. The amount of each award is $3,000 per month for a one- to five-month tenure. Applications must be received by November 15.
*Requirements:* Applicant scholars must have the PhD or equivalent degree or be doctoral candidates at the dissertation stage. Recipients of all fellowships are expected to be in continuous residence at the Huntington and to participate in and make a contribution to its intellectual life.
*Geographic Focus:* All States
*Date(s) Application is Due:* Nov 15
*Amount of Grant:* 3,000 - 15,000 USD
*Contact:* Steve Hindle, Director of Research; (626) 405-3411 or (626) 405-2100; fax (626) 449-5703; shindle@huntington.org or fellowships@huntington.org
*Internet:* huntington.org/WebAssets/Templates/content.aspx?id=566#short
*Sponsor:* Huntington Library, Art Collections, and Botanical Gardens
1151 Oxford Road
San Marino, CA 91108

**Huntington Travel Grants** 1308
The Huntington is an independent research center with holdings in British and American history, literature, art history, and the history of science and medicine. Recipients of the travel grants must be conducting research in a library or archive in Great Britain in any of the fields in which the Huntington collections are strong. The Huntington will reimburse the grantee for economy round-trip airfare before the trip. A stipend of $3,000 will be paid after the grantee submits a detailed report on the research conducted. The travel grants can be taken up as early as June 1, and no later than June 30 of the following year. Applications must be received by November 15.
*Requirements:* Applicant scholars must have the PhD or equivalent degree or be doctoral candidates at the dissertation stage. Recipients of all fellowships are expected to be in continuous residence at the Huntington and to participate in and make a contribution to its intellectual life.

*Geographic Focus:* All States
*Date(s) Application is Due:* Nov 15
*Amount of Grant:* 3,000 USD
*Contact:* Steve Hindle, Director of Research; (626) 405-3411 or (626) 405-2100; fax (626) 449-5703; shindle@huntington.org or fellowships@huntington.org
*Internet:* huntington.org/WebAssets/Templates/content.aspx?id=566#short
*Sponsor:* Huntington Library, Art Collections, and Botanical Gardens
1151 Oxford Road
San Marino, CA 91108

**Huntington Travel Grants and Exchange Fellowships for Study in Great Britain 1309**
The Huntington Library and Art Gallery is an independent research center with holdings in British and American history, literature, art history, the history of science, and photography. The Huntington offers several travel grants in any of the fields in which the Huntington collections are strong and where the research will be carried out in Great Britain. The program also offers exchange fellowships with Corpus Christi, Linacre, Lincoln, and New Colleges, Oxford; and with Trinity Hall, Cambridge. The duration is for one month, and applications should be received by November 15.
*Requirements:* Eligibility is restricted to PhD or equivalent candidates or doctoral candidate at the dissertation stage.
*Geographic Focus:* All States
*Date(s) Application is Due:* Nov 15
*Contact:* Dr. Steve Hindle, Director of Research; (626) 405-3411 or (626) 405-2100; fax (626) 449-5703; shindle@huntington.org or fellowships@huntington.org
*Internet:* huntington.org/WebAssets/Templates/content.aspx?id=566#short
*Sponsor:* Huntington Library, Art Collections, and Botanical Gardens
1151 Oxford Road
San Marino, CA 91108

**Hutchinson Community Foundation Grants 1310**
The Hutchinson Community Foundation provides funding in the counties of Hutchinson and, Reno, Kansas, inluding the immediate area surrounding area. Areas of interest include, but are not, limited to: community needs in areas such as early childhood, youth, education, arts and culutre, and neighborhood development.
*Requirements:* Nonprofit organizations in Kansas are eligible to apply. Special purpose units of government can apply for support of innovative projects located in Reno County.
*Restrictions:* Grant proposals from individuals or non-qualifying organizations will not be considered.
*Geographic Focus:* Kansas
*Contact:* Audrey Abbott Patterson, President; (620) 663-5293; fax (620) 663-9277; aubrey@hutchcf.org
*Internet:* www.hutchcf.org/default.asp?sm=2&si=1
*Sponsor:* Hutchinson Community Foundation
One North Main, Suite 501, P.O. Box 298
Hutchinson, KS 67504-0298

**Hydrogen Student Design Contest 1311**
The annual Hydrogen Student Design Contest challenges university students to design hydrogen energy applications for real-world use. Established in 2004 by the Hydrogen Education Foundation, the Contest showcases the talents of students in many disciplines, including engineering, architecture, marketing, and entrepreneurship. Although the formal rules and guidelines are still under development, this Contest plans to address three key areas: 1) Hydrogen Production—Students will develop a strategy for supplying hydrogen for the community. At least 33% of the community's hydrogen must be produced from renewable sources. In their plans, students must include at least one hydrogen energy station that co-produces power, heat, and hydrogen (also known as a CHHP system). 2) Early Markets—Students will conduct a market analysis to determine near-term customers in the community for their hydrogen supply. Early markets for hydrogen include back-up power, portable power, and materials handling vehicles. Students will identify at least five businesses or other organizations in the community that could be customers for hydrogen in these early markets. 3) Fueling Stations—Students will plan and design a cluster of hydrogen fueling stations for vehicles in the community. Teams will be required to create a plan for the cluster that identifies the number of fueling stations needed to meet passenger vehicle demand; the approximate locations for each station; the station type (retail, private fleet, mixed use, etc.); and total cost estimates for deploying the plan. Students will be required to design the main components of three fueling stations. The Grand Prize Winner will receive the opportunity to present their design in a Keynote Presentation slot in the NHA Annual Hydrogen Conference and Expo, a stipend of up to $5,000 to cover airfare, meals, hotel accommodations, NHA conference registrations, and incidental trip expenses for up to eight team members and their faculty representative, and extensive media recognition. Honorable Mentions will receive the opportunity to give poster presentations of their design at the NHA Annual Hydrogen Conference and Expo, complimentary hotel rooms and registrations to the NHA Conference for up to four team members and their faculty representative.
*Requirements:* The contest is open to current college and university undergraduate and graduate students in the U.S. or abroad. Team members must be enrolled in a college or university at the time of the Contest but do not have to be enrolled full-time. Given the multi-disciplinary nature of the competition, teams are encouraged to include members from any field of study relevant to the team's design. Common disciplines include: engineering, architecture/planning, industrial design, economics, business, environmental science, policy, chemistry, marketing, and education. Registration must be completed by January 22; Entries are due March 24.
*Geographic Focus:* All States
*Date(s) Application is Due:* Jan 22; Mar 24
*Amount of Grant:* 5,000 USD
*Samples:* 2009 Grand Prize: University of Waterloo - Waterloo, Canada; 2009 Honorable Mentions: Wayne State University - Detroit, Michigan.
*Contact:* Kyle Gibeault, Contest Coordinator; (202) 223-5547; fax (202) 223-5537; gibeaultk@hydrogenassociation.org
*Internet:* www.hydrogencontest.org/contest.asp
*Sponsor:* Hydrogen Education Foundation
1211 Connecticut Avenue NW, Suite 600
Washington, D.C. 20036-2701

**IBRO-PERC InEurope Short Stay Grants 1312**
As a result of the successful launch of the FENS/IBRO Program of European Neuroscience Schools, IBRO's Central and Eastern Europe and Western Europe regional committees decided that the collaboration would be further strengthened by the activation of inter-regional networking and mobility of young scientists. InEurope Short Stay Grants fund visits to European institutions by researchers living and working within Europe, so that they can acquire new methods or specific techniques that are necessary for their work. The aim of the programme is to increase intra-European mobility of young researchers by providing grants for short, goal-directed exchanges within European laboratories. Exchanges are limited to no more than four weeks. Funding can be used to cover travel and local expenses. Applications deadlines are in March, June, September and December, and must be submitted through the IBRO online application process. Applicants who are not selected in one round of evaluations will remain in competition for the next round.
*Requirements:* Applicants must be Ph.D. students or postdoctoral Fellows, 35 years of age or less.
*Restrictions:* emailed or hard copy applications are not acceptable.
*Geographic Focus:* Albania, Andorra, Armenia, Austria, Azerbaijan, Belarus, Belgium, Bosnia & Herzegovina, Bulgaria, Croatia, Cyprus, Czech Republic, Denmark, Estonia, Finland, France, Germany, Greece, Hungary, Iceland, Ireland, Italy, Latvia, Liechtenstein, Lithuania, Luxembourg, Macedonia, Malta, Moldova, Monaco, Montenegro, Netherlands, Norway, Poland, Portugal, Romania, Russia, San Marino, Serbia, Slovakia, Slovenia, Spain, Sweden, Switzerland, Turkey, Ukraine, United Kingdom, Vatican City
*Amount of Grant:* Up to 3,000 EUR
*Contact:* Marta Hallak, Senior Director of Grants; 33 (0) 1 46 47 92 92; fax 33 (0) 1 46 47 42 50; mhallak@mail.fcq.unc.edu.ar
*Internet:* ibro.info/professional-development/funding-programmes/regional-grants/europe-grants/
*Sponsor:* International Brain Research Organization
255 rue Saint-Honore
Paris, 75001 France, Metropolitan

**IBRO-PERC Support for European Workshops, Symposia and Meetings 1313**
With the aim to foster cooperation among regional young neuroscientists, IBRO's Western Europe Regional Committee (WERC) offers partial funding of courses, workshops, symposia, and meetings on important topics in neuroscience organized in European countries. Preference will be given to activities that include younger scientists and offer training for scientists from countries with limited resources for research or teaching. Applications must be submitted through the IBRO online application system. emailed or hard copy applications are not acceptable. Applications are accepted until February 15 each year, and awards range up to a maximum of 4,000 Euro.
*Geographic Focus:* Austria, Belgium, Denmark, Finland, France, Germany, Greece, Iceland, Ireland, Italy, Luxembourg, Netherlands, Norway, Portugal, Spain, Sweden, Switzerland, United Kingdom
*Date(s) Application is Due:* Feb 15
*Amount of Grant:* Up to 4,000 EUR
*Contact:* Jochen Pflueger, Chairperson; 33 (0) 1 46 47 92 92; fax 33 (0) 1 46 47 42 50; pflueger@neurobiologie.fu-berlin.de
*Internet:* ibro.info/professional-development/funding-programmes/regional-grants/europe-grants/
*Sponsor:* International Brain Research Organization
255 rue Saint-Honore
Paris, 75001 France, Metropolitan

**IBRO-PERC Support for Site Lectures 1314**
With the aim to facilitate contact between young students and high profile neuroscientists, experience their enthusiasm for brain research and gain first-hand knowledge on modern topics in neuroscience, IBRO-PERC supports the organization of special lectures. These are two one- to two-day long conferences in developing countries, in cooperation with research centres or universities, in which a renowned European neuroscientist lectures about his/her topic of expertise. They carry the title IBRO-PERC Neuroscience Lectures. Applications are to be submitted only by email. The annual application deadline is February 15.
*Requirements:* The application must include in a single document: title of the lecture; dates and venue; applicant's curriculum vitae; and preliminary program and budget.
*Geographic Focus:* Austria, Belgium, Denmark, Finland, France, Germany, Greece, Iceland, Ireland, Italy, Luxembourg, Netherlands, Norway, Portugal, Spain, Sweden, Switzerland, United Kingdom
*Date(s) Application is Due:* Feb 15
*Contact:* Jochen Pflueger, Chairperson; 33 (0) 1 46 47 92 92; fax 33 (0) 1 46 47 42 50; pflueger@neurobiologie.fu-berlin.de
*Internet:* ibro.info/professional-development/funding-programmes/regional-grants/europe-grants/
*Sponsor:* International Brain Research Organization
255 rue Saint-Honore
Paris, 75001 France, Metropolitan

**IBRO/SfN International Travel Grants** 1315

The Society for Neuroscience offers travel fellowships to their annual meeting for neuroscientists under the age of 35 who are currently working and living in a resource-restricted country – those defined by the World Bank as low, lower-middle, or upper-middle income. Registration fees are waived for successful candidates attending the meeting. Awardees will be notified prior to the SfN abstract deadline of May 15 of the current year as to whether they have been selected for the SfN travel grant. However, this award is contingent upon the awardee's abstract being accepted by SfN and proof of attendance at the meeting (in addition to the requirements). Candidates must apply through the IBRO online application system.

*Requirements:* Candidates must be a citizen of and reside in a country that the World Bank has classified low, lower-middle, or upper-middle income. The applicant should be the first author of an abstract to be presented at the SfN annual meeting. A copy of the abstract submitted to SfN for a poster or a platform presentation is required.

*Restrictions:* Candidates from the U.S. and Canada are not eligible to apply.

*Geographic Focus:* All Countries

*Date(s) Application is Due:* Feb 1

*Amount of Grant:* Up to 2,000 EUR

*Contact:* Marta Hallak, Senior Director of Grants; 33 (0) 1 46 47 92 92; fax 33 (0) 1 46 47 42 50; mhallak@mail.fcq.unc.edu.ar

*Internet:* ibro.info/professional-development/funding-programmes/travel-grants/

*Sponsor:* International Brain Research Organization
255 rue Saint-Honore
Paris, 75001 France, Metropolitan

**IBRO Asia Regional APRC Exchange Fellowships** 1316

The IBRO-APRC Exchange Fellowship aims to allow high-quality junior scientists at the level of postdoctoral fellow or junior faculty (under the age of 45) from diverse geographic and scientific areas to broaden the scope of their training in neuroscience by working four to six months abroad in established laboratories. Senior Ph.D. students with good publication records and strong justification for the exchange program will also be considered. Both the applicant and the host laboratory must be within the Asia Pacific Region. Up to $8,500 per Fellowship is available, although funding may vary depending on the country in which the candidate studies. The Fellowship is used along with matching funds from the host laboratory to defray the cost of travel and subsidize living costs. Applications must be submit through the IBRO online application system. emailed or hard copy applications are not acceptable.

*Requirements:* Selection criteria include academic credentials of the applicant and the host, as well as the quality of the research proposal with a structured research plan. Priority is given to applicants from less developed and less well-funded countries. Strong justification must be made by the applicant that he/she will return home after the exchange, bringing with them new knowledge and skills to advance neuroscience in their home countries. A firm commitment from his/her parent institution to this effect will be viewed favorably by the selection committee.

*Geographic Focus:* Afghanistan, Bahrain, Bangladesh, Bhutan, Brunei, Cambodia, China, East Timor, India, Indonesia, Iran, Iraq, Israel, Japan, Jordan, Kazakhstan, Kuwait, Kyrgyzstan, Laos, Lebanon, Malaysia, Maldives, Mongolia, Myanmar (Burma), Nepal, North Korea, Oman, Pakistan, Philippines, Qatar, Russia, Saudi Arabia, Singapore, South Korea, Sri Lanka, Syria, Taiwan, Tajikistan, Thailand, Turkey, Turkmenistan, United Arab Emirates, Uzbekistan, Vietnam, Yemen

*Date(s) Application is Due:* Oct 15

*Amount of Grant:* Up to 8,500 EUR

*Contact:* Marta Hallak, Senior Director of Grants; 33 (0) 1 46 47 92 92; fax 33 (0) 1 46 47 42 50; mhallak@mail.fcq.unc.edu.ar

*Internet:* ibro.info/professional-development/funding-programmes/regional-grants/asiapacific-grants/

*Sponsor:* International Brain Research Organization
255 rue Saint-Honore
Paris, 75001 France, Metropolitan

**IBRO Asia Regional APRC Travel Grants** 1317

The IBRO APRC Travel Grants support young researchers and graduate students working in Asian/Pacific countries to attend and make presentations at international APRC member society meetings, held in English, within the region. Applications from countries of low, lower-middle and upper-middle income in the region (according to The World Bank) are encouraged. Awards range from 1000 to 1500 Euro.

*Requirements:* Graduate students and researchers who have obtained a PhD within the last five years are eligible to apply. For countries in which it is extremely difficult to obtain support for trips to foreign meetings, excellent applications by those aged under 40 are also considered.

*Geographic Focus:* Afghanistan, Bahrain, Bangladesh, Bhutan, Brunei, Cambodia, China, East Timor, India, Indonesia, Iran, Iraq, Israel, Japan, Jordan, Kazakhstan, Kuwait, Kyrgyzstan, Laos, Lebanon, Malaysia, Maldives, Mongolia, Myanmar (Burma), Nepal, North Korea, Oman, Pakistan, Philippines, Qatar, Russia, Saudi Arabia, Singapore, South Korea, Sri Lanka, Syria, Taiwan, Tajikistan, Thailand, Turkey, Turkmenistan, United Arab Emirates, Uzbekistan, Vietnam, Yemen

*Amount of Grant:* 1,000 - 1,500 EUR

*Contact:* Marta Hallak, Senior Director of Grants; 33 (0) 1 46 47 92 92; fax 33 (0) 1 46 47 42 50; mhallak@mail.fcq.unc.edu.ar

*Internet:* ibro.info/professional-development/funding-programmes/regional-grants/asiapacific-grants/

*Sponsor:* International Brain Research Organization
255 rue Saint-Honore
Paris, 75001 France, Metropolitan

**IBRO International Travel Grants** 1318

The IBRO International Travel Grants aim to foster neuroscience research especially in less well-funded countries by providing support to high quality neuroscientists from diverse geographic and scientific areas (US/Canada Region excluded) who wish to participate at international neuroscience meetings. Priority is given to those who have not obtained funding from this programme within the past three years. Applications should be submitted by March 4 for travel to conferences occurring from July to December of that year. Applications should be submitted by September 1 for travel occurring from January to June of the following year. Candidates must apply through the IBRO online application system.

*Restrictions:* Applicants cannot apply for more than one category of funding within travel grants.

*Geographic Focus:* All Countries

*Date(s) Application is Due:* Mar 4; Sep 1

*Amount of Grant:* Up to 1,500 EUR

*Contact:* Marta Hallak, Senior Director of Grants; 33 (0) 1 46 47 92 92; fax 33 (0) 1 46 47 42 50; mhallak@mail.fcq.unc.edu.ar

*Internet:* ibro.info/professional-development/funding-programmes/travel-grants/

*Sponsor:* International Brain Research Organization
255 rue Saint-Honore
Paris, 75001
France, Metropolitan

**IBRO Latin America Regional Diversity Grants** 1319

IBRO Latin America Regional committee awards individual LARC member societies with diversity grants for activities or events that promote diversity and/or the participation of women in neuroscience. The successful society is required to send a program, a report on the proceedings and the numbers and proportions of female plenary lecturers, symposium speakers, session chairs and participants in the meeting after the event has ended. Typically, awards range up to a maximum of 1,000 Euro. The annual deadline for applications is March 31.

*Restrictions:* Applications by email will not be accepted.

*Geographic Focus:* Puerto Rico, Antigua & Barbuda, Argentina, Bahamas, Barbados, Bolivia, Brazil, Chile, Colombia, Costa Rica, Cuba, Dominica, Dominican Republic, Ecuador, El Salvador, Grenada, Guadeloupe, Guyana, Haiti, Honduras, Jamaica, Mexico, Nicaragua, Panama, Paraguay, Peru, Saint Kitts And Nevis, St. Lucia, St. Vincent and the Grenadines, Suriname, Trinidad and Tobago, Uruguay, Venezuela

*Date(s) Application is Due:* Mar 31

*Amount of Grant:* Up to 1,000 EUR

*Contact:* Rebecca Hadid, Director of Professional Development Programs; 33 (0) 1 46 47 92 92; fax 33 (0) 1 46 47 42 50; ibrobecky@gmail.com or ibrocentral@gmail.com

*Internet:* ibro.info/professional-development/funding-programmes/regional-grants/latinamerica-grants/#diversity

*Sponsor:* International Brain Research Organization
255 rue Saint-Honore
Paris, 75001 France, Metropolitan

**IBRO Latin America Regional Funding for Neuroscience Schools** 1320

The Latin America Regional Funding for Neuroscience Schools provides financial support for the training activities of advanced students and young researchers in the field of neuroscience in order to increase the quality of neuroscience education in the region; increase the links between young neuroscientists; create a network of former awardees and senior researchers; and establish a mentorship programme of continuous support of alumni. Candidates must submit applications and all supporting materials through the IBRO online application process. Awards range from 15,000 to 20,000 Euro. The application deadline is March 31.

*Requirements:* The Latin America Regional Committee (LARC) will review applications and grant financial support for Schools of Neuroscience with the following characteristics based on the IBRO School Guidelines (posted on the website): uniform high-quality education with an active involvement of students and teachers during the entire event; three levels of schools (basic, advanced/specialized, and inter-regional); cover broad and topical issues, but also include subjects of general interest, such as guidance on writing papers or grant proposals, and ethical issues; and a well balanced series of lectures and laboratory-type practical activities. Neuroscientists from any country in the region are eligible as prospective school directors. Each school should have no less than 15 and not more than 100 students and young scientists. The level of the school should match the level of selected students. The application should contain the scientific program of the school, a list of local and invited faculty, and the budget. Additional information is available at the IBRO website.

*Restrictions:* Hard copy or emailed applications are not accepted.

*Geographic Focus:* Puerto Rico, Antigua & Barbuda, Argentina, Bahamas, Barbados, Belize, Bolivia, Brazil, Chile, Colombia, Costa Rica, Cuba, Dominica, Dominican Republic, Ecuador, El Salvador, Grenada, Guatemala, Guyana, Haiti, Honduras, Jamaica, Mexico, Nicaragua, Panama, Paraguay, Peru, Saint Kitts And Nevis, St. Lucia, St. Vincent and the Grenadines, Suriname, Trinidad and Tobago, Uruguay, Venezuela

*Date(s) Application is Due:* Mar 31

*Amount of Grant:* 15,000 - 20,000 EUR

*Contact:* Rebecca Hadid, Director of Professional Development Programs; 33 (0) 1 46 47 92 92; fax 33 (0) 1 46 47 42 50; ibrobecky@gmail.com or ibrocentral@gmail.com

*Internet:* ibro.info/professional-development/funding-programmes/regional-grants/latinamerica-grants/#schools

*Sponsor:* International Brain Research Organization
255 rue Saint-Honore
Paris, 75001 France, Metropolitan

**IBRO Latin America Regional Funding for PROLAB Collaborations** 1321

The PROLAB program seeks to foster scientific collaboration and development of human resources among Latin American and Caribbean neuroscience research groups. PROLAB uses the neuroscience scientific network of the region, and provides economic support to joint research projects. Its main goal is to diminish the asymmetries in the development of the neurosciences in the continent, thus allowing established research groups to set up collaboration with emerging groups and open a space for horizontal collaboration among established groups within the region. The scientific collaborations will be supported through grants for research missions (exchange trips by doctoral and post doctoral students from the participating institutions) planned to work on scientific questions specifically proposed in the submitted project. Financial support from PROLAB/IBRO/LARC is exclusively aimed at supporting research missions (three to six months) by doctoral and post-doctoral student members of the research groups for activities within the approved project. Applications must be submitted through the IBRO online application process. There are no hard copy application forms. PROLAB applications are taken until March 31 each year.

*Requirements:* Projects should aim at scientific investigations and training of human resources in research. They should be structured around a scientific research project in neuroscience focused on relevant scientific topics in the forefront of this area of human knowledge. Applications should have a joint coordination by the leaders of the research groups involved. The participant groups will choose one of the research leaders to be the responsible liaison with LARC. The projects should be planned for a two year period, with possibility of extension of financial support for another two years. The application should describe how the collaboration will be organized, and the chronogram of activities for the first two years of collaboration. At the end of each year, the project's liaison to LARC must present a joint report of the collaborative activities that took place among the research groups during that period and an update of the chronogram for the next year. Approval by LARC IBRO is necessary for continuation of the support.
*Geographic Focus:* Puerto Rico, Antigua & Barbuda, Argentina, Bahamas, Barbados, Belize, Bolivia, Brazil, Chile, Colombia, Costa Rica, Cuba, Dominica, Dominican Republic, Ecuador, El Salvador, Grenada, Guatemala, Guyana, Haiti, Honduras, Jamaica, Mexico, Nicaragua, Panama, Paraguay, Peru, Saint Kitts And Nevis, St. Lucia, St. Vincent and the Grenadines, Suriname, Trinidad and Tobago, Uruguay, Venezuela
*Date(s) Application is Due:* Mar 31
*Amount of Grant:* 4,000 EUR
*Contact:* Cecilia Bouzat, Chairperson; 33 (0) 1 46 47 92 92; fax 33 (0) 1 46 47 42 50; inbouzat@criba.edu.ar or ibrocentral@gmail.com
*Internet:* ibro.info/professional-development/funding-programmes/regional-grants/latinamerica-grants/prolab-collaboration-in-latin-america/
*Sponsor:* International Brain Research Organization
255 rue Saint-Honore
Paris, 75001 France, Metropolitan

**IBRO Latin America Regional Funding for Short Research Stays** 1322

The International Brain Research Organization's Latin America Regional Committee (LARC) evaluates applications for funding for short research/training lab visits within the Latin American region. Applications must be submitted through the IBRO online application process. Grant awards range up to a maximum of 1200 Euro. The application deadline is March 31.
*Requirements:* Candidates should not be older than 40 years of age. Priority will be given to doctoral students and postdocs who wish to attend a foreign lab within the region and who plan to link their lab stay with attendance at a course or congress. When the short lab stay is linked to attendance at a course or congress, an application for LARC Travel Grants must also be submitted. Applications for activities outside the region or within the applicant's own country will be considered only if adequately justified. Applicants must supply an activity plan, a letter of support from his or her director - specifically for the proposed activities - and a letter of acceptance from the host institution.
*Restrictions:* emailed applications will not be accepted.
*Geographic Focus:* Puerto Rico, Antigua & Barbuda, Argentina, Bahamas, Barbados, Belize, Bolivia, Brazil, Chile, Colombia, Costa Rica, Cuba, Dominica, Dominican Republic, Ecuador, El Salvador, Grenada, Guatemala, Guyana, Haiti, Honduras, Jamaica, Mexico, Nicaragua, Panama, Paraguay, Peru, Saint Kitts And Nevis, St. Lucia, St. Vincent and the Grenadines, Suriname, Trinidad and Tobago, Uruguay, Venezuela
*Date(s) Application is Due:* Mar 31
*Amount of Grant:* Up to 1,200 EUR
*Contact:* Rebecca Hadid, Director of Professional Development Programs; 33 (0) 1 46 47 92 92; fax 33 (0) 1 46 47 42 50; ibrobecky@gmail.com or ibrocentral@gmail.com
*Internet:* ibro.info/professional-development/funding-programmes/regional-grants/latinamerica-grants/#shortstay
*Sponsor:* International Brain Research Organization
255 rue Saint-Honore
Paris, 75001 France, Metropolitan

**IBRO Latin America Regional Travel Grants** 1323

The International Brain Research Organization's Latin America Regional Committee (LARC) evaluates funding for attending courses/workshops or for presenting results at meetings within the Latin American region. Candidates should not be older than forty (40) years of age. Priority will be given to doctoral students and postdocs who are also planning a short stay in a lab. When the travel grant is linked to a short stay, an application for LARC Short Research Stay Grants must also be submitted. Applications must be submitted through IBRO's online application process. Awards range up to a maximum of 1,200 Euro. The application deadline is March 31.
*Restrictions:* emailed applications are not accepted. Applications for activities outside the region or within the applicant's own country will be considered only if adequately justified.
*Geographic Focus:* Puerto Rico, Antigua & Barbuda, Argentina, Bahamas, Barbados, Belize, Bolivia, Chile, Colombia, Costa Rica, Cuba, Dominica, Dominican Republic, Ecuador, El Salvador, Grenada, Guatemala, Guyana, Haiti, Honduras, Jamaica, Mexico, Nicaragua, Panama, Paraguay, Peru, Saint Kitts And Nevis, St. Lucia, St. Vincent and the Grenadines, Suriname, Trinidad and Tobago, Uruguay, Venezuela
*Date(s) Application is Due:* Mar 31
*Amount of Grant:* 800 - 1,200 EUR
*Contact:* Rebecca Hadid, Director of Professional Development Programs; 33 (0) 1 46 47 92 92; fax 33 (0) 1 46 47 42 50; ibrobecky@gmail.com or ibrocentral@gmail.com
*Internet:* ibro.info/professional-development/funding-programmes/regional-grants/latinamerica-grants/#travelgrant
*Sponsor:* International Brain Research Organization
255 rue Saint-Honore
Paris, 75001 France, Metropolitan

**IBRO Regional Grants for International Fellowships to U.S. Laboratory Summer Neuroscience Courses** 1324

The U.S.-Canada Regional Committee for IBRO (USCRC) provides fellowship support for students from developing nations who are accepted into the eligible high level neuroscience-related summer courses at the Marine Biological Laboratory in Woods Hole, Massachusetts and at the Cold Spring Harbor Laboratory in Cold Spring Harbor, New York. Substantial financial support is available to enable the participation of students from resource-restricted countries in these programs. The Fellowships are an effort to allow the most worthy alumni of IBRO Neuroscience Schools and other outstanding young scientists to gain outstanding experiences in leading training programs. These experiences will benefit scientists who will continue to study and work in less financially advantaged regions such as Latin America, Africa, Central and Eastern Europe, and Asia/Pacific. Candidates should apply directly to MBL or CSHL through their online application process. Each lab will inform candidates if they have been selected. Information about courses offered at each lab is available at the IBRO website.
*Restrictions:* Applicants should contact each lab directly about whether their chosen courses are eligible for the Fellowship.
*Geographic Focus:* All Countries
*Date(s) Application is Due:* Feb 1
*Contact:* Rebecca Hadid, Director of Professional Development Programs; 33 (0) 1 46 47 92 92; fax 33 (0) 1 46 47 42 50; ibrobecky@gmail.com or ibrocentral@gmail.com
Dana Mock-Muñoz de Luna, Research Award Coordinator; (508) 289-7173 or (508) 548-3705; dmock@mbl.edu or researchprograms@mbl.edu
*Internet:* ibro.info/professional-development/funding-programmes/regional-grants/uscanada-grants/
*Sponsor:* International Brain Research Organization
255 rue Saint-Honore
Paris, 75001 France, Metropolitan

**IBRO Research Fellowships** 1325

The IBRO Funding Program aims to foster neuroscience, especially in less well-funded countries, by providing support to high quality neuroscientists from diverse geographic and scientific areas (U.S./Canada region excluded) who wish to broaden the scope of their training in neuroscience by working abroad. IBRO Research Fellowships, which include the John G. Nicholls and Rita Levi-Montalcini Fellowships, support post-doctoral laboratory training to applicants under the age of 45 for up to one year overseas. The funding for a 12-month Fellowship is 35,000 euros. Priority is given to those who have not obtained funding from this programme within the past three years and who, after completion of the Fellowship, are willing to return to their home countries, bringing with them new knowledge and skills to advance neuroscience in their regions.
*Requirements:* Applicants must meet the following *Requirements:* submit a letter of acceptance from the proposed supervisor (host scientist) indicating that he/she will accept the candidate and agrees on the proposed project; a letter of reference from the present supervisor; and, if available, evidence of an offer that the applicant will have a position to return to his/her home country after the Fellowship. Only applications completed through IBRO's online system will be considered.
*Geographic Focus:* All Countries
*Date(s) Application is Due:* Jun 10
*Amount of Grant:* 35,000 EUR
*Contact:* Marta Hallak, Senior Director of Grants; 33 (0) 1 46 47 92 92; fax 33 (0) 1 46 47 42 50; mhallak@mail.fcq.unc.edu.ar
*Internet:* ibro.info/professional-development/funding-programmes/research-fellowships/
*Sponsor:* International Brain Research Organization
255 rue Saint-Honore
Paris, 75001 France, Metropolitan

**IBRO Return Home Fellowships** 1326

The IBRO Fellowships Program aims to foster neuroscience research, especially in less well-funded countries, by providing support to high quality neuroscientists from diverse geographic and scientific areas (US/Canada Region excluded) who wish to broaden the scope of their laboratory training in neuroscience by working abroad. Priority is given to those who have not obtained funding from this program within the past three years and who, after completion of the fellowship, are willing to return to their home countries, bringing with them new knowledge and skills to advance neuroscience in their regions. The objective of the Return Home Program of IBRO is to coordinate efforts with other organizations to both improve the opportunities for productive neuroscience research within the less advantaged regions of the world and to provide more aid to those researchers trained overseas who wish return to their home countries. Three profiles

of scientists will be the main target of this Fellowship: postdoctoral Fellows who have finished research training in neurosciences (including clinical research) in a center of excellence of a developed country; research students who have been trained in a Centre of Excellence in Brain Research (CEBR); and scientists who are developing a successful basic/clinical research career in a developed country and wish to return to their country of origin or to a less developed country for personal or cultural reasons. Only applications completed through the IBRO application system will be considered. The grant amount is up to 20,000 Euros, and the annual deadline for application is September 1.
*Requirements:* Eligibility requirements include: postdoctoral fellows who have finished research training in neuroscience (including clinical research) in a center of excellence of a developed country; research students who have been trained in a Center of Excellence in Brain Research (CEBR); and scientists who are developing a successful basic/clinical research career in a developed country and wish to return to their country of origin or to a less developed country for personal or cultural reasons.
*Geographic Focus:* All Countries
*Date(s) Application is Due:* Sep 1
*Amount of Grant:* Up to 20,000 EUR
*Contact:* Marta Hallak, Senior Director of Grants; 33 (0) 1 46 47 92 92; fax 33 (0) 1 46 47 42 50; mhallak@mail.fcq.unc.edu.ar
*Internet:* ibro.info/professional-development/funding-programmes/return-home-fellowships/
*Sponsor:* International Brain Research Organization
255 rue Saint-Honore
Paris, 75001 France, Metropolitan

**ICC Faculty Fellowships** **1327**
Based on Ernest Boyer's concept that community engagement provides an excellent means by which the three components of faculty professional development of teaching, research, and professional service can support and strengthen each other, Indiana Campus Compact awards Faculty Fellowships as a model for faculty professional development through the scholarship of engagement. Up to 7 fellowships are competitively awarded in each academic year, in addition to one Senior Fellow. The fellowship amount is $3,750. Institutions must match at least $1,250. The fellowship fiscal period is twelve months, from June 1 to May 31.
*Requirements:* Faculty from all disciplines and professional schools at Indiana Campus Compact (ICC) member campuses are eligible to apply. Former Fellows are eligible to apply; however, preference is given to new applicants. Fellows are expected to model the integration of community engagement into all three aspects of their professional development in the following ways: (1) Teaching - Each Fellow will teach a service-learning course within the program period; (2) Research - Each Fellow will design a research, scholarly, or artistic project relating to the service performed with a community organization or to the practice of service-learning, to be initiated during the program period; (3) Professional Service - Each Fellow will provide direct service to a nonprofit organization during the program year to assist in addressing a particular community issue that is related to his/her academic discipline or expertise as an educator. Note: ALL proposals must be properly routed through the fiscal routing system at an applicant's institution prior to electronic submission to Indiana Campus Compact. However, if the institution is an Indiana University campus, routing through Research & Sponsored Programs or Sponsored Research Services is not required.
*Geographic Focus:* Indiana
*Amount of Grant:* 3,750 USD
*Contact:* J.R. Jamison; (317) 274-6500; fax (317) 274-6510; jrjamiso@iupui.edu
*Internet:* www.indianacampuscompact.org/Default.aspx?tabid=86
*Sponsor:* Indiana Campus Compact
620 Union Drive, Suite 203
Indianapolis, IN 46202

**ICC Scholarship of Engagement Faculty Grants** **1328**
Eighteen faculty per year will be selected to choose one of three options related to the Scholarship of Engagement: develop or revise a service-learning course to be taught by the recipient, provide scholarly research on service-engagement, or provide a professional service to a local community agency. The selected faculty members will also encourage the development of service-engagement infrastructure on their respective campuses through trainings and workshops of other faculty. Award amounts are $2,250. Institutions are required to cash match at least $750.
*Requirements:* Adjunct or full-time faculty from all disciplines and professional schools at Indiana Campus Compact (ICC) member campuses are eligible to apply. An adjunct or full-time faculty member applicant must propose a project that focuses on one of three areas: (1) Teaching - Design a new course or revise an existing course to include a service-learning component; (2) Research - Develop and initiate a research project related to his/her discipline that addresses a community issue and advances the field of engagement through service-learning. Preference will be given to projects that are participatory action research; (3) Professional Service - Provide direct service to a nonprofit organization to assist in addressing a particular community issue. This professional service must be related to the faculty member's academic discipline or his/her expertise as an educator. Note: ALL proposals must be properly routed through the fiscal routing system at an applicant's institution prior to electronic submission to Indiana Campus Compact. However, if the institution is an Indiana University campus, routing through Research & Sponsored Programs or Sponsored Research Services is not required.
*Restrictions:* Grants are awarded on a reimbursement basis. Grant funds cannot be used for indirect costs, items of durable equipment costing more than $100 (unless approved in writing by ICC), payment for participants to do service, fundraising activities, entertainment costs, tickets to recreational events, ball games, zoos, etc., clothing such as T-shirts or hats, or cash incentives.
*Geographic Focus:* Indiana
*Date(s) Application is Due:* Nov 6
*Amount of Grant:* 2,250 USD
*Contact:* Lindsay Doucette, Program Director; (317) 274-6500; lgdoucet@iupui.edu
*Internet:* www.indianacampuscompact.org/Default.aspx?tabid=86
*Sponsor:* Indiana Campus Compact
620 Union Drive, Suite 203
Indianapolis, IN 46202

**IDEM Section 319(h) Nonpoint Source Program Grants** **1329**
The Non-point Source (NPS) Section in the Office of Water Quality manages two federal pass-through grant programs aimed at improving water quality in the state: Section 319(h) and Section 205(j); each named after the portion of the Clean Water Act that authorizes the program. The Federal Clean Water Act Section 319(h) provides funding for various types of projects that work to reduce non-point source water pollution. Funds may be used to conduct assessments, develop and implement TMDLs and watershed management plans, provide technical assistance, demonstrate new technology and provide education and outreach. Organizations eligible for funding include nonprofit organizations, universities, and local, State or Federal government agencies. Projects are administered through grant agreements that spell out the tasks, schedule and budget for the project. Projects are normally two to three years long and work to reduce non-point source pollution and improve water quality in the watershed primarily through: education and outreach designed to bring about behavioral changes and best management practice (BMP) implementation that leads to reduced nonpoint source pollution; the development of watershed management plans that meet U.S. EPA's required nine elements; and the implementation of watershed management plans through a cost-share program focusing on BMP implementation that address water quality concerns. IDEM project managers work closely with the project sponsors to help ensure that the project runs smoothly and the tasks of the grant agreement are fulfilled. Site visits are conducted at least quarterly to touch base on the project, provide guidance and technical assistance as needed, and to work with the grantee on any issues that arise to ensure a successful project closeout. Specific ways to address non-point source water pollution include: education and outreach on watershed management; and information gathering activities (such as conducting watershed inventories and water quality assessments for the purpose of developing comprehensive watershed management plans and implementing those plans). A required Notice of Intent (NOI) to apply emails for the current grant solicitation are due June 1, with completed applications due by September 1.
*Requirements:* To be considered for funding, the project sponsor (the entity responsible for the project and its overall success), must be one of the following: municipality; county government; state government; federal government; college or university; or a nonprofit 501(c)3. The program provides funding and technical assistance to groups that work on the watershed level with citizens to develop locally-based solutions to non-point source pollution. IDEM provides sixty percent (60%) of an approved project's total cost with Section 319 funds. A grant recipient must provide the remaining forty percent (40%) of the total project cost as match.
*Restrictions:* The following is a list of activities that cannot be funded with Section 319 funds and cannot be counted as matching funds for a Section 319 grant: permit fees; food for meetings or other events; purchase of agricultural equipment, or other large pieces of equipment (equipment modifications and leasing are allowable); purchase of land or land easements (these activities can be counted as matching funds in some cases); any project which is directed at water quantity rather than water quality, such as dredging, drainage, or flood control; any practices, equipment, or supplies used to fulfill the requirements of any federal permit (NPDES permit, Section 401 Water Quality Certification, permits from the U.S. Army Corps of Engineers, as examples) or to comply with IDEM's Confined Feeding Operation rule or permit requirements, or to meet enforcement requirements; wetland mitigation sites; incentive payments or yield losses; non-point source best management practices not sanctioned by IDEM or not sanctioned by a partner agency of IDEM; practices not installed in accordance with standards and specifications developed by NRCS, IDNR or other recognized standards; office furniture; or sales tax. Additionally, the following will not be funded by this specific grant: septic system pump outs, repairs, rehabilitations, or demonstrations of alternative septic systems; projects whose sole purpose is data collection, research, demonstration of best management practices; or educational outreach. Matching funds cannot come from any federal funding sources.
*Geographic Focus:* Indiana
*Date(s) Application is Due:* Sep 1
*Contact:* Jamie Hosier, Senior Project Manager; (317) 308-3218 or (317) 308-3173; fax (317) 308-3219; jhosier@idem.IN.gov or info@idem.IN.gov
*Internet:* www.in.gov/idem/nps/2524.htm
*Sponsor:* Indiana Department of Environmental Management
Indiana Government Center North, 100 North Senate Avenue
Indianapolis, IN 46204-2251

**IES Faculty Fellowships** **1330**
Founded in 1950, the Institute for the International Education of Students (IES Abroad) is a Chicago-based, not-for-profit academic consortium that offers financial aid to students studying for a semester or full year at one of its centers abroad. IES Abroad operates study-abroad programs on the continents of Europe, Africa, Asia, South America, Australia, and New Zealand. In cooperation with the Newberry Library, the institute is very pleased to offer a one-month research fellowship for an IES Abroad faculty member at the Newberry Library in Chicago, Illinois (USA) during the summer. Recipients will have full access to the Newberry Library's extensive collections in the areas of European discovery, exploration, and settlement of the Americas; American literature and history; the Renaissance; the French Revolution; Portuguese and Brazilian history; American Indian history and literature; the history of cartography; the history

and theory of music; the history of printing; and early philology and linguistics. IES Abroad will cover all international transportation costs, provide a monthly stipend of $1600, and pay housing costs for the duration of the fellowship. The recipient is responsible for any additional expenses. Prospective applicants should contact their respective IES Abroad Center Directors for application instructions and further information. A contact list has been provided at the IES website (click on the "Contact Us" and then on "Advisors and Faculty" links).
*Requirements:* All eligible IES Abroad faculty whose research would benefit from access to the Newberry Library's collections are invited to apply for this fellowship. At the end of the one-month research period, recipients will be asked to present their findings and to report on their experience to the IES Abroad Chicago staff.
*Geographic Focus:* All States, All Countries
*Amount of Grant:* 1,600 USD
*Contact:* Michael Steinberg, Executive, Vice President, Director of Academic Programs, Director of Customized Programs; (800) 995-2300 or (312) 944-1750; fax (312) 944-1448; msteinberg@IESabroad.org or info@IESabroad.org
*Internet:* https://www.iesabroad.org/IES/Advisors_and_Faculty/Faculty_Development/fellowships.html
*Sponsor:* Institute for the International Education of Students
33 North LaSalle Street, 15th Floor
Chicago, IL 60602

**IHI Quality Improvement Fellowships** 1331
The Institute for Healthcare Improvement (IHI), an independent not-for-profit organization based in Cambridge, Massachusetts, is an innovator in health and health care improvement worldwide. The IHI Fellowship Program is a year-long immersion program for mid-career health professionals. The program has two overarching goals, to: (1) develop health care leaders with the drive, skills, and experience to spread improvement in the United States and globally; and (2) build capability within health care organizations to reach dramatically higher levels of performance. Fellows spend one year at IHI's office in Cambridge, Massachusetts, and return to their home organization to lead transformative change. Fellowships are one year in duration, July through June.
*Requirements:* The IHI Quality Improvement Fellowship supports fellows who live and work in the United States. The fellowship is designed for mid-careers professionals who will return to senior leadership positions in their home organizations. IHI will, however, consider a very strong application from a candidate earlier in his/her career. All successful candidates must have full-time employment with an organization who will sponsor them for the fellowship (salary and benefits for the year are paid by the home organization) and who shows commitment to leveraging the fellow's learning appropriately post-fellowship. Applicants are not required to have a M.D; IHI encourages applications from nurse leaders, quality improvement professionals, and allied health professionals.
*Geographic Focus:* All States
*Contact:* Joelle Baehrend, (617) 301-4816; fax (617) 301-4848; jbaehrend@ihi.org
*Internet:* www.ihi.org/offerings/Training/Fellowships/Pages/default.aspx
*Sponsor:* Institute for Healthcare Improvement
20 University Road, 7th Floor
Cambridge, MA 02138

**IIE African Center of Excellence for Women's Leadership Grants** 1332
The African Center of Excellence for Women's Leadership (ACE) aims to develop the capabilities of four women-led organizations in East Africa to be centers of excellence in the delivery of leadership development programs, research and knowledge generation on reproductive health, economic empowerment and girls' education. The ACE for Women's Leadership Program responds to an articulated need by women across the continent for a strategic leadership development program and for a stronger and more coordinated effort to help women advocate for issues that most concern them. The four organizations are intended to serve as places where women can come together for leadership training, strategizing, researching and sharing of best practices across the region.
*Requirements:* The program is based in four countries: Ethiopia, Uganda, Kenya and Rwanda.
*Geographic Focus:* Ethiopia, Kenya, Rwanda, Uganda
*Contact:* Melat Tekletsadik, Program Director; +251 (0) 91 260 9805 or (212) 883-8200; mhaile@iie.org
*Internet:* https://www.iie.org/en/Programs/ACE-for-Womens-Leadership/About
*Sponsor:* Institute of International Education
Room 1905, 19/F, 69 Jervois Street
Sheung Wan, -
Hong Kong

**IIE Ambassador's Distinguished Scholars Program in Ethiopia** 1333
The Ambassador's Distinguished Scholars will be a diverse set of qualified academics advancing scholarship in Ethiopia as full-time teaching faculty members at Ethiopian institutions of higher education. The aim of the program is to strengthen Ethiopian universities' capacity to teach and manage undergraduate and graduate programs and enhance research capacity within Ethiopian universities. This program is supported by the U.S. Embassy in Ethiopia, Bahir Dar University, the University of Gondar, and the Institute of International Education (IIE). Ambassador's Distinguished Scholars will have the opportunity to engage in primary research with doctoral students, work to publish results, teach, design curriculum, mentor students, and collaborate with faculty at their host institution for one academic semester. A living stipend ranging from $1,500 to $3,500 monthly, depending on candidate qualifications. A housing allowance ranging from $175 to $275 monthly, along with round-trip airfare, medical coverage. and other types of support. The annual application deadline is May 15.
*Requirements:* Applications from scholars with a range of experience are welcome, including recent doctoral graduates and professors emeriti. Faculty positions are available in over 60 disciplines. Preferred fields of study include STEM fields, Social Sciences, and Humanities. There is a preference for candidates with academic expertise that matches with the areas of specialization desired by the host institutions.
*Geographic Focus:* All States
*Date(s) Application is Due:* May 15
*Amount of Grant:* 20,000 - 40,000 USD
*Contact:* Saba Gebremedhin; +251-114-700507; shagos@iie.org or adscholars@iie.org
*Internet:* https://www.iie.org/Programs/Ambassadors-Distinguished-Scholars-Program
*Sponsor:* Institute of International Education
P.O. Box 586
Addis Ababa, Code 1110 Ethiopia

**IIE Carnegie African Diaspora All-Disciplines Fellowships** 1334
The Carnegie African Diaspora Fellowship program is a fellowship program for African institutions in Ghana, Kenya, Nigeria, South Africa, Tanzania and Uganda to host an African-born scholar to work in projects in research collaboration, graduate student teaching and mentoring and curriculum co-development. In the area of All-Disciplines Fellowships, the African Diaspora program is open to project requests from any discipline in each of two competition cycle. The Spring competition will open on May 1 and will close on July 1, while the Fall competition will open on October 1 and will close on December 9.
*Requirements:* Anyone born in Africa is eligible to apply (even if they are a citizen of another country).
*Geographic Focus:* All States, All Countries
*Date(s) Application is Due:* Jul 1; Dec 9
*Contact:* Saba Gebremedhin; +251-114-700507; shagos@iie.org or africandiaspora@iie.org
*Internet:* https://www.iie.org/Programs/Carnegie-African-Diaspora-Fellowship-Program/How-to-Apply/Types-of-Fellowships
*Sponsor:* Institute of International Education
P.O. Box 586
Addis Ababa, Code 1110 Ethiopia

**IIE Carnegie African Diaspora Alumni Fellowships** 1335
The Carnegie African Diaspora Fellowship program is a fellowship program for African institutions in Ghana, Kenya, Nigeria, South Africa, Tanzania and Uganda to host an African-born scholar to work in projects in research collaboration, graduate student teaching and mentoring and curriculum co-development. In the area of Alumni Fellowships, the African Diaspora program is open to project requests from any discipline in which an alumni is involved. It provides an opportunity for program alumni to do another stay at their initial host institution. Alumni can also collaborate with new institutions if they choose. The Spring competition will open on May 1 and will close on July 1, while the Fall competition will open on October 1 and will close on December 9.
*Requirements:* Anyone program alumni born in Africa is eligible to apply (even if they are a citizen of another country).
*Geographic Focus:* All States, All Countries
*Date(s) Application is Due:* Jul 1; Dec 9
*Contact:* Saba Gebremedhin, Program Director; +251-114-700507; shagos@iie.org or africandiaspora@iie.org
*Internet:* https://www.iie.org/Programs/Carnegie-African-Diaspora-Fellowship-Program/How-to-Apply/Types-of-Fellowships
*Sponsor:* Institute of International Education
P.O. Box 586
Addis Ababa, Code 1110 Ethiopia

**IIE Carnegie African Diaspora Multi-Institutional Fellowships** 1336
The Carnegie African Diaspora Fellowship program is a fellowship program for African institutions in Ghana, Kenya, Nigeria, South Africa, Tanzania and Uganda to host an African-born scholar to work in projects in research collaboration, graduate student teaching and mentoring and curriculum co-development. In the area of Multi-Institutional Fellowships, the African Diaspora program allows fellows the opportunity to visit and conduct research at multiple institutions. Project requests are specifically encouraged that demonstrate a collaboration between established and newer universities. Collaborations between universities and research organizations are also an option. Multiple-Institutions Fellowship themes with corresponding application dates are as follows: requests in the Spring are accepted from May 1 through July 6 in the areas of science, technology, engineering, arts, mathematics, and sustainable cities and communities; and requests in the Fall are accepted from October 1 through December 9 in the areas of peace, social justice, equality, security (including cyber security), agriculture, industry, and innovation.
*Requirements:* Anyone born in Africa is eligible to apply (even if they are a citizen of another country). Hosting institutions must collaborate on the Project Request.
*Geographic Focus:* All States, All Countries
*Date(s) Application is Due:* Jul 1; Dec 9
*Contact:* Saba Gebremedhin, Program Director; +251-114-700507; shagos@iie.org or africandiaspora@iie.org
*Internet:* https://www.iie.org/Programs/Carnegie-African-Diaspora-Fellowship-Program/How-to-Apply/Types-of-Fellowships
*Sponsor:* Institute of International Education
P.O. Box 586
Addis Ababa, Code 1110 Ethiopia

**IIE Carnegie African Diaspora Thematic Fellowships** 1337

The Carnegie African Diaspora Fellowship program is a fellowship program for African institutions in Ghana, Kenya, Nigeria, South Africa, Tanzania and Uganda to host an African-born scholar to work in projects in research collaboration, graduate student teaching and mentoring and curriculum co-development. In the area of Thematic Fellowships, the African Diaspora program gives preference to different thematic areas over two competition cycles. Though projects in thematic areas will be given priority, many "all-discipline" fellowships will be awarded. Priority Fellowship Themes with corresponding application dates are as follows: requests in the Spring are accepted from May 1 through July 6 in the areas of science, technology, engineering, arts, mathematics, and sustainable cities and communities; and requests in the Fall are accepted from October 1 through December 9 in the areas of peace, social justice, equality, security (including cyber security), agriculture, industry, and innovation.
*Requirements:* Anyone born in Africa is eligible to apply (even if they are a citizen of another country).
*Geographic Focus:* All States, All Countries
*Date(s) Application is Due:* Jul 6; Dec 9
*Contact:* Saba Gebremedhin, Program Director; +251-114-700507; shagos@iie.org or africandiaspora@iie.org
*Internet:* https://www.iie.org/Programs/Carnegie-African-Diaspora-Fellowship-Program/How-to-Apply/Types-of-Fellowships
*Sponsor:* Institute of International Education
P.O. Box 586
Addis Ababa, Code 1110 Ethiopia

**IIE Central Europe Summer Research Institute Summer Research Fellowship** 1338

The Central Europe Summer Research Institute (CESRI) is a fellowship opportunity for U.S. graduate students in science and engineering. With support from the National Science Foundation (NSF) and the German Academic Exchange Service (DAAD), CESRI will provide eight U.S. graduate students with a high-quality international research experience in Austria, the Czech Republic, Germany, Hungary, Poland or Slovakia. Two CESRI grants are specifically for CESRI Fellowships to Germany. The remaining six CESRI grants are funded by NSF for the entire Central Europe region, with preference for grants to countries other than Germany. The program is intended for scientific research projects. The CESRI program will be eight weeks in length, with the first four days spent as a group in Budapest, Hungary in a specially designed cultural and academic orientation to the region. Participants will spend the remaining 7.5 weeks working in individually-arranged placements in university labs or other appropriate sites where they can participate in creative research activities under the supervision of European mentors. In addition to the scholarship, recipients receive round trip airfare to Central Europe, room and board for 8 weeks, health and accident insurance, and a four day academic and cultural orientation in Budapest.
*Requirements:* All applicants must be U.S. citizens or permanent residents in the U.S., and be current Master's or Ph.D. students at a U.S. university in one of the following fields: Biology, Chemistry, Computer Science, Engineering, Environmental Science, or Mathematics. Candidates should submit the online application, along with the online self-evaluation form and two completed professional reference forms.
*Restrictions:* Projects involving field work, policy, ethics, or clinical research on human objects are not eligible for consideration.
*Geographic Focus:* All States
*Date(s) Application is Due:* Feb 1
*Amount of Grant:* 2,000 USD
*Contact:* CESRI Fellowship Contact; (212) 883-8200; fax (212) 984-5452; cesri@iie.org
*Internet:* www.iie.org/Programs/CESRI
*Sponsor:* Institute of International Education
1400 K Street NW, 7th Floor
Washington, D.C. 20005-2403

**IIE China Scholarship Council–East-West Center Visiting Scholarships** 1339

The East-West Center (EWC) welcomes scholars and analysts from China who wish to participate in the "China Scholarship Council – East-West Center Joint Program to consider applying for placement as CSC-EWC Visiting Scholars at the EWC in Honolulu, Hawaii. These placements are designed for scholars who wish to undertake research and writing on topics of relevance to contemporary US-Asia policy that will compliment the Research Program's three broad thematic areas: governance, security and justice; environment, population and health; and innovation, economic integration and growth as well as higher education policy. The program supports up to twelve month residencies at the East-West Center in Honolulu. While in residence, the Scholars will complete an English article or monograph to be considered for publication in an East-West Center publication series. The first full draft of the manuscript should be completed before the end of the residence period. Scholars are also asked to give a public seminar while in residence at the East West Center.
*Requirements:* The CSC-EWC Affiliate Scholar program requires two separate applications. First, an application is submitted to the East-West Center detailing the applicant's plan for dissertation research while in residence at the East-West Center. Successful applicants are then invited to apply for funding from the China Scholarship Council to subsidize their travel, living expenses and health insurance. Scholars are responsible for securing their own housing and medical insurance coverage during their term of residency.
*Restrictions:* For all awards, payment of indirect costs (overhead) is not permitted.
*Geographic Focus:* China
*Date(s) Application is Due:* Nov 15
*Contact:* Diana K. Jahja, Director; +84 4 831 7330; fax +84 4 831 7331; diana.jahja@iief.or.id or info@iief.or.id
Kim Small, Scholarship Program Specialist; (808) 944-7752 or (808) 944-7769; fax (808) 944-7070; fellowships@eastwestcenter.org or smallk@eastwestcenter.org
*Internet:* https://www.eastwestcenter.org/node/35518
*Sponsor:* Institute of International Education
Menara Imperium, 28th Floor, Suite C
Jakarta Selatan, 12980 Indonesia

**IIE David L. Boren Fellowships** 1340

Boren Fellowships provide up to $24,000 funding to U.S. graduate students to add an international and language component to their graduate education through specialization in area study, language study, or increased language proficiency. In addition, Boren Fellowships can provide limited funding for domestic language study that will supplement the overseas component, with a maximum award for a combined overseas and domestic program up to $30,000. Boren Fellowships support study and research in areas of the world that are critical to U.S. interests, including Africa, Asia, Central & Eastern Europe, Eurasia, Latin America, and the Middle East. Boren Fellowships are funded by the National Security Education Program (NSEP), which focuses on geographic areas, languages, and fields of study that are critical to U.S. national security. Applicants should identify how their projects, as well as their future academic and career goals, will contribute to U.S. national security. These award Fellowships are made for a minimum of twelve weeks overseas and a maximum of one year overseas. The annual application submission deadline is January 30, with notification in in mid- to late-April.
*Requirements:* Boren Fellowships promote long term linguistic and cultural immersion. Therefore, applicant preference will be given to those proposing overseas programs of 6 months or longer. However, applicants proposing overseas programs of 3-6 months, especially those in the STEM (science, technology, engineering, and mathematics) fields are encouraged to apply. Boren Fellowships are awarded with preference for countries, languages, and fields of study critical to U.S. national security. Preference is also given to students who will study abroad for longer periods of time, and who are highly motivated by the opportunity to work in the federal government. Applicants should refer to the website for current deadlines and detailed criteria for application.
*Restrictions:* Applicants must commit to a length of study of at last 12 weeks.
*Geographic Focus* All States, District of Columbia, Guam, Marshall Islands, Northern Mariana Islands, Puerto Rico, U.S. Virgin Islands, American Samoa
*Date(s) Application is Due:* Jan 30
*Amount of Grant:* Up to 30,000 USD
*Contact:* Michael Saffle, Boren Outreach and Recruitment Specialist; (212) 984-5357 or (212) 883-8200; fax (212) 984-5401; boren@iie.org
*Internet:* https://www.borenawards.org/fellowships/boren-fellowship-basics
*Sponsor:* Institute of International Education
1400 K Street NW, 7th Floor
Washington, D.C. 20005-2403

**IIE David L. Boren Scholarships** 1341

Boren Scholarships provide undergraduates with up to $20,000 funding to acquire skills and experiences in areas of the world critical to U.S. national security, in exchange for their commitment to seek work in the federal government. Boren Scholarships are funded by the National Security Education Program (NSEP), which focuses on geographic areas, languages, and fields of study critical to U.S. national security. NSEP draws on national security issues such as sustainable development, environmental degradation, global disease and hunger, population growth and migration, and economic competitiveness. Boren Scholarships allow students to study in areas of the world that are underrepresented in study abroad programs, including Africa, Asia, Central and Eastern Europe, Eurasia, Latin America, and the Middle East. Scholars represent a variety of academic backgrounds, but are interested in studying less commonly taught languages, including but not limited to Arabic, Chinese, Korean, Portuguese, Russian, and Swahili. Individual on-campus deadlines will be earlier than February 8.
*Requirements:* Students are eligible to apply if they are U.S. citizens enrolled as undergraduates in a degree-seeking program at a college or university for the duration of the scholarship (they may not graduate until the scholarship is complete); and applying to a study abroad program that meets their home institution standards in a country outside of Western Europe, Canada, Australia, or New Zealand. Boren Scholarships are awarded with preference for countries, languages, and fields of study critical to U.S. national security. Preference is also given to students who will study abroad for longer periods of time, and who are highly motivated by the opportunity to work in the federal government. In exchange for scholarship funding, all Boren scholars must agree to the NSEP Service Requirement posted on the website. Applicants should refer to the website for current information and their individual campus representative for their on-campus deadline.
*Restrictions:* Boren Scholarships are not for study in the U.S.
*Geographic Focus* All States, District of Columbia, Guam, Marshall Islands, Northern Mariana Islands, Puerto Rico, U.S. Virgin Islands, American Samoa
*Date(s) Application is Due:* Feb 8
*Amount of Grant:* 8,000 - 20,000 USD
*Contact:* Lisa Mayorga, Boren Scholarship Program Manager; (800) 618-6737 or (212) 883-8200; fax (212) 984-5401; boren@iie.org
*Internet:* https://www.borenawards.org/scholarships/program-basics/boren-scholarship-basics
*Sponsor:* Institute of International Education
1400 K Street NW, 7th Floor
Washington, D.C. 20005-2403

**IIE East-West Center Asia Pacific Leadership Program Fellowship** 1342

The East-West Center is invites applications annually for the Asia Pacific Leadership Program Fellowship (APLP). The APLP is the center of excellence for leadership

education in the Asia Pacific region and a signature program of the East West Center, coordinated by the Institute if International Education. The APLP seeks outstanding individuals from all walks of life (government, business, NGOs, health sciences, media, monastic orders, and education) to participate in an innovative leadership development program at the East-West Center in Honolulu, Hawaii. The average age of an APLP Fellow is 33, with an average of five to eight years of professional experience prior to entering the program. All fellows live on the East-West Center's campus in Honolulu, Hawaii from August through December for the residential portion of the program. From January through April, the program is run virtually (online) and fellows have the option to return to their home countries or apply for extended fieldwork in Southeast Asia, internships, self-designed projects and employment experiences. Upon acceptance, all participants receive a fellowship valued at approximately $13,000USD to cover a majority of the program costs. Substantial scholarship funding is available for the remainder of the costs of participation, though timing is critical. The earlier an application is submitted, the greater the chance that scholarship funds will be available to support participant costs beyond the fellowship provisions. Scholarship requests and APLP Fellowship applications are due by March 1 each year.
*Requirements:* Applicants should have: at least a three-year bachelor's degree or its equivalent from an accredited U.S. college or university, or from a recognized institution of higher learning abroad (equivalency is determined by the University of Hawaii); at least 5 years of work experience; a demonstrated commitment to promoting peace and prosperity in the Asia Pacific region, and a deep interest in experiential leadership development training. There are no citizenship, age, or ethnicity restrictions.
*Geographic Focus* All States, District of Columbia, Guam, Marshall Islands, Northern Mariana Islands, Puerto Rico, U.S. Virgin Islands, All Countries, American Samoa
*Date(s) Application is Due:* Mar 1
*Amount of Grant:* 13,000 USD
*Contact:* Diana K. Jahja, Director; +84 4 831 7330; fax +84 4 831 7331; diana.jahja@iief.or.id or info@iief.or.id
John N. Hawkins, Co-Director; (808) 944-7769 or (808) 944-7111; fax (808) 944-7070; aplp@eastwestcenter.org or hawkins@gseis.ucla.edu
*Internet:* https://www.eastwestcenter.org/scholarships-fellowships/asia-pacific-leadership-program-fellowship
*Sponsor:* Institute of International Education
Menara Imperium, 28th Floor, Suite C
Jakarta Selatan, 12980 Indonesia

**IIE ExxonMobil Scholarships for Research** **1343**
The ExxonMobil Scholarships for Research (BEI) program, administered by IIE, was created to facilitate the development of research projects in the areas of technology, engineering, science and mathematics. $2,000 dollars will be awarded to outstanding students who are studying their last year of undergraduate studies at the UNAM, to complete their research projects (STEM Fields). In addition, fellows will have the opportunity to participate in a mentoring program with an ExxonMobil executive, a leadership seminar, workshops or conferences presented by academic and professional leaders, a conversation club in English, as well as a visit to the headquarters of ExxonMobil in Houston, Texas. Through this program, we seek to foster a new generation of leaders in the field of mathematics and science in Mexico (STEM Fields).
*Requirements:* Countries of Latin America and the Caribbean are eligible to participate in this program.
*Restrictions:* Applicants must willing and able to reside in the United States for the full duration of study; dependents may not accompany scholarship recipients to the U.S.
*Geographic Focus* Puerto Rico, U.S. Virgin Islands, Anguilla, Antigua & Barbuda, Argentina, Aruba, Bahamas, Barbados, Belize, Bolivia, Brazil, British Virgin Islands, Caribbean, Cayman Islands, Chile, Colombia, Costa Rica, Cuba, Dominica, Dominican Republic, Ecuador, El Salvador, Falkland Islands (Malvinas), French Guiana, Grenada, Guadeloupe, Guatemala, Guyana, Haiti, Honduras, Jamaica, Martinique, Mexico, Montserrat, Nicaragua, Panama, Paraguay, Peru, Saint Kitts And Nevis, St. Lucia, St. Vincent and the Grenadines, Suriname, Trinidad and Tobago, Turks and Caicos Islands, Uruguay, Venezuela
*Contact:* Belén Plascencia, Program Coordinator; +52 55 5592 4817; bplascencia@iie.org or latambecas@iie.org
*Internet:* https://www.iie.org/Programs/ExxonMobil-Scholarship-for-Research-Mexico
*Sponsor:* Institute of International Education
1800 West Loop South, Suite 250
Houston, TX 77027

**IIE General Electric Foundation TECH Awards** **1344**
The GE Foundation TECH Award is designed to stimulate innovation in new technology, design, research methods and applications among university students in China. The Award encourages market-oriented technology innovations in the following areas: energy, water, electronics and electrical engineering, machinery and mechanical processing, chemistry and material science. The Award recognizes innovation and excellence in top Chinese university students in the fields of engineering and the sciences. It also encourages recipients to continue in these fields in China after graduation, and helps to foster the development of local talent and human resources. Recipients also receive a trip to the GE China Technology Center in Shanghai to participate in a recognition ceremony and workshops, as well as the opportunity to intern at GE China.
*Requirements:* Applicants must attend one of the following universities: Shanghai Jiao Tong University; Zhejiang University; China University of Technology; Tsinghua University; Xi'an Jiaotong University; Xi'an University of Electronic Science and Technology; University of Electronic Science and Technology; Harbin Institute of Technology; or Huazhong University of Science and Technology. Along with the online application, students must submit a detailed Chinese plan, its summary, two letters of recommendation, college transcripts, and a copy of the applicant's Master's and/or Ph.D. degree.
*Geographic Focus:* China
*Date(s) Application is Due:* Oct 17
*Amount of Grant:* 800 - 2,000 USD
*Contact:* Award Contact; (8) 52 2103-1505; fax 852 2603-5765; china_awards@iiehongkong.org
*Internet:* www.iie.org/Programs/GE-Foundation-TECH-Award
*Sponsor:* Institute of International Education
1400 K Street NW, 7th Floor
Washington, D.C. 20005-2403

**IIE KAUST Graduate Fellowships** **1345**
The King Abdullah University of Science and Technology (KAUST) engages students, faculty, and researchers in advancing science and technology through collaborative inquiry focused on issues of regional and global significance. Their studies focus on the continued advancement of the global quality of life in all issues associated with energy, water, and food. They also explore environment and the Red Sea, and computational science and engineering (high-performance computing, computing for a small planet) as an enabling technology for their research activities. A KAUST Fellowship includes full tuition, a monthly stipend, housing and a travel benefit.
*Requirements:* Eligibly fields of study include: applied mathematics and computational science; bioscience; chemical and biological engineering; chemical science; computer science; earth science and engineering; electrical engineering; environmental science and engineering; marine science and engineering; materials science and engineering; and mechanical engineering.To qualify for the KAUST fellowship, applicants must meet the following requirements before completing the online application: a bachelor's degree in a KAUST-relevant field of study, pursuing a graduate degree in a similar field; minimum GPA of 3.5; a TOEFL score of 79 or 6.0 on the IELTS (International English Language Testing System; see website for specific requirements where TOEFL is not required); must submit GRE scores; a statement of purpose; three letters of recommendation; and an official university transcript.
*Geographic Focus:* All States, All Countries
*Date(s) Application is Due:* Jan 15
*Contact:* KAUST Fellowship Contact; +966 (0) 2 803 3428
*Internet:* www.kaust.edu.sa/admissions/tokaust/fellowship.html
*Sponsor:* King Abdullah University of Science and Technology
Graduate Affairs, Building 9, Suite 4328
Thuwal, Jeddah, 23955-6900 Saudi Arabia

**IIE Leonora Lindsley Memorial Fellowships** **1346**
The Lindsley Scholarship was established for Leonora Lindsley, an American member of an international all-female combat unit on the Western Front during World War II, who was killed the day before the war officially ended. The scholarship was created in her honor by French citizens who are descendants from Resistance fighters to fund those pursuing educational studies in the United States. An award is granted to one or two students for graduate study for a twelve month period, and is renewable for a second year. The winner(s) are chosen every other year. The amount of the award varies from year to year, but ranges from $10,000 to $12,000.
*Requirements:* The Fulbright Commission in Paris gathers applications and sends them to IIE where they are reviewed for completeness and eligibility. Once this is done, a selection panel is convened and finalists chosen. Any major is eligible.
*Restrictions:* Only descendants of French Resistance fighters are eligible. Grantees must be at an accredited U.S. university.
*Geographic Focus:* All States, France
*Amount of Grant:* Up to 12,000 USD
*Contact:* Leonara Lindsley Fellowship Contact; (212) 984-5552
*Internet:* www.iie.org/Programs/Leonora-Lindsley-Memorial-Fellowship
*Sponsor:* Institute of International Education
1400 K Street NW, 7th Floor
Washington, D.C. 20005-2403

**IIE Rockefeller Foundation Bellagio Center Residencies** **1347**
The Bellagio Center Residency program offers scholars, artists, thought leaders, policymakers and practitioners a serene setting conducive to focused, goal-oriented work, and the opportunity to establish new connections with fellow residents across an array of disciplines and geographies. The Center sponsors three kinds of residencies—for scholars, creative artists and practitioners. The Center offers one-month residency for each category and is able to accommodate a group working on a particular project. The Center is particularly interested in applicants whose work connects in some way to the Foundation's issues areas (basic survival safeguards, global health, climate and environment, urbanization, and social and economic security).
*Requirements:* Applicants must submit an online application, a detailed project proposal of no more than 1,250 words, an abbreviated curriculum vitae, three reference letters, and work samples.
*Geographic Focus:* All States
*Date(s) Application is Due:* Dec 1
*Contact:* Bellagio Center Committee
*Internet:* www.rockefellerfoundation.org/bellagio-center
*Sponsor:* Institute of International Education
1400 K Street NW, 7th Floor
Washington, D.C. 20005-2403

**IIE Whitaker International Fellowships and Scholarships** 1348

The goal of the Whitaker funding is to assist in the development of outstanding professional engineers and scientists who will lead and serve the profession with an international outlook. The term of the Fellowship awards will be one academic year (as defined by the academic calendar of the host country). The term of the Scholarship awards can be for as little as one academic semester or as long as a full academic year. Benefits include maintenance allowance, airfare, accident and sickness insurance, partial tuition reimbursement (Fellows only), and access to grantee events, as well as membership in an elite alumni resource network. Applicants should consult the website for specific information about host institutions in Australia, Brazil, France, Germany, Ireland, Italy, The Netherlands, South Africa, Spain, Switzerland, and the United Kingdom.

*Requirements:* U.S. citizens and non-citizen permanent residents may apply. Fellows must have a BS or MS degree in biomedical engineering or bioengineering, and they will not have a doctorate at the time they start the grant. The Fellows applicant's most recent degree may not have been obtained more than two years prior to the start of the grant. Scholars must have either a degree in biomedical engineering and a doctorate which may not have been obtained more than two years prior to the start of the grant, or must be advanced Ph.D. candidates who have been admitted to candidacy and who will receive the Ph.D. prior to departure.

*Geographic Focus:* All States

*Date(s) Application is Due:* Oct 8

*Amount of Grant:* Up to 35,000 USD

*Contact:* Sabeen Altaf, Program Manager; (212) 984-5442; whitaker@iie.org

*Internet:* www.iie.org/Programs/Whitaker-International-Fellows-and-Scholars-Program

*Sponsor:* Institute of International Education
1400 K Street NW, 7th Floor
Washington, D.C. 20005-2403

**ILA Elva Knight Research Grant** 1349

The International Literacy Association's Elva Knight Research Grant provides $5,000 for research in reading and literacy. Projects should be completed within two years and may be carried out using any research method or approach as long as the focus of the project is on research in reading or literacy. The grant study may be carried out using any research method or approach as long as the focus of the project is on research in reading/writing or literacy or teacher preparation for reading/writing instruction. Proposals must be submitted electronically by March 15.

*Requirements:* To be eligible, all applicants must be members of the International Literacy Association. Specific guidelines and application forms are available at the IRA website. The applicants can complete and submit their applications online using a web-based grant management system. Candidates should not include their name, institutional affiliation, or any other personal information that will identify them other than the Profile page, as the applications are conducted as a blind review.

*Restrictions:* Activities such as developing new programs or instructional materials are not eligible for funding, unless these activities are necessary procedures for the conduct of the research.

*Geographic Focus:* All States, All Countries

*Date(s) Application is Due:* Mar 15

*Amount of Grant:* 5,000 USD

*Contact:* Marcie Craig Post, Executive Director; (302) 731-1600; fax (302) 731-1057; research@reading.org or mpost@reading.org

*Internet:* www.literacyworldwide.org/about-us/awards-grants/ila-elva-knight-research-grant

*Sponsor:* International Literacy Association
800 Barksdale Road, P.O. Box 8139
Newark, DE 19714-8139

**ILA Esther Zolt Academic Research Award** 1350

The ILA Esther Zolt Academic Research Award is given annually for a completed research study on the effective classroom implementation of In2Books that inspires others to improve teaching and learning in innovative ways. This grant was established in memory of Esther Zolt, a life-long elementary classroom teacher. Grants are limited to a maximum of US$2,500 and cover reasonable expenses directly related to the conduct of the research. The annual deadline for application submission is March 15.

*Requirements:* Applicants must be ILA members, and submit the application electronically. In preparing the proposal, the following guidelines must be met: do not include the applicant's name, institutional affiliation, or any other personal information that will identify the applicant on any part of the application other than the Cover Sheet (provided in the guidelines). The completed Cover Sheet must be sent as a separate document to ensure a blind review of all proposals.

*Restrictions:* In preparing an application, do not include the applicant's name, institutional affiliation, or any other personal information that will identify the applicant on any part of the application other than the cover sheet. This will ensure a blind review of all applications. Institutional overhead or indirect costs will not be covered under this grant.

*Geographic Focus:* All States, All Countries

*Date(s) Application is Due:* Mar 15

*Amount of Grant:* 2,500 USD

*Contact:* Marcie Craig Post, Executive Director; (302) 731-1600 or (800) 336-7323; fax (302) 731-1057; research@reading.org or mpost@reading.org

*Internet:* https://www.literacyworldwide.org/about-us/awards-grants

*Sponsor:* International Literacy Association
800 Barksdale Road, P.O. Box 8139
Newark, DE 19714-8139

**ILA Esther Zolt Teacher Research Award** 1351

The International Literacy Association (ILA) Esther Zolt Teacher Research Award is given annually for a completed action research study. The study should be conducted by a teacher who has implemented In2Books or an ePals project, and the Common Core State Standards with exceptional innovation and rigor, leading to the development of curious and independent learners. This grant was established in memory of Esther Zolt, a life-long elementary classroom teacher, and is sponsored by Nina Zolt and Miles Gilburne. Grants are limited to a maximum of $2,500, and cover reasonable expenses directly related to the conduct of the research. The annual deadline for application submission is March 15.

*Requirements:* Applicants must be members of the ILA. In preparing the proposal, these guidelines must be followed: do not include the applicant's name, institutional affiliation, or any other personal information that will identify the applicant on any part of the application other than the Cover Sheet (provided in the guidelines). The completed Cover Sheet should be emailed as a separate document from the proposal to endure a blind review of all proposals.

*Restrictions:* Institutional overhead or indirect costs will not be covered under this grant.

*Geographic Focus:* All States, All Countries

*Date(s) Application is Due:* Mar 15

*Amount of Grant:* 2,500 USD

*Contact:* Marcie Craig Post, Executive Director; (302) 731-1600 or (800) 336-7323; fax (302) 731-1057; research@reading.org or mpost@reading.org

*Internet:* https://www.literacyworldwide.org/about-us/awards-grants

*Sponsor:* International Literacy Association
800 Barksdale Road, P.O. Box 8139
Newark, DE 19714-8139

**ILA Helen M. Robinson Dissertation Research Award** 1352

The Helen R. Robinson Dissertation Research Award, established in memory of Helen M. Robinson, supports doctoral students at the early stages of their dissertation research in the area of reading and literacy. Because Robinson devoted much of her work to developing a better understanding of instructional interventions, the grant in the amount of $1200 will be used to support dissertation research in this area. The annual deadline for submissions is March 15.

*Requirements:* Applicants must be members of the International Literacy Association (ILA) and adhere to the following guidelines: complete and submit a proposal online using a web-based grant management system (Nimble Net) from the IRA website; do not include the applicant's name, institutional affiliation, or other personal information that identifies the applicant on any part of the proposal other than the Profile page (grant competitions will be conducted as a blind review); do not exceed the maximum character requirement (spaces and punctuation are included in the total count; and complete all editing before submitting the proposal, as editing cannot be done after submission. Specified instructions, criteria for proposal review, and the current application is available at the website.

*Restrictions:* Applicants may apply for more than one research grant, but are limited to one research award per year.

*Geographic Focus:* All States, All Countries

*Date(s) Application is Due:* Mar 15

*Amount of Grant:* 1,200 USD

*Contact:* Marcie Craig Post, Executive Director; (302) 731-1600 or (800) 336-7323; fax (302) 731-1057; research@reading.org or mpost@reading.org

*Internet:* www.literacyworldwide.org/about-us/awards-grants/ila-helen-m-robinson-grant

*Sponsor:* International Literacy Association
800 Barksdale Road, P.O. Box 8139
Newark, DE 19714-8139

**ILA Jeanne S. Chall Research Fellowship** 1353

The Jeanne S. Chall Fellowship was established to encourage and support reading research. The Fellowship supports research efforts in the following areas: beginning reading (theory, research, and practice that improves the effectiveness of learning to read); readability (methods of predicting the difficulty of texts); reading difficulty (diagnosis, treatment, and prevention); stages of reading development; the relation of vocabulary to reading; and diagnosing and teaching adults with limited reading ability. The Fellowship is worth $5,000, and the annual deadline is March 15. Additional guidelines, including instructions for the web-based application system and review criteria, are available at the website.

*Requirements:* All applicants must be members of the International Literacy Association. Open to doctoral students who are planning or beginning their dissertations and to university based graduate students who are already pursuing reading research.

*Geographic Focus:* All States, All Countries

*Date(s) Application is Due:* Mar 15

*Amount of Grant:* 5,000 USD

*Contact:* Marcie Craig Post, Executive Director; (302) 731-1600 or (800) 336-7323; fax (302) 731-1057; research@reading.org or mpost@reading.org

*Internet:* www.literacyworldwide.org/about-us/awards-grants/ila-jeanne-s-chall-research-fellowship

*Sponsor:* International Literacy Association
800 Barksdale Road, P.O. Box 8139
Newark, DE 19714-8139

**ILA Leaders Inspiring Readers Award** 1354

The International Literacy Association's Leaders Inspiring Readers Award is presented to a researcher or practitioner who has contributed significantly to advancing the knowledge base and instructional efficacy of educators in the focus area of struggling readers. The annual deadline for applications is March 15.

*Requirements:* All applicants must be members of the International Literacy Association.

*Geographic Focus:* All States, All Countries
*Date(s) Application is Due:* Mar 15
*Contact:* Marcie Craig Post, Executive Director; (302) 731-1600 or (800) 336-7323; fax (302) 731-1057; research@reading.org or mpost@reading.org
*Internet:* https://www.literacyworldwide.org/about-us/awards-grants/ila-leaders-inspiring-readers-award
*Sponsor:* International Literacy Association
800 Barksdale Road, P.O. Box 8139
Newark, DE 19714-8139

**ILA Steven A. Stahl Research Grant** 1355

This $1,000 ILA Steven A. Stahl Research Grant award was established to honor the memory and work of Steven A. Stahl by encouraging and supporting promising graduate students in their research. This award will be given annually to a recipient with at least three years of teaching experience who is conducting classroom research (including action research) focused on improving reading instruction and children's reading achievement. The annual deadline for application submission is March 15.
*Requirements:* The research must be conducted in the classroom(s), focused on improving reading instruction and children's reading achievement, empirically rigorous, and completed within two years.
*Geographic Focus:* All States, All Countries
*Date(s) Application is Due:* Mar 15
*Amount of Grant:* 1,000 USD
*Contact:* Marcie Craig Post, Executive Director; (302) 731-1600 or (800) 336-7323; fax (302) 737-0878; research@reading.org or mpost@reading.org
*Internet:* www.literacyworldwide.org/about-us/awards-grants/ila-steven-a-stahl-research-grant
*Sponsor:* International Literacy Association
800 Barksdale Road, P.O. Box 8139
Newark, DE 19714-8139

**ILA Technology and Literacy Award** 1356

The International Literacy Association Technology and Literacy Award honors educators in grades K through 12 who are making an outstanding and innovative contribution to the use of technology in reading education. All entrants must be educators who work directly with students ages 5 through 18 for all or part of the working day. This award honors one (1) overall winner and up to seven (7) regional winners (including Canada and International). Entries incorporating any brand of computers, operating systems, software applications, and other technologies such as TV or video are acceptable. The annual deadline for applications is March 15.
*Requirements:* Educators who work in public, parochial, or private elementary or secondary schools in the United States, Canada, or other countries (includes U.S. Territories and AP.O. schools) are eligible. Candidates must be: an ILA member; and educators who work directly with student's grades K-12 for all or part of the working day.
*Restrictions:* No programs planned for future implementation will be accepted.
*Geographic Focus:* All States, All Countries
*Date(s) Application is Due:* Mar 15
*Contact:* Marcie Craig Post, Executive Director; (302) 731-1600 or (800) 336-7323; fax (302) 731-1057; mpost@reading.org
*Internet:* www.literacyworldwide.org/about-us/awards-grants/
*Sponsor:* International Literacy Association
800 Barksdale Road, P.O. Box 8139
Newark, DE 19714-8139

**ILA Timothy & Cynthia Shanahan Outstanding Dissertation of the Year Award** 1357

The International Literacy Association's Timothy nd Cynthia Shanahan Outstanding Dissertation of the Year Award is intended for doctoral students who have focused their research in the reading field or who have conducted related research having implications for reading. Studies using any research approach (ethnographic, experimental, historical, survey, etc.) are encouraged. Each study will be assessed by the scholarly qualification of its report and its significant contributions to knowledge within the reading field. The annual deadline for application submission is March 15.
*Requirements:* Dissertations must have been approved by May of the previous application year. Applicants must be current members of the International Reading Association. A forty-page monograph based on the dissertation should be emailed to research@reading.org or mailed to the IRA's address by October 1. The application package should include: title page, abstract, table of contents, dissertation monograph, and references. Applicants must be ILA members.
*Geographic Focus:* All States, All Countries
*Date(s) Application is Due:* Mar 15
*Amount of Grant:* 1,000 USD
*Contact:* Marcie Craig Post, Executive Director; (302) 731-1600 or (800) 336-7323; fax (302) 731-1057; research@reading.org or mpost@reading.org
*Internet:* www.literacyworldwide.org/about-us/awards-grants/ila-outstanding-dissertation-of-the-year-award
*Sponsor:* International Literacy Association
800 Barksdale Road, P.O. Box 8139
Newark, DE 19714-8139

**Illinois Clean Energy Community Foundation Renewable Energy Grants** 1358

The Illinois Clean Energy Community Foundation helps non-profit groups and local government organizations to purchase and install renewable energy technologies, including photovoltaic panels, solar thermal systems, wind turbines and biomass systems. Renewable energy provides heating, cooling, hot water, steam or electricity, in an environmentally sound manner while also lowering or stabilizing utility costs. Before considering renewable energy, make the building as energy efficient as possible so that the installed renewable energy system can be smaller, less costly and able to offset a larger percentage of the energy consumed. The Foundation's goal, through its support of renewable energy technologies, is to improve the environment, help grow the renewable energy sector in Illinois, create jobs and lower energy costs. We also work with colleges and universities, as well as K-12 schools and teachers around the state to help educate and train Illinois' next generation about the benefits and potential of renewable energy.
*Requirements:* 501(c)3 tax-exempt organizations, educational institutions, and state and local government agencies serving Illinois residents are eligible.
*Restrictions:* The foundation will not provide funding for remediation of environmentally impaired properties; technology research; promotion of proprietary products; reoccurring operating costs; political campaigns or lobbying; capital campaigns or support for an organization's endowment; or projects undertaken by individuals.
*Geographic Focus:* Illinois
*Date(s) Application is Due:* Jan 15; Jul 15
*Amount of Grant:* Up to 2,500,000 USD
*Contact:* Bob Romo, Grants Administrator; (312) 372-5191; fax (312) 372-5190; info@illinoiscleanenergy.org
*Internet:* www.illinoiscleanenergy.org/renewable-energy/
*Sponsor:* Illinois Clean Energy Community Foundation
2 N LaSalle Street, Suite 1140
Chicago, IL 60602

**Illinois DCEO Coal Research Grants** 1359

The Illinois Clean Coal Institute (ICCI) Coal Research and Development (R&D) Program is the technical component of the Illinois Department of Commerce and Economic Opportunity (D.C.EO) Office of Coal Development (OCD). The D.C.EO / OCD administers the Illinois Coal Research Program and is dedicated to the development and use of Illinois' extensive coal resources as a fuel source for the 21st century. The main funding cycle of the ICCI is through the annual Request for Proposals (RFP) solicitation. The RFP solicits proposals for research projects up to the commercialization stage. The program provides partial funding to research clean coal technology, coal chemistry, mining productivity and coal combustion byproduct utilization. Subgrants are typically between $60,000 and $250,000 and are issued by Southern Illinois University-Carbondale. Grant periods are limited to 24 months. The annual program cycle begins in April of each year with a Request for Proposals issued by the ICCI. Proposal review and project selection is typically completed by the end of August. Unsolicited proposals may be submitted at any time. All projects under this solicitation are capped at a maximum of $200,000. Project funds requested near the upper range should involve pilot-scale work or other commercial demonstrations, with extensive cost sharing by industrial entities and research partners. ICCI funds may be requested as cost sharing for larger-scale demonstration projects with an Illinois coal emphasis. Allocation of ICCI funds to such projects is of special interest.
*Requirements:* Any entity may apply. Proposals undergo a rigorous peer review. Preference is given to Illinois applicants. Researchers are encouraged to meet with ICCI staff at any time to discuss concepts that may have the potential to further the Illinois coal industry. If the concept appears sound and has the potential to significantly benefit the Illinois coal industry, the ICCI staff will assist the group in the formulation and submission of a Statement of Interest (SOI) document rather than submitting a formal proposal. This document will be sent to researchers after initial discussions with the ICCI. The SOI does not contain all the detail required in a formal proposal, and the scope of the SOI can be as narrow or broad as desired. Another difference between the SOI and a formal proposal is that the SOI allows communications between the ICCI and the prospective proposer prior to preparation of a formal proposal. Usually, the SOI will be between two and four pages. After review of the SOI, the applicant will be notified in writing as to whether or not their project was selected for further consideration. If reviews of the SOI indicate that a formal proposal should be submitted, the proposer will be guided by the ICCI staff to submit an unsolicited proposal. Unsolicited proposals use a similar format as the RFP. Although this program has a coal research emphasis, there are numerous opportunities that do not require experience in hands-on coal research, such as carbon management, materials, catalysis, fundamental and applied chemistry, business practices, mining, and engineering studies.
*Restrictions:* Proposals dealing with the following types of R&D are not of interest: any projects involving the use, classification, and upgrading of coal combustion residues (CCR's). This includes the recovery of special materials or elements from the ash; any projects involving plasma gasification, either as a stand-alone technology or as a supplemental technology; laboratory-scale and small pilot-scale research on CO2 capture and/or separation; or, any projects focusing on fuel cell research. Proposals dealing with these subjects will be deemed nonresponsive and will not be reviewed or considered.
*Geographic Focus:* Illinois
*Amount of Grant:* Up to 250,000 USD
*Contact:* Dr. Francois Botha, (618) 985-3500
Paul Pierre-Louis, (312) 814-3630; fax (312) 814-5247; Paul.Pierre-Louis@illinois.gov
*Internet:* www.icci.org/funding.php
*Sponsor:* Illinois Department of Commerce and Economic Opportunity
500 East Monroe Street
Springfield, IL 62701-1643

**Illinois DCEO Coal Revival Grants** 1360

The Illinois Coal Revival Program provides financial assistance in the form of grants to assist with the development of new, coal-fired electric generation capacity or coal gasification facilities in Illinois. Financial assistance through the program will be

provided in the form of a grant based on State Retail Occupation Taxes that will be paid on Illinois coal purchases for new facilities. Qualifying facilities may be eligible for grants roughly equal to the present value of future sales taxes paid on Illinois-mined coal over a 25-year period, up to a maximum amount of $100 million. Unsolicited proposals may be submitted at any time during the year. Funding of selected proposals may be substantially delayed due to the need for legislative action to establish a funding appropriation.
*Requirements:* Businesses must propose to construct a new electric generating or expand at an existing electric generating facility, including transmission lines and associated equipment to provide baseload electric power or to construct a coal gasification facility. The proposed new facility or facility expansion must: 1) have an aggregate nameplate generating capacity of 400 megawatts or more for all units at one site, will use coal or gases derived from coal as its primary fuel source at the proposed facility, and will support the creation of at least 150 new Illinois coal-mining jobs, or 2) use coal gasification or IGCC to generate chemical feedstocks, transportation fuels or electricity.
*Geographic Focus:* Illinois
*Amount of Grant:* Up to 100,000,000 USD
*Contact:* Dan Wheeler, (217) 785-1671; fax (217) 558-2647; Dan.Wheeler@illinois.gov
*Internet:* www.illinoisbiz.biz/dceo/Bureaus/Coal/Grants/
*Sponsor:* Illinois Department of Commerce and Economic Opportunity
500 East Monroe Street
Springfield, IL 62701-1643

**Illinois DNR Youth Recreation Corps Grants** **1361**
The Illinois Youth Recreation Corps was established for making grants to local sponsors to provide wages to youth operating and instructing in recreational and conservation programs for the benefit of other youth. Such programs shall provide recreational opportunities for children of all age levels and shall include, but not be limited to, the coordination and teaching of physical activities, arts and handicrafts, and learning activities.
*Requirements:* Potential applicants include local sponsors who can provide necessary facilities, materials and management for summer recreational and conservation activities for youth within the community and who desire a grant for hiring eligible youth as supervisors, instructors, instructional aides or maintenance personnel. Local sponsors must be units of local government or not?for?profit entities. Enrollment (hiring) is limited to citizens of the State of Illinois who, at the time of hiring, are 16, 17, 18 or 19 years of age, and who have skills that can be used in the summer recreation or conservation program.
*Restrictions:* Youth currently employed in any manner by the local sponsor are not eligible for inclusion in the program.
*Geographic Focus:* Illinois
*Date(s) Application is Due:* Jun 30
*Contact:* Valerie Keener, Administrator; (217) 524-4126 or (217) 782-7481; fax (217) 782-9599; valerie.keener@illinois.gov or dnr.grants@illinois.gov
*Internet:* www.dnr.illinois.gov/grants/Pages/default.aspx
*Sponsor:* Illinois Department of Natural Resources
One Natural Resources Way
Springfield, IL 62702-1271

**IMLS 21st Century Museum Professionals Grants** **1362**
The 21st Century Museum Professionals program supports a range of activities, including professional training in all areas of museum operations and leadership development. This program provides the museum community with support for a variety of training and personnel development activities for museum staff members across all types of museums, as well as the collection and dissemination of information to museum professionals and the public. Project design could include direct dissemination of information through workshops, seminars, and courses, or indirect communication through publications and Web sites. Projects should benefit multiple institutions or diverse constituencies.
*Requirements:* Museums that fulfill the Eligibility Criteria for Museums may apply. Private not-for-profit museum services organizations or associations that engage in activities designed to advance the well-being of museums and the museum profession also may apply. In addition, institutions of higher education, including public and not-for-profit universities, are eligible. Please see Program Guidelines for specific eligibility criteria.
*Geographic Focus:* All States
*Date(s) Application is Due:* Mar 15
*Amount of Grant:* 15,000 - 500,000 USD
*Contact:* Twinet G. Kimbrough; (202) 653-4703; tkimbrough@imls.gov
*Internet:* www.imls.gov/applicants/grants/21centuryMuseums.shtm
*Sponsor:* Institute of Museum and Library Services
1800 M Street NW, 9th Floor
Washington, D.C. 20036-5802

**IMLS Conservation Project Support Grants** **1363**
The grants provide matching funds to help museums identify conservation needs and priorities and perform activities to ensure the safekeeping of their collections. Projects may assess four types of collections: nonliving, systematics/natural history, living plants, and living animals. Grants are available for different types of conservation projects, such as environmental improvements, research, surveys and long-range plans, training, and treatment. Projects may not exceed two years. IMLS will pay up to one-half the cost of the project. A museum's matching funds must come from cash contributions, earned income, in-kind contributions, or non-federal funds. This program also funds Exceptional Projects. These are projects with broad-reaching effects that benefit multiple institutions. The deadline apples to programs for museums in counties of Alabama, Florida, Louisiana, and Mississippi that have been declared disaster areas by the federal government.
*Requirements:* Museums located in one of the 50 states, the Commonwealth of Puerto Rico, American Samoa, the Virgin Islands, the Northern Mariana Islands, the Trust Territory of the Pacific Islands, Guam, or the District of Columbia may apply. Applicant organizations must have provided museum services, including exhibiting objects to the general public, on a regular basis for at least two years prior to application. A public or private nonprofit organization such as a municipality, college, or university that is responsible for the operation of a museum may, if necessary, apply on behalf of the museum.
*Restrictions:* A museum operated by a department or agency of the federal government is not eligible to apply. Funds may not be used for the acquisition of objects, contributions to endowment funds, costs of social activities, major renovations, new construction, or preaward costs.
*Geographic Focus:* All States
*Date(s) Application is Due:* Oct 1
*Amount of Grant:* 1 - 150,000 USD
*Samples:* University of Wisconsin (Madison, WI)—for conservation project, $24,000.
*Contact:* Mark Feitl, Program Specialist; (202) 653-4635; mfeitl@imls.gov
*Internet:* www.imls.gov/applicants/grants/conservProject.shtm
*Sponsor:* Institute of Museum and Library Services
1800 M Street NW, 9th Floor
Washington, D.C. 20036-5802

**IMLS Grants to State Library Administrative Agencies** **1364**
Through the program, the Institute of Museum and Library Services provides funds to State Library Administrative Agencies (SLAAs) using a population-based formula. State libraries may use the appropriation to support statewide initiatives and services. They also may distribute the funds through subgrant competitions or cooperative agreements to public, academic, research, school, and special libraries in their state.
*Requirements:* State library administrative agencies located in one of the 50 states of the United States, the District of Columbia, the Commonwealth of Puerto Rico, Guam, American Samoa, the U.S. Virgin Islands, the Commonwealth of the Northern Mariana Islands, the Republic of the Marshall Islands, the Federated States of Micronesia, and the Republic of Palau are eligible to submit five-year plans.
*Geographic Focus:* All States
*Date(s) Application is Due:* Apr 1
*Amount of Grant:* 1 - 2,000,000 USD
*Contact:* Laurie C. Brooks; (202) 653-4650; stateprograms@imls.gov
*Internet:* www.imls.gov/programs/programs.shtm
*Sponsor:* Institute of Museum and Library Services
1800 M Street NW, 9th Floor
Washington, D.C. 20036-5802

**IMLS Laura Bush 21st Century Librarian Program Grants** **1365**
The program supports efforts to recruit and educate the next generation of librarians and the faculty who will prepare them for careers in library science. It also supports grants for research related to library education and library staffing needs, curriculum development, and continuing education and training. Program priorities are masters-level programs, doctoral programs, pre-professional programs, research, programs to build institutional capacity, and continuing education and training. The grant period is up to three years, except doctoral programs, which may be up to four years. Applications, guidelines, and examples of successful proposals are available online.
*Requirements:* All applicants for the Laura Bush 21st Century Librarian program are required to file their applications, including all attachments, online through Grants.gov. All types of libraries, except federal and for-profit libraries, may apply. Eligible libraries include public, school, academic, special, private (not-for-profit), archives, library agencies, library consortia, and library associations. In addition, research libraries that give the public access to services and materials suitable for scholarly research not otherwise available to the public and that are not part of a university or college are eligible. Institutions of higher education, including public and not-for-profit universities and colleges, also are eligible. Graduate schools of library and information science may apply as part of an institution of higher education.
*Geographic Focus:* All States
*Date(s) Application is Due:* Dec 15
*Amount of Grant:* 50,000 - 1,000,000 USD
*Contact:* Kevin Cherry, Senior Program Officer; (202) 653-4662; kcherry@imls.gov
*Internet:* www.imls.gov/applicants/grants/21centuryLibrarian.shtm
*Sponsor:* Institute of Museum and Library Services
1800 M Street NW, 9th Floor
Washington, D.C. 20036-5802

**IMLS National Leadership Grants (NLG)** **1366**
National Leadership Grants enable libraries and museums to help people gain the knowledge, skills, attitudes, behaviors, and resources that enhance their engagement in community, work, family, and society. Projects should enable libraries and museums to address current problems in creative ways, develop and test innovative solutions, and expand the boundaries within which cultural heritage institutions operate. The results of these projects will help equip tomorrow's libraries and museums to better meet the needs of a Nation of Learners. Successful proposals will show evidence that they will have national impact and generate results - new tools, research, models, services, practices, or alliances - that can be widely adapted or replicated to extend the benefit of federal support. Proposals will reflect an understanding of current issues and needs, showing the potential for far-reaching impact throughout the museum and/or library community. Projects will provide creative solutions to issues of national importance and provide leadership for other organizations.
*Requirements:* All types of libraries, except federal and for-profit libraries, may apply. Eligible libraries include public, school, academic, special, private (not-for-profit), archives, library agencies, library consortia, and library associations. Research libraries and archives

that give the public access to services and materials suitable for scholarly research not otherwise available and that are not part of a university or college also are eligible. Digital libraries that make library materials publicly available and provide services including selection, organization, description, reference, and preservation under the supervision of at least one permanent professional staff librarian are eligible to apply. Institutions of higher education, including public and not-for-profit universities and colleges, also are eligible. An academic unit, such as a graduate school of library and information science, may apply as part of an institution of higher education. Library applicants may apply individually or as partners. Additionally, all types of museums, large and small, are eligible for funding. Eligible museums include aquariums, arboreta and botanical gardens, art museums, youth museums, general museums, historic houses and sites, history museums, nature centers, natural history and anthropology museums, planetariums, science and technology centers, specialized museums, and zoological parks. Private nonprofit museum services organizations or associations that engage in activities designed to advance the well-being of museums and the museum profession also may apply. In addition, institutions of higher education, including public and nonprofit universities, are eligible.
*Geographic Focus:* All States
*Date(s) Application is Due:* Feb 1
*Amount of Grant:* 50,000 - 1,000,000 USD
*Contact:* Rachel Frick, Program Officer, Libraries; (202) 653-4667; rfrick@imls.gov
*Internet:* www.imls.gov/applicants/grants/nationalLeadership.shtm
*Sponsor:* Institute of Museum and Library Services
1800 M Street NW, 9th Floor
Washington, D.C. 20036-5802

**Independent Institute John M. Templeton Fellowships Essay Contest** **1367**
The Templeton Fellowships Essay Contest gives monetary awards to junior faculty and students who submit an essay based on a particular topic. Essays are submitted from all over the world. First price junior faculty are awarded $10,000, with students awarded first ($2,500), second ($1,500), and third ($1,000) prizes. Applicants should consult the website for current information, as fellowships are not available every year.
*Requirements:* College or university students who are part-time, full-time, undergraduate, graduate, and from all nations and academic disciplines are eligible. (Student must be enrolled on date of essay submission.) Students must be no older than 35 years. Junior faculty members must hold a position of assistant professor or higher, not yet tenured, and may not be older than 35 years.
*Restrictions:* Previous fellowship recipients, judges, and current or past institute employees are ineligible.
*Geographic Focus:* All States, All Countries
*Amount of Grant:* 1,000 - 10,000 USD
*Contact:* Carl Close, Academic Affairs Director; (510) 632-1366; fax (510) 568-6040
*Internet:* www.independent.org/students/essay/winners2010.asp
*Sponsor:* Independent Institute
100 Swan Way
Oakland, CA 94621-1428

**Institute for Agriculture and Trade Policy Food and Society Fellowships** **1368**
The IATP Food and Society Fellows Program provides fellowships to professionals in food and agriculture from across North America, enabling them to use mass media channels to inform and shape the public agenda. The goal of the program is to create sustainable food systems that promote good health, vibrant communities, environmental stewardship, worker justice and accessibility for all. Fellows come from many disciplines: chefs, farmers, nutritionists, activists, public health professionals, fishers, policy experts and academics. Together they form an interdisciplinary team that works to: use communication to influence the issues that reach the public agenda, thereby creating policy changes at the personal, organizational and public policy levels that advance sustainable food and farming systems; increase the mass media communications on issues around sustainable food and farming systems that produce healthy, green, fair and affordable foods; raise the profile of the fellows as food system experts among media and policymakers; and build capacity, leadership, and cohesiveness in a group of experts who collaborate and communicate using mass media channels to bring sustainable food system issues to the public agenda.
*Requirements:* Applicants should be U.S. residents actively involved in a professional career that involves two or more of the following: agriculture, health promotion, youth development, food production, and/or policy analysis.
*Geographic Focus:* All States
*Contact:* Mark Muller, Director; (612) 870-3420; fax (612) 870-4846; mmuller@iatp.org
*Internet:* www.foodandsocietyfellows.org/about/page/about-us
*Sponsor:* Institute for Agriculture and Trade Policy
2105 First Avenue South
Minneapolis, MN 55404

**Institute of Noetic Sciences Research Internships** **1369**
As part of its objective to advance the science of consciousness, IONS is dedicated to training emerging noetic scientists. IONS hosts a limited number of research, education, and communications interns each year. The Institute's interns have been high school and college students wishing to gain research and nonprofit experience, graduate and post-graduate students who engage in more formal training experiences, and those returning to the workforce or making a career change. Internships are unpaid, and interns cannot live on campus.
*Geographic Focus:* All States
*Contact:* Cassandra Vieten, Director of Research; (707) 779-8211 or (707) 779-8212; cvieten@noetic.org or research@noetic.org
*Internet:* www.noetic.org/research/participate/internships/
*Sponsor:* Institute of Noetic Sciences
625 2nd Street, Suite 200
Petaluma, CA 94952-5120

**Intel Corporation Fellowships** **1370**
The Intel Corporation Fellowship program awards fellowships to PhD candidates doing work in fields related to Intel's business and research interests. These fellowships, only available at select U.S. universities, include a cash award (tuition/fees/stipend), an Intel PC, introduction to an Intel Mentor, and the opportunity to conduct research or an internship at Intel. Approximately 35 fellowships are awarded annually. In addition, Intel Foundation collaborates with The National Consortium for Graduate Degrees for Minorities in Engineering and Science (GEM) to help support graduate education for underrepresented minorities at the master's and PhD levels. Contact your school's engineering or computer science department to receive additional information.
*Geographic Focus:* All States
*Date(s) Application is Due:* Nov 15
*Contact:* Fellowship Coordinator; (408) 765-8080; fax (408) 765-9904
*Internet:* www.intel.com/content/www/us/en/education/university/students.html
*Sponsor:* Intel Corporation
2200 Mission College Boulevard
Santa Clara, CA 95054-1549

**Intel Corporation International Internships** **1371**
Several Intel manufacturing facilities, in cooperation with participating international colleges, offer paid internships for students enrolled full-time in specific programs. These internships offer hands-on technical education and experience and valuable networking opportunities. There are a limited number of international internships available, so early application is encouraged. One of the great advantages of these internships is the flexible scheduling, designed to work around a participant's school schedule. The compressed work week offers a three or four day week that gives an Intern time for all other interests and activities. Interns receive: competitive salaries; vacation and holiday time; access to Intel University classes for professional and personal development; networking with Intel managers and executives; relocation assistance; and earned credit toward sabbaticals. Application details are available through the career centers of participating international colleges.
*Requirements:* In order to be eligible for Intel's international internship program, applicants must: have declared in a specific full-time degree program such as semiconductor manufacturing, electronics technology or automation, etc. at a participating college; maintain a 3.0 or higher GPA on a 4.0 scale; be able to work a minimum number of hours as set by their Intel manager; and be enrolled in classes and making significant progress toward a degree. Currently, nearly 300 international internships are offered in the following locations: Israel; Malaysia; and China.
*Geographic Focus:* China, Israel, Malaysia
*Contact:* Internship Coordinator; (408) 765-8080; fax (408) 765-9904
*Internet:* https://intel.taleo.net/careersection/10000/jobsearch.ftl
*Sponsor:* Intel Corporation
2200 Mission College Boulevard
Santa Clara, CA 95054-1549

**International Crane Foundation Internships** **1372**
The International Crane Foundation provides a formal stipend program, with internships being offered in the fields of aviculture and restoration. In the aviculture program, interns will receive intensive hands-on training in the care and management of endangered cranes, including husbandry, handling techniques, behavior, stimulating reproduction, incubation, chick rearing, artificial insemination, health care, and genetic management. The associate also manages the volunteer chick parent program. Each intern is responsible for developing and completing a research project with assistance from the staff. Internships usually run from March through May, June through August, September through November, and December through February. Application deadlines are for spring, summer, fall, and winter internships. Under the restoration program, interns will be involved in the ecosystems restoration program, which directs the restoration of prairie, savanna, and wetland communities on the ICF property. Interns will develop important ecological, hydrological, botanical, and horticultural skills. Depending on the season, interns may be involved in prescribed burns, field studies of wild sandhill care populations; seed collecting, planting, and site preparation for community restorations; and interaction with foreign scientists. The associate also will manage the seed collection program. Internships usually run for six months from March through August and from June through November. Application deadlines are January 1 and April 1 for the spring-summer and summer-fall internships, respectively. The program runs for nine months from March through November.
*Requirements:* Applications may be made by recent college sophomores through college graduates in biology, zoology, or botany who have a willingness to work at manual tasks and can work in a self-directed manner. Individuals seeking graduate training are especially encouraged.
*Geographic Focus:* All States
*Date(s) Application is Due:* Jan 1; Apr 1; Jul 1; Oct 1
*Amount of Grant:* 1,400 USD
*Contact:* Becky Abel, Grants Manager; (608) 356-9462, ext. 144; fax (608) 356-9465; info@savingcranes.org
*Internet:* https://www.savingcranes.org/about-icf/get-involved/intern-opportunities/
*Sponsor:* International Crane Foundation
P.O. Box 447, E-11376 Shady Lane Road
Baraboo, WI 53913-0447

**IREX Mandela Washington Fellowship for Young African Leaders** 1373

Begun in 2014, the Mandela Washington Fellowship is the flagship program of the Young African Leaders Initiative (YALI) that empowers young people through academic coursework, leadership training, and networking. In the summer of each year, the Fellowship will host up to 700 young African leaders. Fellows will travel to U.S. university and college campuses for six-week Academic and Leadership Institutes in one of three tracks: Business and Entrepreneurship, Civic Leadership, or Public Management. For more details on the sector tracks, please review the application instructions. The intensive Institute model includes linkages between academic sessions, leadership development, site visits, community service, networking, and organized cultural and civic activities. Fellows will gain both theoretical knowledge and practical skills, and the Institutes will take place from mid-June to late July. The Institutes will culminate in a high-level, three-day Summit in Washington, D.C. Additionally, up to 100 competitively-selected Fellows will remain in the United States to participate in individually tailored six-week professional development experiences (PDE) at private companies, civil society organizations, or public-sector agencies. The PDE offers Fellows practical training and the opportunity to learn transferable skills, expand their professional networks, and apply concepts learned at the Institutes to real-world situations. The annual deadline for applications is October 11.

*Requirements:* The Mandela Washington Fellowship is open to young African leaders who meet the following criteria: are between the ages of 25 and 35 at the time of application submission, although exceptional applicants younger than 25 will be considered; are eligible to receive a United States J-1 visa; and are proficient in reading, writing, and speaking English. Citizens of one of the following countries are eligible to apply: Angola, Benin, Botswana, Burkina Faso, Burundi, Cameroon, Cape Verde, Central African Republic, Chad, Comoros, Democratic Republic of the Congo (DRC), Republic of the Congo, Cote d'Ivoire, Djibouti, Equatorial Guinea, Eritrea, Ethiopia, Gabon, The Gambia, Ghana, Guinea, Guinea-Bissau, Kenya, Lesotho, Liberia, Madagascar, Malawi, Mali, Mauritania, Mauritius, Mozambique, Namibia, Niger, Nigeria, Rwanda, Sao Tome and Principe, Senegal, Seychelles, Sierra Leone, Somalia, South Africa, South Sudan, Sudan, Swaziland, Tanzania, Togo, Uganda, Zambia, and Zimbabwe.

*Restrictions:* Applicant cannot be: U.S. citizens or permanent residents of the United States; or employees or immediate family members of employees of the U.S. government (including the U.S. Embassy, USAID, and other U.S. government entities).

*Geographic Focus:* Angola, Benin, Botswana, Burkina Faso, Burundi, Cameroon, Cape Verde, Central African Republic, Chad, Comoros, Congo, Congo, Democratic Republic of, Cote d' Ivoire (Ivory Coast), Djibouti, Egypt, Equatorial Guinea, Eritrea, Ethiopia, Gabon, Gambia, Ghana, Guinea, Guinea-Bissau, Kenya, Lesotho, Liberia, Libya, Madagascar, Malawi, Mali, Mauritania, Mauritius, Morocco, Mozambique, Namibia, Niger, Nigeria, Rwanda, Sao Tome & Principe, Senegal, Seychelles, Sierra Leone, Somalia, South Africa, Sudan, Swaziland, Tanzania, Togo, Uganda, Zambia, Zimbabwe

*Date(s) Application is Due:* Oct 11

*Contact:* Amy Bernath, Deputy Director, Education; (202) 628-8188; fax (202) 628-8189; abernath@irex.org or mwfellowship@irex.org

*Internet:* https://www.irex.org/program/mandela-washington-fellowship-application-information

*Sponsor:* International Research and Exchanges Board
1275 K Street, NW, Suite 600
Washington, D.C. 20005

**IREX Mandela Washington Fellowship for Young African Leaders Reciprocal Exchange Component** 1374

The Mandela Washington Fellowship for Young African Leaders, the flagship program of the Young African Leaders Initiative (YALI), was created in 2014 to invest in the next generation of African leaders. The Reciprocal Exchange Component provides Americans with the opportunity to travel to Africa to build upon strategic partnerships and professional connections developed during the Mandela Washington Fellowship in the United States. The program provides funds for American professionals to travel to sub-Saharan African countries to continue collaborative work with Mandela Washington Fellows on the continent. Through these projects, Americans and young African leaders form lasting partnerships, expand markets and networks, and increase mutual understanding between the United States and countries in sub-Saharan Africa. The application submission deadline is April 4.

*Requirements:* American professionals may include those met during the Mandela Washington Fellowship through site visits, networking, the Summit, Professional Development Experiences, or other activities. American professionals participating in the Reciprocal Exchange must be U.S. citizens currently living in the United States and must not have previously received a Reciprocal Exchange award. Americans may not be current U.S. Federal Government employees. Fellows must be considered alumni of the Fellowship by the U.S. Department of State and must not have previously received a Reciprocal Exchange award.

*Geographic Focus:* All States

*Date(s) Application is Due:* Apr 4

*Contact:* Amy Bernath, Deputy Director, Education; (202) 628-8188; fax (202) 628-8189; abernath@irex.org or REmwf@irex.org

*Internet:* https://www.irex.org/program/mandela-washington-fellowship-reciprocal-exchange-opportunity-fellows-and-american

*Sponsor:* International Research and Exchanges Board
1275 K Street, NW, Suite 600
Washington, D.C. 20005

**IREX University Administration Support Program (UASP) Fellowships** 1375

The University Administration Support Program (UASP) benefits university administrators at select state universities from fourteen countries across Eurasia and Africa by providing access to pertinent university management skills and models of administration. The program is further designed to provide training to university administrators in the region while providing more concentrated funding to those university administrators who demonstrate they have the initiative, skills, and institutional support necessary to turn their institutions into models of administrative excellence. To this end, IREX implements a two-tiered program. First, fellowships in University Administration are offered that provide funding for senior level administrators to spend approximately six weeks at host universities in the United States. Second, after completion of their fellowships in the United States, the university administrators apply for small grants to jump-start reform projects at their home universities. Financial provisions include round trip airfare, J-1 visa fees, accident and sickness insurance, housing, daily allowance of $50 per day (meals and incidentals), and a professional development allowance of at least $600 Applications are due by April 15.

*Requirements:* Participating universities from Armenia, Belarus, Ethiopia, Georgia, Ghana, Kenya, Nigeria, Rwanda, Senegal, South Africa, Tanzania, Uganda, and Ukraine are welcome to apply.

*Geographic Focus:* Georgia, Armenia, Belarus, Ethiopia, Ghana, Kenya, Nigeria, Rwanda, Senegal, South Africa, Tanzania, Uganda, Ukraine

*Date(s) Application is Due:* Apr 15

*Amount of Grant:* Up to 4,000 USD

*Contact:* Louisa Manzeske, Senior Program Officer; (202) 628-8188; fax (202) 628-8189; lmanzeske@irex.org or uasp@irex.org

*Internet:* https://www.irex.org/project/university-administration-support-program-uasp

*Sponsor:* International Research and Exchanges Board
1275 K Street, NW, Suite 600
Washington, D.C. 20005

**ISGC Graduate Fellowships** 1376

The ISGC Graduate Fellowships support outstanding graduate students pursuing aerospace, space science, earth system science and other interdisciplinary space-related science, engineering or mathematics fields. Up to ten fellowships of $10,000 each are awarded every academic year. Award winners will be chosen based on student qualifications, scientific and educational relevance of the proposed activity, faculty member's expressed commitment to the project, and the application. Women, under-represented minorities, and persons with disabilities are strongly encouraged to apply. Candidates may apply at the website's online application system. Applications are due by April 12, with award notifications in early May.

*Requirements:* Applicants must meet the following *Requirements:* U.S. citizenship; full-time current enrollment at one of the near twenty ISGC institutions posted on the website; a minimum 3.0 to 4.0 GPA; a good academic record; and research or design experience in a field related to astronomy, aerospace, cosmology, astrophysics, or earth science.

*Geographic Focus:* Illinois

*Date(s) Application is Due:* Apr 12

*Amount of Grant:* 10,000 USD

*Contact:* Diane Jeffers, Associate Director; (217) 244-8048; fax (217) 244-0720; dejeffer@illnois.edu

*Internet:* www.ae.illinois.edu/isgc/funding/grad-fellowship.html

*Sponsor:* Illinois Space Grant Consortium
306 Talbot Laboratory, 104 South Wright Street
Urbana, IL 61801-2935

**ISGC Informal Education Grants** 1377

The Informal Education Grants are designed to increase learning, educate students, educators, and the general public on specific science, technology, aerospace, engineering, or math (STEM) content areas. Innovative activities associated with aerospace, space sciences, and earth system sciences are preferred. The grants are available to faculty members or educators at all ISGC institutions posted at the website. Funds may be used for salary, supplies, and minimal travel. A total of $8,000 has been allocated for one-year programs beginning in May. Programs targeted to women, underrepresented minorities, and persons with disabilities are strongly encouraged. The application is available at the website, and is due February 15.

*Requirements:* Grants must support programs in the STEM content areas. Activities associated with aerospace, physics, astronomy, cosmology, or earth system sciences are preferred. Projects must also include program staff and supplemental materials/handouts. Grants must be matched (at least one-to-one) with funds from non-federal sources. The matching fund's source must be identified and confirmed by a letter from the appropriate authority. Waived indirect costs and faculty effort can qualify as matching funds. Any beneficiary of direct support (salary, wages, travel) from this grant must be a U.S. citizen.

*Restrictions:* Funds cannot be used for the purchase of equipment. Proposal submission for this program is limited to faculty/professional staff (research and professorial) at ISGC affiliate and lead institutions.

*Geographic Focus:* Illinois

*Date(s) Application is Due:* Feb 15

*Amount of Grant:* 8,000 USD

*Contact:* Diane Jeffers; (217) 244-8048; fax (217) 244-0720; dejeffer@illinois.edu

*Internet:* www.ae.illinois.edu/ISGC/funding/informal-ed.html

*Sponsor:* Illinois Space Grant Consortium
306 Talbot Laboratory, 104 South Wright Street
Urbana, IL 61801-2935

**ISGC Multiple Undergraduate Research Grants** 1378

The Multiple Undergraduate Research Grants are designed to fund programs of undergraduate research in aerospace, physics, astronomy, cosmology, earth system science and other interdisciplinary space-related science or engineering fields at ISGC institutions. The program should involve multiple students working with multiple

research advisors. The grants are available to faculty members or research scientists at all ISGC institutions posted on the website. Emphasis will be placed on hands-on activities and projects in areas of interest to NASA. One-year grant periods begin in May. Women, under-represented minorities and persons with disabilities are encouraged to apply. The application is available at the website.
*Requirements:* The grant must support research in an area of interest to NASA, such as aerospace engineering, physics, astronomy, cosmology or earth system science. Grants must be matched (at least one-to-one) with funds from non-federal sources. Prior to the execution of a sub-award agreement, the source of matching funds must be identified and confirmed by a letter from the appropriate authority. Waived indirect costs and faculty effort can qualify as matching funds. Proposals can include requests for salary, supplies, and travel support. Any beneficiary of direct support (salary, wages, travel) from this grant must be U.S. citizen.
*Restrictions:* Funds cannot be used for the purchase of equipment. Participation in this program is limited to faculty (research and professorial) at ISGC affiliate and lead institutions
*Geographic Focus:* Illinois
*Date(s) Application is Due:* Feb 15
*Amount of Grant:* Up to 124,000 USD
*Contact:* Diane Jeffers; (217) 244-8048; fax (217) 244-0720; dejeffer@illinois.edu
*Internet:* www.ae.illinois.edu/ISGC/funding/multi-undergrad.html
*Sponsor:* Illinois Space Grant Consortium
306 Talbot Laboratory, 104 South Wright Street
Urbana, IL 61801-2935

**ISGC Research Seed Grants** 1379

The Research Seed Grants are designed to facilitate the development of research expertise in aerospace engineering/space science at ISGC institutions, especially with smaller research institutions and/or assistant professors. Grants will allow the award recipients to obtain preliminary results needed to support larger proposals to other federal or non-federal funding agencies. Emphasis is placed on research in an area of interest to NASA and new research ideas. A total of $40,000 is allocated for a one-year period beginning in May. The funding level of each seed grant is $10,000. Proposals will be reviewed by ISGC Directors, with selection based on adherence to scientific merit and equitable distribution of resources across the Consortium. Women, under-represented minorities and persons with disabilities are encouraged to apply. The application is available through the online processing system.
*Requirements:* The seed grant must support research in an area of interest to NASA, such as aerospace engineering, physics, astronomy, cosmology or earth system science. Grants must be matched (at least one-to-one) with funds from non-federal sources. Prior to the execution of a sub-award agreement, the source of matching funds must be identified and confirmed by a letter from the appropriate authority. Waived indirect costs and faculty effort can qualify as matching funds. Any beneficiary of direct support (salary, wages, travel) from the grant must be a U.S. citizen. Proposals can include requests for salary, supplies, and travel support.
*Restrictions:* Participation in this program is limited to faculty (research and professorial) at ISGC affiliate and lead institutions that are not participating in an ISGC-funded multiple undergraduate research program. Funds cannot be used for the purchase of equipment.
*Geographic Focus:* Illinois
*Date(s) Application is Due:* Feb 15
*Amount of Grant:* 10,000 USD
*Contact:* Diane Jeffers, Associate Director; (217) 244-8048; fax (217) 244-0720; dejeffer@illinois.edu
*Internet:* www.ae.illinois.edu/ISGC/funding/seedgrant.html
*Sponsor:* Illinois Space Grant Consortium
306 Talbot Laboratory, 104 South Wright Street
Urbana, IL 61801-2935

**ITEEA Maley Scholarship for Technology and Engineering Teacher Professional Development** 1380

The International Technology and Engineering Educators Association (ITEEA) and its Foundation, the Foundation for Technology and Engineering Educators (FTEE), give awards, and scholarships to support the advancement of STEM through technology and engineering education. The Foundation was established in 1986 as a nonprofit 501(c)3 organization, and initiated a program of giving in 1993, in which awards are presented during the ITEEA Annual Conference. FTEE awards support programs that will: make children technologically literate; transfer industrial and corporate research into schools; produce models of excellence in STEM through technology and engineering teaching; create public awareness regarding the nature of technology and engineering education; and help technology and engineering teachers maintain a competitive edge in technology. The Scholarship honors Donald Maley, whose career occurred mostly during the second half of the 20th Century. He is arguably one of the top higher educators in the field of industrial arts and technology. His speeches, articles, and books may be found in ITEEA's Archives at Millersville University in Pennsylvania. The Scholarship's purpose is to support teachers in their preparation to increase the positive outcomes of technology education. Criteria include: evidence of teaching success; plans for action research; recommendations; plans for professional development; and the applicant's financial need. The scholarship is provided directly to the teacher and will be presented during the ITEEA Annual Conference.
*Requirements:* Applicant must be a member of the International Technology Education Association. (Membership may be enclosed with scholarship application). Applicant must be a technology teacher at any grade level who is beginning or continuing graduate study. See the sponsor's website for the list of required items for the application package.
*Geographic Focus:* All States
*Date(s) Application is Due:* Dec 1
*Amount of Grant:* 500 USD
*Contact:* Steven A. Barbato, Executive Director; (703) 860-2100; fax (703) 860-0353; sbarbato@iteea.org@iteea.org or iteea@iteea.org
*Internet:* https://www.iteea.org/Activities/AwardsScholarships/39070/839.aspx
*Sponsor:* International Technology and Engineering Educators Association
1914 Association Drive, Suite 201
Reston, VA 20191-1539

**ITVS Commissioned Funding** 1381

ITVS accepts proposals on an ongoing basis for documentary projects that fall outside of the standing initiatives (DDF, LINCS, Open Call). For development funding, activities may include travel, research, script development, preliminary production for fundraising/work-in-progress reels, or other early-phase activities. For production funding, all production and post-production activities are eligible. Producers should note that ITVS's capacity to fund projects outside of its standing initiatives is extremely limited and highly competitive, but will consider programs for special support and funding on a case-by-case basis. For development funding, ITVS accepts: proposals for documentaries that have not yet begun production; and programs that can become eligible for ITVS production funding initiatives as single programs of standard broadcast length (half-hour or one-hour). For production funding, ITVS accepts proposals for limited series (two to four episodes). There is no deadline for applications. Interested producers may submit their materials at any time. The entire proposal review process takes six to eight weeks.
*Requirements:* Independent producers who are at least 18 years old and are citizens or legal residents of the United States or its external territories are eligible. Applicants must have some previous film or television production experience in a principal role (producer, co-producer, director or co-director) as demonstrated by submitting a sample tape of previously completed work.
*Restrictions:* As with other ITVS initiatives, applicants may only have one project under consideration at any one time. Applicants not eligible to apply include: individuals who do not meet ITVS' definition of an independent producer; students; current signatories of ITVS production contracts who have not completed delivery; producers representing foreign-based production entities; producers regularly employed by a public or commercial broadcast entity, including public access stations; or organizations such as universities, foundations and non-profit media organizations.
*Geographic Focus:* All States
*Contact:* Jonathan Archer, Programming Coordinator; (415) 356-8383, ext. 284; fax (415) 356-8391; jonathan_archer@itvs.org
*Internet:* www.itvs.org/funding/edf
*Sponsor:* Independent Television Service
651 Brannan Street, Suite 410
San Francisco, CA 94107

**J. Larry Landers Student Research Award** 1382

The Gopher Tortoise Council was formed in 1978 by a group of southeastern biologists and other citizens concerned with the decline of the gopher tortoise. The J. Larry Landers Student Research Award is a Gopher Tortoise Council competitive grant program for undergraduate and graduate college students. The amount of the award is variable, but has averaged $1,000.00 over the last few years.
*Requirements:* Proposals can address research concerning gopher tortoise biology or any other relevant aspect of upland habitat conservation and management. A project summary for "The Tortoise Burrow" newsletter or either a presentation or poster at the following year's GTC annual meeting will be required.
*Geographic Focus:* All States
*Date(s) Application is Due:* Sep 15
*Contact:* Jeff Goessling, jeff.goessling@gmail.com
*Internet:* www.gophertortoisecouncil.org/who/grants.php
*Sponsor:* Gopher Tortoise Council
303 Main Street #241
Safety Harbor, FL 34695

**J.N. and Macie Edens Foundation Grants** 1383

Joseph Napoleon (Pole) Edens and Lilac (Macie) Edens were long time residents of Corsicana, Texas. Joseph was a self made man who from a young age managed the family ranch. For years, he bred Herefords, winning numerous prizes throughout the nation. Edens later became bank president of the Corsicana First National Bank. He started 4-H clubs and Future Farmers of America Chapters where his influence is still present today. The Edens' had a strong sense of community and established the J. N. and Macie Edens Foundation to promote the well-being of mankind. Grants are awarded to organizations in Navarro County, Texas for the promotion and extension of religion, education, the alleviation of human suffering, prevention and control of disease, to aid scientific endeavors that contribute to the betterment of mankind, for beautification of public buildings, grounds, or civic betterment and to provide relief to the poor and the indigent. The annual deadline for applications is March 1, with notification of awards on June 30.
*Requirements:* 501(c)3 organizations in, and serving the residents of, Navarro County, Texas, are eligible to apply.
*Geographic Focus:* Texas
*Date(s) Application is Due:* Mar 1
*Samples:* Navarro Council of the Arts, Corsicana, Texas, $10,000 - support for Arts In Education for the children of Navarro County; Viable Options in Community Endeavors, Corsicana, Texas, $10,000 - support for Efforts to Outcomes (performance management software); Family Service Association, Corsicana, Texas, $7,500 - general operating support.
*Contact:* Debra Goldstein Phares, Philanthropic Client Director; (214) 209-1830 or (214) 520-6549; debra.g.phares@ustrust.com or tx.philanthropic@ustrust.com

*Internet:* https://www.bankofamerica.com/philanthropic/foundation.go?fnId=152
*Sponsor:* J.N. and Macie Edens Foundation
901 Main Street, 19th Floor
Dallas, TX 75202-3714

**J.W. Kieckhefer Foundation Grants** **1384**

The Kieckhefer Foundation awards grants to nation-wide 501(c)3 organizations in support of medical research, hospices, and health agencies; family planning services; social services; higher education; youth and child welfare agencies; ecology and conservation; community funds; and cultural programs. Types of support include the following: annual campaigns; building renovation; conferences and seminars; continuing support; emergency funds; endowments; equipment; general operating support; land acquisition; matching and challenge support; program development; publication; and research.
*Requirements:* Applications are not accepted. Organizations should submit a letter of inquiry to the Foundation, with a description of their project and amount requested.
*Restrictions:* Grants are not awarded to individuals.
*Geographic Focus:* All States
*Contact:* John I. Kieckhefer, Trustee; (928) 445-4010
*Sponsor:* J.W. Kieckhefer Foundation
116 East Gurley Street
Prescott, AZ 86301-3821

**Jacob and Valeria Langeloth Foundation Grants** **1385**

The Jacob and Valeria Langeloth Foundation's grant making program is centered on the concepts of health and well-being. The Foundation's purpose is to promote and support effective and creative programs, practices and policies related to healing from illness, accident, physical, social or emotional trauma and to extend the availability of programs that promote healing to underserved populations. The Foundation is particularly interested in funding programs that address the health of individuals who, because of barriers to difficulty accessing care. Its two primary funding priorities are: correctional health; and chronic violence and community health. The required Letter of Intent annual deadline is October 17, with full proposals due December 17.
*Requirements:* 501(c)3 organizations that promote physical and emotional healing, especially to underserved populations, such as: community-based organizations, health care providers and research institutions in the state of New York are eligible to apply.
*Restrictions:* The Foundation will not consider proposals for annual or capital campaigns, for building or renovation projects, for budgetary relief, or preventive medicine. The Foundation does not support projects that focus on children or end-of-life issues. Neither will the Foundation make grants to individuals or for sectarian or religious purposes, or for political activities or lobbying. The Foundation will not fund any organization that discriminates on the basis of age, gender, national origin, race, or sexual preference.
*Geographic Focus:* New York
*Date(s) Application is Due:* Dec 17
*Amount of Grant:* 50,000 - 500,000 USD
*Contact:* Scott Moyer; (212) 687-1133; fax (212) 681-2628; smoyer@langeloth.org
Melissa Houston, Grants and Office Administrator; (212) 687-1133; fax (212) 681-2628; mhouston@langeloth.org
*Internet:* https://www.langeloth.org/content.php?content=60
*Sponsor:* Jacob and Valeria Langeloth Foundation
275 Madison Avenue, Suite 2102
New York, NY 10016

**James Graham Brown Foundation Fellowships** **1386**

The James Graham Brown Foundation's Fellows program is unique in that it is a single initiative that spans two campuses. In partnership with Centre College and the University of Louisville, Brown Fellows receive full tuition, room and board and will participate in an extraordinary summer program featuring international travel as a group before freshman year. Fellows also receive an enhanced academic experience consisting of individual study, field-based experiential learning and community service. Each student will have an on campus mentor, engage in enrichment opportunities and manage a customized Leadership Project. In successive years the program will focus on the individual student to provide an extraordinary summer enrichment experience focused in thematic areas of service, international research and study, and leadership.
*Requirements:* Eligibility and selection criteria include scholastic attainment, intellectual leadership, visionary thinking, and exemplary community involvement. Additional consideration will be given for a demonstrated ability to overcome adversity, as well as a willingness and ability to serve as an ambassador for Kentucky.
*Geographic Focus:* Kentucky
*Date(s) Application is Due:* Mar 4
*Contact:* Andrea Houser, Program Administrator; (502) 896-2440 or (866) 896-5423; fax (502) 896-1774; grants@jgbf.org or info@jgbf.org
*Internet:* www.brownfellows.org/
*Sponsor:* James Graham Brown Foundation
4350 Brownsboro Road, Suite 200
Louisville, KY 40207

**James M. Collins Foundation Grants** **1387**

Established in Texas in 1964 by the late James M. Collins, a Republican who represented the Third Congressional District of Texas from 1968-1983, the Foundation awards grants to Texas nonprofit organizations in its areas of interest, including: the arts; performing arts centers; economic development; health organizations and association; higher education; human services; museums; the Salvation Army; and secondary school. Types of support include: research; program support; and social services. There are no application deadlines or formal applications, and interested partied should begin by contacting the Foundation through an inquiry letter. Most recent funding awards have ranged from $500 up to $50,000.
*Requirements:* Texas nonprofit organizations are eligible to apply.
*Restrictions:* Individuals are not eligible.
*Geographic Focus:* Texas
*Amount of Grant:* 500 - 50,000 USD
*Samples:* Baylor University, Waco, Texas, $50,000 - biomedical research; Crystal Charity Ball, Dallas, Texas, $10,000 - general operating expenses; Our Friends Place, Dallas, Texas, $1,000 - general program support.
*Contact:* Dorothy Dann Collins Torbert, President; (214) 691-2032
*Sponsor:* James M. Collins Foundation
8115 Preston Road, Suite 680
Dallas, TX 75225

**James Marston Fitch Charitable Trust Mid-Career Grants** **1388**

The James Marston Fitch Charitable Trust awards research grants of up to $15,000 to mid-career professionals who have an academic background, professional experience, and an established identity in one or more of the following fields: historic preservation, architecture, landscape architecture, urban design, environmental planning, architectural history, and the decorative arts. Proposals will be considered for the research and/or execution of the preservation-related projects in any of these fields. The foundation endeavors to establish new links between the academic and professional contingents primarily in the field of American historic preservation and to strengthen the connections between theory and practice in conservation technology.
*Requirements:* Applicants must be mid-career professionals with at least ten years experience in historic preservation or related fields, including architecture, landscape architecture, architectural conservation, urban design, environmental planning, archaeology, architectural history, and the decorative arts. Applicants must be legal residents or citizens of the United States.
*Restrictions:* Grants are awarded only to individuals, not organizations. The Foundation does not fund university-sponsored research projects or dissertation research.
*Geographic Focus:* All States
*Date(s) Application is Due:* Oct 21
*Amount of Grant:* Up to 15,000 USD
*Contact:* Cristiana Peña, Executive Director; (212) 252-6809; fax (212) 471-9987; cpena@fitchfoundation.org or info@fitchfoundation.org
*Internet:* fitchfoundation.org/grants/fitch-kress/
*Sponsor:* James Marston Fitch Charitable Trust
232 East 11th Street
New York, NY 10003

**Jane Bradley Pettit Foundation Health Grants** **1389**

The Jane Bradley Pettit Foundation awards grants to initiate and sustain projects in the Greater Milwaukee community, with a focus on programs and projects that serve low-income and disadvantaged individuals, women, children, and the elderly. In the area of Health, the Foundation gives priority to community-based health care and prevention programs which address the physical and mental health needs of families, children, persons at-risk and the elderly. Support and advocacy for victims of abuse and neglect are also a priority. An initial application should be in the form of a letter of intent.
*Requirements:* Milwaukee-area charitable organizations are eligible.
*Geographic Focus:* Wisconsin
*Date(s) Application is Due:* Jan 15; May 15; Sep 15
*Amount of Grant:* 5,000 - 100,000 USD
*Samples:* Mental Health America of Wisconsin, Milwaukee, $10,000 - for general operations; The Healing Center, Milwaukee, Wisconsin, $10,000 - for general operations
*Contact:* Heidi Jones, Director of Administration; (414) 982-2880 or (414) 982-2874; fax (414) 982-2889; hjones@staffordlaw.com
*Internet:* www.jbpf.org/guidelines/index.html
*Sponsor:* Jane Bradley Pettit Foundation
1200 N. Mayfair Road, Suite 430
Wauwatosa, WI 53226-3282

**Jane Coffin Childs Memorial Fund for Medical Research Fellowships** **1390**

The Jane Coffin Childs Memorial Fund for Medical Research Fellowships are intended to further research into the causes, origins, and treatment of cancer. The major effort of the fund, at this time, is directed toward the support of a fellowship program for the training of young men and women at the postdoctoral level for independent research in the same field. Stipend, child dependency allowance, travel for fellow and family, and an institutional allowance are provided. Fellowships are awarded for two to three years. The basic stipend is $52,000 for the first year, $52,500 for the second year, and $53,000 the third year, with an additional $1,000 for each dependent child. An allowance of $2,000 a year toward the cost of the research will be made available to the laboratory sponsoring the fellow. The Fellow will also have access to $1,800 for travel to science meetings. In addition to the travel allowance the Fund provides a travel award for the Fellow and their family to travel to the sponsoring laboratory. In 2008, the Joan A. Steitz Fund of the Community Foundation for Greater New Haven was established as an Endowed Designated Fund for the benefit of Jane Coffin Childs Memorial Fund for Medical Research to provide stipends for child care to women who are recipients of a Jane Coffin Childs post-doctoral fellowship. The online application form becomes available annually on October 1, with an application submission deadline of February 1.
*Requirements:* Applicants should not have more than one year of postdoctoral experience. They must either hold the MD degree or PhD degree in the field in which they propose

to study. Applicants may be citizens of any country, but awards for foreign nationals will be made only for study in the United States. American citizens may hold fellowships either in the United States or in a foreign country.
*Restrictions:* There is no dependency allowance for a spouse. The research allowance is not to be used for travel.
*Geographic Focus:* All States
*Date(s) Application is Due:* Feb 1
*Amount of Grant:* 52,000 - 169,000 USD
*Contact:* Kim Roberts, Administrative Director; (203) 785-4612; fax (203) 785-3301; kim.roberts@yale.edu or info@jccfund.org
*Internet:* www.jccfund.org
*Sponsor:* Jane Coffin Childs Memorial Fund for Medical Research
333 Cedar Street, MC0191, P.O. Box 208000
New Haven, CT 06520-8000

**Japan-United States Friendship Commission-NEH Social Science Fellowships 1391**
The Fellowship Program for Advanced Social Science Research on Japan is a joint activity of the Japan-U.S. Friendship Commission (JUSFC) and the National Endowment for the Humanities. Awards support research on modern Japanese society and political economy, Japan's international relations, and U.S.-Japan relations. The program encourages innovative research that puts these subjects in wider regional and global contexts and is comparative and contemporary in nature. Research should contribute to scholarly knowledge or to the general public's understanding of issues of concern to Japan and the United States. Appropriate disciplines for the research include anthropology, economics, geography, history, international relations, linguistics, political science, psychology, public administration, and sociology. Awards usually result in articles, monographs, books, digital materials, archaeological site reports, translations, editions, or other scholarly resources.
*Requirements:* Citizens and permanent residents of the United States and Japan are eligible. Grants may be made to individual universities or local organizations. Awards made to individuals under all programs normally will be made through academic, professional, artistic, or other appropriate organizations that will examine, recommend, and, in most instances, select the individuals to be supported financially by the commission.
*Restrictions:* Japanese institutions and U.S. academic organizations only may apply. This award is granted twice a year.
*Geographic Focus:* All States, Japan
*Date(s) Application is Due:* Apr 24
*Contact:* Paige Cottingham-Streater, Executive Director; (202) 418-9800; fax (202) 418-9802; grants@jusfc.gov or jusfc@jusfc.gov
Research Programs Team; (202) 606-8200 or (202) 606-8400; questions@neh.gov
*Internet:* www.jusfc.gov/fellowships/
*Sponsor:* Japan-United States Friendship Commission
1201 15th Street NW, Suite 330
Washington, D.C. 20005

**Japan Foundation Center for Global Partnership Intellectual Exchange Grants 1392**
The Japan Foundation Center for Global Partnership (CGP) is dedicated to strengthening the global U.S.-Japan partnership and cultivating the next generation of public intellectuals necessary to sustain this partnership. The grant program supports U.S.-Japan collaborative projects conducted by universities, think-tanks, and other non-profit organizations with 501(c)3 status. Projects should have a foundation in fostering dialogue and promoting partnerships between the two countries through the exploration of policy-oriented issues. This grant program places emphasis on projects with tangible outcomes, and effective dissemination plans that have broad policy impact. The annual deadline for applications is December 1.
*Requirements:* Proposals from U.S. organizations with 501(c)3 status only will be accepted by CGP New York. Appropriate Japan-based institutional collaboration must be demonstrated and secured throughout the course of the project.
*Geographic Focus:* All States
*Date(s) Application is Due:* Dec 1
*Amount of Grant:* Up to 100,000 USD
*Samples:* Carnegie Endowment for International Peace, Washington, D.C., $90,000 - project looked into new and creative ways to strengthen the U.S.-Japan alliance outside of Security alliance strengthening in the coming century; Japan Center for International Exchange, New York, New York, $29,995 - launched a two-year initiative to examine how Japanese and U.S. NGOs can become strategic partners; Southern Illinois University, Carbondale, Illinois, $56,654 - a controlled study that taps into the psychological constructs important in decisions to plead guilty after being accused of wrongdoing.
*Contact:* Ayumi Takita, Intellectual Exchange Program Director; (212) 489-1255, ext. 122; fax (212) 489-1344; ayumi_takita@cgp.org
*Internet:* www.cgp.org/intellectual-exchange/regular-grants
*Sponsor:* Japan Foundation Center for Global Partnership
1700 Broadway, 15th Floor
New York, NY 10019

**Japan Foundation Japanese Studies Fellowships 1393**
In order to promote Japanese Studies overseas, the Japan Foundation's Japanese Studies Fellowships offer preeminent foreign scholars in Japanese studies the opportunity to conduct research in Japan. The foundation offers three types of fellowships: Scholars and Researchers - Long Term (2-12 months); Scholars and Researchers - Short Term (21-59 days); and Doctoral Candidates (4-12 months). Long-term and Doctoral fellows receive the following benefits: a monthly stipend of either 370,000 or 430,000 yen for long-term fellows (amount is determined according to fellow's professional status) or 310,000 yen for Doctoral fellows; a settling-in allowance of 250,000 yen (for fellows not already in Japan); a departure allowance of 50,000 yen; a dependent allowance of 50,000 yen per month (for dependents who reside in Japan continuously for more than six months); coverage of enrollment fees (up to 50,000 yen per month); and a research allowance for Long-term fellows of up to 40,000 yen per month. Short-term fellows receive a per-diem allowance of 17,000 yen per day. All fellows who are not already in Japan receive economy-class airfare to and from Japan and traveler's insurance. U.S. citizens (except for those who are residing in other countries as permanent residents) and permanent U.S. residents should submit their applications online no later than November 1 at the Japan Foundation New York website - and should also follow the application guidelines at that site. All other applicants have a deadline of December 1 and may download their application guidelines and forms from the Japan Foundation website. Applicants who are citizens or permanent residents in countries other than the U.S. or Taiwan should submit their application materials to the nearest Foundation Office or Japanese diplomatic mission. Taiwanese applicants, including permanent residents in Taiwan, should submit their application materials to the Taipei office of the Interchange Association, Japan. Applicants residing in countries in which they hold neither nationality nor permanent residency (including residents in Japan) should submit their applications to the Foundation Headquarters in Tokyo. A list of contact information is included in the downloadable program guidelines (WORD) at the foundation website. Please note that applications received after the deadline will not be considered and that multiple applications from the same applicant will make all applications invalid. The Japan Foundation was established in 1972 by the Japanese legislature and became one of Japan's "Independent Administrative Institutions" in October 2003. The foundation maintains its headquarters in Tokyo and operates through a network of 21 overseas offices in 20 countries worldwide. The foundation's mission is to promote international cultural exchange and mutual understanding between Japan and other countries. To do so it offers three major categories of programming: Arts & Culture; Japanese Studies; and Japanese-Language Education.
*Requirements:* Applicants for long- and short-term fellowships must hold a Ph.D. or equivalent professional experience in research, teaching, and writing in the fields of the humanities or social sciences. Doctoral Candidates must have achieved All-But-Dissertation (ABD) status by the time the fellowship begins. Applicants must hold nationality or permanent residency in countries that have diplomatic relations with Japan (Taiwanese applicants, including permanent residents of Taiwan, are eligible to apply for the long-term and doctoral-candidate fellowships). Applicants must also satisfy the following criteria: be in good health; be proficient in either Japanese or English; secure all affiliation arrangements by the application deadline (excluding the short-term fellowship); and stay continuously in Japan for the term of fellowship.
*Restrictions:* Previous recipients of Japan Foundation long-term or doctoral fellowships are eligible to reapply only if a full three years have elapsed since the termination of their previous fellowship. Previous short-term fellows may reapply after one full year from the termination of their most recent fellowship. Applicants who have received Japan Foundation fellowships in the past on two or more occasions will be given lower priority. Individuals meeting any of the following conditions are not eligible to apply: scholars or researchers in the natural sciences, medicine, or engineering; undergraduates; master's degree candidates; those intending to enroll in undergraduate or master's courses at universities in Japan; graduates intending to pursue Ph.D. courses in Japan; those who intend to use the fellowship for political or religious purposes; those who primarily intend to enhance their Japanese-language ability or technical skills; those who intend to develop and/or produce Japanese-language teaching materials; those who intend to acquire artistic training in the area of Japanese traditional culture (tea ceremony, flower arrangement, etc.); those who plan to receive other grants concurrently with the fellowship; or those who have been in Japan continuously for more than a year at the time of application. Japanese citizens must hold permanent residency in another country to be eligible to apply. Applicants holding nationality of countries not maintaining diplomatic relations with Japan are not eligible to apply, even if they have permanent residency in a country with diplomatic relations with Japan.
*Geographic Focus:* All Countries
*Date(s) Application is Due:* Nov 1; Dec 1
*Amount of Grant:* 357,000 - 4,880,000 JPY
*Contact:* Program Director; 03-5369-6051 or 03-5369-6069; fax 03-5369-6041
*Internet:* www.jpf.go.jp/e/program/intel.html
*Sponsor:* Japan Foundation
4-4-1 Yotsuya, Shinjuku-ku
Tokyo, 160-0004 Japan

**Japan Foundation New York Institutional Support for Japanese Studies Grants 1394**
To encourage the sustained and innovative growth of Japanese Studies in national as well as transnational contexts, the Japan Foundation is expanding its Institutional Support Program (ISP) in the United States. Individual grants will be made annually to colleges, universities (or consortia of colleges and universities), and academic organizations that play a central role in Japanese studies in the U.S. and that demonstrate backing for those proposals through matching funds. Proposals should address the needs, opportunities, and visions of particular institutions. Applicants are encouraged to think practically and inventively about their discrete projects. Projects may engage specific or comprehensive goals and may entail smaller or larger grants. Priority will be given to projects that contribute significantly to infrastructure building for the field, provide opportunities for students and faculty to travel to Japan, and, most desirably, enhance the quality of teaching and research in Japanese Studies as well as the quality of equipment and materials (including those in libraries). Links to application guidelines and forms are available at the website. Application packets must be mailed to the Japanese Foundation New York and must be received by the due date. As application deadlines may vary from year, interested applicants are advised to verify current deadline dates. The Japan

Foundation was established in 1972 by the Japanese legislature and became one of Japan's "Independent Administrative Institutions" in October 2003. The foundation maintains its headquarters in Tokyo and operates through a network of 21 overseas offices in 20 countries worldwide including offices in New York and Los Angeles, and a Center for Global Partnership, also in New York. The foundation's mission is to promote international cultural exchange and mutual understanding between Japan and other countries. To do so, it offers three major categories of programming: Arts & Culture; Japanese Studies; and Japanese-Language Education.
*Requirements:* United States colleges, universities (or consortia of colleges and universities), and academic organizations playing a central role in Japanese Studies are eligible to apply. Small institutions, newly-established institutions, and institutions without an established program of Japanese studies at the graduate level are encouraged to apply and to submit proposals that emphasize lasting impact and sustainability (for example, creation of new faculty positions in Japanese studies, etc.). The foundation requires submission of multi-year proposals (up to three years in length). Applicants may request a maximum of $150,000 per year. Applicants must be prepared to match the grant equally, and optimally, at higher levels. Matching funds can be both cash and in-kind contributions.
*Restrictions:* Previous ISP grantees may not apply through a project whose grant duration overlaps with their current grant project. The ten universities that received sizable Japanese-government donations in 1973 are eligible to apply, but will be given lower priority except in cases in which their proposals are involved in outreach, consortium building, or partnering with institutions and organizations in Japan. Proposed projects must not be used for religious, political, or commercial purposes, or to support specific doctrines or claims. The foundation will not consider the following types of requests: projects in the field of medicine, manual training, or natural sciences; organization of or participation in sporting events and training; projects in the field of art; goodwill exchanges; social-welfare or international-development projects; tourist-oriented programs; and language-education programs (however, professional language training for Japanese-Studies scholars is eligible). The applicant may not request funding for the following items: indirect costs or costs that are not directly needed to carry out the applied projects; overhead costs of the administrative body of the institution; salaries of administrative staff (funds may be requested for assistants directly related to the project, up to a limit of thirty days); equipment such as computers, photocopiers, fax machines, electric appliances, office furniture, or computer supplies; real estate; construction fees; rental fees for facilities that would be used long-term; rental fees for facilities within the applying institution; international airfare that is higher than the estimated economy-class fare for the shortest route from the city of boarding to the destination; or IATA Special Excursion Fares (PEX discount fare).
*Geographic Focus:* All States
*Date(s) Application is Due:* Dec 9
*Amount of Grant:* Up to 450,000 USD
*Samples:* Columbia University, New York, New York, $606,100 - Classic Japanese Music Study and Performance Project; Western Michigan University, Kalamazoo, Michigan, $348,887 - Pre-modern Japanese Culture and Regional Outreach; Furman University, Greenville, South Carolina, $876,869 - Japanese Studies at Furman University for the 21st Century.
*Contact:* Noriko Yamamoto, Program Director, Japanese Studies; (212) 489-0299; fax (212) 489-0409; info@jfny.org
*Internet:* www.jfny.org/japanese_studies/institutionalsupport.html
*Sponsor:* Japan Foundation in New York
1700 Broadway, 15th Floor
New York, NY 10019

**Japan Foundation New York Japanese Studies Fellowships** **1395**

The Japan Foundation New York's Japanese Studies Fellowships give outstanding scholars in the humanities and social sciences in the U.S. the opportunity to conduct research in Japan. The foundation offers three types of fellowships: Scholars and Researchers - Long Term (2-12 months); Scholars and Researchers - Short Term (21-59 days); and Doctoral Candidates (4-12 months). Long-term fellows and Doctoral fellows receive the following benefits: a monthly stipend of either 370,000 or 430,000 yen for long-term fellows (amount is determined according to fellow's professional status) or 310,000 yen for graduate fellows; a settling-in allowance of 250,000 yen (for fellows not already in Japan); a departure allowance of 50,000 yen; a dependent allowance of 50,000 yen per month (for dependents who reside in Japan continuously for more than six months); coverage of enrollment fees (up to 50,000 yen per month); and a research allowance for long-term fellows of up to 40,000 yen per month. Short-term fellows receive a per-diem allowance of 17,000 yen per day. All fellows who are not already in Japan receive economy-class airfare to and from Japan and traveler's insurance. Links to the application guidelines (program instructions) and an online submission system are available at the website. American citizens residing in the U.S. or permanent U.S. residents must submit their applications via this on-line system by the November 1 deadline date. (American citizens with permanent residency abroad have different application requirements and should visit the website and/or contact the foundation for information on how to apply.) All candidates are encouraged to carefully peruse the website and/or email the foundation for more detailed requirements and guidelines. The Japan Foundation was established in 1972 by the Japanese legislature and became one of Japan's "Independent Administrative Institutions" in October 2003. The foundation maintains its headquarters in Tokyo and operates through a network of 21 overseas offices in 20 countries worldwide including offices in New York and Los Angeles, and a Center for Global Partnership, also in New York. The foundation's mission is to promote international cultural exchange and mutual understanding between Japan and other countries. To do so it offers three major categories of programming: Arts & Culture; Japanese Studies; and Japanese-Language Education.
*Requirements:* Applicants must be American citizens, American citizens temporarily residing abroad, or permanent residents of the United States. Non-U.S. citizens are encouraged to visit the Japan Foundation Headquarters website (www.jpf.go.jp/e/) to apply for similar fellowship opportunities. Applicants for long- and short-term fellowships must hold a Ph.D. or equivalent professional experience in research, teaching, and writing in the fields of the humanities or social sciences. Doctoral Candidates must have achieved All-But-Dissertation (ABD) status by the time the fellowship begins. Applicants must also satisfy the following criteria: be in good health; be proficient in either Japanese or English; secure all affiliation arrangements by the application deadline (excluding the short-term fellowship); and stay continuously in Japan for the term of fellowship.
*Restrictions:* The duration of the fellowship may be adjusted by the Foundation. Previous recipients of Japan Foundation fellowships are eligible to reapply only if a full three years have elapsed since the termination of their previous fellowship. Previous Short-Term Research Fellows may reapply after one full year from the termination of their most recent fellowship. Individuals meeting any of the following conditions are not eligible to apply: scholars or researchers in the natural sciences, medicine, or engineering; undergraduates; master's degree candidates; those intending to enroll in undergraduate or master's courses at universities in Japan; graduates intending to pursue Ph.D. courses in Japan; those who intend to use the fellowship for political or religious purposes; those who primarily intend to enhance their Japanese-language ability or technical skills; those who intend to develop and/or produce Japanese-language teaching materials; those who intend to acquire artistic training in the area of Japanese traditional culture (tea ceremony, flower arrangement, etc.); those who plan to receive other grants concurrently with the fellowship; or those who have been in Japan continuously for more than a year at the time of application.
*Geographic Focus:* All States
*Date(s) Application is Due:* Nov 1
*Amount of Grant:* 357,000 - 4,880,000 JPY
*Contact:* Noriko Yamamoto, Program Director, Japanese Studies; (212) 489-0299; fax (212) 489-0409; info@jfny.org
*Internet:* www.jfny.org/japanese_studies/fellowship.html
*Sponsor:* Japan Foundation in New York
1700 Broadway, 15th Floor
New York, NY 10019

**Japan Foundation Six-Month Grants for Specialists in Cultural and Academic Fields** **1396**

The Japan Foundation provides non-native specialists (researchers, postgraduate students, librarians, museum curators, etc.) with an opportunity to take a six-month Japanese-language course specifically designed for their research needs. The program focuses on Japanese-language proficiency necessary for specialized work or research activities and assists participants to acquire language skills necessary to collect and present information in their major fields as well as network with people in their fields. The program also coaches participants as they carry out their research and work activities (visits to other institutions, interviewing professionals, etc.). Training takes place from September through March at the foundation's Japanese-Language Institute (JFJLI, Kansai) in Japan. Exact dates may vary from year to year and are posted at the foundation website. The following expenses will be covered by the training grant: round-trip airfare (discount economy class), airport tax, and foreign-travel tax; accommodation at JFJLI, Kansai; overseas-travel insurance; meals during the training program; and other expenditures necessary for participation in the program. Nationals from the following regions may also receive a set allowance to cover necessary expenses such as public-transportation fees, communication fees, etc.: Asia and Oceania; Central and South America; Eastern Europe; Middle East and North Africa; and Africa. (Applicants should check the guidelines for specifics.) Detailed program information and application guidelines as well as official application forms are downloadable from the foundation website. Applications should be submitted to the applicant's nearest Japan Foundation overseas office or Japanese diplomatic mission no later than December 1. In Taiwan, applications should be submitted to the Interchange Association, Japan, Taipei office. Applicants must notify their respective overseas office, mission, or interchange of their intention to apply prior to submitting an application. Contact information for both the foundation's Japan and overseas offices is available under the "JF Worldwide" tab at the foundation website. Deadlines may vary from year to year. Applicants are encouraged to verify current deadline dates. The Japan Foundation was established in 1972 by the Japanese legislature and became one of Japan's "Independent Administrative Institutions" in October 2003. The foundation maintains its headquarters in Tokyo and operates through a network of 21 overseas offices in 20 countries worldwide. The foundation's mission is to promote international cultural exchange and mutual understanding between Japan and other countries. To do so it offers three major categories of programming: Arts & Culture; Japanese Studies; and Japanese-Language Education.
*Requirements:* This grant is an intensive residential-training course for specialists (e.g. scholars, researchers, full-time postgraduate students, and full-time librarians and museum curators) outside Japan who need a good command of the Japanese language in order to conduct field-related work and research. Applicants must be postgraduate students in institutions outside Japan or other specialists who are employed by (and upon their return continue to work for) applicable organizations outside Japan (e.g. institutions of higher education, research centers, public-sector organizations, research or public libraries, cultural-exchange organizations, or museums which are currently working for international-exchange programs that concern Japan). Permission to participate from the applicant's affiliated institution is an eligibility requirement. Additionally applicants must satisfy the following criteria: be in the fields of Japan-related social sciences or humanities (excluding Japanese-language study and Japanese-language education); have a need for Japanese-language proficiency to successfully carry out their work; have a specific and concrete theme related to Japan in their study or work and a plan to make

a presentation on the achievements; have a minimum proficiency score of Level 4 on the Japanese-Language Proficiency Test (JLPT); be under the age of 40 at time of application; and be in good physical and mental health. All else being equal, priority will be given to the following applicants: specialists who are currently engaged in their profession; postgraduate students who plan to take a job related to Japanese studies; students officially registered in a Ph.D.-level program at a graduate school; applicants who have affiliations with major institutions; and those whose applications are related to major cultural-exchange events announced by the Ministry of Foreign Affairs of Japan.
*Restrictions:* Participants are not permitted to bring family members. The following individuals are not eligible to apply: those who have participated in the training programs for teachers of Japanese language at JFJLI, Urawa in the last five years; those who have participated in the Japanese Language Programs at JFJLI, Kansai for librarians or for specialists in cultural and academic fields; those who have participated in the Japanese Language programs at JFJLI, Kansai for researchers and postgraduate students for more than four-months in the past five years; those who completed their compulsory education in Japan; those who are scheduled for study or training under other programs in Japan; and those who hold nationality in a country that does not have diplomatic relations with Japan (Taiwanese are eligible to apply for this program).
*Geographic Focus:* All States, All Countries
*Date(s) Application is Due:* Dec 1
*Contact:* Program Director, Japanese-Language Teacher and Institutional Support Dept; 048-834-1181-2; fax 048-834-1170
*Internet:* www.jpf.go.jp/e/program/japanese.html#1
*Sponsor:* Japan Foundation
4-4-1 Yotsuya, Shinjuku-ku
Tokyo, 160-0004 Japan

**Jefferson Science Fellowships** **1397**

The Jefferson Science Fellowship program establishes a new model for engaging the American academic science, technology, and engineering communities in the formulation and implementation of U.S. foreign policy. Each fellow will spend one year at the U.S. Department of State for an on-site assignment in Washington, D.C., that may also involve extended stays at U.S. foreign embassies and/or missions. All JSF assignments will be designed in consultation with regional and/or functional bureaus within the U.S. Department of State. Following the fellowship year, the fellow will return to his/her academic career, but will remain available to the U.S. government as an experienced consultant for short-term projects over the following five years. Application and guidelines are available online. The application process opens each year on August 1, with the deadline for submission being October 31.
*Requirements:* Nominees and applicants must be U.S. citizens and must hold a tenured faculty position at a U.S. degree granting academic institution of higher learning. Nominations will be accepted from U.S. academic institutions that have signed a JSF/MOU with the U.S. Department of State.
*Geographic Focus:* All States
*Date(s) Application is Due:* Oct 31
*Amount of Grant:* 50,000 USD
*Contact:* Fellowships; (202) 334-3478 or (202) 334-3560; fax (202) 334-3419; jsf@nas.edu
*Internet:* sites.nationalacademies.org/PGA/Jefferson/index.htm
*Sponsor:* National Academies
500 5th Street, NW, Room 508
Washington, D.C. 20001

**Jeffrey Thomas Stroke Shield Foundation Research Grants** **1398**

The Dr. Jeffrey Thomas Stroke Shield Foundation (JTSSF) welcomes submissions of grant requests for funding covering all aspects of stroke prevention, treatment and awareness. JTSSF is actively seeking to provide seed money grants for projects that are important, achievable, and innovative that will ultimately lead to stroke prevention and a better treatment and quality of life for stroke sufferers. From the Foundation's perspective within clinical stroke treatment and research, it understands the importance of directing targeted research funds to the scientists and investigators who can move the best ideas in stroke research prevention and treatment forward. Letters of Intent (LOIs) are required and must be postmarked by February 28. All LOI submissions will undergo a review and selection process by the JTSSF Scientific Advisory Board. Only those projects deemed to be of suitable scientific merit will be invited to submit a full application. If invited to submit an application, applicants will receive additional instructions and information necessary for submission. Final grant applications are due by May 16.
*Requirements:* Principal investigators who hold appointments/positions in accredited hospitals and clinics, healthcare institutions, academic medical centers, professional associations, member societies and other not?for?profit entities in the U.S. are welcome to apply.
*Geographic Focus:* All States
*Date(s) Application is Due:* May 16
*Contact:* Nancy Weber; (415) 830-6031; info@strokeshieldfoundation.org
*Internet:* strokeshieldfoundation.org/researchers/grant-program/
*Sponsor:* Jeffrey Thomas Stroke Shield Foundation
3053 Fillmore Street, #268
San Francisco, CA 94123

**Jenifer Altman Foundation Grants** **1399**

The foundation is a small foundation working for a socially just and ecologically sustainable future, innovative research and demonstration projects in mind-body health, and improved child care prospects for at-risk children. Small grants are made in two areas: initiatives concerning the impact of endocrine disrupting chemicals on human health and the preservation of the environment and initiatives supporting the global citizens' movement for a just and sustainable future in relation to international governmental decision-making forums. The foundation awards a few grants for innovative research and demonstration programs in mind-body health, especially those focused on cancer. Local initiatives that contribute to the quality of life in Bolinas, CA, are also of interest. Applicants should send a concept letter to determine if their projects align with the foundations areas of interest.
*Requirements:* Creative and innovative nonprofits are eligible to apply for one-time grants.
*Restrictions:* Grants are not made to individuals. The foundation rarely funds programs for the conservation of specific species or habitats and rarely funds research at universities on environmental issues. Grants are not made for civic programs, projects in other countries, or projects that are focused entirely on one city or region in the United States.
*Geographic Focus:* All States
*Amount of Grant:* 500 - 50,000 USD
*Contact:* Marni Rosen, Executive Director; (415) 561-2182 or (415) 561-2187; fax (415) 561-6480; info@jaf.org or mrosen@jaf.org
*Internet:* www.jaf.org/apply/index.html
*Sponsor:* Jenifer Altman Foundation
P.O. Box 29209
San Francisco, CA 94129

**Joe W. and Dorothy Dorsett Brown Foundation Grants** **1400**

The Joe W. and Dorothy Dorsett Brown Foundation awards grants to nonprofit organizations in Louisiana and the Gulf Coast of Mississippi. Areas of interest include medical research; housing for the homeless; support for organizations who care for the sick, hungry or helpless; religious and educational institutions; and organizations and groups concerned with improving the local community. Types of support include: operating budgets; research; and student aid. The foundation also supports service learning, a learn-by-doing approach to the curriculum. Students receive practical, hands-on experience in the subject matter studied by meeting identified community needs through active participation. The listed annual deadline date for application submission is August 31.
*Requirements:* Louisiana and Mississippi nonprofit organizations, with a focus on South Louisiana, the New Orleans area, and the Mississippi Gulf Coast, are eligible. Service Learning grant applications are available yearly to sixth through 12th grades in the following parishes: Orleans, Jefferson, Plaquemines, Saint Bernard, Saint Charles, Tangipahoa, Saint James, Saint John, Saint Tammany, and Washington.
*Geographic Focus:* Louisiana, Mississippi
*Date(s) Application is Due:* Aug 31
*Amount of Grant:* 5,000 - 25,000 USD
*Contact:* Beth Buscher, (504) 834-3433, ext. 200 or (504) 834-3441; bethbuscher@thebrownfoundation.org
*Internet:* www.thebrownfoundation.org
*Sponsor:* Joe W. and Dorothy Dorsett Brown Foundation
320 Hammond Highway, Suite 500
Metairie, LA 70005

**John Carter Brown Library Alexander O. Vietor Memorial Fellowship** **1401**

The John Carter Brown Library, an independently funded and administered center for advanced research in the humanities at Brown University, offers approximately forty residential research fellowships each year, extending from two to four months and carrying a stipend of $2,100 per month. The Alexander O. Vietor Memorial Fellowship is for research in early maritime history. The Fellowship is open to foreign nationals as well as Americans, and to scholars engaged in pre- or postdoctoral, or independent research related to the resources of the library. The Library's resources consist of printed materials (both European and American) related to the discovery, exploration, settlement, and development of North and South America before 1830. Preference may be given to applicants able to take up the fellowship during the course of the academic year, September through May. Application forms may be obtained from the program officer. The library also offers several long-term fellowships of five to nine months, with NEH support and specific eligibility requirements.
*Requirements:* Americans and foreign nationals who are engaged in pre- or postdoctoral or independent research are eligible. Graduate students must have passed their preliminary or general examinations at the time of application.
*Restrictions:* Recipients of all fellowships are expected to be in continuous residence at the library and to participate in the intellectual life of Brown University.
*Geographic Focus:* All States, All Countries
*Date(s) Application is Due:* Dec 1
*Amount of Grant:* 4,200 - 8,400 USD
*Contact:* Neil Safier, Director; (401) 863-2725; fax (401) 863-3477; jcb-director@brown.edu or JCBL_Fellowships@Brown.edu
*Internet:* https://www.brown.edu/academics/libraries/john-carter-brown/fellowships/description-fellowship-program
*Sponsor:* John Carter Brown Library
Box 1894
Providence, RI 02912

**John Carter Brown Library Alice E. Adams Fellowship** **1402**

The John Carter Brown Library, an independently funded and administered center for advanced research in the humanities at Brown University, offers approximately forty residential research fellowships each year, extending from two to four months and carrying a stipend of $2,100 per month. The Alice E. Adams Fellowship is open to foreign nationals as well as Americans, and to scholars engaged in pre- or postdoctoral, or independent research related to the resources of the library. The Library's resources consist of printed materials (both European and American) related to the discovery,

exploration, settlement, and development of North and South America before 1830. Preference may be given to applicants able to take up the fellowship during the course of the academic year, September through May. Application forms may be obtained from the program officer. The library also offers several long-term fellowships of five to nine months, with NEH support and specific eligibility requirements.
*Requirements:* Americans and foreign nationals who are engaged in pre- or postdoctoral or independent research are eligible. Graduate students must have passed their preliminary or general examinations at the time of application.
*Restrictions:* Recipients of all fellowships are expected to be in continuous residence at the library and to participate in the intellectual life of Brown University.
*Geographic Focus:* All States, All Countries
*Date(s) Application is Due:* Dec 1
*Amount of Grant:* 4,200 - 8,400 USD
*Contact:* Neil Safier, Director; (401) 863-2725; fax (401) 863-3477; jcb-director@brown.edu or JCBL_Fellowships@Brown.edu
*Internet:* https://www.brown.edu/academics/libraries/john-carter-brown/fellowships/description-fellowship-program
*Sponsor:* John Carter Brown Library
Box 1894
Providence, RI 02912

**John Carter Brown Library Associates Fellowship** 1403

The John Carter Brown Library, an independently funded and administered center for advanced research in the humanities at Brown University, offers approximately forty residential research fellowships each year, extending from two to four months and carrying a stipend of $2,100 per month. The Library Associates Fellowship is open to foreign nationals as well as Americans, and to scholars engaged in pre- or postdoctoral, or independent research related to the resources of the library. The Library's resources consist of printed materials (both European and American) related to the discovery, exploration, settlement, and development of North and South America before 1830. Preference may be given to applicants able to take up the fellowship during the course of the academic year, September through May. Application forms may be obtained from the program officer. The library also offers several long-term fellowships of five to nine months, with NEH support and specific eligibility requirements.
*Requirements:* Americans and foreign nationals who are engaged in pre- or postdoctoral or independent research are eligible. Graduate students must have passed their preliminary or general examinations at the time of application.
*Restrictions:* Recipients of all fellowships are expected to be in continuous residence at the library and to participate in the intellectual life of Brown University.
*Geographic Focus:* All States, All Countries
*Date(s) Application is Due:* Dec 1
*Amount of Grant:* 4,200 - 8,400 USD
*Contact:* Neil Safier, Director; (401) 863-2725; fax (401) 863-3477; jcb-director@brown.edu or JCBL_Fellowships@Brown.edu
*Internet:* https://www.brown.edu/academics/libraries/john-carter-brown/fellowships/description-fellowship-program
*Sponsor:* John Carter Brown Library
Box 1894
Providence, RI 02912

**John Carter Brown Library Barbara S. Mosbacher Fellowship** 1404

The John Carter Brown Library, an independently funded and administered center for advanced research in the humanities at Brown University, offers approximately forty residential research fellowships each year, extending from two to four months and carrying a stipend of $2,100 per month. The Barbara S. Mosbacher Fellowship is open to foreign nationals as well as Americans, and to scholars engaged in pre- or postdoctoral, or independent research related to the resources of the library. The Library's resources consist of printed materials (both European and American) related to the discovery, exploration, settlement, and development of North and South America before 1830. Preference may be given to applicants able to take up the fellowship during the course of the academic year, September through May. Application forms may be obtained from the program officer. The library also offers several long-term fellowships of five to nine months, with NEH support and specific eligibility requirements.
*Requirements:* Americans and foreign nationals who are engaged in pre- or postdoctoral or independent research are eligible. Graduate students must have passed their preliminary or general examinations at the time of application.
*Restrictions:* Recipients of all fellowships are expected to be in continuous residence at the library and to participate in the intellectual life of Brown University.
*Geographic Focus:* All States, All Countries
*Date(s) Application is Due:* Dec 1
*Amount of Grant:* 4,200 - 8,400 USD
*Contact:* Neil Safier, Director; (401) 863-2725; fax (401) 863-3477; jcb-director@brown.edu or JCBL_Fellowships@Brown.edu
*Internet:* https://www.brown.edu/academics/libraries/john-carter-brown/fellowships/description-fellowship-program
*Sponsor:* John Carter Brown Library
Box 1894
Providence, RI 02912

**John Carter Brown Library Center for New World Comparative Studies Fellowships**

The John Carter Brown Library, an independently funded and administered center for advanced research in the humanities at Brown University, offers approximately forty residential research fellowships each year, extending from two to four months and carrying a stipend of $2,100 per month. The Center for New World Comparative Studies Fellowship are available for scholars engaged in research in the comparative history of the colonial Americas. The Fellowships is open to foreign nationals as well as Americans, and to scholars engaged in pre- or postdoctoral, or independent research related to the resources of the library. The Library's resources consist of printed materials (both European and American) related to the discovery, exploration, settlement, and development of North and South America before 1830. Preference may be given to applicants able to take up the fellowship during the course of the academic year, September through May. Application forms may be obtained from the program officer. The library also offers several long-term fellowships of five to nine months, with NEH support and specific eligibility requirements.
*Requirements:* Americans and foreign nationals who are engaged in pre- or postdoctoral or independent research are eligible. Graduate students must have passed their preliminary or general examinations at the time of application.
*Restrictions:* Recipients of all fellowships are expected to be in continuous residence at the library and to participate in the intellectual life of Brown University.
*Geographic Focus:* All States, All Countries
*Date(s) Application is Due:* Dec 1
*Amount of Grant:* 4,200 - 8,400 USD
*Contact:* Neil Safier, Director; (401) 863-2725; fax (401) 863-3477; jcb-director@brown.edu or JCBL_Fellowships@Brown.edu
*Internet:* https://www.brown.edu/academics/libraries/john-carter-brown/fellowships/description-fellowship-program
*Sponsor:* John Carter Brown Library
Box 1894
Providence, RI 02912

**John Carter Brown Library Charles H. Watts Memorial Fellowship** 1406

The John Carter Brown Library, an independently funded and administered center for advanced research in the humanities at Brown University, offers approximately forty residential research fellowships each year, extending from two to four months and carrying a stipend of $2,100 per month. The Charles H. Watts Memorial Fellowship is open to foreign nationals as well as Americans, and to scholars engaged in pre- or postdoctoral, or independent research related to the resources of the library. The Library's resources consist of printed materials (both European and American) related to the discovery, exploration, settlement, and development of North and South America before 1830. Preference may be given to applicants able to take up the fellowship during the course of the academic year, September through May. Application forms may be obtained from the program officer. The library also offers several long-term fellowships of five to nine months, with NEH support and specific eligibility requirements.
*Requirements:* Americans and foreign nationals who are engaged in pre- or postdoctoral or independent research are eligible. Graduate students must have passed their preliminary or general examinations at the time of application.
*Restrictions:* Recipients of all fellowships are expected to be in continuous residence at the library and to participate in the intellectual life of Brown University.
*Geographic Focus:* All States, All Countries
*Date(s) Application is Due:* Dec 1
*Amount of Grant:* 4,200 - 8,400 USD
*Contact:* Neil Safier, Director; (401) 863-2725; fax (401) 863-3477; jcb-director@brown.edu or JCBL_Fellowships@Brown.edu
*Internet:* https://www.brown.edu/academics/libraries/john-carter-brown/fellowships/description-fellowship-program
*Sponsor:* John Carter Brown Library
Box 1894
Providence, RI 02912

**John Carter Brown Library Environmental Studies Fellowships** 1407

The John Carter Brown Library is keen to welcome applications whose projects have thematic resonance with the study of the environment and its history, lato sensu. In coming years, as part of a multi-year project on The Four Elements in the History of the Americas, the Library will explore the cultural significance of earth, air, fire, and water to the diverse populations of the Americas, from the continents' earliest indigenous inhabitants to the last waves of European scientific explorers at the end of the colonial period. This project will begin with an exhibit and related programming on Subterranean Worlds, and will continue in with a focus on air and climate.
*Requirements:* Americans and foreign nationals who are engaged in pre- or postdoctoral or independent research are eligible. Graduate students must have passed their preliminary or general examinations at the time of application.
*Restrictions:* Recipients of all fellowships are expected to be in continuous residence at the library and to participate in the intellectual life of Brown University.
*Geographic Focus:* All States
*Date(s) Application is Due:* Jan 15
*Contact:* Neil Safier, Director; (401) 863-2725; fax (401) 863-3477; jcb-director@brown.edu or JCBL_Fellowships@Brown.edu
*Internet:* https://www.brown.edu/academics/libraries/john-carter-brown/fellowships/description-fellowship-program
*Sponsor:* John Carter Brown Library
Box 1894
Providence, RI 02912

**John Carter Brown Library Helen Watson Buckner Memorial Fellowship** 1408

The John Carter Brown Library, an independently funded and administered center for advanced research in the humanities at Brown University, offers approximately forty residential research fellowships each year, extending from two to four months and carrying a stipend of $2,100 per month. The Helen Watson Buckner Memorial Fellowship is open to foreign nationals as well as Americans, and to scholars engaged in pre- or postdoctoral, or independent research related to the resources of the library. The Library's resources consist of printed materials (both European and American) related to the discovery, exploration, settlement, and development of North and South America before 1830. Preference may be given to applicants able to take up the fellowship during the course of the academic year, September through May. Application forms may be obtained from the program officer. The library also offers several long-term fellowships of five to nine months, with NEH support and specific eligibility requirements.
*Requirements:* Americans and foreign nationals who are engaged in pre- or postdoctoral or independent research are eligible. Graduate students must have passed their preliminary or general examinations at the time of application.
*Restrictions:* Recipients of all fellowships are expected to be in continuous residence at the library and to participate in the intellectual life of Brown University.
*Geographic Focus:* All States, All Countries
*Date(s) Application is Due:* Dec 1
*Amount of Grant:* 4,200 - 8,400 USD
*Contact:* Neil Safier, Director; (401) 863-2725; fax (401) 863-3477; jcb-director@brown.edu or JCBL_Fellowships@Brown.edu
*Internet:* https://www.brown.edu/academics/libraries/john-carter-brown/fellowships/description-fellowship-program
*Sponsor:* John Carter Brown Library
Box 1894
Providence, RI 02912

**John Carter Brown Library Jeannette D. Black Memorial Fellowship** 1409

The John Carter Brown Library, an independently funded and administered center for advanced research in the humanities at Brown University, offers approximately forty residential research fellowships each year, extending from two to four months and carrying a stipend of $2,100 per month. The Norman Fiering Fund Fellowship is for research in the history of cartography or a closely related area. The Fellowship is open to foreign nationals as well as Americans, and to scholars engaged in pre- or postdoctoral, or independent research related to the resources of the library. The Library's resources consist of printed materials (both European and American) related to the discovery, exploration, settlement, and development of North and South America before 1830. Preference may be given to applicants able to take up the fellowship during the course of the academic year, September through May. Application forms may be obtained from the program officer. The library also offers several long-term fellowships of five to nine months, with NEH support and specific eligibility requirements.
*Requirements:* Americans and foreign nationals who are engaged in pre- or postdoctoral or independent research are eligible. Graduate students must have passed their preliminary or general examinations at the time of application.
*Restrictions:* Recipients of all fellowships are expected to be in continuous residence at the library and to participate in the intellectual life of Brown University.
*Geographic Focus:* All States, All Countries
*Date(s) Application is Due:* Dec 1
*Amount of Grant:* 4,200 - 8,400 USD
*Contact:* Neil Safier, Director; (401) 863-2725; fax (401) 863-3477; jcb-director@brown.edu or JCBL_Fellowships@Brown.edu
*Internet:* https://www.brown.edu/academics/libraries/john-carter-brown/fellowships/description-fellowship-program
*Sponsor:* John Carter Brown Library
Box 1894
Providence, RI 02912

**John Carter Brown Library John M. Monteiro Memorial Fellow in Indigenous Studies Fellowships** 1410

The John Carter Brown Library offers an Indigenous Stusies Fellowship. This two to four-month fellowship is designed for a junior or senior scholar working on the history and/or anthropology of native peoples of the Americas who could benefit from an extended stay and privileged access to the JCB's collection in this field. Tribal historians with non-traditional academic backgrounds are also eligible to apply. The inaugural Indigenous Studies may also help to develop symposia and/or lectures and will be able to access additional funds to bring in speakers and activities to the Library.
*Requirements:* Americans and foreign nationals who are engaged in pre- or postdoctoral or independent research are eligible. Graduate students must have passed their preliminary or general examinations at the time of application.
*Restrictions:* Recipients of all fellowships are expected to be in continuous residence at the library and to participate in the intellectual life of Brown University.
*Geographic Focus:* All States
*Date(s) Application is Due:* Jan 15
*Contact:* Neil Safier, Director; (401) 863-2725; fax (401) 863-3477; jcb-director@brown.edu or JCBL_Fellowships@Brown.edu
*Internet:* https://www.brown.edu/academics/libraries/john-carter-brown/fellowships/description-fellowship-program
*Sponsor:* John Carter Brown Library
Box 1894
Providence, RI 02912

**John Carter Brown Library Long-Term Research Fellowships** 1411

John Carter Brown Library Long-Term Research Fellowships, supported in part by NEH and the Andrew E. Mellon Foundation, are for five to ten months with a monthly stipend of $4,200. The term for these fellowships will typically begin between June 1 and July 15 or between January 15 and March 15. Awards are based on the applicant's scholarly qualifications, the merits of the project, and the appropriateness of the inquiry to the holdings of the library, which is an outstanding collection of primary materials relating to virtually all aspects of the discovery, exploration, settlement, and development of the New World, ranging from the late 15th century to about 1830. Recipients are expected to be in continuous residence at the library and to participate in the intellectual life at Brown University. Application forms may be obtained from the program officer.
*Requirements:* Applicants must be U.S. citizens or have been residents in the United States for the three years immediately preceding the term of the fellowship.
*Restrictions:* Graduate students are not eligible.
*Geographic Focus:* All States
*Date(s) Application is Due:* Dec 15
*Amount of Grant:* 22,000 - 42,000 USD
*Contact:* Neil Safier, Director; (401) 863-2725; fax (401) 863-3477; jcb-director@brown.edu or JCBL_Fellowships@Brown.edu
*Internet:* https://www.brown.edu/academics/libraries/john-carter-brown/fellowships/description-fellowship-program
*Sponsor:* John Carter Brown Library
Box 1894
Providence, RI 02912

**John Carter Brown Library Marie L. and William R. Hartland Fellowship** 1412

The John Carter Brown Library, an independently funded and administered center for advanced research in the humanities at Brown University, offers approximately forty residential research fellowships each year, extending from two to four months and carrying a stipend of $2,100 per month. The Marie L. and William R. Hartland Fellowship is for research related to eighteenth-century maritime social history. The Fellowship is open to foreign nationals as well as Americans, and to scholars engaged in pre- or postdoctoral, or independent research related to the resources of the library. The Library's resources consist of printed materials (both European and American) related to the discovery, exploration, settlement, and development of North and South America before 1830. Preference may be given to applicants able to take up the fellowship during the course of the academic year, September through May. Application forms may be obtained from the program officer. The library also offers several long-term fellowships of five to nine months, with NEH support and specific eligibility requirements.
*Requirements:* Americans and foreign nationals who are engaged in pre- or postdoctoral or independent research are eligible. Graduate students must have passed their preliminary or general examinations at the time of application.
*Restrictions:* Recipients of all fellowships are expected to be in continuous residence at the library and to participate in the intellectual life of Brown University.
*Geographic Focus:* All States, All Countries
*Date(s) Application is Due:* Dec 1
*Amount of Grant:* 4,200 - 8,400 USD
*Contact:* Neil Safier, Director; (401) 863-2725; fax (401) 863-3477; jcb-director@brown.edu or JCBL_Fellowships@Brown.edu
*Internet:* https://www.brown.edu/academics/libraries/john-carter-brown/fellowships/description-fellowship-program
*Sponsor:* John Carter Brown Library
Box 1894
Providence, RI 02912

**John Carter Brown Library Norman Fiering Fund Fellowship** 1413

The John Carter Brown Library, an independently funded and administered center for advanced research in the humanities at Brown University, offers approximately forty residential research fellowships each year, extending from two to four months and carrying a stipend of $2,100 per month. The Norman Fiering Fund Fellowship is open to foreign nationals as well as Americans, and to scholars engaged in pre- or postdoctoral, or independent research related to the resources of the library. The Library's resources consist of printed materials (both European and American) related to the discovery, exploration, settlement, and development of North and South America before 1830. Preference may be given to applicants able to take up the fellowship during the course of the academic year, September through May. Application forms may be obtained from the program officer. The library also offers several long-term fellowships of five to nine months, with NEH support and specific eligibility requirements.
*Requirements:* Americans and foreign nationals who are engaged in pre- or postdoctoral or independent research are eligible. Graduate students must have passed their preliminary or general examinations at the time of application.
*Restrictions:* Recipients of all fellowships are expected to be in continuous residence at the library and to participate in the intellectual life of Brown University.
*Geographic Focus:* All States, All Countries
*Date(s) Application is Due:* Dec 1
*Amount of Grant:* 4,200 - 8,400 USD
*Contact:* Neil Safier, Director; (401) 863-2725; fax (401) 863-3477; jcb-director@brown.edu or JCBL_Fellowships@Brown.edu
*Internet:* https://www.brown.edu/academics/libraries/john-carter-brown/fellowships/description-fellowship-program
*Sponsor:* John Carter Brown Library
Box 1894
Providence, RI 02912

**John Carter Brown Library Paul W. McQuillen Memorial Fellowship Fellowship 1414**

The John Carter Brown Library, an independently funded and administered center for advanced research in the humanities at Brown University, offers approximately forty residential research fellowships each year, extending from two to four months and carrying a stipend of $2,100 per month. The Paul W. McQuillen Memorial Fellowship is open to foreign nationals as well as Americans, and to scholars engaged in pre- or postdoctoral, or independent research related to the resources of the library. The Library's resources consist of printed materials (both European and American) related to the discovery, exploration, settlement, and development of North and South America before 1830. Preference may be given to applicants able to take up the fellowship during the course of the academic year, September through May. Application forms may be obtained from the program officer. The library also offers several long-term fellowships of five to nine months, with NEH support and specific eligibility requirements.

*Requirements:* Americans and foreign nationals who are engaged in pre- or postdoctoral or independent research are eligible. Graduate students must have passed their preliminary or general examinations at the time of application.

*Restrictions:* Recipients of all fellowships are expected to be in continuous residence at the library and to participate in the intellectual life of Brown University.

*Geographic Focus:* All States, All Countries

*Date(s) Application is Due:* Dec 1

*Amount of Grant:* 4,200 - 8,400 USD

*Contact:* Neil Safier, Director; (401) 863-2725; fax (401) 863-3477; jcb-director@brown.edu or JCBL_Fellowships@Brown.edu

*Internet:* https://www.brown.edu/academics/libraries/john-carter-brown/fellowships/description-fellowship-program

*Sponsor:* John Carter Brown Library
Box 1894
Providence, RI 02912

**John Carter Brown Library Ruth and Lincoln Ekstrom Fellowship 1415**

The John Carter Brown Library, an independently funded and administered center for advanced research in the humanities at Brown University, offers approximately forty residential research fellowships each year, extending from two to four months and carrying a stipend of $2,100 per month. The Ruth and Lincoln Ekstrom Fellowship is for research on the history of women and the family in the Americas prior to 1825, including the question of cultural influences on gender formation. The Fellowship is open to foreign nationals as well as Americans, and to scholars engaged in pre- or postdoctoral, or independent research related to the resources of the library. The Library's resources consist of printed materials (both European and American) related to the discovery, exploration, settlement, and development of North and South America before 1830. Preference may be given to applicants able to take up the fellowship during the course of the academic year, September through May. Application forms may be obtained from the program officer. The library also offers several long-term fellowships of five to nine months, with NEH support and specific eligibility requirements.

*Requirements:* Americans and foreign nationals who are engaged in pre- or postdoctoral or independent research are eligible. Graduate students must have passed their preliminary or general examinations at the time of application.

*Restrictions:* Recipients of all fellowships are expected to be in continuous residence at the library and to participate in the intellectual life of Brown University.

*Geographic Focus:* All States, All Countries

*Date(s) Application is Due:* Dec 1

*Amount of Grant:* 4,200 - 8,400 USD

*Contact:* Neil Safier, Director; (401) 863-2725; fax (401) 863-3477; jcb-director@brown.edu or JCBL_Fellowships@Brown.edu

*Internet:* https://www.brown.edu/academics/libraries/john-carter-brown/fellowships/description-fellowship-program

*Sponsor:* John Carter Brown Library
Box 1894
Providence, RI 02912

**John Carter Brown Library William Reese Company Fellowship 1416**

The John Carter Brown Library, an independently funded and administered center for advanced research in the humanities at Brown University, offers approximately forty residential research fellowships each year, extending from two to four months and carrying a stipend of $2,100 per month. The William Reese Company Fellowship is for research in bibliography and the history of printing. The Fellowship is open to foreign nationals as well as Americans, and to scholars engaged in pre- or postdoctoral, or independent research related to the resources of the library. The Library's resources consist of printed materials (both European and American) related to the discovery, exploration, settlement, and development of North and South America before 1830. Preference may be given to applicants able to take up the fellowship during the course of the academic year, September through May. Application forms may be obtained from the program officer. The library also offers several long-term fellowships of five to nine months, with NEH support and specific eligibility requirements.

*Requirements:* Americans and foreign nationals who are engaged in pre- or postdoctoral or independent research are eligible. Graduate students must have passed their preliminary or general examinations at the time of application.

*Restrictions:* Recipients of all fellowships are expected to be in continuous residence at the library and to participate in the intellectual life of Brown University.

*Geographic Focus:* All States, All Countries

*Date(s) Application is Due:* Dec 1

*Amount of Grant:* 4,200 - 8,400 USD

*Contact:* Neil Safier, Director; (401) 863-2725; fax (401) 863-3477; jcb-director@brown.edu or JCBL_Fellowships@Brown.edu

*Internet:* https://www.brown.edu/academics/libraries/john-carter-brown/fellowships/description-fellowship-program

*Sponsor:* John Carter Brown Library
Box 1894
Providence, RI 02912

**John G. Duncan Charitable Trust Grants 1417**

John G. Duncan was born in Sacramento, California, in 1866, and lived a substantial portion of his life in Henderson, Kentucky, where he was actively involved in the philanthropic community. He later relocated to Denver, Colorado, where he also supported a number of different non-profit organizations. Duncan died in 1955 at the age of 89. The John G. Duncan Charitable Trust was established to award grants in the following areas of interest: arts; education; health care; human services; religion; building and renovation; capital campaigns; emergency funds; equipment; program development; seed money; and research. Applications should be submitted through the online grant application form or alternative accessible application designed for assistive technology users. The average award ranges from $5,000 to $10,000. Applications may be submitted year-round, but must be submitted by the following deadlines to be reviewed at the grant meeting held after each date: January 31, April 30, July 31, and October 31.

*Requirements:* Organizations must have 501(c)3 designation, and serve and operate in Colorado. Organizations are eligible to apply once per calendar year. If a grant is awarded, the organization will be ineligible to apply in the following calendar year.

*Restrictions:* The foundation does not fund requests to support: general operating expenses; endowments; organizations located outside of Colorado; other grantmaking organizations; or organizations recognized by the IRS as non-functionally integrated supporting organizations per Section 509(a)3 of the Internal Revenue Code.

*Geographic Focus:* Colorado

*Date(s) Application is Due:* Jan 31; Apr 30; Jul 31; Oct 31

*Amount of Grant:* 5,000 - 10,000 USD

*Contact:* Jason Craig, Trustee; (888) 234-1999; fax (877) 746-5889; grantadministration@wellsfargo.com

*Internet:* https://www.wellsfargo.com/privatefoundationgrants/duncan

*Sponsor:* John G. Duncan Charitable Trust
1740 Broadway, MAC C7300-483
Denver, CO 80274-0001

**John Simon Guggenheim Memorial Foundation Fellowships 1418**

Often characterized as mid-career awards, Guggenheim Fellowships are intended for men and women who have already demonstrated exceptional capacity for productive scholarship or exceptional creative ability in the arts. Fellowships are awarded through two annual competitions: one open to citizens and permanent residents of the United States and Canada, and the other open to citizens and permanent residents of Latin America and the Caribbean. Candidates must apply to the Guggenheim Foundation in order to be considered in either of these competitions. The Foundation receives between 3,500 and 4,000 applications each year. Although no one who applies is guaranteed success in the competition, there is no prescreening: all applications are reviewed. Approximately 220 Fellowships are awarded each year.

*Requirements:* Fellowships are awarded to individuals who have already demonstrated exceptional capacity for productive scholarship or exceptional creative ability in the arts.

*Restrictions:* Guggenheim Fellowships are not scholarships, and they are not available to those who seek to complete their training either as undergraduate, graduate, or part-time students. They are also not intended for immediate postgraduate work.

*Geographic Focus:* All States, Puerto Rico, Anguilla, Antigua & Barbuda, Argentina, Aruba, Bahamas, Barbados, Bolivia, Brazil, British Virgin Islands, Canada, Cayman Islands, Chile, Colombia, Costa Rica, Cuba, Dominica, Dominican Republic, Ecuador, El Salvador, French Guiana, Grenada, Guadeloupe, Guatemala, Haiti, Honduras, Jamaica, Martinique, Mexico, Montserrat, Nicaragua, Panama, Paraguay, Peru, Saint Kitts And Nevis, St. Lucia, St. Vincent and the Grenadines, Trinidad and Tobago, Turks and Caicos Islands, Uruguay, Venezuela, Virgin Islands (U.S.)

*Date(s) Application is Due:* Oct 1; Dec 1

*Amount of Grant:* 40,000 USD

*Contact:* Associate Secretary; (212) 687-4470; fax (212) 697-3248; fellowships@gf.org

*Internet:* www.gf.org/about-the-foundation/the-fellowship/

*Sponsor:* John Simon Guggenheim Memorial Foundation
90 Park Avenue
New York, NY 10016

**John Templeton Foundation Science of Purpose grants Grants 1419**

Sir John Templeton lived in an era of unparalleled scientific and technological progress. The accelerating pace of scientific discovery led Sir John to wonder whether the methods of science might be harnessed to make similar progress in understanding the deepest and most perplexing questions facing humankind. Today, the Foundation that bears his name aspires to fulfill his vision - relentless curiosity in pursuit of infinite discovery. The Templeton Foundation's Natural Sciences Department seeks to support projects that address these and similar big questions concerning the origin and nature of the universe and humankind's place in it. High risk projects that promise substantial rewards are especially encouraged, as are projects unlikely to be supported by standard funding agencies. Hence, as part of the Foundation's commitment to fund 12 Strategic Priorities, the Natural Sciences department will be seeking proposals related to the following priorities: intellectual humility; mathematics and physical sciences; science of purpose; and cultural evolution. In the area of Science of purpose, the Foundation will support

research that seeks to clarify and test notions of apparent agential, goal-directed, and even purposive processes in the sciences. It envisions empirical and theoretical projects that will follow two broad streams: target-oriented behaviors; and long-term directional trends.
*Requirements:* The Foundation strongly prefers to fund projects that are affiliated with an institution (typically a university, research institution or other nonprofit organization.) Whenever possible, individuals are encouraged to link with an appropriate institution that will assist by administering the grant. There are cases, however, when the foundation will make a grant to an individual. Unaffiliated scholars are welcome to propose research or other relevant projects.
*Restrictions:* The Foundation does not accept proposals electronically. Nor does it consider requests for individuals, capital campaigns, scholarships, or endowments.
*Geographic Focus:* All States
*Contact:* Andrea Holt; (610) 941-2828; fax (610) 825-1730; info@templeton.org
*Internet:* https://templeton.org/project/science-of-purpose
*Sponsor:* John Templeton Foundation
300 Conshohocken State Road, Suite 500
West Conshohocken, PA 19428

**Joseph Alexander Foundation Grants** **1420**
Established in 1960 in New York, the Foundation supports primarily in the New York region. Fields of interest include the following: higher education; health organizations; medical research, particularly optic nerve research; social services; and Jewish organizations. Types of support also include the following: annual campaigns; building/renovation; capital campaigns; conferences/seminars; curriculum development; endowments; equipment; exchange programs; general/operating support; program development; research; and scholarship funds. Applicants should submit a letter requesting application guidelines before submitting a proposal. There is no deadline, but the board meets in January, April, July, and October.
*Requirements:* Nonprofit organizations are eligible to apply.
*Restrictions:* Grants are not made to individuals.
*Geographic Focus:* New York
*Amount of Grant:* 5,000 - 50,000 USD
*Samples:* Salk Institute for Biological Studies, La Jolla, California, $40,000, payable over one year; University of Pennsylvania, Philadelphia, Pennsylvania, $50,000, payable over one year; Hebrew Free Burial Association, New York, New York, $3,500, payable over one year.
*Contact:* Robert Weintraub, President; (212) 355-3688
*Sponsor:* Joseph Alexander Foundation
110 East 59th Street
New York, NY 10022-1304

**Joyce Foundation Education Grants** **1421**
The Joyce Foundation is committed to improving education in Midwest cities, especially by eliminating the barriers that prevent low-income students and children of color from reaching their full educational potential. The Foundation is a long-time supporter of Chicago school reform, and has also worked in Milwaukee, Detroit, Cleveland, and currently in Indianapolis and Minneapolis. Over the years, Joyce has explored a variety of strategies, including school finance, parent organizing, educational technology, charter schools, and early childhood education; it has also consistently supported research identifying and evaluating promising new strategies. Because research shows that having a first-rate teacher has an enormous impact on student achievement, the Foundation now concentrates the bulk of its grant making on improving teacher quality. It also continues to support other strategies for addressing the achievement gap, including making sure children learn to read by third grade, and supporting high quality charter schools and other educational innovations.
*Requirements:* Education grant making in K-12 focuses on Chicago, Indianapolis, and Minneapolis. The Education Program works to close the achievement gaps that separate low-income students and communities of color from their peers by improving the quality of teachers they encounter in school, enhancing early reading policies, and exploring such innovations as charter schools. Program priorities are Teacher Quality, Early Reading and Innovation. A small portion of program funds is reserved for other outstanding opportunities, Innovation Grants, to close the achievement gap, especially policy-oriented efforts to expand the supply of high-quality charter schools in Chicago, Indianapolis, and Minneapolis. A letter of interest (LOI) should be submitted first. If acceptable, applicants will be requested to submit a full proposal. Applicants are strongly encouraged to plan their application and proposal submission process for the April or July meetings, since most grant funds will be distributed at those times.
*Restrictions:* The Foundation does not generally support capital proposals, endowment campaigns, religious activities, commercial ventures, direct-service programs, or scholarships.
*Geographic Focus:* Illinois, Indiana, Minnesota
*Date(s) Application is Due:* Apr 8; Aug 14; Dec 3
*Samples:* Advance Illinois, Chicago, IL - $107,085 to organize and train award-winning teachers in Illinois as advocates for policy change. Bellwether Education Partners, Wellesley, MA - $200,000 to support its efforts to advance the national conversation on the sustainability and reform of educator pensions through engaging relevant stakeholder audiences. Center for American Progress, Washington, D.C. - $150,000 to continue development and advancement of policy changes that support improvements in human capital systems throughout the country and access to effective teachers for all students.
*Contact:* Veronica Salter, Grants Manager; (312) 782-2464; fax (312) 595-1350; info@joycefdn.org or vsalter@joycefdn.org
*Internet:* www.joycefdn.org/programs/education/
*Sponsor:* Joyce Foundation
321 North Clark Street, Suite 1500
Chicago, IL 60654

**Joyce Foundation Employment Grants** **1422**
The Joyce Foundation believes that equipping workers to succeed in good-paying jobs benefits individuals, their families and communities, and the Midwest economy as a whole. In the last two decades, the Foundation has tested strategies to boost workers' skills and help women on welfare, low-skilled workers, ex-offenders, and others who face significant barriers in joining the workforce. Through it all, Joyce has consistently funded research to evaluate promising strategies and learn from the results; Joyce-supported research on welfare-to-work and transitional jobs were major contributions to the field. Current evaluation research focuses on adult education programs that prepare adults without high school diplomas or the equivalent with skills needed for careers.
*Requirements:* The Employment Program primarily focuses on federal and state policy grants, but will make some grants to support targeted metro-level progress in Chicago, Indianapolis, and Minneapolis/St. Paul. The current areas of interest are: Basic Foundational Skills - preparing adults with the skills required for jobs in today's market. In particular, the Foundation supports the evaluation and expansion of promising adult education programs that prepare people for post-secondary education and careers; Industry Training Partnerships - The Foundation supports expanding partnerships between industry associations or groups of businesses and educational organizations to create and promote career advancement opportunities. Grants encourage development and promotion of industry-recognized credentials that help low-skilled workers advance in the labor market. The Foundation also supports research into the return on investment relating to employer-sponsored skill development programs; and, Innovation. A letter of interest (LOI) should be submitted first. If acceptable, applicants will be requested to submit a full proposal. Applicants are strongly encouraged to plan their application and proposal submission process for the April or July meetings, since most grant funds will be distributed at those times.
*Restrictions:* The Foundation does not generally support capital proposals, endowment campaigns, religious activities, commercial ventures, direct-service programs, or scholarships.
*Geographic Focus:* Illinois, Indiana, Minnesota
*Date(s) Application is Due:* Apr 8; Aug 14; Dec 3
*Samples:* Center for Energy Workforce Development, Washington, D.C. - $225,000 to build the capacity and sustainability of energy industry workforce consortia in the six Joyce states. Center for Labor and Community Research, Chicago, IL - $150,000 to advocate for expanded and improved advanced manufacturing workforce development in the Chicago area, and to support expansion of such work beyond the Chicago area. EmployIndy, Indianapolis, IN - $250,000 To support phase one of a new business-led workforce development project aimed at closing the skills gap and creating opportunities for low-income workers in the Indianapolis region. (1 yr)
*Contact:* Veronica Salter, Grants Manager; (312) 782-2464; fax (312) 595-1350; info@joycefdn.org or vsalter@joycefdn.org
*Internet:* www.joycefdn.org/programs/employment/
*Sponsor:* Joyce Foundation
321 North Clark Street, Suite 1500
Chicago, IL 60654

**JSA JLab Graduate Fellowships** **1423**
The JLab Graduate Fellowships fund doctoral students at SURA member universities (listed at the website) conducting research related to the theoretical and experimental programs at Jefferson Lab, including the following: nuclear and related particle physics; accelerator physics; associated science and engineering; and free electron laser program. Each Fellowship awards $10,000 toward the normal academic year research assistant stipend, plus a $2,000 supplement. Conference travel support is available with each Fellowship. Selection is based on the following criteria: merit and quality of proposed research; utility and relevance of proposed research to Jefferson Lab and proposed utilization of Lab resources; academic qualifications; letters of recommendation; and the likelihood of accomplishment of proposed research objectives by graduation.
*Requirements:* Eligible students must be enrolled full-time in a relevant doctoral program at a SURA member university. Students may apply at any stage of their graduate careers or while undergraduate seniors. Awards are limited to two years and students seeking second-year support must resubmit a complete application.
*Restrictions:* Awardees may not concurrently hold another major full-time fellowship or internship.
*Geographic Focus:* All States
*Amount of Grant:* 12,000 USD
*Contact:* Hugh Montgomery, JSA President/JLab Director; (757) 269-7100; fax (757) 269-7398; mont@jlab.org
*Internet:* www.jsallc.org/IF/IndexGF.html
*Sponsor:* Jefferson Science Associates
c/o Southeastern Universities Research Association, Inc.
Washington, D.C. 20005

**JSA JLab Sabbatical Research Leave Support Grants** **1424**
The JLab Sabbatical Research Leave Support Grants fund faculty from universities conducting research at the Jefferson Lab during approved sabbatical leave from their home institutions. The Program provides $1,500 per month living expense support for up to twelve months for researchers to relocate to the Lab during their sabbatical. Additional information is available at the website.
*Geographic Focus:* All States
*Amount of Grant:* 18,000 USD
*Contact:* Hugh Montgomery; (757) 269-7100; fax (757) 269-7398; mont@jlab.org
*Internet:* www.jsallc.org/IF/IndexSS.html
*Sponsor:* Jefferson Science Associates
c/o Southeastern Universities Research Association, Inc.
Washington, D.C. 20005

**JSA Junior Scientist Travel Support** 1425

Jefferson Science Associates awards travel support to graduate students and post docs working at the Jefferson Lab for attendance and/or presentations at national and international conferences and workshops. This support enables junior researchers to receive recognition for their work, and helps participants to improve communication skills as they further their careers in the field. Travel funds may cover up to 50% of the travel cost. The traveler is expected to find grants or other support for the remaining cost of travel. Generally, up to $400-$500 may be awarded for local/national meetings and up to $800-$1,000 for international meetings. Advisors/supervisors should complete the request form at the website for matching travel funds. Requests for only attendance to meetings (no presentation or talk expected) are also considered. Requests may be submitted at any time.
*Requirements:* Eligible participants are undergraduate, graduate, or Ph.D. students and post docs who are not employees of the Jefferson Lab and are not receiving other JSA Initiatives Fund support.
*Geographic Focus:* All States, All Countries
*Amount of Grant:* 400 - 1,000 USD
*Contact:* Marco Battaglieri; (757) 269-7100; fax (757) 269-7363; battagli@jlab.org
Douglas Higinbotham; (757) 269-7100; fax (757) 269-7363; doug@jlab.org
*Internet:* www.jsallc.org/IF/IndexJSTF.html
*Sponsor:* Jefferson Science Associates
c/o Southeastern Universities Research Association, Inc.
Washington, D.C. 20005

**Judith and Jean Pape Adams Charitable Foundation ALS Grants** 1426

Jean Pape Adams was interested in supporting charitable organizations actively engaged in discovering the causes of Amyotrophic Lateral Sclerosis (ALS) and in finding a cure. The Foundation encourages research that is collaborative, innovative and aggressive. These initiatives may fill gaps not currently being investigated and may involve novel approaches. The research must be of high scientific merit in finding a cure for ALS. Proposals may be longer than two pages. A lay summary is also required. Applications are reviewed by an Advisory Committee comprised of scientific researchers, neurologists and others who have had close relationships with friends and family members afflicted with ALS. The Committee makes funding recommendations to the Trustees.
*Requirements:* Public agencies and organizations that are classified as a charitable organization described in Section 501(c)3 of the Internal Revenue Code and as a public charity under Section 509(a) of the Internal Revenue Code may apply.
*Restrictions:* Support groups, education and medical equipment are not supported.
*Geographic Focus:* All States
*Date(s) Application is Due:* Sep 13
*Amount of Grant:* 1,000 - 250,000 USD
*Contact:* Shelley L. Carter, Executive Director; (830) 997-7347; fax (918) 591-2433; scarter@adamsfoundation.org
*Internet:* www.adamsfoundation.org/
*Sponsor:* Judith and Jean Pape Adams Charitable Foundation
7030 South Yale Avenue, Suite 600
Tulsa, OK 74136-5749

**Kellogg Institute for International Studies Andrónico Luksic Grants** 1427

The objective of the Kellogg Institute for International Studies Andrónico Luksic Grants program is to encourage the Notre Dame community to find new ways to engage with colleagues from PUC (on research, exchanges, conferences, etc.) and in particular to increase the number of Notre Dame faculty and staff going to Santiago. Notre Dame faculty and staff may apply for grants of up to $10,000. Applicants are encouraged to demonstrate other sources of funding or cost sharing which will complement the Luksic Collaboration fund. Funding is available for a variety of collaborative ventures, including but not limited to the following: initiating collaborative research projects (travel, research costs, etc.); sponsoring or attending research conferences and workshops hosted at PUC; sponsoring educational ventures (faculty and staff exchanges, workshops, etc.); Notre Dame Graduate research assistant and postdoctoral fellow stipends and benefits.
*Restrictions:* Awards will not support clerical or secretarial help. Awards may support expenses incurred only at Notre Dame or by Notre Dame faculty and staff (e.g. hiring of research assistants employed by and working at PUC will generally not be funded through the Notre Dame Luksic grant.
*Geographic Focus:* All States
*Date(s) Application is Due:* Apr 15; Jul 15; Oct 14
*Amount of Grant:* Up to 10,000 USD
*Contact:* Anne Hayes, Associate Director; (574) 631-5527 or (574) 631-6580; fax (574) 631-6717; ahayes2@nd.edu
*Internet:* international.nd.edu/faculty-resources/#Funding
*Sponsor:* Kellogg Institute for International Studies
130 Hesburgh Center
Notre Dame, IN 46556-5677

**Kellogg Institute for International Studies Collaborative Faculty Grants** 1428

Collaborative Faculty Grants promote innovative, collaborative, and interdisciplinary research on issues closely aligned with the Kellogg Institute core research themes, democracy and human development. Collaborative projects might include the expansion of faculty members' current projects or the instigation of new collaborative ventures. The Institute seeks to support scholarly projects that have significant interdisciplinary and collaborative components and an ambitious mix of programmatic elements. Funded projects will transcend those eligible for Kellogg funding in the categories of individual faculty research, working group, or conference support. By providing a strong financial incentive for fellows to undertake major research projects, especially collaborative and longer-term ones, the Institute hopes to foster more, and more significant, innovative research that will make major contributions to scholarship on its core themes. Among other things, Kellogg especially seeks to support research that: crosses disciplinary boundaries and leads to crossfertilization across regions and themes; links important normative and theoretical work with high-level empirical research; unites its core themes of democracy and human development; uses a multidisciplinary approach to address complex problems of development; integrates research on religion and democracy/human development as a transversal theme; and/or contributes to relevant contemporary debates on public policy. Programmatic elements may include but are not limited to: working groups; workshops, roundtables, short-term visitors, and conferences; course buy-outs; team-taught seminars or courses; and graduate student assistance. Project funding may typically run between $10,000 and $70,000 for one or two years.
*Requirements:* Project leaders must be Kellogg faculty fellows, although all participants in the collaborative project need not be. Applicants are strongly encouraged to seek and priority is given to projects with financial support from other sources, both internal and external to Notre Dame.
*Geographic Focus:* All States
*Date(s) Application is Due:* Feb 11; Apr 1; Sep 17; Nov 19
*Amount of Grant:* 10,000 - 70,000 USD
*Contact:* Sharon Schierling, Associate Director; (574) 631-8524 or (574) 631-6580; fax (574) 631-6717; sschierl@nd.edu
*Internet:* kellogg.nd.edu/faculty/grants/collaborative.shtml
*Sponsor:* Kellogg Institute for International Studies
130 Hesburgh Center
Notre Dame, IN 46556-5677

**Kellogg Institute for International Studies Course Development Grants** 1429

New courses that address the questions and concerns of the Kellogg Institute research agenda can serve to enrich and help internationalize the undergraduate curriculum at the University, as well as to foster the intellectual engagement of undergraduates with Kellogg research themes. Funding is available to develop courses during the summer before offering them during the subsequent calendar year. Recipients are expected to offer the new course at least twice within five years following the grant. A maximum of $6,000 is available per grant, including a summer stipend of $2,500 for an individual faculty member, or $1,500 each for two faculty members developing a team-taught course. Up to $3,500 may be budgeted for supplementary funding, which may include but is not limited to teaching materials, books, research assistance, faculty or visitor travel expenses, honoraria, and payment for services.
*Requirements:* All full-time members of the regular faculty (including T&R and SPF) are eligible.
*Restrictions:* Course proposals with an in-country experiential learning component will be considered but Kellogg course development grants may not be used to fund student travel.
*Geographic Focus:* All States
*Date(s) Application is Due:* Feb 11; Apr 1; Sep 17; Nov 19
*Amount of Grant:* Up to 6,000 USD
*Contact:* Sharon Schierling, Associate Director; (574) 631-8524 or (574) 631-6580; fax (574) 631-6717; sschierl@nd.edu
*Internet:* kellogg.nd.edu/faculty/grants/cdev.shtml
*Sponsor:* Kellogg Institute for International Studies
130 Hesburgh Center
Notre Dame, IN 46556-5677

**Kellogg Institute for International Studies Dissertation Year Fellowships** 1430

The Kellogg Institute offers Dissertation Year Fellowships in International Studies to advanced Notre Dame graduate students in the social sciences, history, philosophy, theology, and international human rights law. The fellowships are intended for students to do field research or to complete their dissertations. The dissertation topic must pertain to international studies, addressing one or more of the thematic priorities of the Kellogg Institute: democratization and the quality of democracy; growth and development; religion and society; social movements and organized civil society; public policies for social justice. The awards will be equal to the respective department's academic year stipend for graduate student assistants.
*Requirements:* Applicants must be enrolled and in good standing in the Departments of Economics, Political Science, History, Philosophy, Sociology, or Theology, or in the J.S.D. program in International Human Rights.
*Restrictions:* Students who have other dissertation support equal to a full departmental graduate student stipend from Notre Dame sources for the time period being considered are not eligible. Students applying for this fellowship are not eligible for the Institute's Graduate Research Grant competition.
*Geographic Focus:* All States
*Date(s) Application is Due:* Mar 10
*Contact:* Denise Wright, Program Coordinator; (574) 631-8523 or (574) 631-6580; fax (574) 631-6717; dwright1@nd.edu
*Internet:* kellogg.nd.edu/students/grad/grants/dissert.shtml
*Sponsor:* Kellogg Institute for International Studies
130 Hesburgh Center
Notre Dame, IN 46556-5677

**Kellogg Institute for International Studies Experiencing The World Fellowships** 1431

The Kellogg Institute offers the opportunity for freshmen and sophomores to engage in initial exploratory projects in Africa, Asia or Latin America. The award funds exceptionally qualified and committed undergraduates who seek to undertake innovative projects in Africa, Asia or Latin America. These may include research, non-profit work, study, or other activities that will increase their commitment to and knowledge of one of the

regions. Award recipients may receive up to $5000 to pursue field projects for up to three months in the chosen region. Students interested in this award must justify their project as something that will substantially deepen and enrich their undergraduate experience and must show that this project is something that is otherwise not possible within existing Notre Dame programs. Student projects might include exploratory research for senior theses, area studies essays, or international scholars research; collaborative research with a professor; internship work for a non-profit organization or NGO; enrollment in an academic or language program that would not be possible through existing Notre Dame arrangements; or other creative forms of experiential education.
*Requirements:* Students interested in this award must meet the following criteria: have a GPA of 3.0 or higher; have taken at least one course at Notre Dame related to the proposed region of the project and/or demonstrate prior interest in the region; demonstrate adequate language skills to carry out the proposed project; and have not previously received an ETW Fellowship.
*Geographic Focus:* All States
*Date(s) Application is Due:* Mar 5
*Amount of Grant:* Up to 5,000 USD
*Contact:* Holly Rivers, Assistant Director; (574) 631-6023 or (574) 631-6580; fax (574) 631-6717; hrivers@nd.edu
*Internet:* kellogg.nd.edu/students/etw/index.shtml
*Sponsor:* Kellogg Institute for International Studies
130 Hesburgh Center
Notre Dame, IN 46556-5677

**Kellogg Institute for International Studies Faculty Fellows' Research Grants** 1432
The Kellogg Institute for International Studies offers individual and collaborative research grants to support summer and academic-year research by Kellogg faculty fellows in all disciplines. Funding is available for a variety of project costs, including but not limited to the following: travel, research assistance, or acquisition of research materials, including electronic databases; travel to research meetings and workshops in cases in which other Notre Dame funding does not exist or is not sufficient; communication with colleagues and institutions in other countries for research or other professional purposes; computer software and databases (but not hardware); and collaboration with coauthors and, in collective research projects, with other individuals involved in the project. Proposals will be accepted for most research-related expenses; proposals should justify the relevance and importance of proposed expenses. The committee will consider requests that involve multiple dimensions (e.g., two field trips, one field trip plus research assistance, etc.). Individual grants will not normally exceed $10,000 and proposals will be considered by the Grants Committee for any amount up to this figure.
*Requirements:* Applicants must notify the Kellogg Institute of any funding received from another source; the Institute may adjust the award accordingly. No duplicate funding (that is, funding for the same purpose) is allowed.
*Geographic Focus:* All States
*Date(s) Application is Due:* Feb 11; Apr 1; Sep 17; Nov 19
*Amount of Grant:* Up to 10,000 USD
*Contact:* Sharon Schierling, Associate Director; (574) 631-8524 or (574) 631-6580; fax (574) 631-6717; sschierl@nd.edu
*Internet:* kellogg.nd.edu/faculty/grants/research.shtml
*Sponsor:* Kellogg Institute for International Studies
130 Hesburgh Center
Notre Dame, IN 46556-5677

**Kellogg Institute for International Studies Global Collaboration Initiative Grants** 1433
The Office of the Vice President and Associate Provost for Internationalization, Notre Dame International (NDI), announces the Global Collaboration Initiative (GCI). Under the auspices of the GCI, several grants between $10,000 and $20,000 (with the possibility of renewal for up to three years) will be awarded to fund initiatives in international research, scholarship and collaboration with major foreign universities and research organizations. The GCI is intended to enable through international cooperation the establishment or continuation of programs of research, scholarship and sustained collaboration that will make major contributions in any field of study. These initiatives should contribute to increasing the internationalization of Notre Dame by elevating the university's visibility in the international academic and research community. Successful proposals will demonstrate the potential to create meaningful and sustainable exchange and/or research collaborations with colleagues and students at major foreign research universities and other research institutions, and to elevate the University's visibility in the international academic and research community.
*Requirements:* Eligible applicants include full-time regular members of the Notre Dame faculty, including tenured and tenure-track faculty, research faculty, library faculty and special professional faculty, are eligible to apply. Centers (e.g. NDnano, cSEND) are also eligible to apply provided that at least one faculty member is affiliated with the project.
*Restrictions:* Faculty receiving funding from the GCI Grant Program during the previous fiscal year may not apply for a new grant but may apply for a continuation of the current grant for up to three years. Ineligible project requests include: programs that do not involve collaboration with faculty/researchers outside the U.S; faculty summer salary; faculty academic year course reduction; projects in the final stages of completion; and ongoing research projects funded by another sponsor.
*Geographic Focus:* All States
*Date(s) Application is Due:* Oct 1
*Amount of Grant:* 10,000 - 20,000 USD
*Contact:* Geraldine Meehan, (574) 631-7742 or (574) 631-6580; fax (574) 631-6717; Geraldine.Meehan.19@nd.edu
*Internet:* international.nd.edu/global-collaboration-initiative-gci/
*Sponsor:* Kellogg Institute for International Studies
130 Hesburgh Center
Notre Dame, IN 46556-5677

**Kellogg Institute for International Studies Graduate Research Grants** 1434
The purpose of the Kellogg Institute for International Studies Graduate Research Grants is to support Notre Dame graduate students involved in overseas (or, if demonstrably appropriate, in the United States) research in international studies, addressing the thematic priorities of the Kellogg Institute. The grant is not intended to be a substitute for seeking external support for dissertation research but instead to fill gaps in other funding or to enable exploratory trips that will strengthen students' ability to prepare externally competitive dissertation proposals. Preference is given to students in a doctoral program and with the equivalent of two years of full-time graduate study prior to the beginning date of their proposed research. Funding is available for a variety of project costs, including research materials and assistance as well as travel. Individual grants will not normally exceed $7,000, and proposals will be considered for any amount below this figure.
*Requirements:* Student applicants must be enrolled in graduate school at the University of Notre Dame and be in good standing.
*Restrictions:* Students applying for this award are not eligible to apply for the Institute's Dissertation Year Fellowship in the same academic year.
*Geographic Focus:* All States
*Date(s) Application is Due:* Feb 26; Oct 16
*Amount of Grant:* Up to 7,000 USD
*Contact:* Denise Wright, Program Coordinator; (574) 631-8523 or (574) 631-6580; fax (574) 631-6717; dwright1@nd.edu
*Internet:* kellogg.nd.edu/students/grad/grants/gradresearch.shtml
*Sponsor:* Kellogg Institute for International Studies
130 Hesburgh Center
Notre Dame, IN 46556-5677

**Kellogg Institute for International Studies Grants for Conferences and Workshops** 1435
The Kellogg Institute for International Studies makes grants of up to $35,000 for research conferences organized by its faculty fellows, giving preference to conferences that match major Institute priorities. On top of this amount, the Institute may match up to $15,000 from non-Institute sources for conference funding. Grants up to $10,000 will go through the Grants Committee. The Faculty Committee can supplement those awards with additional amounts for major research conferences addressing Kellogg priorities. Workshops, on the other hand, generally are small and informal, often addressing research issues at an exploratory stage of a project. Workshop grants may also be used to invite scholars and experts to critique a faculty member's manuscript. Grants of up to $5,000 are available for such workshops.
*Requirements:* Faculty members are strongly encouraged to seek other funding sources to supplement and leverage Kellogg support. Kellogg-supported conferences ordinarily are cosponsored by other Notre Dame units or external entities, and need not be held at the University. Conferences often result in published volumes. Applicants' proposals and budgets must include information about other sources of funding and sponsorship.
*Geographic Focus:* All States
*Date(s) Application is Due:* Feb 11; Apr 1; Sep 17; Nov 19
*Amount of Grant:* Up to 35,000 USD
*Contact:* Sharon Schierling, Associate Director; (574) 631-8524 or (574) 631-6580; fax (574) 631-6717; sschierl@nd.edu
*Internet:* kellogg.nd.edu/faculty/grants/events.shtml
*Sponsor:* Kellogg Institute for International Studies
130 Hesburgh Center
Notre Dame, IN 46556-5677

**Kellogg Institute for International Studies Grants for Distinguished Lectureships** 1436
The Kellogg Institute for International Studies will consider proposals from faculty fellows to fund especially distinguished lecturers or performers for periods of up to two weeks. Distinguished lecturers are expected to give a major public lecture or a series of presentations. The maximum award is $10,000. Preference is given to proposals for candidates whose visit would have maximum impact at Notre Dame through such activities as participating in faculty seminars or courses; workshops or other research collaborations; events at residence halls; or extracurricular activities sponsored by student groups.
*Geographic Focus:* All States
*Date(s) Application is Due:* Feb 11; Apr 1; Sep 17; Nov 19
*Amount of Grant:* Up to 10,000 USD
*Contact:* Sharon Schierling, Associate Director; (574) 631-8524 or (574) 631-6580; fax (574) 631-6717; sschierl@nd.edu
*Internet:* kellogg.nd.edu/faculty/grants/events.shtml
*Sponsor:* Kellogg Institute for International Studies
130 Hesburgh Center
Notre Dame, IN 46556-5677

**Kellogg Institute for International Studies Grants for Events** 1437
The Kellogg Institute for International Studies makes available to faculty fellows small grants to support group events such as lectures, seminars, and films that advance understanding of international studies and of Kellogg's major themes in particular. The best proposals will bring together different elements of the Kellogg and/or Notre Dame communities in innovative ways. Proposals may be submitted at any time, but well in

advance of the proposed activity. Proposals will be evaluated by the Grants Committee on a rolling basis. The Grants Committee usually meets only during the academic year, so applicants should plan accordingly. The maximum award is $5,000.
*Geographic Focus:* All States
*Date(s) Application is Due:* Feb 11; Apr 1; Sep 17; Nov 19
*Amount of Grant:* Up to 5,000 USD
*Contact:* Sharon Schierling, Associate Director; (574) 631-8524 or (574) 631-6580; fax (574) 631-6717; sschierl@nd.edu
*Internet:* kellogg.nd.edu/faculty/grants/events.shtml
*Sponsor:* Kellogg Institute for International Studies
130 Hesburgh Center
Notre Dame, IN 46556-5677

**Kellogg Institute for International Studies Intramural Grants** 1438
Kellogg Intramural Grants primarily fund research and educational activities. Requests to support student activities must make explicit the connection between the proposed activity and academic work or educational goals. Proposals by student groups must have a faculty or staff sponsor. The grants are intended to support projects of an institutional or group nature rather than individuals' research. The Institute will provide a maximum of $5,000 to support research conferences organized by Notre Dame faculty members who are not Institute fellows.
*Requirements:* Intramural grants are open to any faculty or staff member, student group, or unit within the ND community. Projects must be within the field of international studies or have a clear international dimension, and preferably relate to one of the Institute's five research themes.
*Restrictions:* Grants are restricted to the University of Notre Dame academic community and are intended primarily (though not exclusively) for on-campus activities and events.
*Geographic Focus:* All States
*Date(s) Application is Due:* Feb 11; Apr 1; Sep 17; Nov 19
*Amount of Grant:* Up to 5,000 USD
*Contact:* Sharon Schierling, Associate Director; (574) 631-8524 or (574) 631-6580; fax (574) 631-6717; sschierl@nd.edu
*Internet:* kellogg.nd.edu/faculty/grants/intramural.shtml
*Sponsor:* Kellogg Institute for International Studies
130 Hesburgh Center
Notre Dame, IN 46556-5677

**Kellogg Institute for International Studies Kroc Undergrad Research Grants** 1439
The Kellogg and Kroc Institutes announce a Research Grants Program for undergraduate students at Notre Dame. The grants are intended for juniors whose interests include a clear international dimension preferably related to the Kellogg Institute's themes. Several grants of up to $5,000 each will be awarded for research abroad or, if demonstrably appropriate, in the United States. Applicants are required to notify the Kellogg Institute of any funding received from another source. If duplicate funding is received, the Institute may adjust the award.
*Requirements:* Students must be in their junior year of studies at Notre Dame and be in good academic standing. They must have the endorsement and pledge of supervision from a faculty member.
*Geographic Focus:* All States
*Date(s) Application is Due:* Mar 1
*Amount of Grant:* Up to 5,000 USD
*Contact:* Holly Rivers, Assistant Director; (574) 631-6023 or (574) 631-6580; fax (574) 631-6717; hrivers@nd.edu
*Internet:* kellogg.nd.edu/students/grants/index.shtml
*Sponsor:* Kellogg Institute for International Studies
130 Hesburgh Center
Notre Dame, IN 46556-5677

**Kellogg Institute for International Studies Quechua Summer Language Fellowships** 1440
The Latin American Studies Program through the Kellogg Institute for International Studies will offer several summer fellowships for the summer to Notre Dame undergraduate and graduate students who would like to pursue the study of Quechua, the language of the Incas and of millions of indigenous peoples living in Bolivia, Colombia, Ecuador, and Peru. Students will attend classes at the Centro Tinku in Cusco, Peru. Classes are available at the beginning, intermediate and advanced levels. Students will participate in a seven week study of the language and culture. Recipients of the award will receive a small stipend as well as funding to cover airfare, program fees, room and board, and health insurance.
*Requirements:* Applicants must meet the following criteria: be able to attend the seven week program; demonstrate interest in Latin American Studies through coursework, work, or living experience, internships, etc; and be proficient in Spanish.
*Geographic Focus:* All States
*Date(s) Application is Due:* Mar 15
*Contact:* Holly Rivers, Assistant Director; (574) 631-6023 or (574) 631-6580; fax (574) 631-6717; hrivers@nd.edu
*Internet:* kellogg.nd.edu/students/lasp/fellowships/quechua.shtml
*Sponsor:* Kellogg Institute for International Studies
130 Hesburgh Center
Notre Dame, IN 46556-5677

**Kellogg Institute for International Studies Summer Internships** 1441
Through the Kellogg Institute for International Studies Summer Internship program, Notre Dame undergraduates are offered a variety of opportunities to gain hands-on experience with policy institutes, non-governmental organizations and various businesses that will better equip them to function in an increasingly globalized world. Students may work with organizations within Africa, Asia, Latin America and the United States. Participation in the program allows students to improve their language skills, learn about major policies and issues facing these regions and gain real world experience in their chosen field. Students spend a minimum of eight weeks with their host institution. Dates of internships vary. Some programs have set requirements while dates for other organizations are arranged between the supervisor and the intern.
*Requirements:* The program seeks students who have the academic interests, skills and initiative that best match the needs of the host organization, ensuring a mutually beneficial relationship for both the students and the organization. Applicants must be able to clearly articulate the relationship between their academic and career goals and the work needed at the host organization. Preference is given to students completing their junior year but sophomores and freshmen may apply. Kellogg internships are available to Notre Dame undergraduate students.
*Restrictions:* Students are not eligible for more than one award.
*Geographic Focus:* All States
*Date(s) Application is Due:* Feb 25
*Contact:* Sharon Schierling, Associate Director; (574) 631-8524 or (574) 631-6580; fax (574) 631-6717; sschierl@nd.edu
*Internet:* kellogg.nd.edu/students/internship/index.shtml
*Sponsor:* Kellogg Institute for International Studies
130 Hesburgh Center
Notre Dame, IN 46556-5677

**Kellogg Institute for International Studies Supplemental Awards for PhD Students** 1442
The Kellogg Institute for International Studies Supplemental Awards for PhD Students aims to attract outstanding graduate students who have already been offered departmental fellowships in the Political Science, Sociology, Economics, or History departments at Notre Dame. The Kellogg Institute will offer the supplemental fellowships for up to five years to qualified students. Recipients will receive $5,000 per year to supplement their departmental stipends. These funds will be disbursed in the fall and spring semesters. Fellowships are renewable annually for four additional academic years, contingent upon satisfactory progress toward the PhD. The annual deadline in in early March.
*Requirements:* To be eligible, a student must intend to devote their major doctoral field to the study of Latin America, Africa, and/or Asia.
*Geographic Focus:* All States
*Date(s) Application is Due:* Mar 3
*Amount of Grant:* Up to 5,000 USD
*Contact:* Denise Wright, Program Coordinator; (574) 631-8523 or (574) 631-6580; fax (574) 631-6717; dwright1@nd.edu
*Internet:* kellogg.nd.edu/students/grad/grants/ccaa.shtml
*Sponsor:* Kellogg Institute for International Studies
130 Hesburgh Center
Notre Dame, IN 46556-5677

**Kellogg Institute for International Studies Visiting Fellowships** 1443
The Kellogg Institute seeks both accomplished and promising scholars whose work and presence will promote interdisciplinary international research on the Institute's core themes. These visiting fellowships afford the time to move a research project forward in an intellectually stimulating environment, share findings, and develop linkages with renowned U.S. and international scholars. As a visiting fellow, you pursue research related to Kellogg Institute themes of democracy and human development, share your research with the Notre Dame scholarly community, and have the opportunity to publish in Kellogg's peer-reviewed Working Paper Series. Visiting Fellowships include: a stipend; travel expenses; medical insurance benefits; and a partial housing subsidy for the Hesburgh Center Residences or other on-campus housing.
*Requirements:* Fellowships are open to candidates from the U.S. and abroad who hold a PhD or equivalent degree in a social science discipline or history (exceptions considered), distinguished senior candidates who do not hold a PhD, and advanced graduate students who will complete their PhD before beginning the fellowship.
*Geographic Focus:* All States
*Date(s) Application is Due:* Nov 1
*Contact:* Denise Wright, Program Coordinator; (574) 631-8523 or (574) 631-6580; fax (574) 631-6717; dwright1@nd.edu
*Internet:* kellogg.nd.edu/vfellowships/index.shtml
*Sponsor:* Kellogg Institute for International Studies
130 Hesburgh Center
Notre Dame, IN 46556-5677

**Kennedy Center Fellowship at the DeVos Institute of Arts Management** 1444
The Fellowship at the DeVos Institute of Art Management is an immersive program in arts management strategy designed to prepare mid-career arts managers for executive positions in today's complex arts environment. Hosted at the nation's cultural center and one of the largest and most dynamic performing arts institutions, the Fellowship offers a structured blend of academic training, hands-on work experience, independent and collective learning opportunities, and personalized mentoring. Fellows attend weekly classes throughout the Fellowship taught by highly experienced performing arts professionals that include case study, independent research, and collaborative group work. During their course of study, Fellows gain hands-on experience in three different Kennedy Center departments working closely with senior Kennedy Center staff. These rotations extend and complement Fellows' experience in their courses and project work. The Seminar Series provides participants with contemporary knowledge essential to leaders of performing arts organizations. Seminar topics include, but are not limited

to: accessibility; arts education; human resource management; law and the arts; media relations; programming; technology and arts management; and union negotiations. Participants receive a $20,000 living allowance, program materials, and reimbursement for health insurance coverage. They also receive tickets to performance at the Kennedy Center and relocation assistance. The application is available at the website.
*Requirements:* Applicants must have a minimum of five years of professional arts management experience, a bachelor's degree in the arts or related discipline, and demonstrated knowledge of issues in arts management.
*Geographic Focus:* All States
*Date(s) Application is Due:* Apr 15
*Amount of Grant:* 20,000 USD
*Contact:* Syrah Gunning, DeVos Fellowship Coordinator; (202) 416-8897; artsmanager@kennedy-center.org
*Internet:* www.kennedy-center.org/education/artsmanagement/fellowships.cfm
*Sponsor:* John F. Kennedy Center for the Performing Arts
2700 F Street NW
Washington, D.C. 20566

**Kennedy Center Summer Fellowships for International Arts Managers** 1445
The DeVos Institute's Summer Fellowships for International Arts Managers offer international arts managers an immersive program of study in arts management strategy. For four weeks each summer, over three consecutive years, international Fellows participate in seminars, group projects, individual mentorships, and practical assignments supervised by the Institute's faculty and Center senior staff. The Fellowship includes the following: seminars and coursework in strategic planning; board development; fundraising; marketing; artistic planning; individual advising from Institute faculty and Center senior staff; air transportation to and from the U.S; lodging during the Fellowship; per diem to cover living expenses; visa sponsorship; and program materials. The application, a list of frequently asked questions, and informational brochure with endorsement comments from previous participants are available at the website.
*Requirements:* Applicants must meet the following criteria: a minimum of five years working experience in an administrative capacity at an arts organization outside North America; currently working as an executive or department head with decision-making authority; ability to commit to the full term of the Fellowship (four weeks in late June/July for three consecutive summers); and excellence verbal and writing skills in business English.
*Geographic Focus:* All Countries
*Contact:* Syrah Gunning, Summer Fellowship Coordinator; (202) 416-8897; segunning@kennedy-center.org or artsmanager@kennedy-center.org
*Internet:* www.kennedy-center.org/education/artsmanagement/international.cfm
*Sponsor:* John F. Kennedy Center for the Performing Arts
2700 F Street NW
Washington, D.C. 20566

**Kenneth T. and Eileen L. Norris Foundation Grants** 1446
The foundation supports Los Angeles County nonprofits in the areas of: medicine—to improve access to health care, increase knowledge through research, and provide facilities for those activities to take place; youth—to provide constructive activities, positive role models, and opportunities for disadvantaged, disabled, and misguided children; community—to support law enforcement agencies, good citizenship, and environmental conservation; culture—to support museums, symphony orchestras, and dance and theater companies; and education and science—to focus on private education, especially secondary and college levels. Types of support include general operating support, continuing support, building construction and/or renovation, equipment acquisition, endowment funds, program development, professorships, scholarship funds, research, and matching funds. Education/science and medicine projects are accepted between May 1 and June 30; youth requests are accepted between February 15 and March 31; cultural (the arts) and community requests are accepted between December 1 and January 31; and medicine proposals are due between May 1 and June 30.
*Requirements:* Grants are awarded to organizations in southern California.
*Geographic Focus:* California
*Date(s) Application is Due:* Jan 31; Mar 31; Jun 30
*Amount of Grant:* 5,000 - 25,000 USD
*Contact:* Lisa D. Hansen; (562) 435-8444; fax (562) 436-0584; grants@ktn.org
*Internet:* www.norrisfoundation.org/grant.html
*Sponsor:* Kenneth T. and Eileen L. Norris Foundation
11 Golden Shore, Suite 450
Long Beach, CA 90802

**Kettering Fund Grants** 1447
Charles F. Kettering (1876-1958) was an American inventor, founder of Delco, and head of the General Motors Research Laboratory for twenty-seven years. He was born in Loudonville, Ohio and lived most of his life in Dayton, Ohio. He was involved in many philanthropic projects during his lifetime, a legacy which the Kettering Family adopted and continues. The mission of the Kettering Fund is to support scientific, medical, social and educational studies and research conducted by nonprofit, charitable organizations that are located and/or provide service in Ohio. The Distribution Committee meets on a semi-annual basis to make funding decisions, generally in May and November. Primary areas of support include: arts, culture and humanities; education; environment; health and medical needs; human services; and public and societal benefit. The Fund committee will consider support for the following purposes: capital needs (such as the purchase of land, new construction, renovations and equipment); seed grants for new organizations or programs; project support; endowments; research; and scholarship funds. Annual deadlines for application submission are July 31 and January 31.
*Requirements:* 501(c)3 nonprofit organizations that provide services in or are located in Ohio may apply.
*Restrictions:* A Request Summary will not be accepted for any of the following purposes: religious organizations for religious purposes; individual public elementary or secondary schools or public school districts; multi-year grants; grants or loans to individuals; tickets, advertising or sponsorships of fundraising events; efforts to carry on propaganda or otherwise attempt to influence legislation; or activities of 509(a)3 Type III Supporting Organizations.
*Geographic Focus:* Ohio
*Date(s) Application is Due:* Jan 31; Jul 31
*Amount of Grant:* 5,000 - 200,000 USD
*Contact:* Judith M. Thompson, Executive Director; (937) 228-1021; fax (888) 719-1185; info@ketteringfamilyphilanthropies.org
*Internet:* https://www.cfketteringfamilies.com/fund/about-us
*Sponsor:* Kettering Fund
40 North Main Street, #1480
Dayton, OH 45423

**Kitty M. Perkins Foundation Grants** 1448
Kitty Shoemaker Perkins was the wife of Edwin Perkins, the inventor of the global packaged soft drink called Kool-Aid. Kitty was Edwinss confidant and partner in this great American story about a young boy who grew up in a small Nebraska town called Hendley, and then went on to make fame and fortune in Chicago, Illinois. The Edwin and Kitty story is the American dream. The Kitty M. Perkins Foundation is dedicated to giving money to charities that help sustain the rural community. Its primary objective is to provide charitable gifts to worthy tax-exempt organizations involving religious, charitable, scientific, literary, artistic, or educational purposes exclusively. Grants may also be made to an eligible state or political subdivision thereof if the grant is made exclusively for public purposes.
*Requirements:* Grants will be made only to such organizations that are tax exempt as defined in Section 501(c)3 of the Internal Revenue Code, and which are not private foundations as defined in Section 509 of the Code.
*Geographic Focus:* Nebraska
*Amount of Grant:* Up to 100,000 USD
*Samples:* Artspirit, Hastings, Nebraska, $7,500 - general operating support; Butler Memorial Library Foundation, Cambridge, Nebraska, $3,000 - general operating support; City of Cambridge, Cambridge, Nebraska, $95,000 - support of golf course operations shortfall and equipment purchase.
*Contact:* Dick Shoemaker; (308) 697-3395; dick.shoemaker@pnpt.com
*Internet:* www.kmpfoundation.com/Guidelines.html
*Sponsor:* Kitty M. Perkins Foundation
304 Nelson Street, P.O. Box 268
Cambridge, NE 69022-3592

**Klarman Family Foundation Grants in Eating Disorders Research Grants** 1449
The Klarman Family Foundation is interested in providing strategic investment in translational research that will accelerate progress in developing effective treatments for anorexia nervosa, bulimia nervosa and binge eating disorder. The program's short-term goal is to support the most outstanding science and expand the pool of scientists whose research explores the basic biology of feeding, anorexia nervosa, bulimia nervosa, and/or binge eating disorder. The long-term goal is to improve the lives of patients suffering from these conditions. Examples of funding areas include but are not limited to: molecular genetic analysis of relevant neural circuit assembly and function; genetic and epigenetic research; animal models created by genetically altering neural circuits; and testing of new chemical entities that might be used in animal models as exploratory treatments. Investigators conducting research in the neuro-circuitry of fear conditioning or reward behavior may also apply but must justify the relevance of their research projects to the basic biology of eating disorders. Two-year awards of $400,000 USD ($200,000 per year inclusive of 10% indirect costs) and one-year pilot studies of up to $150,000 USD (inclusive of 10% indirect costs) are made to investigators with a faculty appointment at a nonprofit academic, medical or research institution in the United States, Canada or Israel.
*Requirements:* At the time of application, investigators must hold a faculty appointment at a nonprofit academic, medical or research institution in the United States, Canada or Israel. Investigators who previously submitted applications or have concluded their Klarman Family Foundation Award remain eligible to apply. Applicants do not need to be nominated by their institutions. Investigators working outside the field of eating disorders research are encouraged to apply. Examples of funding areas include but are not limited to molecular genetic analysis of relevant neural circuit assembly and function; genetic and epigenetic research; animal models created by genetically altering neural circuits; testing of new chemical entities that might be used in animal models as exploratory treatments; and clinical trials involving medications or other non-psychotherapeutic interventions. Applications must be received by The Medical Foundation by February 11, 1:00 pm U.S. Eastern time.
*Restrictions:* Only one application may be submitted per applicant. Clinical psychotherapeutic studies, imaging studies involving humans, and research in the medical complications of these disorders are outside the scope of this program.
*Geographic Focus:* All States, Canada, Israel
*Date(s) Application is Due:* Feb 11
*Amount of Grant:* 400,000 USD
*Contact:* Program Officer; (617) 236-7909; info@klarmanfoundation.org
*Internet:* klarmanfoundation.org/eating-disorders-research/
*Sponsor:* Klarman Family Foundation
P.O. Box 171627
Boston, MA 02117

**Klingenstein-Simons Fellowship Awards in the Neurosciences** 1450

The Klingenstein-Simons Fellowship Awards in the Neurosciences support, in the early stages of their careers, young investigators engaged in basic or clinical research that may lead to a better understanding of epilepsy. The fund recognizes that to accomplish this goal it is necessary to encourage a variety of new approaches. Several areas within the neurosciences are of particular interest to the fund: cellular and molecular neuroscience studies of the mechanisms of neuronal excitability and development, and of the genetic basis of seizure disorders; neural systems studies of the integrative function of the nervous system; and clinical research studies designed to improve the prevention, diagnosis, treatment and the understanding of the causes of epilepsy. The annual deadline for application submission is February 15.

*Requirements:* Applicants must hold the PhD and/or MD degrees, and have completed all research training, including post-doctoral training. U.S. citizenship is not a requirement, but it is expected that candidates will be permanent residents of the U.S. and that their research will be carried out in U.S. institutions.

*Restrictions:* The fund does not contribute to endowments and rarely contributes to buildings or other kinds of capital projects.

*Geographic Focus:* All States

*Date(s) Application is Due:* Feb 15

*Contact:* Kathleen Pomerantz, Vice President; (212) 492-6193 or (212) 492-6181; kathleen.pomerantz@klingenstein.com

*Internet:* www.klingfund.org

*Sponsor:* Esther A. and Joseph Klingenstein Fund
125 Park Avenue, Suite 1700
New York, NY 10017

**Klingenstein Third Generation Foundation Fellowships** 1451

The Klingenstein Third Generation Foundation awards research grants to post-doctoral investigators at esteemed American medical institutions. At its establishment in 1998, the Foundation funded research fellowships in Child and Adolescent depression. In 2007, the Foundation expanded the Program to include funding for research in Child and Adolescent ADHD. In 2015, KTGF expanded to include an access to care fellowship. Each year, the Klingenstein Third Generation Foundation makes up to five fellowship grants to medical institutions that have nominated research projects led by outstanding post-doctoral candidates. The Foundation considers proposals for research projects in the field of child and adolescent ADHD, child and adolescent depression and access to care. The grant awards are for $30,000 per year for two years, and the Primary Investigator is awarded the title of KTGF Fellow in ADHD, KTGF Fellow in Depression or KTGF Fellow in Access to Care. The Foundation desires that the research it supports produce a direct and tangible benefit to children and their families. As such, applicants should be certain to explain, in clear terms designed to be understood by lay people, how their research project will help children and families suffering from these conditions. The Foundation expects institutions to nominate post-doctorial investigators holding a Ph.D. and/or M.D. who have also completed all clinical training. As a general rule, nominated candidates should be at the junior faculty level or on a trajectory for attaining faculty status; applications nominating candidates in T32 or other designated research training programs will be considered. The annual deadline for nominations is November 15.

*Requirements:* The Foundation particularly wishes to support two-year research projects that could lead to larger projects worthy of federal funding and other national research awards. Institutions should, therefore, nominate candidates who have strong evidence of independent research ability potential and who are planning a career in research related to child and adolescent psychiatry or developmental psychopathology.

*Restrictions:* Investigators who have already received R01 awards from NIMH are not eligible for nomination.

*Geographic Focus:* All States

*Date(s) Application is Due:* Nov 14

*Amount of Grant:* 30,000 USD

*Contact:* Kathleen Pomerantz, Vice President; (212) 492-6195 or (212) 492-6181; kathleen.pomerantz@klingenstein.com

*Internet:* www.ktgf.org/fellowship_prog.html

*Sponsor:* Klingenstein Third Generation Foundation
125 Park Avenue Suite 1700
New York, NY 10017-5529

**Knoll-MacDonald Foundation Grants** 1452

Established in 1999, the Knoll-MacDonald Foundation provides grant funding to institutions of higher education in Virginia, Maryland, New York, Rhode Island, and Michigan. The Foundation's primary field of interest is engineering education and biomedical research. Grant amounts are relatively small, most recently ranging from $350 to $2,000. Though there are no annual deadlines, applications should submit a letter of application, which includes: contact information, a detailed description of the project, and amount requested.

*Geographic Focus:* Maryland, Michigan, New York, Rhode Island, Virginia

*Amount of Grant:* 350 - 2,000 USD

*Contact:* Geraldine MacDonald, President; (703) 435-2122

*Sponsor:* Knoll-MacDonald Foundation
12172 Tryton Way
Reston, VA 20190-3281

**Knox County Community Foundation Health Grants** 1453

The Knox County Community Foundation is a nonprofit, public charity created by and for the people of Knox County, Indiana. The Foundation helps nonprofits fulfill their missions by strengthening their ability to meet community needs through grants that assist charitable programs, address community issues, support community agencies, launch community initiatives, and support leadership development. Grant proposals are accepted once each year according to the grant cycle. Proposal requirements may change from year to year; therefore, grant seekers are advised to contact the foundation or see the foundations website, prior to beginning the grant application process. Grants are normally given as one-time support of a project but may be considered for additional support for expansions or outgrowths of an initial project. At the start of each cycle, a notice is mailed to nonprofit organizations that have applied for grants in the past, have received grants in the past, or have otherwise requested notification of the start of each cycle. Grants in the area of health include activities that: improve and promote health outcomes; provide general and rehabilitative health services; offer mental health services; provide crisis intervention programs; strengthen associations or services associated with specific diseases, disorders, and medical disciplines; and support medical research. Samples of previously funded projects are posted on the website.

*Requirements:* The Foundation welcomes proposals from nonprofit organizations that are deemed tax-exempt under sections 501(c)3 and 509(a) of the Internal Revenue Code and from governmental agencies serving the County of Knox, Indiana. Proposals from nonprofit organizations not classified as a 501(c)3 may be considered provided the project is charitable and supports a community need. Proposals submitted by an entity under the auspices of another agency must include a written statement signed by the agency's board president on behalf of the board of directors agreeing to act as the entity's fiscal sponsor, to receive grant monies if awarded, and to oversee the proposed project.

*Restrictions:* Project areas not considered for funding are: religious organizations for strictly religious purposes; political parties or campaigns; endowment creation or debt reduction; operating costs; capital campaigns; annual appeals or membership contributions; travel requests for groups or individuals such as bands, sports teams, or classes. Not more than 20% of any grant request may be for personnel costs, office supplies, or other operating costs. Operating costs for any organization must be directly related to the project or program for which funding is being requested.

*Geographic Focus:* Indiana

*Contact:* Annette Nowaskie, Development and Program Assistant; (812) 886-0093; fax (812) 886-0133; annette@knoxcountyfoundation.org

*Internet:* https://www.communityfoundationalliance.org/knox/program-areas/

*Sponsor:* Knox County Community Foundation
20 North Third Street, Suite 301, P.O. Box 273
Vincennes, IN 47591

**Komen Greater NYC Clinical Research Enrollment Grants** 1454

Women, especially racial and ethnic minorities, are not adequately represented in cancer research efforts. Multiple barriers — cultural, linguistic, financial, and systematic — hinder their participation in clinical research, and particularly in clinical trials. Through Clinical Research Enrollment Grants, Komen Greater NYC seeks to fund projects that employ effective strategies to overcome these barriers to enrollment and retention in breast cancer clinical research. The funding period is two years, based on the Komen Greater NYC fiscal year — April 1st through March 31st. . Programs applying for the CRE grant may request up to $200,000 over this two-year period. Organizations that have previously received Komen Greater NYC funding are welcome to apply for additional funding.

*Requirements:* To apply, the organization must be: a not-for-profit organization (a charitable or educational tax-exempt organization), a government agency, or an Indian tribe; conducting National Cancer Institute (NCI) or Department of Defense (DOD) approved breast cancer clinical research (this includes studies through the Clinical Trials Cooperative Group Program including but not limited to ACOSOG, ECOG, SWOG, NSABP, ACRIN, and RTOG; and located in the Komen Greater NYC service area, which includes the five boroughs of New York City, Long Island, Westchester County, and Rockland County.

*Geographic Focus:* New York

*Date(s) Application is Due:* Oct 9

*Amount of Grant:* Up to 200,000 USD

*Contact:* Michelle Marquez, Director of Development; (212) 461-6186 or (212) 560-9590; fax (212) 560-9598; mmarquez@komennyc.org
Zenia Dacio-Mesina, Grants Program Coordinator; (212) 560-9590; fax (212) 560-9598; zdmesina@komennyc.org or grants@komennyc.org

*Internet:* www.komennyc.org/site/PageServer?pagename=grants_clinicaltrials

*Sponsor:* Susan G. Komen for the Cure Foundation
470 Seventh Avenue, 7th Floor
New York, NY 10018

**Komen Greater NYC Community Breast Health Grants** 1455

The mission of Komen Greater NYC is to eradicate breast cancer as a life-threatening disease by advancing research, education, screening, and treatment. The Greater New York City Affiliate is, therefore, soliciting applications from non-profit organizations that have current or planned breast health programs within the five boroughs of New York City, Long Island, Westchester County, and Rockland County. The purpose of the Community Breast Health Grants Program (formerly known as Screening, Treatment, and Education Program- STEP) is to support community-based programs that provide the medically under-served with access to breast health education, screening coordination, diagnostic, treatment, and support services. The funding period is twelve months, based on the Komen Greater NYC fiscal year — April 1st through March 31st. Programs applying for the CBH grant may request up to $75,000 over this one year period. Organizations that have previously received Komen Greater NYC funding are welcome to apply for additional funding. However, all CBH grant awards are for one year only.

*Requirements:* Applicant organizations must be: a not-for-profit organization (a charitable or educational tax-exempt organization), a government agency, or an Indian tribe; located in the Komen Greater NYC service area, which includes the five boroughs of New York

City, Long Island, Westchester County, and Rockland County; and a current or planned provider of breast health services.
*Geographic Focus:* New York
*Date(s) Application is Due:* Oct 9
*Amount of Grant:* Up to 75,000 USD
*Contact:* Michelle Marquez, Director of Development; (212) 461-6186 or (212) 560-9590; fax (212) 560-9598; mmarquez@komennyc.org
Zenia Dacio-Mesina, Grants Program Coordinator; (212) 560-9590; fax (212) 560-9598; zdmesina@komennyc.org or grants@komennyc.org
*Internet:* www.komennyc.org/site/PageServer?pagename=grants_breasthealthgrants
*Sponsor:* Susan G. Komen for the Cure Foundation
470 Seventh Avenue, 7th Floor
New York, NY 10018

**Kosciuszko Foundation Fellowships for Polish Citizens 1456**
The Kosciuszko Foundation annually awards a number of Non-Teaching Fellowships to Poles for advanced study and research at universities and other institutions of higher learning in the United States. The Foundation provides a cost-of-living stipend which includes: transatlantic travel, housing allowance, health and accident insurance coverage, and, when warranted, domestic travel. Non-Teaching Fellowships are awarded to postgraduate scholars, professionals or artists with doctoral degrees. In determining awards, in addition to overall excellence, consideration is given to number of other factors such as scholarly affiliation or geographic location in order to achieve fair and reasonable diversity among awardees. There are no restrictions as to fields of study or research.
*Requirements:* Only Polish citizens residing permanently in Poland are eligible to apply. This program is not for Poles residing outside of Poland, nor those residing temporarily outside of Poland, whether or not engaged in research or study. Dual citizens (United States and Poland) as well as individuals applying or holding United States permanent residency status (green card) are not eligible. Applicants must have strong English language proficiency to communicate and to carry out proposed projects. The level of English proficiency will be determined during the personal interviews in Warsaw.
*Restrictions:* There is no allowance for dependents (spouse or child) support. Awardees are solely responsible for all expenses of accompanying dependents.
*Geographic Focus:* Poland
*Date(s) Application is Due:* Oct 15
*Amount of Grant:* 7,650 - 25,500 USD
*Contact:* Addy Tymczyszyn, Program Officer, Scholarship & Grants for Americans; (212) 734-2130, ext. 210; fax (212) 628-4552; addy@thekf.org
Program Officer; (48) (22) 621-7067; fax (48) (22) 621-7067
*Internet:* www.thekf.org/scholarships/exchange-us/
*Sponsor:* Kosciuszko Foundation
15 East 65th Street
New York, NY 10065

**Kosciuszko Foundation Graduate Study and Research in Poland Scholarships 1457**
This scholarship supports graduate level research at universities in Poland by American graduate students and university faculty members with funding from the Polish Ministry of Education and the Kosciuszko Foundation. Research projects may be conducted from October through June of the upcoming academic year at Polish universities and institutions which fall under the jurisdiction of the Polish Ministry of Education. The scholarship provides a Ministry stipend in the amount of 1,350 zloty per month for housing and living expenses. Additional funding of $250 per month of approved study/research is awarded by the Kosciuszko Foundation.
*Requirements:* Applicants must have working knowledge of the Polish language and are reviewed based on academic background, motivation for pursuing graduate studies and/or research in Poland, and proposals for studies and research in Poland. Applicants must be U.S. citizens or permanent residents.
*Restrictions:* Transportation to and from Poland is not included. This scholarship does not include tuition to attend classes at Polish universities.
*Geographic Focus:* All States
*Date(s) Application is Due:* Jan 5
*Amount of Grant:* 400 - 4,000 USD
*Contact:* Addy Tymczyszyn, Program Officer, Scholarship & Grants for Americans; (212) 734-2130, ext. 210; fax (212) 628-4552; addy@thekf.org
*Internet:* www.thekf.org/scholarships/exchange-poland/research/
*Sponsor:* Kosciuszko Foundation
15 East 65th Street
New York, NY 10065

**Kosciuszko Foundation Grants for Polish Citizens 1458**
The Kosciuszko Foundation annually awards a number of grants to Poles for study and research at universities and other institutions of higher learning in the United States. The Foundation provides a cost-of-living stipend which includes: transatlantic travel, housing allowance, health and accident insurance coverage, and, when warranted, domestic travel. Grants are awarded to those without doctoral degrees. In determining awards, in addition to overall excellence, consideration is given to number of other factors such as scholarly affiliation or geographic location in order to achieve fair and reasonable diversity among awardees. There are no restrictions as to fields of study or research supported, and each proposal is evaluated individually.
*Requirements:* Only Polish citizens residing permanently in Poland are eligible to apply. This program is not for Poles residing outside of Poland, nor those residing temporarily outside of Poland, whether or not engaged in research or study. Dual citizens (United States and Poland) as well as individuals applying or holding United States permanent residency status (green card) are not eligible. Applicants must have strong English language proficiency to communicate and to carry out proposed projects. The level of English proficiency will be determined during the personal interviews in Warsaw.
*Restrictions:* There is no allowance for dependents (spouse or child) support. Awardees are solely responsible for all expenses of accompanying dependents.
*Geographic Focus:* Poland
*Date(s) Application is Due:* Oct 15
*Amount of Grant:* 7,650 - 25,500 USD
*Contact:* Addy Tymczyszyn, Program Officer, Scholarship & Grants for Americans; (212) 734-2130, ext. 210; fax (212) 628-4552; addy@thekf.org
Program Officer; (48) (22) 621-7067; fax (48) (22) 621-7067
*Internet:* www.thekf.org/scholarships/exchange-us/
*Sponsor:* Kosciuszko Foundation
15 East 65th Street
New York, NY 10065

**Kosciuszko Foundation Teaching Fellowships for Polish Citizens 1459**
The Kosciuszko Foundation annually awards a number of Teaching Fellowships to Poles for teaching assignments at universities and other institutions of higher learning in the United States. The Foundation provides a cost-of-living stipend which includes: transatlantic travel, housing allowance, health and accident insurance coverage, and, when warranted, domestic travel. Teaching Fellowships are for university faculty members with teaching assignments at the United States universities in addition to individual research project. In determining awards, in addition to overall excellence, consideration is given to number of other factors such as scholarly affiliation or geographic location in order to achieve fair and reasonable diversity among awardees. There are no restrictions as to fields of teaching.
*Requirements:* Only Polish citizens residing permanently in Poland are eligible to apply. This program is not for Poles residing outside of Poland, nor those residing temporarily outside of Poland, whether or not engaged in research or study. Dual citizens (United States and Poland) as well as individuals applying or holding United States permanent residency status (green card) are not eligible. Applicants must have strong English language proficiency to communicate and to carry out proposed projects. The level of English proficiency will be determined during the personal interviews in Warsaw.
*Restrictions:* There is no allowance for dependents (spouse or child) support. Awardees are solely responsible for all expenses of accompanying dependents.
*Geographic Focus:* Poland
*Date(s) Application is Due:* Oct 15
*Amount of Grant:* 7,650 - 25,500 USD
*Contact:* Addy Tymczyszyn, Program Officer, Scholarship & Grants for Americans; (212) 734-2130, ext. 210; fax (212) 628-4552; addy@thekf.org
Program Officer; (48) (22) 621-7067; fax (48) (22) 621-7067
*Internet:* www.thekf.org/scholarships/exchange-us/
*Sponsor:* Kosciuszko Foundation
15 East 65th Street
New York, NY 10065

**Kosciuszko Foundation Tuition Scholarships 1460**
Domestic scholarships foster knowledge and appreciation of Polish culture, history, and traditions and serve to strengthen understanding and friendship between citizens of the United States and Poland through educational and cultural ties. Tuition scholarships support full-time study in law, medicine, education, Slavic studies, literature, political science, chemistry, marketing, international relations, journalism, dentistry, nursing, and many other fields. Certain scholarships are available to assist eligible students of law and engineering at DePaul University, Chicago; Bayonne, New Jersey, residents studying nursing, teaching or business; and to eligible students who are residents of Amsterdam, New York, Chicopee, Massachusetts, and the state of New Hampshire. The Tuition Scholarship program also supports certain programs including the one-year Masters program at the Center for European Studies at the Jagiellonian University in Cracow and English Schools of Medicine in Poland. The period of study is one academic year with possible renewal for one additional year. Scholarships range from $1,000 to $7,000.
*Requirements:* Eligible students must be United States citizens of Polish descent, Poles who are permanent residents of the United States, and Americans of non-Polish descent who are pursuing courses of study or research concerning Polish subjects. United States citizens of non-Polish descent are considered when their studies are primarily focused on Polish studies. Only one member per immediate family may receive a Tuition Scholarship during any given academic year.
*Geographic Focus:* All States
*Date(s) Application is Due:* Jan 5
*Amount of Grant:* 1,000 - 7,000 USD
*Contact:* Addy Tymczyszyn, Program Officer, Scholarship & Grants for Americans; (212) 734-2130, ext. 210; fax (212) 628-4552; addy@thekf.org
*Internet:* www.thekf.org/scholarships/tuition/ts/
*Sponsor:* Kosciuszko Foundation
15 East 65th Street
New York, NY 10065

**Kovler Family Foundation Grants 1461**
The Kovler Family Foundation awards grants in the areas of the arts, children/youth services, medical research (particularly diabetes), education, human services, higher education, human services, and Jewish federated giving programs. General operating or research grants are awarded primarily in the Chicago metropolitan area. There are no application forms. Applicants should submit a one to two page written proposal letter with a copy of their IRS determination letter by mid-November. Typical grant awarded is between $1,000-$5,000.

*Requirements:* Illinois nonprofit organizations are eligible to apply.
*Restrictions:* The Foundation does not award grants to individuals.
*Geographic Focus:* Illinois
*Samples:* Columbia College, Chicago, Illinois, $5,000, payable over one year; School of the Art Institute of Chicago, Chicago, Illinois, $1,500, payable over one year; University of Chicago, Chicago, Illinois, $1,000, payable over one year.
*Contact:* Jonathan Kovler, President and Treasurer; (312) 664-5050
*Sponsor:* Kovler Family Foundation
875 North Michigan Avenue
Chicago, IL 60611-1958

**KPMG Foundation Grants** **1462**
The Foundation supports colleges and universities under the following basic programs: employee matching gifts; special grants for specific innovative projects or programs that will create new educational opportunities; KPMG professorships; professional development programs allowing academics to keep abreast of current professional practice by attending KPMG courses; auditing practice and research seminars; and support of educational organizations that aid the Beta Alpha Psi Scholarship program and the minority accounting and information systems doctoral scholarship program. Though unsolicited applications are not encouraged, the Foundation will accept letters of inquiry that demonstrate alignment with its mission. Grants are renewable for up to five years of support.
*Requirements:* Grants must be used for educational purposes related to the firm's functional areas of practice - accounting and auditing, taxation, management consulting, and private business advisory services.
*Geographic Focus:* All States
*Contact:* Jose R. Rodriguez; (201) 307-7932; us-kpmgfoundation@kpmg.com
*Internet:* www.kpmgfoundation.org/funding
*Sponsor:* KPMG Foundation
Three Chestnut Ridge Road
Montvale, NJ 07645

**Lalor Foundation Postdoctoral Fellowships** **1463**
The Lalor Foundation postdoctoral fellowship program supports promising new researchers in establishing scientific and teaching careers. The mission of the program is to support these researchers early in their work so that they can become independently funded in the field of mammalian reproductive biology as related to the regulation of fertility. The individual nominated by the applicant institution for the postdoctoral fellowship for conduct of the work may be a citizen of any country. Fellowships will be $42,000 per year for coverage of fellowship stipend, fringes and institutional overhead.
*Requirements:* U.S. institutions must be exempt from federal income taxes under Section 501(c)3 of the U.S. Internal Revenue Code and must submit a determination letter from the Internal Revenue Service stating that it is not a private foundation. The individual nominated should have training and experience at least equal to the Ph.D. or M.D. level and should not have a faculty appointment (i.e., instructor, lecturer or higher). Potential fellows should not have held the doctoral degree more than two years from receipt of the degree.
*Restrictions:* Institutional overhead may not exceed 10 percent of the total fellowship award.
*Geographic Focus:* All States
*Date(s) Application is Due:* Jan 15
*Amount of Grant:* 42,000 USD
*Contact:* Susan Haff, (617) 426-7080, ext. 323; fax (617) 426-5441; shaff@gmafoundationst.com or fellowshipmanager@gmafoundations.com
*Internet:* www.lalorfound.org/?page_id=13
*Sponsor:* Lalor Foundation
77 Summer Street, 8th Floor
Boston, MA 02110-1006

**LAM Foundation Established Investigator Awards** **1464**
The Established Investigator Awards support research directly related to lymphangioleiomyomatosis (LAM) and is targeted at established investigators who focus on the abnormal proliferation of smooth muscle that occurs in the disease. Awards provide a maximum of $50,000 per year, renewable for up to two additional years. The structure and terms of this award are identical to the LAM Fellowship Award except that with the LAM Established Award, faculty level investigators are eligible to receive funding for technician support and supplies.
*Requirements:* U.S. citizens and foreign nationals with appropriate immigrant visa status are eligible. At the time of application, an applicant must hold a doctoral degree and faculty appointment at the level of assistant or associate professor and be undertaking a project related to LAM.
*Geographic Focus:* All States
*Date(s) Application is Due:* Sep 30
*Contact:* Francis X. McCormack, Director; (513) 558-4831; fax (513) 558-4858; frank.mccormack@uc.edu
*Internet:* www.thelamfoundation.org/research/apply-for-lam-funding
*Sponsor:* LAM Foundation
231 Albert Sabin Way, MSB Room 6053
Cincinnati, OH 45267-0564

**LAM Foundation Pilot Project Grants** **1465**
Pilot project awards are granted for one year to enable investigators to gather sufficient preliminary data for more substantial LAM funding. Funding is intended to support faculty-level investigators, or postdoctoral-level investigators who will be performing LAM research in the laboratory of an established scientist who is an expert in areas that are directly pertinent to LAM. Letters of intent (LOI) are required prior to submission of grant proposals to the LAM Foundation. Application materials and details are available online.
*Requirements:* Applications are accepted from investigators of all nationalities who have MDs or PhDs, but candidates must possess visas that allow for completion of the proposed project in the original laboratory.
*Geographic Focus:* All States
*Date(s) Application is Due:* Sep 30
*Amount of Grant:* Up to 25,000 USD
*Contact:* Francis X. McCormack, Director; (513) 558-4831; fax (513) 558-4858; frank.mccormack@uc.edu
*Internet:* www.thelamfoundation.org/research/apply-for-lam-funding
*Sponsor:* LAM Foundation
231 Albert Sabin Way, MSB Room 6053
Cincinnati, OH 45267-0564

**LAM Foundation Postoctoral Fellowships** **1466**
The foundation offers postdoctoral fellowships for the study of the cellular and molecular basis of the abnormal smooth muscle proliferation that occurs in the disease, Lymphangioleiomyomatosis (LAM). Fellowships are renewable for up to two additional years. The fellowship is intended to support postdoctoral level investigators who will be performing LAM research in the laboratory of an established scientist who is an expert in areas that are directly pertinent to LAM. Letters of intent (LOI) are required prior to submission of grant proposals to the LAM Foundation. Application materials and details are available online.
*Requirements:* Applications are accepted from investigators of all nationalities who have MDs or PhDs, but candidates must possess visas that allow for completion of the proposed project in the original laboratory.
*Restrictions:* The award is generally not available to individuals who have attained faculty rank, or who are seeking transition funds for the establishment of a laboratory in a new setting.
*Geographic Focus:* All States
*Date(s) Application is Due:* Sep 30
*Amount of Grant:* Up to 50,000 USD
*Contact:* Francis X. McCormack, MD, Director; (513) 558-4831; fax (513) 558-4858; frank.mccormack@uc.edu
*Internet:* www.thelamfoundation.org/research/apply-for-lam-funding
*Sponsor:* LAM Foundation
231 Albert Sabin Way, MSB Room 6053
Cincinnati, OH 45267-0564

**LAM Foundation Research Grants** **1467**
The program provides seed money to investigators seeking to understand the abnormal smooth muscle proliferation that occurs in lymphangioleiomyomatosis (LAM). The grant supports clinical, laboratory, epidemiological, or any other kind of research that directly relates to LAM. Grants are subject to annual review and may be granted for up to two years. Scientists who which to apply for LAM Foundation funding must submit a Letter of Intent (LOI). Letters of Intent must reach The LAM Foundation by July 30th.
*Requirements:* U.S. citizens and foreign nationals with appropriate immigrant visa status are eligible. At the time of application, an applicant must hold a doctoral degree and faculty appointment with an academic institution and have completed two years of research training.
*Restrictions:* Residents, interns, postdoctoral fellows, students enrolled in degree-granting programs, and established investigators are ineligible.
*Geographic Focus:* All States
*Date(s) Application is Due:* Sep 30
*Amount of Grant:* Up to 40,000 USD
*Contact:* Francis X. McCormack; (513) 558-4831; frank.mccormack@uc.edu
*Internet:* www.thelamfoundation.org/research/apply-for-lam-funding
*Sponsor:* LAM Foundation
231 Albert Sabin Way, MSB Room 6053
Cincinnati, OH 45267-0564

**Laura B. Vogler Foundation Grants** **1468**
The Vogler Foundation supports innovative programs and projects in New York City and Long Island in the areas of education, health care, and social services. The Foundation is particularly interested in organizations that serve and support children, the elderly, and the disadvantaged. Types of support include general operating support, program development, research grants, and seed money grants. Grants provide one-time, nonrenewable support. Most recent grant awards have ranged from $2,500 to $5,000. The annual deadlines for applications are March 1, July 1, and November 1.
*Requirements:* Nonprofit 501(c)3 organizations in New York City and Long Island, New York, may submit proposals.
*Restrictions:* Grants are not awarded to support building or endowment funds, annual fund-raising campaigns, or matching gifts. Requests for funds for conferences, seminars, or loans are not accepted.
*Geographic Focus:* New York
*Date(s) Application is Due:* Mar 1; Jul 1; Nov 1
*Amount of Grant:* 2,500 - 5,000 USD
*Samples:* Brooklyn Public Library, Brooklyn, New York, $3,000 - support of a creative aging program for adults 55-years-old and older; Cathedral Church of St. John the Divine, New York, New York, $3,500 - fighting poverty; Jewish Guild for the Blind, New York, New York, $5,000 - rehabilitation services to visually impaired/blind children.
*Contact:* Lawrence L. D'Amato, President; (718) 423-3000; fax (631) 251-7162; voglerfound@gmail.com
*Internet:* https://sites.google.com/site/voglerfoundation/
*Sponsor:* Laura B. Vogler Foundation
51 Division Street, P.O. Box 501
Sag Harbor, NY 11963

**Lawrence Fellows Program** 1469

The Lawrence Fellows in the Liberal Arts and Sciences program has two primary goals. First, to help develop the professoriate of the future by providing mentoring, teaching opportunities, and research collaborations for persons who have recently finished their graduate degrees. These activities are designed to better prepare them for careers at selective liberal arts colleges. Second, to bring young scholars to campus who will enrich student learning, share new research directions being pursued at distinguished graduate programs, collaborate with faculty on research and performance, and further strengthen the ability of Lawrence University to offer one-on-one learning experiences for all students. The program was initiated in 2005 with the hiring of eight Fellows for the 2005-06 academic year. Fellows are appointed for two years, during which they teach courses, offer individualized instruction opportunities to students, and continue their professional activities as scholars and/or performers. A unique aspect of the program is its scale and breadth. All fields represented by departments or interdisciplinary areas are eligible to apply for Fellows. For students, the program provides an increase in course offerings and more individualized instruction opportunities due to the presence of additional faculty members, and also helps to maintain small class sizes. For faculty, Fellows are not only valuable by helping to expand departmental curricula, but also as collaborators in research and performances. Fellows may also free Lawrence faculty to teach more specialized courses or to offer more individualized instruction. Further, the connections formed with the premier graduate schools from which Fellows are selected create opportunities for faculty by promoting research collaborations.

*Requirements:* The program is designed for recent PhDs (or M.Mus/MFAs). Applications are accepted from qualified applicants in any discipline and interdisciplinary program offered at the university.

*Geographic Focus:* All States

*Date(s) Application is Due:* Mar 15

*Amount of Grant:* 35,000 - 52,000 USD

*Contact:* Office of the President; (920) 832-7000

*Internet:* www.lawrence.edu/academics/faculty/lawrence_fellows

*Sponsor:* Lawrence University

711 E. Boldt Way

Appleton, WI 54911

**Lawrence S. Huntington Fund Grants** 1470

The Foundation, established in 1997 by its namesake, Lawrence J. Huntington, is dedicated to supporting a historical seaport, as well as for environmental conservation, hospitals and human service programs. Mr. Huntington, who serves as the Chairman of St. Luke's Roosevelt Hospital, Chairman of the Woods Hole Research Center, and Vice Chairman of the Board of the South Street Seaport Museum, is primarily interested in preserving the environment and historical landmarks. There are no specific applications or deadlines with which to adhere, and applicants should approach the Foundation in writing.

*Requirements:* Giving is primarily restricted to the states of New York and Massachusetts, as well as the District of Columbia.

*Geographic Focus:* New York

*Amount of Grant:* Up to 50,000 USD

*Samples:* Beth Israel Medical Center, New York, New York, $15,000 - general operating support; Harvard College, Cambridge, Massachusetts, $10,000 - educational support; New Bedford Whaling Museum, New Bedford, Massachusetts, $10,350 - general operating support.

*Contact:* Lawrence S. Huntington, Director; (212) 717-8633 or (212) 486-2424

*Sponsor:* Lawrence S. Huntington Fund

46 E 70th Street, 4th Floor

New York, NY 10021-4928

**LBI-L Leo Baeck/Studienstiftung Fellowships** 1471

The Leo Baeck Fellowship is an international Fellowship for doctoral candidates pursuing research into the history and culture of German-speaking Jewry. Fellowships fund doctoral students who wish to carry out research in more than one country. Every year, funding is offered for up to 12 junior researchers to spend a year working on research at the location of their choice. In addition to financial support for one year in the form of a monthly stipend and a travel allowance, Fellowships offer regular workshops and a common intranet to encourage Fellows to present their research and discuss it with eminent academics. Students enrolled at a university in Germany may apply for an extension after the first year. The online application form with guidelines is available at the grant website.

*Requirements:* The program is open to students of any nationality at any institution of higher education. Applicants must have one or more university degree(s) with formal qualification for doctoral studies, and a doctoral research project focusing on the history and culture of German-speaking Jewry.

*Restrictions:* Financial support for tuition costs is not included.

*Geographic Focus:* All States, All Countries

*Date(s) Application is Due:* Feb 1

*Amount of Grant:* 13,800 EUR

*Contact:* Prof. Raphael Gross, Director; +44(0) 20 7882 5690 or +44(0) 20 7882 5689; fax +44(0) 20 7882 6901; info@leobaeck.co.uk

Thorsten Klein, Secretary; +49 228 82096-284; thorsten.klein@studienstiftung.de

*Internet:* www.studienstiftung.de/en/leo-baeck.html

*Sponsor:* Leo Baeck Institute London

2nd Floor, Arts Two Building

London, E14NS Great Britain

**LBI-NY/DAAD Fellowships in Germany** 1472

The German Academic Exchange Service (DAAD) and the Leo Baeck Institute (LBI) award one six-month, or two three-month graduate fellowship(s) annually for the purpose of providing financial assistance to doctoral students doing dissertation research or to academics preparing a scholarly essay or book that requires a period of research in libraries, archives or research institutions in the Federal Republic of Germany. The research must be in the field of study served by the LBI (social, communal and intellectual history of German-speaking Jewry). A link to application guidelines is available at the grant website.

*Requirements:* Applicants must be U.S. citizens and PhD candidates or recent PhDs (degree awarded within the last two years).

*Geographic Focus:* All States

*Date(s) Application is Due:* Nov 1

*Amount of Grant:* Up to 2,000 EUR

*Contact:* Carol Kahn Strauss, Executive Director; (212) 744-6400 or (212) 294-8340; fax (212) 988-1305; carol@lbi.cjh.org

Dr. Frank Mecklenburg, Director of Research, Chief Archivist; (212) 744-6400 or (212) 294-8340; fax (212) 988-1305; fmecklenburg@lbi.cjh.org or kim@daad.org

*Internet:* www.lbi.org/about/fellowships/lbidaad-fellowships/

*Sponsor:* Leo Baeck Institute New York

15 West 16th Street

New York, NY 10011

**LBI-NY/DAAD Fellowships in New York** 1473

The German Academic Exchange Service (DAAD) and the Leo Baeck Institute (LBI) award two fellowships annually for research at the LBI in New York on the social, communal, and intellectual history of German-speaking Jewry. The LBI is an international research, study, and lecture center whose library and archives in New York offer comprehensive documentation for the study of the subject area. The DAAD is a German national agency that provides programs and funding for students, faculty, researchers and others in higher education; represents the German higher education system as an academic and research destination; and helps build ties between institutions around the world. The fellowships provide financial assistance to doctoral students for dissertation research and to young academics for the preparation of a scholarly essay or book. Each fellowship consists of a stipend of $2,000, paid in two installments of $1,000, and is normally used within one calendar year. A link to application guidelines and to the application form is provided at the grant website.

*Requirements:* Applicants must be U.S. citizens and Ph.D. candidates or recent Ph.D.s (degree awarded within two years preceding application for the fellowship).

*Restrictions:* Support for travel or family members is not available.

*Geographic Focus:* All States

*Date(s) Application is Due:* Nov 1

*Amount of Grant:* 2,000 USD

*Contact:* Carol Kahn Strauss, Executive Director; (212) 744-6400 or (212) 294-8340; fax (212) 988-1305; carol@lbi.cjh.org or lbaeck@lbi.cjh.org

Dr. Frank Mecklenburg, Director of Research, Chief Archivist; (212) 744-6400 or (212) 294-8340; fax (212) 988-1305; fmecklenburg@lbi.cjh.org

*Internet:* www.lbi.org/about/fellowships/lbidaad-fellowships/

*Sponsor:* Leo Baeck Institute New York

15 West 16th Street

New York, NY 10011

**LBI-NY David Baumgardt Memorial Fellowships** 1474

The David Baumgardt fellowship is intended to assist scholars whose research projects are connected with the writings of Professor David Baumgardt or his scholarly interests in ethics, Wissenschaft des Judentums, and the modern intellectual history of German-speaking Jewry. The fellowship, in principal, requires extensive use of the facilities of the Leo Baeck Institute (LBI) in New York and particularly of the David Baumgardt Collection. The fellowship pays a stipend based on the requirements of the particular project being conducted. Support for travel or family members is not available. The stipend will be paid in monthly installments. Application guidelines and a link to a downloadable application form are available at the grant website. Applications must be submitted in writing no later than November 1.

*Requirements:* Doctoral candidates and academics affiliated with an accredited institution of higher education are eligible to apply. The fellowship should be used by the end of the year in which it was granted.

*Geographic Focus:* All Countries

*Date(s) Application is Due:* Nov 1

*Amount of Grant:* Up to 3,000 USD

*Contact:* Dr. Frank Mecklenburg, Director of Research, Chief Archivist; (212) 744-6400 or (212) 294-8340; fax (212) 988-1305; fmecklenburg@lbi.cjh.org

*Internet:* www.lbi.org/about/fellowships/david-baumgardt-memorial-fellowship/

*Sponsor:* Leo Baeck Institute New York

15 West 16th Street

New York, NY 10011

**LBI-NY Fred Grubel Fellowship** 1475

The Fred Grubel Fellowship is a paid summer internship program for a graduate student who will participate in work on a specific research topic (jointly determined by the candidate and the Leo Baeck Institute) related to Leo Baeck Institute (LBI) collections, which can include archives, and library, photo, and art collections. The research project should pertain to the lives of refugees of the 1930s and 1940s in New York. The fellow will be supervised by the Director of Research and will work on a day-to-day basis with

archives and library staff. The deadline for applications is the fall of the year prior to the summer internship. Decisions will be made in early spring. The compensation is $1,500 per month for three summer months. The application form is available online.
*Requirements:* Ph.D. candidates from history, sociology, literature, or Jewish studies programs are eligible.
*Geographic Focus:* All Countries
*Date(s) Application is Due:* Nov 1
*Amount of Grant:* 1,500 - 4,500 USD
*Contact:* Dr. Frank Mecklenburg, Director of Research, Chief Archivist; (212) 744-6400 or (212) 294-8340; fax (212) 988-1305; fmecklenburg@lbi.cjh.org
*Internet:* www.lbi.org/about/fellowships/fred-grubel-fellowship/
*Sponsor:* Leo Baeck Institute New York
15 West 16th Street
New York, NY 10011

**LBI-NY Fritz Halbers Fellowships** **1476**
The Leo Baeck Institute awards the Fritz Halbers Fellowship to assist scholars whose projects are connected with the culture and history of German-speaking Jewry. The amount of the award will be determined according to the requirements of the project and the availability of funds. The fellowship consists of an award, not exceeding $3,000, to be determined according to the requirements of the project. Application guidelines and an application form are available at the grant website.
*Requirements:* Applicants must be students enrolled in a Ph.D. program at an accredited institution of higher education. The fellowship should be used by the end of the year in which it is granted; support for travel or family members is not available.
*Geographic Focus:* All Countries
*Date(s) Application is Due:* Nov 1
*Amount of Grant:* Up to 3,000 USD
*Contact:* Dr. Frank Mecklenburg, Director of Research, Chief Archivist; (212) 744-6400 or (212) 294-8340; fax (212) 988-1305; fmecklenburg@lbi.cjh.org
*Internet:* www.lbi.org/about/fellowships/fritz-halbers-fellowship/
*Sponsor:* Leo Baeck Institute New York
15 West 16th Street
New York, NY 10011

**LBI-NY Gerald Westheimer Career Development Fellowship** **1477**
The Leo Baeck Institute offers a Career Development Fellowship to a scholar or professional in an early career stage (e.g. before gaining tenure in an academic institution or its equivalent). Proposed work should deal with topics within the Leo Baeck Institute's mission, namely historical or cultural issues of the Jewish experience in German-speaking lands. Funding of up to $20,000 will cover one year and, at the discretion of the reviewing board, may be renewed for a second year. Funding is intended to provide for the following activities and expenses: obtaining scholarly material (e.g. publications); temporary help in research and production needs; membership in scholarly organizations; travel; computer, copying, and communication charges; and a summer stipend for non-tenured academics. Applications outlining the nature and scope of the proposed project including a budget should be submitted, in no more than two pages, by the specified deadline date to the Director of Research. A curriculum vitae, names of three references, and supporting material (e.g. outline of proposed work, draft of chapters, previous publications) should be appended. Submission by email is encouraged.
*Restrictions:* Funding is limited to applicants in early in their career.
*Geographic Focus:* All States, All Countries
*Date(s) Application is Due:* Mar 1
*Amount of Grant:* Up to 20,000 USD
*Contact:* Dr. Frank Mecklenburg, Director of Research, Chief Archivist; (212) 744-6400 or (212) 294-8340; fmecklenburg@lbi.cjh.org
*Internet:* www.lbi.org/about/fellowships/career-development-fellowship/
*Sponsor:* Leo Baeck Institute New York
15 West 16th Street
New York, NY 10011

**LE&RN Postdoctoral Fellowships** **1478**
The Lymphatic Education and Research Network (LE&RN) is a 501(c)3 not-for profit organization whose mission is to fight lymphatic disease and lymphedema through education, research and advocacy. LE&RN seeks to accelerate the prevention, treatment and cure of the disease while bringing patients and medical professionals together to address the unmet needs surrounding lymphatic disorders. The LE&RN Postdoctoral Fellowship program is designed to support postdoctoral scientists in academic, medical or research institutions throughout the world. The awards will support investigators who have recently received their doctorates, a critical point in career development when young scientists choose their lifelong research focus. The goal of the Fellowship Program is to expand and strengthen the pool of outstanding junior investigators in the field of lymphatic research. Two-year fellowships ranging from $87,396 to $98,304 will be awarded to fellows who have completed no more than three years of postdoctoral training by July 1 of the LE&RN funding cycle.
*Requirements:* Each applicant must be working under the supervision of an established investigator who is the designated Mentor. Examples of funding areas include but are not limited to basic, translational or clinical investigations into normal and aberrant lymphatic biology; mechanisms of normal and pathological lymphangiogenesis; imaging and understanding the mechanisms that regulate lymphatic structure and function; animal models of lymphatic system disorders; and the cross-talk between the lymphatic and the immune systems. Applications will be accepted worldwide from postdoctoral fellows who will meet the following eligibility requirements by July 1: received a Ph.D. or equivalent and/or an M.D. or equivalent from an accredited institution; commit at least 90% time for research or, at least 70% time for research if the applicant also has clinical responsibilities; have completed no more than three years of full-time postdoctoral research experience; be a postdoctoral fellow under the supervision of a faculty member; and, conduct the proposed research project at a nonprofit hospital, university or other research institution (there are not institutional limitations on the number of applicants who may submit). Eligibility is not limited to those investigators with past experience in the field. Completed applications must be received by The Medical Foundation by January 10, 1:00 pm, U.S. Eastern time.
*Restrictions:* Projects in lymphocyte biology, leukemia, lymphoma and conditions secondary to lymphedema such as cutaneous infections are outside the scope of this program.
*Geographic Focus:* All States, All Countries
*Date(s) Application is Due:* Jan 10
*Amount of Grant:* 87,396 - 98,304 USD
*Contact:* Erin Johnstone, Program Officer; (617) 279-2240 ext. 710; ejohnstone@hria.org
*Internet:* https://hria.org/tmf/lern/
*Sponsor:* Lymphatic Education and Research Network
261 Madison Avenue
New York, NY 10016

**Leakey Foundation/National Geographic Society Baldwin Fellowsips** **1479**
The Leakey Foundation has partnered with the National Geographic Society to provide additional Baldwin Fellowship funding for paleoanthropology students from South, Southeast, and East Asia. This partnership allows the Foundation to award more Baldwin Fellowships per year and will further help to provide students with the opportunity to get advanced education and training in paleoanthropology. The maximum award is limited to $15,000 per year. Awards are limited to two years. The deadline for new applicants is February 15 and March 1 for returning applicants.
*Requirements:* The countries included in this new opportunity are Bangladesh, Bhutan, Brunei, Cambodia, East Timor, Hong Kong, India, Indonesia, Japan, Laos, Macau, Malaysia, Myanmar (Burma), Nepal, Pakistan, Papua New Guinea, Philippines, Singapore, South Korea, Sri Lanka, Taiwan, Thailand, and Vietnam. Only scholars with citizenship in a developing country are eligible. The applicant must have a Home Sponsor who is a member of the institution where the applicant is affiliated in the home country, as well as a Host Sponsor who is a member of the institution which the candidate plans to pursue training. Should an award be made, the Host Sponsor must be willing to assume responsibility for overseeing the candidate's training. Since the Baldwin Fellowship is intended to be a partnership between the candidate and both the Home and Host Institutions, it is expected that candidates will be offered support by these institutions in the form of financial assistance and, upon completion of training, employment in the home country.
*Geographic Focus:* Bangladesh, Bhutan, Brunei, Cambodia, East Timor, Hong Kong, India, Indonesia, Japan, Laos, Macau, Malaysia, Myanmar (Burma), Nepal, Pakistan, Papua New Guinea, Philippines, Singapore, South Korea, Sri Lanka, Taiwan, Thailand, Vietnam
*Date(s) Application is Due:* Jan 10; Jul 15
*Amount of Grant:* Up to 15,000 USD
*Contact:* Paddy Moore, Grant Officer; (415) 561-4646; fax (415) 561-4647; grants@leakeyfoundation.org
*Internet:* https://leakeyfoundation.org/grants/nat-geo/
*Sponsor:* Leakey Foundation
1003B O'Reilly Avenue, P.O. Box 29346
San Francisco, CA 94129-4646

**Leakey Foundation Franklin Mosher Baldwin Fellowships** **1480**
The Baldwin Fellowships were the brainchild of former Board President Dr. Edwin Munger and Elisabeth G. O'Connor, the widow of Franklin Mosher Baldwin. Baldwin was a distinguished attorney who took a lively interest in the search for human origins. Hoping to honor her late husband's dual interests in early humans and education, O'Connor embraced the idea of a program to educate African scholars in prehistory and paleoanthropology. The first Baldwin Fellowships were awarded in 1978. Today, Franklin Mosher Baldwin Memorial Fellowships are intended for scholars and students with citizenship in a developing country who wish to obtain education or training from an institution outside of the student's home country. This award is for a program of approved, advanced special training or studies leading towards an MA or PhD. It is limited to two years of support. The maximum award is limited to $15,000 per year. Awards are limited to two years. The deadline for new applicants is February 15 and March 1 for returning applicants.
*Requirements:* Only scholars with citizenship in a developing country are eligible. The applicant must have a Home Sponsor who is a member of the institution where the applicant is affiliated in the home country, as well as a Host Sponsor who is a member of the institution which the candidate plans to pursue training. Should an award be made, the Host Sponsor must be willing to assume responsibility for overseeing the candidate's training. Since the Baldwin Fellowship is intended to be a partnership between the candidate and both the Home and Host Institutions, it is expected that candidates will be offered support by these institutions in the form of financial assistance and, upon completion of training, employment in the home country.
*Geographic Focus:* All States, Algeria, Angola, Benin, Botswana, Burkina Faso, Burundi, Cameroon, Cape Verde, Central African Republic, Chad, Comoros, Congo, Congo, Democratic Republic of, Cote d' Ivoire (Ivory Coast), Djibouti, Egypt, Equatorial Guinea, Eritrea, Ethiopia, Gabon, Gambia, Ghana, Guinea, Guinea-Bissau, Kenya, Lesotho, Liberia, Libya, Madagascar, Malawi, Mali, Mauritania, Mauritius, Morocco, Mozambique, Namibia, Niger, Nigeria, Rwanda, Sao Tome & Principe, Senegal, Seychelles, Sierra Leone, Somalia, South Africa, Sudan, Swaziland
*Date(s) Application is Due:* Feb 15; Mar 1
*Amount of Grant:* 15,000 USD

*Contact:* Paddy Moore, Grant Officer; (415) 561-4646; fax (415) 561-4647; grants@leakeyfoundation.org
*Internet:* https://leakeyfoundation.org/grants/baldwins/
*Sponsor:* Leakey Foundation
1003B O'Reilly Avenue, P.O. Box 29346
San Francisco, CA 94129-4646

**Leakey Foundation Primate Research Emergency Fund Grants** **1481**
The Leakey Foundation has emergency funding available to keep long-term primate field sites operational until other funding can be secured. Long-term refers to 10 years or more of data collection. Priority will be given to those sites that have a clear research and conservation focus and that are threatened with loss of long-term datasets. The award maximum is $25,000 for one year. After evaluation, a second year may be funded.
*Geographic Focus:* All States
*Amount of Grant:* Up to 25,000 USD
*Contact:* Paddy Moore, Grant Officer; (415) 561-4646; fax (415) 561-4647; grants@leakeyfoundation.org
*Internet:* https://leakeyfoundation.org/grants/primate-research-fund/
*Sponsor:* Leakey Foundation
1003B O'Reilly Avenue, P.O. Box 29346
San Francisco, CA 94129-4646

**Leakey Foundation Research Grants** **1482**
The Leakey Foundation was established to further research into human origins. Recent priorities include research into the environments, archeology, and human paleontology; into the behavior, morphology, and ecology of the great apes and other primate species; and into the behavioral ecology of contemporary hunter-gatherers. Other areas of study are rarely considered. Today, the Leakey Foundation exclusively funds research related specifically to human origins. Priority of funding is commonly given to exploratory phases of promising new research projects that meet the stated purpose of the Foundation. The majority of Leakey Foundation Research Grants awarded to doctoral students are in the $3,000 to $15,000 range. Larger grants given to senior scientists and post-doctoral students may be funded up to $25,000. The deadlines for research grant applications are January 10 and July 15.
*Requirements:* Advanced doctoral students as well as established scientists are eligible for general research grants. Eligible applicants must either hold a PhD or equivalent qualification in anthropology or a related discipline; or be enrolled in a doctoral program with all degree requirements fulfilled other than the dissertation. There are no citizenship restrictions. There are no citizenship restrictions; however, all applications must be submitted in English.
*Restrictions:* High school and undergraduate college students are not eligible.
*Geographic Focus:* All States, All Countries
*Date(s) Application is Due:* Jan 10; Jul 15
*Amount of Grant:* Up to 25,000 USD
*Contact:* Paddy Moore, Grant Officer; (415) 561-4646; fax (415) 561-4647; grants@leakeyfoundation.org
*Internet:* https://leakeyfoundation.org/grants/research-grants/
*Sponsor:* Leakey Foundation
1003B O'Reilly Avenue, P.O. Box 29346
San Francisco, CA 94129-4646

**Leibniz-DAAD Research Fellowships** **1483**
Leibniz-DAAD Research Fellowships are awarded to highly qualified postdoctoral international researchers who have recently completed their doctoral studies. It provides an opportunity to conduct research at one Leibniz institute of their choice in Germany. Each fellowship can be awarded for up to a maximum of twelve months, and the research must begin between September 1 and January 1 of each year. Fellowship awards include: a monthly installment of 2,000 Euros; a combined health, accident and personal liability insurance as well as long-term care insurance in Germany; a research allowance of 460 Euros per year; and a two-month German language course in Germany (if desired). Fifteen fellowships are awarded each year.
*Requirements:* Applicants for this fellowship must be international postdoctoral researchers.
*Restrictions:* Family-related allowances or support for travel costs are not provided. Projects that do not relate to the activities of the selected institute cannot be considered.
*Geographic Focus:* All States
*Amount of Grant:* 2,000 - 24,000 EUR
*Contact:* Dr. Holger Finken; +49 (228) 882-334; finken@daad.de or daadny@daad.org
*Internet:* www.daad.de/go/en/stipa54654913
*Sponsor:* German Academic Exchange Service (DAAD)
871 United Nations Plaza
New York, NY 10017

**Leo Baeck Essay Prize** **1484**
The Leo Baeck Essay Prize has been established to stimulate new research on the history and culture of German-speaking Jewry and to promote young researchers in the field. The essay can be on any topic on the history and culture of German-speaking, Central-European Jewry from early modern times through the post-war period and up to the present. Along with a cash prize, the winning essayist will receive a free year's print and online subscription to the Leo Baeck Institute Year Book, with the winning essay published in the Year Book's upcoming volume. While there is no runner-up prize, the editors may also publish any entries that are highly commended by the judges. The online application is located at the Baeck website, along with entry guidelines.
*Requirements:* The Leo Baeck Essay Prize is open to researchers who have completed their postgraduate studies no more than five years earlier than the current deadline date. All entrants must grant an exclusive license to the Year Book.
*Restrictions:* Essays submitted for the Leo Baeck Essay Prize must not be under consideration for publication elsewhere.
*Geographic Focus:* All Countries
*Date(s) Application is Due:* Oct 5
*Amount of Grant:* 500 GBP
*Contact:* Professor Raphael Gross, Editor; +44(0) 20 7882 5690 or +44(0) 20 7882 5689; fax +44(0) 20 7882 6901; info@leobaeck.co.uk
*Internet:* www.oxfordjournals.org/our_journals/lbaeck/essay_prize.html
*Sponsor:* Leo Baeck Institute London
2nd Floor, Arts Two Building
London, E14NS Great Britain

**LGA Family Foundation Grants** **1485**
The LGA Family Foundation was established in Honolulu in 1997, with a primary objective to support nonprofit organizations in California and Hawaii. It major areas of interest include: arts and culture; basic and emergency aid; communication media; domesticated animals; elementary and secondary education; family services; health and health care; higher education; historic preservation; human services; museums; natural resources; and nursing care. Its target populations are: academics; Children and youth; economically disadvantaged; low-income and poor people; and students. The Foundation's primary support strategies include: financial sustainability; general operating support; program development; and research. Generally, grant awards range from $1,000 to $50,000, with an average of eighteen awards each year. There are no annual deadlines for application submission.
*Requirements:* Any 501(c)3 organization supporting residents of Hawaii or California are eligible to apply.
*Geographic Focus:* California, Hawaii
*Amount of Grant:* 1,000 - 50,000 USD
*Samples:* University of Colorado Foundation, Denver, Colorado, $325,000 - disease research; Parker School, Waimea, Hawaii, $100,000 - general operating support for elementary and secondary schools; Hawaii Mission Children's Society, Honolulu, Hawaii, $25,000 - historic preservation.
*Contact:* Frank C. Atherton II, President; (808) 694-4540
*Sponsor:* LGA Family Foundation
111 South King Street, P.O. Box 3170, Deptartment 715
Honolulu, HI 96802-3170

**Lil and Julie Rosenberg Foundation Grants** **1486**
The Lil and Julie Rosenberg Foundation was established in Connecticut in 1963. It was founded in memory of Julius Rosenberg, founder of Hartley and Parker in 1941, which grew to become one of the premier wholesalers of fine wines and spirits in the State of Connecticut. Julius started his venture in the alcoholic beverage industry in his late teens at Star Liquor Distributors in New York. Owned by his brother, Abraham, Star Liquor Distributors was the first company to be issued a liquor license in New York State after Prohibition. After several years with Star, Julius and his wife, Lil, moved to Bridgeport, Connecticut, to open what would be called Hartley and Parker, situated on Crescent Avenue. In the mid-1950s, Hartley and Parker moved to a larger warehouse on Front Street and, in 1966, Julius built an even larger warehouse facility located in Stratford. Today, the Foundation has a mission of supporting Jewish organizations and public charities in Connecticut and sometimes nationally. Its primary fields of interest include: human services; Judaism; community nonprofits; education; science; literary works; testing for public safety; fostering national or international amateur sports competition (as long as it doesn't provide athletic facilities or equipment); and the prevention of cruelty to children and animals. There are no annual deadlines for application, and interested parties should submit a detailed description of project and amount of funding requested. Typical awards range from $500 to $5,000.
*Requirements:* 501(c)3 organizations serving the residents of Connecticut are eligible to apply.
*Geographic Focus:* Connecticut
*Amount of Grant:* 500 - 5,000 USD
*Samples:* American Cancer Society, Norwalk, Connecticut, $500 - medical research; Anti Defamation League, New Haven Connecticut, $2,000 - social services support; Fairfield County Community Foundation, $5000 - general operating support.
*Contact:* Jerry Rosenberg, President; (203) 375-5671; fax (203) 378-1463
*Sponsor:* Lil and Julie Rosenberg Foundation
100 Browning Street
Stratford, CT 06615-7130

**Lillian S. Wells Foundation Grants** **1487**
The Lillian S. Wells Foundation established in 1976, primarily supports the Fort Lauderdale, Florida, and Chicago, Illinois regions with funding interests that include: medical research, with emphasis on brain cancer research, women's health, substance abuse, and at-risk youth. Additional funding for education and the arts. Contact the Foundation for an application form, the Board meets quarterly in January, April, July, and October.
*Requirements:* Nonprofit organizations in Chicago, Illinois and Fort Lauderdale, Florida area are eligible to apply.
*Geographic Focus:* Florida, Illinois
*Date(s) Application is Due:* Mar 15; Jun 15; Sep 15; Dec 15
*Amount of Grant:* 1,000 - 1,900,000 USD
*Contact:* Patricia F. Mulvaney; patricia.mulvaney@thewellsfamily.net
*Sponsor:* Lillian S. Wells Foundation
600 Sagamore Road
Fort Lauderdale, FL 33301-2215

**Lilly Endowment Teacher Creativity Fellowships** 1488
One of the longest-standing programs of Lilly Endowment is the Teacher Creativity Fellowship Program. It has evolved to expand eligibility to teachers in private and parochial schools and to principals and assistant principals. The commitment behind this competitive program is to help Indiana become an even more rewarding arena for teachers and other education professionals. The Fellowships have supported creative projects that are personally renewing and intellectually revitalizing. While personal renewal is the main goal of this program, applicants are encouraged to demonstrate how their proposed projects will engage their students in the learning process. The Endowment encourages teams of eligible individuals to submit collaborative proposals. If a collaborative project is funded, each participant may receive an $8,000 fellowship. Proposals will be judged on the substance, clarity, originality and feasibility of the project. Preference will be given to proposals demonstrating that substantial thought and creativity has been given to this renewal opportunity. Proposal criteria and requirements are available at the website, along with tips on writing a proposal. Applicants should also consult the website for detailed information regarding time commitments and reporting for the project.
*Requirements:* Licensed teachers in Indiana public, private and parochial schools (including guidance counselors and library/media specialists), principals and assistant principals in such schools are eligible to apply. They must be currently employed full-time and must be Indiana-certified in grades K-12.
*Restrictions:* Proposals to develop new course outlines or lesson units typically do not receive funding.
*Geographic Focus:* Indiana
*Date(s) Application is Due:* Nov 4
*Amount of Grant:* 8,000 USD
*Contact:* Barbara S. DeHart; (317) 916-7345; teachercreativity@lei.org
*Internet:* www.lillyendowment.org/ed_tc.html
*Sponsor:* Lilly Endowment
2801 N Meridian Street, P.O. Box 88068
Indianapolis, IN 46208-0068

**Lisa Higgins-Hussman Foundation Grants** 1489
The Lisa Higgins-Hussman Foundation was established in Maryland in 2006, with the expressed purpose of supporting medical breakthroughs. Its primary interests include: research efforts to improve scientific knowledge of, and to develop therapies and find cures for conditions such as autism, addiction, alcoholism and neurological disorders; therapeutic care, particularly for hospice care and therapeutic intervention for individuals with debilitating health or neurological disorders; and issues relating to women's health. Most recent awards have ranged from $500 to $125,000, with an average of thirty to thirty-five grants given each year. There is no annual deadline specified, and interested parties should contact the Foundation in writing with a brief overview of their proposal and budgetary needs.
*Requirements:* 501(c)3 organizations throughout the U.S. are eligible to apply.
*Geographic Focus:* All States
*Amount of Grant:* 500 - 125,000 USD
*Samples:* Maryland Food Bank, Baltimore, Maryland, $1,000 - general operations/annual campaign; Organization for Autism Research, Arlington, Virginia, $125,000 - general operations/annual campaign; Wayne State University, Detroit, Michigan, $2,500 - support for programs and services to severely injured service members.
*Contact:* Lisa Marie Higgins, President; (410) 480-8264
*Sponsor:* Lisa Higgins-Hussman Foundation
10215 Tarpley Court
Ellicott City, MD 21042-1681

**LiveStrong Foundation Research Grants** 1490
The LiveStrong Foundation funds and disseminates research that builds on the body of knowledge and services focused on improving the quality of life for cancer survivors across physical, emotional, and practical challenges. The research program also supports research focused on testicular cancer. The foundation funds research that is not readily fundable from traditional sources and that encourages and supports the efforts of both young and established investigators. Young investigators are defined as within eight years of completing a terminal degree or within five years of initiating independent research within a mentored laboratory, while general research awards support new research projects initiated by established investigators. Proposals are accepted from all countries.
*Geographic Focus:* All States, All Countries
*Amount of Grant:* Up to 75,000 USD
*Contact:* Jill Ruiter, Vice-President of Development and Events; (512) 236-8820 or (877) 236-8820; fax (512) 236-8482
*Internet:* https://www.livestrong.org/
*Sponsor:* LiveStrong Foundation
2201 East Sixth Street
Austin, TX 78702

**Lloyd A. Fry Foundation Health Grants** 1491
The Lloyd A. Fry Foundation supports organizations with the strength and commitment to address persistent problems of urban Chicago resulting from poverty, violence, ignorance, and despair. The Foundation's Health Program is committed to increasing access to high quality primary care and reducing health care disparities for Chicago's low-income residents. To accomplish these goals, the Foundation is interested in supporting: efforts to implement medical-home models of care which provide comprehensive integrated primary care services across multi-disciplinary team members in single or multiple settings; high quality primary care services that are not widely available to low-income populations; community outreach to connect hard-to-reach individuals with high-quality primary care; and policy advocacy focused on improving the quality of health care and increasing access to health care for low-income populations in Chicago. Typically, grants in this area range from $18,000 to $100,000.
*Requirements:* Grants are made only to tax-exempt organizations and are rarely made to organizations outside the Chicago metropolitan area.
*Restrictions:* The foundation does not fund government entities, political or religious activities, fundraising events, or medical research.
*Geographic Focus:* Illinois
*Date(s) Application is Due:* Mar 1; Sep 1; Dec 1
*Amount of Grant:* 18,000 - 100,000 USD
*Samples:* Heartland International Health Center, Chicago, Illinois, $100,000 - for Dental Care Services at Senn and Roosevelt High Schools, Hibbard Elementary, and at two community clinics; Lutheran Social Services of Illinois, Des Plaines, Illinois, $40,000 - children and adolescent counseling services; PCC Community Wellness Center, Oak Park, Illinois, $40,000 - for the Maternal and Child Health Services Program.
*Contact:* Unmi Song, President; (312) 580-0310; fax (312) 580-0980; usong@fryfoundation.org
*Internet:* www.fryfoundation.org/grants/health/
*Sponsor:* Lloyd A. Fry Foundation
120 South LaSalle Street, Suite 1950
Chicago, IL 60603-3419

**Louis and Sandra Berkman Foundation Grants** 1492
The Louis and Sandra Berkman Foundation was established in Steubenville, Ohio, in 1952, with giving centered in the States of Ohio and Pennsylvania. The Foundation honors the memory of Louis Berkman, who started in business in the late 1920s in the scrap iron and steel industry. In 1931, he incorporated The Louis Berkman Company in Steubenville. He died in 2013 at the age of 104. The Foundation, which has operated for more than sixty years, has been instrumental in establishing he H.L. Berkman Faculty and Staff Dining Room and the Sandra Weiss Berkman Studio for Ceramic Arts at Bethany College in Bethany, West Virginia. The Foundation also made possible the Learning Resource Center, the Louis Berkman Fireside Lounge in the J.C. Williams Center, and the Science and Technology Building at Franciscan University of Steubenville. The Foundation also played a significant roll in the construction of the Louis and Sandra Berkman Amphitheater at the Old Forth Steuben Project in downtown Steubenville. Over the course of the past several years, the Foundation has contributed extensively to Catholic Central High School in Steubenville, and to the creation of the Berkman Theater at Lanman Hall, which was completed in the Spring of 2012. Today, the foundation's primary interests include support for: art museums; arts and culture; biodiversity; elementary and secondary education; the environment; health and health care; higher education; plant biology; and religion. Typically, funding is provided for: general operating support; program development; research; and research evaluation. The annual deadline for application submission is July 1.
*Requirements:* Any 501(c)3 organization is eligible to apply.
*Geographic Focus:* All States
*Date(s) Application is Due:* Jul 1
*Amount of Grant:* Up to 200,000 USD
*Samples:* Agnes Irwin School, Bryn Mawr, Pennsylvania, $35,000 - general operating support; Cornell University, Ithaca, New York, $180,000 - research; Philadelphia Museum of Art, Philadelphia, Pennsylvania, $10,000 - general operating support.
*Contact:* Robert A. Paul, President; (740) 283-3722
Linda Pirkle, Trustee; (412) 652-9480
*Sponsor:* Louis and Sandra Berkman Foundation
330 North 7th Street, P.O. Box 820
Steubenville, OH 43952-5576

**Louisville Institute Collaborative Inquiry Team Grants** 1493
Louisville Institute's Collaborative Inquiry Team (CIT) program supports teams of four to eight pastors and professors who propose projects to strengthen the life of North American Christian congregations. Teams spend 18 to 36 months exploring together a living question currently confronting church and society. Proposed projects may employ a variety of methodological perspectives, including, but not limited to, historical, systematic and practical theology, the social sciences, history, ethics, or biblical studies. Preference will be given to innovative, interdisciplinary research strategies that investigate adaptive challenges faced by North American congregations. The total grant request may not exceed $45,000. The annual deadline to submit online applications is April 1.
*Requirements:* Collaborative Inquiry Teams are comprised of both academic and pastoral leaders based in the United States or Canada. Applicants must have earned the terminal degree in their chosen vocation. For pastors, this is typically the Master of Divinity degree; for academics it is usually the Ph.D. or Th.D.
*Geographic Focus:* All States, Canada
*Date(s) Application is Due:* Apr 1
*Amount of Grant:* Up to 45,000 USD
*Samples:* Young Lee Hertig, Institute for the Study of Asian American Christianity, Azusa, California, $45,000 - Imagining a More Equal Pulpit: Gender Inclusivity and Asian American and Latino/a Churches; John Winton Morehead, $45,000 - Multi-Faith Matters: Evangelical Multi-faith Engagement, Purity Concerns, and Persuasive Storytelling (2017; Roman Williams, $45,000 - The Interfaith Photovoice Initiative.
*Contact:* Grants Administrator; (502) 992-5432; info@louisville-institute.org
*Internet:* https://louisville-institute.org/programs-grants-and-fellowships/collaborative-inquiry-team/
*Sponsor:* Louisville Institute
1044 Alta Vista Road
Louisville, KY 40205-1798

**Louisville Institute Dissertation Fellowships** 1494

Louisville Institute's Dissertation Fellowship (DF) programs offers up to twelve $22,000 grants to support the final year of Ph.D. or Th.D. dissertation writing. Preference given to students engaged in research pertaining to North American Christianity, especially projects related to Institute mission priorities. This program is also open to outstanding students who may not intend to teach in theological education, but whose dissertation projects have the potential to strengthen the religious life of North American Christians and their institutions. Proposed projects may employ a variety of methodological perspectives, including, but not limited to, history, ethics, the social sciences, biblical studies, and historical, systematic, and practical theology. They may also be interdisciplinary in nature. The annual application online submission deadline is February 1.
*Requirements:* Applicants must be candidates for the Ph.D. or Th.D. degree in an accredited graduate school in the United States or Canada. Applicants must be on track to fulfill all pre-dissertation requirements, including approval of the dissertation proposal, by February 1 of the award year and should expect to complete the dissertation by the end of the following academic year.
*Restrictions:* The institute is interested in funding projects that can contribute in significant ways to the understanding of contemporary religious communities. Consequently, very few dissertation projects funded in recent years have dealt with American religion prior to the mid-nineteenth century. Grantees may not accept other awards that provide a comparable stipend during the tenure of this fellowship nor may they undertake substantial outside employment (more than fifteen hours per week) without written permission.
*Geographic Focus:* All States, Canada
*Date(s) Application is Due:* Feb 1
*Amount of Grant:* Up to 22,000 USD
*Samples:* Kyle Eugene Brooks, Vanderbilt University, Nashville, Tennessee; Leo Guardado, University of Notre Dame, South Bend, Indiana (2017: Lisa Nichols Hickman, Duquesne University, Pittsburgh, Pennsylvania; Nathan Jeremie-Brink, Loyola University Chicago, Chicago, Illinois.
*Contact:* Pamala Collins, Fellowships; (502) 992-5432; info@louisville-institute.org
*Internet:* https://louisville-institute.org/programs-grants-and-fellowships/fellowships/dissertation-fellowship/
*Sponsor:* Louisville Institute
1044 Alta Vista Road
Louisville, KY 40205-1798

**Louisville Institute Doctoral Fellowships** 1495

The Doctoral Fellowship (DOC) program offers up to ten fellowships of $2,000 a year for two years to Ph.D. or Th.D. students to consider theological education as their vocation. In addition, the cohort of ten Doctoral Fellows meets three times during each fellowship year. Proposals should describe the applicant's commitment to the Christian church and theological education and suggest how his or her doctoral work would be an appropriate preparation of a career in theological school teaching. Applicants may represent a variety of academic disciplines, including, but not limited to, history, ethics, the social sciences, biblical studies, and historical, systematic, and practical theology. The annual application online submission deadline is March 1.
*Requirements:* Applicants must be Ph.D. or Th.D. students, currently in their first or second year of doctoral study in an accredited graduate program in the United States or Canada.
*Geographic Focus:* All States, Canada
*Date(s) Application is Due:* Mar 1
*Amount of Grant:* 2,000 USD
*Samples:* Michael Agliardo, Religion and Culture/Sociology of Religion - Reception in North America: A Multi-City Study; Young Lee Lee Hertig, Missiology - Imagining a More Equal Pulpit: Gender Inclusivity and Asian American and Latino/a Churches; Denise E. Thorpe, Practical Theology - Race, Church, and Theological Practices.
*Contact:* Pamala Collins; (502) 992-5432; info@louisville-institute.org
*Internet:* https://louisville-institute.org/programs-grants-and-fellowships/fellowships/doctoral-fellowship/
*Sponsor:* Louisville Institute
1044 Alta Vista Road
Louisville, KY 40205-1798

**Louisville Institute Pastoral Study Project Grants** 1496

The Louisville Institute offers pastoral leaders the gift of time and the resources to investigate issues related to Christian ministry and leadership, North American religious practices and institutions, or major challenges facing wider, contemporary society. The Pastoral Study Project program (PSP) enables pastoral leaders to bracket daily work routines in order to pursue a pressing and significant question for the life of faith. Grants of up to $15,000 support independent or collaborative study projects – projects that privilege pastoral perspectives and rhythms and honor grassroots research conducted by skilled clergy. PSP grantees may use a variety of platforms to share what they learn with a wider audience, extending their leadership in ways that can benefit the broader church and society in North America. The grants supports research periods that range up to eighteen months. The annual application online submission deadline is September 1.
*Requirements:* The Pastoral Study Project program is open to Christian clergy, church staff members, chaplains, denominational staff, nuns/sisters/brothers, members of monastic communities, and others regularly employed in recognized positions of pastoral leadership, ordained and lay. It is also open to ordained ministers who are not currently employed by a religious organization.
*Restrictions:* The PSP grant is restricted to pastoral leaders in the United States and Canada. The Pastoral Study Project does not provide scholarship assistance for Doctor of Ministry or other degree programs. Seminary administrators and faculty members are not eligible for the PSP grant program, nor are members of the Louisville Presbyterian Seminary Board, staff, or student body — including immediate family members (parents, spouse, or children). Pastors enrolled in Ph.D. or Th.D. programs are ineligible for the PSP grant
*Geographic Focus:* All States, Canada
*Date(s) Application is Due:* Sep 1
*Amount of Grant:* Up to 15,000 USD
*Samples:* Liv Larson Andrews, Salem Lutheran Church, Spokane, Washington, $15,000 - Art in Worship for Reconciliation; Moses Berry, Theotokos, Ash Grove, Missouri, $15,000 - Bringing African Americans to Orthodoxy; Rolf T Bouma, Campus Chapel, University of Michigan, Ann Arbor, Michigan, $15,000 - Sacred Place and Sacred Space: Potential Pathways to God.
*Contact:* Grants Administrator; (502) 992-5432; info@louisville-institute.org
*Internet:* https://louisville-institute.org/programs-grants-and-fellowships/grants/pastoral-study-project/
*Sponsor:* Louisville Institute
1044 Alta Vista Road
Louisville, KY 40205-1798

**Louisville Institute Postdoctoral Fellowships** 1497

The Postdoctoral Fellowship (PostDoc) program provides up to six awards of $25,000 (plus insurance, housing & moving benefits) each year to support a two-year visiting professorship in a theological school, college, or university setting. Louisville Postdoctoral Fellows are accompanied by an academic mentor and a pastoral mentor. Fellows also constitute a peer learning cohort that meets a total of six times over two years. Applicants should describe their commitment to the Christian church and theological education and suggest how their doctoral work has prepared them appropriately for a career in theological school teaching. Applicants may represent a variety of academic disciplines, including, but not limited to, history, ethics, the social sciences, biblical studies, and historical, systematic, and practical theology. The annual application online submission deadline is December 1.
*Requirements:* Applicants must have completed all Ph.D. work and received their degree by the time the postdoctoral fellowship begins in September. Preference will be given to recent Ph.D. graduates. Former LI Dissertation Fellows are eligible to apply for the Postdoctoral Fellowship. Scholars from the U.S. and Canada are eligible to apply, and the fellowship can be assigned to theological schools in the U.S. or Canada.
*Restrictions:* Applicants may not submit applications to more than one Louisville Institute grant or fellowship program within the same grant year.
*Geographic Focus:* All States, Canada
*Date(s) Application is Due:* Dec 1
*Amount of Grant:* 25,000 USD
*Samples:* Elizabeth Lynn Anderson, Yale University, New Haven, Connecticut, $25,000 - placement school: Claremont School of Theology; Joung Chul Lee, Claremont School of Theology, Claremont, California, $25,000 - placement school: Pacific School of Religion; Allison Norton, Fuller Theological Seminary, Pasadena, California - placement school: Hartford Seminary.
*Contact:* Pamala Collins; (502) 992-5432; info@louisville-institute.org
*Internet:* https://louisville-institute.org/programs-grants-and-fellowships/fellowships/postdoctoral-fellowship/
*Sponsor:* Louisville Institute
1044 Alta Vista Road
Louisville, KY 40205-1798

**Louisville Institute Project Grant for Researchers** 1498

Louisville Institute's Project Grant for Researchers program (PGR) supports research, reflection, and writing by academics and pastors concerning Christian faith and life, the practice of ministry, and/or religious institutions. Grants of up to $30,000 support a diverse range of projects that may involve independent study, consultations, or collaborative projects involving pastors and academics. Projects that show potential to benefit the church in North America are particularly attractive to the Institute. Proposed projects may employ a variety of methodological perspectives, including, but not limited to, historical, systematic and practical theology, the social sciences, history, ethics, or biblical studies. They may also be interdisciplinary in nature. Research periods supported by this grant may vary. The annual application online submission deadline is October 1.
*Requirements:* The Project Grant for Researchers program is open to both academic and pastoral leaders based in the United States or Canada. Applicants must have earned the terminal degree in their chosen vocation. All applicants must demonstrate a capacity to complete the proposed project in a timely fashion.
*Restrictions:* Awards for sabbatical leaves of less than a full academic year will not be made. Grant funding does not support tuition or degree program expenses. Applicants may only apply to one Louisville Institute grant program within the same grant year.
*Geographic Focus:* All States, Canada
*Date(s) Application is Due:* Oct 1
*Amount of Grant:* Up to 30,000 USD
*Samples:* Randall Balmer, Dartmouth College, Hanover, New Hampshire - Orthodoxy in Alaska: Old World Faith in the New World; Melissa D. Browning, McAfee School of Theology at Mercer University, Atlanta, Georgia - Changing Church in order to Change the World: a study of worshipping congregations practicing deep solidarity; Wendy A. Cadge, Brandeis University, Waltham, Massachusetts - Congregation-Port Connections: Case Studies of Port Chaplaincy in the U.S. and Canada.
*Contact:* Grants Administrator; (502) 992-5432; info@louisville-institute.org
*Internet:* https://louisville-institute.org/programs-grants-and-fellowships/grants/project-grant-for-researchers/
*Sponsor:* Louisville Institute
1044 Alta Vista Road
Louisville, KY 40205-1798

**Louisville Institute Sabbatical Grants for Researchers** **1499**
Louisville Institute's Sabbatical Grant for Researchers program (SGR) enables ecclesially-engaged academics and scholarly religious leaders to conduct a major study that can contribute to the vitality of Christianity in North America. Grants of up to $40,000 support year-long research projects that address Christian faith and life, the practice of ministry, and/or adaptive challenges confronting religious institutions. Proposed projects may employ a variety of methodological perspectives, including, but not limited to, historical, systematic and practical theology, the social sciences, history, ethics, or biblical studies. They may also be interdisciplinary in nature. The annual deadline for the online submission is November 1.
*Requirements:* The Sabbatical Grant for Researchers program is open to both academic and pastoral leaders who are based in the United States or Canada. Preference given to applicants who are fully employed in accredited academic institutions and eligible for a full academic year leave from teaching and administrative duties. Applicants must have earned the terminal degree in their chosen vocation.
*Restrictions:* Seminary administrators and faculty members are not eligible.
*Geographic Focus:* All States, Canada
*Date(s) Application is Due:* Nov 1
*Amount of Grant:* Up to 40,000 USD
*Samples:* Kristin Kobes Du Mez, Calvin College, Grand Rapids, Michigan - Journey of Faith: A Religious History of Hillary Rodham Clinton; Jennie Grillo, Duke Divinity School, Durham, North Carolina - Martyrdom, Idolatry, Liturgy: The Afterlives of the Apocryphal Daniel; Ramon Gutierrez, University of Chicago, Chicago, Illinois - The Millennial Kingdom of the Franciscans in Colonial New Mexico, 1581-1680: Anti-Judaism on New Spain's Frontier.
*Contact:* Grants Administrator; (502) 992-5432; info@louisville-institute.org
*Internet:* https://louisville-institute.org/programs-grants-and-fellowships/grants/sabbatical-grant-for-researchers/
*Sponsor:* Louisville Institute
1044 Alta Vista Road
Louisville, KY 40205-1798

**LSAC Research Grants** **1500**
Law School Admission Council (LSAC) Research Grants fund research on a wide variety of topics related to the mission of LSAC. Specifically included are projects investigating precursors to legal training, selection into law schools, legal education, and the legal profession. LSAC welcomes proposals for research proceeding from any of a variety of methodologies, a potentially broad range of topics, and varying time frames. Proposals will be judged on the importance of the questions addressed, their relevance to the mission of LSAC, the quality of the research designs, and the capacity of the researchers to carry out the project. LSAC's membership includes law schools in the United States, Canada, and Australia. Comparative proposals about topics outside the United States, Canada, and Australia are welcome, but they must include some explicit connection to legal education or the legal profession within those countries. A list of possible topics and topics of particular interest can be found on the website.
*Requirements:* Eligible research projects must inform either the process of selecting law students or legal education itself in a demonstrable way. The program welcomes proposals for research proceeding from any of a variety of methodologies, a potentially broad range of topics, and varying time frames. Proposals will be judged on the importance of the questions addressed, their relevance to the mission of LSAC, the quality of the research designs, and the capacity of the researchers to carry out the project. Eligible investigators need not be members of law school faculties. Proposals from interdisciplinary teams of law faculty and researchers from outside law schools are strongly encouraged. Applicants from all countries are eligible. Principal investigators need not be based in law schools; proposals are welcome from social, behavioral, and educational researchers of all kinds. Collaborations are encouraged between those who know legal education most intimately and those who know most about how to design and conduct empirical research.
*Restrictions:* Grants must be made to an institution or organization, not to individuals. Not eligible to receive grants are members of the LSAC board, members of the LSAC test development and research committee or its grants subcommittee, and persons who had been members of one or more of those bodies within one year prior to application. Grants do not fund legal education tuition.
*Geographic Focus:* All Countries
*Date(s) Application is Due:* Feb 1; Sep 1
*Amount of Grant:* 200,000 USD
*Contact:* Ann Gallagher; (215) 968-1258; fax (215) 504-1408; AGallagher@LSAC.org
*Internet:* www.lsac.org/LSACResources/Grants/lsac-legal-education-grant-program.asp
*Sponsor:* Law School Admission Council
662 Penn Street, P.O. Box 40
Newtown, PA 18940-0040

**Luce/ACLS Collaborative Reading-Workshop Grants in China** **1501**
The Henry Luce Foundation/ACLS Program in China Studies seeks to maintain the vitality of China Studies in North America through fellowships and grants designed primarily for scholars early in their careers. Studies on and in China have developed over the last 30 years in the United States and Canada into a robust field, but current conditions pose daunting problems, especially for scholars just before and just after the dissertation. Collaborative Reading-Workshop Grants are intended for scholars of different disciplines to investigate texts that constitute essential points of entry to Chinese periods, traditions, communities, or events in contemporary or historical times. Applications in all disciplines of the humanities and related social sciences are welcome. The program supports the study of Chinese culture and society in all periods. Research in Hong Kong, Tibet and Taiwan is eligible. Award amounts range up to a maximum of $15,000
*Requirements:* Workshops must be held between June 1 and September of the upcoming academic year. The coordinator of the workshop applies on behalf of all participants. That coordinator must hold a PhD from a university in the United States or Canada; or be a U.S. or Canadian citizen or permanent resident with a PhD from any university. Workshops must be held at a location in the United States or Canada. Research in Hong Kong, Tibet and Taiwan is eligible.
*Geographic Focus:* All States, Canada
*Date(s) Application is Due:* Nov 6
*Amount of Grant:* Up to 15,000 USD
*Contact:* Valerie Popp, Senior Program Officer; (212) 697-1505, ext. 153; fax (212) 949-8058; vpopp@acls.org or chinastudies@acls.org
*Internet:* https://www.acls.org/Competitions-and-Deadlines/Luce-ACLS-Program-in-China-Studies
*Sponsor:* American Council of Learned Societies
633 Third Avenue, 8th Floor
New York, NY 10017-6795

**Luce/ACLS Dissertation Fellowships in American Art** **1502**
The ACLS Dissertation Fellowships in American Art support graduate students in any stage of Ph.D. dissertation research or writing. Ten fellowships are available for a non-renewable, one-year term beginning between June and September for the upcoming academic year. The fellowships may be carried out in residence at the fellow's home institution, abroad, or at another appropriate site for the research. The fellowships may not be used to defray tuition costs or be held concurrently with any other major fellowship or grant. Recently, stipends have been raised to $36,000 and additional funds of up to $4,000 are available for travel and research. The annual deadline for application submission is October 28.
*Requirements:* Applicants must have completed all requirements for the Ph.D. except the dissertation before beginning the one-year tenure. A student must be a Ph.D. candidate in a department of art history, and the candidate's dissertation must be focused on a topic in the history of the visual arts of the United States. Students whose degrees will be granted by another department may be eligible if their principal dissertation adviser is in a department of art history and their dissertation topic is object-oriented. Specific guidelines and the online Fellowship application are available at the website.
*Restrictions:* Fellowship funds may not be used to defray tuition costs. Students who received grants of $5,000 or more from the Henry Luce Program previously administered through university art departments are not eligible, nor are students preparing theses for the Master of Fine Arts degree.
*Geographic Focus* All States, District of Columbia, Guam, Marshall Islands, Northern Mariana Islands, Puerto Rico, U.S. Virgin Islands, American Samoa
*Date(s) Application is Due:* Oct 24
*Amount of Grant:* 36,000 - 40,000 USD
*Contact:* Valerie Popp; (212) 697-1505, ext. 153; fax (212) 949-8058; vpopp@acls.org
*Internet:* https://www.acls.org/Competitions-and-Deadlines/Luce-ACLS-Dissertation-Fellowships-in-American-Art
*Sponsor:* American Council of Learned Societies
633 Third Avenue, 8th Floor
New York, NY 10017-6795

**Luce/ACLS Early Career Fellowships in China** **1503**
The Henry Luce Foundation/ACLS Program in China Studies seeks to maintain the vitality of China Studies in North America through fellowships and grants designed primarily for scholars early in their careers. Studies on and in China have developed over the last 30 years in the United States and Canada into a robust field, but current conditions pose daunting problems, especially for scholars just before and just after the dissertation. Early Career Fellowships are intended for scholars who are preparing their PhD dissertation for publication, or who are embarking on new research projects. The fellowship is in the amount of up to a maximum of $50,000, and tenure is two consecutive semesters released from teaching; fellowship to begin between June to September of the upcoming academic year. Stipends may be used for travel, living expenses, and research costs.
*Requirements:* An applicant must hold a PhD from an institution in the United States or Canada, or be a U.S. or Canadian citizen or permanent resident with a PhD from any institution. An applicant who is not a U.S. or Canadian citizen or permanent resident must have an affiliation, a long-term regular research or teaching appointment, with a university or college in the United States or Canada. A working knowledge of Chinese is required.
*Geographic Focus* All States, District of Columbia, Guam, Marshall Islands, Northern Mariana Islands, Puerto Rico, U.S. Virgin Islands, American Samoa, Canada
*Date(s) Application is Due:* Nov 6
*Amount of Grant:* Up to 50,000 USD
*Contact:* Valerie Popp; (212) 697-1505, ext. 153; fax (212) 949-8058; vpopp@acls.org
*Internet:* https://www.acls.org/Competitions-and-Deadlines/Luce-ACLS-Program-in-China-Studies
*Sponsor:* American Council of Learned Societies
633 Third Avenue, 8th Floor
New York, NY 10017-6795

**Luce/ACLS Predissertation Travel Grants in China** **1504**
The Henry Luce Foundation/ACLS Program in China Studies seeks to maintain the vitality of China Studies in North America through fellowships and grants designed primarily for scholars early in their careers. Studies on and in China have developed over the last 30 years in the United States and Canada into a robust field, but current conditions pose daunting problems, especially for scholars just before and just after the dissertation. Predissertation Travel Grants, for graduate students who wish to conduct preliminary preparations in China prior to beginning basic research for the dissertation. Applications in all disciplines of the humanities and related social sciences are welcome.

The program supports the study of Chinese culture and society in all periods. Research in Hong Kong, Tibet and Taiwan is eligible.
*Requirements:* An applicant must be enrolled in a PhD program at a university in the United States or Canada. The application must contain a letter from the dissertation supervisor stating the need for a summer visit to China prior to dissertation research. Research in Hong Kong, Tibet and Taiwan is eligible.
*Restrictions:* These grants are not intended as extensions of the time devoted to basic research but to support the necessary steps to prepare for it. They are for graduate students—with a PhD prospectus in hand or developing one—seeking to investigate the research currently underway in Chinese archives and field sites, to establish contact with Chinese scholars, and to secure necessary permissions for fieldwork or archival research
*Geographic Focus* All States, District of Columbia, Guam, Marshall Islands, Northern Mariana Islands, Puerto Rico, U.S. Virgin Islands, American Samoa
*Date(s) Application is Due:* Nov 6
*Amount of Grant:* 5,000 USD
*Contact:* Valerie Popp; (212) 697-1505, ext. 153; fax (212) 949-8058; vpopp@acls.org
*Internet:* https://www.acls.org/Competitions-and-Deadlines/Luce-ACLS-Program-in-China-Studies
*Sponsor:* American Council of Learned Societies
633 Third Avenue, 8th Floor
New York, NY 10017-6795

**Luce/ACLS Prpgram in Religion, Journalism and International Affairs Fellowships for Scholars** 1505
Launched in 2015, the Luce/ACLS Program in Religion, Journalism and International Affairs (RJIA) is a two-pronged initiative designed to foster new connections between scholars and journalists covering international affairs. The program offers an interrelated set of awards: programming grants for universities and fellowships for scholars in the humanities and social sciences who study religion in international contexts. Luce/ACLS Religion, Journalism and International Affairs fellowships for Scholars are available to scholars pursuing research on any aspect of religion in an international context. Fellows also will have access to resources to help develop the capacities to relate their work to conversations in public policy and the media. During the fellowship term, fellows may remain at their home institutions or take up residence at one of the grantee institutions. In addition, the program will work with grantee institutions to convene all program-year fellows, together with select journalists and public policy experts at symposia during the fellowship year. The stipend is $55,000, plus up to $3,000 for research costs and related scholarly activities. Awardees electing to take up residence at partnering universities are eligible to receive an additional $5,000 to defray the cost of relocation and residency. The deadline for applications is October 28.
*Requirements:* Fellowships may be held at the awardee's home institution, one of three partnering universities, or any location suitable for the proposed research. Applicants must: be a U.S. citizen or permanent resident as of the application deadline date; have a PhD degree that was conferred by the application deadline; and be able to attend and participate in two program-sponsored events, a workshop in Fall of the academic year (date and location TBD) and a symposium in Spring of the academic year (date and location TBD).
*Geographic Focus* All States, District of Columbia, Guam, Marshall Islands, Northern Mariana Islands, Puerto Rico, U.S. Virgin Islands, American Samoa
*Date(s) Application is Due:* Oct 23
*Amount of Grant:* 55,000 - 63,000 USD
*Contact:* Valerie Popp; (212) 697-1505, ext. 153; fax (212) 949-8058; vpopp@acls.org
*Internet:* https://www.acls.org/Luce-ACLS-in-Religion-Journalism-International-Affairs-Fellowships-for-Scholars
*Sponsor:* American Council of Learned Societies
633 Third Avenue, 8th Floor
New York, NY 10017-6795

**Ludwick Family Foundation Grants** 1506
Founded in 1990 by Arthur and Sarah Ludwick, the Ludwick Family Foundation is a California-based philanthropic organization established exclusively for charitable, scientific, literary, and educational purposes. Grants are awarded to United States or U.S. based international organizations in adherence to the mission of Ludwick Family Foundation and its founding documents. The foundation tends to fund tangible types of items that will remain with and can be used repeatedly by the organization. Since the Foundation only considers invited applications, interested parties should begin by contacting the office with an concept before submitting anything in writing.
*Requirements:* Eligible applicants include U.S. or U.S.-based international organizations, 501(c)3 nonprofit public charities, and government agencies (any level). All grants are to be used exclusively for charitable, public benefit purposes.
*Restrictions:* The foundation does not grant requests for salaries, general operating expenses, scholarships, endowment funds, fundraising events or capital campaigns, feasibility studies, consulting fees, or advertising. The foundation will no longer accept any unsolicited requests for research or from public/private schools (K-12), universities/colleges, child daycare/development centers, hospitals, or libraries. Grant consideration will be given only to those organizations that have been invited to submit formal proposals and that have completed the application process.
*Geographic Focus:* All States
*Amount of Grant:* 5,000 - 50,000 USD
*Contact:* Trista Campbell; (626) 852-0092; fax (626) 852-0776; ludwickfndn@ludwick.org
*Internet:* www.ludwick.org
*Sponsor:* Ludwick Family Foundation
203 South Glendora Avenue, Suite B, P.O. Box 1796
Glendora, CA 91740

**Ludwig Vogelstein Foundation Grants** 1507
The Ludwig Vogelstein Foundation provides one-time grants to individuals working in the fine arts (excluding photography, film, and video) and in the literary arts, including play writing. It also supports certain unaffiliated scholarly and/or research projects. Grants ranging from $1,000 to $3,000 are awarded on the criteria of merit and need; the Foundation is especially concerned with those who have no other source of funding. Though there are no specified deadlines, the ideal contact months for guidelines ans submitted applications are January and February each year.
*Requirements:* Send email or SASE to request current guidelines (email is preferable and quick). The foundation prefers not to receive phone calls. Currently, the sponsor does not have a website.
*Geographic Focus:* All States
*Amount of Grant:* 1,000 - 3,000 USD
*Contact:* Diana Braunschweig, Executive Director; lvf@earthlink.net
*Sponsor:* Ludwig Vogelstein Foundation
4001 Inglewood Avenue, Suite 101-309
Redondo Beach, CA 90278

**Lumina Foundation for Education Grants** 1508
The Lumina Foundation for Education believes that education provides the basis for individual opportunity, economic vitality and social stability. With its partners, Lumina strives to meet workforce demands and close gaps in attainment for groups not historically well-served by higher education. Lumina's overarching goal is to increase the higher education attainment rate of the United States to 60 percent by 2025. This will represent an increase of 23 million graduates above current levels of production. The Foundation's proactive portfolio focuses on: increasing awareness of the benefits of higher education; improving student access to and preparedness for college; improving student success in college; and productivity across the higher education system. Note that Foundation grants vary in size by their scope but that grants resulting from unsolicited inquiries typically range from $50,000-$250,000. The usual duration for a grant is one to three years.
*Requirements:* IRS 501(c)3 or 509(a)(1, 2, or 3) tax-exempt organizations and public school organizations are eligible.
*Restrictions:* The foundation will not award grants that support corporate sponsorships and fundraising events; fund partisan political or lobbying efforts; or provide direct support for individuals, including scholarships or institutional scholarship funds.
*Geographic Focus:* All States
*Amount of Grant:* 50,000 - 250,000 USD
*Contact:* Candace Brandt; (800) 834-5756 or (317) 951-5300; fax (317) 951-5063
*Internet:* www.luminafoundation.org/grants.html
*Sponsor:* Lumina Foundation for Education
30 South Meridian Street, Suite 700, P.O. Box 1806
Indianapolis, IN 46204

**Lumpkin Family Foundation Healthy People Grants** 1509
Historically, the Lumpkin Family Foundation has supported entities that provide services that help keep people in its geographic area healthy. Requests will be evaluated on how well they accomplish one or more of the following: support the creativity of nonprofit organizations by seeding new projects and encouraging experimentation and innovation; support organizations demonstrating outstanding leadership in their field or community; promote the effectiveness of organizations and the nonprofit sector by supporting planning, learning and the professional development of staff and board leaders; facilitate collaboration across traditional organization or sector boundaries for community benefit; and develop public understanding of issues and promote philanthropic support necessary to address issues of community importance.
*Requirements:* Grants are awarded to nonprofit organizations that serve the community without discrimination on the basis of race, sex, or religion. Special consideration will be given to organizations and programs in East Central Illinois.
*Restrictions:* Proposals will not be considered from organizations that are not 501(c)3 tax-exempt; organizations whose primary purpose is to influence legislation; political causes, candidates, organizations or campaigns; individuals; or religious organizations, unless the particular program will benefit a large portion of the community and does not duplicate the work of other agencies in the community.
*Geographic Focus:* All States
*Amount of Grant:* 1,000 - 50,000 USD
*Contact:* Bruce Karmazin, (217) 234-5915 or (217) 235-3361; fax (217) 258-8444; Bruce@lumpkinfoundation.org
*Internet:* www.lumpkinfoundation.org/bWHATbwefund/HealthyPeople.aspx
*Sponsor:* Lumpkin Family Foundation
121 South 17th Street
Mattoon, IL 61938

**Lymphatic Education and Research Network Additional Support Grants for NIH-funded F32 Postdoctoral Fellows** 1510
The Lymphatic Education and Research Network (LE&RN) is a 501(c)3 not for profit organization whose mission is to fight lymphatic disease and lymphedema through education, research and advocacy. LE&RN has retained The Medical Foundation, a division of Health Resources in Action (The Medical Foundation division), to manage the administrative aspects of the LE&RN Awards Program. LE&RN seeks to accelerate the prevention, treatment and cure of the disease while bringing patients and medical professionals together to address the unmet needs surrounding lymphatic disorders. The goal of the Additional Support Awards Program is to help foster career interest in the field of lymphatic research by offering additional funds for the F32 postdoctoral research projects. The additional support may be used for research supplies, equipment, health insurance and travel to scientific meetings.

*Requirements:* Postdoctoral fellows who are supported by the NIHF32 Postdoctoral Fellowship Program, have at least one year remaining of F32 support as of July 1, and conduct research that clearly advances the mission of LE&RN are eligible to apply. Award recipients with an active F32 Fellowship will receive $10,000 annually for up to two years. The additional support award may be used for salary support of personnel who are assisting in the research project, as well as research supplies, equipment, health insurance and travel to scientific meetings. A budget proposal is not required for the application. Online submissions must be received by the Medical Foundation by January 10.
*Geographic Focus:* All States
*Date(s) Application is Due:* Jan 10
*Amount of Grant:* 20,000 USD
*Contact:* Erin Johnstone, Program Officer; (617) 279-2240, ext. 710; fax (617) 423-4619; ejohnstone@hria.org
*Internet:* www.hria.org/tmfservices/tmfgrants/lern32.html
*Sponsor:* Lymphatic Education and Research Network
261 Madison Avenue
New York, NY 10016

**Lynde and Harry Bradley Foundation Fellowships** **1511**
The Bradley Fellowship program is a major part of the Foundation's program interest in strengthening America's intellectual infrastructure at the higher-education level. Begun in 1986, the program is a model for influencing the intellectual framework of national life. It helps fulfill the two basic prerequisites for liberal education — good teaching and good curriculum — by enabling colleges and universities to offer students in the advanced stages of graduate work the opportunity to pursue serious studies with excellent teachers. For the program, the Foundation chooses institutions of higher education each year to support graduate and post-graduate students, who are designated as Bradley Fellows. The Fellows are chosen by each institution upon the recommendation of one or more distinguished professors in the humanities, social sciences, and the law.
*Geographic Focus:* All States
*Contact:* Daniel P. Schmidt; (414) 291-9915; fax (414) 291-9991
*Internet:* www.bradleyfdn.org/bradley_fellows.asp
*Sponsor:* Lynde and Harry Bradley Foundation
1241 North Franklin Place
Milwaukee, WI 53202-2901

**Lynde and Harry Bradley Foundation Grants** **1512**
The foundation aims to encourage projects that focus on cultivating a renewed, healthier, and more vigorous sense of citizenship among the American people and among peoples of other nations. Projects likely to be supported will generally treat free people as self-governing, personally responsible citizens, not as victims or clients; aim to restore the intellectual and cultural legitimacy of common sense, the wisdom of experience, everyday morality, and personal character; seek to reinvigorate and re-empower the traditional, local institutions—families, schools, churches, and neighborhoods—that provide training in and room for the exercise of genuine citizenship, that pass on everyday morality to the next generation, and that cultivate personal character; and encourage decentralization of power and accountability away from centralized, bureaucratic, national institutions back to states, localities, and revitalized mediating structures. Eligible projects may address any arena of public life—economics, politics, culture, or civil society; the problem of citizenship at home or abroad; Milwaukee and Wisconsin community and state projects that aim to improve the life of the community through increasing cultural and educational opportunities, grassroots economic development, and social and health services; the resuscitation of citizenship in the economic, political, cultural, or social realms; policy research and writing about approaches encouraging that resuscitation; academic research and writing that explore the intellectual roots of citizenship; and popular writing and media projects that illustrate for a broader public audience the themes of citizenship. The foundation supports programs that research the needs of gifted children and techniques of providing education for students with superior skills and/or intelligence; research programs investigating how learning occurs in gifted children; and demonstration programs of instruction.
*Requirements:* As an initial step, tax-exempt and nonprofit organizations should prepare a brief letter of inquiry presenting a concise description of their project, its objectives and significance, and the qualifications of the organizations and individuals involved. If the project appears to fall within the foundation's mandate, the applicant will be invited to submit a formal proposal. If invited to submit a formal proposal, the applicant should submit another letter. It should include a more-thorough, yet still concise description of the project, its objectives and significance, and the qualifications of the groups and individuals involved in it. It should also include a project budget, the specific amount being sought from Bradley, and a list of its other sources of support, philanthropic or otherwise.
*Restrictions:* The foundation favors projects that are normally not financed by public funds and will consider requests from religious organizations that are not denominational in character. Grants without significant importance to the foundation's areas of interest will only under special conditions be considered for endowment or deficit financing proposals. Grants will not be made to individuals, for overhead costs, or for fund-raising counsel.
*Geographic Focus:* All States
*Date(s) Application is Due:* Feb 1; May 1; Aug 1; Nov 1
*Amount of Grant:* 25,000 - 300,000 USD
*Samples:* Bel Canto Chorus of Milwaukee, Milwaukee, Wisconsin, $25,000 - to support general operations; Penfield Children's Center, Milwaukee, Wisconsin, $25,000 - to support general operations; Rawhide, Inc., New London, Wisconsin, $20,000 - to support the Residential Care Assistance Program.
*Contact:* Daniel P. Schmidt; (414) 291-9915; fax (414) 291-9991
*Internet:* www.bradleyfdn.org/program_interests.asp
*Sponsor:* Lynde and Harry Bradley Foundation
1241 North Franklin Place
Milwaukee, WI 53202-2901

**M.D. Anderson Foundation Grants** **1513**
The Anderson Foundation funds projects for the improvement of working class conditions among workers, and for the establishment, support and maintenance of hospitals, homes and institutions for the care of the young, sick, the aged, and the helpless. The Foundation also gives for the improvement of general living conditions and for the promotion of health, science, education, and the advancement of knowledge. Funding is given for aging center and services; education; employment; government/public administration; health care; human services; medical specialties public policy and research; and youth services. Types of support include building and/or renovation, equipment, matching/challenge support, research, and seed money.
*Requirements:* Organizations should submit a letter of inquiry with the following information: a copy of their IRS determination letter; a detailed description of their project and amount of funding requested; a copy of the current year's organizational budget and/or project budget; and a listing of additional sources and amount of support. Applicants should submit five copies of the proposal. There are no deadlines. The Board meets once a month, with organizations contacted within four weeks.
*Restrictions:* Funding is given primarily in Texas, with emphasis on the Houston area. Grants are not available to individuals, operating funds, or endowments.
*Geographic Focus:* Texas
*Amount of Grant:* 25,000 - 200,000 USD
*Samples:* Methodist Hospital Foundation, San Antonio, Texas, molecular imaging initiative at the hospital research institute, $200,000 payable over one year; YMCA capital campaign, Houston, Texas, $100,000 payable over one year; Scott and White Memorial Hospital, Temple, Texas, toward support of indigent care program through the development and implementation of the Family Medicine Community Clinic, $25,00 payable over one year.
*Contact:* Karen Jenkins, Grant Contact; (713) 216-1095
*Sponsor:* M.D. Anderson Foundation
P.O. Box 2558
Houston, TX 77252-8037

**M.J. Murdock Charitable Trust College Research Program for Life Sciences** **1514**
The Murdock Trust's College Research Program for Life Science is designed to support research at private, predominantly undergraduate colleges and universities in the Trust's funding region. The primary criterion used in evaluating proposals will be the perceived scientific quality of the research as measured by its likelihood to contribute significantly to fundamental knowledge in the field. A clear commitment of the college or university to support the research will also strengthen a proposal, as will substantial participation of undergraduate students as research associates, rather than as technical assistants. Eligible costs include: minor equipment and supplies needed (with institutional cost sharing); student stipends of up to $3,500 for ten weeks of full time participation; faculty stipends of up to $7,500 for up to ten weeks of full time participation; travel costs; and special/unusual expenses that are essential in the performance of the research. Additional detailed information is available at the Trust website.
*Requirements:* To be eligible, the institution must offer the baccalaureate degree, but no doctoral degree in the sciences. The principal investigator must either hold a faculty appointment in a department of biology, be a member of a department of environmental science doing research on biological systems or organisms, or be a member of a psychology department or its equivalent, doing laboratory neuroscience research. Application forms may be downloaded, but cannot be submitted electronically. The application and all supporting materials must be mailed to the Trust.
*Restrictions:* Faculty members in departments of biochemistry are not eligible. The following are also ineligible for funding: faculty and student academic year stipends, postdoctoral stipends, and graduate student stipends; indirect costs and fringe benefits, except as noted above; secretarial assistance, telephone, routine office supplies; publication costs, books, and journals; service or use charges for equipment housed at the home institution; and travel expenses to scientific meetings.
*Geographic Focus:* Alaska, Idaho, Montana, Oregon, Washington
*Date(s) Application is Due:* Oct 1
*Samples:* Pacific University, Forest Grove, Oregon, study of plant secondary metabolites from "Gaultheria shallon", $48,500; Seattle Pacific University, Seattle, Washington, social brain and vasopressin: regulation of maternal behavior and aggression in the California mouse, $40,000.
*Contact:* Dana Miller, Senior Program Director; (360) 694-8415; fax (360) 694-1819
*Internet:* www.murdock-trust.org/grants/college-research-life-sciences.php
*Sponsor:* M.J. Murdock Charitable Trust
703 Broadway, Suite 710
Vancouver, WA 98660

**M.J. Murdock Charitable Trust College Science Research Program** **1515**
The Murdock College Science Research Program is designed exclusively for private four-year liberal arts colleges in the Trust's grantmaking region. The goal of the program is to strengthen faculty and undergraduate student research efforts in the natural sciences at these colleges. Grants are given to institutions with the capacity for developing division-wide, faculty-student research programs that can be sustained with institutional resources and external funding. To date, the following colleges have been admitted into the program: Gonzaga University, Lewis & Clark College, University of Puget Sound, Whitman College, Linfield College, Northwest Nazarene University, Pacific Lutheran University, Pacific University, Seattle University, Willamette University, and the University of Portland.

*Requirements:* Colleges prepared to enter into and benefit from this program are invited to contact Murdock Trust.
*Restrictions:* Funding is exclusively for private liberal arts colleges and universities in Alaska, Idaho, Montana, Oregon, and Washington State.
*Geographic Focus:* Alaska, Idaho, Montana, Oregon, Washington
*Samples:* Pacific University, Forest Grove, Oregon, for strengthening faculty and student research in the sciences, $325,000; Seattle University, Seattle, Washington, for strengthening faculty and student research, $350,000.
*Contact:* Dana Miller, Senior Program Director; (360) 694-8415; fax (360) 694-1819
*Internet:* www.murdock-trust.org/grants/college-science-research.php
*Sponsor:* M.J. Murdock Charitable Trust
703 Broadway, Suite 710
Vancouver, WA 98660

**M.J. Murdock Charitable Trust Exceptional Opportunity Grants** **1516**
The Murdock Charitable Trust Exceptional Opportunity Grants meets the special program needs of regional graduate research universities. It is appropriate for projects that have special merit, but require special funding due to time constraints, no immediate source of funding, or some other unusual circumstance. The Vice President for Research, or the equivalent, at each of the universities with which arrangements have been made, may authorize an investigator or team to submit a proposal for a maximum amount of $50,000.
*Restrictions:* This grant is open to applicants only upon invitation or special arrangements with the Trust. Any potential applicant or organization interested in this programs should confer with the Trust before serious consideration is given to making application. Proposals may not be submitted to the Trust except at the invitation of the Vice President for Research at the principal investigator's institution. Applicants should contact the Trust for additional information.
*Geographic Focus:* Alaska, Idaho, Montana, Oregon, Washington
*Amount of Grant:* Up to 50,000 USD
*Contact:* Dana Miller, Senior Program Director; (360) 694-8415; fax (360) 694-1819
*Internet:* www.murdock-trust.org/grants/exceptional-opportunity-grants.php
*Sponsor:* M.J. Murdock Charitable Trust
703 Broadway, Suite 710
Vancouver, WA 98660

**M.J. Murdock Charitable Trust General Grants** **1517**
The M.J. Murdock Charitable Trust General Grants provides support for education, arts and culture, and health and human services. The Trust considers educational projects in formal and informal settings, emphasizing enhancement or expansion, as well as new educational approaches. Of special interest to the Trust are the following: performance and visual arts projects which enrich the cultural environment of the region and educational outreach efforts; and programs that emphasize preventative efforts which address physical, spiritual, social, and psychological needs, with a focus on youth. Grants are awarded for capital projects, program initiation, expansion, or for increased organizational capacity. Organizations are encouraged to view the grants awarded link for examples of previously funded grants.
*Requirements:* Before proceeding, interested parties should review the General Grant Application Guidelines to see if their organization is eligible or their project is appropriate for application to the Trust. After determining eligibility and appropriateness, organizations must submit a letter of inquiry before proceeding with the application. Upon approval, the organization will use the Trust's General Grant Application form and procedures.
*Restrictions:* The Trust only funds programs in the Pacific Northwest region: Alaska, Idaho, Montana, Oregon, and Washington. The following requests for funding are not considered: specific individuals and/or their personal benefit; individuals unauthorized to act on behalf of a qualified tax-exempt organization; funds that will ultimately be passed through to other organizations; propagandizing or for influencing legislation and elections; institutions that in policy or practice unfairly discriminate against race, ethnic, origin, sex, creed, or religion; sectarian or religious organizations whose principal activity is for the primary benefit of their own members; or for long-term loans, debt retirement, or operational deficits. The following funding requests are rarely considered: normal ongoing operations or the continuation of existing projects; endowments or revolving funds that act as such; continuation of programs previously financed from other external sources; urgent needs, emergency, or gap funding; organizations organized or operating outside any state or territory of the United States.
*Geographic Focus:* Alaska, Idaho, Montana, Oregon, Washington
*Samples:* Blachet House of Hospitality, Portland, Oregon, $450,000, to serve Portland homeless; Billings District Council of the Society of St. Vincent de Paul, Billings, Montana, to expand client services, $100,000; Whidbey Institute, Clinton, Washington, for program expansion to enhance civic engagement, $150,000.
*Contact:* Marybeth Stewart Goon; (360) 694-8415; fax (360) 694-1819
*Internet:* www.murdock-trust.org/grants/general-grants.php
*Sponsor:* M.J. Murdock Charitable Trust
703 Broadway, Suite 710
Vancouver, WA 98660

**M.J. Murdock Charitable Trust Partners in Science Grants** **1518**
The Murdock Partners in Science Grants are designed to help high school science teachers work with a mentor doing cutting-edge research over the course of two summers. The purpose of this grant is to bring the knowledge from the research lab back into the high school science classroom, promoting hands-on science education. Partners in Science grants provide research opportunities for teachers to work with investigators in academic labs or in labs associated with other non-profit institutions. Teachers are expected to arrange their own partnerships, although the Trust may provide some guidance. Approximately 25 grants are awarded each year to teacher-mentor partnerships in the Pacific Northwest. After completion of the Partners in Science Grants, recipients may then apply for the Partners in Science Supplemental Award of up to $7,000. These awards are made directly to the teacher's high school, and are for the purpose of implementing ideas from the research partnership in their classrooms. The application deadline for these awards is November 1. More information about Supplemental Awards and application forms is available from the Trust.
*Requirements:* Applications are accepted from researchers in the natural sciences at institutions located in areas where the Partners in Science program is available. Applications must be submitted through a college, university or other qualified research institution which accepts responsibility for the administration of the award. To apply, an investigator must hold a Ph.D. and must have an active research program as evidenced by grants and/or publications and hold an appointment in a natural science department at a college, university, or other qualified institution in the Pacific Northwest (Alaska, Idaho, Montana, Oregon, and Washington). High school teachers should possess academic qualifications in appropriate disciplines, in most cases a four year degree or the equivalent in chemistry, physics, biology, or earth science. Applicants must be active in classroom teaching and have a high school appointment. The teacher and high school would normally be within convenient commuting distance of the research host institution so that a research partnership may continue beyond the 8-10 week summer research periods. Qualified faculty members/research scientists (mentors) should request application packets from the Trust (as previously noted awards are made through, and must be endorsed by, a qualified institution of higher education or research institution), or download it from the Trust's website. Applications are in two parts: a short three-page application form to be completed by the mentor; and a two-page teacher candidate statement to be completed by the teacher. Applications must include both parts, for the contributions of each partner will be considered in the review process. There are no restrictions with regard to age, rank, or length of service of either teachers or mentors, but preference may be given to teachers who have not had a previous opportunity to do research, those who have had three or more years of teaching experience, and those who plan to continue teaching for several years, so the benefits of the program can be transferred back to the classroom. Applicants should refer to the website for more information about how mentor/teacher partnerships can be formed.
*Geographic Focus:* Alaska, Idaho, Montana, Oregon, Washington
*Date(s) Application is Due:* Dec 1
*Amount of Grant:* Up to 15,000 USD
*Contact:* Dana Miller, Senior Program Director; (360) 694-8415; fax (360) 694-1819
*Internet:* www.murdock-trust.org/grants/partners-science.php
*Sponsor:* M.J. Murdock Charitable Trust
703 Broadway, Suite 710
Vancouver, WA 98660

**M.J. Murdock Charitable Trust Research Start-up Grants for New Science Faculty** **1519**
The Murdock Start-up Grants offer support to supplement start-up research packages offered to new faculty in the natural sciences departments at private educational institutions. Grants are primarily targeted at undergraduate colleges and universities in the Trust's five-state region (Alaska, Idaho, Montana, Oregon, and Washington).
*Requirements:* Applicants should prepare a two page cover letter that addresses the following items: what is the vision for the department, extending at least over the next five years; how are teaching loads counted in the department, and what is considered a normal load (address at least classroom and also laboratory duties); how many majors have graduated in the department in each of the last five years; what are institutional and departmental expectations for the new hire regarding peer reviewed publications and external funding; how will the new hire fit into the future of the department. The application form is available online and should include the following information: institutional information; departmental information; consultant information; budget summary; department signatures; and a start-up package budget to be offered.
*Geographic Focus:* Alaska, Idaho, Montana, Oregon, Washington
*Date(s) Application is Due:* Aug 1
*Amount of Grant:* Up to 30,000 USD
*Contact:* Dana Miller, Senior Program Director; (360) 694-8415; fax (360) 694-1819
*Internet:* www.murdock-trust.org/grants/start-up-new-science-faculty.php
*Sponsor:* M.J. Murdock Charitable Trust
703 Broadway, Suite 710
Vancouver, WA 98660

**M.J. Murdock Charitable Trust Scientific Research Grants** **1520**
The M.J. Murdock Charitable Trust Scientific Research Grants fund select public research universities and medical institutes located within the Pacific Northwest region. The Trust prefers requests for projects in the natural sciences where the main objective is the acquisition of new knowledge. However, requests for research in engineering and medicine are also eligible. Training students in conducting research is an important consideration. Additional information and the application are located at the Trust website.
*Requirements:* A letter of inquiry is required unless the organization has received a Trust grant for scientific research within the last five years.
*Restrictions:* Only applications from scientific research institutions or major universities will be considered. Applications from individuals will not be considered.
*Geographic Focus:* Alaska, Idaho, Montana, Oregon, Washington
*Samples:* Oregon Health and Science University Foundation, Portland, Oregon, acquisition of a super-resolution microscope for studies of cells and tissues, $450,000; Washington State University Foundation, Pullman, Washington, acquisition of engineering research equipment for studies of biomass processing, $179,000; Western

Washington University Foundation, Bellingham, Washington, multi-disciplinary science research program to support undergraduate research, $138,500.
*Contact:* Dana Miller, Senior Program Director; (360) 694-8415; fax (360) 694-1819
*Internet:* www.murdock-trust.org/grants/general-scientific-research-grants.php
*Sponsor:* M.J. Murdock Charitable Trust
703 Broadway, Suite 710
Vancouver, WA 98660

**Mabel H. Flory Charitable Trust Grants** **1521**
The Mabel H. Flory Charitable Trust was established in Virginia in 1979, and it currently funds organizations in Washington, D.C. metro area and Baltimore, Maryland. Support is available for research designed to develop new cures and methods of treatment for: children with hearing and sight problems; arthritis in both children and adults; and aphasia related problems. Most recent grant awards have ranged from $4,000 to $10,000.
*Requirements:* Your initial approach should be in the form of a letter, prior to the deadline date.
*Restrictions:* No grants are given to individuals.
*Geographic Focus:* District of Columbia, Maryland
*Date(s) Application is Due:* Aug 31
*Amount of Grant:* 4,000 - 10,000 USD
*Samples:* Georgetown University, Washington, D.C., $10,000 - research purposes; Arthritis Foundation, Washington, D.C., $10,000 - research purposes; Foundation Fighting Blindness, Columbia, Maryland, $10,000 - research purposes.
*Contact:* BB&T Trustee; (703) 531-2053 or (888) 575-4586
*Sponsor:* Mabel H. Flory Charitable Trust
P.O. Box 2907
Wilson, NC 27894-2907

**Mabel Y. Hughes Charitable Trust Grants** **1522**
The Mabel Y. Hughes Charitable Trust was established under the last will and testament of Mabel Y. Hughes, a resident of Denver, Colorado, who died on April 9, 1969. Her legacy of giving continues through grants to charitable organizations in the state of Colorado. The Trust awards funding to nonprofit organizations in its areas of interest, including: children and youth services; education; family services; health care; higher education; human services; children's and art museums; performing arts centers; performing arts, opera; and reproductive health and family planning. Types of support include: annual campaigns; continuing support; emergency funds; endowments; equipment; general/operating support; program development; research; and seed money. Approximately fifty awards are given each year, with the average grant size ranging from $5,000 to $25,000. Applications must be submitted by March 1, July 1 or November 1 to be reviewed at the grant meeting that occurs after each deadline.
*Requirements:* Organizations should submit a letter of inquiry to the Trust, and if their project is appropriate for funding, they will be asked to submit a proposal.
*Restrictions:* The Trust does not support funding for individuals, deficit financing, scholarships, fellowships, or loans.
*Geographic Focus:* Colorado
*Date(s) Application is Due:* Mar 1; Jul 1; Nov 1
*Amount of Grant:* 5,000 - 30,000 USD
*Contact:* Peggy Toal, Private Client Services; (720) 947-6725 or (888) 234-1999; fax (877) 746-5889; grantadministration@ wellsfargo.com
*Sponsor:* Mabel Y. Hughes Charitable Trust
1740 Broadway
Denver, CO 80274

**MacArthur Foundation Conservation and Sustainable Development Grants** **1523**
The MacArthur Foundation is concerned about the deteriorating condition of the biosphere that supports life on earth. For the decade 2000-2010, the Foundation pursued a strategy that focused on eight ecological hotspots (places of high biodiversity under threat) in Madagascar, Melanesia, the Andes, Insular Caribbean, Albertine Rift, Eastern Himalayas, and Lower Mekong. In 2011, the Foundation launched a new strategy that is focused on one of the most compelling environmental challenges of the 21st century: the conservation of ecosystems. Ecosystems and their biodiversity underpin human well-being. They provide food and water; regulate floods, drought, and disease; and support soil formation and pollination. They also have intangible value as places of spiritual significance. All these are vital services, essential to economies and nation states. Hence, with the current Conservation and Sustainable Development program, the objective is to slow this degradation of ecosystems and, eventually, to reverse it. The geographical focus will be on three regions: the Great Lakes of East Central Africa, the Greater Mekong and its headwaters, and the watersheds of the Andes. Each is a place of high biodiversity, important freshwater service, and carbon-storage value. MacArthur's conservation and sustainable development program makes grants on a three-year recurring cycle, with one portfolio of grants prepared annually within each region and for the coastal marine initiative.
*Geographic Focus:* All States
*Date(s) Application is Due:* Feb 3; Apr 20; Jun 15
*Amount of Grant:* 50,000 - 2,000,000 USD
*Contact:* Douglas G. Siegel, Grants Administrator; (312) 726-8000; fax (312) 920-6258; 4answers@macfound.org
*Internet:* www.macfound.org/programs/conservation/
*Sponsor:* John D. and Catherine T. MacArthur Foundation
140 S Dearborn Street, Suite 1100
Chicago, IL 60603-5285

**MacDonald-Peterson Foundation Grants** **1524**
Established in 1995 in Texas, the MacDonald-Peterson Foundation offers funding for health and human services programs in Texas. The Foundation's primary fields of interest include: children's services; Christian agencies and churches; community and economic development; food banks; health organizations; higher education; hospitals; human services; medical research; mental health counseling and support groups; performing arts (orchestras); and youth development and community service clubs. There is no specified application format or annual deadlines, so applicants should begin the process by contacting the Foundation directly. Awards generally range from $5,000 to $20,000, and are given for general operations.
*Requirements:* Any 501(c)3 organization in, or serving the residents of, Texas are eligible to apply.
*Geographic Focus:* Texas
*Samples:* 100 Club of Houston, Houston, Texas, $10,000 - general operations; Art and Environmental Architecture, Houston, Texas, $20,000 - general operations; Asian Relief, Riverdale, Maryland, $10,000 - general operations.
*Contact:* Guy Tabor, (713)579-2355
*Sponsor:* MacDonald-Peterson Foundation
1 Riverway, Suite 1000
Houston, TX 77056-1944

**Macquarie Bank Foundation Grants** **1525**
The Foundation is one of Australia's oldest and largest corporate foundations contributing more than $150 million to more than 1500 community organizations world-wide since 1985. The Foundation focuses its resources in five core areas - the arts, education, environment, health, and welfare. The Foundation is also committed to projects specifically aimed at supporting indigenous communities. The Foundation's funding criteria is flexible and open. It welcomes applications from a diverse range of community organizations that are working in innovative ways to provide long-term benefits. Funding levels are flexible and are dictated by the needs of the organization and funding availability. Each application is assessed on its individual merit, with priority given to programs which support a broad section of the community at a regional, state or national level; have the involvement or potential for involvement of Macquarie Bank staff through volunteering, fundraising, pro bono work and board and/or management committee involvement; are located in cities/countries where Macquarie Bank staff are located; and deliver long-term benefits and build community sustainability. Prospective applicants are encouraged to check the Foundation website or contact Foundation Staff for more information on how to apply.
*Geographic Focus:* All States, All Countries
*Amount of Grant:* 100 - 500,000 USD
*Samples:* Po Leung Kuk, Hong Kong, $HK190,000 - a charitable organization supporting disadvantaged children; Oxfam Trailwalker, Melbourne, Australia, $A325,000 (2009-2010); Princess Margaret Hospital Foundation, Toronto, Ontario, Canada, $C350,000 - Ride to Conquer Cancer (2009-2010).
*Contact:* Heather Matwejev, Macquarie Bank Foundation, Asia; +61 2 8232 6951; fax +61 2 8232 0019; heather.matwejev@macquarie.com or foundation@macquarie.com
Lisa George, Macquarie Bank Foundation, Australia; +61 2 8232 6951; fax +61 2 8232 0019; foundation@macquarie.com or lisa.george@macquarie.com
Gail Cunningham, Macquarie Bank Foundation, Canada; +61 2 8232 6951; fax +61 2 8232 0019; foundation@macquarie.com or gail.cunningham@macquarie.com
Rachel Engel, Macquarie Bank Foundation, Europe/Middle East/Africa; +61 2 8232 6951; fax +61 2 8232 0019; foundation@macquarie.com or rachel.engel@macquarie.com
Kathryn O'Neal-Dunham, Macquarie Bank Foundation, United States; +61 2 8232 6951; fax +61 2 8232 0019; foundation@macquarie.com or kathryn.dunham@macquarie.com
*Internet:* www.macquarie.com/mgl/com/foundation/about/application-guidelines
*Sponsor:* Macquarie Bank Foundation
GP.O. Box 4294
Sydney, NSW 1164 Australia

**Maine Community Foundation Keepers Preservation Education Grants** **1526**
The Keepers Preservation Education Fund was established to support aspiring or established historic preservation professionals who wish to increase their professional knowledge or career potential. Awards may be used for tuition, professional meeting attendance, special book or other types of media purchases, domestic and foreign study travel, and other purposes as deemed appropriate. Applications are reviewed on a rolling basis.
*Requirements:* Eligible individuals must be: students enrolled full or part-time in institutions of higher learning with majors in historic preservation or allied fields including American studies, anthropology, archeology, architecture, landscape architecture, art history, architectural history, history, planning, law, economics, or building conservation; volunteer or paid staff of organizations or entities involved in historic preservation; and practicing professionals in historic preservation-related fields. Applicants may be from anywhere in the United States.
*Restrictions:* Individuals must apply through a sponsoring institution or organization, which reviews and forwards the application to the Foundation for funding review. Applicants may not be members of the awards committee or the immediate family of a committee member.
*Geographic Focus:* All States
*Contact:* Cherie Galyean, Grants Administrator; (877) 700-6800; fax (207) 667-0447; cgalyean@mainecf.org)
*Internet:* www.mainecf.org/keepersfund.aspx
*Sponsor:* Maine Community Foundation
245 East Main Street
Ellsworth, ME 04605

**Manitoba Arts Council Aboriginal Arts Creative Development Pilot Projects Grants** 1527

The Aboriginal Arts Creative Development Pilot Projects initiative is designed to enable Aboriginal artists in all arts disciplines, including dance/music, drumming, singing, carving, painting, storytelling, new media, writing, theature or crafts, to create or produce new works over a three month period. Only one Creative Development grant may be received in any given fiscal year (April 1 to March 31). The maximum award is $7,500.
*Requirements:* Professional Aboriginal artists are eligible who: are Canadian citizens or permanent residents of Canada; have lived in Manitoba for at least one year immediately prior to submitting an application; have completed basic training (not necessarily in a formal school, college or university); view the practice of their art as a career and are recognized as professionals in their art form by their peers; have completed at least three full years of professional activity; are able to devote a sufficient time over a six-month period to complete the project for which they are requesting assistance.
*Geographic Focus:* Canada
*Date(s) Application is Due:* Sep 25
*Amount of Grant:* Up to 7,500 CAD
*Contact:* Tracey Longbottom; (204) 945-3180; tlongbottom@artscouncil.mb.ca
*Internet:* artscouncil.mb.ca/2010/02/aboriginal-arts-creative-development/
*Sponsor:* Manitoba Arts Council
525-93 Lombard Avenue
Winnipeg, MB R3B 3B1 Canada

**Manitoba Arts Council Craft Grants** 1528

The Craft Grant is designed to provide assistance to professional Manitoba craft artists in the production of new creative work. The Grant supports contemporary or traditional craft projects based on the artist's original idea. Artists may apply for new work or for work that is part of a series. Craft artists contemplating a major project involving concentrated work over several years may define their proposed project as one phase of this work. Artists may apply for living and project expenses to the following maximums: established artists can receive a maximum of $10,000; and emerging artists can receive a maximum of $5,000. Eligible costs include living costs and direct project costs for new work: research; travel; supplies and materials; the purchase or rental of relevant equipment; documentation; technical assistance; temporary rental of studio space or facilities; and apprenticeship or craft development workshops. Artists are eligible for a maximum of two Manitoba Arts Council Craft Grants every forty-eight months. Artists are encouraged to discuss their eligibility with the program consultant prior to applying.
*Requirements:* Established artists must have a significant record of professional craft activity and have established a national or international profile. Emerging artists must have a modest record of professional craft activity following basic training, including the exhibition or public presentation of their work.
*Restrictions:* The program does not provide for: exhibition expenses which are the responsibility of the exhibition space, such as opening costs, invitations, mailing, printing of catalogs or posters; and the costs of exhibiting existing work.
*Geographic Focus:* Canada
*Date(s) Application is Due:* Sep 15
*Amount of Grant:* Up to 10,000 CAD
*Contact:* Rekha Ramachandran, Program Consultant; (204) 945-2237; fax (204) 945-5925; rramachandran@artscouncil.mb.ca or info@artscouncil.mb.ca
*Internet:* artscouncil.mb.ca/2010/02/craft-grant/
*Sponsor:* Manitoba Arts Council
525-93 Lombard Avenue
Winnipeg, MB R3B 3B1 Canada

**Manitoba Arts Council Dance Creation/Production Grants** 1529

The Dance Creation/Production Grant is available to support the creation of original works by Manitoba choreographers, as well as for dance production projects. Individuals, professional dance artist collectives and professional dance organizations may be eligible. The maximum grant available through this program is $15,000. Assistance may be requested for: research and creation projects (creating new work or developing an existing work); a production project or a series of productions (the public presentation of a choreographic work involving a majority of professional Manitoban artists); a combined creation and production project; and creation of choreography by independent Manitoban choreographers for organizations receiving annual funding by the Manitoba Arts Council (wherein the choreographic fee is the only eligible expense). Eligible costs include: fees to artists (performer, choreographer, mentor, designer, composers, etc.); a monthly living allowance of $2,000 for the primary artist/choreographer (this period of time may include research, creation and/or production); research material; studio expenses; audio-visual expenses; production expenses; and marketing and communication expenses.
*Requirements:* Eligible applicants include professional dance artists, professional dance artist collectives, and professional dance organizations who: do not receive annual/operating support from the Manitoba Arts Council; employ a majority of artists who are Canadian citizens, permanent residents or landed immigrants; employ a majority of artists who have lived in Manitoba for at least one year immediately prior to making an application (exceptional circumstances require a written explanation); and employ a majority of artists who view the discipline as a vocation and are recognized as professionals in their artistic discipline by their peers.
*Restrictions:* This program does not fund: rehearsal costs, dancers' fees or production costs for the creation of choreography by independent Manitoban choreographers for organizations receiving annual funding by the Manitoba Arts Council; projects for Artistic Directors and Choreographers in-residence of organizations receiving annual funding by the Manitoba Arts Council; or amateur or school productions.
*Geographic Focus:* Canada
*Date(s) Application is Due:* Sep 15
*Amount of Grant:* Up to 15,000 CAD
*Contact:* Cathleen Enns, Dance and Music; (204) 945-8631 or (204) 945-2237; fax (204) 945-5925; cenns@artscouncil.mb.ca or info@artscouncil.mb.ca
*Internet:* artscouncil.mb.ca/2010/02/dance-creationproduction-grant/
*Sponsor:* Manitoba Arts Council
525-93 Lombard Avenue
Winnipeg, MB R3B 3B1 Canada

**March of Dimes Agnes Higgins Award** 1530

Established in 1980, the March of Dimes Agnes Higgins Award honors the late Agnes Higgins of the Montreal Diet Dispensary for her innovation and years of service to the cause of improved maternal nutrition. A pioneer in devising methods of nutritional assessment and counseling, Mrs. Higgins greatly advanced the understanding of eating healthy as a crucial factor in healthy pregnancy and prevention of low birthweight. The Agnes Higgins Award is presented in recognition of distinguished achievement in research, education or clinical services in the field of maternal-fetal nutrition. Winners will receive a $5,000 honorarium and will be honored at a presentation and reception at the American Public Health Association Annual Meeting. Applications must be postmarked by March 30.
*Requirements:* Candidates must have: been widely involved in maternal-fetal nutrition through teaching, research, or clinical practice for at least five years; shown a demonstrable effect in raising the quality of maternal-fetal nutritional care through scholarly pursuits, research, education, or practice; and demonstrated the ability to apply maternal-fetal nutritional standards of practice or facilitate their implementation by others.
*Geographic Focus:* All States
*Date(s) Application is Due:* Mar 30
*Amount of Grant:* 5,000 USD
*Contact:* Mary Lavan, Grants Administrator; (914) 997-4609; fax (914) 997-4560; mlavan@marchofdimes.com
*Internet:* www.marchofdimes.org/professionals/agnes-higgins-award-maternal-fetal-nutrition.aspx
*Sponsor:* March of Dimes
1275 Mamaroneck Avenue
White Plains, NY 10605

**March of Dimes Newborn Screening Awards** 1531

The March of Dimes has created four awards to honor state health officials who put policies in place to address this important issue. The Newborn Screening Awards recognize state health officials who establish newborn screening delivery times of 72 hours, 48 hours or 24 hours. The highest award - known as the Robert Guthrie Newborn Screening Award - is presented to the state health official who achieves 95 percent of all newborn screens delivered from the hospital to the lab in 24 hours. An essential component of these awards is transparency - states must monitor their newborn screening transit times and publicly share the results.
*Geographic Focus:* All States
*Contact:* Mary Lavan; (914) 997-4609; fax (914) 997-4560; mlavan@marchofdimes.com
*Internet:* www.marchofdimes.org/professionals/recognizing-state-excellence-in-newborn-screening.aspx
*Sponsor:* March of Dimes
1275 Mamaroneck Avenue
White Plains, NY 10605

**Margaret and James A. Elkins Jr. Foundation Grants** 1532

The Elkins Foundation Grants support nonprofit programs and organizations in the metropolitan Houston, Texas area. Grants are given primarily for charitable, religious, scientific, or educational and literacy programs, including public safety testing, and the prevention of cruelty to children and animals. Fields of interest include: biology/life sciences; child development, education; child development, services; children and youth services; Christian agencies and churches; elementary and secondary education; engineering and technology; health organizations, association; higher education; hospitals (general); medical research, institute; medical education; religion; science; and safety/disasters. The Foundation also supports: building renovation; capital campaigns; emergency funds; endowments; equipment; program development; and research.
*Requirements:* Applications are not required. Applicants should submit a letter of inquiry, along with a copy of their most recent annual report, audited financial statement, or 990 form. Board meetings vary, and there is no deadline.
*Restrictions:* Grants are not made to individuals, or to support annual fundraising campaigns or operating deficits.
*Geographic Focus:* Texas
*Amount of Grant:* 5,000 - 400,000 USD
*Samples:* Saint Thomas Episcopal Church, Houston, Texas, for renovation support, $10,000; Sam Houston State University, Huntsville, Texas, for restoration support, $20,000; University of Texas at Austin, Austin, Texas, for educational fellowship fund, $100,000; Spark School Park Program, Houston, Texas, for program support, $5,000.
*Contact:* Larry Medford, Secretary-Treasurer; (713) 652-2052
*Sponsor:* Margaret and James A. Elkins Jr. Foundation
1001 Fannin Street, 1166 First City Tower
Houston, TX 77002

**Margaret T. Morris Foundation Grants** 1533

The Margaret T. Morris Foundation awards grants, primarily in Arizona, in its areas of interest, including: animal welfare; arts; children and youth, services; education; environment; higher education; homeless service; human services; marine science; medical research and education; mental health and crisis services; museums; performing

arts; reproductive health and family planning; and hospices. The Foundation's types of support include: building renovation; capital campaigns; debt reduction; endowments; general operating support; land acquisition; matching and challenge support; and program development. The Board of Directors meets in August, December, and as needed.
*Requirements:* Applications are not accepted. Applicants should submit a letter of inquiry with their request for funding and a description of the project.
*Geographic Focus:* Arizona
*Contact:* Thomas Polk, Trustee; (928) 445-4010
*Sponsor:* Margaret T. Morris Foundation
P.O. Box 592
Prescott, AZ 86302-0592

**Margoes Foundation Grants** 1534

The Margoes Foundation awards grants to California nonprofits in its areas of interest, including cardiac research, rehabilitation of mentally ill patients, and scholarships for minority students and for Asian and African nationals studying in the United States. Types of funding include general operating support, continuing support, program development, fellowships, scholarship funds, research, and matching funds.
*Requirements:* San Francisco, California, area nonprofits are eligible.
*Restrictions:* Grants are not awarded to agencies supported by federated campaigns or to individuals. Funding for endowments, capital campaigns, arts institutions, and other grants are made at the discretion of the board; proposals are not accepted for these purposes.
*Geographic Focus:* California
*Date(s) Application is Due:* Apr 1; Oct 1
*Amount of Grant:* 10,000 - 20,000 USD
*Contact:* Erin Lingren; (415) 561-6540, ext. 226; fax (415) 561-5477; elingren@pfs-llc.net
*Internet:* www.margoesfoundation.org/for-grantseekers/
*Sponsor:* Margoes Foundation
1660 Bush Street, Suite 300
San Francisco, CA 94109

**Marion Gardner Jackson Charitable Trust Grants** 1535

The Marion Gardner Jackson Charitable Trust was established by Marion Gardner Jackson, the granddaughter of local industrialist, Robert W. Gardner, founder of the Gardner-Denver Company. Before her death in 1976, Marion set the stage for a charitable trust that would keep her wishes alive. What started as a $5 million fund has grown to more than $12 million. In addition, the trust has granted more than $10 million to charitable agencies serving Adams County. Currently, the Trust funds: the arts and humanities; education; health; religion; and human service organizations. Giving is centered around the Quincy, Illinois, area and its surrounding communities in Adams County. The trust supports capital, program, and operating grants. Grants generally range up to a maximum of $25,000. Past awards have included an X-ray system for a dental clinic, several computers for various entities, an upgraded fire suppression system, and a match grant to help build a new home for the Quincy Humane Society. The application is available at the trust website through Bank of America. The annual deadline for submission is August 31. Most recent awards have ranged from $5,000 to $90,000.
*Restrictions:* For program support grants, the yearly request may not be more than 50% of the program's budget. Organizations can submit one application per year and will not receive more than one award from the trust in any given year. Organizations receiving a multi-year award from the trust that continues into the next grant year are not eligible to apply for an additional grant until the end of the grant cycle. The trust does not support requests from individuals, organizations attempting to influence policy through direct lobbying, or any political campaigns.
*Geographic Focus:* Illinois
*Date(s) Application is Due:* Aug 31
*Amount of Grant:* 5,000 - 90,000 USD
*Samples:* West Central Illinois Educational Telecommunications Corporation, Chatham, Illinois, $38,366 - support for public television broadcasting, primarily in educational services, to the West Central Illinois and Northeastern Missouri areas; Adams County Mental Health and Retardation Association, Quincy, Illinois, $37,500 - general operating support; Cheerful Home Association, Quincy, Illinois, $30,000 - program development.
*Contact:* Debra L. Grand, Philanthropic Relationship Manager; (312) 828-2055; ilgrantmaking@ustrust.com
*Internet:* https://www.bankofamerica.com/philanthropic/foundation.go?fnId=60
*Sponsor:* Marion Gardner Jackson Charitable Trust
231 South LaSalle Street, IL1-231-13-32
Chicago, IL 60604

**Marjorie C. Adams Charitable Trust Grants** 1536

The Marjorie C. Adams Charitable Trust, established in New York in 1987, is administered by the JP Morgan Chase Bank of Newark and supports the care and treatment of deaf and blind people, research into the causes and the treatment of deafness and blindness, and K-12 education. Types of support for these programs include: building and renovation projects, capital campaigns, and endowments. There are no specific application forms required or deadlines with which to adhere, and applicants should begin by contacted the Foundation office directly.
*Requirements:* 501(c)3 organizations, primarily serving residents of New York and Maryland, are eligible to apply.
*Geographic Focus:* Maryland, New York
*Amount of Grant:* Up to 150,000 USD
*Samples:* Auditory Oral School New York, Brooklyn, New York, $20,000 - general operating support; Foundation Fighting Blindness, Owings Mill, Maryland, $140,000 - general operating support.
*Contact:* Frank Lemma, Vice President; (212) 464-2439 or (212) 648-1477
*Sponsor:* Marjorie C. Adams Charitable Trust
270 Park Avenue, 16th Floor
New York, NY 10017

**Mary Gates Research Scholarships** 1537

Mary Gates Research Scholarships are competitive scholarships intended to enhance the educational experiences of undergraduate students at the University of Washington while they are engaged in research guided by faculty. Students may pursue research in a variety of disciplines. The application and additional information about the selection criteria are available at the Mary Gates website.
*Requirements:* Applicants must be full time students enrolled at the University of Washington in a four year program. Candidates need not be U.S. citizens. Candidates must already be engaged in research with faculty at the time of application, and continue with their research while full time students.
*Restrictions:* Applicants are not eligible for more than one Mary Gates Scholarship at the same time. Candidates should consult the Gates Endowment office to discuss if the Research Scholarship may interfere with other financial aid they receive.
*Geographic Focus:* All States, All Countries
*Amount of Grant:* 4,000 USD
*Contact:* Jodene Davis; (206) 616-3925; fax (206) 616-4389; mgates@u.washington.edu
*Internet:* www.washington.edu/uaa/mge/apply/research/index.htm
*Sponsor:* Mary Gates Endowment for Students
University of Washington
Seattle, WA 98195-2803

**Mary K. Chapman Foundation Grants** 1538

Mary K. Chapman was born in Oklahoma in 1920. She graduated from the University of Tulsa and worked as a nurse before her marriage to Allen Chapman in 1960. After the death of her husband in 1979, Mary Chapman maintained her own personal charitable giving program. Before her death in 2002, she established The Mary K. Chapman Foundation, a charitable trust founded to perpetuate her own charitable giving program. This foundation was fully funded with a bequest from her estate in 2005. Mary K. Chapman was very interested in supporting education, but as a former nurse and a very compassionate person, much of her charity was directed to health, medical research, and educating and caring for the less fortunate and disadvantaged. There are two steps in the process of applying for a grant. The first is a Letter of Inquiry from the applicant. This letter is used to determine if the applicant will be invited to take the second step of submitting a formal Grant Proposal.
*Requirements:* IRS 501(c)3 non-profits are eligible to apply.
*Restrictions:* Grant requests for the following purposes are not favored: endowments, except as a limited part of a capital project reserved for maintenance of the facility being constructed; deficit financing and debt retirement; projects or programs for which the Chapman Trusts would be the sole source of financial support; travel, conferences, conventions, group meetings, or seminars; camp programs and other seasonal activities; religious programs of religious organizations; project or program planning; start-up ventures are not excluded, but organizations with a proven strategy and results are preferred; purposes normally funded by taxation or governmental agencies; requests made less than nine months from the declination of a previous request by an applicant, or within nine months of the last payment made on a grant made to an applicant; requests for more than one project.
*Geographic Focus:* Oklahoma
*Amount of Grant:* Up to 300,000 USD
*Samples:* Arts and Humanities Council of Tulsa, Tulsa, Oklahoma, $200,000 - Brady District Visual Arts Center capital campaign; Tulsa Zoo Management, Tulsa, Oklahoma, $300,000 - general operating fund; Northeastern Oklahoma A&M College, Miami, Oklahoma, $197,000 - equipment and furnishings for NEO athletic training facility.
*Contact:* Andie Doyle, Program Officer; (918) 496-7882; fax (918) 496-7887; andie@chapmantrusts.com
*Internet:* www.chapmantrusts.org/grants_programs.php
*Sponsor:* Mary K. Chapman Foundation
6100 South Yale, Suite 1816
Tulsa, OK 74136

**Mary Kay Foundation Cancer Research Grants** 1539

In May 2010, the Foundation awarded $1.3 million in grants to select doctors and medical scientists focusing on curing cancers that affect women. These 13 recipients from across the United States received a $100,000 grant to conduct cutting-edge research. Since 1996, the Foundation has given more than $14 million to support this effort. Grants are awarded each year to researchers at medical schools recommended by the The Mary Kay Foundation Research Review Committee, which is composed of prominent doctors who volunteer their time to help the Foundation select the best recipients across the United States. After reviewing these recommendations, the Board of Directors at the Foundation selects the grant recipients. The Mary Kay Foundation accepts grant applications from November to mid-February each year.
*Geographic Focus:* All States
*Amount of Grant:* 100,000 USD
*Samples:* Emily Bernstein, Ph.D., Mount Sinai School of Medicine, $100,000 - Identifying epigenetic drivers of breast cancer; Jeanette Gowen Cook, Ph.D., University of North Carolina at Chapel Hill, $100,000 - Predicting genome instability in breast cancer; Zeng-Quan Yang, Ph.D., Wayne State University, $100,000 - Targeting histone demethylases as a therapeutic strategy in basal breast cancer.

*Contact:* Lana Rowe, (972) 687-4822 or (877) 652-2737; Lana.Rowe@mkcorp.com or MKCares@marykayfoundation.org
Jennifer Cook, (972) 687-5889 or (972) 687-4822; Jennifer.cook@mkcorp.com or MKCares@marykayfoundation.org
*Internet:* www.marykayfoundation.org/Pages/CancerGrantProgram.aspx
*Sponsor:* Mary Kay Foundation
P.O. Box 799044
Dallas, TX 75379-9044

**Mary S. and David C. Corbin Foundation Grants** **1540**
The Corbin Foundation gives primary consideration to charitable organizations and/or local chapters of national charities located in Akron and Summit County, Ohio, although extremely worthy causes outside of this preferred area may be considered. Areas of interest include arts and culture; civic and community; education; health care; housing; social services; medical research; and youth. The Foundation meets in May and November to consider requests. Grant requests must be received no later than March 1 for consideration in May and September 1 for consideration in November. The application and guidelines are available at the Foundation website.
*Requirements:* The Foundation has a general application cover sheet which applicants must complete. Organizations should also send a brief letter on their letterhead, submitting one original and one copy of all application materials.
*Restrictions:* Only written applications will be considered. Telephone and personal interviews are discouraged unless requested by the Foundation. The Foundation does not fund individuals; annual fundraising campaigns; ongoing requests for general operating support (although some repeat grants are made); operating deficits; or organizations which in turn make grants to others.
*Geographic Focus:* Ohio
*Date(s) Application is Due:* Mar 1; Sep 1
*Contact:* Erika J. May; (330) 762-6427; fax (330) 762-6428; corbin@nls.net
*Internet:* foundationcenter.org/grantmaker/corbin/guide.html
*Sponsor:* Mary S. and David C. Corbin Foundation
Akron Central Plaza
Akron, OH 44308-1830

**Massachusetts Cultural Council Artist Fellowships** **1541**
MCC's Artist Fellowships recognize exceptional work by Massachusetts artists across a range of disciplines. These highly competitive awards provide artists crucial validation among their peers and the public. They catalyze artistic advancement and pave the way for creative innovation of enduring cultural value. The Fellowships provide direct, unrestricted support to Massachusetts artists in recognition of exceptional original work, to foster the creation of new art in the Commonwealth. Fellowships in a range of disciplines are awarded to primary creators of original works of art. The Artist Fellowships provide competitive grants of $7,500 and finalist awards of $500. The number of awards per discipline category varies according to the number of applications reviewed, the recommendations of each panel, and program funds available. Drawing, Painting, and Traditional Arts applications are accepted August 15 through October 7. Choreography, Fiction/Creative Nonfiction, and Poetry applications are accepted December 15 through January 30.
*Requirements:* Any artist 18 years or older who has been a legal resident of Massachusetts for the last two years at the time of application is eligible. All Artist Fellows/Finalists must be legal residents of Massachusetts when grants are awarded.
*Geographic Focus:* Massachusetts
*Date(s) Application is Due:* Jan 30; Oct 7
*Amount of Grant:* 500 - 7,500 USD
*Contact:* Kelly Bennett, Program Coordinator; (617) 727-3668, ext. 349 or (800) 232-0960; fax (617) 727-0044; kelly.bennett@art.state.ma.us
*Internet:* www.massculturalcouncil.org/programs/artistfellows.asp
*Sponsor:* Massachusetts Cultural Council
10 St. James Avenue, 3rd Floor
Boston, MA 02116-3803

**Massage Therapy Research Fund (MTRF) Grants** **1542**
The purpose of the MTRF is to fund high quality research that investigates the efficacy, and clinical effectiveness of massage therapy and that contributes to our understanding of how massage therapy achieves its effects. A secondary goal is to increase research knowledge and capacity among massage therapists through their engagement in research practices at all levels. The MTRF invites applications in areas indicated above. For both competition cycles, a total funding amount of $50,000 is offered. Grant applications up to $15,000 will be considered.
*Requirements:* All properly submitted applications will be considered. Studies may be qualitative, quantitative, or mixed method. Initially, funding preference will be given to smaller, pilot, or creative study designs. In this way, the MTRF aims to assist the scientific and massage communities in building a foundation for the eventual construction of more complicated research projects. The MTRF encourages applications that also seek funding from other sources. The principal investigator must be affiliated with a Canadian charitable institution (university or hospital) that is willing to administer the finances of the grant. Funds are awarded to the charitable institution, not to an individual. The research team may include members affiliated with a non-Canadian institution. One member of the research team should be a massage therapist. Students at the masters level and above are encouraged to apply.
*Geographic Focus:* All States, Canada
*Date(s) Application is Due:* Mar 1; Sep 30
*Amount of Grant:* 15,000 USD
*Samples:* 2008 Awards - University of Saskatchewan ($8,929) Donelda Gowan-Moody, RMT and Anne Leis, PhD: Massage Therapists' Research Utilization and Perceptions Towards Research; Carleton University ($15,000) Diane Sliz, MSc (Cand) and Dr. Shawn Hayley: Psychological and Neuro-immune Effects of a Long-term Massage Therapy Treatement in Individuals Suffering from Major Depressive Disorder; Wilfred Laurier University ($14,981) Peter Tiidus, PhD, Kimberley Dawson, PhD and Lance Dawson, RMT: Effectiveness of regular proactive massage therapy for novice recreational runners.
*Contact:* Keren Brown, Executive Director; (416) 778-4443 or (866) 778-4443; fax (416) 778-5438; info@holistichealthresearch.ca
*Internet:* www.holistichealthresearch.ca/massagefund.gk
*Sponsor:* Holistic Health Research Foundation of Canada
80 Carlton Street
Toronto, ON M5B 1L6 Canada

**Matilda R. Wilson Fund Grants** **1543**
Matilda Rausch Dodge Wilson died on September 19, 1967, leaving most of her wealth to the Matilda R. Wilson Fund, a charitable trust she had established in Detroit in 1944. Today, the Matilda R. Wilson Fund awards grants, primarily in southeast Michigan, in support of the arts, youth agencies, higher education, hospitals, and social services. Types of support include building construction/renovation, endowments, equipment acquisition, general operating support, matching/challenge grants, program development, research, and scholarship funds. There are no application forms; initial approach should be a letter of request. The board considers requests at board meetings in January, April, and September.
*Requirements:* Michigan tax-exempt organizations are eligible.
*Restrictions:* Grants or loans are not made to individuals.
*Geographic Focus:* Michigan
*Amount of Grant:* 10,000 - 100,000 USD
*Contact:* David P. Larsen, Grants Administrator; (313) 259-7777; fax (313) 393-7579; roosterveen@bodmanllp.com
*Sponsor:* Matilda R. Wilson Fund
1901 Saint Antoine Street, 6th Floor
Detroit, MI 48226-2310

**Max Bell Foundation Grants** **1544**
The Max Bell Foundation focuses its grant making on innovative projects having substantial impact within the areas of health, Canada and the Asian Pacific, and veterinary sciences and education. Under health care, the foundation directs its support to the areas of health education and preventive health care. It is also interested in assisting health care professionals to update their education on a continuing basis. The foundation also supports research intended to build a better understanding of the Asian Pacific area on the part of Canadian policymakers and business people and assists Canadian scholars to build ties to their Asian counterparts. Also funded are major programs at centers of Asian research in Canadian universities. Within the field of veterinary sciences and education, the foundation seeks to provide broad-based support, although it has developed a particular interest in equine research. Applications are reviewed three times annually, eight weeks before the board meets (usually in May, September, and November).
*Requirements:* Funding is restricted to registered Canadian charities. Applicants should submit a brief letter describing the project for which they are seeking funds, an estimated budget, and the organization's charitable gift registration number. If the letter proves of interest to the foundation, a more detailed proposal will be invited.
*Restrictions:* Support will not be provided for capital projects or building campaigns; continuing operating expenses, funds to cover budget deficits, general fund-raising drives, or annual charitable appeals; academic or sabbatical leaves, internships, bursaries or scholarships; conferences, seminars or workshops, equipment purchases, publication costs, or travel costs.
*Geographic Focus:* Canada
*Amount of Grant:* 20,000 - 400,000 CAD
*Samples:* Alberta Health Services, Edmonton, Alberta, $50,000 - support for All Aboard for Patient Engagement: A Readiness Toolkit for Patients, Providers and Leaders; Canada West Foundation, Calgary, Alberta, $150,000 - to develop and communicate a strategy document aimed at identifying and promoting positions within Canadian debates over water pricing; The Learning Partnership, Toronto, Ontario, $331,060 - support of the Welcome to Kindergarten Program.
*Contact:* Alida White; (403) 215-7310; fax (403) 215-7319; apply@maxbell.org
*Internet:* www.maxbell.org
*Sponsor:* Max Bell Foundation
1201 Fifth Avenue SW, Suite 380
Calgary, AB T2R OY6 Canada

**Maximilian E. and Marion O. Hoffman Foundation** **1545**
The foundation has distributed grants in the areas of medicine, education, the arts, and to various charitable organizations in Connecticut and throughout the country. The principal focus of its philanthropy has been worthy activities in the State of Connecticut.
*Geographic Focus:* Connecticut
*Amount of Grant:* 1,000 - 300,000 USD
*Samples:* Bringham & Women's Hospital, Boston, MA, $115,000—funding for Parkinson's research; Holly Apostles College, Cromwell, CT, $50,000—funding for their capital campaign; Avon Old Farms School, Avon, CT, $40,000—funding to support the Performing Arts Department.
*Contact:* Marion Barrak, President; (860) 521-2949
*Sponsor:* Hoffman Foundation
970 Farmington Avenue, Suite 203
West Hartford, CT 06107-2134

**Mayo Clinic Administrative Fellowship** 1546

The Mayo Clinic Administrative Fellowship Program is a postgraduate fellowship that fosters the development of outstanding individuals committed to careers in health care administration. Since 1983, the Administrative Fellowship Program has provided participants with practical, skill-building experience through a broad variety of administrative rotations. The Administrative Fellowship Program prepares fellows for health care administration leadership opportunities within the Mayo Clinic system through direct participation in activities designed to build and strengthen essential administrative skills. Mayo Clinic Administrative Fellows will build the following skills: acquire and refine managerial competencies, including analytical, persuasive and human relations skills; develop an awareness and understanding of Mayo Clinic organizational structure and the interrelation between departments and personnel; receive individualized mentoring from a wide array of administrators and physicians; and establish a performance record through project work and departmental staff responsibilities. Those interested in the fellowship should review the website for available positions. Deadlines vary, but a general timeline for the fellowship recruitment and selection process is as follows: early August - job posted; mid-September - application deadline; October - phone interviews; early November - candidate on-site interviews; mid-November fellow offers extended; December - acceptance letters mailed; and July - fellowship begins. The application and additional information, including a detailed description of a variety of rotations, informational webinars, and a list of frequently asked questions, are available at the website.

*Requirements:* Diverse candidates from strong master's-level accredited programs are sought for the Administrative Fellowship Program. Qualified candidates are master's-prepared individuals who have completed the degree requirements or are degree candidates in health care administration, business administration, public health, health services administration or related fields of study. Applicants should have experience within the health care industry, such as an internship, externship, fellowship, or applicable employment. Applicants should also possess broad knowledge of current and historical perspectives of the health care industry that can include business management and administration, clinical practice management, clinical research and education, hospital administration and management, finance, human resources and organizational development, information systems, and managed care. Ideal candidates should possess the following traits and behaviors: the need to be challenged at a high level; energetic, hard-working, ambitious and a self-starter; flexibility and willingness to assume a variety of role assignments and deal effectively with ambiguity; objective and insightful; service, team and learning orientation; and understanding and commitment to cultural competency and diversity in the workplace and in the practice of health care.

*Restrictions:* Mayo Clinic is not able to provide visa sponsorship.

*Geographic Focus:* All States

*Date(s) Application is Due:* Sep 15

*Contact:* Stacy M. Fuhrman, ; (507) 266-6599; fuhrman.stacy@mayo.edu
Christopher Benson, Program Director; (480) 301-8000; benson.christopher@mayo.edu
Ajani Dunn, Program Director; (904) 953-2000; dunn.ajani@mayo.edu

*Internet:* www.mayoclinic.org/afp

*Sponsor:* Mayo Clinic
200 First Street SW
Rochester, MN 55905

**Mayo Clinic Administrative Fellowship - Eau Claire** 1547

The Administrative Fellowship Program is a postgraduate fellowship that fosters the development of outstanding individuals committed to careers in health care administration. The Mayo Clinic Health System in Eau Claire has provided participants with practical experience through a broad variety of administrative interactions, committee involvement, and projects leadership opportunities. The Fellowship Program in Healthcare Administration at Mayo Clinic Health System in Eau Claire is designed to foster the development of master's-prepared individuals committed to a career in healthcare administration. Under the guidance of the chief administrative officer, this one-year management training experience is tailored to meet the needs and interests of the individual candidate. This program allows the fellow to experience the major aspects of managing an integrated healthcare organization. The Fellowship Program offers exposure to all levels of management, allowing fellows to gain an applied understanding of the operations of an integrated healthcare system. Fellows are provided with opportunities to attend various executive-level meetings, and encouraged to participate in committees and other team projects. Examples of previous fellowship experiences include a physician need analysis, a community benefit assessment and a supply chain management project. The Fellow will receive a competitive salary and benefits. The application and additional information, including a detailed list of the objectives and expectations of the Fellowship, are posted at the website.

*Requirements:* Diverse candidates from strong master's-level accredited programs are sought for the Administrative Fellowship Program. Qualified candidates are master's-prepared individuals who have completed the degree requirements or are degree candidates in health care administration, business administration, public health, health services administration or related fields of study. Applicants should have experience within the health care industry, such as an internship, externship, fellowship, or applicable employment. Applicants should also possess broad knowledge of current and historical perspectives of the health care industry that can include business management and administration, clinical practice management, clinical research and education, hospital administration and management, finance, human resources and organizational development, information systems, and managed care. Ideal candidates should possess the following traits and behaviors: the need to be challenged at a high level; energetic, hard-working, ambitious and a self-starter; flexibility and willingness to assume a variety of role assignments and deal effectively with ambiguity; objective and insightful; service, team and learning orientation; and understanding and commitment to cultural competency and diversity in the workplace and in the practice of health care.

*Restrictions:* Lodging is the responsibility of the individual. There is no commitment of employment on behalf of the individual or organization beyond the one-year program.

*Geographic Focus:* All States

*Date(s) Application is Due:* Oct 1

*Contact:* Nicole Egan, Administrative Fellow Coordinator; (715) 838-5081; egan.nicole@mayo.edu

*Internet:* mayoclinichealthsystem.org/locations/eau-claire/careers/for-students/administrative-fellowship

*Sponsor:* Mayo Clinic
200 First Street SW
Rochester, MN 55905

**Mayo Clinic Business Consulting Fellowship** 1548

The Business Consulting Fellowship Program at Mayo Clinic is a one-year learning experience designed to develop outstanding individuals committed to a career in health care management. Through this program, recipients will gain practical experience and actively participate in projects within Mayo Clinic. Fellows will assist experienced analysts on large and small projects designed to improve the efficiency, quality, and effectiveness of Mayo Clinic. Sample project topics include the following: business planning; quality improvement; information technology systems implementation; forecasting and access; simulation; patient flow; facilities design; electronic systems usability; and staffing analysis. The Clinic offers competitive compensation and comprehensive benefits for Fellow participants. The program typically begins in June and December. Candidates may apply at the website.

*Requirements:* Applicants must have recently completed a master's degree and have at least one year of relevant internship or work experience, or have a bachelor's degree and at least four years of work or internship experience. Previous health care experience is preferred. Degrees should be in business or health care administration, industrial engineering, operations research, or a closely related field. Successful candidates will have skills and experience in several of these areas: consulting; workflow and operations analysis; staffing analysis; process design, improvement or re-engineering; simulation; electronic system implementation; and project management. Strong candidates should also have the following skills: objectivity; analytical, innovative and results oriented; comfortable with ambiguity; curious and willing to ask questions; highly skilled communicators; and able to work both independently and as a part of team.

*Geographic Focus:* All States

*Contact:* Chad Musolf, Human Resources; (507) 284-4851; musolf.chad@mayo.edu

*Internet:* www.mayoclinic.org/train-sp-rst/

*Sponsor:* Mayo Clinic
200 First Street SW
Rochester, MN 55905

**MAZON Partnership Grants: A Jewish Response to Hunger** 1549

MAZON's domestic partnership grant program makes awards for general support or special projects in four categories. Advocacy education/research grants range from $5,000 to $20,000, and support programs to change public policy or opinion or the operation and administration of federal food assistance programs. Emergency/direct food assistance grants range from $5,000 to $7,000, and support food pantries, soup kitchens, and programs providing meals to special-needs populations. Food bank grants, ranging from $5,000 to $15,000, are awarded to warehouses that collect and distribute food to nonprofit feeding programs. Multi-service organization grants, ranging from $5,000-$15,000, support organizations that provide food and a broad array of services to hungry and low-income people. Deadlines listed are for letters of inquiry, and full applications are by invitation.

*Requirements:* U.S. nonprofit organizations that meet program criteria may apply.

*Restrictions:* MAZON does not make grants to groups that proselytize or place any religious requirements on service; have national name recognition; are principally focused on homelessness; charge individuals for food; or are government entities, professional associations, job training programs, or grantmaking organizations. Capital grants for building projects are not eligible for support. Prepared and perishable food programs are not eligible unless they are in their first two years of operation. MAZON does not fund holiday meal programs.

*Geographic Focus:* All States

*Date(s) Application is Due:* Nov 15

*Amount of Grant:* 5,000 - 20,000 USD

*Contact:* Marla Feldman, Senior Program Director; (310) 442-0020 or (800) 813-0557; fax (310) 442-0030; mfeldman@mazon.org or hello@mazon.org

*Internet:* mazon.org/what-we-do/mazon-grantees

*Sponsor:* MAZON: A Jewish Response to Hunger
10850 Wilshire Boulevard, Suite 400
Los Angeles, CA 90024

**MBL Albert and Ellen Grass Summer Fellowships** 1550

Marine Biological Laboratory (MBL) summer researchers come from all over the world. Visiting researchers find an infrastructure and an informal, interactive scientific community that allows them to launch into research as soon as they arrive. Research awards are available to investigators wishing to do visiting research at the MBL. The Albert and Ellen Grass Fellowship promotes innovative research collaborations, ideally between investigators from different institutions. Preference will be given to junior investigators, who are a required member of any team seeking support from this Fellowship. The MBL Research Awards provide costs for research and housing, and also enable awardees to benefit from the intellectual and interactive environment of the

scientific community at the MBL. Funding requires a minimum stay of six weeks at the MBL. Awards can be used to support independent research, collaborative research, or in affiliation with the discovery courses. Proposals for research award support will be considered in, but are not limited to, the following fields of investigation: cell biology; development biology; ecology; evolution; microbiology; neurobiology; physiology; and tissue engineering. Candidates may apply at the MBL website.
*Requirements:* Research proposed in the resident program or courses requires pre-approval of the host lab and program and the course directors, respectively, in addition to the CASO office.
*Geographic Focus:* All States, All Countries
*Date(s) Application is Due:* Dec 15
*Contact:* Julie Early, Research Award Coordinator; (508) 289-7173; researchawards@mbl.edu or jearly@mbl.edu
*Internet:* hermes.mbl.edu/research/summer/awards_general.html
*Sponsor:* Marine Biological Laboratory
7 MBL Street
Woods Hole, MA 02543

**MBL Ann E. Kammer Memorial Summer Fellowship** 1551
Marine Biological Laboratory (MBL) summer researchers come from all over the world. Visiting researchers find an infrastructure and an informal, interactive scientific community that allows them to launch into research as soon as they arrive. Research awards are available to these visiting researchers. In memory of renowned zoologist, Ann E. Kammer, this Fellowship is available to women investigators working in the neurosciences. The MBL Research Awards provide costs for research and housing, and also enable awardees to benefit from the intellectual and interactive environment of the scientific community at the MBL. Funding requires a minimum stay of six weeks at the MBL. Awards can be used to support independent research, collaborative research, or in affiliation with the discovery courses. Proposals for Research award support will be considered in, but are not limited to, the following fields of investigation: cell biology; development biology; ecology; evolution; microbiology; neurobiology; and physiology. Candidates may fill out applications at the MBL online process.
*Requirements:* Research proposed in the resident program or courses requires pre-approval of the host lab and program and the course directors, respectively, in addition to the CASO office.
*Geographic Focus:* All States, All Countries
*Date(s) Application is Due:* Dec 15
*Contact:* Julie Early; (508) 289-7173; researchawards@mbl.edu or jearly@mbl.edu
*Internet:* hermes.mbl.edu/research/summer/awards_general.html
*Sponsor:* Marine Biological Laboratory
7 MBL Street
Woods Hole, MA 02543

**MBL Associates Summer Fellowships** 1552
Marine Biological Laboratory (MBL) summer researchers come from all over the world. Visiting researchers find an infrastructure and an informal, interactive scientific community that allows them to launch into research as soon as they arrive. Research awards are available to these visiting researchers. The MBL Associates were formed to provide a connection to the Laboratory for friends, both scientists and non-scientists who have an interest in the research, educational and cultural activities of the MBL. The Associates provide fellowship funds to help support the costs of summer research at the MBL. The MBL Research Awards provide costs for research and housing, and also enable awardees to benefit from the intellectual and interactive environment of the scientific community at the MBL. Funding requires a minimum stay of six weeks at the MBL. Awards can be used to support independent research, collaborative research, or in affiliation with the discovery courses. Proposals for Research award support will be considered in, but are not limited to, the following fields of investigation: cell biology; development biology; ecology; evolution; microbiology; neurobiology; physiology; and tissue engineering. Candidates may fill out applications at the MBL online process.
*Requirements:* Research proposed in the resident program or courses requires pre-approval of the host lab and program and the course directors, respectively, in addition to the CASO office.
*Geographic Focus:* All States, All Countries
*Date(s) Application is Due:* Dec 15
*Contact:* Julie Early, Research Award Coordinator; (508) 289-7173; researchawards@mbl.edu or jearly@mbl.edu
*Internet:* hermes.mbl.edu/research/summer/awards_general.html
*Sponsor:* Marine Biological Laboratory
7 MBL Street
Woods Hole, MA 02543

**MBL Baxter Postdoctoral Summer Fellowship** 1553
Marine Biological Laboratory (MBL) summer researchers come from all over the world. Visiting researchers find an infrastructure and an informal, interactive scientific community that allows them to launch into research as soon as they arrive. Research awards are available to these visiting researchers. Baxter International, Inc. has established a Fellowship for postdoctoral investigators in the biological and biomedical sciences. This Fellowship supports a scientist undertaking independent research at the MBL. The MBL Research Awards provide costs for research and housing, and also enable awardees to benefit from the intellectual and interactive environment of the scientific community at the MBL. Funding requires a minimum stay of six weeks at the MBL. Awards can be used to support independent research, collaborative research, or in affiliation with the discovery courses. Candidates may fill out applications at the MBL online process.
*Requirements:* Research proposed in the resident program or courses requires pre-approval of the host lab and program and the course directors, respectively, in addition to the CASO office.
*Geographic Focus:* All States, All Countries
*Date(s) Application is Due:* Dec 15
*Contact:* Julie Early, Research Award Coordinator; (508) 289-7173; researchawards@mbl.edu or jearly@mbl.edu
*Internet:* hermes.mbl.edu/research/summer/awards_general.html
*Sponsor:* Marine Biological Laboratory
7 MBL Street
Woods Hole, MA 02543

**MBL Burr and Susie Steinbach Summer Fellowship** 1554
Marine Biological Laboratory (MBL) summer researchers come from all over the world. Visiting researchers find an infrastructure and an informal, interactive scientific community that allows them to launch into research as soon as they arrive. Research awards are available to these visiting researchers. The Steinbach Fellowship honors the memory of H. "Burr" and Eleanor "Susie" Steinbach. Dr. Steinbach was a scientist and director of the MBL who made important contributions toward the understanding of bioelectric generators, ionic equilibria across cell membranes, and the enzymology of cell organelles. This Fellowship helps cover laboratory fees, housing expenses, and other appropriate personal expenses. The MBL Research Awards provide costs for research and housing, and also enable awardees to benefit from the intellectual and interactive environment of the scientific community at the MBL. Funding requires a minimum stay of six weeks at the MBL. Awards can be used to support independent research, collaborative research, or in affiliation with the discovery courses. Proposals for Research award support will be considered in, but are not limited to, the following fields of investigation: cell biology; development biology; ecology; evolution; microbiology; neurobiology; physiology; and tissue engineering. Candidates may fill out applications at the MBL online process.
*Requirements:* Research proposed in the resident program or courses requires pre-approval of the host lab and program and the course directors, respectively, in addition to the CASO office.
*Geographic Focus:* All States, All Countries
*Date(s) Application is Due:* Dec 15
*Contact:* Julie Early; (508) 289-7173; researchawards@mbl.edu or jearly@mbl.edu
*Internet:* hermes.mbl.edu/research/summer/awards_general.html
*Sponsor:* Marine Biological Laboratory
7 MBL Street
Woods Hole, MA 02543

**MBL E.E. Just Summer Fellowship for Minority Scientists** 1555
Marine Biological Laboratory (MBL) summer researchers come from all over the world. Visiting researchers find an infrastructure and an informal, interactive scientific community that allows them to launch into research as soon as they arrive. Research awards are available to these visiting researchers. The Just Fellowship honors Dr. Just's contributions to developmental biology and experimental embryology. Dr. Just was the first African American to be professionally recognized for his scientific work, much of it accomplished during the early 1900s. This Fellowship will underwrite an outstanding minority scientist's participation in research at the MBL. The MBL Research Awards provide costs for research and housing, and also enable awardees to benefit from the intellectual and interactive environment of the scientific community at the MBL. Funding requires a minimum stay of six weeks at the MBL. Awards can be used to support independent research, collaborative research, or in affiliation with the discovery courses. Proposals for Research award support will be considered in, but are not limited to, the following fields of investigation: cell biology; development biology; ecology; evolution; microbiology; neurobiology; physiology; and tissue engineering. Candidates may fill out applications at the MBL online process.
*Requirements:* Research proposed in the resident program or courses requires pre-approval of the host lab and program and the course directors, respectively, in addition to the CASO office.
*Geographic Focus:* All States, All Countries
*Date(s) Application is Due:* Dec 15
*Contact:* Julie Early, Research Award Coordinator; (508) 289-7173; researchawards@mbl.edu or jearly@mbl.edu
*Internet:* hermes.mbl.edu/research/summer/awards_general.html
*Sponsor:* Marine Biological Laboratory
7 MBL Street
Woods Hole, MA 02543

**MBL Erik B. Fries Summer Fellowships** 1556
Marine Biological Laboratory (MBL) summer researchers come from all over the world. Visiting researchers find an infrastructure and an informal, interactive scientific community that allows them to launch into research as soon as they arrive. Research awards are available to these visiting researchers. The Erik B. Fries Fellowships defray laboratory or library expenses for independent investigators at the MBL. Fries, a long time Woods Hole resident, was a biology professor at City College of New York, and MBL's botany course in the early 1900s. Preference is given to investigator members from Fries' former department at City College. The MBL Research Awards provide costs for research and housing, and also enable awardees to benefit from the intellectual and interactive environment of the scientific community at the MBL. Funding requires a minimum stay of six weeks at the MBL. Awards can be used to support independent research, collaborative research, or in affiliation with the discovery courses. Proposals for Research award support will be considered in, but are not limited to, the following fields

of investigation: cell biology; development biology; ecology; evolution; microbiology; neurobiology; physiology; and tissue engineering. Candidates may fill out applications at the MBL online process.
*Requirements:* Research proposed in the resident program or courses requires pre-approval of the host lab and program and the course directors, respectively, in addition to the CASO office.
*Geographic Focus:* All States, All Countries
*Date(s) Application is Due:* Dec 15
*Contact:* Julie Early; (508) 289-7173; researchawards@mbl.edu or jearly@mbl.edu
*Internet:* hermes.mbl.edu/research/summer/awards_general.html
*Sponsor:* Marine Biological Laboratory
7 MBL Street
Woods Hole, MA 02543

**MBL Eugene and Millicent Bell Summer Fellowships** **1557**
Marine Biological Laboratory (MBL) summer researchers come from all over the world. Visiting researchers find an infrastructure and an informal, interactive scientific community that allows them to launch into research as soon as they arrive. Research awards are available to these visiting researchers. The Eugene and Millicent Bell Fellowships in Tissue Engineering provide funding to visiting graduate students, postdoctoral fellows, or faculty pursuing research in a variety of biomedical questions that advance knowledge of biological composition and function at levels ranging from molecules to cells, to tissue, to organs and whole organisms. This work is grounded in a solid base of chemistry, physics, math and biology and aims to expand the possibilities for tissue and organ reconstitution and replacement. The MBL Research Awards provide costs for research and housing, and also enable awardees to benefit from the intellectual and interactive environment of the scientific community at the MBL. Funding requires a minimum stay of six weeks at the MBL. Awards can be used to support independent research, collaborative research, or in affiliation with the discovery courses. Candidates may fill out applications at the MBL online process.
*Requirements:* Research proposed in the resident program or courses requires pre-approval of the host lab and program and the course directors, respectively, in addition to the CASO office.
*Geographic Focus:* All States, All Countries
*Date(s) Application is Due:* Dec 15
*Contact:* Julie Early; (508) 289-7173; researchawards@mbl.edu or jearly@mbl.edu
*Internet:* hermes.mbl.edu/research/summer/awards_general.html
*Sponsor:* Marine Biological Laboratory
7 MBL Street
Woods Hole, MA 02543

**MBL Evelyn and Melvin Spiegal Summer Fellowship** **1558**
Marine Biological Laboratory (MBL) summer researchers come from all over the world. Visiting researchers find an infrastructure and an informal, interactive scientific community that allows them to launch into research as soon as they arrive. Research awards are available to these visiting researchers. The Spiegal Fellowship support summer research at the MBL in any relevant biomedical discipline. The MBL Research Awards provide costs for research and housing, and also enable awardees to benefit from the intellectual and interactive environment of the scientific community at the MBL. Funding requires a minimum stay of six weeks at the MBL. Awards can be used to support independent research, collaborative research, or in affiliation with the discovery courses. Proposals for Research award support will be considered in, but are not limited to, the following fields of investigation: cell biology; development biology; ecology; evolution; microbiology; neurobiology; physiology; and tissue engineering. Candidates may fill out applications at the MBL online process.
*Requirements:* Research proposed in the resident program or courses requires pre-approval of the host lab and program and the course directors, respectively, in addition to the CASO office.
*Geographic Focus:* All States, All Countries
*Date(s) Application is Due:* Dec 15
*Contact:* Julie Early; (508) 289-7173; researchawards@mbl.edu or jearly@mbl.edu
*Internet:* hermes.mbl.edu/research/summer/awards_general.html
*Sponsor:* Marine Biological Laboratory
7 MBL Street
Woods Hole, MA 02543

**MBL Frank R. Lillie Summer Fellowship** **1559**
Marine Biological Laboratory (MBL) summer researchers come from all over the world. Visiting researchers find an infrastructure and an informal, interactive scientific community that allows them to launch into research as soon as they arrive. Research awards are available to these visiting researchers. The Lillie Fellowship was established in memory of Frank R. Lillie, a distinguished developmental biologist and the laboratory's director for many years. The Fellowship provides a research laboratory and funds for supplies. The MBL Research Awards provide costs for research and housing, and also enable awardees to benefit from the intellectual and interactive environment of the scientific community at the MBL. Funding requires a minimum stay of six weeks at the MBL. Awards can be used to support independent research, collaborative research, or in affiliation with the discovery courses. Proposals for Research award support will be considered in, but are not limited to, the following fields of investigation: cell biology; development biology; ecology; evolution; microbiology; neurobiology; physiology; and tissue engineering. Candidates may fill out applications at the MBL online process.
*Requirements:* Research proposed in the resident program or courses requires pre-approval of the host lab and program and the course directors, respectively, in addition to the CASO office.
*Geographic Focus:* All States, All Countries
*Date(s) Application is Due:* Dec 15
*Contact:* Julie Early; (508) 289-7173; researchawards@mbl.edu or jearly@mbl.edu
*Internet:* hermes.mbl.edu/research/summer/awards_general.html
*Sponsor:* Marine Biological Laboratory
7 MBL Street
Woods Hole, MA 02543

**MBL Frederik B. and Betsy G. Bang Summer Fellowships** **1560**
Marine Biological Laboratory (MBL) summer researchers come from all over the world. Visiting researchers find an infrastructure and an informal, interactive scientific community that allows them to launch into research as soon as they arrive. Research awards are available to these visiting researchers. The Bang Fellowships are offered for the study of the immune capability of marine animals and for the use of marine models for research in molecular biology or biomedicine. Funds primarily support laboratory space and research expenses. Funding requires a minimum stay of six weeks at the MBL. Candidates may fill out applications at the MBL online process.
*Requirements:* Research proposed in the resident program or courses requires pre-approval of the host lab and program and the course directors, respectively, in addition to the CASO office.
*Geographic Focus:* All States, All Countries
*Date(s) Application is Due:* Dec 15
*Contact:* Julie Early, Research Award Coordinator; (508) 289-7173; researchawards@mbl.edu or jearly@mbl.edu
*Internet:* hermes.mbl.edu/research/summer/awards_general.html
*Sponsor:* Marine Biological Laboratory
7 MBL Street
Woods Hole, MA 02543

**MBL Fred Karush Library Readership** **1561**
Marine Biological Laboratory (MBL) summer researchers come from all over the world. Visiting researchers find an infrastructure and an informal, interactive scientific community that allows them to launch into research as soon as they arrive. A variety of research awards are available to these visiting researchers. The Karush Library Readership provides the fees for desk rental, and full privileges in the MBL/WHOI Library on a monthly basis during the summer. The MBL Research Awards provide costs for research and housing, and also enable awardees to benefit from the intellectual and interactive environment of the scientific community at the MBL. Funding requires a minimum stay of six weeks at the MBL. Awards can be used to support independent research, collaborative research, or in affiliation with the discovery courses. Proposals for Research award support will be considered in, but are not limited to, the following fields of investigation: cell biology; development biology; ecology; evolution; microbiology; neurobiology; physiology; and tissue engineering. Candidates may fill out applications at the MBL online process.
*Requirements:* Research proposed in the resident program or courses requires pre-approval of the host lab and program and the course directors, respectively, in addition to the CASO office.
*Geographic Focus:* All States, All Countries
*Date(s) Application is Due:* Feb 15
*Contact:* Julie Early; (508) 289-7173; researchawards@mbl.edu or jearly@mbl.edu
*Internet:* hermes.mbl.edu/research/summer/awards_general.html
*Sponsor:* Marine Biological Laboratory
7 MBL Street
Woods Hole, MA 02543

**MBL Gruss Lipper Family Foundation Summer Fellowship** **1562**
Marine Biological Laboratory (MBL) summer researchers come from all over the world. Visiting researchers find an infrastructure and an informal, interactive scientific community that allows them to launch into research as soon as they arrive. Research awards are available to these visiting researchers. The Gruss Lipper Fellowship underwrites expenses and travel to bring Israeli investigators and advanced students to the MBL to work independently, collaborate with MBL scientists, or to participate in the MBL summer research and short courses. The MBL Research Awards provide costs for research and housing, and also enable awardees to benefit from the intellectual and interactive environment of the scientific community at the MBL. Funding requires a minimum stay of six weeks at the MBL. Awards can be used to support independent research, collaborative research, or in affiliation with the discovery courses. Proposals for Research award support will be considered in, but are not limited to, the following fields of investigation: cell biology; development biology; ecology; evolution; microbiology; neurobiology; physiology; and tissue engineering. Candidates may fill out applications at the MBL online process.
*Requirements:* Research proposed in the resident program or courses requires pre-approval of the host lab and program and the course directors, respectively, in addition to the CASO office.
*Geographic Focus:* Israel
*Date(s) Application is Due:* Dec 15
*Contact:* Julie Early, Research Award Coordinator; (508) 289-7173; fax (508) 289-7931; researchawards@mbl.edu or jearly@mbl.edu
*Internet:* hermes.mbl.edu/research/summer/awards_general.html
*Sponsor:* Marine Biological Laboratory
7 MBL Street
Woods Hole, MA 02543

**MBL H. Keffer Hartline and Edward F. MacNichol, Jr. Fellowships** 1563
Marine Biological Laboratory (MBL) summer researchers come from all over the world. Visiting researchers find an infrastructure and an informal, interactive scientific community that allows them to launch into research as soon as they arrive. Research awards are available to these visiting researchers. The Hartline and MacNichol Fellowships were established in memory of Dr. Hartline and his impact on sensory science. Dr. Hartline came to the MBL as a young scientist, initiated his pioneering studies of the eye, and continued this research at the MBL for many years. These Fellowships support scientists carrying out research in sensory neurobiology. The MBL Research Awards provide costs for research and housing, and also enable awardees to benefit from the intellectual and interactive environment of the scientific community at the MBL. Funding requires a minimum stay of six weeks at the MBL. Awards can be used to support independent research, collaborative research, or in affiliation with the discovery courses. Candidates may fill out applications at the MBL online process.
*Requirements:* Research proposed in the resident program or courses requires pre-approval of the host lab and program and the course directors, respectively, in addition to the CASO office.
*Geographic Focus:* All States, All Countries
*Date(s) Application is Due:* Dec 15
*Contact:* Julie Early, Research Award Coordinator; (508) 289-7173; researchawards@mbl.edu or jearly@mbl.edu
*Internet:* hermes.mbl.edu/research/summer/awards_general.html
*Sponsor:* Marine Biological Laboratory
7 MBL Street
Woods Hole, MA 02543

**MBL Herbert W. Rand Summer Fellowship** 1564
Marine Biological Laboratory (MBL) summer researchers come from all over the world. Visiting researchers find an infrastructure and an informal, interactive scientific community that allows them to launch into research as soon as they arrive. A variety of research awards are available to these visiting researchers. The Rand Fellowship includes a summer stipend, and funds or assists with such expenses as travel, supplies, and laboratory fees. The MBL Research Awards provide costs for research and housing, and also enable awardees to benefit from the intellectual and interactive environment of the scientific community at the MBL. Funding requires a minimum stay of six weeks at the MBL. Awards can be used to support independent research, collaborative research, or in affiliation with the discovery courses. Proposals for Research award support will be considered in, but are not limited to, the following fields of investigation: cell biology; development biology; ecology; evolution; microbiology; neurobiology; physiology; and tissue engineering. Candidates may fill out applications at the MBL online process.
*Requirements:* Research proposed in the resident program or courses requires pre-approval of the host lab and program and the course directors, respectively, in addition to the CASO office.
*Geographic Focus:* All States, All Countries
*Date(s) Application is Due:* Dec 15
*Contact:* Julie Early, Research Award Coordinator; (508) 289-7173; researchawards@mbl.edu or jearly@mbl.edu
*Internet:* hermes.mbl.edu/research/summer/awards_general.html
*Sponsor:* Marine Biological Laboratory
7 MBL Street
Woods Hole, MA 02543

**MBL James E. and Faith Miller Memorial Summer Fellowship** 1565
Marine Biological Laboratory (MBL) summer researchers come from all over the world. Visiting researchers find an infrastructure and an informal, interactive scientific community that allows them to launch into research as soon as they arrive. Research awards are available to these visiting researchers. The Miller Fellowship provides funds for summer research at the MBL in any relevant field of medical research. The MBL Research Awards provide costs for research and housing, and also enable awardees to benefit from the intellectual and interactive environment of the scientific community at the MBL. Funding requires a minimum stay of six weeks at the MBL. Awards can be used to support independent research, collaborative research, or in affiliation with the discovery courses. Proposals for Research award support will be considered in, but are not limited to, the following fields of investigation: cell biology; development biology; ecology; evolution; microbiology; neurobiology; physiology; and tissue engineering. Candidates may fill out applications at the MBL online process.
*Requirements:* Research proposed in the resident program or courses requires pre-approval of the host lab and program and the course directors, respectively, in addition to the CASO office.
*Geographic Focus:* All States, All Countries
*Date(s) Application is Due:* Dec 15
*Contact:* Julie Early; (508) 289-7173; researchawards@mbl.edu or jearly@mbl.edu
*Internet:* hermes.mbl.edu/research/summer/awards_general.html
*Sponsor:* Marine Biological Laboratory
7 MBL Street
Woods Hole, MA 02543

**MBL John M. Arnold Award** 1566
Marine Biological Laboratory (MBL) summer researchers come from all over the world. Visiting researchers find an infrastructure and an informal, interactive scientific community that allows them to launch into research as soon as they arrive. A variety of research awards are available to these visiting researchers. The John M. Arnold Awards funds a retired or senior researcher in any field of study compatible with the mission of the MBL. The MBL Research Awards provide costs for research and housing, and also enable awardees to benefit from the intellectual and interactive environment of the scientific community at the MBL. Funding requires a minimum stay of six weeks at the MBL. Awards can be used to support independent research, collaborative research, or in affiliation with the discovery courses. Proposals for Research award support will be considered in, but are not limited to, the following fields of investigation: cell biology; development biology; ecology; evolution; microbiology; neurobiology; physiology; and tissue engineering. Candidates may fill out applications at the MBL online process.
*Requirements:* Research proposed in the resident program or courses requires pre-approval of the host lab and program and the course directors, respectively, in addition to the CASO office.
*Geographic Focus:* All States, All Countries
*Date(s) Application is Due:* Dec 15
*Contact:* Julie Early; (508) 289-7173; researchawards@mbl.edu or jearly@mbl.edu
*Internet:* hermes.mbl.edu/research/summer/awards_general.html
*Sponsor:* Marine Biological Laboratory
7 MBL Street
Woods Hole, MA 02543

**MBL Laura and Arthur Colwin Summer Fellowships** 1567
Marine Biological Laboratory (MBL) summer researchers come from all over the world. Visiting researchers find an infrastructure and an informal, interactive scientific community that allows them to launch into research as soon as they arrive. Research awards are available to these visiting researchers. The Colwin Summer Fellowships provide support for independent investigators conducting research in the fields of cell and developmental biology. The MBL Research Awards provide costs for research and housing, and also enable awardees to benefit from the intellectual and interactive environment of the scientific community at the MBL. Funding requires a minimum stay of six weeks at the MBL. Candidates may fill out applications at the MBL online process.
*Requirements:* Research proposed in the resident program or courses requires pre-approval of the host lab and program and the course directors, respectively, in addition to the CASO office.
*Geographic Focus:* All States, All Countries
*Date(s) Application is Due:* Dec 15
*Contact:* Julie Early; (508) 289-7173; researchawards@mbl.edu or jearly@mbl.edu
*Internet:* hermes.mbl.edu/research/summer/awards_general.html
*Sponsor:* Marine Biological Laboratory
7 MBL Street
Woods Hole, MA 02543

**MBL Lucy B. Lemann Memorial Summer Fellowship** 1568
Marine Biological Laboratory (MBL) summer researchers come from all over the world. Visiting researchers find an infrastructure and an informal, interactive scientific community that allows them to launch into research as soon as they arrive. Research awards are available to these visiting researchers. Established in memory of Lucy B. Lemann, a Woods Hole resident, this Fellowship helps support the costs of summer research at the MBL. The MBL Research Awards provide costs for research and housing, and also enable awardees to benefit from the intellectual and interactive environment of the scientific community at the MBL. Funding requires a minimum stay of six weeks at the MBL. Awards can be used to support independent research, collaborative research, or in affiliation with the discovery courses. Proposals for Research award support will be considered in, but are not limited to, the following fields of investigation: cell biology; development biology; ecology; evolution; microbiology; neurobiology; physiology; and tissue engineering. Candidates may fill out applications at the MBL online process.
*Requirements:* Research proposed in the resident program or courses requires pre-approval of the host lab and program and the course directors, respectively, in addition to the CASO office.
*Geographic Focus:* All States, All Countries
*Date(s) Application is Due:* Dec 15
*Contact:* Julie Early; (508) 289-7173; researchawards@mbl.edu or jearly@mbl.edu
*Internet:* hermes.mbl.edu/research/summer/awards_general.html
*Sponsor:* Marine Biological Laboratory
7 MBL Street
Woods Hole, MA 02543

**MBL M.G.F. Fuortes Summer Fellowships** 1569
Marine Biological Laboratory (MBL) summer researchers come from all over the world. Visiting researchers find an infrastructure and an informal, interactive scientific community that allows them to launch into research as soon as they arrive. Research awards are available to these visiting researchers. Established in memory of M. G. F. Fuortes, a renowned neurophysiologist, the Fuortes Summer Fellowships are available to young investigators working in the neurosciences. The MBL Research Awards provide costs for research and housing, and also enable awardees to benefit from the intellectual and interactive environment of the scientific community at the MBL. Funding requires a minimum stay of six weeks at the MBL. Awards can be used to support independent research, collaborative research, or in affiliation with the discovery courses. Proposals for Research award support will be considered in, but are not limited to, the following fields of investigation: cell biology; development biology; ecology; evolution; microbiology; neurobiology; physiology; and tissue engineering. Candidates may fill out applications at the MBL online process.
*Requirements:* Research proposed in the resident program or courses requires pre-approval of the host lab and program and the course directors, respectively, in addition to the CASO office.
*Geographic Focus:* All States, All Countries

*Date(s) Application is Due:* Dec 15
*Contact:* Julie Early; (508) 289-7173; researchawards@mbl.edu or jearly@mbl.edu
*Internet:* hermes.mbl.edu/research/summer/awards_general.html
*Sponsor:* Marine Biological Laboratory
7 MBL Street
Woods Hole, MA 02543

**MBL Nikon Summer Fellowship** **1570**
Marine Biological Laboratory (MBL) summer researchers come from all over the world. Visiting researchers find an infrastructure and an informal, interactive scientific community that allows them to launch into research as soon as they arrive. Research awards are available to these visiting researchers. A summer Fellowship at the MBL is available from Nikon, Inc. to a young investigator doing research in an area of biology in which she or he can make extensive use of advanced microscopy or micro-manipulation systems. This Fellowship includes a summer laboratory, housing, and a budget for incidental expenses, equipment rental, and supplies. Requests will also be considered for collaborative research projects at the MBL during the off-season period. The MBL Research Awards provide costs for research and housing, and also enable awardees to benefit from the intellectual and interactive environment of the scientific community at the MBL. Funding requires a minimum stay of six weeks at the MBL. Awards can be used to support independent research, collaborative research, or in affiliation with the discovery courses. Proposals for Research award support will be considered in, but are not limited to, the following fields of investigation: cell biology; development biology; ecology; evolution; microbiology; neurobiology; physiology; and tissue engineering. Candidates may fill out applications at the MBL online process.
*Requirements:* Research proposed in the resident program or courses requires pre-approval of the host lab and program and the course directors, respectively, in addition to the CASO office.
*Geographic Focus:* All States, All Countries
*Date(s) Application is Due:* Dec 15
*Contact:* Julie Early; (508) 289-7173; researchawards@mbl.edu or jearly@mbl.edu
*Internet:* hermes.mbl.edu/research/summer/awards_general.html
*Sponsor:* Marine Biological Laboratory
7 MBL Street
Woods Hole, MA 02543

**MBL Plum Foundation John E. Dowling Fellowships** **1571**
Marine Biological Laboratory (MBL) summer researchers come from all over the world. Visiting researchers find an infrastructure and an informal, interactive scientific community that allows them to launch into research as soon as they arrive. Research awards are available to these visiting researchers. The Dowling Fellowships help underwrite the costs of room and board, supplies, and laboratory space rental for outstanding applicants pursuing research on the neurobiology of vision and retinal research. The MBL Research Awards provide costs for research and housing, and also enable awardees to benefit from the intellectual and interactive environment of the scientific community at the MBL. Funding requires a minimum stay of six weeks at the MBL. Awards can be used to support independent research, collaborative research, or in affiliation with the discovery courses. Proposals for Research award support will be considered in, but are not limited to, the following fields of investigation: cell biology; development biology; ecology; evolution; microbiology; neurobiology; physiology; and tissue engineering. Candidates may fill out applications at the MBL online process.
*Requirements:* Research proposed in the resident program or courses requires pre-approval of the host lab and program and the course directors, respectively, in addition to the CASO office.
*Geographic Focus:* All States, All Countries
*Date(s) Application is Due:* Dec 15
*Contact:* Julie Early; (508) 289-7173; researchawards@mbl.edu or jearly@mbl.edu
*Internet:* hermes.mbl.edu/research/summer/awards_general.html
*Sponsor:* Marine Biological Laboratory
7 MBL Street
Woods Hole, MA 02543

**MBL Robert Day Allen Summer Fellowship** **1572**
Marine Biological Laboratory (MBL) summer researchers come from all over the world. Visiting researchers find an infrastructure and an informal, interactive scientific community that allows them to launch into research as soon as they arrive. Research awards are available to these visiting researchers. The Robert Day Allen Fellowship honors the memory of Professor Allen, a distinguished investigator of cell motility, and one of the developers of video-enhanced optical microscopy with differential interference contrast. The Fellowship funds a young investigator committed to research on cell motility and cytoarchitecture. The MBL Research Awards provide costs for research and housing, and also enable awardees to benefit from the intellectual and interactive environment of the scientific community at the MBL. Funding requires a minimum stay of six weeks at the MBL. Candidates may fill out applications at the MBL online process.
*Requirements:* Research proposed in the resident program or courses requires pre-approval of the host lab and program and the course directors, respectively, in addition to the CASO office.
*Geographic Focus:* All States, All Countries
*Date(s) Application is Due:* Dec 15
*Contact:* Julie Early; (508) 289-7173; researchawards@mbl.edu or jearly@mbl.edu
*Internet:* hermes.mbl.edu/research/summer/awards_general.html
*Sponsor:* Marine Biological Laboratory
7 MBL Street
Woods Hole, MA 02543

**MBL Stephen W. Kuffler Summer Fellowships** **1573**
Marine Biological Laboratory (MBL) summer researchers come from all over the world. Visiting researchers find an infrastructure and an informal, interactive scientific community that allows them to launch into research as soon as they arrive. Research awards are available to investigators wishing to do visiting research at the MBL. The Stephen W. Kuffler Fellowships are intended to encourage the career development of promising young investigators by helping to support them in the intense intellectual atmosphere of the MBL for the summer. It covers part of the costs of laboratory rental, housing, and other personal expenses. The MBL Research Awards provide costs for research and housing, and also enable awardees to benefit from the intellectual and interactive environment of the scientific community at the MBL. Funding requires a minimum stay of six weeks at the MBL. Awards can be used to support independent research, collaborative research, or in affiliation with the discovery courses. Proposals for Research award support will be considered in, but are not limited to, the following fields of investigation: cell biology; development biology; ecology; evolution; microbiology; neurobiology; physiology; and tissue engineering. The application is available at the MBL website.
*Requirements:* Research proposed in the resident program or courses requires pre-approval of the host lab and program and the course directors, respectively, in addition to the CASO office.
*Geographic Focus:* All States, All Countries
*Date(s) Application is Due:* Dec 15
*Contact:* Julie Early; (508) 289-7173; researchawards@mbl.edu or jearly@mbl.edu
*Internet:* hermes.mbl.edu/research/summer/awards_general.html
*Sponsor:* Marine Biological Laboratory
7 MBL Street
Woods Hole, MA 02543

**MBL William Townsend Porter Summer Fellowships for Minority Investigators** **1574**
Marine Biological Laboratory (MBL) summer researchers come from all over the world. Visiting researchers find an infrastructure and an informal, interactive scientific community that allows them to launch into research as soon as they arrive. Research awards are available to these visiting researchers. The Porter Foundation offers fellowship support for young scientists (undergraduates, senior graduate students, and postdoctoral trainees) who are from an under-represented minority group, and are U.S. citizens or permanent residents to do summer research with senior investigators at the MBL. The Fellowship contributes toward the expenses at the MBL during the summer. The MBL Research Awards provide costs for research and housing, and also enable awardees to benefit from the intellectual and interactive environment of the scientific community at the MBL. Funding requires a minimum stay of six weeks at the MBL. Awards can be used to support independent research, collaborative research, or in affiliation with the discovery courses. Proposals for Research award support will be considered in, but are not limited to, the following fields of investigation: cell biology; development biology; ecology; evolution; microbiology; neurobiology; physiology; and tissue engineering. Candidates may fill out applications at the MBL online process.
*Requirements:* Research proposed in the resident program or courses requires pre-approval of the host lab and program and the course directors, respectively, in addition to the CASO office.
*Geographic Focus:* All States, All Countries
*Date(s) Application is Due:* Dec 15
*Contact:* Julie Early; (508) 289-7173; researchawards@mbl.edu or jearly@mbl.edu
*Internet:* hermes.mbl.edu/research/summer/awards_general.html
*Sponsor:* Marine Biological Laboratory
7 MBL Street
Woods Hole, MA 02543

**McCarthy Family Foundation Grants** **1575**
The McCarthy Family Foundation is a small private foundation established in 1988. Organized as a California public benefit nonprofit corporation, it operates from San Diego, California. The foundation's current program interest areas have been funded throughout its almost two decades history. These funding areas are: K-12 science education; HIV/AIDS research, education and direct services; assistance to homeless persons; and child abuse prevention and services for victims and families. The board will occasionally fund a special project outside these categorical areas and/or for regional, national or international charitable purposes. A small portion of the grantmaking budget is provided for director matching grants to encourage and amplify personal philanthropy by the foundation's board of directors. Proposals will be considered in the above general areas or for special project areas established by the Board. Proposals will only be accepted for programs within San Diego County. Multiple year proposals may be considered but it is not the foundation's intention to provide annual support. It is not normally the foundation's desire to be the sole source of funds for a project. Most grants are expected to be small, $5,000-$15,000, reflecting the foundation's limited budget although there are no fixed minimum or maximum amounts. Grant proposals received by March 15 will be considered for decision/funding in June. Grant proposals received by September 15 will be considered for decision/funding in December.
*Requirements:* Proposals will only be accepted for programs within San Diego County, California. All applicants should first submit a letter of inquiry, which can be submitted directly from the Foundation's web site. The letter (1-2 pages) must contain a brief statement describing the applicant and the need for funds with enough information for the foundation to determine whether or not the application falls within its program areas. Proposals submitted without an initial letter of inquiry will not be reviewed by the foundation. Applicants should briefly and clearly provide a statement of their needs and the specific request to the foundation, taking into account other possible sources of funding. Letters of inquiry will be acknowledged upon their receipt, but because the

foundation operates without a professional staff, a more detailed response may be delayed. Applicants who receive a favorable response to their initial inquiry will be invited to submit a grant application. The foundation accepts the Common Grant Application of San Diego Grantmakers, which can be downloaded from the Foundation's web site. Five copies of the application (proposal) should be submitted. Only one copy of required attachments need be submitted. The application should be signed by the organization's board chair or the executive director (or equivalent individuals).
*Restrictions:* The foundation does not consider grants for individuals, scholarship funds, or sectarian religious activities. Normally the foundation does not consider requests for general fundraising drives. It does not make grants intended directly or indirectly to support political candidates or to influence legislation.
*Geographic Focus:* California
*Date(s) Application is Due:* Mar 15; Sep 15
*Amount of Grant:* 500 - 250,000 USD
*Samples:* United Way of San Diego County, $10,000—general operating support; UCLA AIDS Institute, $50,000—stem cell research for HIV prevention; Elementary Institute of Science, $10,000—-science and technology youth education.
*Contact:* Rachel McCarthy Bender, President; (858) 485-0129; mail@mccarthyfamilyfdn.org
*Internet:* www.mccarthyfamilyfdn.org/guide.html
*Sponsor:* McCarthy Family Foundation
P.O. Box 27389
San Diego, CA 92198-1389

**McCombs Foundation Grants** **1576**
The McCombs Foundation, established in 1981, makes grants to Texas nonprofit charitable, philanthropic, educational, and benevolent organizations. Areas of interest include arts, athletics, education and higher education, historic preservation, medical research, philanthropy, recreation and sports, youth services, and voluntarism. There are no application deadlines or forms. Applicants should submit a letter of application stating a brief history of the organization, any available printed support materials, and a budget detail. Final notification takes approximately two weeks.
*Requirements:* Texas nonprofit organizations are eligible. Application form not required.
*Restrictions:* No grants are given to individuals.
*Geographic Focus:* Texas
*Amount of Grant:* 1,000 - 6,000,000 USD
*Samples:* University of Texas, Houston, Texas, $5,000,000 - Anderson Cancer Center; Saint Mary's Hall, San Antonio, Texas, $300,000 - general operations.
*Contact:* Gary Woods, Treasurer; (210) 821-6523
*Sponsor:* McCombs Foundation
755 East Mulberry, Suite 600
San Antonio, TX 78212

**McCune Charitable Foundation Grants** **1577**
The Marshall L. and Perrine D. McCune Charitable Foundation is dedicated to enriching the health, education, environment, and cultural and spiritual life of New Mexicans. The Foundation engages in proactive grantmaking that seeks to foster positive social change. Specifically, the Foundation funds projects that benefit New Mexico in the areas of arts, economic development, education, environment, health, and social services. It is working to stimulate economic diversity, nurture sustainability, and bridge the economic gaps that exist in our communities with the aim of creating wealth for all New Mexicans.
*Requirements:* Grants can be awarded to qualified 501(c)3 nonprofit organizations, federally recognized Indian tribes, public schools, and governmental agencies. Applications are available online via the Foundation's website.
*Restrictions:* Grants are not awarded to individuals or to support endowments.
*Geographic Focus:* New Mexico
*Date(s) Application is Due:* Sep 30
*Amount of Grant:* Up to 150,000 USD
*Contact:* Norty Kalishman, M.D., Program Director; (505) 983-8300; fax (505) 983-7887; mccune@nmmccune.org
*Internet:* www.nmmccune.org/
*Sponsor:* McCune Charitable Foundation
345 East Alameda Street
Santa Fe, NM 87501

**McCune Foundation Civic Grants** **1578**
The McCune Foundation's grants are assigned to one of four program areas, including: education, human services, humanities, and civic. In the area of civic affairs, the Foundation focuses new job growth and generating community wealth, support of core community assets, and quality of life. Community development, public amenities, conservation, and overall regional grants are made in this program area. The civic agenda supports capital projects, research and development, organizational capacity building, and programming with the following strategic priorities: spur investment in Pittsburgh's urban core and metropolitan area; leverage public and private dollars for broad-based support of community assets; revitalize Pittsburgh's neighborhoods by changing blighted market conditions; support broad based community assets improving the livability of the region; and develop a New Economy in Pittsburgh that will return greater prosperity to the region.
*Requirements:* The foundation supports 501(c)3 organizations in southwestern Pennsylvania and throughout the country, with emphasis on the Pittsburgh area. This area includes the following counties: Allegheny, Beaver, Butler, Armstrong, Westmoreland, and Washington. To apply, an organization should send a brief (2 to 3 page) initial inquiry, preferably using the Foundation's website. The letter should contain: project overview - describe what the proposed efforts are intended to achieve for the region as well as for the organization; what activities/actions are planned to meet the stated goals; project timeline; resources required - total cost of the project; anticipated income, including private and public funders; amount of funding requested; IRS 501(c)3 determination letter - attach a copy (either scanned via email or hard copy via regular mail); and a copy of the organization's latest audit (either scanned via email or hard copy via regular mail).
*Restrictions:* Grants are not awarded to individuals or for general operating purposes or loans. Unsolicited proposals from outside the funding area are not accepted.
*Geographic Focus:* All States
*Amount of Grant:* 1,000 - 1,000,000 USD
*Samples:* University of Pittsburgh, Center for Social and Urban Research, Pittsburgh, Pennsylvania, $10,000 - symposium for reducing youth violence; Focus on Renewal Sto-Rox Neighborhood Association, McKees Rocks, Pennsylvania, $175,000 - capacity building; Metro Dallas Homeless Alliance, Dallas, Texas, $300,000 - bridge project.
*Contact:* Henry S. Beukema, Executive Director; (412) 644-8779; fax (412) 644-8059; info@mccune.org
*Internet:* www.mccune.org/foundation:Website,mccune,grants
*Sponsor:* McCune Foundation
750 Sixth PPG Place
Pittsburgh, PA 15222

**McCune Foundation Education Grants** **1579**
The McCune Foundation's grants are assigned to one of four program areas, including: education, human services, humanities, and civic. In the area of education, the Foundation promotes excellence in the institutions of higher education, independent elementary and secondary schools, and ancillary education programs (general education). The education program supports capital projects, research and development, financial aid, programming, and endowment to accomplish the following strategic priorities: leverage university research capacity to enhance the region's competitive advantage in economic development; use universities as anchors for community revitalization strategies; increase access to independent schools and higher education; recruit and retain exceptional faculty; and develop and deliver quality enrichment programs for school aged children, particularly with a focus on math, science and reading.
*Requirements:* The foundation supports 501(c)3 organizations in southwestern Pennsylvania and throughout the country, with emphasis on the Pittsburgh area. This area includes the following counties: Allegheny, Beaver, Butler, Armstrong, Westmoreland, and Washington. To apply, an organization should send a brief (2 to 3 page) initial inquiry, preferably using the Foundation's website. The letter should contain: project overview - describe what the proposed efforts are intended to achieve for the region as well as for the organization; what activities/actions are planned to meet the stated goals; project timeline; resources required - total cost of the project; anticipated income, including private and public funders; amount of funding requested; IRS 501(c)3 determination letter - attach a copy (either scanned via email or hard copy via regular mail); and a copy of the organization's latest audit (either scanned via email or hard copy via regular mail).
*Restrictions:* Grants are not awarded to individuals or for general operating purposes or loans. Unsolicited proposals from outside the funding area are not accepted.
*Geographic Focus:* All States
*Amount of Grant:* 1,000 - 1,000,000 USD
*Samples:* A Wilson Center for African American Culture, Pittsburgh, Pennsylvania, $10,000 - access and services for artists' programs; Airlift Research Foundation, Sharpsburg, Pennsylvania, $10,000 - leadership excellence program; Pennsylvania Institute for Conservation Education, Elysburg, Pennsylvania, $10,000 - wildlife leadership academy.
*Contact:* Henry S. Beukema; (412) 644-8779; fax (412) 644-8059; info@mccune.org
*Internet:* www.mccune.org/foundation:Website,mccune,grants
*Sponsor:* McCune Foundation
750 Sixth PPG Place
Pittsburgh, PA 15222

**McCune Foundation Human Services Grants** **1580**
The McCune Foundation's grants are assigned to one of four program areas, including: education, human services, humanities, and civic. In the area of human services, the Foundation focuses on health, social services, and community improvement. The interdisciplinary approach to grantmaking in this program area works with community-based and regional institutions to address pressing community needs by supporting the critical work of existing programs, as well as new initiatives and programs that seek to find and ameliorate the root causes of community distress and disinvestment. The human services area supports capital projects, research and development, organizational capacity building, and programming with the following strategic priorities: transfer education and research assets into economic opportunities for the region; leverage public and private dollars for broad-based support of community assets; promote self-sufficiency of residents; increase social and economic stability of communities and the region; and test and support effective prevention programs.
*Requirements:* The foundation supports 501(c)3 organizations in southwestern Pennsylvania and throughout the country, with emphasis on the Pittsburgh area. This area includes the following counties: Allegheny, Beaver, Butler, Armstrong, Westmoreland, and Washington. To apply, an organization should send a brief (2 to 3 page) initial inquiry, preferably using the Foundation's website. The letter should contain: project overview - describe what the proposed efforts are intended to achieve for the region as well as for the organization; what activities/actions are planned to meet the stated goals; project timeline; resources required - total cost of the project; anticipated income, including private and public funders; amount of funding requested; IRS 501(c)3 determination letter - attach a copy (either scanned via email or hard copy via regular mail); and a copy of the organization's latest audit (either scanned via email or hard copy via regular mail).
*Restrictions:* Grants are not awarded to individuals or for general operating purposes or loans. Unsolicited proposals from outside the funding area are not accepted.

*Geographic Focus:* All States
*Amount of Grant:* 1,000 - 1,000,000 USD
*Samples:* Hosanna House, Wilkinsburg, Pennsylvania, $500,000 - operations and programs; Mount Lebanon Unites Presbyterian Church, Pittsburgh, Pennsylvania, $7,900 - moving forward action plan; Dallas Children's Advocacy Center, Dallas, Texas, $50,000 - forensic interview wing capital campaign.
*Contact:* Henry S. Beukema; (412) 644-8779; fax (412) 644-8059; info@mccune.org
*Internet:* www.mccune.org/foundation:Website,mccune,grants
*Sponsor:* McCune Foundation
750 Sixth PPG Place
Pittsburgh, PA 15222

**MCEAS Barra Postdoctoral Fellowship** **1581**
Specialists working in any field of early American studies are encouraged to apply, though the center is particularly interested in projects that utilize Philadelphia-area libraries and archives. The fellow will receive a starting stipend of at least $40,000; health insurance; private office space in the Center's building at the northeastern gateway to the University of Pennsylvania's historic campus; library, computer, and other privileges at the university; and access to the Philadelphia area's manuscript, rare book and museum collections. Limited travel and research funding is available. The Barra Postdoctoral Fellow will receive a starting stipend of $43,000, health insurance, and modest funds for travel and research. Barra Sabbatical Fellow will receive a grant of $46,000, with no other salary or benefits, on the assumption that additional support will be provided by his or her home institution during the fellow's period in residence at the McNeil Center. Applications for either fellowship must be submitted on line.
*Requirements:* The competition is open to both junior and senior scholars, but all candidates must have the PhD in hand by the deadline date.
*Geographic Focus:* All States
*Date(s) Application is Due:* Nov 3
*Amount of Grant:* 46,000 USD
*Contact:* Daniel K. Richter, Richard S. Dunn Director; (215) 898-9251; fax (215) 573-3391; drichter@history.upenn.edu
*Internet:* www.mceas.org/postdoctoralfellowship.htm
*Sponsor:* McNeil Center for Early American Studies
3355 Woodland Walk
Philadelphia, PA 19104-4531

**MCEAS Dissertation Fellowships** **1582**
The McNeil Center was established to marshal the vast scholarly and archival resources of ten area institutions in an interdisciplinary research program focused on the history and culture of the United States to 1850, to increase scholarly and public awareness of the heritage of the Delaware Valley. Center offers several pre-doctoral dissertation fellowships each year; most are for a term of nine months, beginning 1 September. Advanced graduate students from any PhD-granting institution who are in the dissertation research or writing stage are eligible to compete for these fellowships, which are open to scholars in any discipline for projects focusing on North America and the Caribbean before 1850. Proposals reliant on research in Philadelphia-area archives and libraries are especially welcome.
*Requirements:* Fellowships are available to advanced graduate students from any PhD-granting institution who are in the dissertation research or writing stage.
*Geographic Focus:* All States
*Date(s) Application is Due:* Feb 2
*Amount of Grant:* 18,000 - 45,000 USD
*Contact:* Daniel K. Richter, Richard S. Dunn Director; (215) 898-9251; fax (215) 573-3391; drichter@history.upenn.edu
*Internet:* www.mceas.org/dissertationfellowships.shtml
*Sponsor:* McNeil Center for Early American Studies
3355 Woodland Walk
Philadelphia, PA 19104-4531

**McGraw-Hill Companies Community Grants** **1583**
McGraw-Hill focuses its charitable efforts on its goal of financial capability for all. The Corporation partners with nonprofit organizations to help individuals gain the necessary knowledge to make smart savings, credit and spending decisions. McGraw-Hill also extends its support to organizations that share the Corporation's dedication to the arts and culture, education and health and human services. The company gives priority consideration to organizations and projects that: promote and support excellence in education and learning, with a primary emphasis on financial literacy; further financial literacy in the communities and markets where the many diverse businesses of The McGraw-Hill Companies operate; utilize unique applications of innovative and developing technologies; extend their reach globally; can be evaluated and can serve as models elsewhere; are staffed and administered by people with demonstrated competence and experience in their fields; have been determined tax-exempt 501(c)3 organizations and qualify as public charities under IRS rules, or are the local country-specific equivalent of a charitable nonprofit organization.
*Requirements:* Proposals must include the following: a program narrative (no more than three pages); a detailed organization overview; a concise description of the grant's purpose; and all required attachments. If the request is considered eligible, a meeting may be arranged with the Corporate Responsibility and Sustainability staff. On-site visits may also be made. Proposals are accepted at any time, with funding decisions made quarterly.
*Restrictions:* McGraw-Hill does not support courtesy advertising, pledges for a walk-a-thon, or similar activities. Funding is also not available for the following: institutions and agencies clearly outside the company's primary geographic concerns and interests; libraries and schools of higher education and K-12; political activities or organizations established to influence legislation; sectarian or religious organizations; member-based organizations, i.e., fraternities, labor, veterans, athletic, social clubs; individuals; endowment funds; or loans. Grants are not renewed automatically. Requests for renewed support must be submitted each year.
*Geographic Focus:* All States
*Contact:* Susan Wallman; (212) 512-6480; fax (212) 512-3611; susan_wallman@mcgraw-hill.com
*Internet:* www.mcgraw-hill.com/site/cr/community/giving#section_2
*Sponsor:* McGraw-Hill Companies
1221 Avenue of the Americas, 20th Floor
New York, NY 10020-1095

**McLean Contributionship Grants** **1584**
Originally established in 1951 as The Bulletin Contributionship for charitable, educational and scientific purposes, the Contributionship became The McLean Contributionship on May 1, 1980 when the association of the McLean family with the Bulletin ended. Independent Publications, also owned by the McLean family, continues as the main financial supporter of the Contributionship. The Contributionship favors projects that stimulate a better understanding of the natural environment and encourage the preservation of its important features; encourage more compassionate and cost-effective care for the ill and aging in an atmosphere of dignity and self-respect; or promote education, medical, scientific, or on occasion, cultural developments that enhance quality of life. In addition, the Trustees from time to time support projects which motivate promising young people to assess and develop their talents despite social and economic obstacles or encourage those in newspaper and related fields to become more effective and responsible in helping people better understand how events in their communities and around the world affect them. The Trustees meet several times a year. The Contribution accepts and processes applications for grants throughout the year. Applications must be received at least six weeks before a meeting date. Interested organizations may view the annual-meeting schedule on the Application Procedure page at the grant website. The Contributionship accepts the common-grant-application form of Delaware Valley Grantmakers Association (link is provided on the Application Procedure web page). Application may also be made by letter (guidelines are included on the Application Procedure web page).
*Requirements:* Applicants must show evidence of tax-exempt status.
*Restrictions:* Geographic area is limited to the following locations: Greater Philadelphia area; Nashua, New Hampshire; Dubois Pennsylvania; and Central Florida. The Contributionship does not fund the costs or expenses of existing staff allocated to a project it is asked to support.
*Geographic Focus:* Florida, New Hampshire, Pennsylvania
*Amount of Grant:* 2,000 - 100,000 USD
*Samples:* A Little Taste of Everything, Philadelphia, Pennsylvania, $1,400 - towards maintaining and enhancing their living roof; Visiting Nurse Association of Greater Philadelphia, Philadelphia, Pennsylvania, $50,000 - toward the installation of the Horizon Enterprise Content Management System; National Constitution Center, Philadelphia, Pennsylvania - towards the upgrade of eight poorly-functioning exhibit cases housing artifacts uncovered during the Center's groundbreaking.
*Contact:* Sandra McLean, Executive Director; (610) 527-6330; fax (610) 527-9733
*Internet:* fdncenter.org/grantmaker/mclean
*Sponsor:* McLean Contributionship
945 Haverford Road, Suite A
Bryn Mawr, PA 19010-3814

**MDA Development Grant** **1585**
The association will consider an application for a research grant from a candidate who may be a member of a research team in the laboratory of a senior investigator under whose guidance the applicant will be given flexibility to conduct a neuromuscular disease research project. Awards are for one to three years. Applications must be requested by the listed application deadlines. Guidelines are available online. Development grants are a maximum of $60,000 per year.
*Requirements:* To be eligible for a Development Grant, an applicant must: hold a Doctor of Medicine (M.D.), Doctor of Philosophy (Ph.D.), Doctor of Science (D.Sc.) or equivalent degree (i.e. D.O.); be a member of a research team at an appropriate institution; be qualified to conduct a program of original research under the supervision of a Principal Investigator; have an acceptable research plan for a specific disease in MDA's program; have access to institutional resources necessary to conduct the proposed research project; and have eighteen (18) months of post-doctoral research laboratory training at the time of application, but no more than 5 years (60 months). A pre-proposal form must be submitted through proposalCENTRAL to formally request an application for an MDA research grant.
*Geographic Focus:* All States
*Date(s) Application is Due:* Jan 15; Jul 15
*Amount of Grant:* Up to 60,000 USD
*Contact:* Program Manager; (520) 529-2000; fax (520) 529-5454; grants@mdausa.org
*Internet:* www.mdausa.org/research/guidelines.html#developmentGrant
*Sponsor:* Muscular Dystrophy Association
3300 E. Sunrise Drive
Tucson, AZ 85718

**MDA Neuromuscular Disease Research Grants** **1586**
MDA supports research aimed at developing treatments for the muscular dystrophies and related diseases of the neuromuscular system. These are the muscular dystrophies (among which are Duchenne and Becker); motor neuron diseases (including ALS and SMA); the peripheral nerve disorders (CMT and Friedreich's ataxia); inflammatory myopathies; disorders of the neuromuscular junction; metabolic diseases of muscle as well as other myopathies. Proposals from applicants outside the United States will be considered for projects of highest priority to MDA and when, in addition to the applicant's having met

the requirements noted above the applicant's country of residence may not have adequate sources of financial support for biomedical research. Funding levels for primary Research Grants are unlimited. Awards are for either one, two or three years for all grant types.
*Requirements:* To be eligible to apply for an MDA research grant, an applicant must: hold a Doctor of Medicine (M.D.), Doctor of Philosophy (Ph.D.), Doctor of Science (D.Sc.) or equivalent degree (i.e. D.O.); be a professional or faculty member (Professor, Associate Professor or Assistant Professor) at an appropriate educational, medical or research institution; be qualified to conduct and mentor a program of original research within their own laboratory; assume both administrative and financial responsibility for the grant; and have access to institutional resources necessary to conduct the proposed research project. A pre-proposal form must be submitted through proposalCENTRAL to formally request an application for an MDA research grant.
*Geographic Focus:* All States
*Date(s) Application is Due:* Jan 15; Jul 15
*Contact:* Program Manager; (520) 529-2000; fax (520) 529-5454; grants@mdausa.org
*Internet:* static.mda.org/research/guidelines.html
*Sponsor:* Muscular Dystrophy Association
3300 E. Sunrise Drive
Tucson, AZ 85718

**MDARD Specialty Crop Block Grant Program-Farm Bill** **1587**
The Specialty Crop Block Grant Program-Farm Bill (SCBGP-FB) provides grants annually to assist in enhancing the competitiveness of specialty crops (fruits, vegetables, tree nuts, and nursery crops). Specialty crops are defined as fruits, vegetables, tree nuts, dried fruits, horticulture, and nursery crops (including floriculture). Examples of project areas that would qualify for funds include, but are not limited to: food safety; food security; nutrition; trade enhancement; education; research; promotion; marketing; plant health programs; "Buy Local" programs; increased consumption; enhanced innovation; improved efficiency of distribution system; environmental concerns and conservation; product development; and cooperative development. This is a cost reimbursement program.
*Requirements:* Eligible applicants must be non-profit organizations, local, state, and federal government entities, for-profit organizations, and universities. Organizations must be legal entities recognized by the IRS and must reside, and/or conduct their business or organization in Michigan and must be in good standing. Universities submitting applications for funding (research) must have the application submitted by the commodity group for which the research is being performed. Applications for grant funds should demonstrate how the project potentially impacts and produces measurable outcomes for the specialty crop industry rather than a single organization, institution, or individual. Multi-state organizations proposals are permitted.
*Restrictions:* Grant funds will not be awarded for projects that directly benefit a particular commercial product or provide a profit to a single organization, institution, or individual, as these projects do not enhance specialty crop industry competitiveness.
*Geographic Focus:* Michigan
*Date(s) Application is Due:* Apr 12
*Contact:* Nancy Nyquist; (517) 241-4381; fax (517) 335-0628; nyquistn@michigan.gov
*Internet:* www.michigan.gov/mdard/0,1607,7-125-1568_51684—-,00.html
*Sponsor:* Michigan Department of Agriculture and Rural Development
P.O. Box 30017
Lansing, MI 48909

**MDEQ Beach Monitoring Grants - Inland Lakes** **1588**
The Beach Monitoring Grants Program provides funding for more effective monitoring of public bathing beaches. With this funding the goals of the program are to: determine levels of bacteria in select bathing beach waters of the state; notify the public of the results; evaluate the effectiveness of state programs in preventing bacterial contamination of surface waters; and develop methods that can differentiate sources of bacterial contamination. Deadlines may vary from year to year based on funding priorities. Applicants will usually have 60 days to prepare and submit a grant application. Grant awards will be announced following department review. The department reserves 90 days for the review process. In the past, a total of $200,000 was available for grants of up to $25,000.
*Requirements:* Local county health departments, universities, and non-profit organizations are eligible to apply. The grants require a minimum of 25 percent local match (can include in-kind services). Grant applications are selected for funding by the department based on program priorities and the following criteria: location and frequency of beach use; history of beach monitoring; history of bacterial contamination; access to lab facilities to analyze samples promptly; ability to communicate results to the public in an efficient manner; ability to respond and take appropriate action in an event of beach contamination; proximity of the beach to known contamination source; availability of matching funds; innovativeness and feasibility of the proposed project; and, ability to reduce time delay between sampling and results.
*Geographic Focus:* Michigan
*Amount of Grant:* Up to 25,000 USD
*Contact:* Shannon Briggs, (517) 335-1214; briggss4@michigan.gov
*Internet:* www.michigan.gov/documents/deq/deq-essd-grantsloans-catalog_210643_7.pdf
*Sponsor:* Michigan Department of Environmental Quality
525 West Allegan Street
Lansing, MI 48909

**MDEQ Local Water Quality Monitoring Grants** **1589**
The water quality monitoring grants are meant to fund water quality monitoring activities to address local issues of concern and to identify new chemicals/issues that may be impacting the quality of Michigan's surface waters. Monitoring activities are limited to surface waters (i.e., rivers, streams, public access lakes, wetlands, the Great Lakes, etc.) and may include ambient chemical, biological, or physical monitoring activities, as well as the development of tools to help with the assessment of such data. Approximately $250,000 is available for water quality monitoring projects (non-beach) under the Clean Michigan Initiative-Clean Water Fund (CMI-CWF). Inland beach grants are available biennially. Disbursement of grant funds is done on a cost-reimbursement basis.
*Requirements:* Local units of government and nonprofit entities are eligible to receive grant funding. Nonprofit entities are those exempt from taxation under Section 501(c)3 of the Internal Revenue Code. Eligible entities generally include county, city, township, and village agencies, watershed and environmental action councils, universities, regional planning agencies, and other incorporated not-for-profit organizations. Applications must be received no later than 5:00 p.m. of the published deadline date. Eligible groups must have undergone a successful financial audit within the 24-month period immediately preceding the application; and, must not have had a grant revoked or terminated, or demonstrated an inability to manage a grant or meet the obligations in a project contract with the Department of Environmental Quality (DEQ), or its predecessor agencies, within the 24-month period immediately preceding the application. There is no maximum dollar amount for these grants. The dollar amount requested should be based upon what is needed to carry out the tasks identified in the project proposal. This grant requires a minimum 25 percent (%) match of public or private funds. Match may not include other CMI funds or Clean Water Act funds awarded as grants by the state. Project contracts cannot exceed two years.
*Restrictions:* Individuals, private lake associations (that are not tax exempt under Section 501(c)3 of the Internal Revenue Code), for-profit organizations, and federal and state agencies are not eligible to apply. Additionally, restoration efforts/education only projects and volunteer monitoring activities are ineligible for funding.
*Geographic Focus:* Michigan
*Date(s) Application is Due:* Mar 22
*Contact:* Dawn Roush, (517) 335-3307; roushd@michigan.gov
*Internet:* www.michigan.gov/deq/0,1607,7-135-3313_3686_3728—-,00.html
*Sponsor:* Michigan Department of Environmental Quality
525 West Allegan Street
Lansing, MI 48909

**MDEQ Scrap Tire Market Development Grants** **1590**
The primary purpose of the MDEQ Scrap Tire Market Development Grants program is to issue grants for projects that will result in the development of increased markets for scrap tires. Overall, the goal is to use available funding to promote increased markets for scrap tire material in order to ensure appropriate markets for this material exist in an effort to reduce the public health and environmental concerns associated with scrap tires. The applicant should submit a Letter of Interest (LOI) for approval prior to submitting a full application. LOIs are due by February 15, with full grants due by April 15.
*Requirements:* The applicant's project must demonstrate new or increased uses of scrap tires in manufactured products, such as modified asphalt, molded or extruded rubber products, or aggregate replacement materials; or the project must be for the research and development of methods to increase the use of scrap tires. The project must be located in Michigan, and the applicant must be registered to do business in Michigan. Scrap tires must be purchased from a Michigan processor or other Michigan generator of scrap tires. The applicant must submit documentation of the property owner's permission for the applicant to do business at that location. The applicant must not be in litigation with the state or any other public entity concerning compliance with the Part 169, Scrap Tires, or Part 115, Solid Waste Management, of the Natural Resources and Environmental Protection Act, 1994, PA 451, as amended. Typical projects must be completed within 18 months of the start date.
*Restrictions:* Grants are available for up to 50% of the cost of purchasing equipment, or for research and development costs. The applicant must commit to at least 50% of the total project amount, in cash. In-kind services are not allowed as part of the match, but will be considered during the application evaluation process.
*Geographic Focus:* Michigan
*Date(s) Application is Due:* Apr 15
*Contact:* Rhonda Oyer, (517) 373-4750; oyerr@michigan.gov
Michael Marshall, (517) 373-4738; marshallm7@michigan.gov
*Internet:* www.michigan.gov/documents/deq/deq-essd-grantsloans-catalog_210643_7.pdf
*Sponsor:* Michigan Department of Environmental Quality
525 West Allegan Street
Lansing, MI 48909

**Meadows Foundation Grants** **1591**
The Meadows Foundation seeks to assist people and institutions of Texas improve the quality and circumstances of life for themselves and future generations. Grants are made in five areas of interest: arts and culture, human services, health, education, and civic and public affairs. Within those three areas, the Foundation seeks to fund programs that: improve educational outcomes of Texas students; preserve and enhance the natural environment; and address the detection and treatment of mental illness.Projects should show the ability to continue and expand after the grant period.
*Requirements:* Grants are made to qualified organizations in Texas or programs benefiting Texas residents. Applicants must prove one or more of the following conditions in their proposal: Foundation support would be vital or catalytic to a proposed project's success; the project is well planned and the agency has the capacity to execute the plan; and financial support from other sources exists to ensure that the project will be implemented and continue after the grant period. Applications can be found at the website and should be submitted online.
*Restrictions:* Grants are not made to: individuals; for church or seminary construction; annual fund-raising events or drives; biomedical research; out-of-state performances or competition expenses; single artistic events or performances; or professional conferences and symposia.

*Geographic Focus:* Texas
*Amount of Grant:* 10,000 - 600,000 USD
*Samples:* Dallas Summer Musicals, Dallas, Texas, $600,000 - emergency working capital to complete the 2011 through 2013 seasons; Faith and Philanthropy Institute, Euless, Texas, $25,000 - for expanding training programs to faith and community-based organizations to improve service delivery; Breakthrough, Austin, Texas, $100,000 - expanding an education-support program that prepares low-income, minority students to graduate from high school and succeed in college.
*Contact:* Cynthia Cass, Grants Administrator; (214) 826-9431 or (800) 826-9431; fax (214) 827-7042; grants@mfi.org
Bruce H. Esterline, Vice President for Grants; (214) 826-9431
*Internet:* www.mfi.org
*Sponsor:* Meadows Foundation of Texas
3003 Swiss Avenue
Dallas, TX 75204-6090

**Medtronic Foundation Fellowships** **1592**
The U.S. based Medtronic Fellows Program works to prepare the next generation of scientific innovators. Through the Foundation's partnerships with preselected universities such as Johns Hopkins, Duke University, Stanford University, and Massachusetts Institute of Technology, it supports college and university scholars and fellows in science, engineering and health-related fields. Applicants should review the website for specific university participants and contact the individual universities for more information.
*Geographic Focus:* All States, United Kingdom
*Contact:* Kris Fortman, U.S. Programs Manager; (763) 514-4000 or (800) 633-8766; fax (763) 505-2648; kris.fortman@medtronic.com
*Internet:* www.medtronic.com/foundation/programs_msp.html
*Sponsor:* Medtronic Foundation
710 Medtronic Parkway
Minneapolis, MN 55432-5604

**Mellon/ACLS Community College Faculty Fellowships** **1593**
ACLS will award up to 26 fellowships in the competition year. The award carries a stipend of up to $40,000, which may be used to support any activities that advance the proposed research project. Examples include: salary replacement during the summer; course buyouts during the academic year; travel costs and registration fees for research and conferences; costs associated with organizing a conference, workshop, or event; fees related to publication or dissemination; stipends for undergraduate research assistants; costs for course materials (if one of the outcomes of the project is curricular); etc. In some cases, a limited portion of the funds may go to a fellow's institution to cover administrative costs associated with the project. Tenure of the fellowship may begin no earlier than July 1 of the year following the deadline and end no later than December 31 of the next year. Fellows will participate in a convening, held in partnership with the Community College Humanities Association, during which they will share their research with one another and connect with other humanities scholars. A stipend of up to $40,000 can be used for any activities that advance the proposed research project. The annual deadline for online submissions is September 25.
*Requirements:* Applicants must: be employed primarily as instructors at a two-year associate's degree-granting college, as defined by the Carnegie Classification of Institutions of Higher Education, and remaining so for the duration of the fellowship (part-time and adjunct instructors are welcome to apply, as long as community college teaching is their primary employment); and have an MA or PhD in the humanities or related social sciences that was conferred by the application deadline. Projects must: address a topic in the humanities or social sciences and/or teaching and learning in those disciplines in a higher education setting; employ predominantly humanistic approaches and qualitative/interpretive methodologies; and include substantial original research.
*Restrictions:* Applicants do not need to be appointed full-time and do not need to be on the tenure-track.
*Geographic Focus:* All States
*Date(s) Application is Due:* Sep 23
*Contact:* Valerie Popp; (212) 697-1505, ext. 153; fax (212) 949-8058; vpopp@acls.org
*Internet:* https://www.acls.org/Competitions-and-Deadlines/Mellon-ACLS-Community-College-Faculty-Fellowships
*Sponsor:* American Council of Learned Societies
633 Third Avenue, 8th Floor
New York, NY 10017-6795

**Mellon/ACLS Dissertation Completion Fellowships** **1594**
The Mellon/ACLS Dissertation Completion Fellowships assist graduate students in the humanities and related social sciences in the last year of Ph.D. dissertation writing. Applicants must be prepared to complete their dissertations within the period of their fellowship tenure or shortly thereafter. ACLS will award 65 Fellowships in this competition for a one-year term beginning between June and September for the following academic year. The Fellowship tenure may be carried out in residence at the Fellow's home institution, abroad, or at another appropriate site for the research. The total award of up to $43,000 includes a stipend plus additional funds of up to $8,000 for university fees and research support. Specific guidelines, the online application, a sample application, and suggestions for writing proposals are available at the website. The annual deadline for online submissions is October 28.
*Requirements:* Applicants must: be Ph.D. candidates in a humanities or social science department in the United States. Applicants from other departments may be eligible if their project is in the humanities or related social sciences, and their principal dissertation supervisor holds an appointment in a humanities field or related social science field. (Students completing master's degrees are not eligible, even if they are the terminal degree in the field); have all requirements for the Ph.D. except the dissertation completed before beginning fellowship tenure; and be no more than six years in the degree program (awardees can hold this Fellowship no later than their seventh year).
*Restrictions:* These Fellowships may not be held concurrently with any other major fellowship or grant.
*Geographic Focus:* All States
*Date(s) Application is Due:* Oct 23
*Amount of Grant:* 35,000 - 43,000 USD
*Contact:* Valerie Popp, Senior Program Officer; (212) 697-1505, ext. 153; fax (212) 949-8058; vpopp@acls.org
*Internet:* https://www.acls.org/Competitions-and-Deadlines/Mellon-ACLS-Dissertation-Completion-Fellowships
*Sponsor:* American Council of Learned Societies
633 Third Avenue, 8th Floor
New York, NY 10017-6795

**Mellon/ACLS Public Fellowships for Recent PhDs** **1595**
The American Council of Learned Societies Public Fellowships for Recent PhDs places up to 21 recent PhDs from the humanities and social sciences in two-year staff positions at partnering organizations in government and the nonprofit sector. Fellows will participate in the substantive work of these organizations and receive professional mentoring. Compensation will be competitive with new professional employees of the hosting organization and will include health insurance for the fellow. The program aims to demonstrate that the capacities developed in the advanced study of the humanities have wide application outside the academic setting. Specific guidelines and the online application are available at the website. The application deadline is March 11.
*Requirements:* ACLS seeks applications from who have received their degrees in the last three years and who aspire to careers in administration, management, and public service by choice rather than circumstance. Competitive applicants will have been successful in both academic and extra-academic experiences. Applicants must possess U.S. citizenship or permanent resident status; have a PhD in the humanities or humanistic social sciences; and not have applied to any other ACLS fellowship programs within the last year. The stipend is equal to $68,000 per year, health insurance coverage for the fellow, a relocation allowance, and up to $3,000 in professional development funds over the course of the fellowship. Prospective applicants should read through all the positions listed at the website and be ready to choose one when beginning the online application process. Applicants may apply to only one position.
*Geographic Focus* All States, District of Columbia, Guam, Marshall Islands, Northern Mariana Islands, Puerto Rico, U.S. Virgin Islands, American Samoa
*Date(s) Application is Due:* Mar 11
*Amount of Grant:* 68,000 - 85,000 USD
*Contact:* Valerie Popp, Senior Program Officer; (212) 697-1505, ext. 153; fax (212) 949-8058; vpopp@acls.org or publicfellows@acls.org
*Internet:* https://www.acls.org/Competitions-and-Deadlines/Mellon-ACLS-Public-Fellows-Program/Mellon-ACLS-Public-Fellows-Competition-for-Recent-PhDs
*Sponsor:* American Council of Learned Societies
633 Third Avenue, 8th Floor
New York, NY 10017-6795

**Mellon Post-Doctoral Fellowship in Philosophy** **1596**
The Philosophy Department invites applications for a two-year post-doctoral fellowship. Applications from candidates who will have their Ph.D. by the start date or who have completed their Ph.D. within the last two years may apply. The areas of specialty and competence are open. The successful candidate will be appointed for a term of two years and will carry a teaching load of one course each semester. Mellon Fellows are included in Department activities as regular junior faculty members. They work closely with a faculty mentor; they participate in Williams' Project in Effective Teaching; and they receive feedback on pedagogical skills and teaching effectiveness through standard evaluation procedures. The baseline stipend is $42,000 per year plus an allowance of $2,250 for research and professional travel expenses. Full benefits including health and dental insurance are provided.
*Requirements:* Applicants must be citizens or permanent residents of the United States, or expect to pursue a teaching career in the United States.
*Geographic Focus:* All States
*Contact:* Steven Gerrard, Philosophy Search; (413) 497-4351; fax (413) 597-3553; communication@williams.edu
*Internet:* dean-faculty.williams.edu/graduate-fellowships/mellon-post-doctoral-fellowship-in-philosophy/
*Sponsor:* Williams College
880 Main Street
Williamstown, MA 01267

**Mellon Post-Doctoral Fellowship in Sustainable Agriculture** **1597**
Williams College Program in Environmental Studies invites applications for a two-year post-doctoral fellowship in Sustainable Agriculture. Candidates who will have their Ph.D. by the start date or who have completed their Ph.D. within the last three years may apply. Candidates with training in the natural or social sciences who can contribute to the curriculum and vitality of an undergraduate program are sought. The position requires vision and initiative and a desire to join Williams' vibrant multidisciplinary community of environmental scholars, practitioners and activists. Relevant fields include the traditional disciplines of agriculture, biology, geography, anthropology, sociology, and economics. Applicants with interdisciplinary training in agrobiodiversity, soil

science, agroecology, or food studies are encouraged. An international focus is highly desirable. Williams places a high value on undergraduate teaching; consequently, strong preference is given to applicants who can demonstrate a capacity for excellent teaching. The successful candidate will be appointed for a term of two years and will carry a teaching load of one course each semester. The baseline stipend is $42,000 per year plus an allowance of $2,250 for research and professional travel expenses. Full benefits including health and dental insurance are provided.
*Requirements:* Applicants may send a cover letter, curriculum vitae, a two-page statement of research project and teaching interests, and three letters of reference.
*Geographic Focus:* All States
*Date(s) Application is Due:* Dec 1
*Contact:* Jennifer L. French; (413) 497-4351; fax (413) 597-3553; jfrench@williams.edu
*Internet:* dean-faculty.williams.edu/graduate-fellowships/mellon-post-doctoral-fellowship-in-sustainable-agriculture/
*Sponsor:* Williams College
880 Main Street
Williamstown, MA 01267

**Mericos Foundation Grants** **1598**
The Mericos Foundation primarily awards grants to organizations based in California with emphasis on Santa Barbara. Fields of interest include aging; animals/wildlife; environment; arts and arts education; child development; children and youth; elementary and secondary education; higher education; hospitals; medical care; medical research; libraries; and museums (art and natural history). Types of support include building/renovation, equipment, fellowships, general/operating support, matching/challenge support, and program development. Grants are usually initiated by the foundation. Interested organizations should contact the Vice-President to discuss their project.
*Requirements:* Nonprofit organizations in California are eligible to apply.
*Restrictions:* Grants are not made to individuals.
*Geographic Focus:* California
*Amount of Grant:* 15,000 - 200,000 USD
*Samples:* Music Academy of the West, Santa Barbara, California, $125,000; Friends of Independent Schools and Better Education, Tacoma, Washington, $25,000; Los Angeles Children's Chorus, Pasadena, California, $25,000.
*Contact:* Linda Blinkenberg, Vice President; (626) 441-5188; fax (626) 441-3672
*Sponsor:* Mericos Foundation
625 South Fair Oaks Avenue, Suite 360
South Pasadena, CA 91030-2630

**Merrick Foundation Grants** **1599**
The Merrick Foundation was established in 1948 by Ward S. Merrick, Sr., as a memorial to his father F.W. Merrick and at the request of F.W.'s wife, Elizabeth B. Merrick. The mission of the Foundation is to enhance the quality of life and the improvement of health for Oklahoma residents and their communities, with a primary emphasis on south central Oklahoma. With this goal in mind, the Foundation trustees are committed to furthering the philanthropic vision of Ward S. Merrick, Sr., by awarding grants to charitable organizations that foster independence and achievement, and that stimulate educational, economic, and cultural growth. Primary fields of interest include: the arts; higher education; human services; medical research; and youth services. The majority of awards are given for general operations. Most recent grants have ranged from $500 to $25,000. The grants committee will review all request letters received through the online grant application process and will email instructions on how to complete an application to qualifying charities. Grant applications will be due October 1 each year.
*Requirements:* The trustees invite proposals from 501(c)3 organizations located in Oklahoma. A letter of request is due by the August 15 deadline, summarizing the project for funding must be submitted through the online grant application process.
*Geographic Focus:* Oklahoma
*Date(s) Application is Due:* Oct 1
*Amount of Grant:* 500 - 25,000 USD
*Samples:* Arbuckle Life Solutions, Ardmore, Oklahoma, $5,000 - general operation support; CASA of Southern Oklahoma, Oklahoma City, Oklahoma, $10,000 - general operating support; Delta Upsilon Fraternity, Oklahoma City, Oklahoma, $5,000 - general operating support.
*Contact:* Randy Macon, Executive Director; 405-755-5571; fax (405) 755-0938; fwmerrick@foundationmanagementinc.com
*Internet:* https://www.foundationmanagementinc.com/foundations/merrick-foundation/
*Sponsor:* Merrick Foundation
2932 NW 122nd Street, Suite D
Oklahoma City, OK 73120-1955

**Meta and George Rosenberg Foundation Grants** **1600**
The Meta and George Rosenberg Foundation was established in California in 1991 in honor of Meta and George Rosenberg. Meta, who would become an Emmy-winning executive producer of the durable television series "The Rockford Files," married talent agent George "Rosey" Rosenberg in 1947, and the two of them soon began representing writers and actors. They also sold innovative series to television networks, including "Julia," starring Diahann Carroll as an African American nurse and single mother; "Hogan's Heroes," a comedy set in a World War II German prison camp; and "Ben Casey," among the first series showcasing medicine as drama. Throughout her career, Meta pursued an interest in photography, collecting the works of internationally known photographers such as Henri Cartier-Bresson, Ansel Adams, and Irving Penn, and aiming her own Leica camera at street scenes from Los Angeles to Paris. Currently, the Foundation's primary fields of interest include: arts and culture; children's diseases; elementary and secondary education; AIDS; and human services. Funding is concentrated in southern California, with some grants being awarded in the State of Iowa. The annual deadline for application submission is August 31. Most recent awards have ranged from $10,000 to $140,000.
*Geographic Focus:* California, Iowa
*Date(s) Application is Due:* Aug 31
*Amount of Grant:* 10,000 - 140,000 USD
*Samples:* Project Grad Los Angeles, North Hollywood, California, $8,400 - general operating support; University of Iowa Foundation, Iowa City, Iowa, $70,000 - general fund contribution; Engage the Art of Aging, Burbank, California, $30,000 - general operating support.
*Contact:* Wallace D. Franson, President and Director; (323) 634-2400
*Sponsor:* Meta and George Rosenberg Foundation
5900 Wilshire Boulevard, Suite 2300
Los Angeles, CA 90036-5050

**Meyer and Pepa Gold Family Foundation Grants** **1601**
Established in 1986 and formerly known as the Gold Family Foundation, the revamped Meyer and Pepa Gold Family Foundation offers grant support in New Jersey and New York. Its primary fields of interest include the arts, health organizations, and medical research agencies. The are no specified application forms required or annual deadlines. Interested parties should begin by contacting the Foundation directly, offering an overview of their program and budgetary needs. Recent grants, which predominately support general operations, range up to $35,000.
*Requirements:* Any 501(c)3 located in, or serving the residents of, New Jersey and/or New York are eligible to apply.
*Restrictions:* No grants are offered to individuals.
*Geographic Focus:* New Jersey, New York
*Amount of Grant:* Up to 35,000 USD
*Samples:* Alzheimer's Foundation of America, New York, New York, $35,000 - general operations; Center for Dementia, New York, New York, $35,000 - general ongoing costs and research; Museum of Jewish Heritage, New York, New York, $5,365 - general operations.
*Contact:* Pepa Gold, Trustee; (732) 229-0569 or (732) 660-1770
*Sponsor:* Meyer and Pepa Gold Family Foundation
P.O. Box 777
Oakhurst, NJ 07755

**Meyer and Stephanie Eglin Foundation Grants** **1602**
The Eglin Foundation supports programs in the Philadelphia area. Fields of interest include: cancer research; higher education; hospitals; Jewish agencies, synagogues, and federated giving programs; museums; music and the performing arts; and United Way programs. Applications are not required, and there are no deadlines. Organizations should submit a letter of intent.
*Requirements:* Along with the letter of intent, organizations should submit an annual report/audited financial statement/990; a 501(c)3 tax-exempt letter; descriptive literature about their organization; and a detailed description of the project, with amount requested.
*Geographic Focus:* Pennsylvania
*Contact:* Stephanie Eglin, President; (215) 496-9381
*Sponsor:* Meyer and Stephanie Eglin Foundation
Eglin Square Garage
Philadelphia, PA 19102

**Meyer Memorial Trust Responsive Grants** **1603**
Responsive Grants are awarded in the areas of human services; health; affordable housing; community development; conservation and environment; public affairs; arts and culture; and education. Funding ranges from $40,000 to $300,000, with grants periods from one to three years in length. Responsive grants help support many kinds of projects, including core operating support, building and renovating facilities, and strengthening organizations. There are two stages of consideration before Responsive Grants are awarded. Initial Inquiries are accepted at any time through MMT's online grants application. Applicants that pass initial approval are invited to submit full proposals. The full two-step proposal investigation usually takes five to seven months. Final decisions on Responsive Grants are made by trustees monthly, except in January, April and August. Additional information about the application process, along with the online application, is available at the website.
*Requirements:* Any 501(c)3 organization supporting the residents of Oregon are welcome to apply. Applicants must fall into one of these categories: a nonprofit agency recognized as tax-exempt by the Internal Revenue Service; a public educational institution; a government or recognized Tribal agency; or an organization that is requesting funding for a project that has a charitable, tax-exempt purpose. Further requirements include that the applicant must: be seeking funding for work that takes place within the state of Oregon; provide equal opportunity in leadership, staffing in service regardless of age, gender, race, ethnicity, sexual orientation, disability, national origin, political affiliation or religious belief; and not require attendance at or participation in religious/faith activities as a condition of service delivery nor require adherence to religious/faith beliefs as a condition of service or employment.
*Restrictions:* The Trust does not fund: work that does not have a significant impact in Oregon; direct grants, scholarships or loans to individuals; endowments; general fund drives, annual appeals, special events or, except in rare cases, conference sponsorships; Program Related Investments (loans and guarantees) to a specific project; elimination of operating deficits; medical research; animal welfare; hospital capital construction; projects/organizations that do not meet the Trust's nondiscrimination policy; earmarks for purposes of influencing legislation; or any expenditures that would violate Meyer's, or a grantee's, tax-exempt status.

*Geographic Focus:* Oregon
*Amount of Grant:* 40,000 - 300,000 USD
*Samples:* Crow's Shadow Institute, Pendleton, Oregon, $165,000 - for organizational support to help boost the programming capacity of this arts organization serving members of the Confederated Tribes of the Umatilla Indian Reservation; Immigration Counseling Service, Portland, Oregon, $66,500 - for outreach, education and legal services to support immigrant integration across Oregon; College Possible, Gresham, Oregon, $135,000 - to support college access and a success program for low-income Portland students.
*Contact:* Chanta Chhay, Grants Manager; (503) 228-5512; chanta@mmt.org
*Internet:* mmt.org/our-portfolios/
*Sponsor:* Meyer Memorial Trust
425 NW 10th Avenue, Suite 400
Portland, OR 97209

**MGM Resorts Foundation Community Grants** **1604**

The MGM Resorts Foundation invites proposals from nonprofit agencies providing direct services to people living in its communities. All of the funds allocated through the Foundation come from employee contributions and their desire to make a difference in the communities where they live and work. Foundation grant allocations are 100% employee-driven. The Foundation empowers MGM Resorts employees to choose to make direct contributions to the agency of their choice, or to contribute to the Community Grant Funds, which provides grants to nonprofits through an annual Request for Proposal (RFP) process. MGM Resorts Foundation grants are for a one-year period and do not automatically renew. Continued or expanded projects and programs (your organization is currently providing these services) can request a per year maximum grant of $65,000 in Southern Nevada; $10,000 in Northern Nevada, Mississippi and the Detroit, Michigan area. New projects and programs (your organization is not currently providing these services) can request a per year maximum grant of $35,000 in Southern Nevada; $5,000 in Northern Nevada, Mississippi and the Detroit, Michigan area.
*Requirements:* To receive a grant from the Foundation, your agency must meet the following *Requirements:* Operate as an IRS 501(c)3 organization and have been doing so for a minimum of 36 months; provide service within the regions MGM Resorts employees live, work and care for their families (Nevada, Mississippi, and the greater Detroit, Michigan area); your organization's administrative costs must be 25% or under; provide a human service; and, meet the MGM Resorts diversity policy: open to all people, without regard to race, color, creed, sex, sexual orientation, religion, disability, or national origin. Agencies must request funding for projects/programs that provide services in the following focus areas: Strengthening Neighborhoods (self-sufficiency, revitalization of communities); Strengthening Children (early childhood development, success in school, prevention / intervention); and, Strength in Difficult Times (recovery and counseling services). Proposals must be received by the Foundation by 5:00 pm of the deadline date.
*Restrictions:* The program does not support the following types of organizations or activities: projects/programs that are exclusively for medical research; public schools or privately funded / tuition-based schools; governmental entities; religious organizations that do not have 501(c)3 status; pass-through agencies (organizations whose staff does not provide direct client services but who allocate funding to subsequent organizations to provide projects/programs and services); sponsorship of special events and/or fundraising activities; capital campaigns or endowment funds; political issues, such as, election campaigns, issue endorsements, bill drafts or legislation reform; organizations that require clients to embrace specific beliefs or traditions; projects/programs that are exclusively recreational or athletic sponsorships; membership-based organizations without a sliding fee scale and scholarship system already in place.
*Geographic Focus:* Michigan, Mississippi, Nevada
*Date(s) Application is Due:* May 2
*Amount of Grant:* 5,000 - 65,000 USD
*Contact:* Shelley Gitomer, Vice President of Philanthropy & Community Engagement; (702) 692-9643; foundation@mgmresorts.com
*Internet:* www.mgmmirage.com/csr/community/foundation.aspx
*Sponsor:* MGM Resorts Foundation
3260 Industrial Road
Las Vegas, NV 89109

**MGN Family Foundation Grants** **1605**

The MGN Family Foundation makes grants to qualified 501(c)3 organizations specializing in the following areas: education; health care and medical research; children in need; armed service personnel. Areas of particular interest include: colleges, universities and private schools, examples would be: to fund a chair to provide lecturers in literature, philosophy or the arts, and provide scholarships; hospitals and clinics that specialize in excellent patient care and continuing medical research such as Memorial Sloan-Kettering Cancer Center and the Mayo and Cleveland Clinics are other examples, also under consideration would be hospice organizations that provide palliative care to the dying; organizations that support children in need whether due to emotional, physical abuse, neglect or disadvantaged circumstances. Support can be for basic necessities such as food, shelter, education, medical as well as, spiritual and emotional counseling; in light of recent events, this Foundation would like to offer assistance to our servicemen/ servicewomen and their families through those charities which give support to their unique needs. The examples provided above reflect the true mission and goals of the MGN Foundation. There should be no exclusions due to race or creed, provided all applicants have a strong moral base and core values in the areas of education, health care, welfare of children and service personnel and their families. The Foundation Board meets semi-annually in May and November. Applications are due by April 1st and October 1st. Application form is available online.
*Requirements:* Qualified 501(c)3 organizations specializing in the following areas: education; health care and medical research; children in need; armed service personnel. To apply, submit seven (7) sets of the following items: Grant application form completed, dated, and signed by the Chief Executive Officer or Chairman of the Board of the organization; list of Board of Directors; Financial Statement (audited if available), for the most recent complete fiscal year; copy of IRS 501(c)3 Determination Letter. If you wish, you may submit your proposal in a narrative format of not more than two pages in addition to the completed application form. Optional materials may be submitted but are not required (such as brochures discussing or depicting the activities of the organization).
*Restrictions:* Do not staple materials or place them in a bound notebook.
*Geographic Focus:* All States
*Date(s) Application is Due:* Apr 1; Oct 1
*Amount of Grant:* 1,000 - 10,000 USD
*Samples:* Agnes Scott College, $3,000—scholarship; Vermont National Guard Charitable Foundation, Inc., $3,000—wounded veteran program; Center for Molecular Medicine and Immunology/Garden State, $5,000—foster research for new technologies in the diagnosis, detection and treatment of cancer.
*Contact:* Pamela Nothstein, Grants Manager; (843) 937-4614; grantinquiries7@wachovia.com
*Internet:* www.mgnfamilyfoundation.org/
*Sponsor:* MGN Family Foundation
16 Broad Street (SC1000)
Charleston, SC 29401

**Microsoft Research Dissertation Research Grants** **1606**

Microsoft recognizes the value of diversity in computing. The Microsoft Research Dissertation Grant aims to increase the pipeline of diverse talent receiving advanced degrees in computing-related fields by providing a research funding opportunity for doctoral students from groups underrepresented in computing (women, African-Americans/Blacks, Latinos, American Indians/Alaskan Natives, Native Hawaiians/ Pacific Islanders, and/or people with disabilities). Grant recipients will receive funding up to $25,000 for the academic year to help them complete research as part of their doctoral thesis work. Microsoft will also arrange and pay for travel and accommodations to grant recipients to attend a two-day Microsoft Research workshop in Redmond, Washington, in the autumn. The annual deadline for applications is March 30.
*Requirements:* Applicants must be enrolled in your fourth year or beyond of PhD studies and self-identify with a group underrepresented in computing. PhD students must be enrolled at a university in the United States or Canada and doing dissertation work that relates to computing topics in which Microsoft Research has expertise.
*Restrictions:* Funding is for use only during their time in the PhD program; it cannot be used for support in a role past graduation, such as a postdoc or faculty position.
*Geographic Focus:* All States
*Date(s) Application is Due:* Mar 30
*Amount of Grant:* 25,000 USD
*Contact:* Dissertation Grant Director; (800) 642-7676; msrgrant@microsoft.com
*Internet:* https://www.microsoft.com/en-us/research/academic-program/dissertation-grant/
*Sponsor:* Microsoft Research
One Microsoft Way, 112/4012
Redmond, WA 98052-6399

**MIT Knight Science Journalism Fellowships** **1607**

The fellowships provide nine-month academic residency at the Massachusetts Institute of Technology, auditing courses, visiting laboratories, and attending twice-weekly seminars with leading scientists. The goal of the program is to increase exposure of the fellows to front-rank researchers and to deepen the fellows' knowledge of current issues in science and technology. The annual online application deadline is March 1. Fellows are selected by a panel of judges - usually five or six judges - after they send a fully completed application.
*Requirements:* Knight Science Journalism Fellowships are for mid-career journalists who demonstrate a high level of journalistic excellence and accomplishment, as well as a long-term commitment to their craft. Journalists from all countries compete on an equal basis.
*Restrictions:* Part-time writers or producers are not eligible.
*Geographic Focus:* All States, All Countries
*Date(s) Application is Due:* Mar 1
*Amount of Grant:* 55,000 USD
*Contact:* Bianca Sinausky; (617) 253-3442; fax (617) 258-8100; singleta@mit.edu
*Internet:* ksj.mit.edu/fellowships
*Sponsor:* Massachusetts Institute of Technology
77 Massachusetts Avenue, Room E19-623
Cambridge, MA 02139-4307

**MLA Aldo and Jeanne Scaglione Prize for a Translation of a Scholarly Study of Literature** **1608**

The Committee on Honors and Awards of the Modern Language Association invites translators and publishers to compete for the biennial Aldo and Jeanne Scaglione Prize for a Translation of a Scholarly Study of Literature. The prize, established by the Aldo and Jeanne Scaglione Endowment Fund, is awarded each odd-numbered year for an outstanding translation into English of a book-length work of literary history, literary criticism, philology, or literary theory. Entries will be accepted until May 1 of odd-numbered years. The winning author will be contacted in September, and a public announcement will be made in early December. The prize, which consists of a cash award and a certificate, will be presented to the winning author at the association's annual convention in January of even numbered years.
*Requirements:* Translators need not be members of the association.
*Geographic Focus:* All States

*Date(s) Application is Due:* May 1
*Amount of Grant:* 2,000 USD
*Samples:* Peimin Ni, Grand Valley State University, for Understanding the Analects of Confucius: A New Translation of Lunyu with Annotations, $2,000 (2017-18); Sylvia Adrian Notini, University of Bologna, for The Venetian Qur'an: A Renaissance Companion to Islam, $2,000 (2017-18); Alexandra Berlina, Düsseldorf, Germany, for Viktor Shklovsky: A Reader, $2,000 (2015-16).
*Contact:* Dennis Looney; (646) 576-5000; fax (646) 458-0030; awards@mla.org
Annie M. Reiser, Coordinator of Book Prizes and Administrative Assistant; (646) 576-5141 or (646) 576-5000; fax (646) 458-0030; awards@mla.org
*Internet:* https://www.mla.org/Resources/Career/MLA-Grants-and-Awards/Award-Submissions-and-Nominations/Competitions-for-MLA-Publication-Awards/Biennial-Prizes-with-Competitions-in-2021/Aldo-and-Jeanne-Scaglione-Prize-for-a-Translation-of-a-Scholarly-Study-of-Litera
*Sponsor:* Modern Language Association
85 Broad Street, Suite 500
New York, NY 10004-2434

**MLA Aldo and Jeanne Scaglione Prize for Comparative Literary Studies** 1609

The Committee on Honors and Awards of the Modern Language Association invites authors to compete for the annual Aldo and Jeanne Scaglione Prize for Comparative Literary Studies, to be awarded for an outstanding scholarly work in the field of comparative literary studies involving at least two literatures. For the current competition, the committee solicits entries of books written by current members of the association and published last year. Works of literary history, literary criticism, philology, and literary theory are eligible, as are works dealing with literature and other arts and disciplines, including cinema; books that are primarily translations will not be considered. The Scaglione Prize for Comparative Literary Studies, which consists of a cash award and a certificate, was established by the Aldo and Jeanne Scaglione Endowment Fund. Publishers may enter more than one title. Entries will be accepted until May 1 each year. The winning author will be contacted in September, and a public announcement will be made in early December. The prize, which consists of a cash award and a certificate, will be presented to the winning author at the association's annual convention in January.
*Requirements:* To enter a book into competition, send four copies and a letter identifying the work and confirming the author's membership in MLA. Entries may be sent at any time but must be received by the listed application deadline date.
*Restrictions:* Books that are primarily translations will not be considered. Open only to members of the association.
*Geographic Focus:* All States
*Date(s) Application is Due:* May 1
*Amount of Grant:* 2,000 USD
*Samples:* Seeta Chaganti, University of California, Davis, for Strange Footing: Poetic Form and Dance in the Late Middle Ages, $2,000 (2018); Siraj Ahmed, Graduate Center and Lehman College, City University of New York, for Archaeology of Babel: The Colonial Foundation of the Humanities, $2,000; Haun Saussy, University of Chicago, for The Ethnography of Rhythm: Orality and Its Technologies, $2,000.
*Contact:* Dennis Looney; (646) 576-5000; fax (646) 458-0030; awards@mla.org
Annie M. Reiser, Coordinator of Book Prizes and Administrative Assistant; (646) 576-5141 or (646) 576-5000; fax (646) 458-0030; awards@mla.org
*Internet:* https://www.mla.org/Resources/Career/MLA-Grants-and-Awards/Award-Submissions-and-Nominations/Competitions-for-MLA-Publication-Awards/Annual-Prizes-with-Competitions-in-2020/Aldo-and-Jeanne-Scaglione-Prize-for-Comparative-Literary-Studies
*Sponsor:* Modern Language Association
85 Broad Street, Suite 500
New York, NY 10004-2434

**MLA Aldo and Jeanne Scaglione Prize for French and Francophone Studies** 1610

The Committee on Honors and Awards of the Modern Language Association invites authors to compete for the annual Aldo and Jeanne Scaglione Prize for French and Francophone Studies, to be awarded for an outstanding scholarly work in the field of French or francophone linguistic or literary studies. For the current year competition, the committee solicits entries of books written by current members of the association and published last year. Works of literary history, literary criticism, philology, and literary theory are eligible for consideration. Entries will be accepted until May 1 each year. The winning author will be contacted in September, and a public announcement will be made in early December. The prize, which consists of a cash award and a certificate, will be presented to the winning author at the association's annual convention in January.
*Requirements:* To enter a book into competition, send four copies and a letter identifying the work and confirming the author's membership in the MLA. Entries may be sent at any time, but all must be received by the listed application deadline date.
*Restrictions:* Books that are primarily translations will not be considered.
*Geographic Focus:* All States
*Date(s) Application is Due:* May 1
*Amount of Grant:* 2,000 USD
*Samples:* Diana Holmes, University of Leeds, for Middlebrow Matters: Women's Reading and the Literary Canon in France since the Belle Epoque, $2,000 (2018); George Hoffmann, University of Michigan, Ann Arbor, for Reforming French Culture: Satire, Spiritual Alienation, and Connection to Strangers, $2,000; Andrew Joseph Counter, University of Oxford, for The Amorous Restoration: Love, Sex, and Politics in Early Nineteenth-Century France, $2,000; Maurice Samuels, Yale University, for The Right to Difference: French Universalism and the Jews, $2,000.
*Contact:* Dennis Looney; (646) 576-5000; fax (646) 458-0030; awards@mla.org
Annie M. Reiser, Coordinator of Book Prizes and Administrative Assistant; (646) 576-5141 or (646) 576-5000; fax (646) 458-0030; awards@mla.org
*Internet:* https://www.mla.org/Resources/Career/MLA-Grants-and-Awards/Award-Submissions-and-Nominations/Competitions-for-MLA-Publication-Awards/Annual-Prizes-with-Competitions-in-2020/Aldo-and-Jeanne-Scaglione-Prize-for-French-and-Francophone-Studies
*Sponsor:* Modern Language Association
85 Broad Street, Suite 500
New York, NY 10004-2434

**MLA Aldo and Jeanne Scaglione Prize for Italian Studies** 1611

The Committee on Honors and Awards of the Modern Language Association invites authors to enter the competition for the eleventh Aldo and Jeanne Scaglione Prize for Italian Studies. The prize, established by the Aldo and Jeanne Scaglione Endowment Fund, is awarded each odd-numbered year to the author of an outstanding scholarly book on any phase of Italian literature or culture or comparative literature involving Italian. This shall include works that study literary or cultural theory, science, history, art, music, society, politics, cinema, and linguistics, preferably but not necessarily relating other disciplines to literature. For the next competition, the committee solicits submission of works published in the current or upcoming even-numbered year. Entries will be accepted until May 1 of odd-numbered years. The winning author will be contacted in September, and a public announcement will be made in early December. The prize, which consists of a cash award and a certificate, will be presented to the winning author at the association's annual convention in January.
*Requirements:* To enter a work into competition, send four copies to the Aldo and Jeanne Scaglione Prize for Italian Literary Studies at MLA. Under separate cover, send a letter indicating the title submitted, the author, the date of publication, and affirmation that the author is a member in good standing of the association.
*Restrictions:* Open only to members of the Association.
*Geographic Focus:* All States
*Date(s) Application is Due:* May 1
*Amount of Grant:* 2,000 USD
*Samples:* Martina Piperno, KU Leuven, Belgium, for Rebuilding Post-revolutionary Italy: Leopardi and Vico's New Science, $2,000 (2018); Diego Pirillo, University of California, Berkeley, for The Refugee-Diplomat: Venice, England, and the Reformation, $2,000 (2018); Serenella Iovino, University of Turin, for Ecocriticism and Italy: Ecology, Resistance, and Liberation, $2,000.
*Contact:* Dennis Looney; (646) 576-5000; fax (646) 458-0030; awards@mla.org
*Internet:* https://www.mla.org/Resources/Career/MLA-Grants-and-Awards/Award-Submissions-and-Nominations/Competitions-for-MLA-Publication-Awards/Biennial-Prizes-with-Competitions-in-2021/Aldo-and-Jeanne-Scaglione-Prize-for-Italian-Studies
*Sponsor:* Modern Language Association
85 Broad Street, Suite 500
New York, NY 10004-2434

**MLA Aldo and Jeanne Scaglione Prize for Studies in Germanic Languages and Literatures** 1612

The Committee on Honors and Awards of the Modern Language Association invites authors to compete for the biennual Aldo and Jeanne Scaglione Prize for Studies in Germanic Languages and Literatures, to be awarded for an outstanding scholarly work on the linguistics or literatures of the Germanic languages, including Danish, Dutch, German, Icelandic, Norwegian, Swedish, and Yiddish. The prize, established by the Aldo and Jeanne Scaglione Endowment Fund, is awarded each even-numbered year. For the current prize, the committee solicits entries of books written by members of the association and published in the two years prior to the upcoming or current even-numbered year. Works of literary history, literary criticism, philology, and literary theory are eligible for consideration. Entries will be accepted until May 1 of even-numbered years year. The winning author will be contacted in September, and a public announcement will be made in early December. The prize, which consists of a cash award and a certificate, will be presented to the winning author at the association's annual convention in January in odd-numbered years.
*Requirements:* The committee solicits entries of books written by current members of the association that were published in the last two years. To enter a book into competition, send four copies and a letter identifying the work and confirming the author's membership in the MLA. Entries may be sent at any time but must be received by the listed application deadline date.
*Restrictions:* Books that are primarily translations will not be considered.
*Geographic Focus:* All States
*Date(s) Application is Due:* May 1
*Amount of Grant:* 2,000 USD
*Samples:* Sabine Hake, University of Texas, Austin, for The Proletarian Dream: Socialism, Culture, and Emotion in Germany, 1863–1933, $2,000 (2016-17); B. Venkat Mani, University of Wisconsin, Madison, for Recoding World Literature: Libraries, Print Culture, and Germany's Pact with Books,, $2,000 (2016-17).
*Contact:* Dennis Looney; (646) 576-5000; fax (646) 458-0030; awards@mla.org
Annie M. Reiser, Coordinator of Book Prizes and Administrative Assistant; (646) 576-5141 or (646) 576-5000; fax (646) 458-0030; awards@mla.org
*Internet:* https://www.mla.org/Resources/Career/MLA-Grants-and-Awards/Award-Submissions-and-Nominations/Competitions-for-MLA-Publication-Awards/Biennial-Prizes-with-Competitions-in-2020/Aldo-and-Jeanne-Scaglione-Prize-for-Studies-in-Germanic-Languages-and-Literature
*Sponsor:* Modern Language Association
85 Broad Street, Suite 500
New York, NY 10004-2434

**MLA Aldo and Jeanne Scaglione Prize for Studies in Slavic Languages and Literatures** 1613

The Committee on Honors and Awards of the Modern Language Association invites authors to compete for the biennial Aldo and Jeanne Scaglione Prize for Studies in Slavic Languages and Literatures, awarded biennially for an outstanding scholarly work on the linguistics or literatures of the Slavic languages. The prize, established by the Aldo and Jeanne Scaglione Endowment Fund, is awarded each odd-numbered year. Works of literary history, literary criticism, philology, and literary theory are eligible for consideration. Entries will be accepted until May 1 of odd-numbered years. The winning author will be contacted in September, and a public announcement will be made in early December. The prize, which consists of a cash award and a certificate, will be presented to the winning author at the association's annual convention in January in even-numbered years.
*Requirements:* Books are eligible if published in the two years preceding the year in which the award is given.
*Restrictions:* Books that are primarily translations will not be considered
*Geographic Focus:* All States
*Date(s) Application is Due:* May 1
*Amount of Grant:* 2,000 USD
*Samples:* Leonid Livak, University of Toronto, for In Search of Russian Modernism, $2,000 (2017-18); Alice Lovejoy, University of Minnesota, Twin Cities, for Army Film and the Avant Garde: Cinema and Experiment in the Czechoslovak Military, $2,000 (2015-16); Emily Van Buskirk, Rutgers University, New Brunswick, for Lydia Ginzburg's Prose: Reality in Search of Literature, $2,000 (2015-16).
*Contact:* Dennis Looney; (646) 576-5000; fax (646) 458-0030; awards@mla.org
Annie M. Reiser, Coordinator of Book Prizes and Administrative Assistant; (646) 576-5141 or (646) 576-5000; fax (646) 458-0030; awards@mla.org
*Internet:* https://www.mla.org/Resources/Career/MLA-Grants-and-Awards/Award-Submissions-and-Nominations/Competitions-for-MLA-Publication-Awards/Biennial-Prizes-with-Competitions-in-2021/Aldo-and-Jeanne-Scaglione-Prize-for-Studies-in-Slavic-Languages-and-Literatures
*Sponsor:* Modern Language Association
85 Broad Street, Suite 500
New York, NY 10004-2434

**MLA Clarivate Analytics / Frank Bradway Rogers Information Advancement Award** 1614

Founded in 1898, the Medical Library Association (MLA) is a nonprofit, educational organization of more than 1,100 institutions and 3,600 individual members in the health sciences information field, committed to educating health-information professionals, supporting health-information research, promoting access to the world's health-sciences information, and working to ensure that the best health information is available to all. The Thomson Reuters/Frank Bradway Rogers Information Advancement Award is presented annually by the Medical Library Association in recognition of outstanding contributions in the following areas: the use of technology to deliver health sciences information; the science of information; and facilitating the delivery of health-sciences information. Originally sponsored by the Institute for Scientific Information (ISI), the award was presented for the first time in 1983 to Frank Bradway Rogers, M.D., for whom it was subsequently renamed. The recipient receives a certificate at the association's annual meeting and a cash award of $500 after the annual meeting. Guidelines and a nomination form are available at the website. In order for the nomination to be considered, all materials and letters of support must be received by MLA via email, fax, or mail (in order of preference) by November 1. The recipient will be notified in March before the annual meeting.
*Requirements:* It is preferred that an individual nominee be a member of MLA, or if a small group is nominated for a combined effort, that at least one member of the group be a member of MLA. Nominations should be for innovative contributions to improved information handling and use. Preference will be given to pioneering efforts (i.e., new theories or applications).
*Geographic Focus:* All States, All Countries
*Date(s) Application is Due:* Nov 1
*Amount of Grant:* 500 USD
*Contact:* Maria Lopez, Awards, Grants, and Scholarships; (312) 419-9094, ext. 15; fax (312) 419-8950; lopez@mail.mlahq.org or awards@mlahq.org
*Internet:* https://www.mlanet.org/p/cm/ld/fid=295
*Sponsor:* Medical Library Association
65 East Wacker Place, Suite 1900
Chicago, IL 60601-7246

**MLA Cunningham Memorial International Fellowship** 1615

Founded in 1898, the Medical Library Association (MLA) is a nonprofit, educational organization of more than 1,100 institutions and 3,600 individual members in the health sciences information field, committed to educating health-information professionals, supporting health-information research, promoting access to the world's health-sciences information, and working to ensure that the best health information is available to all. The Medical Library Association (MLA) Cunningham Memorial International Fellowship was established in 1967 with a bequest from the estate of Eileen R. Cunningham. The fellowship supports the education and training of a health-science librarian from a country outside the United States or Canada and includes a stipend plus approved travel for four months in either country. The fellowship also pays for the cost of attending MLA's Annual Meeting. Guidelines and application materials are available at the website. All materials and letters of support must be received by MLA by December 1.
*Requirements:* Candidates must have both an undergraduate degree and a master's-level library degree (latter requirement may be waived) and must either be working in or preparing to work in a medical library in their country. The fellowship recipient must agree to return to his or her country and work in a medical library for a period of two years. In addition, a satisfactory score must be achieved on the TOEFL English competency examination.
*Restrictions:* Fellowship is not available to U.S. or Canadian citizens. Travel arrangements to and from the United States or Canada must be paid by the fellow. Previous recipients of the Cunningham fellowship are ineligible.
*Geographic Focus:* All Countries
*Date(s) Application is Due:* Dec 1
*Amount of Grant:* 8,000 USD
*Contact:* Maria Lopez, Awards, Grants, and Scholarships; (312) 419-9094, ext. 15; fax (312) 419-8950; lopez@mail.mlahq.org or awards@mlahq.org
*Internet:* https://www.mlanet.org/p/cm/ld/fid=312
*Sponsor:* Medical Library Association
65 East Wacker Place, Suite 1900
Chicago, IL 60601-7246

**MLA David A. Kronick Traveling Fellowship** 1616

Established in 2002, the David A. Kronick Traveling Fellowship grants one $2,000 award each year to cover the expenses involved in traveling to three or more medical libraries in the United States or Canada, for the purpose of studying a specific aspect of health information management. Activities that are operational in nature or have only local usefulness will be considered. Guidelines and application materials are available at the Medical Library Association (MLA) website. Applications and supporting documents must be received by MLA via email, fax, or mail by December 1. email applications and documents will be accepted as PDF or MS Word files only. Founded in 1898, the Medical Library Association (MLA) is a nonprofit, educational organization of more than 1,100 institutions and 3,600 individual members in the health sciences information field committed to educating health-information professionals, supporting health-information research, promoting access to the world's health-sciences information, and working to ensure that the best health information is available to all.
*Requirements:* Eligible candidates should satisfy the following *Requirements:* hold a graduate degree in library science; currently be a practicing health-sciences librarian with at least five years of professional experience; be an individual member of MLA; and be a citizen of or permanent resident in either the United States or Canada. Consideration will be given in exceptional cases to an outstanding candidate not meeting the criteria above. The award winner is required to submit a report to MLA on the results of the project, that may be published in appropriate health information journals.
*Geographic Focus:* All States
*Date(s) Application is Due:* Dec 1
*Amount of Grant:* 2,000 USD
*Contact:* Carla Funk, Executive Director; (312) 419-9094, ext. 14; fax (312) 419-8950; mlapd2@mlahq.org or grants@mlahq.org
*Internet:* www.mlanet.org/awards/grants/index.html
*Sponsor:* Medical Library Association
65 East Wacker Place, Suite 1900
Chicago, IL 60601-7246

**MLA Donald A.B. Lindberg Research Fellowship** 1617

The Medical Library Association (MLA) believes that access to high-quality information improves decision-making by health professionals, scientists, and consumers and is a major determinant in improved health and quality of care both nationally and internationally. MLA awards a $10,000 fellowship annually to support an individual whose research will extend the underlying knowledge-base of health-sciences information management or enhance the practice of the information professions, particularly health-sciences librarianship. The MLA's areas of interest include the organization, delivery, and use of information and knowledge and their impact on health-care access and delivery, public-health services, consumers' use of health information, biomedical research, and education for the health professions. The Lindberg Fellowship is an unrestricted grant that is awarded to the applicant, not to the sponsoring organization, and may be used for salaries, supplies, equipment, travel, fees, insurance, salaries for research assistants, and other research-related costs as specified in the grant application. The award may be used to supplement or extend other awards, including other private or government-supported fellowships, but is not contingent on receiving other awards. The award is not restricted to disbursement in a single year; funding may be disbursed over a period of up to five years depending on the needs of the research fellow. The application form and guidelines are available at the website. The application and supporting documents must be received by MLA via email, fax, or mail by November 15. Applications and documents sent via email will be accepted as PDF or MS Word files only. The Lindberg Research Fellowship is named in honor of Donald A.B. Lindberg, M.D., Director of the National Library of Medicine (NLM), in recognition of his significant national and international achievements at the NLM and its National Center for Biotechnology Information. The Lindberg Research Fellow program of MLA is administered through the Lindberg Research Fellow Jury.
*Requirements:* Health-sciences librarians, health professionals, researchers, educators, and administrators are eligible to apply. Applicants must be sponsored by an institution or organization; one or a combination of the following institutions or organizations may sponsor applicants: MLA's board, committees, sections, and chapters; graduate schools of library and information science; university deans or department chairs in health-professions schools or health-care organizations; library organizations; or scientific academies and societies. Applicants must have a bachelor's, master's, or doctor's degree

or be enrolled in a program leading to such a degree and demonstrate a commitment to the health sciences; additionally applicants must be a citizen or permanent resident of either the United States or Canada, or have been lawfully admitted for permanent residence at the time of appointment.
*Restrictions:* The grant may not be used for institutional overhead, other indirect costs, income tax payments, or tuition. Acceptance of the grant may be subject to institutional rules and regulations and to all applicable tax laws. The fellowship is not designed to support research for a doctoral dissertation or master's thesis. An applicant may resubmit an application to reapply for the Lindberg Fellowship for up to a two-year period.
*Geographic Focus:* All States, Canada
*Date(s) Application is Due:* Nov 15
*Amount of Grant:* 10,000 USD
*Contact:* Carla Funk, Executive Director; (312) 419-9094, ext. 14; fax (312) 419-8950; mlapd2@mlahq.org or grants@mlahq.org
*Internet:* www.mlanet.org/awards/grants/index.html
*Sponsor:* Medical Library Association
65 East Wacker Place, Suite 1900
Chicago, IL 60601-7246

**MLA Erich Meyerhoff Prize** **1618**
Founded in 1898, the Medical Library Association is a nonprofit, educational organization of more than 1,100 institutions and 3,600 individual members in the health sciences information field committed to educating health-information professionals, supporting health-information research, promoting access to the world's health-sciences information, and working to ensure that the best health information is available to all. The Murray Gottlieb Prize was established in 1956 by Ralph and Jo Grimes of the Old Hickory Bookshop (Brinklow, Maryland) in memory of Murray Gottlieb, a New York antiquarian-book dealer. In 2016, it was renamed to honor longtime MLA member Erich Meyerhoff, a legendary figure in medical librarianship with a great devotion to the history of the health sciences. The purpose of the award is to stimulate the health-sciences librarian's interest in the history of medicine. Currently the award is sponsored by the Medical Library Association (MLA) History of the Health Sciences Section and is awarded annually for the best unpublished essay on the history of medicine and allied sciences written by a health-sciences librarian. The author of the winning essay receives complimentary registration to the annual meeting, a certificate at the association's annual meeting, and a cash award of $100 afterwards. The recipient assumes all costs of attending the meeting and the ceremony at which the presentation is made. Guidelines and a submission form are available at the website. All materials must be received by MLA via email, fax, or mail (in order of preference) by November 1. The recipient will be notified in March before the annual meeting.
*Requirements:* Authors of papers submitted must be health-sciences librarians. The submitted paper must treat some aspect of the history of medicine or allied sciences. Entries will be judged according to quality of bibliographic research, quality of the experimental design or arguments developed to support a particular hypothesis, contribution to the study of the history of the health sciences, and originality and style (clarity, appearance, conciseness, and structure).
*Restrictions:* Papers may be under consideration for publication at the time of submission but cannot have been published.
*Geographic Focus:* All States, All Countries
*Date(s) Application is Due:* Nov 1
*Amount of Grant:* 500 USD
*Contact:* Maria Lopez, Awards, Grants, and Scholarships; (312) 419-9094, ext. 15; fax (312) 419-8950; lopez@mail.mlahq.org or awards@mlahq.org
*Internet:* https://www.mlanet.org/p/cm/ld/fid=248
*Sponsor:* Medical Library Association
65 East Wacker Place, Suite 1900
Chicago, IL 60601-7246

**MLA Fenia and Yaakov Leviant Memorial Prize in Yiddish Studies** **1619**
The Committee on Honors and Awards of the Modern Language Association is pleased to invite authors to compete for the ninth and tenth annual Fenia and Yaakov Leviant Memorial Prize. The prize, established in 2000 by the family of Fenia and Yaakov Leviant, is offered each even-numbered year and is awarded alternately to an outstanding translation or an outstanding scholarly work in the field of Yiddish. In 2020, the prize will be awarded to an outstanding scholarly work in English in the field of Yiddish published between 2016 and 2019. Cultural studies, critical biographies, or edited works in the field of Yiddish folklore or linguistic studies are eligible to compete. The 2020 Leviant Memorial Prize, which consists of a cash award and a certificate, will be presented to the winning author at the association's annual convention in January 2021. In the 2022 competition, the prize will be awarded to an English translation of a Yiddish literary work published between 2018 and 2021. For both competitions, the deadline for submission is May 1.
*Requirements:* Publishers may enter more than one title. Provided they meet eligibility requirements, books entered in this competition may also compete for one of the following prizes: the James Russell Lowell Prize, the MLA Prize for a First Book, or the MLA Prize for an Edited Collection.
*Geographic Focus:* All States
*Date(s) Application is Due:* May 1
*Amount of Grant:* 2,000 USD
*Contact:* Dennis Looney; (646) 576-5000; fax (646) 458-0030; awards@mla.org
Annie M. Reiser, Coordinator of Book Prizes and Administrative Assistant; (646) 576-5141 or (646) 576-5000; fax (646) 458-0030; awards@mla.org
*Internet:* https://www.mla.org/Resources/Career/MLA-Grants-and-Awards/Award-Submissions-and-Nominations/Competitions-for-MLA-Publication-Awards/Biennial-Prizes-with-Competitions-in-2020/Fenia-and-Yaakov-Leviant-Memorial-Prize-in-Yiddish-Studies
*Sponsor:* Modern Language Association
85 Broad Street, Suite 500
New York, NY 10004-2434

**MLA Graduate Scholarship** **1620**
The Medical Library Association (MLA) annually awards a $5,000 scholarship to a student who shows excellence in scholarship and potential for accomplishment in health-sciences librarianship. The scholarship is announced at the annual meeting of the association, where the recipient will also receive a one-year student membership in the Medical Library Association and free inclusive student registration at the association's annual meeting. The recipient will assume the other costs of attending the meeting and the ceremony at which the presentation is made. Guidelines and application materials are available at the website. All materials and letters of support must be received by MLA via email, fax, or mail by December 1. email applications and documents will be accepted as PDF or MS Word files only. The recipient will be notified in March before the annual meeting. Founded in 1898, the Medical Library Association is a nonprofit, educational organization of more than 1,100 institutions and 3,600 individual members in the health sciences information field, committed to educating health-information professionals, supporting health-information research, promoting access to the world's health-sciences information, and working to ensure that the best health information is available to all.
*Requirements:* To be eligible applicants must be either citizens or permanent residents of the United States or Canada. Additionally, applicants must either be entering or have completed no more than one-half of the academic requirements of a Masters program at an American Library Association (ALA)-accredited graduate school of library science at the time of the granting.
*Restrictions:* Past recipients of the MLA Scholarship or the MLA Scholarship for Minority Students are not eligible to apply.
*Geographic Focus:* All States, Canada
*Date(s) Application is Due:* Dec 1
*Amount of Grant:* Up to 5,000 USD
*Contact:* Carla Funk, Executive Director; (312) 419-9094, ext. 14; fax (312) 419-8950; mlapd2@mlahq.org or grants@mlahq.org
*Internet:* www.mlanet.org/awards/grants/scholar.html
*Sponsor:* Medical Library Association
65 East Wacker Place, Suite 1900
Chicago, IL 60601-7246

**MLA Graduate Scholarship for Minority Students** **1621**
The Medical Library Association (MLA) annually awards a $5,000 scholarship to a student who shows excellence in scholarship and potential for accomplishment in health-sciences librarianship. The scholarship is announced at the annual meeting of the association, where the recipient will also receive a one-year student membership in the Medical Library Association and free inclusive student registration at the association's annual meeting. The recipient will assume the other costs of attending the meeting and the ceremony at which the presentation is made. Guidelines and application materials are available at the website. All materials and letters of support must be received by MLA via email, fax, or mail (in order of preference) by December 1. email applications and documents will be accepted as PDF or MS Word files only. The recipient will be notified in March before the annual meeting. Founded in 1898, the Medical Library Association (MLA) is a nonprofit, educational organization of more than 1,100 institutions and 3,600 individual members in the health sciences information field, committed to educating health-information professionals, supporting health-information research, promoting access to the world's health-sciences information, and working to ensure that the best health information is available to all.
*Requirements:* To be eligible the applicant must be a member of a minority group (Black or African-American, Hispanic or Latino, Asian, Aboriginal, North-American Indian or Alaskan Native, or Native Hawaiian or other Pacific Islander) and be a citizen or permanent resident of the United States or Canada. Additionally, the applicant must either be entering or have completed no more than one-half of the academic requirements of a Masters program at an American Library Association (ALA)-accredited graduate school of library science at the time of the granting.
*Restrictions:* Past recipients of the MLA Scholarship or the MLA Scholarship for Minority Students are not eligible to apply.
*Geographic Focus:* All States, Canada
*Date(s) Application is Due:* Dec 1
*Amount of Grant:* Up to 5,000 USD
*Contact:* Carla Funk, Program Contact; (312) 419-9094, ext. 14; fax (312) 419-8950; mlapd2@mlahq.org or grants@mlahq.org
*Internet:* www.mlanet.org/awards/grants/minstud.html
*Sponsor:* Medical Library Association
65 East Wacker Place, Suite 1900
Chicago, IL 60601-7246

**MLA Howard R. Marraro Prize** **1622**
The Committee on Honors and Awards of the Modern Language Association invites authors to enter the biennial competition for the Howard R. Marraro Prize. The prize is awarded each even-numbered year to the author of an outstanding scholarly book or essay on any phase of Italian literature or comparative literature involving Italian. For the next competition, the committee solicits submissions of works published in 2019 by current members of the association. Entries will be accepted until May 1 of odd-numbered years. The winning author will be contacted in September, and a public announcement will be

made in early December. The prize, which consists of a cash award and a certificate, will be presented to the winning author at the association's annual convention in January of the next even-numbered year.
*Requirements:* Publishers may submit more than one title. Provided they meet eligibility requirements, books that have been entered in this competition may also compete for the James Russell Lowell Prize, the MLA Prize for a First Book, or the MLA Prize for an Edited Collection.
*Restrictions:* This competition is open to members only.
*Geographic Focus:* All States
*Date(s) Application is Due:* May 1
*Amount of Grant:* 2,000 USD
*Samples:* Gian Maria Annovi, University of Southern California, for Pier Paolo Pasolini: Performing Authorship, $2,000; Marilyn Migiel, Cornell University, for The Ethical Dimension of the Decameron, $2,000.
*Contact:* Dennis Looney; (646) 576-5000; fax (646) 458-0030; awards@mla.org
Annie M. Reiser, Coordinator of Book Prizes and Administrative Assistant; (646) 576-5141 or (646) 576-5000; fax (646) 458-0030; awards@mla.org
*Internet:* https://www.mla.org/Resources/Career/MLA-Grants-and-Awards/Award-Submissions-and-Nominations/Competitions-for-MLA-Publication-Awards/Biennial-Prizes-with-Competitions-in-2020/Howard-R.-Marraro-Prize
*Sponsor:* Modern Language Association
85 Broad Street, Suite 500
New York, NY 10004-2434

**MLA James Russell Lowell Prize** **1623**
The Committee on Honors and Awards of the Modern Language Association invites authors to compete for the fifty-first annual James Russell Lowell Prize, which will be awarded for a scholarly book published in in the past year by a current member of the association. To enter a book into the competition, authors or publishers should send eight copies and confirmation of the author's membership in the MLA. In the case of a born-digital publication, send the relevant media or URL and a letter identifying the work to the email listed. Publishers may enter more than one title. Entries will be accepted until March 1 each year. The winning author will be contacted in September, and a public announcement will be made in early December. The prize, which consists of a cash award and a certificate, will be presented to the winning author at the association's annual convention in January.
*Requirements:* To qualify for the prize, a book must be an outstanding literary or linguistic study or a critical biography. Studies dealing with literary theory, media, cultural history, and interdisciplinary topics are eligible.
*Restrictions:* Books that are primarily translations will not be considered. Books entered in this competition may also compete for one other prize that is not the MLA Prize for a First Book or the MLA Prize for an Edited Collection, provided they meet eligibility requirements.
*Geographic Focus:* All States
*Date(s) Application is Due:* Mar 1
*Amount of Grant:* 1,000 USD
*Samples:* Jonathan P. Eburne, Pennsylvania State University, for Outsider Theory: Intellectual Histories of Unorthodox Ideas (University of Minnesota Press), $1,000, (2018); Deborah L. Nelson, University of Chicago, for Tough Enough: Arbus, Arendt, Didion, McCarthy, Sontag, Weil (University of Chicago Press), $1,000,; Branka Arsic, Columbia University, for Bird Relics: Grief and Vitalism in Thoreau (Harvard University Press), $1,000,.
*Contact:* Dennis Looney; (646) 576-5000; fax (646) 458-0030; awards@mla.org
Annie M. Reiser, Coordinator of Book Prizes and Administrative Assistant; (646) 576-5141 or (646) 576-5000; fax (646) 458-0030; awards@mla.org
*Internet:* https://www.mla.org/Resources/Career/MLA-Grants-and-Awards/Award-Submissions-and-Nominations/Competitions-for-MLA-Publication-Awards/Annual-Prizes-with-Competitions-in-2020/James-Russell-Lowell-Prize
*Sponsor:* Modern Language Association
85 Broad Street, Suite 500
New York, NY 10004-2434

**MLA Katherine Singer Kovacs Prize** **1624**
The Committee on Honors and Awards of the Modern Language Association invites authors to compete for the annual Katherine Singer Kovacs Prize for an outstanding book published in English or Spanish in the field of Latin American and Spanish literatures and cultures. The prize was established in 1990 by a gift from the parents of Katherine Singer Kovacs, a member of the MLA who died in May 1989. A specialist in Spanish and Latin American literature and film, Professor Kovacs taught at Stanford University, the University of Southern California, and Whittier College. Under the terms of the gift, the prize selection committee will be especially interested in original, broadly interpretive books that enhance understanding of the interrelations among literature, the other arts, and society and that offer fresh perspectives on the field. Entries will be accepted until May 1 each year. The winning author will be contacted in September, and a public announcement will be made in early December. The prize, which consists of a cash award and a certificate, will be presented to the winning author at the association's annual convention in January of the following year.
*Requirements:* Publishers may enter more than one title. Provided they meet eligibility requirements, books entered in this competition may also compete for one of the following prizes: the James Russell Lowell Prize, the MLA Prize for a First Book, or the MLA Prize for an Edited Collection.
*Restrictions:* This award competition is open to both members and nonmembers.
*Geographic Focus:* All States
*Date(s) Application is Due:* May 1
*Amount of Grant:* 2,000 USD
*Samples:* Maite Conde, University of Cambridge, for Foundational Films: Early Cinema and Modernity in Brazil, $2,000 (2018); B. Christine Arce, University of Miami, for México's Nobodies: The Cultural Legacy of the Soldadera and Afro-Mexican Women, $2,000; Nancy J. Gates Madsen, Luther College, for Trauma, Taboo, and Truth-Telling: Listening to Silences in Postdictatorship Argentina, $2,000.
*Contact:* Dennis Looney; (646) 576-5000; fax (646) 458-0030; awards@mla.org
Annie M. Reiser, Coordinator of Book Prizes and Administrative Assistant; (646) 576-5141 or (646) 576-5000; fax (646) 458-0030; awards@mla.org
*Internet:* https://www.mla.org/Resources/Career/MLA-Grants-and-Awards/Award-Submissions-and-Nominations/Competitions-for-MLA-Publication-Awards/Annual-Prizes-with-Competitions-in-2020/Katherine-Singer-Kovacs-Prize
*Sponsor:* Modern Language Association
85 Broad Street, Suite 500
New York, NY 10004-2434

**MLA Kenneth W. Mildenberger Prize** **1625**
The Committee on Honors and Awards of the Modern Language Association invites authors to compete for the thirty-sixth Kenneth W. Mildenberger Prize, to be presented for an outstanding scholarly book in the fields of language, culture, literacy, and literature that has a strong application to the teaching of languages other than English. The prize will be awarded in odd years for a book published during the past two years. The selection committee will look for evidence of fresh and effective approaches to teaching and for works that are likely to be useful on a wide scale. Entries will be accepted until May 1 in even-numbered years. The winning author will be contacted in September, and a public announcement will be made in early December. The prize, which consists of a cash award and a certificate, will be presented to the winning author at the association's annual convention in January of odd-numbered years.
*Requirements:* To enter works into competition, send four copies of each work and a letter indicating the titles submitted, the authors, and the dates of publication. Authors of works submitted need not be members of the association.
*Geographic Focus:* All States
*Date(s) Application is Due:* May 1
*Amount of Grant:* 1,000 USD
*Samples:* Claire Kramsch, University of California, Berkeley, and Lihua Zhang, University of California, Berkeley, for The Multilingual Instructor: What Foreign Language Teachers Say about Their Experience and Why It Matters, $1,000 (2017-18); Erin Kearney, University at Buffalo, State University of New York, for Intercultural Learning in Modern Language Education: Expanding Meaning-Making Potentials, $1,000 (2015-16).
*Contact:* Dennis Looney; (646) 576-5000; fax (646) 458-0030; awards@mla.org
Annie M. Reiser, Coordinator of Book Prizes and Administrative Assistant; (646) 576-5141 or (646) 576-5000; fax (646) 458-0030; awards@mla.org
*Internet:* https://www.mla.org/Resources/Career/MLA-Grants-and-Awards/Award-Submissions-and-Nominations/Competitions-for-MLA-Publication-Awards/Biennial-Prizes-with-Competitions-in-2021/Kenneth-W.-Mildenberger-Prize
*Sponsor:* Modern Language Association
85 Broad Street, Suite 500
New York, NY 10004-2434

**MLA Lois Roth Award** **1626**
The Committee on Honors and Awards of the Modern Language Association invites translators and publishers to compete for the annual Lois Roth Award for a translation of a literary work. Since 1999 the MLA has offered the award for an outstanding translation into English of a book-length literary work. To enter a translation into the competition, authors or publishers should send six copies along with brief biographical and professional data about translator, a total of 12–15 pages of the text in its original language taken from the beginning, middle, and end of the work, and a letter identifying the work, the translator, and the date of publication to the Lois Roth Award. Entries will be accepted until May 1 of each year. The winning author will be contacted in September, and a public announcement will be made in early December. The prize, which consists of a cash award and a certificate, will be presented to the winning author at the association's annual convention in January of the following year.
*Requirements:* Publishers may submit more than one title. Works submitted for this prize will also be automatically considered for the Scaglione Prize for a Translation of a Literary Work. Definitions and criteria are the same for both awards.
*Restrictions:* This award competition is open to both members and nonmembers.
*Geographic Focus:* All States
*Date(s) Application is Due:* May 1
*Amount of Grant:* 2,000 USD
*Samples:* Damion Searls, Brooklyn, New York, for Anniversaries: From a Year in the Life of Gesine Cresspahl, by Uwe Johnson, $2,000 (2018); Susan Bernofsky, Columbia University, for Go, Went, Gone, by Jenny Erpenbeck. $2,000.
*Contact:* Dennis Looney, Director of Programs; (646) 576-5000; fax (646) 458-0030; awards@mla.org
Annie M. Reiser, Coordinator of Book Prizes and Administrative Assistant; (646) 576-5141 or (646) 576-5000; fax (646) 458-0030; awards@mla.org
*Internet:* https://www.mla.org/Resources/Career/MLA-Grants-and-Awards/Award-Submissions-and-Nominations/Competitions-for-MLA-Publication-Awards/Annual-Prizes-with-Competitions-in-2020/Lois-Roth-Award
*Sponsor:* Modern Language Association
85 Broad Street, Suite 500
New York, NY 10004-2434

**MLA Matei Calinescu Prize** 1627
The Committee on Honors and Awards of the Modern Language Association invites authors to compete for the annual Matei Calinescu Prize, which will be awarded for a distinguished work of scholarship in twentieth- or twenty-first-century literature and thought in any geographical context. Preference will be given to a first or second work published by an entrant. The prize has been established in honor of Matei Calinescu, a Romanian poet and scholar who taught at Indiana University, Bloomington. Entries will be accepted until May 1 of each year. The winning author will be contacted in September, and a public announcement will be made in early December. The prize, which consists of a cash award and a certificate, will be presented to the winning author at the association's annual convention in January of the following year.
*Requirements:* Publishers may enter more than one title. Provided they meet eligibility requirements, books that have been entered in this competition may also compete for one of the following prizes: the James Russell Lowell Prize, the MLA Prize for a First Book, or the MLA Prize for an Edited Collection.
*Restrictions:* This award competition is open to both members and nonmembers of the association.
*Geographic Focus:* All States
*Date(s) Application is Due:* May 1
*Amount of Grant:* 2,000 USD
*Samples:* Jane Elliott, King's College London, for The Microeconomic Mode: Political Subjectivity in Contemporary Popular Aesthetics, $2,000 (2018); Sharon Luk, University of Oregon, for The Life of Paper: Letters and a Poetics of Living beyond Captivity, $2,000; Jed Rasula, University of Georgia, for History of a Shiver: The Sublime Impudence of Modernism, $2,000.
*Contact:* Dennis Looney; (646) 576-5000; fax (646) 458-0030; awards@mla.org
Annie M. Reiser, Coordinator of Book Prizes and Administrative Assistant; (646) 576-5141 or (646) 576-5000; fax (646) 458-0030; awards@mla.org
*Internet:* https://www.mla.org/Resources/Career/MLA-Grants-and-Awards/Award-Submissions-and-Nominations/Competitions-for-MLA-Publication-Awards/Annual-Prizes-with-Competitions-in-2020/Matei-Calinescu-Prize
*Sponsor:* Modern Language Association
85 Broad Street, Suite 500
New York, NY 10004-2434

**MLA Mina P. Shaughnessy Prize** 1628
The Committee on Honors and Awards of the Modern Language Association invites authors to compete for the thirty-sixth Mina P. Shaughnessy Prize, to be presented for an outstanding scholarly book in the fields of language, culture, literacy, and literature that has a strong application to the teaching of English. The prize will be awarded for a book published in the previous two years. Authors of works competing for the prize need not be members of the association. Entries will be accepted until May 1 in even-numbered years. The winning author will be contacted in September, and a public announcement will be made in early December. The prize, which consists of a cash award and a certificate, will be presented to the winning author at the association's annual convention in January of odd-numbered years.
*Requirements:* To enter works in the competition, send four copies of each work and a letter indicating the titles submitted, the authors, and the dates of publication. Authors of works submitted need not be members of the association.
*Geographic Focus:* All States
*Date(s) Application is Due:* May 1
*Amount of Grant:* 1,000 USD
*Samples:* Bruce Horner, University of Louisville, and Laura Tetreault, University at Albany, State University of New York, editors, for Crossing Divides: Exploring Translingual Writing Pedagogies and Programs, $1,000 (2016-17); Deborah Brandt, University of Wisconsin, Madison, for The Rise of Writing: Redefining Mass Literacy, $1,000 (2014-15).
*Contact:* Dennis Looney; (646) 576-5000; fax (646) 458-0030; awards@mla.org
Annie M. Reiser, Coordinator of Book Prizes and Administrative Assistant; (646) 576-5141 or (646) 576-5000; fax (646) 458-0030; awards@mla.org
*Internet:* https://www.mla.org/Resources/Career/MLA-Grants-and-Awards/Award-Submissions-and-Nominations/Competitions-for-MLA-Publication-Awards/Biennial-Prizes-with-Competitions-in-2020/Mina-P.-Shaughnessy-Prize
*Sponsor:* Modern Language Association
85 Broad Street, Suite 500
New York, NY 10004-2434

**MLA Morton N. Cohen Award for a Distinguished Edition of Letters** 1629
The Committee on Honors and Awards of the Modern Language Association invites editors to compete for the sixteenth Morton N. Cohen Award for a Distinguished Edition of Letters. The award was established in 1989 by a gift from Morton N. Cohen, professor emeritus of English at the City University of New York. The award is presented each odd-numbered year. A multi-volume edition is eligible if at least one volume was published during the two-year period prior to the award ceremony. Editors can apply regardless of the fields they and the authors of the letters represent, and membership in the MLA is not required. Under the terms of the award, the winning collection will be one that provides readers with a clear, accurate, and readable text; necessary background information; and succinct and eloquent introductory material and annotations. The edited collection should be in itself a work of literature. Entries will be accepted until May 1 of odd-numbered years. The winning author will be contacted in September, and a public announcement will be made in early December. The prize, which consists of a cash award and a certificate, will be presented to the winning author at the association's annual convention in January of the next even-numbered year.
*Geographic Focus:* All States
*Date(s) Application is Due:* May 1
*Amount of Grant:* 2,000 USD
*Samples:* Edward M. Burns, emeritus, William Paterson University of New Jersey, for Questioning Minds: The Letters of Guy Davenport and Hugh Kenner, 2 volumes, $2,000 (2017-18); Stephen Bernard, University of Oxford, for The Literary Correspondences of the Tonsons, $2,000 (2015-16).
*Contact:* Dennis Looney; (646) 576-5000; fax (646) 458-0030; awards@mla.org
Annie M. Reiser, Coordinator of Book Prizes and Administrative Assistant; (646) 576-5141 or (646) 576-5000; fax (646) 458-0030; awards@mla.org
*Internet:* https://www.mla.org/Resources/Career/MLA-Grants-and-Awards/Award-Submissions-and-Nominations/Competitions-for-MLA-Publication-Awards/Biennial-Prizes-with-Competitions-in-2021/Morton-N.-Cohen-Award-for-a-Distinguished-Edition-of-Letters
*Sponsor:* Modern Language Association
85 Broad Street, Suite 500
New York, NY 10004-2434

**MLA Prize for a First Book** 1630
The Committee on Honors and Awards of the Modern Language Association invites authors to compete for the annual MLA Prize for a First Book, which will be awarded for an outstanding scholarly work published last year as the first book-length publication by a current member of the association. To enter a book into the competition, authors or publishers should send eight copies and confirmation of the author's membership in the MLA. In the case of a born-digital publication, send the relevant media or URL and a letter identifying the work to the email listed. Publishers may enter more than one title. Entries will be accepted until March each year. The winning author will be contacted in September, and a public announcement will be made in early December. The prize, which consists of a cash award and a certificate, will be presented to the winning author at the association's annual convention in January.
*Requirements:* An author who is not a member of the MLA at the time of publication may become eligible to compete by joining the association prior to May 1 of the following year. To qualify for the prize, a book must be an outstanding literary or linguistic study or a critical biography. Studies dealing with literary theory, media, cultural history, and interdisciplinary topics are eligible.
*Restrictions:* Books that are primarily translations will not be considered. Books entered in this competition may also compete for one other prize that is not the MLA Prize for a First Book or the MLA Prize for an Edited Collection, provided they meet eligibility requirements.
*Geographic Focus:* All States
*Date(s) Application is Due:* Mar 1
*Amount of Grant:* 1,000 USD
*Contact:* Dennis Looney; (646) 576-5000; fax (646) 458-0030; awards@mla.org
Annie M. Reiser, Coordinator of Book Prizes and Administrative Assistant; (646) 576-5141 or (646) 576-5000; fax (646) 458-0030; awards@mla.org
*Internet:* https://www.mla.org/Resources/Career/MLA-Grants-and-Awards/Award-Submissions-and-Nominations/Competitions-for-MLA-Publication-Awards/Annual-Prizes-with-Competitions-in-2020/MLA-Prize-for-a-First-Book
*Sponsor:* Modern Language Association
85 Broad Street, Suite 500
New York, NY 10004-2434

**MLA Prize for an Edited Collection** 1631
The Modern Language Association's Committee on Honors and Awards invites editors to compete for the Modern Language Association Prize for an Edited Collection. Established in 2019 to recognize the intellectual labor involved in producing edited collections, the prize is given without regard to the language of the text presented in the work as long as the work falls within the subject scope of the MLA (e.g., modern languages and literatures, composition theory, folklore, linguistics). Entries will be accepted until May 1 of odd-numbered years. The winner will be contacted in September, and a public announcement will be made in early December. The prize, which consists of a cash award and a certificate, will be presented to the winning author at the association's annual convention in January of the next even-numbered year.
*Restrictions:* Editors must be current MLA members, but contributors to the collections need not be.
*Geographic Focus:* All States
*Date(s) Application is Due:* May 1
*Amount of Grant:* 2,000 USD
*Contact:* Dennis Looney; (646) 576-5000; fax (646) 458-0030; awards@mla.org
Annie M. Reiser, Coordinator of Book Prizes and Administrative Assistant; (646) 576-5141 or (646) 576-5000; fax (646) 458-0030; awards@mla.org
*Internet:* https://www.mla.org/Resources/Career/MLA-Grants-and-Awards/Award-Submissions-and-Nominations/Competitions-for-MLA-Publication-Awards/Biennial-Prizes-with-Competitions-in-2021/MLA-Prize-for-an-Edited-Collection
*Sponsor:* Modern Language Association
85 Broad Street, Suite 500
New York, NY 10004-2434

**MLA Prize for a Scholarly Edition** 1632
The Committee on Honors and Awards of the Modern Language Association invites editors, including editors of digital works, to compete for the biennial MLA Prize for a Scholarly Edition. Established in 1995, the award is presented each odd-numbered year. For the upcoming competition, the committee solicits submissions of editions published during the two previous years. A multi-volume edition is eligible if at least one volume was published during that period. The prize will be given without regard to the field or language either of the editor or of the text presented in the edition. To qualify for the award, an edition should be based on an examination of all available relevant textual

sources; the source texts and the edited text's deviations from them should be fully described; the edition should employ editorial principles appropriate to the materials edited, and those principles should be clearly articulated in the volume; the text should be accompanied by appropriate textual and other historical contextual information; the edition should exhibit the highest standards of accuracy in the presentation of its text and apparatus; and the text and apparatus should be presented as accessibly and elegantly as possible. Entries will be accepted until May 1 of odd-numbered years. The winning author will be contacted in September, and a public announcement will be made in early December. The prize, which consists of a cash award and a certificate, will be presented to the winning author at the association's annual convention in January of the next even-numbered year.
*Requirements:* The editor need not be a member of the MLA. Publishers may enter more than one title. Nominations of a volume from a multivolume edition that has won the award will be accepted only if the principal editor of the current volume was not the principal editor of the volumes that won the award in the past. Substantially revised or updated digital editions may compete once every four years.
*Geographic Focus:* All States
*Date(s) Application is Due:* May 1
*Amount of Grant:* 2,000 USD
*Samples:* Heather Cass White, University of Alabama, Tuscaloosa, for New Collected Poems: Marianne Moore, $2,000 (2017-18); Iman Javadi, Cambridge, United Kingdom; Ronald Schuchard, emeritus, Emory University; and Jayme Stayer, Loyola University Chicago, for Tradition and Orthodoxy, 1934–1939, volume 5 of The Complete Prose of T. S. Eliot: The Critical Edition, $2,000 (2017-18).
*Contact:* Dennis Looney; (646) 576-5000; fax (646) 458-0030; awards@mla.org
Annie M. Reiser, Coordinator of Book Prizes and Administrative Assistant; (646) 576-5141 or (646) 576-5000; fax (646) 458-0030; awards@mla.org
*Internet:* https://www.mla.org/Resources/Career/MLA-Grants-and-Awards/Award-Submissions-and-Nominations/Competitions-for-MLA-Publication-Awards/Biennial-Prizes-with-Competitions-in-2021/MLA-Prize-for-a-Scholarly-Edition
*Sponsor:* Modern Language Association
85 Broad Street, Suite 500
New York, NY 10004-2434

**MLA Prize for Collaborative, Bibliographical, or Archival Scholarship** **1633**
The Committee on Honors and Awards of the Modern Language Association invites submissions for the biennial MLA Prize for Collaborative, Bibliographical, or Archival Scholarship. Established in 1998 for bibliographies, the prize was redefined in 2012 to include archives and digital projects, and expanded in 2019 to recognize collaborative research. It is presented each even-numbered year. For the current award, the committee solicits submissions of enumerative and descriptive bibliographies, archives, digital projects, or collaborative research projects published in the previous two-year period prior to the award year. A multivolume work is eligible if at least one volume was published during that period, and an ongoing digital project is eligible if substantive material was added during that period. The prize will be given without regard to the language of the compiler or of the text presented in the work, as long as it falls within the subject scope of the MLA (e.g., modern languages and literatures, composition theory, folklore, linguistics). Editors or compilers need not be members of the MLA. Entries will be accepted until May 1 of odd-numbered years. The winning author will be contacted in September, and a public announcement will be made in early December. The prize, which consists of a cash award and a certificate, will be presented to the winning author at the association's annual convention in January of the next even-numbered year.
*Requirements:* Criteria for determining excellence include evidence of analytical rigor, meticulous scholarship, intellectual creativity, and subject range and depth. To qualify for the award a work should employ editorial principles appropriate to its materials, and those principles should be clearly articulated in the compilation; the work should provide appropriate contextual information; it should exhibit the highest standards of accuracy in its presentation; and the text should be presented as accessibly and elegantly as possible.
*Restrictions:* This award is open to both members and nonmembers.
*Geographic Focus:* All States
*Date(s) Application is Due:* May 1
*Amount of Grant:* 2,000 USD
*Samples:* Mark Nixon, University of Reading; Dirk Van Hulle, University of Antwerp; Pim Verhulst, University of Antwerp; E. Magessa O'Reilly, Memorial University; and Vincent Neyt, University of Antwerp, for the Beckett Digital Manuscript Project, $2,000 (2016-17).
*Contact:* Dennis Looney; (646) 576-5000; fax (646) 458-0030; awards@mla.org
Annie M. Reiser, Coordinator of Book Prizes and Administrative Assistant; (646) 576-5141 or (646) 576-5000; fax (646) 458-0030; awards@mla.org
*Internet:* https://www.mla.org/Resources/Career/MLA-Grants-and-Awards/Award-Submissions-and-Nominations/Competitions-for-MLA-Publication-Awards/Biennial-Prizes-with-Competitions-in-2020/MLA-Prize-for-Collaborative-Bibliographical-or-Archival-Scholarship
*Sponsor:* Modern Language Association
85 Broad Street, Suite 500
New York, NY 10004-2434

**MLA Prize for Independent Scholars** **1634**
To recognize and further encourage the achievements and contributions of independent scholars, the MLA invites authors to compete for this biennial prize for a scholarly book in the fields of modern languages and literatures, including English. The book must have been published in the preceding calendar year. The award is given to a person who, at the time of publication of the book for which the award is given, has received his/her terminal academic degree no fewer than four years earlier, and does not hold a tenured or tenure-accruing position in a post-secondary educational institution. Entries will be accepted until May 1 of even-numbered years. Normally, part-time instructors and adjunct faculty members are eligible to compete for the prize. The winning author will be contacted in September, and a public announcement will be made in early December. The prize, which consists of a cash award and a certificate, will be presented to the winning author at the association's annual convention in January of odd-numbered years.
*Requirements:* To enter a book into competition, send six copies and a completed application form to MLA Prize for Independent Scholars. Authors of works submitted need not be members of the association.
*Geographic Focus:* All States
*Date(s) Application is Due:* May 1
*Amount of Grant:* 1,000 USD
*Samples:* Margaret Morganroth Gullette, Brandeis University, for Ending Ageism; or, How Not to Shoot Old People, $1,000 (2016-17); Mimi Yiu, Vancouver, British Columbia, for Architectural Involutions: Writing, Staging, and Building Space, c. 1435–1650, $1,000 (2014-15); Joel Des Rosiers, Montreal, for Métaspora: Essai sur les patries intimes, $1,000 (2012-13).
*Contact:* Dennis Looney; (646) 576-5000; fax (646) 458-0030; awards@mla.org
Annie M. Reiser, Coordinator of Book Prizes and Administrative Assistant; (646) 576-5141 or (646) 576-5000; fax (646) 458-0030; awards@mla.org
*Internet:* https://www.mla.org/Resources/Career/MLA-Grants-and-Awards/Award-Submissions-and-Nominations/Competitions-for-MLA-Publication-Awards/Biennial-Prizes-with-Competitions-in-2020/MLA-Prize-for-Independent-Scholars
*Sponsor:* Modern Language Association
85 Broad Street, Suite 500
New York, NY 10004-2434

**MLA Prize for Studies in Native American Literatures, Cultures, & Languages** **1635**
The Modern Language Association's Committee on Honors and Awards invites authors to compete for the annual Modern Language Association Prize for Studies in Native American Literatures, Cultures, and Languages. Established in 2012, the prize is awarded to a member of the association for an outstanding scholarly work in the field of Native American literatures, cultures, and languages. The prize selection committee will consider works that examine and broaden understanding of the cultural expressions of first peoples or nations of the United States, Canada, and Mexico. Works in Indigenous languages, literatures, and cultures focusing on other geographic areas will also be considered if they bring fresh perspectives to the field. Entries will be accepted until May 1 of each year. The winning author will be contacted in September, and a public announcement will be made in early December. The prize, which consists of a cash award and a certificate, will be presented to the winning author at the association's annual convention in January of the following year.
*Requirements:* Publishers may submit more than one title. Provided they meet eligibility requirements, books that have been entered in this competition may also compete for one of the following prizes: the James Russell Lowell Prize, the MLA Prize for a First Book, or the MLA Prize for an Edited Collection.
*Restrictions:* This prize is open only to members of the Association.
*Geographic Focus:* All States
*Date(s) Application is Due:* May 1
*Amount of Grant:* 2,000 USD
*Contact:* Dennis Looney; (646) 576-5000; fax (646) 458-0030; awards@mla.org
Annie M. Reiser, Coordinator of Book Prizes and Administrative Assistant; (646) 576-5141 or (646) 576-5000; fax (646) 458-0030; awards@mla.org
*Internet:* https://www.mla.org/Resources/Career/MLA-Grants-and-Awards/Award-Submissions-and-Nominations/Competitions-for-MLA-Publication-Awards/Biennial-Prizes-with-Competitions-in-2020/MLA-Prize-for-Studies-in-Native-American-Literatures-Cultures-and-Languages
*Sponsor:* Modern Language Association
85 Broad Street, Suite 500
New York, NY 10004-2434

**MLA Prize in United States Latina and Latino and Chicana and Chicano Literary and Cultural Studies** **1636**
The Committee on Honors and Awards of the Modern Language Association is pleased to invite authors to compete for the biennial Modern Language Association Prize in United States Latina and Latino and Chicana and Chicano Literary and Cultural Studies. Established in 2002, the prize will be awarded to a member of the association for an outstanding scholarly study in any language of United States Latina and Latino or Chicana and Chicano literature or culture published during the two years previous to the the award ceremony. Entries will be accepted until May 1 of odd-numbered years. The winning author will be contacted in September, and a public announcement will be made in early December. The prize, which consists of a cash award and a certificate, will be presented to the winning author at the association's annual convention in January of the next even-numbered year.
*Restrictions:* Open only to members of the association. Books that are primarily translations will not be considered.
*Geographic Focus:* All States
*Date(s) Application is Due:* May 1
*Amount of Grant:* 2,000 USD
*Samples:* Stephanie Fetta, University of Massachusetts, Amherst, for Shaming into Brown: Somatic Transactions of Race in Latina/o Literature, $2,000 (2017-18); Licia Fiol-Matta, New York University, for The Great Woman Singer: Gender and Voice in Puerto Rican Music, $2,000 (2017-18).

*Contact:* Dennis Looney; (646) 576-5000; fax (646) 458-0030; awards@mla.org
Annie M. Reiser, Coordinator of Book Prizes and Administrative Assistant; (646) 576-5141 or (646) 576-5000; fax (646) 458-0030; awards@mla.org
*Internet:* https://www.mla.org/Resources/Career/MLA-Grants-and-Awards/Award-Submissions-and-Nominations/Competitions-for-MLA-Publication-Awards/Biennial-Prizes-with-Competitions-in-2021/MLA-Prize-in-United-States-Latina-and-Latino-and-Chicana-and-Chicano-Literary-an
*Sponsor:* Modern Language Association
85 Broad Street, Suite 500
New York, NY 10004-2434

**MLA Research, Development, and Demonstration Project Grant** **1637**
The purpose of the Medical Library Association (MLA)'s Research, Development, and Demonstration Project Grant is to provide support for research, development, or demonstration projects that will help to promote excellence in the field of health-sciences librarianship and information sciences. Grants range from $100 to $1000. The application form and guidelines are available for download at the website. Applications and supporting documents must be received by MLA via email, fax, or mail by December 1. emailed applications and documents will be accepted as PDF or MS Word files only. MLA may offer Research, Development, and Demonstration Project Grants more than once a year. Applicants are encouraged to visit the website or contact the association for information about additional deadlines. Founded in 1898, the Medical Library Association (MLA) is a nonprofit, educational organization of more than 1,100 institutions and 3,600 individual members in the health sciences information field committed to educating health-information professionals, supporting health-information research, promoting access to the world's health-sciences information, and working to ensure that the best health information is available to all.
*Requirements:* Eligible applicants must satisfy the following criteria: hold a graduate degree in library science; be a practicing health-sciences librarian with at least two years of professional experience; be a citizen of or have permanent residence status in either the United States or Canada; and preferably be an individual member of the MLA. Consideration will be given in exceptional cases to an outstanding candidate(s) not meeting the criteria above.
*Restrictions:* Grants will not be given to support an activity that is operational in nature or that has only local usefulness.
*Geographic Focus:* All States, Canada
*Date(s) Application is Due:* Dec 1
*Amount of Grant:* 100 - 1,000 USD
*Samples:* Catherine Mary Boss, "A research study: Does exemplary library services to patients and their family members, friends, and caregivers influence patient satisfaction scores?"; Naomi Broering, "The efficacy of digital conversion of audiotape collection to CD format: measuring change in use and evaluating user satisfaction"; Fern Cheek and Lynda Hartel, "An investigation of electronic book use in an Academic Medical Center".
*Contact:* Carla Funk, Executive Director; (312) 419-9094, ext. 14; fax (312) 419-8950; mlapd2@mlahq.org or grants@mlahq.org
*Internet:* www.mlanet.org/awards/grants/index.html
*Sponsor:* Medical Library Association
65 East Wacker Place, Suite 1900
Chicago, IL 60601-7246

**MLA Rittenhouse Award** **1638**
Founded in 1898, the Medical Library Association is a nonprofit, educational organization of more than 1,100 institutions and 3,600 individual members in the health sciences information field committed to educating health-information professionals, supporting health-information research, promoting access to the world's health-sciences information, and working to ensure that the best health information is available to all. The Rittenhouse Award, initiated in 1967 and sponsored by Rittenhouse Book Distributors, is presented annually by the Medical Library Association (MLA) for the best unpublished paper or web-based project on health-sciences librarianship or medical informatics submitted by a student in an ALA-accredited program of library and information studies or by an internship trainee in medical informatics or health-sciences librarianship. The author of the winning essay or project receives complimentary student registration to MLA's annual meeting, a certificate at the meeting, and a cash award of $500 after the meeting. Winning entries will be submitted to the Journal of the Medical Library Association for consideration. The recipient assumes all costs of attending the meeting and the ceremony at which the presentation is made. Guidelines and a submission form are available at the website. All materials must be received by MLA via email, fax, or mail (in order of preference) by November 1. The recipient will be notified in March before the annual meeting.
*Requirements:* Authors of papers or web-based projects submitted for the Rittenhouse Award must be enrolled in a course for credit in an ALA-accredited graduate program of library and information studies or have recently graduated from such a program and be currently enrolled as an internship trainee in medical informatics or health-sciences librarianship. Submissions must treat some aspect of health-sciences librarianship or medical informatics and may be bibliographical, address a health-sciences issue or topic, or report the results of research. Authors of web-based projects should submit a needs assessment, information on development of the website, evaluation of the website, and a summary outlining the project plan and evaluation results. Entries will be judged according to the following standards: thoroughness and relevance of bibliographic research or needs assessment; importance of topic that would warrant a bibliographic treatment or web-based project; quality of experimental design or arguments developed to support a particular hypothesis or question; originality of argument and positions taken; methodological rigor; quality of website solution; impact on the field of health-sciences librarianship or medical informatics; and style (clarity, appearance, conciseness, organization, and structure).
*Restrictions:* Papers must have been written or web-based project completed during the 18 months preceding the submission deadline and during the student's course work toward a graduate degree in library and information studies or medical informatics. Papers that have been published or that are under consideration for publication are not eligible.
*Geographic Focus:* All States
*Date(s) Application is Due:* Nov 1
*Amount of Grant:* 500 USD
*Contact:* Maria Lopez, Program Contact; (312) 419-9094, ext. 15; fax (312) 419-8950; lopez@mail.mlahq.org or awards@mlahq.org
*Internet:* https://www.mlanet.org/p/cm/ld/fid=289
*Sponsor:* Medical Library Association
65 East Wacker Place, Suite 1900
Chicago, IL 60601-7246

**MLA William Sanders Scarborough Prize** **1639**
The Committee on Honors and Awards of the Modern Language Association invites authors to compete for the annual annual William Sanders Scarborough Prize. A distinguished man of letters and former university president, William Sanders Scarborough was the first African American member of the Modern Language Association. He exemplified the life of the mind combined with community service. Established in 2001, the prize is awarded annually for an outstanding scholarly study of black American literature or culture published the previous year. Entries will be accepted until May 1 each year. The winning author will be contacted in September, and a public announcement will be made in early December. The prize, which consists of a cash award and a certificate, will be presented to the winning author at the association's annual convention in January of the following year.
*Requirements:* Publishers may enter more than one title. Provided they meet eligibility requirements, books entered in this competition may also compete for one of the following prizes: the James Russell Lowell Prize, the MLA Prize for a First Book, or the MLA Prize for an Edited Collection.
*Restrictions:* This award competition is open to both members and nonmembers. Books that are primarily translations will not be considered.
*Geographic Focus:* All States
*Date(s) Application is Due:* May 1
*Amount of Grant:* 2,000 USD
*Samples:* Darius Bost, University of Utah, for Evidence of Being: The Black Gay Cultural Renaissance and the Politics of Violence, $2,000 (2018); Sonya Posmentier, New York University, for Cultivation and Catastrophe: The Lyric Ecology of Modern Black Literature, $2,000; C. Riley Snorton, University of Chicago, for Black on Both Sides: A Racial History of Trans Identity, $2,000.
*Contact:* Dennis Looney; (646) 576-5000; fax (646) 458-0030; awards@mla.org
Annie M. Reiser, Coordinator of Book Prizes and Administrative Assistant; (646) 576-5141 or (646) 576-5000; fax (646) 458-0030; awards@mla.org
*Internet:* https://www.mla.org/Resources/Career/MLA-Grants-and-Awards/Award-Submissions-and-Nominations/Competitions-for-MLA-Publication-Awards/Annual-Prizes-with-Competitions-in-2020/William-Sanders-Scarborough-Prize
*Sponsor:* Modern Language Association
85 Broad Street, Suite 500
New York, NY 10004-2434

**MMA Andrew W. Mellon Fellowships** **1640**
Annual fellowships, funded by the Andrew W. Mellon Foundation, are awarded for training in one or more of the following departments of the Metropolitan Museum: paintings conservation, objects conservation (including sculpture, metalwork, glass, ceramics, furniture, and archaeological objects), musical instruments, arms and armor, paper conservation, textile conservation, the costume institute, and Asian art conservation. Each fellowship will typically be one year in duration with the possibility of renewal for up to two additional years. Fellowship recipients will receive stipends appropriate to their level of training or experience. Additional funds may be available for travel, books, photographs, and other reasonable or necessary expenses incident to the fellowship. The annual deadline for submissions is the first Friday in November.
*Requirements:* Fellowship applicants should already have reached an advanced level of experience or training; recipients will be expected to spend the duration of the fellowships in residence in the departments with which they are affiliated. It is the responsibility of the applicant to discuss his/her application in detail with the department concerned and to obtain that department's approval prior to submitting an application. Such approval should not, however, be construed as assurance that a fellowship will be awarded.
*Geographic Focus:* All States
*Date(s) Application is Due:* Nov 2
*Amount of Grant:* 42,000 - 48,000 USD
*Contact:* Marcie Karp, Senior Managing Educator; (212) 570-3756; fax (212) 650-2763; marcie.karp@metmuseum.org or academic.programs@metmuseum.org
*Internet:* https://www.metmuseum.org/about-the-met/fellowships/art-history-fellowships/endowed-fellowships
*Sponsor:* Metropolitan Museum of Art
1000 Fifth Avenue
New York, NY 10028-0198

**MMAAP Foundation Dermatology Fellowships** **1641**
The mission of the Milstein Medical Asian American Partnership Foundation (MMAAP Foundation) is to improve world health by developing mutually beneficial partnerships between the United States and China, as well as greater Asia. The Foundation invites the submission of applications from Chinese dermatologists and skin biologists for a Fellowship

Award to support one-year training at a prominent sponsoring institution in the U.S. The Foundation will provide the sole support for these initiatives. The aim of the program is to build enduring partnership between the U.S. and Asia through training of future Chinese academic leaders and to encourage long-term collaborations between the two regions. The award will provide support for one fellow in the amount of $60,000 accompanied by a grant of $25,000 to the sponsoring U.S. institution. These funds will be distributed directly to the sponsoring U.S. institution. The annual application deadline is November 1.
*Requirements:* Chinese applicants should come from major medical centers with demonstrated excellence in dermatology and research; at the rank of Instructor or above by the application deadline; an MD or PhD within the specialty of dermatology; research experience of at least five years with a distinguished publication record; commitment to research and academics; demonstrated leadership skills and excellent English speaking proficiency. The ideal candidate should be coming from China at the beginning of the funding period by the MMAAP Foundation; while applicants who are working, studying or training in the U.S. or have plans to visit the U.S. under another program at the time of application may apply, they must disclose such information in the application. After application submission, applicants must provide updates to the MMAAP Foundation if there are any changes in their professional or personal situation.
*Restrictions:* It is not allowed to submit the same application for multiple awards in the same year. Indirect costs are not allowed.
*Geographic Focus:* All States, China
*Date(s) Application is Due:* Nov 1
*Amount of Grant:* 85,000 USD
*Contact:* Mia Wang, (212) 850-4505; fax (212) 889-4959; mia@mmaapf.org
*Internet:* www.mmaapf.org/en/grants/
*Sponsor:* Milstein Medical Asian American Partnership Foundation
335 Madison Avenue, 15th Floor
New York, NY 10017

**MMAAP Foundation Fellowship Award in Hematology** **1642**
The mission of the Milstein Medical Asian American Partnership Foundation (MMAAP Foundation) is to improve world health by developing mutually beneficial partnerships between the United States and China, as well as greater Asia. The Foundation invites the submission of applications from Chinese hematologists and blood researchers for a Fellowship Award to support one-year training at a prominent sponsoring institution in the U.S. The MMAAP Foundation will provide the sole support for these initiatives. The aim of the program is to build an enduring partnership between the U.S. and Asia through training of future Chinese academic leaders and to encourage long-term collaborations between the two regions. The award will provide support for one fellow in the amount of $60,000 accompanied by a grant of $25,000 to the sponsoring U.S. institution. These funds will be distributed directly to the sponsoring U.S. institution.
*Requirements:* Chinese applicants should come from major medical centers with demonstrated excellence in hematology and blood research; at the rank of Instructor or above by the application deadline; an MD or PhD within the specialty of hematology; research experience of at least five years with a distinguished publication record; commitment to research and academics; demonstrated leadership skills; and excellent English speaking proficiency. The ideal candidate should be coming from China at the beginning of the funding period by the MMAAP Foundation; while applicants who are working, studying or training in the U.S. or have plans to visit the U.S. under another program at the time of application may apply, they must disclose such information in the application. After application submission, applicants must provide updates to the MMAAP Foundation if there are any changes in their professional or personal situation.
*Restrictions:* It is not allowed to submit the same application for multiple awards in the same year. Indirect costs are not allowed.
*Geographic Focus:* All States, China
*Amount of Grant:* 85,000 USD
*Contact:* Mia Wang, (212) 850-4505; fax (212) 889-4959; mia@mmaapf.org
*Internet:* www.mmaapf.org/en/grants/
*Sponsor:* Milstein Medical Asian American Partnership Foundation
335 Madison Avenue, 15th Floor
New York, NY 10017

**MMAAP Foundation Fellowship Award in Reproductive Medicine** **1643**
The mission of the Milstein Medical Asian American Partnership Foundation (MMAAP Foundation) is to improve world health by developing mutually beneficial partnerships between the United States and China, as well as greater Asia. The Foundation invites the submission of applications from Chinese investigators and researchers in reproductive medicine for a Fellowship Award to support one-year training at a prominent sponsoring institution in the U.S. The MMAAP Foundation will provide the sole support for these initiatives. The aim of the program is to build an enduring partnership between the U.S. and Asia through training of future Chinese academic leaders and to encourage long-term collaborations between the two regions. The award will provide support for one fellow in the amount of $60,000 accompanied by a grant of $25,000 to the sponsoring U.S. institution. These funds will be distributed directly to the sponsoring U.S. institution. The annual application deadline is November 1.
*Requirements:* Chinese applicants should come from major medical centers with demonstrated excellence in IVF research; at the rank of Instructor or above by the application deadline; an MD or PhD within the specialty of reproductive medicine; significant research experience with a distinguished publication record; commitment to research and academics; demonstrated leadership skills; and excellent English speaking proficiency. The ideal candidate should be coming from China at the beginning of the funding period by the MMAAP Foundation; while applicants who are working, studying or training in the U.S. or have plans to visit the U.S. under another program at the time of application may apply, they must disclose such information in the application. After application submission, applicants must provide updates to the MMAAP Foundation if there are any changes in their professional or personal situation.
*Restrictions:* It is not allowed to submit the same application for multiple awards in the same year. Indirect costs are not allowed.
*Geographic Focus:* All States, China
*Date(s) Application is Due:* Nov 1
*Amount of Grant:* 85,000 USD
*Contact:* Mia Wang, (212) 850-4505; fax (212) 889-4959; mia@mmaapf.org
*Internet:* www.mmaapf.org/en/grants/
*Sponsor:* Milstein Medical Asian American Partnership Foundation
335 Madison Avenue, 15th Floor
New York, NY 10017

**MMAAP Foundation Fellowship Award in Translational Medicine** **1644**
The mission of the Milstein Medical Asian American Partnership Foundation (MMAAP Foundation) is to improve world health by developing mutually beneficial partnerships between the United States and China, as well as greater Asia. The Foundation invites the submission of applications from Chinese medical researchers for a Fellowship Award to support one-year training at a prominent sponsoring institution in the U.S. The MMAAP Foundation will provide the sole support for these initiatives. Translational Medicine refers to the process of converting scientific discoveries into drugs or medical devices to treat patients. The process includes preclinical studies, observational studies, testing treatment methods and the development of best practices of the treatment. The aim of the program is to build an enduring partnership between the U.S. and Asia through training of future Chinese academic leaders and to encourage long-term collaborations between the two regions. The award will provide support for one fellow in the amount of $60,000 accompanied by a grant of $25,000 to the sponsoring U.S. institution. These funds will be distributed directly to the sponsoring U.S. institution. The annual application deadline is November 1.
*Requirements:* Chinese applicants should come from major medical centers with demonstrated excellence in translational medicine; at the rank of Instructor or above by the application deadline; an MD or PhD within any specialty or medical sub-specialties; significant research experience with a distinguished publication record; commitment to research and academics; demonstrated leadership skills; and excellent English speaking proficiency. The ideal candidate should be coming from China at the beginning of the funding period by the MMAAP Foundation; while applicants who are working, studying or training in the U.S. or have plans to visit the U.S. under another program at the time of application may apply, they must disclose such information in the application. After application submission, applicants must provide updates to the MMAAP Foundation if there are any changes in their professional or personal situation.
*Restrictions:* It is not allowed to submit the same application for multiple awards in the same year. Indirect costs are not allowed.
*Geographic Focus:* All States, China
*Date(s) Application is Due:* Nov 1
*Amount of Grant:* 85,000 USD
*Contact:* Mia Wang, (212) 850-4505; fax (212) 889-4959; mia@mmaapf.org
*Internet:* www.mmaapf.org/en/grants/
*Sponsor:* Milstein Medical Asian American Partnership Foundation
335 Madison Avenue, 15th Floor
New York, NY 10017

**MMAAP Foundation Geriatric Project Awards** **1645**
The mission of the Milstein Medical Asian American Partnership Foundation (MMAAP Foundation) is to improve world health by developing mutually beneficial partnerships between the United States and China, as well as greater Asia. The Foundation invites the submission of applications from Chinese geriatricians for a research project award which will have immediate impact on improving skin care in China. The Foundation will be providing the sole support for these initiatives. The aim of the program is to build enduring partnership between the U.S. and Asia through funding a research project and to encourage long term collaborations between the two regions. The award will provide support for the project to be conducted at a Chinese institution in the amount of $50,000 and support for a U.S. partner institution in the amount of $10,000. These funds will be distributed directly to the sponsoring Chinese and U.S. institutions. The annual deadline for applications is November 1.
*Requirements:* Funds may be used for support of a new or ongoing clinical research project with demonstrated potential for immediate impact on improving senior health in China. Applications should come from major Chinese medical centers with demonstrated excellence in dermatology with a prominent U.S. institution as the partner.
*Geographic Focus:* All States, China
*Date(s) Application is Due:* Nov 1
*Amount of Grant:* 60,000 USD
*Contact:* Mia Wang, (212) 850-4505; fax (212) 889-4959; mia@mmaapf.org
*Internet:* www.mmaapf.org/en/grants/
*Sponsor:* Milstein Medical Asian American Partnership Foundation
335 Madison Avenue, 15th Floor
New York, NY 10017

**MMAAP Foundation Hematology Project Awards** **1646**
The mission of the Milstein Medical Asian American Partnership Foundation (MMAAP Foundation) is to improve world health by developing mutually beneficial partnerships between the United States and China, as well as greater Asia. The Foundation invites the submission of applications from Chinese hematologists for a research project award which will have immediate impact on improving blood health in China. The Foundation

will be providing the sole support for these initiatives. The aim of the program is to build enduring partnership between the U.S. and Asia through funding a research project and to encourage long term collaborations between the two regions. The award will provide support for the project to be conducted at a Chinese institution in the amount of $50,000 and support for a U.S. partner institution in the amount of $10,000. These funds will be distributed directly to the sponsoring Chinese and U.S. institutions. The annual deadline for applications is November 1.
*Requirements:* Funds may be used for support of a new or ongoing clinical research project with demonstrated potential for immediate impact on improving blood health in China. Knowledge gained should be applicable to hematological clinical practice with clear generalizability. Applications should come from major Chinese medical centers with demonstrated excellence in hematology with a prominent U.S. institution as the partner.
*Geographic Focus:* All States, China
*Date(s) Application is Due:* Nov 1
*Amount of Grant:* 60,000 USD
*Contact:* Mia Wang, (212) 850-4505; fax (212) 889-4959; mia@mmaapf.org
*Internet:* www.mmaapf.org/en/grants/
*Sponsor:* Milstein Medical Asian American Partnership Foundation
335 Madison Avenue, 15th Floor
New York, NY 10017

**MMAAP Foundation Research Project Award in Reproductive Medicine 1647**
The mission of the Milstein Medical Asian American Partnership Foundation (MMAAP Foundation) is to improve world health by developing mutually beneficial partnerships between the United States and China, as well as greater Asia. The Foundation invites the submission of applications from Chinese investigators in reproductive medicine for a research project award which will have immediate impact on n vitro fertilization (IVF) practices in China. The Foundation will be providing the sole support for these initiatives. The aim of the program is to build enduring partnership between the U.S. and Asia through funding a research project and to encourage long term collaborations between the two regions. The award will provide support for the project to be conducted at a Chinese institution in the amount of $50,000 and support for a U.S. partner institution in the amount of $10,000. These funds will be distributed directly to the sponsoring Chinese and U.S. institutions. The annual deadline for applications is November 1.
*Requirements:* Funds may be used for support of a new or ongoing clinical research project with demonstrated potential for immediate impact on improving IVF practices in China. Applications should come from major Chinese medical centers with demonstrated excellence in reproductive medicine with a prominent U.S. institution as the partner.
*Geographic Focus:* All States, China
*Date(s) Application is Due:* Nov 1
*Amount of Grant:* 60,000 USD
*Contact:* Mia Wang, (212) 850-4505; fax (212) 889-4959; mia@mmaapf.org
*Internet:* www.mmaapf.org/en/grants/
*Sponsor:* Milstein Medical Asian American Partnership Foundation
335 Madison Avenue, 15th Floor
New York, NY 10017

**MMAAP Foundation Research Project Award in Translational Medicine 1648**
The mission of the Milstein Medical Asian American Partnership Foundation (MMAAP Foundation) is to improve world health by developing mutually beneficial partnerships between the United States and China, as well as greater Asia. The Foundation invites the submission of applications from Chinese medical researchers for a research project award which will have immediate impact on improving health in China. The Foundation will be providing the sole support for these initiatives. Translational Medicine refers to the process of converting scientific discoveries into drugs or medical devices to treat patients. The process includes pre-clinical studies, observational studies, testing treatment methods and the development of best practices of the treatment. The aim of the program is to build enduring partnership between the U.S. and Asia through funding a research project and to encourage long term collaborations between the two regions. The award will provide support for the project to be conducted at a Chinese institution in the amount of $50,000 and support for a U.S. partner institution in the amount of $10,000. These funds will be distributed directly to the sponsoring Chinese and U.S. institutions. The annual deadline for applications is November 1.
*Requirements:* Funds may be used for support of a new or ongoing clinical research project with demonstrated potential for immediate impact on improving health in China. Applications should come from major Chinese medical centers with demonstrated excellence in translational medicine with a prominent U.S. institution as the partner.
*Geographic Focus:* All States, China
*Date(s) Application is Due:* Nov 1
*Amount of Grant:* 60,000 USD
*Contact:* Mia Wang, (212) 850-4505; fax (212) 889-4959; mia@mmaapf.org
*Internet:* www.mmaapf.org/en/grants/
*Sponsor:* Milstein Medical Asian American Partnership Foundation
335 Madison Avenue, 15th Floor
New York, NY 10017

**MMAAP Foundation Senior Health Fellowship Award in Geriatrics Medicine and Aging Research 1649**
The mission of the Milstein Medical Asian American Partnership Foundation (MMAAP Foundation) is to improve world health by developing mutually beneficial partnerships between the United States and China, as well as greater Asia. The Foundation invites the submission of applications from Chinese geriatricians for the Irma and Paul Milstein program for Senior Health Fellowship Award to support one year of training at a prominent sponsoring institution in the U.S. for two to four scholars in geriatrics from China. The Foundation will be providing the sole support for these initiatives. The aim of the program is to build enduring partnership between the U.S. and Asia through training of future Chinese academic leaders and to encourage long-term collaborations between the two regions. The award will provide support for the fellow in the amount of $60,000 accompanied by a grant of $25,000 to the sponsoring U.S. institution. These funds will be distributed directly to the sponsoring U.S. institution. The annual application deadline is November 1.
*Requirements:* Chinese applicants should come from major medical centers with demonstrated excellence in geriatrics and research; at the rank of Instructor or above by the application deadline; an MD or PhD within the specialty of geriatrics; research experience of at least five years with a distinguished publication record; commitment to research and academics; demonstrated leadership skills and excellent English speaking proficiency. The ideal candidate should be coming from China at the beginning of the funding period by the MMAAP Foundation; while applicants who are working, studying or training in the U.S. or have plans to visit the U.S. under another program at the time of application may apply, they must disclose such information in the application. After application submission, applicants must provide updates to the MMAAP Foundation if there are any changes in their professional or personal situation.
*Restrictions:* It is not allowed to submit the same application for multiple awards in the same year. Indirect costs are not allowed.
*Geographic Focus:* All States, China
*Date(s) Application is Due:* Nov 1
*Amount of Grant:* 85,000 USD
*Contact:* Mia Wang, (212) 850-4505; fax (212) 889-4959; mia@mmaapf.org
*Internet:* www.mmaapf.org/en/grants/
*Sponsor:* Milstein Medical Asian American Partnership Foundation
335 Madison Avenue, 15th Floor
New York, NY 10017

**MMA Bothmer Fellowship 1650**
Almost 150 years after its founding in 1870, The Metropolitan Museum of Art is as strongly committed as ever to its educational mission of supporting scholarly investigation and research into its encyclopedic collection, and to contributing to broader academic discourses. The annual Bothmer Fellowship is awarded to an outstanding graduate student who has been admitted to a doctoral program of a university in the United States and who has submitted an outline of a thesis dealing with Greek and Roman art. The thesis outline must have already been accepted by the applicant's thesis advisor at the time of the fellowship application. Preference will be given to the applicant who, in the opinion of the grants committee, will profit most from utilizing the resources of the Department of Greek and Roman Art, its collections, library, photographic and other archives, and the guidance of the curatorial staff. The annual deadline for submissions is the first Friday in November.
*Requirements:* It is the responsibility of the applicant, in connection with any project that would require assistance from a particular department, to discuss the project with the department concerned and to obtain that department's approval prior to submitting an application; however, departmental approval should not be construed as assurance of a fellowship.
*Restrictions:* Fellowships generally cannot be given for projects proposing that an exhibition be organized and presented within the fellowship period.
*Geographic Focus:* All States
*Date(s) Application is Due:* Nov 2
*Amount of Grant:* 48,000 - 58,000 USD
*Contact:* Marcie Karp, Senior Managing Educator; (212) 570-3756; fax (212) 650-2763; marcie.karp@metmuseum.org or academic.programs@metmuseum.org
*Internet:* https://www.metmuseum.org/about-the-met/fellowships/art-history-fellowships/endowed-fellowships
*Sponsor:* Metropolitan Museum of Art
1000 Fifth Avenue
New York, NY 10028-0198

**MMA Chester Dale Fellowships 1651**
Almost 150 years after its founding in 1870, The Metropolitan Museum of Art is as strongly committed as ever to its educational mission of supporting scholarly investigation and research into its encyclopedic collection, and to contributing to broader academic discourses. Individuals whose fields of study are related to the fine arts of the Western world and who are, preferably, American citizens under the age of 40 are eligible for Chester Dale fellowships for research at the Metropolitan Museum of Art. Grants typically cover periods from three months to a year. The annual deadline for submissions is the first Friday in November.
*Requirements:* It is the responsibility of the applicant, in connection with any project that may reasonably be expected to require assistance from a particular museum department, to discuss the project with the department concerned and to obtain that department's approval before submitting an application. However, such approval should not be construed as assurance of being awarded a fellowship.
*Restrictions:* Fellowships generally cannot be given for projects proposing that an exhibition be organized and presented within the fellowship period.
*Geographic Focus:* All States
*Date(s) Application is Due:* Nov 2
*Amount of Grant:* 48,000 - 58,000 USD
*Contact:* Marcie Karp, Senior Managing Educator; (212) 570-3756; fax (212) 650-2763; marcie.karp@metmuseum.org or academic.programs@metmuseum.org
*Internet:* https://www.metmuseum.org/about-the-met/fellowships/art-history-fellowships/endowed-fellowships
*Sponsor:* Metropolitan Museum of Art
1000 Fifth Avenue
New York, NY 10028-0198

**MMA Douglass Foundation Fellowship in American Art** 1652

The Metropolitan Museum of Art's Douglass Foundation Fellowship in American Art is awarded in honor of John K. Howat to a promising young scholar for one year's study or research in The American Wing on an aspect of the Museum's collection. Applicants should have been enrolled for at least one year in an advanced degree program in the field of American art or culture. The annual deadline for submissions is the first Friday in November.
*Requirements:* It is the responsibility of the applicant, in connection with any project that may reasonably be expected to require assistance from a particular museum department, to discuss the project with the department concerned and to obtain that department's approval before submitting an application. However, such approval should not be construed as assurance of being awarded a fellowship.
*Restrictions:* Fellowships generally cannot be given for projects proposing that an exhibition be organized and presented within the fellowship period.
*Geographic Focus:* All States
*Date(s) Application is Due:* Nov 2
*Amount of Grant:* 48,000 - 58,000 USD
*Contact:* Marcie Karp, Senior Managing Educator; (212) 570-3756; fax (212) 650-2763; marcie.karp@metmuseum.org or academic.programs@metmuseum.org
*Internet:* https://www.metmuseum.org/about-the-met/fellowships/art-history-fellowships/endowed-fellowships
*Sponsor:* Metropolitan Museum of Art
1000 Fifth Avenue
New York, NY 10028-0198

**MMA Emily K. Rafferty Twelve-Month Internship in Museum Administration** 1653

Emily K. Rafferty is president emerita of The Metropolitan Museum of Art. Recognized nationally and internationally for her leadership and communication expertise, she served The Met for forty years in various positions, including as chief fundraiser, and a decade as President. She oversaw the administration of the institution's 2,300 employees, financial and legal issues, revenue sources, security, and government relations. The Emily K. Rafferty twelve-month internship welcomes applications from students across disciplines who have a strong commitment and passion for leadership and the arts. The internship provides training in many aspects of museum administration, such as education, finance, legal affairs, fundraising, governance, public relations, visitor and market research, government affairs, human resources, merchandising, facilities management, and information systems. With the support and guidance of a staff mentor, the intern rotates through three different departments over the course of twelve months, working on a specific project with a dedicated supervisor in each department. This rotational program will provide maximum exposure to a range of professional experiences and opportunities for a holistic understanding of the operations of the Museum. The ideal applicant will have concentrated their studies in arts administration, art history, or a related field, possess strong interpersonal, communication, and writing skills. The Met strongly encourages applications from members of groups underrepresented in careers related to museums and the visual arts. Compensation is approximately $28,000, plus up to $3,000 for participation in research, conferences, or other professional-development opportunities related to the internship. The annual online application deadline is January 28.
*Requirements:* This internship is open to current undergraduate seniors who will have graduated by the beginning of the program, current graduate students at the master's level, and recent graduates of either an undergraduate or a master's-level program. Recent graduates must have completed their degrees no more than twelve months prior to the application deadline.
*Restrictions:* Current freshmen, sophomores, and juniors are not eligible to apply.
*Geographic Focus:* All States
*Date(s) Application is Due:* Jan 28
*Amount of Grant:* 28,000 - 31,000 USD
*Contact:* Marcie Karp, Senior Managing Educator; (212) 570-3756; fax (212) 650-2763; marcie.karp@metmuseum.org or academic.programs@metmuseum.org
*Internet:* https://www.metmuseum.org/about-the-met/internships/undergraduate-and-graduate-students/long-term
*Sponsor:* Metropolitan Museum of Art
1000 Fifth Avenue
New York, NY 10028-0198

**MMA Hanns Swarzenski and Brigitte Horney Swarzenski Fellowship** 1654

The Metropolitan Museum of Art's Hanns Swarzenski and Brigitte Horney Swarzenski Fellowship is awarded to a promising young scholar for study and research at the Museum. Preference will be given to applicants with a proven interest in museum work or those planning to pursue a museum career in the field of medieval art. Should in any given year the Grants Committee find no suitable candidate for a fellowship in medieval art, the Museum may instead select a suitable candidate in the field of European decorative arts. The annual deadline for submissions is the first Friday in November.
*Requirements:* It is the responsibility of the applicant, in connection with any project that may reasonably be expected to require assistance from a particular museum department, to discuss the project with the department concerned and to obtain that department's approval before submitting an application. However, such approval should not be construed as assurance of being awarded a fellowship.
*Restrictions:* Fellowships generally cannot be given for projects proposing that an exhibition be organized and presented within the fellowship period.
*Geographic Focus:* All States
*Date(s) Application is Due:* Nov 2
*Amount of Grant:* 48,000 - 58,000 USD
*Contact:* Marcie Karp, Senior Managing Educator; (212) 570-3756; fax (212) 650-2763; marcie.karp@metmuseum.org or academic.programs@metmuseum.org
*Internet:* https://www.metmuseum.org/about-the-met/fellowships/art-history-fellowships/endowed-fellowships
*Sponsor:* Metropolitan Museum of Art
1000 Fifth Avenue
New York, NY 10028-0198

**MMA J. Clawson Mills Scholarships** 1655

The J. Clawson Mills Scholarships are awarded for one year of study or research at the Metropolitan Museum in any branch of the fine arts relating to the museum's collections and carry the possibility of renewal for a second year. It is the responsibility of the applicant to discuss the intended project with the department concerned and to obtain that department's approval before submitting an application. The annual deadline for submissions is the first Friday in November.
*Requirements:* Mature scholars of demonstrated ability who are interested in pursuing research projects relevant to the museum's collections are eligible to apply.
*Geographic Focus:* All States
*Date(s) Application is Due:* Nov 2
*Amount of Grant:* 48,000 - 58,000 USD
*Contact:* Marcie Karp, Senior Managing Educator; (212) 570-3756; fax (212) 650-2763; marcie.karp@metmuseum.org or academic.programs@metmuseum.org
*Internet:* https://www.metmuseum.org/about-the-met/fellowships/art-history-fellowships/endowed-fellowships
*Sponsor:* Metropolitan Museum of Art
1000 Fifth Avenue
New York, NY 10028-0198

**MMA Jane and Morgan Whitney Fellowships** 1656

The Jane and Morgan Whitney fellowships are awarded for study, work, or research to students of the fine arts whose fields are related to the museum's collections, with preference given to students in the decorative arts who are under 40 years of age. The fellowship carries the possibility of renewal for one additional year. The annual deadline for submissions is the first Friday in November.
*Requirements:* Applicants must discuss the proposed project with the department concerned and obtain approval before submitting the application.
*Geographic Focus:* All States
*Date(s) Application is Due:* Nov 2
*Amount of Grant:* 48,000 - 58,000 USD
*Contact:* Marcie Karp, Senior Managing Educator; (212) 570-3756; fax (212) 650-2763; marcie.karp@metmuseum.org or academic.programs@metmuseum.org
*Internet:* https://www.metmuseum.org/about-the-met/fellowships/art-history-fellowships/endowed-fellowships
*Sponsor:* Metropolitan Museum of Art
1000 Fifth Avenue
New York, NY 10028-0198

**MMA Lifchez/Stronach Curatorial Internship** 1657

The Lifchez/Stronach Curatorial Internship will be awarded to a student who has recently graduated from college or is enrolled in an art history MA program and would use the internship to continue in a career in art history. The student should come from a background of financial need or other disadvantage that might jeopardize her or his pursuing such a career without this support. The intern will be assigned to a curatorial department to work on projects appropriate to her/his academic background, professional skills, and career goals. The internship is full-time from September to June. Interns are invited to participate in orientation activities during the summer. Annual deadline dates vary; contact the program office for exact dates. Compensation is approximately $21,000, plus up to a $2,000 stipend available for research and educational travel related to the internship. The annual deadline for application submission is January 28.
*Requirements:* Open to current undergraduate seniors, recent graduates from a bachelor's or associate's program, and master's-level graduate students. Recent graduates must have completed their degrees no more than twelve months prior to the application deadline.
*Geographic Focus:* All States
*Date(s) Application is Due:* Jan 28
*Amount of Grant:* 21,000 - 23,000 USD
*Contact:* Marcie Karp, Senior Managing Educator; (212) 570-3756; fax (212) 650-2763; marcie.karp@metmuseum.org or academic.programs@metmuseum.org
*Internet:* https://www.metmuseum.org/about-the-met/internships/undergraduate-and-graduate-students/long-term
*Sponsor:* Metropolitan Museum of Art
1000 Fifth Avenue
New York, NY 10028-0198

**MMA Marica and Jan Vilcek Fellowship** 1658

The Metropolitan Museum of Art's Marica and Jan Vilcek Fellowship is awarded to predoctoral-level scholars from the United States and abroad for one year's study and research. Preference for the Vilcek Fellowship will be given to candidates pursuing doctorate degrees from The Institute of Fine Arts, New York University, but the Museum has the discretion to select other candidates pursuing doctoral degrees if there are more qualified applicants from other institutions. The annual deadline for submissions is the first Friday in November.
*Requirements:* It is the responsibility of the applicant, in connection with any project that may reasonably be expected to require assistance from a particular museum department, to discuss the project with the department concerned and to obtain that department's approval before submitting an application. However, such approval should not be construed as assurance of being awarded a fellowship.

*Restrictions:* Fellowships generally cannot be given for projects proposing that an exhibition be organized and presented within the fellowship period.
*Geographic Focus:* All States
*Date(s) Application is Due:* Nov 2
*Amount of Grant:* 48,000 - 58,000 USD
*Contact:* Marcie Karp, Senior Managing Educator; (212) 570-3756; fax (212) 650-2763; marcie.karp@metmuseum.org or academic.programs@metmuseum.org
*Internet:* https://www.metmuseum.org/about-the-met/fellowships/art-history-fellowships/endowed-fellowships
*Sponsor:* Metropolitan Museum of Art
1000 Fifth Avenue
New York, NY 10028-0198

**MMA Polaire Weissman Fund Fellowships** **1659**
The Polaire Weissman Fund provides fellowships to qualified graduate students who preferably will have completed graduate studies in the fine arts or studies in costume and who are interested in pursuing costume history in museum or teaching careers or other careers including conservation related to the field of art, architecture, cultural studies, design, or costume. These fellowships, generally awarded for nine-month terms, are offered in alternate years. The annual deadline for submissions is the first Friday in November.
*Requirements:* Applicants must discuss the proposed project with the department concerned and obtain approval before submitting the application.
*Geographic Focus:* All States
*Date(s) Application is Due:* Nov 2
*Amount of Grant:* 48,000 - 58,000 USD
*Contact:* Marcie Karp, Senior Managing Educator; (212) 570-3756; fax (212) 650-2763; marcie.karp@metmuseum.org or academic.programs@metmuseum.org
*Internet:* https://www.metmuseum.org/about-the-met/fellowships/art-history-fellowships/endowed-fellowships
*Sponsor:* Metropolitan Museum of Art
1000 Fifth Avenue
New York, NY 10028-0198

**MMA Slifka Foundation Interdisciplinary Fellowship** **1660**
The Metropolitan Museum of Art's Slifka Foundation Interdisciplinary Fellowship is awarded to a completed master's- or PhD-level candidate for training in an interdisciplinary approach, joining art historical research with technical investigation of the Museum's Northern Renaissance paintings. The recipient of this fellowship will conduct research with the curator for the collection catalogue of early Netherlandish paintings. The annual deadline for submissions is the first Friday in November.
*Requirements:* It is the responsibility of the applicant, in connection with any project that may reasonably be expected to require assistance from a particular museum department, to discuss the project with the department concerned and to obtain that department's approval before submitting an application. However, such approval should not be construed as assurance of being awarded a fellowship.
*Restrictions:* Fellowships generally cannot be given for projects proposing that an exhibition be organized and presented within the fellowship period.
*Geographic Focus:* All States
*Date(s) Application is Due:* Nov 2
*Amount of Grant:* 48,000 - 58,000 USD
*Contact:* Marcie Karp, Senior Managing Educator; (212) 570-3756; fax (212) 650-2763; marcie.karp@metmuseum.org or academic.programs@metmuseum.org
*Internet:* https://www.metmuseum.org/about-the-met/fellowships/art-history-fellowships/endowed-fellowships
*Sponsor:* Metropolitan Museum of Art
1000 Fifth Avenue
New York, NY 10028-0198

**MMA Sylvan C. Coleman and Pam Coleman Memorial Fund Fellowship** **1661**
The Sylvan C. Coleman and Pam Coleman Memorial Fund Fellowship is awarded to predoctoral- and postdoctoral-level scholars from the United States and abroad for one year's study and research. This fellowship is reserved for talented individuals with promising art-historical projects that will contribute to the field at large. The annual deadline for submissions is the first Friday in November.
*Requirements:* Competition is open to conservators, art historians, and scientists who are at advanced levels in their training and have demonstrated commitment to the physical examination and treatment of objects of art. It is the responsibility of the applicant to discuss her or his application in detail with the department and to obtain that department's approval to submit the application.
*Geographic Focus:* All States
*Date(s) Application is Due:* Nov 2
*Contact:* Marcie Karp, Senior Managing Educator; (212) 570-3756; fax (212) 650-2763; marcie.karp@metmuseum.org or academic.programs@metmuseum.org
*Internet:* https://www.metmuseum.org/about-the-met/fellowships/art-history-fellowships/endowed-fellowships
*Sponsor:* Metropolitan Museum of Art
1000 Fifth Avenue
New York, NY 10028-0198

**MMA Theodore Rousseau Fellowships** **1662**
Fellowships are for the training of students planning to enter museums as curators of painting, by enabling them to undertake related study in Europe. The purpose of the Theodore Rousseau Fellowship is to develop the skills of connoisseurship by supporting firsthand examination of paintings in major European collections, rather than to support library research for the completion of degree requirements. Short-term fellowships of at least three months will be considered along with 12-month requests. The annual deadline for submissions is the first Friday in November.
*Requirements:* Applicants should have been enrolled for at least one year in advanced degree programs in the field of art history. It is the responsibility of the applicant to discuss the project with the department concerned and to obtain the department's approval before submitting the application.
*Geographic Focus:* All States
*Date(s) Application is Due:* Nov 2
*Amount of Grant:* 48,000 - 58,000 USD
*Samples:* Daniel Schulman, Institute of Fine Arts, New York U (New York, NY)—to support dissertation work on Edgar Degas, $18,000.
*Contact:* Marcie Karp, Senior Managing Educator; (212) 570-3756; fax (212) 650-2763; marcie.karp@metmuseum.org or academic.programs@metmuseum.org
*Internet:* https://www.metmuseum.org/about-the-met/fellowships/art-history-fellowships/endowed-fellowships
*Sponsor:* Metropolitan Museum of Art
1000 Fifth Avenue
New York, NY 10028-0198

**MMA Tiffany and Company Foundation Twelve-Month Curatorial Internship in American Decorative Arts** **1663**
The Tiffany and Company Foundation twelve-month internship is awarded to an individual with a special interest in American decorative arts. Curators in the American Wing work with the Tiffany & Co. Foundation intern to develop their knowledge and skills through hands-on projects researching and documenting the permanent collection, special exhibitions, and/or additional ongoing departmental projects. During their internship, the Tiffany intern also pursues an independent research project culminating in a small gallery installation. Compensation is approximately $28,000, plus up to $3,000 for research and educational travel. The annual online application deadline is January 28.
*Geographic Focus:* All States
*Date(s) Application is Due:* Jan 28
*Amount of Grant:* 28,000 - 31,000 USD
*Contact:* Marcie Karp, Senior Managing Educator; (212) 570-3756; fax (212) 650-2763; marcie.karp@metmuseum.org or academic.programs@metmuseum.org
*Internet:* https://www.metmuseum.org/about-the-met/internships/undergraduate-and-graduate-students/long-term
*Sponsor:* Metropolitan Museum of Art
1000 Fifth Avenue
New York, NY 10028-0198

**MMA Twelve-Month Internship in the Digital Department** **1664**
The Digital Department offers two twelve-month internships for individuals with a background in video editing and production and/or animation, who have completed coursework in any of the following: visual or moving image arts, art education, art history, communications, journalism, history, archaeology or related fields. Individuals will have a strong desire to participate in the creation of new media solutions. The Digital Department leads the creation, production, presentation, and dissemination of the Museum's multimedia content to support the viewing and understanding of The Met collection and exhibitions, both in the galleries and online. The department's Twelve-Month Interns will learn about and assist with all of the work within the department's Creative Production area. Compensation is approximately $28,000. The annual online application deadline is January 28.
*Requirements:* Ideal candidates will have strong and varied multimedia skills and undergraduate coursework in at least one of the areas mentioned above. Strong interest in art history or education, production experience with video or animations, and submission of a portfolio or links to videos is required. This internship is open to current undergraduate seniors who will have graduated by the beginning of the program, and recent graduates from a bachelor's or an associate's program (within twelve months of the application deadline).
*Restrictions:* Graduate students and current undergraduate freshmen, sophomores, and juniors are not eligible to apply.
*Geographic Focus:* All States
*Date(s) Application is Due:* Jan 28
*Amount of Grant:* 28,000 USD
*Contact:* Marcie Karp, Senior Managing Educator; (212) 570-3756; fax (212) 650-2763; marcie.karp@metmuseum.org or academic.programs@metmuseum.org
*Internet:* https://www.metmuseum.org/about-the-met/internships/undergraduate-and-graduate-students/long-term
*Sponsor:* Metropolitan Museum of Art
1000 Fifth Avenue
New York, NY 10028-0198

**MoEF Dr. Salim Ali National Wildlife Fellowship** **1665**
In order to commemorate the memory of the great wildlife conservationist of the country, Dr. Salim Ali to inspire and promote, particularly the younger generation of wildlife managers and scientists for taking up research and experimental projects aimed at conservation of the rich wildlife heritage of the Government of India. The Ministry of Environment and Forests has decided to award the fellowship alternatively each year. The fellowship will be awarded normally for a period of two years at the rate of Rs. 4000 per month. In addition, an amount of Rs. 18,000 per annum will be paid for meeting the contingent expenditure. Preference will be given to the officials of the State and/or Forest and Wildlife Department.

*Requirements:* Only Indian citizens engaged in wildlife management or research are eligible.
*Geographic Focus:* India
*Amount of Grant:* 132,000 INR
*Contact:* Munjurul Hannan Khan, Deputy Secretary; +880 2 71 67 472; fax +880 1 71 344 31 79; dsenv1@moef.gov.bd or munjurulkhan@gmail.com
*Internet:* moef.nic.in/modules/fellowships-and-awards/awards/index.php?f=1
*Sponsor:* Ministry of Environment and Forests
Paryavaran Bhawan, CGO Complex, Lodhi Road
New Delhi, 110 003 India

**MOEF Indira Gandhi Paryavaran Puraskar Award** **1666**

The MOEF Indira Gandhi Paryavaran Puraskar Award was instituted in 1987 by the Ministry of Environment and Forests (MoEF), Government of India. This national prize on environment is named after the former Prime Minister of India late Shrimati Indira Gandhi and shall be known as Indira Gandhi Paryavaran Puraskar. The prize will be given each year on any environment related occasion for significant contributions in the field of environment, which are recognized as having measurable impact on the protection and overall improvement of the environment. From the year 2009 onwards, two prizes of Rs. 5.00 lakh each along with silver lotus trophy under the organization category, and three prizes of Rs. 5.00 lakh, Rs. 3.00 lakh and Rs. 2.00 lakh respectively to individuals in individual category along with silver lotus trophy shall be given annually. Following is an indicative but not exhaustive list of work areas concerned: prevention of pollution; conservation of natural resources; rational use of depletable resources; environmental planning and management; environmental impact assessment; outstanding field work; environmental education; and creating awareness of environmental issues.
*Requirements:* Any citizen of India or Indian organization engaged in the Scientific Work or Field Work for the cause of protection of environment is eligible. There is no age limit for individuals.
*Geographic Focus:* India
*Amount of Grant:* 200,000 - 500,000 INR
*Contact:* S M Munjurul Hannan Khan, Deputy Secretary; +880 2 71 67 472; fax +880 1 71 344 31 79; dsenv1@moef.gov.bd or munjurulkhan@gmail.com
Shri Nilkanth Ghosh, Statistical Advisor; +880 2 43 61 078 or +880 2 43 60 837; neelkanthghosh@nic.in or nilkanta_ghosh@yahoo.com
*Internet:* moef.nic.in/citizen/award/igpp.html
*Sponsor:* Ministry of Environment and Forests
Paryavaran Bhawan, CGO Complex, Lodhi Road
New Delhi, 110 003 India

**Montgomery County Community Foundation Animal Welfare Grants** **1667**

The Montgomery County Community Foundation Grants help fund non-profit organizations and agencies in Montgomery County, Indiana. Significant grants have been awarded to organizations such as the Crawfordsville District Public Library, Boys and Girls Club, and the Family Crisis Shelter. Grants in the area of animal welfare include activities that: improve and promote animal well-being; provide general health services for animals; and support animal research. There are three annual deadlines, including: February 12; July 16; and October 8.
*Requirements:* Grant applications will be accepted from any new or existing nonprofit organization in Montgomery County. These organizations must be a tax-exempt organization whose purposes are described in Section 501(c)3 of the Internal Revenue Service Code. Grant applications will also be accepted from local governmental entities, such as the three Montgomery County school corporations, for charitable purposes. A strong proposal will have several or all of the following characteristics: an estimate of who and how many will benefit; show long term potential; address a community problem of some significance for which funding is not covered by the regular budget; present an innovative and practical approach to solve a community problem or project; identify possible future funding, if needed; give evidence of the stability and qualifications of the organization applying; show cooperation within the organization and avoid duplication effort.
*Restrictions:* The Foundation will usually not fund any of the following: grants to individuals; programs which are religious or sectarian in nature, except when the program is open to the entire community; operating expenses such as salaries and utilities; parades, festivals and sporting events; endowment funds; any propaganda, political or otherwise, attempting to influence legislation or intervene in any political affairs or campaigns; an organization's past debts or existing obligations; or post-event or after-the-fact situations.
*Geographic Focus:* Indiana
*Date(s) Application is Due:* Feb 12; Jul 16; Oct 8
*Amount of Grant:* Up to 50,000 USD
*Contact:* Cheryl Keim, Grants and Community Relations Coordinator; (765) 362-1267; fax (765) 361-0562; cheryl@mccf-in.org
*Internet:* https://www.mccf-in.org/animal-lovers-funds
*Sponsor:* Montgomery County Community Foundation
119 East Main Street
Crawfordsville, IN 47933

**Morgan Adams Foundation Research Grants** **1668**

The Morgan Adams Foundation supports laboratory and clinical research in the area of pediatric cancer, with an emphasis on cancers of the brain and spine. The Foundation's particular interest is to provide seed and bridge grants for viable investigations not yet ripe enough for total funding by larger organizations. Additionally, the Foundation is able to partially fund translational studies, experimental therapeutics studies and/or Phase I and II clinical trials on a case by case basis. The Foundation is seeking out projects that are highly collaborative and/or multi-institutional in nature, hoping to facilitate through the strategic placement of funding, more rapid trials development in order to get the best possible treatment options into circulation with the greatest expediency possible. Through careful consideration and placement of its funding, the Foundation encourages and supports research intended to improve treatment effectiveness, improve treatment outcomes and improve the quality of life for children battling cancer.
*Geographic Focus:* All States
*Contact:* Joan Slaughter, Executive Director; (303) 758-2130; fax (303) 758-2134; joan@morganadamsfoundation.org or info@morganadamsfoundation.org
*Internet:* morganadamsfoundation.hotpressplatform.com/funding-research/how-funds-become-hope
*Sponsor:* Morgan Adams Foundation
5303 East Evans Avenue, Suite 200
Denver, CO 80222

**Motorola Foundation Innovation Generation Collaborative Grants** **1669**

The Innovation Generation Collaborative Grants Support medium to large-scale STEM education collaborations between two or more non-profit organizations, schools and/or school districts, with joint requests ranging from $150,000 to $500,000. Programming must have nationwide and/or online reach or take place near communities where Motorola has an employee presence: Tempe, Arizona; San Diego, California; Miami, Florida; Chicago, Illinois; Lowell, Massachusetts; New York City and Long Island, New York; Philadelphia, Pennsylvania; or Fort Worth, Texas. Program must demonstrate measurable outcomes & impact on at least 250 primary participants. Grants may be for up to two years of project work starting after June of this year. Grant applications are accepted at the Motorola Foundation website, with a deadline date of April 1st.
*Requirements:* Any U.S. non-profit organization, school or school district may apply. Programming must have nationwide and/or online reach or take place near communities where Motorola has an employee presence: Tempe, Arizona; San Diego, California; Miami, Florida; Chicago, Illinois; Lowell, Massachusetts; New York City and Long Island, New York; Philadelphia, Pennsylvania; or Fort Worth, Texas. All requests must be partnerships between two or more non-profit organizations or schools. Only prior Motorola Foundation Innovation Generation grant recipients may apply. Organizations will be expected to interact and share program attributes with other grantees through conferences and online forums. Applications must be accompanied by letters of support from each member of the partnership. Funding priority will be placed on partnerships that: focus on a key STEM interest area such as teacher training, girls programming, out-of-school programming or online programs; engage students and teachers in innovative, hands-on activities; teach not only STEM, but also innovative thinking and creative problem-solving skills; focus on girls and underrepresented minorities; engage Motorola employees as volunteers.
*Restrictions:* Only prior Motorola Foundation Innovation Generation grant recipients may apply.
*Geographic Focus:* All States
*Amount of Grant:* 150,000 - 500,000 USD
*Samples:* Society of Women Engineers, Chicago, IL,—encourages girls in grades K-12 to capture the spirit of innovation and open their eyes and minds to a rewarding profession through a host of events, resources and in-person outreach programs (Out-of-School Programming); Texas Alliance for Minorities in Engineering (TAME), Austin, TX, TAME Statewide Math and Science Competition engages hundreds of female and underrepresented middle school and high school students from across Texas through individual competition and a Team Design Challenge, encouraging teamwork, leadership, academic achievement and an interest in science and math-related careers. (Out-of—School Programming).
*Contact:* Matthew Blakely, (847) 576-7895; fax (847) 576-9440; giving@motorola.com
*Internet:* www.motorola.com/staticfiles/Business/Corporate/US-EN/corporate-responsibility/society/community-investment-education-more-about-innovation-generation-grants.html?globalObjectId=8153
*Sponsor:* Motorola Foundation
1303 E Algonquin Road
Schaumburg, IL 60196

**MSI Alden G. Clayton Dissertation Proposal Competition** **1670**

This annual Alden G. Clayton Dissertation Proposal competition will grant up to five awards of $5,000 each for dissertation proposals. Some preference will be given to proposals that address the current MSI Research Priority Topics. (See the website for current research priorities.) Each winner will receive an unrestricted cash prize of $5,000 designed to aid in the completion of the dissertation. Winners are asked to submit the results to MSI's working paper series. The annual deadline for submissions is August 8.
*Requirements:* Entrants must be candidates for the doctorate in marketing or a related field (e.g., organizational behavior, consumer psychology, economics, strategy, management, finance etc.) in a fully accredited university. The dissertation must be no more than fifty percent completed at the time of submission, and the primary thesis advisor must approve submission of the proposal for the competition. Submissions must be accompanied by a letter from the primary thesis advisor certifying that the dissertation is no more than fifty percent completed at the time of submission.
*Restrictions:* For blind reviewing purposes, omit any identifying information (name, school, etc.) from the proposal itself. Include with the submission a separate file/hard copy sheet containing your name, the title of the dissertation, school, and detailed contact information for both the entrant and his/her thesis advisor(s).
*Geographic Focus:* All States
*Date(s) Application is Due:* Aug 8
*Amount of Grant:* 5,000 USD
*Contact:* Gordon Wyner, Research Director; (617) 491-2060, ext. 231; fax (617) 491-2065; gordon@msi.org or msi@msi.org

*Internet:* www.msi.org/research/awards/clayton-dissertation-proposal-competition/
*Sponsor:* Marketing Science Institute
1000 Massachusetts Avenue
Cambridge, MA 02138-5396

**MSI H. Paul Root Award** **1671**

The H. Paul Root Award is named in honor of H. Paul Root, who was President of MSI from 1990-98. It is given by members of the Journal of Marketing editorial review board to a paper that has made a significant contribution to the advancement of the practice of marketing, and is cosponsored by the American Marketing Association and the Marketing Science Institute. Before 1996, this award was funded by the Alpha Kappa Psi Foundation and was known as the Alpha Kappa Psi Award.
*Geographic Focus:* All States
*Date(s) Application is Due:* Aug 8
*Contact:* Gordon Wyner, Research Director; (617) 491-2060, ext. 231; fax (617) 491-2065; gordon@msi.org or msi@msi.org
*Internet:* www.msi.org/research/awards/msi-h.-paul-root-award/
*Sponsor:* Marketing Science Institute
1000 Massachusetts Avenue
Cambridge, MA 02138-5396

**MSI Research Support Standard Grants** **1672**

MSI funds high-quality research that deals with topics of importance to member companies. The organization supports research with the potential for application by managers as well as more basic or exploratory work. No one approach or methodology is favored over another as long as the form is appropriate to the objectives of the research. Studies may be conceptual or empirical and may involve literature reviews, comparative studies, field or laboratory experiments, model building, or theory development. Cross-disciplinary work building on theories, research results, and methods from disciplines of relevance to marketing are encouraged. MSI and its member companies strongly endorse using actual consumers, customers, and executives rather than student subjects in research projects.
*Requirements:* Academic researchers (faculty members, or doctoral students working collaboratively with faculty advisors) can apply for financial and/or non-financial support for research projects. Most MSI grants are made to cover researchers' out-of-pocket costs for data collection, respondent fees, research assistants, and similar expenses. Both proposals and pre-proposals are accepted, though there is no required standard format. MSI operates on a monthly review cycle, and proposals may be submitted at any time. See website for detailed guidelines.
*Restrictions:* Grants do not provide salary replacement for the principal researcher(s), funds for the purchase of equipment or software, university overhead, tuition, or funds for travel to non-MSI conferences.
*Geographic Focus:* All States
*Amount of Grant:* 5,000 - 20,000 USD
*Contact:* Gordon Wyner, Research Director; (617) 491-2060; fax (617) 491-2065; gordon@msi.org or research@msi.org
*Internet:* www.msi.org/research/obtain-research-support/
*Sponsor:* Marketing Science Institute
1000 Massachusetts Avenue
Cambridge, MA 02138-5396

**Mt. Sinai Health Care Foundation Academic Medicine and Bioscience Grants** **1673**

The Foundation seeks to strengthen Cleveland's prominence as a national leader in medical education and biomedical science, perpetuating Mt. Sinai's leadership role in these areas. Emerging collaborations among Cleveland's academic medical centers present opportunities to build organizational capacity and advance progress. The Board of Directors of the Foundation meets on a quarterly basis to review proposals.
*Requirements:* 501(c)3 nonprofits serving greater Cleveland, Ohio may submit proposals for grant support. The foundation welcomes and encourages an informal conversation with program staff prior to the submission of a grant request.
*Restrictions:* In general, the foundation does not support general operating expenses, direct provision of health services, building or equipment expenses, fund-raising events, projects outside of greater Cleveland, endowment funds, lobbying, program advertising, grants for individuals, or scholarships.
*Geographic Focus:* Ohio
*Date(s) Application is Due:* Jan 1; Apr 1; Jul 1; Oct 1
*Contact:* Shelly Galvin; (216) 421-5500; fax (216) 421-5633; sgg2@case.edu
*Internet:* www.mtsinaifoundation.org/whatwefund_academic.html
*Sponsor:* Mount Sinai Health Care Foundation
11000 Euclid Avenue
Cleveland, OH 44106-1714

**NACC David Stevenson Fellowships** **1674**

The goals of the David Stevenson fellowship are to acknowledge David Stevenson's many contributions to the development of data and information about the nonprofit sector and the important role he played in teaching and working with undergraduate students; and to encourage junior faculty members of color to provide leadership for the field and to assist them in gaining tenure in their respective academic institutions. Two fellowships are awarded annually. Fellowships may be used to support research, travel to present research at conferences, and professional education seminars related to research and teaching. The fellowship must be completed between August 1 and July 31. The award will be made payable to the award recipient's affiliate institution. Guidelines are available online. The Nonprofit Academic Centers Council (NACC) is a membership association comprised of academic centers or programs at accredited colleges and universities that focus on the study of nonprofit organizations, voluntarism and/or philanthropy. Established in 1991, NACC is the first group entirely dedicated to the promotion and networking of centers that provide research and education in philanthropy and the nonprofit sector.
*Requirements:* Applicants must meet the following eligibility criteria: be junior faculty in a tenure track position; be affiliated with an accredited institution of higher education in the United States or Canada (the institution need not be a member of NACC); be members of the African-American, Latino, Asian/Pacific Islander, or American-Indian/Alaskan-Native communities; be committed to excellence in research and teaching in the field of philanthropic or nonprofit-sector studies or nonprofit management from any one of a variety of academic disciplines; and be willing and able to participate in the annual ARNOVA conference (travel expenses will be paid by NACC). Applications will be evaluated by a selection committee of distinguished scholars in the field and colleagues of David Stevenson using the following criteria: significance and quality of proposed research to the field of philanthropic and nonprofit-sector studies; relationship of the topic to the education of undergraduate students or to the development of relevant data and information systems; merit of the proposed project and use of fellowship funds; ability of the applicant to carry out the proposed activity; and the applicant's ability to present a well-written, thoughtfully-prepared application. Priority consideration will be made for nontenured junior faculty. An emphasis on quality and the potential for a candidate to provide leadership in the field will also be a considering factor.
*Restrictions:* Travel expenses to attend the ARNOVA conference are not transferable as additional award money. Travel to attend the ARNOVA conference is limited to the year of the award and up to two years after the award to present fellowship-related research results.
*Geographic Focus:* All States
*Date(s) Application is Due:* Mar 31
*Amount of Grant:* 15,000 USD
*Contact:* Sean W. Shacklett, M.P.A., Executive Director; (216) 368-0969; fax (216) 368-0969; sshacklett-nacc@case.edu
*Internet:* www.naccouncil.org/stevenson.asp
*Sponsor:* Case Western Reserve University
10900 Euclid Avenue
Cleveland, OH 44106-7167

**NACC William Diaz Fellowships** **1675**

The goals of the William Diaz Fellowship are to acknowledge William Diaz's contributions to the development of minority leadership for nonprofit organizations, foundations, and the field of philanthropic and nonprofit management studies and to encourage a better understanding of philanthropy from diverse communities. Two fellowships are awarded annually. Fellowships may be used to support research, travel to present research at conferences, professional education seminars related to research and teaching, and course development. All funds must be expended and accounted for within one year after the end of the fellowship. The award will be made payable to the award recipient's affiliate institution. Guidelines are available online. The Nonprofit Academic Centers Council (NACC) is a membership association comprised of academic centers or programs at accredited colleges and universities that focus on the study of nonprofit organizations, voluntarism and/or philanthropy. Established in 1991, NACC is the first group entirely dedicated to the promotion and networking of centers that provide research and education in philanthropy and the nonprofit sector.
*Requirements:* Applicants must meet the following eligibility criteria: be faculty of any rank (including senior faculty new to philanthropic and nonprofit management studies) whose work focuses on philanthropy, nonprofit organizations, and diversity; be affiliated with an accredited institution of higher education in the United States or Canada (the institution need not be a member of NACC); be members of the African-American, Latino, Asian/Pacific Islander, or American-Indian/Alaskan-Native communities; be committed to excellence in research and teaching in the field of philanthropic or nonprofit-sector studies or nonprofit management from any one of a variety of academic disciplines; be willing and able to participate in the annual ARNOVA Conference (travel expenses will be paid by NACC). Applications will be evaluated by a selection committee of distinguished scholars in the field and colleagues of William Diaz using the following criteria: significance and quality of proposed research; relationship of the proposed research, if any, to the development of minority leadership for nonprofit organizations, foundations, and the field of philanthropic and nonprofit-sector studies; merit of the proposed project and use of fellowship funds; ability of the applicant to carry out the proposed activity; and applicant's ability to present a well-written, thoughtfully-prepared application.
*Restrictions:* Travel expenses to attend the ARNOVA conference are not transferable as additional award money. Travel to attend the ARNOVA conference is limited to the year of the award and up to two years after the award to present fellowship-related research results.
*Geographic Focus:* All States
*Date(s) Application is Due:* Mar 30
*Amount of Grant:* 15,000 USD
*Contact:* Sean W. Shacklett, M.P.A., Executive Director; (216) 368-0969; fax (216) 368-0969; sshacklett-nacc@case.edu
*Internet:* www.naccouncil.org/diaz.asp
*Sponsor:* Case Western Reserve University
10900 Euclid Avenue
Cleveland, OH 44106-7167

**NAEA Research Grants** **1676**

The National Art Education Foundation invites proposals to support research in art education that advances knowledge in the field of art education and that promulgates the goals outlined in 'Creating a Visual Arts Research Agenda Toward the 21st Century.' Funds are awarded to selected art educators whose proposals specifically focus on issues

relating to one of the recommendations identified in this document. NAEF invites proposals to support research in art education that advances knowledge in the field of art education. Grants are awarded to art educators to pursue a broad range of research topics that are aligned with the NAEA Strategic Goals: advocacy, learning, research and knowledge, and organizational vibrancy. NAEF encourages the submissions of proposals that conduct research that supports the impact and importance of arts education in student learning and provides hard data to support the findings of the research. The project year runs from July 1 through June 30.
*Requirements:* NAEF Grants are open to active NAEA members who have been members for at least one year prior to the date of application; to state/province associations of NAEA; and to recognized special issues groups of NAEA. All eligible applicants are encouraged to apply. NAEF encourages submissions of the following proposals: proposals that support the creation of communities of learners, including both researchers and practitioners, working together to explore a research question and/or project; and, proposals that support the identification of best practice and research that leads to further understanding of the impact and importance of arts education to student learning in and through the visual arts in a variety of settings, with an interest in research that provides quantitative data to support its findings. These priorities do not preclude other submissions meeting the eligibility of the current Research Guidelines.
*Restrictions:* The following applicants will not be considered: non-members; nonprofit organizations or institutions; members applying on behalf of the nonprofit organization or institution where they work; relatives of current members of the NAEF Board of Trustees. Applications from individuals who are members of only the state/province association are not considered NAEA members. Individuals must have an NAEA ID# in order to apply.
*Geographic Focus:* All States
*Date(s) Application is Due:* Oct 1
*Amount of Grant:* Up to 10,000 USD
*Contact:* Kathi R. Levin, Program/Development Officer; (703) 860-8000; naef@arteducators.org
*Internet:* www.arteducators.org/grants/national-art-education-foundation
*Sponsor:* National Art Education Foundation
1806 Robert Fulton Drive, Suite 300
Reston, VA 20191

**NAEF Teacher Incentive Grants** **1677**
The National Art Education Foundation (NAEF) invites proposals for scholarship support of projects that promote the improvement of the teaching of art. Funds are awarded to classroom-based action research projects that facilitate development of teachers' capacities to provide visual art instruction that reflects the vision promulgated by the National Art Education Association in areas such as curriculum; student learning; student assessment; classroom behavior, management, or discipline; or other practices relating to instructional interaction and the achievement of student learning. The project year runs from July 1 through June 30. Grants are limited to a maximum of $2,500 per award.
*Requirements:* NAEF Grants are open to active NAEA members who have been members for at least one year prior to the date of application; to state/province associations of NAEA; and to recognized special issues groups of NAEA. All eligible applicants are encouraged to apply.
*Restrictions:* The following applicants will not be considered: non-members; nonprofit organizations or institutions; members applying on behalf of the nonprofit organization or institution where they work; relatives of current members of the NAEF Board of Trustees. Applications from individuals who are members of only the state/province association are not considered NAEA members. Individuals must have an NAEA ID# in order to apply.
*Geographic Focus:* All States
*Date(s) Application is Due:* Oct 1
*Amount of Grant:* Up to 2,500 USD
*Contact:* Kathi R. Levin, Program/Development Officer; (703) 860-8000; naef@arteducators.org
*Internet:* www.arteducators.org/grants/national-art-education-foundation
*Sponsor:* National Art Education Foundation
1806 Robert Fulton Drive, Suite 300
Reston, VA 20191

**NASA Lewis Educational Research and Collaborative Internships** **1678**
The Lewis Educational and Research Collaborative Internship Program (LERCIP), at the NASA Glenn Research Center in Cleveland, Ohio, consists of internships for high school students, college students, secondary school teachers, and NASA scholars. LERCIP offers summer internship opportunities for high school and college students in science, technology, engineering, math (STEM), business and technical areas, as well as internships for high school teachers. The internships are offered under the auspices of LERCIP, a partnership of the NASA Glenn Research Center (GRC) Educational Programs Office and the Ohio Aerospace Institute (OAI). Selected participants work side by side with a GRC mentor on various projects/programs that support NASA mission and are incorporated in to the day-to-day activities of the Center to the fullest extent possible. Students chosen for the program participate in scheduled enrichment activities, GRC Summer Job Fair and Co-op Briefing, and presentations at the Annual LERCIP Student Symposiums as well as opportunities to co-author publications and various papers.
*Requirements:* Teachers must be U.S. citizens employed full-time as secondary school teachers. College students must be U.S. citizens and full-time students pursuing a baccalaureate or higher degree in science, engineering, technology, or professional administration at an accredited college or university in the United States. College-bound high school students or two-year college students are also eligible. Students must have cumulative GPA of 3.0 on a 4.0 scale.
*Geographic Focus:* All States
*Date(s) Application is Due:* Jan 31
*Contact:* Robert LaSalvia; (216) 433-6656; grc-intern@mail.nasa.gov
*Internet:* www.oai.org/education/student/index.html?sc=1
*Sponsor:* Ohio Aerospace Institute
22800 Cedar Point Road
Cleveland, OH 44142

**Nash Avery Foundation Grants** **1679**
The Nash Avery Foundation was established in Minnesota in 2005, and named in honor Nash Wicka, who was diagnosed with Duchenne muscular dystrophy in 2002. The Foundation supports research of muscular dystrophy, primarily in both Minnesota and Massachusetts. Most recent grants have ranged up to approximately $50,000, although a limit has not been established. Though there are no specific annual deadlines for submission, a formal application process is required.
*Requirements:* Any 501(c)3 organization researching Duchenne muscular dystrophy is eligible to apply.
*Geographic Focus:* All States
*Amount of Grant:* Up to 100,000 USD
*Samples:* A.T. Still University of Health Sciences, Kirksville, Missouri - evaluating the utility of anti-inflammatory agents that prevent muscle cell death as potential treatments for DMD; Brown University, Providence, Rhode Island - research of Utrophin, a compensatory protein that can act as a substitute for dystrophin, the missing protein in DMD boys; Children's National Medical Center, Washington, D.C. - investigating four experimental drugs (Celastrol, Resveratrol, Lipoxin A, Cyclosporine A analog) that may prevent muscle degenaeration and increase muscle function.
*Contact:* Thomas C Wicka, Director; (612) 325-5700 or (952) 582-2933; nashaveryfoundation@gmail.com
*Sponsor:* Nash Avery Foundation
4904 Merilane Avenue
Edina, MN 55436-1359

**National Blood Foundation Research Grants** **1680**
The National Blood Foundation (NBF), established in 1983, has a history of supporting research and education that advances transfusion medicine and cellular therapies by funding scientific research that benefits patients and donors. Funds are raised annually from corporations, blood centers, foundations and individuals by the NBF for the National Blood Foundation Research and Education Trust Fund (NBFRET) and the NBF. The NBF and NBFRET, which are 501(c)3 organizations, provide grants for scientific research in the field of transfusion medicine and cellular therapies and support educational initiatives that benefit this community. The National Blood Foundation annually awards grants for one or two-year research projects, with a maximum award per grant of $75,000.
*Requirements:* NBF has also funded research and studies on critical issues that benefit patients and donors in the fields of transfusion medicine and cellular therapies, and are in addition to scientific research. Preference is given to new investigators in the field. There is a $150 application fee.
*Geographic Focus:* All States
*Date(s) Application is Due:* Dec 13
*Amount of Grant:* Up to 65,000 USD
*Contact:* Amy Quiggins, Manager; (301) 215-6552 or (301) 215-6551; fax (301) 215-5751; aquiggins@aabb.org or nbf@aabb.org
*Internet:* www.aabb.org/programs/nbf/Pages/default.aspx
*Sponsor:* National Blood Foundation
8101 Glenbrook Road
Bethesda, MD 20814-2749

**National Brain Tumor Society Research Grants** **1681**
National Brain Tumor Research grants support basic scientific research directed at finding a cure for brain tumors. Grants are awarded for up to a two-year period and may be used for start-up projects or supplementary funding. The Society's research grant application review is a two-step process built on standards set by the National Institutes of Health (NIH). Unless specified, most RFAs require a Letter of Intent (LOI) as a prerequisite to submit a full application. LOIs are then submitted electronically and immediately undergo an administrative review to ensure that application guidelines are met. Members of our Scientific Advisory Council (SAC), led by the Chair and Vice Chair, oversee the review process for each of our grant programs.
*Geographic Focus:* All States
*Amount of Grant:* Up to 100,000 USD
*Contact:* Samantha Masterson, Chief Advancement Officer; (617) 658-7035 or (617) 924-9997; fax (617) 924-9998; smasterson@braintumor.org
*Internet:* braintumor.org/advance-research/research-funding-opportunities/apply-for-a-grant/
*Sponsor:* National Brain Tumor Society
55 Chapel Street, Suite 200
Newton, MA 02458

**National Center for Preservation Technology and Training Internships** **1682**
Paid internships in a variety of projects are available working across disciplines. Interns will have an opportunity to work in one of the following programs: Archeology and Collections; Architecture and Engineering; Historic Landscapes; and, Materials Conservation. Current projects in Historic Landscapes include: Inventory and Mapping of Campus Trees. Current projects in Materials Conservation include: Climate Change and Salts in Masonry #1; Climate Change and Salts in Masonry #2; Evaluation of Paint Primers on Weathered Wood; Study of Freeze Thaw Cycles on Window Glazes; and, Anti-graffiti Coatings on Corten Steel.
*Requirements:* Applicants should demonstrate skills and knowledge in preservation technology. This may include but not be limited to documentation, condition assessments,

literature reviews, laboratory or field research, materials testing, hands-on preservation treatments, or other related skills. Undergraduates, graduate students, or recent graduates will be considered. Studies in disciplines such as architecture, anthropology, conservation, chemistry, engineering, geology, geography, preservation, etc. will be preferred. Most internships last six months unless otherwise noted, beginning and ending dates are negotiable. Interns will become short-term employees of Northwestern State University. Compensation is based on experience, including academic experience. Applications from foreign nationals (non-U.S. citizens) will be accepted. However, these individuals are responsible for obtaining their own visa. Please ensure that application materials are in English.
*Restrictions:* Housing is not provided.
*Geographic Focus:* All States, All Countries
*Date(s) Application is Due:* Jan 31
*Contact:* Dr. Mary F. Striegel, (318) 356-7444; fax (318) 356-9119; mary_striegel@nps.gov
*Internet:* https://www.ncptt.nps.gov/about-us/intern/
*Sponsor:* U.S. Department of the Interior
645 University Parkway
Natchitoches, LA 71457

**National Center for Responsible Gaming Conference Scholarships** **1683**
The National Center for Responsible Gaming (NCRG) is offering the need-based scholarship program for the NCRG Conference on Gambling and Addiction. Insofar it is able, it is the NCRG's wish to provide fair access to the NCRG Conference to all interested stakeholders, especially those in academia and the clinical and public health sectors. Last year, the NCRG supported almost thirty NCRG Conference attendees. This year, the NCRG will award scholarships to more than fifteen individuals, and a portion of these scholarships may also include travel assistance. Additional scholarship funds may be available at a later time. To apply for a scholarship, interested applicants must submit their application and registration form by June 15.
*Geographic Focus:* All States
*Date(s) Application is Due:* Jun 15
*Contact:* Christine Reilly; (978) 338-6610; fax (978) 522-8452; creilly@ncrg.org
*Internet:* www.ncrg.org/public-education-and-outreach/conference/conference-scholarships
*Sponsor:* National Center for Responsible Gaming
900 Cummings Center, Suite 216-U
Beverly, MA 01915

**National Center for Responsible Gaming Large Grants** **1684**
Large Grants provide up to $75,000 for up to two years for discrete, specified, circumscribed research projects related to gambling disorders. The proposed research investigation may focus on a broad range of research that develops and tests psychosocial or pharmacological approaches for prevention, intervention, treatment or relapse prevention of gambling disorders. The NCRG is especially interested in brief interventions targeted at underrepresented populations, such as minorities, young adults and persons with sub-clinical gambling disorders. Other priorities include the following topics: impact of Indian gaming; gambling and minorities; secondary data analysis; and technology and gambling. Applicants may request up to $75,000 in direct costs per year for a period not to exceed 24 months.
*Requirements:* Both public and private nonprofit organizations are eligible to apply. Profit-making organizations should contact Christine Reilly, senior research director (creilly@ncrg.org). Foreign institutions are required to collaborate with a U.S. institution. Applications involving a non-U.S. institution must have a principal investigator and fiscal agent based at a U.S. institution.
*Geographic Focus:* All States
*Date(s) Application is Due:* Aug 3
*Amount of Grant:* Up to 75,000 USD
*Contact:* Christine Reilly; (978) 338-6610; fax (978) 522-8452; creilly@ncrg.org
*Internet:* www.ncrg.org/research-center/apply-ncrg-funding/large-grants
*Sponsor:* National Center for Responsible Gaming
900 Cummings Center, Suite 216-U
Beverly, MA 01915

**National Center for Responsible Gaming Postdoctoral Fellowships** **1685**
The National Center for Responsible Gaming Postdoctoral Fellowship is intended to help ensure that a pool of highly trained scientists is available to address the research needs of the field of gambling disorders. Eligible applicants are within five years of completing their terminal research degree or within five years of completing medical residency. The proposed research investigation may focus on a broad range of research that develops and tests psychosocial or pharmacological approaches for prevention, intervention, treatment or relapse prevention of gambling disorders. The NCRG is especially interested in brief interventions targeted at underrepresented populations, such as minorities, young adults and persons with sub-clinical gambling disorders. Other priorities include the impact of Indian gaming; gambling and minorities; secondary data analysis; and technology and gambling. Applicants may request up to $75,000 in direct costs for a period not to exceed twelve months.
*Requirements:* The Postdoctoral Fellowship is intended for individuals who are within five years of completing their terminal research degree or within five years of completing medical residency (or the equivalent). The candidate for this award may not have been a recipient of an NIH Career Development Award or an NCRG grant. Candidates who are not U.S. citizens but have an appointment at a U.S. institution are eligible to apply. Candidates based at foreign institutions are not eligible to apply.
*Geographic Focus:* All States
*Date(s) Application is Due:* Jun 3
*Amount of Grant:* Up to 75,000 USD
*Contact:* Christine Reilly; (978) 338-6610; fax (978) 522-8452; creilly@ncrg.org
*Internet:* www.ncrg.org/research-center/apply-ncrg-funding/addiction-fellowship
*Sponsor:* National Center for Responsible Gaming
900 Cummings Center, Suite 216-U
Beverly, MA 01915

**National Center for Responsible Gaming Seed Grants** **1686**
National Center for Responsible Gaming Seed Grants support a variety of research activities, including: pilot and feasibility studies; secondary analysis of existing data; small, self-contained research projects; development of research methodology; development of new research technology. The proposed research investigation may focus on a broad range of research that develops and tests psychosocial or pharmacological approaches for prevention, intervention, treatment or relapse prevention of gambling disorders. The NCRG is especially interested in brief interventions targeted at underrepresented populations, such as minorities, young adults and persons with sub-clinical gambling disorders. Other priorities include the following topics: impact of Indian gaming, gambling and minorities, secondary data analysis and technology and gambling. Applicants may request up to $30,000 in direct costs for a period not to exceed 12 months. The annual deadline is September 1.
*Requirements:* Both public and private nonprofit organizations are eligible to receive grants from the NCRG. Applicants from profit-making companies should contact Christine Reilly, senior research director (creilly@ncrg.org). Foreign institutions are required to collaborate with a U.S. institution. Applications involving a non-U.S. institution must have a principal investigator and fiscal agent based at a U.S. institution.
*Geographic Focus:* All States
*Date(s) Application is Due:* Sep 1
*Amount of Grant:* Up to 30,000 USD
*Contact:* Christine Reilly; (978) 338-6610; fax (978) 522-8452; creilly@ncrg.org
*Internet:* www.ncrg.org/research-center/apply-ncrg-funding/seed-grants
*Sponsor:* National Center for Responsible Gaming
900 Cummings Center, Suite 216-U
Beverly, MA 01915

**National Center for Responsible Gaming Travel Grants** **1687**
Travel Grants are intended to support the attendance of postdoctoral investigators at scientific meetings at which they will present a paper or a poster on gambling research. Applicants may request up to $1,500 in support of conference registration fees, airfare, ground transportation, meals and lodging. Eligible meetings include academic conferences that employ rigorous peer review in the selection of presentations and that are not specifically focused on gambling disorder. There are no annual deadlines, and applications are accepted on an ongoing basis.
*Requirements:* Travel Grants are available to scientists who are not more than ten years beyond their doctorate (Ph.D.) or medical residence. The applicant must be the first author of the paper or poster to be presented. Eligible candidates must be based at a U.S. institution.
*Restrictions:* The Principal Investigator may apply for only one Travel Grant per cycle.
*Geographic Focus:* All States
*Amount of Grant:* Up to 1,500 USD
*Contact:* Christine Reilly; (978) 338-6610; fax (978) 522-8452; creilly@ncrg.org
*Internet:* www.ncrg.org/research-center/apply-ncrg-funding/travel-grants
*Sponsor:* National Center for Responsible Gaming
900 Cummings Center, Suite 216-U
Beverly, MA 01915

**National Endowment for the Arts Jazz Masters Fellowships** **1688**
The National Endowment for the Arts Jazz Masters Fellowship is the highest honor bestowed upon a jazz musician. Each year, the program elevates to its ranks a select number of living legends who have made exceptional contributions to the advancement of jazz. For the NEA Jazz Masters, the selection criteria are artistic excellence and significance of the nominees' contributions to the development and performance of jazz. The Arts Endowment will honor musicians who represent a range of styles and instruments. In addition, one fellowship (A.B. Spellman NEA Jazz Masters Fellowship for Jazz Advocacy) will be given to an individual who has made major contributions to the appreciation, knowledge, and advancement of the American jazz art form. Fellowships are awarded to living individuals on the basis of nominations from the public, including the jazz community. The nomination form is available at the NEA website.
*Requirements:* Nominees must have nationally recognized and publicly accessible bodies of work in the field of jazz. The recipients must be citizens or permanent residents of the United States.
*Restrictions:* Posthumous nominations are not eligible for nomination.
*Geographic Focus:* All States
*Date(s) Application is Due:* Oct 1
*Amount of Grant:* 25,000 USD
*Contact:* Katja von Schuttenbach; (202) 682-5711; vonschuttenbach@arts.gov
*Internet:* www.nea.gov/national/jazz/Award.html
*Sponsor:* National Endowment for the Arts
1100 Pennsylvania Avenue NW
Washington, D.C. 20506-0001

**National Endowment for the Arts Literature Fellowships: Creative Writing** **1689**
The Literature Fellowships for Creative Writing offer $25,000 grants in prose (fiction and creative nonfiction) and poetry to published creative writers that enable the recipients to set aside time for writing, research, travel, and general career advancement. The Fellowships operate on a two-year cycle with prose and poetry available in alternating years. Applications are reviewed through an anonymous process in which the only

criteria for review are artistic excellence and artistic merit. To review the applications, the Endowment assembles a different advisory panel every year, each diverse with regard to geography, race and ethnicity, and artistic points of view. Deadlines vary each year according to the writing category. Applications must be submitted through Grants.gov.
*Requirements:* Applicants must be citizens or permanent residents of the U.S. Poetry applicants may apply if they have published within the past ten years. They may also apply if they have published a volume of 48 or more pages; or twenty or more different poems or pages of poetry in five or more literary journals, anthologies, or publications which regularly include poetry as a portion of their format. Up to 16 poems may be in a single volume of poetry of fewer than 48 pages. This volume, however, may count as only one of the required five places of publication. Applicants may use digital, audio, or online publications to establish eligibility, provided that such publications have competitive selection processes and stated editorial policies. If the online publication or website no longer exists, applicants must provide, upon request, sufficient evidence that the work once appeared online. The following may not be used to establish eligibility: pre-publication material, such as galleys, proofs, and advance reader's copies; self-publication including work that has appeared in a publication where the applicant is the editor, publisher, or staff; collaborative work; scholarly writing; instructional writing; journalism; book reviews; editorials/letters to the editor; interviews; student publications and publications which primarily print work by persons who are affiliated with a particular academic institution. In addition, eligibility may not be established by publication in presses that require individual writers to pay for part or all of the publications costs; asks writers to buy or sell copies of the publication; publish the work of anyone who subscribes to the publication or joins the organization through membership fees; publish the work of anyone who buys an advertisement in the publication; publish work without competitive selection; or publish work without professional editing.
*Restrictions:* Individuals may apply only once each year. Applicants must be published authors. The Fellowship is also not available for the following: individuals who previously have received two or more Literature Fellowships (in fiction, creative nonfiction, or poetry) or Translation Fellowships from the NEA; individuals who have received any Literature Fellowship (in fiction, creative nonfiction, or poetry) or Translation Fellowship from the NEA within the past ten years; news reporting; scholarly writing; or work toward academic degrees.
*Geographic Focus:* All States
*Amount of Grant:* 25,000 USD
*Contact:* Amy Stolls, Literature Program Officer; (202) 682-5771 or (202) 682-5034; stollsa@arts.gov or LitFellowships@arts.gov
*Internet:* www.nea.gov/grants/apply/Lit/index.html
*Sponsor:* National Endowment for the Arts
1100 Pennsylvania Avenue NW
Washington, D.C. 20506-0001

**National Endowment for the Arts Literature Fellowships: Translation Projects 1690**
Through Fellowships to exceptionally talented, published translators, the National Endowment for the Arts supports projects for the specific translation of prose, poetry, or drama from other languages into English. The work to be translated should be of interest for its literary excellence and value. Priority will be given to projects that involve work that has not yet been translated into English. Applications must be submitted electronically through Grants.gov.
*Requirements:* Applicants must be U.S. citizens or permanent residents. Individuals are eligible to apply if, alone or in collaboration, they have published a translation into English of a novel or a volume of 48 or more pages of fiction or poetry, or a book-length translation of a work of literary merit; or published a total of 48 pages of translations of creative literature into English in literary magazines, anthologies, or books; or translated into English at least one full-length play that has been produced by a professional theater company.
*Restrictions:* The following is not available for funding: individuals who previously have received three or more Literature Fellowships (in fiction, creative nonfiction, or poetry) or Translation Fellowships; individuals who have received any Literature Fellowship (in fiction, creative nonfiction, or poetry) or Translation Fellowship from the National Endowment for the Arts within the past five years; scholarly writing; or work toward an academic degree.
*Geographic Focus:* All States
*Date(s) Application is Due:* Jan 3
*Amount of Grant:* 12,500 - 25,000 USD
*Contact:* Amy Stolls, Literature Program Officer; (202) 682-5771; stollsa@arts.gov
*Internet:* arts.endow.gov/grants/apply/LitTranslation/index.html
*Sponsor:* National Endowment for the Arts
1100 Pennsylvania Avenue NW
Washington, D.C. 20506-0001

**National Endowment for the Humanities Fellowships at Winterthur 1691**
Founded almost 60 years ago by collector and horticulturist Henry Francis du Pont, Winterthur consists of a prestigious museum of American decorative arts, a research library that serves scholars around the world, and a 1,000 acre preserve located on the grounds of du Pont's childhood home. Winterthur has a National Endowment for the Humanities (NEH)-funded program that offers four- to twelve-month fellowships for scholars wishing to pursue advanced research in the following areas: American studies; American cultural, social, and intellectual history; art and design history; material culture; decorative arts; consumer culture; garden and landscape studies; Shaker studies, travel and tourism; the Atlantic World; and objects in literature. The program encourages the exchange of ideas among research fellows, the Winterthur staff, and graduate students through colloquia, conferences, discussions, and informal gatherings. Housing is available on the Winterthur grounds and is within easy walking distance of the library and museum where fellows enjoy many staff privileges, including 24-hour access to the library. Fellowships stipends range up to $40,000. Detailed guidelines and a downloadable application form are available at Winterthur's website. The preferred method of submission is email; however, applicants may also mail the completed application and required supporting documentation (essay, bibliography, vita, etc.) to the address given. All applicants are strongly encouraged to search Wintercat (the library's online catalog), visit Winterthur, and contact staff members to discuss potential research projects. The suitability of a project to Winterthur's collections is the primary consideration of the fellowship award committee.
*Requirements:* Academic, museum, and independent scholars are invited to apply.
*Restrictions:* This fellowship may not be used in the pursuit of a degree, although fellows with doctorate in hand may revise a dissertation for publication. Applicants must be U.S. citizens or residents for three years prior to application.
*Geographic Focus:* All States
*Date(s) Application is Due:* Jan 15
*Amount of Grant:* Up to 40,000 USD
*Contact:* Rosemary Krill; (302) 888-4637; fax (302) 888-4870; rkrill@winterthur.org
*Internet:* www.winterthur.org/?p=535
*Sponsor:* Winterthur Museum, Garden, and Library
5105 Kennett Pike
Wilmington, DE 19735

**National Gallery of Art CASVA A.W. Mellon Postdoctoral Fellowship 1692**
The Center for Advanced Study in the Visual Arts announces a postdoctoral fellowship supported by a grant from the A.W. Mellon Foundation. The Fellowship, awarded each year for two consecutive academic years, will support research in the history, theory, and criticism of the visual arts of any time period or culture. The A.W. Mellon Postdoctoral Fellow will reside in Washington. During the first year the fellow will carry out research and writing for publication. The fellow will also design and direct an intensive week-long seminar for the seven predoctoral fellows at the Center, focusing on a topic related to the applicant's field of interest and with a special emphasis on methodological issues. In the second academic year, while continuing research and writing in residence, the A.W. Mellon Postdoctoral Fellow will be expected to teach one course (advanced undergraduate or graduate) by arrangement at a neighboring university. The award is $50,000 per year, with the fellow also receiving allowances for travel and research-related materials.
*Requirements:* Applicants for the A. W. Mellon Postdoctoral Fellowship must have received the PhD degree within the five years preceeding the next application deadline.
*Geographic Focus:* All States
*Date(s) Application is Due:* Oct 15
*Amount of Grant:* 50,000 - 60,000 USD
*Contact:* Danielle Horetsky, Fellowship Officer; (202) 842-6482 or (202) 842-6480; fax (202) 789-3026; casva@nga.gov
*Internet:* https://www.nga.gov/research/casva/fellowships/mellon-postdoctoral-fellowships.html
*Sponsor:* National Gallery of Art
2000B South Club Drive
Landover, MD 20785

**National Gallery of Art CASVA Ailsa Mellon Bruce National Gallery of Art Sabbatical Curatorial Fellowship 1693**
The Center for Advanced Study in the Visual Arts announces the Ailsa Mellon Bruce National Gallery of Art Sabbatical Curatorial Fellowship, open to members of the National Gallery of Art curatorial staff. The fellowships are intended to relieve members of the Gallery's curatorial staff of their curatorial and other Gallery responsibilities in order to pursue independent study, research, or publication not directly related to a Gallery project. Fellows will be encouraged to spend their time away from Gallery premises. The Center provides a maximum travel allowance of $8,000, based on need and prorated for the length of the term. As staff members of the National Gallery of Art, recipients will also continue to receive their normal salaries during the period of the fellowship. The fellowship provides for full-time independent research for a minimum of two months to a maximum of four months. The application deadline is November 15.
*Requirements:* The fellowships are reserved for qualified art historians who have worked at least five years in one of the departments of the National Gallery of Art and who have held the appropriate terminal degree in their field for at least five years. In exceptional cases, an equivalent record of professional accomplishment may fulfill the latter requirement.
*Geographic Focus:* All States
*Date(s) Application is Due:* Nov 15
*Amount of Grant:* Up to 8,000 USD
*Contact:* Danielle Horetsky, Fellowship Officer; (202) 842-6482 or (202) 842-6480; fax (202) 789-3026; casva@nga.gov
*Internet:* https://www.nga.gov/research/casva/fellowships/nga-sabbatical-curatorial-fellowship.html
*Sponsor:* National Gallery of Art
2000B South Club Drive
Landover, MD 20785

**National Gallery of Art CASVA Ailsa Mellon Bruce Predoctoral Fellowships for Historians of American Art to Travel Abroad 1694**
The Center for Advanced Study in the Visual Arts offers up to four fellowships to doctoral students in art history who are studying aspects of art and architecture of the United States, including native and pre-Revolutionary America. This fellowship is for a period of four to six weeks of continuous travel abroad in areas such as Africa, Asia, or South America, as well as Europe, to sites of historical and cultural interest, including

museums, exhibitions, collections, and monuments. The travel fellowship is intended to encourage a breadth of art-historical experience beyond the candidate's major field, not for the advancement of a dissertation. Preference will be accorded to those who have had little opportunity for research travel abroad. The amount of the award is dependent on the travel plan, with a maximum of $6,000. A narrative report at the conclusion of the travel period is required. Individuals currently affiliated with the National Gallery of Art are not eligible for this fellowship. Individuals may not apply for other fellowships at the Center while an application is pending or once a fellowship has been awarded. Applications must be submitted on or before November 15.
*Requirements:* Application for a travel fellowship may be made only through nomination by the chair of a graduate department of art history or other appropriate department. Each department may support two candidates. Department chairs must submit one letter of nomination listing all candidates nominated by the department.
*Geographic Focus:* All States
*Date(s) Application is Due:* Nov 15
*Amount of Grant:* Up to 6,000 USD
*Contact:* Danielle Horetsky, Fellowship Officer; (202) 842-6482 or (202) 842-6480; fax (202) 789-3026; casva@nga.gov
*Internet:* https://www.nga.gov/research/casva/fellowships/predoctoral-travel-fellowships.html
*Sponsor:* National Gallery of Art
2000B South Club Drive
Landover, MD 20785

**National Gallery of Art CASVA Andrew W. Mellon Predoctoral Fellowship** 1695
The Center for Advanced Study in the Visual Arts announces its annual program of support for advanced graduate research in the history, theory, and criticism of art, architecture, urbanism, and photographic media. One Andrew W. Mellon Predoctoral Fellowship is awarded annually for 24 months. This fellowship is intended for the advancement and completion of a doctoral dissertation in a period through the 20th century and on a topic other than European or American art. Cross-cultural topics will also be considered, provided that at least one area of focus is a culture outside the European and American traditions. The Andrew W. Mellon Fellow is expected to spend the first year of the fellowship period on dissertation research abroad, and the second year at the Center to complete the dissertation. The stipend for all predoctoral fellowships is $30,000 per year. A housing allowance is available for fellows not in residence. Online applications must be submitted by November 15.
*Requirements:* Applicants must have completed their residence requirements and coursework for the PhD and general or preliminary examinations before the date of application and know two foreign languages related to the topic of the dissertation. Applicants must be either U.S. citizens or enrolled in a university in the United States.
*Geographic Focus:* All States
*Date(s) Application is Due:* Nov 15
*Amount of Grant:* 30,000 - 35,000 USD
*Contact:* Danielle Horetsky, Fellowship Officer; (202) 842-6482 or (202) 842-6480; fax (202) 789-3026; casva@nga.gov
*Internet:* https://www.nga.gov/research/casva/fellowships/predoctoral-dissertation-fellowships.html
*Sponsor:* National Gallery of Art
2000B South Club Drive
Landover, MD 20785

**National Gallery of Art CASVA Chester Dale Predoctoral Dissertation Fellowships** 1696
The Center for Advanced Study in the Visual Arts announces its annual program of support for advanced graduate research in the history, theory, and criticism of art, architecture, urbanism, and photographic media. Two Dale Fellowships are awarded annually: one for 24 months, the other for 12 months. The Twenty-Four-Month Chester Dale Fellowship is intended for the advancement or completion of a doctoral dissertation in any area of Western art, with a preference for modern and contemporary topics. The two-year Dale Fellow is expected to spend the first year on dissertation research either in the United States or abroad; the second year is to be spent in residence at the Center to complete the dissertation. The Twelve-Month Chester Dale Predoctoral Dissertation Fellowship is intended for the advancement or completion of a doctoral dissertation in Western art. The one-year Dale Fellow may use the fellowship to study in the United States or abroad; there are no residence requirements at the Center. The stipend for all predoctoral fellowships is $30,000 per year. A housing allowance is available for fellows not in residence. Online applications must be submitted by November 15.
*Requirements:* Applicants must have completed their residence requirements and coursework for the PhD and general or preliminary examinations before the date of application. Students must know two foreign languages related to the topic of the dissertation. Applicants must be either U.S. citizens or enrolled in a university in the United States.
*Geographic Focus:* All States
*Date(s) Application is Due:* Nov 15
*Amount of Grant:* 30,000 - 35,000 USD
*Contact:* Danielle Horetsky, Fellowship Officer; (202) 842-6482 or (202) 842-6480; fax (202) 789-3026; casva@nga.gov
*Internet:* https://www.nga.gov/research/casva/fellowships/predoctoral-dissertation-fellowships.html
*Sponsor:* National Gallery of Art
2000B South Club Drive
Landover, MD 20785

**National Gallery of Art CASVA David E. Finley Predoctoral Dissertation Fellowship** 1697
The Center for Advanced Study in the Visual Arts announces its annual program of support for advanced graduate research in the history, theory, and criticism of art, architecture, urbanism, and photographic media. One David E. Finley Predoctoral is awarded annually for 36 months. The first two years of the Finley Fellowship are intended for research and travel in Europe to visit collections, museums, monuments, and sites related to a well-advanced dissertation in Western art. During this two-year period, residency abroad is expected, for which travel funds are available. The third year is to be spent in residence at the Center to complete the dissertation and to perform curatorial work. Half of the year in residence will be devoted to a research project, designed to complement the subject of the dissertation, at the Gallery or other Washington-area collections. A primary requirement for this fellowship is that the candidate have a significant interest in curatorial work. The stipend for all predoctoral fellowships is $30,000 per year. A housing allowance is available for fellows not in residence. Online applications must be submitted by November 15.
*Requirements:* Applicants must have completed their residence requirements and coursework for a PhD and general or preliminary examinations before the date of application and know two foreign languages related to the topic of the dissertation. Applicants must be either U.S. citizens or enrolled in a university in the United States.
*Geographic Focus:* All States
*Date(s) Application is Due:* Nov 15
*Amount of Grant:* 30,000 - 35,000 USD
*Contact:* Danielle Horetsky, Fellowship Officer; (202) 842-6482 or (202) 842-6480; fax (202) 789-3026; casva@nga.gov
*Internet:* https://www.nga.gov/research/casva/fellowships/predoctoral-dissertation-fellowships.html
*Sponsor:* National Gallery of Art
2000B South Club Drive
Landover, MD 20785

**National Gallery of Art CASVA Ittleson Predoctoral Fellowship** 1698
The Center for Advanced Study in the Visual Arts announces its annual program of support for advanced graduate research in the history, theory, and criticism of art, architecture, urbanism, and photographic media. One Ittleson Predoctoral Fellowship is awarded annually for 24 months. This fellowship is intended for the advancement and completion of a doctoral dissertation in the visual arts in a field other than European or American art. The Ittleson Fellow is expected to spend the first year of the fellowship period on dissertation research abroad, and the second year at the Center to complete the dissertation. The stipend for all predoctoral fellowships is $30,000 per year. A housing allowance is available for fellows not in residence. Online applications must be submitted by November 15.
*Requirements:* Applicants must have completed their residence requirements and coursework for the PhD and general or preliminary examinations before the date of application and know two foreign languages related to the topic of the dissertation. Applicants must be either U.S. citizens or enrolled in a university in the United States.
*Geographic Focus:* All States
*Date(s) Application is Due:* Nov 15
*Amount of Grant:* 30,000 - 35,000 USD
*Contact:* Danielle Horetsky, Fellowship Officer; (202) 842-6480 or (202) 842-6482; fax (202) 789-3026; casva@nga.gov
*Internet:* https://www.nga.gov/research/casva/fellowships/predoctoral-dissertation-fellowships.html
*Sponsor:* National Gallery of Art
2000B South Club Drive
Landover, MD 20785

**National Gallery of Art CASVA Paul Mellon Predoctoral Dissertation Fellowship** 1699
The Center for Advanced Study in the Visual Arts announces its annual program of support for advanced graduate research in the history, theory, and criticism of art, architecture, urbanism, and photographic media. One Paul Mellon Predoctoral Fellowship is awarded annually for 36 months. This fellowship is intended for the advancement and completion of a doctoral dissertation in Western art and to enable a candidate to reside abroad for two years to develop expertise in a specific city, locality, or region related to the dissertation. The third year is to be spent in residence at the Center to complete the dissertation. The stipend for all predoctoral fellowships is $30,000 per year. A housing allowance is available for fellows not in residence. Online applications must be submitted by November 15.
*Requirements:* Applicants must have completed their residence requirements and coursework for the PhD and general or preliminary examinations before the date of application. Students must know two foreign languages related to the topic of the dissertation. Applicants must be either U.S. citizens or enrolled in a university in the United States.
*Geographic Focus:* All States
*Date(s) Application is Due:* Nov 15
*Amount of Grant:* 30,000 - 35,000 USD
*Contact:* Danielle Horetsky, Fellowship Officer; (202) 842-6482 or (202) 842-6480; fax (202) 789-3026; casva@nga.gov
*Internet:* www.nga.gov/content/ngaweb/research/casva/fellowships/predoctoral-dissertation-fellowships.html
*Sponsor:* National Gallery of Art
2000B South Club Drive
Landover, MD 20785

**National Gallery of Art CASVA Robert H. and Clarice Smith Predoctoral Fellowship** 1700

The Center for Advanced Study in the Visual Arts announces its annual program of support for advanced graduate research in the history, theory, and criticism of art, architecture, urbanism, and photographic media. One Robert H. and Clarice Smith Predoctoral Fellowship is awarded annually for 12 months. This fellowship is intended for the advancement or completion of either a doctoral dissertation or a resulting publication in Northern European art between 1400 and 1700. For a candidate preparing a publication based on the dissertation, the date of graduation may be as early as fall 2014. The Smith Fellow may use the fellowship to study either in the United States or abroad; there are no residence requirements at the Center. The stipend for all predoctoral fellowships is $30,000 per year. A housing allowance is available for fellows not in residence. Online applications must be submitted by November 15.
*Requirements:* Applicants must have completed their residence requirements and coursework for the PhD before the date of application and know two languages related to the topic of the dissertation. Applicants must be either U.S. citizens or enrolled in universities in the United States.
*Geographic Focus:* All States
*Date(s) Application is Due:* Nov 15
*Amount of Grant:* 30,000 - 35,000 USD
*Contact:* Danielle Horetsky, Fellowship Officer; (202) 842-6482 or (202) 842-6480; fax (202) 789-3026; casva@nga.gov
*Internet:* https://www.nga.gov/research/casva/fellowships/predoctoral-dissertation-fellowships.html
*Sponsor:* National Gallery of Art
2000B South Club Drive
Landover, MD 20785

**National Gallery of Art CASVA Samuel H. Kress Predoctoral Dissertation Fellowship** 1701

The Center for Advanced Study in the Visual Arts announces its annual program of support for advanced graduate research in the history, theory, and criticism of art, architecture, urbanism, and photographic media. One Samuel H. Kress Predoctoral Fellowship is awarded annually for 24 months. This fellowship is intended for the advancement and completion of a doctoral dissertation in European art on a topic before the early 19th century. The Fellow is expected to spend the first year of the fellowship period on dissertation research abroad and the second year at the Center to complete the dissertation, devoting half time to Gallery research projects designed to complement the subject of the dissertation and to provide curatorial experience. The stipend for all predoctoral fellowships is $30,000 per year. A housing allowance is available for fellows not in residence. Online applications must be submitted by November 15.
*Requirements:* Applicants must have completed their residence requirements and coursework for the PhD and general or preliminary examinations before the date of application. Students must know two foreign languages related to the topic of the dissertation. Applicants must be either U.S. citizens or enrolled in a university in the United States.
*Geographic Focus:* All States
*Date(s) Application is Due:* Nov 15
*Amount of Grant:* 30,000 - 35,000 USD
*Contact:* Danielle Horetsky, Fellowship Officer; (202) 842-6482 or (202) 842-6480; fax (202) 789-3026; casva@nga.gov
*Internet:* https://www.nga.gov/research/casva/fellowships/predoctoral-dissertation-fellowships.html
*Sponsor:* National Gallery of Art
2000B South Club Drive
Landover, MD 20785

**National Gallery of Art CASVA Senior Fellowships** 1702

Applications will be considered from scholars for study in the history, theory, and criticism of the visual arts (painting, sculpture, architecture, landscape architecture, urbanism, graphics, film, photography, decorative arts, industrial design, etc.) of any geographical area and of any period. Applications are also solicited from scholars in other disciplines whose work examines physical objects or has implications for the analysis and criticism of physical form. In addition, applications are solicited from scholars who are specifically interested in curatorial research related to objects in the painting, sculpture, graphics, and other collections of the gallery. One Paul Mellon Senior Fellowship and four to six Ailsa Mellon Bruce and Samuel H. Kress Senior Fellowships will normally be awarded for an academic year; applications for a single academic term or a quarter are also possible. In exceptional cases, application may be made for a period of two years. The Samuel H. Kress Senior Fellowships are intended to support primarily research related to objects in the collection of the gallery. Senior fellows receive a monthly stipend with additional allowances for research materials, round-trip travel, and local expenses. A senior fellowship award for the academic year is normally limited to one-half of the applicant's salary, up to a maximum of $50,000, depending on individual circumstances. An additional maximum of $12,000 for housing is available.
*Requirements:* Applicants must have held the PhD for five years or more or possess a record of professional accomplishment at the time of application. Senior fellowships are awarded without regard to the age or nationality of the applicant. Fellowships will normally be awarded for an academic year, early fall to late spring, but fellowships for a single term or quarter are also possible. Scholars are expected to reside in Washington, D.C., throughout their fellowship and participate in the activities of the center.
*Restrictions:* Fellowships may not be postponed or renewed.
*Geographic Focus:* All States
*Date(s) Application is Due:* Oct 15
*Amount of Grant:* Up to 62,000 USD
*Contact:* Danielle Horetsky, Fellowship Officer; (202) 842-6482 or (202) 842-6480; fax (202) 842-6733; casva@nga.gov
*Internet:* https://www.nga.gov/research/casva/fellowships/senior-fellowships.html
*Sponsor:* National Gallery of Art
2000B South Club Drive
Landover, MD 20785

**National Gallery of Art CASVA Visiting Senior Fellowships** 1703

Applications will be considered from scholars for short-term fellowships (maximum of 60 days) for study in the history, theory, and criticism of the visual arts (painting, sculpture, architecture, landscape architecture, urbanism, graphics, film, photography, decorative arts, industrial design, etc.) of any geographical area and of any period. Applications are also solicited from scholars in other disciplines whose work examines physical objects or has implications for the analysis and criticism of physical form. In addition, applications are solicited from scholars who are specifically interested in curatorial research related to objects in the painting, sculpture, graphics, and other collections of the gallery. The notable resources represented by the collections of the National Gallery of Art, its library, photographic archives, the Library of Congress, and other specialized research libraries and collections in Washington will be available to the visiting senior fellows; lectures, colloquia, and informal discussions will complement the program. Fellows are provided with a two-month fellowship ranging from $6,000 to $8,000, which includes round-trip travel and local expenses. Fellows may also be eligible for an additional $1,500 per month housing allowance. Application forms must be obtained from the center. This award is granted twice per year.
*Requirements:* Applicants must have held the PhD for five years or more or possess a record of professional accomplishment at the time of application. Fellowships are awarded without regard to age or nationality. All grants are based on individual need. Scholars are expected to reside in Washington, D.C., throughout their fellowship and participate in the activities of the center.
*Restrictions:* Fellowships may not be postponed or renewed.
*Geographic Focus:* All States
*Date(s) Application is Due:* Mar 21; Sep 21
*Amount of Grant:* 6,000 - 11,000 USD
*Contact:* Danielle Horetsky, Fellowship Officer; (202) 842-6482 or (202) 842-6480; fax (202) 842-6733; casva@nga.gov
*Internet:* https://www.nga.gov/research/casva/fellowships/visiting-senior-fellowships.html
*Sponsor:* National Gallery of Art
2000B South Club Drive
Landover, MD 20785

**National Gallery of Art CASVA Wyeth Predoctoral Dissertation Fellowship** 1704

The Center for Advanced Study in the Visual Arts announces its annual program of support for advanced graduate research in the history, theory, and criticism of art, architecture, urbanism, and photographic media. One Wyeth Predoctoral Fellowship is awarded annually for 24 months. This fellowship is intended for the advancement and completion of a doctoral dissertation that concerns aspects of art of the United States, including native and pre-Revolutionary America. The Wyeth Fellow is expected to spend one year of the fellowship period on dissertation research in the United States or abroad, and one year at the Center to complete the dissertation. The stipend for all predoctoral fellowships is $30,000 per year. A housing allowance is available for fellows not in residence. A new initiative of the Wyeth Foundation for American Art provides up to $5,000 in publication subvention for first-time authors. Online applications must be submitted by November 15.
*Requirements:* Applicants must have completed their residence requirements and coursework for the PhD and general or preliminary examinations before the date of application, and know two foreign languages related to the topic of the dissertation. Applicants must be either U.S. citizens or enrolled in a university in the United States.
*Geographic Focus:* All States
*Date(s) Application is Due:* Nov 15
*Amount of Grant:* 30,000 - 35,000 USD
*Contact:* Danielle Horetsky, Fellowship Officer; (202) 842-6482 or (202) 842-6480; fax (202) 789-3026; casva@nga.gov
*Internet:* https://www.nga.gov/research/casva/fellowships/predoctoral-dissertation-fellowships.html
*Sponsor:* National Gallery of Art
2000B South Club Drive
Landover, MD 20785

**National Gallery of Art Graduate Curatorial Internships** 1705

Graduate Curatorial Internships at the National Gallery of Art provide in-depth training for advanced PhD students and recent PhD recipients interested in gaining curatorial experience in a museum setting. Interns work with curators on permanent collection and exhibition projects and attend a weekly seminar that introduces the staff, departments, programs, and functions of the Gallery. The duties and responsibilities are comparable to those of curatorial assistants. Interns receive a stipend of approximately $30,000 that is subject to all applicable taxes. Interns, using an authorized public transportation method, will receive an employer-provided fare subsidy to apply toward their monthly transit costs. The annual deadline for online applications is January 12.
*Requirements:* Eligibility is limited to doctoral students who are in the advanced stage of their dissertations and to recent PhD recipients. Neither previous museum experience nor a stated goal of a museum career is required. Applicants from all backgrounds are encouraged to apply. This is an international program.
*Geographic Focus:* All States
*Date(s) Application is Due:* Jan 12

*Amount of Grant:* Up to 30,000 USD
*Contact:* Danielle Horetsky, Fellowship Officer; (202) 842-6480 or (202) 842-6257; fax (202) 842-6733; dcl@nga.gov
*Internet:* https://www.nga.gov/opportunities/interns-and-fellows/graduate/curatorial-internships.html.html
*Sponsor:* National Gallery of Art
2000B South Club Drive
Landover, MD 20785

**National Gallery of Art Summer Internships** **1706**
Since 1964 the National Gallery of Art has offered professional museum training to candidates from all backgrounds through a variety of internship programs. Nine-week summer internships provide opportunities to work on projects directed by a Gallery curator or department head. Biweekly museum seminars introduce interns to the broad spectrum of museum work, and to Gallery staff, departments, programs, and functions. Interns receive a stipend of approximately $4,500 that is subject to all applicable taxes. Interns using an authorized public transportation method will receive an employer-provided fare subsidy to apply toward their monthly transit costs. The annual deadline for application submission is January 12.
*Requirements:* Eligibility varies according to internship. A few are geared to undergraduates completing their degrees in May and June, but most are for currently enrolled graduate students of all levels and those graduating in May or June with a relevant degree (such as MA, MBA, MFA, M Arch, M Ed, JD, or MLS).
*Geographic Focus:* All States
*Date(s) Application is Due:* Jan 12
*Amount of Grant:* 4,500 USD
*Contact:* Danielle Horetsky, Fellowship Officer; (202) 842-6257 or (202) 842-6480; fax (202) 842-6733; dcl@nga.gov
*Internet:* https://www.nga.gov/opportunities/interns-and-fellows/graduate/summer.html.html
*Sponsor:* National Gallery of Art
2000B South Club Drive
Landover, MD 20785

**National MPS Society Grants** **1707**
The society awards research grants and fellowships to qualified medical researchers to promote medical research in the fields of mucopolyssaccharidosis (MPS) and mucolipidosis (ML). The program supports MPS (I, II, III IV, V, and VI) research, ML II/III research, and general MPS research. In February of each year the Society announces the grants offered, and funding begins July 1. Occasionally a second grant offering is made later in the year. All applications are reviewed and ranked by members of the National MPS Society Scientific Advisory Board. The final decision of award recipients is determined by the Board of Directors of the National MPS Society. Funding for the grants comes from the annual National MPS Society 5k walk/runs in addition to donations and fundraisers. Grants and fellowships are typically awarded for a two-year period.
*Requirements:* Qualified medical researchers are eligible.
*Geographic Focus:* All States
*Date(s) Application is Due:* May 1
*Amount of Grant:* Up to 50,000 USD
*Samples:* Gustavo H.B. Maegawa, MD, PhD, Johns Hopkins School of Medicine, Department of Pediatrics, Baltimore, Maryland, $45,000 - two-year grant for induced-neuronal (iN) cells as tools to study the pathogenesis of neurological manifestations in MPS-II; Shunji Tomatsu, MD, PhD, Nemours Children's Clinic, Delaware Valley of the Nemours Foundation, Wilmington, Delaware, $40,000 - two-year grant for the development of Long Circulating Enzyme Replacement Therapy for MPS IVA.
*Contact:* Barbara Wedehase, Executive Director; (877) 677-1001 or (919) 806-0101; barbara@mpssociety.org or info@mpssociety.org
*Internet:* www.mpssociety.org/research/
*Sponsor:* National MPS Society
P.O. Box 14686
Durham, NC 27709-4686

**National Psoriasis Foundation Bridge Grants** **1708**
The National Psoriasis Foundation (NPF) is offering Bridge Grants to support researchers who have submitted meritorious, but unfunded, K-type (career development) and R-type applications to the National Institutes of Health (NIH), or similar funding body, with a focus on psoriatic disease or related comorbidities. Through this special opportunity, NPF provides a critical year of additional support to near-miss applicants so that they can collect data that strengthens a future application. In line with NPF's mission, Bridge Grants will provide an important life-line to help support long-term careers of psoriatic disease researchers who will go on to make discoveries that bring us closer to a cure for psoriatic disease and improve patients' lives. Applicants may request up to $50,000 (direct costs only) for K-type bridge or up to $100,000 for R-type bridge (direct costs only) per year. The application deadline for cycle one is October 15, with the deadline for cycle two being the following April 15. Award notifications are December 31 and June 30, respectively. As many as two (2) Bridge Grants may be awarded.
*Requirements:* The principal investigator must be personally and actively responsible for the conduct of the proposed research, and eligible to apply for research grants under the guidelines of the applicant's host institution. The project proposed must have been submitted as an unfunded application for either a K-type or R-type (or equivalent) award focused on psoriasis, psoriatic arthritis, or related comorbidities.
*Geographic Focus* All States, District of Columbia, Guam, Marshall Islands, Northern Mariana Islands, Puerto Rico, U.S. Virgin Islands, American Samoa
*Date(s) Application is Due:* Apr 15; Oct 15
*Amount of Grant:* Up to 100,000 USD
*Contact:* Sharlene Vance, Research Programs Coordinator; (503) 244-7404 or (503) 546-8380; fax (503) 245-0626; svance@psoriasis.org or research@psoriasis.org
*Internet:* https://www.psoriasis.org/grants
*Sponsor:* National Psoriasis Foundation
6600 SW 92nd Avenue, Suite 300
Portland, OR 97223-7195

**National Psoriasis Foundation Discovery Grants** **1709**
The National Psoriasis Foundation (NPF) is seeking grant applications that encourage the advancement of innovative research into psoriasis, psoriatic arthritis, and related comorbidities. Discovery Grants fund researchers while they explore preliminary ideas and conduct proof-of-concept experiments. The goal is to stimulate the development of new research programs in the field of psoriatic disease capable of competing for long-term funding from the National Institutes of Health or similar agencies in the future. Applications related to all aspects of psoriatic disease are strongly encouraged, including basic science, translational research, technology, epidemiology, health services, and clinical research. Prospective applicants are encouraged to review recent publications and reports identifying research gaps and priorities in psoriatic disease research. Applicants may request up to $75,000 (direct costs only) per year. Applications must be receive by January 18, with award notifications anticipated by June 30. As many as ten (10) new Discovery Grants may be awarded.
*Requirements:* The principal investigator must be personally and actively responsible for the conduct of the proposed research, and eligible to apply for research grants under the guidelines of the applicant's host institution.
*Restrictions:* The principal investigator need not be a United States citizen and the research need not take place in the United States.
*Geographic Focus* All States, District of Columbia, Guam, Marshall Islands, Northern Mariana Islands, Puerto Rico, U.S. Virgin Islands, American Samoa
*Date(s) Application is Due:* Jan 18
*Amount of Grant:* Up to 75,000 USD
*Contact:* Sharlene Vance, (503) 244-7404 or (503) 546-8380; fax (503) 245-0626; svance@psoriasis.org or research@psoriasis.org
*Internet:* https://www.psoriasis.org/grants
*Sponsor:* National Psoriasis Foundation
6600 SW 92nd Avenue, Suite 300
Portland, OR 97223-7195

**National Psoriasis Foundation Early Career Research Grants** **1710**
The National Psoriasis Foundation (NPF) is seeking grant applications from graduate students, postdoctoral researchers, and researchers at similar trainee-level positions interested in conducting projects focused on psoriatic disease. The goal of this award is to support scientists at this challenging early career stage and to welcome them into the collaborative community of scientists, clinicians, and patients involved with NPF research. Ultimately, recipients of this award are expected to be in a position to compete for future funding through NPF Discovery or Translational Research Grants and go on to establish successful long-term careers conducting psoriatic disease research. Applications related to all aspects of psoriatic disease are strongly encouraged, including basic science, translational research, technology, epidemiology, health services, and clinical research. Prospective applicants are encouraged to review recent publications and reports identifying research gaps and priorities in psoriatic disease research. Applicants may request up to $50,000 (direct costs only) per year. Applications must be received by January 18, with award notifications anticipated by June 30. The project period for this Request for Proposals is one year, from August 1 through July 31. As many as eight (8) new Early Career Research Grants may be awarded.
*Requirements:* The applicant must be personally and actively responsible for the conduct of the proposed research. Eligible applicants include graduate students dedicating at least 50% of their time to research, postdoctoral researchers, fellows, research associates, instructors, and other trainee level positions.
*Restrictions:* The applicant need not be a United States citizen and the research need not take place in the United States.
*Geographic Focus* All States, District of Columbia, Guam, Marshall Islands, Northern Mariana Islands, Puerto Rico, U.S. Virgin Islands, American Samoa
*Date(s) Application is Due:* Jan 18
*Amount of Grant:* Up to 50,000 USD
*Contact:* Sharlene Vance, (503) 244-7404 or (503) 546-8380; fax (503) 245-0626; svance@psoriasis.org or research@psoriasis.org
*Internet:* https://www.psoriasis.org/grants
*Sponsor:* National Psoriasis Foundation
6600 SW 92nd Avenue, Suite 300
Portland, OR 97223-7195

**National Psoriasis Foundation Psoriatic Arthritis Diagnostic Test Grants** **1711**
The National Psoriasis Foundation is currently accepting applications for its Psoriatic Arthritis Diagnostic Test Grant program. The Foundation (NPF) is seeking researchers to submit applications that aim to develop a clinically applicable diagnostic test for psoriatic arthritis that may improve detection of PsA in highly sensitive, specific, and non-invasive manner. The Psoriatic Arthritis Diagnostic Test Grant will fund research initiatives that have the potential to achieve this goal within five years. Projects will occur in two phases, with one year of support to establish proof-of-concept data followed by up to four additional years to complete research, development and validation. The Foundation gives highest priority to proposals that: create, develop or support innovative collaborations among basic, translational, and clinical researchers; target a multi-faceted,

multi-marker approach utilizing a multi-disciplinary team, and/or explain how their approach will have the highest likelihood of developing a diagnostic test for PsA. In year one (current application), applicants may request up to $100,000. Successful applicants will then be required to submit a brief six-month progress report. In year two through year five NPF will provide additional funding for projects that have demonstrated progress during the first year. Renewal will be contingent upon factors, including likelihood of achieving the program goals. Specifically, near the end of year one, applicants will have the opportunity to renew their proposal for up to an additional four years with up to $300,000 annually in funding. As many as six Psoriatic Arthritis Diagnostic Test Grants will be funded during the first year of the program. Researchers interested in applying for this RFP must submit a letter of intent by December 14. Applications are due by April 1, with award notifications anticipated by July 1 of the next year.
*Requirements:* The principal investigator must be personally and actively responsible for the conduct of the proposed research, and eligible to apply for research grants under the guidelines of his/her host institution or organization. The principal investigator need not be a United States citizen and the research need not take place in the United States.
*Geographic Focus* All States, District of Columbia, Guam, Marshall Islands, Northern Mariana Islands, Puerto Rico, U.S. Virgin Islands, American Samoa
*Amount of Grant:* Up to 100,000 USD
*Contact:* Sharlene Vance, Research Programs Coordinator; (503) 244-7404 or (503) 546-8380; fax (503) 245-0626; svance@psoriasis.org or research@psoriasis.org
*Internet:* https://www.psoriasis.org/grants
*Sponsor:* National Psoriasis Foundation
6600 SW 92nd Avenue, Suite 300
Portland, OR 97223-7195

**National Psoriasis Foundation Psoriatic Arthritis Research Grant** 1712
In a collaborative effort, the Arthritis National Research Foundation (ANRF) and the National Psoriasis Foundation (NPF) are seeking grant applications that focus on studying psoriatic arthritis. Applicants may request up to $100,000 (direct cost only) per year. Applications must be receive by the deadline of January 18, with award notifications anticipated by June 30. The total project period for the application submitted in response to this Request for Proposals is one year, August 1 through July 31.
*Requirements:* Applicants must be newer M.D. and/ or Ph.D. applicants at early stages in their career (equivalent to assistant professor/instructor/senior post-docs).
*Restrictions:* The body of the proposal may not exceed nine (9) pages, including figures and tables, excluding references, cover sheet and budget sheet.
*Geographic Focus* All States, District of Columbia, Guam, Marshall Islands, Northern Mariana Islands, Puerto Rico, U.S. Virgin Islands, American Samoa
*Date(s) Application is Due:* Jan 18
*Amount of Grant:* Up to 100,000 USD
*Contact:* Sharlene Vance, Research Programs Coordinator; (503) 244-7404 or (503) 546-8380; fax (503) 245-0626; svance@psoriasis.org or research@psoriasis.org
*Internet:* https://www.psoriasis.org/grants
*Sponsor:* National Psoriasis Foundation
6600 SW 92nd Avenue, Suite 300
Portland, OR 97223-7195

**National Psoriasis Foundation Psoriatic Disease Research Fellowships** 1713
The National Psoriasis Foundation Fellowship program aims to increase the number of clinician-scientists focused on studying and treating psoriatic disease and related conditions. The Foundation is pleased to announce the availability of the Psoriatic Disease Research Fellowships program for the upcoming academic year. The program provides support to eligible institutions to develop and enhance the opportunities for physicians training for research careers in academic dermatology, rheumatology, pediatric dermatology, and pediatric rheumatology. Priority consideration will be given to applications that advance psoriatic disease research. Awardees will receive a maximum of $50,000 to support their fellowship stipend. Institutions are to certify that funds will be applied toward stipend, salary and fringe benefits only, and not to be used for any indirect costs. Applications must be received by January 18, with award notifications anticipated by April 12. The project period for this RFP is from July 1 to June 30. It is anticipated that as many as nine (9) new NPF Psoriatic Disease Research Fellowships will be awarded.
*Requirements:* Psoriatic Disease Research Fellowships may take place at any institution with an eligible fellowship program. Eligibility requirements are as follows: the fellow cannot be expected to be a faculty member during the project period; the research project must be focused on psoriasis and/or psoriatic arthritis; fellowship programs must be non-ACGME-accredited; the fellow cannot be in a residency program during the project period; the fellow must devote at least 75% of their effort to research during the project period; and fellowships are primarily intended for those who already hold an M.D., however applications to support exceptionally promising medical students will be accepted.
*Restrictions:* The fellow need not be a United States citizen and the fellowship need not take place in the United States.
*Geographic Focus* All States, District of Columbia, Guam, Marshall Islands, Northern Mariana Islands, Puerto Rico, U.S. Virgin Islands, American Samoa
*Date(s) Application is Due:* Jan 18
*Amount of Grant:* Up to 50,000 USD
*Contact:* Sharlene Vance, Research Programs Coordinator; (503) 244-7404 or (503) 546-8380; fax (503) 245-0626; svance@psoriasis.org or research@psoriasis.org
*Internet:* https://www.psoriasis.org/fellowship
*Sponsor:* National Psoriasis Foundation
6600 SW 92nd Avenue, Suite 300
Portland, OR 97223-7195

**National Psoriasis Foundation Summer Student Research Grants** 1714
The National Psoriasis Foundation (NPF) is seeking applications from undergraduate and medical students interested in conducting a summer of research focused on psoriatic disease or related comorbidities. The goal is to foster the development of promising young scientists who will go on to become bench researchers or clinician scientists focused on improving the lives of those living with psoriatic disease. Applications related to all aspects of psoriatic disease are strongly encouraged, including basic science, translational research, technology, epidemiology, health services, and clinical research. Prospective applicants are encouraged to review recent publications identifying research gaps and priorities in psoriatic disease research to identify relevant research topics. Successful applicants will receive $4,000 to support salary/stipend and $1,000 to support project related expenses (direct costs only). Applications must be received by January 18, with award notifications anticipated by April 12. As many as four (4) new grants may be awarded. The project period for this Request for Proposals is ten weeks, to take place between May 1 and September 30.
*Requirements:* The applicant must: be a student enrolled as either an undergraduate or medical student for the duration of the award; have the endorsement of an investigator willing to host the student in their lab; and be committed to conducting a summer research project focused on psoriatic disease or related comorbidities.
*Restrictions:* The applicant need not be a United States citizen and the research need not take place in the United States.
*Geographic Focus* All States, District of Columbia, Guam, Marshall Islands, Northern Mariana Islands, Puerto Rico, U.S. Virgin Islands, American Samoa
*Date(s) Application is Due:* Jan 18
*Amount of Grant:* 5,000 USD
*Contact:* Sharlene Vance, Research Programs Coordinator; (503) 244-7404 or (503) 546-8380; fax (503) 245-0626; svance@psoriasis.org or research@psoriasis.org
*Internet:* https://www.psoriasis.org/grants
*Sponsor:* National Psoriasis Foundation
6600 SW 92nd Avenue, Suite 300
Portland, OR 97223-7195

**National Psoriasis Foundation Translational Research Grants** 1715
The National Psoriasis Foundation (NPF) is seeking grant applications that encourage the advancement of innovative research into psoriasis and psoriatic arthritis. Translational Research Grants fund research initiatives that focus on the rapid translation of basic scientific discoveries into clinical applications with a clear benefit for patients with psoriatic disease. Applications related to all aspects of psoriatic disease are strongly encouraged, including basic science, translational research, technology, epidemiology, health services, and clinical research. Prospective applicants are encouraged to review recent publications and reports identifying research gaps and priorities in psoriatic disease research. Applicants may request up to $100,000 (direct costs only) per year. Applications must be receive by January 18, with award notifications anticipated by June 30. Second year funds are contingent upon adequate progress through the first year. The project period for this Request for Proposals is two years. As many as two (2) new Translational Research Grants may be awarded.
*Requirements:* The principal investigator must be personally and actively responsible for the conduct of the proposed research, and eligible to apply for research grants under the guidelines of the applicant's host institution. The principal investigator need not be a United States citizen and the research need not take place in the United States.
*Geographic Focus* All States, District of Columbia, Guam, Marshall Islands, Northern Mariana Islands, Puerto Rico, U.S. Virgin Islands, American Samoa
*Date(s) Application is Due:* Jan 18
*Amount of Grant:* Up to 100,000 USD
*Contact:* Sharlene Vance, Research Programs Coordinator; (503) 244-7404 or (503) 546-8380; fax (503) 245-0626; svance@psoriasis.org or research@psoriasis.org
*Internet:* https://www.psoriasis.org/grants
*Sponsor:* National Psoriasis Foundation
6600 SW 92nd Avenue, Suite 300
Portland, OR 97223-7195

**National Urban Fellows Program** 1716
The National Urban Fellows leadership development program is a rigorous, 14-month, full-time graduate degree program comprising two semesters of academic course work and a nine-month mentorship assignment. Fellows receive a $25,000 stipend, health insurance, a book allowance, relocation and travel reimbursement, and full payment of tuition, in addition to their ongoing personal and professional development. The program culminates in a Master of Public Administration (MPA) degree from the City of New York's Bernard M. Baruch College, School of Public Affairs. During the mentorship, Fellows complete course work via distance learning. An MPA degree is awarded to Fellows upon completion of all academic and program requirements. Application materials are available at the website.
*Requirements:* United States citizens who have bachelor's degrees; have three to five years' full-time work experience in an administrative or management capacity; have demonstrated exceptional ability, maturity, leadership potential, and commitment to the solutions of urban and rural problems; and are willing to relocate for the duration of the fellowship year are eligible. An official transcript of all undergraduate and graduate education will be required.
*Geographic Focus:* All States
*Amount of Grant:* 26,000 USD
*Contact:* Miguel A. Garcia, Jr., Program Director; (212) 730-1700; fax (212) 730-1823
*Internet:* www.nuf.org/fellows-overview
*Sponsor:* National Urban Fellows
989 Avenue of the Americas, Suite 400
New York, NY 10018

**National Wildlife Federation Campus Ecology Fellowships** 1717

National Wildlife Federation's Campus Ecology Fellows confront global warming on their campuses and help to educate and engage the campus community on global warming impacts and solutions. Federation Fellowships allow students to pursue their vision of an ecologically sustainable future through tangible projects to confront global warming on campus and in the community. Fellows gain practical experience in the conservation field and first-hand knowledge of the challenges and opportunities inherent in successful conservation efforts. Fellowship guidelines and the application are available at the Federation website.
*Requirements:* Candidates may apply directly through the Federation's job database.
*Geographic Focus:* All States
*Contact:* Fellowship Coordinator; (800) 822-9919
*Internet:* www.nwf.org/Global-Warming/Campus-Solutions/Get-Involved/Apply-for-a-Fellowship.aspx
*Sponsor:* National Wildlife Federation
11100 Wildlife Center Drive
Reston, VA 20190

**National Wildlife Federation Internships** 1718

Through the internship program, the National Wildlife Federation (NWF) brings students and recent graduates to its offices nationwide, where they can experience many aspects of NWF's conservation mission, including work on endangered species, land stewardship, and protecting our nation's water. Departments benefiting from the skills interns provide include conservation, education, fundraising, communications, legal and publications. Working side by side with knowledgeable professionals, interns receive support from the internship coordinator, while developing a variety of professional and personal life skills. Internship projects support commonsense conservation in communities throughout the nation. Interns receive medical, dental, and holiday benefits. Interns must work a minimum of 20 hours per week and are expected to complete a 24-week internship to be eligible for these benefits. Some departments substitute a paid salary with college credit. Students apply for the internship through the Federation's job page.
*Requirements:* Graduates and graduate students from all fields of study may apply. To ensure a rigorous and meaningful professional, educational, and personal experience, interns must: document and evaluate their experience for the internship coordinator at scheduled intervals throughout the internship. They must also complete the work objectives agreed to at the beginning of the internship. Applicants are selected based on a variety of measures, including professional goals, course work, job experience, resume and cover letter, and grade point average, along with professional/academic references. Candidates should demonstrate commitment to and experience working with multi-cultural communities, environmental justice and diversity. Due to the high volume of applications, only candidates selected for an interview will be contacted.
*Restrictions:* For-credit interns are not eligible for benefits.
*Geographic Focus:* All States
*Amount of Grant:* 500 USD
*Contact:* Internship Coordinator; (800) 822-9919; internopp@nwf.org
*Internet:* www.nwf.org/About/Jobs-at-NWF.aspx
*Sponsor:* National Wildlife Federation
11100 Wildlife Center Drive
Reston, VA 20190

**NCFR Affiliate Councils Outstanding Student Research Papers (Graduate and Undergraduate)** 1719

The National Council on Family Relations, founded in 1938, is the oldest multidisciplinary, nonpartisan professional organization focused solely on family research, practice and education. There are two awards each year, one for a graduate student paper and one for an undergraduate student paper. The graduate student paper winner will receive $250 and a plaque. The undergraduate student paper winner will receive $100 and a certificate.
*Requirements:* While students may submit their papers to NCFR on their own, if multiple submissions are received from within one affiliate, the submissions may be sent back to the local affiliate's board/officers to choose the strongest entry/ies (regional affiliates will be permitted one paper per state in the region). For the Graduate Student Outstanding Paper, authors must be a graduate student, a member of NCFR, and a member of a student, state, or regional affiliate council. For the Undergraduate Student Outstanding Paper, authors must be an undergraduate student and a member of a student, state, or regional affiliate council.
*Restrictions:* Papers already published or presented at a national conference are not eligible.
*Geographic Focus:* All States
*Date(s) Application is Due:* Jun 1
*Amount of Grant:* 100 - 250 USD
*Contact:* Allison Wickler, (763) 231-2881; allisonwickler@ncfr.org
*Internet:* www.ncfr.org/about/awards/affiliate-councils-awards/affiliate-councils-outstanding-student-research-papers-gradua
*Sponsor:* National Council on Family Relations
1201 West River Parkway, Suite 200
Minneapolis, MN 55454-1115

**NCFR Family Therapy Section Best Research Paper Awards** 1720

The National Council on Family Relations, founded in 1938, is the oldest multidisciplinary, nonpartisan professional organization focused solely on family research, practice and education. Family Therapy Section paper awards are intended to foster scholarly activity and assist in identifying future leaders. Winners receive a $250 cash award and a certificate. Two awards will be presented - one to a student and one to a new professional. The awards are presented annually at the Family Therapy Section Annual Meeting at the NCFR Annual Conference. The deadline is concurrent with the NCFR conference submission deadline.
*Requirements:* Upon acceptance of paper for presentation, students are contacted by a section representative and invited to apply and submit their papers for award consideration. The winner must be: a student or new professional within three years of graduation; a current member of NCFR and the FT Section when proposal is submitted; the first author on paper submitted to section; and, ready to present paper at the NCFR Annual Conference.
*Geographic Focus:* All States
*Date(s) Application is Due:* Mar 3
*Amount of Grant:* 250 USD
*Contact:* Jeanne Strand; (763) 781-9331; jeannestrand@ncfr.org
*Internet:* www.ncfr.org/awards/section-awards/family-therapy/best-research-paper-awards-one-student-award-one-new-profession
*Sponsor:* National Council on Family Relations
1201 West River Parkway, Suite 200
Minneapolis, MN 55454-1115

**NCFR Innovation Grants** 1721

The National Council on Family Relations, founded in 1938, is the oldest multidisciplinary, nonpartisan professional organization focused solely on family research, practice and education. NCFR will support up to two $10,000 innovation grants over the next year. The grants program is designed to enhance interdisciplinary collaborations around issues of importance to family well-being. A range of activities could be supported by these funds such as: a) research pilot projects, b) theory development and enhancement, c) policy initiatives, and d) outreach, engagement, and training opportunities. NCFR encourages innovative and groundbreaking proposals that will further the goals of NCFR and, in so doing, strengthen and support families.
*Requirements:* Proposals must involve significant interdisciplinary collaboration and demonstrate how the proposed activities and contributions will be strengthened by the interdisciplinary approach. The proposal must demonstrate how it meets the NCFR Global Ends. The primary applicant (or applicants, in the case of co-PIs) must be an NCFR member who is an academic scholar or other family professional. Postdoctoral fellows may apply as PIs. Students may be involved in the proposed project as collaborators but not as PIs. Applicants may be included on more than one proposal but should be listed as the PI on only on proposal. Priority will be given to proposals with NCFR members from two or more different disciplines. Projects that aim to bridge the areas of research and practice and that build on a stong theoretical framework or advance theory are encouraged. The call for proposals is intentionally broad to encourage a variety of innovative proposals that meet these objectives. Proposals must be received by NCFR by 11:59PM CDT of the deadline date.
*Restrictions:* Funding will not be granted for supporting conferences or workshops.
*Geographic Focus:* All States
*Date(s) Application is Due:* Sep 1
*Amount of Grant:* Up to 10,000 USD
*Contact:* Jeanne Strand; (763) 231-2881; jeannestrand@ncfr.org
*Internet:* www.ncfr.org/awards/ncfr-innovation-grants
*Sponsor:* National Council on Family Relations
1201 West River Parkway, Suite 200
Minneapolis, MN 55454-1115

**NCFR Ruth Hathaway Jewson Award** 1722

The National Council on Family Relations, the NCFR Sections, CFLE Advisory Board, and the Affiliate Councils have established awards to recognize those individuals and groups who further the interests of the organization and/or exemplify the standards to which members aspire. Award winners are distinguished colleagues who have dedicated themselves to working in the family field, and have given outstanding service. Members nominate other members for the awards, send supporting evidence, and list other people who can verify the nominee's qualifications. A selection committee evaluates the applicants and selects the winner(s). This award is given in honor of Ruth Hathaway Jewson, NCFR's second executive director. It funds the the best dissertation proposal submitted by a doctoral candidate in the field of family studies. Dr. Jewson was responsible for expanding NCFR's membership from a small organization to over 4,000 members when she retired. The award winner receives a $500 cash award and a plaque, which are presented at the beginning of a plenary session at the NCFR Annual Conference. This award is presented biannually (even years).
*Requirements:* Awardees must be nominated by NCFR members. The nominee must be a doctoral student and have been a member of NCFR for at least six months. All materials must be received by NCFR by May 1. The winner will be notified by July 31.
*Geographic Focus:* All States
*Date(s) Application is Due:* May 1
*Amount of Grant:* 500 USD
*Contact:* Jeanne Strand; (763) 231-2881; jeannestrand@ncfr.org
*Internet:* www.ncfr.org/awards/ncfr-sponsored-awards/ruth-hathaway-jewson-award-snp
*Sponsor:* National Council on Family Relations
1201 West River Parkway, Suite 200
Minneapolis, MN 55454-1115

**NCHS National Center for Health Statistics Postdoctoral Research Appointments** 1723

The National Center for Health Statistics (NCHS) is the Nation's principal health statistics agency and provides the information needed to develop the programs and policies that will improve the health of the American people. The objective of the NCHS Postdoctoral Research Program is to provide opportunities for postdoctoral candidates of unusual promise and ability to conduct research in problems of their choosing that are compatible

with the interests of NCHS. A postdoctoral researcher is a temporary salaried employee of NCHS analogous to fellows or similar temporary researchers at the postdoctoral level in universities and other organizations. Appointments are two years in length, but extensions are possible with special consideration. Benefits include sick leave, annual leave, thrift savings, and health and life insurance. Flex time and alternate work schedules are available if desired, contingent on the supervisor's approval. Cost of relocation will be determined on an individual basis. An evaluation is conducted after one year to ensure that the postdoctoral researcher is making suitable progress. General areas of interest for research at NCHS include statistical theory, survey methodology, statistical computing, economics, demography, and social and behavioral science. Specific areas of interest are: confidentiality protection and related topics; data linkage; quality control, quality improvement, and its measurement; adjustment for nonresponse in sample surveys; telephone survey methods; data collection in a changing health care system; small-area estimation; cognitive processes and question response; data editing and imputation; statistical computing; sampling and estimation; questionnaire design and measurement research; human factors and usability research; and data mining. To apply, applicants must submit a statement of research interest, a curriculum vitae, official transcripts, and three reference letters. The statement of research interest should describe the purpose and context of the research project, the research approach and methods to be used, key activities to be undertaken during the fellowship period, and the expected contributions of the proposed research. For fullest consideration, applicants should submit the application by the deadline date to the address given. However, applications will be accepted throughout the year until positions are filled. Evaluation of proposals will be conducted annually in March or April by a special panel convened for this purpose. Notification of awards will be made beginning in mid-April. The date on which an appointment would begin is negotiable on an individual basis but is expected to occur between June 1 and December 31. Applicants can find complete information on the postdoctoral research program and on application submission requirements at the program website. Applicants are encouraged to contact NCHS to obtain specific information on current research and available technical facilities and to speak to NCHS researchers working in the specific area of interest.
*Requirements:* Awardees must hold a Ph.D. or other earned research degree recognized in the United States as equivalent to the Ph.D. or must present acceptable evidence of having completed all the formal academic requirements for the degree before appointment. Applicants must have demonstrated ability for creative research. An applicant's training and research experience may be in any appropriate discipline or combination of disciplines required for the proposed research. Postdoctoral researchers must devote their full-time effort to the research program and must be in residence at NCHS during the program. No period of tenure may be spent in residence at another agency or institution.
*Restrictions:* Applicants must be U.S. citizens or legal permanent residents with a work authorization. Postdoctoral research positions are awarded only to persons who have held doctorates for less than 3 years or are in the process of receiving doctorate degrees at the time of application. No additional monetary aid or other remuneration may be accepted from another appointment, fellowship, or similar grant during the period of the program.
*Geographic Focus:* All States
*Date(s) Application is Due:* Feb 28
*Amount of Grant:* 149,744 - 231,484 USD
*Contact:* Cynthia Link, (800) 232-4636 or (301) 458-4344; fax (301) 458-4018; cynthia.link@cdc.hhs.gov
*Internet:* www.cdc.gov/nchs/about/postdoc.htm
*Sponsor:* National Center for Health Statistics
3311 Toledo Road
Hyattsville, MD 20782

**NCI/NCCAM Quick-Trials for Novel Cancer Therapies** **1724**
This PA sponsored by NCI and NCCAM expands the initiative to all cancer sites and provides investigators with rapid access to support for pilot, phase I, and phase II cancer clinical trials testing new agents and patient monitoring and laboratory studies to ensure the timely development of new therapeutic approaches. QUICK-TRIAL will provide a new approach designed to simplify the grant application process and provide a rapid turnaround from application to funding. Features include a modular grant application and award process, inclusion of the clinical protocol within the grant application, and accelerated peer review with the goal of issuing new awards within five months of application receipt. Inclusion of the complete clinical protocol within the PHS 398 grant application is intended to simplify the application process by eliminating the need to duplicate protocol details in the research plan section. Investigators may apply for a maximum of two years of funding support using the exploratory/developmental (R21) grant mechanism for up to $250,000 direct costs per year. Therapeutic trials in human subjects employing new agents and therapeutic approaches, including CAM treatments, whether used as a single agent/modality or in combination, are appropriate. Preference will be given to clinical trials that include laboratory studies to validate mechanistic hypotheses or clinical correlates that can meaningfully guide further clinical development. A letter of intent is due one month prior to application receipt date. PA: PAR-04-155
*Requirements:* Applications may be submitted by foreign and domestic, for-profit and nonprofit organizations, both public and private; units of state and local governments; and eligible agencies of the federal government.
*Geographic Focus:* All States
*Date(s) Application is Due:* Apr 9; Aug 9; Dec 9
*Amount of Grant:* Up to 250,000 USD
*Contact:* Amy Connolly; (301) 496-8786; fax (301) 496-8601; amyconnolly@nih.gov
*Internet:* grants.nih.gov/grants/guide/pa-files/par-04-155.html
*Sponsor:* National Cancer Institute
6130 Executive Boulevard, Executive Plaza N, Suite 3157, MSC 7328
Bethesda, MD 20892-7328

**NCI Cancer Education and Career Development Program** **1725**
This Funding Opportunity Announcement (FOA) represents the continuation of the Cancer Education and Career Development Program (CECDP) established by the National Cancer Institute (NCI). The purpose of the CECDP is to support the development and implementation of institutional curriculum-dependent pre-doctoral and/or postdoctoral programs with specific core didactic and research requirements to train a cadre of scientists in interdisciplinary and collaborative cancer research settings. Examples of interdisciplinary/collaborative areas of cancer research that are particularly applicable to intent of the CECDP are cancer prevention and control, behavioral and population sciences, nutrition, imaging, and molecular diagnosis. This FOA makes an NCI-specific use of the NIH Education R25 grant mechanism.
*Requirements:* Applications may be submitted on behalf of the principal investigator by domestic, nonfederal, public or private organizations. Applications may include more than one institution to create a program through consortium agreements. Predoctoral and postdoctoral candidates must be citizens or noncitizen nationals of the US, or must have been lawfully admitted to the U.S. for permanent residence.
*Restrictions:* No application may request more than $500,000 in direct costs per annum without written approval from NCI.
*Geographic Focus:* All States
*Amount of Grant:* Up to 500,000 USD
*Contact:* Catherine Blount, Grants Administration Branch; (301) 496-3179; fax (301) 496-8601; blountc@mail.nih.gov
*Internet:* grants.nih.gov/grants/guide/pa-files/PAR-06-511.html
*Sponsor:* National Cancer Institute
6130 Executive Boulevard, Executive Plaza N, Suite 3144, MSC 7362
Bethesda, MD 20892-7362

**NCI Centers of Excellence in Cancer Communications Research** **1726**
NCI invites applications for Centers of Excellence in Cancer Communications Research (CECCRs). The purpose is to identify areas of discovery that build upon important recent developments in knowledge and technology and that hold promise for making significant progress against all cancers. The centers must include three or more individual hypothesis-driven research projects, pilot or developmental research projects, shared resources, and a plan for career development. To be effective, the centers' research should integrate cancer communications appropriately into one or more contexts of the cancer continuum—from prevention through treatment to survivorship and end-of-life research. Communications research also is needed about challenging topics such as cancer information seeking, decision making under uncertainty, and genetic testing. Centers' research also should provide insight into mechanisms underlying how people process information. Annual deadline dates may vary; contact program staff for exact dates. RFA: CA-03-007
*Requirements:* Applications may be submitted by domestic for-profit and nonprofit organizations, both public and private; units of state and local governments; and eligible agencies of the federal government.
*Geographic Focus:* All States
*Amount of Grant:* Up to 15,000,000 USD
*Contact:* Crystal Wolfrey, Grants Administration Branch; (301) 496-8634; fax (301) 496-8601; crystal.wolfrey@nih.gov or wolfreyc@mail.nih.gov
*Internet:* grants.nih.gov/grants/guide/rfa-files/RFA-CA-08-004.html
*Sponsor:* National Cancer Institute
6130 Executive Boulevard, Executive Plaza N, Suite 3144, MSC 7362
Bethesda, MD 20892-7362

**NCSS/FASSE/CUFA Inquiry Grant** **1727**
The Fund for the Advancement of Social Studies Education (FASSE) and the College and University Faculty Assembly (CUFA) of the National Council for the Social Studies (NCSS) have established a grant to support inquiry in citizenship education. Grant proposals should affirm social, cultural, and racial diversity and address issues of equality, equity, and social justice. Proposals that address aims for citizen action are preferred. Proposals should be relevant to school, university or community-based educational settings. FASSE and CUFA will award a $10,000 grant to the successful applicant who presents a proposal for a research project that demonstrates potential to inform the educational field about justice oriented, citizenship education. The original and four copies of the proposal must be submitted no later than June 30.
*Requirements:* Assistant, Associate or a full professor, or a graduate student with the demonstrated support of a university mentor/advisor are eligible to apply. Applicant(s) must be an NCSS member. If a graduate student, then the mentor must also be an NCSS member.
*Restrictions:* No fax or email submissions will be accepted.
*Geographic Focus:* All States
*Date(s) Application is Due:* Jun 30
*Amount of Grant:* 10,000 USD
*Contact:* Prema Cordeiro, Program Manager; (301) 588-1800, ext. 106; fax (301) 588-2049; pcordeiro@ncss.org or excellence@ncss.orgexcellence@ncss.org
*Internet:* www.socialstudies.org/fasse
*Sponsor:* National Council for the Social Studies
8555 Sixteenth Street, Suite 500
Silver Spring, MD 20910

**NCSS Exemplary Research in Social Studies Award** **1728**
It is the mission of National Council for the Social Studies to engage and support educators in strengthening and advocating social studies, where social studies is defined as the integrated study of the social sciences and humanities to promote civic competence. In support of this mission, NCSS and the Research Community sponsor this annual research award designed to recognize an exemplary single-study research conception,

model, design, procedure and analysis in social studies and social education. For the purpose of this award, social education includes work in the individual social sciences; women's studies and gender studies; multicultural, citizenship, global, economic, and geographic education; decision making; reflective inquiry; moral education; critical theory; teacher education; teacher thinking; and teaching-learning phenomena. Research is broadly constructed to mean any and all forms of recognized, formal research methodologies, paradigms, and perspectives. Detailed instructions about the nomination process are available at the website.
*Restrictions:* No self-nominations are accepted.
*Geographic Focus:* All States
*Date(s) Application is Due:* May 1
*Contact:* Prema Cordeiro, Program Manager; (301) 588-1800, ext. 106; fax (301) 588-2049; pcordeira@ncss.org or excellence@ncss.org
*Internet:* www.socialstudies.org/awards/research/exemplary/
*Sponsor:* National Council for the Social Studies
8555 Sixteenth Street, Suite 500
Silver Spring, MD 20910

**NCSS Jean Dresden Grambs Distinguished Career Research in Social Studies Award** 1729
The Jean Dresden Grambs Distinguished Career Research Award recognizes professionals who have made extensive contributions to knowledge concerning significant areas of social studies education through meritorious research. The use of the term research in reference to this award shall be broadly construed to mean any and all forms of recognized formal research - methodologies, paradigms. and perspectives. Eligible research is that which has: added significantly to the understanding of the status, impact, characteristics, and activities of social studies education or social education, including several separate and distinct contributions related to curriculum, instruction, evaluation, and the learning and teaching of and in the social studies; stimulated a sense of responsibility among and fostered other professionals to engage in additional research efforts; added significantly to the information bases to those working in the social studies/social education field as evidenced by citations of their work within the social studies literature; and been used by professionals as the basis for significant decisions, rationales, and practices related to one or more areas of social studies curriculum instruction, evaluation, and teaching or learning.
*Requirements:* Applicants are eligible for consideration and receipt of this award when they have met the following conditions: published documentation of research endeavors and findings covering ten or more years; published documentation that research reveals a focus or direction in a significant segment of the research effort, without intending that the individual has spent the entire time on a single or a limited range of research topics; documentation exists that cites in positive ways the person's research contributions; and letters of support on behalf of the researcher are received by not less than five recognized researchers in the field of social studies/social science education. The person does not need to be currently involved in active research and publications in the social studies/social education areas or fields or a current member of NCSS.
*Restrictions:* No self-nominations are accepted.
*Geographic Focus:* All States
*Date(s) Application is Due:* Jul 18
*Contact:* Prema Cordeiro, Program Manager; (301) 588-1800, ext. 106; fax (301) 588-2049; pcordeiro@ncss.org or excellence@ncss.orgexcellence@ncss.orgexcellence@n
*Internet:* www.socialstudies.org/awards/research/grambs/
*Sponsor:* National Council for the Social Studies
8555 Sixteenth Street, Suite 500
Silver Spring, MD 20910

**NCSS Larry Metcalf Exemplary Dissertation Award** 1730
The annual Metcalf Dissertation Award is designed to recognize substantive scholarly inquiry in social education completed in pursuit of the doctoral degree. Social education includes work in the individual social sciences; women studies and gender studies; multicultural, citizenship, global, economic, and geographic education; decision making; reflective theory; moral education; critical theory; teacher education; teacher thinking; and teaching-learning phenomena. Research is broadly defined to include experimental, qualitative, historical, and philosophical work. The award recipient must be willing to present the research at a one-hour session during the annual conference.
*Requirements:* Research in pursuit of the doctoral degree must be completed between March of the year preceding the award and February of the year in which the award is presented.
*Restrictions:* Self nominations are not accepted.
*Geographic Focus:* All States
*Date(s) Application is Due:* Apr 1
*Amount of Grant:* 250 USD
*Contact:* Prema Cordeiro, Program Manager; (301) 588-1800, ext. 106; fax (301) 588-2049; pcordeiro@ncss.org or excellence@ncss.org
*Internet:* www.socialstudies.org/awards/research/metcalf/
*Sponsor:* National Council for the Social Studies
8555 Sixteenth Street, Suite 500
Silver Spring, MD 20910

**NCTM E. Glenadine Gibb 7-12 Classroom Research Grants** 1731
The purpose of this grant is support and encourage classroom-based research in pre-college mathematics education in collaboration with college or university mathematics educators. For the upcoming year, grants with a maximum of $6,000 each will be awarded to classroom teachers currently teaching mathematics at the grades 7–12 level. The research must be a significant collaborative effort involving a college or university mathematics educator (a mathematics education researcher or a teacher of mathematics learning, teaching, or curriculum) and one or more grades 7–12 classroom teachers. The proposal may include, but is not restricted to, research on the following topics: curriculum development and implementation; involvement of at-risk or minority students; students' thinking about a particular mathematics concept or set of concepts; connection of mathematics to other disciplines; focused learning and teaching of mathematics with embedded use of technology (any acquisition of equipment must support the proposed plan but not be the primary focus of the grant); and innovative assessment or evaluation strategies. Involvement of preservice teachers is encouraged but not required. It is hoped that this research will lead to an article suitable for publication in the "Journal for Research in Mathematics Education" or in one of the NCTM school journals.
*Requirements:* Proposals must address the following: research design, the plan for collecting and analyzing data, and the anticipated impact on students' learning. The applicant must be a current Full Individual or E-Member of NCTM or teach at a school having a current NCTM PreK-8 school membership. The college or university mathematics educator must be a member of the NCTM. All grant activities must be completed between June 1 and May 31.
*Restrictions:* Past grantees are not eligible to apply. No person(s) may receive more than one award administered by the Mathematics Education Trust in the same academic year. Past recipients of this grant are not eligible to reapply.
*Geographic Focus:* All States
*Date(s) Application is Due:* Nov 11
*Amount of Grant:* Up to 6,000 USD
*Contact:* Nihad Ziad; (703) 620-9840, ext. 2112; fax (703) 476-2970; infocentral@nctm.org
*Internet:* www.nctm.org/resources/content.aspx?id=22418
*Sponsor:* National Council of Teachers of Mathematics
1906 Association Drive
Reston, VA 20191-1502

**NCTM Edward G. Begle PreK-6 Classroom Research Grants** 1732
The purpose of this grant is to support and encourage classroom-based research in precollege mathematics education in collaboration with college or university mathematics educators. Grants with a maximum of $6,000 each will be awarded to classroom teachers currently teaching mathematics at the grades PreK–6 level. The research must be a significant collaborative effort involving a college or university mathematics educator (a mathematics education researcher or a teacher of mathematics learning, teaching, or curriculum) and one or more grades PreK–6 classroom teachers. The proposal may include, but is not restricted to, research on the following topics: curriculum development and implementation; involvement of at-risk or minority students; students' thinking about a particular mathematics concept or set of concepts; connection of mathematics to other disciplines; focused learning and teaching of mathematics with embedded use of technology (any acquisition of equipment must support the proposed plan but not be the primary focus of the grant); or innovative assessment or evaluation strategies. Involvement of preservice teachers is encouraged but not required. It is hoped that this research will lead to an article suitable for publication in the Journal for Research in Mathematics Education or in one of the NCTM school journals.
*Requirements:* Proposals must address the following: research design, the plan for collecting and analyzing data, and the anticipated impact on students' learning. The applicant must be a current Full Individual or E-Member of NCTM or teach at a school having a current NCTM PreK-8 school membership. The college or university mathematics educator must be a member of the NCTM.
*Restrictions:* No person(s) may receive more than one award administered by the Mathematics Education Trust in the same academic year. Past recipients of this grant are not eligible to reapply.
*Geographic Focus:* All States
*Date(s) Application is Due:* Nov 11
*Amount of Grant:* Up to 6,000 USD
*Contact:* Nihad Ziad, Grants Administrator; (703) 620-9840, ext. 2112; fax (703) 476-2970; infocentral@nctm.org or affiliates@nctm.org
*Internet:* www.nctm.org/resources/content.aspx?id=1330
*Sponsor:* National Council of Teachers of Mathematics
1906 Association Drive
Reston, VA 20191-1502

**NEA National Heritage Fellowships in the Folk and Traditional Arts** 1733
To honor and preserve our nation's diverse cultural heritage, the National Endowment for the Arts (NEA) annually awards eight National Heritage Fellowships to master folk and traditional artists. These fellowships recognize lifetime achievement, artistic excellence, and contributions to our nation's traditional arts heritage. Nominees must be worthy of national recognition and have a record of continuing artistic accomplishment. They must be actively participating in their art form, either as practitioners or as teachers. Fellows are selected according to criteria of artistic excellence, significance within the particular artistic tradition, and contributions to living cultural heritage. In addition, one NEA National Heritage Fellowship will be given to an individual who has made major contributions to the excellence, vitality, and public appreciation of the folk and traditional arts. Nominations are accepted by using the online nomination form located at the NEA website. A list of previous fellowship recipients is also located at the website.
*Requirements:* Recipients are selected on the basis of nominations from the public. Nominations may be individuals or a group of individuals. Nominees must be citizens or permanent residents of the United States. Additional detailed instructions about the nomination process are available at the website.
*Restrictions:* Posthumous or self-nominations are not eligible. Individuals who have previously received a NEA lifetime honor award (National Heritage Fellowship, Jazz Masters Fellowship, or an Opera Honor) are not eligible.

*Geographic Focus:* All States
*Date(s) Application is Due:* Oct 1
*Amount of Grant:* Up to 25,000 USD
*Contact:* Cheryl Schiele, Folk and Traditional Arts Specialist; (202) 682-5587; schielec@arts.gov
*Internet:* www.nea.gov/honors/heritage/index.html
*Sponsor:* National Endowment for the Arts
1100 Pennsylvania Avenue NW
Washington, D.C. 20506-0001

**NEAVS Fellowsips in Women's Health and Sex Differences** **1734**
The American Fund for Alternatives to Animal Research (AFAAR) and NEAVS offer a one-year, $40,000 postdoctoral fellowship grant (with possible renewal) to a woman committed to developing, validating, or using alternatives to animal methods in the investigation of women's health or sex differences. Proposals are reviewed by an award committee with expertise in related fields of the project. The award is not limited to the U.S. – international applicants are welcome. The deadline for applications is December 15, with notice of awards sent by January 15.
*Requirements:* The award is available to postdoctoral female scientists researching women's health or sex differences whose research involves development, validation, or use of non-animal alternatives. Applicants must hold an interest in using or promoting non-animal alternatives in research. Applicants should send a cover letter explaining their interest in alternative research methods and their career goals, along with their CV, research proposal, and three letters of recommendation (including one from their mentor).
*Geographic Focus:* All States, All Countries
*Date(s) Application is Due:* Dec 15
*Amount of Grant:* 40,000 USD
*Contact:* Nina Farley, Project Coordinator; (617) 523-6020; fax (617) 523-7925; nfarley@neavs.org or afaar@neavs.org
*Internet:* alternativestoanimalresearch.org/afaar/programs
*Sponsor:* New England Anti-Vivisection Society
333 Washington Street, Suite 850
Boston, MA 02108

**NEFE Fellowships** **1735**
The National Endowment for Financial Education (NEFE) fellowships are awarded to recent college graduates interested in gaining practical, meaningful professional experience in the personal finance and nonprofit sectors. Fellows work closely with executive staff at the foundation's headquarters near Denver. Fellowships are funded for up to two years and include competitive salary and benefits. Promising students are invited to apply.
*Requirements:* Applicants must hold a bachelor's or master's degree, preferably with an emphasis in family and consumer economics, applied economics, business or personal finance, or journalism or communications.
*Restrictions:* Faxed and email applications are not accepted.
*Geographic Focus:* All States
*Date(s) Application is Due:* Apr 15
*Amount of Grant:* 35,000 USD
*Contact:* Mary Hoch, Manager of Grants and Research; (303) 224-3513 or (303) 741-6333; fax (303) 220-0838; nefegrantrequest@nefe.org
*Internet:* www.nefe.org/
*Sponsor:* National Endowment for Financial Education
1331 17th Street, Suite 1200
Denver, CO 80202

**NEFE Research Grants** **1736**
The National Endowment for Financial Education (NEFE) seeks projects whose outcomes can improve the public's ability to achieve personal and household financial well-being. Of particular interest are pro-active research projects initiated from one of a broad spectrum of scholarly disciplines whose findings may cultivate critical thinking in the financial literacy community. Also of interest are development projects that put research recommendations into action. Project outcomes must be capable of achieving traction and measurable impact with audiences such as financial education intermediaries, researchers, practitioners, decision makers, and others who can achieve effective outreach to a target population with an unmet financial literacy need or to the general public. Funding requests are assessed within the parameters of the following three key grant themes: understanding financial behavior; advance innovative behavior; and assure significance to society. Recent grant awards for research projects have ranged from $100,000 to $150,000. Grant durations generally range from 12 to 36 months. The two annual deadlines are the first Tuesday in June and first Tuesday in December.
*Requirements:* 501(c)3 tax-exempt organizations are eligible. All applicants must submit a completed Letter of Inquiry (LOI) using NEFE's online system as the first step in being considered for funding.
*Restrictions:* NEFE does not fund: individuals; foreign organizations; international programs or projects; an organization's operating and administrative expenses; salaries of permanent staff; capital costs; pass-through funding; budget deficits; projects with a potential conflict of interest; conferences, workshops, seminars and similar activities that fail to provide a long-term solution; distribution methodologies that require ongoing maintenance; projects with commercial or proprietary elements; lobbying, political contributions, fund-raising events, or other similar activities designed to influence legislation; donations, endowments, challenge grants, matching funds, sponsorships, and other similar programs; or direct or matching payments to members of the public.
*Geographic Focus:* All States
*Amount of Grant:* 100,000 - 150,000 USD
*Contact:* Mary Hoch, Manager of Grants and Research; (303) 224-3513 or (303) 741-6333; fax (303) 220-0838; nefegrantrequest@nefe.org
Billy J. Hensley, Senior Director of Education; (303) 224-3505 or (303) 741-6333; fax (303) 220-0838; nefegrantrequest@nefe.org
*Internet:* www.nefe.org/what-we-provide/research-funding.aspx
*Sponsor:* National Endowment for Financial Education
1331 17th Street, Suite 1200
Denver, CO 80202

**NEH Awards for Faculty at Historically Black Colleges and Universities** **1737**
Grants awarded under this program support individual faculty members at Historically Black Colleges and Universities (HBCUs) pursuing research of value to humanities scholars, students, or general audiences. Awards are designed to be flexible, allowing applicants to define the audience, type of research, award periods, and administrative arrangements that best fit their projects. Awards can be used for a wide range of projects that are based on humanities research. Eligible projects include pursuing research in primary and secondary materials; producing articles, monographs, books, digital materials, archaeological site reports, translations, editions, or other scholarly resources; and conducting basic research leading to the improvement of an existing undergraduate course or the achievement of institutional or community research goals.
*Requirements:* This program accepts applications from faculty members and retired faculty members affiliated with Historically Black Colleges or Universities. All U.S. citizens who teach at or have retired from Historically Black Colleges and Universities are eligible to apply. Foreign nationals who have been living in the United States or its jurisdictions for at least the three years prior to the application deadline are also eligible, provided that they teach at or have retired from any of these institutions.
*Restrictions:* NEH Awards for Faculty may not be used for: curricular or pedagogical tools, methods, theories, or surveys; preparation or revision of textbooks; projects that seek to promote a particular political, religious, or ideological point of view; projects that advocate a particular program of social action; works in the creative and performing arts, i.e., painting, writing fiction or poetry, dance performance, etc; or research leading to the improvement of graduate courses.
*Geographic Focus:* All States
*Date(s) Application is Due:* Apr 14
*Amount of Grant:* Up to 50,400 USD
*Contact:* Jane Aikin, Acting Director; (202) 606-8212 or (202) 606-8200; jaikin@neh.gov or facultyawards@neh.gov
Elizabeth Arndt, Senior Program Officer; (202) 606-8276 or (202) 606-8200; earndt@neh.gov or facultyawards@neh.gov
*Internet:* www.neh.gov/grants/guidelines/AF_HBCU.html
*Sponsor:* National Endowment for the Humanities
400 7th Street SW
Washington, D.C. 20506

**NEH Awards for Faculty at Institutions with High Hispanic Enrollment** **1738**
Grants awarded under this program support individual faculty members at Institutions with High Hispanic Enrollment pursuing research of value to humanities scholars, students, or general audiences. Awards are designed to be flexible, allowing applicants to define the audience, type of research, award periods, and administrative arrangements that best fit their projects. Awards can be used for a wide range of projects that are based on humanities research. Eligible projects include pursuing research in primary and secondary materials; producing articles, monographs, books, digital materials, archaeological site reports, translations, editions, or other scholarly resources; and conducting basic research leading to the improvement of an existing undergraduate course or the achievement of institutional or community research goals.
*Requirements:* This program accepts applications from faculty members and retired faculty members affiliated with Institutions with High Hispanic Enrollment. All U.S. citizens who teach at or have retired from Institutions with High Hispanic Enrollment are eligible to apply. Foreign nationals who have been living in the United States or its jurisdictions for at least the three years prior to the application deadline are also eligible, provided that they teach at or have retired from any of these institutions.
*Restrictions:* NEH Awards for Faculty may not be used for: curricular or pedagogical tools, methods, theories, or surveys; preparation or revision of textbooks; projects that seek to promote a particular political, religious, or ideological point of view; projects that advocate a particular program of social action; works in the creative and performing arts, i.e., painting, writing fiction or poetry, dance performance, etc; or research leading to the improvement of graduate courses.
*Geographic Focus:* All States
*Date(s) Application is Due:* Apr 14
*Amount of Grant:* Up to 50,400 USD
*Contact:* Jane Aikin, Acting Director; (202) 606-8212 or (202) 606-8200; jaikin@neh.gov or facultyawards@neh.gov
Elizabeth Arndt, Senior Program Officer; (202) 606-8276 or (202) 606-8200; earndt@neh.gov or facultyawards@neh.gov
*Internet:* www.neh.gov/grants/guidelines/AF_IHHE.html
*Sponsor:* National Endowment for the Humanities
400 7th Street SW
Washington, D.C. 20506

**NEH Awards for Faculty at Tribal Colleges and Universities** **1739**
Grants awarded under this program support individual faculty members at Tribal Colleges and Universities pursuing research of value to humanities scholars, students, or general audiences. Awards are designed to be flexible, allowing applicants to define

the audience, type of research, award periods, and administrative arrangements that best fit their projects. Awards can be used for a wide range of projects that are based on humanities research. Eligible projects include pursuing research in primary and secondary materials; producing articles, monographs, books, digital materials, archaeological site reports, translations, editions, or other scholarly resources; and conducting basic research leading to the improvement of an existing undergraduate course or the achievement of institutional or community research goals.
*Requirements:* This program accepts applications from faculty members and retired faculty members affiliated with Tribal Colleges or Universities. All U.S. citizens who teach at or have retired from Tribal Colleges and Universities are eligible to apply. Foreign nationals who have been living in the United States or its jurisdictions for at least the three years prior to the application deadline are also eligible, provided that they teach at or have retired from any of these institutions.
*Restrictions:* NEH Awards for Faculty may not be used for: curricular or pedagogical tools, methods, theories, or surveys; preparation or revision of textbooks; projects that seek to promote a particular political, religious, or ideological point of view; projects that advocate a particular program of social action; works in the creative and performing arts, i.e., painting, writing fiction or poetry, dance performance, etc; or research leading to the improvement of graduate courses.
*Geographic Focus:* All States
*Date(s) Application is Due:* Apr 14
*Amount of Grant:* Up to 50,400 USD
*Contact:* Jane Aikin, Acting Director; (202) 606-8212 or (202) 606-8200; jaikin@neh.gov or facultyawards@neh.gov
*Internet:* www.neh.gov/grants/guidelines/AF_TCU.html
*Sponsor:* National Endowment for the Humanities
400 7th Street SW
Washington, D.C. 20506

**NEH Collaborative Research Grants** **1740**
Collaborative Research Grants support interpretive research undertaken by a team of two or more scholars, for full-time or part-time activities for periods of a minimum of one year up to a maximum of three years. Support is available for various combinations of scholars, consultants, and research assistants; project-related travel; field work; applications of information technology; and technical support and services. All grantees are expected to communicate the results of their work to the appropriate scholarly and public audiences. Eligible projects include: research that significantly adds to knowledge and understanding in the humanities; conferences on topics of major importance in the humanities that will benefit scholarly research; archaeological projects that include the interpretation and communication of results (projects may encompass excavation, materials analysis, laboratory work, field reports, and preparation of interpretive monographs); and research that uses the knowledge and perspectives of the humanities and historical or philosophical methods to enhance understanding of science, technology, medicine, and the social sciences. These grants support full-time or part-time activities for periods of one to three years.
*Requirements:* Eligibility is limited to: institutional applicants, i.e., U.S. nonprofit organizations with IRS 501(c)3 tax-exempt status, and state and local governments and federally recognized Indian tribal governments; and project directors who are not affiliated with an eligible institution, i.e., U.S. citizens or foreign nationals who have been living in the United States or its jurisdictions for at least the three years immediately preceding the application deadline.
*Restrictions:* Collaborative Research Grants may not be used for: projects undertaken by individuals (without collaboration); specific policy studies or educational or technical impact assessments; projects that seek to promote a particular political, religious, or ideological point of view; projects that advocate a particular program of social action; editions or translations (see the note below); support for the preparation or publication of textbooks intended primarily for classroom use; development of tools, including digital tools, or any materials analysis whose primary goal is refinement of a method, unless the development contributes to the larger interpretive goals of the project; inventories of collections; empirical social science research; projects in cognitive psychology; creation and enhancement of databases unless part of an interpretive humanities project; collection of oral history interviews, unless part of interpretive humanities research; research in educational methods, tests, or measurements, or any projects primarily intended for students in formal learning environments; recurrent meetings of professional organizations, societies, and scholarly organizations; planning or putting on exhibitions; works in the creative and performing arts (e.g., painting, writing fiction or poetry, dance performance); or support for publication costs or publication subvention.
*Geographic Focus:* All States
*Date(s) Application is Due:* Dec 8
*Amount of Grant:* 25,000 - 100,000 USD
*Contact:* Jane Aikin, Acting Director; (202) 606-8212 or (202) 606-8200; jaikin@neh.gov or collaborative@neh.gov
*Internet:* www.neh.gov/grants/guidelines/collaborative.html
*Sponsor:* National Endowment for the Humanities
400 7th Street SW
Washington, D.C. 20506

**NEH Digital Humanities Implementation Grants** **1741**
The NEH Digital Humanities Implementation Grants program is designed to fund the implementation of innovative digital-humanities projects that have successfully completed a start-up phase and demonstrated their value to the field. Such projects might enhance our understanding of central problems in the humanities, raise new questions in the humanities, or develop new digital applications and approaches for use in the humanities. The program can support innovative digital-humanities projects that address multiple audiences, including scholars, teachers, librarians, and the public. Unlike NEH's start-up grant program, which emphasizes basic research, prototyping, experimentation, and potential impact, the Digital Humanities Implementation Grants program seeks to identify projects that have successfully completed their start-up phase and are well positioned to have a major impact. Proposals are welcome for digital initiatives in any area of the humanities. Digital Humanities Implementation Grants may involve: implementation of computationally-based methods or techniques for humanities research; implementation of new digital tools for use in humanities research, public programming, or educational settings; efforts to ensure the completion and long-term sustainability of existing digital resources (typically in conjunction with a library or archive); studies that examine the philosophical or practical implications of the use of emerging technologies in specific fields or disciplines of the humanities, or in interdisciplinary collaborations involving several fields or disciplines; or implementation of new digital modes of scholarly communication that facilitate peer review, collaboration, or the dissemination of humanities scholarship for various audiences. Awards range from $100,000 to $325,000.
*Requirements:* Eligibility is limited to: U.S. nonprofit organizations with IRS 501(c)3 tax-exempt status; and state and local governmental agencies and federally recognized Indian tribal governments.
*Restrictions:* Individuals are not eligible to apply. Digital Humanities Implementation Grants may not be used for: projects that mainly involve digitization; support for recurring publications (such as the creation of a journal); conferences, professional meetings, or workshops; acquisition of computer equipment or software in excess of 20 percent of the grant total; creative or performing arts; work undertaken in the pursuit of an academic degree; projects that seek to promote a particular political, religious, or ideological point of view; or projects that advocate a particular program of social action.
*Geographic Focus:* All States
*Date(s) Application is Due:* Feb 18
*Amount of Grant:* 100,000 - 325,000 USD
*Contact:* Brett Bobley, Chief Information Officer and Director; (202) 606-8401 or (202) 606-8364; bbobley@neh.gov or odh@neh.gov
Perry Collins, Senior Program Officer; (202) 606-8461 or (202) 606-8364; pcollins@neh.gov or odh@neh.gov
*Internet:* www.neh.gov/grants/guidelines/digitalhumanitiesimplementation.html
*Sponsor:* National Endowment for the Humanities
400 7th Street SW
Washington, D.C. 20506

**NEH Digital Humanities Start-Up Grants** **1742**
The National Endowment for the Humanities (NEH) invites applications to the Digital Humanities Start-Up Grants program. This program is designed to encourage innovations in the digital humanities. By awarding relatively small grants to support the planning stages, NEH aims to encourage the development of innovative projects that promise to benefit the humanities. Proposals should be for the planning or initial stages of digital initiatives in any area of the humanities. Digital Humanities Start-Up Grants may involve: research that brings new approaches or documents best practices in the study of the digital humanities; planning and developing prototypes of new digital tools for preserving, analyzing, and making accessible digital resources, including libraries' and museums' digital assets; scholarship or studies that examine the philosophical or practical implications and impact of the use of emerging technologies in specific fields or disciplines of the humanities, or in interdisciplinary collaborations involving several fields or disciplines; innovative uses of technology for public programming and education utilizing both traditional and new media; and new digital modes of publication that facilitate the dissemination of humanities scholarship in advanced academic as well as informal or formal educational settings at all academic levels. Two levels of awards will be made in this program: Level I awards are small grants designed to fund brainstorming sessions, workshops, early alpha-level prototypes, and initial planning; and Level II awards are larger grants that can be used for more fully-formed projects that are ready to begin implementation or demonstrate proofs of concept.
*Requirements:* All applicants must propose an innovative approach, method, tool, or idea that has not been used before in the humanities. Eligibility is limited to: U.S. nonprofit organizations or institutions with IRS 501(c)3 tax-exempt status; and state and local governmental agencies and federally recognized Indian tribal governments.
*Restrictions:* Individuals are not eligible to apply. Digital Humanities Start-Up Grants cannot be used for: projects that mainly involve digitization, unless the applicant is proposing an innovative method for digitization; the creation or conversion of a scholarly journal (however, the exploration of or planning for new models of scholarly publication is allowed); the implementation or assessment of existing digital applications in the humanities (however, exploration of or planning for a new direction or tool for an established project is allowed); recurring or established conferences or professional meetings; acquisition of computer equipment or software in excess of 20 percent of the grant total; creative or performing arts; empirical social scientific research; work undertaken in the pursuit of an academic degree; the preparation or publication of textbooks; projects that seek to promote a particular political, religious, or ideological point of view; or projects that advocate a particular program of social action.
*Geographic Focus:* All States
*Date(s) Application is Due:* Apr 2; Sep 11
*Amount of Grant:* 5,000 - 60,000 USD
*Contact:* Brett Bobley, Director; (202) 606-8401 or (202) 606-8364; bbobley@neh.gov or odh@neh.gov
*Internet:* www.neh.gov/grants/guidelines/digitalhumanitiesstartup.html
*Sponsor:* National Endowment for the Humanities
400 7th Street SW
Washington, D.C. 20506

**NEH Documenting Endangered Languages Grants** 1743

Documenting Endangered Languages is a joint, multi-year funding program of the National Science Foundation and the National Endowment for the Humanities to develop and advance scientific and scholarly knowledge concerning endangered human languages. Made urgent by the imminent death of approximately half of the 6000-7000 currently used human languages, DEL seeks not only to document these endangered languages but to integrate, systematize, and make knowledge concerning them widely available by exploiting advances in information technology. Approximately 6-10 Standard or Continuing Grants ranging from $12,000 to $150,000 per year for one to three years are available. Approximately 12 Fellowships of $4,200 per month for awards lasting from six to twelve months are available; the maximum stipend is $50,400 for a twelve-month tenure period.

*Requirements:* Academic institutions and non-profit, non-academic organizations located in the United States are eligible. U.S. citizens are eligible to apply for fellowships. Foreign nationals who have been living in the United States or its jurisdictions for at least the three years prior to the proposal deadline are also eligible to apply for fellowships.
*Restrictions:* For-profit organizations are not eligible to apply.
*Geographic Focus:* All States
*Date(s) Application is Due:* Sep 20
*Amount of Grant:* 62,400 - 212,400 USD
*Contact:* Barbara Bays, Senior Program Officer; (202) 606-8290 or (202) 606-8269; bbays@neh.gov or publicpgms@neh.gov
Jeff Hardwick, Senior Program Officer; (202) 606-8287 or (202) 606-8269; jhardwick@neh.gov or publicpgms@neh.gov
*Internet:* www.neh.gov/grants/guidelines/del.html
*Sponsor:* National Endowment for the Humanities
1100 Pennsylvania Avenue NW, Room 426
Washington, D.C. 20506

**NEH Fellowships** 1744

Fellowships support individuals pursuing advanced research that is of value to humanities scholars, general audiences, or both. Recipients usually produce articles, monographs, books, digital materials, archaeological site reports, translations, editions, or other scholarly resources in the humanities. Projects may be at any stage of development. Fellowships support continuous full-time work for a period of six to twelve months. NEH encourages submission of Fellowships applications from faculty at Historically Black Colleges and Universities, Institutions with High Hispanic Enrollment, and Tribal Colleges and Universities. Fellowships cover periods lasting from six to twelve months at a stipend of $4,200 per month in outright funds. The maximum stipend is $50,400 for a twelve-month period.

*Requirements:* The Fellowships program accepts applications from researchers, teachers, and writers, whether they have an institutional affiliation or not. All U.S. citizens, whether they reside inside or outside the United States, are eligible to apply. Foreign nationals who have been living in the United States or its jurisdictions for at least the three years prior to the application deadline are also eligible.
*Restrictions:* Fellowships may not be used for: projects that seek to promote a particular political, religious, or ideological point of view; projects that advocate a particular program of social action; specific policy studies; research for doctoral dissertations or theses by students enrolled in a degree program; the preparation or revision of textbooks; curriculum development; the development of pedagogical tools (including teaching methods or theories); educational or technical impact assessments; the creation or enhancement of databases, unless part of a larger interpretive project; empirical social science research, unless part of a larger humanities project; inventories of collections; or works in the creative and performing arts, i.e., painting, writing fiction or poetry, dance performance, etc.
*Geographic Focus:* All States
*Date(s) Application is Due:* Apr 30
*Amount of Grant:* 4,200 - 50,400 USD
*Contact:* Jane Aikin, Director; (202) 606-8212 or (202) 606-8200; jaikin@neh.gov or fellowships@neh.gov
Mary Macklem, Senior Program Officer; (202) 606-8276 or (202) 606-8200; mmacklem@neh.gov or fellowships@neh.gov
*Internet:* www.neh.gov/grants/guidelines/fellowships.html
*Sponsor:* National Endowment for the Humanities
400 7th Street SW
Washington, D.C. 20506

**NEH Fellowships at Independent Research Institutions** 1745

Grants for Fellowship Programs at Independent Research Institutions (FPIRI) support fellowships at institutions devoted to advanced study and research in the humanities. NEH fellowships provide scholars with research time and access to resources that might not be available at their home institutions. Fellowship programs may be administered by independent centers for advanced study, libraries, and museums in the United States; American overseas research centers; and American organizations that have expertise in promoting research on foreign cultures. Individual scholars must apply directly to the institutions themselves. FPIRI grants provide funding to programs for humanities fellowships of four to twelve months. Awards are generally made to support fellowships for three years. Recent grants have ranged from $78,600 to $480,600.

*Requirements:* Any U.S. nonprofit institution with IRS 501(c)3 tax-exempt status is eligible, as are state and local governmental agencies and federally recognized Indian tribal governments. In addition, to apply for a FPIRI grant, institutions must: be financed, governed, and administered independently of institutions of higher education, and have established and maintained fellowship programs with their own or other funding.
*Restrictions:* Individuals are not eligible to apply.
*Geographic Focus:* All States
*Date(s) Application is Due:* Aug 17
*Amount of Grant:* 78,600 - 480,600 USD
*Contact:* Jane Aikin, Acting Director; (202) 606-8212 or (202) 606-8200; jaikin@neh.gov or fpiri@neh.gov
Elizabeth Arndt, Senior Program Officer; (202) 606-8276 or (202) 606-8200; earndt@neh.gov or fpiri@neh.gov
*Internet:* www.neh.gov/grants/guidelines/fpiri.html
*Sponsor:* National Endowment for the Humanities
400 7th Street SW
Washington, D.C. 20506

**NEH Fellowships for Advanced Social Science Research on Japan** 1746

The Fellowship Program for Advanced Social Science Research on Japan is a joint activity of the Japan-U.S. Friendship Commission (JUSFC) and the National Endowment for the Humanities. Awards support research on modern Japanese society and political economy, Japan's international relations, and U.S.-Japan relations. The program encourages innovative research that puts these subjects in wider regional and global contexts and is comparative and contemporary in nature. Research should contribute to scholarly knowledge or to the general public's understanding of issues of concern to Japan and the United States. Appropriate disciplines for the research include anthropology, economics, geography, history, international relations, linguistics, political science, psychology, public administration, and sociology. The fellowships are designed for researchers with advanced language skills whose research will require use of data, sources, and documents in their original languages or whose research requires interviews onsite in direct one-on-one contact. Fellows may undertake their projects in Japan, the United States, or both, and may include work in other countries for comparative purposes. Projects may be at any state of development. Awards usually result in articles, monographs, books, digital materials, archaeological site reports, translations, editions, or other scholarly resources. As tax-supported federal agencies, JUSFC and NEH endeavor to make grant products available to the broadest possible audience. Our goal is for scholars, educators, students, and the American public to have ready and easy access to the wide range of grant products. For projects that lead to the development of websites, all other considerations being equal, preference is given to those that provide free access to the public. Fellowships cover uninterrupted periods lasting from six to twelve months at a stipend of $4,200 per month. The maximum stipend is $50,400 for a twelve-month award period.

*Requirements:* The Fellowships for Advanced Social Science Research on Japan program accepts applications from researchers, teachers, and writers, whether they have an institutional affiliation or not. All U.S. citizens, whether they reside inside or outside the United States, are eligible to apply. Foreign nationals who have been living in the United States or its jurisdictions for at least the three years prior to the application deadline are also eligible.
*Restrictions:* Advanced Social Science Research on Japan Fellowships may not be used for: projects that seek to promote a particular political, religious, or ideological point of view; projects that advocate a particular program of social action; specific policy studies; research for doctoral dissertations or theses by students enrolled in a degree program; the preparation or revision of textbooks; curriculum development; the development of pedagogical tools (including teaching methods or theories); educational or technical impact assessments; the creation or enhancement of databases, unless part of a larger interpretive project; inventories of collections; or works in the creative and performing arts, i.e., painting, writing fiction or poetry, dance performance, etc.
*Geographic Focus:* All States
*Date(s) Application is Due:* May 3
*Amount of Grant:* 25,200 - 50,400 USD
*Contact:* Jane Aikin, Acting Director; (202) 606-8212 or (202) 606-8200; fax (202) 606-8204; jaikin@neh.gov or fellowships@neh.gov
Elizabeth Arndt, Senior Program Officer; (202) 606-8276 or (202) 606-8200; fax (202) 606-8204; earndt@neh.gov or fellowships@neh.gov
*Internet:* www.neh.gov/grants/guidelines/fellowships-japan.html
*Sponsor:* National Endowment for the Humanities
400 7th Street SW
Washington, D.C. 20506

**NEI Clinical Study Planning Grant** 1747

The National Eye Institute (NEI) supports large-scale clinical vision research projects, including randomized clinical trials and epidemiologic studies. At the time of submission, applications requesting support for these activities are expected to provide detailed information regarding the study's rationale, design, analytic techniques, protocols and procedures, facilities and environment, organizational structure, and collaborative arrangements. This information is best conveyed in a well-documented Manual of Procedures (MOP), the development of which represents a costly and time-consuming activity. The Clinical Study Planning Grant is designed to facilitate activities central to the refinement of a study's protocol and procedures and the development of a detailed MOP.

*Requirements:* Applications may be submitted by domestic for-profit and nonprofit organizations, both public and private; units of state and local governments; and eligible agencies of the federal government; domestic and foreign institutions.
*Restrictions:* Foreign institutions are not eligible.
*Geographic Focus:* All States
*Amount of Grant:* Up to 150,000 USD
*Contact:* Donald F. Everett, M.A., (301) 451-2020; fax (301) 402-0528; dfe@nei.nih.gov
*Internet:* grants.nih.gov/grants/guide/pa-files/PAR-05-115.html
*Sponsor:* National Eye Institute
5635 Fishers Lane, Suite 1300, MSC 9300
Bethesda, MD 20892-9300

**Nellie Mae Education Foundation District-Level Change Grants** **1748**

The Nellie Mae Education Foundation District-Level Systems Change Grants support the shift in high school-level education in New England to prepare students to thrive in a global, complex, fast changing society. Grant programs promote student centered models of schooling that offer a path away from a one size fits all approach to a more customized approach to maximize learning for all students. The initiative focuses on three priority areas: developing school and district designs and practices that enable all learners to achieve high standards; creating sustainable policy change to support these new approaches; and generating public will and increasing demand for changes in practice. Examples of allowable activities for funding include: engaging central office, school administrators, and teachers in identifying current initiatives and how they compete with or support a student centered learning frame; surveying student needs and challenges and what motivates them to achieve academically and personally; creating greater building level autonomy and funding streams based on student needs rather than programs; researching feasibility, designing and planning for the appropriate blended learning model including visits to blended learning schools; using technology as an essential tool in learning environments and professional development; or administrator visits from student centered schools. The Foundation will support self-assessment activities, efforts to build, forge or nurture a common vision and mission for the work, and integration of this work with other school/district initiatives. Funds can support purchasing online and hardcopy literacy materials to support complex Common Core State Standards (CCSS) text demands; engaging a team of district stakeholders to conduct an analysis of culture of the system's various dimensions; or administrative visits from student centered schools. Proposal information, including the request for proposal, a list of supplemental materials and instructions, questions to address in the proposal, a list of frequently asked questions and previously funded projects, and an instructional webinar, is located at the Foundation website.
*Requirements:* The Foundation's activities include both making grants to the public charities it supports and providing services to those organizations. The Foundation operates exclusively to promote the charitable and educational purposes of nonprofit educational organizations, including universities, colleges, secondary schools, elementary schools and other educational organizations that are described in the IRS Code Section 501(c)3.
*Restrictions:* Funding for this project is concentrated on urban school districts and learning centers in Connecticut, Massachusetts, and Rhode Island. The Foundation does not fund capital campaigns, endowments, scholarships or fellowships, debt reduction or cash reserves, building construction or renovation, and certain indirect costs.
*Geographic Focus:* Connecticut, Massachusetts, Rhode Island
*Date(s) Application is Due:* Oct 18
*Amount of Grant:* 25,000 - 1,000,000 USD
*Samples:* New England Resource Center for Higher Education, Boston, Massachusetts, Project Compass: a multi-year initiative that supports institutions seeking to systematize changes essential to expanding, sustaining, and integrating rigorous, evidence-based efforts to retain larger numbers of underserved students, $375,000; Sanford School Department, Sanford, Maine, based on the Reinventing Schools Coalition (RISC) model, a model that focuses on a research based approach to student-centered, proficiency based learning where students demonstrate proficiency at commonly understood and meaningful standards, $1.1 million; Hyde Park Task Force, Youth Organizing Project, Jamaica Plain, Massachusetts, project that focuses on developing 50 urban youth leaders who successfully advocate for themselves and their peers in their communities and schools by engaging in community organizing campaigns, $25,000.
*Contact:* Stephanie Cheney, Senior Manager, Grants and Special Programs; (781) 348-4240; fax (781) 348-4299; scheney@nmefdn.org
*Internet:* www.nmefoundation.org/grants/district-level-systems-change
*Sponsor:* Nellie Mae Education Foundation
1250 Hancock Street, Suite 205N
Quincy, MA 02169

**Nellie Mae Education Foundation Public Understanding Grants** **1749**

The Nellie Mae Education Foundation Public Understanding Grants work to promote awareness of the student-centered approaches that lead to deeper learning for all students, and create the public will to demand greater adoption of these approaches. Grants focus on three areas: educating communities on student-centered approaches to learning, what they are, and why they are critical; transforming public perception of how students learn and why it must change; and building public demand for student-centered learning approaches in communities, districts and states. Funding for projects is posted on the Foundation website when it becomes available. A list of frequently asked questions and previously funded projects are located on the website.
*Requirements:* The Foundation's activities include both making grants to the public charities it supports and providing services to those organizations. The Foundation operates exclusively to promote the charitable and educational purposes of nonprofit educational organizations, including universities, colleges, secondary schools, elementary schools and other educational organizations as described in the IRS Code Section 501(c)3. Organizations located in the New England area are eligible to apply (Maine, New Hampshire, Vermont, Massachusetts, Rhode Island, and Connecticut). Additionally, the Foundation supports programs, organizations, research and conferences in other states in order to influence policy, and advance knowledge of public opinion toward education.
*Restrictions:* The Foundation does not fund capital campaigns, endowments, scholarships or fellowships, debt reduction or cash reserves, building construction or renovation, and certain indirect costs. The Foundation also does not support Section 509(a) organizations or individuals.
*Geographic Focus:* All States
*Amount of Grant:* 7,500 - 250,000 USD
*Samples:* Editorial Projects in Education, Inc., Bethesda, Maryland, sponsorship of online stream and archive of Ed Week forum "Managing Your District Creativity," $7,500; KnowledgeWorks Foundation, Cincinnati, Ohio, a day long workshop of Learning 2025, which is designed to expand thinking about the future of learning, $25,500; Rennie Center for Education Research and Policy, Cambridge, Massachusetts, to conduct a public opinion poll in New England to provide the Foundation with information about the public's opinion on the role of K-12 education, beliefs about education reform, and public understanding of student-centered learning, $133,500.
*Contact:* Stephanie Cheney, Senior Manager, Grants and Special Programs; (781) 348-4240; fax (781) 348-4299; scheney@nmefdn.org
*Internet:* www.nmefoundation.org/grants/public-understanding
*Sponsor:* Nellie Mae Education Foundation
1250 Hancock Street, Suite 205N
Quincy, MA 02169

**Nellie Mae Education Foundation Research and Development Grants** **1750**

The Nellie Mae Education Foundation Research and Development Grants work to build and deepen understanding of student-centered learning approaches in order to foster more widespread adoption and implementation. Grants focus on three areas: gathering and promoting existing knowledge of student-centered approaches to learning; creating new research to promote student-centered approaches in order to support overall system change, policy change and community engagement; and developing new tools that support student-centered teaching, learning and assessment. Funding opportunities for this grant are posted on the Foundation website when funding is available. A list of frequently asked questions and additional list of previously funded projects are listed on the website.
*Requirements:* The Foundation's activities include making grants to the public charities it supports and providing services to those organizations. The Foundation operates exclusively to promote the charitable and educational purposes of nonprofit educational organizations, including universities, colleges, secondary schools, elementary schools and other educational organizations that are described in IRS Code Section 501(c)3. Funding is available primarily for the New England area, but the Foundation occasionally supports programs, organizations, research, and conferences outside of the New England area in order to influence policy, and advance knowledge of public opinion toward education.
*Restrictions:* The Foundation does not fund capital campaigns, endowments, scholarships or fellowships, debt reduction or cash reserves, building construction or renovation, and certain indirect costs.
*Geographic Focus:* All States
*Amount of Grant:* 10,000 - 185,000 USD
*Samples:* Council of the Chief State School Officers (CCSSO), Washington, D.C., develop an evaluation framework and design to assess the work of the Innovation Lab Network (ILN), $30,000; Expeditionary Learning Schools Outward Bound, Amherst, Massachusetts, an important step toward developing a solid understanding of underlying practices of student-centered learning as they are implemented in a highly successful student-centered learning environment in an urban context, $235,000; Boston Day and Evening Academy, Roxbury, Massachusetts, funding for the Academy's website for its Responsive Education Alternatives Laboratory, $18,000.
*Contact:* Stephanie Cheney, Senior Manager, Grants and Special Programs; (781) 348-4240; fax (781) 348-4299; scheney@nmefdn.org
*Internet:* www.nmefoundation.org/grants/research-development
*Sponsor:* Nellie Mae Education Foundation
1250 Hancock Street, Suite 205N
Quincy, MA 02169

**Nellie Mae Education Foundation State Level Systems Change Grants** **1751**

The Nellie Mae Education Foundation State-Level Systems Change (SLSC) initiative works to remove barriers and shape state and federal policies that promote deeper learning through student-centered approaches. The SLSC focuses on four areas of priority: state accountability systems that include evaluation of new basic skills such as creative thinking and problem solving; high school graduation requirements based more on students demonstrating competency in target areas than on time spent in the classroom; teacher certification requirements that support teachers and help students develop skills and knowledge through relevant, interest-driven experiences; and state funding formulas that enable and support student-centered learning environments. Request for proposals, with supplementary material and instructions, should be posted on the website if funding is currently available and proposals may be submitted. Previously funded projects and a list of frequently asked questions are located at the Foundation website.
*Requirements:* The Foundation's activities include both making grants to the public charities it supports and providing services to those organizations. The Foundation operates exclusively to promote the charitable and educational purposes of nonprofit educational organizations, including universities, colleges, secondary schools, elementary schools and other educational organizations that are described in IRS Code Section 501(c)3. Funding is concentrated in Connecticut, Maine, New Hampshire, Rhode Island, Vermont, and Massachusetts, but is also available outside the area to support programs, organizations, research, and conference in order to influence policy, and advance knowledge of public opinion toward education.
*Restrictions:* The Foundation does not fund capital campaigns, endowments, scholarships or fellowships, debt reduction or cash reserves, building construction or renovation, and certain indirect costs.
*Geographic Focus:* All States
*Amount of Grant:* 50,000 - 700,000 USD
*Samples:* Massachusetts Afterschool Partnership (MAP), Boston, Massachusetts, to provide resources for programs and families statewide including information and technical assistance, networking, advocacy support, and peer to peer support, $100,000; New England Secondary School Consortium (NESSC) evaluation, University of Massachusetts at Donahue, Hadley, Massachusetts, university as evaluator for the

NESSC educational partnership that works to foster forward-thinking policies and practices regarding the design and delivery of secondary education across New England, $160,000; Maine Coalition for Excellence in Education, Augusta, Maine, developing a two-year campaign to implement a targeted set of policies, structures, and practices to enhance Maine's investments in early childhood, K-12, higher education, and workforce development, $100,000; Maine Department of Education, Augusta, Maine, project to create a virtual "Center for Best Practices and Policies," an online resource center for those working on standards-based education to support services to conduct case studies and a website for sharing information and ideas, $97,000.
*Contact:* Stephanie Cheney, Senior Manager, Grants and Special Programs; (781) 348-4240; fax (781) 348-4299; scheney@nmefdn.org
*Internet:* www.nmefoundation.org/grants/state-level-systems-change
*Sponsor:* Nellie Mae Education Foundation
1250 Hancock Street, Suite 205N
Quincy, MA 02169

**Neotropical Grassland Conservancy Derald G. Langham Memorial Research Grant** 1752
Memorial Research Grants directly promote conservation research. Grants are reviewed by outside scientists and the NGC board. Grants are one-year, unrestricted funds that can be used for direct or indirect costs, including salary. The Derald G. Langham Memorial Research Grant provides a $2,000 grant to scientists in the American tropics featuring grassland plants. The grant is awarded in memory of Derald G. Langham. Derald received the first Ph.D. from Cornell University in plant genetics in 1937. He and his family spent the next 30 years in Venezuela as Derald established plant nurseries and headed the national plant genetics program. His passion for plants, charisma, and support of new ways of thinking of the world are memorialized in this award.
*Geographic Focus:* All States
*Date(s) Application is Due:* Mar 1
*Amount of Grant:* 2,000 USD
*Contact:* Valentina Ferretti; (916) 967-3223; info@conservegrassland.org
*Internet:* www.conservegrassland.org/memorial_research_grants.htm
*Sponsor:* Neotropical Grassland Conservancy
6274 Heathcliff Drive
Carmichael, CA 95608-0909

**Neotropical Grassland Conservancy Memorial Grants** 1753
Memorial Research Grants directly promote conservation research. Grants are reviewed by outside scientists and the NGC board. Grants are one-year, unrestricted funds that can be used for direct or indirect costs, including salary. Specific grants may be restricted to certain taxa or new research initiatives. See individual grants for details. Unless specifically stated, these grants are not restricted to conservation-oriented projects, but you will need to address the issue of how this grant may impact conservation efforts in your area under the section conservation implications of your project. Applications are accepted until the 1st of March each year and are reviewed by outside reviewers and members of the board.
*Geographic Focus:* All States
*Date(s) Application is Due:* Mar 1
*Amount of Grant:* 2,000 - 5,000 USD
*Contact:* Valentina Ferretti; (916) 967-3223; info@conservegrassland.org
*Internet:* www.conservegrassland.org/memorial_research_grants.htm
*Sponsor:* Neotropical Grassland Conservancy
6274 Heathcliff Drive
Carmichael, CA 95608-0909

**Neotropical Grassland Conservancy Remembrance Fund Grants** 1754
The Remembrance Fund honors special people and special events. Many people can contribute to the fund in any amount. Remembrance Fund donations are acknowledged to the person you designate, and are listed on the NGC Web site and in its E-newsletter. A donation on behalf of a person or an occasion is an open expression of thoughtfulness and generosity. Remembrance fund donations have no minimum and can be set-up as one-time or annual contributions. Examples of remembrance events, to which many people can contribute, are graduations, anniversaries, birthdays, memorials, etc. Annual awards are $500 or more, and are named in honor of the person or event (such as the Christopher Miller Award) or can be added to the Remembrance Fund. Grants from the Remembrance Fund are awarded every year and are listed on the web site with a short description.
*Geographic Focus:* All States
*Amount of Grant:* 500 USD
*Contact:* Gary M. Langham, Ph.D.; (916) 967-3223; info@conservegrassland.org
*Internet:* www.conservegrassland.org/remembrance_fund.htm
*Sponsor:* Neotropical Grassland Conservancy
6274 Heathcliff Drive
Carmichael, CA 95608-0909

**Neotropical Grassland Conservancy Student Grants** 1755
The Student Grant Program (SGP) was designed in response to feedback from Venezuelan professors. Graduate student research projects are often hindered by lack of funds for basic needs such as transportation, lodging, meals and consumable field supplies. As little as $500 can launch a lifelong career in conservation. Student researchers in the Neotropical Grassland Conservancy's growing network outline their project and describe its impact on grassland conservation. Applications are accepted on the 1st of April, August, and December and are reviewed by members of the Student Grant Program committee and two outside reviewers. Award decisions are based on the student's demonstrated commitment to conservation, feasibility and merit of the proposed project, and how closely the project meets the mission of the Neotropical Grassland Conservancy.
*Requirements:* Latin American students enrolled in a M.S. or Ph.D. program in a U.S. or South American university are eligible to apply.
*Geographic Focus:* All States
*Date(s) Application is Due:* Apr 1; Aug 1; Dec 1
*Amount of Grant:* 500 - 2,500 USD
*Contact:* Peter B. McIntyre; (916) 967-3223; info@conservegrassland.org
*Internet:* www.conservegrassland.org/student_grant.htm
*Sponsor:* Neotropical Grassland Conservancy
6274 Heathcliff Drive
Carmichael, CA 95608-0909

**Neotropical Grassland Conservancy W. J. and Virginia W. Moorhouse Research Grant** 1756
Memorial Research Grants directly promote conservation research. Grants are reviewed by outside scientists and the NGC board. Grants are one-year, unrestricted funds that can be used for direct or indirect costs, including salary. The W. J. and Virginia W. Moorhouse Research Grant provides a $2,000 grant to scientists doing ornithological research in the grasslands of the American tropics. This grant is awarded in memory of Bill and Jinny Moorhouse.
*Geographic Focus:* All States
*Date(s) Application is Due:* Mar 1
*Amount of Grant:* 2,000 USD
*Contact:* Valentina Ferretti; (916) 967-3223; info@conservegrassland.org
*Internet:* www.conservegrassland.org/memorial_research_grants.htm
*Sponsor:* Neotropical Grassland Conservancy
6274 Heathcliff Drive
Carmichael, CA 95608-0909

**Neotropical Grassland Conservancy Wijdenes Memorial Research Grant** 1757
Memorial Research Grants directly promote conservation research. Grants are reviewed by outside scientists and the NGC board. Grants are one-year, unrestricted funds that can be used for direct or indirect costs, including salary. The Wijdenes Memorial Research Grant provides a $5,000 grant to scientists in the American tropics initiating new grassland conservation research. The grant is awarded in memory of Sue Wijdenes. Sue was an avid birder, traveler, and advocate of conservation.
*Geographic Focus:* All States
*Date(s) Application is Due:* Mar 1
*Amount of Grant:* 5,000 USD
*Contact:* Valentina Ferretti; (916) 967-3223; info@conservegrassland.org
*Internet:* www.conservegrassland.org/memorial_research_grants.htm
*Sponsor:* Neotropical Grassland Conservancy
6274 Heathcliff Drive
Carmichael, CA 95608-0909

**Nestle Foundation Large Research Grants** 1758
The Nestlé Foundation supports research in human nutrition with public health relevance in low-income and lower middle-income countries. The results of the research projects should ideally provide a basis for implementation and action which will lead to sustainable effects in the studied populations as generally applicable to the population at large. They should also enable institutional strengthening and capacity building in a sustainable manner in the host country and further cooperation and collaboration between institutions in developed and developing countries. Categories include: training grants, pilot grants, research grants (small and large), and re-entry grants. Large research grants include a full grant application of a complete research proposal. Applications are accepted all year and evaluated twice a year.
*Requirements:* The Foundation's work is primarily concerned with human nutrition research issues dealing with: maternal and child nutrition, including breastfeeding and complementary feeding; macro- and micronutrient deficiencies and imbalances; interactions between infection and nutrition; and nutrition education and health promotion. Interested scientists should first submit a letter of intent briefly describing the project and the estimated budget. Instructions are available on the Foundation's website. If approved applicants will receive an invitation to submit a full grant proposal. Guidelines and forms are available at the website. The letter of intent and the grant application should include detailed, evidence-based information about the public health relevance of the project as well as its immediate impact and sustainability. Funding is primarily based on the scientific quality, public health relevance in the short and long term, sustainability, capacity-building component, and budget considerations.
*Restrictions:* The Foundation expects research proposals to be primarily the initiative of local researchers from developing countries but considers applications jointly made by scientists from developed countries, provided it is clear that the initiative will result in capacity building and human resource development, and the bulk of the budget is spent in the developing country.
*Geographic Focus:* All Countries
*Date(s) Application is Due:* Jan 10; May 10
*Amount of Grant:* Up to 100,000 USD
*Contact:* Catherine Loeb, Assistant to the Director; +41 21 320 33 51; fax +41 21 320 33 92; nf@nestlefoundation.org
*Internet:* www.nestlefoundation.org/e/research.html
*Sponsor:* Nestle Foundation
4 Place de la Gare
Lausanne, CH-1001 Switzerland

**Nestle Foundation Pilot Grants** 1759

The Nestlé Foundation supports research in human nutrition with public health relevance in low-income and lower middle-income countries. The results of the research projects should ideally provide a basis for implementation and action which will lead to sustainable effects in the studied populations as generally applicable to the population at large. They should also enable institutional strengthening and capacity building in a sustainable manner in the host country and further cooperation and collaboration between institutions in developed and developing countries. Categories include: training grants, pilot grants, research grants (small and large), and re-entry grants. Pilot grants provide support for research with a high potential to lead to a subsequent full research grant. Additional guidelines are available at the Foundation website. Applications are accepted all year and evaluated twice a year.
*Requirements:* The Foundation's work is primarily concerned with human nutrition research issues dealing with: maternal and child nutrition, including breastfeeding and complementary feeding; macro- and micronutrient deficiencies and imbalances; interactions between infection and nutrition; and nutrition education and health promotion. Interested scientists should first submit a letter of intent briefly describing the project and the estimated budget. Instructions are available on the Foundation's website. If approved, applicants will receive an invitation to submit a full grant proposal. Guidelines and forms are available at the website. The letter of intent and the grant application should include detailed, evidence-based information about the public health relevance of the project as well as its immediate impact and sustainability. Funding is primarily based on the scientific quality, public health relevance in the short and long term, sustainability, capacity-building component, and budget considerations.
*Restrictions:* The Foundation expects research proposals to be primarily the initiative of local researchers from developing countries but considers applications jointly made by scientists from developed countries with those from developing countries provided it is clear that the initiative will result in capacity building and human resource development in the latter and the bulk of the budget is spent in the developing country.
*Geographic Focus:* All Countries
*Date(s) Application is Due:* Jan 10; May 10
*Amount of Grant:* Up to 20,000 USD
*Contact:* Catherine Loeb, Assistant to the Director; +41 21 320 33 51; fax +41-21-320 33 92; nf@nestlefoundation.org
*Internet:* www.nestlefoundation.org/e/research.html
*Sponsor:* Nestle Foundation
4 Place de la Gare
Lausanne, CH-1001 Switzerland

**Nestle Foundation Re-entry Grants** 1760

The Nestlé Foundation supports research in human nutrition with public health relevance in low-income and lower middle-income countries. The results of the research projects should ideally provide a basis for implementation and action which will lead to sustainable effects in the studied populations as generally applicable to the population at large. They should also enable institutional strengthening and capacity building in a sustainable manner in the host country and further cooperation and collaboration between institutions in developed and developing countries. Categories include: training grants, pilot grants, research grants (small and large), and re-entry grants. Re-entry grants encourage the return and re-establishment of post-graduate students into careers in their own countries. Applications are accepted all year and evaluated twice a year.
*Requirements:* The Foundation's work is primarily concerned with human nutrition research issues dealing with: maternal and child nutrition, including breastfeeding and complementary feeding; macro- and micronutrient deficiencies and imbalances; interactions between infection and nutrition; and nutrition education and health promotion. Interested scientists should first submit a letter of intent briefly describing the project and the estimated budget. Instructions are available on the Foundation's website. If approved applicants will receive an invitation to submit a full grant proposal. Guidelines and forms are available at the website. The letter of intent and the grant application should include detailed, evidence-based information about the public health relevance of the project as well as its immediate impact and sustainability. Funding is primarily based on the scientific quality, public health relevance in the short and long term, sustainability, capacity-building component, and budget considerations.
*Restrictions:* The host institution will need to guarantee a post for the returnee and ensure career development within the host institution. Contribution of support to the eligible candidate from the host institution is essential, while support and collaboration from the overseas institution where the candidate trained is helpful.
*Geographic Focus:* All Countries
*Date(s) Application is Due:* Jan 10; May 10
*Amount of Grant:* Up to 50,000 USD
*Contact:* Catherine Loeb, Assistant to the Director; +41 21 320 33 51; fax +41 21 320 33 92; nf@nestlefoundation.org
*Internet:* www.nestlefoundation.org/e/research.html
*Sponsor:* Nestle Foundation
4 Place de la Gare
Lausanne, CH-1001 Switzerland

**Nestle Foundation Small Research Grants** 1761

The Nestlé Foundation supports research in human nutrition with public health relevance in low-income and lower middle-income countries. The results of the research projects should ideally provide a basis for implementation and action which will lead to sustainable effects in the studied populations as generally applicable to the population at large. They should also enable institutional strengthening and capacity building in a sustainable manner in the host country and further cooperation and collaboration between institutions in developed and developing countries. Categories include: training grants, pilot grants, research grants (small and large), and re-entry grants. Small research grants provide support of a small research study that represents a continuation of a training grant or a pilot grant. Applications are accepted all year and evaluated twice a year.
*Requirements:* The Foundation's work is primarily concerned with human nutrition research issues dealing with: maternal and child nutrition, including breastfeeding and complementary feeding; macro- and micronutrient deficiencies and imbalances; interactions between infection and nutrition; and nutrition education and health promotion. Interested scientists should first submit a letter of intent briefly describing the project and the estimated budget. Instructions are available on the Foundation's website. If approved applicants will receive an invitation to submit a full grant proposal. Guidelines and forms are available at the website. The letter of intent and the grant application should include detailed, evidence-based information about the public health relevance of the project as well as its immediate impact and sustainability. Funding is primarily based on the scientific quality, public health relevance in the short and long term, sustainability, capacity-building component, and budget considerations.
*Restrictions:* The Foundation expects research proposals to be primarily the initiative of local researchers from developing countries but considers applications jointly made by scientists from developed countries, provided it is clear that the initiative will result in capacity building and human resource development, and the bulk of the budget is spent in the developing country.
*Geographic Focus:* All Countries
*Date(s) Application is Due:* Jan 10; May 10
*Amount of Grant:* Up to 50,000 USD
*Contact:* Catherine Loeb, Assistant to the Director; +41 21 320 33 51; fax +41-21-320 33 92; nf@nestlefoundation.org
*Internet:* www.nestlefoundation.org/e/research.html
*Sponsor:* Nestle Foundation
4 Place de la Gare
Lausanne, CH-1001 Switzerland

**Nestle Foundation Training Grant** 1762

The Nestlé Foundation supports research in human nutrition with public health relevance in low-income and lower middle-income countries. The results of the research projects should ideally provide a basis for implementation and action which will lead to sustainable effects in the studied populations as generally applicable to the population at large. They should also enable institutional strengthening and capacity building in a sustainable manner in the host country and further cooperation and collaboration between institutions in developed and developing countries. Categories include: training grants, pilot grants, research grants (small and large), and re-entry grants. Training grants support a small research project such as a master's degree in science, Ph.D. thesis project, or a training endeavor. Applications are accepted all year and evaluated twice a year.
*Requirements:* The Foundation's work is primarily concerned with human nutrition research issues dealing with: maternal and child nutrition, including breastfeeding and complementary feeding; macro- and micronutrient deficiencies and imbalances; interactions between infection and nutrition; and nutrition education and health promotion. Interested scientists should first submit a letter of intent briefly describing the project and the estimated budget. Instructions are available on the Foundation's website. If approved applicants will receive an invitation to submit a full grant proposal. Guidelines and forms are available on our website. The letter of intent and the grant application should include detailed, evidence-based information about the public health relevance of the project as well as its immediate impact and sustainability. Funding is primarily based on the scientific quality, public health relevance in the short and long term, sustainability, capacity-building component, and budget considerations.
*Restrictions:* The Foundation expects research proposals to be primarily the initiative of local researchers from developing countries but considers applications jointly made by scientists from developed countries with those from developing countries provided it is clear that the initiative will result in capacity building and human resource development in the latter and the bulk of the budget is spent in the developing country.
*Geographic Focus:* All Countries
*Date(s) Application is Due:* Jan 10; May 10
*Amount of Grant:* Up to 20,000 USD
*Contact:* Catherine Loeb; +41-21-320 33 51; fax +41-21-320 33 92; nf@nestlefoundation.org
*Internet:* www.nestlefoundation.org/e/research.html
*Sponsor:* Nestle Foundation
4 Place de la Gare
Lausanne, CH-1001 Switzerland

**Nevada Arts Council Artists Fellowships** 1763

Artists Fellowships of $5,000 and Honorable Mention grants of $500 are awarded awarded in each of three categories – literary, performing and visual arts. They are given to professional individual artists with outstanding talent and demonstrated ability and commitment. The funds may be used for support for the creation of new works, works in progress, and/or professional development. Fellowship support for the categories alternate every other year between creative writing/visual arts and performing arts. Interested artists should first obtain application forms and guidelines from the council.
*Requirements:* Applicants must have been Nevada residents for a full year prior to the application deadline.
*Restrictions:* Individuals who are degree-candidate students are not eligible. Artists who received the fellowships in the past two years are also ineligible.
*Geographic Focus:* Nevada
*Date(s) Application is Due:* May 25
*Amount of Grant:* 5,000 USD

*Contact:* Fran Morrow, Program Coordinator; (775) 687-7106; fax (702) 486-3887; fkmorrow@clan.lib.nv.us
*Internet:* nac.nevadaculture.org/index.php?option=com_content&view=article&id=1043&Itemid=412
*Sponsor:* Nevada Arts Council
716 N Carson Street, Suite A
Carson City, NV 89701

**New America National Fellowships** 1764

New America's Fellows program invests in thinkers—journalists, scholars, filmmakers, and public policy analysts—who offer inventive perspectives on the major challenges facing our society. Fellows advance big ideas through research, reporting, analysis, and storytelling. The big idea can be a sweeping reframing of a familiar subject through new research or a new combination of existing research; a masterful presentation of a case study that advances our understanding of a timeless American theme or stress fracture; an innovative new media or academic project to disseminate knowledge about a shared challenge; or a bold policy prescription for moving domestic and international issues forward. Our goal in the Fellows Program is to find bold, impactful thinkers and to fund them for a year, long enough so that they can make progress in writing a book, develop a series of articles, work on a documentary, or work on another project that would be accessible to a broad audience and long enough to be able to build a real community among the fellows. Fellows benefit from a financial stipend, engagement with each other and with New America's various policy programs, and the expanded audience and exposure from New America and its media partners. Precise terms and stipend levels of fellowships vary widely, as some fellows work full-time at New America in pursuit of their research while many others have other professional commitments during the term of their fellowship.
*Requirements:* Regular fellows show promise in writing for student, academic, or popular publications, but have yet to establish themselves as leading commentators. Senior fellows are generally established academics, policy analysts, or journalists wishing to make departures in their work.
*Geographic Focus:* All States
*Date(s) Application is Due:* Feb 1
*Amount of Grant:* 25,000 - 100,000 USD
*Contact:* Awista Ayub, Deputy Director, Fellows Program; (202) 986-2700; fax (202) 986-3696; ayub@newamerica.org or fellowshipprogram@newamerica.org
*Internet:* https://www.newamerica.org/fellows/how-to-apply/
*Sponsor:* New America
740 15th Street, N.W., Suite 900
Washington, D.C. 20005

**Newberry Library American Musicological Society Short-Term Fellowships** 1765

The Newberry, open to the public without charge, is an independent research library dedicated to the advancement and dissemination of knowledge, especially in the humanities. The library acquires and preserves a broad array of special-collections research materials (spanning six centuries of European and American civilizations) and promotes and provides for these collections' effective use, fostering research, teaching, publication, life-long learning, and civic engagement. Fellowships at the Newberry provide support for researchers wishing to use the Newberry's collections and are either short-term (one to two months) or long-term (four to twelve months). Short-term fellowships are intended to geographically facilitate access for distant scholars to materials essential to a project. Long-term fellowships are intended to support significant works of scholarship that draw on the strengths of the Newberry's collections. The Newberry Library-American Musicological Society (AMS) Fellowship is available to scholars working in music or musicology and demonstrate a specific need for the Newberry's collection. The fellowships provide a stipend of $2,500 per month (for each scholar). At the time of this writing, the current application submission date for most Newberry short-term fellowships is December 15 and awardees are usually notified near the end of April; however deadlines may vary from year to year. Applications must be submitted as email attachments by 11:59 p.m. on the deadline date. Applicants can find guidelines and forms at the grant website. Applicants are also encouraged to verify current deadline dates.
*Requirements:* Applicants must be members of the American Musicological Society at the time of application and, if awarded, through the period of their fellowship. Ph.D. candidates and post-doctoral scholars who live and work outside the Chicago area are eligible to apply. Teams of two or three scholars who plan to collaborate on a single, substantive project are also eligible to apply. All fellowships at the Newberry must be conducted in-residence. The Newberry collections must be necessary to the short-term fellowship project since the purpose of these fellowships is to enable access to particular materials. Applications from former recipients of any fellowship at the Newberry are welcome, but preference is given to applicants who have not held a major fellowship within the past three years. In general, all fellowships at the Newberry must be completed by the end of the academic year for which they are awarded.
*Restrictions:* Most short-term fellowships are restricted to individuals who live outside the Chicago area who need to examine specific items in the Newberry's collections. Applicants may apply for more than one fellowship at the Newberry by designating those for which they wish to be considered on the cover sheet (available at the grant website). Applicants may also apply for both short- and long-term fellowships at the Newberry; however individuals may accept only one fellowship at the Newberry per year. Only scholars who have previously received a short-term fellowship at the Newberry may apply for a long-term fellowship for the same project. No other project extensions are allowable.
*Geographic Focus:* All States, All Countries
*Date(s) Application is Due:* Dec 15
*Amount of Grant:* 2,500 - 7,500 USD
*Contact:* D. Bradford Hunt, Vice President for Research and Academic Programs; (312) 255-3535 or (312) 943-9090; huntb@newberry.org or research@newberry.org
*Internet:* https://www.newberry.org/short-term-fellowships
*Sponsor:* Newberry Library
60 West Walton Street
Chicago, IL 60610-3380

**Newberry Library American Society for Eighteenth-Century Studies Short-Term Fellowships** 1766

The Newberry, open to the public without charge, is an independent research library dedicated to the advancement and dissemination of knowledge, especially in the humanities. The library acquires and preserves a broad array of special-collections research materials (spanning six centuries of European and American civilizations) and promotes and provides for these collections' effective use, fostering research, teaching, publication, life-long learning, and civic engagement. Fellowships at the Newberry provide support for researchers wishing to use the Newberry's collections and are either short-term (one to two months) or long-term (four to twelve months). Short-term fellowships are intended to geographically facilitate access for distant scholars to materials essential to a project. Long-term fellowships are intended to support significant works of scholarship that draw on the strengths of the Newberry's collections. The Newberry American Society for Eighteenth-Century Studies (ASECS) Fellowships are short-term fellowships offered to scholars who wish to pursue studies from the period of 1660 to 1815. These fellowships provide a stipend of $2,500 per month. At the time of this writing, the current application submission date for most Newberry short-term fellowships is December 15 and awardees are usually notified near the end of April; however deadlines may vary from year to year. Applications must be submitted as email attachments by 11:59 p.m. on the deadline date. Applicants can find guidelines and forms at the grant website. Applicants are also encouraged to verify current deadline dates.
*Requirements:* Newberry ASECS Fellowship recipients must be members of the ASECS at the time of the award. Most short-term fellowships are for either Ph.D. candidates or postdoctoral scholars. All fellowships at the Newberry must be conducted in-residence. The Newberry collections must be necessary to the short-term fellowship project since the purpose of these fellowships is to enable access to particular materials. Applications from former recipients of any fellowship at the Newberry are welcome, but preference is given to applicants who have not held a major fellowship within the past three years. In general, all fellowships at the Newberry must be completed by the end of the academic year for which they are awarded.
*Restrictions:* Most short-term fellowships are restricted to individuals who live outside the Chicago area who need to examine specific items in the Newberry's collections. Applicants may apply for more than one fellowship at the Newberry by designating those for which they wish to be considered on the cover sheet (available at the grant website). Applicants may also apply for both short- and long-term fellowships at the Newberry; however individuals may accept only one fellowship at the Newberry per year. Only scholars who have previously received a short-term fellowship at the Newberry may apply for a long-term fellowship for the same project. No other project extensions are allowable.
*Geographic Focus:* All States, All Countries
*Date(s) Application is Due:* Dec 15
*Amount of Grant:* 2,500 - 7,500 USD
*Contact:* D. Bradford Hunt, Vice President for Research and Academic Programs; (312) 255-3535 or (312) 943-9090; huntb@newberry.org or research@newberry.org
*Internet:* www.newberry.org/short-term-fellowships
*Sponsor:* Newberry Library
60 West Walton Street
Chicago, IL 60610-3380

**Newberry Library Andrew W. Mellon Postdoctoral Research Long-Term Fellowships** 1767

The Newberry, open to the public without charge, is an independent research library dedicated to the advancement and dissemination of knowledge, especially in the humanities. The library acquires and preserves a broad array of special-collections research materials (spanning six centuries of European and American civilizations) and promotes and provides for these collections' effective use, fostering research, teaching, publication, life-long learning, and civic engagement. Fellowships at the Newberry provide support for researchers wishing to use the Newberry's collections and are either short-term (one to two months) or long-term (four to twelve months). Short-term fellowships are intended to geographically facilitate access for distant scholars to materials essential to a project. Long-term fellowships are intended to support significant works of scholarship that draw on the strengths of the Newberry's collections. Long-term fellowships support individual research and promote serious intellectual exchange through active participation in the Newberry's scholarly activities, including a biweekly fellows' seminar. Applicants may combine long-term fellowship awards with sabbatical or other stipendiary support. The Andrew W. Mellon Postdoctoral Research Fellowships are long-term fellowships that support postdoctoral scholars in any field relevant to the Newberry's collections for six to nine months of residential research and writing. The fellowship carries a monthly stipend of $4,200. Submission deadlines may vary from year to year. For more information on the program, current deadline dates, and application guidelines, those interested in applying should call or email the Newberry Library using the contact information given on this page.
*Requirements:* Postdoctoral scholars at any stage in their careers are eligible. Applications from former recipients of any fellowship at the Newberry are welcome, but preference is given to applicants who have not held a major fellowship within the past three years. In general, all fellowships must be completed by the end of the academic year for which they are awarded.

*Geographic Focus:* All States, All Countries
*Date(s) Application is Due:* Nov 15
*Amount of Grant:* 16,800 - 37,800 USD
*Contact:* D. Bradford Hunt, Vice President for Research and Academic Programs; (312) 255-3535 or (312) 943-9090; huntb@newberry.org or research@newberry.org
*Internet:* www.newberry.org/long-term-fellowships
*Sponsor:* Newberry Library
60 West Walton Street
Chicago, IL 60610-3380

**Newberry Library Arthur and Janet Holzheimer Short-Term Fellowships in the History of Cartography** 1768

The Newberry, open to the public without charge, is an independent research library dedicated to the advancement and dissemination of knowledge, especially in the humanities. The library acquires and preserves a broad array of special-collections research materials (spanning six centuries of European and American civilizations) and promotes and provides for these collections' effective use, fostering research, teaching, publication, life-long learning, and civic engagement. Fellowships at the Newberry provide support for researchers wishing to use the Newberry's collections and are either short-term (one to two months) or long-term (four to twelve months). Short-term fellowships are intended to geographically facilitate access for distant scholars to materials essential to a project. Long-term fellowships are intended to support significant works of scholarship that draw on the strengths of the Newberry's collections. The Arthur and Janet Holzheimer Short-Term Fellowship in the History of Cartography supports work-in-residence at the Newberry. Projects must be related to the history of cartography and focus on cartographic materials in the Newberry's collection. The fellowship provides a stipend of $2,500 per month. At the time of this writing, the current application submission date for most Newberry short-term fellowships is December 15 and awardees are usually notified near the end of April; however deadlines may vary from year to year. Applications must be submitted as email attachments by 11:59 p.m. on the deadline date. Applicants can find guidelines and forms at the grant website. Applicants are also encouraged to verify current deadline dates.
*Requirements:* Most short-term fellowships are for either Ph.D. candidates or postdoctoral scholars. All fellowships at the Newberry must be conducted in-residence. The Newberry collections must be necessary to the short-term fellowship project since the purpose of these fellowships is to enable access to particular materials. Applications from former recipients of any fellowship at the Newberry are welcome, but preference is given to applicants who have not held a major fellowship within the past three years. In general, all fellowships at the Newberry must be completed by the end of the academic year for which they are awarded.
*Restrictions:* Most short-term fellowships are restricted to individuals who live outside the Chicago area who need to examine specific items in the Newberry's collections. Applicants may apply for more than one fellowship at the Newberry by designating those for which they wish to be considered on the cover sheet (available at the grant website). Applicants may also apply for both short- and long-term fellowships at the Newberry; however individuals may accept only one fellowship at the Newberry per year. Only scholars who have previously received a short-term fellowship at the Newberry may apply for a long-term fellowship for the same project. No other project extensions are allowable.
*Geographic Focus:* All States, All Countries
*Date(s) Application is Due:* Dec 15
*Amount of Grant:* 2,500 - 5,000 USD
*Contact:* D. Bradford Hunt, Vice President for Research and Academic Programs; (312) 255-3535 or (312) 943-9090; huntb@newberry.org or research@newberry.org
*Internet:* www.newberry.org/short-term-fellowships
*Sponsor:* Newberry Library
60 West Walton Street
Chicago, IL 60610-3380

**Newberry Library Arthur and Lila Weinberg Short-Term Fellowships for Independent Scholars and Researchers** 1769

The Newberry, open to the public without charge, is an independent research library dedicated to the advancement and dissemination of knowledge, especially in the humanities. The library acquires and preserves a broad array of special-collections research materials (spanning six centuries of European and American civilizations) and promotes and provides for these collections' effective use, fostering research, teaching, publication, life-long learning, and civic engagement. Fellowships at the Newberry provide support for researchers wishing to use the Newberry's collections and are either short-term (one to two months) or long-term (four to twelve months). Most short-term fellowships are intended to geographically facilitate access for distant scholars to materials essential to a project; however residents in the Chicago area are eligible to apply for Weinberg Fellowships. Long-term fellowships are intended to support significant works of scholarship that draw on the strengths of the Newberry's collections. The Arthur and Lila Weinberg Fellowship is a short-term fellowship awarded to independent scholars or researchers who work outside of traditional academic settings (e.g. scholars, journalists, writers, filmmakers, visual and performing artists, and other humanists) who wish to use the Newberry's collections in the creation of a publication or comparable product. Preference is given to scholars working on historical issues related to social justice or reform. The Weinberg fellowship is intended to support one month-in-residence at the Newberry. Fellows will receive a total stipend of $2,500. At the time of this writing, the current application submission date for most Newberry short-term fellowships is December 15 and awardees are usually notified near the end of April; however deadlines may vary from year to year. Applications must be submitted as email attachments by 11:59 p.m. on the deadline date. Applicants can find guidelines and forms at the grant website. Applicants are also encouraged to verify current deadline dates.
*Requirements:* Both local and non-local residents of the Chicago area are eligible to apply for this fellowship. The Newberry collections must be necessary to the short-term fellowship project since the purpose of these fellowships is to enable access to particular materials. Applications from former recipients of any fellowship at the Newberry are welcome, but preference is given to applicants who have not held a major fellowship within the past three years. In general, all fellowships must be completed by the end of the academic year for which they are awarded.
*Restrictions:* This fellowship is intended to support one month-in-residence at the Newberry; however, local residents who are awarded a Weinberg Fellowship may choose to distribute their time over a period of 9-12 months on a flexible basis to be determined in conversation with Newberry staff.
*Geographic Focus:* All States, All Countries
*Date(s) Application is Due:* Dec 15
*Amount of Grant:* 2,500 USD
*Contact:* D. Bradford Hunt, Vice President for Research and Academic Programs; (312) 255-3535 or (312) 943-9090; huntb@newberry.org or research@newberry.org
*Internet:* www.newberry.org/short-term-fellowships
*Sponsor:* Newberry Library
60 West Walton Street
Chicago, IL 60610-3380

**Newberry Library Associated Colleges of the Midwest/Great Lakes College Association Faculty Long-Term Fellowships** 1770

The Newberry, open to the public without charge, is an independent research library dedicated to the advancement and dissemination of knowledge, especially in the humanities. The library acquires and preserves a broad array of special-collections research materials (spanning six centuries of European and American civilizations) and promotes and provides for these collections' effective use, fostering research, teaching, publication, life-long learning, and civic engagement. Fellowships at the Newberry provide support for researchers wishing to use the Newberry's collections and are either short-term (one to two months) or long-term (four to twelve months). The Associated Colleges of the Midwest (ACM)/Great Lakes College Association (GLCA) Faculty Fellowships support ACM- and GLCA-college faculty as they teach a small group of select undergraduate students in an advanced research seminar while doing their own research. These fellowships are considered long-term fellowships. Fellows can come from any of the ACM or GLCA colleges and from any discipline. At the time of this writing, the current application-submission date for most Newberry long-term fellowships is November 15 and awardees are usually notified in mid- to late-March; however deadlines may vary from year to year. Potential applicants should call or email the ACM contact given for information on how to apply.
*Restrictions:* Applicants may apply for more than one fellowship at the Newberry by designating those for which they wish to be considered on the cover sheet (available at the grant website). Applicants may also apply for both short- and long-term fellowships at the Newberry; however individuals may accept only one fellowship at the Newberry per year. Only scholars who have previously received a short-term fellowship at the Newberry may apply for a long-term fellowship for the same project. No other project extensions are allowable.
*Geographic Focus:* Colorado, Illinois, Indiana, Iowa, Michigan, Minnesota, Ohio, Pennsylvania, Wisconsin
*Date(s) Application is Due:* Nov 15
*Amount of Grant:* 4,200 - 37,800 USD
*Contact:* Joan Gillespie, Vice President; (312) 263-5000; jgillespie@acm.edu
D. Bradford Hunt, Vice President for Research and Academic Programs; (312) 255-3535 or (312) 943-9090; huntb@newberry.org or research@newberry.org
*Internet:* www.newberry.org/long-term-fellowships
*Sponsor:* Newberry Library
60 West Walton Street
Chicago, IL 60610-3380

**Newberry Library Audrey Lumsden-Kouvel Long-Term Fellowships** 1771

The Newberry, open to the public without charge, is an independent research library dedicated to the advancement and dissemination of knowledge, especially in the humanities. The library acquires and preserves a broad array of special-collections research materials (spanning six centuries of European and American civilizations) and promotes and provides for these collections' effective use, fostering research, teaching, publication, life-long learning, and civic engagement. Fellowships at the Newberry provide support for researchers wishing to use the Newberry's collections and are either short-term (one to two months) or long-term (four to nine months). Short-term fellowships are intended to geographically facilitate access for distant scholars to materials essential to a project. Long-term fellowships are intended to support significant works of scholarship that draw on the strengths of the Newberry's collections. Long-term fellowships support individual research and promote serious intellectual exchange through active participation in the Newberry's scholarly activities, including a biweekly fellows' seminar. Applicants may combine long-term fellowship awards with sabbatical or other stipendiary support. The Audrey Lumsden-Kouvel Fellowship is a long-term fellowship for post-doctoral scholars who wish to use the Newberry's extensive holdings in late medieval and early modern history and literature. Preference will be given to projects focusing on Romance cultures, including work that draws on sources from the colonial Americas. Topics in Portuguese, Spanish, and Latin American studies are especially welcome, as are translation projects. The fellowship carries a monthly stipend of $4,200. At the time of this writing, the current application-submission date for most Newberry long-term fellowships is December 12 and awardees are usually notified in mid- to late-March; however deadlines may vary from year to year. Applications must be submitted as email attachments by 11:59 p.m. on the deadline date. Applicants can find guidelines and forms at the grant website. Applicants are also encouraged to verify current deadline dates.

*Requirements:* All long-term fellowships are restricted to postdoctoral scholars. Recent Ph.D. recipients are encouraged to apply. Applicants must request at least four and no more than six months of support. All fellowships at the Newberry must be in-residence. Applications from former recipients of any fellowship at the Newberry are welcome, but preference is given to applicants who have not held a major fellowship within the past three years. In general, all fellowships must be completed by the end of the academic year for which they are awarded.
*Restrictions:* Applicants may apply for more than one fellowship at the Newberry by designating those for which they wish to be considered on the cover sheet (available at the grant website). Applicants may also apply for both short- and long-term fellowships at the Newberry; however individuals may accept only one fellowship at the Newberry per year. Only scholars who have previously received a short-term fellowship at the Newberry may apply for a long-term fellowship for the same project. No other project extensions are allowable.
*Geographic Focus:* All States, All Countries
*Date(s) Application is Due:* Nov 15
*Amount of Grant:* 16,800 - 25,200 USD
*Contact:* D. Bradford Hunt, Vice President for Research and Academic Programs; (312) 255-3535 or (312) 943-9090; huntb@newberry.org or research@newberry.org
*Internet:* www.newberry.org/long-term-fellowships
*Sponsor:* Newberry Library
60 West Walton Street
Chicago, IL 60610-3380

**Newberry Library Charles Montgomery Gray Short-Term Fellowships 1772**

The Newberry, open to the public without charge, is an independent research library dedicated to the advancement and dissemination of knowledge, especially in the humanities. The library acquires and preserves a broad array of special-collections research materials (spanning six centuries of European and American civilizations) and promotes and provides for these collections' effective use, fostering research, teaching, publication, life-long learning, and civic engagement. Fellowships at the Newberry provide support for researchers wishing to use the Newberry's collections and are either short-term (one to two months) or long-term (four to twelve months). Short-term fellowships are intended to geographically facilitate access for distant scholars to materials essential to a project. Long-term fellowships are intended to support significant works of scholarship that draw on the strengths of the Newberry's collections. The Charles Montgomery Gray Fellowship is available to scholars who demonstrate a specific need for the Newberry's collection. This fellowship is open to all fields of study, with preference given to projects in the early modern period or Renaissance, as well as in English history, legal history, or European history. The fellowships provide a stipend of $2,500 per month (for each scholar). At the time of this writing, the current application submission date for most Newberry short-term fellowships is December 15 and awardees are usually notified near the end of April; however deadlines may vary from year to year. Applications must be submitted as email attachments by 11:59 p.m. on the deadline date. Applicants can find guidelines and forms at the grant website. Applicants are also encouraged to verify current deadline dates.
*Requirements:* Ph.D. candidates and post-doctoral scholars who live and work outside the Chicago area are eligible to apply. Teams of two or three scholars who plan to collaborate on a single, substantive project are also eligible to apply. All fellowships at the Newberry must be conducted in-residence. The Newberry collections must be necessary to the short-term fellowship project since the purpose of these fellowships is to enable access to particular materials. Applications from former recipients of any fellowship at the Newberry are welcome, but preference is given to applicants who have not held a major fellowship within the past three years. In general, all fellowships at the Newberry must be completed by the end of the academic year for which they are awarded.
*Restrictions:* Most short-term fellowships are restricted to individuals who live outside the Chicago area who need to examine specific items in the Newberry's collections. Applicants may apply for more than one fellowship at the Newberry by designating those for which they wish to be considered on the cover sheet (available at the grant website). Applicants may also apply for both short- and long-term fellowships at the Newberry; however individuals may accept only one fellowship at the Newberry per year. Only scholars who have previously received a short-term fellowship at the Newberry may apply for a long-term fellowship for the same project. No other project extensions are allowable.
*Geographic Focus:* All States, All Countries
*Date(s) Application is Due:* Dec 15
*Amount of Grant:* 2,500 - 7,500 USD
*Contact:* D. Bradford Hunt, Vice President for Research and Academic Programs; (312) 255-3535 or (312) 943-9090; huntb@newberry.org or research@newberry.org
*Internet:* https://www.newberry.org/short-term-fellowships
*Sponsor:* Newberry Library
60 West Walton Street
Chicago, IL 60610-3380

**Newberry Library Consortium on American Indian Studies Faculty Long-Term Fellowships 1773**

The Newberry, open to the public without charge, is an independent research library dedicated to the advancement and dissemination of knowledge, especially in the humanities. The library acquires and preserves a broad array of special-collections research materials (spanning six centuries of European and American civilizations) and promotes and provides for these collections' effective use, fostering research, teaching, publication, life-long learning, and civic engagement. Fellowships at the Newberry provide support for researchers wishing to use the Newberry's collections and are either short-term (one to two months) or long-term (four to nine months). Short-term fellowships are intended to geographically facilitate access for distant scholars to materials essential to a project. Long-term fellowships are intended to support significant works of scholarship that draw on the strengths of the Newberry's collections. Long-term fellowships support individual research and promote serious intellectual exchange through active participation in the Newberry's scholarly activities, including a biweekly fellows' seminar. Applicants may combine long-term fellowship awards with sabbatical or other stipendiary support. The Newberry Consortium in American Indian Studies (NCAIS) offers long-term fellowships for faculty at participating institutions. The fellowships provide support for one semester in residence at the Newberry and carry a monthly stipend of $4,200. Fellows will work on a project in American Indian Studies requiring research in the Newberry's collections and will receive a research carrel and be accorded the same privileges as other Newberry long-term fellows. At the time of this writing, the current application submission date for most Newberry long-term fellowships is November 15 and awardees are usually notified in mid- to late-March; however deadlines may vary from year to year. Applications must be submitted as email attachments by 11:59 p.m. on the deadline date. Applicants can find guidelines and forms at the grant website. Applicants are also encouraged to verify current deadline dates.
*Requirements:* Fellowship recipients must be faculty at institutions participating in NCAIS. At this writing those institutions include Cornell University, Harvard University, Michigan State University, Princeton University, The University of British Columbia, The University of Illinois at Urbana-Champaign, University of Manitoba, University of Minnesota, University of Montana, University of New Mexico, University of North Carolina at Chapel Hill, University of Winnipeg, University of Wisconsin Milwaukee, University of Wyoming, Vanderbilt University, and Yale University. Preference is given to scholars at an early career stage. Faculty Fellows will present research, participate in both the McNickle Center Seminar in American Indian and Indigenous Studies and the Newberry Library Fellows' Seminar, and be available to NCAIS Graduate Student Fellows for consultation. Applicants must request at least four and no more than six months of support.
*Restrictions:* Applicants may apply for more than one fellowship at the Newberry by designating those for which they wish to be considered on the cover sheet (available at the grant website). Applicants may also apply for both short- and long-term fellowships at the Newberry; however individuals may accept only one fellowship at the Newberry per year. Only scholars who have previously received a short-term fellowship at the Newberry may apply for a long-term fellowship for the same project. No other project extensions are allowable.
*Geographic Focus:* All States, Canada
*Date(s) Application is Due:* Nov 15
*Amount of Grant:* 16,800 - 37,800 USD
*Contact:* D. Bradford Hunt, Vice President for Research and Academic Programs; (312) 255-3535 or (312) 943-9090; huntb@newberry.org or research@newberry.org
*Internet:* www.newberry.org/long-term-fellowships
*Sponsor:* Newberry Library
60 West Walton Street
Chicago, IL 60610-3380

**Newberry Library Consortium on American Indian Studies Graduate Student Short-Term Fellowships 1774**

The Newberry, open to the public without charge, is an independent research library dedicated to the advancement and dissemination of knowledge, especially in the humanities. The library acquires and preserves a broad array of special-collections research materials (spanning six centuries of European and American civilizations) and promotes and provides for these collections' effective use, fostering research, teaching, publication, life-long learning, and civic engagement. The Newberry Consortium on American Indian Studies (NCAIS) offers short-term graduate-student fellowships for doctoral candidates at participating institutions. The fellowships offer support for one to two months of dissertation research and carry stipends of $2,500 per month. NCAIS graduate-student fellows will be offered individual research space at the Newberry and accorded the same privileges as other Newberry short-term fellows. These NCAIS fellowships may also be used to fund research in other libraries, archives, or in the field. At the time of this writing, the current application submission date for most Newberry short-term fellowships is December 15 and awardees are usually notified near the end of April; however deadlines may vary from year to year. Applications must be submitted as email attachments by 11:59 p.m. on the deadline date. Applicants can find guidelines and forms at the grant website. Applicants are also encouraged to verify current deadline dates.
*Requirements:* Fellowship recipients must be doctoral candidates at institutions participating in NCAIS. At this writing those institutions include Cornell University, Harvard University, Michigan State University, Princeton University, The University of British Columbia, The University of Illinois at Urbana-Champaign, University of Manitoba, University of Minnesota, University of Montana, University of New Mexico, University of North Carolina at Chapel Hill, University of Winnipeg, University of Wisconsin Milwaukee, University of Wyoming, Vanderbilt University, and Yale University. Fellowship recipients are expected to present their research at the NCAIS's annual graduate student conference or at a Newberry-sponsored seminar in American Indian and indigenous studies.
*Restrictions:* Applicants may apply for more than one fellowship at the Newberry by designating those for which they wish to be considered on the cover sheet (available at the grant website). Applicants may also apply for both short- and long-term fellowships at the Newberry; however individuals may accept only one fellowship at the Newberry per year. Only scholars who have previously received a short-term fellowship at the Newberry may apply for a long-term fellowship for the same project. No other project extensions are allowable.
*Geographic Focus:* All States, Canada
*Date(s) Application is Due:* Dec 15
*Amount of Grant:* 2,500 - 5,000 USD

*Contact:* D. Bradford Hunt, Vice President of Research and Education; (312) 255-3535 or (312) 943-9090; huntb@newberry.org or research@newberry.org
*Internet:* www.newberry.org/short-term-fellowships
*Sponsor:* Newberry Library
60 West Walton Street
Chicago, IL 60610-3380

**Newberry Library Ecole Nationale des Chartes Exchange Short-Term Fellowship** 1775

The Newberry, open to the public without charge, is an independent research library dedicated to the advancement and dissemination of knowledge, especially in the humanities. The library acquires and preserves a broad array of special-collections research materials (spanning six centuries of European and American civilizations) and promotes and provides for these collections' effective use, fostering research, teaching, publication, life-long learning, and civic engagement. Fellowships at the Newberry provide support for researchers wishing to use the Newberry's collections and are either short-term or long-term. Short-term fellowships are intended to geographically facilitate access for distant scholars to materials essential to a project. Long-term fellowships are intended to support significant works of scholarship that draw on the strengths of the Newberry's collections. Long-term fellowships support individual research and promote serious intellectual exchange through active participation in the Newberry's scholarly activities, including a biweekly fellows' seminar. Applicants may combine long-term fellowship awards with sabbatical or other stipendiary support. The Newberry Library École Nationale des Chartes Exchange Fellowship (considered a short-term fellowship) has existed since 1979 and provides three months of auditor tuition and a monthly stipend of 2,500 Euros for an American or Canadian Ph.D. student to study at the École Nationale des Chartes in Paris, preferably during the fall semester. The École des Chartes is the oldest institution in Europe specializing in the archival sciences, including paleography, codicology, bibliography, diplomatics, textual editing, and the history of the book. Applications must be submitted as email attachments by 11:59 p.m. on the deadline date. At the time of this writing, the current application-submission date for most Newberry long-term fellowships is December 15 and awardees are usually notified in mid- to late-March; however deadlines may vary from year to year. Applicants are encouraged to verify current deadline dates at the grant website. Application forms and more specific guidelines are available at the grant website.
*Requirements:* Applications are especially encouraged from students in the early stages of preparing their dissertations. Applicants must provide specific reference to their need for the training available at the École des Chartes and how it will contribute to their dissertation or future projects. Preference is given to students attending member institutions of the Newberry's Center for Renaissance Studies Consortium which are as follows: DePaul University; Eastern Illinois University; Georgia Institute of Technology; Illinois State University; Indiana University; Kent State University; Loyola University; Marquette University; McGill University; Miami University; Michigan State University; Northern Illinois University; Northwestern University, Ohio University; Oklahoma State University; Purdue University; Queen Mary, University of London; Saint Louis University; Southern Illinois University; Tulane University; University of Aberdeen; University of Akron; University of Arizona; University of Chicago; University of Illinois at Chicago; University of Illinois at Urbana-Champaign; University of Iowa; University of Kansas Lawrence; University of Louisville; University of Michigan; University of Minnesota; University of Missouri; University of Nebraska—Lincoln; University of Nevada; University of North Texas; University of Notre Dame; University of Oklahoma; University of Pittsburgh; University of Tennessee—Knoxville; University of Texas; University of Warwick; University of Waterloo; University of Western Ontario; University of Wisconsin—Madison; University of Wisconsin—Milwaukee; Vanderbilt University; Washington University; Wayne State University.
*Geographic Focus:* All States, All Countries
*Date(s) Application is Due:* Dec 15
*Amount of Grant:* 2,500 - 7,500 USD
*Contact:* D. Bradford Hunt, Vice President for Research and Academic Programs; (312) 255-3535 or (312) 943-9090; huntb@newberry.org or research@newberry.org
*Internet:* www.newberry.org/short-term-fellowships
*Sponsor:* Newberry Library
60 West Walton Street
Chicago, IL 60610-3380

**Newberry Library Frances C. Allen Short-Term Fellowships** 1776

The Newberry, open to the public without charge, is an independent research library dedicated to the advancement and dissemination of knowledge, especially in the humanities. The library acquires and preserves a broad array of special-collections research materials (spanning six centuries of European and American civilizations) and promotes and provides for these collections' effective use, fostering research, teaching, publication, life-long learning, and civic engagement. Fellowships at the Newberry provide support for researchers wishing to use the Newberry's collections and are either short-term (one to two months) or long-term (four to twelve months). Short-term fellowships are intended to geographically facilitate access for distant scholars to materials essential to a project. Long-term fellowships are intended to support significant works of scholarship that draw on the strengths of the Newberry's collections. First awarded in 1983, the Frances C. Allen Fellowships are considered to be short-term fellowships, however their tenure may extend from one month to a full year. Their particular goal is to encourage Native-American women in their studies of any field related to the Newberry's collections. The fellowships provide a stipend of $2,500 per month; however supplemental funding may be available on a case by case basis. At the time of this writing, the current application submission date for most Newberry short-term fellowships is December 15 and awardees are usually notified near the end of April; however deadlines may vary from year to year. Applications must be submitted as email attachments by 11:59 p.m. on the deadline date. Applicants can find guidelines and forms at the grant website. Applicants are also encouraged to verify current deadline dates.
*Requirements:* These fellowships are available to women of Native-American heritage who are pursuing an academic program at any stage beyond the undergraduate degree. Allen fellows are expected to spend a significant part of their tenure in residence at the Newberry's D'Arcy McNickle Center for American Indian History. Applications from former recipients of any fellowship at the Newberry are welcome, but preference is given to applicants who have not held a major fellowship within the past three years. In general, all fellowships must be completed by the end of the academic year for which they are awarded.
*Restrictions:* Most short-term fellowships are restricted to individuals who live outside the Chicago area who need to examine specific items in the Newberry's collections. Applicants may apply for more than one fellowship at the Newberry by designating those for which they wish to be considered on the cover sheet (available at the grant website). Applicants may also apply for both short- and long-term fellowships at the Newberry; however individuals may accept only one fellowship at the Newberry per year. Only scholars who have previously received a short-term fellowship at the Newberry may apply for a long-term fellowship for the same project. No other project extensions are allowable.
*Geographic Focus:* All States, All Countries
*Date(s) Application is Due:* Dec 15
*Amount of Grant:* 2,500 - 7,500 USD
*Contact:* D. Bradford Hunt, Vice President for Research and Academic Programs; (312) 255-3535 or (312) 943-9090; huntb@newberry.org or research@newberry.org
*Internet:* www.newberry.org/short-term-fellowships
*Sponsor:* Newberry Library
60 West Walton Street
Chicago, IL 60610-3380

**Newberry Library Graduate Scholar-in-Residence Opportunities** 1777

The Newberry, open to the public without charge, is an independent research library dedicated to the advancement and dissemination of knowledge, especially in the humanities. The library acquires and preserves a broad array of special-collections research materials (spanning six centuries of European and American civilizations) and promotes and provides for these collections' effective use, fostering research, teaching, publication, life-long learning, and civic engagement. The Graduate Scholar-in-Residence program at the Newberry encourages Ph.D. candidates in the humanities to conduct research in the library's collections and to join its community of scholars. Selected scholars are provided with reserve space, access to the Newberry during extended hours, and opportunities to present their work-in-progress to the Newberry's scholarly community. Applications to become a Graduate Scholar-in-Residence are accepted each year until May 1. Candidates must submit via email a project description along with a filled-in application form (available for download from the website) by 11:59 p.m. on the deadline date. Applicants will be notified of their acceptance in June and will begin their residence in September so they may join the Newberry's new long-term fellows in fall-orientation activities. More detailed guidelines on how to apply are given at the grant website.
*Requirements:* Graduate students with advanced to Ph.D. candidacy are invited to apply for this status, with preference to those whose dissertation projects are well underway. Candidates' need for and intensive use of the Newberry's collections will be a crucial factor in the library's consideration of applications. Graduate scholars are expected to be in residence at least ten hours per week and should be wiling to provide a small amount of service to the Newberry.
*Geographic Focus:* Illinois
*Date(s) Application is Due:* May 1
*Contact:* D. Bradford Hunt, Vice President for Research and Academic Programs; (312) 255-3535 or (312) 943-9090; huntb@newberry.org or research@newberry.org
*Internet:* www.newberry.org/graduate-scholar-residence-program
*Sponsor:* Newberry Library
60 West Walton Street
Chicago, IL 60610-3380

**Newberry Library Herzog August Bibliothek Wolfenbuttel Joint Short-Term Fellowships** 1778

The Newberry, open to the public without charge, is an independent research library dedicated to the advancement and dissemination of knowledge, especially in the humanities. The library acquires and preserves a broad array of special-collections research materials (spanning six centuries of European and American civilizations) and promotes and provides for these collections' effective use, fostering research, teaching, publication, life-long learning, and civic engagement. Fellowships at the Newberry provide support for researchers wishing to use the Newberry's collections and are either short-term (one to two months) or long-term (four to twelve months). Short-term fellowships are intended to geographically facilitate access for distant scholars to materials essential to a project. Long-term fellowships are intended to support significant works of scholarship that draw on the strengths of the Newberry's collections. Applicants for either long- or short-term fellowships at the Newberry may also apply for the Herzog August Bibliothek Wolfenbuttel fellowship which provides an additional two months at the Herzog August library in Wolfenbuttel, Germany. The Herzog award will pay 1,050 EUR per month plus up to 600 EUR for travel expenses. At the time of this writing, the current application submission date for most Newberry short-term fellowships is February 6, and for long-term fellowships, December 15. Short-term awards are usually announced near the end of April; long-term awards, mid- to late-March; however deadlines may vary from year to year. Applications must be submitted as email attachments by 11:59 p.m. on the deadline dates. Applicants can find guidelines and forms at the grant website. Applicants are also encouraged to verify current deadline dates.

*Requirements:* The applicant's proposed project must link the collections of both libraries (the Newberry and the Herzog August); the applicant should plan to hold both fellowships sequentially to ensure continuity of research. Further requirements will depend on the particular Newberry short- or long-term fellowship to which the applicant plans to jointly apply.
*Restrictions:* Restrictions will depend on the particular Newberry short- or long-term fellowship to which the applicant plans to jointly apply.
*Geographic Focus:* All States, All Countries
*Date(s) Application is Due:* Dec 15
*Amount of Grant:* 2,100 - 2,700 EUR
*Contact:* D. Bradford Hunt, Vice President for Research and Academic Programs; (312) 255-3535 or (312) 943-9090; huntb@newberry.org or research@newberry.org
*Internet:* www.newberry.org/short-term-fellowships
*Sponsor:* Newberry Library
60 West Walton Street
Chicago, IL 60610-3380

**Newberry Library John S. Aubrey Short-Term Fellowships 1779**

The Newberry, open to the public without charge, is an independent research library dedicated to the advancement and dissemination of knowledge, especially in the humanities. The library acquires and preserves a broad array of special-collections research materials (spanning six centuries of European and American civilizations) and promotes and provides for these collections' effective use, fostering research, teaching, publication, life-long learning, and civic engagement. Fellowships at the Newberry provide support for researchers wishing to use the Newberry's collections and are either short-term (one to two months) or long-term (four to twelve months). Short-term fellowships are intended to geographically facilitate access for distant scholars to materials essential to a project. Long-term fellowships are intended to support significant works of scholarship that draw on the strengths of the Newberry's collections. The John S. Aubrey Fellowship is available to scholars who demonstrate a specific need for the Newberry's collection. This fellowship is open to all fields of study. The fellowships provide a stipend of $2,500 per month (for each scholar). At the time of this writing, the current application submission date for most Newberry short-term fellowships is December 15 and awardees are usually notified near the end of April; however deadlines may vary from year to year. Applications must be submitted as email attachments by 11:59 p.m. on the deadline date. Applicants can find guidelines and forms at the grant website. Applicants are also encouraged to verify current deadline dates.
*Requirements:* Ph.D. candidates and post-doctoral scholars who live and work outside the Chicago area are eligible to apply. Teams of two or three scholars who plan to collaborate on a single, substantive project are also eligible to apply. All fellowships at the Newberry must be conducted in-residence. The Newberry collections must be necessary to the short-term fellowship project since the purpose of these fellowships is to enable access to particular materials. Applications from former recipients of any fellowship at the Newberry are welcome, but preference is given to applicants who have not held a major fellowship within the past three years. In general, all fellowships at the Newberry must be completed by the end of the academic year for which they are awarded.
*Restrictions:* Most short-term fellowships are restricted to individuals who live outside the Chicago area who need to examine specific items in the Newberry's collections. Applicants may apply for more than one fellowship at the Newberry by designating those for which they wish to be considered on the cover sheet (available at the grant website). Applicants may also apply for both short- and long-term fellowships at the Newberry; however individuals may accept only one fellowship at the Newberry per year. Only scholars who have previously received a short-term fellowship at the Newberry may apply for a long-term fellowship for the same project. No other project extensions are allowable.
*Geographic Focus:* All States, All Countries
*Date(s) Application is Due:* Dec 15
*Amount of Grant:* 2,500 - 7,500 USD
*Contact:* D. Bradford Hunt, Vice President for Research and Academic Programs; (312) 255-3535 or (312) 943-9090; huntb@newberry.org or research@newberry.org
*Internet:* https://www.newberry.org/short-term-fellowships
*Sponsor:* Newberry Library
60 West Walton Street
Chicago, IL 60610-3380

**Newberry Library Lawrence Lipking Short-Term Fellowship at the Newberry 1780**

The Newberry, open to the public without charge, is an independent research library dedicated to the advancement and dissemination of knowledge, especially in the humanities. The library acquires and preserves a broad array of special-collections research materials (spanning six centuries of European and American civilizations) and promotes and provides for these collections' effective use, fostering research, teaching, publication, life-long learning, and civic engagement. Fellowships at the Newberry provide support for researchers wishing to use the Newberry's collections and are either short-term or long-term. Short-term fellowships are intended to geographically facilitate access for distant scholars to materials essential to a project. Long-term fellowships are intended to support significant works of scholarship that draw on the strengths of the Newberry's collections. The annual Lipking-Newberry Fellowship enables a qualified Northwestern graduate student in English to spend a quarter conducting research at the Newberry. A stipend equivalent to Northwestern's University Fellowship stipend ($5,130 in 2009-10) and tuition will be provided. Candidates for the fellowship should supply a short-term cover sheet, a curriculum vitae, one letter of recommendation, and a research plan of no more than 500 words detailing intended use of Newberry resources. The cover sheet and further application guidelines are available at the grant website. Applications must be submitted as email attachments by 11:59 p.m. on the deadline date to the Newberry (and copied to Northwestern). The email addresses to use for both organizations are specified at the grant website. At the time of this writing, the current application submission date for most Newberry short-term fellowships is December 15 and awardees are usually notified near the end of April; however, deadlines may vary from year to year. Candidates are encouraged to verify current deadline dates.
*Requirements:* Applicants must be qualified Northwestern graduate students in the English department. Full-time residence at the Newberry is required.
*Restrictions:* In the unlikely event that no English graduate student should qualify, the fellowship will be opened to other Northwestern departments. Applicants may apply for more than one fellowship at the Newberry by designating those for which they wish to be considered on the cover sheet (available at the grant website). Applicants may also apply for both short- and long-term fellowships at the Newberry; however individuals may accept only one fellowship at the Newberry per year. Only scholars who have previously received a short-term fellowship at the Newberry may apply for a long-term fellowship for the same project. No other project extensions are allowable.
*Geographic Focus:* Illinois
*Date(s) Application is Due:* Dec 15
*Amount of Grant:* 5,500 USD
*Contact:* D. Bradford Hunt, Vice President for Research and Academic Programs; (312) 255-3535 or (312) 943-9090; huntb@newberry.org or research@newberry.org
*Internet:* www.newberry.org/short-term-fellowships
*Sponsor:* Newberry Library
60 West Walton Street
Chicago, IL 60610-3380

**Newberry Library Lester J. Cappon Short-Term Fellowship in Documentary Editing 1781**

The Newberry, open to the public without charge, is an independent research library dedicated to the advancement and dissemination of knowledge, especially in the humanities. The library acquires and preserves a broad array of special-collections research materials (spanning six centuries of European and American civilizations) and promotes and provides for these collections' effective use, fostering research, teaching, publication, life-long learning, and civic engagement. Fellowships at the Newberry provide support for researchers wishing to use the Newberry's collections and are either short-term (generally one to two months) or long-term (four to twelve months). Short-term fellowships are intended to geographically facilitate access for distant scholars to materials essential to a project. Long-term fellowships are intended to support significant works of scholarship that draw on the strengths of the Newberry's collections. The Lester J. Cappon Fellowship in Documentary Editing is considered a short-term fellowship to support scholarly editing projects based on Newberry materials and carried out in residence. Cappon fellowships provide a stipend of $2,500 per month; applicants may request up to two months of support for this fellowship. At the time of this writing, the current application submission date for most Newberry short-term fellowships is December 15 and awardees are usually notified near the end of April; however deadlines may vary from year to year. Applications must be submitted as email attachments by 11:59 p.m. on the deadline date. Applicants can find guidelines and forms at the grant website. Applicants are also encouraged to verify current deadline dates.
*Requirements:* The Cappon Fellowship is restricted to postdoctoral scholars. While most short-term fellowships at the Newberry are restricted to scholars who live outside the Chicago area, applicants for the Cappon fellowship may be local. All fellowships at the Newberry must be conducted in-residence. Applications from former recipients of any fellowship at the Newberry are welcome, but preference is given to applicants who have not held a major fellowship within the past three years. In general, all fellowships at the Newberry must be completed by the end of the academic year for which they are awarded.
*Restrictions:* Applicants may apply for more than one fellowship at the Newberry by designating those for which they wish to be considered on the cover sheet (available at the grant website). Applicants may also apply for both short- and long-term fellowships at the Newberry; however individuals may accept only one fellowship at the Newberry per year. Only scholars who have previously received a short-term fellowship at the Newberry may apply for a long-term fellowship for the same project. No other project extensions are allowable.
*Geographic Focus:* All States, All Countries
*Date(s) Application is Due:* Dec 15
*Amount of Grant:* 2,500 - 5,000 USD
*Contact:* D. Bradford Hunt, Vice President for Research and Academic Programs; (312) 255-3535 or (312) 943-9090; huntb@newberry.org or research@newberry.org
*Internet:* www.newberry.org/short-term-fellowships
*Sponsor:* Newberry Library
60 West Walton Street
Chicago, IL 60610-3380

**Newberry Library Lloyd Lewis Long-Term Fellowships in American History 1782**

The Newberry, open to the public without charge, is an independent research library dedicated to the advancement and dissemination of knowledge, especially in the humanities. The library acquires and preserves a broad array of special-collections research materials (spanning six centuries of European and American civilizations) and promotes and provides for these collections' effective use, fostering research, teaching, publication, life-long learning, and civic engagement. Fellowships at the Newberry provide support for researchers wishing to use the Newberry's collections and are either short-term (one to two months) or long-term (four to nine months). Short-term fellowships are intended to geographically facilitate access for distant scholars to materials essential to a project. Long-term fellowships are intended to support significant works of scholarship that draw on the strengths of the Newberry's collections. Long-term fellowships support individual research and promote serious intellectual exchange through active participation in the

Newberry's scholarly activities, including a biweekly fellows' seminar. Applicants may combine long-term fellowship awards with sabbatical or other stipendiary support. The Lloyd Lewis Fellowships in American History are long-term fellowships for post-doctoral scholars pursuing projects in any area of American history appropriate to the Newberry's collections. The fellowship carries a monthly stipend of $4,200. At the time of this writing, the current application-submission date for most Newberry long-term fellowships is December 12 and awardees are usually notified in mid- to late-March; however deadlines may vary from year to year. Applications must be submitted as email attachments by 11:59 p.m. on the deadline date. Applicants can find guidelines and forms at the grant website. Applicants are also encouraged to verify current deadline dates.
*Requirements:* Applicants must have their Ph.D. or its equivalent and must request at least four but no more than nine months of support. All fellowships at the Newberry must be in-residence. Applications from former recipients of any fellowship at the Newberry are welcome, but preference is given to applicants who have not held a major fellowship within the past three years. In general, all fellowships must be completed by the end of the academic year for which they are awarded.
*Restrictions:* Applicants may apply for more than one fellowship at the Newberry by designating those for which they wish to be considered on the cover sheet (available at the grant website). Applicants may also apply for both short- and long-term fellowships at the Newberry; however individuals may accept only one fellowship at the Newberry per year. Only scholars who have previously received a short-term fellowship at the Newberry may apply for a long-term fellowship for the same project. No other project extensions are allowable.
*Geographic Focus:* All States, All Countries
*Date(s) Application is Due:* Nov 15
*Amount of Grant:* 16,800 - 37,800 USD
*Contact:* D. Bradford Hunt, Vice President for Research and Academic Programs; (312) 255-3535 or (312) 943-9090; huntb@newberry.org or research@newberry.org
*Internet:* www.newberry.org/long-term-fellowships
*Sponsor:* Newberry Library
60 West Walton Street
Chicago, IL 60610-3380

**Newberry Library Midwest Modern Language Association Short-Term Fellowship** **1783**

The Newberry, open to the public without charge, is an independent research library dedicated to the advancement and dissemination of knowledge, especially in the humanities. The library acquires and preserves a broad array of special-collections research materials (spanning six centuries of European and American civilizations) and promotes and provides for these collections' effective use, fostering research, teaching, publication, life-long learning, and civic engagement. Fellowships at the Newberry provide support for researchers wishing to use the Newberry's collections and are either short-term (one to two months) or long-term (four to twelve months). Short-term fellowships are intended to geographically facilitate access for distant scholars to materials essential to a project. Long-term fellowships are intended to support significant works of scholarship that draw on the strengths of the Newberry's collections. A short-term fellowship is available for Ph.D. candidates or post-doctoral scholars who are members of the Midwest Modern Language Association (MMLA) for up to one month's work-in-residence at the Newberry. The fellowship provides a stipend of $2,500 per month. At the time of this writing, the current application submission date for most Newberry short-term fellowships is December 15 and awardees are usually notified near the end of April; however deadlines may vary from year to year. Applications must be submitted as email attachments by 11:59 p.m. on the deadline date. Applicants can find guidelines and forms at the grant website. Applicants are also encouraged to verify current deadline dates.
*Requirements:* MMLA membership must be current at the time of application and through the period of the fellowship. Most short-term fellowships at the Newberry are for Ph.D. candidates and postdoctoral scholars. All fellowships at the Newberry must be conducted in-residence. The Newberry collections must be necessary to the short-term fellowship project since the purpose of these fellowships is to enable access to particular materials. Applications from former recipients of any fellowship at the Newberry are welcome, but preference is given to applicants who have not held a major fellowship within the past three years. In general, all fellowships at the Newberry must be completed by the end of the academic year for which they are awarded.
*Restrictions:* Most short-term fellowships are restricted to individuals who live outside the Chicago area who need to examine specific items in the Newberry's collections. Applicants may apply for more than one fellowship at the Newberry by designating those for which they wish to be considered on the cover sheet (available at the grant website). Applicants may also apply for both short- and long-term fellowships at the Newberry; however individuals may accept only one fellowship at the Newberry per year. Only scholars who have previously received a short-term fellowship at the Newberry may apply for a long-term fellowship for the same project. No other project extensions are allowable.
*Geographic Focus:* All States, Canada
*Date(s) Application is Due:* Dec 15
*Amount of Grant:* 2,500 USD
*Contact:* D. Bradford Hunt, Vice President for Research and Academic Programs; (312) 255-3535 or (312) 943-9090; huntb@newberry.org or research@newberry.org
*Internet:* www.newberry.org/short-term-fellowships
*Sponsor:* Newberry Library
60 West Walton Street
Chicago, IL 60610-3380

**Newberry Library Monticello College Foundation Long-Term Fellowships for Women** **1784**

The Newberry, open to the public without charge, is an independent research library dedicated to the advancement and dissemination of knowledge, especially in the humanities. The library acquires and preserves a broad array of special-collections research materials (spanning six centuries of European and American civilizations) and promotes and provides for these collections' effective use, fostering research, teaching, publication, life-long learning, and civic engagement. Fellowships at the Newberry provide support for researchers wishing to use the Newberry's collections and are either short-term (one to two months) or long-term (four to twelve months). Short-term fellowships are intended to geographically facilitate access for distant scholars to materials essential to a project. Long-term fellowships are intended to support significant works of scholarship that draw on the strengths of the Newberry's collections. Long-term fellowships support individual research and promote serious intellectual exchange through active participation in the Newberry's scholarly activities, including a biweekly fellows' seminar. Applicants may combine long-term fellowship awards with sabbatical or other stipendiary support. The Monticello College Foundation Fellowship for Women is considered a long-term fellowship and is designed primarily for women at early stages in their professional careers whose work gives clear promise of scholarly productivity and whose careers would be significantly enhanced by research, writing, and participation in the intellectual life of the Newberry. The fellowship carries a monthly stipend of $4,200. At the time of this writing, the current application-submission date for most Newberry long-term fellowships is November 15 and awardees are usually notified in mid- to late-March; however deadlines may vary from year to year. Applications must be submitted as email attachments by 11:59 p.m. on the deadline date. Applicants can find guidelines and forms at the grant website. Applicants are also encouraged to verify current deadline dates.
*Requirements:* Applicants' topics should be related to the Newberry's collections; preference will be given to proposals particularly concerned with the study of women. Applicants must have their Ph.D. at the time of application and must request at least four but no more than nine months of support. All fellowships at the Newberry must be in-residence. Applications from former recipients of any fellowship at the Newberry are welcome, but preference is given to applicants who have not held a major fellowship within the past three years. In general, all fellowships must be completed by the end of the academic year for which they are awarded.
*Restrictions:* Applicants may apply for more than one fellowship at the Newberry by designating those for which they wish to be considered on the cover sheet (available at the grant website). Applicants may also apply for both short- and long-term fellowships at the Newberry; however individuals may accept only one fellowship at the Newberry per year. Only scholars who have previously received a short-term fellowship at the Newberry may apply for a long-term fellowship for the same project. No other project extensions are allowable.
*Geographic Focus:* All States
*Date(s) Application is Due:* Nov 15
*Amount of Grant:* 16,800 - 37,800 USD
*Contact:* D. Bradford Hunt, Vice President for Research and Academic Programs; (312) 255-3535 or (312) 943-9090; huntb@newberry.org or research@newberry.org
*Internet:* www.newberry.org/long-term-fellowships
*Sponsor:* Newberry Library
60 West Walton Street
Chicago, IL 60610-3380

**Newberry Library National Endowment for the Humanities Long-Term Fellowships** **1785**

The Newberry, open to the public without charge, is an independent research library dedicated to the advancement and dissemination of knowledge, especially in the humanities. The library acquires and preserves a broad array of special-collections research materials (spanning six centuries of European and American civilizations) and promotes and provides for these collections' effective use, fostering research, teaching, publication, life-long learning, and civic engagement. Fellowships at the Newberry provide support for researchers wishing to use the Newberry's collections and are either short-term (one to two months) or long-term (four to nine months). Short-term fellowships are intended to geographically facilitate access for distant scholars to materials essential to a project. Long-term fellowships are intended to support significant works of scholarship that draw on the strengths of the Newberry's collections. Long-term fellowships support individual research and promote serious intellectual exchange through active participation in the Newberry's scholarly activities, including a biweekly fellows' seminar. Applicants may combine long-term fellowship awards with sabbatical or other stipendiary support. The National Endowment for the Humanities Fellowships are long-term fellowships to support post-doctoral scholars in any field appropriate to the Newberry's collections. The fellowships carries a monthly stipend of $4,200. At the time of this writing, the current application-submission date for most Newberry long-term fellowships is December 12 and awardees are usually notified in mid- to late-March; however deadlines may vary from year to year. Applications must be submitted as email attachments by 11:59 p.m. on the deadline date. Applicants can find guidelines and forms at the grant website. Applicants are also encouraged to verify current deadline dates.
*Requirements:* All long-term fellowships are restricted to postdoctoral scholars. Applicants must be United States citizens or foreign nationals with three years' residence. Applicants must request at least four and no more than nine months of support. All fellowships at the Newberry must be in-residence. Applications from former recipients of any fellowship at the Newberry are welcome, but preference is given to applicants who have not held a major fellowship within the past three years. In general, all fellowships must be completed by the end of the academic year for which they are awarded.

*Restrictions:* Applicants may apply for more than one fellowship at the Newberry by designating those for which they wish to be considered on the cover sheet (available at the grant website). Applicants may also apply for both short- and long-term fellowships at the Newberry; however individuals may accept only one fellowship at the Newberry per year. Only scholars who have previously received a short-term fellowship at the Newberry may apply for a long-term fellowship for the same project. No other project extensions are allowable.
*Geographic Focus:* All States
*Date(s) Application is Due:* Nov 15
*Amount of Grant:* 16,800 - 37,800 USD
*Contact:* D. Bradford Hunt, Vice President for Research and Academic Programs; (312) 255-3535 or (312) 943-9090; huntb@newberry.org or research@newberry.org
*Internet:* www.newberry.org/long-term-fellowships
*Sponsor:* Newberry Library
60 West Walton Street
Chicago, IL 60610-3380

**Newberry Library Northeast Modern Language Assocation Short-Term Fellowship** 1786

The Newberry, open to the public without charge, is an independent research library dedicated to the advancement and dissemination of knowledge, especially in the humanities. The library acquires and preserves a broad array of special-collections research materials (spanning six centuries of European and American civilizations) and promotes and provides for these collections' effective use, fostering research, teaching, publication, life-long learning, and civic engagement. Fellowships at the Newberry provide support for researchers wishing to use the Newberry's collections and are either short-term (one to two months) or long-term (four to twelve months). Short-term fellowships are intended to geographically facilitate access for distant scholars to materials essential to a project. Long-term fellowships are intended to support significant works of scholarship that draw on the strengths of the Newberry's collections. A short-term fellowship is available for Ph.D. candidates or post-doctoral scholars who are members of the Northeast Modern Language Association (NeMLA) for up to one month's work-in-residence at the Newberry. The fellowship provides a stipend of $2,500 per month. Preference will be given to projects focusing on materials written in French, German, Italian, or Spanish. At the time of this writing, the current application submission date for most Newberry short-term fellowships is December 15 and awardees are usually notified near the end of April; however deadlines may vary from year to year. Applications must be submitted as email attachments by 11:59 p.m. on the deadline date. Applicants can find guidelines and forms at the grant website. Applicants are also encouraged to verify current deadline dates.
*Requirements:* Applicants must be members of the NeMLA at the time of the award. Most short-term fellowships at the Newberry are for Ph.D. candidates and postdoctoral scholars. All fellowships at the Newberry must be conducted in-residence. The Newberry collections must be necessary to the short-term fellowship project since the purpose of these fellowships is to enable access to particular materials. Applications from former recipients of any fellowship at the Newberry are welcome, but preference is given to applicants who have not held a major fellowship within the past three years. In general, all fellowships at the Newberry must be completed by the end of the academic year for which they are awarded.
*Restrictions:* Most short-term fellowships are restricted to individuals who live outside the Chicago area who need to examine specific items in the Newberry's collections. Preference is given to residents of the Northeastern United States or Eastern Canada; Applicants may apply for more than one fellowship at the Newberry by designating those for which they wish to be considered on the cover sheet (available at the grant website). Applicants may also apply for both short- and long-term fellowships at the Newberry; however individuals may accept only one fellowship at the Newberry per year. Only scholars who have previously received a short-term fellowship at the Newberry may apply for a long-term fellowship for the same project. No other project extensions are allowable.
*Geographic Focus:* All States, Canada
*Date(s) Application is Due:* Dec 15
*Amount of Grant:* Up to 2,500 USD
*Contact:* D. Bradford Hunt, Vice President for Research and Academic Programs; (312) 255-3535 or (312) 943-9090; huntb@newberry.org or research@newberry.org
*Internet:* www.newberry.org/short-term-fellowships
*Sponsor:* Newberry Library
60 West Walton Street
Chicago, IL 60610-3380

**Newberry Library Rudolph Ganz Long-Term Fellowship** 1787

The Newberry, open to the public without charge, is an independent research library dedicated to the advancement and dissemination of knowledge, especially in the humanities. The library acquires and preserves a broad array of special-collections research materials (spanning six centuries of European and American civilizations) and promotes and provides for these collections' effective use, fostering research, teaching, publication, life-long learning, and civic engagement. Fellowships at the Newberry provide support for researchers wishing to use the Newberry's collections and are either short-term (one to two months) or long-term (four to twelve months). Short-term fellowships are intended to geographically facilitate access for distant scholars to materials essential to a project. Long-term fellowships are intended to support significant works of scholarship that draw on the strengths of the Newberry's collections. Long-term fellowships support individual research and promote serious intellectual exchange through active participation in the Newberry's scholarly activities, including a biweekly fellows' seminar. Applicants may combine long-term fellowship awards with sabbatical or other stipendiary support. The Rudolph Ganz Long-Term Fellowship supports researchers using the Rudolph Ganz Papers and other late nineteenth- and early twentieth-century materials related to Chicago music in that period. Ganz's papers include musical compositions of this world-renowned concert pianist, composer, conductor, and educator, as well as articles, speeches, lectures and essays by him, and two recordings. The collection also includes correspondence to and from prominent musical figures, family correspondence, clippings, photographs, programs, and other artifacts. The fellowship carries a monthly stipend of $4,200. At the time of this writing, the current application-submission date for most Newberry long-term fellowships is November 15 and awardees are usually notified in mid- to late-March; however deadlines may vary from year to year. Applications must be submitted as email attachments by 11:59 p.m. on the deadline date. Applicants can find guidelines and forms at the grant website. Applicants are also encouraged to verify current deadline dates.
*Requirements:* Applicants' topics should be related to the Newberry's collections; preference will be given to proposals particularly concerned with the study of women. Applicants must have their Ph.D. at the time of application and must request at least four but no more than nine months of support. All fellowships at the Newberry must be in-residence. Applications from former recipients of any fellowship at the Newberry are welcome, but preference is given to applicants who have not held a major fellowship within the past three years. In general, all fellowships must be completed by the end of the academic year for which they are awarded.
*Restrictions:* Applicants may apply for more than one fellowship at the Newberry by designating those for which they wish to be considered on the cover sheet (available at the grant website). Applicants may also apply for both short- and long-term fellowships at the Newberry; however individuals may accept only one fellowship at the Newberry per year. Only scholars who have previously received a short-term fellowship at the Newberry may apply for a long-term fellowship for the same project. No other project extensions are allowable.
*Geographic Focus:* All States, Canada
*Date(s) Application is Due:* Nov 15
*Amount of Grant:* 4,200 - 37,800 GBP
*Contact:* D. Bradford Hunt, Vice President for Research and Academic Programs; (312) 255-3535 or (312) 943-9090; huntb@newberry.org or research@newberry.org
*Internet:* www.newberry.org/long-term-fellowships
*Sponsor:* Newberry Library
60 West Walton Street
Chicago, IL 60610-3380

**Newberry Library Scholar-in-Residence Opportunity** 1788

The Newberry, open to the public without charge, is an independent research library dedicated to the advancement and dissemination of knowledge, especially in the humanities. The library acquires and preserves a broad array of special-collections research materials (spanning six centuries of European and American civilizations) and promotes and provides for these collections' effective use, fostering research, teaching, publication, life-long learning, and civic engagement. The Scholar-in-Residence program at the Newberry encourages postdoctoral scholars in the humanities to conduct research in the library's collections and to join its community of scholars. Selected scholars are provided with reserve space, access to the Newberry during extended hours, and opportunities to present their work-in-progress to the Newberry's scholarly community. Applications to become a Scholar-in-Residence are accepted each year until May 1. Candidates must submit via email a project description along with a filled-in application form (available for download from the website) by 11:59 p.m. on the deadline date. Applicants will be notified of their acceptance in June and will begin their residence in September and will join the Newberry's new long-term fellows in fall-orientation activities. Candidates can find more detailed guidelines and information at the grant website.
*Requirements:* Applicants must be postdoctoral scholars in the humanities.
*Geographic Focus:* Illinois
*Date(s) Application is Due:* May 1
*Contact:* D. Bradford Hunt, Executive Vice President for Research and Academic Programs; (312) 255-3535 or (312) 943-9090; huntb@newberry.org or research@newberry.org
*Internet:* www.newberry.org/scholar-residence-program
*Sponsor:* Newberry Library
60 West Walton Street
Chicago, IL 60610-3380

**Newberry Library Short-Term Resident Fellowships for Individual Research** 1789

The Newberry, open to the public without charge, is an independent research library dedicated to the advancement and dissemination of knowledge, especially in the humanities. The library acquires and preserves a broad array of special-collections research materials (spanning six centuries of European and American civilizations) and promotes and provides for these collections' effective use, fostering research, teaching, publication, life-long learning, and civic engagement. Fellowships at the Newberry provide support for researchers wishing to use the Newberry's collections and are either short-term (one to two months) or long-term (four to twelve months). Short-term fellowships are intended to geographically facilitate access for distant scholars to materials essential to a project. Long-term fellowships are intended to support significant works of scholarship that draw on the strengths of the Newberry's collections. Newberry's Short-Term Resident Fellowships provide access to the Newberry's collections for Ph.D. candidates or post-doctoral scholars who live and work outside the Chicago area. Normally, these fellowships are awarded to individual scholars. However, the Newberry will award a small number of fellowships for collaborative projects. The fellowships provide a stipend of $2,500 per month (for each scholar). At the time of this writing, the current application submission date for most Newberry short-term fellowships is December 15 and awardees are usually notified near the end of April; however deadlines may vary from year to year. Applications must be submitted as email attachments by 11:59 p.m. on the deadline date. Applicants can find guidelines and forms at the grant website. Applicants are also encouraged to verify current deadline dates.

*Requirements:* Ph.D. candidates and post-doctoral scholars who live and work outside the Chicago area are eligible to apply. Teams of two or three scholars who plan to collaborate on a single, substantive project are also eligible to apply. All fellowships at the Newberry must be conducted in-residence. The Newberry collections must be necessary to the short-term fellowship project since the purpose of these fellowships is to enable access to particular materials. Applications from former recipients of any fellowship at the Newberry are welcome, but preference is given to applicants who have not held a major fellowship within the past three years. In general, all fellowships at the Newberry must be completed by the end of the academic year for which they are awarded.
*Restrictions:* Most short-term fellowships are restricted to individuals who live outside the Chicago area who need to examine specific items in the Newberry's collections. Applicants may apply for more than one fellowship at the Newberry by designating those for which they wish to be considered on the cover sheet (available at the grant website). Applicants may also apply for both short- and long-term fellowships at the Newberry; however individuals may accept only one fellowship at the Newberry per year. Only scholars who have previously received a short-term fellowship at the Newberry may apply for a long-term fellowship for the same project. No other project extensions are allowable.
*Geographic Focus:* All States, All Countries
*Date(s) Application is Due:* Dec 15
*Amount of Grant:* 2,500 - 7,500 USD
*Contact:* D. Bradford Hunt, Vice President for Research and Academic Programs; (312) 255-3535 or (312) 943-9090; huntb@newberry.org or research@newberry.org
*Internet:* www.newberry.org/short-term-fellowships
*Sponsor:* Newberry Library
60 West Walton Street
Chicago, IL 60610-3380

**Newberry Library Susan Kelly Power and Helen Hornbeck Tanner Short-Term Fellowship** 1790

The Newberry, open to the public without charge, is an independent research library dedicated to the advancement and dissemination of knowledge, especially in the humanities. The library acquires and preserves a broad array of special-collections research materials (spanning six centuries of European and American civilizations) and promotes and provides for these collections' effective use, fostering research, teaching, publication, life-long learning, and civic engagement. Fellowships at the Newberry provide support for researchers wishing to use the Newberry's collections and are either short-term (one to two months) or long-term (four to twelve months). Short-term fellowships are intended to geographically facilitate access for distant scholars to materials essential to a project. Long-term fellowships are intended to support significant works of scholarship that draw on the strengths of the Newberry's collections. The Susan Kelly Power and Helen Hornbeck Tanner Fellowship is a short-term fellowship for Ph.D. candidates and post-doctoral scholars of American Indian heritage. The fellowship supports up to two months of residential research in any field in the humanities using the collections of the Newberry at a stipend of $2,500 per month. At the time of this writing, the current application submission date for most Newberry short-term fellowships is December 15 and awardees are usually notified near the end of April; however deadlines may vary from year to year. Applications must be submitted as email attachments by 11:59 p.m. on the deadline date. Applicants can find guidelines and forms at the grant website. Applicants are also encouraged to verify current deadline dates.
*Requirements:* Most short-term fellowships at the Newberry are for Ph.D. candidates and postdoctoral scholars. All fellowships at the Newberry must be conducted in-residence. The Newberry collections must be necessary to the short-term fellowship project since the purpose of these fellowships is to enable access to particular materials. Applications from former recipients of any fellowship at the Newberry are welcome, but preference is given to applicants who have not held a major fellowship within the past three years. In general, all fellowships at the Newberry must be completed by the end of the academic year for which they are awarded.
*Restrictions:* While most short-term fellowships at the Newberry are restricted to individuals who live outside the Chicago area and who need to examine specific items in the Newberry's collections, applicants for the Susan Kelly Power and Helen Hornbeck Tanner Fellowship may be local. Applicants may apply for more than one fellowship at the Newberry by designating those for which they wish to be considered on the cover sheet (available at the grant website). Applicants may also apply for both short- and long-term fellowships at the Newberry; however individuals may accept only one fellowship at the Newberry per year. Only scholars who have previously received a short-term fellowship at the Newberry may apply for a long-term fellowship for the same project. No other project extensions are allowable.
*Geographic Focus:* All States, Canada
*Date(s) Application is Due:* Dec 15
*Amount of Grant:* 2,500 - 5,000 USD
*Contact:* D. Bradford Hunt, Vice President for Research and Academic Programs; (312) 255-3535 or (312) 943-9090; huntb@newberry.org or research@newberry.org
*Internet:* www.newberry.org/short-term-fellowships
*Sponsor:* Newberry Library
60 West Walton Street
Chicago, IL 60610-3380

**Newberry Library Undergraduate Seminar (NLUS) Faculty Long-Term Fellowship** 1791

The Newberry, open to the public without charge, is an independent research library dedicated to the advancement and dissemination of knowledge, especially in the humanities. The library acquires and preserves a broad array of special-collections research materials (spanning six centuries of European and American civilizations) and promotes and provides for these collections' effective use, fostering research, teaching, publication, life-long learning, and civic engagement. Fellowships at the Newberry provide support for researchers wishing to use the Newberry's collections and are either short-term (one to two months) or long-term (four to nine months). Short-term fellowships are intended to geographically facilitate access for distant scholars to materials essential to a project. Long-term fellowships are intended to support significant works of scholarship that draw on the strengths of the Newberry's collections. The Undergraduate Seminar (NLUS) Faculty Long-Term Fellowships support two professors from any of four participating schools (DePaul University, Loyola University Chicago, Roosevelt University, University of Illinois at Chicago) who are selected to teach a semester-long undergraduate seminar at the Newberry. The fellowship carries a monthly stipend of $4,200. At the time of this writing, the current application-submission date for most Newberry long-term fellowships is November 15 and awardees are usually notified in mid- to late-March; however deadlines may vary from year to year. Applications must be submitted as email attachments by 11:59 p.m. on the deadline date. Applicants can find guidelines and forms at the grant website. Applicants are also encouraged to verify current deadline dates.
*Requirements:* Applicants' topics should be related to the Newberry's collections; preference will be given to proposals particularly concerned with the study of women. Applicants must have their Ph.D. at the time of application and must request at least four but no more than nine months of support. All fellowships at the Newberry must be in-residence. Applications from former recipients of any fellowship at the Newberry are welcome, but preference is given to applicants who have not held a major fellowship within the past three years. In general, all fellowships must be completed by the end of the academic year for which they are awarded.
*Restrictions:* Applicants may apply for more than one fellowship at the Newberry by designating those for which they wish to be considered on the cover sheet (available at the grant website). Applicants may also apply for both short- and long-term fellowships at the Newberry; however individuals may accept only one fellowship at the Newberry per year. Only scholars who have previously received a short-term fellowship at the Newberry may apply for a long-term fellowship for the same project. No other project extensions are allowable.
*Geographic Focus:* All States
*Date(s) Application is Due:* Nov 15
*Amount of Grant:* 4,200 - 37,800 USD
*Contact:* D. Bradford Hunt, Vice President for Research and Academic Programs; (312) 255-3535 or (312) 943-9090; huntb@newberry.org or research@newberry.org
*Internet:* www.newberry.org/long-term-fellowships
*Sponsor:* Newberry Library
60 West Walton Street
Chicago, IL 60610-3380

**Newberry Library Weiss/Brown Publication Subvention Award** 1792

The Newberry, open to the public without charge, is an independent research library dedicated to the advancement and dissemination of knowledge, especially in the humanities. The library acquires and preserves a broad array of special-collections research materials (spanning six centuries of European and American civilizations) and promotes and provides for these collections' effective use, fostering research, teaching, publication, life-long learning, and civic engagement. With support from the Roger W. Weiss and Howard Mayer Brown Fund, the Newberry will award up to $15,000 to subsidize the publication of a scholarly book, monograph, or edition on European civilization before 1700 in one of the following fields: music, theater, French or Italian literature, or cultural studies. This award aspires to enable the publication of works of the highest quality either by publishing them in a particularly appropriate way (with special typography, plates, or appendices that would otherwise be prohibitively expensive, for example) or by significantly reducing the cover price, thereby allowing the publication to reach a wider audience. Once these criteria are met, preference will be given to publications that are unique, that are unusual in concept or execution, that bring into print previously unpublished source material, that represent a departure from the normal habits of a given publishing house or entity, or that promise to reach the broadest possible audience for the type of book envisioned. A link to application guidelines is available at the grant website. Applications are due via email by the deadline date. Awardees will be notified six to eight weeks after the deadline. Deadlines may vary from year to year; applicants are encouraged to verify current deadline dates.
*Requirements:* Applications for the Weiss/Brown Award will require the following documents: the author's curriculum vitae; a copy of the publication contract or other evidence of commitment by the publisher to publish the book; copies of the reader's reports on the manuscript, or equivalent outside evaluations; and a production budget and schedule. The award will be paid to the publisher; as a condition of accepting funds, publishers must include an acknowledgement of the award in the publication and provide the Newberry Library with a gratis copy of the book.
*Restrictions:* Faxed applications will not be accepted.
*Geographic Focus:* All States, All Countries
*Date(s) Application is Due:* Nov 15
*Amount of Grant:* Up to 15,000 USD
*Contact:* D. Bradford Hunt, Vice President for Research and Academic Programs; (312) 255-3535 or (312) 943-9090; huntb@newberry.org or research@newberry.org
*Internet:* www.newberry.org/short-term-fellowships
*Sponsor:* Newberry Library
60 West Walton Street
Chicago, IL 60610-3380

**NFL Charities Medical Grants** 1793

NFL Charities supports medical research and enhancing scientific knowledge in order to benefit all who are actively involved in competitive sports and recreational athletic activities. These grants support research that addresses some of the risk factors that exist

for all athletes and citizens with active lifestyles. The NFL Charities Board considers Medical Grant proposals that place emphasis on scientific merit, clinical relevance and significance to the NFL, especially projects concerning concussion and traumatic brain injury, cardiovascular research and MRSA. Grants will be available in the end of June or early July and will be open for approximately 8 weeks.
*Requirements:* Funding is available for non-profit educational and research institutions within the United States. Applications must be submitted online via the NFL Charities Grant Application Management System (G.A.M.S.) website. Every field in the application must be populated with requested information. Indirect costs should not be included in the project's total budget, and only direct research costs are funded by the grant. Salaries for Principal Investigators (PI's) and Co-Principal Investigators (Co-PI's) (Ph.D's and M.D.'s) must follow NIH guidelines. PI's and Co-PI's must disclose all previous and/or current grant support and any potential overlap must be submitted. Postdoctoral fellows serving as PI or Co-PI must include assurances that they will remain at the same institution for the duration of the grant period; however, PI's and co-PI's may work at different institutions, but the grant must be submitted as associated with the institution of the primary PI.
*Restrictions:* For-profit groups are not eligible for funding. Only submission per PI or Co-PI will be funded, and only one submission per institution can be considered for NFL Charities Medical Grant funding. Prospective applicants first should be vetted through their institutional administrative offices in order to ensure that an institution only endorses and submits its top two proposals. Administrative contacts at the associated institution must sign off on the Compliance Certificate that accompanies each application(s) from a PI's institution, thereby ensuring that such vetting has occurred.
*Geographic Focus:* All States
*Date(s) Application is Due:* Jun 1
*Amount of Grant:* Up to 100,000 USD
*Contact:* Clare Graff, (212) 450-2435 or (917) 816-2885; clare.graff@nfl.com
*Internet:* www.nflcharities.org/grants/medical
*Sponsor:* NFL Charities
280 Park Avenue, 17th Floor
New York, NY 10017

**NFWF Acres for America Grants** **1794**
Acres for America is a partnership between Walmart Stores, Inc. and the National Fish and Wildlife Foundation. The Acres for America program was established to provide funding for projects that conserve important habitat for fish, wildlife, and plants through acquisition of interest in real property. The goal of the Acres for America program is to offset the footprint of Walmart's domestic facilities on at least an acre by acre basis through these acquisitions. Applicants are strongly urged to contact the National Fish and Wildlife Foundation regional director in their area to discuss project ideas prior to submitting pre-proposals. All grant awards require a minimum 1:1 match of cash or contributed goods and services. Federal funds may be considered as match. Higher ratios of matching funds will at times aid in making applications more competitive. All applications to the Acres for America Program will be made through a two-stage process, which includes a pre- and then a full-proposal. Only electronic pre-proposals submitted through the on-line pre-proposal form will be considered. Following a review period of about four weeks, appropriate pre-proposals will be selected and applicants will be invited to submit full proposals.
*Requirements:* Nonprofit conservation organizations, counties and cities, state government agencies, tribes, federal agencies, and schools are eligible.
*Restrictions:* The Foundation does not fund political advocacy or litigation of any kind, shortfalls in government agency budgets, general administrative overhead, or indirect costs.
*Geographic Focus:* All States
*Date(s) Application is Due:* Aug 1
*Amount of Grant:* 200,000 USD
*Contact:* Thomas E. Kelsch, Director; (503) 417-8700, ext. 6008; fax (503) 417-8787; Tom.Kelsch@nfwf.org
*Internet:* www.nfwf.org/AM/Template.cfm?Section=Charter_Programs_List&Template=/TaggedPage/TaggedPageDisplay.cfm&TPLID=60&ContentID=17798
*Sponsor:* National Fish and Wildlife Foundation #6
1133 15th Street, NW, Suite 1100
Washington, D.C. 97205

**NFWF Bird Conservation Initiative Grants** **1795**
The National Fish and Wildlife Foundation and its partners have keystone initiatives which are the core portfolio of multi-year initiatives through which they seek to achieve measurable outcomes. The Bird Conservation Initiative focuses on migratory and resident species and habitats that occur in the United States or its territories and that have been identified as high priorities for the nation. The goal is to find the best conservation investments, fund the best solutions, and deliver measurable results for bird conservation. Under this mission statement the Foundation has selected initiatives that focus on making a measurable impact on the status of specific species and their habitats. Individual initiatives are described on the website. Application is made online from the website.
*Requirements:* Applicants should review the business plans for this initiative and prepare application materials accordingly. Prospective applicants should contact the Director to discuss project ideas prior to submitting proposals to help ensure the relevance of funding requests. The Foundation will consider multi-year funding for important projects, subject to annual performance review of the grantee. A minimum 1:1 match is required for most Foundation grants, and higher ratios of match to Foundation funds are desired and improve the competitiveness of applications.
*Geographic Focus:* All States
*Date(s) Application is Due:* Aug 1; Dec 1
*Amount of Grant:* 50,000 - 375,000 USD
*Samples:* Society of Tympanuchus Cupido Pinnatus, Ltd., Franklin, Wisconsin, $364,350, evaluate the effects of red imported fire ants on abundance of insect prey for Attwater's Prairie-Chickens; Playa Lakes Joint Venture, Lafayette, Colorado, $249,466, continue the restoration of Lesser Prairie-Chicken habitat, connecting already protected areas and unfragmenting grassland habitat in southeastern New Mexico.
*Contact:* Daniel Petit, Director, Bird Conservation; (202) 857-0166; fax (202) 857-0162; daniel.petit@nfwf.org
*Internet:* www.nfwf.org/am/template.cfm?section=Bird_
*Sponsor:* National Fish and Wildlife Foundation #6
1133 15th Street, NW, Suite 1100
Washington, D.C. 97205

**NFWF Freshwater Fish Conservation Initiative Grants** **1796**
The National Fish and Wildlife Foundation and its partners have keystone initiatives which are the core portfolio of multi-year initiatives through which they seek to achieve measurable outcomes. The Freshwater Fish Conservation Initiative focuses on species and habitats species that occur in the United States or its territories and that have been identified as high priorities for the nation. Freshwater fish (anadromous and catadromous) and associated aquatic life such as mussels, crayfish, and other invertebrates are included. The goal is to find the best conservation investments, fund the best solutions, and deliver measurable results for fish conservation. Under this mission statement there are selected initiatives that focus on making a measurable impact on the status of specific species and their habitat. Individual initiatives are listed on the website and are focused on conserving freshwater fish, aquatic organisms, and their habitats. Application is made online from the website.
*Requirements:* Applicants should review the business plans for this initiative and prepare application materials accordingly. Prospective applicants should contact the Director to discuss project ideas prior to submitting proposals to help ensure the relevance of funding requests. The Foundation will consider multi-year funding for important projects, subject to annual performance review of the grantee. A minimum 1:1 match is required for most Foundation grants, and higher ratios of match to Foundation funds are desired and improve the competitiveness of applications.
*Geographic Focus:* All States
*Date(s) Application is Due:* Aug 1; Dec 1
*Amount of Grant:* 50,000 - 375,000 USD
*Samples:* Freshwater Land Trust, Birmingham, Alabama, $14,980, conduct riparian wetland habitat restoration of abandoned residential property; Ecosystem Economics, LLC, Bend, Oregon, $66,924, develop the rationale and integrated framework to design and implement market-oriented strategies and capacities for achieving freshwater conservation and sustainable water management outcomes in Arizona and the Lower Colorado River.
*Contact:* Jim Sedell, Director, Fish Conservation; (503) 417-8700; fax (503) 417-8787
*Internet:* www.nfwf.org/AM/Template.cfm?Section=Fish_Conservation2
*Sponsor:* National Fish and Wildlife Foundation #6
1133 15th Street, NW, Suite 1100
Washington, D.C. 97205

**NFWF Marine & Coastal Conservation Initiative Grants** **1797**
The National Fish and Wildlife Foundation and its partners have keystone initiatives which are the core portfolio of multi-year initiatives through which they seek to achieve measurable outcomes. The Marine and Coastal Conservation Initiative focuses on building conservation partnerships to overcome the most important challenges to the health of the marine and coastal environment. The goal is to find the best conservation investments, fund the best solutions, and deliver measurable results for a broad cross-section of fish and wildlife representing marine and coastal ecosystem health. Under this mission statement, the Foundation has selected initiatives that are focused on making a measureable impact on specific species and their habitat. Individual initiatives are listed on the website but are focused under three themes that are critical for marine and coastal wildlife in America. The themes are promoting sustainable fishers, protecting marine natural heritage, and climate change and adaptation. Application is made online from the website.
*Requirements:* Applicants should review the business plans for this initiative and prepare application materials accordingly. Prospective applicants should contact the Director to discuss project ideas prior to submitting proposals to help ensure the relevance of funding requests. The Foundation will consider multi-year funding for important projects, subject to annual performance review of the grantee. A minimum 1:1 match is required for most Foundation grants, and higher ratios of match to Foundation funds are desired and improve the competitiveness of applications.
*Geographic Focus:* All States
*Date(s) Application is Due:* Aug 1; Dec 1
*Amount of Grant:* 50,000 - 375,000 USD
*Samples:* Seward Association for the Advancement of Marine Science, Seward, Arkansas, $69,263, monitor historic and newly established walrus terrestrial haulout sites to record walrus attendance, duration, and level of disturbance and provide baseline site specific animal counts; California Department of Parks and Recreation, Westport, California, $20,325, design and install interpretive panels and regulatory signage for newly established Marine Protected Areas offshore from California State.
*Contact:* Anthony C. Chatwin, Director, Marine and Coastal Conservation; (202) 857-0166; fax (202) 857-0162; anthony.chatwin@nfwf.org
*Internet:* www.nfwf.org/am/template.cfm?section=Marine_and_Coastal
*Sponsor:* National Fish and Wildlife Foundation #6
1133 15th Street, NW, Suite 1100
Washington, D.C. 97205

**NFWF Wildlife and Habitat Conservation Initiative Grants** 1798

The National Fish and Wildlife Foundation and its partners have keystone initiatives which are the core portfolio of multi-year initiatives through which they seek to achieve measurable outcomes. The Wildlife and Habitat Conservation Initiative addresses conservation needs for mammals, reptiles, amphibians, invertebrates, and plants, as well as landscape-level and issue-based conservation. The goal is to find the best conservation investments, fund the best solutions, and deliver measurable results for wildlife. Under this mission statement, there are selected initiatives focused on making a measureable impact on specific species and their habitats. Individual initiatives are listed on the website and are focused under three themes which are imperiled wildlife species, wildlife movement and migration, and climate change and adaptation. Application is made online from the website.

*Requirements:* Applicants should review the business plans for this initiative and prepare application materials accordingly. Prospective applicants should contact the Director to discuss project ideas prior to submitting proposals to help ensure the relevance of funding requests. The Foundation will consider multi-year funding for important projects, subject to annual performance review of the grantee. A minimum 1:1 match is required for most Foundation grants, and higher ratios of match to Foundation funds are desired and improve the competitiveness of applications.

*Geographic Focus:* All States

*Date(s) Application is Due:* Aug 1; Dec 1

*Amount of Grant:* 50,000 - 375,000 USD

*Contact:* Carly Vynne; (202) 857-0166; fax (202) 857-0162

*Internet:* www.nfwf.org/am/template.cfm?section=Wildlife_and_Habitat12

*Sponsor:* National Fish and Wildlife Foundation #6
1133 15th Street, NW, Suite 1100
Washington, D.C. 97205

**NHLBI Ancillary Studies in Clinical Trials** 1799

This funding opportunity invites research grant applications to conduct time-sensitive ancillary studies related to heart, lung, and blood diseases and sleep disorders in conjunction with ongoing clinical trials and other large clinical studies supported by NIH or non-NIH entities. The program establishes an accelerated review/award process to support the crucial time frame in which these ancillary studies must be performed. Time-sensitive ancillary studies include those that require active longitudinal data collection and thus need to begin recruiting subjects as close as possible to the start of the parent study. The ancillary study can address any research questions related to the mission of NHLBI for which the parent study can provide participants, infrastructure, and data. The parent studies most often will be a clinical trial, but also can be an observational study or registry that can provide a sufficient cohort of well-characterized patients. Each ancillary study application must demonstrate the time-sensitive nature of the application and must explicitly address why an expedited review is essential to its feasibility. Applications are due May 24, September 24, and January 24 by 5:00 pm local time of applicant.

*Requirements:* The following may apply: public/state controlled and private institutions of higher education; nonprofits with or without 501(c)3 IRS status (other than institutions of higher education); small businesses; for-profit organizations (other than small businesses); state, county, city or township governments; special district governments; Indian/Native American Tribal governments; eligible agencies of the Federal government; U.S. territory or possession; independent school districts; public housing authorities/Indian housing authorities; faith-based or community-based organizations; regional organizations; non-domestic (non-U.S.) entities (Foreign Institutions). The following types of Higher Education Institutions are always encouraged to apply for NIH support as public or private institutions of higher education: Hispanic-serving institutions; historically Black colleges and universities; Tribally controlled colleges and universities; Alaska Native and Native Hawaiian serving institutions. Non-domestic (non-U.S.) components of U.S. Organizations are eligible to apply. Any individual(s) with the skills, knowledge, and resources necessary to carry out the proposed research as the Program Director/Principal Investigator (PD/PI) is invited to work with his/her organization to develop an application for support. Individuals from underrepresented racial and ethnic groups as well as individuals with disabilities are always encouraged to apply for NIH support. An application may request a budget for direct costs up to $250,000 (10 modules) per year, excluding subcontractor or consortium facilities and administrative (F&A) costs. The scope of the proposed project should determine the project period. The maximum project period is four years.

*Geographic Focus* All States, Guam, Marshall Islands, Northern Mariana Islands, Puerto Rico, U.S. Virgin Islands, All Countries, American Samoa

*Date(s) Application is Due:* Jan 24; May 24; Sep 24

*Contact:* Suzanne Goldberg, R.N., M.S.N., (301) 435-0532; fax (301) 480-3667; goldbergsh@mail.nih.gov

*Internet:* grants.nih.gov/grants/guide/rfa-files/RFA-HL-14-004.html

*Sponsor:* National Institutes of Health
Building 31, Room 5A48
Bethesda, MD 20892

**NHLBI Bioengineering and Obesity Grants** 1800

This funding opportunity solicits Research Project Grant (R01) applications from institutions/organizations that propose to solicit applications to develop and validate new and innovative engineering approaches to address clinical problems related to energy balance, intake, and expenditure. Novel sensors, devices, imaging, and other technologies, including technologies to detect biochemical markers of energy balance are expected to be developed and evaluated by collaborating engineers, physical scientists, mathematicians, and scientists from other relevant disciplines with expertise in obesity and nutrition. Because the nature and scope of the proposed research will vary from application to application, it is anticipated that the size and duration of each award will also vary. The total amount awarded and the number of awards will depend upon the mechanism numbers, quality, duration, and costs of the applications received.

*Requirements:* The following are eligible to apply: public/state controlled and private institutions of higher education; nonprofit with or without 501(c)3 IRS status (other than institution of higher education); small business; for-profit organization (other than small business); state government; U.S. Territory or Possession; Indian/Native American tribal government (federally recognized); Indian/Native American tribal government (other than federally recognized); Indian/Native American tribally designated organization; Non-domestic (non-U.S.) entity (foreign organization); Hispanic-serving institution; historically Black colleges and universities (HBCUs); tribally controlled colleges and universities (TCCUs); Alaska Native and Native Hawaiian serving institutions; regional organizations. Individuals with the skills, knowledge, and resources necessary to carry out the proposed research are invited to work with their institution/organization to develop an application for support. Individuals from underrepresented racial and ethnic groups as well as individuals with disabilities are always encouraged to apply for NIH support. More than one PD/PI, or multiple PDs/PIs, may be designated on the application. Applicants may submit more than one application, provided each application is scientifically distinct.

*Geographic Focus* All States, Guam, Marshall Islands, Northern Mariana Islands, Puerto Rico, U.S. Virgin Islands, American Samoa

*Date(s) Application is Due:* Jan 25

*Contact:* Abby G. Ershow., (301) 435-0550; fax (301) 480-2858; ErshowA@mail.nih.gov
Tim Baldwin, Ph.D., (301) 435-0513; fax (301) 480-7971; baldwint@nhlbi.nih.gov
Sharon Ross, Ph.D., M.P.H., (301) 594-7547; fax (301) 480-3925; rosssha@mail.nih.gov

*Internet:* grants.nih.gov/grants/guide/pa-files/PA-07-354.html

*Sponsor:* National Institutes of Health
Building 31, Room 5A48
Bethesda, MD 20892

**NHLBI Bioengineering Approaches to Energy Balance and Obesity Grants for SBIR** 1801

This funding opportunity will develop and validate new and innovative bioengineering technology to address clinical problems related to energy balance, intake, and expenditure. Novel sensors, devices, imaging, and other approaches are expected to be developed and evaluated by collaborating engineers, physical scientists, and scientists from other relevant disciplines with expertise in obesity and nutrition. The goal is to increase the number of useful technologies and tools available to scientists to facilitate their research in energy balance and health. Eventually these research tools should facilitate therapeutic advances and behavioral changes to address such problems as weight control and obesity. The total amount awarded and the number of awards will depend upon the mechanism numbers, quality, duration, and costs of the applications received. This FOA will utilize the SBIR (R43/R44) grant mechanisms for Phase I, Phase II, and Fast-Track applications and runs in parallel with a FOA of identical scientific scope, PA-07-436, that solicits applications under the Small Business Technology Transfer (STTR) (R41/R42) grant mechanisms. The total amount awarded and the number of awards will depend upon the mechanism numbers, quality, duration, and costs of the applications received. Program Announcement (PA) Number: PA-07-435.

*Requirements:* Only United States small business concerns (SBCs) are eligible to submit SBIR applications. A small business concern is one that, at the time of award for both Phase I and Phase II SBIR awards, meets all of the following criteria: (1) is independently owned and operated, is not dominant in the field of operation in which it is proposing, has a place of business in the United States and operates primarily within the United States or makes a significant contribution to the U.S. economy, and is organized for profit; (2) is (a) at least 51% owned and controlled by one or more individuals who are citizens of, or permanent resident aliens in, the United States, or (b) for SBIR only, it must be a for-profit business concern that is at least 51% owned and controlled by another for-profit business concern that is at least 51% owned and controlled by one or more individuals who are citizens of, or permanent resident aliens in, the United States; (3) has, including its affiliates, an average number of employees for the preceding 12 months not exceeding. Any individual with the skills, knowledge, and resources necessary to carry out the proposed research is invited to work with his/her organization to develop an application for support. Individuals from underrepresented racial and ethnic groups as well as individuals with disabilities are always encouraged to apply for NIH support.

*Geographic Focus:* All States

*Date(s) Application is Due:* Apr 5; Aug 5; Dec 5

*Contact:* J. Timothy Baldwin, Ph.D., Biomedical Engineer; (301) 435-0513; fax (301) 480-1335; BaldwinT@mail.nih.gov
Abby G. Ershow, Sc.D., (301) 435-0550; fax (301) 480-1338; ErshowA@mail.nih.gov

*Internet:* grants.nih.gov/grants/guide/pa-files/PA-07-435.html

*Sponsor:* National Institutes of Health
Building 31, Room 5A48
Bethesda, MD 20892

**NHLBI Bioengineering Approaches to Energy Balance and Obesity Grants for STTR** 1802

This funding opportunity will develop and validate new and innovative bioengineering technology to address clinical problems related to energy balance, intake, and expenditure. Novel sensors, devices, imaging, and other approaches are expected to be developed and evaluated by collaborating engineers, physical scientists, and scientists from other relevant disciplines with expertise in obesity and nutrition. The goal is to increase the number of useful technologies and tools available to scientists to facilitate their research in energy balance and health. Eventually these research tools should facilitate therapeutic advances and behavioral changes to address such problems as weight

control and obesity. The total amount awarded and the number of awards will depend upon the mechanism numbers, quality, duration, and costs of the applications received. Program Announcement (PA) Number: PA-07-436.
*Requirements:* Only United States small business concerns (SBCs) are eligible to submit STTR applications. Individuals with the skills, knowledge, and resources necessary to carry out the proposed research are invited to work with their organization to develop an application for support. Individuals from underrepresented racial and ethnic groups as well as individuals with disabilities are always encouraged to apply for NIH support. On an STTR application, the PD/PI may be employed with the SBC or the participating non-profit research institution as long as he/she has a formal appointment with or commitment to the applicant SBC, which is characterized by an official relationship between the small business concern and that individual. Applicant SBCs may submit more than one application, provided each application is scientifically distinct. More than one PD/PI, or multiple PDs/PIs, may be designated on the application.
*Geographic Focus:* All States
*Date(s) Application is Due:* Apr 5; Aug 5; Dec 5
*Contact:* Abby G. Ershow., (301) 435-0550; fax (301) 480-1338; ErshowA@mail.nih.gov
J. Timothy Baldwin, Ph.D., Biomedical Engineer; (301) 435-0513; fax (301) 480-1335; BaldwinT@mail.nih.gov
*Internet:* grants.nih.gov/grants/guide/pa-files/PA-07-436.html
*Sponsor:* National Institutes of Health
Building 31, Room 5A48
Bethesda, MD 20892

**NHLBI Biomedical Research Training Program for Individuals from Underrepresented Groups** 1803
The NHLBI established the Biomedical Research Training Program for Individuals from Underrepresented Groups (BRTPUG) to offer opportunities for underrepresented post-baccalaureate individuals to receive training in basic, translational, and clinical research. The purpose of the program is to enhance career opportunities in biomedical sciences, including clinical and laboratory medicine, epidemiology, and biostatistics as applied to the etiology and treatment of heart, blood vessel, lung, and blood diseases. BRTPUG offers each participant the opportunity to work closely with leading research scientists in the Division of Intramural Research and extramural scientists in the Division of Cardiovascular Sciences -Prevention and Population Sciences Program. The program provides participants with hands-on training in a research environment, which will prepare them to continue their studies and advance their careers in clinical and basic research. The trainee's research internship is a one-time appointment of 1-2 year's beginning the summer of selection. Each trainee is assigned to a mentor who is responsible for designing a carefully planned training program.
*Requirements:* The program supports recently completed post-baccalaureates, who have completed academic training in course work relevant to biomedical, behavioral, or statistical research. Applicants must have a cumulative grade point average (GPA) or science course GPA of 3.3 or better on a 4.0 scale, or 4.3 or better on a 5.0 scale, and be a U.S. citizen or permanent resident. Underrepresented post-baccalaureate individuals planning to pursue advanced degrees in the biomedical sciences, who are planning to apply to graduate or professional (medical/dental/veterinary, pharmacy) school are eligible to apply.
*Geographic Focus:* All States
*Date(s) Application is Due:* Jan 31
*Contact:* Dr. Helena O. Mishoe, (301) 451-5081; mishoeh@nhlbi.nih.gov
*Internet:* www.nhlbi.nih.gov/funding/training/redbook/pbbrtu.htm
*Sponsor:* National Institutes of Health
Building 31, Room 5A48
Bethesda, MD 20892

**NHLBI Career Transition Awards** 1804
The purpose of the NHLBI Career Transition Award (K22) program is to provide highly qualified postdoctoral fellows with an opportunity to receive mentored research experience in the NHLBI Division of Intramural Research and then to provide them with funding to facilitate the transition of their research programs as new investigators to extramural institutions. To achieve these objectives, the NHLBI Career Transition Award will support two phases of research: a mentored intramural phase (two years) and an extramural phase (three years), for a total of five years of combined support. Transition from the intramural phase of support to the extramural phase is not automatic. Approval of the transition will be based on the success of the awardees research program as determined by an NHLBI progress review, which will include an evaluation of a research plan to be carried out at the extramural institution. The total period of support is five years (two years intramural, plus three years extramural). Awards are not renewable. Applications are due by 5:00 pm local time of the applicant organization (see dates below).
*Requirements:* The following organizations are eligible to apply: public/state controlled and private institutions of higher education; Hispanic-serving institutions; historically black colleges and universities; tribally controlled colleges and universities; Alaska Native and Native Hawaiian serving institutions; nonprofits with or without 501(c)3 IRS status (other than institutions of higher education); small businesses; for-profit organizations (other than small businesses); state, county, and city/township governments; special district governments; Indian/Native American tribal governments (federally recognized and other than federally recognized); eligible agencies of the Federal government; U.S. territory or possession; independent school districts; public or Indiana housing authorities; Native American tribal organizations (other than federally recognized tribal governments); faith-based or community-based organizations; and regional organizations. Applicant organizations may submit more than one application, provided that each application is scientifically distinct. By the time of award, the individual must be a citizen or a non-citizen national of the United States or have been lawfully admitted for permanent residence. Candidates for this award must have earned a terminal clinical or research doctorate (including PhD, MD, DO, D.C., ND, DDS, DVM, ScD, DNS, PharmD., or equivalent doctoral degree), or a combined clinical and research doctoral degree. The candidate must have postdoctoral research experience, during which the potential for highly productive basic or clinical research was demonstrated. Individuals from underrepresented racial and ethnic groups as well as individuals with disabilities are always encouraged to apply for NIH support. During the intramural phase of the award, the candidate will spend full time on research. The required research experience must be completed in an intramural NIH laboratory. To obtain support for the extramural phase, at the time of the award candidates must have a full time formal tenure-track (or equivalent) appointment offer at the academic institution that is the applicant institution. Candidates who have VA appointments may not consider part of the VA effort toward satisfying the full time requirement at the applicant institution. Candidates with VA appointments should contact the NHLBI staff prior to preparing an application to discuss their eligibility. Support for the extramural phase will be provided to the extramural institution. Total direct costs for the extramural phase cannot exceed $249,000, including fringe benefits, per year. The total costs cannot exceed $747,000 for the three-year period.
*Restrictions:* NIH will not accept any application that is essentially the same as one already reviewed. An individual may not have two or more competing NIH career development applications pending review concurrently. Non-domestic (non-U.S.) entities (foreign institutions) and non-domestic (non-U.S.) components of U.S. organizations are not eligible to apply. Multiple Principal Investigators are not allowed.
*Geographic Focus* All States, Guam, Marshall Islands, Northern Mariana Islands, Puerto Rico, U.S. Virgin Islands, American Samoa
*Date(s) Application is Due:* Feb 12; Jun 11; Oct 12
*Amount of Grant:* Up to 747,000 USD
*Contact:* Herbert M. Geller, Ph.D., Director, Office of Education; (301) 451-9440; fax (301) 594-8133; direducation@nhlbi.nih.gov
Nara Gavini, Ph.D., (301) 451-5081; gavininn@nhlbi.nih.gov
*Internet:* grants.nih.gov/grants/guide/pa-files/PAR-12-137.html
*Sponsor:* National Institutes of Health
Building 31, Room 5A48
Bethesda, MD 20892

**NHLBI Investigator Initiated Multi-Site Clinical Trials** 1805
The support of multi-site clinical trials is one strategy NHLBI uses to improve the understanding of the clinical mechanisms of disease and to improve prevention, diagnosis, and treatment. The purpose of this opportunity is to provide a vehicle for submitting grant applications for investigator-initiated multi-site randomized controlled clinical trials. The trials may address any research question related to the mission and goals of the NHLBI and may test clinical or behavioral interventions. The funding opportunity is appropriate for applications to conduct phase II and phase III randomized clinical trials where participants are recruited from multiple sites. Large-scale pragmatic trials (such as comparative effectiveness trials) as well as trials designed to test efficacy of an intervention are appropriate. The trials may randomize at the individual (patient) level or at a group level (e.g., randomization of clinics, schools, worksites, etc.). Clinical trials involving NHLBI mission-related rare diseases that require coordination across multiple clinical sites are also suitable for submission to this funding opportunity. In any case, the trial should propose the most efficient study design to complete the specific aims. The maximum project period is 5 years.
*Requirements:* The following organizations are eligible to apply: public/state controlled and private institutions of higher education; Hispanic-serving institutions; historically black colleges and universities; tribally controlled colleges and universities; Alaska Native and Native Hawaiian serving institutions; Asian American Native American Pacific Islander serving institutions; nonprofits with or without 501(c)3 IRS status (other than institutions of higher education); small businesses; for-profit organizations (other than small businesses); state, county, and city/township governments; special district governments; Indian/Native American tribal governments (federally recognized and other than federally recognized); eligible agencies of the Federal government; U.S. territory or possession; independent school districts; public or Indiana housing authorities; Native American tribal organizations (other than federally recognized tribal governments); faith-based or community-based organizations; regional organizations; and non-domestic (non-U.S.) entities (foreign institutions) and non-domestic (non-U.S.) components of U.S. organizations. Clinical Coordination Center (CCC) and Data Coordination Center (D.C.C) applications are required to be submitted together when the proposed costs of the clinical trial exceed $500,000 (minus F&A for subcontracts) in any given year. Any individual(s) with the skills, knowledge, and resources necessary to carry out the proposed research as the Program Director(s)/Principal Investigator(s) (PD(s)/PI(s)) is invited to work with his/her organization to develop an application for support. Individuals from underrepresented racial and ethnic groups as well as individuals with disabilities are always encouraged to apply for NIH support.
*Geographic Focus:* All States, All Countries
*Date(s) Application is Due:* Feb 12; Jun 12; Oct 12
*Amount of Grant:* Up to 500,000 USD
*Contact:* David J. Gordon, (301) 435-0564; fax (301) 480-7971; gordond@nhlbi.nih.gov
Gail Weinmann, M.D., (301) 435-0233; weinmang@nhlbi.nih.gov
Ellen M. Werner, Ph.D., (301) 435-0050; wernere@nhlbi.nih.gov
*Internet:* grants.nih.gov/grants/guide/pa-files/PAR-13-128.html
*Sponsor:* National Institutes of Health
Building 31, Room 5A48
Bethesda, MD 20892

**NHLBI Lymphatics in Health and Disease in the Digestive, Cardiovascular and Pulmonary Systems** 1806

This program encourages Exploratory/Developmental Grant (R21) applications for research into aspects of lymphatic vessel physiology and pathophysiology related to health and disease of digestive, cardiovascular and pulmonary system organs and resolution of thromboembolic events, and inflammation and immune responses as they relate to these diseases. However, studies with the major focus on immune mechanisms will not be considered responsive. Studies to understand the factors that control local lymphatic vessel functional anatomy and physiology during health or disease in these organs and systems, and the mechanisms by which alterations of lymphatic vessel function affect organ function, are of interest. Specific areas of interests for the National Heart, Lung, and Blood Institute (NHLBI) include approaches that will identify the genetic, molecular, and cellular defects that contribute to congenital malformation of the lymphatic system; congenital-lymphatic-malformation-induced pulmonary dysfunction; whether and how lymphangiogenesis affects cardiovascular and pulmonary diseases; whether and how lymphatic drainage affects transplant of heart or lungs; whether and how lymphatic vessel hyperplasia in the dermal interstitium is involved in salt-sensitive hypertension; and whether and how the manipulation of platelet-lymphatic endothelial cell interaction may provide clinical benefit during thromboembolic events.

*Requirements:* The following organizations and institutions are eligible to apply: public/state controlled and private institutions of higher education; Hispanic-serving institutions; historically Black colleges and universities (HBCUs); Tribally controlled colleges and universities (TCCUs); Alaska Native and Native Hawaiian serving institutions; Nonprofits with or without 501(c)3 IRS status; small businesses; for-profit organizations; state, county, or city/township governments; special district governments; Indian/Native American tribal governments; Indian/Native American tribally designated organizations; independent school districts; public housing authorities/Indian housing authorities; U.S. territory or possession; Indian/Native American tribal governments (other than federally recognized); regional organizations; non-domestic (non-U.S.) entities (foreign organizations); other eligible agencies of the federal government; and faith-based or community-based organizations. Non-domestic (non-U.S.) components of U.S. Organizations are also eligible to apply. Applicant organizations may submit more than one application, provided that each application is scientifically distinct. Any individual(s) with the skills, knowledge, and resources necessary to carry out the proposed research as the Program Director(s)/Principal Investigator(s) (PD(s)/PI(s)) is invited to work with his/her organization to develop an application for support. Individuals from underrepresented racial and ethnic groups as well as individuals with disabilities are always encouraged to apply for NIH support.

*Restrictions:* NIH will not accept any application that is essentially the same as one already reviewed.

*Geographic Focus* All States, Guam, Marshall Islands, Northern Mariana Islands, Puerto Rico, U.S. Virgin Islands, All Countries

*Date(s) Application is Due:* Feb 16; Jun 16; Oct 16

*Contact:* H. Eser Tolunay, Ph.D.; (301) 435-0560; Eser.Tolunay@nih.gov

Jill L. Carrington, Ph.D., Program Director; (301) 402-0671; carringj@mail.nih.gov

*Internet:* grants.nih.gov/grants/guide/pa-files/PAR-12-260.html

*Sponsor:* National Institutes of Health
Building 31, Room 5A48
Bethesda, MD 20892

**NHLBI Lymphatics in Health and Disease in the Digestive, Urinary, Cardiovascular and Pulmonary Systems** 1807

This program is to encourage Research Project Grant (R01) applications for research into aspects of lymphatic vessel physiology and pathophysiology related to health and disease of digestive system and urinary tract organs, and cardiovascular and pulmonary systems; in resolution of thromboembolic events; and inflammation and immune responses as they relate to these diseases. However, studies with the major focus on immune mechanisms will not be considered responsive. Studies to understand the factors that control local lymphatic vessel functional anatomy and physiology during health or disease in these organs/systems, and the mechanisms by which alterations of lymphatic vessel function affect organ function, are of interest. Applications must be received by 5:00 PM local time of applicant organization.

*Requirements:* The following organizations and institutions are eligible to apply: public/state controlled and private institutions of higher education; Hispanic-serving institutions; historically Black colleges and universities (HBCUs); Tribally controlled colleges and universities (TCCUs); Alaska Native and Native Hawaiian serving institutions; Nonprofits with or without 501(c)3 IRS status; small businesses; for-profit organizations; state, county, or city/township governments; special district governments; Indian/Native American tribal governments; Indian/Native American tribally designated organizations; independent school districts; public housing authorities/Indian housing authorities; U.S. territory or possession; Indian/Native American tribal governments (other than federally recognized); regional organizations; non-domestic (non-U.S.) entities (foreign organizations); other eligible agencies of the federal government; and faith-based or community-based organizations. Non-domestic (non-U.S.) components of U.S. Organizations are also eligible to apply. Applicant organizations may submit more than one application, provided that each application is scientifically distinct. Any individual(s) with the skills, knowledge, and resources necessary to carry out the proposed research as the Program Director(s)/Principal Investigator(s) (PD(s)/PI(s)) is invited to work with his/her organization to develop an application for support. Individuals from underrepresented racial and ethnic groups as well as individuals with disabilities are always encouraged to apply for NIH support.

*Restrictions:* NIH will not accept any application that is essentially the same as one already reviewed.

*Geographic Focus* All States, Guam, Marshall Islands, Northern Mariana Islands, Puerto Rico, U.S. Virgin Islands, All Countries

*Date(s) Application is Due:* Feb 5; Jun 5; Oct 5

*Contact:* H. Eser Tolunay, Ph.D.; (301) 435-0560; Eser.Tolunay@nih.gov

Jill L. Carrington, Ph.D., Program Director; (301) 402-0671; carringj@mail.nih.gov

Deborah K. Hoshizaki, Ph.D., Program Director; (301) 594-7712; dkhosh@nih.gov

*Internet:* grants.nih.gov/grants/guide/pa-files/PAR-07-420.html

*Sponsor:* National Institutes of Health
Building 31, Room 5A48
Bethesda, MD 20892

**NHLBI Mentored Career Award for Faculty at Institutions that Promote Diversity (K01)** 1808

The National Institutes of Health (NIH) recognizes a unique and compelling need to promote diversity in the biomedical, behavioral, clinical and social sciences research workforce. The NIH expects efforts to diversify the workforce to lead to the recruitment of the most talented researchers from all groups, to improve the quality of the educational and training environment, to balance and broaden the perspective in setting research priorities, to improve the ability to recruit subjects from minority and other health disparity populations into clinical research protocols, and to improve the Nation's capacity to address and eliminate health disparities. The objectives of this program are to: (1) Advance the awardee's career development trajectory by strengthening research capacity, publishing and other scholarly activities; (2) Improve success and retention in a research career; (3) Promote scientific collaborations that lead to acquisition of new skills or research in other fields of scholarly interest; and (4) Increase the number of highly trained investigators at institutions that promote diversity whose basic and clinical research interests are grounded in the advanced methods and experimental approaches needed to solve problems related to cardiovascular, pulmonary, and hematologic diseases in general and in populations that suffer disproportionately from these conditions. An award made to a successful applicant to this FOA will be a three-to-five-year non-renewable career development award to support faculty at a non-research intensive institution with an institutional mission focused on serving diverse communities that are not well represented in NIH-funded research, or institutions that have been identified by federal legislation as having specially focused institutional missions. The candidates must have research experience and be committed to developing into independent biomedical investigators in research areas relevant to the mission of the NHLBI (i.e., cardiovascular, pulmonary, hematologic, or sleep disorders research).

*Requirements:* Higher Education Institutions (Public/State Controlled Institutions of Higher Education and Private Institutions of Higher Education) are eligible to apply. The following types of Higher Education Institutions are always encouraged to apply for NIH support as Public or Private Institutions of Higher Education: Hispanic-serving Institutions;Historically Black Colleges and Universities (HBCUs); Tribally Controlled Colleges and Universities (TCCUs); Alaska Native and Native Hawaiian Serving Institutions. The Institution must be a domestic college or university with an institutional mission focused on serving minority and other health disparity populations underrepresented in scientific research, or institutions that have been identified by federal legislation as having specifically focused institutional missions. To demonstrate need for research capacity development in heart, lung, and blood diseases and sleep disorders, applicant institutions should not have received over $1,515,000 in NIH funding (direct costs) from the NHLBI per year in any of the previous five years for research training and education grants in the above NHLBI mission areas. Applicant organizations may submit more than one application, provided that each application is scientifically distinct. Any candidate with the skills, knowledge, and resources necessary to carry out the proposed research as the Program Director/Principal Investigator (PD/PI) is invited to work with his/her mentor and organization to develop an application for support. Individuals from underrepresented racial and ethnic groups as well as individuals with disabilities are always encouraged to apply for NIH support. Multiple Principal Investigators are not allowed. By the time of award, the individual must be a citizen or a non-citizen national of the United States or have been lawfully admitted for permanent residence. Individuals on temporary or student visas are not eligible.

*Restrictions:* Non-domestic (non-U.S.) Entities (Foreign Institutions) and non-domestic (non-U.S.) components of U.S. Organizations are not eligible to apply. Foreign components are not allowed.

*Geographic Focus:* All States

*Date(s) Application is Due:* Feb 7

*Amount of Grant:* Up to 100,000 USD

*Contact:* Henry Chang, M.D., (301) 435-0067; changh@nhlbi.nih.gov

*Internet:* www.nhlbi.nih.gov/funding/training/redbook/newrek01.htm

*Sponsor:* National Institutes of Health
Building 31, Room 5A48
Bethesda, MD 20892

**NHLBI Mentored Career Award For Faculty At Minority Institutions** 1809

The NHLBI Mentored Career Award for Faculty at Minority Institutions is a three-to-five-year non-renewable career development award made to faculty at a minority institution. The candidates are faculty members at a minority institution, must have research experience (length of time may vary), and be committed to developing into independent biomedical investigators in research areas relevant to the mission of the NHLBI (i.e, cardiovascular, pulmonary, hematologic, or sleep disorders research). The award will enable suitable faculty members holding doctoral degrees, such as the Ph.D., M.D., D.O., D.V.M., or an equivalent degree, to undertake special study and supervised research under a mentor who is an accomplished investigator in the research area proposed and has experience in developing independent investigators. Letters of Intent are due on June 21, while full proposals are due by July 19.

*Requirements:* Minority scientists and physicians with limited research experience needing guided coursework and supervised laboratory experiences, as well as minority faculty needing only an intensive research experience under the guidance of an established scientist, are eligible to apply. At the time of award, it is required that at least two years have elapsed since the receipt of the doctoral degree and that the candidate have at least one year of prior documented research experience.
*Restrictions:* Current or past principal investigators of an NIH grant or its equivalent, including the Clinical Investigator Award, Physician Scientist Award, Clinical Investigator Development Award, or Mentored Clinical Scientist Development Award are not eligible for the Mentored Research Scientist Development Award for Minority Faculty. Similarly, individuals serving as responsible investigators or project leaders on large grants, such as a Program Project Grant, are not eligible for the Mentored Research Scientist Development Award for Minority Faculty.
*Geographic Focus:* All States
*Date(s) Application is Due:* Jul 19
*Amount of Grant:* Up to 200,000 USD
*Contact:* Traci Heath Mondoro, (301) 435-0052; fax (301) 480-1060; mondorot@nhlbi.nih.gov
*Internet:* grants.nih.gov/grants/guide/rfa-files/RFA-HL-05-016.html#SectionVII
*Sponsor:* National Institutes of Health
Building 31, Room 5A48
Bethesda, MD 20892

**NHLBI Mentored Career Development Award to Promote Faculty Diversity/ Re-Entry in Biomedical Research** 1810
This program invites applications to increase the number of highly trained investigators, from diverse backgrounds underrepresented in research areas of interest to the NHLBI or from those individuals who wish to re-enter their research careers (e.g., after a hiatus due to family circumstances). Minorities are not the only group underrepresented in the scientific community. Individuals with disabilities are also underrepresented. It is geared toward individuals whose basic and clinical research interests are grounded in the advanced methods and experimental approaches needed to solve problems related to cardiovascular, pulmonary, and hematologic diseases in the general and health disparities populations. NHLBI encourages research training and career development crossing disciplinary boundaries (e.g., biophysics, biostatistics, bioinformatics, bioengineering) to develop a new interdisciplinary work force. Award budgets are composed of salary and other program-related expenses. The total project period may not exceed 5 years. Applications are due by the posted deadline date by 5:00 PM local time of applicant organization.
*Requirements:* The following organizations are eligible to apply: Public/state controlled and private institutions of higher education; nonprofits with and without 501(c)3 IRS Status (other than institutions of higher education); small businesses; for-profit organizations (other than small businesses); state governments; county governments; city or township governments; special district governments; Indian/Native American Tribal governments (federally recognized); Indian/Native American Tribal governments (other than federally recognized); U.S. territory or possession. The following types of higher education institutions are always encouraged to apply for NIH support as public or private institutions of higher education: Hispanic-serving institutions; historically Black colleges and universities; Tribally controlled colleges and universities; Alaska Native and Native Hawaiian serving institutions. Any candidate with the skills, knowledge, and resources necessary to carry out the proposed research as the Program Director/Principal Investigator (PD/PI) is invited to work with his/her mentor and organization to develop an application for support. Individuals from underrepresented racial and ethnic groups as well as individuals with disabilities are always encouraged to apply for NIH support. Multiple Principal Investigators are not allowed. By the time of award, the individual must be a citizen or a non-citizen national of the United States or have been lawfully admitted for permanent residence.
*Restrictions:* Non-domestic (non-U.S.) Entities (Foreign Institutions) are not eligible to apply. Non-domestic (non-U.S.) components of U.S. Organizations are not eligible to apply. Foreign components, as defined in the NIH Grants Policy Statement, are not allowed. Individuals on temporary or student visas are not eligible.
*Geographic Focus* All States, Guam, Marshall Islands, Northern Mariana Islands, Puerto Rico, U.S. Virgin Islands, American Samoa
*Date(s) Application is Due:* Feb 7
*Amount of Grant:* Up to 625,000 USD
*Contact:* Lorraine M. Silsbee, M.H.S., (301) 435-0709; silsbeeL@nhlbi.nih.gov
*Internet:* www.nhlbi.nih.gov/funding/training/redbook/newsck01.htm
*Sponsor:* National Institutes of Health
Building 31, Room 5A48
Bethesda, MD 20892

**NHLBI Mentored Patient-Oriented Research Career Development Awards** 1811
This award supports the career development of investigators who are committed to patient-oriented research. It provides support for supervised study and research for clinically trained professionals who have the potential to develop into productive, clinical investigators focusing on patient-oriented research. This program provides research development opportunities for clinicians with varying levels of research experience. Support is provided for a minimum of 3 years and a maximum of 5 years. New applications are due February 12, June 12, and October 12. Resubmissions are due March 12, July 12, and November 12.
*Requirements:* The following organizations are eligible to apply: Public/state controlled and private institutions of higher education; nonprofits with and without 501(c)3 IRS Status (other than institutions of higher education); small businesses; for-profit organizations (other than small businesses); state governments; county governments; city or township governments; special district governments; Indian/Native American Tribal governments (federally recognized); Indian/Native American Tribal governments (other than federally recognized); U.S. territory or possession. The following types of higher education institutions are always encouraged to apply for NIH support as public or private institutions of higher education: Hispanic-serving institutions; historically Black colleges and universities; Tribally controlled colleges and universities; Alaska Native and Native Hawaiian serving institutions. Applicant organizations may submit more than one application, provided that each application is scientifically distinct. Any candidate with the skills, knowledge, and resources necessary to carry out the proposed research as the Program Director/Principal Investigator (PD/PI) is invited to work with his/her mentor and organization to develop an application for support. Individuals from underrepresented racial and ethnic groups as well as individuals with disabilities are always encouraged to apply for NIH support. Multiple Principal Investigators are not allowed. By the time of award, the individual must be a citizen or a non-citizen national of the United States or have been lawfully admitted for permanent residence. Candidates for this award must have a health-professional doctoral degree. Such degrees include but are not limited to the M.D., D.O., D.D.S., D.M.D., O.D., D.C., Pharm.D., N.D. (Doctor of Naturopathy), as well as a doctoral degree in nursing research or practice. Candidates with Ph.D. degrees are eligible for this award if the degree is in a clinical field and they usually perform clinical duties. Individuals with the Ph.D. or other doctoral degree in clinical disciplines such as clinical psychology, clinical genetics, speech-language pathology, audiology or rehabilitation are also eligible. Individuals holding the Ph.D. in a non-clinical discipline but who are certified to perform clinical duties should contact the appropriate Institute concerning their eligibility for a K23 award. Candidates also must have completed their clinical training, including specialty and, if applicable, subspecialty training prior to receiving an award. However, candidates may submit an application prior to the completion of clinical training.
*Restrictions:* Non-domestic (non-U.S.) Entities (Foreign Organizations) are not eligible to apply. Foreign (non-U.S.) components of U.S. Organizations are not allowed. Individuals on temporary or student visas are not eligible.
*Geographic Focus* All States, Guam, Marshall Islands, Northern Mariana Islands, Puerto Rico, U.S. Virgin Islands, American Samoa
*Date(s) Application is Due:* Feb 12; Mar 12; Jun 12; Jul 12; Oct 12; Nov 12
*Amount of Grant:* Up to 620,000 USD
*Contact:* Sandra Colombini Hatch, MD, Medical Officer, Lung Biology and Disease Program; (301) 435-0222; hatchs@nhlbi.nih.gov
Traci Heath Mondoro, PhD, Program Director; (301) 435-0052; mondorot@nhlbi.nih.gov
*Internet:* www.nhlbi.nih.gov/funding/training/redbook/newk23.htm
*Sponsor:* National Institutes of Health
Building 31, Room 5A48
Bethesda, MD 20892

**NHLBI Mentored Quantitative Research Career Development Awards** 1812
The goal of this program is to foster interdisciplinary collaboration in biomedical and behavioral research by supporting supervised research experiences for scientists with quantitative and engineering backgrounds. This award provides research and career development opportunities for scientists and engineers with little or no biomedical or behavioral research experience who are committed to establishing themselves in careers as independent biomedical or behavioral investigators. Examples of quantitative scientific and technical backgrounds outside of biology and medicine considered appropriate for this award include, but are not limited to: mathematics, statistics, computer science, informatics, physics, chemistry, and engineering. This mechanism is intended for research-oriented investigators from the postdoctoral level to the level of senior faculty. New applications are due February 12, June 12, October 12. Resubmissions are due March 12, July 12, and November 12. Award budgets are composed of salary and other program-related expenses. The total project period may not exceed 5 years.
*Requirements:* Institutions may submit applications on behalf of candidates who hold an advanced degree in a quantitative area of science or engineering. The following organizations are eligible to apply: Public/state controlled and private institutions of higher education; nonprofits with and without 501(c)3 IRS Status (other than institutions of higher education); small businesses; for-profit organizations (other than small businesses); state governments; county governments; city or township governments; special district governments; Indian/Native American Tribal governments (federally recognized); Indian/Native American Tribal governments (other than federally recognized); U.S. territory or possession. The following types of higher education institutions are always encouraged to apply for NIH support as public or private institutions of higher education: Hispanic-serving institutions; historically Black colleges and universities; Tribally controlled colleges and universities; Alaska Native and Native Hawaiian serving institutions. Individuals with an advanced degree in a quantitative area of science or engineering are eligible candidates. Multiple Principal Investigators are not allowed. All candidates must be U.S. citizens, non-citizen nationals, or legal permanent residents of the U.S. The grantee institution must have a strong, well-established research and research training program in the chosen area, accomplished faculty in the basic and clinical sciences, and a commitment to the candidate's research development. The proposed program should include an appropriate mentor.
*Restrictions:* NHLBI has a 6 year limit of cumulative support on institutional and mentored Ks (e.g., K12 or KL2 plus the K25). A candidate for the K25 may not concurrently apply for or have an award pending for any other NIH career development award. Non-domestic (non-U.S.) Entities (Foreign Organizations) are not eligible to apply. Foreign (non-U.S.) components of U.S. Organizations are not allowed.
*Geographic Focus* All States, Guam, Marshall Islands, Northern Mariana Islands, Puerto Rico, U.S. Virgin Islands, American Samoa
*Date(s) Application is Due:* Feb 12; Mar 12; Jun 12; Jul 12; Oct 12; Nov 12

*Contact:* Drew E. Carlson, Ph.D., Program Director, Office of Research Training and Career Development; (301) 435-0535; fax (301) 480-7404; carlsonde@nhlbi.nih.gov
Wayne C. Wang, Ph.D., Program Director, Office of Research Training and Career Development; (301) 435-0535; fax (301) 480-7404; wangwc@nhlbi.nih.gov
*Internet:* www.nhlbi.nih.gov/funding/training/redbook/newk25.htm
*Sponsor:* National Institutes of Health
Building 31, Room 5A48
Bethesda, MD 20892

**NHLBI Microbiome of the Lung and Respiratory Tract in HIV-Infected Individuals and HIV-Uninfected Controls** **1813**

The National Heart, Lung, and Blood Institute (NHLBI) solicits grant applications under this Funding Opportunity Announcement (FOA) to characterize the microbiome of the lung (the airways and airspaces below the glottis and the lung parenchyma) alone or in combination with the nasal and/or oropharyngeal cavities in HIV-infected individuals and matched HIV-uninfected controls, including normal healthy controls, using molecular techniques to identify bacteria and if possible other organisms, e.g., viruses, cell-wall deficient organisms, protozoa, and fungi. Investigators should use high-throughput technology platforms to create a data set of sufficient quality and depth to allow analysis of how changes of microbiota relate to HIV lung disease progression/complications. These data will be used to examine the impact of changes in the respiratory microbiome on the pathogenesis and progression of HIV disease, on HIV-related respiratory complications, and the effects of anti-HIV therapies. The NHLBI also invites applications for four Clinical/Sequencing research sites and one supporting Data Coordinating Center (D.C.C). Each application for a clinical/sequencing research site should propose approaches to characterize the lung microbiome and applicants are encouraged to include at least one hypothesis-driven mechanistic aim. Direct costs up to $525,000 per year for project duration of up to five years may be requested.
*Requirements:* The following organizations/institutions are eligible to apply: Public/State Controlled Institutions of Higher Education; Private Institutions of Higher Education; Hispanic-serving Institutions; Historically Black Colleges and Universities (HBCUs); Tribally Controlled Colleges and Universities (TCCUs); Alaska Native and Native Hawaiian Serving Institutions; Nonprofits with 501(c)3 IRS Status (Other than Institutions of Higher Education); Nonprofits without 501(c)3 IRS Status (Other than Institutions of Higher Education); Small Businesses; For-Profit Organizations (Other than Small Businesses); State Governments; Indian/Native American Tribal Governments (Federally Recognized); Indian/Native American Tribally Designated Organizations; County Governments; City or Township Governments; Special District Governments; Independent School Districts; Public Housing Authorities/Indian Housing Authorities; U.S. Territory or Possession; Indian/Native American Tribal Governments (Other than Federally Recognized); Regional Organizations; Other Eligible Agencies of the Federal Government; and, Faith-based or Community-based Organizations. Any individual with the skills, knowledge, and resources necessary to carry out the proposed research as the PD/PI is invited to work with his/her institution to develop an application for support. Individuals from underrepresented racial and ethnic groups as well as individuals with disabilities are always encouraged to apply for NIH support. More than one PD/PI, or multiple PDs/PIs, may be designated on the application for projects that require a "team science" approach and therefore clearly do not fit the single-PD/PI model. All PDs/PIs must be registered in the NIH eRA Commons prior to the submission of the application.
*Restrictions:* Foreign organizations are not eligible, however, domestic institutions may form consortia and enter into subcontracts with Non-domestic (non-U.S.) Entities (Foreign Organizations) if need can be demonstrated – foreign subcontract use the same criteria as are used for justifying foreign grants.
*Geographic Focus:* All States
*Date(s) Application is Due:* Feb 5; Jun 5; Oct 5
*Amount of Grant:* Up to 525,000 USD
*Contact:* Sandra Colombini Hatch., (301) 435-0222; fax (301) 480-3557; Hatchs@nhlbi.nih.gov
*Internet:* grants.nih.gov/grants/guide/rfa-files/RFA-HL-09-006.html
*Sponsor:* National Institutes of Health
Building 31, Room 5A48
Bethesda, MD 20892

**NHLBI Midcareer Investigator Award in Patient-Oriented Research** **1814**

This award provides protected time to mid-career clinical investigators who are typically at the Associate Professor level or the equivalent and who have their own independent peer-reviewed research support to provide mentoring to junior clinical investigators, particularly K23 grantees, in patient-oriented research and to stabilize the careers of these investigators so that they could continue to conduct patient-oriented research and be available as mentors in patient-oriented research. It is expected that the K24 recipients will obtain new or additional independent peer-reviewed funding for patient-oriented research as a Program Director/Principal Investigator and establish and assume leadership roles in collaborative patient-oriented research programs. In addition, it is expected that there will be an increased effort and commitment to act as a mentor to beginning clinician investigators in patient-oriented research to enhance the research productivity of both the K24 investigator and increase the pool of well-trained clinical researchers of the future. New applications are due on February 12, June 12 and October 12. Renewal applications are due March 12, July 12, and November 12.
*Requirements:* Candidates must be U.S. citizens, non citizen nationals, or legal permanent residents of the U.S. Candidates for this award must have a health-professional doctoral degree or its equivalent. Such degrees include but are not limited to the M.D., D.O., D.D.S., D.M.D., O.D., D.C., Pharm.D., N.D. (Doctor of Naturopathy), as well as a doctoral degree in nursing. Candidates with Ph.D. degrees are eligible for this award if the degree is in a clinical field or they perform patient-oriented research (P.O.R). This may include clinical psychologists, clinical geneticists, speech and language pathologists. Candidates should typically be in the midcareer stage at the Associate Professor level or functioning at that rank in an academic setting or equivalent non-academic setting and must have an established record of independent, peer-reviewed patient-oriented research grant funding including at the time of application and record of publications. This award is intended for individuals who have a record of supervising and mentoring patient-oriented researchers.
*Geographic Focus:* All States
*Date(s) Application is Due:* Feb 12; Mar 12; Jun 12; Jul 12; Oct 12; Nov 12
*Contact:* Sandra Colombini Hatch, M.D., Medical Officer; (301) 435-0222; fax (301) 480-3557; hatchs@nhlbi.nih.gov
*Internet:* www.nhlbi.nih.gov/funding/training/redbook/estk24.htm
*Sponsor:* National Institutes of Health
Building 31, Room 5A48
Bethesda, MD 20892

**NHLBI National Research Service Award Programs in Cardiovascular Epidemiology and Biostatistics** **1815**

The National Research Service Award (NRSA) is one of many training programs sponsored by the National Heart, Lung, and Blood Institute (NHLBI). This program enables research institutions to support pre- and post-doctoral training in areas of particular interest to the Institute and to prepare individuals for careers in biomedical and behavioral research. The cardiovascular epidemiology and biostatistics research training programs offer opportunities in the fields of epidemiology of cardiovascular diseases and statistical analysis. These awards do not support study leading to the M.D., D.D.S. or other similar professional degrees, nor do they support residency training.
*Requirements:* Trainees are selected through review procedures established by the program director at the grantee institution. Students are required to pursue their research training full-time and must be U.S. citizens, noncitizen nationals or legal permanent residents of the United States. At the time of appointment, trainees must have received a baccalaureate degree and must be training at the post-baccalauareate level in a program leading to a Ph.D., Sc.D. or equivalent degree. Students who wish to interrupt their medical, dental, or other professional school studies to engage in full-time research training before completing their professional degrees are also eligible. Attention is given to recruiting individuals from minority groups that are under-represented nationally in the biomedical sciences (African Americans, Hispanics, Native Americans, Alaskan Natives, and Pacific Islanders).
*Restrictions:* Persons on temporary or student visas are not eligible.
*Geographic Focus:* All States
*Date(s) Application is Due:* Jan 25
*Contact:* Lorraine M. Silsbee, (301) 435-0709; fax (301) 480-1667; silsbeeL@nih.gov
*Internet:* www.nhlbi.nih.gov/funding/training/epi-bio/nrsa.htm
*Sponsor:* National Institutes of Health
Building 31, Room 5A48
Bethesda, MD 20892

**NHLBI Pathway to Independence Awards** **1816**

The Pathway to Independence (PI) Award is designed to facilitate a timely transition from a mentored postdoctoral research position to a stable independent research position with independent NIH or other independent research support at an earlier stage than is currently the norm. The PI award will provide up to 5 years of support consisting of two phases. The initial phase will provide 1-2 years of mentored support for postdoctoral research scientists. This phase will be followed by up to 3 years of independent support contingent on securing an independent tenure-track or equivalent research position. New applications are due February 12, June 12, and October 12. Resubmitted applications are due March 12, July 12, and November 12.
*Requirements:* The following organizations are eligible to apply: Public/state controlled and private institutions of higher education; nonprofits with and without 501(c)3 IRS Status (other than institutions of higher education); small businesses; for-profit organizations (other than small businesses); state governments; county governments; city or township governments; special district governments; Indian/Native American Tribal governments (federally recognized); Indian/Native American Tribal governments (other than federally recognized); U.S. territory or possession. The following types of higher education institutions are always encouraged to apply for NIH support as public or private institutions of higher education: Hispanic-serving institutions; historically Black colleges and universities; Tribally controlled colleges and universities; Alaska Native and Native Hawaiian serving institutions. The applicant institution will be the mentored phase K99 institution. All institution/organization types listed above are eligible for both the mentored and independent phase, with one exception: eligible agencies of the Federal government, such as the NIH intramural program, are eligible only for the mentored phase. Any candidate with the skills, knowledge, and resources necessary to carry out the proposed research as the Program Director/Principal Investigator (PD/PI) is invited to work with his/her mentor and organization to develop an application for support. Individuals from underrepresented racial and ethnic groups as well as individuals with disabilities are always encouraged to apply for NIH support. Multiple Principal Investigators are not allowed. At the time of application submission (or resubmission) candidates for this award must have earned a terminal clinical or research doctorate (including Ph.D., M.D., D. O., D.C., N.D., D.D.S., D.V.M., Sc.D., D.N.S., Pharm. D., or equivalent doctoral degree, or a combined clinical and research doctoral degree); and have no more than 5 years (60 months) of research experience since completing the requirements of the doctoral degree (resubmissions must also comply with this requirement). Individuals whose terminal or research doctorate was awarded more than

60 months before submission of the K99/R00 application are strongly urged to contact one of the NHLBI Program staff designated below before submission to determine whether they are eligible to apply. This confirmation of eligibility should be documented in a cover letter to accompany the application. Applications from ineligible individuals will be returned without review. U.S. citizens and non-U.S. citizens are eligible.
*Restrictions:* Non-domestic (non-U.S.) Entities (Foreign Organizations) are not eligible to apply. Foreign (non-U.S.) components of U.S. Organizations are not allowed.
*Geographic Focus* All States, Guam, Marshall Islands, Northern Mariana Islands, Puerto Rico, U.S. Virgin Islands, American Samoa
*Date(s) Application is Due:* Feb 12; Mar 12; Jun 12; Jul 12; Oct 12; Nov 12
*Contact:* Herbert Geller, Ph.D., Director, Office of Education; (301) 451-9440; fax (301) 594-8133; direducation@nhlbi.nih.gov
Helena O. Mishoe, (301) 451-5081; fax (301) 480-0862; mishoeh@nhlbi.nih.gov
*Internet:* grants.nih.gov/grants/guide/pa-files/PA-11-197.html
*Sponsor:* National Institutes of Health
Building 31, Room 5A48
Bethesda, MD 20892

**NHLBI Research on the Role of Cardiomyocyte Mitochondria in Heart Disease: An Integrated Approach** 1817

The National Heart, Lung, and Blood Institute (NHLBI) invites applications for collaborative research projects to develop an integrated understanding of cardiomyocyte mitochondria and its contributions to myocardial adaptations and heart disease progression by combining functional data with information derived from powerful new technologies. Budgets for direct costs of up to $500,000 per year and a project duration of up to four years may be requested for a maximum of $2,000,000 direct costs over a four-year project period. Because the nature and scope of the proposed research will vary from application to application, it is anticipated that the size and duration of each award will also vary.
*Requirements:* The following organizations/institutions are eligible to apply: Public/State Controlled Institutions of Higher Education; Private Institutions of Higher Education; Hispanic-serving Institutions; Historically Black Colleges and Universities (HBCUs); Tribally Controlled Colleges and Universities (TCCUs); Alaska Native and Native Hawaiian Serving Institutions; Nonprofits with 501(c)3 IRS Status (Other than Institutions of Higher Education); Nonprofits without 501(c)3 IRS Status (Other than Institutions of Higher Education); Small Businesses; For-Profit Organizations (Other than Small Businesses); State Governments; Indian/Native American Tribal Governments (Federally Recognized); Indian/Native American Tribally Designated Organizations; County Governments; City or Township Governments; Special District Governments; Independent School Districts; Public Housing Authorities/Indian Housing Authorities; U.S. Territory or Possession; Indian/Native American Tribal Governments (Other than Federally Recognized); Regional Organizations; Non-domestic (non-U.S.) Entities (Foreign Organizations); Other Eligible Agencies of the Federal Government; and, Faith-based or Community-based Organizations. Any individual(s) with the skills, knowledge, and resources necessary to carry out the proposed research as the PD/PI is invited to work with his/her organization to develop an application for support. Individuals from underrepresented racial and ethnic groups as well as individuals with disabilities are always encouraged to apply for NIH support. More than one PD/PI (i.e., multiple PDs/PIs) may be designated on the application for projects that require a "team science" approach and therefore clearly do not fit the single-PD/PI model. All PDs/PIs must be registered in the NIH electronic Research Administration (eRA) Commons prior to the submission of the application. Applicants may submit more than one application, provided each application is scientifically distinct. Request for Applications (RFA) Number: RFA-HL-10-002
*Restrictions:* Applicants are not permitted to submit a resubmission application. Renewals will not be allowed.
*Geographic Focus* All States, Guam, Marshall Islands, Northern Mariana Islands, Puerto Rico, U.S. Virgin Islands, All Countries
*Date(s) Application is Due:* Feb 5; Jun 5; Oct 5
*Amount of Grant:* Up to 2,000,000 USD
*Contact:* Isabella Liang, Ph.D., (301) 435-0504; fax (301) 451-5458; Liangi@mail.nih.gov
*Internet:* grants.nih.gov/grants/guide/rfa-files/RFA-HL-10-002.html
*Sponsor:* National Institutes of Health
Building 31, Room 5A48
Bethesda, MD 20892

**NHLBI Research Supplements to Promote Diversity in Health-Related Research** 1818

This supplement enables principal investigators with eligible NHLBI research grants/contracts to include high school students in their projects. Priority is given to applications requesting support for individuals from underrepresented racial and ethnic groups, individuals with disabilities, and individuals from disadvantaged backgrounds. Nationally, underrepresented groups in biomedical research careers include but are not limited to African Americans, Hispanic Americans, American Indians/Alaska Natives, Native Hawaiians and Pacific Islanders. The research activity proposed for the student must be part of the approved research for the parent grant or contract. The student should be encouraged to participate in ongoing team discussions of research findings and directions. The purpose of this program is to provide high school students with an opportunity to obtain a meaningful experience in various aspects of health-related research to stimulate their interest in careers in biomedical, behavioral, biometric, clinical, nursing and social sciences. Students are expected to devote sufficient effort to the research project and related activities during the period of support to gain insight into the process of scientific discovery. Support for at least three months is encouraged during any one year. Principal Investigators are encouraged to seek high school students who will devote at least two years to this program (i.e., equivalent to two three-month, full-time periods). More than one high school research supplements can be awarded per research grant, subproject of a program project grant (P01), or contract. Applications will be accepted at any time, however, applications should arrive at least three months before the requested start date, to allow time for review.
*Requirements:* All principal investigators at U.S. institutions with eligible NHLBI research grants or contracts may apply. Grants/contracts with adequate time for a summer research experience or one year remaining at the time of award are eligible to apply for high school students. This announcement is for supplements to existing projects. To be eligible, the parent award must be active and the research proposed in the supplement must be accomplished within the competitive segment. All additional costs must be within the scope of the peer-reviewed and approved project. Before submitting an application for a research supplement, applicants are strongly encouraged to contact their program official to discuss the program. Supplemental awards under this program are limited to citizens or non-citizen nationals of the United States or to individuals who have been lawfully admitted for permanent residence in the United States (i.e., in possession of an Alien Registration Receipt Card or some other legal evidence of admission for permanent residence at the time of application). This program may not be used to provide technical support to NIH-supported investigators. Application budgets are limited to no more than the amount of the current parent award, and must reflect actual needs of the proposed project. Direct costs for individual administrative supplements vary from less than $5,000 to more than $100,000 depending on the career level of the candidate. Administrative supplements end with the competitive cycle of the parent grant.
*Restrictions:* Non-domestic (non-U.S.) Entities (Foreign Institutions) and non-domestic (non-U.S.) components of U.S. Organizations are not eligible to apply. Foreign components, as defined in the NIH Grants Policy Statement, are not allowed.
*Geographic Focus:* All States
*Contact:* Nara Gavini, (301) 451-5081; fax (301) 480-0862; gavininn@nhlbi.nih.gov
*Internet:* grants.nih.gov/grants/guide/pa-files/PA-12-149.html
*Sponsor:* National Institutes of Health
Building 31, Room 5A48
Bethesda, MD 20892

**NHLBI Ruth L. Kirschstein National Research Service Awards for Individual Postdoctoral Fellows** 1819

The National Heart, Lung, and Blood Institute is one of many Institutes in the National Institutes of Health that participate in the Ruth L. Kirschstein National Research Service Awards (NRSA) for Individual Postdoctoral Fellows. This program offers health scientists the opportunity to receive full-time research training for up to 3 years in areas that reflect the national need for biomedical, clinical and behavioral research in cardiovascular, pulmonary, hematologic, and sleep disorders. These grants are not intended for study leading to the M.D., D.O., D.D.S., or equivalent professional degrees, nor do they support residency training. The proposed postdoctoral training must offer an opportunity to enhance the applicant's understanding of the health-related sciences, and must be within the broad scope of biomedical, behavioral, or clinical research or other specific disciplines relevant to the research mission of the participating NIH Institutes and Centers. Applicants with a health professional doctoral degree may use the proposed postdoctoral training to satisfy a portion of the degree requirements for a master's degree, a research doctoral degree or any other advanced research degree program.
*Requirements:* Any applicant fellow with the skills, knowledge, and resources necessary to carry out the proposed research as the Project Director/Principal Investigator (PD/PI) is invited to work with his/her sponsor and organization to develop an application for support. Individuals from underrepresented racial and ethnic groups as well as individuals with disabilities are always encouraged to apply for NIH support. Applicants must be U.S. citizens, noncitizen nationals, or legal permanent residents of the U.S. Training can be conducted abroad if the site provides opportunities that are not available in this country. Applicants are cautioned that not all NIH Institutes and Centers (ICs) participate in this program, and that consultation with relevant IC staff prior to submission of an application is strongly encouraged. The participating ICs have different emphases and program requirements for this program. Before submitting a fellowship application, the applicant fellow must identify a sponsoring institution. The sponsoring institution must have staff and facilities available on site to provide a suitable environment for performing high-quality research. Eligible sponsoring institutions include: Public/State Controlled and Private Institutions of Higher Education; Nonprofits with or without 501(c)3 IRS Status (Other than Institutions of Higher Education); Small Businesses; For-Profit Organizations (Other than Small Businesses); Eligible Agencies of the Federal Government; and, Non-domestic (non-U.S.) Entities (Foreign Organizations). The following types of Higher Education Institutions are always encouraged to apply for NIH support as Public or Private Institutions of Higher Education: Hispanic-serving Institutions; Historically Black Colleges and Universities (HBCUs); Tribally Controlled Colleges and Universities (TCCUs); and, Alaska Native and Native Hawaiian Serving Institutions. When the fellowship begins, the applicant must have received a doctoral degree and have arranged to work with a sponsor affiliated with an institution that has the staff and facilities needed for the proposed training.
*Restrictions:* A Kirschstein-NRSA fellowship may not be used to support the clinical years of residency training. However, these awards are appropriate for the research fellowship years of a residency program. Research clinicians must devote full-time to their proposed research training and confine clinical duties to those activities that are part of the research training program.
*Geographic Focus* All States, Guam, Marshall Islands, Northern Mariana Islands, Puerto Rico, U.S. Virgin Islands, All Countries
*Date(s) Application is Due:* Apr 8; Aug 8; Dec 8
*Contact:* Sandra Colombini Hatch, (301) 435-0222; fax (301) 480-3557; hatchs@nhlbi.nih.gov

*Internet:* www.nhlbi.nih.gov/funding/training/redbook/phdf32.htm
*Sponsor:* National Institutes of Health
Building 31, Room 5A48
Bethesda, MD 20892

**NHLBI Ruth L. Kirschstein National Research Service Awards for Individual Predoctoral Fellowships to Promote Diversity in Health-Related Research** 1820

NHLBI is one of many Institutes in the NIH that participate in the Ruth L. Kirschstein National Research Service Awards for Individual Predoctoral Fellowships to Promote Diversity in Health-Related Research. This program encourages students from underrepresented racial and ethnic groups, individuals with disabilities, and individuals from disadvantaged backgrounds to seek research doctoral degrees in the biomedical and behavioral sciences to help increase the number of well-trained scientists from underrepresented groups. The fellowship provides up to 5 years of support for research training leading to the Ph.D. or equivalent research degree, the combined M.D./Ph.D. degree, or other combined degrees in the biomedical or behavioral sciences.
*Requirements:* Individuals with disabilities, or from underrepresented racial and ethnic groups, or individuals from disadvantaged backgrounds pursuing advanced degrees in the biomedical and behavioral sciences are eligible for fellowship awards. The Fellowship applicant must have a baccalaureate degree and be currently enrolled in an eligible doctoral program. At the time of appointment, students must be U.S. citizens, noncitizen nationals, or lawfully admitted to the U.S. for permanent residence. Individuals on temporary or student visas are not eligible. Before submitting a fellowship application, the applicant fellow must identify a sponsoring institution. The sponsoring institution must have staff and facilities available on site to provide a suitable environment for performing high-quality research. Eligible sponsoring institutions include Public/State Controlled and Private Institutions of Higher Education; Nonprofits with or without 501(c)3 IRS Status (Other than Institutions of Higher Education); Small Businesses; For-Profit Organizations (Other than Small Businesses); Eligible Agencies of the Federal Government; Non-domestic (non-U.S.) Entities (Foreign Organizations). The following types of Higher Education Institutions are always encouraged to apply for NIH support as Public or Private Institutions of Higher Education: Hispanic-serving Institutions; Historically Black Colleges and Universities (HBCUs); Tribally Controlled Colleges and Universities (TCCUs); Alaska Native and Native Hawaiian Serving Institutions.
*Restrictions:* NIH will not accept any application that is essentially the same as one already reviewed. An individual may not have two or more competing NIH fellowship applications pending review concurrently.
*Geographic Focus:* All States
*Date(s) Application is Due:* Apr 8; Aug 8; Dec 8
*Contact:* Xenia J. Tigno, PhD, Program Director; (301) 435-0202; tignoxt@mail.nih.gov
*Internet:* www.nhlbi.nih.gov/funding/training/redbook/gradf31.htm
*Sponsor:* National Institutes of Health
Building 31, Room 5A48
Bethesda, MD 20892

**NHLBI Ruth L. Kirschstein National Research Service Awards for Individual Predoctoral MD/PhD Fellows and Other Dual Degree Fellows** 1821

The NHLBI is interested in supporting individual predoctoral fellowships for combined MD/PhD training in research areas relevant to the NHLBI mission. The purpose of the Ruth L. Kirschstein National Research Service Awards (Kirschstein-NRSA) is to provide support to individuals for combined MD/PhD and other dual doctoral degree training (e.g. DO/PhD, DDS/PhD, AuD/PhD). The participating Institutes award this Kirschstein-NRSA individual fellowship (F30) to qualified applicants with the potential to become productive, independent, highly trained physician-scientists and other clinician-scientists, including patient-oriented researchers in their scientific mission areas. This funding opportunity supports individual predoctoral F30 fellowships with the expectation that these training opportunities will increase the number of future investigators with both clinical knowledge and skills in basic, translational or clinical research. Award budgets are composed of stipends, tuition and fees, and institutional allowance.
*Requirements:* Domestic for-profit or non-profit institutions/organizations, or public or private institutions, such as universities, colleges, hospitals and laboratories, eligible agencies of the federal government, and NIH intramural laboratories are eligible to apply. The sponsoring institution must have staff and facilities available on site to provide a suitable environment for performing high-quality research training. The PhD phase of the program may be conducted outside the sponsoring institution, e.g. a Federal laboratory including the NIH intramural laboratories. This training, however, must be part of a combined MD/PhD program. Applications to the program will be accepted from students currently enrolled in a formally combined MD/PhD program. By the time of award, all candidates for the Kirschstein-NRSA F30 award must be citizens or non-citizen nationals of the United States, or must have been lawfully admitted to the United States for Permanent Residence.
*Restrictions:* Foreign institutions are not eligible to apply.
*Geographic Focus:* All States
*Date(s) Application is Due:* Apr 8; Aug 8; Dec 8
*Contact:* Drew E. Carlson, Ph.D., Program Director, Office of Research Training and Career Development; (301) 435-0535; fax (301) 480-7404; carlsonde@nhlbi.nih.gov
*Internet:* www.nhlbi.nih.gov/funding/training/redbook/gradmdphdf30.htm
*Sponsor:* National Institutes of Health
Building 31, Room 5A48
Bethesda, MD 20892

**NHLBI Ruth L. Kirschstein National Research Service Awards for Individual Senior Fellows** 1822

The National Heart, Lung, and Blood Institute is one of many Institutes in the National Institutes of Health that participate in the Ruth L. Kirschstein National Research Service Awards (NRSA) for individual Senior Fellows (Parent F33). This program enables experienced scientists to change the direction of their research careers, broaden their scientific background, acquire new research capabilities, or enlarge their command of an allied research field. The purpose of the senior fellowship (F33) award is to provide senior fellowship support to experienced scientists who wish to make major changes in the direction of their research careers or who wish to broaden their scientific background by acquiring new research capabilities as independent research investigators in scientific health-related fields relevant to the missions of the participating NIH Institutes and Centers. These awards will enable individuals with at least seven years of research experience beyond the doctorate (individuals with PhD, MD, OD, associate professors, full professors, etc.), and who have progressed to the stage of independent investigator, to take time from regular professional responsibilities for the purpose of receiving training to increase their scientific capabilities. In most cases, this award is used to support sabbatical experiences for established independent scientists seeking support for retraining or additional career development. This program is not designed for postdoctoral level investigators seeking to enhance their research experience prior to independence. Award budgets are composed of stipends, tuition and fees, and institutional allowance.
*Requirements:* Applicants must be U.S. citizens, non-citizen nationals, or permanent residents of the U.S. When the fellowship begins, an applicant must have received the Ph.D., M.D., D.O., D.D.S., D.V.M., O.D., D.P.M., Sc.D., D.Eng., DNSc., or equivalent degree. The applicant must arrange to work full-time with a particular sponsor affiliated with an institution that has the staff and facilities needed for the proposed training. Training can be conducted abroad if the site provides opportunities not available in the U.S. The following are eligible sponsoring organizations: public/state controlled and private institutions of higher education; nonprofits with or without 501(c)3 IRS status (other than institutions of higher education); small businesses; for-profit organizations (other than small businesses); eligible agencies of the Federal government; non-domestic (non-U.S.) entities (foreign organizations). The following types of Higher Education Institutions are always encouraged to apply for NIH support as public or private institutions of higher education: Hispanic-serving institutions; historically Black colleges and universities; Tribally controlled colleges and universities; Alaska Native and Native Hawaiian serving institutions.
*Restrictions:* Multiple Principal Investigators are not allowed.
*Geographic Focus* All States, Guam, Marshall Islands, Northern Mariana Islands, Puerto Rico, U.S. Virgin Islands, American Samoa
*Date(s) Application is Due:* Apr 8; Aug 8; Dec 8
*Contact:* Charlotte A. Pratt, Ph.D., RD, Program Director; (301) 435-0382; fax (301) 480-1864; prattc@nhlbi.nih.gov
*Internet:* www.nhlbi.nih.gov/funding/training/redbook/estf33.htm
*Sponsor:* National Institutes of Health
Building 31, Room 5A48
Bethesda, MD 20892

**NHLBI Short-Term Research Education Program to Increase Diversity in Health-Related Research** 1823

The National Heart, Lung, and Blood Institute, National Institutes of Health invites Research Education (R25) applications to promote diversity in undergraduate and health professional participant populations by providing short-term research education support to stimulate career development in cardiovascular, pulmonary, hematologic, and sleep disorders research. The overall goal of the program is to provide research opportunities for individuals from backgrounds underrepresented in biomedical science, including individuals from disadvantaged backgrounds, individuals from underrepresented racial and ethnic groups, and individuals with disabilities that will significantly contribute to a diverse research workforce in the future. The research opportunities should be of sufficient depth to enable the participants, upon completion of the program, to have a thorough exposure to the principles underlying the conduct of research, and help prepare participants interested in research to pursue competitive fellowships, or other research training or career development awards. NHLBI intends to fund three to five new awards corresponding to a total of $900,000 for new grants per fiscal year.
*Requirements:* The following types of organizations are eligible to apply: Higher Education Institutions (Public/State Controlled Institutions of Higher Education, Private Institutions of Higher Education); Nonprofits other than Institutions of Higher Education, with or without 501(c)3 IRS Status; For-Profit Organizations; Small Businesses; State Governments; County Governments; City or Township Governments; Special District Governments; Indian/Native American Tribal Governments (Federally Recognized); Indian/Native American Tribal Governments (Other than Federally Recognized); Eligible Agencies of the Federal Government; U.S. Territory or Possession; Independent School Districts; Public Housing Authorities/Indian Housing Authorities; Native American Tribal Organizations (other than Federally recognized tribal governments); Faith-based or Community-based Organizations; Regional Organizations. The following types of Higher Education Institutions are always encouraged to apply for NIH support as Public or Private Institutions of Higher Education: Hispanic-serving Institutions; Historically Black Colleges and Universities (HBCUs); Tribally Controlled Colleges and Universities (TCCUs); Alaska Native and Native Hawaiian Serving Institutions.Some institutions provide unique opportunities for access to participants and faculty from diverse backgrounds. These institutions can assist NIH in its efforts to recruit the most talented researchers from all groups, to improve the ability to recruit subjects from diverse backgrounds into clinical research protocols, and improve the Nation's capacity to address and eliminate health disparities.

These institutions are therefore encouraged to apply for NIH support. The applicant institution must have a strong research program in the area(s) proposed for research training and must have the requisite staff and facilities to carry out the proposed program. The applicant institution must have the available research facilities, personnel, and support for the program in the areas of cardiovascular, pulmonary, or hematologic diseases or sleep disorders. Institutions with adequate staff and resources in these areas are encouraged to apply. Although the size of the award may vary within the scope of the research education program proposed, it is expected that applications will stay within the following budgetary guidelines: the total institutional annual direct cost should not exceed $319,000. The total project period for an application submitted in response to this funding opportunity may not exceed 5 years.
*Restrictions:* Non-domestic (non-U.S.) Entities (Foreign Institutions) and non-domestic (non-U.S.) components of U.S. Organizations are not eligible to apply. Foreign components, as defined in the NIH Grants Policy Statement, are not allowed.
*Geographic Focus* All States, Guam, Marshall Islands, Northern Mariana Islands, Puerto Rico, U.S. Virgin Islands
*Date(s) Application is Due:* Feb 7
*Amount of Grant:* Up to 2,700,000 USD
*Contact:* Drew Carlson, Ph.D., (301) 435-0535; carlsonde@nhlbi.nih.gov
*Internet:* grants.nih.gov/grants/guide/rfa-files/RFA-HL-13-020.html
*Sponsor:* National Institutes of Health
Building 31, Room 5A48
Bethesda, MD 20892

**NHLBI Summer Internship Program in Biomedical Research** **1824**
The goal of the Summer Internship Program in Biomedical Research at NHLBI is to expose students to research investigation in a highly enriched environment that is devoted exclusively to biomedical research and training. The Program is open to high school and college, graduate or Medical and Dental Students. Participants join a research laboratory for a minimum of ten weeks between June and August and conduct research in selected areas of investigation under the guidance of an NHLBI intramural research scientist. To augment the training experience, students participate in a Seminar Series at the NHLBI, as well as the National Institutes of Health Summer Seminar Series and Summer Research Program Poster Day. While applications may be filed between November 15 and March 1, early applications are encouraged.
*Requirements:* Applicants must be either U.S. citizens or permanent residents and be currently enrolled full-time in high school, college, professional or graduate school with a minimum GPA of 3.0 on a 4.0 scale. College graduates are eligible for the summer between graduation and beginning either graduate or professional school. High school students should be at least 16 years of age prior to June 1. Applications must be submitted online using the link for provided for Summer Internship Applicants.
*Geographic Focus:* All States
*Date(s) Application is Due:* Mar 1
*Contact:* Aurora Taylor, Summer Program Coordinator; (301) 451-9440; fax (301) 594-8133; taylora2@nhlbi.nih.gov or direducation@nhlbi.nih.gov
*Internet:* www.nhlbi.nih.gov/research/intramural/education/summer.htm
*Sponsor:* National Institutes of Health
Building 31, Room 5A48
Bethesda, MD 20892

**NHLBI Training Program Grants for Institutions that Promote Diversity** **1825**
The purpose of this program is to increase the participation of individuals from diverse backgrounds in cardiovascular, pulmonary, hematologic and sleep disorders research across the career development continuum. The NHLBI's T32 Training Program for Institutions That Promote Diversity is a Ruth L. Kirschstein National Research Service Award Program intended to support training of predoctoral and health professional students and individuals in postdoctoral training at non-research intensive institutions with an institutional mission focused on serving diverse communities that are not well represented in NIH-funded research, or identified federal legislation of same, with the potential to develop meritorious training programs in cardiovascular, pulmonary, hematologic, and sleep disorders. The Training Program is designed to expand the capability for biomedical research by providing grant support to institutions that have developed successful programs that promote diversity and that offer doctoral degrees in the health professions or in health-related sciences. These institutions are uniquely positioned to engage health disparities populations in research and translation of research advances that impact health outcomes, as well as provide health care for these populations. The primary goals of the Training Program are to: (1) contribute to the expansion of the future pool of individuals from diverse backgrounds underrepresented in the biomedicine science enterprise, (2) enable trainees to increase their competitiveness for peer-review research funding, (3) strengthen publication record of trainees, and (4) foster institutional environments conducive to professional development in the biomedical science. The application is due by 5:00 pm local time of the applicant organization.
*Requirements:* The following organizations/institutions are eligible to apply: Public/State Controlled Institutions of Higher Education and Private Institutions of Higher Education. The following types of Higher Education Institutions are always encouraged to apply for NIH support as Public or Private Institutions of Higher Education: Hispanic-serving Institutions; Historically Black Colleges and Universities (HBCUs); Tribally Controlled Colleges and Universities (TCCUs); Alaska Native and Native Hawaiian Serving Institutions. The Institution must be a domestic college or university with an institutional mission focused on serving diverse communities or with federal recognition of the same. To demonstrate need for research capacity development in heart, lung, and blood diseases and sleep disorders, applicant institutions should not have received over $1,515,000 in funding (direct costs) from the NHLBI per year in each of the previous five years for research training and education grants in the above NHLBI mission areas. Applicant organizations may submit more than one application, provided that each application is programmatically distinct. Any individual(s) with the skills, knowledge, and resources necessary to carry out the proposed research as the Program Director/Principal Investigator (PD/PI) is invited to work with his/her organization to develop an application for support. Individuals from underrepresented racial and ethnic groups as well as individuals with disabilities are always encouraged to apply for NIH support. The PD/PI should be an established investigator in the scientific area in which the application is targeted and capable of providing both administrative and scientific leadership to the development and implementation of the proposed program. The PD/PI will be expected to monitor and assess the program and submit all documents and reports as required. Although a letter of intent is not required, is not binding, and does not enter into the review of a subsequent application, the information that it contains allows IC staff to estimate the potential review workload and plan the review.
*Restrictions:* Non-domestic (non-U.S.) Entities (Foreign Organizations) are not eligible to apply. Foreign (non-U.S.) components of U.S. Organizations are not allowed.
*Geographic Focus* All States, Guam, Marshall Islands, Northern Mariana Islands, Puerto Rico, U.S. Virgin Islands
*Date(s) Application is Due:* Feb 7
*Amount of Grant:* Up to 250,000 USD
*Contact:* Sandra Colombini Hatch, M.D., (301) 435-0222; hatchs@nhlbi.nih.gov
*Internet:* grants.nih.gov/grants/guide/rfa-files/RFA-HL-12-032.html
*Sponsor:* National Institutes of Health
Building 31, Room 5A48
Bethesda, MD 20892

**NIA Alzheimer's Disease Research Centers Grants** **1826**
The National Institute on Aging funds Alzheimer's Disease Centers (AD.C.s) at major medical institutions across the United States. Researchers at these Centers will work to translate research advances into improved diagnosis and care for people with Alzheimer's disease while focusing on the program's long-term goal—finding a way to cure and possibly prevent Alzheimer's. Areas of investigation range from the basic mechanisms of Alzheimer's to managing the symptoms and helping families cope with the effects of the disease. Center staff will conduct basic, clinical, and behavioral research and train scientists and healthcare providers. Although each center will have its own area of emphasis, a common goal of the AD.C.s is to enhance research on Alzheimer's disease by creating a network that shares new ideas and research results. Collaborative studies draw upon the expertise of scientists from many different disciplines. For patients and families affected by Alzheimer's disease, the AD.C.s will offer: help with obtaining diagnosis and medical management (costs may vary—Centers may accept Medicare, Medicaid, and private insurance); information about the disease, services, and resources; opportunities for volunteers to participate in clinical trials and studies and patient registries; and support groups and other special programs for volunteers and their families. Some AD.C.s will have satellite facilities that offer these services in underserved, rural, and minority communities.
*Requirements:* The following organizations and institutions are eligible to apply: public/state controlled institutions of higher education; private institutions of higher education; Hispanic-serving institutions; historically Black colleges and universities (HBCUs); tribally controlled colleges and universities (TCCUs); Alaska Native and Native Hawaiian serving institutions; nonprofits with 501(c)3 IRS status (other than institutions of higher education); and nonprofits without 501(c)3 IRS status (other than institutions of higher education).
*Restrictions:* At the time of award, the applicant institution cannot have another Alzheimer's Disease Center funded by the NIA.
*Geographic Focus:* All States
*Amount of Grant:* Up to 1,000,000 USD
*Contact:* Creighton H. Phelps, Deputy Director; (301) 496-9350 or (800) 438-4380; fax (301) 496-1494; phelpsc@nia.nih.gov or adear@nia.nih.gov
*Internet:* https://www.nia.nih.gov/alzheimers/alzheimers-disease-research-centers
*Sponsor:* National Institute on Aging
Gateway Building, Suite 350, 7201 Wisconsin Avenue, MSC 9205
Bethesda, MD 20892-9205

**NIA Behavioral and Systems Neuroscience Branch Research Grants** **1827**
The mission of the Behavioral and Systems Neuroscience Branch is to support research aimed at elucidating the age-related alterations in neural systems that support behavioral functions such as cognition, emotion, movement, perception and sensory function. The branch is divided into two sections: Cognitive and Affective Neuroscience and Sensory/Motor Disorders of Aging. Clarification of cognitive, emotional, sensory and motor changes with age and the psychological and/or biological mechanisms underlying them are necessary pre-requisites for disassociating the effects of normal aging from those due to disease. Model systems, including computational models, invertebrates, and mammals, are central to this effort. The ultimate goal of such research is to foster the understanding of individual differences in cognitive, emotional, sensory, and motor function with age; and to elucidate the molecular, cellular and systems level mechanisms underlying age-related change in these behaviors. Better understanding of these mechanisms is important for planning therapeutic interventions, for appropriately using and enhancing skills and expertise of older people, and for specifying the requirements of an aging society.
*Requirements:* Organizations that are eligible to apply include: universities and colleges; medical, dental, and nursing schools; schools of public health; laboratories; hospitals; state and local health departments; other public or private institutions, both nonprofit and for-profit; and individuals. Individuals with professional or scientific degrees are eligible for NRSAs. SBIR grants can be awarded only to domestic small businesses. STTR grants can be awarded only to domestic small business concerns that partner with research institutions in cooperative research and development.

*Geographic Focus:* All States
*Amount of Grant:* 5,000 - 4,000,000 USD
*Contact:* Molly V. Wagster, Branch Chief; (301) 496-9350; wagsterm@nia.nih.gov
*Internet:* https://www.nia.nih.gov/research/dn/research-areas
*Sponsor:* National Institute on Aging
Gateway Building, Suite 350, 7201 Wisconsin Avenue, MSC 9205
Bethesda, MD 20892-9205

**NIA Causes and Effects of Elderly Population Concentrations Grants** 1828

The Causes and Effects of Elderly Population Concentrations program was first published in the NIH Guide for Grants and Contracts on March 26, 1992, and has a current expiration date of January 1, 2019. The intent of the program is to investigate and conduct research in the causes effects that the aged and elderly have upon the environment in which they are concentrated. Applications will be reviewed for scientific and technical merit in accordance with established PHS review procedures. Secondary review will be by the appropriate national advisory council. There are no identified annual deadlines for application submission. The funding ceiling is open.
*Requirements:* Applicants for research grants may be made by public and private, for-profit and non-profit organizations, such as universities college, hospitals, and laboratories. Women and minority investigators, in particular, are encouraged to apply. Foreign institutes are welcome to apply, but are advised to consult NIA staff before applying and are strongly encouraged to apply in collaboration with a U.S. institution. Foreign institutions are only eligible for the research (R01) and conference grant (R13) mechanisms. Applicants for fellowships (F32, F33) must either be U.S. citizens or U.S. permanent residents.
*Restrictions:* This program announcement will not support Career Development Awards.
*Geographic Focus:* All States, All Countries
*Contact:* John Haaga, Ph.D., Director; (301) 496-3131 or (301) 496-3136; jh735i@nih.gov or dbsrquery@nia.nih.gov
*Internet:* grants.nih.gov/grants/guide/notice-files/not92-308.html
*Sponsor:* National Institute on Aging
Gateway Building, Room 533, 7201 Wisconsin Avenue, MSC 9205
Bethesda, MD 20892

**NIA Clarifying the Relationship between Delirium and Alzheimers Disease and Related Dementias Grants** 1829

This Funding Opportunity Announcement (FOA) invites applications that focus on clarifying the relationship between delirium and Alzheimer's disease and related dementias (ADRD). Specifically sought is research focusing on understanding why persons with ADRD are at increased risk to develop delirium, often with a worse prognosis compared to those without antecedent ADRD, and why patients who experience delirium are at higher risk to develop subsequent short- and/or long-term mild cognitive impairment or ADRD, often with an accelerated rate of cognitive decline compared to those without preceding delirium. Relevant research projects may focus on, but are not limited to, those that A) provide insight into possible common, sequential, causative, contributory and/or synergistic pathways underlying both ADRD and delirium, B) elucidate mechanisms that lead to the development of delirium against the background of aging and/or neurodegeneration, with particular emphasis on use of appropriate animal models, C) identify risk factors for the onset and/or progression of delirium in those with ADRD and vice versa, D) diagnose and assess one condition in the setting of the other, E) identify putative phenotypes of patients with co-existing ADRD and delirium, or F) test pharmacologic and/or non-pharmacologic strategies to prevent, treat, or reduce the impact of delirium in patients with ADRD and vice versa. Research supported by this FOA is intended to provide mechanistic insight to improve risk assessment, diagnosis, phenotyping, prevention, and management approaches for both delirium and ADRD.
*Requirements:* Eligible organizations include: public and/or state controlled institutions of higher education; private institutions of higher education; nonprofits other than institutions of higher education; for-profit organizations; state governments; county governments; city or township governments; special district governments; Indian/Native American tribal governments (federally recognized); Indian/Native American tribal governments (other than federally recognized); eligible agencies of the federal government; U.S. territories or possessions; independent school districts; public housing authorities/Indian housing authorities; faith-based or community-based organizations; regional organizations; and non-domestic (non-U.S.) entities (Foreign Institutions).
*Geographic Focus:* All States
*Date(s) Application is Due:* Feb 5; Jun 5; Oct 5
*Contact:* Susan Zieman; (301) 496-6761; fax (301) 496-1494; susan.zieman@nih.gov
*Internet:* https://grants.nih.gov/grants/guide/pa-files/PAR-17-038.html
*Sponsor:* National Institute on Aging
Gateway Building, Suite 350, 7201 Wisconsin Avenue, MSC 9205
Bethesda, MD 20892-9205

**NIA Dementias of Aging Branch Research Grants** 1830

The mission of the Dementias of Aging Branch is to support research on the etiology, epidemiology, diagnosis, treatment, and prevention of Mild Cognitive Impairment (MCI), AD, and other dementias (e.g. Vascular Dementia, Frontotemporal Dementia, Lewy Body Dementia) of older people. This research encompasses basic, clinical, and epidemiological studies as well as clinical trials of these disorders and diseases. The Branch is composed of four sections:Basic Research, Population Studies, Clinical Studies, and Research Centers. There is a close collaboration among the Neurobiology, Behavioral and Systems Neuroscience, and Dementias Branches because of the shared interests involving age-related cognitive decline, MCI, and AD. The ultimate goal of the research supported by the Dementias of Aging Branch is the pre-symptomatic identification of the earliest brain changes that ultimately result in dementia in conjunction with the development of therapeutics which will be able to delay disease onset or progression or to prevent disease entirely. Basic Research supports research on the etiology of MCI, AD and other age-related neurodegenerative disorders, including studies to identify genetic loci associated with inherited and sporadic forms of these diseases; biochemical and molecular genetic analysis of the components of amyloid plaques, neurofibrillary tangles, and other abnormal structures found in the brains of individuals with MCI, AD, and other dementias of aging and the molecular consequences of their accumulation.
*Requirements:* Organizations that are eligible to apply include: universities and colleges; medical, dental, and nursing schools; schools of public health; laboratories; hospitals; state and local health departments; other public or private institutions, both nonprofit and for-profit; and individuals. Individuals with professional or scientific degrees are eligible for NRSAs. SBIR grants can be awarded only to domestic small businesses. STTR grants can be awarded only to domestic small business concerns that partner with research institutions in cooperative research and development.
*Geographic Focus:* All States
*Amount of Grant:* 5,000 - 4,000,000 USD
*Contact:* Laurie M. Ryan, Branch Chief; (301) 496-9350; ryanl@mail.nih.gov
*Internet:* https://www.nia.nih.gov/research/dn/research-areas
*Sponsor:* National Institute on Aging
Gateway Building, Suite 350, 7201 Wisconsin Avenue, MSC 9205
Bethesda, MD 20892-9205

**NIA Failure to Thrive Syndrome Among Older Persons Grants** 1831

The Failure to Thrive Syndrome Among Older Persons program was initiated on November 20, 1992. The geriatric syndrome termed "failure to thrive" has been described, consisting of weight loss, decreased appetite, poor nutrition, and inactivity, often accompanied by dehydration, depressive symptoms, impaired immune function, and low serum cholesterol. Failure to thrive occurs in both acute and chronic forms, leading to impaired functional status, morbidity from infection, pressure sores, and increased mortality. This syndrome has been identified as an aging research priority in the Institute of Medicine's report: Extending Life, Enhancing Life. A National Research Agenda on Aging. The relationships between nutritional, metabolic and other pathophysiologic factors in failure to thrive and their effects on clinical outcomes remain to be clarified. While there is evidence that psychiatric conditions (major and subsyndromal depression and cognitive decline) are frequently associated with failure to thrive, the nature of this association needs to be clarified. The mechanisms for support are: research grants; FIRST awards; career awards (which include Special Emphasis Research Career Awards in Nutritional and Metabolic Factors in Aging, Research Career Development Awards, Clinical Mental Health Academic Award, Clinical Investigator Awards, Physician Scientist Awards, Scientist Development Awards for Clinicians, and Scientist Development Awards.
*Requirements:* Applications may be submitted by foreign and domestic, for-profit and non-profit organizations, public and private, such as universities, colleges, hospitals, laboratories, unit of state and local governments, and eligible agencies of the Federal government. Applications from minority individuals and women are encouraged. Foreign institutions are not eligible for First Independent Research Support and Transition awards (R29) or for the Career Awards (K01, K04, K07, K08, K11, K20, K21).
*Geographic Focus:* All States, All Countries
*Contact:* Heidi Brogdon, Program Specialist; (301) 496-0181 or (301) 496-6402; brogdonh@mail.nih.gov or dabquery@nia.nih.gov
*Internet:* https://grants.nih.gov/grants/guide/pa-files/PA-93-022.html
*Sponsor:* National Institute on Aging
Gateway Building, Suite 2C231, 7201 Wisconsin Avenue, MSC 9205
Bethesda, MD 20892

**NIAF Congressional Fellowships** 1832

The National Italian American Foundation's mission is to promote, preserve and protect Italian American heritage and culture. The NIAF supports funding to organizations, individuals, schools, and communities that provide opportunities to encourage collaboration among students and academics in the United States and Italy and improve the quality of Italian language teaching in the United States. Through the Congressional Fellowship program, the Foundation will place outstanding Italian American college students, graduate students and recent graduates in offices of members of the Italian American Congressional Delegation (IACD) to encourage and support the next generation of Italian American leadership. Five students and/or recent graduates will be selected as NIAF Congressional Fellows and will travel to Washington D.C. to work in a Congressional office. NIAF will award fellows with a stipend of $1000 to aid with travel and living expenses while in Washington. Fellows will also receive training at NIAF headquarters and will be invited to participate in NIAF events throughout the fall, including NIAF Public Policy Forums and the Annual Gala Weekend.
*Requirements:* Students must: be members of the National Italian American Foundation; be a junior or senior in college, or in the first year out of college, a graduate student, or in the first year out of college/graduate school, at beginning of internship date; have at least one Italian grandparent; and provide two references who are either current/former employers or supervisors or current/former professors.
*Geographic Focus:* All States
*Date(s) Application is Due:* Apr 30
*Amount of Grant:* 1,000 USD
*Contact:* Gabriella Mileti, Director of Programs; (202) 939-3114 or (202) 387-0600; fax (202) 387-0800; gmileti@niaf.org or fellowship@niaf.org
*Internet:* www.niaf.org/programs/niaf-congressional-fellowship/
*Sponsor:* National Italian American Foundation
1860 19th Street NW
Washington, D.C. 20009-5501

**NIA Geriatrics Program Grants** 1833

As part of a continuing effort to build and sustain a broad program of research on diseases of older persons and the underlying pathophysiologies, the Geriatrics Program (GP) of the National Institute on Aging encourages investigators to submit applications that propose moderate budgets and durations. GP staff continues to encourage research on aging and physical function and performance, endocrinology, osteoporosis, arthritis, cardiovascular issues, nutrition, metabolism, and related aging research areas. Studies on etiology and pathophysiology of age-related disorders, as well as clinical studies, are encouraged. There are no identified annual deadlines for application submission. No funding ceiling has been determined

*Requirements:* Applicants for research grants may be made by public and private, for-profit and non-profit organizations, such as universities college, hospitals, and laboratories. Women and minority investigators, in particular, are encouraged to apply. Foreign institutes are welcome to apply, but are advised to consult NIA staff before applying and are strongly encouraged to apply in collaboration with a U.S. institution.

*Geographic Focus:* All States, All Countries

*Contact:* Evan C. Hadley; (301) 435-3044 or (301) 496-6761; hadleye@mail.nih.gov

*Internet:* grants.nih.gov/grants/guide/notice-files/not92-112.html

*Sponsor:* National Institute on Aging

Gateway Building, Suite 3C307, 7201 Wisconsin Avenue, MSC 9205

Bethesda, MD 20892-9205

**NIA Institutional Training Grants** 1834

NIA will award new and competing renewal NRSA (NRSA) Institutional Training Grants (T32) to eligible institutions to develop or enhance research training opportunities for individuals, selected by the institution, who are training for careers in specified areas of aging research. The purpose of this announcement is to help ensure a broad cadre of researchers trained in the content and methods of aging research and in the major biomedical, behavioral, and social areas of research most closely related to their field of study within aging. NIA supports both predoctoral and postdoctoral training in aging. The institute also supports short-term research training for students in health- professional programs as part of an overall T32 program. The NIA has a single receipt date each year for applications.

*Requirements:* Applicant must be a citizen or noncitizen national of the United States or must have been lawfully admitted for permanent residence. Predoctoral trainees must have received a baccalaureate degree by the beginning date of their NRSA appointment, and must be training at the postbaccalaureate level and enrolled in a program leading to a PhD in science or in an equivalent research doctoral degree program. Postdoctoral trainees must have received, as of the beginning date of the NRSA appointment, a PhD, MD, or comparable doctoral degree from an accredited domestic or foreign institution. To be eligible for short-term predoctoral research training positions, health-professional students must have completed at least one quarter in a program leading to a clinical doctorate prior to participating in the program.

*Geographic Focus:* All States

*Contact:* Dr. Robin Barr, Director, Office of Extramural Affairs; (301) 496-9322; fax (301) 402-2945; rb42h@nih.gov or barrr@mail.nih.gov

*Internet:* https://www.nia.nih.gov/research/dea/institutional-training-grants

*Sponsor:* National Institute on Aging

Building 31, Room 5C27, 31 Center Drive, MSC 2292

Bethesda, MD 20892-7710

**NIA Minority Dissertation Research Grants in Aging** 1835

Minority Dissertation Research Grants in Aging are small grants to support doctoral dissertation research available for underrepresented minority doctoral candidates. Grant support is designed to aid the research of such investigators and to encourage underrepresented minority individuals from a variety of academic disciplines and programs to conduct research related to aging. It is anticipated that a total of five to six new applications will be funded in each year. The maximum budget request should be limited to $25,000 in direct costs for the initial budget period. No more than $30,000 in total direct costs will be provided across the two year period.

*Requirements:* This research initiative is to provide underrepresented minority students assistance to complete their dissertation research on an aging-related topic and thereby increase their representation in aging research.

*Geographic Focus:* All States

*Date(s) Application is Due:* Mar 15; Nov 15

*Amount of Grant:* Up to 30,000 USD

*Contact:* Laurie M. Ryan, Chief of the Dementias of Aging Branch; (301) 496-9350; fax (301) 496-1494; ryanl@nia.nih.gov

*Internet:* https://grants.nih.gov/grants/guide/rfa-files/RFA-AG-93-003.html

*Sponsor:* National Institute on Aging

Gateway Building, Suite 350, 7201 Wisconsin Avenue, MSC 9205

Bethesda, MD 20892-9205

**NIAMS Pilot and Feasibility Clinical Research Grants in Arthritis and Musculoskeletal and Skin Diseases** 1836

The goal of this initiative is to encourage novel clinical research aimed at improving the understanding of the causes, treatment or prevention of arthritis and musculoskeletal and skin diseases. This Pilot and Feasibility Clinical Research Grants Program is designed to allow initiation of exploratory, short-term clinical studies, intended to facilitate the development of new ideas which may be investigated without stringent requirements for preliminary data. Research aims appropriate for an investigator-initiated R01 grant, such as understanding underlying biological mechanisms of disease, would not be expected in an application under this FOA. Proposed studies should focus on research questions that are likely to gather critical preliminary data in support of a future, planned clinical trial or to benefit clinical research and trials more broadly. Overall goals of these studies are intended to generate data critical to future clinical research or trials such as by demonstrating the feasibility of a recruitment target or approach, obtaining data in support of potential inclusion and exclusion criteria, or to gather preliminary evidence of efficacy, tolerability and/or toxicity of an available drug, biologic or device in a new population relevant to the NIAMS mission. Examples include testing of: new prevention strategies; a new application of existing pharmacologic or non-pharmacologic interventions or their unique combination; a new cost effective or adaptive trial design in a specific population or across populations; and studies intended to expand the science of clinical trials overall. Results generated from this program may lead to critical input into a new clinical trial; trials that inform decision-making in urgent or high priority clinical practice needs for arthritis and musculoskeletal and skin diseases; or preliminary data for trials in understudied and rare diseases. A high priority is the use of such studies to help stimulate the translation of promising research developments from the laboratory into clinical practice. The total project period for an application submitted in response to this funding opportunity may not exceed 2 years. Although the size of award may vary with the scope of research proposed, it is expected that applications will stay within the budgetary guidelines for an exploratory/developmental project; direct costs are limited to $275,000 over an R21 two-year period, with no more than $200,000 in direct costs allowed in any single year.

*Requirements:* The following organizations/institutions are eligible to apply: public/state controlled institutions of higher education; private institutions of higher education; Hispanic-serving institutions; historically Black colleges and universities; tribally controlled colleges and universities; Alaska native and native Hawaiian serving institutions; nonprofits with 501(c)3 IRS status (other than institutions of higher education); nonprofits without 501(c)3 IRS status (other than institutions of higher education); small businesses; for-profit organizations (other than small businesses); state governments; Indian/Native American tribal governments (federally recognized); Indian/Native American tribally designated organizations; county governments; city or township governments; special district governments; independent school districts; public housing authorities/Indian housing authorities; U.S. territory or possession; Indian/Native American tribal governments (other than federally recognized); regional organizations; non-domestic (non-U.S.) entities (foreign organizations); eligible agencies of the federal government; and faith-based or community-based organizations.

*Geographic Focus:* All States

*Date(s) Application is Due:* Mar 1; Jul 1; Nov 1

*Amount of Grant:* Up to 275,000 USD

*Contact:* James Witter, (301) 594-1963; fax (301) 480-4543; witterj@mail.nih.gov

*Internet:* grants.nih.gov/grants/guide/pa-files/PAR-10-282.html#SectionVII

*Sponsor:* National Institute of Arthritis and Musculoskeletal and Skin Diseases

1 AMS Circle

Bethesda, MD 20892-3675

**NIA Neurobiology of Aging Branch Research Grants** 1837

The mission of the Neurobiology of Aging Branch is to support research aimed at elucidating how the nervous system is affected by normal as well as pathological aging. The branch is divided into three sections: Fundamental Neuroscience, Integrative Neurobiology, and Sleep and Biological Rhythms. This Branch encompasses areas of neuroscience exclusive of research focused on the dementias of aging, and fosters research from the most basic levels to the translation of research findings into interventions promoting the health of the public. Several motifs highlight the research questions funded by the Neurobiology of Aging Branch. These are: selective vulnerability of regions of the central nervous system (CNS) to aging effects; molecular genetics of brain aging; oxidative stress, bioenergetic processes and cell death mechanisms; cellular and molecular mechanisms of neural plasticity and brain repair processes; interactions between organ systems (especially the immune, endocrine, and central nervous systems) through common signaling molecules and the potential to uncouple the normal interactions of these systems by age-associated changes, and circadian pacemakers and age-related mechanisms underlying sleep-wakefulness cycles. The ultimate goal of such research is the maintenance of a high quality of life of the older population through a better understanding of the aging nervous system and the application of basic research findings to improve and correct age-associated neuropathologies.

*Requirements:* Organizations that are eligible to apply include: universities and colleges; medical, dental, and nursing schools; schools of public health; laboratories; hospitals; state and local health departments; other public or private institutions, both nonprofit and for-profit; and individuals. Individuals with professional or scientific degrees are eligible for NRSAs. SBIR grants can be awarded only to domestic small businesses. STTR grants can be awarded only to domestic small business concerns that partner with research institutions in cooperative research and development.

*Geographic Focus:* All States

*Amount of Grant:* 5,000 - 4,000,000 USD

*Contact:* Bradley C. Wise, Branch Chief; (301) 496-9350; wiseb@nia.nih.gov

*Internet:* https://www.nia.nih.gov/research/dn/research-areas

*Sponsor:* National Institute on Aging

Gateway Building, Suite 350, 7201 Wisconsin Avenue, MSC 9205

Bethesda, MD 20892-9205

**NIA Neuroscience of Aging Research Grants** 1838

The National Institute on Aging's (NIA) commitment to the systematic study of the aging brain began with its establishment in 1974, and its support of neuroscience research, including research on Alzheimer's disease (AD), has increased dramatically since then. Among the many age-related impairments that lead to institutional care, those related to changes in brain functioning have the most significant implications for public policy and priorities for further research. Changes in the brain, especially those that affect sensory, motor, sleep, cognitive, and emotional functioning, profoundly influence the quality of

life of older individuals. Reduced functional capacity not only limits independence but also influences the attitudes of others toward the aging person, affects the individual's self-image, and often determines the nature and quality of health-care services. Overall, the Division supports a broad spectrum of research aimed at elucidating how the central nervous system and behavior are affected by normal as well as pathological aging. An emerging focus is how the processes of aging and age-related cognitive decline intersect with the development of Alzheimer's disease (AD) and other dementias of aging. The basic theme throughout the Division is to understand the aging nervous system in order to foster the maintenance of health and improve the quality of life of the older population.
*Requirements:* Organizations that are eligible to apply include: universities and colleges; medical, dental, and nursing schools; schools of public health; laboratories; hospitals; state and local health departments; other public or private institutions, both nonprofit and for-profit; and individuals. Individuals with professional or scientific degrees are eligible for NRSAs. SBIR grants can be awarded only to domestic small businesses. STTR grants can be awarded only to domestic small business concerns that partner with research institutions in cooperative research and development.
*Geographic Focus:* All States
*Amount of Grant:* 5,000 - 4,000,000 USD
*Contact:* Creighton H. Phelps, Acting Director; (301) 496-9350; phelpsc@mail.nih.gov
*Internet:* https://www.nia.nih.gov/research/dn/research-areas
*Sponsor:* National Institute on Aging
Gateway Building, Suite 350, 7201 Wisconsin Avenue, MSC 9205
Bethesda, MD 20892-9205

**NIA Oxidative Damage, Antioxidant Defense, and Aging Grants** 1839
The purpose of the Oxidative Damage, Antioxidant Defense, and Aging program is to encourage the submission of applications to support research on the relationship between oxidative damage and antioxidant defense, and how these biological processes are involved in aging and/or change with age. Although the National Institute of General Medical Sciences is not participating in this program announcement, that Institute continues to be interested in receiving applications on basic mechanisms of programmed cell death and oxidative damage when these applications do not focus on aging mechanisms. Modular Grant applications will request direct costs in $25,000 modules, up to a total direct cost request of $250,000 per year.
*Requirements:* The following organizations and institutions may apply: public/state controlled institutions of higher education; private institutions of higher education; Hispanic-serving institutions; historically Black colleges and universities (HBCUs); tribally controlled colleges and universities (TCCUs); Alaska native and native Hawaiian serving institutions; nonprofits with 501(c)3 IRS status; nonprofits without 501(c)3 IRS status; small businesses; for-profit organizations; State governments; regional organizations; U.S. territories or possessions; Indian/Native American tribal governments; Indian/Native American tribally designated organizations; county governments; city or township governments; special district governments; independent school districts; public housing authorities/Indian housing authorities; eligible agencies of the Federal government; and faith-based or community based organizations.
*Geographic Focus:* All States
*Amount of Grant:* Up to 250,000 USD
*Contact:* Heidi Brogdon, Senior Program Specialist; (301) 496-0181 or (301) 496-6402; brogdonh@mail.nih.gov or dabquery@nia.nih.gov
*Internet:* https://grants.nih.gov/grants/guide/notice-files/not93-026.html
*Sponsor:* National Institute on Aging
Gateway Building, Suite 2C231, 7201 Wisconsin Avenue, MSC 9205
Bethesda, MD 20892

**NIA Pilot Clinical Trials for the Spectrum of Alzheimer's Disease and Age-Related Cognitive Decline** 1840
This Funding Opportunity Announcement (FOA) invites applications that propose to develop and implement Phase I or II clinical trials of promising pharmacological and non-pharmacological interventions in individuals with age-related cognitive decline and in individuals with Alzheimer's disease (AD) across the spectrum from pre-symptomatic to more severe stages of disease, as well as to stimulate studies to enhance trial design and methods. Examples of interventions for evaluation include but are not limited to: pharmacological interventions that target eradication or progression of deposition of disease neuropathology; repurposed drugs that have promise for AD treatment such as chemotherapeutic agents or drugs for insulin dysregulation/diabetes; novel cognitive training or cognitive engagement approaches; aerobic exercise and/or other movement therapies, such as Tai Chi; sleep enhancement; mindfulness-based stress reduction; nutritional supplementation or adoption of specific dietary programs; and/or combinations of interventions including the mixture of pharmacological with non-pharmacological therapeutics. NIH intends to fund an estimate of ten to fifteen awards, corresponding to a total of $10 million for the fiscal year. The annual due date for applications is December 16.
*Requirements:* The following organizations and institutions may apply: public and state controlled institutions of higher education; private institutions of higher education; Hispanic-serving institutions; historically Black colleges and universities (HBCUs); tribally controlled colleges and universities (TCCUs); Alaska native and native Hawaiian serving institutions; nonprofits with 501(c)3 IRS status; nonprofits without 501(c)3 IRS status; small businesses; for-profit organizations; State governments; regional organizations; U.S. territories or possessions; Indian/Native American tribal governments; Indian/Native American Tribally designated organizations; county governments; city or township governments; special district governments; independent school districts; public housing authorities/Indian housing authorities; eligible agencies of the Federal government; and faith-based or community based organizations.
*Geographic Focus:* All States
*Date(s) Application is Due:* Dec 16
*Amount of Grant:* 500,000 - 1,500,000 USD
*Contact:* Laurie M. Ryan, Scientific Program Contact; (301) 496-9350; fax (301) 496-1494; ryanl@nia.nih.gov or ryanl@mail.nih.gov
*Internet:* grants.nih.gov/grants/guide/pa-files/PAR-16-365.html
*Sponsor:* National Institute on Aging
Gateway Building, Suite 350, 7201 Wisconsin Avenue, MSC 9205
Bethesda, MD 20892-9205

**NIA Resources Available for Conducting Research on Aging Grants** 1841
The National Institute on Aging (NIA), recognizing that most investigators have neither the facilities nor the fiscal resources needed to develop and maintain colonies of aged animals, has made provision of resources one of its highest priorities. The availability of high quality material for conducting research on aging continues to be a priority area for NIA, with current resources spanning a broad spectrum, including a cell bank, a nematode bank, rodent colonies, and nonhuman primate colonies. Budgets for direct costs of up to $50,000 per year and a project duration of up to two years may be requested for a maximum of $100,000 direct costs over a two-year project period.
*Requirements:* The following organizations and institutions may apply: for-profit organizations; non-profit organizations; public or private institutions, such as universities, colleges, hospitals, and laboratories; units of State government; units of Local government; units of State Tribal government; units of Local Tribal government; eligible agencies of the Federal government; foreign institutions; domestic institutions; faith-based or community-based organizations; Indian/Native American tribal governments (federally recognized); Indian/Native American tribal governments (other than federally recognized); and Indian/Native American tribally designated organizations.
*Geographic Focus:* All States
*Amount of Grant:* Up to 100,000 USD
*Contact:* Heidi Brogdon, Program Specialist; (301) 496-0181 or (301) 496-6402; fax (301) 402-0010; brogdonh@mail.nih.gov or dabquery@nia.nih.gov
*Internet:* grants.nih.gov/grants/guide/notice-files/not93-196.html
*Sponsor:* National Institute on Aging
Gateway Building, Suite 2C231, 7201 Wisconsin Avenue, MSC 9205
Bethesda, MD 20892

**NIA Ruth L. Kirschstein National Research Service Award Individual Postdoctoral Fellowships** 1842
The overall goal of the NIH Ruth L. Kirschstein National Research Service Award (NRSA) program is to help ensure that a diverse pool of highly trained scientists is available in appropriate scientific disciplines to address the Nation's biomedical, behavioral, and clinical research needs. NRSA fellowships support the training of pre-and postdoctoral scientists, dual-degree investigators, and senior researchers. This particular Fellowship supports training for basic and clinical research on aging and the problems of the aging and the aged. The training should be directed toward long-term involvement in research in aging and gerontology. The institute will accept proposals in the behavioral and social sciences related to aging and in the biological and biomedical sciences related to aging. Support may be requested for up to three years. PA: PA-03-067
*Requirements:* Eligible organizations include: higher education institutions, particularly Hispanic-serving Institutions, Historically Black Colleges and Universities (HBCUs), Tribally Controlled Colleges and Universities (TCCUs), Alaska Native and Native Hawaiian Serving Institutions, and Asian American Native American Pacific Islander Serving Institutions (AANAPISIs); nonprofits (other than institutions of higher education); for-profit organizations; government entities; and non-domestic entities.
*Geographic Focus:* All States, All Countries
*Date(s) Application is Due:* Apr 5; Aug 5; Dec 5
*Amount of Grant:* 35,000 - 48,000 USD
*Contact:* Chyren Hunter; (301) 496-9322; fax (301) 402-2945; hunterc@nia.nih.gov
*Internet:* grants.nih.gov/grants/guide/pa-files/PA-16-307.html
*Sponsor:* National Institute on Aging
Building 31, Room 5C27, 31 Center Drive, MSC 2292
Bethesda, MD 20892-7710

**NIAS in the Humanities and Social Sciences Individual Fellowships** 1843
NIAS Individual Fellowships are awarded to senior scholars to carry out advanced research in the humanities and the social sciences. Fellows have at least three years of post-PhD degree academic experience and have already made a considerable contribution to their field. Fellowships are funded directly by NIAS through the Royal Netherlands Academy of Arts and Sciences (KNAW). Awards are by application only, and suit scholars from the Netherlands or scholars from abroad who wish to work on an individual project for one semester (September through January or February through June) or two semesters (September through June) at the Institute.
*Requirements:* Eligibility criteria includes: at least three years of post-Ph.D. degree academic experience; a completed application form with all relevant attachments; and an ffiliation with a university or research institute.
*Restrictions:* NIAS does not provide fellowships for training programs or educational purposes; nor does it offer courses, or financial support to conduct research elsewhere.
*Geographic Focus:* All States, Netherlands
*Contact:* Head of Academic Affairs; +31-(0)20 224 67 03 or +31 (0)20 224 67 00; application@nias.knaw.nl or application@nias.knaw.nl
*Internet:* www.nias.knaw.nl/fellowships/individual-fellowships
*Sponsor:* Netherlands Institute for Advanced Study in the Humanities and Social Sciences
Korte Spinhuissteeg 3, P.O. Box 10855
Amsterdam, 1001 EW Netherlands

**NICHD Studies in Neonatal and Pediatric Resuscitation (R01)** 1844

The purpose of this funding opportunity is to stimulate research on a wide range of topics related to neonatal and pediatric resuscitation. Possible topics may include, but are not limited to: fetal-neonatal transitional cardiovascular and pulmonary physiology, optimizing steps of resuscitation, management of third stage of labor and its effect on the fetus, resuscitation of children with malformations, effect of resuscitation on long-term outcomes and post-resuscitation practices. Applications can include epidemiological studies, studies utilizing fetal-neonatal animal models, computer or other information-technology-based simulations or study designs, clinical observational studies, analyses of pre-existing national or regional datasets, opportunistic studies prospective randomized controlled trials, or any combinations thereof. It is anticipated that the results from well-conducted studies will enable translation of knowledge into evidence-based resuscitation practices ensuring optimal short- and long-term outcomes for all newborn infants and children.

*Requirements:* Eligible applicants include: Higher Education Institutions; Nonprofits Other Than Institutions of Higher Education; For-Profit Organizations; Governments; and Other [Independent School Districts, Public Housing Authorities/Indian Housing Authorities, Native American Tribal Organizations (other than Federally recognized tribal governments), Faith-based or Community-based Organizations, Regional Organizations, and Non-domestic (non-U.S.) Entities (Foreign Institutions)]. The following types of Higher Education Institutions are always encouraged to apply for NIH support as Public or Private Institutions of Higher Education: Hispanic-serving Institutions; Historically Black Colleges and Universities (HBCUs); Tribally Controlled Colleges and Universities (TCCUs); Alaska Native and Native Hawaiian Serving Institutions; and, Asian American Native American Pacific Islander Serving Institutions (AANAPISIs). Application budgets are not limited but need to reflect the actual needs of the proposed project. The scope of the proposed project should determine the project period. The maximum project period is 5 years.

*Restrictions:* Non-domestic (non-U.S.) Entities (Foreign Institutions) and non-domestic (non-U.S.) components of U.S. Organizations are not eligible to apply.

*Geographic Focus:* All States

*Date(s) Application is Due:* Jan 8

*Contact:* Marion Koso-Thomas, M.D, MPH, (301) 496-9485; fax (301) 435-6873; kosomari@mail.nih.gov

*Internet:* https://grants.nih.gov/grants/guide/pa-files/PA-18-485.html

*Sponsor:* National Institutes of Health
P.O. Box 3006
Rockville, MD 20847

**NICHD Studies in Neonatal and Pediatric Resuscitation (R03)** 1845

The purpose of this funding opportunity is to stimulate research on a wide range of topics related to neonatal and pediatric resuscitation. Possible topics may include, but are not limited to: fetal-neonatal transitional cardiovascular and pulmonary physiology, optimizing steps of resuscitation, management of third stage of labor and its effect on the fetus, resuscitation of children with malformations, effect of resuscitation on long-term outcomes and post-resuscitation practices. Applications can include epidemiological studies, studies utilizing fetal-neonatal animal models, computer or other information-technology-based simulations or study designs, clinical observational studies, analyses of pre-existing national or regional datasets, opportunistic studies prospective randomized controlled trials, or any combinations thereof. It is anticipated that the results from well-conducted studies will enable translation of knowledge into evidence-based resuscitation practices ensuring optimal short- and long-term outcomes for all newborn infants and children.

*Requirements:* Eligible applicants include: Higher Education Institutions; Nonprofits Other Than Institutions of Higher Education; For-Profit Organizations; Governments; and Other [Independent School Districts, Public Housing Authorities/Indian Housing Authorities, Native American Tribal Organizations (other than Federally recognized tribal governments), Faith-based or Community-based Organizations, Regional Organizations, and Non-domestic (non-U.S.) Entities (Foreign Institutions)]. The following types of Higher Education Institutions are always encouraged to apply for NIH support as Public or Private Institutions of Higher Education: Hispanic-serving Institutions; Historically Black Colleges and Universities (HBCUs); Tribally Controlled Colleges and Universities (TCCUs); Alaska Native and Native Hawaiian Serving Institutions; and, Asian American Native American Pacific Islander Serving Institutions (AANAPISIs). Application budgets are limited to $50,000 in direct costs per year. The scope of the proposed project should determine the project period. The maximum project period is 2 years.

*Restrictions:* Non-domestic (non-U.S.) Entities (Foreign Institutions) and non-domestic (non-U.S.) components of U.S. Organizations are note eligible to apply.

*Geographic Focus:* All States

*Date(s) Application is Due:* Jan 8

*Amount of Grant:* Up to 100,000 USD

*Contact:* Marion Koso-Thomas, M.D, MPH, (301) 496-9485; fax (301) 435-6873; kosomari@mail.nih.gov

*Internet:* https://grants.nih.gov/grants/guide/pa-files/PA-18-491.html

*Sponsor:* National Institutes of Health
P.O. Box 3006
Rockville, MD 20847

**NICHD Studies in Neonatal and Pediatric Resuscitation (R21)** 1846

The purpose of this funding opportunity is to stimulate research on a wide range of topics related to neonatal and pediatric resuscitation. Possible topics may include, but are not limited to: fetal-neonatal transitional cardiovascular and pulmonary physiology, optimizing steps of resuscitation, management of third stage of labor and its effect on the fetus, resuscitation of children with malformations, effect of resuscitation on long-term outcomes and post-resuscitation practices. Applications can include epidemiological studies, studies utilizing fetal-neonatal animal models, computer or other information-technology-based simulations or study designs, clinical observational studies, analyses of pre-existing national or regional datasets, opportunistic studies prospective randomized controlled trials, or any combinations thereof. It is anticipated that the results from well-conducted studies will enable translation of knowledge into evidence-based resuscitation practices ensuring optimal short- and long-term outcomes for all newborn infants and children.

*Requirements:* Eligible applicants include: Higher Education Institutions; Nonprofits Other Than Institutions of Higher Education; For-Profit Organizations; Governments; and Other [Independent School Districts, Public Housing Authorities/Indian Housing Authorities, Native American Tribal Organizations (other than Federally recognized tribal governments), Faith-based or Community-based Organizations, Regional Organizations, and Non-domestic (non-U.S.) Entities (Foreign Institutions)]. The following types of Higher Education Institutions are always encouraged to apply for NIH support as Public or Private Institutions of Higher Education: Hispanic-serving Institutions; Historically Black Colleges and Universities (HBCUs); Tribally Controlled Colleges and Universities (TCCUs); Alaska Native and Native Hawaiian Serving Institutions; and, Asian American Native American Pacific Islander Serving Institutions (AANAPISIs). Non-domestic (non-U.S.) Entities (Foreign Institutions) and non-domestic (non-U.S.) components of U.S. Organizations are eligible to apply. Application budgets are limited to $275,000 in direct costs over the two-year award period. No more than $200,000 may be requested in any single year. The total project period may not exceed 2 years.

*Geographic Focus:* All States, All Countries

*Date(s) Application is Due:* Jan 8

*Amount of Grant:* Up to 275,000 USD

*Contact:* Marion Koso-Thomas, M.D, MPH, (301) 496-9485; fax (301) 435-6873; kosomari@mail.nih.gov

*Internet:* https://grants.nih.gov/grants/guide/pa-files/PA-14-349.html

*Sponsor:* National Institutes of Health
P.O. Box 3006
Rockville, MD 20847

**Nick Traina Foundation Grants** 1847

The Foundation supports organizations involved in the diagnosis, research, treatment, and/or family support of manic-depression, and other forms of mental illness, suicide prevention, child abuse and children in jeopardy, and provides assistance to struggling musicians in the areas of mental illness. The Foundation may give special consideration to proposals that address manic-depression in children and young adults. There are no deadlines and there is no formal application form. The Board meets four times a year to review proposals.

*Requirements:* Applicants must be 501(c)3 organizations. Proposals should be kept to three pages in addition to attachments and should include: the organization's purpose, including a description of its history and mission and the population served; the reason for the grant request, including the amount requested and how the funds will be used; whether the request is for general organizational support or for a specific project (including the duration of a specific project); a copy of the IRS determination letter; and a copy of the most recent annual report and the last two years' financial statements.

*Restrictions:* Organizations may apply for one grant each calendar year. Requests from individuals are ineligible. Grant funding focuses on the San Francisco Bay area.

*Geographic Focus:* California

*Samples:* At the Crossroads, San Francisco, California, $11,000; Homeless Youth Alliance, San Francisco, California, $27,000; and The Caduceus Society, Abilene, Texas, $1,000.

*Contact:* Danielle Steel, President; (415) 771-4224; info@nicktrainafoundation.org

*Internet:* nicktrainafoundation.com/grants.htm

*Sponsor:* Nick Traina Foundation
P.O. Box 470427
San Francisco, CA 94147-0427

**NIDCR Administrative Supplements for Research on Dietary Supplements Grants** 1848

The purpose of the NID.C.R Administrative Supplements for Research on Dietary Supplements Grants program is to support research in which the supplemental funding would investigate the role of dietary supplements and/or their ingredients in health maintenance and disease prevention. Parent awards need not be focused on dietary supplements; this FOA may provide support to include dietary supplements within the scope of relevant research projects. Research interests of ODS are not limited to specific health conditions, organ systems or population groups. ODS supports all types of research, including pre-clinical, clinical, behavioral, and epidemiological. Additionally, ODS supports research and training programs that build future research capacity for studying the role of dietary supplements in health and disease prevention. Primary consideration for support will be given to applications that stimulate dietary supplement research where it is lacking or lagging, clarify gaps, opportunities and balance between benefits and risks where data are in conflict, target special population groups where additional science on dietary supplements is needed, and focus on the use of dietary supplements in improving or maintaining health and reducing the risk of chronic disease. Budget requests may be for no more than $100,000 in direct costs. The application deadline for submission is April 15.

*Requirements:* Eligible organizations include: higher education institutions; nonprofits (other than higher education institutions); Hispanic-serving institutions; historically Black colleges and universities; tribally-controlled colleges and universities; Alaske native serving institutions; native Hawaiian serving institutions; Asian American serving institutions; Native American serving institutions; Pacific Islander serving institutions;

small businesses; for-profit organizations (other than small businesses); government agencies of all types (including U.S. territories and possessions); independent school districts; faith-based organizations; community-based organizations; regional organizations; and non-domestic (foreign) entities.
*Geographic Focus* All States, District of Columbia, Guam, Marshall Islands, Northern Mariana Islands, Puerto Rico, U.S. Virgin Islands, All Countries, American Samoa
*Date(s) Application is Due:* Apr 15
*Amount of Grant:* Up to 100,000 USD
*Contact:* Dena Fischer, Director, Clinical and Practice-Based Research Program; (301) 594-4876; dena.fischer@nih.gov
*Internet:* https://grants.nih.gov/grants/guide/pa-files/PA-17-307.html
*Sponsor:* National Institute of Dental and Craniofacial Research
6701 Democracy Boulevard, Democracy One, Room 610
Bethesda, MD 20892-4878

**NIDCR Advancing Mechanistic Probiotic/Prebiotic and Human Microbiome Research Grants** 1849
The purpose of the NID.C.R Advancing Mechanistic Probiotic/Prebiotic and Human Microbiome Research Grants program s twofold: to stimulate basic and mechanistic science that facilitates the development of effective probiotics or pre-/probiotic combinations of relevance to human health and disease; and determine biological outcomes for the evaluation of efficacy of pre/probiotics in appropriate test systems and animal models. This FOA encourages basic and mechanistic studies using in vitro, in vivo, ex vivo, and in silico models that focus on prebiotic/probiotic strain selectivity, interaction, and function. It will also encourage inter and multidisciplinary collaborations among scientists in a wide range of disciplines including nutritional science, immunology, microbiomics, genomics, other '-omic' sciences, biotechnology, and bioinformatics. A Letter of Intent is mot required, and there are three cycle deadlines each year: February 5; June 5; and October 5.
*Requirements:* Eligible organizations include: higher education institutions; nonprofits (other than higher education institutions); Hispanic-serving institutions; historically Black colleges and universities; tribally-controlled colleges and universities; Alaske native serving institutions; native Hawaiian serving institutions; Asian American serving institutions; Native American serving institutions; Pacific Islander serving institutions; small businesses; for-profit organizations (other than small businesses); government agencies of all types (including U.S. territories and possessions); independent school districts; faith-based organizations; community-based organizations; regional organizations; and non-domestic (foreign) entities.
*Geographic Focus* All States, District of Columbia, Guam, Marshall Islands, Northern Mariana Islands, Puerto Rico, U.S. Virgin Islands, All Countries, American Samoa
*Date(s) Application is Due:* Feb 5; Jun 5; Oct 5
*Contact:* R. Dwayne Lunsford, Director, Microbiology Program; (301) 827-4635; lunsfordr@nidcr.nih.gov
*Internet:* https://grants.nih.gov/grants/guide/pa-files/PA-15-135.html
*Sponsor:* National Institute of Dental and Craniofacial Research
6701 Democracy Boulevard, Democracy One, Room 610
Bethesda, MD 20892-4878

**NIDCR Biology Of Aging Dental, Oral And Craniofacial Tissues Grants** 1850
The purpose of the Biology Of Aging Dental, Oral And Craniofacial Tissues Grant program is to stimulate collaborative research to understand the biological mechanisms of aging in dental, oral, and craniofacial (DOC) tissues, as they relate to parallel processes in other tissues and organs. The areas of emphasis under this FOA include inflammation, tissue healing and regeneration, and epigenetic regulation. The overarching long-term goal of this effort is to improve oral health in older adults by addressing knowledge gaps in our understanding of the basic biology of age-associated changes in health and disease states of DOC tissues. Application budgets are limited to $250,000 direct costs per year for all years. Although a letter of intent is not required, is not binding, and does not enter into the review of a subsequent application, the information that it contains allows IC staff to estimate the potential review workload and plan the review. The application deadline is January 31, with the AIDS-related application due on February 28.
*Requirements:* Eligible organizations include: higher education institutions; nonprofits (other than higher education institutions); Hispanic-serving institutions; historically Black colleges and universities; tribally-controlled colleges and universities; Alaske native serving institutions; native Hawaiian serving institutions; Asian American serving institutions; Native American serving institutions; Pacific Islander serving institutions; small businesses; for-profit organizations (other than small businesses); government agencies of all types (including U.S. territories and possessions); independent school districts; faith-based organizations; community-based organizations; regional organizations; and non-domestic (foreign) entities.
*Geographic Focus* All States, District of Columbia, Guam, Marshall Islands, Northern Mariana Islands, Puerto Rico, U.S. Virgin Islands, All Countries, American Samoa
*Date(s) Application is Due:* Jan 31
*Amount of Grant:* Up to 250,000 USD
*Contact:* Preethi Chander, Director, Salivary Biology and Immunology Program; (301) 827-4620; preethi.chander@nih.gov
*Internet:* https://grants.nih.gov/grants/guide/rfa-files/RFA-DE-18-009.html
*Sponsor:* National Institute of Dental and Craniofacial Research
6701 Democracy Boulevard, Democracy One, Room 610
Bethesda, MD 20892-4878

**NIDCR Development of Cell and Tissue Platforms to Detect Adverse Biological Consequences of Somatic Cell Gene Editing Grants** 1851
The purpose of the NID.C.R Development of Cell and Tissue Platforms to Detect Adverse Biological Consequences of Somatic Cell Gene Editing Grants program is to solicit applications that propose to develop and validate tissue- and cell-based platforms for assessing potential adverse biological consequences of somatic cell genome editing. Application budgets should not exceed $390,000 direct costs per year for up to five years. A Letter of Intent is due by the deadline of March 3, with the completed application submission deadline of April 3.
*Requirements:* Eligible organizations include: higher education institutions; nonprofits (other than higher education institutions); Hispanic-serving institutions; historically Black colleges and universities; tribally-controlled colleges and universities; Alaske native serving institutions; native Hawaiian serving institutions; Asian American serving institutions; Native American serving institutions; Pacific Islander serving institutions; small businesses; for-profit organizations (other than small businesses); government agencies of all types (including U.S. territories and possessions); independent school districts; faith-based organizations; community-based organizations; regional organizations; and non-domestic (foreign) entities.
*Geographic Focus* All States, District of Columbia, Guam, Marshall Islands, Northern Mariana Islands, Puerto Rico, U.S. Virgin Islands, All Countries, American Samoa
*Date(s) Application is Due:* Apr 3
*Amount of Grant:* Up to 1,950,000 USD
*Contact:* Lillian Shum; (301) 594-0618; lillian.shum@nih.gov
*Internet:* https://grants.nih.gov/grants/guide/rfa-files/RFA-RM-18-015.html
*Sponsor:* National Institute of Dental and Craniofacial Research
6701 Democracy Boulevard, Democracy One, Room 610
Bethesda, MD 20892-4878

**NIDCR Development of Large Animal Reporter Systems for Testing Somatic Cell Genome Editing Tools Grants** 1852
The purpose of the NID.C.R Development of Large Animal Reporter Systems for Testing Somatic Cell Genome Editing Tools Grants opportunity is to create novel large animal reporter models for testing in vivo genome editing technologies. The deliverables from the initiative will be a set of large animal models (pigs and non-human primates), containing one or more reporter genes expressed in all cell types, that will allow facile and quantitative evaluation of genome editing in any cell type of interest. Within the Consortium, these models will be used to test new delivery systems and genome editing tools developed by accompanying components in planned Large Animal Testing Centers. The large animal models will allow investigators to optimize scale up of delivery systems and editing tools, which will be necessary for clinical trials in humans. Application budgets should not exceed $500,000 direct costs per year for up to five years. A Letter of Intent is due by the deadline of March 3, with the completed application due by April 3.
*Requirements:* Eligible organizations include: higher education institutions; nonprofits (other than higher education institutions); Hispanic-serving institutions; historically Black colleges and universities; tribally-controlled colleges and universities; Alaske native serving institutions; native Hawaiian serving institutions; Asian American serving institutions; Native American serving institutions; Pacific Islander serving institutions; small businesses; for-profit organizations (other than small businesses); government agencies of all types (including U.S. territories and possessions); independent school districts; faith-based organizations; community-based organizations; regional organizations; and non-domestic (foreign) entities.
*Geographic Focus* All States, District of Columbia, Guam, Marshall Islands, Northern Mariana Islands, Puerto Rico, U.S. Virgin Islands, All Countries, American Samoa
*Date(s) Application is Due:* Apr 3
*Amount of Grant:* Up to 2,500,000 USD
*Contact:* Lillian Shum; (301) 594-0618; lillian.shum@nih.gov
*Internet:* https://grants.nih.gov/grants/guide/rfa-files/RFA-RM-18-013.html
*Sponsor:* National Institute of Dental and Craniofacial Research
6701 Democracy Boulevard, Democracy One, Room 610
Bethesda, MD 20892-4878

**NIDCR Expanding the Human Genome Editing Repertoire Grants** 1853
The purpose of the NID.C.R Expanding the Human Genome Editing Repertoire Grants program is to solicit applications proposing research on the development of novel and optimized alternatives to existing in vivo genome editing complexes. Application budgets should not exceed $250,000 direct costs per year for up to five years. A Letter of Intent is due by the deadline of March 3, with the completed application due by April 3.
*Requirements:* Eligible organizations include: higher education institutions; nonprofits (other than higher education institutions); Hispanic-serving institutions; historically Black colleges and universities; tribally-controlled colleges and universities; Alaske native serving institutions; native Hawaiian serving institutions; Asian American serving institutions; Native American serving institutions; Pacific Islander serving institutions; small businesses; for-profit organizations (other than small businesses); government agencies of all types (including U.S. territories and possessions); independent school districts; faith-based organizations; community-based organizations; regional organizations; and non-domestic (foreign) entities.
*Geographic Focus* All States, District of Columbia, Guam, Marshall Islands, Northern Mariana Islands, Puerto Rico, U.S. Virgin Islands, Wyoming, American Samoa
*Date(s) Application is Due:* Apr 3
*Contact:* Lillian Shum; (301) 594-0618; lillian.shum@nih.gov
*Internet:* https://grants.nih.gov/grants/guide/rfa-files/RFA-RM-18-017.html
*Sponsor:* National Institute of Dental and Craniofacial Research
6701 Democracy Boulevard, Democracy One, Room 610
Bethesda, MD 20892-4878

**NIDCR Health of Sexual and Gender Minority (SGM) Populations Grants** 1854

The purpose of the NID.C.R Health of Sexual and Gender Minority (SGM) Populations Grants program to support research that will increase scientific understanding of the health status of diverse population groups and thereby improve the effectiveness of health interventions and services for individuals within those groups. Priority is placed on understudied populations with distinctive health risk profiles. This funding opportunity announcement (FOA) focuses on sexual and gender minority (SGM) populations, including lesbian, gay, bisexual, transgender, and intersex populations. Basic, social, behavioral, clinical, and services research relevant to the missions of the sponsoring Institutes and Centers may be proposed. Clinical trial research is optional. The combined budget for direct costs for the two year project period may not exceed $275,000, and no more than $200,000 may be requested in any single year. There are three standard annual deadlines for application submission: February 16; June 16; and October 16.

*Requirements:* Eligible organizations include: higher education institutions; nonprofits (other than higher education institutions); Hispanic-serving institutions; historically Black colleges and universities; tribally-controlled colleges and universities; Alaske native serving institutions; native Hawaiian serving institutions; Asian American serving institutions; Native American serving institutions; Pacific Islander serving institutions; small businesses; for-profit organizations (other than small businesses); government agencies of all types (including U.S. territories and possessions); independent school districts; faith-based organizations; community-based organizations; regional organizations; and non-domestic (foreign) entities.

*Geographic Focus* All States, District of Columbia, Guam, Marshall Islands, Northern Mariana Islands, Puerto Rico, U.S. Virgin Islands, All Countries, American Samoa

*Date(s) Application is Due:* Feb 16; Jun 16; Oct 16

*Amount of Grant:* Up to 275,000 USD

*Contact:* Melissa Riddle, Director, Behavioral and Social Sciences Research; (301) 451-3888; riddleme@nidcr.nih.gov

*Internet:* https://grants.nih.gov/grants/guide/pa-files/PA-18-040.html

*Sponsor:* National Institute of Dental and Craniofacial Research
6701 Democracy Boulevard, Democracy One, Room 610
Bethesda, MD 20892-4878

**NIDCR Human BioMolecular Atlas Program (HuBMAP) Integration, Visualization and Engagement (HIVE) Initiative Grants** 1855

The purpose of the NID.C.R Human BioMolecular Atlas Program (HuBMAP) Integration, Visualization and Engagement (HIVE) Initiative Grants program is to establish the HuBMAP Integration, Visualization and Engagement (HIVE) Collaboratory that will: manage the data generated by the HuBMAP Consortium; coordinate internal and external Consortium activities; develop novel tools for visualizing, searching and modelling data; and build an atlas of tissue maps. The HIVE will be composed of four Components all of which are expected to work closely together to act as the unified backbone for the HuBMAP. The HIVE is expected to work closely with the other funded projects as part of the HuBMAP Consortium to catalyze development of a framework for mapping the human body with high spatial resolution. An application may propose only one Component. However, a coordinated set of separate applications, each describing a separate component, but with a common vision, can be proposed. The annual deadline for letters of intent is February 28, with the submission deadline for completed applications on April 16. The program intends to fund an estimate of four awards, corresponding to $2,000,000 total costs.

*Requirements:* Eligible organizations include: higher education institutions; nonprofits (other than higher education institutions); Hispanic-serving institutions; historically Black colleges and universities; tribally-controlled colleges and universities; Alaske native serving institutions; native Hawaiian serving institutions; Asian American serving institutions; Native American serving institutions; Pacific Islander serving institutions; small businesses; for-profit organizations (other than small businesses); government agencies of all types (including U.S. territories and possessions); independent school districts; faith-based organizations; community-based organizations; regional organizations; and non-domestic (foreign) entities.

*Geographic Focus* All States, District of Columbia, Guam, Marshall Islands, Northern Mariana Islands, Puerto Rico, U.S. Virgin Islands, All Countries, American Samoa

*Date(s) Application is Due:* Apr 16

*Amount of Grant:* Up to 1,000,000 USD

*Contact:* Steven Scholnick; (301) 594-3977; scholnis@mail.nih.gov

*Internet:* https://grants.nih.gov/grants/guide/rfa-files/RFA-RM-18-001.html

*Sponsor:* National Institute of Dental and Craniofacial Research
6701 Democracy Boulevard, Democracy One, Room 610
Bethesda, MD 20892-4878

**NIDCR Innovation Corps at NIH Program for NIH and CDC Translational Research Grants** 1856

The Innovation Corps at NIH Program for NIH and CD.C. Translational Research grant program seeks to develop and nurture a national innovation ecosystem that builds upon biomedical research to develop technologies, products, and services that benefit society. Toward meeting this objective, the I-Corps™ program is being offered. The I-Corps™ at NIH program is focused on educating researchers and technologists on how to translate technologies from the lab into the marketplace. Under this FOA, participating NIH and CD.C. Institutes and Centers will continue providing administrative supplement awards to currently-funded SBIR and STTR Phase I grantees. The program is designed to provide three-member project teams with access to instruction and mentoring in order to accelerate the translation of technologies currently being developed with NIH and Centers for Disease Control and Prevention (CD.C.) SBIR and STTR funding. It is anticipated that outcomes for the I-Corps™ teams participating in this program will include significantly refined commercialization plans and well-informed pivots in their overall commercialization strategies. Prospective applicants are strongly encouraged to contact NIH or CD.C. Scientific/Research staff for more information about the program before applying. The NIH/CD.C. intends to commit up to $1,200,000 to fund up to 24 awards. Application budgets are limited to no more than $50,000 in total direct costs. The current application deadline is March 5.

*Requirements:* Only United States small business concerns (SBCs) are eligible to submit applications for this opportunity.

*Restrictions:* Non-domestic (non-U.S.) Entities (Foreign Institutions) are not eligible to apply.

*Geographic Focus:* All States

*Date(s) Application is Due:* Mar 5

*Amount of Grant:* Up to 50,000 USD

*Contact:* R. Dwayne Lunsford; (301) 827-4635; lunsfordr@nidcr.nih.gov

*Internet:* https://grants.nih.gov/grants/guide/pa-files/PA-18-517.html

*Sponsor:* National Institute of Dental and Craniofacial Research
6701 Democracy Boulevard, Democracy One, Room 610
Bethesda, MD 20892-4878

**NIDCR Innovative Technologies to Deliver Genome Editing Machinery to Disease-relevant Cells and Tissues Grants** 1857

The purpose of the NID.C.R Innovative Technologies to Deliver Genome Editing Machinery to Disease-relevant Cells and Tissues Grants program announcement is to support the development and evaluation of innovative approaches to deliver genome editing machinery into somatic cells, with the ultimate goal of accelerating the development of genome editing therapeutics to treat human disease. Application budgets should not exceed $500,000 direct costs per year for up to five years. A Letter of Intent is due by the deadline of March 3, with the completed application due by April 3.

*Requirements:* Eligible organizations include: higher education institutions; nonprofits (other than higher education institutions); Hispanic-serving institutions; historically Black colleges and universities; tribally-controlled colleges and universities; Alaske native serving institutions; native Hawaiian serving institutions; Asian American serving institutions; Native American serving institutions; Pacific Islander serving institutions; small businesses; for-profit organizations (other than small businesses); government agencies of all types (including U.S. territories and possessions); independent school districts; faith-based organizations; community-based organizations; regional organizations; and non-domestic (foreign) entities.

*Geographic Focus* All States, District of Columbia, Guam, Marshall Islands, Northern Mariana Islands, Puerto Rico, U.S. Virgin Islands, All Countries, American Samoa

*Date(s) Application is Due:* Apr 3

*Amount of Grant:* Up to 2,500,000 USD

*Contact:* Lillian Shum; (301) 594-0618; lillian.shum@nih.gov

*Internet:* https://grants.nih.gov/grants/guide/rfa-files/RFA-RM-18-016.html

*Sponsor:* National Institute of Dental and Craniofacial Research
6701 Democracy Boulevard, Democracy One, Room 610
Bethesda, MD 20892-4878

**NIDCR Limited Competition: Clinical Research Sites for the MACS/WIHS Combined Cohort Study Grants** 1858

The purpose of the NID.C.R Limited Competition: Clinical Research Sites for the MACS/WIHS Combined Cohort Study Grants program is to advance knowledge of the HIV infection, with a focus on HIV-related comorbidities, by supporting current Clinical Research Sites (CRS) of the Multi-Center AIDS Cohort Study (MACS) and the Women's Interagency HIV Study (WIHS) Combined Cohort Study (MACS/WIHS-CCS). Through this FOA, each CRS will implement the full MACS/WIHS-CCS unified science agenda and provide expertise to lead part of the unified science agenda (e.g., protocol development, as a reading center, or laboratory). This FOA will also support recruitment of new participants. A companion cooperative agreement will support the MACS/WIHS-CCS Data Analysis and Coordination Center (DACC). Application budgets may not exceed direct costs of $3,900,000 per year for up to seven years. A letter of intent is due by April 2, with final application submission due by May 2.

*Requirements:* Eligibility is limited to current Clinical Research Site awardees of the Limited Competition Multicenter AIDS Cohort Study: Center for the Coordination, Analysis, and Management of the MACS award (RFA-AI-13-010) and current awardees of the Limited Competition - Women's Interagency HIV Study award (RFA-AI-12-002). Specifically, the eligible awardees are the current MACS and WIHS Clinical Research Sites (WIHS: University of California at San Francisco, University of North Carolina at Chapel Hill, Hektoen Institute for Medical Research, University of Alabama at Birmingham, Montefiore Medical Center, SUNY Downstate Medical Center, Georgetown University, University of Miami, Emory University; MACS: University of California at Los Angeles, Northwestern University, University of Pittsburgh, Johns Hopkins University).

*Geographic Focus:* All States

*Date(s) Application is Due:* May 2

*Amount of Grant:* Up to 27,300,000 USD

*Contact:* Gallya Gannot, Director, Clinical Research and Clinical Technologies; (301) 451-5096; gannotg@mail.nih.gov

*Internet:* https://grants.nih.gov/grants/guide/rfa-files/RFA-HL-19-008.html

*Sponsor:* National Institute of Dental and Craniofacial Research
6701 Democracy Boulevard, Democracy One, Room 610
Bethesda, MD 20892-4878

**NIDCR Limited Competition: Data Analysis and Coordination Center for the MACS/WIHS Combined Cohort Study Grants** 1859

The purpose of the NID.C.R Limited Competition: Data Analysis and Coordination Center for the MACS/WIHS Combined Cohort Study Grants program is to support a Data Analysis and Coordination Center (DACC) for the Multicenter AIDS Cohort Study (MACS) and the Women's Interagency HIV Study (WIHS) Combined Cohort Study (MACS/WIHS-CCS). As an active partner in the scientific endeavor, the DACC will serve as a member of the MACS/WIHS-CCS consortium to advance knowledge of the basic science, clinical science, and epidemiology of HIV infection in the US, with a focus on HIV-related comorbidities. A companion FOA, RFA-HL-19-008, will support the MACS/WIHS-CCS Clinical Research Sites. Application budgets may not exceed direct costs of $2.2 million per year for up to seven years. A Letter of Intent is due by the deadline of April 2, with the completed application due by May 2.
*Requirements:* Eligibility is limited to current awardees of the Limited Competition Multicenter AIDS Cohort Study: Center for the Coordination, Analysis, and Management of the MACS award (RFA-AI-13-010) and current awardees of the Limited Competition - Women's Interagency HIV Study award (RFA-AI-12-002). Specifically, the eligible awardees are the Center for Analysis and Management of MACS Data (CAMACS) and WIHS Data Management and Analysis Center (WDMAC), collocated at Johns Hopkins University.
*Restrictions:* Non-domestic (non-U.S.) entities (Foreign Institutions) and their components are not eligible to apply.
*Geographic Focus:* All States
*Date(s) Application is Due:* May 2
*Amount of Grant:* Up to 15,400,000 USD
*Contact:* Gallya Gannot; (301) 451-5096; gannotg@mail.nih.gov
*Internet:* https://grants.nih.gov/grants/guide/rfa-files/RFA-HL-19-007.html
*Sponsor:* National Institute of Dental and Craniofacial Research
6701 Democracy Boulevard, Democracy One, Room 610
Bethesda, MD 20892-4878

**NIDCR Limited Competition: FaceBase 3 Craniofacial Development and Dysmorphology Data Management and Integration Hub Grants** 1860

The NID.C.R-funded FaceBase data repository and knowledgebase was established to advance craniofacial research by creating comprehensive datasets on craniofacial development and dysmorphologies and disseminating them to the wider craniofacial research community. Since FaceBase's inception, the data management and integration hub has been charged with the critical functions of collecting datasets and analytic tools developed by dataset-generating spoke projects, curating and integrating those resources, and disseminating them. This Request for Applications (RFA) is a limited competition for renewal applications for the Data Management and Integration Hub. As in the past, the FaceBase 3 hub will continue to be responsible for integration and presentation of existing FaceBase datasets, development or adoption of new approaches for data search and visualization, and outreach to the wider craniofacial research community. In addition, the hub will assume an important new mission, that of working with investigators within the craniofacial research community to make their datasets compatible with FaceBase's data models, curating and integrating these new datasets with those already held by FaceBase. The NID.C.R intends to commit up to $2.5 million total costs to fund a single award. A required Letter of Intent is due on June 17, with the application deadline for submission identified as July 17.
*Requirements:* Eligibility for funding through this RFA is limited to the renewal of award(s) currently funded.
*Restrictions:* Non-domestic (non-U.S.) Entities (Foreign Institutions) are not eligible to apply.
*Geographic Focus* All States, District of Columbia, Guam, Marshall Islands, Northern Mariana Islands, Puerto Rico, U.S. Virgin Islands, American Samoa
*Date(s) Application is Due:* Jul 17
*Amount of Grant:* Up to 2,500,000 USD
*Contact:* Steven Scholnick, Director, Developmental Biology and Genetics; (301) 594-3977; scholnis@mail.nih.gov
*Internet:* https://grants.nih.gov/grants/guide/rfa-files/RFA-DE-19-003.html
*Sponsor:* National Institute of Dental and Craniofacial Research
6701 Democracy Boulevard, Democracy One, Room 610
Bethesda, MD 20892-4878

**NIDCR National Research Mentoring Network: The Science of Mentoring, Networking, and Navigating Career Transition Points Grants** 1861

The purpose of the NID.C.R National Research Mentoring Network: The Science of Mentoring, Networking, and Navigating Career Transition Points Grants program is to invite applications for the second phase of the NIH National Research Mentoring Network (NRMN) efforts. The NRMN is a nationwide consortium that was designed to enhance mentoring and career development of individuals from diverse backgrounds, including those from groups underrepresented in the biomedical research workforce. This new cycle of NRMN will expand the scientific scope of the program to encourage grants with varied approaches aimed at the science of mentorship, networking, and navigating critical career transition points. The NRMN will be a set of cooperative agreements in which the program directors/principal investigators will work through a NRMN Coordination Center and the NRMN Resource Center to disseminate effective resources. The NRMN research is expected to align with the hallmarks of success that have been defined by the Diversity Program Consortium and the Center for Evaluation and Coordination (CEC). NIH intends to commit $7.8M total costs per year to fund up to fifteen awards. A Letter of Intent is due May 11, with the completed application deadline on June 11.
*Requirements:* Eligible organizations include: higher education institutions; nonprofits (other than higher education institutions); Hispanic-serving institutions; historically Black colleges and universities; tribally-controlled colleges and universities; Alaske native serving institutions; native Hawaiian serving institutions; Asian American serving institutions; Native American serving institutions; Pacific Islander serving institutions; small businesses; for-profit organizations (other than small businesses); government agencies of all types (including U.S. territories and possessions); independent school districts; faith-based organizations; community-based organizations; and regional organizations.
*Restrictions:* Non-domestic (non-U.S.) Entities (Foreign Institutions) are not eligible to apply.
*Geographic Focus* All States, District of Columbia, Guam, Marshall Islands, Northern Mariana Islands, Puerto Rico, U.S. Virgin Islands, American Samoa
*Date(s) Application is Due:* Jun 11
*Amount of Grant:* Up to 1,000,000 USD
*Contact:* Lynn King, Chief, Research Training and Career Development Branch; (301) 594-5006; lynn.king@nih.gov
*Internet:* https://grants.nih.gov/grants/guide/rfa-files/RFA-RM-18-004.html
*Sponsor:* National Institute of Dental and Craniofacial Research
6701 Democracy Boulevard, Democracy One, Room 610
Bethesda, MD 20892-4878

**NIDCR National Research Mentoring Network Coordination Center Grants** 1862

The National Research Mentoring Network (NRMN) is a nationwide consortium developed to enhance the training and career development of individuals from diverse backgrounds, including those from groups underrepresented in the biomedical research workforce, who are pursuing biomedical research careers by providing networking and mentorship experiences. The NRMN Coordination Center will provide infrastructure and expertise that supports the collection, secure storage, and reporting of data generated by the individual NRMN component projects. Building upon the instruments and processes developed to assess the impact of NRMN activities in the previous Diversity Program Consortium (DPC) funding period, the NRMN Coordination Center will coordinate the early stages of collection of data in an ongoing way, and provide feedback to the consortium to maximize the research benefit of the various NRMN activities. The NRMN Coordination Center will also be responsible for promoting synergistic interactions with the DPC Center for Evaluation and Coordination (CEC) for the long-term collection and storage of data. NIH intends to commit up to $1,100,000 per year in total costs to fund one award. A Letter of Intent is due May 11, with the completed application deadline on June 11.
*Requirements:* Eligible organizations include: higher education institutions; nonprofits (other than higher education institutions); Hispanic-serving institutions; historically Black colleges and universities; tribally-controlled colleges and universities; Alaske native serving institutions; native Hawaiian serving institutions; Asian American serving institutions; Native American serving institutions; Pacific Islander serving institutions; small businesses; for-profit organizations (other than small businesses); government agencies of all types (including U.S. territories and possessions); independent school districts; faith-based organizations; community-based organizations; and regional organizations.
*Restrictions:* Non-domestic (non-U.S.) Entities (Foreign Institutions) are not eligible to apply.
*Geographic Focus* All States, District of Columbia, Guam, Marshall Islands, Northern Mariana Islands, Puerto Rico, U.S. Virgin Islands, American Samoa
*Date(s) Application is Due:* Jun 11
*Amount of Grant:* Up to 1,100,000 USD
*Contact:* Lynn King, Chief, Research Training and Career Development Branch; (301) 594-5006; lynn.king@nih.gov
*Internet:* https://grants.nih.gov/grants/guide/rfa-files/RFA-RM-18-003.html
*Sponsor:* National Institute of Dental and Craniofacial Research
6701 Democracy Boulevard, Democracy One, Room 610
Bethesda, MD 20892-4878

**NIDCR National Research Mentoring Network Resource Center Grants** 1863

NID.C.R National Research Mentoring Network Resource Center Grants announcement is to invite applications from qualified institutions for the NIH National Research Mentoring Network (NRMN) Resource Center. The NRMN is a nationwide consortium intended to enhance the training and career development of individuals from diverse backgrounds, including those from groups underrepresented in the biomedical research workforce, who are pursuing biomedical research careers by providing networking and mentorship experiences. One grant will be awarded for the development and management of a NRMN Resource Center, which will provide a web-based mentoring tool to facilitate real-time mentor/mentee engagement and networking. The awardee will also oversee the management of the NRMN website, report on outputs from NRMN components, and serve as a platform for publicly available mentoring resources and tools. NIH intends to commit $1,100,000 total costs per year to fund one award. A Letter of Intent is due by May 11, with the completed application deadline set on June 11.
*Requirements:* Eligible organizations include: higher education institutions; nonprofits (other than higher education institutions); Hispanic-serving institutions; historically Black colleges and universities; tribally-controlled colleges and universities; Alaske native serving institutions; native Hawaiian serving institutions; Asian American serving institutions; Native American serving institutions; Pacific Islander serving institutions; small businesses; for-profit organizations (other than small businesses); government agencies of all types (including U.S. territories and possessions); independent school districts; faith-based organizations; community-based organizations; and regional organizations.
*Restrictions:* Non-domestic (non-U.S.) Entities (Foreign Institutions) are not eligible to apply.
*Geographic Focus* All States, District of Columbia, Guam, Marshall Islands, Northern Mariana Islands, Puerto Rico, U.S. Virgin Islands, American Samoa
*Date(s) Application is Due:* Jun 11
*Amount of Grant:* Up to 1,000,000 USD
*Contact:* Lynn King, Chief, Research Training and Career Development Branch; (301) 594-5006; lynn.king@nih.gov

*Internet:* https://grants.nih.gov/grants/guide/rfa-files/RFA-RM-18-002.html
*Sponsor:* National Institute of Dental and Craniofacial Research
6701 Democracy Boulevard, Democracy One, Room 610
Bethesda, MD 20892-4878

**NIDCR Opportunities for Collaborative Research at the NIH Clinical Center Grants** 1864

The purpose of the NID.C.R Opportunities for Collaborative Research at the NIH Clinical Center Grants program is to support collaborative translational research projects aligned with NIH efforts to enhance the translation of basic biological discoveries into clinical applications that improve health. It encourages high quality science demonstrating the potential to result in understanding an important disease process or lead to new therapeutic interventions, diagnostics, or prevention strategies within the research interests and priorities of the participating NIH Institutes and Centers. Specifically, the program seeks to broaden and strengthen translational research collaborations between basic and clinical researchers both within and outside NIH to accelerate and enhance translational science by promoting partnerships between NIH intramural investigators (e.g., those conducting research within the labs and clinics of the NIH) and extramural investigators (e.g., those conducting research in labs outside the NIH), and by providing support for extramural investigators to take advantage of the unique research opportunities available at the NIH Clinical Center by conducting research projects in collaboration with NIH intramural investigators. In order to be eligible for this program, the application must include at least one intramural scientist as Program Director/Principal Investigator or collaborator, and at least some of the research must be conducted at the NIH Clinical Center. Through this collaboration, external researchers may gain access to the NIH Clinical Center and leverage the diverse Clinical Center resources, expertise, and infrastructure available to test promising laboratory-and animal-based discoveries with potential for advancing disease diagnosis, treatment and prevention. The special environment of the Clinical Center can support studies that may not be readily supported elsewhere. This may include collaborations that propose targeted increases in new patients enrolled in protocols at the Clinical Center. The maximum amount available per application is $500,000 direct costs per year; this amount includes extramural grantee costs, Clinical Center costs and intramural investigator's costs attributed to the proposed research project. The annual application submission deadline is April 11.
*Requirements:* Eligible organizations include: higher education institutions; nonprofits (other than higher education institutions); Hispanic-serving institutions; historically Black colleges and universities; tribally-controlled colleges and universities; Alaske native serving institutions; native Hawaiian serving institutions; Asian American serving institutions; Native American serving institutions; Pacific Islander serving institutions; small businesses; for-profit organizations (other than small businesses); government agencies of all types (including U.S. territories and possessions); independent school districts; faith-based organizations; community-based organizations; regional organizations; and non-domestic (foreign) entities.
*Geographic Focus* All States, District of Columbia, Guam, Marshall Islands, Northern Mariana Islands, Puerto Rico, U.S. Virgin Islands, All Countries, American Samoa
*Date(s) Application is Due:* Apr 11
*Amount of Grant:* Up to 500,000 USD
*Contact:* Robert Angerer, Scientific Director; (301) 496-1483; rang@mail.nih.gov
Dena Fischer, Director, Clinical and Practice-Based Research Program; (301) 594-4876; dena.fischer@nih.gov
*Internet:* https://grants.nih.gov/grants/guide/pa-files/PAR-18-646.html
*Sponsor:* National Institute of Dental and Craniofacial Research
6701 Democracy Boulevard, Democracy One, Room 610
Bethesda, MD 20892-4878

**NIDCR Rodent Testing Centers for Development of Reporter Systems and Evaluation of Somatic Cell Genome Editing Tools Grants** 1865

The purpose of the NID.C.R Rodent Testing Centers for Development of Reporter Systems and Evaluation of Somatic Cell Genome Editing Tools Grants program is to create rodent reporter models for testing in vivo genome editing technologies, and to use these animal models to test genome editing delivery technologies and new genome editors developed under RFA-RM-18-016 and RFA-RM-18-017. The deliverables from the initiative will be a set of animal models (mice and rat), containing one or more reporter genes expressed in all cell types, that will allow facile and quantitative evaluation of genome editing in any cell type of interest. The Testing Centers will use these models to validate new delivery systems and genome editing tools developed by other investigators in the Somatic Cell Genome Editing program. The creation and testing of these rodent models will accelerate the translation of genome editing technologies into treatments for human diseases. Application budgets should not exceed $225,000 direct costs during the program's initial fiscal year, and $450,000 per fiscal year during the next four years. A Letter of Intent is due by the deadline of March 3, with the completed application due by April 3.
*Requirements:* Eligible organizations include: higher education institutions; nonprofits (other than higher education institutions); Hispanic-serving institutions; historically Black colleges and universities; tribally-controlled colleges and universities; Alaske native serving institutions; native Hawaiian serving institutions; Asian American serving institutions; Native American serving institutions; Pacific Islander serving institutions; small businesses; for-profit organizations (other than small businesses); government agencies of all types (including U.S. territories and possessions); independent school districts; faith-based organizations; community-based organizations; regional organizations; and non-domestic (foreign) entities.
*Geographic Focus* All States, District of Columbia, Guam, Marshall Islands, Northern Mariana Islands, Puerto Rico, U.S. Virgin Islands, All Countries, American Samoa
*Date(s) Application is Due:* Apr 3
*Amount of Grant:* Up to 2,025,000 USD
*Contact:* Lillian Shum, Director, Division of Extramural Research; (301) 594-0618; lillian.shum@nih.gov
*Internet:* https://grants.nih.gov/grants/guide/rfa-files/RFA-RM-18-012.html
*Sponsor:* National Institute of Dental and Craniofacial Research
6701 Democracy Boulevard, Democracy One, Room 610
Bethesda, MD 20892-4878

**NIDCR Somatic Cell Genome Editing Dissemination and Coordinating Center Grants** 1866

The purpose of the NID.C.R Somatic Cell Genome Editing Dissemination and Coordinating Center Grants program is to support the establishment of a Dissemination and Coordinating Center (D.C.C) for the NIH Somatic Cell Genome Editing (SCGE) Consortium. In addition to this D.C.C, the Consortium will include four components (genome editing tools, delivery systems, animal models and biological assays) to produce validated techniques and knowledge through exchange of expertise, information and research tools. The D.C.C is expected to interact with other components of the SCGE Consortium to facilitate collaborative activities within the entire Consortium. The D.C.C will develop an online platform to facilitate information sharing within the Consortium, assemble resources (tools, methods, data and best practices) generated from all Consortium components into a SCGE Toolkit for Therapeutic Genome Editing, and disseminate the SCGE Toolkit and other appropriate resources to researchers across the biomedical research community. Application budgets should not exceed $1,250,000 direct costs for the upcoming fiscal year, $3,250,000 for the following three fiscal years combined, and $1,000,000 in the final year of the program. A Letter of Intent is due by the deadline of March 3, with the completed application due by April 3.
*Requirements:* Eligible organizations include: higher education institutions; nonprofits (other than higher education institutions); Hispanic-serving institutions; historically Black colleges and universities; tribally-controlled colleges and universities; Alaske native serving institutions; native Hawaiian serving institutions; Asian American serving institutions; Native American serving institutions; Pacific Islander serving institutions; small businesses; for-profit organizations (other than small businesses); government agencies of all types (including U.S. territories and possessions); independent school districts; faith-based organizations; community-based organizations; regional organizations; and non-domestic (foreign) entities.
*Geographic Focus* All States, District of Columbia, Guam, Marshall Islands, Northern Mariana Islands, Puerto Rico, U.S. Virgin Islands, All Countries, American Samoa
*Date(s) Application is Due:* Apr 3
*Amount of Grant:* Up to 5,500,000 USD
*Contact:* Lillian Shum, Director, Division of Extramural Research; (301) 594-0618; lillian.shum@nih.gov
*Internet:* https://grants.nih.gov/grants/guide/rfa-files/RFA-RM-18-018.html
*Sponsor:* National Institute of Dental and Craniofacial Research
6701 Democracy Boulevard, Democracy One, Room 610
Bethesda, MD 20892-4878

**NIDCR Time-Sensitive Research on Health Risk and Resilience after Hurricanes Irma & Maria in Puerto Rico and the U.S. Virgin Islands Grants** 1867

The purpose of the NID.C.R Time-Sensitive Research on Health Risk and Resilience after Hurricanes Irma and Maria in Puerto Rico and the U.S. Virgin Islands Grants program is to support time-sensitive research on risk and resilience factors related to short- and long-term health outcomes following Hurricanes Irma and Maria in Puerto Rico and the U.S. Virgin Islands. This initiative will use a shortened submission timeline with expedited peer review, council concurrence and award issuance. The entire cycle, from submission to award, is expected to be within three to four months. The combined budget for direct costs for the two-year project period may not exceed $275,000. No more than $200,000 may be requested in any single year. The application submission deadline is April 16.
*Requirements:* Eligible organizations include: higher education institutions; nonprofits (other than higher education institutions); Hispanic-serving institutions; historically Black colleges and universities; tribally-controlled colleges and universities; Alaske native serving institutions; native Hawaiian serving institutions; Asian American serving institutions; Native American serving institutions; Pacific Islander serving institutions; small businesses; for-profit organizations (other than small businesses); government agencies of all types (including U.S. territories and possessions); independent school districts; faith-based organizations; community-based organizations; regional organizations; and non-domestic (foreign) entities.
*Geographic Focus* All States, District of Columbia, Guam, Marshall Islands, Northern Mariana Islands, Puerto Rico, U.S. Virgin Islands, All Countries, American Samoa
*Date(s) Application is Due:* Apr 16
*Amount of Grant:* Up to 275,000 USD, Health Disparities Research; (301) 594-5394; darien.weatherspoon@nih.gov
*Internet:* https://grants.nih.gov/grants/guide/rfa-files/RFA-MD-18-006.html
*Sponsor:* National Institute of Dental and Craniofacial Research
6701 Democracy Boulevard, Democracy One, Room 610
Bethesda, MD 20892-4878

**NIDCR Training Modules to Enhance the Rigor and Reproducibility of Biomedical Research Grants** 1868

The NIH Research Education Program (R25) supports research education activities in the mission areas of the NIH. The over-arching goal of this R25 program is to support educational activities that complement and/or enhance the training of a workforce to meet the nation's biomedical, behavioral and clinical research needs. To accomplish

the stated over-arching goal, this FOA will support creative educational activities with a primary focus on developing courses for skills development - specifically, exportable training modules designed to enhance the rigor and reproducibility of biomedical research, targeted to trainees and researchers at any level. The NID.C.R will offer a maximum of $1,000,000 for one awardee. The application submission deadline is June 29.
*Requirements:* Eligible organizations include: higher education institutions; nonprofits (other than higher education institutions); Hispanic-serving institutions; historically Black colleges and universities; tribally-controlled colleges and universities; Alaske native serving institutions; native Hawaiian serving institutions; Asian American serving institutions; Native American serving institutions; Pacific Islander serving institutions; small businesses; for-profit organizations (other than small businesses); government agencies of all types (including U.S. territories and possessions); independent school districts; faith-based organizations; community-based organizations; and regional organizations.
*Restrictions:* Non-domestic (non-U.S.) Entities (Foreign Institutions) are not eligible to apply.
*Geographic Focus* All States, District of Columbia, Guam, Marshall Islands, Northern Mariana Islands, Puerto Rico, U.S. Virgin Islands, American Samoa
*Date(s) Application is Due:* Jun 29
*Amount of Grant:* Up to 100,000 USD
*Contact:* Lynn King, Chief, Research Training and Career Development Branch; (301) 594-5006; lynn.king@nih.gov
*Internet:* https://grants.nih.gov/grants/guide/rfa-files/RFA-GM-18-002.html
*Sponsor:* National Institute of Dental and Craniofacial Research
6701 Democracy Boulevard, Democracy One, Room 610
Bethesda, MD 20892-4878

**NIGMS Advancing Basic Behavioral and Social Research on Resilience: an Integrative Science Approach** 1869

This FOA solicits applications that will elucidate mechanisms and processes of resilience within a general framework that emphasizes its dynamics and interactions across both time and scale, multiple contexts, multiple outcomes, and multiple time frames. This framework can be applied to multiple contexts (e.g., acute or chronic stress exposure, including disease, disaster, unemployment, divorce, etc.), multiple outcomes (health, function, psychological well-being, etc.), and multiple time frames (seconds, minutes, days, years). The framework has four features: assessment of a baseline level prior to a challenge; characterization of a specific challenge (acute or chronic); post-challenge measures of outcomes that characterize the response over time; and predictors of outcomes (including predisposing factors at the individual level and environmental moderators). Measurement of response across multiple domains (physiological, psychological etc.) is critical to this holistic framework. Consistent application of this general framework has the potential to reveal underlying principles that describe dynamic trajectories of adaptation to a challenge and to identify potential malleable processes or mechanisms that shape those dynamic response patterns. Typically, it will require repeated measures of outcomes to differentiate resilient patterns of response from non-resilient trajectories. It may also offer insights into individual differences in reactivity and recovery that characterize resilient phenotypes. Previous investigations have identified several genetic, biological and psychological factors that are associated with individuals more successful at adapting to an adverse experience. However, most prior research on resilience has examined processes of adaptation for a single or a limited number of outcome parameters and has addressed the phenomenon from the perspective of a single aspect of functioning, e.g. cardiovascular, mental health, disease progression, substance abuse, etc. In addition, the focus of the majority of the current research has been on the individual; only a few studies have assessed resilience at the level of the family or community. This fragmentation makes it difficult to fully test hypotheses derived from holistic dynamic models of adaptation and recovery. A full model of resilience will require measuring outcomes, processes, and moderators at multiple levels (e.g. genetic, neurobiological, physiological, psychological, behavioral, social, and environmental) as these unfold at different timescales. This poses complex research challenges and may require the development or application of novel methods from other fields, such as engineering and complex systems science. Large sample sizes will be required to address some questions, and integrated expertise is essential. Computational approaches may augment traditional methods of research. This initiative is designed to develop the infrastructure and protocols needed to support well-designed studies of resilience in human subjects (individuals or communities) to examine patterns of response (pre-, during, and post-) to a well characterized challenge as they evolve over time. In order to develop an integrated framework for the concept of resilience, it is critical that investigators examine the concept at multiple levels of analysis and examine interactions, mediators, moderators and potential mechanisms. The goal is to develop comprehensive models of individual and/or social resilience in humans, and to shed light on the mechanisms and processes that account for adaptive or "resilient" response profiles, by leveraging the expertise of investigators from multiple disciplines working in a cooperative and integrated fashion. This FOA seeks teams of investigators who will develop, in the first phase (UG3), the infrastructure and research design for a project that will support testing of hypotheses derived from dynamic models of resilience, incorporating the four key aspects of the general framework articulated above. This approach is intended to address the acknowledged gaps in resilience research in the behavioral and social sciences. Upon completion of the planning phase (UG3), progress will be administratively reviewed for consideration of awarding a second phase (UH3). Letters of intent (not required) are due by November 1, with the application deadline being December 1. UG3 phase support is limited to $250,000 per year with the exception of studies incorporating well-justified pilot studies, in which case there is no budget limit. The budget for the UH3 phase is not limited but needs to reflect the actual needs of the proposed project.
*Requirements:* Eligible organizations include: higher education institutes; nonprofits other than institutions of higher education, either with or without 501(c)3 status; for-profit organizations; governments, including state, county, local, Native American Tribal, and U.S. territories or possessions; independent school districts; public and Indian housing authorities; Native American Tribal organizations; faith-based organizations; community-based organizations; regional organizations; and non-domestic (foreign) entities.
*Geographic Focus* All States, Guam, Marshall Islands, Northern Mariana Islands, Puerto Rico, U.S. Virgin Islands, All Countries, American Samoa
*Date(s) Application is Due:* Dec 1
*Contact:* Stephen Marcus, Program Officer; (301) 451-6446 or (301) 496-7301; marcusst@mail.nih.gov or info@nigms.nih.gov
*Internet:* grants.nih.gov/grants/guide/pa-files/PAR-16-326.html
*Sponsor:* National Institutes of Health
45 Center Drive MSC 6200
Bethesda, MD 20892-6200

**NIGMS Enabling Resources for Pharmacogenomics Grants** 1870

The purpose of this funding opportunity announcement is to support critical enabling resources that will accelerate new research discoveries and/or implementation of research discoveries in pharmacogenomics. A proposed resource must meet an ascertained community demand and benefit the entire scientific field of users. The FOA will support activities that can be clearly and specifically defined, are optimally designed, have evaluative measures built-in, are judiciously staffed, have formed partnerships where appropriate, and ideally have a proven track record and a finite lifetime. The outcome of an enabling resource must be highly impactful in a demonstrable way. Advance consultation with Scientific/Research staff to ensure that a proposed resource fits well with this opportunity is highly encouraged. The current closing date is September 25.
*Requirements:* Eligible applicants include: Native American tribal governments (Federally recognized); private institutions of higher education; special district governments; public and state controlled institutions of higher education; county governments; nonprofits that do not have a 501(c)3 status with the IRS, other than institutions of higher education; for profit organizations other than small businesses; nonprofits having a 501(c)3 status with the IRS, other than institutions of higher education; state governments; small businesses; city or township governments; public housing authorities; Indian housing authorities; independent school districts; Native American tribal organizations (other than Federally recognized tribal governments); Alaska Native and Native Hawaiian serving institutions; Asian American and Native American Pacific Islander serving institutions; eligible agencies of the federal government; faith-based organizations; community-based organizations; Hispanic-serving institutions; historically Black colleges and universities; Indian/Native American tribal governments (other than federally recognized); regional organizations; Tribally controlled colleges and universities (TCCUs); and U.S. territories or possessions.
*Restrictions:* Non-domestic (non-U.S.) entities (foreign institutions) are not eligible to apply.
*Geographic Focus* All States, District of Columbia, Marshall Islands, Northern Mariana Islands, Puerto Rico, U.S. Virgin Islands, American Samoa
*Date(s) Application is Due:* Sep 25
*Contact:* Rochelle M. Long, Research Contact; (301) 594-3827 or (301) 496-7301; rochelle.long@nih.gov or info@nigms.nih.gov
*Internet:* grants.nih.gov/grants/guide/pa-files/PAR-14-185.html
*Sponsor:* National Institutes of Health
45 Center Drive MSC 6200
Bethesda, MD 20892-6200

**NIH Mentored Clinical Scientist Research Career Development Award** 1871

This award enables candidates holding professional degrees (e.g., M.D., D.O., D.V.M., or equivalent degrees) to undertake three to five years of special study and supervised research with the goal of becoming independent investigators. The award also allows awardees to pursue a research career development program suited to their experience and capabilities under a mentor who is competent to provide guidance in the chosen research area. Institutions may submit applications on behalf of candidates who hold professional degrees. At least 2 years must have elapsed since the health professional degree was granted. Candidates can have varying levels of clinical training and research experience. Up to $100,000 per year plus fringe benefits is available, and up to $25,000 per year for research support for new awards.
*Requirements:* Due to the difference in individual Institute and Center program requirements for this FOA, prospective applications should consult the Table of IC-Specific Information, Requirements and Staff Contacts, to make sure that their application is responsive to the requirements of one of the participating NIH Institute and Center programs. The candidate must have an MD degree or its equivalent, must have completed postgraduate clinical training and have secured a faculty appointment in an appropriate research-intensive environment, must identify a mentor with extensive research experience, and must be willing to spend a minimum of 75 percent of full-time professional effort conducting research and research career development. Applications may be submitted on behalf of candidates by domestic, non-federal organizations, public or private, such as medical schools. Minorities and women are encouraged to apply. Candidates must be U.S. citizens or non-citizen nationals, or must have been lawfully admitted for permanent residence and possess Alien Registration Receipt Card (I-551) or some other verification of legal admission as permanent residents.
*Restrictions:* Former principal investigators on NIH research project (R01), sub-projects of program project (P01), or center grants (P50), or the equivalent, are not eligible. A candidate for the award may not concurrently apply for any other PHS award that duplicates the provisions of this award nor have another application pending award. K08 recipients are encouraged to apply for independent research grant support during the period of this award. The K08 recipient would be allowed to maintain the award if other PHS support is procured, as long as the new support does not interfere with the ability to meet the MCSDA requirements.
*Geographic Focus* All States, District of Columbia, Guam, Marshall Islands, Northern Mariana Islands, U.S. Virgin Islands, American Samoa

*Amount of Grant:* Up to 125,000 USD
*Contact:* Alison Cole, Scientific Contact; (301) 594-3827 or (301) 496–7301; fax (301) 480-2802; colea@nigms.nih.gov
Carol Shreffler, Scientific Contact; (919) 541-1445 or (919) 541-3345; fax (301) 480-2978; shreffl1@niehs.nih.gov
Shawn Gaillard, Scientific Contact; 240-627-3857 or 301-496-5717; fax 301-402-3573; gaillardl@niaid.nih.gov
*Internet:* grants.nih.gov/grants/guide/pa-files/PA-16-191.html
*Sponsor:* National Institutes of Health
9000 Rockville Pike
Bethesda, MD 20892

**NIH Ruth L. Kirschstein National Research Service Award Institutional Research Training Grants** 1872

This award enables research institutions to support predoctoral research training in specific areas and fields of shortage. Selection of institutions is by national competition; trainees are selected through local review procedures established by the program director at the grantee institution. Trainees are required to pursue their research training full-time, and trainees in clinical areas are expected to confine their clinical duties to those that are part of their research training. Institutional grants are renewable, through competition, for up to 5 years. The NHLBI will accept all types of competing T32 applications (new, renewal/competing continuation, resubmission/amended) until January 25. Only resubmission/amended T32 applications will be accepted. Because of the difference in individual Institute and Center program requirements for this FOA, prospective applications must consult the Table of IC-Specific Information, Requirements, and Staff Contacts, to make sure that their application is responsive to the requirements of one of the participating NIH Institute and Centers.
*Requirements:* Trainees must be U.S. citizens, noncitizen nationals, or legal permanent residents of the U.S. Persons on temporary or student visas are not eligible. At the time of appointment, trainees must have received a baccalaureate degree and must be training at the graduate level in a program leading to the Ph.D., Sc.D., or other equivalent degree. Trainees who wish to interrupt their medical, dental, veterinary, or other professional school studies to engage in full-time research training before completing their professional degrees are also eligible. Applications requesting $500,000 or more in direct costs in any given year must receive prior approval from the NHLBI before submitting an application.
*Restrictions:* These awards do not support study leading to the M.D., D.O., D.D.S., or other similar professional degrees, nor do they support residency training.
*Geographic Focus* All States, District of Columbia, Guam, Marshall Islands, Northern Mariana Islands, Puerto Rico, U.S. Virgin Islands, American Samoa
*Contact:* Jane Scott, Scientific Program Contact; (301) 435-0535 or (301) 592-8573; scottj2@nhlbi.nih.gov or nhlbiinfo@nhlbi.nih.gov
Alberto L. Rivera-Rentas, Scientific Program Contact; (301) 496-1804 or (301) 496-7243; fax (301) 402-0018; riverara@nidcd.nih.gov or nidcdinfo@nidcd.nih.gov
*Internet:* grants.nih.gov/grants/guide/pa-files/PA-16-152.html
*Sponsor:* National Institutes of Health
9000 Rockville Pike
Bethesda, MD 20892

**NIH Ruth L. Kirschstein National Research Service Awards for Individual Postdoctoral Fellowships** 1873

The purpose of the National Research Service Award Act (NRSA) is to help ensure that diverse pools of highly trained scientists will be available in adequate numbers and in appropriate research areas to carry out the Nation's biomedical, behavioral and clinical research agendas. The National Institutes of Health (NIH) awards individual postdoctoral fellowships (F32) to promising applicants who have the potential to become productive, independent investigators in fields related to the mission of participating NIH Institutes and Centers. The proposed postdoctoral training must be within the broad scope of biomedical, behavioral, or clinical research or other specific disciplines relevant to the research mission of the participating NIH Institutes and Centers. The proposed training must offer an opportunity to enhance the fellow's understanding of the health-related sciences and extend his/her potential for a productive research career. Because of the difference in individual Institute and Center program requirements for this FOA, prospective applications must consult the Table of IC-Specific Information, Requirements, and Staff Contacts, to make sure that their application is responsive to the requirements of one of the participating NIH Institute and Centers.
*Requirements:* Eligible applicants include: county governments; independent school districts; public and state controlled institutions of higher education; for profit organizations other than small businesses; nonprofits that do not have a 501(c)3 status with the IRS, other than institutions of higher education; public housing authorities; Indian housing authorities; private institutions of higher education; city or township governments; special district governments; small businesses; nonprofits having a 501(c)3 status with the IRS, other than institutions of higher education; state governments; Native American tribal governments (federally recognized); Native American tribal organizations (other than Federally recognized); Alaska Native and Native Hawaiian serving institutions; Asian American Native American Pacific Islander serving institutions; eligible agencies of the federal government; faith-based organizations; community-based organizations; Hispanic-serving institutions; Historically Black Colleges and Universities; Indian/Native American tribal governments (other than federally recognized); non-domestic (non-U.S.) entities (foreign organizations); regional organizations; Tribally controlled colleges and universities; and U.S. territories or possessions.
*Geographic Focus* All States, District of Columbia, Guam, Marshall Islands, Northern Mariana Islands, Puerto Rico, U.S. Virgin Islands, American Samoa
*Contact:* Leslie A. Frieden, Scientific Program Contact; (240) 276-5630 or (800) 422-6237; jakowles@mail.nih.gov
Amanda Taylor Boyce, Scientific Program Contact; (301) 594-5585 or (301) 495-4484; boycea@mail.nih.gov or NIAMSinfo@mail.nih.gov
*Internet:* grants.nih.gov/grants/guide/pa-files/PA-16-307.html
*Sponsor:* National Institutes of Health
9000 Rockville Pike
Bethesda, MD 20892

**NIH Ruth L. Kirschstein National Research Service Awards for Individual Predoctoral Fellows** 1874

The objective of this funding opportunity announcement is to help ensure that highly trained scientists will be available in adequate numbers and in appropriate research areas to carry out the Nation's biomedical, behavioral, and clinical research agenda. The participating Institutes of the National Institutes of Health (NIH) provide individual predoctoral research training fellowship awards to promising doctoral candidates who have the potential to become productive, independent investigators in research fields relevant to the missions of these participating NIH Institutes and Centers. Each participating NIH Institute has a unique scientific purview and different program goals and initiatives that evolve over time. Prior to preparing an application, it is critical that all applicants consult the appropriate Institute website for details of research areas supported by that Institute. Applicants should also contact the appropriate Institute scientific/research contact to obtain current information about specific program priorities and policies. Because of the difference in individual Institute and Center program requirements for this FOA, prospective applications must consult the Table of IC-Specific Information, Requirements, and Staff Contacts, to make sure that their application is responsive to the requirements of one of the participating NIH Institute and Centers.
*Requirements:* Eligible applicants include: county governments; independent school districts; public and state controlled institutions of higher education; for profit organizations other than small businesses; nonprofits that do not have a 501(c)3 status with the IRS, other than institutions of higher education; public housing authorities; Indian housing authorities; private institutions of higher education; city or township governments; special district governments; small businesses; nonprofits having a 501(c)3 status with the IRS, other than institutions of higher education; state governments; Native American tribal governments (federally recognized); Native American tribal organizations (other than Federally recognized); Alaska Native and Native Hawaiian serving institutions; Asian American Native American Pacific Islander serving institutions; eligible agencies of the federal government; faith-based organizations; community-based organizations; Hispanic-serving institutions; Historically Black Colleges and Universities; Indian/Native American tribal governments (other than federally recognized); non-domestic (non-U.S.) entities (foreign organizations); regional organizations; Tribally controlled colleges and universities; and U.S. territories or possessions.
*Geographic Focus* All States, District of Columbia, Guam, Marshall Islands, Northern Mariana Islands, Puerto Rico, U.S. Virgin Islands, American Samoa
*Contact:* John Williamson, Scientific Program Contact; (301) 496-2583 or (888) 644-6226; john.williamson@nih.gov
Hua-Chuan Sim, Scientific Program Contact; (301) 594-4882 or (888) 346-3656; simh@mail.nih.gov
*Internet:* grants.nih.gov/grants/guide/pa-files/PA-16-309.html
*Sponsor:* National Institutes of Health
9000 Rockville Pike
Bethesda, MD 20892

**NIH Ruth L. Kirschstein National Research Service Award Short-Term Institutional Research Training Grants** 1875

This program provides funds to research institutions to make awards to individuals in health professional schools for research opportunities that would not be available through their regular course of study. Awards are made to training institutions by national competition. Many of the NIH Institutes and Centers (ICs) use this grant mechanism exclusively to support intensive, short-term research training experiences for students in health professional schools during the summer. In addition, the Short-Term Institutional Research Training Grant may be used to support other types of predoctoral and postdoctoral training in focused, often emerging scientific areas relevant to the mission of the funding IC. Grants may be for project periods up to five years in duration and are renewable. Trainees selected for short-term training are required to pursue research training for 2-3 months on a full-time basis. Because of the difference in individual Institute and Center program requirements for this FOA, prospective applications must consult the Table of IC-Specific Information, Requirements, and Staff Contacts, to make sure that their application is responsive to the requirements of one of the participating NIH Institute and Centers.
*Requirements:* Public and private institutions of higher education; nonprofits with or without 501(c)3 IRS status (other than Institutions of Higher Education); Indian/Native American Tribal Governments; U.S. Territories or Possessions; Native American tribal organizations; and, Faith-based or Community-based Organizations are eligible to apply. The following types of Higher Education Institutions are always encouraged to apply for NIH support as Public or Private Institutions of Higher Education: Hispanic-serving Institutions; Historically Black Colleges and Universities (HBCUs); Tribally Controlled Colleges and Universities (TCCUs); and, Alaska Native and Native Hawaiian Serving Institutions. The proposed training must be in basic, behavioral or clinical research aspects of the health-related sciences. This program is intended to encourage graduate and/or health professional students to pursue research careers by exposure to and short-term involvement in the health-related sciences. The training should be of sufficient depth to enable the trainees, upon completion of the program, to

have a thorough exposure to the principles underlying the conduct of research. Trainees are selected by the grantee institution and must be U.S. citizens, noncitizen nationals, or legal permanent residents of the U.S. Trainees should have successfully completed at least one semester at an accredited school of medicine, optometry, osteopathy, dentistry, veterinary medicine, pharmacy, or public health before entering the program. The award can be used to support individuals who already have an M.S. or Ph.D. and have been accepted to health professional schools.
*Restrictions:* The award cannot be used to support courses that are required for the M.D., D.O., D.D.S., D.V.M., or similar professional degrees. Non-domestic (non-U.S.) Entities (Foreign Organizations) are not eligible to apply. Foreign (non-U.S.) components of U.S. Organizations are not allowed.
*Geographic Focus* All States, Guam, Marshall Islands, Northern Mariana Islands, Puerto Rico, U.S. Virgin Islands, American Samoa
*Contact:* Charlotte Pratt, Ph.D., RD, Scientific Program Contact; (301) 435-0382 or (301) 592-8573; fax (301) 480-1864; prattc@nhlbi.nih.gov or nhlbiinfo@nhlbi.nih.gov
Neeraj Agarwal, Scientific Program Contact; (301) 451-2020 or (301) 496-5248; fax agarwalnee@mail.nih.gov; 2020@nei.nih.gov
Carol Shreffler; (919) 541-1445 or (919) 541-3345; shreffl1@niehs.nih.gov
*Internet:* grants.nih.gov/grants/guide/pa-files/PA-16-151.html
*Sponsor:* National Institutes of Health
9000 Rockville Pike
Bethesda, MD 20892

**NIH Support of Competitive Research (SCORE) Pilot Project Awards** **1876**
The objective of the SCORE program is to foster the development of faculty at minority serving institutions (MSIs) in order to increase their research competitiveness and promote their transition to non-SCORE external sources of funding. This objective is expected to translate into an increase in the number of individuals from groups underrepresented in biomedical and behavioral research professionally engaged in these areas of research, and an enhancement of an institution's research base. SCORE SC grants are offered to eligible MSIs. For the purposes of this program, eligible MSIs are those with more than 50% student enrollment of individuals from groups underrepresented in biomedical and behavioral research. The SC2 awards may not exceed $300,000 (direct costs) for the entire length of the award which may be one to three years maximum.
*Requirements:* Eligible applicants include: public or private post secondary educational institution, such as a university or college awarding associate, undergraduate or graduate degrees located in the United States of America or its territories including Puerto Rico, Guam and Virgin Islands with more than 50 percent student enrollment from groups underrepresented in the biomedical and behavioral sciences (such as African American, Hispanic American, Native American, Alaska Natives or natives of the U.S. Pacific Islands); an Indian tribe that has a recognized governing body and that performs substantial governmental functions, or an Alaska Regional Corporation (ARC), as defined in the Alaska Native Claims Settlement Act (43 U.S.C. 1601 et. Seq.); or an institution that has a significantly high student enrollment from groups underrepresented in the biomedical or behavioral sciences but less than 50% (if the Secretary of the Department of Health and Human Services, through the MBRS Chief, determines that the institution has demonstrated special commitment to the retention and graduation of students from groups underrepresented in the biomedical and behavioral sciences and to the hiring and retention of science faculty from underrepresented groups).
*Restrictions:* Foreign institutions are not eligible to apply.
*Geographic Focus* All States, District of Columbia, Guam, Marshall Islands, Northern Mariana Islands, Puerto Rico, U.S. Virgin Islands, American Samoa
*Date(s) Application is Due:* Jan 25; May 25; Sep 7
*Amount of Grant:* 50,000 - 100,000 USD
*Contact:* Hinda Zlotnik, Research Contact; (301) 594-5132 or (301) 496–7301; fax (301) 480-2753; zlotnikh@nigms.nih.gov
Diane Adger-Johnson, Research Contact; (301) 402-8969 or (301) 496-5717; fax (301) 402-3573; dadger@niaid.nih.gov
Lynn Mertens King, Research Contact; (301) 594-5006 or (301) 496-4261; fax (301) 480-4098; lking@dir.nidcr.nih.gov or nidcrinfo@mail.nih.gov
*Internet:* grants.nih.gov/grants/guide/pa-files/PAR-13-069.html
*Sponsor:* National Institutes of Health
9000 Rockville Pike
Bethesda, MD 20892

**NIH Support of Competitive Research (SCORE) Research Advancement Awards** **1877**
The objective of the SCORE program is to foster the development of faculty at minority serving institutions (MSIs) in order to increase their research competitiveness and promote their transition to non-SCORE external sources of funding. This objective is expected to translate into an increase in the number of individuals from groups underrepresented in biomedical and behavioral research professionally engaged in these areas of research, and an enhancement of an institution's research base. SCORE SC grants are offered to eligible MSIs. For the purposes of this program, eligible MSIs are those with more than 50% student enrollment of individuals from groups underrepresented in biomedical and behavioral research. The anticipated amount of individual SC1 awards is expected to range between $125,000 to $250,000 maximum direct costs per year.
*Requirements:* Eligible applicants include: public or private post secondary educational institution, such as a university or college awarding associate, undergraduate or graduate degrees located in the United States of America or its territories including Puerto Rico, Guam and Virgin Islands with more than 50 percent student enrollment from groups underrepresented in the biomedical and behavioral sciences (such as African American, Hispanic American, Native American, Alaska Natives or natives of the U.S. Pacific Islands); an Indian tribe that has a recognized governing body and that performs substantial governmental functions, or an Alaska Regional Corporation (ARC), as defined in the Alaska Native Claims Settlement Act (43 U.S.C. 1601 et. Seq.); or an institution that has a significantly high student enrollment from groups underrepresented in the biomedical or behavioral sciences but less than 50% (if the Secretary of the Department of Health and Human Services, through the MBRS Chief, determines that the institution has demonstrated special commitment to the retention and graduation of students from groups underrepresented in the biomedical and behavioral sciences and to the hiring and retention of science faculty from underrepresented groups).
*Restrictions:* Foreign institutions are not eligible to apply.
*Geographic Focus* All States, District of Columbia, Guam, Marshall Islands, Northern Mariana Islands, Puerto Rico, U.S. Virgin Islands, American Samoa
*Date(s) Application is Due:* Mar 4; May 25; Sep 25
*Amount of Grant:* 125,000 - 250,000 USD
*Contact:* Hinda Zlotnik, Research Contact; (301) 594-5132 or (301) 496–7301; fax (301) 480-2753; zlotnikh@nigms.nih.gov or info@nigms.nih.gov
Diane Adger-Johnson, Research Contact; (301) 402-8969 or (301) 496-5717; fax (301) 402-3573; dadger@niaid.nih.gov
Lynn Mertens King, Research Contact; (301) 594-5006 or (301) 496-4261; fax (301) 480-4098; lking@dir.nidcr.nih.gov or nidcrinfo@mail.nih.gov
*Internet:* grants.nih.gov/grants/guide/pa-files/PAR-14-019.html
*Sponsor:* National Institutes of Health
9000 Rockville Pike
Bethesda, MD 20892

**NIMH Early Identification and Treatment of Mental Disorders in Children and Adolescents Grants** **1878**
This program supports research on early identification and treatment of mental disorders in children and adolescents. In particular, this announcement intends to encourage research on disorders such as schizophrenia, schizoaffective disorder, bipolar disorder, major depression, obsessive-compulsive disorder, and anorexia nervosa, alone or comorbid with other common mental or substance abuse disorders. PA: PA-00-094
*Requirements:* Applications may be submitted by domestic for-profit and nonprofit organizations, both public and private; units of state or local governments; and eligible agencies of the federal government.
*Geographic Focus:* All States
*Amount of Grant:* 125,000 - 250,000 USD
*Contact:* Shelli Avenevoli, (301) 443-5944; avenevos@mail.nih.gov
*Internet:* grants.nih.gov/grants/guide/notice-files/NOT-MH-08-006.html
*Sponsor:* National Institute of Mental Health
6001 Executive Boulevard
Bethesda, MD 20892

**NLM Express Research Grants in Biomedical Informatics** **1879**
The National Library of Medicine (NLM) offers support for innovative research in biomedical informatics. The scope of NLM's interest in the research domain of informatics is interdisciplinary, encompassing informatics problem areas in the application domains of health care, public health, basic biomedical research, bioinformatics, biological modeling, translational research and health information management in disasters. NLM defines biomedical informatics as the science of optimal organization, management, presentation and utilization of information relevant to human health and biology. Informatics research produces concepts, tools and approaches that advance what is known in the field and have the capacity to improve human health. The number of awards is contingent upon NIH appropriations and the submission of a sufficient number of meritorious applications. The NLM Express Research Grant has a limit of $250,000 per year in direct costs. The scope of the proposed project should determine the project period. The maximum project period is 4 years. (R01; FOA# PAR-13-300)
*Requirements:* Any individual(s) with the skills, knowledge, and resources necessary to carry out the proposed research as the Program Director(s)/Principal Investigator(s) (PD(s)/PI(s)) is invited to work with his/her organization to develop an application for support. Individuals from underrepresented racial and ethnic groups as well as individuals with disabilities are always encouraged to apply for NIH support. Eligible organizations include: higher education institutions (public/state-controlled institutions of higher education, private institutions of higher education); nonprofits other than institutions of higher education (nonprofits with or without 501(c)3 IRS status); for-profit organizations (small businesses, for-profit organizations other than small businesses); governments (state governments, county governments, city or township governments, special district governments, Indian/Native American Tribal Governments (federally recognized), Indian/Native American Tribal Governments (other than federally recognized), eligible agencies of the Federal Government, U.S. Territory or Possession); other (independent school districts, public housing authorities/Indian housing authorities, Native American Tribal Organizations (other than Federally recognized tribal governments), faith-based or community-based organizations, regional organizations, non-domestic (non-U.S.) entities; and, foreign institutions. The following types of Higher Education Institutions are always encouraged to apply for NIH support as public or private institutions of higher education: Hispanic-serving institutions; Historically Black Colleges and Universities (HBCUs); Tribally Controlled Colleges and Universities (TCCUs); Alaska Native and Native Hawaiian Serving Institutions; Asian American Native American Pacific Islander Serving Institutions (AANAPISIs). Applicant organizations must complete and maintain the following registrations as described in the SF 424 (R&R) Application Guide to be eligible to apply for or receive an award: DUNS Number; System for Award Management (SAM); NATP Commercial and Government Entity (NCAGE) Code; eRA Commons; and Grants.gov. All registrations must be completed prior to the

application being submitted. Registration can take 6 weeks or more, so applicants should begin the registration process as soon as possible. Standard NIH deadline dates apply, by 5:00 pm local time of applicant organization.
*Geographic Focus:* All States, All Countries
*Date(s) Application is Due:* Feb 5; Jun 5; Oct 5
*Amount of Grant:* Up to 1,000,000 USD
*Contact:* Dr. Hua-Chuan Sim, Clinical and Public Health Informatics; (301) 594-4882; hua-chuan.sim@nih.gov
*Internet:* www.nlm.nih.gov/ep/GrantResearch.html
*Sponsor:* U.S. National Library of Medicine
8600 Rockville Pike
Bethesda, MD 20894

**NMSF Ernest F. Hollings Ocean Awareness Trust Fund Grants** 1880

The Foundation builds partnerships to promote conservation and develop an ocean literate public who is able to make informed and responsible decisions regarding the ocean and its resources. This fund was established in 2005 as a legacy to the longstanding commitment of former U.S. Senator Ernest F. Hollings to promoting public involvement and understanding of the benefits of the oceans to current and future generations. The Ocean Awareness Trust Fund seeks to continue this important work by ensuring that all Americans understand and value these precious resources, and by promoting opportunities for the people of all ages to become actively interested and involved in ocean stewardship. Special consideration will be given to projects that address marine debris, human interaction with marine mammals, runoff and pollution or public health issues such as red tides or marine pharmaceuticals.
*Requirements:* Eligible applicants are formal or informal education institutions, nonprofits, commercial organizations, and state, local and Indian tribal governments. Federal employees or federal agencies cannot be the primary applicant, but can be a partner in a proposal. A Letter of Inquiry Cover Sheet is required and is available online through the link found at www.nmsfocean.org/our-grants. NMSF will review all Letters of Inquiry, and expects to select approximately 10-15 organizations to move forward with a full proposal. NMSF will send detailed directions for the full proposal submission to those organizations that are selected to move forward. All organizations that submit a Letter of Inquiry will be informed of their status by November. NMSF has added fisheries as a key priority issue this year. The Foundation is especially interested in projects that promote sustainable fisheries in conjunction with place-based management. Projects do not have to include sustainable fisheries. Projects can speak to any of the six priority issues, or any other issue impacting the ocean or Great Lakes. But projects highlighting sustainable fisheries do need to connect to national marine sanctuaries. NMSF is particularly interested in projects where sanctuaries and fisheries issues may intersect, or show promise for cooperation and coordination to meet mutual goals.
*Restrictions:* If you are a current grantee of a Hollings Ocean Awareness Trust Fund Grant in 2005-2014, who has not submitted a final report on your project, then you are not eligible to submit a proposal for funding this grant cycle.
*Geographic Focus:* All States
*Amount of Grant:* 10,000 - 25,000 USD
*Contact:* Allison Alexander, Vice President; (301) 608-3040; fax (301) 608-3044; Allison@nmsfocean.org or grants@NMSFocean.org
*Internet:* nmsfocean.org/hollings-grants
*Sponsor:* National Marine Sanctuary Foundation
8601 Georgia Avenue, Suite 510
Silver Spring, MD 20910

**NOCSAE Research Grants** 1881

Due to the diversity and complexity of potential research proposals, NOCSAE has instituted a two-phase application procedure. Those interested in seeking funding are required to first submit a Preliminary Grant Application. These brief, one page proposals are reviewed by the NOCSAE Board of Directors. The Board votes to invite Full Grant Applications from these preliminary proposals after consideration of the current goals and the available funds. The invited full proposals are assigned and reviewed by an external scientific study section. Applications should describe projects based upon basic and/or applied research bearing a rational relationship toward increasing our understanding of sports injury mechanisms and injury prevention through the use of protective sports equipment. Priority will be given to projects that demonstrate evidence of recurring injury where the injury is either catastrophic, serious, or costly.
*Requirements:* Applications should describe projects based upon basic and/or applied research bearing a rational relationship towards increasing our understanding of sports injury mechanisms and injury prevention through the use of protective sports equipment. Projects related specifically to equipment design are not appropriate. Projects analyzing athletic performance are not appropriate. Any individual and institution residing in North America is eligible. All applications regardless of origin will be evaluated by the same strict scientific guidelines.
*Restrictions:* Projects related specifically to equipment design and development are not appropriate and will not be accepted. Projects analyzing athletic performance are also not appropriate.
*Geographic Focus:* All States
*Date(s) Application is Due:* Oct 1
*Amount of Grant:* 50,000 USD
*Contact:* Mike Oliver; (913) 888-1340; fax (913) 498-8817; mike.oliver@nocsae.org
*Internet:* www.nocsae.org/research/funding.html
*Sponsor:* National Operating Committee on Standards for Athletic Equiipment
11020 King Street, Suite 215
Overland, KS 66210

**Norcliffe Foundation Grants** 1882

The Norcliffe Foundation is a private nonprofit family foundation established in 1952 by Paul Pigott for the purpose of improving the quality of life for all people in the Puget Sound region. The foundation provides grants in the areas of health, education, social services, civic improvement, religion, culture and the arts, the environment, historic preservation and youth programs. Foundation funding types include capital campaigns for building and equipment, certain operating budgets, endowments, challenge/matching grants, land acquisition, new projects, start-up funds, renovation, and research. Requests for publication, videos/films, and website production may be considered as may scholarships, fellowships / chairs, conferences / seminars, social enterprise development, and technical assistance. Multi-year and renewable funding may be considered. Applications are accepted year-round. One copy of a letter proposal and/or common grant application form should be directed to the President in care of the Foundation Manager at the contact information given. An initial phone call is optional. Guidelines, instructions, and a copy of the common grant proposal form are provided at the website. Funding decisions and notification generally occur three to six months after receipt of request.
*Requirements:* 501(c)3 organizations in the Puget Sound region in and around Seattle, Washington are eligible to apply.
*Restrictions:* The Norcliffe Foundation does not provide the following types of assistance: deficit financing, emergency funds, grants to individuals, matches to employee-giving, Program-Related Investments / Loans (PRIs), in-kind services, volunteer / loaned executive, and internships. Applicants may only submit one request per year from date of funding or denial. Applicants who have previously received grants of $50,000 or more from the Foundation must wait two years from final payment before submitting a new application.
*Geographic Focus:* Washington
*Amount of Grant:* 1,000 - 25,000 USD
*Contact:* Arline Hefferline, Foundation Manager; (206) 682-4820; fax (206) 682-4821; arline@thenorcliffefoundation.com
*Internet:* www.thenorcliffefoundation.com/
*Sponsor:* Norcliffe Foundation
999 Third Avenue, Suite 1006
Seattle, WA 98104-4001

**North Dakota Council on the Arts Individual Artist Fellowships** 1883

The Individual Fellowship Program was established to recognize practicing artists residing in North Dakota with a monetary fellowship award. This program is designed to support professional artists with outstanding talent and ability to improve their artistic skills and enhance their career opportunities. Two fellowships of $2,500 are annually awarded. An applicant will be considered professional if they are pursuing their original artwork on an ongoing basis within an artistic discipline, and if they are pursuing it as a means of livelihood, either on a partial or total basis. Works in progress are acceptable with an explanation of the work's goal in the Artist's Statement.
*Requirements:* To be eligible for a fellowship the applicant must: be at least 18 years of age and a resident of the state of North Dakota as of January of the grant year and agree to maintain that residency during the fellowship period; submit a completed application and samples of work on or before the deadline; not be enrolled as an undergraduate or graduate degree-seeking student during the fellowship period; submit samples of work completed within the last three years prior to the application deadline.
*Restrictions:* Collaborative works will not be accepted. Artists may submit collaborative work as evidence of artistic skill only if the applicant's role in the creation of the work is clearly defined.
*Geographic Focus:* North Dakota
*Date(s) Application is Due:* Feb 15
*Amount of Grant:* 2,500 USD
*Contact:* Rebecca Engelman, Arts in Education Director; (701) 328-7593; fax (701) 328-7595; rengelman@nd.gov
*Internet:* www.nd.gov/arts/grants/IAFguidelines.html
*Sponsor:* North Dakota Council on the Arts
1600 E Century Avenue, Suite 6
Bismarck, ND 58503-0649

**Northrop Grumman Corporation Grants** 1884

Northrop Grumman is committed to supporting communities throughout the U.S., especially those where its employees live and work. The Contributions Program seeks to address critical issues and needs by providing financial assistance to accredited schools and 501(c)3 nonprofit organizations. The majority of these contributions address education, services for veterans and the military, health and human services, and the environment.
*Requirements:* The following information must be included in the grant request: Tax Exempt Number; non-profit contact name, title, address, phone number, fax number; brief history of the organization, mission statement, goals and objectives; type and scope of services offered, and the geographical area served; specific details as to how requested funds would be used; project budget and amount requested; demographic impact; current operating budget including latest financial statement; list of other corporate funders; list of directors and/or officers, and their affiliations; Northrop Grumman employee sponsor, if applicable (name and phone number); and contact information.
*Restrictions:* Grants are not awarded to: religious organizations; political groups; fraternal organizations; individuals; athletic groups or activities, including charity-benefit sporting events; charter schools, unless they have open enrollment and hold the same standards as public schools; bands or choirs; capital campaigns; organizations providing services primarily to animals; communities outside of the United States; or organizations whose programs discriminate based on race, color, age, sex, religion, national origin, sexual orientation, disability, veteran status or any other characteristic protected by law.

*Geographic Focus:* All States
*Date(s) Application is Due:* Apr 30; Sep 30
*Contact:* Cheryl Horn, cheryl.horn@ngc.com
*Internet:* www.northropgrumman.com/corporate-responsibility/corporate-citizenship/contribution-guidelines.html
*Sponsor:* Northrop Grumman Corporation
1840 Century Park East
Los Angeles, CA 90067

**Northrop Grumman Foundation Grants** **1885**
Northrop Grumman's priority is to provide opportunities related to science, technology, engineering and mathematics (STEM) for students and teachers. The foundation supports diverse and sustainable NATIONAL-LEVEL programs that enhance the education experience for students and provide teachers with the training and tools they need to be successful in the classroom.
*Requirements:* Grant requests must be submitted online via the Foundation's website. Grant requests must include: tax exempt number; non-profit or institution contact name, title, address, phone number, fax number; brief history of the organization, mission statement, goals and objectives; type and scope of services offered, and the geographical area serviced; specific details as to how requested funds would be used; project budget and amount requested; demographic impacted; current operating budget including latest financial statement; list of other corporate funders; list of directors and/or officers, and their affiliations; Northrop Grumman employee sponsor, if applicable (name and phone number); and contact information.
*Restrictions:* The Foundation will not consider the following requests: individuals; fundraising events such as raffles, walk-a-thons, banquets or dinners; campus student organizations, fraternities, sororities, and honor societies; religious schools or colleges whose primary focus is to promote religious beliefs; athletic teams, support organizations; advertising or underwriting expenses; capital campaigns; endowments; tuition; choirs, bands, or drill teams.
*Geographic Focus:* All States
*Amount of Grant:* 500 - 200,000 USD
*Samples:* University of Nevada, Las Vegas, Nevada, $30,000 - to purchase computer equipment and software for the Transportation Research Center; California State University at Long Beach, Long Beach, Caliornia, $10,000 - for curriculum development in software engineering.
*Contact:* Carleen Beste, (888) 478-5478; carleen.beste@ngc.com
*Internet:* www.northropgrumman.com/corporate-responsibility/corporate-citizenship/foundation-grant-guidelines.html
*Sponsor:* Northrop Grumman Foundation
1840 Century Park E, MS 131/CC
Los Angeles, CA 90067

**Northwest Fund for the Environment Grants** **1886**
The Northwest Fund for the Environment is designated to be spent promoting change in the uses of natural resources to increase their protection and preservation in the state of Washington. Emphasis is placed on the protection of wild fish, native wildlife, natural forests, wetlands and shorelines, and the preservation of pure and free-flowing waters. Special program areas of interest include aquatic ecosystem protection of freshwater and saltwater; growth management to promote smart growth land use policies and to prevent damage to environmentally sensitive areas; and community response in support of grassroots work that feeds into the fund's goals or supports emergent issues and opportunities not covered by previous funding. The Foundation requires grantees to propose a scope of work lasting up to 16 months, complete the entire 16 month grant period, and submit a final report prior to applying for a new standard grant. Standard grants range from $3,000 to $20,000. The Community Response Fund Grants range from $500 to $3,000. Application deadlines vary according to particular funding requested. Samples of previous funding are available at the website.
*Requirements:* Applicants must follow the online application process, which includes calling the Foundation to discuss whether it may be eligible for funding. Unsolicited applications or letters of inquiry will not be accepted. Only nonprofit organizations classified as 501(c)3 public charities serving Washington state are eligible. First time applicants may only apply for project support.
*Restrictions:* The following are not eligible for funding: individuals or for profit organizations; endowments or capital campaigns; debt reduction or loans to organizations; capital projects, such as building acquisition or construction, large equipment purchases or vehicle acquisition; land acquisition; structural restoration, such as tree planting or culvert removal; purely educational activities (except when an integral part but only one component of your application); organization of, scholarships for, or support of public conferences, as distinct from invited meetings or convening of groups engaged in related efforts; youth group support; production of publications and videos for the general public; web page development; museum displays; art projects, including performances; visual art projects and arts education; partisan political activities; academic scholarships or fellowships; academic research; university overhead costs; or field research (except when an integral part but only one component of your application).
*Geographic Focus:* Washington
*Date(s) Application is Due:* Feb 1
*Contact:* Pam Fujita-Yuhas, Foundation Director; (206) 386-7220; fax (206) 386-7223 (call first); staff@nwfund.org
*Internet:* www.nwfund.org
*Sponsor:* Northwest Fund for the Environment
1904 Third Avenue, Suite 615
Seattle, WA 98101

**Norwegian Academy of Science and Letters Abel Prize** **1887**
The Abel Prize is an international prize for outstanding scientific work in the field of mathematics, including mathematical aspects of information technology, mathematical physics, probability theory, numerical analysis and computational science, statistics and applications of mathematics in other sciences. The prize is meant to recognize contributions of extraordinary depth and influence to pure and applied mathematics. Such work may have resolved fundamental problems, created powerful new techniques, introduced unifying principles or opened up major new fields of research. The prize amount is 6 million NOK (about 750,000 Euro or 1,000,000USD) and was awarded for the first time in 2003.
*Requirements:* Candidates may send the committee a nomination letter. The nomination should be accompanied by a description of the work and impact of the nominee/nominees, together with names of distinguished specialists in the field of the nominee/nominees who can be contacted for an independent opinion. Further information and a current online submission form is available online.
*Geographic Focus:* All States, All Countries
*Date(s) Application is Due:* Sep 15
*Amount of Grant:* 6,000,000 NOK
*Contact:* Oyvind Anderson, Secretary General; + 47 22 12 10 90; fax + 47 22 12 10 99; dnva@online.no
*Internet:* www.abelprisen.no/en/abelprisen/nomination.html
*Sponsor:* Norwegian Academy of Science and Letters
Drammensveien 78
Oslo, N-0271 Norway

**Notsew Orm Sands Foundation Grants** **1888**
The Notsew Orm Sands Foundation was established in 1995 via an initial donation from Charles Burnett, III, a British-born race car driver who set the steam-powered land speed record in 2009. Giving is on a national and international basis, with some emphasis on Houston, Texas, and the United Kingdom. Primary fields of interest are: higher education, hospitals, human services, medical research, and Protestant agencies and churches. There are no specific deadlines, and applicants should submit a detailed description of project and amount of funding requested, along with a copy of current year's organizational budget and/or project budget.
*Restrictions:* No grants are given to individuals.
*Geographic Focus:* All States
*Amount of Grant:* 250 - 75,000 USD
*Samples:* Baylor College of Medicine, Houston, Texas, $75,000; University of Saint Thomas, Houston, Texas, $61,000; Yale University, New Haven, Connecticut, $55,636; Centurion Ministries, Princeton, New Jersey, $20,000.
*Contact:* Charles Burnett, III, President and Secretary; (281) 497-0744
*Sponsor:* Notsew Orm Sands Foundation
2470 S. Dairy Ashford Street, Suite 802
Houston, TX 77077-5716

**NPS National Center for Preservation Technology and Training Grants** **1889**
The Preservation Technology and Training (PTT) Grants program provides funding for innovative research that develops new technologies or adapts existing technologies to preserve cultural resources. Grant recipients undertake innovative research and produce technical reports which respond to national needs in the field of historic preservation. NCPTT funds projects within several overlapping disciplinary areas including: archeology, architecture, collections management, engineering, historic landscapes, and materials conservation. PTT Grants will support the following activities: Innovative research that develops new technologies or adapts existing technologies to preserve cultural resources (typically $25,000 to $40,000); Specialized workshops or symposia that identify and address national preservation needs (typically $15,000 to $25,000); How-to videos, mobile applications, podcasts, best practices publications, or webinars that disseminate practical preservation methods or provide better tools for preservation practice (typically $5,000 to $15,000). The maximum grant award is $40,000. The actual grant award amount is dependent on the scope of the proposed activity. Funding Opportunity Number: P18AS00006
*Requirements:* In order to focus research efforts, NCPTT requests innovative proposals that advance the application of science and technology to historic preservation in the following areas: Climate Change Impacts; Disaster Planning and Response; Modeling and Managing Big Data; Innovative Techniques for Documentation; and, Protective Coatings and Treatments. Other research topics may be considered for funding. The maximum grant award amount is dependent on the activity proposed. Eligible applicants are U.S. universities and colleges; U.S. non-profit organizations: museums, research laboratories, professional societies and similar organizations in the U.S. that are directly associated with educational or research activity; and Government agencies in the U.S.: National Park Service and other federal, state, territorial and local government agencies, as well as Hawaiian Natives, Native American and Alaska Native tribes and their Tribal Historic Preservation Offices. Other organizations can participate only as contractors to eligible U.S. partners. Grants funds support only portions of projects that are undertaken or managed directly by U.S. partners and expended in the U.S. and its territories. NCPTT reviews proposals for disciplinary, geographical and institutional distribution. Additionally, a National Park Service (NPS) grants administrator reviews for financial and policy matters. Special consideration will be given to proposals that leverage resources through public and private partnerships.
*Restrictions:* NCPTT does not fund "bricks and mortar" grants.
*Geographic Focus* All States, Guam, Hawaii, Marshall Islands, Northern Mariana Islands, Puerto Rico, U.S. Virgin Islands, American Samoa
*Date(s) Application is Due:* Jan 4
*Amount of Grant:* 5,000 - 40,000 USD

*Samples:* Purdue University, IN - Enhancing Historic Landscape Visualization - $40,000; University of Texas at Austin, TX - Using Eularian Video Magnification to Study the effect of fluctuations in Relative Humidity on cultural heritage materials - $30,500; Cyark, CA - Preserving Heritage with Low Cost Documentation - $31,200.
*Contact:* Tad Britt, Archeology & Collections; (318) 356-7444; fax (318) 356-9119; tad_britt@nps.gov
Andy Ferrell, Architecture & Engineering; (318) 356-7444; fax (318) 356-9119; andrew_ferrell@nps.gov
Todd Wilson, Grant Awarding Officer; 303-358-8403; Todd_wilson@nps.gov
*Internet:* https://www.ncptt.nps.gov/grants/preservation-technology-and-training-grants/
*Sponsor:* U.S. Department of the Interior
645 University Parkway
Natchitoches, LA 71457

**NRA Foundation Grants** 1890

Grant requests must conform to, and foster the purposes set forth in The NRA Foundation's Articles of Incorporation. These purposes are as follows: to promote, advance and encourage firearms and hunting safety; to educate individuals, including the youth of the United States, with respect to firearms and firearms history and hunting safety and marksmanship, as well as with respect to other subjects that are of importance to the well-being of the general public; to conduct research in furtherance of improved firearms safety and marksmanship facilities and techniques; to support activities of the National Rifle Association of America, but only to the extent that such activities are in furtherance of charitable, educational or scientific purposes within the meaning of section 501(c)3 of the Internal Revenue Code; to engage in any other activity that is incidental to, connected with, or in advancement of the foregoing purposes and that is within the scope of allowable purposes under 26 U.S.C. 501(c)3.
*Requirements:* Any organization, association or other entity, whether formally incorporated or not, that has, as a minimum, a unique federal employer identification number (EIN) issued by the Internal Revenue Service. Eligible organizations may apply for an NRA Foundation grant for a qualifying project. Additionally, allowable projects must qualify under IRS 501(c)3 regulations in one of the following categories: charitable; scientific; testing for public safety; literary; educational; fostering national/international amateur sports competition (cannot include the provision of athletic facilities or equipment.)
*Restrictions:* The following organizations are not eligible for grants: political candidates or organizations; labor organizations; state fund committees; friends of NRA committees; private business/private enterprise; other organizations or groups that have not been assigned a federal employer identification number by the Internal Revenue Service. The following activities are not eligible for funding: projects which confer private benefit upon the applying organization/group; deficit financing; projects for commercial ventures; projects which require membership in the NRA or in the applying organization/group; applications from organizations or groups that have not submitted a final report for a previously awarded grant. In addition, the following limitations/restrictions apply to grants which are otherwise eligible for funding: no funding will be awarded to an applicant for payment of administrative fees, office overhead, or other similar charges; the Foundation does not approve multi-year funding of projects; requests must be submitted for consideration each year. Funding cannot be given for competitions requiring NRA or other club or association membership. In addition, although grants may be sought for the purpose of fostering national or international amateur sports competition, grant awards cannot be made for the purpose of providing facilities or equipment to be used in such competitions.
*Geographic Focus:* All States
*Amount of Grant:* 5,000 - 9,000,000 USD
*Contact:* Wayne Sheets; (800) 423-6894; fax (703) 267-3985; nraf@nrahq.org
*Internet:* www.friendsofnra.org/National.aspx?cid=9
*Sponsor:* National Rifle association Foundation
11250 Waples Mill Road
Fairfax, VA 22030

**NSERC Aboriginal Ambassadors in the Natural Sciences and Engineering Award** 1891

The Aboriginal Ambassadors in the Natural Sciences and Engineering award aims to engage Aboriginal students and fellows in promoting interest and participation in the natural sciences and engineering by visiting Canada's Aboriginal communities and schools and sharing their research and education experiences or participating in science promotion events and activities. The award will cover costs associated with the proposed science and engineering promotion activity and Aboriginal community visit. These include costs incurred for: travel (air, rail, car rental, taxi, etc.); accommodation (hotel, etc.); meals; material and supplies needed for demonstrations or experiments; and incidental expenses (photocopying, faxes, etc.). The value of the award will be based on the budget proposed in your application, up to a maximum of $5,000. Applications should be submitted electronically using the Secure Submissions for NSERC's Innovative Collaborations and Science Promotion Programs. There are no fixed deadlines for this award. Applications should be made at least two months before the start of a proposed outreach activity.
*Requirements:* To be eligible for this supplement, you must be: a Canadian citizen or permanent resident of Canada; an Aboriginal person; hold an USRA, Alexander Graham Bell CGS, PGS, PDF, IPS, IRDF or Northern Research Internship. Successful applicants must maintain full-time study, and an active NSERC scholarship or fellowship, during the course of their AANSE supplement funded activities in order to remain eligible. The application must be submitted in Portable Document Format (PDF) as a single document. Keep a copy of the confirmation of receipt.
*Geographic Focus:* All States, Canada
*Amount of Grant:* Up to 5,000 CAD
*Contact:* Coralie Rochefort, Program Officer; (613) 943-7653 or (613) 943-0703; fax (613) 996-2589; ambassadors@nserc-crsng.gc.ca or coralie.rochefort@nserc-crsng.gc.ca
*Internet:* www.nserc-crsng.gc.ca/Students-Etudiants/Aboriginal-Autochtones_eng.asp
*Sponsor:* Natural Sciences and Engineering Research Council of Canada
350 Albert Street, 16th Floor
Ottawa, ON K1A 1H5 Canada

**NSERC Alberta Graduate Student Scholarships in ICT and Nanotechnology** 1892

Alberta Innovates funds scholarships for graduate students engaged in full-time research at Alberta's universities in the platform areas of Information and Communications Technology (ICT), Nanotechnology, and Omics. Our Health sector also offers graduate student scholarships more specific to health — see link to the left on this webpage. Please note that only one scholarship from Alberta Innovates is available to successful applicants at any one time. This funding combines with other programs to create the vibrant and progressive environment Alberta needs to achieve its goals in the areas of environment, energy, food and fiber and health research. As an outcome of these efforts, Alberta is assured the availability of the high-quality graduate student and post-doctoral fellow talent required for Campus Alberta institutions and the multi-disciplinary research teams comprising the Alberta Innovates Centers for Research and Innovation. All applications should be made to the appropriate university.
*Requirements:* The Alberta Innovates Graduate Student Scholarships are open to Canadian, Permanent Resident and International Students. Students in the following categories may apply, provided that they are pursuing graduate studies at an Alberta university in one of the research areas outlined at the website and meet eligibility
*Requirements:* Canadian candidates, whether or not they hold a major (i.e. NSERC) award; and International candidates.
*Geographic Focus:* All Countries, Canada
*Date(s) Application is Due:* Dec 1
*Contact:* Sarah Lee, Team Lead, Graduate Student Scholarships; (780) 450-5553 or (780) 450-5111; sarah.lee@albertainnovates.ca or innotechalberta.ca
*Internet:* www.nserc-crsng.gc.ca/Students-Etudiants/PG-CS/Supplements-Supplements/AICT-ATIC_eng.asp
*Sponsor:* Natural Sciences and Engineering Research Council of Canada
350 Albert Street, 16th Floor
Ottawa, ON K1A 1H5 Canada

**NSERC Alexander Graham Bell Canada Graduate Scholarships and NSERC Postgraduate Scholarships** 1893

The Alexander Graham Bell Canada Graduate Scholarships-Doctoral Program (CGS D) and NSERC Postgraduate Scholarships-Doctoral Program (PGS D) provide financial support to high calibre scholars who are engaged in a doctoral program in the natural sciences or engineering. The CGS D will be offered to the top-ranked applicants and the next tier of meritorious applicants will be offered a PGS D. This support allows these scholars to fully concentrate on their studies and seek out the best research mentors in their chosen fields. Scholarship support for graduate studies through NSERC is limited to a lifetime maximum of four years (48 months) of full-time equivalency. There are different application deadlines to keep in mind. If applying through an institution, there is an internal deadline that that institution's department sets for receiving applications. It would be the earliest of the deadlines, since the department will review applications before passing them along to NSERC. NSERC must receive all applications from Canadian institutions by November 25 of each year. If an applicant is applying directly to NSERC, your application must be submitted electronically to NSERC before 8:00 p.m. (ET) on October 15.
*Requirements:* To be considered eligible for support, as of the application deadline date, an applicant must: be a Canadian citizen or a permanent resident of Canada; intend to pursue, in the following year, full-time graduate studies and research at the doctoral level in an eligible program in one of the areas of the natural sciences and engineering supported by NSERC; and have completed, as of December 31 of the year of application, between zero and 24 months (between four and 36 months if admitted to the doctoral program directly from a bachelor's program) of studies (full-time equivalent) in the doctoral program for which the applicant is requesting funding.
*Restrictions:* Without exception, the CGS D awards are tenable only at eligible Canadian institutions. An applicant may take up his/her PGS D at any eligible Canadian institution. He/she may also take it up at any eligible foreign institution, provided the applicant has received a previous degree from a Canadian institution.
*Geographic Focus:* Canada
*Date(s) Application is Due:* Oct 15; Nov 25
*Amount of Grant:* 21,000 - 35,000 CAD
*Contact:* Coralie Rochefort, Program Officer; (613) 943-0703 or (855) 275-2861; fax (613) 992-5337; schol@nserc-crsng.gc.ca or coralie.rochefort@nserc-crsng.gc.ca
*Internet:* www.nserc-crsng.gc.ca/Students-Etudiants/PG-CS/BellandPostgrad-BelletSuperieures_eng.asp
*Sponsor:* Natural Sciences and Engineering Research Council of Canada
350 Albert Street, 16th Floor
Ottawa, ON K1A 1H5 Canada

**NSERC Banting Postdoctoral Fellowships** 1894

The Banting Postdoctoral Fellowships program is administered by Canada's three research granting councils (the Canadian Institutes of Health Research, the Natural Sciences and Engineering Research Council of Canada, and the Social Sciences and Humanities Research Council of Canada). The program aims to attract and retain top-tier postdoctoral talent, both nationally and internationally, to develop their leadership potential and to position them for success as research leaders of tomorrow, positively contributing to Canada's economic, social and research-based growth through a research-

intensive career. The Fellowship is valued at $70,000 per year (taxable), for two years, with 70 fellowships awarded annually. The annual deadline for applications is September 19.
*Requirements:* Canidates are required to be Canadian citizens or permanent residents of Canada at the time of nomination.
*Geographic Focus:* Canada
*Date(s) Application is Due:* Sep 19
*Amount of Grant:* 70,000 CAD
*Contact:* Coralie Rochefort, Program Officer; (613) 943-0703 or (855) 275-2861; fax (613) 992-5337; banting@cihr-irsc.gc.ca or coralie.rochefort@nserc-crsng.gc.ca
*Internet:* www.nserc-crsng.gc.ca/Students-Etudiants/PD-NP/Banting-Banting_eng.asp
*Sponsor:* Natural Sciences and Engineering Research Council of Canada
350 Albert Street, 16th Floor
Ottawa, ON K1A 1H5 Canada

**NSERC Brockhouse Canada Prize for Interdisciplinary Research in Science and Engineering Grant** 1895

The Brockhouse Canada Prize for Interdisciplinary Research in Science and Engineering recognizes outstanding Canadian teams of researchers from different disciplines who have combined their expertise to produce achievements of outstanding international significance in the natural sciences and engineering. The prize, accompanied by a team research grant of $250,000, reflects NSERC's commitment to supporting Canadian research through strategic investments in people, discovery and innovation. The grant supports the direct costs of university-based research and/or the enhancement of research facilities. The grant may be distributed in one lump sum or up to five instalments, depending on the needs of the recipients.
*Requirements:* Research teams nominated for the Brockhouse Canada Prize must have at least two members who are independent researchers, one of whom must hold an NSERC grant. The team can be part of an international effort, but the majority of the nominated team members must be employed at a Canadian university or public or private organization. NSERC recognizes that teams may change between the time of the specific research achievements and the time of nomination. Nominations will be accepted when changes have occurred but only as long as the core of the team remains intact. Contributions must be primarily in the natural sciences and engineering, and of an interdisciplinary and collaborative nature.
*Restrictions:* Any Canadian citizen may nominate a research team for the Brockhouse Canada Prize. Self nominations will not be accepted. Current NSERC Council members cannot be nominated. Since the prize is for interdisciplinary research, nominators should consult NSERC's Guidelines for the Preparation and Review of Applications in Interdisciplinary Research. Nominators should also consult NSERC's Guidelines for the Preparation and Review of Applications in Engineering and the Applied Sciences. See website for deadline & hyper links to above mentioned guidelines.
*Geographic Focus:* Canada
*Date(s) Application is Due:* Mar 1
*Amount of Grant:* 250,000 CAD
*Contact:* Coralie Rochefort, Program Officer; (613) 943-7653 or (613) 943-0703; fax (613) 992-5337; brockhouse@nserc-crsng.gc.ca or coralie.rochefort@nserc-crsng.gc.ca
*Internet:* www.nserc-crsng.gc.ca/Prizes-Prix/Brockhouse-Brockhouse/Index-Index_eng.asp
*Sponsor:* Natural Sciences and Engineering Research Council of Canada
350 Albert Street, 16th Floor
Ottawa, ON K1A 1H5 Canada

**NSERC Canada Graduate Scholarships Master's (CGS M) Awards** 1896

The NSERC Canada Graduate Scholarships Master's (CGS M) Program provide financial support to high-calibre scholars who are engaged in master's or doctoral programs in the natural sciences or engineering. Its primary objective is to help develop research skills and assist in the training of highly qualified personnel by supporting students who demonstrate a high standard of achievement in undergraduate and early graduate studies. The Canada Graduate Scholarships will be offered to the top-ranked applicants at each level (master's and doctoral) and the next tier of meritorious applicants will be offered an NSERC Postgraduate Scholarship. This support allows these scholars to fully concentrate on their studies and to seek out the best research mentors in their chosen fields. An eligible graduate program must have a significant research component. The annual deadline for application submission is December 1. Awards range up to a maximum of $17,500 for a 12-month period, and are non-renewable. NSERC encourages interested and qualified Aboriginal students to apply.
*Requirements:* To be considered eligible for support, applicants must: be a Canadian citizen or a permanent resident of Canada; hold, or expect to hold (at the time you take up the award), a degree in science or engineering from a university whose standing is acceptable to NSERC; intend to pursue in the following year full-time graduate studies and research at the master's or doctoral level in an eligible program in one of the areas of the natural sciences and engineering supported by NSERC; and have obtained a first-class average (a grade of 'A-') in each of the last two completed years of study (full-time equivalent). Any exceptions to this requirement must be accompanied by supporting comments that justify the submission.
*Restrictions:* Scholarships are not available to students entering graduate studies in a qualifying year. Students who, after receiving an NSERC scholarship, change their field of study and research to a field that is not supported by NSERC, cease to be eligible and their awards will be cancelled. Without exception, CGS M awards are tenable only at eligible Canadian institutions.
*Geographic Focus:* Canada
*Date(s) Application is Due:* Dec 1
*Amount of Grant:* Up to 17,500 CAD
*Contact:* Coralie Rochefort, Program Officer; (613) 943-0703 or (855) 275-2861; fax (613) 996-2589; schol@nserc-crsng.gc.ca or coralie.rochefort@nserc-crsng.gc.ca
*Internet:* www.nserc-crsng.gc.ca/Students-Etudiants/PG-CS/CGSM-BESCM_eng.asp
*Sponsor:* Natural Sciences and Engineering Research Council of Canada
350 Albert Street, 16th Floor
Ottawa, ON K1A 1H5 Canada

**NSERC Canada Graduate Scholarships Michael Smith Foreign Study Supplements** 1897

The Canada Graduate Scholarships Michael Smith Foreign Study Supplements (CGS-MSFSS) program supports high-caliber Canadian graduate students in building global linkages and international networks through the pursuit of exceptional research experiences at research institutions abroad. By accessing international scientific research and training, CGS-MSFSS recipients will contribute to strengthening the potential for collaboration between Canadian and foreign institutions. Supplements of up to $6,000 are available to active CGS (master's or doctoral) or eligible Vanier CGS holders to help offset the costs of undertaking research studies outside Canada for a defined period. A total of 250 awards are available each year and are divided among the three agencies (the Canadian Institutes of Health Research [CIHR], the Natural Sciences and Engineering Research Council of Canada [NSERC] and the Social Sciences and Humanities Research Council of Canada [SSHRC]). This reflects the distribution of CGS and eligible Vanier CGS across the health-related disciplines, the natural sciences and engineering and the social sciences and humanities: 45 for CIHR, 80 for NSERC and 125 for SSHRC. Additional supplements may be awarded should funds become available. There are two annual deadlines for application submission: June 10 and October 10.
*Requirements:* The program is available to Canadian citizens or permanent residents who hold a Joseph-Armand Bombardier, an Alexander Graham Bell, or a Frederick Banting and Charles Best Canada Graduate Scholarship (CGS) at the master's or doctoral level, or a Vanier CGS at the doctoral level. Further, an applicant must: undertake the proposed trip abroad no earlier than the competition deadline date; and not hold, or have held, any other CGS-MSFSS during the course of graduate studies.
*Geographic Focus:* Canada
*Date(s) Application is Due:* Jun 10; Oct 10
*Amount of Grant:* Up to 6,000 CAD
*Contact:* Coralie Rochefort, Program Officer; (613) 943-0703 or (855) 275-2861; schol@nserc-crsng.gc.ca or coralie.rochefort@nserc-crsng.gc.ca
*Internet:* www.nserc-crsng.gc.ca/Students-Etudiants/PG-CS/CGSForeignStudy-BESCEtudeEtranger_eng.asp
*Sponsor:* Natural Sciences and Engineering Research Council of Canada
350 Albert Street, 16th Floor
Ottawa, ON K1A 1H5 Canada

**NSERC Discovery Accelerator Supplements** 1898

The Discovery Accelerator Supplements (DAS) program provides substantial and timely additional resources to accelerate progress and maximize the impact of established, superior research programs. The program provides substantial and timely resources to researchers who have an established, superior research program that is highly rated in terms of originality and innovation, and who show strong potential to become international leaders within their field. These additional resources should enable researchers to capitalize on a research opportunity or a bold idea (e.g., a recent research breakthrough, a paradigm shift, or a new strategy to tackle a scientific problem or research question, etc.). Each award is valued at $120,000 over three years ($40,000 annually) and provides recipients with additional resources to compete with the best in the world. These resources are a supplement to the NSERC Discovery Grant, that may be used to expand the recipient's research group (i.e., students, postdoctoral fellows, technicians), to purchase, or to have access to, specialized equipment, or for other initiatives/resources that would accelerate the progress of their research program, and are only valid during the active period of the Discovery Grant through which the DAS was awarded. No separate application is required. Selection is made in conjunction with the review of Discovery Grant applications each year.
*Requirements:* Applicants to the Discovery Grants Program must present a program of research that is eligible under NSERC's mandate, which is to promote and assist research in the natural sciences and engineering, other than health. The objectives of the research program must be to advance knowledge in the natural sciences or in engineering. The application will be rejected if NSERC determines, at any time during the review cycle, that the subject matter is outside of its mandate. Applicants to the Discovery Grants program must present a program of research that is conceptually distinct from research supported by the Canadian Institutes of Health Research (CIHR) or the Social Sciences and Humanities Research Council (SSHRC).
*Restrictions:* Individuals who are awarded both a DAS and an E.W.R. Steacie Memorial Fellowship will be entitled to receive the two awards, but not concurrently.
*Geographic Focus:* Canada
*Amount of Grant:* 40,000 USD
*Contact:* Coralie Rochefort, Program Officer; (613) 943-0703 or (613) 995-5821; fax (613) 992-5337; das-sad@nserc-crsng.gc.ca or coralie.rochefort@nserc-crsng.gc.ca
*Internet:* www.nserc-crsng.gc.ca/Professors-Professeurs/Grants-Subs/DGAS-SGSA_eng.asp
*Sponsor:* Natural Sciences and Engineering Research Council of Canada
350 Albert Street, 16th Floor
Ottawa, ON K1A 1H5 Canada

**NSERC Discovery Development Grants** 1899

The Discovery Development Grants (DDG) pilot is a complementary program element to the Discovery Grants (DG) program. Its goal is to: promote and maintain a diversified base of high-quality research in small universities across Canada; and provide a stimulating environment for research training in small universities across Canada.

The DDG pilot provides resources to researchers from small universities whose DG applications were deemed to be of appropriate quality to merit research support, but were not funded in the DG Competition. Each award is valued at up to $20,000 over two years (up to $10,000 annually) and provides recipients with resources to build their research program. These resources may be used to expand the recipient's research group (e.g., students, postdoctoral fellows, technicians), to purchase or access specialized equipment, or for other initiatives or resources that would accelerate the recipient's research program. It is expected that the DDG will facilitate recipients' access to additional funding from other sources. There are no annual deadlines.
*Requirements:* In addition to the usual NSERC eligibility criteria for applicants, the pilot is only accessible to researchers at small universities who have been evaluated in the most recent DG Competition. For this purpose, small universities are those that received $4 million or less in NSERC funding on an annual basis (based on a three year average).
*Geographic Focus:* Canada
*Amount of Grant:* Up to 10,000 USD
*Contact:* Coralie Rochefort, Program Officer; (613) 943-0703 or (855) 275-2861; fax (613) 992-5337; ddg-sdd@nserc-crsng.gc.ca or coralie.rochefort@nserc-crsng.gc.ca
*Internet:* www.nserc-crsng.gc.ca/Professors-Professeurs/Grants-Subs/DiscoveryPilot-DecouvertePilote_eng.asp
*Sponsor:* Natural Sciences and Engineering Research Council of Canada
350 Albert Street, 16th Floor
Ottawa, ON K1A 1H5 Canada

**NSERC Discovery Frontiers Grants** **1900**
Discovery Frontiers (DF) grants are initiatives that identify and capitalize on emerging opportunities where Canada can benefit from its world-class capacity to take a leadership role in key areas of research and innovation. They address national research priorities and global challenges by supporting a small number of major new transformative and integrative activities. These initiatives will generate substantial impact from a one-time injection of funding for a defined period, addressing challenges and seizing leadership opportunities in areas of national importance. The initiative will bring together groups of researchers in new ways to address a major research challenge. The groups will incorporate new and emerging ideas, and combine their complementary expertise to conduct transformative, paradigm-changing research. Discovery Frontiers grants support a limited number of large international activities, opportunities or projects that are of high priority in the context of advanced research in Canada. These will be led by world-class Canadian researchers and will comprise teams that will generate substantial impact for the benefit of Canada. There is no annual competition cycle. Calls for proposals will be issued periodically by Natural Sciences and Engineering Research Council of Canada (NSERC). A limited number of proposals for large undertakings with a high level of impact and visibility will be accepted in each competition. Proposals should be for concerted, broad-based, collaborative efforts and not individual research projects. Suitable initiatives include: high risk/high pay-off activities; transformative/disruptive, paradigm-shifting research; new strategies to tackle an important problem of broad interest; research that opens new fields or integrates existing fields in new ways; research that accelerates the establishment of expertise and capacity that is in demand; and research that provides solutions and breaks down barriers at the forefront of significant scientific problems and research questions. Typically, up to $1 million per year for four years is available. The current RFP letter of intent is due March 1, with invited applications due by the July 15 deadline.
*Requirements:* In addition to the usual NSERC eligibility criteria for applicants and co-applicants, each Call for Proposals may identify mandatory eligibility criteria specific to that call. These will be used to screen applications and are pre-requisites for proposals to be accepted, at the Letter of Intent stage, within a specific Discovery Frontiers competition. Discovery Frontiers grants are not tied to the annual Discovery Grants (DG) competition cycle.
*Geographic Focus:* Canada
*Date(s) Application is Due:* Jul 15
*Amount of Grant:* 1,000,000 CAD
*Contact:* Coralie Rochefort, Program Officer; (613) 947-5636 or (613) 943-0703; fax (613) 992-5337; frontiers@nserc-crsng.gc.ca or coralie.rochefort@nserc-crsng.gc.ca
*Internet:* www.nserc-crsng.gc.ca/Professors-Professeurs/Grants-Subs/DF-FD_eng.asp
*Sponsor:* Natural Sciences and Engineering Research Council of Canada
350 Albert Street, 16th Floor
Ottawa, ON K1A 1H5 Canada

**NSERC Discovery Grants** **1901**
The Discovery Grants Program assists in: promoting and maintaining a diversified base of high-quality research capability in the natural sciences and engineering in Canadian universities; fostering research excellence; and providing a stimulating environment for research training. The program supports ongoing programs of research with long-term goals rather than a single short-term project or collection of projects. These grants recognize the creativity and innovation that are at the heart of all research advances. Discovery Grants are considered "grants in aid" of research, as they provide long term operating funds and can facilitate access to funding from other programs, but are not meant to support the full costs of a research program. Recipients of Discovery Grants are not restricted to the specific activities described in their applications, and may pursue new research interests, provided they are within NSERC's mandate. This provides researchers with the flexibility to pursue promising research avenues as they emerge and the opportunity to address higher-risk (higher reward) topics. Researchers can use their grants to participate in collaborative efforts. Applicants are encouraged to increase the inclusion and advancement of women and other under-represented groups in the natural sciences and engineering, as one means to enhance excellence in research and training. Applicants to the Discovery Grant Program are categorized as either Early Career Researchers or Established Researchers. Grants are awarded to individual researchers, normally for five years. The annual deadlines for application submission are: August 1 for intent to apply; and November 1 for completed application.
*Requirements:* Applicants to the Discovery Grants program must present a program of research that is conceptually distinct from research supported by the Canadian Institutes of Health Research (CIHR) or the Social Sciences and Humanities Research Council (SSHRC).
*Restrictions:* Researchers can apply for and hold only one Discovery Grant at a time. Researchers who hold a Discovery Grant cannot reapply for another Discovery Grant until the last year of their current award.
*Geographic Focus:* Canada
*Date(s) Application is Due:* Nov 1
*Contact:* Coralie Rochefort, Program Officer; (613) 943-0703 or (855) 275-2861; fax (613) 992-5337; resgrant@nserc-crsng.gc.ca or coralie.rochefort@nserc-crsng.gc.ca
*Internet:* www.nserc-crsng.gc.ca/Professors-Professeurs/Grants-Subs/DGIGP-PSIGP_eng.asp
*Sponsor:* Natural Sciences and Engineering Research Council of Canada
350 Albert Street, 16th Floor
Ottawa, ON K1A 1H5 Canada

**NSERC Discovery Grants – Northern Research Supplements** **1902**
The Northern Research Supplements (NRS) program has been established to augment and promote Canadian university-based northern research and training. The main purpose of the Northern Research Supplements program is to recognize the added costs unique to conducting research in the Canadian North. The program is aimed at NSERC-funded researchers who intend to conduct Discovery research in Canada's North. Successful applicants will receive a supplement to their Discovery Grant (DG) of up to $25,000 per year, for the duration of their DG. Awards at the upper end of this range will be made to those with higher logistical costs and to recognize outstanding contributions to outreach and engagement with northerners. Recipients of NRS awards are not restricted to the specific activities described in the application, but must pursue research interests in the North. Up to $25,000 per year, for up to five years is available. The annual deadline for application submissions is November 1.
*Requirements:* In addition to the usual NSERC Eligibility Criteria for Faculty, NRS funding is only accessible to researchers who are applying to the current DG competition and whose research endeavors will take them to work in the Canadian North. For the purposes of the program, Canada's North is defined as the land- and ocean-based territory north of the southern limit of the sporadic discontinuous permafrost from northern British Columbia to northern Labrador.
*Restrictions:* Only costs associated with working in the Canadian North are eligible expenses.
*Geographic Focus:* Canada
*Date(s) Application is Due:* Nov 1
*Amount of Grant:* Up to 25,000 CAD
*Contact:* Coralie Rochefort, Program Officer; (613) 943-0703 or (855) 275-2861; fax (613) 992-5337; EnviroSciences-SciencesEnviro@nserc-crsng.gc.ca or coralie.rochefort@nserc-crsng.gc.ca
*Internet:* www.nserc-crsng.gc.ca/Professors-Professeurs/Grants-Subs/DGNRS-SDSRN_eng.asp
*Sponsor:* Natural Sciences and Engineering Research Council of Canada
350 Albert Street, 16th Floor
Ottawa, ON K1A 1H5 Canada

**NSERC E.W.R. Steacie Memorial Fellowships** **1903**
NSERC's E.W.R Steacie Memorial Fellowships honor the memory of Dr. Edgar William Richard Steacie, an outstanding chemist and research leader who made major contributions to the development of science in Canada during, and immediately following, World War II. Steacie believed that young researchers are a great national asset and should be given every opportunity to develop their own ideas. He nurtured Canadian talent and drew many promising scientists to our country through his philosophy that: fundamental research is essential to the development of science; the individual is key to research, and individual ideas are ultimately responsible for important advances in science; there are no national boundaries in science; and complete freedom is required for creative work. And through the Steacie Fellowships, his ideas live on. The Fellowships are awarded to enhance the career development of outstanding and highly promising university faculty who are earning a strong international reputation for original research. Every year, NSERC awards up to six Steacie Fellowships that are held for a two-year period. Successful fellows are relieved of teaching and administrative duties, so that they can devote all their time and energy to research. The Fellowships are held at a Canadian university or affiliated research institution. Each fellow receives a research grant of $250,000 over two years. The Fellowship normally also includes a contribution of up to $90,000 per year to the university toward the fellow's salary. As part of the Fellowship agreement, the university is expected to fund a replacement for the fellow's teaching and administrative responsibilities. The annual deadline for nominations is June 7.
*Requirements:* Candidates should have obtained their doctorate within the last 12 years and hold a grant from NSERC.
*Restrictions:* Candidates may not apply on their own behalf. They must be nominated by senior members of the Canadian science and engineering community, and the nominations must be endorsed by the executive head of the candidate's university.
*Geographic Focus:* All States, Canada
*Date(s) Application is Due:* Jun 7
*Amount of Grant:* 250,000 - 430,000 CAD
*Contact:* Coralie Rochefort, Program Officer; (613) 943-7653 or (613) 943-0703; fax (613) 992-5337; inquiries.steacie@nserc-crsng.gc.ca

*Internet:* www.nserc-crsng.gc.ca/Prizes-Prix/Steacie-Steacie/About-Apropos_eng.asp
*Sponsor:* Natural Sciences and Engineering Research Council of Canada
350 Albert Street, 16th Floor
Ottawa, ON K1A 1H5 Canada

**NSERC Environment Canada Atmospheric and Meteorological Graduate Supplements** **1904**
The purpose of the atmospheric and meteorological graduate supplement is to: promote graduate research in the atmospheric and meteorological sciences in Canadian universities; encourage and support graduate students to carry out research of interest to Environment Canada; increase contacts between Environment Canada researchers and Canadian universities; and promote training of potential candidates for possible future employment at Environment Canada. These supplements provide financial support to high-calibre students working towards a master's or doctoral degree in atmospheric or meteorological sciences, or related fields. Environment and Climate Change Canada offers a maximum of five supplements annually. If an applicant is successful, Environment and Climate Change Canada will supplement the NSERC scholarship by $5,000, provided the applicant holds a scholarship and continues to conduct research in the atmospheric or meteorological sciences.
*Requirements:* This supplement program is available to an applicant only if he/she is awarded an Alexander Graham Bell Canada Graduate Scholarship or an NSERC Postgraduate Scholarship-Doctoral. At the time of application for a scholarship supplement, he/she must be studying or planning to study atmospheric or meteorological sciences in a recognized program at a Canadian university.
*Geographic Focus:* Canada
*Amount of Grant:* 5,000 CAD
*Contact:* Rae Rousseau, Office of the Director General; (416) 739-4639 or (855) 275-2861; rae.rousseau2@canada.ca
*Internet:* www.nserc-crsng.gc.ca/Students-Etudiants/PG-CS/Supplements-Supplements/Atmospheric-Atmospherique_eng.asp
*Sponsor:* Natural Sciences and Engineering Research Council of Canada
350 Albert Street, 16th Floor
Ottawa, ON K1A 1H5 Canada

**NSERC Experience Awards** **1905**
Experience Awards are meant to stimulate interest in research in the natural sciences and engineering. They are also meant to encourage an individual to undertake graduate studies and to pursue a research career in these fields. These awards can provide financial support through a host organization. NSERC encourages qualified Indigenous students to apply for an Experience Award. These awards have a maximum value of $4,500 for a full 16-week period. Award applications may be submitted at any time.
*Requirements:* To be eligible to apply for an award, an applicant must: be a Canadian citizen or permanent resident of Canada; be registered, at the time of application, in a bachelor's degree program in the fields of natural sciences or engineering at an eligible Canadian university or college; ans have obtained, over the previous years of study, a cumulative average of at least second class (a grade of "B" or "B-," if applicable) as defined by the institution.
*Restrictions:* An individual may hold only one Experience Award per fiscal year (April 1 to March 31). Applications are not accepted for an Experience Award if they include: market research or sales promotion; quality control; routine data collection; preparation of routine documentation or procedures (including questionnaires); tech or helpdesk support; or research in non-natural sciences and engineering fields.
*Geographic Focus:* Canada
*Amount of Grant:* Up to 4,500 CAD
*Contact:* Coralie Rochefort, Program Officer; (613) 943-7653 or (613) 943-0703; fax (613) 992-5337; scholfin@nserc-crsng.gc.ca or coralie.rochefort@nserc-crsng.gc.ca
*Internet:* www.nserc-crsng.gc.ca/Students-Etudiants/UG-PC/Experience-Experience_eng.asp
*Sponsor:* Natural Sciences and Engineering Research Council of Canada
350 Albert Street, 16th Floor
Ottawa, ON K1A 1H5 Canada

**NSERC Gerhard Herzberg Canada Gold Medal for Science and Engineering Grants** **1906**
The NSERC Herzberg Medal is awarded annually to an individual who has demonstrated sustained excellence and influence in research, for a body of work conducted in Canada that has substantially advanced the natural sciences or engineering fields. The award is meant to celebrate Canada's most outstanding scientists and engineers and to raise public awareness about the major contributions that Canada's top researchers make to international science and technology, and to bettering people's lives. In addition to the medal, the winner is guaranteed $1 million to use for personal university-based research or to direct in some related way such as the establishment of research scholarships, fellowships or chairs in his or her name at Canadian universities. The monetary award will be distributed over a five-year period. The annual deadline for nominations is March 1.
*Requirements:* To be eligible, a candidate must be a scientist or engineer from a Canadian university, government lab or private firm. Candidates may be nominated by any individual or group; however, the nominators must be Canadian citizens or permanent residents of Canada. Posthumous or self nominations will not be accepted. Current NSERC Council members are not eligible for nomination.
*Restrictions:* An individual may win the NSERC Herzberg Medal only once.
*Geographic Focus:* Canada
*Date(s) Application is Due:* Mar 1
*Amount of Grant:* Up to 1,000,000 CAD
*Contact:* Coralie Rochefort, Program Officer; (613) 943-7653 or (613) 943-0703; fax (613) 995-7753; inquiries.herzberg@nserc-crsng.gc.ca or coralie.rochefort@nserc-crsng.gc.ca
*Internet:* www.nserc-crsng.gc.ca/Prizes-Prix/Herzberg-Herzberg/About-Apropos_eng.asp
*Sponsor:* Natural Sciences and Engineering Research Council of Canada
350 Albert Street, 16th Floor
Ottawa, ON K1A 1H5 Canada

**NSERC Gilles Brassard Doctoral Prize for Interdisciplinary Research** **1907**
Gilles Brassard is a faculty member of the Université de Montréal, where he has been a Full Professor since 1988 and Canada Research Chair since 2001. Brassard received a Ph.D. in Computer Science from Cornell University in 1979, working in the field of cryptography with John Hopcroft as his advisor. He is best known for his fundamental work in quantum cryptography, quantum teleportation, quantum entanglement distillation, quantum pseudo-telepathy, and the classical simulation of quantum entanglement. Some of these concepts have been implemented in the laboratory. In 1984, together with Charles H. Bennett, he invented the BB84 protocol for quantum cryptography. He later extended this work to include the Cascade error correction protocol, which performs efficient detection and correction of noise caused by eavesdropping on quantum cryptographic signals. The NSERC Gilles Brassard Doctoral Prize for Interdisciplinary Research is awarded to an outstanding recipient of an NSERC Vanier Canada Graduate Scholarship who best exemplifies interdisciplinary research. Preference is given to a recipient who plans to pursue doctoral studies at a different university from the one that granted their most recent degree. The award is valued at $10,000 and was established in 2012 by Gilles Brassard, winner of the 2009 Gerhard Herzberg Canada Gold Medal for Science and Engineering.
*Geographic Focus:* All States, Canada
*Date(s) Application is Due:* Mar 1
*Amount of Grant:* 10,000 CAD
*Contact:* Coralie Rochefort, Program Officer; (613) 943-7653 or (613) 943-0703; fax (613) 992-5337; coralie.rochefort@nserc-crsng.gc.ca
*Internet:* www.nserc-crsng.gc.ca/Prizes-Prix/Brassard-Brassard/Index-Index_eng.asp
*Sponsor:* Natural Sciences and Engineering Research Council of Canada
350 Albert Street, 16th Floor
Ottawa, ON K1A 1H5 Canada

**NSERC Industrial Research Chairs** **1908**
Industrial Research Chairs (IRC) are intended to assist universities in building on existing strengths to achieve the critical mass required for a major research endeavour in science and engineering of interest to industry and/or assist in the development of research efforts in fields that have not yet been developed in Canadian universities but for which there is an important industrial need. NSERC offers three types of Industrial Research Chairs: Senior Industrial Research Chairs (for distinguished senior researchers); Associate Industrial Research Chairs (for promising junior faculty researchers); and Executive Industrial Research Chairs (for outstanding R&D professionals). Emphasis is on the development and transfer of new knowledge from the university to industry, and on the training of highly qualified personnel in fields of importance to industry. A reasonable level of industry participation in terms of a monetary, equipment, or personnel commitment is a prerequisite for consideration of NSERC funding. University authorities should contact NSERC concerning application procedures. Applications are accepted at any time.
*Requirements:* Candidates must be researchers of international stature who are at or approaching the most productive period of their careers, who have the necessary qualifications to hold the rank of professor or associate professor and are recognized experts in their fields. The applicant must hold a senior administrative position in a Canadian university, such as department head, dean, vice president or president, with direct or ultimate responsibility for the proposed chair.
*Restrictions:* The applicant cannot be the candidate for the chair.
*Geographic Focus:* Canada
*Contact:* Coralie Rochefort, Program Officer; (613) 995-1111 or (613) 943-0703; fax (613) 992-5337; coralie.rochefort@nserc-crsng.gc.ca
*Internet:* www.nserc-crsng.gc.ca/Professors-Professeurs/CFS-PCP/IRC-PCI_eng.asp
*Sponsor:* Natural Sciences and Engineering Research Council of Canada
350 Albert Street, 16th Floor
Ottawa, ON K1A 1H5 Canada

**NSERC Japan Society for the Promotion of Science Postdoctoral Fellowships** **1909**
The Japan Society for the Promotion of Science (JSPS) has established the JSPS Postdoctoral Fellowships for Foreign Researchers. A limited number of promising early-career researchers are given the opportunity to conduct research in Japanese universities and in designated research institutions and laboratories. NSERC is responsible for recommending Canadian nominees for JSPS Postdoctoral Fellowships. NSERC is acting on the evidence that equity, diversity and inclusion strengthen the scientific and engineering communities and the quality, social relevance and impact of research. Increasing diversity and gender equity in the research enterprise are key priorities in our strategic plan, NSERC 2020, and are highlighted in the strategic goal of "Building a Diversified and Competitive Research Base." The annual deadline for application submission is February 1.
*Requirements:* To be considered eligible for support, as of the application deadline date, an individual must: be a Canadian citizen of a permanent resident of Canada; not have Japanese nationality or be a permanent resident of Japan; hold or expect to hold a doctorate degree in one of the fields of research that NSERC supports prior to the proposed start date of the postdoctoral fellowship; and not hold, or have held, any other JSPS Postdoctoral fellowship.
*Geographic Focus:* Canada
*Date(s) Application is Due:* Feb 1

*Contact:* Coralie Rochefort, Program Officer; (613) 943-0703 or (855) 275-2861; fax (613) 992-5337; schol@nserc-crsng.gc.ca or coralie.rochefort@nserc-crsng.gc.ca
*Internet:* www.nserc-crsng.gc.ca/Students-Etudiants/PD-NP/JSPS-SJPS_eng.asp
*Sponsor:* Natural Sciences and Engineering Research Council of Canada
350 Albert Street, 16th Floor
Ottawa, ON K1A 1H5 Canada

**NSERC John C. Polanyi Awards** 1910

Created in 2006, the NSERC John C. Polanyi Award is given to an individual or team whose research, conducted in Canada, has led to a recent outstanding advance in any NSERC-supported field of the natural sciences or engineering. The award was established in tribute to the excellence in research that John C. Polanyi continues to exemplify. Dr. Polanyi won the 1986 Nobel Prize in Chemistry. The research leading to the advance must have been funded at least partially by an NSERC grant. As part of the award, the winning individual or team will receive a research grant of up to $250,000. The annual deadline for nominations is March 1.
*Requirements:* Candidates may be nominated by any individual or group; however, the nominators must be Canadian citizens or permanent residents of Canada. Self-nominations will not be accepted. Current NSERC Council members are not eligible for nomination. Candidates can be at any stage in their career. At least one of the candidates must hold an NSERC research grant. Also, the research that led to the advance described in the nomination must have been supported by NSERC. NSERC reserves the right to rule on the eligibility of nominations and nominees. In the case of a team nomination, members may be from academia, government or industry, and may be research associates, postdoctoral fellows or students. The team can be part of an international effort, but the majority of the nominated team members must be employed at a Canadian university or public or private organization. NSERC recognizes that teams may change between the time of the specific research achievements and the time of nomination. Nominations will be accepted when changes have occurred, but only where the core of the team remains intact.
*Restrictions:* An individual or team may be nominated for the NSERC John C. Polanyi Award and other major NSERC prizes in the same year, but can receive only one such prize in a given year.
*Geographic Focus:* Canada
*Date(s) Application is Due:* Mar 1
*Amount of Grant:* 250,000 CAD
*Contact:* Coralie Rochefort, Program Officer; (613) 943-7653 or (613) 943-0703; fax (613) 992-5337; polanyi@nserc-crsng.gc.ca or coralie.rochefort@nserc-crsng.gc.ca
*Internet:* www.nserc-crsng.gc.ca/Prizes-Prix/Polanyi-Polanyi/Nomination-Nomination_eng.asp
*Sponsor:* Natural Sciences and Engineering Research Council of Canada
350 Albert Street, 16th Floor
Ottawa, ON K1A 1H5 Canada

**NSERC Postdoctoral Fellowships** 1911

The Postdoctoral Fellowships (PDF) program provides support to a core of the most promising researchers at a pivotal time in their careers. The fellowships are also intended to secure a supply of highly qualified Canadians with leading-edge scientific and research skills for Canadian industry, government, and universities. The value of his fellowship is $45,000 per year for two years. The annual deadline for application submissions is October 15.
*Requirements:* An applicant must be a Canadian citizen or a permanent resident of Canada and hold or expect to hold a doctorate in one of the fields of research that NSERC supports. In addition the applicant must: expect to complete all doctoral requirements by January 1 of the year following the award announcement; have completed all requirements of your first doctoral degree no more than two years before the deadline date of the year in which the application is made.
*Geographic Focus:* Canada
*Date(s) Application is Due:* Oct 15
*Amount of Grant:* 45,000 CAD
*Contact:* Coralie Rochefort, Program Officer; (613) 943-0703 or (855) 275-2861; fax (613) 992-5337; schol@nserc-crsng.gc.ca or coralie.rochefort@nserc-crsng.gc.ca
*Internet:* www.nserc-crsng.gc.ca/Students-Etudiants/PD-NP/PDF-BP_eng.asp
*Sponsor:* Natural Sciences and Engineering Research Council of Canada
350 Albert Street, 16th Floor
Ottawa, ON K1A 1H5 Canada

**NSERC Research Tools and Instruments (RTI) Grants** 1912

Research Tools and Instruments (RTI) Grants foster and enhance the discovery, innovation and training capability of university researchers in the natural sciences and engineering by supporting the purchase of research equipment and installations. RTI Grants are one-year awards that assist in buying or developing research equipment that costs more than $7,000. In exceptional cases, multiple-year requests will be accepted. The onus is on the applicant to clearly explain the rationale. Items costing less than $7,000 each can be purchased with Discovery Grant funds or be included as a complement to a main piece of equipment being requested. RTI grants are divided into three categories according to the total net cost of the equipment, i.e., excluding taxes, shipping, and handling. These categories are: RTI Category 1: $7,001 to $150,000; RTI Category 2: $150,001 to $325,000; and RTI Category 3: more than $325,000. The maximum award amount is $150,000, and the annual deadline for application submission is $October 25.
*Requirements:* To apply, submit an application for a Grant (Form 101) and a Personal Data Form (Form 100) for the applicant and each co-applicant. Only applications from applicants and co-applicants who currently hold or are applying for an NSERC research grant (e.g., Discovery, Strategic, Research Partnership, Canada Research Chairs) at the time of application will be accepted.
*Geographic Focus:* Canada
*Date(s) Application is Due:* Oct 25
*Amount of Grant:* 150,000 USD
*Contact:* Coralie Rochefort, Program Officer; (613) 943-0703 or (855) 275-2861; fax (613) 992-5337; coralie.rochefort@nserc-crsng.gc.ca
*Internet:* www.nserc-crsng.gc.ca/Professors-Professeurs/RTII-OIRI/RTI-OIR_eng.asp
*Sponsor:* Natural Sciences and Engineering Research Council of Canada
350 Albert Street, 16th Floor
Ottawa, ON K1A 1H5 Canada

**NSERC Student Ambassadors Grants** 1913

NSERC has launched an initiative intended to support the promotion of science, technology engineering and mathematics (STEM) to Canadian youth by Canadian youth. This will engage youth directly in becoming mentors and role models for other youth in order to help foster a science and engineering culture in Canada. The purpose of the NSERC Student Ambassadors grant is to encourage post-secondary students and fellows to share their science and engineering experience with youth from groups that are under-represented in STEM. The grant will cover costs associated with the proposed science and engineering promotion outreach activity. These include and are limited to costs incurred for: travel (air, rail, car rental, taxi, etc.) to the venue; accommodation (hotel, etc.); meals; material and supplies needed for demonstrations or experiments; hosting the event (refreshments, renting of facility, audio-visual equipment, etc.); and promotion of the event (photocopying, faxes, etc.). The value of the grant is $1,000. There are no fixed deadlines for this grant, though applications should be submitted at least two months before the start of a proposed outreach activity.
*Requirements:* To be eligible, an applicant must: be registered in a Canadian university, polytechnic, college, or CÉGEP institution in the natural sciences and engineering at the time of submission; be at least 16 years of age; and hold a valid bank account with an institution legally entitled to operate in Canada. An applicant may hold only one NSERC Student Ambassador grant per fiscal year (April 1 – March 31). The outreach activity must: promote science and engineering to youth in elementary school, high school, or the first year of college in Quebec; focus on promoting science and engineering to under-represented groups (for example: girls, Indigenous youth, youth in rural and remote areas, other groups targeted based on gender, ethnicity, language, geographical location or economic status); be organized independently by the applicant (the use of resources or materials developed by other organizations is permitted); and be acknowledged in the outreach activity. The results of the outreach activity must be shared publicly (e.g. online, social media) and reported back to NSERC.
*Restrictions:* Employees of NSERC and members of their immediate families are ineligible to apply.
*Geographic Focus:* Canada
*Amount of Grant:* Up to 1,000 CAD
*Contact:* Coralie Rochefort, Program Officer; (613) 943-7653 or (613) 943-0703; fax (613) 992-5337; ambassadors@nserc-crsng.gc.ca or coralie.rochefort@nserc-crsng.gc.ca
*Internet:* www.nserc-crsng.gc.ca/Students-Etudiants/UG-PC/Ambassadors-Ambassadeurs_eng.asp
*Sponsor:* Natural Sciences and Engineering Research Council of Canada
350 Albert Street, 16th Floor
Ottawa, ON K1A 1H5 Canada

**NSERC Subatomic Physics Discovery Grants** 1914

The Subatomic Physics (SAP) Discovery Individual Grants program assists in: promoting and maintaining a diversified base of high-quality research capability in the natural sciences and engineering in Canadian universities; fostering research excellence; and providing a stimulating environment for research training. The SAP Discovery Grants Program supports ongoing programs of research (with long-term goals) rather than a single short-term project or a collection of unrelated projects that do not contribute to a cohesive program of research. These grants recognize the creativity and innovation that are at the heart of all research advances, whether made individually or in teams. Researchers are free to work in the mode most appropriate for the research area. NSERC recognizes that while being of a grant-in-aid nature, SAP Discovery Grants must be sufficient to support a program of quality research that can have a meaningful impact on the field of study. NSERC is acting on the evidence that equity, diversity and inclusion strengthen the scientific and engineering communities and the quality, social relevance and impact of research. Increasing diversity and gender equity in the research enterprise are key priorities in our current strategic plan, NSERC 2020, and are highlighted in the strategic goal of "Building a Diversified and Competitive Research Base." Applicants are encouraged to increase the inclusion and advancement of women and other under-represented groups in the natural sciences and engineering, as one means to enhance excellence in research and training. Currently, awards are available for individuals (up to five years) and projects (up to three years), with the deadline for letters of intent being August 1 and the final application deadline being November 1 for both types.
*Requirements:* Eligibility guidelines are available at the website for each type of award (individual and project).
*Geographic Focus:* Canada
*Date(s) Application is Due:* Nov 1
*Amount of Grant:* Up to 1,000,000 CAD
*Contact:* Coralie Rochefort; (613) 943-0703 or (855) 275-2861; fax (613) 992-5337; resgrants-subatomicphysics@nserc-crsng.gc.ca or coralie.rochefort@nserc-crsng.gc.ca
*Internet:* www.nserc-crsng.gc.ca/Professors-Professeurs/Grants-Subs/SPDG-SDPS_eng.asp
*Sponsor:* Natural Sciences and Engineering Research Council of Canada
350 Albert Street, 16th Floor
Ottawa, ON K1A 1H5 Canada

**NSERC Synergy Awards for Innovation** 1915

The annual Synergy Awards for Innovation recognize examples of university-industry collaboration that stand as a model of effective partnership. By working together, award-winning companies and universities have proven that effective partnerships are the foundation of achievement. Their success has enriched the academic and research programs within Canadian universities while providing tangible benefits to Canadians. Synergy Awards are offered in four categories: small and Medium-Sized Companies a partnership with a single company of up to 500 employees; large Companies a partnership with a single company of more than 500 employees; two or More Companies a partnership with two or more companies of any size; Leo Derikx Award an established innovative model of long-standing university-industry partnership in pre-competitive R&D that has improved the general well-being of an industry. Up to four awards may be presented annually, one in each category. The annual deadline for nominations is April 15.

*Requirements:* The nominated partnership must be between a Canadian university professor or group of professors and a Canadian-based company or companies. Nominations must be submitted by a senior official from one of the participating organizations. Nominations will be accepted only from the President, CEO or Vice-President of Research (or equivalent) of the industrial partner, or from the President, Rector, Vice-President of Research or Vice-Rector of Research (or equivalent) from the university. Letters of support from the other partner organization(s) must be from a senior executive in the company or executive head of the university, and must be included with the nomination. Refer to Nomination Checklist to ensure that your nomination package is complete. The checklist will be found on the NSERC website.

*Geographic Focus:* Canada

*Date(s) Application is Due:* Apr 15

*Amount of Grant:* Up to 200,000 CAD

*Contact:* Coralie Rochefort, Program Officer; (613) 943-7653 or (613) 943-0703; fax (613) 992-5337; synergy@nserc-crsng.gc.ca or coralie.rochefort@nserc-crsng.gc.ca

*Internet:* www.nserc-crsng.gc.ca/Prizes-Prix/Synergy-Synergie/About-Apropos_eng.asp

*Sponsor:* Natural Sciences and Engineering Research Council of Canada

350 Albert Street, 16th Floor

Ottawa, ON K1A 1H5 Canada

**NSERC Systematics Research Graduate Supplements** 1916

The purpose of the Systematics Research Graduate Supplements (SRGS) is to: promote graduate research in systematics in Canada; encourage and support graduate students to carry out all or part of their research at centers of excellence in systematics research; maintain and strengthen the base of expertise for systematics research in Canada; and provide practical work experience with natural history collections within a professional setting. If an applicant is successful, the sponsoring organization will supplement their scholarship by up to $5,000 per year. The supplement is renewable annually for as long as the successful applicant holds the scholarship. The annual deadline for application submission is April 30.

*Requirements:* This supplement is available to an applicant only if he/she is awarded an Alexander Graham Bell Canada Graduate Scholarship M (CGS M), Postgraduate Scholarship (PGS), or an Industrial Postgraduate Scholarship (IPS). At the time of application for a scholarship supplement, he/she must be studying or planning to study in a university faculty that provides the opportunity to engage in a research project directly related to an area of systematics. In addition, a faculty member or staff research scientist from one of the sponsoring organizations must be on the supervisory committee, and the applicant must carry out all or part of the research in conjunction with one of the participating organizations.

*Geographic Focus:* Canada

*Date(s) Application is Due:* Apr 30

*Amount of Grant:* 5,000 CAD

*Contact:* Mark Graham, Vice-President, Canadian Museum of Nature; (613) 566-4743 or (855) 275-2861; fax (613) 364-4021; mgraham@nature.ca

*Internet:* www.nserc-crsng.gc.ca/Students-Etudiants/PG-CS/Supplements-Supplements/Systematics-Sysematique_eng.asp

*Sponsor:* Natural Sciences and Engineering Research Council of Canada

350 Albert Street, 16th Floor

Ottawa, ON K1A 1H5 Canada

**NSERC Undergraduate Student Research Awards** 1917

Undergraduate Student Research Awards (USRA) are meant to nurture interest and fully develop potential for a research career in the natural sciences and engineering. They are also meant to encourage individuals to undertake graduate studies in these fields. If an applicant would like to gain research work experience that complements studies in an academic setting, these awards can provide financial support through a host university. NSERC encourages qualified Indigenous students to apply for this award. NSERC is acting on the evidence that equity, diversity and inclusion strengthen the scientific and engineering communities and the quality, social relevance and impact of research. Increasing diversity and gender equity in the research enterprise are key priorities in our current strategic plan, NSERC 2020, and are highlighted in the strategic goal of building a diversified and competitive research base. The duration of the award is 16 consecutive weeks on a full-time basis. The value of each award is $4,500, plus organization contribution. Each eligible university sets its own deadline dates for receiving applications.

*Requirements:* To be eligible to apply for an award, an individual must: be a Canadian citizen or permanent resident of Canada; be registered at the time of the application in a bachelor's degree program at an eligible university and have obtained, over the previous years of study, a cumulative average of at least second class (a grade of B or B-, if applicable) as defined by the university; and be engaged on a full-time basis in research and development activities in the natural sciences or engineering during the tenure of the award.

*Restrictions:* To hold an award, an applicant must not have started a program of graduate studies in the natural sciences or engineering at any time. An applicant is not eligible for a USRA if: he/she is currently enrolled in an undergraduate professional degree program in the health sciences (e.g., MD, DDS, BScN); or holds higher degrees in the natural sciences or engineering.

*Geographic Focus:* Canada

*Amount of Grant:* 4,500 CAD

*Contact:* Coralie Rochefort, Program Officer; (613) 943-7653 or (613) 943-0703; fax (613) 992-5337; usra-brpc@nserc-crsng.gc.ca or coralie.rochefort@nserc-crsng.gc.ca

*Internet:* www.nserc-crsng.gc.ca/Students-Etudiants/UG-PC/USRA-BRPC_eng.asp

*Sponsor:* Natural Sciences and Engineering Research Council of Canada

350 Albert Street, 16th Floor

Ottawa, ON K1A 1H5 Canada

**NSERC Vanier Canada Graduate Scholarships** 1918

The Vanier Canada Graduate Scholarships (Vanier CGS) program is administered by the Vanier-Banting Secretariat (Secretariat) on behalf of Canada's three granting agencies (the Canadian Institutes of Health Research [CIHR], the Natural Sciences and Engineering Research Council of Canada, and the Social Sciences and Humanities Research Council of Canada). The Secretariat is housed within CIHR. The Vanier CGS Program aims to attract and retain world-class doctoral students by supporting students who demonstrate both leadership skills and a high standard of scholarly achievement in graduate studies in the natural sciences and engineering, social sciences and humanities, or health sciences. Both Canadian and international students are eligible to be nominated for a Vanier CGS. Areas of research: health research; natural sciences and/or engineering research; or social sciences and humanities research. A Vanier CGS is valued at $50,000 per year for three years. The annual deadline for submissions is October 31.

*Requirements:* To be eligible, an applicant must be a: Canadian citizen; permanent residents of Canada; or foreign citizen.

*Geographic Focus:* Canada

*Date(s) Application is Due:* Oct 31

*Amount of Grant:* Up to 50,000 CAD

*Contact:* Coralie Rochefort, Program Officer; (613) 943-0703 or (855) 275-2861; vanier@cihr-irsc.gc.ca or coralie.rochefort@nserc-crsng.gc.ca

*Internet:* www.nserc-crsng.gc.ca/Students-Etudiants/PG-CS/VanierCGS-VanierBESC_eng.asp

*Sponsor:* Natural Sciences and Engineering Research Council of Canada

350 Albert Street, 16th Floor

Ottawa, ON K1A 1H5 Canada

**NSERC Visiting Fellowships in Canadian Government Laboratories** 1919

The Visiting Fellowships in Canadian Government Laboratories (VF) Program provides promising emerging scientists and engineers with the opportunity to work with research groups or leaders in Canadian government laboratories and research institutions. Fellowships are awarded for one year with the possibility of renewal for a second and third year, at the discretion of the government department concerned. The number of awards varies according to the budgets of participating departments and agencies. The amount of the award is $53,077 per year, renewable for up to two more years. Applications are processed and reviewed as they are received.

*Requirements:* To be eligible an applicant must have received a doctoral degree in the natural sciences or engineering from a recognized university within the past five years. An application will be accepted if the individual is currently enrolled in a doctoral program at a recognized university; however, he/she must expect to complete all requirements for the degree (including the thesis defense) within six months of submitting an application. If an individual has withdrawn from the workforce and active research for maternity leave, or to raise a child for at least one year, after you received a doctorate, NSERC will extend the eligibility period to six years.

*Restrictions:* An individual can apply only twice for a Visiting Fellowship in Canadian Government Laboratories. He/she must not have been barred from participating in NSERC, SSHRC, CIHR, or any other research funding organization programs for reasons of breach of standards of ethics or integrity (including financial or scientific misconduct).

*Geographic Focus:* Canada

*Amount of Grant:* 53,077 USD

*Contact:* Coralie Rochefort, Program Officer; (613) 943-7653 or (613) 943-0703; fax (613) 992-5337; vf-rslgc@nserc-crsng.gc.ca or coralie.rochefort@nserc-crsng.gc.ca

*Internet:* www.nserc-crsng.gc.ca/Students-Etudiants/PD-NP/Laboratories-Laboratoires/index_eng.asp

*Sponsor:* Natural Sciences and Engineering Research Council of Canada

350 Albert Street, 16th Floor

Ottawa, ON K1A 1H5 Canada

**NSF Accelerating Research through International Network-to-Network Collaborations (AccelNet) Grants** 1920

The goals of the Accelerating Research through International Network-to-Network Collaborations (AccelNet) program are to accelerate the process of scientific discovery and prepare the next generation of U.S. researchers for multiteam international collaborations. The AccelNet program supports strategic linkages among U.S. research networks and complementary networks abroad that will leverage research and educational resources to tackle grand scientific challenges that require significant coordinated international efforts. The program seeks to foster high-impact science and engineering by providing opportunities to create new collaborations and new combinations of resources and ideas among linked global networks. This solicitation invites proposals for the creation of international networks of networks in research areas aligned either with one of the NSF

Big Ideas or a community-identified scientific challenge with international dimensions. AccelNet awards are meant to support the connections among research networks, rather than supporting fundamental research as the primary activity. Each network of networks is expected to engage in innovative collaborative activities that promote synergy of efforts across the networks and provide professional development for students, postdoctoral scholars, and early-career researchers. There are two proposal categories covered by this solicitation: Catalytic and Full-Scale Implementation. Letters of Intent (LOI) are due no later than October 30 each year, with full proposals due the following January 31. The number of anticipated annual awards is 7 to 9, with total funding not to exceed $6 million.
*Requirements:* Proposals may only be submitted by the following: Institutions of Higher Education (IHEs) - Two- and four-year IHEs (including community colleges) accredited in, and having a campus located in the US, acting on behalf of their faculty members. Special Instructions for International Branch Campuses of U.S. IHEs: If the proposal includes funding to be provided to an international branch campus of a U.S. institution of higher education (including through use of subawards and consultant arrangements), the proposer must explain the benefit(s) to the project of performance at the international branch campus, and justify why the project activities cannot be performed at the U.S. campus; and non-profit, non-academic organizations: Independent museums, observatories, research labs, professional societies and similar organizations in the U.S. associated with educational or research activities.
*Geographic Focus* All States, District of Columbia, Guam, Marshall Islands, Northern Mariana Islands, Puerto Rico, U.S. Virgin Islands, American Samoa
*Date(s) Application is Due:* Jan 31
*Amount of Grant:* Up to 1,000,000 USD
*Contact:* Claire A. Hemingway; (703) 292-7135 or (703) 292-5111; chemingw@nsf.gov
*Internet:* https://www.nsf.gov/funding/pgm_summ.jsp?pims_id=505584
*Sponsor:* National Science Foundation
2415 Eisenhower Avenue
Alexandria, VA 22314

**NSF Accountable Institutions and Behavior (AIB) Grants** **1921**
The Accountable Institutions and Behavior (AIB) Program supports basic scientific research that advances knowledge and understanding of issues broadly related to attitudes, behavior, and institutions connected to public policy and the provision of public services. Research proposals are expected to be theoretically motivated, conceptually precise, methodologically rigorous, and empirically oriented. Substantive areas include (but are not limited to) the study of individual and group decision-making, political institutions (appointed or elected), attitude and preference formation and expression, electoral processes and voting, public administration, and public policy. This work can focus on a single case or can be done in a comparative context, either over time or cross-sectionally. The Program does not fund applied research. The Program also supports research experiences for undergraduate students and infrastructural activities, including methodological innovations. In addition, we encourage you to examine the websites for the National Science Foundation's Law and Science (LS) and Security and Preparedness (SAP) programs. Full proposals are due annually on August 15 and January 15.
*Requirements:* NSF's mission calls for the broadening of opportunities and expanding participation of groups, institutions, and geographic regions that are underrepresented in STEM disciplines, which is essential to the health and vitality of science and engineering. NSF is committed to this principle of diversity and deems it central to the programs, projects, and activities it considers and supports. NSF also is committed to public access to publications and data, unless there are countervailing interests that prohibit or limit public access to data, including matters of personally identifiable information of research participants, privacy, other issues of vulnerability such as economic, social or other security
*Geographic Focus* All States, District of Columbia, Guam, Marshall Islands, Northern Mariana Islands, Puerto Rico, U.S. Virgin Islands, American Samoa
*Date(s) Application is Due:* Jan 15; Aug 15
*Contact:* Jan E. Leighley; (703) 292-8760 or (703) 292-5111; jleighle@nsf.gov
Mauricia Barnett, Social Scientist; (703) 292-7309 or (703) 292-5111; mbarnett@nsf.gov
Linh Nguyen, Program Assistant; (703) 292-7270 or (703) 292-5111; linguyen@nsf.gov
*Internet:* https://www.nsf.gov/funding/pgm_summ.jsp?pims_id=505715
*Sponsor:* National Science Foundation
2415 Eisenhower Avenue
Alexandria, VA 22314

**NSF Advances in Biological Informatics Grants** **1922**
The Advances in Biological Informatics (ABI) program seeks to encourage new approaches to the analysis and dissemination of biological knowledge for the benefit of both the scientific community and the broader public. The ABI program is especially interested in the development of informatics tools and resources that have the potential to advance- or transform- research in biology supported by the Directorate for Biological Sciences at the National Science Foundation. The ABI program accepts three major types of proposals: Innovation awards that seek to pioneer new approaches to the application of informatics to biological problems, Development awards that seek to provide robust cyberinfrastructure that will enable transformative biological research, and Sustaining awards that seek to support ongoing operations and maintenance of existing cyberinfrastructure that is critical for continued advancement of priority biological research. The annual deadline the second Tuesday in August.
*Geographic Focus:* All States
*Date(s) Application is Due:* Jan 30
*Contact:* Reed S. Beaman, Program Director; (703) 292-7163 or (703) 292-5111; fax (703) 292-9063; rsbeaman@nsf.gov
Peter H. McCartney, Program Director; (703) 292-8470 or (703) 292-5111; fax (703) 292-9063; dbiabi@nsf.gov
*Internet:* https://nsf.gov/funding/pgm_summ.jsp?pims_id=5444
*Sponsor:* National Science Foundation
2415 Eisenhower Avenue
Alexandria, VA 22314

**NSF Advancing Digitization of Biodiversity Collections Grants** **1923**
The Advancing Digitization of Biodiversity Collections program seeks to enhance and expand the national resource of digital data documenting existing vouchered biological and paleontological collections and to advance scientific knowledge by improving access to digitized information (including images) residing in vouchered scientific collections across the United States. The information associated with various collections of organisms, such as geographic, paleogeographic and stratigraphic distribution, environmental habitat data, phenology, information about associated organisms, collector field notes, and tissues and molecular data extracted from the specimens, is a rich resource providing the baseline from which to further biodiversity research and provide critical information about existing gaps in our knowledge of life on earth. The national resource is structured at three levels: a central coordinating organization, a series of thematic networks based on an important research theme, and the physical collections. The national resource builds upon a sizable existing national investment in curation of the physical objects in scientific collections and contributes vitally to scientific research and technology interests in the United States. It will become an invaluable tool in understanding contemporary biological issues and challenges. The annual deadline is the second Friday in October.
*Requirements:* The most frequent recipients are academic institutions and nonprofit research groups. In special circumstances, grants also are awarded to other types of institutions and to individuals.
*Geographic Focus:* All States
*Date(s) Application is Due:* Oct 18
*Contact:* Reed S. Beaman, Program Director; (703) 292-7163 or (703) 292-5111; fax (703) 292-9018; rsbeaman@nsf.gov or biodigit@nsf.gov
Cesar R. Nufio, Program Manager; (703) 292-5113 or (703) 292-5111; fax (703) 292-9018; cnufio@nsf.gov or biodigit@nsf.gov
*Internet:* https://www.nsf.gov/funding/pgm_summ.jsp?pims_id=503559&org=BIO&sel_org=BIO&from=fund
*Sponsor:* National Science Foundation
2415 Eisenhower Avenue
Alexandria, VA 22314

**NSF Advancing Theory in Biology Grants** **1924**
The Biological Sciences Directorate invites submission of proposals that advance our conceptual and theoretical understanding of living systems. The Advancing Theory in Biology (ATB) solicitation supports the development of new theoretical approaches that will improve our understanding of fundamental biological principles that integrate phenomena across levels of biological organization. The anticipated funding amount is $5,000,000 total, of which 10 to 15 grants will be awarded. Awards will not exceed a total of $750,000 over a three year period. Full proposals should be submitted via the FastLane system at NSF or Grants.gov.
*Geographic Focus:* All States
*Date(s) Application is Due:* Jan 30
*Amount of Grant:* Up to 750,000 USD
*Contact:* Michelle Elekonich, Program Director; (703) 292-7202 or (703) 292-5111; fax (703) 292-9061; melekonich@nsf.gov or biotheory@nsf.gov
*Internet:* https://www.nsf.gov/funding/pgm_summ.jsp?pims_id=501066
*Sponsor:* National Science Foundation
2415 Eisenhower Avenue
Alexandria, VA 22314

**NSF Alan T. Waterman Award** **1925**
The annual award recognizes an outstanding young researcher in any field of science or engineering supported by the National Science Foundation. In addition to a medal, the awardee receives a grant of $1,000,000 over a five year period for scientific research or advanced study in the mathematical, physical, biological, engineering, social, or other sciences at the institution of the recipient's choice.
*Requirements:* Institutions may nominate an unlimited number of individuals. Nomination packages consist of a nomination and four letters of reference submitted via FastLane. Candidates must be U.S. citizens, normally 35 years of age or younger, or not more than seven years beyond receipt of the PhD degree by December 31 of the year nominated. Candidates should have demonstrated exceptional individual achievements in scientific or engineering research of sufficient quality to place them at the forefront of their peers. Criteria include originality, innovation, and significant impact on the field.
*Geographic Focus:* All States
*Date(s) Application is Due:* Dec 31
*Amount of Grant:* 1,000,000 USD
*Samples:* David Charbonneau, Professor of Astronomy, Harvard University; Terence Tao, Professor of Mathematics, University of California, Los Angeles; Peidong Yang, Associate Professor of Chemistry, University of California, Berkeley.
*Contact:* Mayra N. Montrose, Program Contact; (703) 292-8040; fax (703) 292-9040; mmontros@nsf
*Internet:* www.nsf.gov/od/waterman/waterman.jsp
*Sponsor:* National Science Foundation
2415 Eisenhower Avenue
Alexandria, VA 22314

**NSF Algebra and Number Theory Grants** 1926

The Algebra and Number Theory program supports research in algebra, algebraic and arithmetic geometry, number theory, and representation theory. For conference proposals with budgets not exceeding $50,000, which in accordance with NSF policy can be reviewed internally at NSF, the following target dates are in effect: for an event that will take place at some time prior to October 1 during a given year, the proposal should be submitted in October of the previous year; for an event that will occur in the period October 1 through December 31 of a given year, the proposal should be submitted in May of that year. A conference proposal with a budget request exceeding $50,000 should be submitted roughly seven months before the event is scheduled to take place, in order to allow time for external review.
*Geographic Focus:* All States
*Date(s) Application is Due:* Oct 12
*Contact:* Andrew D. Pollington, Program Director; (703) 292-4878 or (703) 292-5111; fax (703) 292-9032; adpollin@nsf.gov
*Internet:* https://www.nsf.gov/funding/pgm_summ.jsp?pims_id=5431&org=DMS&sel_org=DMS&from=fund
*Sponsor:* National Science Foundation
2415 Eisenhower Avenue
Alexandria, VA 22314

**NSF Algorithms for Modern Power Systems Grants** 1927

The Algorithms for Modern Power Systems (AMPS) program will support research projects to develop the next generation of mathematical and statistical algorithms for improvement of the security, reliability, and efficiency of the modern power grid. The program is a partnership between the Division of Mathematical Sciences (DMS) at the National Science Foundation (NSF) and the Office of Electricity Delivery & Energy Reliability (OE) at the U.S. Department of Energy (DOE).
*Geographic Focus:* All States
*Contact:* Leland M. Jameson; (703) 292-4883 or (703) 292-5111; ljameson@nsf.gov
*Internet:* https://www.nsf.gov/funding/pgm_summ.jsp?pims_id=505322&org=DMS&sel_org=DMS&from=fund
*Sponsor:* National Science Foundation
2415 Eisenhower Avenue
Alexandria, VA 22314

**NSF Algorithms for Threat Detection Grants** 1928

The Algorithms for Threat Detection (ATD) program will support research projects to develop the next generation of mathematical and statistical algorithms for analysis of large spatiotemporal datasets with application to quantitative models of human dynamics. The program is a partnership between the Division of Mathematical Sciences (DMS) at the National Science Foundation (NSF) and the National Geospatial Intelligence Agency (NGA). The annual deadline is the third Tuesday in February.
*Geographic Focus:* All States
*Date(s) Application is Due:* Feb 19
*Contact:* Leland M. Jameson; (703) 292-4883 or (703) 292-5111; ljameson@nsf.gov
*Internet:* https://www.nsf.gov/funding/pgm_summ.jsp?pims_id=503427&org=DMS&sel_org=DMS&from=fund
*Sponsor:* National Science Foundation
2415 Eisenhower Avenue
Alexandria, VA 22314

**NSF Analysis Grants** 1929

The Analysis program supports basic research in that area of mathematics whose roots can be traced to the calculus of Newton and Leibniz. Given its centuries-old ties to physics, analysis has influenced developments from Newton's mechanics to quantum mechanics and from Fourier's study of heat conduction to Maxwell's equations of electromagnetism to Witten's theory of supersymmetry. More generally, research supported by analysis provides the theoretical underpinning for the majority of applications of the mathematical sciences to other scientific disciplines. Current areas of significant activity include: nonlinear partial differential equations; dynamical systems and ergodic theory; real, complex and harmonic analysis; operator theory and algebras of operators on Hilbert space; mathematical physics; and representation theory of Lie groups/algebras. Emerging areas include random matrix theory and its ties to classical analysis, number theory, quantum mechanics, and coding theory, and development of noncommutative geometry with its applications to modeling physical phenomena. It should be stressed, however, that the underlying role of the Analysis Program is to provide support for research in mathematics at the most fundamental level. Although this is often done with the expectation that the research will generate a payoff in applications at some point down the road, the principal mission of the Program is to tend and replenish an important reservoir of mathematical knowledge, maintaining it as a dependable resource to be drawn upon by engineers, life and physical scientists, and other mathematical scientists, as need arises. Full proposals should be submitted via the FastLane system at NSF or Grants.gov.
*Geographic Focus:* All States
*Date(s) Application is Due:* Oct 2
*Contact:* Kevin Clancey, Program Director; (703) 292-4870 or (703) 292-5111; fax (703) 292-9032; kclancey@nsf.gov
*Internet:* https://www.nsf.gov/funding/pgm_summ.jsp?pims_id=5434&org=DMS&sel_org=DMS&from=fund
*Sponsor:* National Science Foundation
2415 Eisenhower Avenue
Alexandria, VA 22314

**NSF Applied Mathematics Grants** 1930

NSF Applied Mathematics grants support research motivated by or having an effect on problems arising in science and engineering. Mathematical merit and novelty, as well as breadth and quality of impact on applications, are important factors. Proposals to develop critical mathematical techniques from individual investigators as well as interdisciplinary teams are encouraged. The proposal window is November 1 through November 15 each year. To facilitate timely notification of the availability of support: proposals for conferences, workshops, etc., to be held in the U.S. must be submitted eight months in advance of the conference date; proposals to support group travel to meetings outside the U.S. must be submitted twelve months in advance of the meeting date; and proposals for conferences, workshops, etc., whose budget request exceeds $50,000 must be submitted during the annual November 1 through November 15 submission window.
*Restrictions:* Proposals whose primary applications are in the biological sciences are inappropriate for Applied Mathematics, and should be submitted elsewhere.
*Geographic Focus:* All States
*Date(s) Application is Due:* Nov 15
*Contact:* Michael Steuerwalt, Program Director; (703) 292-4860 or (703) 292-5111; fax (703) 292-9032; msteuerw@nsf.gov
*Internet:* https://www.nsf.gov/funding/pgm_summ.jsp?pims_id=5664&org=DMS&sel_org=DMS&from=fund
*Sponsor:* National Science Foundation
2415 Eisenhower Avenue
Alexandria, VA 22314

**NSF Astronomy and Astrophysics Postdoctoral Fellowships (AAPF)** 1931

NSF Astronomy and Astrophysics Postdoctoral Fellowships provide an opportunity for highly qualified young investigators within 3 years of obtaining their PhD to carry out an integrated program of independent research and education. Fellows may engage in research of observational, instrumental, or theoretical nature, in combination with a coherent educational plan for the duration of the fellowship. The program supports researchers for a period of up to three years with fellowships that can be taken to the institution or national facility of their choice. The program is intended to recognize young investigators of significant potential, and provide them with experience in research and education that will establish them in positions of distinction and leadership in the community. The annual deadline is the second Wednesday in October.
*Geographic Focus:* All States
*Date(s) Application is Due:* Oct 15
*Contact:* Harshal Gupta, Program Director; (703) 292-5039 or (703) 292-5111; fax (703) 292-9034; hgupta@nsf.gov
*Internet:* https://nsf.gov/funding/pgm_summ.jsp?pims_id=5291
*Sponsor:* National Science Foundation
2415 Eisenhower Avenue
Alexandria, VA 22314

**NSF Astronomy and Astrophysics Research Grants (AAG)** 1932

The Astronomy and Astrophysics Research Grants program provides individual investigator and collaborative research grants for observational, theoretical, laboratory and archival data studies in all areas of astronomy and astrophysics, including but not limited to the following areas of study: Planetary Astronomy - studies of the detailed structure and composition of the surfaces, interiors and atmospheres of the planets and satellites in the Solar System, the nature of small bodies (asteroids and comets), the inter-planetary medium, and the origin and development of the Solar System; Stellar Astronomy and Astrophysics - studies of the structure and activity of the Sun and other stars, the physical properties and composition of all types of single and multiple stars, compact objects and their interactions, extra-solar system planet formation and detection, star formation and stellar evolution, stellar nucleosynthesis, and the properties of atoms and molecules of relevance to stellar astronomy; Galactic Astronomy - studies on the composition, structure and evolution of the Milky Way galaxy and nearby galaxies (research may focus on the stellar populations in these galaxies, the characteristics of star clusters, the interstellar medium, and the properties of atomic and molecular constituents of the interstellar medium); and Extragalactic Astronomy and Cosmology - studies of the more distant Universe, with research topics including galaxy formation, evolution and interaction, active galaxies, quasars, large-scale structure, and all areas of cosmology.
*Requirements:* Proposals submitted to the AAG Program do not require categorization into one of the study areas identified above. Proposals may span multiple disciplines and/or areas of study and may utilize multiple techniques. Principal Investigators are encouraged to contact one of the Program Officers listed in this announcement prior to submitting a proposal to the AAG Program, particularly if the proposal will include investigators at multiple institutions. Standard NSF eligibility requirements apply.
*Geographic Focus:* All States
*Date(s) Application is Due:* Nov 15
*Contact:* James E. Neff, Coordinator; (703) 292-2475 or (703) 292-5111; fax (703) 292-9034; jneff@nsf.gov
Harshal Gupta, Program Director; (703) 292-5039 or (703) 292-5111; fax (703) 292-9034; hgupta@nsf.gov
*Internet:* https://nsf.gov/funding/pgm_summ.jsp?pims_id=13630
*Sponsor:* National Science Foundation
2415 Eisenhower Avenue
Alexandria, VA 22314

**NSF Atmospheric and Geospace Sciences Postdoctoral Research Fellowships** 1933
The Division of Atmospheric and Geospace Sciences (AGS) awards Postdoctoral Research Fellowships (PRF) to highly qualified early career investigators to carry out an independent research program. The research plan of each Fellowship must address scientific questions within the scope of AGS disciplines. The program supports researchers for a period of up to 2 years with Fellowships that can be taken to the institution of their choice. The program is intended to recognize beginning investigators of significant potential and provide them with experience in research that will broaden perspectives, facilitate interdisciplinary interactions and help establish them in leadership positions within the Atmospheric and Geospace Sciences community. Fellowships are awards to individuals, not institutions, and are administered by the Fellows. Proposals are accepted at any time. However, different programs may have preferred target windows so please contact the relevant program director early in the planning process for submission. Five to eight awards are anticipated each year. Award budgets will be $94,000 in year one of the award and increase to $96,000 in year two.
*Requirements:* This program provides educational opportunities for Postdoctoral Fellows.
*Geographic Focus* All States, District of Columbia, Guam, Marshall Islands, Northern Mariana Islands, Puerto Rico, U.S. Virgin Islands, American Samoa
*Amount of Grant:* 94,000 - 96,000 USD
*Contact:* Sylvia A. Edgerton, Program Director; (703) 292-8522 or (703) 292-5111; fax (703) 292-9022; sedgerto@nsf.gov
Ilia I. Roussev, Program Director; (703) 292-8519 or (703) 292-5111; fax (703) 292-9022; iroussev@nsf.gov or geospaceprf@nsf.gov
*Internet:* https://www.nsf.gov/funding/pgm_summ.jsp?pims_id=12779
*Sponsor:* National Science Foundation
2415 Eisenhower Avenue
Alexandria, VA 22314

**NSF CISE-MPS Interdisciplinary Faculty Program Grants in Quantum Information Science** 1934
The CISE-MPS Interdisciplinary Faculty Program in Quantum Information Science is designed to promote research in the area of Quantum Information Science (QIS) by providing resources to allow QIS researchers and researchers from the CISE or MPS disciplines to actively engage in joint research efforts, addressing problems at the interface between the mathematical and physical sciences and computer and information sciences through long-term visits to a host institution.
*Requirements:* To be eligible for these awards, the principal investigator must hold a tenured or tenure-track position at the academic institution submitting the proposal and have a strong background in one of the disciplines within the preview of the Division of Mathematical Sciences. Proposals may be submitted by academic institutions in support of individual investigators.
*Geographic Focus:* All States
*Contact:* Edward Taylor, Program Director; (703) 292-4872 or (703) 292-5111; fax (703) 292-9032; etaylor@nsf.gov
*Internet:* https://www.nsf.gov/funding/pgm_summ.jsp?pims_id=504743&org=DMS&sel_org=DMS&from=fund
*Sponsor:* National Science Foundation
2415 Eisenhower Avenue
Alexandria, VA 22314

**NSF Collaborative Research in Computational Neuroscience Grants** 1935
Computational neuroscience provides a theoretical foundation and a rich set of technical approaches for understanding the principles and dynamics of the nervous system. Building on the theory, methods, and findings of computer science, neuroscience, biology, psychology, the mathematical and physical sciences, engineering, and other fields, computational neuroscience employs a broad spectrum of analytical and experimental approaches to study function, organization, and computational strategies across all levels of the nervous system. Two classes of proposals will be considered in response to this solicitation: research proposals describing new collaborative research projects, and data sharing proposals to enable sharing of data and other resources. In general, appropriate scientific areas of investigations may be related to any of the participating funding organizations.
*Restrictions:* In response to this solicitation, an investigator may participate as PI or Co-PI in no more than one research proposal per review cycle, and in no more than one data sharing proposal per review cycle.
*Geographic Focus:* All States
*Date(s) Application is Due:* Jan 5; Nov 27
*Amount of Grant:* 300,000 - 1,250,000 USD
*Contact:* Kenneth Whang, Program Director; (703) 292-5149 or (703) 292-5111; fax (703) 292-9073; kwhang@nsf.gov
Jasmine J. Owens, Program Analyst; (703) 292-8377 or (703) 292-5111; fax (703) 292-9073; jowens@nsf.gov
*Internet:* https://www.nsf.gov/funding/pgm_summ.jsp?pims_id=5147&org=BIO&sel_org=BIO&from=fund
*Sponsor:* National Science Foundation
2415 Eisenhower Avenue
Alexandria, VA 22314

**NSF Collections in Support of Biological Research Grants** 1936
The Collections in Support of Biological Research program provides support for biological collection enhancement, computerization of specimen-related data, research to develop better methods for specimen curation and collection management, and activities such as symposia and workshops to investigate support and management of biological collections. Biological collections supported include those housing natural history specimens and jointly curated collections such as frozen tissues and other physical samples, e.g. DNA libraries and digital images. Such collections provide the materials necessary for research in a broad area of biological sciences.
*Requirements:* Proposals are accepted from U.S. organizations, including colleges and universities that maintain research collections, natural history museums including herbaria, and other collections administered by independent organizations or by state, county, or local governments; non-federal and non-profit research organizations that maintain collections; and field stations, marine laboratories, botanical gardens, zoological parks, and aquaria that maintain research collections that document biological diversity. The size of an organization is not a factor in determining eligibility. The maximum that may be requested in a proposal is $500,000. Numbers of awards and average award size/duration are subject to the availability of funds.
*Restrictions:* In any single round of the competition, only one proposal may be submitted from any individual collection within an organization.
*Geographic Focus:* All States
*Amount of Grant:* Up to 500,000 USD
*Contact:* Reed S. Beaman, Program Director; (703) 292-7163 or (703) 292-5111; fax (703) 292-9063; rsbeaman@nsf.gov
Roland P. Roberts, Program Director; (703) 292-7884 or (703) 292-5111; fax (703) 292-9063; rolrober@nsf.gov
*Internet:* https://www.nsf.gov/funding/pgm_summ.jsp?pims_id=503651&org=BIO&sel_org=BIO&from=fund
*Sponsor:* National Science Foundation
2415 Eisenhower Avenue
Alexandria, VA 22314

**NSF Combinatorics Grants** 1937
The NSF Combinatorics program supports research on discrete structures and includes algebraic, enumerative, existential, extremal, geometric, and probabilistic combinatorics, including graph theory. Conference proposals must be submitted at least six months in advance of the conference, and in the same fiscal year (which begins October 1) if possible.
*Requirements:* The foundation supports projects and programs of 501(c)3 tax-exempt organizations located in 10 southwestern Ohio counties, six northern Kentucky counties, and sixty nine Indiana counties.
*Restrictions:* Proposals whose primary applications are in the biological sciences are inappropriate for Applied Mathematics, and should be submitted elsewhere.
*Geographic Focus:* All States
*Contact:* Tomek Bartoszynski, Program Director; (703) 292-4885 or (703) 292-5111; tbartosz@nsf.gov
*Internet:* https://www.nsf.gov/funding/pgm_summ.jsp?pims_id=503570&org=DMS&sel_org=DMS&from=fund
*Sponsor:* National Science Foundation
2415 Eisenhower Avenue
Alexandria, VA 22314

**NSF Computational and Data-Enabled Science and Engineering Grants** 1938
The Computational and Data-Enabled Science and Engineering (CDS&E) program includes all Divisions within the Directorate for Mathematical and Physical Sciences (MPS) and the Directorate for Engineering (ENG), as well as the Office of Advanced Cyberinfrastructure (OAC) in the Directorate for Computer and Information Science and Engineering (CISE). Advanced computational infrastructure and the ability to perform large-scale simulations and accumulate massive amounts of data have revolutionized scientific and engineering disciplines. The goal of the CDS&E program is to identify and capitalize on opportunities for major scientific and engineering breakthroughs through new computational and data analysis approaches. The intellectual drivers may be in an individual discipline or they may cut across more than one discipline in various Directorates. The key identifying factor is that the outcome relies on the development, adaptation, and utilization of one or more of the capabilities offered by advancement of both research and infrastructure in computation and data, either through cross-cutting or disciplinary programs. The CDS&E program welcomes proposals in any area of research supported through the participating divisions that address at least one of the following criteria: promote the creation, development, and application of the next generation of mathematical, computational and statistical theories and tools that are essential for addressing the challenges presented to the scientific and engineering communities by the ever-expanding role of computational modeling and simulation and the explosion and production of digital experimental and observational data; promote and encourage integrated research projects that create, develop and apply novel computational, mathematical and statistical methods, algorithms, software, data curation, analysis, visualization and mining tools to address major, heretofore intractable questions in core science and engineering disciplines, including large-scale simulations and analysis of large and heterogeneous collections of data; encourage adventurous ideas that generate new paradigms and that create and apply novel techniques, generating and utilizing digital data in innovative ways to complement or dramatically enhance traditional computational, experimental, observational, and theoretical tools for scientific discovery and application; and encourage ideas at the interface between scientific frameworks, computing capability, measurements and physical systems that enable advances well beyond the expected natural progression of individual activities, including development of science-driven algorithms to address pivotal problems in science and engineering and efficient methods to access, mine, and utilize large data sets. See the website for all divisional deadlines.
*Restrictions:* The CDS&E program is not intended to replace existing programs that make awards that involve computation and the analysis of large data sets. Rather, the CDS&E program is meant to fund awards that have a significant component of cyber

development or cyber science that goes well beyond what would normally be included in these programs.
*Geographic Focus:* All States
*Contact:* Bogdan Mihaila, Program Director; (703) 292-8235 or (703) 292-5111; fax (703) 292-8235; bmihaila@nsf.gov
Evelyn Goldfield; (703) 292-2173 or (703) 292-5111; egoldfie@nsf.gov
Daryl W. Hess, Program Director; (703) 292-4942 or (703) 292-9035; dhess@nsf.gov
*Internet:* https://www.nsf.gov/funding/pgm_summ.jsp?pims_id=504813&org=DMS&sel_org=DMS&from=fund
*Sponsor:* National Science Foundation
2415 Eisenhower Avenue
Alexandria, VA 22314

**NSF Computational and Data-Enabled Science and Engineering in Mathematical and Statistical Sciences Grants** 1939

The Computational and Data-Enabled Science and Engineering in Mathematical and Statistical Sciences (CDS&E-MSS) program accepts proposals that confront and embrace the host of mathematical and statistical challenges presented to the scientific and engineering communities by the ever-expanding role of computational modeling and simulation on the one hand, and the explosion in production of digital and observational data on the other. The goal of the program is to promote the creation and development of the next generation of mathematical and statistical theories and tools that will be essential for addressing such issues. To this end, the program will support fundamental research in mathematics and statistics whose primary emphasis will be on meeting the aforementioned computational and data-related challenges. This program is part of the wider Computational and Data-enabled Science and Engineering (CDS&E) enterprise in NSF that seeks to address this emerging discipline. The research supported by the CDS&E-MSS program will aim to advance mathematics or statistics in a significant way and will address computational or big-data challenges. Proposals of interest to the program will include a Principal Investigator or co-Principal Investigator who is a researcher in the mathematical or statistical sciences in an area supported by the Division of Mathematical Sciences. The program encourages submission of proposals that include multidisciplinary collaborations or the training of mathematicians and statisticians in CDS&E. The annual proposal window is open from November 25 through December 9.
*Geographic Focus:* All States
*Date(s) Application is Due:* Dec 9
*Contact:* Yong Zeng, Program Director; (703) 292-2301 or (703) 292-5111; fax (703) 292-9032; yzeng@nsf.gov
*Internet:* https://www.nsf.gov/funding/pgm_summ.jsp?pims_id=504687&org=DMS&sel_org=DMS&from=fund
*Sponsor:* National Science Foundation
2415 Eisenhower Avenue
Alexandria, VA 22314

**NSF Computational Mathematics Grants** 1940

This Computational Mathematics program supports mathematical research in areas of science where computing plays a central and essential role, emphasizing algorithms design, numerical methods and their analysis, and symbolic methods. The prominence of computation in the research is a hallmark of the program. Proposals ranging from single-investigator projects that develop and analyze innovative computational methods to interdisciplinary team projects that not only create new mathematical and computational techniques but use them to model, study, and solve important application problems are encouraged. The annual proposal window is November 16 through December 1.
*Geographic Focus:* All States
*Date(s) Application is Due:* Dec 1
*Contact:* Leland M. Jameson, Program Director; (703) 292-4883 or (703) 292-5111; fax (703) 292-9032; ljameson@nsf.gov
*Internet:* https://www.nsf.gov/funding/pgm_summ.jsp?pims_id=5390&org=DMS&sel_org=DMS&from=fund
*Sponsor:* National Science Foundation
2415 Eisenhower Avenue
Alexandria, VA 22314

**NSF Conferences and Workshops in the Mathematical Sciences Grants** 1941

NSF supports research, education, and infrastructure for the mathematical sciences in a variety of modes, including support for symposia, workshops, regional conferences, and special research years. These provide exposure for scholarly work, encourage graduate students and junior scientists early in their careers, and they influence the sense of participation, the concept of a role-model, and the perception of current opportunities in a given scientific field. proposals normally request funding in the range of $5,000 to $25,000, although awards of up to $50,000 have occasionally been made. Their duration is normally for one year. Proposals are accepted at any time.
*Requirements:* Proposals may be submitted by universities, colleges, and nonprofit, nonacademic organizations in support of individual investigators or small groups.
*Geographic Focus:* All States
*Amount of Grant:* 5,000 - 50,000 USD
*Contact:* Tomek Bartoszynski; (703) 292-4885 or (703) 292-5111; tbartosz@nsf.gov
*Internet:* https://www.nsf.gov/funding/pgm_summ.jsp?pims_id=11701&org=DMS&sel_org=DMS&from=fund
*Sponsor:* National Science Foundation
2415 Eisenhower Avenue
Alexandria, VA 22314

**NSF Convergence Accelerator Grants** 1942

With the NSF Convergence Accelerator, NSF's goals are: (i) to pilot a new NSF capability to accelerate use-inspired convergence research in areas of national importance, and (ii) to initiate convergence team-building capacity around exploratory, potentially high-risk proposals in specific convergence topics (tracks). The NSF Convergence Accelerator supports use-inspired, goal-oriented, basic research, encouraging rapid advances through partnerships that include multiple stakeholders (e.g., industry, academic, not-for-profits, government entities, and others). The NSF Convergence Accelerator brings teams together in a cohort that are all focused on a common research goal of national importance, but which may be pursuing many different approaches. As a funder of research and education across all fields of science and engineering and with relationships with universities and funding agencies around the world, NSF is uniquely positioned to pilot this approach to accelerate discovery and innovation. Teams supported by the NSF Convergence Accelerator will focus on grand challenges that require a convergence approach. The teams are multidisciplinary and leverage partnerships; tracks within the NSF Convergence Accelerator relate to a grand challenge problem and have a high probability of resulting in deliverables that will benefit society within a fixed term. The NSF Convergence Accelerator is modeled on acceleration and innovation activities from the most forward-looking companies and universities. Full proposals are accepted at anytime.
*Requirements:* U.S. universities and two- and four-year colleges (including community colleges) are eligible to apply.
*Geographic Focus* All States, District of Columbia, Guam, Marshall Islands, Northern Mariana Islands, Puerto Rico, U.S. Virgin Islands, American Samoa
*Contact:* Douglas Maughan, Program Officer; 703-292-2497 or (703) 292-5111; dmaughan@nsf.gov
Lara A. Campbell; (703) 292-7049 or (703) 292-5111; lcampbel@nsf.gov
Nancy U. Kamei, Program Officer; (703) 292 7236 or (703) 292-5111; nkamei@nsf.gov
*Internet:* https://www.nsf.gov/funding/pgm_summ.jsp?pims_id=505665
*Sponsor:* National Science Foundation
2415 Eisenhower Avenue
Alexandria, VA 22314

**NSF Convergence Accelerator Pilot Phase II Grants** 1943

The goals of NSF's convergence accelerator effort are to support and accelerate use-inspired convergence research in areas of national importance within particular topics (tracks). The NSF Convergence Accelerator supports fundamental research that leads to rapid advances that can deliver useful results to society. Teams are expected to include personnel with the appropriate mix of disciplinary expertise needed to execute their proposed Phase II research and development plan. The team should also include appropriate stakeholders (e.g., industry, Institutions of Higher Education (IHEs), non-profits, government entities, and others), each with a specific role(s) in facilitating the transition of research outputs into practical uses. The proposers should outline a two-year research and development plan in which research transitions to practice. Successful proposals will be funded initially for one year. Each team's progress will be assessed during the year through approximately six virtual and in-person meetings with NSF program staff. The overall progress will be evaluated at the end of one year, based on a report and presentation that the team will make to a panel of reviewers. Teams that show significant progress during the first year, in accordance with the agreed timetable of milestones and deliverables, will receive funding for a second year. Interested teams may request up to $3,000,000 for the first year and up to $5,000,000 in total for the two-year project. Teams should plan on completing the effort within two years; no-cost extensions will be authorized only in extraordinary circumstances. The annual deadline is May 11.
Restrictions: This solicitation is limited to grantees who received a Phase I Award under NSF D.C.L 19-050 and wish to advance to Phase II.
*Geographic Focus* All States, District of Columbia, Guam, Marshall Islands, Northern Mariana Islands, Puerto Rico, U.S. Virgin Islands, American Samoa
*Date(s) Application is Due:* May 11
*Amount of Grant:* 3,000,000 - 5,000,000 USD
*Contact:* Lara A. Campbell, Division Director, Division of Chemistry (MPS/CHE); (703) 292-7049 or 703) 292-5111; lcampbel@nsf.gov
Michael Pozmantier; (703) 292-4475 or (703) 292-5111; mpozmant@nsf.gov
*Internet:* https://www.nsf.gov/funding/pgm_summ.jsp?pims_id=505777
*Sponsor:* National Science Foundation
2415 Eisenhower Avenue
Alexandria, VA 22314

**NSF Critical Techniques, Technologies and Methodologies for Advancing Foundations and Applications of Big Data Sciences and Engineering Grants** 1944

The Critical Techniques, Technologies and Methodologies for Advancing Foundations and Applications of Big Data Sciences and Engineering program seeks novel approaches in computer science, statistics, computational science, and mathematics, along with innovative applications in domain science, including social and behavioral sciences, education, biology, the physical sciences, and engineering that lead towards the further development of the interdisciplinary field of data science. The solicitation invites two categories of proposals: foundations, or those developing or studying fundamental theories, techniques, methodologies, and technologies of broad applicability to big data problems, motivated by specific data challenges and requirements; and innovative applications, or those engaged in translational activities that employ new big data techniques, methodologies, and technologies to address and solve problems in specific application domains. Projects in this category must be collaborative, involving researchers from domain disciplines and one or more methodological disciplines, e.g., computer science, statistics, mathematics, simulation and modeling, etc.

*Requirements:* Proposals in both categories must include a clear description of the big data aspect(s) that have motivated the proposed approach(es), for example: the scalability of methods with increasing data volumes, rates, heterogeneity; or data quality and data bias; etc. Innovative Applications proposals must provide clear examples of the impacts of the big data techniques, technologies and/or methodologies on (a) specific domain application(s).
*Geographic Focus:* All States
*Date(s) Application is Due:* Jan 29
*Contact:* Chaitanya Baru, Senior Advisor for Data Science; (703) 292-4541 or (703) 292-5111; fax (703) 292-9063; cbaru@nsf.gov or bigdata@nsf.gov
*Internet:* https://www.nsf.gov/funding/pgm_summ.jsp?pims_id=504767&org=BIO&sel_org=BIO&from=fund
*Sponsor:* National Science Foundation
2415 Eisenhower Avenue
Alexandria, VA 22314

**NSF Critical Techniques, Technologies and Methodologies for Advancing Foundations and Applications of Big Data Sciences and Engineering Grants** **1945**
The BIGDATA program seeks novel approaches in computer science, statistics, computational science, and mathematics, along with innovative applications in domain science, including social and behavioral sciences, education, biology, the physical sciences, and engineering that lead towards the further development of the interdisciplinary field of data science. The solicitation invites two categories of proposals: foundations - those developing or studying fundamental theories, techniques, methodologies, and technologies of broad applicability to big data problems, motivated by specific data challenges and requirements; and innovative applications - those engaged in translational activities that employ new big data techniques, methodologies, and technologies to address and solve problems in specific application domains. Projects in this category must be collaborative, involving researchers from domain disciplines and one or more methodological disciplines, e.g., computer science, statistics, mathematics, simulation and modeling, etc. Proposals in both categories must include a clear description of the big data aspect(s) that have motivated the proposed approach(es), for example: the scalability of methods with increasing data volumes, rates, heterogeneity; or data quality and data bias; etc. Innovative Applications proposals must provide clear examples of the impacts of the big data techniques, technologies and/or methodologies on (a) specific domain application(s).
*Geographic Focus:* All States
*Contact:* Chaitanya Baru, Program Director; (703) 292-4541 or (703) 292-5111; fax (703) 292-9078; cbaru@nsf.gov
*Internet:* https://www.nsf.gov/funding/pgm_summ.jsp?pims_id=504767&org=DMS&sel_org=DMS&from=fund
*Sponsor:* National Science Foundation
2415 Eisenhower Avenue
Alexandria, VA 22314

**NSF Cultivating Cultures for Ethical STEM Grants** **1946**
Cultivating Cultures for Ethical STEM (CCE STEM) funds research projects that identify factors that are efficacious in the formation of ethical STEM researchers in all the fields of science and engineering that NSF supports. CCE STEM solicits proposals for research that explores the following: What constitutes ethical STEM research and practice, and which cultural and institutional contexts promote ethical STEM research and practice and why? Factors one might consider include: honor codes, professional ethics codes and licensing requirements, an ethic of service and/or service learning, life-long learning requirements, curricula or memberships in organizations (e.g. Engineers without Borders) that stress social responsibility and humanitarian goals, institutions that serve under-represented groups, institutions where academic and research integrity are cultivated at multiple levels, institutions that cultivate ethics across the curriculum, or programs that promote group work, or do not grade. Successful proposals typically have a comparative dimension, either between or within institutional settings that differ along these or other factors.
*Requirements:* Proposals for awards from minority-serving institutions (e.g. Tribal Colleges and Universities, Historically Black Colleges and Universities, Hispanic-Serving Institutions, Alaska Native or Native Hawaiian Serving Institutions), women's colleges, and institutions primarily serving persons with disabilities are strongly encouraged. Proposals including international collaborations are encouraged when those efforts enhance the merit of the proposed work by incorporating unique resources, expertise, facilities or sites of international partners. The U.S. team's international counterparts generally should have support or obtain funding through other sources.
*Geographic Focus:* All States
*Date(s) Application is Due:* Feb 22
*Contact:* John Parker, Program Contact; (703) 292-5034 or (703) 292-5111; fax (703) 292-9068; joparker@nsf.gov
*Internet:* https://www.nsf.gov/funding/pgm_summ.jsp?pims_id=505027&org=BIO&sel_org=BIO&from=fund
*Sponsor:* National Science Foundation
2415 Eisenhower Avenue
Alexandria, VA 22314

**NSF Cyber-Innovation for Sustainability Science and Engineering Grants** **1947**
The Cyber-Innovation for Sustainability Science and Engineering (CyberSEES) program aims to advance interdisciplinary research in which the science and engineering of sustainability are enabled by new advances in computing, and in which computational innovation is grounded in the context of sustainability problems. The program is one component of the National Science Foundation's Science, Engineering, and Education for Sustainability (SEES) activities, a Foundation-wide effort aimed at addressing the challenge of sustainability through support for interdisciplinary research and education. In the SEES context, a sustainable world is one where human needs are met equitably without harm to the environment or sacrificing the ability of future generations to meet their own needs. Computational approaches play a central role in understanding and advancing sustainability. CyberSEES supports research on topics that depend on advances in computational areas including optimization, modeling, simulation, prediction and inference; large-scale data management and analytics; advanced sensing techniques; human computer interaction and social computing; infrastructure design, control and management; and intelligent systems and decision-making. Additionally, the widespread, intensive use of computing technologies also introduces sustainability challenges and motivates new approaches across the lifecycle of technology design and use.
*Requirements:* Proposals may only be submitted by the following: non-profit, non-academic organizations: Independent museums, observatories, research labs, professional societies and similar organizations in the U.S. associated with educational or research activities; or universities and colleges (including community colleges) located and accredited in the U.S., acting on behalf of their faculty members. Such organizations also are referred to as academic institutions. Preliminary proposals are required and must be submitted via the NSF FastLane system, even if full proposals will be submitted via Grants.gov.
*Geographic Focus:* All States
*Contact:* Phillip A. Regalia, Program Director; (703) 292-2981 or (703) 292-5111; fax (703) 292-9059; pregalia@nsf.gov
*Internet:* https://www.nsf.gov/funding/pgm_summ.jsp?pims_id=504829&org=BIO&sel_org=BIO&from=fund
*Sponsor:* National Science Foundation
2415 Eisenhower Avenue
Alexandria, VA 22314

**NSF Designing Materials to Revolutionize and Engineer our Future Grants** **1948**
Designing Materials to Revolutionize and Engineer our Future is the primary program by which NSF participates in the Materials Genome Initiative (MGI) for Global Competitiveness. MGI recognizes the importance of materials science and engineering to the well-being and advancement of society and aims to "deploy advanced materials at least twice as fast as possible today, at a fraction of the cost." MGI integrates materials discovery, development, property optimization, and systems design with a shared computational framework. This framework facilitates collaboration and coordination of research activities, analytical tools, experimental results, and critical evaluation in pursuit of the MGI goals. The MGI Strategic Plan highlights four sets of goals: leading a culture shift in materials science research to encourage and facilitate an integrated team approach; integrating experimentation, computation, and theory and equipping the materials community with advanced tools and techniques; making digital data accessible; and creating a world-class materials science and engineering workforce that is trained for careers in academia or industry. Awards are expected to range from $250,000 to $400,000 per year for a duration of three or four years, thus totaling $750,000 to $1,600,000 over the award period. To cover the breadth of this endeavor, it is expected that proposed projects will be directed by a team of at least two Senior Personnel with complementary expertise.
*Geographic Focus:* All States
*Amount of Grant:* 750,000 - 1,600,000 USD
*Contact:* John A. Schlueter, Program Director; (703) 292-7766 or (703) 292-5111; fax (703) 292-9135; jschluet@nsf.gov
*Internet:* https://www.nsf.gov/funding/pgm_summ.jsp?pims_id=505073&org=DMS&sel_org=DMS&from=fund
*Sponsor:* National Science Foundation
2415 Eisenhower Avenue
Alexandria, VA 22314

**NSF Documenting Endangered Languages Project Grants and Fellowships** **1949**
This funding partnership between the National Science Foundation (NSF) and the National Endowment for the Humanities (NEH) supports projects to develop and advance knowledge concerning endangered human languages. Made urgent by the imminent death of roughly half of the approximately 7,000 currently used languages, this effort aims to exploit advances in information technology to build computational infrastructure for endangered language research. The program supports projects that contribute to data management and archiving, and to the development of the next generation of researchers. Funding can support fieldwork and other activities relevant to the digital recording, documenting, and archiving of endangered languages, including the preparation of lexicons, grammars, text samples, and databases. Funding will be available in the form of one- to three-year project grants as well as fellowships for up to twelve months and doctoral dissertation research improvement grants for up to twenty-four months. Approximately 18 to 22 awards will be chosen, including up to a dozen Fellowships.
*Requirements:* Proposals may only be submitted by: Universities and Colleges [Universities and two- and four-year colleges (including community colleges) accredited in, and having a campus located in the US, acting on behalf of their faculty members. Such organizations also are referred to as academic institutions.]; Non-profit, non-academic organizations (Independent museums, observatories, research labs, professional societies and similar organizations in the U.S. associated with educational or research activities); Unaffiliated Individuals (Scientists, engineers or educators in the U.S. who are U.S. citizens); Tribal organizations and other American Indian, Alaska Native, and Hawaiian serving organizations.
*Geographic Focus:* All States
*Date(s) Application is Due:* Sep 15
*Contact:* Colleen Fitzgerald, Program Director; (703) 292-4381 or (703) 292-5111; fax (703) 292-9068; cfitzger@nsf.gov

*Internet:* www.nsf.gov/funding/pgm_summ.jsp?pims_id=12816
*Sponsor:* National Science Foundation
2415 Eisenhower Avenue
Alexandria, VA 22314

**NSF Earth Sciences Research Instrumentation and Facilities (EAR/IF) Grants 1950**
The Instrumentation and Facilities Program in the Division of Earth Sciences (EAR/IF) supports meritorious requests for infrastructure that promote research and education in areas supported by the Division. EAR/IF will consider proposals for: 1) Acquisition or Upgrade of Research Equipment that will advance laboratory and field investigations and student research training opportunities in the Earth sciences. The maximum request is $500,000. The maximum request for upgrade of research group computing facilities remains $75,000; 2) Development of New Instrumentation, Techniques or Software that will extend current research and research training capabilities in the Earth sciences. The maximum request is $500,000; and 3) Community Facility Support to make complex and expensive instruments, systems of instruments or services broadly available to the Earth science research and student communities. There are no maximum request limitations but potential proposers of new Community Facilities must contact cognizant Program Officers before submission. Support is available through grants or cooperative agreements awarded in response to investigator-initiated proposals. Human resource development and education are expected to be an integral part of all proposals submitted to EAR/IF. All proposers to EAR/IF are encouraged to consider Support of Outreach and/or Broadening Participation Activities. Proposals submitted to the EAR/IF Program may request up to $20,000 for such activities. The maximum allowable request for proposals for Acquisition or Upgrade of Research Equipment and Development of New Instrumentation, Techniques or Software has been changed to $500,000. Proposals are accepted at anytime throughout the year, with an estimated annual awards between 30 and 50.
*Requirements:* Efforts to support participation of underrepresented groups in laboratory and/or field instrument use and training are encouraged. Proposals from early career (tenure track but untenured) lead investigators are also encouraged. Such proposals will be given due consideration as part of the Broader Impacts merit review criterion. Proposals may only be submitted by the following: Institutions of Higher Education (IHEs) - Two- and four-year IHEs (including community colleges) accredited in, and having a campus located in the US, acting on behalf of their faculty members. Special Instructions for International Branch Campuses of U.S. IHEs: If the proposal includes funding to be provided to an international branch campus of a U.S. institution of higher education (including through use of subawards and consultant arrangements), the proposer must explain the benefit(s) to the project of performance at the international branch campus, and justify why the project activities cannot be performed at the U.S. campus; and non-profit, non-academic organizations, including independent museums, observatories, research labs, professional societies and similar organizations in the U.S. associated with educational or research activities.
*Restrictions:* Proposals for Community Facility Support are excluded from the $20,000 maximum for outreach and broadening participation activities.
*Geographic Focus* All States, District of Columbia, Guam, Marshall Islands, Northern Mariana Islands, Puerto Rico, U.S. Virgin Islands, American Samoa
*Amount of Grant:* Up to 500,000 USD
*Contact:* David D. Lambert; (703) 292-8558; fax (703) 292-9025; dlambert@nsf.gov
Russell C. Kelz, 703) 292-4747; fax (703) 292-9025; rkelz@nsf.gov
*Internet:* https://www.nsf.gov/funding/pgm_summ.jsp?pims_id=6186
*Sponsor:* National Science Foundation
2415 Eisenhower Avenue
Alexandria, VA 22314

**NSF Electronic and Photonic Materials (EPM) Grants 1951**
The goal of this program is to advance the field of electronics and photonics through basic, potentially transformative materials science research. The scope of the program encompasses the discovery and understanding of materials and material combinations with potential for major technological advantages. Program focus is on identification and understanding of fundamental atomic and molecular level mechanisms and phenomena associated with synthesis and processing of electronic and photonic materials. High risk, high payoff research is encouraged. Research topics include, but are not limited to, nucleation and growth of thin films and nanostructures; self-assembly; nanostructure definition and etching processes; interface bonding and structure; crystal and interface defects; doping; bulk crystal growth; and interrelationships between synthesis/processing, structure, and properties.
*Requirements:* NSF welcomes proposals on behalf of all qualified scientists, engineers and educators. The Foundation strongly encourages women, minorities and persons with disabilities to participate fully in its programs. Scientists, engineers and educators usually initiate proposals that are officially submitted by their employing organization. Before formal submission, the proposal may be discussed with appropriate NSF program staff. Graduate students are not encouraged to submit research proposals, but should arrange to serve as research assistants to faculty members. NOTE: If the closing date for the submission window falls on a weekend, the closing date moves to the following Monday. The last date of the submission window is an absolute deadline date and proposals must be received by NSF by 5:00 p.m. submitter's local time on that date.
*Restrictions:* The sponsor discourages the submission of more than one proposal from the same Principal Investigator during the proposal-submission window.
*Geographic Focus:* All States
*Date(s) Application is Due:* Oct 31
*Contact:* Z. Charles Ying, EPM Program Director; (703) 292-8428; cying@nsf.gov
*Internet:* www.nsf.gov/funding/pgm_summ.jsp?pims_id=5353&org=MPS&sel_org=MPS&from=fund
*Sponsor:* National Science Foundation
2415 Eisenhower Avenue
Alexandria, VA 22314

**NSF Enabling Access to Cloud Computing Resources for CISE Research and Education (Cloud Access) Grants 1952**
Increasingly, data- and compute-intensive research and education efforts are benefiting from access to cloud computing platforms, which provide robust, agile, reliable, and scalable infrastructure. To better support this growing use of cloud computing resources, the National Science Foundation's (NSF) Directorate for Computer and Information Science and Engineering (CISE) seeks to fund an entity that can serve as a principal interface between the CISE research and education community and public cloud computing providers. Through this solicitation, NSF will support an entity that will have multiple responsibilities, including: 1) establishing partnerships with the various public cloud computing providers; 2) assisting NSF in allocating cloud computing resources to qualifying CISE-funded projects; 3) managing cloud computing accounts and resources allocated to individual CISE projects; 4) providing user training and other support to CISE researchers and educators using cloud computing in their work; and 5) providing strategic technical guidance for CISE researchers and educators interested in using public cloud computing platforms. The maximum award amount is $5,000,000 via a Cooperative Agreement, pending the availability of funds and quality of proposals received. The annual project operating costs are not expected to exceed $850,000 per year for five years, for a total of $4,250,000. The proposed budget may also include a request, not to exceed $750,000, for initial cloud computing resources, as part of the maximum $5,000,000 award amount (see section II, Program Description). Preliminary proposals are due by December 20, with full proposals due by February 19.
*Requirements:* Proposals may only be submitted by the following: Institutions of Higher Education (IHEs) - Two- and four-year IHEs (including community colleges) accredited in, and having a campus located in the US, acting on behalf of their faculty members. Special Instructions for International Branch Campuses of U.S. IHEs: If the proposal includes funding to be provided to an international branch campus of a U.S. institution of higher education (including through use of subawards and consultant arrangements), the proposer must explain the benefit(s) to the project of performance at the international branch campus, and justify why the project activities cannot be performed at the U.S. campus; non-profit, non-academic organizations: Independent museums, observatories, research labs, professional societies and similar organizations in the U.S. associated with educational or research activities.
*Restrictions:* An organization may participate in no more than one proposal submitted in response to this solicitation. Collaborative proposals should be submitted from one organization. If awarded, a single award would be made to the submitting organization, with any collaborators listed as subawards. Collaborative proposals submitted from multiple organizations are not allowed.
*Geographic Focus* All States, District of Columbia, Guam, Marshall Islands, Northern Mariana Islands, Puerto Rico, U.S. Virgin Islands, American Samoa
*Date(s) Application is Due:* Feb 19
*Amount of Grant:* Up to 5,000,000 USD
*Contact:* Kevin Thompson, Program Officer; (703) 292-4220 or (703) 292-5111; kthompso@nsf.gov
Deepankar (Deep) Medhi, Program Officer; (703) 292-8950 or (703) 292-5111; dmedhi@nsf.gov
*Internet:* https://www.nsf.gov/funding/pgm_summ.jsp?pims_id=505591
*Sponsor:* National Science Foundation
2415 Eisenhower Avenue
Alexandria, VA 22314

**NSF Enabling Quantum Leap: Convergent Accelerated Discovery Foundries for Quantum Materials Science, Engineering and Information 1953**
The Division of Materials Research (DMR), the Division of Mathematical Sciences (DMS), the Division of Electrical, Communications and Cyber Systems (ECCS), and the Office of Advanced Cyberinfrastructure (OAC) seek to rapidly accelerate quantum materials design, synthesis, characterization, and translation of fundamental materials engineering and information research for quantum devices, systems, and networks. The new program of Enabling Quantum Leap: Convergent Accelerated Discovery Foundries for Quantum Materials Science, Engineering, and Information (Q-AMASE-i) aims to support these goals by establishing Foundries with mid-scale infrastructure for rapid prototyping and development of quantum materials and devices. The new materials, devices, tools and methods developed by Q-AMASE-i will be shared with the science and engineering communities through a Foundry-operated network. Technology transfer of Foundry activities will be enabled by close cooperation with industrial partners. Six-year awards totaling $20,000,000 to $25,000,000 for the award period are anticipated. Q-AMASE-i Foundries will be awarded as cooperative agreements with an initial commitment of six years, with the possibility of one six-year renewal, subsequent to a rigorous and favorable review by NSF. The annual performance review includes NSF's evaluation of the annual report after the first year, an annual site visit after the second and fourth year of Foundry activities, and a reverse or program management site visit after the third and fifth year. Funding after the second year will depend on the quality of progress and performance documented during the site visits. Letters of Intent are due by September 17, with full applications due November 5.
*Requirements:* Proposals may only be submitted by the following: Institutions of Higher Education (IHEs) - Two- and four-year IHEs (including community colleges) accredited in, and having a campus located in the US, acting on behalf of their faculty members. Special Instructions for International Branch Campuses of U.S. IHEs: If the proposal includes funding to be provided to an international branch campus of a U.S. institution of higher

education (including through use of subawards and consultant arrangements), the proposer must explain the benefit(s) to the project of performance at the international branch campus, and justify why the project activities cannot be performed at the U.S. campus.
*Geographic Focus* All States, District of Columbia, Guam, Marshall Islands, Northern Mariana Islands, Puerto Rico, U.S. Virgin Islands, American Samoa
*Date(s) Application is Due:* Nov 5
*Contact:* Tomasz Durakiewicz; (703) 292-4892 or (703) 292-5111; tdurakie@nsf.gov
Victor Roytburd, Program Officer; (703) 292-8584 or (703) 292-5111; vroytbur@nsf.gov
Amy Walton, Program Officer; (703) 292-4538 or (703) 292-5111; awalton@nsf.gov
*Internet:* https://www.nsf.gov/funding/pgm_summ.jsp?pims_id=505504
*Sponsor:* National Science Foundation
2415 Eisenhower Avenue
Alexandria, VA 22314

**NSF EPSCoR Research Infrastructure Improvement Program: Track-2 Focused Collaborations Grants** 1954

The Established Program to Stimulate Competitive Research (EPSCoR) is designed to fulfill the mandate of the National Science Foundation (NSF) to promote scientific progress nationwide. A jurisdiction is eligible to participate in EPSCoR programs if its level of NSF research support is equal to or less than 0.75 percent of the total NSF research and related activities budget for the most recent three-year period. Through this program, NSF establishes partnerships with government, higher education, and industry that are designed to effect sustainable improvements in a jurisdiction's research infrastructure, Research and Development (R&D) capacity, and hence, its R&D competitiveness. RII Track-2 FEC builds interjurisdictional collaborative teams of EPSCoR investigators in scientific focus areas consistent with NSF priorities. Projects are investigator-driven and must include researchers from at least two RII-eligible jurisdictions with complementary expertise and resources necessary to tackle those projects, which neither party could address as well or rapidly alone. The Science, Technology, Engineering, and Mathematics (STEM) research and education activities should seek to broaden participation through the strategic inclusion and integration of different types of individuals, institutions, and sectors throughout the project. Proposals must describe a comprehensive and integrated vision to drive discovery and build sustainable STEM capacity that exemplifies diversity of all types (individual, institutional, geographic, and disciplinary). The development of diverse early-career faculty is a critical component of this sustainable STEM capacity.
*Requirements:* Individuals and organizations in the following categories may submit proposals: U.S. universities and two-and four-year colleges (including community colleges) acting on behalf of their faculty members; independent museums, observatories, research laboratories, professional societies and similar organizations in the U.S. that are directly associated with educational or research activities; U.S. commercial organizations (especially small businesses with strong capabilities in scientific or engineering research or education); state and local governments; scientists, engineers or educators in the U.S. and U.S. citizens; foreign organizations; and a limited number of other types of Federal agencies.
Restrictions: NSF rarely provides support to foreign organizations. NSF will consider proposals for cooperative projects involving U.S. and foreign organizations, provided support is requested only for the U.S. portion of the collaborative effort. NSF does not normally support research or education activities by scientists, engineers or educators employed by Federal agencies or Federally Funded Research and Development Centers (FFRD.C.s).
*Geographic Focus:* All States
*Date(s) Application is Due:* Jan 26
*Amount of Grant:* Up to 20,000,000 USD
*Contact:* Raffaella Montelli, Program Director; (703) 292-2421 or (703) 292-5111; rmontell@nsf.gov
Andrew V. Ogram; (703) 292-7374 or (703) 292-5111; aogram@nsf.gov
*Internet:* https://www.nsf.gov/funding/pgm_summ.jsp?pims_id=505263&org=DBI&sel_org=DBI&from=fund
*Sponsor:* National Science Foundation
2415 Eisenhower Avenue
Alexandria, VA 22314

**NSF EPSCoR Research Infrastructure Improvement Program Track-1 Grants** 1955

The Established Program to Stimulate Competitive Research (EPSCoR) is designed to fulfill the mandate of the National Science Foundation (NSF) to promote scientific progress nationwide. A jurisdiction is eligible to participate in EPSCoR programs if its level of NSF research support is equal to or less than 0.75 percent of the total NSF research and related activities budget for the most recent three year period (see RII eligibility). Through this program, NSF establishes partnerships with government, higher education, and industry that are designed to effect sustainable improvements in a jurisdiction's research infrastructure, Research and Development (R&D) capacity, and hence, its R&D competitiveness. Research Infrastructure Improvement Track-1 (RII Track-1) awards provide up to $20 million total for 5 years to support improvements to physical and cyber infrastructure and human capital development in research areas selected by the jurisdiction's EPSCoR steering committee as having the best potential to improve future R&D competitiveness of the jurisdiction. The research activities of the project should align with the specific research priorities identified in the approved Science and Technology (S&T) Plan of the jurisdiction.
*Requirements:* Proposals may only be submitted by the following: universities and two- and four-year colleges (including community colleges) accredited in, and having a campus located in, the United States, its territories and possessions (distinct academic campuses (that award their own degrees, have independent administrative structures, admissions policies, alumni associations, etc.) within multi-campus systems qualify as separate institutions); and independent non-profit research museums, independent non-profit research institutions and laboratories, and non-profit research consortia located in the United States, its territories and possessions (these organizations must have an independent administrative and financial structure. Non-profit research consortia must also have 501(c)3 tax status. Any organization can collaborate in a consortium; however, only eligible organizations can receive NSF funds.
Restrictions: Organizations that may not receive funds include foreign organizations (e.g., foreign academic institutions, museums and laboratories); research arms of for-profit firms; and Federally Funded Research and Development Centers (FFRD.C.s). Construction, engineering, architecture and other facility service divisions of for-profit firms that will carry out the work of renovation and repair may receive NSF funds as sub-awards. The American Recovery and Reinvestment Act of 2009 states that funds may not be used by any State or local government, or any private entity, for any casino or other gambling establishment, aquarium, zoo, golf course, or swimming pool.
*Geographic Focus:* All States
*Amount of Grant:* Up to 20,000,000 USD
*Contact:* Sean C. Kennan, Program Director; (703) 292-7575 or (703) 292-5111; fax (703) 292-9047; skennan@nsf.gov
*Internet:* https://www.nsf.gov/funding/pgm_summ.jsp?pims_id=503429&org=DBI&sel_org=DBI&from=fund
*Sponsor:* National Science Foundation
2415 Eisenhower Avenue
Alexandria, VA 22314

**NSF Experimental Program to Stimulate Competitive Research: Workshop Opportunities Grants** 1956

The Experimental Program to Stimulate Competitive Research (EPSCoR) is designed to fulfill the mandate of the National Science Foundation (NSF) to promote scientific progress nationwide. The EPSCoR program is directed at those jurisdictions that have historically received lesser amounts of NSF Research and Development (R&D) funding. Thirty jurisdictions, including twenty-eight states, the Commonwealth of Puerto Rico, and the U. S. Virgin Islands, currently participate in EPSCoR. Through this program, NSF establishes partnerships with government, higher education and industry that are designed to effect sustainable improvements in a jurisdiction's research infrastructure, R&D capacity, and hence, its national R&D competitiveness. The EPSCoR Office welcomes unsolicited proposals from EPSCoR jurisdictions for workshops involving the EPSCoR community. These workshops will focus on innovative ways to address multi-jurisdictional efforts on themes of regional to national importance with relevance to EPSCoR's goals/objectives and NSF's mission. Acceptance of a White Paper by the EPSCoR Office is required before a workshop proposal can be submitted. White Papers may be submitted to the EPSCoR Office via email at any time.
*Geographic Focus:* All States
*Contact:* Denise M. Barnes, Expert; (703) 292-5179 or (703) 292-5111; fax (703) 292-9047; dbarnes@nsf.gov
Sean C. Kennan, Program Director; (703) 292-7575 or (703) 292-5111; fax (703) 292-9047; skennan@nsf.gov
*Internet:* https://www.nsf.gov/funding/pgm_summ.jsp?pims_id=503341&org=DMS&sel_org=DMS&from=fund
*Sponsor:* National Science Foundation
2415 Eisenhower Avenue
Alexandria, VA 22314

**NSF Increasing the Participation and Advancement of Women in Academic Science and Engineering Careers (ADVANCE) Grants** 1957

The goal of the ADVANCE Program is to increase the participation of women in the scientific and engineering workforce through the increased representation and advancement of women in academic science and engineering careers. Through these awards, NSF seeks to support new and systemic approaches to improving the climate for women in U.S. academic institutions and to facilitate women's retention and advancement to the highest ranks of academic leadership. Creative approaches to realize the goal of this program are sought from women and men. Proposals may include international activities that directly support the goals of ADVANCE. This program will support the following types of ADVANCE Projects: Partnerships for Adaptation, Implementation, and Dissemination (PAID) Awards (deadline January 17); Institutional Transformation (IT) Awards (deadline December 6); and Institutional Transformation Planning Grants (deadline December 6).
*Requirements:* Institutional Transformation proposals may be submitted by academic institutions of higher learning in the U.S., its territories or possessions, or the Commonwealth of Puerto Rico, that award degrees in a field supported by NSF. Partnerships for Adaptation, Implementation and Dissemination proposals may be submitted by: academic institutions of higher learning that award degrees in a field supported by NSF; professional societies; or other not-for-profit organizations.
*Geographic Focus:* All States
*Date(s) Application is Due:* Jan 17; Dec 6
*Contact:* Jessie A. DeAro, Program Director; (703) 292-5350 or (703) 292-5111; fax (703) 292-9018; jdearo@nsf.gov or advance@nsf.gov
Sharon Bird, Program Director; (703) 292-5178 or (703) 292-5111; fax (703) 292-9018; sbird@nsf.gov or advance@nsf.gov
*Internet:* https://www.nsf.gov/funding/pgm_summ.jsp?pims_id=5383&org=BIO&sel_org=BIO&from=fund
*Sponsor:* National Science Foundation
2415 Eisenhower Avenue
Alexandria, VA 22314

**NSF Innovative Technology Experiences for Students and Teachers (ITEST)** 1958
ITEST supports the research and development of innovative models for engaging K-12 students in authentic experiences that build their capacity to participate in the science, technology, engineering, and mathematics (STEM) and information and communications technology (ICT) workforce of the future. ITEST projects must include students and may include teachers. ITEST invests in four types of projects: Strategies, Research, Scale-up, and Resource Center projects. 15 to 20 grants will be awarded depending on the availability of funds.
*Requirements:* This program provides educational opportunities for K-12 Educators. This program provides indirect funding for students at this level or focuses on educational developments for this group such as curricula development, training or retention. NSF welcomes proposals on behalf of all qualified scientists, engineers and educators. Before formal submission, the proposal may be discussed with appropriate NSF program staff. The Foundation strongly encourages women, minorities and persons with disabilities to participate fully in its programs. Submission of Letters of Intent is required and are due on September 20 (due by 5 p.m. proposer's local time). Full proposals should be submitted via FastLane or Grants.gov on or before November 13 (due by 5 p.m. proposer's local time).
*Geographic Focus:* All States
*Date(s) Application is Due:* Nov 13
*Contact:* Marie Mapes, Program Manager; (703) 292-2768; mmapes@nsf.gov
*Internet:* www.nsf.gov/funding/pgm_summ.jsp?pims_id=5467&org=EHR&sel_org=EHR&from=fund
*Sponsor:* National Science Foundation
2415 Eisenhower Avenue
Alexandria, VA 22314

**NSF Joint Domestic Nuclear Detection Office-National Science Foundation Academic Research Initiative (ARI) Grants** 1959
The ARI is a joint Domestic Nuclear Detection Office (DNDO) and National Science Foundation (NSF) program seeking novel cross-cutting research that will enhance the nation's ability to detect and interdict nuclear or radiological material outside of regulatory control, and otherwise help prevent nuclear or radiological attacks. This year's solicitation topics will encompass a range of subjects, with an emphasis on unconventional, multidisciplinary approaches to threat detection. A number of small to medium awards are intended in the areas of novel approaches to extremely low-cost threat detection, orthogonal and informatics approaches to threat detection, deterrence analytics, and advanced forensics techniques. A single large award is intended for an integrated, multidisciplinary approach to shielded special nuclear material detection. Primary objectives of the ARI include advancing fundamental knowledge in the above areas and developing intellectual capacity in scientific fields relevant to long-term advances in these areas.
*Requirements:* Organizations eligible to submit proposals under this program include universities and colleges; nonprofit, nonacademic institutions, including museums, professional associations, and community-based organizations; state and local governments, including school districts; and for-profit research institutions and businesses.
*Restrictions:* Proposals outside of the scope described in this solicitation will be returned without review. Research proposals specific to detection of biological, chemical, and conventional weapons are excluded from the scope of this solicitation, however approaches that consider explosives detection and nuclear or radiological materials detection are of interest.
*Geographic Focus:* All States
*Date(s) Application is Due:* Jul 10
*Amount of Grant:* Up to 1,000,000 USD
*Contact:* Gail Dodge; (703) 292-8958 or (703) 292-5111; gdodge@nsf.gov
Bruce Hamilton, Program Officer; (703) 292-8320 or (703) 292-5111; bhamilto@nsf.gov
*Internet:* https://nsf.gov/funding/pgm_summ.jsp?pims_id=503223
*Sponsor:* National Science Foundation
2415 Eisenhower Avenue
Alexandria, VA 22314

**NSF Linguistics Grants** 1960
The Linguistics Program supports basic science in the domain of human language, encompassing investigations of the grammatical properties of individual human languages, and of natural language in general. Research areas include syntax, semantics, morphology, phonetics, and phonology. The program encourages projects that are interdisciplinary in methodological or theoretical perspective, and that address questions that cross disciplinary boundaries, such as (but not limited to): What are the psychological processes involved in the production, perception, and comprehension of language? What are the computational properties of language and/or the language processor that make fluent production, incremental comprehension or rapid learning possible? How do the acoustic and physiological properties of speech inform our theories of language and/or language processing? What role does human neurobiology play in shaping the various components of our linguistic capacities? How does language develop in children? What social and cultural factors underlie language variation and change?
*Requirements:* The Linguistics Program accepts proposals for a variety of project types: research proposals from scholars with PhDs or equivalent degrees, proposals for Doctoral Dissertation Research Improvement Grants, and CAREER proposals. The program will also consider proposals for workshops, conferences, and training activities. NSF welcomes proposals on behalf of all qualified scientists, engineers and educators. The Foundation strongly encourages women, minorities and persons with disabilities to participate fully in its programs. Scientists, engineers and educators usually initiate proposals that are officially submitted by their employing organization. Before formal submission, the proposal may be discussed with appropriate NSF program staff. Graduate students are not encouraged to submit research proposals, but should arrange to serve as research assistants to faculty members. Some NSF divisions accept proposals for Doctoral Dissertation Research Grants when submitted by a faculty member on behalf of the graduate student. Proposal deadlines are January 15 and July 15 annually, or the next federal business day if the target date falls on a weekend or a federal holiday.
*Geographic Focus:* All States
*Date(s) Application is Due:* Jan 15; Jul 15
*Contact:* Dr. Joan Maling; (703) 292-8046; fax (703) 292-9068; jmaling@nsf.gov
*Internet:* nsf.gov/funding/pgm_summ.jsp?pims_id=5408
*Sponsor:* National Science Foundation
2415 Eisenhower Avenue
Alexandria, VA 22314

**NSF Major Research Instrumentation (MRI) Program Grants** 1961
The Major Research Instrumentation (MRI) Program serves to increase access to multi-user scientific and engineering instrumentation for research and research training in our Nation's institutions of higher education and not-for-profit scientific/engineering research organizations. An MRI award supports the acquisition or development of a multi-user research instrument that is, in general, too costly and/or not appropriate for support through other NSF programs. MRI provides support to acquire critical research instrumentation without which advances in fundamental science and engineering research may not otherwise occur. MRI also provides support to develop next-generation research instruments that open new opportunities to advance the frontiers in science and engineering research. Additionally, an MRI award is expected to enhance research training of students who will become the next generation of instrument users, designers and builders. An MRI proposal may request up to $4 million for either acquisition or development of a research instrument. Beginning with the FY 2018 competition, each performing organization may submit in revised "Tracks" as defined below, with no more than two submissions in Track 1 and no more than one submission in Track 2. Track 1 MRI proposals are those that request funds from NSF greater than or equal to $100,000 and less than $1,000,000. Track 2 MRI proposals are those that request funds from NSF greater than or equal to $1,000,000 up to and including $4,000,000. The full proposal application window is from January 1 through the January 19 deadline.
*Requirements:* Proposals may be submitted by U.S. institutions of higher education, independent nonprofit research institutions, research museums, and legally documented incorporated consortia thereof. Consistent with the America COMPETES Act of 2007 (Public Law 110-69), cost sharing of precisely 30% of the total project cost is required for Ph.D.-granting institutions of higher education and for non-degree-granting organizations. Non-Ph.D.-granting institutions of higher education are exempt from the cost-sharing requirement and cannot include it. National Science Board policy prohibits voluntary committed cost sharing.
*Geographic Focus* All States, District of Columbia, Guam, Marshall Islands, Northern Mariana Islands, Puerto Rico, U.S. Virgin Islands, American Samoa
*Date(s) Application is Due:* Jan 19
*Amount of Grant:* 100,000 - 4,000,000 USD
*Contact:* Randy L. Phelps; (703) 292-5049 or (703) 292-5111; rphelps@nsf.gov
*Internet:* https://www.nsf.gov/funding/pgm_summ.jsp?pims_id=5260
*Sponsor:* National Science Foundation
2415 Eisenhower Avenue
Alexandria, VA 22314

**NSF Materials Research (DMR) Grants** 1962
The Division of Materials Research (DMR) supports a wide range of programs that address fundamental phenomena in materials, materials synthesis and processing, structure and composition, properties and performance, and materials education. Research program areas include condensed matter physics; solid-state chemistry and polymers; metals, ceramics, and electronic materials; materials theory; and materials research science and engineering centers. DMR funding modes include support for individual investigators, groups, centers, national facilities, and instrumentation.
*Requirements:* NSF welcomes proposals on behalf of all qualified scientists, engineers and educators. The Foundation strongly encourages women, minorities and persons with disabilities to participate fully in its programs. Scientists, engineers and educators usually initiate proposals that are officially submitted by their employing organization. Before formal submission, the proposal may be discussed with appropriate NSF program staff. Graduate students are not encouraged to submit research proposals, but should arrange to serve as research assistants to faculty members. The window for submitting unsolicited proposals to DMR begins on September 1 and ends on October 31, annually. If the closing date for the submission window falls on a weekend, the closing date moves to the following Monday. The last date of the submission window is an absolute deadline date and proposals must be received by NSF by 5:00 p.m. submitter's local time on that date.
*Geographic Focus:* All States
*Date(s) Application is Due:* Oct 31
*Contact:* Ian Robertson, Division Director, Division of Materials Research; (703) 292-4937; iroberts@nsf.gov
*Internet:* www.nsf.gov/div/index.jsp?div=dmr
*Sponsor:* National Science Foundation
2415 Eisenhower Avenue
Alexandria, VA 22314

**NSF Mathematical Sciences Postdoctoral Research Fellowships** 1963
Thirty to forty fellowships are awarded annually for postdoctoral research in the mathematical sciences. There are two options for awardees: Research Fellowship and Research Instructorship. Awards will support research in areas of mathematics and

statistics, including applications to other disciplines. The fellowships are designed to permit recipients to choose research environments that will have maximum impact on their future scientific development. Awards of these fellowships will be made for appropriate research in pure mathematics, applied mathematics, and statistics. Each applicant will be required to submit a research plan for the tenure period requested. Proposals are due the third Friday in October. The annual deadline for application submission is October 17.
*Requirements:* Fellowships and Instructorships are offered to citizens or nationals of the United States who will have earned a doctoral degree in one of the accepted mathematical sciences in the past five years or have equivalent research training and experience. Applicants must not have held any previous NSF postdoctoral fellowship.
*Restrictions:* The fellowships are not designed to support either the preparation of prior research results for publication or the writing of textbooks.
*Geographic Focus:* All States
*Date(s) Application is Due:* Oct 17
*Amount of Grant:* Up to 128,000 USD
*Contact:* Victor Roytburd, Program Director; (703) 292-8584 or (703) 292-5111; fax (703) 292-9032; vroytburd@nsf.gov or msprf@nsf.gov
*Internet:* https://www.nsf.gov/funding/pgm_summ.jsp?pims_id=5301
*Sponsor:* National Science Foundation
2415 Eisenhower Avenue
Alexandria, VA 22314

**NSF Methodology, Measurement, and Statistics Grants** 1964

The Methodology, Measurement, and Statistics program supports the development, application, and extension of methodology for social, behavioral, and economic research. In general, MMS seeks proposals that are interdisciplinary in nature, methodologically innovative, and grounded in theory. Research includes methods for improving measurement in these sciences; fundamental research on new or existing procedures for data collection; research on the evaluation and comparison of existing databases; and statistical research benefiting one or more of the social, behavioral, and economic sciences. The MMS Program provides support through a number of different funding mechanisms. The following mechanisms are addressed in this solicitation: Regular Research Awards; Awards for conferences, workshops, and community-development activities; Doctoral Dissertation Research Improvement (DDRI) Grants; and Research Experience for Undergraduates (REU) Supplements. There are two annual deadlines for application submission: the last Thursday in January; and the last Thursday in August.
*Geographic Focus:* All States
*Date(s) Application is Due:* Jan 25; Aug 30
*Contact:* Cheryl L. Eavey, Program Director; (703) 292-7269 or (703) 292-5111; fax (703) 292-9068; ceavey@nsf.gov
*Internet:* www.nsf.gov/pubsys/ods/getpub.cfm?ods_key=nsf04504
*Sponsor:* National Science Foundation
2415 Eisenhower Avenue
Alexandria, VA 22314

**NSF Next Generation Networks For Neuroscience Grants** 1965

Understanding the brain is one of the grand scientific challenges at the intersection of experimental, theoretical, and computational investigation in the life, physical, behavioral, and cognitive sciences. Rapid proliferation of advanced measurement instrumentation and techniques has allowed researchers to study the brain and behavior at ever finer physical and temporal scales and in broader social and environmental contexts. At the same time, achieving a comprehensive, transformational understanding of the brain in action and in context will require an increased emphasis on systematic, interdisciplinary collaboration and team science, and the increased use of theoretical frameworks, including evolutionary ones, to explore questions that span organizational levels, scales of analysis, and a wider range of species optimal for experimental exploration of brain function. To catalyze such understanding, NSF announced its intention to support the development of innovative, accessible, and shared capabilities and resources towards the establishment of a coherent national infrastructure for neuroscience research. The goal of this solicitation is to foster the development and dissemination of: innovative research resources, instrumentation, and neurotechnologies; and theoretical frameworks for understanding brain function across organizational levels, scales of analysis, and/or a wider range of species, including humans. This program solicits proposals that will develop and disseminate innovative neurotechnologies and/or theoretical frameworks that will transform our understanding of the linkages between neural activity and cognition and behavior across different systems, environments, and species, while also providing an avenue for widespread dissemination of these technologies and theoretical frameworks as well as broad training opportunities.
*Requirements:* The most frequent recipients of support are academic institutions and nonprofit research groups. In special circumstances, grants also are awarded to other types of institutions and to individuals.
*Geographic Focus:* All States
*Contact:* Reed S. Beaman, Program Director; (703) 292-7163 or (703) 292-5111; fax (703) 292-9063; rsbeaman@nsf.gov
*Internet:* https://www.nsf.gov/funding/pgm_summ.jsp?pims_id=505281&org=DBI&sel_org=DBI&from=fund
*Sponsor:* National Science Foundation
2415 Eisenhower Avenue
Alexandria, VA 22314

**NSF Organizational Change for Gender Equity in STEM Academic Professions (ADVANCE) Grants** 1966

The goals of the ADVANCE program are: to develop systemic approaches to increase the representation and advancement of women in academic STEM careers; to develop innovative and sustainable ways to promote gender equity that involve both men and women in the STEM academic workforce; and to contribute to the research knowledge base on gender equity and the intersection of gender and other identities in STEM academic careers. The ADVANCE program contributes to the development of a more diverse science and engineering workforce because of the focus on equity for STEM academic faculty who are educating, training, and mentoring undergraduate and graduate students and postdoctoral scholars. Members of underrepresented minority groups and individuals with disabilities are especially encouraged to apply. Proposals that address the participation and advancement of women with disabilities and women from underrepresented minority groups are particularly encouraged. There are three program tracks: the Institutional Transformation track; the Adaptation track; and the Partnership track. All projects are expected to build on prior ADVANCE work and gender equity research and literature to broaden the implementation of organizational and systemic strategies to foster gender equity in STEM academic careers.
*Requirements:* Proposals from primarily undergraduate institutions, teaching intensive colleges, community colleges, minority-serving institutions (e.g. Tribal Colleges and Universities, Historically Black Colleges and Universities, Hispanic-Serving Institutions), women's colleges, and institutions primarily serving persons with disabilities are encouraged. Institutional Transformation proposals may be submitted by academic institutions of higher learning in the U.S., its territories or possessions, or the Commonwealth of Puerto Rico, that award degrees in a field supported by NSF. Partnerships involving industry, government, professional societies and other not-for-profit organizations are encouraged but not required; however, in the case of partnerships, the lead partner organization must be an academic institution of higher learning and must accept overall management responsibility for the activity. Government organizations (other than academic institutions of higher learning) are not eligible to apply, but may participate in partnerships when the lead partner organization is an academic institution of higher learning with overall management responsibility for the activity. Partner organizations and institutions must be based in the U.S., its territories or possessions, or the Commonwealth of Puerto Rico. Partnerships for Adaptation, Implementation and Dissemination proposals may be submitted by: academic institutions of higher learning that award degrees in a field supported by NSF; professional societies; or other not-for-profit organizations. Submitting institutions and organizations must be based in the U.S., its territories or possessions, or the Commonwealth of Puerto Rico. Government organizations (other than academic institutions of higher learning) are not eligible to apply, but may participate in partnerships when the lead partner organization is an academic institution of higher learning with overall management responsibility for the activity. Partner institutions and organizations must be based in the U.S., its territories or possessions, or the Commonwealth of Puerto Rico.
*Restrictions:* All ADVANCE proposals are expected to recognize that gender does not exist in isolation from other characteristics, such as race/ethnicity, disability status, sexual orientation, foreign-born and foreign-trained status, faculty appointment type, etc., and should offer strategies to promote gender equity for all faculty. Organizations that have received NSF ADVANCE Institutional Transformation awards are not eligible to apply for another institutional transformation award or for an IT-Start award. Organizations may submit only one Institutional Transformation proposal or one IT-Start proposal. There is no limitation on the number of PAID proposals.
*Geographic Focus:* All States
*Date(s) Application is Due:* Aug 7
*Amount of Grant:* 500,000 USD
*Contact:* Jessie DeAro, Program Director; (703) 292-5350 or (703) 292-5111; fax (703) 292-9018; jdearo@nsf.gov or ADVANCE@nsf.gov
Erika T. Camacho, Program Officer; (703) 292-2834 or 703) 292-5111; fax (703) 292-9018; ecamacho@nsf.gov or ADVANCE@nsf.gov
*Internet:* https://www.nsf.gov/funding/pgm_summ.jsp?pims_id=5383&org=DMS&sel_org=DMS&from=fund
*Sponsor:* National Science Foundation
2415 Eisenhower Avenue
Alexandria, VA 22314

**NSF Perception, Action and Cognition (PAC) Research Grants** 1967

The PAC program funds theoretically motivated research on a wide-range of topic areas related to typical human behavior with particular focus on perceptual, motor, and cognitive processes and their interactions. Central research topics for consideration by the program include (but are not limited to) vision, audition, haptics, attention, memory, written and spoken language, spatial cognition, motor control, categorization, reasoning, and concept formation. Of particular interest are emerging areas, such as the interaction of sleep or emotion with cognitive or perceptual processes, epigenetics of cognition, computational models of cognition, and cross-modal and multimodal processing. The program welcomes a wide range of perspectives, such as individual differences, symbolic and neural-inspired computation, ecological approaches, genetics and epigenetics, nonlinear dynamics and complex systems, and a variety of methodologies spanning the range of experimentation and modeling. The PAC program is open to co-review of proposals submitted to other programs both within the Social, Behavioral, and Economic Sciences Directorate and across other directorates. The proposal window for conference applications is May 15 through June 15; and the proposal windows for research applications are July 15 through August 1 and January 15 through February 1.
*Requirements:* Proposals may be returned without review if the major focus is 1) the organization of neural activity or brain networks; 2) understanding clinical populations;

or 3) non-human animals without a clear and direct impact on our understanding of human perception, action, or cognition. Investigators are encouraged to send the program director a one-page summary of the proposed research before submitting a proposal, in order to determine its appropriateness for the PAC program. Note that a single-copy document that accompanies the proposal but technically is not part of the proposal must be submitted to provide information about the collaborators and other affiliations of all individuals serving as a PI, co-PI, or senior personnel. This information should be prepared in accordance with the PAPPG Chapter II.C.1.
*Geographic Focus* All States, District of Columbia, Guam, Marshall Islands, Northern Mariana Islands, Puerto Rico, U.S. Virgin Islands, American Samoa
*Date(s) Application is Due:* Feb 1; Jun 15; Aug 1
*Amount of Grant:* 1,100 - 5,000,000 USD
*Contact:* Betty K. Tuller; (703) 292-7238 or (703) 292-5111; btuller@nsf.gov
Larry Gottlob, Program Director; (703) 292-4383 or (703) 292-5111; lgottlob@nsf.gov
*Internet:* https://www.nsf.gov/funding/pgm_summ.jsp?pims_id=5686
*Sponsor:* National Science Foundation
2415 Eisenhower Avenue
Alexandria, VA 22314

**NSF Postdoctoral Research Fellowships in Biology (PRFB)** **1968**

The Directorate for Biological Sciences (BIO) awards Postdoctoral Research Fellowships in Biology (PRFB) to recent recipients of the doctoral degree for research and training in selected areas supported by BIO and with special goals for human resource development in biology. The fellowships encourage independence at an early stage of the research career to permit Fellows to pursue their research and training goals in the most appropriate research locations regardless of the availability of funding for the Fellows at that site. For applications under this solicitation, these areas are (1) Broadening Participation of Groups Underrepresented in Biology, (2) Interdisciplinary Research Using Biological Collections, (3) National Plant Genome Initiative (NPGI) Postdoctoral Research Fellowships and (4) Integrative Research Investigating the Rules of Life Governing Interactions Between Genomes, Environment and Phenotypes. The fellowships are also designed to provide active mentoring of the Fellows by the sponsoring scientists who will benefit from having these talented young scientists in their research groups. The research and training plan of each fellowship must address important scientific questions within the scope of BIO and the specific guidelines in this fellowship program solicitation. Because the fellowships are offered to postdoctoral scientists only early in their careers, NSF encourages doctoral students to discuss the availability of these postdoctoral fellowships with their doctoral mentors and potential postdoctoral sponsors early in their doctoral programs to take advantage of this funding opportunity. Fellowships are awards to individuals, not institutions, and are administered by the Fellows. It is estimated that 65 awards will be given annually, with each award being $150,000 to $185,000. The fullproposal annual deadline is November 19.
*Requirements:* Only individuals may apply. NSF postdoctoral fellowships are awards to individuals, and applications are submitted directly by applicants to NSF. However, applications must include sponsoring scientists' statements and the applicants must affiliate with institutions (e.g., colleges and universities, and privately-sponsored nonprofit institutes and museums, government agencies and laboratories, and, under special conditions, for-profit organizations) anywhere in the world. Applicants must: (1) be U.S. citizens (or nationals) or legally admitted permanent residents of the United States (i.e., have a 'green card') at the time of application; (2) earn or plan to earn the doctoral degree in a scientific or engineering field prior to the requested start date of the fellowship; (3) either currently be a graduate student or, at the deadline date, have served in a position requiring the doctoral degree for no more than 12 full time months since earning the degree. There is a one time exception to this criterion for Biological Informatics in November, 2007 only, as described below; (4) must present a research and training plan that falls within the purview of BIO and includes the required information for the specific competition as described below; (5) select a host institution and sponsoring scientist different from the doctoral degree and current position or provide compelling justification for why such a change is not being proposed; (6) not have received Federal funding of more than $20,000 as PI or co-PI (except graduate fellowships and doctoral dissertation improvement grants); (7) not have submitted concurrently the same project to another NSF program; (8) not be a named participant on any other proposal submitted to NSF, including regular research proposals, concurrent with the fellowship application, regardless of who is the named principal investigator; AND (9) meet other limitations on eligibility imposed within the specific competitive areas, if any. There is no limit on the number of applicants that an institution may host. For now, submission can only be made using NSF's FastLane system, given that Grants.gov does not support all required forms. Applications are due the first Monday in November by 5 p.m. proposer's local time.
*Geographic Focus* All States, District of Columbia, Guam, Marshall Islands, Northern Mariana Islands, Puerto Rico, U.S. Virgin Islands, American Samoa
*Amount of Grant:* 150,000 - 185,000 USD
*Contact:* Amanda A. Simcox, Program Director; (703) 292-2532 or (703) 292-5111; fax (703) 292-9461; asimcox@nsf.gov
*Internet:* https://www.nsf.gov/funding/pgm_summ.jsp?pims_id=503622
*Sponsor:* National Science Foundation
2415 Eisenhower Avenue
Alexandria, VA 22314

**NSF Research in Undergraduate Institutions (RUI)** **1969**

The Research in Undergraduate Institutions (RUI) and Research Opportunity Awards (ROA) funding opportunities support research by faculty members at predominantly undergraduate institutions (PUIs). RUI proposals support PUI faculty in research that engages them in their professional field(s), builds capacity for research at their home institution, and supports the integration of research and undergraduate education. ROAs similarly support PUI faculty research, but these awards typically allow faculty to work as visiting scientists at research-intensive organizations where they collaborate with other NSF-supported investigators. All NSF directorates may support RUI and ROA funding activities. Funding for these awards is contained within research and education program allocations and not held as a separate allocation. RUI and ROA proposals are evaluated and funded by NSF programs in the disciplinary areas of the proposed research and are funded at their discretion. Research in Undergraduate Institutions (RUI). An RUI proposal may be: a request to support an individual research project or a collaborative research project involving PUI faculty and students at their own or other institutions; or a request involving shared research instrumentation. The types of ROA opportunities include: a supplement to an existing NSF award to support ROA activities for PUI faculty; requests to rebudget funds in an existing NSF award to support ROA activities for PUI faculty; and submission of a new collaborative proposal between a PUI and another institution(s), with a ROA component as a subaward or as part of a linked collaborative proposal. Prospective PIs should contact disciplinary program officers to identify specific NSF programs and to determine the feasibility and timing of RUI/ROA requests. The number of awards varies across disciplinary research programs. In recent years NSF has made approximately 200 RUI awards and 45 ROA awards per year.
*Requirements:* Eligible PUIs are accredited colleges and universities (including two-year community colleges) that award Associate's degrees, Bachelor's degrees, and/or Master's degrees in NSF-supported fields, but have awarded 20 or fewer Ph.D./D.Sci. degrees in all NSF-supported fields during the combined previous two academic years.
*Geographic Focus* All States, District of Columbia, Guam, Marshall Islands, Northern Mariana Islands, Puerto Rico, U.S. Virgin Islands, American Samoa
*Contact:* Program Contact; (703) 292-5111; info@nsf.gov
*Internet:* https://www.nsf.gov/funding/pgm_summ.jsp?pims_id=5518
*Sponsor:* National Science Foundation
2415 Eisenhower Avenue
Alexandria, VA 22314

**NSF Secondary Satellite Network for the U. S. Academic Research Vessel Fleet (ARF) Grants** **1970**

The Secondary Satellite Network for the U. S. Academic Research Vessel Fleet Grants (ARF) program seeks proposals to support the diverse network-dependent research and operational requirements of vessels in the U.S. Academic Research Fleet (ARF) and associated platforms by providing satellite communication system(s) that can be used as one of multiple options for at-sea access to the Internet. The University National Oceanographic Laboratory System (UNOLS) is an organization of 58 academic institutions involved in oceanographic research. UNOLS institutions operate 5 Global Class ships, 6 Ocean/Intermediate Class ships, and 7 Regional/Coastal/Local Class ships constituting the U.S. Academic Research Fleet (ARF). The UNOLS office coordinates their scheduling and research activities. Productive, successful ocean science research and operations increasingly requires efficient, reliable, and robust connectivity to the Internet. The instrumentation and operating infrastructure have become more complex and demands for stable, reliable off-ship networking have grown. The responsibilities of scientists and vessel operators are increasing, with greater demands on real-time and near real-time data, outreach, shore side IT support and vendor support. These dependencies make functional Internet access mission critical. Letters of Intent are not required, and the annual application deadline is the last week of October
*Requirements:* Proposals may only be submitted by the following: Institutions of Higher Education (IHEs) - Two- and four-year IHEs (including community colleges) accredited in, and having a campus located in the US, acting on behalf of their faculty members. Special Instructions for International Branch Campuses of U.S. IHEs: If the proposal includes funding to be provided to an international branch campus of a U.S. institution of higher education (including through use of subawards and consultant arrangements), the proposer must explain the benefit(s) to the project of performance at the international branch campus, and justify why the project activities cannot be performed at the U.S. campus.
*Geographic Focus* All States, District of Columbia, Guam, Marshall Islands, Northern Mariana Islands, Puerto Rico, U.S. Virgin Islands, American Samoa
*Date(s) Application is Due:* Oct 20
*Contact:* James Holik; (703) 292-7711; fax (703) 292-9034; jholik@nsf.gov
Bauke Houtman; (703) 292-7704; fax (703) 292-9034; bhoutman@nsf.gov
*Internet:* https://www.nsf.gov/funding/pgm_summ.jsp?pims_id=505499
*Sponsor:* National Science Foundation
2415 Eisenhower Avenue
Alexandria, VA 22314

**NSF Social Psychology Grants** **1971**

The Social Psychology program at NSF supports basic research on human social behavior, including cultural differences and development over the life span. Among the many research topics supported are: attitude formation and change, social cognition, personality processes, interpersonal relations and group processes, the self, emotion, social comparison and social influence, and the psycho-physiological and neurophysiological bases of social behavior. The scientific merit of a proposal depends on four important factors: the problems investigated must be theoretically grounded; the research should be based on empirical observation or be subject to empirical validation; the research design must be appropriate to the questions asked; and the proposed research must advance basic understanding of social behavior. Annual full proposal target dates are July 16 and January 16.
*Requirements:* Except where a program solicitation establishes more restrictive eligibility criteria, individuals and organizations in the following categories may submit proposals: universities and colleges; non-profit, non-academic organizations;

for-profit organizations; state and local governments; unaffiliated individuals; foreign organizations; and other Federal agencies.
*Geographic Focus:* All States, All Countries
*Date(s) Application is Due:* Jan 16; Jul 16
*Contact:* Steven Breckler, Program Director; (703) 292-7369 or (703) 292-5111; fax (703) 292-9068; sbreckle@nsf.gov
*Internet:* https://www.nsf.gov/funding/pgm_summ.jsp?pims_id=5712
*Sponsor:* National Science Foundation
2415 Eisenhower Avenue
Alexandria, VA 22314

**NSF Sociology Program Grants** 1972
The Sociology Program supports basic research on all forms of human social organization — societies, institutions, groups and demography — and processes of individual and institutional change. The Program encourages theoretically focused empirical investigations aimed at improving the explanation of fundamental social processes. Included is research on organizations and organizational behavior, population dynamics, social movements, social groups, labor force participation, stratification and mobility, family, social networks, socialization, gender roles, and the sociology of science and technology. The Program supports both original data collections and secondary data analysis that use the full range of quantitative and qualitative methodological tools. The Sociology Program also funds doctoral dissertation research to defray direct costs associated with conducting research, for example, dataset acquisition, additional statistical or methodological training, meeting with scholars associated with original datasets, and fieldwork away from the student's home campus.
*Requirements:* NSF welcomes proposals on behalf of all qualified scientists, engineers and educators. The Foundation strongly encourages women, minorities and persons with disabilities to participate fully in its programs. Before formal submission, the proposal may be discussed with appropriate NSF program staff. Graduate students are not encouraged to submit research proposals, but should arrange to serve as research assistants to faculty members.
*Geographic Focus:* All States
*Date(s) Application is Due:* Feb 15
*Contact:* Dr. Patricia E. White, Sociology Program Director; (703) 292-8762; fax (703) 292-9195; pwhite@nsf.gov
*Internet:* www.nsf.gov/funding/pgm_summ.jsp?pims_id=5369&org=SES&from=home
*Sponsor:* National Science Foundation
2415 Eisenhower Avenue
Alexandria, VA 22314

**NWO Added Value Plus for the Social Sciences Grants** 1973
Advancing knowledge utilization is one of the Netherlands Organization for Scientific Research's (NWO) targets. The policy for encouraging knowledge utilization has various elements. NWO facilitates the awareness of researchers so that third parties can possibly use their research results and can pay due attention to knowledge utilization. Furthermore, NWO is introducing grants that must facilitate the interaction between researchers and potential users of knowledge. Added Value grants, a joint initiative between the Divisions for the Earth and Life Sciences (ALW), Humanities (GW) and Social Sciences (MaGW), has been set up with the aim of supporting researchers who wish to make their research results from current or recently concluded projects (more) suitable and available for third parties. Added Value is a one-off initiative that each of the participating divisions will realize in its own manner appropriate to the field it covers. The aim of Added Value Plus for the Social Sciences is to encourage researchers in the domain of the social and behavioural sciences to make their research results usable in collaboration with their partners and/or users and to increase the involvement of partners and users. Added Value Plus is aimed at research within the so-called Topsectors. Proposals can be submitted if they address the research questions raised in one of the following topsectors: Agro Food, Creative Industry, Energy, Logistics or the Social Infrastructure Agenda. The deadline for application is October 15, and the funding range is 10,000 to 50,000 Euro.
*Requirements:* Applications can be submitted by teams that consist of researchers from NWO-funded research from the open types of support, namely the Open Competition (also TOP and Onderzoekstalent), Innovational Research Incentives Scheme and partners/users. One of the researchers (lecturer, senior lecturer, professor employed by faculties or institutes of a Dutch university or an NWO or KNAW institute) acts as the main applicant. Applications preferably are in Dutch.
*Restrictions:* The project grant is for a maximum period of 12 months.
*Geographic Focus:* All States, Netherlands
*Date(s) Application is Due:* Oct 15
*Amount of Grant:* 10,000 - 50,000 EUR
*Contact:* C.E.A.M. Raat, +31 (0) 70 344 09 08; c.raat@nwo.nl
*Internet:* www.nwo.nl/nwohome.nsf/pages/NWOP_89PH6X_Eng
*Sponsor:* Netherlands Organization for Scientific Research
P.O. Box 93138
The Hague, NL2509 Netherlands

**NWO Aspasia Grants** 1974
Aspasia is designed to alleviate the under-representation of women in the upper echelons of academia. The aim of Aspasia is to encourage the promotion of female academics to senior lecturer or professorial level. The award available for the promotion of each grant recipient is 100,000 Euro. The award offered for the promotion of an unsuccessful, but judged very good or excellent candidate, is 200,000 Euro. This means that the latter group receive no grant under the Innovative Research Incentives Scheme but are awarded a larger subsidy under Aspasia. The award may be used in a number of alternative ways. A condition of grant is that the Board must use at least 50,000 Euro of the award to fund more generic diversity policy measures by the university or faculty to increase the upward movement of female staff within the institution.
*Requirements:* Aspasia is linked to NWO's competitive Vidi and Vici grant schemes. To qualify as an Aspasia candidate, a female applicant for a Vidi or Vici grant must either have been awarded such a grant, or have had her application judged very good or excellent following the interview procedure but been unsuccessful in obtaining one: Executive Boards that promote such applicants to senior lecturer or professor within a year of the Vidi or Vici grant award decision may, subject to certain conditions, qualify for a premium.
*Geographic Focus:* Netherlands
*Amount of Grant:* 100,000 - 200,000 EUR
*Contact:* Aspasia Coordinator; +31 (0) 70 349 40 11; aspasia@nwo.nl
*Internet:* www.nwo.nl/nwohome.nsf/pages/NWOA_4YYAJV_Eng
*Sponsor:* Netherlands Organization for Scientific Research
P.O. Box 93138
The Hague, NL2509 Netherlands

**NWO Athena Grants** 1975
In the Netherlands the percentage of women who work in the natural and technical sciences decreases the higher the position. For example, women are underrepresented in assistant, associate and full professorships. With NWO Athena Grants, the Division for the Chemical Sciences (CW) hopes to increase the flow of women in the large group of post-doctorates with a temporary appointment to a tenured scientific position. This program will encourage the appointment of talented female researchers within chemistry, who have acquired a Veni subsidy, as tenured assistant professors (or a similar permanent position at a research institute). The Athena award is worth 100,000 Euro and is given to the Veni laureate for a maximum period of three years. The grant can be used for the salary of the candidate and for the appointment of additional personnel (PhDs, post-doctorate candidates, and technicians), costs for the acquisition of materials and equipment and/or databases, travel costs and visits to conferences and research institutes.
*Requirements:* The award will be given to female researchers who have been awarded a Veni subsidy from the Division for Chemical Sciences and who, during the course of the Veni project, receive a tenured appointment as assistant professor at a university or a comparable position at a research institute.
*Geographic Focus:* Netherlands
*Amount of Grant:* 100,000 EUR
*Contact:* Dr. Varsha Kapoerchan, +31 (0) 70 349 44 42; v.kapoerchan@nwo.nl
*Internet:* www.nwo.nl/nwohome.nsf/pages/NWOA_6ZXCX3_Eng
*Sponsor:* Netherlands Organization for Scientific Research
P.O. Box 93138
The Hague, NL2509 Netherlands

**NYAM Edward N. Gibbs Memorial Lecture and Award in Nephrology** 1976
In 1901, the widow of Edward N. Gibbs, a patient of Dr. Edward Janeway, established the Edward N.Gibbs Memorial Endowment to award a prize to a physician in practice in the United States for the best original work in the etiology, pathology, and treatment of the diseases of the kidney. Under this program, The New York Academy of Medicine's Edward N. Gibbs Memorial Award and Lecture in Nephrology is presented bi-annually to recognize scientific achievement and outstanding contributions by an individual to basic science or clinical medicine, especially as those contributions relate to nephrology. The awardees' activities may have been accomplished in any nephrology-related area, including research, teaching and education, clinical expertise, medical care programs, and authorship. The distinguished recipient of this honor will present his or her research at a lecture before a broad audience of scientists and clinicians, to be held at The New York Academy of Medicine in the fall. The award recipient will receive the Edward N. Gibbs Memorial Medal and an honorarium of $7,500; all travel expenses associated with the lecture will be paid by the Edward N. Gibbs Memorial Endowment. It is expected that the manuscript from this lecture will be submitted to a scholarly journal for publication.
*Requirements:* Candidates must hold an MD degree and be citizens of the United States to be considered for this award.
*Geographic Focus:* All States
*Date(s) Application is Due:* Jul 30
*Amount of Grant:* 7,500 USD
*Contact:* Donna Fingerhut, Director of Fellowship and Trustee Affairs; (212) 419-3645 or (212) 822-7367; fax (212) 822-7338; gibbs@nyam.org or dfingerhut@nyam.org
*Internet:* https://www.nyam.org/awards-grants/endowed-lectures/
*Sponsor:* New York Academy of Medicine
1216 Fifth Avenue
New York, NY 10029-5202

**NYFA Artist Fellowships** 1977
The New York Foundation for the Arts (NYFA) was founded in 1971 to empower artists at critical stages in their creative lives. Artists' Fellowships are $7,000 cash awards made to individual originating artists living and working in the state of New York for unrestricted use. Grants are awarded in 15 artistic disciplines, with applications accepted in five categories each year in rotating fashion. In the 1212 and 1215 cycles, fellowships will be awarded in painting, folk/traditional arts, interdisciplinary work, fiction, and video/film. In the 1213 and 1216 cycles, fellowships will be awarded in choreography, music and sound, architecture/environmental structures and design, play writing and screenwriting, and photography. In the 1214 and 1217 cycles, fellowships will be awarded in digital/electronic arts, crafts and sculpture, nonfiction literature, poetry, and printmaking/drawing/book arts.
*Requirements:* Applicants must have resided in New York State for two years prior to application. Each recipient will be required to provide public service within the state of New York.

*Geographic Focus:* New York
*Date(s) Application is Due:* Oct 3
*Amount of Grant:* 7,000 USD
*Contact:* Susan Ball, Interim Director of Programs; (212) 366-6900, ext. 217; fax (212) 366-1778; sball@nyfa.org or fellowships@nyfa.org
*Internet:* www.nyfa.org/level2.asp?id=1&fid=1
*Sponsor:* New York Foundation for the Arts
20 Jay Street, 7th Floor
Brooklyn, NY 11201

**NYFA Shelly Pinz Fellowship** **1978**

New York Foundation for the Arts Fellowships are administered by NYFA with leadership support from the New York State Council on the Arts. The Shelly Pinz Fellowship will support one Fellowship in Poetry annually. Shelly Pinz (1944-2004) was a New York songwriter, poet and psychotherapist and is best known for co-writing the Lemon Pipers' 1968 hit "Green Tambourine" as well as her volume of poetry and lyrics titled "Courage to Think." The Shelly Pinz Fellowship was developed in Ms. Pinz's memory to support the careers of poets and composers. The announcement of the Shelly Pinz Fellowship in Poetry will take place at a poetry club, hosted by NYFA, where a reading of selections from "Courage to Think" will occur.
*Requirements:* The Award is open to poets of all nationalities, provided they are residents of New York State. All those who apply to NYFA's Artist Fellowship Program in Poetry will automatically be considered for the Shelly Pinz Fellowship. Applicants must have resided in New York State for two years prior to application. Each recipient will be required to provide public service within the state of New York.
*Geographic Focus:* New York
*Date(s) Application is Due:* Nov 5
*Contact:* Susan Ball, Interim Director of Programs; (212) 366-6900, ext. 321; fax (212) 366-1778; sball@nyfa.org
*Internet:* www.nyfa.org/level3.asp?id=46&fid=1&sid=1
*Sponsor:* New York Foundation for the Arts
20 Jay Street, 7th Floor
Brooklyn, NY 11201

**NYPL Dorothy and Lewis B. Cullman Center for Scholars and Writers Fellowships** **1979**

The Dorothy and Lewis B. Cullman Center for Scholars and Writers is an international fellowship program open to people whose work will benefit directly from access to the collections at the Stephen A. Schwarzman Building - including academics, independent scholars, and creative writers (novelists, playwrights, poets). The Center appoints fifteen Fellows a year for a nine-month term at the Library, from September through May. In addition to working on their own projects, the Fellows engage in an ongoing exchange of ideas within the Center and in public forums throughout the Library. The annual deadline for submission of applications is September 30.
*Requirements:* Applicants must be United States citizens or permanent residents with the legal right to work in the United States.
*Geographic Focus:* All States
*Date(s) Application is Due:* Sep 30
*Contact:* Matthew Knutzen, Director; (212) 930-0829 or (212) 930-0653
*Internet:* https://www.nypl.org/help/about-nypl/fellowships-institutes/center-for-scholars-and-writers
*Sponsor:* New York Public Library
476 Fifth Avenue
New York, NY 10018

**NYPL Lapidus Center Long-Term Fellowships** **1980**

The Lapidus Center for the Historical Analysis of Transatlantic Slavery was funded by a $2.5 million gift from Ruth and Sid Lapidus, matched by The New York Public Library, and it is the only facility of its kind based in a public research library. The Center's mission is to generate and disseminate scholarly knowledge on the slave trade, slavery, and anti-slavery pertaining to the Atlantic World. The Center supports the work of researchers with long-term and short-term fellowships. Given the centrality of Atlantic slavery to the making of the modern world, the Lapidus fellowships ensure that slavery studies are a cornerstone of the Schomburg Center's broader research community and provide a counterbalance to the contemporary direction of scholarship in African American and African Diaspora studies. Currently, two Lapidus Center Long-Term Fellowships are being offered to assist scholars whose research on transatlantic slavery can benefit from extended access to the Schomburg Center's resources. These fellowships will allow recipients to spend six months in residence with an office, a computer, and full access to physical and electronic resources at the Schomburg Center and other research units of The New York Public Library. Fellowships are awarded with a stipend of $30,000. The annual deadline for applications is December 1.
*Requirements:* The Fellowship Program is open to post-doctoral scholars studying the slave trade, slavery, and anti-slavery in the Atlantic World. Only U.S. citizens, permanent residents and foreign nationals who have been resident in the United States for the three years as of December 1 may apply. Fellows are expected to be in continuous residence at the Schomburg Center and to participate in the intellectual life of the program. They must be devoted full time to their research projects and may not be employed during the period except sabbaticals from their own institutions.
*Geographic Focus:* All States
*Date(s) Application is Due:* Dec 1
*Amount of Grant:* 30,000 USD
*Contact:* Sylviane Anna Diouf, Director; (212) 491-2054; lapiduscenter@nypl.org
*Internet:* https://www.nypl.org/help/about-nypl/fellowships-institutes/lapidus-center-fellowships
*Sponsor:* New York Public Library
515 Malcolm X Boulevard
New York, NY 10037

**NYPL Lapidus Center Short-Term Fellowships** **1981**

The Lapidus Center for the Historical Analysis of Transatlantic Slavery was funded by a $2.5 million gift from Ruth and Sid Lapidus, matched by The New York Public Library, and it is the only facility of its kind based in a public research library. The Center's mission is to generate and disseminate scholarly knowledge on the slave trade, slavery, and anti-slavery pertaining to the Atlantic World. The Center supports the work of researchers with long-term and short-term fellowships. Given the centrality of Atlantic slavery to the making of the modern world, the Lapidus fellowships ensure that slavery studies are a cornerstone of the Schomburg Center's broader research community and provide a counterbalance to the contemporary direction of scholarship in African American and African Diaspora studies. Currently, Lapidus Center Short-Term Fellowships are being offered at a duration of three months. These fellowships aim to assist researchers from outside of the New York metropolitan area gain access to collection material at the Schomburg Center not readily available elsewhere. Fellowships are awarded with a stipend of $2,000 per month. The annual deadline for applications is December 1.
*Requirements:* The Fellowship Program is open to doctoral students, post-doctoral scholars, independent researchers, and artists studying the slave trade, slavery, abolition, and anti-slavery in the Atlantic World. Only U.S. citizens, permanent residents and foreign nationals who have been resident in the United States for the three years immediately preceding the application deadline may apply.
*Geographic Focus:* All States
*Date(s) Application is Due:* Dec 1
*Amount of Grant:* 2,000 - 6,000 USD
*Contact:* Sylviane Anna Diouf, Director; (212) 491-2054; lapiduscenter@nypl.org
*Internet:* https://www.nypl.org/help/about-nypl/fellowships-institutes/lapidus-center-fellowships
*Sponsor:* New York Public Library
515 Malcolm X Boulevard
New York, NY 10037

**NYPL Martin Duberman Visiting Fellowships** **1982**

Annually the New York Public Library provides stipends of $15,000 for up to three Martin Duberman Visiting Scholars. The stipends support travel to New York City and related expenses to do research in the Library's premier LGBT (Lesbian, Gay, Bisexual, and Transgender) history collections. The Martin Duberman Visiting Scholars are funded by the support of Martin Duberman and Eli Zal. The awardee will be expected to spend a minimum of three months researching at the Library and at other archives relevant to their topic in the New York City area, to give a public talk on their work, and to write a short piece about their project for the Library's website. The fellowship is open to both academic faculty and independent scholars who have made a significant contribution to the field.
*Requirements:* The program is limited to emerging scholars, those without permanent academic appointments, or those who are unaffiliated with an academic institution. Recipients must supply a written summary of their findings upon completion of their work. Interested applicants should submit a 3–5 page research proposal specifying the relevant collections at the Library for their project, a draft budget and itinerary for their planned trip, a curriculum vitae, and an appropriate letter of recommendation.
*Geographic Focus:* All States
*Date(s) Application is Due:* May 30
*Amount of Grant:* 15,000 USD
*Contact:* Jason Baumann; (212) 930-0887; jasonbaumann@nypl.org
*Internet:* www.nypl.org/node/113221
*Sponsor:* New York Public Library
476 Fifth Avenue
New York, NY 10018

**NYPL Schomburg-Mellon Humanities Summer Institute Fellowships** **1983**

The Schomburg Center for Research in Black Culture and the Andrew W. Mellon Foundation have created the Schomburg-Mellon Humanities Summer Institute to encourage minority students and others with an interest in African-American and African Diasporan Studies to pursue graduate degrees in the humanities. The six-week Institute will, with the help of renowned scholars, develop and nurture the students' interest in the appropriate disciplines and provide them with the requisite intellectual challenges and orientations needed to pursue humanities careers and to reach their full potential. The application deadline is January 10, and selected students will be informed by email by April 5.
*Requirements:* Eligible applicants must: be United States citizens or permanent residents; be enrolled in a college or university in New York, New Jersey, Connecticut or in an Historically Black College or University across the nation; and have a grade point average of 3.0 or higher.
*Geographic Focus:* All States
*Date(s) Application is Due:* Jan 10
*Contact:* Deirdre Lynn Hollman, Director; (917) 275-6975 or (212) 491-2207; sdiouf@nypl.org or schomburged@nypl.org
*Internet:* https://www.nypl.org/help/about-nypl/fellowships-institutes/schomburg-mellon-humanities-summer-institute
*Sponsor:* New York Public Library
515 Malcolm X Boulevard
New York, NY 10037-1801

**NYPL Schomburg Center Scholars-in-Residence Fellowships** 1984

Schomburg Center Scholars-in-Residence Fellowships assist those scholars and professionals whose research in the black experience can benefit from extended access to the Center's resources. Fellowships allow recipients to spend six months or a year in residence with access to resources at the Schomburg Center and other centers of The New York Public Library. The program encourages research and writing on black history and culture, facilitates interaction among participating scholars, and provides widespread dissemination of findings through lectures, publications, and colloquia and seminars. It encompasses projects in African, Afro-American, and Afro-Caribbean history and culture. Funding support for the program has been provided by the Ford Foundation, the Rockefeller Foundation, the Aaron Diamond Foundation, the Irene Diamond Foundation, the Samuel I. Newhouse Foundation, the Andrew W. Mellon Foundation and the National Endowment for the Humanities. Fellowships are awarded for continuous periods of six or twelve months at the Schomburg Center with maximum stipends of $30,000 for six months and $60,000 for twelve months. The application process opens September 1, and the deadline is December 15.
*Requirements:* Fellowships are open to scholars studying the history, literature, and culture of peoples of African descent from a humanistic perspective and to professionals in fields related to the Schomburg Center's collections and program activities. Projects in the social sciences, science and technology, psychology, education, and religion are eligible if they utilize a humanistic approach and contribute to humanistic knowledge.
*Restrictions:* Creative writing (works of poetry and fiction), projects that result in a performance, and dissertation research are not eligible.
*Geographic Focus:* All States
*Date(s) Application is Due:* Dec 15
*Amount of Grant:* Up to 30,000 USD
*Contact:* Deirdre Lynn Hollman; (917) 275-6975 or (212) 491-2207; schomburged@nypl.org
*Internet:* https://www.nypl.org/help/about-nypl/fellowships-institutes/schomburg-center-scholars-in-residency/application
*Sponsor:* New York Public Library
515 Malcolm X Boulevard
New York, NY 10037-1801

**NYPL Short-Term Research Fellowships** 1985

The New York Public Library provides short-term fellowships to support visiting scholars conducting research in the Library's unique research and special collections. Fellowships stipends are available to scholars from outside the New York metropolitan area engaged in graduate-level, post-doctoral, or independent research. Scholars researching in the humanities including but not limited to art history, cultural studies, history, literature, performing arts and photography are welcome to apply. Fellowship stipends are $1,000 per week for a minimum of two and maximum of four weeks. The annual application deadline is January 15, and awarding of the fellowships will be announced on April 1.
*Requirements:* Applicants must be United States citizens or permanent residents with the legal right to work in the United States.
*Geographic Focus:* All States
*Date(s) Application is Due:* Jan 15
*Amount of Grant:* 2,000 - 4,000 USD
*Contact:* Matthew Knutzen; (212) 930-0829 or (212) 930-0653; short.term@nypl.org
*Internet:* www.nypl.org/help/about-nypl/fellowships-institutes/short-term-research-fellowships
*Sponsor:* New York Public Library
476 Fifth Avenue
New York, NY 10018

**NYSCA Dance: Services to the Field Grants** 1986

Services to the Field grants provides support to service organizations that offer professional services for the advancement of dance groups and individuals. Services may include technical assistance in administrative and/or new technological areas, cooperative management, booking and promotional services, research facilities, archival documentation, projects that utilize technology, and performance spaces.
*Requirements:* Applicants must register with the Council before February 22 to be considered for funding. In addition, applicants should demonstrate how Council support would meet the needs of the professional dance field. A paid administrator on staff is required. Applications can be obtained at the website and should be submitted online. First-time applicants are required to call staff before the registration deadline.
*Restrictions:* The request amount should not exceed 50% of the total project budget.
*Geographic Focus:* New York
*Date(s) Application is Due:* Apr 1
*Contact:* Beverly D'Anne, Senior Program Officer; (212) 741-3232 or (212) 741-3331; bdanne@nysca.org or dlim@nysca.org
*Internet:* www.nysca.org/public/guidelines/dance/services.htm
*Sponsor:* New York State Council on the Arts
300 Park Ave South, 10th Floor
New York, NY 10010

**NYSCA Folk Arts: Exhibitions Grants** 1987

Exhibition grants are offered for an exhibition of the work of living folk artists, to explore issues in the folk arts, or to interpret the traditional culture of a specific community through its folk arts. The proposed exhibition should enhance understanding and appreciation of the folk arts for general audiences as well as communities represented in the exhibition. The exhibition may include objects of folk art, photographic representations of folk culture, and/or multimedia kiosks or other installations in the exhibition utilizing new technologies. Requests may include field research expenses for the development of an exhibition and expenses for an exhibition catalog.
*Requirements:* Applicants must register with the Council before February 22 to be considered for funding. The proposed exhibition should focus on traditions currently practiced or within living memory, and may incorporate relevant historical works of folk art. First-time applicants must call program staff before the registration deadline.
*Restrictions:* Funding awarded for exhibitions cannot exceed 50% of project expenses. Support is not available for programming involving artists who appropriate, interpret, or revive the traditions of other communities. Folk Arts emphasizes support for presentations grounded in the traditional modes of practicing folk art, and not programming involving choreography, theatricality, or stylization that significantly alters traditions.
*Geographic Focus:* New York
*Date(s) Application is Due:* Apr 1
*Contact:* Robert Baron, Senior Program Officer; (212) 741-7755 or (212) 741-7143; rbaron@nysca.org or rputnam@nysca.org
*Internet:* www.nysca.org/public/guidelines/folk_arts/exhibitions.htm
*Sponsor:* New York State Council on the Arts
300 Park Ave South, 10th Floor
New York, NY 10010

**NYSCA Folk Arts: Presentation Grants** 1988

The Presentation grants support activities that present folk arts and artists to the public, including concerts, festivals, lecture/demonstrations, residencies, and programs presenting oral narratives. Applications may involve one type of activity or a series of thematically related activities that may involve different modes of presentation. Folk arts events and series are also supported.
*Requirements:* Applicants must register with the Council by February 22 to be considered for funding. Applications should demonstrate a clearly articulated thematic focus for the event or series requested. Also, artists must receive appropriate fees for their presentations, and when travel is involved, adequate per diem. Presentations should also involve personnel and consultants with appropriate cultural expertise and technical skills. These may include, but are not limited to: folklorists, ethnomusicologists, graphic designers, publicists and specialists in lighting, sound reinforcement, stage management, promotion and marketing. In addition, the folk art activity being presented must have its own interpretive materials, and promotion for the program should substantially publicize the folk arts event and the artists involved in the activity. First-applicants must contact program staff before the registration deadline. Applications can be found at the website and should be submitted online.
*Restrictions:* This grant will not fund more than 50% of the project expenses. Applicants may submit no more than one application in this funding category. Folk Arts emphasizes support for presentations grounded in the traditional modes of practicing folk art, and not programming involving choreography, theatricality, or stylization that significantly alters traditions. Support is also not available for programming involving artists who appropriate, interpret, or revive the traditions of other communities.
*Geographic Focus:* New York
*Date(s) Application is Due:* Apr 1
*Contact:* Robert Baron, Senior Program Officer; (212) 741-7755 or (212) 741-7847; rbaron@nysca.org or rputnam@nysca.org
*Internet:* www.nysca.org/public/guidelines/folk_arts/presenting.htm
*Sponsor:* New York State Council on the Arts
300 Park Ave South, 10th Floor
New York, NY 10010

**NYSCA Folk Arts: Regional and County Folk Arts Programs Grants** 1989

Folk Arts, in partnership with regional and local cultural organizations, supports folk arts programs which are regional and county-wide in scope. Support for these programs is intended to generate field research and programming for diverse audiences on an ongoing basis, directed by qualified individuals with experience and expertise in the public folklore field. Grants may be used towards the salaries of staff folklorists employed year-round and expenses associated with their production of public programs, field research on local traditions, and assistance to individual folk artists and other organizations interested in developing folk arts activities.
*Requirements:* Applicants must register with the Council by February 22 to be considered for funding. In addition, an applicant must have: received support from Folk Arts for the past three years; a minimum of four significant public programs that are organized each year; programs to be based to a significant extent upon field research, and work plans that include sufficient time for field research. Also, field research materials collected must be maintained in a locally accessible archive. Staff folklorists supported through this category must have experience organizing folk arts public programs and academic training in folklore or ethnomusicology. The staff folklorist must be employed for a minimum of three days each week. Applicant organizations are also expected to have a long-term commitment to folk arts services and programming. Program income should come from a variety of sources. First-time applicants are required to call staff before the registration deadline.
*Restrictions:* The request amount should not exceed 50% of the total project budget. Folk Arts emphasizes support for presentations grounded in the traditional modes of practicing folk art, and not programming involving choreography, theatricality, or stylization that significantly alters traditions. Support is also not available for programming involving artists who appropriate, interpret, or revive the traditions of other communities.
*Geographic Focus:* New York
*Date(s) Application is Due:* Apr 1
*Contact:* Robert Baron, Senior Program Officer; (212) 741-7755 or (212) 741-7143; rbaron@nysca.org or rputnam@nysca.org
*Internet:* www.nysca.org/public/guidelines/folk_arts/regional.htm
*Sponsor:* New York State Council on the Arts
300 Park Ave South, 10th Floor
New York, NY 10010

**NYSCA Individual Artists: Film, Media & New Technology Production Grants 1990**
Production support is offered annually to an individual artist project for film and video, radio, sound art, or computer-generated art. The project can be a combination of media, including but not limited to, work for tape, installation, single-channel work, or interactive disc formats. Individual Artists supports projects utilizing production methods and tools ranging from the simple to state-of-the-art technologies. The Council extends equal consideration to a wide range of artists and genres. Electronic media artists or filmmakers who wish to collaborate with artists in other disciplines may apply for production support. Priority is given to projects that emphasize the collaborative process and whose goal is to produce a distinctive work resulting from the meeting of disciplines.
*Requirements:* Applicants must register with the Council by February 22 in order to be considered for funding. Also, artists: who receive support for a specific phase of production of a project must complete that phase before re-applying for additional funds for the same project; applying for a documentary project are only eligible for post-production support. A work-in-progress of a minimum of 15 minutes is required. In addition, a sample of a previously completed work is required. First time directors may still apply, but it is recommended that the work sample should represent a project approaching the rough-cut phase. First-time applicants must call program staff before the registration deadline.
*Restrictions:* Production support is not provided for projects that take another discipline as their major focus or aim to provide straightforward documentation of another art form. In addition, organizations may not impose a fee on artists for submitting their applications. Artists may be involved as principal personnel in connection with only one project request. Artists receiving support in this category cannot apply for support in the following two years. (e.g., if funded in FY11, you must wait until FY14 to apply again). This allows the Council to reach a greater number of artists with its limited funds.
*Geographic Focus:* New York
*Date(s) Application is Due:* Apr 1
*Amount of Grant:* 5,000 - 25,000 USD
*Contact:* Helen Cash Jackson, Senior Program Officer; (212) 741-7148; hcash@nysca.org
Arian Blanco, Program Officer; (212) 741-7824; ablanco@nysca.org
*Internet:* www.nysca.org/public/guidelines/individual_artists/film_media.htm
*Sponsor:* New York State Council on the Arts
300 Park Ave South, 10th Floor
New York, NY 10010

**NYSCA Museum: Project Support Grants 1991**
Project Support grants offer an opportunity for museums or related service organizations to seek support for projects or programs which are essential to maintain and/or increase service to their audience/communities. Such arts and cultural projects may include, but are not limited to: exhibitions, education programs, public programs, salary support for a curator or educator, interpretation, collections research, catalogs, audience development, and services to the field. Salary Support prioritizes museums which are poised to move forward in their development by adding an additional staff member for a professional collections-based or education position.
*Requirements:* Applicants must register with the Council by February 22 to be considered for funding. Museums must have been open to the public for a minimum of one year prior to the registration deadline and must provide services at least 120 days per year. Applications can be found at the website and should be submitted online. First-time applicants must contact the Museum staff before the registration deadline.
*Restrictions:* The requested amount shall not exceed 50% of the total project budget. In addition, awards cannot be made for more than three years, and salary support can only partially cover salary and benefits. Organizations applying for or receiving Museum General Operating Support or General Program Support are not eligible for funding in this category. Also, applicants may only submit one Museum request.
*Geographic Focus:* New York
*Date(s) Application is Due:* Apr 1
*Contact:* Kristin Herron, Senior Program Officer; (212) 741-7848; kherron@nysca.org
*Internet:* www.nysca.org/public/guidelines/museums/projects.htm
*Sponsor:* New York State Council on the Arts
300 Park Ave South, 10th Floor
New York, NY 10010

**NYSCA Music: Services to the Field Grants 1992**
Services to the Field grants offer support to organizations that provide professional services to music organizations, artists, and research facilities. Such services may include, but are not limited to, technical assistance in administrative areas, projects that utilize technology, booking and promotional services, and concert and artist management. Support is also available for publications that specifically benefit the professional music field or its audience.
*Requirements:* Applicants must register with the Council by February 22 to be eligible for funding. Also, organizations must have a paid administrator to be eligible for an award. Applications are available at the website and should be submitted online. First-time applicants must contact Music staff before the registration deadline.
*Restrictions:* Request amounts should not exceed 50% of the total project budget. NYSCA also does not fund any projects that are not directed toward programs in New York State, competitions or contests, or out-of-state travel expenses.
*Geographic Focus:* New York
*Date(s) Application is Due:* Apr 1
*Contact:* Beverly D'Anne, Senior Program Officer; (212) 741-3232; bdanne@nysca.org
*Internet:* www.nysca.org/public/guidelines/music/services.htm
*Sponsor:* New York State Council on the Arts
300 Park Ave South, 10th Floor
New York, NY 10010

**NZDIA Lottery Health Research Grants 1993**
Lottery Health Research funding provides grants to organizations for health research projects, scholarships, fellowships and research equipment that will help improve the health of people of New Zealand. Organizations receiving grants are expected to show how their health research will contribute to: better understanding of the causes, prevention and treatment of disorders affecting people of New Zealand; new approaches in health and bio-medical science for the ultimate benefit of all people of New Zealand; and developing and keeping a highly skilled workforce of health and bio-medical researchers in New Zealand. These grants may support: translational research projects; shared research equipment; post-doctoral fellowships; and PhD scholarships. There is one funding round each year for Lottery Health Research. The opening date for the upcoming year is July 11, with final application sdue by the August 22 deadline.
*Requirements:* Citizens or organizations from New Zealand may put in a request for funding.
*Restrictions:* Lottery grants may not be used for any of the following: repaying or servicing debt; refinancing loans, deposits or underwriting projects; commercial, political and/or religious objectives, including employment and/or business initiatives, commercial enterprises, political advocacy or projects which seek to change legislation; fundraisers and projects which seek to raise funds in or for a specific sector, or are involved with the training or employment of fundraisers; projects which seek to redistribute funding to others; overseas aid or disaster relief; alcohol and drug treatment, education and support services; medical expenses, operations, treatments or the purchase of major items of health equipment; capital investment or trust funds; or projects and activities completed (retrospective funding) or items bought before the request. In addition, the Lottery Health Research fund does not support: individuals; operational costs and overheads; computers or tablets, unless required to run health research equipment; research or equipment leading to commercial benefits; project leaders, scholars or fellows who are not New Zealand citizens or permanent residents; conference attendances; requests from overseas institutions or organizations, or for research to be carried out overseas; development of educational materials, unless they have a research component; routine laboratory equipment and equipment ordered before a grant request is made; research where the findings will not be made freely available; relocation expenses; the purchase of land or buildings; or service contracts.
*Geographic Focus:* New Zealand
*Date(s) Application is Due:* Aug 22
*Contact:* Robyn Nicholas, General Manager Community Operations; 0800 824 824 or (03) 546 0904; robyn.nicholas@dia.govt.nz or community.matters@dia.govt.nz
*Internet:* https://www.communitymatters.govt.nz/lottery-health-research/
*Sponsor:* New Zealand Department of Internal Affairs
P.O. Box 805, 147 Lambton Quay
Wellington, 6140 New Zealand

**NZDIA Peace and Disarmament Education Trust Grants 1994**
The Peace and Disarmament Education Trust was established on 3 May 1988, with money received from France in recognition of the events surrounding the sinking of the Greenpeace ship Rainbow Warrior in Auckland on 10 July 1985. The Trust promotes international peace, arms control and disarmament through educational grants. Funded projects are expected to contribute to greater public awareness and understanding of disarmament, arms control, peace and the resolution of international conflict. Grant decisions are based on the likelihood that a completed project, internship, pilot project or thesis will deliver the benefits or outcomes described in the online request. Depending on individual circumstances: up to NZ$14,000 is available for a full year's work for a Masters research thesis; and up to NZ$21,000, plus NZ$5000 tuition fees, is available per year for up to three years for a Doctoral thesis. There are two annual deadlines, including April 4 and July 4.
*Requirements:* Projects must be for charitable purposes and provide a public benefit. There is no limit to the amount an individual or organization may request in support of a project. Based on the number and quality of requests received, however, the Trust may only be able to provide a contribution towards the project costs. These project costs may include salaries directly related to the project's planning or delivery. Preference is given to projects relevant to current New Zealand disarmament and arms control policy.
*Restrictions:* Requests to place interns for a defined period with organizations focused on achieving outcomes that align with the Trust's objectives will only be considered where the intern will be working on disarmament and arms control issues. The Trust will not fund: requests for projects or scholarships that do not relate to or align with the objectives of the Trust; projects to purchase land, buildings, furniture or fittings; projects intended to generate profits, unless any profits are to be reinvested in the project and its sustainability; general running costs or day-to-day administration expenses for organization, such as salaries (unless they are directly related to the project's planning or delivery), stationery, maintenance and office equipment, including computers; retrospective projects or projects that have been completed - project costs incurred before the closing date of the funding round are considered to be retrospective; or projects that are contrary to any policy of the New Zealand government.
*Geographic Focus:* New Zealand
*Date(s) Application is Due:* Apr 4; Jul 4
*Amount of Grant:* 14,000 - 64,500 NZD
*Contact:* Robyn Nicholas, General Manager Community Operations; 0800 824 824 or (03) 546 0904; robyn.nicholas@dia.govt.nz or community.matters@dia.govt.nz
*Internet:* https://www.communitymatters.govt.nz/peace-and-disarmament-education-trust-2/
*Sponsor:* New Zealand Department of Internal Affairs
P.O. Box 805, 147 Lambton Quay
Wellington, 6140 New Zealand

**NZDIA Winston Churchill McNeish Writers' Fellowships** 1995

Sir James McNeish was knighted for services to literature in 2010. He maintained that he would never have become the writer he was without the experience gained from working and living in foreign cultures. In 2012, Sir James and Lady McNeish made a donation to the Trust to encourage promising writers and journalists to travel and live in another culture. The donation enabled the Trust to create its first partnered Fellowship. Winston Churchill McNeish Writer's Fellowships are for promising writers and journalists to travel and live overseas. The intention is for the recipient's horizons to broaden and shift so that their writing is enhanced. Priority is given to applicants who: are an emerging writer or journalist aged between 25 and 40 years; have a track record of publications (one or more publications); and propose travel to non-English speaking, developing countries. The average amount awarded for a fellowship is between $5,000 and $7,000, although requests for up to $10,000 may be made. One funding round for Winston Churchill fellowships is offered every year for overseas travel happening in the following calendar year. These Fellowships are available every second year. The funding round open on March 28 and closes on August 1.

*Requirements:* Fellowship requests must be for: investigation, experience and/or research that will help to advance your field of interest, occupation, trade, industry or profession on your return; overseas travel in the calendar year following the fellowship request; or short-term travel outside New Zealand for up to three months.

*Restrictions:* The Winston Churchill Memorial Trust and the partnered fellowships do not fund requests for: study for academic or professional qualifications; attendance at academic or professional conferences; travel in the calendar year following the fellowship request; or fellowships that do not show how the benefits of what is learned or experienced through overseas travel will be shared with others.

*Geographic Focus:* All States, All Countries, New Zealand

*Date(s) Application is Due:* Aug 1

*Amount of Grant:* 5,000 - 10,000 USD

*Contact:* Robyn Nicholas, General Manager Community Operations; 0800 824 824 or (03) 546 0904; robyn.nicholas@dia.govt.nz or community.matters@dia.govt.nz

*Internet:* https://www.communitymatters.govt.nz/winston-churchill-memorial-trust-fellowships/

*Sponsor:* New Zealand Department of Internal Affairs

P.O. Box 805, 147 Lambton Quay

Wellington, 6140 New Zealand

**NZDIA Winston Churchill Memorial Trust Fellowships** 1996

Winston Churchill Memorial Trust Fellowships help people to travel overseas to learn from others and study topics that will advance their occupation, trade, industry, profession or community and benefit New Zealand. These Fellowships are open to people at all stages in their careers, in any profession, industry, occupation or trade. Fellows return to New Zealand with new knowledge, skills, inspiration and networks to share and support development in their field. All subject areas are considered. There are no qualification requirements. The Trust may also consider requests from people wishing to visit New Zealand for the purpose of completing a fellowship that will contribute to a profession, industry, occupation or trade in New Zealand or benefit New Zealand in general. The average amount awarded for a fellowship is between $5,000 and $7,000, although requests for up to $10,000 may be made. One funding round for Winston Churchill fellowships is offered every year for overseas travel happening in the following calendar year. The funding round open on March 28 and closes on August 1.

*Requirements:* Fellowship requests must be for: investigation, experience and/or research that will help to advance your field of interest, occupation, trade, industry or profession on your return; overseas travel in the calendar year following the fellowship request; or short-term travel outside New Zealand for up to three months.

*Restrictions:* The Winston Churchill Memorial Trust and the partnered fellowships do not fund requests for: study for academic or professional qualifications; attendance at academic or professional conferences; travel in the calendar year following the fellowship request; or fellowships that do not show how the benefits of what is learned or experienced through overseas travel will be shared with others.

*Geographic Focus:* All States, All Countries, New Zealand

*Date(s) Application is Due:* Aug 1

*Amount of Grant:* 5,000 - 10,000 USD

*Contact:* Robyn Nicholas, General Manager Community Operations; 0800 824 824 or (03) 546 0904; robyn.nicholas@dia.govt.nz or community.matters@dia.govt.nz

*Internet:* https://www.communitymatters.govt.nz/winston-churchill-memorial-trust-fellowships/

*Sponsor:* New Zealand Department of Internal Affairs

P.O. Box 805, 147 Lambton Quay

Wellington, 6140 New Zealand

**NZDIA Winston Churchill New Zealand China Friendship Society Fellowships** 1997

In partnership with the New Zealand China Friendship Society, the Winston Churchill Memorial Trust offers a specialized fellowship for travel to China to: gain knowledge, understanding and experience of Chinese culture and values; and share New Zealand culture and values in China. The primary purpose of this Fellowship is the benefit of an occupation, trade or profession, or to benefit New Zealand through the creation or deepening of New Zealand-China relations. The average amount awarded for a fellowship is between $5,000 and $7,000, although requests for up to $10,000 may be made. One funding round for Winston Churchill fellowships is offered every year for overseas travel happening in the following calendar year. The funding round open on March 28 and closes on August 1.

*Requirements:* Fellowship requests must be for: investigation, experience and/or research that will help to advance your field of interest, occupation, trade, industry or profession on your return; overseas travel in the calendar year following the fellowship request; or short-term travel outside New Zealand for up to three months.

*Restrictions:* The Winston Churchill Memorial Trust and the partnered fellowships do not fund requests for: study for academic or professional qualifications; attendance at academic or professional conferences; travel in the calendar year following the fellowship request; or fellowships that do not show how the benefits of what is learned or experienced through overseas travel will be shared with others.

*Geographic Focus:* New Zealand

*Date(s) Application is Due:* Aug 1

*Amount of Grant:* 5,000 - 10,000 USD

*Contact:* Robyn Nicholas, General Manager Community Operations; 0800 824 824 or (03) 546 0904; robyn.nicholas@dia.govt.nz or community.matters@dia.govt.nz

*Internet:* https://www.communitymatters.govt.nz/winston-churchill-memorial-trust-fellowships/

*Sponsor:* New Zealand Department of Internal Affairs

P.O. Box 805, 147 Lambton Quay

Wellington, 6140 New Zealand

**OAH/IEHS John Higham Travel Grants** 1998

The grant is given in memory of John Higham (1920-2003), past president of both organizations and a towering figure in immigration, ethnic, and intellectual history. Travel grants of $500 are awarded to three (3) graduate students each year. Funds are to be used by graduate students toward costs of attending the OAH/IEHS annual meeting. The successful candidates will have a preferred area of concentration in American Immigration and/or American Ethnic and/or American Intellectual history. The committee will seek some balance by gender, region of country, and type of university (e.g., major research university, and second tier).

*Geographic Focus:* All States

*Date(s) Application is Due:* Dec 3

*Amount of Grant:* 500 USD

*Contact:* Kara Hamm; (812) 855-7311; fax (812) 855-0696; khamm@oah.org

*Internet:* www.oah.org/awards/awards.higham.index.html

*Sponsor:* Organization of American Historians

112 N Bryan Avenue, P.O. Box 5457

Bloomington, IN 47408-4141

**OAH Avery O. Craven Award** 1999

The award is given each year for the most original book on the coming of the Civil War, the Civil War years, or the era of Reconstruction, with the exception of works of purely military history. The exception recognizes and reflects Craven's Quaker convictions. Avery O. Craven was president of the OAH 1963-1964. One copy of each entry, clearly labeled Avery O. Craven Award Entry, must be mailed directly to committee members. Guidelines are available online.

*Requirements:* Each entry must be published during the period January 1 through December 31 of the current year.

*Geographic Focus:* All States

*Date(s) Application is Due:* Oct 1

*Amount of Grant:* 500 USD

*Contact:* Kara Hamm; (812) 855-7311; fax (812) 855-0696; khamm@oah.org

*Internet:* www.oah.org/awards/awards.craven.index.html

*Sponsor:* Organization of American Historians

112 N Bryan Avenue, P.O. Box 5457

Bloomington, IN 47408-4141

**OAH Binkley-Stephenson Award** 2000

The Binkley-Stephenson Award has honored scholarly work since 1967. The award recognizes the best article published in the Journal of American History during the preceding calendar year. The journal is published quarterly in March, June, September, and December. It is widely recognized as an important source of research on all aspects of American history. William C. Binkley was president of the OAH 1944-1946 and editor of the Journal of American History 1956-1963.

*Geographic Focus:* All States

*Amount of Grant:* 500 USD

*Contact:* Kara Hamm; (812) 855-7311; fax (812) 855-0696; khamm@oah.org

*Internet:* www.oah.org/awards/awards.binkleystephenson.index.html

*Sponsor:* Organization of American Historians

112 N Bryan Avenue, P.O. Box 5457

Bloomington, IN 47408-4141

**OAH Darlene Clark Hine Award** 2001

The Darlene Clark Hine Award is given annually by the Organization of American Historians to the author of the best book in African American women's and gender history. The award is named for Darlene Clark Hine, a pioneer in African American women's and gender history and past president of the OAH (2001-2002).

*Requirements:* Individuals as well as editors are encouraged to submit nominations. Each entry must be published during the period January 1 through December 31 of the award year.

*Geographic Focus:* All States

*Date(s) Application is Due:* Oct 1

*Amount of Grant:* 500 USD

*Contact:* Kara Hamm; (812) 855-7311; fax (812) 855-0696; khamm@oah.org

*Internet:* www.oah.org/awards/awards.hine.index.html

*Sponsor:* Organization of American Historians

112 N Bryan Avenue, P.O. Box 5457

Bloomington, IN 47408-4141

**OAH David Thelen Award** **2002**
The OAH sponsors this biennial award (formerly the Foreign Language Article Prize through 1998) for the best article on American history published in a foreign language. The winning article will be published in the Journal of American History. An article should be concerned with the past (recent or distant) or with issues of continuity and change. It should also be concerned with events or processes that began, developed, or ended in what is now the United States. It should make a significant and original contribution to the understanding of U.S. history. Comparative and international studies that fall within these guidelines are welcome.
*Requirements:* Entries must have been published during the preceding two calendar years.
*Geographic Focus:* All States
*Date(s) Application is Due:* May 1
*Amount of Grant:* 500 USD
*Contact:* Kara Hamm; (812) 855-7311; fax (812) 855-0696; khamm@oah.org
*Internet:* www.oah.org/awards/awards.thelen.index.html
*Sponsor:* Organization of American Historians
112 N Bryan Avenue, P.O. Box 5457
Bloomington, IN 47408-4141

**OAH Ellis W. Hawley Prize** **2003**
The prize is awarded annually for the best book-length historical study of the political economy, politics, or institutions of the United States, in its domestic or international affairs, from the Civil War to the present. Eligible works include book-length historical studies, written in English, published during a given calendar year. Dissertations completed by October 1 also are eligible. One copy of each entry, clearly labeled Ellis W. Hawley Prize Entry, must be mailed directly to committee members. Guidelines are available online.
*Requirements:* Each entry must be published during the period January 1 through December 31 of the current year,
*Geographic Focus:* All States
*Date(s) Application is Due:* Oct 1
*Amount of Grant:* 500 USD
*Contact:* Kara Hamm; (812) 855-7311; fax (812) 855-0696; khamm@oah.org
*Internet:* www.oah.org/awards/awards.hawley.index.html
*Sponsor:* Organization of American Historians
112 N Bryan Avenue, P.O. Box 5457
Bloomington, IN 47408-4141

**OAH Erik Barnouw Award** **2004**
One or two awards are given annually in recognition of outstanding reporting or programming on network or cable television, or in documentary film, concerned with American history, the study of American history, and/or the promotion of history. The winning film(s) or video program(s) will be screened, and the award(s) will be presented at the annual meeting in March. Films and videos selected as honorable mentions will be shown at the annual meeting, along with the Barnouw winner, as part of the OAH Screening History film series. One copy of each entry, clearly labeled Erik Barnouw Award Entry, must be mailed directly to committee members.
*Requirements:* Films and video programs released between January 1 and December 31 of the current year are eligible for entry.
*Geographic Focus:* All States
*Date(s) Application is Due:* Jan 10
*Amount of Grant:* 1,000 USD
*Contact:* Kara Hamm; (812) 855-7311; fax (812) 855-0696; khamm@oah.org
*Internet:* www.oah.org/awards/awards.barnouw.index.html
*Sponsor:* Organization of American Historians
112 N Bryan Avenue, P.O. Box 5457
Bloomington, IN 47408-4141

**OAH Frederick Jackson Turner Award** **2005**
The award is given annually for an author's first book dealing with some significant phase of American history. The work must be the first book-length study of history published by the author and must have been published or scheduled for publication in the calendar year before the award is given. A copy of each entry should be submitted. Final page proofs may be used for books to be published after September 1 and before January 1 of the following year. The publisher of the winning book will receive a complimentary ad for the book in the Journal of American History and a certificate and the author will receive a medal.
*Requirements:* If the author is a PhD, she/he must have received the degree no earlier than seven years prior to submitting the manuscript for publication. Each entry must be published during the period January 1 through December 31 of the current year.
*Restrictions:* Each press may submit one entry per year.
*Geographic Focus:* All States
*Date(s) Application is Due:* Oct 1
*Amount of Grant:* 1,000 USD
*Contact:* Kara Hamm; (812) 855-7311; fax (812) 855-0696; khamm@oah.org
*Internet:* www.oah.org/awards/awards.turner.index.html
*Sponsor:* Organization of American Historians
112 N Bryan Avenue, P.O. Box 5457
Bloomington, IN 47408-4141

**OAH Friend of History Award** **2006**
First given in 2005, the Friend of History Award recognizes an institution or organization, or an individual working primarily outside college or university settings, for outstanding support for historical research, the public presentation of American history, or the work of the OAH. Recipients will be selected by the OAH Executive Committee from nominations that should include a statement of no more than five pages describing the nominee's distinguished contributions and, if for an individual, a CV of no more than five pages.
*Geographic Focus:* All States
*Amount of Grant:* 500 USD
*Contact:* Kara Hamm; (812) 855-7311; fax (812) 855-0696; khamm@oah.org
*Internet:* www.oah.org/awards/awards.friendsofhistory.index.html
*Sponsor:* Organization of American Historians
112 N Bryan Avenue, P.O. Box 5457
Bloomington, IN 47408-4141

**OAH Germany Residency Program Grants** **2007**
The committee seeks applications from OAH members who are established scholars affiliated with an American or Canadian University interested in leading an advanced undergraduate/graduate student seminar focusing on one aspect of U.S. History. The seminar will be based on a topic in U.S. History of an applicant's design, and all fields and methodologies are welcome. The language of instruction for the seminar is English. The Residency Program will provide round-trip airfare, housing for thirty days, a modest honorarium (1,500 Euros), a graduate assistant, and office space. Tubingen is located in Southwestern Germany, 25 miles south of Stuttgart. It is one of Germany's oldest and most prestigious institutions of higher education and has a vibrant community engaged in the study of the United States.
*Requirements:* Applicants must be members of the OAH, have a Ph.D., and be scholars of American history.
*Geographic Focus:* All States
*Date(s) Application is Due:* Oct 1
*Contact:* Kara Hamm; (812) 855-7311; fax (812) 855-0696; khamm@oah.org
*Internet:* www.oah.org/programs/germany/index.html
*Sponsor:* Organization of American Historians
112 N Bryan Avenue, P.O. Box 5457
Bloomington, IN 47408-4141

**OAH Huggins-Quarles Award** **2008**
The award is given annually to one or two minority graduate students at the dissertation research stage of their PhD program. To apply, a student should submit a brief two-page abstract of his/her dissertation project, along with a one-page budget explaining travel and research plans. Each application must be accompanied by a letter from the dissertation adviser attesting to the student's status and the ways in which the award will facilitate the completion of the dissertation project. Include email addresses for both the applicant and the adviser, if available. One complete copy of each application (including cover letter, abstract, budget, and reference letter) should be clearly labeled Huggins-Quarles Award Entry and received, by the listed deadline date, by each member of the committee. Guidelines are available online.
*Geographic Focus:* All States
*Date(s) Application is Due:* Dec 3
*Amount of Grant:* 1,000 USD
*Contact:* Kara Hamm; (812) 855-7311; fax (812) 855-0696; khamm@oah.org
*Internet:* www.oah.org/awards/awards.hugginsquarles.index.html
*Sponsor:* Organization of American Historians
112 N Bryan Avenue, P.O. Box 5457
Bloomington, IN 47408-4141

**OAH James A. Rawley Prize** **2009**
This prize is given each year for a book dealing with the history of race relations in the United States. Each entry must be published during the period January 1 through December 31 in the year of deadline date. One copy of each entry, clearly labeled James A. Rawley Prize Entry, must be mailed directly to each committee member. The award will be presented at the OAH annual meeting. The publisher will receive a certificate of merit.
*Geographic Focus:* All States
*Date(s) Application is Due:* Oct 1
*Amount of Grant:* 1,000 USD
*Contact:* Kara Hamm; (812) 855-7311; fax (812) 855-0696; khamm@oah.org
*Internet:* www.oah.org/awards/awards.rawley.index.html
*Sponsor:* Organization of American Historians
112 N Bryan Avenue, P.O. Box 5457
Bloomington, IN 47408-4141

**OAH Japanese Residencies Program Grants** **2010**
The OAH sends two American scholars to Japanese universities annually for two-week residencies. Round-trip airfare to Japan, housing, and modest daily expenses are covered by the award. There, in English, the American historians give lectures and seminars in their specialty and provide individual consultation to Japanese scholars, graduate students and sometimes undergraduates studying American history and culture. Visitors also participate in the collegial life of their host institutions and help expand personal scholarly networks between Japan and the U.S. The aim of the program is to contribute to the expansion of personal scholarly networks between the two countries. Round-trip airfare to Japan, housing, and modest daily expenses are covered. Awardees are also encouraged to explore Japan before or after their two-week residency at their own expense.
*Requirements:* Applicants must be members of the OAH, have a Ph.D., and be scholars of American history.
*Geographic Focus:* All States
*Date(s) Application is Due:* Oct 15
*Contact:* Kara Hamm; (812) 855-7311; fax (812) 855-0696; khamm@oah.org

*Internet:* www.oah.org/programs/residencies/index.html
*Sponsor:* Organization of American Historians
112 N Bryan Avenue, P.O. Box 5457
Bloomington, IN 47408-4141

**OAH Lerner-Scott Dissertation Prize** **2011**
This prize is given annually to the best doctoral dissertation in women's history. Each application must contain a letter of support from a faculty member at the degree-granting institution, along with an abstract, table of contents and sample chapter from the dissertation. Also include email addresses for both the applicant and the adviser, if available. One complete copy of each entry must be received by each member of the prize committee by the listed deadline. Finalists will be asked to submit a complete copy of the dissertation at a later date. One copy of each entry, clearly labeled Lerner-Scott Dissertation Prize Entry, must be mailed directly to committee members. Guidelines are available online.
*Requirements:* A dissertation must be completed during the period July 1 of last year through June 30 of the current year.
*Geographic Focus:* All States
*Date(s) Application is Due:* Oct 1
*Amount of Grant:* 1,000 USD
*Contact:* Kara Hamm; (812) 855-7311; fax (812) 855-0696; khamm@oah.org
*Internet:* www.oah.org/awards/awards.lernerscott.index.html
*Sponsor:* Organization of American Historians
112 N Bryan Avenue, P.O. Box 5457
Bloomington, IN 47408-4141

**OAH Liberty Legacy Foundation Award** **2012**
The award will be given for the best book on any historical aspect of the struggle for civil rights in the United States from the nation's founding to the present. One copy of each entry must be received by each committee member by the listed application deadline date. Final page proofs may be used for books to be published after the deadline and before 1 January. If a final page proof is submitted, a bound copy of the entry must be received no later than 7 January. The author of the winning book will receive a monetary prize and a certificate. The publisher of the book will be honored with a certificate of merit. One copy of each entry, clearly labeled 2006 Liberty Legacy Foundation Award, must be mailed directly to committee members. Guidelines are avaialable onlin.
*Requirements:* Each entry must be published during the period 1 January through 31 December.
*Geographic Focus:* All States
*Date(s) Application is Due:* Oct 1
*Amount of Grant:* 2,000 USD
*Contact:* Kara Hamm; (812) 855-7311; fax (812) 855-0696; khamm@oah.org
*Internet:* www.oah.org/awards/awards.liberty.index.html
*Sponsor:* Organization of American Historians
112 N Bryan Avenue, P.O. Box 5457
Bloomington, IN 47408-4141

**OAH Louis Pelzer Memorial Award** **2013**
Candidates for graduate degrees are invited to submit essays for the Louis Pelzer Memorial Award competition. Essays may deal with any period or topic in the history of the United States. The winning essay will be published in the Journal of American History. Significance of the subject matter, literary craftsmanship and competence in the handling of evidence are some of the factors that will be considered in judging the essays. Manuscripts should be clearly labeled (list current year) Louis Pelzer Memorial Award Entry and and addressed to the committee chair). The winner will be notified by the organization and furnished with details of the annual meeting and the awards presentation. Guidelines are available online.
*Requirements:* Essays should be submitted in quintuplicate and should not exceed 7000 words (including endnotes) in length. The endnotes should be triple-spaced. Because manuscripts are judged anonymously, the author's name and graduate program should appear only on a separate cover page.
*Geographic Focus:* All States
*Date(s) Application is Due:* Nov 30
*Amount of Grant:* 500 USD
*Contact:* Kara Hamm; (812) 855-7311; fax (812) 855-0696; khamm@oah.org
*Internet:* www.oah.org/awards/awards.pelzer.index.html
*Sponsor:* Organization of American Historians
112 N Bryan Avenue, P.O. Box 5457
Bloomington, IN 47408-4141

**OAH Merle Curti Award** **2014**
The award is given annually to the best work published in the last two years in the field of American social history (even-numbered years) and intellectual history (odd-numbered years). The author of the winning book will receive a monetary prize, a certificate, and a medal. The publisher(s) of the book(s) will be honored with a certificate(s) of merit. The award will be presented at the annual meeting of the OAH. One copy of each entry, clearly labeled Merle Curti Award Entry, must be mailed directly to the members of the chair. Guidelines are available online.
*Geographic Focus:* All States
*Date(s) Application is Due:* Oct 1
*Amount of Grant:* 2,000 USD
*Contact:* Kara Hamm, ; (812) 855-7311; fax (812) 855-0696; khamm@oah.org
*Internet:* www.oah.org/awards/awards.curti.index.html
*Sponsor:* Organization of American Historians
112 N Bryan Avenue, P.O. Box 5457
Bloomington, IN 47408-4141

**OAH Ray A. Billington Prize** **2015**
The award is given biennially for the best book on American frontier history published during the last two years. American frontier history is defined broadly so as to include the pioneer periods of all geographical areas and comparisons between American frontiers and others. The deadline is for even-numbered years. One copy of each entry, clearly labeled Ray Allen Billington Prize Entry, must be mailed directly to each committee member. Guidelines are available online.
*Geographic Focus:* All States
*Date(s) Application is Due:* Oct 1
*Amount of Grant:* 1,000 USD
*Contact:* Kara Hamm; (812) 855-7311; fax (812) 855-0696; khamm@oah.org
*Internet:* www.oah.org/awards/awards.billington.index.html
*Sponsor:* Organization of American Historians
112 N Bryan Avenue, P.O. Box 5457
Bloomington, IN 47408-4141

**OAH Richard W. Leopold Prize** **2016**
This award is designed to improve contacts and interrelationships within the historical profession where an increasing number of history-trained scholars hold distinguished positions in governmental agencies. This prize recognizes the significant historical work being done by historians outside academe. The award is given every two years for the best book written by a historian connected with federal, state, or municipal government, in the areas of foreign policy, military affairs broadly construed, the historical activities of the federal government, or biography in one of the foregoing areas. One copy of each entry, clearly labeled Richard W. Leopold Prize Entry, must be mailed directly to committee members.
*Requirements:* Each entry must be published during the two-year period between January 1 of the last year and December 31 of the current year. The winner must have been employed in a government position for at least five years.
*Geographic Focus:* All States
*Date(s) Application is Due:* Oct 1
*Amount of Grant:* 1,500 USD
*Contact:* Kara Hamm; (812) 855-7311; fax (812) 855-0696; khamm@oah.org
*Internet:* www.oah.org/awards/awards.leopold.index.html
*Sponsor:* Organization of American Historians
112 N Bryan Avenue, P.O. Box 5457
Bloomington, IN 47408-4141

**OAH Roy Rosenzweig Distinguished Service Award** **2017**
The OAH Roy Rosenzweig Distinguished Service Award is presented each year to an individual or individuals whose contributions have significantly enriched our understanding and appreciation of American history. Beginning in 2008, the award is named in memory of Roy Rosenzweig, who was the Mark and Barbara Fried Chair and founding director of the Center for History and New Media at George Mason University. Rosenzweig, a pioneer in the field of digital humanities, was actively involved in OAH for more than fifteen years, serving on its nominating board, program committees, executive board, and was an OAH Distinguished Lecturer. In 2011, George Mason University renamed the Center for History and New Media to honor his memory. Rosenzweig passed away in 2007 at the age of 57. Those eligible for this award include scholars, teachers, writers, public historians and officeholders who have shaped or deepened our sensitivity to the past.
*Geographic Focus:* All States
*Date(s) Application is Due:* Sep 10
*Amount of Grant:* 2,000 USD
*Contact:* Kara Hamm, Award and Prize Coordinator; (812) 855-7311; fax (812) 855-0696; khamm@oah.org
*Internet:* www.oah.org/awards/awards.distinguished.index.html
*Sponsor:* Organization of American Historians
112 N Bryan Avenue, P.O. Box 5457
Bloomington, IN 47408-4141

**OAH Willi Paul Adams Award** **2018**
The OAH sponsors a biennial award for the best book on American history published in a foreign language. The award was formerly known as the Foreign Language Book Prize. The OAH defines both history and American broadly. To be eligible, a book should be concerned with the past (recent or distant) or with issues of continuity and change. It should also be concerned with events or processes that began, developed, or ended in what is now the United States. Comparative and international studies that fall within these guidelines are welcomed. Authors of eligible books are invited to nominate their work. Under unusual circumstances unpublished manuscripts will be considered.
*Requirements:* Each entry must have been published during the two-year period July 1 through June 30 of the current year.
*Geographic Focus:* All States
*Date(s) Application is Due:* May 1
*Amount of Grant:* 1,000 USD
*Contact:* Kara Hamm; (812) 855-7311; fax (812) 855-0696; khamm@oah.org
*Internet:* www.oah.org/awards/awards.adams.index.html
*Sponsor:* Organization of American Historians
112 N Bryan Avenue, P.O. Box 5457
Bloomington, IN 47408-4141

**Oak Foundation Environmental Grants** **2019**
The foundation awards grants worldwide to address international social and environmental issues, particularly those that have a major impact on the lives of the disadvantaged. In the climate change and marine conservation programs, the Foundation

typically invites organizations to submit letters of enquiry based on annual funding priorities, internal guidelines, budget cycle and timelines. If you have not been contacted by a Program Officer directly to submit a letter of enquiry, the chances are relatively low that an organization will be considered for funding. If you decide to submit an unsolicited letter of inquiry, the Foundation will respond within three months.
*Requirements:* Within this program, Oak funds organizations throughout Europe, the United States, Canada, and Belize.
*Restrictions:* Oak does not provide support to individuals, and does not provide funding for scholarships or tuition assistance for undergraduate or postgraduate studies. The Foundation also does not fund religious organizations for religious purposes, election campaigns, or general fund-raising drives.
*Geographic Focus:* All States, Albania, Andorra, Armenia, Austria, Azerbaijan, Belarus, Belgium, Belize, Bosnia & Herzegovina, Bulgaria, Canada, Croatia, Cyprus, Czech Republic, Denmark, Estonia, Finland, France, Germany, Greece, Hungary, Iceland, Ireland, Italy, Kosovo, Latvia, Liechtenstein, Lithuania, Luxembourg, Macedonia, Malta, Moldova, Monaco, Montenegro, Netherlands, Norway, Poland, Portugal, Romania, Russia, San Marino, Serbia, Slovakia, Slovenia, Spain, Sweden, Switzerland, Turkey, Ukraine, United Kingdom, Vatican City
*Amount of Grant:* 25,000 - 2,000,000 USD
*Samples:* New Economics Foundation, London, England, $249,683 - to promote the economic benefits of sustainable fish stocks and marine ecosystems; Comunidad y Biodiversidad, Guaymas, Mexico, $220,000 - to restore the natural capital of coral reefs for sustainable fisheries in Mexican Caribbean waters; Climate Institute, Sydney, Australia, $478,011 - in support of a program putting a limit and price on carbon pollution.
*Contact:* Leonardo Lacerda; naep@oakfnd.org or eep@oakfnd.ch
*Internet:* www.oakfnd.org/node/22
*Sponsor:* Oak Foundation
Case Postale 115 58, Avenue Louis Casai
Geneva, Cointrin 1216 Switzerland

**Obesity Society Atkinson-Stern Award for Distinguished Public Service** **2020**
The Obesity Society's Atkinson-Stern Award for Distinguished Public Service recognizes an individual or organization whose work has significantly improved the lives of those affected by obesity, whether through research, public policy, patient care or other means. This award includes a plaque and a $1,000 prize. The annual nomination deadline is August 20.
*Requirements:* All award nominations must include a letter of recommendation (signed by two current members of The Obesity Society) stating: reasons for nominating the individual; the nominee's qualifications for the specific award; a summary of the nominee's contributions to the field of obesity; the nominee's complete CV; and other letters of support from individuals (these may be sent under a separate cover letter).
*Restrictions:* Self-nominations are not accepted. Nominees will not be considered for multiple awards in a single year.
*Geographic Focus:* All States
*Date(s) Application is Due:* Aug 20
*Amount of Grant:* 1,000 USD
*Contact:* Regina Williams, Governance and Administrative Coordinator; (301) 563-6526; fax (301) 563-6595; rwilliams@obesity.org or governance@obesity.org
*Internet:* https://www.obesity.org/awards-and-grants/
*Sponsor:* Obesity Society
1110 Bonifant Street, Suite 500
Silver Spring, MD 20910

**Obesity Society Early Career Grants** **2021**
The applicant selected for the Early Career Grant challenge receives a $25,000 research grant based on presentation of a research idea during the Obesity Society's Opening Session at Obesity Week. Four to five finalists will receive a $1,500 travel grant to attend Obesity Week to present their oral arguments to a panel of judges and the audience. Members of the Obesity Society are encouraged to apply by sending in their Letters of Intent by the deadline of May 22, with a full proposal submission deadline of August 20.
*Requirements:* Members of the Obesity Society are encouraged to apply by sending in their Letters of Intent by the deadline of April 14.
*Geographic Focus:* All States
*Date(s) Application is Due:* Aug 20
*Amount of Grant:* Up to 26,500 USD
*Contact:* Regina Williams, Governance and Administrative Coordinator; (301) 563-6526; fax (301) 563-6595; rwilliams@obesity.org or grantsandawards@obesity.org
*Internet:* https://www.obesity.org/awards-and-grants/
*Sponsor:* Obesity Society
1110 Bonifant Street, Suite 500
Silver Spring, MD 20910

**Obesity Society Early Career Research Grants** **2022**
The Obesity Society's Grants program was founded to promote, reward, and encourage research in the field of obesity. The Early Career Research program funds studies proposed by new investigators who have completed their doctorate within the past five years with a PhD or eight years with an MD. One grant will be funded annually up to $25,000 for a one-year pilot study, plus a $1,500 travel grant. Members of the Obesity Society are encouraged to apply by sending in their Letters of Intent by the deadline of May 22, with a full proposal submission deadline of August 20.
*Requirements:* Members of the Obesity Society (domestic and international) are eligible to apply. Individuals must be qualified by training and experience to conduct his or her responsibilities on the research project.
*Geographic Focus:* All States
*Date(s) Application is Due:* Aug 20
*Amount of Grant:* Up to 26,500 USD
*Contact:* Regina Williams, Governance and Administrative Coordinator; (301) 563-6526; fax (301) 563-6595; rwilliams@obesity.org or grantsandawards@obesity.org
*Internet:* https://www.obesity.org/awards-and-grants/
*Sponsor:* Obesity Society
1110 Bonifant Street, Suite 500
Silver Spring, MD 20910

**Obesity Society Early Career Travel Grants** **2023**
The Obesity Society's TOPS Research Achievement Award is made possible by an annual grant from the Take Off Pounds Sensibly Foundation (TOPS) and recognizes a singular achievement or contribution to research in the field of obesity. The award includes a plaque and a $5,000 prize plus a travel grant to the Annual Scientific Meeting. It is presented during a plenary session and the recipient is invited to present the TOPS Research Achievement Award Lecture. The annual nomination deadline is August 20.
*Requirements:* All award nominations must include a letter of recommendation (signed by two current members of The Obesity Society) stating: reasons for nominating the individual; the nominee's qualifications for the specific award; a summary of the nominee's contributions to the field of obesity; the nominee's complete CV; and other letters of support from individuals (these may be sent under a separate cover letter).
*Restrictions:* Self-nominations are not accepted. Nominees will not be considered for multiple awards in a single year.
*Geographic Focus:* All States
*Date(s) Application is Due:* Aug 20
*Amount of Grant:* 6,000 USD
*Contact:* Regina Williams, Governance and Administrative Coordinator; (301) 563-6526; fax (301) 563-6595; rwilliams@obesity.org or governance@obesity.org
*Internet:* https://www.obesity.org/awards-and-grants/
*Sponsor:* Obesity Society
1110 Bonifant Street, Suite 500
Silver Spring, MD 20910

**Obesity Society Early Career Travel Grants** **2024**
The Obesity Society's Early Career Travel Grant supports contributions to the obesity field by early career investigators. Two winners each receive $1,000 for expenses related to ObesityWeek. Members of the Obesity Society are encouraged to apply by sending in their Letters of Intent by the deadline of May 22, with a full proposal submission deadline of August 20.
*Geographic Focus:* All States
*Date(s) Application is Due:* Aug 20
*Amount of Grant:* 1,000 USD
*Contact:* Regina Williams, Governance and Administrative Coordinator; (301) 563-6526; fax (301) 563-6595; rwilliams@obesity.org or grantsandawards@obesity.org
*Internet:* https://www.obesity.org/awards-and-grants/
*Sponsor:* Obesity Society
1110 Bonifant Street, Suite 500
Silver Spring, MD 20910

**Obesity Society Ethan Sims Young Investigator Award** **2025**
Known as the Ethan Sims Young Investigator Award, each year, during the call for abstracts, five young investigators are selected as finalists for this award for excellence in research. All finalists receive a travel grant of $1,000 to cover expenses for ObesityWeek. Each finalist presents his or her abstract during a concurrent ObesityWeek session, with one finalist receiving a $1,000 prize. The recipient is announced at the conclusion of the session. Members of the Obesity Society are encouraged to apply by sending in their Letters of Intent by the deadline of May 22, with a full proposal submission deadline of August 20.
*Requirements:* Members of the Obesity Society are encouraged to apply by sending in their Letters of Intent by the deadline.
*Geographic Focus:* All States
*Date(s) Application is Due:* Aug 20
*Amount of Grant:* 1,000 - 2,000 USD
*Contact:* Regina Williams, Governance and Administrative Coordinator; (301) 563-6526; fax (301) 563-6595; rwilliams@obesity.org or grantsandawards@obesity.org
*Internet:* https://www.obesity.org/awards-and-grants/
*Sponsor:* Obesity Society
1110 Bonifant Street, Suite 500
Silver Spring, MD 20910

**Obesity Society Foster-Schauer International Travel Grants** **2026**
The Obesity Society's Foster-Schauer International Travel Grant is provided to non-U.S. early career or student TOS members who present high scoring oral or poster abstracts at ObesityWeek. The American Society for Metabolic and Bariatric Surgery and TOS each select one recipient. Members of the Obesity Society are encouraged to apply by sending in their Letters of Intent by the deadline of May 22, with a full proposal submission deadline of August 20.
*Requirements:* Members of the Obesity Society are encouraged to apply by sending in their Letters of Intent by the deadline.
*Geographic Focus:* All States
*Date(s) Application is Due:* Aug 20
*Amount of Grant:* Up to 2,000 USD
*Contact:* Regina Williams, Governance and Administrative Coordinator; (301) 563-6526; fax (301) 563-6595; rwilliams@obesity.org or grantsandawards@obesity.org

*Internet:* https://www.obesity.org/awards-and-grants/
*Sponsor:* Obesity Society
1110 Bonifant Street, Suite 500
Silver Spring, MD 20910

**Obesity Society Friends of Albert (Mickey) Stunkard Lifetime Achievement Award** **2027**
The Obesity Society's Friends of Albert (Mickey) Stunkard Lifetime Achievement Award recognizes individuals who, like Mickey Stunkard, have made an outstanding lifetime contribution to the field of obesity in terms of scholarship, mentorship and education. The award includes a plaque and a $1,000 prize and is presented during a plenary session. The recipient is invited to present the Friends of Albert Stunkard Lifetime Achievement Award Lecture.
*Requirements:* All award nominations must include a letter of recommendation (signed by two current members of The Obesity Society) stating: reasons for nominating the individual; the nominee's qualifications for the specific award; a summary of the nominee's contributions to the field of obesity; the nominee's complete CV; and other letters of support from individuals (these may be sent under a separate cover letter).
*Restrictions:* Self-nominations are not accepted. Nominees will not be considered for multiple awards in a single year.
*Geographic Focus:* All States
*Date(s) Application is Due:* Aug 20
*Amount of Grant:* 1,000 USD
*Contact:* Regina Williams, Governance and Administrative Coordinator; (301) 563-6526; fax (301) 563-6595; rwilliams@obesity.org or governance@obesity.org
*Internet:* https://www.obesity.org/awards-and-grants/
*Sponsor:* Obesity Society
1110 Bonifant Street, Suite 500
Silver Spring, MD 20910

**Obesity Society George A. Bray Founders Award** **2028**
The Obesity Society's George A. Bray Founders Award recognizes an individual for significant contributions advancing the scientific or clinical basis for understanding or treating obesity and for extensive involvement with The Obesity Society. The award includes a plaque and a $1,000 prize and is presented during a plenary session. The recipient is invited to present the George A. Bray Founders Award Lecture. The annual nomination deadline is August 20.
*Requirements:* All award nominations must include a letter of recommendation (signed by two current members of The Obesity Society) stating: reasons for nominating the individual; the nominee's qualifications for the specific award; a summary of the nominee's contributions to the field of obesity; the nominee's complete CV; and other letters of support from individuals (these may be sent under a separate cover letter).
*Restrictions:* Self-nominations are not accepted. Nominees will not be considered for multiple awards in a single year.
*Geographic Focus:* All States
*Date(s) Application is Due:* Aug 20
*Amount of Grant:* 1,000 USD
*Contact:* Regina Williams, Governance and Administrative Coordinator; (301) 563-6526; fax (301) 563-6595; rwilliams@obesity.org or governance@obesity.org
*Internet:* https://www.obesity.org/awards-and-grants/
*Sponsor:* Obesity Society
1110 Bonifant Street, Suite 500
Silver Spring, MD 20910

**Obesity Society George A. Bray Master's Thesis and Doctoral Dissertation Awards** **2029**
The Obesity Society's George A. Bray Master's Thesis and Doctoral Dissertation Awards recognize two students for successfully defending either their complete obesity-related master's thesis or doctoral dissertation in the most recent year (from April 1, 2017 to April 30, 2018). Each recipient receives $1,500 to cover travel and registration for the upcoming ObesityWeek plus a $500 prize. Winners are acknowledged at ObesityWeek during the George Bray Founders Award Lecture. The annual nomination deadline is August 20.
*Requirements:* All award nominations must include a letter of recommendation (signed by two current members of The Obesity Society) stating: reasons for nominating the individual; the nominee's qualifications for the specific award; a summary of the nominee's contributions to the field of obesity; the nominee's complete CV; and other letters of support from individuals (these may be sent under a separate cover letter).
*Restrictions:* Self-nominations are not accepted. Nominees will not be considered for multiple awards in a single year.
*Geographic Focus:* All States
*Date(s) Application is Due:* Aug 20
*Contact:* Regina Williams, Governance and Administrative Coordinator; (301) 563-6526; fax (301) 563-6595; rwilliams@obesity.org or governance@obesity.org
*Internet:* https://www.obesity.org/awards-and-grants/
*Sponsor:* Obesity Society
1110 Bonifant Street, Suite 500
Silver Spring, MD 20910

**Obesity Society Rolls-Simons Travel Grants** **2030**
The Rolls-Simons Travel Grant was established by TOS Past President Barbara Rolls after losing her mother, Pat Simons, to obesity-related disease. It provides eight TOS students or post-doctoral members with a travel grant of $1,000 to attend the upcoming ObesityWeek. Members of the Obesity Society are encouraged to apply by sending in their Letters of Intent by the deadline of May 22, with a full proposal submission deadline of August 20.
*Requirements:* Members of the Obesity Society are encouraged to apply by sending in their Letters of Intent by the deadline.
*Geographic Focus:* All States
*Date(s) Application is Due:* Aug 20
*Amount of Grant:* 1,000 USD
*Contact:* Regina Williams, Governance and Administrative Coordinator; (301) 563-6526; fax (301) 563-6595; rwilliams@obesity.org or grantsandawards@obesity.org
*Internet:* https://www.obesity.org/awards-and-grants/
*Sponsor:* Obesity Society
1110 Bonifant Street, Suite 500
Silver Spring, MD 20910

**Obesity Society Thomas A. Wadden Award for Distinguished Mentorship** **2031**
The Obesity Society's Thomas A. Wadden Award for Distinguished Mentorship honors a mid-career or senior TOS member for distinguished mentorship of early career Society members. The award is named for Professor Thomas Wadden, PhD, a long-time mentor of obesity researchers and clinical care professionals. The recipient receives a plaque and a $1,000 prize. The annual nomination deadline is August 20.
*Requirements:* All award nominations must include a letter of recommendation (signed by two current members of The Obesity Society) stating: reasons for nominating the individual; the nominee's qualifications for the specific award; a summary of the nominee's contributions to the field of obesity; the nominee's complete CV; and other letters of support from individuals (these may be sent under a separate cover letter).
*Restrictions:* Self-nominations are not accepted. Nominees will not be considered for multiple awards in a single year.
*Geographic Focus:* All States
*Date(s) Application is Due:* Aug 20
*Amount of Grant:* 1,000 USD
*Contact:* Regina Williams, Governance and Administrative Coordinator; (301) 563-6526; fax (301) 563-6595; rwilliams@obesity.org or governance@obesity.org
*Internet:* https://www.obesity.org/awards-and-grants/
*Sponsor:* Obesity Society
1110 Bonifant Street, Suite 500
Silver Spring, MD 20910

**Obesity Society TOS Scientific Achievement Award** **2032**
The Obesity Society's TOS Scientific Achievement Award recognizes excellence in an obesity-related research career and includes a $5,000 prize, reimbursement of travel expenses up to $3,000 to attend the upcoming ObesityWeek, and a plaque. The award is presented during a plenary session and the recipient is invited to present the TOS Scientific Achievement Award Lecture. The annual nomination deadline is August 20.
*Requirements:* All award nominations must include a letter of recommendation (signed by two current members of The Obesity Society) stating: reasons for nominating the individual; the nominee's qualifications for the specific award; a summary of the nominee's contributions to the field of obesity; the nominee's complete CV; and other letters of support from individuals (these may be sent under a separate cover letter).
*Restrictions:* Self-nominations are not accepted. Nominees will not be considered for multiple awards in a single year.
*Geographic Focus:* All States
*Date(s) Application is Due:* Aug 20
*Amount of Grant:* 5,000 - 8,000 USD
*Contact:* Regina Williams, Governance and Administrative Coordinator; (301) 563-6526; fax (301) 563-6595; rwilliams@obesity.org or governance@obesity.org
*Internet:* https://www.obesity.org/awards-and-grants/
*Sponsor:* Obesity Society
1110 Bonifant Street, Suite 500
Silver Spring, MD 20910

**Obesity Society Weight Watchers Karen Miller-Kovach Research Grants** **2033**
Weight Watchers and The Obesity Society have partnered to create the Karen Miller-Kovach Research Grant. The grant will focus on the development of digital tools that employ evidence-based behavioral weight management strategies. The purpose of the grant is to help investigators collect pilot data to successfully compete for larger extramural funding. One grant will be funded for $50,000 for a one-year period. Members of the Obesity Society are encouraged to apply by sending in their Letters of Intent by the deadline of May 22, with a full proposal submission deadline of August 20.
*Requirements:* Members of the Obesity Society are encouraged to apply by sending in their Letters of Intent by the deadline.
*Geographic Focus:* All States
*Date(s) Application is Due:* Aug 20
*Amount of Grant:* 50,000 USD
*Contact:* Regina Williams, Governance and Administrative Coordinator; (301) 563-6526; fax (301) 563-6595; rwilliams@obesity.org or grantsandawards@obesity.org
*Internet:* https://www.obesity.org/awards-and-grants/
*Sponsor:* Obesity Society
1110 Bonifant Street, Suite 500
Silver Spring, MD 20910

**OISE Muriel and Danny Fung Graduate Scholarships** **2034**
The purpose of this award is to give support to students enrolled in graduate studies in the Department of Sociology and Equity Studies in Education at OISE, University of Toronto. Priority consideration is given to students who are not eligible to receive the University of Toronto's Graduate Funding Package. Specifically, the Muriel and Danny

Fung Graduate Award will be awarded annually to a student who is registered full time in the second year or later year of their program and who's academic and community work promote racial equality and social justice.
*Restrictions:* Student must be registered full-time.
*Geographic Focus:* Canada
*Date(s) Application is Due:* Mar 15
*Contact:* Kathleen Gallagher, (416) 978-0160; fax (416) 926-4744; kathleen.gallagher@utoronto.ca
*Internet:* www.oise.utoronto.ca/sese/Students/Financial_Scholarships/
*Sponsor:* Ontario Institute for Studies in Education
252 Bloor Street West
Toronto, ON M5S 1V6 Canada

**OISE William Waters Scholarships in Urban Education** **2035**
Established in 2005, the Centre for Urban Schooling (CUS) is dedicated to improving the quality of the teaching and learning experiences available to children and youth living in under-served urban communities. The Centre for Urban Schooling conducts research, offers educational programs and professional development opportunities, and supports students, teachers and their communities to develop school reform initiatives with a focus on equity for all students. With more than twenty affiliated professors and researchers from OISE's departments, CU.S. is quickly gaining recognition as Canada's leading research and programming center in issues of urban education. OISE is offering four $30,000 scholarships to promote excellence in teaching in the urban classroom. The successful candidates will be experienced teachers entering a full-time Master's program with an interest in questions of social justice and school success for students from economically disadvantaged neighborhoods attending under-performing schools. In addition to the requirements set out by the graduate program, the successful scholarship recipients will be invited to join the Centre for Urban Schooling and participate in the wide variety of activities related to research, professional development, advocacy and teacher education promoted by the Center over the course of the year.
*Requirements:* Students who are enrolled or will be enrolled full-time in a master's program will be eligible to be considered for assistantships for the first two years of their program. Students who are enrolled or will be enrolled full-time in a doctoral program will be eligible to be considered for assistantships for the first five years of their program.
*Geographic Focus:* Canada
*Date(s) Application is Due:* Jan 16
*Amount of Grant:* 30,000 CAD
*Contact:* Kathleen Gallagher, (416) 978-0160; fax (416) 926-4744; kathleen.gallagher@utoronto.ca
*Internet:* www.oise.utoronto.ca/sese/Students/Financial_Scholarships/
*Sponsor:* Ontario Institute for Studies in Education
252 Bloor Street West
Toronto, ON M5S 1V6 Canada

**Olympus Corporation of Americas Corporate Giving Grants** **2036**
The Olympus Corporation of Americas (OCA) targets corporate giving initiatives designed to benefit and sustain society, making positive contributions. Globally, Olympus strives to play an integral role in helping people around the world to lead safer, healthier and more fulfilling lives. In addition to offering knowledge, expertise and world-renowned products, its employees go further to fulfill the corporate responsibility to its neighbors and to the planet. As a leading medical device company with roots in healthcare going back nearly 100 years, our corporate giving philosophy is aligned with industry guidance and best practices for healthcare companies. Ever cognizant of our duty to comply with laws governing interactions with health care professionals, we adhere to the charitable contributions requirements of the Advanced Medical Technology Association's (AdvaMed) Code of Ethics on Interactions with Health Care Professionals.
*Requirements:* Olympus identifies 501(c)3 charitable organizations* that support the above advancements toward building a sustainable society.
*Restrictions:* Olympus will not make contributions to (a) organizations that discriminate on the basis of race, creed, color, gender, or national origin or (b) individuals. Olympus will not make contributions to denominational religious or political organizations, or similar limited-purpose groups, unless the funds are for purposes that will benefit the community as a whole or further the interests of Olympus.
*Geographic Focus:* All States, Argentina, Bolivia, Brazil, Canada, Caribbean, Chile, Colombia, Costa Rica, Ecuador, El Salvador, Guatemala, Honduras, Mexico, Netherlands Antilles, Nicaragua, Panama, Paraguay, Peru, Uruguay, Venezuela
*Contact:* Manager of Corporate Giving; (484) 896-5000
*Internet:* www.olympusamerica.com/corporate/social_responsibility/community_involvement.asp
*Sponsor:* Olympus Corporation of Americas
3500 Corporate Parkway
Center Valley, PA 18034

**Olympus Corporation of Americas Fellowships** **2037**
The Olympus Fellows program is a full-time, benefits-eligible leadership development program designed for new college graduates who have strong desires to pursue careers with a company whose products and services can help save lives and capture life's most precious moments. It is designed to be educational and challenging while allowing Olympus Fellows to apply analytical and creative skills to make valuable contributions to the business. Based on prior studies and work experience, Olympus has Fellows opportunities in its Medical Systems, Imaging, Scientific Equipment or Corporate Groups. The Fellowship accepts all majors from any Bachelor's or Master's degree program. Graduate and current Fellows have a range of majors, including marketing, finance, engineering, psychology, biochemistry, international relations, supply chain management and information systems, economics, kinesiology, business administration and more. The amount of travel a Fellow can expect will depend on each rotational assignment. Although there is no set deadline, the program encourages students to apply as early in the academic year as possible. Once the next class of Fellows has been identified, the position will be removed from the job posting website until the beginning of the next academic year.
*Geographic Focus:* All States
*Contact:* Manager of Fellowships; (484) 896-5000
*Internet:* www.olympusamerica.com/corporate/recruiting/index.asp
*Sponsor:* Olympus Corporation of Americas
3500 Corporate Parkway
Center Valley, PA 18034

**ONR Maritime Sensing Grants** **2038**
The Maritime Sensing program conducts multidisciplinary science and technology development of acoustic and non-acoustic sensors systems as a means to automatically develop and maintain comprehensive tactical and strategic awareness in the maritime undersea battle space. These systems may be carried as on-board equipment for Navy surface ships, submarines, aircraft, and unmanned vehicles or may deployed from those platforms as mobile, drifting, or fixed ocean surveillance systems operating under autonomous or remote control. The intent of this program is to provide next-generation acoustic and non-acoustic sensing technology for the tactical, surveillance, and undersea warfare missions of the Navy. Project scopes may range from component-level research and development to system-level technology demonstrations. Submission of the Research Proposal should always be preceded by submission of a Preproposal and a discussion of that Preproposal with the assigned Program Officer. A planning letter must be received by close of business on May 23rd in order to receive full consideration.
*Requirements:* Grants are awarded to academic, industrial, and nonprofit organizations, with strong emphasis on university research.
*Restrictions:* New grants or contracts are typically started near the beginning of the fiscal year. Note that the goal is to issue awards early in fiscal year. Planning letters should be received by close of business June 28. New proposals for the next fiscal year should be submitted no later than September 30, unless otherwise directed by the BAA. Competitive renewals should be submitted no later than six months prior to the award termination date after contacting the Program Officer. Conference proposals should be submitted at least four months prior to the conference.
*Geographic Focus:* All States
*Date(s) Application is Due:* Sep 30
*Contact:* Mike Wardlaw, Team Leader; (703) 588-2427 or (703) 588-0615; fax (540) 653-5234; mike.wardlaw@navy.mil
*Internet:* www.onr.navy.mil/Science-Technology/Departments/Code-32/All-Programs/Ocean-Systems-321/Maritime-Sensing.aspx
*Sponsor:* U.S. Department of Defense
One Liberty Center, 875 N. Randolph Street, Suite 1425
Arlington, VA 22203-1995

**Ontario Arts Council Arts Service Project Grants** **2039**
The Ontario Arts Council Arts Service Project Grant program supports collaborative projects, led by Ontario-based ad hoc arts groups/arts collectives or arts organizations that develop or provide unique services for professional arts communities in Ontario. The priorities are to support services developed or delivered for organizations that specifically serve any of OAC's priority groups. Eligible activities include: research; professional development, which may include grant writing, management and human resources, marketing and networking, and other needed skills; equity training services; health and wellness services; employment and fee standards; advocacy initiatives; and other services that promote the purpose of the program. The annual deadline for application submissions is April 10, with notifications approximately four months after the deadline. The maximum award amount is $10,000.
*Requirements:* Eligible applicants include: ad hoc groups/collectives of professional artists or arts professionals with a permanent physical address in Ontario; Ontario-based not-for-profit arts organizations; and organizations must serve professional artists, arts professionals and/or arts organizations eligible to apply to another OAC program.
*Restrictions:* Municipalities, colleges, and universities are ineligible to apply. The program does not fund; ongoing services delivered by any organization funded by an OAC operating program; capital expenses; or projects that serve the applicant themselves. The activities for which funding is requested: cannot start before the deadline; cannot finish before an applicant receives grant results; and must be completed no more than two years after the grant results.
*Geographic Focus:* Canada
*Date(s) Application is Due:* Apr 10
*Amount of Grant:* Up to 10,000 CAD
*Contact:* Nina Charest, Program Administrator; (416) 969-7417 or (800) 387-0058, ext. 7417; ncharest@arts.on.ca
*Internet:* www.arts.on.ca/grants/arts-service-projects
*Sponsor:* Ontario Arts Council
121 Bloor Street East, 7th Floor
Toronto, ON M4W 3M5 Canada

**Ontario Arts Council Chalmers Arts Fellowships** **2040**
The Ontario Arts Council Chalmers Arts Fellowship program provides financial support directly to artists so they may have the opportunity to dedicate themselves to individual creative pursuits. The program aims to provide such support at moments in

artists' careers when a concentration on personal and/or artistic growth or renewal and exploration is most likely to have the greatest impact on their long-term artistic and career development. Fellowships may be awarded to artists at any stage of their career development, from early to mature, and to artists working in a wide range of aesthetics and traditions that reflect Ontario's regional activity, linguistic and cultural diversity and Aboriginal and Franco-Ontarian identity. Artists practicing professionally for ten years or less can receive a maximum of $25,000, while artists practicing professionally for eleven years or more can receive a maximum of $50,000. The annual deadline for application submissions is June 19, with notifications after November 1.
*Requirements:* This program is open to professional Ontario-based artists. In addition, in order to apply to the Chalmers Arts Fellowships program you must: be an individual working in any of the disciplines supported at OAC including crafts, dance, integrated arts, literature, media arts, multi-disciplinary arts, music, theatre or visual arts; have creative and administrative control, as well as overall responsibility for the direction of your project; have had an active professional practice for at least 3 years prior to the application deadline; and have had at least 2 works and/or engagements that include a contract and/or payment (exhibition fee, commission fee, royalties, performance fee, salary, etc.) for the exhibition, distribution, performance, presentation, publishing, etc. of your work. Artistic/music directors of performing arts organizations, arts managers, publishers and curators who are seeking support for a project that may be of benefit to their organization must have no less than 5 years of experience (the grant is awarded to the individual not the organization or performing arts collective/ensemble).
*Restrictions:* Ineligible applicants include: co-applicants, groups, ensembles or collectives; not-for-profit or for-profit producers of creative content; students, including graduate and post graduate (full/part-time), regardless of the program of study; and colleges, universities and municipalities. The fund will not accept applications from artists whose reports for previously awarded OAC or Chalmers program grants are overdue or incomplete.
*Geographic Focus:* Canada
*Date(s) Application is Due:* Jun 19
*Amount of Grant:* Up to 50,000 CAD
*Contact:* Myles Warren, Awards Officer; (416) 969-7422 or (800) 387-0058, ext. 7422; mwarren@arts.on.ca
Rohana Buzamlak, Program Administrator; (416) 969-7416 or (800) 387-0058, ext. 7416; rbuzamlak@arts.on.ca
*Internet:* www.arts.on.ca/grants/chalmers-arts-fellowships
*Sponsor:* Ontario Arts Council
121 Bloor Street East, 7th Floor
Toronto, ON M4W 3M5 Canada

**Ontario Arts Council Craft Project Collaborations Grants** **2041**
The Ontario Arts Council Craft Project Collaborations Grants program provides funding up to $15,000 to help cover the costs of research, experimentation, creation, production and career development. The program's priorities are to support: a range of craft-based practices including work that is functional, conceptual, sculptural or from an art or design perspective; projects that preserve craft traditions or explore new directions; Indigenous artists working with craft materials in contemporary or customary ways; activities that contribute to arts education, public participation and community involvement in craft in Ontario; craft artists at all stages of their careers; and craft artists to create individually or to work with others collaboratively, either within the craft field or in other disciplines. Generally, the program will fund: research activities; creation and production; and development activities. The annual application submission deadline is October 17, with notifications in the middle of the following February.
*Requirements:* The program is open to applications from ad hoc groups, collectives, or collaborative groups with a permanent physical address in Ontario.
*Restrictions:* Ineligible applicants in this category include: undergraduate or graduate students enrolled full-time at an educational institution; curators conducting research or exhibition activities they have been contracted or employed to do for an organization that receives OAC operating grants in the Visual Arts section; public galleries and artist-run centers receiving an operating grant in the Visual Arts section; municipalities, colleges and/or universities; an incorporated for-profit business; or anyone who has received a Chalmers Professional Development grant and not completed and reported on the project. The program will not fund: capital expenses and purchase of equipment; studio and/or business set-up costs; studio renovation, repair or relocation; commissions and student projects; tuition for full-time study; or publication of artist's catalogue.
*Geographic Focus:* Canada
*Date(s) Application is Due:* Oct 17
*Amount of Grant:* Up to 15,000 CAD
*Contact:* Caroline Cotter, Program Administrator; (416) 969-7455 or (800) 387-0058, ext. 7455; ccotter@arts.on.ca
*Internet:* www.arts.on.ca/grants/craft-projects
*Sponsor:* Ontario Arts Council
121 Bloor Street East, 7th Floor
Toronto, ON M4W 3M5 Canada

**Ontario Arts Council Craft Project Grants for Emerging Artists and Curators** **2042**
The Ontario Arts Council Craft Project Grants for Emerging Artists and Curators program provides funding up to $5,000 to help cover the costs of research, experimentation, creation, production and career development. The program's priorities are to support: a range of craft-based practices including work that is functional, conceptual, sculptural or from an art or design perspective; projects that preserve craft traditions or explore new directions; Indigenous artists working with craft materials in contemporary or customary ways; activities that contribute to arts education, public participation and community involvement in craft in Ontario; craft artists at all stages of their careers; and craft artists to create individually or to work with others collaboratively, either within the craft field or in other disciplines. Generally, the program will fund: research activities; creation and production; and development activities. The annual application submission deadline is October 17, with notifications in the middle of the following February.
*Requirements:* The program is open to applications from artists and curators who are Ontario residents and have a permanent physical address in Ontario.
*Restrictions:* Ineligible applicants in this category include: undergraduate or graduate students enrolled full-time at an educational institution; curators conducting research or exhibition activities they have been contracted or employed to do for an organization that receives OAC operating grants in the Visual Arts section; or anyone who has received a Chalmers Professional Development grant and not completed and reported on the project. The program will not fund: capital expenses and purchase of equipment; studio and/or business set-up costs; studio renovation, repair or relocation; commissions and student projects; tuition for full-time study; or publication of artist's catalogue.
*Geographic Focus:* Canada
*Date(s) Application is Due:* Oct 17
*Amount of Grant:* Up to 5,000 CAD
*Contact:* Caroline Cotter, Program Administrator; (416) 969-7455 or (800) 387-0058, ext. 7455; ccotter@arts.on.ca
*Internet:* www.arts.on.ca/grants/craft-projects
*Sponsor:* Ontario Arts Council
121 Bloor Street East, 7th Floor
Toronto, ON M4W 3M5 Canada

**Ontario Arts Council Craft Projects Grants for Mid-Career and Established Artists and Curators** **2043**
The Ontario Arts Council Craft Projects Grants for Mid-Career and Established Artists and Curators program provides funding up to $15,000 to help cover the costs of research, experimentation, creation, production and career development. The program's priorities are to support: a range of craft-based practices including work that is functional, conceptual, sculptural or from an art or design perspective; projects that preserve craft traditions or explore new directions; Indigenous artists working with craft materials in contemporary or customary ways; activities that contribute to arts education, public participation and community involvement in craft in Ontario; craft artists at all stages of their careers; and craft artists to create individually or to work with others collaboratively, either within the craft field or in other disciplines. Generally, the program will fund: research activities; creation and production; and development activities. The annual application submission deadline is October 17, with notifications in the middle of the following February.
*Requirements:* The program is open to applications from artists and curators who are Ontario residents and have a permanent physical address in Ontario.
*Restrictions:* Ineligible applicants in this category include: undergraduate or graduate students enrolled full-time at an educational institution; curators conducting research or exhibition activities they have been contracted or employed to do for an organization that receives OAC operating grants in the Visual Arts section; or anyone who has received a Chalmers Professional Development grant and not completed and reported on the project. The program will not fund: capital expenses and purchase of equipment; studio and/or business set-up costs; studio renovation, repair or relocation; commissions and student projects; tuition for full-time study; or publication of artist's catalogue.
*Geographic Focus:* Canada
*Date(s) Application is Due:* Oct 17
*Amount of Grant:* Up to 15,000 CAD
*Contact:* Caroline Cotter, Program Administrator; (416) 969-7455 or (800) 387-0058, ext. 7455; ccotter@arts.on.ca
*Internet:* www.arts.on.ca/grants/craft-projects
*Sponsor:* Ontario Arts Council
121 Bloor Street East, 7th Floor
Toronto, ON M4W 3M5 Canada

**Ontario Arts Council Deaf and Disability Arts Projects Creation Grants** **2044**
The Ontario Arts Council Deaf and Disability Arts Projects Grants program supports Ontario-based artists and arts professionals who are deaf or have a disability. It also supports ad hoc groups, collectives, and arts organizations mandated to serve and led by artists or arts professionals who are deaf or have a disability. It funds all contemporary and traditional art practices typically supported by the Council. In the category of Creation, funding helps to cover the costs of research and development, exploration and experimentation, and creation of new work. Eligible expenses include: work space rental; equipment rental; living expenses (maximum of $2,000 per month); research materials; artist's fees; transportation and travel; accommodation; and meal allowances. The maximum award in the Creation category is $10,000. Applications are available approximately three months before the annual submission deadline of October 11, with notification of successful awards approximately four months after the deadline.
*Requirements:* Eligible applicants include; professional artists and arts professionals; individuals who are deaf or have a disability; and Ontario residents who have a permanent physical address in Ontario. Applicants can also be: ad hoc groups, collectives, or not-for-profit disability arts organizations that are led by deaf artists or artists with disabilities; ability to demonstrate that the artistic process is led by artists or arts professionals who are deaf or have a disability; and a permanent physical address in Ontario.
*Restrictions:* The Deaf and Disability Arts Projects program does not support: fundraising activities; student projects at all education levels; materials and supplies for workshop activities; capital expenses and equipment purchases; studio renovation, repair or relocation; accredited college or university programs; tuition for full-time study; or business, promotional and marketing expenses. Production expenses are ineligible unless the production element is essential to the development of the work being created.

*Geographic Focus:* Canada
*Date(s) Application is Due:* Oct 11
*Amount of Grant:* Up to 10,000 CAD
*Contact:* Bushra Junaid, Program Officer; (416) 969-7407 or (800) 387-0058, ext. 7407; bjunaid@arts.on.ca
Habibah Haque, Program Administrator; (416) 961-1660, ext. 5081 or (800) 387-0058, ext. 5081; hhaque@arts.on.ca
*Internet:* www.arts.on.ca/grants/deaf-and-disability-arts-projects
*Sponsor:* Ontario Arts Council
121 Bloor Street East, 7th Floor
Toronto, ON M4W 3M5 Canada

**Ontario Arts Council Indigenous Arts Project Grants** **2045**
The Ontario Arts Council Indigenous Arts Project Grants program supports Indigenous (First Nations, Metis, Inuit) artists to research, develop and create new work and engage with their communities, as well as opportunities for Indigenous organizations and communities to strengthen their relationships with the arts. There are two categories: individuals and ad hoc groups, collectives, or organizations. Grants help cover the costs of: research and development that will lead to the creation of new work; creation and production of new work in any art form (e.g. film/video, visual arts, new media, cultural arts, dance, theater, music, pottery, fiction, spoken word and storytelling, etc.); presentations, exhibitions, publications and festivals in which the primary purpose of the project is the presentation, promotion and distribution of art made by Indigenous artists; and connections and partnerships that bring artists together through conferences, workshops, gatherings, delegations, exchanges or other activities. Overall, the program will fund: research and development activities; creation and production of new work; presentation and distribution of art by Indigenous artists; exhibitions; publications; individual and series presentations; and festivals and innovative audience-development projects. The maximum award for an individual is $15,000, while the maximum for ad hoc groups, collectives, or organizations is $30,000. There are two annual deadlines for application submission: March 22 and September 19.
*Requirements:* Eligible applicants can include: professional artists who self-identify as First Nations, Metis, or Inuit, and are Ontario residents; ad hoc groups or collectives, comprised of individuals, of whom at least 50 per cent identify as Indigenous; and not-for-profit Indigenous organizations, centers, and councils. Applicants must have a permanent physical address in Ontario.
*Restrictions:* Ineligible applicants include: students; municipalities; and colleges and/ or universities. The program will not fund: fundraising activities; or capital expenses.
*Geographic Focus:* Canada
*Date(s) Application is Due:* Mar 22; Sep 19
*Amount of Grant:* Up to 30,000 USD
*Contact:* Kateri Gauthier, Program Administrator; (416) 969-7424 or (800) 387-0058, ext. 7424; kgauthier@arts.on.ca
*Internet:* www.arts.on.ca/grants/indigenous-arts-projects
*Sponsor:* Ontario Arts Council
121 Bloor Street East, 7th Floor
Toronto, ON M4W 3M5 Canada

**Ontario Arts Council Indigenous Culture Fund Cultural Research, Documentation, Tools, and Resources Small Grants** **2046**
The Ontario Arts Council Indigenous Culture Fund Cultural Research, Documentation, Tools, and Resources Small Grants program supports status and non-status First Nations, Inuit and Metis community, culture and way-of-life projects in Ontario. These grants are for short-term projects, with applications reviewed with a quick turnaround time. The priorities of this program are to: support the vitality and diversity of First Nations, Inuit and Metis expression and ways of life; create opportunities for Indigenous people to continue community building; support community-based activities that foster well-being and resilience for Indigenous peoples of all ages; support Indigenous peoples to engage in and share cultural practices, knowledge and learning; and support Indigenous language revitalization. Examples of projects that this category might support include: developing a protocols tool-kit that will support archaeological digs in the community; or researching and documenting stories about the local community and land to share with future generations. The program has an open-ended deadline, meaning applications are accepted at any time.
*Requirements:* Ontario-based Indigenous organizations, groups and individuals are eligible. Eligible types of applicant groups include: status and non-status First Nations, Inuit, and Metis individuals; Indigenous ad hoc groups, collectives, and youth councils; and First Nations, Inuit, or Metis organizations, such as incorporated not-for-profit organizations, community centers, cultural centers, friendship centers, libraries, schools, health, social services organizations, band councils, tribal councils, Metis community councils, and provincial organizations. First-time applicants, youth and Elders are encouraged to apply.
*Restrictions:* Ineligible applicants include: non-Indigenous groups, organizations, schools, colleges, universities, and individuals; municipalities; students seeking support for projects related to their student work; and for-profit enterprises.
*Geographic Focus:* Canada
*Contact:* Kateri Gauthier, Program Administrator; (416) 969-7424 or (800) 387-0058, ext. 7424; kgauthier@arts.on.ca
Erika Iserhoff, Indigenous Culture Fund Grants Facilitator; (647) 258-5079 or (800) 387-0058, ext. 5079; eiserhoff@arts.on.ca
*Internet:* www.arts.on.ca/grants/icf-small-grants
*Sponsor:* Ontario Arts Council
121 Bloor Street East, 7th Floor
Toronto, ON M4W 3M5 Canada

**Open Spaces Sacred Places National Awards** **2047**
The initiative offers planning grant support for cross-disciplinary teams to develop proposals for integrated design and research studies to understand the impact of contemplative greenspace on urban dwellers. The awards will directly support the design and implementation of significant new Open Spaces Sacred Places sites designed to address critical human factors related to the physical, mental, and spiritual well-being of individuals living in challenging urban environments. The awards will simultaneously sponsor research on the human impact of encountering and experiencing these open, sacred places in nature in the urban setting.
*Requirements:* Eligible grantees include any 501(c)3 tax-exempt charitable organizations that are classified as public charities and not as private foundations. Such public charities include: a. Institutions that serve a large number of constituents, such as universities, schools, hospitals, medical research organizations, religious institutions, and governmental units and agencies as more specifically described in Section 170 (b)(1)(A)(i-vi) of the Internal Revenue Code (IRC); b. Publicly supported organizations that receive more than one-third of their revenues from modest donors and/or fees and sales of services and goods performed in conducting their charitable activities for a large number of persons as more specifically described in Section 509 (a)(2) of the IRC; c. Organizations created to support and benefit one or more other public charities as more specifically described in Section 509 (a)3 of the IRC. Eligible teams must be cross-disciplinary in makeup and must include an identified Firesoul or Firesouls. Eligible projects must have both an institutional and a community-based partner.
*Geographic Focus:* All States
*Date(s) Application is Due:* Jun 30; Sep 1
*Amount of Grant:* Up to 50,000 USD
*Contact:* Christine Tanabe, Director of Communications & Program Development; (410) 268-1376; fax (410) 268-1379; info@tkffdn.org
*Internet:* www.opensacred.org/grants/request-for-proposal
*Sponsor:* TKF Foundation
410 Severn Avenue, Suite 216
Annapolis, MD 21403

**Oppenstein Brothers Foundation Grants** **2048**
Grants are awarded in the metropolitan area of Kansas, City, Missouri, primarily for social services and early childhood, elementary, secondary, adult-basic, vocational, and higher education; family planning and services; social services and welfare agencies; Jewish welfare organizations; and programs for youth, the handicapped, disadvantaged, mentally ill, homeless, minorities, and elderly. Additional areas of interest include arts/ cultural programs, the performing arts, museums, health care and health organizations, and AIDS research. The foundation considers requests for building and renovation, capital campaigns, curriculum development, emergency funds, equipment, general operating support, program development, seed money, technical support, conferences and seminars, consulting services, and matching funds. Application guidelines are available on request. The board meets every other month. Deadlines are generally three weeks prior to board meetings. Notification of award will take place within two to four months.
*Requirements:* Nonprofit organizations serving the metropolitan area of Kansas City, Missouri are eligible to apply.
*Restrictions:* The foundation primarily supports 501(c)3 nonprofit organizations serving the metropolitan area of Kansas City, Missouri. Grants are not awarded to support individuals or for annual campaigns.
*Geographic Focus:* Missouri
*Amount of Grant:* 5,000 - 15,000 USD
*Samples:* Associated Youth Services, Kansas City, Kansas, $10,000; Child Advocacy Services Center, Kansas City, Missouri, $25,000; Spay and Neuter Kansas City, Kansas City, Missouri, $5,000.
*Contact:* Beth Radtke, Program Officer; (816) 234-2577
*Sponsor:* Oppenstein Brothers Foundation
922 Walnut Street, Suite 200
Kansas City, MO 64106-1809

**Oregon Arts Commission Individual Artist Fellowships** **2049**
The Oregon Arts Commission's Individual Artist Fellowship Program honors Oregon's professional artists and their artistic achievements and supports their efforts to advance their careers. In selecting artists to receive Fellowships, the Commission looks to Oregon artists of outstanding talent, demonstrated ability and commitment to the creation of new work(s). Fellowships are non-matching awards of $3,000. There is no pre-determined number of Fellowships to be awarded in each discipline. The annual deadline for applications is October 31.
*Requirements:* Professional Oregon artists working in any media (with the exception of film/video) may apply. Fellowship applicants must be Oregon residents for one year prior to the application deadline date to be eligible.
*Restrictions:* Full-time students, nonprofessionals who produce art as a hobby, and previous recipients of an individual artist fellowship are ineligible.
*Geographic Focus:* Oregon
*Date(s) Application is Due:* Oct 31
*Amount of Grant:* 3,000 USD
*Contact:* Meagan Atiyeh, Visual Arts/Public Art Coordinator; (503) 986-0084 or (503) 986-0082; fax (503) 986-0260; meagan.atiyeh@oregon.gov or oregon.artscomm@state.or.us
*Internet:* www.oregonartscommission.org/grants/individual-artist-fellowship
*Sponsor:* Oregon Arts Commission
775 Summer Street NE, Suite 200
Salem, OR 97301-1280

**Oregon Arts Commission Media Arts Fellowship** 2050
The Oregon Media Arts Fellowships provide financial and other support for production expenses to enable Oregon film or video artists to create or complete new work or works in progress. The Fellowships recognize and assist media artists whose work shows exceptional promise and who have demonstrated a commitment to film or video art. The program is funded by the Oregon Arts Commission and administered by the Northwest Film Center. Completed applications must be postmarked or hand?delivered to the Northwest Film Center by 5 p.m., January 1. Oregon Media Arts Fellowships range up to $5,000.
*Requirements:* The program is open to both emerging and established individual film and video artists. Applicants must be current Oregon residents and have been an Oregon resident for one full year prior to the application deadline date. ?Only non?commercial, non?instructional, non?promotional projects for which the individual applicant has overall control and primary creative responsibility will be considered. Only one application per individual is permissible and only one application per project.
*Restrictions:* Applications for script development and fundraising clips will not be accepted. Organizations, full?time students and non?professionals who produce art as a hobby are not eligible. Media Arts Fellowships will not fund: capital construction (i.e. studio renovations); tuition assistance for college, university, or other formal courses of study; work that has already been completed; reimbursement of fees already incurred; or promotion of completed work.
*Geographic Focus:* Oregon
*Date(s) Application is Due:* Jan 1
*Amount of Grant:* Up to 5,000 USD
*Contact:* Meagan Atiyeh, Visual Arts/Public Art Coordinator; (503) 986-0084 or (503) 986-0082; fax (503) 986-0260; meagan.atiyeh@oregon.gov
Ben Popp, Filmmaker Services Manager; (503) 221-1156, ext. 13; ben@nwfilm.org
*Internet:* www.oregonartscommission.org/grants/oregon-media-arts-fellowship
*Sponsor:* Oregon Arts Commission
775 Summer Street NE, Suite 200
Salem, OR 97301-1280

**Orentreich Foundation for the Advancement of Science Grants** 2051
Norman Orentreich founded the Orentreich Foundation for the Advancement of Science in 1961. Early investigations sprang naturally from his clinical work as a dermatologist as well as from his interest in the aging and wound healing studies of Alexis Carrell, a pioneer in the field of aging research. In time, OFAS was dedicated to biomedical research that prevents, halts, or reverses those disorders that decrease the quality or length of life. Occasional and limited grants are made for joint or collaborative biomedical research, usually initiated by OFAS, with emphasis in dermatology, aging, endocrinology, and serum markers for human diseases. The amount of the grants varies with joint interest and need. Unsolicited applications may be submitted at any time, since there are no specified deadlines.
*Requirements:* Applicants must be at or above postgraduate level in science or medicine at accredited universities and research institutions in the United States. There is no restriction on citizenship. Submit an outline of proposed program, curriculum vita of principal investigator, and estimated budget.
*Geographic Focus:* All States
*Amount of Grant:* Up to 50,000 USD
*Contact:* Norman Orentreich; (845) 265-4200; fax (845) 265-4210; ofas@orentreich.org
*Internet:* www.orentreich.org/research
*Sponsor:* Orentreich Foundation for the Advancement of Science
Cold Spring-on-Hudson
New York, NY 10516

**Organization for Tropical Studies (OTS) Fellowships** 2052
Organization of Tropical Studies (OTS) fellowships are intended to assist with thesis-related research in tropical biology and similar fields and are available in amounts up to a maximum of $3,650. Among the available fellowships are the following named awards: the William L. Brown Fellowship; the David and Deborah Clark Fellowship; the Rexford Daubenmire Fellowship; the Dole Food Fellowship; the Emily P. Foster Fellowship; the Lillian and Murray Slatkin Fellowship; the Christiane and Christopher Tyson Fellowship (for U.S. and Latin-American graduate students); the GlaxoWellcome Endowment (for Latin-American graduate students only); the Peace Frogs Fellowships (for herpetological research); the Rowe Family Fellowships (for U.S. and Latin-American graduate students); and Water Quality Environmental Outreach. More information about these awards is available from the OTS fellowships webpage. Awards are intended primarily to cover up to fifty days of station fees (food, lodging, and station access) at OTS field stations; current rates are $63 for overnight visits and $31.50 for day visits. Other expenses that may be included in the budget, to a maximum of $500, are research- and export-permit fees, local travel, and reimbursement in whole or in part of one round-trip airline ticket. OTS has a single application process for most of its fellowships; generally applicants do not need to specify the award for which they are applying (with a few exceptions mentioned at the website). Application guidelines and a downloadable cover sheet are available from the OTS Fellowships webpage. The cover sheet and supporting documents must be submitted via email by the deadline date to the Education Director. Notification of decisions will take place about two months after each deadline. OTS is a nonprofit consortium of 64 universities and research institutions in the United States, Costa Rica, Peru, Canada, Mexico, South Africa, and Australia. Founded in 1963, OTS has a mission to provide leadership in education, research and the responsible use of natural resources in the tropics. To this end, graduate and undergraduate field courses are offered in the pure and applied areas of the natural sciences in Costa Rica, Peru, Mexico, the Brazilian Amazon, and South Africa. In addition, OTS operates three research stations in varied ecosystems in Costa Rica: La Selva in the Caribbean lowland rain forest, Palo Verde in the Pacific lowland dry forest, and Las Cruces in a mid-elevation rain forest on Costa Rica's southern Pacific slope. OTS also coordinates and facilitates research, provides environmental-education programs, and works with local agencies on issues of conservation, natural-resource management and land-use practices.
*Requirements:* OTS fellowships are open to graduate students enrolled in degree programs at OTS member institutions and to OTS course alumni. (A list of member institutions is available at the OTS website from the About-Us Quick Link.) Applicants interested in the Water Quality Environmental Outreach program must contact Dr. Catherine Pringle, University of Georgia, cpringle@uga.edu, before submitting an application.
*Restrictions:* Awards will be restricted to OTS field stations (La Selva, Palo Verde, and Las Cruces); proposals for research at other sites will not be considered. Only one research award may be held at a time. Awards do not cover capital equipment; minor amounts of materials and supplies may be considered, but are best covered by other sources of funds. Salaries or stipends for the investigator are not allowed, but salaries for field assistants will be considered if the work cannot be carried out safely or efficiently by one person.
*Geographic Focus:* All States, All Countries
*Date(s) Application is Due:* Mar 1; Oct 1
*Amount of Grant:* Up to 3,650 USD
*Contact:* Education Director; (506) 2524-0607; fax (506) 2524-0608; academic@ots.ac.cr
*Internet:* ots.ac.cr/index.php?option=com_content&task=view&id=170&Itemid=344
*Sponsor:* Organization for Tropical Studies
Duke University, 410 Swift Avenue
Durham, NC 27705

**Organization for Tropical Studies Pilot Awards for Exploratory Research** 2053
Organization for Tropical Studies (OTS) Pilot Awards for Exploratory Research are intended to assist with thesis-related research in tropical biology and similar fields and are available in amounts up to a maximum of $2,075. Awards are intended primarily to cover up to 25 days of station fees (food, lodging and station access) at OTS field stations; current rates are $63 for overnight visits and $31.50 for day visits. Other expenses that may be included in the budget, to a maximum of $500, are research- and export-permit fees, local travel, and reimbursement in whole or in part of one round-trip airline ticket. Application guidelines and a downloadable cover sheet are available from the OTS Fellowships webpage. The cover sheet and supporting documents must be submitted via email by the deadline date to the Education Director. Notification of decisions will take place about two months after each deadline. The Organization for Tropical Studies (OTS) is a nonprofit consortium of 64 universities and research institutions in the United States, Costa Rica, Peru, Canada, Mexico, South Africa, and Australia. Founded in 1963, OTS has a mission to provide leadership in education, research and the responsible use of natural resources in the tropics. To this end, graduate and undergraduate field courses are offered in the pure and applied areas of the natural sciences in Costa Rica, Peru, Mexico, the Brazilian Amazon, and South Africa. In addition, OTS operates three research stations in varied ecosystems in Costa Rica: La Selva in the Caribbean lowland rain forest, Palo Verde in the Pacific lowland dry forest, and Las Cruces in a mid-elevation rain forest on Costa Rica's southern Pacific slope. OTS also coordinates and facilitates research, provides environmental-education programs, and works with local agencies on issues of conservation, natural-resource management and land-use practices.
*Requirements:* Organization for Tropical Studies (OTS) fellowships are open to graduate students enrolled in degree programs at OTS member institutions and to OTS course alumni. (A list of member institutions is available at the OTS website from the About-Us Quick Link.)
*Restrictions:* Only one research award may be held at a time. Awards do not cover capital equipment; minor amounts of materials and supplies may be considered, but are best covered by other sources of funds. Salaries or stipends for the investigator are not allowed, but salaries for field assistants will be considered if the work cannot be carried out safely or efficiently by one person.
*Geographic Focus:* All States, All Countries
*Date(s) Application is Due:* Mar 1; Oct 1
*Amount of Grant:* Up to 2,075 USD
*Contact:* Education Director; (506) 2524-0607; fax (506) 2524-0608; academic@ots.ac.cr
*Internet:* ots.ac.cr/index.php?option=com_content&task=view&id=170&Itemid=344
*Sponsor:* Organization for Tropical Studies
Duke University, 410 Swift Avenue
Durham, NC 27705

**Orthopaedic Trauma Association Kathy Cramer Young Clinician Memorial Fellowships** 2054
The AAOS/OREF/ORS Clinician Scholar Career Development Program (CSCDP) is an annual program seeking applicants in their PGY2-PGY 5 residency years, in fellowships, and Junior Faculty through year three who have the potential/desire to become orthopaedic clinician scientists. The program supports two scholarships for the New Investigators Workshop, as well as two scholarships for the annual Clinician Scholar Career Development program.
*Requirements:* Kathy Cramer Young Clinician Scholarship Award applicants for the AAOS/OREF/ORS Clinician Scholar Career Development Program must be an OTA member in North America ideally in their PGY2-PGY5 residency years, in fellowships and junior faculty through year three who have the potential/desire to become orthopaedic clinician scientists.
*Geographic Focus:* All States
*Contact:* Melanie Hopkins, Fellowship Coordinator; (847) 698-1631; fax (847) 823-0536; hopkins@aaos.org or OTA@aaos.org
*Internet:* ota.org/research/research-studies/
*Sponsor:* Orthopaedic Trauma Association
9400 W. Higgins Road, Suite 305
Rosemont, IL 60018-4975

**Orthopaedic Trauma Association Research Grants** **2055**

The Orthopaedic Trauma Association is accepting pre-proposals for clinical research applications from OTA International Members residing and conducting research. Depending on the year, directed research grants will be given priority evaluation by the research committee. There is no guarantee that these directed grant applications (topics chosen by the OTA Board of Directors) will receive funding, as they will be graded on scientific merit compared to all other grants received that year. The following grants are available for any research issue related to musculoskeletal trauma: Clinical Research Grants ($40,000 per year maximum, for 2 year grant cycle); Basic Research Grants ($50,000 maximum, for 2 year grant cycle); and Resident Research Grant ($20,000 maximum). Full proposal submission deadlines fall during the first two weeks of June.
*Geographic Focus:* All States
*Date(s) Application is Due:* Jun 8
*Amount of Grant:* Up to 50,000 USD
*Contact:* Melanie Hopkins, Fellowship Coordinator; (847) 698-1631; fax (847) 823-0536; hopkins@aaos.org or OTA@aaos.org
*Internet:* ota.org/research/research-studies/
*Sponsor:* Orthopaedic Trauma Association
9400 W. Higgins Road, Suite 305
Rosemont, IL 60018-4975

**OSF Alumni Scholarships** **2056**

The Alumni Scholarship program offers support to Open Society Scholarship alumni to further expand knowledge gained and make a positive contribution to their home country. The competition is open only to Open Society Scholarship Programs alumni of the following programs: Afghan Communications Scholarship Program; American University in Bulgaria; Asia Pacific Scholarship Consortium; Cambridge and Oxford Hospitality Schemes; Civil Service Awards-Moldova; CNOUS-OSI Program; DAAD-OSI Program; Environmental Law/Management Program; Faculty Development Program; Georgian Program for Education Professionals; Global Supplementary Grant Program; Mongolian Professional Fellowship Program; Muskie/FSA Graduate Fellowship Program; OSF Scholarship in European Studies (Maastricht and Aarhus Universities); Palestinian Faculty Development Program; Palestinian Rule of Law Program; Social Work Fellowship Program; Soros Supplementary Grant Program; Supplementary Grant Program Asia (formerly the Supplementary Grant Program–Burma); Supplementary Grant Program – Former Yugoslavia; Undergraduate Exchange Program/Virtual University; and the United Kingdom Scholarship Programs (formerly the Chevening –OSI Scholarships). The application and additional guidelines are available at the website.
*Requirements:* Alumni must be in compliance with the conditions of any past scholarship or Open Society Foundation grant; reside in their home country or region during the project year; and present a project related to or applicable to the home country or region.
*Restrictions:* The program will not support: individual travel grants for participation in conferences, workshops, classes; travel to or from countries not served by the Foundations; expenses taking place outside countries served by the Foundations; additional degrees, courses, or certificates for individual alumnus/a; projects which replicate existing efforts within the country (i.e. local Soros Foundation projects); educational advising centers, or resource centers, for study abroad; or purchasing of office equipment (computers, photocopiers, fax machines).
*Geographic Focus:* All States, All Countries
*Date(s) Application is Due:* May 31
*Contact:* Zarina Usmanova, Program Manager; +44-207-031-0200; fax +44-207-031-0201; zarina.usmanova@opensocietyfoundations.org
*Internet:* www.opensocietyfoundations.org/grants/alumni-grant-program
*Sponsor:* Open Society Foundations
224 West 57th Street
New York, NY 10019

**OSF Aryeh Neier Justice Initiative Fellowships** **2057**

The Aryeh Neier Fellowship Program is a two-year program of practical work experience designed to expand the capacity of lawyers and advocates working internationally on human rights issues. Up to four fellowships will be awarded. During the first year, each Fellow will devote substantial time to one project, while working on other initiatives as they arise involving litigation, research/documentation, advocacy, and/or pilot initiatives designed to enhance the accessibility, fairness, and effectiveness of justice services. During the first year, each Fellow will be based full-time at one of the Justice Initiative's offices in Budapest, London, or New York. During the second year, each Fellow will be based full-time with a local partner organization in Africa, Asia, Europe, or Latin America. Significant attention will be given to ensuring that second-year placements relate to the issues and initiatives that the Fellows worked on during their first year. Several factors will be considered in arranging second-year placements, including the Fellow's preference, visa requirements, budget feasibility, and the organization's ability to support the Fellow's work. Nongovernmental, not-for-profit organizations are preferred, but public, independent offices (such as a government anti-corruption commission) may be considered if a compelling case is made for the potential impact of such a placement. The deadline for application submission is April 24.
*Requirements:* All applicants meeting the criteria are encouraged to apply. The OSF is particularly interested in applicants with experience and interest in the following substantive issues, which reflect current priority areas for the Justice Initiative: Accountability, Liberty and Transparency, focusing on grand corruption, national security and human rights, and freedom of information and expression; Criminal Justice, focusing on accountability for grave crimes and more balanced and effective justice for ordinary crimes, in both national and international systems; Equality & Inclusion, focusing on discrimination, the right to citizenship, and the development and implementation of regional regimes of free movement (eg. the EU, ECOWAS, MERSOCUR); and Rule of Law, focusing on (but not limited to) work to strengthen regional and sub-regional human rights institutions and UN bodies; implementation of rights-based judgments issued by courts and quasi-judicial bodies; legal remedies for economic injustices; and community-based justice services.
*Geographic Focus:* All Countries
*Date(s) Application is Due:* Apr 24
*Contact:* Robert Varenik, Director of Programs; (212) 548-0600; fax (212) 548-4662; robert.varenik@opensocietyfoundations.org or recruitment@opensocietyfoundations.org
*Internet:* https://www.opensocietyfoundations.org/grants/aryeh-neier-open-society-justice-initiative-fellowships-20150121
*Sponsor:* Open Society Foundations
224 West 57th Street
New York, NY 10019

**OSF Baltimore Community Fellowships** **2058**

The Baltimore Community Fellowships program works to meet the challenges facing Baltimore's most needy communities by identifying and supporting social innovators of unusual promise and providing them with the ingredients to ensure their ideas have a stable foundation and long-term viability. The Fellowships program seeks dynamic activists and social entrepreneurs interested in implementing projects that address problems in underserved communities in Baltimore City. Funding of $60,000 is available over eighteen months. The application is available at the OSF-Baltimore website, and candidates may be placed on an application mailing list. The annual application deadline is March 7.
*Requirements:* Fellowships are for initiatives that take place in Baltimore City. Applicants may come from any field, including—but not limited to—business, management, the arts, law, medicine, education, architecture, and engineering. Individuals from underserved communities and people of color are strongly encouraged to apply. Applicants need not be from Baltimore City, but should be knowledgeable about social and economic justice issues affecting Baltimore's communities. Fellows must be willing to participate fully in meetings scheduled for the Community Fellows.
*Restrictions:* Individuals already receiving wages or a salary for the proposed project from a host organization are not eligible. Fellows may not use stipends to supplant funding for activities or projects that the host organization is already implementing. The program does not award fellowships to conduct research or to implement lobbying initiatives.
*Geographic Focus:* All States
*Date(s) Application is Due:* Mar 7
*Amount of Grant:* 60,000 USD
*Contact:* Emily Smith, Baltimore Fellowships and Initiatives; (410) 234-1091; fax (410) 234-2816; emily.smith@opensocietyfoundations.org
*Internet:* www.opensocietyfoundations.org/grants/baltimore-community-fellowships
*Sponsor:* Open Society Foundations
201 North Charles Street, Suite 1300
Baltimore, MD 21201

**OSF Drugs, Security and Democracy Fellowship** **2059**

The Drugs, Security and Democracy (DSD) fellowship is designed to support short-term research that contributes to the literature on drugs in Latin America and the Caribbean on topics and countries that are central to drug policy discussions in the region and beyond. The competition is open to applicants conducting research in Latin America or the Caribbean who are fully embedded in and committed to the region, and whose research focus has a clear and central connection to the field of drugs and to formulating sound drug policy. Successful applicants will be those whose work and interests best match, and who demonstrate a long-term commitment to, these program goals. In addition to conducting individual research, DSD fellows should contribute to the development of a global interdisciplinary network of researchers engaged with drug policy and communicate their findings to relevant audiences over the course of their careers. The annual deadline for applications is March 2.
*Requirements:* Applications are welcome from midcareer and senior researchers/scholars conducting research in Latin America and the Caribbean that addresses issues with a clear and central connection to the field of drugs and to formulating sound drug policy. Eligible applicants must: be fully embedded in and committed to the Latin American and Caribbean region; hold a terminal degree in their field of study or clearly demonstrate equivalent research experience related to the field of drugs, with at least a bachelor's degree in any discipline; focus on one of the topics indicated in the section above; and if proposing to conduct research in a nonnative language, provide evidence of sufficient language proficiency to carry out the project. DSD-funded research must address the primary theme of drugs in Latin America or the Caribbean. Proposals must demonstrate the potential for the research to contribute to a sound and credible knowledge base for informed advocacy and decision-making for drug policy. For the current fellowship competition, applications must address one of the following topics: drug policy and/or legal reform, including different depenalization, decriminalization, legalization, and regulation approaches as well as country-specific obstacles to reform; marijuana, including legalization for medical use; impact of drug laws on prison systems, including costs associated with pretrial detention for drugs; the dynamics and relationships between legal pharmaceutical drug markets and illicit drug production, including barriers to access and incentives/disincentives for producers; drug policy and the peace process in Colombia; analysis of institutional resource distribution between criminal and public health approaches to drug use; or drug economy and its dynamics.
*Geographic Focus:* All States
*Date(s) Application is Due:* Mar 2
*Contact:* Anna Meg Rama, Program Associate; (718) 517-3656 or (212) 377-2700; fax (212) 377-2727; rama@ssrc.org or info@ssrc.org

*Internet:* www.ssrc.org/fellowships/dsd-fellowship/
*Sponsor:* Open Society Foundations
224 West 57th Street
New York, NY 10019

**OSF Eurasia Program Grants** **2060**

The Eurasia Program works to deepen civil society, uphold human rights, and maintain space for independent debate and discussion in Eastern Europe (Moldova, Ukraine), the South Caucasus (Armenia, Azerbaijan, Georgia) and Central Asia (Kazakhstan, Kyrgyzstan, Tajikistan, Turkmenistan, Uzbekistan). The program is advanced through grant making, advocacy and research, regional initiatives, and close engagement with local Open Society foundations. The OSF believes that a vital civil society is fundamental to the promotion of democratic governance and government accountability. Throughout the region, the program works with a wide range of partners from all sectors of civil society, with a particular focus on the following: promoting human rights; combating corruption; strengthening independent, alternative, and social media; securing access to justice for marginalized groups; creating and maintaining platforms for debate and civic engagement; and integrating local actors into the global exchange of ideas.
*Requirements:* Applicants may submit the organization's name and description of the project, list of countries it will benefit, amount requested, and project period. If the funding inquiry meets program funding criteria, the Foundation will contact you to request a full proposal.
*Geographic Focus:* Georgia, Armenia, Azerbaijan, Kazakhstan, Kyrgyzstan, Moldova, Tajikistan, Turkmenistan, Ukraine, Uzbekistan
*Contact:* Dana Geraghty, Senior Program Operations Specialist; (212) 548-0600; fax (212) 548-4600; dana.geraghty@opensocietyfoundations.org
*Internet:* https://www.opensocietyfoundations.org/about/programs/eurasia-program
*Sponsor:* Open Society Foundations
224 West 57th Street
New York, NY 10019

**OSF Fellowships** **2061**

Open Society Fellowships support outstanding individuals from around the world pursuing innovative and unconventional approaches to fundamental open society challenges. The Fellowship enables innovative professionals - including journalists, activists, academics, and practitioners - to work on projects that inspire meaningful public debate, shape public policy, and generate intellectual ferment within the Open Society Foundations. Funding is open to considering a broad range of work products, including: publications (e.g. books, articles, blogs); creative public education projects including online media, pre-production research and post-production outreach for documentary films; and efforts to seed new campaigns or organizations. Stipends range from $80,000 to $100,000. Applicants are required to submit a one- to two-page, single-spaced letter of inquiry that outlines the topic of the project, proposed work product, and relevance. The deadline to submit letters of inquiry on these statements is March 1, with responses to applicants by March 24.
*Requirements:* The Open Society Fellowship accepts proposals from anywhere in the world. Applicants should possess a deep understanding of their chosen subject and a track record of professional accomplishment. Past and current fellows have included journalists, activists, academics, and practitioners in a variety of fields. Successful applicants will be eager to exploit the many resources offered by the Open Society Foundations and be prepared to engage constructively with our global network. Ideal fellows are specialists who can see beyond the basics of their field, and possess the tenacity to complete a project of exceptional merit. Proficiency in spoken English is required. Applicants who are uncertain whether their topic fits within the foundations' focus areas are invited to submit a brief letter of inquiry, accompanied by a CV, before proceeding with the online application process.
*Restrictions:* The fellowship does not fund enrollment for degree or non-degree study at academic institutions, including dissertation research.
*Geographic Focus:* All States, All Countries
*Amount of Grant:* 80,000 - 100,000 USD
*Contact:* Leonard Benardo, Program Coordinator; (212) 548-0600; fax (646) 557-2488; leonard.bernardo@opensocietyfoundations.org
*Internet:* https://www.opensocietyfoundations.org/grants/open-society-fellowship
*Sponsor:* Open Society Foundations
224 West 57th Street
New York, NY 10019

**OSF Hungary Project Grants** **2062**

The Open Society Foundations have been actively supporting Hungarian civil society for the last 30 years. The Hungary Project, part of the Rights and Liberties team of the Open Society Initiative for Europe, is an indirect successor to the national foundation that was transformed into the Hungary Project and operated out of the Budapest office a decade ago. Five years ago, the project staff also directed the emergency assistance that George Soros awarded to the region in the wake of the economic crisis. As such, the Hungary Project has historically been a backbone to support watchdogs and civil society organizations. The overall goal of the Hungary Project is to build broader and deeper constituencies for open society in Hungary. The project doesn't only work with traditional Open Society partners but tries to widen the constituencies for open society values and reach out to new target groups.
*Requirements:* The Hungary Project uses different grant-making methods (open calls, solicited proposals, unsolicited proposals) to select its grantees. In the case of unsolicited proposals the project accepts proposals on a rolling basis throughout the year. Prior to sending a full proposal, potential applicants should submit a brief concept paper of no more than two pages to determine whether projects meet the Hungary Project's current funding priorities and guidelines. The concept paper should include the following: a brief description of the project, planned activities, methodology, and goals; information about the applicant organization; and estimated overall budget and timeframe for the project.
*Geographic Focus:* All States, Hungary
*Contact:* Peter Nizak, Program Manager; +36-1-882-3100; fax +36-1-882-3101; peter.nizak@opensocietyfoundations.org
Eszter Sz?cs, Program Coordinator; +36-1-882-3100; fax +36-1-882-3101; eszter.szucs@opensocietyfoundations.org
*Internet:* https://www.opensocietyfoundations.org/grants/hungary-project-20150610
*Sponsor:* Open Society Foundations
Oktober 6 utca 12
Budapest, H-1051 Hungary

**OSF Internships for Rights and Governance** **2063**

The Open Society Internship for Rights and Governance is a project of the Open Society Foundations that launched in the summer of 2013 in partnership with the School of Public Policy at Central European University. It is designed to inspire a new cohort of practitioners committed to working both in the public interest and at the forefront of global policy. The highly competitive program allows a limited number of students from top public policy schools to immerse themselves in the ideas and practice of open society through a clinical seminar held in Budapest at the School of Public Policy at Central European University, followed by an 8- to 12-week intensive internship at a policy- and rights-oriented nongovernmental organization selected for its outstanding work. Applications are accepted on an ongoing basis.
*Requirements:* Applications will only be accepted from the following universities: Bogaziçi University Department of Political Science and International Relations; Harvard Kennedy School of Government; Jindal School of Government and Public Policy; Koç University Graduate School of Social Sciences and Humanities; Sabanci University Faculty of Arts and Social Sciences; School of Public Policy at Central European University; School of Public Policy and Management of Tsinghua University; Sciences Po Paris School of International Affairs; University of the Andes Alberto Lleras Camargo School of Government; University of Oxford Blavatnik School of Government; University of Pretoria Centre for Human Rights; and University of Toronto Munk School of Global Affairs.
*Geographic Focus:* All States, All Countries
*Contact:* Anne Denes, Program Coordinator; (212) 548-0600; fax (212) 548-4600; anne.denes@opensocietyfoundation.org or osirg@opensocietyfoundations.org
*Internet:* www.opensocietyfoundations.org/grants/open-society-internship-rights-and-governance
*Sponsor:* Open Society Foundations
224 West 57th Street
New York, NY 10019

**OSF Leadership in Government Fellowships** **2064**

The Open Society Foundations announce an opportunity for senior-level government officials and staff in the United States who have recently left public service. Open Society's U.S. Programs will award individual grants to people who have played a significant role in advancing social change from within government in the United States at the local, state, or federal levels. The Foundations are looking for public-sector leaders interested in using this fellowship opportunity to reflect on their time in government and to undertake a project that advances policy ideas and inserts their unique perspectives about driving change from within government into the public dialogue. The fellowship is a grant-making program and is part of a larger effort within the Open Society Foundations to invest in new voices that inform and challenge our collective thinking about advancing open society in the United States. Fellows will receive funding (in the general range of $100,000 to $133,000) depending on work experience, seniority, and percentage of time committed to the project, which will range from 12 to 18 months in duration. The application deadline is July 15.
*Requirements:* All projects must relate to one or more of U.S. Programs' central goals: a more inclusive and accountable American democracy; a fair criminal justice system; full political, economic, and civic participation of communities of color and immigrants; and equitable economic growth. Projects can focus on a particular issue or cut across these broad areas of interest.
*Geographic Focus:* All States
*Date(s) Application is Due:* Jul 15
*Amount of Grant:* 100,000 - 133,000 USD
*Contact:* Leonard Benardo, Program Coordinator; (212) 548-0600; fax (212) 548-4679; leonard.benardo@opensocietyfoundations.org or USPfellowships@opensocietyfoundations.org
*Internet:* https://www.opensocietyfoundations.org/grants/leadership-government-fellows-program-20160525
*Sponsor:* Open Society Foundations
224 West 57th Street
New York, NY 10019

**OSF Leadership in Government Fellowships** **2065**

The Open Society Foundations announce an opportunity for senior-level government officials and staff in the United States who have recently left public service. Open Society's U.S. Programs will award individual grants to people who have played a significant role in advancing social change from within government in the United States at the local, state, or federal levels. The Foundations are looking for public-sector leaders interested in using this fellowship opportunity to reflect on their time in government and to undertake a project that advances policy ideas and inserts their unique perspectives about driving change from within government into the public dialogue. The fellowship is a grant-

making program and is part of a larger effort within the Open Society Foundations to invest in new voices that inform and challenge our collective thinking about advancing open society in the United States. Fellows will receive funding (in the general range of $100,000 to $133,000) depending on work experience, seniority, and percentage of time committed to the project, which will range from 12 to 18 months in duration. The application deadline is July 15.
*Requirements:* All projects must relate to one or more of U.S. Programs' central goals: a more inclusive and accountable American democracy; a fair criminal justice system; full political, economic, and civic participation of communities of color and immigrants; and equitable economic growth. Projects can focus on a particular issue or cut across these broad areas of interest.
*Geographic Focus:* All States
*Date(s) Application is Due:* Jul 15
*Amount of Grant:* 100,000 - 133,000 USD
*Contact:* Leonard Benardo, Program Coordinator; (212) 548-0600; fax (212) 548-4679; leonard.benardo@opensocietyfoundations.org or USPfellowships@opensocietyfoundations.org
*Internet:* https://www.opensocietyfoundations.org/grants/leadership-government-fellows-program-20160525
*Sponsor:* Open Society Foundations
224 West 57th Street
New York, NY 10019

**OSF OECD Thomas J. Alexander Fellowships** **2066**

Fellows will have the opportunity to join one of the world's leading teams in this field at the Organisation for Economic Co-operation and Development (OECD) Headquarters to conduct research and analysis of the extensive OECD educational dataset (from PISA, TALIS, PIAAC, etc.), which is as renowned as it is unique in scale, scope and value. The unique emphasis of the Fellowships on quality and equity in education, particularly in emerging economies, makes them an opportunity to strengthen capacity for quantitative research and analysis on ideas that are crucial for open society. The research will enable an informed response to the growing need for evidence in educational policy and practice. Fellows will work at OECD headquarters in Paris with the team of analysts. They will be expected to spend at least three months and up to a maximum of 12 months in Paris. Fellows are expected to produce a publishable research paper or report at the end of their fellowship.
*Requirements:* Alexander Fellows will typically come from research institutions, governmental agencies, research centers, ministries, or professional organizations. Ideally they will a hold a Ph.D., have experience with quantitative research methods and large-scale surveys, and have had articles published in scientific journals. The program encourages applicants from a variety of disciplines. Proposals focusing on improving equity in education are particularly encouraged. Applicants should explain how their work would benefit the broad public and promote quality and equity in their countries' education systems. A call for proposals will be issued twice a year. Only online applications will be accepted. Applications will be considered by a group of experts in the field of international education assessment. Candidates will be selected on the basis of the originality, potential, and technical quality of the proposals as well as on the qualifications and experience of the applicant and the value added by their time working at OECD headquarters in Paris.
*Geographic Focus:* All States, All Countries
*Date(s) Application is Due:* Dec 21
*Contact:* Program Director; +33 1 45 24 82 00; fax +33 1 45 24 85 00
*Internet:* www.oecd.org/edu/thomasjalexanderfellowship.htm
*Sponsor:* Open Society Foundations
224 West 57th Street
New York, NY 10019

**OSF Palestinian Rule of Law Awards** **2067**

The Palestinian Rule of Law (PROL) Program provides for graduate-level studies in law leading to an LLM (Masters of Laws) degree. The PROL program is predicated upon a belief that the institutionalization of the rule of law will be a key component of political, economic, and social development in the Palestinian territories and underlie any lasting, peaceful resolution of the Palestinian-Israeli conflict. It is a goal of the PROL Program that Fellows, upon their return home, will contribute to the economic and democratic growth of the Palestinian territories and serve as leaders in their fields of specialization. The Program places participants at host law schools in the U.S. and the Central European University in Budapest, an OSF-affiliated institution providing American-style teaching and U.S.-accredited qualifications. Before beginning their degree programs, Fellows attend the intensive OSF Summer School in Istanbul which includes legal writing instruction. During the academic year, Fellow conferences are held to increase the group's cohesiveness as a networked cohort. Fellows who successfully complete the one-year LLM degree can use program funding to undertake an internship/work placement at a nonprofit or nongovernmental institution that furthers their practical training in specialized fields. The Fellowship can be in the U.S., Occupied Palestinian Territories, or elsewhere. However, a primary goal of the Program is that Fellows use their newly acquired skills towards institutionalizing the rule of law and the values of an open society upon their return home. Applications are due no later than February 17.
*Requirements:* Applicants must be: lawyers or law graduates and residents in the Gaza Strip or West Bank, including East Jerusalem; proficient in spoken and written English and Arabic; able to demonstrate professional aptitude and leadership potential; able to begin study in the United States or Hungary in the fall; able to receive and maintain U.S. and/or Hungarian study visas; able to receive and maintain a visa for one-month attendance at a pre-academic summer course in July; able to obtain travel documents to leave and return to the West Bank/Gaza; and committed to returning to the West Bank or Gaza after completion of the program to serve as leaders in their fields of specialization.
*Restrictions:* The Scholarship program does not provide: funding for applicants who are not normally resident in the West Bank or Gaza; funding for study in fields other than law; funding for dependent family members; funding for individuals receiving other OSF grants during the scholarship period; or funding for employees of OSF directly involved in the administration of scholarship programs.
*Geographic Focus:* Egypt, Israel, Jordan, Palestinian Territory
*Date(s) Application is Due:* Feb 17
*Contact:* Joseph Glicksberg, Senior Program Manager; (212) 548-0600; fax (212) 548-4679; joseph.glicksberg@opensocietyfoundations.org or prol@infoscholar.org
*Internet:* https://www.opensocietyfoundations.org/grants/palestinian-rule-law-awards-20161214
*Sponsor:* Open Society Foundations
224 West 57th Street
New York, NY 10019

**OSF Presidential Fellowships** **2068**

The Open Society Presidential Fellowship is awarded yearly to recent JD, LLM, MPA, MPP, and MBA graduates from accredited law, public policy, and business schools. Based in New York City, Fellows pursue work related to human rights, good governance, and justice through an 11-month residence within the Office of the President at the Open Society Foundations. Fellows will work with staff, grantees, and other fellows on special initiatives that address the legal, business, policy, and organizational issues confronted by the Foundations' various programs and projects. They will gain exposure and connections to the Open Society Foundations' leadership and partner organizations, and cultivate knowledge about open society strategies. The term of the Open Society Presidential Fellowship is 11 months. Fellows will receive a salary of $65,000 plus benefits. The annual deadline is December 22.
*Requirements:* The Open Society Presidential Fellowships will be awarded to three (3) students graduating annually from an accredited U.S. law, public policy, or business school.
*Geographic Focus:* All States
*Date(s) Application is Due:* Dec 22
*Amount of Grant:* 65,000 USD
*Contact:* Anna Segelman, Grants and Program Coordinator; (212) 548-0600; fax (212) 548-4600; anna.segelman@osi.hu or humanresources@opensocietyfoundations.org
*Internet:* https://www.opensocietyfoundations.org/grants/open-society-presidential-fellowship
*Sponsor:* Open Society Foundations
224 West 57th Street
New York, NY 10019

**OSF Soros Justice Advocacy Fellowships** **2069**

The Soros Justice Fellowships fund outstanding individuals to implement innovative projects that advance reform and spur debate on a range of issues facing the U.S. criminal justice system. The Fellowships Program is part of a larger effort within the Open Society Foundations' Criminal Justice Fund to reduce the destructive impact of current criminal justice policies on the lives of individuals, families, and communities in the United State by challenging the overreliance on incarceration and harsh punishment, and ensuring a fair and equitable system of justice. Advocacy Fellowships fund outstanding individuals—including lawyers, advocates, grassroots organizers, activist academics, and others with important perspectives—to initiate innovative policy advocacy projects at the local, state, and national levels that will have a measurable impact on one or more of the Open Society Foundations' U.S. criminal justice priorities. Projects may range from litigation to public education to coalition-building to grassroots mobilization to policy-driven research. Advocacy Fellowships are 18 months in duration, may be implemented in conjunction with a host organization, and Fellows are expected to make their projects their full-time work during the term of the Fellowship. The Institute strongly encourages applications for projects that demonstrate a clear understanding of the intersection of criminal justice issues with the particular needs of low-income communities, communities of color, immigrants, gay, lesbian, bisexual and transgender people, women and children, and those disproportionately affected by harsh criminal justice policies; as well as applications for projects that cut across various criminal justice fields and related sectors, such as education, health and mental health, housing and employment. Those with personal experience with these issues are also encouraged to apply. Additional guidelines, along with the application, are available at the website. The annual deadline for application submission is October 12.
*Requirements:* Fellowship projects should address one or more of the following Criminal Justice Fund priorities: expose the excessive and economically destructive costs of incarceration; promote just and effective sentencing practices; foster policies and practices that reduce prison and corrections populations; abolish the privatization of prisons, detention centers, and correctional supervision; combat the criminalization of marginalized populations; eliminate unreasonable barriers to the reintegration of people returning from prison, as well as challenge the stigmatization of people with criminal records; end the prosecution, sentencing and incarceration of children as adults; abolish the death penalty; promote the civic engagement and leadership of people with criminal records; eliminate the use of long-term solitary confinement in prison; challenge harsh immigration enforcement and detention policies and practices; promote new approaches to drug policy; reform discriminatory and abusive police and prosecution practices; improve indigent defense services and systems; and reduce unnecessary pretrial detention.
*Restrictions:* The Fellowships Program does not fund: enrollment for degree or non-degree study at academic institutions, including dissertation research; projects that address criminal justice issues outside the U.S. (applicants themselves, however, can be based outside the U.S., provided their work directly pertains to a U.S. issue); past recipients of a Soros Justice Fellowship; or lobbying activities.

*Geographic Focus:* All States
*Date(s) Application is Due:* Oct 22
*Amount of Grant:* 58,700 - 110,250 USD
*Contact:* Adam Culbreath, Program Coordinator; (212) 548-0600; fax (212) 548-46; adam.culbreath@opensocietyfoundation or sorosjusticefellowships@opensocietyfoundations.org
*Internet:* https://www.opensocietyfoundations.org/grants/soros-justice-fellowships
*Sponsor:* Open Society Foundations
224 West 57th Street
New York, NY 10019

**OSF Soros Justice Media Fellowships** **2070**

The Soros Justice Fellowships fund outstanding individuals to implement innovative projects that advance reform and spur debate on a range of issues facing the U.S. criminal justice system. The Fellowships are part of a larger effort within the Open Society Foundations' Criminal Justice Fund to reduce the destructive impact of current criminal justice policies on the lives of individuals, families, and communities in the U.S. by challenging the overreliance on incarceration and harsh punishment, and ensuring a fair and equitable system of justice. The Soros Media Fellowships support writers, print and broadcast journalists, bloggers, filmmakers, and other individuals with distinct voices proposing to complete media projects for local, regional and national markets that engage the public and spur debate on one or more of the Open Society Foundations' U.S. criminal justice priorities. The Fellowships aim to mitigate the time, space, and market constraints that often discourage individuals from pursuing important but marginalized, controversial or unpopular issues in a comprehensive manner. Media Fellowships are 12 months in duration, and Fellows are expected to make their projects their full-time work during the term of the Fellowship. Open Society strongly encourages applications for projects that demonstrate a clear understanding of the intersection of criminal justice issues with the particular needs of low-income communities, communities of color, immigrants, gay, lesbian, bisexual and transgender people, women and children, and those disproportionately affected by harsh criminal justice policies; as well as applications for projects that cut across various criminal justice fields and related sectors, such as education, health and mental health, housing and employment. The Institute also encourages applications from individuals directly affected by, or with significant direct personal experience with, the policies, practices and systems their projects seek to address. Individuals can apply for one of two awards, depending on the applicant's level of experience. Advocacy Track I comes with an award of $85,750 over 18 months. Advocacy Track II comes with an award of $110,250 over 18 months. Additional guidelines and the application are available at the website. The annual deadline for applications is October 12.
*Requirements:* Applicants who are uncertain whether their proposed project fits with the Fellowship guidelines may submit an email inquiry before submitting the full application. All Fellowship projects should address one or more of the following Criminal Justice Fund priorities: expose the excessive and economically destructive costs of incarceration; promote just and effective sentencing practices; foster policies and practices that reduce prison and corrections populations; abolish the privatization of prisons, detention centers, and correctional supervision; combat the criminalization of marginalized populations; eliminate unreasonable barriers to the reintegration of people returning from prison, as well as challenge the stigmatization of people with criminal records; end the prosecution, sentencing and incarceration of children as adults; abolish the death penalty; promote the civic engagement and leadership of people with criminal records; eliminate the use of long-term solitary confinement in prison; challenge harsh immigration enforcement and detention policies and practices; promote new approaches to drug policy; reform discriminatory and abusive police and prosecution practices; improve indigent defense services and systems; and reduce unnecessary pretrial detention.
*Restrictions:* Funding is not available for enrollment for degree or non-degree study at academic institutions, including dissertation research. Projects that address criminal justice issues outside the U.S. are not eligible, but applicants can be based outside the U.S., if their work pertains to a U.S. issue. Funding is also not available for past recipients of the Social Justice Fellowship, or lobbying activities.
*Geographic Focus:* All States
*Date(s) Application is Due:* Oct 12
*Amount of Grant:* 85,750 - 110,250 USD
*Contact:* Adam Culbreath, Program Manager; (212) 548-0600; fax (212) 548-4600; adam.culbreath@opensocietyfoundation or sorosjusticefellowships@opensocietyfoundations.org
*Internet:* www.opensocietyfoundations.org/grants/soros-justice-fellowships
*Sponsor:* Open Society Foundations
224 West 57th Street
New York, NY 10019

**OSF Think Tank Fund Fellowships** **2071**

Open Society Foundations Organizational Development Grants to Think Tanks aim to strengthen democratic processes by: identifying political, economic, and social problems; researching them in a nonpartisan and scientifically rigorous manner; and providing policy alternatives that enrich public debate. Think tanks are vital not only because they advocate specific remedies, but also because these remedies are informed by carefully accumulated and analyzed data, making them harder to ignore than emotional normative appeals or isolated instances of abuse. The majority of donors operating in the field of policy research provide project funding for analytical materials on specific topics. As a result, both public policy centers and their funders focus almost exclusively on particular policy outputs and their immediate advocacy at the expense of pursuing deeper institutional development of think tanks. This creates a niche for the Think Tank Fund to assist the overall development of grantees in three specific areas: quality of research products, communications and advocacy capacity, and internal development and governance. Our mission is to contribute to building strong think tanks that produce high-quality policy analysis, deliver it to a wide range of audiences to galvanize public interest and debate, and operate as sustainable and internally robust nonprofit organizations. Organizational development grants aim to provide support for improvement in the three specific areas of think tank activity and operation mentioned above. An organizational development grant can cover from one to all three areas for a period between one to two years based on an applicant's demands, merit of the proposal, and the time needed for implementing necessary changes. While applications are no longer accepted on a rolling basis, organizations that would like to have their concepts reviewed should send their concept papers no later than March 18.
*Requirements:* As part of the Open Society Initiative for Europe, the Think Tank Fund for Wider Europe requests proposals for organizational development grants from independent, multi-issue think tanks working in Albania, Bosnia and Herzegovina, Georgia, Kosovo, Macedonia, Moldova, Montenegro, Serbia, and Ukraine. To be considered for an organizational development grant, an organization must meet the following criteria: be a legally registered independent organization in one of the nine countries listed above; be operational for no less than 24 months; be a multi-issue think tank as defined in the complete guidelines section; embrace an inclusive approach to policymaking, which implies making analysis publicly available and targeting a variety of public stakeholders; undertake thorough and rigorous research, and develop thematic expertise, to inform public debates in a responsible and accurate manner and to position itself as a credible interlocutor for the government and other public actors; and be engaged in a broader process of Europeanization in these countries through policy research and advocacy.
*Restrictions:* The Fund does not support: grantees who have received six years of Think Tank Fund core and institutional funding; core funding that does not target one of the organizational development components listed above; thematic project funding or stand-alone policy research; academic research; assistance to public institutions; academic fellowships and scholarships; one-off events (e.g., workshops, seminars, conferences, publications); research in arts, humanities, and natural sciences; or technical assistance projects.
*Geographic Focus:* Georgia, Albania, Bosnia & Herzegovina, Kosovo, Macedonia, Moldova, Montenegro, Serbia, Ukraine
*Date(s) Application is Due:* Mar 18
*Contact:* Jordi Vaquer, Regional Director, Europe; +36 1 882 3171 or +34 93 159 38 38; fax +36 1 882 3101; jordi.vaquer@opensocietyfoundations.org or ttf.applications@opensocietyfoundations.org
*Internet:* https://www.opensocietyfoundations.org/about/programs/think-tank-fund
*Sponsor:* Open Society Foundations
C/Elisabets, 24
Barcelona, 08001 Spain

**OSF Youth Fellowships** **2072**

Launched in 2015, the Open Society Youth Fellowship supports young activists and organizers as thought leaders and allies in crafting solutions and new youth-focused approaches to open society challenges. The fellowship funds work that supports and enriches understanding of youth-led approaches to change, and inspires conversations within the Open Society Foundations and in the world. Some fellows are activists tackling critical open society issues in the regions where they live; others work on projects dedicated to creating content-rich material—like toolkits or studies—that more broadly supports youth work. They may produce reports or blogs, develop innovative education or training resources, launch new campaigns or organizations, or create works of artistic or cultural expression. In all cases, fellows take advantage of the considerable intellectual and logistical resources of the Open Society Foundations and contribute meaningfully to the Foundations' thinking in return.
*Geographic Focus:* All States
*Date(s) Application is Due:* Nov 1
*Contact:* Anna Segelman, Grants and Program Coordinator; (212) 548-0600; fax (212) 548-4600; anna.segelman@osi.hu
*Internet:* https://www.opensocietyfoundations.org/about/programs/open-society-youth-fellowship
*Sponsor:* Open Society Foundations
224 West 57th Street
New York, NY 10019

**Osteosynthesis and Trauma Care Foundation European Visiting Fellowships** **2073**

The Osteosynthesis and Trauma Care Foundation offers Visiting Fellowships in the field of trauma care and orthopaedic surgery. They receive a Travel Grant to visit one of the Host Centers participating in the program. Eligible are surgeons in training or practice in Europe. Objectives of the program are to: enhance professional skills through continuous education; provide experience on latest surgical intervention approaches; and stimulate inter-country exchange between surgeons in Europe. The Visiting Fellowship allows for knowledge acquisition in a variety of fields such as: surgical as well as non-surgical management of musculoskeletal injuries; clinical and radiological musculoskeletal evaluation of the injured patient; algorithmic decision-making with respect to timing and sequencing of multiple injury management; pathophysiology of the multiple-injured patient, including the prophylaxis and/or treatment of complications of musculoskeletal injuries; understanding of advanced technology and instrumentation in treating trauma patients; and ethical, economic, administrative and legal issues as they pertain to orthopaedic trauma care.
*Geographic Focus:* All States
*Contact:* Lee-Ann Finno, Administrator; (212) 305-6392; education@otcfoundation.org
*Internet:* www.otcfoundation.org/fellowships/european-visiting-fellowships-program/
*Sponsor:* Osteosynthesis and Trauma Care Foundation
622 West 168th Street, 11th Floor
New York, NY 10032

**OTS Francis J. Bossuyt Fellowship** 2074

Organization of Tropical Studies (OTS) fellowships in general assist with thesis-related research in tropical biology and similar fields and are available in amounts up to a maximum of $3,650. Awards are intended primarily to cover fees (food, lodging, and station access) at OTS field stations. Other expenses that may be included in the budget, to a maximum of $500, are research- and export-permit fees, local travel, and reimbursement in whole or in part of one round-trip airline ticket. OTS's Bossuyt fellowship was created as a memorial to the life of Francis Bossuyt (1969-2000). At the time of his untimely death, Francis had been collecting his final year of data for his dissertation on Titi monkeys at the Cocha Cashu Biological Station in Manu National Park in Peru. In recognizing Francis' love for the Titi Monkeys and for field-oriented research, the Bossuyt family established the endowed fellowship for the following priority areas: research on animal behavior (preference for Callicebus) in Cocha Cashu and the surrounding areas; research on other subjects in Cocha Cashu; research by Peruvian-students on tropical ecosystems in Peru; and research at other locations in the tropics. Application guidelines and a downloadable cover sheet are available from the OTS Fellowship webpage. The cover sheet and supporting documents must be submitted via email by the deadline date to the Education Director. Notification of decisions will take place about two months after each deadline. The Organization for Tropical Studies (OTS) is a nonprofit consortium of 64 universities and research institutions in the United States, Costa Rica, Peru, Canada, Mexico, South Africa, and Australia. Founded in 1963, OTS has a mission to provide leadership in education, research and the responsible use of natural resources in the tropics. To this end, graduate and undergraduate field courses are offered in the pure and applied areas of the natural sciences in Costa Rica, Peru, Mexico, the Brazilian Amazon, and South Africa. In addition, OTS operates three research stations in varied ecosystems in Costa Rica: La Selva in the Caribbean lowland rain forest, Palo Verde in the Pacific lowland dry forest, and Las Cruces in a mid-elevation rain forest on Costa Rica's southern Pacific slope. OTS also coordinates and facilitates research, provides environmental-education programs, and works with local agencies on issues of conservation, natural-resource management and land-use practices.
*Requirements:* In general, Organization for Tropical Studies (OTS) fellowships are open to graduate students enrolled in degree programs at OTS member institutions and to OTS course alumni. (A list of member institutions is available at the OTS website from the About-Us Quick Link.)
*Restrictions:* Only one OTS research award may be held at a time. Awards do not cover capital equipment; minor amounts of materials and supplies may be considered, but are best covered by other sources of funds. Salaries or stipends for the investigator are not allowed, but salaries for field assistants will be considered if the work cannot be carried out safely or efficiently by one person.
*Geographic Focus:* All States, All Countries, Peru
*Date(s) Application is Due:* Mar 1; Oct 1
*Amount of Grant:* Up to 3,650 USD
*Contact:* Education Director, OTS - Costa Rica Office; (506) 2524-0607; fax (506) 2524-0608; academic@ots.ac.cr
*Internet:* ots.ac.cr/index.php?option=com_content&task=view&id=170&Itemid=344#proposal
*Sponsor:* Organization for Tropical Studies
Duke University, 410 Swift Avenue
Durham, NC 27705

**OTS Hovore-Horn Fellowships** 2075

Organization of Tropical Studies (OTS) fellowships in general assist with thesis-related research in tropical biology and similar fields. Funded through an endowment established in 2005 by biologist Frank T. Hovore in honor of his parents, Alice and Andre Horn, OTS's Hovore fellowship also honors Frank himself, who succumbed to a heart attack September 2006 while doing what he loved, collecting beetles in Ecuador. Interested candidates should propose projects with a strong field component and should emphasize the exploration, discovery and description of biological diversity (e.g., field surveys, faunal inventories, or ecological studies) using insects and other terrestrial arthropods as study organisms. Researchers may work in their own or another Latin-American country. Awards are available in the range of $500 to $1500 and are intended primarily to cover fees (food, lodging, and station access) at Organization for Tropical Studies (OTS) field stations; Other expenses that may be included in the budget, to a maximum of $500, are research- and export-permit fees, local travel, and reimbursement in whole or in part of one round-trip airline ticket. Application guidelines and a downloadable cover sheet are available from the OTS Fellowship webpage. Interested applicants should submit their proposals in Spanish via email along with the other required supporting materials by the deadline date to the contact information given. Notification of decisions will take place about two months after each deadline. OTS is a nonprofit consortium of 64 universities and research institutions in the United States, Costa Rica, Peru, Canada, Mexico, South Africa, and Australia. Founded in 1963, OTS has a mission to provide leadership in education, research and the responsible use of natural resources in the tropics. To this end, graduate and undergraduate field courses are offered in the pure and applied areas of the natural sciences in Costa Rica, Peru, Mexico, the Brazilian Amazon, and South Africa. In addition, OTS operates three research stations in varied ecosystems in Costa Rica: La Selva in the Caribbean lowland rain forest, Palo Verde in the Pacific lowland dry forest, and Las Cruces in a mid-elevation rain forest on Costa Rica's southern Pacific slope. OTS also coordinates and facilitates research, provides environmental-education programs, and works with local agencies on issues of conservation, natural-resource management and land-use practices.
*Requirements:* The Hovore-Horn research fellowships in entomology are available to all Latin-American graduate and advanced-undergraduate students interested in Neotropical entomology. In general applicants should be enrolled in degree programs at OTS member institutions. (A list of member institutions is available at the OTS website from the About-Us Quick Link.)
*Restrictions:* Only one research award may be held at a time. Awards do not cover capital equipment; minor amounts of materials and supplies may be considered, but are best covered by other sources of funds. Salaries or stipends for the investigator are not allowed, but salaries for field assistants will be considered if the work cannot be carried out safely or efficiently by one person.
*Geographic Focus:* Argentina, Belize, Bolivia, Brazil, Caribbean, Chile, Colombia, Costa Rica, Cuba, Dominican Republic, Ecuador, El Salvador, Guatemala, Haiti, Honduras, Mexico, Nicaragua, Panama, Paraguay, Peru, Uruguay, Venezuela
*Date(s) Application is Due:* Mar 1; Oct 1
*Amount of Grant:* 500 - 1,500 USD
*Contact:* Education Director, OTS - Costa Rica Office; (506) 2524-0607; fax (506) 2524-0608; academic@ots.ac.cr
*Internet:* ots.ac.cr/index.php?option=com_content&task=view&id=170&Itemid=344#proposal
*Sponsor:* Organization for Tropical Studies
Duke University, 410 Swift Avenue
Durham, NC 27705

**OTS Stanley Smith Horticultural Fellowships** 2076

Organization for Tropical Studies (OTS) fellowships in general assist with thesis-related research in tropical biology and similar fields and are available in amounts up to a maximum of $3,650. Awards are intended primarily to cover up to fifty days of station fees (food, lodging, and station access) at OTS field stations; current rates are $63 for overnight visits and $31.50 for day visits. Other expenses that may be included in the budget, to a maximum of $500, are research and export permit fees, local travel, and reimbursement in whole or in part of one round-trip airline ticket. OTS's Stanley Smith fellowship is funded by the Stanley Smith Horticultural Trust in memory of Mr. Smith who was a British businessman, horticulturist, and frequent visitor to Las Cruces' Wilson Botanical Garden. The fellowship provides funding for horticultural work and systematics research on tropical plants of interest to the Wilson Botanical Garden. OTS has a single application process for most of its fellowships; generally applicants do not need to specify the award for which they are applying (with a few exceptions which are mentioned at the website). Application guidelines and a downloadable cover sheet are available from the OTS Fellowships webpage. The cover sheet and supporting documents must be submitted via email by the deadline date to the Education Director. Notification of decisions will take place about two months after each deadline. The Organization for Tropical Studies (OTS) is a nonprofit consortium of 64 universities and research institutions in the United States, Costa Rica, Peru, Canada, Mexico, South Africa, and Australia. Founded in 1963, OTS has a mission to provide leadership in education, research and the responsible use of natural resources in the tropics. To this end, graduate and undergraduate field courses are offered in the pure and applied areas of the natural sciences in Costa Rica, Peru, Mexico, the Brazilian Amazon, and South Africa. In addition, OTS operates three research stations in varied ecosystems in Costa Rica: La Selva in the Caribbean lowland rain forest, Palo Verde in the Pacific lowland dry forest, and Las Cruces in a mid-elevation rain forest on Costa Rica's southern Pacific slope. OTS also coordinates and facilitates research, provides environmental-education programs, and works with local agencies on issues of conservation, natural-resource management and land-use practices.
*Requirements:* In general OTS fellowships are open to graduate students enrolled in degree programs at OTS member institutions and to OTS course alumni. (A list of member institutions is available at the OTS website from the About-Us Quick Link.) Curatorial expertise in tropical plant groups is essential to obtain approval for a Stanley Smith Horticultural fellowship.
*Restrictions:* Only one research award may be held at a time. Awards do not cover capital equipment; minor amounts of materials and supplies may be considered, but are best covered by other sources of funds. Salaries or stipends for the investigator are not allowed, but salaries for field assistants will be considered if the work cannot be carried out safely or efficiently by one person.
*Geographic Focus:* All States, All Countries
*Date(s) Application is Due:* Mar 1; Oct 1
*Amount of Grant:* Up to 3,650 USD
*Contact:* Zak Zahawi, Station Director; (506) 2773-4004 ext. 3100; zahawi@ots.ac.cr
*Internet:* ots.ac.cr/index.php?option=com_content&task=view&id=170&Itemid=344#proposal
*Sponsor:* Organization for Tropical Studies
Duke University, 410 Swift Avenue
Durham, NC 27705

**OTS Undergraduate NAPIRE Summer Internships** 2077

With funding from the National Science Foundation (NSF), the Organization for Tropical Studies (OTS) introduces Native-American and Pacific-Islander undergraduate students to ecological research in the tropics. As part of the eight-week program, students will apply scientific methods to ecological inquiry and discovery in a collaborative and team-oriented environment. Students will complete an individual field-research project including experimental design, data collection, analysis and presentation of results. In addition to their original research projects, students will participate in lectures, seminars, and other activities that focus on biodiversity, tropical ecology, conservation, and human ecology as well as the long-term research at the OTS biological stations. Participants will have the opportunity to interact with several Central-American indigenous groups and to understand their efforts towards sustainable livelihoods and tropical-forest conservation. Each student will be assigned a research mentor who will guide their independent project and provide support during the program. The program will be held in Costa Rica in June and July. Students will be in residence mostly at OTS's Las Cruces Biological Station during the eight-week research experience, with some travel to other sites. The award covers the cost of room, board, and travel to and from Costa Rica. Students also receive a $4000 stipend and funds to help cover costs of field equipment. Application

for the internship requires submission of a completed application form (via email), two letters of recommendation, and official transcripts from all universities attended (via standard mail to the address specified at the website). Detailed guidelines and forms are available from the OTS NAPIRE - Application-Instructions webpage. Deadlines may vary from year to year. The Organization for Tropical Studies (OTS) is a nonprofit consortium of 64 universities and research institutions in the United States, Costa Rica, Peru, Canada, Mexico, South Africa, and Australia. Founded in 1963, OTS has a mission to provide leadership in education, research and the responsible use of natural resources in the tropics. To this end, graduate and undergraduate field courses are offered in the pure and applied areas of the natural sciences in Costa Rica, Peru, Mexico, the Brazilian Amazon, and South Africa. In addition, OTS operates three research stations in varied ecosystems in Costa Rica: La Selva in the Caribbean lowland rain forest, Palo Verde in the Pacific lowland dry forest, and Las Cruces in a mid-elevation rain forest on Costa Rica's southern Pacific slope. OTS also coordinates and facilitates research, provides environmental-education programs, and works with local agencies on issues of conservation, natural-resource management and land-use practices.
*Requirements:* U.S. citizens and permanent residents who are undergraduate students enrolled in accredited institutions in the United States (and Pacific Island Territories) may apply to the program.
*Restrictions:* Incoming freshmen and graduating seniors are not eligible. Students must be AMP students at LSAMP institutions. SLAMP is the Louis Stokes Alliance for Minority Participation, an NSF program that was designed to foster achievement by minority students seeking careers in science, technology, engineering and mathematics (STEM).
*Geographic Focus:* All States, Guam, Marshall Islands, Northern Mariana Islands, American Samoa, Micronesia, Palau
*Date(s) Application is Due:* Feb 28
*Amount of Grant:* 4,000 USD
*Contact:* Kattia Mendez, Undergraduate Program Assistant; (506) 2524-0607; fax (506) 2524-0608; kattia.mendez @ ots.ac.cr or napire.program@ots.ac.cr
Wendy R. Townsend, Ph.D., Program Coordinator; wendytownsend@gmail.com
*Internet:* ots.ac.cr/index.php?option=com_content&task=view&id=320&Itemid=450
*Sponsor:* Organization for Tropical Studies
Duke University, 410 Swift Avenue
Durham, NC 27705

**Packard Foundation Fellowships for Science and Engineering 2078**
The Foundation established Fellowships for Science and Engineering to allow the nation's most promising professors to pursue science and engineering research early in their careers with few funding restrictions and limited reporting requirements. Each year the Foundation invites the presidents of 50 universities to nominate two early-career professors each from their institutions. An advisory panel of distinguished scientists and engineers carefully reviews the nominations and selects 16 Fellows to receive individual grants of $875,000, distributed over five years. Requests for nominations will be sent to university presidents by January 31. Nominations are due to the Foundation by March 15. Application materials must be submitted by April 20, with awards announced by October 15.
*Requirements:* Fellows must be faculty members who are eligible to serve as principal investigators engaged in research in the natural and physical sciences or engineering. They must also be within the first three years of their faculty careers. Disciplines that are considered include physics, chemistry, mathematics, biology, astronomy, computer science, earth science, ocean science, and all branches of engineering. Interested applicants should review the Fellowship Guidelines posted on the Foundation's website. If their university is on the "Invited Institutions" list and if they meet the eligibility requirements, the university's Office of Sponsored Research or department chair may be contacted to discuss the selection process.
*Restrictions:* Candidates engaged in research in the social sciences are not considered. Nominees will not be able to access the online application system until they have been identified to the Foundation by the university president.
*Geographic Focus:* All States
*Amount of Grant:* 175,000 USD
*Contact:* Xiao-Wei Wang, Program Specialist; (650) 948-7658 or (650) 917-7275; fax (650) 948-2957; fellows@packard.org or xwang@packard.org
*Internet:* www.packard.org/what-we-fund/conservation-and-science/packard-fellowships-for-science-and-engineering/
*Sponsor:* David and Lucile Packard Foundation
343 Second Street
Los Altos, CA 94022

**Parkersburg Area Community Foundation Arts in Education Special Grants 2079**
Because high quality arts programs aid in the development of competent, creative, and critically thinking adults, the West Virginia Department of Education recommends schools provide more time for the delivery of the art curriculum. The arts in education initiative will make grants for school-level projects and programs that address West Virginia educational content standards and link arts content to science, technology, engineering and/or math (STEM) concepts, as well as help local schools to develop partnerships with arts institutions and higher education. Priority funding consideration will be given to projects that: bring arts programming into schools, using the arts to enhance or introduce science, technology, engineering and/or math (STEM) concepts and that advance state standards and curricular goals; promote collaboration and partnerships to help sustain interest in science, technology, engineering and/or math education and the arts past the period of this grant; offer evidence of a 10% local match for the grant (in-kind or direct support); and projects that did not receive an arts in education award the previous year. Examples of possible projects and additional proposal criteria and instructions are available at the website. Organizations may apply through the online application process.
*Requirements:* To receive consideration, proposals must include all of the following elements: evidence of potential for increased collaboration among educators/schools with arts groups and/or community artists directly in the school environment during the standard school year for K-12 public school students for the current school year (accredited private and parochial schools in the service region with at least 100 students may apply for grant consideration); show potential for creative application/s of all forms of the arts in traditional education, ranging from the visual, to dance, to music, to performance arts; show increased integration of the arts in core disciplines, emphasizing science, technology, engineering and math, to support student advancement and success with clear evidence of the tie between the arts activities to West Virginia content standards that will be pursued; and show evidence that students are engaged in the project and its implementation.
*Restrictions:* A 10% match for the amount of the grant request should be shown in the budget narrative.
*Geographic Focus:* West Virginia
*Date(s) Application is Due:* Oct 1
*Amount of Grant:* Up to 5,000 USD
*Contact:* Jane Winters; (304) 428-4438 or (866) 428-4438; jane.winters@pacfwv.com
*Internet:* www.pacfwv.com/grant-guidelines-and-applications
*Sponsor:* Parkersburg Area Community Foundation
501 Avenuery Street, P.O. Box 1762
Parkersburg, WV 26102-1762

**Parkinson's Foundation ABF Clinical Research Training Fellowship 2080**
The Parkinson's Foundation's collaborative partnerships support top clinician-scientist training, patient-oriented Parkinson's disease (PD) research and medical student PD studies. In 2007 the Foundation and the American Brain Foundation (ABF) began a collaboration to support our mutual interest in training outstanding clinician-scientists in the Parkinson's field. Successful Parkinson's Foundation-ABF Clinical Research Training Fellowship applicants receive a two-year award consisting of a $65,000 annual salary plus $10,000 in educational and research-related expenses per year. The award aims to recognize the importance of good clinical research and encourage young investigators in clinical studies. The full application is due October 1.
*Requirements:* Applicant must be an American Academy of Neurology member interested in an academic career in research, and have completed residency or a post-doctoral fellowship (for a Ph.D.) within the past five years of the start date of the award. Priority will be given to those early in their research careers.
*Restrictions:* Award only funds direct costs.
*Geographic Focus:* All States
*Date(s) Application is Due:* Oct 1
*Amount of Grant:* 200,000 USD
*Contact:* Angela Schlesinger, Executive Director; (763) 317-1308 or (800) 843-2763; fax (305) 243-8144; aschlesinger@parkinson.org or grants@parkinson.org
Kristin Roehl, Grants Program Manager; (612) 928-6082 or (800) 843-2763; fax (305) 243-8144; kroehl@aan.com or grants@parkinson.org
*Internet:* https://www.parkinson.org/research/Grant-Opportunities/Partnership-Awards
*Sponsor:* Parkinson's Foundation
200 Southeast 1st Street, Suite 800
Miami, FL 33131

**Parkinson's Foundation APDA Summer Student Fellowships 2081**
In 2015, the Parkinson's Foundation and the American Parkinson Disease Association (APDA) partnered to expand the foundation's existing Summer Student Fellowships Program, which supports students – from advanced undergraduates to medical students – in their pursuit of Parkinson's-related summer research projects. The goal of the Parkinson's Foundation APDA Summer Student Fellowships is to cultivate students' early interest in helping to solve, treat and end Parkinson's and providing the tools they need to transition into leaders in the field. Fellows work under the close supervision of a sponsor who is an expert in the Parkinson's community and oversees the project. The program offers summer fellowships for ten weeks of clinical or laboratory work with a $4,000 award. The Foundation invites fellows to apply for up to $1,000 in travel funds to attend a related scientific conference to present their research upon project completion. The award supports the student's summer research, defraying tuition costs of research credits or supplementing living or housing costs. This award may be taxable; the Parkinson's Foundation sends an IRS form 1099 to the recipient at the end of the calendar year.
*Requirements:* Undergraduate students and medical students are eligible for summer fellowships. Each applicant must show a mentor with whom he or she will conduct the proposed project. This program is open to national and international applicants.
*Geographic Focus:* All States, All Countries
*Date(s) Application is Due:* Dec 3
*Amount of Grant:* 4,000 - 5,000 USD
*Contact:* Angela Schlesinger, Executive Director; (763) 317-1308 or (800) 473-4636; fax (305) 243-8144; aschlesinger@parkinson.org or grants@parkinson.org
*Internet:* www.parkinson.org/research/Grant-Opportunities/Fellowships-and-Career-Development
*Sponsor:* Parkinson's Foundation
200 Southeast 1st Street, Suite 800
Miami, FL 33131

**Parkinson's Foundation Clinical Research Grants 2082**
The Parkinson Foundation's Clinical Research Fund is designed to support cutting-edge clinical studies of new therapeutic options for Parkinson's patients as well as research comparing the effectiveness of existing treatment options. Conducted by or in collaboration

with Foundation's Centers of Excellence, CRF-funded research aims to discover the best treatment and care options for people living with Parkinson's disease. Broad clinical-research areas funded by the CRF program are as follows: care quality; imaging; cognition; physical therapy; care models; exercise; and biomarkers. CRF Requests for Application are issued as funding becomes available, but no less than once per year. Applicants may request two-year funding (or less) for their projects. An example of the CRF application process is as follows: an RFP is released in August; NPF schedules pre-submission conference calls for September and October; applicants submit their proposals in December; peer-review is conducted January through March; awards are announced in April. Specific deadline dates will vary from year to year and will be detailed in the RFP. More detailed information is generally posted in a listserv to which all applicants are invited to subscribe (address given in the RFP). Applications must be submitted electronically through NPF's online grant submission system. Registration is required. The Parkinson's Foundation Clinical Research Award is for $200,000 total over two years. Up to 10% of the award total, i.e., $20,000, may be used to defray indirect costs. Salary support is permitted but for no more than 40% of the recipient's salary and capped at $50,000 per year. A Letter of Intent is due by January 22, with full applications due by April 8.
*Requirements:* This Award is open to all clinician-scientists who are licensed to care for patients regardless of nationality or location. Physicians must have completed a residency in neurology or be able to justify the relevance of their training pathway is of benefit to people with PD. Allied health professionals must hold an advanced research degree, typically a PhD. All applicants must demonstrate a significant research focus on PD.
*Restrictions:* There are no limits on the number of grants an institution may request but each grant is limited to no more than $200,000 over a two-year period. Postdoctoral fellows, fellows, and students cannot serve as a principal investigator. Grants cannot be used to purchase equipment or to support conference fees or travel (except for travel to a Center of Excellence collaborating on the research project). Indirect costs will not be funded. NPF no longer funds basic laboratory research.
*Geographic Focus:* All States, All Countries
*Date(s) Application is Due:* Apr 8
*Amount of Grant:* Up to 200,000 USD
*Contact:* Angela Schlesinger, Executive Director; (763) 317-1308 or (800) 473-4636; fax (305) 243-8144; aschlesinger@parkinson.org or grants@parkinson.org
*Internet:* www.parkinson.org/research/Grant-Opportunities/clinical-research-awards
*Sponsor:* Parkinson's Foundation
200 Southeast 1st Street, Suite 800
Miami, FL 33131

**Parkinson's Foundation Institutional Movement Disorders Fellowships 2083**
Movement Disorders Neurologists are key specialists who provide expert care to those living with Parkinson's disease and are critical to our efforts to ensure people with Parkinson's have Better Lives. To help those living with Parkinson's receive the care they need today, the Parkinson's Foundation has maintained ample support for this specialized training. The Parkinson's Foundation supported the very first Movement Disorders Fellowship programs in the United States in 1980 at our Centers of Excellence at Columbia University in New York and at Rush University in Chicago. Since then, the Parkinson's Foundation has provided over $15 million dollars to support the training of over 150 neurologists to become experts in treating and helping those with PD. However, the number of people with Parkinson's disease is projected to increase with our aging population; the need for specialized care is greater than ever. Consistent with the Foundation's goal of making the lives of those with PD better, the Parkinson's Foundation is widening the reach of our training program for physicians to become experts in Parkinson's care. Institutions in North America may now apply to the Parkinson's Foundation Movement Disorders Training Award to support the training of neurologists to become specialists in Movement Disorders. The award provides an institution with the funds to support the two-year long training of an incoming movement disorders fellow, covering their salary and associated costs. The award provides annual support of up to $75,000 for salary, matching the institution's PGY-5/6 pay scale. An additional 25% of salary will be included for other costs such as fringe benefits, malpractice insurance, research allowance, and travel. A stipend of $10,000 is provided for dedicated mentoring support. The potential total award is $103,750 per year for up to two years. The annual application deadline is January 7.
*Requirements:* The award is open to North American institutions with an established movement disorders fellowship program. Potential fellows may be of any nationality and must have completed a residency in neurology. Individuals may not apply.
*Geographic Focus:* All States, American Samoa, Canada
*Date(s) Application is Due:* Jan 22
*Amount of Grant:* Up to 103,750 USD
*Contact:* Angela Schlesinger, Executive Director; (763) 317-1308 or (800) 473-4636; fax (305) 243-8144; aschlesinger@parkinson.org or grants@parkinson.org
*Internet:* www.parkinson.org/research/Grant-Opportunities/Parkinsons-Foundation-Institutional-Movement-Disorders-Training-Award
*Sponsor:* Parkinson's Foundation
200 Southeast 1st Street, Suite 800
Miami, FL 33131

**Parkinson's Foundation Parkinson's Advocates in Research Leadership Awards 2084**
The Parkinson's Foundation PAIR Leadership Awards foster patient engagement in Parkinson's research. These grants support scientists working to formally engage Parkinson's Foundation Research Advocates and other patient advocates (people with Parkinson's and care partners) in their research. The Foundation bases this award on its philosophy that when it brings together two kinds of experts – scientists who study PD and individuals who live with it – it is likely to find answers more quickly. The award is an extension of the Parkinson's Foundation Parkinson's Advocates in Research (PAIR) program. Some possible patient advocate activity examples include: identifying research priorities of the Parkinson's community; developing specific study design and protocols; and serving as co-investigators on specific clinical studies. Up to $10,000 in funding is available for outstanding project ideas. The annual application submission deadline is June 1.
*Requirements:* A successful PAIR Leadership Award recipient is a Parkinson's researcher in the U.S. who has: established pre-existing relationship(s) with Parkinson's Foundation Research Advocate(s) or other patient advocates within her/his community; already identified a specific project that will engage Parkinson's Foundation Research Advocate(s) or other patient advocates and in which all parties (advocates and professionals) believe there is potential to accelerate PD research; received the support of at least one Parkinson's Foundation Research Advocate(s) or another patient advocate who is collaborating on the project; and demonstrated a commitment to patient engagement to improve research.
*Geographic Focus:* All States
*Date(s) Application is Due:* Jun 1
*Amount of Grant:* Up to 10,000 USD
*Contact:* Angela Schlesinger, Executive Director; (763) 317-1308 or (800) 473-4636; fax (305) 243-8144; aschlesinger@parkinson.org or grants@parkinson.org
*Internet:* https://www.parkinson.org/research/information-for-researchers/awards-for-independent-investigators
*Sponsor:* Parkinson's Foundation
200 Southeast 1st Street, Suite 800
Miami, FL 33131

**Parkinson's Foundation Postdoctoral Fellowships for Basic Scientists 2085**
The Postdoctoral Fellowships for Basic Scientists are two-year fellowships for young scientists, fresh from their Ph.D. training, to study at major research institutions. This program grants $110,000 over two years. This consists of $50,000 annually (to be used as salary) as well as a $5,000 annual research allowance to be used at the Fellow's discretion, with sponsor/mentor approval, to pay for items such as books, training courses, travel costs (up to $2,000) or a computer. There is no provision for the deduction of postdoctoral taxes, institutional overhead or fees. The Foundation may approve $3,000 of the research allowance to defray the health insurance costs paid by the Fellow if the Fellow sends a written request. All applications must be submitted online through the Parkinson's Foundation website; no paper copies will be accepted. The application period typically opens in December and closes in early February. A Letter of Interest (LOI) is due each year on November 15, with the full application due by February 19.
*Requirements:* The applicant must be within five years of receiving his or her Ph.D. at the time the award starts, and must name an individual who will serve as his or her research mentor and supervisor. This program is open to both national and international applicants. If awarded, a Postdoctoral Fellow must commit 90 percent of his or her effort to the Parkinson's Foundation-funded project.
*Geographic Focus:* All States
*Date(s) Application is Due:* Feb 19
*Amount of Grant:* 110,000 USD
*Contact:* Angela Schlesinger, Executive Director; (763) 317-1308 or (800) 473-4636; fax (305) 243-8144; aschlesinger@parkinson.org or grants@parkinson.org
*Internet:* https://www.parkinson.org/research/information-for-researchers/early-career-fellowships/postdoctoral?_ga=2.148920498.549293731.1548697453-56726270.1548697453
*Sponsor:* Parkinson's Foundation
200 Southeast 1st Street, Suite 800
Miami, FL 33131

**Parkinson's Foundation Postdoctoral Fellowships for Clinical Neurologists 2086**
These Fellowships are for young clinicians who have completed their neurology residency and are seeking clinical research experience, the Parkinson's Foundation offers the Postdoctoral Fellowships for Clinical Neurologists, two-year awards in the amount of $130,000. This award includes salary support of $60,000 per year. Similar to the Fellowship for Basic Scientists, the Fellow, with approval of the mentor, may use an additional research allowance of $5,000 per year to pay for such items as books, training courses, travel costs (up to $2,000) or a computer. There is no provision for the deduction of postdoctoral taxes, institutional overhead or fees. However, if the Fellow sends a written request, we may approve up to $3,000 of the research allowance to defray the cost of health insurance paid by the Fellow. All applications must be submitted online through the Parkinson's Foundation website; no paper copies will be accepted. A Letter of Interest (LOI) is due each year on November 15, with the full application due by February 19.
*Requirements:* Applicants seeking a Postdoctoral Fellowship for Neurologists must possess an M.D. or equivalent and be within three years of having completed a residency in neurology at the time the award starts. Applicants may not have their own lab and must name an individual who will serve as his or her mentor and supervisor of their research. This program is open to national and international applicants. If awarded, the Foundation requires a Clinical Fellow to commit 80 percent of his or her effort to the Parkinson's Foundation-funded project.
*Geographic Focus:* All States
*Date(s) Application is Due:* Feb 19
*Amount of Grant:* Up to 130,000 USD
*Contact:* Angela Schlesinger, Executive Director; (763) 317-1308 or (800) 473-4636; fax (305) 243-8144; aschlesinger@parkinson.org or grants@parkinson.org
*Internet:* www.parkinson.org/research/Grant-Opportunities/Fellowships-and-Career-Development
*Sponsor:* Parkinson's Foundation
200 Southeast 1st Street, Suite 800
Miami, FL 33131

**Parkinson's Foundation PSG Mentored Clinical Research Award** **2087**

The grant from the Parkinson's Foundation to the Parkinson Study Group funds the Mentored Clinical Research Award for new investigators. The grant supports new investigators for a one-year project in patient-oriented research in Parkinson's or other parkinsonian disorders under the mentorship of an experienced investigator. Funding an investigator with the potential to become an independent researcher is the award goal. The Foundation anticipates an award of $50,000 for the current application period. This is a collaborative fellowship and for this program, applicants cannot use the Parkinson's Foundation online grant management system. Instead, applicants should submit an electronic copy of the proposal in Microsoft Word of PDF format to the Parkinson Study Group. The annual application deadline is January 18.

*Requirements:* Applicants for the Mentored Clinical Research Award should be clinicians and scientists who are within five years of having completed formal training. Fellows may apply. If during formal training unusual circumstances occur delaying completion, time out periods will be allowed. Applicants must name a mentor with extensive research experience. Either the applicant or the mentor must be a member of the Parkinson Study Group. The applicant may have co-mentors.

*Restrictions:* This award only funds direct costs.

*Geographic Focus:* All States

*Date(s) Application is Due:* Jan 18

*Amount of Grant:* Up to 50,000 USD

*Contact:* Angela Schlesinger, Executive Director; (763) 317-1308 or (800) 843-2763; fax (305) 243-8144; aschlesinger@parkinson.org or grants@parkinson.org

Roseanna Battista, (585) 273-2862 or (585) 275-1642; Roseanna.Battista@chet.rochester.edu

*Internet:* www.parkinson.org/research/Grant-Opportunities/Partnership-Awards

*Sponsor:* Parkinson's Foundation

200 Southeast 1st Street, Suite 800

Miami, FL 33131

**Parkinson's Foundation Research Centers of Excellence Awards** **2088**

The goal of the Parkinson's Foundation Research Centers of Excellence is to provide funding for innovative team science at institutions working on a thematic area of Parkinson's disease. The Foundation is interested in proposals that tackle critical issues or problems that are impeding progress in the field of PD research. The Foundation's vision and mission stand on three pillars: understanding Parkinson's through research, ensuring better care for people with Parkinson's, and educating and empowering the community. As such, proposals that touch upon these pillars (for example, studies unraveling the pathophysiology of PD, preclinical work towards disease-modifying therapies, or symptomatic therapy for unmet clinical needs) will be given priority. Awards are $500,000 a year for 4 years, for a total award of $2,000,000. Up to 10% ($200,000) of the total award may be used towards indirect costs.

*Requirements:* This award program is open to institutions that have or can recruit a team of scientists working in the field of Parkinson's disease. The team may all reside in the same geographical area and work at the same institution, or through a virtual center with scientists spread across regions and institutions. Preference will be given to applications which demonstrate cross-departmental or cross-institutional collaboration, bringing new approaches and perspectives to solving PD. Institutions that hold a Parkinson's Foundation Center of Excellence designation are also welcome to apply.

*Restrictions:* Students cannot enroll in a combined medical, dental or veterinary/Ph.D. program (e.g., M.D./Ph.D.) or Ph.D., or Sc.D. program, or have a Ph.D. or Sc.D. in a laboratory-based biological science.

*Geographic Focus:* All States

*Date(s) Application is Due:* Apr 1

*Amount of Grant:* 2,000,000 USD

*Contact:* Angela Schlesinger, Executive Director; (763) 317-1308 or (800) 473-4636; fax (305) 243-8144; aschlesinger@parkinson.org or grants@parkinson.org

*Internet:* https://www.parkinson.org/research/Grant-Opportunities/Parkinsons-Foundation-Research-Centers-of-Excellence

*Sponsor:* Parkinson's Foundation

200 Southeast 1st Street, Suite 800

Miami, FL 33131

**Parkinson's Foundation Visiting Scholar Award** **2089**

The goal of the Parkinson's Foundation Visiting Scholar Award is to provide graduate students with opportunities for expanding their existing skillset to benefit their Parkinson's research. Gone are the days when research is done in silos. Collaboration and learning from one another are key aspects of research conducted today, regardless of the field. The Visiting Scholar Program provides this opportunity so that future Parkinson's research continues to be innovative, creative, and multi-disciplinary. This award provides a $4,000 stipend to support the travel and housing of scholar while visiting a host laboratory to learn new techniques that will benefit their Parkinson's research. Examples include microscopy, molecular biology, surgical techniques, computer modelling, and others. While the goal is to learn a new technique, it should not be done in isolation. A small project incorporating the new technique is required, ideally incorporating the Scholars current research project. Scholars will work under the close supervision of the principal investigator of the host laboratory who will oversee the project. The award is to be used at the Scholar's discretion to supplement research costs or to supplement travel and housing costs while visiting the host laboratory. Upon completion of their project, Scholars are invited to apply for up to $1,000 in travel funds to attend a related scientific conference to present their research. All applications must be submitted online through the Parkinson's Foundation website; no paper copies will be accepted. The application period typically opens in October and closes in early December.

*Requirements:* Ph.D. students conducting Parkinson's research for their thesis are eligible for this program. This program is open to applicants regardless of nationality or location. Each applicant must identify a mentor with whom he or she will conduct the proposed project. Students must be in their second year or later, and have an expected graduate date no sooner than 6 months after the end of the summer project. Host mentors and their laboratory are not required to conduct Parkinson's research, but the technique learned must be applicable to Parkinson's research. The host laboratory must be a significant distance from the Fellow's Institution and must not be an affiliate of the Fellow's Institution.

*Geographic Focus:* All States

*Date(s) Application is Due:* Dec 3

*Amount of Grant:* Up to 4,000 USD

*Contact:* Angela Schlesinger, Executive Director; (763) 317-1308 or (800) 473-4636; fax (305) 243-8144; aschlesinger@parkinson.org or grants@parkinson.org

*Internet:* www.parkinson.org/research/Grant-Opportunities/Fellowships-and-Career-Development

*Sponsor:* Parkinson's Foundation

200 Southeast 1st Street, Suite 800

Miami, FL 33131

**Patron Saints Foundation Grants** **2090**

The Patron Saints Foundation is a private foundation that provides grants to public charities that improve the health of individuals residing in the West San Gabriel Valley through health care programs which are consistent with the moral and religious teachings of the Roman Catholic Church. Grants will be made to qualified public charities to sponsor charitable, scientific or educational health care programs which are not inconsistent with the moral and religious teachings of the Roman Catholic Church, in the following health care categories: community health services (direct health care services); community health care education; capital expenditures; equipment and supplies; medical research. Grants typically range from $5,000 to $15,000. Larger or smaller grants are at the discretion of The Patron Saints Foundation Board of Directors based on the merits of the project.

*Requirements:* California 501(c)3 tax-exempt organizations serving the West San Gabriel Valley are eligible, including: the cities of Alhambra, Arcadia, Duarte, El Monte, La Canada Flintridge, Monrovia, Monterey Park, Pasadena, Rosemead, San Gabriel, San Marino, Sierra Madre, South El Monte, South Pasadena, Temple City and the unincorporated areas of Los Angeles County known as Altadena and South San Gabriel along with the unincorporated portions of Los Angeles County within the West San Gabriel Valley.

*Restrictions:* Grants are not awarded for endowment funds, political activities, travel, surveys or fund raising activities.

*Geographic Focus:* California

*Date(s) Application is Due:* Mar 1; May 1

*Amount of Grant:* 5,000 - 15,000 USD

*Contact:* Kathleen T. Shannon, Executive Director; (626) 564-0444; fax (626) 564-0444; patronsaintsfdn@sbcglobal.net

*Internet:* www.patronsaintsfoundation.org/index.php?nav=Grant%20Guidelines

*Sponsor:* Patron Saints Foundation

260 S Los Robles Avenue, Suite 201

Pasadena, CA 91101-3614

**Paul Marks Prizes for Cancer Research** **2091**

The Paul Marks Prizes for Cancer Research recognize outstanding young investigators who have made significant contributions to increase the understanding of cancer or improve the treatment of the disease through basic or clinical research. The winners will present their work at the center, be honored at a dinner, and share a cash award of $150,000. Nomination packets must include a letter from the nominator outlining the significance of the accomplishments for which the candidate should be recognized; a one-page scientific biography of the candidate; a list of up to eight of the candidate's significant published papers with a brief (fewer than 100 words) explanation of the importance of each one; and the candidate's curriculum vita. Up to three supporting letters may also be submitted. The bi-annual deadline date is April 30 of odd years.

*Requirements:* Nominees are required to be age 45 or younger at the time of the submission deadline.

*Geographic Focus:* All States

*Date(s) Application is Due:* Apr 30

*Amount of Grant:* 150,000 USD

*Contact:* Program Contact; (212) 639-6561 or (212) 639-2000

*Internet:* www.mskcc.org/mskcc/html/53983.cfm

*Sponsor:* Memorial Sloan-Kettering Cancer Center

1275 York Avenue

New York, NY 10021

**PC Fred H. Bixby Fellowships** **2092**

The Population Council offers a Fellowship program to expand training opportunities for social scientists and biomedical researchers in the health and population fields. This Fellowship is geared toward developing-country nationals in the early stages of their careers, and to those with a demonstrated commitment to remaining in their home countries to build capacity in local institutions or returning home after working/studying abroad. The Council's Fred H. Bixby Fellowship is a ten-year program which offers a limited number of Fellowships each year. The Bixby Fellowship allows candidates to work with experienced mentors in the Council's network of offices all over the world. Fellows work on projects in the areas of HIV and AIDS, poverty, gender, and youth, and reproductive health. The term of appointment is two years, with the second year contingent upon a successful evaluation and review of performance in the first year.

The Fellowship is a full-time commitment, meaning no additional research, studies, or work may be pursued for the duration of the Fellowship. Awards consist of a monthly stipend, health insurance, an allowance for relocation expenses, and attendance and travel to one international professional meeting per year. The Fellow is responsible for all local accommodations and living costs. Fellows will be provided with office space in the Council's office, plus access to office support facilities and a notebook computer.
*Requirements:* Candidates must have recently completed a Ph.D. or equivalent degree within the last five years in the social sciences, public health, or biomedical sciences. All applicants should have previous direct experience with either biomedical research, program research, or policy-relevant social science research. Applicants must be legal citizens of a developing country and proficient in English. Candidates interested in applying for a Fred H. Bixby Fellowship must first email the Fellowship Coordinator with a letter indicating professional objectives and goals, a short statement about research interests and plans for the proposed Fellowship period, as well as the name(s) of the mentor(s) (up to three) with whom they would like to work. They should also include a copy of their curriculum vitae. The Fellowship Coordinator will then initiate contact with the appropriate mentor(s) and instruct each candidate on how to proceed. Additional information about the application process is available at the Council website.
*Restrictions:* Candidates with master's-level degrees in public health or a related field may be eligible only if they have considerable skills, background, experience, and peer-reviewed publications, including 2–3 years of relevant job experience post-master's.
*Geographic Focus:* All Countries
*Date(s) Application is Due:* Jan 31
*Contact:* Coordinator; (212) 339-0500; fax (212) 755-6052; bixbyfellowship@popcouncil.org
*Internet:* www.popcouncil.org/what/bixby.asp
*Sponsor:* Population Council
One Dag Hammarskjold Plaza
New York, NY 10017

**Peabody Foundation Grants** **2093**
The Peabody Foundation was established in 1894 in Massachusetts. Grants provide care, treatment, rehabilitation, education, and assistance to children with physical disabilities as well as encouraging and supporting medical research in the causes of crippling disease, particularly in children. The majority of recipient organizations will be located in the Boston, Massachusetts area. Proposals may be submitted at any time and are reviewed in the spring. Funding ranges from $20,000 to $200,000. Interested organizations should contact the Administrative Director for application procedures and eligibility information.
*Restrictions:* Grants are not made to individuals. Giving is limited to Massachusetts, with emphasis on the Boston area.
*Geographic Focus:* Massachusetts
*Date(s) Application is Due:* Feb 1
*Amount of Grant:* 20,000 - 200,000 USD
*Samples:* Agassiz Village, Lexington, Massachusetts, $35,000 - to benefit disabled children and/or related research; Brigham and Women's Hospital, Boston, Massachusetts, $75,000 - to benefit disabled children and/or related research; Cotting School for Handicapped Children, Lexington, Massachusetts, $41,000.
*Contact:* Judi Mullen; (508) 728-8780; jemullen12@comcast.net
Jonathan Bashein, Office Administrator; (617) 345-1000 or (617) 310-4100; fax (617) 345-1300; jbashein@nixonpeabody.com
*Sponsor:* Peabody Foundation
100 Summer Street
Boston, MA 02110-2131

**Pediatric Brain Tumor Foundation Early Career Development Grants** **2094**
The Pediatric Brain Tumor Foundation (PBTF) is a 501(c)3 nonprofit charitable organization whose mission is as follows: to find the cause of and cure for childhood brain tumors by supporting medical research; to increase public awareness about the severity and prevalence of childhood brain tumors; to aid in the early detection and treatment of childhood brain tumors; to support a national database on all primary brain tumors; and to provide educational and emotional support for children and families affected by this life-threatening disease. The goal of the PBTF's Early Career Development grants program is to support talented researchers in the field of pediatric brain tumors who are in the first five years of their first faculty position. The expectation is that this funding will facilitate awardees' transition to becoming fully independent investigators.
*Requirements:* Applicant organizations must be 501(c)3 institutions located in the U.S., Canada, and Australia.
*Geographic Focus:* All States
*Amount of Grant:* Up to 400,000 USD
*Samples:* Dana-Farber Cancer Institute, Boston, Massachusetts, $300,000 - resistance to BET-bromodomain inhibitors in MYC-amplified medulloblastoma; Heinrich-Heine University, Dusseldorf, Germany, $300,000 - unraveling medulloblastoma biology by proteomics; University of California, San Francisco, California, $300,000 - genetic susceptibility to ependymoma and interaction with perinatal risk factors.
*Contact:* Robin Boettcher, President and CEO; (800) 253-6530 or (828) 665-6891; fax (828) 665-6894; rboettcher@curethekids.org or pbtfus@pbtfus.org
*Internet:* www.curethekids.org/research/what-we-fund/research-funding/#.WMAQXYVExWa
*Sponsor:* Pediatric Brain Tumor Foundation
302 Ridgefield Court
Asheville, NC 28806

**Pediatric Brain Tumor Foundation Early Career Development Grants** **2095**
The Pediatric Brain Tumor Foundation (PBTF) is a 501(c)3 nonprofit charitable organization whose mission is as follows: to find the cause of and cure for childhood brain tumors by supporting medical research; to increase public awareness about the severity and prevalence of childhood brain tumors; to aid in the early detection and treatment of childhood brain tumors; to support a national database on all primary brain tumors; and to provide educational and emotional support for children and families affected by this life-threatening disease. The goal of the PBTF's Early Career Development grants program is to support talented researchers in the field of pediatric brain tumors who are in the first five years of their first faculty position. The expectation is that this funding will facilitate awardees' transition to becoming fully independent investigators.
*Requirements:* Applicant organizations must be 501(c)3 institutions located in the U.S., Canada, Germany, and Australia.
*Geographic Focus:* All States, Australia, Canada, Germany
*Amount of Grant:* Up to 400,000 USD
*Samples:* Dana-Farber Cancer Institute, Boston, Massachusetts, $300,000 - resistance to BET-bromodomain inhibitors in MYC-amplified medulloblastoma; Heinrich-Heine University, Dusseldorf, Germany, $300,000 - unraveling medulloblastoma biology by proteomics; University of California, San Francisco, California, $300,000 - genetic susceptibility to ependymoma and interaction with perinatal risk factors.
*Contact:* Robin Boettcher, President and CEO; (800) 253-6530 or (828) 665-6891; fax (828) 665-6894; rboettcher@curethekids.org or pbtfus@pbtfus.org
*Internet:* www.curethekids.org/research/what-we-fund/research-funding/#.WMAQXYVExWa
*Sponsor:* Pediatric Brain Tumor Foundation
302 Ridgefield Court
Asheville, NC 28806

**Pediatric Brain Tumor Foundation Institute Grants** **2096**
The Pediatric Brain Tumor Foundation (PBTF) is a 501(c)3 nonprofit charitable organization whose mission is as follows: to find the cause of and cure for childhood brain tumors by supporting medical research; to increase public awareness about the severity and prevalence of childhood brain tumors; to aid in the early detection and treatment of childhood brain tumors; to support a national database on all primary brain tumors; and to provide educational and emotional support for children and families affected by this life-threatening disease. PBTF Institute grants concentrate investment in centers with well-recognized expertise in pediatric brain tumor research. This enables the Foundation to be flexible with project selection and to reprioritize during the funding period based on project performance. Other advantages of the Institute model include: enhanced coordination of discovery and translational research efforts; greater sharing and leveraging of core resources between recipient centers; and enhanced exchange of experimental findings and approaches between principal investigators.
*Requirements:* Institute projects must be hypothesis-driven basic or translational scientific research on pediatric brain tumors. Applicant organizations must be 501(c)3 institutions located in the U.S., Canada, and Australia.
*Geographic Focus:* All States, Australia, Canada
*Amount of Grant:* Up to 1,000,000 USD
*Samples:* Hospital for Sick Children, Toronto, Canada, $1,000,000 - epigenetic alterations define lethal CMP-positive ependymomas of infancy; Duke University, Durham, North Carolina, $400,000 - development of a peptide vaccine for DIPG; University of California, San Francisco, California, $300,000 - oligodendrocyte developmental methylome to characterize progenitors for pediatric glioma.
*Contact:* Robin Boettcher, President and CEO; (800) 253-6530 or (828) 665-6891; fax (828) 665-6894; rboettcher@curethekids.org or pbtfus@pbtfus.org
*Internet:* www.curethekids.org/research/what-we-fund/research-funding/#.WMANToVExWY
*Sponsor:* Pediatric Brain Tumor Foundation
302 Ridgefield Court
Asheville, NC 28806

**Pediatric Brain Tumor Foundation Opportunity Grants** **2097**
The Pediatric Brain Tumor Foundation (PBTF) is a 501(c)3 nonprofit charitable organization whose mission is as follows: to find the cause of and cure for childhood brain tumors by supporting medical research; to increase public awareness about the severity and prevalence of childhood brain tumors; to aid in the early detection and treatment of childhood brain tumors; to support a national database on all primary brain tumors; and to provide educational and emotional support for children and families affected by this life-threatening disease. Research initiatives that the Foundation judges to be critical to filling in the gaps and enhancing the pediatric brain tumor research landscape are eligible for opportunity grants. Discussions with the Research Advisory Network and the broader scientific community helps to identify candidates for funding. Applications are accepted by invitation only. However, the Foundation welcomes discussion of research ideas that fit with its strategic priorities, particularly if the ideas involve multi-institutional efforts.
*Requirements:* Institute projects must be hypothesis-driven basic or translational scientific research on pediatric brain tumors. Applicant organizations must be 501(c)3 institutions located in the U.S., Canada, and Australia.
*Geographic Focus:* All States, Australia, Canada
*Amount of Grant:* Up to 500,000 USD
*Samples:* University of California, San Francisco, California, $500,000 - support for the Pacific Pediatric Neuro-Oncology Consortium, Operations Center; University of California, San Diego, California, $234,000 - Immunosignature Strategy for Development of Clinical Biomarkers and Identification of New Drug Target Candidates

for Pediatric Brain Cancer; Children's Oncology Group, Philadelphia, Pennsylvania, $100,000 - support for Project: Every Child.
*Contact:* Robin Boettcher, President and CEO; (800) 253-6530 or (828) 665-6891; fax (828) 665-6894; rboettcher@curethekids.org or pbtfus@pbtfus.org
*Internet:* www.curethekids.org/research/what-we-fund/research-funding/#.WMApJ4VExWb
*Sponsor:* Pediatric Brain Tumor Foundation
302 Ridgefield Court
Asheville, NC 28806

**Pediatric Cancer Research Foundation Grants** **2098**
Since its establishment in 1982 as a grass-roots organization, the Pediatric Cancer Research Foundation (PCRF) has focused its efforts upon improving the care, quality of life, and survival rate of children with malignant diseases. With these ends in mind, PCRF concentrates on laboratory research that will translate into immediate treatment for children with cancer. PCRF primarily funds research that brings innovative new drugs and treatment regimens to children. Proposals should be based on molecular, cellular, or integrated systems; be conceptually innovative; and have a clear plan for study and future clinical use of the factual data obtained. Current areas of foundation funding include stem-cell transplantation, stem-cell biology, molecular oncology, and molecular and cellular genetics. Awards will generally be made for a one- to three-year period at the Board's discretion. Renewals for additional years may be made for promising projects as recognized by the Scientific Review Committee. Prospective applicants can download the application form at the PCRF website. Completed applications must be received at the PCRF office by 5:00 PM on May 1st. If a deadline falls on a Saturday or Sunday, the applications are due the next business day.
*Requirements:* Applications may be submitted by individuals holding a M.D., Ph.D., or equivalent degree and working in domestic non-profit organizations such as universities, colleges, hospitals, or laboratories (the "sponsoring institution"). Applications may be multi-institutional in nature. Applicants currently holding a grant may re-apply for extension of similar research by submitting a Renewal Application.
*Restrictions:* No overhead or indirect costs to institutions will be funded.
*Geographic Focus:* All States
*Date(s) Application is Due:* May 1
*Samples:* Columbia University, New York, New York, $740,548; Children's Hospital of Los Angeles, Los Angeles, California, $64,250; MD Anderson, Houston, Texas, $37,500.
*Contact:* Joseph M. Galosic, Director of Scientific Affairs; (949) 859-6312; fax (949) 859-6323; admin@pcrf-kids.org
*Internet:* www.pcrf-kids.org/
*Sponsor:* Pediatric Cancer Research Foundation
9272 Jeronimo Road, Suite 122
Irvine, CA 92618

**Penn Graduate School of Education Dean's Fellowships** **2099**
The Dean's Fellowship provides funding to doctoral level students and includes, but is not limited to, a tuition scholarship, a living stipend, and a research and teaching apprenticeship. Fellowships are taxable and include a service requirement in the form of the apprenticeship, which promotes the development of the student's research and teaching skills. There is no application for Dean's Fellowships; award decisions are made by the academic division based on the application and communicated at the time of the admissions decision. All incoming Ph.D. students receive a fellowship package, which includes tuition and fee support, health insurance, a living stipend, and a research and teaching apprenticeship. The fellowship covers four years of study and requires that students maintain full-time enrollment.
*Requirements:* A limited number of Dean's Fellowships are awarded to incoming full-time, traditional Ed.D. students who show high academic potential.
*Geographic Focus:* All States
*Contact:* Dr. Andrew Porter, Dean; (215) 898-7014; fax (215) 746-6884; andyp@gse.upenn.edu
*Internet:* www.gse.upenn.edu/admissions_financial/scholarships#fellowships
*Sponsor:* Penn Graduate School of Education
3700 Walnut Street
Philadelphia, PA 19104-6216

**Penn Graduate School of Education Foreign Language Area Studies Fellowship** **2100**
The Foreign Language and Area Studies (FLAS) Fellowship is a tuition scholarship and a living stipend awarded to meritorious undergraduate students and graduate students undergoing training in modern foreign languages and related area or international studies, or with the international aspects of professional or fields of study. FLAS fellowships are funded by the U.S. Department of Education and administered by the University of Pennsylvania's Title VI National Resource Centers. FLAS awards are available only for specific languages, and are contingent on federal funding. Awards are made for the Academic Year or Summer. Academic Year Fellows must be admitted to or enrolled in undergraduate, graduate, or professional programs at the University of Pennsylvania. FLAS awards may be used in some cases for students participating in official overseas language programs and in very limited cases for dissertators.
*Requirements:* Applicants must be U.S. citizens or permanent residents of the United States. Students receiving FLAS Fellowships must be enrolled in full-time study for the duration of the FLAS award and must take one language course and one related area or international studies course each semester.
*Geographic Focus:* All States
*Contact:* Dr. Andrew Porter; (215) 898-7014; fax (215) 746-6884; andyp@gse.upenn.edu
*Internet:* www.gse.upenn.edu/admissions_financial/scholarships#fellowships
*Sponsor:* Penn Graduate School of Education
3700 Walnut Street
Philadelphia, PA 19104-6216

**Penn Graduate School of Education Pre-Doctoral Training in Interdisciplinary Methods for Field-Based Research in Education** **2101**
Penn GSE offers a Pre-Doctoral Training Program designed to support and enhance the Ph.D. training of students in different disciplines who are committed to careers in applied, field-based education research. The initiative involves six University of Pennsylvania schools and multiple school districts in Pennsylvania and New Jersey and is supported by the Institute of Education Sciences (IES) in the U.S. Department of Education. Fellows receive financial support, access to specialized resources and experts in the field of study, and become part of an emerging national network of students, faculty and practitioners conducting interdisciplinary research in the education sciences.
*Geographic Focus:* All States
*Contact:* Dr. Andrew Porter; (215) 898-7014; fax (215) 746-6884; andyp@gse.upenn.edu
*Internet:* www.gse.upenn.edu/admissions_financial/scholarships#fellowships
*Sponsor:* Penn Graduate School of Education
3700 Walnut Street
Philadelphia, PA 19104-6216

**Penn Graduate School of Education Research Assistantships** **2102**
Research Assistantships provide a partial or full living stipend and/or tuition support for doctoral students in exchange for assisting faculty with eligible research projects. Research Assistantships provide valuable mentored research experiences for doctoral students and are often offered in conjunction with other forms of aid to create a financial assistance package.
*Geographic Focus:* All States
*Contact:* Dr. Andrew Porter; (215) 898-7014; fax (215) 746-6884; andyp@gse.upenn.edu
*Internet:* www.gse.upenn.edu/admissions_financial/scholarships#assistantships
*Sponsor:* Penn Graduate School of Education
3700 Walnut Street
Philadelphia, PA 19104-6216

**Penny Severns Breast, Cervical and Ovarian Cancer Research Grants** **2103**
The Penny Severns Breast, Cervical and Ovarian Cancer Research Fund is a special fund within the state treasury that is used for breast, cervical and ovarian cancer research grants. Revenue sources include general revenue funds, income tax contributions and gifts, as well as grants and awards from private foundations, nonprofit organizations and other governmental entities or persons. Grants support research in areas related to breast, cervical and ovarian cancer prevention, etiology, pathogenesis, early detection, treatment and behavioral sciences. Research also may include clinical trials. One-year grants are available with the possibility of two subsequent 12-month renewals. Standard research grants are intended to develop and advance the understanding, techniques and modalities effective in early detection, prevention, cure, screening and treatment of breast, cervical and ovarian cancers and may include clinical trials. The maximum award for standard research grants will not exceed $75,000. Fellowship research grants support supervised post-doctoral research training. These grants are intended to further develop the skills necessary for a career in breast, cervical and ovarian cancer research. Fellowship research grants will receive maximum annual funding of $35,000. Special consideration is given to single-year applications and to applications for pilot projects that have the potential for subsequent funding from other sources. Continuation grants, with a maximum of two additional years, are available for those grantees that upon peer review demonstrate progress toward stated goals.
*Requirements:* Although a majority of the applications submitted are bio-medical in nature, researchers in the fields of behavioral and social sciences also are encouraged to apply. Funding is granted to institutions, not individuals. Research is conducted by an individual(s) under the authority of an institution.
*Geographic Focus:* Illinois
*Amount of Grant:* 20,000 - 75,000 USD
*Contact:* Grant Coordinator; 217-524-6088; fax 217-557-3326; dph.owhline@illinois.gov
*Internet:* www.idph.state.il.us/about/womenshealth/grants/penny.htm
*Sponsor:* Illinois Department of Public Health
535 W. Jefferson St., First Floor
Springfield, IL 62761-0001

**Percy B. Ferebee Endowment Grants** **2104**
Grants from the Percy B. Ferebee Endowment are awarded to support charitable, scientific and literary projects, and in particular, governmental and civic projects designed to further the cultural, social, economic and physical well-being of residents of Cherokee, Clay, Graham, Jackson, Macon and Swain Counties of North Carolina and the Cherokee Indian Reservation. Grants are also awarded in the form of scholarships to assist worthy and talented young men and women who reside in said counties in pursuing their college/university degree education within the state of North Carolina. The deadline for the submission of a grant application is September 30. The deadline for scholarship applications is January 31. Application forms are available online. Applicants will receive notice acknowledging receipt of the grant request, and subsequently be notified of the grant declination or approval.
*Requirements:* 501(c)3 non-profits in the Cherokee, Clay, Graham, Jackson, Macon and Swain Counties of North Carolina are eligible to apply for grants. Proposals should be submitted in the following format: completed Common Grant Application Form; an original Proposal Statement; an audited financial report and a current year operating budget; a copy of your official IRS Letter with your tax determination; a listing of your Board of Directors. Proposal Statements (second item in the above Format) should answer these questions: what are the objectives and expected outcomes of this program/project/request; what strategies will be used to accomplish your objective; what is the timeline for completion; if this is part of an on-going program, how long has it been in operation; what criteria will you use to measure success; if the request is not fully funded, what other sources can you engage; an Itemized budget should be included;

please describe any collaborative ventures. Prior to the distribution of funds, all approved grantees must sign and return a Grant Agreement Form, stating that the funds will be used for the purpose intended. Progress reports and Completion reports must also be filed as required for your specific grant. All current grantees must be in good standing with required documentation prior to submitting new proposals to any foundation. Scholarship recipients must be a resident of these areas and must attend a college or university in the state of North Carolina. Contact the Foundation for additional application requirements.
*Restrictions:* Grants are not made for political purposes, nor to organizations which discriminate on the basis of race, ethnic origin, sexual or religious preference, age or gender.
*Geographic Focus:* North Carolina
*Date(s) Application is Due:* Jan 31; Sep 30
*Amount of Grant:* 3,000 - 15,000 USD
*Samples:* Nantahala Regional Library, $15,000—general operating support; Macon County Historical Society, $7,500—general operating expenses; Reach, Inc. Resources Education Assistance Counseling & Housing, $4,000—general operating support.
*Contact:* Wachovia Bank, N.A., Trustee; grantinquiries6@wachovia.com
*Internet:* https://www.wellsfargo.com/private-foundations/ferebee-endowment
*Sponsor:* Percy B. Ferebee Endowment
Wachovia Bank, NC6732, 100 North Main Street
Winston Salem, NC 27150

**Perkin Fund Grants** **2105**
Established in 1967 in New York, the Perkin Fund is a charitable family foundation that provides grants to established institutions in the fields of astronomy, medicine and scientific research. In recent years, the Fund has also supported leading institutions in the arts, education and social services. The Fund reflects the broad interests of the founders, Richard S. and Gladys T. Perkin, and their family. Prior to his death in 1969, Mr. Perkin had been a founder and Chairman of the Board of the Perkin-Elmer Corporation, a leading manufacturer of scientific instruments and optical systems. Following Perkin's death, and until her death in 2001, Mrs. Perkin served as Chairman of the Board of Trustees of the Perkin Fund, lending her support to the advancement of scientific endeavors. Today, the Perkin Fund continues to be led by members of the Perkin family who, along with non-family Trustees, serve the Perkin Fund without compensation. The Fund continues to be primarily interested in scientific research, education, medicine and related disciplines. However, over the past twenty years, the Fund has grown in size and broadened its mandate. This has enabled the Trustees to consider a limited number of requests in the areas of human services and the arts. The Fund bases its grants on written requests from qualified institutions for support of worthwhile projects and programs. The Trustees evaluate these proposals to determine which are deemed especially important and fall within the Fund's guidelines. Types of support have included: annual campaigns; capital and infrastructure; continuing support; endowments; equipment purchase and rental; fellowships; general operating support; program development; and research and research evaluation. Recent award amounts have ranged up to a maximum of $50,000.
*Requirements:* Giving is primarily limited to 501(c)3 organizations that support the residents in Connecticut, Massachusetts, and New York.
*Geographic Focus:* Connecticut, Massachusetts, New York
*Date(s) Application is Due:* Mar 15; Sep 15
*Amount of Grant:* Up to 50,000 USD
*Samples:* Blue Ocean Society for Marine Conservation, Portsmouth, New Hampshire, $25,000 - general operating support; Columbia Business School, New York, New York, $20,000 - general operating support; Harvard Fencing Team, Cambridge, Massachusetts, $50,000 - general operating support.
*Contact:* Winifred P. Gray; (978) 468-2266; theperkinfund@verizon.net
*Sponsor:* Perkin Fund
176 Bay Road, P.O. Box 2220
South Hamilton, MA 01982-2232

**Perpetual Trust for Charitable Giving Grants** **2106**
The Perpetual Trust for Charitable Giving was established in 1957 to support and promote quality educational, human-services, and health-care programming for underserved populations. Special consideration is given to medical aid and medical research organizations, as well as to institutions of higher learning. Grant requests for general operating support are strongly encouraged. Program support will also be considered. Small, program-related capital expenses may be included in general operating or program requests. To better support the capacity of nonprofit organizations, multi-year funding requests are strongly encouraged. Applicants must apply online at the grant website. Applicants are strongly encouraged to do the following before applying: review the downloadable state application procedures for additional helpful information and clarifications; review the downloadable online-application guidelines at the grant website; review the trust's funding history (link is available from the grant website); review the online application questions in advance; and review the list of required attachments. These will generally include: a list of board members, financial statements (audited, reviewed, or compiled by independent auditor); an organization summary; a list of other funding sources; an IRS Determination letter; and other required documents. All attachments must be uploaded in the online application as PDF, Word, or Excel files. The application deadline for the Perpetual Trust is 11:59 p.m. on September 1. Applicants will be notified of grant decisions before November 30.
*Restrictions:* The trust does not support requests from individuals, organizations attempting to influence policy through direct lobbying, or any political campaigns.
*Geographic Focus:* Massachusetts
*Date(s) Application is Due:* Sep 1
*Samples:* Forsyth Institute's Forsyth Dental Infirmary for Children, Boston, Massachusetts, $60,000, ForsythKids Program; Bridge Over Troubled Waters, Boston, Massachusetts, $30,000, general operating support; Codman Academy Foundation, Dorchester, Massachusetts, $25,000, general operating support.
*Contact:* Miki C. Akimoto, Vice President; (866) 778-6859; miki.akimoto@baml.com
*Internet:* https://www.bankofamerica.com/philanthropic/fn_search.action
*Sponsor:* Perpetual Trust for Charitable Giving
225 Franklin Street, 4th Floor, MA1-225-04-02
Boston, MA 02110

**Perry County Community Foundation Health Grants** **2107**
The Perry County Community Foundation is a nonprofit, public charity created for the people of Perry County, Indiana. The Foundation connects donors with the causes they care about. The Foundation will accept grant proposals from all charitable organizations seeking funding; however, special attention will be given to projects that promote healthy living and that offer training and/or support to potential small business entrepreneurship. The Foundation considers proposals for grants on a yearly cycle which begins each July 7 and runs through the annual deadline of September 14. At the start of each cycle, a notice is mailed to nonprofit organizations that have applied for or received grants in the past, or have otherwise requested notification of the start of each cycle. The grant committee makes its recommendations on funding to the Foundation's Board of Trustees, which will make final funding recommendations to the board of directors of the Community Foundation Alliance. All organizations who submit proposals are notified of committee's decision no later than December 1. Grants in the area of health include activities that: improve and promote health outcomes; provide general and rehabilitative health services; offer mental health services; provide crisis intervention programs; strengthen associations or services associated with specific diseases, disorders, and medical disciplines; and support medical research. Samples of previously funded projects are posted on the website.
*Requirements:* The Foundation welcomes proposals from nonprofit organizations that are 501(c)3 and 509(a) tax-exempt under the IRS code and from governmental agencies serving the County of Perry, Indiana. Proposals from nonprofit organizations not classified as a 501(c)3 public charity may be considered if the project is charitable and supports a community need. Proposals submitted by an entity under the auspices of another agency must include a written statement signed by the agency's board president on behalf of the board of directors agreeing to act as the entity's fiscal sponsor, to receive grant monies if awarded, and to oversee the proposed project.
*Restrictions:* Project areas not considered for funding: religious organizations for religious purposes; political parties or campaigns; endowment creation or debt reduction; operating costs; capital campaigns; annual appeals or membership contributions; travel requests for groups or individuals such as bands, sports teams, or classes.
*Geographic Focus:* Indiana
*Date(s) Application is Due:* Sep 14
*Contact:* Renate Warner, Regional Director of Community Engagement and Impact; (812) 547-3176; rwarner@communityfoundationalliance.org or renate@perrycommunityfoundation.org
*Internet:* https://www.communityfoundationalliance.org/perry/program-areas/
*Sponsor:* Perry County Community Foundation
817 12th Street, P.O. Box 13
Tell City, IN 47586-1785

**Peter F. McManus Charitable Trust Grants** **2108**
The Peter F. McManus Charitable Trust was established in Pennsylvania in 2000, and designated to support mental health research on the causes of alcoholism and other substance abuses. Proposed projects may examine basic, clinical, social, and environmental causes of substance abuse. Grants may be requested in an amount up to a maximum of $50,000. Since a formal application is not required, interested parties should forward a brief summary proposal describing the project and the amount of funding requested, along with investigator's biosketch, proposed budget for the project, and a copy of the organization's IRS determination letter. The Board meets once annually, in either October or November. The annual application submission deadline is August 31
*Requirements:* The Trust gives primarily to 501(c)3 organizations in California, Connecticut, Maryland, and Massachusetts.
*Restrictions:* No funding is available for treatment of alcohol or substance abuse. No more than 10% of the grant amount may be used for indirect costs.
*Geographic Focus:* California, Connecticut, Maryland, Massachusetts
*Date(s) Application is Due:* Aug 31
*Amount of Grant:* Up to 50,000 USD
*Samples:* John Hopkins University, Baltimore, Maryland, $50,000 - dopamine system research project; Boston University School of Medicine, Boston, Massachusetts, $50,000 - neurological basis of alcohol addiction; Stanford University, Stanford, California, $50,000 - exploring physician opioid.
*Contact:* Katharine G. Lidz, Trust Administrator; (610) 647-4974; fax (610) 647-8316
*Sponsor:* Peter F. McManus Charitable Trust
31 Independence Place
Chesterbrook, PA 19087-5824

**Pfizer Appropriate Immunisations in Adult Patients with Immune-Mediated Inflammatory Conditions Awards** **2109**
The mission of Pfizer Independent Grants for Learning and Change (IGLC) is to partner with the global healthcare community to improve patient outcomes in areas of mutual interest through support of measurable learning and behavior change strategies. The intent of this document is to encourage organizations with a focus in healthcare professional quality improvement and/or healthcare education to submit a letter of intent (LOI) in

response to a Request for Proposal (RFP) that is related to quality improvement or education in a specific disease state, therapeutic area, or broader area of unmet need. The RFP model is a two-stage process. Stage 1 is the submission of the LOI. After review of the LOI, an interested party may be invited to submit a full grant proposal. Stage 2 is the submission of the full grant proposal. The annual deadline for LOI submissions is April 12, with the full grant proposal deadline being July 18. Individual projects of all scope and size will be considered. For instance, small-scale single-institution initiatives may submit grant requests ranging from $20,000 to $150,000. Large-scale multi-country collaborative projects may request grants of up to $750,000. The total available budget related to this RFP is $750,000.
*Requirements:* The following may apply: medical, nursing, allied health, and/or pharmacy professional schools; healthcare institutions (both large and small); professional associations; government agencies; patient organizations and other entities with a mission related to healthcare improvement. Collaborations within institutions (e.g., between departments and/or inter-professional), as well as between different institutions, organizations, and associations, are encouraged. Applicants must reside in: Albania, Andorra, Armenia, Australia, Austria, Azerbaijan, Belarus, Belgium, Bosnia and Herzegovina, Bulgaria, Croatia, Cyprus, Czech Republic, Denmark, Estonia, Finland, France, Georgia, Germany, Greece, Hungary, Iceland, Ireland, Israel, Italy, Kazakhstan, Kosovo, Latvia, Liechtenstein, Lithuania, Luxembourg, Macedonia, Malta, Moldova, Monaco, Montenegro, New Zealand, Netherlands, Norway, Poland, Portugal, Romania, Russia, San Marino, Serbia, Slovakia, Slovenia, Spain, Sweden, Switzerland, Turkey, Ukraine, United Kingdom, and Vatican City.
*Geographic Focus:* Albania, Andorra, Armenia, Australia, Austria, Azerbaijan, Belarus, Belgium, Bosnia & Herzegovina, Bulgaria, Croatia, Cyprus, Czech Republic, Denmark, Estonia, Finland, France, Germany, Greece, Hungary, Iceland, Ireland, Israel, Italy, Kazakhstan, Kosovo, Latvia, Liechtenstein, Lithuania, Luxembourg, Macedonia, Malta, Moldova, Monaco, Montenegro, Netherlands, New Zealand, Norway, Poland, Portugal, Romania, Russia, San Marino, Serbia, Slovakia, Slovenia, Spain, Sweden, Switzerland, Turkey, Ukraine, United Kingdom, Vatican City
*Date(s) Application is Due:* Jul 18
*Amount of Grant:* 20,000 - 750,000 GBP
*Contact:* Mikael Dolsten, President of Worldwide Research and Development; (212) 733-2323; mikaeldolsten@pfizer.com
*Internet:* https://www.pfizer.com/purpose/medical-grants/request-proposals
*Sponsor:* Pfizer Corporation
235 East 42nd Street
New York, NY 10017-5755

**Pfizer ASPIRE Awards in Endocrinology Research** 2110

The mission of the ASPIRE Awards in Endocrinology Research is to advance the medical knowledge of the fundamental mechanisms of endocrine disease and the mechanisms of action of current and potential future treatments to improve the care of patients with endocrine disorders. Pfizer invites basic-science, innovative-translational, and clinical-research proposals that evaluate endocrine disease, disease mechanisms, and outcomes, primarily in the growth-hormone insulin-like growth-factor field. Funding will be provided to research proposals that address understanding disease mechanisms, genetic factors involved in disease development, factors affecting response to therapeutic agents, identification of novel biomarkers, and identification of novel therapeutic targets or therapeutic approaches. Pfizer will fund up to four awards of 50,000 euros each for one year. Applicants must submit their proposals through the Integrated System for Pfizer Investigator Initiated Research (INSPIIRE) web portal, the link to which is available from the competitive grant announcement on the ASPIRE website. Pfizer invites interested investigators in Europe, Australia and New Zealand to apply. Applications must be received by 11:59 p.m. eastern daylight time on June 15. Interested applicants can download an application guide from this location for more detailed information about application requirements.
*Requirements:* Applicants must hold a professional degree (M.D., Ph.D., Pharm.D., or equivalent), be qualified to carry out the proposed investigation, and still be in the early stages of their career (preferably within ten years of having completed a post-graduate degree). ASPIRE review committee members will consider the following criteria in evaluating proposals: scientific merit of the research proposal; qualifications of the applicant; relevance of the proposal to the ASPIRE program's mission; evidence of the applicant's commitment to an academic research career, evidence of a suitable research environment; and evidence of suitable mentors.
*Restrictions:* No other government, non-governmental, or industry-sponsored funding may cover the same work scope as the grant application to the ASPIRE program; however the proposal may be related to other funding as long as there is no direct overlap. Due to the Foreign Corrupt Practices Act (FCPA), submissions selected for awards must go through an internal review at Pfizer to ensure provision of the grant is appropriate and that relevant information regarding potentially influencing government official status, as well as beneficiaries and controllers information, is captured, when applicable.
*Geographic Focus:* Albania, Armenia, Australia, Belgium, Bosnia & Herzegovina, Bulgaria, Denmark, Estonia, Finland, France, Germany, Great Britain, Greece, Hong Kong, Hungary, Ireland, Italy, Liechtenstein, Luxembourg, Monaco, Netherlands, New Zealand, Portugal, Spain, Sweden, Switzerland, The Netherlands, Turkey, United Arab Emirates, Vatican City
*Amount of Grant:* Up to 50,000 EUR
*Contact:* Mikael Dolsten, President; (212) 733-2323; mikaeldolsten@pfizer.com
*Internet:* https://www.aspireresearch.org/endocrinologyregional/index.html
*Sponsor:* Pfizer Corporation
235 East 42nd Street
New York, NY 10017-5755

**Pfizer ASPIRE Breast Cancer Research awards** 2111

The ASPIRE Breast Cancer Research Awards program underscores Pfizer's commitment to supporting investigators in advancing their academic research in the field of breast cancer. The primary purpose of the program is to support continuous medical research to maximize the available therapies by improving medical knowledge and clinical management of the disease. Pfizer is interested in supporting clinical studies in breast cancer, focusing in the following areas: hormone receptor positive (HR+) advanced breast cancer, safety and efficacy profile of palbociclib/hormone therapy given concurrently with radiation in patients with bone disease; HER2-advanced breast cancer, safety and efficacy profile of talazoparib novel combination (except CDK4/6 inhibitors) in patients with HRD/DDR deficiency; HER2-advanced breast cancer, safety and efficacy profile of gedatolisib plus talazoparib combination in appropriate patients; and early stage breast cancer setting, translational end point driven study to explore potential effect of gedatolisib on immune system in regard to PD-1, PD-L1, TILs, and other appropriate markers, with or without an immunotherapy agent. The program intends to fund four to six clinical studies within research scope, for a total of approximately four million dollars. Selection of research proposals will be performed by an independent external review panel of experts.
*Requirements:* The program is open to U.S. investigators only.
*Geographic Focus* All States, District of Columbia, Guam, Marshall Islands, Northern Mariana Islands, Puerto Rico, U.S. Virgin Islands, American Samoa
*Amount of Grant:* Up to 1,500,000 USD
*Contact:* Mikael Dolsten, President of Worldwide Research and Development; (212) 733-2323; mikaeldolsten@pfizer.com
*Internet:* https://www.aspireresearch.org/onc/index.html
*Sponsor:* Pfizer Corporation
235 East 42nd Street
New York, NY 10017-5755

**Pfizer ASPIRE Dermatology Research Awards** 2112

The purpose of the Pfizer ASPIRE Dermatology Research Awards is to support basic science, translational, clinical and health services research through a competitive grant program that increases medical knowledge in inflammatory and immune-mediated skin diseases. Pfizer is interested in supporting studies in the following areas: clinical research related to inflammatory or immune-mediated skin diseases; basic research in the etiology of inflammatory or immune-mediated skin diseases including clarification of the intracellular immune cascade in the skin; approaches researching all aspects of atopic dermatitis; and PDE4 and crisaborole. ASPIRE Dermatology Disease Research Awards are up to a maximum of $150,000 each. Research projects are expected to complete within two years of study start. Up to five awards may be given.
*Geographic Focus* All States, District of Columbia, Guam, Marshall Islands, Northern Mariana Islands, Puerto Rico, U.S. Virgin Islands, American Samoa
*Amount of Grant:* Up to 150,000 USD
*Contact:* Mikael Dolsten, President of Worldwide Research and Development; (212) 733-2323; mikaeldolsten@pfizer.com
*Internet:* https://www.aspireresearch.org/dermatology/index.html
*Sponsor:* Pfizer Corporation
235 East 42nd Street
New York, NY 10017-5755

**Pfizer ASPIRE Emerging Markets Thrombosis Investigator-Initiated Research Awards** 2113

Pfizer invites investigators to submit emerging markets thrombosis-initiated research proposals from researchers across Africa, Asia, Latin America and the Middle East. The key objective is to encourage the generation of new data to improve patient care for those requiring oral anticoagulation treatment in relevant disease areas of unmet medical needs or special interest. Funds can be used to cover the awardees' salary and fringe benefits, as well as other direct expenses incurred during the research project. Direct expenses may include the salary of technical associates, the purchase of relevant laboratory supplies, and direct expenses necessary for proper conduct of research (e.g. subject-related costs, study-related personnel costs, diagnostic fees/services, data management expenses). Indirect costs may include travel expenses to scientific meetings, software licensing fees, publication costs, etc. Prior to submitting an application, applicants are requested to have a medical to medical conversation with their Pfizer representative. Applicants can email iir@pfizer.com to obtain the appropriate Pfizer contact information for their region. Downloadable application guidelines and online submission instructions are available at the grant website. Applications must be submitted to Pfizer through an online submission website. From there they will be routed to Pfizer's global investigator-initiated research (INSPIIRE) web portal. Applications must be received by the annual deadline date of March 22. The maximum award will be $100,000 USD.
*Requirements:* Applicants must have a professional degree (M.D., Ph.D., Pharm.D., or equivalent) and reside in select countries of Africa, Asia, Latin America, or the Middle East. Countries of Africa include: Algeria, Angola, Benin, Botswana, Burkina Faso, Burundi, Cameroon, Central African Republic, Chad, Congo, Cote D' Ivoire, Djibouti, Egypt, Equatorial Guinea, Eritrea, Ethiopia, Gabon, Gambia, Ghana, Guinea, Kenya, Liberia, Libya, Madagascar, Malawi, Mali, Mauritania, Mauritius Islands, Morocco, Mozambique, Nambia, Niger, Nigeria, Rwanda, Senegal, Sierra Leone, Somalia, South Africa, Sudan, Tanzania, Togo, Tunisia, Uganda, Western Sahara, Zaire (Democratic Republic of the Congo), Zambia, and Zimbabwe. Countries of Asia include: Bangladesh, Bhutan, Brunei, Cambodia, Hong Kong, India, Indonesia, Laos, Macau, Malaysia, Maldives, Myanmar, Nepal, Pakistan, Philippines, Singapore, Taiwan, Thailand, and Vietnam. Countries of Latin America include: Argentina, Belize, Bolivia, Brazil, Chile, Colombia, Costa Rica, Cuba, Dominican Republic, Ecuador, El Salvador, Guatemala, Haiti, Honduras, Jamaica, Mexico, Nicaragua, Panama, Paraguay, Peru, Uruguay, and

Venezuela. Countries of the Middle East include: Saudi Arabia, Bahrain, Iran, Iraq, Jordan, Kuwait, Lebanon, Oman, Qatar, Syria, United Arab Emirates, and Yemen.
*Geographic Focus:* Afghanistan, Algeria, Angola, Bahrain, Bangladesh, Belize, Benin, Bhutan, Bolivia, Botswana, Brazil, Brunei, Burkina Faso, Burundi, Cambodia, Cameroon, Central African Republic, Chad, Chile, Colombia, Congo, Costa Rica, Cote d' Ivoire (Ivory Coast), Cuba, Djibouti, Dominican Republic, Ecuador, Egypt, El Salvador, Equatorial Guinea, Eritrea, Ethiopia, Gabon, Gambia, Ghana, Guatemala, Guinea, Haiti, Honduras, Hong Kong, India, Indonesia, Iran, Iraq, Jamaica, Jordan, Kenya, Kuwait, Laos, Lebanon, Liberia, Libya, Macau, Madagascar, Malawi, Malaysia, Maldives, Mali, Mauritania, Mauritius, Mexico, Morocco, Mozambique, Myanmar (Burma), Namibia, Nepal, Nicaragua, Niger, Nigeria, Oman, Pakistan, Panama, Paraguay, Peru, Philippines, Qatar, Rwanda, Saudi Arabia, Senegal, Sierra Leone, Singapore, Somalia, South Africa, Sudan, Syria, Taiwan, Tanzania, Thailand, Tonga, Tunisia, Uganda, United Arab Emirates, Uruguay, Venezuela, Vietnam, Western Sahara, Yemen, Zaire, Zambia, Zimbabwe
*Date(s) Application is Due:* Mar 22
*Amount of Grant:* Up to 100,000 USD
*Contact:* Mikael Dolsten, President; (212) 733-2323; mikaeldolsten@pfizer.com
*Internet:* https://www.aspireresearch.org/emrista/index.html
*Sponsor:* Pfizer Corporation
235 East 42nd Street
New York, NY 10017-5755

**Pfizer ASPIRE Gene and anti-TFPI Therapies for Hemophilia Global Research Awards** 2114

The Advancing Science through Pfizer: Investigator Research Exchange (ASPIRE) Gene and anti-TFPI Therapies for Hemophilia Global Research Awards program is a competitive grants program that reflects the commitment of Pfizer Hemophilia Rare Disease group to support ongoing basic science and clinical investigation in gene and anti-TFPI therapies for hemophilia. Ongoing basic science research and clinical research are critical to deepen understanding of disease mechanisms and to ensure advancement of management strategies for hemophilia and related comorbidities. In an era of increased competition for research funding, the ASPIRE Gene and anti-TFPI Therapies in Hemophilia Global Research Awards Program is designed to support laboratory and clinical research in pathogenesis, complications, management, and clinical outcomes of these therapies for hemophilia. Areas of research focus include: basic science of gene therapy for hemophilia; basic science of TFPI and anti-TFPI monoclonal antibodies; and patients with MILD hemophilia A or B. Selection of research proposals will be performed by an independent, external expert panel comprised of nationally known academic clinicians. Project duration should be one to two years and should be approximately $125,000 per year, inclusive of overhead costs (capped at 28%). It is anticipated two to three awards will be made.
*Geographic Focus* All States, District of Columbia, Guam, Marshall Islands, Northern Mariana Islands, Puerto Rico, U.S. Virgin Islands, American Samoa
*Amount of Grant:* Up to 125,000 USD
*Contact:* Mikael Dolsten, President of Worldwide Research and Development; (212) 733-2323; mikaeldolsten@pfizer.com
*Internet:* https://www.aspireresearch.org/genetherapyforhemophilia/index.html
*Sponsor:* Pfizer Corporation
235 East 42nd Street
New York, NY 10017-5755

**Pfizer ASPIRE Global Transthyretin (TTR) Amyloidosis Competitive Research Grants** 2115

The Global ASPIRE Transthyretin Amyloidosis Competitive Research Grant program underscores Pfizer's commitment to supporting investigators with an interest in advancing their academic research in TTR amyloidosis. The primary mission of the program is to support specific research in basic science and broad clinical research through a competitive grants program that advances medical knowledge in the understanding, diagnosis and treatment of TTR amyloidosis. In-scope research submissions along the entire clinical spectrum of TTR amyloidosis (from TTR familial amyloid polyneuropathy (TTR-FAP) to TTR cardiomyopathy) and including mixed phenotypes are highly encouraged. Within the specified areas of research interest, investigators are expected to generate data to better understand important diagnostic and clinical management issues relevant to treating patients with TTR amyloidosis. Proposals may include post-hoc exploratory analyses and well-designed retrospective or prospective analyses to generate and/or test hypotheses or studies to validate new assessments. Awards will range from $25,000 USD to a maximum of $50,000 USD. The annual application deadline is May 14.
*Requirements:* Pfizer invites junior investigators from around the world with a terminal degree (MD and/or PhD and/or PharmD or equivalent) who are developing their research careers in TTR Amyloidosis to apply through submission of innovative research proposals designed to enhance our ability to understand the epidemiology, basic science and early diagnosis and treatment of patients with TTR amyloidosis.
*Restrictions:* No other government, non-governmental, or industry-sponsored projects may cover the same work scope as the grant application to the EUROPE ASPIRE Program. However, a EUROPE ASPIRE grant may be related to other funding from foundations or government agencies, as long as there is no direct overlap. The institutional overhead costs cannot exceed 28% of the direct costs. Pfizer does not pay overhead on indirect costs.
*Geographic Focus:* All Countries
*Date(s) Application is Due:* Mar 14
*Amount of Grant:* 25,000 - 50,000 USD
*Contact:* Mikael Dolsten, President; (212) 733-2323; mikaeldolsten@pfizer.com
*Internet:* https://www.aspireresearch.org/TTRFAP/index.html
*Sponsor:* Pfizer Corporation
235 East 42nd Street
New York, NY 10017-5755

**Pfizer ASPIRE North America Hemophilia Research Awards** 2116

Pfizer's ASPIRE Hemophilia Research awards are intended to support basic-science, translational, and clinical research that advances medical knowledge, pathogenesis, and treatment of hemophilia and to support academic research as well as the career development of promising young and established scientists. Projects are expected to yield results that will advance the medical knowledge of Hemophilia A or Hemophilia B. It is expected that results will be presented at scientific meetings and published in peer reviewed journals. Proposed projects should last one to two years and cost approximately $$50,000 to $70,000 annually, inclusive of overhead costs. Funds may be used to cover the awardees' salary and fringe benefits, as well as other direct expenses incurred during the research project. Direct expenses may include the salary of technical associates; the purchase of relevant laboratory supplies; direct expenses necessary for proper conduct of research (e.g. subject-related costs, study-related personnel costs, diagnostic fees/services, data management expenses). Indirect costs may include travel expenses to scientific meetings, software licensing fees, publication costs, etc. Research proposals will be evaluated and selected by an independent, external expert panel comprised of nationally known academic clinicians. Applicants must submit their proposals through the Integrated System for Pfizer Investigator Initiated Research (INSPIIRE) web portal, the link to which is available from the competitive grant announcement on the ASPIRE website. Applications must be received by 11:59 p.m. eastern daylight time on the deadline date of April 13. Deadlines may vary from year to year. Interested applicants are advised to verify the current deadline date at the website or by contacting Pfizer.
*Requirements:* Applicants must have a professional degree (M.D., Ph.D., Pharm.D., or equivalent) and reside in the United States. Applicants should consider the following components when developing their research proposals: in vitro, in vivo, and clinical investigations; evaluation of disease pathogenesis and host response; and comparison of host response and clinical impact. Proposals will be evaluated on the following criteria: scientific merit of the research proposal; qualifications of the applicant; relevance of the proposed research to Pfizer's hematology program mission; evidence of the applicant's commitment to an academic research career; and evidence of a suitable research environment.
*Restrictions:* ASPIRE Hemophilia award winners may apply for awards in the future; awards are based upon scientific merit commensurate with funding that is available. No other government, non-governmental, or industry-sponsored funding may cover the same work scope as the grant application to the ASPIRE program; however the proposal may be related to other funding as long as there is no direct overlap.
*Geographic Focus* All States, District of Columbia, Guam, Marshall Islands, Northern Mariana Islands, Puerto Rico, U.S. Virgin Islands, American Samoa
*Date(s) Application is Due:* Apr 13
*Amount of Grant:* Up to 70,000 USD
*Contact:* Mikael Dolsten, President of Worldwide Research and Development; (212) 733-2323; mikaeldolsten@pfizer.com
*Internet:* https://www.aspireresearch.org/ASPIRE-Hemophilia/index.html
*Sponsor:* Pfizer Corporation
235 East 42nd Street
New York, NY 10017-5755

**Pfizer ASPIRE North America Rheumatology and Inflammatory Bowel Disease Research Awards** 2117

Pfizer's ASPIRE Rheumatology Research awards are designed to increase the understanding of health disparities as they relate to rheumatoid arthritis in Hispanic and other underrepresented minority populations. Specifically, proposals should address the following areas: epidemiology and current trends in diagnosis and management; clinical phenotypic differences; and genomic research into the heterogeneity of and within the targeted population. Pfizer will provide up to a maximum of $150,000 to fund research conducted over a two-year period. Applicants must submit their proposals through the Integrated System for Pfizer Investigator Initiated Research (INSPIIRE) web portal, the link to which is available from the competitive grant announcement on the ASPIRE website. Applications must be received by 11:59 p.m. eastern daylight time on July 10.
*Requirements:* Applicants must have a professional degree (M.D., Ph.D., Pharm.D., or equivalent) and reside in the United States. Investigators are expected to generate exploratory data to better understand the variance in clinical presentation, severity of illness, response to therapy, and predictors of efficacy and tolerability in the targeted population. ASPIRE review committee members will consider the following criteria in evaluating proposals: scientific merit of the research proposal; qualifications of the applicant; relevance of the proposal to the ASPIRE program's mission; evidence of the applicant's commitment to an academic research career, evidence of a suitable research environment; and evidence of a mentorship program.
*Restrictions:* ASPIRE Rheumatology Research awards are limited to one award per investigator per lifetime. No other government, non-governmental, or industry-sponsored funding may cover the same work scope as the grant application to the ASPIRE program; however the proposal may be related to other funding as long as there is no direct overlap.
*Geographic Focus:* All States
*Date(s) Application is Due:* Jul 10
*Amount of Grant:* Up to 150,000 USD
*Contact:* Mikael Dolsten, President of Worldwide Research and Development; (212) 733-2323; mikaeldolsten@pfizer.com
*Internet:* https://www.aspireresearch.org/Rheumatology/index.html

*Sponsor:* Pfizer Corporation
235 East 42nd Street
New York, NY 10017-5755

**Pfizer ASPIRE Sickle Cell Disease Research Awards** **2118**

Pfizer ASPIRE Sickle Cell Disease Research Awards program offers a new opportunity for researchers in the United States and Canada, underscoring Pfizer's commitment to supporting investigators with an interest in advancing innovation in Sickle Cell Disease. The primary purpose is to support: basic science, translational and clinical research through a competitive grants program that advances medical knowledge in the pathogenesis and treatment of Sickle Cell Disease; innovative pilot studies focused on improving patient outcomes in Sickle Cell Disease; and academic research as well as the career development of promising young and established scientists. Examples of research topics are listed for guidance below (although other applications will be considered); biomarkers of sickle cell disease, including VOC; novel potential drug targets in sickle cell disease; phenotypic differences in clinical course and/or response to therapy; clinical trial endpoint validation in sickle cell disease; patient-reported outcomes, health-related quality of life; improving care of specific complications of sickle cell disease; improving care delivery models; and improving acute pain management. Review of applications will be performed by an independent, external panel of experts. Project duration should be one or two years, and the approximate total award will not exceed $55,000 USD total, inclusive of overhead costs (capped at 28%). The annual application submission date is March 19.
*Requirements:* Sickle Cell Disease ASPIRE program is open to investigators in the U.S. and Canada.
*Geographic Focus:* All States, Canada
*Date(s) Application is Due:* Mar 19
*Amount of Grant:* Up to 55,000 USD
*Contact:* Mikael Dolsten, President of Worldwide Research and Development; (212) 733-2323; mikaeldolsten@pfizer.com
*Internet:* https://www.aspireresearch.org/ASPIRE-Sicklecell/index.html
*Sponsor:* Pfizer Corporation
235 East 42nd Street
New York, NY 10017-5755

**Pfizer ASPIRE United States Investigator Research Awards in Endocrinology** **2119**

As competition for research funding opportunities increases, the ASPIRE program underscores Pfizer's commitment to supporting exceptional investigators with an interest in advancing their academic research careers in Endocrinology. The mission of the ASPIRE United States Investigator Research Awards in Endocrinology is to advance the medical knowledge of the fundamental mechanisms of disease, mechanisms of action of current and potential future treatments, and to improve the care of patients with endocrine disorders. Scientific research continues to advance our knowledge and understanding of endocrine disorders and their treatments. The 2018 U.S. ASPIRE Investigator Research Awards in Endocrinology is designed to: support exceptional investigators working in the field of endocrinology; and advance the understanding of disease mechanisms or treatments through translational basic science or clinical research in order to enhance the clinical care of adult patients with endocrine diseases. Currently, funding will be provided to research proposals that seek to explore translational or clinical research in the following endocrinology areas acromegaly in adults. The program will fund two awards of up to $75,000 each over two years to conduct basic science, translational or clinical research generating data that will help advance the science of acromegaly in adults. Applications must be received by April 27.
*Requirements:* The interested applicant must meet the following criteria to be eligible to apply: be a medical graduate currently residing, working, and conducting research within the United States; be registered to practice medicine in the United States; and be a current trainee in Adult Endocrinology in the United States. Pfizer grants are awarded with the expectation that timely submissions for presentation will be made at appropriate medical meetings or research forums and that papers will be submitted for publication in appropriate peer-reviewed medical journals. The grant recipient will be required to provide Pfizer with an opportunity to review any proposed publications.
*Restrictions:* Awardees must remain in United States for the duration of the study.
*Geographic Focus* All States, District of Columbia, Guam, Marshall Islands, Northern Mariana Islands, Puerto Rico, U.S. Virgin Islands, American Samoa
*Amount of Grant:* Up to 75,000 AUD
*Contact:* Mikael Dolsten, President of Worldwide Research and Development; (212) 733-2323; mikaeldolsten@pfizer.com
*Internet:* https://www.aspireresearch.org/endocrineglobal/index.html
*Sponsor:* Pfizer Corporation
235 East 42nd Street
New York, NY 10017-5755

**Pfizer Australasian Paediatric Endocrine Care (APEC) Research Grants** **2120**

The Australasian Pediatric Endocrine Care (APEC) Research Grants were established in 2007 to support and advance research in the field of Pediatric Endocrinology. The program has been designed to promote excellence in medical research throughout Australasia. It seeks to encourage and help develop the best of our junior clinical researchers to enter or continue in the field of Pediatric Endocrinology and extend Pfizer Australia's commitment and contribution to medicine and the health of our community. APEC Grants are designed to encourage junior researchers to enter or continue in the field of Pediatric Endocrinology. Pfizer Australia has allocated AUD $100,000 to this initiative as part of its continued commitment to Pediatric Endocrinology research. Individual grants will be available up to AUD $50,000 (excluding GST). Each grant is for a period of twelve months. The annual application deadline is April 3.
*Requirements:* Applications are welcome from medical graduates who are currently residing, working, and conducting research within Australasia. The applicant must be currently registered to practice medicine in Australasia; the majority of the research must be conducted in Australasia, and as of 01 January of next year, applicants must be a current trainee in pediatric endocrinology in Australasia, or within five years of being awarded FRACP, or within five years of having completed pediatric endocrinology training as determined by the RACP (Royal Australasian College of Physicians) Endocrinology SAC.
*Restrictions:* The applicant must not have held an APEC Grant within the past five years.
*Geographic Focus:* Australia
*Date(s) Application is Due:* Apr 3
*Amount of Grant:* Up to 50,000 AUD
*Contact:* Maureen Surian, Grants Administrator; +61 2 9850 3973 or +61 2 9850 3333; maureen.surian@pfizer.com
Rosie Moffat; +61 2 9850 3973 or +61 2 9850 3333; rosie.moffat@pfizer.com
*Internet:* https://www.pfizergrants.com.au/grants/PaediatricEndocrine.aspx
*Sponsor:* Pfizer Corporation
235 East 42nd Street
New York, NY 10017-5755

**Pfizer Awards Antimicrobial Stewardship in the Asia-Pacific Region** **2121**

The Joint Commission, in collaboration with Pfizer Independent Grants for Learning and Change, is initiating this Request for Proposal (RFP) for research projects focused on the implementation and/or assessment of antimicrobial stewardship programs in the Asia-Pacific region. The mission of Pfizer Independent Grants for Learning and Change (IGLC) is to partner with the global healthcare community to improve patient outcomes in areas of mutual interest through support of measurable learning and behavior change strategies. In order to reduce the development and spread of resistant bacteria and deliver better patient outcomes, healthcare organizations must implement measures to ensure optimal use of antibiotics. Implementation of an antimicrobial stewardship program will help organizations reach the goal of providing patients requiring antibiotic treatment with the right antibiotics, at the right time, at the right dose, and for the right duration. The program is seeking quality improvement research, health services research and/or educational research applications that identify and evaluate practices for successfully implementing key elements of antimicrobial stewardship programs within hospitals and other healthcare organizations. Individual projects requesting up to $250,000 will be considered. The total available budget related to this RFP is $1,000,000. Letters of intent are due by the deadline of May 4, with full proposals having a deadline of August 3.
*Requirements:* Hospitals, healthcare systems, healthcare institutions (both large and small), professional associations, societies, and others with a mission related to healthcare improvement may apply. Collaborations within institutions (e.g., between departments and/or inter-professional), as well as between different institutions, organizations, and associations, are encouraged. Applicants must serve residents in: Australia, Bangladesh, Bhutan, Brunei, Burma, Cambodia, Canada, China, Chile, Cook Islands, Fiji, French Polynesia, Guam, India, Indonesia, Japan, Kiribati, Laos, Malaysia, Maldives, Marshall Islands, Micronesia, Mongolia, Nauru, Nepal, New Caledonia, New Zealand, Niue, North Korea, Pakistan, Palau, Papua New Guinea, Peru, Philippines, Russia, Samoa, Singapore, Solomon Islands, South Korea, Sri Lanka, Thailand, Timor-Lester, Tonga, Tuvalu, Vanuatu, Vietnam, and the United States.
*Geographic Focus:* All States, Guam, Marshall Islands, American Samoa, Australia, Bangladesh, Bhutan, Brunei, Burma (Myanmar), Cambodia, Canada, Chile, China, Cook Islands, Fiji, French Polynesia, India, Indonesia, Japan, Kiribati, Laos, Malaysia, Maldives, Micronesia, Mongolia, Nauru, Nepal, New Caledonia, New Zealand, Niue, North Korea, Pakistan, Palau, Papua New Guinea, Peru, Philippines, Russia, Singapore, Solomon Islands, South Korea, Sri Lanka, Thailand, Timor-Lester, Tonga, Tuvalu, Vanuatu, Vietnam
*Date(s) Application is Due:* Aug 3
*Amount of Grant:* Up to 250,000 USD
*Contact:* Mikael Dolsten, President of Worldwide Research and Development; (212) 733-2323; mikaeldolsten@pfizer.com
*Internet:* https://www.pfizer.com/purpose/medical-grants/request-proposals
*Sponsor:* Pfizer Corporation
235 East 42nd Street
New York, NY 10017-5755

**PFizer Compound Transfer Agreements** **2122**

Pfizer has a pure-powder-transfer program encompassing all of the company's publicly-known compounds and routinely enters Compound Transfer Agreements (CTAs) that provide compounds free of charge to third parties to conduct agreed-upon studies. To request a Pfizer pure-substance compound, applicants should complete the submission form located on the Integrated System for Pfizer Investigator Initiated Research (INSPIIRE) web portal, the link to which is available on the Compound-Transfer-Program web page. Processing time from initial submission to shipment of the approved compound(s) takes approximately twelve weeks. Applicants will receive an auto-reply email confirming their submission has been completed and can track the status of that submission on the INSPIIRE portal.
*Requirements:* An attachment with a full description of the specific research study proposed should be included with the applicant's on-line submission. Pfizer reviews each request for content, value, and consistency with Pfizer's direction for the requested compound and reserves the right to deny any request the reviewers feel does not advance the compound(s) or add value to human or animal health. Pfizer requires a manuscript or report of the conclusions to bring closure to the study.
*Restrictions:* Studies involving humans or requesting funding support should be submitted through Pfizer's Investigator Initiated Research program.
*Geographic Focus:* All States, All Countries

*Contact:* Mikael Dolsten, President; (212) 733-2323
*Internet:* https://www.pfizer.com/science/compound-transfer-program
*Sponsor:* Pfizer Corporation
235 East 42nd Street
New York, NY 10017-5755

**Pfizer Dermatology Fellowships** 2123

The mission of Pfizer Independent Grants for Learning and Change (IGLC) is to partner with the global healthcare community to improve patient outcomes in areas of mutual interest through support of measurable learning and behavior change strategies. The Pfizer Fellowship program is a reflection of Pfizer's commitment to advancing research, innovation, clinical expertise and medical education for medical, nursing, and allied health professionals. This program provides career-building opportunities for practicing clinicians and academic researchers in outcomes and patient-oriented research that is vital to improving healthcare. The intent of this Request for Proposal (RFP) is to encourage organizations with an active fellowship program relevant to inflammatory and immune -mediated skin diseases fellowship program to submit a proposal. The purpose of the Fellowship program grant is to support the continued training of fellows in any two years of a program. Organizations seeking funding must submit proposals describing the strength of their program in supporting clinical training and research. Individual requests for fellowship grants up to $80,000 per year for up to two years will be considered. The annual application deadline is March 23.
*Requirements:* The following requirements are in place for the Fellowship: funding is to be used as a salary supplement for a fellow at the recipient organization; funding will be awarded to the organization, which will select the recipient; funding must be used primarily as salary and fringe support for the selected fellow, not for overhead or indirect costs; and the U.S. Physician Payments Sunshine Act and other transparency law reporting requirements must be satisfied with respect to meals, payments, travel and other transfers of value.
*Geographic Focus* All States, District of Columbia, Guam, Marshall Islands, Northern Mariana Islands, Puerto Rico, U.S. Virgin Islands, American Samoa
*Date(s) Application is Due:* Mar 23
*Amount of Grant:* 80,000 - 160,000 GBP
*Contact:* Mikael Dolsten, President of Worldwide Research and Development; (212) 733-2323; mikaeldolsten@pfizer.com
Amanda Solis, Program Officer; (212) 733-2323; amanda.solis@pfizer.com
*Internet:* https://www.pfizer.com/purpose/medical-grants/request-proposals
*Sponsor:* Pfizer Corporation
235 East 42nd Street
New York, NY 10017-5755

**Pfizer Developing Clinical Research and Publication Skills Award in Africa, Europe, and the Middle East** 2124

The mission of the British Medical Journal is to publish rigorous, accessible, and entertaining material that will help doctors and medical students in their daily practice, lifelong learning, and career development. As well as publishing research, the British Medical Journal has developed the Research to Publication program which is part of the Journal's commitment to building medical research capabilities around the world. Research to Publication is a gold standard online program of self- study courses to enable doctors and healthcare researchers to develop clinical research skills, write the best possible papers, and succeed in getting papers published. The mission of Pfizer Independent Grants for Learning and Change (IGLC) is to partner with the global healthcare community to improve patient outcomes in areas of mutual interest through support of measurable learning and change strategies. The intent of this program is to encourage organizations with a focus in healthcare professional education and/or quality improvement to submit an application for funding to purchase access to Research to Publication for their organization. The total available budget related to this RFP is £50,000. The annual application submission deadline is April 30.
*Requirements:* The following organizations may apply: medical, dental, nursing, allied health, and/or pharmacy professional schools; healthcare institutions (both large and small); academic institutions and other entities with a mission related to healthcare improvement. Applicants must reside in select countries of Africa, Europe, or the Middle East. Countries of Africa include: Algeria, Angola, Benin, Botswana, Burkina Faso, Burundi, Cameroon, Central African Republic, Chad, Congo, Cote D' Ivoire, Djibouti, Egypt, Equatorial Guinea, Eritrea, Ethiopia, Gabon, Gambia, Ghana, Guinea, Kenya, Liberia, Libya, Madagascar, Malawi, Mali, Mauritania, Mauritius Islands, Morocco, Mozambique, Nambia, Niger, Nigeria, Rwanda, Senegal, Sierra Leone, Somalia, South Africa, Sudan, Tanzania, Togo, Tunisia, Uganda, Western Sahara, Zaire (Democratic Republic of the Congo), Zambia, and Zimbabwe. Countries of Europe include: Albania, Andorra, Armenia, Austria, Azerbaijan, Belarus, Belgium, Bosnia and Herzegovina, Bulgaria, Croatia, Cyprus, Czech Republic, Denmark, Estonia, Finland, France, Georgia, Germany, Greece, Hungary, Iceland, Ireland, Italy, Kazakhstan, Kosovo, Latvia, Liechtenstein, Lithuania, Luxembourg, Macedonia, Malta, Moldova, Monaco, Montenegro, Netherlands, Norway, Poland, Portugal, Romania, Russia, San Marino, Serbia, Slovakia, Slovenia, Spain, Sweden, Switzerland, Turkey, Ukraine, United Kingdom, and Vatican City. Countries of the Middle East include: Saudi Arabia, Bahrain, Iran, Iraq, Jordan, Kuwait, Lebanon, Oman, Qatar, Syria, United Arab Emirates, and Yemen.
*Geographic Focus:* Albania, Algeria, Andorra, Angola, Armenia, Austria, Azerbaijan, Bahrain, Belarus, Belgium, Benin, Bosnia & Herzegovina, Botswana, Bulgaria, Burkina Faso, Burundi, Cameroon, Central African Republic, Chad, Congo, Cote d' Ivoire (Ivory Coast), Croatia, Cyprus, Czech Republic, Denmark, Djibouti, Egypt, Equatorial Guinea, Eritrea, Estonia, Ethiopia, Finland, France, Gabon, Gambia, Germany, Ghana, Great Britain, Greece, Guinea, Hungary, Iceland, Iran, Iraq, Ireland, Italy, Jordan, Kazakhstan, Kenya, Kosovo, Kuwait, Latvia, Lebanon, Liberia, Libya, Liechtenstein, Lithuania, Luxembourg, Macedonia, Madagascar, Malawi, Mali, Malta, Mauritania, Mauritius, Moldova, Monaco, Montenegro, Morocco, Mozambique, Namibia, Netherlands, Niger, Nigeria, Norway, Oman, Poland, Portugal, Qatar, Romania, Russia, Rwanda, San Marino, Saudi Arabia, Senegal, Serbia, Sierra Leone, Slovakia, Slovenia, Somalia, South Africa, Spain, Sudan, Sweden, Switzerland, Tanzania, Togo, Tunisia, Turkey, Uganda, Ukraine, United Arab Emirates, United Kingdom, Vatican City, Western Sahara, Yemen, Zaire, Zambia, Zimbabwe
*Date(s) Application is Due:* Apr 30
*Amount of Grant:* Up to 50,000 GBP
*Contact:* Mikael Dolsten, President of Worldwide Research and Development; (212) 733-2323; mikaeldolsten@pfizer.com
*Internet:* https://www.pfizer.com/purpose/medical-grants/request-proposals
*Sponsor:* Pfizer Corporation
235 East 42nd Street
New York, NY 10017-5755

**Pfizer Developing Clinical Research and Publication Skills Award in the United Kingdom** 2125

The mission of the British Medical Journal is to publish rigorous, accessible, and entertaining material that will help doctors and medical students in their daily practice, lifelong learning, and career development. As well as publishing research, the British Medical Journal has developed the Research to Publication program which is part of the Journal's commitment to building medical research capabilities around the world. Research to Publication is a gold standard online program of self- study courses to enable doctors and healthcare researchers to develop clinical research skills, write the best possible papers, and succeed in getting papers published. The mission of Pfizer Independent Grants for Learning and Change (IGLC) is to partner with the global healthcare community to improve patient outcomes in areas of mutual interest through support of measurable learning and change strategies. The intent of this program is to encourage organizations with a focus in healthcare professional education and/or quality improvement to submit an application for funding to purchase access to Research to Publication for their organization. The total available budget related to this RFP is £50,000. The annual application submission deadline is April 30.
*Requirements:* The following organizations may apply: medical, dental, nursing, allied health, and/or pharmacy professional schools; healthcare institutions (both large and small); academic institutions and other entities with a mission related to healthcare improvement.
*Geographic Focus:* United Kingdom
*Date(s) Application is Due:* Apr 30
*Amount of Grant:* Up to 50,000 GBP
*Contact:* Mikael Dolsten, President; (212) 733-2323; mikaeldolsten@pfizer.com
*Internet:* https://www.pfizer.com/purpose/medical-grants/request-proposals
*Sponsor:* Pfizer Corporation
235 East 42nd Street
New York, NY 10017-5755

**Pfizer Foundation Healthcare Charitable Contributions** 2126

The Pfizer Foundation's Healthcare Charitable Contributions support the following types of requests and projects: patient education (including health screening); patient advocacy for disease awareness; and improving patent access to care. Funding is currently available in the following clinical areas to support healthcare charitable contributions: arthritis and pain management; Gaucher disease; growth disorders; hemophilia; infectious disease (bacterial, pneumococcal disease prevention); neurology; amyloidosis; dementia; diabetic peripheral neuropathy; fibromyalgia; multiple sclerosis; oncology; breast cancer; gastrointestinal stromal tumors; leukemia (CML and ALL); lung cancer; non-Hodgkin's lymphoma; pancreatic neuroendocrine tumors; renal cell carcinoma; pulmonary arterial hypertension; respiratory (COPD, smoking cessation); rheumatoid arthritis; transplantation (kidney transplant); urology (overactive bladder); and women's health (menopause, cardiovascular risk, depression, bladder health, and sexual health). Applicants have four opportunities during the year to apply for Pfizer's Healthcare Charitable Contributions: from December 1 through January 15 for activities starting in April; March 1 through April 15 for activities starting in July; June 1 through July 15 for activities starting in October; and September 1 through October 15 for activities starting in January. Interested applicants must apply through the Pfizer Grants Application System. Applicants can register and log into the system from the grant webpage.
*Requirements:* 501(c)3 organizations are eligible to apply.
*Restrictions:* Pfizer must not receive any significant value in terms of good or services in return for their donation.
*Geographic Focus:* All States, All Countries
*Date(s) Application is Due:* Jan 15; Apr 15; Jul 15; Oct 15
*Contact:* Caroline Roan; (866) 634-4647 or (212) 209-8997; healthcharitables@pfizer.com
*Internet:* https://www.pfizer.com/responsibility/grants_contributions/description_funding_recipient_organizations
*Sponsor:* Pfizer Foundation
235 East 42nd Street
New York, NY 10017-5755

**Pfizer Foundation Knowledge Gap Track Two Grants** 2127

The mission of the Pfizer Knowledge Gap Track Two grants program is to cooperate with health-care delivery organizations and professional associations to narrow professional practice gaps through support of continuing education, learning, and continuous-improvement strategies that result in measurable improvement in competence, performance, or patient outcomes. Applicants may apply for two types of medical-

education grants: healthcare quality improvement (track one) grants; and annual meetings for purposes of emerging science/knowledge exchange (track two) grants. The purpose of track-two grants is to support the live presentation of original research and the exchange of emerging clinical information. Eligible providers can request funding to support live (fact-to-face) activities at major national and regional conferences and congresses in the following clinical areas: amyloidosis (familial amyloid polyneuropathy); cardiovascular metabolic risk (lipids, diabetes); chronic obstructive pulmonary disease; cough and cold; diabetic peripheral neuropathy; dietary supplements; fibromyalgia; Gaucher disease; growth disorders; hemophilia; infectious disease; multiple sclerosis; oncology (breast cancer, gastrointestinal stromal tumors, leukemia, lung cancer, non-Hodgkin's lymphoma, pancreatic neuroendocrine tumors, and renal cell carcinoma); pain; pulmonary arterial hypertension; rheumatoid arthritis; smoking cessation; transplantation (kidney); urology (overactive bladder); women's health (menopause); and support for the Continuing Medical Education (CME)/Continuing Education (CE)/Continuing Professional Development (CPD) profession. Track two grants have five submission cycles annually: January 1 through February 15 for activities starting in April; March 1 through April 15 for activities starting in June; May 1 through June 15 for activities starting in August; July 1 through August 15 for activities starting in October; and September 1 through October 15 for activities starting in December. Regional meeting requests are capped at $25,000 and national meeting requests have a $50,000 cap. Application must be made through the Medical Education Group's online Grant Management System (GMS) which is also available at the website. Applicants must first register themselves in the GMS to create an account.
*Requirements:* Eligibility is based on the following criteria: whether or not the activity aligns with Pfizer's posted areas of interest (see Description); whether the activity is a live (face-to-face) annual activity which serves as a platform for the exchange of new clinical and scientific information and reaches a national or regional audience; whether the activity was developed by the applicant organization specifically for its professional members or constituents; whether the activity is an established part of the applicant organization's ongoing educational program; and whether the activity (if new) is clearly based on an assessment of the educational needs of the applicant's target audience.
*Restrictions:* Only one request per annual meeting will be accepted.
*Geographic Focus:* All States, All Countries
*Date(s) Application is Due:* Feb 15; Apr 15; Jun 15; Aug 15; Oct 15
*Amount of Grant:* 25,000 - 50,000 USD
*Contact:* Caroline Roan; (866) 634-4647 or (212) 209-8997; healthcharitables@pfizer.com
*Internet:* https://www.pfizer.com/purpose/medical-grants/process
*Sponsor:* Pfizer Foundation
235 East 42nd Street
New York, NY 10017-5755

**Pfizer Global Investigator Sponsored Research Grants 2128**
Pfizer supports Investigator Sponsored Research (ISR) projects that advance medical and scientific knowledge about our therapies. An ISR is a type of grant that supports an independent research study where the investigator or organization is the sponsor of the study and where Pfizer provides financial and/or non-financial support for the development or refinement of specific and defined medical knowledge relating to a Pfizer asset. This global program is open to all researchers who are interested in conducting their own research. This grant type is used as support for pre-clinical and clinical studies (including interventional and non-interventional), that involve a Pfizer asset (e.g., commercial drug, investigational drug, pure compound).
*Requirements:* Following are types of research eligible for support: clinical studies of approved and unapproved uses involving approved or unapproved Pfizer drugs; observational studies, such as epidemiology studies and certain outcomes research studies where the primary focus is the scientific understanding of disease; other types of independent research on disease states, including novel diagnostic screening tools and surveys where Pfizer has no direct commercial interest; and in vitro or animal studies which include a request for funding. Applicants should note that Pfizer requires the following documents before it can initiate support (either drug and/or funding) of an IIR request: internal review board (IRB)/ ethics committee approval; regulatory response documentation - Investigational New Drug (IND) or Compound Transfer Agreement (CTA) documentation if applicable, a fully executed IIR agreement, and a final study protocol.
*Restrictions:* Pfizer will not support the following types of requests: requests for ongoing or new research without an associated study protocol or synopsis; general education and training activities; support for ongoing clinical programs that are part of an organization's routine operations; start-up funds to establish new clinical or research programs or to expand existing programs; purchases of capital equipment unrelated to the study or that would generate revenue; construction funds to build new facilities; and hiring of staff that are not dedicated to the study.
*Geographic Focus:* All States, All Countries
*Contact:* Mikael Dolsten, President of Worldwide Research and Development; (212) 733-2323; mikaeldolsten@pfizer.com or GlobalMedicalGrants@pfizer.com
*Internet:* https://www.pfizer.com/purpose/independent-grants/investigator-sponsored-research
*Sponsor:* Pfizer Corporation
235 East 42nd Street
New York, NY 10017-5755

**Pfizer Global Research Awards for Nicotine Dependence 2129**
The mission of Pfizer's Global Research Awards for Nicotine Dependence (GRAND) Program is to advance the understanding of the mechanisms of tobacco and nicotine dependence and its treatment. Projects should yield results that merit first submission as abstracts to scientific meetings and, subsequently, publication in peer-reviewed journals. Pfizer intends to fund at least five grants in amounts between $50,000 and $200,000. Funds from the GRAND program may be used to support the applicant's salary and fringe benefits, technical salaries, and supplies. Applicants must submit their proposals through the Integrated System for Pfizer Investigator Initiated Research (INSPIIRE) web portal, the link to which is available from the competitive-grant-announcement web page. Applications must be received by 11:59 p.m. eastern daylight time on July 3. Interested applicants can also download an application guide from the announcement web page for more detailed information about application requirements.
*Requirements:* Applicants holding an M.D., a Ph.D., or equivalent degree are eligible to apply. Research projects should aim to provide information that could directly improve the use of pharmacotherapy in clinical practice. The GRAND review committee will consider the following criteria in evaluating proposals: scientific merit of the research proposal; qualifications of the applicant; relevance of the proposed research to the advancement of the treatment of nicotine dependence; evidence of a suitable research environment; and feasibility of the project. Prior to submitting an application via the INSPIIRE portal, applicants should notify the GRAND coordinator via the current email address given in the guidelines so that the GRAND coordinator can expedite and track the processing of the application.
*Restrictions:* The following topics fall outside of the scope of the GRAND Program in the current year: animal studies; genetics; epidemiology (unrelated to pharmacotherapy, e.g. harm and prevalence of dependence); tobacco policy; public health; community interventions; economic and social factors; and educational programs. No other government, non-governmental, or industry-sponsored funding may cover the same work scope as the grant application to the GRAND program; however the proposal may be related to other funding as long as there is no direct overlap. If the conduct of the research results in any invention or discovery by the awardee that relates to a Pfizer product, the awardee will grant to Pfizer a perpetual, royalty-free worldwide, non-exclusive license to each such invention.
*Geographic Focus:* All States, All Countries
*Date(s) Application is Due:* Jul 3
*Amount of Grant:* 50,000 - 200,000 USD
*Contact:* Mikael Dolsten, President; (212) 733-2323; mikaeldolsten@pfizer.com
*Internet:* www.grandawardsprogram.org/
*Sponsor:* Pfizer Corporation
235 East 42nd Street
New York, NY 10017-5755

**Pfizer Improving the Clinical Management of Uncontrolled Acromegaly Grants 2130**
The mission of Pfizer Independent Grants for Learning and Change (IGLC) is to partner with the global healthcare community to improve patient outcomes in areas of mutual interest through support of measurable learning and behavior change strategies. The intent of this program is to encourage organizations with a focus in healthcare professional education and/or quality improvement to submit a proposal in response to a Request for Proposal (RFP) that is related to education in a specific disease state, therapeutic area, or broader area of educational need. The program will support projects that focus on addressing challenges relating to the clinical management of adults with uncontrolled acromegaly, a hormonal disorder that develops when the pituitary gland produces too much growth hormone during adulthood. When this happens, bones increase in size, including those of in the hands, feet and face. Acromegaly usually affects middle-aged adults. Individual projects requesting up to $120,000 will be considered. The annual deadline for full proposal submission is May 31.
*Requirements:* The following may apply: medical, nursing, allied health, and/or pharmacy professional schools; healthcare institutions (both large and small); professional associations; government agencies; patient organizations and other entities with a mission related to healthcare improvement. Collaborations within institutions (e.g., between departments and/or inter-professional), as well as between different institutions, organizations, and associations, are encouraged. Applicants must serve residents in: Albania, Armenia, Australia, Azerbaijan, Bosnia and Herzegovina, Georgia, Israel, Japan, Kazakhstan, Kosovo, Macedonia, Montenegro, New Zealand, Russia, Serbia, Turkey, Ukraine, and Uzbekistan,
*Geographic Focus:* All States, Albania, Armenia, Australia, Azerbaijan, Bosnia & Herzegovina, Israel, Japan, Kazakhstan, Kosovo, Macedonia, Montenegro, New Zealand, Russia, Serbia, Turkey, Ukraine, Uzbekistan
*Date(s) Application is Due:* May 31
*Amount of Grant:* Up to 120,000 USD
*Contact:* Mikael Dolsten, President of Worldwide Research and Development; (212) 733-2323; mikaeldolsten@pfizer.com
*Internet:* https://www.pfizer.com/purpose/medical-grants/request-proposals
*Sponsor:* Pfizer Corporation
235 East 42nd Street
New York, NY 10017-5755

**Pfizer Rheumatology and IBD Research Grants 2131**
Pfizer is committed to research in the Inflammation area and supports medical innovation through this Competitive Investigator Research Awards program. The Pfizer Inflammation ASPIRE programme is designed to advance clinical and scientific research and generate data that will help inform clinical practice and treatment decisions in rheumatoid arthritis (RA), psoriatic arthritis (PsA) and ulcerative colitis (UC). This special program has increased funding available for all three therapy areas. The opportunity will allow investigators to conduct clinically relevant investigations, providing an enhanced understanding of the disease area with potential benefits translated into clinical management. The primary purpose of these Research Awards is to support clinical research in RA and UC, as well as translational and clinical research in PsA through a competitive grants program that advances medical knowledge in

understanding the role of the Janus Kinase (JAK)–Signal Transducer and Activator of Transcription (STAT) pathways in Immune-Mediated Inflammatory Diseases (IMID). Of particular interest is the effect of targeted inhibition of the JAK–STAT pathways as related to specific clinical effects examining either efficacy and/or safety outcomes and the role of the JAK–STAT pathways in PsA. Within the proposed topics, investigators are expected to generate clinical data on treatment optimisation with JAK inhibitors in RA, UC and PsA. Investigators are also expected to generate in-vitro data with the primary objective of further understanding the role of the JAK–STAT pathways in PsA. Project duration should not exceed three years. Grants will be awarded based upon: scientific merit of the research proposal; qualifications of the applicant; relevance of proposed research to the programme's mission; evidence of the applicant's commitment to an academic research career; and evidence of a suitable research environment. Pfizer is funding awards up to €1.5-2.0 million for exceptional proposals in multicentre, multinational clinical studies in RA. All other awards in RA will vary depending on the research question. Funding in PsA and UC will also depend on the research question, with maximum awards of up to €200,000 for clinical proposals and up to €80,000 for translational research in PsA. It is expected that the results will be presented at scientific meetings and published in peer-reviewed journals. The annual application deadline is March 23, with award announcements in July.
*Geographic Focus:* Albania, Armenia, Australia, Azerbaijan, Bosnia & Herzegovina, Israel, Japan, Kazakhstan, Kosovo, Macedonia, Montenegro, New Zealand, Russia, Serbia, South Korea, Turkey, Ukraine, Uzbekistan
*Amount of Grant:* Up to 2,000,000 EUR
*Contact:* Mikael Dolsten, President of Worldwide Research and Development; (212) 573-2323; mikaeldolsten@pfizer.com
*Internet:* https://www.aspireresearch.org/europerheumatology/index.html
*Sponsor:* Pfizer Corporation
235 East 42nd Street
New York, NY 10017-5755

**Pfizer Rheumatology Fellowships** **2132**
The mission of Pfizer Independent Grants for Learning and Change (IGLC) is to partner with the global healthcare community to improve patient outcomes in areas of mutual interest through support of measurable learning and behavior change strategies. The Pfizer Fellowship program is a reflection of Pfizer's commitment to advancing research, innovation, clinical expertise and medical education for medical, nursing, and allied health professionals. This program provides career-building opportunities for practicing clinicians and academic researchers in outcomes and patient-oriented research that is vital to improving healthcare. The intent of this Request for Proposal (RFP) is to encourage organizations with an active Rheumatology fellowship program to submit a proposal. The purpose of the Fellowship program grant is to support the continued training of fellows in any two years of a program. Organizations seeking funding must submit proposals describing the strength of their program in supporting clinical training and research. Individual requests for fellowship grants up to $80,000 per year for up to two years will be considered. The annual application deadline is March 23.
*Requirements:* The following requirements are in place for the Fellowship: funding is to be used as a salary supplement for a fellow at the recipient organization; funding will be awarded to the organization, which will select the recipient; funding must be used primarily as salary and fringe support for the selected fellow, not for overhead or indirect costs; and the U.S. Physician Payments Sunshine Act and other transparency law reporting requirements must be satisfied with respect to meals, payments, travel and other transfers of value.
*Geographic Focus:* All States
*Date(s) Application is Due:* Mar 23
*Amount of Grant:* 80,000 - 160,000 USD
*Contact:* Mikael Dolssten, President of Worldwide Research and Development; (212) 733-2323; mikaeldolsten@pfizer.com
Amanda Solis, Program Officer; (212) 733-2323; amanda.solis@pfizer.com
*Internet:* https://www.pfizer.com/purpose/medical-grants/request-proposals
*Sponsor:* Pfizer Corporation
235 East 42nd Street
New York, NY 10017-5755

**Phi Kappa Phi Foundation Love of Learning Awards** **2133**
The Phi Kappa Phi Love of Learning Awards help fund post-baccalaureate studies and/or career development for active Phi Kappa Phi members to include (but not be limited to) graduate or professional studies, doctoral dissertations, continuing education, career development, travel related to teaching/studies, etc. Submission guidelines and a current application are available at the website.
*Requirements:* The awards are open to all active Phi Kappa Phi members who have completed their baccalaureate studies in any major by the application submission deadline, and have not received a Phi Kappa Phi Fellowship or Award of Excellence grant.
*Restrictions:* Recipients of the Fellowship award are not eligible to apply.
*Geographic Focus:* All States
*Date(s) Application is Due:* Jun 30
*Amount of Grant:* 500 USD
*Contact:* Maria Davis; (800) 804-9880; LoveofLearning@PhiKappaPhi.org
*Internet:* www.phikappaphi.org/Web/Awards/Love_of_Learning.html
*Sponsor:* Phi Kappa Phi Foundation
7576 Goodwood Boulevard
Baton Rouge, LA 70806

**Phi Kappa Phi Graduate Fellowships** **2134**
The Phi Kappa Phi Fellowships help fund graduate work for students entering graduate school within one year of their baccalaureate degree. However, persons of any age having delayed graduate study may apply through the chapter of their initiation or chapter of current membership for consideration as the chapter nominee. Any major is eligible as long as the applicant is a PKP member in good standing. Applications must be filed with the secretary of the local chapter by the established deadline on special forms available at the PKP website. Each chapter selects one applicant to compete for the award. Five fellowships of $15,000 each, and 50 fellowships of $5,000 each are available. Specific guidelines, a current application, advice for applicants, and personal statement samples are available at the PKP website.
*Requirements:* Recipients must be active members of Phi Kappa Phi and have initiated plans to enroll for an advanced degree in a recognized graduate or professional school. Applications will be accepted from individuals selected for membership but not yet initiated. Applications from international students and schools are eligible as long as the student's PKP membership is in good standing.
*Restrictions:* Applicants who have earned ten or more graduate credits before the current academic year are not eligible. The deadline for the application is usually spring, but dates vary for individual chapters.
*Geographic Focus:* All States, All Countries
*Amount of Grant:* 5,000 - 15,000 USD
*Contact:* Chapter Officer; (800) 804-9880; fax (225) 388-4900; awards@phikappaphi.org
*Internet:* www.phikappaphi.org/Web/Awards/Fellowship.html
*Sponsor:* Phi Kappa Phi Foundation
7576 Goodwood Boulevard
Baton Rouge, LA 70806

**Phi Kappa Phi Scholar Award** **2135**
The Phi Kappa Phi Scholar Award honors those who demonstrate the ideals of the Society through their activities, achievements, and scholarship. The Award recognizes excellence in teaching, research, and public service. The Award is given every three years. Chapters and active members are encouraged to nominate an individual for this award; self-nominations are encouraged. Although nominees are often faculty members at colleges and universities, a chapter or any active member may nominate an active Society member who is not institution-affiliated provided all criteria are met. The Award recipient will be presented with a check for $1,000, a one year active membership in Phi Kappa Phi, a recognition certificate, and an invitation to the annual Convention. Travel, lodging and meal expenses incurred by the recipient are reimbursed. Specific criteria and a nomination form are located at the website.
*Requirements:* Applicants must be Society members in good standing; hold a distinguished record of past scholarly achievements; possess documented evidence of national or international prominence and visibility; demonstrated consistency of excellence in teaching and/or service; and a distinguished record of honors, publications, patents, inventions or other types of recognition for excellence and scholarly achievement.
*Geographic Focus:* All States, All Countries
*Date(s) Application is Due:* Feb 15
*Amount of Grant:* 1,000 USD
*Contact:* Scholar Award Contact; (800) 804-9880; Scholar@PhiKappaPhi.org
*Internet:* www.phikappaphi.org/Web/Awards/PKP_Scholar_and_Artist.html
*Sponsor:* Phi Kappa Phi Foundation
7576 Goodwood Boulevard
Baton Rouge, LA 70806

**Phil Hardin Foundation Grants** **2136**
The Phil Hardin Foundation targets its funding to education in the State of Mississippi. Specific priorities are to strengthen the capacity of communities to nurture and educate young children; the capacity of higher education institutions to renew communities and their economies; the capacity of communities for locally initiated educational improvement and economic development; and state- and local-level policy and leadership initiatives that fit with foundation goals. Types of support include general operating support, continuing support, building construction/renovation, equipment acquisition, endowment funds, program development, conferences and seminars, professorships, publication, seed grants, fellowships, scholarship funds, research, and matching funds. The foundation also operates four K-12 fellowship programs. Deadlines vary between programs; contact program staff for specific fellowship deadlines.
*Requirements:* Applicants must either be based in Mississippi or the project must benefit Mississippi, depending on the program. Contact program staff for eligibility.
*Geographic Focus:* Mississippi
*Samples:* Covington County Vocational Technical Center, Collins, Mississippi, $2,000 - support of the Click, Click, Boom program; Meridian Public School District, Meridian, Mississippi, $100,000 - recruitment, retention and incentive program for highly-effective educators; Trinity Dyslexia Education Center, Meridian, Mississippi, $15,000 - general operating costs.
*Contact:* Lloyd Gray, (601) 483-4282; fax (601) 483-5665; info@philhardin.org
*Internet:* www.philhardin.org/application-and-instructions.cfm
*Sponsor:* Phil Hardin Foundation
2750 North Park Drive
Meridian, MS 39305

**Phi Upsilon Omicron Alumni Research Grant** **2137**
Phi Upsilon Omicron National Honor Society was founded in 1910 at the University of Minnesota in the area of Family and Consumer Sciences. The purpose of the society's Alumni Research Grant is to support post-graduate research in family and consumer sciences. Proposed research should contribute to the knowledge base of family and consumer sciences or further the purposes of Phi Upsilon Omicron, especially in regard to academic

excellence, leadership development, and/or professional service. The grant is awarded on even-numbered years. Application forms may be obtained from the Society's National Office (see contact section on this page). The application packet must be postmarked no later than November 1. Applicants will be notified of the disposition of their application following the meeting of the Alumni Research Grant Advisory Committee.
*Requirements:* One or more of the following entities are eligible to apply: individual society alumni; society alumni committees or task forces; and alumni chapters.
*Geographic Focus:* All States
*Date(s) Application is Due:* Nov 1
*Amount of Grant:* 2,500 USD
*Contact:* Susan Rickards; (304) 368-0612; rickards@phiu.org or info@phiu.org
*Internet:* phiu.org/awards.htm
*Sponsor:* Phi Upsilon Omicron, Inc.
P.O. Box 329
Fairmont, WV 26555

**Phi Upsilon Omicron Candle Fellowships** **2138**
Phi Upsilon Omicron is a national honor society in Family and Consumer Sciences that was founded in 1910 at the University of Minnesota. The society awards two Candle fellowships annually for study leading to a master's degree in family and consumer sciences. The application guidelines can be downloaded from the fellowship website. The application packet must be postmarked no later than February 1. The society offers numerous graduate fellowship opportunities to its members. Applicants may apply for more than one fellowship opportunity with the same application packet and should list the name(s) of the fellowships for which application is being made.
*Requirements:* Applicant must be a member of Phi Upsilon Omicron who is enrolled in or planning to enroll in graduate school. Selection will be based upon the following criteria: applicant's scholastic record; honors and recognitions; participation in honor society as well as professional, community, and other organizations; scholarly work; statement of professional goals; and recommendations.
*Geographic Focus:* All States
*Date(s) Application is Due:* Feb 1
*Amount of Grant:* 750 - 1,500 USD
*Contact:* Susan Rickards; (304) 368-0612; rickards@phiu.org or info@phiu.org
*Internet:* phiu.org/fellowships.htm#
*Sponsor:* Phi Upsilon Omicron, Inc.
P.O. Box 329
Fairmont, WV 26555

**Phi Upsilon Omicron Diamond Anniversary Fellowships** **2139**
Phi Upsilon Omicron is a national honor society in Family and Consumer Sciences that was founded in 1910 at the University of Minnesota. Two Diamond Anniversary fellowships will be awarded per year for study at the master's or doctoral level in family and consumer sciences or a related area. The application guidelines can be downloaded from the fellowship website. The application packet must be postmarked no later than February 1. The society offers numerous graduate fellowship opportunities to its members. Applicants may apply for more than one fellowship opportunity with the same application packet and should list the name(s) of the fellowships for which application is being made.
*Requirements:* Applicant must be a member of Phi Upsilon Omicron and have been accepted into (or be currently enrolled in) a graduate program in consumer and family sciences. Selection will be based upon the following criteria: applicant's scholastic record; honors and recognitions; participation in honor society as well as professional, community, and other organizations; scholarly work; statement of professional goals; and recommendations.
*Geographic Focus:* All States
*Date(s) Application is Due:* Feb 1
*Amount of Grant:* 750 - 1,500 USD
*Contact:* Susan Rickards; (304) 368-0612; rickards@phiu.org or info@phiu.org
*Internet:* phiu.org/fellowships.htm#
*Sponsor:* Phi Upsilon Omicron, Inc.
P.O. Box 329
Fairmont, WV 26555

**Phi Upsilon Omicron Florence Fallgatter Distinguished Service Award** **2140**
Phi Upsilon Omicron National Honor Society was founded in 1910 at the University of Minnesota in the area of Family and Consumer Sciences. The society's Florence Fallgatter Distinguished Service Award is granted biennially to a qualified alumni member of the society for outstanding achievement in family and consumer sciences. The recipient's expenses to attend the society's biennial meeting will be paid, and the member will be recognized with a plaque at the meeting. The nomination deadline is November 1. Interested applicants should contact the society for information on nomination procedures.
*Requirements:* A nominee must exhibit the following characteristics: be an alumni member of the society; have finished his or her bachelor's degree no less than ten years ago; have demonstrated excellence in business, dietetics/nutrition, education, cooperative extension; journalism; research; social services, or volunteer services; and have had a significant positive impact on the family and consumer sciences profession.
*Geographic Focus:* All States
*Date(s) Application is Due:* Nov 1
*Contact:* Susan Rickards; (304) 368-0612; rickards@phiu.org or info@phiu.org
*Internet:* phiu.org/awards.htm
*Sponsor:* Phi Upsilon Omicron, Inc.
P.O. Box 329
Fairmont, WV 26555

**Phi Upsilon Omicron Founders Fellowship** **2141**
Phi Upsilon Omicron is a national honor society in Family and Consumer Sciences that was founded in 1910 at the University of Minnesota. The society's Founders Fellowship supports advanced graduate study in the field of consumer and family sciences. The application guidelines can be downloaded from the fellowship website. The application packet must be postmarked no later than February 1. The society offers numerous graduate fellowship opportunities to its members. Applicants may apply for more than one fellowship opportunity with the same application packet and should list the name(s) of the fellowships for which application is being made.
*Requirements:* Applicant must be a member of Phi Upsilon Omicron, have completed at least half the credit-hour requirements toward the doctorate in some area of family and consumer sciences, and have had several years of successful employment in the profession. Selection will be based upon the following criteria: applicant's scholastic record; honors and recognitions; participation in honor society as well as professional, community, and other organizations; scholarly work; statement of professional goals; and recommendations.
*Geographic Focus:* All States
*Date(s) Application is Due:* Feb 1
*Amount of Grant:* 750 - 1,500 USD
*Contact:* Susan Rickards; (304) 368-0612; rickards@phiu.org or info@phiu.org
*Internet:* phiu.org/fellowships.htm#
*Sponsor:* Phi Upsilon Omicron, Inc.
P.O. Box 329
Fairmont, WV 26555

**Phi Upsilon Omicron Geraldine Clewell Doctoral Fellowship** **2142**
Phi Upsilon Omicron is a national honor society in Family and Consumer Sciences that was founded in 1910 at the University of Minnesota. The society's Geraldine Clewell Doctoral Fellowship is awarded annually to a doctoral student in family and consumer sciences or a related area. The application guidelines can be downloaded from the fellowship website. The application packet must be postmarked no later than February 1. Applicants should note that one of the three required recommendations must come from their major advisor. The society offers numerous graduate fellowship opportunities to its members. Applicants may apply for more than one fellowship opportunity with the same application packet and should list the name(s) of the fellowships for which application is being made.
*Requirements:* Applicants must be members of Phi Upsilon Omicron. Preference is given to a student who desires to teach at the college/university level. Selection will be based upon the following criteria: applicant's scholastic record; honors and recognitions; participation in honor society as well as professional, community, and other organizations; scholarly work; statement of professional goals; and recommendations.
*Geographic Focus:* All States
*Date(s) Application is Due:* Feb 1
*Amount of Grant:* 750 - 1,500 USD
*Contact:* Susan Rickards; (304) 368-0612; rickards@phiu.org or info@phiu.org
*Internet:* phiu.org/fellowships.htm#
*Sponsor:* Phi Upsilon Omicron, Inc.
P.O. Box 329
Fairmont, WV 26555

**Phi Upsilon Omicron Geraldine Clewell Masters Fellowship** **2143**
Phi Upsilon Omicron is a national honor society in Family and Consumer Sciences that was founded in 1910 at the University of Minnesota. The society offers numerous graduate fellowship opportunities to its members. The society's Geraldine Clewell Masters Fellowship is awarded annually to a student pursuing a master's degree in family and consumer sciences or related area. The application guidelines can be downloaded from the fellowship website. The application packet must be postmarked no later than February 1. Applicants should note that one of the three required recommendations must come from their major advisor. The society offers numerous graduate fellowship opportunities to its members. Applicants may apply for more than one fellowship opportunity with the same application packet and should list the name(s) of the fellowships for which application is being made.
*Requirements:* Applicants must be members of Phi Upsilon Omicron. Preference is given to a student majoring in family and consumer-sciences education and who desires to teach at the elementary/secondary level. Selection will be based upon the following criteria: applicant's scholastic record; honors and recognitions; participation in honor society as well as professional, community, and other organizations; scholarly work; statement of professional goals; and recommendations.
*Geographic Focus:* All States
*Date(s) Application is Due:* Feb 1
*Amount of Grant:* 750 - 1,500 USD
*Contact:* Susan Rickards; (304) 368-0612; rickards@phiu.org or info@phiu.org
*Internet:* phiu.org/fellowships.htm#
*Sponsor:* Phi Upsilon Omicron, Inc.
P.O. Box 329
Fairmont, WV 26555

**Phi Upsilon Omicron Jean Dearth Dickerscheid Fellowship** **2144**
Phi Upsilon Omicron is a national honor society in Family and Consumer Sciences that was founded in 1910 at the University of Minnesota. The society's Jean Dearth Dickerscheid Fellowship is awarded annually to a doctoral student who is pursuing his or her degree in family and consumer sciences or a related area and has earned at least one other degree in consumer and family sciences. The application guidelines can be downloaded from the fellowship website. The application packet must be postmarked no later than February 1. The society offers numerous graduate fellowship opportunities to its members. Applicants

may apply for more than one fellowship opportunity with the same application packet and should list the name(s) of the fellowships for which application is being made.
*Requirements:* Applicants must be members of Phi Upsilon Omicron and intend to pursue a career in academia. Selection will be based upon the following criteria: applicant's scholastic record; honors and recognitions; participation in honor society as well as professional, community, and other organizations; scholarly work; statement of professional goals; and recommendations.
*Geographic Focus:* All States
*Date(s) Application is Due:* Feb 1
*Amount of Grant:* 750 - 1,500 USD
*Contact:* Susan Rickards, Executive Director; (304) 368-0612; rickards@phiu.org or info@phiu.org
*Internet:* phiu.org/fellowships.htm#
*Sponsor:* Phi Upsilon Omicron, Inc.
P.O. Box 329
Fairmont, WV 26555

**Phi Upsilon Omicron Margaret Drew Alpha Fellowship** 2145
Phi Upsilon Omicron is a national honor society in Family and Consumer Sciences that was founded in 1910 at the University of Minnesota. The society's Margaret Drew Alpha Fellowship supports graduate study in a field of family and consumer sciences with preference given to dietetics or food and nutrition. The application guidelines can be downloaded from the fellowship website. The application packet must be postmarked no later than February 1. The society offers numerous graduate fellowship opportunities to its members. Applicants may apply for more than one fellowship opportunity with the same application packet and should list the name(s) of the fellowships for which application is being made.
*Requirements:* Applicant must be a member of Phi Upsilon Omicron. Selection will be based upon the following criteria: applicant's scholastic record; honors and recognitions; participation in honor society as well as professional, community, and other organizations; scholarly work; statement of professional goals; and recommendations.
*Geographic Focus:* All States
*Date(s) Application is Due:* Feb 1
*Amount of Grant:* 750 - 1,700 USD
*Contact:* Susan Rickards, Executive Director; (304) 368-0612; rickards@phiu.org or info@phiu.org
*Internet:* phiu.org/fellowships.htm#
*Sponsor:* Phi Upsilon Omicron, Inc.
P.O. Box 329
Fairmont, WV 26555

**Phi Upsilon Omicron Presidents Research Fellowship** 2146
Phi Upsilon Omicron is a national honor society in Family and Consumer Sciences that was founded in 1910 at the University of Minnesota. The society's President Research Fellowship supports graduate research in family and consumer sciences at the master's, doctoral, or postdoctoral level. The application guidelines can be downloaded from the fellowship website. The application packet must be postmarked no later than February 1. The society offers numerous graduate fellowship opportunities to its members. Applicants may apply for more than one fellowship opportunity with the same application packet and should list the name(s) of the fellowships for which application is being made.
*Requirements:* Applicant must be a member of Phi Upsilon Omicron. Selection will be based upon the following criteria: the organization and need for the research as stated in the applicant's research prospectus; applicant's scholastic record; honors and recognitions; participation in honor society as well as professional, community, and other organizations; scholarly work; statement of professional goals; and recommendations.
*Geographic Focus:* All States
*Date(s) Application is Due:* Feb 1
*Amount of Grant:* 750 - 1,500 USD
*Contact:* Susan Rickards; (304) 368-0612; rickards@phiu.org or info@phiu.org
*Internet:* phiu.org/fellowships.htm#
*Sponsor:* Phi Upsilon Omicron, Inc.
P.O. Box 329
Fairmont, WV 26555

**Pike County Community Foundation Health Grants** 2147
The Pike County Community Foundation is a charitable organization formed to strengthen the Pike County, Indiana community by awarding grants to local nonprofits, by bringing individuals together to address community needs, and by offering personalized charitable gift planning services to donors. At the start of each cycle, a notice is mailed to nonprofit organizations that have previously applied for and received grants, or have otherwise requested notification of each cycle. Proposals are accepted from July 6 through the September 8 deadline. The grants committee will make its recommendations on funding to the Foundation's Board of Trustees, which will make final funding recommendations to the board of directors of the Community Foundation Alliance. All organizations that have submitted grant proposals are notified of the final outcome no later than December 1. Grants in the area of health include activities that: improve and promote health outcomes; provide general and rehabilitative health services; offer mental health services; provide crisis intervention programs; strengthen associations or services associated with specific diseases, disorders, and medical disciplines; and support medical research. Samples of previously funded projects are posted on the website.
*Requirements:* The Foundation welcomes proposals from nonprofit organizations that are deemed tax-exempt under sections 501(c)3 and 509(a) of the Internal Revenue Code and from governmental agencies serving Pike County. Proposals from nonprofit organizations not classified as a 501(c)3 public charity may be considered provided the project is charitable and supports a community need. Proposals submitted by an entity under the auspices of another agency must include a written statement signed by the agency's board president on behalf of the board of directors agreeing to act as the entity's fiscal sponsor, to receive grant monies if awarded, and to oversee the proposed project.
*Restrictions:* Project areas not considered for funding include: religious organizations for religious purposes; political parties or campaigns; endowment creation or debt reduction; operating costs; capital campaigns; annual appeals or membership contributions; travel requests for groups or individuals such as bands, sports teams, or classes.
*Geographic Focus:* Indiana
*Date(s) Application is Due:* Sep 8
*Contact:* Cindy Gaskins; (812) 354-6797; director@pikecommunityfoundation.org
*Internet:* https://www.communityfoundationalliance.org/pike/program-areas/
*Sponsor:* Pike County Community Foundation
714 Main Street, P.O. Box 587
Petersburg, IN 47567

**Pinkerton Foundation Grants** 2148
The Pinkerton Foundation is an independent grantmaking foundation established in 1966 by Robert Allan Pinkerton with the broad directive to reduce the incidence of crime and to prevent juvenille delinquency. The Foundation's principal program interests are focused on economically disadvantaged children, youth and families, and severely learning disabled children and adults of borderline intelligence. The Foundation supports efforts to strengthen and expand community-based programs for children, youth and families in New York City. The Foundation also occasionally funds research, demonstration and evaluation projects in its principal program areas. While grants for direct service projects are usually limited to New York City, those with potential for national impact or replication may go beyond this geographic limitation. The Foundation's Board of Directors have two grantmaking meetings per year, in May and in December. Letters of inquiry are welcome throughout the year. Additional guidelines are available at the Foundation's website.
*Requirements:* Grants are awarded primarily to New York City 501(c)3 nonprofit public charitable organizations.
*Restrictions:* The foundation does not grant requests for emergencies, medical research, direct provision of health care, religious education, conferences, publications, or capital projects.
*Geographic Focus:* All States
*Date(s) Application is Due:* Feb 1; Sep 1
*Amount of Grant:* 25,000 - 250,000 USD
*Samples:* Reel Works Teen Filmmaking, Brooklyn, NY, $35,000—operating support for filmmaking program that serves up to 180 high school students; Madison Square Boys & Girls Club (New York, NY) $250,000—for after school and summer program operations at Madison's clubhouses; New York Cares, Inc., New York, NY, $20,000—in support of their SAT Prep Program.
*Contact:* Joan Colello; (212) 332-3385; fax (212) 332-3399; pinkfdn@pinkertonfdn.org
*Internet:* fdncenter.org/grantmaker/pinkerton
*Sponsor:* Pinkerton Foundation
610 Fifth Avenue, Suite 316
New York, NY 10020

**Piper Trust Healthcare and Medical Research Grants** 2149
The Piper Trust's grantmaking focuses on Virginia Galvin Piper's commitment to improving the quality of life for residents of Maricopa County. Piper Trust's particular interest lies with projects that benefit young children, adolescents and older adults in Maricopa County. The Trust makes grants to faith-based organizations that serve these target populations in a manner consistent with program guidelines. For Healthcare and Medical Research grants, the trust is most interested in proposals that address improved facilities for children, adolescents and older adults; better trained healthcare workforce; increased access to basic healthcare; and, centers for advancement in personalized medicine.
*Requirements:* Piper Trust makes grants to actively operating Section 501(c)3 organizations in Maricopa County. These organizations must have been in operation for at least three years from the effective date of their IRS ruling. Special rules apply to private foundations and 509(a)3 (Type III) organizations. There are no deadlines on initial proposals, and letters of inquiry throughout the year are reviewed throughout the year. If the Trust asks for a full proposal, its disposition depends on its completeness and the meeting schedule of the Piper trustees.
*Geographic Focus:* Arizona
*Contact:* Terri Leon, Program Officer; (480) 556-7121; tleon@pipertrust.org
*Internet:* pipertrust.org/our-grants/healthcare-medical-research/
*Sponsor:* Virginia G. Piper Charitable Trust
1202 East Missouri Avenue
Phoenix, AZ 85014

**Playboy Foundation Grants** 2150
The Playboy Foundation seeks to foster social change by confining its grants and other support to projects of national impact and scope involved in fostering open communication about, and research into, human sexuality, reproductive health and rights; protecting and fostering civil rights and civil liberties in the United States for all people, including women, people affected and impacted by HIV/AIDS, gays and lesbians, racial minorities, the poor and the disadvantaged; and eliminating censorship and protecting freedom of expression and First Amendment rights. The Foundation does not accept unsolicited proposals, but welcomes letters of inquiry via post for the areas of interest noted above. Grants awarded by the Foundation are typically up to $10,000.
*Restrictions:* The foundation will not consider religious programs, individual needs, capital campaigns, endowments, scholarships, or fellowships; social services, including residential

care, clinics, treatment, or recreation programs; national health, welfare, educational, or cultural organizations, or their state affiliates; or government agencies or projects.
*Geographic Focus:* All States
*Amount of Grant:* 5,000 - 10,000 USD
*Contact:* Executive Director; (312) 373-2437 or (312) 751-8000; fax (312) 751-2818; giving@playboy.com
*Internet:* www.playboyenterprises.com/foundation
*Sponsor:* Playboy Foundation
680 North Lake Shore Drive
Chicago, IL 60611

**Posey County Community Foundation Health Grants** 2151

The Posey County Community Foundation is a nonprofit, public charity created by and for the people of Posey County, Indiana. Grant proposals are accepted once each year as a one-time project support. At the beginning of each cycle in January, notices are sent to nonprofit organizations that have previously applied for grants, have received grants in the past, or have otherwise requested notification. Applicants are encouraged to schedule a meeting with the Foundation's director to receive an overview of the grant process. Proposals are accepted from January through August 1. The grants committee will make its recommendations on funding to the Foundation's Board of Trustees, which will make final funding recommendations to the board of directors of the Community Foundation Alliance. All organizations that have submitted grant proposals are notified of the final outcome no later than December 1. Grants in the area of health include activities that: improve and promote health outcomes; provide general and rehabilitative health services; offer mental health services; provide crisis intervention programs; strengthen associations or services associated with specific diseases, disorders, and medical disciplines; and support medical research. Samples of previously funded projects are posted on the website.
*Requirements:* The Foundation welcomes funding requests from nonprofit organizations that are 501(c)3 or 509(a) tax exempt, and from governmental agencies serving Posey County. For those organizations not tax exempt, requests may be considered if the project is charitable and supports a community need. Proposals submitted by an entity under the auspices of another agency must include a written statement signed by the agency's board president on behalf of the board of directors agreeing to act as the entity's fiscal sponsor, to receive grant monies if awarded, and to oversee the proposed project.
*Restrictions:* Project areas not considered for funding include: religious organizations for religious purposes; political parties or campaigns; endowment creation or debt reduction; operating costs; capital campaigns; annual appeals or membership contributions; travel requests for groups or individuals such as bands, sports teams, or classes.
*Geographic Focus:* Indiana
*Date(s) Application is Due:* Aug 1
*Amount of Grant:* Up to 5,000 USD
*Contact:* Monica Spencer, Regional Director; (812) 838-0288; fax (812) 838-8009; monica@poseycommunityfoundation.org
*Internet:* https://www.communityfoundationalliance.org/posey/program-areas/
*Sponsor:* Posey County Community Foundation
402 Main Street, P.O. Box 746
Mt. Vernon, IN 47620

**Price Family Charitable Fund Grants** 2152

Established in 1983, the Price Family Charitable Fund serves the San Diego, Carlsbad and, San Marcos, California region. The foundation is interested in supporting the economically disadvantaged, giving primarily for education and philanthropy purposes. The types of support include: annual campaigns; fellowships; program evaluation; scholarship funds; scholarships to individuals and; general operating support. There's no formal application to submit. Potential grantees much first submit a letter of Inquiry including: project goals, objectives and expected results (maximum of one page); project narrative (maximum of five pages); project budget (maximum of one page); a list of grants, if any, received by organization in the last 12 months for this program/project (sources and amounts); a list of the organization's current board of directors, including each member's name, profession, and office help on the board, if any.
*Requirements:* Non-profit organizations with 501(c)3 status or governmental units such as public schools or city departments in are available for funding. The Foundation gives primarily in the following region of: San Diego; Carlsbad; San Marcos, California.
*Restrictions:* Grants are not made: to organizations whose primary purpose is religious, or for propagandizing, influencing legislation and/or elections, promoting voting registration, for political candidates, political campaigns or organizations engaged in political activities; or to federal appeals or to organizations the collect funds for redistribution to other non-profit groups. Unsolicited requests for funds are not accepted.
*Geographic Focus:* California
*Samples:* George G. Glenner Alzheimer's Family Centers, Inc., San Diego, CA, $33,000—Dementia Care Training & Public Scholarships; Episcopal Community Services, San Diego, CA, $60,000—general support; A Reason to Survive, San Diego, CA, $100,000—capacity building.
*Contact:* Terry Malavenda, (858) 551-2330
*Sponsor:* Price Family Charitable Fund
7979 Ivanhoe Avenue, Suite 520
La Jolla, CA 92037-4513

**Princess Grace Foundation Theater, Dance, and Film Grants** 2153

The Princess Grace Foundation Grants for Theater, Dance, and Film are awarded to individuals, through their affiliation with nonprofit organizations, in the areas of dance, theater, and film. The grants are made on an annual basis by application for scholarships, fellowships, and apprenticeships. This unique collaborative process fills voids in the artistic community. Scholarships, apprenticeships, and fellowships give emerging artists the financial assistance and moral encouragement to focus on artistic excellence, while monetary support for the nominating organizations eases fund raising challenges, directing resources toward the creative process. The March 31 application deadline is for theater and playwriting, April 30 for dance, and June 1 for film. The application and specific guidelines for each category of funding are available at the Foundation website.
*Requirements:* Applicants must be U.S. citizens or have permanent resident status. They must also be studying with a company in the United States. All applicants except playwrights must be nominated by a school department chair or dean or company artistic director. The nominating organization must be a registered 501(c)3 company.
*Restrictions:* Short training programs, intensives, and film festivals are not eligible for funding.
*Geographic Focus:* All States
*Date(s) Application is Due:* Mar 31; Apr 30; May 15; Jun 1
*Contact:* Toby Boshak; (212) 317-1470; fax (212) 317-1473; pgfusa@pgfusa.com
*Internet:* www.pgfusa.com/awards/grants/index.html
*Sponsor:* Princess Grace Foundation - USA
150 East 58th Street, 25th Floor
New York, NY 10155

**Princeton University Center for Human Values Graduate Prize Fellowships** 2154

The Graduate Prize Fellowship program is intended to recognize and support post-generals graduate students with distinguished academic records in any discipline whose dissertation research centrally involves the critical study of human values. In addition to providing tangible support, a main purpose is to enable people working on topics in human values to engage with each other both informally and in a faculty-led fellows' seminar, where they present their work in progress. The principal considerations in the selection of fellows will be the promise of excellence in an applicant's dissertation work, the applicant's capacity to contribute constructively to the fellows' group and the likelihood that participation in the program will help to shape and improve the applicant's dissertation. Accordingly, preference will be given to applicants who are in the relatively early stages of work on their dissertations. Graduate Prize Fellowships provide a stipend of $32,000 for twelve months.
*Requirements:* The fellowships are for Princeton University graduate students only. Fellows must be enrolled for the duration of the fellowship and may not take on any teaching or other employment. Except in extraordinary cases, we do not expect to support students who would be beyond their fifth year of enrollment during the fellowship year.
*Geographic Focus:* All States
*Date(s) Application is Due:* Mar 6
*Contact:* Erin Graham, Program Coordinator; (609) 258-1460 or (609) 258-4798; fax (609) 258-2729; egraham@princeton.edu
*Internet:* uchv.princeton.edu/fellowships_awards/graduate_prize_fellowships.php
*Sponsor:* Princeton University
5 Ivy Lane, 304 Louis Marx Hall
Princeton, NJ 08544

**Princeton University Center for Human Values Graduate Student Top-Up Prizes** 2155

The University Center for Human Values offers a limited number of top-up prizes to admitted graduate students who show promise in the area of ethics, political theory/philosophy, and human values. The prize is a one-time only award that is added to the stipend during the first year of study. Each department may nominate several students for this award. All applicants who have been offered admission are eligible.
*Geographic Focus:* All States
*Contact:* Erin Graham, Program Coordinator; (609) 258-1460 or (609) 258-4798; fax (609) 258-2729; egraham@princeton.edu
*Internet:* uchv.princeton.edu/fellowships_awards/topup.php
*Sponsor:* Princeton University
5 Ivy Lane, 304 Louis Marx Hall
Princeton, NJ 08544

**Princeton University Center for Human Values Harold T. Shapiro Postdoctoral Research Associate in Bioethics** 2156

The Harold T. Shapiro Postdoctoral Research Associate position supports outstanding scholars studying ethical issues arising from developments in medicine or the biological sciences. We particularly encourage proposals focusing on problems of practical importance that have broader theoretical interest. The Shapiro Research Associate spends from one to three years at Princeton conducting research and teaching the equivalent of one course each year. He or she participates in the Ira W. DeCamp Seminar in Bioethics and is invited to participate in other activities of the University Center for Human Values, including a year-long research seminar for visiting fellows and Center faculty. The term of the position is one year with the possibility of extension for up to two further years. Princeton University offers competitive salary and employee benefits.
*Requirements:* Applicants must have completed all the requirements for the Ph.D., M.D., or other equivalent doctoral degree by the start of the appointment and must not have held the degree for more than three years. An applicant may have a background in any area of ethical studies, not necessarily in bioethics. However, the proposed research plan should be related to the field of bioethics.
*Geographic Focus:* All States
*Contact:* Erin Graham, Program Coordinator; (609) 258-1460 or (609) 258-4798; fax (609) 258-2729; egraham@princeton.edu
*Internet:* uchv.princeton.edu/fellowships_awards/hts_bioethics.php
*Sponsor:* Princeton University
5 Ivy Lane, 304 Louis Marx Hall
Princeton, NJ 08544

**Princeton University Center for Human Values Laurance S. Rockefeller Visiting Fellowships** 2157

The University Center for Human Values invites applications for Laurance S. Rockefeller Visiting Faculty Fellowships for the academic year. Fellows devote an academic year in residence at Princeton to do research and writing about topics involving human values in public and private life. The program is open to scholars in all disciplines provided their research plans qualify. In recent years fellows have been drawn from fields including philosophy, political theory, literature, history, classics, economics and law, but this list is not meant to be exhaustive.
*Requirements:* Fellows are expected to reside in or around Princeton and to be active contributors to the intellectual life of the Center. This includes participating in a weekly seminar attended by fellows and Center faculty to discuss work in progress and in various other seminars, colloquia and lectures sponsored by the Center. Fellows enjoy access to Firestone Library and to a wide range of activities throughout the University. Candidates should have a doctorate or equivalent professional degree and a strong record of research publications appropriate to their career stage.
*Geographic Focus:* All States
*Date(s) Application is Due:* Nov 5
*Contact:* Erin Graham, Program Coordinator; (609) 258-1460 or (609) 258-4798; fax (609) 258-2729; egraham@princeton.edu
*Internet:* uchv.princeton.edu/fellowships_awards/lsr_visiting_fellowships.php
*Sponsor:* Princeton University
5 Ivy Lane, 304 Louis Marx Hall
Princeton, NJ 08544

**Princeton University Center for Human Values Political Philosophy Travel Grants** 2158

The University Center for Human Values, along with the Program in Political Philosophy, offers Political Philosophy Research and Travel grants, supported by a fund established by Amy Gutmann, former provost of the University and founding director of the University Center for Human Values. Individual grants of up to $2,000 are awarded to graduate students working on topics in political philosophy and enrolled in the departments that participate in the program (Classics, History, Philosophy, Politics, Religion) for summer travel.
*Requirements:* The graduate student must be enrolled at the university during the time of travel and grants are to be used in support of research and travel related to research. The intent is to increase the overall support provided by the University for excellent work in the field of political philosophy. The terms of the fund specify that grants must be used to augment resources already available at Princeton from other sources.
*Geographic Focus:* All States
*Amount of Grant:* Up to 2,000 USD
*Contact:* Erin Graham, Program Coordinator; (609) 258-1460 or (609) 258-4798; fax (609) 258-2729; egraham@princeton.edu
*Internet:* uchv.princeton.edu/fellowships_awards/graduate_student_travel_grants.php
*Sponsor:* Princeton University
5 Ivy Lane, 304 Louis Marx Hall
Princeton, NJ 08544

**Princeton University Center for Human Values Postdoctoral Research Associate in Values and Public Policy** 2159

The University Center for Human Values and the Woodrow Wilson School of Public and International Affairs invite applications for postdoctoral positions in Values and Public Policy. The program aims to support highly promising scholars trained in moral and political philosophy, political theory, normative economics and related areas to develop a research agenda in the ethical dimensions of public policy. Candidates selected will undertake a research project exploring a normative problem arising in an area of public policy in which the Woodrow Wilson School sponsors research. They will be affiliated with a unit of the School that can inform and support their work. They will also be invited to participate in programs of the University Center for Human Values. Candidates will be expected to contribute one course each year to the School's undergraduate program on a topic related to ethical issues in public policy, subject to sufficient enrollment and approval by the Dean of the Faculty. The term of appointment is one year, normally beginning September 1, with the expectation of renewal for one further year assuming good performance. Applicants are expected to be in residence for the duration of the appointment. Princeton offers competitive salary and employee benefits. Candidates should submit an online application.
*Requirements:* Applicants must have completed the requirements for the Ph.D. by the start of the appointment and must not have held the degree for more than three years by that date. They may not be employed by another institution during the term of their Princeton appointment.
*Geographic Focus:* All States
*Date(s) Application is Due:* Dec 3
*Contact:* Erin Graham, Program Coordinator; (609) 258-1460 or (609) 258-4798; fax (609) 258-2729; egraham@princeton.edu
*Internet:* uchv.princeton.edu/fellowships_awards/vpp_postdoc.php
*Sponsor:* Princeton University
5 Ivy Lane, 304 Louis Marx Hall
Princeton, NJ 08544

**Princeton University Center for Human Values Short Movie Prize** 2160

The UCHVFest is to be an annual gathering of the Center community to celebrate the achievements and participation of the Center's undergraduate students in their engagement with human values. The Center's undergraduate activities include the Undergraduate Certificate Program in Values and Public Life, the Film Forum, the Environmental Affairs Forum, and the Human Values Forum. The UCHVFest adds another way for undergraduates to engage with human values through the Center, namely, by making a short movie addressing the current year's theme. Selected movies will be screened and a $1,000 Short Movie Prize will be awarded. Undergraduate students enrolled at Princeton during academic year are invited to submit entries no later than February 15.
*Requirements:* The movie should be no more than seven minutes and should address the current year's theme.
*Geographic Focus:* All States
*Date(s) Application is Due:* Feb 15
*Amount of Grant:* 1,000 USD
*Contact:* Erum Syed, Assistant Director; (609) 258-2739 or (609) 258-4798; fax (609) 258-2729; erum@princeton.edu
*Internet:* uchv.princeton.edu/fellowships_awards/short_movie_prize.php
*Sponsor:* Princeton University
5 Ivy Lane, 304 Louis Marx Hall
Princeton, NJ 08544

**Princeton University Lewis Center for the Arts Hodder Fellowships** 2161

The Hodder Fellowship will be given to writers and other artists of exceptional promise to pursue independent projects at Princeton University during the academic year. Hodder Fellows may be writers, composers, choreographers, visual artists, performance artists, or other kinds of artists or humanists who have much more than ordinary intellectual and literary gifts; they are selected more for promise than for performance. Given the strength of the applicant pool, most successful Fellows have published a first book or have similar achievements in their own fields; the Hodder is designed to provide Fellows with the studious leisure to undertake significant new work. Hodder Fellows spend an academic year at Princeton, but no formal teaching is involved. A $68,000 stipend is provided. Apply by November 1 through the Princeton Human Resources website.
*Requirements:* One need not be a U.S. citizen to apply.
*Restrictions:* Fellowships cannot fund work leading to the Ph.D. degree.
*Geographic Focus:* All States
*Date(s) Application is Due:* Nov 1
*Amount of Grant:* 68,000 USD
*Contact:* Michael Cadden, Acting Chair; (609) 258-1500; fax (609) 258-2230; mcadden@princeton.edu or oconnorm@princeton.edu
*Internet:* www.princeton.edu/arts/lewis_center/society_of_fellows/
*Sponsor:* Princeton University
185 Nassau Street
Princeton, NJ 08544

**Princeton University Library Research Grants** 2162

The Friends of the Princeton University Library offers short-term research grants to promote scholarly use of the research collections of the library. The Program in Hellenic Studies supports a limited number of library fellowships in Hellenic Studies, and the Cotsen Children's Library supports research in its collection on aspects of children's books. These Library Research Grants are meant to help defray expenses incurred in traveling to and residing in Princeton during the tenure of the grant. The length of the grant will depend on the applicant's research proposal, but is usually up to one month. Application information and guidelines are available at the website.
*Requirements:* Applicants are asked to read the FAQ at the website, then complete an online application form, and submit a Word or PDF file (preferred) containing a budget form, a curriculum vitae or resume, and a research proposal not exceeding one thousand words in length. Applicants must also arrange for two confidential letters of recommendation to be sent directly to the Library Research Grants Committee. The proposal should address specifically the relevance to the proposed research of unique resources found in the Princeton University Library collections. Applications will be considered for scholarly use of archives, manuscripts, rare books, and other rare and unique holdings of the Department of Rare Books and Special Collections, the Mudd Library, the Marquand Library of Art and Archaeology, and the East Asian Library (Gest Collection). Prospective grantees are urged to consult the Library's home page for detailed descriptions of the collections, especially those in the Department of Rare Books and Special Collections. Applicants should have specific Princeton resources in mind as they prepare their proposals. A committee consisting of members of the faculty, the library staff, and the Friends of the Library will award the grants on the basis of the relevance of the proposal to unique holdings of the library, the significance of the project, and the applicant's scholarly qualifications.
*Restrictions:* Recommendation letters must be received from an institutional email address, and not a commercial provider such as hotmail. The general circulating collections and electronic resources of the Princeton University Library are not relevant for purposes of these grants.
*Geographic Focus:* All States
*Date(s) Application is Due:* Jan 15
*Amount of Grant:* Up to 3,500 USD
*Contact:* Research Grants Contact; (609) 258-3184; fax (609) 258-2324; rbsc@princeton.edu or pulgrant@princeton.edu
*Internet:* www.princeton.edu/rbsc/fellowships/f_ships.html
*Sponsor:* Princeton University Library
One Washington Road
Princeton, NJ 08544

**Prudential Foundation Education Grants** 2163

The Prudential Foundation's areas of interest are ready-to-learn programs, ready-to-work programs, and ready-to-live programs. In order to promote sustainable communities and improve social outcomes for community residents, the Foundation focuses its strategy in the following educational areas: education leadership to support reform in public education by increasing the capacity of educators, parents, and community residents to

implement public school reform; and youth development to build skills and competencies needed for young people to be productive citizens (this includes expanding arts education opportunities and supporting effective out-of-school-time programs for young people). Finally, the Foundation also funds organizations whose efforts influence policy that adapts promising practices and evidence-based approaches to instruction and learning in schools. Types of support include operating support, continuing support, annual campaigns, seed money, matching funds and employee matching gifts, consulting services, technical assistance, employee-related scholarships, research, capital campaigns, conferences and seminars, and projects/programs. The Foundation is especially interested in proposals that anticipate and address potential major problems. Funds are targeted to areas where Prudential has a strong presence. Applicant should make initial contact with a brief letter to determine whether a more detailed proposal would be acceptable.
*Requirements:* The Prudential Foundation supports nonprofit, charitable organizations, and programs whose mission and operations are broad and non-discriminatory. The Foundation focuses its resources to support organizations whose activities address social needs or benefit underserved groups and communities. Priority in order of preference goes to programs in Newark, New Jersey, and surrounding communities; Los Angeles, California; Hartford, Connecticut; New York, New York; Chicago, Illinois; Jacksonville, Florida; Atlanta, Georgia; Minneapolis, Minnesota; Philadelphia and Scranton, Pennsylvania; Houston and Dallas, Texas; Dubuque, Iowa; Phoenix, Arizona; Honolulu, Hawaii; and New Orleans, Louisiana.
*Restrictions:* The Foundation does not fund: organizations that are not tax-exempt under paragraph 501(c)3 of the U.S. Internal Revenue Code; labor, religious, political, lobbying, or fraternal groups—except when these groups provide needed services to the community at large; direct grants or scholarships to individuals; support for single-disease health groups; or good will advertising.
*Geographic Focus:* Arizona, California, Connecticut, Florida, Georgia, Hawaii, Illinois, Iowa, Louisiana, Minnesota, New Jersey, New York, Pennsylvania, Texas
*Amount of Grant:* Up to 1,000,000 USD
*Contact:* Lata N. Reddy, Vice President and Head of Corporate Social Responsibility; (973) 802-4791; community.resources@prudential.com
*Internet:* https://www.prudential.com/links/about/corporate-social-responsibility
*Sponsor:* Prudential Foundation
751 Broad Street, 15th Floor
Newark, NJ 07102-3777

**PSG Mentored Clinical Research Awards** **2164**
The Mentored Clinical Research Award (MCRA) for new investigators is funded by a grant from the Parkinson's Disease Foundation (PDF) to the Parkinson Study Group (PSG). The PDF has partnered with the PSG to encourage the professional and scientific development of young investigators on their path to independence. To this end, this grant will support a new investigator for a one year project in patient oriented research in Parkinson's disease or other parkinsonian disorders under the mentorship of an experienced investigator. The training should lead a junior investigator to gain skills in clinical research. The research plan should address unmet needs of people living with PD, have the potential for broad application among the PD community, and lead to advances in clinically relevant treatment options. An award of $50,000 will be available each year. The annual deadline for application submission is January 12, with successful applicants being notified by March 16.
*Requirements:* Appropriate applicants for the MCRA are clinicians and scientists who are within 5 years of having completed formal training (this includes PhD professionals working in the field of Parkinson's disease). Fellows may apply. See website for additional guidelines.
*Geographic Focus:* All States
*Date(s) Application is Due:* Jan 12
*Amount of Grant:* 50,000 USD
*Contact:* Donna Moszkowicz, PSG Mentoring Committee Administrator; (585) 273-2862; fax (585) 275-1642; donna.moszkowicz@chet.rochester.edu
*Internet:* www.parkinson-study-group.org/resources
*Sponsor:* Parkinson Study Group
MassGeneral Institute for Neurodegenerative Disease
Boston, MA 02129

**PSG Retrospective Datamining and Planning Project Awards** **2165**
The Parkinson Study Group (PSG) has funding available for investigators seeking initial support for Parkinson disease research. Candidates propose a hypothesis driven, one year research plan for a retrospective datamining research project in Parkinson disease using PSG databases. Funds have been provided by the Parkinson's Disease Foundation to the PSG to support retrospective datamining projects. An award of $50,000 will be available each year. The annual deadline for application submission is December 1, with successful applicants being notified by February 9.
*Geographic Focus:* All States
*Date(s) Application is Due:* Dec 1
*Contact:* Donna Moszkowicz, Development and Event Manager; (585) 273-2862; fax (585) 275-1642; donna.moszkowicz@chet.rochester.edu
*Internet:* www.parkinson-study-group.org/resources
*Sponsor:* Parkinson Study Group
MassGeneral Institute for Neurodegenerative Disease
Boston, MA 02129

**PSG Utilizing Datatop Biospecimens Grants** **2166**
The Utilizing Datatop Biospecimens Grant program is in collaboration with The Michael J. Fox Foundation to support the identification and validation of novel biomarkers through use of the DATATOP biospecimen and clinical data resource. The DATATOP (Deprenyl And Tocopherol Antioxidative Therapy Of Parkinsonism) intervention trial, conducted by the PSG in the late 1980s, also collected serum, cerebrospinal fluid, urine and DNA from trial participants to be banked for subsequent analyses. The biologic specimens paired with the clinical data from this study are a unique resource for identifying and verifying objective markers that change with clinical features of Parkinson disease. Awards have ranged up to a maximum of $25,000.
*Geographic Focus:* All States
*Amount of Grant:* Up to 25,000 USD
*Samples:* Anne-Marie Wills, Massachusetts General Hospital, Charlestown, Massachusetts, $22,950 - support to study Body Mass Index and PD Survival; Rachel Biemiller, University of Rochester Medical Center, Rochester, New York, $25,000 - support to study PD phenotype and response to dopaminergic therapy: a secondary analysis of the CALM-PD trial.
*Contact:* Donna Moszkowicz, PSG Mentoring Committee Administrator; (585) 273-2862; fax (585) 275-1642; donna.moszkowicz@chet.rochester.edu
*Internet:* www.parkinson-study-group.org/resources
*Sponsor:* Parkinson Study Group
MassGeneral Institute for Neurodegenerative Disease
Boston, MA 02129

**PVA Research Foundation Grants** **2167**
The foundation, established by the Paralyzed Veterans of America, supports innovative research and fellowships that improve the lives of those with spinal cord injury and disease (SCI/D). The Foundation funds five research categories: laboratory research in the basic sciences to find a cure for SCI/D; clinical and functional studies of the medical, psychosocial and economic effects of SCI/D, and interventions to alleviate these effects; design and development of assistive technology for people with SCI/D, which includes improving the identification, selection and utilization of these devices; fellowships for postdoctoral scientists, clinicians and engineers to encourage training and specialization in the field of spinal cord research; and conferences and symposia that provide opportunities for collaboration and interaction among scientists, health care providers and others involved in the SCI/D research community. The PVA Research Foundation online application site will open on May 1 to receive new applications. The annual deadline is September 1.
*Requirements:* Eligible grantee institutions must be located in the United States or Canada. However, investigators and fellows are not required to be U.S. or Canadian citizens. All grant applicants must have a professional degree: Ph.D. or M.D. preferred, master's degree acceptable under some circumstances. Senior fellows are encouraged to apply as principal investigators. Post-doctoral individuals are eligible to apply for fellowship support within four years of receiving a Ph.D. or completing M.D. residency. Graduate students can participate in Foundation-related research and be paid from a Foundation award.
*Restrictions:* Graduate students cannot apply for a Foundation grant either as a fellow or as a principal investigator.
*Geographic Focus:* All States, Canada
*Date(s) Application is Due:* Sep 1
*Amount of Grant:* 50,000 - 75,000 USD
*Contact:* Barbara Zupnik, Grants Portfolio Manager; (202) 416-7652 or (800) 424-8200; fax (202) 416-7641; barbaraz@pva.org or foundations@pva.org
*Internet:* www.pva.org/site/c.ajIRK9NJLcJ2E/b.6305827/k.7268/PVA_Research_Foundation.htm
*Sponsor:* Paralyzed Veterans of America Research Foundation
801 Eighteenth Street, NW
Washington, D.C. 20006-3517

**R.C. Baker Foundation Grants** **2168**
The R.C. Baker Foundation makes grants to U.S. nonprofit organizations for projects and programs that support social services for youth and the elderly, crime prevention, education, religion (Christian, Episcopal, Friends, Jewish, Methodist, and Presbyterian), scientific research, culture, and health. Support will be provided for fellowships and scholarships, general operating grants, challenge/matching grants, emergency funds, building funds, equipment, continuing support, annual campaigns, capital campaigns, renovation projects, and special projects. A cover letter should be submitted with a one- to two-page proposal. Annual deadlines are May 1 and October 1. The board meets in June and November to consider requests.
*Requirements:* 501(c)3 nonprofit organizations are eligible to apply.
*Geographic Focus:* All States
*Date(s) Application is Due:* May 1; Oct 1
*Amount of Grant:* 1,000 - 280,000 USD
*Samples:* A Bridge for Kids, San Diego, California, $2,750 - general operating support; Adventure Cycling Association, Missoula, Montana, $3,000 - general operating support; Friends of the Children's Museum, La Habra, California, $2,500 - general operating support.
*Contact:* Frank Laurence Scott, Jr., President; (760) 632-3600
*Sponsor:* R.C. Baker Foundation
330 Encinitas Boulevard, Suite 101
Encinitas, CA 92024-3723

**RACGP Foundation Adelaide North East Division of General Practice Research Grant** **2169**
The RACGP Foundation and the former Adelaide North East Division of General Practice (ANEDGP) have established a research grant for the purpose of conducting primary care research related to the boundaries of the former Adelaide North East Division of General Practice. The objective is to fund a researcher to conduct innovation and quality improvement research that advances the profession of general practice. The grant is available for a period of one year to provide GP researchers with an opportunity to conduct research that advances the profession of general practice. Funding of up to $20,000 (excluding GST) is available. Applications for this research grant will close on May 7 each year.

*Requirements:* To be eligible for the RACGP Foundation Adelaide North East Division of General Practice research grant, the principal investigator must be a general practitioner or general practice registrar practicing within the boundaries of the former Adelaide North East Division of General Practice.
*Restrictions:* By applying for the award each applicant will be taken to have consented to permit RACGP to publicize their name, their institution, and the title of their research project.
*Geographic Focus:* Australia
*Date(s) Application is Due:* May 7
*Amount of Grant:* 20,000 AUD
*Contact:* Sarah Maguire, Awards Secretariat; +61 3 8699 0550 or +61 3 8699 0488; fax +61 3 8699 0560; sarah.maguire@racgp.org.au or foundation@racgp.org.au
*Internet:* https://racgpfoundation.smartygrants.com.au/ANE2018
*Sponsor:* Royal Australian College of General Practitioners Foundation
100 Wellington Parade
East Melbourne, VIC 3002 Australia

**RACGP Foundation Alan Chancellor Award** **2170**

Dr. Alan Chancellor, who passed away in 1996, donated this award in 1990 for new researchers in the field of general practice. A recipient of the Rose Hunt Award for a lifetime of service to the RACGP, Dr. Chancellor was Chairman of the National Research Committee, a founding supervisor of the RACGP Training Program, and a foundation member of the RACGP National Research and Evaluation Ethics Committee. The Award is presented each year to the GP or GP registrar considered to be the best first time presenter of a research paper at the RACGP Annual Conference. The prize is $500 (excluding GST) and an engraved medallion. This Award will be presented at the annual RACGP Foundation Gala Dinner, and is made available by donations from the RACGP and RACGP members. The annual deadline for submissions is August 31.
*Requirements:* To be eligible for the Award, an individual must be: a GP or GP registrar; a member of the RACGP; in the early stages of their research career; a first-time presenter of a research paper at the RACGP Annual Scientific Conference, and; a major contributor to the research study being presented.
*Restrictions:* By applying for the award each applicant will be taken to have consented to permit RACGP to publicise their name, their institution, and the title of their research project.
*Geographic Focus:* Australia
*Date(s) Application is Due:* Aug 31
*Amount of Grant:* 500 AUD
*Contact:* Sarah Maguire, Awards Secretariat; +61 3 8699 0550 or +61 3 8699 0488; fax +61 3 8699 0560; sarah.maguire@racgp.org.au or awards@racgp.org.au
*Internet:* https://racgpfoundation.smartygrants.com.au/ACA18
*Sponsor:* Royal Australian College of General Practitioners Foundation
100 Wellington Parade
East Melbourne, VIC 3002 Australia

**RACGP Foundation BOQ Specialist Research Grant** **2171**

BOQ Specialist and the RACGP Foundation have established a research grant to encourage innovation in preventative healthcare in general practice. BOQ Specialist has been Australia's leading financial services partner for medical professionals for over 25 years. By offering a distinctive range of personal and business banking products, BOQ Specialist helps its niche clients to achieve their professional and personal ambitions. Together, BOQ Specialist and the RACGP Foundation are committed to supporting medical professionals and recognize the pivotal role that general practitioners play in promoting good health, providing preventative care and facilitating early detection and intervention. One grant of up to $20,000 (excluding GST) is available for a period of one year to provide a GP researcher or GP registrar with an opportunity to conduct research into preventive care in general practice. Applications for this research grant will close on May 7 each year.
*Requirements:* To be eligible for the RACGP Foundation BOQ Specialist research grant, the principal investigator must be a general practitioner or general practice registrar.
*Restrictions:* Unless otherwise indicated, Grants must be used in the year for which they are awarded. Grants will be paid only to incorporated bodies that have an ABN and ACN and not to individuals. By applying for a Grant each applicant will be taken to have consented to permit the RACGP to publicise their name, institution, and title of their research project.
*Geographic Focus:* Australia
*Date(s) Application is Due:* May 7
*Amount of Grant:* 20,000 AUD
*Contact:* Sarah Maguire, Awards Secretariat; +61 3 8699 0550 or +61 3 8699 0488; fax +61 3 8699 0560; sarah.maguire@racgp.org.au or foundation@racgp.org.au
*Internet:* https://racgpfoundation.smartygrants.com.au/BOQ18
*Sponsor:* Royal Australian College of General Practitioners Foundation
100 Wellington Parade
East Melbourne, VIC 3002 Australia

**RACGP Foundation Charles Bridges-Webb Memorial Award** **2172**

The RACGP has established an award in memory of one of its most dedicated researchers, Professor Charles Bridges-Webb AO. Charles became a member of the RACGP in 1959, and was an inspiration and a wise mentor for generations of general practice researchers in Australia. He was awarded both the Rose Hunt Award and Life Fellowship of the RACGP and was Director of the RACGP NSW&ACT research unit until June of 2010. Professor Bridges-Webb was also recognized by the Australian Association for Academic Primary Care through Life Membership and by the World Health Organization of Family Doctors (WONCA) through Fellowship, the most prestigious award of the international organization awarded to individuals who have rendered outstanding service to primary care and family medicine at a global level. The award of up to $1,000 (excluding GST) is available to pay for direct research costs associated with the conduct of any study that has ethics approval and is deemed to have relevance to general practice, health care, health outcomes or health professional education. Direct research costs include any item or service that materially contributes to the collection and analysis of data relevant to the aims of the research project. For instance, it can include payments to practices for participation in a research project, payment for equipment and disposables, payment for paper or photocopying, telephone calls, data entry and statistical help. Applications for this research grant will close on May 7 each year.
*Requirements:* General practice registrars who have recently undertaken, will undertake, or are currently undertaking an academic term in a University Department or Discipline of General Practice or Rural Health are eligible to apply.
*Geographic Focus:* Australia
*Date(s) Application is Due:* May 7
*Amount of Grant:* Up to 1,000 AUD
*Contact:* Sarah Maguire, Awards Secretariat; +61 3 8699 0550 or +61 3 8699 0488; fax +61 3 8699 0560; sarah.maguire@racgp.org.au or foundation@racgp.org.au
*Internet:* https://racgpfoundation.smartygrants.com.au/CBW18
*Sponsor:* Royal Australian College of General Practitioners Foundation
100 Wellington Parade
East Melbourne, VIC 3002 Australia

**RACGP Foundation Chris Silagy Research Scholarship** **2173**

This scholarship is named in honor of Professor Christopher Allen Silagy, AO MBBS PhD FRACGP FAFPHM. At the age of 31 Chris was awarded a PhD by Monash University for his research into the prevention of cardiovascular disease in the elderly. The same year he received the Sir Robert Menzies Memorial Scholarship in Medicine; the first general practitioner to receive this prestigious award. Chris worked in the Department of Public Health & Primary Care at the University of Oxford for two years, during which time he became involved with the Cochrane Collaboration, dedicated to the development of evidence-based medicine. The Chris Silagy Research Scholarship is designed to support a twelve month research project into an aspect of evidence-based primary health care by a general practitioner researcher. The scholarship is to the value of $15,000 per annum which includes a return economy airfare (booked by the College) to either attend a meeting of the Cochrane Collaboration or visit a Cochrane Research Centre during the year of their scholarship. Scholarships are awarded by the Chris Silagy Research Scholarship Committee, a subcommittee of the RACGP National Research Awards Committee. Applications for this research grant will close on May 7 each year.
*Requirements:* Applicants for the Chris Silagy Research Scholarship must be: a General Practitioner or General Practice Registrar; a member of the RACGP; and currently undertaking a higher degree (Masters, PHD, MD) or have completed such a course in the past five years.
*Restrictions:* Unless otherwise indicated, the Scholarship must be used in the year for which it is awarded. The Scholarship will be paid only to incorporated bodies that have an ABN and ACN and not to individuals. By applying for the Scholarship each applicant will be taken to have consented to permit the RACGP to publicise their name, institution, and title of their research project.
*Geographic Focus:* Australia
*Date(s) Application is Due:* May 7
*Amount of Grant:* 15,000 AUD
*Contact:* Sarah Maguire, Awards Secretariat; +61 3 8699 0550 or +61 3 8699 0488; fax +61 3 8699 0560; sarah.maguire@racgp.org.au or foundation@racgp.org.au
*Internet:* https://racgpfoundation.smartygrants.com.au/SIL18
*Sponsor:* Royal Australian College of General Practitioners Foundation
100 Wellington Parade
East Melbourne, VIC 3002 Australia

**RACGP Foundation Diabetes Australia Research Grant** **2174**

Diabetes is one of the most frequently managed chronic conditions in general practice, and its prevalence has been increasing in recent years. General practitioners provide most of the medical care to people with diabetes. Diabetes Australia was established in 1984 as the national peak body for diabetes in Australia providing a single, powerful, collective voice for people living with diabetes, their families and carers. A non-profit organization, Diabetes Australia works in partnership with diabetes consumer organizations, health professionals, educators and researchers to minimize the impact of diabetes in the Australian community. Diabetes Australia is committed to turning diabetes around through awareness, prevention, detection, management and the search for a cure. The RACGP is a key contributor to the improvement of Australian health through general practice. Together, the RACGP and Diabetes Australia have established a grant to support research into diabetes in general practice. One grant of up to $60,000 is available for a period of one year to fund research into diabetes management, prevention, and/or clinical care in general practice. Applications for this research grant will close on May 7 each year.
*Requirements:* To be eligible for the RACGP Foundation Diabetes Australia research grant, the principal investigator must be a general practitioner or general practice registrar.
*Geographic Focus:* Australia
*Date(s) Application is Due:* May 7
*Amount of Grant:* Up to 60,000 AUD
*Contact:* Sarah Maguire, Awards Secretariat; +61 3 8699 0550 or +61 3 8699 0488; fax +61 3 8699 0560; sarah.maguire@racgp.org.au or foundation@racgp.org.au
*Internet:* https://racgpfoundation.smartygrants.com.au/DIA2018
*Sponsor:* Royal Australian College of General Practitioners Foundation
100 Wellington Parade
East Melbourne, VIC 3002 Australia

**RACGP Foundation Education Research Grants** 2175

Royal Australian College of General Practitioners Foundation Education Research Grants (ERGs) aim to continue to build Regional Training Organization research capability in medical education and training with a specific focus on education improvement. The overarching objective is to develop evidence-based education initiatives that increase training capacity and improve educational outcomes. ERGs are awarded via a competitive process to Regional Training Organizations (RTOs), that partner with academic institutions to undertake and complete a research project. The application window for the RACGP Education Research Grant round is open from each year through February 5.
*Requirements:* To be eligible for the Education Research grants, the chief investigator must be a GP or GP registrar who is a member of the RACGP and must have a project that relates clearly to general practice.
*Restrictions:* Unless otherwise indicated, the Grant must be used in the year for which it is awarded. The Grant will be paid only to incorporated bodies that have an ABN and ACN and not to individuals. By applying for the Grant each applicant will be taken to have consented to permit the RACGP to publicise their name, institution, and title of their research project.
*Geographic Focus:* Australia
*Date(s) Application is Due:* Feb 5
*Amount of Grant:* Up to 20,000 AUD
*Contact:* Sarah Maguire, Awards Secretariat; +61 3 8699 0550 or +61 3 8699 0488; fax +61 3 8699 0560; sarah.maguire@racgp.org.au or foundation@racgp.org.au
*Internet:* https://www.racgp.org.au/becomingagp/australian-general-practice-training-program-(agpt)/education-research-grants/
*Sponsor:* Royal Australian College of General Practitioners Foundation
100 Wellington Parade
East Melbourne, VIC 3002 Australia

**RACGP Foundation Family Medical Care Education and Research Grant** 2176

The Family Medical Care Education and Research (FMCER) grants have been offered by the RACGP since 1980. The aim of the grant is to encourage and support general practitioners in the early stages of their research career to conduct research into an aspect of primary health care. The objective of the grant is to advance education in, and research into, medical knowledge and science. These grants are available for a period of one year to provide new and emerging GP researchers with an opportunity to conduct research. Funding of up to $20,000 (excluding GST) is available for research that benefits primary health care. Two grants are available. Applications for this research grant will close on May 7 each year.
*Requirements:* To be eligible for an FMCER grant, the principal investigator must be: a general practitioner or general practice registrar; and an early career researcher. Additionally, the project must relate directly to general practice.
*Restrictions:* Unless otherwise indicated, the Fellowship must be used in the year for which it is awarded. The Fellowship will be paid only to incorporated bodies that have an ABN and ACN and not to individuals. By applying for the Fellowship each applicant will be taken to have consented to permit the RACGP to publicise their name, institution, and title of their research project.
*Geographic Focus:* Australia
*Date(s) Application is Due:* May 7
*Amount of Grant:* 20,000 AUD
*Contact:* Sarah Maguire, Awards Secretariat; +61 3 8699 0550 or +61 3 8699 0488; fax +61 3 8699 0560; sarah.maguire@racgp.org.au or foundation@racgp.org.au
*Internet:* https://racgpfoundation.smartygrants.com.au/FMC18
*Sponsor:* Royal Australian College of General Practitioners Foundation
100 Wellington Parade
East Melbourne, VIC 3002 Australia

**RACGP Foundation Hospitals Contribution Fund Research Foundation Grants** 2177

The Hospitals Contribution Fund (HCF) Research Foundation and the RACGP Foundation have established a research grant for the purpose of funding research into health services issues of relevance to general practice. The HCF Research Foundation is a not-for-profit charitable trust established to encourage health service research for the benefit of all Australians, with objectives to assist in the improvement of prevention, treatment and cure of diseases. The focus of the HCF Research Foundation is achieving implementation of research activities. Health services research examines how people get access to health care, how much health care costs, and what happens to patients as a result of this care. The main goals of health services research are considered to be to identify the most effective ways to organize, manage, finance, and deliver high quality care; reduce medical errors; and improve patient safety. Two grants of up to $60,000 each (excluding GST) are available for a period of one year to fund research into health services issues of relevance to general practice. This might include access to care, effectiveness of care, quality of care, efficiency of care, or cost of care. Applications for this research grant will close on May 7 each year.
*Requirements:* To be eligible for the RACGP Foundation Hospitals Contribution Fund Research Foundation grant, the principal investigator must be a general practitioner or general practice registrar.
*Geographic Focus:* All States
*Date(s) Application is Due:* May 7
*Amount of Grant:* Up to 60,000 AUD
*Contact:* Sarah Maguire, Awards Secretariat; +61 3 8699 0550 or +61 3 8699 0488; fax +61 3 8699 0560; sarah.maguire@racgp.org.au or foundation@racgp.org.au
*Internet:* https://racgpfoundation.smartygrants.com.au/HCF2018
*Sponsor:* Royal Australian College of General Practitioners Foundation
100 Wellington Parade
East Melbourne, VIC 3002 Australia

**RACGP Foundation Indigenous Health Award** 2178

The Royal Australian College of General Practitioners Foundation supports research which makes a difference to health outcomes, policy and practice, and the capacity of general practice to engage Aboriginal and Torres Strait Islanders' health issues. It particularly encourages a systematic approach to strategic and policy-driven research focusing on primary care, developing collaborative approaches and the building of research capacity within Indigenous populations and communities. The Award is valued at $2500 and is available for a 12 month period to provide support for: indigenous medical students; medical students to gain experience in indigenous health; indigenous medical graduates undertaking GP training; registrars and GPs undertaking research and educational projects in indigenous health, or; other activities supporting the provision of high quality health care to indigenous people. One or more RACGP Indigenous Health Awards, valued at up to $9,000 in total (excluding GST), are available for a period of one year. Applications for this research grant will close each year on May 7.
*Requirements:* Applicants for the award must be either a medical student or a GP or GP registrar who is a member of the RACGP.
*Restrictions:* Unless otherwise indicated, the Award must be used in the year for which it is awarded. The Award will be paid only to incorporated bodies that have an ABN and ACN and not to individuals. By applying for the Award each applicant will be taken to have consented to permit the RACGP and APHCRI to publicise their name, institution, and title of their research project.
*Geographic Focus:* Australia
*Date(s) Application is Due:* May 7
*Amount of Grant:* 9,000 AUD
*Contact:* Sarah Maguire, Awards Secretariat; +61 3 8699 0550 or +61 3 8699 0488; fax +61 3 8699 0560; sarah.maguire@racgp.org.au or foundation@racgp.org.au
*Internet:* https://racgpfoundation.smartygrants.com.au/IHA2018
*Sponsor:* Royal Australian College of General Practitioners Foundation
100 Wellington Parade
East Melbourne, VIC 3002 Australia

**RACGP Foundation IPN Medical Centres Research Grant** 2179

IPN Medical Centres and the RACGP Foundation have established a research grant to encourage innovation in chronic disease management in general practice. IPN manages over 220 multidisciplinary medical centers throughout Australia which deliver over ten million patient consultations each year. Chronic diseases are the leading cause of illness, disability and death in Australia. IPN has a focus on delivering caring medicine, and given that the majority of chronic disease is managed in the general practice setting, supports innovation that will improve patient care. One grant of up to $10,000 (excl. GST) is offered for a period of one year to fund projects on innovation in the area of chronic disease management in the general practice setting. Applications for this research grant will close on May 7 each year.
*Requirements:* To be eligible for the RACGP Foundation IPN Medical Centres research grant, the principal investigator must be a general practitioner or general practice registrar.
*Restrictions:* Unless otherwise indicated, the Fellowship must be used in the year for which it is awarded. The Fellowship will be paid only to incorporated bodies that have an ABN and ACN and not to individuals. By applying for the Fellowship each applicant will be taken to have consented to permit the RACGP to publicise their name, institution, and title of their research project.
*Geographic Focus:* Australia
*Date(s) Application is Due:* May 7
*Amount of Grant:* Up to 10,000 AUD
*Contact:* Sarah Maguire, Awards Secretariat; +61 3 8699 0550 or +61 3 8699 0488; fax +61 3 8699 0560; sarah.maguire@racgp.org.au or foundation@racgp.org.au
*Internet:* https://racgpfoundation.smartygrants.com.au/IPN2018
*Sponsor:* Royal Australian College of General Practitioners Foundation
100 Wellington Parade
East Melbourne, VIC 3002 Australia

**RACGP Foundation Peter Mudge Medal** 2180

The RACGP Peter Mudge Medal is an award established to honor the work and commitment of Professor Peter Mudge in general practice and research. Professor Mudge is currently the Patron of the RACGP Foundation and holds a Rose-Hunt Medal, the RACGP's most prestigious honor, for his remarkable contribution to the RACGP both as Chair of the RACGP Tasmania Board and as Chair of RACGP Council. The Medal is awarded once a year to a presenter at the annual RACGP Annual Conference who has advanced the discipline of general practice and the goals of the RACGP and whose original research has the most potential to significantly influence daily general practice. The prize is $500 (excluding GST) and an engraved medallion. This Award will be presented at the annual RACGP Foundation Gala Dinner, and is made available by donations from the RACGP and RACGP members. The annual deadline for application submission is August 31.
*Requirements:* To be eligible for the Peter Mudge Medal, an individual must be: an RACGP Fellow; and a major contributor to the research study being presented.
*Restrictions:* By applying for the award each applicant will be taken to have consented to permit RACGP to publicise their name, their institution, and the title of their research project.
*Geographic Focus:* Australia
*Date(s) Application is Due:* Aug 31
*Amount of Grant:* 500 AUD
*Contact:* Sarah Maguire, Awards Secretariat; +61 3 8699 0550 or +61 3 8699 0488; fax +61 3 8699 0560; sarah.maguire@racgp.org.au or awards@racgp.org.au
*Internet:* https://racgpfoundation.smartygrants.com.au/PMM18
*Sponsor:* Royal Australian College of General Practitioners Foundation
100 Wellington Parade
East Melbourne, VIC 3002 Australia

**RACGP Foundation Rose-Hunt Award** 2181

The Rose-Hunt Award is the highest accolade awarded by the RACGP. The award is a gift from the Royal College of General Practitioners to the RACGP, commemorating the founding members of the Royal College of General Practitioners, Dr. Fraser Rose and Lord Hunt of Fawley. The Rose-Hunt Award is awarded to one person who is a Fellow or a member, who - in the opinion of the Council - has rendered outstanding service in the promotion of the objectives of the RACGP, either by individual patient care, organization, education, research or any other means. Outstanding service may be over a period of ten years or a substantial and significant contribution to general practice over a shorter period. Nominations can be made by any two Fellows of the RACGP, and must be received by the annual deadline of May 18.
*Geographic Focus:* Australia
*Date(s) Application is Due:* May 18
*Contact:* Sarah Maguire, Awards Secretariat; +61 3 8699 0550 or +61 3 8699 0488; fax +61 3 8699 0560; sarah.maguire@racgp.org.au or awards@racgp.org.au
*Internet:* https://www.racgp.org.au/awards
*Sponsor:* Royal Australian College of General Practitioners Foundation
100 Wellington Parade
East Melbourne, VIC 3002 Australia

**RACGP Foundation Vicki Kotsirilos Integrative Medicine Grants** 2182

An increasing number of doctors are transcending the "alternative vs conventional" medicine paradigm by incorporating scientifically-proven natural therapies into their practices. Instead of a pill they prescribe a herb or a walk in the park. Integrative medicine has multiple benefits. It allows doctors to use the best of both worlds, traditional and conventional, while affording the freedom of choice to patients. Dr. Vicki Kotsirilos has embraced this form of medicine as a health practitioner and an educator. Such is her belief in this form of medicine that she had established the Vicki Kotsirilos Integrative Medicine Grant endorsed by the Royal Australian College of General Practitioners. The Grant is intended to encourage research that assists the integration of safe and ethical forms of natural and complementary medicines or therapies into mainstream general practice. The Grant is designed to encourage research that assists the integration of safe and ethical forms of natural and complementary medicines or therapies into mainstream general practice. An award of up $5,000 is offered to applicants conducting general practice research in the following areas: clinical research that contributes to the knowledge base concerning the efficacy of natural and complementary medicines or therapies; any project that supports research in behavioral and lifestyle factors impacting on health (such as diet, exercise, and stress management); and research that contributes to the efficient dissemination of the evidence regarding integrative medicine. The annual deadline for application submissions is May 7.
*Requirements:* Applicants must be: a General Practitioner or General Practice Registrar; and a member of the RACGP.
*Restrictions:* Unless otherwise indicated, the Grant must be used in the year for which it is awarded. The Grant will be paid only to incorporated bodies that have an ABN and ACN and not to individuals. By applying for the Grant each applicant will be taken to have consented to permit the RACGP to publicise their name, institution, and title of their research project.
*Geographic Focus:* Australia
*Date(s) Application is Due:* May 7
*Amount of Grant:* 5,000 AUD
*Contact:* Sarah Maguire, Awards Secretariat; +61 3 8699 0550 or +61 3 8699 0488; fax +61 3 8699 0560; sarah.maguire@racgp.org.au or foundation@racgp.org.au
*Internet:* foundation.racgp.org.au/grants
*Sponsor:* Royal Australian College of General Practitioners Foundation
100 Wellington Parade
East Melbourne, VIC 3002 Australia

**RACGP Foundation Walpole Grieve Memorial Award** 2183

The RACGP Foundation has established this award in memory of Dr. Rex Outhwaite Walpole and Dr. PWH Grieve. Dr. Rex Outhwaite Walpole was a general practitioner in Melbourne for over 30 years, until appointed as the foundation Medical Director of the Victorian Academy of General Practice (1976-1984). He was chairman of the Medical Education Committee in Victoria and was the first to use public television as a medium for the education of doctors. He was editor of the Australian Family Physician from 1972 to 1974, chairman of the Accreditation Committees in the Victoria Faculty and Council, chair of the Board of the Victoria Faculty and medical editor of the check Program. One of his major contributions to the training of future doctors was the establishment of the Academy Fellowship Program. Dr. Grieve was the second Secretary General of the RACGP and made outstanding contributions to his profession. Peter Grieve joined the RACGP in 1963 and became a member of the Faculty Board of NSW, Chairman of the Medical Education Committee, Chairman of the Faculty Board, Chairman of the Organizing Committee for 'Convention 1975' and the Faculty Representative on Council. He was appointed as Secretary General of the RACGP in 1978. The award, to the value of $5,000 (excluding GST), is designed to support travel for research or educational purposes.
*Requirements:* Applicants for the award must be: a general practitioner or general practice registrar; and an early career researcher.
*Restrictions:* Unless otherwise indicated, the Award must be used in the year for which it is awarded. The Award will be paid only to incorporated bodies that have an ABN and ACN and not to individuals. By applying for the Award each applicant will be taken to have consented to permit the RACGP to publicise their name, institution, and title of their research project.
*Geographic Focus:* Australia
*Date(s) Application is Due:* May 7
*Amount of Grant:* 5,000 AUD
*Contact:* Sarah Maguire, Awards Secretariat; +61 3 8699 0550 or +61 3 8699 0488; fax +61 3 8699 0560; sarah.maguire@racgp.org.au or awards@racgp.org.au
*Internet:* https://racgpfoundation.smartygrants.com.au/WGA2018
*Sponsor:* Royal Australian College of General Practitioners Foundation
100 Wellington Parade
East Melbourne, VIC 3002 Australia

**Radcliffe Institute Carol K. Pforzheimer Student Fellowships** 2184

The Arthur and Elizabeth Schlesinger Library on the History of Women in America invites Harvard undergraduates to make use of the library's collections with competitive awards of amounts up to $2,500 for relevant research projects. Preference will be given to applicants pursuing research in the history of work and the family, community service and volunteerism, the culinary arts, or women's health. The research may be, but is not required to be, in connection with a project for academic credit. Applications will be evaluated on the significance of the research and the project's potential contribution to the advancement of knowledge as well as its creativity in drawing on the library's holdings. The awards may be used to cover photocopying, other incidental research expenses, or living expenses in lieu of term time or summer employment. See website for current deadlines and application.
*Requirements:* Each application must include: the Schlesinger Library Grants and Fellowships cover page; a project description (no longer than four double-spaced pages in twelve-point font) indicating the purpose of the research, the Schlesinger Library holdings to be consulted, and the significance of these holdings to the project overall; a one-page bibliography of principal secondary sources consulted in designing the research; a résumé no longer than two pages; the name of one reference who has agreed to send a supporting letter; and a proposed budget indicating how the funds requested will be spent.
*Restrictions:* Only Harvard undergraduates are eligible. Funding does not cover the purchase of durable equipment.
*Geographic Focus:* All States
*Amount of Grant:* Up to 2,500 USD
*Contact:* (617) 496-3048; science@radcliffe.edu
*Internet:* www.radcliffe.edu/schles/pforzheimer_grant.aspx
*Sponsor:* Radcliffe Institute for Advanced Study
10 Garden Street
Cambridge, MA 02138

**Radcliffe Institute Dissertation Grants** 2185

The Arthur and Elizabeth Schlesinger Library on the History of Women in America invites scholars whose dissertation research requires use of the library's collections to apply for research support. Applications are evaluated on the significance of the research and the project's potential contribution to the advancement of knowledge as well as its creativity in drawing on the library's holdings. Grants cover travel expenses, living expenses, photocopying, and other incidental research expenses.
*Requirements:* Applicants must be a doctoral candidate in a relevant field and have an approved dissertation topic. Priority is given to those whose projects require use of materials available only in the Schlesinger Library. A brief report about work accomplished under the grant will be due no later than one year from the date of the award.
Restrictions: Non-U.S. citizens are eligible but should contact the library regarding visas and other required paperwork prior to applying.The grant does not cover the purchase of durable equipment.
*Geographic Focus:* All States, All Countries
*Amount of Grant:* Up to 3,000 USD
*Contact:* Grants Administrator; (617) 495-8647; slgrants@radcliffe.edu
*Internet:* www.radcliffe.edu/schles/dissertation_grant.aspx
*Sponsor:* Radcliffe Institute for Advanced Study
10 Garden Street
Cambridge, MA 02138

**Radcliffe Institute Fellowships for Creative Artists** 2186

The Radcliffe Institute is unique among centers for advanced study across the country in its commitment to encouraging and developing the creative arts along with all other disciplines. Through fellowships, public lectures and exhibits, the Institute offers artists not only a quiet place to create or compose, but also an opportunity to showcase their new work.
*Requirements:* Scholars in any field with a doctorate or appropriate terminal degree at least two years prior to appointment (by December of the prior year) in the area of the proposed project are eligible to apply. Only scholars who have published at least two articles or monographs are eligible to apply. Artists and writers need not have a PHD or an MFA to apply; however, they must meet other specific eligibility requirements. Fellows are expected to be free of their regular commitments so they may devote themselves full time to the work outlined in their proposal. Since this is a residential fellowship, fellows are expected to reside in the Boston area during that period and to have their primary office at the Institute so that they can participate fully in the life of the community. Applicants are strongly encouraged to use the online application system.
*Restrictions:* Former fellows of the Radcliffe Institute Fellowship Program (1999 to present) are not eligible to apply.
*Geographic Focus:* All States
*Date(s) Application is Due:* Oct 1
*Amount of Grant:* 70,000 USD
*Contact:* Science Office; (617) 496-3048; science@radcliffe.edu
*Internet:* www.radcliffe.edu/research/arts.aspx
*Sponsor:* Radcliffe Institute for Advanced Study
10 Garden Street
Cambridge, MA 02138

**Radcliffe Institute Individual Residential Fellowships** 2187

The Radcliffe Institute Fellowship Program is a scholarly community where individuals pursue advanced work across a wide range of academic disciplines, professions, and creative arts. Radcliffe Institute fellowships are designed to support scholars, scientists, artists, and writers of exceptional promise and demonstrated accomplishment who wish to pursue work in academic and professional fields and in the creative arts. In recognition of Radcliffe's historic contributions to the education of women and to the study of issues related to women, the Radcliffe Institute sustains a continuing commitment to the study of women, gender, and society. Applicants need not focus on gender, however. Women and men from across the United States and throughout the world are encouraged to apply. email fellowships@radcliffe.edu for questions regarding humanities, social sciences, or creative arts fellowships. email sciencefellowships@radcliffe.edu for questions regarding natural science and mathematics fellowships.

*Requirements:* Scholars in any field with a doctorate or appropriate terminal degree at least two years prior to appointment (by December of the prior year) in the area of the proposed project are eligible to apply. Only scholars who have published at least two articles or monographs are eligible to apply. Artists and writers need not have a PHD or an MFA to apply; however, they must meet other specific eligibility requirements. Fellows are expected to be free of their regular commitments so they may devote themselves full time to the work. Applicants must reside in the Boston area, with their primary office at the Institute. Applications are available online, but applicants should refer to website for current deadlines and application requirements for each academic discipline.

*Restrictions:* Former fellows of the Radcliffe Institute Fellowship Program (1999 to present) are not eligible to apply.

*Geographic Focus:* All States, All Countries
*Amount of Grant:* 35,000 - 70,000 USD
*Contact:* (617) 496-1324
*Internet:* www.radcliffe.edu/fellowships/apply.asp
*Sponsor:* Radcliffe Institute for Advanced Study
10 Garden Street
Cambridge, MA 02138

**Radcliffe Institute Oral History Grants** 2188

The Arthur and Elizabeth Schlesinger Library invites scholars who are conducting oral history interviews relevant to the history of women or gender in the United States to apply for funding support. This grant stipulates that the interviews take place in accordance with guidelines of the Oral History Association, that consent is obtained from interviewees for their words to be viewed by researchers worldwide, and that true copies or transcripts of the original recording of the oral interviews, as well as copies of the consent forms, be deposited in the Schlesinger Library. Applicants may request support to cover travel expenses, living expenses, transcription services, and incidental research materials. Applications will be evaluated on the significance of the research and documentation planned and the project's potential contribution to the advancement of knowledge.

*Requirements:* Each application must include: the Schlesinger Library Grants and Fellowships cover page; a project description (no longer than four double-spaced pages) indicating the purpose of the research, the persons to be interviewed, and the interview arrangements; a curriculum vitae of no more than two pages; the name of one reference who has agreed to send a supporting letter; and a proposed budget indicating how the requested funds will be spent. A brief report about work accomplished under the grant will be due no later than one year from the date of the award.

Restrictions: The grant does not cover funding for durable equipment. Non-U.S. citizens are eligible but should contact the library regarding visas and other required paperwork prior to applying.

*Geographic Focus:* All States, All Countries
*Amount of Grant:* Up to 3,000 USD
*Contact:* Administrator; (617) 495-8647; fax (617) 496-4640; slgrants@radcliffe.edu
*Internet:* www.radcliffe.edu/schles/oral_history_grant.aspx
*Sponsor:* Radcliffe Institute for Advanced Study
10 Garden Street
Cambridge, MA 02138

**Radcliffe Institute Research Support Grants** 2189

The Arthur and Elizabeth Schlesinger Library on the History of Women in America invites scholars to apply for research support. Applications are evaluated on the significance of the research and the project's potential contribution to the advancement of knowledge as well as its creativity in drawing on the library's holdings. Grants cover travel and living expenses, photocopying, and other incidental research expenses.

*Requirements:* Each application must include: the Schlesinger grants and fellowships cover page; a detailed project description; a curriculum vitae; name of one reference; and a detailed proposed budget. Applicants must have a doctoral degree or equivalent research and writing experience. Priority is given to those who demonstrate research productivity and whose projects require use of materials available only at the Schlesinger Library. Applicants should refer to the grant website for requirement details and current deadlines.

Restrictions: Non-U.S. citizens are eligible but should contact the library regarding visas and other required paperwork prior to applying. The grant does not cover the purchase of durable equipment such as laptop computers.

*Geographic Focus:* All States, All Countries
*Amount of Grant:* Up to 3,000 USD
*Contact:* Grants Administrator; (617) 495-8647; slgrants@radcliffe.edu
*Internet:* www.radcliffe.edu/schles/research_grant.aspx
*Sponsor:* Radcliffe Institute for Advanced Study
10 Garden Street
Cambridge, MA 02138

**Radcliffe Institute Residential Group Fellowships** 2190

The Radcliffe Institute Fellowship Program is a scholarly community where individuals pursue advanced work across a wide range of academic disciplines, professions, and creative arts. Radcliffe Institute fellowships are designed to support scholars, scientists, artists, and writers of exceptional promise and demonstrated accomplishment who wish to pursue work in academic and professional fields and in the creative arts. In recognition of Radcliffe's historic contributions to the education of women and to the study of issues related to women, the Radcliffe Institute sustains a continuing commitment to the study of women, gender, and society. Applicants' projects need not focus on gender, however. Women and men from across the United States and throughout the world are encouraged to apply.

*Requirements:* Scholars in any field with a doctorate or appropriate terminal degree at least two years prior to appointment (by December of the prior year) in the area of the proposed project are eligible to apply. All Humanists, Mathematicians, and Social and Natural Scientists applicants must have earned their doctorate at least one year prior to application. Creative Artists applicants need not have a PHD or an MFA to apply; however, they must meet other specific eligibility requirements. Applicants should refer to the website's eligibility guidelines for further details and current application deadlines.

*Restrictions:* Former fellows of the Radcliffe Institute Fellowship Program (1999 to present) are not eligible to apply. Only scholars who have published at least two articles or monographs are eligible to apply.

*Geographic Focus:* All States, All Countries
*Amount of Grant:* 35,000 - 70,000 USD
*Contact:* Judith Vichniac, Associate Dean, Fellowship Program; (617) 495-8212 or (617) 496-1324; fax (617) 495-8136
*Internet:* radcliffe.onlineapplicationportal.com/misc/guidelines/
*Sponsor:* Radcliffe Institute for Advanced Study
10 Garden Street
Cambridge, MA 02138

**Rathmann Family Foundation Grants** 2191

The foundation's main funding areas are education, with priority given to science and math; the arts; children and youth health organizations; and preservation of the environment. Types of support include general operating support, continuing support, capital campaigns, equipment acquisition, endowment funds, program development, conferences and seminars, seed grants, curriculum development, fellowships, internships, scholarship funds, research, and matching funds. There are no specific deadlines or application forms, and interested parties should begin by contacting the Foundation directly.

*Requirements:* Grants are awarded to organizations in the San Francisco Bay, California area; the Annapolis, Maryland area; the Seattle, Washington area; the Philadelphia, Pennsylvania area: and metropolitan Minneapolis/Saint Paul, Minnesota.

*Restrictions:* Grants are not awarded to/for private foundations; religious organizations for religious activities; civil rights; social action; advocacy organizations; fraternal groups; political purposes; mental health counseling; individuals; or fundraisers, media events, public relations, annual appeals, or propaganda.

*Geographic Focus:* California, Maryland, Minnesota, Pennsylvania, Washington
*Amount of Grant:* 1,000 - 100,000 USD
*Contact:* Rick Rathmann, (410) 349-2376; fax (410) 349-2377
*Sponsor:* Rathmann Family Foundation
1290 Bay Dale Drive, P.O. Box 352
Arnold, MD 21012

**RCPSC/AMS CanMEDs Research and Development Grants** 2192

The primary objective of the AMS CanMEDs Research and Development grant is to support research, development, and/or implementation of projects to enhance specialty education in the promotion of the key roles for specialists as defined by CanMEDS. Applications may be individual or joint, involving one or more departments, disciplines, or faculties of medicine. It is not necessary to be a Fellow of the RCPSC or a physician; however, a Fellow of the Royal College must contribute to the project. Eligible projects may include faculty development, course development and/or medical education research that both apply and promote the CanMEDS roles. Budgets may include requests to defray the costs for travel, accommodation and meals for investigators to disseminate the project findings. One or more grants of up to $25,000 will be awarded each year. The annual deadline for applications is March 3.

*Requirements:* Applicants must be involved in teaching and/or medical education research in Canada. Applications that include research must meet contemporary standards of educational research, which include the relevance of the research question, the quality of the research design, ethical review of the submission, accountability for funds received, and the ability to complete the project during the funding period. Applications that include project development must include an evaluative component assessing the impact of the project.

*Restrictions:* Budgets should not include salary or honoraria for investigators. Ineligible costs include tuition fees, capital equipment costs (e.g. computer hardware, office equipment), overhead costs, university or other organization administrative fees.

*Geographic Focus:* All States
*Date(s) Application is Due:* Mar 3
*Amount of Grant:* Up to 25,000 USD
*Contact:* Allison Montpetit, Grant Administrator; (800) 668-3740, ext. 291 or (613) 730-8177; fax (613) 730-8830; researchgrants@royalcollege.ca or awards@royalcollege.ca
*Internet:* www.royalcollege.ca/portal/page/portal/rc/awards/grants/medical_education/ams_grant
*Sponsor:* Royal College of Physicians and Surgeons of Canada
774 Echo Drive
Ottawa, ON K1S 5N8 Canada

**RCPSC Detweiler Traveling Fellowships** 2193
Fellowships are made to improve the quality of medical and surgical practice in Canada. The Detweiler Traveling Fellowship will enable the recipients to visit medical centers in Canada or abroad to study or gain experience in the use or application of new knowledge or techniques in their fields or to further the pursuit of a fundamental or clinical research problem. Five or more fellowships offered annually to a maximum of $21,000 will be reserved for junior applicants. Two additional fellowships offered annually will be reserved for senior applicants: one will be granted to a Fellow who has held Royal College certification for five years or more and has been in private practice for at least three years, and one will be granted to a Fellow who has held Royal College certification for five years or more and has been in academic practice for at least three years. Each award will be to a maximum of $10,500.
*Requirements:* Applicants must be Fellows of the Royal College in good standing, or residents in the final year of their training who will be admitted to Fellowship in the Royal College prior to December 31st.
*Restrictions:* Travel fellowships will not be awarded retroactively.
*Geographic Focus:* Canada
*Date(s) Application is Due:* Sep 30
*Amount of Grant:* 1,750 - 21,000 USD
*Contact:* Kimberly Ross, Awards and Grants Section; (800) 668-3740, ext. 355 or (613) 730-8177; fax (613) 730-8830; kross@royalcollege.ca or awards@royalcollege.ca
*Internet:* www.royalcollege.ca/portal/page/portal/rc/awards/grants/travelling_fellowships/detweiler_travelling_grant
*Sponsor:* Royal College of Physicians and Surgeons of Canada
774 Echo Drive
Ottawa, ON K1S 5N8 Canada

**RCPSC Harry S. Morton Traveling Fellowship in Surgery** 2194
The Harry S. Morton Traveling Fellowship in Surgery, established by The Royal College of Physicians and Surgeons of Canada in 2013, is funded from the investment income of the H. S. Morton Exchange Fellowship Fund of Canada. Dr. Morton obtained his medical degrees from the Royal London Hospital Medical School in England and joined the Royal Canadian Navy in 1938. Following the Second World War, he taught at McGill and practiced as Chief Surgeon at Queen Mary's Veterans Hospital in Montreal. He was made a Patron of the Royal College of Surgeons in 1999. The purpose of this Fellowship is to advance and to promote the study of and research in surgical training, and the practice of surgery, and to support the building of capacity and additional expertise within Canada's academic surgical programs. The Fellowship provides funds for Canadian surgeons and surgical residents travelling to the United Kingdom and may be used to support surgery-related clinical, scientific, or medical education studies or research. The Travelling Fellowship will provide extended training support (3 months or more) for individuals with academic surgical career plans when they return to practice in Canada. Fellowships are supported by a grant of $50,000 per year, pro-rated monthly, and are contingent upon a financial commitment from a University/Department of Surgery to provide funding support for salary/living expenses (amount should be reflected on application form). The annual deadline for applications is September 2.
*Requirements:* The Traveling Fellowship is open to Royal College Fellows certified in any surgical discipline recognized by the Royal College, and to residents in the final year of their surgical training who will be admitted to Fellowship in the Royal College prior to December 31 of the year following receipt of the Fellowship.
*Geographic Focus:* Canada
*Date(s) Application is Due:* Sep 2
*Amount of Grant:* Up to 50,000 USD
*Contact:* Kimberly Ross, Grant Administrator; (800) 668-3740, ext. 355 or (613) 730-8177; fax (613) 730-8830; kross@royalcollege.ca or awards@royalcollege.ca
*Internet:* www.royalcollege.ca/portal/page/portal/rc/awards/grants/travelling_fellowships/morton_travelling_fellowship
*Sponsor:* Royal College of Physicians and Surgeons of Canada
774 Echo Drive
Ottawa, ON K1S 5N8 Canada

**RCPSC Medical Education Research Grants** 2195
The primary objective of the Medical Education Research Grant is to support quality research in Canada relevant to the field of postgraduate medical education or continuing professional development. Applications may be individual or joint, involving one or more departments, disciplines, or faculties of medicine. It is not necessary to be a Fellow of the RCPSC or a physician; however, a Fellow of the Royal College must contribute to the research project. The following areas of research are encouraged: curriculum design and implementation; instructional methods; assessment; program evaluation; continuing education; professional development; and faculty development. Up to $25,000 per year of study to a maximum of $50,000 per project is available. The annual deadline for applications is March 2.
*Requirements:* Applicants must be involved in teaching and/or medical education research in Canada. Applications must meet contemporary standards of educational research, which include the relevance of the research question, the quality of the research design, ethical review of the submission, accountability for funds received, and the ability to complete the project during the funding period.
*Restrictions:* Ineligible costs include tuition fees, capital equipment costs (e.g. computer hardware, office equipment), overhead costs, university or other organization administrative fees.
*Geographic Focus:* All States, Canada
*Date(s) Application is Due:* Mar 2
*Amount of Grant:* Up to 25,000 USD
*Contact:* Allison Montpetit, Grant Administrator; (800) 668-3740, ext. 291 or (613) 730-8177; fax (613) 730-8830; researchgrants@royalcollege.ca or awards@royalcollege.ca
*Internet:* www.royalcollege.ca/portal/page/portal/rc/awards/grants/medical_education/meded_research_grant
*Sponsor:* Royal College of Physicians and Surgeons of Canada
774 Echo Drive
Ottawa, ON K1S 5N8 Canada

**RCPSC Peter Warren Traveling Scholarship** 2196
Dr. Charles Peter Warren was the inaugural Chair of the Royal College's History and Heritage Advisory Committee, and provided leadership to the Royal College's history and heritage activities until his sudden passing in May of 2011. This traveling scholarship was introduced in 2013 in recognition of Dr. Warren's significant contributions to the History of Medicine in Canada. It is intended to encourage post-graduate trainees in medicine to undertake research projects in the history of specialty medicine, the history of the Royal College of Physicians and Surgeons of Canada, and/or the history of Post-Graduate Medical Education in Canada; and to encourage the use of the Royal College historical and archival collections as a resource for scholarly work within the field of the history of medicine. Up to $1,500, based on the budget provided with the application package, is available. Applications should be submitted electronically by the annual September 2 deadline.
*Requirements:* The Peter Warren Traveling Scholarship is open to applicants from all medical and surgical disciplines recognized by the Royal College. Applicants must be post-graduate trainees in a Canadian medical school who are undertaking a clearly defined project in the history of specialty medicine including but not limited to the history of the Royal College and/or the history of Post-Graduate Medical Education in Canada.
*Geographic Focus:* Canada
*Date(s) Application is Due:* Sep 2
*Amount of Grant:* Up to 1,500 CAD
*Contact:* Kimberly Ross, Grant Administrator; (800) 668-3740, ext. 355 or (613) 730-8177; fax (613) 730-8830; kross@royalcollege.ca or awards@royalcollege.ca
*Internet:* www.royalcollege.ca/portal/page/portal/rc/awards/grants/travelling_fellowships/warren_scholarship
*Sponsor:* Royal College of Physicians and Surgeons of Canada
774 Echo Drive
Ottawa, ON K1S 5N8 Canada

**RCPSC Robert Maudsley Fellowship for Studies in Medical Education** 2197
Dr. Robert Maudsley, MD, FRCSC was a passionate mentor and innovator in the field of medical education. He was instrumental in the development of the original CanMEDS Framework, and at the time of his passing was Associate Director, Internationally Educated Health Professionals at the Royal College. Dr. Maudsley passed away suddenly in October 2012. he goal of this one-year fellowship is to increase the number and quality of professionally trained medical educators in Canada by providing training in the science of medical education to selected promising candidates. The program is to help specialists acquire knowledge and skills in the field of medical education in order to develop educational programs, evaluation methods and medical education research that apply to any phase of the continuum of medical education. Up to $40,000 for residents and Fellows per year of study is available. The annual deadline for applications is September 2.
*Requirements:* Applicants must: register in a Masters- or PhD-level university program leading to a degree in Education, or a subject field/program of study closely related to medical education; be registered and attending university as a full-time or part-time student, either within Canada or abroad; be Canadian citizens or permanent residents; be Fellows of the Royal College in good standing or residents who, at the time of the application, are registered in a program accredited by the Royal College; and provide a letter of support from a Canadian Faculty of Medicine (residents must also provide a letter of support from their RCPSC Residency Program Director indicating that the Program Director has approved the resident's intended course of study in education).
*Geographic Focus:* Canada
*Date(s) Application is Due:* Sep 2
*Amount of Grant:* Up to 40,000 CAD
*Contact:* Allison Montpetit, Grant Administrator; (800) 668-3740, ext. 291 or (613) 730-8177; fax (613) 730-8830; researchgrants@royalcollege.ca
*Internet:* www.royalcollege.ca/portal/page/portal/rc/awards/grants/medical_education/fellowship_grant
*Sponsor:* Royal College of Physicians and Surgeons of Canada
774 Echo Drive
Ottawa, ON K1S 5N8 Canada

**RCY - NSF Ocean Sciences Division Research Grants** 2198
This division supports research to improve understanding of the ocean and ocean basins. Basic research programs support individual scientists, small groups of cooperating scientists, and some large coordinated projects. Ocean sciences also supports efforts to develop, acquire, and operate the instrumentation and facilities needed to carry out these research programs. Programs include ocean sciences instrumentation development, oceanographic instrumentation, shipboard scientific support equipment, and ocean sciences research support and its subprograms of physical oceanography, chemical oceanography, marine geology and geophysics, and biological oceanography; oceanographic centers and facilities; and the ocean drilling program.
*Geographic Focus:* All States
*Date(s) Application is Due:* Feb 15; Aug 15
*Contact:* David L. Garrison, (703) 292-7588; fax (703) 292-9085; dgarriso@nsf.gov
*Internet:* www.geo.nsf.gov/oce/ocefund.htm
*Sponsor:* National Science Foundation
2415 Eisenhower Avenue
Alexandria, VA 22314

**Reinberger Foundation Grants** 2199

Clarence T. Reinberger was born in 1894 on Cleveland's west side, and began his business career in the 1920's as a pioneer in the automobile replacement parts field. Starting as a clerk in the Cleveland retail store of the National Automotive Parts Association's (NAPA) Automotive Parts Company, he became president of that company in 1948. In the 1960's the Automotive Parts Company merged with the Genuine Parts Company of Atlanta. Mr. Reinberger held the position of Chairman of the Board of Genuine Parts Company until his death in 1968. Louise Fischer Reinberger was born in Germany. After graduation from high school in the United States, she was employed by the Halle Brothers Company, a large Cleveland department store. The Reinberger Foundation was established by Mr. and Mrs. Reinberger in 1966. Mr. Reinberger left a substantial bequest to the Foundation at his death in 1968. Upon Mrs. Reinberger's death in 1984, the major portion of her estate was also bequeathed to the Foundation. Although the Reinbergers had no children, the foundation continues to be managed by several generations of Mr. Reinberger's family. Since its inception, the foundation has distributed over $91,000,000 to the non-profit community. The foundation divides its support among the following program areas: Arts, Culture, and Humanities; Education; Human Service - Health; and Human Service - Other. Categories supported under Arts, Culture and Humanities include museums, visual arts, performing arts, media and communication, arts education, zoos, and public recreation. Categories supported under Education include K-12 schools, early childhood education, adult education/literacy, libraries, and workforce development. Categories supported under Human Service - Health include hospitals and clinics, substance abuse prevention/treatment, medical and disease research, disease prevention, and speech and hearing. Categories supported under Human Service - Other include children and youth services, residential and home care, emergency food programs, youth development, domestic violence, and temporary housing/homeless shelters. Letters of Inquiry for grants of any type may be submitted according to the following program-area schedule: Education - March 1; Human Service (Health) - June 1; Human Service (Other) - September 1; and Arts, Culture, and Humanities - December 1. Letters of Inquiry may either be sent by U.S. Mail or emailed and are due in the foundation office by the deadline date. Full proposals are accepted only at the request of the foundation.

*Requirements:* Applicants must be 501(c)3 organizations. Preferential consideration is given to organizations serving Northeast Ohio, or the greater-Columbus area.

*Restrictions:* No loans are made, nor are grants given to individuals. The Reinberger Foundation does not make more than one grant to a particular organization during a given calendar year, nor will new proposals be considered until existing multi-year commitments have been paid.

*Geographic Focus:* Ohio

*Date(s) Application is Due:* Mar 1; Jun 1; Sep 1; Dec 1

*Samples:* Cleveland Zoological Society, Cleveland, Ohio, $100,000; Cuyahoga Community College Foundation, Cleveland, Ohio, $37,500; The Free Medical Clinic of Greater Cleveland, Cleveland, Ohio, $50,000; Community Kitchen, Inc., Columbus, Ohio, $5,000.

*Contact:* Karen R. Hooser, President; (216) 292-2790; fax (216) 292-4466; info@reinbergerfoundation.org

*Internet:* reinbergerfoundation.org/apply.html

*Sponsor:* Reinberger Foundation
30000 Chagrin Boulevard #300
Cleveland, OH 44122

**Research Corporation Cottrell Scholars Awards** 2200

Cottrell Scholar Awards (CSA) originate from the Foundation's concern with the apparent separation of teaching and research in Ph.D. institutions. The main goal is to promote and support the university scholar model. University scholars are faculty members who have both excellent research programs and excellent approaches to student learning at the undergraduate level. Desired outcomes include: a cultural shift in Ph.D.-granting institutions toward valuing the university scholar; increased attraction and retention of undergraduates in science; and increased undergraduates from Ph.D.-granting institutions pursuing graduate degrees. A key objective is to build a community of outstanding scholar-educators who are dedicated to becoming leaders in both research and teaching and who collectively have the potential to change the way science is taught nationally.

*Requirements:* Eligible applicants must be tenure-track faculty members at United States institutions whose primary appointment is in a Bachelor's and Ph.D.-granting department of astronomy, biochemistry, biophysics, chemistry, or physics, but not in a school of medicine or engineering. Applicants must demonstrate a strong commitment to lower division undergraduate education and high prospects for becoming a national leader in their field of research. Potential applicants begin the online submission process by completing an eligibility quiz. If eligible, applicants gain access to the application packet and instructions for electronic submission.

*Geographic Focus:* All States

*Date(s) Application is Due:* Aug 1

*Amount of Grant:* 75,000 - 75,000 USD

*Samples:* The State University of New York, Buffalo, New York, field of chemistry; University of Maryland, College Park, New York, field of astronomy; and University of Notre Dame, Notre Dame, Indiana, field of physics.

*Contact:* Silvia Ronco, Ph.D., Program Officer; (520) 571-1111; fax (520) 571-1119; sronco@rescorp.org or csa@rescorp.org

*Internet:* www.rescorp.org/grants-and-awards/cottrell-scholar-awards/about-the-program

*Sponsor:* Research Corporation for Science Advancement
4703 East Camp Lowell Drive, Suite 201
Tucson, AZ 85712

**Research Corporation Multi-Investigator Cottrell College Science Awards** 2201

Multi-Investigator Cottrell College Science Awards (CCSA) are a new initiative aimed at helping start sustainable, collaborative programs of research by cross-disciplinary teams of faculty from science departments in primarily undergraduate institutions. Grants are focused on early career faculty. The research project should be one that could not be effectively attacked by an individual researcher or a group of researchers within the same discipline. Rather it is aimed at projects that, by their complexity and interdisciplinary nature, require a cross-disciplinary team approach to achieve significant progress and sustainability. Funding provides $75,000 for a two-investigator team and $100,000 for a three-investigator team. The award duration is two years with a single, one-year, extension available on request. An institutional match of $25,000 is required.

*Requirements:* An eligible team will consist of at least two, and no more than three, faculty from at least two distinct science departments, all from within the same institution. One of the team members must be from a department of chemistry, physics or astronomy. Only team members from non-Ph.D.-granting departments are eligible. While one team member may be a tenured faculty, the remaining team members must be tenure-track faculty within six years of their first faculty appointment at the time of an award. An on-line eligibility quiz for both institutional and team eligibility is available. Eligible teams will be linked to a pre-proposal application form to be submitted by September 15. Pre-proposals most closely meeting the review criteria and considered to have the highest prospects for establishing a sustainable program of cross-disciplinary collaborative research will be invited to submit the full application with an application target date of November 15.

*Restrictions:* Team members may have previously held a Single Investigator CCSA, but cannot have an active award at the time of application. Team members may not have both a Single-Investigator and Multi-Investigator CCSA application pending at the same time. Applications containing members from a Department or School of Engineering or Medicine will not be accepted.

*Geographic Focus:* All States

*Contact:* Silvia Ronco, Ph.D.; (520) 571-1111; fax (520) 571-1119; sronco@rescorp.org

*Internet:* www.rescorp.org/grants-and-awards/cottrell-college-science-awards/multi-investigator

*Sponsor:* Research Corporation for Science Advancement
4703 East Camp Lowell Drive, Suite 201
Tucson, AZ 85712

**Research Corporation Scialog Grants** 2202

Scialog® Grants seek to accelerate the work of 21st-century transformational science through funding research, intensive dialog and community building. Each multi-year initiative will promote scientific innovation in the face of a complex research challenge that serves as a driver in contemporary science. Successful grantees will be asked to address a few narrowly focused issues on a particular research initiative and to communicate with one another in an annual closed conference for the purpose of sharing insights and building further collaborations. Recent funding was for Scialog: Solar Energy Conversion. This funding cycle has ended. Please check the website for the announcement of a new round of funding. Funding is $100,000 for an individual or $250,000 for a team.

*Requirements:* The application process begins with completion of the online eligibility quiz. Applicants who pass the eligibility quiz will be given access to necessary forms to submit a proposal. However, final determination of eligibility will be made by a Program Officer.

*Restrictions:* Multiple proposals from the same investigator are not accepted. There is, however, no limit on the number of proposals submitted from the same institution.

*Geographic Focus:* All States

*Samples:* University of Florida, Gainesville, Florida, field of chemistry; University of Oregon, Eugene, Oregon, field of physics; and University of Wisconsin, Madison, Wisconsin, field of chemistry.

*Contact:* Silvia Ronco, Ph.D.; (520) 571-1111; fax (520) 571-1119; sronco@rescorp.org

*Internet:* www.rescorp.org/grants-and-awards/scialog/solar-energy-conversion

*Sponsor:* Research Corporation for Science Advancement
4703 East Camp Lowell Drive, Suite 201
Tucson, AZ 85712

**Research Corporation Single-Investigator Cottrell College Science Awards** 2203

Single-Investigator Cottrell College Science Awards (CCSA) support research in astronomy, chemistry, physics and closely related fields that significantly overlap these three disciplines at public and private, predominantly undergraduate institutions. Proposed projects are judged on the basis of scientific merit, originality, significance, feasibility, overlap with the three core disciplines and the ability of the institutional environment to sustain the activity. The involvement of undergraduate students in the research is expected, and is an important factor in most awards. The overarching goal is providing "seed" funding for significant research that will enhance the professional and scholarly development of beginning faculty at predominantly undergraduate institutions, along with their students. Funding is expected to help early career faculty establish sustainable programs of research. Faculty are expected to explore new areas of science, to make new discoveries that contribute to their discipline, and to initiate new research programs that can be sustained by other extramural funding sources, as well as with institutional support. Successful programs are not summer research projects, but rather those undertaken with continuity throughout the academic year. A key element is the involvement of undergraduate students in meaningful ways and thereby raising the quality of undergraduate education. Two-year awards are $35,000, and an institutional match of $10,000 is required.

*Requirements:* The principal investigator must have a faculty appointment at a United States college or university in a department of astronomy, chemistry or physics; or, if from another department, propose research that significantly overlaps with research in astronomy, chemistry or physics. The applicant's home department must offer at least baccalaureate, but not doctoral, degrees. At the time of application the applicant must be within the first three years of her/his first tenure-track appointment. The potential of a

proposed research project to add to fundamental scientific knowledge is a prime criterion in its evaluation, as is its potential for developing into a long-term viable program capable of attracting future support from other agencies. Other factors considered are the environment for research at the institution, student participation, and the contribution the research will make to the college's science programs. An on-line eligibility quiz is available. Applicants meeting the basic eligibility requirements will be directed to submit a brief pre-proposal by September 15.
*Restrictions:* Faculty with a previous CCSA award or whose appointment is in a department or school of engineering or medicine are not eligible. Expenses are limited to five budget categories: equipment, supplies, student stipends, faculty stipends and travel away from the home institution to conduct research. No indirect costs or fringe benefits (other than FICA at 7.65 percent) are allowed.
*Geographic Focus:* All States
*Date(s) Application is Due:* Nov 15
*Contact:* Silvia Ronco, Ph.D., Program Officer; (520) 571-1111; fax (520) 571-1119; sronco@rescorp.org or ccsa@rescorp.org
*Internet:* www.rescorp.org/grants-and-awards/cottrell-college-science-awards/single-investigator
*Sponsor:* Research Corporation for Science Advancement
4703 East Camp Lowell Drive, Suite 201
Tucson, AZ 85712

**RFF Gilbert F. White Postdoctoral Fellowships** 2204
Resources for the Future (RFF) will award resident fellowships for the upcoming academic year in honor of Gilbert F. White, retired chairman of the RFF Board of Directors, distinguished geographer, and statesman of science. The fellowships are intended for researchers who have a doctorate degree and wish to devote a year to scholarly work in areas related to natural resources, energy, or the environment. Social scientists, as well as natural scientists, interested in policy-relevant interdisciplinary research are encouraged to apply. Because fellows will interact closely with RFF researchers, selection criteria will include the nature of the applicant's proposed research program and how it fits with RFF work in progress. Fellows receive an annual stipend based upon their current salary and also receive research support, office facilities at RFF, and an allowance of up to $1,000 for moving or living expenses. The deadline for applications is February 19.
*Requirements:* The program is open to both U.S. and non-U.S. citizens, provided the latter have proper work and residency documentation. The award is open to individuals in any discipline who have completed their doctoral requirements by the beginning of the current academic year. Teaching and/or research experience at the postdoctoral level is preferred, though not essential. Individuals holding positions in government, as well as at academic institutions, are eligible.
*Geographic Focus:* All States, All Countries
*Date(s) Application is Due:* Feb 19
*Contact:* Robert C. Williams, Director of Academic Programs; (202) 328-5031 or (202) 328-5000; fax (202) 939-3460; white-award@rff.org or williams@rff.org
*Internet:* www.rff.org/about/gilbert-f-white-postdoctoral-fellowship-program
*Sponsor:* Resources for the Future
1616 P Street NW
Washington, D.C. 20036-1400

**RFF John V. Krutilla Research Stipend** 2205
Resources for the Future (RFF) will award one research grant for the academic year in honor of the late John V. Krutilla, a renowned environmental economist who served as a senior fellow at RFF for most of his career. The grant is intended for young scholars who have been awarded a doctoral degree in the past five years. The focus of the award is research related to environmental and resource economics. Special attention will be given to applications that seek to pursue research into one or more of the areas pioneered by John Krutilla. The awardee may be invited to present the results of the research activities at RFF in the year after completion of the work. One award for $5,500 will be made each year. It can be used for summer salary support, to pay for research assistance, or for any other legitimate research expenses. The deadline for applications is February 19.
*Requirements:* The award is open to young scholars who have received their doctoral degrees since June 30, five years earlier.
*Geographic Focus:* All States
*Date(s) Application is Due:* Feb 19
*Amount of Grant:* 5,500 USD
*Contact:* Robert C. Williams, Director of Academic Programs; (202) 328-5031 or (202) 328-5000; fax (202) 939-3460; krutilla-award@rff.org or williams@rff.org
*Internet:* www.rff.org/about/john-v-krutilla-research-stipend
*Sponsor:* Resources for the Future
1616 P Street NW
Washington, D.C. 20036-1400

**RFF Joseph L. Fisher Dissertation Fellowships** 2206
In honor of the late Joseph L. Fisher, president of Resources for the Future (RFF) from 1959–1974, RFF will award fellowships for the coming academic year in support of doctoral dissertation research on issues related to the environment, natural resources, or energy. RFF's primary research disciplines are economics and other social sciences. Proposals originating in these fields will have the greatest likelihood of success. Proposals from the physical or biological sciences must have an immediate and obvious link to environmental policy matters. Awardees will be invited to present their research at RFF, and RFF will pay reasonable travel expenses for the trip. The fellowships carry a stipend of $18,000 for the academic year. The fellowship is intended to be a student's principal source of support. The deadline for applications is February 19.
*Requirements:* This fellowship supports graduate students in the final year of their dissertation research. Fellowship applicants must expect to complete all requirements for the doctorate by the end of summer. Applicants must also have completed the preliminary examinations for their doctorate prior to the application deadline. This requirement will be strictly enforced.
*Restrictions:* The fellowship stipend will be reduced dollar-for-dollar by the amount of any financial assistance (other than tuition support) from other sources (however, the fellowship will still pay a minimum stipend of $2,000 to awardees whose stipends would otherwise be reduced below that amount).
*Geographic Focus:* All States
*Date(s) Application is Due:* Feb 19
*Amount of Grant:* 18,000 USD
*Contact:* Robert C. Williams, Director of Academic Programs; (202) 328-5031 or (202) 328-5000; fax (202) 939-3460; fisher-award@rff.org or williams@rff.org
*Internet:* www.rff.org/about/joseph-l-fisher-doctoral-dissertation-fellowships
*Sponsor:* Resources for the Future
1616 P Street NW
Washington, D.C. 20036-1400

**RFF Summer Research Internships** 2207
Resources for the Future (RFF), an independent nonprofit organization specializing in research, policy analysis, and public education on environmental, energy, and natural resource issues, offers summer internships for outstanding undergraduate and graduate students. Priority consideration is given to graduate students. Internships are available beginning in June and ending in late August for a 10-week assignment. To meet individual needs, some flexibility in the beginning and ending dates is possible. Each intern can expect to work with one or more RFF researchers on a variety of ongoing projects or assist in the development of entirely new areas of research and policy analysis. Highly motivated candidates in the social and natural sciences with training in economics and quantitative methods, or with a keen interest in public policy, are encouraged to apply. Students will earn a stipend ($375 per week for graduate students; $350 per week for undergraduates). Housing assistance is not provided.
*Requirements:* The program is open to both U.S. and non-U.S. citizens, provided the latter have proper work and residency documentation. Candidates should display strong writing skills and a desire to analyze complex environmental policy problems amenable to interdisciplinary methods. The ability to work without supervision in a careful and conscientious manner is essential.
*Geographic Focus:* All States, All Countries
*Date(s) Application is Due:* Mar 4
*Amount of Grant:* 3,500 - 3,750 USD
*Contact:* Robert C. Williams, Director of Academic Programs; (202) 328-5031 or (202) 328-5000; fax (202) 939-3460; IC@rff.org or williams@rff.org
*Internet:* www.rff.org/about/rff-summer-research-internship-program
*Sponsor:* Resources for the Future
1616 P Street NW
Washington, D.C. 20036-1400

**RFF Walter O. Spofford, Jr., Memorial Internships** 2208
Resources for the Future (RFF), an independent, nonprofit organization specializing in research, policy analysis, and public education on environmental, energy, and natural resource issues, has established the Walter O. Spofford, Jr., Memorial Internship. With Spofford's help, the RFF China Program was launched in 1989, and his name became synonymous with RFF's work in China. The intern can expect to work with one or more RFF researchers on a variety of ongoing projects or assist in the development of entirely new areas of research and policy analysis. RFF seeks highly motivated candidates in their first or second year of graduate training in the social or natural sciences. Students will earn a stipend in the amount of $375/week. No housing assistance is provided. All application materials must be received by February 19.
*Requirements:* Candidates should also have outstanding policy analysis and writing skills. The program is open to both U.S. and non-U.S. citizens, provided the latter have proper work and residency documentation.
*Geographic Focus:* All States, All Countries
*Date(s) Application is Due:* Feb 19
*Contact:* Robert C. Williams, Director of Academic Programs; (202) 328-5031 or (202) 328-5000; fax (202) 939-3460; spofford-award@rff.org or williams@rff.org
*Internet:* www.rff.org/about/walter-o-spofford-jr-memorial-internship
*Sponsor:* Resources for the Future
1616 P Street NW
Washington, D.C. 20036-1400

**RGk Foundation Grants** 2209
RGK Foundation awards grants in the broad areas of Education, Community, and Medicine/Health. The foundation's primary interests within Education include programs that focus on formal K-12 education (particularly mathematics, science and reading), teacher development, literacy, and higher education. Within Community, the foundation supports a broad range of human services, community improvement, abuse prevention, and youth development programs. Human service programs of particular interest to the foundation include children and family services, early childhood development, and parenting education. The foundation supports a variety of Community Improvement programs including those that enhance non-profit management and promote philanthropy and voluntarism. Youth development programs supported by the foundation typically include after-school educational enrichment programs that supplement and enhance formal education systems to increase the chances for successful

outcomes in school and life. The foundation is also interested in programs that attract female and minority students into the fields of mathematics, science, and technology. The foundation's current interests in the area of Medicine/Health include programs that promote the health and well-being of children, programs that promote access to health services, and Foundation-initiated programs focusing on ALS. Annual deadlines for applicants are March 6, June 12, and September 16.
*Requirements:* Although there are no geographic restrictions to the foundation's grantmaking program, the foundation no longer accepts unsolicited requests for international agencies or programs. All applicants must complete an electronic Letter of Inquiry (LOI) from the website as the first step. RGK Foundation will entertain one electronic LOI per organization in a twelve-month period. While the foundation occasionally awards grants for operating expenses, capital campaigns, endowments, and international projects, such grants are infrequent and usually initiated by the foundation. Grants are made only to nonprofit organizations certified as tax exempt under Sections 501(c)3 or 170(c) of the IRS code. Hospitals, educational institutions, and governmental institutions meeting these requirements are eligible to apply.
*Restrictions:* The foundation refrains from funding annual funds, galas or other special-event fundraising activities; debt reduction; emergency or disaster relief efforts; dissertations or student research projects; indirect and/or administrative costs; sectarian religious activities, political lobbying, or legislative activities; institutions that discriminate on the basis of race, creed, gender, or sexual orientation in policy or in practice; or loans, scholarships, fellowships, or grants to individuals; unsolicited requests for international organizations or programs or for ALS research projects.
*Geographic Focus:* All States
*Amount of Grant:* 1,000 - 200,000 USD
*Samples:* Chicago Youth Programs, Chicago, Illinois, $15,000 - support to expand early literacy services in four high-need Chicago neighborhoods; Menninger Clinic Foundation, Houston, Texas, $13,860 - support for genetic analysis of adolescent blood samples for the McNair Initiative for Neuroscience Discovery; Colorado Humanities, Greenwood Village, Colorado, $25,000 - support for Motheread/Fatheread Early Childhood and Family Literacy Program.
*Contact:* Suzanne Haffey, Grants Associate; (512) 474-9298; fax (512) 474-7281; shaffey@rgkfoundation.org
*Internet:* www.rgkfoundation.org/public/guidelines
*Sponsor:* RGk Foundation
1301 West 25th Street, Suite 300
Austin, TX 78705-4236

**Rhode Island State Government Internships** **2210**
The Rhode Island State Government Internships give undergraduate and graduate students the opportunity to participate in state government operations by interning in executive, legislative, and judicial departments. A spring internship and a summer internship are offered. Under the spring program, students receive academic credit toward their major field of study. Students are required to complete papers, work eight to 10 hours per week within a field placement, attend weekly seminars, and complete a take-home midterm and final exam. During the eight-week summer internship program, July and August, interns work 35 hours per week within their chosen field placement, are responsible for time sheets, and are responsible for mandatory student progress reports and final assessment. The summer program has a Rhode Island residency clause. A stipend is awarded on a monthly basis for the two-month period. Work study students are welcome to apply. Deadlines to apply vary according to semester appointment of the internship.
*Requirements:* The Internship Program is open to currently enrolled college students, graduate students, and law school students. Students are required to be currently enrolled in an accredited institution and have a minimum 2.5 grade point average. While all academic majors are accepted, student experience, academic background and educational goals are taken into consideration for placement.
*Restrictions:* Financial aid and grants are not awarded.
*Geographic Focus:* Rhode Island
*Date(s) Application is Due:* May 10; Aug 31; Dec 7; Dec 21
*Contact:* Bea Lanzi; (401) 222-6782; fax (401) 222-4447; intern@rilegislature.gov
*Internet:* www.rilegislature.gov/internship/pages/Home.aspx
*Sponsor:* Rhode Island State Internship Program
State House, Room 8AA
Providence, RI 02903

**Richard Davoud Donchian Foundation Grants** **2211**
The Richard Davoud Donchian Foundation was founded after Mr. Donchian's death in April of 1993. Extending Mr. Donchian's lifetime passion, its mission is to help others meet their potential and achieve high degrees of personal and professional success. The Foundation's Board of Directors is continuous in its efforts to preserve the Donchian legacy of leadership and integrity, as well as its commitment to learning and personal growth. Consequently, the Foundation concentrates its primary giving activities in the areas of ethical leadership in business and community affairs; education, personal development and literacy; and moral, ethical and spiritual advancement in all areas of life. Applications are accepted throughout the year and are reviewed in the order they are received. The Foundation encourages pilot initiatives that test new program models.
*Requirements:* The majority of the Foundation's grantmaking is focused in the northeastern United States, although, occasionally, grants may be made in other regions of the country and/or abroad. All applicants must have tax-exempt 501(c)3 status as a non-profit organization as defined by the Internal Revenue Service. The applicant must have an active board of directors with policy-making authority. Grants may range from a few thousand dollars up to $50,000. In unique circumstances, the Foundation does consider a more significant grant for a program having a major impact in one or more of our areas of interest. Of particular interest to the Foundation are organizations that promote partnerships and collaborative efforts among multiple groups and organizations. Priority will be given to requests that show specific plans for funding beyond the present and program innovation and tangible outcomes, with an emphasis on opportunities for significant and lasting social improvement.
*Restrictions:* The Foundation generally will not provide grants for the following: organizations not determined to be tax-exempt under section 501(c)3 of the Internal Revenue Code; individuals; general fundraising drives; endowments; government agencies; or organizations that subsist mainly on third party funding and have demonstrated no ability or expended little effort to attract private funding.
*Geographic Focus:* Connecticut, Delaware, District of Columbia, Florida, Maine, Maryland, New Jersey, New York, North Carolina, South Carolina, Vermont, Virginia
*Amount of Grant:* Up to 50,000 USD
*Contact:* Donchian Administrator, (203) 629-8552; fax (203) 547-6112; rdd@fsllc.net
*Internet:* www.foundationservices.cc/RDD2/grantrequests.htm#Guidelines
*Sponsor:* Richard Davoud Donchian Foundation
640 W. Putnam Avenue 3rd Floor
Greenwich, CT 06830

**Robert Bowne Foundaion Edmund A. Stanley, Jr. Research Grants** **2212**
The foundation provides research grants as part of the Afterschool Matters Initiative to foster high quality cutting-edge research that has a lasting impact on the field. Research should support either original empirical research in or about community-based youth programs during the non-school hours or research syntheses or policy analyses of community-based youth programs.
*Requirements:* This is a national grant competition, and research may be conducted in any state. Contact foundation for application details.
*Geographic Focus:* All States
*Amount of Grant:* 10,000 USD
*Contact:* Anne Lawrence; (212) 792-6250; anne.lawrence@bownefoundation.org
*Internet:* www.bownefoundation.org/index.php?option=com_content&view=category&layout=blog&id=74&Itemid=96
*Sponsor:* Robert Bowne Foundation
6 East 39th Street, 10th Floor
New York, NY 10016

**Robert G. Cabell III and Maude Morgan Cabell Foundation Grants** **2213**
The Robert G. Cabell III and Maude Morgan Cabell Foundation supports capital projects of Virginia nonprofit organizations in the areas of arts and culture, community development, higher education, historic preservation, religion (Christian, Episcopal, Protestant, and Roman Catholic), social services, and community welfare. Preference will be given to the Richmond, Virginia, area. There are no application forms. Applicants may submit a written proposal that includes a brief description of the overall mission of the organization, a brief description of how the specific project supports the overall mission, a project budget, the level and type of support for the project from the local area, applicant's current operating budget and most recent budget, list of governing board and its officers, evidence of 501(c)3 status, and cover letter. Proposals are considered at grant review meetings each spring and fall, with deadlines being March 1 and September 1.
*Requirements:* Virginia 501(c)3 tax-exempt organizations that are not classified as private foundations or private operating foundations are eligible.
*Restrictions:* The Foundation does not make grants to individuals, primary or secondary schools, or state-supported organizations; nor will grants be made for debt reduction, endowments (except for faculty development), general operating expenses, research, or scholarships.
*Geographic Focus:* District of Columbia, Maryland, Virginia
*Date(s) Application is Due:* Mar 1; Sep 1
*Amount of Grant:* 25,000 - 750,000 USD
*Samples:* Bath County Historical Society, Warm Springs, Virginia, $85,000 - support for an expansion of public meeting space and exhibit and artifact storage at the Historical Society; Goochland Free Clinic and Family Services, Goochland, Virginia, $500,000 - support of the Goochland Cares: All.Here.Now. campaign; Randolph College, Lynchburg, Virginia, $250,000 - support of campus modernization efforts.
*Contact:* Jill A. McCormick, Executive Director; (804) 780-2050; fax (804) 780-2198; cabell.foundation@gmail.com
*Internet:* www.thecabellfoundation.org/
*Sponsor:* Robert G. Cabell III and Maude Morgan Cabell Foundation
901 E Cary Street, Suite 1402
Richmond, VA 23219-4037

**Roche Organ Transplant Research Foundation Grants** **2214**
The Roche Organ Transplantation Research Foundation (ROTRF) is committed to funding a broad range of transplantation-related research, with the ultimate aim of improving long-term outcomes for people with organ transplants. The ROTRF supports important and innovative projects in clinical research applicable to the understanding and care of human organ transplant recipients and related tissue transplants, such as islet transplantation. The ROTRF gives priority to clinically oriented research projects, such as observational clinical studies, or studies that use human transplant samples for laboratory examinations, and focus on understanding the pathogenesis and prevention of human transplantation diseases, including organ injury and rejection, viral infection, and late graft deterioration. Furthermore, the ROTRF may consider funding research that addresses related clinical issues in non-transplant patients, including hepatitis C infection and other viral infections, and inflammatory processes relevant to human organ transplantation. These related clinical issues should demonstrate their relevance to disease phenotypes in organ transplant patients. The Trustees will also consider

funding studies that investigate transplant populations, ethics, organ preservation and allocation, and healthcare delivery. Moreover, the ROTRF will welcome research in new emerging technologies that examine the pathogenesis of human disease states in organ transplantation. If animal models are used to complement the clinical research, a higher priority will be given to those that simulate pathological states and mechanisms operating in human organ transplantation. Application is made through the online letter of intent. Applicants are given a submission ID number which allows them to access and edit the letter of intent an any time until the next submission deadline.
*Requirements:* To be considered for a grant award, applications should have demonstrable applicability to human organ transplantation. Investigators working in areas such as antibody-mediated rejection and antibody formation, graft pathology during rejection events, histocompatibility, infectious agents, and disease phenotypes in transplant patients are encouraged to apply. The Trustees will also consider funding studies that investigate transplant populations, ethics, organ preservation and allocation, and health care delivery. Furthermore, the ROTRF will welcome research in new emerging technologies that examine the pathogenesis of human disease states in organ transplantation.
*Restrictions:* Applications must be submitted in English. Only established, faculty member of academic staff at universities, transplant centers and research institutes are eligible as Principal Applicants. Every applicant can submit only one Letter of Intent for each funding cycle. Every ROTRF grantee (including co-investigator) can only hold one ROTRF grant at a time.
*Geographic Focus:* All States, All Countries
*Date(s) Application is Due:* Oct 1
*Contact:* Grants Administrator; (41) 41 377 53 53; fax (41) 41 377 53 34; admin@rotrf.org
*Internet:* www.rotrf.org
*Sponsor:* Roche Organ Transplant Research Foundation
Postfach 222
Meggen, CH-6045 Switzerland

**Rockford College Faculty Fellowships** **2215**

Faculty Fellows use funds to explore how active citizenship can be applied to their curriculum and/or research interests. In some cases, this is a continuation of work already being done. For other faculty members, the fellowship affords an opportunity to explore a new dimension of their discipline. The overall goals of the program include the following: supporting the practice of the scholarship of engagement; developing as faculty leaders in modeling best practices, recruiting other faculty, and institutionalizing socially responsive knowledge to complement each student's education; encouraging Faculty Fellows to build a strong and productive social and intellectual community of scholars—coming together across disciplinary and community boundaries, reading and sharing work, discussing ideas, and grounding that work in new theoretical and conceptual contexts; and providing Fellows opportunities to conduct research that builds knowledge about civic engagement or is "civically engaged" and developing curricular approaches that prepare students to be effective citizens.
*Requirements:* All full-time and adjunct faculty members of Rockford College are eligible to apply for these fellowships.
*Geographic Focus:* Illinois
*Contact:* Ryan Kessler, Student Advocate; (815) 394-4262 or (815) 226-4062; fax (815) 394-5174; SAS@rockford.edu
*Internet:* www.rockford.edu/?page=JACCEScholarshipProg
*Sponsor:* Rockford College
5050 E State Street
Rockford, IL 61108-2393

**Ronald McDonald House Charities Grants** **2216**

Ronald McDonald House Charities provides grants to nonprofit, tax-exempt organizations whose national or global programs help improve the health and well being of children under 21. Types of support include capital grants, challenge/matching grants, formula grants, professorships, program grants, research grants, scholarships, seed money grants, and visiting scholars. The board meets quarterly to review, select, and award grants to organizations that have demonstrated an ability to respond to the needs of specific groups of children in a definitive, hands-on manner that yields measurable results.
*Requirements:* To be considered for funding, an applicant must be designated a not-for-profit, tax-exempt charitable organization. U.S. charities must have a current 501(c)3 tax-exempt status letter on file with the Internal Revenue Service. Organizations seeking funding should have: a broad base of funding support; management and staff capacity to effectively execute the project; a longer-term organizational strategic plan (evidence to ensure program and organization sustainability); and a clear, concise plan for project evaluation with outcome measurement that shows potential for meaningful change. Organizations seeking funding should have a specific program that: directly improves the health and well being of children; addresses a significant funding gap or critical opportunity; has long-term impact in terms of replication or reach; produces measurable results; and is sustainable without relying on RMHC funding. Prior to submitting a full proposal, organizations must submit a letter of inquiry which will be reviewed by the Board. If interested, the Board will invite organizations to submit a formal grant proposal.
*Restrictions:* The charity does not fund to projects/programs that are local in scope and that do not benefit children under 21; to reduce debt; for annual fund appeals; for medical research; for for-profit organizations; for fundraising sponsorships; for programs/projects administered by activist groups; for projects that have already been completed; for sectarian or religious purposes; for scholarships and fellowships to individuals; to propagandize or influence elections or legislation; for advertising or fundraising drives; to intermediary funding agencies; for endowment campaigns; for individuals.
*Geographic Focus:* All States
*Date(s) Application is Due:* Mar 5; Sep 7
*Samples:* 2008 Recipients: Andean Health & Development, Big Shoulders Fund, Childrens Hospital Los Angeles, Foundation for the Children of the Californias, Hospice St. Joseph for the Diocese of Norwich, ORBIS International, RMHC China.
*Contact:* Michael Singer, Executive Director; (630) 623-7048; fax (630) 623-7488; grants@rmhc.org
*Internet:* rmhc.org/what-we-do/grants/how-to-apply/
*Sponsor:* Ronald McDonald House Charities
1 Kroc Drive
Oak Brook, IL 60523

**Rose Community Foundation Health Grants** **2217**

The Rose Community Foundation recognizes that improving access to quality care requires well-informed, visionary leaders. For this reason, the Foundation promotes initiatives that develop health-policy and public-health leadership. While the Foundation invests in health leadership that can improve access to quality care over the long term, it also works to effect more immediate change. Committed to improving access to care for low-income children, youth and families, the Foundation continues to support efforts to enroll them in programs such as Child Health Plan Plus (CHP+), a publicly supported health insurance program. Because at least half of all premature deaths are the result of lifestyle choices that individuals may be able to change, the Foundation also supports health promotion and disease prevention programs that encourage healthy choices and discourage behaviors that lead to illness and injury. In addition, the Foundation funds community organizations that provide information about the steps individuals can take to prevent such conditions as diabetes, teen pregnancy, HIV, heart disease and cancer. There are no annual deadlines.
*Requirements:* Colorado 501(c)3 tax-exempt organizations serving the residents of Adams, Arapahoe, Boulder, Denver, Douglas, and Jefferson Counties are eligible.
*Restrictions:* The Foundation will generally not support: grants to individuals or endowments, including academic chairs and scholarships; grants to one organization to be passed to another; annual appeals or membership drives; fundraisers and other one-time events; or financial support for political candidates. An application will not be accepted if: reports for a prior grant to the organization/program in question are past due; or the same applicant has other applications pending.
*Geographic Focus:* Colorado
*Contact:* Whitney Gustin Connor, Program Officer; (303) 398-7410 or (303) 398-7400; fax (303) 398-7430; wconnor@rcfdenver.org
*Internet:* https://rcfdenver.org/nonprofits-and-grants/what-we-fund/health/
*Sponsor:* Rose Community Foundation
600 South Cherry Street, Suite 1200
Denver, CO 80246-1712

**Royal Belgian Numismatic Society Research Grants** **2218**

Each year the society awards a number of small research grants, available to fellows and nonfellows, for any field of numismatic research. There is no standard application form, but applicants should set out the purpose of their project; what they plan to achieve; detailed costings; and what other sources of finance they have available, have applied for, or intend to apply for. Applications in writing, with the names of two referees, should be submitted; the awards should be announced within two months of the closing date.
*Requirements:* Contenders must be no more than 35 years old as of the deadline date; participation is not restricted to holders of diplomas, titles, nor to nationality.
*Geographic Focus:* All States, All Countries
*Date(s) Application is Due:* Dec 31
*Amount of Grant:* 635 EUR
*Contact:* Department of Coins and Medals; jm@bvdmc.com
*Internet:* www.numisbel.be/
*Sponsor:* Royal Belgian Numismatic Society
Boulevard de l'Empereur 4 à 1000 Bruxelles
Brussels, Belgium

**Royal Caribbean Cruises Ocean Fund** **2219**

The Royal Caribbean Cruises Ocean Fund supports efforts to restore and maintain a healthy marine environment, minimize the impact of human activity on the environment, and promote awareness of ocean and coastal issues and respect for marine life. Ocean Fund grants are made to a variety of nonprofit groups and institutions whose activities are directly related to marine conservation, including initiatives in research, education and innovative technologies. Previous funding for organizations around the world include include projects that relate to ocean science, climate change, key marine species, education and innovative technologies. Previously funded projects are available at the Ocean Fund website.
*Requirements:* Organizations seeking grant funding from the Ocean Fund are welcome to email throughout the year to introduce their organization, mission, programs and potential projects relevant to the mission of the Ocean Fund within the body of the email. Due to the volume of email inquiries, only those organizations who are requested to submit a letter of interest will receive a personalized response to their email. The Ocean Fund grant selection process is by invitation only.
*Restrictions:* Any attachments to emails will not be reviewed.
*Geographic Focus:* All States
*Amount of Grant:* 20,000 - 40,000 USD
*Contact:* Ocean Fund Administrator; (305) 539-6573; oceanfund@rccl.com
*Internet:* www.theoceanfund.com/about
*Sponsor:* Royal Caribbean Cruises
1050 Caribbean Way
Miami, FL 33132-2096

**Royal Norwegian Embassy Kavli Prizes** 2220

The Kavli Prize is awarded every two years and includes three separate prizes awarded for outstanding scientific research in the fields of astrophysics, neuroscience and nanoscience. The Prize, which consists of a cash award, a scroll, and a medal, is intended not only as an accolade for the individual scientists concerned, but also as a recognition of important research efforts. It aims to promote international cooperation between researchers and to increase awareness among the general public of the importance of research. The Norwegian Academy of Science and Letters has full responsibility for selecting the prize winners to ensure that this process is independent of the funding organization. The Academy has set up three selection committees made up of top international scientists within the fields for which the prize is awarded. Four of the committee members are Nobel Laureates. The committee members have been chosen in close cooperation with the corresponding institutions in the USA, China, France, the UK and Germany.

*Requirements:* The nomination process is open to all who wish to nominate candidates. The prize can be awarded to a single person for profound scientific achievements or shared for closely related fundamental contributions. Specific guidelines and a current nomination form, which must be completed online in one session, are available at the website.

*Restrictions:* Self nominations are not accepted, and deceased persons may not be nominated.

*Geographic Focus:* All States, All Countries

*Date(s) Application is Due:* Dec 1

*Amount of Grant:* 1,000,000 USD

*Contact:* Eirik Lislerud; +47 221236640; eirik.lislerud@dnva.no

*Internet:* www.norway.org/studywork/research/prizes/Kavli/

*Sponsor:* Royal Norwegian Embassy in Washington

2720 34th Street NW

Washington, D.C. 20008

**Roy J. Carver Charitable Trust Higher Education Grants** 2221

The Roy J. Carver Charitable Trust supports Iowa and western Illinois colleges and universities. The Trust's interests in this area extend to those projects that strengthen institutions and offer innovative learning opportunities to students enrolled in baccalaureate and graduate programs throughout the grant-making region of Iowa and parts of Illinois. Typically, funding is awarded to capital or programmatic initiatives that seek to substantively improve instruction or increase access to educational resources, with the goal of better preparing students for future academic and professional challenges. Projects incorporating interdisciplinary approaches and demonstrating the potential for lasting impact among a diverse range of students are especially encouraged. A majority of awards have centered on teaching and research in the natural sciences.

*Requirements:* Grants are made to 501(c)3 tax-exempt colleges and universities located in Iowa and western Illinois.

*Restrictions:* The trust does not support annual campaigns or ongoing operations, direct grants to individuals, religious activities, organizations without 501(c)3 status, fund-raising benefits or program advertising, or political parties or candidates.

*Geographic Focus:* Illinois, Iowa

*Date(s) Application is Due:* Feb 15; May 15; Aug 15; Nov 15

*Amount of Grant:* 10,000 - 1,250,000 USD

*Samples:* Augustana College, Rock Island, Illinois, $1,250,000 - support expansion of the science building; Grand View University, Des Moines, Iowa, $200,000 - to acquire scientific instrumentation; Grinnell College, Grinnell, Iowa, $200,000 - to support development of a new academic program in data science.

*Contact:* Dr. Troy K. Ross; (563) 263-4010; fax (563) 263-1547; info@carvertrust.org

*Internet:* https://www.carvertrust.org/program-areas/higher-education/

*Sponsor:* Roy J. Carver Charitable Trust

202 Iowa Avenue

Muscatine, IA 52761-3733

**Roy J. Carver Charitable Trust Medical and Science Research Grants** 2222

With its Medical and Science Research program, the Roy J. Carver Charitable Trust supports Iowa and western Illinois nonprofit organizations and projects in the area of medical and scientific research. The goal of the program in medical and scientific research is to provide support for innovative investigation that may hold great promise for advancing scientific knowledge and improving human health. In this context, the Trust has recognized a variety of university-based scientific endeavors, most notably in the fields of medicine, engineering and the natural sciences. Although grants are occasionally awarded to address institutional capital needs, the focus of this program area is typically on the work of specific laboratories, often emphasizing multidisciplinary approaches and the development of untried, inherently risky research directions. Applicants are encouraged to contact staff prior to the development of a proposal. The board considers requests at quarterly meetings in January, April, July, and October. Approved grant payments are made semiannually in April and October.

*Requirements:* Grants are made to 501(c)3 tax-exempt scientific organizations in Iowa and western Illinois.

*Restrictions:* The trust does not support annual campaigns or ongoing operations, direct grants to individuals, religious activities, organizations without 501(c)3 status, fund-raising benefits or program advertising, or political parties or candidates.

*Geographic Focus:* Illinois, Iowa

*Date(s) Application is Due:* Feb 15; May 15; Aug 15; Nov 15

*Amount of Grant:* 10,000 - 3,500,000 USD

*Samples:* Iowa State University, Ames, Iowa, $98,196 - to develop a novel system for early-stage detection of oral cancer; University of Illinois, Urbana, Illinois, $124,528 - to fund an investigation into the replication mechanism of the influenza virus; University of Iowa, Iowa City, Iowa, $3,100,000 - to provide major program and endowment support at the University of Iowa Carver College of Medicine–Neuroscience.

*Contact:* Dr. Troy K. Ross, Executive Administrator; (563) 263-4010; fax (563) 263-1547; info@carvertrust.org

*Internet:* https://www.carvertrust.org/program-areas/medical-and-scientific-research/

*Sponsor:* Roy J. Carver Charitable Trust

202 Iowa Avenue

Muscatine, IA 52761-3733

**Rumsfeld Foundation Graduate Fellowships** 2223

The Rumsfeld Foundation Graduate Fellowships provide funding for promising students with an interest in serving the nation through policy-relevant fields or by pursuing a career in public service. Depending on the circumstances of each selected student, he or she can receive a stipend for the academic year and tuition assistance for a combined total of up to $25,000 per student. Fellowships are awarded on the basis of recommendations from fourteen advisers at the following institutions: Georgetown University; Princeton University; University of Pennsylvania; Harvard University; Stanford University; University of Virginia; Indiana University; University of Chicago; Yale University; Johns Hopkins University; University of Notre Dame; George Washington University; and Missouri State University. These fellows include students pursuing a Master's, Ph.D., or Juris Doctorate with concentrations including political and theological ethics, public management, economics, international affairs, philosophy, strategic studies, Middle East Politics, and East Asian civilizations. Once nominations are submitted, the Foundation trustees meet to review and determine the fellowships to be granted.

*Requirements:* Fellowship recipients must be nominated and attend one of the 14 universities listed.

*Geographic Focus:* All States

*Amount of Grant:* Up to 25,000 USD

*Contact:* Lawrence Di Rita, Trustee; contact@rumseldfoundation.org

*Internet:* www.rumsfeldfoundation.org/focus/our/fellowships

*Sponsor:* Rumsfeld Foundation

1718 M Street NW, #366

Washington, D.C. 20036

**Ruth and Vernon Taylor Foundation Grants** 2224

The foundation awards grants to nonprofit organizations in the areas of arts and humanities, civic and public affairs, secondary schools, higher education, environment, hospitals, human services, health, youth services, and social services. Types of support include general operating support, building construction and renovation, endowment funds, and research. The foundation suggests that initial contact be made in writing, since unsolicited requests for funds are not accepted. The Board meets in May and September.

*Requirements:* Organizations located in Colorado, Illinois, Montana, New Jersey, New York, Pennsylvania, Texas, or Wyoming are eligible.

*Restrictions:* Grants are not awarded to individuals.

*Geographic Focus:* Colorado, Illinois, Montana, New Jersey, New York, Pennsylvania, Texas, Wyoming

*Amount of Grant:* 1,000 - 20,000 USD

*Contact:* Douglas Taylor, Trustee; (303) 893-5284; fax (303)893-8263

*Sponsor:* Ruth and Vernon Taylor Foundation

518 17th Street, Suite 1670

Denver, CO 80202

**Ruth Estrin Goldberg Memorial for Cancer Research Grants** 2225

The Ruth Estrin Goldberg Memorial for Cancer Reearch Grants are intended to support cancer research projects being conducted by an investigator at a recognized hospital, university, or other research institution. Grants are awarded for one year with possible renewal. Application forms and guidelines are available upon request.

*Requirements:* The principal investigator must be affiliated with an accredited institution.

*Restrictions:* Focus is limited to New Jersey, New York, and Pennsylvania.

*Geographic Focus:* New Jersey, New York, Pennsylvania

*Amount of Grant:* 10,000 - 15,000 USD

*Contact:* Rhoda Goodman, Grants Contact; (908) 686-5508 or (973) 467-3838

*Sponsor:* Ruth Estrin Goldberg Memorial for Cancer Research

c/o Greenberg and Company

Springfield, NJ 07081

**RWJF Childhood Obesity Grants** 2226

Through its Childhood Obesity Grant program, RWJF has developed three integrated strategies to reverse the childhood obesity epidemic: funding local organizations to make positive changes at the community level, advocating for healthier policies in the public and private sectors, and providing grants to researchers and evaluators to strengthen the evidence about what works. The Foundation's grant making is centered on advancing these strategies in ways that make it easier for all children to lead healthy lives. The program is intended to improve understanding of school, community, state and national policies and environmental factors affecting youth diet, physical activity, obesity, and tobacco, alcohol and drug use, and to evaluate the effectiveness of interventions to prevent youth obesity.

*Requirements:* Eligible applicants for research grants are scholars associated with educational institutions, research organizations, health care providers or other public or nonprofit organizations. Preference will be given to public entities or nonprofit organizations tax exempt under Section 501(c)3 of the Internal Revenue Code. Applicants may be independent of or associated with organizations currently receiving grants.

*Geographic Focus:* All States

*Contact:* C. Tracy Orleans, Senior Program Officer/Senior Scientist; (609) 627-5962 or (877) 843-7953; cto@rwjf.org

*Internet:* www.rwjf.org/en/about-rwjf/program-areas/childhood-obesity/programs-and-grants.html
*Sponsor:* Robert Wood Johnson Foundation
Route 1 and College Road East, P.O. Box 2316
Princeton, NJ 08543-2316

**RWJF Clinical Scholars Grants** 2227
RWJF Clinical Scholars Grants foster the development of leaders who will transform U.S. health and health care. These future leaders will be equipped to work with communities and to conduct innovative research as a result of the didactic and experiential learning they receive at each of four chosen training sites (University of California, Los Angeles; University of Michigan; University of Pennsylvania; and Yale University). The program's focus emphasizes community-based research and leadership training. The goal of the program is the integration of scholars' clinical training with skills in program development and research methodology to help them find solutions for issues in public policy, community health, and health services research. The program offers graduate-level study and research as part of a university-based post-residency training program. Up to 29 scholars will be selected for appointments beginning the following year. Under the program, the universities provide Scholars with their stipends and health insurance. Stipends currently range from $67,000 to $75,500 per year, depending on the training site. Limited financial support is also available at each university for expenses associated with Scholars' research, including travel.
*Requirements:* To be eligible, physicians must be: U.S. citizens; be committed to a career consonant with the program's purposes and priorities; be highly regarded by those responsible for their clinical training; and plan to complete the clinical requirements of their residency/fellowship training by the date of entry into the program (except for surgeons). Scholars may not hold appointments as sub-specialty fellows during their tenure in the program.
*Geographic Focus:* All States
*Date(s) Application is Due:* Feb 28
*Amount of Grant:* 67,000 - 75,500 USD
*Contact:* Kristin Siebenaler; (919) 962-8025; fax (919) 843-2666; ksieben@med.unc.edu
*Internet:* rwjcsp.unc.edu/
*Sponsor:* Robert Wood Johnson Foundation
Route 1 and College Road East, P.O. Box 2316
Princeton, NJ 08543-2316

**RWJF Disparities Research for Change Grants** 2228
The RWJF Disparities Research for Change program was created to support rigorous evaluations of practical and replicable solutions designed to reduce and eliminate disease specific racial and ethnic health care disparities. It seeks to improve the quality of health care provided to patients affected by racial and ethnic disparities by: Granting funds to discover and evaluate practical and replicable solutions designed to reduce and eliminate disease specific racial and ethnic health care disparities; Focusing on interventions aimed at health care delivery for one or more of the following health concerns: cardiovascular disease, depression, or diabetes; Conducting systematic reviews of the literature regarding racial and ethnic health care disparities interventions; and Disseminating results from these research efforts and systematic reviews to encourage health care systems to address racial and ethnic gaps in care.
*Requirements:* For-profit organizations are eligible to apply, however preference will be given to applicants that are public agencies or are tax-exempt under Section 501(c)3 of the Internal Revenue Code and are not private foundations as defined under Section 509(a). Proposals must focus on innovative interventions that reduce racial and ethnic disparities in the treatment of cardiovascular disease, depression, and diabetes. There are no age restrictions for the target populations of the interventions proposed for evaluation using program funds. Although there are no restrictions on the age of target populations, proposals targeting pediatric and geriatric populations are encouraged. Applications are accepted in two phases. The first phase is a brief proposal submission open to all interested research teams. The second phase is a full proposal submission and will be accepted by invitation only. All applications, both the brief and full proposals, must be submitted through the Robert Wood Johnson Foundation's Grantmaking Online system.
*Restrictions:* Individuals are not eligible to apply. Stand-alone prevention interventions such as tobacco cessation or exercise programs are not eligible unless they are one part of a larger, more comprehensive, intervention. The program will only directly fund health care delivery institutions. However, eligible institutions may partner/sub-contract with academic or research groups.
*Geographic Focus:* All States
*Amount of Grant:* 100,000 - 275,000 USD
*Contact:* Marshall H. Chin, Director; (866) 344-9800 or (773) 702-4769; fax (773) 702-4620; mchin@medicine.bsd.uchicago.edu or info@solvingdisparities.org
*Internet:* www.solvingdisparities.org/
*Sponsor:* Robert Wood Johnson Foundation
Route 1 and College Road East, P.O. Box 2316
Princeton, NJ 08543-2316

**RWJF Enhancing Coastal Community Resilience and Well-being in the Gulf of Mexico Region Grants** 2229
Enhancing Coastal Community Resilience and Well-being in the Gulf of Mexico Region is a funding opportunity of the Gulf Research Program of the National Academies of Sciences, Engineering, and Medicine developed and funded in collaboration with the Robert Wood Johnson Foundation. RWJF collaboration builds upon its common interest in: fostering the development of scientifically-sound research and practices that improve people's lives and the environments in which they live; and understanding the many complex factors that influence health and well-being, such as where people live, how they work, the safety of their surroundings, and the strength and resilience of families and communities. A total of $10 million is available for an anticipated three to six projects of up to 36 months in duration. Grants will support the development of research and practices that focus on the human dynamics of resilience. Specifically, RWJF is interested in projects that enhance community resilience and well-being by accounting for the influence of social, cultural, and health factors on a community's capacity to adapt and thrive as part of efforts to mitigate and respond to the adverse impacts of climate change, severe weather, and major environmental disasters. Projects should result in a stronger evidence-base for the kinds of strategies and approaches that can enhance community resilience in ways that also improve well-being. Projects should produce information that is actionable –in that it can be used immediately to benefit the residents of the U.S. Gulf of Mexico region, which includes the states of Alabama, Florida, Louisiana, Mississippi, and Texas. Projects should seek to develop sustained capacity at the community level for enhancing resilience. Letters of Intent are due by August 11, with final proposals due by October 5 each year.
*Requirements:* Proposed projects should bring together researchers, practitioners, and individuals from communities, including those from community organizations, state and local government, industry, and local businesses, so that all relevant perspectives inform possible approaches to advancing the science and practice of resilience. Involvement of local leaders and community members, particularly those substantially affected, is strongly encouraged.
*Geographic Focus:* All States
*Date(s) Application is Due:* Oct 5
*Amount of Grant:* Up to 5,000,000 USD
*Contact:* Chris Elfring, Executive Director; (202) 334-1973; fax (202) 334-1539; celfring@nas.edu or gulfgrants@nas.edu
*Internet:* www.nas.edu/gulf/grants/rwjf/index.htm
*Sponsor:* Robert Wood Johnson Foundation
Route 1 and College Road East, P.O. Box 2316
Princeton, NJ 08543-2316

**RWJF Harold Amos Medical Faculty Development Research Grants** 2230
The Harold Amos Medical Faculty Development Research program was established to increase the number of faculty from historically disadvantaged backgrounds who can achieve senior rank in academic medicine and dentistry and who will encourage and foster the development of succeeding classes of such physicians and dentists. Four-year postdoctoral research awards are offered to universities, schools of medicine and dentistry and research institutions to support the research and career development of physicians and dentists from historically disadvantaged backgrounds who are committed to developing careers in academic medicine and dentistry and to serving as role models for students and faculty of similar background. The program defines the term historically disadvantaged to mean the challenges facing individuals because of their race, ethnicity, socioeconomic status or similar factors. The program will fund up to ten four-year awards up to $420,000 each, with scholars to receive an annual stipend of up to $75,000 each, complemented by a $30,000 annual grant toward support of research activities.
*Requirements:* Eligible to apply are minority physicians who are U.S. citizens at the time of application, have excelled in their post-secondary education within the United States, are now completing or will have completed formal clinical training, and are committed to academic careers.
*Geographic Focus:* All States
*Date(s) Application is Due:* Mar 15
*Amount of Grant:* 75,000 - 420,000 USD
*Contact:* Nina Ardery, Deputy Director; (317) 278-0500; amfdp@starpower.net
*Internet:* www.rwjf.org/en/library/funding-opportunities/2016/harold-amos-medical-faculty-development-program—amfdp—.html
*Sponsor:* Robert Wood Johnson Foundation
Route 1 and College Road East, P.O. Box 2316
Princeton, NJ 08543-2316

**RWJF Harold Amos Medical Faculty Development Scholars** 2231
The Harold Amos Medical Faculty Development Program supports basic/biomedical, clinical, and health services/epidemiology research. The purpose of the AMFDP award is to facilitate the transition of the newly-trained clinician who wishes to develop into an independent investigator. A completed application consists of two parts: the completion of the online proposal process, and submission of other (paper) supporting documents. The maximum amount of support is $420,000 over a four-year period. This includes an annual $75,000 stipend, plus $30,000 per year for research support.
*Requirements:* To be eligible, applicants must be physicians from historically disadvantaged backgrounds (ethnic, financial, or educational) who are U.S. citizens or permanent residents at the time of application deadline; have excelled in their education; are now completing or have completed their formal clinical training; are prepared to devote four consecutive years to research; are committed to pursuing academic careers; and are committed to improving the health status of the under-served, decreasing health disparities, or serving as role models for students and faculty from historically disadvantaged backgrounds Preference will be given to physicians who have recently completed their formal clinical training. In order to pursue the advanced research training required by this program, applicants must identify faculty mentors with whom a research plan will be developed. The mentor's experience in the supervision of trainees and the adequacy of the mentor's research environment and support will be strongly considered.
*Geographic Focus:* All States
*Date(s) Application is Due:* Mar 15
*Amount of Grant:* Up to 105,000 USD

*Contact:* Nina Ardery, Deputy Director; (301) 565-4080 or (317) 278-0500; fax (301) 565-4088; nardery@erols.com or amfdp@starpower.net
*Internet:* www.amfdp.org/for-applicants
*Sponsor:* Robert Wood Johnson Foundation
Route 1 and College Road East, P.O. Box 2316
Princeton, NJ 08543-2316

**RWJF Health Policy Fellowships** 2232

The purpose of RWJF Health Policy Fellowships is to give outstanding mid-career health professionals in academic and community-based settings an opportunity to learn more about the legislative process through direct participation. Fellows actively contribute to the formulation of national health policies and accelerate their careers as leaders in health policy. Up to six grants of up to $165,000 each will be made. Each fellow will receive up to $94,000 for the Washington stay in salary plus fringe benefits or fellowship stipend. Fellows will receive an additional allowance for relocation subject to limitations provided in detail on the program's Web site. November 15 is the deadline for receipt of applications submitted online.
*Requirements:* Applicants may have backgrounds in the following disciplines: allied health professions; biomedical sciences; dentistry; economics or other social sciences; health services organization and administration; medicine; nursing; public health; or social and behavioral health. Applicants must be U.S. citizens or have permanent residency status at the time of application. Potential applicants must obtain documented sponsorship from an eligible institution to support their application. Nonprofit health care organizations and academic centers are eligible sponsors. Eligible institutions may sponsor more than one candidate, but no more than one individual will be selected from any one sponsoring organization.
*Restrictions:* No additional direct costs (such as laptops or PDAs) or indirect costs are paid for during the first year of this program.
*Geographic Focus:* All States
*Date(s) Application is Due:* Nov 15
*Amount of Grant:* Up to 165,000 USD
*Contact:* Marie Michnich, Director; (202) 334-1506; fax (202) 334-3862; mmichnich@nas.edu or info@healthpolicyfellows.org
*Internet:* www.healthpolicyfellows.org/
*Sponsor:* Robert Wood Johnson Foundation
Route 1 and College Road East, P.O. Box 2316
Princeton, NJ 08543-2316

**RWJF Healthy Eating Research Grants** 2233

Healthy Eating Research is a national program that supports research to identify, analyze, and evaluate environmental and policy strategies that can promote healthy eating among children and prevent childhood obesity. Special emphasis is given to research projects that benefit children in the low-income and racial/ethnic populations at highest risk for obesity. The program funds two types of research grants: studies to identify and/or evaluate promising school food environment and policy changes (12- to 18-month awards up to $100,000 each, and 18- to 36-month awards up to $400,000 each); and analyses of macro-level policy or system determinants of school food environments and policies (12- to 18-month awards up to $75,000 each). Applications must be submitted through the online process in response to future calls for proposals. Calls for Proposals and their guidelines are available online.
*Requirements:* Preference will be given to those applicants that may be either public entities or nonprofit organizations that are tax-exempt under Section 501(c)3 of the Internal Revenue Code. Applicant organizations must be based in the United States or U.S. Territories. The focus of this program is the United States; studies of policies in other countries will be considered only to the extent that they may directly inform U.S. policy. The experience and qualifications of the research team is one of the primary criteria for proposal review. A doctorate or terminal degree is preferred for the principal investigator.
*Restrictions:* This program does not fund demonstration projects. The foundation does not award grants to private individuals or applicants from outside the U.S. or U.S. Territories. No faxed, emailed or mailed proposals will be accepted.
*Geographic Focus:* All States
*Amount of Grant:* Up to 400,000 USD
*Contact:* Kathy Kosiak, Research Coordinator; (800) 578-8636; fax (612) 624-9328; healthyeating@umn.edu
*Internet:* www.healthyeatingresearch.org/
*Sponsor:* Robert Wood Johnson Foundation
Route 1 and College Road East, P.O. Box 2316
Princeton, NJ 08543-2316

**RWJF Increasing Openness and Transparency in Research Grants** 2234

The purpose of the Increasing Openness and Transparency in Research Grant solicitation is to encourage and support a series of projects aimed at increasing transparency and accountability throughout the research life cycle, with a special emphasis on open access, to increase the speed of the flow of information from funded research to action. RWJF wants to use these grants to help increase our understanding, and the understanding of the research community, about the potential for these approaches to accelerate the discovery process and expand access to research findings. Up to $1.5 million has been allocated for this work, and the Foundation anticipates making six–eight individual awards for projects to be completed within 12–24 months. Letters of Intent are due by October 5, with full invited proposals due by December 9.
*Requirements:* The following criteria are required to apply to this solicitation: all projects must have implications and impact in the United States; lead organizations must be based in the United States or its territories (international individuals and organizations can partner with U.S.-based institutions in order to apply; and previous work in open access and/or open science.
*Geographic Focus:* All States
*Date(s) Application is Due:* Nov 16
*Amount of Grant:* Up to 300,000 USD
*Contact:* Debra Joy Perez, Senior Program Officer; (215) 732-2200; fax (215) 732-8123; djperez@fas.harvard.edu or info@rwjf-newconnections.org
*Internet:* www.rwjf.org/en/library/funding-opportunities/2016/increasing—openness—and—transparency—in—research.html
*Sponsor:* Robert Wood Johnson Foundation
Route 1 and College Road East, P.O. Box 2316
Princeton, NJ 08543-2316

**RWJF Interdisciplinary Nursing Quality Research Initiative Grants** 2235

The Interdisciplinary Nursing Quality Research Initiative has most recently awarded two-year grants to support the generation, dissemination, and translation of research to improve the contributions of nursing to the quality of hospital care. It will continue to support interdisciplinary teams of nurse scholars and scholars from other disciplines to address gaps in knowledge about the relationship between nursing and health care quality. Topic areas may include: identifying, developing, and testing new measures to advance quality nursing research; and testing the effects of implementing the nursing-sensitive performance measures endorsed by the National Quality Forum. Investigator-initiated ideas that are consistent with the broader set of program goals also will be considered. The two-phase application process begins with online submittal of a brief proposal; full proposals are by invitation, and are due approximately four months after invitations to submit are given. All current and future calls for proposals are detailed at the web site.
*Requirements:* Teams of researchers who are U.S. citizens or permanent residents are eligible. Each team must have at least one nursing researcher and at least one researcher from another discipline. Preference will be given to 501(c)3 applicants that are not 509(a) private foundations.
*Geographic Focus:* All States
*Amount of Grant:* Up to 300,000 USD
*Contact:* Lori A. Melichar, Senior Program Officer; (301) 571-8161 or (800) 978-8309; INQRIhelpdesk@constellagroup.com
*Internet:* www.inqri.org/grants
*Sponsor:* Robert Wood Johnson Foundation
Route 1 and College Road East, P.O. Box 2316
Princeton, NJ 08543-2316

**RWJF Investigator Awards in Health Policy Research** 2236

The Investigator Awards in Health Policy Research program supports researchers whose cross-cutting and innovative ideas promise to contribute meaningfully to improving health and health policy in America. Each year, the program awards 24- to 36-month grants of up to $335,000 to highly qualified individuals to fund approximately ten research projects that have national policy relevance. The program funds projects with a broader scope than traditional NIH or NSF grants. Data-driven health services research focusing on a narrowly defined research question is not appropriate for this program, nor are pilot projects that are designed to evaluate clinical care in a specific geographic area or to demonstrate service delivery programs or financing strategies.
*Requirements:* Individuals in the United States who are affiliated either with a US-based educational institution or a 501(c)3 tax-exempt organization are eligible. Applications are welcomed from investigators in fields such as anthropology, business, demography, economics, engineering, ethics, genetics, health and social policy, history, journalism, law, medicine, nursing, political science, psychology, public health, science policy, social work, sociology and others. The program seeks a diverse group of applicants including minorities, researchers early in their careers, and individuals in nonacademic settings.
*Restrictions:* Past awardees and teams of more than two researchers are not eligible to apply.
*Geographic Focus:* All States
*Amount of Grant:* Up to 335,000 USD
*Contact:* Alan Cohen; (617) 353-9220, ext. 1; fax (617) 353-9227; rwjfihp@bu.edu
*Internet:* www.investigatorawards.org
*Sponsor:* Robert Wood Johnson Foundation
Route 1 and College Road East, P.O. Box 2316
Princeton, NJ 08543-2316

**RWJF New Connections: Increasing Diversity of RWJF Programming** 2237

New Connections provides funding for research that aims to expand the diversity of perspectives that inform Robert Wood Johnson Foundation (RWJF) programming and introduce new researchers and scholars to RWJF, while simultaneously helping to meet staff needs for data analysis. Through grantmaking, mentorship, career development and networking, New Connections enhances the research capacity of its grantees and network members. The researchers in this program transcend disciplines (health; health care; social sciences; business; urban planning; architecture and engineering); work to build the case for a Culture of Health with strong qualitative and quantitative research skills; and produce and translate timely research results. Grants of up to $50,000 each will be given, with a total of up to 17 grants awarded each year. Of the 17 grants awarded in this funding round, up to two grants will be designated for New Connections–Policies for Action awards. The deadline for receipt of full proposals is May 4.
*Requirements:* This program seeks scholars who have historically been underrepresented in research activities supported by RWJF. This includes people from ethnic or racial minority groups, first-generation college graduates and people from low-income communities. The initiative will support research activities that address specific questions identified by RWJF programming teams and portfolios. Contact the sponsor for the programming focus. There are two stages in the application process.
*Geographic Focus:* All States

*Date(s) Application is Due:* May 4
*Amount of Grant:* Up to 50,000 USD
*Contact:* Renée Byng Yancey, National Program Officer; (215) 732-2200; fax (215) 732-8123; info@rwjf-newconnections.org
*Internet:* www.rwjf.org/en/library/funding-opportunities/2016/new-connections—increasing-diversity-of-rwjf-programming—-juni.html
*Sponsor:* Robert Wood Johnson Foundation
Route 1 and College Road East, P.O. Box 2316
Princeton, NJ 08543-2316

**RWJF Nurse Faculty Scholars** 2238

The goal of the Robert Wood Johnson Foundation Nurse Faculty Scholars (NFS) program is to develop the next generation of national leaders in academic nursing through career development awards for outstanding junior nursing faculty. The program aims to strengthen the academic productivity and overall excellence of nursing schools by providing mentorship, leadership training, and salary and research support to junior faculty. The Nurse Faculty Scholars program is part of the Foundation's Human Capital Portfolio, which aims to ensure that the nation has a diverse, well-trained workforce to meet the health care needs of all Americans. Up to 12 awards of up to $350,000 each over three years will be available. February 12 is the deadline for receipt of applications.
*Requirements:* Accredited colleges and universities within the United States are eligible to apply. To be eligible, candidates must meet the following criteria: be a registered nurse with a research doctorate in nursing or a related discipline; be a junior faculty member in an accredited school of nursing with an academic position that could lead to tenure; after completing the doctorate, be in an academic position that could lead to tenure for at least two years and no more than five years at the start of the program; identify at least one senior leader in the school of nursing to serve as a primary (nursing) mentor for academic career planning, and to provide access to organizations, programs and colleagues helpful to the candidate's work as a scholar; identify at least one senior researcher in the university with similar or complementary research interests to serve as a research mentor; and be a U.S citizen or permanent resident at the time of application.
*Geographic Focus:* All States
*Date(s) Application is Due:* Feb 12
*Amount of Grant:* Up to 350,000 USD
*Contact:* Maryjoan D. Ladden; (877) 738-0737; fax (410) 614-8285; rwjfnfs@jhu.edu
*Internet:* www.rwjf.org/applications/solicited/npo.jsp?FUND_ID=55121
*Sponsor:* Robert Wood Johnson Foundation
Route 1 and College Road East, P.O. Box 2316
Princeton, NJ 08543-2316

**RWJF Pioneer Portfolio Grants** 2239

The Pioneer Portfolio Grant program aims to support novel, high-return ideas that may have far-reaching impact on people's health, the quality of care they receive and the systems through which that care is provided. It seeks ideas not only from the mainstream of health and health care but also looks to sources outside of these fields for innovations that might have transformative impact. Several projects apply approaches from diverse sectors such as finance, design and entertainment to forge new solutions in health and health care arenas.
*Requirements:* Preference is given to public agencies, public charities or are tax-exempt under section 501(c)3 of the Internal Revenue Code. Organizations must be located in the United States or one of its territories.
*Restrictions:* The organization does not fund any of the following: Ongoing general operating expenses or existing deficits; Endowment or capital costs; Basic biomedical research; Research on drug therapies or devices; International programs; Direct support of individuals; Lobbying of any kind. Additionally, RWJF no longer funds projects in the following areas: End-of-life care; Long-term care; Specific chronic conditions (that are not part of the strategies of other Program Areas); Physical activity for adults age 50 or older.
*Geographic Focus:* All States
*Contact:* Deborah H. Bae; (609) 627-5812 or (877) 843-7953; dbae@rwjf.org
*Internet:* www.rwjf.org/en/grants/what-we-fund.html
*Sponsor:* Robert Wood Johnson Foundation
Route 1 and College Road East, P.O. Box 2316
Princeton, NJ 08543-2316

**RWJF Policies for Action: Policy and Law Research to Build a Culture of Health Grants** 2240

Policies for Action: Policy and Law Research to Build a Culture of Health (P4A) was created to help build an evidence base for policies that can lead to a Culture of Health. P4A seeks to engage long-standing health and health care researchers, as well as experts in fields like housing, education, transportation, and the built environment, to name a few, who have not worked in health before. The goal is to develop research that generates actionable evidence—the data and information that can guide legislators and other policymakers, public agencies, educators, advocates, community groups, and individuals. The research may examine established laws, regulations, and policies as well as potential new policies and approaches. The research funded under this call for proposals (CFP) should inform the significant gaps in our knowledge regarding what policies can serve as levers to improve population health and well-being, and achieve greater levels of health equity. Approximately $1.5 million will be awarded through this CFP. Each grant will award up to $250,000 for a maximum funding period of 24 months. The deadline for required Letters of Intent is March 15, with full invited proposals due June 17.
*Requirements:* Applicants may be either 501(c)3 public entities or nonprofit organizations and not classified as a private foundation under Section 509(a).
*Restrictions:* In-kind services and money to pay for capital costs or renovation may not be used to match funds.
*Geographic Focus:* All States
*Date(s) Application is Due:* Jun 17
*Amount of Grant:* Up to 250,000 USD
*Contact:* Bethany Saxon, Director of Communications; (215) 204-2134
*Internet:* www.rwjf.org/en/library/funding-opportunities/2016/policies-for-action—policy-and-law-research-to-build-a-culture-.html
*Sponsor:* Robert Wood Johnson Foundation
Route 1 and College Road East, P.O. Box 2316
Princeton, NJ 08543-2316

**S. D. Bechtel, Jr. Foundation / Stephen Bechtel Fund Preventive Healthcare and Selected Research Grants** 2241

The Foundation?fund believe that access to quality care, preventive health care and research programs can help individuals of all ages increase life expectancy and improve their quality of life. The directors have a particular interest in promoting nutritional health, fitness, and wellness in order to increase positive health outcomes. The Foundation/Fund support organizations working to achieve the following objectives: enabling San Francisco Bay Area community clinics and school-based health centers to improve the quality and delivery of medical care; enhancing and expand school-based efforts in the San Francisco Bay Area that improve the nutritional health and fitness of children; and advancing research related to Alzheimer's, aging, nutrition or other Board-determined specific research initiatives.
*Requirements:* The primary geographic focus of the Foundation/Fund is the San Francisco Bay area. The Foundation/Fund: support non-profit organizations providing quality programs in science, technology, engineering and math (STEM) education, environment, environmental education, character and citizenship development and preventive healthcare and selected research; provide capital support; provide operational support; and provide project support.
*Restrictions:* The Foundation/Fund do not provide endowment funding, international grants, or grants for individuals.
*Geographic Focus:* California
*Date(s) Application is Due:* Oct 1
*Amount of Grant:* 5,000 - 25,000 USD
*Contact:* Coordinator; (415) 284-8675; fax (415) 284-8571; sdbjr@sdbjrfoundation.org
*Internet:* www.sdbjrfoundation.org/program_areas.htm#health
*Sponsor:* S. D. Bechtel, Jr. Foundation / Stephen Bechtel Fund
P.O. Box 193809
San Francisco, CA 94119-3809

**SAH-HABS Sally Kress Tompkins Architectural History Fellowship** 2242

The Society of Architectural Historians (SAH) promotes the study, interpretation, and conservation of architecture, design, landscapes, and urbanism worldwide for the benefit of all. The Sally Kress Tompkins Fellowship, a joint program of the Society of Architectural Historians (SAH) and the National Park Service's Historic American Buildings Survey (HABS), gives a graduate student architectural historian the opportunity to work on a 12-week HABS history project during the summer. The Fellow will prepare a written history that focuses on a significant U.S. building or site for the permanent HABS collection at the Library of Congress. The Fellow's research interests and goals will inform the building or site selected for documentation by HABS staff. The Fellow is usually stationed in the HABS Washington, D.C., office. award includes a $10,000 stipend and will be presented during the Society's Annual Conference. The award will also be announced in the Journal of the Society of Architectural Historians and the SAH Newsletter after the presentation. Applications will be accepted from August 3 through December 31.
*Requirements:* Applicants should be pursuing graduate studies in architectural history or other related fields.
*Geographic Focus:* All States
*Date(s) Application is Due:* Dec 31
*Amount of Grant:* 10,000 USD
*Contact:* Lisa Davidson; (202) 354-2179; fax (202) 371-6473; lisa_davidson@nps.gov
*Internet:* www.sah.org/jobs-and-careers/sah-fellowships-and-grants/research-fellowships
*Sponsor:* Society of Architectural Historians
1365 North Astor Street
Chicago, IL 60610-2144

**SAH Annual Conference Independent Scholar Fellowships** 2243

The Society of Architectural Historians (SAH) promotes the study, interpretation, and conservation of architecture, design, landscapes, and urbanism worldwide for the benefit of all. Fellowships of up to $1,000 each supports the travel of an independent scholar whose paper has been accepted for delivery at the SAH Annual International Conference. An independent scholar is anyone who conducts scholarly research outside universities and traditional academia. Online applications will be accepted from August 1 through September 7.
*Requirements:* Applicant must have been a member of SAH for at least one year before applying for this award and must be engaged in advanced graduate study (normally beyond the master's level) that involves some aspect of the history of architecture or of one of the fields closely allied to it (e.g., city planning, landscape architecture, decorative arts, or historic preservation).
*Geographic Focus:* All States
*Date(s) Application is Due:* Sep 7
*Amount of Grant:* Up to 1,000 USD
*Contact:* Chris Higgins; (312) 573-1365; fax (312) 573-1141; chiggins@sah.org or info@sah.org
*Internet:* www.sah.org/jobs-and-careers/sah-fellowships-and-grants/annual-conference-fellowships

*Sponsor:* Society of Architectural Historians
1365 North Astor Street
Chicago, IL 60610-2144

**SAH Annual Conference Senior Scholar Fellowships** 2244

The Society of Architectural Historians (SAH) promotes the study, interpretation, and conservation of architecture, design, landscapes, and urbanism worldwide for the benefit of all. The annual Senior Scholar Fellowships of up to $1,000 each support the travel of senior-level, international scholars whose papers have been accepted for delivery at the SAH Annual International Conference. Research areas include Europe in the nineteenth and twentieth centuries as well as built environments world wide (other than Europe) from ancient times to the present. Online applications will be accepted from August 1 through September 7.
*Requirements:* Applicant must have been a member of SAH for at least one year before applying for this award and must be engaged in advanced graduate study (normally beyond the master's level) that involves some aspect of the history of architecture or of one of the fields closely allied to it (e.g., city planning, landscape architecture, decorative arts, or historic preservation).
*Geographic Focus:* All States
*Date(s) Application is Due:* Sep 7
*Amount of Grant:* Up to 1,000 USD
*Contact:* Chris Higgins; (312) 573-1365; fax (312) 573-1141; chiggins@sah.org or info@sah.org
*Internet:* www.sah.org/jobs-and-careers/sah-fellowships-and-grants/annual-conference-fellowships
*Sponsor:* Society of Architectural Historians
1365 North Astor Street
Chicago, IL 60610-2144

**SAH Edilia and Francois-Auguste De Montequin Fellowship** 2245

The Society of Architectural Historians (SAH) promotes the study, interpretation, and conservation of architecture, design, landscapes, and urbanism worldwide for the benefit of all. The Edilia and Francois-Auguste De Montequin Fellowship provides support for travel related to research on Spanish, Portuguese, or Ibero-American architecture. The awards consist of a $2,000 stipend for a junior scholar and a $6,000 award for a senior scholar. The Fellowship will be presented at the Society's Annual International Conference and will be announced in the Journal of the Society of Architectural Historians and the SAH Newsletter after the presentation. Online applications will be accepted from August 1 through September 5.
*Requirements:* The fellowship is intended to support the research of junior scholars (usually scholars engaged in doctoral dissertation research) annually, and senior scholars (scholars who have completed their Ph.D. or equivalent terminal degree) every other year in even-numbered years. The research to be supported must focus on Spanish, Portuguese, or Ibero-American architecture, including colonial architecture produced by the Spaniards in the Philippines and what is today the United States. Applicants must be a current member of SAH.
*Geographic Focus:* All States
*Date(s) Application is Due:* Sep 5
*Amount of Grant:* 2,000 - 6,000 USD
*Contact:* Chris Higgins, Program Coordinator; (312) 573-1365; fax (312) 573-1141; chiggins@sah.org or info@sah.org
*Internet:* www.sah.org/jobs-and-careers/sah-fellowships-and-grants/research-fellowships
*Sponsor:* Society of Architectural Historians
1365 North Astor Street
Chicago, IL 60610-2144

**Saint Mary's College Fellowships** 2246

The Center for Women's InterCultural Leadership (CWIL) seeks to increase awareness of the importance of intercultural dialogue in today's world, to highlight the leadership roles women have adopted in such dialogue, and to foster the next generation of women leaders. In pursuit of these goals, CWIL offers visiting and faculty Fellowships. All appointments are for one year with the possibility of a renewal for a second year. CWIL Fellows have three general types of responsibilities: research; teaching courses relating to cultural or women's leadership issues; and assisting the CWIL and/or the department in its programmatic and community activities. These activities will vary depending on the type of Fellowship and the needs of the sponsoring department.
*Geographic Focus:* All States
*Contact:* Linda Biggins; (574) 284-4051; fax (574) 284-4141; cwil@saintmarys.edu
*Internet:* www.centerforwomeninleadership.org/research-and-scholarship/fellows
*Sponsor:* Saint Mary's College
Center for Women's Intercultural Leadership
Notre Dame, IN 46556-5001

**Samuel Freeman Charitable Trust Grants** 2247

The Samuel Freeman Charitable Trust was given charitable status in March of 1982, and is located in Providence, Rhode Island. The Trust has broad funding interests in supporting U.S. charitable, scientific, literary, and education programs. The Trust awards grants to support: cancer research and treatment; the preservation, exhibition, and operation of historical railway equipment; and secondary schools and universities. Types of support include: general operating support; continuing support; annual campaigns; and seed money. There are no application deadlines, and proposals are reviewed throughout the year following an invitation generated by a Letter of Interest.
*Requirements:* U.S. nonprofit organizations are eligible to apply.
*Geographic Focus:* All States
*Amount of Grant:* 1,000 - 150,000 USD
*Samples:* Statue of Liberty, Ellis Island, New York, $45,000 - general operating support; Citadel Foundation, Charleston, South Carolina, $50,000 - general operating support; Storm King Art Center, Mountainville, New York, $50,000 - general operating support.
*Contact:* Catherine Murray Smith, Trustee; (888) 866-3275
*Sponsor:* Samuel Freeman Charitable Trust
P.O. Box 1802
Providence, RI 02901-1802

**Samuel H. Kress Foundation Conservation Fellowships** 2248

The purpose of the Kress Conservation Fellowship program is to provide a wide range of post-graduate fellowship opportunities that will help develop the skills of emerging conservators. The Kress Conservation Fellowships provide competitive grants to museums and other conservation facilities which sponsor supervised internships in the conservation of specific objects and onsite training. Nine $32,000 Fellowships are expected to be awarded each year for one-year post-graduate internships in advanced conservation at a museum or conservation facility. Typically, $27,000 is allocated as a fellowship stipend, and $5,000 toward host institution administrative costs, benefits for the Fellow, and other direct costs of hosting the Fellowship. Most Fellowships begin in late summer or early fall, and run for a term of 9 to 12 months. The application must be received by January 22.
*Requirements:* Applications must be made by the museum or conservation facility at which the fellowship will be based. Prior to beginning the fellowship, fellows should have completed a masters-level degree in conservation. The fellowship candidate may be identified in advance of application by the host institution or recruited subsequently.
*Restrictions:* Fellowships will not be awarded to complete a degree program.
*Geographic Focus:* All States
*Date(s) Application is Due:* Jan 22
*Amount of Grant:* 32,000 USD
*Contact:* Wyman Meers; (212) 861-4993; fax (212) 628-3146; info@kressfoundation.org
*Internet:* www.kressfoundation.org/fellowships/conservation/
*Sponsor:* Samuel H. Kress Foundation
174 East 80th Street
New York, NY 10075

**Samuel H. Kress Foundation History of Art Institutional Fellowships** 2249

Advanced training in European art history requires direct exposure to the object of study, prolonged access to key information resources such as libraries and photographic archives, the development of professional relationships with colleagues abroad, and sustained immersion in European cultures. These related needs are often best satisfied by extended engagement with a European art research center. The Kress History of Art: Institutional Fellowships are intended to provide promising young art historians with the opportunity to experience just this kind of immersion. Six pre-doctoral Kress Institutional Fellowships in the History of European Art will be awarded each year. Kress Institutional Fellowships are being administered by the host institutions, which include: Kunsthistorisches Institut in Florence, Italy; Kunsthistorisch Instituut der Rijksuniversiteit in Leiden, the Netherlands; Courtauld Institute of Art & Warburg Institute of Art in London, England; Zentralinstitut für Kunstgeschichte in Munich, Germany; Institut national d'histoire de l'art in Paris, France; and Bibliotheca Hertziana in Rome, Italy. Applications must be received by the November 30 deadline. The fellowship award is $30,000 per annum.
*Requirements:* Dissertation research must focus on European art from antiquity to the early 19th century. Candidates must be nominated by their academic department.
*Restrictions:* These Fellowships are restricted to pre-doctoral candidates in the history of art and related disciplines (such as archaeology, architecture, or classics). Nominees must be U.S. citizens or individuals matriculated at an American university.
*Geographic Focus:* All States
*Date(s) Application is Due:* Nov 30
*Amount of Grant:* 30,000 USD
*Contact:* Wyman Meers; (212) 861-4993; fax (212) 628-3146; info@kressfoundation.org
*Internet:* www.kressfoundation.org/fellowships/institutional/
*Sponsor:* Samuel H. Kress Foundation
174 East 80th Street
New York, NY 10075

**Samueli Foundation Integrative Health Grants** 2250

The Samueli Foundation considers grants to agencies, primarily in Orange County, California, that serve the community, and whose programs meet the guidelines listed. Grants are usually approved for a defined period of time, but may be paid over a multi-year period. In the area of integrative health, Foundation goals are to: support programs promoting alternative and integrative medicine; and utilize creative and innovative approaches to education, prevention, treatment and research of disease and illness.
*Requirements:* 501(c)3 tax-exempt organizations throughout the U.S. are eligible to apply.
*Restrictions:* The Foundation does not fund umbrella fund raising organizations, political campaigns, or grants to individuals.
*Geographic Focus:* All States
*Amount of Grant:* 1,000 - 50,000 USD
*Contact:* Gerald R. Solomon; (949) 760-4400; fax (949) 760-4110; Info@samueli.org
*Internet:* https://www.samueli.org/funding-priorities/integrative-health/
*Sponsor:* Samueli Foundation
2101 East Coast Highway, 3rd Floor
Corona del Mar, CA 92625

**San Antonio Area Foundation Biomedical Research Grants** 2251

The San Antonio Area Foundation's Biomedical Research grants are awarded on an annual basis to programs in the areas of cancer, diabetes and heart disease. Applications are evaluated on: originality and importance of the research; likelihood that the project can gain competitive funding nationally; the project's benefit to the San Antonio and South Texas community; and capability of the investigator to complete the study. Priority is given to: junior investigators initiating pilot projects which, if funded, can be leveraged to gain additional national funding; senior investigators seeking bridge funding or defining new directions for their laboratories; and laboratories encompassing teaching components, inter-institutional collaborations or scientific inquiries of particular relevance to our South Texas population. Annual funding is announced each April.

*Requirements:* Eligible organizations to submit applications include nonprofit organizations exempt from federal income tax under section 501(c)3 of the Internal Revenue Code, schools, churches or governmental bodies that serve Bexar and the following counties: Frio, Atascosa, Karnes, Wilson, Guadalupe, Gonzales, Comal, Blanco, Kendall, Gillespie, Kerr, Bandera and Medina.

*Restrictions:* The Area Foundation does not fund: endowments; debt reduction; operating deficits; salaries; or travel and printing costs related to publication.

*Geographic Focus:* Texas

*Amount of Grant:* Up to 50,000 USD

*Samples:* St. Mary's University, San Antonio, Texas, $18,500 - support of research in coaptation depth in tricuspid annulus cinching; Texas Biomedical Research Institute, San Antonio, Texas, $33,700 - research in identifying causes of increased fracture risk in individuals with diabetes; University of Texas at San Antonio, San Antonio, Texas, $30,000 - support for quantifying cancer molecular signatures with a double-clad fiber optic probe.

*Contact:* Catherine Garcia; (210) 228-3757 or (210) 225-2243; cgarcia@saafdn.org

*Internet:* www.saafdn.org/Nonprofits/Grants/BiomedicalResearch.aspx

*Sponsor:* San Antonio Area Foundation
303 Pearl Parkway, Suite 114
San Antonio, TX 78215

**Sandia National Laboratories Harry S. Truman Fellowship in National Security Science and Engineering** 2252

Sandia established the President Harry S. Truman Fellowship in National Security Science and Engineering to attract the best nationally recognized new Ph.D. scientists and engineers. This three-year fellowship is named for President Truman who charged Sandia in 1949 with providing "an exceptional service in the national interest" - a motto that leads Sandia to excel to this day. Truman Fellows conduct independent ground-breaking research that supports Sandia's national-security mission. Fellows choose their own research topics and benefit by having access to Sandia's state-of-the-art facilities and collaborating with some of the nation's best scientists and engineers. Fellows may work at either of Sandia's principal locations in New Mexico and California. Sandia's competitive wage and benefits package includes an annual salary of $111,200; flexible work arrangements; 11 paid holidays; three weeks of vacation; health, vision, and dental insurance; and a 401(k) savings plan with company match. The annual deadline to apply is November 1. Fellows may work at either of the laboratories' principal locations, in Albuquerque, New Mexico, and Livermore, California. The fellowship is a three-year appointment normally commencing on October 1, although exceptions may be made to accommodate special circumstances.

*Requirements:* Eligibility criteria include: U.S. citizenship, the ability to obtain a Department of Energy clearance; research in areas of interest to national security; the candidate must have been awarded a PhD within the past three years at the time of application or will have completed all PhD requirements by commencement of appointment; and, candidates seeking their first national laboratory appointment. A minimum GPA of 3.5/4.0 (undergraduate) and 3.7/4.0 (graduate) is preferred.

*Geographic Focus:* All States

*Date(s) Application is Due:* Nov 1

*Amount of Grant:* 111,200 USD

*Contact:* Yolanda V. Moreno, (505) 284-2106; fax (505) 845-0580

*Internet:* www.sandia.gov/careers/students_postdocs/fellowships/truman_fellowship.html

*Sponsor:* Sandia National Laboratories
P.O. Box 5800, MS-1351
Albuquerque, NM 87185-1351

**Sandia National Laboratories John von Neumann Fellowship** 2253

The John von Neumann Postdoctoral Research Fellowship in Computational Science offers an exceptional opportunity to conduct innovative research in computational mathematics and scientific computing on advanced computing architectures. As members of either the Center for Computing Research at Sandia/New Mexico or the Computer Sciences and Information Systems Center at Sandia/California, von Neumann Fellows apply their research findings to a broad range of science and engineering problems of national importance. Sandia's computational science research programs span computational and discrete mathematics, computational physics and engineering, and systems software and tools. As a world leader in large-scale parallel computer systems, algorithms, software, and applications, Sandia provides a collaborative and highly multidisciplinary environment for solving computational problems at extreme scales. Sandia's state-of-the-art parallel-computing environment includes advanced architectures, such as the Cielo petascale machine, and numerous large-scale clusters and visualization servers, such as the 615-TFlop Sky Bridge cluster and the 409-TFlop Chama cluster. The John von Neumann Fellowship is a one-year appointment with a possible renewal for a second year. The fellowship offers a highly competitive salary, moving expenses, and a generous professional travel allowance. Von Neumann Fellows are supported by the Applied Mathematics program, which is part of the U.S. Department of Energy's Office of Advanced Scientific Computing Research.

*Requirements:* Applicants must have: a PhD, received prior to commencement of the fellowship, in Applied or Computational Mathematics, or related computational science and engineering discipline; a maximum of three years of postdoctoral experience; and a strong record of research publications. U.S. Citizenship not required.

*Geographic Focus:* All States, All Countries

*Date(s) Application is Due:* Dec 1

*Contact:* Yolanda V. Moreno, (505) 284-2106; fax (505) 845-0580

*Internet:* www.sandia.gov/careers/students_postdocs/fellowships/johnvonneumann_fellowship.html

*Sponsor:* Sandia National Laboratories
P.O. Box 5800, MS-1351
Albuquerque, NM 87185-1351

**Sandler Foundation Asthma Research Grants** 2254

Herb Sandler, along with his late wife Marion, co-founded and ran Golden West Financial Corporation for 43 years. The company reportedly achieved the best long-term earnings record of any company other than Warren Buffett's Berkshire Hathaway. While Marion's own struggle with asthma originally put asthma on the Sandlers' philanthropic radar, Herb says, "from the beginning [we knew] that we would not make any breakthroughs during our lifetime." Indeed, the Sandlers recognized asthma as a pervasive and critical health issue. The Sandler Foundation is a charitable foundation formed in 1991 and, in 2006, the Sandlers made a contribution of $1.3 billion to the Foundation, which was the second largest American charitable contribution in that year. Only invited applications are welcome, so interested parties should begin by contacting the Foundation with a query. With that in mind, the mission of the Sandler Foundation is to develop important new pathways of investigation in basic research regarding asthma. The program, which is open to investigators from all fields, particularly encourages applications from investigators not currently studying asthma. Innovation and risk are strongly encouraged. The program especially (but not exclusively) seeks applications in the following areas: development of new animal models for asthma; studies in humans or of human materials (excluding clinical trials); genetics of asthma; and technology development relevant to asthma. Three-year senior investigator awards and three-year junior investigator awards will be made. Awards to junior investigators are intended to support investigators who have demonstrated exceptional early accomplishment as independent investigators, permitting the expansion of their work into the field of asthma; they are not start-up funds. The proposed work should be directed toward uncovering basic mechanisms in the pathogenesis of asthma.

*Requirements:* Senior investigators will have well-established research programs and an international reputation for their research. They will usually hold a full-time academic appointment as professor, tenured associate professor, or the equivalent. Junior investigators will have already established an independent research program. They will usually hold a full-time academic appointment as an assistant professor, early associate professor, or the equivalent. Investigators may apply from any U.S. or Canadian nonprofit organization that can and will provide the necessary facilities and infrastructure for the research, and that will assure compliance with guidelines for animal and human studies established by the U.S. Department of Health and Human Services. Applicants from Canadian institutions must also provide evidence that the institution has been designated as a public charity by the U.S. Internal Revenue Service.

*Geographic Focus:* All States

*Amount of Grant:* 150,000 - 250,000 USD

*Contact:* Herbert Sandler, President; (415) 777-0411; fax (415) 514-0734

*Internet:* www.sandlerresearch.org

*Sponsor:* Sandler Foundation
4 Embarcadero Center, Suite 3150
San Francisco, CA 94111-4113

**Sara Elizabeth O'Brien Trust Grants** 2255

The Sara Elizabeth O'Brien Trust was established in 1981 to support charitable organizations that provide treatment or care to those who are blind or have cancer. The O'Brien Trust also supports medical research in the areas of blindness or cancer. The majority of grants from the O'Brien Trust are one year in duration; on occasion, multi-year support is awarded. Applicants must apply online at the grant website. Applicants are strongly encouraged to do the following before applying: review the downloadable state application procedures for additional helpful information and clarifications; review the downloadable online-application guidelines at the grant website; review the trust's funding history (link is available from the grant website); review the online application questions in advance; and review the list of required attachments. These will generally include: a list of board members, financial statements (audited, reviewed, or compiled by independent auditor); an organization summary; a list of other funding sources; an IRS Determination letter; and other required documents. All attachments must be uploaded in the online application as PDF, Word, or Excel files. The application deadline for the Sara Elizabeth O'Brien Trust is 11:59 p.m. on October 15. Applicants will be notified of grant decisions before December 31.

*Requirements:* Applicants must have 501(c)3 tax-exempt status.

*Restrictions:* The trust does not support requests from individuals, organizations attempting to influence policy through direct lobbying, or any political campaigns.

*Geographic Focus:* Massachusetts

*Date(s) Application is Due:* Oct 15

*Samples:* Foundation of the Massachusetts Eye and Ear Infirmary, Boston, Massachusetts, $60,000, to support research aimed at discovering genes associated with diabetic retinopathy; Dana-Farber Cancer Institute, Boston, Massachusetts, $33,000, fellowships in pediatric palliative care; MAB Community Services, Brookline, Massachusetts, $47,000, Seeing is Not Falling Project.

*Contact:* Michealle Larkins; (866) 778-6859; michealle.larkins@baml.com
*Internet:* https://www.bankofamerica.com/philanthropic/fn_search.action
*Sponsor:* Sara Elizabeth O'Brien Trust
225 Franklin Street, 4th Floor, MA1-225-04-02
Boston, MA 02110

**SARE Farmer/Rancher Research and Education Grants-Western Region 2256**
The Farmer/Rancher Research and Education Grants are conducted by agricultural producers with support and guidance from a technical advisor. Individual farmers or ranchers may apply for up to $15,000, and a group of three or more producers may apply for up to $25,000. Producers typically use their grants to conduct on-site experiments that can improve their operations and the environment, and can be shared with other producers. Grant recipients may also focus on marketing and organic production. Funding may be used for the following purposes: mini-grants to producers to compensate for field research and educational activities; cost of field sampling, crop analysis, and educational surveys; materials and supplies need for the project; small tools and equipment; outreach expenses such as a field day, tours, making a brochure, or creating a webpage; local or out of town travel need for the project; hired labor for heavy equipment; and producer labor for project activities. Specific guidelines, the application, and instructions about contacting state officers are available at the website.
*Requirements:* All proposals must have at least one producer. Applicants are producers if their primary occupation is farming or ranching; they have a farm tax number, or; they are a part-time producer with at least $1,000 documented annual income from their operation. The producer is fiscally responsible for the project and, if funded, provides his or her Social Security number. All proposals must include the following: yearly plan for outreach to other producers; a list of the educational materials they plan to produce and use in that outreach; and expected changes in these producers.
*Geographic Focus:* Alaska, Arizona, California, Colorado, Guam, Hawaii, Idaho, Montana, Nevada, New Mexico, Northern Mariana Islands, Oregon, Utah, Washington, Wyoming, American Samoa, Micronesia
*Date(s) Application is Due:* Dec 2
*Amount of Grant:* Up to 25,000 USD
*Contact:* Robert Newhall, Deputy Director; (435) 797-2183 or (435) 797-2257; fax (435) 797-3344; bob.newhall@usu.edu or wsare@usu.edu
*Internet:* www.westernsare.org/Grants/Types-of-Grants
*Sponsor:* Sustainable Agriculture Research and Education
University of Maryland
College Park, MD 20742-6715

**SARE Graduate Student Grants - North Central Region 2257**
The North Central Region SARE (NCR-SARE) Graduate Student Grants support projects by graduate students that address sustainable agriculture issues and are part of the student's degree program. At least 15 projects of up to $10,000 each are funded each year. Projects may last up to 36 months. Funded proposals have contributed to farmer or rancher profitability, environmental quality, and the enhancement of the quality of life of farmers or ranchers, rural communities, and society as a whole. NCR-SARE strongly encourages students to involve farmers and ranchers in their grant projects. NCR-SARE usually issues a call for proposals in the fall, and graduate student grant proposals are due in January. Usually, the Council announces the awards in the spring. Funds are typically available in the fall. Specific guidelines and information, including a video and list of previously funded projects, are available at the website.
*Requirements:* Graduate students may only receive one NCR-SARE Graduate Student award during their graduate student career. Projects must benefit agriculture in the North Central Region.
*Geographic Focus:* Illinois, Indiana, Iowa, Kansas, Michigan, Minnesota, Missouri, Nebraska, North Dakota, Ohio, South Dakota, Wisconsin
*Date(s) Application is Due:* May 5
*Amount of Grant:* Up to 10,000 USD
*Contact:* Beth Nelson; (612) 626-4436; fax (612) 626-3132; bethnelson@umn.edu
*Internet:* www.northcentralsare.org/Grants/Our-Grant-Programs/Graduate-Student-Grant-Program
*Sponsor:* Sustainable Agriculture Research and Education
University of Maryland
College Park, MD 20742-6715

**SARE Graduate Student Grants - Northeast Region 2258**
The SARE Graduate Student Grants are available for graduate students who are researching topics in sustainable agriculture that will serve the interests of farmers and agricultural service providers, such as extension staff, NRCD personnel, and others in the agricultural community. Successful Graduate Student Grant proposals will address inquiries in sustainable agriculture that are important to farmers, Cooperative Extension staff, and agricultural researchers. Awards may encompass a wide range of topics including, but not limited to cropping systems, pest management, livestock health, farm energy production, farm labor, soil quality, and the marketing of local food. SARE funds can be used for supplies, labor, equipment specific to the project (which must then become the property of the college or university), farm equipment rental or operating charges specific to the project, and travel expenses specific to the project and consistent with institutional mileage and per-diem rates. The grant can also be used for indirect costs up to 10% of your project's direct costs. Specific guidelines, examples of previously funded projects, and application submission instructions are available at the website.
*Requirements:* Grants are open to graduate students enrolled at an accredited college, university, or veterinary school in the Northeast region. Applicants must have a faculty advisor who will endorse the proposal, oversee the research, and act as the official principal investigator. Applications must come from one qualified individual. Groups or two-person teams may not participate. Collaboration is encouraged, but one person must be in charge of the project, ensuring that the work is done and reports are filed.
*Restrictions:* Grant funds cannot be used for any of the following: graduate student tuition remission; tuition payment; curriculum fees; travel to scholarly meetings (unless essential to the project, such as for a presentation of project results); travel outside the country; costs associated with preparing your thesis; the purchase of classroom texts; or capital costs such as the purchase of land, buildings, livestock, orchards, greenhouses, or other major farm improvements.
*Geographic Focus:* Connecticut, Delaware, District of Columbia, Maine, Maryland, Massachusetts, New Hampshire, New Jersey, New York, Pennsylvania, Rhode Island, Vermont, West Virginia
*Date(s) Application is Due:* May 25
*Amount of Grant:* 15,000 USD
*Contact:* Judy Gifford; (410) 648-5753; fax (443) 480-1853; gifford@baybroadband.net
*Internet:* www.nesare.org/Grants/Get-a-Grant/Graduate-Student-Grant
*Sponsor:* Sustainable Agriculture Research and Education
University of Maryland
College Park, MD 20742-6715

**SARE Graduate Student Grants - Southern Region 2259**
Sustainable Agriculture Research and Education Graduate Student Grants support sustainable agriculture research funding for Master's and Ph.D. students. Funds may be used for the following purposes: supplies, including software; equipment (any equipment obtained becomes the property of the university that the funded graduate student attends); labor; special texts not readily available; farm equipment rental or operating charges; and travel and per diem necessary for the project. The application and additional guidelines are available at the website.
*Requirements:* Full time Master's or Ph.D. students at accredited institutions in the southern region, with research projects that address issues of sustainable agriculture of current and potential importance to the southern region, may apply for funding. The grant will fund a project maximum of $11,000 for up to three years work on a student's project.
*Restrictions:* Only graduate students enrolled in academic programs in the southern region may apply for funding. Funds may not be used for the following purposes: graduate student stipends; travel to scholarly meetings or any international travel; preparation of thesis copy; publication costs; purchase of classroom books or payment of tuition; and permanent capital improvements. Applicants may receive only one SARE Graduate Student Grant during their graduate student career.
*Geographic Focus* Arkansas, Florida, Georgia, Kentucky, Louisiana, Mississippi, North Carolina, Oklahoma, Puerto Rico, South Carolina, Tennessee, Texas, U.S. Virgin Islands, Virginia
*Date(s) Application is Due:* Jun 1
*Amount of Grant:* Up to 11,000 USD
*Contact:* Paige Patton; (770) 412-4787; fax (770) 412-4789; ppatton@uga.edu
*Internet:* www.southernsare.org/Grants/Types-of-Grants/Graduate-Student-Grants
*Sponsor:* Sustainable Agriculture Research and Education
University of Maryland
College Park, MD 20742-6715

**SARE Graduate Student Grants in Sustainable Agriculture - Western Region 2260**
The Western SARE Graduate Student Grants in Sustainable Agriculture offer grants of up to $25,000 that may last up to two years. Grants are awarded to the applicant's university, with the funds to be assigned to the graduate student's project. An initial 90% of the award will be paid by reimbursement of allowable project expenses. The final 10% will be reimbursed after Western SARE has received and approved a final project report. Funds may be used for: outreach expenses such as holding a field day, tours, making a brochure, or creating a webpage; supplies, including software; small tools and equipment (this becomes property of the graduate student's university) not to exceed $5,000 per piece of equipment; hourly labor for documented work on the Western SARE project; farm equipment rental or operating charges; out-of-town travel and per diem necessary for the project (as prescribed by your university rules); local travel (where personal auto is used and reimbursed) needed for project; or publication costs (not to exceed $1,000. Specific guidelines and additional information about how to apply and proposals will be judged are available at the website.
*Requirements:* Applicants may receive only one grant during his or her graduate program. Western SARE will consider proposals from master's and Ph.D. students enrolled at accredited colleges or universities in the Western region. The student must be considered full-time, according to his or her institution's requirements, at the time the proposal is submitted. Proposals must address issues in sustainable agriculture of current and potential importance to the Western Region. Proposals are requested within the following guidelines: plant health and production and plant products; animal health and production and animal products; food safety, nutrition, and health; renewable energy, natural resources and environment; agriculture systems and technology; and agriculture economics and rural communities. Western SARE requires annual progress reports, a final report and one copy each of all theses, dissertations, and all other publications that are developed through the funded project. Project reports are compiled each year and shared with producers, interested policymakers and leaders at community, state, regional and national levels.
*Restrictions:* Funds may not be used for: preparation of thesis copy; purchase of classroom books or payment of tuition; permanent improvements to a farm or ranch, such as constructing or remodeling a building; graduate student stipends; or any single piece of equipment costing more than $5,000.
*Geographic Focus:* Alaska, Arizona, California, Colorado, Guam, Hawaii, Idaho, Montana, Nevada, New Mexico, Northern Mariana Islands, Oregon, Utah, Washington, Wyoming, American Samoa, Micronesia

*Date(s) Application is Due:* May 13
*Amount of Grant:* Up to 25,000 USD
*Contact:* Robert Newhall, Deputy Coordinator; (435) 797-2257 or (435) 797-2183; fax (435) 797-3344; bob.newhall@usu.edu or wsare@usu.edu
*Internet:* www.westernsare.org/Grants/Types-of-Grants
*Sponsor:* Sustainable Agriculture Research and Education
University of Maryland
College Park, MD 20742-6715

**SARE Large Systems Research Grants - Southern Region** 2261
The SARE Large Systems Research Grants are designed to assure research capacity in addition to gaining a near-term result to a specific research question. They are also designed to improve the opportunities for teams to engage in interdisciplinary activities. This ongoing process of research design followed by research activities, analysis, and redesign does not fit standard short-term modes of funding common in many grant programs. The Southern Region SARE Program hopes to address that funding gap. Project maximum is $100,000 per year for three years with opportunity for additional three-year renewals. Additional guidelines and the application is available at the website.
*Requirements:* Large Systems Grants are for teams who already have successful agricultural systems research in place but need support to accomplish additional long-term research goals. Research activities that not only keep the research team and infrastructure together over time but that also allow the teams to analyze and assess results over time, integrating new questions that arise into the long-term research protocol.
*Geographic Focus* Alabama, Arkansas, Georgia, Kentucky, Louisiana, Mississippi, North Carolina, Oklahoma, Puerto Rico, South Carolina, Tennessee, Texas, U.S. Virgin Islands, Virginia
*Amount of Grant:* Up to 100,000 USD
*Contact:* Sandra Young, Administrative Associate; (770) 229-3212; fax (770) 412-4790; syoung@griffin.uga.edu
*Internet:* www.southernsare.org/Grants/Types-of-Grants/Large-Systems-Research-Grants
*Sponsor:* Sustainable Agriculture Research and Education
University of Maryland
College Park, MD 20742-6715

**SARE On-Farm Research Grants - Southern Region** 2262
The Southern Region SARE program recognizes the value and importance of on-farm research in developing sustainable solutions to agricultural production problems by offering funding opportunities to Extension, Natural Resources Conservation Service personnel, university researchers, and non-governmental organizations who currently work with farmers and conduct on-farm research with at least one farmer/rancher cooperator. The On-Farm Research Grant program assists these individuals and organizations in conducting on-farm research or marketing projects related to sustainable agriculture to help the farmers they work with find solutions to various agricultural production issues. Successful proposals include: cooperators — on-farm research projects must include at least one farmer and/or rancher project cooperator; an outreach plan — to provide producers, researchers, and Extension personnel with the ability to learn from project results (outreach plans may include workshops, field days, fact sheets, or brochure); and clear goals — clear and simple goals help ensure a successful project. On-Farm Research Grant funds may be used for the following purposes: costs of sampling and sample analysis; materials and supplies needed for the project; outreach expenses such as holding a field day; travel (maximum 51 cents/mile) needed for the project; hired labor, including farmer/rancher labor beyond their normal farming duties; and refreshments at field days. Specific guidelines and the application are available at the website.
*Requirements:* Cooperative Extension, NRCS, university personnel, governmental and non-governmental organizations who work with farmers and ranchers, and are interested in conducting on-farm research or marketing projects related to sustainable agriculture are eligible to apply for the grant. On-farm research and marketing projects that promote sustainable agriculture are eligible to apply. Projects should be innovative, generate results that are useful beyond one year, and produce information that many farmers can use.
*Restrictions:* Funding may not be used for the following purposes: starting or expanding a farm, farming operation or NGO; equipment; permanent improvements to a farm or ranch, (e.g. planting an orchard, buying a herd, permanent fencing, permanent irrigation, permanent greenhouses or buildings); lunches or other full meals at field days or large gatherings; testing of commercial products.
*Geographic Focus* Alabama, Arkansas, Florida, Georgia, Kentucky, Mississippi, North Carolina, Oklahoma, Puerto Rico, South Carolina, Tennessee, Texas, U.S. Virgin Islands, Virginia
*Date(s) Application is Due:* Nov 1
*Amount of Grant:* Up to 15,000 USD
*Contact:* Paige Patton; (770) 412-4787; fax (770) 412-4789; ppatton@uga.edu
*Internet:* www.southernsare.org/Grants/Types-of-Grants/On-Farm-Research-Grants
*Sponsor:* Sustainable Agriculture Research and Education
University of Maryland
College Park, MD 20742-6715

**SARE Professional + Producer Research and Education Grants - Western Region** 2263
With a Professional + Producer Research and Education Grant, an agricultural professional, working with at least five producers, develops a proposal to conduct research and education on a sustainable agricultural topic and incorporate such items as: on-farm demonstrations, farmer-to-farmer educational outreach, and other items to assist in producer adoption in an area of sustainable agriculture. The goal is to achieve results that can be communicated to producers and professionals – information that can improve income, the environment, communities, and quality of life for all citizens. Producers will also have the opportunity to expand their knowledge and expertise in sustainable agriculture. Sustainable agriculture is defined as an integrated system of plant and animal production practices having a site-specific application that will, over the long term: satisfy human food and fiber needs; enhance environmental quality and the natural resource base upon which the agriculture economy depends; make the most efficient use of nonrenewable resources and on-farm resources and integrate, where appropriate, natural biological cycles and controls; sustain the economic viability of farm operations; and enhance the quality of life for farmers and society as a whole. On-farm education and experience of FFA, 4-H and other similar aged youth groups dealing with sustainable agricultural techniques and practices are encouraged. The SARE Council typically selects proposals diverse in subject matter and geography and that demonstrate outcomes that farmers and ranchers in the region can successfully adopt. Specific guidelines and instructions for application are available at the website.
*Requirements:* A minimum of five producers (separate operations) are required to participate in the development of on-farm research and demonstration with educational outreach. Each producer must be an independent and separate operation. Any agriculture/natural resource professional in the Western region may apply. These professionals are cooperative extension educators/agents and specialists, USDA-NRCS field staff, agricultural consultants, nonprofit staff members or other agricultural professionals assisting producers at the local level. Congress mandates that the SARE grant program depart from "business as usual." To that end, the Administrative Council requires that producers are involved from start to finish in the planning, design, implementation, and educational outreach of any SARE-funded project. Projects should: relate ideas to the definition of sustainable agriculture; address the changes that could come from the adoption of the results by other producers; and detail creative educational outreach plans that deliver this knowledge to other producers in the Western Region.
*Geographic Focus:* Alaska, Arizona, California, Colorado, Guam, Hawaii, Idaho, Montana, Nevada, New Mexico, Northern Mariana Islands, Oregon, Utah, Washington, Wyoming, American Samoa, Micronesia
*Date(s) Application is Due:* Dec 2
*Amount of Grant:* Up to 50,000 USD
*Contact:* Robert Newhall, Deputy Director; (435) 797-2257 or (435) 797-2183; fax (435) 797-3344; bob.newhall@usu.edu or wsare@usu.edu
*Internet:* www.westernsare.org/Grants/Types-of-Grants
*Sponsor:* Sustainable Agriculture Research and Education
University of Maryland
College Park, MD 20742-6715

**SARE Professional Development Grants - Northeast Region** 2264
The Professional Development (PD) Grants fund outcome-based projects that train Cooperative Extension educators and other agricultural service providers in sustainable techniques and concepts. Projects must be directed toward increasing the skill and understanding of these service providers and consistent with SARE's larger goal of broad farmer adoption of sustainable practices. The program emphasizes training the trainers, disseminating sustainable practices and technologies, advancing new content, and seeking measurable, verifiable change that leads to the improved and widespread practice of agricultural sustainability. Professional Development projects can cover a wide range of topics, such as ecological weed management, labor, whole-farm nutrient planning, tillage, season extension, carbon trading, and the needs of new farmers. Grants focus on training extension and NRCS staff, agricultural consultants, and other service providers who work directly with farmers and advise them on their farm and business practices. Projects should seek a cascade effect so that the people who took the training are positioned to teach others. SARE funds have also been used to produce books, videos, webcasts, and other media tools to improve participant understanding of sustainable techniques. Specific guidelines and application instructions are available at the website.
*Requirements:* Anyone can apply for a PD grant, but applicants should have direct and substantial experience delivering agricultural education to extension and other agricultural service providers. All applicants should demonstrate a firm grasp of current barriers and issues in sustainable agriculture. Proposals typically come from extension staff and the agricultural nonprofit sector, as well as college and university faculty and staff, and people who work on research farms and experiment stations. Northeast SARE uses outcome funding to select and follow through on its PD projects. All PD proposals should address the Northeast SARE outcome statement ("agriculture in the Northeast will be diversified and profitable, providing healthful products to its customers; it will be conducted by farmers who manage resources wisely, who are satisfied with their lifestyles, and have a positive influence on their communities and the environment.). Applicants not familiar with outcome funding should read the instruction guides and handouts on the website. The outcome approach permeates the entire application process, uses a distinct vocabulary, and operates on specific assumptions. Applicants must be conversant in outcome funding to successfully apply for a PD grant.
*Restrictions:* Applicants must be located in the Northeast SARE region.
*Geographic Focus:* Connecticut, Delaware, District of Columbia, Maine, Maryland, Massachusetts, New Hampshire, New Jersey, New York, Pennsylvania, Rhode Island, Vermont, West Virginia
*Date(s) Application is Due:* Oct 20
*Amount of Grant:* 60,000 - 200,000 USD
*Contact:* Tom Morris, Program Coordinator; (860) 486-0637; fax (860) 486-0682; thomas.morris@uconn.edu
*Internet:* www.nesare.org/Grants/Get-a-Grant/Professional-Development-Grant
*Sponsor:* Sustainable Agriculture Research and Education
University of Maryland
College Park, MD 20742-6715

**SARE Research and Education Grants - North Central Region** **2265**

The North Central Region SARE (NCR-SARE) Research and Education Grants award funding to researchers and educators involved in projects that explore and promote environmentally sound, profitable, and socially responsible food and/or fiber systems. Research and Education projects include a strong outreach component and significant farmer/rancher or other end user involvement from inception of the idea through implementation of the project. Many projects are interdisciplinary and/or multi-institutional, involving a broad range of agricultural interests. Project coordinators in the past have explored sustainable agriculture under the following topics: biocontrol, crop production, education and extension, networking, livestock production, marketing, quality of life, soil quality, value-added marketing, waste management, water quality, and weed control. Awards range from $10,000 to $200,000, with projects lasting up to 36 months. Coordinators are encouraged to explore social, environmental, and economic aspects of whole systems, keeping in mind that projects must have a strong outreach component. Specific guidelines, previously funded projects, and application instructions are available at the website.

*Requirements:* Farmers and ranchers help develop and then become the end-users of NCR-SARE research and educational products.

*Geographic Focus:* Illinois, Indiana, Iowa, Kansas, Michigan, Minnesota, Missouri, Nebraska, North Dakota, Ohio, South Dakota, Wisconsin

*Date(s) Application is Due:* Apr 5

*Amount of Grant:* Up to 175,000 USD

*Contact:* Beth Nelson, Grant Program Coordinator; (612) 626-4436 or (612) 626-3113; fax (612) 626-3132; bethnelson@umn.edu or ncrsare@umn.edu

*Internet:* www.northcentralsare.org/Grants/Our-Grant-Programs/Research-and-Education

*Sponsor:* Sustainable Agriculture Research and Education
University of Maryland
College Park, MD 20742-6715

**SARE Research and Education Grants - Northeast Region** **2266**

The Research and Education Grants fund outcome-based projects offering research, education, and demonstration projects that benefit farmers and explore new sustainable farm practices. Projects should be directed toward research results that will translate quickly into farmer benefit. Farmers, extension staff, and other stakeholders should be involved in planning project content. Research and Education projects can cover a wide range of topics. Northeast SARE has recently funded cover crop trials, farm-based energy, new approaches to livestock health, nutrient management and recycling, agritourism, and new directions in integrated crop and pest management. Whatever the topic, the project outcome must focus on a measurable change in farm practices and management that improves sustainability. SARE funds have also been used to produce books, videos, webcasts, and other media tools to improve the understanding of sustainable techniques. Specific guidelines, instructions for the application process, and a list of previously funded projects are available at the website.

*Requirements:* Anyone can apply for the grant, but applicants should have experience doing agricultural research, preferably research in cooperation with commercial farmers and applied research that will translate quickly into improved practices on the ground. All applicants must demonstrate a firm grasp of current barriers and issues in sustainable agriculture. Proposals typically come from university and extension staff, the agricultural nonprofit sector, and from those working on research farms and experiment stations. There is no requirement that applicants have a specific affiliation, but there is a clear expectation that applicants are credible and experienced in the science of agriculture and agricultural sustainability and understand the importance of profitability, good stewardship, and sustainable communities.

*Geographic Focus:* Connecticut, Delaware, District of Columbia, Maine, Maryland, Massachusetts, New Hampshire, New Jersey, New York, Pennsylvania, Rhode Island, Vermont, West Virginia

*Date(s) Application is Due:* Oct 20

*Contact:* Judy Gifford; (410) 648-5753 or (443) 480-1853 cell; gifford@baybroadband.net

*Internet:* www.nesare.org/Grants/Get-a-Grant/Research-and-Education-Grant

*Sponsor:* Sustainable Agriculture Research and Education
University of Maryland
College Park, MD 20742-6715

**SARE Research and Education Grants - Southern Region** **2267**

Research and Education projects are conducted by teams of interdisciplinary researchers specializing in sustainable agricultural systems. At the core of sustainable agriculture research is the application of a systems approach to research that aims to understand how a complex system functions as a whole, often beginning with a conceptual model. This approach recognizes that agricultural systems are complex and that interaction among components determines characteristics of the system. Research and Education Grants encourage that systems research approach in sustainable agriculture. Successful proposals include systems orientation and farmer or end-user involvement. Specific guidelines, application procedures, and samples of previous funded projects are available at the website.

*Requirements:* Research-based activities with an educational/outreach component that focus on developing sustainable agricultural systems or moving existing systems toward sustainable agriculture are eligible to apply. The following entities may apply for funding: teams of researchers from public and private institutions such as 1862 and 1890 land-grant universities, environmental agencies, historically African American universities, USDA-ARS labs, non-governmental organizations, and community based organizations. All Research and Education Grant applicants must propose systems oriented research that fits one of three grant categories: production research — focused on actual production methods, this kind of research has made up the bulk of SARE's project portfolio and has developed techniques that have become common tools for farmers; postharvest/food systems research — these projects examine what happens past the farm gate such as in the markets, distribution systems and policy making. This category can serve as a funding path for social science researchers to also make a difference in our farm and food systems; and a combination of production and postharvest/food systems research — the ultimate in systems research connects what goes on in the ground with everything that happens after a crop is harvested, including adding value, marketing, infrastructure for processing and transportation, as well as policy making.

*Geographic Focus:* Alabama, Arkansas, Florida, Georgia, Kentucky, Louisiana, Mississippi, North Carolina, Oklahoma, Puerto Rico, South Carolina, Tennessee, Texas, Virginia, Virgin Islands (U.S.)

*Amount of Grant:* Up to 300,000 USD

*Contact:* Sandra Young; (770) 229-3212; fax (770) 412-4790; syoung@griffin.uga.edu

*Internet:* www.southernsare.org/Grants/Types-of-Grants/Research-Education-Grants

*Sponsor:* Sustainable Agriculture Research and Education
University of Maryland
College Park, MD 20742-6715

**SARE Research and Education Grants - Western Region** **2268**

The SARE Research and Education Grants involve scientists, producers, and others using interdisciplinary approaches to address issues related to sustaining agriculture. The program encourages projects that: recognize the importance of whole-farm or whole-ranch systems approaches; utilize multiple disciplines (project cooperators) to assist in the project; produce impacts and outcomes that can be identified, measured and disseminated; address weak links or information gaps in a whole-farm or whole-ranch system; include educational outreach to producers and educational products to state extension and the Natural Resource Conservation Service (NRCS) field offices; incorporate research and education efforts with on-farm/ranch projects; use both research methods and educational outreach components to accomplish the SARE goal of producer adoption on the ground. The project's primary force should always: conduct both research and extension/education outreach on the sustainable agriculture project with producers; produce scholarly and educational materials; conduct outreach and disseminate the findings to producers and agricultural professionals (extension and USDA-NRCS, sustainable and conventional agricultural organizations, nonprofits, and NGOs); and assist with producer adoption of the findings. Specific guidelines and information about state offices and contacts is available at the website.

*Requirements:* Applicants must show relevance to the national definition of sustainable agriculture; include a minimum of five separate producers for on-farm testing; provide educational outreach plans to producers and agricultural producers; and include quantifiable scholarly and educational products with distribution techniques to producers and extension/NRCS field office staffs.

*Geographic Focus:* Alaska, Arizona, California, Colorado, Guam, Hawaii, Idaho, Montana, Nevada, New Mexico, Northern Mariana Islands, Oregon, Utah, Washington, Wyoming, American Samoa, Micronesia

*Date(s) Application is Due:* Jun 3

*Contact:* Robert Newhall, Deputy Coordinator; (435) 797-2183 or (435) 797-2257; fax (435) 797-3344; bob.newhall@usu.edu or wsare@usu.edu

*Internet:* www.westernsare.org/Grants/Types-of-Grants

*Sponsor:* Sustainable Agriculture Research and Education
University of Maryland
College Park, MD 20742-6715

**Sarkeys Foundation Grants** **2269**

Governed by a dedicated Board of Trustees, the foundation that bears SJ Sarkeys' name is deeply committed to furthering his vision to improve the quality of life in Oklahoma. The Foundation provides grants to a diverse group of nonprofit organizations and institutions, almost all of which are located in Oklahoma. Major areas of foundation support include education, arts and cultural endeavors, scientific research, animal welfare, social service and human service needs, and cultural and humanitarian programs of regional significance. Grant proposals are considered at the April and October meetings of the board of trustees.

*Requirements:* Most organizations classified by the IRS as being a 501(c)3 that is not a private foundation and is headquartered and offering services in Oklahoma may apply. Preference is given to organizations that have been in operation at least 3 years. Organizations are required to submit a Letter of Inquiry to determine whether they meet the criteria and priorities for funding. Representatives are encouraged to speak with a program officer for more information and to ask any questions about the process. An organization may submit one request in a twelve month period.

*Restrictions:* The Sarkeys Foundation will not fund: local programs appropriately financed within the community; direct mail solicitations and annual campaigns; out of state institutions; hospitals; operating expenses; purchase of vehicles; grants to individuals; responsibility for permanent financing of a program; programs whose ultimate intent is to be profit making; start-up funding for new organizations; feasibility studies; grants which trigger expenditure responsibility by Sarkeys Foundation; direct support to government agencies; individual public or private elementary or secondary schools, unless they are serving the needs of a special population which are not being met elsewhere; and religious institutions and their subsidiaries.

*Geographic Focus:* Oklahoma

*Date(s) Application is Due:* Feb 3; Aug 1

*Amount of Grant:* Up to 50,000 USD

*Contact:* Susan Frantz, Senior Program Officer; (405) 364-3703; susan@sarkeys.org or sarkeys@sarkeys.org

*Internet:* www.sarkeys.org/grant_guidelines.html

*Sponsor:* Sarkeys Foundation
530 East Main Street
Norman, OK 73071

**SASS Birgit Baldwin Fellowship** 2270

The Society for the Advancement of Scandinavian Study (SASS) offers the Birgit Baldwin Fellowship in Scandinavian Studies, a grant to encourage the research and writing of dissertations for the Ph.D. in topics concerned with Scandinavian (Nordic) literature or film that can most effectively be pursued in the archives and libraries of the Nordic countries. The stipend is $20,000 payable in two installments for travel to and living expenses in the Nordic region during one academic year. Specific guidelines and a current application are available online. Applicants submit materials to sass.baldwin.fellowship@gmail.com.

*Requirements:* Applicants must be members in good standing of the Society for the Advancement of Scandinavian Study, both during the year in which they submit applications and in the fellowship year. After completion of the fellowship year, the Baldwin Fellow is invited to give a short report at the next annual meeting of SASS. Applicants should submit a dissertation proposal that includes research plans and evidence of competency (including linguistic competency) to carry them out, supported by letters from two scholars. The successful applicant will demonstrate the significance of the subject, the need to work in Scandinavia (the Nordic region), and the feasibility of accomplishing the objectives of the research plans.

*Restrictions:* Applicants must be earning their degrees in North American universities. The Fellowship recipient is required to enroll in an international medical insurance policy for the duration of their stay overseas. SASS will issue the first fellowship check only after receiving a copy of the medical insurance policy and approving it.

*Geographic Focus:* All States

*Date(s) Application is Due:* Nov 1

*Contact:* Mabel Williams; (801) 422-8326; mabelwilliams@hotmail.com

*Internet:* www.scandinavianstudy.org/site/index.php?option=com_content&view=article&id=57&Itemid=62

*Sponsor:* Society for the Advancement of Scandinavian Study
Brigham Young University
Provo, UT 84602-6702

**SASS Einar and Eva Lund Haugen Memorial Scholarship** 2271

The Society for the Advancement of Scandinavian Study (SASS) offers the Einar and Eva Lund Haugen Memorial Scholarship. The scholarship supports outstanding graduate students whose doctoral dissertations examine a Scandinavian or Scandinavian-American topic in History and the Social Sciences. Scandinavia is defined as the five countries of the Nordic region: Norway, Sweden, Finland, Denmark, and Iceland. The grant can be used for all aspects of dissertation work. Preference will be given to those who need to conduct research in archives and libraries or to pursue fieldwork or excavation in places remote from their home institutions. The current application and specific guidelines, including a list of all materials required, are available at the website.

*Requirements:* Applicants must be enrolled in a doctoral program in North America. They must be members in good standing of the Society for the Advancement of Scandinavian Study, both during the year in which they submit applications and in the year of funding. A successful applicant must prove completion of course work and preliminary requirements for a doctoral degree before receiving the grant. After completion of the period funded by the scholarship, the recipient is invited to give a short report at the annual meeting of SASS.

*Geographic Focus:* All States, Denmark, Finland, Iceland, Norway, Sweden

*Date(s) Application is Due:* Jan 15

*Amount of Grant:* 5,000 USD

*Contact:* Mabel Williams; (801) 422-8326; mabelwilliams@hotmail.com

*Internet:* www.scandinavianstudy.org/site/index.php?option=com_content&view=article&id=88&Itemid=70

*Sponsor:* Society for the Advancement of Scandinavian Study
Brigham Young University
Provo, UT 84602-6702

**Schramm Foundation Grants** 2272

The foundation awards grants to Colorado nonprofit organizations in its areas of interest, including arts and culture, civic affairs, community development, education (elementary, secondary, and higher), health care, housing, humanities, medical research, science, social services delivery, technology, women's issues, and youth. Types of support include building construction/renovation, continuing support, equipment acquisition, general operating support, matching/challenge grants, program development, and scholarship funds. Applications are accepted from July 1 through August 31 (postmarked).

*Requirements:* Colorado 501(c)3 nonprofit organizations are eligible. Preference is given to requests from the Denver area. Applications must clearly express the reason(s) for the request, attach financial statements and copy of exemption letter.

*Restrictions:* Grants do not support advertising, advocacy organizations, individuals, international organizations, political organizations, religious organizations, school districts, special events, or veterans organizations.

*Geographic Focus:* Colorado

*Date(s) Application is Due:* Aug 31

*Amount of Grant:* 500 USD

*Contact:* Gary Kring, President; (303) 861-8291

*Sponsor:* Schramm Foundation
800 Grant Street, Suite 330
Denver, CO 80203-2944

**Scleroderma Foundation Established Investigator Grants** 2273

The Scleroderma Foundation is seeking applications from promising established investigators both inside and outside the field of SSc who wish to propose pilot studies to obtain preliminary data dealing with a highly innovative and/or highly relevant theme related to SSc. Applications along with the instructions are available at the SF website.

*Requirements:* Applications may be submitted by domestic and foreign nonprofit organizations, public and private, such as universities, colleges, hospitals, and laboratories. Applicants must have a doctorate degree in medicine, osteopathy, veterinary medicine, or one of the sciences; must have completed a postdoctoral fellowship by the grant award date; and have been principal investigator on grants from the foundation or other national, private, or government agencies. Applicants may request up to $75,000. Up to 8% of the direct costs per year of award may be set aside as indirect costs. The indirect cost amount will be subtracted from the yearly total [up to $75,000) of the award) per year for up to two years (total for 2 years not to exceed $150,000).

*Restrictions:* Funding are not renewable. Before completion of this grant, investigators are encouraged to seek more substantial continuing support for research through other grant mechanisms through private or government agencies. Replacement of the principal investigator on this award is not permitted. There will be no routine escalation for future years. Awards are contingent on the availability of funds and the receipt of sufficiently meritorious applications meeting the stated eligibility requirements.

*Geographic Focus:* All States, All Countries

*Date(s) Application is Due:* Sep 15

*Amount of Grant:* Up to 75,000 USD

*Contact:* Tracey Sperry, Director of Development and Research; (800) 722-4673 or (978) 463-5843; fax (978) 463-5809; tsperry@scleroderma.org

*Internet:* www.scleroderma.org/site/PageServer?pagename=prof_research_types#.UCz0vXAZzbg

*Sponsor:* Scleroderma Foundation
300 Rosewood Drive, Suite 105
Danvers, MA 01923

**Scleroderma Foundation New Investigator Grants** 2274

The Scleroderma Foundation is seeking applications from promising new investigators who hold faculty or equivalent positions and who wish to pursue a career in research related to systemic sclerosis (scleroderma). This grant will support promising research that is likely to lead to individual research project grants. Awards are contingent on the availability of funds and the receipt of sufficiently meritorious applications meeting the state eligibility requirements. The application is available at the SF website.

*Requirements:* Applications may be submitted by domestic non-profit organizations, public and private such as universities, colleges, hospitals and laboratories. Applicants must have a doctoral degree in Medicine, Osteopathy, Veterinary Medicine or one of the sciences and must have completed a postdoctoral fellowship by the grant award date. Applicants may request up to $50,000. Up to 8% of the direct costs per year of award may be set aside as indirect costs. The indirect costs will be subtracted from the yearly total (up to $50,000 of the award) per year for up to three years (total for three years not to exceed $150,000).

*Restrictions:* Foreign organizations and institutions are not eligible. These new investigator grants may not be used to support thesis or dissertation research or fellowship training. Applicants who have been a Principal Investigator on grants from the Scleroderma Foundation or other national, private or government agencies other than fellowship grants are not eligible for this award. Awards are not renewable. Before completion of this grant, investigators are encouraged to seek continuing support for research through other grant mechanisms through private or government agencies. Replacement of the Principal Investigator on this award is not permitted. There will be no routine escalation for future years.

*Geographic Focus:* All States

*Date(s) Application is Due:* Sep 15

*Amount of Grant:* Up to 50,000 USD

*Contact:* Tracey Sperry, Director of Development and Research; )978) 463-5843; fax (978) 463-5809; tsperry@scleroderma.org

*Internet:* www.scleroderma.org/site/PageServer?pagename=prof_research_types#.UCzz-3AZzbg

*Sponsor:* Scleroderma Foundation
300 Rosewood Drive, Suite 105
Danvers, MA 01923

**Searle Freedom Trust Grants** 2275

The Searle Freedom Trust was founded by Daniel C. Searle in 1998 to foster research and education on public policy issues that affect individual freedom and economic liberty. Through its grantmaking, the foundation seeks to develop solutions to the country's most important and challenging domestic policy issues. Recognizing that good policymaking relies on the availability of high-quality research, the foundation invests primarily in scholarship that results in the publication of books, journal articles, and policy papers. Funding is typically provided in the form of research grants, fellowships, and other types of targeted project support. The Searle Freedom Trust also provides funding for public interest litigation and supports outreach to the public through a variety of forums, including sponsorship of research conferences and seminars, film and journalism projects, and new media initiatives. With funding help from the Trust, university and think tank scholars investigate a wide range of issues such as tax and budget policy; cost-benefit analysis of regulatory practices and proposals; the workings of the legal system; environmental policy; social welfare reform; and K-12 and higher education policy. The Board meets three times a year. Proposals should be submitted through the Trust's online system. Proposals are accepted on a rolling basis, with deadlines of December 12 for the Spring meeting; April 16 for the Summer meeting; and August 22 for the Winter meeting.

*Requirements:* In addition to the application, organizations must submit a project description; the project budget; the organization's annual budget; any supporting materials; a list of key supporters; and evidence of the organization's non-profit status.

*Restrictions:* The Trust will typically not support: building projects; individuals; for profit organizations; organizations primarily depend on government funds; capital campaigns;

or endowments. Funding is also not available to support unspecified overhead costs associated with university research projects.
*Geographic Focus:* All States
*Date(s) Application is Due:* Apr 16; Aug 22; Dec 12
*Contact:* Kimberly O. Dennis, President and CEO; (202) 375-7822 or (202) 375-7820; kdennis@searlefreedomtrust.org
*Internet:* https://www.searlefreedomtrust.org/application-guidelines/
*Sponsor:* Searle Freedom Trust
1055 Thomas Jefferson Street, NW, Suite L26
Washington, D.C. 20007

**Searle Scholars Program Grants** **2276**
The aim of the Searle Scholars Program (administered by the Kinship Foundation) is to identify junior faculty members who have demonstrated a potential for doing innovative research in medicine, chemistry, and the biological sciences. Approximately 15 scholars are selected each year to receive the award. Each scholar receives $100,000 per year to support their research for three years. Submissions are restricted to selected candidates from invited institutions only and are not open to individuals who have not been selected by their institution to participate in the competition. The list of invited institutions can be found on the Participating Institutions page on the Program's website. Individuals from an invited institution who are considering submitting an application should consult with the individual at their institution who handles outside awards to determine how many applications their institution has been allowed to submit, as well as how selections will be made within their institution.
*Requirements:* Applicants will be expected to be pursuing independent research careers in biochemistry, cell biology, genetics, immunology, neuroscience, pharmacology, related areas in chemistry, medicine, and the biological sciences. Candidates should be in their first or second year at the assistant professor level. This appointment must be a tenure-track position and must be in an academic department of an invited, degree-granting institution. Potential applicants whose institutions do not have tenure-track appointments should consult with the scientific director of the program prior to preparing an application.
*Restrictions:* Applications from any institution not invited to participate will not be considered. No more than two applications will be accepted from any one institution.
*Geographic Focus:* All States
*Date(s) Application is Due:* Sep 28
*Amount of Grant:* 100,000 USD
*Contact:* Jennifer Nesler, Administrative and Grants Coordinator; (312) 803-6200; fax (312) 803-6201; jennifer.nesler@kinshipfoundation.org
*Internet:* www.searlescholars.net/go.php?id=5
*Sponsor:* Searle Scholars Program
303 West Madison Street, Suite 1800
Chicago, IL 60606

**Seattle Foundation Medical Funds Grants** **2277**
The Seattle Foundation administers the Medical Funds program to support medical research of potential benefit to the community and to address specific healthcare needs. In the area of medical research, grants are available in the fields of cancer, cardio-pulmonary disease, multiple sclerosis and diabetes. Special consideration will be given to research projects related to immunology, oncology, neurology, molecular biology and genetics. Grants will also be given to requests for specific equipment. Grants are also available to organizations that are administering projects addressing the healthcare needs of low-income children. In this area, grants will be given to requests for specific equipment and support for capital campaigns or facility renovation projects. A total amount of $200,000 is available for grants annually in this entire program. Typically no more than $50,000 is disbursed to any one organization.
*Requirements:* Preference will be given to organizations/institutions with the capacity to disseminate research findings and who receive regular support from other recognized funding sources (including the federal government).
*Restrictions:* These funds are not to be used for patient care. Research activities can include patient care, but the primary purpose of these grants is to support the purchase of equipment used in medical research.
*Geographic Focus:* Washington
*Samples:* Benaroya Research Institute at Virginia Mason, Seattle, Washington, $50,000 - to support the purchase of BD FACSVerse flow cytometer with 3-laser and 8-color capability; Fred Hutchinson Cancer Research Center, Seattle, Washington, $18,879 - to purchase a QuantaMaster 40 Spectrofluorometer suite; Puget Sound Blood Center, Seattle, Washington, $27,558 - to purchase a Perkin Elmer Micro Beta 2 Plate Reader to detect the biological profile of samples in small wells known as micro titer plates.
*Contact:* Ceil Erickson, Grantmaking Director; (206) 515-2131 or (206) 515-2109; fax (206) 622-7673; grantmaking@seattlefoundation.org or c.erickson@seattlefoundation.org
*Internet:* www.seattlefoundation.org/nonprofits/medicalfunds/Pages/MedicalFunds.aspx
*Sponsor:* Seattle Foundation
1200 Fifth Avenue, Suite 1300
Seattle, WA 98101-3151

**Seaver Institute Grants** **2278**
The Institute's focuses its giving in four primary areas: scientific and medical research, education, public affairs, and the cultural arts. Seed grants are awarded to nonprofit organizations for projects that offer the potential for a significant advancement in their fields. The institute seeks proposals that incorporate state-of-the-art scientific research, exceptional educational development, unique approaches to the creative arts, and the ramifications of shifting societal realities. There are no application deadlines.
*Requirements:* U.S. 501(c)3 nonprofits are eligible. Interested applicants should submit an inquiry letter of two or three pages and include the goals of the project, a preliminary budget, and the qualifications of the organization and principal investigator. If interested, the institute will invite a full proposal.
*Restrictions:* Grants are not made for construction, endowments, ongoing projects, operating budgets, deficits, or individual scholarships.
*Geographic Focus:* All States
*Amount of Grant:* 1,000 USD
*Contact:* Victoria Seaver Dean, President; (310) 979-0298; vsd@theseaverinstitute.org
*Sponsor:* Seaver Institute
12400 Wilshire Boulevard
Los Angeles, CA 90025

**SeaWorld and Busch Gardens Conservation Fund Grants** **2279**
The fund's grants support wildlife conservation projects conducted by recognized charitable organizations and noted scientists throughout the world. Grants support projects in the areas of species research, terrestrial; species research, aquatic; habitat protection; animal rescue and rehabilitation; and conservation education. Grants are awarded on a semiannual basis. Both guidelines and application are available online.
*Requirements:* Charitable organizations and scientists worldwide are eligible.
*Geographic Focus:* All States
*Date(s) Application is Due:* Jun 1; Dec 1
*Amount of Grant:* 5,000 - 25,000 USD
*Contact:* Fred Jacobs, Senior Director; (314) 613-6077 or (877) 792-4332; fax (314) 613-6079; fred.jacobs@anheuser-busch.com or mailbox@swbgfund.org
*Internet:* www.seaworld.org/conservation-matters/conservation-fund/index.htm
*Sponsor:* SeaWorld and Busch Gardens Conservation Fund
423 Lynch Street, Building 260-2
Saint Louis, MO 63118

**Sensient Technologies Foundation Grants** **2280**
The foundation supports organizations involved with arts and culture; children/youth, services; community/economic development; education; education, fund raising/fund distribution; education, research; family services; food services; general charitable giving; health care; higher education; homeless, human services; hospitals (general); human services; medical research, institute; mental health/crisis services; nutrition; performing arts; residential/custodial care, hospices; United Ways and Federated Giving Programs; Urban/community development; and voluntarism promotion. The foundation's types of support include: annual campaigns; capital campaigns; emergency funds; endowments; general operating support; matching/challenge support; program development; research; and scholarship funds.
*Requirements:* An application form is not required. Submit a letter of inquiry as an initial approach. The advisory board meets in January and June/July or as needed, with deadline for proposal review one month prior to board meetings.
*Restrictions:* The foundation gives primarily to its areas of interest in Indianapolis, Indiana; St. Louis, Missouri; and Milwaukee, Wisconsin. It does not support sectarian religious, fraternal, or partisan political organizations. It does not give grants to individuals.
*Geographic Focus:* Indiana, Missouri, Wisconsin
*Contact:* Douglas L. Arnold, (414) 271-6755; fax (414) 347-4783
*Sponsor:* Sensient Technologies Foundation
777 E Wisconsin Avenue
Milwaukee, WI 53202-5304

**SfN Chapter-of-the-Year Award** **2281**
The Chapter-of-the-Year Award recognizes outstanding chapters for their efforts and accomplishments across a broad range of activities that are in line with the mission and strategic initiatives of SfN. Recipients receive a $1,000 grant to support chapter goals and programs. Awards are presented during the Chapters Workshop at the SfN annual meeting.
*Requirements:* For award consideration, chapters must: submit an annual chapter report; and meet reporting requirements for all other programs through which funding from SfN was received (e.g., chapter grants). Additionally, a representative from the chapter leadership must be present at the Chapter's Workshop to accept the award. If requested, the chapter will give a presentation on a topic selected by SfN at the Chapter's Workshop.
*Geographic Focus:* All States, All Countries
*Date(s) Application is Due:* Jun 10
*Amount of Grant:* 1,000 USD
*Contact:* Program Director; (202) 962-4000; chapters@sfn.org
*Internet:* www.sfn.org/Awards-and-Funding/Chapter-Grants-and-Awards/Chapter-of-the-Year-Award
*Sponsor:* Society for Neuroscience
1121 14th Street, NW, Suite 1010
Washington, D.C. 20005

**SfN Chapter Grants** **2282**
SfN's offers annual Chapter Grants of $500 to $2,000 to support chapter activities that advance SfN's mission, including outreach, advocacy, and professional development programs. For award consideration, chapters must have submitted at least one annual report during the previous two calendar years and must have provided reports from all other programs through which funding was received. Previous grants have funded activities such as: student-oriented regional conferences; beginning, continuing, or expanding Brain Awareness Week activities, including increased community involvement or the purchase of educational supplies; hosting a local political representative to promote neuroscience; hosting a lecturer from outside the chapter's geographic region (previously funded by the

Distinguished Traveling Scientists program); collaborating with local organizations on professional development initiatives; and supporting a chapter annual meeting.
*Requirements:* Chapters must spend grant money on the approved budget (submitted with the application) within one year of being awarded. A final financial and project report, with return of unused funds, is due to SfN 30 days after the event.
*Geographic Focus:* All States, All Countries
*Amount of Grant:* 500 - 2,000 USD
*Contact:* Program Coordinator; (202) 962-4000; chapters@sfn.org
*Internet:* www.sfn.org/Awards-and-Funding/Chapter-Grants-and-Awards/Chapter-Grants
*Sponsor:* Society for Neuroscience
1121 14th Street, NW, Suite 1010
Washington, D.C. 20005

**SfN Donald B. Lindsley Prize in Behavioral Neuroscience** 2283
The Donald B. Lindsley Prize in Behavioral Neuroscience recognizes a young neuroscientist's outstanding Ph.D. thesis in the general area of behavioral neuroscience (defined as neuroscience research involving behavioral variables or oriented toward the solution of behavioral problems). The recipient receives a $2,500 prize; and air or ground transportation, two nights hotel, and complimentary registration for the Society's annual meeting. The Society's President presents the prize at a lecture at the annual meeting. All nomination materials must be submitted through the Society's award nomination link at the Society's website. Nominations open on March 7 and close on May 20.
*Requirements:* Nominees must have a thesis submitted and approved between January 1 and December 31 of the preceding calendar year. Nominations must be made or endorsed by SFN members, although nominees do not need to be SFN members.
*Restrictions:* Self nomination is not permitted. No person may nominate more than one candidate. Current Society officers and councilors are ineligible for nomination.
*Geographic Focus:* All States, Puerto Rico, Argentina, Australia, Brazil, Canada, China, Denmark, India, Israel, Malaysia, Mexico, New Zealand, Nigeria, Norway, Singapore, South Korea, Spain, Switzerland, Turkey, Ukraine, United Kingdom
*Date(s) Application is Due:* May 20
*Amount of Grant:* 2,500 USD
*Contact:* Steven E Petersen; (314) 362-3317; fax (314) 362-2186; sep@npg.wustl.edu
*Internet:* www.sfn.org/Awards-and-Funding/Individual-Prizes-and-Fellowships/Young-Scientists-Achievements-and-Research/Donald-B-Lindsley-Prize-in-Behavioral-Neuroscience
*Sponsor:* Society for Neuroscience
1121 14th Street, NW, Suite 1010
Washington, D.C. 20005

**SfN Federation of European Neuroscience Societies Forum Travel Awards** 2284
The SfN offers up to fifteen (15) travel awards of $2,000 each to support the participation of graduate students from the U.S., Canada, and Mexico at the Federation of European Neuroscience Societies (FENS) Forum, held every two years. Recipients of the travel awards, selected by the SfN International Affairs Committee, are based on the scientific merit of their abstract submitted to the Forum, and the strength of related application materials. Applicants will be emailed about the status of their application. Recipients must attend the Forum to receive their award checks.
*Requirements:* The applicant must be a U.S., Canadian, or Mexican citizen or permanent resident and graduate student who has advanced to candidacy for a Ph.D. The nominee must be the first author of an abstract to be presented at the Forum. Both the applicant and his or her advisor must be SfN members, and the applicant may not be a recipient of another SfN travel award in the same year. In addition to the online application form, applicants must submit a paragraph describing what they hope to achieve by attending the Forum; a copy of the abstract that would be submitted for the Forum; a one page letter of recommendation from their advisor; and a one page CV which includes their education, honors/awards, and publications (abstracts and manuscripts).
*Geographic Focus:* All States, Canada, Mexico
*Date(s) Application is Due:* Feb 2
*Amount of Grant:* 2,000 USD
*Contact:* Kelly McCabe, Global Programs Assistant; (202) 962-4000; fax (202) 962-4941; kmccabe@sfn.org or globalaffairs@sfn.org
*Internet:* www.sfn.org/Awards-and-Funding/Individual-Prizes-and-Fellowships/Professional-Development-Awards/FENS-Forum-Travel-Awards
*Sponsor:* Society for Neuroscience
1121 14th Street, NW, Suite 1010
Washington, D.C. 20005

**SfN Jacob P. Waletzky Award** 2285
The Jacob P. Waletzky Award is presented to a scientist who has conducted research or plans to conduct research in the area of substance abuse and the brain and nervous system. Recipients receive a $25,000 prize and complimentary registration, transportation (economy air or ground), and two nights hotel accommodations for the SfN annual meeting. The SfN president presents the prize at a lecture at the meeting. Nominations open on March 7 and close on June 3.
*Requirements:* Nominations must be made or endorsed by a Society member, although nominees do not need to be Society members. All materials are submitted using the Society's award nomination process located at the Society's website.
*Geographic Focus:* All States, Puerto Rico, Argentina, Australia, Brazil, Canada, Denmark, India, Israel, Malaysia, Mexico, New Zealand, Nigeria, Norway, Singapore, South Korea, Spain, Switzerland, Turkey, Ukraine, United Kingdom
*Date(s) Application is Due:* Jun 3
*Amount of Grant:* 25,000 USD
*Contact:* Daniel H Geschwind; (310) 206-6814; fax (310) 267-2401; dhg@mednet.ucla.edu
*Internet:* www.sfn.org/Awards-and-Funding/Individual-Prizes-and-Fellowships/Outstanding-Research-and-Career-Awards/Jacob-P-Waletzky-Award
*Sponsor:* Society for Neuroscience
1121 14th Street, NW, Suite 1010
Washington, D.C. 20005

**SfN Janett Rosenberg Trubatch Career Development Award** 2286
The Janett Rosenberg Trubatch Award recognizes individuals who have demonstrated originality and creativity in research. The Award also promotes success during academic transitions prior to tenure. The recipient receives a $2,000 award and complimentary registration to the Society's annual meeting. Two recipients are selected each year. The Awards are presented during the Society's annual meeting. All materials must be submitted using the Society's online nomination site. Nominations open on March 7 and close on May 20.
*Requirements:* Nominees with a Ph.D. or M.D. (or equivalent) are eligible until they have completed ten years of full-time work. Candidates working in non-academic environments are eligible if their work is published and meets academic standards. Nominees must not have received university tenure by the nomination deadline. Nominations must be made or endorsed by a Society member, although nominees do not need to be Society members.
*Restrictions:* Self-nomination is not permitted. No person may nominate more than one candidate. Current Society officers and councilors are ineligible for nomination.
*Geographic Focus:* All States, Puerto Rico, Argentina, Australia, Brazil, Canada, China, Denmark, India, Israel, Malaysia, Mexico, New Zealand, Nigeria, Norway, Singapore, South Korea, Spain, Switzerland, Turkey, Ukraine, United Kingdom
*Date(s) Application is Due:* May 20
*Amount of Grant:* 2,000 USD
*Contact:* Nancy Y Ip, Ph.D.; 85-223-587-267; fax 85-223-581-464; boip@ust.hk
*Internet:* www.sfn.org/Awards-and-Funding/Individual-Prizes-and-Fellowships/Young-Scientists-Achievements-and-Research/Janett-Rosenberg-Trubatch-Career-Development-Award
*Sponsor:* Society for Neuroscience
1121 14th Street, NW, Suite 1010
Washington, D.C. 20005

**SfN Japan Neuroscience Society Meeting Travel Awards** 2287
The Society for Neuroscience (SfN) and the Japan Neuroscience Society (JNS) support a joint travel award exchange program for graduate students or postdoctoral investigators to their respective annual meetings. The Society sponsors five eligible members from Canada, Mexico, and the U.S. who intend on participating in the JNS meeting in Japan. The JNS will also sponsor five members to attend the SfN annual meeting in the U.S. (locations vary). Recipients are selected on the merit of their abstract, CV, and letters of recommendation. Students are selected by the International Affairs Committee. Award checks are distributed to selectees following the meeting. Awardees will need to provide a certificate of attendance and airline boarding passes to receive their award.
*Requirements:* Applicants must be Ph.D. candidates or postdoctoral scientists studying or working in Canada, Mexico, or the U.S. They must also be citizens or permanent residents of those locations. Candidates must be SfN members at the time of application and first author on an abstract to be presented at the JNS meeting. Candidates may not have been selected for another SfN travel award within the last calendar year. Preference is given to those who have not previously had extensive experience living or working in Japan. In addition to the online application, candidates must submit a paragraph describing what the applicant hopes to achieve by attending the JNS meeting; a copy of the abstract to be submitted; a one page recommendation from their advisor; and the applicant's CV including information on education, date of Ph.D. candidacy, honors/awards, and publications.
*Geographic Focus:* All States, Canada, Japan, Mexico
*Date(s) Application is Due:* Feb 3
*Contact:* Kelly McCabe, Global Programs Assistant; (202) 962-4000; fax (202) 962-4941; kmccabe@sfn.org or globalaffairs@sfn.org
*Internet:* www.sfn.org/Awards-and-Funding/Individual-Prizes-and-Fellowships/Professional-Development-Awards/JNS-Meeting-Travel-Awards
*Sponsor:* Society for Neuroscience
1121 14th Street, NW, Suite 1010
Washington, D.C. 20005

**SfN Julius Axelrod Prize** 2288
The Julius Axelrod Prize, supported by the Eli Lilly and Company Foundation, was established to honor a scientist with distinguished achievements in the broad field of neuropharmacology (or a related area) and exemplary efforts in mentoring young scientists. The recipient receives a $25,000 prize. They also attend the SfN annual meeting with complimentary air or ground transportation, hotel, and registration. The SfN president presents the prize at a lecture at the meeting. Nominations open on March 7 and close on June 3.
*Requirements:* Nominees must have accomplishments in the field of neuropharmacology or a related area. They also need to have shown dedication to mentoring young scientists. Nominations are made or endorsed by SfN members, although nominees do not need to be SfN members. All nomination materials are submitted using the Society's award nomination site.
*Restrictions:* Self nomination is not permitted, and no person may nominate more than one candidate. Current SfN officers and councilors are ineligible for nomination.
*Geographic Focus:* All States, Puerto Rico, Argentina, Australia, Brazil, Canada, China, Denmark, India, Israel, Malaysia, Mexico, New Zealand, Nigeria, Norway, Singapore, South Korea, Spain, Switzerland, Turkey, Ukraine, United Kingdom

*Date(s) Application is Due:* Jun 3
*Amount of Grant:* 25,000 USD
*Contact:* Guoping Feng, Ph.D.; (617) 715-4898; fax (617) 324-6752; fengg@mit.edu
*Internet:* www.sfn.org/Awards-and-Funding/Individual-Prizes-and-Fellowships/Outstanding-Research-and-Career-Awards/Julius-Axelrod-Prize
*Sponsor:* Society for Neuroscience
1121 14th Street, NW, Suite 1010
Washington, D.C. 20005

**SfN Louise Hanson Marshall Specific Recognition Award 2289**
The Louise Hanson Marshall Special Recognition Award recognizes an individual who has significantly promoted the professional development of women in neuroscience through teaching, organizational leadership, public advocacy, or other efforts that are not necessarily related to research. The recipient receives the Marshall Award, along with complimentary transportation, hotel, and registration for the SfN annual meeting. The Award is presented at the Celebration of Women in Neuroscience luncheon held during the SfN annual meeting. All nominating materials should be submitted using SfN's award nomination site. Nominations for the award open on March 7 and close on June 10.
*Requirements:* Nominees must have served the professional through SfN and/or related organizations. They must also have attained recognition at national or international levels as a scientist, educator, businessperson, or administrator in neuroscience. In addition, nominees must have demonstrated a high degree of imagination, innovation, and initiative in the pursuit of neuroscience, and facilitated mentoring or the advancement of young women in neuroscience. Nominees may only be made by SfN members, but the nominees do not need to be SfN members.
*Restrictions:* Self nomination is not permitted. No one may nominate more than one candidate.
*Geographic Focus:* All States, Puerto Rico, Argentina, Australia, Brazil, Canada, China, Denmark, India, Israel, Malaysia, Mexico, New Zealand, Nigeria, Norway, Singapore, South Korea, Spain, Switzerland, Turkey, Ukraine, United Kingdom
*Date(s) Application is Due:* Jun 10
*Contact:* Nancy Y Ip, Ph.D.; 85-223-587-267; fax 85-223-581-464; boip@ust.hk
*Internet:* www.sfn.org/Awards-and-Funding/Individual-Prizes-and-Fellowships/Promotion-and-Mentoring-of-Women-in-Neuroscience/Louise-Hanson-Marshall-Special-Recognition-Award
*Sponsor:* Society for Neuroscience
1121 14th Street, NW, Suite 1010
Washington, D.C. 20005

**SfN Mika Salpeter Lifetime Achievement Award 2290**
The Mika Salpeter Lifetime Achievement Award recognizes an individual with outstanding career achievements in neuroscience who has also significantly promoted the professional advancement of women in neuroscience. The recipient receives a $5,000 award, in addition to complimentary air or ground transportation, hotel, and registration to the SfN annual meeting. The nominee must have exhibited a dedication to facilitating the mentoring and entry of young women into neuroscience or to the advancement of women in neuroscience; sustained exceptional achievements in neuroscience as evidenced by publications, inventions, and/or awards; served the profession through SfN and/or related organizations; recognized at national or international levels as a scientist, educator, businessperson, or administrator in neuroscience; and demonstrated a high degree of imagination, innovation, and initiative in the pursuit of neuroscience. Nominations for the award open on March 7 and close on June 10.
*Requirements:* Nominations must be made or endorsed by a Society member, although nominees do not need to be Society members. All nomination materials are submitted using SfN's award nomination site.
*Restrictions:* Self nomination is not permitted, and no person may nominate more than one candidate. Current SfN officers and councilors are ineligible for nomination.
*Geographic Focus:* All States, Puerto Rico, Argentina, Australia, Brazil, Canada, China, Denmark, India, Israel, Malaysia, Mexico, New Zealand, Nigeria, Norway, Singapore, South Korea, Spain, Switzerland, Turkey, Ukraine, United Kingdom
*Date(s) Application is Due:* Jun 10
*Amount of Grant:* 5,000 USD
*Contact:* Nancy Y Ip, Ph.D.; 85-223-587-267; fax 85-223-581-464; boip@ust.hk
*Internet:* www.sfn.org/Awards-and-Funding/Individual-Prizes-and-Fellowships/Outstanding-Research-and-Career-Awards/Mika-Salpeter-Lifetime-Achievement-Award
*Sponsor:* Society for Neuroscience
1121 14th Street, NW, Suite 1010
Washington, D.C. 20005

**SfN Neuroscience Fellowships 2291**
The Neuroscience Scholars Program (NSP) is a two-year fellowship administered by the Society for Neuroscience to enhance career development and professional networking opportunities for underrepresented diverse undergraduate and graduate students, and postdoctoral fellows in neuroscience. The program, funded by the National Institute of Neurological Disorders and Stroke, provides scholars with: annual stipend for enrichment activities outside the scholar's home institution; support for and annual meeting travel expenses; access to annual meeting workshops, courses, and events; complimentary SfN meeting registration and abstract fee waivers; complimentary SfN membership and online subscription to The Journal of Neuroscience; and mentoring. Other benefits include opportunities to network, expand professional contacts, and acquire professional skills. The application and samples of the biosketch required are located at the SfN website.
*Requirements:* Applicants must be citizens or permanent residents of the U.S., and enrolled in a degree-granting program or postdoctoral fellowship. Candidates must be from a group recognized as underrepresented in the biomedical, behavioral, clinical, and social sciences. Additional information is available at the website.
*Restrictions:* Previous fellowship recipients are ineligible.
*Geographic Focus:* All States
*Contact:* Mona Miller, Director of Programs; (202) 962-4000; mmiller@sfn.org
*Internet:* www.sfn.org/Awards-and-Funding/Individual-Prizes-and-Fellowships/Fellowships/Neuroscience-Scholars-Program
*Sponsor:* Society for Neuroscience
1121 14th Street, NW, Suite 1010
Washington, D.C. 20005

**SfN Patricia Goldman-Rakic Hall of Honor 2292**
The Patricia Goldman-Rakic Hall of Honor is a posthumous award for a neuroscientist who pursued career excellence and exhibited dedication to the advancement of women in neuroscience. The recipient is recognized at the Celebration of Women in Neuroscience Luncheon held during the SfN annual meeting and the family of the deceased honoree receives an engraved Tiffany and Company crystal bowl. Nominations for the award open on March 7 and close on June 10.
*Requirements:* Nominees must have: exhibited dedication to facilitating the mentoring and entry of young women into neuroscience or to the advancement of women in neuroscience; sustained exceptional achievements in neuroscience as evidenced by publications, inventions, and/or awards; served the profession through SfN and/or related organizations; been recognized at a national or international level as a scientist, educator, businessperson, or administrator in neuroscience; and demonstrated a high degree of imagination, innovation, and initiative in the pursuit of neuroscience.
*Geographic Focus:* All States, All Countries
*Date(s) Application is Due:* Jun 10
*Contact:* Nancy Y Ip, Ph.D.; 85-223-587-267; fax 85-223-581-464; boip@ust.hk
*Internet:* www.sfn.org/Awards-and-Funding/Individual-Prizes-and-Fellowships/Promotion-and-Mentoring-of-Women-in-Neuroscience/Patricia-Goldman-Rakic-Hall-of-Honor
*Sponsor:* Society for Neuroscience
1121 14th Street, NW, Suite 1010
Washington, D.C. 20005

**SfN Peter and Patricia Gruber International Research Award 2293**
The Peter and Patricia Gruber International Research Award in Neuroscience recognizes two young neuroscientists for outstanding research and educational pursuit in an international setting. If the nominee is a U.S. citizen or permanent resident, the nominee must be studying or working at an institution located outside the U.S; if a non-U.S. citizen or permanent resident, the nominee must be studying or working at an institution that is located in the U.S. Nominees must be active in research at the time the award is given. Successful candidates are graduate students, postdoctoral fellows, or research associates. Each recipient receives a $25,000 prize, in addition to air or ground transportation to the SfN annual meeting with two nights hotel accommodations, and complimentary registration. All nominating materials must be submitted using the SfN's award nomination site. Nominations for the award open on March 7 and close on May 20.
*Requirements:* Although nominees do not need to be SfN members, they must be nominated or endorsed by an SfN member.
*Restrictions:* Self-nomination is not permitted. No person may nominate more than one candidate. Current SfN officers and councilors are ineligible for nomination.
*Geographic Focus:* All States, Puerto Rico, Argentina, Australia, Brazil, Canada, China, Denmark, India, Israel, Malaysia, Mexico, New Zealand, Nigeria, Norway, Singapore, South Korea, Spain, Switzerland, Turkey, Ukraine, United Kingdom
*Date(s) Application is Due:* May 20
*Amount of Grant:* 25,000 USD
*Contact:* Roberta Diaz Brinton, Ph.D., Chairperson; (323) 442-1430; fax (323) 442-1740; awards@sfn.org or rbrinton@usc.edu
*Internet:* www.sfn.org/Awards-and-Funding/Individual-Prizes-and-Fellowships/Young-Scientists-Achievements-and-Research/Peter-and-Patricia-Gruber-International-Research-Award
*Sponsor:* Society for Neuroscience
1121 14th Street, NW, Suite 1010
Washington, D.C. 20005

**SfN Ralph W. Gerard Prize in Neuroscience 2294**
The Ralph W. Gerard Prize in Neuroscience honors an outstanding scientist who has made significant contributions to neuroscience throughout his or her career. The Gerard Prize is the highest recognition conferred by the Society, and open to all scientists with contributions to the field of neuroscience. The recipient receives $25,000, in addition to complimentary air or ground transportation, hotel, and registration for the SfN annual meeting. Nominations for the award open on March 7 and close on June 3.
*Requirements:* Nominations must be made or endorsed by Society member, although nominees do not need to be Society members. All nomination materials need to be submitted using SfN's award nomination site.
*Restrictions:* Self nominations are not permitted, and no person may nominate more than one candidate. Current SfN officers and councilors are ineligible for nomination.
*Geographic Focus:* All States, Puerto Rico, Argentina, Australia, Brazil, Canada, China, Denmark, India, Israel, Malaysia, Mexico, New Zealand, Nigeria, Norway, Singapore, Spain, Switzerland, Turkey, Ukraine, United Kingdom
*Date(s) Application is Due:* Jun 3
*Amount of Grant:* 25,000 USD
*Contact:* Nicholas C Spitzer, Ph.D.; (858) 534-3896; fax (858) 534-7309; nspitzer@ucsd.edu

*Internet:* www.sfn.org/Awards-and-Funding/Individual-Prizes-and-Fellowships/Outstanding-Research-and-Career-Awards/Ralph-W-Gerard-Prize-in-Neuroscience
*Sponsor:* Society for Neuroscience
1121 14th Street, NW, Suite 1010
Washington, D.C. 20005

**SfN Swartz Prize for Theoretical and Computational Neuroscience** 2295
The Swartz Prize is presented annually to an individual whose activities have produced a significant cumulative contribution to theoretical models or computational methods in neuroscience, or who has made a noteworthy advance in theoretical or computational neuroscience. The recipient receives $25,000, in addition to complimentary transportation, hotel, and registration to the SfN annual meeting, where the Prize is presented. The recipient also receives a lecture slot at the Dynamical Neuroscience Satellite Event, sponsored by the NIMH. All nomination materials must be submitted using SfN's award nomination site. Nominations for the award open on March 7 and close on June 3.
*Requirements:* Nominations are open to all individuals who have contributed to theoretical or computational neuroscience. Nominations should be made or endorsed by SfN members, although nominees do not need to be SfN members.
*Restrictions:* Self nomination is not permitted, and no person may nominate or write a letter of recommendation for more than one candidate. Current SfN officers and councilors are not eligible for nomination.
*Geographic Focus:* All States, Puerto Rico, Argentina, Australia, Brazil, Canada, China, Denmark, India, Israel, Malaysia, Mexico, New Zealand, Nigeria, Norway, Singapore, South Korea, Spain, Switzerland, Turkey, Ukraine, United Kingdom
*Date(s) Application is Due:* Jun 3
*Amount of Grant:* 25,000 USD
*Contact:* Thomas D. Albright, Ph.D.; (858) 453-4100; fax (858) 546-8526; tom@salk.edu
*Internet:* www.sfn.org/Awards-and-Funding/Individual-Prizes-and-Fellowships/Outstanding-Research-and-Career-Awards/Swartz-Prize-for-Theoretical-and-Computational-Neuroscience
*Sponsor:* Society for Neuroscience
1121 14th Street, NW, Suite 1010
Washington, D.C. 20005

**SfN Travel Awards for the International Brain Research Organization World Congress** 2296
The Society for Neuroscience (SfN) offers up to 15 travel awards of $2,000 each to support the participation of U.S., Canadian, and Mexican graduate students at the IBRO World Congress held every four years. Recipients of the travel awards are selected by the SfN International Affairs Committee, based on the scientific merit of the abstract submitted to the Congress and the strength of related application materials. Award checks are presented at the conference. Applicants are notified by email of the status of their application. Deadlines vary according to the date of the World Congress.
*Requirements:* Applicants must be U.S., Canadian, or Mexican citizens or permanent residents, and graduate students who have advanced to candidacy for a Ph.D. Nominees must be first authors of an abstract to be presented at the conference. Applicants and their advisors must be SfN members. Applicants may not be recipients of another SfN travel award in the same award year. In addition to the online application form, applicants must submit: a paragraph describing what they hope to achieve by attending the conference; a copy of the abstract submitted for the conference; a one page letter of recommendation from their advisor; and a one page CV which includes their education, honors/awards, and publications (abstracts and manuscripts).
*Geographic Focus:* All States, Canada, Mexico
*Amount of Grant:* 2,000 USD
*Contact:* Chairperson; (202) 962-4000; globalaffairs@sfn.org
*Internet:* www.sfn.org/Awards-and-Funding/Individual-Prizes-and-Fellowships/Professional-Development-Awards/IBRO-World-Congress-Travel-Awards
*Sponsor:* Society for Neuroscience
1121 14th Street, NW, Suite 1010
Washington, D.C. 20005

**SfN Young Investigator Award** 2297
The Young Investigator Award recognizes the outstanding achievements and contributions by a young neuroscientist who has recently received his or her advanced professional degree. The recipient receives a $15,000 award, along with complimentary ground or air transportation, hotel, and registration for the SfN annual meeting, where the awards are presented. Nominations must be submitted using SfN's award nomination site. Nominations for the award open on March 7 and close on June 3.
*Requirements:* Nominees must be active in research at the time the award is given, and have received a Ph.D. or M.D. (or equivalent) within the past ten years. Nominations must be made or endorsed by an SfN members, but nominees do not need to be SfN members.
*Restrictions:* Self-nomination is not permitted. No person may nominate more than one candidate. Current SfN officers and councilors are ineligible for nomination.
*Geographic Focus:* All States, Puerto Rico, Argentina, Australia, Brazil, Canada, China, Denmark, India, Israel, Malaysia, Mexico, New Zealand, Nigeria, Norway, Singapore, South Korea, Spain, Switzerland, Turkey, Ukraine, United Kingdom
*Date(s) Application is Due:* Jun 3
*Amount of Grant:* 15,000 USD
*Contact:* Catherine G Dulac, Ph.D.; (617) 495-7893; fax (617) 495-1819; dulac@fas.harvard.edu
*Internet:* www.sfn.org/Awards-and-Funding/Individual-Prizes-and-Fellowships/Young-Scientists-Achievements-and-Research/Young-Investigator-Award
*Sponsor:* Society for Neuroscience
1121 14th Street, NW, Suite 1010
Washington, D.C. 20005

**Shastri Indo-Canadian Institute Arts Fellowships** 2298
Arts fellowships are for visual and performing artists who are recognized and emerging practitioners of an Indian art form and wish to undergo training or engage in related activity in India to improve their skills or expand their repertoire; or are not practitioners of an Indian art form but wish to undergo training in India in an Indian art form to develop their own work in new, interesting, and creative ways. Arts Fellowships are for a minimum of four months and a maximum of twelve months. Artists at all stages of their career are welcome to apply. All fellowships include: one round-trip excursion airfare from the place of affiliation in Canada to the place of affiliation in India by the most direct and least expensive route; a living allowance of Rs 21,600 per month; an accountable research allowance of Rs 5,400 per month (receipts and expenditure statements required); and a reimbursement of affiliation fee up to a maximum of Rs 8,000 (to be reimbursed by the Shastri India Office on presentation of receipts).
*Restrictions:* Eligibility is restricted to citizens and permanent residents of Canada who are a minimum of 18 years of age.
*Geographic Focus:* Canada
*Date(s) Application is Due:* Jun 30
*Amount of Grant:* 81,000 - 324,000 INR
*Contact:* Mahmuda Aldeen; (403) 220-7467; fax (403) 289-0100; sici@ucalgary.ca
*Internet:* www.sici.org/programmes/details/arts-fellowship-2012-2013/
*Sponsor:* Shastri Indo-Canadian Institute
1402 Education Tower, 2500 University Drive NW
Calgary, AB T2N 1N4 Canada

**Shastri Indo-Canadian Institute Collaborative Research Project Grants** 2299
The Shastri Indo-Canadian Institute with funding support from the Department of Education and Ministry of Human Resource Development, invites proposals for Collaborative Research Projects. Applications from Indian and Canadian Shastri member institutions are welcome. The purpose of these projects is to fill information gaps for policy makers and others who influence policy-making processes. Grant applicants should demonstrate their ability to bring forward innovative ideas, as well as how they plan to share their findings with stake holders and at the grass roots level. The intention behind the projects should be to promote awareness among students and the public in the areas of: environment management; economic development and competitiveness; trade; democracy; rule of law; human rights; managing diversity; energy; public health; poverty alleviation; sustainable development; India-Canada relations; or gender equality.
*Requirements:* Indian and Canadian Shastri member institutions are eligible to apply. Applicants are required to name a partner institution in the other country. The partner institution may or may not be a Shastri member.
*Geographic Focus:* Canada, India
*Date(s) Application is Due:* Nov 1
*Contact:* Mahmuda Aldeen; (403) 220-7467; fax (403) 289-0100; sici@ucalgary.ca
*Internet:* www.sici.org/programmes/details/collaborative-research-projects-2011-2013/
*Sponsor:* Shastri Indo-Canadian Institute
1402 Education Tower, 2500 University Drive NW
Calgary, AB T2N 1N4 Canada

**Shastri Indo-Canadian Institute Doctoral Student Research Fellowships** 2300
Doctoral Student Research Fellowships are awarded to students enrolled in Doctoral programs outside India with a significant India Studies focus, to enable them to conduct research in India towards their degree. Students must be affiliated with an Indian institution. All fellowships include: one round-trip excursion airfare from the place of affiliation in Canada to the place of affiliation in India by the most direct and least expensive route; a living allowance of Rs 21,600 per month; an accountable research allowance of Rs 5,400 per month (receipts and expenditure statements required); and a reimbursement of affiliation fee up to a maximum of Rs 8,000 (to be reimbursed by the Shastri India Office on presentation. The Institute may offer support for more than one year provided satisfactory standing and progress are maintained. Student Graduate Degree Fellowships are for a minimum of three months to a maximum of twelve months. The awards are made for the April 1st through March 31st period. Fellowships for successful applicants are subject to approval of their research topics/areas by the Indian Consulate or High Commission closest to their city of residence.
*Requirements:* Students must be registered in a graduate program. Canadian citizens enrolled in graduate programs outside Canada are eligible for student fellowships.
*Restrictions:* Eligibility is restricted to citizens and permanent residents of Canada who are a minimum of 18 years of age. Applicants who have accepted a grant or award from the Shastri Institute within the last 12 months are not eligible to apply. Individuals who have received previous grants or awards from the Shastri Institute and have yet to submit their final reports are not eligible to apply.
*Geographic Focus:* Canada
*Date(s) Application is Due:* Jun 30
*Amount of Grant:* 81,000 - 324,000 INR
*Contact:* Mahmuda Aldeen; (403) 220-7467; fax (403) 289-0100; sici@ucalgary.ca
*Internet:* www.sici.org/programmes/details/doctoral-student-research-fellowship-2012-2013/
*Sponsor:* Shastri Indo-Canadian Institute
1402 Education Tower, 2500 University Drive NW
Calgary, AB T2N 1N4 Canada

**Shastri Indo-Canadian Institute Faculty Research Fellowships** 2301
Faculty fellowships support research and training in India by Canadian scholars with full- or part-time appointments in a Canadian institution of higher education. Retired faculty who maintain affiliation with a Canadian institution of higher education are eligible to apply if, upon return to Canada, they will continue to contribute to India

studies through either teaching or research. Fellowships are awarded on the basis of detailed proposals, which are peer reviewed. Faculty research fellowships are for scholars who have demonstrated excellence in a specific field of India studies. Faculty training fellowships are for scholars who have had little or no previous involvement with India studies and wish to undertake an intensive program of study at an Indian institution; wish to change their disciplinary focus within India studies; wish to develop or enhance their knowledge of an Indian language; or wish to carry out comparative research in India. All fellowships include: one round-trip excursion airfare from the place of affiliation in Canada to the place of affiliation in India by the most direct and least expensive route; a living allowance of Rs 21,600 per month; an accountable research allowance of Rs 5,400 per month (receipts and expenditure statements required); and a reimbursement of affiliation fee up to a maximum of Rs 8,000 (to be reimbursed by the Shastri India Office on presentation of receipts).
*Requirements:* Eligibility is restricted to citizens and permanent residents of Canada who are a minimum of 18 years of age. Eligible Scholars hold full or part-time appointments in an institution of higher education. Retired faculty who maintain affiliation with an institution of higher education are eligible to apply if upon return to Canada they will continue to contribute to India Studies through either teaching or research.
*Restrictions:* Applicants who have accepted a grant or award from the Shastri Institute within the last 12 months are not eligible to apply. Individuals who have received previous grants or awards from the Shastri Institute and have yet to submit their final reports are not eligible to apply.
*Geographic Focus:* Canada
*Date(s) Application is Due:* Jun 30
*Amount of Grant:* 81,000 - 324,000 INR
*Contact:* Mahmuda Aldeen; (403) 220-7467; fax (403) 289-0100; sici@ucalgary.ca
*Internet:* www.sici.org/programmes/details/faculty-research-fellowship-2012-2013/
*Sponsor:* Shastri Indo-Canadian Institute
1402 Education Tower, 2500 University Drive NW
Calgary, AB T2N 1N4 Canada

**Shastri Indo-Canadian Institute Faculty Training Fellowships** **2302**
Faculty Training Fellowships are for scholars: who have had little or no previous involvement with India Studies and wish to undertake an intensive program of study at an Indian institution; who wish to change their disciplinary focus within India Studies; who wish to develop or enhance their knowledge of an Indian Language; and who wish to carry out comparative research in India. Academic Fellowships are for a minimum of three to a maximum of twelve months. All fellowships include: one round-trip excursion airfare from the place of affiliation in Canada to the place of affiliation in India by the most direct and least expensive route; a living allowance of Rs 21,600 per month; an accountable research allowance of Rs 5,400 per month (receipts and expenditure statements required); and a reimbursement of affiliation fee up to a maximum of Rs 8,000 (to be reimbursed by the Shastri India Office on presentation of receipts). Faculty Training Fellowships are for a minimum of three months to a maximum of twelve months. The awards are made for the April 1st through March 31st period. Fellowships for successful applicants are subject to approval of their research topics/areas by the Indian Consulate or High Commission closest to their city of residence.
*Requirements:* Faculty Fellowships support research and training in India by Canadian scholars with full- or part-time appointments in a Canadian institution of higher education. Retired faculty who maintain affiliation with a Canadian institution of higher education are eligible to apply if upon return to Canada they will continue to contribute to India studies through either teaching or research. They are awarded on the basis of detailed proposals, which are peer reviewed.
*Restrictions:* Eligibility is restricted to citizens and permanent residents of Canada who are a minimum of 18 years of age. Applicants who have accepted a grant or award from the Shastri Institute within the last 12 months are not eligible to apply. Individuals who have received previous grants or awards from the Shastri Institute and have yet to submit their final reports are not eligible to apply.
*Geographic Focus:* Canada
*Date(s) Application is Due:* Jun 30
*Amount of Grant:* 81,000 - 324,000 INR
*Contact:* Mahmuda Aldeen; (403) 220-7467; fax (403) 289-0100; sici@ucalgary.ca
*Internet:* www.sici.org/programmes/details/faculty-training-fellowship-2012-2013/
*Sponsor:* Shastri Indo-Canadian Institute
1402 Education Tower, 2500 University Drive NW
Calgary, AB T2N 1N4 Canada

**Shastri Indo-Canadian Institute Graduate Degree Fellowships from an Indian Institution** **2303**
Student Graduate Degree Fellowships are given to students who wish to enroll in an Indian institution of higher learning to obtain a graduate degree. Applicants should seek confirmation from the India Office that the Government of India approves the proposed institution as a degree-granting body and that their current Canadian university supports their proposed Indian program and will allow transfer of credits. The applicant must have the support of the current chairperson of the department of study in Canada. Students should note that a Canadian university might not accept an Indian degree and courses as equivalent. All fellowships include: one round-trip excursion airfare from the place of affiliation in Canada to the place of affiliation in India by the most direct and least expensive route; a living allowance of Rs 21,600 per month; an accountable research allowance of Rs 5,400 per month (receipts and expenditure statements required); and a reimbursement of affiliation fee up to a maximum of Rs 8,000 (to be reimbursed by the Shastri India Office on presentation. The Institute may offer support for more than one year provided satisfactory standing and progress are maintained. Student Graduate Degree Fellowships are for a minimum of six months to a maximum of twelve months (extension could be offered under special circumstances). The awards are made for the April 1st through March 31st period. Fellowships for successful applicants are subject to approval of their research topics/areas by the Indian Consulate or High Commission closest to their city of residence.
*Requirements:* Eligibility is restricted to citizens and permanent residents of Canada who are a minimum of 18 years of age. Students must be registered in a graduate program (Canadian citizens enrolled in graduate programs outside Canada are eligible for student fellowships).
*Restrictions:* Applicants who have accepted a grant or award from the Shastri Institute within the last 12 months are not eligible to apply. Individuals who have received previous grants or awards from the Shastri Institute and have yet to submit their final reports are not eligible to apply.
*Geographic Focus:* Canada
*Date(s) Application is Due:* Jun 30
*Amount of Grant:* 81,000 - 324,000 INR
*Contact:* Mahmuda Aldeen; (403) 220-7467; fax (403) 289-0100; sici@ucalgary.ca
*Internet:* www.sici.org/graduate-degree-fellowships-degree-from-an-indian-institution-2012-2013/
*Sponsor:* Shastri Indo-Canadian Institute
1402 Education Tower, 2500 University Drive NW
Calgary, AB T2N 1N4 Canada

**Shastri Indo-Canadian Institute Librarian Fellowships** **2304**
Librarian Fellowships are awarded for Canadian university librarians to undertake: an intensive program of study at an Indian Institution; enhancement of knowledge of and Indian Language; or enhancement of research support for south Asian studies. Academic Fellowships are for a minimum of three to a maximum of twelve months. Fellowships for successful applicants are subject to approval of their research topics/areas by the Indian Consulate or High Commission closest to their city of residence. All fellowships include: one round-trip excursion airfare from the place of affiliation in Canada to the place of affiliation in India by the most direct and least expensive route; a living allowance of Rs 21,600 per month; an accountable research allowance of Rs 5,400 per month (receipts and expenditure statements required); and a reimbursement of affiliation fee up to a maximum of Rs 8,000 (to be reimbursed by the Shastri India Office on presentation of receipts). Librarian Fellowships are for a minimum of three months to a maximum of twelve months. The awards are made for the April 1st to March 31st period. Fellowships for successful applicants are subject to approval of their research topics/areas by the Indian Consulate or High Commission closest to their city of residence.
*Requirements:* Faculty Fellowships support research and training in India by Canadian scholars with full- or part-time appointments in a Canadian institution of higher education. Retired faculty who maintain affiliation with a Canadian institution of higher education are eligible to apply if upon return to Canada they will continue to contribute to India studies through either teaching or research. They are awarded on the basis of detailed proposals, which are peer reviewed.
*Restrictions:* Eligibility is restricted to citizens and permanent residents of Canada who are a minimum of 18 years of age. Applicants who have accepted a grant or award from the Shastri Institute within the last 12 months are not eligible to apply. Individuals who have received previous grants or awards from the Shastri Institute and have yet to submit their final reports are not eligible to apply.
*Geographic Focus:* Canada
*Date(s) Application is Due:* Jun 30
*Amount of Grant:* 81,000 - 324,000 INR
*Contact:* Mahmuda Aldeen; (403) 220-7467; fax (403) 289-0100; sici@ucalgary.ca
*Internet:* www.sici.org/programmes/details/librarian-fellowship-2012-2013/
*Sponsor:* Shastri Indo-Canadian Institute
1402 Education Tower, 2500 University Drive NW
Calgary, AB T2N 1N4 Canada

**Shastri Indo-Canadian Institute Master's Student Research Fellowships** **2305**
Master's Student Research Fellowships are awarded to students enrolled in Master's programs outside India with a significant India Studies focus, to enable them to conduct research in India towards their degree. Students should be affiliated with an Indian institution.
*Requirements:* Eligible are Canadian citizens and landed immigrants with full- or part-time appointments in Canadian institutions who plan to return to Canada following the tenure of the fellowship. All fellowships include: one round-trip excursion airfare from the place of affiliation in Canada to the place of affiliation in India by the most direct and least expensive route; a living allowance of Rs 21,600 per month; an accountable research allowance of Rs 5,400 per month (receipts and expenditure statements required); and a reimbursement of affiliation fee up to a maximum of Rs 8,000 (to be reimbursed by the Shastri India Office on presentation of receipts).
*Restrictions:* Eligibility is restricted to citizens and permanent residents of Canada who are a minimum of 18 years of age.
*Geographic Focus:* Canada
*Date(s) Application is Due:* Jun 30
*Amount of Grant:* 81,000 - 324,000 INR
*Contact:* Mahmuda Aldeen; (403) 220-7467; fax (403) 289-0100; sici@ucalgary.ca
*Internet:* www.sici.org/programmes/details/masters-student-research-fellowship-2012-2013/
*Sponsor:* Shastri Indo-Canadian Institute
1402 Education Tower, 2500 University Drive NW
Calgary, AB T2N 1N4 Canada

**Shastri Indo-Canadian Institute Millennium Development Research Grants** **2306**
Funds will be provided to a Canadian or Indian member university on behalf of a binational research teams to undertake public policy relevant research toward one of the eight United Nations Millennium Development Goals. Grant recipients will be required

to disseminate their findings, both to the academic and to the broader community. Acceptable expenses include: travel and accommodation for participants, meals, communications costs (audio, video conferencing, promotional materials), organizational support, translation, copying and printing, dissemination of research, research assistant allowance, and reasonable out of pocket expenses. Travel expenses may either be for Canadians going to India or for Indians coming from India or domestic travel for research purposes. Total awards are $80,000 maximum.
*Requirements:* The lead applicant may be an individual researcher/faculty member or research team affiliated with a Canadian or Indian member institution of the Shastri Indo-Canadian Institute. Applicant must have a partner in the other country. Grant recipients will be required to integrate consideration of gender equality and environmental sustainability in their research model. (Preferences will be given to: applicants who partner with business, and or government; and applicants who also partner with business, civil society organizations and/or government.
*Restrictions:* The Shastri Institute will not pay institutional overhead.
*Geographic Focus:* Canada
*Date(s) Application is Due:* Aug 31
*Amount of Grant:* Up to 80,000 CAD
*Contact:* Mahmuda Aldeen; (403) 220-7467; fax (403) 289-0100; sici@ucalgary.ca
*Internet:* www.sici.org/programmes/details/shastri-millennium-development-research-grants/
*Sponsor:* Shastri Indo-Canadian Institute
1402 Education Tower, 2500 University Drive NW
Calgary, AB T2N 1N4 Canada

**Shastri Indo-Canadian Institute Partnership Development Seed Grants 2307**
The purpose of the grant is to be a catalyst for scholarly activities which will foster the creation of new knowledge and contribute to formal institutional research agreements or other long-term partnerships. Funds may be used to support a binational seminar, conference, meeting, or other activities that: build awareness of potential collaboration opportunities with Indian institutions; promote sharing of knowledge, information, resources and expertise; showcase existing Canada-India research collaborations and their impact on society; and assist researchers to take advantage of Canada-India collaborative opportunities. The Shastri Institute will provide five grants to Canadian or Indian institutions of higher learning to support catalyst activities such as workshops, seminars, conferences, scholarly discussions and planning meetings that are expected to lead to significant research collaborations with universities or research institutions from the other country. Acceptable expenses include: travel and accommodation for speakers and participants, meals, communication costs (audio, video conferencing, promotional materials), organizational support, translation, copying and printing. Travel expenses may either be for Canadians going to India or for Indians coming to Canada. Applications which promote partnerships with business, civil society organizations and/ or government are especially welcome.
*Requirements:* Individual researcher/faculty member or research teams affiliated with a Canadian or Indian university are eligible. The lead applicant must be employed or affiliated with a Canadian or Indian member institution.
*Restrictions:* The Shastri Institute will not pay institutional overhead.
*Geographic Focus:* Canada
*Amount of Grant:* 10,000 USD
*Contact:* Mahmuda Aldeen; (403) 220-7467; fax (403) 289-0100; sici@ucalgary.ca
*Internet:* www.sici.org/programmes/details/partnership-development-seed-grant-2011-2012/
*Sponsor:* Shastri Indo-Canadian Institute
1402 Education Tower, 2500 University Drive NW
Calgary, AB T2N 1N4 Canada

**Shastri Indo-Canadian Institute Post-Doctoral Research Fellowships 2308**
Post-doctoral Research Fellowships are open to scholars who have completed a Ph.D. and wish to do research in India. Applicants need not have an academic position. These awards will not normally be made if it has been greater than five years since completion of a doctoral degree. All fellowships include: one round-trip excursion airfare from the place of affiliation in Canada to the place of affiliation in India by the most direct and least expensive route; a living allowance of Rs 21,600 per month; an accountable research allowance of Rs 5,400 per month (receipts and expenditure statements required); and a reimbursement of affiliation fee up to a maximum of Rs 8,000 (to be reimbursed by the Shastri India Office on presentation. The Institute may offer support for more than one year provided satisfactory standing and progress are maintained. Student Graduate Degree Fellowships are for a minimum of three months to a maximum of twelve months (extension could be offered under special circumstances). The awards are made for the April 1st through March 31st period. Fellowships for successful applicants are subject to approval of their research topics/areas by the Indian Consulate or High Commission closest to their city of residence.
*Requirements:* Eligibility is restricted to citizens and permanent residents of Canada who are a minimum of 18 years of age.
*Restrictions:* Applicants who have accepted a grant or award from the Shastri Institute within the last 12 months are not eligible to apply. Individuals who have received previous grants or awards from the Shastri Institute and have yet to submit their final reports are not eligible to apply.
*Geographic Focus:* Canada
*Date(s) Application is Due:* Jun 30
*Amount of Grant:* 81,000 - 324,000 INR
*Contact:* Mahmuda Aldeen; (403) 220-7467; fax (403) 289-0100; sici@ucalgary.ca
*Internet:* www.sici.org/programmes/details/post-doctoral-fellowship-2012-2013/
*Sponsor:* Shastri Indo-Canadian Institute
1402 Education Tower, 2500 University Drive NW
Calgary, AB T2N 1N4 Canada

**Shastri Indo-Canadian Institute Research Collaboration Development Grants 2309**
The Shastri Indo Canadian Institute announces a new institutional grant program for its Canadian and Indian Member Institutions, the Institutional Research Collaboration Development Grant (IRCDG). The objective of IRCDG is to support the Institute's research collaboration programs in the area of socio-economic development with focus on sustainable development. Themes of particular interests to the Institute are food security, child and youth education and employment, environment and energy, gender and health, and good governance. The program, made possible by the funding that the Institute receives from the Canadian International development Agency (CIDA), is to encourage new institutional investments in these areas of research. Small matching grants of up to CAD $10,000 are available to selected member institutions with commitment of equivalent amount of funds to initiate and develop research projects or programs that address CIDA's development priorities, are based on bi-national collaboration, and will lead to future research in these broad areas.
*Requirements:* The initial phase of the collaboration will have to be completed within three months. Canadian and Indian member institutions are eligible to apply for this grant. They may submit only one application (whether they are the lead institution or not). Other academic or development partners may be involved in a project but do not qualify as co-applicants. Either applicant, in India or in Canada, can be the lead applicant.
*Geographic Focus:* Canada
*Date(s) Application is Due:* Sep 30
*Amount of Grant:* Up to 10,000 USD
*Contact:* Mahmuda Aldeen; (403) 220-7467; fax (403) 289-0100; sici@ucalgary.ca
*Internet:* www.sici.org/programmes/details/shastri-institutional-research-collaboration-development-grant/
*Sponsor:* Shastri Indo-Canadian Institute
1402 Education Tower, 2500 University Drive NW
Calgary, AB T2N 1N4 Canada

**Shastri Indo-Canadian Institute Student Language Training Fellowships 2310**
Student Language Training Fellowships are awarded to students enrolled in graduate degree programs outside India, who have taken training in an Indian language in Canada and wish to pursue advanced studies in that language in India. The award is tenable only at recognized educational institutions that award appropriate diplomas or certificates. Student Language Training Fellowships are for a minimum of three months to a maximum of twelve months. The awards are made for the April 1st to March 31st period. Fellowships for successful applicants are subject to approval of their research topics/areas by the Indian Consulate or High Commission closest to their city of residence. All fellowships include: one round-trip excursion airfare from the place of affiliation in Canada to the place of affiliation in India by the most direct and least expensive route; a living allowance of Rs 21,600 per month; an accountable research allowance of Rs 5,400 per month (receipts and expenditure statements required); and a reimbursement of affiliation fee up to a maximum of Rs 8,000 (to be reimbursed by the Shastri India Office on presentation of receipts).
*Requirements:* Eligibility is restricted to citizens and permanent residents of Canada who are a minimum of 18 years of age. Students must be registered in a graduate program (Canadian citizens enrolled in graduate programs outside Canada are eligible for student fellowships).
*Restrictions:* Applicants who have accepted a grant or award from the Shastri Institute within the last 12 months are not eligible to apply. Individuals who have received previous grants or awards from the Shastri Institute and have yet to submit their final reports are not eligible to apply.
*Geographic Focus:* Canada
*Date(s) Application is Due:* Jun 30
*Amount of Grant:* 8,100 - 324,000 INR
*Contact:* Mahmuda Aldeen; (403) 220-7467; fax (403) 289-0100; sici@ucalgary.ca
*Internet:* www.sici.org/programmes/details/student-language-training-2012-2013/
*Sponsor:* Shastri Indo-Canadian Institute
1402 Education Tower, 2500 University Drive NW
Calgary, AB T2N 1N4 Canada

**Shastri Indo-Canadian Institute University of Ottawa Overseas India Summer Course Grants 2311**
Shastri Indo-Canadian Institute and The University of Ottawa in affiliation with the University of Mumbai announce the University of Ottawa Overseas India Summer Course, with a focus on the religious and cultural tradition of Jainism. This unique program will provide students from all over Canada an opportunity to participate in a cultural and educational immersion in India. In India, the program will be based at the University of Mumbai with students residing on the U of M campus. Students will also spend several days in Varanasi, as well as visit the Ajanta-Ellora caves. The dates for the program are: May 25-June 17. Students in the program are eligible to receive 3 credits (graduate or undergraduate credits, depending on their degree programs) through the accrediting host institution, the University of Ottawa. The reduced cost rate of the program is the cost of registration for the University of Ottawa 3-credit course (approx. $650), plus $2500 for the India portion which includes application fees, airfare to and from India, health insurance, accommodation, three vegetarian meals a day, flights/ trains/buses within India, course materials, field trip costs.
*Requirements:* Graduate Students with an interest in India Studies and advanced Undergraduate Students who have completed 3 years of undergraduate study, including some component of India Studies coursework (in Religious Studies, History, or Philosophy etc.) or have a particular academic interest in India Studies are eligible.
*Geographic Focus:* Canada
*Date(s) Application is Due:* Dec 15
*Contact:* Mahmuda Aldeen; (403) 220-7467; fax (403) 289-0100; sici@ucalgary.ca

*Internet:* www.sici.org/programmes/details/shastri-institute-university-of-ottawa-study-jainism-in-india-summer-co/
*Sponsor:* Shastri Indo-Canadian Institute
1402 Education Tower, 2500 University Drive NW
Calgary, AB T2N 1N4 Canada

**Shelley and Donald Rubin Foundation Grants** **2312**
The foundation's mission is to support innovative efforts to transform society's institutions and to empower individuals to develop their full potential. The foundation is primarily interested in supporting the inclusion of art from non-Western European cultures into the mainstream of scholarship and display. In addition, the foundation supports research, action, and other projects designed to reveal and understand barriers to the full access of all people to American society and the larger international community. Areas of particular interest include, but are not limited to: access to health care; AIDS and its effects on society's institutions; the celebration of ethnic and cultural diversity that simultaneously encourages intergroup understanding; and cultural and arts programs that encourage individual and community identity. Projects in other areas that are most often funded are those that in themselves serve as catalysts for social change, addressing emerging problems as well as better known, long-standing problems in new ways. On a limited basis, programs eligible for funding may also include evaluation, technical assistance, and demonstration projects leading to the development of innovative models. Some examples of technical assistance of interest are the establishment of websites, the development of web/TV productions, and related outreach activities. Most recent awards have ranged from $2,500 to $100,000.
*Requirements:* Only organizations with tax-exempt status under U.S. Internal Revenue Code, or those with qualifying fiscal sponsors, will be considered. Organizations must be active in the five boroughs of New York City. Potential grantees must be organizations that supply one or more of the following to audiences: arts education; public art; art in community and service centers; artistic activism; community-based museums; art in the service of social justice or change; and promotion of under-recognized artistic practice.
*Restrictions:* Grants do not support individuals, capital projects, endowments, scholarships, fellowships, or fund raising activities.
*Geographic Focus:* New York
*Amount of Grant:* 2,500 - 100,000 USD
*Contact:* Alexander Gardner, Executive Director; (212) 780-2035; erich@sdrubin.org
*Internet:* www.sdrubin.org/guidelines.htm
*Sponsor:* Shelley and Donald Rubin Foundation
17 West 17th Street, 9th Floor
New York, NY 10011

**Shell Oil Company Foundation Environment Grants** **2313**
The mission of the Shell Oil Company Foundation is to help foster the general well being of communities where Shell Oil Company employees live and work and to provide educational opportunities that prepare students and faculty to succeed while meeting the needs of the ever-changing workplace. Focus areas for funding include: community development; education; and the environment. In the area of the environment, the focus is on biodiversity initiatives with support to programs that restore critical eco systems, address water, air quality research, preserve wetlands and sponsor wetlands initiatives. As part of the Foundation's commitment to environmental stewardship, it supports projects that restore and protect critical ecosystems. In addition to restoration and preservation efforts, the Foundation funds research projects for threatened wildlife and/or habitats.
*Requirements:* Shell will consider charitable contributions to eligible nonprofit organizations with priority consideration given to organizations serving in or near U.S. communities where Shell has a major presence.
*Restrictions:* Shell Oil Company and the Shell Oil Company Foundation will not consider contributions for the following purposes: individuals; private foundations; non-profit organizations without a current 501(c)3 exempt status; conferences or symposia; endowment funds; fraternal and labor organizations; capital campaigns; conferences, workshops, or seminars not directly related to Shell business interests; religious organizations that do not serve the general public on a non-denominational basis; organizations located in or benefiting nations other than U.S. and its territories; or organizational operating expenses.
*Geographic Focus:* All States
*Amount of Grant:* 10,000 - 50,000 USD
*Contact:* Nancy Tootle, External Relations Social Investment Manager; (832) 337-0635 or (832) 337-2034; nancy.tootle@shell.us
*Internet:* www.shell.us/sustainability/request-for-a-grant-from-shell.html
*Sponsor:* Shell Oil Company Foundation
One Shell Plaza 910 Louisiana, Suite 4478A
Houston, TX 77002

**SHOT Abbott Payson Usher Prize** **2314**
The SHOT Abbott Payson Usher Prize was established to encourage the publication of original research of the highest standard. It is awarded annually to the author of the best scholarly work published during the preceding three years under the auspices of the Society for the History of Technology. A certificate is included with the $400 prize.
*Geographic Focus:* All States
*Amount of Grant:* 400 USD
*Contact:* Bernie Carlson; (434) 987-6230; fax (434) 975-2190; shot@virginia.edu
*Internet:* www.historyoftechnology.org/awards/usher.html
*Sponsor:* Society for the History of Technology
P.O. Box 400744
Charlottesville, VA 22904-4744

**SHOT Brooke Hindle Postdoctoral Fellowships** **2315**
THIS Fellowship in the History of Technology honors the contribution of Brooke Hindle to the work of the Society for the History of Technology. The fellowship is for $10,000 and may be used for any purpose connected with research or writing in the history of technology for a period of not less than four months during the year following the award. Applicants should be intending either to prepare a dissertation for publication as articles or as a monograph, whether or not this involves fresh primary research, or to develop a new project based on primary research.
*Requirements:* Applicants must hold a doctorate in the history of technology or a related field, normally awarded within the preceding four years, or expect to have graduated by the time of the award. Applications must be in English.
*Geographic Focus:* All States
*Date(s) Application is Due:* Apr 15
*Amount of Grant:* 10,000 USD
*Contact:* shot@virginia.edu, Secretary; (434) 987-6230; fax (434) 975-2190; shot@virginia.edu or shot@virginia.edu
*Internet:* www.historyoftechnology.org/awards/hindle.html
*Sponsor:* Society for the History of Technology
P.O. Box 400744
Charlottesville, VA 22904-4744

**SHOT Dibner Award for Excellence in Museum Exhibits** **2316**
The Dibner Award for Excellence in Museum Exhibits was established in 1985 to recognize excellence in museums and museum exhibits that interpret the history of technology, industry, and engineering to the general public. Winning exhibits, in addition to being well designed and produced, should raise pertinent historical issues. Artifacts and images should be used in a manner that interests, teaches, and stimulates both the general public and historians. The award consists of a plaque and up to $1,000 to cover expenses for a member of the design team to accept the award at the SHOT awards banquet. The Society especially encourages nominations from local and regional historical societies. Virtual exhibits are not eligible for the award.
*Requirements:* Anyone, including the institution or individual responsible for its creation, may nominate an exhibit for the Dibner Award.
*Restrictions:* Exhibits are eligible for this award if they have been open to the public for no more than 24 months before the deadline for nominations.
*Geographic Focus:* All States
*Date(s) Application is Due:* May 1
*Amount of Grant:* 1,000 USD
*Contact:* Bernie Carlson; (434) 987-6230; fax (434) 975-2190; shot@virginia.edu
*Internet:* www.historyoftechnology.org/awards/dibner.html
*Sponsor:* Society for the History of Technology
P.O. Box 400744
Charlottesville, VA 22904-4744

**SHOT Eugene S. Ferguson Prize** **2317**
The Eugene S. Ferguson Prize is awarded by SHOT for outstanding and original reference work that will support future scholarship in the history of technology. The Ferguson Prize recognizes work that is in the tradition of scholarly excellence established by Eugene S. Ferguson (1916-2004), SHOT's pioneering bibliographer, a founding member of the Society (President, 1977-78; da Vinci Medalist, 1977), museum curator and exhibit catalog author, editor, annotator, university professor, and scholar of the history of engineering and technology. The prize consists of a plaque and a cash award of $2,500.
*Requirements:* Works published in the four calendar years prior to the year of the award are eligible for consideration. Works that extend beyond the printed text, including other media, such as CDs, World Wide Web sites, and electronic data bases and tools, keeping in mind that the principal criterion of support [of] future scholarship for judging excellence requires that such works exhibit a degree of permanence similar to that of books.
*Geographic Focus:* All States
*Date(s) Application is Due:* Apr 1
*Amount of Grant:* 2,500 USD
*Contact:* Bernie Carlson; (434) 987-6230; fax (434) 975-2190; shot@virginia.edu
*Internet:* www.historyoftechnology.org/awards/ferguson.html
*Sponsor:* Society for the History of Technology
P.O. Box 400744
Charlottesville, VA 22904-4744

**SHOT IEEE Life Members Prize in Electrical History** **2318**
This prize is awarded annually to the best paper in electrical history published during the previous year. The prize is sponsored by the life members of the Institute of Electrical and Electronics Engineers (IEEE). The competition is open to any historical paper published in a learned journal or magazine if it treats the art or engineering aspects of electrotechnology and its practitioners. Electrotechnology encompasses power, electronics, telecommunications, and computer science. The cash portion of the prize is shared among joint authors; individual certificates are presented to each author. The prize consists of a cash award of $500 and a certificate.
*Requirements:* The article must be written in English.
*Geographic Focus:* All States
*Date(s) Application is Due:* Apr 15
*Amount of Grant:* 500 USD
*Contact:* Bernie Carlson; (434) 987-6230; fax (434) 975-2190; shot@virginia.edu
*Internet:* www.historyoftechnology.org/awards/ieee.html
*Sponsor:* Society for the History of Technology
P.O. Box 400744
Charlottesville, VA 22904-4744

**SHOT International Scholars** 2319

The SHOT International Scholars program is intended to foster a stronger international community for the study of the history of technology by identifying and helping non-U.S. historians of technology to participate in the meetings and governance of the society and receive recognition from their respective governments and national academic communities. Appointments will be for two-year terms beginning each January. International scholars will be made full members of the society and receive complimentary subscriptions to the society's journal. They will be strongly encouraged to attend the annual meetings during their terms. In addition, they will be asked to prepare at least one report or essay review on the history of technology scene in their country for possible presentation to, or publication by, the society. Up to seven scholars are appointed each year.
*Requirements:* Scholars must reside and work in underrepresented countries outside the United States; they may be both junior and more advanced scholars. An individual may be nominated, or may nominate her/himself, by sending a letter containing a brief statement about the individual (current status, employment, field of interest and research, etc.), a brief curriculum vita, a list of publications, and a fax number or email address.
*Geographic Focus:* All Countries
*Date(s) Application is Due:* May 30
*Contact:* Bernie Carlson; (434) 987-6230; fax (434) 975-2190; shot@virginia.edu
*Internet:* www.historyoftechnology.org/international_scholars.html
*Sponsor:* Society for the History of Technology
P.O. Box 400744
Charlottesville, VA 22904-4744

**SHOT Joan Cahalin Robinson Prize** 2320

This prize is awarded annually for the best paper presented at the SHOT annual meeting by a historian under 30 years of age or by a person past 30 who is an accredited graduate student or candidate for a higher degree and who is presenting his or her first paper at a SHOT annual meeting. Candidates for the award are judged not only on the quality of the historical research and scholarship demonstrated in the written papers submitted to the Robinson Prize Committee before the meeting, but also on the effectiveness of their oral presentations. A certificate is included with the prize.
*Geographic Focus:* All States
*Date(s) Application is Due:* Jul 1
*Amount of Grant:* 350 USD
*Contact:* Bernie Carlson; (434) 987-6230; fax (434) 975-2190; shot@virginia.edu
*Internet:* www.historyoftechnology.org/awards/robinson.html
*Sponsor:* Society for the History of Technology
P.O. Box 400744
Charlottesville, VA 22904-4744

**SHOT Leonardo da Vinci Medal** 2321

The highest recognition from the Society for the History of Technology is the Leonardo da Vinci Medal, presented to an individual who has made an outstanding contribution to the history of technology, through research, teaching, publications, and other activities. Andras Beck (formerly of the Hungarian Academy of Arts) designed the medal, the face of which shows Leonardo's head modeled after the artist's self-portrait. The reverse design shows (in the words of the sculptor) "the basic sources of energy: water, wind, and fire." The prize consists of a medal and a certificate. SHOT invites members to nominate candidates for the DaVinci Medal. To put forward a person's name for consideration, send a letter of nomination, describing the person's outstanding contributions to the history of technology.
*Geographic Focus:* All States
*Contact:* Bernie Carlson; (434) 987-6230; fax (434) 975-2190; shot@virginia.edu
*Internet:* www.historyoftechnology.org/awards/davinci.html
*Sponsor:* Society for the History of Technology
P.O. Box 400744
Charlottesville, VA 22904-4744

**SHOT Melvin Kranzberg Dissertation Fellowship** 2322

The Melvin Kranzberg Dissertation Fellowship is presented annually to a doctoral student engaged in the preparation of a dissertation on the history of technology, broadly defined. This award is in memory of the co-founder of the Society and honors Melvin Kranzberg's many contributions to developing the history of technology as a field of scholarly endeavor and SHOT as a professional organization. The $4,000 award is unrestricted and may be used in any way that the winner chooses to advance the research and writing of his or her dissertation. Possible uses include underwriting the costs of travel to archival collections; photocopying or microfilming; translation of documents; and so on.
*Requirements:* Students from institutions of higher learning anywhere in the world who are working on projects in the history of technology are eligible to apply; doctoral candidates from outside the United States are especially encouraged to submit application materials. Applicants must have completed all requirements for their doctorate except for the dissertation by September 1 of the year the award is made.
*Restrictions:* The award may not be used for university tuition or fees.
*Geographic Focus:* All States, All Countries
*Date(s) Application is Due:* Apr 15
*Amount of Grant:* 4,000 USD
*Contact:* Bernie Carlson; (434) 987-6230; fax (434) 975-2190; shot@virginia.edu
*Internet:* www.historyoftechnology.org/awards/kranzberg.html
*Sponsor:* Society for the History of Technology
P.O. Box 400744
Charlottesville, VA 22904-4744

**SHOT NASA Fellowship** 2323

The NASA Fellowship in the History of Space Technology, offered by SHOT and supported by the National Aeronautics and Space Administration (NASA) History Division, will fund one predoctoral or postdoctoral fellow for up to one academic year to undertake a research project related to the history of space technology. The fellowship may support advanced research related to all aspects of space history, leading to publications on the history of space technology broadly considered, including cultural and intellectual history, institutional history, economic history, history of law and public policy, and history of engineering and management. The fellowship carries a stipend of $21,500, paid quarterly. Fellows will carry out their research projects using personal office space, equipment, and supplies.
*Requirements:* Applicants must possess a doctorate in history of technology or in a closely related field, or be enrolled as a student in a doctoral degree program and have completed all requirements for the Ph.D., except the dissertation, in history of technology or a related field. Eligibility is not limited to U.S. citizens or residents.
*Restrictions:* Funds may not be used to support tuition or fees.
*Geographic Focus:* All States
*Date(s) Application is Due:* Apr 1
*Amount of Grant:* 21,500 USD
*Contact:* David Lucsko; (334) 844-6770; fax (334) 844-6673; shotsec@auburn.edu
*Internet:* www.historyoftechnology.org/awards/nasa.html
*Sponsor:* Society for the History of Technology
P.O. Box 400744
Charlottesville, VA 22904-4744

**SHOT Samuel Eleazar and Rose Tartakow Levinson Prize** 2324

This Samuel Eleazar and Rose Tartakow Levinson Prize is awarded annually by the Society for the History of Technology for an original essay in the history of technology that is the author's first work intended for publication. The essay must explicitly examine in some detail a technology or technological device within the framework of social or intellectual history. To nominate an essay, all materials should be forwarded by email in MS Word or PDF format.
*Requirements:* Manuscripts already published or accepted for publication are not eligible. In order to be considered, manuscripts must be in English and of a length suitable for publication as an article in Technology and Culture-approximately 7,500 words.
*Geographic Focus:* All States
*Date(s) Application is Due:* Apr 15
*Amount of Grant:* 450 USD
*Contact:* Bernie Carlson; (434) 987-6230; fax (434) 975-2190; shot@virginia.edu
*Internet:* www.historyoftechnology.org/awards/levinson.html
*Sponsor:* Society for the History of Technology
P.O. Box 400744
Charlottesville, VA 22904-4744

**SHRM Foundation Research Grants** 2325

The Society for Human Resource Management (SHRM) Foundation seeks to fund high-impact research on all aspects of human resource management with a focus on addressing current challenges or understanding emerging trends. Any topic may be funded however the research must be aimed at an academic audience while also having direct, actionable implications for HR practice. It is crucial that there is continuity between the research questions proposed and the methods used. The research should also be able to reasonably generalize across people and settings. Investigators should include a statement in the proposal regarding the individuals, groups, industry sectors or countries for which their findings are expected to generalize, given the sample and study design.
*Requirements:* The Foundation welcomes proposals from any country. Submissions must be written in English and applications will be read by reviewers whose first language is English. There is no requirement that researchers be based in the United States. In recent years, the funding rate has been similar for researchers both inside and outside the U.S. The SHRM Foundation encourages submitting a formal letter of inquiry to find out whether or not your proposal will be a good fit with SHRM Foundation funding criteria. Overhead up to 15% of direct costs may be requested when required by the PI's institution. However the maximum overall grant budget must not exceed $200,000 USD.
*Restrictions:* Employees or national board members of SHRM, HRCI or the SHRM Foundation and members of their immediate family are not eligible to apply for SHRM Foundation grants, awards or scholarships.
*Geographic Focus:* All States, All Countries
*Date(s) Application is Due:* Apr 1; Oct 1
*Amount of Grant:* Up to 200,000 USD
*Contact:* Lynn McFarland, Ph.D., Research Coordinator; (803) 767-6992; lynnmcfarland@bellsouth.net
*Internet:* www.shrm.org/about/foundation/research/pages/applyforgrant.aspx
*Sponsor:* Society for Human Resource Management Foundation
1800 Duke Street
Alexandria, VA 22314

**Sick Kids Foundation New Investigator Research Grants** 2326

Sick Kids Foundation New Investigator Research Grants are jointly sponsored by SickKids Foundation and the Canadian Institutes of Health Research Institute of Human Development, Child and Youth Health. Grant recipients may obtain up to three years of support for research in the biomedical, clinical, health systems and services, population and public health sectors. The program provides support to child health researchers early in their careers. It is intended that New Investigator Research Grants will enhance the grant recipient's capacity to compete with more senior investigators for research grants from

other funders. The Foundation funds research that has the potential to have a significant impact on child health outcomes. Multi-year grants range up to $300,000.
*Requirements:* The applicant must be a new investigator as of the application deadline. For the purpose of this award, a new investigator is defined as an individual who as a principal or co-principal investigator has not received combined operating grant funding of $500,000 or more (Canadian dollars) and is within five years of their first academic appointment. Academic appointment is defined as an appointment which allows an individual to apply for research grants as an independent investigator.
*Geographic Focus:* Canada
*Amount of Grant:* Up to 300,000 CAD
*Samples:* Dr. Jacob Jaremko, University of Alberta, Edmonton, Alberta, Canada, $257,245 - 3D Ultrasound in Hip Dysplasia; Dr. Alexander Beristain, University of British Columbia, Vancouver, British Columbia, Canada, $299,385 - Effect of Obesity-Associated Inflammation on the Maternal-Fetal Interface in Early Pregnancy; Dr. Sarah Fraser, University of Montreal, Montreal, Quebec, Canada, $283,288 - Collaborative Mental Health Care from Communities, to Professionals, to Policy Change for Inuit of Canada.
*Contact:* Ted Garrard, President; (416) 813-6166 or (800) 661-1083; fax (416) 813-5024; ted.garrard@sickkidsfoundation.com
*Internet:* www.sickkidsfoundation.com/about-us/grants/new-investigator-research-grants
*Sponsor:* Sick Kids Foundation
525 University Avenue, 14th Floor
Toronto, ON M5G 2L3 Canada

**Sidgmore Family Foundation Grants** 2327
The Sidgmore Family Foundation honors the legacy of John W. Sidgmore by taking a proactive approach to helping others succeed. The Foundation desires to use its resources to find creative and innovative solutions so that people may achieve their full potential and become responsible, healthy and productive members of society. In recognition that an impoverished environment limits the possibilities for people to develop and thrive, the Sidgmore Family Foundation is particularly interested in funding organizations that: improve the quality of education and teacher training; further the advancement of knowledge in the field of medicine with a special emphasis on hearing and cardiology; utilize entrepreneurial skills to explore and develop creative, scalable, and sustainable solutions to critical social problems; and, provide support and services to those in need in the Washington D.C. area. Grants are awarded to organizations that have a clear, replicable plan for success, measured sustainable results, and high approval ratings from charity evaluator organizations, such as Charity Navigator. The Foundation also awards multi-year grants that can sustain a program or project.
*Requirements:* Nonprofit 501(c)3 organizations are eligible. Preference is given to organizations that serve residents in the Washington, D.C. metropolitan area, Maryland, and Virginia. Applicants must begin the process with an initial Letter of Inquiry (LOI). LOIs will receive a response if the Foundation wishes to receive a proposal from your organization.
*Restrictions:* The Sidgmore Family Foundation does not make grants to individuals, national health organizations, government agencies, or political and public policy advocacy groups.
*Geographic Focus:* All States
*Contact:* M. Gelbwaks, Director; (516) 541-2713; SidgmoreFound@aol.com
*Internet:* www.sidgmorefoundation.com/#application_process
*Sponsor:* Sidgmore Family Foundation
71 Leewater Avenue
Massapequa, NY 11758

**Sid W. Richardson Foundation Grants** 2328
Grants are provided to tax-exempt organizations in Texas in the areas of education (museums, learning centers, day schools, K-12 schools, higher education institutions, and business and economic education), health (medical schools, organ donor registries, hospitals, disease prevention, health science centers, and nursing associations), arts (arts councils, visual and performing arts festivals, museums, ballet, symphony orchestra, and arts education programs), and human services (boys' and girls' clubs, united funds, the elderly, crime prevention, the disabled, housing opportunities, food programs, and drug and alcohol abuse prevention). Types of support include operating budgets, seed grants, building construction funds, equipment acquisition, endowment funds, research, publications, conferences and seminars, matching funds, continuing support, and projects/programs. Award amounts vary depending on proposed projects. Applications must be received no later than January 15.
*Requirements:* Grant requests will be considered only from tax-exempt organizations described in Section 501(c)3 of the Internal Revenue Code of 1986 and classified as other than a private foundation within the meaning of Section 509(a) of the Code, or from a qualified public entity described in Section 170 of the Code.
*Restrictions:* Grants are not made to individuals, or for the support of school trips, testimonial dinners, fundraisers, or marketing events. An organization is limited to one application per calendar year.
*Geographic Focus:* Texas
*Date(s) Application is Due:* Jan 15
*Amount of Grant:* 10,000 - 100,000 USD
*Samples:* Back to School Roundup, Fort Worth, Texas, $10,000 - general support for the 7th annual Tarrant County Back to School Roundup; Baylor College of Medicine Department of Internal Medicine, Houston, Texas, $100,000 - support for Dr. James L. Pool's program in Internal Medicine; Christ's Haven for Children, Keller, Texas, $7,500 - general support for the program assisting foster children.
*Contact:* Carolyn Johns, (817) 336-0494; fax (817) 332-2176; cjohns@sidrichardson.org or info@sidrichardson.org
*Internet:* www.sidrichardson.org/grants/
*Sponsor:* Sid W. Richardson Foundation
309 Main Street
Fort Worth, TX 76102

**Sigma Theta Tau International /American Nurses Foundation Grant** 2329
Sigma Theta Tau International (STTI), the only international nursing honor society worldwide, is a global community of nurse leaders dedicated to using knowledge, scholarship, service, and learning to improve the health of the world's people. STTI members belong to 475 chapters in 90 countries. STTI and the American Nurses' Foundation (ANF) have joined forces to provide funding for the STTI/ANF research grant. The purpose of the grant is to encourage the research career development of nurses through support of research conducted by beginning nurse researchers or experienced nurse researchers who are entering a new field of study. Applications must be submitted directly to STTI via their online submission system by the May 1 deadline date. Funding will begin in October. Links to an FAQ, application-review criteria, and help with grant-writing are available at the website.
*Requirements:* Eligible applicants must meet the following criteria: be a registered nurse with a current license; have a master's or doctoral degree or be enrolled in a doctoral program; have a clinical nursing research topic; submit a completed research application packagbe and signed research agreement; and be ready to implement the research project when funding is received. Preference will be given to Sigma Theta Tau members, other qualifications being equal. Allocation of funds is based on the quality of the proposed research, the future promise of the applicant, and the applicant's research budget.
*Restrictions:* Funds for this grant do not cover expenses incurred prior to the funding date.
*Geographic Focus:* All States, All Countries
*Date(s) Application is Due:* May 1
*Amount of Grant:* Up to 7,500 USD
*Samples:* Amy M. Sawyer, Ph.D., R.N. (USA) - "A Tailored Intervention to Promote CPAP Adherence"; Anne M. Fink, R.N., M.S., C.C.R.N. (USA) - "Cytokines and Fatique in Chronic Heart Failure"; Laurie Beth Hartjes, M.S., P.N.P.-B.C. (USA) - "Development and Evaluation of an Innovative Malaria-Risk Reduction Tool for Travelers".
*Contact:* Research Services; (888) 634-7575 or (317) 634-8171; fax (317) 634-8188; research@stti.iupui.edu
*Internet:* www.nursingsociety.org/Research/Grants/Pages/grant_anf.aspx
*Sponsor:* Sigma Theta Tau International
550 W. North Street
Indianapolis, IN 46202-2156

**Sigma Theta Tau International / Council for the Advancement of Nursing Science Grants** 2330
Sigma Theta Tau International (STTI), the only international nursing honor society worldwide, is a global community of nurse leaders dedicated to using knowledge, scholarship, service, and learning to improve the health of the world's people. STTI members belong to 475 chapters in 90 countries. STTI and the Council for the Advancement of Nursing Science (CANS) have joined forces to provide funding for the STTI/CANS research grant. The purpose of the grant is to encourage qualified nurses to improve the health of the world's people through research. Proposals for clinical, educational or historical research, including plans for broadly disseminating the research findings, may be submitted for the grant. Applications must be submitted directly to STTI via their online submission system by the July 1 deadline date. Funding will begin in November. Links to an FAQ, application-review criteria, and help with grant-writing are available at the website.
*Requirements:* Eligible applicants must meet the following criteria: be a member in good standing of both STTI and CANS; hold at minimum a Master's degree or equivalent; submit a completed research application via STTI's on-line submission system; be ready to implement research project when funding is received; submit a final report; submit a completed abstract to Virginia Henderson International Nursing Library; submit the completed project abstract for presentation at the Sigma Theta Tau or Council for Advancement of Nursing Science research meeting; and credit research-grant partners in all publications and presentations of the research.
*Restrictions:* Funds for this grant do not cover expenses incurred prior to the funding date.
*Geographic Focus:* All States, All Countries
*Date(s) Application is Due:* Jul 1
*Amount of Grant:* Up to 5,000 USD
*Samples:* Carla Fry, M.S.N. (USA) - "Addressing Pneumococcal Vaccine Uptake Disparities among African-American Adults in the U.S"; Therea M. Wadas, M.S.N. (USA) - "Relationship among AP.O.E genotypes, inflammatory markers, and risk factors in African Americans with ischemic stroke"; Carolyn Herrington, R.N., Ph.D., N.N.P.-B.C. (USA) "Comparison of Skin Conductance and Salivary Cortisol as Indicators of Pain Response in Premature Infants Undergoing Heel Stick for Necessary Blood Sampling".
*Contact:* Research Services; (888) 634-7575 or (317) 634-8171; fax (317) 634-8188; research@stti.iupui.edu
*Internet:* www.nursingsociety.org/Research/SmallGrants/Pages/CANS.aspx
*Sponsor:* Sigma Theta Tau International
550 W. North Street
Indianapolis, IN 46202-2156

**Sigma Theta Tau International / Oncology Nursing Society Grant** 2331
Sigma Theta Tau International (STTI), the only international nursing honor society worldwide, is a global community of nurse leaders dedicated to using knowledge, scholarship, service, and learning to improve the health of the world's people. STTI members belong to 475 chapters in 90 countries. STTI and the Oncology Nurses Society (ONS) have joined forces to provide funding through ONS's Small Research Grants

program for this grant. From the perspectives of both organizations, the purpose of the grant is to stimulate clinically-related oncology-nursing research, increase the knowledge base for oncology-nursing practice, and to prepare future oncology-nurse researchers. Application guidelines are available through ONS; applications must be submitted directly to ONS. Contact information and a link to the ONS website is available at the grant website.
*Requirements:* Eligible applicants must meet the following criteria: be a registered nurse actively involved in some aspect of cancer-patient care, education or research and have earned a master's degree. Other qualifications being equal, preference will be given to Sigma Theta Tau International members.
*Geographic Focus:* All States, All Countries
*Date(s) Application is Due:* Oct 1
*Amount of Grant:* 10,000 USD
*Samples:* Kathleen Ruccione, M.P.H., R.N., C.P.O.N.®, F.A.A.N. (USA), $10,000 - "Iron, Anthracyclines and Cardiac Outcomes among Childhood Cancer Survivors"; Shu-Feng Sung, Ph.D., R.N. (USA), $10,000 - " Translaton Cardiovascular Risk and Breast Cancer Outcomes"; Sharon Kozachik, R.N., Ph.D. (USA), $10,000 - "Sleep, HPA Axis Activity and Paclitaxel-Induced Neuropathic Pain".
*Contact:* Debbie Kubiak, Grants Coordinator; (412) 859-6224 or (866) 357-4667; fax (877) 369-5497; dkubiak@ons.org
Kathi Dandracs, Grants Specialist; (412) 859-6233 or (866) 357-4667; fax (877) 369-5497; kkandracs@ons.org
Carlton G. Brown, R.N., Ph.D., A.O.C.N., President; (866) 357-4667 or (412) 859-6298; fax (877) 369-5497
*Internet:* www.nursingsociety.org/Research/Grants/Pages/grant_ons.aspx
*Sponsor:* Sigma Theta Tau International
550 W. North Street
Indianapolis, IN 46202-2156

**Sigma Theta Tau International Doris Bloch Research Award** 2332
Sigma Theta Tau International (STTI), the only international nursing honor society worldwide, is a global community of nurse leaders dedicated to using knowledge, scholarship, service, and learning to improve the health of the world's people. STTI members belong to 475 chapters in 90 countries. The purpose of STTI's Doris Bloch Research Award is to encourage nurses to contribute to the advancement of nursing through research. Allocation of funds is based on the quality of the proposed research, the future promise of the applicant, and the applicant's research budget. Applications from novice researchers who have received no other national research funds are encouraged and will receive preference for funding, other aspects being equal. Preference will be given to Sigma Theta Tau International members, other qualifications being equal. Applications must be submitted via STTI's online submission system. A link to the submission system will be available in July. Funding will begin in June. Links to an FAQ, application-review criteria, and help with grant-writing are available at the website. Funding is provided by a gift to the honor society's Research Endowment from the Doris Bloch estate.
*Requirements:* Eligible applicants must meet the following criteria: be a registered nurse with a current license; hold a master's or doctoral degree or be enrolled in a doctoral program; submit a completed research application package and signed research agreement; be ready to implement the research project when funding is received; complete the project within one (1) year of funding; submit to STTI a final report; and submit a completed abstract to STTI's Virginia Henderson International Nursing Research Library.
*Geographic Focus:* All States, All Countries
*Date(s) Application is Due:* Dec 1
*Amount of Grant:* Up to 5,000 USD
*Samples:* Carina Katigbak, M.S., B.Sc.N. (USA) - "Exploring the role of Community Health Workers upon Hypertension Related Behaviors with Filipino Americans in New York City"; Hsiu-Chin Chen, R.N., Ph.D., Ed.D. (USA) - "The significance of student input to nursing program evaluation: Part II"; Jennifer Hobbs, R.N., Ph.D. (USA) - "Standardizing the Particular in Clinical Information Systems: Nursing's Role in the Development of Clinical Standards during the Late 20th Century".
*Contact:* Research Services; (888) 634-7575 or (317) 634-8171; fax (317) 634-8188; research@stti.iupui.edu
*Internet:* www.nursingsociety.org/Research/Grants/Pages/grant_bloch.aspx
*Sponsor:* Sigma Theta Tau International
550 W. North Street
Indianapolis, IN 46202-2156

**Sigma Theta Tau Small Grants** 2333
Sigma Theta Tau International (STTI), the only international nursing honor society worldwide, is a global community of nurse leaders dedicated to using knowledge, scholarship, service, and learning to improve the health of the world's people. STTI members belong to 475 chapters in 90 countries. The purpose of STTI's Small Grants Program is to encourage nurses to contribute to the advancement of nursing through research. Ten to fifteen grants are awarded annually. Allocation of funds is based on the quality of the proposed research, the future promise of the applicant, and the applicant's research budget. Applications from novice researchers who have received no other national research funds are encouraged and will receive preference for funding, other aspects being equal. Preference will be given to Sigma Theta Tau members, other qualifications being equal. Applications must be submitted via STTI's online submission system. A link to the submission system will be available in July. Funding will begin in June. Links to an FAQ, application-review criteria, and help with grant-writing are available at the website.
*Requirements:* Eligible applicants must meet the following criteria: be a registered nurse with a current license; have a master's or doctoral degree or be enrolled in a doctoral program; have a clinical-nursing research topic; submit a completed research application package and signed research agreement; be ready to implement the research project when funding is received; complete the project within one (1) year of funding; submit a final report; and submit a completed abstract to STTI's Virginia Henderson International Nursing Research Library.
*Restrictions:* Funds for this grant do not cover expenses incurred prior to the funding date.
*Geographic Focus:* All States, All Countries
*Date(s) Application is Due:* Dec 1
*Amount of Grant:* Up to 5,000 USD
*Samples:* Tara Albrecht, Ph.D., M.S.N., B.S.N. (USA) - "Symptom Management and Psychosocial needs in Patients with Acute Myeloid Leukemia"; Kathleen J. Finlayson, Ph.D., M.S.N. (Australia) - "Identifying Effective New Interventions to Prevent Recurrence of Venous Leg Ulcers in Adults with Chronic Venous Insufficiency"; Orla Marie Smith, M.N. (USA) - "Prevalence and Predictors of Anxiety, Depression and Risk for Post-Traumatic Stress in Family Members of Patients in the Intensive Care Unit: The ICU-Adapts Survey".
*Contact:* Research Services; (888) 634-7575 or (317) 634-8171; fax (317) 634-8188; research@stti.iupui.edu
*Internet:* www.nursingsociety.org/Research/Grants/Pages/small_grants.aspx
*Sponsor:* Sigma Theta Tau International
550 W. North Street
Indianapolis, IN 46202-2156

**Silicon Valley Creates Artist Laureate Grants** 2334
Silicon Valley Creates recognizes the region's finest creative artists who are seriously committed to developing their art forms and to enriching the community with their exhibits, performances, presentations, and service. Artist Laureate awards provide unrestricted cash prizes to support artists while they pursue their creative work. Through Artist Laureate awards, SVC seeks to: recognize artistic excellence and impact of our finest local artists; provide monetary support to encourage continuing development of their work; and spotlight the important role individual artists play in contributing to a vibrant and creative community. Artists may apply on their own (self-nominate) or apply with the support of a nominator, to compete in one of five different categories: On the Wall-art that is affixed to, hung on, or projected on a wall or flat panel/screen; Off the Wall-free-standing or suspended art that can be viewed from all angles; On Stage-art that is performed by the artist for an audience, on a formal or informal stage; Off Stage-art that is created or developed by the artist offstage, backstage, or pre-performance; and Emerging Artist-early career artists in all genres. One $5,000 laureate prize is awarded in each of the five categories. Complete applications must be received by November 7.
*Requirements:* Applicants in the Emerging Artist category must be nominated (they may not self-nominate for this category). Applicant/nominee must: be at least 18 years of age at the time of the application deadline; spend at least ten hours a week on and derive income from the practice/creation of art; have exhibited, performed, presented and/or published artistic work in a public context within the last three years (to demonstrate activities are current); be the primary or principal creator/performer or the sole author of work(s) submitted; and be a resident of Santa Clara County or in one of several immediately bordering counties (Alameda, San Mateo, and Santa Cruz Counties) for at least one year prior to the application deadline, AND can demonstrate through professional history that artistic activity has had significant impact on residents of Santa Clara County.
*Restrictions:* Applicant/nominee may not be: enrolled as an undergraduate student in a degree program of any kind; enrolled as a graduate student in a program related to the specific discipline of this application; a prior recipient of an Artist Fellowship or Laureate from Silicon Valley Creates/Arts Council Silicon Valley; a current or former employee or board member of Silicon Valley Creates/Arts Council Silicon Valley/1stACT Silicon Valley. This program does not include architecture, landscaping, industrial/graphic/computer systems design, journalism/photojournalism, or other commercial activities.
*Geographic Focus:* California
*Date(s) Application is Due:* Nov 7
*Amount of Grant:* 5,000 USD
*Contact:* Audrey Struve; (408) 998-2787, ext. 204; fax (408) 971-9458; audreys@svcreates.org
*Internet:* www.svcreates.org/page/grants_artist_laureate
*Sponsor:* Silicon Valley Creates
4 North Second Street, Suite 500
San Jose, CA 95113-1324

**Sioux Falls Area Community Foundation Community Fund Grants** 2335
The purpose of the Sioux Falls Area Foundation (SFACF)'s grant making program is to provide support across a wide spectrum of charitable needs and interests. Grant making categories include arts and humanities (e.g., theatre, music, arts, dance, cultural development, historic preservation, library programs, and museums); community affairs and development (e.g., citizen participation, public use of parks and recreation, administration of justice, economic development, employment, and training); education (e.g., lifelong-learning activities in formal educational settings, support of educational facilities and systems, and scholarships); environment (e.g., protection of natural areas, conservation of energy, prevention and elimination of pollution or hazardous waste, wildlife protection, and water quality); health (e.g., improvement of healthcare, prevention of substance abuse; support of mental-health needs, and medical research); human services (e.g., assistance to families, youth, the elderly, disabled, special groups, social service providers, and those who stand in need); and religion (e.g., support for churches, religious institutions, and religion programs). SFACF offers two grant programs from its unrestricted funds: Spot Grants for projects up to $3,000 and Community Fund Grants for projects over $3,000. The majority of Community Fund Grants are made in the range of $5,000 to $10,000. Proposals for Community Fund Grants must be submitted using a standard application form (available from the SFACF office or downloadable from the website). Applications are accepted anytime and will

be reviewed by the Grants Committee at their next scheduled meeting. Meetings take place six times a year: January, March, May, July, September, and November (a schedule is posted at the SFACF website).
*Requirements:* Nonprofit organizations serving residents in the Sioux Falls, South Dakota area (Minnehaha, Lincoln, McCook, and Turner counties) are eligible to apply. SFACF considers grant requests for programs that require start-up funds to address important community needs or opportunities, expansion of programs that meet important community needs or opportunities, assistance to organizations weathering unforeseen or unusual financial crises; programs that increase an organization's capacity to advance its mission more efficiently or effectively; and programs or studies that inform the community's understanding of needs or opportunities. Requests are evaluated by the following criteria: comparative benefit to the community; the organization's capacity to achieve the stated objectives; the amount of support requested versus the number of people benefited; a well-planned approach to achieving stated objectives; a reasonable expectation that the program can be sustained over time (where applicable); the organization's history of working collaboratively to address community needs and opportunities; and when applicable, the organization's past SFACF grant performance.
*Restrictions:* SFACF does not consider grant requests for individuals, national fundraising efforts, political advocacy, and sectarian religious programs. The following types of requests are discouraged: large capital improvements or construction drives; ongoing operational support; reduction or elimination of organizational deficits; reimbursement of expenses undertaken prior to submission of a grant application; computer hardware and software, unless these are the focus of a new or enhanced program; public art for which approval and placement has not yet been secured; and multi-year requests.
*Geographic Focus:* South Dakota
*Date(s) Application is Due:* Jan 1; Mar 1; May 1; Jul 1; Sep 1; Nov 1
*Contact:* Andy Patterson, President; (605) 336-7055, ext. 15; fax (605) 336-0038; apatterson@sfacf.org
Patrick Gale; (605) 336-7055, ext. 20; fax (605) 336-0038; pgale@sfacf.org
*Internet:* www.sfacf.org/applying-for-grants/
*Sponsor:* Sioux Falls Area Community Foundation
200 North Cherapa Place
Sioux Falls, SD 57103-2205

**Sioux Falls Area Community Foundation Spot Grants** **2336**
The purpose of the Sioux Falls Area Foundation (SFACF)'s grant making program is to provide support across a wide spectrum of charitable needs and interests. Grant making categories include arts and humanities (e.g., theatre, music, arts, dance, cultural development, historic preservation, library programs, and museums); community affairs and development (e.g., citizen participation, public use of parks and recreation, administration of justice, economic development, employment, and training); education (e.g., lifelong-learning activities in formal educational settings, support of educational facilities and systems, and scholarships); environment (e.g., protection of natural areas, conservation of energy, prevention and elimination of pollution or hazardous waste, wildlife protection, and water quality); health (e.g., improvement of healthcare, prevention of substance abuse; support of mental-health needs, and medical research); human services (e.g., assistance to families, youth, the elderly, disabled, special groups, social service providers, and those who stand in need); and religion (e.g., support for churches, religious institutions, and religion programs). SFACF offers two grant programs from its unrestricted funds: Community Fund Grants for projects over $3,000 and Spot Grants for projects up to $3,000. Spot Grant proposals may be submitted at any time and do not require SFACF's standard application form. Applicants should include the following components in their requests: a typed summary of their program in two pages or fewer; signatures of the organization's executive director and board chair; a board of directors roster; a copy of the organization's IRS tax determination letter; and a project budget. In most cases, SFACF will review and respond to Spot Grant requests within two weeks of receipt. SFACF has provided complete guidelines and an informative FAQ at their website.
*Requirements:* Nonprofit organizations serving residents in the Sioux Falls, South Dakota area (Minnehaha, Lincoln, McCook, and Turner counties) are eligible to apply. SFACF considers grant requests for programs that require start-up funds to address important community needs or opportunities, expansion of programs that meet important community needs or opportunities, assistance to organizations weathering unforeseen or unusual financial crises; programs that increase an organization's capacity to advance its mission more efficiently or effectively; and programs or studies that inform the community's understanding of needs or opportunities. Requests are evaluated by the following criteria: comparative benefit to the community; the organization's capacity to achieve the stated objectives; the amount of support requested versus the number of people benefited; a well-planned approach to achieving stated objectives; a reasonable expectation that the program can be sustained over time (where applicable); the organization's history of working collaboratively to address community needs and opportunities; and when applicable, the organization's past SFACF grant performance.
*Restrictions:* SFACF does not consider grant requests for individuals, national fundraising efforts, political advocacy, and sectarian religious programs. The following types of requests are discouraged: large capital improvements or construction drives; ongoing operational support; reduction or elimination of organizational deficits; reimbursement of expenses undertaken prior to submission of a grant application; computer hardware and software, unless these are the focus of a new or enhanced program; public art for which approval and placement has not yet been secured; and multi-year requests.
*Geographic Focus:* South Dakota
*Amount of Grant:* Up to 3,000 USD
*Contact:* Andy Patterson, President; (605) 336-7055, ext. 15; fax (605) 336-0038; apatterson@sfacf.org
Patrick Gale; (605) 336-7055, ext. 20; fax (605) 336-0038; pgale@sfacf.org
*Internet:* www.sfacf.org/applying-for-grants/
*Sponsor:* Sioux Falls Area Community Foundation
200 North Cherapa Place
Sioux Falls, SD 57103-2205

**Skaggs Foundation Grants** **2337**
The Skaggs Foundation, supported by the Samuel D. Skaggs Investment firm, was established in 1962 in California, and currently provides funding throughout Alaska. The Foundation's primary areas of interest include: arts and culture; animals and wildlife; preservation; the environmental protection; children's services; special education; environmental education; natural resources; and marine science. Types of support available with these funds include: continuing support; endowments; equipment; general operating support; internship funds; land acquisition; matching or challenge funding; program evaluation; research; and technical assistance. Typical awards range up to a maximum of $15,000.
*Requirements:* Nonprofit organizations are eligible, preference is given to requests from Alaska. The initial approach should be a one or two page letter of inquiry. There is no formal application form. Applicants should submit the following: copy of IRS Determination Letter; copy of current year's organizational budget and/or project budget; one copy of the proposal.
*Restrictions:* No grants are given directly to individuals.
*Geographic Focus:* Alaska
*Date(s) Application is Due:* May 1
*Amount of Grant:* Up to 15,000 USD
*Samples:* Alaska Arts Southeast, Sitka, Alaska, $15,000 - general operating support; Sitka Sound Science Center, Sitka, Alaska, $8,000 - support of the annual fellowship program; Preservation Theater, Douglas, Alaska, $2,500 - support of the Theater in the Wild program; Rivers Without Borders, Clinton, Alaska, $5,000 - general operating support.
*Contact:* Samuel D. Skaggs Jr., President; (907) 463-4843 or (907) 463-5511
*Sponsor:* Skaggs Foundation
119 Seward Street, Suite 7
Juneau, AK 99801

**Smith Richardson Foundation Domestic Public Research Fellowships** **2338**
The Domestic Public Policy program supports projects that will help the public and policy makers understand and address critical challenges facing the United States. An overarching goal of the Foundation's grant making is to support projects that help stimulate and inform important public policy debates. To that end, the Foundation supports research on and evaluation of existing public policies and programs, as well as projects that inject new ideas into public debates. The Foundation is interested in a wide range of topics. Education and school reform have been central to the Foundation's grant making in recent years, including efforts to assess the effectiveness of various school reform initiatives, such as charter schools. The Foundation has expanded on its interest in public finance issues by beginning to explore the future of the Medicare program and the larger challenge of coping with expected increases in the costs of federal entitlement programs. The Foundation also provides support for projects that assess the impact of immigration on American society, examine the effectiveness of our regulatory policies, and explore the interaction between the policy making process and the political process through support for projects on congressional redistricting and the conduct and financing of political campaigns. The Foundation continues to support projects that inform the debate over policies designed to assist disadvantaged families.
*Requirements:* Fellowships will only be awarded to individual researchers and not to research teams. He or she must hold a position as a full-time faculty member at a college or university or as a full-time fellow at a public policy think tank or research organization in the United States. An applicant should explain how he or she meets these requirements in the cover letter to his or her research proposal.
*Geographic Focus:* All States
*Amount of Grant:* 60,000 USD
*Contact:* Marin J. Strmecki; (203) 222-6222; fax (203) 222-6282; jhollings@srf.org
*Sponsor:* Smith Richardson Foundation
60 Jesup Road
Westport, CT 06880-4311

**Smith Richardson Foundation Strategy and Policy Fellowship** **2339**
The Smith Richardson Foundation's International Security and Foreign Policy program announces the Strategy and Policy Fellows program, its annual grant competition to support young scholars and policy thinkers on American foreign policy, international relations, international security, military policy, and diplomatic and military history. The purpose of the program is to strengthen the U.S. community of scholars and researchers conducting policy analysis in these fields. The Foundation will award at least three research grants of $60,000 each to enable the recipients to research and write a book. Within the academic community, this program supports junior or adjunct faculty, research associates, and post-docs who are engaged in policy-relevant research and writing. Within the think tank community, the program supports members of the rising generation of policy thinkers who are focused on U.S. strategic and foreign policy issues. The Foundation must receive all Strategy and Policy and Fellows proposals by 5:00 p.m., EDT, June 14.
*Requirements:* An applicant must have a Ph.D. by the time of the deadline, preferably in Political Science, Public Policy, Policy Analysis, International Political Economy, or History. The proposal must be submitted by the institution the applicant has an affiliation and be accompanied by a letter from the institution's representative who is authorized to accept the grant on behalf of the institution. The applicant should also explain how he or she meets all of these requirements in a cover letter to the proposal. The recipients are likely to be under the age of 40. There is, however, no formal age requirement.

*Geographic Focus:* All States
*Date(s) Application is Due:* Jun 14
*Amount of Grant:* 60,000 USD
*Contact:* Marin J. Strmecki, Senior Vice President; (203) 222-6222; fax (203) 222-6282; jhollings@srf.org
*Sponsor:* Smith Richardson Foundation
60 Jesup Road
Westport, CT 06880-4311

**Smith Richardson Foundation World Politics and Statecraft Fellowship** 2340
The Smith Richardson Foundation announces the World Politics and Statecraft Fellowship program, its annual grant competition to support Ph.D. dissertation research on American foreign policy, international relations, international security, strategic studies, area studies, and diplomatic and military history. The fellowship's objective is to support the research and writing of policy-relevant dissertations through funding of fieldwork, archival research, and language training. In evaluating applications, the Foundation will accord preference to those projects that could directly inform U.S. policy debates and thinking, rather than dissertations that are principally focused on abstract theory or debates within a scholarly discipline. The Foundation will award up to twenty grants of $7,500 each. The Foundation must receive all World Politics and Statecraft Fellowship proposals by 5:00 p.m., EDT, October 15.
*Geographic Focus:* All States
*Amount of Grant:* 7,500 USD
*Contact:* Marin J. Strmecki, Senior Vice President; (203) 222-6222; fax (203) 222-6282; jhollings@srf.org
*Sponsor:* Smith Richardson Foundation
60 Jesup Road
Westport, CT 06880-4311

**Smithsonian Biodiversity Genomics and Bioinformatics Postdoctoral Fellowships** 2341
The Smithsonian Institution (SI) Postdoctoral Fellowships in Biodiversity Genomics and Bioinformatics promote collaborative research in these fields (60%), plus a well-defined outreach component (40%) oriented toward building genomics expertise in the greater SI research community. Research should involve comparative genomic approaches such as phylogenomics, population genomics, metagenomics or transcriptomics, and have a component that involves significant bioinformatics analysis. Your proposal should also detail in at least one page your bioinformatics outreach plan, which can include collaborative work with other SI projects and teams, training workshops or development of software, pipelines or tutorials. We plan to fill three or more Fellowships in the area of Biodiversity Genomics and are especially interested in recruiting a team of Fellows who will work together to advance bioinformatics at the Institution. Coordinated applications dealing with multiple genomic approaches are strongly encouraged. Proposals are due April 1 each year.
*Requirements:* Applicants must propose to conduct research in-residence for a period of 12 to 24 months. Applicants must have completed or be near completion of the Ph.D. Recipients who have not completed the Ph.D. at the time of application must provide proof of completion of the degree before the fellowship begins.
*Geographic Focus:* All States
*Date(s) Application is Due:* Apr 1
*Contact:* Bryan T. Fair, Program Assistant; (202) 275-0655 or (202) 633-7070; fax (202) 275-0489; siofg@si.edu
*Internet:* www.si.edu/ofg/fell.htm
*Sponsor:* Smithsonian Institution
470 L'Enfant Plaza SW, Suite 7102, MRC 902 P.O. Box 37012
Washington, D.C. 20013-7012

**Smithsonian Environmental Research Center Graduate Student Fellowships** 2342
Graduate Student Fellowships are offered to students formally enrolled in a graduate program of study, who have completed at least one semester, and not yet been advanced to candidacy if in a PhD program. Applicants must submit a proposal for research in a discipline which is pursued at the Smithsonian Institution. The term is 10 weeks with a stipend of $6,500. Fellows are selected in a single, annual competition. The annual application deadline is January 15 of the proposed year of sponsorship.
*Geographic Focus:* All States
*Date(s) Application is Due:* Jan 15
*Amount of Grant:* 6,500 USD
*Contact:* Dan Gustafson, Internship, Fellowship, and Volunteer Coordinator; (443) 482-2217 or (443) 482-2200; fax (443) 482-2380; gustafsond@si.edu
*Internet:* www.serc.si.edu/pro_training/fellowships/fellowships.aspx
*Sponsor:* Smithsonian Institution
647 Contees Wharf Road, P.O. Box 28
Edgewater, MD 21037

**Smithsonian Environmental Research Center Internships** 2343
The Smithsonian Environmental Research Center's (SERC) Internship program offers undergraduate and beginning graduate students a unique opportunity to gain hands-on experience in the fields of environmental research and education. This program enables students to work on specific projects under the direction of SERC's professional staff and is tailored to provide the maximum educational benefit to each participant. Projects include terrestrial, atmospheric and estuarine environmental research within the disciplines of ecology, biology, chemistry, microbiology, botany, zoology, mathematics and physics. Projects are also offered in environmental education and environmental information management. Although students will become familiar with all the research being conducted by a particular staff member and with much of the research of SERC in general, individuals will devote most of their time to a project and content area of particular interest to them. Students will have the opportunity to expand their knowledge in a selected field of study and to learn a variety of research techniques through firsthand experience. At the conclusion of the internship, student participants will be expected to present the findings of their independent projects in a formal seminar to the SERC community. As an additional educational benefit, SERC maintains an ongoing seminar program that is designed to bring in noted outside speakers on a variety of topics in the environmental sciences. During the program's spring and summer seasons, field trips to research facilities and Smithsonian museums conducting similar programs are also coordinated. Selected candidates will receive a stipend of $475.00 per week. Limited on-site dormitory space may be available for $95.00 per week.
*Requirements:* The Smithsonian Environmental Research Center will consider applications from currently enrolled undergraduate and beginning graduate students, or students who have recently graduated from an undergraduate or Masters program. Applicants must be in a position to commit fully to the completion of a project. U.S. citizenship is not a requirement to participate in this program. Selection is based upon the student's academic credentials, extent of relevant training or experience, letters of recommendation, and the congruence of the student's expressed goals with those of the internship program.
*Restrictions:* SERC does not supply board, although cooking facilities and utensils are available.
*Geographic Focus:* All States
*Date(s) Application is Due:* Feb 1; Jun 1; Nov 15
*Amount of Grant:* 475 - 5,700 USD
*Contact:* Dan Gustafson, Internship, Fellowship, and Volunteer Coordinator; (443) 482-2217 or (443) 482-2200; fax (443) 482-2380; gustafsond@si.edu
*Internet:* www.serc.si.edu/pro_training/internships/internships.aspx
*Sponsor:* Smithsonian Institution
647 Contees Wharf Road, P.O. Box 28
Edgewater, MD 21037

**Smithsonian Environmental Research Center Postdoctoral/Senior Fellowships** 2344
Smithsonian Environmental Research Center Postdoctoral and Senior Fellowships are offered to scholars who have held their degree or equivalent for less than seven years. Senior fellowships are offered to scholars who held their degree or equivalent for seven years or more. Applicants must submit a detailed proposal including a justification for conducting research in-residence at the Smithsonian. The term is 12-24 months at SERC with the second year of funding contigent upon satisfactory progress and availability of funds. Both fellowships have annual stipend of $45,000 plus allowances. Fellows are selected in a single, annual competition. The annual application deadline is January 15 of the proposed year of sponsorship.
*Geographic Focus:* All States
*Date(s) Application is Due:* Jan 15
*Amount of Grant:* 45,000 - 90,000 USD
*Contact:* Dan Gustafson, Internship, Fellowship, and Volunteer Coordinator; (443) 482-2217 or (443) 482-2200; fax (443) 482-2380; gustafsond@si.edu
*Internet:* www.serc.si.edu/pro_training/fellowships/fellowships.aspx
*Sponsor:* Smithsonian Institution
647 Contees Wharf Road, P.O. Box 28
Edgewater, MD 21037

**Smithsonian Environmental Research Center Predoctoral Fellowships** 2345
Predoctoral Fellowships are offered to doctoral candidates who have completed preliminary course work and examinations. The applicant must submit a detailed proposal including a justification for conducting their research in-residence at the Institution. Candidates must have the approval of their universities to conduct doctoral research at the Smithsonian. The term is 3 to 12 months with annual stipend of $30,000 plus allowances. Fellows are selected in a single, annual competition. The annual application deadline is January 15 of the proposed year of sponsorship.
*Geographic Focus:* All States
*Date(s) Application is Due:* Jan 15
*Amount of Grant:* 7,500 - 30,000 USD
*Contact:* Dan Gustafson, Internship, Fellowship, and Volunteer Coordinator; (443) 482-2217 or (443) 482-2200; fax (443) 482-2380; gustafsond@si.edu
*Internet:* www.serc.si.edu/pro_training/fellowships/fellowships.aspx
*Sponsor:* Smithsonian Institution
647 Contees Wharf Road, P.O. Box 28
Edgewater, MD 21037

**Smithsonian Graduate Student Fellowships** 2346
Students actively engaged in graduate study at any level may apply to study and conduct research for 10-week periods under the guidance of Smithsonian staff members. Applicants must propose to conduct research in disciplines that are pursued at the Smithsonian (American history, art, and culture; history of science and technology; history of art, design, crafts, and decorative arts; anthropology, archaeology, linguistics, and ethnic studies; evolutionary, systematic, behavioral, and environmental biology and conservation; earth, mineral, and planetary sciences; and materials characterization and conservation) and must plan to spend their entire tenure in residence. These fellowships allow students to conduct research for periods of three to twelve months. Applications are posted to the Web site in September.
*Requirements:* Applicants must have completed coursework and preliminary examinations for the doctoral degree, and must be engaged in dissertation research. In addition, candidates must have the approval of their universities to conduct their doctoral research at the Smithsonian.
*Geographic Focus:* All States

*Date(s) Application is Due:* Jan 15
*Amount of Grant:* 6,500 USD
*Contact:* Bryan T. Fair; (202) 275-0655 or (202) 633-7070; fax (202) 275-0489; siofg@si.edu
*Internet:* www.si.edu/ofg/fell.htm
*Sponsor:* Smithsonian Institution
470 L'Enfant Plaza SW, Suite 7102, MRC 902 P.O. Box 37012
Washington, D.C. 20013-7012

**Smithsonian Latino Studies Fellowships** 2347
Smithsonian Latino Studies Fellowships provide opportunities to U.S. Hispanic predoctoral students and postdoctoral and senior scholars to pursue research topics that relate to Latino art, culture, and history. Interdisciplinary subjects are encouraged and can be undertaken at more than one of the Smithsonian museums and/or research units, and advised by one or more of the Smithsonian research staff members. These types of fellowships are available at both the senior and postdoctoral level. A research and travel allowance is made available to cover additional costs of spending up to one-third of the fellowship tenure away from the Smithsonian, if appropriate and necessary, but not at the fellow's home institution. Applications are posted to the Web site in September.
*Requirements:* Postdoctoral fellowships are offered to scholars who have held a PhD or equivalent for less than seven years. Senior fellowships are offered to scholars who have held a PhD or equivalent for seven years or more. Predoctoral fellowships are offered to doctoral candidates who have completed preliminary course work and examinations, and have been advanced to candidacy.
*Geographic Focus:* All States
*Date(s) Application is Due:* Jan 15
*Amount of Grant:* 30,000 - 45,000 USD
*Contact:* Bryan T. Fair, Program Assistant; (202) 275-0655 or (202) 633-7070; fax (202) 275-0489; siofg@si.edu
*Internet:* www.si.edu/ofg/fell.htm
*Sponsor:* Smithsonian Institution
470 L'Enfant Plaza SW, Suite 7102, MRC 902 P.O. Box 37012
Washington, D.C. 20013-7012

**Smithsonian Minority Visiting Student Awards** 2348
Through the Minority Student Awards Program the Office of Fellowships offers internships and visiting student awards to increase participation of U.S. minority groups who are underrepresented in Smithsonian scholarly programs, in the disciplines of research conducted at the Institution, and in the museum field. Visiting Student Awards are available for currently enrolled advanced graduate students. Visiting Students pursue independently designed research projects in association with Smithsonian staff. Students should contact the Office of Fellowships for application information. The stipend amount is $550 per week for up to ten weeks.
*Requirements:* U.S. minority undergraduate and graduate students are invited to apply.
*Geographic Focus:* All States
*Date(s) Application is Due:* Feb 1; Oct 15
*Amount of Grant:* 550 - 5,500 USD
*Contact:* Bryan T. Fair, Program Assistant; (202) 275-0655 or (202) 633-7070; fax (202) 275-0489; siofg@si.edu
*Internet:* www.si.edu/ofg/fell.htm
*Sponsor:* Smithsonian Institution
470 L'Enfant Plaza SW, Suite 7102, MRC 902 P.O. Box 37012
Washington, D.C. 20013-7012

**Smithsonian Molecular Evolution Fellowships** 2349
Postdoctoral Fellowships in Molecular Evolution are available to support research that uses the resources and research opportunities offered at the National Zoological Park (Washington, D.C.), the National Museum of Natural History (Washington, D.C.) or the Smithsonian Tropical Research Institute (Panama). The stipend being offered is currently $45,000 for one year (12 months) to $90,000 for two years (24 months). The annual deadline is January 15 for awards to begin on or after June 1.
*Geographic Focus:* All States
*Date(s) Application is Due:* Jan 15
*Amount of Grant:* 45,000 - 90,000 USD
*Contact:* Bryan T. Fair, Program Assistant; (202) 275-0655 or (202) 633-7070; fax (202) 275-0489; siofg@si.edu
*Internet:* www.si.edu/ofg/fell.htm
*Sponsor:* Smithsonian Institution
470 L'Enfant Plaza SW, Suite 7102, MRC 902 P.O. Box 37012
Washington, D.C. 20013-7012

**Smithsonian Museum Conservation Institute Analytical Studies Group Internships** 2350
The Smithsonian Museum Conservation Institute in conjunction with the NSF REU Sites Program is offering short-term opportunities for undergraduates to work on research projects with staff members of its technical studies and research group. The Smithsonian's Museum Conservation Institute (MCI) is the center for specialized technical collections research and conservation for all Smithsonian museums, collections, and research centers. MCI combines state-of-the-art instrumentation and scientific techniques to provide technical research and characterization of objects. Internships will take place during the summer. Acceptable majors include, but are not limited to: museum studies, art history, archaeology, paleontology, chemistry, biochemistry, biology, ecology, earth science, chemical engineering, and materials science engineering. A stipend will be offered for a period of up to 10 weeks.
*Requirements:* Applicants must be currently enrolled as an undergraduate in a university program or scheduled to matriculate in the fall of the current year. Basic chemistry and mathematical coursework is preferred; laboratory experience is not required. Applicants must be U.S. citizens or permanent residents.
*Geographic Focus:* All States
*Date(s) Application is Due:* Mar 30
*Contact:* Paula T. DePriest, Deputy Director; (301) 238-1206 or (301) 238-1240; fax (301) 238-3709; DepriestP@si.edu or MCIweb@si.edu
*Internet:* si.edu/MCI/english/professional_development/2012TechInternships.html
*Sponsor:* Smithsonian Institution
4210 Silver Hill Road
Suitland, MD 20746

**Smithsonian Museum Conservation Institute Internships for Research Projects** 2351
The Smithsonian's Museum Conservation Institute is offering short-term opportunities to work on research projects with staff members. The internships are intended mainly for the summer, but consideration will be given to projects at other times during the year. Preference will be given to applicants currently enrolled in conservation training programs or graduate students in related subjects. A stipend may be offered for a period of up to 10 weeks. Projects will be supervised by a number of staff members (listed at the website), whom applicants are urged to contact in advance of application. Applications should be made online prior to the annual March 30 deadline, and selection notifications will be sent by April 15.
*Geographic Focus:* All States
*Date(s) Application is Due:* Mar 30
*Contact:* Paula T. DePriest, Deputy Director; (301) 238-1206 or (301) 238-1240; fax (301) 238-3709; DepriestP@si.edu or MCIweb@si.edu
*Internet:* si.edu/MCI/english/professional_development/2012ResearchInternships.html
*Sponsor:* Smithsonian Institution
4210 Silver Hill Road
Suitland, MD 20746

**Smithsonian Museum Conservation Institute Research Post-Doctorate/Post Graduate Fellowships** 2352
Fellowships at the Smithsonian Museum Conservation Institute offers fellowship opportunities involving a variety of artifact analysis, preservation, and conservation treatment specialties. MCI's commitment is to enhance the experience of the fellow or intern, providing specialized technical and scientific training. Such training contributes significantly to the qualifications of the students in their subsequent professional employment. Applications are currently invited for a post-doctoral or a post-graduate fellowship position at the mass spectrom'try and proteomics laboratory at the Smithsonian's Museum Conservation Institute. The aim of the laboratory is to develop mass spectrometry and proteomics technologies relevant to museums' specimens. Projects include but are not limited to: proteomics, biological dating using various MS techniques such as amino acid racemization; analysis of insoluble proteinaceous or polymeric materials; analysis of paints, inks, etc. using surface ionization techniques; and development of portable separation-mass spectrometry devices for onsite chemical/biological analysis. Qualified applicants should send a letter of interest and resume, including lists of publications and references to Dr. Mehdi Moini (moinim@si.edu).
*Requirements:* The successful applicant will have a Ph.D. or Master's degree in the area of separation, microfabrication, mass spectrometry, or proteomics. Working experience with Thermo LTQ Orbitrap Veloes, ABI Qstar and 4700 TOF/TOF, as well as with capillary electrophoresis and nano-LC is highly desirable.
*Geographic Focus:* All States
*Contact:* Paula T. DePriest, Deputy Director; (301) 238-1206 or (301) 238-1240; fax (301) 238-3709; DepriestP@si.edu or MCIweb@si.edu
*Internet:* si.edu/MCI/english/professional_development/2012ResearchPostDoctorate.html
*Sponsor:* Smithsonian Institution
4210 Silver Hill Road
Suitland, MD 20746

**Smithsonian National Air and Space Museum Aviation Space Writers Foundation Grant** 2353
The National Air and Space Museum is offering a grant of $5,000, to be awarded in even-numbered years (2012, 2014, 2016 etc.), to support research on aerospace topics. The product created as a result of the grant must be in any form suitable for potential public dissemination in print, electronic, or broadcast or other visual medium, including, but not limited to a book manuscript, a video or film script, or monograph. Potential topics might be contemporary aviation or space events that are of interest to the general public; significant persons, historical events or trends that illuminate the history of human flight in air and space; records or compendia of aerospace source material. Funds may be used to support research travel and expenses, or to produce a product of the research as outlined.
*Requirements:* Recipients of the award need not be in residence at the National Air and Space Museum. In addition, award winners are required to provide a summary report in the form of a memorandum.
*Geographic Focus:* All States
*Date(s) Application is Due:* Jan 15
*Amount of Grant:* 5,000 USD
*Contact:* Collette Williams, Grant Manager; (202) 633-2648; fax (202) 786-2447; williamsce@si.edu
*Internet:* airandspace.si.edu/getinvolved/fellow/writer_grant.cfm
*Sponsor:* Smithsonian Institution
Sixth and Independence Avenue SW, Room 3313, MRC 312
Washington, D.C. 20560

**Smithsonian National Air and Space Museum Charles A. Lindbergh Chair in Aerospace History** 2354

Charles A. Lindbergh (1902-1974) was one of the most significant figures in the history of aviation. On May 20-21, 1927, he made a historic nonstop solo flight in the Ryan NY-P Spirit of St. Louis from New York to Paris that changed the face of aviation for all time. As a result, he was given international honors and acclaimed a hero. The Charles A. Lindbergh Chair in Aerospace History is a competitive twelve-month fellowship open to senior scholars with distinguished records of publication who are at work on, or anticipate being at work on, books in aerospace history. Support is available for replacement of salary and benefits up to a maximum of $100,000 a year. Tenure is typically for an academic year (September through August) and applications are due January 15 of the year preceding the award of the fellowship.

*Requirements:* The Lindbergh Chair is open to established and recognized senior scholars with distinguished records of publication who are at work on, or anticipate being at work on, books in aerospace history. Residency in the Washington, D.C. area is a requirement. Interaction with Museum professional staff is highly desired and expected, including attendance at selected professional seminars and colloquia.

*Geographic Focus:* All States

*Date(s) Application is Due:* Jan 31

*Contact:* Collette Williams, Fellowship Coordinator; (202) 633-2648; fax (202) 786-2447; williamsce@si.edu

*Internet:* airandspace.si.edu/getinvolved/fellow/lindfellow.cfm

*Sponsor:* Smithsonian Institution

Sixth and Independence Avenue SW, Room 3313, MRC 312

Washington, D.C. 20560

**Smithsonian National Air and Space Museum Guggenheim Fellowships** 2355

The Daniel and Florence Guggenheim Foundation was established in 1924 to advance the "well-being of mankind throughout the world." The Foundation supported Robert Goddard's work on the development of liquid-propelled rockets and established many research centers and scholarly activities concerned with the aerospace sciences and exploration. The National Air and Space Museum Guggenheim Fellowships are designed to promote research into, and writing about, the history of aviation and space flight. Candidates are encouraged to pursue programs of research and writing that support publication of works that are scholarly in tone and substance and intended for publication as articles in peer-reviewed journals or in book form from a reputable publisher (in the case of post postdoctoral applicants) or in a doctoral dissertation (in the case of pre-docs). Each fellow will work closely with staff members who have similar interests. Staff members of the Aeronautics Division currently conduct research into aviation history including such themes as the growth and impact of aeronautics on society, the evolution of aircraft technology, and the development of air transport and military aviation. The Division emphasizes both U.S. and international aspects of aviation history. Members of the Space History Division conduct research in the history of post-war science and technology and the history of space flight, and have written major works on the history of rocketry and the origins of space science.

*Requirements:* The Guggenheim Fellowships are a competitive three- to twelve-month in-residence fellowship for pre- or postdoctoral research in aviation and space history. Predoctoral applicants should have completed preliminary course work and examinations and be engaged in dissertation research. Postdoctoral applicants should have received their Ph.D. within the past seven years. An annual stipend of $30,000 for predoctoral candidates and $45,000 for postdoctoral candidates will be awarded, with limited additional funds for travel and miscellaneous.

*Geographic Focus:* All States

*Date(s) Application is Due:* Jan 15

*Amount of Grant:* 30,000 - 45,000 USD

*Contact:* Collette Williams; (202) 633-2648; fax (202) 786-2447; williamsce@si.edu

*Internet:* airandspace.si.edu/getinvolved/fellow/gfellow.cfm

*Sponsor:* Smithsonian Institution

Sixth and Independence Avenue SW, Room 3313, MRC 312

Washington, D.C. 20560

**Smithsonian National Air and Space Museum Postdoctoral Earth and Planetary Sciences Fellowship** 2356

The National Air and Space Museum has established the Postdoctoral Earth and Planetary Sciences Fellowship to support scientific research in this area. Scientists in the Center for Earth and Planetary Studies concentrate on geologic and geophysical research of the Earth and other terrestrial planets, using remote sensing data obtained from Earth-orbiting and interplanetary spacecraft. Research also focuses on global environmental change. Appointments can be made for one or more years. Stipends are compatible with NRC postdoctoral fellowships in the applicant's field. Applications for fellowships at the National Air and Space Museum are typically posted in November. The application deadline is January 15.

*Geographic Focus:* All States

*Date(s) Application is Due:* Jan 15

*Contact:* Collette Williams, Fellowship Coordinator; (202) 633-2648; fax (202) 786-2447; williamsce@si.edu or watterst@si.edu

*Internet:* airandspace.si.edu/getinvolved/fellow/cepsfell.cfm

*Sponsor:* Smithsonian Institution

Sixth and Independence Avenue SW, Room 3313, MRC 312

Washington, D.C. 20560

**Smithsonian National Air and Space Museum Verville Fellowships** 2357

Alfred V. Verville (1890-1970), an innovative designer, made numerous contributions to civilian and military aviation during his 47 years in the aerospace industry. The Verville Fellowship is designed to promote research into, and writing about, the history of aviation and space flight. Candidates are encouraged to pursue programs of research and writing professional in tone and substance, but addressed to an audience with broad interests. Each fellow will work closely with staff members who have similar interests. Staff members of the Aeronautics Division currently conduct research into aviation history including such themes as the growth and impact of aeronautics on society, the evolution of aircraft technology, and the development of air transport and military aviation. The Division emphasizes both U.S. and international aspects of aviation history. Members of the Space History Division conduct research in the history of post-war science and technology and the history of space flight, and have written major works on the history of rocketry and the origins of space science.

*Requirements:* The Verville Fellowship is a competitive nine- to twelve-month in-residence fellowship intended for the analysis of major trends, developments, and accomplishments in the history of aviation or space studies. The fellowship is open to all interested candidates with demonstrated skills in research and writing. An advanced degree in history, engineering, or related fields is not a requirement. An annual stipend of $55,000 will be awarded for a 12-month fellowship, with limited additional funds for travel and miscellaneous expenses.

*Geographic Focus:* All States

*Date(s) Application is Due:* Jan 15

*Amount of Grant:* 55,000 USD

*Contact:* Collette Williams, Fellowship Coordinator; (202) 633-2648 or (202) 357-2515; fax (202) 786-2447; williamsce@si.edu

*Internet:* airandspace.si.edu/getinvolved/fellow/vfellow.cfm

*Sponsor:* Smithsonian Institution

Sixth and Independence Avenue SW, Room 3313, MRC 312

Washington, D.C. 20560

**Smithsonian National Museum of American History Lemelson Center Fellowships** 2358

The Lemelson Center Fellows Program supports projects that present creative approaches to the study of invention and innovation in American society. These include, but are not limited to, historical research and documentation projects, resulting in publications, exhibitions, educational initiatives, and multimedia products. The Lemelson Center invites application covering a broad spectrum of research topics that resonate with its mission to foster a greater understanding of invention and innovation, broadly defined. However, the Center especially encourages project proposals that will illuminate the role of women inventors; inventors with disabilities; inventors from diverse backgrounds; or any inventions and technologies associated with groups that are traditionally under-represented in the historical record. Fellowship tenure is based upon the applicants' stated needs (and available funding) up to a maximum of ten weeks. The term is up to ten weeks in length, with pre-doctoral fellows earning a $575 stipend per week, and post-doctoral and professionals earning an $870 stipend per week.

*Requirements:* The Center offers fellowships to pre-doctoral graduate students, postdoctoral scholars, and other professionals who have completed advanced training. Fellows are expected to reside in the Washington, D.C. area, to participate in the Center's activities, and to make presentations on their work to colleagues at the museum.

*Geographic Focus:* All States

*Date(s) Application is Due:* Jan 15

*Amount of Grant:* 575 - 8,700 USD

*Contact:* Omar Eaton-Martínez, Internship and Fellowship Manager; (202) 633-3556; fax (202) 786-2851

Eric Hintz, Historian; (202) 633-3734; fax (202) 357-4517; hintze@si.edu

*Internet:* americanhistory.si.edu/getinvolved/fellowindex.cfm

*Sponsor:* Smithsonian Institution

14th Street and Constitution Avenue, NW

Washington, D.C. 20013

**Smithsonian National Museum of National History Lemelson Center Travel to Collections Awards** 2359

The Travel to Collections Awards Program supports research on the history of invention and innovation based on the holdings of the Museum's Archives Center and curatorial divisions. The Archives Center holds more than 20,000 feet of archival materials. The collections are particularly strong in personal papers and business records documenting the history of American enterprise and technology. To encourage use of its invention-related collections, the Lemelson Center offers five short-term travel awards per year. Awards are $150 per day for ten business days maximum. The travel award may be used to cover transportation and living expenses, and copying pertinent archival resources. Applications are accepted from June 1 to July 1 each year.

*Requirements:* Scholars, graduate students, and independent researchers not residing or attending school within commuting distance of the National Museum of American History may apply for this program. Recipients must commence their research at the National Museum of American History within one year of being notified of the award.

*Restrictions:* Awards may not be used to extend other Smithsonian appointments. Only one award can be offered to a visitor within a twelve-month period.

*Geographic Focus:* All States

*Date(s) Application is Due:* Jul 1

*Amount of Grant:* 150 - 1,500 USD

*Contact:* Omar Eaton-Martínez; (202) 633-3556; fax (202) 786-2851

*Internet:* invention.smithsonian.org/resources/research_travel.aspx

*Sponsor:* Smithsonian Institution
14th Street and Constitution Avenue, NW
Washington, D.C. 20013

**Smithsonian Native American Community Scholars Awards** 2360
The Office of Fellowships offers Community Scholars Awards to Native Americans who are formally or informally related to a Native American community to undertake individually designed research projects related to Native American topics and using Native American resources at the Smithsonian. The stipend is equal to $150 per day for up to 21 days, with a travel allowance and small research allowance also being offered. Deadlines are February 1 (to start after June 1), June 1 (to start after October 1), and October 1 (to start after January 1).
*Requirements:* Native American undergraduate and graduate students are eligible.
*Geographic Focus:* All States
*Date(s) Application is Due:* Feb 1; Jun 1; Oct 1
*Amount of Grant:* Up to 3,500 USD
*Contact:* Bryan T. Fair; (202) 275-0655 or (202) 633-7070; fax (202) 275-0489; siofg@si.edu
*Internet:* www.si.edu/ofg/fell.htm
*Sponsor:* Smithsonian Institution
470 L'Enfant Plaza SW, Suite 7102, MRC 902 P.O. Box 37012
Washington, D.C. 20013-7012

**Smithsonian Native American Visiting Student Awards** 2361
Appointments are available for currently enrolled advanced Native American graduate students. Visiting students pursue independent research in association with Smithsonian staff. Stipends are per day or per week and include a travel allowance and small research allowance. This award is granted three times a year, depending upon when the funds are requested. The stipend is $150 per day for up to a maximum of 21 days, and $550 per week for three to ten weeks. A travel allowance and a small research allowance is also offered. Deadlines are February 1 (to start after June 1), June 1 (to start after October 1), and October 1 (to start after January 1).
*Requirements:* Native American graduate students are eligible.
*Geographic Focus:* All States
*Date(s) Application is Due:* Feb 1; Jun 1; Oct 1
*Amount of Grant:* 150 - 5,500 USD
*Contact:* Bryan T. Fair, Program Assistant; (202) 275-0655 or (202) 633-7070; fax (202) 275-0489; siofg@si.edu
*Internet:* www.si.edu/ofg/fell.htm
*Sponsor:* Smithsonian Institution
470 L'Enfant Plaza SW, Suite 7102, MRC 902 P.O. Box 37012
Washington, D.C. 20013-7012

**Smithsonian Postdoctoral Fellowships** 2362
Postdoctoral Fellowships of three to twelve months are available for scholars who have held the doctoral degree or equivalent for fewer than seven years as of the application deadline. Proposals for research may be offered in fields of study conducted at the institute (American history, art, and culture; history of science and technology; history of art, design, crafts, and decorative arts; anthropology, archaeology, linguistics, and ethnic studies; evolutionary, systematic, behavioral, and environmental biology and conservation; earth, mineral, and planetary sciences; and materials characterization and conservation). The fellowship term is three to 12 months. Applications are posted on the Web site each September.
*Requirements:* Applicants must have a PhD or equivalent degree for less than seven years.
*Geographic Focus:* All States
*Date(s) Application is Due:* Jan 15
*Amount of Grant:* 45,000 USD
*Contact:* Bryan T. Fair; (202) 275-0655; fax (202) 275-0489; siofg@si.edu
*Internet:* www.si.edu/ofg/fell.htm
*Sponsor:* Smithsonian Institution
470 L'Enfant Plaza SW, Suite 7102, MRC 902 P.O. Box 37012
Washington, D.C. 20013-7012

**Smithsonian Postgraduate Fellowships in Conservation of Museum Collections** 2363
Smithsonian Postgraduate Fellowships in Conservation of Museum Collections are offered to recent graduates of masters programs in art conservation or the equivalent or conservation scientists, including those at the postdoctoral level, who wish to conduct research and gain further training in Smithsonian conservation laboratories for a period of one year. Additional facilities may be available to museum or archives fellows for analytical work at the Museum Conservation Institute (MCI). Fellowship duration in this category is three to twelve months. A stipend of $35,000 is being offered, plus allowances. The deadline is January 15 for awards to begin on or after June 1. Applicants are urged to apply concurrently to all other Smithsonian programs for which they may be eligible.
*Geographic Focus:* All States
*Date(s) Application is Due:* Jan 15
*Amount of Grant:* 35,000 USD
*Contact:* Bryan T. Fair, Program Assistant; (202) 275-0655 or (202) 633-7070; fax (202) 275-0489; siofg@si.edu
*Internet:* www.si.edu/ofg/fell.htm
*Sponsor:* Smithsonian Institution
470 L'Enfant Plaza SW, Suite 7102, MRC 902 P.O. Box 37012
Washington, D.C. 20013-7012

**Smithsonian Predoctoral Fellowships** 2364
Fellowships support independently conceived and designed research study by predoctoral students to work with Smithsonian staff members and use Smithsonian collections, facilities, and other special resources. Proposals are accepted for research in the areas of study conducted at the institute (American history, art, and culture; history of science and technology; history of art, design, crafts, and decorative arts; anthropology, archaeology, linguistics, and ethnic studies; evolutionary, systematic, behavioral, and environmental biology and conservation; earth, mineral, and planetary sciences; and materials characterization and conservation). The fellowship term is three to 12 months. Applications are posted to the Web site in September.
*Requirements:* Candidates must have completed all coursework and examinations and be ready to engage in dissertation research.
*Geographic Focus:* All States
*Date(s) Application is Due:* Jan 15
*Amount of Grant:* 30,000 USD
*Contact:* Bryan T. Fair, Program Assistant; (202) 275-0655 or (202) 633-7070; fax (202) 275-0489; siofg@si.edu
*Internet:* www.si.edu/ofg/fell.htm
*Sponsor:* Smithsonian Institution
470 L'Enfant Plaza SW, Suite 7102, MRC 902 P.O. Box 37012
Washington, D.C. 20013-7012

**Smithsonian Senior Fellowships** 2365
Senior Fellowships of three to twelve months are available for scholars who have held the doctoral degree or equivalent for more than seven years as of the application deadline. Applications for these fellowships may be made up to eighteen months in advance. Stipends for senior fellowships are the same as for the postdoctoral program, but the Smithsonian's stipend may be matched by other sources of funding such as a sabbatical salary. Each year, appointments are made to distinguished senior scientists for research in the fields of anthropology, natural history, environmental science and ecological processes, tropical biology, African art and culture, oriental art, astrophysics and related topics, history of science and technology, American cultural and social history, or art history. Prior to submitting applications, interested senior scientists should consult with Smithsonian staff in their chosen fields. Applications are posted to the Web site in September.
*Requirements:* Applicants must have held the PhD or equivalent degree for at least seven years.
*Geographic Focus:* All States
*Date(s) Application is Due:* Jan 15
*Amount of Grant:* 45,000 USD
*Contact:* Bryan T. Fair, Program Assistant; (202) 275-0655 or (202) 633-7070; fax (202) 275-0489; siofg@si.edu
*Internet:* www.si.edu/ofg/fell.htm
*Sponsor:* Smithsonian Institution
470 L'Enfant Plaza SW, Suite 7102, MRC 902 P.O. Box 37012
Washington, D.C. 20013-7012

**SOBP A.E. Bennett Research Award** 2366
The Society of Biological Psychiatry offers an annual award of $5,000 each in basic science and in clinical science for the purpose of stimulating international research in biological psychiatry by young investigators. Candidates must be actively engaged in the research for which the award is sought. Applicants need only write a brief description of their research, submit two to three published or submitted papers and arrange for two letters of support (at least one of the letters needs to be from a member of the Society). Although the research is not to be judged in comparison with the work of the more senior investigators, special consideration will be given to the originality of the approach and independence of thought evident in the submission. Nominations should be submitted on-line on or before September 30.
*Requirements:* Submissions are welcomed from laboratory researchers in any country who have not passed their 45th birthday or have not been engaged in research for greater than 10 years following award of their terminal degree or the end of formal clinical/fellowship training, whichever is later by deadline date; candidates need not be current members of the society.
*Geographic Focus:* All States, All Countries
*Date(s) Application is Due:* Sep 30
*Amount of Grant:* 5,000 USD
*Contact:* Maggie Peterson, Executive Director; (904) 953-2842 or (615) 432-0096; fax (904) 953-7117; maggie@mayo.edu or sobp@sobp.org
*Internet:* https://sobp.org/a-e-bennett-research-award/
*Sponsor:* Society of Biological Psychiatry
5034A Thoroughbred Lane
Brentwood, TN 37027

**SOBP Domestic Travel Fellowship Award for Early Career Investigator** 2367
The training of a new generation of psychiatric academicians is a major objective of the Society of Biological Psychiatry (SOBP). Participation in professional meetings, where new information is exchanged and contacts are made, can have a critical impact on the career of a developing clinician-scientist. In recognition of the importance of academic exchange, SOBP annually selects distinguished Early Career Investigators from North American institutions to be a part of our Travel Fellowship Award program. Travel Awards recognize excellence in scholarly activity by junior researchers by providing the opportunity to attend the annual meeting and participate in special programming for the awardees. These awards are supported by the Society of Biological Psychiatry and in part by Elsevier, publisher of the Society's journals, Biological Psychiatry and Biological Psychiatry: Cognitive Neuroscience and Neuroimaging. The award is equal to $2,000, which can be used to help defray the travel and living cost to attend the annual meeting.

The deadline to submit the online application is on or before September 30. Up to 34 candidates may be selected for the awards.
*Requirements:* Individuals from any university, department or institute in the U.S., Canada or Mexico may submit an application. Applicant must be within ten years of completing their terminal research degree or within ten years of completing their medical residency at the time of application.
*Restrictions:* Academic faculty with rank higher than Assistant Professor (or equivalent) are not eligible to apply.
*Geographic Focus* All States, District of Columbia, Guam, Marshall Islands, Northern Mariana Islands, Puerto Rico, U.S. Virgin Islands, American Samoa
*Date(s) Application is Due:* Sep 30
*Amount of Grant:* Up to 2,000 USD
*Contact:* Maggie Peterson, Executive Director; (904) 953-2842 or (615) 432-0096; fax (904) 953-7117; maggie@mayo.edu or sobp@sobp.org
*Internet:* https://sobp.org/travel-fellowship-award-early-career-investigator-domestic/
*Sponsor:* Society of Biological Psychiatry
5034A Thoroughbred Lane
Brentwood, TN 37027

**SOBP George N. Thompson Award for Distinguished Service** **2368**
The Society of Biological Psychiatry established the George N. Thompson Award for Distinguished Service in 1982. The award recognized Dr. George N. Thompson's role as one of the founders of the Society in 1944 and his contributions over the years to the growth of the Society. The award is intended to honor members of the Society who have given outstanding service to promote the welfare of the organization. The recipient may be honored for work in the areas of leadership, governance, development of publications, special projects, financial contributions, or other service recognized as distinguished by the Council. The award will be advertised yearly, but Council reserves the right to make no award in any year. Members of the Society should submit a letter describing the distinguished work of the member being nominated. The award committee will review all nominations and recommend a recipient to Council for approval. All applications must be submitted on-line.
*Requirements:* Nominations must be submitted on-line before or on September 30.
*Geographic Focus* All States, District of Columbia, Guam, Marshall Islands, Northern Mariana Islands, Puerto Rico, U.S. Virgin Islands, American Samoa
*Date(s) Application is Due:* Sep 30
*Contact:* Maggie Peterson, Executive Director; (904) 953-2842 or (615) 432-0096; fax (904) 953-7117; maggie@mayo.edu or sobp@sobp.org
*Internet:* https://sobp.org/george-n-thompson-award-for-distinguished-service/
*Sponsor:* Society of Biological Psychiatry
5034A Thoroughbred Lane
Brentwood, TN 37027

**SOBP Gold Medal Award** **2369**
The Gold Medal Award was established by the Society of Biological Psychiatry to honor pioneering contributions to the field and for significant and sustained work that advances and extends knowledge in biological psychiatry. Nominees need not be members of the Society of Biological Psychiatry. The decision of the award committee shall be based on the nominee's scientific accomplishments, their independence, originality, and impact; their development of new concepts, theories and techniques, that constitute a definitive, mature and significant contribution to knowledge sustained over a period of years. An award may be made every year, but the award committee and Council reserve the right to make no award in any year. Members of the Society or members of other national or international scientific associations may submit nominations. Nominations must be submitted on-line before or on September 30.
*Requirements:* All scientists working in the field of biological psychiatry including but not limited to individuals from academic institutions, foundations, governmental, industrial and research organizations regardless of age, sex, race or nationality shall be eligible.
*Restrictions:* Members of the award committee are not eligible to receive the award during their tenure on the award committee.
*Geographic Focus* All States, District of Columbia, Guam, Marshall Islands, Northern Mariana Islands, Puerto Rico, U.S. Virgin Islands, American Samoa
*Date(s) Application is Due:* Sep 30
*Contact:* Maggie Peterson, Executive Director; (904) 953-2842 or (615) 432-0096; fax (904) 953-7117; maggie@mayo.edu or sobp@sobp.org
*Internet:* https://sobp.org/society-of-biological-psychiatry-gold-medal-award/
*Sponsor:* Society of Biological Psychiatry
5034A Thoroughbred Lane
Brentwood, TN 37027

**SOBP Humanitarian Award** **2370**
The Society of Biological Psychiatry Humanitarian Award is intended to recognize those who have demonstrated service and advocacy for mental illness. Nomination materials should include a statement or letter describing the the nominee's service and advocacy for mental illness. An award may be made every year, but the award committee and Council reserve the right to make no award in any year. Members of the Society or members of other national or international scientific associations may submit nominations. Nominations must be submitted on-line on or before September 30.
*Restrictions:* Members of the award committee are not eligible to receive the award during their tenure on the award committee.
*Geographic Focus* All States, District of Columbia, Guam, Marshall Islands, Northern Mariana Islands, Puerto Rico, U.S. Virgin Islands, American Samoa
*Date(s) Application is Due:* Sep 30
*Contact:* Maggie Peterson, Executive Director; (904) 953-2842 or (615) 432-0096; fax (904) 953-7117; maggie@mayo.edu or sobp@sobp.org
*Internet:* https://sobp.org/society-of-biological-psychiatry-humanitarian-award/
*Sponsor:* Society of Biological Psychiatry
5034A Thoroughbred Lane
Brentwood, TN 37027

**SOBP International Travel Fellowship Award for Early Career Investigators** **2371**
In recognition of the importance of academic exchange, the Society of Biological Psychiatry (SOBP) annually selects distinguished early career investigators from outside of North American and from one of the following five IBRO Regions (Africa, Asian/Pacific, Central and Eastern Europe, Latin America, Western Europe) who would benefit from attending the annual meeting of the Society of Biological Psychiatry. Priority will be given to applicants from less developed and less well-funded countries. This program will provide exposure of the fellowship recipients to the best science the Society has to offer and facilitate the professional contacts that may yield enduring, collaborative and collegial relationships. Travel Awards recognize excellence in scholarly activity by junior researchers by providing the opportunity to attend the annual meeting and participate in special programming for the awardees. These awards are supported by the Society of Biological Psychiatry and in part by Elsevier, publisher of the Society's journals, Biological Psychiatry and Biological Psychiatry: Cognitive Neuroscience and Neuroimaging. The award is equal to $2,000, which can be used to help defray the travel and living cost to attend the annual meeting. The deadline to submit the online application is on or before September 30.
*Requirements:* This award is designed for Predoctoral Scholars currently enrolled in an academic program (MD, MD/PhD, PsyD, PharmD, MA and MS Students, etc) from any university, graduate study department or institute in the following five IBRO regions: Africa, Asian/Pacific, Pan European, Latin America, or Middle East/Northern Africa.
*Geographic Focus:* All Countries
*Date(s) Application is Due:* Sep 30
*Amount of Grant:* Up to 2,000 USD
*Contact:* Maggie Peterson, Executive Director; (904) 953-2842 or (615) 432-0096; fax (904) 953-7117; maggie@mayo.edu or sobp@sobp.org
*Internet:* https://sobp.org/travel-fellowship-awards-early-career-investigator-domestic/
*Sponsor:* Society of Biological Psychiatry
5034A Thoroughbred Lane
Brentwood, TN 37027

**SOBP Somerfeld-Ziskind Research Awards** **2372**
The Society of Biological Psychiatry offers an annual award of $5,000 in either basic or clinical research for the purpose of stimulating investigations in biological psychiatry by senior investigators who are members in good standing of the Society of Biological Psychiatry. The Ziskind-Somerfeld Research Foundation makes this award possible. No submission is required to be considered for this award. The research will be judged in comparison with the work of other senior investigators. Special consideration is given to the originality of the approach and independence of thought evident in the submission. The monetary value of this award is $5,000.
*Requirements:* Candidates will be at least thirty-five years of age at the time of submission. When more than one investigator is listed as an author on the submitted work, the principal investigators (first and last authors) must be over the age of 45.
*Geographic Focus* All States, District of Columbia, Guam, Marshall Islands, Northern Mariana Islands, Puerto Rico, U.S. Virgin Islands, American Samoa
*Amount of Grant:* 5,000 USD
*Contact:* Maggie Peterson, Executive Director; (904) 953-2842 or (615) 432-0096; fax (904) 953-7117; maggie@mayo.edu or sobp@sobp.org
*Internet:* https://sobp.org/ziskind-somerfeld-research-award/
*Sponsor:* Society of Biological Psychiatry
5034A Thoroughbred Lane
Brentwood, TN 37027

**SOBP Travel Fellowship Award for Predoctoral Scholars** **2373**
The training of a new generation of psychiatric academicians is a major objective of the Society of Biological Psychiatry. Participation in professional meetings, where new information is exchanged and contacts are made, can have a critical impact on the career of a developing clinician-scientist. In recognition of the importance of academic exchange, the SOBP annually selects Predoctoral Scholars for travel fellowship awards to attend the Annual Meeting. Travel Awards recognize excellence in scholarly activity by junior researchers by providing the opportunity to attend the annual meeting and participate in special programming for the awardees. These awards are supported by the Society of Biological Psychiatry and in part by Elsevier, publisher of the Society's journals, Biological Psychiatry and Biological Psychiatry: Cognitive Neuroscience and Neuroimaging. The award is equal to $2,000, which can be used to help defray the travel and living cost to attend the annual meeting. The deadline to submit the online application is on or before September 30.
*Requirements:* This award is designed for Predoctoral Scholars currently enrolled in an academic program (MD, MD/PhD, PsyD, PharmD, MA and MS Students, etc) from any university, graduate study department or institute in the U.S., Canada, and Mexico, or outside of North America from the following five IBRO regions: Africa, Asian/Pacific, Pan European, Latin America, or Middle East/Northern Africa.
*Geographic Focus* All States, District of Columbia, Guam, Marshall Islands, Northern Mariana Islands, Puerto Rico, U.S. Virgin Islands, American Samoa, Canada, Mexico
*Date(s) Application is Due:* Sep 30
*Amount of Grant:* Up to 2,000 USD

*Contact:* Maggie Peterson, Executive Director; (904) 953-2842 or (615) 432-0096; fax (904) 953-7117; maggie@mayo.edu or sobp@sobp.org
*Internet:* https://sobp.org/travel-fellowship-award-predoctoral-scholars/
*Sponsor:* Society of Biological Psychiatry
5034A Thoroughbred Lane
Brentwood, TN 37027

**Sony Corporation of America Grants** 2374

Sony Corporation of America consists of three operating companies as well as the corporate headquarters, which is based in New York City: Sony Electronics, Inc. (headquarters in San Diego, California); Sony Pictures Entertainment, Inc. (headquarters in Culver City, California); and Sony Music Entertainment, Inc. (headquarters in New York City). Each company, as well as the overall corporation, has its own philanthropic priorities and resources (e.g. grants, product donations, and recordings and screenings) that benefit a multitude of causes. Taken together the corporation's areas of interest cover arts and culture; health and human services; civic and community outreach, education; the environment; disaster response; and volunteerism. The core of Sony's various corporate philanthropy programs are their contributions to the communities in which Sony employees work and live; however, the corporation and subsidiaries contribute to national nonprofits as well. In the past, types of support have included general operating budgets, continuing support, annual campaigns, seed grants, building construction, equipment acquisition, endowment funds, employee matching gifts, internships, and employee-related scholarships. Sony Corporation of America and its subsidiaries welcome requests for support throughout the year. There is no grant application form. Requests must be submitted in writing to the corporation or its operating companies. Contact information is given on this page and at the website. Guidelines for what to include in the application as well as more information on types of programs the corporation and/or its subsidiaries have supported are available at the grant website. Notification of grant-request approval or rejection will be made in writing within one month of receipt of all proposed materials.
*Requirements:* U.S. nonprofits, including schools and school districts, are eligible.
*Restrictions:* The corporation does not consider multi-year requests for support. The following types of organizations will not be funded: organizations that discriminate on the basis of race, color, creed, gender, religion, age, national origin, or sexual orientation; partisan political organizations, committees, or candidates and public office holders; religious organizations in support of their sacramental or theological functions; labor unions; endowment or capital campaigns of national origin; organizations whose prime purpose is to influence legislation; testimonial dinners in general; for-profit publications or organizations seeking advertisements or promotional support; individuals seeking self-advancement; foreign or non-U.S.-based organizations; and organizations whose mission is outside of the U.S.
*Geographic Focus:* All States
*Amount of Grant:* 1,000 - 100,000 USD
*Contact:* Janice Pober, Senior Vice President, Corporate Social Responsibility; (310) 244-7737; SPE_CSR@spe.sony.com
Karen E. Kelso, Vice-President; (212) 833-8000; SCA_CSR@sonyusa.com
Julie Wenzel, Senior Manager, Community Relations; (858) 942-2400; SELCommunityAffairs@am.sony.com
*Internet:* www.sony.com/SCA/philanthropy/guidelines.shtml
*Sponsor:* Sony Corporation of America
550 Madison Avenue, 33rd Floor
New York, NY 10022-3211

**Special Olympics Eunice Kennedy Shriver (EKS) Fellowships** 2375

The Eunice Kennedy Shriver (EKS) Program aims to develop leadership on the field of intellectual disability through development of cross-sector bonds among professionals working in the field of intellectual disability in developing countries. Fellowships will be granted to candidates with a commitment to NGO or governmental efforts that support people with intellectual disabilities. Participants will receive an intense, hands-on learning experience that will benefit them both professionally and personally. Fellows will receive Mentors in their Host Programs that will support their learning. Likewise, the Special Olympics Program in their home country will provide aid in their six-month volunteer project upon their return. Fellows will also receive programmatic support from a dedicated team of professionals based in the Washington, D.C Special Olympics headquarters office. All housing, food, transportation to and from the host country and other travel as directed by Special Olympics, Inc. will be covered at the U.S. government per diem rate. In addition, a stipend for other living expenses will be awarded to each Fellow in the Program.
*Requirements:* Fellows will be financially responsible for obtaining their passports and visas.
*Geographic Focus:* All States
*Contact:* Lisa Dietz, EKS Fellowship Program Specialist; (202) 824-0228; fax (202) 824-0200; ldietz@specialolympics.org
*Internet:* www.specialolympics.org/EKS_fellowship.aspx
*Sponsor:* Special Olympics
1133 19th Street, NW
Washington, D.C. 20036-3604

**Special Olympics Health Professions Student Grants** 2376

The purpose of this program is to engage health professions students to work with persons with intellectual disabilities as a way of filling in a gap that exists in most health program curricula. The program promotes short-term projects exploring issues that impact the health and well-being of all persons with intellectual disabilities, including, but not limited to, Special Olympics athletes. Projects may include: data collection and analysis on issues impacting persons with intellectual disabilities; measurement of attitudes, opinions and behaviors of health professionals, coaches, family/caregivers and athletes; follow-up assessments of existing programs; or health promotion projects. Projects that involve collaborations with Special Olympics Programs or other CD.C. grant recipients (e.g., state and local health departments) are encouraged. Special Olympics is not currently prioritizing specific research areas for funding, as the program's primary goal is to engage health professions students to work with the population of persons with intellectual disabilities. All awards will be made to the institution on behalf of the student grantee. Funding must include all expenses and no indirect costs may be covered by this grant. Applications are accepted year-round. Application and guidelines are available online.
*Requirements:* Health professions students at the bachelor, master, doctorate and post-doctorate levels from accredited institutions or programs in the United States or other countries are eligible. Fields or disciplines may include, audiology, dentistry, pre-med, medicine, nursing, optometry, physical therapy, podiatry, public health, social work, law, public administration, or other related health areas. All students??? projects will require a faculty advisor.
*Geographic Focus:* All States
*Amount of Grant:* Up to 5,000 USD
*Contact:* Darcie Mersereau, Manager, Research and Evaluation; (202) 628-3600 or (800) 700-8585; fax (202) 824-0200; 2006hastudentgrant@specialolympics.org
*Internet:* www.specialolympics.org/content.aspx?id=6336&terms=student
*Sponsor:* Special Olympics
1133 19th Street, NW
Washington, D.C. 20036-3604

**Special Olympics Youth Fan Grants** 2377

Special Olympics is the leader in cutting-edge research and evaluation to better understand the many challenges faced by people with intellectual disabilities and the significant impact of Special Olympics on their lives. The program's independent research is also a driving force for realizing improved policies, laws and rights for people with intellectual disabilities around the world. Projects may be research-based or programmatic (conferences, health promotion programs, etc), but if they are programmatic they must still contain an evaluation component.
*Geographic Focus:* All States
*Contact:* Kathleen Palermo, Grants Coordinator – Healthy Athletes; (202) 628-3630 or (202) 824-0238; fax (202) 628-3926; kpalermo@specialolympics.org
*Internet:* www.specialolympics.org/research_studies.aspx
*Sponsor:* Special Olympics
1133 19th Street, NW
Washington, D.C. 20036-3604

**Spencer County Community Foundation Health Grants** 2378

The Spencer County Community Foundation is a nonprofit, public charity created for the people of Spencer County, Indiana. The Foundation helps nonprofits fulfill their missions by strengthening their ability to meet community needs through grants that assist charitable programs, address community issues, support community agencies, launch community initiatives, and support leadership development. Grant proposals are accepted year round, and encompasses three different request levels: level one requests of $2,500 and below, with an online grant application required to apply); level two requests of $2,501 to $10,000, with an online grant application also required to apply; and level three requests over $10,000, with an online letter of inquiry and face-to-face meeting with the grants committee required to apply). Grants are typically given as one time support of a project, but may be considered for additional support for expansions or outgrowths of an initial project. Funding requests will be reviewed throughout the year as follows: level one on September 15, December 15, March 15, and June 15; level two on December 15 and June 15; and level three is open-ended. Grants in the area of health include activities that: improve and promote health outcomes; provide general and rehabilitative health services; offer mental health services; provide crisis intervention programs; strengthen associations or services associated with specific diseases, disorders, and medical disciplines; and support medical research. Samples of previously funded projects are posted on the website.
*Requirements:* The Spencer County Community Foundation welcomes proposals from nonprofit organizations that are tax exempt under sections 501(c)3 and 509(a) and from governmental agencies serving the County of Spencer, Indiana. Proposals from nonprofit organizations not classified as a 501(c)3 public charity may be considered if the project is charitable and supports a community need. Proposals submitted by an entity under the auspices of another agency must include a written statement signed by the agency's board president on behalf of the board of directors agreeing to act as the entity's fiscal sponsor, to receive grant monies if awarded, and to oversee the proposed project. School administrators and teachers planning to apply for a grant must contact the Foundation prior to beginning the application process to discuss guidelines pertaining to funding requests from school personnel and school corporation oversight of the application process.
*Restrictions:* Project areas not considered for funding: religious organizations for religious purposes; political parties or campaigns; endowment creation or debt reduction; or annual appeals or membership contributions.
*Geographic Focus:* Indiana
*Date(s) Application is Due:* Mar 15; Jun 15; Sep 15; Dec 15
*Amount of Grant:* Up to 20,000 USD
*Contact:* Laura Harmon, Regional Director; (812) 649-5724; laura@spencercommunityfoundation.org or lharmon@communityfoundationalliance.org
*Internet:* https://www.communityfoundationalliance.org/spencer/program-areas/
*Sponsor:* Spencer County Community Foundation
Lincoln Commerce Center
Rockport, IN 47635

**Spencer Foundation Conference Grants** 2379

The Spencer Foundation's Conference Grant program is designed to provide support to scholars for small research conferences and focused symposia. The conference grant program intends to bring together researchers whose substantive knowledge, theoretical insight, and methodological expertise can be assembled in ways that build upon and advance best practices in education research. Through this grant program, the Foundation rotates the area of focus periodically to generate fresh ideas and perspectives on pressing educational challenges. This grant program supports conference proposals with budgets of $50,000 or less. The Foundation is currently seeking proposals from scholars whose interests are related to measuring educational quality and improvement. The deadlines for proposals is 4:00 pm CDT, May 31 and December 1.
*Restrictions:* The Foundation does not fund: scholarships; curriculum development; program development and implementation; program/curriculum evaluation; or professional development programs.
*Geographic Focus:* All States
*Date(s) Application is Due:* May 31; Dec 1
*Amount of Grant:* Up to 50,000 USD
*Contact:* Judy Klippenstein, Associate Accountant; (312) 274-6536 or (312) 337-7000; fax (312) 337-0282; jklippenstein@spencer.org or conferences@spencer.org
*Internet:* https://www.spencer.org/grant_types/conference-grants
*Sponsor:* Spencer Foundation
625 North Michigan Avenue, Suite 1600
Chicago, IL 60611

**Spencer Foundation Dissertation Fellowships** 2380

Offered in collaboration with the National Academy of Education, the Spencer Foundation Education Dissertation Fellowships program seeks to encourage a new generation of scholars from a wide range of disciplines and professional fields to undertake research relevant to the improvement of education. These $25,000 fellowships support individuals whose dissertations show potential for bringing fresh and constructive perspectives to the history, theory, or practice of formal or informal education anywhere in the world. This highly competitive program aims to identify the most talented researchers conducting dissertation research related to education. Like all Spencer Foundation programs, the Dissertation Fellowship program receives many more applications than it can fund. This year, up to 600 applications are anticipated and up to 25 fellowships will be awarded.
*Requirements:* Grants are made only to institutions, not directly to individuals. Applicants nominated by their institutions must be candidates for the doctoral degree in any field of study at a graduate school in the United States. Dissertation topics must concern education. All predissertation requirements must be completed by June 1 of the upcoming year.
*Restrictions:* Awards are not designed for individuals in the fields of guidance or curriculum development, nor for those planning to be school administrators.
*Geographic Focus:* All States
*Date(s) Application is Due:* Nov 1
*Amount of Grant:* 27,500 USD
*Contact:* Annie Brinkman, Program Contact; (312) 274-6526 or (202) 334-2341; fax (312) 337-0282; abrinkman@spencer.org or info@naeducation.org
*Internet:* https://www.spencer.org/grant_types/dissertation-fellowship
*Sponsor:* Spencer Foundation
625 North Michigan Avenue, Suite 1600
Chicago, IL 60611

**Spencer Foundation Journalism Fellowships** 2381

The Spencer Education Journalism Fellowship, headquartered at Columbia University Graduate School of Journalism, is dedicated to supporting long-form reporting that enhances the public's understanding of education. This fellowship allows journalists to spend the academic year at Columbia University studying with scholars from a variety of disciplines and with mentors in the Journalism School. The program contributes to the Foundation's goal of engaging with the public in order to make education research more accessible. Projects by prior fellows have resulted in books, radio documentaries and major pieces in national magazines and other outlets. Each fellow receives a $75,000 stipend for living expenses and an additional $7,500 for research support. Tuition and fees for courses at Columbia University are included in the fellowship. The application deadline is in the winter for the following academic year, which typically runs from August to mid-June.
*Restrictions:* The Foundation does not fund: scholarships; curriculum development; program development and implementation; program/curriculum evaluation; or professional development programs.
*Geographic Focus:* All States
*Amount of Grant:* 82,500 USD
*Contact:* Maricelle Garcia, Program Contact; (312) 274-6515 or (312) 337-7000; fax (312) 337-0282; mgarcia@spencer.org or midcareer@spencer.org
*Internet:* https://www.spencer.org/grant_types/journalism-program
*Sponsor:* Spencer Foundation
625 North Michigan Avenue, Suite 1600
Chicago, IL 60611

**Spencer Foundation Large Research Grants** 2382

The Spencer Foundation was established in 1962 by Lyle M. Spencer. The Foundation received its major endowment upon Spencer's death in 1968 and began formal grant making in 1971. The Foundation is intended, by Spencer's direction, to investigate ways in which education, broadly conceived, can be improved around the world. From the first, the Foundation has been dedicated to the belief that research is necessary to the improvement in education. The Foundation is thus committed to supporting high-quality investigation of education through its research programs and to strengthening and renewing the educational research community through its fellowship and training programs and related activities. The Foundation has identified the following five areas of inquiry: education and social opportunity; organizational learning; purposes and values of education; teaching, learning, and instructional resources; and field-initiated proposals. Small grants are those $50,000 or less. The Large Research Grants on Education Program supports education research projects that will contribute to the improvement of education, broadly conceived, with budgets ranging from $125,000 up through $500,000 and for projects durations of one to five years. This program is field-initiated in that proposal submissions are not in response to a specific request for a particular research topic, discipline, design, or method. Our goal for this program is to support rigorous, intellectually ambitious and technically sound research that is relevant to the most pressing questions and compelling opportunities in education.
*Requirements:* Initial proposals are accepted at any time, regardless of the level of funding requested. Applicants must possess an earned doctorate and be affiliated with a school, college, university, or cultural institution. Educators currently employed in K-12 settings who are not pursuing a graduate degree are also eligible.
*Restrictions:* Principal investigators are not permitted to receive two research grants simultaneously. Projects may not extend beyond a year in duration, and no extensions will be granted. Grants do not support indirect costs. The Foundation does not fund: scholarships; curriculum development; program development and implementation; program/curriculum evaluation; or professional development programs.
*Geographic Focus:* All States
*Date(s) Application is Due:* Jun 20
*Contact:* Marcelle Garcia, Program Contact; (312) 337-7000; fax (312) 337-0282; largegrants@spencer.org
*Internet:* https://www.spencer.org/grant_types/large-research-grant
*Sponsor:* Spencer Foundation
625 North Michigan Avenue, Suite 1600
Chicago, IL 60611

**Spencer Foundation Lyle Spencer Research Awards** 2383

The Lyle Spencer Research Awards program was introduced in 2014 as the successor to the Foundation's long-standing Major Research Grants program. Funded research projects vary widely, ranging from medium-sized studies that can be completed within a year by an individual researcher to more extensive collaborative studies that last several years. There are no funding priorities for subjects of research; projects originate from research ideas initiated in the field by scholars and other researchers. The Foundation's aim is to support intellectually ambitious, large-scale education research projects with budgets between $525,000 and $1,000,000. Investigators of projects with budgets over $525,000 are asked to submit a preliminary proposal. If the Foundation is interested in the project, the Principal Investigator will be invited to submit a full proposal. Normally, applicants will be informed about the status of the preliminary proposal within two to four weeks of receipt.
*Requirements:* Ordinarily, principal investigators applying for a Major Research Grant must be affiliated with a school district, a college or university, a research facility, or a cultural institution. Researchers must also have an earned doctorate in an academic discipline or professional field or appropriate experience in the teaching profession.
*Restrictions:* Grantees may not receive two research grants simultaneously from the foundation. The foundation does not fund direct interventions or evaluations of programs. The foundation does not pay government-approved overhead rates on research grants; overhead requests of more than $75,000 may not exceed 15 percent of the requested direct costs. The foundation does not pay indirect costs on research grants of $75,000 or less. The Foundation does not fund: scholarships; curriculum development; program development and implementation; program/curriculum evaluation; or professional development programs.
*Geographic Focus:* All States
*Date(s) Application is Due:* May 22
*Amount of Grant:* 525,000 - 1,000,000 USD
*Contact:* Annie Brinkman, Grants Manager; (312) 274-6511 or (312) 337-7000; fax (312) 337-0282; abrinkman@spencer.org or lyles@spencer.org
*Internet:* https://www.spencer.org/grant_types/lyle-spencer-research-awards
*Sponsor:* Spencer Foundation
625 North Michigan Avenue, Suite 1600
Chicago, IL 60611

**Spencer Foundation Mentor Award** 2384

The Spencer Foundation was established in 1962 by Lyle M. Spencer. The Foundation received its major endowment upon Spencer's death in 1968 and began formal grant making in 1971. The Foundation is intended, by Spencer's direction, to investigate ways in which education, broadly conceived, can be improved around the world. From the first, the Foundation has been dedicated to the belief that research is necessary to the improvement in education. The Foundation is thus committed to supporting high-quality investigation of education through its research programs and to strengthening and renewing the educational research community through its fellowship and training programs and related activities. The Foundation has identified the following five areas of inquiry: education and social opportunity; organizational learning; purposes and values of education; teaching, learning, and instructional resources; and field-initiated proposals. The Mentor Award is a grant intended to support the ongoing mentorship of the winner. This award will provide a $25,000 grant to a nominated mentor who has demonstrated extraordinary contributions to supporting the research training and career trajectories of graduate students and junior scholars that engage in education research, broadly construed. The annual deadline for applications is March 25.
*Geographic Focus:* All States
*Date(s) Application is Due:* Mar 25
*Amount of Grant:* 25,000 USD

*Contact:* Jasmine Janicki, Program Contact; (312) 274-6511 or (312) 337-7000; fax (312) 337-0282; jjanicki@spencer.org or mentoraward@spencer.org
*Internet:* https://www.spencer.org/grant_types/mentorship-award
*Sponsor:* Spencer Foundation
625 North Michigan Avenue, Suite 1600
Chicago, IL 60611

**Spencer Foundation Postdoctoral Fellowships** **2385**
Administered by the National Academy of Education, the postdoctoral fellowships are designed to promote scholarship in the United States and abroad on matters relevant to the improvement of education in all its forms. The program supports up to 30 early career scholars working in critical areas of education research. These $70,000 fellowships support non-residential postdoctoral proposals that make significant scholarly contributions to the field of education. The program also develops the careers of its recipients through professional development activities involving National Academy of Education members. Scholars anywhere in the world who have completed their doctorates within the last five years and who wish to conduct research related to education may apply.
*Geographic Focus:* All States
*Amount of Grant:* 70,000 USD
*Contact:* Annie Brinkman, Grants Manager; (312) 274-6526 or (202) 334-2341; fax (312) 337-0282; abrinkman@spencer.org or info@naeducation.org
*Internet:* https://www.spencer.org/grant_types/post-doctoral-fellowship
*Sponsor:* Spencer Foundation
625 North Michigan Avenue, Suite 1600
Chicago, IL 60611

**Spencer Foundation Research-Practice Partnership Grants** **2386**
The Research-Practice Partnership Grant program accepts proposals with budgets up to $400,000 seeking to strengthen existing research-practice partnerships (RPPs). RPPs are a novel model for conducting relevant and useful research and facilitating the long-term accumulation of knowledge about education policy and practice. They enable researchers and practitioners to jointly study problems in specific settings over time. This program is intended to provide intermediate-term stability to existing RPPs by supporting a range of activities vital to their operation. The program has identified three common types of RPPs: place-based research alliances that work with either school districts or youth-serving organizations that cross multiple sectors in addition to education; design research teams that seek to simultaneously build and study solutions in real world contexts; and networked improvements communities (NICs) that seek to leverage diverse experiences in multiple settings to advance understandings about what works where, when, and under what conditions.
*Requirements:* Initial proposals are accepted at any time, regardless of the level of funding requested.
*Restrictions:* The Foundation does not fund: scholarships; curriculum development; program development and implementation; program/curriculum evaluation; or professional development programs.
*Geographic Focus:* All States
*Date(s) Application is Due:* Jun 19
*Amount of Grant:* Up to 400,000 USD
*Contact:* Maricelle Garcia, Program Contact; (312) 274-6511 or (312) 337-7000; fax (312) 337-0282; research-practice-partnership@spencer.org
*Internet:* www.spencer.org/research-practice-partnership-program
*Sponsor:* Spencer Foundation
625 North Michigan Avenue, Suite 1600
Chicago, IL 60611

**Spencer Foundation Small Research Grants** **2387**
The Spencer Foundation was established in 1962 by Lyle M. Spencer. The Foundation received its major endowment upon Spencer's death in 1968 and began formal grant making in 1971. The Foundation is intended, by Spencer's direction, to investigate ways in which education, broadly conceived, can be improved around the world. From the first, the Foundation has been dedicated to the belief that research is necessary to the improvement in education. The Foundation is thus committed to supporting high-quality investigation of education through its research programs and to strengthening and renewing the educational research community through its fellowship and training programs and related activities. The Foundation has identified the following five areas of inquiry: education and social opportunity; organizational learning; purposes and values of education; teaching, learning, and instructional resources; and field-initiated proposals. Small grants are those $50,000 or less. The majority of small grant proposals that are funded by the Foundation are field-initiated, in the sense that they are not submitted in response to a Request for Proposal (RFP). The annual deadline for applications is July 1.
*Requirements:* Initial proposals are accepted at any time, regardless of the level of funding requested. Applicants must possess an earned doctorate and be affiliated with a school, college, university, or cultural institution. Educators currently employed in K-12 settings who are not pursuing a graduate degree are also eligible.
*Restrictions:* Principal investigators are not permitted to receive two research grants simultaneously. Projects may not extend beyond a year in duration, and no extensions will be granted. Grants do not support indirect costs. The Foundation does not fund: scholarships; curriculum development; program development and implementation; program/curriculum evaluation; or professional development programs.
*Geographic Focus:* All States
*Date(s) Application is Due:* Jul 1
*Amount of Grant:* Up to 50,000 USD
*Contact:* Jasmine Janicki, Program Contact; (312) 274-6511 or (312) 337-7000; fax (312) 337-0282; smallgrants@spencer.org
*Internet:* www.spencer.org/small-research-grants
*Sponsor:* Spencer Foundation
625 North Michigan Avenue, Suite 1600
Chicago, IL 60611

**SPIRE Postdoctoral Fellowships** **2388**
The Seeding Postdoctoral Innovators in Research and Education (SPIRE) Postdoctoral Fellowship Program was one of the first grantees to receive funding in 1999 from the special Institutional Research and Academic Career Development Award (IRACDA) program, which is now part of the division of Training, Workforce Development, & Diversity under the National Institute for General Medical Sciences (NIGMS). The goal of SPIRE was to address the growing crisis in science education by investing in tomorrow's science scholars and educators and providing them with a holistic approach to postgraduate professional development. SPIRE strives to blend an outstanding research environment with broad professional development activities and hands-on classroom experiences for those scientists who want to combine research with undergraduate and graduate education. Fellowship stipends will be competitive with federal guidelines as set by the National Institutes of Health (NIH), with annual increases as allowed by these guidelines. Health insurance will also be provided. In addition, each fellowship includes an allowance for research and teaching activities, travel, support for mentoring an undergraduate student, workshops on workplace skills, and participation in annual SPIRE events. Each fellowship is three years.
*Requirements:* Applicants can be either new PhDs or postdocs with up to two years postgraduate training at the time of appointment. Candidates from groups underrepresented in the biological, biomedical, chemical, and related STEM fields in sciences are strongly urged to apply and applicants from diverse backgrounds are welcome. According to the NIH regulations for postdoctoral training grant eligibility: Applicants must be U.S. citizens or non-citizen nationals, or have been lawfully admitted for permanent residence and possess an Alien Registration Receipt Card or some other verification of legal status as a permanent resident. Non-citizen nationals, although not U.S. citizens, owe permanent allegiance to the US. They are generally born in lands, which are not states but are under U.S. sovereignty, jurisdiction, or administration. Individuals on temporary or student visas are not eligible for SPIRE fellowships.
*Geographic Focus:* All States
*Contact:* Brian Rybarczyk, Ph.D., (919) 962-2505; brybar@unc.edu
*Internet:* spire.unc.edu/apply/
*Sponsor:* University of North Carolina at Chapel Hill
CB#4010
Chapel Hill, NC 27599

**SPJ Eugene C. Pulliam Fellowship for Editorial Writing** **2389**
The Pulliam Fellowship is awarded to an outstanding editorial writer to help broaden his or her journalistic horizons and knowledge of the world. The award can be used to cover the cost of study, research, and/or travel in any field. In some cases, the fellowship results in editorials and other writings, including books. The Fellowship has no formal application form. Each applicant should submit the following materials: a cover letter stating the purpose and nature of the proposed study, a time line for accomplishing the work, and a plan for how the stipend will be used. The cover letter should contain complete contact information for the applicant including mailing address, telephone number and email address; the editor's endorsement; a one-page professional biography and summary of professional experience; and five samples of editorials or columns.
*Requirements:* To be eligible for a Pulliam fellowship, a candidate must: hold a position as a part-time or full-time editorial writer or columnist at a news publication located in the U.S; have at least three years experience as an editorial writer or columnist; demonstrate outstanding writing and analytical abilities; secure assurances by the editor or publisher that the applicant will be allowed sufficient time to pursue the fellowship without jeopardizing employment. (fellows do not have to leave their jobs.); demonstrate ability and intent to publish work within 18 months of selection. (If selected, work must be published within 18 months of receiving the fellowship). The selected applicant must provide a post-fellowship written report on how funds were used. Each Fellowship recipient will become a mentor to the following year's recipient.
*Geographic Focus:* All States
*Date(s) Application is Due:* Jun 22
*Amount of Grant:* 75,000 USD
*Contact:* Sue Kopen Katcef, Vice Chair, Awards Committee; (301) 405-7526 or (301) 405-2399; fax (301) 314-9166; skatcef@spg.org
*Internet:* www.spj.org/a-pulliamfellow.asp
*Sponsor:* Society of Professional Journalists
3909 North Meridian Street
Indianapolis, IN 46208

**SPJ Pulliam/Kilgore Freedom of Information Internships** **2390**
The Society of Professional Journalists (SPJ) annually awards two Pulliam/Kilgore Freedom of Information Internships. One intern works in the offices of the Society's First Amendment legal counsel in Washington, D.C. The other intern works at the Society's National Headquarters in Indianapolis. The interns research and write about freedom of information issues while assisting the Society's Freedom of Information Committee and First Amendment legal counsel in preparing SPJ's annual Freedom of Information Report. Other responsibilities may include working on the SPJ website and assisting with SPJ programs and initiatives. Interns receive a stipend of $400 per week and work during a mutually agreeable ten week period during the summer. Recipients are notified in February, and should arrange to start their internships in late May or early June. The application and specific guidelines for each internship are available at the SPJ website.

*Requirements:* Candidates must be senior or graduate journalism students, or law students with journalism backgrounds. To be considered, applicants must include the following: a completed application (located at the website); resume; a one page essay illustrating their understanding of Freedom of Information issues and what they expect to gain from the internship; and no more than five writing samples.
*Geographic Focus:* All States
*Date(s) Application is Due:* Jan 16
*Amount of Grant:* 4,000 USD
*Contact:* Joe Skeel; (317) 927-8000, ext. 216; fax (317) 920-4789; jskeel@spj.org
*Internet:* www.spj.org/i-pullkilgore.asp
*Sponsor:* Society of Professional Journalists
3909 North Meridian Street
Indianapolis, IN 46208

**SRC AMD/Mahboob Khan Fellowship** **2391**

The Mahboob Khan Fellowship was created in memory of Dr. Khan and in recognition of his support of SRC students, especially the Fellows. Advanced Micro Devices (AMD) partially funds this Fellowship. Succeeding Fellowships will be awarded as the current recipient completes doctoral studies. AMD selects this Fellow from the applications received during the regular Graduate Fellowship Program (GFP) application process. In addition to the academic and citizenship criteria, AMD seeks to identify a student with technical interest specifically aligned with that of AMD. As part of the GFP, this Fellowship provides full tuition and fees and a competitive stipend for up to five years of doctoral study. Fellowships also include an unrestricted annual gift of $2,000 for use by the student's faculty advisor in support of the student. The Fellowships will continue to completion of the degree program, but for no more than five years. Continuation will be contingent only upon satisfactory progress in the proposed course of study and research as determined by the faculty advisor and GRC. The Graduate Fellowships are offered through a national competition and are awarded based on outstanding academic achievement in the field of microelectronics and the relevance of the proposed research to SRC/semiconductor industry goals. Application materials are typically available in November, and completed applications are due mid-February of the following year with awards made in April. All named Fellowships will be awarded consecutively with the next award being made at the completion of the current Fellowship. Notice of Fellowship competition is through universities having GRC-sponsored research contracts (see the website for a complete list of universities and contacts), and other sources as appropriate.
*Requirements:* This Fellowship is open to students who are U.S. citizens or have permanent resident, refugee, or political asylum status in the U.S; are pursuing or planning to pursue a Ph.D. degree with research relevant to microelectronics under the guidance of a GRC-approved faculty member and with at least two years to completion of the doctoral degree; will be performing research under a sponsored research program approved by GRC; and are willing to provide a copy of their Ph.D. dissertation to GRC for publication to the SRC website.
*Geographic Focus:* All States
*Date(s) Application is Due:* Feb 15
*Contact:* Mahboob Khan; (919) 941-9400; fax (919) 941-9450; apply@src.org
*Internet:* www.src.org/student-center/fellowship/#gfp
*Sponsor:* Semiconductor Research Corporation
P.O. Box 12053
Research Triangle Park, NC 27709-2053

**SRC Aristotle Award** **2392**

The Aristotle Award was created by the SRC Board of Directors to recognize SRC-supported faculty whose deep commitment to the educational experience of SRC students has had a profound and continuing impact on their professional performance and a significant impact for members over a long period of time. The Award is intended to acknowledge outstanding teaching in its broadest sense, emphasizing student advising and teaching during the research project, thereby contributing to the maturation of the student. The following equally-weighted selection criteria will be considered by the Awards Committee: innovation in student advising; provision of learning experiences for students in sound execution of semiconductor technology (device physics, process, design, etc.), and deployment of statistical methodology such as design of experiments; creation of an environment that encourages development in leadership skills; and student exposure to and interaction with industry.
*Requirements:* Nominations are accepted by email from the SRC community including the SRC Board of Directors, Advisory Boards, Industrial Liaisons, university researchers (self-nominations will not be accepted), government agencies that participate in the SRC research program, and alumni of the SRC student programs. Nominations must include: the nomination form (located at the website); a nomination narrative covering each of the four selection criteria; and supporting documentation.
*Geographic Focus:* All States
*Date(s) Application is Due:* Apr 18
*Contact:* Lee Ann Clewell; (919) 941-9444; fax (919) 941-9450; clewell@src.org
*Internet:* www.src.org/award/aristotle/
*Sponsor:* Semiconductor Research Corporation
P.O. Box 12053
Research Triangle Park, NC 27709-2053

**SRC Company-Named Fellowships** **2393**

The Company-Named Fellowships were created to recognize outstanding academic and research achievement and to increase the number of Graduate Fellows in research areas of special interest to the co-sponsoring companies. These Fellowships are named for the companies providing matching funding. Fellows are chosen from the applicants for the Graduate Fellowship Program (GFP) based on the eligibility requirements used to select the GFP Fellows, and on the match between the student's proposed research and the co-sponsoring company's research interests. The recipient will be mentored for the term of the Fellowship by an industry advisor from the co-sponsoring company, and will be awarded internships and other support from the co-sponsoring company as appropriate. As a part of the Graduate Fellowship Program, this Fellowship provides full tuition and fees and a competitive stipend for up to five years of doctoral study. Fellowships also include an unrestricted annual gift of $2,000 for use by the student's faculty advisor in support of the student. The Fellowships will continue to completion of the degree program, but for no more than five years. The Graduate Fellowships are offered through a national competition and are awarded based on outstanding academic achievement in the field of microelectronics and the relevance of the proposed research to SRC/semiconductor industry goals. Application materials are typically available in November, and completed applications are due mid-February of the following year with awards made in April. Additional information is available at the SRC website.
*Requirements:* Fellowships are open to students who: are U.S. citizens or have permanent resident, refugee, or political asylum status in the U.S; are pursuing or planning to pursue a Ph.D. degree with research relevant to microelectronics under the guidance of a GRC-approved faculty member and with at least two years to completion of the doctoral degree; will be performing research under a sponsored research program approved by GRC; and are willing to provide a copy of his/her Ph.D. dissertation to GRC for publication to the SRC website.
*Geographic Focus:* All States
*Date(s) Application is Due:* Feb 15
*Contact:* Coordinator for Company-Named Fellowships; (919) 941-9400; fax (919) 941-9450; apply@src.org
*Internet:* www.src.org/student-center/fellowship/
*Sponsor:* Semiconductor Research Corporation
P.O. Box 12053
Research Triangle Park, NC 27709-2053

**SRC Company-Named Scholarships** **2394**

Company-Named scholarships were created to increase the number of Master's Scholarships in research areas of special interest to the co-sponsoring companies. These scholarships are named for the companies providing matching funding. The recipient of a company-named scholarship will be mentored for the term of the scholarship by an industry advisor from the co-sponsoring company, and will be awarded internships and other support from the co-sponsoring company as appropriate. As part of the Master's Scholarship program, these scholarships provide full tuition and fees and a competitive stipend for up to two years of master's level study. Scholarships also include an unrestricted gift of $2,000 for use by the student's faculty advisor in support of the student. The Scholarships will continue to completion of the degree program, but for no more than two years. Continuation will be contingent only upon satisfactory progress in the proposed course of study and research as determined by the faculty advisor and Global Research Collaboration (GRC). Scholarships are offered through a national competition and are awarded based on minority or female status and outstanding academic achievement. Students in BS/MS degree programs are encouraged to apply for the Master's Scholarships at the beginning of their senior year. Application materials are typically available in November, and completed applications are due mid-February of the following year with awards made in April. Notice of scholarship competition is through universities having GRC-sponsored research contracts, and other sources as appropriate. Candidates may email or write directly to SRC for an application.
*Requirements:* Scholarships are open to students who are women or are members of an underrepresented minority category (African-American, Hispanic, Native American); meet admission requirements for graduate school at a GRC-participating university (see website for a complete list of universities and contacts); are U.S. citizens or have permanent resident, refugee, or political asylum status in the U.S; are planning to pursue a master's degree with research relevant to microelectronics under the guidance of a GRC-approved faculty member and under a GRC-approved research program; and are willing to provide a copy of their master's thesis where appropriate for publication to the SRC secure website.
*Geographic Focus:* All States
*Date(s) Application is Due:* Feb 15
*Contact:* Company-Sponsored Scholarship Coordinator; (919) 941-9400; fax (919) 941-9450; apply@src.org
*Internet:* www.src.org/student-center/fellowship/
*Sponsor:* Semiconductor Research Corporation
P.O. Box 12053
Research Triangle Park, NC 27709-2053

**SRC Core Research Grants** **2395**

Global Research Collaboration (GRC) supports two kinds of research: core and custom. Core research is selected by representatives of the entire GRC membership. Currently, core research is about 80% of the total research budget. GRC, as a mission-oriented research organization, funds research directly responsive to its defined agenda. The funding terms for contracts vary; however, the GRC trend is toward multi-year research efforts (typically three years) with an initial 12-month funding term. At the end of the multi-year period, researchers may re-compete for new support.
*Requirements:* Proposals should provide sufficient information to permit an evaluation of the intended research, its technical merits, innovative approach, and relevance to the GRC research agenda. The author should presume general technical knowledge in the field by reviewers, but should not assume that the merits of the proposal are obvious. The capability of the institution to adequately support the proposed research and the

technical expertise of the researchers to conduct this research should be justified. Refer to the Proposal Submission Guide, located at the website. for specific authoring and submission instructions.
*Geographic Focus:* All States
*Contact:* Core Research Coordinator; (919) 941-9400; fax (919) 941-9450
*Internet:* www.src.org/program/grc/about/core-research/
*Sponsor:* Semiconductor Research Corporation
P.O. Box 12053
Research Triangle Park, NC 27709-2053

**SRC Cross-Disciplinary Research (CSR) Program Grants** **2396**
The goal of the Cross-Disciplinary Semiconductor Research Program is to foster exploratory, multi-disciplinary, high-risk university research leading to novel high-payoff solutions for the technology challenges faced by the semiconductor industry at and beyond the time horizons of the SIA International Technology Roadmap for Semiconductors (ITRS). Successful projects will involve synergistic research across normal disciplinary boundaries that offers innovative/disruptive solutions supporting the continuation of the Moore's Law cost/performance improvement cadence. Possible targets for the research can address semiconductor technologies ranging from factory systems through device technologies to packaged systems. Multi-disciplinary projects leading to divergent applications of semiconductor technologies will also be considered. The role of the Program is to stimulate non-traditional thinking about the issues facing the industry. The Program is also intended to complement the traditional GRC science area research programs. Cross-Disciplinary Semiconductor Research projects are exploratory in nature and should lead to research programs that address challenges identified by ITRS. Funding for a project is provided on a non-renewable grant basis, with maximum funding not to exceed $40,000.
*Requirements:* International university faculty are eligible. Participants in the program are expected to participate in an annual workshop to describe the results of their study, and to develop a proposal for follow-on funding for expanded programs by the GRC Core Research Program. Follow-on program funding will depend on the availability of science area funds and alignment with the science area strategic plans.
*Geographic Focus:* All Countries
*Amount of Grant:* Up to 40,000 USD
*Contact:* Victor Zhirnov; (919) 941-9400; fax (919) 941-9450; victor.zhirnov@src.org
*Internet:* www.src.org/program/grc/csr/
*Sponsor:* Semiconductor Research Corporation
P.O. Box 12053
Research Triangle Park, NC 27709-2053

**SRC Franco Cerrina Fellowship** **2397**
The Cerrina Fellowship encourages, sponsors, and supports young scientists and technologists whose work integrates leading-edge research into new application areas, crossing boundaries between traditional or established disciplines, with original ideas and solutions. The Franco Cerrina Fellowship will be awarded to students who best represent the spirit of Cerrina by taking research in new directions in areas such as silicon/biological interfaces and applications, beyond-CMOS computational devices, nanotechnology and nanofabrication, and opening new frontiers in electrical and computer engineering.
*Geographic Focus:* All States
*Contact:* Fellowship Administrator; (919) 941-9400; fax (919) 941-9450
*Internet:* www.src.org/program/srcea/cerrina/
*Sponsor:* Semiconductor Research Corporation
P.O. Box 12053
Research Triangle Park, NC 27709-2053

**SRC Graduate Fellowships** **2398**
The Graduate Fellowship Program (GFP) addresses the issues of improving educational opportunities at the doctoral level and supplying a relevantly educated workforce for the semiconductor industry. The objectives of the program are to encourage academically gifted U.S./permanent resident students to pursue doctoral degrees in research areas consistent with SRC program goals; and to develop high quality doctoral graduates for member companies and U.S. universities. The GFP was created to attract exceptionally talented students with U.S. citizenship to academic areas of interest to SRC members. The program has since been opened to students holding permanent resident, refugee, or political asylum status in the U.S. While the Fellows are not required to take employment within the SRC community upon graduation, they are strongly encouraged to do so, and assistance is provided in finding appropriate employment in an SRC member company, U.S. government agency, or U.S. university. Graduating Fellows entering member companies facilitate the transfer of new science and technology from the participating universities to the supporting organizations. Fellows joining faculties at universities worldwide carry with them the expertise to stimulate new research activities and to encourage additional student interest in semiconductor-related fields. Students who are supported by the GFP are actively involved in GRC-approved sponsored research; however, the scope of the student's work may be beyond the scope of the research defined by the research sponsor. The Fellows are encouraged to conduct research leading to novel, high-payoff solutions for the technology challenges faced by the semiconductor industry at and beyond the time horizons of the International Technology Roadmap for Semiconductors. The Graduate Fellowships are offered through a national competition and are awarded based on outstanding academic achievement in the field of microelectronics and the relevance of the proposed research to SRC/semiconductor industry goals. Application materials are typically available in November, and completed applications are due mid-February of the following year with awards made in April. Notice of Fellowship competition is through universities having GRC-sponsored research contracts, and other sources as appropriate. A list of those universities and their contacts is posted on the SRC website.
*Requirements:* The Fellowships are open to students who are U.S. citizens or have permanent resident, refugee, or political asylum status in the U.S; are pursuing or planning to pursue a Ph.D. degree with research relevant to microelectronics under the guidance of a GRC-approved faculty member and with at least two years to completion of the doctoral degree; will be performing research under a sponsored research program approved by GRC; and are willing to provide a copy of his/her Ph.D. dissertation to GRC for publication to the SRC website. It is expected that all Fellows will complete their Ph.D. degrees in an area relevant to microelectronics and that upon completion of the program, secure employment with an SRC member or U.S. government agency or a faculty position in an accredited four-year U.S. college or university.
*Geographic Focus:* All States, All Countries
*Date(s) Application is Due:* Feb 15
*Contact:* Fellowship Coordinator; (919) 941-9400; fax (919) 941-9450; apply@src.org
*Internet:* www.src.org/student-center/fellowship/#gfp
*Sponsor:* Semiconductor Research Corporation
P.O. Box 12053
Research Triangle Park, NC 27709-2053

**SRC GRC and SEMATECH Joint Projects Grants** **2399**
GRC and SEMATECH, in order to expedite the transition of research results into applications with even more value benefiting members, have formed a partnership to conduct applied research at jointly funded centers. Through the collaborative process GRC funds research concepts, develops research options, and narrows the research options. GRC and SEMATECH jointly fund a transition period to develop the feasibility of the research, and transition of the research concepts is made to technology hardening which is funded by SEMATECH. Additional information about currently funded projects is available at the SRC website.
*Geographic Focus:* All States
*Contact:* Robert Havemann, ESH Thrust Manager; (919) 941-9400; fax (919) 941-9450; bob.havemann@src.org
*Internet:* www.src.org/program/grc/about/sematech-joint-projects/
*Sponsor:* Semiconductor Research Corporation
P.O. Box 12053
Research Triangle Park, NC 27709-2053

**SRC Hans J. Coufal Fellowship** **2400**
The SRC/NRI Hans J. Coufal Fellowship was created in memory of Hans J. Coufal, the founding director of the Nanoelectronics Research Initiative (NRI), for the study of nanoelectronics-related disciplines. The Coufal Fellowship encourages non-traditional thinking and exploratory, high-risk research leading to novel, high-payoff solutions for challenges faced by the semiconductor industry at and beyond the time horizons of the International Technology Roadmap for Semiconductors. The recipient is expected to work on an NRI-funded research program, but the student's work is encouraged to go beyond the scope of the research defined in the research contract. The Coufal Fellowship is administered by NRI and provides full tuition and fees, a competitive stipend, and a $2,000 annual unrestricted gift to the recipient's faculty advisor. The Fellowship will continue to completion of the degree program, but for no more than five years. Continuation will be contingent only upon satisfactory progress in the proposed course of study and research as determined by the faculty advisor and NRI.
*Requirements:* The Fellowship will be awarded to a graduate student who has completed a bachelor's degree and is pursuing, or is planning to pursue, a doctoral program in an academic area appropriate to the Fellowship. Although this Fellowship is not restricted to any particular area of nanoelectronics, preference will be given to proposed work in Dr. Coufal's primary areas of interest (nanoelectronic device technologies, physics, and materials). The recipient will hold citizenship or permanent resident status in the U.S.
*Geographic Focus:* All States
*Date(s) Application is Due:* Feb 1
*Contact:* Fellowship Administrator; (919) 941-9400; fax (919) 941-9450
*Internet:* www.src.org/student-center/fellowship/#coufal
*Sponsor:* Semiconductor Research Corporation
P.O. Box 12053
Research Triangle Park, NC 27709-2053

**SRC International Fellowship** **2401**
The International Fellowship was first awarded in recognition of GRC's international status. As part of the Graduate Fellowship Program, this fellowship provides full tuition and fees and a competitive stipend for up to five years of doctoral study. Fellowships also include an unrestricted annual gift of $2,000 for use by the student's faculty advisor in support of the student. Applicants may contact the SRC about the status of this Fellowship, which is offered as funding becomes available.
*Geographic Focus:* All Countries
*Date(s) Application is Due:* Feb 15
*Contact:* Fellowship Coordinator; (919) 941-9400; fax (919) 941-9450; apply@src.org
*Internet:* www.src.org/student-center/fellowship/
*Sponsor:* Semiconductor Research Corporation
P.O. Box 12053
Research Triangle Park, NC 27709-2053

**SRC Landmark Innovation Award** 2402

The SRC Intellectural Assets Study Group established the Landmark Innovation Award to be given to a SRC university researcher/researchers when an SRC-supported invention proves to have a long-term, significant benefit for the industry. Given the nature of this award, the selection criteria is intentionally general in nature. The impact of the innovation may be technological or economic in nature, and preferably should have created a broad or globally recognized benefit for the industry. Examples that could support an award are significant cost savings or revenue generation for SRC member companies, compelling reasons for SRC membership, proven commercialization and implementation in industry, direct/indirect beneficial impact on the industry, and long term benefits to the industry and/or society. The cash award is shared equally among key contributors who develop innovative patented technology that significantly enhances the productivity and competitiveness of the semiconductor industry. There intentionally has been no specific dollar amount of this award, thus the impact of the landmark innovation can be correspondingly rewarded. Nominations may be submitted to the SRC Intellectual Property Office for consideration by the award committee and the SRC Board of Directors. The nomination form is available at the SRC website.
*Requirements:* Nominations may be submitted at any time, directed to SRC for the consideration of the SRC Board. Nominations are accepted from selected members of the SRC community, including the SRC Board of Directors, SRC Technical Advisory Boards, mentors, member company employees, university principal investigators, and research task leaders. Nominations may be sent via email or directly to the SRC address, and will be confirmed by receipt. No particular form or format is required, and any additional supporting documentation (letters, publication bibliographical information, etc.) that helps distinguish a nomination is welcome.
*Geographic Focus:* All States
*Contact:* Director of Corporate Legal Affairs and IP Counsel; (919) 941-9400; fax (919) 941-9450; legal@src.org
*Internet:* www.src.org/award/landmark/
*Sponsor:* Semiconductor Research Corporation
P.O. Box 12053
Research Triangle Park, NC 27709-2053

**SRC Mahboob Khan Outstanding Industry Liaison Award** 2403

The SRC Industry Liaison Program is one of the most effective methods developed by SRC to provide its members with early access to key technologies, substantial leverage in the explicit direction of technology development, and ultimate transfer of semiconductor technology from the research base to the supporting industry. Each year SRC recognizes individuals who, in serving as SRC Industry Liaisons, have had significant impact on the technical contributions of the research, are deeply committed to the education of graduate students, and have been instrumental in transferring the research results to industry. The following selection criteria is used by the committee: the liaison's role in developing the workfoce; the liaison's effectiveness and involvement in aligning research to industry needs, as well as guidance and material help during all phases of the research process; and the liaison's role ion technology transfer from the university community to the industrial community. The nomination form is available at the website.
*Requirements:* Nominations for individuals serving as SRC Industry Liaisons for a period of at least one year are accepted from members of the SRC community. Past winners may only receive the award once for an individual task.
*Geographic Focus:* All States
*Date(s) Application is Due:* Apr 18
*Contact:* Lee Ann Clewell; (919) 941-9444; fax (919) 941-9450; leeann.clewell@src.org
*Internet:* www.src.org/award/khan-liaison/call/
*Sponsor:* Semiconductor Research Corporation
P.O. Box 12053
Research Triangle Park, NC 27709-2053

**SRC Master's Scholarships** 2404

The Master's Scholarship Program (MSP) addresses issues of improving educational opportunities at the master's level for students in underrepresented minority categories. The objectives of the program are to encourage academically gifted U.S./permanent resident students in these populations to pursue graduate research in areas consistent with Global Research Collaboration (GRC) goals; and to develop a cadre of highest quality minority candidates for doctoral study and hire by GRC companies. The MSP Scholarships provide full tuition and fees and a competitive stipend for up to two years of master's level study. Scholarships also include an unrestricted gift of $2,000 for use by the student's faculty advisor in support of the student. The Scholarships will continue to completion of the degree program, but for no more than two years. Continuation will be contingent only upon satisfactory progress in the proposed course of study and research as determined by the faculty advisor and GRC. The Master's Scholarships are offered through a national competition and are awarded based on minority or female status and outstanding academic achievement. Students in BS/MS degree programs are encouraged to apply for the Master's Scholarships at the beginning of their senior year. Application materials are typically available in November, and completed applications are due mid-February of the following year with awards made in April. Notice of scholarship competition is through universities having GRC-sponsored research contracts, and other sources as appropriate. Applications are available by email request or by writing directly to the SRC.
*Requirements:* The MSP is open to students who are women or are members of an underrepresented minority category (African-American, Hispanic, Native American); meet admission requirements for graduate school at a GRC-participating university (see website for a complete list of universities and contacts); are U.S. citizens or have permanent resident, refugee, or political asylum status in the U.S; are planning to pursue a master's degree with research relevant to microelectronics under the guidance of a GRC-approved faculty member and under a GRC-approved research program; and are willing to provide a copy of his/her master's thesis where appropriate for publication to the SRC secure website.
*Geographic Focus:* All States
*Date(s) Application is Due:* Feb 15
*Contact:* Scholarship Coordinator; (919) 941-9400; fax (919) 941-9450; apply@src.org
*Internet:* www.src.org/student-center/fellowship/
*Sponsor:* Semiconductor Research Corporation
P.O. Box 12053
Research Triangle Park, NC 27709-2053

**SRC Peter Verhofstadt Fellowship** 2405

The Verhofstadt Fellowship was created by SRC in memory of Peter Verhofstadt, SRC Chief Scientist and Co-Executive Director of MARCO, an SRC subsidiary that manages the Focus Center Research Program. The Verhofstadt Fellowship funds study in the physical sciences, specifically chemistry, physics, mathematics, or biophysics. This Fellowship strives to stimulate non-traditional thinking and encourage exploratory, high-risk research leading to novel, high-payoff solutions for challenges faced by the semiconductor industry at and beyond the time horizons of the International Technology Roadmap for Semiconductors. The recipient is expected to work on a GRC-funded research program, but the student's work is encouraged to be beyond the scope of the research defined in the research program. As part of the Graduate Fellowship Program, the Verhofstadt Fellowship provides full tuition and fees and a competitive stipend for up to five years of doctoral study. Fellowships also include an unrestricted annual gift of $2,000 for use by the student's faculty advisor in support of the student. The Graduate Fellowships are offered through a national competition and are awarded based on outstanding academic achievement in the field of microelectronics and the relevance of the proposed research to SRC semiconductor industry goals. Application materials are typically available in November, and completed applications are due mid-February of the following year with awards made in April.
*Requirements:* The Fellowship is open to students who: are U.S. citizens or have permanent resident, refugee, or political asylum status in the U.S; are pursuing or planning to pursue a Ph.D. degree with research relevant to microelectronics under the guidance of a GRC-approved faculty member and with at least two years to completion of the doctoral degree; will be performing research under a sponsored research program approved by GRC; and are willing to provide a copy of his/her Ph.D. dissertation to GRC for publication to the SRC website.
*Geographic Focus:* All States, All Countries
*Date(s) Application is Due:* Feb 15
*Contact:* Fellowship Coordinator; (919) 941-9400; fax (919) 941-9450; apply@src.org
*Internet:* www.src.org/student-center/fellowship/
*Sponsor:* Semiconductor Research Corporation
P.O. Box 12053
Research Triangle Park, NC 27709-2053

**SRC Research Customization Program (RCP) Grants** 2406

The Research Customization Program (RCP) was launched to increase member benefits by providing the flexibility for eligible members to directly select a few university projects that are of particular interest to that member. The projects that a member selects must be consistent with the Global Research Collaboration (GRC) mission, responsive to the needs articulated by the International Technology Roadmap for Semiconductors (ITRS), and suitable for university research. Unless a unique situation exists, the nominal commitment is for three years of at least $50,000 per year, the minimum for a project with a Ph.D. candidate to achieve meaningful results. Additional information about the selection process is available at the SRC website.
*Requirements:* Eligibility requirements are specific and must be followed. A member company ascertains the amount of funding available for RCP. The Finance Department provides quarterly reports outlining the available funding to the Executive TAB (ETAB) representative and Board of Director of eligible members. A member company makes a preliminary selection after reviewing a short written description of the planned research. The ETAB representative for the member company makes the final selection of the project after receiving inputs from within the company. The ETAB representative then verifies the nature of the proposed research with the Principal Investigator (PI), provides the PI with the name of the Industry Liaison(s), obtains a Proposal ID from SRC, and requests the PI to submit a formal proposal in the standard format to GRC. The ETAB representative provides an official authorization for funding to GRC. When the GRC has received both the proposal and the member company authorization, the project is reviewed by the Research Management Committee (RMC). Projects meeting the RCP project guidelines will be approved for funding. A standard GRC contract is made with the university (a list of supporting universities is posted on the website). The appropriate Science Area implements and manages the project in a manner similar to the other projects in the research portfolio. Additional information is available at the website.
*Geographic Focus:* All States
*Contact:* Customization Coordinator; (919) 941-9400; fax (919) 941-9450; legal@src.org
*Internet:* www.src.org/program/grc/about/custom-research/
*Sponsor:* Semiconductor Research Corporation
P.O. Box 12053
Research Triangle Park, NC 27709-2053

**SRC Robert M. Burger Fellowship** 2407

The Robert M. Burger Fellowship was created by SRC in recognition of Dr. Burger's ten years of service to SRC as Vice President and Chief Scientist. This Fellowship supports doctoral study in non-technology topics related to microelectronics and the semiconductor industry. Relevant topics include public policy (e.g. regulatory, tax/

fiscal, immigration, or federal investment in research and education), management, technology policy, fiscal, innovation, workforce education, or economic issues. As part of the Graduate Fellowship Program, the Burger Fellowship provides full tuition and fees and a competitive stipend for up to five years of doctoral study. Fellowships also include an unrestricted annual gift of $2,000 for use by the student's faculty advisor in support of the student. The Graduate Fellowships are offered through a national competition and are awarded based on outstanding academic achievement in the field of microelectronics and the relevance of the proposed research to SRC/semiconductor industry goals. Application materials are typically available in November, and completed applications are due mid-February of the following year with awards made in April.
*Requirements:* The Burger Fellowship is awarded to a graduate student who is a U.S. citizen, has completed a master's degree, and who is pursuing or planning to pursue a doctoral program.
*Geographic Focus:* All States
*Date(s) Application is Due:* Feb 15
*Contact:* Fellowship Coordinator; (919) 941-9400; fax (919) 941-9450; apply@src.org
*Internet:* www.src.org/student-center/fellowship/
*Sponsor:* Semiconductor Research Corporation
P.O. Box 12053
Research Triangle Park, NC 27709-2053

**SRC Simon Karecki Fellowship** **2408**
The Simon Karecki Fellowship Fund was created in Simon's memory to continue his research and to provide other outstanding young people with the opportunity to make their contributions to environmental research. The Fund will make an annual cash award or fellowship for outstanding contributions to environmental research. The Karecki Fellowship is awarded to a doctoral student who is performing or is planning to perform research through the Center. The Fellowship will be awarded competitively on the basis of quality, relevance, and timeliness of the proposed research; demonstrated academic and leadership ability, and the appropriateness of the institutional environment for the research. The Awards are based on academic excellence, demonstrated concern for the environment, including significant contribution to research in environmentally benign semiconductor manufacturing, demonstrated ability for technical communication, and demonstrated ability to work with others and foster cooperation. The level of annual funding is determined by the Education Alliance and the Advisory Board.
*Geographic Focus:* All States
*Contact:* Fellowship Administrator; (919) 941-9400; fax (919) 941-9450; srcea@src.org
*Internet:* www.src.org/program/srcea/karecki/
*Sponsor:* Semiconductor Research Corporation
P.O. Box 12053
Research Triangle Park, NC 27709-2053

**SRC Technical Excellence Award** **2409**
The Technical Excellence Award was established by the Board of Directors as an incentive and recognition program for research of exceptional value to SRC members. The $5,000 award is shared equally among key contributors for innovative technology that significantly enhances the productivity/competitiveness of the semiconductor industry. Nominations are accepted from all members of the SRC community for work performed during previous years. Most nominations are from industry, indicating the extent to which results from SRC-funded programs are being applied or likely will be applied by the industry. The award is presented during the fall of each year. Nomination forms and a description of the narrative required with the form are posted at the website.
*Requirements:* In reviewing each nomination for the award, the Awards Review Committee considers creativity, relevance to SRC/industry research objectives, value to industry, and the researcher's role in technology transfer. The Awards Committee presents its recommendations to the Board of Directors; the Board approves and confirms all award recipients.
*Geographic Focus:* All States
*Date(s) Application is Due:* Apr 18
*Amount of Grant:* 5,000 USD
*Contact:* Lee Ann Clewell, Awards Coordinator; (919) 941-9400 or (919) 941-9444; fax (919) 941-9450; clewell@src.org
*Internet:* www.src.org/award/tech-excellence/
*Sponsor:* Semiconductor Research Corporation
P.O. Box 12053
Research Triangle Park, NC 27709-2053

**SRC Undergraduate Research Opportunities** **2410**
Undergraduate Research Opportunities (URO) at the Semiconductor Research Corporation offer an innovative program co-sponsored by SRC's Education Alliance and the Intel Foundation that provides undergraduates with valuable research experience and mentoring. Participating undergraduates not only gain confidence in their ability to perform hands-on research, but they also come to appreciate the doors that are opened to those with an advanced degree. The objectives of the SRC URO program are to increase academic retention among undergraduates who express interest in physical science and engineering disciplines, and encourage them to continue their education beyond their four-year degree. The SRC URO program seeks to attract a diverse student population, including women and under-represented minorities. The program also offers workshops and resources to assist students with the graduate school application process. In addition to working on research projects that address the needs of SRC member companies, students have meaningful interactions with industry scientists and engineers, as well as with researchers from a broad academic community.
*Requirements:* Applicants should contact the URO Coordinator at their college for further information about applying. The following universities host SRC URO programs: Carnegie Mellon University; Cornell University; Georgia Institute of Technology; Howard University; North Carolina A&T; Oregon State University; Portland State University; Purdue University; University of California at Berkeley; University of California at Los Angeles; University of Illinois at Urbana-Champaign; University of Michigan; University of Texas at Austin; and the University of Washington.
*Geographic Focus:* All States
*Contact:* Undergraduate Research Coordinator; (919) 941-9400; fax (919) 941-9450; educationalliance@src.org
*Internet:* www.src.org/program/srcea/uro/
*Sponsor:* Semiconductor Research Corporation
P.O. Box 12053
Research Triangle Park, NC 27709-2053

**SSHRC-APC Automotive Partnership Canada** **2411**
Automotive Partnership Canada (APC) provides $145 million in research funding over five years to support significant, collaborative research and development activities to benefit the entire Canadian automotive industry. This initiative is a partnership between five federal research and granting agencies under the Industry Canada umbrella, which has identified collaborative, industry-driven automotive research as a priority. APC's mission is to support research and development that will move the Canadian automotive sector to a greater level of innovation. APC's emphasis will be on transformative, integrated projects or programs that will give Canadian industry and academia the resources required to further their automotive research and development ambitions. Smaller projects are also eligible for support through APC. However, the urgent requirement to boost automotive research will require big ideas and wide-ranging resources. All research to be supported must fall under at least one of the ten priority areas: improving the automobile's environmental performance and impact - lighter weight and/or sustainable materials; advanced powertrain; energy storage; and application of alternative fuels; the cognitive car - vehicular software; electronics and mechatronics for safety and performance enhancement; wireline and wireless communications for vehicular application; next generation manufacturing - manufacturing processes for mass reduction, cost reduction, and quality improvement; and improving manufacturing flexibility and efficiency. APC will consider proposals in social sciences and humanities that fall outside the priority areas if they follow these two *Requirements:* they must maintain and enhance the competitiveness of the Canadian automotive sector; and the research must be carried out with a non-academic partner(s) that can apply and use the research results. APC proposals are accepted on an ongoing basis. This allows proposals to come forward as industry needs arise and also allows for proposals to be uniquely crafted according to the requirements of the research and industry challenges at hand. There are no minimum or maximum dollar amounts that can be requested. Projects can range from six months to five years in duration. However, all projects should first be defined by the scope of research and development that needs to be carried out, after which an appropriate project budget and duration can be determined. Additional information and guidelines, including possible research partners, are posted on the APC website.
*Requirements:* Researchers must first contact the APC Project Office to discuss their project aims, scope, industrial involvement, and other parameters. If their idea fits the mandate of the APC, they are asked to prepare a written overview of the proposed research. If the project still fits within the APC parameters, a full proposal is requested. The Project Office will provide advice and feedback, review drafts, and work with the applicant until a satisfactory full proposal is completed.
*Restrictions:* Unsolicited full proposals are not accepted. The APC definition of "automotive" includes vehicles licensed for on-road driving (cars, light trucks, heavy trucks, buses and recreational vehicles). The definition does not include agricultural vehicles, military vehicles, tracked vehicles, off-road and rail vehicles.
*Geographic Focus:* All States, Canada
*Contact:* John Wood, Executive Director, Automotive Projects; (905) 403-0600; fax (905) 403-1053; john.wood@nserc-crsng.gc.ca
Jeffery Nerenberg, Senior Manager, Special Initiatives; (613) 947-9316; fax (613) 992-5337; jeffery.nerenberg@nserc-crsng.gc.ca
*Internet:* www.apc-pac.ca/About-Renseignements/Background-Contexte_eng.asp
*Sponsor:* Social Sciences and Humanities Research Council of Canada
350 Albert Street, P.O. Box 1610
Ottawa, ON K1P 6G4 Canada

**SSHRC-CIHR-SC Research Data Centres Joint Initiative** **2412**
Decision-makers need an up-to-date and in-depth understanding of Canadian society to help them respond not only to today's needs, but to anticipate tomorrow's as well. This need is underlined by a growing demand for analytical output from the rich sources of data collected by Statistics Canada. The Canadian initiative on social statistics access to research data centres (RD.C.) is a strategic joint initiative of SSHRC, the Canadian Institutes of Health Research and Statistics Canada. This initiative strives to promote research and training in the application of social statistics. RD.C.s provide researchers with access, in a secure university setting, to microdata from population and household surveys. The centres are staffed by Statistics Canada employees. They are operated under the provisions of the Statistics Act in accordance with all the confidentiality rules and are accessible only to researchers with approved projects who have been sworn in under the Statistics Act as 'deemed employees. These RD.C.s are located throughout the country, so researchers do not need to travel to Ottawa to access Statistics Canada microdata. Applications may be submitted at any time. Results will be announced approximately eight weeks after submission. Additional information is posted on the Statistics Canada website.

*Requirements:* The application process and guidelines are based on the affiliation of the Principal Investigator and the type of research being conducted at a Research Data Centre (RD.C.). Proposals depend on the type of research academics involved. Complete descriptions of the proposal process by category of researcher (faculty, masters or doctoral student, by academic course, those working for academic institutions who are not members of the Canadian Research Data Centre Network, or type of government research funded) are posted on the Statistics Canada website.
*Geographic Focus:* All States, Canada
*Contact:* Mika Oehling, Program Officer, Partnerships Portfolio; (613) 992-4227; fax (613) 947-0223; mika.oehling@sshrc-crsh.gc.ca
*Internet:* www.statcan.gc.ca/rdc-cdr/index-eng.htm
*Sponsor:* Social Sciences and Humanities Research Council of Canada
350 Albert Street, P.O. Box 1610
Ottawa, ON K1P 6G4 Canada

**SSHRC-FHSS Awards to Scholarly Publications** **2413**
The Awards to Scholarly Publications (ASPP) provides funding to help authors and academic presses publish scholarly books that make a significant original contribution to knowledge in the humanities and social sciences in Canada, but would not be published for commercial purposes. Up to 180 subsidies of $8,000 are available each year. Five subsidies of $12,000 are also available for the translation of scholarly works. Detailed eligibility criteria for authors and publishers and a list of frequently asked questions are posted on the FHSS website.
*Restrictions:* Funding is not available for the publication of textbooks, technical reports, original works of poetry, fiction and drama, scholarly journals and articles, and conference proceedings. If the manuscript is a "trade" or popular publication (self-help book, volume of personal reflections, etc.), it is not eligible for funding.
*Geographic Focus:* Canada
*Amount of Grant:* 8,000 CAD
*Contact:* Jessica Clark; (613) 238-6112, ext. 352; fax (613) 238-6114; ideas-idees.ca
*Internet:* www.ideas-idees.ca/aspp
*Sponsor:* Social Sciences and Humanities Research Council of Canada
350 Albert Street, P.O. Box 1610
Ottawa, ON K1P 6G4 Canada

**SSHRC-IDRC International Partnerships for Sustainable Societies** **2414**
The SSHRC-IDRC joint initiative supports international partnerships to produce high-quality research to inform academic, public, and policy debates in ways that can help create just, inclusive, and sustainable social and economic development. Thematic areas include information and networks; inclusive growth; governance, security, and justice; and green growth. Partnerships must be led by a project director based in an eligible low- or middle-income country (LMIC), according to the World Bank listing, and a project director based in Canada. At the letter of intent stage, up to 10 applicants will be awarded a $20,000 grant to prepare a formal application. However, SSHRC may consider requests of up to $30,000 if well justified. Requests for support at the formal application stage would typically be in the range of $500,000 to $2,500,000 over four to seven years. The project director based in an eligible LMIC may request up to $1,250,000 from IDRC and the project director based in Canada may request up to $1,250,000 from SSHRC.
*Requirements:* The Canadian project director must submit a proposal on behalf of the Canadian and LMIC partnership to SSHRC's Partnership Grants – Letter of Intent funding opportunity. This proposal will describe in its entirety the research program, the partnership, and the budget request from SSHRC and IDRC. In addition, the LMIC project director must submit the IDRC Supplementary Information Form to IDRC. SSHRC and IDRC will combine the application forms to conduct the merit review process. Applicants who are successful at the LOI stage will be invited to submit, again jointly, a formal application. They will be given instructions on how to complete the application at that time.
*Geographic Focus:* Canada
*Date(s) Application is Due:* Feb 15
*Amount of Grant:* 20,000 - 1,250,000 CAD
*Contact:* David O'Brien; (613) 236-6163, ext 2052; fax (613) 238-7230; dobrien@idrc.ca
Partnership Grants Coordinator; (613) 943-1007; partnershipgrants@sshrc-crsh.gc.ca
*Internet:* www.idrc.ca/EN/Funding/Competitions/Pages/CompetitionDetails.aspx?CompetitionID=48
*Sponsor:* Social Sciences and Humanities Research Council of Canada
350 Albert Street, P.O. Box 1610
Ottawa, ON K1P 6G4 Canada

**SSHRC-NSERC College and Community Innovation Enhancement Grants** **2415**
The objective of the Innovation Enhancement Grants is to increase innovation at the community and/or regional level by enabling Canadian colleges to increase their capacity to work with local small and medium-sized companies. They support applied research and collaborations that facilitate commercialization, as well as technology transfer, adaptation and adoption of new technologies. Grants will stimulate applied research that brings together expertise from diverse fields such as natural sciences and engineering, social sciences and humanities, and/or health sciences to address business-driven challenges and opportunities. Grants provide funding to colleges on a competitive basis to enhance their applied research capacity, and carry out applied research and technology transfer activities in collaboration with, and to the benefit of small to medium size companies in particular. These grants will focus on an area where the college has recognized expertise, that meet local or regional needs, and where there is the potential to increase economic development of the community. Grants are awarded for either a two-year or a five-year period. For the two-year grant, $100,000 per year for each of two years is provided. For the five-year grant, a base grant of up to $500,000 per year is provided for the first three years. During the course of the grants, colleges are expected to increase their effective collaboration with local or regional companies, and other existing community resources. To ensure increased and sustained commitment and involvement from college partners in years four and five of the five-year grants, grants will provide four-fifths of the average annual base funding received in the first three years. It is expected that other sources of support will be available from business and other partners (either cash or in-kind contributions). Additional guidelines, including entry-level and five year grants requirements and descriptions, are available at the NSERC website.
*Requirements:* Canadian colleges and universities are eligible to apply. These institutions must offer programs in the natural sciences, engineering, social sciences, humanities and/or health sciences. Faculty members involved in the grants must be involved in these disciplines. Institutions must also provide the space, facilities and services to enable its natural sciences, engineering, social sciences, humanities and/or health sciences faculty members to conduct research. Additional requirements and guidelines for businesses are available at the NSERC website.
*Geographic Focus:* Canada
*Contact:* Eric Bergeron, Program Officer, Partnerships Portfolio; (613) 996-1422 or (613) 944-5802; fax (613) 947-0223; eric.bergeron@sshrc-crsh.gc.ca
*Internet:* www.nserc-crsng.gc.ca/Professors-Professeurs/RPP-PP/CCI-ICC_eng.asp
*Sponsor:* Social Sciences and Humanities Research Council of Canada
350 Albert Street, P.O. Box 1610
Ottawa, ON K1P 6G4 Canada

**SSHRC-Public Safety Canada Kanishka Project Initiative** **2416**
The Kanishka Project Initiative funds research on pressing questions for Canada on terrorism and preventing violent extremism. The Project strives to understand what terrorism means in the Canadian context, how it is changing over time, and what can be done to support effective policies and programs to counter terrorism and violent extremism in Canada. Funding is available to researchers from a variety of disciplines to support a range of activities, including research, workshops, and publications. Research themes include ideological extremism and violence; perception and emotion; collective dynamics and resilience; and organization and effectiveness. An extensive list of guidelines and funding opportunities is posted on the Public Safety Canada website. Descriptions of ongoing projects are also included.
*Geographic Focus:* Canada
*Amount of Grant:* Up to 2,500,000 CAD
*Contact:* Brett Kubicek, Manager of Research and Academic Relations; (800) 830-3118 or (613) 944-4875; fax (613) 943-5186; Kanishka@ps-sp.gc.ca
Kanishka Project Coordinator; (613) 943-1007; partnershipgrants@sshrc-crsh.gc.ca
*Internet:* www.publicsafety.gc.ca/prg/ns/ai182/kpcp/knsh-eng.aspx
*Sponsor:* Social Sciences and Humanities Research Council of Canada
350 Albert Street, P.O. Box 1610
Ottawa, ON K1P 6G4 Canada

**SSHRC Aid to Scholarly Journals** **2417**
The objective of the Aid to Scholarly Journals is to increase access to, and readership for, original research results in the social sciences and humanities through Canadian scholarly journals. Through this funding opportunity, SSHRC will contribute to the broad dissemination and impact of original research results in social science and humanities scholarship through Canadian scholarly journals. Funding will also assist journals as they seek to take advantage of advances in communication technologies. Funds will be awarded to help defray the costs of publishing scholarly articles, to assist with distribution costs, and to support journal organizations in their transitions to digital media. A list of eligible expenses is available at the SSHRC website. Funding is available for three years, with up to $30,000 per year. Up to $5,000 a year may also be requested for: distribution costs, or those costs directly related to the circulation of the journal, be they either print or electronic (printing and mailing costs, fees paid to electronic-service providers, or support of existing websites); and transition costs, or those costs directly related to the transition to an electronic medium (the purchase of new computers, including software, costs related to the launch of a website, and staff training). Detailed estimates of cost must be provided for these items. For this competition, a scholarly journal is defined as a peer-reviewed academic publication that disseminates the results of original scholarship. Subscription-based journals, journals using an open-access business model, and journals using a "moving wall" arrangement to provide delayed open access are eligible for support. Also, for this competition, an article is defined as a record of previously unpublished, high-quality, original research or scholarship that has been accepted for publication after a process of peer review. Eligible article types include: research or theoretical articles and/or essays, and literature reviews, provided they meet the other criteria of being original, peer-reviewed and high-quality. Applicants will be required in the application to justify their choices of peer-reviewed articles. The application and additional guidelines are available at the website.
*Requirements:* To be eligible for support under this program, a journal must meet the definition of a scholarly journal; submit each article to be published to a rigorous, independent peer-review process; founded at least two years prior to the application deadline and published within that timeframe at least four issues, each having at least three peer-reviewed articles, or, for electronic journals not publishing in issue format, at least 12 peer-reviewed articles; a minimum of 250 regular readers, as demonstrated through a detailed web-usage report or other verifiable documentation, such as a list of subscribers; an editor-in-chief who is a faculty member of a Canadian postsecondary institution; an editorial board responsible for the intellectual direction and content of the journal of which at least one-third of the board's members are scholars affiliated with a Canadian postsecondary institution; and published primarily in English and/or French.

*Restrictions:* Article types that are not eligible for funding include: introductions; individual book, article or event reviews; letters to the editor; or editorials. A journal is not eligible if it disseminates previously published research results, or mainly publishes material assembled by non-specialists from secondary sources; is a professional or trade magazine, bulletin, newsletter, or advocacy publication, or is instructional in nature; is sponsored or published by, or on behalf of, a government department, agency, or commission; is edited and published solely by students; any member of its editorial board is under SSHRC sanction for financial or research misconduct; has an operating surplus equal to or greater than one year's operating expenses; or a non-Canadian publisher owns the exclusive copyright to its contents.
*Geographic Focus:* Canada
*Date(s) Application is Due:* Jun 30
*Amount of Grant:* Up to 35,000 CAD
*Contact:* Julie Lafontaine; (613) 947-9654; fax (613) 992-7635; julie.lafontaine@sshrc.ca
*Internet:* www.sshrc-crsh.gc.ca/funding-financement/programs-programmes/scholarly_journals-revues_savantes-eng.aspx
*Sponsor:* Social Sciences and Humanities Research Council of Canada
350 Albert Street, P.O. Box 1610
Ottawa, ON K1P 6G4 Canada

**SSHRC Aid to Small Universities** **2418**
The objective of this program is to enable small universities to develop and strengthen focused research capacity in the social sciences and humanities. Examples of activities that the ASU program supports: start-up costs/partial funding of research centers; stipends to doctoral students, provided the program of studies is related to the ASU grant; agenda-setting seminars; visiting scholars (travel and stipend); organization of colloquia or symposia; and seed funding for collaborative research or the development of partnerships.
*Requirements:* Eligible institutions must have active degree-granting status in the SSHRC fields at the undergraduate level or beyond, hold membership in the Association of Universities and Colleges of Canada, have fewer than 250 full-time faculty in the social sciences and humanities, and be independent of the federal government for the purpose of faculty employment status.
*Geographic Focus:* Canada
*Date(s) Application is Due:* Dec 1
*Amount of Grant:* Up to 30,000 CAD
*Contact:* Lorraine Anderson, Team Leader, Research Portfolio; (613) 947-4233; fax (613) 992-7635; lorraine.anderson@sshrc-crsh.gc.ca
*Internet:* www.sshrc-crsh.gc.ca/funding-financement/programs-programmes/aid_small_universities-aide_petites_universites-eng.aspx
*Sponsor:* Social Sciences and Humanities Research Council of Canada
350 Albert Street, P.O. Box 1610
Ottawa, ON K1P 6G4 Canada

**SSHRC Banting Postdoctoral Fellowships** **2419**
The Banting Postdoctoral Fellowships strive to attract and retain top-tier postdoctoral talent, nationally and internationally; develop their leadership potential; and position them for success as research leaders, contributing to Canada's economic, social, and research based growth through a research-intensive career. Selection criteria is based on research excellence and leadership in the research domain; quality of applicant's proposed research program; and institutional commitment and demonstrated synergy between applicant and institutional strategic priorities. The Fellowships are distributed equally across Canada's three federal granting agencies: the Canadian Institutes of Health Research (CIHR), the Natural Sciences and Engineering Research Council (NSERC) and the Social Sciences and Humanities Research Council (SSHRC). Up to 70 new two-year awards are given annually, with up to 140 awards active at any given time. Candidates may apply through the ResearchNet application process at the Banting website. Additional information and frequently asked questions are also posted on the website.
*Requirements:* Canadian citizens, permanent residents, and international citizens may apply. Applicants must fulfill or have fulfilled all degree requirements for a Ph.D., its equivalent, or health professional degree (a complete list of eligible health professionals is posted on the website). Applicants must contact the hot research institution whose endorsement they wish to receive, and continue contact to discuss benefits and their rights and responsibilities.
*Restrictions:* Applicants outside Canada may only hold their Banting Fellowship at a Canadian institution. Canadian or permanent resident applicants who receive their Ph.D. from an international institution may only hold the Banting Fellowship at a Canadian institution. Awards must be taken up no earlier than April 1 and no later than October 1 following the year of application.
*Geographic Focus:* All States, All Countries, Canada
*Date(s) Application is Due:* Nov 1
*Amount of Grant:* 70,000 USD
*Contact:* Banting Fellowship Coordinator; (613) 943-7777; banting@researchnet-recherchenet.ca or fellowships@sshrc-crsh.gc.ca
*Internet:* www.sshrc-crsh.gc.ca/funding-financement/programs-programmes/fellowships/banting-eng.aspx
*Sponsor:* Social Sciences and Humanities Research Council of Canada
350 Albert Street, P.O. Box 1610
Ottawa, ON K1P 6G4 Canada

**SSHRC Bora Laskin National Fellowship in Human Rights Research** **2420**
The purpose of this fellowship is to support research, preferably of a multidisciplinary or interdisciplinary nature, and to develop Canadian expertise in the field of human rights, with emphasis on themes and issues relevant to Canadian human rights. The Fellowship supports research in all areas of humanities and social sciences, including journalism and law. Preference is given to multidisciplinary or interdisciplinary research projects. One non-renewable fellowship, tenable for 12 months in Canada, is awarded every year. The award consists of a $45,000 stipend plus $10,000 for research and research-related travel expenses.
*Requirements:* Applicants must be affiliated with a Canadian postsecondary institution. Preference will be given to applicants with at least five years of proven research experience in their field. Awards must be used between April 1 and October 1 of the year for which they are awarded.
*Restrictions:* The Fellowship is not intended for scholars just beginning their research careers, and may not be used to complete the requirements for a degree program.
*Geographic Focus:* All States, Canada
*Amount of Grant:* 55,000 CAD
*Contact:* Fellowships and Institutional Grants Contact; (613) 943 7777; fax (613) 943 1329; fellowships@sshrc-crsh.gc.ca
*Internet:* www.sshrc-crsh.gc.ca/funding-financement/programs-programmes/fellowships/bora_laskin-eng.aspx
*Sponsor:* Social Sciences and Humanities Research Council of Canada
350 Albert Street, P.O. Box 1610
Ottawa, ON K1P 6G4 Canada

**SSHRC Digging into Data Challenge Grants** **2421**
The Social Sciences and Humanities Research Council of Canada and several other organizations launch the next round of the Digging into Data Challenge, a competition for humanities and social sciences research on cyberscholarship. As the world becomes increasingly digital, new techniques will be needed to search, analyze, and understand these everyday materials. Digging into Data challenges the research community to help create the new research infrastructure for 21st century scholarship. In recognition of the international nature of cyberinfrastructure/e-science, the purpose of the program is to bring together international research teams to advance research and to share their results openly, so that others may learn from them. Along with SSHRC, the following research funders are also involved: the Arts and Humanities Research Council (United Kingdom), the Economic and Social Research Council (United Kingdom), the Institute of Museum and Library Services (United States), the Joint Information Systems Committee (United Kingdom), the National Endowment for the Humanities (United States), the National Science Foundation (United States), and the Netherlands Organisation for Scientific Research (Netherlands). Specific objectives are: to promote the development and deployment of innovative research techniques in large-scale data analysis; to foster interdisciplinary collaboration among scholars in the humanities, social sciences, computer sciences, information sciences, and other fields, around questions of text and data analysis; to promote international collaboration; and to work with data repositories that hold large digital collections to ensure efficient access to these materials for research. Applicants will form international teams from at least two of the participating countries. Winning teams will receive grants from two or more of the funding agencies and, two years later, will be invited to show off their work at a special conference sponsored by the eight funders. Total project funding is approximately U.S. $4.8 million.
*Geographic Focus:* All States, Canada, Netherlands, United Kingdom
*Amount of Grant:* Up to 4,800,000 CAD
*Contact:* Gail Zboch; (613) 943-1148; gail.zboch@sshrc-crsh.gc.ca
Michael Adams; (613) 944-1758; michael.adams@sshrc-crsh.gc.ca
*Internet:* www.sshrc.ca/web/apply/program_descriptions/digging_into_data_e.asp
*Sponsor:* Social Sciences and Humanities Research Council of Canada
350 Albert Street, P.O. Box 1610
Ottawa, ON K1P 6G4 Canada

**SSHRC Doctoral Fellowships** **2422**
Doctoral fellowships are offered to assist in developing research and scholarly resources in the social sciences and humanities. Fellowships are for one, two, three, or four years for universities in Canada or abroad. Direct applications are due November 7, while university applications are due January 9. Applicants are also eligible for the Bombardier Doctoral Scholarship worth $35,000 per year, although the Bombardier Scholarship may only be used at Canadian universities. Applicants apply for both awards through one application form, located at the website. Doctoral Fellowship holders are also eligible for department-awarded Queen's Fellowship of $15,000, William Taylor Fellowship of $5,000, and Aileen Ross Fellowship of $10,000.
*Requirements:* Candidates must be Canadian citizens or have permanent resident (landed immigrant) status at the time of application and must have completed one year of doctoral study or a master's degree by the date of taking the award. Applicants must be pursuing doctoral studies in the social sciences or humanities.
*Restrictions:* Applicants must not have already received a scholarship or fellowship from SSHRC, NSERC, or CIHR, and not be applying to NSERC or CIHR.
*Geographic Focus:* Canada
*Date(s) Application is Due:* Jan 9; Nov 7
*Amount of Grant:* 20,000 CAD
*Contact:* Division of Fellowships and Institutional Grants; (613) 943-7777; fax (613) 943-1329; fellowships@sshrc.ca
*Internet:* www.sshrc-crsh.gc.ca/funding-financement/programs-programmes/fellowships/doctoral-doctorat-eng.aspx#a5
*Sponsor:* Social Sciences and Humanities Research Council of Canada
350 Albert Street, P.O. Box 1610
Ottawa, ON K1P 6G4 Canada

**SSHRC Graduate Scholarships-Michael Smith Foreign Study Supplements** **2423**
The Canada Graduate Scholarships-Michael Smith Foreign Study Supplements (CGS-FSS) program will support high-calibre Canadian graduate students in building global linkages and international networks through the pursuit of exceptional research

experiences in Canada. By accessing international scientific research and training, CGS-FSS recipients will contribute to strengthening the potential for collaboration between Canadian universities and affiliated research institutions and universities, or other research institutions outside of Canada. The defined research study period abroad to be covered must be no less than three months and no more than six months. Supplements of up to $6,000 will be awarded to Canadian citizens or permanent residents who hold an active Joseph-Armand Bombardier, Alexander Graham Bell or Frederick Banting and Charles Best Canada Graduate Scholarship (CGS) at the master's or doctoral level, or a Vanier CGS at the doctoral level. A total of 250 CGS-MSFSS awards are available each year, and are divided among the three granting agencies to reflect the distribution of CGS awards across the social sciences and humanities, natural sciences and engineering, and health-related disciplines: 45 for the Canadian Institutes of Health Research (CIHR); 80 for the Natural Sciences and Engineering Research Council (NSERC); and 125 for SSHRC.
*Requirements:* To be eligible to apply for this program, candidates must be a Canadian citizen or permanent resident of Canada, and be registered at a Canadian institution at the time of application. Eligible institutions abroad include foreign universities, international or national research institutes, or other institutions that actively carry out research.
*Geographic Focus:* Canada
*Date(s) Application is Due:* Jun 10; Oct 10
*Amount of Grant:* Up to 6,000 CAD
*Contact:* Research Training Portfolio Coordinator; (613) 995-2694; scholfin@sshrc-crsh.gc.ca
*Internet:* www.sshrc-crsh.gc.ca/funding-financement/programs-programmes/fellowships/cgs_foreign_study-besc_etudes_etranger-eng.aspx
*Sponsor:* Social Sciences and Humanities Research Council of Canada
350 Albert Street, P.O. Box 1610
Ottawa, ON K1P 6G4 Canada

**SSHRC Insight Development Grants** **2424**

Insight Development Grants support research in its initial stages. The grants enable the development of new research questions, as well as experimentation with new methods, theoretical approaches and/or ideas. Funding is provided for short-term research development projects of up to two years, proposed by individuals or teams. The objectives of the Insight program are to build knowledge and understanding from disciplinary, interdisciplinary and/or cross-sector perspectives through support for the best researchers; support new approaches to research on complex and important topics; provide a high-quality research training experience for students; fund research expertise that relates to societal challenges and opportunities; and mobilize research knowledge, to and from academic and non-academic audiences, with the potential to lead to intellectual, cultural, social and economic influence, benefit and impact. Proposed projects may involve case studies, pilot initiatives, and critical analyses of existing research. Projects may also involve national and international research collaboration, and the exploration of new ways of producing, structuring and mobilizing knowledge within and across disciplines and sectors. Funding is available for emerging and established scholars. Emerging scholar funding is used to develop new research questions and/or approaches. Funding for established scholars provides support to explore new research questions and/or approaches that are distinct from the applicant's previous research. The SSHRC also welcomes applications involving research-creation. Insight Development Grants are adjudicated by academic and/or other experts drawn from, but not limited to, the following research fields: Group 1 - history; medieval studies; classics; literature; fine arts; philosophy; religious studies; and related fields: Group 2 - anthropology; archaeology; linguistics; translation; political science; public administration; law; criminology; geography; urban planning and environmental studies; and related fields: Group 3 - business and management; economics; and related fields: Group 4 - sociology; demography; communication studies; journalism; media studies; gender studies; cultural studies; library and information science; and related fields: Group 5 - education; psychology; social work; career guidance; and related fields. In the application form, applicants will be asked to indicate which of the five groups they consider most appropriate for their proposal, and to select a primary discipline and area of research from a detailed list on the application form. The application form and Canadian Common CV are available at the SSHRC website.
*Requirements:* Applications may be submitted by an individual researcher or team of researchers. Applicants must be affiliated with a Canadian university or college at the time of application.
*Restrictions:* Insight Development Grants are not intended to support large-scale initiatives. Long-term support for research is offered through SSHRC's Insight Grants. Research projects should be clearly defined and in the early stages of the research process. Insight Development Grant funding is not intended to support ongoing research for established scholars.
*Geographic Focus:* All States, Canada
*Date(s) Application is Due:* Feb 1
*Amount of Grant:* 7,000 - 75,000 CAD
*Contact:* Fatima Bidas, Program Officer, Research Portfolio; (613) 943-1560; fax (613) 992-7635; fatima.bidas@sshrc-crsh.gc.ca
Joanne Kay, Senior Research and Policy Analyst; (613) 294-3036; fax (514) 939-1918; joanne.kay@pch.gc.ca
*Internet:* www.sshrc-crsh.gc.ca/funding-financement/programs-programmes/insight_development_grants-subventions_de_developpement_savoir-eng.aspx
*Sponsor:* Social Sciences and Humanities Research Council of Canada
350 Albert Street, P.O. Box 1610
Ottawa, ON K1P 6G4 Canada

**SSHRC Insight Grants** **2425**

Insight Grants support research excellence in the social sciences and humanities. Funding is available to both emerging and established scholars for long-term research initiatives of three to five years. The Insight Grants funding opportunity supports research proposed by scholars and judged worthy of funding by their peers and/or other experts. Insight Grants research initiatives may be undertaken by an individual researcher or a team of researchers working in collaboration. Insight Grants are adjudicated by committees of academic and/or other experts drawn from, but not limited to, the following research fields: Group 1 - history; medieval studies; classics; literature; fine arts; philosophy; religious studies; and related fields: Group 2 - anthropology; archaeology; linguistics; translation; political science; public administration; law; criminology; geography; urban planning and environmental studies; and related fields: Group 3 - business and management; economics; and related fields: Group 4 - sociology; demography; communication studies; journalism; media studies; gender studies; cultural studies; library and information science; and related fields: and Group 5 - education, psychology; social work; and related fields. In the application form, applicants will be asked to indicate which of the five groups they consider most appropriate for their proposal, and to select a primary discipline and area of research from a detailed list available on the application form. The maximum value of a grant is $500,000, over three to five years. A minimum request of $7,000 is required in at least one of the years. A maximum of $200,000 is available in a single year. Additional guidelines are available at the website.
*Requirements:* Applications may be submitted by an individual researcher or a team of researchers. All applicants must be affiliated with a Canadian university, college, or research institute. Before the full application, applicants must first submit a notice of intent to the SSHRC.
*Geographic Focus:* All States, Canada
*Date(s) Application is Due:* Oct 15
*Amount of Grant:* 7,000 - 500,000 CAD
*Contact:* Fatima Bidas, Program Officer, Research Portfolio; (613) 943-1560; fax (613) 992-7635; fatima.bidas@sshrc-crsh.gc.ca
Joanne Kay, Senior Research and Policy Analyst; (613) 294-3036; fax (514) 939-1918; joanne.kay@pch.gc.ca
*Internet:* www.sshrc-crsh.gc.ca/funding-financement/programs-programmes/insight_grants-subventions_savoir-eng.aspx
*Sponsor:* Social Sciences and Humanities Research Council of Canada
350 Albert Street, P.O. Box 1610
Ottawa, ON K1P 6G4 Canada

**SSHRC Institutional Grants** **2426**

Under this program, SSHRC provides annual block grants for eligible Canadian postsecondary institutions in order to fund small-scale research activities by their faculty in the social sciences and humanities. The program assists Canadian universities in developing, increasing, or strengthening research excellence in social sciences and humanities; assisting researchers beginning their research career to become competitive in grant competitions at the national level; helping established researchers with modest funding requirements or those wishing to reorient and strengthen their research programs; and supporting national and international dissemination and collaboration. The application and additional guidelines, including the terms of the funding, are posted on the SSHRC website.
*Requirements:* To be eligible for funding, a postsecondary institution must be: eligible to manage SSHRC funds (see SSHRC's list of eligible institutions); an institutional member of the Association of Universities and Colleges of Canada (AUCC) and/or the Association of Canadian Community Colleges (ACCC); and have active degree-granting status in at least one of the social sciences and humanities disciplines; or the institution must be eligible to manage SSHRC funds (see SSHRC's list of eligible institutions); be an institutional member of the AUCC and/or the ACCC; affiliated or federated with an eligible institution, but grant their own degrees, receive their own operating budget directly from the provincial government, and have their own board of directors; and offer programs in at least one of the social sciences and humanities disciplines.
*Restrictions:* Institutions must adhere to the following *Restrictions:* applicants must maintain an affiliation with the postsecondary institution for the duration of the grant period; visiting fellows and scholars are not eligible, nor are the costs incurred for their travel and residency (although postdoctoral fellows affiliated with the institution are eligible to receive support provided they meet the other eligibility criteria); students are not eligible to hold their own award (but SSHRC encourages the employment of students in SIG-funded research); and support must be for scholarly activity in the humanities and the social sciences.
*Geographic Focus:* Canada
*Date(s) Application is Due:* Dec 1
*Amount of Grant:* 5,000 CAD
*Contact:* Lorraine Anderson, Team Leader, Research Portfolio; (613) 947-4233; fax (613) 992-7635; lorraine.anderson@sshrc-crsh.gc.ca
*Internet:* www.sshrc-crsh.gc.ca/funding-financement/programs-programmes/institutional_grants-subventions_institutionnelles-eng.aspx
*Sponsor:* Social Sciences and Humanities Research Council of Canada
350 Albert Street, P.O. Box 1610
Ottawa, ON K1P 6G4 Canada

**SSHRC Joseph-Armand Bombardier CGS Doctoral Scholarships** **2427**

The J.-Armand Bombardier CGS Doctoral Scholarships programs aim to develop research skills and assist in the training of highly-qualified personnel by supporting students who demonstrate a high standard of scholarly achievement in undergraduate and graduate studies in the social sciences and humanities. Scholarships are valued at $35,000 per annum for 36 months. Scholarship applicants are also eligible for the SSHRC

Doctoral Fellowships at $20,000 per annum for 12, 24, 36, or 48 months. Holders of the Bombardier Scholarship are also eligible for consideration of the department-awarded Queen's Fellowship of $15,000, William Taylor Fellowship of $5,000, and the Aileen Ross Fellowship of $10,000. Direct applications are due November 7, while university applications are due January 9.
*Requirements:* To qualify to apply for the J.-Armand Bombardier CGS Doctoral Awards programs, applicants must: be a citizen or permanent resident of Canada; be applying for support to pursue a first PhD (or equivalent); be pursuing doctoral studies (or a combined MA/PhD) in the social sciences or humanities; not have already received a scholarship or fellowship from SSHRC, NSERC or CIHR to undertake or complete a doctoral degree or combined MA/PhD; not be applying in the upcoming academic year to NSERC or CIHR; or not be pursuing a degree program that combines undergraduate and graduate degrees.
*Restrictions:* The Scholarship may only be used at an eligible university in Canada. A detailed list of those universities is located at the website.
*Geographic Focus:* Canada
*Date(s) Application is Due:* Jan 9; Nov 7
*Amount of Grant:* 35,000 CAD
*Contact:* Doctoral Awards Coordinator; (613) 943-7777 or (613) 995-4273; fax (613) 943-1329; fellowships@sshrc.ca
*Internet:* www.sshrc-crsh.gc.ca/funding-financement/programs-programmes/fellowships/doctoral-doctorat-eng.aspx#a5
*Sponsor:* Social Sciences and Humanities Research Council of Canada
350 Albert Street, P.O. Box 1610
Ottawa, ON K1P 6G4 Canada

**SSHRC Jules and Gabrielle Leger Fellowship** 2428
This fellowship is offered in odd-numbered years to assist in developing research and scholarly resources in the social sciences and humanities. The fellowship is awarded for research and writing on the historical and contemporary contribution of the Crown and its representatives, federal and provincial, in addition to the political, constitutional, cultural, intellectual, and social life of the country. One non-renewable fellowship is offered every second year and is tenable for 12 months. There are no restrictions on the place of tenure. The award consists of a $40,000 stipend plus $10,000 for research and research-related travel expenses. This award may not be used to complete the requirements for a degree program. Awards must be taken up between April 1 and October 1 of the year for which they are awarded. Additional guidelines are available at the SSHRC website.
*Requirements:* Applicants must be affiliated with a Canadian postsecondary institution. This fellowship is not intended for scholars at the beginning of their research careers. Preference will be given to applicants with at least five years of proven research experience in their field. Please note that all applicants, as well as anyone else participating in any capacity in the proposed project, must not be under SSHRC sanction for financial or research misconduct.
*Geographic Focus:* All States, Canada
*Date(s) Application is Due:* Oct 1
*Amount of Grant:* 50,000 CAD
*Contact:* Leger Program Officer; (613) 943-7777; fax (613) 943-1329; fellowships@sshrc-crsh.gc.ca
*Internet:* www.sshrc-crsh.gc.ca/funding-financement/programs-programmes/fellowships/leger-eng.aspx
*Sponsor:* Social Sciences and Humanities Research Council of Canada
350 Albert Street, P.O. Box 1610
Ottawa, ON K1P 6G4 Canada

**SSHRC Metropolis Project** 2429
The purpose of the annual Metropolis National Research Competition is to select and finance a major national-level policy research project in the field of immigration and diversity. The findings of the research will be of direct relevance to policy-makers within the Government of Canada and, more specifically, to the main funding partners of the Metropolis Project. At a broader level, the objectives of the competition are to produce a pan-Canadian research study that informs federal policy on issues surrounding immigration and diversity; produce knowledge that affects federal policy and programs related to immigration and diversity; widely disseminate, in collaboration with the Metropolis secretariat, the results of this research through a range of knowledge transfer/mobilization activities targeting federal policy-makers and the research community; and create knowledge that informs and expands the broader academic discourse on immigration and diversity. The maximum value of the grant is $125,000 for 12 to 18 months. Extensions are not available. The application and detailed guidelines are posted at the website.
*Requirements:* Researchers affiliated with Canadian postsecondary educational institutions are invited to submit a proposal. Successful applicants must seek affiliation with one of the five Metropolis centres (an up-to-date list is found at the website). Individuals who wish to become affiliated with a Metropolis centre can apply by completing the appropriate application forms, available on the Metropolis website, or by contacting one of the centres directly.
*Geographic Focus:* Canada
*Date(s) Application is Due:* Oct 11
*Amount of Grant:* Up to 125,000 CAD
*Contact:* Mika Oehling, Program Officer, Partnerships Portfolio; (613) 992-4227; fax (613) 947-0223; mika.oehling@sshrc-crsh.gc.ca
*Internet:* www.sshrc-crsh.gc.ca/funding-financement/programs-programmes/metropolis-eng.aspx
*Sponsor:* Social Sciences and Humanities Research Council of Canada
350 Albert Street, P.O. Box 1610
Ottawa, ON K1P 6G4 Canada

**SSHRC Postdoctoral Fellowships** 2430
The Postdoctoral Fellowships support the most promising new scholars in the social sciences and humanities and assists them in establishing a research base at an important time in their research careers. Postdoctoral Fellowships provide stipendiary support to recent PhD graduates who are: undertaking original research; publishing research findings; developing and expanding personal research networks; broadening teaching experience; and preparing to become competitive in national research grants competitions. Posdoctoral Fellowships are valued at approximately $38,000 per year, plus a research allowance of up to $5,000 per award. These are non-renewable fellowships, tenable for a minimum of 12 months and a maximum of 24 months. Fellowships will normally be awarded to candidates affiliated with a university other than that which awarded the PhD. Recipients of this Fellowship are also eligible for department consideration of the Postdoctoral Prize of $10,000. Additional guidelines and the application are available at the SSHRC website.
*Requirements:* To apply to this program, applicants must be a citizen or permanent resident of Canada; have earned a first doctorate from a recognized university within three years of the application date or have completed a first doctorate no earlier than five years before the application date, but have had their career interrupted or delayed for the purpose of child rearing, illness, or health-related family responsibilities; not hold a tenure or tenure-track faculty position; have finalized arrangements for affiliation with a recognized university or research institution; and have applied not more than twice before for a SSHRC Postdoctoral Fellowship.
*Restrictions:* Applicants who obtain their PhD from a foreign university are eligible to apply only if they wish to hold their award at a Canadian university. Applicants are eligible to apply to hold their award at a foreign university only if their PhD was earned at a Canadian university.
*Geographic Focus:* All States, Canada
*Date(s) Application is Due:* Oct 4
*Amount of Grant:* Up to 43,000 CAD
*Contact:* Fatima Bidas, Research Training Portfolio Contact; (613) 943-7777; fax (613) 943-1329; fatima.bidas@sshrc-crsh.gc.ca or fellowships@sshrc-crsh.gc.ca
*Internet:* www.sshrc-crsh.gc.ca/funding-financement/programs-programmes/fellowships/postdoctoral-doctorat-eng.aspx
*Sponsor:* Social Sciences and Humanities Research Council of Canada
350 Albert Street, P.O. Box 1610
Ottawa, ON K1P 6G4 Canada

**SSHRC Sport Participation Research Initiative: Doctoral Award Supplements** 2431
Applicants for a SSHRC doctoral award who indicate their interest on the application form and who are working in one or more of the target areas will be considered for a Sport Canada Doctoral Award Supplement. Those target areas are the following: identification and overcoming of barriers to participation in sport; training of participants, volunteers, coaches and administrators in sport; development, monitoring and evaluation of policies designed to enhance participation in sport; development of capacity and infrastructure that will enhance participation in sport; and identification and assessment of the benefits and outcomes of participation in sport. Those who already hold a SSHRC doctoral award and are working in one or more of the target areas are also eligible to apply for this supplement. For applicants registered at a Canadian postsecondary institution, the postsecondary institution sets the deadline; for all others, November 7 is the deadline.
*Requirements:* Applicants for, and holders of, a SSHRC doctoral award who are working in one or more of the target areas of interest, and who are affiliated with an eligible Canadian university are eligible to apply.
*Geographic Focus:* Canada
*Amount of Grant:* Up to 10,000 CAD
*Contact:* Fatima Bidas, Program Officer, Research Portfolio; (613) 943-1560; fax (613) 992-7635; fatima.bidas@sshrc-crsh.gc.ca
Joanne Kay; (613) 294-3036; fax (514) 939-1918; joanne.kay@pch.gc.ca
*Internet:* www.sshrc-crsh.gc.ca/funding-financement/programs-programmes/sport_can-eng.aspx#atarget
*Sponsor:* Social Sciences and Humanities Research Council of Canada
350 Albert Street, P.O. Box 1610
Ottawa, ON K1P 6G4 Canada

**SSHRC Sport Participation Research Initiative: Postdoctoral Fellowship Supplements** 2432
Applicants for a SSHRC Postdoctoral Fellowship who indicate their interest on the application form and who are working in one or more of the following target areas will be considered for a Sport Canada Postdoctoral Fellowship Supplement. Those areas of research are: identification and overcoming of barriers to participation in sport; training of participants, volunteers, coaches and administrators in sport; development, monitoring and evaluation of policies designed to enhance participation in sport; development of capacity and infrastructure that will enhance participation in sport; and identification and assessment of the benefits and outcomes of participation in sport. Sport Participation Research Initiative Postdoctoral Fellowship Supplements are tenable for 12 months and worth $10,000, in addition to the annual value of the SSHRC Postdoctoral Fellowship. Additional guidelines are posted on the SSHRC website.
*Requirements:* Applicants for a SSHRC Postdoctoral Fellowship who are working in one or more of the target areas of interest, and who are affiliated with an eligible Canadian postsecondary institution, are eligible to apply.
*Restrictions:* Only applications that are recommended for funding at SSHRC are referred to Sport Canada.
*Geographic Focus:* All States, Canada
*Date(s) Application is Due:* Oct 4
*Amount of Grant:* 10,000 CAD

*Contact:* Fatima Bidas, Program Officer, Research Portfolio; (613) 943-1560; fax (613) 992-7635; fatima.bidas@sshrc-crsh.gc.ca
Joanne Kay, Senior Research and Policy Analyst; (613) 294-3036; fax (514) 939-1918; joanne.kay@pch.gc.ca
*Internet:* www.sshrc-crsh.gc.ca/funding-financement/programs-programmes/sport_can-eng.aspx
*Sponsor:* Social Sciences and Humanities Research Council of Canada
350 Albert Street, P.O. Box 1610
Ottawa, ON K1P 6G4 Canada

**SSHRC Therese F. Casgrain Fellowship** **2433**
The Therese F. Casgrain Fellowship supports research on women and social change in Canada. This Fellowship supports research in the field of social justice, particularly in defense of individual rights and the promotion of the economic and social interests of Canadian women. The Foundation is particularly interested in research that includes a discussion of public policy options and makes concrete recommendations for change. The non-renewable Fellowship, tenable for 12 months and only in Canada, is available every fourth year. The award consists of a $40,000 stipend, paid in three installments, of which up to $10,000 may be used for travel and research expenses. The successful applicant must take up the award between April 1 and October 1 following the announcement of awards.
*Requirements:* Applicants must be Canadian citizens or permanent residents of Canada, hold a doctorate degree or equivalent advanced professional degree, have at least five years of proven research experience, and not be under SSHRC sanction for financial or research misconduct. Affiliation with a university or an appropriate research institute or similar organization is desirable, but is not a condition of the award.
*Geographic Focus:* Canada
*Amount of Grant:* Up to 40,000 CAD
*Contact:* Fellowships and Institutional Grants Contact; (613) 943-7777; fax (613) 943-1329; fellowships@sshrc-crsh.gc.ca
*Internet:* www.sshrc-crsh.gc.ca/funding-financement/programs-programmes/fellowships/casgrain-eng.aspx
*Sponsor:* Social Sciences and Humanities Research Council of Canada
350 Albert Street, P.O. Box 1610
Ottawa, ON K1P 6G4 Canada

**SSRC-Van Alen Fellowships** **2434**
The SSRC-Van Alen New York Prize Fellowships support work on the intersection between social dynamics and the built environment. Since its inception, two rounds of Fellowships have been offered in collaboration with the Van Alen Institute, an independent nonprofit architectural organization, on issues of urban sustainability and the 'spatialization' of information policy. Grants are awarded to partnerships between social scientists and architectural/design specialists. Investigations may take a range of formats, including installations, demonstrations, performances, and workshops, but are by their involvement of the general public. Additional information is available at the SSRC website.
*Geographic Focus:* All States
*Contact:* Joe Karaganis, Program Director; (718) 517-3673 or (212) 377-2700; fax (212) 377-2727; karaganis@ssrc.org
*Internet:* www.ssrc.org/fellowships/ssrc-van-alen-fellows-program/
*Sponsor:* Social Science Research Council
One Pierrepont Plaza, 15th Floor
Brooklyn, NY 11201

**SSRC Abe Fellowship for Journalists** **2435**
The Abe Fellowship for Journalists is designed to encourage in-depth coverage of topics of pressing concern to the U.S. and Japan through individual short-term policy-related projects. Applicants are invited to submit proposals on one of the following three themes: traditional and non-traditional approaches to security and diplomacy (transnational terrorism, internal ethnic and religious strife, infectious diseases, food safety, climate change, and non-proliferation, as well as the role of cultural initiatives in peace building); global and regional economic issues (regional and bilateral trade arrangements, international financial stability, globalization and the mitigation of its adverse consequences, sustainable urbanization, and environmental degradation); and social and cultural issues (demographic change, immigration, the role of civil society and media as champions of the public interest, social enterprise, corporate social responsibility, and revitalization of multi-cultural urban areas). Fellows are expected to produce an analytical article or feature story that will inform public debate or a policy community. The application and additional requirements are available at the SSRC website.
*Requirements:* The AFJ competition is open to citizens of the U.S. and Japan with at least five years of professional journalistic experience with newspapers, newsmagazines, wire services, and on-line news organizations. Freelancers are also eligible. Nationals of other countries must be permanent residents of the U.S. or Japan, or have a long-term affiliation with the American or Japanese journalistic communities. U.S.-based applicants with no previous journalistic employment in Japan have priority. Japan-based applicants with no previous journalistic employment in the U.S. have priority.
*Restrictions:* Fellows may receive salary from their employers, but cannot carry out assignments while on the award.
*Geographic Focus:* All States, Japan
*Date(s) Application is Due:* Sep 15
*Amount of Grant:* Up to 26,100 USD
*Contact:* Fernando Rojas, U.S; (202) 377-2700; fax (212) 377-2727; abe@ssrc.org
Takuya Toda-Ozaki, Program Manager, Tokyo; +81-3-5369-6072 or +81-3-5369-6085; fax +81-3-5369-6042; ssrcABE@gol.com or tozaki@gol.com
*Internet:* www.ssrc.org/fellowships/abe-fellowship-for-journalists/
*Sponsor:* Social Science Research Council
One Pierrepont Plaza, 15th Floor
Brooklyn, NY 11201

**SSRC Abe Fellowships** **2436**
The Abe Fellowship is designed to encourage international multidisciplinary research on topics of pressing global concern. The program seeks to foster the development of a new generation of researchers who are interested in policy-relevant topics of long-range importance and who are willing to become key members of a bilateral and global research network built around such topics. Rather than seeking to promote greater understanding of a single country, the Abe Fellowship encourages research on the experiences and future challenges of the U.S. and Japan in a comparative or global perspective. The program committee seeks applications for research focusing explicitly on policy-relevant and contemporary issues that have a comparative or transnational perspective and that draw the study of the U.S. and Japan into wider disciplinary or theoretical debates. Applicants are invited to submit proposals for research in the social sciences or the humanities relevant to any one or combination of the following three themes: global issues; problems common to industrial and industrializing societies; or issues that pertain to U.S.-Japan relations. Applications are welcome from scholars and non-academic research professionals. Funding for the program is provided by the Japan Foundation Center for Global Partnership. The application and additional information are available at the SSRC website.
*Requirements:* Applications are welcome from scholars and non-academic research professionals. Competition is open to citizens of the U.S. and Japan as well as to nationals of other countries who can demonstrate strong and serious long-term affiliations with research communities in Japan or the U.S. Applicants must hold a Ph.D. or the terminal degree in their field, or have attained an equivalent level of professional experience.
*Geographic Focus:* All States
*Date(s) Application is Due:* Sep 1
*Contact:* Fernando Rojas; (212) 377-2000; fax (212) 377-2727; abe@ssrc.org
Takuya Toda-Ozaki, Program Manager, Tokyo; +81-3-5369-6072 or +81-3-5369-6085; fax +81-3-5369-6042; ssrcABE@gol.com or tozaki@gol.com
*Internet:* www.ssrc.org/fellowships/abe-fellowship/
*Sponsor:* Social Science Research Council
One Pierrepont Plaza, 15th Floor
Brooklyn, NY 11201

**SSRC African Peacebuilding Network Research Grants** **2437**
The African Peacebuilding Network (APN) of the Social Science Research Council (SSRC) invites research grant applications from African researchers, policy analysts, and practitioners working on conflict and peacebuilding at universities and research institutions or regional governmental and non-governmental organizations in Africa. A core component of the APN, the Research Grants program is a vehicle for enhancing the quality and visibility of African peacebuilding research regionally and globally and for making such knowledge accessible to key policymakers and research centers of excellence in Africa and around the world. Substantial support is available for research and analysis on issues such as: conflict prevention, mediation, management, resolution, and transformation; environmental change and conflict; post-conflict democratization, governance, and reconstruction; the relationship between peacebuilding and statebuilding, including state-society relations and state reconstruction; transitional justice, reconciliation, social and economic justice; peacekeeping and peace support operations; and disarmament, demobilization, and reinsertion. The APN is interested in innovative projects that demonstrate strong potential for producing high-quality research and analysis that can lead to practical action on peacebuilding and/or facilitate inter-regional collaboration and networking among African researchers and practitioners. Additional proposal guidelines are available at the SSRC website. Grants are awarded on a competitive, peer-reviewed basis and are intended to support six to seven months of research. Up to fifteen grants of a maximum of $15,000 will be awarded. The application is available through the SSRC's application portal at its website.
*Requirements:* Applicants must be African citizens currently residing in a Sub-Saharan African country. Academics must hold a faculty or research position at an African university or research organization and have either a Ph.D. or a master's, with at least two years of research experience. Policy analysts and practitioners must be based in Africa at a regional or sub-regional institution, a government agency, or a non-governmental, media, or civil society organization and have a master's with at least two years of work experience.
*Geographic Focus:* Angola, Benin, Botswana, Burkina Faso, Burundi, Cameroon, Cape Verde, Central African Republic, Chad, Comoros, Congo, Democratic Republic of, Cote d' Ivoire (Ivory Coast), Djibouti, Equatorial Guinea, Eritrea, Ethiopia, Gabon, Gambia, Ghana, Guinea, Guinea-Bissau, Kenya, Lesotho, Liberia, Madagascar, Malawi, Mali, Mauritania, Mauritius, Mozambique, Namibia, Niger, Nigeria, Rwanda, Sao Tome & Principe, Senegal, Seychelles, Sierra Leone, South Africa, Sudan, Swaziland, Tanzania, Togo, Uganda, Zambia, Zimbabwe
*Date(s) Application is Due:* Jun 15
*Amount of Grant:* Up to 15,000 USD
*Contact:* Kirin Bains, Research Coordinator; (718) 517-3679 or (212) 377-2700; fax (212) 377-2727; kbains@ssrc.org
*Internet:* www.ssrc.org/fellowships/apn-research-grants/
*Sponsor:* Social Science Research Council
One Pierrepont Plaza, 15th Floor
Brooklyn, NY 11201

**SSRC Collaborative Research Grants on Environment and Health in China** 2438

The collaborative grants program was begun to stimulate new research on environment and health in China that is directly linked to policy and practice, to encourage collaboration across disciplines, and to help develop a network of scholars working in this field. Each year seven to eight grants of $10,000 to 15,000 each are given to Chinese institutes for projects that: integrate environment or health perspectives into research currently focused on a single domain by sharing existing data or conducting limited additional data collection; develop a new focus on environment and health within the context of ongoing research on development; broaden the scope of analysis through collaboration with partners who bring additional disciplinary perspectives or international experience; or work with NGO or government partners to make existing research more accessible and useful. Additional information is available at the SSRC website.

*Requirements:* Each project must address an environmental health problem that affects a substantial number of people and it must be clearly useful in generating or disseminating knowledge that will inform policy or the work of government or civil society actors. In order to ensure feasibility, applicants must demonstrate the necessary access to data, or relationships with local communities or government agencies, to conduct the project successfully. If the project is being carried out at the local level, researchers must either live and work in the same province, or otherwise demonstrate how the project will build local capacity to work on environment and health issues.

*Geographic Focus:* China

*Amount of Grant:* 10,000 - 15,000 USD

*Contact:* Chao Cai, Communications Officer, China; (212) 377-2700; fax (212) 377-2727; chao@ssrc.org

Xiaofang Liu, Program Assistant; (212) 377-2700; fax (212) 377-2727; liu@ssrc.org

*Internet:* www.ssrc.org/fellowships/grants-for-collaborative-research-on-environment-and-health-in-china/

*Sponsor:* Social Science Research Council

One Pierrepont Plaza, 15th Floor

Brooklyn, NY 11201

**SSRC Dissertation Proposal Development Faculty Field Fellowships** 2439

The Dissertation Proposal Development Fellowship (DPDF) Faculty Field Competition is open to tenured humanities and social sciences faculty interested in creating or reinvigorating interdisciplinary fields of study through the training of the next generation of researchers. Selected research directors guide the development of effective doctoral dissertation proposals within innovative fields by helping fellows sharpen the focus of their research and identify appropriate methods of investigation and analysis. Annually, the DPDF program selects up to five research fields, each proposed by two senior faculty with different institutional affiliations and, as relevant, different disciplinary specializations. Faculty selected as field research directors design and lead two workshops for student fellows, in coordination with DPDF staff, and serve as mentors to the fellows during the course of their summer pre-dissertation research. Proposed research fields should be domains of inquiry centered on a core set of issues and questions that can be valuably addressed by diverse scholarship. Fields should address topics of broad public concern that can be approached from multiple intellectual, societal, and geographic orientations. Fields should also address topics that encompass broad geographic areas rather than focus on one specific country. Research fields proposed by pairs of faculty who bring distinct perspectives from any different disciplines are welcome, but fields stemming jointly from the humanities and social sciences are particularly encouraged. Additional information, frequently asked questions, and a sample application are available at the SSRC website.

*Requirements:* Applicants must apply in pairs, and be based at different university, with at least one applicant based in the U.S. Both applicants must be available to attend the spring and fall workshops in their entirety (workshop dates found at website). Both applicants must be tenured at the time of the application and trained in different disciplines or bring different methodological frameworks to their proposed research field. At least one applicant must be based at a doctoral-degree granting university. However, both applicants must demonstrate records of carrying out research activities and mentoring graduate students.

*Restrictions:* Research fields proposed to be led by more than two people will not be accepted.

*Geographic Focus:* All States

*Date(s) Application is Due:* Sep 30

*Amount of Grant:* Up to 13,000 USD

*Contact:* Marika Dunn, Program Officer; (718) 517-3645 or (212) 377-2700; fax (212) 377-2727; dunn@ssrc.org

*Internet:* www.ssrc.org/fellowships/dpdf-faculty-fields-competition/

*Sponsor:* Social Science Research Council

One Pierrepont Plaza, 15th Floor

Brooklyn, NY 11201

**SSRC Dissertation Proposal Development Student Fellowships** 2440

The Dissertation Proposal Development Fellowship (DPDF) is organized to help early-stage graduate students in the humanities and social sciences formulate effective doctoral dissertation proposals. Each year, the program offers training within different interdisciplinary fields of study under the leadership of pairs of tenured senior faculty, who define the fields and serve as research directors for groups of 12 graduate students. The students participate in two workshops: one in the spring, to prepare students to undertake summer research that will inform the design of their dissertation research; the other in the fall, to help students apply their summer research experiences to writing dissertation and funding proposals. Students may apply for up to $5,000 to cover summer research costs. Working together, research directors and graduate students design research that will help to shape evolving fields in the humanities and social science. Current research fields include: Black Atlantic Studies; The Political Economy of Redistribution; Visual Culture; Critical Studies of Science and Technology Policy; Animal Studies; Human Dimensions of Global Environmental Change; Muslim Modernities; Science/Art Studies; Global Indigenous Politics; and Migration and Gender Studies. Research fields change every year. The application and additional information are available at the SSRC website.

*Requirements:* The DPDF program is open to doctoral students in social sciences or humanities disciplines who have completed their major course requirements, and are beginning to design research proposals. Students who have completed their comprehensive, general, or qualifying exams are eligible to apply as long as their dissertation proposals will be formally approved by their department after the fall DPDF workshop. Typically such students will be second and third year graduate students, but first and fourth year student may, under exceptional circumstances, be eligible. Applicants must be enrolled full-time in a Ph.D. program within an accredited university in the U.S., unless indicated otherwise for a particular field.

*Geographic Focus:* All States

*Date(s) Application is Due:* Sep 30

*Amount of Grant:* Up to 5,000 USD

*Contact:* Schuyler Marquez, Program Assistant; (718) 517-3704 or (212) 377-2700; fax (212) 377-2727; marquez@ssrc.org

Brian Paragas, Program Assistant, Migration and Human Rights; (718) 517-3695; fax (212) 377-2727; paragas@ssrc.org

*Internet:* www.ssrc.org/fellowships/dpdf-fellowship/

*Sponsor:* Social Science Research Council

One Pierrepont Plaza, 15th Floor

Brooklyn, NY 11201

**SSRC Drugs, Security and Democracy Fellowships** 2441

The Drugs, Security and Democracy (DSD) Fellowships support research on organized crime, drug policy, issues of governance, and associated topics across the social sciences and related disciplines. The fellowship seeks to develop a concentration of researchers who are interested in policy-relevant outcomes and membership in a global interdisciplinary network. Funded research must address the relationship among at least two of the following three themes: drugs — potential topics include, but are not limited to, international and regional drug policy, drug trafficking, organized crime, drug production, and impact on communities including youth delinquency and gangs; security — potential topics include, but are not limited to, issues of traditional and non-traditional security, public safety, the pluralization of security actors, the role of law enforcement, the accountability of police forces, formal and informal strategies to increase security, violence, instability, immigration and border security; and/or democracy — potential topics include, but are not limited to, issues of governance, state responses to organized violence, civil society networks and how they mobilize against organized crime and drugs, the framing of incentives to develop appropriate policies, freedom of the press, impunity, corruption, and the relationship between state and non-state actors. Examples of possible research topics that address the relationship among two of the three themes include: substance control in border regions, violence and electoral campaigns, human rights and security policy, and media and violence and global drug flows. The program encourages interdisciplinary projects and those that address transnational and transregional issues. Applications from researchers in professions outside of academia are welcome. Additional information and requirements are available at the SSRC website.

*Requirements:* The DSD Program features two competitions: Dissertation Fellowship — open to PHD and JSD candidates worldwide who have an approved dissertation prospectus by July 1, but have not completed writing for final submission; and Research Fellowship — applications are accepted from PHD recipients worldwide who have completed their PHD within seven years of the application deadline and researchers in Latin America or the Caribbean without a PHD but with a master's or the terminal degree in their field or equivalent professional experience.

*Geographic Focus:* All States, Puerto Rico, Antigua & Barbuda, Argentina, Aruba, Bahamas, Barbados, Bolivia, Brazil, British Virgin Islands, Cayman Islands, Chile, Colombia, Costa Rica, Cuba, Dominican Republic, Ecuador, El Salvador, Grenada, Guadeloupe, Guatemala, Haiti, Honduras, Jamaica, Martinique, Mexico, Nicaragua, Panama, Paraguay, Peru, Saint Kitts And Nevis, St. Lucia, Trinidad and Tobago, Turks and Caicos Islands, Uruguay, Venezuela, Virgin Islands (U.S.)

*Date(s) Application is Due:* Jan 20

*Contact:* Jessica Mack, Program Associate; (212) 377-2700 or (718) 517-3656; fax (212) 377-2727; mack@ssrc.org

*Internet:* www.ssrc.org/fellowships/dsd-fellowship/

*Sponsor:* Social Science Research Council

One Pierrepont Plaza, 15th Floor

Brooklyn, NY 11201

**SSRC Eurasia Program Dissertation Development Awards** 2442

The Eurasia Program offers financial and academic support to graduate students in the early stages of dissertation development, Ph.D. candidates near completion of their doctoral programs in the social sciences and related humanities, and young scholars within five years of the completion of their Ph.D. Dissertation Development Awards (DDAs) are intended to provide one academic year of support to enable the prompt completion of a Ph.D dissertation. Fellows will participate in professionalization activities and a spring conference, and contribute to the Eurasia Program's new working paper and policy brief series. The application is available through the Council's online application system. Frequently asked questions are also available at the website.

*Requirements:* Applicants must be citizens or permanent residents of the U.S., and have completed all requirements for the Ph.D. except for the dissertation at the time the fellowship begins. Applicants must have obtained ABD status (must have completed all

of the requirements for the Ph.D degree except for the dissertation) by the application submission deadline. Applicants should demonstrate the ability to complete their dissertation by the end of the fellowship support period, and demonstrate evidence of not less than 10% University cost-sharing. This may be an in-kind contribution, or may come in the form of a tuition remission, a waiver of required fees, or health coverage.
*Restrictions:* Only the following countries are eligible for funding under the Eurasian Program: Armenia, Azerbaijan, Belarus, Georgia, Kazakhstan, Kyrgyzstan, Moldova, Russia, Tajikistan, Turkmenistan, Ukraine and Uzbekistan.
*Geographic Focus:* All States
*Amount of Grant:* Up to 20,000 USD
*Contact:* Denise Mishiwiec, Program Coordinator; (212) 377-2700; fax (212) 377-2727; mishiwiec@ssrc.org or eurasia@ssrc.org
*Internet:* www.ssrc.org/fellowships/eurasia-fellowship/
*Sponsor:* Social Science Research Council
One Pierrepont Plaza, 15th Floor
Brooklyn, NY 11201

**SSRC Eurasia Program Pre-Dissertation Awards** 2443
The Eurasia Program provides financial and academic support to graduate students in the early stages of dissertation development; Ph.D. candidates near completion of their doctoral programs in the social sciences and related humanities; and young scholars within five years of the completion of their Ph.D. Pre-Dissertation Awards (PDAs) enable early stage graduate students to perform initial field assessments of up to 4 weeks for archival exploration, preliminary interviews, and other forms of feasibility studies related to their dissertations. Proposals should reflect a clear plan for initial field assessment, require a budget of less than $4,000, and clearly articulate the policy relevance of the proposed project. The Council anticipates that six scholars will be able to gain firsthand knowledge of their proposed field sites, establish contacts within local communities, meet with local scholars, and gain insight into how their dissertation topic resonates with regional intellectual, political and social currents. Frequently asked questions are available at the Council website, in additional to the online application process.
*Requirements:* Applicants must be citizens or permanent residents of the United States, and have completed all requirements for the Ph.D. except for the dissertation at the time the fellowship commences. Applicants must have obtained ABD status (must have completed all of the requirements for the Ph.D degree except for the dissertation) by the application submission deadline. Applicants should demonstrate the ability to complete their dissertation by the end of the fellowship support period, and demonstrate evidence of not less than 10% University cost-sharing.
*Restrictions:* The following countries are eligible for funding under the SSRC Eurasia Fellowship Program: Armenia, Azerbaijan, Belarus, Georgia, Kazakhstan, Kyrgyzstan, Moldova, Russia, Tajikistan, Turkmenistan, Ukraine and Uzbekistan.
*Geographic Focus:* All States
*Date(s) Application is Due:* Dec 1
*Amount of Grant:* Up to 4,000 USD
*Contact:* Denise Mishiwiec, Program Coordinator; (212) 377-2700; fax (212) 377-2727; mishiwiec@ssrc.org or eurasia@ssrc.org
*Internet:* www.ssrc.org/fellowships/eurasia-fellowship/
*Sponsor:* Social Science Research Council
One Pierrepont Plaza, 15th Floor
Brooklyn, NY 11201

**SSRC International Dissertation Research Fellowships** 2444
The International Dissertation Research Fellowship (IDRF) offers nine to twelve months of support for up to 80 graduate students in the humanities and social sciences who are enrolled in doctoral programs in the U.S. and conducting dissertation research outside of the U.S. IDRF promotes research that is situated in a specific discipline and geographical region but is also informed by interdisciplinary and cross-regional perspectives. Research topics may address all periods in history, but applicants should be alert to the broader implications of their research as it relates to contemporary issues and debates. The program invites proposals for empirical and site-specific dissertation research outside the U.S. It will consider applications for dissertation research grounded in a single site, informed by broader cross-regional and interdisciplinary perspectives, as well as applications for multi-sited, comparative, and transregional research. Fellowship amounts vary depending on the research plan, with a per-fellowship average of $20,000. The fellowship includes participation in an SSRC-funded interdisciplinary workshop upon the completion of IDRF-funded research. The application and additional information are available at the SSRC website.
*Requirements:* The program is open to graduate students in the humanities and social sciences - regardless of citizenship - enrolled in doctoral programs in the U.S. Applicants must complete all Ph.D. requirements except on-site research by the time the fellowship begins or by December, whichever comes first.
*Restrictions:* Proposals that identify the U.S. as a case for comparative inquiry are welcome; however, proposals that require no field research outside the U.S. are not eligible. The IDRF program expects fellows to remain at their research site(s) for the full nine- to twelve-month funding period. The IDRF program will not support study at foreign universities, conference participation, or dissertation write-up. The program does not accept applications from Ph.D. programs in law, business, medicine, nursing or journalism, nor does it accept applications in doctoral programs that do not lead to a Ph.D.
*Geographic Focus:* All States
*Date(s) Application is Due:* Nov 3
*Amount of Grant:* Up to 19,000 USD
*Contact:* Elsa Ransom, Program Manager; (718) 517-3642 or (212) 377-2700; fax (212) 377-2727; ransom@ssrc.org
*Internet:* www.ssrc.org/fellowships/idrf-fellowship/
*Sponsor:* Social Science Research Council
One Pierrepont Plaza, 15th Floor
Brooklyn, NY 11201

**SSRC Islamic Traditions and Muslim Societies in World Contexts Grants** 2445
The Islamic Traditions and Muslim Societies in World Contexts Grants support public engagement by academic experts on Islamic traditions and Muslim societies. The grants program seeks to encourage projects permitting the dissemination of academic research on Islamic traditions and Muslim societies to targeted constituencies, with particular emphasis on media and policymakers. With this program the SSRC looks to nurture the public relevancy of area studies research by promoting strategic partnerships that will break down barriers and build connections between academic researchers, journalists, policymakers, and practitioners. The SSRC also seeks to assist faculty in developing the necessary skills they require to be able to engage specialized nonacademic audiences on the topic of Islam in the world. Funds are available to innovative projects that will effect change by creating robust and sustainable capacities for public and policy engagement on university campuses. Projects which include strong faculty participation and leadership will be given funding priority as will projects that disseminate expert knowledge born of context specific research. Successful applications will demonstrate usage of adequate distribution channels for all deliverables intended to reach targeted constituencies. The application and additional information are available at the SSRC website.
*Requirements:* Grants are available to interdisciplinary research centers on U.S. university campuses, including but not limited to Title VI National Resource Centers. Research centers are eligible to apply individually for grants up to $50,000 to be used over a period of fifteen months. Extraordinary projects involving multiple collaborating centers or universities may be considered for grants up to $100,000.
*Geographic Focus:* All States
*Date(s) Application is Due:* Dec 3
*Amount of Grant:* Up to 100,000 USD
*Contact:* Thomas Asher, Program Director; (718) 517-3661 or (212) 377-2700; fax (212) 377-2727; asher@ssrc.org
*Internet:* www.ssrc.org/fellowships/islam-and-muslims-in-world-contexts/
*Sponsor:* Social Science Research Council
One Pierrepont Plaza, 15th Floor
Brooklyn, NY 11201

**SSRC Japan Society for the Promotion of Science (JSPS) Fellowships** 2446
The Japan Society for the Promotion of Science (JSPS) Fellowship Program provides recent PhD recipients and ABDs (please see program eligibility requirements) with opportunities to conduct research in Japan under the leadership of a host researcher. Fellows are encouraged to advance their own research and at the same time closely collaborate with young Japanese researchers and contribute to Japanese research communities. Fellows are selected based on nominations made by the SSRC Japan Advisory Board. The SSRC Japan Advisory Board recommends to JSPS up to 20 candidates annually from the social sciences and humanities. Award offers are made by JSPS in early-summer for fellowships that must commence between April 1 and November 30 (long-term) or April 1 and March 31 (short-term). Applications are welcome from all social science and humanities disciplines and need not be explicitly related to the study of Japan.
*Requirements:* Applicants must be U.S. citizens or permanent residents at the time of application and submit proof of affiliation with an eligible host research institution in Japan as part of the application packet. Permanent residents must provide a copy of a permanent resident card (green card). Citizens of other countries are eligible for short-term fellowships if they have completed a master's or PhD course at an institution of higher education in the U.S. and, upon completing the course, have for at least three continuous years conducted high-level research at a university in the U.S. Priority will be given to researchers who are not Japanese citizens.
*Restrictions:* JSPS does not extend fellowships to scholars employed in the institutions under jurisdiction of the U.S. Department of Defense. Scholars who have previously received funding from JSPS for 12 months or longer are not eligible to apply for JSPS fellowships.
*Geographic Focus:* All States
*Date(s) Application is Due:* Dec 1
*Contact:* Fernando Rojas, Fellowships Associate; (212) 377-2700 or (718) 517-3639; fax (212) 377-2727; japan@ssrc.org or rojas@ssrc.org
*Internet:* www.ssrc.org/fellowships/jsps-fellowship/
*Sponsor:* Social Science Research Council
One Pierrepont Plaza, 15th Floor
Brooklyn, NY 11201

**SSRC Mellon Mays Predoctoral Research Grants** 2447
The SSRC-Mellon Mays Graduate Initiatives Program offers three grants (Graduate Studies Enhancement, Predoctoral Research Development and Dissertation Completion) designed to meet specific needs during the doctoral degree process. The requirements of each category are described in detail at the website. The application period for each is from May 1 through November 1.
*Requirements:* Applicants must have been selected as Mellon Mays Fellows as undergraduates. Fellows may apply for one grant per year and must be enrolled in a doctoral program in a Mellon-designated field of study, or have filed a petition for inclusion of another field.
*Geographic Focus:* All States
*Date(s) Application is Due:* Nov 1
*Amount of Grant:* Up to 5,000 USD

*Contact:* Adam Radwan, Program Associate; (212) 377-2700 or (718) 517-3617; fax (212) 377-2727; radwan@ssrc.org or mellonmays@ssrc.org
*Internet:* www.ssrc.org/fellowships/mellon-mays-predoctoral-research-grants/
*Sponsor:* Social Science Research Council
One Pierrepont Plaza, 15th Floor
Brooklyn, NY 11201

**SSRC New Directions in the Study of Prayer Research Grants** 2448
The New Directions in the Study of Prayer program invites proposals from scholars in all disciplines for studies that will enhance knowledge of the social, cultural, psychological, and cognitive dimensions of prayer, and of its origins, variations, and correlations in human life. Supported with funding from the John Templeton Foundation, and developed in conjunction with the SSRC's program on Religion and the Public Sphere, the project is led by a multi-disciplinary SSRC advisory committee. Approximately 25 research grants, ranging from $50,000 to $200,000, are awarded. Both individual and collaborative projects will be considered, and a small number of journalism grants of up to $50,000 each will also be awarded. All grantees will be asked to participate in a series of interdisciplinary workshops, conferences, and online initiatives organized in conjunction with the project.
*Requirements:* Candidates must submit a letter of intent before pursuing a project.
*Geographic Focus:* All States
*Amount of Grant:* 50,000 - 200,000 USD
*Contact:* Taline Cox, Program Assistant; (718) 517-3682 or (212) 377-2700; fax (212) 377-2727; cox@ssrc.org
*Internet:* www.ssrc.org/programs/new-directions-in-the-study-of-prayer/
*Sponsor:* Social Science Research Council
One Pierrepont Plaza, 15th Floor
Brooklyn, NY 11201

**SSRC Next Generation Social Sciences in Africa: Doctoral Dissertation Completion Fellowships** 2449
The Social Science Research Council offers Fellowships to support the advancement of social science faculty toward completion of doctoral degrees and to promote next generation social science research in Ghana, Nigeria, South Africa, Tanzania, and Uganda. The Fellowships support dissertations and research on peace, security, and development topics. The Doctoral Dissertation Completion Fellowship supports a one-year leave from teaching responsibilities and a stipend up to $15,000 to permit the completion of a dissertation that advances research on peace, security, and development topics.
*Requirements:* All applicants must be citizens of and reside in a sub-Saharan African country while holding a current faculty position at an accredited college or university in Ghana, Nigeria, South Africa, Tanzania, or Uganda. Applicants for any of the funding opportunities offered through this program must have a master's degree, and be working toward completion of the doctoral degree. The program seeks to promote diversity and encourages women to apply. To be eligible for the dissertation completion fellowship, applicants must be in the final year of writing and have completed at least one dissertation chapter.
*Geographic Focus:* Ghana, Nigeria, South Africa, Tanzania, Uganda
*Date(s) Application is Due:* Dec 1
*Amount of Grant:* Up to 15,000 USD
*Contact:* Kimberly Greenberg, Program Associate; (718) 517-3663 or (212) 377-2700; fax (212) 377-2727; greenberg@ssrc.org or nextgenafrica@ssrc.org
*Internet:* www.ssrc.org/fellowships/next-generation-social-sciences-in-africa-doctoral-dissertation-completion-fellowship/
*Sponsor:* Social Science Research Council
One Pierrepont Plaza, 15th Floor
Brooklyn, NY 11201

**SSRC Next Generation Social Sciences in Africa: Doctoral Dissertation Proposal Fellowships** 2450
The Social Science Research Council offers approximately 40 Fellowships each year to support the advancement of social science faculty toward completion of doctoral degrees and to promote next generation social science research in Ghana, Nigeria, South Africa, Tanzania, and Uganda. The Fellowships support dissertations and research on peace, security, and development topics. The Doctoral Dissertation Proposal Fellowship supports short-term research costs of up to $3,000 to develop a doctoral dissertation proposal. The application and additional information are available at the SSRC website.
*Requirements:* All applicants must be citizens of and reside in a sub-Saharan African country while holding a current faculty position at an accredited college or university in Ghana, Nigeria, South Africa, Tanzania, or Uganda. Applicants for any of the funding opportunities offered through this program must have a master's degree, and be working toward completion of the doctoral degree. The program seeks to promote diversity and encourages women to apply. To be eligible for the Dissertation Proposal Fellowship, applicants must be admitted to a graduate program but have yet to undertake dissertation research.
*Geographic Focus:* Ghana, Nigeria, South Africa, Tanzania, Uganda
*Date(s) Application is Due:* Dec 1
*Amount of Grant:* Up to 3,000 USD
*Contact:* Kimberly Greenberg, Program Associate; (718) 517-3663 or (212) 377-2700; fax (212) 377-2727; greenberg@ssrc.org or nextgenafrica@ssrc.org
*Internet:* www.ssrc.org/fellowships/next-generation-social-sciences-in-africa-proposal-fellowship/
*Sponsor:* Social Science Research Council
One Pierrepont Plaza, 15th Floor
Brooklyn, NY 11201

**SSRC Next Generation Social Sciences in Africa: Doctoral Dissertation Research Fellowships** 2451
The Social Science Research Council offers approximately 40 Fellowships each year to support the advancement of social science faculty toward completion of doctoral degrees and to promote next generation social science research in Ghana, Nigeria, South Africa, Tanzania, and Uganda. The Fellowships support dissertations and research on peace, security, and development topics. The doctoral dissertation research Fellowship supports 9-12 months of dissertation research costs of up to $15,000. The application and additional information are available at the SSRC website.
*Requirements:* All applicants must be citizens of and reside in a sub-Saharan African country while holding a current faculty position at an accredited college or university in Ghana, Nigeria, South Africa, Tanzania, or Uganda. Applicants for any of the funding opportunities offered through this program must have a master's degree, and be working toward completion of the doctoral degree. The program seeks to promote diversity and encourages women to apply. To be eligible for the $15,000 dissertation research fellowship, applicants must have an approved dissertation proposal, but will not yet have undertaken research of nine months or more.
*Geographic Focus:* Ghana, Nigeria, South Africa, Tanzania, Uganda
*Date(s) Application is Due:* Dec 1
*Amount of Grant:* Up to 15,000 USD
*Contact:* Kimberly Greenberg, Program Associate; (718) 517-3663 or (212) 377-2700; fax (212) 377-2727; greenberg@ssrc.orc or nextgenafrica@ssrc.org
*Internet:* www.ssrc.org/fellowships/next-generation-social-sciences-in-africa-research-fellowship/
*Sponsor:* Social Science Research Council
One Pierrepont Plaza, 15th Floor
Brooklyn, NY 11201

**SSRC Our Shared Past Grants** 2452
Our Shared Past grants promote the development of international scholarly communities committed to analyzing history curriculum, and reframing the teaching of world history through the identification of new scholarship and the development of new curricular content that illustrate shared cultural, economic, military, religious, social, and scientific networks and practices as well as shared global norms and values that inform world history and society. The project will encourage both the synthesis of existing scholarship on these topics and the exploration of concrete ways that this reframing can be successfully introduced into teaching curriculum in European, Middle Eastern, North African, or North American contexts. Our Shared Past grants may support workshops that convene scholars and advisory boards as well as partnerships between regional educational organizations and advocates. Additionally, projects may use a grant to support public-facing events that allow for engagement with curriculum reform around world history. Funds are available to projects that will develop robust networks of faculty who possess the expertise to study existing world history curriculum (K–14, i.e., at pre-collegiate and collegiate levels) in a specific national or regional setting and who are positioned to propose alternative curriculum design based on that scholarship. Successful projects will make use of relevant distribution channels for all project deliverables. Five grants are expected to be awarded. Organizations must apply through the application portal at the SSRC website. Hard copy applications will not be accepted.
*Requirements:* Grants will be made to research teams or research centers on U.S. university campuses, including, but not limited to, Title VI National Resource Centers. Research teams/consortia may include partner institutions from the U.S. and/or Middle East/North Africa region, but the principal investigator must be based at a U.S. university.
*Geographic Focus:* All States
*Date(s) Application is Due:* May 31
*Amount of Grant:* Up to 75,000 USD
*Contact:* Kimberly Greenberg, Program Associate; (212) 377-2700; fax (212) 377-2727; greenburg@ssrc.org or osp@ssrc.org
*Internet:* www.ssrc.org/fellowships/our-shared-past/
*Sponsor:* Social Science Research Council
One Pierrepont Plaza, 15th Floor
Brooklyn, NY 11201

**SSRC Postdoctoral Fellowship for Transregional Research: Inter-Asian Contexts and Connections** 2453
The Postdoctoral Fellowship for Transregional Research: Inter-Asian Contexts and Connections program is a pilot postdoctoral fellowship program to support transregional research under the rubric Inter-Asian Contexts and Connections. Its purpose is to strengthen the understanding of issues and geographies that may not fit neatly into existing divisions of academia or the world and to develop new approaches, practices, and opportunities in international, regional, and area studies in the U.S. Funded by the Andrew W. Mellon Foundation, these fellowships will help junior scholars (those at the postdoctoral stage, one to seven years out of the Ph.D.) complete first books and/or undertake second projects. In addition to funding research, the program creates networks and shared resources to support Fellows well beyond the grant period. The Postdoctoral Fellowship for Transregional Research will provide promising scholars important support at critical junctures in their careers. Fifteen fellowships are awarded over the two-year course of the pilot program.
*Geographic Focus:* All States
*Contact:* Holly Danzeisen, Associate Director, Eurasia and Inter-Asia; (718) 517-3662 or (212) 377-2700; fax (212) 377-2727; danzeisen@ssrc.org or transregional@ssrc.org
*Internet:* www.ssrc.org/fellowships/transregional-research-fellowship/
*Sponsor:* Social Science Research Council
One Pierrepont Plaza, 15th Floor
Brooklyn, NY 11201

**SSSP Arlene Kaplan Daniels Paper Award** 2454

The Arlene Kaplan Daniels Paper Award for the best paper on Women and Social Justice was established in 2016 and will be awarded annually with a stipend of $500. Arlene was the first female editor of Social Problems (1975-1978) and became the Society's President in 1986. Her scholarship addressed women's exclusion in the labor market and examined women's career contingencies, union involvement, feminist networking, and the salience of women's voluntary work. This scholarship found its way into her 1987 SSSP Presidential Address "Invisible Work" and her 1988 book, Invisible Careers. She supported other women sociologists across the discipline and professional associations through mentoring, preparing references, and editing papers. The annual deadline for nominations is April 15. All nominations are conducted online at the website. SSSP will honor Dr. Daniels for her service with this award. Her scholarship has made an important contribution to the discipline through its unparalleled awareness of women's contribution to society.
*Requirements:* Eligible papers must have been published in the past two calendar years. Members of the Society are urged to nominate one or more persons for the award.
*Geographic Focus:* All States
*Date(s) Application is Due:* Apr 15
*Amount of Grant:* 500 USD
*Contact:* Michele Smith Koontz, Administrative Officer; (865) 689-1531; fax (865) 689-1534; mkoontz3@utk.edu or sssp@utk.edu
Dr. Tanya L. Saunders, Chairperson; (865) 689-1531; fax (865) 689-1534; saunderstanya@gmail.com
*Internet:* https://www.sssp1.org/index.cfm/m/654/Arlene_Kaplan_Daniels_Paper_Award/
*Sponsor:* Society for the Study of Social Problems
901 McClung Tower, University of Tennessee
Knoxville, TN 37996-0490

**SSSP C. Wright Mills Award** 2455

C. Wright Mills wrote in The Power Elite that: "Only when mind has an autonomous basis, independent of power, but powerfully related to it, can mind exert its force in the shaping of human affairs. This is democratically possible only when there exists a free and knowledgeable public, to which [people] of knowledge may address themselves, and to which [people] of power are truly responsible." The Society for the Study of Social Problems established the C. Wright Mills Award in 1964. The SSSP annually gives its prestigious C. Wright Mills Award to the author of what the committee considers to be the most outstanding book written in the tradition of C. Wright Mills and his dedication to a search for a sophisticated understanding of the individual and society. Past winners have included Patricia Hill Collins, William Julius Wilson, Pun Ngai, Pierrette Hondagneu-Sotelo, Charles Tilly, Theda Skocpol, Rosabeth Moss Kanter, Harry Braverman, Frances Fox Piven-Richard Cloward, Jacqueline P. Wiseman, Laud Humphreys, and Jerome H. Skolnick. The books nominated must have been published in the year preceding the year of the award.
*Geographic Focus:* All States
*Date(s) Application is Due:* Dec 15
*Amount of Grant:* 1,000 USD
*Contact:* Michele Smith Koontz, Administrative Officer; (865) 689-1531; fax (865) 689-1534; mkoontz3@utk.edu or sssp@utk.edu
Dr. Carla Shedd; (865) 689-1531; fax (865) 689-1534; cshedd@gc.cuny.edu or sssp@utk.edu
*Internet:* https://www.sssp1.org/index.cfm/m/259/C_Wright_Mills_Award/
*Sponsor:* Society for the Study of Social Problems
901 McClung Tower, University of Tennessee
Knoxville, TN 37996-0490

**SSSP Doris Wilkinson Faculty Leadership Award** 2456

The Doris Wilkinson Faculty Leadership Award will be given annually to an outstanding faculty member who has exercised an extensive leadership role within the SSSP and other professional societies and within the larger community. The award was established in 2016 and will be awarded annually with a stipend of $500. In her 2001 Social Problems article, "Making History: A Personal Narrative of My Presidential Year," Doris chronicles her activities in the Society during the early 1970s, when SSSP's programs and members offered her a refuge during a restless period within the American Sociological Association (ASA). She was a participant-observer of the events that occurred when Sociologists for Women in Society and the Association of Black Sociologists were seen as "outsiders within" and sought meaningful entry and participation in the ASA. Doris turned her observation into action and became one of the leading voices in bringing change to the society and sociology as a discipline. Her leadership was consistent with the Society's mission, and the Doris Wilkinson Faculty Leadership Award will be given to a scholar who effectively: addresses social issues of contemporary public importance; brings a deepened public understanding to these issues; advances social understanding of leadership for change within the larger society; and proposes courses of action to effect that change. The annual deadline for award nominations is April 15.
*Geographic Focus:* All States
*Date(s) Application is Due:* Apr 15
*Amount of Grant:* 500 USD
*Contact:* Michele Smith Koontz, Administrative Officer; (865) 689-1531; fax (865) 689-1534; mkoontz3@utk.edu or sssp@utk.edu
Dr. Shirley A. Jackson, Awards Committee Chairperson; (865) 689-1531; fax (865) 689-1534; shja2@pdx.edu or sssp@utk.edu
*Internet:* https://www.sssp1.org/index.cfm/m/801/Doris_Wilkinson_Faculty_Leadership_Award/
*Sponsor:* Society for the Study of Social Problems
901 McClung Tower, University of Tennessee
Knoxville, TN 37996-0490

**SSSP Joseph B. Gittler Award** 2457

Joseph B. GittlerJoseph B. Gittler was a sociologist by training who received his doctorate from the University of Chicago in 1941. For more than 60 years, he was a member of the faculty at many leading universities in this country and abroad, including Duke University, George Mason University, the University of Rochester, Iowa State University, the University of Georgia, Yeshiva University, Cardozo Law School, Hiroshima University in Japan and Ben-Gurion University in Israel. A member of the American Sociological Association and a fellow of the New York Academy of Sciences, he wrote numerous scholarly articles and books, including "Understanding Minority Groups"; "Ethnic Minorities in the United States: Perspectives from the Social Sciences"; "Ethnic and Racial Conflict"; and "Ideas of Concord and Discord in Religious Systems," among many others. Established in 2007 at the bequest of Joseph B. Gittler, this award is made in recognition of the significant scholarly achievements that a SSSP member has made in contributing to the ethical resolution of social problems. Any member of the Society may nominate one or more persons for the award. Members of the Joseph B. Gittler Award Committee are encouraged to nominate. The award will be presented at the Annual Meeting in Denver, Colorado. Nominations and supporting documents should be submitted online no later than April 15.
*Requirements:* The nominee must have been an active member of the Society for the Study of Social Problems for at least three years prior to receiving the award. The nominee must have produced and disseminated scholarship promoting ethical solutions to social problems over the preceding three or more calendar years. All nominations must be accompanied by supporting evidence sufficiently detailed for the committee to render a decision (e.g., a resume; media accounts of activist activities inspired by the nominee's scholarly efforts, testimonials from grass roots organizations or advocacy agencies; or additional supporting description of the nominee's work, demonstrating that the contributions meet the criteria for nomination).
*Geographic Focus:* All States
*Date(s) Application is Due:* Apr 15
*Amount of Grant:* 500 USD
*Contact:* Dr. Paul D. Steele, Awards Committee Chairperson; (865) 689-1531; fax (865) 689-1534; steelepauld@gmail.com
Michele Smith Koontz, Administrative Officer; (865) 689-1531; fax (865) 689-1534; mkoontz3@utk.edu or sssp@utk.edu
*Internet:* https://www.sssp1.org/index.cfm/m/294/Joseph_B_Gittler_Award/
*Sponsor:* Society for the Study of Social Problems
901 McClung Tower, University of Tennessee
Knoxville, TN 37996-0490

**SSSP Kathleen S. Lowney Mentoring Award** 2458

Established in 2016, the Kathleen S. Lowney Mentoring Award will be given annually to an outstanding faculty member or community activist. The award will be presented at the awards ceremony during the annual meeting. The award recipient will receive a plaque and up to three of the award recipient's mentees will receive a one-year SSSP membership as well as annual meeting registration fees for the year the award is presented. The award was established to recognize the value of quality mentoring relationships between mentor and mentee and/or mentoring programs especially those for undergraduate or graduate students and/or for social activists, particularly for younger scholars and activists. The goals of effective mentoring should include: fostering sensitivity and understanding of significant problems of social life; encouraging understandings of institutional inequalities and strategies for overcoming them; and promoting problem-centered social research and/or social activism. The annual deadline for nominations is April 15.
*Geographic Focus:* All States
*Date(s) Application is Due:* Apr 15
*Contact:* Dr. Ivy Ken; (202) 994-1886; ivyken@gwu.edu
Michele Smith Koontz, Administrative Officer; (865) 689-1531; fax (865) 689-1534; mkoontz3@utk.edu or sssp@utk.edu
*Internet:* https://www.sssp1.org/index.cfm/m/684/Kathleen_S_Lowney_Mentoring_Award_/
*Sponsor:* Society for the Study of Social Problems
901 McClung Tower, University of Tennessee
Knoxville, TN 37996-0490

**SSSP Racial/Ethnic Minority Graduate Fellowship** 2459

The Society for the Study of Social Problems (SSSP), in keeping with its philosophy of active engagement with social problems, participation in social problem solutions, and advancement of knowledge through study, service and critical analysis, established the Racial/Ethnic Minority Graduate Scholarship at its annual meeting in August 1993. The purpose of the scholarship is: to identify and support developing minority scholars who exemplify and give fresh voice to the SSSP history and commitment to scholar activism; to give renewed energy and wider lenses to diversity in scholarship; to increase the pool of minority social and behavioral scientists; and to establish a formal commitment to diversity through support of a minority doctoral student in the social and/or behavioral sciences inclusive of course work or dissertation research support who demonstrates a commitment, through his or her scholarly examination, of any aspect of inequality, injustice and oppression. A $15,000 scholarship will be funded to two students, with an additional $500 awarded for attendance at the annual meeting.
*Requirements:* All applicants must be current SSSP members when applying. Eligibility is any person identified as either Black/African American, Hispanic/ Latino, Asian/ Asian-American, Native Hawaiian or Other Pacific Islander, or American Indian or Alaska Native accepted into an accredited doctoral program in any one of the social and/ or behavioral sciences so as to expand their perspectives in the pursuit and investigations into social problems.
*Geographic Focus:* All States

*Date(s) Application is Due:* Feb 1
*Amount of Grant:* 15,000 - 15,500 USD
*Contact:* Michele Smith Koontz, Administrative Officer; (865) 689-1531; fax (865) 689-1534; mkoontz3@utk.edu or sssp@utk.edu
Dr. France Winddance Twine, Chairperson; (805) 893-4398 or (805) 893-8000; winddance@soc.ucsb.edu
*Internet:* https://www.sssp1.org/index.cfm/m/261/Racial/Ethnic_Minority_Graduate_Fellowship/
*Sponsor:* Society for the Study of Social Problems
901 McClung Tower, University of Tennessee
Knoxville, TN 37996-0490

**Stackpole-Hall Foundation Grants** **2460**
The Stackpole-Hall Foundation awards grants to nonprofit organizations and institutions designed to enhance the social welfare of the Pennsylvania area. The Foundation concentrates on the areas of education, health care, cultural, youth development, social welfare, environmental, and community development. Priority is given to Elk County. Grants are awarded in support of matching, seed, partnership, and under certain circumstances, operational grants. The Foundation trustees meet four times a year to award grants. Proposals are due a month prior to the meeting for which they will be considered. Application forms are not required. Detailed instructions about information to include with the letter of inquiry are located at the Foundation website.
*Requirements:* The Foundation considers requests from organizations which qualify as churches; governmental organizations; or 501(c)3 organizations; or are classified as not being a private foundation under section 509(a) of the federal tax code.
*Restrictions:* Unless usual circumstances exist, requests for operating or endowment grants are given low priority. Funding is not available for individuals.
*Geographic Focus:* Pennsylvania
*Amount of Grant:* Up to 50,000 USD
*Contact:* William Conrad, Executive Director; (814) 834-1845; fax (814) 834-1869; stackpolehall@windstream.net
*Internet:* www.stackpolehall.org/application.html
*Sponsor:* Stackpole-Hall Foundation
44 South St. Marys Street
St. Marys, PA 15857

**Stanford Center on Adolescence Youth Purpose Research Grants** **2461**
Small grants for dissertation and postdoctoral research will be awarded in March for research to be completed by September. The Stanford Center on Adolescence aims to spread the recognition of youth purpose as an important area of scientific research by encouraging all inquiries that shed light on topics related to purpose among young people, such as adolescent motivation, personal meaning, or positive youth development, in an array of settings found in contemporary society. The Center operationally defines purpose as a stable and generalized intention to accomplish something that is meaningful to the self and that leads to productive engagement with some aspect of the world beyond the self. Through its own research, the Center seeks to address questions about how purpose develops, the antecedents of purpose, the difference that purpose makes in the lives of young people, and how to best cultivate purpose in youth. The Center encourages researchers from all disciplines to submit proposals for projects that will contribute to the scientific study of this research topic. Applicants may request up to $10,000 in funding for research-related expenses.
*Requirements:* Applicants must be affiliated with an accredited college or university and should be U.S citizens or permanent residents. Postdoctoral applicants must have received their Ph.D. within the past five years.
*Geographic Focus:* All States
*Amount of Grant:* Up to 10,000 USD
*Samples:* Lisa M. Dorner, Assistant Professor of Educational Psychology, Research and Evaluation, University of Missouri, St. Louis, MO, Project: Purposeful Work: Civic Development from the Perspective of Adolescent Language Brokers; Ruth H. Kim, Doctoral Candidate, Education Department, University of California, Santa Cruz, CA, Project: Poetry as Purpose: Extending Beyond the Self through Youth Spoken Word; Laura E. Welfare, and Elysia V. Clemens, Assistant Professor of Counselor Education, Virginia Polytechnic Institute and State University, Blacksburg, VA, Project: Facilitating the Development of Purpose: Evidence for the Effectiveness of a Group Counseling Intervention with Middle School Adolescents; Michael F. Steger, with Brian J. Dyk, and Bill Peisner, Assistant Professor, Department of Psychology Colorado State University, Fort Collins, CO, Project: The Development and Exploration of a Youth Purpose Program for Middle Schools.
*Contact:* Lisa Staton, Administrative Associate; (650) 725-8205; fax (650) 725-8207; stanfordcoa@stanford.edu
*Internet:* www.stanford.edu/group/adolescent.ctr/researchawards
*Sponsor:* Stanford Center on Adolescence
505 Lasuen Mall, Stanford University
Stanford, CA 94305-3083

**Stanford Humanities Center Arts Writer/Practitioner Fellowship** **2462**
The Stanford Humanities Center and the Stanford Institute for Creativity and the Arts (SiCa) are offering one year-long fellowship to an arts practitioner who is also a writer, scholar, or critic pursuing a research or critical project in the arts. Fellows are expected to remain in residence, living in a ten-mile radius of the university, during the regular academic year (mid-September through mid-June). The arts practitioner/writer fellow will be in residence with other fellows at the Humanities Center and is expected to make an intellectual contribution to Stanford University by participating in the scholarly life of the Humanities Center, e.g. regularly attending lunch and participating in weekly research presentations. Applicants should be interested in engaging with the intellectual life at Stanford and exploring connections with departments and other centers on campus. Fellows are awarded a stipend of up to $60,000 and a housing and moving allowance of up to $15,000. Medical insurance is not provided. Applications are submitted via the center's online application system which becomes accessible at the center's website each August. All applications must be in English. Annual deadline dates may vary. The center notifies applicants of the outcome in late March. Application guidelines are available at the website.
*Requirements:* Applicants may practice in any field of art (visual, performance, composition, mixed- and inter-media, architecture and design) but must demonstrate significant professional accomplishment as arts practitioners and as critics or scholars. Fellowships will be awarded on the basis of a scholarly or critical project in the arts and not on the basis of art production. Applicants may, but need not, be affiliated with an academic institution; independent practitioners are welcome to apply. Applications will be judged on the promise of the specific research project being proposed, the research project's potential interest to scholars in different fields of the humanities, the originality and intellectual distinction of the candidate's previous work, and the applicant's perceived ability to engage in collegial interaction and to contribute to the discussion of presentations.
*Geographic Focus:* All States, All Countries
*Amount of Grant:* Up to 75,000 USD
*Contact:* Robert Barrick, Fellowship Program Administrator; (650) 723-3052 ext. 3-3054; fax (650) 723-1895; rbarrick@stanford.edu or humanities-center@stanford.edu
*Internet:* shc.stanford.edu/fellowships/arts/#arts_answer_6
*Sponsor:* Stanford University
424 Santa Teresa Street
Stanford, CA 94305-4015

**Stanford Humanities Center External Faculty Fellowships** **2463**
Founded in 1980, the Stanford Humanities Center is a multidisciplinary research institute at Stanford University dedicated to advancing knowledge about culture, philosophy, history, and the arts. The center's external-faculty fellowships are offered to non-Stanford scholars (both to members of humanities departments as traditionally defined and to other scholars seriously interested in humanistic issues). Fellowship duration is one full academic year. Fellows are expected to remain in residence, living in a ten-mile radius of the university, during the regular academic year (mid-September through mid-June). Additionally fellows are expected to make contributions to the intellectual life of the Stanford community by taking part in a Geballe Research Workshop, regularly attending lunches, and participating in weekly research presentations. Fellows are awarded a stipend of up to $60,000 and a housing and moving allowance of up to $15,000. Medical insurance is not provided. Applications are submitted via the center's online application system which becomes accessible at the center's website each August. All applications must be in English. Annual deadline dates may vary. The center notifies applicants of the fellowship competition outcome in late March. Application guidelines are available at the website.
*Requirements:* External fellowships are intended primarily for non-Stanford scholars who are currently teaching in or affiliated with an academic institution; however, independent non-Stanford scholars may also apply. All applicants must have a Ph.D. External fellowships fall into two categories: senior fellowships for well-established scholars, and junior fellowships for scholars who during the fellowship year will be at least three but no more than 10 years beyond receipt of their Ph.D. Research projects must be in the humanities. The center is open to projects employing information technology in humanities research. Applications will be judged on the promise of the specific research project being proposed, the research project's potential interest to scholars in different fields of the humanities, the originality and intellectual distinction of the candidate's previous work, and the applicant's perceived ability to engage in collegial interaction and to contribute to the discussion of presentations.
*Restrictions:* Creative arts projects are not eligible for funding. Former external faculty fellows are not eligible to apply. Scholars who have received major research grants within the last three years generally will not be as competitive as other applicants.
*Geographic Focus:* All States, All Countries
*Date(s) Application is Due:* Oct 15
*Amount of Grant:* Up to 75,000 USD
*Contact:* Robert Barrick, Fellowship Program Administrator; (650) 723-3052 ext. 3-3054; fax (650) 723-1895; rbarrick@stanford.edu or humanities-center@stanford.edu
*Internet:* shc.stanford.edu/fellowships/non-stanford-faculty/
*Sponsor:* Stanford University
424 Santa Teresa Street
Stanford, CA 94305-4015

**Stanley Smith Horticultural Trust Grants** **2464**
The Stanley Smith Horticultural Trust was created in 1970 by May Smith, in honor of her late husband. The trust typically makes grants to botanical gardens, arboreta, and universities primarily in North and South America to support education and research in ornamental horticulture. Specifically, the trust is interested in funding organizations pursuing the following activities: the advancement of research in ornamental horticulture and the publication of the results of such research; assisting in the creation, development, preservation, and maintenance of gardens accessible to the public for educational purposes; promotion of the environmentally responsible introduction, cultivation, and distribution of plants which have ornamental horticultural value; assisting in the publication of books or other works relating to the science of horticulture; and informal and/or formal educational activities which further ornamental horticulture. The trust typical awards grants in the amount of $20,000. Organizations seeking funding

should first confirm that their projects meet the trust's funding interests and eligibility requirements by taking the eligibility quiz which is accessible from the grantseeker's webpage at the trust's website. Grant seekers meeting the qualifications should send an email to the Grants Director with a brief description of their project. Organizations with projects that appear to be qualified and that are of interest to the trust will be sent current application guidelines along with an invitation to submit a full proposal. Proposals are accepted until August 15 and are reviewed by the trustees once a year in November. Funding of approved grants occurs in December for use in the following calendar year.
*Requirements:* Organizations that are recognized as 501(c)3 public charities by the United States Internal Revenue Service and non-U.S. organizations that can demonstrate that they would meet the requirements for such status are eligible to apply. The trust provides support for projects in North and South America in the following areas: development of programs and projects; salaries; physical improvements; signage; access; equipment; publications; and, under some circumstances, general operations.
*Restrictions:* The trust does not fund endowments, indirect costs, or projects concerned primarily with agriculture, environmental issues, science education, or horticultural therapy.
*Geographic Focus:* All States, All Countries
*Date(s) Application is Due:* Aug 15
*Amount of Grant:* Up to 20,000 USD
*Samples:* Cedar Valley Arboretum & Botanical Garden, Waterloo, Iowa, $20,000 - toward the salary of the Director of Horticulture; Transition Zone Horticultural Institute, Flagstaff, Arizona, $10,000 - to fund a horticulture internship at the Arboretum; Missouri Botanical Garden, St. Louis, Missouri, $47,000 - for staff salaries related to the research and publication project "Flora of China".
*Contact:* Dr. Thomas F. Daniel, Grants Director, Stanley Smith Horticultural Trust; (415) 379-5350; tdaniel@calacademy,org
Janet Ferriaolo, Grants Manager; (415) 332-0166; jferraiolo@adminitrustllc.com
*Internet:* www.adminitrustllc.com/stanley-smith-horticultural-trust/
*Sponsor:* Stanley Smith Horticultural Trust
c/o Adminitrust LLC
Sausalito, CA 94965

**Stella and Charles Guttman Foundation Grants** **2465**

The majority of grants are made to organizations providing services to people in the New York City metropolitan area. Beginning in 2014, the Stella and Charles Guttman Foundation intends to direct a substantial portion of its grantmaking to programs that serve low income infants, toddlers and preschoolers as they transition to kindergarten. Special emphasis will be placed on programs that improve quality, expand services and create a strong continuum of care for children ages 0 to 3 in high-need neighborhoods. Systemic investments in early childhood programs may include the expansion of evidence-based home visiting programs, infant health and mental health programs and professional development for center-based teachers, as well as home-based caregivers. The Foundation is also committed to fund programs in neighborhoods with high levels of poverty and a large concentration of public housing. In addition to early childhood programs, the Foundation will support programs that work to build a network of education, health and social services for children from birth through college graduation.
*Requirements:* Charitable 501(c)3 or 170(b)1 organizations are eligible with a strong emphasis on New York City and Israel.
*Restrictions:* The foundation does not make grants directly to individuals or to organizations not qualified as charitable, for foreign travel or study, to initiate or defend public interest litigation, to support anti-vivisectionist causes, or to religious organizations for religious observances.
*Geographic Focus:* New York
*Amount of Grant:* 25,000 - 100,000 USD
*Samples:* Citizens' Committee for Children of New York, New York, New York, $50,000 - support of educational, policy and advocacy initiatives focused on early childhood education; Good Shepherd Services, Brooklyn, New York, $50,000 - support of the Bronx Opportunity Network, a collaboration of seven community-based organizations; Henry Street Settlement, New York, New York, $100,000 - support of the Henry Street capital campaign and the renovation of the Charles and Stella Guttman Building.
*Contact:* Elizabeth Olofson, Executive Director; (212) 371-7082; fax (212) 371-8936; eolofson@guttmanfoundation.org
*Internet:* www.guttmanfoundation.org
*Sponsor:* Stella and Charles Guttman Foundation
122 East 42nd Street, Suite 2010
New York, NY 10168

**Strake Foundation Grants** **2466**

The Foundation supports hospitals, schools, colleges, and Catholic charities, as well as projects focusing in adult basic education and literacy, museums, and arts and culture. Support is considered for operating budgets, capital campaigns, special projects, research, matching funds and general purposes.
*Requirements:* Awards are made to 501(c)3 organizations located only in the United States, primarily in Texas. Organizations may submit only one request per calendar year. There are no set amounts for requests, however awards generally range between $2,000 and $20,000 with a few exceptions as high as $50,000.
*Restrictions:* Awards are not made to support individuals, nor for deficit financing, consulting services, technical assistance, publications, or loans.
*Geographic Focus:* All States
*Samples:* Corporation for Educational Radio and Television, New York, NY, $2,500–to support the 'Black American Conservatism' companion website; El Centro de Corazon, Houston, TX, $$7,500–operating support; Fund for American Studies, Washington, D.C., $5,000–scholarship support.
*Contact:* George Strake, Jr., President; (713) 216-2400; foundation@strake.org
*Sponsor:* Strake Foundation
712 Main Street, Suite 3300
Houston, TX 77002

**Streilein Foundation for Ocular Immunology Visiting Professorships** **2467**

The SFOI Visiting Professorship is intended to stimulate the spread of knowledge concerning the immunological causes and approaches for managing ocular diseases. The SFOI Visiting Professorship should provide a clear scientific and educational benefit for the host institution. The Professorship will support a short visit of a senior scientist or clinician with a track record of ocular immunology research to a host institution. An important feature of the Visiting Professorship is to encourage and mentor young investigators. The visiting professor shall contribute one full day of teaching and interacting with students, residents, postdoctoral fellows, and faculty. The host institution will receive a $5,000 award to pay for coach airfare, and/or standard ground transportation, hotel (up to two nights), meals, honorarium, and expenses associated with this visit (e.g., refreshments, advertising). Transportation by personal car will be reimbursed at $0.51/mile. Meal allowance will be set by the host institution. Also included is an honorarium of $1,500 paid at the time expense reimbursement is submitted.
*Requirements:* Any institution of higher education or research institute is eligible to apply. The application for a SFOI Visiting Professorship must be submitted by a departmental Chairman of an educational institution, the Director of a research institute, or an official of similar stature from the host organization. The application must be received by the SFOI by July 1st of the calendar year in which the professorship is intended.
*Geographic Focus:* All States
*Date(s) Application is Due:* Jul 1
*Amount of Grant:* 6,500 USD
*Contact:* Joan Stein-Streilein, Ph.D., Treasurer; (617) 912-7494; fax (617) 912-0105; jstein@vision.eri.harvard.edu
*Internet:* www.streilein-foundation.org/fellowships.html
*Sponsor:* Streilein Foundation for Ocular Immunology
P.O. Box 6104
Boston, MA 02114

**Streisand Foundation Grants** **2468**

Established in 1986 in New York, the Foundation was founded by Barbra Streisand, singer-songwriter and actress. Streisand has won two Academy Awards, eight Grammy Awards, five Emmy Awards, a Special Tony Award, and is one of very few entertainers to have won all of these honors. In 2001, Streisand received the AFI Life Achievement Award, as well as she is the highest ranking female artist on the Recording Industry Association of America's (RIAA) Top Selling Artists list. Today, the Foundation awards grants for national programs affecting women and at-risk youth. Areas of support include programs for children and youth, with a focus on the economically disadvantaged; civil rights, with emphasis on race relations between African Americans and Jews; AIDS research, advocacy, service, and litigation; nuclear disarmament; and environmental preservation. Types of support include general operating grants, technical assistance, and project-specific support. Applicants are asked to submit one- to two-page letters of inquiry. Invited applications are accepted between September 1 and the December 1 deadline.
*Requirements:* National 501(c)3 nonprofits and California projects in the Los Angeles area are eligible.
*Restrictions:* Grants are not made to individuals nor do they support start-up organizations, endowments, or capital campaigns.
*Geographic Focus:* All States
*Date(s) Application is Due:* Dec 1
*Amount of Grant:* Up to 30,000 USD
*Samples:* Union of Concerned Scientists, Cambridge, Massachusetts - general operating support; Brennan Center for Justice, New York, New York - general operating support; Iraq And Afghanistan Veterans Of America (IAVA), New York, New York - general operating support; Mother Jones, San Francisco, California - general operating support.
*Contact:* Margery Tabankin, Executive Director; fax (310) 314-8396; stfnd@aol.com
*Internet:* www.barbrastreisand.com/streisand-foundation/
*Sponsor:* Streisand Foundation
1327 Ocean Avenue, Suite H
Santa Monica, CA 90401

**Sun Academic Excellence Grants** **2469**

The Sun Academic Excellence Grant program has helped numerous schools kick-start their projects by providing equipment grants and access to free training and certifications. Funding for grants focus on areas such as: curriculum integration—the integration of Sun Technology in undergraduate and graduate level courses through the development and use of model curricula; community development—the sponsorship of a Sun or Open Source community development projects that align with the academic and research interests of your students, faculty and researchers and result in put-backs to the code; localization—the localization of the code or development tools and resources to languages other than English; community outreach—to promote broad awareness of Sun technologies on campus and participation of faculty, students and other members of the academic community in Sun communities through events, workshops and end-user communications; Java software development—NetBeans IDE plug-ins that result in put backs to the Net Beans community; application development—development of an application, technology or solution using the Sun platform which can be deployed in the Education, Life Science, Healthcare or Government industry; and thin client—deployment of Sun Ray thin clients; ideally 100 or more seats. Grants are made on a published quarterly calendar and are for hardware donations only.

*Requirements:* In the United States, higher education eligible institutions must have: 3304(f) or a 501(c)3 or Section 115 tax exempt status; and be a degree granting institution or be involved in basic research. Outside of the United States, higher education institutions must either be degree-granting or involved in basic research and be recognized as a non-commercial enterprise by their national government. To be eligible as a primary or secondary educational institution, the following guidelines must be met: public or private formal full-time educational institution offering classes in the range or equivalence of grades Primary and Secondary (K12); or licensed, certified or accredited to operate as a school or school administrative unit in the state or country of operation.
*Restrictions:* Grants do not include maintenance, service or technical support. Non-profit institutions with 501(c)3 status which are neither degree-granting nor directly involved with basic research may be eligible if they directly support the academic and research communities in their mission to create and disseminate knowledge, and if they can demonstrate financial need.
*Geographic Focus:* All States
*Date(s) Application is Due:* Feb 26; May 31; Aug 31; Nov 30
*Contact:* Linda Rogers, Senior Director of Strategy; (650) 336-0487 or (650) 960-1300; fax (650) 856-2114; corpaffrs@corp.sun.com
*Internet:* www.sun.com/solutions/landing/industry/education/aeg.xml
*Sponsor:* Sun Microsystems Corporation
4150 Network Circle
Santa Clara, CA 95054

**Sun Grid Research Grants** **2470**

Sun fosters close relationships with researchers in academia, national labs and nonprofit organizations to create the future of computer and information technology. Sun Grid Research Grants award hours of central processing units (CPUs) on the Sun Grid Compute Utility. Sun Grid helps educators and researchers derive immediate benefits from an open, grid-based computing infrastructure on a utility basis providing choice and control over how IT is purchased. The Sun External Research Office collaborates with the academic community to define and create the future of computer and information technology. Sun and the university work to transform key research outcomes into future products.
*Requirements:* In the United States, higher education eligible institutions must have: 3304(f) or a 501(c)3 or Section 115 tax exempt status; and be a degree granting institution or be involved in basic research. Outside of the United States, higher education institutions must either be degree-granting or involved in basic research and be recognized as a non-commercial enterprise by their national government.
*Geographic Focus:* All States
*Date(s) Application is Due:* Feb 26; May 31; Aug 31; Nov 30
*Contact:* Linda Rogers, Senior Director of Strategy; (650) 336-0487 or (650) 960-1300; fax (650) 856-2114; corpaffrs@corp.sun.com
*Internet:* be.sun.com/teleweb/grantprog/pdf/STAR_DS_new_brandV5.pdf
*Sponsor:* Sun Microsystems Corporation
4150 Network Circle
Santa Clara, CA 95054

**Sun Microsystems Foundation Summer Academic Enrichment Grants** **2471**

The Sun Foundation donates dollars to provide access to academic enrichment programs for low-income and under represented students. Through the Sun Foundation's Summer Academic Enrichment Program, cash grants have been awarded to non-profit organizations in the Greater Boston area, Denver Metro Area, California Bay Area, and U.K. (Scotland, England). The grant recipients' innovative programs help youth develop skills through hands-on collaborative learning. Additionally, the programs encourage students to pursue further studies in math, science, technology and engineering. The Sun Foundation's educational grants support the early development of talent necessary to meet the high-tech workforce needs of the future.
*Requirements:* U.S. 3304(f) or 501(c)3 tax-exempt organizations that serve students in the metropolitan regions of: Boston, Massachusetts; Denver, Colorado; and California Bay are eligible to apply, as well as the international areas of the United Kingdom and Scotland.
*Geographic Focus:* California, Colorado, Massachusetts, United Kingdom
*Samples:* Girls For A Change, San Jose, California—to help over one thousand girls become critical thinkers, innovators and leaders; InternalDrive Foundation, Campbell, California—to fund the Hands-On Tech Outreach Program, a summer computer camp in Santa Clara; UMass Lowell Summer DESIGNCAMP, Lowell, Massachusetts——to improve critical thinking and problem solving skills of 360 students.
*Contact:* Linda Rogers, Senior Director of Strategy; (650) 336-0487 or (650) 960-1300; fax (650) 856-2114; corpaffrs@corp.sun.com
*Internet:* www.sun.com/aboutsun/comm_invest/grants.html
*Sponsor:* Sun Microsystems Foundation
901 San Antonio Road, MS MPK15-204
Palo Alto, CA 94303

**Sunny and Abe Rosenberg Foundation Grants** **2472**

Abe Rosenberg, a retired liquor importer and distributor who was active in many philanthropies, helped to establish Star Liquor Dealers in Long Island City shortly after the sale of liquor became legal again in 1933 with the repeal of Prohibition. Incorporated in New York in 1966, the Rosenberg Foundation awards grants to New York City area nonprofit organizations. Funding priorities include programs that promote excellence in public education; artistic and cultural education for public school students; and activities that promote learning for disadvantaged children and youth. The Foundation also awards support to supplementary services for children currently in foster care and young adults moving out of the foster system. Typically, an awarded grant ranges from $5,000 to $50,000. The Foundation does not accept unsolicited proposals, so interested parties should begin with either a letter of inquiry or a telephone contact to the Foundation office.
*Requirements:* New York City-area nonprofit organizations are eligible. Applicants should contact the Foundation to see if their project may be eligible for funding.
*Restrictions:* The Foundation does not support advocacy, litigation, or political causes. Funding is also not available for individuals or other private foundations.
*Geographic Focus:* New York
*Amount of Grant:* 5,000 - 50,000 USD
*Samples:* Bay Street Theater, Sag Harbor, New York, $25,000 - general operating support; Caridad Center, Boynton Beach, Florida, $12,000 - general operating support; American Ballet Theater, New York, New York, $5,000 - general operating support.
*Contact:* Maryanne Passafiume, (212) 993-6230 or (212 ) 993-6232; mpassafiume@rosenbergfoundation.org
*Sponsor:* Sunny and Abe Rosenberg Foundation
Ansonia Station
New York, NY 10023-0033

**Surdna Foundation Arts Teachers Fellowships** **2473**

The Surdna Foundation announces the annual national initiative to support the artistic revitalization of outstanding arts teachers in public arts high schools. Surdna recognizes that arts teachers often lack the time and resources to reconnect with other arts professionals and with the artistic processes they teach. Through the Surdna Arts Teachers Fellowship Program (SATF), Fellows design individualized courses of study that provide both immersion in their own creative work and the opportunity to interact with other professional artists in their fields. Surdna believes that this approach to professional development will enhance the effectiveness of arts teachers and will directly benefit the young people they teach. All permanently assigned full- and part-time arts faculty in public arts high schools are eligible. Twenty awards of $5,500 each, with a complementary grant of $1,500 to the Fellow's school to support post-fellowship activities, will be made. The fellowship award may be used to defray the costs of tuition and other fees, room and board, travel, purchase of materials and/or equipment for personal art-making, childcare, and other relevant expenses.
*Requirements:* All permanently assigned full- and part-time arts faculty in public arts high schools are eligible. Applicants must be minimally in their fifth year of teaching arts in high school, and plan to continue as an arts teacher in their public arts high school during the upcoming school year.
*Geographic Focus:* All States
*Date(s) Application is Due:* Feb 10
*Amount of Grant:* Up to 6,500 USD
*Contact:* Kimberly Bartosik, SATF Program Director; (212) 557-0010, ext. 256; fax (212) 557-0003; artsfellowship@surdna.org
*Internet:* www.surdna.org/what-we-fund/thriving-cultures/109.html
*Sponsor:* Surdna Foundation
330 Madison Avenue, 30th Floor
New York, NY 10017-5001

**Surdna Foundation Sustainable Environments Grants** **2474**

The Surdna Foundation seeks to create just and sustainable communities where consumption and conservation are balanced and innovative solutions to environmental problems improve people's lives. It works from a sustainable development perspective to demonstrate that a healthy environment is the backbone of a healthy economy and a democratic society. The Foundation funds three key related priority areas - Climate Change, Green Economy, and Transportation and Smart Growth - that aim to transform how Americans work, consume, and move. Together these will help make the theory of a carbon free society into a practical and achievable reality for communities across the United States. The Surdna Foundation accepts applications on an ongoing basis. However, grants are approved three times per year: in February, May and September. Applicants should use the online letter of inquiry to apply for funding.
*Requirements:* 501(c)3 organizations may apply. The foundation urges applicants to send two- to three-page letters of intent before sending proposals. IRS nonprofit status certification, recent audited financial statements, and project budget should also be included.
*Restrictions:* Generally, the Surdna Foundation does not support: programs addressing toxics, hazardous waste, land and habitat conservation; animal welfare, biodiversity and ocean management; individuals; or academic fellowships.
*Geographic Focus:* All States
*Amount of Grant:* 15,000 - 1,000,000 USD
*Samples:* Global Philanthropy Partnership, Chicago, Illinois, $600,000 - to deepen connections and sharing among 90 leading North American municipal sustainability directors through a professional network and build new collaborative initiatives for policy development, professional development and sustainability innovation; Transportation Equity Network, Chicago, Illinois, $500,000 - continued support for a national grassroots campaign promoting fair and sustainable transportation policies through reforms at the national, state and regional level.
*Contact:* Edward Skloot, Executive Director; (212) 557-0010; fax (212) 557-0003; grants@surdna.org or executivedirector@surdna.org
*Internet:* www.surdna.org/what-we-fund/sustainable-environments.html
*Sponsor:* Surdna Foundation
330 Madison Avenue, 30th Floor
New York, NY 10017-5001

**Susan G. Komen Breast Cancer Foundation Brinker Awards for Scientific Distinction** 2475

Established through a partnership between Susan G. Komen for the Cure and Brinker International in 1992, the Brinker Award for Scientific Distinction is a marquee award that recognizes pioneering work in breast cancer research and treatment. The Brinker Award honors leading scientists for their significant achievements and contributions to basic and translational science and clinical practice that have advanced the fight to save lives and realize our vision of a world without breast cancer. Since 1992, the roster of Komen Brinker Award laureates has grown to include names of researchers who have made the most significant advances in breast cancer research and medicine. The Brinker Award for Scientific Distinction is given annually in two categories. The Basic Science Award is presented to a researcher whose scientific discoveries or novel technologies have added substantively to our understanding of breast cancer and the intrinsic molecular processes that drive the disease, and/or whose work has bridged the gap between basic research and patient care. The Clinical Research Award is presented to a clinical or translational researcher who has advanced the identification of new prevention, detection or treatment approaches for breast cancer and promoted their incorporation into clinical care.
*Requirements:* Breast cancer researchers who are affiliated with nonprofit institutions (academic, industry, government, or other) in the United States and abroad are eligible.
*Geographic Focus:* All States
*Date(s) Application is Due:* Jun 30
*Amount of Grant:* 20,000 USD
*Contact:* Award Administrator; (972) 855-1600; fax (972) 855-1605
*Internet:* ww5.komen.org/ResearchGrants/BrinkerAward.html
*Sponsor:* Susan G. Komen Breast Cancer Foundation
5005 LBJ Freeway, Suite 250
Dallas, TX 75244

**Susan G. Komen Breast Cancer Foundation Career Catalyst Research Grants** 2476

Career Catalyst Research (CCR) Grants are intended to foster promising breast cancer researchers who are in the early stages of their faculty careers by providing support for up to three years of "protected time" for research career development under the guidance of a mentor committee. These Grants provide funding for research projects that have significant potential to advance our understanding of breast cancer and lead to reductions in breast cancer incidence and/or mortality within the next decade. It is expected that awardees will launch independent research careers and successfully compete for subsequent research project funding in breast cancer following the successful completion of a Career Catalyst Research Grant. CCRs provide support for hypothesis-driven research projects that have significant potential to advance our understanding of breast cancer and lead to reductions in breast cancer incidence and/or mortality within the next decade. Studies focusing on quality of life or survivorship issues are not appropriate for this mechanism. Applicants may request funding of up to $150,000 per year (combined direct and indirect costs) for up to three years.
*Requirements:* Applicants and Institutions must conform to the eligibility criteria to apply for a CCR Grant (eligibility requirements must be met at the time of full application submission). Applicants must: have a doctoral degree, including MD, PhD, DrPH, DO, or equivalent; currently hold a faculty appointment or have a formal offer letter at the time of application, verified by the Letter of Institutional Support; not have held any faculty appointment, including non-tenure and tenure track appointments combined, for more than a total of 6 years at the time of application, verified by the Letter of Institutional Support; not currently hold or simultaneously apply for a Komen Investigator-Initiated Research Grant; not currently be a Principal Investigator on an existing R01 research grant; have adequate space to conduct proposed research and protected time for research, verified by the Letter of Institutional Support; and ensure that all past and current Komen-funded grants are up to date and in compliance with all Komen requirements; e.g. Applicants are not required to be U.S. citizens or residents. Institutions must be a nonprofit institution or organization anywhere in the world.
*Geographic Focus:* All States, All Countries
*Date(s) Application is Due:* Dec 17
*Amount of Grant:* Up to 450,000 USD
*Contact:* Director, Grants and Sponsored Programs; (972) 855-1600 or (866) 921-9678; fax (972) 855-1605; helpdesk@komengrantsaccess.org
*Internet:* ww5.komen.org/uploadedFiles/Content/ResearchGrants/GrantPrograms/new-CareerCatalystRFA2012-2.pdf
*Sponsor:* Susan G. Komen Breast Cancer Foundation
5005 LBJ Freeway, Suite 250
Dallas, TX 75244

**Susan G. Komen Breast Cancer Foundation Challege Grants: Breast Cancer and the Environment** 2477

Major advances have been made in understanding the biology and diversity of breast cancer, but much more remains to be discovered about the many causes of breast cancer – particularly what contributions a diverse array of environmental factors may be making – and how to prevent it. The challenges are many: The scientific community has been presented with conflicting and inconclusive results from past studies. With increased knowledge of the complexity of breast cancer biology, it has become apparent that future research into environmental influences will need to focus on early-life exposures, associations with specific tumor types, and gene-environment interactions. Adding to the complexity of this task is the fact that for a wide array of exposures the assessment methodologies, tools and resources are limited. Through its Challenge Grants: Breast Cancer and the Environment program, the Foundation seeks to address key research needs in the field of environmental contributions to breast cancer risk. Susan G. Komen for the Cure and its Scientific Advisory Board have requested that the Institute of Medicine (IOM) review the current evidence on environmental risk factors for breast cancer, consider gene-environment interactions in breast cancer, explore evidence-based actions that might reduce the risk of breast cancer, and recommend research needed in these areas. At the conclusion of their report, the IOM issued 13 recommendations for further research, of which three have been chosen by the Komen Scientific Advisory Board as the subject of Challenge Grants: Studies of Occupational Cohorts and Other Highly Exposed Populations; New Exposure Assessment Tools; and Minimizing Exposure to Ionizing Radiation.
*Requirements:* U.S. nonprofit organizations and international institutions will be eligible to apply.
*Geographic Focus:* All States
*Amount of Grant:* Up to 250,000 USD
*Contact:* Director, Grants and Sponsored Programs; (972) 855-1600; fax (972) 855-1605; grants@komen.org
*Internet:* ww5.komen.org/ResearchGrants/FundingOpportunities.html
*Sponsor:* Susan G. Komen Breast Cancer Foundation
5005 LBJ Freeway, Suite 250
Dallas, TX 75244

**Susan G. Komen Breast Cancer Foundation Challege Grants: Investigator Initiated Research** 2478

Susan G. Komen Breast Cancer Foundation Challenge Grants seek to stimulate exploration of new ideas and novel approaches to research on breast cancer and the environment, through a life course approach, that have significant potential to lead to reductions in breast cancer incidence and/or mortality within the next decade. In the area of Studies of Occupational Cohorts and Other Highly-Exposed Populations, the research seeking to study populations exposed to potentially cancer-causing agents or external toxins, chemicals, hormones or radiation, such as occupational cohorts, persons with event-related high exposures, or patient groups receiving high-dose or long-term medical treatments, including medical radiation. In the area of New Exposure Assessment Tools, the research is seeking to improve methodologies for measuring, across the life course, personal exposure to and biologically effective doses of environmental factors that may alter risk for or susceptibility to breast cancer. In the area of Minimizing Exposure to Ionizing Radiation, comparative effectiveness research is seeking to assess the relative benefits and harms of medical radiation, including imaging procedures and diagnostic/follow-up radiologic tests using algorithms in common practice. Applicants may request either three or four years of funding as follows: up to a total of $750,000 over three years (combined direct and indirect costs); or up to a total of $1,000,000 over four years (combined direct and indirect costs). Reasonable compensation of advocates is allowed when advocates perform services that would otherwise be a contracted expense. Compensation may be in the form of salary, per-hour compensation, or honoraria. Additionally, grant funds can be used for advocate participation in scientific conferences that would enhance their knowledge and skills related to the research project.
*Requirements:* A wide range of nonprofits, including research groups and health and human service agencies, are eligible to apply. Budgets are not required to be equivalent across each year of the Grant, but rather should reflect the costs appropriate to support the research project each year.
*Restrictions:* Personnel on the project are limited to a base salary at or below $250,000 per year, and PIs must provide a 10% minimal level of effort. Equipment costs are limited to no more than 30% of total direct costs. Indirect costs cannot exceed 25% of total direct costs (including any indirect costs paid through subcontracts or consortia). Costs that are not allowed include: travel costs; publication costs and meeting-related poster printing costs are allowed; graduate and postdoctoral fellow tuition; visa costs; or professional membership dues.
*Geographic Focus:* All States
*Date(s) Application is Due:* Dec 20
*Amount of Grant:* Up to 1,000,000 USD
*Contact:* Director, Grants and Sponsored Programs; (972) 855-1600 or (866) 921-9678; fax (972) 855-1605; helpdesk@komengrantaccess.org
*Internet:* ww5.komen.org/uploadedFiles/Content/ResearchGrants/GrantPrograms/new-Challenge-InvestInitiatedRFA2012-2.pdf
*Sponsor:* Susan G. Komen Breast Cancer Foundation
5005 LBJ Freeway, Suite 250
Dallas, TX 75244

**Susan G. Komen Breast Cancer Foundation Challege Grants: Career Catalyst Research** 2479

Challenge Grants-Career Catalyst Research are intended to foster promising breast cancer researchers who are in the early stages of their faculty careers by providing support for up to three years of protected time for research career development under the guidance of a mentor committee. It is expected that awardees will launch independent research careers and successfully compete for subsequent research project funding in breast cancer following the successful completion of the Grant. These Grants provide funding for research projects that have significant potential to advance our understanding of breast cancer and the environment, through a life course approach, that have significant potential to lead to reductions in breast cancer incidence and/or mortality within the next decade. Applications will be accepted within the following three focus areas: studies of occupational cohorts and other highly-exposed populations; new exposure assessment tools; and minimizing exposure to ionizing radiation. The pre-application deadline is September 10 annually, and there is a limit of $450,000 combined direct and indirect costs ($150,000 per year for up to 3 years).
*Requirements:* CCR grants are open to scientists who have held faculty positions for no more than six years at the time of full application to achieve research independence. Applicants and Institutions must conform to the eligibility criteria to apply for a CCR

Grant (eligibility requirements must be met at the time of full application submission). Applicants must: have a doctoral degree, including MD, PhD, DrPH, DO, or equivalent; currently hold a faculty appointment or have a formal offer letter at the time of application, verified by the Letter of Institutional Support; not have held any faculty appointment, including non-tenure and tenure track appointments combined, for more than a total of 6 years at the time of application, verified by the Letter of Institutional Support; not currently hold or simultaneously apply for a Komen Investigator-Initiated Research Grant; not currently be a Principal Investigator on an existing R01 research grant; have adequate space to conduct proposed research and protected time for research, verified by the Letter of Institutional Support; and ensure that all past and current Komen-funded grants are up to date and in compliance with all Komen requirements; e.g. Applicants are not required to be U.S. citizens or residents. Institutions must be a nonprofit institution or organization anywhere in the world.
*Geographic Focus:* All States, All Countries
*Date(s) Application is Due:* Dec 17
*Amount of Grant:* Up to 450,000 USD
*Contact:* Director, Grants and Sponsored Programs; (972) 855-1600 or (866) 921-9678; fax (972) 855-1605; helpdesk@komengrantsaccess.org
*Internet:* ww5.komen.org/uploadedFiles/Content/ResearchGrants/GrantPrograms/Challenge-CareerCatalystRFA2012-2.pdf
*Sponsor:* Susan G. Komen Breast Cancer Foundation
5005 LBJ Freeway, Suite 250
Dallas, TX 75244

**Susan G. Komen Breast Cancer Foundation Investigator Initiated Research Grants** **2480**
Susan G. Komen Breast Cancer Foundation Investigator Initiated Research Grants seek to stimulate exploration of new ideas and novel approaches in breast cancer research and clinical practice that have significant potential to lead to reductions in breast cancer incidence and/or mortality within the next decade. Applications addressing topics other than those described below will be administratively withdrawn from consideration, and will not be reviewed or scored: Novel Therapeutics and/or Resistance - Therapeutic Implications of Tumor Genomics; Biology of Breast Cancer - Implications of the Immune System in Breast Cancer Biology; Prevention/Early Detection - New Strategies for Early Detection; and Disparities in Breast Cancer Outcomes - Outcomes of Specific Populations after Diagnosis. The pre-application submission deadline is September 10, with the full application due December 17. $1,000,000 in combined direct and indirect costs are allowable, which equals $250,000 per year for up to 4 years.
*Requirements:* Grants will be awarded to a single Principal Investigator (PI) or one PI and one Co-Principal Investigator (Co-PI). PIs and Co-PIs: must have a doctoral degree, including MD, PhD, DrPH, DO, or equivalent; must have a full time faculty appointment at the time of application; may only apply in the final year of their funding and may not hold both grants simultaneously; must ensure that all past and current Komen-funded Grants are up to date and in compliance with all Komen requirements; and are not required to be U.S. citizens or residents. Primary Institution must be a nonprofit institution or organization anywhere in the world.
*Geographic Focus:* All States, All Countries
*Date(s) Application is Due:* Dec 17
*Amount of Grant:* Up to 1,000,000 USD
*Contact:* Director, Grants and Sponsored Programs; (972) 855-1600 or (866) 921-9678; fax (972) 855-1605; helpdesk@komengrantsaccess.org
*Internet:* ww5.komen.org/uploadedFiles/Content/ResearchGrants/GrantPrograms/InvestInitiatedRFA2012-2-new2a.pdf
*Sponsor:* Susan G. Komen Breast Cancer Foundation
5005 LBJ Freeway, Suite 250
Dallas, TX 75244

**SWE Scholarships** **2481**
The SWE Scholarship Program provides financial assistance to women admitted to accredited baccalaureate or graduate programs, in preparation for careers in engineering, engineering technology and computer science. Currently, SWE disbursed more than 180 new and renewed scholarships valued at $470,000. Grants are announced in the summer (sophomore, junior, senior and graduate students) and late summer/early fall (freshmen and reentry students) for use during the following academic year. Grant payments are made in the fall for corporate-sponsored scholarships and in both the fall and spring for endowed scholarships, upon proof of registration. Scholarships and fellowships range from $1,000 to $10,000 each.
*Requirements:* All SWE-administered scholarships are open only to women majoring in engineering in a school, college, or university with an accredited engineering program, and who will be in a specified year of study during the academic year following presentation of the grant. Reentry scholarships are limited to women who have been out of the engineering job market for a minimum of two years and who will return to school for an engineering program, undergraduate or graduate.
*Geographic Focus:* All States
*Date(s) Application is Due:* Feb 15
*Amount of Grant:* 1,000 - 10,000 USD
*Contact:* Reena Singhall Lee, (312) 596-5223; reenasinghallee@gmail.com or scholarshipapplication@swe.org
*Internet:* societyofwomenengineers.swe.org/index.php?option=com_content&task=view&id=222&Itemid=229
*Sponsor:* Society of Women Engineers
120 S La Salle Street, Suite 1515
Chicago, IL 60603

**T.R. Higgins Lectureship Award** **2482**
The T.R. Higgins Lectureship Award recognizes an outstanding lecturer and author whose technical research paper or papers may be identified as an outstanding contribution to engineering literature on fabricated structural steel published during the eligibility period. The award is named for Theodore R. Higgins, former AISC Director of Engineering and Research, who was widely acclaimed for his many contributions to the advancement of engineering technology related to fabricated structural steel. Winners will receive: a framed certificate, which is presented at the annual AISC North American Steel Construction Conference (NASCC); and a $15,000 cash award.
*Requirements:* The author must be a permanent resident of the United States. The paper or papers must have been published in a professional journal that provides for a discussion of papers by the engineering profession within the recent five-year period.
*Geographic Focus:* All States
*Date(s) Application is Due:* Aug 1
*Amount of Grant:* 10,000 USD
*Contact:* Janet T. Cummins, Engineering and Research Coordinator; (312) 670-2400; fax (312) 670-5403; cummins@aisc.org
*Internet:* https://www.aisc.org/content.aspx?id=560
*Sponsor:* American Institute of Steel Construction
One East Wacker Drive, Suite 700
Chicago, IL 60601-1802

**Talbert and Leota Abrams Foundation** **2483**
The Talbert and Leota Abrams Foundation awards grants to organizations in Michigan in its fields of interest, including higher education, adult basic education and literacy programs, libraries and reading, and community funds. Grants support research and program development. The board meets in May to consider requests.
*Requirements:* Nonprofit organizations in Michigan, primarily central Michigan, are eligible. Applicants submit a two page letter of inquiry, along with a copy of their 501(c)3 exemption. Churches may apply for nonsectarian, nonreligious grants.
*Restrictions:* Grants do not support athletic activities, churches for sectarian use, individuals, operating expense, loans, or traveling expenses. The Foundation does not fund the maintenance of existing programs or services.
*Geographic Focus:* Michigan
*Date(s) Application is Due:* May 31
*Contact:* Joe Foster Jr., Secretary; (517) 706-5791
*Sponsor:* Talbert and Leota Abrams Foundation
400 Lake Lansing Road, Suite F
Lansing, MI 48912-3674

**Tandem Launch Technologies Grants** **2484**
Tandem Launch Technologies is currently accepting proposals for their latest round of multimedia invention development funding. This unique acceleration program provides entrepreneurial university inventors with up to $1 million worth of funding, business and technical development resources, and connections in the multimedia industry. Multimedia projects appropriate for the program are scalable to the consumer electronics market, fit with a licensing model, and require a budget of less than $1 million to develop an initial prototype. Projects in the following fields can be submitted: video processing and encoding; display technologies; camera and image capture; audio processing and encoding; natural human computer interaction; and natural speech synthesis and recognition.
*Requirements:* The TandemLaunch model is only suitable for technologies with the following characteristics: protectable intellectual property, ideally through patents; scalable to the broader consumer electronics and not limited to niche markets; amenable to a licensing business model; development of an initial demonstration prototype with an overall budget of less than $1M requiring no restricted capital equipment; and support from dedicated and highly skilled inventors with a keen interest in commercializing the technology.
*Geographic Focus:* All States, All Countries, Canada
*Date(s) Application is Due:* Aug 15
*Amount of Grant:* Up to 1,000,000 USD
*Contact:* Katie Young Morris, Project Manager; (438) 380-5435, ext. 209; Catherine.young-morris@tandemlaunchtech.com or projects@tandemlaunchtech.com
David Zangwill, (438) 380-5435, ext. 204; david.zangwill@tandemlaunchtech.com
*Internet:* www.tandemlaunchtech.com/proposals.php
*Sponsor:* TandemLaunch Technologies
245 Victoria Avenue, Suite 200
Westmount, QC H3Z 2M6 Canada

**Tau Beta Pi Association Fellowships** **2485**
The Fellowship Program is Tau Beta Pi's single most important project for the advancement of engineering education and the profession. It is a time-proven idea which was pioneered by Tau Beta Pi in the honor-society field. The plan was adopted by the 1928 Convention as a result of the recommendation and labors of President A. D. Moore. Since the beginning, the purpose has been to finance, for a select group of members chosen for merit and need, a year of graduate study at the colleges of their choice. A distinguishing feature of Tau Beta Pi fellowships has always been that they are as free of binding rules as possible. Tau Beta Pi fellows are free to do graduate work in any field which will enable them to contribute to the engineering profession. The only specific duty of awardees is to write a report at the completion of the fellowship year summarizing their work. The annual application deadline is February 1.
*Requirements:* Tau Beta Pi Constitution and Bylaws specifically restrict these fellowships to initiated members of the Tau Beta Pi Association on the date the fellowship is awarded.
*Geographic Focus:* All States
*Date(s) Application is Due:* Feb 1

*Amount of Grant:* 10,000 USD
*Contact:* Steve Pierre; (865) 546-4578; fax (865) 546-4579; fellowship@tbp.org
*Internet:* www.tbp.org/fellowships.cfm
*Sponsor:* Tau Beta Pi Association
P.O. Box 2697
Knoxville, TN 37901-2697

**Taubman Foundation for the Arts Grants** **2486**
The Taubman Foundation for the Arts Grants was established by Nicholas F. Taubman in Texas in 2010. Today, its primary fields of interest include: art museums; general arts and culture; elementary and secondary education; graduate and professional education; higher education; historic preservation; historical activities; history museums; in-patient medical care; Judaism; libraries; museums; nonprofits; opera; and performing arts. Types of support often include: leadership; professional development; research; and general operating funds. An application letter should be submitted by interested organizations, and should include an overall program description and budgetary needs. Grants typically range from $1,000 to $100,000, with twelve to fifteen awards given each year. There are no specified annual deadlines.
*Requirements:* Though giving is primarily centered around Virginia and New York, grants are also awarded across the United States.
*Geographic Focus:* District of Columbia, Indiana, New York, Texas, Virginia
*Amount of Grant:* 1,000 - 100,000 USD
*Samples:* Smithsonian Institution, Washington, D.C., $10,000 - contribution to general operations; George W. Bush Presidential Center, Dallas, Texas, $100,000 - contribution to general operations; Virginia Historical Society, Richmond, Virginia, $10,000 - contribution to general operations.
*Contact:* Nicholas F. Taubman, President; (540) 777-4506
*Sponsor:* Taubman Foundation for the Arts
2965 Colonnade Drive, Suite 300
Roanoke, VA 24018-3557

**TD4HIM Foundation Grants** **2487**
The San Francisco 49ers Foundation is the non-profit community funding extension of the San Francisco 49ers. Now in its 16th year, the 49ers Foundation supports development programs for underserved youth that keep them safe, on track and in school. A significant portion of its funding goes toward family violence prevention programs and activities that teach youth leadership and respect.
*Requirements:* 501(c)3 organizations and programs in the greater Bay Area that are in alignment with the mission of the 49ers Foundation: keeping kids safe, on track and in school, are eligible to apply.
*Restrictions:* Please be advised that the 49ers Foundation does not fund the following areas: land, building or vehicle purchase; endowments; political campaigns; religious organizations (projects to promote religion); loans of any type; personal scholarships; travel; drives; specific individuals; organizations based outside the greater San Francisco Bay Area.
*Geographic Focus:* All States
*Contact:* TD4HIM Program Coordinator; (408) 829-3381
*Internet:* www.49ers.com/community/foundation.html
*Sponsor:* TD4HIM Foundation
60 Shoreline Circle
Incline Village, NV 89451-9529

**Teagle Foundation Grants** **2488**
The Teagle Foundation's commitment to promoting and strengthening liberal education grounds their grant funding. Awards generally encourage collaboration among institutions, seeking to generate new knowledge on issues of importance to higher education. The Foundation is committed to widely disseminating this knowledge. The Foundation's signature Outcomes and Assessment initiative grows from their conviction that nothing has as much potential to affect students' educational experience as a sustained and systematic assessment of what and how they learn. The College-Community Connections program provides grants to community-based organizations in New York City that help disadvantaged young people prepare for and succeed in college, and works to build closer ties between these organizations and area colleges and universities. Requests for grant support will be considered following a two-stage application process. First, the Foundation asks that prospective grantees share brief concept papers. After review of the concept papers, a limited number of applicants will then be invited to submit full proposals. Concept papers are considered on a rolling basis. The Teagle Board of Directors reviews all grant requests when it meets in February, May, and November. If a proposal is invited, program staff will confer with the applicant to determine the appropriate timeline for submitting a full proposal in time for potential review by the Board.
*Requirements:* The Teagle Foundation invites proposals from two- and four-year colleges, universities, higher education consortia, and non-profit entities that share its commitment to the liberal arts and aim to pursue programs under one of its current initiatives.
*Restrictions:* The Foundation does not accept unsolicited proposals.
*Geographic Focus:* All States
*Contact:* Loni Bordoloi Pazich; (212) 373-1972; info@teaglefoundation.org
*Internet:* www.teaglefoundation.org/Grants-Initiatives/How-We-Grant
*Sponsor:* Teagle Foundation
570 Lexington Avenue, 38th Floor
New York, NY 10022

**Terra Foundation-Yale University Press American Art in Translation Book Prize** **2489**
The Terra Foundation for American Art, in partnership with Yale University Press, offers a prize for an unpublished manuscript or previously published manuscript in a language other than English written by a non-U.S. author. The manuscript should make a significant contribution to scholarship on the historical visual arts of what is now the geographic United States. In helping to overcome the language barrier that often divides scholars and deters international research and collaboration, the prize aims to advance and internationalize scholarship on American art and seeks to recognize original and thorough research, sound methodology, and significance in the field. The award is especially intended to encourage authors who take the field of American art history into new historical and interpretive terrain, or who establish connections among the work of scholars within and outside the United States, providing a model of international exchange important to sustaining relevance and academic rigor for the future of the field. Topics of interest include the following aspects of the visual arts of the United States (dating from circa 1500 to 1980): painting, sculpture, graphic arts, photography, decorative arts, craft, and folk art. Books focusing on methodology, criticism, patronage, iconography, and social history may also be nominated. The winner receives a $5,000 cash prize; the Terra Foundation funds production of the book, which is published (in print and electronic form) in English by Yale University Press. In addition, Yale University Press invites the winner to present a lecture on the book, upon publication, at Yale University. The annual Letter of Inquiry deadline is August 1, with invitations for formal proposals received by September 7. The annual deadline for final proposals is October 15.
*Requirements:* Single-author, book-length academic manuscripts—previously unpublished or published within the past five calendar years—in the field of American art history are eligible.
*Geographic Focus:* All Countries
*Date(s) Application is Due:* Oct 15
*Amount of Grant:* 5,000 USD
*Contact:* Amy Gunderson, Grants Manager; (312) 654-2275 or (312) 664-3939; gunderson@terraamericanart.org or grants@terraamericanart.org
*Internet:* https://www.terraamericanart.org/what-we-offer/grant-fellowship-opportunities/terra-foundation-yale-university-press-american-art-in-translation-book-prize/
*Sponsor:* Terra Foundation
120 East Erie Street
Chicago, IL 60611

**Terra Foundation Affiliated Fellowship in Rome** **2490**
The Terra Foundation Affiliated Fellowship in Rome offers support for a postdoctoral or senior scholar to conduct research on American art and visual culture (not including architecture) prior to 1980 while in residence at the American Academy in Rome for a period of four to six months. During that time, the fellowship recipient will pursue their work in an atmosphere conducive to intellectual and artistic freedom, interdisciplinary exchange, and innovation. Topics dealing with Italian-American exchanges and requiring firsthand exposure to collections and archives in Rome and its vicinity will be given priority. This award is administered by the American Academy in Rome. The Fellowship includes housing (single room), meals, a private study, a stipend of $2,500 per month, and privileged access to Rome. Topics dealing with Italian-American exchanges and requiring first-hand exposure to collections and archives in Rome and its vicinity will be given priority. The annual deadline for application submission is November 15.
*Requirements:* Applicants may be of any nationality, provided their research would be facilitated by residence at the Academy. Applicants must hold a PhD at the time of application. Senior scholars are welcome to apply.
*Geographic Focus:* All States, All Countries
*Date(s) Application is Due:* Nov 15
*Amount of Grant:* 10,000 - 15,000 USD
*Contact:* Amy Gunderson, Grants Manager; (312) 654-2275 or (312) 664-3939; gunderson@terraamericanart.org or grants@terraamericanart.org
*Internet:* https://www.terraamericanart.org/what-we-offer/grant-fellowship-opportunities/terra-foundation-affiliated-fellowship-rome/
*Sponsor:* Terra Foundation
120 East Erie Street
Chicago, IL 60611

**Terra Foundation Fellowships at the Smithsonian American Art Museum** **2491**
These one-year residential fellowships at the Smithsonian American Art Museum, in Washington, D.C., support full-time independent and dissertation research by scholars from abroad researching historical American art (circa 1550–1980) or by U.S. scholars, particularly those investigating international contexts for American art. These awards are administered by the Smithsonian American Art Museum. Three twelve-month fellowships are awarded annually, one each at the predoctoral, postdoctoral, and senior levels, to scholars from abroad who are researching American art or to U.S. scholars with an exceptionally strong international component to their study. The annual stipend for Terra Foundation Fellows is $36,000 at the predoctoral level and $50,400 at the postdoctoral and senior levels, with an allowance of up to $4,000 available for short research trips. The annual deadline for application submission is November 1.
*Requirements:* Applications that meet the criteria for the Terra Foundation Fellowships will be evaluated on the quality of the proposed project and the applicant's academic standing, scholarly qualifications, and experience.
*Geographic Focus:* All States
*Date(s) Application is Due:* Nov 1
*Amount of Grant:* 36,000 - 54,400 USD
*Contact:* Amy Gunderson, Grants Manager; (312) 654-2275 or (312) 664-3939; gunderson@terraamericanart.org or grants@terraamericanart.org
*Internet:* https://www.terraamericanart.org/what-we-offer/grant-fellowship-opportunities/terra-foundation-fellowships-at-the-smithsonian-american-art-museum/
*Sponsor:* Terra Foundation
120 East Erie Street
Chicago, IL 60611

**Terra Foundation International Curatorial Travel Grants for Travel to the United States** 2492

International Curatorial Travel Grants for Travel to the United States allow international curators to travel to the United States to research and develop specific exhibition ideas about any aspect of historical American art (except architecture, conservation, or film) by visiting public and private art collections, archives, and libraries, and meeting with U.S.-based specialists. To encourage and enrich these international partnerships between art museums and to deepen research and dialogue, for the next two years the foundation will offer these awards. An institution may apply for a travel grant even if it has applied (or plans to apply) separately for general exhibition support for the same project. Proposals for travel grants and general exhibition support are each considered on their own merits; grant approval in one program does not guarantee grant approval in the other. Funding is available for travel, lodging, meals, and research fees and expenses. Grants range from $3,000 to $7,000 each. There are three annual deadlines for applications: January 15; May 15; and September 15.
*Requirements:* The foundation only accepts proposals from institutions with United States 501(c)3 status or the international equivalent. International Curatorial Travel Grants for Travel to the United States are not made to individuals. Only curators permanently employed by a museum outside the United States will be considered at this time. Travel must take place within 12 months from the time of award.
*Geographic Focus:* All States, All Countries
*Date(s) Application is Due:* Jan 15; May 15; Sep 15
*Amount of Grant:* 3,000 - 7,000 USD
*Contact:* Amy Gunderson, Grants Manager; (312) 654-2275 or (312) 664-3939; gunderson@terraamericanart.org or grants@terraamericanart.org
*Internet:* https://www.terraamericanart.org/what-we-offer/grant-fellowship-opportunities/international-curatorial-travel-grants-for-travel-to-the-united-states/
*Sponsor:* Terra Foundation
120 East Erie Street
Chicago, IL 60611

**Terra Foundation International Essay Prize** 2493

The Terra Foundation for American Art has awarded the Smithsonian American Art Museum a grant to administer a biennial prize which recognizes excellent scholarship by a non-U.S. citizen working in the field of historical American art. The aim of the award is to stimulate and actively support non-U.S. scholars working on American art topics, foster the international exchange of new ideas, and create a broad, culturally comparative dialogue on American art. Manuscripts should advance the understanding of American art, demonstrating new findings and original perspectives. The prize winner will be given the opportunity to work toward publication in American Art, the Smithsonian American Art Museum's scholarly journal. The winner will also receive a $1,000 cash award and a travel stipend of up to $3,500 to give a presentation in Washington, D.C., and meet with museum staff and fellows. Submission must be by the annual deadline of January 15.
*Requirements:* Manuscripts previously published in a foreign language are eligible if released within the last two years.
*Restrictions:* Essays that have been published in English will not be considered. Architecture and film studies are not eligible.
*Geographic Focus:* All States, All Countries
*Date(s) Application is Due:* Jan 15
*Amount of Grant:* 1,000 - 4,500 USD
*Contact:* Amy Gunderson, Grants Manager; (312) 654-2275 or (312) 664-3939; gunderson@terraamericanart.org or TerraEssayPrize@si.edu
*Internet:* https://www.terraamericanart.org/what-we-offer/grant-fellowship-opportunities/terra-foundation-international-essay-prize-smithsonian-american-art-museum/
*Sponsor:* Terra Foundation
120 East Erie Street
Chicago, IL 60611

**Terra Foundation International Research Travel Grants for U.S.-Based Scholars** 2494

Terra Foundation International Research Travel Grants offer US-based scholars working on American art and visual culture prior to 1980 the opportunity to conduct research abroad. Grant funding is available for projects that require study of materials outside the United States and that will enable scholars to: discover new primary source material; experience works of art first-hand in museums and private collections; make contact with artists, critics, art dealers, archivists, curators, and university scholars; consult archives and library collections outside the U.S; and establish professional networks for future research. Up to ten doctoral, postdoctoral, and senior grants are awarded annually as follows: maximum of $6,000 per grant for doctoral students; and maximum of $9,000 per grant for postdoctoral and senior scholars. The annual deadline for applications is January 15.
*Requirements:* For this program, the foundation accepts proposals from U.S.-based doctoral students, postdoctoral, and senior scholars. Travel should be undertaken within one calendar year (June through May) after the announcement of the selection results in April. The final report must be submitted within three months of completion of travel. Recipients must acknowledge Terra Foundation support in any publication that results from research conducted during the grant.
*Restrictions:* Travel may not take place prior to notification of the grant award.
*Geographic Focus:* All States
*Date(s) Application is Due:* Jan 15
*Amount of Grant:* Up to 9,000 USD
*Contact:* Amy Gunderson, Grants Manager; (312) 654-2275 or (312) 664-3939; gunderson@terraamericanart.org or grants@terraamericanart.org
*Internet:* https://www.terraamericanart.org/what-we-offer/grant-fellowship-opportunities/terra-foundation-for-american-art-international-research-travel-grant-for-us-based-scholars/
*Sponsor:* Terra Foundation
120 East Erie Street
Chicago, IL 60611

**Terra Foundation Paul Mellon Centre Fellowship** 2495

Beginning in 2018, the Terra Foundation-Paul Mellon Centre Fellowship will be offered annually to support a non-U.S. scholar pursuing research on an aspect of Anglo-American artistic exchange. Instituted as part of a growing collaboration between the Paul Mellon Centre and the Terra Foundation, this four-month research fellowship is designed to facilitate original, rigorous, and exciting new scholarship on an aspect of the artistic relationship between Britain and the United States from 1500 to 1980. The Fellowship will provide an honorarium of £9,500, and can be used towards replacement teaching costs at the fellow's home institution, or for replacement curatorial staff. The annual application submission deadline is January 31, with awards being announced in the spring.
*Requirements:* The fellowship is for an individual employed at a college, university, museum or archive, or for an independent scholar. The award is open to a range of individuals, from early-career postdoctoral scholars to those who are internationally recognized in their field of expertise. This fellowship must be taken over a four-month period between September and September.
*Restrictions:* U.S. citizens are not eligible to apply for this award. The fellowship cannot be used to support research in the fields of archaeology, the current practice of architecture, or the performing arts. Funding cannot be deferred to a later academic year nor can it run concurrently with a fellowship awarded by another institution.
*Geographic Focus:* All Countries
*Date(s) Application is Due:* Jan 31
*Amount of Grant:* 9,500 GBP
*Contact:* Amy Gunderson, Grants Manager; (312) 654-2275 or (312) 664-3939; gunderson@terraamericanart.org or grants@terraamericanart.org
*Internet:* https://www.terraamericanart.org/what-we-offer/grant-fellowship-opportunities/terra-foundation-for-american-art-paul-mellon-centre-fellowship/
*Sponsor:* Terra Foundation
120 East Erie Street
Chicago, IL 60611

**Terra Foundation Postdoctoral Research Fellowship in Paris** 2496

The Terra Foundation Postdoctoral Research Fellowship in Paris is designed to build capacity and professional expertise in American art history by supporting the research of postdoctoral scholars from outside the United States. This residential fellowship at the Institut national d'histoire de l'art (INHA), in Paris, offers a postdoctoral scholar the opportunity to pursue his or her own work while in residence for 12 months. The Terra Foundation fellow will be able to consult local libraries and archives, including the INHA library, the largest art history library in the world. Additionally, the Terra Foundation fellow will benefit from a travel allowance to visit museum and archival collections in the United States and attend the College Art Association conference or another American scholarly event. These awards are administered by the Institut national d'histoire de l'art. The annual application deadline is January 15.
*Requirements:* The Postdoctoral Research Fellowship in Paris is open to candidates having successfully completed a PhD dissertation on American art before 1980 at a Francophone academic institution anywhere in the world.
*Geographic Focus:* All States
*Date(s) Application is Due:* Jan 15
*Contact:* Amy Gunderson, Grants Manager; (312) 654-2275 or (312) 664-3939; gunderson@terraamericanart.org or grants@terraamericanart.org
*Internet:* https://www.terraamericanart.org/what-we-offer/grant-fellowship-opportunities/postdoctoral-research-fellowship-in-american-art-paris/
*Sponsor:* Terra Foundation
120 East Erie Street
Chicago, IL 60611

**Terra Foundation Research Travel Grants to the United States** 2497

Terra Foundation Research Travel Grants provide support for research on topics concerning American art and visual culture prior to 1980. These grants enable scholars outside the United States to consult resources and visit collections within the United States. Grant funding is available for short-term travel that gives scholars: an opportunity to discover new source material; experience works of art first-hand in museums and private collections; make contact with artists, art dealers, archivists, curators, university scholars, etc; consult local archives and library collections; and establish professional networks for future research. Up to twenty grants are awarded annually (up to five of them are specifically dedicated to scholars from Asia, Africa, and Latin America) as follows: up to $6,000 per grant for doctoral students from Europe, Canada, Australia, and New Zealand; up to $7,000 per grant for doctoral students from Asia, Africa, and Latin America; and up to $9,000 per grant for all postdoctoral and senior scholars. Funds can be used for related transportation, lodging, meals, and research fees and expenses of the Grantee only (not of family, travel companions, etc). The annual deadline for applications is January 15.
*Requirements:* The foundation only accepts proposals from doctoral students, postdoctoral scholars, and senior scholars outside the United States. Travel should be undertaken within one calendar year (June through May) after the announcement of the selection results in April. A final report must be submitted within three months of completion of travel. Recipients must acknowledge the support of the Terra Foundation in any publication that results from research conducted during the grant.
*Restrictions:* Funds cannot be used for the purchase of computers or other equipment.
*Geographic Focus:* All Countries

*Date(s) Application is Due:* Jan 15
*Amount of Grant:* Up to 9,000 USD
*Contact:* Amy Gunderson, Grants Manager; (312) 654-2275 or (312) 664-3939; gunderson@terraamericanart.org or grants@terraamericanart.org
*Internet:* https://www.terraamericanart.org/what-we-offer/grant-fellowship-opportunities/research-travel-grants-to-the-united-states/
*Sponsor:* Terra Foundation
120 East Erie Street
Chicago, IL 60611

**Terra Foundation U.S. Curatorial Travel Grants for Travel Outside the United States** **2498**
To encourage and enrich international partnerships between art museums and to deepen research and dialogue, for the next two years the foundation offers U.S. Curatorial Travel Grants for Travel outside the United States. An institution may apply for a travel grant even if it has applied (or plans to apply) separately for general exhibition support for the same project. Proposals for travel grants and general exhibition support are each considered on their own merits; grant approval in one program does not guarantee grant approval in the other. Funding is available for travel, lodging, meals, and research fees and expenses. Grants range from $3,000 to $7,000 each. There are three annual deadlines for application submission: January 15; May 15; and September 15.
*Requirements:* The foundation only accepts proposals from institutions with United States 501(c)3 status. Travel must take place within 12 months from the time of award.
*Restrictions:* U.S. Curatorial Travel Grants for Travel outside the United States are not made to individuals.
*Geographic Focus:* All States
*Date(s) Application is Due:* Jan 15; May 15; Sep 15
*Amount of Grant:* 3,000 - 7,000 USD
*Contact:* Amy Gunderson, Grants Manager; (312) 654-2275 or (312) 664-3939; gunderson@terraamericanart.org or grants@terraamericanart.org
*Internet:* https://www.terraamericanart.org/what-we-offer/grant-fellowship-opportunities/us-curatorial-travel-grants-for-travel-outside-the-united-states/
*Sponsor:* Terra Foundation
120 East Erie Street
Chicago, IL 60611

**Texas Tech College of Visual and Performing Arts Fellowships** **2499**
The College of Visual and Performing Arts administers Graduate Recruitment Fellowships distributed to the college by the Graduate School and two graduate fellowships funded through generous endowments from the Helen Jones and The CH Foundations. These include the Helen DeVitt Jones Graduate Fellowship and The CH Foundation Graduate Fellowship. To be considered for these awards, outstanding prospective and continuing students must be nominated by the faculty and Graduate Advisor/Coordinator of the degree program they are or will be pursuing. The Graduate Recruitment Fellowship is equal to $4,000 annually, renewable for up to two years for Master's, up to three years for terminal Master's (MFA-ART and MFA-THA), and up to four years for Doctoral. The Helen DeVitt Jones Graduate Fellowship is equal to $4,500 annually, renewable for up to five years. The CH Foundation Graduate Fellowship is equal to $5,000 annually, renewable for up to five years.
*Requirements:* Recipients will be full-time graduate students enrolled in the CVPA, selected based on merit by award committees at the unit and college levels. Fellowships may be awarded along with teaching or graduate assistantships. Preference will be given to applicants who have a strong record of community service. To renew, recipients must maintain a 3.5 GPA status, make satisfactory progress toward completion of their degree as defined by the requirements of the college, and demonstrate their involvement in community service activities.
*Geographic Focus:* All States
*Date(s) Application is Due:* Feb 1
*Amount of Grant:* 4,000 - 5,000 USD
*Contact:* Andrew Martin; (806)-742-3825 or (806) 742.0700; fax (806) 742-1971
*Internet:* www.depts.ttu.edu/cvpa/students/Fellowships.php
*Sponsor:* Texas Tech University
1011 Boston Avenue
Lubbock, TX 79409-2081

**Textron Corporate Contributions Grants** **2500**
Textron Inc. was founded in 1923 as a small textile company and has since become one of the world's best known multi-industry companies. Textron focuses philanthropic giving in the following areas: workforce development and education; healthy families/vibrant communities; and sponsorships. In the area of workforce development and education, the company focuses on job-training and employment development (eg., school-to-work, welfare-to-work, job-training for underserved-audiences, literacy, and English-as a-Second-Language programs). In the area of healthy families/vibrant communities, the company focuses on arts and culture (with emphasis on outreach programs that enhance learning and target low- and moderate-income individuals), community revitalization (eg., affordable housing and economic development in low-income areas), and health and human-service organizations (eg., food pantries, homeless shelters, and services for low-income residents). In the area of sponsorships, the company encourages volunteerism and sponsors worthwhile events that benefit the communities where employees live and work. Textron's grant history has included funding of general-operating costs, capital campaigns, building construction/renovation, equipment acquisition, program development, conferences and seminars, publication, seed money, fellowships, scholarship funds, research, technical assistance, consulting services, and matching funds. Downloadable guidelines (PDF) and application (Word document) are available online. The completed application and required accompanying documentation must be received via mail by the deadline date.
*Requirements:* Textron targets its giving to nonprofit agencies located in its headquarters state of Rhode Island and those locations where the company has divisional operations. Organizations outside of Rhode Island should contact the Textron company in their area; a listing of Textron businesses along with their contact information can be accessed by clicking the "Contact Us" link at the Textron website.
*Restrictions:* Textron will review only one request per organization during a 12-month period. Contributions will not be made to the following types of organizations: organizations without 501(c)3 status as defined by the Internal Revenue Service; individuals; political, fraternal or veterans organizations; religious institutions when the grant would support sectarian activities; and organizations that discriminate by race, creed, gender, ethnicity, sexual orientation, disability, age or any other basis prohibited by law.
*Geographic Focus:* Georgia, Illinois, Kansas, Louisiana, Maryland, Massachusetts, New York, North Carolina, Pennsylvania, Rhode Island, Texas, Germany, Great Britain
*Date(s) Application is Due:* Mar 1; Sep 1
*Samples:* The Providence Center, Providence, Rhode Island, $10,000 - to provide training opportunities in receptionist, production, driver, and computer vocations.
*Contact:* Karen Warfield, Community Affairs Manager; (401) 421-2800; fax (401) 457-2225
*Internet:* www.textron.com/about/commitment/corp-giving/
*Sponsor:* Textron Charitable Trust
40 Westminster Street
Providence, RI 02903

**Thermo Fisher Scientific Corporate Contributions Grants** **2501**
Thermo Fisher Scientific supports the next generation of thinkers and leaders with in-kind, financial and volunteer support programs that encourage students to explore and develop their interest in STEM (science, technology, engineering and math). In addition, the Corporation's mission as a company spurs it to share knowledge and skills with the communities in which it operates. Support for education in science, technology, engineering, and math is needed more than ever. However, many communities and schools lack the budget or the tools to effectively conduct classroom programs in STEM. Thermo Fisher believes exposure to hands-on learning in these subjects is essential to its shared future. Its customer, South Dakota Biotech, shares a similar view and in 2017, Thermo Fisher partnered to build 800 STEM-credible safety kits at their annual summit. Its teams helped to source the materials included in the kits and collaborated with the summit attendees for their final assembly. Members of a local Boys and Girls of Club of America were invited to participate in hands-on science activities where they received a donation of the safety kits to take back to their club.
*Requirements:* 501(c)3 nonprofit organizations are eligible to apply.
*Restrictions:* The corporation does not provide funding for advertising journals or booklets; fundraising events such as telethons, walkathons, and races; specific performances or concerts; sporting events; endowment campaigns; film, video, television, or radio projects; grants to individuals; political causes or candidates; or organizations that practice discrimination or limit membership on the basis of race, creed, gender, age, sexual orientation, or national origin.
*Geographic Focus:* All States
*Contact:* Contributions Manager; 847) 674-5917
*Internet:* https://www.thermofisher.com/us/en/home/about-us/corporate-social-responsibility/philanthropic-giving.html
*Sponsor:* Thermo Fisher Scientific
8025 Lamon Ave # 125
Skokie, IL 60077

**Third Millennium Foundation Fellowships** **2502**
Through the Third Millennium Foundation's Fellowship programs, college students, graduate students and activists receive financial and professional support that enables them to build their capacity across a full range of human rights issues. In doing so, the Foundation aims to develop a cadre of young activists from different countries whose understanding of, commitment to, and effectiveness in human rights work has been enhanced. These Fellowships provide a unique opportunity for in-depth exposure to and reflection on a broad range of perspectives, situations and experiences related to human rights work.
*Geographic Focus:* All States
*Contact:* Fellowships Manager; (212) 421-5244; fax (212) 421-5243; tmf.usa@verizon.net
*Internet:* www.seedsoftolerance.org/initiative-fellowships.html
*Sponsor:* Third Millennium Foundation
340 West 12th Street
New York, NY 10014

**Third Millennium Foundation Grants** **2503**
The Third Millennium Foundation's principal goal is to support programs and initiatives designed to promote tolerance, particularly among the young. Its work is focused on developing young children's understanding of and respect for the differences that exist among themselves, especially those related to culture, ethnicity, gender, race, and socioeconomic status. The Foundation emphasizes art in its grant making and continues to explore creative new avenues for unlearning intolerance. Foundation programs build on research findings demonstrating that gaining self-control and the ability to relate positively to others is developed in early childhood and that family, educators, and child caregivers play significant roles in shaping lifelong attitudes. The Foundation makes grants in the United States and internationally, and supports arts and human rights

programs. Its programmatic emphasis is on education, children's rights and capacity building for grassroots organizations.
*Geographic Focus:* All States
*Contact:* Grants Manager; (212) 421-5244; fax (212) 421-5243; tmf.usa@verizon.net
*Internet:* www.seedsoftolerance.org/initiative-grantmaking.html
*Sponsor:* Third Millennium Foundation
340 West 12th Street
New York, NY 10014

**Thomas Austin Finch, Sr. Foundation Grants** **2504**
The purpose of the Finch Foundation is the improvement of the mental, moral and physical well-being of the inhabitants of Thomasville, North Carolina, with emphasis on improving education, improving health related facilities and attracting new business to the community. However, grants are made for a wide variety of charitable causes throughout the greater Thomasville area. In general, grants are awarded for seed money, matching funds and general purposes. Primary consideration is given to projects of a non-recurring nature or to start up funding of limited duration. The foundation meets twice a year to review grant requests. Requests must be submitted by February 15th for the Spring Meeting and October 15th for the Fall Meeting. Application forms are available online. Applicants will receive notice acknowledging receipt of the grant request, and subsequently be notified of the grant declination or approval.
*Requirements:* Eligible applicants are IRS 501(c)3 non-profit organizations located within the corporate limits of the City of Thomasville, North Carolina or, organizations where the benefits of the grant will inure primarily to the residents of Thomasville. Proposals should be submitted in the following format: completed Common Grant Application Form; an original Proposal Statement; an audited financial report and a current year operating budget; a copy of your official IRS Letter with your tax determination; a listing of your Board of Directors. Proposal Statements (second item in the above Format) should answer these questions: what are the objectives and expected outcomes of this program/project/request; what strategies will be used to accomplish your objective; what is the timeline for completion; if this is part of an on-going program, how long has it been in operation; what criteria will you use to measure success; if the request is not fully funded, what other sources can you engage; an Itemized budget should be included; please describe any collaborative ventures. Prior to the distribution of funds, all approved grantees must sign and return a Grant Agreement Form, stating that the funds will be used for the purpose intended. Progress reports and Completion reports must also be filed as required for your specific grant. All current grantees must be in good standing with required documentation prior to submitting new proposals to any foundation.
*Restrictions:* Grants are generally not made for typical operational or maintenance-oriented purposes, political purposes, nor to organizations which discriminate on the basis of race, ethnic origin, sexual or religious preference, age or gender.
*Geographic Focus:* North Carolina
*Date(s) Application is Due:* Feb 15; Oct 15
*Amount of Grant:* 1,500 - 100,000 USD
*Samples:* Memorial United Methodist Church, $30,000—general operating funds; Hospice of Davidson County, Inc., $60,000—construction of inpatient facility; Thomasville City Schools, $24,400—active classroom project.
*Contact:* Wachovia Bank, N.A., Trustee; grantinquiries6@wachovia.com
*Sponsor:* Thomas Austin Finch, Sr. Foundation
Wachovia Bank, NC6732, 100 North Main Street
Winston Salem, NC 27150

**Thomas C. Ackerman Foundation Grants** **2505**
The Thomas C. Ackerman Foundation was founded in 1991. By Spring, 1992, substantial assets had been received from the Thomas C. Ackerman Trust, funded as a result of Thomas Ackerman's death on February 13, 1991; the Board of Directors was fully constituted; and the guidelines and objectives for grant making were adopted. The Foundation's areas of interest include: arts and culture; education; and health and human services. To be considered for a grant from the Foundation, an organization must initially submit a Letter of Intent online. Directions for completing the grant application will be provided to an organization, if it is invited to submit one. Awards generally range from $1,000 to $40,000. There are three annual deadlines for full application submission: February 28, September 5, and November 28.
*Requirements:* California nonprofits, primarily in San Diego County, are eligible to apply. To be considered for a grant from the Thomas C. Ackerman Foundation, an organization must initially submit a Letter of Intent online.
*Restrictions:* It is a policy of the Foundation not to provide continuous support for any project to the extent that the project becomes dependent upon the Foundation for its continued existence. The Foundation may make multi-year commitments on occasion. Grants are not made to individuals. Generally, the Foundation does not support conferences or symposia. While occasional grant support is given to religious organizations, those grants are made for direct support of nonsectarian educational or service projects and not for projects which are of primary benefit to members of a particular religion or belief or which primarily promote a particular religion or belief. The Foundation will not consider grants relating to human medical or biomedical research.
*Geographic Focus:* California
*Date(s) Application is Due:* Feb 28; Sep 5; Nov 28
*Amount of Grant:* 1,000 - 40,000 USD
*Samples:* San Diego Public Library, San Diego, California, $35,000 - community development; San Diego Food Bank, San Diego, California, $15,000 - in support of the Food 4 Kids program; Elementary Institute of Science, San Diego, California, $5,000 - educational expansion.
*Contact:* Lynne Newman; (619) 741-0113; info@AckermanFoundation.org
*Internet:* www.ackermanfoundation.org/
*Sponsor:* Thomas C. Ackerman Foundation
3755 Avocado Boulevard, #518
La Mesa, CA 91941-7301

**Thomas F. and Kate Miller Jeffress Memorial Trust Grants** **2506**
The Jeffress Memorial Trust was established under the will of Robert M. Jeffress, a business executive and philanthropist of Richmond, Virginia. Jeffress had a deep interest in his native state, and he made large contributions during his life in support of scientific research in the Commonwealth of Virginia. The Trust provides $60,000 awards as seed funding to support one-year pilot studies that encourage the development of innovative interdisciplinary strategies that integrate computational and quantitative scientific methodologies across a broad range of scientific disciplines. Eligible research areas include astronomy, biosciences, chemistry, computer sciences, engineering, environmental sciences, material science, mathematics and physics. The application deadline is generally in January of each year and the notification period is in March.
*Requirements:* The Jeffress Memorial Trust grant review process is administered by The Medical Foundation, a division of HRiA.
*Restrictions:* Research in social, economic and behavioral sciences, including psychology, are currently outside the scope of Jeffress funding.
*Geographic Focus:* Virginia
*Amount of Grant:* 60,000 USD
*Samples:* University of Virginia, Charlottesville, Virginia, $60,000 - research support for Biophysical, Dynamical-systems Models of Neurons that Recognize Complex Acoustic Patterns; College of William and Mary Foundation, Williamsburg, Virginia, $60,000 - research support for Exploring the Evolutionary and Molecular Dynamics of Mixed Biofilms to Disrupt their Stability; George Mason University Foundation, Fairfax, Virginia, $60,000 - research support for Applying the Ontology-guided Machine Learning to Anazlyze the Surveillance, Epidemiology, and End Results Programs - Medicare Health Outcomes Survey (SEER-mHOS) Linked Database.
*Contact:* Sarah Kay, Vice President; (804) 887-8773; sarah.kay@ustrust.com
*Internet:* https://www.bankofamerica.com/philanthropic/foundation.go?fnId=124
*Sponsor:* Thomas F. and Kate Miller Jeffress Memorial Trust
1111 E. Main Street, VA2-300-12-92
Richmond, VA 23219

**Thomson Reuters / MLA Doctoral Fellowships** **2507**
The purpose of the Thomson Reuters/MLA Doctoral Fellowship is to foster and encourage superior students to conduct doctoral work in health-sciences librarianship or information science by providing support to individuals who have been admitted to candidacy. The award was originally established in 1986 by the Institute of Scientific Information (ISI, which is now a part of Thomson Reuters) and is administered by the Medical Library Association (MLA). The award is granted bi-annually in even-numbered years. The fellowship pays $2,000 in one payment toward the following types of expenses: project expenses; related travel expenses; or augmentation of another larger, separately-funded project relevant to health-science librarianship and which is part of the requirements of a doctoral degree. Guidelines and application materials are available at the website. All materials and letters of support must be received by MLA via email, fax, or mail by December 1. email applications and documents will be accepted as PDF or MS Word files only. Founded in 1898, the Medical Library Association (MLA) is a nonprofit, educational organization of more than 1,100 institutions and 3,600 individual members in the health sciences information field committed to educating health-information professionals, supporting health-information research, promoting access to the world's health-sciences information, and working to ensure that the best health information is available to all.
*Requirements:* Applicants must satisfy the following eligibility criteria: be graduates of an American Library Association (ALA)-accredited school of library science; be candidates in a Ph.D. program with emphsis on biomedical and health-related information science; and be citizens of or have permanent residence status in the United States or Canada. Preference will be given to applicants who have at least 75% of their course work completed and dissertation prospectus either approved or in the approval process.
*Restrictions:* Past recipients of the Thomson Reuters / MLA Doctoral Fellowship are ineligible. The following types of expenses are not supported: tuition fees; tuition-related expenses; and living expenses.
*Geographic Focus:* All States, Canada
*Date(s) Application is Due:* Dec 1
*Amount of Grant:* Up to 2,000 USD
*Contact:* Carla Funk, Program Contact; (312) 419-9094, ext. 14; fax (312) 419-8950; mlapd2@mlahq.org or grants@mlahq.org
*Internet:* www.mlanet.org/awards/grants/index.html
*Sponsor:* Medical Library Association
65 East Wacker Place, Suite 1900
Chicago, IL 60601-7246

**Thoracic Foundation Grants** **2508**
The major emphasis of the Thoracic Foundation of Boston is to offer support for medical research in thoracic diseases. Contributions are also made to hospitals providing service to thoracic and cardiac cases. Fields of interest include: asthma research; cancer research; children and youth services; heart and circulatory diseases; specialty hospitals; medical school education; and smoking cessation. In addition, funds are awarded to assist with publications, conferences, and seminars in the same field. Grants are awarded thoughout Massachusetts, but primarily in the Boston metropolitan area. A letter of application

may be submitted at any time, offering a detailed description of the project and amount of funding requested. Awards range from $5,000 to $75,000.
*Geographic Focus:* District of Columbia, Massachusetts
*Samples:* Action on Smoking and Health, Washington, D.C., $5,000 - funding for non-smokers' rights; Boston Biomedical Research Institution, Boston, Massachusetts, $75,000 - research and development related to thoracic disease; Beth Israel Deaconess Medical Center, Boston, Massachusetts, $35,000 - experimental radiology program.
*Contact:* Bradley R. Cook, Trustee, c/o Taylor, Ganson, and Perrin, LLP; (617) 951-2777; fax (617) 951-0989
*Sponsor:* Thoracic Foundation
160 Federal Street, 20th Floor
Boston, MA 02110-1700

**Thrasher Research Fund Early Career Awards** **2509**
The Fund recognizes that young investigators may find it difficult to remain in pediatric research because of a lack of funding. Therefore, the purpose of this program is to encourage the development of medical research in child health by awarding small grants to new researchers, helping them gain a foothold in this important area. The Fund is open to a variety of research topics important to children's health. Both incidence and severity are considered when determining the significance of a problem being studied. The program is particularly interested in applicants that show great potential to impact that field of children's health through medical research. Both an applicant's aptitude and inclination toward research are considered. The quality of the mentor and the mentoring relationship are also considered to be important predictors of success. There are three funding cycles per year for Early Career Award Program grants.
*Requirements:* Eligible applicants are: physicians who are in a residency/fellowship training program, or who completed that program no more than one year before the date of submission of the concept paper; post-doctoral researchers who received the doctoral level degree no more than three years prior to the date of submission of the concept paper. Each project needs to be under the guidance of a mentor. The qualifications and experience of the mentor will be considered in the evaluation of the application. A mentor may have only one Thrasher Research Fund Early Career Awardee at a time. There are no restrictions with regard to citizenship. The Fund is open to applications from institutions both inside and outside the United States.
*Restrictions:* A new investigator who holds a National Institutes of Health (NIH) K award or a Clinical and Translational Science Award (CTSA) is not eligible to apply for the Thrasher Early Career Award.
*Geographic Focus:* All States, All Countries
*Date(s) Application is Due:* Mar 30; Jul 27; Nov 30
*Contact:* Megan Duncan, MSPH; (801) 240-4720; DuncanME@thrasherresearch.org
Aaron V. Pontsler, M.S., MBA; (801) 240-6385; PontslerAV@thrasherresearch.org
*Internet:* www.thrasherresearch.org/sites/www_thrasherresearch_org/Default.aspx?page=231
*Sponsor:* Thrasher Research Fund
68 S. Main Street, Suite 400
Salt Lake City, UT 84101

**Thrasher Research Fund Grants** **2510**
The Thrasher Research Fund provides grants for pediatric medical research. Because significant solutions for many children's health problems remain undiscovered, the Fund invites a broad array of applications. The Thrasher Research Fund seeks to foster an environment of creativity and discovery aimed at finding solutions to children's health problems. The Fund awards grants for research that offers substantial promise for meaningful advances in prevention and treatment of children's diseases, particularly research that offers broad-based applications.
*Requirements:* Principal Investigators must be qualified in terms of education and experience to conduct research. A doctoral-level degree is required. There are no citizenship or residency requirements. The Fund is open to applications from institutions both inside and outside the United States.
*Restrictions:* The fund does not award grants for educational programs; general operating expenses; general bridge funding for incomplete projects; construction or renovation of buildings or facilities; loans, student aid, scholarships, tuition; or, support of other funding organizations. Grants generally are not awarded for conferences, workshops, or symposia. Proposals in the areas of research on human fetal tissue or behavioral science research will not be considered.
*Geographic Focus:* All States, All Countries
*Amount of Grant:* 100,000 - 350,000 USD
*Contact:* Megan Duncan, MSPH, Research Manager; (801) 240-4720; fax (801) 240-1625; DuncanME@thrasherresearch.org
Aaron V. Pontsler, M.S., MBA, Research Manager; (801) 240-6385; fax (801) 240-1625; PontslerAV@thrasherresearch.org
*Internet:* www.thrasherresearch.org/sites/www_thrasherresearch_org/Default.aspx?page=28
*Sponsor:* Thrasher Research Fund
68 S. Main Street, Suite 400
Salt Lake City, UT 84101

**TIAA-CREF Grants** **2511**
TIAA-CREF Institute grants help identify real-world solutions to pressing issues related to lifetime financial security, organizational success and the business of higher education. Grants from the Institute support research on new business models in higher education, including such topics as innovation and financial sustainability, new workforce models and new governance models. Supported small grant projects are expected to be completed within one year, while larger grants may have a multi-year timetable. The annual application submission date is July 18.
*Geographic Focus:* All States
*Date(s) Application is Due:* Jul 18
*Contact:* Marilyn Grande, Program Contact; (516) 650-9683; mgrande@tiaa-cref.org
*Internet:* https://www.tiaa-crefinstitute.org/public/institute/grants
*Sponsor:* Teachers Insurance and Annuity Association College Retirement Equities Fund
730 Third Avenue
New York, NY 10017-3206

**TIAA-CREF Paul A. Samuelson Award** **2512**
The Samuelson Award recognizes outstanding scholarly writing on issues related to lifelong financial security. It is given each year in recognition of an outstanding research publication containing ideas that the public and private sectors can use to maintain and improve Americans financial well being. The work must be theoretical or empirical in nature and must cover a subject directly relevant to lifelong financial security, such as: retiree healthcare issues; the design of public and private programs providing income or health benefits in retirement; the functioning of insurance markets; the design of investment products and strategies for pension funds and other long-term savings plans; the impact of tax and regulatory policies on savings, retirement, and insurance products and programs; and the changing nature of consumption needs and labor market activities over the life cycle. The winner(s) will announced late in the year and is honored at a reception during the annual meeting of the Allied Social Science Association. The application is available at the TIAA-CREF website.
*Requirements:* Individuals who have published research, in either book or article form, between January 1 of the previous year and June 20 of the current year are eligible.
*Restrictions:* Publications previously submitted will not be considered. Publications authors by current employees and trustees of TIAA-CREF, judges of the award, or their close relatives will not be considered. Collected paper in an edited volume are not eligible as a submission.
*Geographic Focus:* All States
*Date(s) Application is Due:* Sep 10
*Amount of Grant:* 10,000 USD
*Contact:* David P. Richardson, Ph.D., Award Coordinator; (800) 842-2252
*Internet:* https://www.tiaa-crefinstitute.org/public/institute/awards/samuelson
*Sponsor:* Teachers Insurance and Annuity Association College Retirement Equities Fund
730 Third Avenue
New York, NY 10017-3206

**TIAA-CREF Theodore M. Hesburgh Award for Leadership Excellence** **2513**
The Theodore Hesburgh Award for Leadership Excellence seeks to strengthen the teaching tradition at U.S. undergraduate colleges and universities by acknowledging and rewarding exceptional programs. The Award recognizes leadership and commitment to higher education and contributions to the greater good. It is presented to a current college or university president or chancellor who embodies the spirit of Father Hesburgh, his commitment and contributions to higher education and society. The Award is administered by the TIAA-CREF Institute and presented at the American Council on Education annual meeting.
*Requirements:* Any U.S. undergraduate public or private, two and four year colleges and universities are eligible for nomination. The winners are chosen through a rigorous process by a panel of higher education leaders representing a full range of institutions.
*Restrictions:* While more than one program may be submitted, no institution will receive more than one award.
*Geographic Focus:* All States
*Amount of Grant:* 5,000 - 30,000 USD
*Contact:* Anne Ollen; (866) 842-2442 or (212) 913-2803; aollen@tiaa-cref.org
*Internet:* www.tiaa-crefinstitute.org/institute/awards/hesburgh/index.html
*Sponsor:* Teachers Insurance and Annuity Association College Retirement Equities Fund
730 Third Avenue
New York, NY 10017-3206

**Tides Foundation Antonio Pizzigati Prize for Software in the Public Interest** **2514**
The Tides Foundation Antonio Pizzigati Prize for Software in the Public Interest annually awards a $10,000 cash grant to one individual who has created or led an effort to create an open source software product of significant value to the nonprofit sector and movements for social change. The Pizzigati Prize honors the brief life of Tony Pizzigati, an early advocate of open source computing. Nominations may be made by an individual for their own work, or by a peer or colleague an behalf of a nominee. All application materials are due by December 15.
*Requirements:* The Pizzigati Prize welcomes applications from—and nominations for—single individuals. Those nominated for the prize should have developed a software product that is open source, as defined by the Open Source Initiative, and easily and widely available. This software must have already demonstrated its value to at least one nonprofit organization. Additionally, it should be a product that can be a value to multiple nonprofit organizations. Applicants will be evaluated on a range of criteria by an advisory panel that includes past winners of the Prize.
*Geographic Focus:* All States
*Date(s) Application is Due:* Dec 15
*Amount of Grant:* 10,000 USD
*Contact:* Program Associate; (415) 561-6400 or (415) 561-6300; fax (415) 561-6401; info@tides.org
*Internet:* https://www.tides.org/impact/awards-prizes/pizzigati-prize/
*Sponsor:* Tides Foundation
P.O. Box 29198
San Francisco, CA 94129-0198

**Tiffany and Company Foundation Coral Conservation Grants** 2515

The Tiffany and Company Foundation values healthy oceans and the important role that corals play in these ecosystems. The Foundation believes that precious corals cannot be sustainably removed from the oceans for use in jewelry or home décor. The Foundation supports organizations that work to improve the health of oceans through research, preservation and management of coral reefs. Strategies include: to promote awareness and education of the importance of corals and marine ecosystems through outreach to targeted constituencies such as consumers, ocean enthusiasts and select marine-tourism providers; and to support key research and reef management strategies to directly contribute to saving reef ecosystems.

*Requirements:* To qualify for a grant, a prospective grantee must be a U.S.-based, tax-exempt organization with 501(c)3 status. A prospective grantee must also be engaged in work that fits within the Foundation's program guidelines.

*Restrictions:* The Tiffany and Company Foundation does not offer grants to: organizations that do not have tax-exempt status under Section 501(c)3 of the Internal Revenue Code, or the equivalent; religious, political, lobbying, social or fraternal organizations; fundraising benefits; individuals, including those with a 501(c)3 organization willing to accept funding on their behalf; or organizations that discriminate or limit their membership, support or services based on race, color, sex, religion, national origin, sexual orientation, gender identification, alien or citizenship status, age, disability or any other characteristic protected by law.

*Geographic Focus:* All States

*Amount of Grant:* 20,000 - 3,000,000 USD

*Samples:* Foundation Center, New York, New York, $20,000 - support for the Ocean Web Portal, a dynamic, web-based knowledge portal to track and inspire ocean conservation efforts around the world; Give2Asia, San Francisco, California, $500,000 - support for Reef Aid, a project of Greening Australia, to improve water quality in the Great Barrier Reef and build the resilience of coral in the face of climate; Nature Conservancy, Arlington, Virginia, $1,000,000 (over two years) - support for a regional coral restoration strategy in the Caribbean as a global model for marine conservation.

*Contact:* Rebecca Burton, Program Manager; (212) 230-6591; fax (212) 230-5341; foundation@tiffany.com

*Internet:* www.tiffanyandcofoundation.org/guidelines/coral.aspx

*Sponsor:* Tiffany and Company Foundation
200 Fifth Avenue
New York, NY 10010

**TIRF Doctoral Dissertation Grants** 2516

Since 2002, The International Research Foundation for English Language Education (TIRF) has supported students completing their doctoral research on topics related to the foundation's priorities. Each year applicants who have been advanced to candidacy in legitimate PhD or EdD programs are invited to submit proposals for Doctoral Dissertation Grants (DDGs). Grants will be awarded in amounts up to $5,000 USD per awardee. Each year the highest rated Doctoral Dissertation Grant proposal is awarded in the name of the late Russell N. Campbell. Dr. Campbell, a Past President of TESOL, was one of the founding fathers of the Foundation.

*Requirements:* Applicants must be enrolled in a legitimate doctoral program; be advanced to candidacy and have had a research plan approved by a faculty committee at their university; write a proposal which is clearly related to TIRF's research priorities; follow specific instructions located in the call for proposals; and have their research supervisor submit an official letter of support attesting to the applicant's readiness to complete the doctoral dissertation. Please note that DDG applicants do not need to be U.S. citizens. Each year the research priority for competition is announced. Check the website for this year's priority.

*Geographic Focus:* All States

*Date(s) Application is Due:* May 14

*Amount of Grant:* Up to 5,000 USD

*Samples:* 2012: Patsy Vinogradov (Russell N. Campbell Award Winner), Little Desks, Big Ideas: The Unlikely Meeting of Adult Educators in Kindergarten to Explore Early Literacy Instruction; Lixia Cheng, Effects of Pragmatic Task Features, English Proficiency, and Learning Setting on Chinese ESL/EFL Learners' Spoken Performance of Requests; Tasha L. Darbes, Hard to Place: Multilingual Immigrant Youth in Community Colleges; Joyce Kling, Teacher Cognition: English-medium Instruction at LIFE; Derya Kulavuz-Onal, ESL/EFL Teachers' Learning to Teach with Technology through Participation in an Online Community of Practice: A Netnography of "Webheads in Action"; Hyojung Lim, Investigating the Cognitive Validity of a L2 Reading Test: An Eye-tracking Study; Briana Ronan, Making Meaning across Modes: English Language Learners and their Academic Writing within a Digital Space; Weiwei Yang, The Role of Cognitive Complexity of Independent Writing Tasks in L2 Writing Quality and in Accuracy, Complexity, and Fluency of Language Production in L2 Writing

*Contact:* Ryan Damerow, fax 831-647-6650; info@tirfonline.org

*Internet:* www.tirfonline.org/research-grants/doctoral-dissertation-grants/

*Sponsor:* International Research Foundation for English Language Education
177 Webster St., #220
Monterey, CA 93940

**Todd Brock Family Foundation Grants** 2517

Established in Texas in 2007, the Todd Brock Family Foundation offers support for educational programs, children, athletics, research, and community projects, as well Protestant agencies and churches. A formal application is required, and applicants should forward the entire proposal to the office. The Foundation rarely offers funding outside of Texas. There are no deadlines for submitting a completed proposal, and grants have most recently ranged from $300 to $250,000.

*Geographic Focus:* Texas

*Samples:* Memorial High School Boosters, Houston, Texas, $1,000 - funding for annual gold tournament; Legacy Christian Academy, Beaumont, Texas, $250,000 - for general operations; Yellowstone Academy, Houston, Texas, $25,000 - support for low-income children and families.

*Contact:* Todd O. Brock, President; (409) 833-6226; fax (409) 832-3019

*Sponsor:* Todd Brock Family Foundation
1670 E Cardinal Drive, P.O. Box 306
Beaumont, TX 77704-0306

**Tourette Syndrome Association Post-Doctoral Fellowships** 2518

The Tourette Syndrome Association is interested in supporting any research that has the potential to contribute to the understanding of Tourette Syndrome, including genetics, pathogenesis, pathophysiology, clinical treatments and animal model development. Post-doctoral research fellowships of up to $40,000 for one year are awarded. Applications are evaluated on the following information: candidate's qualifications and objectives; candidate's experience relevant to the project; methodology; significance and/or relevance to the TS field; percentage of time to be devoted to project; adequacy and availability of research facilities and other project support; and the ability to complete the project in a timely manner. The application and deadline information are available at the website.

*Requirements:* Candidates must have a M.D., Ph.D. or equivalent. Previous experience in the field of movement disorders is desirable, but not essential. International applicants are welcome.

*Geographic Focus:* All States, All Countries

*Amount of Grant:* Up to 40,000 USD

*Contact:* Heather Cowley, Fellowship Administrator; (718) 224-2999, ext. 247 or (718) 224-2999, ext. 222; fax (718) 279-9596; heather.cowley@tsa-usa.org

*Internet:* tsa-usa.org/research.html

*Sponsor:* Tourette Syndrome Association
42-40 Bell Boulevard
Bayside, NY 11361-2820

**Tourette Syndrome Association Research Grants** 2519

The Tourette Syndrome Association (TSA) is interested in receiving applications from any research study that may contribute to the understanding of Tourette Syndrome, including genetics, pathogenesis, pathophysiology, clinical treatments and animal model development. Research grants of up to $75,000 for one year or up to $150,000 for two years are awarded. The TSA will consider reasonable requests for the following: salaries (NIH salary caps apply); laboratory supplies and other research-related expenses; travel expenses for investigators (up to $1,500 per year); research equipment up to $5,000; and indirect costs of up to 10% may be included. Applicants are evaluated on the following information: candidate's qualifications and objectives; their relevant experience to the project; methodology; significance/relevance to the TS field; percentage of time to be devoted to the project; adequacy and availability of research facilities and other project support; and the ability to complete the project in a timely manner. For preliminary screening, a pre-proposal briefly describing the scientific basis and relevance of the proposed project is required. Preproposal applications are available at the website. Applicants for grants involving clinical treatment are encourage to read the posted clinical studies documents at the website. Payment schedules will vary according to one year or two year awards. Applicants invited to submit a full research proposal will be contacted with instructions.

*Requirements:* Investigators are required to have an advanced degree, such as a Ph.D., M.D., or equivalent. International applicants are welcome to apply. Investigators for nonprofit and for-profit organizations may apply. Previous experience in the field of movement disorders is desirable, but not essential.

*Geographic Focus:* All States, All Countries

*Date(s) Application is Due:* Nov 1

*Amount of Grant:* 75,000 - 150,000 USD

*Contact:* Heather Cowley, TSA Research Grant Coordinator; (718) 224-2999, ext. 247 or (718) 224-2999, ext. 222; fax (718) 279-9596; heather.cowley@tsa-usa.org

*Internet:* tsa-usa.org/research.html

*Sponsor:* Tourette Syndrome Association
42-40 Bell Boulevard
Bayside, NY 11361-2820

**TRDRP California Research Awards** 2520

This program is focused on research projects that address questions specific to tobacco-related disease or tobacco control issues in California. The awards will provide funding for average annual direct costs of $140,000; $170,000 per year for research involving human subjects. Full indirect costs are paid to eligible institutions. Maximum duration is 3 years.

*Requirements:* Principal Investigators from California not-for-profit organizations are eligible for funding. Research undertaken with TRDRP funds must be conducted primarily in California. The Principal Investigator must conduct and supervise the research project directly and in person. Underrepresented racial and ethnic investigators are encouraged to submit applications. To be competitive, proposals must present a compelling justification, including evidence if available, supporting the California-specific criterion. The proposal should be fully developed, scientifically rigorous, and include sound background information, hypothesis, and promising preliminary studies or preliminary data.

*Restrictions:* All submitted applications will be screened for their direct relevance to tobacco use or tobacco-related disease. Proposed research that is not relevant will not undergo peer review for scientific merit. It is incumbent on the applicant to make a compelling case that the proposed work is directly relevant to the mission of TRDRP.

*Geographic Focus:* California

*Date(s) Application is Due:* Jan 15

*Amount of Grant:* 140,000 - 170,000 USD
*Contact:* M.F. Bowen, (510) 987-9811; mf.bowen@ucop.edu
*Internet:* www.trdrp.org/fundingopps/callawardmechs.asp#calrchawd
*Sponsor:* Tobacco-Related Disease Research Program
Office of the President, 300 Lakeside Drive, 6th Floor
Oakland, CA 94612-3550

**TRDRP Dissertation Research Awards** **2521**
This award is intended to support the dissertation research of a doctoral candidate pursuing tobacco-related research. The award is designed for students advanced to candidacy no later than July 1, and initiating their dissertation research. Maximum of $20,000 annual direct costs averaged over the duration of the award (maximum 2 years) for stipend, supplies, and domestic travel. This award also provides an additional maximum of $10,000 per year for tuition/enrollment fee remission, fringe benefits, and health insurance.
*Requirements:* Applications in all relevant research areas are welcomed, but applications in the social/behavioral sciences and in public policy are encouraged. The applicant and principal mentor must be affiliated with an academic research institution. U.S. citizenship is not a requirement. The candidate must commit a minimum of 80 percent time to the research project.
*Restrictions:* No equipment purchases are allowed. No indirect costs are allowed.
*Geographic Focus:* California
*Date(s) Application is Due:* Jan 15
*Amount of Grant:* 1 - 30,000 USD
*Contact:* M.F. Bowen, (510) 987-9811; mf.bowen@ucop.edu
*Internet:* www.trdrp.org/fundingopps/callawardmechs.asp
*Sponsor:* Tobacco-Related Disease Research Program
Office of the President, 300 Lakeside Drive, 6th Floor
Oakland, CA 94612-3550

**TRDRP Exploratory/Developmental Research Awards** **2522**
The purpose of these grants is to gather preliminary data or demonstrate proof-of-principle (i.e., pilot projects), or to conduct a research project within the specified limits of money and time. The ultimate goal of these awards is to provide the foundation for proposals for fully-developed research project awards from other funding programs. Total direct costs of $250,000 for the entire project duration (maximum of 2 years). Indirect costs are paid to eligible institutions.
*Requirements:* Principal Investigators from California not-for-profit organizations are eligible for funding. Research undertaken with TRDRP funds must be conducted primarily in California. The Principal Investigator must conduct and supervise the research project directly and in person. Underrepresented racial and ethnic investigators are encouraged to submit applications. Although Principal Investigators may submit as many proposals for these awards as they wish for separate projects, no more than one Exploratory/Developmental Research Award will be made to any individual PI in the same year. Only proposals that address one of TRDRP's Primary research area will be eligible and will undergo peer review. Exploratory/Developmental Research Awards are for new projects only; that is, competing renewal applications will not be accepted.
*Restrictions:* All submitted applications will be screened for their direct relevance to tobacco use or tobacco-related disease. Proposed research that is not relevant will not undergo peer review for scientific merit. It is incumbent on the applicant to make a compelling case that the proposed work is directly relevant to the mission of TRDRP.
*Geographic Focus:* All States
*Date(s) Application is Due:* Jan 15
*Amount of Grant:* 250,000 USD
*Contact:* M.F. Bowen, (510) 987-9811; mf.bowen@ucop.edu
*Internet:* www.trdrp.org/fundingopps/callawardmechs.asp#explordev
*Sponsor:* Tobacco-Related Disease Research Program
Office of the President, 300 Lakeside Drive, 6th Floor
Oakland, CA 94612-3550

**TRDRP New Investigator Awards** **2523**
This award mechanism is specifically designed to enable new investigators to initiate an independent research program in a tobacco-related field. These are awards for investigators who have a doctoral degree, have not yet served as a principal investigator of a research project grant (e.g., an NIH R01-type grant), and have had no more than five years experience as an independent investigator, i.e., no more than five years since completing formal postdoctoral training, or since the doctoral degree if no postdoctoral training. Direct costs of $90,000 per year can be averaged over the duration of the award. The indirect cost rate is capped at 8% for eligible institutions.
*Requirements:* New investigators must commit to a minimum of 50 percent time to the research project for the award duration.
*Restrictions:* Current or past new investigator awardees from TRDRP, other UC Special Research Programs, or other agencies (for example, current or past recipients of NIH K-series career development awards) are not eligible.
*Geographic Focus:* California
*Date(s) Application is Due:* Jan 15
*Amount of Grant:* 90,000 USD
*Contact:* M.F. Bowen, (510) 987-9811; mf.bowen@ucop.edu
*Internet:* www.trdrp.org/fundingopps/callawardmechs.asp
*Sponsor:* Tobacco-Related Disease Research Program
Office of the President, 300 Lakeside Drive, 6th Floor
Oakland, CA 94612-3550

**TRDRP Participatory Research Awards: CARA/SARA** **2524**
The purpose of the Community Academic Research Awards (CARA) and the School Academic Research Awards (SARA) is to stimulate and support collaborations between community-based organizations/ schools with academic investigators. These awards are for a collaborative partnership to perform scientifically rigorous research into tobacco control issues that are identified as important and meaningful to specific communities/schools in the state.
*Requirements:* The applicant partners must demonstrate the use of methods that are relevant, culturally sensitive, and appropriate in terms defined and accepted by the participating communities/schools. Applicants are encouraged to consider TRDRP's Primary research areas and and the TEROC recommended research strategies that lend themselves to a participatory research approach.
*Restrictions:* All submitted applications will be screened for their direct relevance to tobacco use or tobacco-related disease. Proposed research that is not relevant will not undergo peer review for scientific merit. It is incumbent on the applicant to make a compelling case that the proposed work is directly relevant to the mission of TRDRP.
*Geographic Focus:* California
*Date(s) Application is Due:* Jan 15
*Amount of Grant:* 100,000 - 300,000 USD
*Contact:* M.F. Bowen, (510) 987-9811; mf.bowen@ucop.edu
*Internet:* www.trdrp.org/fundingopps/callawardmechs.asp
*Sponsor:* Tobacco-Related Disease Research Program
Office of the President, 300 Lakeside Drive, 6th Floor
Oakland, CA 94612-3550

**TRDRP Postdoctoral Fellowship Awards** **2525**
These are awards for individuals to obtain postdoctoral research training under a designated mentor. The application must be prepared and submitted exclusively by the fellow and must outline an original research project (separate from the project of a mentor). Letters of support addressing the candidate's training, potential, and the commitment of the mentor and the department to the candidate's career development are essential.
*Requirements:* Principal Investigators from California not-for-profit organizations are eligible for funding. Research undertaken with TRDRP funds must be conducted primarily in California. The Principal Investigator must conduct and supervise the research project directly and in person. Underrepresented racial and ethnic investigators are encouraged to submit applications. To be eligible, the candidate must be recognized by the applicant institution as a postdoctoral fellow no later than July 1 of the year intended for the fellowship. U.S. citizenship is not a requirement. The fellow must commit a minimum of 75 percent time to the research project. Direct costs of $45,000 per year averaged over the duration of the award. No indirect costs are allowed. Maximum duration is 3 years.
*Restrictions:* All submitted applications will be screened for their direct relevance to tobacco use or tobacco-related disease. Proposed research that is not relevant will not undergo peer review for scientific merit. It is incumbent on the applicant to make a compelling case that the proposed work is directly relevant to the mission of TRDRP.
*Geographic Focus:* California
*Date(s) Application is Due:* Jan 15
*Amount of Grant:* 45,000 USD
*Contact:* M.F. Bowen, (510) 987-9811; mf.bowen@ucop.edu
*Internet:* www.trdrp.org/fundingopps/callawardmechs.asp#postdoc
*Sponsor:* Tobacco-Related Disease Research Program
Office of the President, 300 Lakeside Drive, 6th Floor
Oakland, CA 94612-3550

**TRDRP Research Grants** **2526**
TRDRP will consider unsolicited requests to support research-related activities that build research infrastructure, disseminate research findings, or stimulate new research directions. Support can be requested for scientific conferences to assess tobacco's impact on California's population; or to allow tobacco investigators to evaluate, in a timely manner, new and breaking trends in tobacco control or tobacco-related disease research. The TRDRP Scientific Advisory Committee will make recommendations regarding funding. These opportunities will be limited in number, scope, cost, and duration. Contact a TRDRP research administrator regarding the appropriateness of your proposal prior to submission.
*Requirements:* In order to qualify for funding, the planned activities must be directly related to one or more of TRDRP's Priority research areas. The activity must primarily take place in California, involve California investigators, and include, where applicable, discussants and speakers funded by TRDRP. Unsolicited requests may be submitted at any time. Requests will be evaluated expeditiously by ad-hoc peer review when appropriate.
*Restrictions:* All submitted applications will be screened for their direct relevance to tobacco use or tobacco-related disease. Proposed research that is not relevant will not undergo peer review for scientific merit. It is incumbent on the applicant to make a compelling case that the proposed work is directly relevant to the mission of TRDRP.
*Geographic Focus:* California
*Contact:* M.F. Bowen, (510) 987-9811; mf.bowen@ucop.edu
*Internet:* www.trdrp.org/fundingopps/callawardmechs.asp
*Sponsor:* Tobacco-Related Disease Research Program
Office of the President, 300 Lakeside Drive, 6th Floor
Oakland, CA 94612-3550

**Turtle Conservation Fund Grants** **2527**
The mission of Turtle Conservation Fund (TCF) is to ensure that no species of tortoise or freshwater turtle becomes extinct and that sustainable populations of all species persist and thrive in the wild. TCF provides the critical funding necessary for facilitating and coordinating global turtle conservation initiatives. As one of its main functions, TCF

administers a turtle conservation and research grants program. Awards are granted to organizations or individuals for specific conservation or research projects dealing with highly endangered tortoises or freshwater turtles, but not marine turtles, with either partial or full support as funding allows. TCF Awards at the present time are approximately in the $2,000 to $5,000 range per project, with two funding cycles annually. Total disbursement has averaged between $30,000 to $50,000 per year to date, and is growing as the TCF grows.
*Requirements:* Priority for funding is given to projects that focus on tortoises or freshwater turtles that are already highly threatened according to the IUCN Red List criteria, either Critically Endangered or Endangered, or which may be highly threatened but not yet officially classified as such on the Red List. Award recipients enter into contractual agreement with either Conservation International or Chelonian Research Foundation to produce the proposed work (these two non-profit organizations serve as TCF's financial umbrella organizations). Award recipients are also encouraged to publish at least partial results of the supported research in the international scientific turtle journal, Chelonian Conservation and Biology, published by Chelonian Research Foundation. Applications must be submitted electronically and must include the required TCF Proposal Cover Sheet
*Restrictions:* TCF does not provide support for indirect organizational overhead expenses associated with administration of grants, and will not fund requests for such indirect costs.
*Geographic Focus:* All States
*Date(s) Application is Due:* May 1; Nov 1
*Amount of Grant:* 2,000 - 5,000 USD
*Samples:* Attum, Omar, 2009, A local community and science based approach to conserving the critically endangered Egyptian tortoise. Caccone, Adalgisa and Flanagan, Joe, 2009, Rescuing two giant Galapagos tortoise species from extinction. Giustiniano, Danielle, 2009, Paternity, sperm storage and sperm competition in the critically endangered western swamp tortoise, Pseudemydura umbrina. Som, Sitha; Chey, Koulang; Sun, Yoeung; Kim, Chamnan; Kheng, Sokhorn; and Sitha, Prum, 2009, Community-based nest protection for Cantor's giant softshell turtle in the Mekong River, Cambodia. Rashid, S.M.A, 2009, Survey of Batagur baska in the Sunderbans, Bangladesh. Ihlow, Flora and Böhme, Wolfgang, 2009, Release study of Indotestudo elongata in the Kulen Promtep Wildlife Sanctuary in the Northern Plains of Cambodia with regards to the species' ecology and natural history. Walker, Ryan and Rakotondrainy, Riana, 2009, Range-wide population density survey of the mid range Madagascar spider tortoise (Pyxis arachnoides arachnoides ); incorporating GIS application into conservation planning. McCormack, Tim; Ha, Hoang Van; and Nhan, Nguyen Chi, 2009, Ensuring a future for the Vietnamese pond turtle: Establishing the Mauremys annamensis conservation project (MAP), in central Vietnam. Kholis, Munawar, 2009, Painted terrapin (Callagur borneoensis) distribution and status in northern Sumatra, Indonesia.
*Contact:* Hugh R. Quinn, Co-Chair; 605-342-0583; DoubleHQ@aol.com
Anders G.J. Rhodin; 978-582-9668; fax 978-582-6279; RhodinCRF@aol.com
*Internet:* www.turtleconservationfund.org/grants/
*Sponsor:* Turtle Conservation Fund
2011 Crystal Drive, Suite 500
Arlington, VA 22202

**U.S. Department of Education Innovation and Research Grants** **2528**
The Education Innovation and Research (EIR) Program, established under section 4611 of the Elementary and Secondary Education Act (ESEA), as amended by Every Student Succeeds Act (ESSA), provides funding to create, develop, implement, replicate, or take to scale entrepreneurial, evidence-based, field-initiated innovations to improve student achievement and attainment for high-need students; and rigorously evaluate such innovations. The EIR program is designed to generate and validate solutions to persistent educational challenges and to support the expansion of effective solutions to serve substantially larger numbers of students. There are three types of grants under this program: Early-phase grants, Mid-phase grants, and Expansion grants. These grants differ in terms of the level of prior evidence of effectiveness required for consideration for funding, the expectations regarding the kind of evidence and information funded projects should produce, the level of scale funded projects should reach, and, consequently, the amount of funding available to support each type of project. Depending upon Congressional funding, the annual application submission date has typically fallen during the second week of April.
*Requirements:* Eligible applicants include: a local educational agency; a State educational agency; the Bureau of Indian Education; a consortium of State educational agencies or local educational agencies; a nonprofit organization; and a State educational agency, an local educational agency, a consortium, or the Bureau of Indian Education, in partnership with a nonprofit, a business, an educational service agency, or an institution of higher education.
*Geographic Focus* All States, District of Columbia, Guam, Marshall Islands, Northern Mariana Islands, Puerto Rico, U.S. Virgin Islands, American Samoa
*Samples:* Metropolitan Nashville Public Schools, Nashville, Tennessee, $7,683,178 - mid-phase support of the Scaling Up Pyramid Model (PM) Implementation in Preschool and Kindergarten Classrooms; Niswonger Foundation, Tusculum, Tennessee, $7,814,288 - mid-phase support of the Rural Literacy Initiative Focused on Effectiveness program; DuPage Regional Office Of Education, DuPage, Illinois, $3,907,144 - early-phase support of the Partners To Lead – A Consortium of 26 Rural and/or high need Local Education Agencies.
*Contact:* Kelly Terpak, Program Contact; (202) 453-7122 or (202) 453-7200; kelly.terpak@ed.gov or eir@ed.gov
*Internet:* https://innovation.ed.gov/what-we-do/innovation/education-innovation-and-research-eir/
*Sponsor:* U.S. Department of Education
400 Maryland Avenue, SW
Washington, D.C. 20202-5960

**U.S. Department of Education Jacob K. Javits Gifted and Talented Students Education Grants** **2529**
The U.S. Department of Education Jacob K. Javits Gifted and Talented Students Education Grants program conducts research for the purpose of carrying out activities described in Sec. 5464(b) of the statute including research on methods and techniques for identifying and teaching gifted and talented students and for using gifted and talented programs and methods to serve all students. It also conducts program evaluations and surveys. As part of its work, the center collects, analyzes, and develops information about gifted and talented education. Emphasis is given to the identification of and services for students traditionally not included in gifted and talented education, including individuals with limited English proficiency (LEP), individuals with disabilities, and individuals living under economically disadvantaged conditions. The estimated range of awards will be up to a maximum of $500,000. Depending upon Congressional funding, the annual application submission date has typically fallen during the third week of June.
*Requirements:* Institutions of Higher Education, State Education Agencies and a consortium of IHEs and SEAs may apply.
*Geographic Focus* All States, District of Columbia, Guam, Marshall Islands, Northern Mariana Islands, Puerto Rico, U.S. Virgin Islands, American Samoa
*Samples:* Arizona Department of Education, Phoenix, Arizona, $410,202 - to improve the equitable access, participation and performance of high ability and high potential students; Colorado Department of Education, Denver, Colorado, $283,780 - to use various resources and implement innovative, locally defined programs so strong sustainable gifted programs exist in every part of the state; Kentucky Department of Education, Frankfurt, Kentucky, $363,853 - to provide professional learning to school leaders on the Excellence Gap and increase state capacity to identify and serve more underrepresented students.
*Contact:* Sylvia Lyles, Director; (202) 260-1979 or (202) 453-7200; fax (202) 260-8969; anne.sweet@ed.gov or oese@ed.gov
Theda Zawaiza, (202) 205-3783; fax (202) 260-8969; theda.zawaiza@ed.gov or oese@ed.gov
*Internet:* https://www2.ed.gov/programs/javits/index.html
*Sponsor:* U.S. Department of Education - 4
400 Maryland Avenue, SW
Washington, D.C. 20202-6200

**UCR Graduate Student in the Humanities Fellowship** **2530**
The College of Humanities, Arts and Social Sciences and the Center for Ideas and Society are pleased to announce a new one-year fellowship for exceptional graduate students in the humanities who have advanced to candidacy and are in the final stages of completing their dissertations. This program is designed to support advanced doctoral students and to encourage the collaboration, interdisciplinary dialogue, and innovation that are fundamental to research in the arts and humanities in the University of California. Along with Faculty Research Fellows, Graduate Fellows will be part of a new UC Society of Fellows in the Humanities. During the fellowship year, UC Graduate Fellows will be affiliated with the Center for Ideas and Society, and are expected to give a lecture sometime during their fellowship year. They will also be asked to take part in the activities of the Center, and will be eligible for workspace within it. Graduate Fellows will be invited to participate in an annual meeting of UC Faculty Research Fellows and UC Graduate Fellows and to submit a brief essay for inclusion in an e-publication disseminated by the UC Humanities Research Institute. Information about the application is available at the website.
*Requirements:* Graduate students need not apply separately for the UC Graduate Fellowship in the Humanities. The Center will select Fellowship holders from among those who apply for the Graduate Division's Dissertation Year Program Awards (graduate.ucr.edu/fin_aid.html).
*Restrictions:* Applicants must be humanities graduate students at the UCR.
*Geographic Focus:* All States
*Contact:* Laura Lozon, Associate Director; (951) 827-1555; laura.lozon@ucr.edu
*Internet:* www.ideasandsociety.ucr.edu/fellowships_grants/UC_grad_fellows/index.html
*Sponsor:* University of California, Riverside
900 University Avenue
Riverside, CA 92521

**UCR Humanities Graduate Student Research Grants** **2531**
The Humanities Graduate Student Research Grants assist graduate students involved in original research or creative projects in the humanities, as broadly defined by the National Endowment for the Humanities. The bill creating the National Endowment for the Humanities indicates "...the term humanities includes, but is not limited to, the study of the following: language; linguistics; literature; history; jurisprudence; philosophy; archaeology; comparative religion; ethics; the history, criticism, theory and practice of the arts; those aspects of the social sciences which have humanistic content and employ humanistic methods; and the study and application of the humanities to the human environment with particular attention to the relevance of the humanities to the current condition of national life..." The Endowment supports projects involving additional humanities-based disciplines not explicitly named in the law, such as the history of science and cultural anthropology. These awards are primarily intended for research proposals that relate to a student's dissertation work. However, if funds permit, other proposals will be considered. The application and specific guidelines regarding research protocol and expenditures are available at the website.
*Requirements:* Applicants must be graduate students enrolled in the humanities at UCR.
*Restrictions:* Expenditures must be within the categories budgeted in the application. Written permission is required for budget adjustments prior to spending funds. Major purchases of non-expendable equipment will not be funded. Reusable supplies and equipment purchased from these funds remain the property of UCR. Research grants

are neither for stipends nor for the actual preparation of thesis copy such as typing, charts, or photographs.
*Geographic Focus:* All States
*Date(s) Application is Due:* Mar 15
*Amount of Grant:* Up to 1,500 USD
*Contact:* Laura Lozon, Associate Director; (951) 827-1555; laura.lozon@ucr.edu
*Internet:* www.ideasandsociety.ucr.edu/fellowships_grants/grad_research_grants/index.html
*Sponsor:* University of California, Riverside
900 University Avenue
Riverside, CA 92521

**UCR Resident Fellowships** **2532**
Each quarter four UCR faculty are awarded Resident Fellowships to support collaborative research. The purpose of this program is to promote multi-disciplinary and cross-disciplinary humanistic scholarship by assembling groups of scholars from different disciplines to examine areas of common interest. Normally fellowships are awarded to groups, not to individuals, and should include at least some members from the College of Humanities, Arts, and Social Sciences. Preference will ordinarily be given to those who have not been Resident Fellows. Home departments are provided with $5,000 funding for release time for the Faculty Fellows. In exchange, the fellow recipients are relieved of departmental and teaching obligations during their residence, which is normally during the quarter with the lightest teaching load. Fellows are provided with work space at the Center for Ideas and Society and meet in seminar at least once a week. The format of the meeting as well as the duration is at the discretion of the Fellows. Resident Fellows are expected to take part in the Center's activities, such as attending Center sponsored talks. Faculty who receive Fellowships become eligible again in four years. The application package is available at the website.
*Requirements:* Applicants must be faculty members from the College of Humanities, Arts, and Social Sciences.
*Restrictions:* Preference will be given to those who have not been Resident Fellows.
*Geographic Focus:* All States
*Date(s) Application is Due:* Dec 5
*Amount of Grant:* 5,000 USD
*Contact:* Laura Lozon, Assistant Director; (951) 827-1555; fax (951) 827-6377; laura.lozon@ucr.edu or ideassoc@citrus.ucr.edu
*Internet:* www.ideasandsociety.ucr.edu/fellowships_grants/resident_fellows/index.html
*Sponsor:* University of California, Riverside
900 University Avenue
Riverside, CA 92521

**UNC Carolina Postdoctoral Program for Faculty Diversity** **2533**
As part of a continuing commitment to building a culturally diverse intellectual community and advancing scholars from underrepresented groups in higher education, The University of North Carolina at Chapel Hill (UNC) Carolina Postdoctoral Program for Faculty Diversity makes available postdoctoral research appointments for a period of two years. The purpose of the program is to develop scholars from underrepresented groups for possible tenure track appointments at UNC and other research universities. Postdoctoral scholars will be engaged full-time in research and may teach only one course per fiscal year. The program supports 10 postdoctoral scholars, engaged full-time in research and writing for a two-year term starting July 1. Fellows are provided in the program with research funds, program laptop, as well as department mentor, office, computer, health benefits and $39,874 annual salary. Meetings and workshops are structured for the purposes of networking, social interaction and supporting professional development of diversity postdoctoral scholars within the program.
*Requirements:* Applicants who will have completed their doctoral degree no later than July 1 of the current academic year and no earlier than July 1 of four years prior to the current academic year are eligible to apply. The primary criterion for selection is evidence of scholarship potentially competitive for tenure track appointments at UNC and other research universities. An important secondary criterion is the support of prospective departments. Preference will be given to U.S. citizens and permanent residents. UNC strongly encourages applications from African American, Native American and Hispanic scholars.
*Geographic Focus:* All States
*Date(s) Application is Due:* Jan 4
*Amount of Grant:* 39,874 USD
*Contact:* Judson H. Fraley; (919) 962-9982; jhfraley@email.unc.edu
*Internet:* research.unc.edu/carolina-postdocs/
*Sponsor:* University of North Carolina at Chapel Hill
301 Bynum Hall, CB #4100
Chapel Hill, NC 27599-4100

**UNC School of Information and Library Science Doctoral Fellowships** **2534**
The School of Information and Library Science (SILS) at the University of North Carolina seeks applicants interested in earning a doctoral degree in digital curation. The fellowship offers the following: a 20 hour a week position as a Research Fellow in digital curation; an annual two year stipend; health coverage with in-state tuition; and extensive opportunities to meet leaders in the digital curation research and practice areas with workshops and symposia.
*Requirements:* Applicants should follow the regular application procedures found on the SILS Ph.D. admissions page. Students are encouraged to apply early, as this ensures consideration for the greatest amount of university funding. In addition to the required written statement of the intended research focus, applicants will write a separate essay elaborating on their educational goals and how they are related to the goals of DigCCurr II. See the DigCCurr II website and promotional flyer for more details, including current deadlines and application.
*Restrictions:* Applicants must be U.S. citizens.
*Geographic Focus:* All States
*Amount of Grant:* 38,000 USD
*Contact:* Dr. Helen Tibbo, Fellowship Contact; (919) 962-8063; tibbo@email.unc.edu
Dr. Cal Lee, Fellowship Contact; (919) 962-7024; callee@ils.unc.edu
*Internet:* ils.unc.edu/digccurr/aboutII.html
*Sponsor:* University of North Carolina-Chapel Hill
Campus Box 3360, Manning Hall
Chapel Hill, NC 27599-3360

**UNESCO Co-Sponsored Fellowships** **2535**
UNESCO has adopted a proactive policy towards the encouragement of partnerships with governments, institutions of higher education and foundations in order to increase training opportunities for other Member States. Under this policy, Member States, foundations, institutions and private donors may offer fellowships (contributions-in-kind), or may finance fellowships (cash contributions, or either through cost-sharing arrangements and through total funding) in fields within UNESCO's competence. The terms for each program under the Co-sponsored Fellowships Scheme are negotiated and agreed upon with each donor. UNESCO cooperates with the donor country or institution in administering the training scheme jointly. The Organization announces the scheme to the targeted Member States, solicits and collects applications, and prepares a short-list of candidates, which is then submitted to the donor to make the final selection. The Fellowships Section then grants the fellowships to the selected beneficiaries. Deadlines and Fellowship amounts vary by country and particular Fellowship. Additional information is available at the website.
*Geographic Focus:* All Countries
*Contact:* Irina Bokova, Director-General; +33 (0) 1 45 68 10 00; fax +33 (0) 1 45 66 94 29
*Internet:* portal.unesco.org/en/ev.php-URL_ID=14133&URL_DO=DO_TOPIC&URL_SECTION=201.html
*Sponsor:* United Nations Educational, Scientific, and Cultural Organization
7 place de Fontenoy
Paris 07 SP, 75352 France

**UNESCO Fellowships** **2536**
The UNESCO Fellowships are designed to provide an opportunity for study and training to qualified persons who are seeking to pursue advance studies in the fields of education, natural sciences, social and human sciences, culture, communication, information and informatics. Deadlines and placement of Fellows vary by country and fields of discipline. The application and detailed guidelines are available at the website.
*Requirements:* All applications must be channeled through the National Commissions of the candidate's country and must be addressed to the Director-General of UNESCO. Applicants must have exceptional merit, and demonstrate that they possess outstanding potential to contribute to the social, cultural and/or economic development of their country to be considered for a UNESCO fellowships. Candidates are usually persons who have already begun their careers, are at an advanced stage of their university education, and are pursuing short term post graduate studies of six months or less. Priority is given to women, youth, the disadvantaged, underdeveloped countries, and member states of Africa. Candidates should be in good health and no more than 45 years of age.
*Restrictions:* The UNESCO Secretariat cannot accept requests directly from individuals. Undergraduate students who wish to pursue studies in their own country cannot be beneficiaries of UNESCO fellowships.
*Geographic Focus:* All States, All Countries
*Contact:* Irina Bokova, Director-General; +33 (0) 1 45 68 24 96; fax +33 (0) 1 45 68 55 70; fellowships@unesco.org
*Internet:* portal.unesco.org/en/ev.php-URL_ID=17616&URL_DO=DO_TOPIC&URL_SECTION=201.html
*Sponsor:* United Nations Educational, Scientific, and Cultural Organization
7 place de Fontenoy
Paris 07 SP, 75352 France

**Union Pacific Foundation Community and Civic Grants** **2537**
The Foundation has a strong interest in promoting organizational effectiveness among nonprofits. To that end, this Foundation dedicates the majority of their grants to help nonprofit organizations build their capacity, increase their impact, and operate more efficiently and effectively. Grants are made primarily to proposals in the areas of community and civic, and health and human services. The community and civic grants category focuses on assisting community-based organizations and related activities that improve and enrich the general quality of life in the community. This category includes organizations such as aquariums, botanical gardens, children's museums, history/science museums, public libraries, public television and radio, and zoos.
*Requirements:* The Foundation will accept only online applications; printed copies of the application are not available and will not be accepted. Grants are made to institutions located in communities served by Union Pacific Corporation and its operating company Union Pacific Railroad Company.
*Restrictions:* The Foundation will not consider a request from or for: individuals; organizations/projects/programs that do not fit within the Foundation's funding priorities; organizations without a Section 501(c)3 public charity determination letter from the Internal Revenue Service; organizations that channel grant funds to third parties; organizations whose dominant purpose is to influence legislation or participate/intervene in political campaigns on behalf of or against any candidate for public office; organization/projects/programs for which the Foundation is asked to serve as the sole funder; organizations that already have an active multi-year Union Pacific Foundation grant; religious organizations for non-secular programs (i.e. programs which promote religious doctrine); organizational

deficits; local affiliates of national health/disease-specific organizations; non-U.S.-based charities; organizations whose program activities are mainly international; elementary or secondary schools; athletic programs or events; donations of railroad equipment; conventions, conferences or seminars; fellowships or research; loans; labor organizations; or organizations whose programs have a national scope.
*Geographic Focus:* Arizona, Arkansas, California, Colorado, Idaho, Illinois, Iowa, Kansas, Louisiana, Minnesota, Missouri, Montana, Nebraska, Nevada, New Mexico, Oklahoma, Oregon, Tennessee, Texas, Utah, Washington, Wisconsin, Wyoming
*Date(s) Application is Due:* Aug 15
*Contact:* Darlynn Myers, Director; (402) 271-5600; fax (402) 501-2291; upf@up.com
*Internet:* www.up.com/found/grants.shtml
*Sponsor:* Union Pacific Foundation
1400 Douglas Street, Stop 1560
Omaha, NE 68179

**Union Pacific Foundation Health and Human Services Grants 2538**
The Foundation has a strong interest in promoting organizational effectiveness among nonprofits. To that end, this foundation dedicates the majority of their grants to help nonprofit organizations build their capacity, increase their impact, and operate more efficiently and effectively. Grants are made primarily to proposals in the areas of community and civic, and health and human services. The health and human services category assists organizations dedicated to improving the level of health care or providing human services in the community. Types of support include general operating support, continuing support, capital campaigns, building construction and renovation, curriculum development, equipment acquisition, program development, and matching funds.
*Requirements:* The Foundation will accept only online applications; printed copies of the application are not available and will not be accepted. Grants are made to institutions located in communities served by Union Pacific Corporation and its operating company Union Pacific Railroad Company.
*Restrictions:* The Foundation generally will not consider a request from or for: individuals; organizations/projects/programs that do not fit within the Foundation's funding priorities; organizations without a Section 501(c)3 public charity determination letter from the Internal Revenue Service; organizations that channel grant funds to third parties; organizations whose dominant purpose is to influence legislation or participate/intervene in political campaigns on behalf of or against any candidate for public office; organization/projects/programs for which the Foundation is asked to serve as the sole funder; organizations that already have an active multi-year Union Pacific Foundation grant; religious organizations for non-secular programs; organizational deficits; local affiliates of national health/disease-specific organizations; non U.S.-based charities; organizations whose program activities are mainly international; elementary or secondary schools; athletic programs or events; donations of railroad equipment; conventions, conferences or seminars; fellowships or research; loans; labor organizations; or organizations whose programs have a national scope.
*Geographic Focus:* Arizona, Arkansas, California, Colorado, Idaho, Iowa, Kansas, Louisiana, Minnesota, Missouri, Montana, Nebraska, Nevada, New Mexico, Oklahoma, Oregon, Tennessee, Texas, Utah, Washington, Wisconsin, Wyoming
*Date(s) Application is Due:* Aug 15
*Contact:* Darlynn Myers, Director; (402) 271-5600; fax (402) 501-2291; upf@up.com
*Internet:* www.up.com/found/grants.shtml
*Sponsor:* Union Pacific Foundation
1400 Douglas Street, Stop 1560
Omaha, NE 68179

**United Methodist Church Asian, Black, Hispanic and Native American History Research Grants 2539**
The purpose of the Asian, Black, Hispanic and Native American History Research award is to promote excellence in research and writing in the history of these four groups in the United Methodist Church or its antecedents. Grants totaling $3,000 will be available with each grant being at least $1,000 awarded at the discretion of the committee of judges. Selection is made by a committee consisting of three persons: one historian who is a member of The United Methodist Church, one member of the General Commission on Archives and History and one member chosen at large by the General Secretary of the General Commission. Submitted proposals should be postmarked no later than December 31. Grant recipients will be announced by May 1 of the following year.
*Requirements:* The candidate must submit an application in English and include biographical information; a description of the project, including its significance, what resources will be used, and how it will be made available to the public; the expected date of completion; and an itemized budget. The candidate must submit three letters of recommendation in English from persons who can provide evidence of his/her capability. A reference copy of the final product will be donated to the General Commission on Archives and History of the United Methodist Church. Submit materials to the office.
*Geographic Focus:* All States
*Date(s) Application is Due:* Dec 31
*Amount of Grant:* 1,000 - 3,000 USD
*Contact:* Robert J. Williams, General Secretary; (973) 408-3189; fax (973) 408-3909; rwilliams@gcah.org or research@gcah.org
*Internet:* www.gcah.org/AwardsGrants/RacialEthnic.htm
*Sponsor:* United Methodist Church
36 Madison Avenue, P.O. Box 127
Madison, NJ 07940

**United Methodist Church Center for the Evangelical UBH Audrie E. Reber Memorial Award 2540**
The Center for the Evangelical United Brethren Heritage is offering an annual award to an undergraduate, seminary, or graduate student enrolled in an institution of higher education who submits the best paper on some aspect of Evangelical United Brethren Church history, theology or church life. The Evangelical United Brethren heritage includes the history of its predecessor bodies, the Evangelical Church and the Church of the United Brethren in Christ, as well as the wider Pietist heritage from which these bodies derive. The award is in memory of Audrie E. Reber, former missionary, author and active church-woman who was a Founding Member of the Center and a Member of its Advisory Board. The winning paper may be published in Methodist History, and a cash award of $700 will be given.
*Requirements:* Three clean copies of each paper must be submitted accompanied by a letter of nomination from the instructor under whose supervision it was written. The paper must not be longer than 20 double-spaced typewritten or computer processed pages. It must be properly footnoted and bear evidence of thorough and reliable handling of sources using the annotation standards accepted at the school. A bibliography also must be included. By April 1st each year, all competing papers must be in the hands of the Director of The Center for the Evangelical United Brethren Heritage, United Theological Seminary, 4501 Denlinger Road, Dayton, OH 45426.
*Restrictions:* The decision of the judges is final. No award will be made if the judges determine that no paper submitted merits the award.
*Geographic Focus:* All States
*Date(s) Application is Due:* Apr 1
*Amount of Grant:* 700 USD
*Contact:* Robert J. Williams, General Secretary; (973) 408-3189; fax (973) 408-3909; rwilliams@gcah.org or research@gcah.org
*Internet:* www.gcah.org/site/c.ghKJI0PHIoE/b.4619069/k.552C/EUB_Audrie_E_Reber_Award.htm
*Sponsor:* United Methodist Church
36 Madison Avenue, P.O. Box 127
Madison, NJ 07940

**United Methodist Church John Harrison Ness Memorial Award 2541**
The General Commission on Archives and History of The United Methodist Church offers an annual award of $500 (first prize) and $300 (second prize) to the students enrolled in the M. Div. program (or its equivalent) in United Methodist or other seminaries accredited by the Association of Theological Schools who submit the best papers on some aspect of United Methodist history broadly conceived. Guidelines are available online.
*Requirements:* The paper must have been judged by its author's professor as one of quality. Once the paper has passed his or her critical judgment, the professor will submit it to the General Commission on Archives and History. The paper must be properly footnoted and bear evidence of thorough and reliable handling of sources using the annotation standards accepted at the seminary. A bibliography also must be included. The paper should normally range in length from 4,000 to 8,000 words. Three copies of the paper are to be submitted for judging. The winning paper(s) may be published in 'Methodist History', at the discretion of the editor.
*Restrictions:* The decision of the judges will be final. No prize will be awarded if the judges decide that no paper submitted has the required quality of scholarship for an award.
*Geographic Focus:* All States
*Date(s) Application is Due:* Feb 1
*Amount of Grant:* 300 - 500 USD
*Contact:* Robert J. Williams, General Secretary; (973) 408-3189; fax (973) 408-3909; rwilliams@gcah.org or research@gcah.org
*Internet:* www.gcah.org/site/c.ghKJI0PHIoE/b.3526379/k.8DA6/John_Ness_Award.htm
*Sponsor:* United Methodist Church
36 Madison Avenue, P.O. Box 127
Madison, NJ 07940

**United Methodist Church Josephine Forman Scholarship 2542**
The scholarship is named for Josephine Forman, archivist for eighteen years of the Southwest Texas Conference of The United Methodist Church and author of We Finish to Begin: A History of Travis Park United Methodist Church, 1846–1991. As Conference Archivist, Ms. Forman collecteded records from closed churches along with administrative records for the Conference. For many years she was also the Church Historian at Travis Park United Methodist Church in San Antonio. The purpose of the Josephine Forman Scholarship is to provide financial support to minority students pursuing graduate education in archival science, to encourage students to pursue a career as an archivist, and to promote the diversification of the American archives profession. The scholarship is given to applicants who demonstrate excellent potential for scholastic and personal achievement and who manifest a commitment both to the archives profession and to advancing diversity concerns within it. The selection process is managed by the General Commission on Archives and History in partnership with the Society of American Archivists.
*Requirements:* The applicant must be: a citizen or permanent resident of the United States; of American Indian/Alaska Native, Asian, Black/African American, Hispanic/Latino, or Native Hawaiian/Other Pacific Islander descent; currently enrolled in a graduate program or a multi-course program in archival administration, or have applied to such a program for the next academic year; and enrolled in a graduate program and begin school no later than September 1 or the fall semester/quarter immediately following the award. The applicant shall have completed no more than half of the credit requirements toward her/his graduate degree at the time of the award
*Geographic Focus:* All States
*Amount of Grant:* 10,000 USD
*Contact:* Robert J. Williams; (973) 408-3189; fax (973) 408-3909; rwilliams@gcah.org
Selection Cpmmittee; (973) 408-3189; fax (973) 408-3909

*Internet:* www.gcah.org/site/c.ghKJI0PHIoE/b.6364989/k.C50B/Josephine_Forman_Scholarship.htm
*Sponsor:* United Methodist Church
36 Madison Avenue, P.O. Box 127
Madison, NJ 07940

**United Methodist Church Women in History Research Grant** 2543
The purpose of the Women in United Methodist History Research Grant is to provides seed money (travel, secretarial services, etc.) for research projects relating specifically to the history of women in the United Methodist Church or its antecedents. Grants totaling $2,500 will be available with each grant being at least $1,000 awarded at the discretion of the committee of judges. A copy of the final product will be deposited with the General Commission on Archives and History of The United Methodist Church.
*Requirements:* Selection is made by a committee consisting of three persons who are historians of women in United Methodism. Proposals on women of color and on history at the grass roots level are especially encouraged. Application guidelines are available online. The candidate must submit an application which includes the following: vita or resume; description of the project, including the significance of the research; description of the final product - applicants are encouraged to consider a variety of formats (written, audiovisual, oral history, bibliographies, archival guides, etc.); timetable for the project, including the expected date of completion; budget; indication of how the research results will be disseminated. The candidate must submit three letters of recommendation from persons who can provide evidence of his/her scholarly capability.
*Restrictions:* The grant money is not to be used for equipment, publications costs, or researcher's salary.
*Geographic Focus:* All States
*Date(s) Application is Due:* Dec 31
*Amount of Grant:* 1,000 - 2,500 USD
*Contact:* Robert J. Williams, General Secretary; (973) 408-3189; fax (973) 408-3909; rwilliams@gcah.org or research@gcah.org
*Internet:* www.gcah.org/site/c.ghKJI0PHIoE/b.3526385/k.D947/Women_in_UM_History_Research.htm
*Sponsor:* United Methodist Church
36 Madison Avenue, P.O. Box 127
Madison, NJ 07940

**United Methodist Church Women in History Writing Award** 2544
One Women in History Writing Award is given each year for an original manuscript on the history of women in the United Methodist Church or its antecedents. Consideration will be given to completed, original manuscripts of article length. The winning manuscript is considered for publication in Methodist History. A copy of the winning entry will be deposited with the General Commission on Archives and History of the United Methodist Church. Guidelines are available online.
*Requirements:* Manuscripts should not be longer than twenty double-spaced, typewritten pages. Footnotes and bibliography should follow the Kate Turabian Manual of Style. Selection is made by a committee consisting of three persons who are historians of women in United Methodism.
*Geographic Focus:* All States
*Date(s) Application is Due:* Dec 31
*Amount of Grant:* 500 USD
*Contact:* Robert J. Williams, General Secretary; (973) 408-3189; fax (973) 408-3909; rwilliams@gcah.org or research@gcah.org
*Internet:* www.gcah.org/site/c.ghKJI0PHIoE/b.3526381/k.7DAA/Women_in_UM_History_Award.htm
*Sponsor:* United Methodist Church
36 Madison Avenue, P.O. Box 127
Madison, NJ 07940

**United Methodist Church World is My Parish United Methodist History Research Grant** 2545
The World is My Parish United Methodist History Research Grant was created to enable the collection of written and oral history and encourage research regarding the development of The United Methodist Church and its antecedents in any of the conferences not in the United States. Grants totaling $3,000 will be available with each grant being at least $1,000 awarded at the discretion of the committee of judges. A reference copy of the final product will be donated to the General Commission on Archives and History of the United Methodist Church. The Commission will send copies to other United Methodist general agencies as appropriate.
*Requirements:* The applicant must submit an application which includes the following: (if at all possible the application should be in English but the applicant's language will be accepted) Biographical information; description of the project, including its significance, what resources will be used, and how it will be made available to the public; expected date of completion; and, an itemized budget. The applicant must submit three letters of recommendation from persons who can provide evidence of his/her capability.
*Geographic Focus:* All States
*Date(s) Application is Due:* Dec 31
*Amount of Grant:* Up to 4,000 USD
*Contact:* Robert J. Williams, General Secretary; (973) 408-3189; fax (973) 408-3909; rwilliams@gcah.org or research@gcah.org
*Internet:* www.gcah.org/site/c.ghKJI0PHIoE/b.4619067/k.B158/World_is_My_Parish_Research.htm
*Sponsor:* United Methodist Church
36 Madison Avenue, P.O. Box 127
Madison, NJ 07940

**United Methodist Committee on Relief Global Health Grants** 2546
UMCOR Global Health programs work internationally with more than 300 United Methodist hospitals and clinics, using education as well as preventative and curative measures to confront major health issues such as malnutrition, maternal and child mortality, HIV and AIDS, and malaria. Our programs also seek to increase access to clean water and better sanitation. In the United States, UMCOR Global Health programs connect people to resources and support networks and advocate for people with disabilities. UMCOR's holistic approach to health empowers people to take charge of their own lives. This strategy emphasizes education and the development of local resources so that improvements in public health are sustained over time. UMCOR's programs reach entire communities, especially the most vulnerable, and value local cultural practices as sources of preventative and curative health care. The goals are to: end poverty and hunger; attain universal education; achieve gender equality; promote child health; promote maternal health; combat HIV/AIDS and other disease of poverty; attain environmental sustainability; and build a global partnership.
*Restrictions:* This program is only for United Methodist and Methodist-related projects outside the U.S.
*Geographic Focus:* All States, All Countries
*Contact:* Shannon Trilli, Director of Global Health Initiatives; (212) 870-3870 or (212) 870-3951; strilli@umcor.org or umcor@umcor.org
*Internet:* new.gbgm-umc.org/umcor/www.umcor.org/UMCOR/Programs/Global-Health/Global-Health-and-Developmentwork/health/malaria/
*Sponsor:* United Methodist Committee on Relief
475 Riverside Drive, Room 1520
New York, NY 10115

**United Methodist Communications Stoody-West Fellowship** 2547
United Methodist Communications gives two $6,000 Stoody-West Fellowships per year; one of these goes to a United Methodist student. The fellowships are given to students who intend to pursue careers in religious journalism in their post-graduate studies at an accredited U.S. college or university. The purpose of the fellowship is to enhance the recipient's professional competence and thereby help to perpetuate the standards exemplified by Ralph Stoody and Arthur West. Religious journalism is broadly interpreted to include audiovisual, electronic, and print journalism. Half of the grant is paid in August or September after the recipient is enrolled in the graduate program; the second half is paid in December. Applications are available on the Web site.
*Requirements:* Applicant should submit three letters of recommendation, an official academic transcript mailed directly by the school, a statement about interest in religious journalism and future plans, four examples of journalistic work, a list of published articles, and a recent photograph.
*Restrictions:* Funds are not paid for summer sessions.
*Geographic Focus:* All States
*Date(s) Application is Due:* Mar 15
*Amount of Grant:* 6,000 USD
*Contact:* Scholarship Administrator; (615) 742-5769 or (615) 742-5400; fax (615) 742-5485; scholarships@umcom.org or foundation@umcom.org
*Internet:* archives.umc.org/interior.asp?ptid=1&mid=2664
*Sponsor:* Foundation for United Methodist Communications
P.O. Box 320
Nashville, TN 37202-0320

**United States Institute of Peace - Jennings Randolph Peace Dissertation Scholarships** 2548
The Jennings Randolph Program for International Peace awards nonresidential Peace Dissertation Scholarships to students at U.S. universities who are writing doctoral dissertations on topics related to peace, conflict and international security. Each year the program awards approximately ten scholarships. Funding lasts for ten months starting in September. Doctoral students from any county may apply, and projects in all disciplines are welcome. Scholars receive $20,000 for 10 months. Candidates may apply through the online process system at the USIP website.
*Requirements:* Citizens of any country may apply. Applicants must be enrolled in recognized doctoral programs in accredited universities in the United States. Successful candidates must have completed all course work and examinations towards their doctoral degrees by the time their scholarship funding begin. Proposals should be consistent with the Institute's mandate and present a research agenda with clear relevance to policy issues. Historical topics are appropriate if they promise to shed light on contemporary issues. Area studies projects and single case studies will be competitive if they focus on conflict and its resolution, apply to other regions and cases, or both.
*Restrictions:* Peace Scholar awards may not be made for projects that constitute policymaking for a government agency or private organization, focus to any substantial degree on conflicts within U.S. domestic society, or adopt a partisan, advocacy or activist stance. Peace Scholar awards may not be deferred. They generally may not be combined with any other major award or fellowship except in special circumstances and with the written approval of the Institute.
*Geographic Focus:* All States, All Countries
*Date(s) Application is Due:* Dec 14
*Amount of Grant:* 20,000 USD
*Contact:* Steven Riskin, Senior Program Officer, Fellowships; (202) 429-3869 or (202) 457-1700; fax (202) 429-6063; interviews@usip.org
*Internet:* www.usip.org/grants-fellowships/jennings-randolph-peace-scholarship-dissertation-program
*Sponsor:* United States Institute of Peace
2301 Constitution Avenue, NW
Washington, D.C. 20037

**United States Institute of Peace - Jennings Randolph Senior Fellowships** 2549

The Jennings Randolph Senior Fellowships provide scholars, policy analysts, policy makers, journalists, and other experts with opportunities to spend time in residence at the Institute, reflecting and writing on pressing international peace and security challenges. Applications are invited from all disciplines and professions. Senior Fellowships usually last for ten months, starting in October, but shorter-term fellowships are also available. Fellowships are open to citizens of any country. The Institute awards between 10 and 12 fellowships per year. The program attempts to match the recipient's earned income during the year preceding the fellowships, up to a maximum of $100,000 for 10 months. The Institute will provide coverage of 80% of health premiums for the Fellow and his/her eligible dependents, with a cap of $500 per month. The Institute will also cover travel to and from Washington, D.C., for Fellows and their dependents. Each Fellow is provided with a part-time research assistant during his/her fellowship. The Institute does not provide housing in Washington D.C., but it provides information on housing, schools and daycare.
*Requirements:* Citizens of any country may apply. Non-U.S. citizens without permanent resident status must obtain a J-1 exchange visitor visa to participate in the Fellowship Program. This status requires recipients to reside in their home country for two years following the fellowship before applying for the H or L visa, or for permanent residency in the United States. There is no specific educational degree requirement for candidates. Fellows come from a variety of professional backgrounds and from early, middle and late stages of their careers. Priority is given to proposals deemed likely to make timely and significant contributions to the understanding and resolution of ongoing and emerging conflicts and other challenges to international peace and security. Historical topics are appropriate if they promise to shed light on contemporary issues. Area studies projects and single-case studies will be competitive if they focus on conflict and its resolution, apply to other regions and cases, or both. Fellows are expected to be at the Institute and participate in its daily life. Fellows are expected to devote full attention to their Fellowship work in order to complete their projects within the period of residency. The Institute requires first right of review for manuscripts produced as a result of Fellowship support.
*Restrictions:* Senior Fellow awards may not be granted for projects that constitute policymaking for a government agency or private organization, focus to any substantial degree on conflicts within U.S. domestic society, or adopt a partisan, advocacy, or activist stance. Joint applications (two or more applicants for a single project) will not be accepted. An Institute Fellowship may not be deferred or combined with any other major award.
*Geographic Focus:* All States, All Countries
*Date(s) Application is Due:* Sep 10
*Amount of Grant:* Up to 100,000 USD
*Contact:* Elizabeth Cole, Senior Program Officer; (202) 429-3869 or (202) 457-1700; fax (202) 429-6063; interviews@usip.org
*Internet:* www.usip.org/grants-fellowships/jennings-randolph-senior-fellowship-program
*Sponsor:* United States Institute of Peace
2301 Constitution Avenue, NW
Washington, D.C. 20037

**United States Institute of Peace - Transatlantic Post-Doc Fellowship for International Relations and Security** 2550

The Transatlantic Post-Doc Fellowship for International Relations and Security (TAPIR) is open to candidates who have recently received their doctorate in social and political sciences or economics, and whose research focuses on topics of international relations and/or international peace and security issues. Up to five Fellowships are awarded each year. Fellowships are granted for a duration of 24 months to prepare Fellows for a career in policy-oriented and international research at renowned think tanks and political consulting research institutes. Fellows spend three eight-month stays at institutions participating in the program - at least one on the Eastern and one on the Western side of the Atlantic. During their stay, Fellows are supervised by a specialist at the Institute, while being integrated into the relevant research units or projects. Fellows will have the opportunity to work on issues of major and immediate political significance, and to publish their work in journals or paper series that will be read by decision-makers at national and international levels. The Fellowship aims to involve participants actively in the work and research of leading international think tanks. It is not intended to primarily fund the completion of major individual publications or of academic peer-reviewed journal articles. The Fellowship consists of a monthly stipend; a monthly allowance for health insurance; and a one-time travel allowance for travel to and from the host institutions. The TAPIR Fellowship is administered by the Stiftung Wissenschaft und Politik in Germany. A list of participating institutes is posted on their website.
*Requirements:* To be eligible for the Transatlantic Post-Doc Fellowship, candidates must successfully defend their Ph.D. by the beginning of their Fellowship. Candidates must have a strong interest in working in a policy-oriented international environment and must demonstrate their ability to do so by previous international academic experience. Applicants must either be citizens of the U.S., Canada or Europe (in a broader geographical definition) OR must have resided in the U.S., Canada or Europe at least for two-and-a-half years in the five years before the application deadline.
*Restrictions:* Fees for residence permits are not included.
*Geographic Focus:* All States, Albania, Andorra, Armenia, Austria, Azerbaijan, Belarus, Belgium, Bosnia & Herzegovina, Bulgaria, Canada, Croatia, Cyprus, Czech Republic, Denmark, Estonia, Finland, France, Germany, Greece, Hungary, Iceland, Ireland, Italy, Kazakhstan, Latvia, Liechtenstein, Lithuania, Luxembourg, Macedonia, Malta, Moldova, Monaco, Montenegro, Netherlands, Norway, Poland, Portugal, Romania, Russia, San Marino, Serbia, Slovakia, Spain, Sweden, Switzerland, Turkey, Ukraine, United Kingdom, Vatican City
*Date(s) Application is Due:* Feb 1
*Amount of Grant:* Up to 51,500 EUR
*Contact:* Tobias von Lossow, TAPIR Coordinator; +49 (0)30 88007-106; fax +49 (0)30 88007-5106 / -100; tobias.vonlossow@swp-berlin.org or transatlanticfellowship@swp-berlin.org
*Internet:* www.usip.org/grants-fellowships
*Sponsor:* United States Institute of Peace
2301 Constitution Avenue, NW
Washington, D.C. 20037

**United States Institute of Peace Annual Grants** 2551

The Institute's Annual Grants support innovative peacebuilding projects involving research, the identification of promising models and effective practices, the development of practitioner resources and tools, the development and delivery of education, training and dialogue programs, and the production of films, radio programs, and other media. It funds projects focused on preventing, managing, and resolving violent conflict and promoting post-conflict peacebuilding in settings outside the borders of the United States. Awards support activities that apply across a broad range of relevant disciplines, skills, and approaches. The Institute welcomes proposals of an interdisciplinary or multidisciplinary nature. Topic areas of interest include, but are not limited to: conflict analysis and prevention; mediation and conflict resolution; postconflict peace and stability operations; religion and peacemaking; women and girls in conflict and peacebuilding; rule of law and transitional justice; economies and conflict; social, psychological, and physical impacts of war and conflict; and media and conflict. Candidates may access the online application process at the Institute website. Additional guidelines and a detailed list of previously funded projects are also available at the website.
*Requirements:* The Institute may provide grant support to domestic and international nonprofit organizations and individuals including the following: institutions of post-secondary, community, and secondary education; public and private education, training or research institutions, and libraries. Individuals whose proposals are funded will be required to identify a nonprofit organization to receive and manage the grant monies. When applicants are employed by an eligible institution, such as a college or university, the Institute requires that grants be made to the institution rather than to the individual.
*Restrictions:* Support for degree work is not eligible in this program. Only one application per competition may be submitted by the same project director. Unsuccessful applicants of previous grant program competitions may not submit the same application unless it has been substantially revised. The Institute will not accept applications that list Institute participants, consultants, or project personnel members of the Institute's Board of Directors or staff. In addition, any application that lists the Institute as a collaborator in the project will not be accepted. Individuals who are currently working as Institute contractors may not be eligible to apply for grant support.
*Geographic Focus:* All States, All Countries
*Date(s) Application is Due:* Oct 1
*Amount of Grant:* 50,000 - 120,000 USD
*Samples:* Asylum Access, San Francisco, California, $47,000 - to develop, test, refine, and disseminate a refugee rights toolkit to enable local advocates to set up refugee rights projects in their home country; Center for a New American Security, Washington, D.C., $112,000 - a study of the likely effects of a nuclear Iran and present specific policy options for the U.S. to mitigate these consequences and deter a nuclear-armed Iran should international prevention efforts fail; Foundation for Information Technology Education and Development, Quezon City, Philippines, $64,160 - to develop and produce audio materials for primary and secondary students in history, civics, and social studies classes, with accompanying teachers guides for presenting and integrating the audio materials into existing curriculum.
*Contact:* Steven Riskin; (202) 429-3869 or (202) 457-1700; fax (202) 429-6063
*Internet:* www.usip.org/grants-fellowships/annual-grant-competition
*Sponsor:* United States Institute of Peace
2301 Constitution Avenue, NW
Washington, D.C. 20037

**University of Chicago Career Advancement PRISM Research Grants** 2552

PRISM Research Grants offer undergraduate students the funding to develop independent academic research projects. Students must develop a research proposal, which stresses not only the personal interest of the student, but how the project is linked to their academic and career interests. The average grant amount is $1,500, but each proposal is considered individually for amounts up to $2,500. The annual deadline for applications is April 15, and applications should be submitted through UChicago Handshake. On average, eight to ten grants are awarded each year.
*Requirements:* Only currently enrolled first- through third-year undergraduate students (including International Students) with declared majors in the PRISM fields (Art History, Anthropology, Comparative Human Development, English Language and Literature, History Philosophy/Allied Fields, Political Science, Psychology, and Sociology) are eligible to apply.
*Restrictions:* The grants cannot be used to support work as a faculty research assistant or unpaid internships experiences. A student may only receive the grant once while at the University of Chicago. Applicants who have received a Metcalf Internship may not apply for a PRISM Research Grant for the same position or project.
*Geographic Focus:* Illinois
*Date(s) Application is Due:* Apr 15
*Amount of Grant:* Up to 2,500 USD
*Contact:* Jeannette Daly; (773) 702-7040; fax (773) 834-2277; jdaly@uchicago.edu
*Internet:* https://careeradvancement.uchicago.edu/jobs-internships-research/prism-grant-guidelines
*Sponsor:* University of Chicago
1212 E. 59th Street
Chicago, IL 60637

**University of Chicago Career Advancement Seidel Scholars PRISM Grants 2553**
The Seidel Scholars PRISM Grant gives three students the opportunity to spend 8-10 weeks during the summer exploring how they can connect their academic passions with a professional occupation. The three students selected as Seidel Scholars will each receive a $4,000 stipend for a career-related, individual project. This project should creatively put some aspect of your PRISM major (methodologies, research interests, etc.) into practice, but the grant does not fund internships or projects that are exclusively academic. Some previous examples of Seidel projects include writing novels, curating museum exhibits, filming documentaries, and researching family histories. The annual deadline for applications is April 15, and applications should be submitted through UChicago Handshake.
*Geographic Focus:* Illinois
*Date(s) Application is Due:* Apr 15
*Amount of Grant:* 4,000 USD
*Contact:* Jeannette Daly; (773) 702-7040; fax (773) 834-2277; jdaly@uchicago.edu
*Internet:* https://careeradvancement.uchicago.edu/jobs-internships-research/seidel-prism-grants
*Sponsor:* University of Chicago
1212 E. 59th Street
Chicago, IL 60637

**University of Chicago Chapin Hall Doris Duke Fellowships 2554**
Chapin Hall at the University of Chicago offers the Doris Duke Fellowships for the Promotion of Child Well-Being. These fellowships are designed to identify and develop a new generation of leaders interested in and capable of creating practice and policy initiatives that will enhance child development and improve the nation's ability to prevent all forms of child maltreatment. The fellows receive an annual stipend of $30,000 for up to two years to support the completion of their dissertation and related research at their academic institution. Up to 15 fellowships are awarded annually. Fellows are guided by an academic mentor whom they select; fellows also identify a policy or practice mentor to assist them in better understanding how to frame their research questions with an eye toward maximizing policy and practice relevance. Applicants may be enrolled in any discipline. The formal application period will run from August 1 through December 1.
*Requirements:* Applicants must be enrolled in an accredited U.S. doctoral program and have substantially completed the coursework required to be advanced to candidacy. They are expected to complete or make significant progress on their dissertation within the two-year fellowship period. Applicants must be U.S. citizens or permanent residents in the U.S.
*Geographic Focus:* All States
*Date(s) Application is Due:* Dec 1
*Amount of Grant:* 30,000 USD
*Contact:* Emily Siegel, Fellowship Administrator; (773) 256-5227; fax (773) 256-5351; ddfellowship@chapinhall.org or esiegel@chapinhall.org
*Internet:* www.chapinhall.org/fellowships/doris-duke-fellowships
*Sponsor:* University of Chicago
1313 East 60th Street
Chicago, IL 60637

**University of Chicago Chapin Hall Harold A. Richman Postdoctoral Fellowship 2555**
Named for Chapin Hall's founding director and former Dean at the School of Social Service Administration, the Harold A. Richman Postdoctoral Fellowship is an in-residence, one-year term appointment, renewable for up to one additional year. The position offers early-career scholars the opportunity to strengthen their research portfolio. Selected candidates will work with senior researchers at Chapin Hall and faculty at the School of Social Service Administration, and may make connections with other University of Chicago faculty in order to help them launch careers in applied social policy research. This program is aimed at strengthening the intellectual and research qualifications of outstanding scholars interested in applied research careers focusing on child, youth, and family well-being. The Fellow will receive $52,000 per calendar year and an opportunity to participate in the University of Chicago's benefit program.
*Requirements:* Researchers interested in improving programs and policies that impact children, youth, their families, and communities are invited to apply for the Richman Fellowship. The Fellowship is open to individuals who have received their doctorate within the last five years or expect to receive it by August 31 of the current academic year. The degree must be in a relevant discipline, such as public policy, social work, economics, education policy, human development, public health, statistics, epidemiology, etc. A minimum of three years of research experience (dissertation-related research is acceptable) and experience conducting empirical research in the context of child, youth, or family policy strongly preferred. The successful candidate will have demonstrated a potential for success in publishing and obtaining research grants.
*Geographic Focus:* All States
*Date(s) Application is Due:* Dec 15
*Amount of Grant:* 52,000 USD
*Contact:* Sylvie Bendier Decety, Fellowship Manager; (773) 256-5125 or (773) 256-5100; fax (773) 256-5351; sbendier@chapinhall.org or RichmanPostdoc@uchicago.edu
*Internet:* www.chapinhall.org/fellowships/harold-richman-fellowship
*Sponsor:* University of Chicago
1313 East 60th Street
Chicago, IL 60637

**University of Chicago Study Abroad African Studies Research Grants 2556**
African Studies Research Grants in the amount of $4,000 each will be awarded by CLAS to students in their third year who will spend a minimum of eight weeks conducting research in Africa. Applications need to indicate a related topic for the BA paper and specific research goals for the summer. Research grants are primarily for third-year students in the College, but exceptional second-year students and fourth-year students who plan to spend part or all of a fifth year in the College are also eligible to apply. All applications must be submitted online and supporting hard-copy documents submitted to the Study Abroad Office by the February 12 deadline. Those offered awards will be asked to either accept or decline their awards by April 10.
*Requirements:* Eligible students are those registered at the University of Chicago for the spring quarter before the summer they go abroad and will be registered the following fall quarter.
*Geographic Focus:* All States
*Date(s) Application is Due:* Feb 12
*Amount of Grant:* 4,000 USD
*Contact:* Elana Kranz, Associate Director of Study Abroad; (773) 834-5424; fax (773) 702-1755; ekranz@uchicago.edu or sitg@uchicago.edu
*Internet:* study-abroad.uchicago.edu/programs/african-studies-research-grant
*Sponsor:* University of Chicago
Harper Memorial 203, 1116 E 59th Street
Chicago, IL 60637

**University of Chicago Study Abroad Nassef Sawiris Research Grants 2557**
The Nassef Sawiris Research Grant program is a $4,000 award intended to underwrite the costs of conducting undergraduate summer research in Egypt. Research grants are primarily for third-year students in the College, but exceptional second-year students and fourth-year students who plan to spend part or all of a fifth year in the College are also eligible to apply. The annual deadline for application submission is February 12. Those offered awards will be asked to either accept or decline their awards by April 10.
*Requirements:* Eligible students are those registered at the University of Chicago for the spring quarter before the summer they go abroad and will be registered the following fall quarter. Students taking a leave of absence before or after the grant period should contact CLAS office to discuss whether their case merits an exception to the spring- or fall-quarter registration requirement.
*Geographic Focus:* All States
*Date(s) Application is Due:* Feb 12
*Amount of Grant:* 4,000 USD
*Contact:* Elana Kranz, Associate Director of Study Abroad; (773) 834-5424; fax (773) 702-1755; ekranz@uchicago.edu or sitg@uchicago.edu
*Internet:* study-abroad.uchicago.edu/programs/nassef-sawiris-research-grant-sawiris
*Sponsor:* University of Chicago
Harper Memorial 203, 1116 E 59th Street
Chicago, IL 60637

**University of Chicago Study Abroad Third Year International Travel Grants 2558**
The F. Champion Ward Third Year International Travel Grant program provides $3,000 to support students conducting summer research outside the U.S. In most cases these awards support research leading to a BA paper, however, other academic research projects may also be considered. Research grants are primarily for third-year students in the College, but exceptional second-year students and fourth-year students who plan to spend part or all of a fifth year in the College are also eligible to apply. The annual application deadline has been identified as February 12. Those offered awards will be asked to either accept or decline their awards by April 10.
*Requirements:* Eligible students are those registered at the University of Chicago for the spring quarter before the summer they go abroad and will be registered the following fall quarter. Students taking a leave of absence before or after the grant period should contact the CLAS office to discuss whether their case merits an exception to the spring- or fall-quarter registration requirement.
*Geographic Focus:* All States
*Date(s) Application is Due:* Feb 12
*Amount of Grant:* 3,000 USD
*Contact:* Elana Kranz, Associate Director of Study Abroad; (773) 834-5424; fax (773) 702-1755; ekranz@uchicago.edu or sitg@uchicago.edu
*Internet:* study-abroad.uchicago.edu/programs/third-year-international-travel-grant
*Sponsor:* University of Chicago
Harper Memorial 203, 1116 E 59th Street
Chicago, IL 60637

**University of Southern Mississippi H.A. & Margret Rey Translation Fellowships 2559**
This fellowship program, established by the Rey Foundation and the de Grummond Children's Literature Collection at the University of Southern Mississippi, awards grants to afford the hiring of translators (i.e. German, Portuguese, French, Russian) for the purpose of translating the H. A. and Margret Rey documents into English to further facilitate use in research. Grants totaling up to $1,664 may be made to applicants who demonstrate experience in translating.
*Requirements:* Recipients will be required to produce completed translations of papers from the H.A. and Margret Rey collection in the de Grummond Children's Literature Collection.
*Geographic Focus:* All States
*Date(s) Application is Due:* Oct 31
*Amount of Grant:* Up to 1,664 USD
*Contact:* Ellen Ruffin; (601) 266-4349; fax (601) 266-6269; ellen.ruffin@usm.edu
*Internet:* www.lib.usm.edu/legacy/degrum/public_html/html/research/re-fellowships.shtml
*Sponsor:* University of Southern Mississippi
118 College Drive, #5148
Hattiesburg, MS 39406-5148

**University of Southern Mississippi Keats/Domanska/de Grummond Children's Literature Research Fellowships** 2560

This program, established by the Ezra Jack Keats Foundation, the Janina Domanska Literary Estate, and the de Grummond Children's Literature Research Collection of the University of Southern Mississippi, awards grants to scholars engaged in projects based substantially on the holdings of the de Grummond collection. The purpose of the program is to encourage imaginative and productive research in children's literature. The de Grummond Collection focuses on American and British children's literature, historical as well as contemporary. Original manuscripts and illustrations are a special emphasis of the collection. Grants of up to $1,582 may be made to applicants of any discipline who engage in projects which require substantial use of the de Grummond Collection. Awards are based upon the applicant's scholarly qualifications, the merits of the proposal submitted, and the appropriateness of the proposal to the resources of the de Grummond Collection. Funding can be used for travel, lodging, and meals. Applications are available at the website.
*Requirements:* Applicants may be researchers in any discipline who engage in projects that require substantial use of the de Grummond collection. Preference is given to projects that have a good likelihood of being published.
*Geographic Focus:* All States
*Date(s) Application is Due:* Oct 31
*Amount of Grant:* Up to 1,582 USD
*Contact:* Ellen Ruffin; (601) 266-4349; fax (601) 266-6269; ellen.ruffin@usm.edu
*Internet:* www.lib.usm.edu/legacy/degrum/public_html/html/research/re-fellowships.shtml
*Sponsor:* University of Southern Mississippi
118 College Drive, #5148
Hattiesburg, MS 39406-5148

**USA Funds Access to Education Scholarships** 2561

USA Funds is a nonprofit organization that helps students from groups underrepresented in higher education pursue a college degree. USA awards scholarships to low income students, with special consideration for students who are members of an ethnic minority group or who have a documented physical disability. The scholarship recipients must either plan to enroll or be enrolled in full- or half-time undergraduate or full-time graduate- or professional-degree coursework at an accredited two- or four-year college, university or vocational/technical school. See website for current guidelines and application. The recipients were selected based on additional factors, including the student's past academic performance and future potential, leadership and participation in school and community activities, work experience, career and educational aspirations and goals, and any unusual personal or family circumstances.
*Requirements:* Eligibility includes high-school seniors, currently enrolled college students, or incoming college students who plan to enroll or are enrolled in full- or half-time undergraduate or full-time graduate coursework at an accredited two- or four-year college, university, or vocational/technical school; have an annual adjusted gross family income of $35,000 or less, as documented on a federal income-tax return on which the applicant is claimed as a federal tax exemption; and are U.S. citizens or eligible non-citizens (permanent U.S. residents with a Green Card, or refugees with a Departure Record (Form I-94) from the U.S. Immigration and Naturalization Service.
*Restrictions:* Applicants may not be in default on a previous federal education loan.
*Geographic Focus:* All States
*Amount of Grant:* 1,500 USD
*Contact:* Scholarship Management Services; (800) 537-4180; contact@usafunds.org
*Internet:* www.usafunds.org/about_usa_funds/index.html
*Sponsor:* USA Funds
P.O. Box 6028
Indianapolis, IN 46206-6028

**USAID Comprehensive District-Based Support for Better HIV/TB Patient Outcomes Grants** 2562

The United States Government, as represented by the United States Agency for International Development (USAID) Mission to Southern Africa, is seeking applications from organizations interested in implementing a five-year comprehensive district-based HIV-related services support program for better patient outcomes, as fully described in this Request for Applications (RFA). The purpose of this program is to strengthen SAG systems in order to improve patient outcomes and prevent HIV by supporting comprehensive clinic-based HIV-related services in select districts. All support will be in line with SAG national, provincial and district policies, standards, guidelines, and implementation plans as well as the Partnership Framework between the USG and SAG. The program will build and capitalize on the accomplishments and lessons learned since the national ARV rollout began in 2004. It will focus on institutionalization of routine and consistent use of systems designed to improve efficiencies in district and sub-district management that ultimately improve patient and public health outcomes. The award ceiling is $20,000,000, with an estimated four grants to be funded. The closing date will be determined at the time of requesting full applications.
*Requirements:* Any local South African non-governmental organizations (NGOs), private voluntary organizations (PVOs), Faith-based and community organizations, and for-profit companies willing to forego profit will be eligible for award under this RFA. In support of the Agency's interest in fostering a larger assistance base and expanding the number and sustainability of development partners, USAID encourages applications from potential new partners.
*Geographic Focus* All States, District of Columbia, Guam, Marshall Islands, Northern Mariana Islands, Puerto Rico, U.S. Virgin Islands, American Samoa, South Africa
*Amount of Grant:* Up to 20,000,000 USD
*Contact:* Beatrice Lumanade, Acquisitions and Assistance Specialist; +27 12 452 2377 or +27 12 452 2000; blumande@usaid.gov or applications4@usaid.gov
*Internet:* www.grants.gov/web/grants/view-opportunity.html?oppId=169154
*Sponsor:* U.S. Agency for International Development
Ronald Reagan Building, 1300 Pennsylvania Avenue, NW
Washington, D.C. 20523-1000

**USAID Development Innovation Accelerator Broad Agency Announcement for Global Forestry and Biodiversity Conservation Grants** 2563

This Broad Agency Announcement (BAA) seeks opportunities to co-create, co-design, co-invest, and collaborate in the research, development, piloting, testing, and scaling of innovative, practical and cost-effective interventions to address critical biodiversity conservation challenges. The United States Agency for International Development (USAID) invites organizations and companies to participate with USAID in response to a Critical Biodiversity Conservation Challenge Addendum issued under this BAA, to create more strategic, focused and results-oriented biodiversity programming and apply scientific and research-motivated approaches to support sustainable biodiversity conservation and development outcomes.
*Requirements:* Public, private, for-profit, and non-profit organizations, as well as institutions of higher education, public international organizations, non-governmental organizations, U.S. and non-U.S. government organizations, and international donor organizations are eligible under this BAA, unless otherwise stated in the individual BAA Addendum.
*Geographic Focus* All States, District of Columbia, Guam, Marshall Islands, Northern Mariana Islands, Puerto Rico, U.S. Virgin Islands, American Samoa
*Contact:* Agreement Specialist; DIAglobalscale@usaid.gov
*Internet:* www.grants.gov/web/grants/view-opportunity.html?oppId=271284
*Sponsor:* U.S. Agency for International Development
Ronald Reagan Building, 1300 Pennsylvania Avenue, NW
Washington, D.C. 20523-1000

**USAID Development Innovation Ventures Grants** 2564

As part of its commitment to increase investments and engagement in cost-efficient innovations, USAID launched Development Innovation Ventures (DIV) as a way of producing development outcomes more effectively and cost-efficiently while managing risk and obtaining leverage. Through DIV, USAID seeks to identify and rigorously test promising projects with the potential to significantly (rather than incrementally) improve development outcomes, and help replicate and scale projects that are proven successful. USAID/DIV expects its most successful of investments will have an accelerated growth path to reach tens of millions of beneficiaries worldwide within 10 years. DIV Grants will support projects at three stages: Stage 1, the development needed to support proof of concept and feasibility; Stage 2, implementing the project at large scale with rigorous impact testing; and Stage 3, transitioning innovations to widespread adoption throughout one country and/or additional adoption in other countries. For Stage 1 applicants, a Concept Paper and its attachments are the only application documents required. Stage 2 and 3 applicants may submit a Concept Paper and wait to be invited (or not) to submit a Full Application. Alternatively Stage 2 and 3 applicants may elect to forego the Concept Paper and submit a Full Application. Interested applicants must first submit an Letters of Interest using the LOI form posted. Only invited applicants will receive the Full Application form.
*Requirements:* USAID/DIV welcomes applications from many types of organizations, including: U.S. non-governmental organizations (NGOs); non-U.S. NGOs; faith-based organizations; U.S. private businesses; non-U.S private businesses; business and trade associations; international organizations; U.S. colleges and universities (public and private); non-U.S. colleges and universities (public and private); civic groups; regional organizations; and foundations.
*Restrictions:* All applicants must be legally recognized organizational entities under applicable law. Individuals and governments are not eligible to apply directly at this time. The participation of foreign government organizations under this APS is possible only through an approved subaward agreement with a prime recipient.
*Geographic Focus* All States, District of Columbia, Guam, Marshall Islands, Northern Mariana Islands, Puerto Rico, U.S. Virgin Islands, All Countries, American Samoa
*Date(s) Application is Due:* Apr 15
*Amount of Grant:* 100,000 - 15,000,000 USD
*Contact:* Tameka J. Laws, Contracting Officer; (202) 567-5066 or (202) 712-4810; fax (202) 216-3524; divapplication@usaid.gov or tlaws@usaid.gov
*Internet:* https://www.usaid.gov/div
*Sponsor:* U.S. Agency for International Development
Ronald Reagan Building, 1300 Pennsylvania Avenue, NW
Washington, D.C. 20523-1000

**USAID Family Planning and Reproductive Health Methods Grants** 2565

The United States Agency for International Development (USAID) is seeking concept papers from qualified U.S. nonprofit non-Governmental Organizations (NGOs) for a program titled Family Planning and Reproductive Health Methods to Address Unmet Need, for funding of Cooperative Agreements. The purpose of this APS is to support the research, development, and introduction of technologies and approaches that better meet the needs of women and girls as their sexual and reproductive health concerns change over time. The General Objectives of the APS are to: refine existing FP methods to address method-related reasons for non-use; respond to product-related issues about currently available FP methods that arise at purchase and/or from the field, and that may affect provider/user perceptions and/or the supply chain; develop new FP methods that address method-related reasons for non-use, and/or fill gaps in the existing method mix; conduct research to foster the introduction and uptake of new and/or underutilized woman-initiated methods, particularly non-hormonal barriers, contraceptive vaginal

rings, and fertility awareness methods based on knowledge and monitoring of the menstrual cycle; and develop multipurpose prevention technologies (MPTs) that address the simultaneous risks of unintended pregnancy, HIV, and other sexually transmitted infections (STIs) – particularly Herpes Simplex Virus (HSV) and Human Papillomavirus (HPV). The funding range of awards is $500,000 to $80,000,000.
*Requirements:* Eligible organizations include registered U.S. and non-U.S. private non-governmental organizations, non-profit organizations, and for-profit organizations willing to forego profit.
*Geographic Focus* All States, District of Columbia, Guam, Marshall Islands, Northern Mariana Islands, Puerto Rico, U.S. Virgin Islands, All Countries, American Samoa
*Date(s) Application is Due:* Jan 14
*Amount of Grant:* 500,000 - 80,000,000 USD
*Contact:* Marcus Moon; (202) 567-4406 or (202) 712-4810; fax (202) 216-3524; mmoon@usaid.gov
*Internet:* www.grants.gov/web/grants/view-opportunity.html?oppId=215873
*Sponsor:* U.S. Agency for International Development
Ronald Reagan Building, 1300 Pennsylvania Avenue, NW
Washington, D.C. 20523-1000

**USAID Fighting Ebola Grants** **2566**
This Broad Agency Announcement (BAA) seeks opportunities to co-create, co-design, co-invest, and collaborate in the development, testing, and scaling of practical and cost-effective innovations that can help healthcare workers on the front lines provide better care and stop the spread of Ebola. The United States Agency for International Development (USAID) invites organizations and companies to participate with USAID, in cooperation with its partners, in response to Fighting Ebola Challenge Addenda issued under this BAA, to provide innovations and technologies that further the U.S. Government's commitment to addressing the Ebola epidemic. No annual deadline for submission of applications has been identified. Multiple awards are anticipated.
*Requirements:* Public, private, for-profit, and non-profit organizations, as well as institutions of higher education, public international organizations, non-governmental organizations, U.S. and non-U.S. government organizations, and international donor organizations are eligible under this BAA.
*Geographic Focus* All States, District of Columbia, Guam, Marshall Islands, Northern Mariana Islands, Puerto Rico, U.S. Virgin Islands, All Countries, American Samoa
*Contact:* Lorraine Sherman, Program Specialist; (202) 712-4810; fax (202) 216-3524; lsherman@usaid.gov or submitebolaEOI@usaid.gov
*Internet:* www.grants.gov/web/grants/view-opportunity.html?oppId=268349
*Sponsor:* U.S. Agency for International Development
Ronald Reagan Building, 1300 Pennsylvania Avenue, NW
Washington, D.C. 20523-1000

**USAID Limodzi Program Grants** **2567**
The United States Agency for International Development in Malawi is issuing this Broad Agency Announcement (BAA) to seek participants to co-create, co-design, co-invest, and collaborate on research and development interventions to strengthen citizen engagement and foster agency for change by testing new approaches to Malawi's most critical development challenges. USAID/Malawi invites organizations and companies to submit an Expression of Interest. The intent of the BAA is to allow co-creation and co-design to the maximum extent to create high quality, effective partnerships with great efficiency in time and resources. USAID/Malawi will invite selected for-profit and non-profit, public and private organizations, as detailed below, to co-create research and development (R&D) solutions to the Problem and Challenge Statements stated in this BAA, including those organizations that have ideas, expertise, resources, and/or funding to add to potential solutions. The current closing date for application submission is May 25.
*Requirements:* The application is open to any type of agency, without restrictions.
*Geographic Focus* All States, District of Columbia, Guam, Marshall Islands, Northern Mariana Islands, Puerto Rico, U.S. Virgin Islands, All Countries, American Samoa
*Date(s) Application is Due:* May 25
*Amount of Grant:* 35,000,000 - 45,000,000 USD
*Contact:* Desiree Savoy, Agreement Officer; (202) 712-4656; dsavoy@usaid.gov
Paul Martin, Agreement Officer; +265 17 72 455; pmartin@usaid.gov
*Internet:* www.grants.gov/web/grants/view-opportunity.html?oppId=304133
*Sponsor:* U.S. Agency for International Development
Ronald Reagan Building, 1300 Pennsylvania Avenue, NW
Washington, D.C. 20523-1000

**USAID Microbicides Research, Development, and Introduction Grants** **2568**
Clinical trials of vaginal microbicide have been tested recently in South Africa to determine efficacy of reduction of HIV infection. Microbicide research, development, and introduction activities which could potentially be supported through this APS mechanism are expected to contribute substantially to the Principles of the Global Health Initiative (GHI), particularly in promoting research and innovation; implementing a woman- and girl-centered approach; strengthening and leveraging inputs from multilateral organizations, global health partnerships, and the private sector; and encouraging country ownership and leadership. Potential activities could also contribute significantly to strengthening the health system building blocks related to commodities and procurement, service delivery, and leadership and governance. It is expected that gender analyses and other gender considerations will continue to inform the design and implementation of activities implemented as part of the Agency's microbicide program. All awards will also be subject to USAID environmental requirements, including completion of an initial environmental examination. The award floor is $100,000, with a maximum grant award of $2,250,000.
*Requirements:* To qualify for funding, applicants must: be either U.S. or non-U.S. organizations registered and working in USAID priority countries; and have substantial experience (5 or more years) working directly with relevant HIV/AIDS-related activities in multiple countries in sub-Saharan Africa. Relevancy of experience will be determined by USAID.
*Restrictions:* Non-eligible entities include other U.S. Government agencies and departments. Other U.S. Government agencies and departments may not apply for USAID funding under this APS.
*Geographic Focus* All States, District of Columbia, Guam, Marshall Islands, Northern Mariana Islands, Puerto Rico, U.S. Virgin Islands, All Countries, American Samoa
*Amount of Grant:* 100,000 - 2,250,000 USD
*Contact:* Kathleen Bumpass, Contracts Specialist; (202) 567-5008; kbumpass@usaid.gov
*Internet:* www.grants.gov/web/grants/view-opportunity.html?oppId=187913
*Sponsor:* U.S. Agency for International Development
Ronald Reagan Building, 1300 Pennsylvania Avenue, NW
Washington, D.C. 20523-1000

**USAID Middle East Regional Cooperative (MERC) Project Grants** **2569**
USAID's MERC program funds peer-reviewed collaborative research projects between Israel and its Arab neighbors. Established as a result of the Camp David Peace Accords to promote Arab-Israeli cooperation, MERC has expanded beyond its original participants, Egypt and Israel, and also funded Arab-Israeli projects with researchers and institutions in Jordan, Morocco, West Bank/Gaza, Lebanon, and Tunisia. Other Arab countries are also eligible, including those in the Maghreb and Gulf regions. MERC projects are not restricted to a particular technical sector. The program seeks applied research and technology development proposals within the broad areas of the natural, physical, and social sciences and engineering. Investigators are also encouraged to integrate social science approaches, economics, and the use of information technology into the research where appropriate. As a pre-proposal, researchers should first submit four-to-six pages to describe the proposed work. Pre-proposals are judged by technical experts within USAID. Experts in the proposed topic will score the pre-proposal on Arab-Israeli cooperation, relevance to regional development, technical merit and innovation. The MERC Committee consisting of USAID technical staff and State Department representatives vote whether or not to invite the investigators to move to the next step by submitting a full proposal. MERC receives between 40-100 pre-proposals yearly. Usually 15-20 pre-proposals are invited to submit full proposals. Full proposals contain detailed information about the experimental plan, investigators and outreach of the project. USAID calls peer-review panels of scientists and engineers active in the fields of research proposed. Several topically related proposals are discussed at one meeting and between three and five outside reviewers read each proposal. After a discussion, the panel votes to recommend if the proposal should be funded or not. The reviewers also suggest changes to the proposal to help improve the work. There is no set minimum amount. Recently awarded grants ranged from approximately $100,000 to $900,000 with an average size of about $500,000. Larger projects approaching $1 million are not discouraged. The annual deadline for pre-proposal submission is December 15.
*Requirements:* Eligibility to apply is unrestricted.
*Restrictions:* MERC does not fund purely basic research, but rather, projects that will contribute to development in an estimated time frame of ten years or less.
*Geographic Focus:* Egypt, Israel, Jordan, Lebanon, Morocco, Tunisia
*Amount of Grant:* Up to 1,000,000 USD
*Contact:* Maria Longi; (202) 712-0084 or (202) 712-0300; merc.copies@gmail.com
*Internet:* https://www.usaid.gov/where-we-work/middle-east/merc-funding-process
*Sponsor:* U.S. Agency for International Development
Ronald Reagan Building, 1300 Pennsylvania Avenue, NW
Washington, D.C. 20523-1000

**USAID Palestinian Community Infrastructure Development Grants** **2570**
The Palestinian Community Infrastructure Development (PCID) Program aims to increase Palestinian access to water, sanitation and other small and medium scale community infrastructure in the West Bank. This focus will be complemented by a minimum stand-by capability to respond rapidly to needs in Gaza should political and security conditions require. Any interventions in Gaza will be identified and discussed by USAID and the implementing partner on a case by case basis after award. No interventions in Gaza are contemplated at this time. Accordingly, PCID will meet the following objectives: increase access to water and sanitation systems, reduce water losses and enhance water and sanitation system sustainability; increase the impact of USAID-funded programs in health, democracy and governance, education and the private sector by addressing underlying infrastructure needs which may be limiting the impact of those programs; and respond rapidly to urgent water, sanitation and other public infrastructure needs. The estimated ceiling for this RFA is $100 million.
*Requirements:* U.S.-based non-governmental organizations are eligible to submit applications. Applicants must have established financial management, monitoring and evaluation, internal control systems, and policies and procedures that comply with established U.S. Government standards, laws, and regulations.
*Restrictions:* Applications from non-U.S. organizations as well as Public International Organizations will not be considered.
*Geographic Focus:* All States
*Amount of Grant:* Up to 100,000,000 USD
*Contact:* Claudia Koziol, Agreements Officer; 972-3-519-8507 or 972-3-511-4868; fax ckoziol@usaid.gov; WBG-OCM-packages@usaid.gov
*Internet:* www.grants.gov/web/grants/view-opportunity.html?oppId=148373
*Sponsor:* U.S. Agency for International Development
Ronald Reagan Building, 1300 Pennsylvania Avenue, NW
Washington, D.C. 20523-1000

**USAID Policy, Advocacy, and Communication Enhanced for Population and Reproductive Health Grants** 2571

The United States Agency for International Development (USAID) seeks concept papers from qualified applicants. This APS publicizes the intention of the United States Government (USG) to fund up to three (3) awards through the USAID, Global Health Bureau to address the overarching APS Program Purpose of assuring family planning and population issues are included in policies and programs as key to sustainable and equitable economic growth and development. The award(s) under this APS will advance the Agency's goals of ending extreme poverty and ending preventable child and maternal deaths by ensuring that decision makers understand the health and economic benefits of voluntary family planning and in turn, implement policies and programs to help meet unmet demand for family planning. Awards will range up to a maximum of $55,000,000. The annual closing date for applications is April 15.

*Requirements:* To qualify for funding, organizations must be: a U.S. organization or non-U.S. organization registered and working in USAID-assisted countries; a U.S. or non-U.S. academic institution (university, professional school, or two- or four-year college, including community colleges); or a research institute, professional society or similar organization.

*Geographic Focus* All States, District of Columbia, Guam, Marshall Islands, Northern Mariana Islands, Puerto Rico, U.S. Virgin Islands, All Countries, American Samoa

*Date(s) Application is Due:* Mar 12

*Amount of Grant:* Up to 55,000,000 USD

*Contact:* Samantha Corey; (202) 567-4517; fax (202) 216-3524; scorey@usaid.gov

*Internet:* www.grants.gov/web/grants/view-opportunity.html?oppId=275168

*Sponsor:* U.S. Agency for International Development
Ronald Reagan Building, 1300 Pennsylvania Avenue, NW
Washington, D.C. 20523-1000

**USAID Research and Innovation for Health Supply Chain Systems and Commodity Security Grants** 2572

The purpose of this APS is to publicize the United States Government's (USG) plan to fund a limited number of awards through the U.S. Agency for International Development's (USAID) Washington Global Health Bureau (GH) to support research on a focused set of health supply chain systems and related commodity security issues in low and middle income countries. Funds may support: formative, analytical research on challenges to improving supply chain systems and the enabling environments for commodity security; and/or development and testing of interventions in the field based on formative research conducted under this APS or from other sources. The awards under this APS will contribute to advancing and supporting health programs worldwide, contributing to ending preventable child and maternal deaths, creating an AIDS-free generation, and protecting communities from infectious diseases. In addition, this APS will address several key focus areas of the Global Health Initiative (GHI), including building sustainability through health systems strengthening and accelerating results through research and innovation. The objectives of the APS are listed below: improve health supply chain systems through transformative changes that use industry best practice; improve the quantity, reliability, and efficiency of financing for health commodities and supply chain systems; and improve governance and accountability for commodity security.

*Requirements:* To qualify for funding, organizations must be: U.S. organizations or non-U.S. organizations registered and working in USAID-assisted countries; and academic organizations (universities, professional schools, or two- or four-year colleges, including community colleges), or research institutes, professional societies or similar organizations that are directly associated with research activities.

*Geographic Focus* All States, District of Columbia, Guam, Marshall Islands, Northern Mariana Islands, Puerto Rico, U.S. Virgin Islands, American Samoa

*Date(s) Application is Due:* May 29

*Amount of Grant:* 500,000 - 2,000,000 USD

*Contact:* Lesley Stewart; (202) 567-4762 or (202) 567-4664; lestewart@usaid.gov

*Internet:* www.federalgrants.com/Research-and-Innovation-for-Health-Supply-Chain-Systems-and-Commodity-Security-46262.html

*Sponsor:* U.S. Agency for International Development
Ronald Reagan Building, 1300 Pennsylvania Avenue, NW
Washington, D.C. 20523-1000

**USCD/NOAA Illinois-Indiana Sea Grant Core Proposals** 2573

The Illinois-Indiana Sea Grant (IISG) program funds two to five projects during each two-year cycle. Requests for proposals are usually issued in December of even years for funding beginning 14 months later, in February of the following even year. IISG begins the proposal selection process by requesting preproposals. Preproposals are screened by an internal committee appointed by the IISG director for their applicability to IISG topic areas, potential to impact management decisions, appearance of scientific rigor, and funding level. Preproposal selection is the only opportunity the IISG committee has to independently select proposal topics of most interest. An IISG internal committee meets to make final funding recommendations based on the expert panel ranking and the research funds available. Recommendations are submitted to the National Sea Grant Office for approval. Upon approval, researchers are asked to submit any necessary final paperwork. This information is included in an Omnibus Proposal submitted to the National Sea Grant Office in November of each year. Projects start March 1 of the following year.

*Geographic Focus:* Illinois, Indiana

*Contact:* Tomas Hook, Associate Director of Research; (765) 496-6799 or (217) 333-6444; fax (217) 333-8046; thook@purdue.edu or iisg@illinois.edu

*Internet:* www.iisgcp.org/funding/core.html

*Sponsor:* U.S. Department of Commerce
1401 Constitution Avenue, NW, Room 5128
Washington, D.C. 20230

**USDA 1890 Institution Teaching, Research, and Extension Capacity Building Grants** 2574

The 1890 Institution Teaching, Research, and Extension Capacity Building Grants (CBG) are intended to strengthen teaching, research and extension programs in the food and agricultural sciences by building the institutional capacities of the 1890 land-grant institutions, and Tuskegee University. The CBG program supports projects that strengthen teaching programs in the food and agricultural sciences in the need areas of curriculum design, materials development, and faculty development. CBG also supports projects that strengthen research and extension programs in need areas of studies and experimentation, and extension program development support systems. The CBG also support integrated project grants. Applications submitted to CBG must address at least one of the following NIFA strategic goals: sustainable bioenergy; food security; childhood obesity prevention; climate change; or food safety. The electronic applications, abstracts of funded projects, and specific guidelines are available at the website.

*Requirements:* Applicants must be 1890 land-grant institutions to be eligible for funding.

*Geographic Focus:* Alabama, Arkansas, Delaware, Georgia, Kentucky, Louisiana, Maryland, Mississippi, Nebraska, North Carolina, Oklahoma, South Carolina, Tennessee, Texas, Virginia, West Virginia

*Date(s) Application is Due:* Mar 27

*Amount of Grant:* 150,000 - 600,000 USD

*Contact:* Edwin Lewis; (202) 690-4565; fax (202) 720-2030; elewis@nifa.usda.gov

*Internet:* www.nifa.usda.gov/fo/1890capacity.cfm

*Sponsor:* U.S. Department of Agriculture
1400 Independence Avenue, SW, Stop 2201
Washington, D.C. 20250-2201

**USDA Agriculture and Food Research Initiative Competitive Grants - Food Security** 2575

The USDA Agricultural and Food Research Initiative Competitive Grants focuses on the societal challenge to keep American agriculture competitive while ending world hunger. In the Food Security Challenge Area, specific program areas are designed to achieve the long-term outcome of increasing food availability and decreasing the number of food insecure individuals. Project types supported include single-function extension projects, multi-function integrated research, education and/or extension projects, and Food and Agricultural Science Enhancement (FASE) Grants. Applicants should refer to the online request for application to find the specific contact person for their project. Information about the application, criteria, and particular guidelines for each grant area are located at the website. Funding is based on grant categories.

*Requirements:* All proposals submitted to this challenge area must be multidisciplinary in which two or more disciplines collaborate to address a common problem. These collaborations, where appropriate, may involve an integration of the biological, physical, chemical, or social sciences.

*Geographic Focus:* All States

*Date(s) Application is Due:* Mar 1

*Contact:* Margo Holland, Standard, Conference, and FASE Grants Contact; (202) 401-5044; mholland@nifa.usda.gov
Peter Johnson, Integrated Projects Contact; (202) 401-1896; pjohnson@nifa.usda.gov
Martin Draper, Extension Projects Contact; (202) 401-1990; mdraper@nifa.usda.gov

*Internet:* www.csrees.usda.gov/funding/rfas/afri.html

*Sponsor:* U.S. Department of Agriculture
1400 Independence Avenue, SW, Stop 2201
Washington, D.C. 20250-2201

**USDA Alaska Native-Serving and Native Hawaiian-Serving Institutions Education Competitive Grants** 2576

The purpose of the program is to promote and strengthen the ability of Alaska Native-Serving Institutions and Native Hawaiian-Serving Institutions to carry out education, applied research, and related community development programs. NIFA intends this program to address educational needs, as determined by each institution, within a broadly defined arena of food and agricultural sciences-related disciplines. Priority will be given to those projects that enhance educational equity for under represented students; strengthen institutional educational capacities; prepare students for careers related to the food, agricultural, and natural resource systems of the United States; and maximize the development and use of resources to improve food and agricultural sciences teaching programs. Samples of previous projects are available at the website. Applications must be filed electronically through Grants.gov.

*Requirements:* Private and state controlled institutions of higher education are eligible to apply. Individual public or private, nonprofit Alaska Native-serving and Native Hawaiian-serving institutions of higher education that meet the definitions of Alaska Native-Serving Institution or Native Hawaiian-Serving Institution established in Title III, Part A of the Higher Education Act of 1965, are also eligible to apply.

*Geographic Focus:* Alaska, Hawaii

*Date(s) Application is Due:* Mar 1

*Amount of Grant:* 150,000 - 1,500,000 USD

*Contact:* Saleia Afele-Faamuli, National Program Leader; (202) 720 - 0384; fax (202) 720 - 4924; sfaamuli@nifa.usda.gov

*Internet:* www.nifa.usda.gov/funding/rfas/alaska.html

*Sponsor:* U.S. Department of Agriculture
1400 Independence Avenue, SW, Stop 2201
Washington, D.C. 20250-2201

**USDA Animal Health and Disease Research Grants** 2577

The purpose of this funding is to increase animal health and disease research activities at accredited State veterinary schools or colleges or agricultural experiment stations that conduct animal health and disease research. To support animal health and disease research at eligible Schools and Colleges of Veterinary Medicine, and State Agricultural Experiment Stations whose purpose is to improve the health and productivity of food animals and horses through effective prevention, control or treatment of disease, reduction of losses from transportation and other hazards, and protect human health through control of animal diseases transmissible to people.

*Requirements:* Eligibility is restricted to schools and colleges of veterinary medicine and state agricultural experiment stations.
*Geographic Focus:* All States
*Amount of Grant:* Up to 100,000 USD
*Contact:* Peter Johnson; (202) 401-1896; fax (202) 205-3641; pjohnson@nifa.usda.gov
*Internet:* https://www.cfda.gov/?s=program&mode=form&tab=step1&id=1ddd2def6f16faee9d008a7ed30a8a7b
*Sponsor:* U.S. Department of Agriculture
1400 Independence Avenue, SW, Stop 2201
Washington, D.C. 20250-2201

**USDA Basic and Applied Research Grants** 2578

USDA Basic and Applied Research Grant program has as its primary mission to make agricultural research discoveries, evaluate alternative ways of attaining research goals, and provide scientific technical information. In conducted these project grants, research is conducted that is in cooperation with and is correlated with the Agricultural Research Service's in-house research programs and projects. Limited discretionary research funds are periodically made available. A peer review panel considers each proposal, evaluates the qualifications of applicants in line with research to be undertaken and determines priority for final negotiations of the grant. The grant range is $1,000 to $25,000, with an average award being $3,000. There are no identified deadlines.

*Requirements:* Nonprofit institutions of higher education or other nonprofit research organizations, whose primary purpose is conducting scientific research, are eligible to apply. Preapplication coordination is required. Environmental impact information is not required for this program.
*Geographic Focus:* All States
*Amount of Grant:* 1,000 - 25,000 USD
*Contact:* Kathleen S. Townson, Grants Management Specialist; (301) 504-1702; kathleen.townson@ars.usda.gov
*Internet:* www.ars.usda.gov/Research/Research.htm
*Sponsor:* U.S. Department of Agriculture
Jamie L. Whitten Building
Washington, D.C. 20250

**USDA Biotechnology Risk Assessment Grants** 2579

The purpose of the USDA Biotechnology Risk Assessment Grants (BRAG) program is to support the generation of new information that will assist Federal regulatory agencies in making science-based decisions about the effects of introducing into the environment genetically engineered organisms (GE), including plants, microorganisms (including fungi, bacteria, and viruses), arthropods, fish, birds, mammals and other animals excluding humans. The BRAG program accomplishes its purpose by providing Federal regulatory agencies with scientific information relevant to regulatory issues. Although investigators are not required to perform actual risk assessments in the research they propose, they should design studies that will provide information useful to regulators for making science-based decisions in their assessments of genetically-modified organisms. Organizations are required to submit the Grants.gov online application available on its website. Specific guidelines for the individual proposals and Grants.gov are located at the NIFA website and within the request for application.

*Requirements:* Eligible applicants include any public or private research or educational institution or organization.
*Restrictions:* Submission of a full proposal with a corresponding letter of intent will not be considered for review. Awards will not be made for food safety risk assessment or risk management, health risk assessment or risk management of humans or domestic food animals exposed to GE organisms, social or economic research, methods for seed storage, clinical trials, commercial product development, product marketing strategies, product stewardship, or other research deemed inappropriate to risk assessment or risk management relative to the environment.
*Geographic Focus:* All States
*Date(s) Application is Due:* Feb 1
*Amount of Grant:* 25,000 - 1,000,000 USD
*Contact:* Shing Kwok; (202) 401-6060; skwok@nifa.usda.gov
Mark Mirando; (202) 401-4336; mmirando@nifa.usda.gov
Jack Okamuro, Program Leader for Crop Production and Protection; (301) 504-5912 or (202) 285-9520 cell; jack.okamuro@ars.usda.gov
*Internet:* www.nifa.usda.gov/funding/rfas/biotech_risk.html
*Sponsor:* U.S. Department of Agriculture
1400 Independence Avenue, SW, Stop 2201
Washington, D.C. 20250-2201

**USDA Cochran Fellowships** 2580

The Cochran Fellowship Program (CFP) provides U.S.-based agricultural training opportunities for senior and mid-level specialists and administrators from public and private sectors who are concerned with agricultural trade, agribusiness development, management, policy, and marketing. The program offers short-term training opportunities, most ranging from 1-3 weeks depending on the objectives of the program. Participants meet with professionals in their fields, participate in field observations and industry visits, experience on-the-job training, attend university courses and seminars, or participate in a combination of these experiences. CFP works closely with U.S. agricultural trade and market development associations to identify potential areas of training for countries participating in the Program. Recruitment typically occurs between December and March each year. Most training activities are then conducted between March and October. However, most Eurasian recruitment takes place from June to August, and most Eurasian training activities are conducted January to June. Applicants should refer to the website for guidelines and contact information for their specific country.

*Requirements:* Program participation is open to the staff of agribusinesses, government departments, universities, and other agricultural organizations. In their own countries, applicants may be managers, technicians, scientists, professors, administrators, or policy makers. All participants must be in good health. Applicants are required to have a thorough physical exam prior to traveling to the U.S., as well as a medical letter stating he/she is fit to travel. Participants who are proficient in the oral and written usage of the English language are preferred. Applications are required to be completed in English (preferably typed). The officer conducts an initial screening of applications to determine whether the proposed study program is appropriate and is consistent with the country strategy. The application must include a detailed description of the specific training requested by the applicant.
*Restrictions:* The Cochran Program does not fund international travel to and from the participating country, except in the following countries: Armenia, Azerbaijan, Bulgaria, Georgia, Kazakhstan, Kyrgyzstan, Moldova, Tajikistan, Turkmenistan, and Uzbekistan.
*Geographic Focus:* All Countries
*Contact:* Scott Lewis, Chief, Cochran Fellowship Branch; (202) 720-9033 or (202) 690-1734; fax (202) 690-0349; CochranFellowships@fas.usda.gov or scott.lewis@fas.usda.gov
*Internet:* www.fas.usda.gov/icd/cochran/cochran.asp
*Sponsor:* U.S. Department of Agriculture
1400 Independence Avenue SW
Washington, D.C. 20250

**USDA Cooperative Extension System Education Grants** 2581

The program seeks to help people improve their lives and communities through an educational process that uses scientific knowledge focused on issues critical to the economic, agricultural, societal, health/safety, and environmental progress of all Americans. In addition, the program seeks to identify and solve farm, home, and community problems through the practical application of research findings of the USDA and land-grant colleges and universities. The cooperative extension system is a future-oriented, self-renewing, national educational network providing excellence in programs that focus on contemporary issues and needs of people.

*Requirements:* Grants are available to designated 1862, 1890, and 1994 land-grant institutions in the 50 states and Puerto Rico, Guam, the Virgin Islands, American Samoa, Micronesia, Northern Marianas, and the District of Columbia.
*Geographic Focus:* All States
*Amount of Grant:* 890,000 - 20,000,000 USD
*Contact:* Andrea L. Brandon; (202) 720-9181 or (202) 720-4423; abrandon@nifa.usda.gov
*Internet:* www.csrees.usda.gov/nea/education/education.cfm
*Sponsor:* U.S. Department of Agriculture
1400 Independence Avenue, SW, Stop 2201
Washington, D.C. 20250-2201

**USDA Decadal and Regional Climate Prediction Using Earth System Models Grants** 2582

The USDA Decadal and Regional Climate Prediction Using Earth System Models (EaSM) Grants seek to improve and expand current modeling capabilities in order to substantively contribute to the advancement of reliable regional and decadal climate predictions. This solicitation will capitalize on the synergy between development of climate models, their use in both the assessment and attribution of climate variability and impacts, and the development of approaches to effectively inform adaptation policy. This interdisciplinary scientific challenge calls for the development and application of next-generation Earth System Models that include coupled and interactive representations of such things as ocean and atmospheric currents, human activities, agricultural working lands and forests, urban environments, biogeochemistry, atmospheric chemistry, the water cycle and land ice. This solicitation seeks to attract scientists from the disciplines of geosciences, social sciences, agricultural and biological sciences, mathematics and statistics, physics, and chemistry. Successful proposals will develop intellectual excitement in the participating disciplinary communities and engage diverse interdisciplinary teams with sufficient breadth to achieve the scientific objectives. The project encourages proposals that have strong broader impacts, including public access to data and other research products of general interest, as well as educational, diversity, or societal impacts. The following are specific areas of interest to the funding agencies for EaSM: (i) research that has the potential to dramatically improve predictive capabilities; (ii) prediction and attribution studies; (iii) research that addresses critical issues linking relevant earth system processes over a variety of spatial and temporal scales; (iv) research that examines the relationships between climate variability and change to human and natural environments from the human perspective; (v) development and applications of metrics, methods, and tools for testing and evaluating climate and climate impact predictions and their uncertainty characterization. The electronic application and specific guidelines are available at the website. The submission of the application is through the National Science Foundation.

*Requirements:* Proposals may only be submitted by the following: universities and colleges, including community colleges; non-profit and non-academic organizations; for-profit organizations; and federal agencies and federally funded research and development centers.
*Geographic Focus:* All States
*Date(s) Application is Due:* May 11

*Amount of Grant:* 300,000 - 1,000,000 USD
*Contact:* Nancy Cavallaro, National Program Leader; (202) 401-5176 or (202) 401-6070; ncavallaro@nifa.usda.gov or easm2@nsf.gov
*Internet:* www.nifa.usda.gov/fo/earthsystemmodeling.cfm
*Sponsor:* U.S. Department of Agriculture
1400 Independence Avenue, SW, Stop 2201
Washington, D.C. 20250-2201

**USDA Distance Education Grants for Higher Education Institutions in Insular Areas** **2583**
The Distance Education Grants Program for Higher Education Institutions in Insular Areas (DEG) is a NIFA-administered competitive grants program focused on improving formal, postsecondary agricultural sciences education. The USDA program strives to strengthen resident instruction, curriculum, and teaching programs in the food and agricultural sciences in these insular areas through distance education technology. The purpose of the DEG program is to promote and strengthen the ability of insular area institutions to: (1) acquire the equipment, instrumentation, networking capability, hardware and software, digital network technology, and infrastructure necessary to teach technology in the classroom; (2) develop and provide educational services to prepare students or faculty seeking a degree or certificate; (3) provide teacher education, library and media specialist training, and preschool and teacher aid certification to individuals who seek to acquire or enhance technology skills; (4) implement a joint project to provide education regarding technology in the classroom; or (5) provide leadership development to administrators, board members, and faculty of eligible institutions with institutional responsibility for technology education. Funded projects improve the economic health and viability of rural communities through the development of degree programs emphasizing new and emerging employment opportunities in agriscience and agribusiness. Projects should also address the national challenge to increase the number and diversity of students entering other agriculture related science, technology, engineering, and mathematics (STEM) disciplines (i.e., having an agricultural sciences workforce representative of the nation's population). Funding varies according to project goals and objectives. The electronic application and specific guidelines are available at the website.
*Requirements:* Applications must state how the funded project will address the two distance education program goals: to increase the number and diversity of students who will pursue the complete a 2- or 4-year postsecondary degree in the food and agricultural sciences, or other STEM fields closely related to the food and agricultural sciences, and encourage study in areas that contribute to any of the five NIFA priority areas (global food security and hunger, climate change, sustainable bioenergy, childhood obesity, and food safety); and to enhance the quality of postsecondary instruction in order to help meet current and future national food and agricultural sciences workplace needs. Applications must also give a compelling educational challenge, indicating how the methodology is both unique and with merit; offer significant promise of adoption by others; and the expectation that impacts will continue once grant funds end. Individual land-grant colleges and universities and other institutions that have secured land-grant status, and are located in insular areas are eligible for awards under this program, either as direct applicants or as parties to a consortium agreement. Eligible insular areas include the Commonwealth of Puerto Rico, Guam, American Samoa, the Commonwealth of the Northern Mariana Islands, the Federated States of Micronesia, the Republic of the Marshall Islands, the Republic of Palau, and the Virgin Islands of the United States.
*Geographic Focus* Guam, Puerto Rico, U.S. Virgin Islands, American Samoa
*Date(s) Application is Due:* Mar 1
*Amount of Grant:* 70,000 - 750,000 USD
*Contact:* Saleia Afele-Faamuli, National Program Leader; (202) 720 - 0384; fax (202 720-4924; sfaamuli@nifa.usda.gov
*Internet:* www.nifa.usda.gov/fo/distanceeducationgrant.cfm
*Sponsor:* U.S. Department of Agriculture
1400 Independence Avenue, SW, Stop 2201
Washington, D.C. 20250-2201

**USDA Emerging Markets Grants** **2584**
The purpose of the program is to assist U.S. organizations, public and private, to improve the market access and to develop and promote U.S. agricultural products and/or processes in low- to middle-income countries that offer promise of emerging market opportunities in the near to medium term. This is to be accomplished by providing U.S. technical assistance through projects and activities in those emerging markets consistent with U.S. foreign policy interests. The program funds technical assistance activities to leverage the export and marketing of U.S. agricultural products to emerging markets. The emphasis is on marketing opportunities where there are risks that the private sector would not normally undertake alone, with funding provided on a project-by-project basis. The program is intended to support primarily small- to medium-sized U.S. firms that may need federal assistance in realizing or maintaining access in overseas markets. Application deadline is announced in the Federal Register.
*Requirements:* Any U.S. agricultural and/or agribusiness organization, university, or state department of agriculture is eligible to participate in the program. Priority will be given to those proposals that include significant support and involvement by private industry. Proposals from research and consulting organizations will be considered if they provide evidence of substantial participation by U.S. industry. U.S. market development cooperators may seek funding to address priority, market-specific issues and to undertake activities not already serviced by or unsuitable for funding under other FAS marketing programs.
*Restrictions:* No proposal will be considered without the element of cost-sharing.
*Geographic Focus:* All States
*Date(s) Application is Due:* May 20
*Amount of Grant:* Up to 500,000 USD
*Contact:* Maureen Quinn; (202) 720-7115; fax (202) 720-1727; maureen.quinn@usda.gov
Sally Klusaritz; (202) 720-3448; fax (202) 720-1727; sally.klusaritz@usda.gov
*Internet:* www.fas.usda.gov/mos/em-markets/em-markets.asp
*Sponsor:* U.S. Department of Agriculture
1400 Independence Avenue SW
Washington, D.C. 20250

**USDA Food and Agricultural Sciences National Needs Graduate Fellowship Grants** **2585**
The Food and Agricultural Sciences National Needs Graduate Fellowship Grants support training students for master's and doctoral degrees in food, agricultural, and natural resource sciences. It also supports special international study or thesis/dissertation research travel allowances for eligible USDA NNF beneficiaries. The objectives of the NNF program, working collaboratively with eligible institutions, are: supporting the training of students at the master's and doctoral levels in order to fill the nationally identified expertise shortage areas in the food and agricultural sciences; encouraging and recruiting outstanding students, including those from traditionally underrepresented groups, to enter and complete graduate training in the food and agricultural sciences; increasing the quality and diversity of students completing graduate degrees in the food and agricultural sciences; increasing the number of students with graduate degrees in the food, agricultural and related sciences awarded by at least 50 per annum; and supporting at least 10 eligible National Needs Fellows to build science competency and have international exposure within the Targeted Expertise Shortage Area (TESA) in food and agricultural sciences through the IRTA awards. Applications are accepted from institutions that offer a graduate degree in at least one of the following targeted expertise shortage areas: animal and plant production; forest resources; agricultural educators and communicators; agricultural management and economics; food science and human nutrition; sciences for agricultural biosecurity; and training in integrative biosciences for sustainable food and agricultural systems. Specific guidelines and additional information, including abstracts for previously funded projects, are located at the website. The application is available at grants.gov.
*Requirements:* Applications may be submitted by institutions having the capacity to carry out the teaching of food and agricultural sciences, including land grant institutions, public and private colleges and institutions, and colleges and universities with significant minority enrollments. Applications also may be submitted by research foundations maintained by eligible colleges or universities. NNF stipend funds for fellowships must be used to: (1) support the same Fellow at the master's degree level who maintains satisfactory progress and is eligible to receive grant support for a maximum of 24 months during a 36 period, and/or (2) support the same Fellow at the doctoral degree level who maintains satisfactory progress and is eligible to receive grant support for a maximum of 36 months within a 42 month period.
*Restrictions:* Indirect costs are not eligible for funding. NIFA provides a per Fellow cost-of-education allowance each year in lieu of indirect costs.
*Geographic Focus:* All States
*Date(s) Application is Due:* Jun 1
*Amount of Grant:* 4,500 - 262,000 USD
*Contact:* Ray A. Ali; (202) 720-2727; fax (202) 690-0062; rali@nifa.usda.gov
*Internet:* www.nifa.usda.gov/fo/graduateandpostgraduatefellowshipsserd.cfm
*Sponsor:* U.S. Department of Agriculture
1400 Independence Avenue, SW, Stop 2201
Washington, D.C. 20250-2201

**USDA Foreign Market Development Grants** **2586**
The Foreign Market Development Program (FMD), also known as the cooperator program, is administered by the Foreign Agricultural Service (FAS) of the U.S. Department of Agriculture (USDA). The goal of the program is to develop, maintain, and expand long-term export markets for U.S. agricultural products. The Cooperator program is designed to create, expand, and maintain foreign markets for U.S. agricultural commodities and products through cost-share assistance. Financial assistance under the Cooperator program will be made available on a competitive basis and applications will be reviewed against the evaluation criteria contained herein and in the Cooperator program regulations. All U.S. agricultural commodities, except tobacco, are eligible for consideration. Under the Cooperator program, the FAS enters into agreements with eligible nonprofit U.S. trade organizations to share the cost of certain overseas marketing and promotion activities.
*Requirements:* To participate in the Cooperator program, an applicant must: be a nonprofit U.S. agricultural trade organization; and agree to contribute resources to its proposed promotional activities.
*Geographic Focus:* All States
*Contact:* Maureen Quinn, Director; (202) 720-7115 or (202) 720-4327; fax (202) 720-1727; maureen.quinn@usda.gov or podadmin@fas.usda.gov
Sally Klusaritz, Deputy Director; (202) 720-3448 or (202) 720-4327; fax (202) 720-1727; sally.klusaritz@usda.gov or podadmin@fas.usda.gov
*Internet:* www.fas.usda.gov/mos/programs/fmdprogram.asp
*Sponsor:* U.S. Department of Agriculture
1400 Independence Avenue SW
Washington, D.C. 20250

**USDA Hazardous Fuels Woody Biomass Utilization Grants** **2587**
The U.S. Forest Service, State and Private Forestry, Technology Marketing Unit, located at the Forest Products Laboratory, requests proposals for wood energy projects that require engineering services. These projects will use woody biomass, such as material removed from forest restoration activities, wildfire hazardous fuel treatments, insect and disease mitigation, forest management due to catastrophic weather events, and/or thinning overstocked stands.

The woody biomass shall be used in a bioenergy facility that uses commercially proven technologies to produce thermal, electrical or liquid/gaseous bioenergy. Grant funding must be used to further the planning of such facilities by funding the engineering services necessary for final design and cost analysis. Examples of projects might include the following: engineering design of a woody biomass boiler for steam at a sawmill, hospital or school; non-pressurized hot water system for various applications; and biomass power generation facility. Specific guidelines and current application information are available at the website.
*Requirements:* The following entities are eligible for funding: state and county governments; city or township governments; special district governments; independent school districts; public, private, or state controlled institutions of higher education; federally recognized tribal governments; public housing authorities; nonprofits with or without 501(c)3 status; for profit organizations; and small businesses.
*Geographic Focus:* All States
*Date(s) Application is Due:* Mar 1
*Amount of Grant:* 50,000 - 250,000 USD
*Contact:* Susan LeVan-Green, Program Manager; (608) 231-9518; slevan@fs.fed.us
*Internet:* www.forestsandrangelands.gov/Woody_Biomass/opportunities.shtml
*Sponsor:* U.S. Department of Agriculture
1400 Independence Avenue SW
Washington, D.C. 20250

**USDA Higher Education Challenge Grants** **2588**
The USDA National Institute of Food and Agriculture requests applications for the Higher Education Challenge Grants Program (HEC) to stimulate and enable colleges and universities to provide the quality of education necessary to produce baccalaureate or higher degree level graduates capable of strengthening the Nation's food and agricultural, scientific, and professional workforce. Projects supported by the USDA Higher Education Challenge Grants will: address a state, regional, national, or international educational need; involve a creative or non-traditional approach toward addressing the need that can serve as a model to others; encourage and facilitate better working relationships in the university science and education community, as well as between universities and the private sector; and result in benefits that will likely transcend the project duration and USDA support. Electronic application information is available at the website.
*Requirements:* Applications may be submitted by: U.S. public or private nonprofit colleges and universities offering a baccalaureate or first professional degree in at least one discipline or area of the food and agricultural sciences; land-grant colleges and universities; colleges and universities having significant minority enrollments and the ability to carry out the teaching of food and agricultural sciences; and other colleges and universities having a demonstrable capacity to carry out the teaching of food and agricultural sciences.
*Restrictions:* Organizations are required to match 25% of the USDA funds awarded from non-federal sources. Funds may not be used for the renovation or reburbishment of research, education, or extension space; the purchase or installation of fixed equipment in such space, or to plan, repair, rehabilitate, acquire, construct building or facilities (i.e., greenhouses, laboratories, barns, or other structures). Promotional items such as t-shirts and other give-a-ways, and food functions such as cookouts or other social meal gatherings are considered entertainment, and are also not allowable expenses. Federal Cost Principles governing this grant program state that items not eligible for federal funding may also not be used as an applicant's contribution toward meeting matching requirements.
*Geographic Focus:* All States, All Countries
*Date(s) Application is Due:* Mar 30
*Amount of Grant:* 30,000 - 750,000 USD
*Contact:* Gregory Smith, National Program Leader; (202) 720-1973; gsmith@csrees.usda.gov or HEC@nifa.usda.gov
*Internet:* www.csrees.usda.gov/fo/fundview.cfm?fonum=1082
*Sponsor:* U.S. Department of Agriculture
1400 Independence Avenue, SW, Stop 2201
Washington, D.C. 20250-2201

**USDA Hispanic-Serving Institutions Education Grants** **2589**
The Hispanic-Serving Institutions Education Grants are intended to promote and strengthen the ability of Hispanic-Serving Institutions to carry out higher education programs in the food and agricultural sciences. The programs aim is to attract outstanding students and produce graduates capable of enhancing the nation's food and agricultural scientific and professional work force. Award recipients may subcontract to organizations not eligible to apply provided such organizations are necessary for the conduct of the project. The budget for subcontractors must be included with the corresponding budget narrative. Applicants are encouraged to collaborate with a USDA agency. Regular project applications must be received by February 9, while collaborative applications are due March 8. Application information, including samples of application packages and frequently asked questions, are available at the website.
*Requirements:* Only public or other non-profit Hispanic-Serving Institutions are eligible to apply for this program. To qualify as an Hispanic-Serving Institution, applicants must at the time of application, have an enrollment of undergraduate full-time equivalent students that is at least 25 percent. By submitting an application, the organization certifies that it is eligible to receive funding under this program.
*Geographic Focus:* All States
*Amount of Grant:* Up to 1,100,000 USD
*Contact:* Irma Lawrence, National Program Leader; (202) 720-2082 or (202) 720-1793; fax (202) 720-2030; ilawrence@nifa.usda.gov
*Internet:* www.csrees.usda.gov/fo/fundview.cfm?fonum=1094
*Sponsor:* U.S. Department of Agriculture
1400 Independence Avenue, SW, Stop 2201
Washington, D.C. 20250-2201

**USDA Integrated Pest Management Competitive Grants - Western Region** **2590**
The USDA Regional Integrated Pest Management Competitive Grants (RIPM) support the continuum of research and extension efforts needed to increase the implementation of IPM methods. The RIPM program supports projects that develop individual pest control tactics, integrate individual tactics into an IPM system, and develop and implement extension and education programs. The program is administered by the land-grant university system's four regional IPM Centers (North Central, Northeastern, Southern, Western) in partnership with NIFA. Funding will support projects that promote cooperative efforts across appropriate disciplines, with linkages between research and extension efforts, and components of existing or emerging pest management systems. The proposal should describe the role of each member of the multi-disciplinary team and their responsibilities to the project. The electronic application, specific guidelines, and abstracts of previously funded projects are available at the website.
*Requirements:* Organizations eligible to receive funding include the following: state agricultural experiment stations; 1862 and 1890 land-grant colleges and universities, including Tuskegee University and West Virginia State University; 1994 land-grant colleges and universities; research foundations established by land-grant colleges and universities; colleges and universities receiving funds under the Act of October 10, 1962 (16 USC 582a et seq.); accredited schools or colleges of veterinary medicine; and the University of the District of Columbia. Organizations eligible to receive Extension awards are 1862 and 1890 land-grant colleges and universities, including Tuskegee University and West Virginia State University, and the University of the District of Columbia.
*Restrictions:* Funding may not be used for renovation or refurbishment of research, education, or extension space; the purchase or installation of fixed equipment in such space; or the planning, repair, rehabilitation, acquisition, or construction of buildings or facilities.
*Geographic Focus:* Alaska, Arizona, California, Colorado, Guam, Hawaii, Idaho, Montana, Nevada, New Mexico, Northern Mariana Islands, Oregon, Utah, Washington, Wyoming, American Samoa, Micronesia, Palau
*Date(s) Application is Due:* Feb 28
*Amount of Grant:* 180,000 USD
*Contact:* Robert Nowierski; (202) 401-4900; fax (202) 401-4888; rnowierski@nifa.usda.gov
*Internet:* www.nifa.usda.gov/funding/rfas/ipm_western.html
*Sponsor:* U.S. Department of Agriculture
1400 Independence Avenue, SW, Stop 2201
Washington, D.C. 20250-2201

**USDA Integrated Research Education and Extension Grants - Organic Transitions** **2591**
The Integrated Research Education and Extension Grants - Organic Transitions are soliciting proposals to assess the impact of organic systems on environmental outcomes in soil conservation, land stewardship, soil carbon sequestration, and greenhouse gas emissions. Fieldwork to set up treatments or collect data on organic practices must be done on certified organic land, or land in transition to organic certification. Practices and systems to be addressed include those associated with organic crops, organic animal production (including dairy), and organic systems integrating plant and animal production. Applicants are encouraged to develop content suitable for delivery through eXtension. Funding may be requested to: establish an eXtension Community of Practice (CoP); or exhance an existing CoP, such as eOrganic, to endure that information and technology transfer reaches potential adopters as quickly as possible. Projects are expected to use field and modeling information to demonstrate the benefits or challenges to environmental services and climate change litigation posed by implementing certified organic practices and/or combinations of practices. Specific guidelines, abstracts of funded projects, workshops, proposal examples, and grant writing tips are available at the website. The application is available at grants.gov.
*Requirements:* The following entities are eligible to apply: 1862, 1890, and 1994 land grant institutions; Hispanic serving institutions; and private and public institutions of higher education. Applicants are expected to describe stakeholder involvement in problem identification, planning, implementation and evaluation. Applications describing multi-state, multi-institutional, multidisciplinary, multifunctional activities and combinations are encouraged. S single university demonstrating significant collaboration with various agencies or organizations within the host state, as appropriate to project goals, may also be competitive. Applicants are strongly encouraged to assemble project teams that include those with expertise in research, education, extension, and evaluation. Projects should plan to deliver applied production information to producers and/or students. Applicants are also encouraged to describe how results at the field and farm scale might be extrapolated beyond the experimental boundaries of the proposed project.
*Restrictions:* If a grant provides a particular benefit to a specific agricultural commodity, the grant recipient is required to provide funds awarded on a dollar-for-dollar basis from non-federal sources with cash and/or in-kind contributions. NIFA may waive the matching funds requirement for a grant if it is determined that: the results of the project, while of particular benefit to a specific agricultural commodity, are likely to be applicable to agricultural commodities generally; or the project involves a minor commodity, the project deals with scientifically important research, and the grant recipient is unable to satisfy the matching funds requirement. Projects in organic agriculture research typically qualify for a waiver of the matching requirement.
*Geographic Focus:* All States
*Date(s) Application is Due:* Apr 25
*Amount of Grant:* Up to 750,000 USD
*Contact:* Steven Smith, National Program Leader; (202) 401-6134; fax (202) 401-1782; sismith@nifa.usda.gov
*Internet:* www.csrees.usda.gov/fo/organictransitionsprogram.cfm
*Sponsor:* U.S. Department of Agriculture
1400 Independence Avenue, SW, Stop 2201
Washington, D.C. 20250-2201

**USDA International Science and Education Competitive Grants** 2592

The International Science and Education Competitive Grants (ISE) supports activities that will improve the capabilities of American colleges and universities to conduct international collaborative research, extension and teaching. The grants are expected to enhance the international content of curricula; ensure that faculty work beyond the U.S. and bring lessons learned back home; promote international research partnerships; enhance the use and application of foreign technologies in the U.S; and strengthen the role that colleges and universities play in maintaining U.S. competitiveness. Multi-year proposals of up to four years may be submitted. Partnerships with minority serving institutions and/or community college are encouraged. Current applications, specific guidelines, and abstracts of previously funded projects are available at the website.

*Requirements:* Public and private colleges and universities are eligible to apply, in addition to 1862, 1890, and 1994 land-grant institutions.

*Restrictions:* Internationalization of U.S. campus-based programs is the primary focus of the ISE program. Proposals that do not show benefit to the U.S. will not be reviewed. Since the ISE program is an integrated program, proposals should focus on at least two of the three main components (agricultural research, extension and teaching). Single function proposals will not be reviewed.

*Geographic Focus:* All States

*Date(s) Application is Due:* Jan 26

*Contact:* Patricia Fulton, National Program Leader; (202) 690-3852 or (202) 720-3801; fax (202) 690-2355; pfulton@nifa.usda.gov

*Internet:* www.nifa.usda.gov/fo/internationalscienceandeducation.cfm

*Sponsor:* U.S. Department of Agriculture

1400 Independence Avenue, SW, Stop 2201

Washington, D.C. 20250-2201

**USDA Interregional Research Project Minor Crop Pest Management Grants** 2593

The primary goal of the IR-4 program is to provide safe, effective, and economical pest management solutions for growers of minor/specialty crops. IR-4 facilitates crop protection by providing expert assistance with product development and registration. IR-4 positively impacts the ability of research and extension personnel, producers, and grower organizations to provide a continuous supply of affordable minor/specialty crops to the public. Successful applicants will demonstrate the capacity and commitment required to accelerate the registration of newer, reduced risk pest management tools; expedite access to these pest management tools for minor/specialty crop growers; and conduct efficacy research to identify new and more effective minor/specialty crop pest management solutions. The FY 2012 IR-4 Program will fund four geographically-based IR-4 Regional Centers and an IR-4 National Headquarters. NIFA does not require matching support for this program, and matching resources will not be factored into the review process. The electronic application, specific guidelines, and abstracts of previously funded projects are available at the website.

*Requirements:* Land-grant colleges and universities (1862, 1890, and 1994) and state agricultural experiment stations are eligible to apply. Applications may be submitted by state agricultural experiment stations, research foundations established by land grant colleges and universities, colleges and universities receiving funds under the McIntire-Stennis Act, and accredited schools or colleges of veterinary medicine for the purpose of facilitating or expanding ongoing state-federal food and agricultural research programs. Applicants should clearly describe a plan for achieving three core objectives: to maintain regulatory clearances of effective crop protection agents for high-value, minor/specialty food crops and for minor uses (special circumstance needs) on major food crops with special emphasis on lower risk chemicals and uses that are compatible with integrated pest management (IPM) programs; to support research to enhance the development and registration of bio-pesticides for use in food and non-food pest management programs; and to support research on crop protection products that will expand their uses on ornamental crops (nursery, floral, turf, and other non-food crop systems) to allow management of new and important pest species.

*Restrictions:* Indirect costs and tuition remissions are not allowable costs. Costs that are a part of an institution's indirect cost pool (e.g., administrative or clerical salaries) may not be reclassified as direct costs for the purpose of making them allowable. Funding may not be used for the renovation or refurbishment of research, education, or extension space; the purchase or installation of fixed equipment in such space; or the planning, repair, rehabilitation, acquisition, or construction of buildings or facilities.

*Geographic Focus:* All States

*Date(s) Application is Due:* Mar 20

*Amount of Grant:* 1,000,000 - 4,800,000 USD

*Contact:* Monte Johnson; (202) 401-1108; fax (202) 401-1782; mpjohnson@nifa.usda.gov

*Internet:* www.nifa.usda.gov/funding/rfas/minor_crop.html

*Sponsor:* U.S. Department of Agriculture

1400 Independence Avenue, SW, Stop 2201

Washington, D.C. 20250-2201

**USDA Market Access Program Grants** 2594

The Market Access Program (MAP), formerly the Market Promotion Program, uses funds from the U.S. Department of Agriculture's (USDA) Commodity Credit Corporation (CCC) to help U.S. producers, exporters, private companies, and other trade organizations finance promotional activities for U.S. agricultural products. The MAP encourages the development, maintenance, and expansion of commercial export markets for agricultural commodities. Activities financed include consumer promotions, market research, technical assistance, and trade servicing.

*Requirements:* To participate in the MAP, an applicant must be a nonprofit U.S. agricultural trade organization, a nonprofit SRTG, a nonprofit U.S. agricultural cooperative, or a State government agency. A small-sized U.S. commercial entity may participate through a MAP participant.

*Geographic Focus:* All States

*Contact:* Maureen Quinn, Director; (202) 720-7115 or (202) 720-4327; fax (202) 720-1727; maureen.quinn@usda.gov or podadmin@fas.usda.gov

Sally Klusaritz, Deputy Director; (202) 720-3448 or (202) 720-4327; fax (202) 720-1727; sally.klusaritz@usda.gov or podadmin@fas.usda.gov

*Internet:* www.fas.usda.gov/mos/programs/map.asp

*Sponsor:* U.S. Department of Agriculture

1400 Independence Avenue SW

Washington, D.C. 20250

**USDA McIntire-Stennis Cooperative Forestry Research Grants** 2595

The USDA McIntire-Stennis Grants assist states in continuing programs of research at forestry schools and in developing a trained pool of scientists to conduct forestry research. Categories of forestry research supported include reforestation and management of land for the production of crops of timber and other related products of the forest; management of forest and related watershed lands to improve conditions of waterflow and to protect resources against flood and erosion; management of forest and related rangeland for production of forage, domestic livestock and game, and improvement of food and habitat for wildlife; management of forest lands for outdoor recreation; protection of forest resources against fire, insects, diseases, and other destructive agents; utilization of wood and other forest products; development of sound policies for the management of forest lands, and the harvesting and marketing of forest products; and other studies necessary to obtain the fullest and most effective use of forest resources. The application package must be obtained from grants.gov, with the appropriate funding number entered. Current specific instructions are available at the website.

*Requirements:* Applicants must be state or territorial institutions.

*Geographic Focus:* All States

*Date(s) Application is Due:* Aug 29

*Amount of Grant:* 33,000 - 1,200,000 USD

*Contact:* Linda Toppin, Administrative Assistant; 202-690-2563; ltoppin@nifa.usda.gov

*Internet:* www.csrees.usda.gov/business/awards/formula/mcintire_stennis.html

*Sponsor:* U.S. Department of Agriculture

1400 Independence Avenue, SW, Stop 2201

Washington, D.C. 20250-2201

**USDA National Integrated Food Safety Initiative Grants** 2596

The National Integrated Food Safety Initiative Grants support food safety projects that demonstrate an integrated approach to solving problems in applied food safety research, education, or extension. Special emphasis is given to research describing multi-functional activities, such as research and extension, research and education, or extension and education. Applications describing multi-state, multi-institutional, multidisciplinary, and multifunctional activities are encouraged. If international activities (e.g. partnerships, exchanges, travel, etc) are proposed, then applicants should describe indicators that will be used to assess those activities. Project teams should include university, government, and industry collaborators. Specific criteria (standard vs. special emphasis projects, conferences, and bridge funding), samples of previous proposals, the electronic application, and workshop presentations are available at the website.

*Requirements:* The following entities are eligible to apply: 1862, 1890, and 1994 land grant institutions; Hispanic serving institutions (four year and two year, if the institution offers an associate degree in an agriculture-related field); and private and public institutions of higher education, particularly institutions that serve high-risk, under-served, or hard-to-reach populations. If a grant provides a particular benefit to a specific agricultural commodity, the grant recipient is required to match the USDA funds awarded on dollar-for-dollar basis from non-federal sources with cash and in-kind contributions.

*Geographic Focus:* All States

*Date(s) Application is Due:* Jun 27

*Amount of Grant:* Up to 2,000,000 USD

*Contact:* Jan Singleton; (202) 401-1954; fax (202) 401-4888; jsingleton@nifa.usda.gov

*Internet:* www.nifa.usda.gov/fo/nationalintegratedfoodsafetyinitiative.cfm

*Sponsor:* U.S. Department of Agriculture

1400 Independence Avenue, SW, Stop 2201

Washington, D.C. 20250-2201

**USDA Norman E. Borlaug International Agricultural Science and Technology Fellowships** 2597

The Norman E. Borlaug International Agricultural Science and Technology Fellowships promote food security and economic growth by increasing scientific knowledge and collaborative research to improve agricultural productivity. It accomplishes this by: providing training opportunities for early and mid-career researchers and policymakers from developing and middle-income countries; fostering collaborative research to improve agricultural productivity; facilitating the transfer of new science and agricultural technologies to strengthen agricultural practices; and addressing obstacles to the adoption of technology, such as inadequate policies and regulations. Fellowships range from six to 12 weeks depending on the research topic and funding availability. Areas of training and research can be in any agriculture-related field including plant pathology, entomology, veterinary sciences, microbiology, agricultural economics, food safety, sanitary and phytosanitary topics, natural resources management, agricultural biotechnology, global climate change, and agricultural policy. Allowable training topics vary by country and year and can be found in the country-specific application announcements posted on the website.

*Requirements:* Candidates must be female citizens of the county in which the Fellowship is offered, hold a minimum of a Master's degree, and be in the early to middle stage of their professional career, with a good working knowledge of the English language. They must also have at least two years of practical work experience, with current employment in a university,

government, non-profit, or private agricultural research entity in their county, with the intention to continue working in their home country after the Fellowship is completed.
*Restrictions:* Passports must be valid for at least six months from the date the applicant enters the U.S.
*Geographic Focus:* Bangladesh, Burundi, Comoros, Djibouti, Eritrea, Ethiopia, Ghana, Guatemala, Haiti, Honduras, Kenya, Liberia, Madagascar, Malawi, Mali, Mauritius, Mayotte, Mozambique, Nigeria, Reunion, Rwanda, Senegal, Seychelles, Somalia, Tanzania, Uganda, Zambia, Zimbabwe
*Contact:* Karina Ramos-Hides; (202) 720-3736; karina.ramos@fas.usda.gov
*Internet:* www.fas.usda.gov/icd/borlaug/Borlaug.asp
*Sponsor:* U.S. Department of Agriculture
1400 Independence Avenue SW
Washington, D.C. 20250

**USDA Organic Agriculture Research and Extension Initiative 2598**

The USDA Organic Agriculture Research and Extension Initiative (OREI) seeks to solve critical organic agriculture issues, priorities, or problems through the integration of research and extension activities. The purpose of this program is to fund projects that will enhance the ability of producers and processors who have already adopted organic standards to grow and market high quality organic agricultural products. Priority concerns include biological, physical, and social sciences, including economics. The OREI is particularly interested in projects that emphasize research and outreach that assist farmers and ranchers with whole farm planning. Projects should plan to deliver applied production information to producers. Fieldwork must be done on certified organic land or on land in transition to organic certification, as appropriate to project goals and objectives. Funding amounts are based on specific project categories.
*Requirements:* The following are eligible to apply for the initiative: public and state controlled institutions of higher education; nonprofits having a 501(c)3 status with the IRS, other than institutions of higher education; nonprofits that do not have a 501(c)3 status with the IRS, other than institutions of higher education; private institutions of higher education; individuals; for profit organizations other than small businesses; and small businesses. The following entities are also eligible to apply: state agricultural experiment stations; colleges and universities; university research foundations; other research institutions and organizations; federal agencies; national laboratories; private organizations or corporations; individuals who are United States citizens or nationals; or a group consisting of two or more of the above entities.
*Restrictions:* Recipients are required to provide funds or in-kind support from non-federal sources in an amount at least equal to the amount provided by the funding agency.
*Geographic Focus:* All States
*Date(s) Application is Due:* Mar 9
*Amount of Grant:* 50,000 - 2,000,000 USD
*Contact:* Steven Smith; (202 401-6134; fax (202) 401-1782; sismith@nifa.usda.gov
*Internet:* www.nifa.usda.gov/funding/rfas/OREI.html
*Sponsor:* U.S. Department of Agriculture
1400 Independence Avenue, SW, Stop 2201
Washington, D.C. 20250-2201

**USDA Plant and Animal Disease, Pest Control, and Animal Care Grants 2599**

The Plant and Animal Disease, Pest Control, and Animal Care Grants help protect U.S. agriculture from economically injurious plant and animal diseases and pests, guarantee the safety and potency of veterinary biologics, and ensure the humane treatment of animals. Project and training grants are awarded to eligible organizations to carry out these objectives. Possible projects include conducting surveys, inspections to detect and appraise infestations, eradication and control activities, and carrying out regulatory actions to prevent interstate spread of infestations and diseases. Additional guidelines and application instructions are available at the website for the Catalog of Federal Domestic Assistance (project number 10.025). Applicants may contact their state Animal and Plant Health Inspection Service office.
*Requirements:* Eligible to apply are foreign, state, local, and U.S. territorial government agencies; nonprofit institutions of higher education; and nonprofit associations or organizations requiring federal support to eradicate, control, or assess the status of injurious plant and animal diseases and pests that are a threat to regional or national agriculture and to conduct related demonstration projects. A letter from applicants seeking support or cooperation with the Animal and Plant Health Inspection Service is required. Applicants must indicate the nature and severity of the specific problem and explain how they will address the problem. Applications are approved upon determination that the project will contribute toward accomplishment of the Agency's overall mission and meet any established project evaluation and/or selection criteria.
*Restrictions:* Grant recipients share some project or program costs. The cost-sharing arrangements are developed between the USDA and the recipients in advance. Individuals are not eligible for funding.
*Geographic Focus:* All States
*Contact:* Eileen M. Berke; (301) 734-8330; eileen.m.berke@aphis.usda.gov
*Internet:* https://www.cfda.gov/index?s=program&mode=form&tab=step1&id=4eef76daa8c24580301ca56617a3462e
*Sponsor:* U.S. Department of Agriculture
4700 River Road
Riverdale, MD 20737

**USDA Plant Feedstock Genomics for Bioenergy: A Joint Research Grant 2600**

The National Institute of Food and Agriculture's (NIFA) Competitive Programs and the Department of Energy's Office of Science, Office of Biological and Environmental Research (OBER) announce the inter-agency program to support genomics-based research that will lead to the improved use of biomass and plant feedstocks for the production of fuels such as ethanol or renewable chemical feedstocks. Specifically, applications are sought for fundamental research on plants that will improve biomass characteristics, biomass yield, or sustainability. Systems biology approaches to identify genetic indicators enabling plants to be efficiently bred or manipulated, or research to predict phenotype from underlying genotype leading to improved feedstock characterization and sustainability are also encouraged. Applicants may request up to three years of funding. Please note that applicants will submit applications through DOE and a pre-application is requested rather than a letter of intent. The electronic application, specific guidelines, and abstracts for previously funded projects are available at the website.
*Requirements:* The following are eligible to apply for funding: 1862, 1890, and 1994 land grant institutions; public and private institutions; small businesses and all other for profit businesses; Hispanic serving institutions; nonprofits with or without 501(c)3; and state agricultural experiment stations. Federal agencies, national laboratories, private organizations or corporations, and individuals are also eligible to apply.
*Restrictions:* Research that seeks to increase starch content for improved nutrient qualities or to facilitate the digestion and fermentation of starch to produce sugars and other bio-based products or bio-fuels is not eligible for funding. Research that seeks to increase grain yield or seed-oil productions is also not eligible.
*Geographic Focus:* All States
*Date(s) Application is Due:* Feb 24
*Amount of Grant:* 200,000 - 500,000 USD
*Contact:* Dr. Ed Kaleikau, National Program Leader; (202) 401-1931; fax (202) 401-1782; ekaleikau@nifa.usda.gov
Dr. Catherine Ronning; (301) 903-9549; Catherine.ronning@science.doe.gov
*Internet:* www.nifa.usda.gov/fo/plantfeedstockgenomicsforbioenergy.cfm
*Sponsor:* U.S. Department of Agriculture
1400 Independence Avenue, SW, Stop 2201
Washington, D.C. 20250-2201

**USDA Policy Research Center Grants 2601**

The USDA Policy Research Center (PRC) Grants are available for centers to conduct research and education programs that are objective, operationally independent, and external to the federal government. They should also study the effect of public policies and trade agreements in the following areas: the farm and agricultural sectors (including commodities, livestock, dairy, and specialty crops, such as fruits and vegetables, tree nuts, and plants used for medicinal purposes); the environment; rural families, households, and economies; and consumers, food, and nutrition. Research may be either disciplinary or interdisciplinary and include the following activities: quantify implications of public policies and regulations; develop theoretical and research methods; collect, analyze, and disseminate data for policy makers, analysts, and individuals; and develop programs to train future analysts. A PRC is defined as having the analytical capacity to either perform policy analysis across multiple sets of public policy issues or have a specialized capacity in a single policy area. A center may reside in a single institution or be a collaborative effort across multiple institutions. Matching support is not required. Instructions for the Grants.gov application, specific qualifications guidelines, and evaluation criteria are available at the website.
*Requirements:* Applications may be submitted by state agricultural experiment stations, colleges and universities, other research institutions and organizations, private organizations, corporations, and individuals. Award recipients may subcontract to organizations not eligible to apply provided such organizations are necessary for the conduct of the projec.
*Geographic Focus:* All States
*Date(s) Application is Due:* Feb 13
*Amount of Grant:* 450,000 - 960,000 USD
*Contact:* Robbin Shoemaker, National Program Leader; (202) 720-5468; fax (202) 401-5179; rshoemaker@nifa.usda.gov
*Internet:* www.nifa.usda.gov/funding/rfas/prc.html
*Sponsor:* U.S. Department of Agriculture
1400 Independence Avenue, SW, Stop 2201
Washington, D.C. 20250-2201

**USDA Rangeland Research Grants 2602**

The primary purpose of the Rangeland Research Grants is to provide U.S. agricultural producers, rural landowners, and land managers with integrated science strategies to make informed land management decisions with an emphasis on enhancing the restoration and sustainable integrity of U.S. rangelands. Rangeland research should employ a methodology that integrates approaches to discover, understand, and address phenomena that are too broad to be explained by standard disciplinary science. Priority consideration will be given to the following areas: rangeland restoration, management of rotational grazing, and management of co-grazing. Successful applicants will: document the experience level of key personnel working within a collaborative, integrated rangeland research project; demonstrate the ability to participate in a comprehensive nationally coordinated effort that shares data and its analysis, and derived conclusions; involve novel or non-traditional partners (e.g., non-governmental organizations, public or private foundations, social scientists or psychologists) to move programs toward improved rangeland management and behavior change in stakeholders; demonstrate capacity to conduct efficient and successful integrated rangeland research projects; demonstrate ability to develop new scientific methodologies and innovations that transcend standard disciplinary science; develop multidisciplinary research projects which result in an enhanced land and resource management knowledge base; and conceptualize and implement research projects that support the purpose and goals of the research grants. Specific guidelines, including abstracts of previously funded projects, are available at the website. The application is available at grants.gov.
*Requirements:* Land-grant colleges and universities, state agricultural experiment stations, in addition to colleges, universities, and federal laboratories having a demonstrable capacity

in rangeland research are eligible to apply for funding. Award recipients may subcontract to organizations not eligible to apply provided such organizations are necessary for the conduct of the project. Applicants are required to provide 50% matching funds from non-federal sources for all proposed federal funds being sought. Non-federal matching contributions, such as cash and third party in kind, are accepted under this program.
*Restrictions:* Grant funds may not be used for the following: renovation or refurbishment of research, education, or extension space; the purchase or installation of fixed equipment in such space; or the planning, repair, rehabilitation, acquisition, or construction of buildings or facilities. Funding is also not available for indirect costs, tuition remission, or time extensions beyond five years.
*Geographic Focus:* All States
*Date(s) Application is Due:* Jul 15
*Amount of Grant:* 350,000 - 500,000 USD
*Contact:* James Dobrowolski, National Program Leader; (202) 401 - 5016; fax (202) 401 - 1706; jdobrowolski@nifa.usda.gov
*Internet:* www.nifa.usda.gov/fo/rangelandresearchprogram.cfm
*Sponsor:* U.S. Department of Agriculture
1400 Independence Avenue, SW, Stop 2201
Washington, D.C. 20250-2201

**USDA Regional Integrated Pest Management Grants - North Central Region 2603**
The USDA Regional Integrated Pest Management Competitive Grants (RIPM) support the ongoing research and extension efforts needed to increase the implementation of IPM methods. The RIPM program supports projects that develop individual pest control tactics, integrate individual tactics into an IPM system, and develop and implement extension and education programs. The program is administered by the land-grant university system's four regional IPM Centers (North Central, Northeastern, Southern, Western) in partnership with NIFA. Funding will support projects that promote cooperative efforts across appropriate disciplines, with linkages between research and extension efforts, and components of existing or emerging pest management systems. The proposal should describe the role of each member of the multi-disciplinary team and their responsibilities to the project. The electronic application, specific guidelines, and abstracts of previously funded projects are available at the website.
*Requirements:* The following organizations are eligible for funding: state agricultural experiment stations; 1862 and 1890 land-grant colleges and universities, including Tuskegee University and West Virginia State University; 1994 land-grant colleges and universities; research foundations established by land-grant colleges and universities; colleges and universities receiving funds under the Act of October 10, 1962 (16 USC 582a et seq.); accredited schools or colleges of veterinary medicine; and the University of the District of Columbia. Organizations eligible to receive extension awards are 1862 and 1890 land-grant colleges and universities, including Tuskegee University and West Virginia State University, and the University of the District of Columbia.
*Restrictions:* Funding may not be used for renovation or refurbishment of research, education, or extension space; the purchase or installation of fixed equipment in such space; or the planning, repair, rehabilitation, acquisition, or construction of buildings or facilities.
*Geographic Focus:* Illinois, Indiana, Iowa, Kansas, Michigan, Minnesota, Missouri, Nebraska, North Dakota, Ohio, South Dakota, Wisconsin
*Date(s) Application is Due:* Feb 28
*Amount of Grant:* Up to 175,000 USD
*Contact:* Robert Nowierski; (202 401-4900; fax (202) 401-4888; rnowierski@nifa.usda.gov
*Internet:* www.nifa.usda.gov/fo/regionalintegratedpestmgtnorthcentral.cfm
*Sponsor:* U.S. Department of Agriculture
1400 Independence Avenue, SW, Stop 2201
Washington, D.C. 20250-2201

**USDA Regional Integrated Pest Management Grants - Northeastern Region 2604**
The USDA Regional Integrated Pest Management Competitive Grants (RIPM) support the continuum of research and extension efforts needed to increase the implementation of IPM methods. The RIPM program supports projects that develop individual pest control tactics, integrate individual tactics into an IPM system, and develop and implement extension and education programs. The program is administered by the land-grant university system's four regional IPM Centers (North Central, Northeastern, Southern, Western) in partnership with NIFA. Funding will support projects that promote cooperative efforts across appropriate disciplines, with linkages between research and extension efforts, and components of existing or emerging pest management systems. The proposal should describe the role of each member of the multi-disciplinary team and their responsibilities to the project. The electronic application, specific guidelines, and abstracts of previously funded projects are available at the website.
*Requirements:* The following organizations are eligible to receive funding: state agricultural experiment stations; 1862 and 1890 land-grant colleges and universities, including Tuskegee University and West Virginia State University; 1994 land-grant colleges and universities; research foundations established by land-grant colleges and universities; colleges and universities receiving funds under the Act of October 10, 1962 (16 USC 582a et seq.), accredited schools or colleges of veterinary medicine; and the University of the District of Columbia. Organizations eligible to receive extension awards are: 1862 and 1890 land-grant colleges and universities, including Tuskegee University and West Virginia State University, and the University of the District of Columbia.
*Restrictions:* Funding may not be used for renovation or refurbishment of research, education, or extension space; the purchase or installation of fixed equipment in such space; or the planning, repair, rehabilitation, acquisition, or construction of buildings or facilities.
*Geographic Focus:* Connecticut, Delaware, District of Columbia, Maine, Maryland, Massachusetts, New Hampshire, New Jersey, New York, Pennsylvania, Rhode Island, Vermont, West Virginia
*Date(s) Application is Due:* Feb 28
*Amount of Grant:* Up to 180,000 USD
*Contact:* Robert Nowierski, National Program Leader; (202) 401-4900; fax (202) 401-4888; rnowierski@nifa.usda.gov
*Internet:* www.nifa.usda.gov/fo/regionalintegratedpestmgtnortheastern.cfm
*Sponsor:* U.S. Department of Agriculture
1400 Independence Avenue, SW, Stop 2201
Washington, D.C. 20250-2201

**USDA Regional Integrated Pest Management Grants - Southern Region 2605**
The USDA Regional Integrated Pest Management Competitive Grants (RIPM) support the ongoing research and extension efforts needed to increase the implementation of IPM methods. The RIPM program supports projects that develop individual pest control tactics, integrate individual tactics into an IPM system, and develop and implement extension and education programs. The program is administered by the land-grant university system's four regional IPM Centers (North Central, Northeastern, Southern, Western) in partnership with NIFA. Funding will support projects that promote cooperative efforts across appropriate disciplines, with linkages between research and extension efforts, and components of existing or emerging pest management systems. The proposal should describe the role of each member of the multi-disciplinary team and their responsibilities to the project. The electronic application, specific guidelines, and abstracts of previously funded projects are available at the website.
*Requirements:* Organizations eligible to receive funding include the following: state agricultural experiment stations; 1862 and 1890 land-grant colleges and universities, including Tuskegee University and West Virginia State University; 1994 land-grant colleges and universities; research foundations established by land-grant colleges and universities; colleges and universities receiving funds under the Act of October 10, 1962 (16 USC 582a et seq.); accredited schools or colleges of veterinary medicine; and the University of the District of Columbia. Organizations eligible to receive Extension awards are 1862 and 1890 land-grant colleges and universities, including Tuskegee University and West Virginia State University, and the University of the District of Columbia.
*Restrictions:* Funding may not be used for renovation or refurbishment of research, education, or extension space; the purchase or installation of fixed equipment in such space; or the planning, repair, rehabilitation, acquisition, or construction of buildings or facilities.
*Geographic Focus* Alabama, Arkansas, Florida, Georgia, Kentucky, Louisiana, Mississippi, North Carolina, Oklahoma, Puerto Rico, South Carolina, Tennessee, Texas, U.S. Virgin Islands, Virginia
*Date(s) Application is Due:* Feb 28
*Amount of Grant:* Up to 171,000 USD
*Contact:* Robert Nowierski, National Program Leader; (202) 401-4900; fax (202) 401-4888; rnowierski@nifa.usda.gov
*Internet:* www.nifa.usda.gov/fo/regionalintegratedpestmgtsouthern.cfm
*Sponsor:* U.S. Department of Agriculture
1400 Independence Avenue, SW, Stop 2201
Washington, D.C. 20250-2201

**USDA Resident Instruction Grants for Higher Education in Insular Areas 2606**
The purpose of this program is to promote and strengthen the ability of insular area institutions to carry out teaching and education programs within a broadly defined arena of food and agricultural sciences-related disciplines. Insular areas include the Commonwealth of Puerto Rico, Guam, American Samoa, the Commonwealth of the Northern Mariana Islands, the Federated States of Micronesia, the Republic of the Marshall Islands, the Republic of Palau, and the Virgin Islands of the United States. By strengthening institutional educational capacities in instruction and curriculum, and enhancing the quality of teaching and learning, NIFA intends that this program will help insular area institutions meet their unique needs. This program also will assist insular area institutions to make more efficient use of existing educational funds by providing resources for partnerships between faculties at insular area and mainland institutions. The electronic application and specific guidelines are available at the website.
*Requirements:* Eligible applicants include higher education institutions located in an insular area with a capacity to carry out teaching and extension programs in the food and agricultural sciences. Individual land-grant colleges and universities, and other institutions that have secured land-grant status through federal legislation, and located in insular areas are automatically eligible for awards under this program, either as direct applicants or as parties to a consortium agreement.
*Geographic Focus* Guam, Marshall Islands, Northern Mariana Islands, Puerto Rico, U.S. Virgin Islands, American Samoa, Micronesia, Palau
*Date(s) Application is Due:* Mar 1
*Amount of Grant:* 70,000 - 850,000 USD
*Contact:* Saleia Afele-Faamuli, National Program Leader; (202) 720-0384; fax (202) 720-4924; sfaamuli@nifa.usda.gov
*Internet:* www.nifa.usda.gov/fo/residentinstructionforinsularareas.cfm
*Sponsor:* U.S. Department of Agriculture
1400 Independence Avenue, SW, Stop 2201
Washington, D.C. 20250-2201

**USDA Scientific Cooperation Exchange Program 2607**
The Scientific Cooperation Exchange Program (SCEP) with the People's Republic of China supports international exchanges to promote bilateral scientific exchanges designed to promote agricultural cooperation, development, and trade. The program offers opportunities for U.S. teams of up to five members to initiate linkages with potential long-term collaborators at Chinese institutions throughout the People's Republic of China (PRC). Under a special USDA Scientific Technological Exchange

Agreement with the PRC, the Chinese Ministry of Agriculture offers two-week visits to China designed to meet the scientific objectives of each team. Priority will be given to innovative proposals that address food security, climate change, food safety, animal and plant health, biotechnology and other emerging technologies, food processing and product development, non-food product development, and sustainable plant and animal management. Specific guidelines, including funding and cost share requirements, program purpose and priorities, focus areas and recommended topics, application and submission information, and frequently asked questions are available at the website.
*Requirements:* Proposals may be received from universities and colleges, including community colleges, accredited in and having a campus located in the United States, acting on behalf of their institutions. U.S. non-profit research organizations, including research laboratories, professional societies, or similar organizations in the U.S. that are directly associated with education, extension, research, or capacity building activities; private organizations; federal and state agencies; or consortia led by the eligible organizations listed here are also eligible. A proposal from a consortium of organizations must be submitted as a single proposal with one U.S. organization serving as the lead and all other organizations as team members and sub-awardees, when applicable. The Principal Investigator (PI) must hold a position at an eligible U.S. institution. Especially encouraged are proposals from smaller academic institutions, as well as minority serving institutions, researchers early in their career, extension agents, and women that hold positions at an eligible institution.
*Restrictions:* The U.S. team is responsible for the international airfare between the U.S. and China. MOA will cover all in-country program related expenses, including lodging, transportation, meals, translations, and cultural activities. USDA does not provide any funding to support U.S. teams.
*Geographic Focus:* All States
*Date(s) Application is Due:* Jul 8
*Contact:* Khaliaka Meardry; (202) 690-1866; Khaliaka.Meardry@fas.usda.gov
Nora Banks, Program Contact; (202) 720-3904; Nora.Banks@fas.usda.gov
*Internet:* www.fas.usda.gov/icd/grants/scep/scep.asp
*Sponsor:* U.S. Department of Agriculture
1400 Independence Avenue SW
Washington, D.C. 20250

**USDA Scientific Cooperation Research Program** **2608**

The Scientific Cooperation Research Program (SCRP) supports joint research, extension, and education projects among domestic and international agricultural professionals. These projects address issues including agricultural trade and market access, animal and plant health, biotechnology, food safety and security, and sustainable natural resource management in the U.S. and internationally. The objective of SCRP is to reduce global poverty and hunger by supporting applied scientific research, extension, or education projects that aim to address challenges faced by smallholder farmers in emerging economies. For the purposes of this announcement, smallholders, as defined by the World Bank, are farmers that own or lease less than 124 acres of land (less than 50 hectares). Since 1980, the program has supported more than 400 projects with approximately 95 partnering countries, enhancing the technical skills of more than 1,000 agricultural professionals. Specific guidelines, along with application and submission information, countries defined by the World Bank as eligible emerging economies, and frequently asked questions are available at the website.
*Requirements:* Proposals may be received from academic institutions such as universities and colleges, including community colleges, accredited in and having a campus located in the U.S., acting on behalf of their institution. Non-profit research organizations in the U.S., including research laboratories, professional societies, or similar organizations that are directly associated with education, extension, research, or capacity building activities; private organizations; federal and state agencies; or consortia led by these eligible organizations. A proposal from a consortium of organizations must be submitted as a single proposal with one U.S. organization serving as the lead and all other organizations as team members and sub-awardees when applicable. All proposals must include foreign collaborations with institutions in emerging economies that have diplomatic relations with the U.S. All proposals should address one of the three focus areas: improving agricultural productivity, creating sustainable agricultural systems, or building regional or global trade capacities. Priority will be given to: minority serving institutions; countries defined by the World Bank as emerging economies; and projects that address challenges to women. Proposals should also utilize the knowledge and technologies of the scientific communities to help aid in developing practical solutions to these challenges. Projects must include foreign collaborations and may not exceed two years. Funding may be allocated to foreign collaborators through sub-awards. All proposals must match at least 25% of the requested funds; third party cost sharing is allowed.
*Restrictions:* Projects may not exceed two years. Funding will be on a cost reimbursable basis. FAS reserves the right to allocate partial funding or terminate funding at any time. Scholarships, stipends, sabbaticals, tuition, tuition remission, general purpose equipment (no particular scientific, technical, or programmatic purpose), entertainment, capital improvements, thank you gifts, and other expenses not directly related to the project are not eligible for funding.
*Geographic Focus:* All States
*Date(s) Application is Due:* Jul 5
*Amount of Grant:* Up to 45,000 USD
*Contact:* Khaliaka Meardry, Program Contact; (202) 690-1866; Khaliaka.Meardry@fas.usda.gov or SCRPProposals@fas.usda.gov
Nora Banks, Program Contact; (202) 720-3904; Nora.Banks@fas.usda.gov
*Internet:* www.fas.usda.gov/icd/SCRP/default.asp
*Sponsor:* U.S. Department of Agriculture
1400 Independence Avenue SW
Washington, D.C. 20250

**USDA Small Business Innovation Research Grants - Phase I** **2609**

The Small Business Innovation Research Grant (SBIR) - Phase I provides funding to qualified small businesses to support high quality, advanced concepts research related to important scientific problems and opportunities in agriculture that could lead to significant public benefit if successful. Objectives of the SBIR program include the following: stimulating technological innovation in the private sector; strengthening the role of small businesses in meeting federal research and development needs; increasing private sector commercialization of innovations derived from USDA-supported research and development efforts; and fostering and encouraging participation by women-owned and socially and economically disadvantaged small business firms in technological innovation. Selected areas for research are forests and related resources; plant production and protection; animal production and protection; air, water, and soils; food science and nutrition; rural and community development; aquaculture; and industrial applications. Phase I awards will be made for periods normally not to exceed six months to determine the scientific or technical feasibility of ideas in the selected research areas. Phase II awards will be made to firms with approaches that appear sufficiently promising as a result of Phase I studies for a period not to exceed 24 months. The electronic application, specific guidelines, and abstracts from previously funded projects are available on the website.
*Requirements:* Small, for-profit businesses are eligible to apply. Each organization submitting a proposal must qualify as a small business for research or research and development purposes. Participation by university faculty or government scientists in SBIR projects is strongly encouraged.
*Restrictions:* Phase I winners who have previously been rejected for Phase II funding are not eligible to apply. Funding is not available for loans or establishing a business.
*Geographic Focus:* All States
*Date(s) Application is Due:* Jul 13
*Amount of Grant:* 70,000 - 100,000 USD
*Contact:* Dr. Charles Cleland, National Program Leader; (202) 401-4002 or (202) 401-5852; fax (202) 401-6070; ccleland@nifa.usda.gov
*Internet:* www.csrees.usda.gov/fo/fundview.cfm?fonum=1220
*Sponsor:* U.S. Department of Agriculture
1400 Independence Avenue, SW, Stop 2201
Washington, D.C. 20250-2201

**USDA Small Business Innovation Research Grants - Phase II** **2610**

The USDA Small Business Innovation Research Grants (SBIR) - Phase II provide funding to qualified small businesses to support high quality, advanced concepts research related to important scientific problems and opportunities in agriculture that could lead to significant public benefit if successful. Objectives of the SBIR program include the following: stimulating technological innovation in the private sector; strengthening the role of small businesses in meeting Federal research and development needs; increasing private sector commercialization of innovations derived from USDA-supported research and development efforts; and fostering and encouraging participation by women-owned and socially and economically disadvantaged small business firms in technological innovation. The electronic application, specific guidelines, and abstracts from previously funded projects are available on the website.
*Requirements:* To be eligible for Phase II, applicants must be Phase I winners as described in the Request for Application. Small businesses and small proprietorships that are in business for profit are eligible to submit applications to this program. Each organization submitting a proposal must qualify as a small business concern for research or research and development purposes. Participation by university faculty or government scientists in SBIR projects is strongly encouraged.
*Restrictions:* Phase I winners who have previously been rejected for Phase II funding are not eligible to apply. Funding is not available for loans or establishing a business.
*Geographic Focus:* All States
*Date(s) Application is Due:* Mar 1
*Amount of Grant:* Up to 450,000 USD
*Contact:* Charles Cleland, National Program Leader; (202) 401-5852 or (202) 401-4002; fax (202) 401-6070; ccleland@nifa.usda.gov
*Internet:* www.nifa.usda.gov/fo/sbir.cfm
*Sponsor:* U.S. Department of Agriculture
1400 Independence Avenue, SW, Stop 2201
Washington, D.C. 20250-2201

**USDA Special Research Grants** **2611**

The program supports research to facilitate or expand promising breakthroughs in areas of the food and agricultural sciences of importance to the nation and to facilitate or expand ongoing state-federal food and agricultural research programs. Areas of basic and applied research are generally limited to high-priority problems of a regional or national scope. Areas currently considered are water quality, integrated pest management, potato breeding, and range land research. Application deadlines are announced each fiscal year in the Federal Register and may also be obtained by calling the Office of Extramural Programs.
*Requirements:* State agricultural experiment stations, all colleges and universities, other research institutions and organizations, federal agencies, private organizations or corporations, and individuals having a demonstrable capacity to conduct research to facilitate or expand promising breakthroughs in areas of the food and agricultural sciences of importance to the United States are eligible to apply.
*Geographic Focus:* All States
*Amount of Grant:* Up to 95,000,000 USD
*Contact:* Andrea L. Brandon; (202) 720-9181; abrandon@nifa.usda.gov
*Internet:* www.csrees.usda.gov/funding/rfas/pmap.html
*Sponsor:* U.S. Department of Agriculture
1400 Independence Avenue, SW, Stop 2201
Washington, D.C. 20250-2201

**USDA Technical Agricultural Assistance Research Grants** **2612**

The Technical Agricultural Assistance Research Grants identify and apply the most appropriate solutions to international agricultural problems, and increase the capabilities of U.S. educational institutions and nonprofit agencies in agricultural technical assistance and research. Technical assistance provided through these reimbursable agreements benefit foreign governments and related agricultural institutions in their countries. Examples of previous projects include the following: development of effective management concepts and methodologies for implementation of technical agricultural and environmental protection projects; hurricane rehabilitation activities in Central America and the Caribbean, emphasizing watershed rehabilitation, food safety, and food security; sound natural resources management, agricultural and environmental policy reform; micro-finance for economic and rural development in developing countries; and strengthening of host government institutions in post-conflict environments. Programs currently funded include provision of technical assistance for expanding the use of effective program management concepts and technologies in developing countries to promote food security; guidance on natural resources management policy, procedures and techniques; trade capacity building; post-disaster and post-conflict reconstruction; and integrated approaches to international agricultural development and environmental protection. Cooperative agreements are generally funded for one to two years. Specific guidelines, including how to submit requests for proposal packages, are available at the USDA/FAS website.

*Requirements:* U.S. institutions of higher learning, and public and private/nonprofit organizations whose primary purpose is agriculture, natural resources management, and/or rural development (including those located in U.S. territories).

*Geographic Focus:* All States

*Amount of Grant:* 30,000 - 600,000 USD

*Contact:* Grant Pettrie; (202) 690-1924; grant.pettrie@usda.gov

*Internet:* https://www.cfda.gov/index?s=program&mode=form&tab=core&id=3e3bdecd95f214083e79af9ad9a83de3

*Sponsor:* U.S. Department of Agriculture
1400 Independence Avenue SW
Washington, D.C. 20250

**USDA Tribal College Research Grants** **2613**

The USDA Tribal College Research Grants are designed to assist 1994 Land-Grant Institutions (Tribal Colleges) in building institutional research capacity through applied projects that address student educational needs and solve community, reservation or regional problems. Awards are available to conduct applied research that enhances educational or career prospects of American Indians through student-developed research. Collaboration with 1862 or 1890 Land Grant Institutions is a requirement. Eligible institutions may propose projects in any discipline of the food and agricultural sciences. Examples include but are not limited to: developing new techniques for increased/more efficient bio-energy and biomass conversion through implementing sustainable use of natural resources; research on sustainable agriculture, sustainable forestry, biotechnology, agribusiness management and marketing, or aquaculture; research on human nutrition, including health and obesity prevention; research on natural resource issues including phyto-remediation of soils; water quality and quantity; alternative and sustainable energy sources; waste management; and ecology of grasslands, forest and animal systems; and research on the interventions and approaches that promote the participation of American Indian students in the food and agricultural sciences. The electronic application, specific guidelines (including tribal colleges eligible for funding), evaluation criteria, and abstracts from previously funded projects are available at the website.

*Requirements:* Applicants must be 1994 land grant institutions. Grant applications submitted under this program must certify that the research will be performed in collaboration with at least one 1862 or 1890 Land-Grant college or university.

*Restrictions:* Colleges must collaborate with an 1862 or 1890 land grant institution, but as the lead institution, the 1994 land grant institution must retain at least 60% of the award funds. The use of grant funds to plan, acquire, or construct a building or facility is not allowed under this program. With prior approval, some grant funds may be used for minor alterations, renovations, or repairs deemed necessary to retrofit existing teaching or research spaces in order to carry out a funded project. However, requests to use grant funds for such purposes must demonstrate that the alterations, renovations, or repairs are essential to achieving the major purpose of the project. Grant funds may not be used for endowment investing.

*Geographic Focus:* Arizona, California, Kansas, Michigan, Minnesota, Montana, Nebraska, New Mexico, North Dakota, South Dakota, Washington, Wisconsin

*Date(s) Application is Due:* Mar 14

*Amount of Grant:* 35,000 - 200,000 USD

*Contact:* Tim Grosser; (202) 690-0402; fax (202) 720-3945; tgrosser@nifa.usda.gov

*Internet:* www.nifa.usda.gov/fo/tribalcollegesresearchtcrgp.cfm

*Sponsor:* U.S. Department of Agriculture
1400 Independence Avenue, SW, Stop 2201
Washington, D.C. 20250-2201

**USDA Wildlife Services Grants** **2614**

The objectives of the Wildlife Services Grants are to reduce damage caused by mammals and birds and those mammal and bird species that are reservoirs for zoonotic diseases. Wherever feasible, humane methods will be employed. Recipients are to work closely with state departments of fish and game, agriculture, health, and counties to alleviate wild animal damage. Those who receive funding are expected to conduct surveys and campaigns to reduce wild animal damage, develop methods to control wild animal damage, and provide technical advice and assistance. There are no deadlines and funding is available year round. Specific guidelines and additional information is available at the CFDA website (project 10.028). For direct technical assistance, state fish and game departments should be contacted.

*Requirements:* State and local governments, federally recognized tribal governments, public/private nonprofit organizations, nonprofit institutions of higher education, and individuals are eligible to apply.

*Geographic Focus:* All States

*Contact:* Eileen M. Berke; (301) 734-8330; eileen.m.berke@aphis.usda.gov

*Internet:* https://www.cfda.gov/index?s=program&mode=form&tab=step1&id=12c2396f3a0d6463d2d9132339523995

*Sponsor:* U.S. Department of Agriculture
4700 River Road
Riverdale, MD 20737

**USDC/NOAA Anadromous Fish Conservation Act Grants** **2615**

These projects are conducted for the conservation, development, and enhancement of anadromous fishery resources (those that migrate from salt to fresh water for spawning) including similar species in the Great Lakes and Lake Champlain. All projects must be cleared with the State fishery agency of the State that the work is carried out in. Many projects funded under this program are critical elements of larger multi-funded programs to manage, restore or enhance United States anadromous species such as Pacific and Atlantic salmon, Atlantic and shortnose sturgeon, and American Shad and river herrings. Funds may be used for spawning area improvement, installment of fishways, data collection, construction of fish protection devices and hatcheries, and research to improve management and increase anadromous fish resources. This program is also supported by the Fish and Wildlife Service of the DOI. Applications are accepted at any time for one- to three-year projects. The project application should be submitted 90 days in advance of desired effective date.

*Requirements:* Any interested person or organization may propose a cooperative undertaking; however, all proposals must be coordinated with the state fishery agency having responsibility for the resource to be affected by the proposal.

*Restrictions:* Funds may not be used for law enforcement, public relations, or construction of facilities and vessels, whose primary purpose is to commercially harvest, handle, and process fishery products and may not be used for projects in the Columbia River Basin with the exception of the state of Idaho. The federal share of a project cost shall usually not exceed 50 percent. Federal funds for projects supporting an interstate fishery management plan may be up to 90 percent.

*Geographic Focus:* All States

*Amount of Grant:* 2,000 - 400,000 USD

*Contact:* Michelle Winowitch, (301) 713-3155; fax (301) 713-0596; Michelle.Winowitch@noaa.gov or grants.information@noaa.gov

*Internet:* www.nmfs.noaa.gov/sfa/state_federal/State-Federal-WEB/afca.htm

*Sponsor:* U.S. Department of Commerce
1401 Constitution Avenue, NW, Room 5128
Washington, D.C. 20230

**USDC/NOAA Coastal and Marine Fish Habitat Protection Grants** **2616**

NOAA Fisheries and the National Fish and Wildlife Foundation are soliciting projects to support the protection of coastal and marine fish habitats through NFHP Fish Habitat Partnerships. Successful proposals will: provide protection of coastal and marine fish habitats through cooperative, non-regulatory approaches as a primary benefit of the project; demonstrate how proposed actions will result in quantifiable fish habitat protection; address a strategic habitat protection priority of one or more of the eight coastally-focused Fish Habitat Partnerships, as identified in a Partnership's strategic plan; support knowledge transfer and lessons learned to other entities protecting fish habitat; and be supported by the relevant NFHP Fish Habitat Partnership, local constituencies and affected landowners. The total funding available for this program is $135,000. Awards considered for funding under this initiative are likely to fall within the $15,000 to $50,000 range and should meet project selection criteria above while demonstrating a strong, on-the ground benefit to coastal or marine habitat(s). Applicants are encouraged to provide non-federal match of $2 for every $1 of grant funds requested. Eligible non-federal matching sources can include cash, in-kind donations, and/or volunteer labor.

*Requirements:* All persons, organizations, and agencies (excluding U.S. Federal Government) working on projects to protect coastal and marine fish habitats are eligible to apply for funding.

*Restrictions:* Federal agencies may not receive funds, but may be listed as partners on an application. No part the grant funds or the non-federal match may be used to: support litigation expenses or lobbying activities; cover permanent federal employee salary expenses; supplement shortfalls in government agency budgets; support basic research; support basic planning, outreach, or education projects without an on-the-ground component.

*Geographic Focus:* All States

*Date(s) Application is Due:* Jun 29

*Amount of Grant:* 15,000 - 50,000 USD

*Contact:* Cara Rose, Assistant Director, Western Partnership Office; (503) 417-8700, ext. 6008; cara.rose@nfwf.org

*Internet:* www.nfwf.org/AM/Template.cfm?Section=Search&TEMPLATE=/CM/ContentDisplay.cfm&CONTENTID=24847

*Sponsor:* U.S. Department of Commerce
1401 Constitution Avenue, NW, Room 5128
Washington, D.C. 20230

**USDC/NOAA Community-Based Habitat Restoration Program Grants** **2617**

The Nature Conservancy (TNC) and the National Oceanic and Atmospheric Administration (NOAA) Community-based Restoration Program (CRP) is intended to bring together interested groups, public, private, tribal and non-profit organizations to implement habitat restoration projects to benefit NOAA trust resources (coastal and

marine species and their habitats). This innovative program recognizes the significant role that partnerships can play in making habitat restoration happen within communities, and acknowledges that habitat restoration is often best implemented through technical and monetary support provided at a community level. The program funds grass-roots marine and coastal habitat restoration projects that will benefit andromous fish species, commercial and recreational resources, and endangered and threatened species. The program invites the public to submit proposals for available funding to implement grass-roots habitat restoration projects that will benefit living marine resources, including andromous fish. The Federal Funding Opportunity describes the conditions under which applications (project proposals) will be accepted under the CRP and describes criteria under which applications will be evaluated for funding consideration. Projects funded through the CRP will be expected to have strong on-the-ground habitat restoration components that provide educational and social benefits for people and the communities in addition to long-term ecological habitat improvements for NOAA trust resources. Proposals selected for funding through this solicitation will be implemented through a project grant or cooperative agreement mechanism. Preference will be given to projects at priority sites identified through Marine Ecoregional Assessments and other TNC priority setting approaches at the state and territory level.
*Requirements:* Eligible applicants are institutions of higher education, hospitals, other non-profits, commercial (for profit) organizations, organizations under the jurisdiction of foreign governments, international organizations, and state, local and Indian tribal governments whose projects have the potential to benefit NOAA trust resources. Non-federal match is required at the rate of 1:1. The 1:1 match can come from a variety of (non-federal) public and private sources and can include in-kind goods and services from project partners.
*Restrictions:* Applications from federal agencies or employees of Federal agencies will not be considered.
*Geographic Focus:* All States
*Date(s) Application is Due:* Apr 13
*Amount of Grant:* 25,000 - 250,000 USD
*Contact:* Boze Hancock; (401) 874-6121; bhancock@tnc.org
*Internet:* www.habitat.noaa.gov/funding/northeast.html
*Sponsor:* U.S. Department of Commerce
1401 Constitution Avenue, NW, Room 5128
Washington, D.C. 20230

**USDC/NOAA Community-Based Marine Debris Removal Grants** 2618
The Program provides funding to catalyze the implementation of locally driven, community-based marine debris prevention and removal projects that will benefit coastal habitat, waterways, and NOAA trust resources including diadromous fish. Projects funded through the MDP have strong on-the-ground habitat components involving the removal of marine debris and derelict fishing gear that will provide educational and social benefits for people and their communities in addition to long-term ecological habitat improvements for NOAA trust resources. The principal objective of the Program is to provide federal financial and technical assistance to grass-roots, community-based activities that improve living marine resource habitats through the removal of marine debris and the prevention of debris entering the marine environment, and that promote stewardship and a conservation ethic for NOAA trust resources. Generally, costs that are allowable include salaries and equipment and supplies (as long as these are necessary and reasonable specifically for the purpose of the award). Typical project awards will range from $15,000 to $150,000. NOAA will not accept proposals for less than $15,000 or proposals for more than $250,000 under this solicitation.
*Requirements:* Eligible applicants are institutions of higher education, other non-profits, commercial (for profit) organizations, organizations under the jurisdiction of foreign governments, international organizations, and state, local and Indian tribal governments whose projects have the potential to benefit NOAA trust resources. Applicants are encouraged to demonstrate a minimum 1:1 non-Federal match for MDP funds requested to conduct the proposed project.
*Restrictions:* Applications from federal agencies or employees of Federal agencies will not be considered.
*Geographic Focus:* All States
*Date(s) Application is Due:* Nov 1
*Amount of Grant:* 15,000 - 250,000 USD
*Contact:* Tom Barry, (301) 427-8653; tom.barry@noaa.gov
*Internet:* www.habitat.noaa.gov/funding/marinedebris.html
*Sponsor:* U.S. Department of Commerce
1401 Constitution Avenue, NW, Room 5128
Washington, D.C. 20230

**USDC/NOAA Marine Fisheries Initiative (MARFIN) Grants** 2619
The Marine Fisheries Initiative (MARFIN) is a competitive Federal assistance program that funds projects seeking to optimize research and development benefits from U.S. marine fishery resources through cooperative efforts involving the best research and management talents to accomplish priority activities. Projects funded under MARFIN provide answers for fishery needs covered by the NMFS Strategic Plan, available from National Marine Fisheries Service (NMFS), particularly those goals relating to: rebuilding over-fished marine fisheries, maintaining currently productive fisheries, and integrating conservation of protected species and fisheries management. Funding priorities for MARFIN are formulated from recommendations received from non-federal scientific and technical experts and from NMFS research and operations officials. The NMFS Southeast Regional Office anticipates awarding approximately ten projects that will range from $25,000 to $175,000 per year for each project (not to exceed $175,000 per year). The total federal amount that may be requested shall not exceed $175,000 for a one-year project, $350,000 for a two-year project, and $525,000 for a three-year project.
*Requirements:* Eligible applicants may be institutions of higher education, nonprofits, commercial organizations, individuals, state, local and Indian tribal governments.
*Restrictions:* Federal agencies or institutions are not eligible. Foreign governments, organizations under the jurisdiction of foreign governments, and international organizations are excluded for purposes of this solicitation
*Geographic Focus:* All States
*Date(s) Application is Due:* Aug 9
*Amount of Grant:* 25,000 - 525,000 USD
*Contact:* Ellie Francisco Roche; (727) 824-5324; Ellie.Roche@noaa.gov
*Internet:* sero.nmfs.noaa.gov/grants/marfin.htm
*Sponsor:* U.S. Department of Commerce
1401 Constitution Avenue, NW, Room 5128
Washington, D.C. 20230

**USDC/NOAA National Marine Aquaculture Initiative Grants** 2620
NOAA launched the National Marine Aquaculture Initiative (NMAI) in 1998 to encourage scientific research and development to support a robust, sustainable domestic marine aquaculture industry. The competition fosters dynamic partnerships that channel resources toward the development of sustainable aquaculture technologies. Projects often involve partnerships among commercial companies, research institutions, universities, state governments, and coastal communities. Currently, NOAA Sea Grant's National Marine Aquaculture Initiative is making $3.2 million available through a Sea Grant Aquaculture Research competitive grants program. Funds will support marine aquaculture research projects as part of the overall plan to support the development of environmentally and economically sustainable ocean, coastal, and Great Lakes aquaculture. Priorities for this competition include: research to inform specific regulatory decisions; research that supports multi-use spatial planning; and socio-economic research targeted to understand aquaculture in a larger context. The deadline for receipt of preproposals via electronic mail at the National Sea Grant Office is 4:00 p.m. Eastern Time on February 7.
*Requirements:* Institutions of higher education, nonprofit organizations, commercial organizations, State, local and Indian tribal governments and individuals are eligible.
*Restrictions:* Federal agencies and their personnel are not permitted to receive federal funding under this competition; however, federal scientists can serve as uncompensated partners or co-Principal Investigators on research proposals. Directors of the state Sea Grant Programs are not eligible to compete for funds under this announcement, although for administrative purposes, they will be considered to be the Principal Investigator for all awards made to their state programs.
*Geographic Focus:* All States
*Date(s) Application is Due:* Feb 7
*Amount of Grant:* 50,000 - 500,000 USD
*Contact:* Dr. Gene Kim, (301) 734-1281; oar.hq.sg.aquaculture@noaa.gov
*Sponsor:* U.S. Department of Commerce
1401 Constitution Avenue, NW, Room 5128
Washington, D.C. 20230

**USDC/NOAA National Marine Fisheries Service Sea Grant Joint Graduate Fellowships** 2621
NOAA National Sea Grant Office and NOAA Fisheries established a Graduate Fellowship Program in two specialized areas: population dynamics and marine resource economics. Population dynamics is the study of fish populations as affected by fishing mortality, growth, recruitment and natural mortality. Ph.D. candidates interested in the population dynamics of living marine resources and the development and implementation of quantitative methods for assessing their status can receive up to three years of funding. Ph.D. students in marine resource economics, concentrating on the conservation and management of living marine resources, can receive two years of funding. The four main goals of the NOAA Fisheries/Sea Grant Fellowship Program are: to encourage qualified applicants to pursue careers in either population dynamics and stock assessment or in marine resource economics; to increase available expertise related to these fields; to foster closer relationships between academic scientist and NOAA Fisheries; and to provide real-world experience to graduate students and accelerate their career development.
*Requirements:* The Fisheries Fellowship is available to U.S. citizens who are graduate students enrolled in PhD degree programs in academic institutions in the United States and its territories.
*Geographic Focus:* All States
*Date(s) Application is Due:* Jan 20
*Contact:* Terry Smith, Program Manager; (301) 734-1077, ext. 1075; fax (301) 713-0799; terry.smith@noaa.gov
*Internet:* www.seagrant.noaa.gov/funding/rfp.html#nmfs
*Sponsor:* U.S. Department of Commerce
1401 Constitution Avenue, NW, Room 5128
Washington, D.C. 20230

**USDC/NOAA Nature Conservancy Community-based Matching Grants** 2622
The objectives of TNC and NOAA's Community-based Restoration Program (CRP) are to bring together interested groups, public, private, tribal and non-profit organizations to implement habitat restoration projects to benefit NOAA trust resources (coastal and marine species and their habitats). This innovative program recognizes the significant role that partnerships can play in making habitat restoration happen within communities, and acknowledges that habitat restoration is often best implemented through technical and monetary support provided at a community level. NOAA and TNC are looking to support community-based restoration of a diversity of habitat types and no habitats are excluded. The office will consider any innovative restoration project nationwide that supports NOAA trust resources, and particularly those projects that have a multi-species

benefit or emphasize Ecosystem-Based Management. A focal area of particular interest, though not exclusive or limiting, is native shellfish (bivalve) restoration projects. Typical grants will be in the range of $25,000 to $85,000.
*Geographic Focus:* All States
*Date(s) Application is Due:* Apr 13
*Contact:* Boze Hancock; (401) 874-6121; bhancock@tnc.org
*Internet:* www.habitat.noaa.gov/funding/northeast.html
*Sponsor:* U.S. Department of Commerce
1401 Constitution Avenue, NW, Room 5128
Washington, D.C. 20230

**USDC/NOAA Restoration Center National and Regional Partnership Grants 2623**
For more than 20 years, the NOAA Restoration Center has worked to maintain strong relationships with a number of non-profit organizations focused on restoring coastal habitat nationwide. These partnerships help the Restoration Center leverage greater funds and effort for restoration through match and in-kind contributions such as funding, land easements, staff expertise, volunteer support, and other services. In return, the NOAA Restoration Center offers technical and permitting assistance in restoring habitats from coast to coast. The office offers funding for multi-year national and regional habitat restoration Partnerships that will result in implementation of a wide-range of habitat restoration projects - from locally-driven, hands-on projects that emphasize stewardship, to mid-scale, watershed-scale projects that yield significant ecological and socioeconomic benefits. Grants range from $500,000 to $1,000,000.
*Requirements:* Eligible applicants are institutions of higher education, non-profits, commercial (for profit) organizations, U.S. Territories, and state, local and Indian tribal governments.
*Restrictions:* Applications from Federal agencies or employees of Federal agencies will not be considered. Federal agencies are strongly encouraged to work with states, non-governmental organizations, municipal and county governments, conservation corps organizations and others that are eligible to apply.
*Geographic Focus:* All States
*Date(s) Application is Due:* Sep 30
*Amount of Grant:* 500,000 - 1,000,000 USD
*Contact:* Robin Bruckner, (301) 713-0174; Robin.Bruckner@noaa.gov
Melanie Gange, (301) 713-0174; Melanie.Gange@noaa.gov
*Internet:* www.habitat.noaa.gov/funding/partnerships.html
*Sponsor:* U.S. Department of Commerce
1401 Constitution Avenue, NW, Room 5128
Washington, D.C. 20230

**USDC/NOAA Saltonstall-Kennedy Grants 2624**
The Saltonstall-Kennedy Act established a fund (known as the S-K fund) that the Secretary of Commerce uses to provide grants or cooperative agreements for fisheries research and development projects addressing aspects of U.S. fisheries, including, but not limited to, harvesting, processing, marketing, and associated infrastructures. U.S. fisheries include any fishery, commercial or recreational, that is, or may be, engaged in by citizens or nationals of the United States, or citizens of the Northern Mariana Islands (NMI), the Republic of the Marshall Islands, Republic of Palau, and the Federated States of Micronesia.
*Requirements:* Any U.S. citizen or national, or any citizen of the Northern Mariana Islands, the Republic of the Marshall Islands, Republic of Palau, or the Federated States of Micronesia; or group including state and local governments may apply.
*Restrictions:* Employees of federal agencies and regional fishery management councils and their employees are not eligible.
*Geographic Focus:* All States
*Date(s) Application is Due:* Oct 1
*Amount of Grant:* 100,000 - 300,000 USD
*Contact:* Daniel Namur, Program Manager; (301) 713-2259; Dan.Namur@noaa.gov
*Internet:* www.nmfs.noaa.gov/mb/financial_services/skhome.htm
*Sponsor:* U.S. Department of Commerce
1401 Constitution Avenue, NW, Room 5128
Washington, D.C. 20230

**USDC Afghanistan and Iraq Investment and Reconstruction Task Forces Internships 2625**
The Afghanistan and Iraq Investment and Reconstruction Task Forces of the U.S. Department of Commerce serve as clearinghouses of information for U.S. companies interested in doing business in Iraq and Afghanistan. The Task Force works closely with other U.S. government agencies, Afghan and Iraqi government agencies, and international organizations to provide U.S. companies counseling on the latest developments and business opportunities in Iraq and Afghanistan. The Task Forces seek unpaid international trade interns for one or two competitive positions each semester (Spring, Summer, Fall). This is an excellent opportunity for students to learn about how the U.S. government operates and contributes to Afghanistan and Iraq reconstruction efforts. Interns work on both Task Forces.
*Requirements:* Qualifications include: concurrently enrolled undergraduate or graduate students interested in learning how the U.S. government operates and contributes to Afghanistan and Iraq reconstruction efforts; majoring in international affairs, international trade, economics, or business; excellent research, writing, and computer skills; available at least 20 hours per week; and U.S. citizenship required.
*Geographic Focus:* All States
*Contact:* Tracy M. Rollins; (202) 482-0073; fax (202) 482-2443; SABITApply@mail.doc.gov
*Internet:* www.trade.gov/afghanistan/tg_aftf_003400.asp
*Sponsor:* U.S. Department of Commerce
1401 Constitution Avenue, NW
Washington, D.C. 20230

**USDC Dissertation Fellowships 2626**
The Census Bureau Dissertation Fellowship program seeks to award a limited number of contracts each year to Ph.D. -granting accredited institutions of higher education in the United States to fund the dissertation research of doctoral candidates who propose and investigate solutions to problems on the list of research topics of primary interest to the Census Bureau. When fully implemented, up to ten dissertation fellowships will be awarded annually for up to $50,000 each. Dissertation Fellowships will be awarded to institutions of higher education for doctoral candidates of unusual promise and ability whose general fields of research relate to statistics, economics, and demography. Direct connections to the cited listing of research topics is essential. Exceptionally strong candidates writing master's theses that address these problems will also be considered.
*Requirements:* The application materials and supporting documents for each Dissertation Fellow must be submitted by his or her university. A Dissertation Fellow must: be a doctoral candidate or exceptionally strong master's candidate; be funded by a contract awarded to institution for tuition, research materials, travel, and salary; devote full-time effort, be enrolled full-time, and in residence at Fellow's Institution; report progress biannually; collaborate on-site with Census Bureau staff for one to two months; and present research results in a Census Bureau seminar. United States Citizenship is required.
*Geographic Focus:* All States
*Date(s) Application is Due:* Feb 28
*Amount of Grant:* Up to 50,000 USD
*Contact:* Dr. Tommy Wright; (301) 763-1702; fax (301) 763-8399; tommy.wright@census.gov
Carol A. Gentry, carol.a.gentry@census.gov
*Internet:* www.census.gov/research/opportunities/
*Sponsor:* U.S. Department of Commerce
4600 Silver Hill Road
Washington, D.C. 20233

**USDC Low-Power Television and Translator Upgrade Grants 2627**
The Low-Power Television and Translator Upgrade program reimburses the eligible costs to upgrade analog low power television broadcast stations, Class A television stations, television translator stations, and television booster stations in rural communities to digital transmission. Applications are received the first business day of each month, and the final closing date for receipt of Upgrade program applications will be July 2.
*Requirements:* Eligible stations must be broadcasting a licensed digital signal before applying for a grant.
*Geographic Focus:* All States
*Date(s) Application is Due:* Jul 2
*Contact:* Dr. Bernadette McGuire-Rivera, Associate Administrator; (202) 482-5802; fax (202) 501-8009; bmcguire-rivera@ntia.doc.gov
*Internet:* www.ntia.doc.gov/category/low-power-television-and-translator-program
*Sponsor:* U.S. Department of Commerce
1401 Constitution Avenue, NW
Washington, D.C. 20230

**USDC Market Development Cooperator Program (MDCP) Grants 2628**
The goal of the Market Development Cooperator Program is to develop, maintain, and expand foreign markets for non-agricultural goods and services produced in the United States. The intended beneficiaries of the program are U.S. producers of non-agricultural goods or services that seek to export such goods or services. The International Trade Administration (ITA) encourages applicants to propose activities that would be most appropriate to the market development needs of their U.S. industry or industries. Examples of activities might include: opening an overseas office or offices to perform a variety of market development services for companies joining a consortium to avail themselves of such services; detailing a private-sector representative to a US&FCS post in accordance with 15 USC 4723(c); commissioning overseas market research, participating in overseas trade exhibitions, and trade missions to promote U.S. exports, and/or hosting reverse trade missions; and conducting U.S. product demonstrations abroad. The applicant must contribute at least two dollars for each federal dollar provided. Full guidelines are available on the website or may be obtained by contacting the office listed. No award will exceed $300,000. ITA anticipates concluding seven (7) to ten (10) awards.
*Requirements:* U.S. trade associations; nonprofit industry organizations; state trade departments and their regional associations, including centers for international trade development; and private industry firms or groups of firms in cases where no entity described above represents that industry are eligible.
*Geographic Focus:* All States
*Date(s) Application is Due:* Feb 21
*Amount of Grant:* Up to 300,000 USD
*Contact:* Brad Hess; (202) 482-2969; fax (202) 482-5828; Brad.Hess@trade.gov
*Internet:* ita.doc.gov/td/mdcp/
*Sponsor:* U.S. Department of Commerce - 1
14th Street and Constitution Avenue NW, Herbert C Hoover Building
Washington, D.C. 20230

**USDC Planning and Local Technical Assistance Grants 2629**
The Economic Development Administration announces general policies and application procedures for grant-based investments under the Planning and Local Technical Assistance Programs. These programs will help communities develop the planning and technical expertise to support communities and regions in their comprehensive, entrepreneurial, and innovation-based economic development efforts. Resulting in increased private investment and higher-skill, higher-wage jobs in areas experiencing substantial and persistent economic distress, these programs are designed to enhance the competitiveness of regions. EDA's Planning and Local Technical Assistance Programs are two of the six economic development

assistance programs the agency operates. Under the Planning Program, EDA provides assistance to eligible recipients to create regional economic development plans in order to stimulate and guide the economic development efforts of a community or region. EDA's Local Technical Assistance Program helps eligible recipients fill the knowledge and information gaps that may prevent leaders in the public and nonprofit sectors in economically distressed regions from making optimal decisions on local economic development issues.
*Requirements:* Eligible applicants include: district organizations; Indian tribes or a consortium of Indian tribes; states, cities or other political subdivisions of a state, including special purpose units of a state or local government engaged in economic or infrastructure development activities, or a consortium of political subdivisions; institutions of higher education or a consortium of institutions of higher education; or public or private non-profit organizations or associations acting in cooperation with officials of a political subdivision of a state.
*Restrictions:* EDA is not authorized to provide grants directly to individuals or to for-profit entities seeking to start or expand a private business.
*Geographic Focus:* All States
*Contact:* Lindsey Pangretic, Director, Research and National Technical Assistance Division; (202) 482-4085 or (202) 482-2900; grants@eda.doc.gov
*Internet:* www.grants.gov/search/search.do;jsessionid=kDW2PsLT1zdv3HLW1Bpwx3yQyQbpJPt1XnmTfyM1yGJpBP99tt2g!-757993493?oppId=58876&mode=VIEW
*Sponsor:* U.S. Department of Commerce
1401 Constitution Avenue, NW, Suite 7800
Washington, D.C. 20230

**USDC Postdoctoral Research Grants** **2630**
The objectives of the Census Bureau Postdoctoral Research Program are to provide postdoctoral candidates of unusual promise and ability opportunities for research on problems, largely of their own choice, that are compatible with the interests of the Census Bureau, and to contribute to the overall efforts of federal statistical agencies. A Postdoctoral Researcher is a resident researcher and a temporary employee of the Census Bureau. Postdoctoral Researchers are analogous to fellows or similar temporary researchers at the postdoctoral level in universities and other organizations. Postdoctoral Research positions are not intended to be, or to compete with, permanent professional career positions. Awardees are offered two-year appointments, but extensions are possible with special considerations. An evaluation is conducted after one year to ensure that the Postdoctoral Researcher is making suitable progress. An applicant's training and research experience may be in any appropriate discipline or combination of disciplines required for the proposed research. The annual base salary for a Census Bureau Postdoctoral Researcher starts from $74,872 to $97,333 with adjustments depending on qualifications and experience. For fullest consideration, submit application by January 31.
*Requirements:* No commitment on the part of a Postdoctoral Researcher or the Census Bureau with regard to later employment is implied or should be inferred by the offer or acceptance of an award as a Postdoctoral Researcher. Postdoctoral Researchers must devote their full-time effort to the research program and must be in residence at the Census Bureau during the program. No period of tenure may be spent in residence at another agency or institution. No additional monetary aid or other remuneration may be accepted from another appointment, fellowship, or similar grant during the period of the program. Postdoctoral Research positions are awarded only to persons who have held the doctorate less than six years or are in the process of receiving a doctorate degree at the time of application. If a doctorate has been held for more than six years, refer to the ASA/NSF/Census Bureau Fellowship Program for another research opportunity. Awardees must hold the Ph.D. or other earned research degree recognized in the U.S. as equivalent to the Ph.D. or must present acceptable evidence of having completed all the formal academic requirements for the degree before appointment. Applicants must have demonstrated ability for creative research.
*Geographic Focus:* All States
*Amount of Grant:* 74,872 - 97,333 USD
*Contact:* Chris A. Cowan; (301) 480-3538; chris.cowan@hhs.gov
*Internet:* www.census.gov/hrd/www/jobs/prp.html
*Sponsor:* U.S. Department of Commerce
4600 Silver Hill Road
Washington, D.C. 20233

**USDC Presidential Management Fellowships** **2631**
For the past 27 years, the Presidential Management Fellows (PMF) Program and its predecessor, the Presidential Management Intern (PMI) Program, have been attracting outstanding masters, law, and doctoral-level students to the Federal service. The PMF Program seeks candidates who display excellent communication, interpersonal, planning and analytical skills, decision-making ability, and the potential to perform in leadership positions. This rigorous two-year paid program includes 80-hours of training each year, rotational opportunities, challenging assignments, and opportunities to network between agencies. The Census Bureau is supportive of the PMF program and has a strong commitment to training and professional development. Current and former PMFs work in numerous positions throughout the Bureau and engage in a variety of diverse tasks. As a PMF at Census, you could have the opportunity to work with current data on topics such as unemployment, business productivity, poverty, and health insurance. This data has a profound impact on policy formation, evaluation, and decision making in both the public and private sectors.
*Geographic Focus:* All States
*Contact:* Chris A. Cowan; (301) 480-3538; chris.cowan@hhs.gov
*Internet:* www.census.gov/hrd/www/jobs/pmip.html
*Sponsor:* U.S. Department of Commerce
4600 Silver Hill Road
Washington, D.C. 20233

**USDC State Broadband Initiative Grants** **2632**
The National Telecommunications and Information Administration's State Broadband Initiative implements the purposes of both the American Recovery and Reinvestment Act and the Broadband Data Improvement Act, which envisioned a comprehensive program — led by state entities or non-profit organizations working at their direction — to help integrate broadband and information technology into state and local economies. Through the State Broadband Initiative (formerly called the State Broadband Data and Development grant program), NTIA awarded a total of $293 million to 56 grantees, one from each of the 50 states, five territories, and the District of Columbia, or their designees. Grantees are using this funding to support the efficient and creative use of broadband technology to better compete in the digital economy. These state-created efforts vary depending on local needs but include programs to assist small businesses and community institutions in using technology more effectively, conduct research on barriers to broadband adoption, implement innovative applications that increase access to government services and information, and support state and local task forces to expand broadband access and adoption. Since accurate data is critical for broadband planning, another purpose of the State Broadband Initiative is to assist states in gathering data twice a year on the availability, speed, and location of broadband services, as well as the broadband services that community institutions use. The data will be used by NTIA to update its publicly searchable, interactive National Broadband Map.
*Geographic Focus:* All States
*Contact:* Dr. Bernadette McGuire-Rivera, Associate Administrator; (202) 482-5802; fax (202) 501-8009; bmcguire-rivera@ntia.doc.gov
*Internet:* www.ntia.doc.gov/category/state-broadband-initiative
*Sponsor:* U.S. Department of Commerce
1401 Constitution Avenue, NW
Washington, D.C. 20230

**USDD Acoustic Transduction Materials and Devices Grants** **2633**
Research on innovative materials and device structures is sought that enhances the performance of the electro-mechanical transducers used by the Department of Navy to generate, detect and suppress undersea sound waves. The present Office of Naval Research program has two thrusts. The first aims to devise and validate first-principles quantum-mechanical methods to evaluate the properties of materials with structural phase transitions; this will enable the exploration of the properties of new materials systems in advance of their synthesis. The second focuses on the new high-coupling, high-strain relaxor piezoelectric single crystals; efforts underway aim to produce and characterize these piezocrystals and to design/fabricate innovative transducers from them for Navy SONAR systems. Applications include: Mine Identification SONAR; UUV Acoustic Modem; Vector Sensor Towed Arrays; Echocardiography, Medical Ultra-Sound; and 4D Realtime Volumetric Harmonic Imaging. Submission of the Research Proposal should always be preceded by submission of a Preproposal and a discussion of that Preproposal with the assigned Program Officer. Applications for the upcoming fiscal year should be received by September 30.
*Requirements:* Grants are awarded to academic, industrial, and nonprofit organizations, with strong emphasis on university research. Investigators are encouraged to contact the staff prior to formally submitting a proposal.
*Restrictions:* New grants or contracts are typically started near the beginning of the fiscal year. New proposals for the next fiscal year should be submitted no later than September 30, unless otherwise directed. Competitive renewals should be submitted no later than six months prior to the award termination date after contacting the Program Officer. Conference proposals should be submitted at least four months prior to the conference.
*Geographic Focus:* All States
*Date(s) Application is Due:* Sep 30
*Contact:* Dr. Wallace Smith, Program Officer; (703) 696-0284; fax (703 )696-0934; wallace.smith1@navy.mil
*Internet:* www.onr.navy.mil/Science-Technology/Departments/Code-33/All-Programs/332-naval-materials/Acoustic-Transduction-Materials-Devices.aspx
*Sponsor:* U.S. Department of Defense
One Liberty Center, 875 N. Randolph Street, Suite 1425
Arlington, VA 22203-1995

**USDD Active Aperture Array Grants** **2634**
The Office of Naval Research (ONR) Active Aperture Array Program seeks to develop innovative electronics technologies that will enable the revolutionary architectures and capabilities of future electronically scanned array (ESA) apertures. The emphasis is on enabling affordable solutions without sacrificing advanced performance at RF frequencies from microwave through millimeter wave. Research efforts are currently focused on: improvement of solid state device parameters including: switching speed,bandwidth, noise, dynamic range, linearity and robustness to dense signal environments; increasing the performance and efficiency of component technologies via advanced semiconductor III-V materials such as InP, GaN, SiC and others such as evolving nanotechnology materials; demonstrating size reduction of components needed to fit within an array lattice; and mixed analog and digital conversion circuits and signal processing architectures. Proposers are encouraged to contact the program officer to discuss their research interest prior to the submission of formal proposals. Grants for the upcoming fiscal year must be received by September 30.
*Requirements:* Grants are awarded to academic, industrial, and nonprofit organizations, with strong emphasis on university research.
*Restrictions:* New grants or contracts are typically started near the beginning of the fiscal year (October 1). New proposals for the next fiscal year should be submitted no later than September 30, unless otherwise directed by the BAA. Competitive renewals should be submitted no later than six months prior to the award termination date after

contacting the Program Officer. Conference proposals should be submitted at least four months prior to the conference.
*Geographic Focus:* All States
*Date(s) Application is Due:* Sep 30
*Contact:* Dr. Deborah Van Vechten, Program Officer; (703) 696-4219 or (703) 696-5031; fax (703) 696-5940; deborah.vanvechten@navy.mil
*Internet:* www.onr.navy.mil/Science-Technology/Departments/Code-31/All-Programs/312-Electronics-Sensors/Active-Aperture-Array.aspx
*Sponsor:* U.S. Department of Defense
One Liberty Center, 875 N. Randolph Street, Suite 1425
Arlington, VA 22203-1995

**USDD Active Authentication Grants** **2635**
DARPA is soliciting innovative research proposals in support of the development of new software-based biometric modalities. Proposed research should investigate innovative approaches that enable revolutionary advances in science, devices, or systems. The Active Authentication Grants program seeks to change the current focus from user proxies (e.g., passwords and CACs) when validating identity on the Department of Defense IT systems to a focus on the individual. Within this program, the intention is to focus on the unique factors that make up the individual, also known as their biometrics, without requiring the deployment of additional hardware sensors. Research resulting from these Grants will support that overall program intent by investigating novel software-based biometric modalities that can be used to provide meaningful and continual authentication when later integrated into a cyber security system. Application information is available on the website. Multiple awards not exceeding $500,000 per effort are expected.
*Requirements:* All responsible sources capable of satisfying Government requirements are eligible. Historically black colleges and universities, small businesses, small disadvantaged businesses and minority institutions are encouraged to apply.
*Restrictions:* Specifically excluded is research that primarily results in evolutionary improvements to the existing state of practice.
*Geographic Focus:* All States
*Date(s) Application is Due:* Mar 6
*Contact:* Richard Guidorizzi; (703) 526-6630; ActiveAuthentication@darpa.mil
*Internet:* www.grants.gov/search/search.do;jsessionid=Kk1NP54Tt0jpnMLC6J7vB0Sh53T9jH1YXXSfy5LnVN1hCmPD7TKp!-1993000306?oppId=138033&mode=VIEW
*Sponsor:* U.S. Department of Defense
3701 North Fairfax Drive
Arlington, VA 21702-1714

**USDD Advances in Bioscience for Airmen Performance Grants** **2636**
The USDD Air Force Research Laboratory Advances in Bioscience for Airmen Performance Grants are intended for innovative science and technology projects to support advanced bioscience research. Specifically, unique and innovative research concepts are sought in the following technical areas: applied biotechnology; biobehavioral performance; biomechanics; and counterproliferation. The purpose is to augment core research activities and facilities to provide radically new military capabilities that improve warfighter performance and combat effectiveness. The research will provide for a broad range of revolutionary and evolutionary technologies to meet the critical needs of the warfighter for full dimensional protection, survivability, and performance enhancement. Grants will be offered for multiple years, and applicants may submit white papers at any time. Applicants whose white papers are determined to meet the needs of the Air Force will be invited to submit a full proposal. Total annual funding is expected to be more than $8,000,000 with approximately ten awards being made per year.
*Requirements:* This is an unrestricted solicitation. Small businesses are encouraged to propose on all or any part of this solicitation.
*Geographic Focus:* All States
*Date(s) Application is Due:* Sep 30
*Contact:* Angela G. Lauofo; (937) 255-5937; Angela.Lauofo@wpafb.af.mil
Mark Fagan, Technical Specialist; (937) 904-9546; Mark.Fagan@wpafb.af.mil
*Internet:* www.grants.gov/search-grants.html?agencies%3DDOD|Department%20of%20Defense
*Sponsor:* U.S. Department of Defense
1400 Defense Pentagon
Washington, D.C. 20301-1400

**USDD Aerospace Medicine, Clinical Research, Human Performance Research, and Expeditionary Medicine Grants** **2637**
The USDD Air Force Research Laboratory is offering Aerospace Medicine, Clinical Research, Human Performance Research, and Expeditionary Medicine Grants. The Grants will support technology research primarily focused in the following research and technology areas. For aerospace and expeditionary medicine and clinical research the areas are: advanced diagnostics; therapeutics; occupational toxicology; expeditionary ground medicine and aeromedical evacuation; product developmental requirements; general acceleration and hypobaric test and evaluation; aircrew and battlefield airman health and performance; and clinical and translational research. For human performance and systems integration, research is needed in order to optimize warfighter effectiveness in all environments including, but not limited to: hot and humid, cold and dry, and high altitude. Interested applicants may submit white papers at any time. Applicants submitting white papers assessed as meeting Air Force needs will be asked to submit a technical and cost proposal. This is a five-year, open-ended solicitation. Five or six awards annually are anticipated.
*Requirements:* This is an unrestricted solicitation. Small businesses are encouraged to propose on all or any part of this solicitation.
*Geographic Focus:* All States
*Date(s) Application is Due:* Dec 15
*Contact:* Nicholas J. Voiles; 937-656-9808; nicholas.voiles@wpafb.af.mil
Nicholas Voiles, Contract Specialist; (937) 656-9808; nicholas.voiles@wpafb.af.mil
Kimberly Rhoads, Contract Negotiator; (937) 656-9763; kimberly.rhoads@wpafb.af.mil
*Internet:* www.grants.gov/search-grants.html?agencies%3DDOD|Department%20of%20Defense
*Sponsor:* U.S. Department of Defense
1400 Defense Pentagon
Washington, D.C. 20301-1400

**USDD Air Force Defense Research and Development Rapid Innovation Fund Grants** **2638**
Air Force Research Laboratory Contracting Directorate (AFRL/PK) is soliciting white papers for the Air Force Defense Research and Development Rapid Innovation Fund (RIF) Program to facilitate the rapid insertion of innovative technologies into military systems or programs meeting critical national security needs. This BAA is open to any technologies, products, or processes at the component, subsystem, or system level demonstrating near-term potential to address Air Force requirements or improve existing Air Force acquisition programs addressing specific areas outlined above. Proposal invitation will be determined based on evaluation of white papers using the criteria published in this BAA. Total program value is $50 million; however, no individual award is anticipated to exceed $3 million. The Government may provide additional funds to be used for new awards. The Air Force anticipates selecting 16 or more awards from this announcement. White papers are due no later than 3:00 pm EST, November 27.
*Requirements:* This is an unrestricted solicitation. Small businesses are encouraged to propose on all or any part of this solicitation. Cost share is encouraged but not required.
*Geographic Focus:* All States
*Date(s) Application is Due:* Nov 27
*Amount of Grant:* Up to 3,000,000 USD
*Contact:* Kristina J. Croake; (937) 255-4758; kristina.croake@wpafb.af.mil
Robert Matthews, Program Manager; (937) 255-8986; robert.matthews@wpafb.af.mil
*Sponsor:* U.S. Department of Defense
1400 Defense Pentagon
Washington, D.C. 20301-1400

**USDD Air Force Medical Support Agency/Modernization Directorate Research/Development and Innovations Grants** **2639**
The USDD Air Force Research Lab, on behalf of the Air Force Medical Support Agency (AFMSA)/Modernization Directorate is offering Research/Development and Innovations Grants. The Grants will ensure the AFMSA identifies and develops new knowledge and technologies to fulfill critical mission requirements. The intent of the Grants is to fund research targeted towards Air Force specific needs. AFMSA is soliciting white papers (and later technical/cost proposals) on the following research effort: (1) force health protection; (2) enroute care; (3) occupational medicine/in garrison; (4) expeditionary medicine; and (5) human performance. Interested applicants may submit white papers by the published due date. Selected applicants will be invited to submit a full proposal. Ten to twenty awards are anticipated per year with an estimated value of $100,000 to $1,000,000 annually, with total contract values of up to $3,000 over a three-year period. This is a five-year open ended solicitation. It is anticipated that updated requests for white papers will be posted on a yearly basis. Updated requests will likely occur in the first quarter of each calendar year. The continuation of requests for white papers and the submission dates should be verified from the website. Instructions for white paper submission can be found on the website.
*Requirements:* This is an unrestricted solicitation. Small businesses are encouraged.
*Geographic Focus:* All States
*Contact:* Kimberly Rhoads; (937) 255-0908; Kimberly.Rhoads@wpafb.af.mil
Nicholas Voiles, Contract Specialist; (937) 656-9808; Nicholas.Voiles@wpafb.af.mil
*Sponsor:* U.S. Department of Defense
1400 Defense Pentagon
Washington, D.C. 20301-1400

**USDD Air Warfare and Naval Weapons Applications Grants** **2640**
Naval unique or essential projects in this division are involved with applied research and advanced technology aligned with current and future naval capability gaps and innovative naval prototypes. The portfolio of current future naval capability programs includes: Discriminate and Provide Terminal Guidance, Enhanced Weapon Systems Technologies, Helicopter Low-Level Operations, Laser-Based Landing Aids, and Increased Capabilities Against Moving and Stationary Targets. The Innovative Naval Prototype portfolio consists of the Electromagnetic Railgun and the Free Electron Laser programs. Submission of the Research Proposal should always be preceded by submission of a Pre-proposal and a discussion of that Pre-proposal with the assigned Program Officer.
*Requirements:* Grants are awarded to academic, industrial, and nonprofit organizations, with strong emphasis on university research.
*Restrictions:* New grants or contracts are typically started near the beginning of the fiscal year (October 1). New proposals for the next fiscal year should be submitted no later than June 1, unless otherwise directed by the BAA. Competitive renewals should be submitted no later than six months prior to the award termination date after contacting the Program Officer. Conference proposals should be submitted at least four months prior to the conference.
*Geographic Focus:* All States
*Contact:* Roger Ellis, (703) 696-9504 or (703) 588-2427; fax (703) 696-5940; roger.ellis@navy.mil

*Internet:* www.onr.navy.mil/Science-Technology/Departments/Code-35/All-Programs/air-warfare-352.aspx
*Sponsor:* U.S. Department of Defense
One Liberty Center, 875 N. Randolph Street, Suite 1425
Arlington, VA 22203-1995

**USDD Alloys and Joining Grants** **2641**
Structural metals are the basis of the modern Navy. Ship construction depends critically upon the assembly of metal plates bonded to each other and to an underlying metal support framework. The physical and mechanical properties of these metals limit the designs of ships, aircraft and ground vehicles. The chemistry and solid-state physics of alloy formation certainly plays an important role in determining the properties of any particular alloy. This Office of Naval Research program aims to design an alloy to possess optimal mechanical properties for a particular application, and the highest corrosion resistance, with the lightest weight. The joining and welding of metals also pervades the military. The integrity of a welded joint, under all possible environmental conditions, is critically important in ships and aircraft. Research supported in this program focuses closely on the physics, chemistry and materials science of welds to understand the origins of bonding strength and adhesion between different joined substances. Submission of the Research Proposal should always be preceded by submission of a Preproposal and a discussion of that Preproposal with the assigned Program Officer. Applications for the upcoming fiscal year should be received by September 30.
*Requirements:* Grants are awarded to academic, industrial, and nonprofit organizations, with strong emphasis on university research. Investigators are encouraged to contact the staff prior to formally submitting a proposal.
*Restrictions:* New grants or contracts are typically started near the beginning of the fiscal year. New proposals for the next fiscal year should be submitted no later than September 30, unless otherwise directed. Competitive renewals should be submitted no later than six months prior to the award termination date after contacting the Program Officer. Conference proposals should be submitted at least four months prior to the conference.
*Geographic Focus:* All States
*Date(s) Application is Due:* Sep 30
*Contact:* Dr. William Mullins, Program Officer; (703) 696-0707; fax (703) 696-0934; william.m.mullins@navy.mil
*Internet:* www.onr.navy.mil/Science-Technology/Departments/Code-33/All-Programs/332-naval-materials/Alloys-Joining.aspx
*Sponsor:* U.S. Department of Defense
One Liberty Center, 875 N. Randolph Street, Suite 1425
Arlington, VA 22203-1995

**USDD Antifouling/Fouling Release Coatings Grants** **2642**
A significant limit to the performance of any marine platform arises from surface fouling by marine organisms. The response to this is to develop nontoxic approaches to antifouling and fouling release coatings. The state-of-the-art antifouling coatings are copper rosin or copper self polishing coatings, while fouling-release coatings are silicone-based. The overall program is broad with contributions from surface and adhesion science, analytical chemistry, and marine biology. Portions of this program are allied with the polymer science programs and are focused on the synthesis of new coatings. Current approaches include improved silicones, low surface energy coatings, amphiphilic, zwitterionic-containing and textured/patterned coatings with complex topographies. The long-term goals of this Office of Naval Research (ONR) program are to develop new approaches to antifouling or foul-release coatings with the following attributes: non-toxic or environmentally benign; durable in order to allow 12 -year dry-dock intervals; improved anti-fouling performance over existing copper-based paints. As marine fouling involves living organisms, there is a close cooperation between this program and programs maintained by Dr. Linda Chrisey of the ONR Human and Bioengineered Systems Division. ONR welcomes research ideas and suggestions from a variety of scientific and technical disciplines. Submission of the Research Proposal should always be preceded by submission of a Preproposal and a discussion of that Preproposal with the assigned Program Officer. Applications for the upcoming fiscal year should be received by September 30.
*Requirements:* Grants are awarded to academic, industrial, and nonprofit organizations, with strong emphasis on university research. Investigators are encouraged to contact the staff prior to formally submitting a proposal.
*Restrictions:* New grants or contracts are typically started near the beginning of the fiscal year. New proposals for the next fiscal year should be submitted no later than September 30, unless otherwise directed. Competitive renewals should be submitted no later than six months prior to the award termination date after contacting the Program Officer. Conference proposals should be submitted at least four months prior to the conference.
*Geographic Focus:* All States
*Date(s) Application is Due:* Sep 30
*Contact:* Dr. Stephen McElvany, Program Officer; (703) 696-1449; fax (703) 696-6887; steve.mcelvaney@navy.mil
*Internet:* www.onr.navy.mil/Science-Technology/Departments/Code-33/All-Programs/332-naval-materials/Antifouling-Fouling-Release-Coating.aspx
*Sponsor:* U.S. Department of Defense
One Liberty Center, 875 N. Randolph Street, Suite 1425
Arlington, VA 22203-1995

**USDD Applications and Transitions Grants** **2643**
The Office of Naval Research (ONR) Applications and Transitions Division supports programs in surface and aerospace surveillance, communications, and electronic combat. Research areas in surface and aerospace surveillance include sensors, primarily radar and electro-optic/infrared, and associated signal and image processing methods. Areas of interest in communications include military radio communications with an emphasis on anti-jam and low-probability-of-intercept techniques, radio networks, and dynamic internetworking. Electronic combat research areas include threat warning systems, electronic support measures, decoys and electronic countermeasures. Navigation research includes GPS anti-jam and associated inertial navigation techniques. Applications for the upcoming fiscal year should be received by September 30.
*Requirements:* Grants are awarded to academic, industrial, and nonprofit organizations, with strong emphasis on university research.
*Restrictions:* New grants or contracts are typically started near the beginning of the fiscal year. New proposals for the next fiscal year should be submitted no later than September 30, unless otherwise directed by the BAA. Competitive renewals should be submitted no later than six months prior to the award termination date after contacting the Program Officer. Conference proposals should be submitted at least four months prior to the conference.
*Geographic Focus:* All States
*Date(s) Application is Due:* Sep 30
*Contact:* Michael Pollock, Division Director; (703 )696-4202 or (703) 696-4217; fax (703) 696-3098; michael.a.pollock@navy.mil
*Internet:* www.onr.navy.mil/Science-Technology/Departments/Code-31/All-Programs/313-applications-transitions.aspx
*Sponsor:* U.S. Department of Defense
One Liberty Center, 875 N. Randolph Street, Suite 1425
Arlington, VA 22203-1995

**USDD Applied and Computational Analysis Grants** **2644**
The Office of Naval Research (ONR) Applied Computational Analysis Program develops and exploits methods in modern and classical mathematical analysis with emphasis on mathematical and computational models of physical phenomena. Naval interest in waves, flows, materials, structures, and information processing motivates research in the areas of multi-phase, multi-physics, multi-material problems; fluid transport and mixing; predictability of models for nonlinear dynamics; and inverse problems arising from acoustic and electromagnetic wave propagation and scattering. The Applied Computational Analysis Program currently has three thrusts: Multi-Scale/Multi-Physics Modeling; Dynamical Systems and Oceanic and Atmospheric Modeling; and Inverse Methods. Proposers are encouraged to contact the program officer to discuss their research interest prior to the submission of formal proposals.
*Requirements:* Grants are awarded to academic, industrial, and nonprofit organizations, with strong emphasis on university research.
*Restrictions:* New grants or contracts are typically started near the beginning of the fiscal year (1 October). New proposals for the next fiscal year are ongoing. Competitive renewals should be submitted no later than six months prior to the award termination date after contacting the Program Officer. Conference proposals should be submitted at least four months prior to the conference.
*Geographic Focus:* All States
*Contact:* Dr. Reza Malek-Madani; (703) 696-0195; reza.malekmadani@navy.mil
*Internet:* www.onr.navy.mil/Science-Technology/Departments/Code-31/All-Programs/311-Mathematics-Computers-Research/Applied-Computational-Analysis.aspx
*Sponsor:* U.S. Department of Defense
One Liberty Center, 875 N. Randolph Street, Suite 1425
Arlington, VA 22203-1995

**USDD Arctic and Global Prediction Grants** **2645**
The Office of Naval Research (ONR) Arctic and Global Prediction Program is motivated by the rapid decline in summer ice extent that has occurred in recent years. This has prompted renewed Department of Navy interest in understanding and predicting the Arctic physical environment at a variety of time and space scales. The ability to predict the location of the ice edge, the space-time evolution of the ice cover, and the ice thickness will be particularly critical to safe naval operations in the Arctic. To achieve this, models integrating the ocean, waves, ice and atmosphere must be able to represent the physical processes, interactions and feedbacks involved in the seasonal evolution of ice extent, area, thickness and volume. Consequently, the program has three focus areas: developing integrated ocean-ice-wave-atmosphere Earth System models for improved prediction at a variety of time scales; improving understanding of the physical environment and processes in the Arctic Ocean; and investigating new technologies, e.g., sensors, platforms and communications, for sustained operation and observation in the challenging Arctic environment. This announcement will remain open until September 30.
*Requirements:* The Arctic and Global Prediction Program is committed to contributing to Navy STEM (Science, Technology and Mathematics) education efforts, particularly the education of the next generation of scientists specializing in the physics of the Arctic Ocean system. Investigators are encouraged to include students (undergraduate and graduate) and post-doctoral researchers in their research plans. As part of this commitment, the program will also consider requests for Early Student Support (ESS) to assist beginning graduate students working with ONR-funded investigators.
*Geographic Focus:* All States
*Contact:* Dr. Martin Jeffries; (703) 696-7825; martin.jeffries@navy.mil
*Internet:* www.onr.navy.mil/Science-Technology/Departments/Code-32/All-Programs/Atmosphere-Research-322/Arctic-Global-Prediction.aspx
*Sponsor:* U.S. Department of Defense
One Liberty Center, 875 N. Randolph Street, Suite 1425
Arlington, VA 22203-1995

**USDD ARO Broad Agency Announcement Grants** 2646

The United States Army Research Office (ARO) solicits proposals for basic and scientific research in mechanical sciences, environmental sciences, mathematical sciences, computer sciences, electronics, computational and information sciences, physics, chemistry, life sciences, and materials science. Interested applicants are strongly encouraged to make preliminary inquiries as to the general need for the type of research effort contemplated before preparing a proposal. Each technical area has a point of contact identified in the full announcement at the website. This is a multiple year continuously open Broad Agency Announcement (BAA). Amendments may be posted. Applicants are encouraged monitor FedBizOpps.gov (https://www.fbo.gov/) and/or Grants.gov (grants.gov/applicants/find_grant_opportunities.jsp) for updates. The ARO prefers proposals for a 3-year period which include a brief summary of work contemplated for each 12-month period so that awards may be negotiated for an entire 3-year program or for individual 1-year increments of the total.

*Requirements:* Proposals may be submitted by degree-granting universities, nonprofit organizations, or industrial concerns. Proposals are encouraged from historically black colleges and universities and from minority institutions.

*Geographic Focus:* All States

*Date(s) Application is Due:* Mar 31

*Contact:* Kevin Bassler; (919) 549-4295; kevin.j.bassler.civ@mail.mil

*Internet:* www.grants.gov/search/search.do;jsessionid=T23pP9VJLR1BL3prQrvr7Rq3JwPh12ckQPsJ17XClKWQcZh8ScJj!-2099600874?oppId=41952&mode=VIEW

*Sponsor:* U.S. Department of Defense
4400 Martin Road, Redstone Arsenal
Huntsville, AL 35898

**USDD Atomic, Molecular and Quantum Physics Grants** 2647

The research supported under this program seeks primarily to understand and exploit the interactions of light, atoms, ions, and molecules. Areas supported may include: laser cooling and trapping of both ions and neutral atoms; Bose-Einstein condensation; atom laser concepts; coherent control, e.g., of chemical reactions; femtosecond pulse shaping; quantum computing; and quantum state teleportation. Also supported are novel theoretical methods for computational electronic structure. Challenges addressed through the atomic and molecular physics program include: obtaining ultra-high precision spectroscopy; overcoming limitations in atomic time and frequency standards, leading, for example, to the pursuit of the physics of ultra-cold collisions; the ability to control chemical dynamics and branching ratios; and the ability to predict properties of materials, and devise novel new materials (e.g., through a better understanding of strong electron correlations). Many of these subprograms and goals are interrelated and interdependent. Submission of the research proposal should always be preceded by submission of a preproposal and a discussion of that preproposal with the assigned Program Officer. Proposals are preferred by April 1 of each year.

*Requirements:* Grants are awarded to academic, industrial, and nonprofit organizations, with strong emphasis on university research.

*Restrictions:* New grants or contracts are typically started near the beginning of the fiscal year (1 October). New proposals for the next fiscal year should be submitted no later than April 1, unless otherwise directed. Competitive renewals should be submitted no later than six months prior to the award termination date after contacting the Program Officer. Conference proposals should be submitted at least four months prior to the conference.

*Geographic Focus:* All States

*Date(s) Application is Due:* Apr 1

*Contact:* Dr. Charles W. Clark, (703) 696-5267 or (703) 696-5031; fax (703) 696-5940; charles.clark3@navy.mil or 312_AM@onr.navy.mil

*Internet:* www.onr.navy.mil/sci_tech/31/312/atomic_molecular_physics.asp

*Sponsor:* U.S. Department of Defense
One Liberty Center, 875 N. Randolph Street, Suite 1425
Arlington, VA 22203-1995

**USDD Basic Research Challenge: Carbon Molecular Electronics Grants** 2648

The Office of Naval Research is interested in receiving proposals on carbon-based molecular electronics. The objective are to encourage research and innovation in bottom-up chemical synthesis and assembly of carbon, particularly graphene, based electronic devices and circuits with atomic precision and Angstrom resolution. The program will address the challenges and limitations in top-down fabricated graphene nanoelectronic devices where structures are too large (limited by lithography) and intrinsic graphene properties are often masked by edge roughness and hard to measure. The long term vision of the program is to enable a new kind of molecular electronics, a single giant molecule made of carbon or other related materials that functions as electronic, magnetic or optical devices or circuits. White paper submission is not required but strongly encouraged. White papers are accepted electronically. Full proposals must be submitted through Grants.gov. Two to six awards are anticipated. The target budget for single investigator awards is $150,000 or less. The target budget for teams of no more than four is $600,000 or less. The award period will be three years plus two additional years of option period.

*Requirements:* All responsible sources from academia and industry are eligible. Historically black colleges and universities and minority institutions are encouraged to apply and to join others in submitting proposals.

*Geographic Focus:* All States

*Date(s) Application is Due:* May 1

*Contact:* Jennifer Brown, Contract Specialist; (703) 588-2432; jennifer.brown4@navy.mil
Dr. Chagaan Baatar, Technical Specialist; chagaan.baatar@navy.mil

*Internet:* www.onr.navy.mil/Contracts-Grants/Funding-Opportunities/Special-Notices.aspx

*Sponsor:* U.S. Department of Defense
One Liberty Center, 875 N. Randolph Street, Suite 1425
Arlington, VA 22203-1995

**USDD Basic Research Challenge: Decentralized Online Optimization Grants** 2649

The Office of Naval Research is offering a Basic Research Challenge: Decentralized Online Optimization. The objective is to develop scientific principles and algorithms for solving decentralized, online optimization problems. Proposals should scientifically address efficient, robust computational techniques and have the following characteristics: (1) The techniques can be applied to problems possessing a high degree of decentralization, and in which there is limited communication between system components; (2) The techniques can be applied to problems in which not all relevant information of the environment is known a priori, and is revealed incrementally to individual system components; information spreads through the system as communications become available and with potentially substantial time delays; and (3) The techniques are capable of producing high-quality solutions in a reasonable amount of time. Solution quality is measured against optimality, and solution time is measured against the time scale of changes in the environment. Analysis of these measures is expected to be mathematically rigorous. White papers are not required but are strongly encouraged for all applicants. White papers may be submitted electronically, and proposals may be submitted through Grants.gov. Eight to ten awards are anticipated, and the project period may be from one to five years. Questions regarding this announcement may be submitted electronically no later than two weeks prior to the due date. Answers will be addressed in the form of an amendment and will be posted to the following web pages: Federal Business Opportunities (https://www.fbo.gov/); Grants.gov (www.grants.gov/); and ONR Broad Agency Announcement (www.onr.navy.mil/Contracts-Grants/Funding-Opportunities/Special-Notices.aspx).

*Geographic Focus:* All States

*Date(s) Application is Due:* May 15

*Amount of Grant:* 100,000 - 150,000 USD

*Contact:* Gordon Jaquith, Contract Specialist; 703) 696-7831; gordon.jaquith@navy.mil
Donald Wagner, Technical Specialist; (703) 696-4257; Don.Wagner@navy.mil

*Internet:* www.onr.navy.mil/Contracts-Grants/Funding-Opportunities/Special-Notices.aspx

*Sponsor:* U.S. Department of Defense
One Liberty Center, 875 N. Randolph Street, Suite 1425
Arlington, VA 22203-1995

**USDD Basic Research for Combating Weapons of Mass Destruction Broad Agency Announcement Grants** 2650

The Defense Threat Reduction Agency (DTRA) seeks Basic Research for Combating Weapons of Mass Destruction (C-WMD) Grants. Grants are focused on soliciting basic research projects that support the DTRA mission to safeguard America and its allies from WMD (e.g., chemical, biological, radiological, nuclear, and high-yield explosives) by providing capabilities to reduce, eliminate, and counter the threat and mitigate its effects. To this end DTRA seeks research across broad strategic thrust areas: (1) science of WMD sensing and recognition; (2) cognitive and information science; (3) science for protection; (4) science to defeat WMD; and (5) science to secure WMD. Pre-application white papers are due January 13, and invitation-only proposals are accepted from April 2 through 20 only. Single scope awards will average $150,000 per year, and multidisciplinary awards will average $350,000 per year. This is a Broad Agency Announcement (BAA) which will have multiple, consecutive opportunities/periods not yet determined. It is anticipated that there will be at least one opportunity per year announced in the December timeframe. There may be additional opportunities/periods in any year. Subsequent submission opportunities/periods with corresponding topics, application packages, and applicable deadlines, will be published via amendments to this BAA. Applicants are encouraged to monitor Grants.gov Grant Opportunities (grants.gov/applicants/find_grant_opportunities.jsp) and the DTRA Basic and Fundamental Research Portal (www.dtrasubmission.net/portal) for announcements and updates.

*Requirements:* Applicants eligible to submit pre-application white papers and proposals include the following United States and foreign-based equivalents: accredited degree-granting colleges and universities; industrial/commercial basic research centers, including small businesses with a portfolio predominantly in basic research proposing a project with significant participation (minimum 30% of total effort value on the proposed project) by one or more accredited degree-granting colleges and/or universities; and not-for-profit organizations/basic research centers with a portfolio predominantly in basic research proposing a project with significant participation (minimum 30% of total effort value on the proposed project) by one or more accredited degree-granting colleges and/or universities. Proof of 501(c)3 status may be required for not-for-profits. For foreign-based establishments entirely based outside the United States and/or its territories, proof of not-for-profit status may be required.

*Geographic Focus:* All States, All Countries

*Contact:* Gena Leake; (703) 767-5870; HDTRA1-BRCWMD-A@dtrasubmission.net

*Internet:* www.dtra.mil/Research.aspx

*Sponsor:* U.S. Department of Defense
6200 Meade Road
Fort Belvoir, VA 22060-5264

**USDD Basic Research Initiative Grants** 2651

The Air Force Office of Scientific Research (AFOSR) manages the basic research investment for the United Stares Air Force (USAF). As a part of the Air Force Research Laboratory (AFRL), AFOSR's technical experts foster and fund research within the AFRL, universities, and industry laboratories to ensure the transition of research results to support USAF needs. The Basic Research Initiative Grants are intended to initiate a basic research program that begins to build the foundational understanding of the factors that influence human reliance on autonomous systems and offer both practical guidance for the Air Force and future research avenues for applied research. The research encompassed will focus on a set of studies for targeting the psychological, neurological, or contextual elements of human reliance on autonomous systems. The intent is to

identify general principles of human trust, and the results should not be bound to a particular system. Suggested research areas may include: (1) empirical studies to reveal the psychological, organizational, technological, and group-level factors that shape reliance on automation; (2) neuroscience and psychophysiology research to establish the markers for said reliance; (3) case studies of Department of Defense users with experience using autonomous vehicles to identify practical lessons learned and realistic perspectives for use or disuse of such systems; (4) anthropological/cross-cultural science to understand the barriers to optimal use of automation in other countries; (5) cross-service studies to reveal potential differences in perceptions/use of automated systems; (6) harnessing and understanding relevant and practical metrics for establishing trustworthiness of autonomous systems (e.g., verification and validation); and (7) understanding the impact of different generations (i.e., age groups) on reliance. Three to four awards of $100,000 to $250,000 for three years are expected. White papers are strongly encouraged prior to submitting a full proposal. There is no deadline for submission. This announcement remains open until superseded.
*Requirements:* All responsible potential applicants from academia and industry are eligible.
*Geographic Focus:* All States
*Date(s) Application is Due:* Mar 30
*Amount of Grant:* Up to 750,000 USD
*Contact:* Dr. Joseph Lyons; (703) 696-6207; Joseph.Lyons@afosr.af.mil
*Internet:* www07.grants.gov/search/search.do;jsessionid=cTNwP5vW3XRK6sZQhKynnCW1wJYNK5n2LXd28SkRTWH312MZ1Q9D!-1993000306?oppId=129053&mode=VIEW
*Sponsor:* U.S. Department of Defense
875 North Randolph Street, Suite 325, Room 3112
Arlington, VA 22203-1768

**USDD Battlespace Environment Division Research Interests Grant** **2652**
The USDD Air Force Research Laboratory Space Vehicles Research Directorate, Battlespace Environment Division, is interested in receiving proposals to advance the state-of-the-art and scientific knowledge in space and missile technology. Proposals that either enable future space systems to meet performance, reliability, maintainability, supportability and affordability goals, or which enhance performance or significantly reduce cost, schedule, or risk of existing technologies, are of particular interest. This is a multiple year open-ended Broad Agency Announcement (BAA). Periodically proposal call announcements (CALLS) may be issued to request proposals for specific research efforts under technical topic areas. These subsequent CALLS will contain specific descriptions of the research effort to be addressed, anticipated period of performance, information peculiar to the specific research technical topic area, and the expected dollar range for proposals received under the CALL. No proposals should be submitted against this baseline BAA. Proposal due dates and times will be specified in each CALL. A minimum of one award per specific CALL is anticipated. Anticipated funding for this multiple-year program (not per award) is approximately $9 million to $10 million per fiscal year.
*Geographic Focus:* All States
*Date(s) Application is Due:* Nov 10
*Amount of Grant:* Up to 58,000,000 USD
*Contact:* Jean Barnes, Contract Specialist; (505) 846-4695; jean.barnes@kirtland.af.mil. Dr. Joel B. Mozer, Technical Specialist; (505) 853-0059; Joel.Mozer@kirtland.af.mi
*Internet:* www.grants.gov/search-grants.html?agencies%3DDOD|Department%20of%20Defense
*Sponsor:* U.S. Department of Defense
1400 Defense Pentagon
Washington, D.C. 20301-1400

**USDD Breast Cancer Clinical Translational Research Grants** **2653**
Applications for the Breast Cancer Research Program (BCRP) are being solicited by the Assistant Secretary of Defense for Health Affairs, Defense Health Program. The BCRP was established in 1992 to support innovative research focused on ending breast cancer. The BCRP challenges the scientific community to design research that will address the urgency of the vision to end breast cancer. Specifically, the BCRP seeks to accelerate high-impact research, encourage innovation and stimulate creativity, bring new investigators into the breast cancer field, and facilitate multidisciplinary collaborations. The BCRP Clinical Translational Research (CTR) Grants are intended to promote significant improvements over current approaches to breast cancer prevention and therapy, such as studies that will prevent primary or recurrent breast cancer and studies that may result in a new treatment to prevent breast cancer progression to metastasis. CTR Grants support research projects that are likely to have a major impact on breast cancer by applying promising research findings to patients with, or populations at risk for, breast cancer. Principal Investigators (PIs) and their collaborators may have originated projects in their laboratories that will form the basis for 1-2 years of advanced translational research leading to a prospective clinical trial to be conducted during this award period. Alternatively, PIs may leverage partnerships with industry. The ability to conduct the required translational research and early phase clinical trial during the award period must be demonstrated. Investigators must demonstrate availability of and access to an appropriate patient population that will support meaningful clinical outcomes. Submission is a multi-step process requiring both pre-application submission through the CDMRP eReceipt System (https://cdmrp.org/) and application submission through Grants.gov (www.grants.gov/). The maximum period of performance is 5 years. The maximum allowable direct costs for the entire period is $12M plus indirect costs.
*Requirements:* PIs must be at or above the level of assistant professor (or equivalent).
*Restrictions:* Applicants are encouraged to sign up on Grants.gov for "send me change notification emails" by following the link on the Synopsis page for the Program Announcement/Funding Opportunity. If the application package is updated or changed, the original version of the application package may not be accepted by Grants.gov.
*Geographic Focus:* All States
*Date(s) Application is Due:* Aug 15
*Contact:* Grants Officer; (301) 682-5507; help@cdmrp.org
*Sponsor:* U.S. Department of Defense - 1
1400 Defense Pentagon
Washington, D.C. 20301-1400

**USDD Breast Cancer Era of Hope Scholar Grants** **2654**
The Era of Hope Scholar Grants support individuals who are early in their careers and have high potential for innovation in breast cancer research. Since the intent is to recognize creative and innovative individuals rather than projects, the central feature of the award is the innovative contribution that the Principal Investigator (PI) can make toward ending breast cancer. The PI should articulate a vision that challenges current dogma and demonstrates an ability to look beyond tradition and convention. Experience in breast cancer research is not required; however, the application must focus on breast cancer and the PI must commit at least 50% of his/her full-time professional effort during the award period to breast cancer research. Five awards are anticipated. The maximum period of performance is 5 years, and the maximum direct costs for the entire period is $2.5M plus indirect costs.
*Requirements:* Individuals should be exceptionally talented scientists who have demonstrated that they are the "best and brightest" in their field(s) through extraordinary creativity, vision, and productivity. They should have demonstrated experience in forming effective partnerships and collaborations and should exhibit strong potential for leadership in the breast cancer research community. The PI must be an independent investigator within 6 years of his/her last mentored position as of the deadline.
*Restrictions:* Postdoctoral fellows, clinical fellows (including residents and interns), and other researchers in training positions are not eligible for this award.
*Geographic Focus:* All States
*Date(s) Application is Due:* Apr 19
*Contact:* Grants Officer; (301) 682-5507; help@cdmrp.org
*Sponsor:* U.S. Department of Defense - 1
1400 Defense Pentagon
Washington, D.C. 20301-1400

**USDD Breast Cancer Idea Grants** **2655**
The Idea Award is designed to promote new ideas that are still in the early stages of development and have the potential to yield highly impactful data and new avenues of investigation. This mechanism supports conceptually innovative, high-risk/high-reward research that could lead to critical discoveries or major advancements that will accelerate progress toward ending breast cancer. Applications should include a well-formulated, testable hypothesis based on strong scientific rationale. Innovation and impact are the most important aspects of the Idea Award. Applications that demonstrate exceptional scientific merit but lack innovation and high potential impact do not meet the intent of the Idea Award. Presentation of preliminary data is not consistent with the intent of the Idea Award mechanism. While the inclusion of preliminary data is not prohibited, the strength of the application should not rely on preliminary data, but on the innovative idea and the potential impact it will provide. Submission is a multi-step process requiring both pre-application submission through the CDMRP eReceipt System (https://cdmrp.org/) and application submission through Grants.gov (www.grants.gov/). Approximately fifty awards are anticipated. The maximum period of performance is 3 years. The maximum allowable direct costs for the entire period is $375,000 plus indirect costs.
*Requirements:* Investigators at all academic levels (or equivalent) are eligible.
*Restrictions:* Applicants are encouraged to sign up on Grants.gov for "send me change notification emails" by following the link on the Synopsis page for the Program Announcement/Funding Opportunity. If the application package is updated or changed, the original version of the application package may not be accepted by Grants.gov.
*Geographic Focus:* All States
*Date(s) Application is Due:* Aug 15
*Contact:* Grants Officer; (301) 682-5507; help@cdmrp.org
*Sponsor:* U.S. Department of Defense - 1
1400 Defense Pentagon
Washington, D.C. 20301-1400

**USDD Breast Cancer Impact Grants** **2656**
The Breast Cancer Research Program (BCRP) Impact Grants support research projects (from small to largescale) that specifically focus on scientific and clinical breast cancer issues, which, if successfully addressed, will revolutionize the understanding, prevention, and/or treatment of breast cancer and make major advances toward the goal of ending the disease. The BCRP particularly encourages applications that focus on poorly understood issues that are of critical significance in breast cancer, such as those related to prevention, susceptibility, recurrence, or metastasis. The proposed work must be based on sound overall research and fully supported by preliminary data and/or published reports. The most important aspect is the potential of the proposed research to have an unprecedented impact on reducing breast cancer incidence, recurrence, and/or mortality. The research project may be from any discipline or combination of disciplines, including basic, translational, clinical (clinical trials are allowed), behavioral, and/or epidemiological research. The potential impact may be near-term or long-term, but it must be significant and non-incremental. Applications are encouraged that include meaningful and productive collaborations between investigators. The Partnering PI Option is structured to accommodate two PIs, who will each receive a separate award. One partner is identified as the Initiating PI and the other partner is identified as the Partnering PI. The Initiating and Partnering PIs have different submission requirements; however, both PIs should

contribute to the preparation of a single application. The collaborative partners may have expertise in similar or disparate scientific disciplines, but each partner is expected to bring different strengths to the application. New collaborations are encouraged but not required. It is the responsibility of the collaborating investigators to describe how their combined expertise in the collaboration will better address the research question and explain why the work should be done together rather than through separate efforts. Submission is a multi-step process requiring both pre-application submission through the CDMRP eReceipt System (https://cdmrp.org/) and application submission through Grants.gov (www.grants.gov/). Five awards are anticipated. The maximum period of performance is 5 years. The maximum allowable direct costs for the entire period is $2M plus indirect costs.
*Requirements:* PIs must be at or above the level of Assistant Professor (or equivalent).
*Restrictions:* Applicants are encouraged to sign up on Grants.gov for "send me change notification emails" by following the link on the Synopsis page for the Program Announcement/Funding Opportunity. If the application package is updated or changed, the original version of the application package may not be accepted by Grants.gov.
*Geographic Focus:* All States
*Date(s) Application is Due:* Aug 15
*Contact:* Grants Officer; (301) 682-5507; help@cdmrp.org
*Sponsor:* U.S. Department of Defense - 1
1400 Defense Pentagon
Washington, D.C. 20301-1400

**USDD Breast Cancer Innovation Grants** **2657**
The Breast Cancer Research Program (BCRP) Innovator Grants support visionary individuals who have demonstrated creativity, innovative work, and leadership in any field including, but not limited to, breast cancer. The Innovator Grants will provide these individuals with the funding and freedom to pursue their most novel, visionary, high-risk ideas that could ultimately lead to ending breast cancer. Since the intent of the Innovator Grant mechanism is to recognize creative and innovative individuals rather than projects, the central feature of the award is the innovative contribution that the Principal Investigator (PI) can make toward ending breast cancer. Experience in breast cancer research is not required; however, the application must focus on breast cancer, and the PI must commit at least 50% of his/her full-time professional effort during the award period to breast cancer research. Submission is a multi-step process requiring both pre-application submission through the CDMRP eReceipt System (https://cdmrp.org/) and application submission through Grants.gov (www.grants.gov/). Three awards are anticipated. The maximum period of performance is 5 years. The maximum allowable direct costs for the entire period is $5M plus indirect costs.
*Requirements:* PIs must be at or above the level of associate professor (or equivalent). The PI should have a past record of creativity, promise for continued innovation in future work, and a vision that challenges current dogma and demonstrates an ability to look beyond tradition and convention. The PI is also expected to have demonstrated success at forming and leading effective partnerships and collaborations.
*Restrictions:* Previous recipients of the BCRP Innovator Grants and BCRP Integration Panel members are ineligible. Applicants are encouraged to sign up on Grants.gov for "send me change notification emails" by following the link on the Synopsis page for the Program Announcement/Funding Opportunity. If the application package is updated or changed, the original version of the application package may not be accepted by Grants.gov.
*Geographic Focus:* All States
*Date(s) Application is Due:* Aug 2
*Contact:* Grants Officer; (301) 682-5507; help@cdmrp.org
*Sponsor:* U.S. Department of Defense - 1
1400 Defense Pentagon
Washington, D.C. 20301-1400

**USDD Breast Cancer Research Postdoctoral Fellowships** **2658**
The Postdoctoral Fellowships support exceptionally talented recent doctoral or medical graduates in their pursuit of innovative, high-impact breast cancer research during their postdoctoral training and allow them to obtain the necessary experience for an independent career at the forefront of breast cancer research. These individuals should be exceptionally talented scientists who have demonstrated that they are the "best and brightest" of their peers. Applicants must exhibit a strong desire to pursue a career in breast cancer research. Under this award mechanism, the postdoctoral fellow is considered the Principal Investigator (PI) and, as such, should write the project narrative with appropriate direction from the mentor. While the PI is not required to have previous experience in breast cancer research, the application must focus on breast cancer research. Submission is a multi-step process requiring both pre-application submission through the CDMRP eReceipt System (https://cdmrp.org/) and application submission through Grants.gov (www.grants.gov/). Approximately 22 awards are anticipated. The maximum period of performance is 3 years. The maximum allowable direct costs for the entire period of performance are $300,000 plus indirect costs.
*Requirements:* At the time of application, all eligible PIs must have: successfully completed the requirements for a doctoral or medical degree; been in the laboratory or clinical research setting in which the proposed research is to be performed for no more than 2 years as a postdoctoral researcher; and a total of less than 4 years of postdoctoral research experience (excluding clinical residency or fellowship training).
*Restrictions:* Only postdoctoral fellows are eligible for this award. Faculty members, instructors, and all other non-postdoctoral positions are not eligible to apply. Applicants are encouraged to sign up on Grants.gov for "send me change notification emails" by following the link on the Synopsis page for the Program Announcement/Funding Opportunity. If the application package is updated or changed, the original version of the application package may not be accepted by Grants.gov.
*Geographic Focus:* All States
*Date(s) Application is Due:* Apr 19
*Contact:* Fellowship Coordinator; (301) 682-5507; help@cdmrp.org
*Sponsor:* U.S. Department of Defense - 1
1400 Defense Pentagon
Washington, D.C. 20301-1400

**USDD Breast Cancer Transformative Vision Grant** **2659**
The Breast Cancer Research Program (BCRP) Transformative Vision Grant supports research projects that will fulfill an extraordinary vision for dramatically affecting the prevention or treatment of breast cancer. The Grant requires a plan that will test and achieve the vision as quickly as possible through the translation of the research ideas to individuals with, and/or those at risk for, breast cancer. The scope of the effort may include a broad spectrum of research spanning from basic to clinical studies. There are two critical components. The first is vision and impact. A vision for a new approach that will have a revolutionary impact on the prevention or treatment of breast cancer must be articulated. The time to the final impact may vary, but the success of the vision must be transformative and significantly advance the goal of ending breast cancer. The second is implementation. The vision must be supported by a detailed plan that identifies critical milestones, outlines the innovations and technical solutions that will be implemented to accomplish the milestones, and explains how these solutions will be translated to individuals with, and/or those at risk for, breast cancer. It is expected that the proposed plan will present an exceptional level of innovation and creativity and that the Principal Investigator (PI) will assemble the team necessary to realize the vision. Submission is a multi-step process requiring both pre-application submission through the CDMRP eReceipt System (https://cdmrp.org/) and application submission through Grants.gov (www.grants.gov/). One award is anticipated. The maximum period of performance is 5 years. The maximum allowable direct costs for the entire period is $12M plus indirect costs.
*Requirements:* PIs must be at or above the level of Assistant Professor (or equivalent).
*Restrictions:* Applicants are encouraged to sign up on Grants.gov for "send me change notification emails" by following the link on the Synopsis page for the Program Announcement/Funding Opportunity. If the application package is updated or changed, the original version of the application package may not be accepted by Grants.gov.
*Geographic Focus:* All States
*Date(s) Application is Due:* Aug 15
*Contact:* Grants Officer; (301) 682-5507; help@cdmrp.org
*Sponsor:* U.S. Department of Defense - 1
1400 Defense Pentagon
Washington, D.C. 20301-1400

**USDD Broad Agency Announcement for the Naval Medical Research and Development Center-Fredrick Grants** **2660**
The Naval Medical Research and Development Center-Frederick has a broad role in competitive funding of meritorious research across a spectrum of science and engineering disciplines. This Broad Agency Announcement (BAA) is to solicit proposals for innovative approaches in infectious disease research, non-communicable disease research, and combat casualty care research. Amendments will be posted to one or more of the following web pages: FedBizOpps (https://www.fbo.gov/); Grants.gov (www.grants.gov/), or Naval Medical Logistics Command Research and Grants (www.nmlc.med.navy.mil/DBU-RnG.asp). Applicants are encouraged to monitor these sites. Applicants are strongly encouraged to contact the Grants Officer prior to submitting a proposal. Funding and the number of awards is indeterminate. Selection of research projects is based on the evaluation of the proposal, programmatic review and the availability of funds.
*Requirements:* All responsible sources from academia and industry are eligible. Historically black colleges and universities and minority institutions are encouraged to apply.
*Geographic Focus:* All States
*Date(s) Application is Due:* Dec 31
*Contact:* Christopher Sherman, Supervisory Contract Specialist; (301) 619-2157; NMLC-Research@med.navy.mil
*Internet:* www.grants.gov/search/search.do;jsessionid=vKBrPDydsD287nDBQ7lpKh0LBwgGkYc3QTGFG2XGDQMHCF5L2fG1!-757993493?oppId=138693&mode=VIEW
*Sponsor:* U.S. Department of Defense
693 Neiman Street
Fort Detrick, MD 21702-9203

**USDD Broad Agency Announcement Grants** **2661**
The Department of Defense HIV/AIDS Prevention Program (DHAPP) has successfully engaged over 80 countries in efforts to combat HIV/AIDS among their respective military services. DHAPP is a partner collaborating with the United States State Department, Health and Human Services, United States Agency for International Development, and Centers for Disease Control and Prevention, in the President's Emergency Plan for AIDS Relief (PEPFAR). This BAA is intended to solicit existing partners and establish new partners in order to expand the DHAPP program. Proposals should focus on rapidly extending HIV/AIDS services. Applicants are encouraged to target specific needs with a practical business plan, using small grass-roots organizations to provide community-based services as a way to enhance organic capabilities and sustainability. Concept Papers may be submitted at any time during this period. Early submission is strongly encouraged due to the possible lack of funding for the later evaluation periods. The period of performance is generally one year, with up to two additional phases subject to availability of funding. Applicants are encouraged to propose additional phases if appropriate. This is a multiple year BAA which may be revised as needed. Changes will be posted at Grants.gov (grants.gov/applicants/find_grant_opportunities.jsp). Applicants are encouraged to sign up on Grants.gov for "send me change notification emails" by following the link on the synopsis page for the program announcement/funding opportunity.

*Requirements:* All responsible sources from academia, industry, and non-governmental organizations are eligible. Applicants respondents must demonstrate the active support of the in-country military and the Department of Defense representative in the corresponding United States Embassy in the planning and execution of proposals.
*Restrictions:* Funding is available for Angola, Botswana, Caribbean Regional, Central America Regional, Cote d' Ivoire, Democratic Republic of the Congo, Dominican Republic, Ethiopia, Ghana, Guyana, India, Indonesia, Lesotho, Malawi, Mozambique, Namibia, Rwanda, South Africa, Swaziland, Uganda, Ukraine, Vietnam and Zambia. A description of the work needed, along with the program areas and estimated funding is available in the BAA found at the website. All questions should be submitted to the points of contact by email.
*Geographic Focus:* All States
*Date(s) Application is Due:* Sep 30
*Amount of Grant:* 300,000 - 700,000 USD
*Contact:* Latrice Rubenstein; (619) 532-4357; Latrice.Rubenstein@navy.mil
Cynthia Simon-Arndt; Cynthia.SimonArndt@med.navy.mil
*Internet:* www.grants.gov/search/search.do;jsessionid=1GhLPGQSVtvhVkKQYJgRnyFGWgBT8SQyFXjYv31mW6Ky7klFQ1Vx!-757993493?oppId=67733&mode=VIEW
*Sponsor:* U.S. Department of Defense
5450 Carlistle Park, P.O. Box 2050
Mechanicsburg, PA 17055-0791

**USDD Broad Agency Announcement HIV/AIDS Prevention Grants** **2662**
The Department of Defense HIV/AIDS Prevention Program (DHAPP) has successfully engaged over 80 countries in efforts to combat HIV/AIDS among their respective military services. DHAPP is a partner collaborating with the United States State Department, Health and Human Services, United States Agency for International Development, and Centers for Disease Control and Prevention, in the President's Emergency Plan for AIDS Relief (PEPFAR). This Broad Agency Announcement (BAA) allows DHAPP to assist countries where militaries are not funded under PEPFAR by soliciting existing partners and establishing new partners in furtherance of DHAPP and partner military program goals. Proposals should focus on rapidly extending HIV/AIDS services. Applicants are encouraged to target specific needs with a practical business plan, using small grass-roots organizations to provide community-based services as a way to enhance organic capabilities and sustainability. Concept papers are required and full proposals are by invitation only. Early submission is strongly encouraged. The period of performance is generally one year, with up to two additional phases subject to availability of funding. Applicants are encouraged to propose additional phases if appropriate. This is a multiple year BAA which may be revised as needed. Based on the availability of funds, up to three evaluation cycles will be conducted. Changes will be posted at Grants.gov (grants.gov/applicants/find_grant_opportunities.jsp). Applicants are encouraged to sign up on Grants.gov for "send me change notification emails" by following the link on the synopsis page for the program announcement/funding opportunity.
*Requirements:* All responsible sources from academia, industry, and non-governmental organizations are eligible. Applicants must demonstrate the active support of the in-country military and the USDD representative in the corresponding United States Embassy in the planning and execution of proposals.
*Restrictions:* Funding is available for Ghana and El Salvador. A description of the work needed, along with the program areas and estimated funding is available in the BAA found at the website. All questions should be submitted to the points of contact by email.
*Geographic Focus:* All States
*Amount of Grant:* 60,000 - 500,000 USD
*Contact:* Latrice Rubenstein; (619) 556-5276; Latrice.Rubenstein@navy.mil
Cynthia Simon-Arndt; Cynthia.SimonArndt@med.navy.mil
*Sponsor:* U.S. Department of Defense
5450 Carlistle Park, P.O. Box 2050
Mechanicsburg, PA 17055-0791

**USDD Broad Agency Announcement Research and Development Grants** **2663**
The United States Army Engineer Research and Development Center (ERD.C.) has issued a Broad Agency Announcement (BAA) for various research and development topic areas. This BAA solicits proposals for basic and applied research and that part of development not related to the development of a specific system or hardware procurement. To be eligible the technology or methodology must be either basic research, applied research, advanced technology development not for a specific system/hardware, or demonstration and validation. This BAA will only be used when meaningful proposals with varying technical/scientific approaches can be reasonably anticipated. The ERD.C. consists of the Coastal and Hydraulics Lab, the Geotechnical and Structures Lab, the Environmental Lab, and the Information Technology Lab in Vicksburg, Mississippi; the Cold Regions Research and Engineering Lab in Hanover, New Hampshire; the Construction Engineering Research Lab in Champaign, Illinois; and the Topographic Engineering Center in Alexandria, Virginia. The ERD.C. is responsible for conducting research in the broad fields of hydraulics, dredging, coastal engineering, instrumentation, oceanography, remote sensing, geotechnical engineering, earthquake engineering, soil effects, vehicle mobility, self-contained munitions, military engineering, geophysics, pavements, protective structures, aquatic plants, water quality, dredged material, treatment of hazardous waste, wetlands, physical/mechanical/ chemical properties of snow and other frozen precipitation, infrastructure and environmental issues for installations, computer science, telecommunications management, energy, facilities maintenance, materials and structures, engineering processes, environmental processes, land and heritage conservation, and ecological processes. Applicants are encouraged to make preliminary inquiries about the research considered before preparing a proposal. Points of contact are listed for each research areas for each laboratory. The proposal should present the proposed research effort in sufficient detail to allow ERD.C. to evaluate the scientific merit and relevance of the proposed research and to determine funding availability. The ERD.C. also supports conferences and symposia in special areas of science that bring experts together to discuss recent research or educational findings or to expose other researchers or advanced graduate students to new research and educational techniques. The ERD.C. encourages the convening, in the United States, of major international conferences, symposia, and assemblies of international alliances. Conference support proposals should be submitted a minimum of six months prior to the date of the conference.
*Requirements:* Eligible applicants are education institutions, non-profit organizations, and private industries.
*Geographic Focus:* All States
*Date(s) Application is Due:* Jan 31
*Contact:* Lisa Luckett, Grants Officer; (601) 634-2633; Lisa.K.Luckett@usace.army.mil
*Sponsor:* U.S. Department of Defense
3909 Halls Ferry Road
Vicksburg, MS 39180-6199

**USDD Bulk Nanostructured Materials Grants** **2664**
This program takes advantage of the unusual properties of nanomaterials to build hybrid structural materials as combinations or composites of both nano and regular materials in order to obtain new materials with enhanced properties compared to the original constituents. For example, there is interest in developing materials with greatly increased strength over that of the components without, at the same time, increasing the composite's brittleness or fragility. In addition, work is underway to develop thermal sprayed nanostructured coatings. Such coatings may offer greatly increased resistance to wear and corrosion especially in moving/rotating parts. Submission of the Research Proposal should always be preceded by submission of a Preproposal and a discussion of that Preproposal with the assigned Program Officer. Applications for the upcoming fiscal year should be received by September 30.
*Requirements:* Grants are awarded to academic, industrial, and nonprofit organizations, with strong emphasis on university research. Investigators are encouraged to contact the staff prior to formally submitting a proposal.
*Restrictions:* New grants or contracts are typically started near the beginning of the fiscal year. New proposals for the next fiscal year should be submitted no later than September 30, unless otherwise directed. Competitive renewals should be submitted no later than six months prior to the award termination date after contacting the Program Officer. Conference proposals should be submitted at least four months prior to the conference.
*Geographic Focus:* All States
*Date(s) Application is Due:* Sep 30
*Contact:* Lawrence Kabacoff, Program Officer; (703) 696-0283 or (703) 696-3073; fax (703) 796-0934; lawrence.kabacoff@navy.mil
*Internet:* www.onr.navy.mil/Science-Technology/Departments/Code-33/All-Programs/332-naval-materials/Bulk-Nanostructured-Materials.aspx
*Sponsor:* U.S. Department of Defense
One Liberty Center, 875 N. Randolph Street, Suite 1425
Arlington, VA 22203-1995

**USDD Capacitors for Pulsed Power Applications Grants** **2665**
The Capacitors for Pulsed Power Applications program's goal is to increase the energy density of large dielectric capacitors suitable for handling large amounts of power in conditioners, converters and pulsed power application. The current state of the art capacitors are based on metallized biaxially oriented polypropylene and feature high-breakdown strength, low dielectric losses and graceful failure. Approaches currently in the program include improved synthesis and characterization of olefinic polymers, PVDF and other polar polymers, and composite and nanocomposite approaches. Within the program, there is collaboration between synthetic, physical, and theoretical chemists, materials scientists, and engineers. Submission of the Research Proposal should always be preceded by submission of a Preproposal and a discussion of that Preproposal with the assigned Program Officer. Applications for the upcoming fiscal year should be received by September 30.
*Requirements:* Grants are awarded to academic, industrial, and nonprofit organizations, with strong emphasis on university research. Investigators are encouraged to contact the staff prior to formally submitting a proposal.
*Restrictions:* New grants or contracts are typically started near the beginning of the fiscal year. New proposals for the next fiscal year should be submitted no later than September 30, unless otherwise directed. Competitive renewals should be submitted no later than six months prior to the award termination date after contacting the Program Officer. Conference proposals should be submitted at least four months prior to the conference.
*Geographic Focus:* All States
*Date(s) Application is Due:* Sep 30
*Contact:* Dr. Paul Armistead, Program Officer; (703) 696-4315; fax (703) 696-0001; paul.armistead@navy.mil
*Internet:* www.onr.navy.mil/Science-Technology/Departments/Code-33/All-Programs/332-naval-materials/Capacitors-Pulsed-Power-Applications.aspx
*Sponsor:* U.S. Department of Defense
One Liberty Center, 875 N. Randolph Street, Suite 1425
Arlington, VA 22203-1995

**USDD Care for the Critically Injured Burn Patient Grants** **2666**
The Assistant Secretary of Defense for Health Affairs, Defense Health Program Medical Research and Development Office is soliciting proposals for the Defense Medical Research and Development Program (DMRDP). The Telemedicine and Advanced Technology Research Center (TATRC) is administering the application process, and the United States Army Medical Research Acquisition Activity (USAMRAA) is issuing this program

announcement and will be negotiating all resulting awards. The goal is to advance the state of medical science in those areas of most pressing need and relevance to today's battlefield experience, in this case burn injuries. The objectives of the DMRDP are to discover and explore innovative approaches to protect, support, and advance the health and welfare of military personnel, families, and communities; to accelerate the transition of medical technologies into deployable products; and to accelerate the translation of advances in knowledge into new standards of care for injury prevention, treatment of casualties, rehabilitation, and training systems that can be applied in theater or in the clinical facilities of the Military Health System. The research results are expected to increase the body of knowledge available to professionals and practitioners in health, medical science and related fields. The research impact is expected to benefit the military community and civilian community, particularly as it relates to the following priority areas: (1) multicenter development and assessment of the impact of checklists on standardization of burn care and improvement of outcomes in burn patients; (2) multicenter studies of ICU-based rehabilitation outcomes, i.e., evaluation of physical and rehabilitation technologies/therapies directed at short-term ICU outcomes, understanding the effect of rehabilitation on pathophysiology in the ICU burn patient, and documenting the efficacy of these immediate ICU interventions on long-term quality-of-life outcomes such as the return-to-work/duty rate; (3) device/drug development and validation for inhalation lung injury; and (4) research into the formation of hypertrophic scarring and the development of therapies to manage the condition. Pre-applications are due by December 2, and applications are by invitation only. Three to six awards are anticipated.
*Requirements:* Principal investigators must be independent investigators at any academic level from academia, research institutions, industry, local, state or federal governments, and private foundations that possess the skills, knowledge, and resources necessary to carry out the proposed research. Applications from investigators within the military services and other federal agencies are encouraged, as are applications involving multidisciplinary collaborations among academia, industry, the military services, the Department of Veterans Affairs, and other federal government agencies.
*Geographic Focus:* All States
*Date(s) Application is Due:* Mar 9
*Amount of Grant:* 1,500,000 USD
*Contact:* Mary Rico, Grants Officer; (703) 518-4726; mary.rico@amedd.army.mil
*Sponsor:* U.S. Department of Defense - 1
1400 Defense Pentagon
Washington, D.C. 20301-1400

**USDD Cellular Materials Grants** 2667
Structural cellular materials, either as stochastic foams or integrated into various periodic configurations, are highly porous structures with 20 percent or less of their interior volume occupied by material. These materials, therefore, are useful in a number of naval applications. To enhance our scientific understanding of these materials, the Office of Naval Research (ONR) supports basic research into the behavior of cellular materials under the influence of dynamic deformation processes. An objective is to integrate different materials with different structures and orientations in order to get superior thermal conductance in complex structures. In addition, ONR is very interested in improved protection from fragments or blast under static and dynamic load conditions. ONR seeks to explore further the use of different materials and materials combinations to enhance overall protection to the warfighter. Experimental characterization of material structures under blast and/or fragmentation threats should yield physics-based models and validated algorithms that, in turn, should enable optimal design of blast and fragmentation resistant structures. Finally, the use of light-weight structural cellular materials for thermal management in non-electronic systems is of interest. Submission of the Research Proposal should always be preceded by submission of a Preproposal and a discussion of that Preproposal with the assigned Program Officer. Applications for the upcoming fiscal year should be received by September 30.
*Requirements:* Grants are awarded to academic, industrial, and nonprofit organizations, with strong emphasis on university research. Investigators are encouraged to contact the staff prior to formally submitting a proposal.
*Restrictions:* New grants or contracts are typically started near the beginning of the fiscal year. New proposals for the next fiscal year should be submitted no later than September 30, unless otherwise directed. Competitive renewals should be submitted no later than six months prior to the award termination date after contacting the Program Officer. Conference proposals should be submitted at least four months prior to the conference.
*Geographic Focus:* All States
*Date(s) Application is Due:* Sep 30
*Contact:* Dr. David Shifler; (703) 696-0285 or (703) 696-0934; david.shifler@navy.mil
*Internet:* www.onr.navy.mil/Science-Technology/Departments/Code-33/All-Programs/332-naval-materials/Cellular-Materials.aspx
*Sponsor:* U.S. Department of Defense
One Liberty Center, 875 N. Randolph Street, Suite 1425
Arlington, VA 22203-1995

**USDD Chemical and Biological Defense Innovations and Technologies Grants** 2668
The Space and Naval Warfare Systems Center (SPAWAR) is soliciting proposals for basic research, applied research and advanced research in areas relating to the advancement of technology to meet USDD requirements for chemical, biological, and pandemic disease defense. This Broad Agency Announcement seeks white papers and full proposals for proof-of-concept research to develop an in vitro platform of human organ constructs that could recapitulate the human response to threat agents and accurately predict human safety, efficacy, and pharmacokinetics of candidate Medical Countermeasures. Proposed research should investigate unique and innovative approaches for: rapid detection and diagnosis of chemical and biological threats (to include chemical, viral, bacterial, and protozoan); dissemination of information to local, state, and federal authorities for action; and defining and developing next generation capabilities in chemical, biological and pandemic disease defense.
*Geographic Focus:* All States
*Date(s) Application is Due:* Dec 11
*Contact:* David Hill, Contract Specialist; (619) 553-4349; david.l.hill3@navy.mil
*Sponsor:* U.S. Department of Defense
P.O. Box 190022
North Charleston, SC 29419-9022

**USDD Clinical Functional Assessment Investigator-Initiated Grants** 2669
The United States Army Medical Research and Materiel Command (USAMRMC) is seeking health information technology research that may lead to outcomes and products that may qualify as advanced development activities within the Department of Defense (USDD). The results of this research will support the Military Health System in the adoption and implementation of clinical functional disability assessments and standards, by leveraging the skills and talents of United States based businesses and teaching/research academic institutions for success. The intent is to create enduring initiatives and collaborations that will develop and support sustained business and enterprise solutions for the USDD as well as organizations such as the Veterans Benefits Administration, Centers for Medicare and Medicaid Services, and the general public. Proposals are invited that develop products and systems and generate new knowledge that supports medical protection and treatment for USDD personnel and beneficiaries. Military relevance will be a primary criterion for evaluation in all areas. Pre-proposals are due by November 30, and full proposals are by invitation only. One or two awards are anticipated.
*Requirements:* Eligible applicants include United States based businesses, teaching research academic institutions, and multi-center collaborations, to include international partners, that develop products and systems and generate new knowledge that supports medical protection and treatment for USDD personnel and beneficiaries. Principal investigators must be independent investigators at any academic level from academia, research institutions, government, industry, and private foundations that possess the skills, knowledge, and resources necessary to carry out the proposed research.
*Geographic Focus:* All States
*Date(s) Application is Due:* Mar 15
*Amount of Grant:* 1,500,000 USD
*Contact:* Mary Rico, Grants Officer; (703) 518-4726; mary.rico@amedd.army.mil
*Sponsor:* U.S. Department of Defense - 1
1400 Defense Pentagon
Washington, D.C. 20301-1400

**USDD Collaborative Research in Structural Sciences** 2670
The focus of this program will be on the following three technical thrust areas: adaptive, spatial and temporal, multi-scaling strategies for long duration path-dependent combined-environment structural simulation; damage-integrated, structural-thermal, reduced-order-models for realistic structural components; and high-fidelity, continuum-level, constitutive and material damage evolution models. A major goal within AFRL is to establish cooperative relationships with external researchers in order to increase the agility and responsiveness of our research efforts. We wish to pursue a collaborative effort in Structural Science. The objective of the collaboration is to provide research in aerospace structural design through a broad and agile base of expertise. The collaborative research is expected to further new inter-disciplinary research areas of unique interest to the Air Force; for example, faculty members at universities can hire top graduate students and formulate research projects of long-term interest to AFRL. By conducting this collaborative research, AFRL can tap into the required expertise to help accelerate innovative research in structures in a variety of areas. The Air Force anticipates awarding one (1) award for this announcement, but reserves the right to award all, part, or none of the proposals received and reserves the right to award grants or cooperative agreements. The base value is estimated to be $2,500,000, with a $1,000,000 priced Option, for a total program value of $3,500,000.
*Requirements:* This is an unrestricted solicitation. Small businesses are encouraged to propose on all or any part of this solicitation.
*Geographic Focus:* All States
*Date(s) Application is Due:* Oct 31
*Amount of Grant:* Up to 3,500,000 USD
*Contact:* Noreen Bennett; (937) 656-9671; noreen.bennett@wpafb.af.mil
*Sponsor:* U.S. Department of Defense
1400 Defense Pentagon
Washington, D.C. 20301-1400

**USDD Communications and Networking Discovery and Invention Grants** 2671
The goal of the Communications and Networking Program within the Office of Naval Research is to support the FORCEnet vision by developing measurable advances in technology that can directly enable and enhance end-to-end connectivity and quality-of-service for mission-critical information exchange among such widely dispersed naval, joint, and coalition forces. The vision is to provide high throughput robust communications and networking to ensure all warfighters, from the operational command to the tactical edge, have access to information, knowledge, and decision-making necessary to perform their assigned tasks. Proposals are sought under the following focus areas: (1) Back-end electronics devices/sub-systems/architectures/algorithms for ultra-wideband apertures with simultaneous low noise, low distortion, and very high instantaneous dynamic range; (2) Innovative approaches for spectrum co-existence underlay/overlay, interference cancellation, etc.) of military waveforms with commercial communications; (3) High pulse power (variable pulse rep frequency) fiber laser emission in the blue-green (450-520 nm) using all-fiber power-scaling and combining techniques. Fully-passive (or minimal

electrical/thermal control) filter designs with high transmissivity, narrow optical bandpass (<0.5 nm), and large field-of-view in the blue-green region. High speed (Gbps), single-photon counting, blue-green detectors and high performance receiver waveforms/protocols with relaxed sync/timing accuracies; (4) Machine learning techniques and other novel approaches for autonomous network management; and (5) Robust tools/approaches for de-confliction of traffic prioritization policies, according to Commander's Intent, over a distributed, tactical network. The Office of Naval Research is also receptive to highly innovative ideas in other general communications and networking areas that are not within the designated focus areas specified above but nonetheless are important to the Navy/Marine Corps. White papers are accepted by mail with delivery confirmation or by commercial carrier. An expanded oral presentation will be requested from applicants whose proposals are identified as being of "particular value" to the Navy. Following oral presentations, applicants whose technology is still considered as having "particular value" to the Navy will be encouraged to submit detailed technical and cost proposals. Average awards are $500,000 per year. The award period may be from one to three years.
*Requirements:* All responsible sources from academia and industry are eligible. Historically black colleges and universities and minority institutions are encouraged to apply. Teams are encouraged and may submit proposals in any and all areas.
*Geographic Focus:* All States
*Date(s) Application is Due:* Jun 28
*Contact:* Sheila Richardson; (703) 696-1575; sheila .richardson@navy.mil
Dr. Santanu Das, Contract Specialist; (703) 696-1575; Santanu.Das@navy.mil
*Internet:* www.onr.navy.mil/Contracts-Grants/Funding-Opportunities/Broad-Agency-Announcements.aspx
*Sponsor:* U.S. Department of Defense
One Liberty Center, 875 N. Randolph Street, Suite 1425
Arlington, VA 22203-1995

**USDD Computer-Aided Materials Design Grants** **2672**
Advances in theory (e.g., classical, quantum and statistical mechanics), along with computing hardware and software, now make it possible to suggest molecules, their structures and assemblies (viz., materials) and to compute expected properties, in some cases, even before actual synthesis and processing have been carried out. With computer-aided design, the Office of Naval Research (ONR) believes that fundamental science can be explored, the cost of research and development can be substantially lowered, and, in the process, optimally designed materials can be derived. This Computer-Aided Materials Design (CAMD) program currently focuses on alloys, superconducting materials, new polymer composites, high-temperature materials, enhanced capacitors and batteries, and anti-fouling materials. Several topic areas reflect cooperation with other programs in the division and in ONR. General computer-aided organic or inorganic chemical synthesis is not an element of this program. Materials and systems of emphasis are alloys, polymer composites and materials relevant to power and energy. Submission of the Research Proposal should always be preceded by submission of a Preproposal and a discussion of that Preproposal with the assigned Program Officer. Applications for the upcoming fiscal year should be received by September 30.
*Requirements:* Grants are awarded to academic, industrial, and nonprofit organizations, with strong emphasis on university research. Investigators are encouraged to contact the staff prior to formally submitting a proposal.
*Restrictions:* New grants or contracts are typically started near the beginning of the fiscal year. New proposals for the next fiscal year should be submitted no later than September 30, unless otherwise directed. Competitive renewals should be submitted no later than six months prior to the award termination date after contacting the Program Officer. Conference proposals should be submitted at least four months prior to the conference.
*Geographic Focus:* All States
*Date(s) Application is Due:* Sep 30
*Contact:* Dr. Kenny Lipkowitz, Program Officer; (703) 696-0487 or (703) 696-0707; fax (703) 696-0934; kenny.lipkowitz@navy.mil
*Internet:* www.onr.navy.mil/Science-Technology/Departments/Code-33/All-Programs/332-naval-materials/Computer-Aided-Materials-Design.aspx
*Sponsor:* U.S. Department of Defense
One Liberty Center, 875 N. Randolph Street, Suite 1425
Arlington, VA 22203-1995

**USDD Crowd Sourced Formal Verification Grants** **2673**
DARPA-Information Innovations Office is soliciting innovative research proposals in the area of crowd sourced formal verification of software. Proposed research should investigate innovative approaches that enable revolutionary advances in science, devices, or systems. The goal of the Crowd Sourced Formal Verification (CSFV) program is to make formal verification of software more cost-effective by enabling non-specialists to participate productively in the formal verification process. The approach is to transform the verification task into a more accessible task by creating a game that reflects a relevant abstraction of the program for verification purposes, is intuitively understandable by ordinary people, and is fun to play. Completion of the game effectively helps the formal verification tool complete the corresponding formal program verification proof. The scope of the game-based formal verification technical area is limited to automatic game construction, bidirectional interfaces to formal verification tools, and necessary modifications to those tools. As the CSFV program goals involve certifying the absence of certain security vulnerabilities in code, logically sound verification techniques are required. The CSFV program seeks improved compiler optimization by reusing annotations useful for verification to provide more precise control and data flow information. Finally, a research integrator role will bring the verification tools and games together in a coherent system and website. Multiple awards are anticipated. Annual funding is expected to range from $4.7 to $11.05 million. The level of funding for individual awards has not been determined. Application instructions can be found on the website.
*Requirements:* All responsible sources capable of satisfying the government's needs are eligible. Historically black colleges and universities, small businesses, small disadvantaged businesses and minority institutions are encouraged to submit proposals and team with others to submit proposals.
*Restrictions:* Specifically excluded from CSFV is the construction of new formal verification tools. Also excluded is research that primarily results in evolutionary improvements to the existing state of practice.
*Geographic Focus:* All States
*Date(s) Application is Due:* Mar 20
*Contact:* Dean Drew, Program Manager; (703) 526-6630; CSFV@darpa.mil
*Sponsor:* U.S. Department of Defense
3701 North Fairfax Drive
Arlington, VA 21702-1714

**USDD Defense Sciences Research and Technology Grants** **2674**
The mission of the Defense Advanced Research Projects Agency (DARPA) Defense Sciences Office (DSO) is to pursue and exploit fundamental science and innovation for national defense. The Defense Sciences Reach and Tecnology Grants solicit proposal abstracts and full proposals for advanced research and development in a variety of enabling technical areas. Proposal abstracts and full proposals may be either basic or applied research. Applicants should demonstrate that their proposed effort is aimed at high-risk/high-payoff technologies that have the potential for making, in the five- to ten-year timeframe, revolutionary rather than incremental improvements to national security, including emerging threats and operational challenges. While there is no specific requirement for cost and duration of the proposed effort, it is recommended that proposers include a timeline that plots the most critical issues on the path to success. Topic areas include: physical sciences; material science; biology; neuroscience; mathematics; and testing and demonstration. Each area is described in detail at the website and should be reviewed by applicants. Abstracts/proposals that are not within the topic areas described may be considered out of scope and may not be evaluated. Proposal abstracts and full proposals may be submitted at any time until the deadline.
*Requirements:* All responsible sources capable of satisfying the government's needs are eligible. Historically black colleges and universities, small businesses, small disadvantaged businesses and minority institutions are encouraged to apply.
*Restrictions:* Abstracts/proposals for the integration of existing technologies or development of systems will be considered out of scope and will not be evaluated. The preferred method of communication is email.
*Geographic Focus:* All States
*Contact:* Dr. Jon Mogford, Technical Specialist; DARPA-BAA-11-65@darpa.mil
*Sponsor:* U.S. Department of Defense
3701 North Fairfax Drive
Arlington, VA 22203-6630

**USDD Developer Tools for Medical Education Public Physiology Research Platform Grant** **2675**
The Assistant Secretary of Defense for Health Affairs, Defense Health Program is soliciting proposals for a Developer Tools for Medical Education Public Physiology Research Platform (DTME-PRP) Grant. The DTME-PRP Grant will be issued and awarded by the United States Army Medical Research Acquisition Activity (USAMRAA) and managed by the Army's Telemedicine and Advanced Technology Research Center. This initiative is primarily focused on the creation of publicly accessible, free, open-license, and/or open-source tools for creation of medical training and education content and technology. Principally, the DTME initiative seeks to support the strategic goals of the Joint Program Committee 1 – Medical Training and Health Information Sciences Research Program through the leveraging of technologies created by this effort, finding cost efficiencies through utilization of free tools and allowing more commonality for military medical training technology systems. It is envisioned that expanding a research effort to create this capability to an open public trial and experimentation basis will allow for feedback from a larger part of the medical training community than would otherwise be possible from a single group. Feedback regarding novel uses of the evolving research platform and the data created by this research will lead to improvements in the science regarding simulated physiology for use in public and private sector medical education. Proposals should address research that will do the following: (1) create a public physiology research platform for use by the medical simulation development and education community; (2) enable accurate and consistent simulated physiology across multiple training applications and platforms; (3) allow for community contributions to this effort; (4) encourage greater use of scientifically-based physiological and pharmacological predictive technologies; (5) lower barriers to creating medical training content through availability of this resource; (6) serve the training needs of the military; and (7) expand the corpus of knowledge regarding the use of simulated physiology for medical education. Approximately $7 million is available, and one award is anticipated. The maximum period of performance is five years. Pre-proposals are due by January 10, and full proposal submissions are by invitation only.
*Requirements:* Eligible organizations must be extramural and include nonprofit, public, and private organizations, such as universities, colleges, hospitals, laboratories, and commercial firms. Historically black colleges and universities/minority institutions are encouraged to apply.
*Geographic Focus:* All States
*Date(s) Application is Due:* Apr 26
*Contact:* Mary Rico, Grants Officer; (703) 518-4726; mary.rico@amedd.army.mil
*Sponsor:* U.S. Department of Defense - 1
1400 Defense Pentagon
Washington, D.C. 20301-1400

**USDD Development of Quantum Computing Algorithms Grants** 2676
The United States Army Research Office (ARO) together with the National Security Agency is soliciting proposals to develop new quantum computing algorithms for hard computational problems, develop insights into the power of quantum computation, and consider issues of quantum complexity and computability. Proposals for research in quantum algorithms should primarily be to devise novel quantum algorithms for solving mathematically and computationally hard problems from such diverse fields as algebra, number theory, geometry, analysis, optimization, graph theory, differential equations, combinatorics, topology, logic, and simulation. Quantum algorithms that are developed should focus on constructive solutions for specific tasks and on general methodologies for expressing and analyzing algorithms tailored to specific problems. Complexity analysis such as upper and lower bounds on algorithms, including developing new methodologies for deriving such bounds, is encouraged. Analysis of the robustness of an algorithm against uncertainty (e.g. small perturbations in the problem's defining parameters or input states) is also encouraged. Investigators should presuppose the existence of a fully functional quantum computer and consider what algorithmic tasks are particularly well suited to such a machine. A necessary component of this research will be to compare the efficiency of the quantum algorithm to the best existing classical algorithm for the same problem. Although quantum algorithm proposals may consider general architectural constraints (e.g. nearest neighbor only gates) for implementing algorithms, they should otherwise concentrate on developing the algorithm. Quantum algorithm proposals may consider computational models other than the circuit model (e.g. the adiabatic model), however proposals should assiduously avoid tying algorithms to specific physical models. Applicants are strongly encouraged to submit white papers. Full proposals are required to apply. This is a multiple year Broad Agency Announcement (BAA). Applicants are encouraged to periodically check the following websites for updates and/or amendments: FedBizOpps.gov (https://www.fbo.gov/), Grants.gov (grants.gov/applicants/find_grant_opportunities.jsp), and the ARO website (www.arl.army.mil/www/default.cfm?Action=6&Page=8).
*Requirements:* Degree-granting universities (foreign and domestic), nonprofit organizations, and industrial concerns (large and small businesses) are eligible. Proposals are encouraged from historically black colleges and universities and from minority institutions.
*Geographic Focus:* All States, All Countries
*Date(s) Application is Due:* May 8
*Amount of Grant:* 1,000,000 USD
*Contact:* Jessica Leach, Grants Officer; (919) 549-4387; jessica.e.leach.civ@mail.mil
Dr. T.R. Govindan, Technical Specialist; (919) 549-4387; t.r.govindan@civ@mail.mil
*Sponsor:* U.S. Department of Defense
4400 Martin Road, Redstone Arsenal
Huntsville, AL 35898

**USDD Dielectric Films for Capacitors Grants** 2677
Dielectric films for capacitors are critical for the development of energy- and power-dense systems for shipboard use impulse power, power conditioning and back-up power applications. The Office of Naval Research program focuses on the materials and structures that have the capacity to increase energy density, reducing weight and volume allocation for capacitor banks while extending service life in extreme conditions. Current efforts are focused on multi-scale models of dielectric charge storage and transfer including failure mechanisms, synthesis of novel polymer and composite dielectric materials, optimization for rapid, controlled discharge rates and development of high temperature (150 – 200 C) polymer dielectrics. Applications include: conditioners; capacitors; rail guns; auxiliary power; and power electronic inverters and converters. Submission of the Research Proposal should always be preceded by submission of a Preproposal and a discussion of that Preproposal with the assigned Program Officer. Applications for the upcoming fiscal year should be received by September 30.
*Requirements:* Grants are awarded to academic, industrial, and nonprofit organizations, with strong emphasis on university research. Investigators are encouraged to contact the staff prior to formally submitting a proposal.
*Restrictions:* New grants or contracts are typically started near the beginning of the fiscal year. New proposals for the next fiscal year should be submitted no later than September 30, unless otherwise directed. Competitive renewals should be submitted no later than six months prior to the award termination date after contacting the Program Officer. Conference proposals should be submitted at least four months prior to the conference.
*Geographic Focus:* All States
*Date(s) Application is Due:* Sep 30
*Contact:* Dr. Paul Armistead, Program Officer; (703) 696-4315; fax (703) 696-0001; paul.armistead@navy.mil
*Sponsor:* U.S. Department of Defense
One Liberty Center, 875 N. Randolph Street, Suite 1425
Arlington, VA 22203-1995

**USDD Electro-Optical (EO)/Infrared (IR) Sensor Processing Grants** 2678
The Office of Naval Research (ONR) Electro-Optical (EO) / Infrared (IR) Sensors and Sensor Processing Program objective is to detect, classify/identify, and localize/geolocate air, sea-surface, and ground targets by developing innovative sensor technology and by improving the performance of the signal and image processing techniques associated with electro-optical (EO) and infrared (IR) sensors (passive and active) that operate from the visible through longwave infrared bands. Proposers are encouraged to contact the program officer to discuss their research interest prior to the submission of formal proposals.
*Requirements:* Grants are awarded to academic, industrial, and nonprofit organizations, with strong emphasis on university research.
*Restrictions:* New grants or contracts are typically started near the beginning of the fiscal year (October 1). New proposals for the next fiscal year should be submitted no later than September 30, unless otherwise directed by the BAA. Competitive renewals should be submitted no later than six months prior to the award termination date after contacting the Program Officer. Conference proposals should be submitted at least four months prior to the conference.
*Geographic Focus:* All States
*Date(s) Application is Due:* Sep 30
*Contact:* Ravindra Athale, Program Officer; (703) 696-2237; fax (703) 696-2206; ravindra.athale@navy.mil
*Internet:* www.onr.navy.mil/Science-Technology/Departments/Code-31/All-Programs/312-Electronics-Sensors/EO-IR-Sensor-Processing.aspx
*Sponsor:* U.S. Department of Defense
One Liberty Center, 875 N. Randolph Street, Suite 1425
Arlington, VA 22203-1995

**USDD Electrochemical Materials Grants** 2679
The Office of Naval Research Electrochemical Materials Program is focused on developing a fundamental understanding of charge (electron and ion) storage, transport and transfer mechanisms and applying that knowledge to develop materials, materials architectures and devices that address unique Navy and Marine Corps application needs, such as current and future logistics fueled devices and air-independent propulsion for unmanned undersea vehicles (UUVs). Current foci in this program include in-situ probes for determining reaction mechanisms in high temperature electrochemical systems, multi-scale modeling to meaningfully span atomistic to cell level behavior, challenges unique to three-dimensional battery architectures and fundamental understanding of lithium-based battery catastrophic failure modes. Submission of the Research Proposal should always be preceded by submission of a Preproposal and a discussion of that Preproposal with the assigned Program Officer. Applications for the upcoming fiscal year should be received by September 30.
*Requirements:* Grants are awarded to academic, industrial, and nonprofit organizations, with strong emphasis on university research. Investigators are encouraged to contact the staff prior to formally submitting a proposal.
*Restrictions:* New grants or contracts are typically started near the beginning of the fiscal year. New proposals for the next fiscal year should be submitted no later than September 30, unless otherwise directed. Competitive renewals should be submitted no later than six months prior to the award termination date after contacting the Program Officer. Conference proposals should be submitted at least four months prior to the conference.
*Geographic Focus:* All States
*Date(s) Application is Due:* Sep 30
*Contact:* Dr. Michele Anderson, Program Officer; (703) 696-1938; fax (703) 696-6887; michele.anderson1@navy.mil
*Internet:* www.onr.navy.mil/Science-Technology/Departments/Code-33/All-Programs/332-naval-materials/Electrochemical-Materials.aspx
*Sponsor:* U.S. Department of Defense
One Liberty Center, 875 N. Randolph Street, Suite 1425
Arlington, VA 22203-1995

**USDD Electromagnetic Materials Grants** 2680
The Office of Naval Research (ONR) Electromagnetic Materials Program develops advanced electronic and magnetic materials critical to the advancement of electronic and magnetic devices. Included are III-V semiconductors, oxides, multi-ferroic and multi-functional materials, magnetic films and materials that address micro-scale thermal management, integration and/or novel functionality. Focused experimental and theoretical work is supported. This program currently has three major thrust areas: growth and integration of multi-ferroic materials and their hetero-structures to achieve novel functional devices including, but not limited to, uncooled magnetic sensors, microwave circulators, isolators and tunable filters and phase shifters; III-V device technologies to maximize RF performance metrics including power, linearity, bandwidth, efficiency and noise by improving materials to overcome fundamental limitations such as thermal resistance or capacitance; and emerging materials research that offers the potential for unprecedented functionality such as bandwidth, power handling, linearity, tunability or other new capabilities. Investigators are encouraged to contact the staff prior to formally submitting a proposal. Submissions for the upcoming fiscal year must be received by September 30.
*Requirements:* Grants are awarded to academic, industrial, and nonprofit organizations, with strong emphasis on university research.
*Geographic Focus:* All States
*Date(s) Application is Due:* Sep 30
*Contact:* Dr. Daniel Green; (703) 696-4220; daniel.s.green1@navy.mil
*Internet:* www.onr.navy.mil/Science-Technology/Departments/Code-31/All-Programs/312-Electronics-Sensors/Electronic-Materials.aspx
*Sponsor:* U.S. Department of Defense
One Liberty Center, 875 N. Randolph Street, Suite 1425
Arlington, VA 22203-1995

**USDD Entry-Level Faculty Awards in Ocean Acoustics** 2681
The purpose of the USDD Entry-Level Faculty Awards in Ocean Acoustics is to encourage research careers in ocean acoustics or the related disciplines of undersea signal processing, marine structural acoustics and transducer materials science by supporting the research of outstanding young university faculty members. Applicants must be U.S. citizens and must apply within the first four years following the start of his or her first tenure-track appointment. Awardees will be selected based on past performance and quality of proposal. These awards are 36 months in duration and support young faculty scientists at $100,000 per year. ONR expects to grant up to two new entry-level faculty awards per year.

*Requirements:* A detailed technical proposal in the format described in Planning Letter/ Pre-Proposal and Proposal Process is required.
*Geographic Focus:* All States
*Date(s) Application is Due:* Aug 1
*Amount of Grant:* 100,000 USD
*Contact:* Dr. Ellen Livingston, Program Officer; (703) 696-4203 or (703) 696-5031; fax (703) 696-5940; ellen.livingston@navy.mil
*Internet:* www.onr.navy.mil/Science-Technology/Departments/Code-32/All-Programs/ Atmosphere-Research-322/Ocean-Acoustics/Awards-Ocean-Acoustics.aspx
*Sponsor:* U.S. Department of Defense
One Liberty Center, 875 N. Randolph Street, Suite 1425
Arlington, VA 22203-1995

**USDD Extended Solids Grants** **2682**
The Defense Advanced Research Projects Agency (DARPA) is soliciting innovative multidisciplinary research proposals to develop and demonstrate synthesis methods of materials with superior properties (>1.5X over SOA) that currently can only be made using ultrahigh pressure techniques, such that processing under conditions amenable to production scales (<2000 K and 0.5 GPa) and stability under ambient conditions is achieved. Extended solids are polymorphs/phases of simple molecules that are currently formed under ultrahigh pressure conditions where strong intermolecular bonding and tight crystal packing can be induced, leading to dramatic changes in physical, mechanical, and functional properties. Abstracts are encouraged in advance of full proposals in order to provide potential applicants with a rapid response to minimize unnecessary effort. Multiple awards are anticipated.
*Requirements:* All responsible sources capable of satisfying the government's needs are eligible. Historically black colleges and universities, small businesses, small disadvantaged businesses and minority institutions are encouraged to apply and to join others in submitting proposals.
*Geographic Focus:* All States
*Date(s) Application is Due:* May 22
*Contact:* Dr. Judah Goldwasser, Technical Specialist; DARPA-BAA-12-20@darpa.mil
*Sponsor:* U.S. Department of Defense
3701 North Fairfax Drive
Arlington, VA 22203-6630

**USDD Fundamental Research to Counter Weapons of Mass Destruction Grants** **2683**
This is a Broad Agency Announcement (BAA) by the Defense Threat Reduction Agency (DTRA). It is an extramural endeavor that combines the fundamental research, educational program, or other effort needs of DTRA and other Department of Defense interests. DTRA safeguards America and its allies from weapons of mass destruction (WMD) and provides capabilities to reduce, eliminate and counter the threat and effects from chemical, biological, radiological, nuclear, and high yield explosives. DTRA seeks to identify, adopt, and adapt emerging, existing and revolutionary sciences that may demonstrate high payoff potential to Counter-WMD (C-WMD) threats. This BAA solicits ideas and topic-based pre-application white papers for long-term challenges that offer a significant contribution in the following: the current body of knowledge; the understanding of phenomena and observable facts; advancement in revolutionary technology; and new concepts for technology application or that may have impact on future C-WMD threat reduction or capabilities. A portion of this effort is expected to be devoted to awards for science, technology, engineering, and mathematics education programs with a C-WMD focus; such as, but not limited to, postdoctoral fellowships, stipends, degrees, visiting scientist programs, student exchange programs, and development of accredited C-WMD curricula. There are 6 thrust areas: (1) science of WMD sensing and recognition; (2) cognitive, information and network science; (3) science for protection; (4) science to defeat WMD; (5) science to secure WMD; and (6) cooperative counter WMD research with global partners. All non-topic based proposed efforts must be coordinated with the relevant technical point of contact prior to submission. Contact information is listed in the BAA on the website. On average 15 to 30 individual awards are expected per year. There are two categories of awards. Single scope awards (averaging $150K per year) are research projects or efforts that focus on exploratory aspects of a unique problem, a high risk approach, or innovative research in subjects with potential for high impact to C-WMD science. Multidisciplinary awards (averaging $350K per year) are research projects or efforts that involve a comprehensive program of innovative research in an interdisciplinary area with potential for high impact. Application packages and instructions vary by topic and thrust area. Applicants should carefully review the instructions found in the BAA at the website. Generally there are no deadlines for pre-application white papers, and the deadline for invited proposals will be provided in the letter of invitation. Proposal invitations are limited to available funds. Timeline details are available at the DTRA Basic and Fundamental Research Community Portal (www.dtrasubmission.net/portal). Applicants are responsible for checking the site for changes and/or updates.
*Requirements:* Applicants eligible to submit pre-application white papers and proposals include the following United States and foreign-based equivalents: accredited degree-granting colleges and universities; industrial/commercial basic research centers, including small businesses with a portfolio predominantly in basic research proposing a project with significant participation (minimum 30% of total effort value on the proposed project) by one or more accredited degree-granting colleges and/or universities; and not-for-profit organizations/basic research centers with a portfolio predominantly in basic research proposing a project with significant participation (minimum 30% of total effort value on the proposed project) by one or more accredited degree-granting colleges and/or universities. Proof of 501(c)3 status may be required for not-for-profits. For foreign-based establishments entirely based outside the United States and/or its territories, proof of not-for-profit status may be required.
*Geographic Focus:* All States, All Countries
*Contact:* Gena Leake; (703) 767-5870; HDTRA1-09-14-FRCWMD@dtra.mil
*Internet:* www.dtra.mil/Research.aspx
*Sponsor:* U.S. Department of Defense
6200 Meade Road
Fort Belvoir, VA 22060-5264

**USDD Future Naval Fuels Grants** **2684**
During the 21st Century, the availability of petroleum-derived fuels will decline, causing a global shift towards alternative fuels. Additionally, because the United States currently imports approximately 60 percent of our current oil needs, our energy security is potentially subject to uncertainties that could impact the operations of the Navy and Marine Corps assets. The Navy uses two specified fuels for aircraft and shipboard propulsion: JP-5 and NATO F-76. They are compatible with seawater compensation and have a higher flash point that meets international requirements for shipboard usage. A number of key system parameters must be investigated in determining the impact of alternative fuel changes on the combustions systems (gas turbines and diesel engines) that power Navy and Marine Corps assets. Alternative fuels aboard Navy ships must be compatible with current Navy fuels, tolerant to seawater compensation, have flash point characteristics equivalent to current Navy fuels, have long-term storage capabilities, resistance to biocontamination, and will not negatively impact the current Navy fuels logistics. Research needs to address a methodology for fuel characterization and combustion qualification for acceptable operation of current and proposed gas turbines and diesel engines that may be deployed by the Navy and Marine Corps. We need to understand the impact of varying fuel stock and/or fuel processing on fuel properties and corresponding effects on engine operations, reliability and life. Research must also explore the long-term impact, if any, of alternative fuels on the engine materials and components. Investigations will include validated models that can predict the combustion dynamics, flame stability and emissions of varying fuel constituents. By understanding petroleum-based fuel characteristics and combustion dynamics, opportunities exist to design synthetic fuels with higher heat specific heat content that could extend operational range of assets. Submission of the Research Proposal should always be preceded by submission of a Preproposal and a discussion of that Preproposal with the assigned Program Officer. Applications for the upcoming fiscal year should be received by September 30.
*Requirements:* Grants are awarded to academic, industrial, and nonprofit organizations, with strong emphasis on university research. Investigators are encouraged to contact the staff prior to formally submitting a proposal.
*Restrictions:* New grants or contracts are typically started near the beginning of the fiscal year. New proposals for the next fiscal year should be submitted no later than September 30, unless otherwise directed. Competitive renewals should be submitted no later than six months prior to the award termination date after contacting the Program Officer. Conference proposals should be submitted at least four months prior to the conference.
*Geographic Focus:* All States
*Date(s) Application is Due:* Sep 30
*Contact:* Sharon Beermann-Curtin, Program Officer; (703) 588-2358; fax (703) 696-0934; sharon.beermanncurti@navy.mil
*Internet:* www.onr.navy.mil/Science-Technology/Departments/Code-33/All-Programs/332-naval-materials/Future-Naval-Fuels.aspx
*Sponsor:* U.S. Department of Defense
One Liberty Center, 875 N. Randolph Street, Suite 1425
Arlington, VA 22203-1995

**USDD Graduate Traineeship Awards in Ocean Acoustics** **2685**
USDD Graduate Traineeship Awards in Ocean Acoustics, for study and research leading to a doctoral degree, will be given to individuals who have demonstrated a special aptitude and desire for advanced training in ocean acoustics or the related disciplines of undersea signal processing, marine structural acoustics and transducer materials science. Applicants must have received their baccalaureate or master degrees within three years of the year in which they apply and must plan to attend an institution with doctoral granting programs in the ocean sciences or related fields. Special consideration will be given to students who have passed doctoral candidate qualifying exams. These awards are 36 months in duration and cover the student's full tuition, salary and required fees. Applications should include a letter of intent (describing applicant's research interests and academic institution), three reference letters, and official transcripts. ONR expects to grant up to five new graduate traineeships per year.
*Geographic Focus:* All States
*Date(s) Application is Due:* Aug 1
*Contact:* Dr. Ellen Livingston, Program Officer; (703) 696-4203 or (703) 696-5031; fax (703) 696-5940; ellen.livingston@navy.mil
*Internet:* www.onr.navy.mil/Science-Technology/Departments/Code-32/All-Programs/ Atmosphere-Research-322/Ocean-Acoustics/Awards-Ocean-Acoustics.aspx
*Sponsor:* U.S. Department of Defense
One Liberty Center, 875 N. Randolph Street, Suite 1425
Arlington, VA 22203-1995

**USDD High Assurance Cyber Military Systems Grants** **2686**
The Defense Advanced Research Projects Agency (DARPA) is soliciting innovative research proposals in the area of the clean-slate development of software for high-assurance cyber-physical systems. Proposed research should investigate innovative approaches that enable revolutionary advances in science or systems. The goal of the High-Assurance Cyber Military Systems (HACMS) Grants program is to create technology for the construction of high-assurance, cyber-physical systems, where high assurance is defined to mean functionally correct and satisfying appropriate safety and

security properties. Achieving this goal requires a fundamentally different approach from what the software community has taken to date. Consequently, HACMS will adopt a clean-slate, formal methods–based approach that enables semi-automated code synthesis from executable, formal specifications. In addition to generating code, such a synthesizer will produce a machine-checkable proof that the generated code satisfies the functional specification as well as security and safety policies. A key technical challenge is the development of techniques to ensure that such proofs are composable, allowing the construction of high-assurance systems out of high-assurance components. The HACMS program is structured as a 4.5-year effort, split into three 18-month phases. The five technical areas are: military vehicle experts; operating systems; control systems; research integration and red team. Multiple awards are anticipated.
*Requirements:* All responsible sources capable of satisfying the government's needs are eligible. Historically black colleges and universities, small businesses, small disadvantaged businesses and minority institutions are encouraged to apply and to team with others to submit proposals.
*Restrictions:* Specifically excluded is research that primarily results in evolutionary improvements to the existing state of practice.
*Geographic Focus:* All States
*Date(s) Application is Due:* Jul 10
*Contact:* Kathleen Fisher, Program Manager; (703) 526-6630; HACMS@darpa.mil
*Sponsor:* U.S. Department of Defense
3701 North Fairfax Drive
Arlington, VA 21702-1714

**USDD High School Apprenticeship Program and Undergraduate Research Apprenticeship Program Grants** 2687

The Army Research Office (ARO) intends to conduct the High School Apprenticeship Program (HSAP) and Undergraduate Research Apprenticeship Program (URAP) for another year. HSAP/URAP will fund the science, technology, engineering, or math (STEM) apprenticeship of promising high school juniors, seniors, and undergraduate college students to work in a university structured research environment under the direction of existing ARO-sponsored principal investigators (PI) serving as mentors. The overall objectives include: ensuring a high quality and high quantity STEM workforce in the United States; fostering student interest in STEM subject areas; facilitating follow-on student opportunities including university enrollment and Department of Defense scholarships/internships; and ultimately exposing the apprentices to the attractive, meaningful, and challenging military and civilian career STEM opportunities. The PI/mentor is principally responsible for identifying meaningful and significant duties and responsibilities under the auspices of the current grant and will provide the day-to-day guidance and oversight for the apprentice. Awards will not exceed $3,000 per apprentice. Each modification can have no more than two apprentices for grants and no more than six apprentices for UARCs. Application instructions are on the website.
*Requirements:* Awards will be made as add-ons to existing single-investigator grants, Multidisciplinary University Research Initiative grants, and University Affiliated Research Centers that have at least 12 months' period of performance remaining from the date of proposal submission. An apprentice must be at least a junior in high school, i.e., started the fall semester of their junior year or be an undergraduate college student.
*Geographic Focus:* All States
*Date(s) Application is Due:* Feb 29
*Contact:* Ashley Wade, ARO Youth Science Program Manager; (919) 549-4205; ashley.e.wade.civ@mail.mil
*Sponsor:* U.S. Department of Defense
4400 Martin Road, Redstone Arsenal
Huntsville, AL 35898

**USDD HIV/AIDS Prevention: Military Specific HIV/AIDS Prevention, Care, and Treatment for Non-PEPFAR Funded Countries Grants** 2688

The Department of Defense HIV/AIDS Prevention Program (DHAPP) has successfully engaged over 80 countries in efforts to combat HIV/AIDS among their respective military services. DHAPP is a partner collaborating with the United States State Department, Health and Human Services, United States Agency for International Development, and Centers for Disease Control and Prevention, in the President's Emergency Plan for AIDS Relief (PEPFAR). This Broad Agency Announcement (BAA) allows DHAPP to assist countries where militaries are not funded under PEPFAR and to solicit existing partners and establish new partners in furtherance of DHAPP and partner military program goals. Proposals should focus on rapidly extending HIV/AIDS services. Applicants are encouraged to target specific needs with a practical business plan, using small grass-roots organizations to provide community-based services as a way to enhance organic capabilities and sustainability. Concept papers are required and full proposals are by invitation only. Submission within the first cycle, ending February 21, is strongly encouraged. The period of performance is generally one year, with up to two additional phases subject to availability of funding. Applicants are encouraged to propose additional phases if appropriate. This is a multiple year BAA which may be revised as needed. Based on the availability of funds, up to three evaluation cycles will be conducted. Changes will be posted at Grants.gov (grants.gov/applicants/find_grant_opportunities.jsp). Applicants are encouraged to sign up on Grants.gov for "send me change notification emails" by following the link on the synopsis page for the program announcement/funding opportunity.
*Requirements:* All responsible sources from academia, industry, and non-governmental organizations are eligible. Applicants respondents must demonstrate the active support of the in-country military and the Department of Defense representative in the corresponding United States Embassy in the planning and execution of proposals.
*Restrictions:* Funding is available for Chad, Niger, Mali, and Burkina Faso. A description of the work needed, along with the program areas and estimated funding is available in the BAA found at the website. All questions should be submitted to the points of contact by email.
*Geographic Focus:* All States
*Date(s) Application is Due:* Sep 30
*Amount of Grant:* 30,000 - 290,000 USD
*Contact:* Latrice Rubenstein; (619) 556-5276; Latrice.Rubenstein@navy.mil
Cynthia Simon-Arndt, Technical/Progrommatic Officer; Cynthia.SimonArndt@med.navy.mil
*Sponsor:* U.S. Department of Defense
5450 Carlistle Park, P.O. Box 2050
Mechanicsburg, PA 17055-0791

**USDD HIV/AIDS Prevention Program Information Systems Development Grants** 2689

The Department of Defense HIV/AIDS Prevention Program (DHAPP) has successfully engaged over 80 countries in efforts to combat HIV/AIDS among their respective military services. DHAPP is a partner collaborating with the United States State Department, Health and Human Services, United States Agency for International Development, and Centers for Disease Control and Prevention, in the President's Emergency Plan for AIDS Relief. This BAA is intended to solicit existing partners and establish new partners to support the development of a DHAPP.org web site. This site will serve as a source for program updates, with both a public side and a password protected intranet side. A secure online database is needed and needs to aggregate all individual partners' submissions into a single file for download by the DHAPP operations staff in a common format, such as XL and CSV and TXT. Other site capacities should include a file structure to upload and download files, internet and email messaging, and database functions to support information gathering and analysis. Proposed work should include an infrastructure plan and a plan for support and/or turnover of the site. The period of performance is generally one year, with additional phases at the discretion of the government. Applicants are encouraged to propose additional phases if appropriate. This is a multiple year BAA which may be revised as needed. Based on the availability of funds, up to three evaluation cycles will be conducted. Changes will be posted at Grants.gov (grants.gov/applicants/find_grant_opportunities.jsp). Applicants are encouraged to sign up on Grants.gov for "send me change notification emails" by following the link on the synopsis page for the program announcement/funding opportunity.
*Requirements:* All responsible sources from academia, industry, and non-governmental organizations are eligible. Applicants must demonstrate the active support of the in-country military and the USDD representative in the corresponding United States Embassy in the planning and execution of proposals. Applicants should have prior experience with development and hosting of websites preferably for funding multiple partners, from many countries and cultures and languages, and experience in training staff to use and maintain web based databases. Applicants should also have experience with development and hosting of USDD websites and be familiar with the most recent Information Assurance requirements.
*Geographic Focus:* All States
*Date(s) Application is Due:* Sep 30
*Amount of Grant:* 80,000 - 100,000 USD
*Contact:* Latrice Rubenstein; (619) 556-5276; Latrice.Rubenstein@navy.mil
Cynthia Simon-Arndt; Cynthia.SimonArndt@med.navy.mil
*Internet:* www.afosr.af.mil/pdfs/BAA2005-1.pdf
*Sponsor:* U.S. Department of Defense
5450 Carlistle Park, P.O. Box 2050
Mechanicsburg, PA 17055-0791

**USDD Hull Grooming/Advanced Hull Cleaning Technologies Grants** 2690

While all marine operations hope eventually to be able to prevent bio-fouling at low cost and no maintenance, at present it is periodically necessary to mechanically clean a ship's hull. This requires periodic maintenance for the removal of marine biofouling. Because this is a labor-intensive operation, the Navy is developing autonomous robotic devices that can continuously "groom" the hull surface. Some of the challenges, apart from building an autonomous device, are associated with the development of sensor systems (detectors and analysis-and-control software) that can provide both navigation and path planning and discrimination between clean and fouled surfaces. While large diver-operated aggressive cleaning methods currently exist, a more proactive continuous grooming of the hull that occurs at the onset of biofouling settlement/colonization by autonomous vehicles should reduce maintenance costs and maintain a clean hull that will minimize fuel consumption and greenhouse gas emissions. Approaches to grooming methods that minimize surface damage or are non-contact in nature are of extreme interest to the development of this technology. Submission of the Research Proposal should always be preceded by submission of a Preproposal and a discussion of that Preproposal with the assigned Program Officer. Applications for the upcoming fiscal year should be received by September 30.
*Requirements:* Grants are awarded to academic, industrial, and nonprofit organizations, with strong emphasis on university research. Investigators are encouraged to contact the staff prior to formally submitting a proposal.
*Restrictions:* New grants or contracts are typically started near the beginning of the fiscal year. New proposals for the next fiscal year should be submitted no later than September 30, unless otherwise directed. Competitive renewals should be submitted no later than six months prior to the award termination date after contacting the Program Officer. Conference proposals should be submitted at least four months prior to the conference.
*Geographic Focus:* All States
*Date(s) Application is Due:* Sep 30
*Contact:* Dr. Stephen McElvany, Program Officer; steve.mcelvaney@navy.mil

*Internet:* www.onr.navy.mil/Science-Technology/Departments/Code-33/All-Programs/332-naval-materials/Hull-Grooming-Cleaning-Technologies.aspx
*Sponsor:* U.S. Department of Defense
One Liberty Center, 875 N. Randolph Street, Suite 1425
Arlington, VA 22203-1995

**USDD Human Social, Cultural and Behavior Modeling Thrust Grants** 2691
The Office of Naval Research (ONR) Human Social, Culture and Behavior Modeling Program invests in research on building capability through the development of a knowledge base, building models and creating training capacity in order to understand, predict and shape human behavior cross-culturally. The program seeks to: understand the human, social, cultural and behavioral factors that influence human behavior; improve our ability to model these influences and understand their impact on human behavior at the individual-, group- and society-level of analysis; improve computational modeling and simulation capabilities, visualization software toolsets, and training/mission rehearsal systems that provide forecasting capabilities for socio-cultural responses; and develop and demonstrate an integrated set of model description data (metadata), information systems and procedures that will facilitate assessment of the software engineering quality of sociocultural behavior models, their theoretical foundation and the translation of theory into model constructs.
*Requirements:* Grants are awarded to academic, industrial, and nonprofit organizations, with strong emphasis on university research.
*Geographic Focus:* All States
*Contact:* Dr. Ivy Estabrooke, Thrust Manager; (703) 588-2396; fax (703) 696-1211; ivy.estabrooke@navy.mil
*Internet:* www.onr.navy.mil/Science-Technology/Departments/Code-30/All-Programs/Human-Behavioral-Sciences.aspx
*Sponsor:* U.S. Department of Defense
One Liberty Center, 875 N. Randolph Street, Suite 1425
Arlington, VA 22203-1995

**USDD Image Optimization for Small Focal Plane Arrays Grants** 2692
While the Office of Naval Research (ONR) continues to develop larger format Focal Plane Arrays (FPAs) to provide wide area coverage sensors, it has been recognized that this approach will lead to ever increasing FPA costs. A more cost effective approach could be defined by an investigation of how to achieve sensor performance using smaller, more affordable FPAs by optimizing the imagery from these systems to meet detection, tracking and identification requirements. This Broad Agency Announcement (BAA) seeks to develop techniques to optimize imagery from passive and active Electro-Optics Infrared (EO-IR) imaging sensors based on smaller, less expensive FPAs. Some past and on-going ONR efforts include coded sensor processing, compressive sensing, and "Lucky Imaging" algorithms. There may also be techniques for adding processing to read out integrated circuits (ROICs) that will enable better utilization of smaller FPAs. Although this effort is not intended to cover the development of new EO-IR sensor hardware, including FPAs and ROICs, the impact of this investigation may lead to the proposal and implementation of new sensor hardware architectures. The overall intent of the BAA is the development of image enhancement techniques and methodologies that can be incorporated in processing of imagery from a variety of different small-format EO-IR sensor architectures. This approach can greatly improve the Navy's ability to achieve Information Dominance with a broader spectrum of lower-cost sensors. Work funded may include basic research, applied research and some advanced technology development. All applicants must submit white papers which are accepted electronically only. An expanded oral presentation will be requested from applicants whose proposals are identified as being of "particular value" to the Navy. Following oral presentations, applicants whose technology is still considered as having "particular value" to the Navy will be encouraged to submit full proposals. Two or three awards are anticipated. The award period may be from one to three years.
*Requirements:* All responsible sources from academia and industry are eligible. Historically black colleges and universities and minority institutions are encouraged to apply.
*Geographic Focus:* All States
*Date(s) Application is Due:* May 1
*Contact:* Rebecca Foster, Contract Specialist; (703) 696-2972; rebecca.d.foster@navy.mil
Dr. Antti Makinen, Technical Specialist; (703) 696-2972; antti.makinen@navy.mil
*Internet:* www.onr.navy.mil/Contracts-Grants/Funding-Opportunities.aspx
*Sponsor:* U.S. Department of Defense
One Liberty Center, 875 N. Randolph Street, Suite 1425
Arlington, VA 22203-1995

**USDD Information Innovation Office Grants** 2693
The Defense Advanced Research Projects Agency (DARPA) is soliciting innovative research proposals of interest to the Information Innovation Office (I2O). Proposed research should investigate innovative approaches that enable revolutionary advances in science, devices, or systems. I2O seeks unconventional approaches that are outside the mainstream, undertaking directions that challenge assumptions and have the potential to radically change established practice. I2O is interested in submissions related to the focus areas of understanding, empowering, and connecting. I2O seeks to enable understanding through the development of computational and analytic techniques that can process and assimilate the huge volumes of data generated by modern sensor, information, and communication systems, and by incorporating these techniques and new enabling sensor technologies in intelligence, surveillance, and reconnaissance exploitation and language processing systems that can classify behaviors, correlate events, identify trends, detect anomalies, and generate alerts. I2O works to empower warfighters at all echelons to meet challenges by developing advanced computing technologies, including modeling and simulation, visualization, knowledge management, geographic information systems, semantic web techniques, social networks, crowd-sourced, and Web 2.0/3.0 techniques, and implementing these in command and control systems that automate and improve the application of information to military decision-making, planning, training, mission rehearsal, and operations support. I2O aims to ensure connection by developing more resilient computing and communication technologies and by creating new approaches to ensure the security of defense networks and computing systems. These technologies will enable more rapid detection, analysis, and mitigation of cyber threats; more efficient utilization of network resources; and more effective monitoring and control of computing and communications infrastructure. Multiple awards are anticipated. The level of funding for individual awards has not been determined and will depend on the quality of the proposals received and the availability of funds. Applicants may submit an abstract in advance of a proposal to reduce potential effort in proposal preparation. DARPA will respond to abstracts indicating whether there is interest for the proposed work. Proposals may be submitted and will be reviewed irrespective of comments or feedback received in response to the abstract. Abstracts and proposals may be submitted until the deadline and will be reviewed on a rolling basis through the submission due date.
*Requirements:* All responsible sources capable of satisfying the government's needs are eligible. Historically black colleges and universities, small businesses, small disadvantaged businesses and minority institutions are encouraged to submit proposals and team with others to submit proposals.
*Restrictions:* Specifically excluded is research that results primarily in evolutionary improvements to the existing state of the art.
*Geographic Focus:* All States
*Date(s) Application is Due:* May 30
*Contact:* I2O BAA Coordinator; (703) 526-6630; DARPA-BAA-11-34@darpa.mil
*Sponsor:* U.S. Department of Defense
3701 North Fairfax Drive
Arlington, VA 21702-1714

**USDD Innovative Systems for Military Missions - Tactical Technology Office Grants** 2694
The Tactical Technology Office (TTO) of the Defense Advanced Research Projects Agency (DARPA) is soliciting executive summaries, white papers and proposals for advanced research and development of Innovative Systems for Military Missions. Innovative Systems are integrated systems or critical systems components, which often incorporate emerging advanced technologies, and which enable revolutionary improvements to the capability, efficiency and effectiveness of the military. TTO seeks responses relating to three mission thrust areas: (1) advanced weapons systems, (2) advanced platforms, and (3) advanced space systems. TTO's solicitation focuses on the high risk/high payoff development, integration, demonstration and evaluation of innovative systems or critical systems components enabled by, and incorporating, new or emerging technologies. Proposed efforts must also show significant promise to provide the United States military with revolutionary new mission capabilities, and/or enable significant increases in mission effectiveness. Innovative system concepts of interest to TTO typically address emerging technical opportunities, advanced systems concepts, emergent threats, and/or new technology-enabled concepts of operation. TTO strongly encourages respondents to adopt a complete systems engineering approach to the problems. Executive summaries, white papers, and proposals are accepted. Applicants are strongly encouraged to submit a one page executive summary describing the technology area and new ideas of interest prior to submitting a white paper or proposal. Executive summaries will be reviewed as they are received. Based on a favorable review of an executive summary, applicants will be invited to submit a six page white paper describing the proposer's proposed concept prior to submitting a proposal. A DARPA Program Manager may contact applicants to discuss the executive summary or white paper. This procedure is intended to minimize unnecessary effort in proposal preparation and review. Submission of an executive summary and/or white paper is not required prior to submitting a full proposal. Executive summaries are due by December 16. White papers are due by January 6. Full proposals are due by February 17. Multiple awards are anticipated. The number of awards made will depend on the quality of proposals and the availability of funds. In general awards are anticipated to be for less than $1 million and less than 18 months duration.
*Requirements:* All responsible sources capable of satisfying the government's needs are eligible. Historically black colleges and universities, small businesses, small disadvantaged businesses and minority institutions are encouraged to submit proposals and join others in submitting proposals.
*Geographic Focus:* All States
*Date(s) Application is Due:* Feb 17
*Contact:* David Neyland, Director; (703) 526-6630; DARPA-BAA 11-13@darpa.mil
*Sponsor:* U.S. Department of Defense
3701 North Fairfax Drive
Arlington, VA 22203-6630

**USDD Ionospheric Modification Grants** 2695
The Office of Naval Research, Ionospheric Modifications program investigates the interaction of high power, high frequency electromagnetic radiation with the ionosphere for the purpose of generating low frequency communication signals. The approach is to observe, diagnose, and interpret: naturally occurring, highly variable phenomena associated with the high latitude ionosphere; and the relatively small but useful perturbations that can be induced for communication applications. Emphasis is given to research to be conducted at the High-Frequency Active Auroral Research Program Observatory in Gakona, Alaska, but research at other facilities may also be considered, if supportive of the program goals. Average awards range at $50,000. Applications for the upcoming fiscal year should be received by September 30.

*Requirements:* Grants are awarded to academic, industrial, and nonprofit organizations, with strong emphasis on university research.
*Restrictions:* New grants or contracts are typically started near the beginning of the fiscal year. New proposals for the next fiscal year should be submitted no later than September 30, unless otherwise directed by the BAA. Competitive renewals should be submitted no later than six months prior to the award termination date after contacting the Program Officer. Conference proposals should be submitted at least four months prior to the conference.
*Geographic Focus:* All States
*Date(s) Application is Due:* Sep 30
*Amount of Grant:* 25,000 - 75,000 USD
*Contact:* Craig Selcher, Program Officer; Craig_Selcher@onr.navy.mil
*Internet:* www.onr.navy.mil/Science-Technology/Departments/Code-31/All-Programs/312-Electronics-Sensors/Ionospheric-Modification.aspx
*Sponsor:* U.S. Department of Defense
One Liberty Center, 875 N. Randolph Street, Suite 1425
Arlington, VA 22203-1995

**USDD Littoral Geosciences and Optics Grants** **2696**
The Office of Naval Research Littoral Geosciences and Optics program combines the former Coastal Geosciences and Environmental Optics programs and will leverage the combined expertise of our investigators to better address the multidisciplinary nature of many new basic research areas. We will continue to support basic and applied research efforts directed at solving nearshore, geoscience, optical and remote sensing problems in addition to new collaborative projects. Main areas of interest include predicting the 4D coastal environment and predicting seafloor characteristics in the global ocean margins. Areas of interest are motivated by on-going and future Navy and Marine Corps needs, including littoral operations in nearshore, estuarine, and riverine environments; prediction of sediment transport phenomena across a broad spectrum of coastal environments; and assessment of sources of sonar and optical clutter, among many others. Basic research interests include: nearshore, estuarine and riverine processes; remote sensing of coastal and riverine environments; and sediment transport processes. Studies by individual investigators and collaborative teams are often undertaken in conjunction with other programs at ONR, including the Physical Oceanography, Ocean Engineering and Marine Systems, and Ocean Acoustics programs. Focused, field-oriented research programs are undertaken as Departmental Research Initiatives (DRIs) or as Multidisciplinary University Research Initiatives (MURIs), each of which typically has a five-year timeframe and substantial funding for field studies. Submission of the Research Proposal should always be preceded by submission of a Preproposal and a discussion of that Preproposal with the assigned Program Officer.
*Requirements:* Grants are awarded to academic, industrial, and nonprofit organizations, with strong emphasis on university research.
*Restrictions:* New grants or contracts are typically started near the beginning of the fiscal year. New proposals for the next fiscal year should be submitted no later than September 30, unless otherwise directed by the BAA. Competitive renewals should be submitted no later than six months prior to the award termination date after contacting the Program Officer. Conference proposals should be submitted at least four months prior to the conference.
*Geographic Focus:* All States
*Date(s) Application is Due:* Sep 30
*Contact:* Dr. Thomas Drake, Division Director; (703) 696-1206 or (703) 588-2427; fax (703) 696-5940; tom.drake@navy.mil or 321_CG@onr.navy.mil
*Internet:* www.onr.navy.mil/Science-Technology/Departments/Code-32/All-Programs/Atmosphere-Research-322/Littoral-Geosciences-Optics.aspx
*Sponsor:* U.S. Department of Defense
One Liberty Center, 875 N. Randolph Street, Suite 1425
Arlington, VA 22203-1995

**USDD Long Range Broad Agency Announcement for Navy and Marine Corps Science and Technology Grants** **2697**
The Office of Naval Research is interested in receiving proposals for long-range science and technology projects which offer potential for advancement and improvement of Navy and Marine Corps operations. This is a Broad Agency Announcement intended for proposals related to basic research, applied research, or advanced technology development. White papers and full proposals are expected to be unclassified; however, classified proposals are permitted. Proposals may be submitted at any time until the closing of this BAA, which is September 30.
*Requirements:* All responsible sources from academia and industry are eligible to apply. Historically black colleges and universities and minority institutions are encouraged to apply and to join others in submitting proposals.
*Geographic Focus:* All States
*Date(s) Application is Due:* Sep 30
*Contact:* Ganesh Krish; (703) 588-2784 or (703) 696-4046; Ganesh.Krish@navy.mil
*Internet:* www.onr.navy.mil/Contracts-Grants/Funding-Opportunities/Broad-Agency-Announcements.aspx
*Sponsor:* U.S. Department of Defense
One Liberty Center, 875 N. Randolph Street, Suite 1425
Arlington, VA 22203-1995

**USDD Marine Mammals and Biological Oceanography Grants** **2698**
ONR's Marine Mammals and Biological Oceanography program was created late in 2007 and replaced two previous ONR programs (the Marine Mammals and Biology Program in Code 34 and the biological component of the Ocean Optics & Biology Program in Code 32). The Marine Mammals and Biological Oceanography program supports basic and applied research and technology development related to understanding the effects of sound on marine mammals, including physiological, behavioral, ecological effects and population-level effects. Our current program thrusts include, but are not limited to: Monitoring and Detection; Integrated Ecosystem Research; Effects of Sound on Marine Life; and Models and Databases for Environmental Compliance. Pre-proposals are required, and are due April 15. An invitation for full proposals will be sent out in either June or July, with final drafts due by the end of August.
*Requirements:* Grants are awarded to academic, industrial, and nonprofit organizations, with strong emphasis on university research.
*Geographic Focus:* All States
*Date(s) Application is Due:* Aug 30
*Contact:* Michael Weise; (703) 696-4533; fax (703) 696-2007; michael.j.weise@navy.mil
*Internet:* www.onr.navy.mil/sci_tech/32/322/marine_mammals_biology.asp
*Sponsor:* U.S. Department of Defense
One Liberty Center, 875 N. Randolph Street, Suite 1425
Arlington, VA 22203-1995

**USDD Medical Practice Initiative Breadth of Medical Practice and Disease Frequency Exposure Grants** **2699**
USDD Medical Practice Initiative Breadth of Medical Practice and Disease Frequency Exposure (MPI-BMP) Grants are being solicited by the Assistant Secretary of Defense for Health Affairs, Defense Health Program (DHP). The MPI is primarily focused on the development of medical training systems and competency assessment tools for the sustainment of military medical readiness. The preliminary focus will be on determining the viability of extracting relevant metrics from existing data for use in the determination of skills degradation and using this data to assist in identifying gaps. The MPI-BMP does not only concern itself with degradation from being removed from practice, but also looks at the breadth of clinical exposures and looks across each physician's expected clinical knowledge domain. The data collected in the course of this research will aid both public and private sector organizations in their efforts to improve the quality of care provided in the outpatient setting. The primary, but not exclusive, purpose is research that will: (1) identify when and why degradation of cognitive clinical skills occurs; (2) create validated analytical tools that can predict the probable onset of cognitive skills degradation and determine, with specificity, when skill or knowledge areas have degraded or will be likely to degrade; and (3) propose methods and tools which will enable physicians to preemptively refresh knowledge and maintain familiarity and fluency across the expected competencies for their specialty. Pre-proposals are due December 22, and proposals are by invitation only. Two awards are anticipated. The maximum period of performance is three years.
*Requirements:* Eligible organizations must be extramural and include nonprofit, public, and private organizations, such as universities, colleges, hospitals, laboratories, and commercial firms. Historically black colleges and universities/minority institutions are encouraged to apply.
*Geographic Focus:* All States
*Date(s) Application is Due:* Apr 19
*Amount of Grant:* 4,000,000 USD
*Contact:* Mary Rico, Grants Officer; (703) 518-4726; mary.rico@amedd.army.mil
*Sponsor:* U.S. Department of Defense - 1
1400 Defense Pentagon
Washington, D.C. 20301-1400

**USDD Medical Practice Initiative Procedural Skill Decay and Maintenance Grants** **2700**
USDD Medical Practice Initiative Procedural Skill Decay and Maintenance (MPI-PSD) Grants are awarded by the United States Army Medical Research Acquisition Activity (USAMRAA) and managed by the Army's Telemedicine and Advanced Technology Research Center. The MPI is primarily focused on the development of medical training systems and competency assessment tools for the sustainment of military medical readiness. The preliminary focus will be on determining the viability of extracting relevant metrics from existing data for use in the determination of skills degradation and using this data to assist in identifying gaps. The data collected in the course of this research will aid both public and private sector organizations in their efforts to improve the quality of care provided in the outpatient setting. The primary, but not exclusive, purpose is research that will: (1) identify when and why degradation of psychomotor and procedural clinical skills occurs; (2) create validated analytical tools that can predict the probable onset of skills2 degradation and determine, with specificity, when skills have degraded or will be likely to degrade; and (3) propose methods and tools which will enable physicians or surgeons to preemptively refresh the expected competencies for their specialty. The study must focus on one of more of the following physician practices: (1) surgical; (2) intensive care medicine; and (3) outpatient invasive procedures. Pre-proposals are due December 22, and proposals are by invitation only. Two awards are anticipated. The maximum period of performance is three years. The maximum period of performance is three years.
*Requirements:* Eligible organizations must be extramural and include nonprofit, public, and private organizations, such as universities, colleges, hospitals, laboratories, and commercial firms. Historically black colleges and universities/minority institutions are encouraged to apply.
*Geographic Focus:* All States
*Date(s) Application is Due:* Apr 19
*Amount of Grant:* 2,000,000 USD
*Contact:* Mary Rico, Grants Officer; (703) 518-4726; mary.rico@amedd.army.mil
*Sponsor:* U.S. Department of Defense - 1
1400 Defense Pentagon
Washington, D.C. 20301-1400

**USDD Metamaterials for Radion Frequency and Optical Applications Grants 2701**
This is a multiple year Broad Agency Announcement (BAA). The objective is to encourage a flow of supplementary and/or complementary technologies for electromagnetic metamaterials. Such technologies may include active electronic matching networks, frequency selective structures, and periodic structure with engineered dispersion, including photonic band-gap materials, acoustic metamaterials. Innovative advancements in using metamaterials and/or complementary technologies that leverage commercial economies of scale are highly desired. The areas of focus are radio frequency, optical, acoustic and bio-related metamaterials. White papers only are being accepted. White papers will be reviewed to determine if full proposals will be requested. The award period is expected to be between 9 and 24 months with individual awards ranging from $50K to $1M. Ten awards are anticipated.
*Requirements:* This is an unrestricted solicitation. Small businesses are encouraged to apply.
*Geographic Focus:* All States
*Contact:* Dawn M. Dalhamer, Contracting Officer; (937) 255-5372; dawn.dalhamer@wpafb.af.mil
*Sponsor:* U.S. Department of Defense
1400 Defense Pentagon
Washington, D.C. 20301-1400

**USDD Mixed Signal, Signal Processing Devices, Circuits and Architecture Grants 2702**
The Office of Naval Research Electronic Circuits Program seeks new approaches to achieving 100+ GHz logic and circuits, which includes investigations in devices, architecture, power dissipation, materials and related all-encompassing aspects. This work is expected to lead to the ability to efficiently digitize and synthesize RF signals at frequencies up to and beyond 50 GHz as well as to amplitude, phase and frequency modulate these signals. Additional objectives of this work are expected to lead to digital beam-forming and steering of electronic antenna arrays and the use of analog and digital signal processing and noise shaping techniques to improve the dynamic range of RF transmit/receive systems. Techniques to control amplitude and phase jitter throughout RF systems are also relevant. Proposal submissions are ongoing, but proposals are preferred by April 1 each year.
*Requirements:* Grants are awarded to academic, industrial, and nonprofit organizations, with strong emphasis on university research.
*Restrictions:* New grants or contracts are typically started near the beginning of the fiscal year (October 1). New proposals for the next fiscal year should be submitted no later than April 1, unless otherwise directed by the BAA. Competitive renewals should be submitted no later than six months prior to the award termination date after contacting the Program Officer. Conference proposals should be submitted at least four months prior to the conference.
*Geographic Focus:* All States
*Date(s) Application is Due:* Apr 1
*Contact:* Dr. Stephen Pappert, Program Officer; (703) 696-5268 or (619) 553-5704; stephen.pappert@navy.mil
*Internet:* www.onr.navy.mil/Science-Technology/Departments/Code-31/All-Programs/312-Electronics-Sensors/RF-Signal-Circuits-Architecture.aspx
*Sponsor:* U.S. Department of Defense
One Liberty Center, 875 N. Randolph Street, Suite 1425
Arlington, VA 22203-1995

**USDD Multidisciplinary Research Program of the University Research Initiative Grant 2703**
The MURI program supports basic research in science and engineering at U.S. institutions of higher education (hereafter referred to as "universities") that is of potential interest to DoD. The program is focused on multidisciplinary research efforts where more than one traditional discipline interact to provide rapid advances in scientific areas of interest to the DoD. As defined by the DoD, basic research is systematic study directed toward greater knowledge or understanding of the fundamental aspects of phenomena and of observable facts without specific applications towards processes or products in mind. It includes all scientific study and experimentation directed toward increasing fundamental knowledge and understanding in those fields of the physical, engineering, environmental, and life sciences related to long-term national security needs. It is farsighted high payoff research that provides the basis for technological progress. The DoD's basic research program invests broadly in many specific fields to ensure that it has early cognizance of new scientific knowledge.
*Requirements:* Proposals from a team of university investigators are warranted when the necessary expertise in addressing the multiple facets of the topics may reside in different universities, or in different departments in the same university. By supporting multidisciplinary teams, the program is complementary to other DoD basic research programs that support university research through single-investigator awards. Proposals shall name one Principal Investigator (PI) as the responsible technical point of contact.
*Geographic Focus:* All States
*Date(s) Application is Due:* Dec 10
*Contact:* Dr. Bill Lukens, Program Manager; (703) 696-4668; william.lukens1@navy.mil
Paula Barden, AFRL/RX Contact; (703) 696-4111; paula.barden.ctr@navy.mil
*Sponsor:* U.S. Department of Defense
875 North Randolph Street, Suite 325, Room 3112
Arlington, VA 22203-1768

**USDD Nano-Bio Manufacturing Institute Grants 2704**
The objective of the NBMI is to accelerate risk assessment and the development of robust mitigation approaches for critical nano-bio materials and fabrication innovations by developing common technology and manufacturing platforms through public-private partnerships between federal laboratories, DoD acquisition centers, academia, U.S. manufacturing enterprises, and commercial industries. To achieve this objective, the institute is envisioned to create a community of large and small companies, academia, federal laboratories, and the states to accelerate innovation by investing in industrially-relevant nano-bio manufacturing technologies. Ultimately, a fully functioning NBMI will have both virtual and physical entities that will enable partners to bridge the gap between basic research and product transition by developing verified and validated open architecture processing-design-performance, forecasting, and risk assessment tools for integration of new nano-bio materials and process technologies. Additionally, it will provide shared assets to help companies' access cutting-edge capabilities and equipment, and create an unparalleled environment to educate and train students and workers in advanced manufacturing skills. A successful institute will serve as a technical center of excellence, providing the infrastructure to support manufacturing enterprises at a prototype level that are relevant and scalable; and ensuring that the key capabilities for DoD needs is available.
*Requirements:* This is a restricted solicitation limited to a non-profit 501(c) designated] organization.
*Geographic Focus:* All States
*Date(s) Application is Due:* Nov 13
*Amount of Grant:* Up to 4,400,000 USD
*Contact:* Randall Wallace, Contracts Negotiator; (937) 255-0816; randall.wallace@wpafb.af.mil
*Sponsor:* U.S. Department of Defense
1400 Defense Pentagon
Washington, D.C. 20301-1400

**USDD National Consortium for Measurement and Signature Intelligence Research Grants 2705**
Measurement and Signature Intelligence (MASINT) is an intelligence discipline that employs a broad range of scientific developments to gather foreign intelligence. MASINT oversight and direction is provided by the National MASINT Office. The National Consortium for MASINT Research as the focal point for MASINT innovation provides cutting-edge MASINT research for the Intelligence Community. A formal consortium of laboratories, academia, and industry forms the core of the advanced MASINT research and development program and creates a dynamic forum for diverse research partners to the benefit of MASINT. Applicants are invited to present related work, on-going research activities and proposed future activities associated with the following areas: (1) remote assessment of missile performance characteristics such as location, thrust, throw weight, warhead accuracy, and defensive capabilities; (2) remote assessment and detection of weapons of mass destruction such as nuclear, biological, chemical and radiological weapons (excluding improvised explosive devices utilizing standard explosives such as dynamite, TNT, C4, and ANFO); and (3) remote assessment and detection of directed energy weapons (which include all lasers primarily designed as weapons as well as high-powered microwave and electromagnetic pulse weapons). Applicants are encouraged to first submit white papers for evaluation. Feedback will be provided. White papers are due February 1. Although proposals will be accepted through the published deadline, to ensure timely consideration they must be received by April 4. Proposals submitted after April 4 will be reviewed and considered if additional funds become available.
*Requirements:* All responsible sources from academia (colleges and universities) are eligible, including academia from the British Commonwealth nations (Canada, Great Britain, Australia and New Zealand). Historically black colleges and universities and minority institutions are encouraged to apply.
*Geographic Focus:* All States, Australia, Canada, Great Britain, New Zealand
*Date(s) Application is Due:* Dec 31
*Contact:* Janet Norton; (808) 473-7669; janet.norton@navy.mil
*Sponsor:* U.S. Department of Defense
5450 Carlistle Park, P.O. Box 2050
Mechanicsburg, PA 17055-0791

**USDD National Geospatial-Intelligency Agency Academic Research Grants 2706**
The National Geospatial-Intelligence Agency (NGA) welcomes all innovative ideas for path-breaking research that may advance the geospatial intelligence (GEOINT) mission. The NGA mission is to provide timely, relevant, and accurate GEOINT in support of national security objectives. GEOINT is the exploitation and analysis of imagery and geospatial information to describe, assess and visually depict physical features and geographically referenced activities on the earth. GEOINT consists of imagery, imagery intelligence, and geospatial information. NGA offers a variety of critical GEOINT products in support of United States national security objectives and natural disasters, including aeronautical, geodesy, hydrographic, imagery, targeting information and topographical. The NGA Academic Research Program (NARP) is focused on innovative, far-reaching basic and applied research in science, technology, engineering and mathematics that has the potential to advance the GEOINT mission. The objective of the NARP is to support innovative, high-payoff research that provides the basis for revolutionary progress in areas of science and technology affecting the needs and mission of NGA. This research also supports the National System for Geospatial Intelligence, which is the combination of technology, systems and organizations that gather, produce, distribute and consume geospatial data and information. The end result is aimed at advancing GEOINT capabilities by improving analytical methods, enhancing and expanding systems capabilities, and leveraging resources for common NSG goals. Amendments will be posted to Grants.gov (www.grants.gov/) and to the Intelligence Community ARC (https://arc.westfields.net). Applicants are encouraged to periodically check these websites for updates and amendments and to set up alerts. White papers may be submitted at any time, but no later than April 30, 2013. Proposals may be submitted at any time, but no later than September 30, 2013. Applicants are highly encouraged to submit white papers prior to submitting proposals in order for NGA to

determine the level of interest in the proposed research. Awards will be made on a rolling basis subject to availability of funds.
*Requirements:* United States institutions of higher education are eligible. Proposals may be submitted in any of the following four categories for which the applicant is eligible: (1) university research initiatives; (2) historically black colleges and universities and minority institutions; (3) outstanding new science and technical innovative researcher; and (4) research collaboration forum.
*Geographic Focus:* All States
*Date(s) Application is Due:* Sep 13
*Amount of Grant:* 300,000 - 4,800,000 USD
*Contact:* Eric Rauch, Grants Officer; (571) 577-2462; Eric.Rauch@nga.mil
*Sponsor:* U.S. Department of Defense
7500 GEOINT Drive
Springfield, VA 22150-7500

**USDD Naval Research Facilities Grants** **2707**
The Research Facilities program manages, under a charter party agreement with selected universities, the operation and maintenance of all oceanographic research vessels owned by ONR for Department of the Navy. As part of the management of the ONR fleet, the program actively supports fleet modernization to maintain the level of excellence that is required to support cutting edge oceanographic research. Other oceanographic platforms such as unmanned underwater vehicles, aerial unmanned vehicles may be used to collect field data through the Research Facilities Program. Submission of the Research Proposal should always be preceded by submission of a Preproposal and a discussion of that Preproposal with the assigned Program Officer. Applications for the upcoming fiscal year should be received by September 30.
*Requirements:* Grants are awarded to academic, industrial, and nonprofit organizations, with strong emphasis on university research. Investigators are encouraged to contact the staff prior to formally submitting a proposal.
*Restrictions:* New grants or contracts are typically started near the beginning of the fiscal year. New proposals for the next fiscal year should be submitted no later than September 30, unless otherwise directed by the BAA. Competitive renewals should be submitted no later than six months prior to the award termination date after contacting the Program Officer. Conference proposals should be submitted at least four months prior to the conference.
*Geographic Focus:* All States
*Date(s) Application is Due:* Sep 30
*Contact:* Robert Schnoor, Team Leader; tim.schnoor@navy.mil
*Internet:* www.onr.navy.mil/Science-Technology/Departments/Code-32/All-Programs/Ocean-Systems-321/Navy-Research-Facilities.aspx
*Sponsor:* U.S. Department of Defense
One Liberty Center, 875 N. Randolph Street, Suite 1425
Arlington, VA 22203-1995

**USDD Navigation and Timekeeping Grants** **2708**
The Office of Naval Research (ONR) Navigation and Timekeeping Program seeks new and innovative navigation technologies that will provide more accurate, reliable, maintainable and affordable systems for Naval air, surface, subsurface and ground platforms and forces.Precision navigation and timekeeping are essential for many modem naval and maritime systems, and it is essential that navigation and timekeeping services be made available to platforms and weapons at the highest level of accuracy and with the highest possible confidence at reasonable cost. Lack of precise navigation and timekeeping technologies may jeopardize the success of military operations. In the past six years, ONR has been conducting a series of Science and Technology Projects In the following three technology areas; GPS Anti-Jam Technology, Precision Time and Time Transfer Technology, and Non-GPS Navigation Technology. Proposers are encouraged to contact the program officer to discuss their research interest prior to the submission of formal proposals. Applications for the upcoming fiscal year should be received by September 30.
*Requirements:* Grants are awarded to academic, industrial, and nonprofit organizations, with strong emphasis on university research.
*Restrictions:* New grants or contracts are typically started near the beginning of the fiscal year. New proposals for the next fiscal year should be submitted no later than September 30, unless otherwise directed by the BAA. Competitive renewals should be submitted no later than six months prior to the award termination date after contacting the Program Officer. Conference proposals should be submitted at least four months prior to the conference.
*Geographic Focus:* All States
*Date(s) Application is Due:* Sep 30
*Contact:* John C. Kim, Program Officer; (703) 696-4214; john.c.kim1@navy.mil
*Internet:* www.onr.navy.mil/Science-Technology/Departments/Code-31/All-Programs/312-Electronics-Sensors/Navigation-Timekeeping.aspx
*Sponsor:* U.S. Department of Defense
One Liberty Center, 875 N. Randolph Street, Suite 1425
Arlington, VA 22203-1995

**USDD Non-Destructive Evaluation and Prognostics: Advanced Sensors a Technologies Grants** **2709**
Integrated system health management (SHM) for platforms systems and subcomponents is now an integral part of planning the design and construction of future assets (e.g., ships, aircraft, vehicles, buildings). Health management of critical assets depends on two equally important processes. The first is health diagnostics; the second is health and lifetime prediction, or prognostics. The diagnostic step certainly feeds vital information to the prognostic arm. A variety of sensors have been developed, and continue to be developed or enhanced, to detect and analyze vibrations, current surges, stress and strain, as well as to see and identify defects such as cracks and corrosion pits. An example: spectroscopic oil analysis can be a valuable tool to identify potential stressors to engines and engine lifetime. Sensor development focuses on efficiency improvements through miniaturization, sensitivity enhancement and energy harvesting to reduce power requirements. The prognostic part of the program is perhaps the more difficult at this time. Prognosis involves prediction, and prediction can only be reliably made if one knows the physics, chemistry and allied engineering of the underlying failure mechanisms. This Office of Naval Research program supports work in a variety of fields to address these issues. Submission of the Research Proposal should always be preceded by submission of a Preproposal and a discussion of that Preproposal with the assigned Program Officer. Applications for the upcoming fiscal year should be received by September 30.
*Requirements:* Grants are awarded to academic, industrial, and nonprofit organizations, with strong emphasis on university research. Investigators are encouraged to contact the staff prior to formally submitting a proposal.
*Restrictions:* New grants or contracts are typically started near the beginning of the fiscal year. New proposals for the next fiscal year should be submitted no later than September 30, unless otherwise directed. Competitive renewals should be submitted no later than six months prior to the award termination date after contacting the Program Officer. Conference proposals should be submitted at least four months prior to the conference.
*Geographic Focus:* All States
*Date(s) Application is Due:* Sep 30
*Contact:* Dr. Ignacio Perez, Program Officer; (703) 696-0688; fax (703) 696-0934; ignacio.perez1@navy.mil
*Internet:* www.onr.navy.mil/Science-Technology/Departments/Code-33/All-Programs/332-naval-materials/NDE-Sensors-Prognostics.aspx
*Sponsor:* U.S. Department of Defense
One Liberty Center, 875 N. Randolph Street, Suite 1425
Arlington, VA 22203-1995

**USDD Non-Destructive Evaluation and Prognostics: Fatigue and Fracture Damage Protection Grants** **2710**
The Non-Destructive Evaluation and Prognostics: Fatigue and Fracture Damage Protection program focuses on fatigue and damage accumulation of structural metals and alloys, including nanostructured materials. Fatigue focuses on quantifying the damage parameters in terms of ?K (cyclic damage) and Kmax (static damage), which are key measures of damage in materials. Because damage frequently arises from a combination of mechanical and corrosion effects, these mechanisms are the primary areas of focus in the program, with the goal of a reliable fatigue damage life prediction capability. Because they are important metals to the Navy, there is a focus on aluminum, steel and titanium alloys. In nanostructured alloys, research involves bi-modal and tri-modal microstructures of 5xxx aluminum alloys and their cumulative effects, in order to characterize fracture and fatigue as a function of processing variables. Submission of the Research Proposal should always be preceded by submission of a Preproposal and a discussion of that Preproposal with the assigned Program Officer. Applications for the upcoming fiscal year should be received by September 30.
*Requirements:* Grants are awarded to academic, industrial, and nonprofit organizations, with strong emphasis on university research. Investigators are encouraged to contact the staff prior to formally submitting a proposal.
*Restrictions:* New grants or contracts are typically started near the beginning of the fiscal year. New proposals for the next fiscal year should be submitted no later than September 30, unless otherwise directed. Competitive renewals should be submitted no later than six months prior to the award termination date after contacting the Program Officer. Conference proposals should be submitted at least four months prior to the conference.
*Geographic Focus:* All States
*Date(s) Application is Due:* Sep 30
*Contact:* Dr. A.K. Vasudevan; (703) 696-8181; vasu.vasudevan@navy.mil
*Internet:* www.onr.navy.mil/Science-Technology/Departments/Code-33/All-Programs/332-naval-materials/NDE-Fatigue-Fracture.aspx
*Sponsor:* U.S. Department of Defense
One Liberty Center, 875 N. Randolph Street, Suite 1425
Arlington, VA 22203-1995

**USDD NRL Broad Agency Announcement Grants** **2711**
The Naval Research Laboratory (NRL) is the Navy's coporate laboratory. NRL conducts basic and applied research for the Navy in a variety of scientific and technical disciplines. The basic research program is driven by perceptions about future requirements of the Navy. NRL conducts most of its research program at its own facilities but also funds some related research such as anticipated by this announcement. The Navy's operational effectiveness depends on its ability to keep pace with rapidly developing technologies. Proposals may range from theoretical studies to proof-of-concept to include fabrication and delivery of a prototype. This is a multiple year Broad Agency Announcement (BAA). Amendments may be published at Grants.gov (grants.gov/applicants/find_grant_opportunities.jsp) and/or at FedBizOpps (https://www.fbo.gov/). Applicants are encouraged to monitor these sites or to sign up for notification of changes. White papers may be submitted and will be reviewed within a reasonable period of time, generally 30 days. Instructions for formal proposals, as well as classified proposals, can be found on the website.
*Requirements:* Eligible applicants include any non-governmental entity, including commercial firms, institutions of higher education with degree-granting programs in science or engineering (universities), or by consortia led by such concern. NRL encourages participation by small businesses, veteran-owned small businesses, service-disabled veteran-owned small businesses, small disadvantaged businesses, HUBZone small businesses, woman-owned small businesses, and historically black colleges and universities and minority institutions.

*Geographic Focus:* All States
*Contact:* Mary Johnson, Contract Specialist; (202) 767-2021 or (301) 342-3744; nrlproposals@nrl.navy.mil
*Sponsor:* U.S. Department of Defense
4555 Overlook Avenue Southwest
Washington, D.C. 20375

**USDD Ocean Acoustics Research Grants** **2712**
The Ocean Acoustics Research program supports research that addresses an understanding of the physics of the generation, propagation, and scattering of narrowband and broadband acoustic (and elastic) waves in the temporally and spatially varying ocean environment. Though research in acoustical signal processing is no longer a significant thrust of this program, support is granted to research in collaboration with the Undersea Signal Processing Program, which also manages development and demonstration efforts. The Ocean Acoustics program contains three primary thrusts and a miscellaneous thrust. The Shallow-Water Acoustics thrust seeks to understand the propagation and scattering of low frequency (10Hz to a few kHz) acoustic energy within the shallow-water ocean environment. Areas of research include investigations of the dominant shallow-water scattering mechanisms, the conversion of seafloor-incident acoustic energy into elastic body waves and interface waves, and acoustic propagation through linear and nonlinear internal waves. The High-Frequency Acoustics thrust seeks to understand the interaction of high frequency (few kHz to thousand kHz) sound with the ocean environment. Components of this thrust include the propagation of sound through an intervening turbulent or stochastic medium; scattering from rough surfaces, biologics, and bubbles; and penetration/propagation within the porous seafloor. The Long-Range Propagation thrust seeks to understand the behavior of sound as it propagates over very long ranges (several hundred kilometers to several thousand kilometers) in the ocean. The main area of interest is understanding the effect of ocean internal waves and mesoscale on transmitted broadband acoustic signals. Submission of the Research Proposal should always be preceded by submission of a Preproposal and a discussion of that Preproposal with the assigned Program Officer. Applications for the upcoming fiscal year should be received by September 30.
*Requirements:* Grants are awarded to academic, industrial, and nonprofit organizations, with strong emphasis on university research. Investigators are encouraged to contact the staff prior to formally submitting a proposal.
*Restrictions:* New grants or contracts are typically started near the beginning of the fiscal year. New proposals for the next fiscal year should be submitted no later than September 30, unless otherwise directed. Competitive renewals should be submitted no later than six months prior to the award termination date after contacting the Program Officer. Conference proposals should be submitted at least four months prior to the conference.
*Geographic Focus:* All States
*Date(s) Application is Due:* Sep 30
*Contact:* Dr. Robert Headrick; (703 )696-4135; fax (703) 696-2007; bob.headrick@navy.mil
*Internet:* www.onr.navy.mil/Science-Technology/Departments/Code-32/All-Programs/Atmosphere-Research-322/Ocean-Acoustics.aspx
*Sponsor:* U.S. Department of Defense
One Liberty Center, 875 N. Randolph Street, Suite 1425
Arlington, VA 22203-1995

**USDD Ocean Engineering and Marine Systems Grants** **2713**
The goals of the Ocean Engineering and Marine Systems program are to improve the knowledge base of fluid-structure interactions for engineering designs and to develop and demonstrate new technologies for expeditionary operations and special warfare. The investigators conduct multidisciplinary science and technology efforts in the behavior of ocean systems in general as well as specifically for the Marine Corps, Navy, Special Forces, and Navy Explosive Ordnance Disposal. The basic research program focuses on fluid forcing and structural response mechanisms and the coupled nonlinear interactions of fluids with generic structural components and unmanned platform systems. Technology development efforts are primarily in the areas of detection and imaging technologies applicable to underwater and buried objects, technologies for neutralization of explosive devices, underwater life support technologies, surface and subsurface transport systems for Navy Special Forces, mobile autonomous platform system technologies for very shallow water, surf zone and beach zone reconnaisance, and technologies leading to capability for rapid wide-area clearance of mines and obstacles from the surf zone. Submission of the Research Proposal should always be preceded by submission of a Preproposal and a discussion of that Preproposal with the assigned Program Officer. Applications for the upcoming fiscal year should be received by September 30.
*Requirements:* Grants are awarded to academic, industrial, and nonprofit organizations, with strong emphasis on university research. Investigators are encouraged to contact the staff prior to formally submitting a proposal.
*Restrictions:* New grants or contracts are typically started near the beginning of the fiscal year. New proposals for the next fiscal year should be submitted no later than September 30, unless otherwise directed. Competitive renewals should be submitted no later than six months prior to the award termination date after contacting the Program Officer. Conference proposals should be submitted at least four months prior to the conference.
*Geographic Focus:* All States
*Date(s) Application is Due:* Sep 30
*Contact:* Dr. Thomas Swean, Team Leader; (703) 696-4025 or (703) 696-5031; fax (703) 696-5940; tom.swean@navy.mil
*Internet:* www.onr.navy.mil/Science-Technology/Departments/Code-32/All-Programs/Ocean-Systems-321/Ocean-Engineering-Marine-Systems.aspx
*Sponsor:* U.S. Department of Defense
One Liberty Center, 875 N. Randolph Street, Suite 1425
Arlington, VA 22203-1995

**USDD Office-Wide Broad Agency Announcement Microsystems Technology Office Grant** **2714**
The mission of the Defense Advanced Research Projects Agency's (DARPA) Microsystems Technology Office (MTO) is to exploit breakthroughs in materials, devices, circuits, and mathematics to develop beyond leading edge microsystems components with revolutionary performance and functionality to enable new platform capability for the Department of Defense. To execute this mission, MTO supports revolutionary research in electronics, photonics, MEMS, algorithms, and combined microsystems technology to deliver new capabilities to sense, communicate, energize, actuate, and process data and information for the war fighter. These Grants seek revolutionary research ideas for topics not being addressed by ongoing MTO programs or other published solicitations. These Grants are primarily, but not solely, intended for early stage research (studies) that may lead to larger, focused, MTO programs in the future. Studies are defined as single phase efforts of short duration (less than 12 months) costing less than $1,000,000. Whereas proposers are strongly encouraged to submit studies, multi-phase efforts will also be considered. multi-phase efforts are efforts with more than one phase that may span longer than 12 months and that may cost more than $1,000,000. Multi-phase efforts must also cover discernable research topic areas that lead to clearly defined, quantitative technical metrics. Multiple awards are anticipated. Funding awarded will depend on the quality of the proposals received and the availability of funds. Submission of abstracts in advance of a full proposal is strongly encouraged. This procedure is intended to minimize unnecessary effort in proposal preparation and review. Early submission of abstracts and proposals is strongly encouraged because selections may be made at any time during the period of solicitation. Abstracts and proposals may be submitted until the deadline.
*Requirements:* All responsible sources capable of satisfying the government's needs are eligible. Historically black colleges and universities, small businesses, small disadvantaged businesses and minority institutions are encouraged to submit proposals and join others in submitting proposals.
*Geographic Focus:* All States
*Date(s) Application is Due:* Sep 1
*Contact:* Dr. Thomas Lee, Director; (703) 526-6630; fax (703) 741-0087; DARPA-BAA-10-35@darpa.mil
*Sponsor:* U.S. Department of Defense
3701 North Fairfax Drive
Arlington, VA 22203-6630

**USDD Organic Photovoltaics Grants** **2715**
The Office of Naval Research Organic Photovoltaics program's purpose is to develop low-cost, light-weight, rugged thin-film organic photovoltaics with overall power conversion efficiencies of 15 percent or greater. This technology would enable the in-field implementation of flexible, disposable photovoltaic devices such as sensors, solar tent canopies and battery rechargers – greatly reducing the logistical burden of maintaining power generation capabilities in extensible positions. Current focus areas include the development of advanced materials for separate functionalities (hole/electron transport, sensitizing), investigating the effects of interfaces and constituent morphology on cell performance, developing tools to optimize device efficiency and develop reproducible material processing capabilities for affordability. Applications include: man-portable sunlight-powered devices; disposable sensors; and battery rechargers.
*Requirements:* Grants are awarded to academic, industrial, and nonprofit organizations, with strong emphasis on university research. Investigators are encouraged to contact the staff prior to formally submitting a proposal.
*Restrictions:* New grants or contracts are typically started near the beginning of the fiscal year. New proposals for the next fiscal year should be submitted no later than September 30, unless otherwise directed. Competitive renewals should be submitted no later than six months prior to the award termination date after contacting the Program Officer. Conference proposals should be submitted at least four months prior to the conference.
*Geographic Focus:* All States
*Date(s) Application is Due:* Sep 30
*Contact:* Dr. Paul Armistead, Program Officer; (703) 696-4315; fax (703) 696-0001; paul.armistead@navy.mil
*Internet:* www.onr.navy.mil/Science-Technology/Departments/Code-33/All-Programs/332-naval-materials/Organic-Photovoltaics.aspx
*Sponsor:* U.S. Department of Defense
One Liberty Center, 875 N. Randolph Street, Suite 1425
Arlington, VA 22203-1995

**USDD Physical Oceanography Grants** **2716**
The Physical Oceanography program supports process-oriented and hypothesis-driven science and technology in the area of physical oceanography and in the applications of oceanic research tools and techniques for fleet usage. In response to post-Cold War naval strategy and tactics, increased emphasis is given to the littoral, defined here as the oceanographic region encompassing the continental shelf and slope and the adjacent deep water. The approach is principally on field observations with theory and modeling expected to be closely integrated. The following thrusts are supported: air-sea interaction: impact on mixed layer dynamics for improved parameterizations in oceanic and atmospheric models; internal waves/turbulence: development of a global littoral model and a low level of open ocean work; marginal seas/straits: identify key physics, develop archetypes, generalize results; shelf/slope dynamics: shelf/slope coupling, parameterizations for improved slope/shelf modeling; boundary current/slope interaction: quantify key physics for enhancement of improved regional modeling; fleet meteorological and oceanographic support: develop tools, techniques, and observations of direct impact to activities at sea. An overarching objective of the program is to

foster transition of research products such as numerical and theoretical models, analysis algorithms, in situ data, sea-going instrumentation, and platforms into operational naval systems. In addition, support is also available for critical evaluation of the impact of the environment on fleet exercises that test and evaluate operational naval systems. Regular planning letters are dure by June 30, and applications for the upcoming fiscal year should be received by September 30.
*Requirements:* Grants are awarded to academic, industrial, and nonprofit organizations, with strong emphasis on university research.
*Restrictions:* New grants or contracts are typically started near the beginning of the fiscal year. New proposals for the next fiscal year should be submitted no later than September 30, unless otherwise directed by the BAA. Competitive renewals should be submitted no later than six months prior to the award termination date after contacting the Program Officer. Conference proposals should be submitted at least four months prior to the conference.
*Geographic Focus:* All States
*Date(s) Application is Due:* Sep 30
*Contact:* Dr. Terri Paluszkiewicz; (703) 696-6680; terri.paluszkiewicz@navy.mil
*Internet:* www.onr.navy.mil/Science-Technology/Departments/Code-32/All-Programs/Atmosphere-Research-322/Physical-Oceanography.aspx
*Sponsor:* U.S. Department of Defense
One Liberty Center, 875 N. Randolph Street, Suite 1425
Arlington, VA 22203-1995

**USDD Polymeric Organic Materials: Conducting/Semiconducting Polymers Grants** **2717**
The Office of Naval Research Functional Polymeric/Organic Materials program supported the research that led to the discovery of high-electrical conductivity in doped conjugated polymers. In the past, the program supported research towards improving the stability of conducting polymers and developing conducting polymers for several anti-static applications. As semiconducting properties emerged, the research was focused on organic light-emitting diodes and non-linear optical applications. At this time, the primary focus is on photovoltaic devices and the design, synthesis, assembly, fabrication and characterization of the active materials in these devices. Research is primarily conducted on organic and hybrid thin-film approaches including bulk heterojunction and planar device architectures. Developing low-cost, potentially disposable, light-weight flexible organic and/or polymeric thin-film photovoltaics is critical for military use in distributed operations. Possible applications include the ability to recharge batteries in-field as well as helping to power sensors and unmanned undersea vehicles. Closely related to this program is work on thin-film dielectric capacitors for energy storage. Applications include: flexible organic and/or polymeric thin-film photovoltaics; nonlinear optical films; data processing; light-emitting diodes; and piezoelectric and electrostrictive materials for acoustic sensing or transduction. Submission of the Research Proposal should always be preceded by submission of a Preproposal and a discussion of that Preproposal with the assigned Program Officer. Applications for the upcoming fiscal year should be received by September 30.
*Requirements:* Grants are awarded to academic, industrial, and nonprofit organizations, with strong emphasis on university research. Investigators are encouraged to contact the staff prior to formally submitting a proposal.
*Restrictions:* New grants or contracts are typically started near the beginning of the fiscal year. New proposals for the next fiscal year should be submitted no later than September 30, unless otherwise directed. Competitive renewals should be submitted no later than six months prior to the award termination date after contacting the Program Officer. Conference proposals should be submitted at least four months prior to the conference.
*Geographic Focus:* All States
*Date(s) Application is Due:* Sep 30
*Contact:* Dr. Paul Armistead, Program Officer; (703) 696-4315; fax (703) 696-0001; paul.armistead@navy.mil
*Internet:* www.onr.navy.mil/Science-Technology/Departments/Code-33/All-Programs/332-naval-materials/Functional-Polymeric-Organic-Materials.aspx
*Sponsor:* U.S. Department of Defense
One Liberty Center, 875 N. Randolph Street, Suite 1425
Arlington, VA 22203-1995

**USDD Polymer Matrix Composites Grants** **2718**
The Office of Naval Research Polymer Matrix Composites program focuses on advancing understanding to exploit research on carbon nanotube fibers. These fibers have been found to possess remarkable tensile strength compared to ordinary carbon and other fibers. Fiber re-enforcement of polymer matrices has been used for decades to increase the strength of polymeric materials. This program supports efforts to increase the yield and decrease the cost of producing sufficiently large quantities of carbon nanotube fibers to build and test polymer matrix components. In addition, research is encouraged to find optimum blends of polymeric materials to use with the fibers. The concomitant testing of materials is supported. Submission of the Research Proposal should always be preceded by submission of a Preproposal and a discussion of that Preproposal with the assigned Program Officer. Applications for the upcoming fiscal year should be received by September 30.
*Requirements:* Grants are awarded to academic, industrial, and nonprofit organizations, with strong emphasis on university research. Investigators are encouraged to contact the staff prior to formally submitting a proposal.
*Restrictions:* New grants or contracts are typically started near the beginning of the fiscal year. New proposals for the next fiscal year should be submitted no later than September 30, unless otherwise directed. Competitive renewals should be submitted no later than six months prior to the award termination date after contacting the Program Officer. Conference proposals should be submitted at least four months prior to the conference.
*Geographic Focus:* All States
*Date(s) Application is Due:* Sep 30
*Contact:* Dr. Ignacio Perez, Program Officer; (703) 696-0688; fax (703) 696-0934; ignacio.perez1@navy.mil
*Internet:* www.onr.navy.mil/Science-Technology/Departments/Code-33/All-Programs/332-naval-materials/Polymer-Matrix-Composites.aspx
*Sponsor:* U.S. Department of Defense
One Liberty Center, 875 N. Randolph Street, Suite 1425
Arlington, VA 22203-1995

**USDD Postdoctoral Fellowship Awards in Ocean Acoustics** **2719**
USDD Postdoctoral Fellowship Awards in Ocean Acoustics will be given to outstanding doctoral degree holders who wish to continue research in ocean acoustics or the related disciplines of undersea signal processing, marine structural acoustics and transducer materials science. Applicants must identify and provide a letter of recommendation from an ocean scientist or acoustician at an academic institution or naval laboratory with whom they propose to work. Special consideration will be given to applications that include collaboration with a scientist at a naval research laboratory or naval warfare center division. These postdoctoral awards are 24 months in duration and cover salary and benefits. A detailed planning letter is required. ONR expects to grant up to three such awards per year.
*Geographic Focus:* All States
*Date(s) Application is Due:* Aug 1
*Contact:* Dr. Ellen Livingston, Program Officer; (703) 696-4203 or (703) 696-5031; fax (703) 696-5940; ellen.livingston@navy.mil
*Internet:* www.onr.navy.mil/Science-Technology/Departments/Code-32/All-Programs/Atmosphere-Research-322/Ocean-Acoustics/Awards-Ocean-Acoustics.aspx
*Sponsor:* U.S. Department of Defense
One Liberty Center, 875 N. Randolph Street, Suite 1425
Arlington, VA 22203-1995

**USDD Power and Thermal Management Technology Development Grants** **2720**
The objective of the Power and Thermal Management Technology Development Grants is to obtain cutting edge technology from industry and academia that the Air Force Research Laboratory, Propulsion Directorate, Wright Research Site, has not previously considered for funding. Contracts will be awarded that provide a broad range of highly unique evolutionary and revolutionary technology advances in three application areas: integrated power and thermal subsystems; special purpose power and thermal; and battlespace fuels. White papers may be submitted at any time. Review is anticipated to take approximately 45 working days. Applicants whose white papers meet Air Force needs will be asked to submit a technical and cost proposal within 30 working days. Instructions and the format for white paper and proposal submission can be found on the website. Awards will normally be made for up to three years. This is an initial Broad Agency Announcement. Please see the website for updates.
*Requirements:* This is an unrestricted solicitation. Small businesses are encouraged to propose on all or any part of this solicitation.
*Geographic Focus:* All States
*Amount of Grant:* 200,000 - 3,000,000 USD
*Contact:* Lori Shirdon, Contract Negotiator; (937) 255-8077; lori.shirdon@wpafb.af.mil
Frank Gulczinski, Technical Specialist; (937) 255-0375; frank.gulczinski@wrafb.af.mil
Sarah E.P. Chaffee, Contract Specialist; (937) 255-5499; Sarah.Chaffee@wrafb.af.mil
*Sponsor:* U.S. Department of Defense
1400 Defense Pentagon
Washington, D.C. 20301-1400

**USDD Power Efficiency Revolution for Embedded Computing Technologies Grants2721**
The Defense Advanced Research Projects Agency (DARPA) is soliciting proposals for innovative research and development in the area of embedded power efficient computing. The goal of the Power Efficiency Revolution for Embedded Computing Technologies (PERFECT) Grants program is to provide the technologies and techniques to overcome the power efficiency barriers that currently constrain embedded computing systems capabilities and limit the potential of future embedded systems. This goal will be achieved by taking a revolutionary approach to processing power efficiency. This approach includes near threshold voltage operation and massive heterogeneous processing concurrency, combined with techniques to effectively utilize the resulting concurrency and tolerate the resulting increased rate of soft errors. Multiple awards are anticipated. Funding awarded will depend on the quality of the proposals received and the availability of funds. Submission of abstracts in advance of a full proposal is strongly encouraged. This procedure is intended to minimize unnecessary effort in proposal preparation and review. Early submission of abstracts is encouraged. Abstracts will be reviewed in the order received. Detailed instructions for submitting abstracts and proposals are provided on the website. Abstracts are due on or before 4:00 pm eastern time February 23, and proposals are due on or before 4:00 pm eastern time April 16.
*Requirements:* All responsible sources capable of satisfying the government's needs are eligible. Historically black colleges and universities, small businesses, small disadvantaged businesses and minority institutions are encouraged to submit proposals and join others in submitting proposals.
*Geographic Focus:* All States
*Contact:* Dr. Charlie Holland; (703) 526-6630; DARPA-BAA-12-24@darpa.mil
*Sponsor:* U.S. Department of Defense
3701 North Fairfax Drive
Arlington, VA 22203-6630

**USDD President's Emergency Plan for AIDS Relief Military Specific HIV Seroprevalence and Behavioral Epidemiology Risk Survey Grants** 2722

The Department of Defense HIV/AIDS Prevention Program (DHAPP) has successfully engaged over 80 countries in efforts to combat HIV/AIDS among their respective military services. DHAPP is a partner collaborating with the United States State Department, Health and Human Services, United States Agency for International Development, and Centers for Disease Control and Prevention, in the President's Emergency Plan for AIDS Relief (PEPFAR). This Broad Agency Announcement (BAA) is to solicit existing partners and establish new partners to assist select military troops with performing an independent HIV Seroprevalence and Behavioral Epidemiology Risk Survey. Proposals should focus on rapidly extending HIV/AIDS services. Applicants are encouraged to target specific needs with a practical business plan, using small grass-roots organizations to provide community-based services as a way to enhance organic capabilities and sustainability. Concept papers are required and full proposals are by invitation only. Concept papers may be submitted at any time. Early submission is strongly encouraged due to the possible lack of funding for the later evaluation periods. The period of performance is generally one year, with up to two additional phases subject to availability of funding. Applicants are encouraged to propose additional phases if appropriate. This is a multiple year BAA which may be revised as needed. Based on the availability of funds, up to three evaluation cycles will be conducted. Changes will be posted at Grants.gov (grants.gov/applicants/find_grant_opportunities.jsp). Applicants are encouraged to sign up on Grants.gov for "send me change notification emails" by following the link on the synopsis page for the program announcement/funding opportunity.

*Requirements:* All responsible sources from academia, industry, and non-governmental organizations are eligible. Applicants must demonstrate the active support of the in-country military and the USDD representative in the corresponding United States Embassy in the planning and execution of proposals.

*Restrictions:* Countries covered by this BAA are: Democratic Republic of Congo, Liberia, Malawi, Mali, Sierra Leone, South Africa, and Vietnam.

*Geographic Focus:* All States

*Date(s) Application is Due:* Sep 30

*Amount of Grant:* 100,000 - 700,000 USD

*Contact:* Latrice Rubenstein; (619) 556-5276; Latrice.Rubenstein@navy.mil
Cynthia Simon-Arndt; Cynthia.SimonArndt@med.navy.mil

*Sponsor:* U.S. Department of Defense
5450 Carlistle Park, P.O. Box 2050
Mechanicsburg, PA 17055-0791

**USDD Propulsion Materials Grants** 2723

Reliable high-temperature propulsion materials are crucial to improved engine efficiency, reduced fuel costs and decreased maintenance/total life cycle costs. The useful life of a material in any application depends on diverse factors such as the marine environment, temperature and cyclic activities, and mechanical stress. The materials also need to resist oxidation, corrosion or alternating cycles of oxidation and corrosion. The research in this Office of Naval Research (ONR) topic area involves, in part, the thermodynamics and kinetics of materials interactions and materials stability under marine operating environments and temperatures. In order to eventually provide optimal materials for application, research is needed to develop models to assist the development of new materials and to help predict the life of both existing and new materials. Basic research in this area should help ONR understand how to reduce and manage materials instabilities and degradation. The research to investigate the mechanisms that lead to materials degradation should also explore how the mechanisms fundamentally relate to or depend on mechanics, interdiffusion, coatings and materials chemistry, as well as the environmental parameters of temperature, pressure and humidity. Computational approaches to creating new materials and material processes or improving the performance of current materials are encouraged. Current materials include Ni-base single-crystal superalloys, ceramic matrix composites, ceramics, Mo-based superalloys, thermal barrier coatings, environmental coatings and other coatings resistant to oxidation and corrosion environments. Submission of the Research Proposal should always be preceded by submission of a Preproposal and a discussion of that Preproposal with the assigned Program Officer. Applications for the upcoming fiscal year should be received by September 30.

*Requirements:* Grants are awarded to academic, industrial, and nonprofit organizations, with strong emphasis on university research. Investigators are encouraged to contact the staff prior to formally submitting a proposal.

*Restrictions:* New grants or contracts are typically started near the beginning of the fiscal year. New proposals for the next fiscal year should be submitted no later than September 30, unless otherwise directed. Competitive renewals should be submitted no later than six months prior to the award termination date after contacting the Program Officer. Conference proposals should be submitted at least four months prior to the conference.

*Geographic Focus:* All States

*Date(s) Application is Due:* Sep 30

*Contact:* Dr. David Shifler; (703) 696-0285; fax (703) 696-0934; david.shifler@navy.mil

*Internet:* www.onr.navy.mil/Science-Technology/Departments/Code-33/All-Programs/332-naval-materials/Structural-Propulsion-Materials.aspx

*Sponsor:* U.S. Department of Defense
One Liberty Center, 875 N. Randolph Street, Suite 1425
Arlington, VA 22203-1995

**USDD Radar and Signal Processing Grants** 2724

RADAR, an acronym for radio detection and ranging, is a method of locating distant targets by sending bursts of electromagnetic radiation and measuring their refl ections. It can be used for determining both the location and speed of an object. RADAR senses the environment by detecting and characterizing electromagnetic radiation that is reflected from objects within the environment. The Office of Naval Research (ONR) RADAR and Signal Processing Program has the objective of developing an affordable, fully automated network of time-coordinated mono-static, bi-static and passive surveillance sensors providing real-time tracking, identification, and engagement information with persistent wide area awareness. Electromagnetic propagation and scattering with emphasis on antennas; target, clutter and background modeling; and sensor performance prediction. Applications for the upcoming fiscal year must be received by September 30.

*Requirements:* Grants are awarded to academic, industrial, and nonprofit organizations, with strong emphasis on university research.

*Restrictions:* New grants or contracts are typically started near the beginning of the fiscal year. New proposals for the next fiscal year should be submitted no later than September 30, unless otherwise directed by the BAA. Competitive renewals should be submitted no later than six months prior to the award termination date after contacting the Program Officer. Conference proposals should be submitted at least four months prior to the conference.

*Geographic Focus:* All States

*Date(s) Application is Due:* Sep 30

*Contact:* Michael Pollock, Division Director; (703) 696-4202 or (703 )696-4217; fax (703) 696-3098; michael.a.pollock@navy.mil

*Internet:* www.onr.navy.mil/Science-Technology/Departments/Code-31/All-Programs/312-Electronics-Sensors/Radar-Systems-Signal-Processing.aspx

*Sponsor:* U.S. Department of Defense
One Liberty Center, 875 N. Randolph Street, Suite 1425
Arlington, VA 22203-1995

**USDD Radio Frequency-Field Programmable Gate Arrays Grants** 2725

The Defense Advanced Research Projects Agency (DARPA) is soliciting innovative research and development proposals in the area of Radio Frequency Field Programmable Gate Arrays (RF-FPGAs), a thrust within the DARPA Microsystems Technology Office's Adaptive RF Technology program. The goal of RF-FPGA is to enable a common hardware architecture that facilitates reutilization of the same set of RF front-end components across disparate applications through programmability of the transceiver chain. RF-FPGAs will impact the areas of communications, electronic warfare, and signal intelligence by eliminating redundant and costly hardware development required for the adoption or recognition of a new wireless function or waveform. Multiple awards are anticipated. The funding amount of awards will depend on the quality of the proposals received and the availability of funds. Applicants are strongly encouraged to submit an abstract in advance of a full proposal. This procedure is intended to minimize unnecessary effort in proposal preparation and review. Instructions for submission of abstracts and proposals can be found at the website. Abstracts are due on or before 4:00 pm eastern time December 7. Proposal are due on or before 4:00 pm eastern time February 16.

*Requirements:* All responsible sources capable of satisfying the government's needs are eligible. Historically black colleges and universities, small businesses, small disadvantaged businesses and minority institutions are encouraged to submit proposals and join others in submitting proposals.

*Geographic Focus:* All States

*Contact:* Dr. William J. Chappel, Program Manager; (703) 526-6630 or (571) 218-4507; William.Chappel@darpa.mil or DARPA-BAA-12-13@darpa.mil

*Sponsor:* U.S. Department of Defense
3701 North Fairfax Drive
Arlington, VA 22203-6630

**USDD Research and Analytical Support for the 711th Human Performance Wing's Human Effectiveness Directorate Grants** 2726

The goal of the UDDD's Air Force Research Laboratory Grants is to obtain technical and analytical support for the conduct of research at the 711th Human Performance Wing's (HPW) Human Effectiveness Directorate. This includes all locations (Wright-Patterson Air Force Base, Ohio; Brooks City-Base, Texas; and Mesa Research Site, Arizona). Support may take the following forms: provide appropriately training and experienced undergraduate and graduate students, post-doctoral fellows and faculty for direct work on 711 HPW's research, analyses, and studies; provide faculty to supervise students to perform advanced research analysis and studies; match faculty expertise to designated research and study needs with 711 HPW; and provide a pool of subjects for research studies. This is a multi-year offering and applicants may submit white papers at any time. Applicants whose white paper are considered to meet the needs of the Air Force will be invited to submit a full proposal. Early white paper submission is encouraged. Individual Grants will normally be for 12 months or less.

*Requirements:* This is an unrestricted solicitation. Small businesses are encouraged to propose on all or any part of this solicitation.

*Geographic Focus:* All States

*Date(s) Application is Due:* Sep 30

*Contact:* Pamela J. Covey; (937) 255-8958; pamela.covey@wpafb.af.mil
Linda Lange, Technical Specialist; (937) 255-0406; Helen.Williams@wpafb.af.mil

*Internet:* www.grants.gov/search-grants.html?agencies%3DDOD|Department%20of%20Defense

*Sponsor:* U.S. Department of Defense
1400 Defense Pentagon
Washington, D.C. 20301-1400

**USDD Research and Studies for the Office of Net Assessment Grants** 2727

The Department of Defense's Office of Net Assessment conducts and sponsors analyses that compare the standing, trends, and future prospects of United States and foreign military capability and military potential. Priority is given to assessing aspects of the security environment and parts of the world that are likely to change in the future

and that may present threats or opportunities to the United States. Assessments may focus on specific theaters, regions, functions, mission areas, major weapons categories, doctrine, etc., as well as on demographic, economic, and political developments that may affect the power and strategies of nations. Proposals will be considered that undertake comprehensive approaches to one or more of the following areas of interest or that investigate one or more particular aspect(s) or case(s) that belong to the broader areas listed. There are five specific areas of interest and one area for additional research topics. They include: (1) future naval warfare; (2) proliferated world; (3) space, (4) sustaining current areas of United States advantage; (5) history (military); and (6) additional research topics (other germane topics may be of interest). The content and format for applications can be found on the website. Proposals should be submitted as early as possible and are valid for a period of one year after receipt. Awards will be made for a period of five years on a bi-monthly basis. The period of performance typically ranges from one to five years.
*Requirements:* All profit and non-profit organizations as well as institutions of higher education are eligible.
*Geographic Focus:* All States
*Date(s) Application is Due:* Jun 15
*Contact:* Sherry Goodman; 703-588-1138; Sherry.Goodman.ctr@whs.mil
*Sponsor:* U.S. Department of Defense
1400 Defense Pentagon
Washington, D.C. 20301-1400

**USDD Research Initiatives at the Naval Postgraduate School Grants** 2728
The Naval Postgraduate School (NPS) is interested in receiving proposals for research initiatives which offer potential for advancement and improvement in the NPS core mission of graduate education and research. This includes research initiatives across a spectrum of science and engineering, business and policy, operational and informational sciences, and interdisciplinary disciplines that support the NPS' graduate education and research mission. Work funded may include basic research, applied research and some advanced research. Applicants are strongly encouraged to contact an NPS point of contact whose program best matches their field of interest. See the website to locate the appropriate point of contact. Applicants are encouraged to submit a white paper prior to submitting a full proposal. The funding amount and award period will vary depending on the research area and the technical approach pursued by the applicant.
*Requirements:* All responsible sources from academia and industry are eligible. Historically black colleges and universities and minority institutions are encouraged to apply.
*Geographic Focus:* All States
*Date(s) Application is Due:* Mar 31
*Contact:* Janet Norton; (808) 473-7669; janet.norton@navy.mil
*Sponsor:* U.S. Department of Defense
5450 Carlistle Park, P.O. Box 2050
Mechanicsburg, PA 17055-0791

**USDD Research Interest of the Air Force Office of Scientific Research Grants** 2729
The Air Force Office of Scientific Research (AFOSR) plans, coordinates, and executes the Air Force Research Laboratory's (AFRL) basic research program in response to technical guidance from AFRL and requirements of the Air Force; fosters, supports, and conducts research within Air Force, university, and industry laboratories; and ensures transition of research results to support USAF needs. The focus of AFOSR is on research areas that offer significant and comprehensive benefits to our national warfighting and peacekeeping capabilities. These areas are organized and managed in three scientific directorates. First is Aerospace, Chemical and Material Sciences (RSA), which focuses on meeting the basic research challenges related to future air and space power by leading the discovery and development of fundamental science and engineering across three integrated research focus areas of: aero-structure interaction and control; complex materials and structures; and energy, power and propulsion. Second is Physics and Electronics, which generates the fundamental knowledge needed to advance Air Force operational capabilities and is categorized into the following three broad areas of: complex electronics and fundamental quantum processes; plasma physics and high energy density nonequilibrium processes; and optics, electromagnetics, communication, and signal processing. Third is Mathematics, Information and Life Sciences, which supports a wide range of fundamental mathematics, information and computer science, biological sciences, and human behavioral/performance research to provide novel options that increase operational flexibility and performance of systems and environments of relevance to the Air Force. Research is primarily encompassed in the following four focus areas: information and complex networks research; decision-making research; dynamical systems, optimization, and control research; and natural materials and systems research. This announcement remains open until superseded. Proposals may be submitted at any time, and various contacts are listed in the proposal.
*Geographic Focus:* All States
*Contact:* Contracting Officer; afosr.baa@afosr.af.mil
*Sponsor:* U.S. Department of Defense
875 North Randolph Street, Suite 325, Room 3112
Arlington, VA 22203-1768

**USDD Research Interests of the U.S. Air Force Academy Grants** 2730
The United States Air Force Academy (USAFA) invests in an active research program for three main reasons. First and foremost, research significantly enhances the cadet learning experience. Research is done by, for and with cadets who work alongside fellow cadets and faculty mentors. Secondly, research programs provide opportunities for essential faculty development. Research broadens and deepens the experience base of the faculty and infuses current, relevant, state-of-the-art and cutting-edge applications and examples into the curriculum. Third, USAFA strives to conduct research to enhance the ability of the Air Force to perform its mission. There are currently ongoing research projects spanning topics as diverse as super cooled cesium atoms, cyber security, spatial disorientation and homeland defense. This is a Broad Agency Announcement (BAA) by USAFA to solicit proposals for basic and applied research from business and academia. This strategy is an acquisition tool for USAFA with the flexibility to solicit proposals and make awards to develop technologies to meet present and future Air Force research needs as technology issues are identified. Proposals are invited for research in many broad areas which are described in detail in the BAA on the website. This BAA will remain open until superseded. White papers may be submitted at any time and are reviewed as received. Proposal call announcements (Calls) may be issued against this BAA. Calls will contain specific instruction including further technical details, application instructions and due dates.
*Requirements:* All responsible, potential applicants from academia and industry are eligible. Proposals are particularly encouraged from small businesses, historically black colleges and universities, minority institutions and minority researchers.
*Restrictions:* Questions about a technical research area should be directed to the technical program manager for each specific research topic listed in the BAA.
*Geographic Focus:* All States
*Amount of Grant:* Up to 90,000,000 USD
*Contact:* Jessica Niebuhr; (719) 333-3907; Jessica.Niebuhr@us.af.mil
*Sponsor:* U.S. Department of Defense
4102 Pinion Drive
Colorado Springs, CO 80840

**USDD RF Semiconductors, RF Solid State Amplifiers and Wide Bandgap Materials Grants** 2731
The Office of Naval Research (ONR) RF Semiconductor Devices, RF Solid State Amplifiers and Wide Bandgap Materials Program's objective is to develop solid state (e.g., wide bandgap semiconductor) low noise, high speed, and power devices and circuits for use in surveillance, electronic warfare, multifunctional RF systems communications, and power applications. Emphasis is on device advancements that enable a significant improvement in broadband performance, high linearity, and/or high (e.g., 60%) power added efficiency. Power output of at least five times that of GaAs-based devices. Emerging applications for wide bandgap materials and devices in the areas of transmitter and receiver components, mixed signal and digital circuits are also included. The program investigates electronic properties of wide bandgap, III-V and silicon-based semiconductors and novel devices; focusing on novel high power microwave and switching devices for broadband radar, electronic warfare, communications, and multifunctional systems applications and for power control applications. Proposers are encouraged to contact the program officer to discuss their research interest prior to the submission of formal proposals.
*Requirements:* Grants are awarded to academic, industrial, and nonprofit organizations, with strong emphasis on university research.
*Restrictions:* New grants or contracts are typically started near the beginning of the fiscal year (October 1). New proposals for the next fiscal year should be submitted no later than September 30, unless otherwise directed by the BAA. Competitive renewals should be submitted no later than six months prior to the award termination date after contacting the Program Officer. Conference proposals should be submitted at least four months prior to the conference.
*Geographic Focus:* All States
*Date(s) Application is Due:* Sep 30
*Contact:* Dr. Paul Maki, Program Officer; (703) 696-3060 or (703) 588-2427; fax (703) 696-8496; paul.maki@navy.mil
*Internet:* www.onr.navy.mil/Science-Technology/Departments/Code-31/All-Programs/312-Electronics-Sensors/Semiconductor-Devices.aspx
*Sponsor:* U.S. Department of Defense
One Liberty Center, 875 N. Randolph Street, Suite 1425
Arlington, VA 22203-1995

**USDD RF Superconducting Technologies Grants** 2732
The Office of Naval Research (ONR) RF Superconducting Technologies Program's objective is to develop new and improved electronic devices and circuits using superconducting materials that are relevant to RF systems applications. Digital and mixed signal circuits clocking above 20 GHz are the focus. AD.C. and interference cancellation are specific topics of interest. Improvements in packaging and memory technologies are also sought. Basic research on the fundamental physics of phonons as they interact with light and of heat transport and cooling are of interest. Proposers are encouraged to contact the program officer to discuss their research interest prior to the submission of formal proposals. Basic research proposals are preferred by June 1 of each year, while applied research proposals should be submitted in approximately March in response to an annual specific broad agency announcement from the ONR electronics program.
*Requirements:* Grants are awarded to academic, industrial, and nonprofit organizations, with strong emphasis on university research.
*Geographic Focus:* All States
*Date(s) Application is Due:* Jun 1
*Contact:* Deborah Van Vechten, Program Officer; (703) 696-4219; fax (703) 696-8496; deborah.vanvechten@navy.mil
*Internet:* www.onr.navy.mil/Science-Technology/Departments/Code-31/All-Programs/312-Electronics-Sensors/RF-Superconducting-Technologies.aspx
*Sponsor:* U.S. Department of Defense
One Liberty Center, 875 N. Randolph Street, Suite 1425
Arlington, VA 22203-1995

**USDD Sabbatical Leave Fellowships** 2733

The Sabbatical Leave program provides fellowship appointments for a minimum of one semester to a maximum of one year in length. The purpose of sabbatical leave is to provide an opportunity for faculty members to engage in scholarly, creative, professional, research, or other academic activities that will enhance the faculty member's further contributions to their institution. There are three levels of appointment: summer faculty fellow, senior summer faculty fellow and distinguished summer faculty fellow. Participants in the Sabbatical Leave program receive stipends ranging from $1,400 to $1,900 per week. Each fellow will be reimbursed for expenses incurred on an optional pre-program visit to the sponsoring laboratory and one round-trip encompassing travel to the sponsoring laboratory at the beginning of the program and travel back to their home residence at the end of the program. Relocation assistance is provided to qualifying participants.
*Requirements:* Sabbatical leave appointments are contingent on the availability of sabbatical leave positions. Participants must conduct research on-site. Prospective applicants should contact the research facility at which they are interested in working to jointly develop a suitable research proposal. Also required is an estimate of base salary and percentage of salary provided by the home institution while on sabbatical leave.
*Restrictions:* Faculty members from Historically Black Colleges and Universities, Minority Institutions, American Indian Tribal Colleges and Universities, and Hispanic Serving Institutions (HBCU/MI/TCU/HIS), as designated by the U.S. Department of Education, are especially encouraged to apply.
*Geographic Focus:* All States
*Amount of Grant:* 14,000 - 19,000 USD
*Contact:* V. Nadine Daniel, Point of Contact; (202) 289-9806, ext. 126; daniel.vontilla.ctr@navy.mil or ndaniel@aghe.org
*Internet:* onr.asee.org/about_the_sabbatical_leave_program
*Sponsor:* U.S. Department of Defense
One Liberty Center, 875 N. Randolph Street, Suite 1425
Arlington, VA 22203-1995

**USDD Science, Technology, Engineering and Mathematics (STEM) Grants** 2734

The United States Army Engineer Research and Development Center (ERD.C.) seeks proposals to stimulate young pupils in the sciences, technology, engineering and mathematics (STEM). Depending upon the availability of appropriated funds, ERD.C. may: (1) make multiple awards; and (2) consider options exercisable for multi-year performance. Funding is limited and proposals are primarily sought in the not-to-exceed $30,000 range; however, larger awards may be considered when appropriate.
*Requirements:* All responsible sources from educational institutions and non-profit organizations are eligible. Historically black colleges and universities and minority institutions are encouraged to submit proposals and join others in submitting proposals. The target students attend primary, middle, and high schools in the following locations; Warren County, Mississippi; Champaign County, Illinois; Grafton County, New Hampshire; Orange County, Vermont; Windsor County, Vermont; Sullivan County, New Hampshire; Arlington County, Virginia; and Alexandria, Virginia. The target students may attend public or private schools.
*Geographic Focus:* All States
*Date(s) Application is Due:* May 15
*Amount of Grant:* 30,000 USD
*Contact:* Kimberly Dulaney, Contract Specialist; (601) 634-3954 or (601) 634-3188; kimberly.l.dulaney@usace.army.mil
*Sponsor:* U.S. Department of Defense
3909 Halls Ferry Road
Vicksburg, MS 39180-6199

**USDD Science and Tecnology for Warfighter Training and Aiding Grants** 2735

The 711th Human Performance Wing (HPW), Warfighter Readiness Research Division, conducts pioneering basic, applied, and advanced research and development in support of a variety of research and operational needs. The objectives of the Grants are to research, demonstrate, evaluate, and transition human performance methods and technology to enable warfighters to have the right skills, knowledge, experience and capabilities at the right time to make the right decisions. The objective will be met through the following three science and technology areas: cognitive modeling; immersive environments; and continuous learning. This is a multi-year offering and applicants may submit white papers at any time. Applicants whose white paper are considered to meet the needs of the Air Force will be invited to submit a full proposal. Early white paper submission is encouraged. Estimated funding over five years is $49 million. Three to five awards per year are anticipated.
*Requirements:* This is an unrestricted solicitation. Small businesses are encouraged to propose on all or any part of this solicitation.
*Geographic Focus:* All States
*Date(s) Application is Due:* Sep 30
*Contact:* Angela G. Lauofo; (937) 255-5091; Angela.Lauofo@wpafb.af.mil
Helen Williams, Contracts Negotiator; (937) 656-9833; helen.williams@wpafb.af.mil
*Internet:* www.grants.gov/search-grants.html?agencies%3DDOD|Department%20of%20Defense
*Sponsor:* U.S. Department of Defense
1400 Defense Pentagon
Washington, D.C. 20301-1400

**USDD Sensor Innovative Research Grants** 2736

The Electromagnetics Technology Division (RYH), Sensors Directorate (RY), Air Force Research Laboratory (AFRL), is looking to conduct research in the following areas: Antenna Technology, Opto-Electronic Technology, Electromagnetic Scattering, and Infrared Sensor Technology. The Air Force anticipates awarding up to 25 awards for this announcement. Potential applicants should see the full proposal for specific details concerning the description of efforts and primary objectives. Typical grant awards will range from $50,000 to $1,000,000.
*Requirements:* This is an unrestricted solicitation. Small businesses are encouraged to propose on all or any part of this solicitation.
*Geographic Focus:* All States
*Amount of Grant:* 50,000 - 1,000,000 USD
*Contact:* Brian Lowe, Contract Officer; (937) 255-3585; brian.lowe@wpafb.af.mil
*Sponsor:* U.S. Department of Defense
1400 Defense Pentagon
Washington, D.C. 20301-1400

**USDD Ship Systems and Engineering Research Grants** 2737

The division is focused on providing technologically superior warfighting capabilities at reduced total ownership costs for surface and subsurface platforms through investments in basic and applied research, and advanced technology development of programs in the following areas: Advanced Sea Platforms; Sea Platform Survivability; and Surface Ship Hydrodynamics. Research topic areas of interest that contribute to these focus areas include: electrical power generation systems and equipment; energy conversion, distribution and storage; electro-mechanics; thermoelectrics; phototonics; thermal energy management; stability and control of electrical networks; ship design and analysis; systems engineering; ship systems integration; hydromechanics; maneuvering and seakeeping; hydroacoustics; structural mechanics; solid mechanics; thermo-mechanics; structural acoustics; structural dynamics; computational mechanics; advanced structural systems; protection systems and design methodology; automation; control of dynamic systems; damage control and firefighting; control of acoustic and non-acoustic signatures; high-speed platforms; sense and respond logistics; and inter- and intra-ship material movement and tracking. Submission of the Research Proposal should always be preceded by submission of a Preproposal and a discussion of that Preproposal with the assigned Program Officer. Applications for the upcoming fiscal year should be received by September 30.
*Requirements:* Grants are awarded to academic, industrial, and nonprofit organizations, with strong emphasis on university research.
*Restrictions:* New grants or contracts are typically started near the beginning of the fiscal year. New proposals for the next fiscal year should be submitted no later than September 30, unless otherwise directed. Competitive renewals should be submitted no later than six months prior to the award termination date after contacting the Program Officer. Conference proposals should be submitted at least four months prior to the conference.
*Geographic Focus:* All States
*Date(s) Application is Due:* Sep 30
*Contact:* Dr. John Pazik, Division Director; (703) 696-4404; john.pazik@navy.mil
*Internet:* www.onr.navy.mil/Science-Technology/Departments/Code-33/All-Programs/331-ship-systems-engineering.aspx
*Sponsor:* U.S. Department of Defense
One Liberty Center, 875 N. Randolph Street, Suite 1425
Arlington, VA 22203-1995

**USDD Signal Processing for Networked Sensing Grants** 2738

A majority of performance improvements in sensors, networking and communication connectivity are expected to come from conceptualization of new systems and innovations in signal processing. The Office of Naval Research Signal Processing for Networked Sensing program's goal is to leverage developments in signal processing techniques to bring about orders of magnitude improvements in sensing, target resolution, small-target detectability, multi-target tracking and to address hard problems in sensor data fusion, track fusion and communications. The program addresses hard problems such as: secure error free data communication in networked sensors; processing tracks for highly maneuvering and dense target scenarios; tracking and identifying clustered, grouped, disintegrating targets; achieving a reliable conflict free track per target view in a multi-target scenario across networked sensors; and maintaining tracks and a consistent target picture in the presence of clutter, out-of-order measurements, electronic interference and miscellaneous disrupting challenges. The submission deadline is ongoing.
*Requirements:* Grants are awarded to academic, industrial, and nonprofit organizations, with strong emphasis on university research.
*Geographic Focus:* All States
*Contact:* Dr. Rabinder Madan; (703) 696-4217; rabinder@madan@navy.mil
*Internet:* www.onr.navy.mil/Science-Technology/Departments/Code-31/All-Programs/311-Mathematics-Computers-Research/Signal-Processing-Networked-Sensing.aspx
*Sponsor:* U.S. Department of Defense
One Liberty Center, 875 N. Randolph Street, Suite 1425
Arlington, VA 22203-1995

**USDD Small University Grants** 2739

The USDD's Air Force Research Laboratory Space Vehicles Directorate is offering Small University Grants in space related basic, applied, and advanced research projects. The scope of the research includes the entire spectrum of space that is applicable to the Air Force including all tangentially space-related research. Research includes, but is not limited to, fundamental research in: space system components and technologies; space environment and surveillance; and space system integration. Proposals may be submitted at any time. However, applicants must call the technology points of contact to make sure funding is available for the technology area of consideration before submitting a proposal. Each technology area has a specific contact and the names and telephone numbers can be found in the full announcement at the website. Proposals are reviewed and evaluated as they are received. Multiple Grants of various values are anticipated with a period of performance from one to two years. At minimum one Grant per fiscal year

is planned. Anticipated annual funding in total is expected to be $790,000 to $990,000. It is anticipated that Grants will remain open for a period of five years. Any updates or possible cancellation will be issued as an amendment and can be found on the website.
*Requirements:* Only United States and United States territory educational institutions are eligible.
*Geographic Focus:* All States
*Date(s) Application is Due:* Aug 26
*Amount of Grant:* Up to 4,950,000 USD
*Contact:* LaDonna Nivens; (505) 853-2279; amanda.qualls@kirtland.af.mil
*Internet:* www.grants.gov/search-grants.html?agencies%3DDOD|Department%20of%20Defense
*Sponsor:* U.S. Department of Defense
1400 Defense Pentagon
Washington, D.C. 20301-1400

**USDD Software and Computer Systems Grants** 2740

The Office of Naval Research Software and Computer Systems program focuses on developing the fundamental principles and models for the design and construction of complex software systems that meet required assurances for security, safety, reliability and performance. It builds upon foundations of mathematical logics and computational complexity but takes into account the uncertainties and constraints of the operational environment. The overall goal is to measurably improve the ability of complex software-intensive systems to meet the information-processing challenges of future naval systems. The program has several thrust areas that build around the theme of information as a computational resource to be managed, secured and shared in different contexts: Principles for Correctness and Security Properties; Trusted Network Computing; Secure Information Management, Sharing and Interaction; Information Assurance of Cyber-Physical Systems; and Quantum Information Sciences for Secure Computing and Communication. The annual deadline for the next fiscal year is September 30.
*Requirements:* Grants are awarded to academic, industrial, and nonprofit organizations, with strong emphasis on university research.
*Restrictions:* New grants or contracts are typically started near the beginning of the fiscal year. New proposals for the next fiscal year should be submitted no later than September 30, unless otherwise directed by the BAA. Competitive renewals should be submitted no later than six months prior to the award termination date after contacting the Program Officer. Conference proposals should be submitted at least four months prior to the conference.
*Geographic Focus:* All States
*Date(s) Application is Due:* Sep 30
*Contact:* Dr. J. Sukarno Mertoguno; (703) 696-0107; sukarno@mertoguno@navy.mil
*Internet:* www.onr.navy.mil/Science-Technology/Departments/Code-31/All-Programs/311-Mathematics-Computers-Research/Software-Computing-Systems.aspx
*Sponsor:* U.S. Department of Defense
One Liberty Center, 875 N. Randolph Street, Suite 1425
Arlington, VA 22203-1995

**USDD Solid and Liquid Waste Treatment Grants** 2741

The Office of Naval Research Solid and Liquid Waste Treatment program seeks to enhance the science and technology base in order to assist the development of sensors, materials, methods and, ultimately, systems for the treatment or destruction of solid and liquid wastes both on board ships and at shore-side support activities. Areas of interest include: novel approaches to on-board treatment of both oily and non-oily wastewaters to meet discharge standards; reliable sensor and control systems to automate shipboard waste treatment and reduce the manpower burden; novel approaches to reduce the threat of the transport of non-indigenous species in ballast operations; and alternative approaches to treat or remediate solid and hazardous wastes. Submission of the Research Proposal should always be preceded by submission of a Preproposal and a discussion of that Preproposal with the assigned Program Officer. Applications for the upcoming fiscal year should be received by September 30.
*Requirements:* Grants are awarded to academic, industrial, and nonprofit organizations, with strong emphasis on university research. Investigators are encouraged to contact the staff prior to formally submitting a proposal.
*Restrictions:* New grants or contracts are typically started near the beginning of the fiscal year. New proposals for the next fiscal year should be submitted no later than September 30, unless otherwise directed. Competitive renewals should be submitted no later than six months prior to the award termination date after contacting the Program Officer. Conference proposals should be submitted at least four months prior to the conference.
*Geographic Focus:* All States
*Date(s) Application is Due:* Sep 30
*Contact:* Dr. Stephen McElvany, Program Officer; (703) 696-1449; fax (703) 696-6887; steve.mcelvaney@navy.mil
*Internet:* www.onr.navy.mil/Science-Technology/Departments/Code-33/All-Programs/332-naval-materials/Solid-Liquid-Waste-Treatment.aspx
*Sponsor:* U.S. Department of Defense
One Liberty Center, 875 N. Randolph Street, Suite 1425
Arlington, VA 22203-1995

**USDD Solid Mechanics Grants** 2742

The field of solid mechanics bridges research in materials science and structural mechanics and design. The Office of Naval Research Solid Mechanics program supports research in mechanics of deformation, damage evolution and failure of composite hull structural materials in the naval environment, when subjected to complex three-dimensional static, cyclic and dynamic loading conditions. The program aims to establish physically based models for material behavior, with predictive capabilities. The current focus is on marine composites and sandwich structures. Areas of interest include: 3D failure modes, mode coupling and failure criteria; effects of moisture, sea water, hydrostatic pressure and temperature; durability and life prediction; effects of dynamic loading (e.g. impact, shock, blast implosion); mechanics of nanocomposites and multifunctional composites; concepts for mitigation of damage and failure, and life extension; concepts for the enhancement of thermo-mechanical properties and structural performance; and materials systems of interest include composites and sandwich structures for naval ship hulls. Submission of the Research Proposal should always be preceded by submission of a Preproposal and a discussion of that Preproposal with the assigned Program Officer. Applications for the upcoming fiscal year should be received by September 30.
*Requirements:* Grants are awarded to academic, industrial, and nonprofit organizations, with strong emphasis on university research. Investigators are encouraged to contact the staff prior to formally submitting a proposal.
*Restrictions:* New grants or contracts are typically started near the beginning of the fiscal year. New proposals for the next fiscal year should be submitted no later than September 30, unless otherwise directed. Competitive renewals should be submitted no later than six months prior to the award termination date after contacting the Program Officer. Conference proposals should be submitted at least four months prior to the conference.
*Geographic Focus:* All States
*Date(s) Application is Due:* Sep 30
*Contact:* Dr. Yapa Rajapakse, Program Officer; (703) 696-4405; yapa.rajapakse@navy.mil
*Internet:* www.onr.navy.mil/Science-Technology/Departments/Code-33/All-Programs/332-naval-materials/Solid-Mechanics.aspx
*Sponsor:* U.S. Department of Defense
One Liberty Center, 875 N. Randolph Street, Suite 1425
Arlington, VA 22203-1995

**USDD Strategic Technologies Grants** 2743

The Defense Advanced Research Projects Agency's (DARPA) Strategic Technology Office (STO) is soliciting innovative proposals for the performance of research, development, design, and testing that directly supports STO. This includes finding difficult targets; communications, networks and electronic warfare; shaping the environment; and foundational technologies that support multiple STO focus areas. The recommended order of submissions is executive summary, abstract, and proposal; however, it is not mandatory to submit an executive summary and/or abstract before submitting a full proposal. Applicants are strongly encouraged to submit an executive summary prior to abstracts or proposals. This procedure is intended to minimize unnecessary effort in proposal preparation and review. Instructions for submitting executive summaries, abstracts, and proposals can be found on the website. Multiple awards are possible. The funding awarded will depend on the quality of the proposals received and the availability of funds.
*Requirements:* All responsible sources capable of satisfying the government's needs are eligible. Historically black colleges and universities, small businesses, small disadvantaged businesses and minority institutions are encouraged to submit proposals and join others in submitting proposals.
*Geographic Focus:* All States
*Date(s) Application is Due:* Jan 12
*Contact:* BAA Coordinator; (703) 526-6630; fax DARPA-BAA-12-09@DARPA.mil
*Sponsor:* U.S. Department of Defense
3701 North Fairfax Drive
Arlington, VA 22203-6630

**USDD U.S. Army Medical Research and Materiel Command Broad Agency Announcement for Extramural Medical Research Grants** 2744

The United States Army Medical Research and Materiel Command's mission is to provide solutions to medical problems of importance to the American warfighter at home and abroad. The scope of this effort and the priorities attached to specific projects are influenced by changes in military and civilian medical science and technology, operational requirements, military threat assessments, and national defense strategies. Research areas of interest for the Extramural Medical Research Grants include the following: military infectious diseases; combat casualty care; military operational medicine; clinical and rehabilitative medicine; medical biological defense; medical chemical defense; telemedicine and advanced technology; and special programs (including health care delivery; detection, diagnosis, control or eradication of specified diseases, conditions, or syndromies; and the advancement of military medical interests). This Broad Agency Announcement (BAA) is intended to solicit extramural research and development ideas. Funding has not been set aside specifically for this announcement and the number of awards has not been determined. Applicants are encouraged to submit proposals that span their entire research project up to five years. Pre-proposals and full proposals may be submitted at any time until the deadline.
*Requirements:* Eligible applicants include the following: private/public/state controlled institutions of higher education; Hispanic-serving institutions; historically black colleges and universities/minority institutions; tribally controlled colleges and universities; Alaska Native and Native Hawaiian serving institutions; nonprofits with 501(c)3 status (other than institutions of higher education; small businesses; for-profit organizations (other than small businesses); Indian/Native American tribal governments; Indian/Native American tribally designated organizations; non-domestic (non-United States) entities (foreign organizations); and federal, state, and local government agencies.
*Geographic Focus:* All States
*Date(s) Application is Due:* Sep 30
*Contact:* Grants Officer; (703) 518-4726; (QA.BAA@amedd.army.mil),
*Sponsor:* U.S. Department of Defense - 1
1400 Defense Pentagon
Washington, D.C. 20301-1400

**USDD Ultra-High Temperature Materials Grants** **2745**

This Office of Naval Research program seeks to develop materials, structures and computational tools that will enable the development of expendable systems that operate above 2500°C with no ablation or erosion. The primary application areas are naval hypersonics technologies, including advanced missile systems and Electromagnetic Railgun projectile. This program is also developing high-temperature structural insulator materials that mechanically support and thermally isolate internal systems (e.g., sensors, guidance mechanisms, weapons payload) from high-heat loads generated by burning fuel and/or aerothermal heating during launch/firing/re-entry of advanced naval weapons systems. Understanding the fundamental relationships between composition, microstructure and physical response (thermal and mechanical) at the micro- and meso-scale is of high importance, as is the potential for low-cost manufacturing techniques. Primary applications include: high-temperature structural insulators; electromagnetic railgun projectiles; missile systems; leading edges; and rocket nozzles. Submission of the Research Proposal should always be preceded by submission of a Preproposal and a discussion of that Preproposal with the assigned Program Officer. Applications for the upcoming fiscal year should be received by September 30.

*Requirements:* Grants are awarded to academic, industrial, and nonprofit organizations, with strong emphasis on university research. Investigators are encouraged to contact the staff prior to formally submitting a proposal.

*Restrictions:* New grants or contracts are typically started near the beginning of the fiscal year. New proposals for the next fiscal year should be submitted no later than September 30, unless otherwise directed. Competitive renewals should be submitted no later than six months prior to the award termination date after contacting the Program Officer. Conference proposals should be submitted at least four months prior to the conference.

*Geographic Focus:* All States

*Date(s) Application is Due:* Sep 30

*Contact:* Dr. Eric Wuchina; (301) 227-3949; fax (301) 227-4732; eric.wuchina@navy.mil

*Internet:* www.onr.navy.mil/Science-Technology/Departments/Code-33/All-Programs/332-naval-materials/Ultra-High-Temperature-Materials.aspx

*Sponsor:* U.S. Department of Defense
One Liberty Center, 875 N. Randolph Street, Suite 1425
Arlington, VA 22203-1995

**USDD Undersea Signal Processing Grants** **2746**

The USDD Undersea Signal Processing program's goal is to develop signal processing algorithms that improve the Navy's ability to detect, identify and locate submarines in shallow and deep ocean environments. To this end, we fund basic and applied research projects at universities, federal laboratories and industry. The program is organized into the following technical thrusts: Active Sonar Signal Processing - projects focus on active sonar systems for wide area surveillance, distributed search and battle group defense; Passive Sonar Signal Processing - projects focus on signal processing algorithms that detect, classify and locate modern diesel-electric submarines at tactically useful ranges; Data Fusion Technology - the program's objective is to develop new ways to fuse sonar data from multiple sensors; and Fundamental Research Initiatives - this program supports basic research in statistical and ocean acoustic signal processing projects conducted at universities throughout North America. Submissions for fiscal year planning letters are due by the close of business, June 14. Applications for the upcoming fiscal year should be received by September 30.

*Requirements:* Grants are awarded to academic, industrial, and nonprofit organizations, with strong emphasis on university research.

*Restrictions:* New grants or contracts are typically started near the beginning of the fiscal year. Fiscal year planning letters are due by the close of business, June 14. New proposals for the next fiscal year should be submitted no later than September 30, unless otherwise directed by the BAA. Competitive renewals should be submitted no later than six months prior to the award termination date after contacting the Program Officer. Conference proposals should be submitted at least four months prior to the conference.

*Geographic Focus:* All States

*Contact:* John Tague, Team Leader; (703) 696-4399; john.a.tague@navy.mil

*Internet:* www.onr.navy.mil/Science-Technology/Departments/Code-32/All-Programs/Ocean-Systems-321/Undersea-Signal-Processing.aspx

*Sponsor:* U.S. Department of Defense
One Liberty Center, 875 N. Randolph Street, Suite 1425
Arlington, VA 22203-1995

**USDD Vacuum Devices Grants** **2747**

The Office of Naval Research (ONR) Vacuum Devices Program develops techniques and technologies for compact high-power, high-frequency vacuum electronic amplifiers exhibiting high gain, bandwidth, and efficiency. Of particular interest is the development of compact high power amplifiers that can be applied to the millimeter and sub-millimeter wave frequency range. Geometry and physics-based computationally efficient algorithms, models and simulation based design are included. Applications for the upcoming fiscal year should be received by September 30.

*Requirements:* Grants are awarded to academic, industrial, and nonprofit organizations, with strong emphasis on university research.

*Restrictions:* New grants or contracts are typically started near the beginning of the fiscal year. New proposals for the next fiscal year should be submitted no later than September 30, unless otherwise directed by the BAA. Competitive renewals should be submitted no later than six months prior to the award termination date after contacting the Program Officer. Conference proposals should be submitted at least four months prior to the conference.

*Geographic Focus:* All States

*Date(s) Application is Due:* Sep 30

*Contact:* Ingham Mack, Program Officer; (703) 696-4825; ingham.mack@navy.mil

*Internet:* www.onr.navy.mil/Science-Technology/Departments/Code-31/All-Programs/312-Electronics-Sensors/Vacuum-Devices.aspx

*Sponsor:* U.S. Department of Defense
One Liberty Center, 875 N. Randolph Street, Suite 1425
Arlington, VA 22203-1995

**USDD Warfighter Interface Technologies Advanced Research Programs Grants** **2748**

The 711th Human Performance Wing, Human Effectiveness Directorate is soliciting white papers for innovative technologies to support the development of Warfighter Interface Technologies Advanced Research Programs. The objective is to address highly unique and innovative research and development concept areas in Warfighter Interface Technologies. Proposed work should address the innovative and strategic thought of the 21st century adversaries, and develop new concepts to counter these adversaries with innovative warfighter interface capabilities. The technology research will be primarily focused in the following four areas: battlespace acoustics; battlespace visualizations; human role in autonomous systems; and distributed decision making. This is a multi-year offering and applicants may submit white papers at any time. Applicants whose white paper are considered to meet the needs of the Air Force will be invited to submit a full proposal. Early white paper submission is encouraged. Anticipated annual funding ranges from $5,900,000 to $9,900,000. Ten awards are expected per year.

*Requirements:* This is an unrestricted solicitation. Small businesses are encouraged to propose on all or any part of this solicitation.

*Geographic Focus:* All States

*Date(s) Application is Due:* Sep 30

*Contact:* Angela G. Lauofo; (937) 656-5937; Angela.Lauofo@wpafb.af.mil
Ronald Yates, Technical Specialist; (937) 255-7587; Ronald.Yates@wpafb.af.mil

*Internet:* www.grants.gov/search-grants.html?agencies%3DDOD|Department%20of%20Defense

*Sponsor:* U.S. Department of Defense
1400 Defense Pentagon
Washington, D.C. 20301-1400

**USDD Water Desalination and Purification Grants** **2749**

The Office of Naval Research (ONR) Water Desalination/Purification program is currently an applied research and development program focusing on maturing promising technologies for incorporation into prototype desalination systems. The goal is to improve the Navy desalination capability in terms of increased operational availability in littoral waters, reduced maintenance, decreased size and weight, and reduced energy usage. The systems must be designed for the operational constraints of Navy ships as detailed in ONR BAA 09-013: Component Development for Advanced Shipboard Desalination. The fiscal year will consist of two-year efforts to mature promising component level technologies followed by development of complete shipboard desalination systems. Though the program is quite applied now, ONR involvement derives from prior basic research on the development of novel membranes for shipboard wastewater treatment (solid and liquid waste treatment). Submission of the Research Proposal should always be preceded by submission of a Preproposal and a discussion of that Preproposal with the assigned Program Officer. Applications for the upcoming fiscal year should be received by September 30.

*Requirements:* Grants are awarded to academic, industrial, and nonprofit organizations, with strong emphasis on university research. Investigators are encouraged to contact the staff prior to formally submitting a proposal.

*Restrictions:* New grants or contracts are typically started near the beginning of the fiscal year. New proposals for the next fiscal year should be submitted no later than September 30, unless otherwise directed. Competitive renewals should be submitted no later than six months prior to the award termination date after contacting the Program Officer. Conference proposals should be submitted at least four months prior to the conference.

*Geographic Focus:* All States

*Date(s) Application is Due:* Sep 30

*Contact:* Dr. Paul Armistead, Program Officer; (703) 696-4315; fax (703) 696-0001; paul.armistead@navy.mil

*Internet:* www.onr.navy.mil/Science-Technology/Departments/Code-33/All-Programs/332-naval-materials/Water-Desalination-Purification.aspx

*Sponsor:* U.S. Department of Defense
One Liberty Center, 875 N. Randolph Street, Suite 1425
Arlington, VA 22203-1995

**USDD Workplace Violence in the Military Grants** **2750**

The Defense Health Program is seeking applications for the Psychological Health and Traumatic Brain Injury (PH/TBI) Research Program for Workplace Violence in the Military. The PH/TBI Research Program was established to complement ongoing Department of Defense (USDD) efforts towards promoting optimal care for PH (including workplace violence across the USDD service areas) and TBI in the areas of prevention, detection and intervention. United States warfighters endure increasingly demanding and high tempo conditions both in garrison and in the combat field of operations in order to keep pace with ongoing wartime mission requirements. In addition to the loss of lives overseas, warfighters have lost their lives on an Army post stemming from acts of violence in the workplace. The tragic shooting of military personnel at Fort Hood in November 2009 highlighted the need for the USDD to thoroughly review its approach to force health protection and to broaden their historical focus on hostile external threats. The USDD force protection policies, programs, and procedures need to be improved and integrated into a cohesive and comprehensive approach. This Grant program stems directly from recommendations yielded from the Independent Review Related to Fort Hood, which determined that USDD programs, policies, process and procedures that address identification of indicators for violence and radicalization are

outdated, incomplete, and fail to include key indicators of potentially violent behaviors. It was also determined that currently the USDD has no risk assessment system available to supervisors and commanders to assist them in the identification, mitigation, and tracking of internal threats. The Independent Review outlined actions needed to identify behavioral indicators of violence. Applications are sought to support a retrospective study and a prospective study across USDD service areas to develop a set of empirical behavioral-based predictors of potential violence. Applications may include a combined, two-tiered approach that would first involve the retrospective study with those results informing the design of the prospective study. Submitters may also choose to propose only a retrospective or a prospective study. Approximately $8M is available, and five awards are anticipated. Pe-applications should be submitted by February 27.
*Requirements:* Applications from investigators within the military services and other federal agencies are encouraged, as are applications involving multidisciplinary collaborations among academia, industry, the military services, the Department of Veterans Affairs, and other federal government agencies. A principal investigator must be an independent investigator at any academic level from government, academia, research institutions, industry, and private foundations that possess the skills, knowledge, and resources necessary to carry out the proposed research.
*Geographic Focus:* All States
*Date(s) Application is Due:* Apr 26
*Amount of Grant:* 3,000,000 USD
*Contact:* Grants Manager; (703) 674-2500; programannouncements@tatrc.org.
*Sponsor:* U.S. Department of Defense - 1
1400 Defense Pentagon
Washington, D.C. 20301-1400

**USDoD Applied and Computational Analysis Grants** **2751**
The Office of Naval Research (ONR) Applied Computational Analysis Program develops and exploits methods in modern and classical mathematical analysis with emphasis on mathematical and computational models of physical phenomena. Naval interest in waves, flows, materials, structures, and information processing motivates research in the areas of multi-phase, multi-physics, multi-material problems; fluid transport and mixing; predictability of models for nonlinear dynamics; and inverse problems arising from acoustic and electromagnetic wave propagation and scattering. The Applied Computational Analysis Program currently has three thrusts: Multi-Scale/Multi-Physics Modeling; Dynamical Systems and Oceanic and Atmospheric Modeling; and Inverse Methods. Proposers are encouraged to contact the program officer to discuss their research interest prior to the submission of formal proposals. The annual deadline is September 30.
*Requirements:* Grants are awarded to academic, industrial, and nonprofit organizations, with strong emphasis on university research.
*Restrictions:* New grants or contracts are typically started near the beginning of the fiscal year (October 1). New proposals for the next fiscal year should be submitted no later than June 1, unless otherwise directed by the BAA. Competitive renewals should be submitted no later than six months prior to the award termination date after contacting the Program Officer. Conference proposals should be submitted at least four months prior to the conference.
*Geographic Focus:* All States
*Date(s) Application is Due:* Jun 1
*Contact:* Dr. Reza Malek-Madani, Program Officer; (703) 696 4314 or (703) 696-5031; fax (703) 696-5940; reza.malekmadani@navy.mil
*Internet:* www.onr.navy.mil/Science-Technology/Departments/Code-31/All-Programs/311-Mathematics-Computers-Research/Applied-Computational-Analysis.aspx
*Sponsor:* U.S. Department of Defense
One Liberty Center, 875 N. Randolph Street, Suite 1425
Arlington, VA 22203-1995

**USDoD Assured Cyber Effects Grants** **2752**
The Office of Naval Research (ONR) Assured Cyber Effects program focuses on developing new approaches in cyber and information technology to support tactical decision makers with agile and responsive battlespace applications that can be achieved via rapidly evolving cyberspace. This is achieved by producing full-spectrum cyber effects that encompass defensive and other capabilities to assure access to and integrity of all information in the battlespace. Specific research foci are centered around tactical and real-time cyber capabilities that can be delivered from both manned and unmanned Naval platforms. Technologies sought are ones that support both real-time hardware and software, sensing, monitoring, analysis and technical applications that protect our information enterprise and also enable our warfighters to dominate adversary information systems. Research ideas, building upon fundamental research in computer and information sciences for ensuring the integrity of the U.S. Naval information enterprise including the protection of our hull, mechanical and electrical systems are being sought. The cyber space domain that U.S. Naval Forces operate in requires advances in cyber that enable our warfighters to remain persistently ahead of the adversary. Research in methods to assure access to cyberspace and the integrity of command and control advancing research on the convergence of cyber and electronic maneuver are desired.
*Requirements:* Grants are awarded to academic, industrial, and nonprofit organizations, with strong emphasis on university research.
*Restrictions:* New grants or contracts are typically started near the beginning of the fiscal year. New proposals for the next fiscal year should be submitted no later than September 30, unless otherwise directed by the BAA. Competitive renewals should be submitted no later than six months prior to the award termination date after contacting the Program Officer. Conference proposals should be submitted at least four months prior to the conference.
*Geographic Focus:* All States
*Date(s) Application is Due:* Sep 30
*Contact:* Gary Toth, Program Officer; (703) 696 4961 or (703) 696-5031; fax (703) 696-5940; gary.toth@navy.mil
*Internet:* www.onr.navy.mil/Science-Technology/Departments/Code-31/All-Programs/311-Mathematics-Computers-Research/Assured-Cyber-Effects.aspx
*Sponsor:* U.S. Department of Defense
One Liberty Center, 875 N. Randolph Street, Suite 1425
Arlington, VA 22203-1995

**USDoD Automated Image Understanding Thrust Grants** **2753**
The objective of Automated Image Understanding thrust is to further develop promising computational methods based on principled approaches for building reliable vision systems. The primary focus is on high-level vision, namely, recognizing objects and activities, and inferring intentions and threats in surveillance imagery, as well as for semantic search of large image and video databases. Suggested research topics of interest include the following: methods for integration of high-level knowledge and image data, in particular hierarchical and compositional models for representation and processing, and reasoning; learning complex visual concepts, in particular learning relationships among objects, activities involving interactions of people and objects/spaces, etc; using non-visual information to provide additional context for recognition and inference; reconstruction of "scenes and events" from ad hoc networks of heterogeneous cameras, where cameras may be still or video, stationary or mobile, with different resolutions and image quality, and with images taken from widely different views; and methods for analyzing images from other modalities and fusion of multi-modal imagery for enhanced recognition.
*Requirements:* Grants are awarded to academic, industrial, and nonprofit organizations, with strong emphasis on university research.
*Restrictions:* New grants or contracts are typically started near the beginning of the fiscal year. New proposals for the next fiscal year should be submitted no later than September 30, unless otherwise directed by the BAA. Competitive renewals should be submitted no later than six months prior to the award termination date after contacting the Program Officer. Conference proposals should be submitted at least four months prior to the conference.
*Geographic Focus:* All States
*Date(s) Application is Due:* Sep 30
*Contact:* Dr. Behzad Kamgar-Parsi, Program Officer; (703) 696-5754 or (703) 696-5031; fax (703) 696-2611; behzad.kamgarparsi@navy.mil
*Internet:* www.onr.navy.mil/Science-Technology/Departments/Code-31/All-Programs/311-Mathematics-Computers-Research/Computational-Methods-Decision-Making.aspx
*Sponsor:* U.S. Department of Defense
One Liberty Center, 875 N. Randolph Street, Suite 1425
Arlington, VA 22203-1995

**USDoD Command and Control Grants** **2754**
The Office of Naval Research (ONR) Command and Control program focuses on developing new approaches to support tactical decision makers and information superiority. This is performed by producing distributed situation assessment of a commander's environment through development of a common tactical picture; developing collaborative, distributed mission plans; monitoring and dynamically replanning mission execution as required; understanding their environment by being able to access distributed data bases intelligently through use of software agents; effectively using sensor information and making resource allocation decisions and information operations. Specific research involves real-time computing, decision aids and collaborative workspaces; secure distributed architecture; and information warfare techniques to protect secure information. The Command and Control program has several focus areas that build upon the central theme of providing distributed fusion, information processing and information management within a service-oriented architecture, including: Management of Coordinated Electronic Surveillance; Management of Coordinated Electronic Surveillance Higher Level Situation and Knowledge-Based Services; Data Workflow Management Across Globally Netted Maritime Headquarters With Maritime Operations Center; and Dynamic Management Across Wide Variety of Networks. The annual deadline is ongoing.
*Requirements:* Grants are awarded to academic, industrial, and nonprofit organizations, with strong emphasis on university research.
*Restrictions:* New grants or contracts are typically started near the beginning of the fiscal year (October 1). New proposals for the next fiscal year should be submitted no later than September 30, unless otherwise directed by the BAA. Competitive renewals should be submitted no later than six months prior to the award termination date after contacting the Program Officer. Conference proposals should be submitted at least four months prior to the conference.
*Geographic Focus:* All States
*Date(s) Application is Due:* Sep 30
*Contact:* Dr. Gary Toth, Program Officer; (703) 696-4961 or (703) 696-5031; fax (703) 696-2611; gary.toth@navy.mil
*Internet:* www.onr.navy.mil/Science-Technology/Departments/Code-31/All-Programs/311-Mathematics-Computers-Research/Command-Control.aspx
*Sponsor:* U.S. Department of Defense
One Liberty Center, 875 N. Randolph Street, Suite 1425
Arlington, VA 22203-1995

**USDoD Communications and Networking Grants** **2755**
The Office of Naval Research (ONR) Communications and Networking Program supports basic and applied research and development of antennas, radio communications and wireless networking relevant to naval applications. Interest areas in antenna technology include electrically small antennas, wideband multifunction antennas, compatibility of phased array antennas with naval platforms and marine environments, directional beam forming/steering techniques, and special-purpose submarine communication antenna

systems. Areas of interest in radio communications include anti-jam and low-probability-of-intercept techniques, SATCOM performance enhancements, interference mitigation, adaptive equalization, bandwidth efficient modulation, cognitive radio for dynamic spectrum management, and high data rate tactical communications techniques including communications at speed and depth (for submarines). Areas of interest in wireless networks include mobile ad-hoc wireless networking algorithms/protocols, end-to-end Quality-of-Service, joint/coalition interoperability, service oriented tactical networking, mission-based policy and network controls and management. Submission of the Research Proposal should always be preceded by submission of a Pre-proposal and a discussion of that Pre-proposal with the assigned Program Officer.
*Requirements:* Grants are awarded to academic, industrial, and nonprofit organizations, with strong emphasis on university research.
*Restrictions:* New grants or contracts are typically started near the beginning of the fiscal year (October 1). New proposals for the next fiscal year should be submitted no later than June 1, unless otherwise directed by the BAA. Competitive renewals should be submitted no later than six months prior to the award termination date after contacting the Program Officer. Conference proposals should be submitted at least four months prior to the conference.
*Geographic Focus:* All States
*Date(s) Application is Due:* Jun 1
*Contact:* Santanu Das, Program Officer; (703) 588-1036 or (703) 588-2427; fax (703) 696-5940; santanu.das@navy.mil
*Internet:* www.onr.navy.mil/Science-Technology/Departments/Code-31/All-Programs/311-Mathematics-Computers-Research/Communications-Networking.aspx
*Sponsor:* U.S. Department of Defense
One Liberty Center, 875 N. Randolph Street, Suite 1425
Arlington, VA 22203-1995

**USDoD Cyber Security and Complex Software Systems Grants** 2756
The Office of Naval Research Software and Computer Systems program focuses on developing the fundamental principles and models for the design and construction of complex software systems that meet required assurances for security, safety, reliability and performance. It builds upon foundations of mathematical logics and computational complexity but takes into account the uncertainties and constraints of the operational environment. The overall goal is to measurably improve the ability of complex software-intensive systems to meet the information-processing challenges of future naval systems. Cyber Security sub-program investigates and develops wide-ranging principles and techniques for continuously maintaining confidentiality, integrity, and availability of information and information infrastructures, focusing on the software, the hardware and the network. The program heavily emphasizes on automation and autonomy in the cyber environment, preferring human-on-the-loop paradigm over as human-in-the-loop, whenever possible. The program seeks to establish an autonomic, secure and dependable informational infrastructure toward achieving Information Dominance. Complex Software sub-program investigates principles, algorithms and methods, and develops software engineering tools for achieving efficient, timely, robust and secure software executables, focusing on science for software construction, correctness and efficiency by revisiting software development and deployment methodology. Efficient, timely, robust and secure software is a requirement for secure information infrastructure toward achieving Information Dominance. Proposers are encouraged to contact the program officer to discuss their research interest prior to the submission of formal proposals. The annual deadline for each upcoming fiscal year is September 30.
*Requirements:* Grants are awarded to academic, industrial, and nonprofit organizations, with strong emphasis on university research.
*Restrictions:* New grants or contracts are typically started near the beginning of the fiscal year (October 1). New proposals for the next fiscal year should be submitted no later than September 30, unless otherwise directed by the BAA. Competitive renewals should be submitted no later than six months prior to the award termination date after contacting the Program Officer. Conference proposals should be submitted at least four months prior to the conference.
*Geographic Focus:* All States
*Date(s) Application is Due:* Sep 30
*Contact:* Dr. J. Sukarno Mertoguno, Program Officer; (703) 696-0107 or (703) 696-5031; fax (703) 696-5940; sukarno.mertoguno@navy.mil
*Internet:* www.onr.navy.mil/Science-Technology/Departments/Code-31/All-Programs/311-Mathematics-Computers-Research/Cyber-Security-Software.aspx
*Sponsor:* U.S. Department of Defense
One Liberty Center, 875 N. Randolph Street, Suite 1425
Arlington, VA 22203-1995

**USDoD Defense University Research Instrumentation Program Grants** 2757
The Defense University Research Instrumentation Program (DURIP) is a multi-agency DoD program within the University Research Initiative designed to improve the capabilities of U.S. institutions of higher education to conduct research and to educate scientists and engineers in areas important to national defense by providing funds for the acquisition of research equipment. Each year that DURIP is in effect, a complete announcement will provide detailed program information, including the deadline date for submitting proposals. DURIP proposals submitted to ONR should facilitate research in an area of interest to ONR, as described in the Science and Technology section of the ONR Website. Potential proposers may contact the appropriate program managers, listed with each area of interest, to explore possible mutual interests before submitting proposals. Proposals are evaluated competitively. The evaluation criteria are listed in the current program announcement. Proposals may request $50,000 to $1,000,000. Full proposals will be accepted until July 22.
*Requirements:* Grants are awarded to academic, industrial, and nonprofit organizations, with strong emphasis on university research.
*Geographic Focus:* All States
*Date(s) Application is Due:* Jul 22
*Amount of Grant:* 50,000 - 1,000,000 USD
*Contact:* Paula Barden, (703) 696-5031; fax (703) 696-5940; paula.barden.ctr@navy.mil
*Internet:* www.onr.navy.mil/Science-Technology/Directorates/office-research-discovery-invention/Sponsored-Research/University-Research-Initiatives/DURIP.aspx
*Sponsor:* U.S. Department of Defense
One Liberty Center, 875 N. Randolph Street, Suite 1425
Arlington, VA 22203-1995

**USDoD Electronic Warefare Grants** 2758
The Office of Naval Research (ONR) Electronic Warfare Program goal is to control the electromagnetic spectrum (EMS) by exploiting, deceiving, or denying enemy use of the spectrum while ensuring its use by friendly forces. To that end, ONR's EW Technology Program provides naval forces (including Navy and Marine Corps) with improved threat warning systems; electronic warfare support (ES); decoys and countermeasures against weapon tracking and guidance systems; electronic attack (EA) against adversary command, control, communications, computers, intelligence, surveillance, and reconnaissance (C4ISR); and electronic protection (EP) of our own weapons and C4ISR from intentional and unintentional interference. Proposers are encouraged to contact the program officer to discuss their research interest prior to the submission of formal proposals. The annual deadline for each upcoming fiscal year is September 30.
*Requirements:* All responsible sources from academia and industry are eligible. Historically black colleges and universities and minority institutions are encouraged to apply. Teams are encouraged and may submit proposals in any and all areas.
*Restrictions:* New grants or contracts are typically started near the beginning of the fiscal year (October 1). New proposals for the next fiscal year should be submitted no later than September 30, unless otherwise directed by the BAA. Competitive renewals should be submitted no later than six months prior to the award termination date after contacting the Program Officer. Conference proposals should be submitted at least four months prior to the conference.
*Geographic Focus:* All States
*Date(s) Application is Due:* Sep 30
*Contact:* Dr. Bradley Binder, Program Officer; (703) 696-5031; fax (703) 696-5940; bradley.binder@navy.mil
*Internet:* www.onr.navy.mil/Science-Technology/Departments/Code-31/All-Programs/312-Electronics-Sensors/Electronic-Warfare.aspx
*Sponsor:* U.S. Department of Defense
One Liberty Center, 875 N. Randolph Street, Suite 1425
Arlington, VA 22203-1995

**USDoD Faculty Sabbatical Leave Program Grants** 2759
The Faculty Sabbatical Leave Program provides science and engineering faculty the opportunity to conduct research at Navy laboratories while on sabbatical leave. Participants in the Sabbatical Leave Program will receive a monthly stipend making up the difference between salary and sabbatical leave pay from their home institution. In addition, participants will receive reimbursement for travel to and from the laboratory site and a relocation allowance for those who must relocate their residence during their sabbatical leave tenure. Sabbatical leave appointments are contingent on the availability of sabbatical leave positions. Appointments are for a minimum of one semester and a maximum of one year. This is a residential sabbatical - participants must conduct research on site.
*Requirements:* Faculty should submit applications at least six months prior to their proposed sabbatical leave starting date.
*Geographic Focus:* All States
*Contact:* Maria E. Honeycutt, (703) 696-5031; maria.e.honeycutt.ctr@navy.mil
*Internet:* www.onr.navy.mil/Education-Outreach/Summer-Faculty-Research-Sabbatical.aspx
*Sponsor:* U.S. Department of Defense
One Liberty Center, 875 N. Randolph Street, Suite 1425
Arlington, VA 22203-1995

**USDoD Fires Thrust Grants** 2760
The Office of Naval Research (ONR) Fires thrust seeks to enable warfighters employed in small, distributed units with tools to locate and decisively destroy larger enemy forces by applying timely, reliable, precise and accurate fires from a myriad of platforms. Products will equip warfighters with integrated, light-weight optics and sensors to see through all battlefield conditions (day, night, low-light, obscuration) and light-weight, organic, advanced weapons for the rapid, accurate, effective application of firepower. Productions would also equip tactical units frequently operating well beyond conventional parameters of direct fire mutual support. The increased intelligence capabilities delivered by company intelligence cells will generate more potential targets in the future. Technology investment areas include: developing targeting technologies (detection, locating, identification, designation, and tracking) for faster, more precise engagements, while simplifying tasks for the warfighter (specific interests include: precision location of targets, reduction of ballistic flight error, and eye-safe, low-power, designation technology); providing more capable, lighter weight ammunition across the spectrum of lethality (to include contained collateral damage situations); with increased reliability, range and precision; for direct and indirect fire weapons, small arms through major caliber; with low costs and logistics burden, and increased operating, transportation and storage safety; and increased shelf life with reduced storage maintenance requirements (specific interests include affordable guided flight control, improved first round Probability of Kill (Pk) and novel seeker technologies); increasing the capabilities and reducing the weight of the Marine Ground

Combat Element (GCE) through Marine Rifle Company organic weapons systems and Navy Expeditionary Combat Command organic weapons systems (specific interests include: reducing weight and cost, increasing range and safety, and extending the service life); and investigating the fundamental science of energetic materials that will lead to the development of ammunition propellants with improved performance and minimized safety hazards. Specific interests include: improving timing and initiation behavior; reducing shock, impact, storage and transportation hazards; furthering the understanding of material properties at the nanoscale. The Expeditionary Maneuver and Combating Terrorism Department welcomes innovative ideas and white papers in response to the ONR long-range broad agency announcement.
*Requirements:* Grants are awarded to academic, industrial, and nonprofit organizations, with strong emphasis on university research.
*Restrictions:* New grants or contracts are typically started near the beginning of the fiscal year (October 1). New proposals for the next fiscal year should be submitted no later than September 30, unless otherwise directed by the BAA. Competitive renewals should be submitted no later than six months prior to the award termination date after contacting the Program Officer. Conference proposals should be submitted at least four months prior to the conference.
*Geographic Focus:* All States
*Contact:* Dan Simons, Thrust Manager; (703) 696 4840 or (703) 696-5031; fax (703) 696-5940; dan.simons@navy.mil
*Internet:* www.onr.navy.mil/Science-Technology/Departments/Code-30/All-Programs/Fires.aspx
*Sponsor:* U.S. Department of Defense
One Liberty Center, 875 N. Randolph Street, Suite 1425
Arlington, VA 22203-1995

**USDoD Force Protection Thrust Grants** **2761**
The Office of Naval Research (ONR) Force Protection thrust seeks to develop and mature technologies that provide protection from myriad modes of enemy attack throughout the spectrum of warfare, including concepts such as asymmetric and irregular warfare and distributed operations which concentrate on the small unit (battalion and below) and individual warfighter levels. End products will include protective systems expeditionary in nature, light-weight and capable of providing a far greater degree of performance than any comparable system currently available. The functional areas of investigation are explosive hazard defeat through detection, breaching and neutralization of all explosive hazards, counter sniper, counter rocket, artillery, and mortar (CRAM), counter-bomber and personal protective equipment. Technology investment areas include detection, neutralization and mitigation. The Expeditionary Maneuver and Combating Terrorism Department welcomes innovative ideas and white papers in response to the ONR long-range broad agency announcement.
*Requirements:* Grants are awarded to academic, industrial, and nonprofit organizations, with strong emphasis on university research.
*Restrictions:* New grants or contracts are typically started near the beginning of the fiscal year. New proposals for the next fiscal year should be submitted no later than September 30, unless otherwise directed by the BAA. Competitive renewals should be submitted no later than six months prior to the award termination date after contacting the Program Officer. Conference proposals should be submitted at least four months prior to the conference.
*Geographic Focus:* All States
*Contact:* Lee Mastroianni, Thrust Manager; (703) 696-3073 or (703) 696-5031; fax (703) 696-5940; lee.mastroianni@navy.mil
*Internet:* www.onr.navy.mil/Science-Technology/Departments/Code-30/All-Programs/Force-Protection.aspx
*Sponsor:* U.S. Department of Defense
One Liberty Center, 875 N. Randolph Street, Suite 1425
Arlington, VA 22203-1995

**USDoD Human Performance Training and Education Grants** **2762**
The Office of Naval Research (ONR) Human Performance Training and Education (HPT&E) thrust seeks to develop, evaluate and deliver scientifically proven methodologies and technologies that enable the cognitive and physical superiority of Marines. This thrust seeks to develop technologies, architectures and systems that accelerate the training at the individual and team level by enhancing physical and cognitive preparedness. Efforts are focused on providing tools, technologies and methodologies to support the overall goal of either the physical or cognitive technology areas of investment by advancing the delivery of their science into systems for use by the warfighter. In the area of physical performance, ONR's HPT&E thrust is focused on technologies that provide protection against disease and injury as well as greater resilience and optimized physical and physiological performance across the spectrum of conflict at both individual and team levels. This technology area of investment includes: Warfighter Physiology - conducting assessments of the physiological impact of extended full-spectrum conflicts in order to develop methods for mitigating the physiological decline inherent to these types of operations; and Physical Preparedness - pursuing technologies and methodologies for improving and sustaining the warfighter's physical readiness. In the area of cognitive performance, HPT&E's thrust is concentrated on developing human performance training technologies that can train warriors faster and to a higher standard, and resulting in increased retention. It also pursues technologies that improve the ability of Marines to perform cognitively demanding mission-specific tasks, including: Adaptive (Dynamic) Learning - assessing the development of the trainee's cognition and behavior in real time in order to adapt the training environment to the individual and team; Advanced Language Learning - expanding ONR's understanding of how the brain develops during language learning in order to leverage this knowledge to create better training methods, enhanced skill retention and to incorporate socioeconomic knowledge; Experiential Learning - developing first-person immersive training that can provide the visual, auditory, tactile and olfactory stimuli of a complex battlefield within all spectrum of warfare, including culturally realistic, reactive, dynamic and synthetic entities that allow realistic interaction with civilians and opposing forces; and Perception and Perceptuomotor - developing tactically relevant perceptual skills and control interfaces in order to enhance situational awareness.
*Requirements:* Grants are awarded to academic, industrial, and nonprofit organizations, with strong emphasis on university research.
*Restrictions:* New grants or contracts are typically started near the beginning of the fiscal year (October 1). New proposals for the next fiscal year should be submitted no later than June 1, unless otherwise directed by the BAA. Competitive renewals should be submitted no later than six months prior to the award termination date after contacting the Program Officer. Conference proposals should be submitted at least four months prior to the conference.
*Geographic Focus:* All States
*Contact:* Peter Squire, Thrust Manager; (703) 696-0407 or (703) 696-5031; fax (703) 696-5940; peter.squire@navy.mil
*Internet:* www.onr.navy.mil/Science-Technology/Departments/Code-30/All-Programs/Human-Performance-Training.aspx
*Sponsor:* U.S. Department of Defense
One Liberty Center, 875 N. Randolph Street, Suite 1425
Arlington, VA 22203-1995

**USDoD Image Understanding Thrust Grants** **2763**
The goal of Image Understanding thrust is to develop theory and algorithms for understanding surveillance imagery, for semantic search of visual datasets, and for autonomous agent perception. The main focus is on reconstructing 3D scenes, recognizing object classes and specific objects, recognizing activities and events, inferring intentions, as well as succinct natural language descriptions of images and video. Of particular interest is developing visual representations, methods for building visual knowledge bases optimized for inference, and methods for integrating reasoning with high-level knowledge and image data. Proposers are encouraged to contact the program officer to discuss their research interest prior to the submission of formal proposals. The annual deadline for each upcoming fiscal year is September 30.
*Requirements:* Grants are awarded to academic, industrial, and nonprofit organizations, with strong emphasis on university research.
*Restrictions:* New grants or contracts are typically started near the beginning of the fiscal year. New proposals for the next fiscal year should be submitted no later than September 30, unless otherwise directed by the BAA. Competitive renewals should be submitted no later than six months prior to the award termination date after contacting the Program Officer. Conference proposals should be submitted at least four months prior to the conference.
*Geographic Focus:* All States
*Date(s) Application is Due:* Sep 30
*Contact:* Dr. Behzad Kamgar-Parsi, Program Officer; (703) 696-5754 or (703) 696-5031; fax (703) 696-2611; behzad.kamgarparsi@navy.mil
*Internet:* www.onr.navy.mil/Science-Technology/Departments/Code-31/All-Programs/311-Mathematics-Computers-Research/Machine-Learning-Reasoning-Intelligence.aspx
*Sponsor:* U.S. Department of Defense
One Liberty Center, 875 N. Randolph Street, Suite 1425
Arlington, VA 22203-1995

**USDoD Information Integration Thrust Grants** **2764**
The Office of Naval Research (ONR) Information Integration program is concerned with basic research in probability (a mathematical framework to represent the likelihood of phenomena with uncertain outcomes) and statistics (the process of collecting data and extracting and visualizing the information contained in the data). Recently, connection with several areas of mathematics such as analysis, geometry, algebra, combinatorics and theoretical computer science has generated a number of new scientific ideas with important applications. Such synergism will likely be required to develop innovative techniques for solving the challenging problems discussed below. Several focus areas build around the central themes of information representation, discovery, processing and integration, including: Information Representation and Discovery; Information Integration and Data Fusion; Decision Making Under Uncertainty; and Image Analysis for Intelligence, Surveillance, Reconnaissance, Targeting. The annual deadline is September 30.
*Requirements:* Grants are awarded to academic, industrial, and nonprofit organizations, with strong emphasis on university research.
*Restrictions:* New grants or contracts are typically started near the beginning of the fiscal year. New proposals for the next fiscal year should be submitted no later than September 30, unless otherwise directed by the BAA. Competitive renewals should be submitted no later than six months prior to the award termination date after contacting the Program Officer. Conference proposals should be submitted at least four months prior to the conference.
*Geographic Focus:* All States
*Date(s) Application is Due:* Sep 30
*Contact:* Dr. Predrag Neskovic, Program Officer; (703) 696 4304 or (703) 696-5031; fax (703) 696-5940; predrag.neskovic@navy.mil
*Internet:* www.onr.navy.mil/Science-Technology/Departments/Code-31/All-Programs/311-Mathematics-Computers-Research/Computational-Methods-Decision-Making.aspx
*Sponsor:* U.S. Department of Defense
One Liberty Center, 875 N. Randolph Street, Suite 1425
Arlington, VA 22203-1995

**USDoD Integrated Topside Innovative Naval Prototype Grants** 2765

Integrated Topside (InTop) is an Innovative Naval Prototype (INP) program that will develop a scalable family of Electronic Warfare (EW), radar, Signals Intelligence (SIGINT) and communications capability to support multiple classes of ships and other Navy platforms. InTop is utilizing a modular / open RF design to facilitate best-of-breed technology and cost effective upgrades. The InTop vision is to: dominate the RF spectrum; enable innovation through a RF Open Architecture (hardware and software); and create affordable systems that are scalable across platforms. InTop plans to reduce the number of topside apertures present on Navy ships through the use of integrated, multi-function, multi-beam arrays. In the past, the topside design approach was based on developing separate systems and associated antennas for each individual RF function which led to a significant increase in topside antennas. This increase has led to problems with Electro-Magnetic Interference (EMI), Radar Cross Section (RCS) and the overall performance of critical ship EW, sensing and communication functions. InTop builds off of a past ONR effort, the Advanced Multi-Function RF Concept (AMRFC), which was a proof-of-principle demonstration that showed the feasibility of multi-beam, multi-function apertures capable of simultaneously supporting EW, communications and radar functions in two arrays (transmit and receive). The AMRFC test bed also provided a modular architecture where interfaces were sufficiently defined between units (arrays, signal processing, resource manager, etc.) that systems developed by different companies and government labs could be integrated and work together in one overall system.
*Requirements:* Grants are awarded to academic, industrial, and nonprofit organizations, with strong emphasis on university research.
*Restrictions:* New grants or contracts are typically started near the beginning of the fiscal year (October 1). New proposals for the next fiscal year should be submitted no later than September 30, unless otherwise directed by the BAA. Competitive renewals should be submitted no later than six months prior to the award termination date after contacting the Program Officer. Conference proposals should be submitted at least four months prior to the conference.
*Geographic Focus:* All States
*Date(s) Application is Due:* Sep 30
*Contact:* Betsy DeLong, Program Officer; (703) 588-0069 or (703) 696-5031; fax (703) 696-5940; betsy.delong@navy.mil
*Internet:* www.onr.navy.mil/Science-Technology/Departments/Code-31/All-Programs/312-Electronics-Sensors/Integrated-Topside.aspx
*Sponsor:* U.S. Department of Defense
One Liberty Center, 875 N. Randolph Street, Suite 1425
Arlington, VA 22203-1995

**USDoD Intelligence, Surveillance and Reconnaissance Thrust Grants** 2766

The Office of Naval Research (ONR) Intelligence, Surveillance and Reconnaissance (ISR) thrust seeks to develop and leverage advanced technologies for applications in future intelligence, surveillance and reconnaissance systems. It also seeks to enhance situational awareness to enable real-time tactical decision making for distributed operations and provide proactive and predictive capabilities for asymmetric and irregular warfare. ISR's technology investment areas include: Sensor Fields; Relevant and Situational Information on Demand; and Actionable Intelligence for Expeditionary and Irregular Warfare. The Expeditionary Maneuver and Combating Terrorism Department welcomes innovative ideas and white papers in response to the ONR long-range broad agency announcement.
*Requirements:* Grants are awarded to academic, industrial, and nonprofit organizations, with strong emphasis on university research.
*Restrictions:* New grants or contracts are typically started near the beginning of the fiscal year (October 1). New proposals for the next fiscal year should be submitted no later than September 30, unless otherwise directed by the BAA. Competitive renewals should be submitted no later than six months prior to the award termination date after contacting the Program Officer. Conference proposals should be submitted at least four months prior to the conference.
*Geographic Focus:* All States
*Date(s) Application is Due:* Sep 30
*Contact:* Martin Kruger, Thrust Manager; (703) 696-5349 or (703) 696-5031; fax (703) 696-4274; martin.kruger1@navy.mil
*Internet:* www.onr.navy.mil/Science-Technology/Departments/Code-30/All-Programs/Intelligence-Surveillance-ISR.aspx
*Sponsor:* U.S. Department of Defense
One Liberty Center, 875 N. Randolph Street, Suite 1425
Arlington, VA 22203-1995

**USDoD Intelligent for Autonomous Agents Grants** 2767

The Office of Naval Research (ONR) Intelligent and Autonomous Agents program focuses on developing the intelligence needed for agents to function autonomously in a variety of situations. There are three areas of particular interest, including: Building Blocks of Machine Intelligence; Teams of Unmanned Vehicles; and Human-Agent Collaboration. Some suggested topics of interest for Building Blocks of Machine Intelligence are: methods for building knowledge bases from diverse sources; learning complex concepts and tasks from examples, instructions, and demonstrations; reasoning with uncertain and qualitative information, as well as methods for meta-reasoning for self-assessment; planning in large domains in partially known environments and incompletely modeled goals and domains; and intelligent architectures that seamlessly integrate knowledge-bases, learning, reasoning, and planning, for decision-making. Some suggested topics of interest for Teams of Unmanned Vehicles are: computational methods for building decentralized collaborating teams of autonomous agents, in particular agents that are fairly capable in terms of sensing, communication and computational resources; and mathematical theories of swarm control, particularly engineered swarms with desired behaviors. Some suggested topics of interest for Human-Agent Collaboration are: multi-modal, multi-participant, human-agent dialogue systems for seamless interactions that are natural to humans; and computational models of human behavior and decision-making for use by autonomous agents. Proposers are encouraged to contact the Program Officer to discuss their research interest prior to the submission of formal proposals. The annual deadline for each upcoming fiscal year is September 30.
*Requirements:* Grants are awarded to academic, industrial, and nonprofit organizations, with strong emphasis on university research.
*Restrictions:* New grants or contracts are typically started near the beginning of the fiscal year (October 1). New proposals for the next fiscal year should be submitted no later than September 30, unless otherwise directed by the BAA. Competitive renewals should be submitted no later than six months prior to the award termination date after contacting the Program Officer. Conference proposals should be submitted at least four months prior to the conference.
*Geographic Focus:* All States
*Date(s) Application is Due:* Sep 30
*Contact:* Dr. Behzad Kamgar-Parsi, Program Officer; (703) 696-5754 or (703) 588-2427; fax (703) 696-5940; behzad.kamgar-parsi@.navy.mil or 311_IS@onr.navy.mil
*Internet:* www.onr.navy.mil/Science-Technology/Departments/Code-31/All-Programs/311-Mathematics-Computers-Research/Machine-Learning-Reasoning-Intelligence.aspx
*Sponsor:* U.S. Department of Defense
One Liberty Center, 875 N. Randolph Street, Suite 1425
Arlington, VA 22203-1995

**USDoD Logistics Thrust Grants** 2768

The Office of Naval Research (ONR) Logistics thrust seeks to provide Marines of the future with a precisely tailored level of sustained logistic support from sea-based platforms to rapidly maneuver forces ashore. Logistics delivery systems of the future will be more responsive and flexible, enabling Marines to out-pace rapidly changing operational scenarios. Likewise, delivered logistic commodities will provide more operational value per unit weight, enhancing combat unit self-sufficiency and maneuverability. Finally, operational units will benefit from technologies that maximize equipment readiness by minimizing both down-time and maintenance requirements. Technology areas of investment include: Asset Visibility; Logistic Transport; Operational Sustainment; and Maintenance Reduction. Proposers are encouraged to contact the program officer to discuss their research interest prior to the submission of formal proposals.
*Requirements:* Grants are awarded to academic, industrial, and nonprofit organizations, with strong emphasis on university research.
*Restrictions:* New grants or contracts are typically started near the beginning of the fiscal year (October 1). New proposals for the next fiscal year should be submitted no later than September 30, unless otherwise directed by the BAA. Competitive renewals should be submitted no later than six months prior to the award termination date after contacting the Program Officer. Conference proposals should be submitted at least four months prior to the conference.
*Geographic Focus:* All States
*Date(s) Application is Due:* Sep 30
*Contact:* Billy Short, Program Officer; (703) 696-6993 or (703) 696-5031; fax (703) 696-5940; billy.short@onr.navy.mil
*Internet:* www.onr.navy.mil/Science-Technology/Departments/Code-30/All-Programs/Logistics.aspx
*Sponsor:* U.S. Department of Defense
One Liberty Center, 875 N. Randolph Street, Suite 1425
Arlington, VA 22203-1995

**USDoD Maneuver Thrust Grants** 2769

The Office of Naval Research (ONR) Maneuver thrust explores technologies to increase the warfighting capabilities and effectiveness of the Marine Corps Air Ground Task Force with emphasis on improving survivability and enhancing maneuver and maneuver-enabling systems in both decentralized and asymmetric warfare. Marine forces of the future will be significantly more agile, lethal, mobile and survivable. This thrust's technology investment areas include: Survivability - providing survivability within the battlefield in order to support expeditionary maneuver warfare; Advanced Mobility - increasing operational tempo and gaining temporal dominance in order to achieve positional advantage; and Maneuver Enablers - providing maneuver-enabling technologies to separate warfighters from potentially hazardous missions and to increase efficiency and economy of force. Proposers are encouraged to contact the program officer to discuss their research interest prior to the submission of formal proposals.
*Requirements:* Grants are awarded to academic, industrial, and nonprofit organizations, with strong emphasis on university research.
*Restrictions:* New grants or contracts are typically started near the beginning of the fiscal year (October 1). New proposals for the next fiscal year should be submitted no later than September 30, unless otherwise directed by the BAA. Competitive renewals should be submitted no later than six months prior to the award termination date after contacting the Program Officer. Conference proposals should be submitted at least four months prior to the conference.
*Geographic Focus:* All States
*Date(s) Application is Due:* Sep 30
*Contact:* Jeff Bradel; (703) 588-2552 or (703) 696-5031; fax (703) 696-5940; jeff.bradel@navy.mil
*Internet:* www.onr.navy.mil/Science-Technology/Departments/Code-30/All-Programs/Maneuver.aspx
*Sponsor:* U.S. Department of Defense
One Liberty Center, 875 N. Randolph Street, Suite 1425
Arlington, VA 22203-1995

**USDoD Mathematical Data Science Program Grants** 2770

The Office of Naval Research (ONR) Mathematical Data Science program is concerned with basic research in mathematics, probability, statistics, signal processing, machine learning, data engineering, and information theory. The program aims to develop rigorous mathematical and algorithmic answers to questions that are currently addressed using heuristics or non-principled approaches. Program focus areas include: big data; small data; representation, learning, and inference; complex networks; multi-modal, multi-scale information integration; and decision making under uncertainty. Proposers are encouraged to contact the Program Officer to discuss their research interest prior to the submission of formal proposals. The annual deadline for each upcoming fiscal year is September 30.
*Requirements:* Grants are awarded to academic, industrial, and nonprofit organizations, with strong emphasis on university research.
*Restrictions:* New grants or contracts are typically started near the beginning of the fiscal year (October 1). New proposals for the next fiscal year should be submitted no later than September 30, unless otherwise directed by the BAA. Competitive renewals should be submitted no later than six months prior to the award termination date after contacting the Program Officer. Conference proposals should be submitted at least four months prior to the conference.
*Geographic Focus:* All States
*Date(s) Application is Due:* Sep 30
*Contact:* Dr. Predrag Neskovic, Program Officer; (703) 696-5031; fax (703) 696-5940; predrag.neskovic@navy.mil
*Internet:* www.onr.navy.mil/Science-Technology/Departments/Code-31/All-Programs/311-Mathematics-Computers-Research/Mathematics-Data-Science.aspx
*Sponsor:* U.S. Department of Defense
One Liberty Center, 875 N. Randolph Street, Suite 1425
Arlington, VA 22203-1995

**USDoD Mathematical Optimization and Operations Research Grants** 2771

The Office of Naval Research (ONR) Mathematical Optimization and Operations Research program supports basic and applied research in mathematical optimization. The primary focus of the program's basic-research portion is linear, nonlinear, integer and combinatorial optimization. Theoretical development, algorithm design and analysis, computational techniques, and software prototypes for solving large-scale problems are of interest. Techniques that produce provably optimal or near-optimal solutions, as well as techniques applicable to large problem domains are favored. This includes, but is not limited to, cutting-plane and polyhedral techniques for mixed-integer programming and interior-point and first-order algorithms for conic/convex optimization. Innovative techniques for dealing with uncertainty from stochastic optimization, robust optimization and simulation-based optimization are of growing interest. The primary focus of the applied-research portion of the program is on the development of innovative models and solution techniques for Navy-relevant problems in resource allocation, such as those arising in mission planning, command and control, and logistics. Current application areas of interest are maritime mission planning and unmanned sensor management and allocation. New proposals for the next fiscal year should be submitted no later than September 30.
*Requirements:* Grants are awarded to academic, industrial, and nonprofit organizations, with strong emphasis on university research.
*Restrictions:* New grants or contracts are typically started near the beginning of the fiscal year. New proposals for the next fiscal year should be submitted no later than September 30, unless otherwise directed by the BAA. Competitive renewals should be submitted no later than six months prior to the award termination date after contacting the Program Officer. Conference proposals should be submitted at least four months prior to the conference.
*Geographic Focus:* All States
*Date(s) Application is Due:* Sep 30
*Contact:* Dr. Donald Wagner, Program Officer; (703) 696-4313 or (703) 696-5031; fax (703) 696-5940; don.wagner@navy.mil
*Internet:* www.onr.navy.mil/Science-Technology/Departments/Code-31/All-Programs/311-Mathematics-Computers-Research/Mathematical-Optimization.aspx
*Sponsor:* U.S. Department of Defense
One Liberty Center, 875 N. Randolph Street, Suite 1425
Arlington, VA 22203-1995

**USDoD Multidisciplinary Research Program of the University Research Initiatives Grants** 2772

The Multidisciplinary Research Program of the University Research Initiatives (MURI) is a multi-agency DoD program that supports research teams whose efforts intersect more than one traditional science and engineering discipline. Multidisciplinary team effort can accelerate research progress in areas particularly suited to this approach. Multidisciplinary research also can help to hasten the transition of research findings to practical application. MURI awards are made in research topics specified by the participating defense agencies each year that the program is in force. Specified topics change each year. Awards are typically for a period of three years (funded incrementally or as options) with two additional years possible as options to bring the total award to five-years, and at a funding level ranging from half a million to about a million dollars per year, with the size of the award dependent upon the topic, technical goals, and availability of appropriations. White papers are due on August 1, and full proposals are due on November 15. Submission of the Research Proposal should always be preceded by submission of a pre-proposal and a discussion of that pre-proposal with the assigned Program Officer.
*Requirements:* Grants are awarded to academic, industrial, and nonprofit organizations, with strong emphasis on university research.
*Geographic Focus:* All States
*Date(s) Application is Due:* Nov 15
*Contact:* Paula Barden, (703) 696-5031; fax (703) 696-5940; paula.barden.ctr@navy.mil
*Internet:* www.onr.navy.mil/Science-Technology/Directorates/office-research-discovery-invention/Sponsored-Research/University-Research-Initiatives/MURI.aspx
*Sponsor:* U.S. Department of Defense
One Liberty Center, 875 N. Randolph Street, Suite 1425
Arlington, VA 22203-1995

**USDoD Nanoscale Computing Devices and Systems Grants** 2773

The Office of Naval Research (ONR) Nanoscale Computing Devices and Systems (Nanoelectronics) Program objective is to study the electronic, optical and magnetic properties of, and potential device functionalities in, nanometer scale materials and structures. In order to accomplish this objective, the program will address science and technology challenges at multiple levels: material level - synthesis, characterization and understanding of novel nanomaterials with desired functionalities is a challenging task; device and circuit level - need to invent and demonstrate entirely new classes of devices and circuits that fully utilize the properties of nanomaterials; architecture level - conventional von Neumann architecture is not scalable; and fabrication and manufacturing - as we approach sub-10nm scale, current fabrication techniques have difficulty with resolution and/or throughput Potential proposers are encouraged to communicate with the program officer prior to white paper submission. Proposers are encouraged to contact the program officer to discuss their research interest prior to the submission of formal proposals. White papers are preferred by April 30 of each year. Typically awards range from $50,000 to $400,000.
*Requirements:* Grants are awarded to academic, industrial, and nonprofit organizations, with strong emphasis on university research.
*Restrictions:* Competitive renewals should be submitted no later than six months prior to the award termination date after contacting the Program Officer.
*Geographic Focus:* All States
*Date(s) Application is Due:* Apr 30
*Amount of Grant:* 50,000 - 400,000 USD
*Contact:* Dr. Chagaan Baatar, Program Officer; (703) 696-0483 or (703) 588-2427; fax (703) 696-5940; chagaan.baatar@navy.mil
*Internet:* www.onr.navy.mil/Science-Technology/Departments/Code-31/All-Programs/312-Electronics-Sensors/Nanoscale-Electronics.aspx
*Sponsor:* U.S. Department of Defense
One Liberty Center, 875 N. Randolph Street, Suite 1425
Arlington, VA 22203-1995

**USDoD National Defense Science and Engineering Graduate Fellowships** 2774

The National Defense Science and Engineering Graduate Fellowship program is a joint program of the United States Army, Navy and Air Force within the University Research Initiative (URI), designed to increase the number of U.S. citizens trained in disciplines of science and engineering important to defense goals. DoD awards approximately 100 to 150 new three-year graduate fellowships each year to individuals for study and research leading to doctoral degrees in, or closely related to, the disciplines of aeronautical and astronautical engineering; biosciences, chemical engineering; chemistry; cognitive, neural, and behavioral sciences; electrical engineering; geosciences; civil engineering; computer and computational sciences; materials science and engineering; mathematics; mechanical engineering; naval architecture and ocean engineering; oceanography; and physics. The Fellowships are intended for students at or near the beginning of graduate study. The tenure of an NDSEG Fellowship is 36 months. Consult the current program announcement for stipend levels. In addition to stipends, the DoD pays the Fellow's full tuition and required fees. The deadline for application is in January.
*Requirements:* Applicants must be citizens or nationals of the United States who intend to pursue a Ph.D. in one of the designated fields. Applications are encouraged from women, underrepresented minorities and persons with disabilities. Recipients of new three-year fellowships must begin tenure not later than the fall following award.
*Geographic Focus:* All States
*Contact:* Maria E. Honeycutt, Point of Contact; (703) 696-5031; fax (703) 696-5940; maria.e.honeycutt.ctr@navy.mil or 363_NDSEG@onr.navy.mil
*Internet:* www.onr.navy.mil/Education-Outreach/undergraduate-graduate/NDSEG-graduate-fellowship.aspx
*Sponsor:* U.S. Department of Defense
One Liberty Center, 875 N. Randolph Street, Suite 1425
Arlington, VA 22203-1995

**USDoD Naval Research Enterprise Internships** 2775

The 10-week Naval Research Enterprise Internship program (NREIP) is designed to provide opportunities for undergraduate and graduate students to participate in research, under the guidance of an appropriate research mentor, at a participating Navy laboratory. Interns will be selected based upon academic achievement, personal statements, recommendation, and career and research interests. NREIP stipends are $5,400 for undergraduate sophomores and $8,100 for undergraduate juniors and seniors. Graduate level students receive $10,800. Online application will be open from September 15 to October 23.
*Requirements:* Applicants must be: U.S. citizens (Permanent Resident Alien status is considered at some labs); junior, senior, or graduate students; and enrolled at a participating university.
*Geographic Focus:* All States
*Date(s) Application is Due:* Oct 23
*Amount of Grant:* 5,400 - 10,800 USD
*Contact:* Dr. Michelle Skoorka, Program Manager; (703) 696-5031; fax (703) 696-5940; michelle.skoorka@navy.mil

*Internet:* www.onr.navy.mil/sci_tech/3t/corporate/nreip.asp
*Sponsor:* U.S. Department of Defense
One Liberty Center, 875 N. Randolph Street, Suite 1425
Arlington, VA 22203-1995

**USDoD Office of Naval Research (ONR) Command, Control, Computers and Communication (C4) Thrust Grants** 2776

The Office of Naval Research (ONR) Command, Control, Computers and Communication (C4) Thrust seeks to improve C4 capabilities for naval war fighters with an emphasis on mall units, asymmetric and irregular warfare, information analysis and communication, and istributed operations. Its technology investment areas include network-centric warfare and interoperability, over-the-horizon communications gateways, and small-unit technologies. In the area of network-centric warfare and interoperability, ONR seeks to nearly ubiquitous communications and availability of information for naval warfighters, particularly in austere environments. This technology area of investment includes: mobility management; and information assurance application tolerance. In the area of over-the-horizon communications, ONR seeks to provide long-reach communications enabling technologies that allow lower-echelon warfighters to exploit the global network. This would include beyond-line-of-sight (BLoS) communications, which provide reach-back to higher echelons by all users, and networked RF distribution systems, which would provide access to the Global Information Grid services by appropriate echelons. In the area of small-unit technologies, ONR seeks to close technology gaps to allow small-unit warfighters to gain timely, accurate, and nearly complete situational awareness.
*Requirements:* Grants are awarded to academic, industrial, and nonprofit organizations, with strong emphasis on university research.
*Restrictions:* New grants or contracts are typically started near the beginning of the fiscal year (October 1). New proposals for the next fiscal year should be submitted no later than June 1, unless otherwise directed by the BAA. Competitive renewals should be submitted no later than six months prior to the award termination date after contacting the Program Officer. Conference proposals should be submitted at least four months prior to the conference.
*Geographic Focus:* All States
*Contact:* Captain Ben Pimentel, Thrust Manager; (703) 696-2760 or (703) 696-5031; fax (703) 696-5940; benjamin.a.pimentel1@navy.mil
*Internet:* www.onr.navy.mil/Science-Technology/Departments/Code-30/All-Programs/C4.aspx
*Sponsor:* U.S. Department of Defense
One Liberty Center, 875 N. Randolph Street, Suite 1425
Arlington, VA 22203-1995

**USDoD Science and Engineering Apprentice Program (SEAP)** 2777

The Science and Engineering Apprentice Program (SEAP) places academically talented high school students with interest and ability in science, technology, engineering and mathematics (STEM) as apprentices in Department of Defense (DoD) laboratories for eight weeks during the summer. There are more than 25 Navy laboratories across the country currently participating in the Navy SEAP program. While working closely with research personnel, who act as mentors, the students will experience scientific and engineering practice not available in their school environment. The program offers students a unique and positive experience in their fields of interest and encourages them to pursue STEM related careers. Exposure to the science and engineering activities at DoD laboratories may also encourage a positive image toward the defense community and foster an interest in future employment as a researcher within the DoD. Participation in this program also helps these students serve as positive role models for their peers by encouraging other students to take more science and math courses. Online application will be open from September 15 to October 23.
*Requirements:* Applicants must be U.S. citizens. In some laboratories, security clearance is required.
*Geographic Focus:* All States
*Date(s) Application is Due:* Oct 23
*Amount of Grant:* 3,300 - 3,800 USD
*Contact:* Dr. Michelle Skoorka, Research Directorate Leadership; (703) 696-5031; fax (703) 696-5940; michelle.skoorka@navy.mil
*Internet:* www.onr.navy.mil/Education-Outreach/K-12-Programs/SEAP.aspx
*Sponsor:* U.S. Department of Defense
One Liberty Center, 875 N. Randolph Street, Suite 1425
Arlington, VA 22203-1995

**USDoD Small Business Innovation Research (SBIR) Grants** 2778

The ONR Small Business Innovation Research (SBIR) program is an integral part of the Naval Science and Technology program. SBIR topics address naval needs and should ultimately transition to the fleet through enabling technologies for future naval capabilities or directly to naval acquisition programs. Topics are derived from ONR Program Officers via periodic calls which are issued by the SBIR Program Manager to the ONR Department Heads. Topics are published and solicited with those of other Naval components in the Department of Defense SBIR Program Solicitations. Qualified small businesses may submit proposals for these topics which are then competitively evaluated by Program Officers. Awards are made to the best proposing companies within allotted funding. Funding for this program is not appropriated, but provided annually under the statutory 2-1/2% set-aside from ONR extramural research, development, testing and evaluation (RDT&E) funds. DoD goals are for all Phase Is to be under contract within 4 months of receipt and for all Phase IIs to be under contract within 6 months following the end of the Phase I contract.
*Requirements:* A business may apply that is independently owned and operated; is not dominant in the field of operation in which it is proposing; has its principal place of business located in the United States; is organized for profit; is at least 51 percent owned, or in the case of a publicly owned business, at least 51 percent of its voting stock is owned by United States citizens or lawfully admitted permanent resident aliens; and including its affiliates, has a number of employees amounting to 500 or less.
*Geographic Focus:* All States
*Contact:* Lore-Anne Ponirakis, Program Manager; (703) 696-5031 or (703) 588-2427; fax (703) 696-5940; loreanne.ponirakis@navy.mil
*Internet:* www.onr.navy.mil/Science-Technology/Directorates/Transition/SBIR-STTR.aspx
*Sponsor:* U.S. Department of Defense
One Liberty Center, 875 N. Randolph Street, Suite 1425
Arlington, VA 22203-1995

**USDoD Small Business Technology Transfer (STTR) Program Grants** 2779

Small Business Technology Transfer (STTR) is a mission-oriented science and technology program seeking to produce commercializable products through scientific research areas. The program requires that small businesses and non-profit research institutions submit a joint proposal responsive to broad topics defined in the Navy section of the annual DOD STTR solicitation. Unlike the Navy's SBIR program, the Navy STTR program focuses on a different science or technology area each year. Interested parties may register to receive the DoD Solicitation on-line via the DoD SBIR Program.
*Requirements:* A business may apply that is independently owned and operated; is not dominant in the field of operation in which it is proposing; has its principal place of business located in the United States; is organized for profit; is at least 51 percent owned, or in the case of a publicly owned business, at least 51 percent of its voting stock is owned by United States citizens or lawfully admitted permanent resident aliens; and including its affiliates, has a number of employees amounting to 500 or less.
*Geographic Focus:* All States
*Contact:* Lore-Anne Ponirakis, Program Manager; (703) 696-5031 or (703) 588-2427; fax (703) 696-5940; loreanne.ponirakis@navy.mil
*Internet:* www.onr.navy.mil/Science-Technology/Directorates/Transition/SBIR-STTR.aspx
*Sponsor:* U.S. Department of Defense
One Liberty Center, 875 N. Randolph Street, Suite 1425
Arlington, VA 22203-1995

**USDoD Summer Faculty Research Program Grants** 2780

The Summer Faculty Research program provides science and engineering faculty members from institutions of higher education the opportunity to participate in research at Navy laboratories for a ten week period during the summer. Participants work with professional peers in the Navy laboratories on research tasks of mutual interest. Information about the research interests of the participating laboratories is available via the ASEE website. The research for the summer is defined in advance through correspondence and an optional pre-program visit to the research site, for which funds are provided. Three levels of appointment are available: Summer Faculty Fellow, Senior Summer Faculty Fellow, and Distinguished Summer Faculty Fellow. Stipends vary by level of appointment and are set each year. In addition, each Fellow will be reimbursed for his/her personal travel for an optional pre-program visit to the research site. Stipends range from $1,400 to $1,900 per week for the summer program. Applications will be accepted from September 14 through December 14.
*Requirements:* The ONR Summer Faculty Research Program is open to United States citizens or nationals and to legal permanent residents holding a teaching or research position at a U.S. college or university. Some restrictions may apply to permanent residents regarding eligibility for a DoD security clearance.
*Geographic Focus:* All States
*Date(s) Application is Due:* Dec 14
*Amount of Grant:* 14,000 - 19,000 USD
*Contact:* Maria E. Honeycutt, (703) 696-5031; fax (703) 696-5940; maria.e.honeycutt.ctr@navy.mil
*Internet:* www.onr.navy.mil/Education-Outreach/Summer-Faculty-Research-Sabbatical.aspx
*Sponsor:* U.S. Department of Defense
One Liberty Center, 875 N. Randolph Street, Suite 1425
Arlington, VA 22203-1995

**USDoD Young Investigator Program Grants** 2781

The objectives of the Young Investigator Program (YIP) are to attract to naval research outstanding new faculty members at institutions of higher education, to support their research, and to encourage their teaching and research careers. ONR anticipates making about 24 new awards annually. Awards of up to $170,000 per year for three years, with the possibility of additional support for capital equipment or collaborative research with a Navy laboratory, are made, based on research proposals and supporting materials. Special attention will be given to proposals in naval priority research areas, listed in the announcement.
*Requirements:* Awards under this announcement will be made only to U.S. institutions of higher education that award degrees in science, engineering and/or mathematics. Historically black colleges and universities (HBCUs) and minority institutions (MIs) are encouraged to submit proposals and join others in submitting proposals
*Geographic Focus:* All States
*Date(s) Application is Due:* Dec 1
*Amount of Grant:* 170,000 USD
*Contact:* Reginald G. Williams, Program Manager; (703) 696-5031; fax (703) 696-5940; reginald.g.williams@navy.mil
*Internet:* www.onr.navy.mil/Science-Technology/Directorates/office-research-discovery-invention/Sponsored-Research/YIP.aspx
*Sponsor:* U.S. Department of Defense
One Liberty Center, 875 N. Randolph Street, Suite 1425
Arlington, VA 22203-1995

**USDOE Next Generation Electric Machines: Enabling Technologies Grant** 2782

The DOE Office of Energy Efficiency and Renewable Energy's Advanced Manufacturing Office (AMO) makes research, development, and demonstration (RD&D) investments in advanced manufacturing process and materials technologies. These technologies are foundational, pervasive, and promise crosscutting industrial applications and impact in reducing industry's energy footprint and greenhouse gas (GHG) emissions, as well as supporting the global competitiveness of clean energy products. By targeting the development of energy-related advanced manufacturing technology, AMO's work can create completely new supply chains and stimulate significant economic growth and job creation. AMO's Next Generation Electric Machines (NGEM) program is an RD&D effort leveraging recent technology advancements in power electronics and electric motors to develop a new generation of energy efficient, high power density, high speed, integrated MV drive systems for a wide variety of critical energy applications. Improvements to these systems can be realized through the application of key enabling technologies, such as wide bandgap devices, advanced magnetic materials, improved insulation materials, aggressive cooling techniques, high speed bearing designs, and improved conductors or superconducting materials. Through this development program, NGEM will facilitate a step-change that enables more efficient use of electricity, as well as reduced drive system size and weight, developing lasting capabilities for motor material development, design, and analysis that are cost-shared with industry stakeholders. This specific Financial Opportunity Announcement (FOA) is focused on advancing three key enabling technologies that can take the development of electric machines a giant step further by addressing significant reductions in three major categories of energy losses in an electric machine. The FOA leverages on recent technical advancements made in nanomaterial research, 6.5wt% siliconizing process in electrical steel manufacturing, and improved in-field performance of high temperature superconductor technology to realize the complete vision of developing a new generation of energy efficient, high power density, high speed electric machines for a wide variety of critical energy applications. The expected number of awards is twelve, ranging from $1,000,000 to $6,500,000. Concept papers are due by April 12, with the final application deadline of June 7.

*Requirements:* The eligibility for application is unrestricted.
*Geographic Focus:* All States
*Date(s) Application is Due:* Jun 7
*Amount of Grant:* 1,000,000 - 6,500,000 USD
*Contact:* Isaac Chan, Program Manager; (202) 586-9488 or (720) 356-1800; fax (720) 356-1750; Isaac.Chan@ee.doe.gov or AMONGEM@ee.doe.gov
*Internet:* www.grants.gov/web/grants/view-opportunity.html?oppId=282056
*Sponsor:* U.S. Department of Energy
15013 Denver West Parkway
Golden, CO 80401

**USDOT Aviation Research and Development Grants** 2783

The objectives of Aviation Research and Development are to encourage and support: innovative and advanced research in the areas of potential benefit to the long-term technical needs of the National Airspace System; areas related to research on the prevention of catastrophic failures; and areas related to research, development, and implementation of technologies and procedures to counteract terrorist acts against civil aviation. The grants may be used to support research and development projects in the following areas: communications, navigation, and surveillance; capacity and air traffic control technology; aviation weather; airport technology; aircraft safety technology; system security technology; human factors and aviation medicine; environment and energy; and operations research. The program is ongoing through December of 2019, with the maximum award being $5,000,000.

*Requirements:* Colleges, universities, and nonprofit institutions are eligible, as well as nonprofits having a 501(c)3 status with the IRS, other than institutions of higher education. Profit-making organizations may be eligible in the area of aviation security.
*Geographic Focus:* All States
*Amount of Grant:* 25,000 - 5,000,000 USD
*Contact:* Robert Boggi, Grants Specialist; (609) 485-6911 or (866) 835-5322; fax (609) 485-6509; robert.boggi@faa.gov
*Internet:* www07.grants.gov/search/search.do?&mode=VIEW&oppId=134953
*Sponsor:* U.S. Department of Transportation
800 Independence Avenue, SW
Washington, D.C. 20591

**USDOT Dwight David Eisenhower Transportation Fellowships** 2784

The Dwight David Eisenhower Transportation Fellowship program's goals are to attract people to the field of transportation, to enhance the careers of transportation professionals by encouraging them to seek advanced degrees, and to retain top talent in the transportation community. The Fellowship awards are: the Eisenhower Graduate Fellowships to allow students to pursue master's degrees or doctorates in transportation-related fields; the Eisenhower Grants for Research (GRF) Fellowship to acquaint undergraduates and graduate students with transportation research, development, and technology transfer activities at U.S. Department of Transportation facilities; the Eisenhower Historically Black Colleges and Universities Fellowships to provide HBCU students with additional opportunities to enter careers in transportation; the Eisenhower Hispanic Serving Institutions Fellowships to provide HSI students with opportunities to enter transportation careers; the Eisenhower Tribal Colleges Initiatives, which provide fellowships for Native American students and faculty at tribal colleges; the Eisenhower People with Disabilities (PWD) Fellowship to provide additional opportunities for people with disabilities to pursue education that prepares them to enter careers in transportation; the Eisenhower Intern (EIF) Fellowship to provide students with opportunities to perform a wide range of significant transportation-related activities at public and private-sector transportation organizations; and the Eisenhower Community College (CC) Fellowship to provide students at community colleges with opportunities to enter careers in transportation. The annual closing date is listed as April 14, with up to eighty awards ranging from $5,000 to $35,500.

*Requirements:* Individuals are welcome to apply. Applicants must: possess a baccalaureate degree from an accredited U.S. Institution; be enrolled in an accredited U.S. Institution of higher education; be pursuing a degree in a full-time program in a transportation-related discipline; have at least one full academic year remaining in program of study; conduct ongoing research in one or more transportation-related disciplines; and plan to enter the transportation profession after completing their higher level education.
*Geographic Focus:* All States
*Date(s) Application is Due:* Apr 14
*Amount of Grant:* 5,000 - 35,500 USD
*Contact:* Fawn Thompson, Program Manager; (404) 562-3917 or (703) 235-0538; fawn.thompson@dot.gov
*Internet:* https://www.fhwa.dot.gov/tpp/ddetfp.htm
*Sponsor:* U.S. Department of Transportation
1200 New Jersey Avenue, SE
Washington, D.C. 20590

**USDOT Research on Clarus System Data Grants** 2785

Similar to the Use Case effort, in early 2010 FHWA issued a Broad Agency Announcement (BAA) soliciting proposals to use the Clarus System in new/transformative ways. The goal of the BAA was to foster collaboration between transportation engineering, computer science, and atmospheric science disciplines and to support research and scientific study on the use of Clarus System data to improve surface transportation weather management/operations, create innovative interfaces, and/or develop new applications including weather-responsive traffic management tools. The tools and applications that are being developed from this research must address a specific surface transportation weather management problem and be implementable by agencies that draw data from the Clarus system. The tools and applications that will be developed from research must address a specific surface transportation weather management problem and be implementable by agencies that draw data from the Clarus system.

*Requirements:* Application eligibility is unrestricted.
*Geographic Focus:* All States
*Contact:* Freida Byrd; (202) 366-6547; fax (202) 366-3705; freida.byrd@dot.gov
*Internet:* www.its.dot.gov/clarus/clarus_baa.htm
*Sponsor:* U.S. Department of Transportation
1200 New Jersey Avenue, SE
Washington, D.C. 20590

**USFWS African Elephant Conservation Grants** 2786

The program objectives are to provide financial assistance to support projects that will enhance sustainable conservation programs to ensure effective, long-term conservation of African elephants (Loxodonta africana). Funds may be used for approved conservation projects. Funds are matched by non-federal funds or in-kind support, which must be equal to or exceed the amount of federal funds provided. U.S. Fish and Wildlife Service (USFWS) also supports African elephant conservation projects in Central Africa through the Africa Regional Program (formerly the Wildlife Without Borders - Africa program). Funding Opportunity Number: F18AS00055

*Requirements:* Qualified individuals; non-profit, non-governmental organizations; public and private institutions of higher education; multi-national secretariats, and range state government agencies with demonstrated experience in elephant conservation are eligible to apply. U.S. non-profit, non-governmental organizations with 501(c)3 Internal Revenue Status (IRS) must provide a copy of their status determination letter received from the IRS. Applicants must ensure that activities occurring outside the United States are coordinated as necessary with appropriate U.S. and foreign government authorities and that any necessary licenses, permits, or approvals are obtained prior to undertaking proposed activities.
*Restrictions:* The USFWS will not fund the purchase of firearms or ammunition; buying of intelligence information or paying informants; gathering information by persons who conceal their true identity; law enforcement operations that prompt suspects to carry out illegal activities so they may be arrested (entrapment); or any activity that would circumvent sanctions, laws or regulations of either the U.S. or the country in which the activity would occur.
*Geographic Focus:* All States, All Countries
*Date(s) Application is Due:* Feb 13
*Amount of Grant:* 200,000 - 400,000 USD
*Contact:* African Elephant Conservation Fund; (703) 358-1754; fax (703) 358-2849; MSCF_africanelephant@fws.gov
*Internet:* https://www.fws.gov/international/grants-and-reporting/how-to-apply.html
*Sponsor:* U.S. Department of the Interior
5275 Leesburg Pike
Falls Church, VA 22401

**USFWS Central Africa Regional Grants** 2787

Central Africa is a globally important region for forest and biodiversity conservation. The U.S. Fish and Wildlife Service (USFWS) works closely with national governments, U.S. agencies, and a range of other partners to ensure a strategic, results-based approach to wildlife stewardship in the region. In collaboration with the U.S. Agency for International Development's (USAID) Central Africa Regional Program for the Environment (CARPE), USFWS is providing this funding opportunity to reduce pressures on key wildlife populations, and to provide training to develop the requisite individual and institutional capacity for durable, long-term conservation success. Funding Opportunity Number: F18AS00060

*Requirements:* Applicants under this program can be government agencies, non-governmental organizations, institutions of higher education, and individuals. Applicants must ensure that activities occurring outside the United States are coordinated as necessary with appropriate U.S. and foreign government authorities and that any necessary licenses, permits, or approvals are obtained prior to undertaking proposed activities. Funding will only be considered for projects that impact wildlife populations in the 10 Commission of Central African Forests (COMIFAC) countries: Burundi, Cameroon, Chad, Central African Republic, Democratic Republic of the Congo, Equatorial Guinea, Gabon, Republic of Congo, Rwanda, and São Tomé and Príncipe. A pre-proposal concept of up to four pages should be submitted, including a Statement of Need and one-page summary budget. Applicants will be notified of the opportunity to submit a full application for either a grant or cooperative agreement by invitation only.
*Restrictions:* The Division of International Conservation will not fund the purchase of firearms or ammunition; buying or purchasing of intelligence, evidence or information or paying informants; gathering information by persons who conceal their true identity or any activity that would circumvent sanctions, laws, or regulations of either the U.S. or the country in which the activity would occur.
*Geographic Focus:* All States, All Countries
*Date(s) Application is Due:* May 4
*Contact:* Dirck Byler, (703) 358-2337; Dirck_Byler@fws.gov
Nancy Gelman, Pre-proposal Questions; nancy_gelman@fws.gov
*Internet:* https://www.fws.gov/international/grants-and-reporting/how-to-apply.html
*Sponsor:* U.S. Department of the Interior
5275 Leesburg Pike
Falls Church, VA 22401

**USFWS Combating Wildlife Trafficking Grants** **2788**
The Combating Wildlife Trafficking Strategy and Partnerships Program provides financial assistance to projects that address the illegal trade and trafficking of priority species. Wildlife trafficking is defined in the as "the poaching or other taking of protected or managed species and the illegal trade in wildlife and their related parts and products." For the purposes of this funding opportunity, the term "wildlife" includes terrestrial and aquatic animal species and plant species subject to illegal trade. Funding will support multiple, one-year to two-year grants, with one-year proposals receiving up to $100,000. Two-year proposals demonstrating strong justification for multi-year support will be considered for funding of up to $200,000 for the entire project period. Funding Opportunity Number: F18AS00087
*Requirements:* Applicants under this program can be: individuals, multi-national secretariats, foreign governments, U.S. and foreign non-profits, non-governmental organizations, community and indigenous organizations, U.S. and foreign public and private institutions of higher education, and U.S. and foreign for-profit organizations. Applicants must ensure that activities occurring outside the United States are coordinated as necessary with appropriate U.S. and foreign government authorities and that any necessary licenses, permits, or approvals are obtained prior to undertaking proposed activities.
*Restrictions:* The USFWS International Affairs will not fund the purchase of firearms or ammunition; buying of intelligence information or paying informants; gathering information by persons who conceal their true identity; law enforcement operations that prompt suspects to carry out illegal activities so they may be arrested (entrapment); or any activity that would circumvent sanctions, laws or regulations of either the U.S. or the country in which the activity would occur.
*Geographic Focus:* All States, All Countries
*Date(s) Application is Due:* May 10
*Amount of Grant:* Up to 100,000 USD
*Contact:* Tatiana Hendrix, (703) 358-2655; Tatiana_Hendrix@fws.gov
*Internet:* https://www.fws.gov/international/grants-and-reporting/how-to-apply.html
*Sponsor:* U.S. Department of the Interior
5275 Leesburg Pike
Falls Church, VA 22401

**USFWS Great Ape Conservation Grants** **2789**
In response to the decline of ape populations in Africa and Asia, the U.S. Government enacted the Great Ape Conservation Act of 2000. The Act provides for the conservation and protection of apes by supporting conservation programs in countries within their ranges and the projects of persons and organizations with demonstrated expertise in ape conservation. The U.S. Fish and Wildlife Service (USFWS) works closely with national governments, U.S. agencies, and a range of other partners to ensure a strategic, results-based approach to ape conservation in Africa and Asia. The Great Ape Conservation Fund (GACF) welcomes project proposals for the conservation of gorillas (Gorilla spp.), chimpanzees (Pan troglodytes), bonobos (Pan paniscus), orangutans (Pongo spp.), and gibbons (family Hylobatidae) throughout their ranges. Proposed project work should occur within the ape range or, if work is to be conducted outside of the range, the proposal should demonstrate a clear relevance to ape conservation. Applied research projects should address specific management needs and actions, and/or threats from emerging issues, such as disease, that affect the species and their habitats. Funding Opportunity Number: F18AS00091
*Requirements:* Applicants under this program can be government agencies, non-governmental organizations, institutions of higher education, and individuals. Applicants must ensure that activities occurring outside of the United States are coordinated as necessary with appropriate U.S. and foreign government authorities and that any necessary licenses, permits, or approvals are obtained prior to undertaking proposed activities.
*Restrictions:* The Division of International Conservation will not fund: the purchase of firearms or ammunition; gathering information by persons who conceal their true identity; buying or purchasing of intelligence, evidence or information, or paying informants; law enforcement operations that, to arrest suspects, prompt them to carry out illegal activities (entrapment); any activity that would circumvent sanctions, laws or regulations of either the U.S. or the country of proposed activity; projects in countries determined by the U.S. Department of State to have repeatedly provided support for acts of international terrorism; projects in countries included in comprehensive sanction programs administered by the U.S. Department of Treasury, Office of Foreign Asset Controls, without proper licenses.
*Geographic Focus:* All States, All Countries
*Date(s) Application is Due:* Apr 30
*Contact:* Dr. Kenneth Cameron; (703) 358-2331; fax (703) 358-2115; kenneth_cameron@fws.gov
Dr. Meenakshi Nagendran, Contact for Asia; (703) 358-2315; fax (703) 358-2115; meenakshi_nagendran@fws.gov
*Internet:* https://www.fws.gov/international/grants-and-reporting/how-to-apply.html
*Sponsor:* U.S. Department of the Interior
5275 Leesburg Pike
Falls Church, VA 22401

**USFWS Marine Turtle Conservation Grants** **2790**
The Marine Turtle Conservation Fund (MTCA) is soliciting proposals for the conservation of marine turtles (Chelonia mydas, Caretta caretta, Dermochelys coriacea, Eretmochelys imbricata, Lepidochelys olivacea, Lepidochelys kempii) throughout their range outside of the United States and its territories. The goal of this program is to reduce threats to marine turtles in their natural habitat. Proposals should identify specific conservation actions that have a high likelihood of creating lasting benefits. Project activities that emphasize data collection and status assessment should describe a direct link to management action, and explain how lack of information has been a key limiting factor for management action in the past. Proposed project work should occur within the species range, outside of the United States. If work is to be conducted outside of the species range, the proposal should show a clear relevance to its conservation. Proposals are typically funded at or less than $50,000 USD for every year of project activities. The amount of funding requested must match scope of the activities, the anticipated results, and the length of the project period. Higher amounts may be requested with appropriate justification.
*Requirements:* Applicants can be individuals; multi-national secretariats; state and local government agencies; non-profit, non-governmental organizations; for-profit organizations; public and private institutions of higher education, and foreign entities.
*Restrictions:* MTCA will not fund the purchase of firearms or ammunition; buying of intelligence, evidence or information or paying informants; gathering information by persons who conceal their true identity or any activity that would circumvent sanctions, laws, or regulations of either the U.S. or the country in which the activity would occur.
*Geographic Focus:* All States, All Countries
*Date(s) Application is Due:* Jan 22; Apr 2
*Amount of Grant:* Up to 50,000 USD
*Contact:* Earl Possardt; (703) 358-2277; fax (703) 358-2115; earl_possardt@fws.gov
*Internet:* https://www.fws.gov/international/grants-and-reporting/how-to-apply.html
*Sponsor:* U.S. Department of the Interior
5275 Leesburg Pike
Falls Church, VA 22401

**USFWS Rhinoceros and Tiger Conservation Grants - Africa** **2791**
The Rhinoceros and Tiger Conservation Fund (RTCF) is soliciting proposals for the conservation of African rhinoceroses. Administered by the U.S. Fish & Wildlife Service (USFWS), its purpose is to provide financial assistance to support projects that work to conserve populations of rhinos and tigers in the wild. Proposals for Africa and for Asia are handled through different program offices. The amount of funding requested must match scope of the activities, the anticipated results, and the length of the project period. Funding Opportunity Number: F18AS00092
*Requirements:* Any organization responsible for the conservation and protection of rhinoceroses in the wild and any other organization, multi-national secretariat, public or private university, or individual with demonstrated experience in rhinoceros conservation may submit proposals to this fund. U.S. non-profit, non-governmental organizations with 501(c)3 Internal Revenue Status (IRS) must provide a copy of their status determination letter received from the IRS. Proposed project work should occur within rhinoceros range, or, if work is to be conducted outside of the range, the proposal should show clear relevance to rhinoceros conservation. Proposals must identify immediate threats to rhino survival and support direct conservation activities to eliminate or reduce those threats. If the project includes research, the applicant must provide a convincing argument that the research addresses priority threats and that the results are likely to result in management actions. Priority will be given to projects involving indigenous subspecies within natural range. Ex situ populations are not eligible.
*Geographic Focus:* All States
*Date(s) Application is Due:* May 16
*Contact:* African Rhino Conservation Program; (703) 358-1754; fax (703) 358-2849; africanrhino@fws.gov
*Internet:* https://www.fws.gov/international/wildlife-without-borders/rhino-and-tiger-conservation-fund.html
*Sponsor:* U.S. Department of the Interior
5275 Leesburg Pike
Falls Church, VA 22401

**USFWS Rhinoceros and Tiger Conservation Grants - Asia** **2792**
The Rhinoceros and Tiger Conservation Fund (RTCF) is soliciting proposals for the conservation of rhinoceroses and/or tigers throughout their ranges. Administered by the U.S. Fish & Wildlife Service (USFWS), its purpose is to provide financial assistance to support projects that work to conserve populations of rhinos and tigers in the wild.

Proposals for Africa and for Asia are handled through different program offices. The amount of funding requested must match scope of the activities, the anticipated results, and the length of the project period. Funding Opportunity Number: F18AS00094
*Requirements:* Any organization responsible for the conservation and protection of rhinoceroses in the wild and any other organization, multi-national secretariat, public or private university, or individual with demonstrated experience in rhinoceros conservation may submit proposals to this fund. U.S. non-profit, non-governmental organizations with 501(c)3 Internal Revenue Status (IRS) must provide a copy of their status determination letter received from the IRS. Proposed project work should occur within the rhinoceros or tiger range, or, if work is to be conducted outside of the range, the proposal should show clear relevance to rhinoceros or tiger conservation. Proposals must identify immediate threats to rhino survival and support direct conservation activities to eliminate or reduce those threats. If the project includes research, the applicant must provide a convincing argument that the research addresses priority threats and that the results are likely to result in management actions. Priority will be given to projects involving indigenous subspecies within natural range. Ex situ populations are not eligible.
*Geographic Focus:* All States, All Countries
*Date(s) Application is Due:* Apr 30
*Contact:* Dr. Meenakshi Nagendran; (703) 358-2315; Meenakshi_nagendran@fws.gov
*Internet:* https://www.fws.gov/international/wildlife-without-borders/rhino-and-tiger-conservation-fund.html
*Sponsor:* U.S. Department of the Interior
5275 Leesburg Pike
Falls Church, VA 22401

**USFWS South America Regional Grants** **2793**
South America is home to an unparalleled number of plant and animal species, ecosystem diversity, and species endemism making it unique among continents. South America's wildlife and ecosystems are among the most threatened with some of the highest land conversion and deforestation rates in the world spurred by unsustainable resource and land management, illegal wildlife trade, human-wildlife conflict, and limited conservation incentives. The U.S. Fish and Wildlife Service (USFWS) works with national governments, U.S. agencies, civil society, and a range of other partners to ensure a results-based approach to conserving priority species, habitats and ecological processes across landscapes with high biodiversity value in South America. USFWS is providing this funding opportunity to reduce threats to key species and ecosystems in South America and to strengthen the requisite local individual and institutional capacity to sustain conservation programs in the long-term through strategic partnerships with key stakeholders at the local, national, regional, and/or international levels. Project activities should take place in the focal countries in South America. If work is to be conducted in the United States, the proposal must show a clear impact on biodiversity conservation in South America to be eligible. Funding Opportunity Number: F18AS00090
*Requirements:* The USFWS seeks to support conservation of the large-scale terrestrial ecosystems that are viable for long-term wildlife conservation and critical to protecting regional watersheds and environmental services. Because existing protected areas alone are not sufficient to maintain ecosystem functions and large scale processes, this funding opportunity seeks landscape-scale approaches to conserve the region's unique ecosystems and iconic wildlife that can also serve as umbrella species for the rest of the region's biodiversity. Consideration will be given to projects that address one or more of the programmatic funding priorities in the following countries: Argentina, Bolivia, Chile, Colombia, Ecuador, Paraguay, Peru.
*Geographic Focus:* All States, Argentina, Bolivia, Chile, Colombia, Ecuador, Paraguay, Peru
*Date(s) Application is Due:* May 11
*Contact:* Jenny Martínez; (703) 358-2041; fax (703) 358-2115; Jenny_Martinez@fws.gov
*Internet:* https://www.fws.gov/international/grants-and-reporting/how-to-apply.html
*Sponsor:* U.S. Department of the Interior
5275 Leesburg Pike
Falls Church, VA 22401

**USGS Development of Kalman Filter Optimized for Earthquake Ground Motions Cooperative Agreement** **2794**
The U.S. Geological Survey (USGS) provides support for research that will assist in achieving the goals of the earthquake ground motion program. The work in this addendum will require leveraging the hardware and software previously developed to stand-up an operational network in Costa Rica covered under the existing "Innovative Geodetic Sensors" cooperative agreement. The primary objectives are: the development of Kalman filter combining smartphone accelerometers and GNSS data, optimized for earthquake ground motions. The fundamentals of this were done already in Minson et al. but for smartphones that were 5 years older than they are now (moreover, the release of the Android-N operating system warrants development of a Kalman filter specific to that operating system, testing and algorithm development; and help software contractor modify QED software for Android-N and to incorporate Kalman Filter developed in #1.
*Requirements:* Colleges, universities, for-profit and nonprofit organizations, and state and local governments may submit applications for support by a named principal investigator. It is strongly recommended that the applicant contact the appropriate regional or topical coordinator and other USGS points of contact to learn whether the proposed work duplicates work being done internally, and how their proposed work can complement and help support the goals and objectives of internal efforts. Research proposals may be accepted and approved out of cycle (after the closing date) only in cases where there is compelling circumstance or emergency (e.g., seismic event), which must be acted on before the next competitive review cycle. Proposers should contact the appropriate Regional or Topical Coordinator prior to submitting out-of-cycle proposals.
*Geographic Focus:* All States, All Countries
*Date(s) Application is Due:* Sep 20
*Amount of Grant:* 58,687 - 117,374 USD
*Contact:* Donald J. Shannon, Contracts Specialist; (916) 278-9342; dshannon@usgs.gov
*Internet:* https://earthquake.usgs.gov/research/external/
*Sponsor:* U.S. Department of the Interior
12201 Sunrise Valley Drive
Reston, VA 22092

**USGS Numerical Modelling in Rivers, Estuaries and Coastal Waters Cooperative Agreement** **2795**
The U.S. Geological Survey (USGS) provides support for research that will assist in achieving the goals of the earthquake coastal research program. This cooperative agreement is planned as a 3-year research project between the U.S. Geological Survey and Deltares USA that builds on several past cooperative agreements starting in 2001. Since 2001, Deltares and the U.S. Geological Survey (USGS) have been collaborating intensively on coastal morphodynamic research. This collaboration is linked to the Coastal and Marine Geology program that is operated from science centers in Santa Cruz, California, St. Petersburg, Florida, and Woods Hole, Massachusetts. Deltares is uniquely qualified in this collaboration considering the technical expertise and decades-long continuous development of state-of-the-science software and methods for coastal and estuarine modeling. Deltares is a world-renowned independent institute with more than 800 specialists, many whom are well-decorated scientists in their fields. This cooperative agreement follows on several similar agreements and builds on more than a decade of collaborative research. At present, collaborative research and products are focused on improving the fidelity of coastal change, groundwater, hydrodynamic, and morphodynamic models, establishing robust methods to support operational and early warning systems, and communicating hazards and risks. Deltares' research interests are strongly aligned with those of the USGS Coastal Marine Geology Program, particularly with respect to extreme weather events, and coastal ecosystems and hazards. As such, and in support of the requirements in 43 U.S.C. § 1457b, this cooperative agreement is designed to benefit both USGS and Deltares. Deltares supplies cutting-edge models and software as well technical expertise in model development and coastal processes. The USGS provides observation data, model testing beds, and expertise in the said topics. The collaboration between USGS and Deltares is crucial for furthering scientific advancement, understanding, and accurate representation of coastal ecosystems and hazards, a corner-stone of the USGS Natural Hazards Mission.
*Requirements:* Colleges, universities, for-profit and nonprofit organizations, and state and local governments may submit applications for support by a named principal investigator. It is strongly recommended that the applicant contact the appropriate regional or topical coordinator and other USGS points of contact to learn whether the proposed work duplicates work being done internally, and how their proposed work can complement and help support the goals and objectives of internal efforts. Research proposals may be accepted and approved out of cycle (after the closing date) only in cases where there is compelling circumstance or emergency (e.g., seismic event), which must be acted on before the next competitive review cycle. Proposers should contact the appropriate Regional or Topical Coordinator prior to submitting out-of-cycle proposals.
*Geographic Focus* All States, District of Columbia, Guam, Marshall Islands, Northern Mariana Islands, Puerto Rico, U.S. Virgin Islands, All Countries, American Samoa
*Date(s) Application is Due:* Sep 21
*Amount of Grant:* 308,000 - 2,105,000 USD
*Contact:* Donald J. Shannon, Contracts Specialist; (916) 278-9342; dshannon@usgs.gov
*Internet:* https://earthquake.usgs.gov/research/external/
*Sponsor:* U.S. Department of the Interior
12201 Sunrise Valley Drive
Reston, VA 22092

**UUA Minesterial Fellowships** **2796**
The Office of Ministerial Credentialing has limited funds available to students enrolled full-time in a Masters of Divinity degree program leading to ordination as a Unitarian Universalist minister. Aspirant status or candidate status is required. Applicants for financial aid through the UUA are automatically considered for a number of named fellowships; no additional application materials are required. Once ministers are granted Preliminary Fellowship by the Ministerial Fellowship Committee (MFC), they continue their journey towards Final Fellowship by meeting the MFC's credentialing requirements outlined in the Rules and Policies of the Ministerial Fellowship Committee. The Ministries and Faith Development Staff Group provides support to the MFC and can also offer guidance on preparation and materials required for the renewal process, waiver requests, changes in fellowship status, appropriate forms and deadlines for renewal applications.
*Geographic Focus:* All States
*Amount of Grant:* 2,500 - 11,000 USD
*Contact:* Alicia LeBlanc, Ministerial Credentialing Administrator; (617) 948-4371 or (617) 742-2100; fax (617) 742-0321; aleblanc@uua.org
*Internet:* https://www.uua.org/careers/ministers/support/fellowship
*Sponsor:* Unitarian Universalist Association
24 Farnsworth Street
Boston, MA 02210-1409

**Vancouver Foundation Grants and Community Initiatives Program** **2797**
Vancouver Foundation's Grants and Community Initiatives program provides Community Impact Grants in nine fields of interest: Animal Welfare; Arts and Culture; Children, Youth and Families; Education; Environment; Health and Social Development; Health and Medical Research; Youth Homelessness; Youth Philanthropy. The Vancouver Foundation has a two-stage application process. The first stage is a

letter of intent. This consists of a brief proposal to determine basic suitability. The second stage is submission of a full grant application by a specific date. The letter of intent may be submitted online. You will be notified, if you are invited to submit a full grant application. The following is a list of deadline dates for the letters of intent and applications. Animal Welfare, Arts and Culture, Children, Youth and Families, Education, Environment, and Health and Social Development grants have a deadline date of: Letter of Intent, June 18, Application, August 6. Youth Homelessness grants: Letter of Intent, August 6, Application, September 24. Youth Philanthropy grants: Letter of Intent, September 17, Application, October 1. Health and Medical Research grants deadline dates are undetermined, at this time.
*Requirements:* Under our Grants and Community Initiatives program, Vancouver Foundation funds eligible applicants, which include: registered charities and qualified donees under the Income Tax Act; First Nations that may be considered a public body performing a function of government.
*Restrictions:* The Foundation does not fund the following: 100% of a proposal's costs; an organization's ongoing operational or core expenses; retroactive funding, or for any expenses to be incurred prior to the Foundation's decision date; debt retirement or reserves; mortgage pay-downs; conferences, competitions, symposia, annual events, or travel to/attendance at such events; office equipment and furniture; activities of religious organizations that serve primarily their membership and/or their direct religious purposes, unless the community at large will benefit significantly; sabbatical leaves, student exchanges; medical facilities or equipment; league-based sports and recreation programs; library acquisitions and construction; school construction; publication of books; academic or dissertation research; research with human subjects; endowment matching grants; capital requests, with the exceptions noted above; activities previously supported through government funding; bursaries, scholarships and awards; school trips, annual events or equipment; grants to individuals or businesses; funding requests if reports on any previous grants are overdue.
*Geographic Focus:* All States, Canada
*Contact:* Mark Gifford, Director; (604) 629-5362; markg@vancouverfoundation.ca
*Internet:* www.vancouverfoundation.ca/grants/communityinitiatives.htm
*Sponsor:* Vancouver Foundation
555 West Hastings Street, Suite 1200
Vancouver, BC V6B 4N6 Canada

**Vanderburgh Community Foundation Health Grants** **2798**
The Vanderburgh Community Foundation is a nonprofit, public charity created for the people of Vanderburgh County, Indiana. The Foundation allows nonprofits to fulfill their missions by strengthening their ability to meet community needs through grants that assist charitable programs, address community issues, support community agencies, launch community initiatives, and support leadership development. The grant cycle is open year-round, and funding requests will be reviewed throughout the year with quarterly deadlines as follows: March 1; June 1; September 1; and December 1. Grants in the area of health include activities that: improve and promote health outcomes; provide general and rehabilitative health services; offer mental health services; provide crisis intervention programs; strengthen associations or services associated with specific diseases, disorders, and medical disciplines; and support medical research. Samples of previously funded projects are posted on the website.
*Requirements:* The Foundation gives primarily to 501(c)3 tax-exempt organizations in Vanderburgh County, Indiana. Organizations must first submit a preliminary proposal to determine their eligibility. Proposals will be reviewed within thirty days, then select applicants will receive a formal invitation to submit a full proposal.
*Restrictions:* The following project areas are not considered for funding: religious organizations for religious purposes; political parties or campaigns; endowment creation or debt reduction; capital campaign; operating costs not directly related to the proposed project; annual appeals or membership contributions; and travel requests for groups or individuals such as bands, sports teams, or classes.
*Geographic Focus:* Indiana
*Date(s) Application is Due:* Mar 1; Jun 1; Sep 1; Dec 1
*Amount of Grant:* Up to 10,000 USD
*Contact:* Sarah Wagner, Regional Director of Development; (812) 422-1245; fax (812) 429-0840; swagner@communityfoundationalliance.org
*Internet:* https://www.communityfoundationalliance.org/vanderburgh/program-areas/
*Sponsor:* Vanderburgh County Community Foundation
401 South East 6th Street, Suite 203
Evansville, IN 47713

**Vernon K. Krieble Foundation Grants** **2799**
The Foundation was established in 1984 by the family of Professor Vernon K. Krieble, scientist, educator, inventor, and entrepreneur. Recognizing that the Foundation's assets are the product of a free and democratic society, the founders considered it fitting that those assets be used "to further democratic capitalism and the preserve and promote a society of free, educated, healthy and creative individuals." The Foundation offers support to non-profit charitable and educational organizations that demonstrate leadership in furthering the original objectives, so that future generations can aspire to and achieve their full potential in a free society. Awards range from $2,500 to $50,000. There are no deadlines.
*Requirements:* Nonprofit 501(c)3 organizations are eligible. Funding is provided only for those organizations and projects which involve public policy research and education on issues supporting the preservation, and in some cases the restoration, or freedom and democracy in the United States, according to the principles of the Founding Fathers. Written proposals should include a summary of the project, the project budget, the amount requested, the qualifications of individuals involved, and a copy of the organization's Internal Revenue Service determination letter.
*Geographic Focus:* All States
*Samples:* Foundation for West Hartford Schools, West Hartford, Connecticut, $2,500 - general support; Leadership Program of the Rockies, Denver, Colorado, $50,000 - general support; and Atlas Economic Research Foundation, Arlington, Virginia, $25,000 - general support.
*Contact:* Helen E. Krieble, President; (303) 758-3956; fax (303) 488-0068
*Internet:* www.krieble.org/grants
*Sponsor:* Vernon K. Krieble Foundation
1777 S Harrison Street, Suite 807
Denver, CO 80210

**Victor E. Speas Foundation Grants** **2800**
The Victor E. Speas Foundation was established to provide medical care for the needy, to further medical research, and to support and promote quality educational, cultural, human-services, and health-care programming. In the area of arts, culture, and humanities, the foundation supports programming that: fosters the enjoyment and appreciation of the visual and performing arts; strengthens humanities and arts-related education programs; provides affordable access; enhances artistic elements in communities; and nurtures a new generation of artists. In the area of education, the foundation supports programming that: promotes effective teaching; improves the academic achievement of, or expands educational opportunities for, disadvantaged students; improves governance and management; strengthens nonprofit organizations, school leadership, and teaching; and bolsters strategic initiatives of area colleges and universities. In the area of health, the foundation supports programming that improves the delivery of health care to the indigent, uninsured, and other vulnerable populations and addresses health and health-care problems that intersect with social factors. In the area of human services, the foundation funds programming that: strengthens agencies that deliver critical human services and maintains the community's safety net and helps agencies respond to federal, state, and local public policy changes. In the area of community improvement, the foundation funds capacity-building and infrastructure-development projects including: assessments, planning, and implementation of technology for management and programmatic functions within an organization; technical assistance on wide-ranging topics, including grant writing, strategic planning, financial management services, business development, board and volunteer management, and marketing; and mergers, affiliations, or other restructuring efforts. Grant requests for general operating support and program support will be considered. Grants from the foundation are one year in duration. Application materials are available for download at the grant website. There are no application deadlines for the Victor E. Speas Foundation. Proposals are reviewed on an ongoing basis.
*Requirements:* Applicants must have 501(c)3 tax-exempt status and serve the residents of Kansas City, Missouri. Two copies of the completed application must be mailed.
*Restrictions:* Grant requests for capital support will not be considered. The trust does not support requests from individuals, organizations attempting to influence policy through direct lobbying, or any political campaigns.
*Geographic Focus:* Missouri
*Samples:* Childrens Mercy Hospital Foundation, Kansas City, Missouri, $100,000, to fund research and clinical trials of reformulated adult drugs to benefit pediatric patients with rare and neglected blood and other cancers; Open Options, Kansas City, Missouri, $25,000, United Cerebral Palsy of Greater Kansas City technology upgrade; Metropolitan Community Colleges Foundation, Kansas City, Missouri, $50,000, for medical equipment for the Health Science Institute training facility.
*Contact:* Spence Heddens; (816) 292-4301; Spence.heddens@baml.com
*Internet:* https://www.bankofamerica.com/philanthropic/fn_search.action
*Sponsor:* Victor E. Speas Foundation
1200 Main Street, 14th Floor, P.O. Box 219119
Kansas City, MO 64121-9119

**Victor Grifols i Lucas Foundation Research Grants** **2801**
The Víctor Grífols i Lucas Foundation awards six grants for research in bioethics. The grants, each worth 5000 euros, are awarded to research projects to be undertaken on any issue related to bioethics and its practical application. Awards will be paid as follows: one-third of the amount when the grant is initially awarded; two-thirds of the award upon completion of the project research. Projects may be proposed individually or by teams. The annual deadline is May 31.
*Requirements:* If a project is not completed, the grant recipient must return the initial one-third of the award given.
*Geographic Focus:* All States, Spain
*Date(s) Application is Due:* May 31
*Amount of Grant:* 5,000 EUR
*Contact:* Grant Director; +34 935 710 410; fundacio.grifols@grifols.com
*Internet:* www.fundaciongrifols.org/portal/en/2/becas_investigacion
*Sponsor:* Victor Grifols i Lucas Foundation
Carrer Jesús i Maria, 6
Barcelona, 08022 Spain

**Virginia Foundation for the Humanities Discretionary Grants** **2802**
The Virginia Foundation for the Humanities (VFH) was established in 1974 as a nonprofit organization dedicated to promoting the humanities, and to using the humanities to address issues of broad public concern. Since 1974, the VFH has awarded more than 3,000 grants, supporting tens of thousands of separate activities, and serving audiences in every city and county in Virginia. Goals of the VFH grant program are as follows: to encourage the development of high-quality educational programs in the humanities; to support accessible programs that reach the broadest possible audience in Virginia; to support the work of humanities institutions - museums, libraries, historical societies, colleges, and universities - as well as other non-profit organizations working within the

humanities; to explore the stories that define Virginia and its people; and to address the issues that are most important to communities in Virginia. Projects supported include exhibits, public forums and discussions, media programs (film, video, radio, and digital media), publications, research, teachers' institutes and seminars, oral history projects, lectures and conferences, and other kinds of programs that draw on the resources of the humanities, address important issues, and enrich the cultural life of the state. The program names six key-priority areas of interest: Books, Reading, and Literacy; Rights and Responsibilities; Media and Culture; Violence and Community; Science, Technology, and Society; Virginia History. Other identified areas of interest are teacher education programs (especially those related to Virginia's Standards of Learning); African American history and culture; Native American history and culture; history and culture of minority communities in Virginia; Virginia's folklife and traditional culture(s); and the future of rural Virginia. The foundation's interest is not confined to only these topics. Discretionary Grants allow VFH to make smaller grants (up to $3,000) at any time of the year. These grants may be used to plan larger projects or to implement programs which require only a modest amount of funding. Decisions are normally made within four weeks following receipt of the application. Applications should be submitted via the VFH online system (available at the foundation's website) which requires as a first step, creation of an account and password. Grant guidelines are available at the website in downloadable Word format. Applicants should contact VFH staff in advance before submitting a Discretionary Grant proposal. Draft proposals are strongly encouraged and are recommended to be submitted at least three weeks prior to the deadline to allow time for a thorough staff review.
*Requirements:* Any incorporated non-profit organization in Virginia is eligible to apply. Incorporated non-profit organizations based outside of Virginia are also eligible if their project deals with a subject directly related to Virgina and a significant audience within the state is anticipated. All VFH grants must be matched with at least an equal amount of cost share, which can be in the form of cash or in-kind contributions from non-Federal sources. Sources and amounts of anticipated cost share should be indicated at the time of the proposal.
*Restrictions:* VFH Grants may not be used to support the following: advocacy or political action programs that promote a particular solution or point of view; creative or performing arts, unless they are used in a supporting role to enhance discussion of issues or interpretation; research or writing unless these are integral to programs having a direct public audience; subvention of publications; courses for credit, except those designed especially for teachers; acquisition of equipment; building construction, restoration, or preservation; meals, other than necessary travel expenses for program personnel; international travel; indirect costs; and projects whose primary audience is children or youth.
*Geographic Focus:* All States
*Amount of Grant:* Up to 3,000 USD
*Contact:* David Bearinger, Director of Grants and Public Programs; (434) 924-3296; fax (434) 296-4714; dab@virginia.edu
Jeanne Nicholson Siler, (434) 924-3296; fax (434) 296-4714; vhfgrants@virginia.edu
*Internet:* virginiahumanities.org/grants/
*Sponsor:* Virginia Foundation for the Humanities
145 Ednam Drive
Charlottesville, VA 22903-4629

**Virginia Foundation for the Humanities Open Grants** **2803**
The Virginia Foundation for the Humanities (VFH) was established in 1974 as a nonprofit organization dedicated to promoting the humanities, and to using the humanities to address issues of broad public concern. Since 1974, the VFH has awarded more than 3,000 grants, supporting tens of thousands of separate activities, and serving audiences in every city and county in Virginia. Goals of the VFH grant program are as follows: to encourage the development of high-quality educational programs in the humanities; to support accessible programs that reach the broadest possible audience in Virginia; to support the work of humanities institutions - museums, libraries, historical societies, colleges, and universities - as well as other non-profit organizations working within the humanities; to explore the stories that define Virginia and its people; and to address the issues that are most important to communities in Virginia. Projects supported include exhibits, public forums and discussions, media programs (film, video, radio, and digital media), publications, research, teachers' institutes and seminars, oral history projects, lectures and conferences, and other kinds of programs that draw on the resources of the humanities, address important issues, and enrich the cultural life of the state. The program names six key-priority areas of interest: Books, Reading, and Literacy; Rights and Responsibilities; Media and Culture; Violence and Community; Science, Technology, and Society; and Virginia History. Other identified areas of interest are teacher-education programs (especially those related to Virginia's Standards of Learning); African-American history and culture; Native-American history and culture; history and culture of minority communities in Virginia; Virginia's folklife and traditional culture(s); and the future of rural Virginia. The foundation's interest is not confined to only these topics. Open Grant applications are usually considered in three annual grant cycles with deadlines of October 15, February 1, and May 1. Applications should be submitted via the VFH online system (available at the foundation's website) which requires as a first step, creation of an account and password. Grant guidelines are available at the website in downloadable Word format. All applicants are urged to contact the staff prior to submission of their proposals. Draft proposals are strongly encouraged and are recommended to be submitted at least three weeks prior to the deadline to allow time for a thorough staff review.
*Requirements:* Any incorporated non-profit organization in Virginia is eligible to apply. Incorporated non-profit organizations based outside of Virginia are also eligible if their project deals with a subject directly related to Virgina and a significant audience within the state is anticipated. All VFH grants must be matched with at least an equal amount of cost share, which can be in the form of cash or in-kind contributions from non-Federal sources. Sources and amounts of anticipated cost share should be indicated at the time of the proposal.
*Restrictions:* VFH Grants may not be used to support the following: advocacy or political action programs that promote a particular solution or point of view; creative or performing arts, unless they are used in a supporting role to enhance discussion of issues or interpretation; research or writing unless these are integral to programs having a direct public audience; subvention of publications; courses for credit, except those designed especially for teachers; acquisition of equipment; building construction, restoration, or preservation; meals, other than necessary travel expenses for program personnel; international travel; indirect costs; and projects whose primary audience is children or youth.
*Geographic Focus:* All States, Canada
*Date(s) Application is Due:* Feb 1; May 1; Oct 15
*Amount of Grant:* Up to 10,000 USD
*Samples:* Southern Memorial Association, Lynchburg, Virginia - Remembering Tinbridge Hill publication project; Living Archives, New York, New York - The Loving Story film documentary project; Eastern Shore of Virginia Historical Society, Onancock, Virginia - Stories from the Shore oral history project.
*Contact:* David Bearinger, Director of Grants and Public Programs; (434) 924-3296; fax (434) 296-4714; dab@virginia.edu
Jeanne Nicholson Siler, (434) 924-3296; fax (434) 296-4714; vhfgrants@virginia.edu
*Internet:* virginiahumanities.org/grants/
*Sponsor:* Virginia Foundation for the Humanities
145 Ednam Drive
Charlottesville, VA 22903-4629

**Virginia Foundation for the Humanities Residential Fellowships** **2804**
The Virginia Foundation for the Humanities (VFH) supports scholarly work on the humanities and in the public interest. The foundation's residential fellowship program is located at its facility in Charlottesville, Virginia, in the foothills of the Blue Ridge Mountains. The fellowship offers time, space, and resources to scholars and writers bringing the humanities to visibility - applying the tools of history, philosophy, ethics, cultural studies, and literary criticism to matters of public concern. Fellowships are awarded for one semester or for a full year. Stipends range up to $15,000 per semester. Fellows have visiting-faculty status at the nearby University of Virginia as well as the opportunity to get to know VFH staff members and become involved, formally or informally, in a variety of public humanities projects. Fellows also appear on With Good Reason, a VFH Radio program. Candidates must apply through an online application system at the foundation website. Applicants will need to create a login account to use this system. Letters of recommendation must be mailed separately to the contact information given.
*Requirements:* The foundation seeks applications that are intellectually stimulating, imaginative, and accessible to the public. There are no restrictions on topic, and applications are invited from across the broad spectrum of the humanities. Candidates may be faculty members in the humanities, independent scholars, or others working on projects in the humanities and need not have advanced degrees. However, the foundation generally does not support dissertation work and strongly encourages postdoctoral applicants to apply for projects other than dissertation revisions.
*Restrictions:* The foundation does not support work toward a degree. Former VFH Fellows must wait three years before applying for another fellowship. Former Fellows are welcome to apply for space (without stipend) at any time.
*Geographic Focus:* All States, All Countries
*Date(s) Application is Due:* Dec 1
*Amount of Grant:* Up to 30,000 USD
*Contact:* Ann Spencer, Program Associate, Research and Education; (434) 924-3296; fax (434) 296-4714; aspencer@virginia.edu or VFHfellowships@virginia.edu
*Internet:* virginiahumanities.org/fellowships/
*Sponsor:* Virginia Foundation for the Humanities
145 Ednam Drive
Charlottesville, VA 22903-4629

**W.C. Griffith Foundation Grants** **2805**
The foundation awards grants, primarily in Indiana, to nonprofit organizations in its areas of interest, including arts and culture, cancer and medical research, children and youth services, Christian churches and organizations, community development, education, environmental programs, health organizations and hospitals, higher education, homeless services, human reproduction and fertility, library science and libraries, minorities, museums, music and performing arts, secondary education, and social services delivery. Types of support include building construction/renovation, capital campaigns, and continuing support. There are no application forms or deadlines. The board meets in June and November; proposals should be received in May and October.
*Requirements:* Nonprofit organizations are eligible. Preference is given to requests from Indianapolis, Indiana.
*Geographic Focus:* All States
*Amount of Grant:* 500 - 50,000 USD
*Contact:* Curt Farran, c/o National City Bank of Indiana; (317) 267-7262
*Sponsor:* W.C. Griffith Foundation
101 West Washington Street, Suite 600 East
Indianapolis, IN 46255

**W.K. Kellogg Foundation Healthy Kids Grants** **2806**
The Foundation believes that children need nutrition, stimulation, healthy living conditions and access to quality health care. It helps many of them get all four, by funding organizations that improve birth outcomes and first food experiences, create access to healthy foods, improve health services, and educate families and communities about the inter-related factors that determine well-being. The Foundation especially focuses on children who are disadvantaged by multiple societal factors, a disproportionate percentage of whom are children of color.

*Requirements:* To be eligible for support, your organization or institution, as well as the purpose of the proposed project, must qualify under regulations of the United States Internal Revenue Service. The Kellogg Foundation does not have any submission deadlines. Grant applications are accepted throughout the year and are reviewed at its headquarters in Battle Creek, Michigan, or in its regional office in Mexico.
*Restrictions:* In general, the Foundation does not provide funding for: operational phases of established programs; capital requests (which includes the construction, purchase, renovation, and/or furnishing of facilities); equipment; conferences and workshops; films, television and/or radio programs; endowments; development campaigns; or research/studies (unless they are an integral part of a larger program budget being considered for funding).
*Geographic Focus:* All States
*Contact:* Deborah A. Rey; (269) 969-2133 or (269) 968-1611; fax (269) 968-0413
*Internet:* www.wkkf.org/what-we-support/healthy-kids.aspx
*Sponsor:* W.K. Kellogg Foundation
1 Michigan Avenue East
Battle Creek, MI 49017-4012

**W.M. Keck Foundation Medical Research Grants** 2807

The W.M. Keck Foundation concentrates on strengthening studies and programs in accredited universities and colleges and major, independent medical research institutions. Medical Research grants seek to advance the frontiers of the life sciences by supporting basic research that is high-risk and has the potential to transform its field. Successful projects are distinctive and novel in their approach to problems, push the edge of their field, or question the prevailing paradigm. Past grants have been awarded to major universities, medical schools, and independent research institutions to support pioneering biological research, including the development of new technologies, instrumentation, or methodologies. Historically, grants range from $500,000 to $5 million, and are typically $2 million or less. Applicants are strongly urged to contact Foundation staff well in advance of submitting a Phase I Application. The best times for these contacts are between January 1 and February 15 leading up to a May 1 submittal, or between July 1 and August 15 leading up to a November 1 submittal.
*Requirements:* Research universities, medical colleges and major private independent scientific and medical research institutes are eligible. Organizations must be exempt from federal taxation as defined by Section 501(c)3 of the Internal Revenue Code and be designated as a public charity (and not a private foundation) as defined by Section 509(a)1 or 509(a)2 or 170(b) of the Internal Revenue Code. If the institution is located in the State of California, the organization must also be exempt from California State Franchise or Income Tax under Section 23701(d) of the Revenue and Taxation Code.
*Restrictions:* Grants are not considered for: general operating expenses, endowments or deficit reduction; general and federated campaigns, including fundraising events, dinners or mass mailings; individuals; conference or seminar sponsorship; book publication and film or theater productions; public policy research; institutions that are located outside the United States; institutions that do not have at least three consecutive full, certified, audited financial statements; conduit organizations, unified funds or organizations that use grant funds from donors to support other organizations or individuals; or institutions that are subsidiaries or affiliates of larger entities that do not have a separate board of directors and independent audited financial statements.
*Geographic Focus:* All States
*Date(s) Application is Due:* May 1; Nov 1
*Amount of Grant:* 500,000 - 5,000,000 USD
*Samples:* Dirk Englund, Jonathan Owen, & Rafael Yuste, Columbia University, New York, New York, $1,000,000 - to develop a novel method for direct, real-time imaging of neuronal voltage signals; Laura Ranum & Maurice Swanson, University of Florida, Gainesville, Florida, $1,000,000 - to pursue critical questions: How does RAN translation work? Is RAN translation a key, previously unrecognized cellular process? Are repetitive sequences throughout the genome translated into proteins and if so, what is their function?.
*Contact:* Margie Antonetti, Grants and Database Specialist; (213) 680-3833; fax (213) 614-0934; mantonetti@wmkeck.org or MedRsch@wmkeck.org
*Internet:* www.wmkeck.org/grant-programs/medicalresearch.html
*Sponsor:* W.M. Keck Foundation
550 S Hope Street, Suite 2500
Los Angeles, CA 90071

**W.M. Keck Foundation Science and Engineering Research Grants** 2808

The W.M. Keck Foundation concentrates on strengthening studies and programs in accredited universities and colleges and major, independent medical research institutions. Science and Engineering Research grants seek to benefit humanity by supporting projects that are distinctive and novel in their approach, question the prevailing paradigm, or have the potential to break open new territory in their field. Past grants have been awarded to major universities and independent research institutions to support pioneering science and engineering research, the development of new technologies, and to facilitate the purchase of advanced instruments where such instruments would further specific research ventures. Historically, grants range from $500,000 to $5 million, and are typically $2 million or less. Applicants are strongly urged to contact Foundation staff well in advance of submitting a Phase I Application. The best times for these contacts are between January 1 and February 15 leading up to a May 1 submittal, or between July 1 and August 15 leading up to a November 1 submittal.
*Requirements:* Research universities, medical colleges and major private independent scientific and medical research institutes are eligible. Organizations must be exempt from federal taxation as defined by Section 501(c)3 of the Internal Revenue Code and be designated as a public charity (and not a private foundation) as defined by Section 509(a)1 or 509(a)2 or 170(b) of the Internal Revenue Code. If the institution is located in the State of California, the organization must also be exempt from California State Franchise or Income Tax under Section 23701(d) of the Revenue and Taxation Code.
*Restrictions:* Grants are not considered for: general operating expenses, endowments or deficit reduction; general and federated campaigns, including fundraising events, dinners or mass mailings; individuals; conference or seminar sponsorship; book publication and film or theater productions; public policy research; institutions that are located outside the United States; institutions that do not have at least three consecutive full, certified, audited financial statements; conduit organizations, unified funds or organizations that use grant funds from donors to support other organizations or individuals; or institutions that are subsidiaries or affiliates of larger entities that do not have a separate board of directors and independent audited financial statements.
*Geographic Focus:* All States
*Date(s) Application is Due:* May 1; Nov 1
*Amount of Grant:* 500,000 - 5,000,000 USD
*Samples:* Alexander Fridman, Drexel University, Philadelphia, Pennsylvania, $1,000,000 - to develop a fundamental understanding of bio-molecular interaction pathways between strongly non-equilibrium electrical plasma and living cells; Jules S. Jaffe, University of California, San Diego, California, $1,000,000 - to build the world's first high resolution, 3D, in situ, underwater microscope.
*Contact:* Margie Antonetti, Grants and Database Specialist; (213) 680-3833; fax (213) 614-0934; mantonetti@wmkeck.org or SciEng@wmkeck.org
*Internet:* www.wmkeck.org/grant-programs/science-engineering.html
*Sponsor:* W.M. Keck Foundation
550 S Hope Street, Suite 2500
Los Angeles, CA 90071

**W.M. Keck Foundation Undergraduate Education Grants** 2809

The W.M. Keck Foundation concentrates on strengthening studies and programs in accredited universities and colleges and major, independent medical research institutions. The Foundation's Undergraduate Education program promotes distinctive learning and research experiences in science, engineering, and the liberal arts at four-year undergraduate colleges only in Foundation designated states, or through national organizations that address undergraduate needs. Historically, grants range from $200,000 to $1 million, and are typically under $500,000. On a case-by-case basis, the Foundation may consider the award of smaller seed grants for planning and piloting. Applicants are strongly urged to contact Foundation staff well in advance of submitting a Phase I Application. The best times for these contacts are between January 1 and February 15 leading up to a May 1 submittal, or between July 1 and August 15 leading up to a November 1 submittal.
*Requirements:* Public institutions and research universities located in the designated states of Alaska, Arizona, California, Colorado, Hawaii, Idaho, Kansas, Louisiana, Mississippi, Montana, Nevada, New Mexico, Oklahoma, Oregon, Texas, Utah, Washington, and Wyoming may apply.
*Restrictions:* Grants are not considered for: general operating expenses, endowments or deficit reduction; general and federated campaigns, including fundraising events, dinners or mass mailings; individuals; conference or seminar sponsorship; book publication and film or theater productions; public policy research; institutions that are located outside the United States; institutions that do not have at least three consecutive full, certified, audited financial statements; conduit organizations, unified funds or organizations that use grant funds from donors to support other organizations or individuals; or institutions that are subsidiaries or affiliates of larger entities that do not have a separate board of directors and independent audited financial statements.
*Geographic Focus:* Alaska, Arizona, California, Colorado, Hawaii, Idaho, Kansas, Louisiana, Mississippi, Montana, Nevada, New Mexico, Oklahoma, Oregon, Texas, Utah, Washington, Wyoming
*Date(s) Application is Due:* May 1; Nov 1
*Amount of Grant:* 200,000 - 1,000,000 USD
*Contact:* Margie Antonetti, Grants and Database Specialist; (213) 680-3833; fax (213) 614-0934; mantonetti@wmkeck.org or UGProgram@wmkeck.org
*Internet:* www.wmkeck.org/grant-programs/undergraduate-program.html
*Sponsor:* W.M. Keck Foundation
550 S Hope Street, Suite 2500
Los Angeles, CA 90071

**W.W. Smith Charitable Trust Grants for Medical Research in Cancer, AIDS, and Heart Disease** 2810

Concentration of the W.W. Smith Charitable Trust Grants for Medical Research program in this area is toward basic scientific research that can help uncover and solve some of the mysteries of cancer, heart disease, and AIDS. Types of support include challenge/matching grants, general operating grants, program development grants, research grants, and scholarships. Funding is directed to projects that are new or headed by little-known investigators who are not yet qualified for large government grants. The trust also supports promising research outside the conventional pathways. Grant applications should be submitted no later than July 15 for cancer and AIDS research and no later than December 15 for cardiovascular research. Each application should be accompanied by an audited financial statement for the institution of the researcher. Projects requested should be one to three years in length and should not request funds for equipment alone. A special application form is required.
*Requirements:* Institutions in the general geographic area of New Jersey, eastern Pennsylvania, and Maryland may submit proposals. Specialized application form and specific guidelines should be obtained from the trust prior to submitting a proposal.
*Restrictions:* Only 10 percent of indirect costs are allowed on all grant proposals. Funds to support students, fellows, and postdoctoral candidates should be supported by curriculum vita. Applications should be submitted for the work of individual investigators and not for general work of research centers.

*Geographic Focus:* New Jersey, Pennsylvania
*Date(s) Application is Due:* Jun 15; Dec 15
*Samples:* Thomas Jefferson U, Jefferson Medical College (Philadelphia, PA)—to endow a chair in cardiac research, $1.5 million; Coriell Institute for Medical Research—to support cancer research, $165,000; Fox Chase Cancer Ctr—to support AIDS research, $166,645.
*Contact:* Louise L. Smith, Medical Grant Advisor; (610) 397-1844; fax (610) 397-1680; info@wwsmithcharitabletrust.org
*Internet:* www.wwsmithcharitabletrust.org/medical.html
*Sponsor:* W.W. Smith Charitable Trust
200 Four Falls Corporate Center, Suite 300
West Conshohocken, PA 19428

**Wallace Foundation Grants** **2811**

The foundation has three current objectives: strengthen education leadership to improve student achievement; improve after-school learning opportunities; and expand participation in arts and culture. Specifically, the fund seeks to develop effective school leaders to improve student learning, create new standards for community participation in the arts, and provide high-quality informal learning opportunities in communities for children and families. Most grants are awarded as part of funds-initiated programs, and unsolicited proposals are rarely funded. Nevertheless, organizations may submit a one- to two-page letter of inquiry describing the project, the organization, the estimated total for the project, and the portion requiring funding.
*Requirements:* U.S. nonprofit organizations are eligible. In most cases, the foundation identifies and evaluates prospective grantees through the issuance of Requests for Proposals or other careful screening processes. This also means that unsolicited proposals are rarely funded. Nevertheless, you may submit an inquiry by email briefly describing the project, your organization, the estimated total for the project and the portion requiring funding to: The Wallace Foundation, grantrequest@wallacefoundation.org.
*Restrictions:* Grants do not support religious or fraternal organizations; international programs; conferences; historical restoration; health, medical, or social service programs; environmental/conservation programs; capital campaigns; emergency funds or deficit financing; private foundations; or individuals.
*Geographic Focus:* All States
*Amount of Grant:* 75,000 - 4,000,000 USD
*Samples:* Advance Illinois, Chicago, IL - $300,000; Jefferson County Public Schools, Louisville, KY - $3,800,000; Education Development Center, Inc., Newton, MA - $700,000; Learning Matters, New York, NY - $500,000; Rand Corporation, Santa Monica, CA - $1,490,000.
*Contact:* Holly Dodge, Grants Administrator; (212) 251-9700; fax (212) 679-6990
*Internet:* www.wallacefoundation.org/GrantsPrograms/GrantApproach/Pages/FundingGuidelines.aspx
*Sponsor:* Wallace Foundation
5 Penn Plaza, 7th Floor
New York, NY 10001

**Walmart Foundation Manufacturing Innovation Grants** **2812**

The Walmart U.S. Manufacturing Innovation Fund is a partnership between Walmart, the Walmart Foundation and the United States Conference of Mayors (USCM). The Fund is focused on the development of U.S. manufacturing, with a specific goal of making it both easier and more competitive to make household goods in the U.S. Funding research in the manufacturing space in focus areas identified as major challenges is another way Walmart can facilitate and accelerate U.S. manufacturing. The Fund will achieve this goal through grants that directly support applied research projects advancing innovative solutions to key challenges that, once addressed, can lower the cost of making consumer products in the U.S. The Fund looks primarily at the following criteria: impact on consumer product manufacturing; stage of development and commercial viability; degree of innovation; and ability of the organization and team to successfully carry out the proposed project. Projects must have a budget exceeding $100,000 per year. Currently, the program is seeking projects that address: weaving; fabric dyeing; and cut and sew.
*Requirements:* Only U.S. 501(c)3 organizations and public universities that are instrumentalities of a state government are eligible for funding. At least 50% of project teams must be based in the U.S.
*Restrictions:* The Fund is currently not awarding grants for building or capital projects, supplier agreements, or other non-research projects. Individuals, municipalities and other government entities, political organizations and current or potential Walmart suppliers are ineligible for funding.
*Geographic Focus:* All States
*Contact:* Julie Gehrki; (479) 273-4000 or (800) 530-9925; fax (479) 273-6850
*Internet:* corporate.walmart.com/global-responsibility/us-manufacturing/walmart-us-manufacturing-innovation-fund
*Sponsor:* Walmart Foundation
702 SW 8th Street, Department 8687, No. 0555
Bentonville, AR 72716-0555

**Warrick County Community Foundation Health Grants** **2813**

The Warrick County Community Foundation is a nonprofit, public charity created for the people of Warrick County, Indiana. The Foundation helps nonprofits fulfill their missions by strengthening their ability to meet community needs through grants that assist charitable programs, address community issues, support community agencies, launch community initiatives, and support leadership development. Funding requests are accepted once each year according to the Foundation's grant cycle. Application requirements may change from year to year; therefore, grant seekers are advised to revisit the web page prior to beginning the grant application process. Grants are normally given as one-time support of a project, but may be considered for additional support and for expansions or outgrowths of an initial project. Applications are accepted from December 1 through the February 1 deadline. Grants in the area of health include activities that: improve and promote health outcomes; provide general and rehabilitative health services; offer mental health services; provide crisis intervention programs; strengthen associations or services associated with specific diseases, disorders, and medical disciplines; and support medical research. Samples of previously funded projects are posted on the website.
*Requirements:* The Foundation welcomes proposals from nonprofit organizations that are deemed tax-exempt under sections 501(c)3 and 509(a) of the IRS code and from governmental agencies serving Warrick County, Indiana. Proposals from nonprofit organizations not classified as a 501(c)3 public charity may be considered if the project is charitable and supports a community need.
*Restrictions:* Project areas not considered for funding include: religious organizations for religious purposes; political parties or campaigns; endowment creation or debt reduction; operating costs; capital campaigns; annual appeals or membership contributions; or travel requests for groups or individuals such as bands, sports teams, or classes.
*Geographic Focus:* Indiana
*Date(s) Application is Due:* Feb 1
*Amount of Grant:* Up to 10,000 USD
*Contact:* Melinda Waldroup, Director of Programs and Strategic Engagement; (812) 429-1191, Ext. 3 or (812) 455-8220; mwaldroup@communityfoundationalliance.org
*Internet:* https://www.communityfoundationalliance.org/warrick/program-areas/
*Sponsor:* Warrick County Community Foundation
224 West Main Street, P.O. Box 215
Boonville, IN 47601

**Washington Area Women's Foundation Strategic Opportunity and Partnership Fund Grants** **2814**

In response to the findings in "A Portrait of Women & Girls in the Washington Metropolitan Area," the Women's Foundation created Stepping Stones to help build the economic security and financial independence of low-income, women-headed families. Stepping Stones is the first multi-year initiative of its kind for the region and will be the major focus of grants and program work. One area of funding under the Stepping Stones umbrella is known as Strategic Opportunity and Partnership Fund Grants. This Fund provides grants to nonprofit organizations engaged in research, policy advocacy, grassroots efforts, capacity building or communications and media activities that support the overall goals of Stepping Stones.
*Requirements:* Nonprofit organizations in the Washington metropolitan area are eligible to apply.
*Geographic Focus:* District of Columbia
*Contact:* Nicole Cozier, Chief of Strategic Operations; (202) 347-7737, ext. 203; fax (202) 347-7739; ncozier@wawf.org
Jennifer Lockwood-Shabat, Vice-President; (202) 347-7737, ext. 206; fax (202) 347-7739; jlockwoodshabat@wawf.org
*Internet:* thewomensfoundation.org/our-work/stepping-stones/strategic-opportunity-and-partnership-fund/
*Sponsor:* Washington Area Women's Foundation
1331 H Street, NW, Suite 1000
Washington, D.C. 20005

**Watson-Brown Foundation Grants** **2815**

The Watson-Brown Foundation awards grants to qualifying organizations that have an abiding interest in the history and culture of the south and southeast. Grants take many forms, including: book subventions; museum and archive support; preservation awards; educational programming; and research support. Typical recipients include: colleges and universities; scholarly societies; and other public charities that collect or advance knowledge of the region. The Foundation supports historic preservation in part through its Junior Board of Trustees, a talented group of high school students that makes grant awards to worthy preservation projects in east central Georgia. The majority of grants awarded support southern studies and related programming at the college level. Academic disciplines and areas of interest also include: history; English; literature; law; agricultural education; and historic preservation. Types of support the Foundation will consider to advance its mission include: operating support; publication subventions; research grants; and program grants. The annual deadline for a particular year is April 15 of that year.
*Requirements:* 501(c)3, 509(a)1, 509(a)2, and 509(a)3 nonprofit organizations are eligible. Generally, the foundation supports only organizations, institutions, or programs located in the southeastern region of the United States.
*Restrictions:* Grants do not support individuals (except for scholarships awarded through the foundation's internal scholarship program); religious programs, or political/lobbying activities; debt reduction; advertising or marketing efforts; social events such as sports tournaments or galas; or individual projects for more than one year.
*Geographic Focus:* Alabama, Arkansas, Florida, Georgia, Kentucky, Louisiana, Mississippi, Missouri, North Carolina, South Carolina, Tennessee, Virginia, West Virginia
*Date(s) Application is Due:* Apr 15
*Contact:* Tad Brown; (706) 595-8886; fax (706) 595-3948; tbrown@watson-brown.org
*Internet:* www.watson-brown.org/grants/index.html
*Sponsor:* Watson-Brown Foundation
310 Tom Watson Way
Thomson, GA 30824

**Web We Want Small Grants** **2816**

The Web We Want campaign exists to support global, national and regional efforts to make the Web universal, free and open. The campaign is focused on creating a world where everyone, everywhere is free to collaborate and create without fear of censorship or

persecution. A limited number of small grants are available to support national advocacy and campaigning efforts to make the web free, open, socially relevant and accessible to all. Applications will be accepted in either the Rapid Response or Surveillance Action Research categories. The Rapid Response has a rolling deadline and are micro-grants designed to help when you face an unexpected crisis or strategic opportunity that demands more resources fast. Surveillance Action Research grants will enable national groups or coalitions to investigate their own country's surveillance policies, laws and practices. Funds are available for action research to provide solid evidence to underpin your own local advocacy, media work and campaigning. The results of the country research will also be published under your group's name in a special report on global surveillance. Applications will be accepted until September 30.
*Requirements:* Applicants must join the Web We Want mailing list (signup at the website) prior to applying. Applications are encouraged from a broad and diverse range of groups and organizations, around the world. Priority will go to non-profit organisations in the global South and to those who show that they have the partnerships and follow-up mechanisms needed to reach, influence and mobilise others in their country. Rapid Response applicants may apply for any amount between $1,000 and $2,000 and can submit an application at any time. Decisions will be made within three (3) working days. If the application is successful funds will be administered within five (5) working days of approval decision. Surveillance Action Research applications can apply for any amount between $1,000 and $3,000 towards the costs of research, report writing and follow-up media work and advocacy based on your findings. All successful grant applicants will be listed publicly, with a brief description of project the grant is supporting on webwewant.org, unless the grant applicant shows that this would put the project at risk.
*Restrictions:* Grants cannot be used to fund the development or implementation of applications, software, computer equipment, computer literacy training, technology start-ups, hubs etc. These grants are only for advocacy, campaigning and influencing activities.
*Geographic Focus:* All States, All Countries
*Date(s) Application is Due:* Sep 30
*Amount of Grant:* 1,000 - 3,000 USD
*Contact:* Program Manager
*Internet:* webwewant.org/about/
*Sponsor:* World Wide Web Foundation
1889 F Street NW, 2nd Floor
Washington, D.C. 20006

**Welch Foundation Award in Chemistry** 2817

The purpose of the Robert A. Welch Award in Chemistry is to foster and encourage basic chemical research and to recognize, in a substantial manner, the value of chemical research contributions for the benefit of mankind. Any person can be considered for the award. The award is intended to recognize contributions that have not previously been rewarded in a similar manner. All nominations must include a brief biographical sketch of the proposed nominee, a list of 25 of the nominee's most important scientific publications, a statement of the research areas in chemistry on which the nomination is based, and not more than three seconding nomination letters. Nominations with supporting documents should be sent to the Foundation office by the January 31 deadline. The monetary amount of the award is $500,000.
*Requirements:* Nominations will be solicited from selected scientific organizations and individuals. Nominations received from other organizations and individuals will also be considered, though no person may nominate oneself.
*Restrictions:* Directors and officers of the foundation and members of the scientific advisory board of the foundation are not eligible.
*Geographic Focus:* All States
*Date(s) Application is Due:* Jan 31
*Amount of Grant:* 500,000 USD
*Contact:* Carla J. Atmar, Grant Programs Coordinator; (713) 961-9963, ext. 13 or (713) 961-9884; fax (713) 961-5168; welchaward@welch1.org
*Internet:* www.welch1.org/awards/welch-award-in-chemistry
*Sponsor:* Welch Foundation
5555 San Felipe, Suite 1900
Houston, TX 77056-2730

**Welch Foundation Departmental Grants** 2818

The purpose of the Welch Foundation's Departmental grant is to support chemical research by members of the chemistry department faculty, and to provide an opportunity for students to study chemistry in a less structured way. The grant is also designed to provide students with the tools they need to gain and develop an interest in pursuing a career in chemistry. This funding also ensures that students will have scholarships, lab equipment and chemicals and can travel to industry meetings and conferences. These grants are awarded to smaller and medium-sized educational institutions in Texas. Proposal submissions are by invitation only and must be submitted through proposalCentral by September 15 of each year.
*Restrictions:* Applications are restricted to smaller and medium-sized educational institutions in Texas.
*Geographic Focus:* Texas
*Date(s) Application is Due:* Sep 15
*Amount of Grant:* 150,000 USD
*Contact:* Carla J. Atmar, Grant Programs Coordinator; (713) 961-9963, ext. 13 or (713) 961-9884; fax (713) 961-5168; info@welch1.org
*Internet:* www.welch1.org/grants-programs/departmental-grants
*Sponsor:* Welch Foundation
5555 San Felipe, Suite 1900
Houston, TX 77056-2730

**Welch Foundation Endowed Chairs** 2819

The purpose of the Welch Foundation Endowed Chair program is to provide a faculty position for an eminent scientist, thus further increasing the basic scientific research in chemistry and other sciences in Texas. Historically, the institution must be willing to match the endowment funds dollar for dollar. Applications must be submitted by the last business day of January to be considered for the next grant year. The official online application on proposalCENTRAL must be used.
*Requirements:* To be considered for an endowed chair, a Texas educational institution must have a Ph.D. program in chemistry.
*Geographic Focus:* Texas
*Date(s) Application is Due:* Jan 31
*Contact:* Carla J. Atmar, Grant Programs Coordinator; (713) 961-9963, ext. 13 or (713) 961-9884; fax (713) 961-5168; info@welch1.org
*Internet:* www.welch1.org/grants-programs/endowed-chairs
*Sponsor:* Welch Foundation
5555 San Felipe, Suite 1900
Houston, TX 77056-2730

**Welch Foundation Norman Hackeman Award in Chemical Research** 2820

The purpose of the Welch Foundation Norman Hackeman Award in Chemical Research is to encourage and recognize young chemical scientists in Texas for their past research endeavors. The award is meant not only to recognize the work of a scientist, but to encourage those who are embarking on careers dedicated to increasing fundamental understanding of chemistry. The award is not intended to finance any further research or activities of the recipient, nor is the recipient required to render any future services as a condition of receiving the award. The award will normally be presented at a function on the campus of the recipient's institution. The current monetary amount of the award is $100,000.
*Requirements:* A candidate for the award must be 40 years of age or younger on June 30 for the year in which nominated. Any Texas college or university may nominate two candidates for any given year. The nomination should normally be by letter from the institution in which the individual is a faculty member. To be eligible, a nominee should have spent the most recent three years ending June 30 doing chemical research in Texas.
*Restrictions:* Directors and officers of the foundation and members of the scientific advisory board of the foundation are not eligible. Self-nominations are not accepted.
*Geographic Focus:* Texas
*Date(s) Application is Due:* Sep 15
*Amount of Grant:* 100,000 USD
*Contact:* Carla J. Atmar, Grant Programs Coordinator; (713) 961-9963, ext. 13 or (713) 961-9884; fax (713) 961-5168; nhacr@welch1.org
*Internet:* www.welch1.org/awards/norman-hackerman-award-in-chemical-research
*Sponsor:* Welch Foundation
5555 San Felipe, Suite 1900
Houston, TX 77056-2730

**Welch Foundation Research Grants** 2821

The Welch Foundation supports fundamental chemical research at educational institutions within the state of Texas. Each research grant provides a minimum of $150,000 in funding over a three year period to support research in chemistry by a full-time tenured or tenure-track faculty member who serves as principal investigator. Each research grant may utilize its funds on stipends, fellowships, scholarships, permanent scientific equipment, expendable scientific items, services, equipment maintenance, publication expenses, and travel. A research grant provides a minimum of $65,000 per year in funding to support research in chemistry by a full-time regular faculty member with tenure or on tenure-track who serves as principal investigator.
*Restrictions:* Applications are restricted to universities, colleges, or other educational institutions located in the state of Texas. Only one proposal may be submitted naming a particular principal investigator.
*Geographic Focus:* Texas
*Date(s) Application is Due:* Jan 31
*Amount of Grant:* 65,000 - 200,000 USD
*Contact:* Carla J. Atmar, Grant Programs Coordinator; (713) 961-9963, ext. 13 or (713) 961-9884; fax (713) 961-5168; info@welch1.org
*Internet:* www.welch1.org/grants-programs/research-grants
*Sponsor:* Welch Foundation
5555 San Felipe, Suite 1900
Houston, TX 77056-2730

**Wellcome Trust Biomedical Science Grants** 2822

Wellcome Trust supports research into all aspects of science: from molecules and cells vital to life, through the spread of diseases or the vectors of disease across the globe, to clinical and public health research to improve the quality of healthcare. Research can be based in the laboratory, the clinic or the field, and may involve experimental or theoretical approaches. The Trust not only funds scientists but also supports undergraduate and doctoral students, clinicians, dentists and veterinarians. It funds the best researchers with the most innovative and exciting ideas in the form of personal and research support, collaborations and support for symposia, conferences and workshops. A significant proportion of the funding is devoted to support those working overseas.
*Requirements:* Overseas applicants should fulfill the eligibility criteria for principal applicants. Principal applicants must be established researchers who are expected to apply from an eligible organization, able to sign up to the Wellcome Trust's Grant Conditions, and should normally: hold an academic or research post (or equivalent); have at least five years' postdoctoral (or equivalent) research experience (ndividuals who hold an established lectureship or a well-recognized fellowship, but who have less than five

years postdoctoral experience are also eligible; other researchers with less than five years' experience should contact the Grants Information Desk for further advice); be in receipt of salary funding for the duration of the grant requested or, if this is not in place, have a position that is underwritten by their employing institution for the duration of the grant.
*Geographic Focus:* All States, All Countries
*Contact:* Kevin Moses, Director of Science; +44 (0)20 7611 5757 or +44 (0)20 7611 8888; fax +44 (0)20 7611 8258; sciencegrants@wellcome.ac.uk
*Internet:* www.wellcome.ac.uk/Funding/Biomedical-science/index.htm
*Sponsor:* Wellcome Trust
Gibbs Building, 215 Euston Road
London, GLONDON NW1 2BE United Kingdom

**Wellcome Trust New Investigator Awards** **2823**
The Wellcome Trust is particularly interested in receiving applications from university lecturers within the first five years of their independent research careers. New Investigator Awards are intended to support strong researchers who are in the early stages of their independent research careers and have already shown that they can innovate and drive advances in their field of study. Candidates should be no more than five years from appointment to their first academic position. These Awards provide flexible support at a level and length appropriate to enable the best researchers to address the most important questions about health and disease. Awards may equally be small or large, but candidates should be able to articulate a compelling vision for their research, while ensuring that their proposal and requested funding is appropriate to their research experience to date. The research should contribute to the Trust's vision to achieve extraordinary improvements in human and animal health, and the proposal should describe how it will address one or more of our five major challenges. Prospective applicants are encouraged to contact the Trust if they wish to discuss their proposal or their suitability for this scheme.
*Requirements:* To be eligible for a New Investigator Award you should be based in the UK, Republic of Ireland or a low- or middle-income country, and you should: have an established academic post at an eligible higher education or research institution (you are employed on a permanent, open-ended or long-term rolling contract, salaried by your host institution); and be no more than five years from appointment to your first established academic post on the date you submit your application. More experienced researchers who have obtained significant, independent research support prior to their first academic appointment will be advised to apply for a Senior Investigator Award. Applicants are also eligible to apply if you have a written guarantee of an
*Geographic Focus:* Afghanistan, Albania, Algeria, American Samoa, Angola, Argentina, Armenia, Azerbaijan, Bangladesh, Belarus, Belize, Benin, Bhutan, Bolivia, Bosnia & Herzegovina, Botswana, Brazil, Bulgaria, Burkina Faso, Burundi, Cambodia, Cameroon, Cape Verde, Central African Republic, Chad, China, Colombia, Comoros, Congo, Congo, Democratic Republic of, Costa Rica, Cote d' Ivoire (Ivory Coast), Cuba, Djibouti, Dominica, Dominican Republic, Ecuador, Egypt, El Salvador, Eritrea, Ethiopia, Fiji, Gabon, Gambia, Ghana, Grenada, Guatemala, Guinea, Guinea-Bissau, Guyana, Haiti, Honduras, India, Indonesia, Iran, Iraq, Ireland, Jamaica, Jordan, Kazakhstan, Kenya, Kiribati, Korea, Kosovo, Kyrgyzstan, Laos, Lebanon, Lesotho, Liberia, Libya, Macedonia, Madagascar, Malawi, Malaysia, Maldives, Mali, Mauritania, Mauritius, Mexico, Micronesia, Moldova, Mongolia, Montenegro, Morocco, Mozambique, Myanmar (Burma), Namibia, Nepal, Nicaragua, Niger, Nigeria, Pakistan, Palau, Panama, Papua New Guinea, Paraguay, Peru, Philippines, Romania, Rwanda, Sao Tome & Principe, Senegal, Serbia, Seychelles, Sierra Leone, Solomon Islands, Somalia, South Africa, South Sudan, Sri Lanka, St. Lucia, St. Vincent and the Grenadines, Suriname, Swaziland, Syrian Arab Republic, Tajikistan, Tanzania, Thailand, Timor-Lester, Togo, Tonga, Tunisia, Turkey, Turkmenistan, Tuvalu, Uganda, Ukraine, United Kingdom, Uzbekistan, Vanuatu, Venezuela, Vietnam, Yemen, Zambia, Zimbabwe
*Date(s) Application is Due:* Mar 14; Jul 14; Nov 14
*Contact:* Kevin Moses, Director of Science; +44 (0)20 7611 5757 or +44 (0)20 7611 8888; fax +44 (0)20 7611 8258; sciencegrants@wellcome.ac.uk
*Internet:* www.wellcome.ac.uk/Funding/Biomedical-science/Funding-schemes/Investigator-Awards/WTX059284.htm
*Sponsor:* Wellcome Trust
Gibbs Building, 215 Euston Road
London, GLONDON NW1 2BE United Kingdom

**Weyerhaeuser Company Foundation Grants** **2824**
The foundation's mission is to improve the quality of life in communities where Weyerhaeuser has a major presence and to provide leadership that increases public understanding of issues where society's needs intersect with the interests of the forest products industry. Community service grants include awards in the fields of education and youth, health and welfare, civic and community improvement, and culture and the arts. In addition to community grants, industry-related awards are made to educational institutions, environmental groups, and professional organizations that promote further understanding of how the forest products industry responds to a changing society. Types of support include seed money, building construction/renovation, equipment acquisition, employee-related scholarships, publication, conferences and seminars, fellowships, lectureships, operating budgets, research, program development, employee matching gifts, and technical assistance. Applicants should send a short letter that introduces the project and sponsoring organization and provide tax-exempt status evidence. If further consideration is warranted, the foundation may ask for additional information or a formal proposal. Proposals may be submitted at any time.
*Requirements:* Applying organizations must have nonprofit, tax-exempt status. To be considered for funding, an organization must: serve a community within a 50-mile radius of a major Weyerhaeuser facility; support a state-wide issue of interest to the Foundation and Weyerhaeuser in the key states of Alabama, Arkansas, Louisiana, Mississippi, North Carolina, Oklahoma, Oregon or Washington; or support a selected, high-priority national or international initiative directly related to the sustainability and importance of working forests. A limited number of smaller awards are also made to other locales where fewer employees are based.
*Restrictions:* Grants are not awarded to individuals or for political campaigns, activities that influence legislation, religious organizations seeking funds for theological purposes, or funds to purchase tickets or tables at fundraising benefits.
*Geographic Focus:* Alabama, Arizona, Arkansas, California, Colorado, District of Columbia, Georgia, Idaho, Illinois, Kentucky, Louisiana, Maryland, Michigan, Minnesota, Mississippi, Missouri, New Hampshire, New Jersey, New Mexico, North Carolina, Ohio, Oklahoma, Oregon, Pennsylvania, South Carolina, Texas, Utah, Virginia, Washington, West Virginia, Wisconsin
*Date(s) Application is Due:* Aug 31
*Contact:* Anne Levya, Team Coordinator; (253) 924-3159; fax (253) 924-3658; foundation@weyerhaeuser.com or anne.leyva@weyerhaeuser.com
*Internet:* www.weyerhaeuser.com/Sustainability/Foundation
*Sponsor:* Weyerhaeuser Company Foundation
P.O. Box 9777
Federal Way, WA 98063-9777

**WHOI Geophysical Fluid Dynamics Fellowships** **2825**
The Geophysical Fluid Dynamics Fellowship involve an intense ten-week summer research experience and vigorous discussions of concepts that span different disciplines. Participants include graduate students and researchers from a variety of fields who share a common interest in the nonlinear dynamics of rotating, stratified fluids. Under the supervision of staff, Fellows pursue a research project and present a lecture and a written report for a proceedings volume. Participants receive a stipend of $6,284 and a $650 allowance for travel expenses within the U.S. Up to ten Fellowships are available. Applications and a list of required materials are available at the website.
*Requirements:* Applicants should be graduate students in any field sharing a common interest in the nonlinear dynamics of rotating, stratified fluids. These fields include classical fluid dynamics, physical oceanography, meteorology, astrophysics, planetary atmospheres, geological fluid dynamics, hydromagnetics, physics and applied mathematics. Fellows will be expected to attend the full ten weeks of the Program, complete a research project under the supervision of the staff, and present a lecture and a written report for a proceedings volume at the end of the program.
*Geographic Focus:* All States
*Date(s) Application is Due:* Feb 15
*Amount of Grant:* 6,284 - 6,934 USD
*Contact:* Janet Fields, GFD Program Coordinator; (508) 289-2950; fax (508) 457-2115; jfields@whoi.edu or gfd@whoi.edu
*Internet:* www.whoi.edu/main/gfd/fellowships
*Sponsor:* Woods Hole Oceanographic Institution
266 Woods Hole Road
Woods Hole, MA 02543-1050

**WHOI Guest Student Internships** **2826**
The Woods Hole Oceanographic Institution offers Guest Student Internships for a limited number of graduate and undergraduate students. Guest Student positions are extremely limited, and the ability to place a Guest Student depends on whether the student's interests and academic background coincide with available openings at the time the application is received. Scientists may invite their own Guest Students, so it is strongly recommended that a prospective student contact the scientist(s) with whom he or she is interested in working. Applications are accepted year-round and evaluated on a rolling basis, and are available at the admissions website.
*Requirements:* Applicants must be current undergraduate or graduate students. Advanced high school students enrolled in Upper Cape and Island schools may be eligible.
*Restrictions:* Guest Students are responsible for arranging their own room and board, and are expected to arrive with their own health insurance. Woods Hole does not issue academic credit for Guest Student work, but a description and evaluation of a student's work is available upon request.
*Geographic Focus:* All States
*Contact:* Kama Thieler, Undergraduate Programs Coordinator; (508) 289-3478; fax (508) 457-2188; kthieler@whoi.edu
*Internet:* www.whoi.edu/gueststudent/
*Sponsor:* Woods Hole Oceanographic Institution
266 Woods Hole Road
Woods Hole, MA 02543-1050

**WHOI Marine Policy Fellowships** **2827**
Qualified individuals in the social sciences are invited to apply their training and expertise to the economic, legal and policy issues that arise from use of the world's oceans. At the Institution's Marine Policy Center, emphasis is placed on multidisciplinary research to advance the conservation and management of coastal and marine resources. The work of Marine Policy Fellowship scholars integrates law, policy analysis, and statistics with WHOI's basic strengths in ocean sciences. Current areas of research concentration include fisheries and aquaculture; the economics of coastal communities; marine spatial planning and ecosystem-based management; sea level rise and shoreline change; and water resources management, with a particular focus on coastal aquifers and the effects of climate change. The main objectives of the program are the following: provide support and experience to Research Fellows interested in marine policy issues; offer opportunities for interdisciplinary application of social sciences and natural sciences to marine policy problems; and to conduct research and convey information necessary for the development of effective local, national, and international ocean policy. The Fellowship Program also

provides successful applicants the opportunity to participate in Woods Hole seminars and study groups and to develop collaborative research with other members of the scientific and technical staff at the Institution. Workshops and conferences involving senior policymakers and scholars are occasionally held as part of the Center's program. Recipients will receive a stipend of $57,000 for a period of one year and are eligible for group health and dental insurance. In addition, modest research and travel funds will be made available. Recipients are appointed for one year; however, depending upon the nature and scope of the recipient's effort, opportunities may be available to continue a professional affiliation with the Center. The application and information about additional materials needed are located at the website.
*Requirements:* Applicants must have completed a doctoral level degree or possess equivalent professional qualifications through career experience. The Center also welcomes experienced professionals who can arrange a leave or sabbatical.
*Geographic Focus:* All States
*Date(s) Application is Due:* Jan 5
*Amount of Grant:* 57,000 USD
*Contact:* Janet Fields, Fellowship Coordinator; (508) 289-2950; fax (508) 457-2115; jfields@whoi.edu or postdoc@whoi.edu
*Internet:* vishnu.whoi.edu/devel/webshare/whoi-redesign-final/education/postdoc_prgms_mpfellowship.html
*Sponsor:* Woods Hole Oceanographic Institution
266 Woods Hole Road
Woods Hole, MA 02543-1050

**WHOI Ocean Science Journalism Fellowship** **2828**
The Ocean Science Journalism Fellowship is designed to introduce science journalists to the interdisciplinary and wide-ranging fields of oceanography and ocean engineering. Through seminars, laboratory visits, and brief field expeditions, Fellows gain access to new research findings. They also gain fundamental background information in engineering, marine biology, geology and geophysics, marine chemistry and geochemistry, and physical oceanography. Topics include the following: harmful algal blooms to deep-sea hydrothermal vents; seafloor earthquakes to ice-sheet dynamics; the ocean's role in climate change to the human impact on fisheries and coastline change; and ocean instruments and observatories to underwater robots. The one-week residential experience offers travel allowance and room and board. International journalists are welcome to apply. The application and a list of required materials are available at the website. Applications are due by May 13, but acceptance begins in early January.
*Requirements:* The Fellowship is open to professional writers, producers, and editors working for print, broadcast, radio, and Internet media. Ideal candidates will have at least two years of writing, producing, or editing experience for a general-interest audience.
*Restrictions:* Freelancers, book authors, and writers from non-traditional media are also considered, but should be prepared to show a substantial portfolio of work. The application may be emailed, but required materials should be mailed.
*Geographic Focus:* All States, All Countries
*Date(s) Application is Due:* May 13
*Contact:* Erin Koenig, Media Relations Coordinator; (508) 289-2270; ekoenig@whoi.edu or media@whoi.edu
*Internet:* www.whoi.edu/page.do?pid=8029
*Sponsor:* Woods Hole Oceanographic Institution
266 Woods Hole Road
Woods Hole, MA 02543-1050

**WHOI Postdoctoral Fellowships** **2829**
Postdoctoral Fellows with national or international competitively awarded fellowships may be appointed at the Woods Hole Oceanographic Institution, provided they are able to locate a scientist on the staff willing to act as a sponsor. Once a sponsor has been located, and the appointment has been approved by the sponsor's Department chair and the Dean, the recipient will be appointed the appropriately named Postdoctoral Fellow at the Institution. An official letter of notification of the award from the national or international awarding organization must also be received. There is no deadline. The appointment time period will coincide with the fellowship award time period.
*Requirements:* Fellows must find a scientist on staff with the Institution who is willing to act as their sponsor.
*Geographic Focus:* All States
*Contact:* Janet Fields, Postdoctoral Coordinator; (508) 289-2950; fax (508) 457-2188; jfields@whoi.edu or postdoc@whoi.edu
*Internet:* www.whoi.edu/page.do?pid=8029
*Sponsor:* Woods Hole Oceanographic Institution
266 Woods Hole Road
Woods Hole, MA 02543-1050

**WHOI Postdoctoral Scholar Awards** **2830**
Postdoctoral Scholar awards are offered annually for 18 months to recipients of new or recent doctorates in the fields of chemistry, engineering, geology, geophysics, mathematics, meteorology, physics, and biology as well as oceanography. The competitive awards are designed to further the education and training of the applicant with primary emphasis placed on the individual's research promise. The 18 month appointment includes $58,000 per year, plus $28,500 for the additional six months, plus a relocation allowance. All appointees are eligible for group health and dental insurance. Limited support is available for travel expenses, equipment supplies, and special services. Additional guidelines are posted on the website. The application and a description of additional required materials are available at the website.
*Requirements:* In order to be eligible for one of these awards, applicants must have received their doctoral degree within the three years. As a condition of the award, successful applicants must be able to present a certified copy of their diploma, or a certified letter from the appropriate college or university official stating that they have completed all of the requirements for their degree including the completion of their final thesis, before they begin their scholar appointment. However, applicants do not need to have already received their degree prior to applying.
*Restrictions:* MIT/WHOI Joint Program Students, and those holding any type of WHOI appointment at the post-Ph.D. level during the 12 months prior to April 1st of the Scholar award year are not eligible for the Scholar competition.
*Geographic Focus:* All States
*Date(s) Application is Due:* Jan 5
*Amount of Grant:* 58,000 - 88,500 USD
*Contact:* Janet Fields, Postdoctoral Coordinator; (508) 289-2950; fax (508) 457-2115; jfields@whoi.edu or postdoc@whoi.edu
*Internet:* www.whoi.edu/apo/postdoctoral/scholarship/
*Sponsor:* Woods Hole Oceanographic Institution
266 Woods Hole Road
Woods Hole, MA 02543-1050

**WHOI Summer Student Fellowships** **2831**
The Summer Student Fellowships bring bright, talented students to Woods Hole to learn more about ocean science, attend lectures by Woods Hole scientists, and conduct independent research working alongside world-class scientists. All Fellows are expected to work on a project selected in collaboration with their sponsor(s) that will provide meaningful results in one summer's work. Project topics span the vast spectrum of research in ocean sciences and engineering conducted in WHOI's science departments and the Woods Hole Field Station of the U. S. Geological Survey (USGS). Fellowship recipients have the opportunity to attend and participate in a busy schedule of talks, seminars and a hands-on, one-day, at-sea practicum cruise focusing on data collection and sampling methods with advanced oceanographic technology and instruments. Fellows also have opportunities to interact with current MIT/WHOI Joint Program graduate students with a question and answer session designed specifically for that purpose with the Ethics in Science Workshop. Fellows receive $585 per week for the ten to twelve week program. A stipend of $650 for round-trip travel to Woods Hole is also available. International students are welcome to apply. Additional guidelines and program requirements are posted on the website. The application is available at the Admissions page.
*Requirements:* Students who have completed their junior year at colleges or universities by the start of the Fellowship period are eligible to apply. Preference is given to students studying in any of the fields of science or engineering including but not limited to the fields of biology, chemistry, engineering, geology, geophysics, mathematics, meteorology, physics, oceanography, and marine policy. Students must have at least a tentative interest in the ocean sciences, oceanographic engineering, mathematics, or marine policy.
*Geographic Focus:* All States, All Countries
*Date(s) Application is Due:* Feb 15
*Amount of Grant:* 6,500 - 7,670 USD
*Contact:* Kama Thieler, Undergraduate Programs Coordinator; (508) 289-3478; fax (508) 457-2188; kthieler@whoi.edu or education@whoi.edu
*Internet:* www.whoi.edu/main/summer-student-fellowship
*Sponsor:* Woods Hole Oceanographic Institution
266 Woods Hole Road
Woods Hole, MA 02543-1050

**Wilhelm Sander Stiftung Foundation Grants** **2832**
The Wilhelm Sander Stiftung Foundation was established in Germany in 1973. The purpose of the Foundation is to support research in the field of human medicine, with a strong focus on cancer research, both clinical and clinical-experimental. Applications are accepted electronically (online) or via hard copy. There are no annual deadlines specified, and interested applicants should begin by contacting the Foundation directly.
*Restrictions:* The support by the Wilhelm Sander Stiftung Foundation is limited to Germany and Switzerland.
*Geographic Focus:* Germany, Sweden
*Amount of Grant:* Up to 100,000 EUR
*Contact:* Bernhard Knappe; +49 (089) 544-1870; fax +49 (089) 544-18720; info@sanst.de
*Internet:* wilhelm-sander-stiftung.de/
*Sponsor:* Wilhelm Sander-Stiftung Foundation
Goethe route 74
Munich, 80336 Germany

**Willard and Pat Walker Charitable Foundation Grants** **2833**
The Willard and Pat Walker Charitable Foundation awards grants to Arkansas nonprofit organizations in its areas of interest, including: the arts; children and youth services; health organizations; higher education; residential and custodial hospice care; and social services. Types of support include: building construction and renovation; capital campaigns; continuing support; endowments; equipment; general operating grants; matching grants; program development; and scholarship endowment funds. There are no application forms, so applicants should provide a detailed description of their project and the amount of funding requested. There are two annual deadlines of March 1 and October 1. The Board meets twice per year, in April and November
*Requirements:* Arkansas nonprofit organizations are eligible to apply.
*Restrictions:* Individuals are not eligible.
*Geographic Focus:* Arkansas
*Date(s) Application is Due:* Mar 1; Oct 1

*Amount of Grant:* Up to 1,000,000 USD
*Samples:* Crystal Bridges Museum of American Art, Bentonville, Arkansas, $1,000,000 - endowment contribution; New School, Fayetteville, Arkansas, $1,000,000 - challenge grant; Razorback Foundation, Fayetteville, Arkansas, $625,000 - capital campaign.
*Contact:* John M. Walker; (479) 582-2310; fax (479) 582-2292; walkerfamily1@sbcglobal.net
*Sponsor:* Willard and Pat Walker Charitable Foundation
P.O. Box 10500
Fayetteville, AR 72703-2857

**Willary Foundation Grants** **2834**
The Willary Foundation supports projects that are interesting, creative, and imaginative and that benefit communities in northeastern Pennsylvania. The Foundation is particularly interested in projects that support leadership and the development of leadership in business, the economy, education, human services, government, the arts, media, and research. The Foundation gives preference to efforts that have a ripple effect in the community and those that are conducted in conjunction with other sources of funding. Willary seeks to foster both individuals and groups with unique, innovative, or unusual ideas and efforts. The application and samples of previously funded projects are available at the Foundation website.
*Requirements:* Individuals and nonprofits in Lackawanna and Luzerne counties, Pennsylvania, are eligible to apply.
*Restrictions:* Generally, Willary will not consider applications for capital campaigns and annual drives.
*Geographic Focus:* Pennsylvania
*Date(s) Application is Due:* Mar 10; Aug 27
*Amount of Grant:* 1,200 - 50,000 USD
*Samples:* Center for Cancer Wellness, Forty Fort, Pennsylvania, an exercise program geared toward improving the health of the participants to increase their physical activity: $20,000; United Neighborhood Centers, Scranton, Pennsylvania, photography and creative writing program for low income youth between the ages of 11 and 18: $25,000.
*Contact:* Linda Donovan, Administrative Director; (570) 961-6952; fax (570) 961-7269; info@willary.org
*Internet:* www.willary.org
*Sponsor:* Willary Foundation
P.O. Box 283
Scranton, PA 18501-0937

**William and Flora Hewlett Foundation Education Grants** **2835**
The William and Flora Hewlett Foundation has been making grants since 1967 to solve social and environmental problems at home and around the world. Two types of Education Grants are funded. The first is deeper learning which delivers the skills and knowledge students need to succeed in a world that is changing at an unprecedented pace. Deeper learning prepares students to: master core academic content; think critically and solve complex problems; work collaboratively; communicate effectively; and learn how to learn (e.g., self-directed learning). The second is open educational resources which are high-quality, openly licensed, online educational materials that offer an extraordinary opportunity for people everywhere to share, use, and reuse knowledge. Open educational resources demonstrate great potential as a mechanism for instructional innovation as networks of teachers and learners share best practices.
*Requirements:* Applicants whose project or organization aligns closely with the Foundation's goals and strategies, may submit a brief on-line letter of inquiry. The Foundation will provide Education Grants supporting deeper learning to organizations that: promote policies or strategies that create incentives for K-12 schools and community colleges to focus on deeper learning; build the capacity of the education system and teaching practices both online and in the classroom to reach large numbers of students with deeper learning; support proof points, including model K-12 schools and community colleges, and fund research that promotes deeper learning as an attainable and necessary goal for all students; and develop new, innovative models to increase access for all students and to improve deeper learning. The Foundation will provide Education Grants supporting open educational resources primarily by supporting the infrastructure needed to sustain a well-functioning open educational ecosystem and demonstration projects that strengthen student access to deeper learning in K-12 and community colleges.
*Restrictions:* The Foundation does not consider requests to fund student aid, individual scholarships, construction, equipment and computer purchases, health research, or health education programs. Few, if any, letters of inquiry are funded. A full proposal should not be submitted unless expressly requested. The majority of awards are made to organizations by invitation. Due to the volume of submissions the Foundation does not respond to phone calls or emails regarding letters of inquiry. Submissions via the online letter of inquiry form receive a response within thirty days or as soon thereafter as possible.
*Geographic Focus:* All States
*Contact:* Barbara Chow; (650) 234-4500; fax (650) 234-4501; info@hewlett.org
*Internet:* www.hewlett.org/Grantseekers
*Sponsor:* William and Flora Hewlett Foundation
2121 Sand Hill Road
Menlo Park, CA 94025

**William and Flora Hewlett Foundation Environmental Grants** **2836**
The William and Flora Hewlett Foundation has been making grants since 1967 to solve social and environmental problems at home and around the world. The Foundation's Environmental Grants goals are to: conserve the Western United States and Canada for wildlife and people; slow global climate change by reducing greenhouse gas emissions; ensure that the United States energy supply is clean and consumption is efficient; and address environmental problems that disproportionately affect disadvantaged communities in the San Francisco Bay Area. Support for western conservation support four key strategies: land (protecting large open spaces); water (restoring river flows and conserve areas near rivers and streams); energy (reducing fossil fuel development and increase energy efficiency and renewable energy sources; and broad-based support (building broad-based support for land, water, and energy goals among key stakeholders). Support for energy and climate pursue strategies in three areas: global climate policy; national energy policy; and sustainable transportation. Support for the San Francisco Bay Area is provided to regional organizations partnering with residents of these communities to: improve and expand urban parks; support outdoor recreation programs for youth; and improve public transportation available to these communities.
*Requirements:* Applicants whose project or organization aligns closely with the Foundation's goals and strategies, may submit a brief on-line letter of inquiry. Western conservation is supported in twelve states and three Canadian provinces: Alaska, Arizona, California, Colorado, Idaho, Montana, Nevada, New Mexico, Oregon, Washington, Wyoming, British Columbia, Alberta, and Yukon.
*Restrictions:* The Foundation does not consider requests to fund student aid, individual scholarships, construction, equipment and computer purchases, health research, or health education programs. Few, if any, letters of inquiry are funded. A full proposal should not be submitted unless expressly requested. The majority of awards are made to organizations by invitation. Due to the volume of submissions the Foundation does not respond to phone calls or emails regarding letters of inquiry. Submissions via the online letter of inquiry form receive a response within thirty days or as soon thereafter as possible.
*Geographic Focus:* All States, Canada
*Contact:* Tom Steinbach; (650) 234-4500; fax (650) 234-4501; info@hewlett.org
*Internet:* www.hewlett.org/programs/environment-program
*Sponsor:* William and Flora Hewlett Foundation
2121 Sand Hill Road
Menlo Park, CA 94025

**William Andrews Clark Memorial Library Ahmanson-Getty Postdoctoral Fellowships** **2837**
Theme-based, residential fellowships are awarded for periods of two consecutive academic quarters for participation in the interdisciplinary, cross-cultural programs of the Center for 17th- & 18th-Century Studies/William Andrews Clark Memorial Library. The theme for each academic year is announced the preceding fall. An organized research unit of the University of California, the center provides a forum for the discussion of central issues in the study of the 17th and 18th centuries, facilitates research and publication, and encourages the creation of programs that advance understanding of this time period. The William Andrews Clark Memorial Library, which is administered by the center, is known for its collections on 17th- and 18th-century Britain, Oscar Wilde and the 1890s, the history of printing, and certain aspects of Western Americana. Application materials are available upon request. The stipend for this fellowship is equal to $43,692 for the three-quarter period including paid medical benefits for scholar and dependents.
*Requirements:* Scholars who have received a Ph.D. in the last six years and are engaged in research pertaining to the announced theme are eligible to apply. Fellows are expected to make a substantive contribution to the Center's workshops and seminars. Awards are for one full academic year in residence at the Clark Library.
*Geographic Focus:* All States
*Date(s) Application is Due:* Feb 1
*Amount of Grant:* 43,692 USD
*Contact:* Myrna Ortiz, Fellowship Coordinator; (310) 206-8552 or (323) 731-8529; fax (323) 731-8617; ortiz@humnet.ucla.edu or clark@humnet.ucla.edu
*Internet:* www.1718.ucla.edu/research/postdoctoral/
*Sponsor:* William Andrews Clark Memorial Library
2520 Cimarron Street
Los Angeles, CA 90018

**William Andrews Clark Memorial Library ASECS/Clark Fellowships** **2838**
One-month resident fellowships at the William Andrews Clark Memorial Library are available to postdoctoral scholars with projects in the Restoration or the 18th century. The library, which is administered by the Center for 17th- and 18th-Century Studies, is known for its collections on 17th- and 18th-century Britain, Oscar Wilde and the 1890s, the history of printing, and certain aspects of the American West. The application deadline is for fellowships to be held between July 1 of the current year and June 30 of the next. Applications are available on the Web site. The stipend is a total of $3,000 for the one-month residency.
*Requirements:* Scholars who have received a PhD in the last six years and are engaged in research pertaining to the announced theme are eligible to apply.
*Geographic Focus:* All States
*Date(s) Application is Due:* Feb 1
*Amount of Grant:* 3,000 USD
*Contact:* Myrna Ortiz, Program Contact; (310) 206-8552 or (323) 731-8529; fax (323) 731-8617; ortiz@humnet.ucla.edu or clark@humnet.ucla.edu
*Internet:* www.1718.ucla.edu/research/postdoctoral/
*Sponsor:* William Andrews Clark Memorial Library
2520 Cimarron Street
Los Angeles, CA 90018

**William Andrews Clark Memorial Library Clark-Huntington Joint Bibliographical Fellowships** **2839**
Sponsored jointly by the William Andrews Clark Memorial Library and the Huntington Library, this two-month fellowship provides support for bibliographical research in early modern British literature and history as well as other areas where the two libraries have common strengths. The Application form is an online submission, and the applicant

should forward a description of research plans, specifying any material required at the Clark and the Huntington Libraries (maximum of 1,000 words, double-spaced). The stipend is equal to $6,000 for the two-month residency.
*Requirements:* Applicants should hold a PhD degree or have appropriate research experience.
*Geographic Focus:* All States
*Date(s) Application is Due:* Feb 1
*Amount of Grant:* 6,000 USD
*Contact:* Myrna Ortiz, Fellowship Coordinator; (310) 206-8552 or (323) 731-8529; fax (323) 731-8617; ortiz@humnet.ucla.edu or clark@humnet.ucla.edu
*Internet:* www.1718.ucla.edu/research/postdoctoral/
*Sponsor:* William Andrews Clark Memorial Library
2520 Cimarron Street
Los Angeles, CA 90018

**William Andrews Clark Memorial Library Dissertation Fellowships 2840**
One or two fellowships are awarded each year to UCLA doctoral candidates whose dissertation involves extensive research in the Clark Library's holdings. The award of $18,000 (plus fixed graduate fees, excluding nonresident tuition) is for nine months during the academic year. Information on the library's holdings, application, and guidelines are available on the Web site.
*Geographic Focus:* All States
*Date(s) Application is Due:* Feb 1
*Amount of Grant:* 18,000 USD
*Contact:* Myrna Ortiz, Fellowship Manager; (310) 206-8552 or (323) 731-8529; fax (323) 731-8617; ortiz@humnet.ucla.edu or clark@humnet.ucla.edu
*Internet:* www.1718.ucla.edu/research/graduate/
*Sponsor:* William Andrews Clark Memorial Library
2520 Cimarron Street
Los Angeles, CA 90018

**William Andrews Clark Memorial Library Kanner Fellowship in British Studies 2841**
The Kanner Fellowship in British Studies is a three-month fellowship that supports research at the Clark Library in any area pertaining to British history and culture. The fellowship is open to both postdoctoral and pre-doctoral scholars. Information on the library's holdings, application, and guidelines are available on the Web site. The stipend is equal to $9,000 for the three-month tenure.
*Requirements:* The fellowship is open to both postdoctoral and pre-doctoral scholars. Scholars holding PhD or equivalent degrees involved in a project suitable to the collection are eligible to apply.
*Geographic Focus:* All States
*Date(s) Application is Due:* Feb 1
*Contact:* Myrna Ortiz, Fellowship Manager; (310) 206-8552 or (323) 731-8529; fax (310) 206-8577; ortiz@humnet.ucla.edu or clark@humnet.ucla.edu
*Internet:* www.humnet.ucla.edu/humnet/c1718cs/applic5.htm
*Sponsor:* William Andrews Clark Memorial Library
2520 Cimarron Street
Los Angeles, CA 90018

**William Andrews Clark Memorial Library Predoctoral Fellowships 2842**
Three-month residential Predoctoral Fellowships are awarded to advanced doctoral candidates at the William Andrews Clark Memorial Library whose dissertations concern any area represented in the Clark's collections or are linked to Center/Clark programs supported by these collections or by others at UCLA. The Center for 17th- and 18th-Century Studies, an organized research unit of UCLA, provides a forum for the discussion of central issues in the study of the 17th and 18th centuries, facilitates research and publication, supports scholarship, and encourages the creation of interdisciplinary, cross-cultural programs that advance the understanding of this period. The application deadline of February 1 is for fellowships to be carried out between July 1 of one year and June 30 of the next. The stipend is equal to $3,000 per month for the one-to-three-month residency.
*Requirements:* UCLA graduate students are eligible.
*Geographic Focus:* All States
*Date(s) Application is Due:* Feb 1
*Amount of Grant:* 3,000 - 9,000 USD
*Contact:* Myrna Ortiz, Fellowship Coordinator; (310) 206-8552 or (323) 731-8529; fax (323) 731-8617; ortiz@humnet.ucla.edu or clark@humnet.ucla.edu
*Internet:* www.1718.ucla.edu/research/graduate/
*Sponsor:* William Andrews Clark Memorial Library
2520 Cimarron Street
Los Angeles, CA 90018

**William Andrews Clark Memorial Library Short-Term Resident Fellowships 2843**
The Short-Term Fellowships offered by the William Andrews Clark Memorial Library facilitate access to the Department of Special Collections for researchers and scholars residing outside of the Los Angeles area. Fellowships are held April through December for periods of one to three months. The department collects primary resources in the humanities and social sciences. Application requirements include a cover letter, a curriculum vita, a brief outline of the research and specific collections to be used (two-page maximum), dates to be spent in residence, and three letters of recommendation. Information on the library's holdings, application, and guidelines are available on the Web site. The stipend for this fellowship is equal to $3,000 per month.
*Requirements:* The fellowship is open to both postdoctoral and pre-doctoral scholars. Scholars holding PhD or equivalent degrees involved in a project suitable to the collection are eligible to apply.
*Geographic Focus:* All States
*Date(s) Application is Due:* Feb 1
*Amount of Grant:* 3,000 - 9,000 USD
*Contact:* Myrna Ortiz, Fellowship Contact; (310) 206-8552 or (323) 731-8529; fax (323) 731-8617; ortiz@humnet.ucla.edu or clark@humnet.ucla.edu
*Internet:* www.1718.ucla.edu/research/postdoctoral/
*Sponsor:* William Andrews Clark Memorial Library
2520 Cimarron Street
Los Angeles, CA 90018

**William and Sandy Heitz Family Foundation Grants 2844**
The William and Sandy Heitz Family Foundation was established in Milwaukee, Wisconsin, in 1994. The Foundation's primary fields of interest include: cancer research; the study of diseases and conditions; graduate and professional education; health and health care; heart and circulatory system diseases; human services; independent living for people with disabilities; medical education; public health; and special population support. Population groups targeted are: academics; children and youth; the economically; people with disabilities; hearing impairments; vision impairments; and students. Types of funding include: equipment purchase; general operating support; program development; and research and evaluation. Applicants should submit a detailed description of project and the amount of funding requested. Most recent awards have ranged from $250 to $75,000. There are no specified annual deadlines for application submission.
*Requirements:* Any 501(c)3 organization serving the Foundation's primary fields of interest are eligible to apply.
*Geographic Focus:* All States
*Amount of Grant:* 250 - 75,000 USD
*Samples:* Clemson University Foundation, Clemson, South Carolina, $250 - support of education; Medical College of Wisconsin, Milwaukee, Wisconsin, $75,000 - research support; Lukemia Lymphoma Foundation, Port Chester, New York, $2,000 - research support.
*Contact:* Sandra Heintz, President; (262) 377-3970
*Sponsor:* William and Sandy Heitz Family Foundation
10800 N. Haddonstone Place
Mequon, WI 53092

**William Blair and Company Foundation Grants 2845**
Contributing to the community is an important part of the culture of William Blair and Company. The Foundation was officially established in Illinois in 1980, with giving primarily centered around metropolitan Chicago. All partners of the firm contribute part of their individual share of profits to the Foundation. Donation requests are made to the Foundation by partners and employees. The Foundation supports a broad range of causes including: civic affairs; public safety; arts and culture; higher education; youth-oriented activities; healthcare research; cultural affairs; and civic charities. Types of support include: annual campaigns; building and renovation; capital campaigns; general operating support; endowments; fellowships; internship programs; and scholarship funding. There are no specific deadlines or applications forms required. Funding typically ranges from $500 to $25,000.
*Requirements:* Requests can be made by sending a letter with a general description of the organization and its special purpose. Activities should have a significant impact on the Chicago metropolitan area.
*Geographic Focus:* Illinois
*Samples:* Doctors without Borders, Chicago, Illinois, $20,500 - general operating support; Department of Prints & Drawing at the Art Institute of Chicago, Illinois, $5,000 - general operating support; Greater Chicago Food Depository, Chicago, Illinois, $10,000 - general operating support.
*Contact:* E. David Coolidge III, Vice President; (312) 236-1600
*Sponsor:* William Blair and Company Foundation
222 W Adams Street, 28th Floor
Chicago, IL 60606

**William G. and Helen C. Hoffman Foundation Grants 2846**
Helen C. Hoffman resided in the Village of South Orange, New Jersey. Her foundation was established in 1998 in memory of herself and her husband after the death of their daughter Corinne Blair. Her testamentary wish was to establish this foundation for charitable, religious, scientific, literary, and educational purposes. Her preference was to support blindness and its cure. Approximately 90% of the grants will provide support to the blind and, to medical research for the prevention of blindness. The remaining 10% will fund annual grants in the following areas of interest: education, the arts, environment and, social/civic causes. Requests must be received by January 15 for the March meeting and, August 22 for the October meeting. Application forms are available online. Applicants will receive notice acknowledging receipt of the grant request, and subsequently be notified of the grant declination or approval.
*Requirements:* Any U.S. 501(c)3 non-profit organizations may apply. Proposals should be submitted in the following format: completed Common Grant Application Form; an original Proposal Statement; an audited financial report and a current year operating budget; a copy of your official IRS Letter with your tax determination; a listing of your Board of Directors. Proposal Statements (second item in the above Format) should answer these questions: what are the objectives and expected outcomes of this program/project/request; what strategies will be used to accomplish your objective; what is the timeline for completion; if this is part of an on-going program, how long has it been in operation; what criteria will you use to measure success; if the request is not fully funded, what other sources can you engage; an Itemized budget should be included; please describe any collaborative ventures. Prior to the distribution of funds, all approved grantees must sign and return a Grant Agreement Form, stating that the funds will be used for the purpose intended. Progress reports and Completion reports must also be

filed as required for your specific grant. All current grantees must be in good standing with required documentation prior to submitting new proposals to any foundation.
*Restrictions:* Grants are not made for political purposes, nor to organizations which discriminate on the basis of race, ethnic origin, sexual or religious preference, age or gender.
*Geographic Focus:* All States
*Date(s) Application is Due:* Jan 15; Aug 22
*Amount of Grant:* 5,000 - 15,000 USD
*Samples:* Tri-County Scholarship Fund, $5,000—scholarships for economically disadvantaged students to receive a private elementary/secondary education; New Eyes for the Needy, Inc., $20,000— eyeglasses for poor Kentuckians; Christian Health Care Center Foundation, $10,000— support senior subsidies for adult day services.
*Contact:* Wachovia Bank, N.A., Trustee; grantinquiries2@wachovia.com
*Internet:* https://www.wellsfargo.com/private-foundations/hoffman-foundation
*Sponsor:* William G. and Helen C. Hoffman Foundation
190 River Road, NJ3132
Summit, NJ 07901

**William G. Rohrer Charitable Foundation Grants** **2847**
William G. Rohrer (1909-1989) was an American Republican Party politician, businessman and philanthropist who served as the Mayor of Haddon Township, New Jersey, for 36 years. The Foundation founded in his honor was established in New Jersey in 1990. Aside from contributing more $12 million to Rowan University, where the College of Business was named the William G. Rohrer College of Business, the Foundation currently awards grants to New Jersey nonprofit organizations in its areas of interest, including: higher education; health care; health associations; and human services. There are no application deadlines or forms with which to adhere, and interested parties should contact the office directly.
*Geographic Focus:* New Jersey
*Amount of Grant:* Up to 200,000 USD
*Samples:* Coriell Institute for Medical Research, Camden, New Jersey, $200,000 - general operating support; Haddon Township School District, Haddon Township, New Jersey, $15,000 - general operating support; Exponent Philanthropy, Washington, D.C., $2,725 - general operating support.
*Contact:* Grants Administrator; (856) 638-4876 or (856) 638-4862
*Sponsor:* William G. Rohrer Charitable Foundation
1600 Market Street, 4th Floor
Philadelphia, PA 19103-7240

**William H. Adams Foundation for ALS Grants** **2848**
Established in 1999, the William H. Adams Foundation researches and funds programs specializing in the treatment of Amyotrophic Lateral Sclerosis (ALS) or Lou Gehrig's disease. Since there are no formal applications required, interested parties should contact the Foundation directly. Most recent grant amounts have ranged from $10,000 to $250,000, and are either in the form of general donations or research specific. There are no specific annual deadlines for making application.
*Geographic Focus:* All States
*Amount of Grant:* 10,000 - 250,000 USD
*Samples:* John Hopkins University, Baltimore, Maryland, $250,000 - medical research; San Francisco Film Society, San Francisco, California, $10,000 - general donation; University of California, San Francisco Foundation, San Francisco, California, $20,000 - research for ALS.
*Contact:* Ellen Adams, President; (415) 592-8151
*Sponsor:* William H. Adams Foundation for ALS
1918 43rd Avenue
San Francisco, CA 94116-1025

**William O. Seery Foundation Grants** **2849**
The William O. Seery Foundation was established in Connecticut, and was named in honor of Reverend William Owen Seery, Parochial Vicar at SS Peter and Paul Church and Saint Leo the Great Church in Waterbury. He graduated from Marianapolis Preparatory School and the University of Hartford, with a Masters Degree in Organizational Behavior. He served in the USMC and prior to entering the seminary he owned William O. Seery Associates in Hartford. The Foundation primary fields of interest include: breast cancer; other types of cancer; education; health; and research. A formal application is not required, and interested parties should begin by contacting the Foundation directly. There are no specified application deadlines.
*Geographic Focus:* All States
*Amount of Grant:* Up to 50,000 USD
*Samples:* Health Research in Action, Boston, Massachusetts, $45,000 - to support cancer research; Multiple Myeloma Research for Jim Grazick, Norwalk, Connecticut, $45,000 - to support Myeloma research at Washington University Medical.
*Contact:* Co-Trustee; (888) 866-3275
*Sponsor:* William O. Seery Foundation
P.O. Box 1802
Providence, RI 02901-1802

**William T. Grant Foundation Scholars Program** **2850**
The William T. Grant Scholars Program supports promising early-career researchers from diverse disciplines, who have demonstrated success in conducting high-quality research and are seeking to further develop and broaden their expertise. Candidates are nominated by a supporting institution and must submit five-year research plans that demonstrate creativity, intellectual rigor, and a commitment to continued professional development. Every year, four to six William T. Grant Scholars are selected and each receives $350,000 distributed over a five-year period.
*Requirements:* To be eligible for consideration, applicants must: be employed at a nonprofit institution, either in the United States or abroad; submit a project that is consistent with the Foundation's Current Research Interests; address issues that have compelling relevance for theory, and policies or practices, affecting the settings of youth ages 8 to 25 in the United States or a vulnerable subpopulation of those youth; and have received their terminal degree within seven years of submitting their application.
*Restrictions:* The Foundation rarely funds international work. The William T. Grant Scholars award must not replace the institution's current support of the applicant's research nor should it duplicate any other funding, received or pending.
*Geographic Focus:* All States
*Date(s) Application is Due:* Jul 6
*Amount of Grant:* 350,000 USD
*Samples:* Elizabeth Oltmans Ananat, Ph.D., Duke University, North Carolina, $350,000 - "Economic and Social Determinants of the Educational Choices of Young Adults"; Phillip Atiba Goff, Ph.D., University of California, Los Angeles, $350,000 - "Broken Windows, Broken Youth: The Effect of Law Enforcement Activity on non-White Adolescent Male Development"; Patrick Sharkey, Ph.D., New York University, New York, $350,000 - "The Impact of Acute Violence and Other Environmental Stressors on Cognitive Functioning and School Performance".
*Contact:* Irene Williams, Grants Coordinator; (212) 752-0071; fax (212) 752-1398; iwilliams@wtgrantfdn.org or info@wtgrantfdn.org
*Internet:* www.wtgrantfoundation.org/funding_opportunities/fellowships/william_t__grant_scholars/william_t_grant_scholars
*Sponsor:* William T. Grant Foundation
570 Lexington Avenue, 18th Floor
New York, NY 10022-6837

**Wilton and Effie Hebert Foundation Grants** **2851**
The Wilton and Effie Hebert Foundation was established in Texas in 1992, with funding aimed primarily at supporting programs that serve children and youth, as well as hospitals and health organizations. Funding is also available for elementary and secondary education, higher education, social services of all types, and Catholic organizations and churches. Support typically takes the form of capital and infrastructure, general operating funding, individual development programs, research and evaluation, and scholarship funds. There are no annual deadlines for application submissions, and awards generally range up to a maximum of $40,000.
*Requirements:* Any 501(c)3 organization serving the residents of Texas is eligible to apply.
*Restrictions:* Funding is not given directly to individuals.
*Geographic Focus:* Texas
*Amount of Grant:* Up to 40,000 USD
*Contact:* James M. Black, Attorney at Law; (409) 982-9433 or (409) 727-2345
*Sponsor:* Wilton and Effie Hebert Foundation
P.O. Box 908
Port Neches, TX 77619

**Winn Feline Foundation/AVMF Excellence In Feline Research Award** **2852**
The Winn Foundation has partnered with the American Veterinary Medical Foundation (AVMF) to recognize the veterinary research scientist whose work mirrors Winn's efforts to improve the lives of "Every Cat, Every Day". The Winn/AVMF Excellence in Feline Research Award is accompanied by a $2,500 cash award, a crystal cat sculpture, plus up to $1,000 to offset round trip airfare and other travel expenses. The Research Award is paired with a matching scholarship award by the AVMF for a veterinary student interested in feline medicine. For nomination forms and additional information related to supporting materials, award criteria, and the selection process, visit www.avma.org/awards.
*Requirements:* Applicants should include a cover letter, a list of significant research achievements, and a concise curriculum vita (CV) including relevant publications. The cover letter must include the name of the award; the name, mailing address, telephone number(s), email address, and college and year of graduation of the nominee; the nature of the nominee's professional activity (type of practice or type of salaried work); organizational memberships (professional and scientific) of the nominee; a narrative sketch of the nominee's professional background; and a statement pertaining to the nominee's qualifications for the award. To be considered, nominees must have demonstrated Winn's mission to advance feline health and welfare through research; have a pattern of following good research study protocols; complete progress reports in a timely manner; and have their research published and acknowledge Winn as a donor of grants when appropriate.
*Restrictions:* No self-nominations will be considered.
*Geographic Focus:* All States
*Date(s) Application is Due:* Feb 1
*Contact:* Janet Wolf, Executive Director; (856) 447-9798; winnfeline@aol.com
Eileen Hoblit, Award Contact; (800) 248-2862, ext. 6778; fax (847) 925-1329
*Internet:* www.winnfelinehealth.org/Pages/Researchers.html
*Sponsor:* Winn Feline Foundation
390 Amwell Road, Suite 402
Hillsborough, NJ 08844

**Winn Feline Foundation Grants** **2853**
The foundation supports research studies into medical problems affecting cats with the Winn Feline Foundation Grant and the Miller Trust Grant. The Foundation supports research on feline asthma, anemia, inflammatory bowel disease, hyperthyroidism, FIP, megacolon, polycystic kidney disease, transdermal drug delivery, liver disease, and vaccine-associated fibrosarcoma. Studies general in scope are primarily encouraged; however, the foundation is also interested in projects that address problems unique

to individual breeds. Continuation of grants awarded will be considered. Application guidelines and current deadlines are available online.
*Requirements:* Applicants may be faculty veterinarians, postdoctoral fellows, practicing veterinarians, or veterinary students.
*Restrictions:* The foundation does not fund salaries of principal investigators, major equipment expenditures, travel or indirect costs.
*Geographic Focus:* All States
*Amount of Grant:* Up to 35,000 USD
*Samples:* DNA Array Analyses for Cat Diseases, University of California, Davis, $25,000; A Reproducible Protocol to Isolate a Characterized Population of Adult Feline Progenitor Cell, Louisiana State University School of Veterinary Medicine, $9,995; Evolution of Feline Infectious Peritonitis Virus Within FIP Cats and Tissue-specific Adaptation of the Virus to Activating Proteases, Cornell University, $23,986.
*Contact:* Janet Wolf, Grants Administrator; (908) 359-1184; fax (908) 359-7619; winn@winnfelinehealth.org
*Internet:* www.winnfelinehealth.org
*Sponsor:* Winn Feline Foundation
390 Amwell Road, Suite 402
Hillsborough, NJ 08844

**Winston Churchill Foundation Scholarships** **2854**
The program offers American citizens of exceptional ability and outstanding achievement the opportunity to pursue graduate studies in engineering, mathematics, or the sciences at the University of Cambridge. The foundation awards at least thirteen scholarships and hopes to increase that number to fifteen soon. The one-year awards lead to the Masters of Philosophy (MPhil), the Certificate of Post-Graduate Study (CPGS in different fields), the Certificate of Advanced Study (CAS in theoretical or applied mathematics), and a Diploma (in Computer Science). On occasion the foundation offers the award to an individual who has made outstanding contributions exemplifying Churchill's attributes and ideals. The Scholarship is worth between $44,000 and $50,000, and covers all University and College tuition and fees. In addition, students receive a living allowance of 10,000 GBP if enrolled in a nine-month academic program and 12,000 GBP if enrolled in a full-year academic program. They also receive an allowance of up to $1,000 for travel to and from the United Kingdom. The foundation also offers the possibility of a Special Research Grant of up to $2,000; this grant may cover travel for presentations at international conferences, short stays at another university or institute for special research, and other activities. Married students should consult with the Foundation about the possibility of additional support.
*Requirements:* Applicants for a Churchill Scholarship must be citizens of the United States and must be a senior who is enrolled in one of the institutions participating in the program competition or a student who has recently graduated from one of those institutions. Upon taking up a scholarship, a Churchill Scholar must be between the ages of 19 and 26, hold a bachelor's degree or its equivalent, and may not have attained a doctorate. Instructions, forms and applications are available at the sponsor's website.
*Geographic Focus:* All States
*Date(s) Application is Due:* Nov 1
*Amount of Grant:* 44,000 - 50,000 USD
*Contact:* Michael Morse, Executive Director; (212) 752-3200; fax (212) 246-8330; info@winstonchurchillfoundation.org
*Internet:* www.winstonchurchillfoundation.org/Scholarships.html
*Sponsor:* Winston Churchill Foundation of the United States
600 Madison Avenue, Suite 1601
New York, NY 10022-1615

**Winterthur Museum Dissertation Fellowships** **2855**
Founded almost 60 years ago by collector and horticulturist Henry Francis du Pont, Winterthur consists of a prestigious museum of American decorative arts, a research library that serves scholars around the world, and a 1,000 acre preserve located on the grounds of du Pont's childhood home. The organization offers various opportunities to both the public and the academic community. Winterthur's Dissertation Fellowships support doctoral candidates in conducting research and writing their dissertations in areas of social and cultural history, including material culture, architecture, decorative arts, design, consumer culture, garden and landscape studies, Shaker studies, travel and tourism, the Atlantic World, and objects in literature. Dissertation fellowships last four to nine months. Fellows receive a stipend of up to $7,000 per semester. Housing is available on the Winterthur grounds and is within easy walking distance of the library and museum where fellows enjoy many staff privileges, including 24-hour access to the library. Detailed guidelines and a downloadable application form are available at the website. The preferred method of submission is email; however, applicants may also mail the completed application and required supporting documentation (essay, bibliography, vita, etc.) to the address given. All applicants are strongly encouraged to search Wintercat (the library's online catalog), visit Winterthur, and contact staff members to discuss potential research projects. The suitability of a project to Winterthur's collections is the primary consideration of the fellowship award committee.
*Requirements:* Applicants must be doctoral candidates researching or writing their dissertation on a topic that fits well with Winterthur's collections.
*Restrictions:* Candidates applying for a Dissertation Fellowship or a National Endowment for the Humanities Fellowship may ask to also be considered for a short-term research fellowship in the event that the primary application is not accepted.
*Geographic Focus:* All States, All Countries
*Date(s) Application is Due:* Jan 15
*Amount of Grant:* Up to 7,000 USD
*Contact:* Rosemary Krill; (302) 888-4637; fax (302) 888-4870; rkrill@winterthur.org
*Internet:* www.winterthur.org/?p=535
*Sponsor:* Winterthur Museum, Garden, and Library
5105 Kennett Pike
Wilmington, DE 19735

**Winterthur Research Fellowships** **2856**
Founded almost 60 years ago by collector and horticulturist Henry Francis du Pont, Winterthur consists of a prestigious museum of American decorative arts, a research library that serves scholars around the world, and a 1,000 acre preserve located on the grounds of du Pont's childhood home. The organization offers various opportunities to both the public and the academic community. Winterthur offers one- to three-month research fellowships to encourage research in America's artistic, cultural, intellectual, and social history. The stipend is $1,500 per month. While applicants need not apply for a specific named fellowship, Winterthur will designate certain awards, for instance: the Faith Andrews Fellowships (for the study of Shaker life and material culture); the Robert Lee Gill Fellowships (for research on American decorative arts, painting, architecture, or historic preservation); and the Neville McD. Thompson Fellowships (for the study of domestic life, late 19th- and early 20th-century design and material culture). Grants are available to academic, museum, and independent scholars, and to support dissertation research. Resources include a library of approximately one-half million imprints, manuscripts, visual materials, and printed ephemera supporting interdisciplinary study of American life into the early 20th century. Housing is available on the grounds. Fellows are provided with a library carrel and 24-hour access to the library's circulating collections. Detailed application guidelines and a downloadable form are available at the website. The preferred method of submission is email; however, applicants may also mail the completed application and required supporting documentation (essay, bibliography, vita, etc.) to the address given. All applicants are strongly encouraged to search Wintercat (the library's online catalog), visit Winterthur, and contact staff members to discuss potential research projects. The suitability of a project to Winterthur's collections is the primary consideration of the fellowship award committee.
*Requirements:* Applications are welcome from academic, museum, and independent scholars, including graduate students.
*Geographic Focus:* All States, All Countries
*Date(s) Application is Due:* Jan 15
*Amount of Grant:* 1,500 - 5,500 USD
*Contact:* Rosemary Krill, Senior Lecturer; (302) 888-4637; rkrill@winterthur.org
*Internet:* www.winterthur.org/?p=535
*Sponsor:* Winterthur Museum, Garden, and Library
5105 Kennett Pike
Wilmington, DE 19735

**Wisconsin American Legion Auxiliary Scholarships** **2857**
The American Legion Auxiliary in Wisconsin offers scholarships each year to high school students, college students, and graduate students to provide financial support for education to students with outstanding scholastic ability who are in financial need. Various named scholarships are available as well as scholarships for special areas such as nursing, health careers, and special education. When requesting an application form, the applicant should specify if he or she is a high school student, college student, or graduate student; there are three different forms. Some of the special area scholarships have additional requirements; ask for guidelines when requesting the forms.
*Requirements:* Students must be residents of Wisconsin; be the daughter, son, wife, or widow of a veteran; have a minimum 3.2 GPA on a 4.0 scale; and be in need of financial aid to continue education. The granddaughter or great-granddaughter may apply if she is an American Legion Auxiliary member.
*Geographic Focus:* Wisconsin
*Date(s) Application is Due:* Mar 15
*Amount of Grant:* 750 - 2,000 USD
*Contact:* Grants Administrator; (608) 745-0124 or (866) 664-3863; fax (608) 745-1947; alawi@amlegionauxwi.org
*Internet:* www.amlegionauxwi.org
*Sponsor:* Wisconsin American Legion Auxiliary
P.O. Box 140
Portage, WI 53901-0140

**WMU Department of Spanish Travel Study Abroad Award** **2858**
The Department of Spanish Travel Study Abroad Award is intended for Spanish majors or minors. The purpose of these awards is to encourage students to further develop their Spanish language skills and to broaden their understanding of other cultures and peoples. These awards may be used for travel and/or study abroad during any part of the academic year or summer. Preference will be given to those who enroll in WMU sponsored programs.
*Requirements:* Applicants must have: a declared major or minor in the Department of Spanish; completed a minimum of two courses in the Spanish major/minor by the end of the semester of application; and a minimum overall GPA of 3.25, plus a GPA of 3.5 in the Spanish major or minor.
*Restrictions:* No student may receive a Travel/Study Abroad Award more than once. Recipients of this award are not eligible for the President's Award or Scholarship nor are the recipients of the President's Award/Scholarships eligible for a Travel/Study Abroad Award.
*Geographic Focus:* All States
*Contact:* Irma Lopez; (269) 387-3023; fax (269) 387-3103; spanish@wmich.edu
*Internet:* www.wmich.edu/spanish/scholarships-awards/travstud.htm
*Sponsor:* Western Michigan University
420 Sprau Tower
Kalamazoo, MI 49008-5338

**Wolfsonian-FIU Fellowships** 2859

The Wolfsonian is a museum, library, and research center donated by Mitchell Wolfson, collector of decorative and propaganda arts, to Florida International University in 1997. The museum explores, through its exhibitions and special programs, the critically important role of design at the height of the industrial age (1885-1945) in the context of social, political, and technological issues. The collections comprise approximately 120,000 objects from the period in a variety of media including furniture; industrial-design objects; works in glass, ceramics, and metal; rare books; periodicals; ephemera; works on paper; paintings; textiles; and medals. The Wolfsonian-FIU Fellowship program promotes scholarly research on The Wolfsonian's collections. Since its inception in 1995, the program has supported projects from a wide range of academic fields. Fellowships are awarded for full-time research at The Wolfsonian, generally for periods of three to four weeks. Fellowships include a stipend, accommodations, and round-trip travel. The timing of dates will be negotiated with individual awardees. The annual submission deadline is December 31 for a residency beginning no earlier than July 1 of the following year. Applicants are strongly encouraged to discuss their project with the Fellowship Coordinator prior to submitting their application. Applications are submitted via an online form accessible from the fellowship's webpage at the Wolfsonian website.

*Requirements:* The program is open to holders of master's or doctoral degrees, Ph.D. candidates, and to others who have a significant record of professional achievement in relevant fields. Scholars from outside of the United States are eligible. Fellowships are awarded without regard to race, color, gender, religion, national origin, sexual orientation, age, or disability.

*Geographic Focus:* All States, All Countries

*Date(s) Application is Due:* Dec 31

*Contact:* Coordinator; (305) 535-2623 or (305) 531-1001; research@thewolf.fiu.edu

*Internet:* www.wolfsonian.org/research-library/fellowships

*Sponsor:* Florida International University

1001 Washington Avenue

Miami Beach, FL 33139

**Women's International League for Peace and Freedom Internships** 2860

The Women's International League for Peace and Freedom (WILPF) works to achieve through peaceful means world disarmament, full rights for women, racial and economic justice, an end to all forms of violence, and to establish those political, social, and psychological conditions which can assure peace, freedom, and justice for all. Internships, which provide a monthly stipend of 900 CHF, are available to women throughout the year, including the Jane Addams Internship, in Philadelphia, Pennsylvania; the United Nations Internship, in New York, New York; and the WILPF International Internship, in Geneva, Switzerland. Internships are available from January 1 to March 30; June 12 to August 30; and September 15 to December 15. To apply, send a resume writing sample, internship goals, and date available with desired internship placement.

*Requirements:* Internships are for women, preferably between the ages of 25 and 30. Preference is given to WILPF members.

*Restrictions:* The internship does not provide for health or other insurance, or health-related expenses.

*Geographic Focus:* All States, Switzerland

*Date(s) Application is Due:* Apr 1; Jul 15; Nov 15

*Contact:* Ashish Mahajan; (212) 682-1265; fax (212) 682-1265; amahajan@wilpf.ch

*Internet:* https://wilpf.org/get-involved/work-with-us/

*Sponsor:* Women's International League for Peace and Freedom

777 UN Plaza, 6th Floor

New York, NY 10017

**WWF Russell E. Train Education for Nature Alumni Grants** 2861

Education for Nature (EFN) is committed to ensuring that our alumni continue to advance in their careers and become a voice for conservation. By providing these ongoing education opportunities, EFN Alumni can gain up-to-date training and knowledge needed to advance conservation efforts in their home countries and regions. EFN Alumni Grants support ongoing education and training for former Train Fellows or Scholars who have earned a degree through EFN. Many Alumni use these opportunities to present a paper at a conference, attend a workshop or short-term training courses offered by universities and other institutions, or to conduct research within a WWF priority place.

*Requirements:* To be eligible for an EFN Alumni Grant, applicants must: be a former Russell E. Train Fellow or Scholar; have submitted all reports required by their grant contracts; have completed their two-year work requirement in their home country; have a bank account in their own name that can receive U.S. dollars; not have received an EFN Alumni Grant from EFN in the past three years.

Restrictions: Grant requests may not exceed U.S. $3,500 and alumni must observe a three-year waiting period before requesting another grant.

*Geographic Focus:* All States, All Countries

*Contact:* Mercedes Gonzalez, Senior Program Officer; (202) 778-9612; efn@wwfus.org or mercedes.gonzalez@wwfus.org

*Internet:* worldwildlife.org/initiatives/efn-alumni-grants

*Sponsor:* World Wildlife Fund

P.O. Box 97180

Washington, D.C. 20090-7180

**WWF Russell E. Train Fellowships** 2862

Russell E. Train Fellowships support individuals pursuing a master's or doctoral degree in conservation. Each year, WWF supports a small number of committed conservationists from target countries to receive financial support for their studies and field research. Applicants can apply to attend any university around the world and must return to their home countries to work in conservation for at least two years after completing their degree. Applicants from select WWF-U.S. priority countries must meet all of the eligibility criteria to be considered for a grant. Fellows will receive financial support for tuition, books, travel to and from the training institution, and room and board for an academic program for a maximum of $20,000 USD (per year) for up to two years; limited costs for field research required by the academic institution to obtain the degree; long-term professional development support through the EFN Alumni Program.

*Requirements:* To be eligible for a Train Fellowship, applicants must: be a citizen and legal resident of a participating country; have a minimum of two year's work experience in conservation; be enrolled, admitted, or have applied to an institution of higher education; and, begin their studies within one year from the application deadline for their respective countries. Participating countries: Cameroon, Central African Republic, Democratic Republic of Congo, Gabon, Kenya, Malawi, Mozambique, Nepal, Republic of Congo, Tanzania, Thailand, and Vietnam. Eligibility criteria may vary slightly among countries.

*Restrictions:* Individuals who have received a fellowship from EFN in the past and who wish to pursue a second degree are not eligible.

*Geographic Focus:* Cameroon, Central African Republic, Congo, Congo, Democratic Republic of, Gabon, Kenya, Malawi, Mozambique, Nepal, Tanzania, Thailand, Vietnam

*Date(s) Application is Due:* Feb 28

*Amount of Grant:* Up to 20,000 USD

*Contact:* Mercedes Gonzalez, Senior Program Officer; (202) 778-9612; efn@wwfus.org or mercedes.gonzalez@wwfus.org

*Internet:* worldwildlife.org/initiatives/russell-e-train-fellowships

*Sponsor:* World Wildlife Fund

P.O. Box 97180

Washington, D.C. 20090-7180

**Xerox Foundation Grants** 2863

Xerox Foundation Grants assist a variety of social, civic and cultural organizations that provide broad-based programs and services in cities where our employees work and live. The Foundation also remains committed to a program of grants to colleges and universities to prepare qualified men and women for careers in business, science, government, and general education. The Foundation seeks to further advance knowledge in science and technology, and to enhance learning opportunities for minorities and the disadvantaged. Worldwide, Xerox philanthropy tries to engage national leadership in addressing major social problems and to support programs in education, employability and cultural affairs. Other areas of particular focus include programs responsive to the national concern for the environment and the application of information technology. Large grants may be approved for more than one year (multi-year grants). All organizations that have previously received support on an annual basis from the Foundation, must re-submit a request each year to be evaluated for continued support.

*Requirements:* Grants are made only to 501(c)3 and 509(a) organizations. No specific application form is used. Requests for grants/funding should be submitted in letter form describing the project or program. This request should contain the legal name of the organization, the official contact person, its tax- exempt status, a brief description of its activities and programs, the purpose for which the grant is being requested, the benefits expected, the plans for evaluation, the projected budget, and the expected sources and amount of needed funds.

*Restrictions:* The foundation declines requests to support individuals; capital grants (new construction or renovation); endowments or endowed chairs; organizations supported by United Way, unless permission has been granted by United Way to a member agency to conduct a capital fund drive or a special benefit; political organizations or candidates; religious or sectarian groups; or municipal, county, state, federal, or quasi-government agencies.

*Geographic Focus:* All States

*Amount of Grant:* 500 - 50,000 USD

*Contact:* Joseph M. Cahalan, President; (203) 968-2453 or (203) 968-3000

*Internet:* www.xerox.com/about-xerox/citizenship/xerox-foundation/enus.html

*Sponsor:* Xerox Foundation

45 Glover Avenue, P.O. Box 4505

Norwalk, CT 06856-4505

**Yale University Beinecke Library Fellowships for Visiting Scholars** 2864

The Yale University Beinecke Rare Book and Manuscript Library offers short-term Fellowships to support visiting scholars pursuing post-doctoral or equivalent research in its collections. The Fellowships are designed to provide access to the library for scholars living outside the greater New Haven area. Fellowships pay for travel to and from New Haven, and include a living allowance of $4,000 per month. Granted for one month, Fellowships must be taken between September 1 and May 31. Recipients are expected to be in residence during the period of their award, and are encouraged to participate in the activities of Yale University. The application is located at the Beinecke website, along with a list of frequently asked questions about the application process.

*Requirements:* Applicants must explain in specific detail the relationship of the Beinecke collections to their project and its significance within the larger field of scholarly work. Along with the online application form, candidates must submit the following items to the selection committee: their curriculum vitae; a brief research proposal; a detailed list of specific research materials to be consulted at the library during their Fellowship; and two letters of recommendation, either sealed, or sent directly from the recommenders.

*Restrictions:* Students enrolled in degree programs are not eligible.

*Geographic Focus:* All States

*Date(s) Application is Due:* Dec 7

*Amount of Grant:* 4,000 USD

*Contact:* Jennifer Castellon, Senior Administrative Assistant; (203) 432-2977 or (203) 432-7325; fax (203) 432-4047; jennifer.castellon@yale.edu or beinecke.library@yale.edu

*Internet:* www.library.yale.edu/beinecke/brbleduc/brblapplyvisiting.html

*Sponsor:* Yale University Library
Sterling Memorial Library
New Haven, CT 06520-8240

**Yale University Beinecke Library Research Fellowships for Yale Graduate and Professional School Students** 2865

The Beinecke Library offers Research Fellowships for Yale Graduate and Professional School students who wish to use Beinecke collections as a primary resource for their dissertations or culminating projects. All awarded Fellows are expected to dedicate a minimum of 25 hours of research in the Beinecke collections per week throughout the Fellowship period (attendance is monitored and stipends are adjusted to reflect shortened research periods). At the end of the research period, Fellows are also asked to provide a brief statement indicating how this Fellowship support was used to accomplish research goals. Graduate students seeking these Fellowships must have completed their course work, passed their qualifying examinations, and be prepared to pursue research based upon an approved prospectus. Professional school students seeking these Fellowships must be engaged in research for an approved culminating project required for their degree, with the expectation that their project will be undertaken in the final year of their program. Short-term and long-term Fellowships are available, depending on the needs of the applicant. Short-term Fellowships for $3,000 per month are available for one, two, or three months, and may be held at any time. Long-term Fellowships for $13,650 are for one full semester. The application is available at the Beineke website.
*Requirements:* Along with the application, the following materials must be submitted: a curriculum vitae; a proposal explaining in detail the specific relationship between the Beinecke Collectons and the applicant's research; a detailed list of specific research materials to be consulted at the Beineke; and two sealed letters of recommendation. Graduate students must submit an approved or pending prospectus. Professional School students must submit a statement explaining how their project fulfills the culminating requirement in their school.
*Restrictions:* Only students whose proposals demonstrate a need for extensive use of Beinecke materials will be considered for these awards. Awarded Fellows may not hold concurrent teaching positions or any concurrent Fellowships (other than, when appropriate, the standard Graduate School support). Applications from previous recipients of these Fellowships will be considered only in exceptional circumstances.
*Geographic Focus:* All States
*Date(s) Application is Due:* Mar 8; Oct 19
*Contact:* Jennifer Castellon; (203) 432-7325; jennifer.castellon@yale.edu
*Internet:* www.library.yale.edu/beinecke/brbleduc/brblapplystudent_research.html
*Sponsor:* Yale University Library
Sterling Memorial Library
New Haven, CT 06520-8240

**Z. Smith Reynolds Foundation Fellowships** 2866

The Z. Smith Reynolds Foundation offers one two-year, paid fellowship, beginning July 1 of each year, to any individual who demonstrates an interest in philanthropy, public policy, community service and/or the nonprofit sector. These positions are salaried and include benefits. The fellow acquires increasing responsibility over the course of two years in evaluating grant proposals, visiting with grant applicants, assisting with grantmaking administration, working on projects of importance to the Foundation, and providing additional support to Foundation staff as needed. Through the grant process and related programs, the fellow gains experience with nonprofit organizations, public policy, Foundation trustees and other leaders across North Carolina. Each fellow has the opportunity to assist in furthering projects related to the current focus areas of the Foundation. The fellowship is structured to enable the fellow to further explore areas that are of interest to his or her professional development and that align with the Foundation's goals, mission and vision. The fellow is also responsible for coordinating the Foundation's internship programs. Applications are due by January 3 each year.
*Requirements:* Eligible candidates must be natives or residents of North Carolina, and/or graduates of an accredited North Carolina college or university. Further qualifications include: strong communication skills, both written and verbal; the ability to multi-task; the desire to contribute to an ever-changing work environment; a sense of responsibility and personal initiative; confidence and poise; willingness to work as a team player; and respect for diverse communities and leadership. Preference is given to recent college/ university graduates. People of color are encouraged to apply.
*Geographic Focus:* North Carolina
*Date(s) Application is Due:* Jan 3
*Contact:* Khaetlyn Grindell, Fellow; (336) 725-7541, ext. 114; fax (336) 725-6069; khaetlyng@zsr.org or info@zsr.org
*Internet:* https://www.zsr.org/grants-programs/zsr-fellowship
*Sponsor:* Z. Smith Reynolds Foundation
102 West Third Street, Suite 1110
Winston-Salem, NC 27101

**Z. Smith Reynolds Foundation Non-Profit Internships** 2867

Summer internships offer a crucial pathway to future employment for the next generation of nonprofit leaders. However, the universality of unpaid internships in the nonprofit sector create financial hurdles for students to participate in experiential learning. In an effort to diversify the pool of young nonprofit professionals in North Carolina, the Z. Smith Reynolds Foundation established the Non-Profit Internship Program (NPIP) in 2011. Since 2011, the program has awarded paid internship opportunities to 20 Pell grant-eligible college and university students enrolled in four-year institutions across the state. This year, the program will expand through a collaboration of six funders and will offer 36 paid internships for the summer of 2017. Funders involved in the collaboration include: Blue Cross and Blue Shield of North Carolina Foundation; the Foundation for a Healthy High Point; Kate B. Reynolds Charitable Trust; Weaver Foundation; the Winston-Salem Foundation; and Z. Smith Reynolds Foundation. Interns receive a taxable stipend of $3,200 of which the funder provides two-thirds ($2,133) in a lump sum to organizations. Host organizations must provide the remaining one-third ($1,067) and are responsible for paying the stipend to the intern. Each Host Organization is eligible for only one intern per summer.
*Requirements:* Eligible students must be: a rising sophomore, junior, or senior enrolled full-time in a four-year North Carolina college or university, or a rising sophomore, junior or senior North Carolina resident enrolled full-time in a four-year accredited college outside the state; and Pell grant-eligible in the academic year in which you apply for NPIP.
*Geographic Focus:* North Carolina
*Date(s) Application is Due:* Jan 23
*Amount of Grant:* 3,200 USD
*Contact:* Allie Garrett, Associate Program Officer; (336) 725-7541, ext. 113; fax (336) 725-6069; allieg@zsr.org or info@zsr.org
*Internet:* https://www.zsr.org/grants-programs/non-profit-internship-program
*Sponsor:* Z. Smith Reynolds Foundation
102 West Third Street, Suite 1110
Winston-Salem, NC 27101

**Z. Smith Reynolds Foundation Summer Internship** 2868

The Z. Smith Reynolds Foundation offers one eight-week paid summer internship to an undergraduate or graduate student interested in philanthropy, public policy, community service and/or the nonprofit sector. The Foundation provides a stipend for the internship as well as an additional housing stipend if the student does not live within commuting distance of Winston-Salem. Interns have the opportunity to assist staff in researching projects related to the long-term goals of the Foundation, particularly those that align with its current focus areas. They also have the opportunity to shadow Foundation staff, attend conferences and meetings, and/or pursue personal projects of interest. Interested candidates should complete the online application by January 3.
*Requirements:* Eligible candidates must be either currently enrolled in an accredited North Carolina undergraduate or graduate program or a North Carolina resident. Applicants must demonstrate an interest in philanthropy, public policy, community service and/or the nonprofit sector. Applicants must also demonstrate strong communication skills, personal initiative and the willingness to support Foundation staff through intensive research projects.
*Geographic Focus:* North Carolina
*Date(s) Application is Due:* Jan 3
*Contact:* Khaetlyn Grindell, Fellow; (336) 725-7541, ext. 114; fax (336) 725-6069; khaetlyng@zsr.org or info@zsr.org
*Internet:* https://www.zsr.org/grants-programs/zsr-summer-internship
*Sponsor:* Z. Smith Reynolds Foundation
102 West Third Street, Suite 1110
Winston-Salem, NC 27101

**Zabrowski Family Foundation Grants** 2869

The Zabrowski Family Foundation was established by Dr. Randal Zabrowski in 2007, a radiologist in Sheboygan, Wisconsin, who is affiliated with Aspirus Grand View Hospital. He received his medical degree from University of Wisconsin School of Medicine and Public Health and has been in practice for more than 20 years. The Foundation's primary fields of interest are to support: Catholicism; education; and human service programs. This funding is generally given for: general operating support; policy, advocacy, and systems reform; research; and research evaluation. There are no application forms or annual deadlines, so interested parties should begin by contacting the Foundation directly. Most recently, awards have ranged up to a maximum of $2,000.
*Requirements:* Any 501(c)3 organization serving residents of Wisconsin or Virginia are eligible to apply.
*Geographic Focus:* Virginia, Wisconsin
*Amount of Grant:* Up to 2,000 USD
*Samples:* Army Emergency Relief, Alexandria, Virginia, $2,000 - general operating support; Foundation of Religious Retirement, St. Francis, Wisconsin, $125 - general operating support; Lake Country Academy, Sheboygan, Wisconsin, $2,000 - general operating support.
*Contact:* Dr. Randal Zabrowski, President; (920) 457-0211
*Sponsor:* Zabrowski Family Foundation
4541 North Evergreen Drive
Sheboygan, WI 53081

**Zayed Sustainability Prize-Energy** 2870

The Zayed Sustainability Prize is a tribute to the legacy of the late founding father of the UAE, Sheikh Zayed bin Sultan Al Nahyan–a champion of environmental stewardship and global sustainability. This annual award celebrates achievements that are driving impact, innovation and inspiration across five distinct categories: Health, Food, Energy, Water and Global High Schools. This prize recognizes organizations and programs that empower people by providing access to energy and energy efficiency solutions.
*Requirements:* To be eligible for this prize, you should be a small or medium-sized enterprise, a social enterprise, a non-governmental organization (NGO), a non-profit organization (NP.O.), or a similar organization (with revenues of less than $50 million USD per year). Organizations from the Americas, Europe and Central Asia, Sub-Saharan Africa, East Asia and the Pacific, South Asia, or MENA are eligible to apply.
*Geographic Focus:* All States, All Countries
*Date(s) Application is Due:* Aug 9
*Amount of Grant:* 600,000 USD
*Contact:* Prize Coordinator; +971 2 653 0005; fax +971 2 653 6002; info@zayedsustainabilityprize.com

*Internet:* https://zayedsustainabilityprize.com/en/prize-categories/energy/
*Sponsor:* Zayed Sustainability Prize
P.O. Box 54115
Abu Dhabi, 00000 United Arab Emirates

**Zayed Sustainability Prize-Water** **2871**
The Zayed Sustainability Prize is a tribute to the legacy of the late founding father of the UAE, Sheikh Zayed bin Sultan Al Nahyan–a champion of environmental stewardship and global sustainability. This annual award celebrates achievements that are driving impact, innovation and inspiration across five distinct categories: Health, Food, Energy, Water and Global High Schools. This prize recognizes organizations and programs that empower people by providing access to safe and affordable drinking water. Also of interest are organizations that are active in improving water quality, sanitation, water use efficiency, protection of water-related ecosystems and community participation.
*Requirements:* To be eligible for this prize, you should be a small or medium-sized enterprise, a social enterprise, a non-governmental organization (NGO), a non-profit organization (NP.O.), or a similar organization (with revenues of less than $50 million USD per year). Organizations from the Americas, Europe and Central Asia, Sub-Saharan Africa, East Asia and the Pacific, South Asia, or MENA are eligible to apply.
*Geographic Focus:* All States, All Countries
*Date(s) Application is Due:* Aug 9
*Amount of Grant:* 600,000 USD
*Contact:* Prize Coordinator; +971 2 653 0005; fax +971 2 653 6002; info@zayedsustainabilityprize.com
*Internet:* https://zayedsustainabilityprize.com/en/prize-categories/water/
*Sponsor:* Zayed Sustainability Prize
P.O. Box 54115
Abu Dhabi, 00000 United Arab Emirates

**Zonta International Foundation Amelia Earhart Fellowships** **2872**
The Zonta International Foundation awards 35 Fellowships of $10,000 each annually to assist women in advanced study and research in a suitable area of aerospace science or engineering. These Amelia Earhart Fellowships are open to women of any nationality and may be used at any institution of higher learning offering accredited graduate courses and degrees in aerospace studies. The Fellowship may be used for tuition, books and fees, or living expenses. Requests for a second award by current fellows undergo the same evaluation procedures as the initial application; a third award may be made in exceptional cases. The successful candidate identifies the relevance of her field of study to aerospace-related science or aerospace-related engineering, demonstrates outstanding academic credentials, and specifically outlines her professional or career goals as well as detailing an original focus in her research project. Applications for the Fellowships must be received or post-marked by November 15.
*Requirements:* Women of any nationality pursuing a Ph.D./doctoral degree who demonstrate a superior academic record in the field of aerospace-related sciences and aerospace-related engineering are eligible.
*Geographic Focus:* All States, All Countries
*Date(s) Application is Due:* Nov 15
*Amount of Grant:* 10,000 USD
*Contact:* Martina Gamboa, Programs Senior Coordinator; (630) 928-1400; fax (630) 928-1559; mgamboa@zonta.org or programs@zonta.org
*Internet:* https://foundation.zonta.org/Our-Programs/Educational-Programs/Amelia-Earhart-Fellowship
*Sponsor:* Zonta International Foundation
1211 West 22nd Street, Suite 900
Oak Brook, IL 60523

# SUBJECT INDEX

**Armenia**



**Art Appreciation**



**Art Conservation**



**Art Criticism**



**Art Education**



**Art History**

**Arts, Fine**



**Arts, General**

**Biology Education**



**Biology, Behavioral**



**Biology, Cellular**



**Biology, Conservation**



**Biology, Developmental/Evolutionary**



**Biology, Molecular**

**Biomedical Research Resources**



**Biomedical Research Training**

**Children's Literature**



**Children's Museums**



**Chile**



**China**



**Chinese Art**



**Chinese Language/Literature**



**Chinese Studies**



**Choreography**



**Chromatography**



**Chronic Illness**



**Churches**

**Clinical Psychology**



**Clinical Research**



**Clinical Sociology**



**Clinical Trials**



**Clinics**

**Community Education**



**Community Outreach**



**Community Services**

**Conferences, Travel to**



**Conflict/Dispute Resolution**



**Conservation**



**Conservation Education**



**Conservation, Agriculture**

**Dentistry**



**Depression**



**Dermatology**



**Deserts and Arid Zones**



**Design Arts**



**Developing/Underdeveloped Nations**



**Developmental Psychology**



**Developmentally Disabled**



**Diabetes**



**Diabetic Retinopathy**



**Diagnosis, Medical**

**Education**

**Engineering Administration/Management**



**Engineering Design**



**Engineering Education**

**Feminism**



**Fencing**



**Fetus Research**



**Fiber Optics**



**Fiction**



**Film Production**



**Films**



**Finance**



**Financial Aid (Scholarships and Loans)**

**Genomics**



**Geochemistry**



**Geography**



**Geography Education**



**Geology**



**Geology, Marine**



**Geometry**



**Geophysics**



**Geosciences**



**Geriatrics**



**German Language/Literature**



**German Studies**



**Germany**



**Gifted/Talented Education**



**Girls**



**Global Change**



**Global Issues**

**Health Care**

**Health Promotion**



**Health Records**



**Health Research**



**Health Sciences**

**Higher Education Administration**



**Higher Education Studies**

**Humanities**

**Humanities Education**



**Hungary**



**Hunger**



**Hurricanes**

**International Justice**



**International Organizations**



**International Planning/Policy**



**International Programs**



**International Relations**



**International Students**

**Mathematics Education**



**Mathematics and Statistics**



**Mathematics, Applied**



**Mechanical Engineering**

**Mechanics**



**Media**



**Media Arts**



**Medical Devices Engineering**



**Medical Education**



**Medical Ethics**



**Medical Informatics**



**Medical Physics**

**Medical Programs**



**Medical Research**

**Medical Sciences**



**Medical Technology**



**Medical/Diagnostics Imaging**



**Medicinal Chemistry**



**Medicine**

**Minorities, Ethnic**



**Minority Education**



**Minority Employment**



**Minority Health**



**Minority Schools**



**Minority/Woman-Owned Business**



**Molecular Probes**



**Mongolia**



**Montana**



**Morphology**



**Multiculturalism**



**Multidisciplinary Arts**

**Native Americans**



**Natural History**



**Natural Resources**



**Natural Sciences**

**Physical Disability**



**Physical Medicine and Rehabilitation**



**Physical Sciences**



**Physical Sciences Education**



**Physics**



**Physics Education**



**Physics, Applied**



**Physics, Atomic**



**Physics, Computational**



**Physics, High Energy**



**Physics, Molecular**



**Physics, Nuclear**

**Rehabilitation/Therapy**



**Rehabilitation/Therapy, Emotional/Social**



**Rehabilitation/Therapy, Occupational/Vocational**



**Rehabilitation/Therapy, Physical**



**Religion**



**Religious Pluralism**



**Religious Studies**

**Research Resources (Health/Safety/Medical)**



**Respiratory Diseases**



**Respiratory System**



**Restoration and Preservation**



**Restoration and Preservation, Art Works/Artifacts**



**Restoration and Preservation, Manuscripts/Books/Music Scores**



**Restoration and Preservation, Structural/Architectural**



**Rheology**



**Rheumatic Diseases**



**Rheumatology**



**Rhode Island**



**Risk Factors/Analysis**



**Roma**



**Roman Catholic Church**

**Science Education**

**Seafood**



**Secondary Education**



**Security**



**Sedimentology**



**Seismology**



**Semiconductors**



**Seminars**



**Senile Dementia**

**Social Services**



**Social Services Delivery**



**Social Stratification/Mobility**



**Social Work**

# PROGRAM TYPE INDEX

NOTE: Numbers refer to entry numbers

**Basic Research**

**Building Construction and/or Renovation**



**Capital Campaigns**

**Centers: Research/Demonstration/Service**



**Citizenship Instruction**



**Community Development**

**Consulting/Visiting Personnel**

**Cultural Outreach**

**Curriculum Development/Teacher Training**



**Demonstration Grants**

**Educational Programs**

**Emergency Programs**



**Endowments**



**Environmental Programs**

**Fellowships**

**General Operating Support**

**Graduate Assistantships**



**Grants to Individuals**

**International Exchange Programs**



**International Grants**

**Job Training/Adult Vocational Programs**

**Land Acquisition**



**Matching/Challenge Funds**



**Materials/Equipment Acquisition (Computers, Books, Videos, etc.)**

**Preservation/Restoration**

**Professorships**



**Publishing/Editing/Translating**



**Religious Programs**

**Scholarships**



**Seed Grants**

**Service Delivery Programs**

**Training Programs/Internships**



**Travel Grants**

**Vocational Education**

# GEOGRAPHIC INDEX

Note: This index lists grants for which applicants must be residents of or located in a specific geographic area. Numbers refer to entry numbers.

## United States

**All States**

**Alabama**



**Alaska**



**American Samoa**

**Arizona**



**Arkansas**

**Florida**



**Georgia**

**Guam**

**Hawaii**



**Idaho**



**Illinois**



**Indiana**

**Maryland**

**Ohio**



**Oklahoma**



**Oregon**



**Pennsylvania**



**Puerto Rico**

## Foreign Countries

**Cape Verde**



**Caribbean**

www.ingramcontent.com/pod-product-compliance
Ingram Content Group UK Ltd.
Pitfield, Milton Keynes, MK11 3LW, UK
UKHW051420290926
15471UKWH00009B/1706

9 781940 750507